A FLORA OF SOUTHERN CALIFORNIA

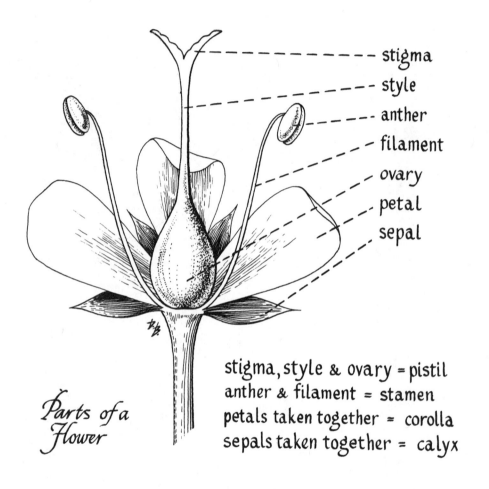

A FLORA
OF SOUTHERN
CALIFORNIA

PHILIP A. MUNZ

DIRECTOR EMERITUS
RANCHO SANTA ANA BOTANIC GARDEN
CLAREMONT, CALIFORNIA

UNIVERSITY OF CALIFORNIA PRESS
BERKELEY, LOS ANGELES, LONDON

UNIVERSITY OF CALIFORNIA PRESS
BERKELEY AND LOS ANGELES, CALIFORNIA
UNIVERSITY OF CALIFORNIA PRESS, LTD.
LONDON, ENGLAND

COPYRIGHT © 1974, BY
THE REGENTS OF THE UNIVERSITY OF CALIFORNIA
ISBN-0-520-02146-0
LIBRARY OF CONGRESS CATALOG CARD NUMBER: 73-174462
PRINTED IN THE UNITED STATES OF AMERICA
2 3 4 5 6 7 8 9 0

To the Memory of

SUSANNA BIXBY BRYANT

this book is respectfully dedicated. Loyal daughter of California, lover of its out-of-doors and its vegetation, and creator of the Rancho Santa Ana Botanic Garden.

CONTENTS

INTRODUCTION — 1

- BOUNDARIES — 1
- CLIMATE — 1
- THE VEGETATION — 1
- PLANT DISTRIBUTION — 5
- GEOLOGICAL HISTORY — 6
- INSULAR FLORA — 6
- DISCONTINUOUS DISTRIBUTION — 7
- SUMMARY — 7
- ACKNOWLEDGMENTS — 7
- ABBREVIATIONS — 8

CLASSIFICATION OF SOUTHERN CALIFORNIA VASCULAR PLANTS — 11

- KEY TO MAJOR GROUPS — 11
- SUBDIVISION LYCOPSIDA — 11
- SUBDIVISION SPHENOPSIDA — 15
- SUBDIVISION PTEROPSIDA — 16
    - CLASS FILICAE — 16
    - CLASS CONIFERAE — 35
    - CLASS GNETAE — 43
    - CLASS ANGIOSPERMAE — 45
        - SUBCLASS DICOTYLEDONES — 45
        - SUBCLASS MONOCOTYLEDONES — 862

ADDITIONS AND CORRECTIONS — 1017

GLOSSARY — 1019

INDEX — 1033

INTRODUCTION

BOUNDARIES

The area designated as southern California in this book extends from its northern boundary of Point Conception, Santa Barbara County, eastward along the crest of the Santa Ynez Mountains to the Mt. Pinos region in Ventura County, Fort Tejon in Kern County, the Tehachapi and Piute mountains, then northward to Little Lake in Inyo County and along the east slopes of the Inyo and White mountains to the Deep Springs region. It does not cover the northern part of Santa Barbara County, the southern end of the San Joaquin Valley, nor any of the Owens Valley, since each of those areas includes many elements of a more northern flora. As here defined, southern California comprises a *cismontane* area between the sea and the mountains, a *montane* area which in some cases reaches considerable elevation (Mt. Pinos, San Gabriel, San Bernardino, San Jacinto, Santa Rosa, Palomar, Cuyamaca and Laguna ranges), and a *transmontane* or desert area which is constituted largely of the Mojave and Colorado deserts. A fourth area consists of the islands off the coast: a northern group or Channel Islands (San Miguel, Santa Rosa, Santa Cruz and Anacapa islands) and a southern group (Santa Catalina, San Clemente, Santa Barbara and San Nicolas islands).

CLIMATE

(Sprague, M., Climate of California, U.S. Yearbook of Agriculture 1941: 783–797.)

Owing to the wide range of topography and the influence of coast and mountains, there are several unlike climates, marked differences in temperatures and precipitation occurring within a very few miles. The major factor in temperature is the isotherms that tend to follow topographic contours. The immediate coast tends to be cool and without the extreme fluctuation in temperature found in inland valleys and deserts with a more continental climate. Most of the region has a wet and a dry season, the cooler months usually the wetter. In the montane region snow falls mostly above 5000 ft., and many days have frost, which is rare at the extreme coast. Average annual precipitation tends to increase with altitude, varying from less than 2 inches in Death Valley to perhaps 50 in the wettest parts of the mountains. In such an arid climate torrential rains may cause much washing. Proximity to the coast, and fog, are important elements moderating the climate.

THE VEGETATION

In a climate like those described above, with a long dry season and rather mild winters, our typical "California vegetation" is largely Mediterranean in type, with harsh-leaved evergreens (sclerophylls) that are ± woody and with herbs in between. The brushy types are often spoken of as *chaparral*. Where the rainfall is greater, arborescent types occur and we get *woodland*, then at still higher elevations which are cooler and wetter, we generally pass into coniferous vegetation or *forests*. Beginning at the coast, the principal Plant Communities are as follows:

(1) **Coastal Strand.** Sandy beaches and dunes, with many succulent plants. Rainfall running from 10 to 25 inches; temperature rather even; little or no frost. Characteristic genera: Abronia, herbaceous species of Atriplex, Camissonia, Convolvulus.

(2) **Coastal Salt Marsh.** Tidelands, etc., with wet saline places and same climate as in 1. Much succulence. Distichlis; chenopods like Suaeda, Salicornia; Limonium.

(3) **Freshwater Marsh.** At various elevations mostly below 500 ft., in wet or semi dry places, with varying climates, but wet substratum. Typha, Scirpus, Eleocharis.

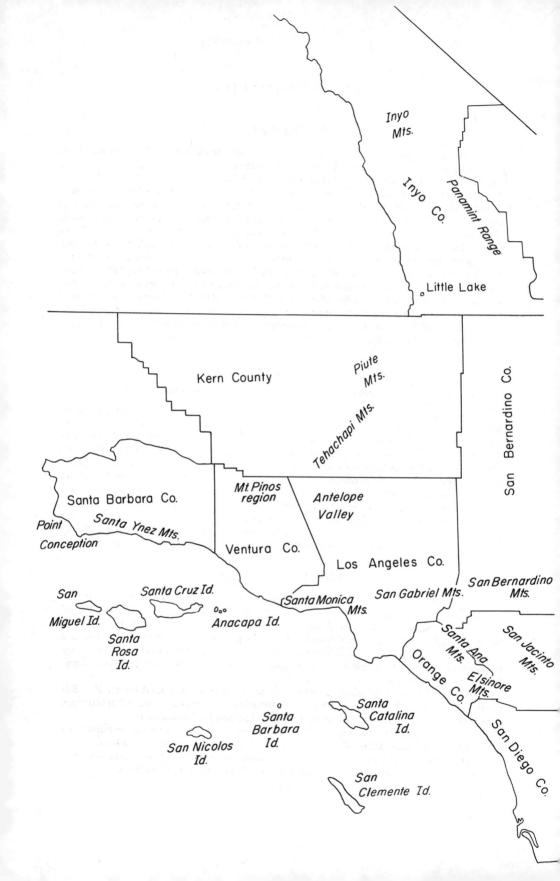

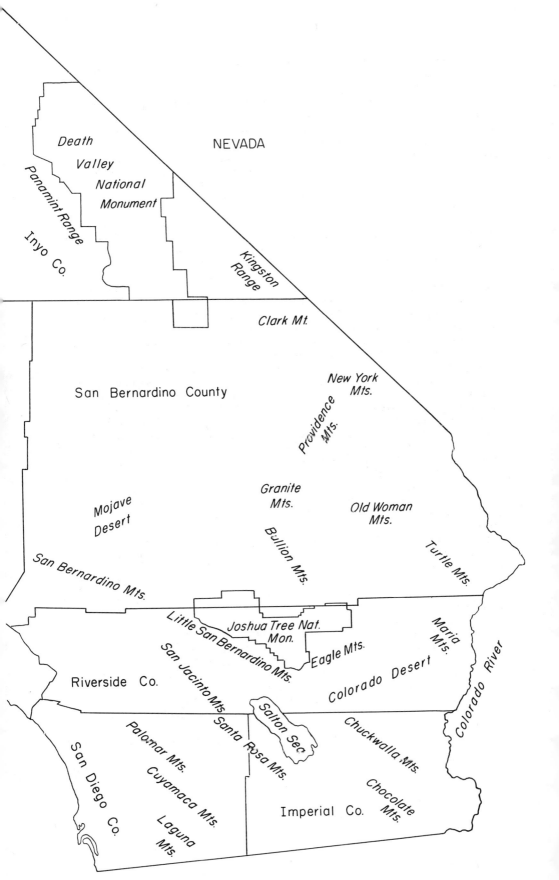

(4) **Coastal Sage Scrub.** Half-woody shrubs; cismontane from the coast to ca. 3000 ft. Precipitation 10–20 inches, usually not much frost. Artemisia, various species of Salvia, *Eriogonum fasciculatum*, Haplopappus species.

(5) **Chaparral.** Woodier shrubs, with well-developed broad-leaved sclerophylls; partly developed in the past by fire and many species are stump-sprouters after fire. Rainfall 14–25 inches, winters cooler; from above Coastal Sage Scrub to Yellow Pine Forest; cismontane. Heteromeles, Adenostoma, Cercocarpus, Arctostaphylos, shrubby Quercus, Ceanothus, Rhamnus.

(6) **Valley Grassland.** Coastal hot interior valleys below 4000 ft.; rainfall 6–20 inches, mixed in with the two preceding communities, but often on the south-facing slopes. Many wildflowers, including Ranunculus and Delphinium; originally largely perennial bunch-grasses, now largely weedy annual introduced grasses.

(7) **Southern Oak Woodland.** Open woodland, with ca. 15–25 inches of rain; oaks, walnut, sycamore, elderberry, sugar bush. Northward running into the woodland of the Sierran foothills and the Coast Ranges west of the San Joaquin Valley.

(8) **Montane Coniferous Forest.** In California as a whole this community is divisible into a lower one (Yellow Pine Forest), then a layer (Red Fir Forest) with more snow, cooler weather and somewhat different species of trees. However, in our southern mountains with a comparatively small area at these higher altitudes as compared to the much larger one in the Sierra Nevada, these communities are somewhat telescoped, and I am not attempting a division between about 5000 and 9000 ft. Precipitation runs from perhaps 35–50 or to 60 inches and growing season from 3 to 6 or 7 months. *Pinus ponderosa, P. jeffreyi, Abies concolor, P. coulteri, Calocedrus decurrens, Quercus kelloggii*, various species of Arctostaphylos and Ceanothus and of Cercocarpus are abundant. In its upper parts *Pinus murrayana* is found. Above 9500 ft. in a few places, such as in the San Bernardino Mountains, there are more Limber Pine (*P. flexilis*) and Lodgepole, a poor representation of the so-called Subalpine Forest of the Sierra but without Whitebark Pine or Hemlock.

(9) **Bristle-cone Pine Forest.** On the upper parts of the higher mountains of the northern Mojave Desert (Panamint, Inyo and White mountains), mostly above 8000 ft. and with perhaps 15–20 inches annual precipitation. The dominant tree is *Pinus longaeva* (Bristle-cone) and such shrubs as *Cercocarpus ledifolius*, Chrysothamnus and *Artemisia nova*.

(10) **Sagebrush Scrub.** Descending toward the desert from Bristle-cone Pine and even at its elevation, on drier more exposed slopes, are great expanses of low gray species of woody Artemisia (*A. cana, A. nova*, and farther down, *A. tridentata*), with Rabbit-Brush (Chrysothamnus), Coleogyne and Purshia. Average precipitation 8–15 inches; elevations, 4000–9000 ft.

(11) **Shadscale Scrub.** In heavy soil, often with underlying hardpan, and at about 3000–6000 ft. is another low shrubby vegetation, with perhaps 6–15 inches of rain and often near Joshua Tree Woodland, but apparently with different soils. Largely in the extreme northern Mojave Desert, it includes such species as *Atriplex confertifolia, Grayia spinosa, Eurotia lanata, Artemisia spinescens, Menodora spinescens, Gutierrezia sarothrae, Coleogyne ramosissima*.

Pinyon-Juniper Woodland and Joshua Tree Woodland have about the same climatic conditions as the two preceding communities, but often granitic and less saline soils.

(12) **Pinyon-Juniper Woodland.** Includes Pinyon (*Pinus monophylla*), Juniper (*Juniperus californica* or *J. osteosperma*), Scrub Oak (*Quercus turbinella*), *Yucca schidigera*, etc. It occurs commonly at 5000–8000 ft., receives 12–20 inches of precipitation, with some snow. Shrubs are Purshia, Cowania, Fallugia, Cercocarpus.

(13) **Joshua Tree Woodland.** At 2500–4000 ft., with 6–15 inches of rain, partly as summer showers, extends from extreme west to extreme east of the Mojave Desert. *Yucca brevifolia, Y. schidigera*, Salazaria, Lycium, Salvia, Eriogonum.

(14) **Creosote Bush Scrub.** The great mass of the floor of the deserts and their lower slopes are covered by a scrub vegetation of Larrea or Creosote Bush. Rainfall 2–8 inches, partly in summer. Burr weed (*Ambrosia dumosa*), Ocotillo (Fouquieria), Incienso (Encelia), etc.

(15) **Alkali Sink.** A term used for alkaline fl ıts and low places with no or poor drainage, largely at the elevations of the Creosote Bush, i.e., below 4000 ft. The plants are largely fleshy halophytes (Allenrolfea, Salicornia, Atriplex, Suaeda, with Salt Grass (Distichlis) and Mesquite (Prosopis) and some introduced plants like Tamarisk.

These Plant Communities are useful since they indicate for a given species something as to its plant associates and its ecology (altitudes, rainfall, etc.), and thus often enable a student to be more certain that he has identified a plant correctly. Plant Communities are often rather indefinable; for example, sometimes a north-facing slope holds its moisture longer than the hotter south side opposite it and accommodates a different lot of plants, such as chaparral on the north-facing slope and coastal sage scrub on the south. Naturally where the two meet there is confusion. This can be true also as the result of burning, deforestation, agriculture. After a given plant cover has been badly despoiled, it becomes weedy and may recover something of its earlier nature only after many years.

PLANT DISTRIBUTION

One cannot be long in southern California and observe its vegetation without being impressed by certain distribution patterns, for example, the likeness of the Pine belt to the Sierra Nevada. There is a series of "stepping stones" between the southern Sierra and our more southern mountains, notably the Tehachapi Mountains, Mt. Pinos region (8826 ft.), Sawmill and Liebre mountains, then the San Gabriel Mountains (10,080), San Bernardino Range (nearly 11,500 ft.), San Jacinto Mountains (ca. 10,800 ft.), Santa Rosa Mountains (8716 ft.), Palomar Mountains (6126 ft.), Cuyamaca Mountains (6515 ft.), and Laguna Mountains (over 5000 ft.). With so many high spots in a more or less broken series, it is not surprising to find that a large portion of the plant species of the pine belt have a Sierran affinity, with perhaps 100 such species as far south as the mountains of San Diego County and many more in the San Jacinto and San Bernardino mountains, in meadows and cool areas. Furthermore, the southern climate was formerly much cooler than now. There were cirques and glacial moraines in the San Bernardino Mountains, where now little or no snow persists until August.

In the same way marked distribution patterns exist at lower elevations. There is similarity between the Mojave Desert and Great Basin; many species in the eastern Mojave of California extend into Utah, the Colorado Desert and northwestern Mexico, often combining in the Sonoran Desert with quite a different makeup from the Mojave, such as trees like Olneya and Cercidium, of a thorny forest type. Our cismontane region is in many ways the typical "California" flora, having the sclerophyll or broad-leaved evergreen groups which often arose in the highlands of Mexico. These migrated north in a time more moist than now, then west to the coast, but since have been cut off from their Mexican relatives by the intervening deserts. Examples are manzanitas (Arctostaphylos), Fremontodendron, Ceanothus, evergreen oaks and Cupressus.

One would expect much similarity between our higher mountains (including the Sierra Nevada) and the Rocky Mountains, but this is not so great as it might be. Undoubtedly the summer dryness of our area as compared with that of the Rocky Mountains means that many northern plants, which in many cases are circumpolar and have worked their way south, have disappeared in our dry region and have survived in the Rockies with more summer rain. Of course, some species, like species of Juncus, and plants that grow in mountain meadows, have survived.

A rather unexpected phenomenon is the presence in our area of such Eurasian genera as Aquilegia, Anemone, Delphinium and Crepis. Some that have been carefully monographed originated deep in Asia and radiated out from there to Europe, northern Africa, and North America, by way of what is now the Bering Strait.

GEOLOGICAL HISTORY

Eocene. Seventy million or more years ago, there were three major vegetation units in North America:

(1) *Neotropical Tertiary* extended north to southeastern Alaska and to Canada. The climate was uniformly warm, with a rainfall of 80 inches or more. This unit included many broad-leaved evergreens of tropical families such as Cinnamomum, Cycas, Persea (avocado) and Sabal. As yet there were no extensive mountain ranges.

(2) *Arcto-Tertiary* was a northern unit during the Eocene period and comprised Abies, Acer, Alnus, Chamaecyparis, Cornus, Lithocarpus, Picea, Populus, relatives of Sequoia and Thuja. Many of these are still here, but more of them, especially broad-leaved and often deciduous species, have persisted in our wetter eastern climate.

(3) *Madro-Tertiary* (named for Sierra Madre of Mexico) originated in southwestern North America. Smaller-leaved drought-resistant sclerophylls existed by Middle Eocene and spread with increasing dryness. Example: Arbutus, Juglans, some pines (Pinyon, Digger), Quercus (live oaks), Lyonothamnus, Umbellularia, Cupressus, certain species of Prunus, Cercocarpus, Purshia.

Later than Eocene. Elevation of mountain ranges and their accompanying rain-shadows and development of drier summers; development of deserts. Tropical flora pushed south. Many genera migrated to the coast. A montane-coniferous forest developed and invaded higher mountains in late Miocene and in Pliocene; it reached southern California in Pleistocene and persisted as our pine belt, so that some 325 such species are found in the San Bernardino Mountains.

The Madro-Tertiary contributed largely to our California Woodland element such as the digger-pine woodland, oak-walnut woodland, and to our pinyon-juniper woodland. Chaparral ancestry is quite old (manzanita, Ceanothus, Cercocarpus, Dendromecon, Fremontodendron) and was restricted to cismontane California in late Pliocene.

Pleistocene. Temperate vegetation showed further advancement southward than at any other time, the redwood reaching Carpenteria in Santa Barbara County. The climate was cooler and more moist than now. Death Valley was a lake 90 miles long and 600 feet deep. With glaciation in some of our mountains, the pine belt plants came far south; later with more modern warmth and dryness, many were pushed north or survived here and there as relicts (example, Male Fern in Holcomb Valley, found once in the 1880s; *Anaphalis margaritacea* in a single canyon in the San Bernardino Mountains).

Present. At present we have a mixture of genera and species of northern ancestry: *Artemisia tridentata*, Salsola, many grasses, Chrysothamnus, Eurotia, Aquilegia, Delphinium, Crepis; some of these are from the old Arcto-Tertiary origin and include many circumpolar species: ferns, grasses, sedges, etc. We also have plants of southern derivation, often coming down from Madro-Tertiary sources: Larrea, Fremontodendron, Eriogonum, Grayia. The third big element is of local derivation (especially true of species) and represents much endemism. Local endemics (correlated with tremendous variation in topography and climate, summer dryness, etc.) may occur: (a) mountain range by mountain range (*Sedum niveum, Lesquerella bernardina, Arabis parishii, Horkelia wilderae, Sidalcea pedata, Taraxacum californicum*); (b) valley by valley (*Ribes hesperium, Ribes parishii, Lathyrus laetiflorus, Collinsia parryi, Calochortus plummerae*). Others are general southern California endemics: Matilija Poppy, Eriogonum species, *Adenostoma sparsifolium*, Forbes cypress, Lathyrus species. Another large group is island endemics, some rather general, others confined to one or few islands: *Salvia brandegei, Eriogonum giganteum, E. arborescens*, Lyonothamnus, *Ceanothus arboreus*.

INSULAR FLORA

For a discussion of our insular flora see R. Thorne, The California Islands, Ann. Mo. Bot. Gard. 56: 391–408, 1969. The chain of islands off the southern California coast represents, at least in part, the westerly extension of the Santa Monica Mountains. Plants from the mainland migrated west and had almost disappeared from the mainland by the end of the Pliocene. Speciation developed when the offshore lands broke up into further separate islands.

Two general groups are a northern group (Santa Rosa, Santa Cruz, San Miguel, Anacapa) with pines, *Quercus tomentella, Rhamnus pirifolius, Ceanothus arboreus, Prunus lyonii*, and a southern group (San Nicolas, Santa Barbara, Santa Catalina and San Clemente) lacking pines, but having Lyonothamnus, *Quercus tomentella, Rhamnus pirifolia, Prunus lyonii, Ceanothus arboreus*. Raven says that there are 76 endemic taxa on the islands, 23 in the northern group, 38 in the southern group, and 15 common to both groups (P. Raven, The floristics of the California Islands *in* R. N. Philbrick (ed.), Proceedings of the Symposium on the Biology of the California Islands, p. 63, 1967 [published by Santa Barbara Botanic Garden]). San Clemente Island has several species restricted otherwise to Guadalupe Island: *Phacelia floribunda, Camissonia guadalupensis*. Other Guadalupe Island species occur on both Catalina and San Clemente: *Scrophularia villosa, Galvesia speciosa, Crossosoma californica*.

DISCONTINUOUS DISTRIBUTION

Some striking cases probably represent relict distribution, with certain areas maintaining old species because of special soil and precipitation factors, for example species in the mountains of San Diego County and cismontane central California: *Thermopsis californica* as var. *semota, Viola lobata* (Santa Lucia Mountains north), *Ceanothus foliosus* as *C. austromontanus* in region about Julian; *Salvia sonomensis*; *Zigadenus venenosus, Downingia concolor* as var. *brevior*.

California (largely cismontane) with the other states of our west coast and extreme southern South America have over 100 species closely related or perhaps identical: *Phacelia magellanica* group, Gilia, Larrea, Prosopis, Apiaceae like *Bowlesia incana, Sanicula crassicaulis* and *S. graveolens*. So far, sporadic long-distance transtropical dispersal seems to be a hypothesis for explanation of this discontinuity; most such species are self-compatible so that a single introduction might suffice.

SUMMARY

The geographic area here called southern California, with perhaps an extent of 30,000 square miles, ranges from below sea level in Death Valley and Salton Sea to over 11,000 ft. in the San Bernardino Mountains. It has an annual precipitation from about one inch to perhaps 50 inches. The flora has had a long history, with northern, Mexican and at first tropical elements early represented; with later an increasing elevation of mountains and diversity of topography accompanied by a drying process and by rain-shadows. The elements from the various ancient floras have been highly modified. With the Pleistocene came a great migration southward, thus taxa from various areas met for the first time, allowing much hybridization. In the Rancho Santa Ana Botanic Garden this same situation has occurred as species from various parts of the state have for the first time been brought to grow near each other; so much hybridization has occurred that for genera like Ceanothus, seeds harvested in the Garden cannot be used to represent the species with any certainty. Thus in nature new forms arose and found various ecological niches. Many species moved north again or to higher altitudes as the more recent drier and warmer climate evolved. The result is that now we have a very complex flora with many not very ancient species. It is characterized by many large genera: Eriogonum, Lupinus, Astragalus, Penstemon, Trifolium, Cryptantha, Phacelia, Mimulus, Lotus, Gilia.

ACKNOWLEDGMENTS

In compiling a manual of the plants of any region, the author naturally must, in addition to the study of plant specimens themselves, turn to recently published works for changes in concept and literature. For the most part I cite such references used herein immediately after the generic description or sometimes briefly in a discussion of a given species.

Then too, one finds it helpful to consult friends who may be specialists or students of various groups and get from them opinions not yet published. In this particular I have depended primarily on Robert Thorne, first for his concept of groups above families and of families themselves and for the names he employs, although I have not always followed his usage. I have also profited greatly by his knowledge of certain aquatics, as well as of sedges, grasses, etc. and of certain geographical areas such as Santa Catalina Island and San Gabriel Mountains.

I have had help from Lee W. Lenz particularly with the Brodiaea complex, from Lyman Benson with Cactaceae, from James Henrickson with Fouquieria and other plants, from his former student Dieter H. Wilken with *Baccharis viminea* and the genus Hulsea, from James Reveal with that ever-puzzling genus Eriogonum. Others to whom I have turned are Stephen Tillett on Abronia, Alva Day in Gilia, John Tucker in Quercus, Reid Moran in Crassulaceae and plants of Lower California, especially those of Guadalupe Island, Reed Rollins in Brassicaceae, Wallace Ernst in Papaveraceae, and Dr. Wahl's determinations of Chenopodium as left in our herbarium. Still others who have contributed largely are John Thomas Howell, Ernest Twisselmann, and Percy C. Everett.

Harry Thompson and Joyce Roberts have kindly prepared the manuscript for Loasaceae. Lawrence Heckard has loaned me notes on Orobanche; he and Rimo Bacigalupi have helped with Castilleja. Ella May Dempster has loaned me unpublished manuscript on Galium, A. Michael Powell on Laphamia and Perityle, Delbert Wiens on Arceuthobium and prepared a key for the southern California species of the genus. David Dunn has sent notes on Lupinus.

Others who have been particularly helpful with specimens, often showing new distribution records or exemplifying new taxa, have been Tom C. Fuller, Ernest Twisselmann, R. W. Townsend, Louis B. Ziegler, John and Lucille Roos, Janice Beatley.

I especially wish to express my gratitude to my wife Alice M. Munz for her many hours spent on mounting drawings that make up the plates illustrating over 600 species.

The illustrations in the book represent the work of several persons. Those unsigned are largely the work of Stephen Tillett, with a few, such as for orchids, by Richard Shaw. On those signed, the letter J means Jeanne Janish, who has illustrated western plant species more widely than any other artist; DB stands for Dick Beasley, at present a professor of art. Both of these artists have an unusual ability to throw into three dimensions a flattened specimen mounted on an herbarium sheet. The other drawings were largely by former students: RC or Cross by the late Rodman Cross, Craig by the late Tom Craig, Z by Milford Zornes, SHL by Shue-Huei Liao, a student from Taiwan.

I am grateful to Art Gibson in helping prepare some of the last pages of manuscript.

ABBREVIATIONS

acc.–according to
adv.–adventive
Afr.–Africa
ak., aks.–akene, akenes
Alta.–Alberta
Am.–America, American
Ariz.–Arizona
Ark.–Arkansas
auth.–authors
B.C.—British Columbia
C.–Canyon
ca.–circa (about or approximately)
Calif.–California
Can.–Canada
caps.–capsule
cent.–central

cm.–centimeter, centimeters
co.–county, counties
Colo.–Colorado
cult.–cultivated, cultivation
diam.–diameter
dm.–decimeter, decimeters
e.–east, eastward, eastern
elev., elevs.–elevation, elevations
Eu.–Europe
F.–forest
f.–forma
fil., fils.–filament, filaments
fl., fls.–flower, flowers
Fla.–Florida
fld.–flowered
fr., frs.–fruit, fruits

Abbreviations

ft. – foot, feet
hemis. – hemisphere
id., ids. – island, islands
Ida. – Idaho
Ill. – Illinois
Ind. – Indiana
infl., infls. – inflorescence, inflorescences
introd. – introduction, introduced
invol., invols. – involucre, involucres
Kans. – Kansas
Ky. – Kentucky
La. – Louisiana
L. Calif. – Lower California
lf. – leaf
lft., lfts. – leaflet, leaflets
loc., locs. – locality, localities
lvd. – leaved
lvs. – leaves
m. – meter, meters
Man. – Manitoba
Medit. – Mediterranean
Mex. – Mexico, Mexican
Mich. – Michigan
Minn. – Minnesota
Miss. – Mississippi
mm. – millimeter, millimeters
Mont. – Montana
mt., mts. – mountain, mountains
n. – north, northern, northward
N. Am. – North America, North American
nat. – native
natur. – naturalized
N. Dak. – North Dakota
ne. – northeast, northeastern
n. Mex. – northern Mexico
Nebr. – Nebraska
Nev. – Nevada
New Mex. – New Mexico
Nfld. – Newfoundland
no. – number
nw. – northwest, northwestern
Okla. – Oklahoma
Ore. – Oregon
R. – river
ref., refs. – reference, references
s. – south, southern, southward
S. Am. – South America
Sask. – Saskatchewan
S. Dak. – South Dakota
se. – southeast, southeastern
segm., segms. – segment, segments
Son. – Sonora
sp., spp. – species (singular and plural)
ssp., sspp. – subspecies (singular and plural)
sw. – southwest, southwestern
temp. – temperate
Tex. – Texas
trop. – tropical, tropics
U.S. – United States
V. – valley
Va. – Virginia
var., vars. – variety, varieties
w. – west, western, westward
Wash. – Washington
wd., wds. – woodland, woodlands
W.I. – West Indies
Wis. – Wisconsin
Wyo. – Wyoming

SIGNS USED

± – more or less
♂ – staminate
♀ – pistillate
× – sign of hybrid, used before the species name
μ – micron, the millionth part of a meter

NOTE TO THE READER

In a book where one group has to come after another in a linear series, it is practically impossible to reveal relationship or a true phylogeny. For that reason and for convenience, I have simply used an alphabetic arrangement for families within major groups such as dicotyledons, genera within families, and species within genera.

The reader may notice that in many instances word divisions at the ends of lines do not follow common American practice. This is due partly to the fact that the book was printed in England and partly to restrictions of line justification.

VASCULAR PLANTS OF SOUTHERN CALIFORNIA
Division **TRACHEOPHYTA**
KEY TO MAJOR GROUPS
1. Plants without seeds or fls., reproducing by 1-celled spores borne in sporangia.
    2. Plants not fernlike and not free-floating, the lvs. mostly minute or ± scalelike (except in *Isoetes*), 1-veined; the fertile lvs. with 1 sporangium on the adaxial surface.
        3. Stems not jointed; lvs. not whorled and not forming a sheath at the nodes
        Subdivision I. *Lycopsida* p. 11
        3. Stems of hollow joints; lvs. whorled and forming a sheath at the solid nodes
        Subdivision II. *Sphenopsida* p. 15
    2. Plants fernlike and with large lvs. (fronds) or free-floating aquatics and with small or scalelike overlapping lvs.; mostly elaborately veined; sporangia of spreading lvs. usually abaxial .... Subdivision III. *Pteropsida* Class I. *Filicae* p. 16
1. Plants with seeds usually produced by cones or fls.
    4. Seeds not inclosed in ripened pistils, but naked and usually borne on the surface of a scale; the crowded or overlapping scales usually forming a cone or strobilus; plants not bearing typical fls.
        5. Stems not jointed; lvs. needlelike or linear, sometimes scalelike and then closely overlapping; plants mostly with resin. Mostly trees or arborescent
        Class II. *Coniferae* p. 35
        5. Stems jointed; lvs. scalelike, in 2's or 3's in widely separated whorls; plants without resin; cones small, with thin scales. Desert shrubs ,.... Class III. *Gnetae* p. 43
    4. Seeds inclosed in ripened pistils; plants producing true fls. . Class IV. *Angiospermae* p. 45
        5a. Cotyledons 2; stems usually exogenous (increasing in diam. by cambial activity between the xylem and phloem); lvs. mostly net-veined; fls. usually on the plan of 4 or 5 .... Subclass I. *Dicotyledones* p. 45
        5a. Cotyledon 1; stem mostly endogenous, or if with cambial activity, this adding whole vascular bundles; lvs. mostly parallel-veined; fls. mostly on the plan of 3
        Subclass II. *Monocotyledones* p. 862

Subdivision I. **LYCOPSIDA**
Class **LYCOPODIAE**

Plant-body a sporophyte with slender stems and roots and small mostly spiral or paired or whorled 1-veined lvs. or with cormlike stem and longer lvs. Sporangium solitary on the adaxial face and near the base of the sporangium-bearing lf. (sporophyll); sporophylls often segregated from the sterile foliage lvs. and aggregated into ± loose strobili. Spores developing into minute or microscopic gametophytes producing antheridia and archegonia, the ♂ and ♀ sex-organs. Fusion of gametes results in a new sporophyte.

KEY TO ORDERS
Stems above ground, ± elongate and branched, covered with many minute overlapping scalelike lvs. less than 1 cm. long... 2. *Selaginellales*
Stems underground, cormlike, bearing tufted lvs. 2.5–25 cm. long............. 1. *Isoetales*

Order 1. ISOETALES

Characters as given for the family *Isoetaceae*

1. Isoetàceae. QUILLWORT FAMILY

Small, aquatic, amphibious or terrestrial herbs, with a 2–3-lobed corm crowned with many sedgelike elongate-subulate lvs. with large basal sporangia. Sporangia solitary, on upper surface of lf.-base, of 2 kinds: micro- and macro-sporangia, the former producing many small spores each of which forms an antheridium, the latter with large spores forming ♀ gametophytes. One genus of worldwide distribution.

1. *Isòetes* L. QUILLWORT

Corm fleshy, bearing roots below. Lvs. with an upper part that is sterile and septate, with 4 longitudinal series of air-spaces and with or without peripheral strands, with or without stomates. Lvs. also with a fertile basal part forming a large plano-convex imbedded sporangium partly covered apically by a thin membrane (velum) above which is a small green triangular structure (ligule). Ca. 60 spp. (Greek, *isos*, ever, and *etas*, green, of uncertain application.)

Plants mostly entirely submersed; lvs. lacking peripheral strands; stomata few or lacking; corm usually 2-lobed .... 1. *I. bolanderi*
Plants terrestrial or amphibious, on wet soil or about margins of ponds or streams; stomata numerous.
    Corm usually 2-lobed; velum not completely covering the sporangium ...... 2. *I. howellii*
    Corm usually 3-lobed; velum usually nearly or quite covering the sporangium .. 3. *I. orcuttii*

1. **I. bolánderi** Engelm. [*I. californica* Engelm.] Plate 1, Fig. A. Submersed; corm deeply 2-lobed; lvs. 6–25, tapering to a very fine point, bright green, soft, 6–15 (–25) cm. long, with some stomates but usually no peripheral strands; ligule small, cordate; sporangia 3–4 mm. long, ca. ⅓ covered by the velum; macrospores ca. 0.3–0.4 mm. in diam., with low tubercles.—Mountain lakes and ponds, 5000–11,000 ft.; Montane Coniferous F.; San Bernardino Mountains, Sierra Nevada to Wash., Rocky Mountains.

2. **I. howéllii** Engelm. [*I. melanopoda* var. *californica* A. A. Eat.] Amphibious; corm 2-lobed; lvs. mostly 10–30, bright green, spreading, 5–28 cm. long, with many stomates and usually 4 peripheral strands; ligule narrow, elongate-triangular; velum covering ⅓ of the sporangium; macrospores white (tan when wet), 0.25–0.6 mm. in diam., usually with tubercles and anastomosing and distinct crests.—In water and on mud, below 9000 ft.; many Plant Communities; mesas near San Diego to San Bernardino Mts. and Mirror Lake, Ventura Co.; to Sierra Nevada and Wash., Mont.

3. **I. orcúttii** A. A. Eat. [*I. nuttallii* var. *o.* Clute.] Corm 3-lobed; lvs. 3–25, spreading, 3–6 cm. long, with many stomates and no peripheral strands; velum complete; ligule small, triangular; macrospores gray (or brown when wet), 0.2–0.4 mm. in diam., smooth, glossy, sometimes ± tubercled.—In water of vernal pools and on mud, at low elevs.; Chaparral, V. Grassland, etc.; San Diego Co. to cent. Calif.

Order 2. SELAGINELLALES

Characters as given for the family *Selaginellaceae*

Family 1. Selaginellàceae

Low terrestrial plants, freely branched, with numerous very small, usually imbricate lvs. that are spirally arranged and all alike or in 4 longitudinal rows and of 2 kinds. Sporangia in terminal quadrangular sessile spikes or strobili of slightly modified lvs. (sporo-

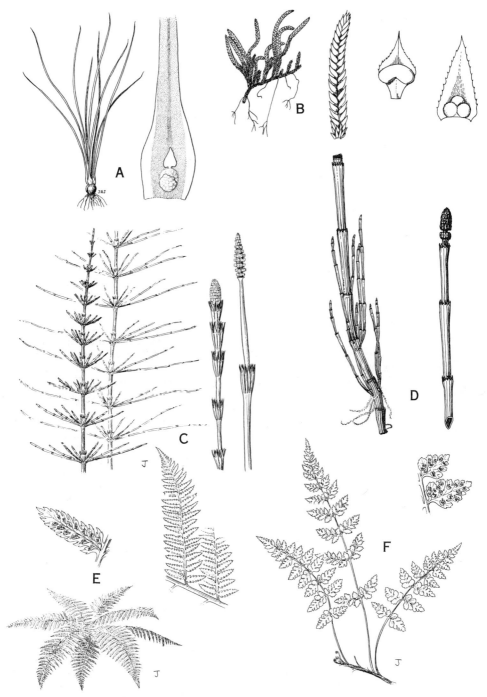

Plate 1. Fig. A, *Isoetes bolanderi*, adaxial view of separate lf. with basal sporangium, partial velum, triangular ligule. Fig. B, *Selaginella bigelovii*, separate strobilus, smooth microsporangium, lobed macrosporangium. Fig. C, *Equisetum arvense*, early fertile stem and later branched sterile stem. Fig. D, *Equisetum laevigatum*, cones on vegetative stem. Fig. E, *Athyrium filix-femina*, elongate indusia. Fig. F, *Cystopteris fragilis*, rounded sori.

phylls), each strobilus containing both microsporophylls and macrosporophylls, the former with many minute reddish or orange microspores, the latter with 1–4 rather large macrospores. One genus.

1. Selaginélla Beauv. SPIKE-MOSS. LITTLE CLUB-MOSS

Characters of the family. Ca. 600 spp., widely distributed, especially in the tropics. (Name, a diminutive of *Selago*, a classical name for some sp. of *Lycopodium*.)

(Tryon, R. M. S. S. rupestris and its allies. Ann. Mo. Bot. Gard. 42: 1–99. 1955).
1. Stems erect or ascending, rooting only at or near the base .... 2. *S. bigelovii*
1. Stems prostrate to decumbent, rooting at or near the apex.
   2. Lvs. on the underside of the main stems not decurrent, lacking a terminal seta; stems forming an ashy carpet-like growth very near the ground. San Diego region...... 3. *S. cinerascens*
   2. Lvs. on the under side of the main stems strongly decurrent.
      3. Lvs. lacking a terminal seta at maturity; plants strongly dorsi-ventral. Desert
                                                                                                           4. *S. eremophila*
      3. Lvs. with a terminal seta; stems radially symmetrical.
         4. Setae from 1/3–2/3 the length of the lf.-blade .... 1. *S. asprella*
         4. Setae less than 1/3 the length of the lf.-blade.
            5. The setae yellowish, smooth; cilia 2–7 on each side of the lf. or wanting. Mostly from above 7500 ft. .... 6. *S. watsonii*
            5. The setae white, ± scabrous; cilia more numerous. Panamint and Providence Mountains, mostly from below 7500 ft. .... 5. *S. leucobryoides*

1. **S. asprélla** Maxon. Stems loosely matted, 3–6 cm. long, creeping, with laxly ascending branches 1–2 cm. long; lvs. ascending, somewhat imbricate, deltoid-subulate, glaucous, white-margined, eciliate or ciliate near the apex, 2.8–3.2 mm. long; the setae 0.7–0.9 mm. long, white-hyaline, forming tufts at the ends of the branches; strobili lax, 1–2 cm. long, arcuate, the sporophylls 2.5–3 mm. long, long-acuminate.—Dry rocky places, 5000–8900 ft.; Montane Coniferous F.; San Gabriel, San Bernardino, San Jacinto, Santa Ana and Laguna mts.; s. Sierra Nevada; n. L. Calif.

2. **S. bigelòvii** Underw. Plate 1, Fig. B. Stems slender, shortly branched, ascending to erect, 5–20 cm. long, from widely creeping rhizomes; lvs. appressed-imbricate, gray-green, the blades 1.2–1.8 mm. long, ciliate, narrow-lanceolate, each with a short white seta; strobili 4–15 mm. long, erect at tips of short lateral branches, the sporophylls ovate, ca. 2 mm. long.—Dry rocky banks below 6000 ft.; Chaparral, Coastal Sage Scrub, etc.; cismontane from n. L. Calif. to Sonoma Co. and s. Sierra Nevada. Channel Ids.

3. **S. cineráscens** A. A. Eat. [*S. bryoides* Underw.] Plants forming a close ashen carpet on the ground; stems wiry, mostly 5–12 cm. long, branched; lvs. appressed-imbricate, ashy gray, linear to broadly subulate, ciliate, not setigerous, 1.2–1.5 mm. long, with deep dorsal groove; strobili 2–4 mm. long; sporophylls broadly cordate, short-acuminate, ciliate.—Dry slopes and mesas; Coastal Sage Scrub, Chaparral; sw. San Diego Co.; adjacent L. Calif.

4. **S. eremóphila** Maxon. [*S. parishii* of Calif. refs.] Plants prostrate, dorsiventral, the stems 5–12 cm. long, with spreading branches; lvs. crowded, broadly lanceolate, those of the lower side 2 mm. long, thin, acutish, not setigerous when mature, with ca. 25 cilia on each side; lvs. of upper side subvertical, 1–1.4 mm. long, with 6–12 cilia on each side; strobili many, curved, 6–10 mm. long, the sporophylls with 12–18 cilia on each side.—Sheltered places among rocks, below 3000 ft.; Creosote Bush Scrub; canyons along w. edge of Colo. Desert, Chuckwalla Mts.; Ariz., L. Calif.

5. **S. leucobryoìdes** Maxon. Plants cushionlike, the stems 1–2 cm. long, readily fragmenting when dry, with thick short erect branches; lvs. uniform, appressed-imbricate, glaucous, linear-subulate, 2.5–3.2 mm. long, short-setigerous, with 8–16 cilia on each side; strobili many, 5–10 mm. long, erect; sporophylls rigidly appressed-imbricate, deltoid-ovate, ciliate and toothed.—Rock-crevices and slopes, 2000–7500 ft.; Shadscale Scrub to Pinyon-Juniper Wd.; Panamint and Providence mts., e. Mojave Desert.

6. **S. wátsonii** Underw. Cespitose, the stems prostrate, short, 2–8 cm. long, with numerous branches 1–2 cm. long, ascending; lvs. densely appressed-imbricate, thick, linear-oblong, 2–2.4 mm. long, each with broad dorsal groove and stout smooth seta, with few or no cilia; strobili 1–2.5 cm. long, sharply quadrangular, the ovate sporophylls ciliate on basal half.—Locally frequent in dry rocky places, 7500–11,000 ft.; Montane Coniferous F., Alpine Fell-fields; Santa Rosa and San Jacinto mts. n. through the mts. to Sierra Nevada and White Mts.; to Ore., Mont., Utah.

Subdivision II. SPHENOPSIDA

Class 1. EQUISETAE

With a single living order (Equisetales) and family. Characters as given for that family.

1. Equisetàceae. HORSETAIL FAMILY

Rushlike, often branching plants, with perennial creeping branching rhizomes rooting at the nodes. Aerial stems perennial or annual, cylindrical, fluted, simple or with whorled branches at the solid sheathed nodes, the internodes generally hollow. Sheaths made up of the united minute whorled lvs. which may be free at the tips. Surfaces of the stems overlaid with silica; stomates arranged in bands or rows in the grooves. Strobili or cones terminal, formed of whorls of stalked peltate structures around a cent. axis, each with a circle of sporangia beneath. Spores uniform, green, the outer layer forming 4 hygroscopic bands. One genus.

1. *Equisètum* L. HORSETAIL. SCOURING-RUSH

Characters of the family. Ca. 25 spp., widely distributed. (Latin, *equus*, horse, and *seta*, bristle.)

(Hauke, R. A taxonomic monograph of the genus Equisetum subgenus Hippochaete. Beihefte zur Nova Hedwigia 8: 1–123. 1963)
Aerial stems all green, not dimorphous; stomates in regular rows.
  Sheaths much longer than broad, ± funnel-shaped, green, normally with a narrow black band at the top .... 3. *E. laevigatum*
  Sheaths nearly or quite as broad as long, cylindrical, usually ashy with 2 black bands.
    2. *E. hyemale*
Aerial stems of 2 kinds, the fertile ones not green or branched, the sterile green and with many slender branches; stomates scattered.
  Sterile stems stout, generally over 3 mm. thick and 6–12 dm. tall; sheaths with 20–30 teeth.
    4. *E. telmateia*
  Sterile stems slender, generally less than 3 mm. thick and less than 6 dm. tall; sheaths with 8–12 teeth .... 1. *E. arvense*

1. **E. arvénse** L. COMMON HORSETAIL. Plate 1, Fig. C. Fertile stems 5–25 cm. tall, flesh-colored, soon dying, the pale sheaths with 8–12 lanceolate brownish teeth; sterile stems 1–6 dm. tall, green, 6–14-ridged, roughish, the sheaths with hyaline-margined teeth; branches slender, in dense whorls; cones lance-ovoid, 2–3 cm. long; $n=$ ca. 108 (Manton, 1950).—Wet places below 10,000 ft., many Plant Communities; cismontane Calif., Inyo Co. to Modoc Co.; to Alaska, Labrador, Nfld., most of the U.S.; Eurasia. Many growth forms have been given names.

2. **E. hyemàle** L. var. **affìne** (Engelm.) A. A. Eat. [*E. robustum* var. *affine* Engelm. *E. praealtum* Raf. *E. h.* var. *californicum* Milde. *E. h.* var. *robustum* A. A. Eat. *E. r.* A. Br. ex Engelm.] Stem normally unbranched, upright, persisting for at least 2 years, 5–20 dm. tall, 4–12 mm. thick, with 16–48 ridges, these very rough with 1 or 2 rows of transverse bands of silica; sheaths but little longer than broad, cylindrical and ashy with black bands

at both ends, the short teeth quite persistent; lvs. 3-4-keeled, the cent. keel plainly or faintly grooved; cones ovoid, 1-2.5 cm. long, sharply apiculate; $n=$ ca. 108 (Manton, 1950).—Occasional as colonies in moist places below 8500 ft.; many Plant Communities; cismontane and montane Calif.; to Alaska, Quebec, most of U.S.

3. **E. laevigàtum** A. Br. [*Hippochaete l.* Farw. *A. funstoni* A. A. Eat. *E. kansanum* Schaffn. *E. fontinale* Copel.] Plate 1, Fig. D. Aerial stems of 1-2 years duration, normally unbranched, upright, smooth, tufted, normally 2-8 dm. tall, smoothish to rough with crossbands of silica, 15-30-ridged; sheaths longer than broad, usually dilated upward, the lower in old stems becoming girdled with brown, 7-15 mm. long; cones brown to yellow obtuse or with a slight apiculum, containing well formed spores.—In damp or moist places below 8000 feet; many Plant Communities; cismontane Calif., Owens V.; to B.C., L, Calif., Ontario, Tex., Mex.

4. **E. telmatèia** Ehrh. var. **bráunii** Milde. [*E. maximum* Lam.] GIANT HORSETAIL. Fertile stems erect, short-lived, whitish or brownish, 2-6 dm. tall, smooth, with loose membranous sheaths 2-5 cm. long and with 20-30 teeth; sterile stems 5-25 dm. tall, 5-20 mm. thick, pale green, 20-40 ridged, the sheaths cylindrical, the teeth broadly hyaline-margined; branches solid, in dense whorls; cones stout-pedunculate, 4-8 cm. long; $n=$ ca. 108 (Manton, 1950).—Occasional in swampy places and along streams, below 4500 ft.; several Plant Communities; cismontane Calif.; to B.C. also in Mich.

E. × ferrisii Clute. [*E. hyemale* var. *affine* × *E. laevigatum*. *E. h.* var. *intermedium* A. A. Eat.] Plants occasionally occur intermediate between spp. 1 and 4 and with sterile spores which do not shed from the cones. Reported from Monterey, Kern cos., etc.

Subdivision III. **PTEROPSIDA.** FERNS
Class 1. **FILICAE**

Terrestrial, sometimes epiphytic or aquatic plants, usually with evident alternation of generations. The sporophyte generation, the fern plant, usually differentiated into stems, roots and lvs. In the homosporous ferns all or certain lvs. (sporophylls) bear the minute sporangia which produce the 1-celled spores that develop into the small usually independent gametophyte generation. This forms antheridia and archegonia and fertilization results in a new sporophyte. Heterosporous ferns aquatic and floating, or on mud; producing the sporangia in special structures (sporocarps) that contain micro- and macrosporangia and result in microscopic ♂ and ♀ gametophytes.

KEY TO ORDERS

Plants fernlike, terrestrial or on mud, not floating; producing 1 kind of spore in sporangia.
    Sporangia large, globular, without an annulus, borne in a stalked spike or panicle from the base of the green blade which appears lateral; rootstock almost none; frond usually 1, seeming to arise from a cluster of fleshy roots. (Eusporangiatae)
        3. *Ophioglossales*   p. 33
    Sporangia minute, stalked, with an annulus of thick-walled cells on 1 side and borne on the underside of the frond; rootstock developed; fronds more than 1; roots not fleshy. (Leptosporangiatae) .... 1. *Filicales*   p. 17
Plants not particularly fernlike, but floating aquatics or creeping on mud, producing 2 kinds of spores in round to ovoid bony sporocarps near the base of lf.-stalks or on the underside of the branches.
    Lvs. circinate, petioled, the blades sometimes lacking; plants mostly creeping on and rooting in the mud .... 2. *Marsileales*   p. 33
    Lvs. not circinate, sessile, overlapping; plants mostly floating on water . 4. *Salviniales*   p. 34

Order 1. **FILICALES**

Herbaceous to arborescent ferns, the stem varying from a creeping rhizome to an erect trunk. Lvs. of circinate vernation, several to many, sometimes differentiated into sterile and fertile types or portions. Sporangia on the underside of the blades, scattered or in separate clusters (sori). Sporangia leptosporangiate (developed from a single epidermal cell), with wall 1 cell thick and bearing an annulus. Gametophyte mostly a separate minute thallus. Several families and many genera with several thousand spp.

KEY TO FAMILIES

Sporangia borne near or at the margin of the lvs., i.e. near the apex of the veins, or decurrent on the veins or completely covering them .... 5. *Pteridaceae* p. 24
Sporangia not marginal, but borne in separate sori away from the margins.
   Sori round to oval.
      Indusia wanting, the sori naked.
         Stipes jointed to the rhizome; blades pinnatifid .... 4. *Polypodiaceae* p. 23
         Stipes continuous with the rhizome; blades 2-pinnate, then pinnatifid
                                                       1. *Aspidiaceae* (*Athyrium*) p. 17
      Indusia present, although sometimes soon concealed by the maturing sporangia; stipes not jointed to the rhizome .... 1. *Aspidiaceae* p. 17
   Sori oblong, linear to lunate, or horseshoe-shaped.
      Fronds 6–20 dm. tall; indusium often ± curved.
         Venation partly areolate; sori 2–6 mm. long .... 3. *Blechnaceae* p. 22
         Venation entirely open; sori to ca. 1 mm. long .... 1. *Aspidiaceae* p. 17
      Fronds 0.3–2.5 dm. tall, pinnate; indusium straight .... 2. *Aspleniaceae* p. 22

1. **Aspidiàceae**

Mostly terrestrial ferns, with creeping or ascending or erect rhizome, rarely forming a short trunk or scandent, paleate. Stipe rarely articulate; fronds pinnate in plan, simple to decompound, mostly uniform. Sori typically away from the margin, round, sometimes elongate, or the sporangia extending indefinitely along the veins and even over the surface; indusium usually present, fixed beneath the sorus, and opening around the margin, but sometimes peltate, or opening over the sorus, or elongate, sometimes wanting. Annulus longitudinal, of 10–40 thickened cells, interrupted by the pedicel. Spores bilateral, epispore usually present, often conspicuous. Ca. 60 genera, largely trop.

Indusia wanting, the sorus roundish; ultimate segms. of lvs. very narrow. Alpine
                                                                                 1. *Athyrium* p. 18
Indusia present, peltate or reniform.
   The indusium round, centrally attached; lvs. once pinnate, the basal pinnae sometimes pinnate; texture coriaceous.
      Veins free. Native .... 5. *Polystichum* p. 19
      Veins anastomosing. Occasional escape from cult .... 2. *Cyrtomium* p. 18
   The indusium not centrally attached; lvs. at least bipinnate throughout; texture not coriaceous.
      Stipes slender, less than 1.5 mm. in diam.; blade of frond 1–1.5 dm. long and 5–8 cm. wide.
         Indusium under the sorus, with stellate divisions .... 7. *Woodsia* p. 21
         Indusium hoodlike, fixed on 1 side, with a broad base .... 3. *Cystopteris* p. 18
      Stipes coarser, 2–4 mm. in diam.; blade of frond 2.5–5 or more dm. long and 1–2 dm. wide.
         Indusium distinctly reniform and quite circular in outline, attached only along the sinus.
            Blades with distinctly toothed segms., if only 1–2 times pinnate, sometimes 3 times pinnate, veins 2–several times forked .... 4. *Dryopteris* p. 19
            Blades with entire or subentire segms.; veins simple or once forked 6. *Thelypteris* p. 21
         Indusium merely curved, elongate rather than round, attached along the inner side .... 1. *Athyrium* p. 18

1. Athýrium Roth. LADY FERN

Medium to large upright ferns, the rhizome commonly upright or long-creeping, with membranous thin-walled scales. Roots mostly stout. Fronds usually large, the stipe short to long. Blades usually pinnately decompound, sometimes pinnate or simple, mostly glabrous except on the axes; veins typically free. Sori on the surface, typically elongate along 1 or both sides of the veins, the indusium typically curved across the distal end of the sorus, sometimes wanting. Sporangia with slender pedicels, the annulus of 12–20, commonly 16, thickened cells. A large widely distributed genus. (Greek, *a*, without, and *thurium*, shield.)

Indusium lacking; pinnules mostly distant. White Mts., Sierra Nevada .... 1. *A. alpestre*
Indusium mostly crescent-shaped, broadly hooked or horseshoe-shaped; pinnules mostly close. San Jacinto Mts. and Santa Cruz Id. n. .... 2. *A. filix-femina*

1. A. alpéstre (Hoppe) Rylands var. **cyclosòrum** (Ledeb.) Moore. [*A. americanum* Maxon. *Phegopteris alpestris* var. *americana* Jeps.] Rhizomes branched, making massive rounded tufts; scales many, brown, thin, to ca. 1 cm. long; fronds clustered, 2–9 dm. high; stipes short, straw-colored, from a darker base, sparsely chaffy; blades lance-oblong or narrower, 2–6 dm. long, 0.4–2.5 dm. wide, usually twice pinnate, then pinnatifid; pinnae ascending-spreading, acuminate; pinnules delicate, stalked, separated, ± oblong, incised; sori many, round, without indusia.—Meadows and moist places, 5500–12,000 ft.; Montane Coniferous F.; White Mts., Sierra Nevada, n. to Alaska, e. to Colo.

2. A. filix-fémina (L.) Roth var. **califórnicum** Butters. Plate 1, Fig. E. Rhizome stout, with dark brown lanceolate scales to 1 cm. long; fronds 6–15 dm. high, erect-arching, mostly 1–2 dm. broad; stipes short, straw-colored, paleaceous at the dark base; blades 2–3 times pinnate, the pinnae lance-oblong, sessile, ascending-spreading, puberulent beneath along the rachises, not paleaceous; pinnules lance-oblong, crenate-dentate to pinnatifid, largely 0.5–2 cm. long; sori oblong to lunate, usually less than 1 mm. long; indusia toothed or ciliate on the free edge.—Along streams and in meadows, mostly 4000–9500 ft.; Yellow Pine F. to Subalpine F.; San Jacinto and San Bernardino mts., Sierra Nevada, N. Coast Ranges; to Ida., Colo., n. Mex.

Var. **sitchénse** Rupr. [*A. f.* ssp. *cyclosorum* (Rupr.) C. Chr.] Pinnae longer; fronds wider with scattered scales on underside of rachises, not puberulent; pinnules largely 2–3 cm. long.—Wet places near the coast; Santa Cruz Id., San Luis Obispo Co. to Alaska, Quebec; Eurasia.

2. Cyrtòmium Presl.

Near to *Polystichum*, but differing in its anastomosing veins. Several spp., Old World. (Name, Greek, a *bow*.)

1. C. falcàtum (L.f.) Presl. [*Polypodium f.* L. f.] HOLLY FERN. Stiff, erect, the stipes shaggy; fronds dark green, 3–6 dm. long, 1–2 dm. wide, pinnate; pinnae alternate, 7–10 cm. long, ± ovate; $n=82$ (Mitui, 1965).—Natur. at La Jolla, San Diego, Big Dalton C. (San Gabriel Mts.); native of E. Asia, S. Afr., Polynesia.

3. Cystópteris Bernh.

Rather delicate ferns with creeping rhizomes. Fronds erect or ± spreading, the stipes slender, not jointed to the rhizome; blades 1–4-pinnate, delicate, the fertile commonly less leafy and longer-stalked than the sterile. Sori roundish, on the veins, separate; indusium attached to the base of the receptacle and partly under the sorus, soon pushed back by the sporangia and partly hidden, withering, the sori in age seemingly naked. Sporangium roundish, on a slender pedicel, the annulus longitudinal, of 14–16 thickened cells; spores reniform, smooth or muriculate. A small, widely distributed genus. (Greek, *cystis*, bladder, and *pteris*, fern.)

1. C. frágilis (D.) Bernh. [*Polypodium f.* L. *Filix f.* Gilib.] BRITTLE FERN. Plate 1, Fig. F. Rhizome with thin ovate-acuminate scales toward the tip; fronds few to several, the slender stipes 0.5–2 dm. long, the blades 1–2.5 dm. long, mostly lance-oblong to -ovate, acuminate, nearly or quite 2-pinnate, thin; pinnules decurrent on the margined rachis; sori small; indusia roundish or ovate-acuminate, deeply convex; $n=84$ (Manton, 1950). —Mostly moist, ± rocky places, frequently ± shaded; mostly above 3500 ft.; many Plant Communities, cismontane and montane Calif., occasional in desert mts.; to Alaska, Atlantic Coast; Eurasia, trop. Am.

4. Dryópteris Adans. WOOD FERN. SHIELD FERN

Mainly woodland ferns of moderate or large size; rhizome short and stout, ascending to erect or creeping, with broad paleae that are entire or glandular-margined. Fronds borne singly or in a crown, the elongate stipes commonly scaly. Blades bipinnatifid to decompound, uniform, firm in texture, mostly glabrous; veins free, forked. Sori normally on the veins, round, indusiate or not, the indusia round-reniform and attached by the inner end of the sinus. Annulus of 14 or more cells. Spores tuberculate or echinulate. A genus of ca. 150 spp., mostly trop. (Greek, *drys*, oak, and *pteris*, fern).

Pinnae sessile, lance-oblong, the lower basal pinnule with a semicordate base, this overlying the primary rachis; veinlets all ending in salient spinelike teeth. Common .. 1. *D. arguta*
Pinnae mostly short-stalked, deltoid-lanceolate, the basal pinnules symmetrical, not semicordate, not overlying the primary rachis; veinlets usually ending in curved teeth. San Bernardino Mts .... 2. *D. filix-mas*

1. **D. argùta** (Kàulf.) Watt. [*Aspidium a.* Kaulf. *D. rigida* var. *a.* Underw.] Plate 2, Fig. A. Rhizomes stout, short-creeping, woody, with thin attenuate bright brown scales; fronds several, close, erect, 3–8 dm. high; stipes stout, scaly, shorter than the blades; blades lance-ovate to -oblong, acuminate, 2.5–6 dm. long, 1–3 dm. broad, mostly twice-pinnate; pinnules lance-oblong, subcoriaceous, rounded-obtuse, serrate to incise, the teeth often spinelike; sori in 2 rows, large, close; indusia firm, with a deep narrow sinus; $n=41$ (Löve, 1964). Common on shaded slopes and in open woods, mostly below 5000 ft.; many Plant Communities; cismontane from San Diego Co. to Wash ; Santa Barbara Ids.

2. **D. filix-más** (L.) Schott. [*Polypodium f.-m.* L. *Aspidium f.-m.* Sw.] MALE FERN. Fronds 3–10 dm. long; stipes scaly; blades 2.5–8 dm. long, nearly twice-pinnate, the basal pinnules not subcordate, marginal teeth curved, not spinelike; $n=82$ (Manton, 1950).— Known for Calif. from a single collection in 1882, at 8000 ft., Holcomb V., San Bernardino Mts.; Ariz. to B.C., Atlantic Coast; Eurasia.

5. Polýstichum Roth

Rather coarse ferns with mostly short ascending paleate rhizomes, the scales generally lacerate. Fronds several, rigidly ascending or recurved; stipes paleate; blades uniform or dimorphous, pinnate to decompound, the ultimate divisions or teeth usually mucronate, the segms. of harsh texture; veins free. Sori on the veins, round, the indusium peltate or rarely none. Annulus mostly of 18 or more cells. Spores oblong to roundish, usually tuberculate or echinulate. Ca. 175 spp., mainly of temp. and boreal regions. (Greek, *polus*, many, and *stichos*, row, some spp. having many rows of sori.)

Fronds once pinnate, the pinnae variously incised or serrate, never lobed or pinnatifid
1. *P. munitum*
Fronds with at least the lower pinnae pinnately lobed or divided. .... 2. *P. scopulinum*

1. **P. munìtum** (Kaulf.) Presl. subsp. **munìtum.** [*Aspidium m.* Kaulf.] SWORD FERN. Coarse evergreen fern from strong woody suberect very scaly rhizomes; fronds many, sometimes 75–100, rigidly ascending in heavy crowns or clumps, 6–14 dm. high; stipes stout, 1–3 (–6) dm. long, conspicuously paleaceous with large lanceolate chestnut-brown scales mixed with shorter lance-linear ciliate ones; blades pinnate, lanceolate, rather short-acuminate to slender-subcaudate at tip, dark lustrous green above, paler beneath,

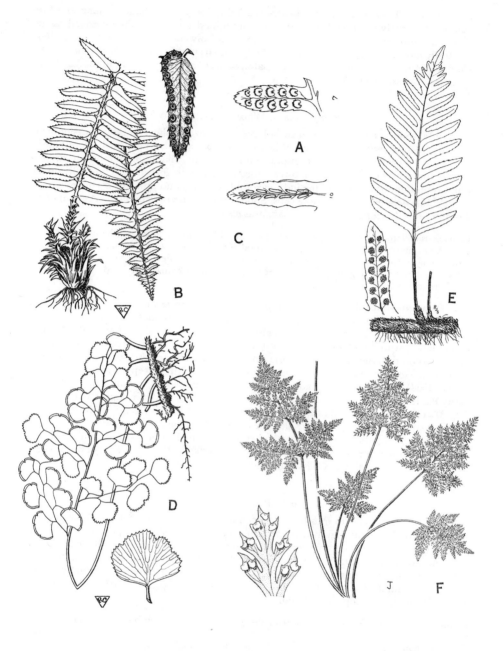

Plate 2. Fig. A, *Dryopteris arguta*, fragment with ± reniform indusia. Fig. B, *Polystichum munitum*, ± peltate indusia. Fig. C, *Woodwardia fimbriata*, indusia-like chain links. Fig. D, *Adiantum jordanii*, reflexed indusium-like margin. Fig. E, *Polypodium californicum*, no indusia. Fig. F, *Cheilanthes californica*, marginal sori.

3-6 (-10) dm. long, 9-16 (-25) cm. broad; pinnae densely and evenly placed, very many, narrow-lanceolate, auriculate, straight or falcate, pungently toothed or incised, the teeth short, firm, bristle-tipped; sori 1-1.5 mm. broad, usually in 2 rows, submarginal; indusia roundish, fringed, irregularly and tardily deciduous; $n = 82$ (Wagner, 1963).—Santa Cruz Id.; Monterey Co. to Alaska, Mont.

Subsp. **ímbricans** (D. C. Eat.) Munz, comb. nov. [*Aspidium m.* var. *i.* D. C. Eat., Ferns N. Am. 1: 188. 1878. *Polystichum m.* var. *i.* Maxon.] Plate 2, Fig. B. Fronds 3-5 dm. long, the stipes with a basal tuft of scales; pinnae crowded, obliquely imbricate, mostly 2-3 cm. long.—Canyon slopes, San Diego Co. to B.C.

Subsp. **cúrtum** Ewan. Rhizomes shorter, very chaffy; fronds fewer, 4-8 dm. long, the more slender stipes 2-3 dm. long, densely paleaceous near base, sparingly so above; blades gradually tapering to tip, 2.5-4 (-5) dm. long, 6-9 cm. wide, usually with lance-oblong pinnae.—Canyon slopes, 1500-8500 ft.; Chaparral to Montane Coniferous F.; Santa Lucia Mts. to mts. of San Diego Co.

2. **P. scopulìnum** (D. C. Eat.) Maxon. [*Aspidium aculeatum* var. *s.* D. C. Eat. *P. mohrioides* var. *s.* Fern.] Rhizome stout, erect or decumbent, paleaceous with light brown or linear to oblong-ovate denticulate scales; fronds 6-10, erect-spreading, 1.5-4 (-6) dm. long; stipes 0.3-1.4 dm. long, stout, grooved, paleaceous especially toward the base; blades linear to lance-oblong, coriaceous, fibrillose-paleaceous beneath, 1-3 dm. long, 2.5-6 cm. wide, pinnate; pinnae many, deltoid-ovate to -oblong, ± pinnately divided at the base, the lobes and teeth oblique, pungent; sori many, large, close, near the middle, in 2 confluent rows; indusia erose-dentate.—Dry crevices and rocky places, 5000-10,500 ft.; Montane Coniferous F.; San Bernardino and San Gabriel mts.; Sierra Nevada to Wash., Utah, Quebec.

6. *Thelýpteris* Schott

Moderate-sized ferns; rhizome short- or long-creeping, or ascending to erect, paleate, the paleae rarely dense, often pubescent. Lf.-blades typically bipinnatifid and usually narrowed toward both ends, rarely more compound, often hairy with simple unicellular hairs; veins free and usually simple and reaching the margin. Sori on lower surface on veins, small, round or rarely elongate. Indusium round-reniform, if present. A large widely distributed genus. (Greek, *thelys*, ♀, and *pteris*, fern.)

Lowest veins truly united and sending an excurrent vein out to the sinus, the next distal
vein also connivent to the sinus; lf.-surface not pubescent .... 1. *T. acuminata*
Lowest veins merely connivent to the sinus; lf.-surface short-pubescent beneath   2. *T. puberula*

1. **T. acumináta** (Houtt.) Morton. [*Polypodium a.* Houtt.] Blades gradually narrowed toward the base, the lowest pinnae scarcely shorter than the middle ones, stipes elongate. —An oriental fern reported by *Morton* as natur. in a nursery (Rosecroft Gardens) near San Diego and to be looked for outside.

2. **T. pubérula** (Baker) Morton. [*Aspidium p.* Baker. *Dryopteris p.* Kuntze. *Lastrea augescens* J. Sm. *D. feei* C. Chr. *T. f.* Moxley. *T. augescens* (Link) M. & J.] Rhizome woody, slender, long-creeping, with few apical rusty linear-attenuate, short-hirsute scales; fronds few, 3-10 dm. long; stipes stout, naked; blades broadly oblong, 3-7 dm. long, 1.5-4 dm. wide, pinnate-pinnatifid, freely puberulous beneath; pinnae spreading, linear-attenuate, cut to ca. 2-3 mm. from the stout midrib; segms. oblique, subfalcate; veins 8-11 pairs, the 2 lower pairs running to the hyaline sinus; sori small, close, supramedial; indusia firm, pilose. —Occasional in wet shaded canyons below 3000 ft.; Chaparral, Creosote Bush Scrub; about Santa Barbara, Santa Monica Mts., s. face of San Gabriel Mts., e. base of San Jacinto Mts.; to L. Calif., Mex.

7. *Woódsia* R. Br.

Rhizome erect, clothed with broad thin paleae. Fronds pinnate or bipinnate, hairy, sometimes also scaly, or glabrescent, numerous, densely clustered; veins free. Sori roundish, separate or ± confluent in age, the indusia basal, breaking irregularly into lacerate divisions, or stellate, the filiform divisions often concealed by sporangia. Sporangia small, globose, the annulus longitudinal, of 18–20 thickened cells; spores ± reticulate. Ca. 40 spp., best developed in China. (Named for J. *Woods*, 1776–1864, English botanist.)

Blades glabrous; indusia divided into hairlike segms.; fronds 1–2 dm. high ..... 1. *W. oregana*
Blades glandular-puberulent; indusia divided into lanceolate segms.; fronds 1–4 dm. high .... 2. *W. scopulina*

1. **W. oregàna** D. C. Eat. Rhizomes short-creeping, rather slender, tufted, densely paleaceous with pale brown often dark-striped scales; fronds many, 0.5–2.5 dm. high; stipes 2–12 cm. high, straw-colored with a brownish base; blades lance-oblong to linear, 5–12 cm. long, glabrous, bright green, delicately herbaceous, with 6–12 pairs of pinnatifid deltoid-oblong pinnae with toothed lobes; sori submarginal.—Rare in dry rocky places, 4000–11,000 ft.; Pinyon-Juniper Wd. to Subalpine F.; Santa Rosa and San Jacinto mts., higher mts. of Mojave Desert; to B.C., S. Dak., Quebec, New Mex.

W. plúmmerae Lemmon differing from *W. oregana* in the indusium having non-ciliate instead of ciliate lobes. It was reported by Dodge (Nova Hedwigia 16: 107. 1964) from Colo. Desert, San Diego Co., *Orcutt*. Since Dodge cited a number of Orcutt collections from mountains of n. L. Calif., one wonders whether this specimen has a faulty label.

2. **W. scopulìna** D. C. Eat. Similar to *W. oregana*, larger, the blades with numerous light green glandular-puberulent pinnae and few to many flat septate hairs; indusial segms. lanceolate.—Occasional in exposed rocky places, 4000–12,000 ft.; San Bernardino Mts., Sierra Nevada, White Mts.; to Alaska, Quebec, Colo., N. Car.

2. Asplenìàceae

Terrestrial, sometimes epiphytic ferns with creeping or suberect paleaceous rhizomes. Stipes not articulate. Fronds simple to decompound. small to large, mostly firm in texture; veins forking, free or anastomosing. Sori elongate, along the veinlets, with elongate indusia attached to the veinlets. Annulus longitudinal, incomplete, commonly of ca. 20 cells. Spores bilateral. Ca. 9 genera

1. *Asplènium* L. SPLEENWORT

Ours smallish ferns, with erect or ascending rhizomes and rigid scales. Fronds pinnate. Stipes and rachises dark. Veins free. Sori oblong or linear, straight, oblique to midrib. Indusium fixed lengthwise by one edge. Ca. 700 spp., cosmopolitan. (Greek, *a*, without, and *spleen*, spleen, once used medicinally.)

1. **A. vespertìnum** Maxon. [*A. trichomanes* var. *v.* Jeps.] Tufted evergreen with ascending or erect rhizome 1–2 cm. long, with linear-acicular scales 2–2.5 mm. long, dark brown; fronds numerous, 5–28 cm. long; stipes purplish-brown, shining; blades linear or lance-oblong, 1–2.5 cm. wide, the rachis fibrillose; pinnae 20–30 pairs, oblong to linear-oblong, deeply crenate, 6–10 mm. long; sori 4–12 on a pinna, with narrow firm crenulate indusium.—Moist shaded rocky places below 3000 ft.; Chaparral, Coastal Sage Scrub, S. Oak Wd.; from Santa Monica and s. face of San Gabriel mts. to San Diego Co.; n. L. Calif.

3. Blechnàceae

Terrestrial ferns, with creeping or erect paleaceous rhizomes. Stipes not articulate. Fronds commonly large and coarse, mostly pinnate or pinnatifid, sometimes more compound. Veinlets branching and anastomosing. Sori on the secondary veins. discrete or united. Indusium open on the costal side, rarely wanting. Sporangia large; annulus longitudinal, interrupted. Spores usually without epispore; bilateral. Eight genera

1. **Woodwárdia** Sm. CHAIN FERN

Coarse rather large ferns with stout erect to short-creeping rhizomes. Fronds several to many in a crown, firm, typically bipinnatifid, with entire or serrulate margins; veinlets anastomosing. Sori borne on the outer horizontal veins of a continuing series of elongate areoles, the indusium elongate, arched. Annulus of 18–24 cells. Spores smooth or flocculose. Ca. 5 spp. of N. Am. and Eurasia. (Named for T. J. *Woodward*, English botanist.)

1. **W. fimbriàta** Sm. in Rees. [*W. chamissoi* Brack. *W. radicans* var. *americana* Hook.] Plate 2, Fig. C. Rhizome woody, the scales lance-attenuate, 1–3 cm. long, glossy, bright brown, entire; fronds almost erect, in a circle, 1–2 m. high; stipes short, straw-colored from a brown base; blades 2–5 dm. wide, ± oblong, pinnate, narrow at base; pinnae deeply pinnatifid, 1–2.5 dm. long, the segms. lanceolate, spinulose-serrate, firm-herbaceous; indusia almost straight, 1.5–6 mm. long; $n=34$ (Manton & Sledge, 1954).—Springy and boggy places in canyons below 5000 (8000) ft.; many Plant Communities; Santa Cruz Id. and cismontane Calif., occasional on desert edge; to B.C., Ariz.

4. Polypodiàceae

Epiphytic, sometimes terrestrial fern with creeping or ascending rhizome bearing broad to setiform scales. Stipes usually articulate. Fronds usually simple to pinnate, usually firm in texture, glabrous to hairy or scaly; venation free or usually reticulate. Indusia lacking, the sori typically round, sometimes elongate along the veins and sporangia sometimes spread over the laminar surface. Annulus usually of 12–14 cells; stomium well developed. Spores bilateral, sometimes tetrahedral, with or without a thin epispore. Ca. 65 genera.

1. *Polypòdium* L. POLYPODY

Fronds uniform, pinnatifid to compound, glabrous or paleate, rarely pubescent. Sori on back of frond, terminal or nearly so on the lowest free veinlet included in an areole, round to elliptical. Ca. 75 spp., mostly of N. Hemis. (Greek, *polus*, many, and *pous*, foot, some spp. having many knoblike places on the rhizome.)

Blades coriaceous or leathery; rhizome glaucous; sori 3–4 mm. broad. Coastal .... 3. *P. scouleri*
Blades not leathery; rhizome not glaucous; sori 1–3 mm. broad.
  Pinnae usually more than 3 cm. long, pointed, with serrate margins. From below 4000 ft. .... 1. *P. californicum*
  Pinnae usually less than 2.5 cm. long, rounded at tips, with entire or crenate margins. At 5000–8500 ft. .... 2. *P. hesperium*

1. **P. califórnicum** Kaulf. [*P. intermedium* H. & A. *P. vulgare* var. *i.* Fern.] Plate 2, Fig. E. Creeping rhizome 5–10 mm. thick, with deciduous deltoid-ovate rusty brown scales 3–7 mm. long; fronds not evergreen, 1–3.5 dm. high; stipes stout, straw-colored, naked, mostly shorter than the blades; blades oblong to narrow-ovate, pinnatifid nearly to the rachis, mostly 0.5–3 dm. long, 0.5–1.5 dm. wide, the segms. membranous, linear-oblong, 3–7 cm. long, veins mostly dark, opaque, 3–5-times forked, ± casually joined and often forming an irregular series of areoles; sori oval, slightly inframedial; $n=37$ or 74 (Lloyd, 1963).—Common winter and spring fern on rocky ledges and moist banks below 4000 ft.; Chaparral, Coastal Sage Scrub, etc.; cismontane s. Calif. n. through the Coast Ranges and Sierra Nevada; Santa Barbara Ids.; L. Calif. Unusually coriaceous specimens from exposed sea bluffs, rocks, etc. are var. **kaulfussii** D. C. Eat.

P. austràle Fée, with ovate fronds with lower pinnae inflexed and 2–8 indurated annulus cells (not 10–21), otherwise like *P. californicum.*—Reported from San Clemente Id.

2. **P. hespérium** Maxon. [*P. vulgare* L. var. *columbianum* Gilbert.] Rhizome ca. 5 mm. thick, densely paleaceous with ovate acuminate scales 3–5 mm. long; fronds rather close, mostly 1–2 dm. long; stipes ca. as long as blade, straw-colored, naked; blades deltoid-

oblong to linear-oblong, pinnatifid nearly to the naked rachis; segms. narrow-oblong to oval, rounded-obtuse, crenate to crenate-serrulate; veins mostly twice forked, translucent; sori round-oval, medial; $n=37$ or 74 (Lloyd & Lang, 1964).—Rare, rock-ledges and crevices, 5000–8500 ft.; Montane Coniferous F.; San Jacinto Mts., San Bernardino Mts., Sierra Nevada; to Alaska, S. Dak., New Mex., L. Calif.

Kiefer reports hybrids between this sp. and both of the others here given.

3. **P. scoùleri** Hook. & Grev. [*P. pachyphyllum* D. C. Eat. *P. carnosum* Kell.] Rhizome woody, 6–10 mm. thick, loosely chaffy, white-pruinose and naked in age, the scales dark brown, ca. 1 cm. long, denticulate; fronds few, 1.5–7 dm. long; stipes stout, naked, shorter than the blade; blades deltoid-ovate, ca. 1–4 dm. long, pinnatisect into 2–14 pairs of linear to narrow-oblong obtuse, spreading, coriaceous crenate segms. 8–20 mm. broad; midribs scaly beneath; veins joined in a series of areoles; sori crowded against the midrib, mostly on the upper segms.; $n=37$ (Manton, 1951).—Mossy logs, cliffs, etc., along the Coast, Santa Cruz Id.; Santa Cruz Co. to B.C.

5. Pteridàceae

Terrestrial ferns with a creeping rhizome or ascending to erect stem, the indument of hairs or paleae. Fronds pinnate in plan, decompound to simple and entire, not articulate to the rhizome. Sori typically marginal and protected by an indusium opening toward the margin, or by a reflexed margin, or elongate along the veins and without an indusium, or covering the whole fertile surface. Annulus sometimes oblique and interrupted, or mostly longitudinal and interrupted, the sporangia opening through a definite stomium. Spores almost always tetrahedral. Ca. 60 genera of temp. and trop. regions.

1. Sporangia following the veins throughout, hence sporangia on the back of the fronds
       6. *Pityrogramma*   p. 31
1. Sporangia borne at or near the apex of the veins, hence near margin of fronds.
    2. Fronds dimorphous, the fertile pinnae very narrow. High montane
       3. *Cryptogramma*   p. 28
    2. Fronds uniform or nearly so, sterile and fertile pinnae almost alike.
        3. Plants large coarse, the fronds mostly 3–15 dm. high; stipes light-colored, not brown; indusium double, the inner minute concealed.
            4. Fronds usually 3 times pinnate in lower part. Common native . 7. *Pteridium*   p. 32
            4. Fronds once pinnate. Natur. in canyons on s. face of San Gabriel Mts.
               8. *Pteris*   p. 32
        3. Plants mainly small, less than 3 dm. high (see *Pellaea*); stipes mostly brown or purplish; indusium single or none.
            5. Reflexed lf.-margin not continuous, appearing as separate large indusia; ultimate segms. of frond at least 1 cm. broad; maidenhair ferns
               1. *Adiantum*   p. 25
            5. Reflexed lf.-margin continuous, or if discontinuous, the ultimate segms. of the frond 1–5 mm. broad.
                6. Foliage nearly or quite glabrous, the inrolled margins quite continuous.
                    7. Lf.-blades deltoid, not much longer than broad.
                        8. Sori solitary, with separate short round-lunate indusia; ultimate lf.-segments 3–7 mm. long .... 2. *Cheilanthes*   p. 25
                        8. Sori contiguous, with a common narrowly linear indusium formed by the inrolled lf.-margin .... 5. *Pellaea*   p. 29
                    7. Lf.-blades narrower, much longer than broad .... 5. *Pellaea*   p. 29
                6. Foliage mostly hairy or scaly, if subglabrous, then with margins not inrolled or the inrolling discontinuous; lf.-blades much longer than wide.
                    9. Sori covered with the inrolled lf.-margin.
                        10. Fronds lanceolate to lance-oblong, not waxy-granular . 2. *Cheilanthes*   p. 25
                        10. Fronds deltoid-pentagonal, waxy-granular beneath . 4. *Notholaena*   p. 28
                    9. Sori naked, the lf.-margin scarcely if at all inrolled ..... 4. *Notholaena*   p. 28

1. Adiántum L. MAIDENHAIR

Delicate ferns with long-creeping or short, paleate rhizomes, the scales usually brown or black, narrow. Fronds distichous or in several ranks, ascending to drooping, with firm dark usually shining stipes; blades simple, 1-3 times pinnate or decompound, mostly broad, mostly firm-herbaceous, the veins free or rarely anastomosing. Sori appearing marginal, the sporangia borne along and sometimes between the ends of the free, forking veins, on the lower side of the reflexed indusiform marginal lobes of the pinnules. Sporangia with an annulus of ca. 18 thickened cells; spores dark, smooth. Ca. 200 spp., largely of trop. Am. (Greek, *a*, without, and *diaine*, unwetted, referring to the shedding of rain drops.)

Blade much longer than wide, not forked, but with a continuous main rachis.
  Indusia distinct, ca. 2 mm. long; segms. wedge-shaped, deeply lobed and with irregular outline .... 1. *A. capillus-veneris*
  Indusia nearly continuous, becoming 8 mm. long; segms. with ± rounded, scarcely lobed margin and regular outline... 2. *A. jordani*
Blade at least as wide as long, divided at base into 2 equal parts, each with several pinnate branches .... 3. *A. pedatum*

1. **A. capíllus-véneris** L. [*A. modestum* Underw.] VENUS-HAIR FERN. Rhizome creeping, slender, the scales thin, light brown, lance-linear, entire; fronds ± spaced, ascending to pendent, 2-7 dm. long, the stipes slender, almost black, to ca. as long as the blades; blades 1-4 dm. long, 2-3 times pinnate at base, the upper third once pinnate, pinnules stalked, obovate to rhombic, etc., 5-30 mm. long, cuneate at base, the outer edge lobed or incised, with toothed margins; sori mostly oblong-lunate, solitary on the lobes, 1-2 mm. long; $n=30$ (Manton, 1950).—Calcareous seeps on rocky walls, etc., mostly below 4000 ft.; many Plant Communities; widely scattered in cismontane s. Calif., occasional on the desert; Santa Cruz and Santa Catalina ids.; warmer regions of both hemis.

2. **A. jórdani** K. Müll. [*A. emarginatum* D. C. Eat.] CALIF. MAIDENHAIR. Plate 2, Fig. D. Slender rhizome rather densely paleaceous with dark brown rigid attenuate scales; fronds several, rather close, 2-5 dm. long; blades ovate, the pinnules rounded, 6-25 mm. broad, entire below, shallowly 2-5-lobed at outer edge, the sori close, 4-8 mm. long.—Damp shaded banks at base of rocks and trees, mostly below 3500 ft.; several Plant Communities; Coast Ranges from San Diego Co. n. to Ore. and foothills of Sierra Nevada; on ids. off s. Calif. coast.

3. **A. pedàtum** L. spp. **aleùticum** (Rupr.) Calder & Taylor. [*A. p.* var. *a.* Rupr.] FIVE-FINGER FERN. Rhizome rather short and thick, the scales brown, lance-oblong to deltoid; fronds close, erect, 2-8 dm. high, stipe dark and chaffy, blades mostly reniform-orbicular in outline, 1-5 dm. broad, forked at base, the ± recurved branches bearing on the outer side several slender spreading pinnate divisions 1-4 dm. long; pinnules close, many, ± oblong, cleft on upper margin; sori linear to oblong-lunate; $2n=58$ (Löve, 1964).—Moist shaded rock-crevices, swampy woods and canyons, sea level to 10,000 ft.; many Plant Communities; San Bernardino and San Gabriel mts.; n. to Alaska, Utah, Quebec; Santa Cruz Id.

2. Cheilánthes Sw. LIP FERN

Low ferns with glandular-pubescent, tomentose, scaly or rarely waxy or glabrous lvs. Fronds uniform, 2-4-pinnate, the segms. usually minute and often beadlike. Sori rounded, distinct or nearly confluent, borne at the enlarged tips of the free veinlets, covered by the inrolled margin of the lf. A rather large genus of warmer parts of the world. (Greek, *cheilos*, margin, and *anthus*, fl.)

1. Indusia not continuous; fronds glabrous or, if hairy, ± glandular-viscid.
  2. Fronds deltoid, not at all viscid .... 1. *C. californica*
  2. Fronds narrowly elongate, ± glandular.
    3. The fronds conspicuously hairy at base of stipe only. Desert species ...... 8. *C. viscida*
    3. The fronds hairy throughout. Cismontane .... 3. *C. cooperae*

1. Indusia continuous; fronds scaly or woolly.
    4. Fronds without scales or coarse fibers. E. Mojave Desert.................. 5. *C. feei*
    4. Fronds with scales or fibers.
        5. Segms. hairy on upper surface.
            6. Rachis with scales; pinnules loosely woolly beneath. W. desert edge .... 7. *C. parishii*
            6. Rachis with coarse fibrils; pinnules densely woolly beneath. W. base of San Jacinto Mts.
        5. Segms. without hairs on upper surface. 6. *C. fibrillosa*
            7. Fronds few, 5–10 mm. apart; segms. mostly subcordate-orbicular, fertile nearly to the base, the subentire margin closely revolute. Cismontane............. 2. *C. clevelandii*
            7. Fronds many, close; segms. oval or irregularly roundish, fertile distally, the ± crenate margins deeply recurved.
                8. Rhizome scales narrow, rigid, with strongly sclerotic walls; scales of blade with deeply cordate base with overlapping basal lobes. San Gabriel, Santa Ana and Laguna mts. to Ariz. .... 4. *C. covillei*
                8. Rhizome scales broader, thinner, pale brown or only partly sclerotic-walled; scales of blade rounded at base or merely subcordate. E. Mojave Desert to Ariz.
                    9. *C. wootoni*

1. **C. califórnica** Mett. [*Hypolepis c.* Hook. *Aspidotis c.* Nutt. ex Copel.] Plate 2, Fig. F. CALIFORNIA LACE FERN. Rhizome scales 2–2.5 mm. long; fronds 0.5–3.5 dm. long, many; stipes dark brown, glossy, wiry, 0.5–2.5 dm. long, nearly naked; blades ± 4 times pinnate, the segms. bright green, glabrous, linear to elliptic, decurrent, 2–3 mm. long; sori solitary at enlarged vein-tips, with round-lunate, ample false indusium adherent to slender saccate marginal tooth at each side.—Common on dry shaded slopes and cliffs below 2500 ft.; many Plant Communities; Coast Ranges, San Diego to Humboldt cos., Kern Co. to Butte Co., Santa Barbara Ids.; n. L. Calif.

2. **C. clevelándii** D. C. Eat. Rhizome slender, creeping, brown-scaly; fronds scattered, 1–4 dm. tall; stipes stout, light brown, scaly; blades oblong to lanceolate, 3–4-pinnate, the segms. close, beadlike, suborbicular, glabrous and green above, with numerous light brown ciliate scales beneath; sporangia in the strongly revolute border.—Frequent in dry rocky places below 5000 ft.; Chaparral, Coastal Sage Scrub; cismontane San Diego Co. to Banning; near Santa Barbara; Santa Cruz and Santa Rosa ids.; n. L. Calif.

3. **C. coóperae** D. C. Eat. Rootstock short, thick, with tufted scales; lvs. tufted, 0.5–2 dm. high; stipes dark brown, they and the blades hairy and ± glandular; blades oblong to wider, 2-pinnate, the pinnules elliptic, cleft, 4–6 mm. long, grayish-green; sori 1 or 2 to a lobe.—Occasional in limestone clefts below 2000 ft.; Coastal Sage Scrub, Chaparral, Foothill Wd.; Slover Mt. near Colton, Piru, Santa Inez Mts.; to Santa Cruz Co., Tulare Co. to Shasta Co.

4. **C. covíllei** Maxon. Plate 3, Fig. A. Rhizome short, creeping, appressed-paleaceous; fronds few to many, 0.5–3 dm. tall; stipes brown or purplish, with small paler scales; blades oblong to deltoid, 3-pinnate, the rachises and lower surface of the segms. densely scaly, the scales imbricate, white to brownish, not ciliate except at base; lf.-segms. green and glabrous above, beadlike, strongly recurved on margins.—Common in rocky places, 1500–9000 ft.; Chaparral, Yellow Pine F., Pinyon-Juniper Wd., etc.; deserts and cismontane s. Calif. to Marin and Mendocino cos., Tulare Co.; to Nev., Ariz., n. L. Calif.

Plants with a few minute stellate scales on upper surface of lvs. and the scales beneath small, many, dark, often reduced to tangled cilia, occur sparingly between 4000 and 8000 ft., in the San Bernardino Mts. They seem referrable to **C. intertéxta** (Maxon) Maxon [*E. covillei* var. *i.* Maxon.] which is known in Coast Ranges from Monterey Co. n. and in the Sierra Nevada from Tulare Co. n.

5. **C. feèi** T. Moore. [*Myriopteris gracilis* Fee. *C. g.* Riehl, not Kaulf.] Rhizome with short scaly branches bearing tufted linear scales 5–7 mm. long; fronds tufted, numerous, 0.5–2 dm. long; stipes slender, brown, ± pilose; blades lance-oblong to narrow-ovate, thinly villous and beneath with pale brown tomentum, usually 3-pinnate, the ultimate segms. gray-green, oval to rounded, simple or lobed, the margins narrowly recurved.—Limestone crevices, 4000–6500 ft.; Pinyon-Juniper Wd., Joshua Tree Wd.; Providence, New York, Clark and Inyo-White mts. of Mojave Desert; to B.C., Tex., n. Mex.

Plate 3, Fig. A, *Cheilanthes covillei*, beadlike scaly segms. Fig. B, *Cryptogramma crispa*, fertile and sterile fronds. Fig. C, *Notholaena californica*, reflexed indusium-like margin. Fig. D, *Notholaena parryi*, beadlike woolly segms.

6. **C. fibrillòsa** Davenp. ex Underw. [*C. lanuginosa* var. *f.* Davenp.] Much like *C. parishii*, the rachis fibrillose, not scaly; blades thrice pinnate, the segms. woolly beneath, subglabrate in age.—Known from a single collection in Chaparral, San Jacinto Canyon, Riverside Co. One wonders whether it might not have been a hybrid.

7. **C. parishii** Davenp. Rhizomes short, with tufted scales 3–4 mm. long; fronds 0.5–1.5 dm. long, the stipes dark brown, fibrillose-paleaceous; blades 3–4-pinnate, lance-oblong, villous above and beneath with long flexuous hyaline hairs, paleaceous on rachises beneath with light brown elongate scales; ultimate lf.-segms. minute, gray-green, with ± recurved margins.—About rocks, Creosote Bush Scrub; Quail Springs (Little San Bernardino Mts.), Andreas C. near Palm Springs, and Sentenac C., e. San Diego Co. Pray finds the spores sterile and believes this to be a hybrid.

8. **C. víscida** Davenp. Rhizome thick, with subglobose divisions and red-brown lanceolate scales; fronds tufted, 1–2 dm. high; stipes slender, brittle, dark brown, viscid-glandular; rachises glandular and hairy; blades viscid, 2–3-pinnate, lanceolate in outline, the lower pinnae distant; pinnules or ultimate segms. incised or toothed, minute, green, not hairy; sori solitary on the ultimate lobes.—Occasional about rocks below 4000 ft.; Creosote Bush Scrub, w. edge of Colo. Desert, Twentynine Palms, Granite Mts. near Victorville, El Paso Range, Darwin, Panamint Mts.; L. Calif.

9. **C. woótoni** Maxon. Rhizome creeping, slender, pale brown, with loosely imbricate oblong-ovate to lance-oblong scales; fronds several, 1–3 dm. long; stipe slender, castaneous; blades oblong, thrice pinnate, the rachises and under surface imbricately brown-scaly, the scales conspicuously long-ciliate especially toward the base.—Dry rocky places, 4000–8000 ft.; Joshua Tree Wd., Pinyon-Juniper Wd.; Panamint, Inyo-White, New York and Providence mts., e. Mojave Desert; to New Mex.

3. *Cryptográmma* R. Br. in Richards.

Rather small ferns with stout ascending rhizomes clothed with thin brown paleae. Fronds crowded, glabrous green, dimorphous, the fertile usually with narrower and longer pinnules than the sterile. Blades 2–3-pinnate; veins free. Sorus submarginal, covering the branches of forked veins, protected by a continuous reflexed margin. Annulus of 20–24 thickened cells; spores usually tetrahedral, hyaline, tuberculate. Five spp., Chile, Eurasia, N. Am. (Greek, *cryptos*, hidden, and *gramme*, line, because of the concealed sori.)

1. **C. críspa** (L.) R. Br. var. **acrostichoides** (R. Br. in Richards.) Clarke [*C. crispa* ssp. *a.* Hult.] ROCK-BRAKE. PARSLEY FERN. Plate 3, Fig. B. Rhizomes in tufts, the scales lance-ovate, 4–5 mm. long; fronds 1–3 dm. high; stipes of fertile lvs. straw-colored, of sterile green; sterile blades 3–12 cm. long, 2–3-pinnate, ovate to lance-ovate, the pinnae few, close, with ultimate segms. crowded, ovate to oblong; fertile blades simpler, the ultimate segms. linear-oblong, 6–12 mm. long, ca. 2 mm. wide, with revolute margins.—Rocky ledges, cliffs, etc., 6000–11,000 ft.; upper Montane Coniferous F.; San Jacinto and San Bernardino mts., Sierra Nevada and White Mts. to Alaska, Labrador.

4. *Notholaèna* R. Br.

Ours small, xerophytic ferns with short or elongate rhizomes. Lvs. glandular, waxy, hairy or tomentose beneath. Blades 1–4-pinnate and variously lobed, narrow to broad. Sori on the veins, near or at their tips, ± or less confluent laterally. Indusia lacking, the margin revolute and ± covering the sporangia. Ca. 60 spp., largely of trop. Am. (Greek, *nothos*, spurious, and Latin, *laena*, cloak, referring to the false indusia.)

(Tryon, R. A revision of the American spp. of Notholaena. Contr. Gray Herb. 179: 1–106. 1956.) Fronds not at all hairy or scaly.
    The fronds covered with a yellowish or white powder, especially beneath; blade deltoid; pinnae close .... 1. *N. californica*
    The fronds not at all powdery; blade lanceolate; pinnae remote .... 2. *N. jonesii*

Fronds hairy or scaly.
  The fronds scaly, once pinnate or bipinnatifid .... 5. *N. sinuata*
  The fronds woolly, bipinnate or tripinnatifid.
    Lvs. averaging 10–15 cm. long; tomentum close; plant not viscid. Cismontane
        .... 3. *N. newberryi*
    Lvs. averaging 7–12 cm. long; tomentum very loose; plant ± viscid. Deserts .. 4. *N. parryi*

1. **N. califórnica** D. C. Eat. [*N. cretacea* auth., not Liebm. *Aleuritopteris c.* auth., not Fourn. *N. californica* ssp. *nigrescens* Ewan.] Plate 3, Fig. C. Tufted plants, the rhizomes dark paleaceous, the scales rigidly acicular, denticulate; fronds 4–12 cm. high, few to many; stipes brown; blades deltoid-pentagonal, 2–5 cm. long, 2–4 cm. wide, 3-pinnate on the large basal pinnae, segms. with in-curved margins, oblong, obtuse, yellowish- or whitish-waxy beneath, the sporangia not concealed.—About rocks and cliffs below 2500 ft.; Coastal Sage Scrub, Creosote Bush Scrub; mts. from n. base of San Gabriel Mts., Victorville, Colton, Palm Springs, Temescal Canyon, Yaqui Well and Beale's Well (Colo. Desert); to n. L. Calif., Ariz.

2. **N. jònesii** Maxon. [*Cheilanthes j.* Munz.] Tufted, with short oblique light brown paleaceous rhizome, the scales linear, long-attenuate; fronds few to many, 3–15 cm. long, with curved red-brown stipes; blades 2–8 cm. long, mostly twice pinnate, the 3–7 pairs of pinnae distant; pinnules entire or crenately lobed, rounded, 2–5 mm. long, glabrous, not at all pulverulent; sporangia in a submarginal band.—Crevices of limestone cliffs, 3500–6000 ft.; Joshua Tree Wd., Pinyon-Juniper Wd., etc.; scattered locations Mojave Desert and Palm Springs region to White Mts., Tulare Co. and Santa Barbara Co.; Utah, Ariz.

3. **N. newbérryi** D. C. Eat. [*Cheilanthes n.* Domin.] COTTON FERN. Rhizomes slender, appressed-scaly; fronds erect, clustered, 0.8–2.2 dm. high, feltlike or white tomentose or at length tawny on the under surface; stipes wiry, purplish; blades 3-pinnate, oblong-lanceolate; pinnae ca. 10 pairs, the lower longest and distant; pinnules crowded, their segms. minute, ca. 1 mm. long, round-ovate; sporangia protruding from the tomentum in age.—Common on dry slopes and walls below 2500 ft.; Chaparral, Coastal Sage Scrub; Ventura Co. and s. base of San Gabriel Mts. to San Bernardino and San Diego cos., San Clemente Id.

4. **N. párryi** D. C. Eat. [*Cheilanthes p.* Domin.] PARRY CLOAK FERN. Plate 3, Fig. D. Rhizomes slender, branched, creeping, appressed-scaly; fronds erect, clustered, 0.8–2.2 dm. long, the stipes wiry, purplish-brown; blades linear- to ovate-oblong, 3–8 cm. long, thrice pinnate, the pinnae 5–9 pairs, ultimate segms. round to ovate-oblong, 2–3 mm. long, closely enveloped in dense light gray or brownish wool; sporangia dark, partly evident in age.—Common under overhanging rocks, etc. below 7000 ft.; Creosote Bush Scrub to Pinyon-Juniper Wd.; deserts from White Mts. s. to Colo. Desert; to Utah, Ariz.

5. **N. sinuàta** (Sw.) Kaulf. var. **cochisénsis** (Goodd.) Weath. [*N. c.* Goodd. *Cheilanthes s.* var. *c.* Munz.] Erect plants 1–3 dm. high, the rootstock short, woody, with reddish, linear, sparsely ciliate scales; fronds tufted, simply pinnate, grayish-green above, grayish-brown and scaly beneath, the pinnae subquadrate, 5–9 mm. long, almost as broad, with 1–2 (–3) pairs of lobes.—Dry limestone slopes and crevices at 3200–5500 ft.; Joshua Tree Wd., Pinyon-Juniper Wd.; Providence and Clark mts., e. San Bernardino Co.; to Ariz., n. Mex.

5. Pellaèa Link. CLIFF-BRAKE

Ours mostly rather small ferns with short or elongate rhizomes and erect mostly glabrous, persistent lvs. Blades 1–4 times pinnate, the rachises usually dark, lustrous; segms. round-oval to linear, small to larger, ± articulate; veins free, not thickened at the tips of the branches. Sori terminal and subterminal, rounded or oblong, often confluent in a submarginal band, ± concealed by the strongly revolute indusiform margin. Ca. 80 spp., mostly temp. (Greek, *pellos*, dusky, referring to the stipes.)

(Tryon, Alice F. A revision of the fern genus Pellaea section Pellaea. Ann. Mo. Bot. Gard. 44: 125–193. 1957.)

Blades once pinnate, the pinnae mostly 2-parted with the upper lobe larger ...... 2. *P. breweri*

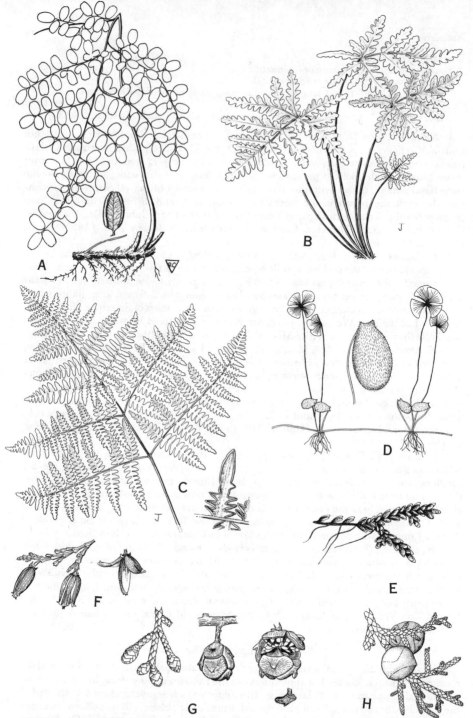

Plate 4. Fig. A, *Pellaea andromedaefolia*, frond segm. with reflexed margin. Fig. B, *Pityrogramma triangularis*. Fig. C, *Pteridium aquilinum*, reflexed indusium-like margin. Fig. D, *Marsilea vestita*, bean-shaped sporocarp. Fig. E, *Azolla filiculoides*, floating, small-lvd. Fig. F, *Calocedrus decurrens*, small ♀ cones. Fig. G, *Cupressus forbesii*, terminal ♂ cones, ♀ cone with peltate scales. Fig. H, *Juniperus californica*, ♀ cone formed by union of fleshy scales.

Blades 2-4 times pinnate.
  Pinnules obtuse, not at all mucronulate, 4-8 mm. wide; rhizomes slender, long-creeping
                                                                          1. *P. andromedæfolia*
  Pinnules mucronulate, narrower; rhizomes stout, short.
    Segms. with opaque whitish borders, usually membranaceous; sporangia long stalked. E.
    Mojave Desert .... 3. *P. longimucronata*
    Segms. entirely green, the border undifferentiated or narrow, lutescent green; sporangia
    short stalked. General .... 4. *P. mucronata*

  1. **P. andromedæfòlia** (Kaulf.) Fee var. **andromedæfòlia**. [*Pteris a.* Kaulf. *Pellaea a.* var. *rubens* D. C. Eat. *P. rafaelensis* Mox.] Coffee Fern. Plate 4, Fig. A. Rhizome slender, long-creeping, with imbricate, narrow, hair-pointed scales; fronds distichous, 1.5-7 dm. long, the stipes glaucous, straw-colored; blades 0.5-2 dm. broad, 3 times pinnate, with glabrous slender rachis; pinnules remote, oblong, many with 5-7 segms., the sterile ones flat, the fertile strongly revolute; sporangia partly hidden; $n=64$, occasionally 32 (Pray, 1968).—Dry rather stony places, mostly below 4000 ft.; many Plant Communities; cismontane s. Calif., n. to Mendocino and Butte cos., n. L. Calif.

  Var. **pubéscens** D. C. Eat. Lvs. less than 3.5 dm. long, bipinnate, a few pinnules ternate, stipe and rachis puberulent; $n = 32, 64$ (Pray, 1968).—Coastal and insular s. Calif.; to L. Calif.

  2. **P. brèweri** D. C. Eat. [*Allosurus b.* O. Ktze.] Rhizome short-creeping, with tufted twisted scales 7-10 mm. long; fronds 0.5-2 dm. long, the stipes glossy, bright brown, transversely corrugate; blades 1.5-3.5 cm. broad, with 6-12 pairs of pinnae that are mostly 2-parted with the upper lobe larger; sporangia terminal and sub-terminal, confluent, nearly hidden by the reflexed margin.—Exposed dry rocky places, 7000-12,000 ft.; Montane Coniferous F.; San Bernardino, Panamint and White mts., Sierra Nevada; to Wash., Ida., Utah.

  3. **P. longimucronàta** Hook. [*P. wrightiana* var. *l.* Davenp.] Rhizome decumbent, multicipital, with appressed brown scales; fronds 12-35 cm. long, stiff, the lower pinnae often sterile, the upper fertile; stipe and rachis glabrous, castaneous; blade 8-22 cm. long, triangular, bipinnate, rarely tripinnate, gray-green, the lower pinnae pinnate with 9-21 segms., these with a narrow, white, crenulate or dentate border, mucronulate; sporangia usually long stalked.—Rocky places, 3000-8000 ft.; Piñyon Juniper Wd., etc.; mts. of e. Mojave Desert; to Rocky Mts.

  4. **P. mucronàta** (D. C. Eat.) D. C. Eat. var. **mucronàta**. [*Allosurus m.* D. C. Eat. *P. ornithopus* Hook.] Bird's Foot Fern. Rhizome thick, woody, with closely tufted castaneous scales; fronds mostly fertile, 1.5-5 dm. long, stiff; stipes dullish, purplish-brown; blades 1-3 dm. long, 2-3 times pinnate, the pinnae distant, nearly perpendicular to the rachis and divided into 6-20 pairs of usually ternate pinnules, or these with as many as 11 linear-oblong to elliptical gray-green segms. 2-6 mm. long, wrinkled, revolute to the middle; sporangia mostly hidden.—Dry rocky places mostly below 6000 ft.; many Plant Communities; nearly throughout cismontane Calif., sparingly on the deserts; n. L. Calif.

  Var. **califórnica** (Lemmon) M. & J. [*P. wrightiana* var. *c.* Lemmon. *P. bella* Baker. *P. w.* var. *compacta* Davenp. *P. compacta* Maxon. *Platyloma bella* T. Moore.] Pinnae ascending at acute angles to the rachis, becoming imbricate toward the apical portion of the frond; pinnules less than 10 pairs per pinna, usually entire.—Mostly above 6000 ft.; Montane Coniferous F.; mts. of s. Calif., Sierra Nevada.

6. *Pityrográmma* Link

  Small to medium-sized ferns with short-creeping to oblique rhizomes covered with dark rigid lance-linear paleae. Fronds crowded, uniform, the stipe scaly at base, dark, polished; blades 1-3 times pinnate, linear to ovate, ± densely waxy-powdered beneath, sometimes glandular above, usually without scales. Sori borne along the veins, confluent, non-indusiate. Annulus of 20-24 thickened cells; spores globose-tetrahedral, dark, irregularly reticulate-ribbed. A rather small genus, largely from trop. Am. (Greek, *pityron*, bran, and *gramme*, line, because of the sori.)

(Weatherby, C. A. Varieties of P. triangularis. Rhodora 22: 113–120. 1920. Alt, K. S. and V. Grant, Cytotaxonomic observations on the goldback fern. Brittonia 12: 153–170. 1960.)

1. **P. triangulàris** (Kaulf.) Maxon var. **triangulàris**. [*Gymnogramme t.* Kaulf. *Gymnopteris t.* Underw. *Ceropteris t.* Underw.] GOLDENBACK FERN. Plate 4, Fig. B. Rhizome stout, with brownish or blackish scales 2–2.5 mm. long; fronds many, 1–4 dm. high, the stipes red-brown when young, older in age, twice as long as the blades, these deltoid-pentagonal, 6–18 cm. long, almost as broad, pinnate or the basal pinnae again pinnate; pinnules oblong, lobed to subentire, yellow-powdery beneath, glabrous above; sori ± covering the backs of the pinnules; $n=30$ or 60 (Alt & Grant, 1960).—Common in ± rocky shaded places below 5000 ft.; many Plant Communities; most of cismontane Calif.; from n. L. Calif. to B.C.

Var. **máxoni** Weath. Stipes red-brown, glossy; blades nearly 3-pinnate below, sparsely yellowish-glandular on upper surface, light yellow or whitish powdery on lower.—Rocky places, Creosote Bush Scrub to Pinyon-Juniper Wd.; e. Mojave Desert, Colo. Desert; to L. Calif., Ariz., Son.

Var. **víridis** Hoover. Backs of lvs. green, with sparse waxy granules.—Santa Cruz Id.; San Luis Obispo Co.

Var. **viscòsa** (Nutt.) ex D. C. Eat. Weath. [*Gymnogramme t.* var. *v.* Nutt. *Ceropteris v.* Underw. *P. v.* Maxon.] SILVERBACK FERN. Stipes red-brown; blades viscid above, with white powder beneath; $n=30$ (Alt & Grant, 1960).—Dry slopes, Coastal Sage Scrub, Chaparral, along the coast, San Diego Co. and n. L. Calif. to Santa Barbara Co.; Channel Ids.

7. *Pteridium* Scop. BRAKE or BRACKEN

Coarse ferns with long-creeping, branched underground rhizomes clothed with hairs. Fronds stout, erect to reclining, with stout stipes having a feltlike covering near the base. Blades pinnately compound, coriaceous, ± densely hairy, triangular to elongate, the ultimate segms. entire to lobed, the veins free except for a marginal strand. Sorus continuous along the lf.-margin, with a double indusium, the outer (false) formed by the reflexed margin, the inner (true) nearly concealed by the slender-stalked sporangia; annulus of ca. 13 thickened cells; spores tetrahedral or nearly so, smooth. Ca. 6–8 widely distributed spp. (Diminutive of *Pteris*, a fern genus.)

1. **P. aquilìnum** (L.) Kuhn var. **pubéscens** Underw. [*P. a.* var. *lanuginosum* of Calif. auth.] Plate 4, Fig. C. Lvs. erect or ascending, 3–15 dm. tall; stipes straw-colored; blades 2–10 or more dm. long, usually 3 times pinnate in lower part; segms. linear-oblong, pubescent to tomentose beneath, sometimes hairy above, mostly 0.5–2.5 cm. long, entire or ± sinuate; indusia narrow, villous; $n=52$ (Manton, 1950).—In moist places at lower elevs., common groundcover in forests at higher ones, up to 10,000 ft.; many Plant Communities; widely distributed in mts., occasional to lower cismontane elevs.; Santa Rosa and Santa Cruz ids.; to Alaska, Mont., Mex.

8. *Ptèris* L.

A large genus of the warmer parts of the world, the plants of medium or large size. Sori on a narrow receptacle connected in a marginal line under a simple indusium formed of the revolute margin of the frond, mostly connecting the ends of the free veins. Our sp. with once-pinnate lvs. (Greek, *pteris*, fern.)

1. **P. vittàta** L. LADDER-BRAKE. Rootstock stout; lvs. dark green, erect or nearly so, clustered, 2–5 dm. long; petioles green, scaly; lfts. firm, lanceolate, ± acuminate; $n=58$ (Kurita, 1963).—Natur. from cult. at ca. 1500 ft., Eaton Canyon, San Dimas Canyon, Big Dalton Canyon, San Gabriel Canyon (along s. base of San Gabriel Mts.). Native of China.

Order 2. MARSILEALES

Characters as given for the family *Marsileaceae*

1. Marsileàceae. MARSILEA FAMILY

Small perennial herbs with slender creeping rhizomes rooting in the mud. Lvs. petioled and with 2- to 4-foliolate blades or filiform and bladeless. Sori borne within hard round to bean-shaped sporocarps which arise from the base of the petioles or near them and bear microspores which produce antheridia and macrospores which produce ♀ gametophytes. Three genera.

Lf.-blades present; plants hairy .... 1. *Marsilea*
Lf.-blades absent; plants glabrous .... 2. *Pilularia*

1. *Marsilea* L.

Plants on mud or in water with floating lvs. Lfts. 4. Sporocarps ovoid or bean-shaped, usually with 2 teeth near the base, 2-chambered vertically and with many transverse partitions, each compartment being a sorus and containing both microsporangia and macrosporangia. Ca. 70 spp., in most parts of the world. (Named for A. *Marsigli*, Italian botanist of the 18th century.)

1. **M. vestìta** Hook. & Grev. Plate 4, Fig. D. Rhizomes long-creeping, densely hairy at the nodes; petioles 2–18 cm. long; lfts. broadly cuneate, hairy, 5–15 mm. long; peduncles distinct from the petiole, short; sporocarps solitary, 4–8 mm. long, at first densely hairy.—Muddy banks, edges of ponds, etc., especially about vernal pools, below 5000 ft.; mostly in Coastal Sage Scrub and V. Grassland; scattered stations from San Diego Co. n.; to B.C.

2. *Pilulària* L. PILL-WORT

Rhizome slender, creeping; lvs. minute, filiform, bladeless. Sporocarps globose, short-peduncled, axillary, longitudinally 2–4 loculed, each locule a sorus with microsporangia and macrosporangia. Ca. 6 widely distributed spp. (Latin, *pilula*, a little ball, referring to the sporocarps.)

1. **P. americàna** A. Br. Lvs. 2–5 cm. long, 1 or more at a node; sporocarps ca. 2 mm. in diam.—Occasional in heavy soil, largely of vernal pools, below 5500 ft.; San Diego region, Elsinore, Menifee V., Upland, Santa Barbara; to Ore., Nebr., Ark., Ga.; Chile.

Order 3. OPHIOGLOSSALES

Characters as given for the family *Ophioglossaceae*

1. Ophioglossàceae. ADDER'S-TONGUE FAMILY

Perennial herbs with a short fleshy rhizome, ± fleshy fibrous roots and 1–several lvs. with a bud containing the undeveloped succeeding lvs. commonly enclosed in the sheathing base of the stalk of the last expanded lf. Lvs. erect (in ours), simple to variously compound, commonly with a sterile blade near the base of which arises a stalked sporebearing spike or panicle. Sporangia naked, developed partly from subepidermal tissue (eusporangiate). Spores all alike (plants homosporous), yellowish. Gametophytes tuberlike, underground. Three genera, widely distributed.

Sterile lf.-blade lobed or divided; fertile portion usually a panicle; venation dichotomous, open
      1. *Botrychium*
Sterile lf.-blade simple; fertile portion a spike; venation reticulate .... 2. *Ophioglossum*

1. Botrýchium Sw. GRAPE FERN. MOONWORT

Rootstock small, erect; bud for following year near base of naked stalk. Fronds 1-3; sterile portion erect or bent down in vernation, stalked or sessile when developed, 1-4 times pinnate; fertile portion a spike to a panicle, the large globose sporangia distinct, borne in 2 rows on the ultimate divisions. Ca. 25 spp., mostly in temp. regions of both hemis. (Greek, *botrus*, a cluster, referring to the groups of sporangia.)

(Clausen, R. T. Mem. Torrey Bot. Club 19: 22-107. 1938.)

Common stalk of lf. almost entirely above ground; sterile blade with the apex decurved in vernation .... 1. *B. lunaria*
Common stalk of lf. ca. ½ to entirely underground; sterile blade erect in vernation (to be sought in the bud for next season) .... 2. *B. simplex*

1. B. lunària (L.) Sw. [*Osmunda l.* L.] Plants rather stout, 4-25 cm. high; common stalk 4-25 cm. high; sterile blade commonly sessile, 1-6 cm. long, half or less than half as wide, pinnately divided, with flabellate sometimes overlapping lobes; fruiting spike 1-8 cm. long; sporangia 0.5-1 mm. in diam.; $n=45$ (Manton, 1950).—Moist grassy places, 7000-10,300 ft.; Montane Coniferous F.; San Bernardino and San Gabriel mts.; Sierra Nevada; to Alaska, Atlantic Coast.

Variable, Calif. plants often largely referred to var. **minganénse** (Victorin) Dole, with narrow sterile blade having remote segms., but the variation not geographic and our plants not closely fitting the variety.

2. B. símplex E. Hitchc. Plants slender, 3-20 cm. high; sterile blade simple, entire or once pinnate, inserted either basally or suprabasally, distinctly stalked, 1-5 cm. long, 0.5-2.5 cm. wide, the segms. cuneate or flabelliform, 0.8-1.2 mm. in diam.; sporangia 0.8-1.2 mm. in diam.; $n=45$ (Wagner, 1955).—Open meadows and damp places, 5000-10,000 ft.; Red Fir F., Lodgepole F.; San Bernardino Mts., Mt. Pinos, Sierra Nevada, to B.C., Atlantic Coast; Eurasia.

Again, variable, some of our material having been referred to var. **compósitum** (Lasch) Milde, with a tendency to ternate blades.

2. Ophioglóssum L. ADDER'S-TONGUE FERN

Rootstock erect, fleshy, short; the bud for the following year by the side of the base of the naked stalk; sterile portion of lf.-blades fleshy, simple; fertile portion a spike with 2 rows of connate sporangia. Perhaps 40 spp., widely distributed in both hemis. (Greek, *ophis*, snake, and *glossa*, tongue.)

1. O. califórnicum Prantl. [*O. lusitanicum* ssp. *c.* Clausen.] Plant 3-13 cm. high, the sterile blade ± fleshy, 0.5-1.5 cm. wide, tending to be acute; fertile part of spike 0.5-2 cm. long, with 8-15 pairs of sporangia.—Vernal pools; near Dan Diego, cent. Calif.; n. L. Calif., reported by Beauchamp from Lee Valley and Jamal Mt., San Diego Co.

Order 4. SALVINIALES

Characters as given for the family *Salviniaceae*

1. Salviniàceae. SALVINIA FAMILY

Floating aquatic plants, with ± elongate and sometimes branching axis bearing apparently 2-ranked lvs. Sporocarps soft and thin-walled, 1-loculed, each containing a cent. receptacle bearing microsporangia or macrosporangia. Two genera.

1. Azólla Lam.

Small mosslike plants with pinnately branched stems covered with minute imbricate 2-lobed lvs. producing rootlets beneath. Sporocarps of 2 kinds, borne in pairs beneath the stem, the small ones ovoid and bearing 1 macrospore, the larger globose and containing masses of microspores. Six spp., widely distributed. (Greek name of doubtful origin.)

1. **A. filículoides** Lam. DUCKWEED FERN. Plate 4, Fig. E. Plants compact, often reddish, 1–2.5 cm. long, easily breaking apart; lvs. ovate, 1 mm. long, the dorsal lobe papillate-hairy, the ventral submersed and forming the sporocarps.—Frequent in sluggish water at low elevs., cismontane Calif., occasional on the desert as on Mojave R.; to Wash., Ariz., S. Am.

Subdivision III. PTEROPSIDA

Class 2. CONIFERAE. CONE-BEARING PLANTS

Woody plants with scaly to needlelike or broader lvs. Monoecious or occasionally dioecious, the ♂ structures (microsporophylls) usually in short-lived strobili and bearing the sporangia on their abaxial surface. Ovulate structures in cones or not, the ovule or ovules borne terminally and singly, or more commonly adaxially on cone-scales, hence no development of a stigma. Seeds thus not inclosed in a seed-vessel, but released when the cone-scales pull apart. An ancient division with some extinct fossil groups and ca. 500 living spp.

Order 1. CONIFERALES

Ovulate cone usually dry, ± woody, more fleshy in *Juniperus*, but the constituent ± fused ovuliferous scales evident even there. Several families.

KEY TO FAMILIES

Lvs. scalelike, mostly decurrent, thickly clothing the branches .... 1. *Cupressaceae*
Lvs. needlelike or linear, not scalelike .... 2. *Pinaceae*

1. Cupressàceae. CYPRESS FAMILY

Monoecious or dioecious trees or shrubs, with opposite or whorled scalelike or sometimes linear lvs. that are decurrent and thickly clothing the branchlets or sometimes are jointed at the base. Staminate cones small, terminal on the branchlets or axillary; stamens with mostly 3–6 pollen-sacs (microsporangia) attached to the lower half of the thin shield-like expanded portion. Ovuliferous scales paired or whorled with 1–many erect ovules near the base. Cones woody or fleshy, the scales shield-shaped or imbricate; seeds often angled or winged; cotyledons 2–several. Ca. 20 genera of both hemis.

Fr. a woody cone; lvs. scalelike, decussate.
    Cones and their 6 scales oblong, imbricated or valvate, maturing the first year, seeds 2 to each scale .... 1. *Calocedrus*
    Cones subglobose, their scales shield- or wedge-shaped, maturing in 1 or 2 years; seeds many on each scale .... 2. *Cupressus*
Fr. berrylike, formed by the fusion of the scales; ovules 1–2; lvs. in 2's or 3's .... 3. *Juniperus*

1. *Calocèdrus* Kurz. INCENSE-CEDAR

Evergreen trees with distichous strongly compressed branchlets forming flat sprays. Lvs. scalelike, 4-ranked, closely imbricate, decussate, decurrent, the pairs of ca. equal length. Monoecious, the ♂ cones of 6 decussate scales each with 4 pollen-sacs on the underside. Ovulate cones oblong, ± truncate, maturing the first year, the scales in 3 pairs, woody, imbricate, mucronulate on the back near the tip, the lower pair small, sterile, connate into a flat woody plate. Seeds 2 on each fertile scale, compressed, with 2 unequal lateral oblong wings; cotyledons 2. Ca. 3 spp. of e. Asia, w. N. Am. (Greek, *kalos*, beautiful, and *kedros*, cedar.)

1. **C. decúrrens** (Torr.) Florin. [*Thuja d.* Torr. *Libocedrus d.* Torr. *Heyderia d.* K. Koch.] Plate 4, Fig. F. Forest tree, 25–35 (–50) m. high, aromatic, with a straight conical trunk from a broad base, the lower branches curved downward, the upper erect; crown conical; bark 1–2.5 cm. thick, cinnamon-brown, fibrous; branchlets flattened, often vertically placed; lvs. light green, 3–10 mm. long; ♂ catkins yellow, 5–6 mm. long; ♀ cones pendulous, 2–2.5 cm. long; seeds 8–10 mm. long.—Mt. slopes and canyons, 4500–8200 ft.; Yellow Pine F.; mts. from n. L. Calif. to Ore., w. Nev. Wood light, soft, close-grained, durable in the ground, usually with cavities due to dry rot; used for shingles, posts, lead pencils, railroad ties.

2. Cupréssus L. Cypress

Evergreen trees, sometimes shrubby, ± resinous, with fibrous or exfoliating bark, stout branches and naked buds. Lvs. scalelike, ovate, acuminate to blunt, opposite, thickened, rounded, often glandular on the back, with appressed or spreading slender tips. Staminate and ♀ cones on separate branchlets of same tree, the former mostly 3–5 mm. long, with 10–20 triangular opposite scales, each with 3–10 pollen-sacs. Ovulate cones solitary at tips of short branchlets, numerous and appearing clustered, maturing at end of second season, globose to oblong, with much thickened peltate scales with cent. boss or umbo, separating at maturity or remaining closed for years. Seeds many on all but upper and lower pairs of scales, often 100 per cone, irregular in shape, with narrow thin wing; cotyledons 3–5. Perhaps 25–30 spp. of w. N. Am., s. Eu., w. Asia. Widely cult. (Classical name of the cypress.)

(Wolf, C. B. Taxonomic and distributional studies of the New World Cypresses. Aliso 1: 1–250. 1948.)

Dorsal surface of mature lvs. usually lacking an active gland or pit; foliage green, not grayish
                        1. *C. forbesii*
Dorsal surface of mature lvs. usually with an active gland or pit; foliage blue-gray or gray-green
                        2. *C. stephensonii*

1. **C. fórbesii** Jeps. TECATE CYPRESS. Plate 4, Fig. G. Small tree, usually less than 10 m. tall, with an irregular spreading crown and exfoliating mahogany-brown or cherry-red bark; foliage light rich green to ± dull green; lvs. 1.5 mm. long, rounded or ridged on back; ♂ cones 3–4 mm. long, mostly with 12–14 scales; ♀ cones dull brown or gray, to 3 cm. in diam., the umbos inconspicuous, the cones not opening up for many years; seeds rich dark brown, 5–6 mm. long.—Dry slopes, 1500–5000 ft.; Chaparral; Claymine and Gypsum canyons, Santa Ana Mts. in Orange Co., Otay Mt. and Mt. Tecate, San Diego Co.

2. **C. stephensònii** C. B. Wolf. [*C. arizonica* var. *s.* Little.] CUYAMACA CYPRESS. Tree 10–16 m. high, the thin bark cherry-red, smooth, exfoliating; foliage blue-gray or gray-green; lvs. ca. 1 mm. long, with active gland; ♂ cones 2–4 mm. long; ♀ cones to ca. 2.5 cm. in diam.; seeds 100–125, 5–8 mm. long.—Dry slopes at 3000–4500 ft.; Chaparral; headwaters of King Creek, sw. slope of Cuyamaca Peak, San Diego Co.

3. Juníperus L. Juniper

Aromatic trees or shrubs with thin shredding bark and opposite or whorled (in 3's) lvs. that are needlelike and spreading or scalelike and appressed. Plants dioecious or sometimes monoecious. Staminate cones minute, solitary or clustered, with ovate or peltate scales bearing 3–6 pollen-sacs. Ovulate cones ovoid to subglobose, berrylike by the union of the 2–6 fleshy scales, some or all of which bear 1–3 erect ovules. Seeds wingless, ovoid, terete or angled, brown, with a large 2-lobed hilum; cotyledons 2–6. Ca. 60 spp. of N. Hemis. (Ancient classical name.)

Mature fr. red or reddish brown, with sweet dry pulp; shrubby 1–6 m. tall.
 Lvs. mostly in 3's, glandular-pitted on the back, rounded at apex; berry 12–18 mm. long
                            1. *J. californica*
 Lvs. mostly in 2's; berry 6–9 mm. long . . . . 3. *J. osteosperma*
Mature fr. blue or blue-black, with resinous pulp; tree 8–20 m. tall . . . . 2. *J. occidentalis*

1. **J. califórnica** Carr. [*Sabina c.* Antoine.] CALIFORNIA JUNIPER. Plate 4, Fig. H. Arborescent shrub mostly 1–4 m. high, with stout irregular stems and a broad erect but fairly open head, rarely a small tree to ca. 10 m. tall; bark ashy-gray; branchlets stiff; lvs. scalelike, closely appressed, slightly keeled and conspicuously glandular-pitted on back, cartilaginously fringed on margins, bluntly pointed, 3–4 mm. long; berries at first bluish with a dense bloom, later reddish-brown beneath the bloom, oblong-ovoid, 12–18 mm. long, nearly smooth; seeds 1–2, sharp-pointed, angled, 6–9 mm. long, light brown and shining except for the dull yellowish base; cotyledons 4–6.—Dry slopes and flats mostly below 5000 ft.; Pinyon-Juniper Wd., Joshua Tree Wd., Chaparral; desert slopes from w. edge of Colo. Desert and Joshua Tree National Monument to Kern Co.; interior cismontane s. Calif., to s. Sierra Nevada and inner Coast Ranges.

2. **J. occidentàlis** Hook. subsp. **austràlis** Vasek. WESTERN JUNIPER. Tree mostly 8–20 m. tall, with a well defined trunk 1 (–2) m. thick, with red-brown shreddy bark; branches often large and spreading; lvs. in 3's, closely appressed, rounded and gland-pitted on the back; mostly dioecious; berries rounded to oblong-ovoid, 6–8 mm. long, blue-black at maturity, very glaucous; seeds 2–3, ovoid, acute, rounded and grooved or pitted on the back, ca. 6 mm. long; cotyledons 2–4.—Dry slopes and flats, to 10,000 ft.; Montane Coniferous F.; San Bernardino Mts. and e. San Gabriel Mts. n. through the Sierra Nevada to Lassen Co. Var. **occidentàlis** occurs from there to Wash.

3. **J. osteospérma** (Torr.) Little. [*J. tetragona* var. *o.* Torr. *Sabina o.* Antoine. *J. californica* var. *utahensis* Engelm. *J. u.* Lemmon.] UTAH JUNIPER. Arborescent shrub or small tree, mostly 3–6 m. tall, forming rounded clumps or crowns; branchlets stiff; lvs. without glands or inconspicuously glandular, acute to acuminate, 2–3 mm. long; fr. 6–9 mm. long, rounded or oblong, red-brown beneath the bloom when mature; seeds 1–2, ovoid, strongly angled, ca. 7 mm. long; cotyledons 4–6.—Dry slopes and flats, 4800–8500 ft.; Pinyon-Juniper Wd.; mts. of e. Mojave Desert to Bridgeport region, Mono Co.; to sw. Ida., sw. Wyo., and w. New Mex.

2. Pinàceae. PINE FAMILY

Resinous, mostly evergreen trees, sometimes shrubs, with needlelike lvs. spirally arranged, solitary or fascicled. Stamens and ovules in different cones on the same tree. Staminate cones with many spirally arranged stamens (microsporophylls), each having 2 pollen-sacs beneath, the cones drying and soon dropping after anthesis. Ovulate cones with spirally arranged scales, each of which is subtended by a bract, the cones maturing in 1 or 2 or 3 years and becoming ± woody; ovules naked, 2 at the base of each scale on its upper side, developing into seeds usually bearing a terminal wing made of tissue from the surface parts of the ovuliferous scale. Ca. 9 genera and 200 spp.

Cones erect on the branch, the scales falling separately at maturity; lvs. solitary, blunt, flat in cross section; terminal buds blunt, enveloped in resin .... 1. *Abies*
Cones reflexed or pendulous, the scales persistent; lvs. solitary or fascicled, acute to sharp at apex; terminal buds mostly pointed, not enveloped in resin.
    Cone maturing in 2–3 seasons; lvs. fascicled, needle-like, in axillary 1–5-lvd. clusters, inclosed at the base in a membranous sheath .... 2. *Pinus*
    Cone maturing in 1 season; lvs. scattered, linear .... 3. *Pseudotsuga*

1. Àbies Mill. FIR

Pyramidal evergreen trees, the old bark thick and furrowed. Lvs. entire, sessile, flattened and usually grooved above, keeled beneath with 2 pale stomatic bands, spirally arranged but often appearing 2-ranked, or frequently curving upward; abscission scar smooth. Cones from axillary winter buds, the ♂ borne on under side of the rather stiff branches of upper half of tree. Ovulate cones borne on topmost branches, erect, with numerous 2-ovuled, imbricated scales falling away separately. Seeds with large thin wing. Ca. 40 spp. of N. Hemis. (Classical Latin name.)

1. A. cóncolor (Gord. & Glend.) Lindl. [*Picea c.* Gord. & Glend. *P. lowiana* Gord. *A. c.* var. *l.* Lemmon.] WHITE FIR. Tree 15–40 or more m. tall, with rather narrow spirelike crown and short stiff branches; branchlets nearly glabrous; lvs. 3–6 cm. long, bluish green and stomatiferous above, with pale stomatiferous bands beneath separated by a median keel; ♂ cones usually reddish; ♀ cones 7–12 cm. long, greenish or becoming brown; seeds ca. 1–12 mm. long; $n=12$ (Sax & Sax, 1933).—Common on dry slopes mostly above 6000 ft., mostly Yellow Pine F.; mts. of San Diego Co. to n. Calif., Ore., s. Rocky Mts., L. Calif. Wood of second grade quality, used for packing cases, etc.

2. Pinùs L. PINE

Ours trees with furrowed or scaly bark. Lvs. needlelike, borne in clusters of 1–several on deciduous spurs in the axils of scalelike primary lvs. and enclosed in the bud by numerous scales lengthening and forming a sheath at the base of each cluster. Staminate cones of many overlapping stamens. Ovulate cones becoming woody, composed of many ovuliferous scales with 2 ovules on each upper surface. Each scale having a ± thickened exposed part (the apophysis) with a terminal or dorsal brown protuberance or elevation (umbo). Seeds usually obovoid, usually winged at tip; cotyledons 3–18. Ca. 80 spp. of N. Hemis. (Classical Latin name.)

(Shaw, G. R. The genus Pinus. Pub. Arnold Arboretum No. 5: 1–96. 1914.)
Sheaths of the lf.-clusters deciduous in the first year; base of the bracts of the lf.-bearing spurs not decurrent; needles with 1 cent. vascular strand. (Subgenus Haploxylon.)
  Lvs. 5 in a cluster.
    The cones with dorsal umbos armed with slender prickles, the cones 7–9 cm. long; needles 25–35 mm. long. Panamint and Inyo-White mts. .... 7. *P. longaeva*
    The cones with terminal unarmed umbos on the back of the scales.
      Cones with short or almost no stalks, the cone-body 8–25 cm. long; needles 2.5–7 cm. long. Mts. bordering the deserts .... 4. *P. flexilis*
      Cones with stalks 5–8 cm. long, the cone-body 25–40 cm. long; needles 5–10 cm. long. Montane .... 6. *P. lambertiana*
  Lvs. 1–4 in a cluster.
    The lvs. commonly in 1's or 2's and more than 1 mm. wide, ±grayish-green, not very glaucous.
      Lvs. mostly single, terete. Common in desert mts. .... 8. *P. monophylla*
      Lvs. mostly in 2's, semiterete, deeply channeled. Rare, Little San Bernardino and New York mts. .... 3. *P. edulis*
    The lvs. commonly in 4's, ± glaucous, scarcely 1 mm. wide. W. edge of Colo. Desert
                                                                                    12. *P. quadrifolia*
Sheaths of the lf.-clusters persistent; base of the spur-bracts decurrent; needles with 2 vascular strands. (Subgenus Diploxylon.)
  Lvs. mostly 2 in a cluster; cones 3–7 cm. long.
    The cones subterminal, 3–5 cm. long, almost symmetrical, opening when mature, deciduous; needles 3–5 cm. long. High montane .... 10. *P. murrayana*
    The cones lateral, 5–8 cm. long, asymmetrical, persistent and remaining closed for some years; needles 10–15 cm. long. Coastal.
      Scale-tips of young ♀ cones reflexed; mature cones with tips of scales produced into a strongly recurved hook; cones reflexed .... 9. *P. muricata*
      Scale-tips of young ♀ cones erect; mature cones with tips of scales plane or slightly rounded; cones at right angles to the branch .... 13. *P. remorata*
  Lvs. more than 2 in a cluster.
    Needles 5 in a cluster, gray-green, 20–30 cm. long; cones 10–15 cm. long. S. coast
                                                                                    15. *P. torreyana*
    Needles mostly 3 in a cluster.
      Cones subterminal, symmetrical, opening at maturity, usually deciduous above the basal scales persistent on the branch.
        Lvs. dull gray-green; branchlets glaucous; cones 15–35 cm. long .... 5. *P. jeffreyi*
        Lvs. yellow-green; branchlets not glaucous; cones 7–15 cm. long ...... 11. *P. ponderosa*

Cones lateral.
The cones asymmetrical (the outer scales much enlarged), remaining closed and attached to the branches for years; needles 8–17 cm. long .... 1. *P. attenuata*
Cones ± symmetrical, opening at maturity, deciduous; needles 15–30 cm. long.
    The cones oblong-ovoid, the seeds longer than their wings; lvs. gray-green, drooping 14. *P. sabiniana*
    The cones oblong-conic, the seeds shorter than their wings; lvs. blue-green, erect 2. *P. coulteri*

1. **P. attenuàta** Lemmon. [*P. tuberculata* Gord. *P. californica* Hartw., not H. & A.] KNOBCONE PINE. Tree, 2–12 (–15) m. tall, forming a straggling crown in age; old bark thin, with low ridges and loose scales; lvs. in 3's, slender, stiff, pale yellow-green, 8–17 cm. long, with evident rows of stomates; ♂ cones ca. 12 mm. long; ♀ cones short-stalked, often whorled, remaining closed and persisting for years, light brown, narrow-ovoid, asymmetrical, the scales on the outer side enlarged into prominent knobs with thick flattened incurved prickles; seeds dark, 5–7 mm. long.—Dry barren or rocky places below 4000 ft.; Chaparral, Foothill Wd.; widely scattered from Santa Ana Mts., San Bernardino Mts., Santa Ynez Mts. etc. to n. Calif.; s. Ore.

2. **P. coùlteri** D. Don. [*P. macrocarpa* Lindl.] COULTER PINE. Tree 12–25 m. tall, with broad pyramidal or asymmetrical crown; bark blackish brown, deeply broad-ridged, with thin scales; branchlets stout, rough, becoming dark in age; lvs. in 3's, stiff, stout, deep blue-green, 15–30 cm. long, with many stomatal bands, persisting for 3–4 years; ♂ cones ca. 2.5 cm. long; ♀ cones oblong-ovoid, 20–30 cm. long, hanging on the trees several years after releasing the seeds; scales with prominent stout incurved flattened spurs; seeds dark chestnut-brown, 12–18 mm. long, the wings 25–30 mm. long.—Dry rocky slopes, 4000–7000 ft.; Yellow Pine F., Foothill Wd.; mts. from n. L. Calif. to inner Coast Ranges of Contra Costa Co. Wood light, soft, not strong, brittle, coarse-grained, used for second-class lumber.

3. **P. édulis** Engelm. [*Caryopitys e.* Small. *P. cembroides* var. *e.* Voss.] PINYON. Differing from *P. monophylla* chiefly in 2- or 3-lvd. clusters and perhaps scarcely specifically distinct.
    At 4200–7000 ft., with *P. monophylla*, apparently as relict stands; Little San Bernardino Mts., New York Mts.; to Wyo., Tex., n. Mex. Seeds an important food.

4. **P. fléxilis** James. [*Apinus f.* Rydb.] LIMBER PINE. Tree 10–20 m. high, with a short trunk and ultimately broad crown; bark at first smooth, grayish white, later dark and broken into brown plates covered with thin scales; branchlets, at first finely pubescent; lvs. in 5's, stout, erect, stiff, dark green, with 1–4 rows of stomates on each side; ♂ cones reddish, ca. 1 cm. long; ♀ cones subcylindric, 8–25 cm. long, light brown, ± purplish-tinged, the scales thickened and often incurved at apex; seeds dark red-brown, mottled with black, 10–12 mm. long, the narrow wings usually remaining on the scale; $n=12$ (Sax & Sax, 1933).—Dry slopes and ridges, 7500–11,000 ft.; Lodgepole F., Subalpine F., Bristlecone F.; Santa Rosa and San Jacinto mts., n. through s. Calif. mts. (including desert ranges) to B.C. and Rocky Mts.

5. **P. jéffreyi** Grev. & Balf. in A. Murr. [*P. ponderosa* var. *j.* Balf. ex Vasey. *P. deflexa* Torr.] JEFFREY PINE. Tree 20–60 m. tall, with long symmetrical crown and darker bark than in *P. ponderosa*, often in narrow plates, the inner surface of the scales creamy pinkish- or chocolate-brown, the scales hard, without resin pits; odor in the deep furrows rather strong, pleasant, ± vanilla-like; lvs. in 3's, ± blue-green, 12–28 cm. long, persisting 5–8 years; with plainly marked rows of stomates; ♂ cones 20–35 mm. long; ♀ cones subterminal, short-stalked, long-oval, 15–25 cm. long, the scales stout, almost vertical to the cone-axis, with long mostly deflexed prickles, the points seldom protruding outward; seeds 10–13 mm. long, mottled, the wings to 3 cm. long; $n=12$ (Sax & Sax, 1933).—Dry slopes, mostly 6000–9000 ft. Montane Coniferous F.; n. L. Calif. though the main (non-desert) ranges to s. Ore. Hybridizing with *P. ponderosa* and *P. coulteri*.

6. **P. lambertiàna** Dougl. [*Strobus l.* Mold.] SUGAR PINE. The largest of our pines, 20–75 m. tall, the crown open and narrow when young, later flat-topped; bark smooth, dark green when young, later divided into platelike ridges with loose reddish-brown

scales; branchlets pubescent when young; lvs. in 5's, slender, 7–10 cm. long, with several rows of stomates on both sides; ♂ cones yellow, 8–10 mm. long; ♀ cones cylindric, 25–45 cm. long, on stalks 5–8 cm. long, the scales with thin tips and terminal scarlike umbos; seeds dark, 9–12 mm. long, the wings twice as long; $2n = 24$ (Mehra & Khoshoo, 1948).—Common forest tree, 4500–9000 ft.; Montane Coniferous F.; mts. from n. L. Calif. to Ore. Wood light, soft, straight-grained, of high commercial value.

7. **P. longaèva** Engelm. [*P. aristata* Engelm. *P. balfouriana* var. *a.* Engelm.] BRISTLECONE PINE. Heavy bushy trees, 6–16 m. tall, sometimes with very thick trunks; bark thin, red-brown, with flat irregular ridges and appressed scales; branchlets light orange color, glabrous or at first puberulous; lvs. in 5's, slender, 2.5–3.5 cm. long; ♂ cones 10–12 mm. long; ♀ cones almost sessile, ovoid, 7–9 cm. long, dark purplish-brown, the scales thickened and ridged, with a slender incurved prickle ca. 6 mm. long; seeds light brown, with darker mottling, ca. 8 mm. long, the wings slightly longer.—Dry rocky slopes, 7500–11,500 ft.; Bristlecone Pine F.; Inyo-White, Panamint, Funeral, Grapevine mts.; to Colo., New Mex.

8. **P. monophýlla** Torr. & Frém. [*P. fremontiana* Endl. *P. edulis* var. *m.* Torr. *P. cembroides* var. *m.* Voss. *Caryopitys m.* Rydb.] ONE-LVD. PINYON. A low tree, 5–15 m. tall, usually with divided trunk, ± flat-topped in age; bark with narrow flat ridges; lvs. mostly in 1's, rigid, incurved, pale gray-green, 2.5–3.5 cm. long; ♂ cones ca. 6 mm. long; ♀ cones subglobose to broadly ovoid, 3.5–5.5 cm. long, brown, the scales 4-sided, knobbed at tip; seeds ca. 15 mm. long, the wings somewhat shorter.—Dry rocky slopes and ridges, 3500–9000 ft.; Pinyon-Juniper Wd.; mts. bordering and in the Mojave Desert, to Mono Co. and w. base of Sierra Nevada; to Utah, Ariz., n. L. Calif. Seeds important for food.

9. **P. muricàta** D. Don. [*P. edgariana* Hartw.] BISHOP PINE. Plate 5, Fig. A. Slender tree, 10–20 m. high; old bark with narrow rounded ridges; lvs. in 2's, dark yellow-green, 8–15 cm. long; cones asymmetrical, deflexed, in whorls, the scales prolonged into prominent knobs, each armed with a stout elongate spine.—On Santa Cruz Id. and along coast from Santa Barbara Co. to Humboldt Co. The wood is light, strong, hard, coarse-grained; occasionally used for lumber.

10. **P. murrayàna** Grev. & Balf. in A. Murr. [*P. contorta* var. *m.* Engelm. *P. c.* ssp. *m.* Critchfield.] LODGEPOLE PINE. TAMARAC PINE. A tree 15–40 m. tall, slender and straight when in close stands, heavier and more open in older and spaced trees; bark thin, close and firm, light brown, with small thin scales; branchlets slender; lvs. in 2's, yellow-green, 3–6 cm. long; cones ovoid, 3–5 cm. long, brown, opening at maturity, the scales thin, concave, with slender prickles.—Borders of meadows and on ridges and slopes above 7000 ft.; San Jacinto, San Bernardino and San Gabriel mts.; n. to Alaska; n. L. Calif. Wood light, pitchy, soft, close- and straight-grained. Questionably distinct as a sp. from *P. contorta* Dougl. of the north coast.

11. **P. ponderòsa** Laws. [*P. benthamiana* Hartw.] YELLOW PINE. PONDEROSA PINE. Long-lived tree 15–70 m. high, the branches short, often pendulous, generally turned up at the ends; bark on old trees often 6–10 cm. thick and separated into broad yellow-brown to reddish plates with small soft concave scales whose smooth inner surface is yellow and with small dark resin-pits; odor in the furrows resinous; branchlets of season yellow-green, not at all glaucous; lvs. yellow-green, in 3's, lasting ca. 3–5 years, 12–25 cm. long, ± scabrous on margins, the rows of stomates scarcely distinguishable; ♂ cones 2–3 cm. long; ♀ cones subterminal, subhorizontal, oval, mostly 7–15 cm. long, the spreading scales with short prickles protruding outward; seeds 6–7 mm. long, dark purple, ± mottled, the wings 4–5 times as long; $n = 12$ (Sax & Sax, 1933).—Forming large parklike forests at 4500–8000 ft.; Yellow Pine F.; mts. from San Diego Co. n. to the Coast Ranges and Sierra Nevada; to B.C. The wood is hard, strong, rather fine-grained and much used in construction.

12. **P. quadrifòlia** Parl. ex Sudw. [*P. parryana* Engelm., not Gord. *P. cembroides* var. *p.* Voss.] FOUR-LVD. PINYON. Much like *P. monophylla* in habit, the young branchlets pubescent; lvs. mostly 4, pale blue-green on back, with conspicuous rows of stomates on inner surfaces; seed 15 mm. long, the wings ca. 3 mm.—Dry slopes, 3500–5500 ft.; Pin-

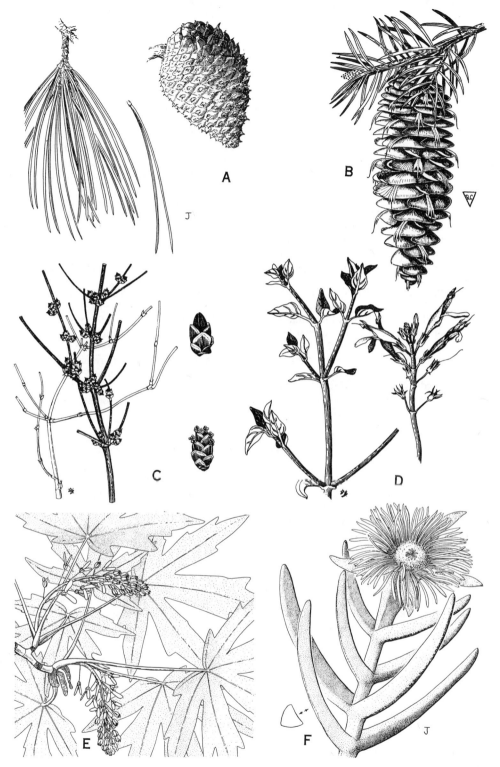

Plate 5. Fig. A, *Pinus muricata*, 2-needled pine. Fig. B, *Pseudotsuga macrocarpa*, seedbearing scales in axils of thin toothed scales. Fig. C, *Ephedra nevadensis*, ♂ and ♀ cones. Fig. D, *Beloperone californica*, young lvs. and tubular fls. Fig. E, *Acer macrophyllum*, young samaras and fls. Fig. F, *Carpobrotus edulis*.

yon-Juniper Wd.; w. edge of Colo. Desert from Santa Rosa and San Jacinto mts. into n. L. Calif. Seeds edible.

13. **P. remoràta** Mason. SANTA CRUZ ID. PINE. Much like *P. muricata*; lvs. dark green; cones ovoid, more symmetrical, scarcely or not deflexed, 5–8 cm. long, each scale plane or slightly rounded, armed with a minute often deciduous prickle.—Associated with *P. muricata* on Santa Cruz Id. and sparingly so on the mainland near La Purissima Mission and in Marin Co. It has been stated by H. L. Mason that this sp. has been distinct since early Pleistocene and is now a remnant of a past flora.

14. **P. sabiniàna** Dougl. DIGGER PINE. Tree mostly 10–20 m. high, the trunk dividing and supporting an open leafy crown; bark dark brown with purplish tinge; branchlets at first glaucous; lvs. in 3's, stout, gray-green, 20–30 cm. long, with many bands of stomates and lasting 3–4 years; ♂ cones 3–4 mm. long; ♀ cones long-stalked, deflexed, broadly ovoid to subglobose, mostly 15–25 cm. long, light reddish-brown, the scales ending in stout, flattened, downwardly projecting hooks; seeds blackish-brown, 20–24 mm. long, with very short wings.—Dry slopes and ridges below 4500 ft.; Foothill Wd.; n. Los Angeles Co. away from the coast in hills bordering the Cent. V. Wood of little use. Seeds sweet and edible.

15. **P. torreyàna** Parry ex Carr. [*P. lophosperma* Lindl.] TORREY PINE. Tree usually 10–15 m. high in nature with broad open crown, but much taller in cult.; bark 2–3 cm. thick, red brown, with broad irregular ridges; branchlets stout; lvs. in 5's, forming large terminal tufts, dull green, stout, 20–30 cm. long, with many rows of stomates on all 3 surfaces; ♂ cones yellowish, 2–3.5 cm. long; ♀ cones on elongate stalks, broadly ovoid, 10–15 cm. long, chestnut-brown, the scales thickened apically into a low pyramidal knob with a minute prickle; seeds oval, ± angled, 2–2.5 cm. long, brown, ± mottled, nearly surrounded by the long wings.—Restricted endemics on dry slopes below 500 ft. near Del Mar on the San Diego Co. coast and Santa Rosa Id. Wood light, soft, coarse-grained; now cult. widely in Kenya, New Zealand, etc.

3. *Pseudotsùga* Carr.

Pyramidal trees with deeply furrowed bark, slender usually horizontal irregularly whorled branches and slender drooping branchlets. Lvs. linear, flat, solitary, appearing 2-ranked, short-petiolate, dark green above, paler beneath. Cones ovoid-oblong, acute, pendulous, maturing in one season, with rounded concave, rigid, persistent scales which are shorter than the prominent, less woody, mostly 3-toothed bracts. Ca. 5 spp. of w. N. Am. and e. Asia. (Greek, *pseudos*, false, and *tsuga*, hemlock.)

1. **P. macrocárpa** (Vasey) Mayr. [*Abies m.* Vasey. *P. douglasii* var. *m.* Engelm.] BIG-CONE SPRUCE. Plate 5, Fig. B. Tree 12–20 m. tall, the trunk with deep rounded ridges; lvs. blue-green, 2–3 cm. long, pointed; ♂ cones pale yellow; ♀ cones 10–15 cm. long, cylindrical, short-stalked, the scales thick and rigid, the bracts ca. 8 mm. wide, conspicuous; seeds 12 mm. long, equal to the wings.—Dry slopes and canyons, 2000–6000 ft.; Chaparral, Yellow Pine F.; mts. from San Diego Co. to Santa Barbara and Kern cos.

Subdivision III. PTEROPSIDA

Class 3. GNETAE

Order 2. EPHEDRALES

Characters of the family *Ephedraceae*

1. Ephedràceae. EPHEDRA FAMILY

Shrubs or small trees with jointed grooved greenish stems and scalelike opposite or whorled lvs. Plants usually dioecious, the axillary cones with basal decussate scalelike bracts. Staminate cones with 2-8 stamens free or united into a column and each with 1-8 terminal usually bilocular pollen-sacs. Ovulate cones with 1-3 cent. ovules enclosed in an urn-shaped envelope formed by union of a lower pair of bracts to form an outer integument and of an upper pair to form an elongate tubular inner integument. Seed enclosed in the indurate envelope; cotyledons 2. One genus of arid parts of N. Hemis.

1. *Ephèdra* L. EPHEDRA. MORMON- or MEXICAN-TEA

Broomlike shrubs, the branches somewhat resembling an Equisetum. Perhaps 40 spp. of N. Am., N. Afr., Eurasia. (Ancient name used by Pliny for the horsetails.)

Lvs. and bracts in 3's; cones sessile.
  Lvs. persisting, becoming shredded and gray in age, 5-13 mm. long; tips of twigs spinose
                                                                                  6. *E. trifurca*
  Lvs. falling off in age or remaining firm, 2-6 mm. long; tips of twigs not spinose.
    Seeds brown, smooth, subglobose; young stems yellow-green, almost smooth, except for the longitudinal furrows. Widely distributed .... 2. *E. califoinica*
    Seeds cream to light brown, rough, angular, less than half as thick as long; young stems pale gray-green, slightly roughened and with longitudinal furrows. Death V. region.
                                                                                  4. *E. funerea*
Lvs. and bracts in 2's; cones often pedunclled.
  Seeds solitary, or if more than 1, light in color.
    The seeds smooth, brown to chestnut in color; lf.-bases brown and persistent; stems usually rough .... 1. *E. aspera*
    The seeds furrowed or scabrous, light brown to gray-green; lf.-bases gray and deciduous; stem usually smooth .... 3. *E. fasciculata*
  Seeds paired, brown to almost black.
    Lf.-bases brown; seeds lightly furrowed longitudinally; branchlets green not glaucous, 1-1.5 mm. in diam., numerous, tending to erect and ± parallel .... 7. *E. viridis*
    Lf.-bases gray; seeds smooth; branchlets gray-green, ± glaucous, ca. 2 mm. in diam., fewer and divergent .... 5. *E. nevadensis*

1. **E. áspera** Engelm. ex Wats. [*E. nevadensis* var. *a.* L. Benson.] Erect, 3-12 dm. high, with rigid terete branchlets to 3 mm. thick, the internodes 1-5 cm. long, pale to dark green, mostly scabrous, becoming yellow; lvs. opposite, 1.2-5 mm. long, the basal sheath subpersistent; ♂ cones mostly paired, obovoid, 4-7 mm. long, the bracts in 6-10 pairs, obovate, 3 mm. long, membranaceous, yellow to red-brown; staminal column 4-5 mm. long, with 4-6 subsessile anthers; ♀ cones ovoid, 6-10 mm. long, with 5-7 pairs of bracts, these round, 2-5 mm. long, thickened, red-brown with membranaceous margins; seeds 1, smooth to slightly rough, light brown to chestnut, 5-8 mm. long.—Occasional on dry rocky slopes below 5000 ft.; Creosote Bush Scrub, Joshua Tree Wd.; Mojave and Colo. deserts; to Tex., n. Mex.

2. **E. califórnica** Wats. Erect or spreading, 3-10 dm. high, with semiflexible to rigid branchlets to 4 mm. thick and internodes 3-6 cm. long, at first yellow-green, glaucous, almost smooth; lvs. in 3's, 2-6 mm. long, with a green-brown dorsi-median thickening, the

hard brown sheath subpersistent; ♂ cones 1-several, ovoid, 6-7.5 mm. long, the bracts in 3's, in 8-12 whorls, ovate, 2.5-3 mm. long, membranaceous, light yellow with hyaline margins; staminal column 3-5 mm. long, with 3-7 subsessile anthers; ♀ cones ovoid, 7-10 mm. long, the bracts in 4-6 whorls, round, 5-7 mm. long, pale yellow-translucent except for the orange- or green-yellow center and base; seeds mostly 1, subglobose, 7-9 mm. long, brown to chestnut.—Dry slopes and fans largely below 3000 ft.; Creosote Bush Scrub, both deserts; in Chaparral, San Diego Co.; V. Grassland along inner Coast Ranges; to Merced Co., L. Calif.

3. **E. fasciculàta** A. Nels. Low, often prostrate, the branches flexuous, up to 3.5 mm. thick, the internodes 1-5 cm. long, pale green, smooth to ± roughened, becoming yellowed; lvs. opposite, 1-3 mm. long, with a hyaline subpersistent white sheath; ♂ cones 2-several, narrow-ellipsoid, 4-8 mm. long, sessile, with 4-8 pairs of obovate bracts 2-3 mm. long; staminal column 3-9 mm. long, with 6-10 sessile or short-stipitate anthers; ♀ cones ellipsoidal, 6-13 mm. long, with 4-7 whorls of elliptic bracts 3-7 mm. long, with light brown to green bracts and hyaline margins; seed usually 1, longitudinally furrowed, light brown, 8-13 mm. long.—Occasional in sandy places, Creosote Bush Scrub, as at Kelso and Palm Springs; to Ariz.

Var. **clòkeyi** (Cutler) Clokey. [*E. c.* Cutler.] Erect; ♀ cones obovoid, 6-10 mm. long; seed 5-8 mm. long.—Dry slopes and washes; Creosote Bush Scrub; both deserts.

4. **E. funèrea** Cov. & Mort. [*E. californica* var. *f.* L. Benson.] Erect, 3-15 dm. high, with stiff hard branches to 3.5 mm. thick and internodes 2-6 cm. long, pale, gray-green, glaucous, slightly roughened; lvs. in 3's, 3-6 mm. long; ♂ cones elongate-elliptic, 5-8 mm. long, with 6-9 whorls of ovate bracts 3-4 mm. long, membranaceous, yellowish; ♀ cones lance-obovate, 8-13 mm. long, the bracts in 6-9 whorls, 4-8 mm. long, yellow-translucent except for the green cent.; seeds mostly 1, tetragonal, pale green to light brown, 6-9 mm. long, smooth to scabrous.—At ca. 2000-5000 ft.; Creosote Bush Scrub; Death V. region.; sw. Nev.

5. **E. nevadénsis** Wats. Plate 5, Fig. C. Spreading-erect, to ca. 12 dm. high, the young stems pale green, glaucous, almost smooth, later yellow or gray; lvs. in 2's, 2-4 (-7) mm. long, with a dorsi-median thickening, splitting and falling off, leaving gray bases; ♂ cones 1-several, ellipsoid, 4-8 mm. long, with 5-9 pairs of obovate bracts 3-4 mm. long, yellow to light brown; staminal column 3-5 mm. long; ♀ cones roundish, 5-11 mm. long, pedunculate, with 3-5 pairs of bracts 4-8 mm. long, light brown to yellow-green; seeds paired, smooth, brown, 6-9 mm. long.—Common on dry slopes and hills mostly below 4500 ft.; Creosote Bush Scrub, Joshua Tree Wd.; Colo. and Mojave deserts to Owens V.; to Utah, Ariz.

6. **E. trifúrca** Torr. Erect, 5-20 dm. high, with rigid terete branchlets to 3.5 mm. thick and internodes 3-9 cm. long, at first pale green, almost smooth, later yellow, then gray-green; lvs. in 3's, spinosely tipped, their sheath becoming fibrous, grayish, persistent; ♂ cones 1-several, obovoid, 6-9 mm. long, the bracts in 3's, red-brown; staminal column 4-5 mm. long, with 4-5 anthers; ♀ cones obovoid, 10-14 mm. long, the bracts in 6-9 whorls of 3, orbicular, 8-12 mm. long, translucent except for the red-brown cent.; seed 1, sometimes 2-3, light brown, smooth, 9-14 mm. long.—Dry sandy and rocky places, below 2000 ft.; Creosote Bush Scrub; Colo. Desert; to w. Tex., nw. Mex., L. Calif.

7. **E. víridis** Cov. [*E. nevadensis* var. *v.* Jones.] Erect, 5-15 dm. high, with very numerous, broomlike yellow-green scabrous slender branchlets; lvs. opposite, 1.5-4 mm. long, setaceously tipped from a dorsi-median thickening, deciduous and leaving a thickened persistent brown base; ♂ cones 2 or more, obovoid, 5-7 mm. long, with 6-10 pairs of bracts 2-4 mm. long, light yellow; staminal column 2-4 mm. long, with 5-8 sessile anthers; ♀ cones obovoid, 6-10 mm. long, with 4-8 pairs of ovate bracts 4-7 mm. long; seeds paired, brown, trigonal, smooth, 5-8 mm. long; $n=14$ (Raven, Kyhos & Hill, 1965).—Frequent on dry rocky slopes, canyon walls, etc., 3000-7000 ft.; Creosote Bush Scrub to Pinyon-Juniper Wd.; w. edge of Colo. Desert through the Mojave Desert, to Mono and Lassen cos. to w. Colo., Utah, Ariz.

Class 4. **ANGIOSPERMAE.** FLOWERING PLANTS

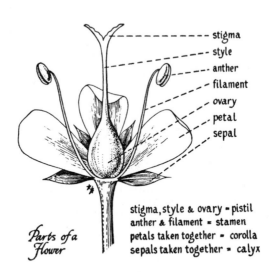

stigma, style & ovary = pistil
anther & filament = stamen
petals taken together = corolla
sepals taken together = calyx

Parts of a Flower

Subclass I. **DICOTYLEDONES**

Stem exogenous, with pith, xylem, cambium, phloem and cortex, and if of more than short duration, increasing in diam. by tissue formed by the cambium. Lvs. net-veined, usually pinnately or palmately so. Fls. mostly with parts in 4's or 5's. Embryo generally with 2 cotyledons, the first lvs. of the germinating plantlet opposite. Ca. 3/4 of the flowering plants belong here.

KEY TO FAMILIES

1. Petals lacking or not evident (sometimes simulated by the petallike calyx, as in *Fremontodendron* and *Nyctaginaceae*).
   2. Plants definitely woody; well developed trees and shrubs .... **Group 1** p. 45
   2. Plants herbaceous, sometimes slightly woody at the base only .... **Group 2** p. 46
1. Petals present, evident.
   3. Corolla of separate petals.
      4. Stamens numerous, more than twice as many as the petals .... **Group 3** p. 48
      4. Stamens few, not more than twice as many as the petals .... **Group 4** p. 49
   3. Corolla of ± connate petals, these often markedly grown together .... **Group 5** p. 52

Group 1. WOODY PLANTS WITHOUT PETALS, OR PETALS NOT EVIDENT

1. Lvs. compound, the blade divided into separate parts.
   2. Plants woody vines climbing by twisting petioles; lvs. opposite; stamens and pistils many .... *Ranunculaceae* (Clematis) p. 721
   2. Plants trees or shrubs, not climbing.
      3. Lvs. alternate; fr. a nut .... *Juglandaceae* p. 521

3. Lvs. opposite; fr. a samara.
　　　　4. Lfts. 3; samaras pubescent .... *Aceraceae* (Acer negundo)　p. 54
　　　　4. Lfts. 3–7; samaras glabrous .... *Oleaceae* (Fraxinus)　p. 586
1. Lvs. simple, the blade entire, toothed or lobed.
　　5. Plants parasitic on branches of trees and shrubs and not connected with the ground
　　　　　　　　　　　　　　　　　　　　　　　　　　　Viscaceae　p. 856
　　5. Plants growing normally on the ground.
　　　　6. Fls. (at least the ♂) in catkins.
　　　　　7. Lvs. alternate; plants often monoecious.
　　　　　　8. Calyx present in ♂ fls.
　　　　　　　9. Fr. a nut inclosed in a leafy invol., or fr. conelike; plants deciduous
　　　　　　　　　　　　　　　　　　　　　　　　　　　Betulaceae　p. 245
　　　　　　　9. Fr. an acorn with a basal cup, or a spiny bur inclosing 1–3 nuts; plants
　　　　　　　　　mostly evergreen .... *Fagaceae*　p. 478
　　　　　　8. Calyx none, the ♂ fls. subtended by bracts.
　　　　　　　9a. Fr. a caps. containing comose seeds; ♀ fls. in aments ..... *Salicaceae*　p. 771
　　　　　　　9a. Fr. 1-seeded, without coma; ♀ fl. 1, in axil of bract ..... *Myricaceae*　p. 576
　　　　　7. Lvs. opposite; plants dioecious.
　　　　　　10. Catkins erect; maritime subshrub with linear-oblanceolate lvs.
　　　　　　　　　　　　　　　　　　　　　　　　　　　Batidaceae　p. 242
　　　　　　10. Catkins drooping; montane shrub with broader lvs. ...... *Garryaceae*　p. 484
　　　　6. Fls. not in catkins.
　　　　　11. Plants normally trees with a well defined cent. trunk.
　　　　　　12. Lvs. entire, lanceolate to ± ovate, strongly aromatic when crushed.
　　　　　　　13. Ovary superior, maturing into a drupe; stamens in 3 series of 3 each, not
　　　　　　　　　highly colored. Native .... *Lauraceae*　p. 544
　　　　　　　13. Ovary inferior, maturing into a woody caps.; stamens many, white to
　　　　　　　　　yellow or red. Introd. .... *Myrtaceae* (Eucalyptus)　p. 577
　　　　　　12. Lvs. ± toothed to lobed, ovate to roundish, not strongly aromatic.
　　　　　　　14. Lvs. 12–25 cm. broad, deeply 5-lobed; fls. in round heads . *Platanaceae*　p. 636
　　　　　　　14. Lvs. 2–5 cm. broad, ± serrate; fls. in small clusters or solitary . *Ulmaceae*　p. 845
　　　　　11. Plants normally shrubs with several stems from near base.
　　　　　　15. Fls. yellowish, 2.5–6 cm. broad, the wide sepals or calyx-lobes petaloid
　　　　　　　　　　　　　　　　　　　　Sterculiaceae (Fremontodendron)　p. 841
　　　　　　15. Fls. smaller.
　　　　　　　16. Ovary 3–4-loculed; styles 3–4.
　　　　　　　　17. Fr. a lobed caps.; lvs. often alternate .... *Euphorbiaceae*　p. 405
　　　　　　　　17. Fr. a smooth cylindrical caps.; lvs. opposite .... *Buxaceae*　p. 308
　　　　　　　16. Ovary 1-celled (or if 2–3-celled, the style single); styles 1–3.
　　　　　　　　18. Fr. an ak. or utricle.
　　　　　　　　　19. Styles 2–3; ovary not surrounded by a floral tube.
　　　　　　　　　　20. Fls. subtended by an invol. of ± united bracts; ak. triangular or
　　　　　　　　　　　　lenticular .... *Polygonaceae*　p. 670
　　　　　　　　　　20. Fls. without an invol. of bracts; fr. mostly a utricle or depressed-
　　　　　　　　　　　　globose ak. .... *Chenopodiaceae*　p. 351
　　　　　　　　　19. Style 1, ± plumose in fr.; ovary surrounded by floral tube
　　　　　　　　　　　　　　　　　　Rosaceae (Cercocarpus, Coleogyne)　p. 742
　　　　　　　　18. Fr. a samara, drupe or berry.
　　　　　　　　　21. Lvs. opposite.
　　　　　　　　　　22. Plants with silvery-scurfy branchlets and lvs.; fr. red, fleshy
　　　　　　　　　　　　　　　　　　　　　　　　　　Elaeagnaceae　p. 395
　　　　　　　　　　22. Plants glabrous; fr. a samara or blackish drupe ..... *Oleaceae*　p. 586
　　　　　　　　　21. Lvs. alternate; fr. a drupe or berrylike .... *Rhamnaceae*　p. 731

　　　　Group 2. HERBACEOUS PLANTS WITHOUT PETALS, OR PETALS NOT EVIDENT
1. Aquatic plants growing ± submerged, or, as water dries up, on wet mud.
　2. Lvs. largely dissected, whorled.
　　3. Ovary superior; fr. an ak.; lvs. 3-times forked .... *Ceratophyllaceae*　p. 350

```
        3. Ovary inferior; fr. a nutlet or splitting into 4 nutlets; lvs. ±pinnatifid
                                                                    Haloragaceae   p. 493
     2. Lvs. entire, often opposite and often crowded into terminal rosettes.
        2a. Stamen 1.
           3a. Ovary superior, splitting into 4 parts when ripe; lvs. 3–8 mm. long
                                                                    Callitrichaceae  p. 321
           3a. Ovary inferior; fr. indehiscent; lvs. 5–30 mm. long ........ Hippuridaceae  p. 495
        2a. Stamens 4–5; ovary inferior, forming a caps. ........ Onagraceae (Ludwigia)  p. 611
1. Land plants, sometimes growing in wet places.
  5. Ovary superior, free from the calyx, although sometimes surrounded by it.
     6. Perianth lacking entirely.
        7. Fls. white, perfect, borne in a bracted spike which is subtended by a conspi-
           cuous invol. ........................................ Saururaceae  p. 776
        7. Fls. imperfect, either ♂ or ♀.
           8. Fls. borne in clusters often surrounded by an invol. resembling a perianth;
              caps. 3-lobed; sap often milky; plants monoecious ....... Euphorbiaceae  p. 405
           8. Fls. borne in catkins disposed in terminal spikes; plants dioecious. Seashore
                                                                    Batidaceae  p. 242
     6. Perianth present as a calyx (often petaloid).
        9. Pistils more than 1; stamens 10–many .................. Ranunculaceae  p. 719
        9. Pistil 1.
           10. Fls. perigynous, the ovary inclosed in or seated in a floral tube.
              11. Stipules present; lvs. alternate; fr. an ak. ................ Rosaceae  p. 739
              11. Stipules none; lvs. opposite.
                 12. Stamens many; fr. circumscissile .......... Aizoaceae (Sesuvium)  p. 58
                 12. Stamens 3–5; fr. an ak. ......................... Nyctaginaceae  p. 577
           10. Fls. hypogynous; the ovary not so inclosed.
              13. Style and stigma single.
                 14. Calyx not tubular.
                    15. Lvs. subulate, squarrose-spreading, 3–6 mm. long
                                                          Caryophyllaceae (Loeflingia)  p. 340
                    15. Lvs. ovate, the blades 10–20 mm. long.
                       16. Calyx 5–6-parted; stamens 6–7 .. Euphorbiaceae (Eremocarpus)  p. 409
                       16. Calyx largely 4-parted; stamens 4 .............. Urticaceae  p. 846
                 14. Calyx tubular, corollalike, subtended by bracts often forming a calyx-
                    like invol. ..................................... Nyctaginaceae  p. 577
              13. Styles and stigmas more than 1.
                 17. Lvs. mostly deeply palmately 5–7-lobed or with 5–7 lfts.; fls. imper-
                    fect ........................................... Moraceae  p. 575
                 17. Lvs. mostly entire or shallowly lobed; fls. perfect or imperfect.
                    18. Ovary 1-loculed (except sometimes in Silene which has a tubular
                       gamosepalous calyx).
                       19. Ovary or seed solitary; fr. mostly an ak. or utricle; lvs. often
                          alternate.
                          20. Fls. borne in a tubular to campanulate invol. ... Polygonaceae  p. 670
                          20. Fls. not borne in an invol.
                             21. Lvs. with evident stipular sheath above each node
                                                                    Polygonaceae  p. 670
                             21. Lvs. lacking such stipular sheath.
                                22. Calyx mostly 6-cleft; stamens 3, 6, or 9 ... Polygonaceae  p. 670
                                22. Calyx-lobes or sepals 1, 4, or 5.
                                   23. Bracts subtending fls. not scarious; plants mostly mealy,
                                      scurfy or fleshy .................... Chenopodiaceae  p. 351
                                   23. Bracts subtending fls. scarious; plants not mealy,
                                      scurfy or fleshy ................... Amaranthaceae  p. 58
                       19. Ovules or seeds more than 1; fr. a caps.; lvs. opposite.
                          24. Lvs. of each pair very unequal ................ Aizoaceae  p. 55
                          24. Lvs. of each pair normally ca. equal ....... Caryophyllaceae  p. 334
                    18. Ovary more than 1-loculed; fr. a berry .......... Phytolaccaceae  p. 632
```

48   *Dicotyledones*

    5. Ovary ± inferior, partly or wholly adnate to the floral tube.
       25. Fls. imperfect; stamens 8–many; fr. a 1-loculed caps.
          27. Lvs. divided; large green herb .... *Datiscaceae* p. 393
          27. Lvs. reduced to minute bracts; very small parasite on Dalea . . *Rafflesiaceae* p. 719
       25. Fls. perfect.
          28. Fr. a bony nut, indehiscent; lvs. rhombic-ovate, entire
                                                       Aizoaceae (Tetragonia) p. 58
          28. Fr. a caps.
             29. Lvs. entire, oblong, glabrous; stamens 3–6. Root-parasite . . *Santalaceae* p. 776
             29. Lvs. toothed or crenate to lobed, roundish; stamens 4, 5 or 8. *Saxifragaceae* p. 778

       Group 3. PLANTS WITH SEPARATE PETALS; STAMENS NUMEROUS, MORE THAN TWICE AS
                                 MANY AS PETALS

1. Ovary superior.
    2. Plants woody, i.e. shrubs or trees.
       3. Plants with evident stipules.
          4. Plants with many exserted stamens; lvs. pinnate or the pinnules suppressed
             and the petiole flattened into a lf.-like vertical structure; pistil 1, forming a
             legume .... *Fabaceae* (Acacia) p. 420
          4. Plants not as above.
             5. Stamens united into a tube around the pistil; lvs. rounded, often stellate-
                pubescent .... *Malvaceae* p. 561
             5. Stamens not united into a tube; lvs. various, not stellate-pubescent
                                                              Rosaceae p. 739
       3. Plants without stipules.
          6. Desert plants with conspicuous thorny branches .... *Simaroubaceae* (Castela) p. 829
          6. Cismontane or desert plants, not thorny.
             7. Pistil 1; sepals 2, rarely 3 or 4, often caducous .... *Papaveraceae* p. 622
             7. Follicles 2–9, separate; calyx of 5 persistent lobes ...... *Crossosomataceae* p. 389
    2. Plants herbaceous, or woody at base only; not definite shrubs.
       8. Sepals 2, rarely 3 or 4 (joined into a pointed cap in *Eschscholzia*).
          9. Plants fleshy; sepals persistent .... *Portulacaceae* p. 708
          9. Plants not fleshy; sepals caducous at anthesis .... *Papaveraceae* p. 622
       8. Sepals normally more than 2.
          10. Stamens united into a tube around the pistil .... *Malvaceae* p. 561
          10. Stamens not so united.
             11. Maturing ovary at summit exposing the contained seeds; fls. irregular
                                                             Resedaceae p. 730
             11. Maturing ovary closed at summit; fls. mostly regular.
                12. Stamens hypogynous.
                   13. Lvs. opposite.
                      14. Lvs. punctate with pellucid dots; stamens usually in 3 or 5 bundles
                                                             Hypericaceae p. 519
                      14. Lvs. not so punctate; stamens separate .... *Cistaceae* p. 370
                   13. Lvs. alternate.
                      15. Lvs. reduced to small scales; petals 4; thorny shrubs *Koeberliniaceae* p. 521
                      15. Lvs. well developed; plants not thorny shrubs.
                         16. Ovary 3–5-celled.
                            17. Styles 3, bilobed; stamens united in ranks of 5
                                                  Euphorbiaceae (Ditaxis) p. 408
                            17. Style 1, with 3 stigmas; stamens not so united
                                                   Cistaceae (Helianthemum) p. 370
                        16. Ovary or ovaries 1-celled; stamens spirally arranged.
                            18. Sepals persistent; anthers maturing in centrifugal sequence
                                                             Paeoniaceae p. 622
                            18. Sepals deciduous; anthers maturing centripetally *Ranunculaceae* p. 719
                12. Stamens perigynous; stipules usually well developed ...... *Rosaceae* p. 739
1. Ovary at least partly inferior.
    19. Ovary only partly inferior.
       20. Fr. a dry caps. .... *Saxifragaceae* p. 778

Key to Families   40

    20. Fr. a fleshy pome .... *Rosaceae* p. 739
  19. Ovary wholly inferior.
    21. Petals numerous (technically may be modified stamens); plant succulent.
      22. Plants not spiny; lvs. well developed .... *Aizoaceae* p. 55
      22. Plants spiny, with lvs. much reduced and caducous, or wanting . *Cactaceae* p. 308
    21. Petals few; plants not succulent
      23. Trees or shrubs; fr. fleshy .... *Rosaceae* p. 739
      23. Herbs; fr. a caps. .... *Loasaceae* p. 549

Group 4.   Plants with separate petals: stamens not more than twice as many as petals

1. Pistils more than 1, nearly or quite separate.
  2. Plants fleshy, at least the lvs. so; lvs. simple .... *Crassulaceae* p. 382
  2. Plants not succulent.
    3. Insertion of the stamens on the floral tube; fls. perigynous.
      4. Stipules usually absent; pistils mostly fewer than sepals ...... *Saxifragaceae* p. 778
      4. Stipules usually present; pistils mostly as many as sepals .... *Rosaceae* p. 739
    3. Insertion of stamens hypogynous; small herbs with long taillike spike of pistils
                                         Ranunculaceae (Myosurus)   p. 725
1. Pistil 1, of 1 or more ± united carpels.
  5. Plants climbing by means of tendrils; lvs. palmately veined.
    6. Fr. a berry; frs. borne in clusters. Grapes .... *Vitaceae* p. 858
    6. Fr. a pepo, the frs. borne singly. Gourds, melons, etc. .... *Cucurbitaceae* p. 389
  5. Plants not climbing by means of tendrils.
    7. Styles 2–5, separate to near base.
      8. Plants definitely woody, well developed shrubs or trees.
        9. Lvs. small, scalelike, appressed; fls. minute, in large clusters . *Tamaricaceae* p. 843
        9. Lvs. well developed.
          10. Ovary superior.
            11. Fr. a samara; lvs. palmately veined, opposite .... *Aceraceae* p. 54
            11. Fr. not a samara; lvs. pinnately veined, alternate ... *Anacardiaceae* p. 64
          10. Ovary appearing ± inferior.
            12. Lvs. alternate, with stipules.
              13. Fr. a pome, the dry pistils technically superior and surrounded by a fleshy floral tube; lvs. ± oblong, pinnately veined
                                       Rosaceae (Heteromeles)   p. 747
              13. Fr. a berry, inferior; lvs. mostly palmately lobed
                                       Saxifragaceae (Ribes)   p. 784
            12. Lvs. opposite, without stipules .... *Saxifragaceae* (Jamesia)   p. 781
      8. Plants herbaceous.
        14. The plant a submerged aquatic or on wet mud flats and rooting at nodes.
          15. Lvs. ± finely dissected .... *Haloragaceae* p. 493
          15. Lvs. entire, opposite .... *Elatinaceae* p. 396
        14. The plant a normal terrestrial plant.
          16. Ovary superior.
            17. Lvs. compound with 3 lfts.; stamens 10 .... *Oxalidaceae* p. 620
            17. Lvs. simple.
              18. The lvs. largely basal.
                19. Plants fleshy, not maritime .... *Portulacaceae* p. 708
                19. Plants not fleshy, maritime .... *Plumbaginaceae* p. 636
              18. The lvs. mostly cauline.
                20. Lvs. for the most part alternate.
                  21. Sepals 2, or if more, the lvs. fleshy.
                    22. Sepals 2; plants not large vines; stamens 1–3. *Portulacaceae* p. 708
                    22. Sepals 5; plants large vines; stamens 5 ...... *Basellaceae* p. 242
                  21. Sepals 3 several; plants not fleshy.
                    23. Stamens 10; plants ± gray-pubescent
                                   Euphorbiaceae (Ditaxis)   p. 408
                    23. Stamens 3–8; plants not gray-pubescent.
                      24. Fls. regular, 5-merous; caps. closed until mature.

             25. Ovary 1-celled .... *Caryophyllaceae* p. 334
             25. Ovary 2–5-celled .... *Linaceae* p. 547
          24. Fls. irregular, 2–7-merous; caps. open at top before
                  maturity .... *Resedaceae* p. 730
      20. Lvs. opposite.
          26. Stamens 5 or 10; caps. 3–10-valved or- toothed; placenta
                  largely cent. .... *Caryophyllaceae* p. 334
          26. Stamens 4–7; caps. 2–4-valved; placenta basal but parietal
                                                      Frankeniaceae p. 484
   16. Ovary ± inferior.
      27. Ovules several in each cavity of the ovary; fr. a caps. or many-seeded
              berry .... *Saxifragaceae* p. 778
      27. Ovules solitary in each cavity of the ovary; fls. in umbels.
          28. Fr. a berry; lvs. 3–15 dm. long .... *Araliaceae* p. 90
          28. Fr. dry, splitting into 2 carpels; lvs. mostly smaller .... *Apiaceae* p. 67
7. Style 1, sometimes divided toward the apex.
   29. Ovary inferior.
      30. Plant scabrous with short barbed hairs; fr. dry .. *Loasaceae* (Petalonyx) p. 558
      30. Plant not scabrous with short barbed hairs.
          31. Plants well developed shrubs or arborescent; fr. fleshy.
              32. Fls. racemose or solitary; lvs. alternate; stamens usually 5
                                                  Saxifragaceae (Ribes) p. 784
              32. Fls. in heads, cymes or umbels; lvs. mostly opposite; stamens 4
                                                      Cornaceae p. 379
          31. Plants herbaceous, sometimes suffrutescent; fr. dry; fls. mostly 4-merous
                                                      Onagraceae p. 589
   29. Ovary superior, but sometimes appearing inferior because inclosed in but
          not adnate to floral tube.
      33. Plants well developed shrubs or trees.
          34. Fls. irregular, the petals not all alike.
              35. Petals 3 (2 forming a pair, the 3d hooded above); sepals 5, the lateral
                      petaloid .... *Polygalaceae* p. 669
              35. Petals 4–5; calyx not petaloid.
                  36. Ovary 3-loculed; lvs. opposite, palmately compound; fls. not
                      papilionaceous, white in a large thyrse; fr. a 1-seeded caps
                                                      Hippocastanaceae p. 495
                  36. Ovary 1-loculed (sometimes divided into 2 parts by a false par-
                      tition).
                      37. Fls. papilionaceous; fr. a legume; lvs. mostly compound; sta-
                              mens more than 4 .... *Fabaceae* p. 419
                      37. Fls. not papilionaceous; fr. a spiny 1-seeded pod; lvs. simple;
                              stamens 4 .... *Krameriaceae* p. 521
          34. Fls. regular, the petals essentially alike.
              38. Lvs. compound, consisting of 2 or more lfts. (separately deciduous).
                  39. Fr. a samara; wing of fr. terminal; lfts. 3–9; lvs. opposite
                                                  Oleaceae (Fraxinus) p. 586
                  39. Fr. not a samara.
                      40. The fr. fleshy; petals 3 plus 3; lfts. mostly more than 3
                                                      Berberidaceae p. 242
                      40. The fr. dry; petals in 1 series.
                          41. Ovary 1-loculed, becoming a legume; stamens mostly 10
                                                      Fabaceae p. 419
                          41. Ovary more than 1-loculed.
                              42. Lfts. 13–25; fls. 5-merous .... *Burseraceae* p. 307
                              42. Lfts. 2–3.
                                  43. Sepals and petals 5; fr. not inflated ..... *Zygophyllaceae* p. 859
                                  43. Sepals and petals 4; fr. inflated . *Capparaceae* (Isomeris) p. 330
              38. Lvs. simple, but sometimes divided, looking like lfts., but the parts
                      not separately deciduous.
                  44. Fr. stipitate, 2-locular; fls. yellow ..... *Brassicaceae* (Stanleya) p. 302
                  44. Fr. sessile.

Key to Families 51

   45. The fr. a legume, oblong, flat; stamens 10; fls. red-purple
                     Fabaceae (Cercis)   p. 438
   45. The fr. not a legume.
    46. Lvs. pellucid-punctate (gland-dotted); petals 4, white; lvs. linear; stamens 8 .... *Rutaceae* (Cneoridium)   p. 770
    46. Lvs. not as above; stamens not 8; petals not 4.
     47. Trees with large palmately lobed lvs.; fls. imperfect, reduced, in spherical clusters .... *Platanaceae*   p. 639
     47. Trees or shrubs, the lvs. not large and palmately lobed.
      48. Lvs. small, scalelike, linear, caducous; branchlets rigid, spine-tipped. Colo. Desert ..... *Koeberliniaceae*   p. 521
      48. Lvs. large, pinnately veined.
       49. Stamens 4–5 (sometimes 6–10 in *Forsellesia*, a small spiny desert shrub).
        50. The stamens opposite the petals ... *Rhamnaceae*   p. 731
        50. The stamens alternate with the petals or more numerous.
         50a. Ovules 1–2 in each ovary-cell. Natives
                    Celastraceae   p. 349
         50a. Ovules several to many in each ovary-cell. Introduced .... *Pittosporaceae*   p. 632
       49. Stamens 10. Cismontane chaparral
                    Rosaceae (Adenostoma)   p. 741
33. Plants herbaceous.
 51. Sepals 2 or 3.
  52. Plants fleshy, succulent; sepals 2, persistent .... *Portulacaceae*   p. 708
  52. Plants not succulent.
   53. Sepals caducous, 2 or 3; stamens 6–12; lvs. entire ...*Papaveraceae*   p. 622
   53. Sepals not early deciduous; lvs. dissected; stamens in 2 sets of 3 each; petals 4, in 2 dissimilar pairs .... *Papaveraceae*   p. 622
 51. Sepals 4 or 5, sometimes more.
  54. Fls. nearly or quite regular.
   55. Lvs. compound, the lfts. separately deciduous.
    56. Fr. 1-loculed, forming a legume; lvs. pinnate *Fabaceae* (Cassia)   p. 437
    56. Fr. 2–5-loculed; fr. not a legume.
     57. Lfts. 3; lvs. alternate.
      58. Stamens 10; foliage sour to taste, not ill-smelling
                    Oxalidaceae   p. 620
      58. Stamens 2–6; foliage ± acrid, ill-smelling .... *Capparaceae*   p. 328
     57. Lfts. 10–16; lvs. opposite.
   55. Lvs. simple and entire or divided, but not truly compound.
    59. Ovary appearing inferior, actually superior and free in the floral tube; fls. 4–6-merous .... *Lythraceae*   p. 560
    59. Ovary appearing superior.
     60. Sepals 4; fils. 6, 4, rarely 2, distinct .... *Brassicaceae*   p. 267
     60. Sepals mostly 5; fils. 5, 8 or 10, often united at base.
      61. Carpels in mature fr. separating from each other as 1-seeded structures.
       62. Stipules scarious; carpels tailed in mature fr.; lower lvs. usually opposite .... *Geraniaceae*   p. 489
       62. Stipules none; carpels not tailed; lvs. alternate
                    Limnanthaceae   p. 547
      61. Carpels when mature forming a many-seeded caps. .... *Ericaceae*   p. 397
  54. Fls. quite definitely irregular, the petals unlike.
   62. The fls. papilionaceous; fr. a legume .... *Fabaceae*   p. 419
   62. The fls. not papilionaceous; fr. not a legume.
    63. Lvs. not peltate.
     64. Fr. a 1-loculed, 3-valved caps.; plants not over 3 dm. high .... *Violaceae*   p. 852

　　　　　64. Fr. a 5-loculed, elastically dehiscent, explosive caps.; plants
　　　　　　　to 1 m. high .... *Balsaminaceae* p. 241
　　　　　63. Lvs. peltate; fr. 3-lobed, 3-loculed, each locule 1-seeded
　　　　　　　　　　　　　　　　　　　　　　　　　　　Tropaeolaceae p. 845

Group 5. PLANTS WITH PETALS ± CONNATE

1. Ovary superior.
　2. Stamens more than 5.
　　3. Corolla not markedly sympetalous (not urn-shaped or tubular) the petals united
　　　only near the base.
　　　4. Pistils 4–5; stamens 10; plants succulent .... *Crassulaceae* p. 382
　　　4. Pistil 1.
　　　　5. Corolla regular.
　　　　　6. Fls. minute, in dense heads or spikes; stamens 10 to many; fr. a legume.
　　　　　　Shrubs or trees .... *Fabaceae* p. 419
　　　　　6. Fls. not as above; fr. not a legume.
　　　　　　7. Stamens many, united in a tube about the pistil; lvs. simple, pal-
　　　　　　　mately veined or lobed .... *Malvaceae* p. 561
　　　　　　7. Stamens 10, united at very base only; lvs. 3-foliolate .... *Oxalidaceae* p. 620
　　　　5. Corolla irregular, the petals not all alike.
　　　　　8. Petals 4, in 2 unlike pairs; sepals 2; lvs. dissected .... *Papaveraceae* p. 622
　　　　　8. Petals 3, appearing to be 5 because of 2 petaloid sepals; other 3 sepals
　　　　　　not petaloid; lvs. entire .... *Polygalaceae* p. 669
　　3. Corolla urn-shaped or tubular or with petals markedly united.
　　　9. Stamens nearly or quite free from the corolla.
　　　　10. Style 1; stigma 1; plants not thorny .... *Ericaceae* p. 397
　　　　10. Styles 3, united to middle; anthers opening lengthwise; plants thorny
　　　　　desert shrubs .... *Fouquieriaceae* p. 483
　　　9. Stamens inserted in corolla-throat; root-parasites .... *Lennoaceae* p. 545
　2. Stamens not more than 5.
　　10a. Plants parasitic, lacking chlorophyll.
　　　11. Twining or trailing vines with small haustoria and minute fls.; no connec-
　　　　tion with ground when mature .... *Convolvulaceae* (Cuscuta) p. 376
　　　11. Root-parasites.
　　　　12. Corolla regular; stamens 5 .... *Lennoaceae* p. 545
　　　　12. Corolla 2-lipped; stamens 4 .... *Orobanchaceae* p. 617
　　10a. Plants not or only partially parasitic and having chlorophyll.
　　　13. Corolla regular.
　　　　14. Pistils 2 (ovaries distinct, but styles or stigmas united); plants with milky
　　　　　juice.
　　　　　15. Styles united; stamens connivent around the stigma ..... *Apocynaceae* p. 87
　　　　　15. Styles distinct below; stamens monadelphous and adnate to the stylar
　　　　　　column .... *Asclepiadaceae* p. 91
　　　　14. Pistil 1.
　　　　　16. Ovary 1-loculed, 1-ovuled; style and stigma 1; fr. dry, hard; the seem-
　　　　　　ingly colored corolla really a calyx inside a 1-fld. invol. ... *Nyctaginaceae* p. 577
　　　　　16. Ovary and fr. not as above.
　　　　　　17. Stamens as many as the corolla-lobes and opposite them, or with an
　　　　　　　additional series of staminodia; placentae cent. or basal.
　　　　　　　18. Fr. a several-seeded caps.; style 1 .... *Primulaceae* p. 715
　　　　　　　18. Fr. 1-seeded; styles 5 .... *Plumbaginaceae* p. 636
　　　　　　17. Stamens as many as or fewer than the corolla-lobes and alternate with
　　　　　　　them.
　　　　　　　17a. Foliage punctate with translucent dots .... *Myoporaceae* p. 576
　　　　　　　17a. Foliage not punctate with translucent dots.
　　　　　　　　19. Corolla small, dry-scarious, veinless; stamens 2 or 4
　　　　　　　　　　　　　　　　　　　　　　　　　　　　　Plantaginaceae p. 633
　　　　　　　　19. Corolla not dry-scarious, veiny; caps. usually not as above.
　　　　　　　　　20. Ovary 4-celled, commonly 4-lobed, each lobe forming a nut-
　　　　　　　　　　let when mature (although some may abort); infl. usually a
　　　　　　　　　　scorpioid cyme .... *Boraginaceae* p. 246

20. Ovary 1-, 2- or 3-celled.
   21. Stamens 2, opposite each other; shrub; caps. circumscissile
                                                                   Oleaceae (Menodora)   p. 588
   21. Stamens 4–7; fr. not opening by a lid.
      22. Anthers opening by terminal pores; fls. large, white or
          yellowish .... *Ericaceae* (Rhododendron)   p. 404
      22. Anthers opening by longitudinal slits.
         23. Style 3-cleft; ovary 3-loculed; caps. 3-valved
                                                                        Polemoniaceae   p. 637
         23. Style not 3-cleft; ovary 1–2-celled.
            24. Calyx 4–5-toothed or -cleft; style 1, entire.
               25. Ovary 1-celled; stigmas 2; lvs. opposite
                                                                          Gentianaceae   p. 485
               25. Ovary 2-celled; stigma usually 1.
                  26. Lvs. opposite; fls. in dense axillary clusters,
                      these forming interrupted leafy spikes
                                                                 Loganiaceae   p. 559
                  26. Lvs. alternate; infl. never spicate ... *Solanaceae*   p. 829
            24. Calyx of 5 distinct sepals, or sepals united only at
              very base; styles 2 or 1, usually partly divided.
               27. Plants twining or trailing; infl. not coiled;
                  corolla plaited in bud .... *Convolvulaceae*   p. 371
               27. Plants erect or diffuse; infl. cymose, often
                  coiled; corolla not plaited in bud . *Hydrophyllaceae*   p. 495
13. Corolla ± irregular.
   28. Fr. of 2–4 nutlets; lvs. opposite.
      29. Ovary not lobed; style apical, entire .... *Verbenaceae*   p. 850
      29. Ovary 4-lobed; style arising between the ovary-lobes, cleft at apex
                                                                         Lamiaceae   p. 522
   28. Fr. a caps.
      30. Ovary 1-celled
         31. Plants aquatic, often with dissected lvs.; corolla spurred; stamens 2
                                                                       Lentibulariaceae   p. 545
         31. Plants terrestrial, the lvs. not dissected; corolla gibbous; stamens 4
                                                                              Martyniaceae   p. 574
      30. Ovary 2-celled.
         32. Seeds not winged; placentae axile.
            33. Calyx with a pair of bractlets at base; desert shrub with red corolla;
                stamens 2 .... *Acanthaceae*   p. 54
            33. Calyx without bractlets; herbs to shrubs; stamens 5, 4, or rarely 2
                                                                        Scrophulariaceae   p. 789
         32. Seeds winged; placentae parietal; caps. linear, 15–25 cm. long.
            Desert shrub .... *Bignoniaceae*   p. 246
1. Ovary inferior or partly so.
   34. Stamens more than 5.
      35. Anthers opening by terminal pores or chinks; fr. fleshy .... *Ericaceae*   p. 397
      35. Anthers opening by longitudinal slits; fr. dry .... *Styraceae*   p. 843
   34. Stamens 5 or fewer.
      36. Stamens distinct.
         37. Lvs. alternate; fls. regular (± irregular in *Nemacladus* and *Parishella*); stamens
            5 .... *Campanulaceae*   p. 322
         37. Lvs. opposite or whorled.
            38. Stamens 1–3; fls. irregular; fr. 1-seeded .... *Valerianaceae*   p. 847
            38. Stamens 4–5, rarely 2; fls. often regular.
               39. Ovary 1-celled; fls. in short spikes or involucral heads; fr. an ak. Introd.
                  herbs .... *Dipsacaceae*   p. 395
               39. Ovary 2–5-celled; fls. not so arranged; fr. usually not a solitary ak.
                  40. Lvs. opposite and evidently stipulate, or whorled and not stipulate
                                                            Rubiaceae   p. 761
                  40. Lvs. opposite or perfoliate, rarely stipulate and when so, the stipules
                    minute .... *Caprifoliaceae*   p. 331

36. Stamens united by the anthers.
    41. Plants bearing tendrils; lvs. palmate; stamens 5, with 2 pairs united and appearing to be 3 .... Cucurbitaceae p. 389
    41. Plants not tendril-bearing.
        42. Fls. not in involucrate heads; stamens free from the corolla . Campanulaceae p. 322
        42. Fls. in involucrate heads; stamens adnate to the corolla ..... Asteraceae p. 95

1. Acanthàceae. ACANTHUS FAMILY

Ours low shrubs with spreading often leafless branches. Fls. axillary, forming short 4-rowed racemes. Calyx 5-parted. Corolla dull scarlet, tubular, bilabiate, the upper lip 2-, the lower 3-lobed. Stamens 2 in ours. Ovary 2-celled. Fr. a loculicidal caps. Ca. 200 genera and 2000 spp., mostly of warmer regions. Many are prized ornamentals.

1. *Beloperòne* Nees.

Ours a subshrub with spreading often leafless branches. Fls. axillary, forming short 4-rowed racemes. Anther-cells somewhat unequal and oblique. Caps. clavate, the locules 2-seeded. Ca. 30 spp., of trop. Am. (Greek, *belos*, an arrow, and *perone*, something pointed.)

    1. **B. califórnica** Benth. CHUPAROSA. Plate 5, Fig. D. Branches greenish-canescent, 3–15 dm. long, ± arched, slender; lvs. ovate, 1–1.5 cm. long, short-petioled, deciduous; calyx canescent, 4–5 mm. long; the segms. lanceolate; corolla red, straight, 3–3.5 cm. long, the lips ca. 12 mm. long, oblong, truncate; stamens exserted; $2n = 28$ (Grant, 1956). —Common in sandy places along water courses below 2500 ft.; Creosote Bush Scrub; w. and n. edges of Colo. Desert; to L. Calif., Ariz., Son. March–June and later.

2. Aceràceae. MAPLE FAMILY

Shrubs to trees, with opposite, simple or compound lvs. Plants polygamous to dioecious, the fls. small, borne in racemes, corymbs or fascicles. Calyx generally 5-parted (sometimes 4–9). Petals as many as sepals or 0. Stamens 4–9, often 8, hypogynous or borne on the edges of a disk; fils. filiform. Pistil 1 with a 2- (rarely to 10-) lobed ovary and 2 styles. Fr. of 2 winged carpels, united below (samaras). Two genera and ca. 125 spp. of N. Hemis.

1. *Acer* L. MAPLE

Fl.-clusters drooping. Ca. 120 spp. of N. Temp. (Latin name of the maple.)

Lvs. simple, palmately lobed; petals present.
    Lvs. 1.5–2.5 cm. wide; fls. in corymbs; frs. glabrous .... 1. *A. glabrum*
    Lvs. 10–15 cm. wide; fls. many, in racemes; frs. hairy .... 2. *A. macrophyllum*
Lvs. pinnate with 3 lfts.; petals absent .... 3. *A. negundo*

    1. **A. glàbrum** Torr. var. **diffùsum** (Greene) Smiley. [*A. d.* Greene. *A. bernardinum* Abrams.] Shrub or small tree to 6 m. high; twigs whitish-gray; lf.-blades 1.5–2.5 cm. long, 1.2–2.8 cm. wide, with few blunt teeth on the lobes; peduncle plus pedicel 1–2 cm. long; samara-pairs ca. 3–6, each samara 2–3 cm. long.—At 6500–9000 ft.; mostly Montane Coniferous F.; San Jacinto and San Bernardino mts., Panamint Mountains, Inyo-White Range, Clark Mt.; e. slope of Sierra Nevada; to Utah. April–May.

    2. **A. macrophýllum** Pursh. [*A. dactylophyllum, A. leptodactylon*, and *A. politum* Greene.] BIG-LEAF MAPLE. CANYON MAPLE. Plate 5, Fig. E. Broad tree, 5–25 m. tall; lvs. roundish in outline, darker green above, mostly 1–2 dm. wide, palmately 3–5-parted, pubescent beneath, few- and remote-toothed; racemes many-fld.; petals ca. 3 mm. long; stamens 7–9, villous below; samara hispid on the body, wing 2–2.5 cm. long; $2n = 26$ (Wright, 1957).—Common on stream banks and in canyons, below 5000 ft.; Chaparral, Coastal Sage Scrub, etc.; most of cismontane s. Calif. except valley plains; to Alaska. April–May. Very variable.

    3. **A. negúndo** L. ssp. **califórnicum** (T. & G.) Wesmael. [*Negundo c.* T. & G. *A. n.*

var. *c.* Sarg.] BOX ELDER. Round-headed tree 6–15 m. tall; twigs rather slender, pubescent, greenish; lvs. pinnately 3-foliolate, the terminal lft. largest, 3–5-lobed, ovate, 5–12 cm. long, coarsely serrate, long-petiolulate, the lateral smaller, short-petiolulate, all densely pubescent especially beneath and when young; petioles 2–8 cm. long; fls. unisexual, the ♂ fascicled, the ♀ drooping-racemose, appearing slightly before the lvs., greenish; pedicels filiform; stamens 4–5; samara strawcolor when mature, finely pubescent, 2.5–3 cm. long.—Along streams and bottom lands, below 6000 ft., lower edge of Yellow Pine F. and lower; local in San Jacinto and San Bernardino mts., Santa Barbara and Kern cos.; n. to n. Calif. March–April.

3. Aizoàceae. CARPET-WEED FAMILY

Herbs or low shrubs, erect or prostrate, often fleshy. Lvs. alternate or opposite or whorled, with or without stipules. Fls. usually perfect, regular. Fl.-tube free or adnate to the ovary. Sepals 4–8, imbricate. Petals 0 to many, linear. Stamens perigynous, many in several series, few in 1 series, or 1; anthers small, 2-celled, opening lengthwise. Ovary superior or inferior, 3–several-loculed; styles 3–many. Fr. a caps. or drupaceous or nutlike, often clasped by the persistent calyx. Seeds usually many, embryo annular; endosperm copious to scant. A large family, mainly trop. and of S. Hemis.

Petals present, numerous; ovary inferior.
    Lvs. largely alternate.
        Plants annual; herbage with shining colorless papillae; fls. white to reddish; placentation axile .... 5. *Gasoul*
        Plants perennial; herbage smooth; fls. bright yellow; placentation parietal ...... 6. *Herrea*
    Lvs. largely opposite; perennials; herbage often smoothish.
        Lf.-blades flat, expanded; fls. red, 1 cm. wide .... 1. *Aptenia*
        Lf.-blades semiterete or 3-sided, linear.
            Lvs. 3-angled; stigmas and ovary-locules 8–20 .... 2. *Carpobrotus*
            Lvs. mostly semiterete; stigmas and ovary-locules 4–6.
                Fls. pale pink, less than 2 cm. in diam........................... 4. *Drosanthemum*
                Fls. more intensely colored, larger.
                    The fls. orange-red, to 5 cm. diam........................... 4. *Drosanthemum*
                    The fls. not orange-red, to 4 cm. diam.
                        Lvs. 2.5–3.5 cm. long, 5 mm. wide; fls. rose-red .... 3. *Disphyma*
                        Lvs. 1.5–2.5 cm. long, 2 mm. wide; fls. intense-red .... 7. *Lampranthus*
Petals none; calyx sometimes colored within and petaloid.
    Lvs. alternate; fl.-tube united to ovary; lvs. triangular-ovate .... 10. *Tetragonia*
    Lvs. opposite or whorled; fl.-tube free from ovary.
        The lvs. whorled; fl.-tube nearly or quite wanting; caps. loculicidal .... 8. *Mollugo*
        The lvs. opposite; fl.-tube evident; caps. circumscissile.
            Stipules present, scarious; stamens 5–10; lvs. obovate .... 11. *Trianthema*
            Stipules absent; stamens 5–many; lvs. spatulate .... 9. *Sesuvium*

1. Aptènia N. E. Br.

Perennials, suffrutescent, prostrate. Lvs. petioled, flat, cordate-ovate, acute, opposite. Fls. terminal or lateral, short-stalked. Fl.-tube obconic; sepals unequal, the 2 larger flat, the others subulate. Petals purple. One sp. (Greek, *apten*, wingless.)

1. **A. cordifòlia** (L.f.) N. E. Br. [*Mesembryanthemum c.* L. f.] Stems 3–6 or more dm. long; lvs. 1–3 cm. long; peduncles 8–15 mm. long; petals ca. 5 mm. long; caps. 1.3–1.5 mm. high; seeds 1 mm. long or more; $2n=18$ (Snoad, 1951).—Occasional escape from cult. as at Santa Catalina Id., La Jolla, San Diego, Ventura; native of S. Afr. April–May.

2. Carpobròtus N. E. Br.

Subshrubs with long, 2-angled, prostrate stout branches. Lvs. large, opposite, united at base, straight or curved saber-shaped, 3-angled, thick, with entire or dentate margins.

Fls. large, solitary. Stigmas 10–16, plumose; fr. fleshy, indehiscent. Ca. 25 spp., largely of S. Hemis.

Lvs. straight, not serrate, 3–5 cm. long; fls. rose-magenta, 3–5 cm. broad ..... 1. *C. aequilateris*
Lvs. curved, serrate on lower angle, becoming 7–10 cm. long; fls. yellow or purple, 8–10 cm. broad .... 2. *C. edulis*

1. **C. aequilaterus** (Haw.) N. E. Brown. [*Mesembryanthemum a.* Haw.] [*C. chilénsis* (Mol.) N. E. Br. *Mesembryanthemum c.* Mol.] SEA-FIG. Perennial with stems 1 m. or so long, forming extensive mats; lvs. 3–5 cm. long, 3-sided; fls. terminal, sessile or short-peduncled, 3–5 cm. broad; fl.-tube turbinate, 2–3 cm. long; sepals unequal, the larger foliaceous; petals magenta; ovary 8–10-loculed; seeds obovoid, somewhat compressed.—Sand dunes and bluffs along the coast; Coastal Strand, Coastal Sage Scrub, etc.; L. Calif. to Ore.; Chile. April–Sept.

2. **C. édulis** (L.) Bolus. [*Mesembryanthemum e.* L.] HOTTENTOT-FIG. Plate 5, Fig. F. Like *C. aequilateris*, but the lvs. somewhat curved, serrate on lower angle, 6–10 cm. long; fls. 8–10 cm. in diam., yellow, drying pink; fr. becoming yellow, edible.—Much planted along highways and banks to control erosion; natur. on dunes along coast; from S. Afr. April–Oct.

3. *Disphỳma* N. E. Br.

Creeping caespitose herbs, the branches often rooting at the nodes. Lvs. opposite, little connate at the base, semicylindric at the base, triquetrous at apex; peduncles solitary, subcompressed, thickened upward; calyx subturbinate, the 5 segms. unequal; petals 2-seriate, ca. twice as long as calyx-segms. Ca. 3 spp. of S. Afr.

1. **D. crassifòlium** (L.) L. Bol. [*Mesembryanthemum c.* L.] Prostrate, the stems 3 or more dm. long; lvs. bright green, 2.5–3.5 cm. long; peduncles 2.5–3.5 cm. long; $2n=36$ (Snoad, 1951).—Escape as at Ojai, Ventura; native of S. Afr. Aug.–Oct.

4. *Drosánthemum* Schwant.

Succulent perennials, woody at base, the stems branching, rough with small pustules; lvs. compressed, 3-angled or terete, with glistening papillae; fls. 1–3, usually pedicelled; stigmas 4–6; caps. 4–6-celled, the valves with winged keels. Ca. 100 spp. of S. Afr.

Fls. orange-red; lvs. 2–4 cm. long .... 2. *D. speciosum*
Fls. rose; lvs. 1–1.5 cm. long .... 1. *D. floribundum*

1. **D. floribúndum** (Haw.) Schwant. [*Mesembryanthemum f.* Haw.] Stems slender, subdecumbent, 1–2 dm. long; lvs. cylindrical, somewhat curved, 1–1.5 cm. long, obtuse, glittering from minute papillae; peduncles solitary, 2–3 cm. long; fls. 2–2.5 cm. in diam.; calyx-segms. equal, linear, obtuse; petals 1–2-seriate, rose.—Established near the beach, Santa Barbara and Ventura cos.; native of S. Afr. Summer.

2. **D. speciòsum** (Haw.) Schwant. [*Mesembryanthemum s.* Haw.] Suffruticose; stems to 6 dm. long; lvs. semiterete, 2–4 cm. long; fls. solitary, terminal, peduncled, orange-red, ca. 4 cm. across.—Natur. along coastal bluffs in Orange Co.; native S. Afr. Oct.

5. *Gasoùl* Adans.

Annual or biennial tall herbs, usually erect, sometimes prostrate, often thick and fleshy, branched; lvs. alternate or opposite, rather large; fls. solitary or numerous, in ± elongate inflorescences, white, pink or red, rarely yellow. Petals often filiform. Whole plant or parts of it, glistening with papillae.

Lvs. ovate to broadly spatulate, flat .... 1. *G. crystallinum*
Lvs. linear, semiterete .... 2. *G. nodiflorum*

1. **G. crystallìnum** (L.) Rotm. [*Mesembryanthemum c.* L. *Cryophytum c.* N. E. Br.] ICE-PLANT. Very succulent annual, prostrate, much-branched, the branches 2–6 dm. long,

with large vescicles; lvs. ovate or spatulate, 2–10 cm. long, narrowed to a short amplexicaul base or somewhat petioled; fl.-tube campanulate, 8–12 mm. long; petals white to reddish, 6–8 mm. long; caps. 5-loculed; seeds brown, round-papillate, ca. 0.8 mm. long; $2n = 18$ (Sugiura, 1936).—More or less saline places, Coastal Strand, Coastal Sage Scrub; along the coast from L. Calif. to cent. Calif. Apparently natur. from S. Afr. March–Oct.

2. **G. nodiflòrum** (L.) Rotm. [*Mesembryanthemum n.* L. *Cryophyton n.* L. Bolus.] Annual, freely branched from base, the branches 0.5–2 dm. long, suberect to procumbent, covered with fine colorless vescicles; lvs. linear, subterete, 1–2 cm. long, 1–2 mm. thick; fls. solitary in axils, subsessile or on short stout pedicels; fl.-tube turbinate, 4–5 mm. long; sepals 4–5 mm. long; petals white, ca. 4 mm. long; ovary 5-loculed; seeds finely papillate; $n = 18$ (Reese, 1957).—Coastal Strand, Coastal Sage Scrub; Santa Barbara Co. to L. Calif.; Santa Barbara Id., Catalina Id.; probably introd. here from S. Afr. April–Nov.

6. *Hérrea* Schwant.

Decumbent, glabrous, perennial, succulent herbs. Lvs. opposite to alternate, terete to semiterete. Peduncles straight to sigmoidally bent. Fls. expanding in afternoon, closing at sunset. Petals very numerous, in several series, often pallid in lower part, sulphur or golden in upper part. Ca. 25 spp. of S. Afr.

1. **H. elongàta** (Haw.) L. Bol. [*Mesembryanthemum e.* Haw. *M. pugioniforme* DC.] Stems 3 dm. long, dying off after flowering, little branched; lvs. loosely arranged, alternate, spreading, curved, ± amplexicaul at base, 10–15 cm. long, 6 mm. wide, slightly grooved on upper surface; pedicels to 15 cm. long; petals shining sulphur-yellow, ca. 2.5 cm. long. —Becoming natur. from Santa Barbara Co. n.; native of S. Afr. May–Oct.

7. *Lampránthus* N. E. Br.

Perennial branching herbs, somewhat shrubby at base; lvs. slightly united at base, very narrow; fls. solitary or sometimes 3, large and showy, in various colors; stigmas usually 5, subulate or plumose, often inconspicuous; caps. 5 celled, with valves slightly winged on the keels. Over 100 spp. of S. Afr. (Greek, *lampro-*, and *anthos*, glossy fl.)

1. **L. coccíneus** (Haw.) N. E. Br. [*Mesembryanthemum c.* Haw.] Stems erect, 4–6 dm. tall, lvs. 3-angled, 1.2–2.5 cm. long, ca. 2 mm. wide, grayish green and punctate; fls. on pedicels 5–10 cm. long, scarlet, with 2 lf.-like bracts near the middle, 3.5–4 cm. in diam.— Natur. along the coast as in San Diego and Los Angeles cos.; native of S. Afr.

8. *Mollùgo* L. Indian-Chickweed

Low glabrous, much-branched annuals. Lvs. whorled or alternate. Fls. axillary on slender pedicels. Sepals 5, white inside. Petals 0. Stamens hypogynous, 5 and alternate with the sepals, or 3 and alternate with the locules of the ovary. Stigmas 3. Caps. 3-loculed, 3-valved, loculicidal, the partitions breaking away from the many-seeded axis. Seeds without a strophiole. Ca. 12 spp., mostly trop. (Old name for Galium *Mollugo*, transferred to this genus, perhaps because of whorled lvs.)

Lvs. linear; plant erect or ascending .... 1. *M. cerviana*
Lvs. spatulate; plant prostrate .... 2. *M. verticillata*

1. **M. cerviàna** (L.) Ser. [*Pharnaceum c.* L.] Stems very slender, 4–15 cm. high; lvs. glaucous, linear, 5–10 in a whorl, the basal linear-spatulate, 3–15 mm. long; fls. whorled; pedicels filiform; sepals 1.5 mm. long; caps. subglobose; seeds 0.4–0.5 mm. long; $2n = 18$ (Sugiura, 1936).—Sandy flats below 5000 ft.; Creosote Bush Scrub, Joshua Tree Wd., Yellow Pine F.; scattered stations, desert and cismontane; Sept.–March. Natur. from Old World.

2. **M. verticillàta** L. Prostrate, forming mats 1–3 (–5) dm. across; lvs. spatulate, 5–6 in a whorl, unequal, 5–20 mm. long, short-petioled; pedicels filiform, several at a node;

sepals oblong, 2 mm. long; caps. slightly exceeding sepals, ovoid; seeds reniform, shining or slightly granular, brown, ca. 0.6 mm. long; $2n=64$ (Sugiura, 1936).—Waste places and fields, becoming widely natur. in Calif.; from trop. Am. April–Nov.

9. Sesùvium L. Sea-Purslane

Usually prostrate or decumbent fleshy herbs or subshrubs. Lvs. opposite, exstipulate. Fls. solitary, axillary. Fl.-tube turbinate. Sepals 5, persistent, purplish or rose-pink inside. Petals 0. Stamens 5–many, inserted on the fl.-tube. Styles 3–5, separate. Caps. 3–5-locular, circumscissile at middle. Seeds minute, smooth, several to many in each locule. Ca. 5 spp., largely maritime. (Name unexplained.)

1. **S. verrucòsum** Raf. [*S. sessile* auth., not Pers.] Plate 6, Fig. A. Perennial, glabrous, freely branched; stems 1–5 dm. long or more, papillose; lvs. broadly spatulate, 1–4 cm. long; fls. almost sessile, 8–10 mm. long; sepals scarious-margined, short-horned near apex; caps. conic, ca. 5 mm. high; seeds black, shining, 1 mm. long.—Low, ± saline places; V. Grassland, Coastal Sage Scrub, Alkali Sink, etc.; cismontane Calif. to L. Calif. and Sacramento V., deserts; to Tex., Mex. April–Nov.

10. Tetragònia L. Sea-Spinach

Annual or perennial herbs or subshrubs. Lvs. alternate, often fleshy. Fls. axillary, solitary or few, sessile or not. Fl.-tube adnate to ovary, fleshy. Sepals 3–5. Petals 0. Stamens 1–many, perigynous. Ovary inferior when mature, 3–9-loculed; styles 3–9. Ovule 1 in each locule. Fr. nutlike, indehiscent, in ours 2–5-horned. Seeds subreniform. Ca. 60 spp. of S. Hemis., especially S. Afr. (Greek, *tetra*, four, and *gonu*, knee, or angle, referring to fr.)

1. **T. tetragònioides** (Pallas) Kuntze. [*T. expansa* Murr.] New-Zealand-Spinach. Succulent annual with many spreading or procumbent branches 3–several dm. long; herbage with small crystalline vescicles; lvs. triangular-ovate, entire or undulate, 2–5 cm. long, abruptly contracted at base into a cuneately winged petiole; fls. subsessile, solitary in axils, yellow-green; sepals spreading, 1.5–2.5 mm. long; fr. 8–10 mm. long, 2–5-horned; seeds brownish, ca. 2 mm. long; $2n=32$ (Sugiura, 1936).—Natur. along beaches and near salt marshes, along the coast to Ore; native of se. Asia; Australasia. April–Sept. Santa Cruz and San Miguel ids.

11. Triánthema L. Horse-Purslane

Prostrate herbs or subshrubs, branched from base. Lvs. opposite, those of each pair unequal; stipules present. Fl. solitary, axillary. Fl.-tube short. Sepals 5. Petals 0. Stamens 5–10, perigynous. Ovary truncate, 1–2-locular; styles 1–2. Caps. short-cylindric or turbinate, 1–5-seeded, tardily circumscissile, the upper part thick and tough, usually with 2 marginal crests. Seeds reniform, roughened. Ca. 15 spp., mostly trop. (Greek, *treis*, three, and *anthemon*, fl.)

1. **T. portulacástrum** L. Diffusely branched herb, annual, glabrous except for the rows of soft hairs, the branches rather succulent, 2–6 dm. long; lvs. obovate, 1–2 cm. long, smaller on axillary branchlets; petioles short, dilated at base into bidentate stipular expansions; fls. small, purplish within, solitary in axils; sepals 2.5 mm. long; caps. cylindrical, 4 mm. long; seeds black, rough, ca. 2 mm. in diam.—Occasional in saline spots; Alkali Sink, etc.; Imperial V. to Kern and Tulare cos.; Loma Linda, San Bernardino Co.; to Tex., trop. Am. May–Nov.

4. Amaranthàceae. Amaranth Family

Herbs or shrubs. Lvs. alternate or opposite, exstipulate, simple, entire. Fls. small, perfect or unisexual, inconspicuous, in ours congested in clusters or spikes, with 3 dry, scarious, persistent, often colored bracts. Calyx commonly of 5 sepals, sometimes 1–4, persistent, usually scarious. Petals 0. Stamens usually as many as the sepals. Ovary superior, 1-celled,

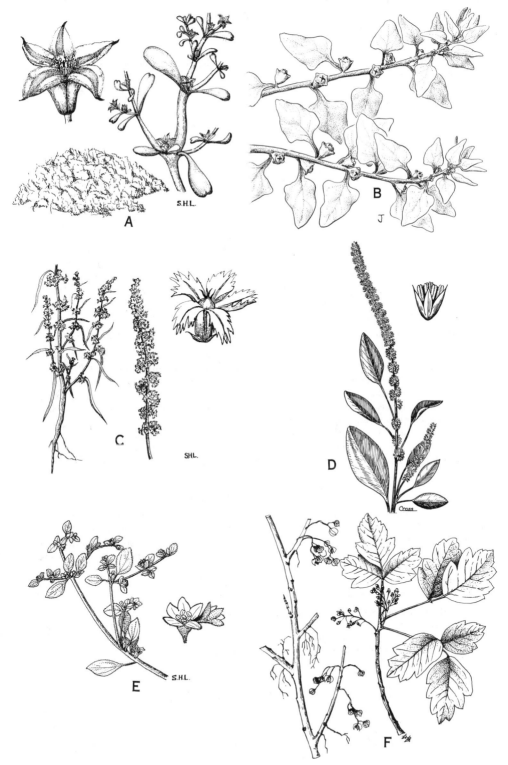

Plate 6. Fig. A, *Sesuvium verrucosum*. Fig. B, *Tetragonia expansa*. Fig. C, *Amaranthus fimbriatus*. Fig. D, *Amaranthus palmeri*. Fig. E, *Tidestromia oblongifolia*. Fig. F, *Toxicodendron diversilobum*.

usually with 2–3 stigmas. Fr. a membranous or fleshy utricle, indehiscent, irregularly dehiscent, or circumscissile. Seed 1, with copious endosperm. Often weedy plants, a few grown for ornament, as pot-herbs, or in Latin Am. for their edible seed. Ca. 40 genera and 500 spp., largely of warm climates.

Lvs. alternate; anthers 4-celled; plants nearly or quite glabrous .... 2. *Amaranthus*
Lvs. largely opposite; anthers 2-celled; plants white stellate-woolly or villous.
    Fls. glomerate, with an invol. of upper lvs.
        Stamens hypogynous. Mostly desert plants .... 4. *Tidestromia*
        Stamens perigynous. Occasional cismontane introd. .... 3. *Brayulinea*
    Fls. in axillary headlike spikes, without invol. Cismontane introd. .... 1. *Alternanthera*

1. Alternánthera Forsk.

Herbs or subshrubs, erect to prostrate. Lvs. opposite, sessile or petioled. Fls. bracteate and bibracteolate, perfect, in short axillary spikes. Calyx of 5 often unlike sepals. Stamens mostly 5, sometimes 2 or 3; the fils. united into a tube with staminodia alternating with stamens. Utricle membranaceous, indehiscent; stigma mostly capitate. Seeds inverted, smooth. Ca. 170 spp. of warmer parts of New World. (Latin, *alternus*, alternate, referring to alternate stamens and staminodia.)

Heads sessile; staminodia shorter than the fils. .... 1. *A. peploides*
Heads long-peduncled; staminodia not shorter than fils. .... 2. *A. philoxeroides*

    1. A. peploìdes (H. & B.) Urban. [*Illecebrum p.* H. & B. *A. repens* auth.] Perennial with prostrate stems and thick woody vertical root; stems branched, pubescent, 2–5 dm. long; lvs. many, rhombic-ovate, green, those of the pair very unequal, the blades 1–3 cm. long, petioles shorter; heads short-cylindric or ovoid, 5–15 mm. long, sessile, whitish; bracts ovate, ciliate; sepals unequal, ovate, villous on the 3 nerves, 3–5 mm. long.—Occasional as an escape at low elevs., cismontane; to s. Atlantic Coast, Mex. Aug.–Dec.

    2. A. philoxeroìdes (Mart.) Griseb. ALLIGATOR WEED. Subglabrous aquatic plant with lvs. 3–11 cm. long; heads on long peduncles.—Introd. in Rio Hondo, Los Angeles Co.; in Tulare Co., etc.; sometimes choking irrigation canals; se. U.S.; S. Am.

2. Amaránthus L. AMARANTH

Annual, usually coarse herbs, erect to prostrate, usually branched. Lvs. alternate, entire, petioled. Fls. 3-bracted, small, green to purplish, in glomerules in axillary or terminal spiked clusters, commonly with ♂ and ♀ fls. in same cluster. Calyx glabrous. Stamens 5, 3 or 2, separate. Stigmas 2 or 3. Fr. a 1-seeded ovoid utricle, 2–3-beaked at apex and usually exceeding calyx. Seed erect, compressed, smooth; embryo coiled into a ring around the endosperm. Ca. 50 spp., widespread except in cold regions. (Greek, *amarantos*, unfading, because of the dry persistent calyx and bracts.)

1. Sepals of ♀ fls. lacking; fls. dioecious; plants erect, stout, to 1.5 m. tall; lvs. narrow, to 1 dm. long. Rare introd. .... 15. *A. tamariscanus*
1. Sepals present on ♂ fls. and perfect fls.
    2. Sepals of ♀ fls. spatulate, the claw narrow, the calyx ± urceolate.
        3. Plants dioecious.
             4. Bracts 2–3 times as long as the pistillate calyx.
                  5. Infl. not leafy; bracts rigid and spinose; plants not viscid-pubescent . 10. *A. palmeri*
                  5. Infl. leafy at least below; bracts not spinose; plants viscid-pubescent
                                                                                      16. *A. watsonii*
            4. Bracts not longer than pistillate calyx .... 2. *A. arenicola*
        3. Plants monoecious.
            6. Bracts equaling or exceeding sepals of ♀ fls., these sepals not fimbriate . 12. *A. pringlei*
            6. Bracts shorter than sepals of ♀ fls., these sepals fimbriate .... 8. *A. fimbriatus*
    2. Sepals of ♀ fls. oblong or obovate, not with a narrow claw, the calyx not urceolate; plants monoecious.

7. Axils of many lvs. bearing pair of sharp spines .... 14. *A. spinosus*
7. Axils of lvs. without spines.
  8. Utricle indehiscent, fleshy; spikes terminal; prostrate plant with slender stems
                                                                                                                           7. *A. deflexus*
  8. Utricle dehiscent, the top falling away like a lid.
    9. Infl. of terminal or axillary spikes; sepals 5.
      10. Style-branches recurved; lateral branches of infl. few to none.
        11. Base of style-branches slender, forming shallow saddle; midrib of bract very slender, rather long-excurrent; sepals recurved; lateral spikes of infl. few to none
                                                5. *A. caudatus*
        11. Bases of style-branches stout, forming cleft at summit of broad tower; midrib of bract very thick, excurrent; sepals straight; lateral spikes of infl. long, few, widely spaced .... 11. *A. powellii*
      10. Style-branches erect; lateral spikes of infl. numerous, crowded.
        12. Sepals oblong, acute, straight; midrib of bract long, excurrent; style branches with slender bases.
          13. Lateral spikes of infl. long; sepals 1.5 mm. long; bracts to 1 and ½ times as long as sepals, usually red or purple .... 6. *A. cruentus*
          13. Lateral spikes of infl. short; sepals 1.5–2 mm. long; bracts twice as long as sepals, usually green or pale reddish .... 9. *A. hybridus*
        12. Sepals narrowly obovate, emarginate, recurved; midrib of bracts barely excurrent; style-branches with moderately stout bases .... 13. *A. retroflexus*
    9. Infl. wholly of axillary glomerules.
      13a. Sepals 4–5 in both ♂ and ♀ fls.; stems prostrate; seeds 1.5 mm. broad
                                                                           3. *A. blitoides*
      13a. Sepals 1–3; seeds less than 1 mm. broad.
        14. The sepals of ♀ fls. 2–3, but only 1 well developed; stems slender, prostrate
                                                                      4. *A. californicus*
        14. The sepals of ♀ fls. 3, subequal; stems ascending .... 1. *A. albus*

1. **A. álbus** L. [*A. graecizans* many auth.] TUMBLEWEED. Stems erect or ascending, bushy-branched, pale, 2–15 dm. long; lvs. elliptic to spatulate or obovate, obtuse or retuse, 1–6 cm. long, the petioles short; monoecious, the fls. greenish, in small axillary clusters; bracts subulate, green, rigid, 2–4 times as long as sepals; sepals 3, scarcely 1 mm. long; stamens 3; utricle rugose, exceeding calyx; seeds round, dark red-brown, shining, ca. 0.8 mm. in diam.; $2n=32$ (Heiser & Whitaker, 1948), 34 (Sharma & Banik, 1965).—Common weed in cult. and waste places through much of N. Am.; native of trop. Am. May–Oct. Blowing about as a tumbleweed when dry.

2. **A. arenícola** Jtn. Differing from *A. palmeri* and *A. watsonii* (both of which have bracts 2–3 times as long as the ♀ calyx) in the bracts not exceeding the ♀ calyx which is 2.5–3 mm. long.—Reported as adventive in Santa Barbara Co.; native Iowa to Colo. and Texas.

3. **A. blitoìdes** Wats. [*A. graecizans* Am. auth., not L.] Stems prostrate, 1–5 dm. long, branched from base, glabrous, often purplish; lvs. many, often crowded, the petioles 2–15 mm. long, the blades spatulate to obovate, 8–25 mm. long, pale green; monoecious; fls. in dense axillary clusters usually shorter than petioles; bracts ovate-oblong, ca. as long as sepals; sepals 4–5, oblong, acute or acuminate, 2–3 mm. long; utricle scarcely rugose, equaling calyx; seeds round, dull black, 1.5 mm. broad; $2n=32$ (Grant, 1959).—Occasional native weed in waste and cult. places below 5000 ft.; many Plant Communities; widely scattered in Calif.; to Wash., Rocky Mts., Mex. May–Nov.

4. **A. califórnicus** (Moq.) Wats. [*Mengea c.* Moq. *A. carneus* Greene. *A. albomarginatus* Uline & Bray.] Stems prostrate, glabrous, much-branched, 5–20 cm. long, forming whitish or somewhat reddish mats; lvs. many, crowded, the petioles 2–10 mm. long, the blades obovate to oblong, obtuse or rounded, 5–15 mm. long; monoecious, the fls. in small axillary clusters; bracts lanceolate, subulate-tipped, ca. 1 mm. long; sepals 2–3 in ♂ fls., usually only 1 well developed in ♀ fls. and barely 1 mm. long; utricle smooth, bursting irregularly, subglobose, often red or purple; seeds round, dark red-brown, 0.7 mm. in diam.—On dried mud, moist flats, etc.; often at low elevs., but reaching 8000 ft.; many

Plant Communities; cismontane and in Coachella V.; San Diego Co. to Wash., Alta, etc. July–Oct.

5. **A. caudàtus** L. Infl. thick and pendulous, terminal spike long, laterals few and short or absent; bract short or medium, with slender, rather long-excurrent midrib; sepals recurved, broadly obovate or spatulate, obtuse to emarginate, 1.5–2 mm. long.—Reported from San Antonio Canyon, San Gabriel Mts. at 4400 ft. and occasional in Ariz.; to cent. Am. July.

6. **A. cruéntus** L. Infl. lax, the terminal spike short, the laterals long, very numerous and crowded; bracts extremely short, not much longer than the sepals, with long-excurrent midrib; sepals straight, 1.5 mm. long, oblong, acute.—To be looked for in Calif.; widely introd. in N. Am. Native in Cent. and S. Am.

7. **A. defléxus** L. Stems slender, much branched, ascending or decumbent, subglabrous or pubescent above; lf.-blades rhombic-ovate to lanceolate, obtuse, 5–20 mm. long, on slender petioles ca. as long; bracts ovate, acute, cuspidate, ca. 2 mm. long; sepals 2–3, oblong, ca. as long as bracts; utricle oblong, indehiscent, much exceeding calyx, 3–5-nerved, with fleshy walls; seeds dark red-brown, oval, 1 mm. long; $2n=34$ (Grant, 1959).—Weed in gardens, along streets, etc. at widely scattered stations, cismontane s. Calif.; natur. from Old World. May–Nov.

8. **A. fimbriàtus** (Torr.) Benth. [*Sarratia berlandieri* var. *f.* Torr. *Amblogyna f.* Gray.] Plate 6, Fig. C. Stems slender, erect, 1–5 dm. high, simple or branched above, glabrous or slightly puberulent above; lvs. narrowly lanceolate to linear, bright green, the blades 2–6 cm. long, the petioles 1–2 cm.; monoecious; fls. in loose clusters, these scattered or in loose spikes; bracts ovate, half as long as calyx; sepals 2.5–3 mm. long, fimbriate, often rose or lavender, obtuse in ♀, acute in ♂ fls.; stamens 3; utricle circumscissile; seeds roundish, 0.8 mm. diam.—Occasional, dry gravelly places, below 5000 ft.; Creosote Bush Scrub, Joshua Tree Wd., etc.; Colo. and e. Mojave deserts; to Utah, Mex. Aug.–Nov.

9. **A. hýbridus** L. Much like *A. retroflexus*, the stems more slender, often reddish at base, 5–15 dm. high, glabrous or pubescent; lvs. rather dark green, 2–12 cm. long, on somewhat shorter petioles; infl. very many-fld., of few to mostly numerous spikes in panicles, the spikes often crowded, tawny-green, the lateral ± ascending, 2–6 cm. long, 6–12 mm. thick, the terminal usually longer and thicker; bracts almost twice as long as sepals; ♀ fls. 1.5–2 mm. long, acute; utricle not exceeding calyx; seed round, black, 1 mm. in diam.; $2n=32$ (Covas & Schnock, 1946), (Grant, 1959).—Waste and cult. ground at low elevs.; through much of cismontane Calif.; semicosmopolitan weed. June–Nov.

10. **A. pálmeri** Wats. Plate 6, Fig. D. Stems stout, erect, 2–15 dm. high, pale green, glabrous, or villous about the spikes, simple or with short erect branches; lvs. broadly ovate to rhombic or lanceolate, the blades 1.5–5 cm. long on petioles 2–6 cm. long; dioecious; the fls. in long, erect, nearly continuous spikes 1–3 dm. long; bracts usually 2–3 times as long as fls., linear to ovate, with stout rigid spinose tips; calyx 2–3 mm. long, the ♂ sepals acute, the ♀ obtuse; stamens 5; utricle subglobose, circumscissile, rugose at summit; seeds 1.3 mm. in diam.; $2n=34$ (Grant, 1959).—Weed in waste places and fields, at low elevs.; deserts and interior valleys; Creosote Bush Scrub, V. Grassland, Chaparral, etc.; s. Calif. to cent. U.S., Mex. June–Oct.

11. **A. powéllii** Wats. Much like *A. retroflexus*, but with the terminal panicle of few erect stiff spikes, the cent. one much elongate, 1–2.5 dm. long, the lateral 4–12 cm. long; bracts 2–3 times the length of the calyx; sepals of ♀ fls. ca. 3 mm. long, acute, tapering gradually at apex; stamens usually 3; seeds 1.1–1.3 mm. in diam.; $2n=34$ (Grant, 1959).—Occasional weed in waste places, cismontane Calif.; to Ore., Wyo., Mex. July–Oct.

12. **A. prínglei** Wats. Stems stout, erect, glabrous, much branched, 1.5–2 dm. high; petioles 3–25 mm. long; blades elliptic, oblong to linear, 1–4.5 cm. long; monoecious; fls. in dense axillary glomerules and short terminal spikelike panicles; bracts equal to or longer than fls., acute, rigidly spinose; sepals 5, those of ♂ fls. oblong, acute; of ♀ fls. broadly spatulate, 1.5–2 mm. long, often purplish; stamens 3; utricle ovoid, circumscissile near middle; seed orbicular, lenticular, 1 mm. in diam., black, shining.—Dry places, at

Amaranthus

5300 ft.; Pinyon-Juniper Wd.; Clark Mt., e. Mojave Desert; to Nev., Tex., Mex. Aug.–Sept.

13. **A. retrofléxus** L. Stout, usually branched, erect, rough-pubescent, 3–15 dm. tall, with erect or ascending branches; lf.-blades ovate or oblong-ovate, 3–10 cm. long, the uppermost narrower; petioles 1–7 cm. long; monoecious; fls. in dense spikes, 8–20 mm. thick and crowded into terminal lobulate panicles with crowded lateral spikes 1–5 cm. long; bracts ovate, subulate, ca. twice as long as sepals; sepals 5, linear-oblong, often mucronate, ca. 3 mm. long; stamens 5; utricle rugulose, shorter than sepals; seeds compressed, rounded or obovate, black, ca. 1 mm. long; $2n=32$ (Heiser & Whitaker, 1948; Grant, 1959).—Garden and orchard weed, also in waste places; native of trop. Am. June–Nov.

14. **A. spinòsus** L. Glabrous, bushy, with reddish stems 2–12 dm. long; lvs. lance-ovate to rhombic-ovate, 2–10 cm. long, dull green, with pair of axillary spines; fertile clusters round, axillary; upper clusters sterile, spicate; fls. yellow-green, 1.5 mm. long; utricle thin, bursting irregularly; seeds smooth, black, shining, 0.8–1 mm. in diam.; $2n=34$ (Takagi, 1933; Sharma & Banik, 1965).—Occasional, as weed in San Fernando and Antelope valleys, Los Angeles Ca.; native of Old World. Fall.

15. **A. tamariscànus** Nutt. [*Acnida t.* Wood.] Stout, erect, much-branched, to 1.5 m. tall; lf.-blades oblong to lanceolate, to 1 dm. long; dioecious; spikes mostly naked (bractless), very slender, stiff and straight, single or panicled, 2–4 dm. long; stamens 5; utricle definitely circumscissile.—Along a roadside, Pala, San Diego Co., *Townsend*; native of cent. states. July–Sept.

16. **A. watsònii** Standl. Erect, glabrous, 2–10 dm. high, much branched; petioles stout, 5–20 mm. long; lf.-blades oblong to elliptic, 2–4 cm. long, yellow-green; fls. in glomerules in stout spikelike terminal panicles; bracts oblong, acuminate, not exceeding fls.; sepals of ♂ fls. oblong, acute, 2–3 mm. long, of ♀ broadly spatulate, 2–2.5 mm. long; stamens 5; utricle subglobose, circumscissile, seed ca. 1 mm. across, black.—Mostly in Creosote Bush Scrub; se. deserts; to L. Calif., Son. April–Aug.

3. *Brayulìnea* Small

Prostrate to decumbent perennial herbs; stems branched at base, often zigzag. Fls. perfect, subtended by bracts and in axillary clusters. Sepals 5, pubescent. Stamens 5, perigynous; fils. broad; anthers 1-celled. Ovary flattened, 1-celled; style short; stigma notched. Utricle membranous, indehiscent. (Named for W. L. *Bray* and E. B. *Uline*, students of N. Am. Amaranthaceae.)

1. **B. dénsa** (Willd.) Small. [*Illecebrum d.* Willd.] Prostrate perennial, densely lanate; stems to 3 dm. long; cauline lvs. wing-petioled, elliptical to broadly oval, 3–15 mm. long; fls. densely glomerate; bracts ovate, scarious, white; calyx 2–2.5 mm. long; utricle included.—Occasional introd. from se. states; to S. Am. July–Oct.

4. *Tidestròmia* Standl.

Annual or perennial herbs or small shrubs, closely grayish stellate-tomentose or villous. Lvs. largely opposite, petioled. Fls. minute, perfect, axillary, mostly glomerate, subtended by leafy involucral bracts and 3 small bractlets. Sepals 5, equal, thin, pubescent. Stamens 5, the fils. united below into a short cup, with 5 short teeth alternating with them. Utricle subglobose, 1-seeded, indehiscent. Three spp. of sw. U.S. and Mex. (Named for I. *Tidestrom*, Am. botanist.)

Plant annual; staminodia almost lacking. W. San Diego Co. . . . . 1. *T. lanuginosa*
Plant perennial; staminodia almost half as long as fils. Deserts . . . . 2. *T. oblongifolia*

1. **T. lanuginòsa** (Nutt.) Standl. [*Achyranthes l.* Nutt. *Alternanthera l.* Moq.] Prostrate or procumbent annual, much branched, the branches 1–5 dm. long, stellate-pubescent; lf.-blades round to round-ovate, 5–30 mm. long, stellate-pubescent, on slender petioles as

long or shorter; glomerules few-fld.; perianth 1–3 mm. long, 3 times the bracts in length; staminodia minute or wanting; seed 0.5 mm. long.—Established on banks of Otay Lake, San Diego Co.; native from Utah to Kans. and Mex. July–Oct.

2. **T. oblongifòlia** (Wats.) Standl. [*Cladothrix o.* Wats.] Plate 6, Fig. E. Perennial from a stout woody taproot, erect to decumbent, much-branched, forming low broad plants 1.5–3 dm. high and ca. twice as wide; lf.-blades oblong to almost round, 1–3 cm. long, obtuse, prominently veined, on somewhat shorter petioles; invol. of 3 broad partly united bracts 3–4 mm. long; calyx 1–2 mm. long; seeds 0.5 mm. long.—Dry sandy places, such as washes, mostly below 2000 ft.; Creosote Bush Scrub; Colo. Desert through e. Mojave Desert to Death V.; Nev., Ariz. April–Dec.

Wiggins recognizes ssp. *cryptantha* (Wats.) Wiggins [*Cladothrix c.* Wats.] for plants about the Salton Sea, as having smaller lvs. (2–10 mm. long); invol. deeper (3–4 mm.); and stems very slender.

5. **Anacardiàceae.** Sumac Family

Trees or shrubs, with acrid, resinous or milky sap and alternate lvs. Polygamo-dioecious or with perfect fls., these small, regular, in axillary or terminal panicles. Calyx commonly 5-cleft, a glandular ring or cuplike disk lining its base. Petals commonly 5, the stamens as many or twice as many. Ovary 1-celled, 1-ovuled, free from calyx and disk. Styles 3, terminal. Fr. a berrylike dry drupe; seed without endosperm. Ca. 70 genera and 600 spp., mostly from warm regions.

Stamens 5; native shrubs, mostly with entire or 3-foliolate lvs.
  Lfts. 3- (rarely 5-) foliolate, deciduous.
    Fls. whitish, in loose axillary panicles; fr. glabrous; whitish; lvs. subglabrous, shining; branches stiff, erect or climbing .... 3. *Toxicodendron*
    Fls. yellowish in sessile spikes; fr. hairy, red; lvs. usually pubescent; branches tending to be arched .... 1. *Rhus*
  Lfts. simple, persistent, coriaceous .... 1. *Rhus*
Stamens 10; introd. trees mostly with pendent pinnate lvs. .... 2. *Schinus*

1. *Rhus* L. Sumac

Polygamous shrubs or small trees. Lvs. simple or compound. Fls. small, the sepals 4–6, mostly 5, persistent. Petals spreading in anthesis. Disk annular. Stamens 5. Drupe small, glabrous or pubescent, with persistent or deciduous exocarp. Seed 1, inverted. Ca. 120 spp., widely distributed. (Greek, *rhous*, ancient name for sumac.)

(Barkley, F. A. A monographic study of Rhus and its immediate allies in N. and Cent. Am. Ann. Mo. Bot. Gard. 24: 265–498. 1937.)

Lvs. 3- (rarely 5-) foliolate, deciduous .... 4. *R. trilobata*
Lvs. simple, persistent, coriaceous.
  Panicle with numerous slender branchlets; fr. glabrous, whitish; lvs. oblong-lanceolate, almost 3 times longer than wide .... 2. *R. laurina*
  Panicle with fewer stout branchlets; fr. pubescent red.
    Lfts. oblong-elliptic .... 1. *R. integrifolia*
    Lfts. ovate, somewhat folded along midrib .... 3. *R. ovata*

1. **R. integrifòlia** (Nutt.) Benth. & Hook. [*Styphonia i.* Nutt. *Schmaltzia i.* Barkl.] Lemonadeberry. Plate 7, Fig. A. Rounded aromatic shrub, 1–3 m. high, with finely pubescent, ± reddish stoutish twigs; lvs. coriaceous, flat, subentire or with shallow sharp teeth, oblong-ovate, rounded-obtuse at both ends, subglabrous, 2.5–5 cm. long, 2–3 cm. wide; sometimes ± lobed; petioles 3–6 mm. long; fls. in close panicles, rose to white, subtended by roundish hairy bracts; petals ciliolate, ca. 3 mm. long; drupe viscid, acid, pubescent, reddish, flattened, ca. 10 mm. in diam.; stone flat, ca. 7 mm. long.— Ocean bluffs and canyons, dry places under 2600 ft.; inland to w. Riverside Co.; Coastal Sage Scrub, Chaparral; Santa Barbara Co. to L. Calif. Feb.–May. Channel Ids.

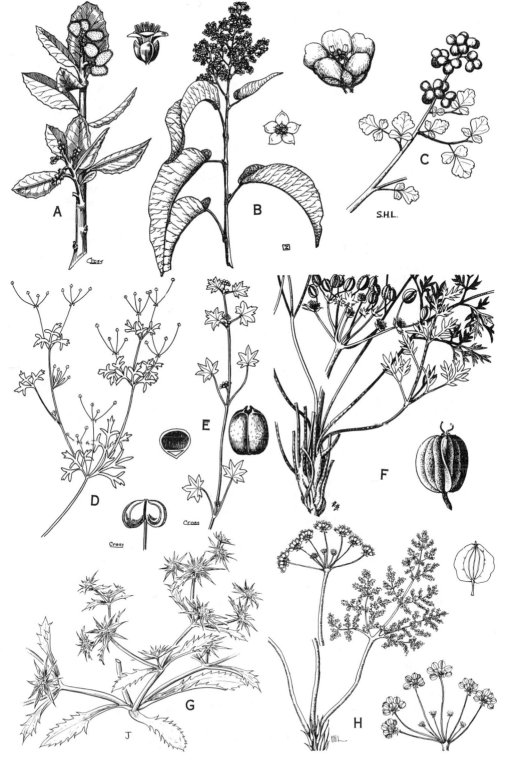

Plate 7. Fig. A, *Rhus integrifolia*. Fig. B, *Rhus laurina*. Fig. C, *Rhus trilobata*. Fig. D, *Apiastrum angustifolium*. Fig. E, *Bowlesia incana*. Fig. F, *Cymopterus panamintensis*. Fig. G, *Eryngium armatum*. Fig. H, *Lomatium mohavense*.

2. **R. laurìna** Nutt. in T. & G. [*Malosma l.* Nutt. ex Abrams.] LAUREL SUMAC. Plate 7, Fig. B. Large rounded evergreen shrub or small tree, 2–5 m. tall, glabrous, aromatic; twigs and veins reddish; lvs. lance-oblong, bicolored, 4–10 cm. long, 2–4 cm. wide, entire, mucronate, somewhat folded along the midrib; petioles 1–3 cm. long; panicle dense, intricately branched, 5–15 cm. long; fls. white, ca. 1 mm. long; drupe whitish, glabrous, 2–3 mm. in diam.; stone smooth, ca. 2 mm. long, flattish.—Dry slopes below 3000 ft.; Chaparral, Coastal Sage Scrub; cismontane slopes from L. Calif. to Santa Barbara Co.; inland to desert edge. June–July. Quite susceptible to heavy frost. Catalina Id.

3. **R. ovàta** Wats. [*Neostyphonia o.* Abrams. *Schmaltzia o.* Barkl. and var. *traskiae* Barkl.] SUGAR BUSH. Evergreen shrub to small tree, 1.5–3 or more m. high, with stout reddish glabrous twigs; lvs. coriaceous, ovate, acute, 4–8 cm. long, ± folded along midrib; infl. dense; bracts ovate, 1.5 mm. long; sepals and petals ciliate, the former 2.5, the latter 5 mm. long; drupes glandular, viscid, ± acid, reddish, 7–8 mm. in diam.; stone flat, smooth, ca. 4 mm. long.—Dry bluffs and slopes, below 4000 ft.; Coastal Sage Scrub, Chaparral, usually away from the coast, but also there; L. Calif. to Santa Barbara Co. Santa Cruz and Santa Catalina ids. March–May.

4. **R. trilobàta** Nutt. ex T. & G. var. **pilosíssima** Engler in DC. [*Schmaltzia malacophylla* Greene. *R. t.* var. *m.* Munz.] SQUAW BUSH. Plate 7, Fig. C. Diffusely branched shrub, 8–14 dm. tall, the branches spreading, often turned down and rooting at tips, strong-scented when crushed; ultimate branchlets rather coarse, 2 mm. in diam., heavily pubescent; lfts. rhombic-ovate to cuneate-obovate, obtuse, crenate, the terminal one 1–3 cm. long, ca. as wide, larger than the 2 lateral; fls. yellowish, in clustered spikes; petals ca. 2 mm. long; fr. viscid, pilose, 4–5 mm. in diam., reddish; seeds ± striate, 4–5 mm. long.—Common in canyons and washes especially of interior valleys, mostly below 3500 ft.; Coastal Sage Scrub, Chaparral, S. Oak Wd., etc.; cismontane s. Calif. to Little San Bernardino Mts. and n. to Butte Co. March–April.

New branchlets densely pubescent; lvs. tomentose at least beneath .... var. *pilosissima*
New branchlets and lvs. mostly glabrate, at least not densely pubescent.
    Terminal lft. not deeply lobed; lfts. ± glabrate. Deserts .... var. *anisophylla*
    Terminal lft. deeply lobed; lfts. mostly somewhat pubescent. Uncommon, cismontane.
                                                                                                                               var. *quinata*

Var. **anisophýlla** (Greene) Jeps. [*Schmaltzia a.* Greene.] Twigs finely puberulent; lvs. green, subglabrous, the lfts. small, the terminal one 1–1.5 cm. long, not lobed, the lateral pair less than 1 cm. long and often unequal; frs. bright crimson.—Dry slopes, 3500–5500 ft.; Creosote Bush Scrub to Juniper-Pinyon Wd.; mts. of Mojave and n. Colo. deserts; to Utah, Ariz. Spring.

Var. **quinàta** (Greene) Jeps. [*Schmaltzia q.* Greene. *S. cruciata* Greene.] Ultimate twigs slender, 1–1.5 mm. thick, very finely puberulent; lvs. sparsely and finely pubescent in maturity, the terminal lft. usually 3-lobed, so that the lf. as a whole appears ± 5-lobed or -foliolate.—Dry slopes and thickets; Chaparral, S. Oak Wd.; San Diego Co. and L. Calif. throughout cismontane Calif.; to Ore.

2. *Schìnus* L. PEPPER-TREE

Dioecious trees with alternate odd-pinnate resinous lvs. Fls. in bracteate panicles, small, whitish. Calyx short, 5-parted. Petals 5, imbricated. Ovary sessile, forming a globose drupe. Ca. 15 spp., mostly from S. Am. (Greek name for *Pistacia*.)

1. **S. mólle** L. Evergreen, 5–15 m. high, with slender pendulous twigs; lvs. 2–3 dm. long, pendulous, with numerous lance-linear lfts. 3–6 cm. long; fr. reddish, 6–8 mm. in diam.; $2n = 28$ (Schnack & Covas, 1947).—The Peruvian Peppertree, commonly cult. in Calif., frequently becomes natur., especially in canyons.

3. Toxicodéndron Tourn.

Deciduous trees, shrubs or woody vines with poisonous properties. Lvs. alternate, palmately ternate or imparipinnate, estipulate, thin, glabrous to tomentose beneath, glabrous to sericeous above; lfts. opposite except for the terminal one. Fls. in axillary, lateral, thyrsoid panicles or racemes, ultimate clusters determinate, of 3 or 4 fls.; bracts of the infl. lanceolate, deciduous; fruiting infl. usually pendent. Sepals 5, veined, glabrous, ovate, usually with reflexed tips, smaller in ♀ fls. than in ♂. Ovary 1-celled by abortion; style terminal, stigma 3-parted; ovule raised on an elongate basal funicle. Drupes subglobular or laterally flattened, whitish to yellow; seeds bony, often ridged. Stamens 5, inserted below a lobed disk, reduced in ♀ fls. (Latin, *toxico*, poisonous, and *dendron*, tree).

(Gillis, W. T. The systematics and ecology of poison-ivy and the poison-oaks. Rhodora 73: 72–159; 161–237; 465–540. 1971.)

1. **T. diversilòbum** (T. & G.) Greene. [*Rhus d.* T. & G. *T. comarophyllum* Greene. *T. isophyllum* Greene. *T. radicans* ssp. *diversilobum* Thorne.] POISON-OAK. Plate 6, Fig. F. Erect shrub, 1–2 (–3) m. tall, stiffly branched, sparingly pubescent or glabrous; lvs. pinnately 3-foliolate, the lfts. obtuse, usually crenulate or even lobed, ovate to suborbicular, 2–7 cm. long, bright green and shining above, paler beneath; panicles axillary, racemose; petals of ♂ fls. 3–4 mm. long, of ♀ 2–3 mm.; fr. whitish, glabrous, subglobose, 4–7 mm. thick; seeds flattened, 3–6 mm. long, irregularly roughened; $2n=15$ pairs (Raven et al., 1965).—Common in low places and thickets and wooded slopes below 5000 ft.; Coastal Sage Scrub, Chaparral, etc.; cismontane; to Wash., L. Calif. Causing a painful dermatitis for human beings. Often conspicuous in the dry season by its red foliage. April–May. Variable, a climbing form is forma *radicans* and f. *quinquifolia*, both of these forms by McNair in the genus *Rhus*.

6. Apiàceae. CARROT FAMILY

Aromatic herbs, commonly with hollow stems and alternate compound or simple lvs., the petioles frequently dilated at the base. Fls. small, epigynous, perfect or sometimes imperfect, in compound or simple umbels or rarely a head, the umbels usually with an invol. of bracts and the umbellets with an involucel of bractlets. Sepals usually 5, sometimes obsolete. Petals 5, the tips often inflexed. Stamens 5, alternate with the petals; fls. filiform, anthers versatile. Ovary inferior, bicarpellate, 2-loculed; styles 2, usually borne on a stylopodium (swollen base). Ovule 1 in each locule. Fr. dry, usually ribbed or winged, the 2 carpels separating at maturity along the plane of their contiguous faces (commissure), either flattened laterally (at right angles to the commissure) or dorsally (parallel with the commissure), sometimes terete. The 2 mericarps attached to a carpophore; pericarp usually containing oil-tubes between the ribs (in the intervals) and on the commissure. Seeds mostly adnate to the pericarp; endosperm cartilaginous; embryo small. Ca. 250 genera and 2000 spp., widely distributed and many of economic importance for food, flavoring, etc. Some are poisonous.

(Mathias, Mildred E. and L. Constance, *in* N. Am. Flora 28B: 43–295. 1944–1945.)
1. Infl. capitate, not umbellate.
    2. Fr. winged, not squamose .... 13. *Cymopterus*
    2. Fr. not winged, ribless, variously scaly .... 15. *Eryngium*
1. Infl. a distinct umbel, not capitate.
    3. Ovary and fr. bearing prickles, bristles, tubercles or scales.
        4. Lvs. simple, entire, usually the basal with parallel veining and the upper broad and perfoliate .... 9. *Bupleurum*
        4. Lvs. compound.
            5. Ovary and fr. armed with spines, hooked bristles, or tubercles.
                6. Plants perennial or biennial; fls. perfect and ♂ .... 28. *Sanicula*
                6. Plants annual; fls. all perfect.

Apiaceae

      7. The plants glabrous; lf.-divisions ± elongate, filiform .... 5. *Apiastrum*
      7. The plants ± pubescent; lf.-divisions shorter.
         8. Invol. of conspicuous foliaceous bracts; lvs. 3–4-pinnatisect; fr. bristly on ribs only .... 10. *Caucalis*
         8. Invol. none or of linear bracts; lvs. pinnate to 3-pinnatisect; fr. bristly or tubercled throughout .... 32. *Torilis*
   5. Ovary and fr. armed with bristles which are never hooked.
      9. The fr. several times longer than broad; oil-tubes absent or obscure.
         10. Plants annual; the fr. with a beak longer than the body .... 28. *Scandix*
         10. Plants perennial; fr. not beaked or if so, the beak shorter than the body
                                                                                    22. *Osmorhiza*
      9. The fr. not more than twice as long as broad; oil-tubes evident.
         11. Foliage glabrous; fr. armed with unequal subulate bristles ..... 2. *Ammoselinum*
         11. Foliage ± pubescent; fr. armed with barbed bristles .... 14. *Daucus*
3. Ovary and fr. not bearing prickles, bristles, tubercles or scales.
   12. Lvs. simple; umbels simple or proliferous.
      13. Ovary and fr. glabrous; lvs. glabrous; plants ± aquatic .... 18. *Hydrocotyle*
      13. Ovary and fr. covered with stellate hairs; lvs. ± stellate-pubescent; plants of dry land
                                                                                    8. *Bowlesia*
   12. Lvs. and umbels variously compound.
      14. Ribs of the fr. not prominently winged; fr. terete in cross section or flattened somewhat laterally.
         15. Fls. yellow.
            16. Involucel absent; plants with anise odor .... 16. *Foeniculum*
            16. Involucel present; plants lacking anise odor .... 31. *Tauschia*
         15. Fls. not yellow.
            17. Fr. elongate, several times longer than wide .... 22. *Osmorhiza*
            17. Fr. roundish to oblong, not more than twice as long as broad.
               18. Plants annual; with celery odor and taste if perennial .... 6. *Apium*
               18. Plants perennial or biennial, not with celery odor and taste.
                  19. Plants caulescent, mostly rather tall; invol. usually present.
                     20. Bracts 3-parted to the middle into filiform divisions, closely reflexed
                                                                                           1. *Ammi*
                     20. Bracts entire or toothed, mostly spreading, sometimes none.
                        21. Stems purple-dotted; oil-tubes none or obscure; lvs. decompound into small divisions .... 12. *Conium*
                        21. Stems not purple-dotted; oil-tubes present; lvs. pinnately or ternate-pinnately divided into mostly larger divisions.
                           22. The lvs. all once-pinnate .... 7. *Berula*
                           22. The lvs. pinnately or ternate-pinnately divided or the upper once-pinnate.
                               23. The stylopodium prominent; ribs not corky; roots tuberous or fusiform .... 25. *Perideridia*
                               23. The stylopodium none or low; ribs corky.
                                    24. Styles ⅕–⅓ as long as the fr.; fr. broadly ovoid to roundish
                                                                              11. *Cicuta*
                                  24. Styles ca. ½ as long as the fr.; fr. subcylindric ... 20. *Oenanthe*
                19. Plants acaulescent, low; invol. absent.
                   25. Plants pubescent; sepals evident.
                      26. Pedicels of the fls. subequal; sepals not rigid .... 26. *Podistera*
                      26. Pedicels of the ♂ fls. longer than or equaling the fr.; sepals rigid
                                                                        21. *Oreonana*
                   25. Plants glabrous; sepals minute or evident .... 31. *Tauschia*
     14. Ribs of the fr. (all or some of them) winged; fr. ± dorsally compressed.
        27. Lateral ribs winged; dorsal ribs filiform.
           28. Corollas all alike; oil-tubes mostly as long as the fr.
              29. Lvs. simply pinnate into ovate divisions.
                  30. Fls. white; dorsal and intermediate ribs apparently 5. Native aquatic plants
                                                                      23. *Oxypolis*
                    30. Fls. yellow; dorsal and intermediate ribs 3. The garden Parsnip escaped
                                                                      24. *Pastinaca*

29. Lvs. pinnately or ternate-pinnately divided into mostly linear or filiform divisions.
    31. Plants annual, with leafy stems; lf.-divisions filiform; plants with anise odor ..... 3. *Anethum*
    31. Plants perennial, acaulescent or short-stemmed; lf.-divisions often wider; no anise odor .... 19. *Lomatium*
28. Corollas of marginal fls. of the umbel larger than of cent.; oil-tubes reaching only half way to base of fr.; tall, coarse, ± woolly plants .... 17. *Heracleum*
27. Lateral, dorsal and intermediate ribs winged or prominent.
  32. Stems leafy, rather tall.
    33. Umbellets not capitate .... 4. *Angelica*
    33. Umbellets capitate .... 30. *Sphenosciadium*
  32. Stems scarcely or not developed above ground.
    34. Sepals prominent; bractlets usually inconspicuous .... 27. *Pteryxia*
    34. Sepals not prominent; bractlets usually conspicuous .... 13. *Cymopterus*

1. Ámmi L. BISHOP'S WEED

Slender, erect, caulescent, branching, essentially glabrous herbs. Lvs. petioled, ternate-pinnately or pinnately dissected into filiform or lanceolate ultimate segms. Infl. of loose compound umbels. Invol. of many entire to divided bracts. Rays many, spreading-ascending. Involucel of many entire bractlets. Fls. white. Sepals minute. Stylopodium depressed-conic; styles slender, elongate. Fr. oblong to ovoid, laterally flattened, glabrous; ribs acute; oil-tubes solitary in the intervals, 2 on the commissure. Six spp. of Medit. region. (Ancient Latin name.)

Umbels spreading in fr.; infl. not borne on a discoid receptacle .... 1. *A. majus*
Umbels compact in fr.; infl. borne on a discoid receptacle .... 2. *A. visnaga*

    1. **A. màjus** L. Two–8 dm. high, scabrous in the infl.; lvs. oblong in outline, the blades 5–20 cm. long, ternate or pinnate, the lfts. lanceolate, 1–1.5 cm. long, setulose-serrate; petioles 3–12 cm. long; bracts exceeding rays; bractlets of involucel linear; rays 50–60, 2–7 cm. long; pedicels 1–10 mm. long; fr. oblong, 1.5–2 mm. long; $n=11$ (Håkansson, 1953).—Occasional weed in fields and waste places, as at Ojai and n.; introd. from Eurasia. May–June.
    2. **A. visnága** (L.) Lam. [*Daucus v.* L.] Glabrous, 2–8 dm. high; lvs. deltoid in outline, the blades 5–20 cm. long, with filiform or linear ultimate divisions 5–30 mm. long; petioles ca. 10 cm. long; peduncles 8–12 cm. long; bracts exceeding the rays; rays 60–100, 2–5 cm. long; pedicels 3–12 mm. long; fr. oblong-ovoid to ovoid, 2–2.5 mm. long; $2n=20$ (Gardé & Gardé, 1951).—Occasional weed in waste places, as near mouth of Santa Ana R., near Goleta, etc.; native of Eurasia. June–July.

2. *Ammoselìnum* T. & G.

Low mostly branched annuals, ± roughened. Lvs. ternately dissected into linear or spatulate ultimate segms. Infl. of loose compound umbels. Invol. usually wanting. Involucel of several narrow entire or toothed bractlets. Rays few. Fls. white. Sepals obsolete. Stylopodium low-conic; styles short. Fr. oblong-ovoid to ovoid, flattened laterally; ribs prominent, glabrous to coarsely scabrous; oil-tubes 1–3 in intervals, 2–4 on the commissure. Three spp. of sw. N. Am. (Greek, *ammos*, sand, and *Selinum*, an Old World genus of Apiaceae.)

    1. **A. gigantèum** Coult. & Rose. [*A. occidentale* M. & J.] Stems 1–several, 1–2 dm. high; lvs. 1.5–2.5 cm. long, glabrous, ternately-pinnately dissected into linear segms., 4–12 mm. long; umbels axillary and terminal; bractlets few, linear-lanceolate, 1–3 mm. long; pedicels 1–8 mm. long, unequal; fr. 3–5 mm. long, scabrous; oil-tubes 3 on the intervals, 2 on the commissure.—Occasional in flat basins; Creosote Bush Scrub; near Borrego and Hay-fields, Colo. Desert; to Ariz., Coahuila. March–April.

3. *Anèthum* L. DILL

Tall branching annuals. Lvs. pinnately dissected into filiform divisions. Fls. yellow, in compound umbels. Invol. and involucels wanting. Rays many. Pedicels slender, spreading. Sepals obsolete. Stylopodium conic; styles short, reflexed. Fr. ovate, strongly flattened dorsally, glabrous; ribs narrowly winged, the lateral broader than the dorsal; oil-tubes solitary in intervals. Two spp., Old World. (The ancient Latin name.)

1. **A. gravèolens** L. Plants 4–15 dm. tall; lvs. oblong to obovate in outline, the blades 1–3 dm. long; petioles 5–6 cm. long; peduncles 7–15 cm. long; rays 10–45, spreading, 3–10 cm. long; pedicels 20–50, 6–10 mm. long; fr. ovoid, ca. 4 mm. long; $2n = 22$ (Tamamschjan, 1933).—Waste places, as a garden escape, as near Long Beach; June–Aug. Used in pickling.

4. *Angélica* L.

Stout or slender, mostly erect perennials, caulescent, from stout taproots. Lvs. ternate-pinnately or pinnately compound, the lfts. broad, distinct, serrate, toothed or lobed. Petioles sheathing, the cauline sheaths often inflated and bladeless. Infl. of loose compound umbels. Invol. usually none; involucel of many entire bractlets or none. Sepals minute or obsolete. Fls. white, pink or purplish. Fr. strongly flattened dorsally, the carpels with filiform to narrowly or corky-winged dorsal ribs, the lateral broadly thin- or corky-winged; stylopodium low-conical; oil-tubes few to many. Ca. 50 spp., circumboreal. (Latin, *angelica*, angelic, because of medicinal properties.)

Lf.-divisions linear to linear-oblong, less than 1 cm. broad. Panamint and White mts. n.
2. *A. lineariloba*
Lf.-divisions lanceolate to oval, broader.
Rays 7–14; fr. 4–5 mm. long, 2–3 mm. broad; lvs. oblong .... 1. *A. kingii*
Rays 25–40; fr. 8–10 mm. long, 6–7 mm. broad; lvs. deltoid in outline .... 3. *A. tomentosa*

1. **A. kíngii** (Wats.) Coult. & Rose. [*Selinum k.* Wats.] Stout, 3–9 dm. tall; lf.-blades oblong in outline, 1.5–4 dm. long, ternate-pinnate, the lfts. lanceolate to lance-ovate, entire to ± serrate, 4–15 cm. long, mostly 1–3.5 cm. broad; petioles 1.5–3 dm. long; involucel wanting; rays 7–14, unequal, 1–10 cm. long; pedicels 1–6 mm. long, webbed; fls. white, the petals pubescent on back; ovaries hispid; dorsal ribs of fr. narrowly winged, the lateral much narrower than the body; oil-tubes 1–2 in intervals, 2 on commissure.—Damp banks, 7000–9500 ft.; Pinyon-Juniper Wd.; White Mts.; to Nev., Ida. June–Aug.

2. **A. lineariloba** Gray. [*A. l.* var. *culbertsonii* Jeps.] Stout, subglabrous to scabrous, 5–15 dm. tall; lf.-blades ternate-pinnately decompound, 1–3.5 dm. long; the segms. linear to linear-oblong, entire, 2–8 mm. broad; petioles 5–25 cm. long; involucel wanting; rays 20–40, subequal, 3–7 cm. long; pedicels 3–10 mm. long; fls. white or pinkish; ovaries glabrous to scabrous; fr. oblong to cuneate, 10–13 mm. long, 5–7 mm. wide, the dorsal ribs narrowly winged, the lateral broader, ca. equal to the body; oil-tubes 1–2 in intervals, 4 on commissure.—Moist places and gravelly or talus slopes, 7500–10,500 ft.; Pinyon-Juniper Wd.; Bristle-cone Pine F.; Panamint and White mts.; to Sierra Nevada. June–Aug.

3. **A. tomentòsa** Wats. [*A. californica* Jeps.] Stout, 6–18 dm. tall, the lvs. glaucous beneath and stiff-villous with occasional forked hairs; infl. villous; lf.-blades deltoid in outline, 1.5–4.5 dm. long, ternate-pinnately divided, the lfts. oval or oblong to lanceolate, 3–15 cm. long, 1–8 cm. broad; petioles 2–3 dm. long; involucel of several narrow villous bractlets, 2–5 mm. long; rays 25–40, unequal, 3–12 cm. long; pedicels 2–12 mm. long; petals white, villous on back; ovaries densely villous; fr. oblong-ovale, the dorsal ribs narrowly winged, the lateral ca. as wide as the body; oil-tubes 1 in intervals, 4 on commissure. Rather moist and ± shaded places, below 6000 ft.; Chaparral, Yellow Pine F.; mts. from San Diego Co. n.; to n. Calif. June–Aug.

5. Apiástrum Nutt. in T. & G.

Smooth branching annuals with slender stems and finely dissected lvs. Umbels naked, unequally few-rayed. Sepals obsolete. Petals white. Fr. ellipsoid-cordate, with obscure or obsolete ribs, ± papillose-roughened. Stylopodium depressed; styles short. Oil-tubes solitary in the intervals and beneath the ribs, 2 on the ± concave commissural side. One sp. (Latin, *apium*, celery, and *aster*, wild.)

1. **A. angustifòlium** Nutt. in T. & G. WILD-CELERY. Plate 7, Fig. D. Stems 0.5–5 dm. long, erect; lvs. 2–5 cm. long, ternately finely dissected; umbels sessile; rays unequal, 1–5 cm. long; pedicels 0–1.5 cm. long; fr. 1–1.5 mm. long, the ribs inconspicuous.—Common in dry sandy valleys and on slopes below 3000 ft.; Coastal Sage Scrub, Chaparral, etc.; cismontane from San Diego Co. n.; L. Calif.; occasional at desert edge; on Santa Catalina, San Clemente, Santa Barbara, Santa Rosa, Santa Cruz ids. March–April.

6. Àpium L.

Glabrous annual to perennial caulescent herbs with pinnate to ternate-pinnately decompound lvs. Fls. white or greenish-yellow, in compound umbels. Invol. and involucel wanting to conspicuous. Sepals obsolete. Stylopodium short-conic to depressed; styles short. Fr. oblong-oval to subglobose or ellipsoid, mostly glabrous, ribs prominent, filiform; oil-tubes 1 in the intervals, 2 on the commissure. Ca. 30 spp. of Eurasia and S. Hemis. (The ancient Latin name.)

Plants perennial; lf.-divisions ovate to suborbicular or cuneate .... 1. *A. graveolens*
Plants annual; lf.-divisions linear to filiform .... 2. *A. leptophyllum*

1. **A. gravèolens** L. CELERY. Erect to ascending, 5–12 dm. tall; basal lvs. pinnate, 1–6 dm. long, petioled, the upper much reduced; lfts. 5–9, 2.5–7 cm. long; petioles 3–25 cm. long; umbels sessile or short-peduncled; invol. and involucels none; rays 7–16; pedicels 1–6 mm. long; fr. subglobose to ellipsoid, ca. 1.5 mm. long; $2n=22$ (Shah, 1953).—Common in wet places at low elevs.; natur. from Eu. May–July. San Nicolas Id.

2. **A. leptophýllum** (Pers.) F. Muell. ex Benth. & Muell. [*Pimpinella l.* Pers.] MARSH-PARSLEY. Prostrate to suberect, 1–6 dm. high; basal lvs. 3–4-pinnately decompound, 3–10 cm. long, long-petioled, the upper smaller, short-petioled; ultimate divisions 4–30 mm. long, linear to subfiliform; rays 3–5; bracts and bractlets none; fr. ovoid, 1.3–3 mm. long; $n=7$ (Bell & Constance, 1957).—Occasional as weed in gardens, lawns, etc.; Riverside and Los Angeles cos. n.; e. and s. to W. I., S. Am. April–Aug.

7. Bérula Hoffm. in Bess. WATER-PARSNIP

Erect, caulescent, branched, glabrous perennial with petioled pinnately variously cut lvs. and small white fls. Infl. of loose compound umbels. Invol. of conspicuous narrow entire or toothed bracts. Involucel of conspicuous narrow bractlets. Rays rather few. Sepals minute, subulate. Stylopodium conic; styles short. Fr. oval to orbicular, glabrous, laterally flattened; ribs filiform, obscure in the thick corky pericarp; oil-tubes many. One sp. of N. Temp. (Latin name of some aquatic plant.)

1. **B. erécta** (Huds.) Cov. [*Sium e.* Huds.] Stems 2–8 dm. high, stoutish; lfts. in 5–9 pairs, oblong, subentire to serrate or laciniate, 1.5–4 cm. long; rays 6–15; pedicels 2–5 mm. long; fr. scarcely 2 mm. long, with inconspicuous ribs; $2n=18$ (Scheerer, 1940); $n=6$ (Bell & Constance, 1957).—Marshes and sluggish water, mostly below 3000 ft.; Freshwater Marsh, Chaparral, Coastal Sage Scrub, Creosote Bush Scrub, etc.; cismontane and somewhat scattered on the Mojave desert; to B.C., Atlantic Coast, L. Calif.; Eurasia. June–Sept. San Miguel Id.

8. Bòwlesia R. & P.

Slender branching annuals with stellate pubescence. Lvs. opposite, simple, lobed, the stipules scarious, lacerate. Umbels on axillary pedicels, simple, few-fld. Fls. white, minute. Sepals rather prominent. Fr. broadly ovoid, stellate-pubescent, with narrow commissure and lacking ribs or oil-tubes, the dorsal part of each carpel turgid. Ca. 15 spp., mostly of Latin Am. (W. Bowles, Irish Naturalist of the 18th century.)

(Mathias, M. E. and L. Constance. A revision of the genus Bowlesia Ruiz & Pavon and its relatives. Univ. Calif. Pub. Bot. 38: 1–73. 1965.)

1. **B. incàna** R. & P. [*B. lobata* auth. *B. septentrionalis* Coult. & Rose.] Plate 7, Fig. E. Delicate, with weak trailing stems 1–5 dm. long, dichotomously branched; petioles slender, 2–8 cm. long; lf.-blades thin, reniform to cordate in outline, 5–30 mm. broad, 5–7-lobed, the lobes entire or toothed; umbels 1–6-fld., on short peduncles; fr. 1–1.5 mm. long, sessile or subsessile; $n = 16$ (Bell & Constance, 1960), $n = 8$ (Bell & Constance, 1966).— Shaded places below 2500 ft.; S. Oak Wd., Coastal Sage Scrub, Chaparral, etc.; n. L. Calif. through cismontane Calif. to cent. Calif.; desert edge; Santa Catalina, San Clemente, Santa Rosa, Santa Cruz ids.; to La. March–April.

9. Bupleùrum L.

Annual to perennial herbs, rather low, erect or spreading, glabrous. Basal lvs. petioled, entire; cauline usually sessile and clasping, auriculate or perfoliate. Infl. of loose compound umbels. Invol. of conspicuous bracts or wanting. Involucel of broad foliaceous often connate bractlets often longer than the fls. and frs. Rays few. Fls. yellow. Sepals obsolete. Stylopodium depressed-conic; styles short. Fr. oblong to subglobose, slightly flattened laterally and constricted at the commissure; ribs filiform; oil-tubes many to wanting or obscure. Rather a large genus, chiefly of the Old World. (Greek, *bous*, ox, and *pleuron*, a rib.)

1. **B. subovàtum** Link. Annual, 2–4 dm. high, glaucous and glabrous, divaricately branched; cauline lvs. broadly ovate, perfoliate, 2–10 cm. long; peduncles 1–8 cm. long; invol. wanting; involucel of 5–6 suborbicular ± united bractlets 1–1.5 cm. long; pedicels 10–20, half as long as fr.; fr. ovoid-globose, 3–5 mm. long, dark, tuberculate or rugose, the ribs prominent; $n = 14$ (Bell & Constance, 1960).—Occasional weed in gardens as at Loma Linda, Twentynine Palms; introd. from Medit. region. May–June.

10. Caùcalis L.

Hispid annuals with pinnately dissected decompound lvs. Umbels irregularly compound. Fls. white. Sepals evident. Fr. flattened laterally, oblong to ovoid. Carpel with 5 filiform bristly ribs and 4 prominently winged secondary ones with barbed or hooked bristles. Oil-tubes solitary in the intervals (under the secondary ribs), 2 on the commissural face. Stylopodium conical. Seed-face deeply grooved. Five spp., N. Temp. Zone. (Ancient classical name.)

1. **C. microcárpa** H. & A. [*Daucus brachiatus* Torr.] Erect, slender-stemmed, 1–3.5 dm. high; lvs. 2–3 times ternate, much dissected; umbels unequally 1–9-rayed; bracts foliaceous, pinnate; pedicels unequal; fr. oblong, 3–7 mm. long, armed with rows of hooked bristles; $n = 11$ (Bell & Constance, 1960 and 1966).—Occasional, open and shaded slopes, below 5000 ft., largely Coastal Sage Scrub, Chaparral, S. Oak Wd., Pinyon-Juniper Wd., etc.; cismontane and Mojave Desert; Santa Catalina, San Clemente and Santa Cruz ids.; to B.C., Utah, Ariz., Mex. April–June.

11. Cicùta L. WATER-HEMLOCK

Glabrous branching caulescent perennials from a tuberous base with fibrous to tuberous roots. Lvs. petiolate, broad, 1–3-pinnate or ternate-pinnate, the divisions serrate to incised. Petioles sheathing. Infl. of loose compound umbels. Invol. wanting or inconspicuous. Involucel of several narrow bractlets. Rays many, slender, spreading-ascending. Pedicels slender, spreading. Fls. white or greenish. Sepals evident. Stylopodium depressed or low-conic. Fr. oval to ovoid or subglobose, flattened laterally, often constricted at the commissure; ribs usually prominent, corky; oil-tubes solitary in the intervals, 2 on the commissure. Ca. 8 spp.; circumboreal. (Ancient Latin name of poisonous hemlock.)

1. **C. doúglasii** (DC.) Coult. & Rose. [*Sium d.* DC.] Stout, 0.5–2 m. high; lf.-blades oblong to ovate in outline, 1.2–3.5 dm. long, 1–3-pinnate, the lfts. linear-lanceolate to broader, 3–10 cm. long, serrate to incised; petioles 1–8 cm. long; peduncles 5–15 cm. long; bracts of invol. 1–several or none; bractlets several, 2–15 mm. long; rays 2–6 cm. long; pedicels 3–8 mm. long; fr. ovoid to subglobose, 2–4 mm. long; ribs low, corky; $n=22$ (Bell & Constance, 1957).—Mostly freshwater wet places below 8000 ft.; Freshwater Marsh, to Yellow Pine F.; most of cismontane and montane Calif.; to Alaska, Alta., Ariz., n. Mex. June–Sept.

12. Conium L. Poison-Hemlock

Tall biennial glabrous herbs with spotted stems and pinnately decompound lvs. Infl. of compound, many-rayed umbels. Bracts of invol. many, lanceolate, inconspicuous. Bractlets of involucel many, shorter than the pedicels. Rays many, spreading-ascending. Sepals obsolete. Fls. white. Stylopodium depressed-conic. Fr. broadly ovoid, flattened laterally, glabrous; ribs prominent, obtuse, undulate; oil-tubes obscure, irregular. Two spp., 1 Eurasian, 1 S. African. (Greek, *coneion*, the ancient name.)

1. **C. maculàtum** L. Stems 5–30 dm. high; lower lvs. petioled, upper sessile, all finely dissected into segms. ovate in outline, dentate or incised; lower lf.-blades 15–30 cm. long; rays 1.5–4.5 cm. long; fr. 2–2.5 mm. long, the ribs prominent when dry; $2n=22$ (Gardé & Gardé, 1949).—Common in low waste places below 5000 ft., especially in cismontane Calif.; natur. from Eu. San Nicolas Id. April–Sept.

13. Cymópterus Raf.

Perennial from long thickened or fusiform taproots, the stems mostly subterranean (pseudoscapes) bearing the tuft of lvs. and peduncles at the surface of the ground. Lvs. variously lobed, divided or compound, petioled, thin to subcoriaceous. Umbels congested and globose, or spreading; invol. present or absent; involucels usually present, with conspicuous herbaceous to ± hyaline bractlets. Fls. yellow, purple or white. Sepals small or obsolete; stylopodium wanting. Fr. ovoid to oblong, somewhat flattened dorsally, the ribs conspicuously winged or the dorsal sometimes wingless, the wings thin or thick and corky toward the outer edge; oil-tubes small, 1–many in intervals, 2–many on commissure. Ca. 30 spp. of w. N. Am. (Greek, *kuma*, wave, and *pteron*, wing, some spp. having undulate wings.)

Lvs. hirtellous to scaberulous.
    Umbels not globose, the rays evident; invol. wanting, or of a few linear bracts
        1. *C. aboriginum*
    Umbels congested and globose; invol. conspicuous .... 2. *C. cinerarius*
Lvs. glabrous.
    Umbels congested and globose, the rays obsolete; fls. purple .... 3. *C. deserticola*
    Umbels not globose, the rays evident.
        Involucral bracts mostly wanting, not scarious; bractlets of involucel inconspicuous or if conspicuous, not hyaline.
            Lvs. simply ternate or pinnate, the divisions deltoid, confluent; fls. purplish . 4. *C. gilmanii*
            Lvs. ternate, then 2–3-pinnate, the divisions linear, distinct; fls. greenish-yellow
                6. *C. panamintensis*
        Involucral bracts evident to conspicuous, scarious; bractlets conspicuous, hyaline.
            Bractlets purple or greenish-white, many-nerved; pedicels less than 1 mm. long
                5. *C. multinervatus*
            Bractlets white or whitish, few-nerved; pedicels 3–12 mm. long ...... 7. *C. purpurascens*

1. **C. aboríginum** Jones. [*Aulospermum a.* Mathias.] Acaulescent, the lvs. and peduncles from a short rootstock covered with persistent lf.-bases, 1–3.5 dm. high; lf.-blades oblong in outline, 3–10 cm. long, ternate-bipinnate or tripinnate, gray-green, hirtellous, ultimate divisions crowded, linear, acute, 2–8 mm. long; petioles 3–12 cm. long; invol. wanting or of few linear bracts; involucel-bractlets several, linear; rays 3–10, spreading, 4–35 mm. long; fls. white; fr. ovoid to oblong, 6–11 mm. long, 5–8 mm. wide, the wings linear.—Dry rocky places, 4000–9300 ft.; Joshua-Tree Wd., Pinyon-Juniper Wd.; Inyo-White Range to Panamint Mts., Grapevine Mts.; w. Nev. April–June.

2. **C. cineràrius** Gray. Acaulescent, the lvs. arising directly from the root-crown, to 7 or 8 cm. high; lf.-blades oblong-ovate in outline, 1.5–2.5 cm. long, cinereous-hirtellous, 2-pinnate; petioles 3–5 cm. long; umbel small, discoid; bracts united below the middle, triangular-lanceolate, scarious-margined; fls. white; fr. narrowly cuneate, 6 mm. long, glabrous, the wings a little constricted at the base, subacute at apex, the dorsal and lateral similar.—Dry open places, 9000–11,500 ft.; Bristle-cone Pine F., Lodgepole F., etc.; Inyo-White Mts. and Pilot Knob, to Sierra Nevada and w. Nev. June–July.

3. **C. deserticola** Bdg. Acaulescent, the lvs. arising from the root-crown, glabrous, 1–1.5 dm. high; lf.-blades broadly oblong-ovate, 2–6 cm. long, glaucous, ternate-bipinnate, the ultimate divisions 1–4 mm. long; petioles ca. as long as blade; umbel globose, compact; invol. none; bractlets paleaceous, mostly aborted; fls. purple; frs. pubescent, 5–7 mm. long, oblong-ovoid to cuneate, the lateral wings narrower than the body, the dorsal absent or reduced; $n = 11$ (Bell & Constance, 1960).—Creosote Bush Scrub, Joshua Tree Wd.; Mojave Desert from e. of Victorville to Muroc and Kramer. April. Very rare.

4. **C. gílmanii** Mort. Subacaulescent, glabrous, 1–2 dm. high; lf.-blades round-reniform in outline, 2.5–4.5 cm. long, ternate, with deltoid confluent laciniately lobed lfts.; rays ca. 8, 1–2 cm. long; invol. none; bractlets several, distinct, linear-subulate, exceeding the purplish fls.; fr. broadly oval, 7–8 mm. long, the wings broader than the body, narrowed or broadened at the base.—Dry rocky slopes, 3500–6500 ft.; Creosote Bush Scrub; Inyo Co. from Last Chance Mts. to Death V. region; w. Nev. April–May.

5. **C. multinervàtus** (Coult. & Rose) Tides. [*Phellopterus m.* Coult. & Rose.] Acaulescent or with a pseudoscape, glabrous, 0.5–2 dm. high; lf.-blades ovate-oblong in outline, 1–9 cm. long, pinnate to ternate-pinnate, pallid, the lfts. entire or lobed, the lobes acute to obtusish, 1–6 mm. long; petioles 2–7 cm. long; invol. a low scarious sheath or of 1–2 conspicuous nerved bracts or a purplish connate lobed cup; bractlets conspicuous, ± ovate, many-nerved, subconnate; umbels compact, the fertile rays mostly 1–5, glabrous, 5–25 mm. long; fls. purplish; fr. ovoid to ovoid-oblong, 8–17 mm. long and broad, often purplish, the wings long, 2–3 times as broad as body.—Dry slopes and plains, 3500–6000 ft.; Joshua Tree Wd., Pinyon-Juniper Wd., n. base of San Bernardino Mts., New York Mts., e. Mojave Desert; to Utah, w. Tex., Son. March–April.

6. **C. panaminténsis** Coult. & Rose var. **acutifòlius** (Coult. & Rose) Munz. [*Aulospermum p.* var. *a.* Coult. & Rose.] Like var. *panamintensis*, but the lf.-divisions more remote, the ultimate ones 3–20 mm. long; $n = 11$ (Bell & Constance, 1960).—At 2000–4700 ft.; Creosote Bush Scrub; Mojave Desert from Sheephole Mts. and Kelso to Barstow and Eagle Mts. (n. Colo. Desert). March–April.

Var. **panaminténsis**. [*Aulospermum p.* Coult. & Rose.] Plate 7, Fig. F. Glabrous, acaulescent, 0.5–4 dm. high; lf.-blades broadly ovate-oblong, 3–12 cm. long, ternate, then 2–3-pinnate, the divisions linear, spinulose, distinct, 1–5 mm. long; rays 5–15, 1–6 cm. long; invol. wanting; bractlets several, ± united, linear-attenuate; fls. greenish-yellow; fr. oblong-ovoid, 6–10 mm. long, the wings equaling or exceeding the body, thin, enlarged at base.—Dry rocky places, 3000–8300 ft.; Creosote Bush Scrub to Pinyon-Juniper Wd.; Argus Mts. and El Paso Range of Mojave Desert to mts. about Death V. and to Mono Co. March–May.

7. **C. purpuráscens** (Gray) Jones. [*C. montanus* var. *p.* Gray. *Phellopterus p.* Coult. & Rose.] Acaulescent or subacaulescent, 3–15 cm. high; lf.-blades glabrous, ovate-oblong in outline, 1.5–5 cm. long, bipinnate or pinnate, pale, the ultimate lobes rounded to

acute; 1–8 mm. long, confluent; petioles 1–4 cm. long; bracts of invol. white, mostly connate below the middle; bractlets white with 1–5 green or white nerves; umbels globose, compact; fertile rays 3–5, only 4–10 mm. long; fls. purplish; fr. usually broadly ovoid, 8–18 mm. long, 8–16 mm. wide, the wings thin, 2–3 times as broad as body.—Rocky limestone hills, 4500–7000 ft., largely Pinyon-Juniper Wd.; Cottonwood Mts., Death V. region, New York Mts., Clark Mt.; to Ida., Utah, Ariz. March–May.

14. Daùcus L.

Pubescent or bristly caulescent annuals or biennials. Lvs. pinnately decompound into small narrow ultimate divisions. Invol. of leafy pinnately divided bracts. Involucels of many entire or divided bractlets. Infl. compact in fr. Fls. usually white. Sepals present or obsolete. Fr. oblong to ovoid, flattened dorsally; primary ribs slender, bristly; secondary ribs winged and with a row of barbed bristles. Oil-tubes solitary in intervals, 2 on the commissure. Ca. 25 spp., widely distributed. (The ancient Greek name.)

Bracts pinnately divided into short linear or lanceolate divisions; rays usually 0.4–4 cm. long; carpel usually broadest below the middle; cent. fl. of the umbellet white .... 2. *D. pusillus*
Bracts pinnately divided into elongate filiform divisions; rays 3–7 cm. long; carpel broadest at middle; cent. fl. of the umbellet usually rose or purple .... 1. *D. carota*

1. **D. caròta** L. WILD CARROT. QUEEN ANNE'S LACE. Biennial, with erect bristly stems 2–10 dm. high, from a fusiform root; lvs. finely dissected; rays many, unequal, the infl. compact, concave in fr.; pedicels unequal, 3–10 mm. long; fr. ovoid, 3–4 mm. long; $2n = 18$ (Heiser & Whitaker, 1948).—Occasionally natur., especially in s. Calif. from cult.; native of Eu. May–Sept.

2. **D. pusíllus** Michx. RATTLESNAKE WEED. Annual, 3–8 dm. high, simple or few-branched, retrorsely papillate-hispid; lf.-blades 3–10 cm. long; peduncles 1–4 dm. long; rays unequal; fr. 3–5 mm. long, the commissure with 2 rows of stiff bristles.—Common on dry slopes, especially after fire and disturbance, below 5000 ft.; Coastal Sage Scrub, Chaparral, S. Oak Wd., etc.; cismontane and occasional on desert; to B.C., Atlantic Coast, n. Mex. April–June. Channel Ids.

15. Erýngium L.

Biennials or perennials, usually glabrous, from taproots or clustered fibrous roots. Lvs. often prickly, entire to lobed or divided, the blades somewhat obsolete; petioles sheathing, sometimes septate. Infl. capitate, the heads solitary, racemose or cymose; invol. of entire or lobed bracts subtending the head. Bractlets of involucel entire or lobed. Fls. white, blue or purple. Sepals evident, entire to spinescent. Rays and pedicels lacking. Stylopodium lacking. Fr. globose to ovoid, slightly flattened laterally if at all, variously covered with thin scales or with tubercles, the ribs obsolete. Oil-tubes none or obscure. Ca. 200 spp. of temp. and trop. regions. (Greek name used by Dioscorides.)

Bracts and bractlets ± callous-margined; petals 0.5–1 mm. long .... 2. *E. armatum*
Bracts and bractlets not callous-margined; petals 1.5 mm. long .... 1. *E. aristulatum*

1. **E. aristulàtum** Jeps. var. **aristulàtum.** [*E. jepsoni* Coult. & Rose.] Stems erect to prostrate, 1–8 dm. long; basal lvs. oblanceolate to obovate, 3.5–25 cm. long, 0.5–2.5 cm. broad, spinulose-serrate to incised or lobed, or blades obsolete; petioles slender, 5–25 cm. long; cauline lvs. reduced upward; infl. cymose, the heads few to many, 5–12 mm. in diam.; bracts rigid, spreading, linear to ± lanceolate, 5–25 mm. long, spinose on margins, ± scarious-winged at base; bractlets 5–15 mm. long, with 1–3 pairs of lateral spines or spineless; sepals lanceolate, 1.5–3.5 mm. long; petals oblanceolate, 1.5 mm. long; fr. ovoid, 1.5–2.5 mm. long, covered with lanceolate scales, the upper to 1 mm. long.—Salt marshes and vernal pools below 3000 ft.; Coastal Salt Marsh to Coastal Sage Scrub; 5 mi. n. of Ventura to San Benito and Plumas cos. May–Aug.

Var. **párishii** (Coult. & Rose) Math. & Const. [*E. p.* Coult. & Rose. *E. jepsonii* var. *p.*

Jeps.] Sepals ovate, 1.5–2 mm. long, usually puberulent.—Vernal pools; Chaparral; w. of Murrieta, Riverside Co., San Diego region; n. L. Calif.

2. **E. armàtum** (Wats.) Coult. & Rose. [*E. petiolatum* var. *a.* Wats.] Plate 7, Fig. G. Stems 0.5–4 dm. long; basal lf.-blades oblanceolate, 0.5–3 dm. long, 0.5–3 cm. wide, remotely serrate to coarsely spinose-incised; petioles short, broad; cauline lvs. reduced; heads many, in a cyme, short-pedunculate, yellowish, sometimes bluish, many-fld.; bracts 8–10, rigid, lanceolate, 10–20 mm. long, mostly entire; bractlets 5–10 mm. long, scarious-winged at base; sepals ovate-lanceolate, 1–2 mm. long, acuminate, entire; petals 0.5–1 mm. long, oblong to oblanceolate; fr. 1.5–3 mm. long, scaly; $n = 16$ (Bell & Constance, 1957).—Vernal pools, near coast, near Goleta, Santa Barbara Co.; to Humboldt Co. May–Aug.

16. *Foeniculum* Adans.

Erect, caulescent biennial or perennial herbs, with anise odor. Lvs. decompound into linear or capillary divisions. Fls. yellow, in large compound umbels. Invol. and involucels none. Sepals obsolete. Stylopodium conical. Fr. oblong, slightly flattened laterally, glabrous; ribs prominent; oil-tubes solitary in the intervals. Four spp., Old World. (Diminutive of Latin, *foenum*, hay, because of odor.)

1. **F. vulgàre** Mill. [*Anethum foeniculum* L.] SWEET FENNEL. Perennial, 1–2 m. high, with striate branching stems, glaucous; lf.-blades ovate to deltoid in outline, to ca. 3 dm. long, pinnately decompound into filiform divisions 4–10 mm. long; petioles broadly sheathing; rays 15–40, ± unequal, 1–6 cm. long; frs. oblong, 3.5–4 mm. long, the ribs acute; $2n = 22$ (Gardé & Gardé, 1949).—Common in waste places especially in s. and cent. Calif.; natur. from Eu. May–Sept.

17. *Herácleum* L. COW-PARSNIP

Erect, tall, stout, biennial or perennial, from taproots or fascicled fibrous roots. Lvs. petioled, ternately or pinnately compound, the lfts. broad, serrate to lobed. Petioles sheathing, mostly dilated. Infl. of loose compound umbels on terminal and axillary peduncles. Invol. usually none. Involucel of many entire narrow bractlets. Rays many. Fls. usually white; sepals obsolete; petals oval to obcordate, the outer of the marginal fls. radiant and often 2-cleft. Stylopodium conical. Fr. orbicular to obovate to elliptic, strongly flattened dorsally; dorsal ribs filiform, the lateral broadly thin-winged. Oil tubes large, solitary in intervals, 2–4 on commissure, extending only part way to the base of the fr. Ca. 60 spp., circumboreal. (Named for *Hercules*, who was supposed to have first used it for medicine.)

1. **H. sphondýlium** L. ssp. **montànum** (Schleicher ex Gaudin) Briq. [*H. lanatum* Michx. *H. douglasii* DC.] Perennial, woolly; 1–3 m. high; lf.-blades 2–5 dm. long, the lfts. 1.5–4 dm. long, ovate to roundish, cordate, coarsely serrate and ± lobed; peduncles 1–4 dm. long; involucral bracts 5–10, 0.5–2 cm. long; bractlets similar; rays 15–30; pedicels 8–20 mm. long; petals white; fr. broad, 8–12 mm. long; $n = 11$ (Bell & Constance, 1957).—Moist places below 9000 ft.; Montane Coniferous F.; San Jacinto and San Bernardino mts.; to Alaska, Atlantic Coast. April–July.

18. *Hydrocótyle* L. MARSH PENNYWORT

Low perennials with slender creeping stems or rootstocks, rooting at the nodes. Lvs. petioled, entire or parted to base. Petioles slender, not sheathing. Peduncles axillary, obsolete to longer than lvs. Invol. wanting or present. Fls. white, greenish to yellow; petals ovate, plane; calyx-teeth obsolete or minute. Fr. orbicular or ellipsoid, strongly flattened laterally. Carpel with 5 primary ribs. Ca. 75 spp., widely distributed. (Greek, *hudor*, water, and *cotule*, a low vessel or cup.)

Lvs. peltate.
    Infl. a simple umbel .... 3. *H. umbellata*

Infl. an interrupted spike .... 4. *H. verticillata*
Lvs. not peltate.
    Fr. pedicellate .... 1. *H. ranunculoides*
    Fr. sessile .... 2. *H. sibthorpioides*

    1. **H. ranúnculoìdes** L. f. Floating or creeping on mud; lvs. 0.5–8 cm. broad, round-reniform with cordate base, 3–7-cleft with crenate lobes, on equal or longer petioles; peduncles shorter, recurved in fr.; umbels simple, capitate, 5–10-fld.; pedicels 1–3 mm. long; fr. suborbicular, 2–3 mm. long, with thick pericarp and obscure filiform ribs.—Ponds and slow streams, below 5000 ft.; many Plant Communities; cismontane Calif. and desert edge (Borrego, Victorville); to Wash., Penn., S. Am. Eu. March–Aug.

    2. **H. sibthórpioides** Lam. Lvs. 5–10 mm. in diam.; fr. sessile, not pedicelled, otherwise much as in *H. ranunculoides*.—Rare, as at Westwood, Los Angeles; native of Asia, Africa.

    3. **H. umbellàta** L. Branches of rootstock descending, with round tubers; petioles and peduncles 3–12 cm. long; lf.-blades round-peltate, crenate, mostly 0.5–5 cm. wide; umbels many-fld., simple, the pedicels 2–25 mm. long; fr. 1–2 mm. long, 2–3 mm. broad, strongly notched at base and apex; pericarp thin between thick corky ribs; $2n = 48$ (Wanscher, 1932).—Frequent in wet places and sluggish water, below 5000 ft.; many Plant Communities; cismontane Calif.; to Ore., Mex., Atlantic Coast, S. Am. March–July.

    4. **H. verticillàta** Thunb. var. **triràdiata** (A. Rich.) Fern. [*H. polystachya* var. *t.* A. Rich. *H. prolifera* Kell.] Resembling var. *verticillata*, but fr. with pedicels 1–10 mm. long, the spike mostly simple.—Sluggish water, Ventura Co.; to cent. Calif., Atlantic states, S. Am.

    Var. **verticillàta** [*H. cuneata* Coult. & Rose. *H. v.* var. *c.* Jeps.] With habit of *H. umbellata*, but the infl. an interrupted spike, simple or bifurcate; fr. sessile or subsessile, shallowly notched at apex, narrowly rounded to abruptly cuneate at base, 1–3 mm. long; $2n = 88 \pm 4$ (C. B. Smith, 1968).—Occasional as in Orange and San Bernardino cos.; to cent. Calif., Atlantic states, Mex., W. I. April–Sept.

19. *Lomàtium* Raf.

    Acaulescent or short-caulescent perennials with slender or thick subfusiform roots. Lvs. ternate, pinnate or decompound. Fls. in compound umbels, yellow to white or purple. Invol. mostly lacking; involucels present or sometimes none. Sepals small. Fr. strongly flattened dorsally; carpels with filiform dorsal ribs, the lateral winged, thin to corky; stylopodium lacking; oil-tubes 1–several in the intervals, sometimes obsolete, 2–10 on the commissure. Seed flattened dorsally, the face plane or ± concave. Ca. 80 spp., w. N. Am. (Greek, *loma*, a border, because of the wings on the fr.)

(Mathias, M. E. A revision of the genus Lomatium. Ann. Mo. Bot. Gard. 25: 225–297. 1938. Theobald, W. L. Lomatium dasycarpum-mohavense-foeniculaceum complex. Brittonia 18: 1–18. 1966.)

  1. Fr. notched at each end, the wings distinct on each side of the body; lfts. cuneate to obovate.
    2. Lf.-divisions not pinnatifid, merely toothed or sometimes 3-lobed; wings twice as broad as body of fr. Santa Barbara to San Diego cos. .... 7. *L. lucidum*
    2. Lf.-divisions pinnatifid; wings ca. as broad as body of fr. San Nicolas Id. ... 6. *L. insulare*
  1. Fr. not or scarcely notched; lvs. mostly dissected into numerous narrow segms.
    3. Bractlets broad, rounded to obovate; fls. yellow.
      4. Plants usually with several cauline lvs.; wings of fr. broader than the body, the dorsal ribs obsolete.
        5. Fr. ovate to obovate, 9–15 mm. long; sepals prominent in young frs. ... 12. *L. vaseyi*
        5. Fr. oblong to ovate, 5–11 mm. long; sepals obsolete .... 11. *L. utriculatum*
      4. Plants with not more than 1 stem lf.; wings of fr. narrower than body, or if broader, the dorsal ribs evident.
        6. Fls. yellow; plants glabrous or slightly pubescent. Insular .... 2. *L. caruifolium*
        6. Fls. white; plants scaberulous to densely pubescent. Largely from deserts
                                                           9. *L. nevadense*

3. Bractlets narrow, linear to lanceolate; fls. white, purple or yellow.
7. Lvs. with mostly few or large divisions, ternately or pinnately divided, the divisions mostly remote; fls. yellow to creamy white.
8. Lf.-blades ovate to obovate in outline, ternate-pinnately or quinate-pinnately divided. Ventura and Kern cos. n. .... 1. *L. californicum*
8. Lf.-blades narrow-oblong in outline, 2–3 pinnate, the ultimate segms. linear. Mts. of Mojave Desert .... 10. *L. parryi*
7. Lvs. decompound, dissected into many minute divisions.
9. Ovaries and young fr. variously pubescent or roughened.
10. Plants short-caulescent; petals densely pubescent over entire upper surface. Cismontane .... 3. *L. dasycarpum*
10. Plants acaulescent; petals glabrous or sometimes pubescent mostly along the margins. Largely desert.
11. Petioles longer than lf.-blades; plants hoary-pubescent, never villous; wings of fr. more than half the width of the body .... 8. *L. mohavense*
11. Petioles usually shorter than lf.-blades; plants glabrate to villous; wings of fr. mostly less than half the width of the body .... 5. *L. foeniculaceum*
9. Ovaries and young fr. glabrous; fr. with corky-thickened wings. Desert slopes San Bernardino and San Gabriel mts. n. .... 4. *L. dissectum*

1. **L. califórnicum** (Nutt.) Math. & Const. [*Leptotaenia c.* Nutt. ex T. & G.] Caulescent, 3–12 dm. high, glabrous, glaucous; lf.-blades ovate to deltoid in outline, 1–3 dm. long, ternate or biternate, the segms. cuneate to obovate, usually 3-cleft and coarsely toothed, 2–5 cm. long; petioles 0.5–2.5 dm. long, sheathing at base; bractlets linear, scarious; rays many, subequal, 3–8 cm. long; pedicels 4–12 mm. long; fls. many, yellow; fr. oblong-oval, 10–15 mm. long, glabrous, the wings ± corky, narrower than body; oil-tubes 3–4 in intervals, 6–10 on commissure, or obscure; $n=11$ (Bell & Constance, 1957). —Wooded or brushy slopes below 5500 ft.; Chaparral, Foothill Wd., Yellow Pine F., etc.; Ventura and Kern cos. n. to Ore. April–June.

2. **L. caruifòlium** (H. & A.) Coult. & Rose. [*Ferula c.* H. & A.] Acaulescent or short-caulescent, 1.5–4.5 dm. high, glabrous to pubescent, from very long slender taproot; lf.-blades ovate to obovate in outline, 5–30 cm. long, 1–3-ternate or 1-ternate, then bipinnate, the ultimate segms. linear, 3–15 mm. long; petioles 4–7 cm. long; bractlets entire or toothed, scarious-margined; rays 6–15, unequal, 2–12 cm. long; pedicels mostly 2–8 mm. long; fls. yellow; fr. narrowly ovate to obovate, 8–12 mm. long, usually obtuse at both ends, glabrous, the wings thickish, narrower than body; oil-tubes obscure.—Santa Cruz Id.; Coast Ranges to Mendocino Co. March–May.

3. **L. dasycárpum** (T. & G.) Coult. & Rose. subsp. **dasycárpum**. [*Peucedanum d.* T. & G. *Cogswellia d.* Jones.] Plants 1–4.5 dm. high, the lower stem horizontal for 3–8 cm.; lf.-blades 3–14 cm. long, the ultimate divisions linear, 0.5–2.5 mm. long; invol. wanting or inconspicuous; rays 10–23, 0.7–9 cm. long; involucel of linear-lanceolate bractlets; pedicels 5–24 mm. long, many; fls. greenish to greenish-yellow, appearing white because of the pubescent petals; fr. orbicular to ovate-oblong, 7–20 mm. long, 6–15 mm. broad, the wings at least as wide as body, body sparsely pubescent to subglabrous; oil-tubes 1–3 in intervals, 2–4 on commissure; $n=11$ (Theobald, 1966).—Common on dry ridges below 5000 ft.; Chaparral, etc.; n. L. Calif., through cismontane Calif. to Humboldt Co. March–June.

Subsp. **tomentòsum** (Benth.) Theobald. [*Peucedanum. t.* Benth. *L. t.* Coult. & Rose.] Plants 1–5 dm. high; lf.-blades 4–16 cm. long, the ultimate divisions crowded, linear to filiform, 0.5–3 mm. long; pedicels 3–24 mm. long, many; fr. ovate-oblong, 7–21 mm. long, 8–18 mm. broad, the wings ± tomentose, narrower than or equal in width to body, body densely tomentose; oil-tubes 1–4 in intervals, 2–4 on commissure; $n=11$ (Theobald, 1966).—Tehachapi Mts. n. to Shasta Co.

4. **L. disséctum** (Nutt.) Math. & Const. var. **multífidum** (Nutt.) Math. & Const. [*Leptotaenia m.* Nutt. ex T. & G.] Mostly caulescent, 8–14 dm. tall, puberulent to subglabrous, from a stout caudex and thickened root; lf.-blades deltoid-roundish in outline, 1.5–3.5 dm. long, ternate, then 2–4-pinnate, the segms. linear-oblong, 2–22; petioles

broadly sheathing at the stem; rays many, 3–13 cm. long; bractlets few, entire; pedicels 4–20 mm. long; fls. purple or yellow; fr. oblong-oval, 12–16 mm. long, glabrous, the wings much narrower than the body, thick and corky; oil-tubes obscure; $n=11$ (Bell & Constance, 1960).—Occasional on rocky slopes, 3500–8000 ft.; Pinyon-Juniper Wd., Yellow Pine Wd.; n. face of San Bernardino and San Gabriel mts. to Sierra Nevada and n. to B.C., Rocky Mts. May–July.

5. **L. foeniculàceum** (Nutt.) Coulter & Rose ssp. **fimbriàtum** Theobald. Plants 0.7–3 dm. high, villous; lvs. broadly ovate to ovate-oblong in outline, the ultimate divisions linear, 1–5 mm. long, 0.5–1 mm. broad, apiculate; petioles shorter than the blades, usually sheathing, purplish; peduncles 1–3 dm. long; rays 2–14, 0.5–6 cm. long, usually unequal; involucel of linear bractlets; pedicels 2–10 mm. long, many; fls. usually yellow, or with some purple; petals pubescent along margins and sometimes with a few hairs on upper surface; fr. ovate-oblong to suborbicular, 5–9 mm. long, 4–7 mm. broad, the body and wings pubescent, the wings less wide than the body; oil-tubes 1–6 in intervals, 3–8 on commissure; $n=11$ (Theobald, 1966).—Dry places, 7000–10,400 ft.; Pinyon-Juniper Wd., Bristle-cone Pine F.; Inyo-White Mts.; w. Nev. June–July.

Subsp. **inyoénse** (Math. & Const.) Theobald. [*L. i.* Math. & Const.] Petals glabrous; petioles sheathing, purplish; fls. cream white to pale yellow; fr. body and wings puberulent.—At 9000–11,100 ft.; Inyo Mts.; to Nev., Ida. July–Aug.

6. **L. insulàre** (Eastw.) Munz. [*Peucedanum i.* Eastw.] Acaulescent, 1.5 dm. tall; lvs. biternate to bipinnate; lfts. cuneate, dentate or incised; rays 10–18; bractlets filiform; pedicels 6–12 mm. long; fls. yellow; fr. oblong-ovate, 12–15 mm. long, emarginate at both ends; wings thick, ca. as broad as body; oil-tubes 2 in intervals, 4 on commissure.—Sea bluffs; San Nicolas Id.; Guadalupe Id. Feb.–April.

7. **L. lùcidum** (Nutt.) Jeps. [*Eurypterа l.* Nutt. ex T. & G. *Peucedanum e.* Gray. *P. hassei* Coult. & Rose. *Cogswellia l.* and *h.* Jones.] Short-caulescent, 2–6 dm. tall, glabrous, from long slender taproot; lf.-blades ± ovate in outline, 5–9 cm. long, 1–2-ternate, the ultimate divisions deltoid to cuneate, 1.5–7 cm. long, dentate to lobed; petioles 3–13 cm. long; rays 10–20, 2–8 cm. long; bractlets of involucel lance-linear, ca. as long as fls.; pedicels 7–17 mm. long; fls. yellow; fr. suborbicular to broadly elliptic, 6–15 mm. long, emarginate especially at base, the wings thick, broader than body; oil-tubes solitary in intervals, 2–4 on commissure; $n=11$ (Bell & Constance, 1960).—Occasional on brushy slopes below 5000 ft.; mostly Chaparral; away from immediate coast, San Diego Co. to Santa Barbara Co. Jan.–April.

8. **L. mohavénse** (Coult. & Rose) Coult. & Rose subsp. **mohavénse**. [*Peucedanum m.* Coult. & Rose. *P. argense* Jones.] Plate 7, Fig. H. Acaulescent, 1–3 dm. tall, short hoary-pubescent; lvs. ovate to oblong-ovate in outline, the ultimate divisions obovate, 2–4 mm. long, rarely more than 2–3 times longer than broad; fls. purple, yellow or yellow with a purplish tinge; pedicels 1–10 mm. long; bracts linear; rays 10–16, subequal, 1–4 cm. long; fr. 4.5–9 mm. broad, ovate to orbicular, the body and wings densely pubescent; $n=11$ (Theobald, 1966).—Abundant on dry plains and slopes, 2000–7000 ft.; Creosote Bush Scrub to Pinyon-Juniper Wd.; Mojave Desert from Barstow and Antelope V. to Inyo Co. and Santa Barbara Co. April–May.

Ssp. **longilòbum** Theobald. Ultimate divisions of lvs. oblong to oblong-ovate, usually more than 3 times longer than broad; fls. usually yellow, sometimes with purplish tinge; fr. 5–11 mm. broad; $n=11$ (Theobald, 1966).—At 2000–5000 ft.; Joshua Tree Wd., Pinyon-Juniper Wd.; n. base San Gabriel Mts. to Joshua Tree, s. along w. Colo. Desert to n. L. Calif.

9. **L. nevadénse** (Wats.) Coult. & Rose var. **párishii** (Coult. & Rose) Jeps. [*Peucedanum parishii* Coult. & Rose.] Acaulescent or nearly so, 1–2 dm. tall, glaucous, closely pubescent; lvs. decompound, the divisions pinnately divided into acute segms. 4–10 mm. long, or less divided and linear, to 2 cm. long; umbel few- and equally-rayed; rays mostly 1.5–5.5 cm. long; pedicels 3–12 mm. long; fls. white; fr. glabrous, 7–10 mm. long, the wings usually narrower than the body; ribs filiform or obsolete, oil-tubes often obscure, 1–4 in intervals, 4–7 on commissure; $n=11$ (Bell & Constance, 1957).—Dry slopes, 5000–

9000 ft.; Pinyon-Juniper Wd., to Montane Coniferous F.; desert slopes of San Bernardino and San Gabriel mts. to e. Ore., New Mex. April-June.

Var. **pseudorientàle** (Jones) Munz. [*Cogswellia n.* var. *p.* Jones. *L. n.* var. *holopterum* Jeps.] Petioles with more prominent scarious margins; wings of fr. ca. as broad as body, the dorsal ribs evident.—Dry slopes, 3000–6000 ft.; Creosote Bush Scrub to Pinyon-Juniper Wd.; mts. of e. Mojave Desert; to Nev., Ariz.

10. **L. párryi** (Wats.) Macbr. [*Peucedanum p.* Wats.] Acaulescent, glabrous, 2–4 dm. tall, from a long stout taproot; lf.-blades narrow-oblong in outline, 1–2 dm. long, 2–3-pinnate, the segms. linear, 2–9 mm. long; petioles 6–9 cm. long, sheathing below; bractlets several, linear, acute, subscarious; rays ca. 15, 2–4 cm. long; pedicels 10–17 mm. long; fls. yellow, ca. 10 in number; fr. oblong, 9–12 mm. long, the wings ca. as broad as body; oil-tubes 2–3 in intervals, 4 on commissure.—Occasional on rocky slopes, 4000–8500 ft.; mostly Pinyon-Juniper Wd.; Providence, New York, Panamint mts., etc. of ne. Mojave Desert; to Utah. May–June.

11. **L. utriculàtum** (Nutt.) Coult. & Rose. [*Peucedanum u.* Nutt. ex T. & G. *Cogswellia u.* Jones. *C. chandleri* Jones.] Caulescent, 1–5 dm. tall, purplish below, glabrous to pubescent; lf.-blades oblong in outline, 3–16 cm. long, tripinnate into linear segms. 4–25 mm. long; petioles 1–10 cm. long; bractlets obovate, entire to cleft, ± scarious on margins; rays 5–13, 2–12 cm. long; pedicels 2–9 mm. long; fls. ca. 20, yellow; fr. ovate to oblong, 5–11 mm. long, puberulent when young, the wings thin, usually broader than body; oil-tubes 1–3 in dorsal intervals, 1–4 in lateral, 2–6 on commissure; $n=11$ (Bell & Constance, 1957).—Grassy places below 5000 ft.; V. Grassland, S. Oak Wd., etc.; cismontane s. Calif. to desert edge (Argus Mts., etc.; Santa Rosa and Santa Cruz ids.; to B.C. March–April.

12. **L. vàseyi** (Coult. & Rose) Coult. & Rose. [*Peucedanum v.* Coult. & Rose. *Cogswellia v.* Coult. & Rose.] Resembling *L. utriculatum*, the ultimate lf.-segms. oblong, 3–17 mm. long; rays 10–20, 2–7 cm. long; umbellets ca. 30-fld.; fls. yellow; fr. 9–15 mm. long, the thin wings usually broader than the body; oil-tubes 1 in intervals, 4 on commissure.—Dry slopes and mesas, below 5000 ft.; Coastal Sage Scrub, Chaparral, etc.; interior cismontane s. Calif. to San Luis Obispo and Inyo cos. March–April.

20. Oenánthe L.

Decumbent to erect, branching, caulescent, mostly glabrous perennials. Lvs. petioled, pinnate to pinnately decompound. Petioles sheathing. Infl. of loose compound umbels. Invol. wanting or reduced. Rays many, spreading, subequal. Involucels of many narrow bractlets. Fls. white. Sepals lanceolate. Stylopodium conical. Fr. in ours oblong, subterete, glabrous; ribs low, obtuse, subequal, corky, often ± confluent; oil-tubes usually solitary in intervals, 2 on commissure. Ca. 30 spp., mostly of Old World. (Ancient Greek name of some thorny plant.)

1. **O. sarmentòsa** Presl. [*Helosciadium californicum* H. & A. *O. c.* Wats.] Plate 8, Fig. A. Stems succulent, 5–15 dm. long, decumbent and ascending; lf.-blades oblong to ovate in outline, 1–3 dm. long, bipinnate, the lfts. ovate, 1–5 cm. long, coarsely dentate to incised; petioles 1–3 dm. long; peduncles 5–13 cm. long; bracts few, linear, 5–15 mm. long; bracteoles 4–5 mm. long; rays 10–20, 1.5–3 cm. long; fr. oblong, 2.4–3.5 mm. long, the prominent ribs broader than the intervals; $n=22$ (Bell & Constance, 1957).—Marshes and sluggish water, below 5000 ft.; several Plant Communities; cismontane s. Calif. and to desert edge as at Victorville; to B.C. May–Sept.

21. Oreonàna Jeps.

Low tufted acaulescent plants from thickened taproot. Lvs. petioled, pinnately or ternately decompound, the segms. small, oblong, crowded. Infl. of subcapitate compound umbels. Invol. none. Rays few to many, short, stout. Pedicels of sterile fls. longer than fr. and rysa, those of fertile fls. obsolete. Fls. white or purplish. Sepals evident. Fr. ovoid,

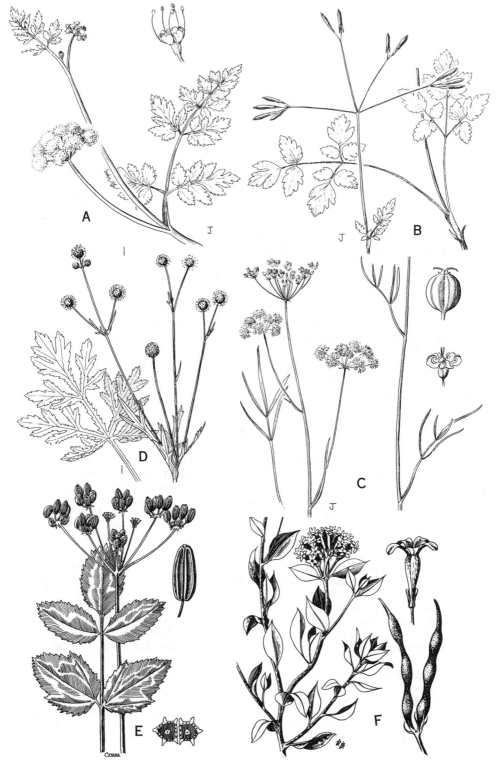

Plate 8. Fig. A, *Oenanthe sarmentosa*. Fig. B, *Osmorhiza chilensis*. Fig. C, *Perideridia gairdneri*. Fig. D, *Sanicula bipinnatifida*. Fig. E, *Tauschia arguta*. Fig. F, *Amsonia brevifolia*.

somewhat laterally compressed; ribs filiform; oil-tubes several in the intervals and on the commissure. Two spp. (Greek, *oreos*, mountain, and *nannos*, dwarf.)

1. **O. vestìta** (Wats.) Jeps. [*Deweya v.* Wats.] Plants 4–15 cm. high, densely white-woolly throughout; umbels dense, mostly above the lvs.; bractlets many; sterile pedicels 10–25 mm. long; fr. subsessile, 5–6 mm. long; oil-tubes 3–4 in the intervals, 3 on the commissure.—Dry gravel or talus, 6500–11,000 ft.; upper Montane Coniferous F.; San Gabriel and San Bernardino mts. June–July.

22. *Osmorhìza* Raf. SWEET-CICELY

Slender to stoutish caulescent perennials from thick fascicled roots. Lvs. petiolate, ternate to ternate-quinate, the lfts. lanceolate to roundish, serrate to pinnatifid, with mucronate teeth or lobes. Petioles sheathing. Infl. of loose compound umbels, on terminal or lateral peduncles. Invol. wanting or present. Involucels of several narrow reflexed bractlets or wanting. Rays few, slender, unequal. Fls. white, purple or greenish-yellow. Sepals obsolete. Stylopodium conic. Fr. linear to oblong, cylindric to clavate, obtuse, tapering, beaked or constricted at apex, rounded or tapering at base; ribs filiform, acute, often bristly. Oil-tubes obscure or wanting. Seed subterete in cross section, the face concave or grooved. Ca. 11 spp. of N. Am., w. S. Am., e. Asia. (Greek, *osme*, odor, and *rhiza*, root.)

(Constance, L. C. and R. H. Shan. The genus O. Univ. Calif. Pub. Bot. 23: 111–156. 1948.)

Involucels of several bractlets, conspicuous .... 1. *O. brachypoda*
Involucels lacking or rudimentary .... 2. *O. chilensis*

1. **O. brachýpoda** Torr. [*Washingtonia b.* Heller. *O. b.* var. *fraterna* Jeps.] Erect, usually pubescent, aromatic, 4–6 dm. tall, branched; lvs. 1–2 dm. long; lfts. 2–3 cm. long, coarsely laciniate-cleft and serrate; umbel 1- to 6-rayed, with invol. mostly absent, and involucels of linear bracts 4–5 mm. long; fr. 12–20 mm. long, short-attenuate at base, rough-bristly on ribs.—Shaded woods, below 5500 ft.; S. Oak Wd., Yellow Pine F.; San Diego Co. n. to cent. Calif.; to Ariz. March–May.

2. **O. chilénsis** H. & A. [*O. nuda* Torr. *Washingtonia brevipes* Coult. & Rose.] Plate 8, Fig. B. Stems glabrous or pubescent, 3–8 dm. tall; lvs. 5–10 cm. long, on longer petioles, the lfts. ovate, 3-lobed or -cleft, coarsely serrate; umbels on long peduncles; rays 2–4; involucels lacking; fr. 10–20 mm. long, tapering toward the slender beak, narrowed and densely hispid at base; $n=11$ (Bell & Constance, 1957).—Woods, mostly between 5000 and 9000 ft.; Montane Coniferous F.; mts. from San Diego Co. to Alaska, Atlantic Coast; in S. Am. April–July.

23. *Oxýpolis* Raf.

Smooth erect, caulescent swamp herb from fascicled tubers. Lvs. petioled, simply pinnate or ternate. Petioles sheathing. Infl. of loose compound umbels. Invol. of few slender bracts or none. Involucel of lfts. like the bracts or none. Rays few to many. Pedicels slender; sepals prominent or minute. Stylopodium conic; styles slender. Fr. oblong to obovoid, glabrous, strongly flattened laterally; ribs filiform, the lateral broadly thin-winged, the nerves resembling ribs and making the carpel appear 5-ribbed; oil-tubes large, solitary in the intervals, 2–6 on the commissure. Ca. 7 spp., of N. Am. (Greek, *oxys*, sharp, and *polis*, city; of uncertain application.)

1. **O. occidentàlis** Coult. & Rose. Plants 6–10 dm. high; lvs. 1–3 dm. long, the lfts. 5–13, lanceolate to oblong-ovate, crenate to incised, 3.5–6 cm. long; petioles 1–3 dm. long; stem-lvs. few, reduced; peduncles to 3 dm. long; rays 12–24, ± unequal, 2–8 cm. long; bracts mostly 1–2, 5–25 mm. long; bractlets 5–15 mm. long; fls. white; sepals evident; fr. 5–6 mm. long; $n=18$ (Bell & Constance, 1957).—Shallow water, 4000–8500 ft.; Montane Coniferous F.; San Bernardino Mts.; Sierra Nevada to Ore. July–Aug.

24. Pastináca L.

Stout biennial with thick root and angular leafy stems. Lvs. pinnately compound. Invols. and involucels small or none. Fls. yellow or red. Sepals obsolete. Stylopodium depressed-conical. Fr. flattened dorsally, oval to obovate. Carpels with winged lateral and filiform dorsal ribs; oil-tubes solitary in intervals, 2–4 on commissure. Ca. 14 spp., Eurasian. (Latin name from *pastino*, to prepare the ground for planting of the vine.)

1. **P. satìva** L. COMMON PARSNIP. Glabrous, 3–10 dm. high; lvs. oblong to ovate, the blades 1.5–2.5 dm. long, the lfts. 5–10 cm. long; petioles 1–1.5 cm. long; peduncles stout, 7–15 cm. long; rays 15–25, unequal, 2–10 cm. long; fls. yellow; fr. 5–6 mm. long; $2n=22$ (Ogawa, 1929).—Localized as natur. plant, as at Julian, in San Jacinto Mts., San Gabriel Mts.; escaping from gardens. June–July.

25. Perideridia Rchb.

Erect, caulescent branching herbs, from tuberous or fusiform fascicled roots. Lvs. petioled, ternately, pinnately or ternate-pinnately compound, the ultimate divisions linear to ovate. Infl. of loose compound umbels. Invol. of few to many entire narrow ± scarious bracts. Involucel of usually scarious or colored bractlets. Sepals evident. Fls. white to pinkish. Fr. flattened laterally; carpels with filiform ribs; stylopodium ± conical; oil-tubes 1–5 in intervals, 2–8 on commissure. Ca. 9 spp. of N. Am. The tuberous roots were important as food for the Indians. (Greek, *peri*, around, and *derris*, a leather coat.)

(Chuang, T. and L. Constance. A systematic study of Perideridia. Univ. Calif. Pub. Bot. 55: 1–74. 1969.)

Petioles and rachis not dilated, the ultimate divisions not dimorphic.
    Bractlets usually setaceous; fr. ca. as broad as long .... 1. *P. gairdneri*
    Bractlets scarious or with scarious margins; fr. longer than broad .... 2. *P. parishii*
Petioles and rachis dilated, the ultimate lf.-divisions usually dimorphic .... 3. *P. pringlei*

1. **P. gáirdneri** (H. & A.) Mathias. [*Ataenia g.* H. & A.] SQUAW ROOT. Plate 8, Fig. C. Plants slender, 3–12 dm. tall; lvs. mostly pinnate into linear or lanceolate, mostly entire divisions 2–12 cm. long; bracts mostly obsolete; bractlets several, linear, green or scarious, 1–4 mm. long; rays usually 8–20, 1.5–6 cm. long; pedicels 3–7 mm. long; fr. ± subglobose, 2–3 mm. in diam.; oil-tubes solitary in the intervals, 2 on the commissure; $n=17$ (Bell & Constance, 1960).—Wet places at rather low elevs.; San Diego Co. n. to B.C., Rocky Mts.; in the mts. farther n. June–July.

2. **P. párishii** (Coult. & Rose) Nels. & Macbr. ssp. **párishii** [*Pimpinella p.* Coult. & Rose. *Eulophus p.* Coult. & Rose.] Slender, 2–8 dm. tall; lvs. lanceolate to ovate in outline, the blades 5–15 cm. long, ternate or simple or biternate, the lfts. linear to lanceolate, 2–10 cm. long; petioles 3–7 cm. long; peduncles 6–15 cm. long, bracts mostly obsolete; involucel of several linear to obovate, scarious to colored bractlets 2–4 mm. long; rays 6–11, ± unequal, 1–4 cm. long; pedicels 3–8 mm. long; fr. oblong, 3–5 mm. long; oil-tubes 2–4 in intervals, 6 on commissure; $n=17, 18, 19$ (Bell & Constance, 1960).—Damp meadows, etc., 4000–7500 ft.; largely Montane Coniferous F.; San Bernardino Mts.; to New Mex.

Ssp. **latifolia** (Gray) Chuang & Constance. Rays usually 12–14, subequal; bractlets subequal to pedicels; fr. subglobose to ovoid.—At 7000–8000 ft.; mts. of s. Calif. not including San Bernardino Mts.; to n. Calif.

3. **P. prínglei** (Coult. & Rose) Nels. & Macbr. [*Eulophus p.* Coult. & Rose.] Resembling *P. parishii*, but lf.-blades ovate to deltoid in outline, 5–10 cm. long, pinnately dissected into linear divisions 0.2–8 cm. long; bractlets subulate, 2–4 mm. long, scarious; rays 5–8; fr. oblong, 4–6 mm. long.—Reported from n. Los Angeles Co., thence to Kern and San Luis Obispo cos. April–June.

26. Podístera Wats.

Low cespitose scabrous herbaceous perennial with long thickened roots. Lvs. pinnate to bipinnate, the ultimate divisions 1–3 mm. long. Infl. compact; invols. absent; involucels subscarious. Rays few, pedicels short or obsolete. Calyx prominent. Fls. yellow to purplish. Fr. ovoid to ± oblong, slightly flattened laterally, glabrous; ribs filiform; oil-tubes 2–several in intervals and on commissure. Three spp. of w. N. Am. (Greek, *podos*, foot, and *stereos*, solid, because of compactness.)

1. **P. nevadénsis** (Gray) Wats. [*Cymopterus n.* Gray.] Plant 1–5 cm. tall; lvs. 8–15 mm. long; peduncles solitary, 1–2.5 cm. long; umbels subcapitate; fr. 2–3 mm. long; Subalpine, reported from Sugarloaf Peak, San Bernardino Mts.; White Mts., cent. Sierra Nevada. July–Sept.

27. *Pterýxia* Nutt. ex Coult. & Rose

Low cespitose perennials from a deep-seated root, caulescent or acaulescent. Lvs. bipinnate or pinnately or ternate-pinnately decompound, with ultimate divisions linear, oblong or subcuneate. Petioles sheathing. Infl. of compound umbels. Invol. usually wanting. Involucels of narrow herbaceous bractlets. Fls. yellow to white or purple. Sepals evident. Fr. narrowly oblong to ovoid, flattened dorsally; lateral ribs winged, thin, some or all of the dorsal winged. Stylopodium none. Oil-tubes 1 to several in the intervals, several on the commissure. Five spp., w. N. Am. (Greek, *pteris*, fern, and *ixia*, the chameleon plant.)

1. **P. petraèa** (Jones) Coult. & Rose. [*Cymopterus p.* Jones.] Stems slender, 1.5–3 dm. tall; lf.-blades pale green, narrow-oblong in outline, 3–15 cm. long, ternate-bipinnate to tripinnate; the ultimate linear divisions 1–8 mm. long; petioles 5–12 cm. long; involucel with linear bractlets 1–3 mm. long; rays 3–7, unequal, to 5 cm. long; pedicels 1–6 mm. long; fls. yellow; fr. ovoid to ± oblong, 4.5–7 mm. long, 2–4 mm. broad, the wings plane, not broader than the body; oil-tubes 1–3 in intervals, 5–15 on commissure.—Rocky places 5000–7500 ft.; Pinyon-Juniper Wd., Inyo-White Mts.; to 11,000 ft. to the n. May–June.

28. *Sanícula* L. SANICLE. SNAKEROOT.

Biennial or perennial herbs with nearly naked or few-lvd. stems. Lvs. palmately or pinnately divided, rarely entire. Umbels irregularly compound, few-rayed, bearing invols. and involucels. Sepals evident, persistent. Corolla greenish-yellow or purple. Fr. subglobose, densely covered with tubercles or hooked bristles. Carpels not ribbed; oil-tubes usually several to many. Ca. 40 spp., widely distributed in n. temp. regions, S. Afr. and S. Am. (Diminutive, derived from Latin, *sanare*, to heal.)

(Shan & Constance. The genus Sanicula in the Old World and New. Univ. Calif. Pub. Bot. 25: 1–78. 1951. Bell, C. B. The S. crassicaulis complex. Univ. Calif. Pub. Bot. 27: 133–230. 1954.)

1. Basal lvs. ternately or palmately divided, rarely entire.
    2. Frs. prickly on upper part only, pedicellate or stipitate .... 1. *S. arguta*
    2. Frs. prickly to the base.
        3. Involucel bractlets 1–2 mm. long; calyx-lobes of ♂ fls. 0.5–0.7 mm. long; styles twice as long as calyx; frs. distinctly pedicellate .... 4. *S. crassicaulis*
        3. Involucel bractlets 3–5 mm. long; calyx-lobes of ♂ fls. 1–2.3 mm. long; styles ca. as long as calyx; frs. subsessile .... 6. *S. hoffmannii*
1. Basal lvs. pinnatifid or pinnately or ternate-pinnately dissected or decompound.
    4. Lvs. once or twice pinnatifid with a toothed rachis .... 3. *S. bipinnatifida*
    4. Lvs. 1- to 3-pinnate or pinnately or ternate-pinnately decompound, without a margined or toothed rachis.
        5. Stems from fusiform taproots.
            6. Lvs. 1- to 3-pinnate; ♂ fls. few and inconspicuous .... 2. *S. bipinnata*
            6. Lvs. ternate-pinnately decompound; ♂ fls. several, conspicuous ..... 5. *S. graveolens*
        5. Stems from globose or irregular tuberous roots .... 7. *S. tuberosa*

1. **S. argùta** Greene ex Coult. & Rose. Stems erect from a taproot, glabrous, 1.5–5 dm. tall, ± branched; lvs. mainly basal, blades deltoid in outline, 3–10 cm. long, palmately 3–5-parted, the middle divisions longer, all spinose-serrate to sublaciniate, decurrent with a broad toothed wing on the rachis, ± glandular-roughened above; bracts reduced, lf.-like; involucel with entire or 3-lobed bractlets; fertile rays 3–5, the umbellets globose; fls. yellow; fr. 4–6 mm. long, obovoid, stipitate, bristly above, almost naked basally; $n=8$

(Bell, 1954).—Dry ± grassy slopes, below 2500 ft.; V. Grassland, Coastal Sage Scrub, Chaparral, etc.; cismontane, n. L. Calif. to Monterey Co. Santa Rosa, Santa Cruz, San Nicolas, Santa Clemente, Santa Catalina ids. March–April.

2. **S. bipinnàta** H. & A. [*S. pinnatifida* Torr.] POISON SANICLE. Taproot vertical; stems slender, erect, 1–6 dm. high, solitary, branched from near base; lvs. largely near base, the blades 4–10 cm. long, 2–3-pinnate, the divisions not decurrent, ovate to oblong, toothed; cauline lvs. reduced; umbels 2–5-rayed; bracts 1–2 cm. long; bractlets 6–8, entire, 1–2 mm. long; fls. pale yellow; frs. 3–8, 2–3 mm. long, subsessile; $n=8$ (Bell, 1954).—Shaded or open slopes below 5000 ft.; V. Grassland, Chaparral, etc.; interior cismontane, from Los Angeles Co. n. April–May.

3. **S. bipinnatífida** Dougl. ex Hook. PURPLE SANICLE. Plate 8, Fig. D. Stems rather stout, 1.5–7 dm. high, from a thickened often branched taproot-crown; basal lvs. several, polymorphic, the blades 3–7-parted, 5–12 cm. long, the divisions cleft to lobed, decurrent on the rachis forming a toothed wing; umbel 3–5-rayed, the bracts leaflike; bractlets 2.5 mm. long, lanceolate; fls. purplish-red to yellow; frs. 5–10, sessile, ovoid to subglobose, 3–6 mm. long; $n=8$ (Bell, 1954).—Frequent on open slopes, below 4500 ft.; V. Grassland, Yellow Pine F., Chaparral, etc.; cismontane from n. L. Calif. to B.C. March–May.

4. **S. crassicáulis** Poepp. ex DC. [*S. menziesii* H. & A.] Stem solitary, erect, branched above, 3–8 dm. tall, from stout taproot; lvs. near base, round-cordate in outline, 3–9 cm. broad, palmately and deeply 3–5-lobed, the segms. broad, obovate, sharply lobed or incised; umbels in open panicle, rays few; invol. of 2–3 small bracts; bractlets 6–8, small, entire; fls. yellow; fr. subglobose, 2–4 mm. long, stipitate, $n=8$, 12, or 16 (Bell, 1954).—Frequent on shaded and wooded slopes below 4500 ft.; S. Oak Wd., Chaparral, etc.; cismontane, n. to B.C., also in S. Am. Santa Catalina and Santa Cruz ids. March–May.

5. **S. gravèolens** Poepp. ex DC. [*S. nevadensis* Wats.] Root vertical, mostly fusiform; stems slender and erect or low and spreading, 0.5–4.5 dm. high; lvs. ternate, ± ovate, 1.5–4 cm. long, the primary divisions oblong-ovate, petiolulate, 3–5-lobed or 1–2-pinnatisect, the segms. ± ovate, incised or lobed; invol. bracts pinnatifid, 5–10 mm. long; rays 3–5; bractlets 6–10, 1 mm. long; fls. yellow; frs. 2–5, ovoid-globose, 3–5 mm. long; $n=8$ (Bell, 1954).—Open mostly coniferous forests, mts. San Diego Co., Santa Ana Mts., etc. n. to B.C. April–June.

6. **S. hoffmánnii** (Munz) Shan & Const. [*S. bipinnatifida* var. *h.* Munz.] Stout, erect, 3–9 dm. high, from a taproot; stem mostly branched above; basal lvs. many, glaucous, rounded-deltoid in outline, 5–13 cm. broad, deeply 3–5-lobed, the segms. obovate, ± lobed, serrate; upper lvs. reduced; umbels in loose panicle, 3–4-rayed; bractlets lanceolate, 3–5 mm. long; fls. greenish-yellow; frs. 4–10, ovoid, 3–5 mm. long, subsessile; $n=8$ (Bell, 1954).—Coastal Sage Scrub, etc. near coast of San Luis Obispo and Santa Barbara cos.; Santa Rosa, Santa Cruz and San Nicolas ids. March–May.

7. **S. tuberòsa** Torr. Stems simple or divided near base, 1–8 dm. high, from a small globose tuber; lvs. mostly in lower part, the blades 3–12 cm. long, ternate then pinnatifid or tripinnatisect; the primary divisions petiolulate, remote, the segms. entire to dissected, narrow; bracts 2, compound; rays mostly 3; bractlets 6–10, ca. 1–3 mm. long; fls. yellow; frs. 3–5, 1.5–2 mm. long, mostly subsessile; $n=8$ (Bell, 1954).—Open to wooded places below 5000 ft.; S. Oak Wd., Chaparral, etc.; L. Calif. n. to sw. Ore. March–June.

29. *Scándix* L.

Annual, with pinnately decompound lvs. and sheathing petioles. Infl. of loose compound or simple umbels. Invol. usually none. Bractlets several, usually dissected. Sepals minute or obsolete. Petals white, mostly unequal. Stylopodium depressed. Fr. linear or linear-oblong, laterally flattened, beaked, prominently ribbed; oil-tubes solitary in intervals or obsolete. Ca. 10 spp. of Old World. (Ancient Greek name.)

1. **S. pécten-véneris** L. SHEPHERD'S NEEDLE. Plants 1.5–3.5 dm. high, ± hispid, mostly branched from the base; lvs. oblong in outline, the blades 3–12 cm. long, finely dissected; bractlets lanceolate, ciliate; pedicels to 8 mm. long; fr. 6–15 mm. long, with

an additional 20-70 mm. in the beak; $2n=16$ (Wanscher, 1931).—Occasional in waste places, especially from Ventura and Santa Barbara cos. n.; natur. from Eurasia. April–June.

30. *Sphenosciàdium* Gray

Tall subsimple, stout, thick-rooted perennials, glabrous below the tomentose infl. Lvs. 1-2-pinnate or ternate-pinnate. Petioles inflated, sheathing. Infl. of loose compound umbels. Invol. wanting. Involucel of many linear-setaceous tomentose bractlets. Rays many, rather long. Fls. white or purplish, sessile in capitate umbellets. Stylopodium small, conical. Fr. strongly flattened dorsally, cuneate-obovate, ribbed at base, winged above. Oil-tubes 1 in intervals, 2 on commissure. One sp. (Greek, *sphenos*, wedge, and *sciadios*, umbrella, referring to the umbel.)

1. **S. capitellàtum** Gray. [*Selinum eryngifolium* Greene.] RANGER'S BUTTON. WHITE HEADS. Scabrous, 5-16 dm. high; lf.-blades ovate to oblong in outline, 1-4 dm. long, the lfts. linear-oblong to lance-ovate, 1-12 cm. long, 0.5-5 cm. broad, paler and scabrous beneath; petioles 1-4 dm. long; rays 4-18, subequal, 1.5-9 cm. long, tomentose; fr. 5-8 mm. long, 3-5 mm. broad, tomentose, the dorsal ribs winged, the lateral more broadly winged; $n=11$ (Bell & Constance, 1960).—Wet places, 3000-10,000 ft.; Yellow Pine F. to Subalpine F.; San Jacinto and San Bernardino mts., Mt. Pinos, White Mts.; to Ore., L. Calif. July–Aug.

31. *Táuschia* Schlecht.

Acaulescent or short-caulescent perennials from taproots or tubers. Lvs. petioled, entire, pinnate to ternately decompound. Infl. of loose compound umbels. Invol. usually wanting. Involucels dimidiate, usually of prominent bractlets. Fertile rays few to many; fertile pedicels short. Fls. yellow, white or purplish. Sepals prominent to obsolete. Stylopodium obsolete. Fr. oblong to round or ellipsoid, slightly flattened laterally, glabrous; ribs prominent to filiform, not winged; oil-tubes large to small, solitary to several in intervals, 2-several on commissure. Ca. 20 spp., w. and c. N. Am. (I. F. *Tausch*, 19th-cent. botanist.)

Lvs. simply pinnate; fr. with conspicuous ribs .... 1. *T. arguta*
Lvs. ternate-pinnate or ternately or pinnately decompound.
    Ultimate lf.-divisions 6-12 mm. long; plants glabrous .... 3. *T. parishii*
    Ultimate lf.-divisions 20-60 mm. long; plants minutely scabrous .... 2. *T. hartwegii*

1. **T. argùta** (T. & G.) Macbr. [*Deweya a.* T. & G. *Velaea a.* Coult. & Rose.] Plate 8, Fig. E. Short-caulescent, glabrous, 3-7 dm. high; lf.-blades oblong to ovate in outline, 8-16 cm. long, the lfts. oblong to oval, distinct, 3-8 cm. long, sharply serrate; petioles 6-20 cm. long; peduncles 1.5-4 dm. long; involucel of several entire to lobed bractlets 2-10 mm. long; rays 15-25, 2-12 cm. long; sepals evident; fr. oblong, 6-9 mm. long, ribs prominent; oil-tubes 3-5 in intervals, 4-6 on commissure; $n=11$ (Bell & Constance, 1957).—Locally common on dry fans and slopes, below 6000 ft.; Coastal Sage Scrub, Chaparral, etc.; L. Calif. to Santa Barbara Co., inland to desert edge. April–June.

2. **T. hartwégii** (Gray) Macbr. [*Deweya h.* Gray.] Acaulescent, 3-10 dm. high, minutely scabrous; lf.-blades 1-2-ternate-pinnate, 1-2 dm. long, the lfts. oblong to ovate, confluent, 2-6 cm. long, mucronulate-serrulate; petioles 5-25 cm. long; peduncles 2-8 dm. long; bractlets 5-12 mm. long; sepals minute; rays 2-13 cm. long; pedicels 2-7 mm. long; fls. yellow; fr. suborbicular, 4-7 mm. long; oil-tubes 3-5 in intervals, 6-8 on commissure; $n=11$ (Bell & Constance, 1957).—Occasional on wooded or brushy slopes, below 5000 ft.; Chaparral, Foothill Wd.; from Santa Monica Mts., Ft. Tejon, Ventura Co., etc. to Butte Co. March–May.

3. **T. párishii** (Coult. & Rose) Macbr. [*Velaea p.* Coult. & Rose.] Acaulescent, 1-4 dm. high, glabrous and glaucous; lf.-blades ovate to oblong in outline, ternate-pinnate or bipinnate, 8-15 cm. long, the ultimate segms. ovate, toothed, 6-12 mm. long; petioles 5-15 cm. long; peduncles 1-3 dm. long; bractlets few, 5-12 mm. long; rays subequal, 3-6 cm. long; pedicels 2-7 mm. long; sepals evident; fls. yellow; fr. oblong to oval, 5-8 mm.

long, the ribs filiform; oil-tubes 4–5 in intervals, 8–10 on commissure; $n=11$ (Bell & Constance, 1957).—Frequent on dry slopes, 3000–8000 ft.; Chaparral, Yellow Pine F., Pinyon-Juniper Wd.; mts. from San Diego Co. to Mt. Pinos and s. Sierra Nevada. May–July.

32. *Tórilis* Adams. HEDGE-PARSLEY

Hispid or pubescent annuals with pinnately compound lvs. Fls. white, in compound umbels. Invol. of few small bracts or wanting. Involucels of several linear bractlets. Sepals triangular, acute, sometimes obsolete. Stylopodium thick, conic. Fr. ovoid or oblong, tuberculate or prickly; the primary ribs filiform, setulose, the lateral displaced on to the commissural surface, the secondary hidden by the many barbed or hooked bristles or tubercles; oil-tubes solitary under the secondary ribs, 2 on the commissural side. Ca. 20 spp., N. Hemis. (Derivation not known.)

1. **T. nodòsa** (L.) Gaertn. [*Tordylium n.* L.] Erect, few-branched, 1–7 dm. tall, the stems retrorsely scabrous; lvs. pinnately decompound, 0.5–2 dm. long, the lfts. bipinnately dissected; umbels short-peduncled, solitary, opposite the lvs.; invol. none or of 1 bract; involucel-bractlets longer than the pedicels; rays few; fr. ovoid, 3–5 mm. long, the outer carpel hooked-bristly, the inner smooth or with tubercles; $2n=22$ (Gardé & Gardé, 1949).—Occasional on open hills below 2000 ft., natur. from Eu. April–June. San Miguel, Santa Cruz and Santa Catalina ids.

7. Apocynàceae. DOGBANE FAMILY

Perennial herbs, shrubs or vines, some trop. forms arboreal. Juice milky. Lvs. entire, chiefly opposite, sometimes whorled or opposite, estipulate. Fls. regular, bisexual, solitary and axillary, or cymose, or panicled. Calyx free from ovary, persistent, 5-parted, imbricated in bud. Corolla 5-lobed, convolute and often twisted in bud. Stamens 5, alternate with corolla lobes, inserted on the tube or throat; anthers often produced at base into a sterile appendage, connivent around the stigma. Carpels 2, distinct or united and making a 1-celled ovary; styles simple or divided; stigma simple. Fr. of 2 follicles, or drupaceous. Seeds often bearing a coma; endosperm fleshy. Ca. 150 genera and 1000 spp.; widely distributed, especially in warmer regions where many are very ornamental (Oleander, Frangipani, etc.)

Lvs. alternate; stamens inserted at summit of corolla-tube; seeds without coma .... 1. *Amsonia*
Lvs. opposite.
  Plant creeping or trailing; fls. blue; seeds naked .... 4. *Vinca*
  Plant not trailing; fls. not blue; seeds comose.
    Corolla 1.5–2 cm. long; style filiform with membranous reflexed collar under stigma
        3. *Cycladenia*
    Corolla not over 6 mm. long; style very short, not appendaged .... 2. *Apocynum*

1. *Amsònia* Walt.

Perennial herbs; lvs. many, alternate. Fls. in a terminal compound cyme. Calyx small, 5-parted. Corolla with narrow-funnelform tube and spreading or reflexed lobes, somewhat bluish, the tube reflexed-hairy within. Stamens included. Style filiform; stigma in ours subtended by a globose thickening. Follicles slender, torulose, several-seeded. Seeds cylindric or oblong, without coma. Ca. 17 spp. of N. Am. and Japan. (Dr. Charles *Amson*, 18th-century resident of Virginia.)

(Woodson, R. Monograph of the genus Amsonia. Ann. Mo. Bot. Gard. 15: 379–434. 1928.)

1. **A. brevifòlia** Gray. Plate 8, Fig. F. Stems several from branching crown, 2–4 dm. tall, glabrous throughout, bright green; lvs. ovate to lanceolate, subsessile, acuminate, entire, 2–4 cm. long; calyx 3–4 mm. long; corolla 12–15 mm. long, the lobes half as long as the tube; follicles 5–9 cm. long, glabrous; seeds largely 10–13 mm. long, ca. 3 mm.

wide, tapering to each end.—Rather common on dry slopes and banks, 2500–6000 ft.; Creosote Bush Scrub, Joshua Tree Wd., Pinyon-Juniper Wd.; Mojave Desert from n. base of San Bernardino Mts. e. along n. edge of Colo. Desert and n. to Eureka V.; to Utah, Ariz. March–May.

A. tomentòsa Torr. & Frém. [*A. brevifolia* var. *t.* Jeps.] Grayish-tomentose throughout; seeds largely 8–10 mm. long, 3–4 mm. wide, subtruncate to oblique at the ends. Occurs with *A. brevifolia* for the most part and doubtfully distinct as a sp., although so maintained by Woodson.

2. Apócynum L. Dogbane. Indian-Hemp

Perennial herbs with upright branching stems, opposite lvs., tough fibrous bark, and small pale cymose fls. on short pedicels. Calyx deeply 5-cleft. Corolla campanulate or cylindrical, 5-lobed, with 5 small triangular appendages opposite the lobes. Stamens inserted on very base of corolla; fils. short and broad; anthers connivent, sagittate. Style none or very short; stigma ovoid. Follicles terete, slender. Seeds many, comose. Ca. 7 spp. of N. Am. (Greek, *apo*, from, and *kuon*, dog, ancient name of the dogbane of the Old World.)

(Woodson, R. N. Am. Fl. 29: 188–192. 1938.)
Corolla 4–6 mm. long, 2–3 times as long as calyx.
    Lvs. drooping; corolla at least 3 times as long as calyx.
        Corolla campanulate, the orifice of the tube at least twice the width of the base; follicles normally pendulous .... 1. *A. androsaemifolium*
        Corolla cylindric, the orifice of the tube ca. as great as the width of the base; follicles suberect
            4. *A. pumilum*
    Lvs. spreading; corolla ca. twice the length of the calyx .... 3. *A. medium*
Corolla 2–3 mm. long; lvs. ascending.
    Bracts of the infl. scarious and aristate; lvs. evidently petiolate; follicles 9–15 cm. long
        2. *A. cannabinum*
    Bracts of the infl. semifoliaceous or blade-bearing; lvs. nearly or quite sessile; follicles 4–10 cm. long .... 5. *A. sibiricum*

1. **A. androsaèmifolium** L. Plate 9, Fig. A. Stems erect, 2–4 dm. tall, glabrous, diffusely branched; lvs. drooping, ovate to oblong, pale and tomentulose beneath, 2–5 cm. long, subsessile; calyx 1.5–3 mm. long; corolla campanulate, white, usually with pinkish veins, 4–7 mm. long; follicles pendulous at maturity, 4–12 cm. long; seeds linear, ca. 1 mm. long, with a pale tawny coma; $2n=16$? (Schürhoff et al., 1937).—Dry flats and slopes, 5000–9000 ft.; Montane Coniferous F.; San Jacinto and San Bernardino mts.; Ariz. to B.C. and Atlantic Coast. June–Aug.

2. **A. cannábinum** L. var. **glabérrimum** A. DC. Plate 9, Fig. B. Glabrous, 3–6 dm. tall; lvs. petioled or the lower subsessile, lance-ovate, 4–10 cm. long, mucronate; calyx-lobes scarious, glabrous, 1.5–2 mm. long; corolla 2–3 mm. long, almost as wide, greenish to whitish, cylindric to urn-shaped; follicles pendulous, 12–20 cm. long; seeds ca. 4 mm. long, the coma whitish, much longer.—Occasional in damp places below 6000 ft.; many Plant Communities, cismontane and occasional on the desert; to Can. and Atlantic Coast. June–Aug.

3. **A. mèdium** Greene var. **floribúndum** (Greene) Woodson. [*A. f.* Greene. *A. lividum* Greene.] Stems 3–5 dm. tall; lvs. spreading, ovate-lanceolate, subsessile, glabrous; calyx-lobes 1.5–3 mm. long, slightly ciliate-erose; corolla campanulate, 4–5 mm. long; follicles 7–10 cm. long, pendulous; coma pale tawny, 2 cm. long.—Rare in San Bernardino Mts. at 6000–8800 ft.; to Wash., Rocky Mts. June–Aug.

4. **A. pùmilum** (Gray) Greene var. **pùmilum.** [*A. androsaemifolium* var. *p.* Gray.] Like *A. androsaemifolium*, but mostly less than 3 dm. tall; lvs. more oblong-oval, subglabrous, more obtuse; follicles erect at maturity, coma of seeds white.—Occasional, at 5000–7000 ft.; Yellow Pine F.; San Bernardino Mts., n. slopes San Gabriel Mts.; to Wash., Mont. June–Aug.

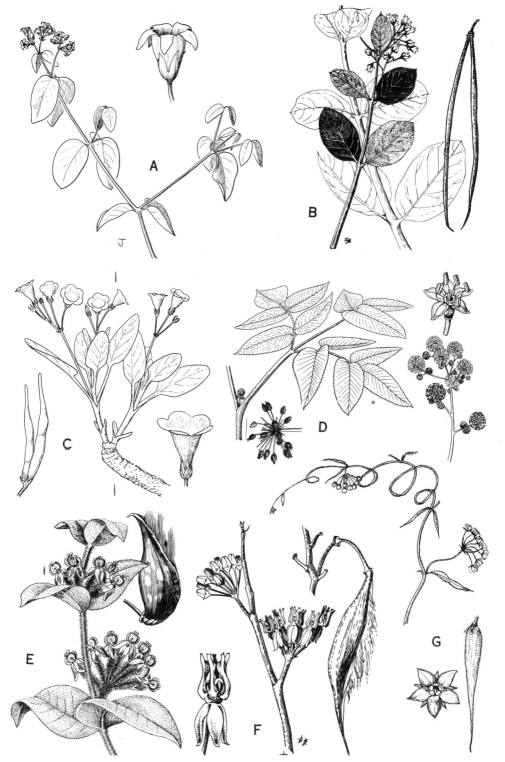

Plate 9. Fig. A, *Apocynum androsaemifolium*. Fig. B, *Apocynum cannabinum*. Fig. C, *Cycladenia humilis*. Fig. D, *Aralia californica*. Fig. E, *Asclepias californica*. Fig. F, *Asclepias subulata*. Fig. G, *Sarcostemma hirtellum*, twining.

Var. **rhomboideum** (Greene) Bég & Bel. [*A. r.* Greene.] The plant, or at least, the lower surface of the lvs. pubescent.—From Palomar Mts. (San Diego Co.) to n. Calif.

5. **A. sibíricum** Jacq. var. **salígnum** (Greene) Fern. [*A. s.* Greene.] Glabrous, 4–7 dm. tall; lvs. sessile, even amplexicaul, oblong-ovate, mostly 4–8 cm. long; calyx 1.5–2 mm. long, scarious; corolla cylindric to urn-shaped, 3–4 mm. long, white or greenish, the lobes erect or slightly spreading; follicles glabrous, pendulous at maturity, 4–9 cm. long; seeds 3.5–4 mm. long, the coma 1.5–2 cm. long.—Moist places to 7500 ft.; many Plant Communities; cismontane and montane, to desert edge; to B.C., Minn. and Tex. June–Aug.

3. Cycladènia Benth.

Low perennials from fleshy roots, with 1–several stems. Lvs. opposite, in 2 or 3 pairs. Fls. 2 or 3 on axillary peduncles. Calyx 5-lobed. Corolla funnelform, 5-lobed and with 5 small appendages alternate with the lobes. Stamens borne near base of tube. Style filiform with membranous collar. Follicles slightly fleshy, terete, rather stout; seeds many, comose at apex, narrowly urn-shaped, compressed. One polymorphous sp. (Greek, *kuklos*, ring, and *aden*, a gland, referring to the annular disk.)

1. **C. hùmilis** Benth. var. **venústa** (Eastw.) Woodson ex Munz. [*C. v.* Eastw.] Plate 9, Fig. C. Plant glabrous, 1–1.5 dm. tall; lvs. round-ovate, thick, 3–5 cm. long, on shorter petioles; calyx and corolla conspicuously pilose; corolla 15–20 mm. long, rose-purple; follicles ca. 8 cm. long.—High gravelly, talus-like slopes, 7000–9000 ft.; Red Fir F.; San Gabriel Mts., Santa Lucia Mts. and in Inyo Co. May–July.

4. Vínca L. Periwinkle

Trailing perennial herbs, with fertile erect stems. Corolla salverform, with callous constriction at throat. Style slender; stigma annular. Follicles narrow, terete. Seeds many, compressed, truncate at both ends. Ca. 12 spp. of Old World. (Ancient Latin name.)

1. **V. màjor** L. Sterile stems trailing, a m. or so long, flowering ones shorter; lvs. ovate-orbicular, dark green, 2–3 cm. long, on petioles 0.5–2 cm. long; fls. solitary in axil of every other lf., violet or blue, 2.5–3 cm. long; follicles 4–5 cm. long, somewhat torulose; $2n=92$ (Rutland, 1941).—Occasional escape about old dwelling-sites and becoming natur. especially in shaded places below 3000 ft.; native of Eu. March–July. Santa Catalina and Santa Cruz ids.

8. Araliàceae. Ginseng Family

Herbs, shrubs or trees, sometimes climbing. Lvs. alternate to whorled, entire to compound or decompound. Fls. small, greenish or whitish, perfect or imperfect, mostly in umbels or umbellate heads. Fl.-tube adnate to the ovary. Calyx small, often absent. Petals usually 5, valvate or imbricate. Stamens as many as petals and alternate with them. Ovary 1, inferior, 1- to several-loculed, with 1 ovule in each locule; styles as many as the locules. Fr. a berry or drupelike. Seeds flattened or 3-angled; testa thin; embryo small in copious endosperm. Ca. 50 genera and 500 spp.; widely distributed in temp. and trop. regions.

Plant large perennial herb, with large compound lvs. . . . . 1. *Aralia*
Plant sprawling or climbing by aerial roots; lvs. simple, ± lobed . . . . 2. *Hedera*

1. Aràlia L. Spikenard

Lvs. large, pinnately or ternately decompound; lfts. serrate. Umbels 2 or more in an infl., radiating or laxly corymbose or in large compound racemes, with small bracts and bractlets. Pedicels articulate with calyx. Plants polygamo-monoecious, the fls. 5–6-merous, mostly glabrous. Fils. short. Disk small. Styles 4–6, free above or at the base. Fr. baccate, subglobose, sharply angled. Ca. 30 spp. of N. Am. and Asia. (Name unexplained.)

Aralia

1. **A. califórnica** Wats. Plate 9, Fig. D. Roots large, with milky juice; stems 1–3 m. high; lvs. glabrous, ternate then pinnately 3–5-foliolate; petioles to 3 dm. long; lfts. ovate or ± oblong, serrate, subcordate at base, 0.5–2.5 dm. long; infl. ample, 3–4 dm. long, with numerous glandular-tomentulose many-fld. umbels; pedicels 1–2 cm. long; sepals minute; petals ca. 2 mm. long; fr. 3–5 mm. in diam., dark when mature, seeds almost 3 mm. long, light in color; $2n=48$ (Bowden, 1945).—In ± moist and shaded spots, below 6000 ft.; Chaparral, Yellow Pine F., etc.; cismontane from Orange Co. n.; to Ore. June–Aug.

2. *Hédera* L. Ivy

Plants with a sterile juvenile form which is climbing and with aerial roots and lobed lvs. and an adult form fertile, nonclimbing, and with lvs. not lobed. Lvs. persistent, alternate. Fls. bisexual, greenish, in umbels arranged in terminal racemes or panicles. Sepals 5; petals 5; styles united. Fr. a 3–5-seeded berry. Ca. 6 spp. of Eurasia, commonly cult. in Calif. as English and Algerian Ivy. (The classical name.)

1. **H. hèlix** L. English Ivy. Juvenile lvs. usually 3–5-lobed, triangular-ovate to subreniform, mostly 2–10 cm. wide, cordate to truncate at base; $2n=48$ (Jacobsen, 1954).—Occasional escape from cult. and persisting for some time; native of Eurasia.

9. Asclepiadàceae. Milkweed Family

Perennial, herbs, vines or shrubs with milky juice. Lvs. opposite, whorled, rarely alternate, without stipules. Fls. perfect, regular, mostly umbellate. Calyx deeply 5-lobed, the lobes usually imbricate. Corolla 5-lobed, the lobes commonly valvate in the bud. A 5-lobed crown is usually present between the corolla and stamens, adnate to either or to both. Stamens 5, inserted on the corolla-tube, usually near base; fils. monadelphous or sometimes distinct; anthers united and tipped by a scarious membrane inflexed on the summit of the stylar disk. Pollen grains united into waxlike or granular pollinia. Carpels 2, with distinct superior ovaries and styles, but united above by the stigmas. Fr. of 2 follicles; seeds many, compressed, with long coma. Ca. 200 genera and 2500 spp., widely distributed, but most frequent in warmer regions.

Stems not twining .... 2. *Asclepias*
Stems twining.
  Crowns wanting; corolla urn-shaped or campanulate; pollinia strictly pendulous . 3. *Cynanchum*
  Crowns present; corolla rotate or open-campanulate; pollinia pendulous or horizontal.
    Fls. solitary; pollinia horizontal .... 4. *Matelea*
    Fls. borne in umbels or racemes.
      The fls. white. Escape from cult. .... 1. *Araujia*
      The fls. not white. Native .... 5. *Sarcostemma*

1. *Aráujia* Brot.

Corolla-tube inflated at the base; lobes 5, overlapping in the bud; crown with 5 scales attached at or below the middle of the tube; stigma often 2-beaked at apex. Ca. 5 spp. native in Brazil and Argentina. (*Araujia*, native S. Am. name.)

1. **A. sericófera** Brot. Bladder-Flower. Vigorous climber; stem covered with pale down when young; lvs. ovate-oblong, 5–10 cm. long, pale green, minutely pitted beneath; fls. white, salverform, 2–3 cm. across, the tube 12 mm. long; pod grooved, 12 cm. long, 5–7 cm. wide.—Reported as an escape from cult., Riverside to Placer cos.; native of S. Am.

2. *Asclèpias* L. Milkweed

Herbs or shrubs from deep-seated roots. Lvs. opposite, alternate or whorled. Fls. perfect, regular, in axillary or terminal umbels. Calyx small, usually with small glands at the base of the 5 lobes. Corolla rotate, deeply 5-cleft, the lobes usually reflexed in anthesis.

Crown usually with a column, the lobes 5, concave and hoodlike, each bearing a hornlike or toothlike projection within, sometimes hornless. Fils. connate into a tube, the anthers winged, the wings broadened below the middle. Pollinia solitary in the sacs, pendulous. Follicles fusiform or narrower, mostly acuminate. Seeds compressed, comose. A rather large genus of the New World. (The Greek name of *Aesculapias*, for whom the genus is named.)

(Woodson, R. E. Jr. The N. Am. spp. of Asclepias. Ann. Mo. Bot. Gard. 41: 1–211. 1954.)
1. Corolla-segms. spreading during anthesis; hoods of crown joined to each other by a lobed disk. E. Mojave Desert .... 2. *A. asperula*
1. Corolla-segms. reflexed during anthesis; hoods distinct from each other or lacking.
    2. The purplish hoods without horns on inner surface; herbage densely tomentose
    3. *A. californica*
    2. The hoods with horns on inner surface.
        3. The hoods 2–3 times as long as stamens and stigma; fil.-column almost lacking; pedicels deflexed in fr.
            4. Lvs. ovate; stems herbaceous; plant 1–2 dm. high .... 7. *A. nyctaginifolia*
            4. Lvs. filiform or lacking; stems woody below, 1–1.5 m. tall .... 8. *A. subulata*
        3. The hoods ca. as long as stamens and stigma; fil.-column well developed.
            5. Lvs. 1 cm. or less wide, often early deciduous; pedicels erect in fr.
                6. Plants herbaceous, rarely 1 m. high, with persistent green lvs.; stems green
                6. *A. fascicularis*
                6. Plants shrubby, 1–2 m. high, the lvs. soon deciduous; stems white .. 1. *A. albicans*
            5. Lvs. mostly 2–7 cm. wide; pedicels deflexed in fr.
                7. Lateral umbels sessile or subsessile; rare, on Mojave Desert .... 9. *A. vestita*
                7. Lateral umbels well peduncled.
                    8. Lvs. usually 3 or more at a node; horns mostly included within the hoods. Cismontane .... 4. *A. eriocarpa*
                    8. Lvs. opposite; horns exserted. Desert .... 5. *A. erosa*

1. **A. albicáns** Wats. Shrubby, 1–2 m. tall, the stems whitish; lvs. in 2's or 3's, filiform, 1–2 cm. long, early deciduous; umbels somewhat woolly; corolla-lobes greenish-white with some brown or pink, 5–6 mm. long; hoods yellowish, shorter than the anthers; horns incurved, barely exserted; follicles on erect pedicels, tomentulose when young, later glabrate and whitish, smooth, lanceolate in shape, ca. 1 dm. long; seeds narrow, ca. 6 mm. long; coma ca. 1.5 cm. long.—Occasional, dry rocky places below 2500 ft.; Creosote Bush Scrub; Colo. Desert, e. Mojave Desert from Sheephole Mts. to Whipple Mts., etc.; Ariz., L. Calif. March–May.

2. **A. aspérula** (Dcne.) Woodson. [*Acerates a.* Dcne. *A. capricornu* Woodson ssp. *occidentalis* Woodson. *Ananthrix decumbens* Nutt. *Asclepiodora d.* Gray.] Stems several, herbaceous, decumbent or ascending, 3–6 dm. long, puberulent; lvs. green, scabrous-puberulent, lanceolate, 4–12 cm. long, mostly alternate, short-petioled, sometimes in 3's; umbels many fld., largely solitary and terminal, peduncled; pedicels, calyx and outer surface of corolla-lobes puberulent; corolla-lobes widely spreading, greenish-white, ca. 8 mm. long; hoods purplish, 5–6 mm. long; follicles 7–8 cm. long, erect on recurved pedicels; seeds ca. 6 mm. long.—Dry open places, 3000–6500 ft.; largely Pinyon-Juniper Wd.; New York, Providence and Clark mts., of e. Mojave Desert; to Colo., Texas. May–July and in Sept.

3. **A. califórnica** Greene. [*Gomphocarpus tomentosus* Gray. *G. t.* var. *xanti* Gray. *G. torreyi* Macbr.] Plate 9, Fig. E. Herbaceous, soft white-woolly, the stems mostly 1.5–5 dm. long; lvs. mostly opposite, ovate to lanceolate, 5–15 cm. long, short-petioled, acute to acuminate at tip, obtuse to cordate at base; lateral umbels if present, tending to be sessile; fls. 6–12; calyx and outer surface of corolla white-woolly; corolla-lobes purplish, 8–10 mm. long; hoods dark maroon, broadly ovoid, centrally attached to column, shorter than the anthers; horns lacking; follicles ovoid, hoary, 5–8 cm. long, acuminate; seeds ca. 1 cm. long, the coma 3.5 cm. long.—Frequent, on dry slopes below 7500 ft.; Yellow Pine F., Pinyon-Juniper Wd., Chaparral; n. L. Calif. through cismontane s. Calif. to Kern and Inyo cos. April–July.

4. **A. eriocárpa** Benth. [*A. fremontii* Torr. ex Gray.] Heavy-tomentose throughout,

Asclepias

the herbaceous stems simple, erect, 4–9 dm. tall; lvs. usually at least in part in whorls of 3 or 4, some opposite, elongate-oblong, 6–15 (–20) cm. long, on very short petioles, truncate to subcordate at base; umbels few to several, many-fld., stout-peduncled; pedicels 2–5 cm. long; calyx 2.5–3 mm. long; corolla-lobes cream, oblong-ovate, 4–5 mm. long; hoods shorter than stamens, cream or with purplish tinge; horns barely exserted; follicles hoary, 6–9 cm. long; seeds ca. 8 mm. long.—Frequent in dry barren places below 7000 ft.; many Plant Communities; L. Calif. to Mendocino and Shasta cos. June–Aug. Poisonous to sheep and sometimes in baled hay to rabbits and other animals. A form with shorter follicles (3–4 cm.) occasional in the San Bernardino Mts. is the var. *microcarpa* M. & J.

5. **A. eròsa** Torr. [*A. e.* var. *obtusata* Gray. *A. o.* and *rothrockii* Greene.] Finely woolly, later glabrate, the herbaceous stems 5–8 dm. tall; lvs. opposite, oblong-ovate to lance-ovate, 8–15 cm. long, coriaceous, short-petioled, acute to acuminate; umbels peduncled, many-fld.; pedicels slender, 1.5–2 cm. long; calyx woolly, ca. 2.5 mm. long; corolla greenish-white, ± tomentose on outside, 5–6 mm. long; hoods broadly obovoid, little exceeding stamens; follicles tomentulose, short-acuminate, 5–8 cm. long; seeds ca. 9 mm. long, pubescent.—Dry slopes and washes, below 5000 ft.; Creosote Bush Scrub, Joshua Tree Wd., Pinyon-Juniper Wd.; deserts n. to Inyo Co., upper San Joaquin V.; to Utah, Ariz. May–July.

6. **A. fasciculàris** Dcne. in A. DC. [*A. mexicana* auth., not Cav.] Herbaceous, the stems several, erect, 6–8 dm. high, glabrous; lvs. linear to linear-lanceolate, usually in whorls of 3–6, short-petioled, 3–10 mm. wide, 4–10 cm. long, commonly folded along midrib; umbels in terminal corymbose clusters, many-fld.; the peduncles exceeding the pedicels; calyx pubescent, ca. 2 mm. long; corolla greenish-white, often tinged purple, the oblong lobes 4–5 mm. long; hoods ca. as long as stamens; horns slender, exserted, incurved; follicles smooth, narrow, 6–9 cm. long; seeds ca. 6 mm. long.—Frequent in colonies occurring in dry places mostly below 7000 ft.; many Plant Communities, mostly in cismontane areas away from the coast, occasional in the desert; Catalina and Santa Cruz ids.; to Wash., Ida., Utah, L. Calif. June–Aug.

7. **A. nyctaginifòlia** Gray. Herbaceous, with several ascending stems 1–1.5 dm. long, green, puberulent, lvs. ovate, acute, somewhat crisped, 4–7 cm. long, on petioles 1–2 cm. long; umbels nearly or quite sessile; corolla thin, greenish-white, 12–14 mm. long; hoods narrowly oblong, twice as long as stamen-column; follicles 5–6 cm. long.—Dry slopes, 4000–5000 ft.; Pinyon-Juniper Wd.; Providence and New York mts., e. Mojave Desert; to Ariz. May–June.

8. **A. subulàta** Dcne. in A. DC. Plate 9, Fig. F. Almost shrubby, the greenish-white stems in clusters, erect, rushlike, 1–1.5 m. tall, leafless or with few filiform lvs. 2–5 cm. long; umbels few to several, at and near summit of branches, rather short-peduncled; pedicels rather stout, ascending, somewhat tomentulose, mostly 1.5–2.5 cm. long; calyx tomentose, ca. 3 mm. long; corolla thin, greenish-white, 7–8 mm. long; hoods ca. as long, twice as long as stamens; horns scarcely exserted; follicles slender, 6–12 cm. long, smoothish; seeds 6–7 mm. long; $2n = 22$ (Lewis, 1961).—Occasional in desert washes and sandy places below 2000 ft.; Creosote Bush Scrub; Colo. Desert, e. Mojave Desert; L. Calif., Ariz. April–Dec.

9. **A. vestìta** H. & A. Stems herbaceous, simple, ascending, 2–5 dm. tall, white woolly, finally glabrate; lvs. opposite, ovate to oblong-lanceolate, subacuminate, 4–10 cm. long, short-petioled; umbels 1–4, the lateral ones sessile; corolla largely purplish, sometimes greenish-white, 4–6 mm. long; hoods nearly erect, truncate at summit and entire, not exceeding stamens; horns not exserted, blunt, slightly incurved; follicles 5–6.5 cm. long, hoary in youth; seeds ca. 10 mm. long.—Dry flats and slopes below 5000 ft.; s. Mojave Desert (Cajon Pass) to Monterey Co. and Inyo Co. A poorly marked form is the var. *parishii* Jeps.; lvs. more glabrous and infl. more solitary and less arachnoid; occurs about the w. part of the Mojave Desert. April–June.

3. Cynánchum L.

Ours shrubs or suffrutescent plants with slender twining stems. Lvs. opposite, slender or reduced. Fls. small, in axillary umbels or small cymes. Calyx 5-parted, the segms. acute. Corolla campanulate to urn-shaped, 5-lobed. Crowns wanting. Pollen-masses solitary in each pollen-sac. Follicles long-acuminate, smooth, terete. As here understood, a large genus of Old and New Worlds. (Greek, *cyon*, dog, and *anchein*, to strangle; ancient name of some plant supposed to poison dogs.)

1. **C. utahénse** (Engelm.) Woodson. [*Astephanus u.* Engelm.] Perennial from a branched crown, the stems slender, glabrous, 2–5 dm. tall; lvs. linear, acuminate, 2–3 cm. long, spreading or reflexed; umbels short-peduncled, 3–10-fld., with a few subulate bracts; corolla dull yellow, 2 mm. wide, the lobes ovate, somewhat hooded; anthers unappendaged at apex; follicles 4–6 cm. long; seeds rough-granulate.—Occasional, dry sandy places, below 3000 ft.; Creosote Bush Scrub; both deserts; to Utah, Ariz. April–June.

4. Matelèa Aubl.

As here understood a rather large genus of herbs, vines and shrubs mostly with both long eglandular hairs and short bulbose emergences. Ours with petioled opposite, usually cordate lvs. Fls. in axillary peduncled cymelike fascicles or umbels. Calyx deeply 5-cleft, glandular within. Corolla rotate to campanulate, 5-parted, the corona consisting of a unit enation of the anther-fil., subtending an additional inner enation. Stamens with connate fils. forming a tube; anthers tipped with a small scarious inflexed membrane in ours. Pollinia solitary in each sac, horizontal. Follicles thick, smooth or tuberculate or angled. Seeds compressed, with coma. Widely distributed. (Name not explained.)

1. **M. parvifòlia** (Torr.) Woodson. [*Gonolobus p.* Torr. *G. hastulatus* Gray. *Vincetoxicum h.* Heller. *G. californicus* Jeps.] Somewhat woody at base, ± twining, 1–4 dm. high, puberulent; lvs. cordate-sagittate, 5–20 mm. long, on petioles half as long; fls. 1, in the axil, sometimes 2; calyx 1–1.5 mm. long; corolla greenish, 3–4 mm. long; crown corolla-like, 5-lobed, attached to the stamen-column by 5 thin vertical plates; follicles 5–7 cm. long, tapering; seeds 4.5 mm. long.—Rare, in dry rocky places, 2000–3000 ft.; Creosote Bush Scrub; Colo. Desert (Corn Springs, Cottonwood Springs, Yaqui Well), Mojave Desert (Kelso); to Tex., L. Calif. March–May.

5. Sarcostémma R. Br.

Suffrutescent, trailing or twining vines. Lvs. usually foliose, opposite, usually with 1 or more glands on the ventral surface of the midrib near the base. Fls. 1–30, in cymes or umbels. Calyx deeply 5-lobed. Corolla subrotate to campanulate or salverform, 5-lobed. Stamens 5, the fils. coherent into a column, each fil. bearing an inflated vescicular segm. (corona-vescicle) just below the anther; anthers 2-celled, the membranous dorsal appendage ovate to deltoid; pollinia solitary in each anther-sac, pendulous. Follicle fusiform to clavate. Seeds flattened or unequally biconvex, with a micropylar coma. A rather large genus of warmer parts of the world. (Greek, *sarx*, flesh, and *stemma*, crown, referring to the fleshy inner corona.)

(Holm, R. W. The Am. species of Sarcostemma. Ann. Mo. Bot. Gard. 37: 477–560. 1950.)

Corolla purplish; plant glabrous or with subappressed hairs .... 1. *S. cynanchoides*
Corolla greenish-yellow; plant with short spreading hairs .... 2. *S. hirtellum*

1. **S. cynanchoìdes** Dcne. ssp. **hartwégii** (Vail) R. Holm. [*Philibertella h.* Vail. *Funastrum h.* Schlechter. *S. heterophyllum* auth.] Stems 1–2 m. long; lvs. glabrous to puberulent, green, linear to lanceolate, 3–4 cm. long, short-petioled; umbels several-fld., on peduncles 1–5 cm. long; pedicels 7–14 mm. long; calyx-lobes 2–3 mm. long; corolla rotate-subcampanulate, the lobes 5–7 mm. long; apical appendages of anthers orbicular; follicles slender, fusiform, 7–10 cm. long; seeds 6 mm. long.—Fairly frequent in dry places, below 2000 ft., Coastal Sage Scrub, Chaparral, Creosote Bush Scrub, etc., both deserts away from the

coast; San Diego and Imperial cos. to Ventura and San Bernardino cos.; to Utah, Ariz. April-July.

2. **S. hirtéllum** (Gray) R. Holm. [*S. heterophyllum* var. *h.* Gray. *Philibertella h.* Vail. *Funastrum h.* Schlechter.] Plate 9, Fig. G. Much like the preceding but more canescent; corolla-lobes 4–5 mm. long; apical appendage of anthers elliptic-ovate; follicles 4–4.5 cm. long; seeds 8 mm. long.—Frequent in washes, below 3500 ft.; Creosote Bush Scrub; Colo. Desert and e. Mojave Desert; Ariz., Nev. March–May.

10. Asteràceae. Sunflower Family

Herbs or shrubs; in the tropics sometimes trees. Lvs. opposite or alternate, rarely whorled, entire to dissected, estipulate. Fls. borne in a head, on a common receptacle, surrounded by an invol. of phyllaries (bracts), perfect, polygamous, monoecious or dioecious. Heads usually many-fld., sometimes few-fld., or even 1-fld. Receptacle usually flattened, sometimes conical or even columnar, with bracts (*chaff*), scales or bristles subtending the fls., or without (*naked*). Corolla gamopetalous, either regular, tubular, with a usually 5-toothed limb, valvate in bud, the veins bordering the margins of the lobes, or bilabiate, or ligulate, with the limb (*ray*) strap-shaped and toothed at apex, the corolla rarely wanting in the ♀ fls. Stamens 5 (rarely 4 or 3), usually united by the elongated anthers into a tube (*syngenesious*), inserted on the corolla. Style normally 2-branched, the branches stigmatiferous inside, in hermaphrodite (functional ♂) fls. usually entire; ovary inferior, 1-celled, maturing into an ak., with a single seed without endosperm, usually bearing a persistent pappus (representing the calyx-limb) of bristles, scales or paleae, or even a crown or ring. The tubular ♂ or hermaphrodite fls. are called disk-fls., and the head is discoid when composed only of these. The ligulate (ray) fls. are commonly ♀, sometimes perfect or neutral, and the head is called *radiate* when composed of cent. tubular fls. and marginal ray-fls.; if all are strap-shaped, it is called *ligulate*. The largest family of vascular plants, with possibly 950 genera and 20,000 spp., chiefly herbaceous and world-wide in distribution. Plate 10, Fig. A.

KEYS TO THE GENERA OF ASTERACEAE

1. Anthers tailed or sagittate at their base.
    2. Some or all of the corollas appearing definitely bilabiate .... **Group A**
    2. Corollas not at all or obscurely bilabiate.
        3. Style without ring of hairs or distinct thickened ring below the branches; plants not prickly; receptacle naked .... **Group B**
        3. Style with a ring of hairs or with a thickened ring below the papillate branches; plants usually prickly; receptacle usually densely bristly .... **Group C**
1. Anthers not tailed or sagittate at their base.
    4. Heads with fls. all perfect and corollas strap-shaped, 5-toothed at apex (true ligules)
        Group D
    4. Heads with fls. tubular when perfect, and ± regular; marginal fls. often ♀ or neutral and often strap-shaped, 2–3-toothed (ray-fls. or rays).
        5. Pappus none or vestigial, sometimes a mere crown.
            6. Rays none or vestigial .... **Group E**
            6. Rays evident .... **Group F**
        5. Pappus evident on some or all of the aks.
            7. Rays none or vestigial.
                8. Aks. with a pappus of scales or awns or both, these sometimes united into a low crown
                    Group G
                8. Aks. with a pappus of capillary bristles, rarely with additional outer scales
                    Group H
            7. Rays evident, present.
                9. Aks. with a pappus of firm awns or chaffy scales.
                    10. Rays yellow to orange or brown .... **Group I**

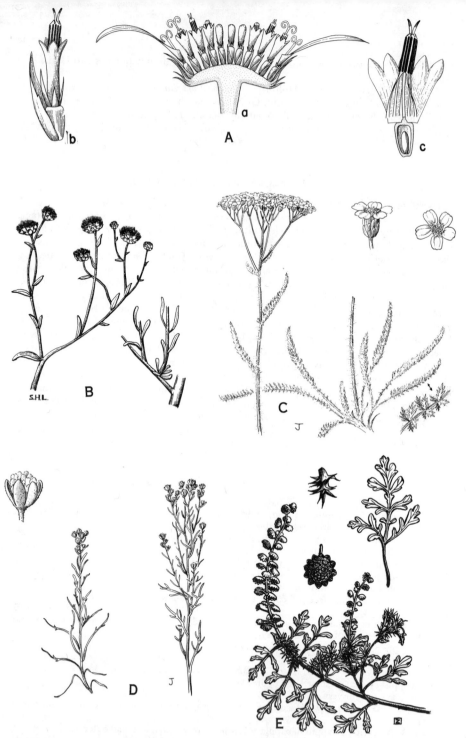

Plate 10. Fig. A, Diagram through an asteraceous head, with tubular central disk-fls., outer row of petallike ray-fls., outer invol. of overlapping bracts; separately disk-fls. with subtending bract (chaff), pappus (calyx) of 4 scales, inferior ovary, anthers connate, pistil with 2 stigmas. Fig. B, *Acamptopappus sphaerocephalus*. Fig. C, *Achillea millefolium* var. *californica*. Fig. D, *Amblyopappus pusillus*. Fig. E, *Ambrosia chamissonis*, ♂ heads in terminal racemes, ♀ heads burlike.

Asteraceae 97

  10. Rays white to pink, red or purplish .... **Group J**
 9. Aks. with a pappus of capillary bristles, rarely with additional outer scales
                                    Group K

Group A. (Corollas bilabiate; anthers tailed)

1. Heads 1-fld., capitate-clustered and surrounded by spiny-toothed semi-translucent bracts
                                    70. *Hecastocleis*
1. Heads many-fld., not capitate-clustered.
 2. Plants herbaceous; corollas pink or whitish; invol. well imbricated .... 109. *Perezia*
 2. Plants shrubby; corollas yellow; invol. in only 2 distinct rows .... 143. *Trixis*

Group B. (Heads discoid except in *Pulicaria*, anthers tailed; plants not prickly)

1. Heads with short yellow rays; pappus 2-ranked. Introd. .... 121. *Pulicaria*
1. Heads rayless.
 2. Receptacle naked; phyllaries many, not scalelike; often perennial.
  3. Phyllaries dry but not scarious; plants not tomentose; large herbs or shrubs . 116. *Pluchea*
  3. Phyllaries scarious; herbage woolly.
   4. Fls. all fertile, perfect and ♀ in the same head; taprooted .... 65. *Gnaphalium*
   4. Fls. dioecious or nearly so; fibrous-rooted.
    5. Pappus-bristles of ♀ fls. united at base; basal lvs. tufted, persistent; strictly dioecious
                                   11. *Antennaria*
    5. Pappus-bristles distinct; basal lvs. soon deciduous; ♀ plants commonly with few central functionally ♂ fls. .... 9. *Anaphalis*
 2. Receptacle chaffy, at least near the margin; phyllaries several or none, when present saclike (open in *Evax*) bearing ♀ (fertile) fls.; small woolly annuals.
  6. Fr.-bearing bracts open, merely subtending the ♀ fls., persistent .... 57. *Evax*
  6. Fr.-bearing bracts conduplicate or saclike, enclosing the ♀ fl. falling with the ak.
   7. Aks. gibbous, the style lateral; true invol. below the fertile bracts of ca. 5 scarious scales
                                     101. *Micropus*
   7. Aks. straight, the style apical or nearly so.
    8. Lvs. alternate; receptacle convex to almost linear.
     9. Fertile ♀ fls. all epappose, completely enclosed in woolly hyaline-appendaged phyllaries; receptacle chaffy throughout; true invol. none .... 133. *Stylocline*
     9. Fertile ♀ fls. of 2 sorts, the outer epappose and bract-enclosed, the inner pappose but not bract-subtended; true invol. scanty .... 58. *Filago*
    8. Lvs. opposite; receptacle subglobose to obpyriform; true invol. and pappus none; inner fls. bractless .... 119. *Psilocarphus*

Group C. (Anthers tailed; style with a ring of hairs or thickened ring below the papillate branches; plants usually prickly)

1. Aks. basifixed; marginal fls. not enlarged.
 2. Plants not thistlelike or spiny; phyllaries hooked at tip .... 13. *Arctium*
 2. Plants thistlelike, often spiny.
  3. Fils united below; lvs. white-mottled along the veins; pappus several-seriate, not plumose
                                     128. *Silybum*
  3. Fils. separate; lvs. not white-mottled.
   4. Pappus of short scales in several series; outer phyllaries leaflike; yellow- or orange-fld. annuals .... 32. *Carthamus*
   4. Pappus of elongate slender bristles, these sometimes united at base; outer phyllaries not leaflike; fls. not bright yellow or orange.
    5. Phyllaries enlarged at base, broad; fls. violet; receptacle fleshy ...... 48. *Cynara*
    5. Phyllaries narrower; fls. white, pink, or purplish; receptacle not fleshy.
     6. Pappus-bristles plumose (or those of marginal fls. sometimes merely barbellate)
                                         41. *Cirsium*
     6. Pappus-bristles merely barbellate .... 31. *Carduus*
1. Aks. obliquely inserted on the receptacle; marginal fls. often enlarged.
 7. Lvs. not prickly on the margin; heads not subtended by the upper lvs.; fls. in most spp. not yellow .... 39. *Centaurea*
 7. Lvs. prickly on the margin; heads subtended by the upper lvs.; fls. yellow .... 42. *Cnicus*

Group D. (Corollas all strap-shaped)

1. Herbage usually coarsely hairy; fls. blue; aks. short, truncate at apex; pappus paleaceous or none; pollen pale yellow; style branches mostly elongate .... 40. *Cichorium*
1. Not in all respects as above.
    2. Low, acaulescent, glabrous; pollen orange; style-branches short and blunt.
        3. Aks. not beaked; pappus (at least in part) broad at base, tapering into bristlelike awns

102. *Microseris*
        3. Aks. beaked; pappus of white capillary bristles .... 5. *Agoseris*
    2. Not in all respects as above.
        4. Annuals to shrubs, mostly glabrous, some appressed-tomentose, none hirsute with spreading hairs; fls. pink, white or sometimes yellow; aks. cylindrical, fusiform, or beaked, never flattened; pappus setose (none in *Atrichoseris*); style-branches mostly short.
            5. Aks. pappose.
                6. Plants nonscapose, usually branched.
                    7. Plants shrubs 1–2 m. high; lvs. 6–13 cm. long, in tufts at ends of branches, irregularly sinuate-toothed. San Clemente Id. .... 105. *Munzothamnus*
                    7. Plants herbaceous, or at least lower in height.
                        8. Pappus not plumose.
                            9. Plants perennial or annual; heads relatively slender.
                                10. The pappus of 5 rigid tapering awns .... 35. *Chaetadelphia*
                                10. The pappus of many capillary bristles .... 94. *Lygodesmia*
                            9. Plants annual; heads subglobose .... 97. *Malacothrix*
                        8. Pappus plumose at least above.
                            9a. Aks. beakless; fls. rose or flesh-colored; annual or usually perennial

132. *Stephanomeria*
                            9a. Aks. beaked; fls. white or nearly so; fistulous-stemmed annual

122. *Rafinesquia*
                6. Plants scapose or subscapose; heads many-fld.; annuals except in a few spp. of *Malacothrix*.
                    10. Pappus of plumose bristles .... 10. *Anisocoma*
                    10. Pappus of capillary bristles, never plumose, quickly deciduous.
                        11. Aks. beakless, truncate .... 97. *Malacothrix*
                        11. Aks. beaked.
                            12. Infl. not glandular; aks. abruptly short-beaked, truncate .. 64. *Glyptopleura*
                            12. Infl. with tack-shaped glands; aks. tapering into the shallowly cup-tipped beak .... 30. *Calycoseris*
            5. Aks. epappose; glaucous scapose annual with broad spinulose-denticulate lvs.

18. *Atrichoseris*
        4. Not in all respects as above.
            13. Herbs with grasslike clasping lvs.; heads large, solitary, yellowish or purplish; pappus of ± flattened plumose bristles. Weeds .... 140. *Tragopogon*
            13. Not in all respects as above.
                14. Herbs with coarse spreading hirsute pubescence; fls. mostly yellow; aks. mostly fusiform or beaked; pappus mostly of coarse plumose setae.
                    15. Receptacle chaffy-bracted; at least the inner aks. beaked; lvs. mostly basal

83. *Hypochoeris*
                    15. Receptacle naked.
                        16. Aks. terete, truncate, the outer ones enveloped by the hardening phyllaries

71. *Hedypnois*
                        16. Aks. beaked.
                            17. Stems leafy; outer phyllaries broader than the inner .... 114. *Picris*
                            17. Stems scapose: outer phyllaries very small .... 90. *Leontodon*
                14. Herbs of various habit; pappus of numerous nonplumose setae.
                    18. Aks. flattened (obscurely so in *Sonchus arvensis*); leafy-stemmed herbs; pappus of soft and numerous bristles.
                        19. Invol. campanulate; aks. not beaked .... 131. *Sonchus*
                        19. Invol. cylindrical; aks. beaked .... 86. *Lactuca*
                    18. Aks. not obviously flattened; lvs. mostly basal or subbasal; pappus various or none.

Asteraceae 99

    19. Aks. truncate; pappus fragile, usually sordid or fuscous; caulescent
                                                                                   76. *Hieracium*
    19. Aks. beaked; pappus mostly bright white.
      20. Heads in panicles or cymes; pappus early deciduous .... 47. *Crepis*
      20. Heads solitary on fistulous naked scapes; pappus not early deciduous
                                                                                        137. *Taraxacum*

Group E. (Rays and pappus none or vestigial)

1. Heads unisexual; both ♂ and ♀ on the same plant; invol. of ♀ heads closed and ± burlike; ♂ heads in a raceme or spike, with open invols.
    2. Phyllaries of ♂ heads united.
      3. Fruiting invol. winged with broad scarious scales; lvs. or their divisions filiform
                                                                                      79. *Hymenoclea*
      3. Fruiting invol. not so winged; lvs. toothed or pinnatifid .... 7. *Ambrosia*
    2. Phyllaries of ♂ heads distinct; invol. of ♀ heads becoming a stout spiny bur . 149. *Xanthium*
1. Heads not unisexual; ♀ invol. not becoming a bur.
    4. Phyllaries in a single series or lacking.
      5. Lvs. opposite.
        6. Phyllaries 1–3; head usually 1-fld. Introd. .... 59. *Flaveria*
        6. Phyllaries several; head several-fld. Native .... 88. *Lasthenia*
      5. Lvs. mostly alternate.
        6a. Fls. all perfect; low small-lvd. perennial. Ne. Mojave Desert .... 111. *Perityle*
        6a. Fls. imperfect, ♂ and ♀ in same head.
          7. Lvs. or their lobes linear-filiform; aks. long-villous; plant woody .... 84. *Iva*
          7. Lvs. or their lobes not linear-filiform; aks. not villous; plant not woody.
            8. Aks. obovoid or pyriform, not winged; perennial herbs; phyllaries ca. 5 . 84. *Iva*
            8. Aks. flattened, pectinate-winged; annual herbs; phyllaries 6–7, the 1 or 2 enlarged
                                                                        49. *Dicoria*
    4. Phyllaries in 2 or more usually imbricated series.
      9. Heads in spikes, racemes or panicles .... 16. *Artemisia*
      9. Heads solitary or in capitate clusters or in corymbs.
        10. Aks., especially the marginal, stipitate .... 46. *Cotula*
        10. Aks. sessile.
          11. Receptacle convex or hemispheric, not chaffy .... 99. *Matricaria*
          11. Receptacle flat.
            12. Receptacle chaffy; plant woody, 4–6 dm. high, with tomentose pinnately divided lvs. .... 125. *Santolina*
            12. Receptacle not chaffy.
              13. Perennials with capitate to corymbiform infl. .... 136. *Tanacetum*
              13. Annuals with burlike heads sessile in forks of stems .... 130. *Soliva*

Group F. (Rays evident; pappus none or vestigial)

1. Receptacle chaffy with bracts at base of at least some of the fls. or with stiff bristles.
    2. Phyllaries usually imbricated, or at least in more than 1 series.
      3. Lvs. opposite. Introd. weeds.
        4. Ray-ak. completely enclosed by the subtending inner phyllary .... 100. *Melampodium*
        4. Ray-ak. not so enclosed.
          5. Heads (including rays) barely 1 cm. in diam.; rays 1–5, white ...... 52. *Eclipta*
          5. Heads (including rays) ca. 3 cm. in diam.; rays yellow .... 67. *Guizotia*
      3. Lvs alternate (at least in upper $\frac{2}{3}$ of plant) or largely basal.
        6. Rays white.
          7. Heads relatively large, solitary or few .... 12. *Anthemis*
          7. Heads very small, many, in dense flattish or rounded panicles .... 2. *Achillea*
        6. Rays yellow.
          8. Invol. of 2 very unlike series of phyllaries; herbs or ± woody and with fleshy stems
                                                                         44. *Coreopsis*
          8. Invol. of 2 or more series of similar phyllaries, often ± suffrutescent.
            9. Disk-aks. thickened, angled, not or scarcely compressed; large-lvd. perennial herb
                                                                      22. *Balsamorhiza*

9. Disk-aks. strongly flattened.
   10. Plants scapose, herbaceous, with very large solitary heads; aks. with a cartilaginous border .... 54. *Enceliopsis*
   10. Plants leafy-stemmed, ± shrubby, the heads several, medium-sized; aks. ciliate
   53. *Encelia*
2. Phyllaries in a single series.
   11. Receptacle bristly, honeycombed; phyllaries not enclosing the ray-aks.; plant ± tomentose, annual.
      12. Rays orange. Garden escape .... 145. *Venidium*
      12. Rays yellow. Native .... 56. *Eriophyllum*
   11. Receptacle not bristly.
      13. Ray-aks. enclosed in the obcompressed phyllaries, of which the infolded lateral margins meet.
         14. Disk-fls. 6; ray-aks. obcompressed .... 87. *Lagophylla*
         14. Disk-fl. 1 and the ray-ak. obcompressed, or more than 1 and ray-aks. not obcompressed .... 96. *Madia*
      13. Ray-aks. not completely enclosed, the phyllaries not obcompressed.
         15. Rays 3-toothed or 3-lobed, the lobes subparallel; herbage without tack-shaped glands.
            16. Upper lvs. and phyllaries not terminated by open pit glands .... 74. *Hemizonia*
            16. Upper lvs. and phyllaries terminated by open pit glands. ...... 77. *Holocarpha*
         15. Rays 3-cleft or -parted into palmately spreading lobes; herbage mostly with tack-shaped glands .... 29. *Calycadenia*
1. Receptacle naked, without bristles or chaffy bracts.
   17. Phyllaries graduated in several unequal series, closely imbricate, at least the inner ± scarious.
      18. Rays well developed, the heads solitary to corymbose .... 37. *Chrysanthemum*
      18. Rays short, inconspicuous, the heads in capitate to crowded-corymbose infls.
      136. *Tanacetum*
   17. Phyllaries in 1–2 series, or at least not imbricate.
      19. Rays white or pink. Low introd. daisy .... 24. *Bellis*
      19. Rays yellow or orange.
         20. Phyllaries scarious-margined, incurved; ray-aks. strongly incurved; lvs. entire. Garden escape .... 28. *Calendula*
         20. Phyllaries and ray-aks. not strongly incurved.
            21. Lvs. opposite.
               22. Phyllaries in 1 series; plants mostly annual .... 88. *Lasthenia*
               22. Phyllaries in 2 or more series; plants perennial, succulent .... 85. *Jaumea*
            21. Lvs. alternate.
               23. Phyllaries broad, in 2 distinct series (the outer ± reflexed, the inner erect); sunflowerlike perennial .... 144. *Venegasia*
               23. Phyllaries not as above.
                  24. Rays persistent and becoming papery; phyllaries many; woolly perennials of the deserts .... 21. *Baileya*
                  24. Rays not as above.
                     25. Rays with a tooth or lobe on the inner side of the throat opposite the ligule
                     103. *Monolopia*
                     25. Rays without a lobe on the inside.
                        26. Plants rather succulent, glabrous except for a microscopic glandular pubescence and very sparse flocs of wool in the leaf-axils and below the heads .... 26. *Blennosperma*
                        26. Plants not as above.
                           27. Herbage not white-woolly; annual herbs.
                              28. Lvs. 1–3-times ternately divided; plants 3–6 dm. high; San Bernardino and Santa Rosa mts. .... 20. *Bahia*
                              28. Lvs. coarsely sinuate-dentate; plants 1–3 dm. high. Garden escape
                              50. *Dimorphotheca*
                           27. Herbage white-woolly at least when young.
                              29. Rays yellow or white. Widely distributed .... 56. *Eriophyllum*
                              29. Rays ± purplish. Desert slopes, San Gabriel and San Bernardino mts.
                              134. *Syntrichopappus*

Asteraceae

Group G. (Rays none or vestigial; pappus of paleae or flattened scales or stiff bristles or awns or both)

1. Receptacle chaffy.
   2. Pappus of plumose awns; plant a low almost leafless shrub .... 23. *Bebbia*
   2. Pappus not of plumose awns.
      3. Aks. with pectinately toothed wings. Annuals .... 49. *Dicoria*
      3. Aks. not with pectinately toothed wings.
         4. Awns or teeth of pappus not retrorsely barbed. Shrubby .... 53. *Encelia*
         4. Awns or teeth retrorsely barbed. Herbaceous .... 25. *Bidens*
1. Receptacle not chaffy.
   5. Lvs., at least all the lower, opposite.
      6. Plant woody; lvs. deltoid, small, on long petioles. Deserts .... 115. *Pleurocoronis*
      6. Plants herbaceous.
         7. Perennial; lvs. hastate-triangular, 3–7 cm. long; plant 1–1.5 m. tall .... 110. *Pericome*
         7. Annuals; lvs. mostly narrower, or if broad, shorter; plants smaller.
            8. Lvs. not auricled at base; phyllaries ca. 5–15 .... 88. *Lasthenia*
            8. Lvs. auricled at base, ± oblong; phyllaries ca. 15–30 .... 141. *Trichocoronis*
   5. Lvs. mostly alternate.
      9. Pappus a mere crown; phyllaries imbricate, ± scarious or scarious-margined.
         10. Perennials with heads in capitate or corymbiform infl. .... 136. *Tanacetum*
         10. Annuals with heads solitary or somewhat corymbed .... 99. *Matricaria*
      9. Pappus of well developed paleae or awns or both.
         11. Lvs. and invol. punctate with translucent oil-glands.
            12. Pappus of 2–6 entire paleae with or without awns; phyllaries united into a cup or tube .... 135. *Tagetes*
            12. Pappus of 10–15 paleae or these dissected into bristles; phyllaries ± 2-seriate
                                                        51. *Dyssodia*
         11. Lvs. and invol. not punctate with translucent oil-glands.
            13. Pappus of bristles or these alternating with minute paleae or awns.
               14. The pappus of 2–8 deciduous awns; plants perennial, usually strongly viscid. Mostly cismontane .... 66. *Grindelia*
            11. The pappus of ca. 4 bristles alternating with ca. 4 minute paleae; annual, not glutinous. Colo. Desert .... 98. *Malperia*
            13. Pappus of flat or awnlike paleae.
               15. Pappus of 20–30 silvery awns; plant woody. Deserts .... 1. *Acamptopappus*
               15. Pappus of mostly fewer parts.
                   16. Phyllaries in a single series.
                      17. Pappus-paleae erose or fimbriate; lvs. 3-lobed at apex
                                                    56. *Eriophyllum pringlei*
                      17. Pappus-paleae nearly or quite entire.
                         18. Pappus of subulate awns; slender annuals with entire linear lvs.
                                                    124. *Rigiopappus*
                         18. Pappus not of subulate awns.
                            19. Fls. whitish or purplish.
                               20. Pappus-paleae linear, with a strong midrib; lvs. entire 107. *Palafoxia*
                               20. Pappus-paleae nerveless; lvs. toothed to pinnatifid .. 34. *Chaenactis*
                            19. Fls. yellow.
                               21. Pappus-paleae 4; lvs. dissected .... 34. *Chaenactis glabriuscula*
                               21. Pappus-paleae more numerous.
                                   22. Plants annual; viscid; lvs. mostly linear, entire. Coastal
                                                         6. *Amblyopappus*
                                 22. Plants perennial, not viscid; lvs. mostly dissected. Interior
                                                         56. *Eriophyllum*
                 16. Phyllaries in 2 or more series.
                     23. Pappus-paleae 5, dissected into bristles; white-woolly annual. Colo. Desert
                                                         142. *Trichoptilium*
                     23. Pappus-paleae 10–22, not so dissected.
                          24. The pappus-paleae obtuse; lvs. in a basal rosette ..... 80. *Hymenopappus*
                        24. The pappus-paleae lanceolate, each ending in an awn; lvs. both basal and cauline .... 81. *Hymenothrix*

Group H. (Rays none or vestigial; pappus of capillary bristles)

1. Heads unisexual; plants dioecious, ± woody at base .... 19. *Baccharis*
1. Heads not strictly unisexual.
   2. Plants well developed shrubs.
      3. Phyllaries proper, 4–7, in a single series, of equal length; lvs. narrow, entire
         138. *Tetradymia*
      3. Phyllaries more numerous.
         4. Lvs. scalelike, at least on flowering branches.
            5. Aks. sericeous; heads solitary at ends of branchlets .... 17. *Aster intricatus*
            5. Aks. glabrous; heads closely placed in larger clusters .... 91. *Lepidospartum*
         4. Lvs. not scalelike.
            6. Phyllaries arranged in ± distinctly vertical ranks .... 39. *Chrysothamnus*
            6. Phyllaries not so arranged.
               7. Fls. whitish.
                  8. Pappus double, the inner of bristles, the outer of short paleae; petioles longer than the small lf.-blades .... 115. *Pleurocoronis*
                  8. Pappus simple, of capillary bristles only; petioles mostly shorter than lf.-blades
                     27. *Brickellia*
               7. Fls. yellow.
                  9. Phyllaries merely of 1 series, subequal, subulate; desert shrub with terete resinous-punctate lvs. .... 113. *Peucephyllum*
                  9. Phyllaries of more than 1 series, unequal; widely distributed .. 69. *Haplopappus*
   2. Plants herbaceous.
      10. Plant annual.
         11. Herbage white-woolly, aromatic; lvs. broad, rounded; heads solitary in forks of branches. Deserts .... 118. *Psathyrotes*
         11. Herbage and lvs. not as above.
            12. Phyllaries in a single series, equal or nearly so; lvs. dentate to pinnatifid. Garden weed .... 127. *Senecio vulgaris*
            12. Phyllaries in more than 1 series, unequal.
               13. Outer corollas enlarged, more deeply cleft on inner side; herbage often strong-smelling .... 92. *Lessingia*
               13. Outer corollas not enlarged, very slender; herbage not usually strong-smelling
                  43. *Conyza*
      10. Plants biennial to perennial.
         14. Principal phyllaries essentially equal and in 1 series, although some basal outer reduced bracts may be present.
            15. Lvs. opposite. Yellow-fld. plants of mt. moist places .... 15. *Arnica*
            15. Lvs. alternate or basal.
               16. Pappus-bristles plumose; plants ± scapose .... 123. *Raillardella*
               16. Pappus-bristles barbellate, not truly plumose; fls. yellow to orange . 127. *Senecio*
         14. Principal phyllaries ± imbricate in 2–several series.
            17. Phyllaries and lvs. with conspicuous translucent oil-glands ...... 117. *Porophyllum*
            17. Phyllaries and lvs. not having translucent oil-glands.
               18. Fls. not yellow.
                  19. Aks. 5-angled or -ribbed; lvs. often opposite .... 4. *Ageratina*
                  19. Aks. 10-ribbed; lvs. alternate .... 27. *Brickellia*
               18. Fls. yellow.
                  20. Phyllaries not in vertical ranks.
                     21. Style-appendages very short, 0.5 mm. long or less; phyllaries scarcely imbricate .... 55. *Erigeron*
                     21. Style-appendages mostly 0.7 mm. or more long; phyllaries plainly imbricate.
                        22. Plant a biennial or tap-rooted perennial with squarrose phyllaries
                           95. *Machaeranthera*
                        22. Plant not as above.
                           23. Aks. ± flattened; invol. 9–12 mm. high; plants herbaceous 38. *Chrysopsis*
                           23. Aks. terete or angled; invol. mostly 12 mm. or higher; plants mostly definitely woody at base .... 69. *Haplopappus*
                  20. Phyllaries in vertical ranks; invol. 11–12 mm. high; suffrutescent herb with narrow lvs. .... 112. *Petradoria*

Asteraceae 103

Group I. (Rays well developed, yellow to orange or brown, at least when fresh; pappus of well developed paleae or scales or of stiff awns)

1. Receptacle chaffy or bristly throughout or with a circle of chaffy bracts surrounding the disk-fls.
    2. Invol. of 1–several series of phyllaries, none enfolding the ray-aks.
        3. Receptacle bristly .... 60. *Gaillardia*
        3. Receptacle chaffy.
            4. Phyllaries in 2 distinct dissimilar series; aks. flattened at right angles to the radius of the head.
                5. Pappus of 2–6 firm mostly retrorsely barbed awns; lvs. entire to pinnately compound; aks. flattened .... 25. *Bidens*
                5. Pappus of 2 minute teeth; lvs. pinnately dissected .... 44. *Coreopsis*
            4. Phyllaries in 2 or more similar series.
                6. Rays sessile and persistent on their aks., becoming papery; disk-aks. 4-angled. E. Mojave Desert .... 126. *Sanvitalia*
                6. Rays not persistent or papery.
                    7. Plants scapose or nearly so, with broad entire ± silvery-pubescent lvs. and large long-peduncled heads .... 54. *Enceliopsis*
                    7. Plants not scapose.
                        8. Rays ♀.
                            9. Disk-aks. strongly compressed, 2-winged; lvs. mostly opposite . 146. *Verbesina*
                            9. Disk-aks. compressed-quadrangular; lvs. alternate .... 148. *Wyethia*
                        8. Rays neutral, the aks. not developing.
                            10. Aks. ± thickened.
                                11. Pappus of 2 awns and several short squamellae .... 147. *Viguiera*
                                11. Pappus of 2, rarely many, deciduous paleaceous awns ... 73. *Helianthus*
                            10. Aks. very flat.
                                12. Aks. with 2 white wings; lvs. opposite, at least below .... 146. *Verbesina*
                                12. Aks. not winged; lvs. usually alternate.
                                    13. Plants annual; awns 2, strong; ak.-body conspicuously white-margined 63. *Geraea*
                                    13. Plants perennial, forming low rounded shrubs; awns 1–2, weak 53. *Encelia*
    2. Invol. with 1 series of phyllaries, each partly or completely enclosing an ak.
        14. Ray-aks. completely enfolded by the phyllaries.
            15. Fls. red-brown; pappus silvery-scarious, the longer paleae longer than the aks.; aks. 10-costate and tuberculate-scabrous .... 3. *Achyrachaena*
            15. Fls. yellow; pappus and aks. not as above .... 89. *Layia*
        14. Ray-aks. half-enclosed by the phyllaries .... 74. *Hemizonia*
1. Receptacle not chaffy or bristly.
    15a. Invol. and lvs. with translucent oil-glands.
        16. Phyllaries uniseriate, united almost to apex; pappus of 5–6 unequal paleae . 135. *Tagetes*
        16. Phyllaries ± 2-seriate, if 1-seriate, then quite free; pappus not as above.
            17. Aks. pubescent; lvs. lacking stiff spreading basal bristles; phyllaries ± 2-seriate 51. *Dyssodia*
            17. Aks. glabrous; lvs. with a few stiff spreading bristles at base; phyllaries 1-seriate 108. *Pectis*
    15a. Invol. and lf. lacking translucent oil-glands.
        18. Invol. of several series of graduated phyllaries.
            19. Pappus of 2–8 slender deciduous awns; heads rather large, gummy; coarse herbs 66. *Grindelia*
            19. Pappus not as above; heads small.
                20. Pappus of disk-fls. of 1 series of straight paleae; lvs. mostly linear .. 68. *Gutierrezia*
                20. Pappus of disk-fls. of many ± twisted narrow paleae or flat bristles; lvs. elliptic to obovate .... 8. *Amphipappus*
        18. Invol. of ca. 1–2 equal or subequal but not graduated series of phyllaries.
            21. Lvs. opposite; mostly low annuals .... 88. *Lasthenia*
            21. Lvs. mostly alternate.
                22. Rays persistent and becoming papery, few, broad, conspicuous; herbage white-woolly. Deserts .... 120. *Psilostrophe*

22. Rays not persistent or becoming papery.
　23. Phyllaries in 1 series .... 56. *Eriophyllum*
　23. Phyllaries in more than 1 series.
　　24. Lvs. entire to pinnately cut, green above, white-woolly beneath; heads 3–8 cm. across; subacaulescent escape from gardens. .... 62. *Gazania*
　　24. Lvs. various; heads mostly smaller. Natives.
　　　25. Phyllaries reflexed; rays short, broad .... 72. *Helenium*
　　　25. Phyllaries not reflexed.
　　　　26. Receptacle ± rounded; aks. turbinate, mostly 5-angled; pappus-paleae mostly 5 .... 82. *Hymenoxys*
　　　　26. Receptacle flat; aks. linear, flattened or somewhat 3-angled; pappus-paleae mostly 4 .... 78. *Hulsea*

Group J. (Rays well-developed, white to pink, red or purplish; pappus of paleae or scales or of stiff awns)

1. Heads 5–7 cm. across; peduncles 1.5–3 dm. long; white-woolly perennial. Garden escape
　　　　　　　　　　　　　　　　　　　　　　　　　　　　　　　　　　14. *Arctotis*
1. Heads smaller.
　2. Outermost aks. ± enfolded by their subtending phyllary.
　　3. Ray-aks. half enclosed by the phyllaries; disk-aks. sterile, undeveloped, glabrous, epappose
　　　　　　　　　　　　　　　　　　　　　　　　　　　　　　　　　　74. *Hemizonia*
　　3. Ray-aks. completely enclosed by the phyllaries .... 89. *Layia*
　2. Outermost aks. not so enfolded.
　　4. Lvs. opposite; rays 4–5, short; aks. black. Introd. weed .... 61. *Galinsoga*
　　4. Lvs. alternate or basal.
　　　5. Phyllaries dotted or striped with oil-glands; rays purplish; perennial .. 106. *Nicolletia*
　　　5. Phyllaries not bearing conspicuous oil-glands; mostly annuals.
　　　　6. Aks. compressed, 2-edged or 2-nerved.
　　　　　7. Pappus of several to many paleae or flattened bristles; phyllaries scarious-margined .... 139. *Townsendia*
　　　　　7. Pappus of 1–2 bristlelike awns or a crown of scales or both; phyllaries not scarious-margined .... 111. *Perityle*
　　　　6. Aks. ± thickened, 4–5-angled.
　　　　　8. Herbage hispid-pilose; upper lvs. subtending the heads .... 104. *Monoptilon*
　　　　　8. Herbage woolly; upper lvs. not subtending the heads .... 56. *Eriophyllum*

Group K. (Rays well developed; pappus of soft capillary bristles, mostly without any paleae)

1. Rays yellow.
　2. Phyllaries ca. 5, partly enclosing the ray-aks.; pappus of 35–40 white bristles. Mojave Desert
　　　　　　　　　　　　　　　　　　　　　　　　　　　　　　　134. *Syntrichopappus*
　2. Phyllaries not at all enclosing the ray-aks.
　　3. Phyllaries many, usually ± imbricated; style-tips not truncate.
　　　4. Plants annual.
　　　　5. Lvs. entire, narrow, not bristle-tipped .... 36. *Chaetopappa*
　　　　5. Lvs. pinnatifid, the lobes bristle-tipped .... 69. *Haplopappus gracilis*
　　　4. Plants perennial.
　　　　6. Ray-aks. without pappus; tall leafy herbs with broad lvs. .... 75. *Heterotheca*
　　　　6. Ray-aks. with pappus.
　　　　　7. Pappus in 1 series.
　　　　　　8. Pappus-bristles unequal; invol. mostly 6–18 mm. high, if shorter, mostly on shrubby plants .... 69. *Haplopappus*
　　　　　　8. Pappus-bristles equal; invol. mostly 4–5 mm. high, or if larger, on herbaceous plants.
　　　　　　　9. Heads with phyllaries in ± distinct vertical ranks; lvs. coriaceous
　　　　　　　　　　　　　　　　　　　　　　　　　　　　　　　　112. *Petradoria*
　　　　　　　9. Heads with phyllaries not in vertical ranks; lvs. not coriaceous . 129. *Solidago*
　　　　　7. Pappus in 2 series, the outer much shorter than the inner and of short linear scales or bristles.

    10. Lvs. oblong-spatulate to oblanceolate or elliptic, well distributed on the stems
     38. *Chrysopsis*
     10. Lvs. linear, mostly basal .... 55. *Erigeron*
  3. Phyllaries fewer, or if more, at least not strongly imbricated, mostly subequal and in 1 series.
    11. Lvs. mostly opposite .... 15. *Arnica*
    11. Lvs. alternate or basal.
     12. Pappus plumose .... 123. *Raillardella*
     12. Pappus not plumose .... 127. *Senecio*
 1. Rays white to pink, blue or purple.
   13. Pappus of ray-fls. much reduced.
    14. Style-appendages of disk-fls. ± comose; perennial. Montane and cismontane
     45. *Corethrogyne*
    14. Style-appendages of disk-fls. not comose; annual. Deserts .... 95. *Machaeranthera*
   13. Pappus well developed in both ray- and disk-fls.
    15. Heads nodding in bud; lvs. narrow; annuals with slender stems .... 36. *Chaetopappa*
    15. Heads erect in bud.
     16. Rays very numerous, filiform, scarcely exceeding the disk-fls.; annual weeds with numerous sublinear lvs. .... 43. *Conyza*
     16. Rays much surpassing the disk.
      17. Phyllaries subequal or ± imbricate, neither leafy nor with chartaceous base and herbaceous green tip; style-appendages 0.5 mm. long or less, lanceolate or broader
       55. *Erigeron*
      17. Phyllaries either subequal and the outer leafy, or usually imbricate, with chartaceous and green tips; style-appendages lanceolate or narrower, usually more than 0.5 mm. long.
       18. Plants annual or biennial or perennial from a distinct taproot; lvs. mostly toothed; phyllaries chartaceous or coriaceous toward base, often spreading or recurved at the herbaceous tips .... 95. *Machaeranthera*
       18. Plants mostly perennial and if so, rhizomatous or fibrous-rooted; phyllaries mostly herbaceous.
        19. Style-appendages lanceolate to subulate, acute; lvs. mostly 3–20 cm. long; plants often tallish. Mostly montane and cismontane .... 17. *Aster*
        19. Style-appendages ovate or oblong, obtuse; lvs. to ca. 1 cm. long; plants low, tufted. Of desert mts. .... 93. *Leucelene*

1. *Acamptopáppus* Gray. GOLDENHEAD

Low much branched desert shrubs with white bark. Lvs. alternate, small, usually spatulate or oblanceolate, entire, 1-nerved. Heads yellow, radiate or discoid, subglobose, solitary or cymosely arranged at branch-tips, the fls. all fertile. Invol. broad, ca. 4-seriate, strongly graduate, the phyllaries broad, rounded, whitish with greenish tip and thin scarious margin. Receptacle convex, alveolate-fimbrillate. Style-branches linear, aks. subturbinate, short, densely villous. Pappus persistent, of ca. 30–40 silvery flattened paleae and bristles of different widths, the broader ones usually somewhat dilated apically. Two known spp. (Greek, *akamptos*, stiff, and *pappos*, pappus.)

Heads discoid, small, gathered in small cymes .... 2. *A. sphaerocephalus*
Heads radiate, large, solitary at tips of branches .... 1. *A. shockleyi*

 1. **A. shóckleyi** Gray. Rounded shrub 1.5–5 dm. high, spinescent-branched, the herbage finely hirtellous or hispidulous; lvs. spatulate, oblanceolate or elliptic, 5–15 mm. long, usually mucronulate; heads globose, radiate, 1.5–3 cm. wide, solitary at ends of nearly naked peduncles; invol. 8–11 mm. high; rays 8–13, the oblong ligule ca. 1 cm. long; $n=9$ (Raven et al., 1960).—Desert plains and washes, 3000–6200 ft.; Creosote Bush Scrub, Joshua Tree Wd.; White Mts., Inyo Co. to Clark Mt., e. San Bernardino Co.; s. Nev. April–June.

 2. **A. sphaerocéphalus** (Harv. & Gray) Gray. [*Aplopappus s.* Harv. & Gray.] Plate 10, Fig. B. Round-topped shrub 2–9 dm. high, corymbosely branched, often densely

twiggy, glabrous throughout or sparsely hispidulous on lf.-margins; lvs. linear to spatulate, 5–20 mm. long, 1.5–5 mm. wide, obtuse to acute, mucronulate, sessile; heads 7–10 mm. high, mostly solitary but approximate at ends of cymosely arranged branchlets; invol. 5–6 mm. high; $n=9$ (Raven et al., 1960).—Open desert, 200–4200 ft.; Creosote Bush Scrub, Joshua Tree Wd.; e. Mojave Desert and w. border of Colo. Desert; to Utah, Ariz. May–June. A form with stems and lvs. densely hirtellous is var. *hirtéllus* Blake, distributed from the region of Hesperia and Adelanto to Lone Pine, Inyo Co.

2. *Achillèa* L. Yarrow

Perennial aromatic herbs with alternate subentire to pinnately dissected lvs. Heads several to many, rather small, in terminal corymbs, radiate or rarely discoid. Rays mostly 3–12, ♀ and fertile, white to pink or yellow, short, broad. Phyllaries dry, imbricate in several series, with scarious or hyaline margins and often greenish midrib. Receptacle conic or convex, chaffy throughout. Disk-fls. ca. 10–75, perfect and fertile. Aks. compressed parallel to the phyllaries, callous-margined, glabrous. Pappus none. Ca. 75 spp., N. Hemis. (In honor of *Achilles*.)

1. **A. millefòlium** L. The treatment is here used that was proposed by Malcolm A. Nobs in Abrams, Illus Fl. 4: 390–391. 1960, who wrote that this is a highly polymorphic circumboreal sp. of most diverse habitats and wide distribution. In N. Am. it is represented by 2 chromosome forms; the tetraploid ($n=18$, *A. lanulosa* Nutt.) and the hexaploid ($n=27$, *A. borealis* Bong.). These types show parallel variation, are indistinguishable on morphological criteria, and intergrade into European forms of *A. millefolium*. Dr. Nobs considered them as units in a large polymorphic sp. with 10 recognizable forms in the w. states. Many individual plants cannot definitely be assigned to one of these units but are intermediate.

Lf.-segms. thick and fleshy, the terminal ones ± ovate, densely congested. San Miguel and San Nicolas ids. .... 1. var. *arenicola*
Lf.-segms. thin, not fleshy, the terminal ones ± oblanceolate to linear.
  Herbage gray, villous to woolly; margins of phyllaries light brown to straw-colored. Yellow Pine F. to Lodgepole F. .... 3. var. *lanulosa*
  Herbage usually green, moderately villous.
    Lf.-segms. oblanceolate, all segms. tending to be oriented in same plane. Possible escape in lawns, etc. .... 4. var. *millefolium*
    Lf.-segms. linear, spreading and ascending. Native.
      Invol. generally 5–6 mm. high. Yellow Pine F. .... 5. var. *pacifica*
      Invol. generally 6–9 mm. high. Mostly below 2500 ft., cismontane .... 2. var. *californica*

Var. **arenícola** (Heller) Nobs in Ferris. [*A. a.* Heller. *A. m.* var. *maritima* Jeps.] Stems stout, 1–6 dm. high, ascending to decumbent, arising from well developed rhizomes; root-system mainly fibrous from the rhizomes but retaining a strong cent. taproot; herbage gray, densely white-villous throughout; lvs. oblanceolate, 1–2 dm. long, tripinnately dissected, the primary pinnae perpendicular to the rachis, secondary and tertiary crowded, ovate to oblanceolate, fleshy, the tips mucronate; heads 5–6 mm. high, densely packed in round-topped corymbs, the phyllaries with light brown margins; $n=27$.—Sandy soil, Coastal Strand; San Miguel and San Nicolas ids.; Monterey Co. to Del Norte co. March–Nov.

Var. **califórnica** (Pollard) Jeps. [*A. c.* Pollard. *A. borealis* ssp. *c.* Keck.] Plate 10, Fig. C. Stems stout, 5–12 dm. high, from thick well developed rhizomes with a well developed taproot; herbage green, occasionally grayish, moderately villous; lvs. 1–3 dm. long, linear-lanceolate, finely dissected into linear, spine-tipped segms.; heads numerous, to 9 mm. high, in loose corymbs, the phyllaries with light brown margins; $n=27$.—Mostly below 2500 ft., occasional higher, Coastal Sage Scrub, Chaparral; cismontane s. Calif. to Wash. San Clemente, Santa Catalina, Santa Barbara, Santa Cruz and Santa Rosa ids. March–July.

Var. **lanulòsa** (Nutt.) Piper. [*A. l.* Nutt.] Stems 3–5 dm. high; herbage gray, often woolly; lvs. 0.5–1.5 dm. long, linear-lanceolate, finely dissected into linear segms., the ultimate ones with prominent spine-tips; heads 5–7 mm. high, in dense flat-topped corymbs, the margins with light brown edges; $n = 18$.—Largely in Yellow Pine F., mts. of San Diego Co. n. to B.C. and e. to Rocky Mts. May–Aug. S. Calif. material is difficult to distinguish from var. *californica*.

Var. **millefòlium**. Lf.-segms. oblanceolate and all oriented in the same plane; plant to 1 m. high; invol. 4–5 mm. high; phyllaries with light brown margins; $n = 27$.—To be expected in lawns, etc. natur. in e. U.S.; native of Old World.

Var. **pacífica** (Rydb.) G. N. Jones. [*A. p.* Rydb.] Much like var. *californica*, less grayish herbage than in var. *lanulosa*; heads with invols. 5–6 mm. high, the phyllaries hyaline to light straw-color; $n = 18$.—Said to occur in mts. of San Diego Co., along the e. side of the Sierra Nevada and n. to Alaska. May–Sept.

3. *Achyrachaèna* Schauer

Spring annuals with opposite lvs. clasping below, alternate above, linear. Heads solitary, long-peduncled, terminating the stems and few ascending branches, heterogamous, radiate, all fls. fertile. Invol. oblong-campanulate, nearly equal to the inconspicuous fls. Ray-fls. 1-seriate; ligules yellow, becoming orange or crimson. Receptacular bracts in 1 row between ray and disk, free. Aks. 10-ribbed, tuberculate-scabrous; pappus of silvery scales, spreading on ripe aks., to form a round head. One sp. (Greek, *achuron*, chaff, and Latin, *achaenium*, an ak., referring to the chaffy pappus.)

1. **A. móllis** Schauer. [*Lepidostephanus madioides* Bartl.] BLOW-WIVES. Simple or few-branched, 1–4 dm. high, villous, moderately glandular above; lvs. entire or remotely serrulate, to 13 cm. long, to 7 mm. wide, mostly much less; heads in fl. 1.5–2 cm. high, in fr. globose, 3 cm. across; phyllaries closely investing the ray-aks.; ray-fls. ca. 8, the ligules inconspicuous; disk-fls. ca. 15–35; aks. clavate, black, ca. 5 mm. long, those of ray without pappus and smooth-ribbed, those of disk with pappus and scabrous with brown teeth; pappus of 10 oblong blunt scales; $n = 8$ (Johansen, 1933).—Frequent in moist grassy fields with heavy soil and below 1500 ft.; V. Grassland, S. Oak Wd.; cismontane valleys from n. L. Calif.; n. to Ore. April–May. San Clemente, Santa Rosa and Santa Cruz ids.

4. *Ageratìna* Spach

Perennial herbs or shrubs. Lvs. mostly opposite, sometimes alternate, usually serrate, sometimes entire to crenulate. Heads 10–40-fld., mostly in corymbose infl. of cymes. Fls. all tubular, perfect, on a flat, glabrous or sparsely hairy receptacle. Phyllaries in 1–2 series, appressed, herbaceous. Corolla white to pink or purple. Anthers with large often truncate appendage. Pollen spherical, tricolpate. Aks. 5- or 10-angled; pappus of ± numerous capillary bristles. A segregate genus from *Eupatorium*.

Lf.-blades mostly 4–10 cm. long; petioles mostly 2–4 cm. long. Escape from cult. . 1. *A. adenophora*
Lf.-blades mostly 1.5–3.5 cm. long; petioles mostly 0.5–1 cm. long. Native, e. Mojave Desert
2. *E. herbacea*

1. **A. adenóphora** (Spreng.) King & Robins. [*Eupatorium a.* Spreng. *E. pasadenense* McClat.] Stems purplish, erect, simple or few-branched, 3–5 dm. high, glandular-puberulent; lvs. deltoid-ovate, serrate, 2–4 cm. long, on somewhat shorter petioles; cymes small, compact, in open panicles, with numerous heads 5 mm. high; invol. campanulate, glandular-puberulent; fls. white, sometimes aging pink.—Occasional escape from gardens, as about San Diego, Los Angeles, Pasadena, mouth of San Gabriel R., Santa Barbara; native of Mex. Most months.

2. **A. herbàcea** (Gray) King & Robins. [*Eupatorium h.* (Gray) Greene. *E. ageratifolium* var? *h.* Gray] From a woody caudex; stems several, 5–7 dm. high, leafy, branching above; herbage light green, cinereous-puberulent; lvs. mostly opposite, ovate, commonly cordate, crenate-serrate, acute; heads 6–8 mm. high, numerous, in dense ± corymbose

cluster of cymes; invol. puberulent; corolla white; aks. black, ca. 2 mm. high; $n = 17$ (Turner & Flyr, 1966).—Among rocks, 5000–6400 ft.; Pinyon-Juniper Wd.; New York Mts., Clark Mt. in e. Mojave Desert; to Utah, New Mex. Sept.–Oct.

5. *Agóseris* Raf. MOUNTAIN DANDELION

Annual or perennial herbs with strong taproots; vertical caudex short or long, single or multicipital. Lvs. mostly in radical tufts or a few scattered on lower stem, glabrous to villous. Infl. scapose, the head erect; invol. campanulate to subcylindric, the phyllaries subequal or imbricate. Fls. yellow to orange, often drying darker, usually longer than phyllaries at anthesis. Aks. narrowed into a stout or slender beak $\frac{1}{6}$ to almost 4 times as long as body; pappus of 50 or more white capillary bristles, not plumose. Ca. 8–9 spp., of w. N. Am. and 1 in S. Am. (Greek, *aix*, goat, and *seris*, chicory.)

Plants annual; ak.-beak mostly 2–3 times as long as ak.-body; invol. 12–18 mm. high
                                           3. *A. heterophylla*
Plants perennial; ak.-beak variable; invol. mostly higher.
 Ak.-beak mostly not more than ½ as long as ak.-body, stout, ± striate. Inyo-White Mts.
                                          1. *A. glauca*
 Ak.-beak more than twice as long as ak.-body, slender, not plainly striate.
  Ak.-body tapering at apex, gradually beaked; lf.-segms. not retrorse; phyllaries mostly in 3 sets . . . . 2. *A. grandiflora*
  Ak.-body truncate at apex, abruptly beaked; lf.-segms. retrorse; phyllaries in 2 quite unlike sets . . . . 4. *A. retrorsa*

1. A. glaùca (Pursh) Greene var. **montícola** (Greene) Q. Jones. [*A. m.* Greene.] Perennial, with 1 to several stems from the short branches of the root-crown, the scapes stout, mostly 1–3 dm. high, pubescent just below the heads and toward the base; lvs. largely basal, linear to oblanceolate, 5–20 cm. long, acute or acuminate, glabrous or short-hairy, not arachnoid; invol. 15–25 mm. high, subglabrous, the phyllaries in ca. 3 lengths, the outer conspicuously shorter; fls. yellow; aks. 5–10 mm. long, with a short stout striate beak; pappus 10–12 mm. long; $2n = 18, 36$ (Stebbins et al., 1953).—In rather dryish to moist places, 9000–10,500 ft.; Sagebrush Scrub, Montane Coniferous F.; Inyo-White Mts.; to Wash., Nev. June–Aug.

2. A. grandiflòra (Nutt.) Greene. [*Stylopappus g.* Nutt. *Troximon g.* Gray. *T. plebeium* Greene.] Scapes robust, few to several from the branched root-crown, stout, 1.5–6 dm. high; herbage hirsute-pubescent to glabrate, or the invol. and lower parts ± tomentose; lvs. oblanceolate to lance-elliptic, 10–25 cm. long, entire to laciniate or with spreading lobes; heads 2.5–4 cm. high, the invol. with broad short outer phyllaries and narrow much longer inner ones; ligules exserted, yellow; ak.-body ca. 4–7 mm. long, tapering to a beak 2–4 times as long; $2n = 18$ (Stebbins et al., 1953).—Dry or ± moist places below 8000 ft.; Chaparral, Montane Coniferous F.; most of cismontane s. Calif. mainland; reported by C. Smith from Santa Rosa Id.; ranging n. to B.C., Rocky Mts. May–July.

3. A. heteróphylla (Nutt.) Greene. [*Macrorhyncus h.* Nutt. *Troximon h.* Greene.] Subglabrate to pubescent annual 0.3–4 dm. tall, often with several scapes from the base; lvs. oblong to spatulate or linear, entire to denticulate or sinuate, 5–15 cm. long, in a ± basal cluster; invol. 5–20 mm. long, sparsely villous, the phyllaries lance-acuminate; ligules scarcely to conspicuously exserted, yellow; ak.-body 3–5 mm. long, several-ribbed, the beak ca. twice as long, slender; $2n = 18, 36$ (Stebbins et al., 1953).—Open grassy places and flats below 7500 ft.; V. Grassland, Coastal Sage Scrub, etc.; cismontane; to B.C., Rocky Mts. April–July. Variable, some forms having received names. Santa Cruz and Santa Rosa ids.

4. A. retrórsa (Benth.) Greene. [*Macrorhynchus r.* Benth. *Troximon r.* Gray.] Perennial, the stout scapes 1.5–5 dm. tall; woolly pubescent when young, later ± glabrate; lvs. lanceolate to oblong-lanceolate, 8–30 cm. long, pinnately parted into linear or lanceolate retrorse and rather scattered segms., the terminal one very long; invol. 2.5–4 cm. high, the phyllaries in 2 sets; ligules somewhat longer than invol., yellow, but sometimes drying pinkish; ak.-body 5–7 mm. long, ± truncate at apex and narrowed suddenly into the slender nonstriate beak 2–4 times as long; pappus-bristles 10–16 mm. long.—Common on dry slopes and ridges, 2500–8000 ft.; Chaparral, Yellow Pine F.; mts. of s. Calif.; to Wash., Nev. May–Aug.

6. *Amblyopáppus* H. & A.

Small annual herbs. Lvs. narrowly linear, all but the lowermost alternate. Heads small, apparently discoid, cymose-paniculate. Invol. broadly obconic, 1–2-seriate, the phyllaries obovate to obovate-oblong, herbaceous, rather thin, concave. Receptacle short-conic. Marginal fls. ♀; corolla tubular, 2–3-toothed. Disk-fls. perfect, fertile; corolla 5-lobed. Aks. short-clavate, 3–4-angled. Pappus of 7–12 irregular oblong obtuse paleae. One sp. (Greek, *amblus*, blunt, and *pappos*, pappus.)

1. **A. pusíllus** H. & A. [*Aromia tenuifolia* Nutt. *Infantea chilensis* Remy.] Plate 10, Fig. D. Plant yellow-green, balsamic glandular-granuliferous; stems 1–4 dm. high, leafy, erect, corymbosely but strictly branching above; lvs. entire or the lower pinnately divided into 3–5 lobes; heads numerous, short-peduncled; invol. 3–5 mm. high, ca. equal to the 10–30 inconspicuous yellowish fls.; aks. 1.5–2 mm. long, black, sparsely glandular and hispidulous; pappus-paleae whitish or reddish, 0.5 mm. long.—Coastal bluffs, old dunes, etc., below 500 ft.; Coastal Strand, Coastal Salt Marsh, Coastal Sage Scrub; L. Calif. to cent. Calif.; Peru, Chile. All the major Channel Ids.

7. *Ambròsia* L.

Annual or perennial monoecious herbs or subshrubs. Lvs. opposite or alternate, mostly lobed or dissected. Staminate heads nodding in bractless terminal racemes or spikes; invols. bowl-shaped or turbinate; paleae of the receptacle subtending at least the outer fls.; the corollas funnelform. Pistillate (fertile) heads borne in lf.-axils below the ♂ racemes, 1-fld., their invols. ± turbinate, at maturity armed with a single row of prickles above the middle or with several series of spines or prickles and then becoming a bur, which may represent more than 1 fl., each fl ending in a beak. Ca. 40 spp., largely of warmer parts of Am. (Ancient Greek, also Latin, name of several plants.)

1. Fruiting invol. unarmed or with a single row of short prickles below the single beak.
    2. Plants with lvs. once-pinnatifid into coarse lobes, greenish, rough-pubescent 9. *A. psilostachya*
    2. Plants with lvs. twice-pinnatifid into minute obtuse lobes 1 mm. wide, very canescent with soft hairs .... 10. *A. pumila*
1. Fruiting invol. armed with several rows of spines below the 1–4 beaks.
    3. Plants herbaceous.
        4. Bur 2–4 mm. long, its spines mostly hooked .... 5. *A. confertiflora*
        4. Bur 7–10 mm. long, its spines flattened and channeled, not hooked.
            5. Plants decumbent littoral perennial herbs .... 3. *A. chamissonis*
            5. Plants erect weedy annuals .... 1. *A. acanthicarpa*
    3. Plants shrubby or at least woody at the base.
        6. Lvs. sessile and cordate-clasping, coarsely spinose-toothed; fr. ovoid, 10–20 mm. long
            8. *A. ilicifolia*
        6. Lvs. petioled, the blades not spinose.
            7. Bur with straight spines.
                8. Lvs. deeply toothed to pinnatifid; fr. fusiform, 8 mm. long, 1-beaked, the spines densely long-villous .... 7. *A. eriocentra*
                8. Lvs. once to thrice pinnately divided into very small obtuse lobes; fr. globular, 4–6 mm. long, 2-beaked, the spines not villous .... 6. *A. dumosa*
            7. Bur with hooked spines; lvs. not pinnatifid or divided.
                9. Body of the bur lanate like the base of the spines .... 4. *A. chenopodiifolia*
                9. Body of the bur essentially glabrous, the spines prominently glandular-puberulent
                    2. *A. ambrosioides*

1. **A. acanthicárpa** Hook. [*Franseria a.* Cov. *F. californica* Gand. *F. palmeri* Rydb.] SAND-BUR. Rather strict or corymbosely branched annual 1–7 dm. high, rather densely

appressed-canescent and sparingly hispid; lvs. bipinnatifid, the blade ovate in outline, with oblong obtusish divisions, 2–6 cm. long, petiolate; ♂ heads very numerous, 2–4 mm. wide; fr. 1-fld., glabrous or sparsely villous, 5–10 mm. long, the spines 9–18, much flattened, shallowly sulcate; $n=18$ (Payne et al., 1964).—Common weed of sandy plains, stream bottoms, etc., mostly below 2500 ft., occasional to 6500 ft.; V. Grassland, Foothill Wd., Coastal Sage Scrub, etc.; cismontane s. Calif., and desert edge; to Wash., Tex. Aug.–Nov. Santa Catalina Id.

2. **A. ambrosioìdes** Cav. [*Gaertneria a.* Kuntze.] Spreading shrubby perennial 6–15 dm. high; stems hispid-hirsute and glandular; lvs. gray-green, hirsutulous and scabrous, simple, lanceolate, acuminate, ± truncate at base, 4–12 cm. long, 1.5–3 cm. wide, ascending, on petioles 1–2 cm. long, irregularly dentate; ♂ heads few, 6–8 mm. wide; fr. 2–4-fld., 10–15 mm. long, the 70–90 slender subulate hook-tipped spines glandular-puberulent, 2–4 mm. long; $n=18$ (Payne et al., 1964).—Hillsides and waste places, below 1000 ft.; Coastal Sage Scrub, Chaparral; near San Diego; s. Ariz., n. Mex. March–June.

3. **A. chamissònis** Less. [*Gaertneria c.* Kuntze. *Franseria cuneifolia* Nutt.] Plate 10, Fig. E. Perennial herb with radiating procumbent branching stems, forming loose mats 1–3 m. across and 1.5–3 dm. high; herbage silvery canescent with silky hairs, the stems more hirsute; lvs. simple, the blades ovate, rhombic, or oval-oblanceolate, usually obtuse, crenate-serrate to bluntly toothed or lobed or even incised, 2–5 cm. long, tapering to a petiole nearly as long; ♂ heads in congested terminal spikes, 7–8 mm. wide; fr. 1-fld., sparsely hirsute and glandular-atomiferous, 8–10 mm. long, the spines 15–25, dilated at base, boat-shaped; $n=18$ (Payne et al., 1964).—Coastal Strand, n. to B.C. July–Nov. San Clemente, Santa Catalina, San Miguel and Santa Cruz ids.

Growing along with the form above described are plants with lvs. once to thrice pinnatifid into oblong or obovate segms., less silvery; fr. glandular but not hairy, the spines often more slender and less sulcate; $n=18$ (Payne et al., 1964).—These two forms are said to hybridize freely, yet the two main type persist side by side. This second form has been treated taxonomically in various ways: as a var. in *Franseria* as *bipinnatisecta* Less.; as a subsp. by Wiggins & Stockw., also under the name *bipinnatifida*, as a sp. in *Gaertneria* by Kuntze, in *Ambrosia* by Greene; as a var. *dubia* under *F. bipinnatifida* by Eastw.; and as a forma *bipinnatisecta* in *Franseria* by Calder & Taylor. Payne (J. Arn. Arb. 45: 417–418. 1964) gives it no taxonomic recognition.

4. **A. chenopòdiifolia** (Benth.) Payne. [*Gaertneria c.* Abrams. *Franseria c.* Benth.] Rounded shrub 3–10 dm. high, the many stems white-tomentose when young, the herbage gray yellow-green; lvs. greenish on upper surface, tomentose beneath, deltoid-ovate, petiolate, the blade 1–6 cm. long, dentate and often obscurely 3–5-lobed; ♂ heads spicate, rather few, the invol. rotate, shallowly 7–10-lobed, tomentulose; fr. globose, 2–3-beaked, lanate between the 20–30 glandular spines, these subulate, flattened and sulcate below, hook-tipped, 2–3 mm. long; $2n=36$ (Payne et al., 1964).—Dry sunny hillsides below 600 ft.; Coastal Sage Scrub; sw. San Diego Co.; Ariz., Mex. March–June.

5. **A. confertiflòra** DC. [*Franseria c.* Rydb.] Herbaceous perennial, with 1–several erect stems 3–12 dm. high from slender rootstocks; herbage dark gray-green, strigose-hispidulous; lvs. bipinnately parted into oblong-linear segms., 4–10 cm. long, lance-oblong to obovate in outline; ♂ heads narrowly racemose, campanulate, 2–2.5 mm. wide; fr. obovoid, glomerate, 1–2-beaked, 2.5–4 mm. long; reticulate-ridged between the 10–20 broad-based, hook-tipped spines.—Dry places below 1000 ft.; Coastal Sage Scrub, S. Oak Wd.; cismontane s. Calif.; to Kans., Tex., Mex. May–Nov.

6. **A. dumòsa** (Gray) Payne. [*Franseria d.* Gray. *F. albicaulis* Torr. *Gaertneria d.* Kuntze.] BURRO-WEED. BUR-SAGE. Plate 11, Fig. A. Low intricately branched rounded shrub 2–6 dm. high, stems white, becoming spinescent; herbage densely cinereous-strigose; lvs. mostly fascicled, 5–20 mm. long, ovate in outline, 1–3-pinnatifid with short obtuse lobes; ♂ heads spicate-racemose, rather few, with ♀ heads often scattered between, ♂ invol. shallow, cinereous and glandular; fr. obovoidal, 2-beaked, glandular-puberulent or glabrate, the spines 20–30, lance-subulate, flattened, channeled, scattered, 2–2.5 mm. long; $n=18, 36, 54, 63$ (Payne, 1964).—Abundant on well drained soils below 3500 ft. (occa-

Plate 11. Fig. A, *Ambrosia dumosa*. Fig. B, *Ambrosia eriocentra*. Fig. C, *Ambrosia psilostachya* var. *californica*, ♀ head with teeth not spines. Fig. D, *Antennaria rosea*. Fig. E, *Arnica chamissonis*. Fig. F, *Artemisia douglasiana*.

sional to 5500) throughout the desert; Creosote Bush Scrub, Joshua Tree Wd.; Inyo Co. to Imperial Co. and San Diego Co.; to Son., L. Calif., Utah. Feb.–June, Sept.–Nov.

7. **A. eriocéntra** (Gray) Payne. [*Franseria e.* Gray. *Gaertneria e.* Kuntze.] Plate 11, Fig. B. Rigidly branched spreading shrub 3–8 dm. high, minutely tomentose-canescent; lvs. glabrate above, round-ovate, obtuse, sinuate-crenate to dentate, petioled, the blades 1.5–3 cm. long; staminate heads 7–8 mm. broad, ± glomerate, the invol. shallow, glandular and villous; ♀ heads 1-fld., sessile; fr. fusiform, 8 mm. long, the hairs tufted toward the ends of the 12–20 flattened-subulate straight spines; $n = 18$ (Payne et al., 1964).—Dry gravelly and rocky places, 2500–5000 ft.; Creosote Bush Scrub, Joshua Tree Wd., Pinyon-Juniper Wd.; mts. of e. Mojave Desert, San Bernardino Co. to Utah, Ariz. March–May.

8. **A. ilicifòlia** (Gray) Payne. [*Franseria i.* Gray. *Gaertneria i.* Kuntze.] Shrubby, 8–10 dm. high, the branches rigid, leafy, hispid-hirsute; lvs. rigidly coriaceous, 2–5 cm. long, sessile, ovate, coarsely dentate with spine-tipped teeth; ♂ heads 10–12 mm. broad; bur 1 cm. long, 2-beaked, glandular-pubescent, the spines 40–70, shallowly grooved but subterete, prominently hooked; $n = 18$ (Payne et al., 1964).—Rare, forming mats in sandy canyons and washes, below 1000 ft.; Creosote Bush Scrub; Colo. Desert e. of Salton Sea, Riverside and Imperial cos.; w. Ariz., n. L. Calif. Feb.–April.

9. **A. psilostàchya** DC. var. **califórnica** (Rydb.) Blake. [*A. c.* Rydb.] WESTERN RAGWEED. Plate 11, Fig. C. Perennial with running underground rootstocks; herbage rather harshly pubescent with short stiff appressed or spreading hairs, glandular, aromatic when rubbed; stems simple, 3–12 dm. high; lvs. mostly only once pinnatifid, 4–12 cm. long, thickish, often subsessile; ♂ invols. scabrous-hirtellous; ♀ invols. merely tuberculate above or quite unarmed; $n = 36, 54, 72$ (Payne et al., 1964).—One of the very common weeds of roadsides and uncultivated lands, sea level to 2500 ft.; Coastal Sage Scrub, Chaparral, etc.; cismontane and in Colo. Desert; n. to Wash. July–Nov. Santa Catalina Id.

10. **A. pùmila** (Nutt.) Gray. [*Franseria p.* Nutt.] Herbaceous, from a small branching rhizomelike caudex, 0.5–3 (–5) dm. high; herbage grayish silky-canescent; lvs. crowded, ± oblanceolate in outline, 2–3-pinnatifid into narrowly oblong-oblanceolate crowded obtuse lobes, with long slender petioles; ♀ invols. in fr. unarmed, pubescent, obovoid, 2 mm. long.—Dry, sunny places, along roadsides, etc., 100–600 ft.; V. Grassland; sw. San Diego Co., Lake Hodges to National City; L. Calif. June–Sept.

8. *Amphipáppus* T. & G. CHAFF-BUSH

Low shrub with divaricate spinescent branchlets. Lvs. alternate, small, entire, oval to elliptic or obovate, short-petioled. Heads radiate, few-fld., small, glomerate at tips of branchlets. Invol. obovoid, ca. 3-seriate, strongly graduate, pale, the 7–12 broad rounded phyllaries thin, with scarious erose margin. Receptacle fimbrillate. Ray-fls. 1–2, small, pale yellow, ♀, fertile. Disk-fls. 4–7, perfect, sterile. Anthers narrowly lance-tipped. Style branches thick. Ray-aks. broadly oblanceolate, pilose, their pappus of 15–20 short, basally united unequal white paleae. Disk-aks. undeveloped, their pappus of 25 tortuous white paleae of unequal width. One sp. (Greek, *amphi*, both kinds of, and *pappos*, pappus.)

1. **A. fremóntii** T. & G. ssp. **fremóntii**. [*Amphiachyris f.* Gray.] Much-branched white-barked shrub 3–6 dm. high, with yellow-green cast, glabrous throughout, slightly glutinous around heads; lvs. 5–12 mm. long, 2–4 mm. wide, 1-nerved; heads 4–5 mm. high, the phyllaries closely appressed; $n = 9$ (Raven et al., 1960).—Open desert and semialkaline flats, 500–5200 ft.; Creosote Bush Scrub; Death V. region from Inyo and Argus ranges to Funeral Mts. and Pahrump V., Inyo and n. San Bernardino cos.; Nev. April–May.

Ssp. **spinòsus** (A. Nels.) Keck. [*A. s.* A. Nels. *A. f.* var. *s.* C. L. Porter. *Amphiachyris f.* var. *s.* Nels.] Herbage densely scabro-hispidulous.—E. Mojave Desert in s. San Bernardino Co. (Goffs, Kelso, Essex, etc.); to Utah, Ariz.

9. **Anáphalis** DC. PEARLY EVERLASTING

White-woolly perennials with simple erect equably leafy stems. Lvs. entire, alternate. Heads mostly numerous, in terminal compound corymbs. Plants dioecious or polygamo-dioecious, the ♀ heads sometimes with a few cent. ♂ fls. Receptacle naked, flat or convex. Heads discoid or disciform, many-fld. Phyllaries imbricate in several series. Staminate fls. tubular, generally with undivided style; ♀ fls. tubular-filiform, with bifid style. Pappus of capillary bristles, not conspicuously barbellate. Ca. 25 spp. of N. Temp. (Ancient Greek name of some Everlasting.)

1. **A. margaritàcea** (L.) Benth. ex C. B. Clarke. [*Gnaphalium m.* L. *A. m.* var. *occidentalis* Greene.] With slender running rootstocks; stems commonly 2–9 dm. high, loosely white-woolly or somewhat rusty in age; lvs. lanceolate to linear or linear-oblong, 2–8 (–12) cm. long, sessile, obtuse to acuminate, often greener above than beneath, sometimes ± revolute; heads to 1 cm. wide; invol. ca. 5–7 mm. high; phyllaries pearly white, ovate, sometimes with a basal dark spot; aks. papillate; $2n=28$ (Maude, 1939).—Shaded slopes at ca. 6000 ft.; Montane Coniferous F.; a small side canyon off Mill Creek Canyon, San Bernardino Mts.; Sierra Nevada to Alaska, Atlantic Coast, Eurasia. June–Aug.

10. **Anisocòma** T. & G. SCALE BUD

Low scapose annual with broad rosette of pinnately parted or toothed lvs. and several ascending 1-headed scapes. Invol. cylindric, strongly graduated, the inner phyllaries linear and acute, the outer successively shorter and rounded-obtuse, appressed, all with brownish-green to purplish midrib and broad scarious margins. Receptacle flat, with linear bracts. Ligules pale yellow. Aks. oblong, truncate, 10–15-nerved, pubescent. Pappus bright white, of 10–12 plumose bristles in 2 unequal series. One sp. (Greek, *anisos*, unequal, and *kome*, tuft of hair, because of the 2 unlike sets of pappus-bristles.)

1. **A. acáulis** T. & G. Lvs. many, ± tomentulose, 3–5 cm. long, oblong in outline, with toothed segms.; scapes glabrous, 5–20 cm. long; invol. 2–3 cm. high, the phyllaries often edged with red toward the tips and with reddish dots; aks. 4 mm. long; longer pappus-bristles ca. 2–3 cm. long, the others much shorter; $2n=14$ (Stebbins & others, 1953).—Common in washes and sandy places, 2000–7300 ft.; Creosote Bush Scrub, Joshua Tree Wd., etc.; Mojave and Colo. deserts; Nev., Ariz. April–June.

11. **Antennària** Gaertn. PUSSYTOES

Low woolly perennials with simple basal and alternate lvs., the former mostly tufted and conspicuous, the cauline mostly ± reduced. Plants dioecious; heads 1–many in a usually congested infl., disciform or discoid. Fls. many. Phyllaries imbricate in several series, scarious at least in upper part, often whitish or colored. Receptacle flattish or convex, naked. The ♂ fls. with filiform corolla, entire or merely notched style, and a scanty pappus of usually clavate bristles. The ♀ fls. with tubular 5-toothed corolla, 2-cleft style and a copious pappus of capillary naked sometimes barbellate bristles. Aks. terete or slightly compressed. A confusing genus, some spp. having both bisexual and parthenogetic races. Perhaps 25–30 spp., largely of N. Am., some circumpolar, some S. Am. (Latin, *antenna*, because of the resemblance in ♂ fls. to insect antennae.)

(Sharsmith, C. W. in Abrams, Illus. Fl. 4: 474–485. 1960.)

Heads solitary, terminal, or at most 2; cespitose from a multicipital caudex ..... 2. *A. dimorpha*
Heads several to many.
    Basal lvs. distinctly less pubescent on their upper surface than beneath. ...... 3. *A. marginata*
    Basal lvs. as densely pubescent on upper surface as on lower.
        Distal scarious portion of phyllaries blackish green .... 4. *A. media*
        Distal scarious portion of phyllaries white or roseate.
            Phyllaries with conspicuous dark or blackish spot at the base .... 1. *A. corymbosa*
            Phyllaries greenish or merely brownish at the base, mostly roseate .... 5. *A. rosea*

1. **A. corymbòsa** E. Nels. [*A. dioica* var. *c.* Jeps.] Loosely mat-forming from a slender branching caudex, with rosulate basal lvs. and slender, leafy stolons up to 7 cm. long and erect leafy stems 7–25 cm. high; basal and stolon lvs. thinnish, mostly oblanceolate, 1.5–4 cm. long 2–5 mm. wide, rather thinly subsericeous-tomentose and mostly greenish; cauline lvs. gradually reduced; heads several, glomerate or in a corymbiform cyme; ♀ invols. woolly from base to the middle, 3–5 mm. high, the phyllaries with mostly greenish base, a conspicuous dark brown or blackish medial spot, and firm ivory-white tips; aks. sparsely puberulous.—At 12,000 or more ft.; Alpine Fell-fields; White Mts.; to Ore., Rocky Mts. June–Aug.

2. **A. dimórpha** (Nutt.) T. & G. [*Gnaphalium d.* Nutt.] Cespitose from a low multicipital caudex, forming small mats, gray-tomentose throughout except on invols.; with erect basal lvs. and few-lvd., 1-headed stems mostly 1–4 cm. long; basal lvs. 0.8–2.5 cm. long, oblanceolate or wider; ♀ invol. cylindric-turbinate, 10–16 mm. long, thinly woolly at base, the outer phyllaries ± ovate, with brownish center, the inner lance-linear, with pale brown center; ♂ heads broader, their invols. 5–9 mm. high; aks. with persistent crown of pappus.—Dry slopes and ridges, 4800–9000 ft.; Sagebrush Scrub to Lodgepole F., Pinyon-Juniper Wd.; San Bernardino and San Gabriel mts., Mt. Pinos, Panamint Mts.; to B.C., Rocky Mts. April–July.

3. **A. marginàta** Greene. [*A. dioica* var. *m.* Jeps.] Gray-tomentose throughout except on the upper surface of the lvs., with leafy stolons to 1 dm. long and erect sparsely leafy stems to 2.5 dm. long; lvs. ± spatulate, 1–3 cm. long, quickly glabrate and bright green on upper surface; heads 3–6, ± glomerate; ♀ invols. woolly at base, 6–8 mm. high, the phyllaries with pale green base, often darker medial spot and white or roseate tips; aks. glandular.—Occasional at 6000–7000 ft.; Yellow Pine F.; San Bernardino F.; to New Mex. and Colo. May–July.

4. **A. mèdia** Greene. [*A. alpina* var. *m.* Jeps.] Mat-forming, stoloniferous, with a branching root-crown and stems to 1.5 dm. high; lower lvs. oblanceolate to spatulate, white-tomentose; heads mostly 3–6 in a subcapitate group; ♀ invol. 4–7 mm. high, woolly below, mostly blackish-green at the tips; aks. glandular.—Mostly above 11,000 ft.; Alpine Fell-fields; San Bernardino Mts., White Mts.; to B.C., Rocky Mts. July–Aug.

5. **A. ròsea** Greene. [*A. dioica* var. *r.* D. C. Eat.] Plate 11, Fig. D. Cespitose, forming leafy mats, closely and persistently gray-tomentose; stems mostly 0.5–2.5 dm. high, scatteringly leafy; basal lvs. oblanceolate or spatulate, mostly ca. 1–3 cm. long; stem-lvs. reduced; heads several, in cymes; ♀ invol. 4–7 mm. high, woolly below, the scarious part of the phyllaries ± oblong, deep pink to bright or dull white, ± striate under magnification; $2n = 56$ (Löve & Löve, 1964).—Dryish to moist, ± wooded places, 6500–10,000 ft.; Yellow Pine F., Lodgepole F., Bristle-cone Pine F.; San Jacinto and San Bernardino mts., White Mts.; to Alaska, Rocky Mts. June–Aug.

12. *Ánthemis* L.

Annual or perennial herbs, usually aromatic, with alternate incised-dentate to pinnately dissected lvs. Heads medium-sized, radiate or rarely discoid. Rays elongate, white or yellow, ♀ or neutral. Phyllaries subequal or imbricate, dry, ± hyaline on margins. Disk-fls. many, perfect. Receptacle convex to conic, ± chaffy, at least near summit. Aks. terete, or 4–5-angled, sometimes ± compressed. Pappus a short crown or none. Ca. 60 spp. of Old World. (Ancient Greek name of the Chamomile.)

1. **A. cótula** L. MAYWEED. Ill-smelling annual, 1–5 dm. high, ± branched, subglabrous; lvs. mostly 2–6 cm. long, 2–3-pinnatifid into very narrow segms.; heads rather many, short-peduncled, 1.5–2 cm. across; phyllaries ± pointed; rays white, 5–11 mm. long, sterile; aks. subterete, ca. 10-ribbed; pappus none; $2n = 18$ (Harding, 1950).—Common weed in waste places, fields, etc.; much of Calif.; native of Eu. April–Aug.

13. Arctium L. BURDOCK

Coarse biennials, branched, with large unarmed, rounded or ovate, mostly cordate petioled lvs. Heads several to many, many-fld. Fls. all tubular, perfect, similar, pink or purplish. Invol. subglobose. Phyllaries imbricate, coriaceous, appressed at base, attenuate to long stiff points with hooked tips. Receptacle bristly. Aks. oblong, flattened, transversely wrinkled. Pappus short, of many rough bristles, separate and deciduous. Four spp., Eurasian. (Greek, *arction*, a plant from *arctos*, bear, because of the rough invol.)

1. **A. láppa** L. Plant 1–2.5 m. tall; petioles to 4–5 dm. long, woolly; lf.-blades 1–5 dm. long, 0.5–3 dm. wide, subglabrous above, white-tomentose beneath; infl. corymbiform, with leafy bracts; heads 3–4 cm. wide; invol. glabrous, green, the outer phyllaries 1 cm. long; aks. 6–7 mm. long; $2n=32$ (Sigura, 1936).—Occasional weed in scattered stations, cismontane; natur. from Eu. June–Aug.

14. Arctòtis L.

Perennial herbs with ± white-woolly herbage. Lvs. alternate. Heads of large or medium size, solitary, long-peduncled, with rays. Phyllaries imbricate in several series, the inner scarious. Receptacle honeycombed, mostly bristly. Aks. grooved, usually villous, ours with a tuft of silky hairs near the base; pappus crownlike or composed of scales. Ca. 30 spp., mostly of S. Afr.; a few grown as ornamentals. (Greek, *arktos*, bear, and *otis*, ear, referring to the pappus-scales.)

1. **A. stoechadifòlia** Berg. var. **grándis** (Thunb.) Less. [*A. g.* Thunb.] AFRICAN DAISY. Perennial but often grown as an annual, bushy, 4–7 dm. high; stems leafy; lvs. 8–15 cm. long, obovate-oblong, toothed, concolorous; peduncles 1.5–3 dm. long; heads 5–7 cm. across; outer phyllaries tomentose; rays white, but violet on the outside; $2n=18$ (Bilquez, 1951).—Occasional escape from cult.; native of S. Afr. June–Aug.

15. Arnica L.

Perennial herbs, ± glandular or aromatic, from a rhizome or caudex and with fibrous roots. Lvs. simple, opposite or the uppermost sometimes alternate. Heads rather large, turbinate to hemispheric, solitary to rather many, radiate or discoid. Phyllaries green, ± biseriate, subequal and connivent. Receptacle flat or convex, naked. Rays when present, yellow or orange, rather few and broad. Disk-fls. perfect and fertile, yellow or orange. Anthers entire to minutely sagittate. Style-branches ± flattened, truncate. Aks. subterete, 5–10-nerved; pappus white to tawny, the bristles many, capillary, barbellate to subplumose. Ca. 30 spp., circumboreal, but running s. in the mts. Some spp. with apomixis and some of hybrid origin, hence confused. (Origin of name obscure.)

(Maguire, B. A. A monograph of the genus Arnica. Brittonia 4: 386–510. 1943.)
Heads characteristically radiate, sometimes with rayless plants growing with rayed ones.
    Cauline lvs. mostly 5–10 pairs. San Bernardino and White mts. . . . . 1. *A. chamissonis*
    Cauline lvs. mostly 2–4 pairs. White Mts.. . . . . 3. *A. mollis*
Heads characteristically discoid. Santa Ana and Santa Ynez mts. . . . . 2. *A. discoidea*

1. **A. chamissònis** Less. ssp. **foliòsa** (Nutt.) Maguire. [*A. f.* Nutt.] Plate 11, Fig. E. Perennial from long nearly naked rhizomes; stems solitary, 3–8 dm. high; herbage ± villous-puberulent to -hirsute, ± glandular or viscid above; cauline lvs. mostly 5–10 pairs, not much reduced upwards, lanceolate to oblanceolate, 5–20 cm. long, including petioles, uppermost sessile, blades mostly entire to denticulate; heads 5–15, hemispheric-campanulate, 15–18 mm. high; invol. 8–12 mm. high; phyllaries acutish, with hairs at base ± septate; ligules pale yellow, 12–18 mm. long; disk-corollas 7.5–9 mm. long; aks. 4–5 mm. long; pappus stramineous, barbellate, the bristles 0.2 mm. long; aks. 4–5 mm. long; $n=53$–54 (Ornduff et al., 1963).—Moist places, at ca. 10,000 ft.; Bristle-cone Pine F.; Panamint Mts.; to Alaska, Mont. Plants with phyllaries more blunt and rounded at tips

are var. *bernardìna* (Greene) Maguire. [*A. b.* Greene.] At ca. 7000–9100 ft.; Montane Coniferous F.; San Bernardino Mts., White Mts. July–Aug.

   2. **A. discoìdea** Benth. Rhizomes extensive, nearly naked or forming an approximate crown; stems mostly solitary, 3–6 dm. tall, glandular-puberulent and often also long-hairy, commonly branched especially above; cauline lvs. 3–several pairs, mostly crowded toward base, there opposite, and long-petioled; lvs. ± hairy, the blade ovate to deltoid, 3–8 cm. long, sharply or undulately dentate; upper lvs. sessile, reduced; heads several, discoid; invol. 9–13 mm. high, glandular and spreading-villous; phyllaries mostly obtusish, 9–11 mm. long; aks. glandular, and ± hairy; pappus strongly barbellate; $n=38$ (Ornduff et al., 1963).—Open woods at ca. 3500–4500 ft.; Chaparral, Yellow Pine F.; Santa Ana and Santa Ynez mts.; to n. Calif. May–July.

   3. **A. móllis** Hook. Rhizomes dark brown, thick, short; stems 2–6 dm. high, simple, ± scabrid-puberulent to pilose and glandular; cauline lvs. mostly 3–4 pairs, entire to denticulate; heads 1–few; disk to ca. 3 cm. wide; invol. 10–16 mm. high; phyllaries ± acuminate, long-hairy at base; rays mostly 12–18, yellow, 15–25 mm. long; pappus sub-plumose; aks. hirsute; $n=54-57$ (Taylor, 1967).—At ca. 10,150 ft., White Mts.; to B.C., Rocky Mts. July–Sept.

16. *Artemísia* L. SAGEBRUSH. WORMWOOD

Herbs or shrubs, usually aromatic. Lvs. alternate, entire to toothed or dissected. Infl. spiciform, racemiform, or paniculate. Heads small, discoid or disciform, sometimes only with perfect fls., sometimes the outer ♀ and the cent. then sometimes sterile. Phyllaries dry, imbricate, at least the inner scarious or scarious-margined. Receptacle flat or hemispheric, naked or with long hairs. Style-branches flattened, truncate, penicillate. Aks. obovoid or oblong, usually glabrous; pappus a very short crown or more often none. Over 100 spp., of N. Hemis. and S. Am. (*Artemisia*, wife of Mausolus, king of Caria.)

(Keck, D. D. A revision of the A. vulgaris complex in N. Am. Proc. Calif. Acad. Sci. IV. 25: 421–468. 1946. Ward, G. H. Artemisia, section Seriphidium, in N. Am. Contr. Dudley Herb. 4: 155–205. 1953. Beetle, A. A. New names within the section Tridentatae. Rhodora 61: 82–85. 1959. A study of sagebrush. Univ. Wyo. Bull. 368. 1960.)

1. Fls. all perfect; shrubs except *A. palmeri*.
   2. Ray-fls. present in at least some of the heads. Mojave Desert .... 2. *A. bigelovii*
   2. Ray-fls. absent; all fls. perfect and fertile.
      3. Receptacle naked; plants shrubby, mostly ± canescent.
         4. Inner phyllaries usually canescent; plants mostly more than 4 dm. high; fls. 3–6 (–12) in a head; aks. resinous-granuliferous .... 13. *A. tridentata*
         4. Inner phyllaries usually glabrous; plants mostly less than 4 dm. high.
            5. Fls. mostly 10–16; aks. granuliferous; lvs. 0.5–5 cm. long .... 11. *A. rothrockii*
            5. Fls. mostly 3–5; aks. glabrous; lvs. 0.5–1.5 cm. long .... 9. *A. nova*
      3. Receptacle with chaff; lvs. green above. San Diego region .... 10. *A. palmeri*
1. Fls. not all perfect, the marginal ♀, the cent. ones fertile or sterile.
   6. Plants definitely woody at base, not rhizomatous.
      7. Plant with long naked spines; stems 1–3 dm. long; lvs. with spatulate lobes. Deserts
         .... 12. *A. spinescens*
      7. Plant not spiny; stems taller; lvs. with linear lobes.
         8. Lvs. with linear-filiform segms. less than 1 mm. wide, ± revolute; ray-fls. 6–10; disk-fls. 15–30. Widely distributed .... 3. *A. californica*
         8. Lvs. with segms. 1–3 mm. wide, not revolute; ray-fls. to 15; disk-fls. to 40. Insular
            .... 7. *A. nesiotica*
   6. Plants not woody at base.
      9. Herbage glabrous, rarely pubescent, never tomentose.
         10. Invol. 2–3 mm. high.
            11. Annual or biennial; lvs. (at least the lower) 2–3-pinnatifid. Weed in waste places
               .... 1. *A. biennis*
            11. Perennial; lvs. linear, mostly entire. Native .... 5. *A. dracunculus*
         10. Invol. 4–7 mm. high. White Mts. .... 8. *A. norvegica*
      9. Herbage ± pubescent, mostly white-tomentose beneath.

11. Fls. all fertile; plants with branched caudex .... 8. *A. norvegica*
11. Fls. of center of disk sterile; plants from horizontal rhizomes.
  12. Principal lvs. narrow, not more than 1 cm. wide (exclusive of lobes when present), tomentose on both sides or green above; stems rarely over 1 m. high
  6. *A. ludoviciana*
  12. Principal lvs. 1–5 cm. wide exclusive of lobes when present, frequently entire; lvs. greener above than beneath; stems to 3 m. high .... 4. *A. douglasiana*

1. **A. biénnis** Willd. Erect glabrous annual or biennial, 3–8 dm. high, leafy; lvs. 2-pinnately parted or the upper pinnatifid, the lobes linear, acute, mostly cut-toothed; heads subglobose, crowded, erect, subsessile; invol. 2–3 mm. high; larger glabrous phyllaries rounded, broadly scarious on margin, with green midrib; disk-fls. 15–40, fertile; corolla campanulate, ca. 1 mm. long; aks. glabrous, 4–5-nerved.—Occasional weed in widely scattered places, up to 7000 ft.; cismontane and to desert edge as at Oro Grande; native in nw. U.S. Aug.–Oct.

2. **A. bigelòvii** Gray in Torr. Low evergreen shrub 2–4 dm. high, many-stemmed, with grayish-brown bark; twigs and lvs. canescent; lvs. of vegetative shoots sessile or short-petioled, narrowly cuneate, 1–2 cm. long, 2–5 mm. wide, with a truncate sharply 3-toothed apex, sometimes entire; lvs. of infl. mostly entire; infl. narrowly paniculate, dense, with several heads on each short recurved panicle-branch; invol. turbinate, 2–4 mm. high; phyllaries densely tomentose, 8–12, the outer ovate, the inner oblong; rays 0–2, ♀, the tubular corolla 2-toothed, 1–2 mm. long; disk-fls. 1–3, perfect, fertile, funnelform, 5-toothed, 1.5–3 mm. long; aks. ellipsoid, 4–5-ribbed, glabrous; $2n=18$ (Ward, 1953).—Dry limestone slopes, 5000–6000 ft.; mostly Pinyon-Juniper Wd.; Inyo, Clark, and New York mts., e. Mojave Desert; to Colo., w. Texas. Aug.–Sept.

3. **A. califórnica** Less. Coastal Sagebrush. Grayish shrub 6–15 dm. high, the lvs. numerous, strigulose, the lower 1–5 cm. long, palmately once or twice parted into linear-filiform segms. less than 1 mm. wide, the upper sometimes entire and fascicled; heads many, in long racemose panicles; invol. 2–3 mm. long; fls. rather numerous, the rays 6–10 and disk 15–30; aks. with a minute squamellate crown; $2n=18$.—Lower slopes and fans below 2500 ft.; Coastal Sage Scrub, Coastal Strand, etc.; cismontane, L. Calif. n. to cent. Calif. Aug.–Dec. San Clemente, Santa Catalina, Santa Cruz, San Miguel and Santa Rosa ids.

4. **A. douglasiàna** Bess. in Hook. [*A. vulgaris* var. *californica* Bess. *A. v.* var. *heterophylla* Jeps.] Plate 11, Fig. F. Stout rhizomatous perennial, 5–15 (–30) dm. high; stems simple or sometimes branched above; lvs. commonly 7–15 cm. long, lanceolate to elliptic and entire, or oblanceolate to obovate in outline and coarsely few-lobed or -toothed toward apex, the lobes entire, mostly lanceolate, strongly discolored, glabrous and green to sparsely tomentulose above, densely gray-tomentose beneath, mostly plane; infl. leafy, open or dense, elongate, paniculate; heads erect or nodding; invol. campanulate, 3–4 mm. long, ± tomentose; phyllaries 8–14; ray-fls. 6–10; disk-fls. 10–25; $n=27$ (Clausen et al., 1940).—Mostly low waste places, at elevs. up to 6000 ft.; many Plant Communities; cismontane and to desert edge as at Victorville; to Wash., Ida., L. Calif. June–Oct.

5. **A. dracúnculus** L. [*A. glauca* Pall. *A. dracunculoides* Pursh.] Almost odorless perennial with horizontal rhizome; stem erect, 5–15 dm. tall, glabrous to villous-puberulent; lvs. linear or nearly so, 3–8 cm. long, entire or sometimes cleft, the lower mostly deciduous; panicle with elongate leafy ascending branches; heads many, soon spreading or nodding; invol. subglobose, 2–3 mm. broad; phyllaries 7–12, scarious-margined; fls. ca. 20–30, the outer ♀ and fertile, the inner sterile; aks. ellipsoid, glabrous, not ribbed; $2n=18$ (Weinedel, 1928).—Dry ± disturbed places, up to 11,000 ft.; Coastal Sage Scrub to Montane Coniferous F. and above; cismontane s. Calif. to cent. Calif.; Joshua Tree Wd. and above, Mojave Desert; to B.C., Wis., Tex. Aug.–Oct.

6. **A. ludoviciàna** Nutt. ssp. **álbula** (Woot.) Keck. [*A. a.* Woot. *A. l.* var. *a.* Shinners.] Like ssp. *ludoviciana* but with lvs. mostly 1–2 cm. long, obovate to elliptic, and with forward-projecting teeth or lobes, or lance-linear and entire, often ± revolute; invol. ca. 3 mm. high; phyllaries 11–16; ray-fls. 8–11; disk-fls. 8–13.—Mostly dry slopes below

7000 ft.; Creosote Bush Scrub to Pinyon-Juniper Wd., Yellow Pine F.; w. edge of Colo. Desert, Mojave Desert; to Rocky Mts., Tex., L. Calif. May–Oct.

Ssp. **incómpta** (Nutt.) Keck. [*A. i.* Nutt.] Stems usually entirely herbaceous; herbage commonly ± green, the lvs. glabrate above, white-tomentose beneath; lower lvs. 2–8 cm. long, parted into linear or lanceolate lobes, some of these again toothed or lobed; upper lvs. less cut to entire; invol. 3–3.5 mm. high, silky-tomentose to shining-glabrate; phyllaries 9–14; ray-fls. 6–10; disk-fls. 15–30; $n=18$ (Clausen et al., 1940).—Dry places, 5000–8000 ft.; Joshua Tree Wd., Pinyon-Juniper Wd., Montane Coniferous F.; Santa Rosa Mts. to San Gabriel Mts.; Kingston Mts.; to Modoc Co. and Rocky Mts. July–Sept.

Ssp. **ludoviciàna** [*A. gnaphalodes* Nutt. *A. l.* var. *g.* T. & G.] Perennial rhizomatous herb; stems mostly 3–10 dm. tall, simple to infl., ± white-tomentose at least above; lvs. many, linear to lanceolate, oblanceolate or elliptic, sometimes cuneate, mostly 3–11 cm. long, entire or few-toothed or -lobed especially apically, the lobes mostly entire, permanently and densely white-tomentose on both sides or loosely floccose to green and glabrate above; infl. ample or narrow, an elongate usually compact panicle 1.5–5 dm. broad; heads erect or nodding, often in glomerules; invol. 3–4 mm. high, usually tomentose; phyllaries 7–13; ray-fls. 5–12; disk-fls. 6–20; $n=18$ (Clausen et al., 1940).—Dry open places, below 8000 ft.; Sagebrush Scrub, Montane Coniferous F., Joshua Tree Wd., etc.; San Jacinto Mts. to San Gabriel Mts.; to Wash., Ontario, Ark., New Mex. July–Sept.

7. **A. nesiótica** Raven. [*Crossostephium insulare* Rydb. *A. californica* var. *i.* Munz.] Resembling *A. californica* in habit and deeply divided lvs., but the lf.-lobes not or scarcely revolute, 1–3 mm. wide, fleshy; ray-fls. to 15; disk-fls. to 40. Foggy hillsides, Coastal Sage Scrub; San Clemente Id., Santa Barbara Id., San Nicolas Id. April–Sept.

8. **A. norvègica** Fries var. **saxátilis** (Bess. in Hook.) Jeps. [*A. chamissoniana* var. *s.* Bess.] Perennial from a branched caudex; stems 2–6 dm. tall, mostly loosely villous; lvs. ovate in outline, the basal with dissected blades 2–10 cm. long and ultimate acute narrow segms.; cauline lvs., reduced upward, becoming sessile; infl. loosely racemose or racemose-paniculate; heads many-fld.; invol. 4–7 mm. high; phyllaries ± woolly, villous to glabrous, dark-margined; disk-corollas long-hairy near the junction with the mostly glabrous aks.—Rocky places, Subalpine Fell-fields; Inyo-White range; to the Arctic, Rocky Mts. July–Sept.

9. **A. nòva** A. Nels. [*A. arbuscula* ssp. *n.* Ward. *A. tridentata* ssp. *n.* Hall & Clements.] Low spreading evergreen shrub mostly 1–3 dm. high; lvs. canescent, 1–2 cm. long, apically 3-toothed; panicle on very slender stem, narrow, strict, with many heads; phyllaries 8–12, the inner mostly glabrous; invol. 3–3.5 mm. high; fls. mostly 3–5, perfect; aks. glabrous; $2n=18, 36$ (Ward, 1953).—Dry rocky places, 5000–11,000 ft.; Sagebrush Scrub, Pinyon-Juniper Wd., etc.; Hemet V., San Jacinto Mts.; desert slopes of San Bernardino Mts., New York, Clark, Panamint, Kingston, Inyo-White mts.; to Rocky Mts. Sept.–Nov.

10. **A. pálmeri** Gray. [*Artemisiastrum p.* Rydb.] A ± suffrutescent perennial with long wandlike stems 5–8 dm. high, grayish-pubescent over a reddish color; lvs. 5–12 cm. long, pinnately parted into 3–9 linear or lance-linear lobes, or linear and entire, subglabrous above, finely tomentose beneath; infl. broadly paniculate, 1.5–5 dm. long; heads numerous; invol. hemispheric, 2.5–4 mm. long; phyllaries 8–12, ovate-acutish, sparingly pubescent or glabrous, the inner little longer than the outer, mostly subglabrous; fls. 12–30, perfect, usually each subtended by a bract; corolla 1.5–2.5 mm. long; aks. 4-angled, granuliferous; $2n=18$ (Ward, 1953).—At low elevs.; Coastal Sage Scrub; sw. San Diego Co.; n. L. Calif. July–Sept.

11. **A. rothróckii** Gray. [*A. tridentata* ssp. *r.* Hall & Clem.] Low spreading evergreen shrub 2–6 dm. high, often root-sprouting; twigs densely tomentose when young; lvs. of vegetative shoots 0.5–5 cm. long, mostly broadly cuneate or flabelliform, 3-toothed or -lobed, sometimes lanceolate, entire, densely canescent, sometimes glabrate in age; infl. lvs. entire; infl. spicate or narrowly paniculate, 0.5–4 dm. long; invol. campanulate, 4–5.5 mm. long; phyllaries 10–14, the outer broadly ovate, acute or acuminate, ± tomentose, the inner elliptic, less hairy; fls. mostly 10–16, perfect; corolla 5-toothed, 2.3–3.5 mm. long; aks. granuliferous; $2n=36, 54, 72$ (Ward, 1953).—Rocky slopes, 2500–11,000 ft.;

Joshua Tree Wd. to Bristle-cone Pine F.; n. base San Bernardino Mts., Panamint Mts., White Mts.; to cent. Sierra Nevada. Aug.–Sept.

12. **A. spinéscens** D. C. Eat. Woody, intricately branched, 1–3 dm. high, spiny, white-tomentose on young branches and ± so or stiff-pubescent on the grayish foliage; lvs. crowded, 5–8 mm. long, pedately 5–7-parted and the divisions 3-lobed; heads few, in short lateral spikes which become the spines the next year; fls. 5–12; invol. globose, 3 mm. long; phyllaries 5–8, obovate, tomentose; corolla and ak. villous-cobwebby.—Semialkaline flats, 2000–6000 ft.; Shadscale Scrub, Creosote Bush Scrub, etc.; w. Mojave Desert from Lancaster region n. and e.; to Ore., Rocky Mts. April–May.

13. **A. tridentàta** Nutt. ssp. **párishii** (Gray) Hall & Clem. [*A. p.* Gray.] Plants erect; lvs. mostly linear, entire or shallowly notched; infl. often with drooping branches; aks. glandular and sparingly short-villous to arachnoid-hairy; $2n = 36$ (Ward, 1953).—Dry valleys, 1000–2500 ft.; Coastal Sage Scrub, Joshua Tree Wd., etc.; Santa Clara River V., n. Los Angeles Co. to w. Mojave Desert (Antelope V.). Oct.–Nov.

Ssp. **tridentàta** [*A. t.* var. *angustifolia* Gray.] BASIN SAGEBRUSH. Plate 12, Fig. A. Rounded evergreen shrub mostly 0.5–3 m. high, with a short thick trunk or few branches from base, ± silvery-canescent throughout; lvs. of vegetative shoots sessile or short-petioled, cuneate with 3 blunt teeth or sometimes 4–9-toothed or entire, 1–4 cm. long, 0.2–1.3 cm. wide, those of infl. mostly entire; infl. mostly with erect branches and 1.5–4 dm. long; invol. 3–3.8 mm. long; phyllaries 8–15, the outer short, round-ovate, the inner elliptic; fls. 3–6 (–12) in a head, perfect; corolla narrow-funnelform, 2–3 mm. long; aks. resinous-granuliferous, rarely sparsely short-villous; $2n = 18, 36$ (Ward, 1953).—Dry slopes and plains, 1500–10,600 ft.; Sagebrush Scrub, Yellow Pine F., Chaparral, Pinyon-Juniper Wd., etc.; n. L. Calif., along w. edge of Colo. Desert and w. Mojave Desert, Panamint Mts., Inyo-White Mts.; to B.C., Rocky Mts. Aug.–Oct.

17. **Áster** L. ASTER

Summer- or fall-flowering herbs, usually rhizomatous or fibrous-rooted perennials, rarely shrubs. Lvs. alternate, entire or toothed. Heads usually numerous and radiate, in panicles, corymbs or racemes, rarely solitary. Invol. turbinate, campanulate or hemispheric; phyllaries usually herbaceous and imbricated, sometimes coriaceous with green tips. Ray-fls. usually present, occasionally without ligule, ♀, shades of blue or purple, rarely white. Disk-fls. mostly yellow or reddish-purple. Style-branches flattened, with lanceolate to subulate appendages. Aks. ± compressed, hairy or glabrous; pappus of subequal ± numerous capillary bristles, occasionally with a few short outer bristles or scales. Some 250 or more spp., centred in N. Am., but widely distributed in temp. regions. (Greek, *aster*, a star, from the radiate heads of the fls.)

(Cronquist, A. W. N. Am. spp. of Aster centering about Aster foliaceus Lindl. Am. Midl. Nat. 29: 429–468. 1943; Revision of the Oreastrum group of Aster. Leafl. W. Bot. 5: 73–82. 1948.)
1. Plants annual.
    2. Rays evident, surpassing the pappus; phyllaries distinctly graduate, acuminate with a prominent scarious margin .... 6. *A. exilis*
    2. Rays present but very inconspicuous, equaling or shorter than the mature pappus, several-seriate or more numerous than the disk-fls. .... 7. *A. frondosus*
1. Plants perennial.
    3. Plants woody toward the base; stems rushlike, almost leafless, the lvs. inconspicuous at anthesis.
        4. Plant shrubby, glaucous, not spiny; heads discoid, yellowish; aks. pubescent
        .... 10. *A. intricatus*
        4. Plant essentially herbaceous, not glaucous, often spinose in the lf.-axils; heads radiate, white; aks. glabrous .... 15. *A. spinosus*
    3. Plants herbaceous; stems leafy or subscapose, the lvs. obvious.
        5. Pappus distinctly double, the outer series of bristles very short; heads solitary; lvs. numerous, narrow, uniform; stems 5–12 cm. high. White Mts. .... 14. *A. scopulorum*
        5. Pappus simple or occasionally ± double; habit various.

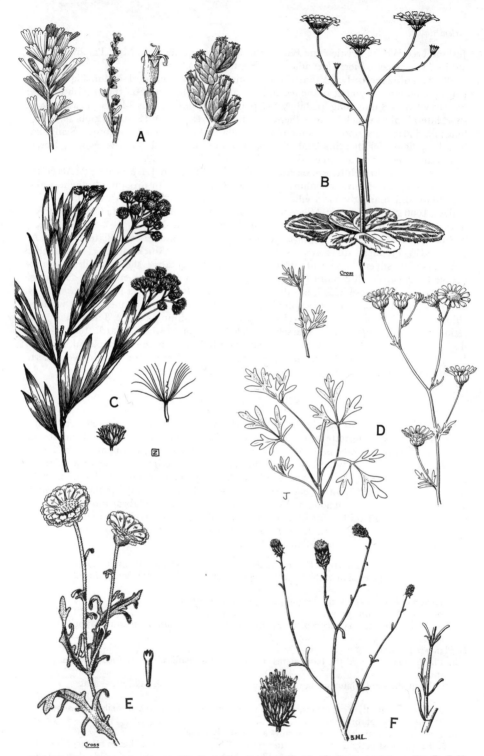

Plate 12. Fig. A, *Artemisia tridentata*. Fig. B, *Atrichoseris platyphylla*. Fig. C, *Baccharis glutinosa*. Fig. D, *Bahia dissecta*. Fig. E, *Baileya pleniradiata*. Fig. F, *Bebbia juncea*.

6. Plants with taproot or erect taprootlike caudex, without rhizomes or numerous fibrous roots; low plants with grasslike lvs. and found in meadows, San Jacinto Mts.
.... 2. *A. alpigenus*
6. Plants with creeping rhizomes or if with woody caudex, this ± horizontal and with numerous fibrous roots.
   7. Phyllaries often with purple tip or margin; disk-corollas with tube equaling or surpassing the limb. Santa Cruz Id. .... 13. *A. radulinus*
   7. Phyllaries sometimes purple-tipped but not purple-margined; disk-corollas with tube distinctly shorter than the limb.
      8. Pubescence of stem in lines below the lf.-bases, commonly neither uniform under the heads nor confined to the infl. .... 9. *A. hesperius*
      8. Pubescence of stem uniform, or if in lines, uniform under the heads or scanty and confined to the infl.
         9. Invol. not strongly graduated, or, if so, the phyllaries markedly acute; phyllaries acute, or if obtuse, enlarged and foliaceous.
            10. Infl. a narrow leafy panicle, usually with numerous heads; rays usually pink or white. White Mts .... 5. *A. eatonii*
            10. Infl. shorter, more open, cymose-paniculate, often with much reduced lvs.
               11. Middle stem-lvs. mostly less than 7 cm. wide and more than 7 times as long as wide; plant rhizomatous .... 11. *A. occidentalis*
               11. Middle stem-lvs. 2 or more cm. wide and less than 7 times as long; lvs. frequently toothed .... 8. *A. greatai*
         9. Invol. strongly graduated, at least the outer phyllaries obtuse, markedly shorter than the inner, not foliaceous.
            12. Infl. conspicuously bracteate with small lvs.; innermost phyllaries usually obtusish; mostly cismontane.
               13. Herbage, including phyllaries, glabrate, or if pubescent, the hairs short or appressed; infl. an open often divergent panicle .... 4. *A. chilensis*
               13. Herbage, including phyllaries, cinereous with short spreading hairs; infl. a close panicle or raceme .... 3. *A. bernardinus*
            12. Infl. not closely beset with small uniform bracts; innermost phyllaries usually acuminate; mostly montane or transmontane .... 1. *A. adscendens*
               14. Lower lvs. oblanceolate; heads in cymes or panicles, Montane Coniferous F .... 1. *A. adscendens*
               14. Lower lvs. linear; heads solitary at ends of branchlets. E. Mojave Desert
.... 12. *A. pauciflorus*

1. **A. adscéndens** Lindl. in Hook. [*A. chilensis* ssp. *a*. Cronq.] Stems slender, 2–8 dm. high, with erect or ascending branches above, uniformly pubescent at summit of peduncles, often in lines and becoming glabrous below; lvs. usually entire, with scabrous margin, otherwise glabrous to ± pubescent, the lower narrowly oblanceolate, the middle cauline lanceolate to linear, amplexicaul, mostly less than 1 cm. wide; heads few to many, in a nearly naked cyme or cymose panicle; invol. 4–7 mm. high, strongly graduate; phyllaries ciliolate, usually glabrous on back, green-tipped, linear to linear-oblong, the innermost subacuminate; rays 20–35, violet or purplish, 6–10 mm. long; $n=8$ (Clausen et al., 1940). —In meadows, etc., 100–7500 ft.; mostly Montane Coniferous F.; San Jacinto to San Gabriel mts.; to Wash., Rocky Mts. July–Oct.

2. **A. alpígenus** (T. & G.) Gray ssp. **andersònii** (Gray) Onno. [*Erigeron andersonii* Gray. *A. a.* Gray.] Subscapose herb from a short caudex surmounting a fleshy taproot, 5–40 cm. high, thinly tomentulose below, more densely so upward; basal lvs. tufted, grasslike, linear or oblance-linear, entire, 4–20 cm. long, 2–10 mm. wide; cauline lvs. few, much reduced, sessile; heads solitary, showy; invol. broadly hemispheric, 2–3-seriate, 6–10 mm. high; phyllaries subequal, or ± graduate, the inner subscarious and often purplish on the margin; rays 7–15 mm. long, purple; aks. pilose to base; $n=18$ (Raven et al., 1960).—Meadows, 7000–9000 ft.; Montane Coniferous F.; San Jacinto Mts., White Mts. and Sierra Nevada to Ore. June–Sept.

3. **A. bernardínus** Hall. [*A. deserticola* Macbr. Possibly *A. defoliatus* Parish.] Stems erect from a woody rhizome, 4–12 dm. high, leafy with fascicles developing in the axils; herbage densely pubescent with soft straight hairs; primary lvs. linear or narrowly lanceolate, the lower ones promptly deciduous, 3–7 cm. long, 3–6 mm. wide; heads numerous,

medium-sized, terminating densely bracteate twigs in racemose panicles; invol. 5.5–6.5 mm. high; phyllaries pubescent on back and ciliate, the outer usually very obtuse or rounded.—Damp meadows, 100–3500 (–7000) ft.; Freshwater Marsh, Coastal Sage Scrub, etc.; cismontane s. Calif., to desert edge as at Victorville. July–Nov.

4. **A. chilénsis** Nees. [*A. menziesii* Lindl. in Hook.] Erect or ascending, 4–10 dm. high, paniculately branched above, uniformly pubescent above, often in lines below; lvs. usually entire, with scabrous margin and sometimes rough above, glabrous beneath, the lower 8–12 cm. long, oblanceolate or broader, on a winged petiole; the mid-cauline linear-lanceolate, sessile, the upper crowded, becoming bractlike; infl. mostly paniculate; heads usually numerous, invol. 5–7 mm. high, 4–5-seriate; phyllaries ciliolate, glabrous on back, green-tipped, oblong; rays 20–35, violet or purple to whitish, 6–12 mm. long.— Dry banks, grassy fields, etc., sea-level to 5000 ft.; many Plant Communities; mts. from San Diego Co. to Santa Barbara Co.; Santa Cruz Id. June–Oct.

5. **A. eatònii** (Gray) Howell. [*A. foliaceus* var. *e.* Gray.] Stem 4–10 dm. high, from a stout creeping rhizome, many-branched, minutely pubescent and with scattered hairs; lvs. lance-linear, numerous, usually entire, scabrous-margined, the lower petioled, early deciduous, the middle and upper sessile, 5–15 cm. long, 4–18 mm. wide; heads many, in a narrow leafy panicle; invol. 5–7 mm. high; phyllaries subequal or loosely graduate, sometimes subtended by leafy bracts, ciliolate; rays 20–35, lavender or violet; 5–10 mm. long.—At ca. 5000–8000 ft.; Sagebrush Scrub; White Mts.; to B.C., New Mex. July–Sept.

6. **A. éxilis** Ell. Slender erect glabrous annual 3–15 dm. high, openly much branched above; lvs. linear to oblanceolate, 5–12 cm. long, mostly entire, 2–10 mm. wide, 1-nerved, the uppermost bractlike; panicle diffuse, rather narrow; heads numerous; invol. 4–6 mm. high, 3–4-seriate; phyllaries lance-linear, attenuate, with translucent midrib and hyaline margin; rays light pink to purple, many, scarcely exceeding the scanty pappus; $n=5$ (Solbrig et al., 1964).—Common in wet, often alkaline places below 500 ft.; many Plant Communities; Imperial Co., cismontane s. Calif., Santa Cruz Id.; to n. Calif. and Atlantic Coast. July–Oct.

7. **A. frondòsus** (Nutt.) T. & G. [*Tripolium f.* Nutt. *Brachyactis f.* Gray.] Erect or decumbent annual, 1–4 dm. high, subglabrous, simple to diffusely branched, leafy; lvs. linear to lance- or oblance-oblong, hispid-ciliate, 1–3 cm. long, the lower wing-petioled; heads crowded in narrow leafy panicles; invol. hemispherical, the bracts linear-oblong to subspatulate, obtuse, loose; rays pinkish, scarcely exceeding the disk; aks. strigose.— Occasional, moist alkaline flats, marshes and lake borders, 2500–7000 ft.; Alkali Sink, Sagebrush Scrub, etc.; Lake Elsinore, Baldwin L., Rabbit Springs, Death V., Victorville; to e. Wash., Wyo., New Mex. May–Oct.

8. **A. greàtai** Parish. Perennial, the stems slender, 5–12 dm. high, leafy, sparsely but rather uniformly coarse-pubescent above; lvs. ample, thin, entire or sometimes serrulate, obovate to oblanceolate, 2–4 cm. wide, 6–12 cm. long, the upper reduced and sessile; infl. paniculate with leafy bracts; invols. 5.5–6.5 mm. high; phyllaries subequal to graduate, linear-lanceolate, ciliolate, with green acuminate tip; rays 25–50, light purple, 8–12 mm. long; $n=8$ (Solbrig et al., 1969).—Moist or dry places in canyons, 2000–4000 ft.; Chaparral, S. Oak Wd.; s. face of San Gabriel Mts., Verdugo Mts. Aug.–Oct.

9. **A. hespérius** Gray. Perennial from a creeping rhizome, 5–12 dm. tall, with numerous flowering branches showing lines of coarse pubescence; lvs. firm, linear-lanceolate, sessile by a broad base, acuminate, glabrous on the surfaces, ciliate, 5–10 cm. long, sharply shallowly serrate, the upper reduced so that the panicle is leafy-bracted; heads many; invols. 5–8 mm. high; phyllaries 4–5-seriate, linear, acuminate, with green tip; rays 21–35, white or pink to bluish-purple, 6–12 mm. long.—Stream banks and meadows below 5500 ft.; Chaparral, S. Oak Wd., Sagebrush Scrub; cismontane s. Calif. n. through Owens V. to Mono Co., Alta., Wis., etc. Aug.–Oct.

10. **A. intricàtus** (Gray) Blake. [*Bigelovia i.* Gray. *A. carnosus* Gray, not Gilib.] Rounded bush 5–9 dm. high, the slender glaucescent almost leafless stems rigidly and divaricately much-branched, woody below, essentially glabrous; lower lvs. linear, fleshy, 1–2 cm. long, those of the branches reduced to scales; heads many, few-fld., small, discoid,

yellowish, solitary at tips of branches; invol. 5.5–6.5 mm. high, turbinate, glabrous; phyllaries 4–5-seriate; aks. slender, striate, ± sericeous, pappus copious, reddish.—Occasional, saline soils, below 3500 ft.; Alkali Sink, Coastal Sage Scrub, Creosote Bush Scrub; San Bernardino and Los Angeles cos. to Mono Co.; to Nev., Ariz. June–Oct.

11. **A. occidentàlis** (Nutt.) T. & G. [*Tripolium o.* Nutt.] Stems from a slender branching caudex, 2–4 or –5 dm. high, often reddish, scantily pubescent above, often in lines, but uniformly toward summit of peduncles; lvs. entire, persistent, linear-oblanceolate, usually less than 10 cm. long, the basal with narrowly winged ciliate petiole, the middle cauline 3–10 mm. wide, sessile; heads 1–several, ± cymose; invol. 5–7 mm. high, scarcely graduate; phyllaries linear, often purple-tipped, ciliolate; rays 25–35, lavender or violet, 7–10 mm. long; $n=16$ (Clausen et al., 1940).—Moist places, 4000–10,000 ft.; Montane Coniferous F.; mts. from n. L. Calif. to n. Calif. July–Sept. Variable; the s. Calif. plants sometimes referred to var. *párishii* (Gray) Ferris. [*A. fremontii* var. *p.* Gray.] They tend to be ca. 3 dm. high; lvs. slightly clasping.

12. **A. pauciflòrus** Nutt. Stems erect from creeping rootstocks, wiry, simple or branched, 3–12 dm. high; lvs. ± fleshy, linear, acuminate, sessile, entire, 5–10 cm. long, 3–6 mm. wide, bractlike in infl.; heads solitary at ends of branchlets; invol. 5–7 mm. high, viscid; phyllaries mostly herbaceous, 2–3-seriate, rather loose; rays 5–8 mm. long, lavender to whitish.—Most saline places, 800–2300 ft.; Alkali Sink; Amargosa Desert region (Tecopa), se. Inyo Co.; to Tex., Sask. June–Oct.

13. **A. radulìnus** Gray. Perennial, 1–5 dm. high, scabrous-pubescent; lvs. obovate to oblong, 6–10 cm. long, sharply and deeply serrate; infl. cymose; invol. turbinate, 6–9 mm. high; phyllaries 5–7-seriate; rays 10–15, white to pale violet, 6–12 mm. long.—Dry woods; Santa Cruz Id.; San Luis Obispo Co. to Vancouver Id. July–Oct.

14. **A. scopulòrum** Gray. Stems numerous from a branching woody caudex, 5–12 cm. high; herbage subtomentose; lvs. crowded, sessile, entire, 5–15 mm. long, 1–3 mm. wide; peduncles 1–4 cm. long; invol. 7–11 mm. high; phyllaries 3–4-seriate; rays 8–15, blue or deep violet, 7 12 mm. long.—Dry rocky places 5000–10,000 ft.; Sagebrush Scrub, Pinyon-Juniper Wd.; White Mts. to Ore., Wyo. May–July.

15. **A. spinòsus** Benth. [*Leucopsis s. Greene.*] MEXICAN DEVIL-WEED. Broomlike clumps of much-branched almost leafless thorny stems 6–25 dm. high; lvs. thickish, narrow, 2–4 cm. long, the upper bracteate; heads small, solitary at ends of branches; invol. 4–6 mm. high; phyllaries 3–5-seriate, imbricate, with scarious margins; rays 15–30, white, short; aks. glabrous; $n=9$ (Raven et al.).—Troublesome weed in low subsaline places below 500 ft.; Alkali Sink, Coastal Sage Scrub, Creosote Bush Scrub; s. San Diego Co., Colo. River V. n. to Needles; to La., Mex., Cent. Am. June–Dec.

18. *Atrichóseris* Gray. TOBACCO-WEED

Low scapose annual with basal rosette of broad thick lvs. and 1 to few scapes cymosely branched above. Heads on slender peduncles. Invol. of ca. 15 equal linear acute phyllaries and several small outer ones. Receptacle scrobiculate. Ligules white. Aks. oblong with corky-thickened ribs. Pappus none. One sp. (Greek, *athrix*, without hair, and *seris*, a cichoriaceous genus.)

1. **A. platyphýlla** Gray. Plate 12, Fig. B. Lvs. flat on the ground, light green, often spotted, somewhat glaucous, round-oblong, 3–10 cm. long, short-petioled, spinulose-denticulate; scapes 3–7 dm. high, minutely bracteate; invol. 6 mm. high, the phyllaries with scarious margins; ligules oblong, 1–2 cm. long; fls. fragrant; aks. white, 4 mm. long; $2n=18$ (Stebbins et al., 1953).—Common in sandy washes; Creosote Bush Scrub; Colo. Desert e. of Mecca and Imperial V., Mojave Desert e. of Ord Mts. and in Death V. region; to Utah, Ariz. Mostly March–May.

19. *Báccharis* L.

Dioecious shrubs or undershrubs, or some perennial herbs; often resinous or glutinous. Lvs. alternate, entire or toothed. Heads numerous, corymbose or paniculate, many-fld., discoid. Invol. imbricated, of chartaceous whitish phyllaries. Staminate heads composed of tubular slightly dilated 5-toothed ♂ fls. with abortive ovary and scabrous, often tortuous and scanty pappus of clavellate bristles. Pistillate heads of tubular-filiform truncate or obscurely toothed ♀ fls. with copious pappus of capillary bristles. Aks. small, somewhat compressed, 5–10-ribbed. Complex and diverse Am. genus of some 300 spp., best developed in e. S. Am. (After the god Bacchus.)

1. Herbage puberulent or pubescent; aks. puberulent.
   2. Lvs. entire, 0.5–1.2 cm. long, mostly lost before anthesis .... 1. *B. brachyphylla*
   2. Lvs. acutely serrate, 2–5 cm. long, persistent .... 6. *B. plummerae*
1. Herbage not hairy.
   3. Stems herbaceous to base; lvs. ovate-lanceolate; phyllaries thin, viscid-ciliate; aks. puberulent .... 2. *B. douglasii*
   3. Stems woody below; plants shrubby; phyllaries not viscid-ciliate; aks. glabrous.
     4. Plants persistently leafy, not broomlike (or sometimes somewhat so in *B. emoryi*).
       5. Lvs. linear or linear-lanceolate, entire or evenly toothed, 5–15 cm. long; aks. 5-nerved
                                                                                       4. *B. glutinosa*
       5. Lvs. cuneate to oblong-lanceolate, few-toothed above the middle or entire; aks. 10-nerved.
         6. Lvs., at least the upper, narrowly oblong; infl. sparsely leafy ...... 3. *B. emoryi*
         6. Lvs. all ovate or broadly cuneate; infl. densely leafy .... 5. *B. pilularis*
     4. Plants with lvs. mostly deciduous before anthesis, broomlike with numerous sulcate fastigiate branches; heads scattered; aks. 10-nerved.
       7. Pistillate pappus very short, ca. 3 mm. long in fr.; larger lvs. obovate
                                                                                                      8. *B. sergiloides*
       7. Pistillate pappus to 10 mm. or more in fr.; larger lvs. linear .... 7. *B. sarathroides*

1. **B. bráchyphylla** Gray. Woody at base, much branched, 6–10 dm. high, sparsely leaved, puberulent, the lvs. linear, acute, entire, sessile, scabrous, the lower 1–1.5 cm. long, others mostly scalelike on the branches; heads loosely paniculate; invol. of ♀ heads 5–6 mm. high, 3–4-seriate, the phyllaries with green hispidulous midrib and whitish scarious margin; ripe aks. ca. 2.5 mm. long, puberulent, the brownish pappus ca. 7 mm. long.—Local in dry rocky washes below 3500 ft.; Creosote Bush Scrub; Morongo Wash, Eagle Mts., Little San Bernardino Mts., Crucero; to San Diego Co.; e. to Ariz., Son. Aug.–Nov.

2. **B. doúglasii** DC. [*B. haenkei* DC.] Stems several from a suffrutescent base, 1–2 m. high, simple or with ascending branches above, green, glutinous; lvs. narrowly to broadly lanceolate, short-petioled, entire or serrulate, glandular-punctate, 3–10 cm. long, 6–25 mm. wide, 3-nerved; heads many, in terminal often compound corymbose clusters; invol. of ♂ and ♀ heads ca. 5 mm. high, 4–5-seriate, the phyllaries lanceolate, acuminate, thin, with narrow greenish center, viscid-ciliate toward apex; ripe aks. 0.8 mm. long, viscid-puberulent, 4–5-nerved; $n=9$ (Solbrig et al., 1964).—Moist places near streams, below 2500 ft.; Coastal Salt Marsh, Coastal Sage Scrub, etc.; San Gabriel Mts. and e. San Diego Co. to the coast and n. to Ore. Santa Catalina and Santa Cruz ids. July–Oct.

3. **B. émoryi** Gray. Erect, loosely branched shrub mostly 1–3 m. high with striate, ± glutinous branchlets; lvs. cuneate, or oblong, obtuse or acute, tapering to base, commonly few-toothed in distal half, 2–5 cm. long, 0.5–2 cm. wide, 3-nerved, the upper linear, 1-nerved; heads many, in large pyramidal panicles; invol. of ♀ heads 7–9 mm. high, ca. 6-seriate, the outer phyllaries ovate, the inner lance-linear, acuminate; invol. of ♂ heads broader, ca. 6 mm. high, 5-seriate; ripe aks. 1.5–2 mm. long, glabrous, 1-nerved, the pappus 1 cm. long.—Mostly along streams below 2000 (–4000) ft.; Coastal Sage Scrub, Creosote Bush Scrub, etc.; n. L. Calif. to Los Angeles, Ventura, Bakersfield, Death V. region; to Utah, w. Tex. Santa Catalina Id. Aug.–Dec.

4. **B. glutinòsa** Pers. [*B. coerulescens* DC. *B. viminea* DC.] MULE FAT. Plate 12, Fig. C. Willowlike shrubs, 1–4 m. high, the virgate stems simple to infl. or branched, lvs. ± glutinous, entire or denticulate, 5–15 cm. long, 7–18 mm. wide, lance-linear, acuminate, short-petioled, the midrib stronger than the 2 laterals; heads many, in terminal compound

corymbs or close cymose clusters on ends of short lateral branches; invol. ca. 4 mm. high, 3–5-seriate, the phyllaries ovate to lance-oblong, with obscure green midrib, ± ciliolate; $n = 9$ (Turner et al., 1961; Solbrig et al., 1964).—Mostly below 3500 ft., along watercourses; Coastal Sage Scrub, Chaparral, etc.; n. L. Calif. to n. cent. Calif., coast to Colo. Desert and Kelso, Darwin, etc.; to Colo., Tex., Mex. S. Am. Santa Catalina Id., Santa Barbara, Santa Cruz and Santa Rosa ids. Most of the year. Generally *B. viminea* is recognized as a separate sp. by having laterally placed cymes, but there is some evidence (presented by D. H. Wilken) that the terminal corymbose infl. of *B. glutinosa* comes first and the side branches may develop later.

5. **B. pilulàris** DC. ssp. **consangúinea** (DC.) C. B. Wolf. [*B. c.* DC.] COYOTE BRUSH. Much-branched erect or rounded shrub 1–4 m. high; lvs. very numerous, oval or obovate, 1.5–4 cm. long, 5–15 mm. wide, usually with 5–9 teeth, resinous, 1-nerved; heads numerous, in small axillary and terminal glomerules on the leafy branchlets; invol. 3–5 mm. high, ca. 5-seriate; phyllaries ovate (outer) to lance-oblong (inner), obtuse, stramineous, scurfy-glandular, indurate, with narrow scarious fimbrillate margin; ripe aks. 1.3–1.5 mm. long, glabrous, 10-nerved, the pappus 6–10 mm. long; $n = 9$ (Raven et al., 1960).— Hillsides and canyons below 2500 ft.; Coastal Strand, Coastal Sage Scrub; along the coast inland to Fallbrook, etc.; San Diego Co. to Ore. Santa Catalina, San Clemente, Santa Barbara, Santa Cruz, Santa Rosa ids. Aug.–Dec.

6. **B. plúmmerae** Gray. Rounded bush 6–10 dm. high, herbaceous above the woody base, loosely much-branched, viscid-pubescent especially above; lvs. linear-oblong, obtuse, numerous, sessile, serrate, 2–5 cm. long, early deciduous; heads in short panicles; staminate invols. 4–5 mm. long, the ♀ 6–8 mm. high, ca. 5-seriate, the linear-lanceolate phyllaries green except for the ciliate scarious margin; aks. 2.7–3.2 mm. long, viscid-puberulent.—Bushy canyons and slopes near the sea, below 1000 ft.; Coastal Sage Scrub; coast of Santa Barbara, Ventura and Los Angeles cos.; Santa Cruz Id. Aug.–Oct.

7. **B. sarathroìdes** Gray. BROOM BACCHARIS. Erect, glabrous, glutinous green shrub 2–4 m. high, nearly leafless, with broomlike, angular-sulcate branches; lvs. all sublinear, entire, rigid, up to 2 cm. long; heads mostly solitary at the tips of the numerous branchlets; invol. of ♀ heads 6–8 mm. high, ca. 6-seriate, cream-color, the outer phyllaries broadly ovate, the inner linear-oblong, obtuse, indurate; ♂ heads with invol. 3–4 mm. high, the broad blunt phyllaries with small green apical spot; aks. 1.7–2.2 mm. long, glabrous, 10-nerved, the pappus 6–11 mm. long; $2n = 18$ (DeJong & Montgomery, 1963). —Sandy washes below 1200 ft.; Coastal Sage Scrub, Creosote Bush Scrub; San Diego and Riverside regions, Colo. Desert; to L. Calif., Ariz., New Mex., Sinaloa. June–Oct.

8. **B. sergiloìdes** Gray. SQUAW WATERWEED. Green glabrous, rounded, often nearly leafless shrub 7–20 dm. high, the broomlike branches striate-angled; larger lvs. spatulate or obovate, entire or rarely few-toothed, 1–2.5 cm. long, 3–10 mm. wide, obtuse, obscurely punctate; heads many, in dense panicles; invol. 3 mm. high, 4–5-seriate, stramineous; phyllaries oval to lance-oblong, obtuse, firm, scurfy-glandular, with narrow whitish margin; ripe aks. 1.5 mm. long, glabrous, 10-nerved, the pappus 2.5–3 mm. long; $2n = 18$ (DeJong & Montgomery, 1963).—Washes and canyon-bottoms, below 4500 ft.; Creosote Bush Scrub to Pinyon-Juniper Wd.; both deserts from Death V. region s., interior cismontane areas in Chaparral, S. Oak Wd.; to L. Calif., Utah, Ariz., Son. April–Sept.

20. *Bahía* Lag.

Annuals or perennials, rarely suffrutescent, pubescent. Lvs. alternate or opposite, entire to dissected. Heads radiate or rarely discoid, yellow, terminating the branches. Invol. obconic to hemispheric; phyllaries not punctate, 1–2-seriate, subequal, broader above the middle. Receptacle flat. Ray-fls. when present, ♀, fertile. Disc-fls. perfect, fertile. Aks. narrow, tapering to base, 4-angled. Pappus paleaceous, usually with callous-thickened base or midrib. Ca. 15 spp. of sw. U.S., Mex., and w. S. Am. (Honoring J. F. *Bahi*, Barcelona botany professor.)

Biennial; ray-fls. present; pappus wanting .... 1. *B. dissecta*
Annual; ray-fls. wanting; pappus present .... 2. *B. neomexicana*

1. **B. disśecta** (Gray) Britton. [*Amauria d.* Gray. *Villanova chrysanthemoides* Gray. *B. c.* Gray.] Plate 12, Fig. D. Biennial or short-lived perennial 3–8 dm. high; stems minutely hirtellous and glandular, often anthocyanous, openly branching above; lvs. once to thrice ternately divided into linear or oblong rounded lobes, 2–7 cm. long, strigillose; peduncles 1–6 cm. long; densely glandular-pubescent; invol. hemispheric, 5–6 mm. high, the phyllaries ± glandular-hairy, narrowly obovate, abruptly narrowed to a short caudate tip; rays mostly 10–13, ca. 6 mm. long; aks. black, striate, hirtellous, epappose; $n = 36$ (Turner & Flyr, 1966).—Dry open slopes and ridges, 6000–8600 ft.; Yellow Pine F.; San Bernardino Mts., Santa Rosa Mts.; to Wyo., Tex., n. Mex. Aug.–Sept.

2. **B. neomexicàna** (Gray) Gray. [*Schkuhria n.* Gray ex Rydb. *Amblyopappus n.* Gray.] Slender, often many-branched annual 1–2 dm. high, hirtellous; lvs. opposite below, alternate within the corymbose infl., linear-filiform, usually 3-divided, 2–3 cm. long, impressed-punctate; heads broadly obconic, discoid; invol. short-pilose and glandular-puberulent, ca. 6 mm. high, the phyllaries oblanceolate, obtuse, the thin outer margin purplish; fls. pale yellow or whitish, small; aks. quadrangular, slender, tapering to the sericeous base, strigillose above, 2–5 mm. long; pappus of 8 obovate paleae 1–1.5 mm. long.—Sandy slopes and washes, ca. 5000 ft.; Pinyon-Juniper Wd.; Clark Mts., e. Mojave Desert; to Colo., Tex., n. Mex.; w. S. Am. Sept.

21. *Bàileya* Harv. & Gray. DESERT-MARIGOLD

Densely white-woolly annual to perennial herbs, usually rather freely branched. Lower lvs. pinnatifid, the upper entire, all alternate. Heads solitary, radiate, peduncled; invol. of numerous distinct woolly phyllaries. Receptacle naked. Ray-fls. ♀, numerous; ligules large, persistent, accrescent, becoming deflexed and papery in age. Disk-fls. many, fertile. Style-branches short, obtuse. Aks. clavate or oblong, truncate and obscurely toothed apically, many-striate, glabrous. Pappus none. Ca. 3 or 4 spp. of sw. U.S. and adjacent Mex. (Honoring J. W. *Bailey*, early Am. microscopist.)

Ray-fls. 5–7; heads relatively small, loosely cymose; invol. ca. 6 mm. broad ... 2. *B. pauciradiata*
Ray-fls. 20–50; heads larger, mostly solitary on elongate peduncles; invol. 10 or more mm. broad.
  Stems leafy to above the middle or nearly to apex, the medium-sized heads on peduncles 10 cm. or less long .... 3. *B. pleniradiata*
  Stems leafy only below the middle, the large heads on peduncles 10–20 cm. long.
.... 1. *B. multiradiata*

1. **B. multiràdiata** Harv. & Gray. [*B. m.* var. *nudicaulis* Gray.] Biennial or perennial, much like no. 3, but the basal lvs. persisting usually as a rosette and in the lower half of the stem; peduncles bractless, 1–3 dm. long, often stout; invol. 7–8 mm. high, 11–16 mm. wide; ligules 25–30, 11–15 mm. long; $n = 16$ (Carlquist, 1956).—Sandy plains and rocky slopes, 2000–5200 ft.; Creosote Bush Scrub, Joshua Tree Wd.; e. San Bernardino Co.; to Utah, Tex., n. Mex. April–July; Oct.

2. **B. pauciradiàta** Harv. & Gray. Densely floccose much-branched annual from a taproot, sometimes persisting, 2–6 dm. high, leafy throughout; lvs. linear to linear-lanceolate and nearly all entire, or the lower irregularly pinnatifid or bipinnatifid with 2–5 pairs of short linear lobes, 3–10 cm. long; peduncles 2–5 cm. long; invol. 5–6 mm. high; ligules 5–7, lemon yellow, 5–8 mm. long; aks. pale, muriculate; $n = 16$ (Raven & Kyhos, 1961).—In sand, below 2200 (3500) ft.; Creosote Bush Scrub; common in Colo. Desert, less so in e. Mojave Desert; Ariz., adjacent Mex. Feb.–June; Oct.

3. **B. pleniradiàta** Harv. & Gray. [*B. multiradiata p.* Cov. *B. nervosa* Jones.] Plate 12, Fig. E. Annual or sometimes perennial, white-floccose with ± appressed wool, with many branching stems, 2–5 dm. high, leafy to above the middle; lvs. spatulate to linear-oblong, the lower ones 4–7 cm. long and soon withering, pinnately few-lobed, petioled, the upper smaller, sessile, mostly entire; peduncles mostly less than 10 cm. long, slender; invol.

6–8 mm. high, 8–14 mm. wide; ligules 20–40, golden to pale yellow, 8–10 mm. long; aks. pale, sparsely glandular-atomiferous; $n=16$ (Raven & Kyhos, 1961).—Sandy places, below 5000 ft.; Creosote Bush Scrub, Joshua Tree Wd.; common on Mojave and ne. Colo. deserts, Inyo Co. to Riverside co.; to Utah, w. Tex., n. Mex. March–June; Oct.

22. *Balsamorhìza* Hook. ex Nutt. BALSAM ROOT

Low perennials with thick fusiform rough-barked taproot and a usually multicipital caudex. Lvs. large, in a basal tuft. Stems naked or few-lvd., bearing 1–few heads of yellow fls. Invol. broad, in ours with outer phyllaries foliaceous. Ray-fls. ♀, fertile, showy. Disk-fls. fertile. Aks. without pappus, those of the disc 4-sided or 3-sided. Ca. 12 spp. of w. U.S. (Greek, *balsamos*, balsam, and *rhiza*, root.)

(Sharp, Ward M. A critical study of certain epappose genera of the Heliantheae–Verbesininae of the natural family Compositae. Ann. Mo. Bot. Gard. 22: 51–153. 1935.)

1. **B. deltoìdea** Nutt. Green scabrous; stems 2–8 dm. high, scapiform, but usually with several greatly reduced lvs.; blades of basal lvs. triangular-hastate to cordate-ovate, 1–3 dm. long, 0.5–2 dm. wide, moderately hispidulous and reticulate-coriaceous in age, usually entire or crenate; heads solitary or few, the summit of peduncle and base of invol. often densely hirsute and glandular; ligules 1.5–3 cm. long; aks. glabrous.—Dry banks usually 2000–7000 ft.; Chaparral, Foothill Wd., Yellow Pine F.; n. Los Angeles Co. (Bouquet C., Elizabeth Lake region, etc.) to Ventura and Kern cos. and to B.C. April–June.

23. *Bébbia* Greene

Low, rounded, many-stemmed odorous half-shrub with green nearly leafless branches. Heads solitary or loosely cymose, discoid, yellow. Invol. hemispheric, ca. 3-seriate, the phyllaries imbricate, striate, shorter than the disc. Receptacular bracts partly enfolding the aks. Corolla linear, the tube glandular, the limb hairy. Aks. slender, strigose. Pappus of 15–20 plumose bristles much longer than the ak. Two spp. of the N. Am. deserts. (Honoring M. S. *Bebb*, Illinois student of w. willows.)

1. **B. jùncea** (Benth) Greene. [*Carphephorus j.* Benth. *B. j.* var. *aspera* Greene.] SWEET BUSH. Plate 12, Fig. F. Diffuse, 5–10 dm. high, glabrate or minutely hispid, or towards the heads somewhat canescent, the often leafless slender branches junciform; lvs. remote, early deciduous, mostly linear, 1–5 cm. long, often bearing 1–2 salient teeth; invol. 4–8 mm. high; phyllaries lance-ovate to lanceolate, the outer usually canescent, the inner often somewhat anthocyanous.—Gravelly fans, rocky washes, etc., below 4000 ft.; Creosote Bush Scrub; both deserts, but only in the e. part of the Mojave, Inyo Co. to L. Calif.; cismontane s. Calif. from San Bernardino s., largely in Coastal Sage Scrub; to Nev., New Mex., Son. April–July.

24. *Béllis* L. DAISY

Low annual or perennial herbs, with rosulate lvs. and medium-sized solitary heads on scapelike peduncles. Invol. hemispheric; phyllaries herbaceous, blunt, biseriate, equal. Receptacle conic, naked. Ray-fls. ♀, white to pink or purple. Disc-fls. perfect, greenish-yellow; style-branches flattened, with short ovate puberulent appendage scarcely longer than broad. Aks. compressed, mostly 2-nerved; pappus none. Ca. 6 spp., native to Eu. (Latin, *bellus*, pretty.)

1. **B. perénnis** L. ENGLISH DAISY. Moderately hirsute, except for the silky-strigose upper part of the scapes; lvs. broadly oblanceolate to obovate or orbicular, the blade thin, dentate or denticulate, ca. as long as the petiole, the two combined 2–7 cm. long, 7–20 mm. wide; scape 5–20 cm. high; invol. 4–7 mm. high; rays 30–80, 1 cm. or less long; $2n=18$ (Negodi, 1935).—Natur. in lawns or perhaps sometimes introd. in cheap seed or intentionally; more common northward; introd. from Eu. April–Sept.

25. Bìdens L. Bur-Marigold

Annual or perennial herbs with opposite simple to ternately or pinnately compound lvs. Heads radiate or discoid, many-fld., usually yellow, solitary or paniculate. Invol. mostly 2-seriate, the inner phyllaries membranous, often striate, broader than the herbaceous outer ones. Ray-fls. mostly neutral; disc-fls. perfect. Aks. obcompressed or 3–4-angled, usually with a pappus of 2–4 rigid retrorsely barbed persistent awns. Ca. 200 spp., of all warm regions, chiefly Am. (Latin, *bidens*, 2-toothed, referring to the pappus.)

(Sherff, Earl E. The genus Bidens. Field Mus. Nat. Hist., Bot. Ser. 16: 1–709. 1937.)
Aks. flat, cuneate to obovate.
    Lvs. compound; rays inconspicuous .... 1. *B. frondosa*
    Lvs. simple; rays 1.5–3 cm. long .... 2. *B. laevis*
Aks. narrow, linear-tetragonal; rays inconspicuous or lacking .... 3. *B. pilosa*

1. B. frondòsa L. Stick-Tight. Annual, erect, branching, 2–12 dm. high, nearly glabrous; lvs. slenderly petiolate, sparsely hispidulous, with 3–5 lanceolate acuminate serrulate lfts. up to 8 cm. long and 25 mm. wide, the terminal one slightly petiolulate; heads discoid or nearly so, ca. 1 cm. in diam. at anthesis; outer phyllaries commonly 8, ± foliaceous, ciliate toward the base; disk orange; aks. cuneate, sparsely hispidulous, papillate-rugose in age, 2-awned; $2n=48$ (Löve, 1964).—Occasional, damp ground in waste places below 5200 ft.; natur. in Calif.; Riverside, Victorville, Lake Arrowhead, Los Angeles, Santa Monica Mts., etc.; to Wash., Atlantic Coast. Aug.–Oct.

2. B. laèvis (L.) BSP. [*Helianthus l.* L. *B. speciosa* Parish, not Gardn.] Bur-Marigold. Glabrous perennial 5–20 dm. high; stem stout, smooth; lvs. narrowly to broadly lanceolate, sessile, ± connate at base, 7–16 cm. long, serrate; heads rather few, large, radiate; invol. with spreading oblong ciliolate outer phyllaries and broadly ovate thin brownish yellow inner ones; rays deep yellow, 1.5–3 cm. long; aks. 2–4-awned; $n=11$ (Torres, 1958).—Wet lowlands, river-bottoms, etc.; below 2600 ft.; Freshwater Marsh, etc.; cismontane s. Calif.; to Cent. Calif., Atlantic Coast. Aug.–Nov.

3. B. pilòsa L. [*B. californica* DC.] Beggar-Ticks. Rather weak-stemmed, branching annual, 3–15 dm. high, the stem ± pilose; lvs. pinnate with 3–5 lfts., these ovate, serrate, 1–3 cm. long; heads discoid or inconspicuously radiate, the rays 2–3 mm. long, yellowish-white; invol. 5–7 mm. high, the inner phyllaries hyaline-margined; aks. linear, tetragonal, tuberculate-strigose, with 2–4 awns; $2n=72$ (Harvey, 1966), $n=12, 14$ (Powell & Turner, 1963).—Natur. weed in lowlands of s. Calif.; native Am. tropics. May–Nov.

26. *Blennospérma* Less.

Low annual herbs with alternate usually pinnately parted lvs. Heads radiate, many-fld., terminating the branches. Invol. hemispheric or depressed, the phyllaries uniseriate, thin, with membranous margin, united at base. Receptacle flat, naked. Ray-fls. fertile. Disc-fls. perfect but functionally ♂; tube slender; throat broadly campanulate. Aks. obovate, densely covered with minute papillae which become mucilaginous when wet; epappose. Three spp., of Calif. and Chile. (Greek, *blenna*, mucus, and *sperma*, seed.)

1. B. nànum (Hook.) Blake. [*Chrysanthemum? n.* Hook. *Coniothele californica* DC. *B. c.* T. & G.] Stems 8–20 cm. high, glabrous; lvs. 1–3 cm. long, with linear remote lobes; invol. depressed-hemispheric, 5–6 mm. high, the 5–12 phyllaries elliptic, purple-tipped and with an apical tuft of hairs; rays of same number, oblong, entire, yellow with purplish-brown backs, 4–6 mm. long; aks. 2.5–3 mm. long; $n=7$ (Heiser, 1947).—Local at scattered mud-flats (Murietta region, Julian, Cuyamaca, Wilmington, Gardena); to n. Calif. Feb.–April.

27. *Brickèllia* Ell.

Perennial to annual or subshrubs. Lvs. alternate or opposite, simple, veiny, mostly resinous-dotted. Fls. white to creamy or pink-purple, all perfect and tubular, 3–many in discoid mostly narrow heads. Invol. cylindric to campanulate, the phyllaries striate, chartaceous to membranous, imbricated. Receptacle naked, mostly flat. Anthers minutely round-sagittate at base; style-branches with short stigmatic lines and an elongate papillate appendage. Aks. 10-nerved; pappus of 10–80 barbellate or nearly smooth to rarely subplumose bristles. Almost 100 spp., of N. and S. Am., chiefly in warm regions. (Dr. J. Brickell, early physician and botanist of Georgia.)

(Robinson, B. L. A monograph of the genus Brickellia. Mem. Gray Herb. 1: 1–151. 1917.)
1. The heads clustered, usually racemose-paniculate, not in corymbs.
  2. Heads 3–7-fld.
    3. Phyllaries 10–12; lvs. linear .... 7. *B. longifolia*
    3. Phyllaries ca. 20; lvs. lanceolate to ovate.
      4. Lvs. entire; phyllaries subglabrous .... 9. *B. multiflora*
      4. Lvs. dentate-serrate; phyllaries puberulent .... 6. *B. knappiana*
  2. Heads 8–26-fld.
    5. Phyllaries with recurved or spreading tips; lvs. sessile or nearly so.
      6. Intermediate phyllaries entire; lvs. white-lanate .... 10. *B. nevinii*
      6. Intermediate phyllaries mostly 3-toothed; lvs. greenish.
        7. Stems glandular-villous; aks. 4–4.5 mm. long .... 8. *B. microphylla*
        7. Stems finely woolly; aks. 3.5 mm. long .... 12. *B. watsonii*
    5. Phyllaries erect; lvs. petioled, the petioles at least ⅓ the length of the blades.
      8. Heads mostly 9–10 mm. high; invol. puberulous .... 3. *B. desertorum*
      8. Heads mostly 12–14 mm. high; invol. essentially glabrous .... 2. *B. californica*
1. The heads solitary or at the ends of corymbosely arranged branchlets.
  9. Plants green; heads not over 17 mm. high.
    10. Outer phyllaries broadly ovate, leaflike; lvs. ovate, usually sharply toothed   1. *B. arguta*
    10. Outer phyllaries linear to narrowly oblong, not leaflike.
      11. Lvs. spatulate; those of the divaricate branches not over 8 mm. long . 4. *B. frutescens*
      11. Lvs. linear-oblong, those of the ascending branchlets 12–20 mm. long
                                                                                   11. *B. oblongifolia*
  9. Plants white-tomentose; heads ca. 20 mm. high .... 5. *B. incana*

1. **B. argùta** Rob. [*B. atractyloides* var. *a.* Jeps.] Plate 13, Fig. A. Much-branched shrub 2–4 dm. high, glandular-pilose; stems zigzag; lvs. bright green, rigid-coriaceous, alternate, ovate, saliently toothed to entire, acute or acuminate, granular-scabrous, 1–2 cm. long, subsessile; heads solitary, slender-pedunculate, 13–15 mm. high, the invol. campanulate; outer phyllaries lance-ovate, entire or nearly so, the inner linear-attenuate; aks. 4 mm. long.—Frequent in rocky places below 7500 ft.; Creosote Bush Scrub, Joshua Tree Wd.; Darwin and Last Chance Range (Inyo Co.) to n. L. Calif. April–May. Var. *odontolepis* Rob. Outer phyllaries conspicuously dentate.—W. Colo. Desert.

2. **B. califórnica** (T. & G.) Gray [*Bulbostylis c.* T. & G. *Coleosanthus c.* Kuntze.] Woody at base, 5–10 dm. high, much branched, puberulent to thinly tomentose; lvs. alternate, deltoid-ovate, crenate-serrate, 1–4 cm. long, subtruncate or subcordate at base, short-petioled, the uppermost reduced; panicle leafy, the heads in small terminal clusters on lateral branchlets; heads 8–18-fld., 12–14 mm. high, cylindrical; phyllaries 3–5-striate, green or purplish, obtusish, subglabrous; aks. 3 mm. high; $n=9$ (Gaiser, 1953).—Common in washes and on dry slopes, below 8000 ft.; Coastal Sage Scrub, Chaparral, cismontane areas and occasional on desert as in Inyo Mts. (Creosote Bush Scrub, Pinyon-Juniper Wd.); to n. Calif., Colo., Tex., n. Mex. Aug.–Oct. Santa Catalina Id., Anacapa and Santa Cruz ids.

3. **B. desertòrum** Cov. [*B. californica* var. *d.* Parish. *Coleosanthus d.* Cov.] Shrubby, intricately branched, 1–1.5 m. tall, puberulent; lvs. opposite or alternate, ovate, crenate-serrate, obtuse, 3–12 mm. long, short-petioled, tomentulose; heads in glomerules at ends of short lateral branchlets, 8–9 mm. high, 8–12 fld.; phyllaries 3–4-striate, greenish or yellowish, tomentulose, acutish; aks. 2.3 mm. long, pubescent; $n=9$ (Gaiser, 1953).—Occasional in dry rocky places, 800–7600 ft.; Creosote Bush Scrub, Joshua Tree Wd., Coastal Sage Scrub; w. edge of Colo. Desert, cismontane Riverside and San Bernardino cos. to Eagle and Panamint mts.; Nev., Ariz. Aug.–Nov.

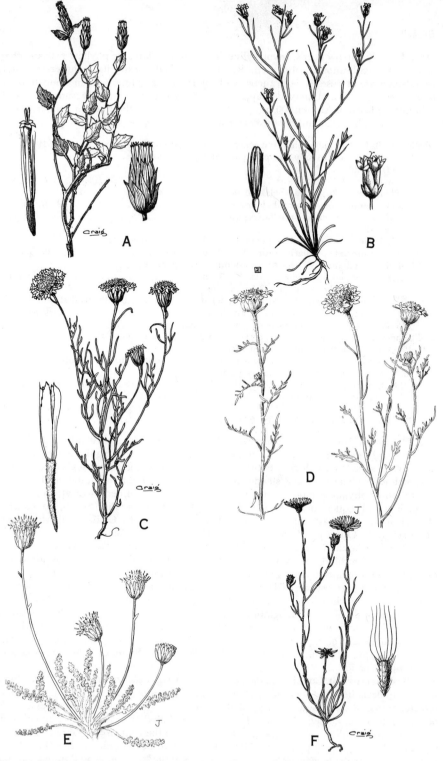

Plate 13. Fig. A, *Brickellia arguta*. Fig. B, *Calycadenia tenella*. Fig. C, *Chaenactis fremontii*. Fig. D, *Chaenactis glabriuscula*. Fig. E, *Chaenactis santolinoides*. Fig. F, *Chaetopappa aurea*.

4. **B. frutéscens** Gray. [*Coleosanthus f.* Kuntze.] Intricately branched aromatic rigid shrub 3–6 dm. high, with divaricate ± spiny cinereous-pubescent slender branches; lvs. alternate, spatulate-oblong, obtuse, entire, pale green, 3–12 mm. long, 1–3 mm. wide, short-petioled; heads solitary, terminal, 13–14 mm. high; phyllaries oblong-linear, 4-striate, obtusish; aks. hispidulous-scabrous, ca. 3.3 mm. long.—Occasional, dry rocky slopes, 2000–3500 ft.; Creosote Bush Scrub; w. edge Colo. Desert; L. Calif. April–June.

5. **B. incàna** Gray. [*Coleosanthus i.* Kuntze.] Rounded white-tomentose bush 4–10 dm. high, woody at base; lvs. alternate, sessile, ovate, serrulate to entire, obtuse to acute, 1–3 cm. long; heads solitary, peduncled, ca. 22 mm. high; invol. campanulate; phyllaries imbricated, ovate-oblong, tomentose, the inner narrower; aks. 1 cm. long, cinereous; $n=9$ (Gaiser, 1953).—Sandy washes and flats below 5000 ft.; Creosote Bush Scrub, Shadscale Scrub, Joshua Tree Wd.; n. Colo. Desert (Coachella V., etc.), Mojave Desert n. to Tecopa; w. Nev., Ariz. April–Oct.

6. **B. knappiàna** E. Drew. [*Coleosanthus k.* Greene.] Slender, willowlike shrub 1–2 m. tall, somewhat viscid and hispidulous; lvs. alternate, lanceolate or narrow-ovate, 2.5–3.5 cm. long, 1–1.5 cm. wide; petioles 4–5 mm. long; panicle leafy with a few heads at the tip of each branch; heads ca. 7 mm. high, 5–7-fld., the phyllaries obtuse, 4-nerved; aks. 2.5 mm. long, minutely hispidulous.—Reported from ca. 2500–3500 ft.; Joshua Tree Wd.; Mohave R. and Panamint Mts.

7. **B. longifòlia** Wats. [*Coleosanthus l.* Kuntze.] Glabrous shrub, branched, 1–1.5 m. high; lvs. alternate, linear, entire, attenuate, 3–10 cm. long, 2–10 mm. wide, almost sessile; heads 3–5-fld., in small cymes which are racemosely arranged; outer phyllaries ovate, obtuse, the inner lance-oblong, 3-nerved; aks. 1.8 mm. long.—Rare, dry stream-terraces, etc., 3000–5000 ft.; Creosote Bush Scrub, Joshua Tree Wd.; Inyo Co. (Argus Mts., Panamint Mts. etc.) to Utah, Ariz.

8. **B. microphýlla** (Nutt.) Gray. [*Bulbostylis m.* Nutt. *Coleosanthus m.* Kuntze.] Shrubby, branched at base, glandular-villous, 3–6 dm. high, the stems paniculately much branched above, the short branchlets with 1–3 heads; lvs. green, round-ovate, 7–20 mm. long, subentire to denticulate, short-petioled; heads ca. 22-fld., 10–11 mm. high; phyllaries green-tipped; aks. 4–4.5 mm. long, hispidulous; $n=9$ (Gaiser, 1953).—Occasional, dry rocky places, 3000–8000 ft.; Creosote Bush Scrub, Joshua Tree Wd., Pinyon-Juniper Wd.; San Gabriel Mts., w. Mojave Desert, Inyo Mts.; to Ore. Aug.–Nov.

9. **B. multiflòra** Kell. [*Coleosanthus m.* Kuntze.] Erect branched shrub 1–2 m. tall; lvs. glabrous, alternate, lanceolate, acute, subentire, 3-nerved, gummy, 3–8 cm. long, 1–2.5 cm. wide; heads 3–5-fld., ca. 7 mm. high; phyllaries 3-nerved; aks. 1.7 mm. long.—Common, dry washes and stony places, 2000–7000 ft.; Creosote Bush Scrub to Pinyon-Juniper Wd.; mts. of Inyo Co., Clark Mt. (e. San Bernardino Co.); w. Nev. Sept.–Nov.

10. **B. nevínii** Gray. [*Coleosanthus n.* Heller.] Dense white-tomentose shrub, the stems several, erect, 3–5 dm. tall, loosely branched; lvs. alternate, ovate, sometimes subcordate, sessile, 0.6–1.5 cm. long; panicle open, with 1–few heads at ends of short branchlets; heads ca. 23-fld., 15 mm. tall; phyllaries woolly, 3–4-nerved, the outer with spreading or recurved tips; aks. ca. 4 mm. long, hispidulous; $n=9$ (Gaiser, 1953).—Dry slopes and washes, 800–5500 ft.; Coastal Sage Scrub, Chaparral; from Santa Monica Mts. and s. face of San Gabriel Mts. to Santa Barbara Co. and sw. Kern Co. Sept.–Nov.

11. **B. oblongifòlia** Nutt. var. **linifòlia** (D. C. Eat.) Rob. [*B. l.* D. C. Eat. *B. mohavensis* Gray. *Coleosanthus l.* Kuntze.] Stems many from a branching woody base and forming rounded clumps 2–4 dm. high; plant cinereous-pubescent and somewhat glandular; lvs. ovate-oblong to linear, acute, entire or 1–2-toothed, 10–22 mm. long, sessile or nearly so; heads solitary, 12–16 mm. high, cylindrical, pedunculate; phyllaries linear, 2–4-nerved, imbricate, acute, the outer short and lance-oblong; aks. pilose, eglandular.—Occasional in dry stony places, below 8800 ft.; Joshua Tree Wd., Pinyon-Juniper Wd.; w. Colo. Desert, Mojave Desert n. to Mono Co.; to Colo., New Mex. May–July.

12. **B. watsònii** Rob. Shrubby at base, intricately branched, 2–3 dm. high, aromatic, tomentulose-puberulent; lvs. ovate, 3–10 mm. long, sparingly dentate, light green; heads 1–3 on ends of branchlets of rather a small panicle, 15–18-fld., 9–11 mm. high; phyllaries

green-tipped; aks. 3.5 mm. long.—Dry rocky places, 3800–7000 ft.; largely Pinyon-Juniper Wd.; Clark, Funeral, Panamint, and Inyo mts.; to Utah. Sept.–Oct.

28. Caléndula L.

Annual or perennial herbs with simple alternate lvs. Heads mostly large. Invol. broad; phyllaries usually scarious-margined, incurved, in 1–2 rows. Receptacle naked, plane. Ray-fls. yellow or orange; ray-aks. glabrous, incurved. Disk-fls. infertile. Pappus none. Ca. 15 spp., Medit. region to Iran, Canary Ids. (Latin, *calendae*, throughout the months.)

Lvs. linear-lanceolate; heads nodding in fr.; 1–2 cm. broad in fl. .... 1. *C. arvensis*
Lower lvs. spatulate, the upper lanceolate to linear; fruiting heads erect; flowering heads 2–5 cm. broad .... 2. *C. officinalis*

1. C. arvénsis L. FIELD MARIGOLD. Annual; stems slender, 1–3 dm. long, leafy throughout, finely glandular-pubescent; lower lvs. short-petioled, the middle and upper sessile, lance-oblong to lanceolate, 2–7 cm. long; heads peduncled, solitary at ends of branches; invol. broadly campanulate, ca. 8–10 mm. high, the phyllaries uniseriate, narrow; ligules 7–12 mm. long, 3-toothed; aks. strongly muricate on the incurved back; $2n=36$ (Negodi, 1936).—Reported occasionally as natur., as at Santa Barbara; native of Eurasia. March–April.

2. C. officinàlis L. POT-MARIGOLD. Annual, 3–6 dm. high, much-branched; lvs. oblong to oblong-obovate, 5–15 cm. long, entire or remotely denticulate, ± clasping. Heads solitary on stout peduncles; phyllaries lanceolate to ovate, 12–20 mm. high, ciliate; rays 2–3.5 cm. long; aks. thorny-muricate; $2n=28$ (Negodi, 1936), 32 (Weedle, 1941).—Occasional waif escaping from gardens where in common cult.; from Eu. March–May.

29. Calycadènia DC. ROSIN WEED

Xerophytic annuals with linear to filiform entire revolute grasslike lvs., those at the base of the rigid stem crowded into an erect rosette, more scattered above and often fasciculate, those of the fascicles and the uppermost usually with a prominent gland at the tip and often similar glands along the margins, these bractlike lvs. also with pectinate cilia. Ray-fls. few, 1–5 (–8), with broad palmately 3-lobed or -parted ligules. Receptacular bracts united into a cup surrounding the few disk-fls. Ray-aks. with nearly cent. terminal areola. Disk-fls. also fertile, angular, usually with a paleaceous pappus. Self-sterile. Eleven spp., mostly only in Calif. (Greek, *kalux*, cup, and *adenos*, gland, in reference to the glands of the infl.)

1. C. tenélla (Nutt.) T. & G. [*Osmadenia t.* Nutt. *Hemizonia t.* Gray.] Plate 13, Fig. B. Stem 1.5–5 dm. high, divaricately branched, the ultimate branchlets subcapillary, leafy and sparsely villous; lvs. 1.5–5 cm. long, scabrous and ± hirsute, the floral bracts white-ciliate; heads scattered; invol. 3.5–4.8 mm. high, ovoid; ray-fls. 3–5, the ligules white or with a crimson blotch; disk-fls. 3–10, white, fading roseate; ray-aks. 1.8–2.7 mm. long, the cent. areola strongly beaked; pappus of 4–5 red-flecked aristiform-tipped paleae 3.5 mm. long; $n=9$ (Clausen et al., 1934).—Abundant in light soils below 3250 ft.; Coastal Sage Scrub, S. Oak Wd.; cismontane Los Angeles Co. to L. Calif.

30. Calycóseris Gray

Annual, branched from base, glabrous below, sprinkled with tack-shaped glands above. Lvs. mostly basal, pinnately parted into narrow divisions. Heads rather showy, peduncled. Invol. many-fld., of many narrow scarious-margined equal phyllaries and of an outer series of very short loose ones. Receptacle with capillary bristles. Aks. fusiform, with 5 or 6 ribs, tapering into a short beak, this expanded apically into a shallow denticulate cup. Pappus abundant, white, of hispidulous white bristles falling away together. Two spp. (Greek, *kalux*, cup, and *seris*, a cichoriaceous genus.)

Fls. yellow; aks. smooth, ca. 8 mm. long including the beak .... 1. *C. parryi*
Fls. white; aks. rugulose, ca. 6 mm. long including the beak .... 2. *C. wrightii*

1. **C. párryi** Gray. Plant mostly 1–3 dm. high, subsimple to divaricately branched; lvs. 3–12 cm. long, pinnately parted into short linear lobes or the upper subentire; stipitate glands dark; invol. 10–15 mm. high; phyllaries linear, attenuate; ligules 15–25 mm. long; pappus longer than ak.—Common on desert flats and open slopes, 300–6000 ft.; Creosote Bush Scrub, Joshua Tree Wd.; Calif. deserts; to Utah, Ariz. March–May.

2. **C. wrìghtii** Gray. [*C. w.* var. *californica* Bdg.] Habit of *C. parryi*; glands pale; ligules white with rose or purplish dots or streaks on back; aks. rugulose on sides; $2n = 14$ (Stebbins et al., 1953).—Less frequent, below 6200 ft.; mostly Creosote Bush Scrub; w. edge Colo. Desert, e. Mojave Desert to Eureka V. and Death V.; to Utah, w. Tex. March–May.

31. *Cárduus* L. PLUMELESS THISTLE

Much like *Cirsium*, but pappus-bristles merely barbellate or smoothish. Plants biennial. Lvs. conspicuously decurrent, spiny. Large genus of Eurasia and n. Afr. (Ancient Latin name.)

Heads 4–6 cm. broad; invol. 2.5–3 cm. high .... 1. *C. nutans*
Heads 1–2 cm. broad; invol. 1.5–2 cm. high.
  The heads usually 1–5 at the ends of branches; phyllaries with small rough hairs on margin and back .... 2. *C. pycnocephalus*
  The heads usually 5–20 at the ends of branches; phyllaries glabrous or subciliate
                                                                    3. *C. tenuiflorus*

1. **C. nùtans** L. MUSK THISTLE. Biennial, 4–10 dm. high; lvs. bipinnately lobed, to 2 dm. long; heads solitary, on long naked peduncles; outer phyllaries reflexed; fls. purplish, aks. shining, 5 mm. long; pappus ca. 2 cm. long.—As a weed in a few stations (Walnut in e. Los Angeles Co.; near Victorville, etc.); native of Eu. June–July.

2. **C. pycnocéphalus** L. ITALIAN THISTLE. Annual, 3–18 dm. high, the stems slender, narrowly spiny-winged especially below; lvs. pinnatifid, to ca. 12 cm. long, the lobes and teeth spine-tipped, white-tomentose beneath, green but ± arachnoid above; heads subcylindric; phyllaries not membranous-margined, ± persistently floccose-tomentose; corolla-lobes ca. 3 times as long as the throat, rose-purple; aks. light tan or buff, ca. 20-nerved, 5–6 mm. long; pappus 15–20 mm. long, sordid; $2n =$ ca. 54 (Moore & Franklin, 1962).—Occasional weed along roadsides, etc.; mostly to the north of our area; introd. from Eu. May–July.

3. **C. tenuiflòrus** Curt. Like the preceding, but the stems more definitely spiny-winged in the upper parts; phyllaries ± membranous-margined, very scantily tomentose; corolla-lobes 1.5–2.5 times as long as the throat; aks. gray-brown, usually 10–13-nerved; pappus 1–1.5 cm. long.—Occasional weed, as at Covina, Los Angeles Co.; s. Eu. May–July.

32. *Carthàmus* L.

Annuals with alternate spinose lvs. Heads terminal, solitary or corymbose. Invol. with spreading leafy outer phyllaries and ± spiny inner ones. Receptacle chaffy. Corolla expanded above the tube. Aks. glabrous, mostly 4-ribbed; pappus scalelike or none. Ca. 20 spp., Eurasia. One cult. for dye from the fls. (Arabic name alluding to fl. color.)

Fls. yellow; aks. brownish .... 1. *C. baeticus*
Fls. orange; aks. white .... 2. *C. tinctorius*

1. **C. baèticus** (Boiss. & Reut.) Nym. [*Kentrophyllum b.* Boiss. & Reut. *C. nitidus* Calif. refs.] Leafy spiny plant 4–10 dm. high; lvs. rigid, pinnatifid, clasping at the base, to ca. 1 dm. long, with long stout marginal spines; outer phyllaries ending in leaflike appendage, much longer than the fls.; fls. yellow, with red veins; aks. 5–6 mm. long, straw color with

some brown.—Occasional weed reported from San Diego, Brea (Orange Co.), etc.; w. Medit. July-Aug.

2. **C. tinctòrius** L. SAFFLOWER. Lvs. oblong to lance-ovate, the upper clasping, minutely and softly spinose-toothed; fls. red-orange; aks. white, shining, ca. 6 mm. long, with a slightly notched scar near base; pappus of numerous narrow scales; $2n = 24$ (Poddubnaja, 1931).—Reported in Antelope V. and from Kern Co. n.; Eurasia. Cult. as an oil-seed crop.

33. *Centáurea* L. STAR THISTLE

Annual to perennial herbs with alternate entire dentate incised or pinnatifid lvs. and large or middle-sized heads of tubular purple, violet, pinkish, white or yellow fls., the outer sometimes enlarged. Invol. ovoid, globose to subcylindric, the phyllaries appressed, imbricated, entire to fimbriate or dentate, sometimes spine-tipped. Receptacle flat, bristly. Marginal fls. usually sterile; cent. perfect and fertile. Aks. ± compressed, usually smooth and shining, attached at or near the base. Pappus setose or partly chaffy or none. Ca. 500 spp., mostly of Old World, many cult. for the fls. (*Centaurie*, ancient Greek name, without clear application.)

1. Phyllaries not spinose at the tips.
    2. Plants annual.
        3. Plants glabrous; phyllaries entire, rounded at tip .... 7. *C. moschata*
        3. Plant with flocculent pubescence; phyllaries fringed .... 3. *C. cyanus*
    2. Plants perennial.
        4. Plant with creeping root; fls. blue to pink or white .... 9. *C. repens*
        4. Plant not with creeping root; fls. purple .... 2. *C. cineraria*
1. Phyllaries ending in definite terminal, but sometimes short, spines.
    5. Terminal spine of phyllary less than 5 mm. long; fls. purplish to white.
        6. Phyllaries entire, not fringed; invol. villous .... 8. *C. muricata*
        6. Phyllaries deeply fringed toward apex; invol. glabrous .... 4. *C. diluta*
    5. Terminal spine of phyllary more than 5 mm. long.
        7. Fls. purplish or pinkish; stems wingless.
            8. Aks. with definite pappus .... 5. *C. iberica*
            8. Aks. without pappus or with merely a vestige .... 1. *C. calcitrapa*
        7. Fls. yellow; stems winged with decurrent lf.-bases.
            9. Spines slender, purplish, 1 cm. long or less; plants branched mostly above the base; corolla glandular .... 6. *C. melitensis*
            9. Spines stout, yellow, 1–2 cm. long; plants branched from the base; corolla without glands
               10. *C. solstitialis*

1. **C. calcítrapa** L. Annual or biennial, divaricately branched, 3–6 dm. high; lvs. pubescent, the basal rosulate, pinnately divided into linear-lanceolate toothed or incised segms.; cauline reduced, simpler; heads sessile or nearly so, on leafy branches; invol. ovoid, smooth, ca. 1.5 cm. long without the spines; several of the subcoriaceous phyllaries tipped with divergent to reflexed stiff spines 1–2.5 cm. long; fls. purplish; aks. ca. 3 mm. long, straw color, mottled with dark brown, without pappus; $2n = 20$ (Vignoli, 1945).—In waste places and uncult. lands, at scattered stations from San Diego Co. n.; native of Eu. July-Oct.

2. **C. cinerària** L. DUSTY MILLER. Erect branching perennial 3–10 dm. high, closely white-tomentose throughout; lvs. pinnately parted into narrow obtuse lobes; heads rather large, the invol. round-ovoid, ca. 1.5–2 cm. long; phyllaries with a membranous black margin, long-ciliate, the apical bristle thicker than the others; fls. purple, the margin slightly enlarged; pappus copious, white, ca. 7–8 mm. long; $2n = 18$ (Larsen, 1956).—Occasional escape from gardens as Santa Catalina Id., Ojai, Santa Barbara; from Eu. June-Sept.

3. **C. cyànus** L. BACHELOR'S BUTTON. CORNFLOWER. Slender-stemmed annual 3–6 dm. high, with long ascending branches ending in solitary heads; lightly flocculent-tomentose when young; invol. ovoid, ca. 1.5 cm. long, of ca. 4 unequal series of pale scabrous-fimbriate phyllaries; fls. deep purplish-blue to pink or white, the marginal enlarged and

raylike; aks. metallic pale blue, ca. 4–5 mm. long; $2n=24$ (Fritsch, 1935).—Escaped from gardens and natur. at widely scattered stations, cismontane; native of Eu. May–Aug.

4. **C. dilùta** Ait. Lvs. with scabrous margins, the lower oblong, dentate, the upper lance-oblong, entire, decurrent; heads on well formed peduncles; invol. conic-ovoid, 1.5–2 cm. long, subglabrous; phyllaries lacerate on margins, the outer and middle with a stout terminal spine 1–2 mm. long; fls. pinkish or pale violet, ca. 2 cm. long, or the outer much larger; pappus white, capillary.—Occasionally natur. as at East Whittier, Watts, Santa Barbara, Vista (San Diego Co.); from Medit. region. May–July.

5. **C. ibèrica** Trev. Perennial 5–10 dm. high, ± cobwebby-pubescent; basal lvs. deeply sinuate-lobed, 1.5–2.5 dm. long, the cauline remote, reduced, few- and narrow-lobed; heads in open corymbose panicle; invol. ovoid, ca. 1.5 cm. high, the phyllaries green-chartaceous, ovate at base, abruptly narrowed into widespread stout spines 1–2 cm. long, with traces of short spines near their base; fls. purplish-pink near their tips, occasionally whitish; aks. 3–4 mm. long, straw color to grayish or mottled with dark brown, with ca. 3 rows of barbellulate bristles; $2n=16$ (Poddubnaja, 1931).—Occasionally natur. as at Ramona (San Diego Co.), Santa Barbara Co.; from Asia Minor. Aug.–Nov.

6. **C. meliténsis** L. TOCALOTE. Annual 3–8 dm. high, erect, commonly much-branched, grayish-pubescent, the stems winged by the decurrent lvs.; basal lvs. lyrate, 5–12 cm. long, with obtuse lobes; upper lvs. narrow, entire; heads solitary or 2 or 3 together; invol. ovoid, ca. 1 cm. high, arachnoid, the phyllaries rigid, the outer with palmatifid spining, the inner and middle with a rigid spine 4–8 mm. long; fls. yellow; aks. ca. 2.5 mm. long, grayish, with pappus-bristles in ca. 3 rows; $2n=22$ (Covas & Schnack, 1947).—Common weed in grain fields, pasture, along roadsides, etc., much of Calif. including the ids.; introd. from Eu. May–June.

7. **C. moschàta** L. SWEET SULTAN. Green glabrous annual, erect, branching below, to 2 ft.; lvs. toothed or pinnatifid and the lobes dentate; fls. white, yellow or purple, in long-peduncled solitary heads 5 cm. across, the marginal fls. enlarged; phyllaries entire, the innermost with scarious margins; pappus equaling the hairy ak.—Persisting in the wild about Santa Barbara, etc.; Old World. Summer.

8. **C. muricàta** L. Two 8 dm. high, ± villous below, subglabrous above; basal lvs. lyrate-pinnatifid into dentate lobes, upper lvs. entire; stem-branches long, slender, loosely paniculate, ending in single heads; invol. ovoid, ca. 15 mm. long, villous-pubescent; phyllaries imbricate in many ranks, ovate, entire, the terminal spine 3–4 mm. long; fls. violet to white; aks. dark, ca. 3 mm. long, striate and with transverse pits between, the pappus of outer imbricate scales and inner white bristles.—Occasional escape at Santa Barbara; native of Spain. May–June.

9. **C. rèpens** L. RUSSIAN KNAPWEED. Perennial, 3–10 dm. high, from creeping rootstocks, arachnoid-tomentose; lvs. firm, the basal oblong, sinuate-pinnatifid, 4–10 cm. long, the cauline numerous, linear-lanceolate to -oblong, entire, 2–3 cm. long, heads peduncled, oblong-cylindrical; invol. 10–14 mm. high, the phyllaries entire, roundish, scarious-margined, the inner hairy at the tip; fls. blue to pink or white, the marginal enlarged; pappus-bristles in several ranks, deciduous, barbellulate; $2n=26$ (Heiser & Whitaker, 1948).—Rather widely natur. Imperial Co. and cismontane cos.; from the region of the Caucasus. May–Sept.

10. **C. solstitiàlis** L. BARNABY'S THISTLE. Cottony-pubescent annual, branched from the base, 3–7 dm. high; basal lvs. 5–8 cm. long, deeply lobed; the upper to ca. 3 cm. long, entire, narrow, decurrent on the thus winged stems; invol. globose-ovoid, 14–18 mm. high; lower phyllaries with 3-pronged spines, middle with simple stout spines 6–20 mm. long, uppermost spineless; fls. bright yellow; aks. light-colored and with white pappus-bristles or dark and without pappus; $2n=16$ (Heiser & Whitaker, 1948).—Widely distributed weed in Calif.; from Eu. May–Oct.

34. *Chaenáctis* DC.

Annual, biennial or low perennial herbs, usually ± floccose, at least when young. Lvs. alternate, entire to compoundly dissected. Heads discoid, the fls. all perfect, but the marginal ones often enlarged and subligulate, peduncled. Invol. campanulate, turbinate or hemispheric; phyllaries herbaceous, in 1–3 series, free. Receptacle flat, alveolate, naked, or in 1 sp. with some bristles. Fls. yellow, white or pink, perfect; limb 5-lobed. Aks. linear-clavate, terete or compressed. Pappus of 4–20 hyaline erose-denticulate scales, or obsolete. Ca. 25 spp. of w. U.S. and adjacent borders. (Greek, *chaino*, to gape, and *aktis*, ray, the marginal fls. in many spp. flaring and raylike.)

(Stockwell, Palmer. A revision of the genus Chaenactis. Contr. Dudley Herb. 3: 89–168. 1940.)
1. Pappus-paleae 10 or more in 2 series; perennials or biennials.
 2. Lvs. linear-oblong, with 10–15 pairs of small crispate pinnae. Mts. of s. Calif.
                                                 8. *C. santolinoides*
 2. Lvs. broader with fewer segms.
  3. Lf.-segms. linear, mostly entire; suffruticose. San Jacinto and Santa Rosa mts.
                                                      7. *C. parishii*
  3. Lf.-segms. broader, often again toothed or lobed, perennial. From Panamint Mts. n.
                                                         3. *C. douglasii*
1. Pappus-paleae 4–5 in 1 series or with outer reduced paleae, or obsolete; annuals.
 4. Fls. yellow .... 5. *C. glabriuscula*
 4. Fls. white or tinged with purple.
  5. Plants 3–15 dm. high; pappus none or rudimentary; aks. flattened ... 1. *C. artemisiaefolia*
  5. Plants mostly lower than 3 dm.; pappus of persistent entire or erose paleae; aks. subterete.
   6. Stamens included; corollas pink, hoary, much exceeding the invol. ... 6. *C. macrantha*
   6. Stamens exserted.
    7. Phyllaries attenuate into a bristle-tip .... 2. *C. carphoclinia*
    7. Phyllaries obtuse or acute.
     8. Lvs. entire or once-parted into few linear lobes, green; plants soon glabrate.
      9. Phyllaries 8–10 mm. long, essentially glabrous including the short straight tip; marginal corollas much enlarged; pappus uniseriate .... 4. *C. fremontii*
      9. Phyllaries 12–16 mm. long, at least the curving tip densely puberulent; marginal corollas scarcely enlarged; pappus biseriate .... 10. *C. xantiana*
     8. Lvs. bipinnatifid with short thick segms., grayish; plants ± persistently tomentose; phyllaries 6–9 mm. long, glandular-puberulent .... 9. *C. stevioides*

  1. **C. artemisiaefòlia** (Harv. & Gray) Gray. [*Acarphaea a.* Harv. & Gray.] Annual, simple below, paniculately branched above, 3–15 dm. high; herbage farinose below, glandular-hirsute above; lvs. 5–15 cm. long, deltoid to ovate in outline, bi- or tri-pinnatifid into linear or oblong, irregular divisions; heads hemispheric, the invol. 9–12 mm. high, densely glandular-pubescent; phyllaries linear-lanceolate, acute; corollas nearly uniform, white or pinkish, the tube densely glandular-puberulent; aks. compressed, subglabrous; pappus none or rudimentary; $n=8$ (Stockwell, 1940, Raven & Kyhos, 1961).—Dry canyon sides and disturbed places like burns, below 5600 ft.; Coastal Sage Scrub, Chaparral; near the coast, from Ventura Co. to San Gabriel Mts., along the foothills to L. Calif. April–July.

  2. **C. carphoclínia** Gray. Diffusely branched annual, the branching tending to be zigzag, the stems very slender, 1–4 dm. high, farinose; lvs. on lower stems, 2–5 cm. long, 1–2-pinnatifid into filiform divisions, gradually reduced up the stem; peduncles very slender, 2–6 cm. long; invol. campanulate, 6–8 mm. high, the phyllaries 15–30, tapering into bristlelike awns; fls. whitish, the marginal not much enlarged; pappus-paleae of cent. fls. 4, ovate-lanceolate, almost as long as their corollas, those of outer fls. sometimes shorter and broader.—Common on hot desert pavements, Creosote Bush Scrub; below 2500 ft.; s. Inyo Co. across both deserts to n. L. Calif., e. Utah, Ariz. March–May. A variable sp., plants from the Death V. region and about head of San Joaquin V., and to w. Colo. Desert up to 4800 ft., with pappus-paleae not more than ⅓ the length of the corolla, constitute the var. **attenuàta** (Gray) Jones. [*C. a.* Gray.] At the e. end of the Santa Rosa Mts. have been found plants with the lvs. mostly gathered into a dense basal cluster, many tripinnatifid and named var. **peirsònii** (Jeps.) Munz. [*C. p.* Jeps.]

  3. **C. doúglasii** (Hook.) H. & A. [*Hymenopappus d.* Hook.] Biennial or short-lived perennial with basal rosette and erect stem 2–4 dm. high, floccose when young, usually

simple below, corymbosely branched above; lvs. glandular-punctate, loosely tomentose, later glabrate, 3–10 cm. long, 1–4 cm. wide, 2–3-pinnatifid with 4–8 pairs of pinnae; invol. broadly turbinate, 10–14 mm. high, densely glandular-puberulent; corolla whitish or pinkish; pappus of 10 usually linear-oblong palcae $\frac{1}{2}$ $\frac{3}{4}$ as long as corolla.—Dry open places and woods, 9000–11,000 ft., Sagebrush Scrub, Pinyon-Juniper Wd., Bristle-cone Pine F.; Panamint Mts., Inyo Mts., etc.; to n. Our form is sometimes referred to *C. panamintensis* Stockwell, to *C. d.* var. *achilleaefolia* (H. & A.) Nels., and to var. *rubricaulis* (Rydb.) Ferris.

4. **C. fremóntii** Gray. Plate 13, Fig. C. Annual, usually branched at base, 1–3 dm. high, subglabrous; lvs. somewhat fleshy, scattered, 2–5 cm. long, entire and linear or the lower once pinnate into linear lobes; heads few, pedunculate; invol. 8–10 mm. high; phyllaries acute, essentially glabrous; fls. white, 5–6 mm. long; pappus-paleae of disk-fls. 4, equal, lance-acuminate, those of outer fls. shorter; $2n = 10$ (Stockwell, 1940).—Common on sandy mesas and open sandy slopes, 100–4000 (–7000) ft.; Creosote Bush Scrub, Joshua Tree Wd., Chaparral; Mojave and Colo. deserts, to Cent. V.; Nev., Ariz. March–May.

5. **C. glabriúscula** DC. Plate 13, Fig. D. Annual, 1–4 dm. high, openly branched, thinly floccose to woolly; leafy only below or mostly above, the lvs. ± scattered, mostly 1–2-pinnate into linear lobes usually 2–8 mm. long, uppermost lvs. entire; peduncles usually 3–10 cm. long; invol. 5–10 mm. high, the phyllaries plane, 1-nerved; corollas yellow, the marginal with flaring palmate limb surpassing the disk; pappus-paleae 4–5; $n = 6$ (Stockwell, 1940).—A widespread and variable sp., of which the following variants may be mentioned:

Peduncles very long, subscapose and slender or from short leafy stem and stout.
 Heads on long delicate scapose peduncles; plants branching only at base, woolly; lvs. in a basal rosette .... var. *lanosa*
 Heads on stout often fistulous peduncles; plant branching in lower half, glabrate, leafy below
                                 var. *denudata*
Peduncles relatively short, usually much less than half the length of the stem.
 Phyllaries oblong, plane, with thick obtuse tips; marginal corollas mostly with palmate limb.
  Phyllaries 8–10 mm. high, the heads medium large. Lowland .... var. *glabriuscula*
  Phyllaries 5–7 mm. high, the heads small. Montane .... var. *curta*
 Phyllaries boat-shaped, narrowly linear; marginal corollas only slightly enlarged, seldom with palmate limb.
  Lf.-segms. very narrow, rather short and recurved .... var. *tenuifolia*
  Lf.-segms. succulent, short, obtuse .... var. *orcuttiana*

Var. **cúrta** (Gray) Jeps. [*C. heterocarpha* var. *c.* Gray.] Heads small, the invols. 5–7 mm. high; phyllaries narrow; pappus paleae of cent. fls. 4–5, usually half as long as corolla, sometimes reduced more; $n = 6$ (Raven & Kyhos, 1961).—Sandy or gravelly soils, up to 6500 ft.; V. Grassland, Chaparral, etc.; mts. from Santa Barbara and Kern cos. to San Diego Co. April–June.

Var. **denudàta** (Nutt.) Munz. [*C. d.* Nutt.] Rather large and coarse, 3–4 dm. high, the numerous peduncles becoming fistulous; invol. 9–10 mm. high; marginal corollas prominently enlarged; pappus-paleae of inner fls. 4–5, nearly as long as corolla; $n = 6$ (Raven & Kyhos, 1961).—Old dunes and sandy places usually near the coast and below 1000 ft.; Coastal Sage Scrub, S. Oak Wd.; San Luis Obispo Co. to Orange Co. and San Bernardino V. April–July.

Var. **glabriúscula.** One to 4 dm. high, thinly floccose, soon glabrate; lvs. scattered throughout, 3–8 cm. long, with few to several pairs of narrowly linear lobes usually 2–8 mm. long; peduncles 3–8 cm. long; invol. 7–10 mm. high, the phyllaries plane, 1-nerved; pappus-paleae 4, those of cent. fls. $\frac{3}{4}$ as long as corolla; $n = 6$ (Stockwell, 1940).—Sandy valleys and foothills, below 4000 ft.; Chaparral, V. Grassland; Los Angeles and Santa Barbara cos. to Monterey Co. March–May.

Var. **lanòsa** (DC.) Hall. [*C. l.* DC.] Leafy only at the branching base, 2–3 dm. high, with many scapose peduncles; herbage floccose with tardily deciduous wool; lvs. mostly pinnatifid with few narrowly linear lobes or the upper entire; invol. ca. 8 mm. high; marginal corollas not very ampliate; pappus-paleae of cent. fls. 4 (or 5), subequal, acutish, nearly as long as corolla; $n=6$ (Stockwell, 1940).—Common in dry sandy places, below 2000 ft.; V. Grassland, Chaparral, Los Angeles Co. to San Francisco Bay. March–June.

Var. **orcuttiàna** (Greene) Hall. [*C. tenuifolia* var. *o.* Greene *C. o.* Parish.] Two–6 dm. high, stoutish, ± succulent; lvs. sometimes crowded near the base, 2–3-pinnatifid, fleshy, the ultimate lobes very short and blunt; infl. viscid; invol. 6–7 mm. high; pappus-paleae of inner fls. 4, ca. ¾ as long as corolla; $n=6$ (Stockwell, 1940).—Dunes near the coast; Coastal Strand, San Diego Co.; n. L. Calif. April–July.

Var. **tenuifòlia** (Nutt.) Hall. [*C. t.* Nutt. *C. filifolia* Gray.] Tall, slender, 2–6 dm. high, simple or much branched; lf.-divisions filiform, short or much elongated; ± succulent and viscid, not very lanose; heads small; invol. 5–6 mm. high; phyllaries ± keeled, narrow; pappus-paleae of inner fls. uniseriate, somewhat shorter than the corolla; $n=6$ (Stockwell, 1940).—Open sandy places up to 5000 ft.; Coastal Sage Scrub, Chaparral, S. Oak Wd.; cismontane s. Calif., Los Angeles Co. to n. L. Calif. April–June.

6. **C. macrántha** D. C. Eat. Annual, 5–20 cm. high, commonly widely branched from base, floccose-tomentose when young; lvs. once or twice pinnatifid with linear or oblong lobes; invol. campanulate, 12–15 mm. high, ± tomentose; corollas vespertine, white to flesh-color, hoary, much exceeding the invol., the marginal ones not much enlarged; stamens included; pappus of 4 linear-oblong paleae half as long as the corolla and 2–4 very short outer ones, or these absent; $n=6$ (Raven & Kyhos, 1961).—Gravelly desert plains or hills, 2000–5000 ft.; Creosote Bush Scrub, Shadscale Scrub, Sagebrush Scrub, Pinyon-Juniper Wd.; s. Mono Co. through the Mojave Desert; to Ariz., Utah, sw. Ida. April–May.

7. **C. párishii** Gray. Suffruticose perennial 2–4 dm. high; stems several, erect, subscapose, from a branched decumbent woody base, simple or erectly branched above to bear 2–4 long-peduncled heads, the sterile shoots short, pannose; lvs. lanulose and punctate, 2–5 cm. long, oblong in outline, pinnate with relatively few linear obtuse, mostly entire lobes, petiolate; invol. 11–13 mm. high, minutely glandular; phyllaries linear, loose, unequal; corolla cream-white or pinkish, 7–9 mm. long, at least the lobes hispidulous; aks. densely hirsute; pappus of 14–18 unequal linear-lanceolate scales shorter than the corolla; $n=6$ (Raven & Kyhos, 1961).—Infrequent, dry rocky slopes, 5000–7000 ft.; Yellow Pine F., Chaparral; San Jacinto and Santa Rosa mts.; n. L. Calif. June–July.

8. **C. santolinoìdes** Greene. [*C. s.* var. *indurata* Stockw.] Plate 13, Fig. E. Perennial with a cespitose often branching crown on a woody taproot; stems subscapose, sometimes branched, 1–3.5 dm. high, ± glandular; lvs. principally in dense basal rosettes, permanently white-tomentose and punctate, 3–10 cm. long, 4–7 mm. wide, linear-oblong, pinnatisect with many short oblong or rounded crispate segms.; invol. 8–13 mm. high, densely glandular; phyllaries oblong, obtuse; corollas cream-white or pink, 6–7 mm. long, ± pilose; aks. densely hirsute, equal to corollas; pappus of 12–16 very unequal oblanceolate scales shorter than the corolla; $n=6$ (Raven & Kyhos, 1961).—Open pine woods and dry ridges, 4500–8000 ft.; Yellow Pine F.; San Bernardino Mts., San Gabriel Mts., Mt. Pinos; Greenhorn Mts. June–July.

9. **C. stevioìdes** H. & A. [*C. latifolia* Stockw.] Freely branched, 1–2.5 dm. high, floccose when young, somewhat glabrate later, glandular-puberulent above; lvs. 3–8 cm. long, scattered, grayish, twice pinnatifid into short lobes; peduncles slender, 1–4 cm. long, glandular-puberulent; invol. 6–9 mm. high, the phyllaries linear, obtusish, glandular-puberulent; corollas white, 5 mm. long, the marginal moderately enlarged; pappus-paleae 4, lance-oblong, ca. ⅔ the length of the corolla, acute; $n=5$ (Raven & Kyhos, 1961).—Sandy desert floors and slopes, below 5000 ft.; Creosote Bush Scrub; both deserts, from White Mts. to San Diego Co.; to New Mex., Wyo., Ida. March–May. Plants with pappus not more than ⅓ the length of the corolla are the var. **brachypáppa** (Gray) Hall. [*C. b.* Gray.]

10. **C. xantiàna** Gray. [*C. x.* var. *integrifolia* Gray.] Rather stout, simple or branched above, annual, 1–4 dm. high, glabrate; lvs. somewhat fleshy, 2–6 cm. long, entire or pin-

natifid into linear divisions; peduncle fistulose, leafy; invol. 12–16 mm. high; phyllaries linear, acute, floccose when young, the tips somewhat foliaceous and hirsutulous; corollas white or flesh-color; pappus-paleae double, 4 long surrounded by 4 very short ones; $n=7$ (Stockwell, 1940, Raven & Kyhos, 1961).—Desert slopes, 1400–7000 ft.; largely Sagebrush Scrub, Pinyon-Juniper Wd., some Chaparral; n. base of San Gabriel Mts. and Mojave R. w. to Mt. Pinos; n. to Ore., w. Nev. April–June.

35. *Chaetadélphia* Gray

Diffuse much-branched broomlike perennial herb from a heavy underground branching rootstock. Lvs. alternate, entire, linear to lance-linear, thickish, acuminate, scarious-margined. Heads solitary on ends of branches; invol. with 5 principal phyllaries and a few short outer ones. Fls. 5 to a head, pale lavender, almost whitish. Aks. 5-angled, narrow; pappus of 5 tapering awns bearing on each side toward the base 3–5 shorter slender rigid bristles. One sp. (Greek, *chaete*, bristle, and *adelphe*, sister, referring to the united bristles in the pappus.)

1. **C. wheèleri** Gray. Glabrous, 1–3 dm. high, with ridged branches; lvs. 2–5 cm. long, deciduous; invol. 12–15 mm. high; aks. 8–10 mm. long; pappus light brown; $2n=18$ (Stebbins et al., 1953).—Sand dunes, etc., 3000–5000 ft.; Creosote Bush Scrub, Sagebrush Scrub; base of Inyo-White Range; w. Nev., s. Ore. May–Sept.

36. *Chaetopáppa* DC.

Ours low very slender annuals with simple or diffusely branched stems and alternate chiefly linear lvs. Heads small, few- to many-fld., terminating very slender peduncles, radiate, disciform or discoid, all florets potentially fertile, yellow, white or reddish. Invol. turbinate to hemispheric, the phyllaries 2–5-seriate, graduate or equal, thin, green-centered, scarious-margined, usually setulose-tipped, persistent. Receptacle convex, naked. Pistillate florets 1–3-seriate, ligulate or tubular; hermaphroditic fls. slender, 3–5-toothed. Aks. linear fusiform, often compressed, pubescent. Pappus of 3–many fragile slender bristles, sometimes dilated and ± joined at base, or wanting. Ca. 15 spp. of sw. U.S. and Mex. (Greek, *chaete*, bristle, and *pappos*, pappus.)

Ray-fls. inconspicuous; invol. turbinate. From Santa Barbara Co. n. .... 1. *C. alsinoides*
Ray-fls. conspicuous; invol. broadly hemispheric.
    Invol. pubescent. Santa Catalina Id., coastal Los Angeles Co. .... 3. *C. lyonii*
    Invol. glabrous. General cismontane distribution .... 2. *C. aurea*

1. **C. alsinoìdes** (Greene) Keck. [*Pentachaeta a.* Greene.] Diffuse, 3–12 cm. high and wide, ± villous; lvs. filiform or nearly so; invol. 2.5–3 mm. high, glabrous or somewhat hirsute, the phyllaries subequal, 6–7, oblong or nearly so, green, lacerate toward apex; ♀ fls. ca. 4–6, capillary, tubular or with minute involute ligule; disk-fls. reddish, commonly with imperfect anthers; aks. brownish, villous; pappus of 3 capillary bristles; $n=9$ (Solbrig et al., 1964).—Grassy places; Santa Ynez Mts., Santa Barbara Co. to n. Calif. April–May.

2. **C. áurea** (Nutt.) Keck. [*Pentachaeta a.* Nutt.] Plate 13, Fig. F. Usually diffuse, 8–30 cm. high, glabrous or with lf.-margins ciliate; lvs. linear, the lower 1–3.5 cm. long, to 2 mm. wide, the upper shorter and narrower; heads 1–2.5 cm. wide; invol. broad, 4–7 mm. high; phyllaries ca. 4–5-ranked, outer lance-ovate to oblong, acuminate to obtuse, the greenish cent. portion scarcely wider than the scarious margins; ray-fls. 0–70, the ligule 5–12 mm. long; disk-fls. usually many; aks. red-brown, 1 mm. long, short-strigose; $n=9$ (Solbrig et al., 1964).—Dry open and grassy places, below 6000 ft.; V. Grassland, Coastal Sage Scrub to Yellow Pine F.; n. L. Calif. to Los Angeles and San Bernardino cos. April–July.

3. **C. lyònii** (Gray) Keck. [*Pentachaeta l.* Gray.] Like *C. aurea*, lightly pubescent; invol. pubescent, ca. 5 mm. high; phyllaries subequal, acuminate; pappus-bristles 10–12, fili-

form, fragile, flared at very base and forming a rudimentary corona.—V. Grassland, etc.; Santa Catalina Id., coastal Los Angeles Co. March–April.

37. *Chrysánthemum* L.

Annual or perennial herbs with alternate entire, toothed or pinnatifid lvs. Heads 1 to many, small to quite large, radiate or rarely discoid. Rays when present, ♀ and fertile, white to yellow or pink. Phyllaries ± imbricate in 2–4 series, dry, becoming scarious or hyaline on at least the margin and tip, sometimes with greenish midrib. Receptacle flat or convex, naked. Disk-fls. tubular and perfect, the corolla 5 (4)-lobed. Aks. subterete or angular, 5–10-ribbed, or those of rays with 2–3 wing-angles. Pappus a short crown or none. Ca. 100 spp., chiefly of N. Hemis, and Old World. (Greek, *chrusos*, gold, and *anthemon*, fl.)

Plants annual.
    Heads with a tricolored effect, the disk purple, the rays white or red or purple with yellow or purple ring .... 1. *C. carinatum*
    Heads not tricolored, light yellow .... 2. *C. coronarium*
Plants perennial; rays white.
    Heads borne singly .... 3. *C. leucanthemum*
    Heads borne in dense flat-topped clusters .... 4. *C. parthenium*

1. C. carinàtum L. TRICOLOR CHRYSANTHEMUM. Glabrous, simple or forked annual 5–9 dm. high; lvs. ± succulent, remotely twice pinnatifid into linear lobes; heads solitary, long-peduncled, 4–6 cm. across; outer phyllaries keeled; rays differently colored at base so as to make a ring in the fl.-head; disk purple; aks. flat, winged; $n=9$ (Rana, 1965).—Escape from cult., as near San Diego, El Segundo dunes, etc.; native of Morocco. April–June.

2. C. coronàrium L. GARLAND C. Ca. 2–10 dm. tall, annual, stiff; lvs. not succulent, twice pinnatifid into oblong to subovate divisions; heads ca. 2–4 cm. across, yellow or yellowish-white; phyllaries not keeled; aks. ± prismatic, angled but not winged, with minor ribs between; $2n=18, 36$ (Shimotomai & Huziwara, 1935).—Occasionally natur., as at San Diego, El Segundo, Redondo, Monterey Park, Ontario, Ventura, Santa Barbara; native of Medit. region. April–Aug.

3. C. leucánthemum L. [*L. vulgare* Lam.] OX-EYE DAISY. Perennial with ± of a rhizome; stems mostly 2–8 dm. high, ± simple, glabrous or sparsely hairy; lower lvs. spatulate to oblanceolate, petioled, 4–15 cm. long, ± deeply crenate, the cauline reduced, subsessile, subentire to blunt-toothed; heads solitary, long-pedunculate, 2.5–5 cm. across; disk 1–2 cm. in diam.; phyllaries narrow, with a brown submarginal area; rays white, 1–2 cm. long; aks. terete, ca. 10-ribbed; pappus none; $2n=36, 54$ (Dowrick, 1952).—Occasionally natur. as at Julian, Loma Linda, more abundant in n. Calif.; native of Old World. June–Aug.

4. C. parthènium (L.) Bernh. Perennial with a caudex; stems leafy, subglabrous, 3–8 dm. high; lvs. pinnatifid with rounded to pinnate segms., blade to 7 cm. long; heads several to many, corymbose; disk 5–9 mm. wide; rays white, ca. 10–20, 4–8 mm. long; aks. 8–10-nerved; $2n=18$ (Harling, 1951).—Occasional as in San Antonio Canyon, Santa Barbara, Ventura; from Eu. June–Sept.

38. *Chrysópsis* (Nutt.) Ell. GOLDEN-ASTER

Low mostly perennial pubescent herbs, sometimes suffrutescent. Lvs. alternate, usually entire. Heads radiate or discoid, yellow, medium-sized, terminating the stems and branches. Invol. campanulate to hemispheric, the phyllaries numerous, narrow and imbricated. Receptacle low-convex, foveolate. Ray-fls. fertile, ♀; ligules narrow. Disk-fls. perfect, fertile. Style-branches flattened, with long hairy appendages much longer than the stigmatic portion. Aks. ± flattened; pappus usually double, brownish, the inner of numerous capillary bristles, the outer, when present, of short linear scales or bristles. Ca.

20 spp. of temp. N. Am. (Greek, *chrysos*, golden, and *opsis*, appearance, from the color of the heads.)

Heads discoid; outer pappus none or indistinct .... 1. *C. breweri*
Heads radiate; outer pappus linear-squamellate .... 2. *C. villosa*

(At least three workers have recently indicated the desirability of merging *Chrysopsis* with *Heterotheca* under the latter name. I have not done so here; since the California forms are still so unsatisfactorily worked out, I do not wish to make the necessary new combinations.)

1. **C. bréweri** Gray. [*C. wrightii* Gray. *Heterotheca b.* Shinners.] Few to many stems from a woody caudex, 2–8 dm. high, mostly corymbosely branched, each leafy branch ending in a single head; herbage moderately hirtellous, glandular-puberulent; lvs. lanceolate to narrowly oblong to lance-ovate, 1–3 cm. long; heads campanulate, discoid; invol. 7–11 mm. high, 2–3-seriate; outer pappus of fine bristles.—Collected on Mt. San Gorgonio, San Bernardino Mts. in 1882 by W. G. Wright, not again; Sierra Nevada to Shasta Co. July–Sept.

2. **C. villòsa** (Pursh) Nutt. [*Amellus v.* Pursh.] var. **echioìdes** (Benth.) Gray. [*C. e.* Benth. *Heterotheca e.* Shinners.] Stems erect, 3–8 dm. high, virgate, simple or branching throughout; lvs. crowded, spreading, hirsute-canescent, firm, the upper 1–2 cm. long, sessile; heads many, paniculate to cymose; phyllaries many, slender, hispid-hirsute; outer pappus of short narrow squamellae.—Dry sandy open places, below 6500 ft.; Coastal Sage Scrub, V. Grassland, Chaparral, Yellow Pine F.; mostly away from the immediate coast except about San Diego; n. L. Calif. to cent. Calif. July–Nov.

KEY TO VARIETIES

Heads rather many, cymose or paniculate, not subtended by leafy bracts. Mostly away from immediate coast.
    Lvs. pubescent with spreading hairs.
        Herbage densely canescent and hispid-hirsute .... var. *echioides*
        Herbage rarely canescent, moderately hispid .... var. *hispida*
    Lvs. pubescent with silky appressed hairs .... var. *fastigiata*
Heads large, solitary or few in a cluster, closely subtended by leafy bracts; herbage obviously glandular. Mostly coastal .... var. *sessiliflora*

Var. **fastigiàta** (Greene) Hall. [*C. f.* Greene.] Stems 3–10 dm. high, subsimple to many-branched, usually with many small ascending linear-oblong to elliptic mucronate lvs., these silky-tomentose, especially dorsally, 1–2 (–3) cm. long; outer pappus absent, or present as short bristles, not as scales; $n = 9$ (Raven et al., 1960).—Dry places below 6500 ft.; Coastal Sage Scrub, Chaparral, etc.; Orange Co. to Ventura Co.; w. Sierran foothills. July–Nov.

Var. **híspida** (Hook.) Gray ex D. C. Eat. [*Diplopappus h.* Hook.] Stems slender, virgate, 2–4 dm. high; herbage moderately hispid or villous, somewhat glandular; lvs. linear-oblanceolate, 8–20 mm. long, 2–5 mm. wide; heads many, rather small; phyllaries minutely glandular and sparsely hispid; outer pappus of short narrow scales; $n = 9$ (Raven et al., 1960).—Rocky places, 3000–6400 ft.; Joshua Tree Wd.; Little San Bernardino Mts.; n. to B.C., Rocky Mts. April–May; Oct.–Nov.

Var. **sessiliflòra** (Nutt.) Gray. [*C. s.* Nutt. *Heterotheca s.* Shinners.] Herbage mostly grayish, villous-canescent and glandular; lvs. crowded, oblong or spatulate, 1–2 cm. long; heads mostly large and solitary, foliose-bracteate; outer pappus evident, squamellate.— Washes, brushy places, etc. below 5500 ft.; Coastal Sage Scrub, Chaparral; n. L. Calif.; San Diego Co. (Torrey Pines, Palomar Mt.); Riverside and Los Angeles cos, etc. to n. Calif. July–Sept.

39. *Chrysothámnus* Nutt. RABBIT-BRUSH

Shrubs or subshrubs, usually much-branched with erect stems. Lvs. alternate, entire, narrow, not fascicled, sometimes glandular-punctate. Heads many, discoid, narrow, mostly 5-fld., in cymes, racemes or panicles. Invol. cylindraceous; phyllaries strongly imbricate, usually in 5 vertical ranks, chartaceous or coriaceous, sometimes green-tipped, mostly carinate and sometimes with a costal gland. Appendages of style-branches usually longer than the stigmatic portion. Aks. slender, terete or angled or flattened, densely sericeous to glabrous. Pappus copious, of soft capillary bristles. Thirteen spp. of w. N. Am. (Greek, *chrysos*, gold, and *thamnos*, shrub, the infl. mostly very showy.)

(Hall, H. M. and F. E. Clements. Carnegie Inst. Wash. Pub. 326: 157–234. 1923.)
Lvs. resinous-punctate, terete.
    Fls. white .... 1. *C. albidus*
    Fls. yellow.
        Phyllaries without an apical gland or greenish spot .... 4. *C. paniculatus*
        Phyllaries with a prominent gland at apex .... 6. *C. teretifolius*
Lvs. not resinous-punctate.
    Twigs densely pannose-tomentose.
        Infl. mostly cymose; phyllaries obtuse to moderately attenuate .... 3. *C. nauseosus*
        Infl. mostly racemose or spicate; phyllaries very attenuate .... 5. *C. parryi*
    Twigs not pannose-tomentose.
        Fls. white; phyllaries with slender tapering tip .... 1. *C. albidus*
        Fls. yellow; phyllaries obtuse to acute.
            Invol. 8–15 mm. high, the phyllaries acuminate, keeled .... 2. *C. depressus*
            Invol. mostly 5–7 mm. high, the phyllaries usually obtuse, not strongly keeled
                                                                               7. *C. viscidiflorus*

1. **C. álbidus** (Jones) Greene. [*Bigelovia a.* Jones ex Gray.] Shrubs 3–15 dm. high, white-barked, glabrous, resinous-viscid, aromatic; lvs. filiform, 1.5–3 cm. long, 0.5–1 mm. wide, terete, impressed-punctate, crowded, with axillary fascicles; heads in small compact terminal cymes, 4–6-fld.; invol. 7–9 mm. high; phyllaries ca. 4-seriate, graduate, in obscure vertical ranks, the outer lanceolate to ovate, herbaceous-thickened in outer half, with subulate curved tip, the inner oblong, acuminate-tipped, the narrow hyaline margin ± erose; corolla white, 7–8 mm. long, the lobes ca. 2 mm. long; aks. pilose; $n=9$ (Anderson, 1966).—Dry alkaline plains, below 6500 ft.; Alkali Sink, Shadscale Scrub; rare in Calif., Saline V., Amargosa Desert se. of Death V.; to Utah. Aug.–Nov.

2. **C. depréssus** Nutt. [*Linosyris d.* Torr. *Bigelovia d.* Gray.] Depressed subshrub with many erect herbaceous stems from a woody spreading crown, 1–3 dm. high, cinereous with a dense scabrid puberulence; lvs. oblanceolate or spatulate, the lowermost rounded or obtuse, the upper sharply apiculate, 7–20 mm. long, 1.5–4 mm. wide; heads 5-fld., in compact cymes; invol. 9–12 mm. high; phyllaries 5-seriate, in 5 well defined vertical ranks, lance-acuminate, strongly keeled, the outer green and minutely puberulent, the inner broader, scarious, with hyaline margins; aks. glabrate; pappus brownish-white; $n=9$ (Anderson, 1966).—Dry canyons, 3500–6000 ft.; Joshua Tree Wd., Pinyon-Juniper Wd.; mts. of ne. Mojave Desert; to Colo., New Mex. Aug.–Oct.

3. **C. nauseòsus** (Pall.) Britton. [*Chrysocoma n.* Pall.] The typical form not in Calif.

KEY TO SUBSPECIES OF CALIFORNIA

Aks. densely pubescent.
    Phyllaries, at least the outer, ± pubescent or tomentose, sometimes only ciliate.
        Corolla-lobes lanceolate, 1.3–2.5 mm. long; style-appendage longer than the stigmatic portion; invol. 10–13 mm. high .... ssp. *bernardinus*
        Corolla-lobes ovate, 0.5–1 mm. long; style-appendage shorter than the stigmatic portion; invol. 6–7 mm. high .... ssp. *hololeucus*
    Phyllaries not hairy, but sometimes glandular or viscid; lvs. not white-tomentose.
        Invol. 6.5–9 mm. high, not sharply angled, the phyllaries slightly if at all keeled   ssp. *consimilis*
        Invol. 9–10 mm. long, sharply angled, the strongly keeled phyllaries in very distinct vertical rows .... ssp. *mohavensis*
Aks. glabrous or with merely a few hairs on the prominent ribs.
    Mts. of e. Mojave Desert .... ssp. *leiospermus* (Gray) Hall & Clements

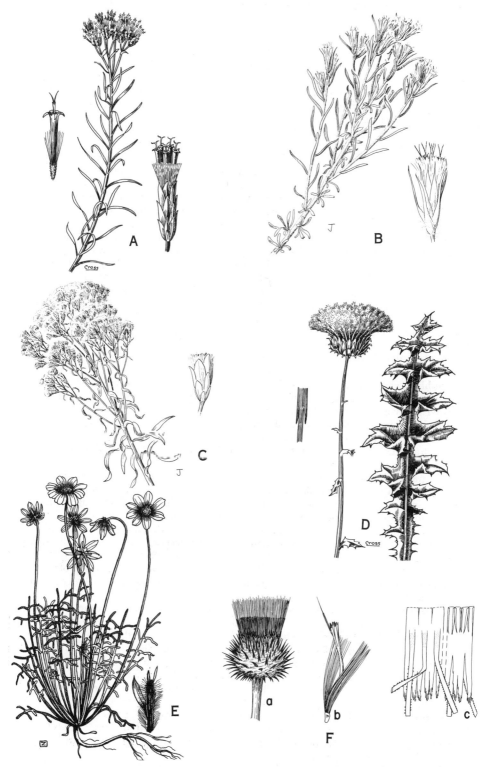

Plate 14. Fig. A, *Chrysothamnus nauseosus bernardinus*. Fig. B, *Chrysothamnus parryi*. Fig. C, *Chrysothamnus viscidiflorus*. Fig. D, *Cirsium californicum*. Fig. E, *Coreopsis bigelovii*. Fig. F, *Cirsium occidentale*, with tubular corolla and connate anthers caudate at their base.

Ssp. **bernardínus** (Hall) Hall & Clem. [*C. n.* var. *b.* Hall.] Plate 14, Fig. A. Rounded, 3–5 dm. high, twigs with white tomentum; lvs. usually green, linear, 3–5 cm. long, 1–2 mm. wide; invol. 10–13 mm. high; phyllaries sharply acuminate, keeled, with margin broadly hyaline and cut-fimbriate or merely ± ciliate, the outer often puberulent or rarely tomentose, together with the peduncles ± glandular-atomiferous; corolla 9–12 mm. long, the lobes 1.5–2.5 mm. long, the tube puberulent.—Dry benches, 6000–9500 ft.; Montane Coniferous F.; San Gabriel, San Bernardino and San Jacinto mts. Aug.–Oct.

Ssp. **consímilis** (Greene) Hall & Clem. [*C. c.* Greene. *C. n.* var. *c.* Hall. *C. n.* var. *viridulus* Hall.] Shrub 5–30 dm. high; tomentum gray, greenish-yellow or whitish; lvs. mostly linear-filiform, 2.5–5 cm. long, less than 1 mm. wide, green or gray-tomentulose; infl. tending to be narrow and elongate rather than flat-topped; invol. 6.5–8.5 mm. high, glabrous; phyllaries acute or obtuse, not sharply keeled; corolla 7–9.5 mm. long, the lobes 1–2.5 mm. long; $n = 9$ (Anderson, 1966).—Occasional in open alkaline valleys, Sagebrush Scrub, skirting the deserts from San Diego Co. to e. of the Sierra Nevada to e. Ore., Ida., Utah.

Ssp. **hololeùcus** (Gray) Hall & Clem. [*Bigelovia graveolens* var. *h.* Gray. *C. gnaphalodes* Greene. *C. n.* vars. *h.* and *g.* Hall.] Shrub 5–20 dm. high, the twigs white, gray or yellowish-green; herbage fragrant; lvs. 1–3 cm. long, 0.5–1.5 mm. wide, gray- or white-tomentose; invol. 6–7 mm. high; phyllaries rather obtuse, keeled, woolly but not ciliate; corolla 6.5–8 mm. long, the lobes 0.5–1 mm. long; aks. densely pubescent; $n = 9$ (Anderson, 1966).—Sandy neutral soils of high deserts, 3000–9000 ft.; Alkali Sink, Sagebrush Scrub, Pinyon-Juniper Wd.; w. ends of Colo. and Mojave deserts, e. to Nev., n. to Mono Co. Sept.–Nov.

Ssp. **mohavénsis** (Greene) Hall & Clem. [*Bigelovia m.* Greene. *C. m.* Greene. *C. n.* var. *m.* Hall.] Shrub 6–20 dm. high, often fastigiately branched, the branches often leafless and rushlike, closely gray- or greenish-yellow-tomentose; lvs. narrowly linear, 2–3 cm. long, less than 1 mm. wide, green or gray-tomentulose; infl. a rounded or somewhat elongate thyrse; invol. narrow, 8–10.5 mm. high, glabrous, sharply 5-angled; phyllaries keeled, mostly obtuse, in very distinct vertical ranks, the costa usually dilated at apex into an oblong gland; corolla 8–10 mm. long, the lobes 1.5–2.5 mm. long; $n = 9$ (Anderson, 1966). —Well drained, scarcely alkaline soils, 2500–6000 ft.; Joshua Tree Wd., Creosote Bush Scrub; w. Mojave Desert to cent. Calif.; e. to Nev. Sept.–Oct.

4. **C. paniculàtus** (Gray) Hall. [*Bigelovia p.* Gray. *Ericameria p.* Rydb.] Loosely branched, broadly rounded shrub 6–20 dm. high, the herbage glabrous, resinous, strongly glandular-punctate, ± glaucescent; lvs. terete, mucronate-tipped, 1–3 cm. long, 0.5 mm. wide; heads in profuse panicles, 5–8-fld.; invol. subcylindric, 5.5–6.5 mm. high; phyllaries 4-seriate, graduate, the vertical ranks not sharply defined, oblong, obtuse, whitish, indurate, the costa narrow, scarcely if at all glandular-thickened above; aks. appressed-villous or sericeous; pappus brownish-white; $2n = 18$ (DeJong & Montgomery, 1963).—Stony open places mostly below 4000 ft.; Creosote Bush Scrub; w. and n. borders of Colo. Desert, local across Mojave Desert to Tehachapi, e. Inyo and e. San Bernardino cos.; to Utah. Ariz. June–Dec.

5. **C. párryi** (Gray) Greene ssp. **ásper** (Greene) Hall & Clem. [*C. a.* Greene. *C. p.* var. *a.* Munz.] Plate 14, Fig. B. Low shrub 1.5–3 dm. high, with erect or spreading stems, the younger white-tomentose; lvs. 2–4 cm. long, 1–3 mm. wide, 1-nerved, green, with short-stalked resin-glands; heads few, in short racemes or subpaniculate; invol. 11–15 mm. long; phyllaries 9–13, usually glandular-puberulent as well as arachnoid-ciliate below; fls. 5–10; aks. appressed-pilose; $n = 9$ (Anderson, 1966).—Occasional under pines, 7000–8600 ft.; Montane Coniferous F.; San Bernardino Mts., Mt. Pinos; s. Sierra Nevada; Nev. Autumn.

Ssp. **ímulus** Hall & Clem. [*C. p.* var. *i.* Jeps.] Spreading at base, the shoots ca. 1 dm. long; lvs. narrowly spatulate, 1–1.5 cm. long, 2–3 mm. wide, gray-tomentose; heads few, 11–15-fld.; phyllaries ca. 16, obscurely ranked, the outer tomentose; $n = 9$ (Anderson, 1966).—Montane Coniferous F.; San Bernardino Mts.

6. **C. teretifòlius** (Dur. & Hilg.) Hall. [*Linosyris t.* Dur. & Hilg. *Bigelovia t.* Gray.] Fastigiately branched globose or spreading shrub 2–15 dm. high, the herbage glabrous,

glandular-punctate, resinous, dark green; lvs. terete, obtuse, not mucronate, 1–2.5 cm. long, 0.5–1 mm. wide; heads in short terminal spikes, 4–6-fld.; invol. subcylindric, 6–8 mm. high; phyllaries 4–5-seriate, strongly graduate in vertical ranks, oblong, obtuse, indurate, strawcolor, tipped with a conspicuous green glandular spot; aks. appressed-villous; pappus stramineous; $n=9$ (Anderson, 1966).—Canyon walls and rocky slopes, 3200–8000 ft.; Sagebrush Scrub, Creosote Bush Scrub, Joshua Tree Wd., Pinyon-Juniper Wd.; deserts from Santa Rosa Mts. to Tehachapi and Mono Co.; to Utah, Ariz. Sept.–Nov.

7. **C. viscidiflòrus** (Hook.) Nutt. [*Crinitaria v.* Hook. *Bigelovia v.* DC.] ssp. **pùmilus** (Nutt.) Hall & Clem. [*C. p.* Nutt. *C. v.* var. *p.* Jeps.] Shrub 1–4 dm. high, with brittle twigs; lvs. linear, pungently acute, 2–4 cm. long, 1–2.5 mm. wide, plane or tortuous, bright green, viscidulous or slightly glandular and the margins slightly scabrous otherwise glabrous; cyme small but sometimes lax, with glabrous branches; invol. 5–6 mm. high; phyllaries oblong, not keeled, the outer acute, the inner obtuse or submucronate, all without a subapical spot; aks. ± strigose; $n=9$ (Anderson, 1966).—Frequent on dry benches, 5000–11,000 ft.; Sagebrush Scrub, Pinyon-Juniper Wd.; San Bernardino Mts., mts. of Inyo Co.; to Wash., Ida. Sept.–Nov.

Ssp. **pubérulus** (D. C. Eat.) Hall & Clem. [*Linosyris v.* var. *p.* D. C. Eat.] Low shrub 2–5 dm. high; lvs. narrowly linear, acute, 1.5–4 cm. long, rarely over 1 mm. wide, often twisted or revolute, pale grayish-green, densely puberulent; cyme mostly small and compact; the branches densely puberulent; invol. ca. 6 mm. high; phyllaries oblong, not keeled, obtuse or the outer almost acute; aks. densely strigose or silky; $n=9$ or 18 (Anderson, 1966).—Sagebrush-covered plains and slopes below 11,000 ft.; Sagebrush Scrub, Pinyon-Juniper Wd.; San Bernardino Mts., Inyo-White Range; e. flank of Sierra Nevada to Wash., Wyo., Nev.

Ssp. **stenophýllus** (Gray) Hall & Clem. [*Bigelovia douglasii* var. *s.* Gray.] Shrub 1–3 dm. high; lvs. paler, less than 1 mm. wide, rigid; invol. 4–6 mm. high.—E. end of Bear V., San Bernardino Mts.; Inyo-White Range; to Ore.

Ssp. **viscidiflòrus.** Rounded white-barked shrub less than 1 m. high; lvs. 2–5 cm. long, 2–5 mm. wide, 1–3-nerved, flat or twisted, spreading or reflexed, glabrous, viscid, sometimes scabro-ciliate; heads ca. 5 fld., in broad terminal cymes; invol. 5–7 mm. high; phyllaries lanceolate to linear-oblong, obtuse to acute, not keeled, strongly graduate; pappus brownish-white; $n=9$, 18 (Anderson, 1966).—Dry open places, 4000–7500 ft.; Yellow Pine F.; San Jacinto and San Bernardino mts.; Sierra Nevada to B.C., New Mex. and Mont. Aug.–Sept. (Plate 14, Fig. C.)

40. *Cichòrium* L. CHICORY

Annual or perennial herbs, with mostly basal lvs. Stems branching, with reduced lvs. Fls. mostly blue, in sessile heads, all ligulate and perfect. Invol. with 2 series of phyllaries, the outer shorter. Receptacle naked. Aks. glabrous, striate-nerved, 5-angled, truncate, beakless. Pappus in 2–3 series of short blunt paleae. Ca. 9 spp., largely Medit. (Name altered from the Arabic.)

1. **C. íntybus** L. Stems erect from a deep perennial taproot, glabrous or hairy, 3–10 dm. high; lower lvs. oblanceolate, toothed or pinnatifid, 1–2 dm. long, upward reduced and sessile; heads 1–3 in axils of much reduced upper lvs., the branches nearly naked, racemiform; heads ca. 4 cm. broad in anthesis, blue, occasionally white; invol. 10–15 mm. high, the outer phyllaries loose, fewer and ca. half as long as the inner; aks. 2–3 mm. long; pappus-paleae minute, narrow; $2n=18$ (Stebbins et al., 1953).—Natur. in waste places in much of cismontane Calif.; from Eu. June–Oct. The root used in many regions as an adulterant or substitute for coffee.

41. *Cirsium* L. THISTLE

Annual, biennial or perennial herbs, spiny, with alternate toothed or more usually pinnatifid lvs. and solitary to clustered discoid heads. Receptacle flat to subconic, densely bristly. Invol. with several series of phyllaries, these subequal or usually ± imbricate, generally some or most of them spine-tipped, in some spp. with a thickened glutinous dorsal ridge. Corolla tubular, its segms. linear, white or yellowish to red or purple. Fils. usually papillose-hairy; anthers caudate at base. Style with a thickened minute hairy ring, the branches nearly connate; stigmatic lines marginal, extending nearly to the free tips. Aks. flattened or quadrangular, glabrous, 4–many-nerved; pappus of bristles, mostly plumose, sometimes barbellate, deciduous in a ring. Ca. 200 spp., of N. Hemis. The spp. hybridize freely and much introgression occurs. (Greek, *kirsion*, a kind of thistle.)

1. Heads usually unisexual, the plants partly dioecious; perennial from creeping rootstocks; invol. 1–2.5 cm. high. Introd. weed .... 1. *C. arvense*
1. Heads with bisexual fls.
  2. Stem spinose-winged by decurrent lf.-bases almost as long as internodes; lvs. scabrous-hispid above. Introd. weed .... 13. *C. vulgare*
  2. Stems not spinose-winged by decurrent lf.-bases or only shortly so; lvs. not scabrous-hispid above.
    3. Middle or outer phyllaries with a conspicuous glutinous dorsal ridge.
      4. Heads many, rather small, the invol. mostly 2–2.5 cm. high. Mojave Desert
                                                                           5. *C. mohavense*
      4. Heads mostly solitary, larger, the invol. 2.5–4 cm. high.
        5. Stems 1–4 dm. high; spines of the phyllaries not more than 5–6 mm. long. Occasional introduction .... 12. *C. undulatum*
        5. Stems 5–15 dm. high; spines of the phyllaries ca. 1 cm. long. At 5000–10,000 ft., n. Inyo Co. .... 9. *C. ochrocentrum*
    3. Middle or outer phyllaries not bearing a conspicuous glutinous dorsal ridge (except possibly in *C. neomexicanum* and *C. californicum*).
      6. Heads leafy-bracteate (the spiny foliaceous subtending bracts usually more than 3), usually ± clustered.
        7. Plants acaulescent, to ca. 3 dm. high. Damp places.
          8. Heads 2.5–3.3 cm. long; corollas purple; outer phyllaries flexible, membranous; plants acaulescent .... 4. *C. congdonii*
          8. Heads 2.5–4.5 cm. high; corollas ochroleucous or pink; outer phyllaries stiff; plants acaulescent or stemmed .... 11. *C. tioganum*
        7. Plants caulescent, 5–20 dm. high. Dry places .... 2. *C. brevistylum*
      6. Heads not subtended by leafy bracts, usually not much clustered.
        9. Phyllaries essentially glabrous or arachnoid along margins only, not squarrose or reflexed. E. Mojave Desert .... 7. *C. nidulum*
        9. Phyllaries densely arachnoid-tomentose on back, sometimes ± glabrate in age, the middle ones ± squarrose, the outer reflexed.
          10. Stems coarse, the peduncles mostly 3–5 mm. in diam.
            11. Heads often as broad as long or broader; phyllaries generally straight, almost acicular above the appressed base; corollas scarcely exceeding the invol.; corollas usually purplish-red, rarely white .... 8. *C. occidentale*
            11. Heads generally longer than broad; phyllaries generally curved downward, outward or inward, linear to ovatish, but not acicular; corollas usually crimson-red, but varying to purplish-red .... 10. *C. proteanum*
          10. Stems more slender, the peduncles mostly 2–2.5 mm. in diam.
            12. Fls. mostly pinkish or lavender; lvs. mostly not strongly decurrent. Cismontane
                                                                                  3. *C. californicum*
            12. Fls mostly whitish; lvs. often strongly decurrent. Mojave Desert
                                                                                  6. *C. neomexicanum*

    1. **C. arvénse** (L.) Scop. [*Serratula a.* L.] CANADA THISTLE. Perennial with creeping rootstocks; stems 3–9 dm. high, slender; lvs. oblong or lanceolate, subrigid, smooth or slightly woolly beneath, at last green on both sides, strongly sinuate-pinnatifid, prickly-margined, 5–12 cm. long; heads imperfectly dioecious, rather numerous, the invol. 1.5–2 cm. high; outer phyllaries appressed, ovate, subulate-tipped, the inner with thin attenuate tips; fls. mostly pinkish-purple, not much exserted; $2n = 34$ (Ehrenberg, 1945).—Occasionally natur. in low places; native of Eu. June–Sept.
    2. **C. brevístylum** Cronq. [*C. edule* Calif. refs., not Nutt.] Plant robust short-lived perennial from taproot; stem mostly 1–2 m. high, ± crisp-arachnoid, leafy to the top; lvs. green above, ± arachnoid beneath, mostly 5–15 cm. long, shallowly sinuate-pinnatifid,

the upper auriculate-clasping, all ± weakly spinulose-ciliate; heads clustered at summit of branches, often exceeded by subtending lvs.; invol. 2–4 cm. high, ± arachnoid, not much imbricate; phyllaries slender, tapering gradually, all but the innermost with short erect spines; fls. dull purple-red or sometimes whitish, not much exserted; aks. 3.5–4 mm. long, yellowish at tip.—Brushy and wooded slopes, Coastal Sage Scrub, Chaparral; occasional, cismontane to B.C. April–Aug.

3. **C. califórnicum** Gray. [*Carduus c.* Greene.] Plate 14, Fig. D. Mostly biennial, slender, 5–18 dm. high, leafy near the base, cymosely branched above, white-woolly, glabrate in age; lvs. narrow, 1–3.5 dm. long, sinuately to deeply pinnatifid, moderately spinose; heads mostly solitary on the long slender peduncles; invol. hemispheric, ± arachnoid-woolly, 2–3 cm. high; outer phyllaries with spreading terminal spines, the inner erect and more herbaceous; fls. lavender to whitish, well exceeding invol.—Common on dry slopes and fans, below 7000 ft.; Coastal Sage Scrub, Chaparral, Yellow Pine F.; cismontane s. Calif.; to cent. Calif. April–July. Santa Catalina Id. Rather low less spinose plants, more densely white-tomentose have been called var. *bernardinum* (Greene) Petrak.

4. **C. cóngdònii** Moore & Frankton. [*C. drummondii* Calif. refs. in part.] Acaulescent; lvs. usually ca. 12 cm. (–20) long, 6 cm. wide, pinnatifid to ⅔ the width of the blade, with slender marginal spines, blades light green above, gray-green beneath, in a prostrate rosette and often with purple pigment on veins and margins; heads 3–9, sessile in center of rosette, 2.5–3.5 cm. high; invol. 2–2.5 cm., outer phyllaries lanceolate, 2–4 mm. wide, green or brownish-green, eglandular, with marginal arachnoid pubescence, the terminal spine slender, 2–3 (–4) mm. long, inner phyllaries longer, innermost unarmed, often with some purple, chartaceous, curled; corolla purple or reddish-purple, 22–30 mm. long; pappus 17–18 mm. long; aks. 4–4.5 mm. long, dark brown with yellow apical rim; $2n = 34$ (Moore & Frankton, 1967).—Meadowy and damp places, 5000–10,000 ft.; Yellow Pine F., Bristle-cone Pine F.; San Bernardino Mts., White Mts.; to Sierra Nevada. June–Aug.

5. **C. mohavénse** (Greene) Petr. [*Carduus m.* Greene.] Short-lived perennial or biennial, the stem rather stout, 6–15 dm. high, simple or paniculately branched above; herbage white-tomentose or the wool deciduous on upper surface of lvs.; lvs. rather narrow, 1–2.5 dm. long, evenly sinuate-pinnatifid, the lobes tipped with numerous long yellow spines; upper lvs. rather strongly decurrent; invol. oblong, becoming hemispheric, 1.5–3 cm. high, glabrate; phyllaries with dorsal glutinous ridge, strawcolor and ending in stout spines; fls. pinkish or light red.—Moist alkaline places at 1400–7000 ft.; Creosote Bush Scrub, Alkali Sink, Pinyon-Juniper Wd.; Mojave Desert. July–Oct.

6. **C. neoméxicanum** Gray. Biennial, stout, 1.5–2 m. tall, white-woolly; basal lvs. narrow to ± elliptic, deeply and regularly sinuate-pinnatifid, strongly spiny, to ca. 4 dm. long; cauline lvs. gradually reduced up the stem, quite strongly decurrent; heads 4–5 cm. broad; invol. 2.5–4 cm. high, subglobose, sparsely arachnoid, glabrate; outer phyllaries reflexed, the middle squarrose, strongly spinose, innermost attenuate, scarcely spine-tipped; fls. whitish, well exserted.—Dry rocky places, 3000–6000 ft.; Joshua Tree Wd., Pinyon-Juniper Wd.; mts. of e. Mojave Desert; to Colo., New Mex. April–May.

7. **C. nìdulum** (Jones) Petr. [*Cnicus n.* Jones.] Perennial, with stems erect, branched above, ca. 5–8 dm. high, arachnoid-tomentose; basal lvs. 2–3.5 dm. long, deeply and regularly sinuate-pinnatifid, the lobes short, lanceolate, with very long yellow spines; upper lvs. sessile, reduced, very spiny; heads solitary, ovoid; invol. ca. 3.5 cm. high, the phyllaries glabrous except for the arachnoid margins, with a rather faint glutinous ridge, the lower phyllaries spreading-reflexed, they and the middle ending in yellow spines to 2.5 cm. long, the innermost acuminate, not spine-tipped; fls. light red-purple, well exserted.—Rocky places, below 9000 ft.; Creosote Bush Scrub to Pinyon-Juniper Wd.; New York Mts., Inyo-White Range; to n. Ariz. July–Oct.

8. **C. occidentàle** (Nutt.) Jeps. [*Carduus o.* Nutt. *Cirsium coulteri* Harv. & Gray.] Plate 14, Fig. E. Biennial or winter annuals, erect, leafy, usually 0.5–1 m. tall, arachnoid- or lanate-tomentose; lower lvs. 1–3 dm. long, sinuately rather deeply lobed, the lobes deltoid, toothed, spine-tipped; stems ± leafy, the uppermost lvs. spiny, reduced; heads soli-

tary at ends of the stems, 3–5 cm. long; invols. campanulate to broadly hemispheric, the phyllaries interconnected by a webbing of fine whitish filmy tomentum, acicular, the outer tending to be reflexed and downward-pointing, the middle ± spreading horizontally, the inner flattened toward the tip, sharply acute or weakly spine-tipped; fls. purplish-red, rarely white, scarcely exceeding invol.; corollas 2.5–3.5 cm. long; aks. 5–6 mm. long, brownish.—Sandy to grassy or brushy places, mostly at fairly low elevs.; Coastal Strand, Coastal Sage Scrub, Chaparral, etc.; cismontane s. Calif. to cent. Calif.; on all the major ids. April–July.

9. **C. ochrocéntrum** Gray. Stem stout, 5–15 dm. high, from a perennial base, tomentose throughout; lvs. pinnatifid with rather crowded segms. and armed with long yellowish spines; upper lvs. quite decurrent; heads mostly solitary at ends of branches, 4–6 cm. high; invol. of many imbricate lanceolate phyllaries with conspicuous glutinous ridges and yellow spines 1–2 cm. long; fls. red-purple, well exserted.—A weedy introduction, as at Beaumont, Escondido; native from Ariz. to Tex. and Nebr. May–Oct.

10. **C. proteànum** J. T. Howell. Much like *C. occidentale*, but the heads longer than broad, generally 4–5 cm. long; phyllaries not acicular, often subglabrate; fls. usually crimson-red but varying to more purplish; aks. 6–7 mm. long, truncate at base.—From Kern and Ventura cos. n. to Shasta Co. April–Aug.

11. **C. tiogànum** (Congd.) Petr. [*Cnicus t.* Congd.] Acaulescent to 1.5 dm. high; stem thick, fleshy, ribbed, simple or few-branched; lvs. oblanceolate, to 3 dm. long, the lower generally lobed with marginal spines, light green, lightly arachnoid, the uppermost subentire; heads 3–5 (–9), in a close terminal group exceeded by the upper cauline lvs., mostly 2.5–3.5 (–4.5) cm. high, the invol. of 5–7 rows of phyllaries, the outer ovate to ovate-lanceolate, the terminal spine mostly 2–5 mm. long; inner phyllaries longer, membranous at tip, pointed; corolla ochroleucous, white to pinkish, 25–35 cm. long; aks. 5–6 mm. long; $2n = 34$ (Moore & Frankton, 1967).—Meadows, 3000–7000 ft.; Chaparral, Yellow Pine F.; Palomar and San Bernardino mts.; to Sierra Nevada, Colo., New Mex. June–Aug.

12. **C. undulàtum** (Nutt.) Spreng. [*Carduus u.* Nutt.] Plants with a well developed underground rootstock with tuberous thickenings; stems 1–4 dm. high, mostly simple, persistently white-tomentose, as are lower surfaces of lvs.; lvs. coarsely toothed to pinnatifid with ovoid or deltoid lobes; heads mostly several, at branch-ends; invol. 2.5–4 cm. high, the phyllaries ± tomentose on margins, with a dorsal glutinous ridge, well imbricate, the inner attenuate, and often crisped at tip; fls. mostly pink-purple, well exserted; $2n = 26$ (Frankton & Moore, 1961).—Apparently an occasional introduction in fields as at Escondido, W. Covina; Ariz. to B.C., Minn. July–Sept.

13. **C. vulgàre** (Savi) Ten. [*Carduus v.* Savi. *Cirsium lanceolatum* auth., not Hill.] BULL THISTLE. Coarse spreading biennial, 6–12 dm. tall; rosette-lvs. oblanceolate to elliptic, coarsely toothed; cauline lvs. lanceolate, to ca. 3 dm. long, deeply pinnatifid into lanceolate lobes, green and hirtellous above, armed with long fierce prickles, tomentose beneath; lf.-bases decurrent on stem as long interrupted prickly wings; heads 1–few; invol. ovoid to subglobose, 3–4 cm. long, the phyllaries mostly lanceolate to linear, attenuate to subulate-acerose, spreading; fls. purple, well exserted; $2n = 68$ (Poddubnaja, 1931).—An aggressive weed becoming common in waste places below 5000 ft.; from Eu. June–Sept.

42. Cnìcus L. BLESSED THISTLE

Annual loosely pubescent herb with sinuate-dentate or pinnatifid lvs. having spiny teeth or lobes. Heads large, sessile, of yellow tubular fls. and solitary, terminal, subtended by the upper lvs. Phyllaries imbricated in several series, the outer ovate, the inner lanceolate, tipped by long pinnately branched spines. Receptacle bristly. Aks. terete, striate, laterally attached, 10-toothed at apex. Pappus of 2 series of awns, the outer long, naked, yellow, the inner white, hispidulous. One sp. (Latin name of the Safflower, from the Greek *cnecos*.)

1. **C. benedíctus** L. Branched, 2–6 dm. high; lvs. oblong-lanceolate in outline, con-

spicuously veiny, 6–15 cm. long; invol. 3–4 cm. long; $2n=22$ (Varama, 1947).—Occasional weed in waste places, fields, etc.; cismontane; natur. from Eu. April–July.

43. Con*y*za L., emend. Less.

Annual or perennial herbs. Lvs. alternate, subentire to bipinnatifid. Heads small, several to many, disciform or minutely radiate. Invol. few-seriate, the subequal phyllaries narrow, subherbaceous, strongly reflexed in age. Pistillate fls. very numerous, several-seriate, their corollas tubular-filiform, in ours usually with an inconspicuous white or purplish ligule barely equaling the style and pappus. Disk-fls. perfect, few, their narrowly tubular corollas (in ours) enlarged above. Style-branches with short ovate hispidulous appendages. Aks. small, oblong, compressed; pappus of rather few capillary bristles. More than 50 spp., chiefly of warmer regions. (Name used by Dioscoridary and Pliny for some kind of Fleabane, presumably from the Greek, *konops*, a flea.)

Upper and lower lvs. coarsely dentate; ray-fls. without ligules; herbage glandular-viscid
                                                               3. *C. coulteri*
Upper lvs., at least, entire; ray-fls. with ligules.
    Invol. copiously hairy, at least the outer phyllaries herbaceous . . . . 1. *C. bonariensis*
    Invol. subglabrous, the phyllaries with scarious margins . . . . 2. *C. canadensis*

1. **C. bonariénsis** (L.) Cronq. [*Erigeron b.* L. *E. linifolius* Willd.] Annual, up to 1 m. high, subsimple or with erect leafy branches; herbage grayish green, ± densely strigose and hirsute; lvs. as in *C. canadensis*; heads rather many; invol. 4–5 mm. high, the phyllaries densely hirsutulous; rays 125–180; disk-fls. 10–20; pappus whitish or stramineous; $2n=54$ (Holmgren, 1919).—Frequent weed in waste ground; to Atlantic Coast; introd. from S. Am. June–Aug.

2. **C. canadénsis** (L.) Cronq. [*Erigeron c.* L.] Horseweed. Strict leafy annual up to 2 m. high, sometimes with erect branches in upper half, green, subglabrous to hirsute; lvs. many, the lower oblanceolate, to 1 dm. long, entire or serrate, the upper narrower, sessile, entire; heads small, numerous, panicled; invol. 3–4 mm. high, the phyllaries linear and glabrous or nearly so, with conspicuous oil-filled cent. area and narrow subscarious margin; rays 25–40, white, very inconspicuous; disk-fls. 7–12; $2n=18$ (Okabe, 1934).—Common weed of waste ground; throughout the U.S. and s. Canada; trop. and S. Am. June–Sept.

3. **C. còulteri** Gray. [*Erigeron discoidea* Kell. *Conyzella c.* Greene.] Erect annual 2–10 dm. high, glandular-pubescent and villous or hirsute throughout, the rigid very leafy stem simple below, paniculately branched above; lvs. narrowly oblong, coarsely toothed, the lower petioled, the cauline sessile and 2–6 cm. long; heads small, numerous; invol. 2–3 mm. high, the linear-attenuate phyllaries hirsute and glandular; ray-fls. 125–250, without ligules; disk-fls. 5–15; pappus soft, whitish.—Occasional in moist flats below 1000 ft.; Coastal Sage Scrub, etc.; cismontane and Imperial V.; Santa Catalina and San Clemente ids.; Santa Cruz Id. May–Oct. E. to Colo., Tex., Mex.

44. *Coreópsis* L. Coreopsis. Tickseed

Annual or perennial herbs, or sometimes shrubby. Lvs. usually opposite, entire, dentate, tripartite or 1–4 times pinnately dissected. Heads showy, long-peduncled, radiate. Invol. 2-seriate in ours, the outer phyllaries 5–8, spreading, herbaceous, the inner 8–12, erect, membranous. Paleae of the flat receptacle thin, scarious. Aks. obcompressed, oblong to orbicular, the margin smooth or ciliate, thin-winged or thickened, with pappus a small cup, or of 2 teeth or narrow paleae over the angles, or obsolete. Ca. 100 spp. of warm and temp. regions, chiefly Am. (Greek, *koris*, a tick, and *opsis*, resemblance, from the form of the ak).

(Sharsmith, Helen K. The native Calif. spp. of the genus Coreopsis. Madroño 4: 209–231. 1938.)
Lvs. entire; perennial occasionally escaping from cult . . . . 6. *C. lanceolata*
Lvs. 1–2-pinnate. Native spp.

Perennials; stems stout; heads 5–8 cm. across. Coastal.
    Heads numerous, cymosely clustered on leafy peduncles. Los Angeles Co. n. 5. *C. gigantea*
    Heads few, on naked peduncles. San Diego Co. .... 7. *C. maritima*
Annuals; stems slender; heads 2–5 cm. across. Not maritime.
    Aks. nonciliate; pappus-paleae none.
        The aks. dull, beset with clavellate hairs, the corky wing thick and rugose
            2. *C. californica*
        The aks. shining, smooth or nearly so, the corky wings thin .... 4. *C. douglasii*
    Aks. of disk ciliate, with 2 pappus-paleae; ray-aks. glabrous, epappose.
        Outer phyllaries linear, obtuse; stems essentially scapose .... 1. *C. bigelovii*
        Outer phyllaries broadly ovate, acute; stems leafy below .... 3. *C. calliopsidea*

1. **C. bigelòvii** (Gray) Hall. [*Pugiopappus b.* Gray.] Plate 14, Fig. F. Erect glabrous annual, with few 1-headed scapes 1–5 dm. high; lvs. basal or nearly so, 4–12 cm. long, ovate in outline, 1–2-pinnate into linear lobes as wide as the 1–2 mm. rachis; heads 2–4.5 cm. across, golden; outer phyllaries linear, shorter or longer than the dark oblong-ovate inner ones, glabrous; ray-aks. oblong-obovate, glabrous, narrowly winged, without pappus; disk-aks. oblanceolate, dorsally glabrous, the pappus of 2 bright white linear-lanceolate fimbriate paleae ca. 2.5 mm. long.—Dry gravelly hillsides below 6000 ft.; Creosote Bush Scrub, Joshua Tree Wd., Chaparral, Pinyon-Juniper Wd.; San Gorgonio Pass to Santa Monica Mts. and across the desert to Death V. region and to Monterey and Tulare cos. March–May.

2. **C. califórnica** (Nutt.) H. K. Sharsmith. [*Leptosyne c.* Nutt. *L. douglasii* auth., not DC.] Habit of *C. bigelovii*; heads 1–3.5 cm. across, yellow; aks. obovate, dull and minutely papillate, with stubby hyaline clavellate or capitate hairs on wing and body, brownish, the lighter-colored wing conspicuously spongy-thickened, ± ciliolate; pappus reduced to a cupule.—Sandy or gravelly flats and slopes below 3000 ft.; Creosote Bush Scrub, V. Grassland, S. Oak Wd.; Imperial Co. to s. Inyo Co., cismontane Los Angeles Co. to San Diego Co.; to Ariz., L. Calif. March–May.

3. **C. calliopsídea** (DC.) Gray. [*Agarista c.* DC. *Pugiopappus c.* Gray.] Stout erect glabrous annual, the stems monocephalous, 1–5 dm. high; lvs. on lower half of the stem, the lower 4–8 cm. long, ovate in outline, 1–2-pinnate into linear lobes 0.5–2 mm. wide; heads 2–9 cm. across, golden; outer phyllaries deltoid-ovate, shorter than the ovate inner ones; palea adherent to the disk-fl.; ray-aks. oval, glabrous, broad-winged, epappose; disk-fls. lance-oblong, dorsally glabrous and shining, ventrally rather copiously silky, the margin densely ciliate with longer hairs, their pappus of 2 yellowish fimbriate awns ca. 4 mm. long.—Dry open gravelly ground, below 3200 ft.; Creosote Bush Scrub, Joshua Tree Wd.; w. Mojave Desert from San Bernardino Co. w. and n. to Alameda Co. March–May.

4. **C. doúglasii** (DC.) Hall. [*Leptosyne d.* DC.] Much like *C. californica*, to 2.5 dm. high, the lvs. entire or few-lobed, nearly filiform, grooved above, remotely glandular-ciliolate; heads 1–2.5 cm. across; outer phyllaries narrowly lanceolate, inner lance-ovate; aks. obovate, concavo-convex, essentially glabrous, brown, the wing yellowish, not rugose, scarcely thickened; pappus a mere cupule.—Below 2000 ft.; Chaparral, etc.; Santa Barbara Co. to Santa Clara Co. March–April.

5. **C. gigántea** (Kell.) Hall. [*Leptosyne g.* Kell. *Tuckermannia g.* Jones.] Stout erect fleshy arborescent few-branched shrub 4–15 (–30) dm. high, glabrous, the main trunk to 12 cm. thick; lvs. in dense tufts at ends of primary branches, alternate, 5–25 cm. long, 3–4-pinnate into nearly filiform segms.; lvs. deciduous in dry season; heads in cymose clusters on long scapifòrm peduncles, 4–8 cm. across, yellow; aks. oblong to obovate, plane, narrowly winged; pappus none.—Rocky cliffs and exposed dunes, immediate coast; Coastal Strand, Coastal Sage Scrub; Los Angeles Co. to San Luis Obispo Co. Ids. from Santa Catalina n. March–May.

6. **C. lanceolàta** L. Perennial herb with 1–several erect or ascending few-branched stems 2–6 dm. high; lvs. several in the lower half of the stem, linear-lanceolate to oblanceolate, the lowermost slender-petioled, all entire; heads 3–6 cm. across, solitary on long

slender peduncles, yellow; phyllaries all broad; aks. orbicular, 2.5–3 mm. diam., black, with broad thin wing.—Occasional escape from gardens; native in e. U.S. May–July.

7. **C. marítima** (Nutt.) Hook. f. [*Tuckermannia m.* Nutt.] Robust erect glabrous perennial, with fleshy tubers and fistulous stems 2–5 dm. high; lvs. fleshy, slender, petiolate, 5–25 cm. long, 2–3-pinnate into widely divaricate linear segms. 2–3 mm. wide; heads few, on naked stout peduncles 15–50 cm. long, 6.5–8 cm. across, yellow; phyllaries all broad, the inner scarcely exceeding the outer; rays 15–20; style-branches triangular-acute; aks. oblong to obovate, plane, the thin wing to 1 mm. wide, rarely with 1–2 awns.—Coastal bluffs and dunes below 200 ft.; Coastal Strand, Coastal Sage Scrub; s. San Diego Co.; n. L. Calif. and adjacent ids. March–June.

45. *Corethrógyne* DC.

Perennial herbs, sometimes suffrutescent at base, clothed with a soft white, at length ± deciduous wool; resembling *Aster*. Heads many-fld., solitary or cymose, or paniculate. Invol. turbinate to hemispheric; phyllaries imbricate in several series, narrow, with erect to squarrose green tips. Ray-fls. neutral, numerous, 1-seriate; ligule linear, violet. Disk-fls. yellow. Style-branches linear, the short blunt appendage with a dense tuft of rigid yellow hairs. Aks. of ray none or rudimentary, those of disk cuneiform or linear-turbinate, pubescent. Pappus of numerous fulvous bristles. Three spp. (Greek, *korethron*, a brush for sweeping, and *gune*, style, from the brushlike style appendages.)

1. **C. filáginifolia** (H. & A.) Nutt. [*Aster? f.* H. & A.] Suffrutescent or herbaceous, slender, erect or ascending, mostly paniculately few- to several-branched above, 2–8 dm. high; herbage with ± tardily deciduous tomentum; lvs. lanceolate to oblanceolate, or spatulate, often sharply serrate above, 2–8 cm. long, the upper reduced and sessile; invol. 5–11 mm. high.

KEY TO VARIETIES

Invol. glandular at anthesis.
  Invol. 10–12 mm. high, hemispheric, with long-stalked glands. Coastal San Diego Co.
                                                                                         3. var. *incana*
  Invol. under 9 mm. high, the glands sessile or short-stipitate.
    Plants ± green and glandular from base to apex, very obscurely if at all tomentose. Santa Cruz Id. . . . . 11. var. *viscidula*
    Plants ± tomentose, at least below.
      Stems usually under 4 dm. high.
        Tomentum only about the base, not more than half-way up the stem, the upper glandular portion bright green. San Gabriel Mts. . . . . 7. var. *pinetorum*
        Tomentum at least half-way up the stem.
          Invol. 7–9 mm. high; stems very stout, ± depressed. Santa Barbara Ids. 8. var. *robusta*
          Invol. 6–7 mm. high; stems fairly slender, quite erect. Montane . . . 2. var. *glomerata*
      Stems usually over 4 dm. high.
        Tomentum extending up to invol.; invol. turbinate, 5–7 mm. high. Inland valleys
                                                                        1. var. *bernardina*
        Tomentum not extending up to invol., but upper parts glandular.
          Invol. turbinate to broadly campanulate. Islands . . . . 10. var. *virgata*
          Invol. cylindric, 6–8-seriate; phyllaries squarrose, very glandular. San Fernando V. to Mt. Pinos, Tehachapi Mts. . . . . 6. var. *peirsonii*
Invol. tomentose at anthesis.
  Lvs. linear. San Diego region . . . . 5. var. *linifolia*
  Lvs. not linear.
    Invol. 7–8 mm. high, turbinate; lvs. mostly broadly oblong. Coastal . . . . . . 4. var. *latifolia*
    Invol. 9–14 mm. high, campanulate; lvs. ovate to spatulate. San Bernardino Mts.
                                                                              9. var. *sessilis*

1. Var. **bernardína** (Abrams) Hall. [*C. virgata* var. *b.* Abrams.] Stems slender, mostly 4–9 dm. high, rather persistently tomentose except on invols. and upper parts of peduncles; lvs. commonly narrow, 1–7 cm. long; heads terminating the long divaricate branches; invol. turbinate, 5–7 mm. high, the phyllaries often squarrose; $n=5$ (Raven et al., 1960).

—Sandy plains and rocky canyons, 1000–3500 ft.; cismontane, Ventura Co. to Riverside Co. July–Oct.

2. Var. **glomeràta** Hall. [*C. brevicula* Greene. *C. f.* var. *b.* Canby.] Stems erect, quite herbaceous, 2–4 (–6) dm. high, the tomentum deciduous from the uppermost parts; lvs. yellow-green; heads few; invol. turbinate, 6–7 mm. high, with generally recurved phyllaries; $n=5$ (Raven et al., 1960).—Frequent under pines, largely at 5000–8000 ft.; Laguna Mts. to s. Sierra Nevada. July–Oct.

3. Var. **incàna** (Nutt.) Canby. [*C. i.* Nutt. *C. f.* var. *pacifica* Hall.] Stout, erect, 5–8 dm. high, with tomentum deciduous from upper parts; lvs. linear to narrowly lanceolate or oblanceolate, mostly entire, to 7 or 8 cm. long; infl. an open panicle with very large heads; invol. 10–12 mm. high, densely beset with stout-stalked prominent glands.—Common on sandy slopes facing the sea; Coastal Sage Scrub; sw. San Diego Co. June–Aug.

4. Var. **latifòlia** Hall. [*C. flagellaris* Greene.] Suffrutescent, stout, 3–6 dm. high, tomentose throughout including the invols.; principal lvs. broadly oblong, 1–4 cm. long, 5–12 mm. wide; heads few; invols. 7–8 mm. long, ca. 5-seriate; phyllaries with slightly spreading tips; rays showy, 10 mm. long.—Slopes overlooking the coast; Coastal Sage Scrub; Redondo, Los Angeles Co. to Santa Barbara Co. Anacapa Id. Aug.–Dec.

5. Var. **linifòlia** Hall. Erect, 2–4 dm. high, permanently hoary even on the invols.; lvs. narrow, 1–2 (–5) mm. wide; invols. 8–10 mm. high, 6–7-seriate; rays lavender, 6–7 mm. long.—Common on bluffs and brushy slopes near the sea; Coastal Sage Scrub, Chaparral; sw. San Diego Co. July–Sept.

6. Var. **peirsònii** Canby. Stout, 3–9 dm. high, tomentose except in the paniculate to dark green glandular infl.; lvs. oblanceolate to obovate, 1–5 cm. long; invol. cylindric, dark-granular, 7–8 mm. high, the phyllaries numerous, prominently squarrose.—Open places and canyons below 5000 ft.; Coastal Sage Scrub, Chaparral, S. Oak Wd.; mts. of Ventura Co. to Tehachapi Mts. and w. end of San Gabriel Mts. Aug.–Nov.

7. Var. **pinetòrum** Jtn. Herbaceous, 1–5 dm. high; lower lvs. and lower parts of stem permanently tomentose, the upper half of plant oily-green, with dense stipitate glands; heads few; invol. turbinate, 5–7 mm. high; phyllaries squarrose; $n=5$ (Raven et al., 1960).—Dry rocky places 4000–5500 ft. or somewhat lower; Chaparral, Yellow Pine F.; San Gabriel Mts. July–Oct.

8. Var. **robústa** Greene. Stems stout, not erect, tomentose except on upper parts; lvs. numerous, broadly obovate, to spatulate, 2–4 cm. long; panicle dense, corymbose; invol. hemispheric, 7–9 mm. high, scarcely glandular.—Coastal Sage Scrub; San Miguel and Santa Rosa ids.; n. Santa Barbara Co. to San Luis Obispo Co.

9. Var. **séssilis** (Greene) Canby. [*C. s.* Greene.] Stout erect, 2–7 dm. high, densely and rather permanently tomentose including the invols.; lvs. spatulate to ovate, rather equably distributed; heads many-fld.; invol. campanulate, 9–14 mm. high, 7–9-seriate; phyllaries broad; rays 7–11 mm. long.—Common among pines, 4000–7500 ft.; Yellow Pine F.; San Bernardino Mts. July–Oct.

10. Var. **virgàta** (Benth.) Gray. [*C. v.* Benth. *C. f.* var. *rigida* Gray.] Slender, erect, 4–12 dm. high, ± gray- or white-tomentose below, becoming green and short-stipitate-glandular above; lvs. oblanceolate to linear-lanceolate, entire or with serrate tips; infl. a diffuse panicle; invol. 5–8 mm. high, the phyllaries sometimes squarrose; $n=5$ (Raven et al., 1960).—Our commonest form; open or brushy places below 2000 ft.; Coastal Strand, Coastal Sage Scrub, S. Oak Wd.; Catalina Id.; along the coast and outer Coast Ranges, L. Calif. to cent. Calif. July–Oct.

11. Var. **viscídula** (Greene) Keck. [*C. v.* Greene.] Erect, 4–7 dm. high; tomentose when young, the tomentum soon lost, the plant yellowish-green and very viscid almost throughout at anthesis, often much branched; Coastal Sage Scrub, Chaparral; Santa Cruz Id.; Monterey region. June–Dec.

46. *Cótula* L.

Low mostly diffuse herbs with alternate toothed, lobed or dissected lvs. Heads pedunculate, disciform. Fls. yellow. Receptacle flat or nearly so. Phyllaries greenish, subequal, in 1 or 2 ranks. Outer series of fls. ♀, fertile. Disk-fls. tubular, 4-toothed, fertile or not. Mature aks. ± pedicelled, compressed in ours and spongy-margined or narrow-winged. Pappus none. Ca. 50 spp. of wide distribution in S. Hemis. (Greek, *kotule*, a small cup, referring to a hollow at the base of the amplexicaul lvs.)

Annual; lvs. finely pinnate; heads 2–5 mm. broad .... 1. *C. australis*
Perennial; lvs. entire or pinnatifid; heads 8–10 mm. broad .... 2. *C. coronopifolia*

1. **C. austràlis** (Sieber) Hook. f. [*Anacyclus a.* Sieber.] Slender-stemmed, branched from base, spreading, 0.3–2 dm. high, sparsely spreading-pubescent; lvs. 1–3 cm. long, 1–2-pinnate into linear lobes; phyllaries brownish at apex, round-oblong, with scarious margins; marginal aks. minutely glandular on both surfaces, but glabrous on margins, slightly over 1 mm. long.—Very common troublesome weed about gardens, city lots, etc.; from Australia. Jan.–May.

2. **C. coronópifolia** L. Brass-Buttons. Perennial, decumbent or repent, fleshy, glabrous, branched; stems 2–3 dm. long; lvs. 2–7 cm. long, linear-oblong to -oblanceolate, entire to coarsely and deeply few-toothed, with broad rachis; heads depressed, bright yellow; phyllaries oblong, 3–5-veined, ± anthocyanous; ♀ fls. in 1 row, on pedicels as long as invol., without corolla; disk-fls. on much shorter pedicels, numerous; aks. winged, almost 2 mm. long; $2n=20$ (Castro & Fontes, 1946).—Common on mud and moist banks, about salt marshes, etc., many Plant Communities; cismontane and occasionally to desert edge; natur. from S. Afr. March–Dec. Santa Catalina, Santa Cruz and Santa Rosa ids.

47. Crèpis L. Hawksbeard

Annual to perennial herbs, usually ± caulescent. Lvs. largely ± basal, entire to toothed or deeply lobed. Heads in panicles or cymose clusters. Invol. cylindric or campanulate, the inner phyllaries equal, in a single series, with ± thickened midribs; outer calyculate phyllaries small or wanting. Receptacle naked. Fls. yellow, all ligulate. Aks. columnar or fusiform, narrowed toward the summit, 10–35-ribbed. Pappus of many soft mostly white bristles. Many polyploid apomicts occur that are ± intermediate between the diploid spp. and cannot be accounted for in a brief treatment. Ca. 200 spp. widely distributed in N. Hemis. (Greek, *krepis*, a sandal, an ancient plant name.)

(Babcock, E. B. The genus Crepis. Univ. Calif. Publ. Bot. 21–22, pp. 1–1030. 1947.)
Plants native perennials.
  The plants glabrous, densely tufted, 2–7 cm. high; lvs. entire .... 3. *C. nana*
  The plants ± hairy or tomentose, larger; lvs. toothed or pinnatifid.
    Largest heads of the infl. with 5–7 inner phyllaries; heads 5–10-fld. .... 1. *C. acuminata*
    Largest heads of infl. with 8–13 inner phyllaries; heads 9–40-fld.
      Invols. 5–9 mm. wide at anthesis, mostly 12–25-fld. .... 4. *C. occidentalis*
      Invols. 3–5.5 mm. wide at anthesis, mostly 8–10-fld. .... 2. *C. intermedia*
Plants introduced annuals or biennials, weedy .... 5. *C. vesicaria*

1. **C. acuminàta** Nutt. Stems 2–6 dm. high, with gray-green foliage; basal lvs. 1–4 dm. long, 1–10 cm. wide, pinnately lobed with 5–10 pairs of lateral entire or dentate segms., the apical part acuminate, 3–8 cm. long; cauline lvs. few, remote, reduced upward; infl. corymbiform, usually ca. 30–100-headed; heads 5–10-fld.; invol. cylindric-campanulate, 9–15 mm. long; outer phyllaries 5–7, ciliate; inner 5–8, lanceolate, ciliate at apex; corollas 10–18 mm. long; aks. pale yellow or brownish, 6–9 mm. long, ca. 12-ribbed; $2n=22, 33, 44, 55?, 88?$ (Babcock, 1947).—Dry flats, Montane Coniferous F.; at 7000–10,300 ft.; San Bernardino Mts., Mt. Pinos, San Gabriel Mts., White Mts.; to Wash., Rocky Mts. June–Aug.

2. **C. intermèdia** Gray. [*C. acuminata* var. *i.* Jeps.] Mostly 3–7 dm. high, canescent-tomentose; lower lvs. 1.5–4 dm. long, 2–9 cm. wide, elliptic-lanceolate, pinnatifid, the apical part long-attenuate, the lateral lobes lanceolate, entire or dentate; infl. corymbiform, with 10–60 heads, these 7–12 (–16)-fld.; invol. 10–16 mm. high, tomentose to glabrous; outer phyllaries 6–8, lance-deltoid, the inner mostly 7–8, becoming carinate;

corollas 14–30 mm. long; aks. yellow to brown, 6–9 mm. long, 10–12-ribbed; pappus 7–10 mm. long; $2n = 33?$, 44?, 55, 88? (Babcock, 1947).—Dry slopes, 6000–9500 ft.; Sagebrush Scrub, Montane Coniferous F.; Grapevine Mts., Last Chance Mts., Inyo-White Mts., all of Inyo Co.; to Wash., Rocky Mts. June–Aug.

3. **C. nàna** Richards. Plants tufted, 0.2–0.7 dm. high, the stems many, slender, arising from branches from a root-crown; lvs. mostly basal, glabrous, obovate-spatulate to ± elliptical, the blades 1–5 cm. long, on petioles ca. as long, ± purplish, entire or nearly so; branches congested, 2–4-headed, the heads borne among the lvs.; invol. 10–13 mm. long, cylindrical; outer phyllaries 5–8, unequal; inner 10, equal, with a purplish ciliate apex; corolla 7–8 mm. long, yellow with purplish tinge on outer face; aks. 4–6 mm. long, golden-brown, 10–13-ribbed; $2n = 14$ (Babcock, 1947).—Stony or gravelly scree or talus, 8000–12,000 ft.; Lodgepole Pine F., Bristle-cone Pine F.; San Gabriel, Panamint and more n. mts.; to Alaska, Labrador; Asia. July–Aug.

4. **C. occidentàlis** Nutt. subsp. **occidentàlis**. Mostly 1.5–3 dm. high, with a close gray tomentum and infl. ± glandular; lvs. 1–3 dm. long, sinuately dentate or runcinately or deeply pinnatifid with linear or lanceolate toothed lobes; lvs. gradually reduced up the stems; stems erect, stout, several-branched, each main branch with ca. 10–30 heads; larger heads 18–30-fld.; outer phyllaries 6–8, to ca. half as long as inner; inner 8–13, with some glandular pubescence and ± carinate toward outer base; corolla ca. 22 mm. long; aks. brown, 6–10 mm. long, 10–18-ribbed; $2n = 22$, 33, 44, 55?, 66?, 77, 88? (Babcock, 1947).—Dry rocky and stony places, mostly 4000–8000 ft.; Montane Coniferous F., Sagebrush Scrub; San Bernardino Mts., Inyo Mts.; to Wash., Rocky Mts. June–Aug.

Subsp. **pùmila** (Rydb.) Babcock & Stebbins. [*C. p.* Rydb.] Lacking glandular pubescence on the tomentose phyllaries.—With the typical ssp. from Kern and Ventura cos. n.

5. **C. vesicària** L. ssp. **taraxacifòlia** (Thuill.) Thell. Annual or biennial, 1–8 dm. high; stems 1–few, paniculately branched above; herbage subglabrous, but with short scattered bristles; lower lvs. pinnately parted to dentate, 4–10 cm. long; heads many-fld.; invol. 8–12 mm. high; outer phyllaries 6–12, inner 9–13; corolla 10–12 mm. long; aks. 6–8 mm. long, pale brown; $2n = 8$, 16 (Babcock, 1947).—Introd. weed which has been reported from Los Angeles Co.; native of Eu. June–Aug.

48. *Cynàra* L.

Stout thistlelike perennial herbs, mostly coarse and prickly, with large pinnatifid or bipinnatifid lvs. and very large solitary globose heads. Invol. broad or nearly globose; phyllaries in many series and ± enlarged at the base. Receptacle fleshy, plane, bristly. Corolla slender-tubed, violet, blue or white. Aks. thick, glabrous, compressed or 4-angled, with truncate apex. Pappus of many series of plumose bristles. Ca. 12 spp. of Medit. region and Canary Ids.; 2 grown as vegetables. (Greek, *kuon*, dog, the invol.-spines likened to dog's teeth.)

1. **C. cardúnculus** L. Cardoon. Robust, freely branched, 5–10 (–20) dm. high; stems and under side of lvs. white-tomentose, the upper side of lvs. grayish-green, ± arachnoid; lvs. 3–6 dm. long, deeply pinnatifid, prominently spiny; invol. 3–4 cm. high; phyllaries spine-tipped; heads purple-fld.; $2n = 34$ (Covas & Schnack, 1947).—Occasionally natur. in low places, as in Orange Co. and cent. Calif.; from Medit. region. May–July.

49. *Dicòria* T. & G.

Diffusely branched annual herbs. Upper lvs. alternate, petioled, entire or toothed. Heads small, numerous, nodding, spicately arranged in loose panicles, heterogamous and also some ♂. Phyllaries strongly dimorphic, the outer ca. 5, small, herbaceous, the inner subtending the 1–2 ♀ fls. subscarious, accrescent, much surpassing the outer at maturity. Paleae narrow, tardily deciduous. Pistillate fls. without corolla. Staminate fls. with funnelform corolla, the stamens with coherent fils. and free anthers. Aks. plano-convex,

Plate 15. Fig. A, *Dicoria canescens*, inner phyllaries enlarged and subtend ♀ fls. Fig. B, *Dyssodia porophylloides*, glandular phyllaries. Fig. C, *Encelia farinosa*. Fig. D, *Erigeron foliosus*. Fig. E, *Eriophyllum confertiflorum*. Fig. F, *Filago californica*, outer fl. ♀, enclosed by bract.

keeled on each face, pectinately wing-margined, epappose. Ca. 4 spp. of arid sw. U.S. (Greek, *dis*, twice, and *koris*, a bug, from the aspect of the 2 aks. of the original sp.)

1. **D. canéscens** T. & G. ssp. **canéscens**. Plate 15, Fig. A. Widely branching plants 3–9 dm. high, white-strigose-canescent or -hirsute and often also somewhat hispid and glandular; juvenile lvs. deltoid-lanceolate, dentate, 3–5 cm. long, floral lvs. broadly ovate to subrotund, denticulate to subentire, ca. 1 cm. long; outer phyllaries elliptic, 2–3 mm. long, reflexed in age, the inner ones glandular-puberulent, orbicular, becoming deeply concave and accrescent in fr., to 6–8 mm. long; ak. 4.5–5.5 mm. long, the wing shallowly to deeply toothed; $n = 18$ (Payne et al., 1964).—Open sandy places, Creosote Bush Scrub; Colo. and Mojave deserts; Ariz., Son. Sept.–Jan.

Ssp. **clárkae** (Kenn.) Keck. [*D. c.* Kenn.] Like ssp. *canescens*, but inner phyllaries strongly accrescent, 10–13 mm. long in fr.; ak. 5.5–6 mm. long, its wings with shorter ± corneous teeth.—Creosote Bush Scrub, Joshua Tree Wd.; more common in Mojave Desert, less so in Colo.; w. Ariz. Sept.–Dec.

Ssp. **hispídula** (Rydb.) Keck. [*D. h.* Rydb.] Stems white-strigose-canescent, also with scattered spreading pustulate-based hispid hairs; inner phyllaries scarcely accrescent, 4–5 mm. long; ak. 3.5–4.5 mm. long, the narrow margin with ± remote teeth.—Creosote Bush Scrub, Joshua Tree Wd.; mostly in w. Mojave Desert, uncommon on Colo. Desert.; w. Ariz. Sept.–Dec.

50. *Dimorphothèca* Vaill. CAPE-MARIGOLD

Herbs or undershrubs; lvs. alternate, entire to pinnatifid. Heads solitary on terminal peduncles; invol. broadly campanulate; phyllaries in 1 series, with scarious margins; receptacle naked. Ray-fls. ♀, fertile in 1 row, the style divided into 2 long stigmatic branches. Disk-fls. bisexual, fertile; style with 2 short branches. Aks. of ray-fls. 3-angled to subterete, usually wrinkled or tuberculate, or disk-fls. smooth and with thickened margins. Ca. 7 spp. of S. Afr.

Rays white above, purplish beneath .... 1. *D. ecklonis*
Rays orange-yellow .... 2. *D. sinuata*

1. **D. ecklònis** DC. Shrubby perennial with narrow toothed lvs.; peduncles bearing large daisy-like heads having white rays that are purplish beneath; head closing at night. —Occasional escape as at Santa Barbara.

2. **D. sinuàta** DC. Annual, 1–3 dm. high, loosely branched, glandular-pubescent; lvs. oblong-lanceolate, to 9 cm. long, coarsely sinuate-dentate; heads 3.5 cm. across, the rays orange-yellow.—Reported from San Diego, Riverside, San Bernardino, Ventura, Santa Barbara and Kern cos.

51. *Dyssòdia* Cav.

Strong-scented annual or perennial herbs or subshrubs, the herbage marked with conspicuous translucent oil-glands, the stems striate. Lvs. opposite or alternate, entire to pinnatisect. Heads small to rather large, terminal, radiate, yellow or orange. Invol. of 2 equal series of quite distinct to ± united phyllaries, these usually subtended by an outer series of much shorter calyculate bracts. Receptacle convex, puberulent. Ray-fls. ♀, fertile. Disk-fls. fertile. Style-branches with a short conic appendage. Aks. terete or angled, striate. Pappus of bristle-tipped paleae or these dissected into numerous bristles. Possibly 40 spp., natives of sw. U.S. and adjacent Mex. (Greek, *dysodia*, a disagreeable odor.)

(Strother, J. L. Systematics of Dyssodia Cav. Univ. Calif. Publ. Bot. 48: 1–88. 1969.)
Invol. 12–16 mm. high; lvs. alternate, inconspicuous.
    Stems scabridulous; lvs. merely spinose-dentate .... 1. *D. cooperi*
    Stems glabrous; lvs. 3–5-parted into narrow lobes .... 3. *D. porophylloides*
Invol. 5–7 mm. high; lvs. opposite, mostly longer than the internodes.
    Pappus-paleae each dissected into 5–12 capillary bristles; heads sessile or subsessile
                                                                     2. *D. papposa*
    Pappus-paleae tipped with only 1–3 bristles; heads on long slender peduncles   4. *D. pentachaeta*

1. **D. coòperi** Gray. Perennial from a woody base, 3–5 dm. high; stems rather stout, puberulous or hispidulous; lvs. lanceolate to ovate, 1–2 cm. long, spinulose-dentate, rarely with small lobes; invol. turbinate, 14–18 mm. high, the principal phyllaries 20–30, linear-lanceolate, nerved, subulate-acuminate, the glands if present terminal only; rays 8–12, bright orange to yellow, often turning saffron, 1 cm. long; $n = 13$ (Raven & Kyhos, 1961).—Open places, 2000–5000 ft.; Creosote Bush Scrub, Joshua Tree Wd.; e. and s. Mojave Desert from Death V. region to Victorville and n. Colo. Desert; Nev., Ariz. May–June, Sept.–Nov.

2. **D. pappòsa** (Vent.) Hitchc. [*Tagetes p.* Vent.] Much-branched annual 1–4 dm. high; lvs. 2–5 cm. long, pinnatifid into linear divisions; heads subsessile; invol. 6–7 mm. high, the principal phyllaries 8–10, hyaline-margined, with brownish glands; rays yellow, short.—Occasional weed as at Loma Linda; native, Ariz. to La., Ohio, Mont. May–June.

3. **D. porophylloìdes** Gray. Plate 15, Fig. B. Very glabrous perennial from a woody base, 3–6 dm. high, with numerous virgate glaucescent stems; lvs. narrow, thick, 1–2 cm. long, the lower petioled, parted into lance-linear acerose-tipped divisions, the upper often entire or incised; invol. turbinate, 12–15 mm. high, the principal phyllaries 14–20, indurate, with oval terminal gland and linear marginal ones; rays orange-yellow, inconspicuous or none; $n = 13$ (Raven & Kyhos, 1961).—Sandy washes, slopes and mesas, below 3500 ft.; Creosote Bush Scrub, s. Mojave and Colo. deserts, from the w. border of the latter to Ariz., Son., L. Calif. March–June.

4. **D. pentachaèta** (DC) Rob. ssp. **p.** var. **benenídium** (DC) Strother. [*D. thurberi* (Gray) Nels. *Hymenatherum t.* Gray.] Low pubescent perennial 1–2 dm. high, diffusely branched; lvs. mostly opposite, sessile, 1–3 cm. long, pinnatifid into 5–7 filiform spinulose-tipped divisions; peduncles slender, 4–8 cm. long; invol. turbinate-campanulate, 4–5 mm. high, naked or with 1–3 minute accessory bracts, each longer bract with 3–9 glands near apex, the free margins ciliate; ligules 3 mm. long; pappus of 10 lanceolate, awn-tipped bristles.—Washes or rocky places, often on limestone, 3000–5600 ft.; Creosote Bush Scrub, Joshua Tree Wd.; from San Jacinto Mts. to mts. of e. Mojave Desert; to Tex., Mex. April–June.

52. *Eclipta* L.

Low annual herbs with leafy procumbent or ascending stems. Lvs. opposite, narrow, entire or toothed. Heads radiate but inconspicuous, white-fld., on peduncles from the upper axils. Invol. hemispheric, 2-seriate, the outer phyllaries somewhat longer and broader than the inner. Receptacle flat, with bristlelike bracts. Rays short. Disk-fls. perfect, fertile. Aks. short, thick, 3- or 4-angled, becoming corky-rugose on the angles, hairy at summit. Pappus an obscure crown or none. Small genus of riparian herbs, chiefly trop. (Greek, *ekleipo*, to be deficient because of lack of pappus.)

1. **E. álba** (L.) Hassk. [*Verbesina a.* L.] Stem 1–10 dm. long; lvs. lanceolate, 2–10 cm. long, sessile, ± short-strigose; rays equalling disk.; $n = 11$ (Turner & Flyr, 1966).—Shores and wet banks, Freshwater Marsh; scattered localities; adv. from e. U.S. All months.

53. *Encèlia* Adans.

Low branching shrubs. Lvs. alternate, entire or remotely toothed. Heads medium-sized, solitary or panicled, peduncled, radiate or discoid, yellow or the disk purplish. Invol. 2–3-seriate. Bracts of the convex receptacle soft and scarious, embracing the aks. and falling with them. Ray-fls. neutral. Disk-fls. flat, obovate, villous-ciliate, ± pubescent on the faces; pappus 0 or of 2 slender awns. Ca. 14 spp. from sw. U.S. to Mex., Peru, Chile, Galapagos Ids. (Christopher *Encel*, who published on oak-galls, in 1577.)

(Blake, S. F. A revision of Encelia and some related genera. Proc. Am. Acad. 49: 358–376. 1913.)
Heads in cymose panicles; peduncles glabrous; lvs. white-tomentulose . . . . 2. *E. farinosa*
Heads solitary at the tips of pubescent peduncles; lvs. not tomentulose.
    Disks purple; rays 18–30 mm. long. Cismontane . . . . 1. *E. californica*
    Disks yellow. Interior and deserts.

Rays none; lvs. scabrous with scattered pustulate-based hairs. E. deserts ... 3. *E. frutescens*
Rays present, 5–15 mm. long; lvs. finely appressed-pubescent, sometimes with slender scabrous hairs between .... 4. *E. virginensis*

1. **E. califórnica** Nutt. Rounded bush, 6–12 dm. high and often broader, much branched; stems and peduncles canescent; lvs. green, appressed-pubescent, ovate to lanceolate, entire or repand-dentate, petioled; invol. 10–12 mm. high, tomentose; rays 1.5–3 cm. long, yellow; disk-corollas purplish; aks. 5–6 mm. high.—Coastal bluffs and open or bushy slopes below 2000 ft.; Coastal Sage Scrub, Chaparral; Santa Barbara Co. to L. Calif., inland to w. Riverside and w. San Bernardino cos. Santa Catalina, San Clemente, Santa Cruz ids. Feb.–June.

2. **E. farinòsa** Gray ex Torr. BRITTLE-BUSH. INCIENSO. Plate 15, Fig. C. Roundish bush 3–8 (–16) dm. high, fragrant, from a woody trunk and bearing dense clusters of lvs. of the season; lvs. narrowly to broadly ovate, obtuse to acute, entire or ± repand-toothed, silvery-tomentose, 3–8 cm. long, on shorter petioles; panicle almost naked, with whitish-yellow branches, quite glabrous except under the heads; invol. 5–8 mm. high; rays 10–15 mm. long; pappus none; $n=17$ (Turner & Flyr, 1966), 18 (Kyhos, 1967).—Dry stony slopes below 3000 ft.; Creosote Bush Scrub, Coastal Sage Scrub; Death V. region s. through e. Mojave Desert to Colo. Desert; w. Riverside and San Bernardino cos., coastal San Diego Co.; to Utah, L. Calif., Mex. March–May. A form with purple disk-corollas is in the Colo. Desert and to Ariz. and L. Calif. and has been named var. **phenicodónta** (Blake) Jtn.

3. **E. frutéscens** Gray. [*Simsia f.* Gray.] Round-topped bushy shrub 8–15 dm. tall, with short scabrous pubescence on younger parts, the hairs pustulate-based; lvs. 1–2.5 cm. long, oblong to ovate, short-petioled, ± scabrous, greenish; peduncles naked, monocephalous; invol. 6–10 mm. high, scabrous; rays usually absent; aks. 7–8 mm. long; $n=17$ (Jackson, 1960).—Dry slopes and mesas, Creosote Bush Scrub; e. and s. Mojave Desert, Colo. Desert to e. San Diego Co.; Ariz. Feb.–May.

4. **E. virginénsis** A. Nels. ssp. **áctoni** (Elmer) Keck. [*E. a.* Elmer. *E. frutescens* var. *a.* Blake.] Like ssp. *virginensis*, but lvs. often larger, mostly broadly ovate, up to 4 cm. long and 3 cm. wide, ± densely canescent or velutinous, often yellowish-green; rays 10–16 mm. long.—Desert slopes, below 5000 ft.; Creosote Bush Scrub; w. Colo. and Mojave deserts n. to Death V. region and White Mts.; interior cismontane as at San Francisquito Creek, San Jacinto region, Palomar, etc. Feb.–July.

Ssp. **virginénsis**. [*E. frutescens* var. *v.* Blake.] Like *E. frutescens*; stems scabrous; lvs. green, finely canescent and with sessile or stalked glands, with stouter tuberculate-based hairs between, usually 1–2.5 cm. long, 6–16 mm. wide; rays 5–10 mm. long.—Washes, gravelly places, etc. below 5000 ft.; Creosote Bush Scrub, e. Mojave Desert, Providence and Kingston mts., Needles, etc.; to Utah, Ariz. April–May; Dec.

54. *Enceliópsis* (Gray) A. Nels.

Scapose xerophytic perennial herbs with stout woody taproot and often much-branched caudex, the herbage silvery-velutinous or canescent. Lvs. in basal tufts, thick, oval, or rhombic, entire. Heads large, many-fld., solitary on scapiform peduncles, yellow, radiate in ours. Invol. 2–3-seriate. Bracts of the ± convex receptacle soft and scarious, embracing the aks. with which they fall. Ray-fls. ♀, sterile. Disk-fls. strongly compressed, blackish, with a narrow white cartilaginous border and crown. Pappus of 2 short subulate awns, with or without squamellae between, or lacking. Three or four spp. of sw. U.S. (*Encelia*, and Greek, *opsis*, likeness, habitally similar to *Encelia*.)

Lvs. acute, rhombic-ovate, the blade usually longer than the winged petiole; pubescence silvery
1. *E. covillei*
Lvs. obtuse, obovate or suborbicular, the blade usually shorter than the slender petiole; pubescence dull .... 2. *E. nudicaulis*

1. **E. covíllei** (Nels.) Blake. [*Helianthella c.* A. Nels. *E. grandiflora* A. Nels. *E. argophylla* var. *g.* (Jones) Jeps.] Stems short, woody below, with tufted basal lvs. on the several

branches, 3–6 dm. high; lvs. rhombic-obovate, 5–8 cm. long, greenish-white with appressed hairs, narrowed into margined petioles; peduncles very long (3–5 dm.); invol. ca. 18 mm. high; rays many, 3.5–4 cm. long, light yellow; aks. 1 cm. long.—Clayey or rocky subalkaline canyon sides and sandy washes, 1200–4000 ft.; Creosote Bush Scrub; w. side of Panamint Mts., Inyo Co. April–June.

2. **E. nudicáulis** (Gray) A. Nels. [*Encelia n.* Gray. *Helianthella n.* Gray.] Cespitose, 1–4 dm. high, densely tomentose-canescent, often with flocs of wool in the lf.-axils; lvs. ovate to orbicular, the blade 2–6 cm. long, the petiole 1–3 times as long; disks 2–3.5 cm. across; invol. hemispheric; phyllaries 1–2 cm. long, lance-subulate; rays ca. 21, 2–2.5 cm. long; aks. silky-villous.—Sand, rocky clays etc., below 6000 ft.; Sagebrush Scrub, Shadscale Scrub, Creosote Bush Scrub; Death V. region, Inyo Co. to Utah, Ida. May.

55. Erigeron L. FLEABANE

Annual to perennial herbs with leafy or subscapose stems and entire or toothed, alternate, generally sessile lvs. Heads corymbose and solitary, rarely paniculate; many-fld. Invol. campanulate to hemispheric; phyllaries only slightly or not graduated, narrow, not herbaceous-tipped, but herbaceous throughout to scarcely herbaceous throughout. Receptacle flat, naked. Ray-fls. ♀, fertile, numerous, bearing evident often narrow, white, pink, purple, or bluish ligules, or the ligules or even the ray-fls. sometimes lacking. Disk-fls. ± numerous, yellow, rarely reddish; style-branches with lanceolate and acute or usually triangular and obtuse appendages. Aks. flattened, usually pubescent and 2-nerved, sometimes 4–10-nerved; pappus of capillary often fragile bristles, sometimes with an outer series of short minute bristles or scales. Largely Am. genus of ca. 200 spp., some in the Old World, nearly all of temp. and boreal regions. (Greek, *eri*, early, and *geron*, old man, the ancient name of an early-fl. plant with hoary pubescence.)

(Cronquist, A. Revision of the N. Am. spp. of Erigeron, n. of Mex. Brittonia 6: 121–302. 1947.)
1. Pistillate corollas very numerous, filiform, with very narrow, short, inconspicuous erect rays 2–3 mm. long, peduncles erect. Inyo White and San Bernardino mts. .... 10. *E. lonchophyllus*
1. Pistillate corollas few to numerous, rarely absent, the tube generally cylindrical; rays mostly well developed and spreading, sometimes reduced or absent, but not short, narrow and erect.
   2. Internodes very numerous and short; lvs. linear or narrowly oblong, uniform from base to near top of plant; phyllaries markedly graduate.
      3. Root-crown subterranean, giving rise to slender rhizomatous stems rarely more than 1.5 mm. thick, which become aerial stems; heads solitary or few on slender stems 3. *E. breweri*
      3. Root-crown superficial; base of stem, if rhizomatous, very short or stout or both; heads often numerous .... 8. *E. foliosus*
   2. Internodes not excessively numerous or usually very short; lvs. variously shaped, sometimes linear, but then the basal obviously larger than the cauline; phyllaries equal or imbricate.
      4. Plants maritime, 1–4 dm. high, scapose; heads large, hemispheric, the disk 14–35 mm. wide; disk-corollas 4.5–7 mm. long .... 9. *E. glaucus*
      4. Plants not maritime; heads smaller except sometimes in some tall spp.
         5. Cauline lvs. ample, sessile and ± clasping, 8–15 cm. long; rays 150–400, 5–10 mm. long. Moist places below 4000 ft. .... 12. *E. philadelphicus*
         5. Cauline lvs. mostly not very well developed, commonly linear or narrow, not as above.
            6. Lvs., or many of them, trilobed or 2–4 times ternate .... 6. *E. compositus*
            6. Lvs. all entire or nearly so.
               7. Lvs. silvery-strigose; aks. 4–8 (–10)-nerved. Desert plants.
                  8. Basal lvs. forming a persistent tuft; aks. mostly 6–8-nerved. Inyo Co.
                                                                              2. *E. argentatus*
                  8. Basal lvs. not forming a persistent tuft; aks. mostly 4-nerved.
                     9. Stems silvery-pubescent; outer pappus of evident narrow scales. N. base of San Bernardino Mts. .... 11. *E. parishii*
                     9. Stems merely gray-green; outer pappus of inconspicuous setae. Providence Mts. and e. .... 16. *E. utahensis*
               7. Lvs. not silvery-strigose; aks. mostly 2-nerved.
                  10. Pubescence of stem widely spreading, sometimes scanty. Inyo Co.

11. Plants alpine or subalpine with 1, rarely 2 heads, radiate.
    12. Stems ± foliose; lower lvs. narrowly oblanceolate .... 4. *E. clokeyi*
    12. Stems scapose or nearly so; basal lvs. broadly oblanceolate to rounded
                                                                                      15. *E. uncialis*
11. Plants of low elevs., or if higher in mts., discoid; heads 1–many.
    13. Heads evidently radiate.
        14. Pappus simple or nearly so; disk-corollas 5–6 mm. long. Santa Rosa Id.;
            Santa Ynez Mts. and n. .... 14. *E. sanctarum*
        14. Pappus distinctly double. Interior.
            15. Biennial or nearly so; disk-corollas 2–3 mm. long ... 7. *E. divergens*
            15. Perennial; disk-corollas 3.5–5 mm. long .... 13. *E. pumilus*
    13. Heads disciform, the ♀ fls. rayless.
        16. Biennial; disk-corollas 2–3 mm. long .... 7. *E. divergens*
        16. Perennial; disk-corollas 3–5 mm. long .... 1. *E. aphanactis*
10. Pubescence of stem ± closely appressed; pulvinate-cespitose perennial with densely clustered linear lvs. 0.4–2 cm. long. Inyo Mts. ..... 5. *E. compactus*

1. **E. aphanáctis** (Gray) Greene. [*E. concinnus* var. *a*. Gray.] Cespitose perennial with a taproot and branching caudex; stems 1–3 dm. high, ± branched, the herbage densely short-hirsute; basal lvs. crowded, linear-oblanceolate to spatulate, long-petioled, up to 8 cm. long, 6 mm. wide; the cauline many but reduced; heads several or solitary, yellow, often brownish in age; invol. 4–6 mm. high, short-hirsute; ♀ fls. eradiate, or sometimes with very short rays; inner pappus of 7–20 bristles, the outer of evident scales; $n = 9$ (Solbrig et al., 1964).—Dry stony places, 5000–8000 ft.; Sagebrush Scrub, Yellow Pine F., Joshua Tree Wd.; San Bernardino Mts., Providence, Grapevine, Argus and Inyo mts.; to Sierra Nevada. May–Sept. An essentially scapose plant, with corolla-lobes turning reddish in age is var. **congéstus** (Greene) Cronquist [*E. c.* Greene], the common form in the San Bernardino Mts. at 6000–11,400 ft.

2. **E. argentátus** Gray. Densely cespitose perennial; stems many, sparsely leafy, 1–4 dm. high; herbage silvery-strigose; basal lvs. densely tufted, linear-oblanceolate, up to 6 cm. long, 6 mm. wide; heads usually solitary, the disk 10–18 mm. wide; invol. 6–9 mm. high; phyllaries strongly imbricate, the outer silvery-strigose, pointed; rays 20–50, 9–15 mm. long, lavender-purple or paler; pappus of 25–40 slender white bristles and as many short outer setae.—Rocky places, 6000–8500 ft.; Pinyon-Juniper Wd.; Last Chance Mts., Inyo-White Mts.; to Utah. June–July.

3. **E. bréweri** Gray var. **bréweri**. Root-crown usually deeply seated, producing slender rhizomes and trailing and erect stems 1–3 dm. high; herbage retrorse-hirtellous or partly spreading-hirsute to glabrate; lvs. many, rather uniform, linear to oblong-oblanceolate, 0.5–4 cm. long, 1–6 mm. wide; heads mostly several, the disk 7–15 mm. wide; invol. 4–6 mm. high, rather densely glandular-atomiferous; phyllaries relatively broad, pointed, usually with greenish tips; rays 10–45, 4–10 mm. long, generally blue; pappus of 20–50 slender bristles and a few inconspicuous outer setae.—Dry rocky places 5000–10,500 ft.; Yellow Pine F. to Lodgepole F. San Bernardino and San Gabriel mts.; to n. Calif., w. Nev. July–Aug.

Var. **jacínteus** (Hall) Cronq. [*E. j.* Hall.] Stems 1 dm. long or shorter; pubescence longer and finer; lvs. crowded, not over 12 mm. long.—At 6000–11,480 ft.; Pinyon-Juniper Wd., Yellow Pine F., Lodgepole F., Subalpine F.; San Jacinto, San Gabriel, San Bernardino mts. July–Aug.

Var. **porphyréticus** (Jones) Cronq. [*E. p.* Jones.] Invol. generally hispidulous as well as glandular, 5–8 mm. high, phyllaries without evident green tips; lvs. rarely more than 2 mm. wide.—At 5000–9500 ft.; Pinyon-Juniper Wd., Yellow Pine F.; San Bernardino, Kingston, Inyo-White mts., Sierra Nevada, Sweetwater Mts. May–Aug.

4. **E. clòkeyi** Cronq. Cespitose perennial with woody taproot; stems slender, 3–20 cm. long; herbage moderately hirsute with curved hairs; lvs. crowded at base, narrowly oblanceolate, entire, slender-petiolate, up to 8 cm. long and 6 mm. wide, the cauline mostly linear; heads mostly solitary, the disk 8–11 mm. wide; invol. 4–7 mm. high, glandular and hirsute; rays 20–50, lavender, 6–10 mm. long.—Sandy or rocky places, 8000–12,000

ft.; Sagebrush Scrub, Bristle-cone Pine F., Subalpine F.; Panamint Mts., Last Chance Mts.; Inyo Mts.; Sierra Nevada; to Utah. June–Aug.

5. **E. compáctus** Blake. Pulvinate-cespitose perennial with a short multicipital caudex on a stout taproot; lvs. basal, densely clustered, white-strigose, linear, 4–20 mm. long; scapes 2–6 cm. high; heads solitary, the disk 7–14 mm. wide; invol. 5–6 mm. high, strigose and glutinous; phyllaries subequal, or ± imbricate, pallid with dark midrib, firm, acute; rays 15–30, 6–9 mm. long, white or pink; pappus of 30–40 slender firm sordid bristles and a few slender outer setae.—Dry gravelly soils, 6000–7500 ft.; Pinyon-Juniper Wd.; Inyo Mts.; to Utah. June.

6. **E. compósitus** Pursh var. **discoìdeus** Gray. [*E. trifidus* Hook.] Commonly compact and small; lvs. mostly only once ternate; otherwise similar to var. *glabratus*.—Range and environments of var. *glabratus* but much less common in Calif.

Var. **glabràtus** Macoun. [*E. multifidus* Rydb.] Dwarf cushionlike plant from a compactly branched caudex; herbage ± hispid-hirsute and glandular; lvs. crowded at base, 2–6 cm. long, long-petioled, the fanlike blade 2–3-times ternate, the lobes commonly crowded and linear-oblong; stems 5–15 cm. high, 1-headed, subscapose, with few reduced, often entire lvs.; invol. 5–8 mm. high, the thin slender phyllaries purple-tinged; ♀ fls. 20–50, the ligules white, pink or bluish, to 10 mm. long, sometimes essentially lacking.—Rocky places, 8000–11,500 ft.; Lodgepole F. to Alpine Fell-fields; San Gorgonio Peak, White Mts. to Alaska, Greenland, Rocky Mts. July–Aug.

7. **E. divérgens** T. & G. [*E. californicus* Jeps.] Annual to biennial with several to many leafy ascending stems 1–4 dm. high, with short, scarcely spreading, hispid pubescence; basal lvs. oblanceolate to obovate, entire or rarely toothed or pinnatifid, 1–3 cm. long, narrowed into a winged petiole; cauline becoming sessile and reduced; heads solitary at the ends of the slender branches, hence in open corymbs; invol. 4 mm. high, the phyllaries linear, acuminate, somewhat membranous-margined, pubescent, sometimes glandular; rays very numerous, filiform, 5–7 mm. long, violet to white; pappus of inner scanty slender bristles and outer short scales.—Dry sandy places, 2000–8000 ft.; Montane Coniferous F., Pinyon-Juniper Wd., etc.; L. Calif. and Laguna Mts. to San Bernardino and San Gabriel mts., occasional on both deserts, to n. Calif.; B.C., Mont., Tex. March–Nov.

8. **E. foliòsus** Nutt. var. **foliòsus**. Plate 15, Fig. D. Perennial with a stout taproot and rather superficial root-crown; stems erect, 2–10 dm. high, glabrous to ± strigose; lvs. numerous, ± crowded, rather uniform, linear to narrowly oblong, 2–6 cm. long, the larger ones mostly over 1.5 mm. wide, hairy, often with pustulate hairs; heads arranged in a usually naked corymbiform infl., the disk 10–18 mm. wide; invol. 4–7 mm. high, glabrate to strigose or spreading-hairy or glandular or both; phyllaries imbricate, acuminate, the inner strawcolor with brown midrib; rays 20–60, 5–12 mm. long, blue or purple; pappus of 20–30 slender tawny bristles and a few inconspicuous outer setae; aks. 2-nerved, sparsely hairy.—Grassy or brushy places below 5000 ft.; Chaparral, Yellow Pine F., S. Oak Wd.; Santa Catalina, Santa Cruz ids.; cismontane s. Calif. and n. L. Calif. to cent. Calif. May–July.

A variable sp., of which the other principal forms in s. Calif. are:

Var. **covíllei** (Greene) Compton. [*E. c.* Greene.] Grayish with a very stiff dense pubescence of coarse almost lanceolate hairs.—Below 6000 ft.; Joshua Tree Wd., Pinyon-Juniper Wd., etc.; Mojave Desert from n. base of San Bernardino Mts. to Inyo Co. and e. side of Sierra Nevada. May–Aug.

Var. **stenophýllus** (Nutt.) Gray. [*E. s.* Nutt. *E. tenuissimus* Greene.] More glabrous than the usual var. *foliosus* and with lvs. filiform, 1–2 mm. wide.—With var. *foliosus* in s. Calif.

9. **E. gláucus** Ker. SEASIDE-DAISY. Stems erect from a decumbent base, 1–4 dm. high, leafy, hirsute and ± viscid, arising from a basal rosette of lvs. on the fleshy caudex; lvs. rather thick, finely puberulent, light green, spatulate or obovate, mostly entire, 3–9 cm. long, the upper reduced; heads solitary at the ends of the stems or the few branches, large, the disk 1.5–3.5 cm. wide; invol. 8–12 mm. high; phyllaries equal, densely shaggy to long-villous; rays ca. 100, 9–15 mm. long, pale violet to lavender.—Coastal bluffs; Coastal Sage Scrub; Santa Barbara Ids.; San Luis Obispo to Ore. April–Aug.

10. **E. lonchophýllus** Hook. [*E. racemosus* Nutt.] Perennial or biennial, with crowded erect leafy stems, 1–4 dm. high, subglabrous to sparingly hirsute, yellow-green; lvs. spatulate to linear, the blades entire, 2–6 cm. long, the lower long-petioled; heads few, in simple panicles; invol. hemispherical, 6–7 mm. high, with lanceolate pubescent phyllaries; rays filiform, 25–50, 2–3 mm. long, white or pale lavender, inconspicuous; pappus of 20–30 long slender bristles, sometimes with few inconspicuous outer setae.—Wet meadows, 7000–8000 ft.; Yellow Pine F.; upper Santa Ana River system, San Bernardino Mts., at higher elevs.; Bristle-cone Pine F.; Inyo-White Range; to Alaska, Que. July–Aug.

11. **E. párishii** Gray. Perennial, with branched woody root-crown, 1–3.5 dm. high, the stems few-branched, silvery with soft appressed hairs; lvs. linear to oblance-linear, 2–5 cm. long, 2–4 mm. wide; heads solitary at ends of branches; invol. greenish, puberulent, 6–7 mm. high; phyllaries lance-linear, imbricate; rays 30–50, rose to lavender, 6–13 mm. long; pappus of 18–26 firm sordid bristles and several conspicuous white outer setae. —Dry slopes, 3500–5000 ft.; Joshua Tree Wd.; about Cushenbury Canyon, s. Mojave Desert, n. base of San Bernardino Mts. May–June.

12. **E. philadélphicus** L. Biennial or short-lived perennial, the stem simple, 5–9 dm. high; herbage hispid-hirsute; lvs. oblong-obovate to spatulate, serrate or with few coarser teeth; basal lvs. petioled, 1 dm. or longer, upper smaller and auriculate-clasping, uppermost reduced to leafy bracts; heads few, in small corymbs; invol. 5–6 mm. high; phyllaries subequal, ± hirsute; rays 150–400, deep pink to white, 5–10 mm. long, 0.2–0.6 mm. wide; $2n=18$ (Mulligan, 1957); $n=9$ (Taylor & Brockman, 1966).—Moist places, stream-banks, etc. below 4000 ft.; many Plant Communities; scattered localities, cismontane; to Can. and e. U.S. April–July.

13. **E. pùmilus** Nutt. ssp. **concinnoìdes** Cronq. [*Distasis concinnus* H. & A. *E. c.* T. & G.] Perennial, with thickened, usually branched root-crown and several erect simple or few-branched stems 1–3 dm. high; herbage canescent-hispid; lower lvs. linear to spatulate, acute, 1.5–3 cm. long, petioled, the upper reduced; peduncles slender; heads solitary; invol. hemispheric, ca. 5 mm. high; rays numerous, narrow, violet or rose to nearly white, 5–6 mm. long; inner pappus of bristles, the outer of relatively broad scales. —Occasional on dry slopes, 3000–8000 ft.; Creosote Bush Scrub, Joshua Tree Wd., Pinyon-Juniper Wd.; e. Mojave Desert, from e. San Bernardino Co. to s. Inyo Co. May–June.

14. **E. sanctàrum** Wats. Stems erect or ± decumbent from a very slender branching rootstock, 5–30 cm. high, sparsely pubescent with spreading or retrorse hairs; lvs. entire, minutely rough-hairy and ± ciliate, oblanceolate, 2–5 cm. long, 3–10 mm. wide, reduced up the stem; heads usually solitary on a subnaked peduncle 2–10 cm. long, the disk 12–17 mm. wide; invol. 6–9 mm. high; phyllaries densely hirsute; rays 45–90, 8–12 mm. long, rose-purple.—Hills near the coast, below 1000 ft.; Coastal Sage Scrub, Chaparral; Santa Rosa Id., Santa Ynez Mts.; to San Luis Obispo Co. March–June.

15. **E. unciàlis** Blake. Cespitose perennial, with heavy, much-branched, deep-seated root-crown; lvs. tufted, oblanceolate, obtuse, entire, the blades 4–9 mm. long, 3–4 mm. wide, on somewhat longer slender petioles, rather soft-pubescent; invol. 4 mm. high, the phyllaries lance-linear, villous, green in center, hyaline-margined; rays ca. 30, oblong-linear, 3–4 mm. long, whitish to pink; pappus scant, simple.—Crevices of limestone cliffs, 7000–9500 ft.; Sagebrush Scrub, Bristle-cone Pine F.; Pinyon-Juniper Wd.; Clark Mts., e. San Bernardino Co.; Inyo Mts., Tin Mt., Inyo Co. June–July.

16. **E. utahénsis** Gray. Habit of *E. parishii* Gray, from a heavy root, the erect stems 1–5 dm. high, sometimes branching above, silvery strigose, densely so at base, ± viscidulous; lvs. well distributed, strigose and gray-green, the lower linear-oblanceolate, to 10 cm. long and 6 mm. wide, commonly much narrower, the upper gradually reduced; heads 1–10, the disk 8–15 mm. wide; invol. 4–6 mm. high; phyllaries strongly imbricate; rays 10–40, lavender or whitish, 9–18 mm. long; pappus of 35–45 firm bristles, the outer smaller setae often obscure.—Dry limestone slopes, ca. 5000 ft.; Pinyon-Juniper Wd.; Providence Mts., Mojave Desert; to Utah, Colo., Ariz. May–June.

56. Eriophýllum Lag.

Annual to perennial, herbaceous to shrubby, ± permanently floccose or tomentose plants, with alternate, entire to pinnately lobed or dissected lvs. Heads radiate, rarely discoid, solitary and peduncled at the ends of the branches or in open or condensed corymbs. Invol. hemispheric to oblong, the phyllaries in 1–2 series, mostly distinct, rigid, permanently erect, concave and partly enveloping the marginal aks. Receptacle ± convex. Rays 4–15, ♀, fertile, sometimes lacking. Disk-corollas commonly with a glandular and hairy tube. Aks. usually 4–5-angled, narrow; pappus of 4–14 hyaline paleae, which may be laciniate-cleft, sometimes wanting. (Greek, *erion*, wool, and *phyllon*, lf., the herbage woolly.)

(Constance, L. A systematic study of the genus Eriophyllum. Univ. Calif. Publ. Bot. 18: 69–136. 1937.)

1. Perennial (or biennials), somewhat woody at the base.
  2. Heads large, the invol. 7–10 mm. high, solitary on long peduncles; lvs. often entire
                                                                                        3. *E. lanatum*
  2. Heads smaller, the invol. 4–7 mm. high, clustered; lvs. mainly divided or lobed.
    3. Rays 5–6; peduncles supporting infls. slender .... 2. *E. confertiflorum*
    3. Rays 8–12; peduncles supporting infls. stout.
      4. Lvs. white-tomentose on both surfaces; pappus-paleae 4–6; aks. not glandular
                                                                                          7. *E. nevinii*
      4. Lvs. green above; pappus-paleae 8–12; aks. glandular .... 9. *E. staechadifolium*
1. Annuals, 2–20 cm. high.
  5. Heads mostly sessile in leafy-bracteate terminal clusters; pappus-paleae lacerate-fimbriate.
    6. Rays present; pappus-paleae shorter than corolla-tube .... 6. *E. multicaule*
    6. Rays lacking; pappus-paleae equaling the corolla-tube .... 8. *E. pringlei*
  5. Heads pedunculate; pappus-paleae if present not lacerate-fimbriate.
    7. Rays absent; plants 1–2 cm. high. Barstow region .... 5. *E. mohavense*
    7. Rays present; plants 3–20 cm. high; receptacle conical or convex.
      8. Anthers with acute to obtuse tips; peduncles usually 2–10 cm. long; ligules yellow; pappus less than 0.4 mm. long .... 1. *E. ambiguum*
      8. Anthers with linear-subulate tips.
        9. Rays white; pappus of unequal paleae. E. Mojave Desert .... 4. *E. lanosum*
        9. Rays yellow, rarely white to reddish. More widely distributed ..... 10. *E. wallacei*

1. **E. ambíguum** (Gray) Gray. [*Lasthenia a.* Gray.] Annual, few- to many-branched, ascending to erect, 5–25 cm. tall; tomentum tardily deciduous from lvs. and stems; lvs. alternate, oblanceolate to spatulate, entire or few-toothed, 1–3 cm. long; peduncles 1–5 cm. long, monocephalous; invol. 5–6 mm. high; phyllaries ca. 8, ± united; rays 5–10, yellow, 3–8 mm. long; pappus a crown of erose scales, sometimes conspicuous and lacerate, sometimes quite wanting; $n=7$ (Carlquist, 1956).—Gravelly slopes and washes, below 7500 ft.; Creosote Bush Scrub to Pinyon Juniper Wd.; fairly frequent in Palm Springs area, more so in n. and w. Mojave Desert; to Inyo Co. and Nev. April–June. Variable.

2. **E. confertiflòrum** (DC.) Gray var. **confertiflòrum**. [*Bahia c.* DC.] GOLDEN-YARROW. Plate 15, Fig. E. Perennial, somewhat woody at base to quite shrubby, the stems slender, simple or few-branched, finely white-tomentose, 3–6 dm. tall; lvs. 2–3 cm. long, pinnatifid or bipinnatifid into linear almost filiform divisions not over 1 mm. wide, bright green above, white-tomentose beneath, with revolute margins; heads almost sessile in small corymbose clusters; invol. oblong, 3–4 mm. long, floccose-glabrate, the phyllaries ca. 5, ovate; rays 4–6, yellow, 3 mm. long; paleae 8–12, subequal, ca. 0.5 mm. long; $n=8$, 16 (Carlquist, 1956).—Dry slopes and washes near the coast; Coastal Sage Scrub, Chaparral; n. L. Calif. to n. Calif. April–Aug. Exceedingly variable and intergrading freely with a number of poorly marked variants which may be keyed out as follows:

Ligules wanting. Upper San Antonio Canyon, San Gabriel Mts. . var. *discoideum* (Rydb.) Munz
Ligules present.
  Plant herbaceous, scarcely woody, 1–3 dm. high. Dry slopes under pines; Yellow Pine F.
                                                                                                   var. *tridactylum* (Rydb.) Munz

Plant woody at base, 3-6 dm. high.
Ultimate lf.-divisions 1-2 mm. wide, few in number, mostly 3-5.
Cismontane at low elevs.; Coastal Sage Scrub, Chaparral. Santa Catalina Id.
var. *trifidum* (Nutt.) Gray
Ultimate lf.-divisions not over 1 mm. wide, more numerous.
Lvs. green above; heads in compact corymbose clusters . . . . var. *confertiflorum*
Lvs. gray-tomentose above; heads in ± open corymbs. Dry slopes, mts. along w. edge Colo. Desert to Little San Bernardino Mts., then to Mint Canyon, Mt. Pinos, etc.; cent. Calif.
var. *laxiflorum* Gray

3. **E. lanàtum** (Pursh) Forbes var. **hállii** Const. Stems numerous, 3-4 dm. high, herbaceous, from a perennial base; herbage thinly floccose; lvs. broad and thin, 2.5-5 cm. long, ovate in outline, pinnately incised or pinnatifid, the oblong divisions entire or toothed; heads solitary or few, relatively small; invol. campanulate, 8-12 mm. high; rays ca. 8, yellow, 10-13 mm. long; tube of disk corollas glabrous; aks. 4-5 mm. long, sparsely hispidulous; pappus of 8-12 unequal or subequal oblong and/or lanceolate erose paleae up to 2 mm. long.—Known only from the type region, 3500 ft.; Foothill Wd.; near Ft. Tejon, Kern Co. June-July.

Var. **obovàtum** (Greene) Hall. [*E. o.* Greene.] Stems several, slender, leafy, 2-4 dm. high; herbage moderately white-tomentose; lvs. oblanceolate, entire or shallowly toothed, 1.5-4 cm. long; invol. 7-10 mm. high; rays 10-13, 8-10 mm. long; aks. 3-4 mm. long, glabrous; pappus of 8-10 lance-oblong erose paleae up to 1 mm. long.—Openings under pines, 4000-7600 ft.; Yellow Pine F.; San Bernardino Mts.; Kern Co. to Tulare Co. June-July.

4. **E. lanòsum** (Gray) Gray. [*Burrielia l.* Gray. *Actinolepis l.* Gray.] Annual, loosely floccose, few-branched, ascending or erect, 5-12 cm. high; lvs. linear or somewhat oblanceolate, 1-2 cm. long, 1-3 mm. wide; peduncles few, 1-4 cm. long; invol. narrow, 5-6 mm. high; rays ca. 10, the ligules white, 5 mm. long; pappus of 4-5 paleae produced into scabrous awns and equaling the corolla, and of as many alternating obtuse paleae half as long; $n=4$ (Carlquist, 1956).—Sandy deserts, 500-3500 ft.; Creosote Bush Scrub; e. Mojave and Colo. deserts; to Utah, Ariz., L. Calif. Feb.-May.

5. **E. mohavénse** (Jtn.) Jeps. [*Eremonanus m.* Jtn.] Woolly-villous cespitose dwarf annuals 1-2.5 cm. high, 2-3 cm. across; lvs. spatulate, entire and mucronate, or sharply 3-toothed at apex, attenuate to base, 5-10 mm. long; heads solitary at ends of the branches, discoid, 3-4-fld.; invol. cylindric, 3-4 mm. high; phyllaries 3-4, linear-oblong, obtuse, concave; disk-fls. ca. 2 mm. long; aks. terete, 2-2.5 mm. long; pappus of 12-14 oblong to obovate subequal paleae ca. 1.5 mm. long.—Sandy or rocky places, 2000-3000 ft.; Creosote Bush Scrub; known only from within 30 miles of Barstow, Mojave Desert. April-May.

6. **E. multicáule** (DC.) Gray. [*Actinolepis m.* DC.] Annual, diffusely branched from the base, 2-15 cm. high, loosely woolly; lvs. 5-10 mm. long, broadly spatulate, usually with 2-3 rounded teeth at the apex, green and subglabrous above, white-woolly beneath; heads in leafy-bracteate close terminal clusters; invol. 3 mm. high; phyllaries 5-7, ovate-oblong, distinct, boat-shaped; rays 3-7, yellow, 2 mm. long; pappus of 10-15 narrow, somewhat fimbriate paleae continued above into subulate awns; $n=7$ (Carlquist, 1956). —Sandy places and chaparral openings below 2000 ft.; Coastal Sage Scrub, Chaparral; San Diego Co. to cent. Calif. March-May.

7. **E. nevínii** Gray. Shrubby branching perennial, 5-20 dm. high; stems stout, tomentose, the old lvs. deciduous, the shoots of the season densely leafy, terminated by a glabrate leafless portion bearing the corymbiform much-branched many-headed compact infl.; lvs. 5-25 cm. long, broadly ovate in outline, 2-3-pinnatifid into many linear-oblong segms., tomentose above and beneath; invol. narrow-campanulate, 6-7 mm. high; phyllaries 8-12, oblong, obtuse; rays 4-8, inconspicuous, 2 mm. long; aks. 3-4 mm. long; pappus-paleae 4-6, often ± fused, erose, unequal; $n=16$ (Carlquist, 1956).—Rocky coastal bluffs; Coastal Sage Scrub; Santa Catalina and San Clemente ids. April-Aug.

8. **E. prínglei** Gray. [*Actinolepis p.* Greene.] Annual, branching, forming white-woolly

tufts 1–5 cm. high; lvs. and heads much as in *E. multicaule*, but without ligules; pappus-paleae ca. 10, silvery, lanceolate, cleft into several short bristle-like divisions; $n=7$ (Carlquist, 1956).—Sandy places, 1200–6600 ft.; Creosote Bush Scrub; Sagebrush Scrub; w. borders of both deserts, San Diego Co. to Mono Co.; Ariz. April–June.

9. **E. staechádifolium** Lag. var. **artemisiaefòlium** (Less.) Macbr. [*Bahia a.* Less.] Shrubby much-branched perennial 3–15 dm. high, tomentose but often glabrate in age; lvs. 1–2-pinnatifid, with linear to oblong lobes, subcoriaceous, persistently tomentose beneath, 3–7 cm. long, the margin revolute, with linear to oblong lobes and rather broad rachis; heads numerous, short-peduncled, in rather dense corymbs; invol. campanulate, 5–7 mm. high; phyllaries 8–11, oblong, obtuse, carinate-thickened at base; rays (0–) 6–9, yellow, 3–5 mm. long; aks. 3–4 mm. long; pappus of 8–12 oblong or oblanceolate unequal paleae up to 1 mm. long; $n=16$ (Carlquist, 1956).—Shores and fields near the coast; Coastal Strand, Coastal Sage Scrub; Santa Barbara Ids.; Point Conception to Ore. March–Aug.

Var. **depréssum** Greene. Stems 1–2 dm. high, stout, depressed; lvs. with rachis to 1 cm. wide, densely clothing the stem and sometimes overtopping the heads.—Santa Rosa, Santa Cruz and Anacapa ids.; about Surf. March–July.

10. **E. wállacei** Gray. Annual, freely branched from base, the stems spreading-ascending; copiously and rather permanently white-woolly throughout; lvs. obovate or spatulate, obtuse, mostly entire, 5–14 mm. long; heads very short-peduncled; invol. 5 mm. high, the phyllaries overlapping, not united; rays 8–10, the ligules yellow, 4 mm. long; aks. ± pubescent; pappus of 6–10 oblong paleae 0.5 mm. long or lacking; $n=5$ (Carlquist, 1956).—Sandy places, below 6000 ft.; Creosote Bush Scrub, Joshua Tree Wd., Chaparral; both deserts, San Diego Co. to Mono Co.; San Bernardino, San Jacinto and San Fernando valleys; n. L. Calif., Utah, Ariz. March–May. A form with ligules white to rose, along the w. edge of the Colo. Desert, is recognized as var. **rùbellum** (Gray) Jeps. [*Antheropeas r.* Gray.]

57. *Èvax* Gaertn.

Ours low woolly annuals with entire alternate lvs. and small discoid heads with circles of bractlike lvs. beneath. Phyllaries very closely imbricate, covering but not enclosing the ♀ fls. and forming a short-cylindric head. Receptacle columnar, villous, tipped with a cuplike whorl of mostly 5 bracts at summit, the cup bearing mostly 2–4 ♂ fls. Phyllaries and bracts of ♀ fls. indurate, persistent, the upper or sterile bracts deciduous. The ♀ fls. with filiform corollas. Pappus none. A genus of some size, sometimes perennial in the Old World. (Supposedly the name of an Arabian chief.)

Plants mostly acaulescent; lf.-blades ca. one third as wide as long. Near Murrieta, Riverside Co.
    1. *E. acaulis*
Plants mostly caulescent; lf.-blades ca. half as wide as long. Near the coast ...... 2. *E. sparsiflora*

1. **E. acáulis** (Kell.) Greene. [*Stylocline a.* Kell.] Much like the next sp., but mostly quite acaulescent, sometimes with prostrate stems 1–3 cm. long; lvs. spatulate, the blades mostly 2–6 mm. long and 1–3 mm. wide.—S. Oak Wd. at 6 mi. w. of Murrieta, Riverside Co.; w. base of Sierra Nevada. March–May.

2. **E. sparsiflòra** (Gray) Jeps. [*E. caulescens* var. *s.* Gray.] Erect, simple or branched, mostly 3–10 cm. high, grayish; lf.-blades spatulate, 1–1.5 cm. long, on slender petioles ca. as long; heads axillary, often ± crowded toward top; receptacle-bracts woolly on back, long-hirsute at base; cent. fls. ca. 4.—Dry open places below 2500 ft.; many Plant Communities; w. San Diego Co., Santa Rosa Id.; to Ore. March–May. Plants with shorter stems below 3 cm. long have been called var. *brevifòlia* (Gray) Jeps.

58. *Filàgo* L.

White woolly annual with alternate lvs. Heads small, discoid, in small capitate clusters. Invol. scanty, the phyllaries resembling the bracts of the cylindric to convex receptacle.

Outer fls. ♀, fertile, with tubular-filiform corolla, placed in several series, the outer epappose, and subtended by concave partly enclosing bracts, the inner bractless and with pappus of capillary bristles; cent. fls. 2–5, often sterile, bractless, with capillary pappus. Aks. subterete. Ca. 12 spp. of temp. and warm-temp. Eurasia, Afr., Am. (Latin, *filum*, thread, referring to the hairs.) The four spp. treated here are placed in a genus *Oglifa* by Chrtek & Holub.

Uppermost lvs. much longer than the heads.
    Lvs. linear; receptacle ± obconic or convex .... 1. *F. arizonica*
    Lvs. subulate with broadish base; receptacle low, nearly flat .... 4. *F. gallica*
Uppermost lvs. scarcely or not longer than the heads.
    Plants erect, 6–25 cm. high .... 2. *F. californica*
    Plants depressed, spreading, the stems 3–10 cm. long .... 3. *F. depressa*

1. **F. arizónica** Gray. Stem branched at or above the base, erect or diffuse, 4–15 cm. long, with slender naked internodes and few lvs. except at upper part; lvs. linear to ± oblong, mostly ca. 5–15 mm. long, the uppermost involucrate around and exceeding the heads; marginal ♀ fls. 10–15, their bracts firm, ovate, those within the inner circle of paleae ca. 4–5, all perfect; aks. slightly curved, smooth.—Occasional in dry and disturbed places below 3000 ft.; Coastal Sage Scrub, Chaparral, Creosote Bush Scrub; cismontane and deserts. Calif.; San Clemente and Santa Catalina ids.; to Ariz., L. Calif. March–May.

2. **F. califórnica** Nutt. Plate 15, Fig. F. Erect, simple or branched, 0.5–3.5 dm. high, leafy; lvs. 0.8–2 cm. long, oblong-linear to subspatulate, sessile; heads ovoid, 3–4 mm. high, scarcely exceeded by the involucrate lvs.; bracts of outer ♀ fls. 8–10, woolly, boat-shaped, with hyaline tip, the inner ones thinner and less woolly, the inner florets ca. 12–20, only ca. 2–4 often perfect; inner aks. papillose.—Common in dry open places, on burns, etc.; cismontane s. Calif. below 3500 ft.; Coastal Sage Scrub, Chaparral, etc.; occasional on deserts in Creosote Bush Scrub; to Utah, Ariz., L. Calif. March–June. On several of the s. Calif. islands.

3. **F. depréssa** Gray. Branched from base, depressed or spreading, the stems 4–12 cm. long; lvs. oblong to narrow-obovate, 3–9 mm. long; heads 2–3 mm. high, the involucrate lvs. short; outer bracts with a hyaline appendage ca. as long as the body, the marginal ♀ fls. 5–6; aks. all smooth.—Open places below 2500 ft.; Creosote Bush Scrub, Joshua Tree Wd.; both deserts; to Ariz. March–May.

4. **F. gállica** L. Erect, simple or branched, 2–3 dm. high; lvs. appressed, subulate, 1–2.5 cm. long, those below the heads involucrate, divaricate; receptacle almost flat; heads 2–5, obconic, ca. 4 mm. high; marginal aks. completely enclosed in the subtending triangular pointed woolly bract; aks. 0.5 mm. long.—Burns and waste places, San Diego Co. to n. Calif.; natur. from Eu. Santa Catalina Id. April–June.

59. *Flavèria* Juss.

Glabrous to pubescent annuals, ± succulent. Heads individually inconspicuous. Lvs. opposite, entire or toothed, sessile, sometimes connate. Invol. narrow, prismatic; phyllaries 1–8, subequal. Ray mostly 1 and yellow, or 0. Disk-fls. 1–15. Ak. narrow, 8–10-ribbed. Pappus wanting or rarely of 2–4 scales. Ca. 10 spp., mostly American. (Latin, *flavus*, yellow.)

1. **F. trinérvia** (Spreng.) C. Mohr. Stem 2–12 dm. tall, widely branched, subglabrous; lvs. linear to linear-elliptic, 3–10 cm. long, serrate, 3-ribbed; heads usually 1-fld., in axillary or involucrate clusters; corolla of ♀ fls. 1.5 mm. long, the ligule oblique; corolla of perfect fl. 2 mm. long; ak. 2 mm. long.—Ala. to Ariz. and S. Am. Natur. at Calimesa, Riverside Co., *Fuller*.

60. *Gaillárdia* Foug. BLANKET FLOWER

Annual or perennial herbs. Lvs. alt. or all basal, entire to pinnatifid, sometimes resinous-punctate. Heads solitary, radiate, rather large, mostly long-pedunculate. Invol. 2–3-seriate, herbaceous, reflexed in fr. Receptacle convex to subglobose, with soft or firm setae that do not individually subtend the disk-fls. Ray-fls. neutral, or sometimes ♀ and fertile; ligules yellow or partly or wholly reddish-purple. Disk-fls. perfect, fertile. Aks. broadly obpyramidal, villous at least at base. Pappus of 5–10 scarious often awned paleae. Ca. a dozen spp., chiefly e. N. Am., one S. Am. (For *Gaillard* de Merentonneau, French botanist.)

Plant perennial; lvs. 5–15 cm. long; ligules 25–30 mm. long .... 1. *G. aristata*
Plant annual; lvs. 4–8 cm. long; ligules 12–20 mm. long .... 2. *G. pulchella*

1. **G. aristàta** Pursh. Perennial; lvs. 5–15 cm. long; ligules 25–30 mm. long, purple at base, the remainder yellow; $n=34$ (Taylor & Brockman, 1966).—Reported from w. of Tehachapi, Kern Co., *Fuller*; and in more n. cos. Native from Ore. to B.C. and N. Dak.

2. **G. pulchella** Foug. Leafy-stemmed annual 2–4 dm. high, harsh-puberulent; lower lvs. oblanceolate, short-petioled, 4–8 cm. long, entire, dentate or sinuately pinnatifid, the upper sessile; invol. ca. 10 mm. high; phyllaries lanceolate, hirsute, chartaceous at base; $n=17, 18$ (Biddulph, 1944).—Locally natur. as escape from gardens as in Ventura Co.; native from Ariz. to Colo., Nebr., La. June–Aug.

61. *Galinsòga* R. & P.

Annuals with opposite lvs. and small cymose heads. Phyllaries few, relatively broad, rather membranous, with several green nerves, each subtending a ray. Receptacle conic, the paleae membranous, flat and narrow. Ray-fls. 4–5, only slightly surpassing the disk, fertile, white in ours. Disk-fls. perfect, yellow. Aks. 4-angled, the outer somewhat obcompressed. Pappus-paleae fimbriate, sometimes reduced or none in the rays. Ca. 6 spp. from U.S. to Argentina, some widely distributed as weeds. (For M. M. *Galinsoga*, Spanish physician and botanist.)

1. **G. parviflòra** Cav. Slender annual 2–7 dm. high, leafy and often freely branching throughout, sparsely hairy; lvs. ovate, acute, short-petiolate, 2–5 cm. long, 1–3 cm. wide, crenulate or entire, thin; heads broadly campanulate, 3–4 mm. high; pappus of the disk-fls. of 8–14 fimbriate blunt paleae, that of the ray-fls. obsolete; $n=16$ (Turner & Flyr, 1966).—Locally common weed in irrigated orchards, etc., cismontane Orange, San Bernardino and Los Angeles cos., Inyo Co.; native from Mex. to S. Am., but now a cosmopolitan weed. May–Nov.

62. *Gazània* Gaertn.

Mostly perennial herbs with short leafy stems or lvs. crowded in a basal tuft, entire or pinnatifid. Heads radiate, fairly large, solitary on long peduncles. Phyllaries in 2–several rows, united and cuplike at base. Rays white, yellow, orange or scarlet, closing at night. Receptacle pitted. Aks. villous, the pappus of 2 series of delicate scarious toothed scales, often hidden in the ak.-wool. Ca. 25 spp. of S. Afr. (Named for T. *Gaza*, 15th-century Italian scholar.)

1. **G. longiscàpa** DC. Subacaulescent; lvs. variable, lanceolate, entire or pinnately cut, subglabrous and green above, white-tomentose beneath, ± revolute; heads 3–8 cm. across, on glabrous peduncles exceeding lvs.; rays yellow, often with black spot at base; invol. glabrous, the phyllaries finely pointed, equaling or exceeding the tube, the outer ones ciliate.—Occasional escape from gardens, as in Santa Barbara Co.; native of S. Afr. Many months.

63. *Geraèa* T. & G.

Herbs with alternate dentate lvs. Heads rather few, paniculate, showy, yellow-fld. Invol. hemispheric, 2–3-seriate. Bracts of the low-convex receptacle softly scarious, clasping the aks. and falling away with them. Ray-fls. neutral. Disk-aks. flat, narrowly cuneate, villous-ciliate, the sides villous medianly, black with whitish narrow margin and thickened crown produced into 2 slender persistent awns. Two spp. (Greek, *geraios*, old, the aks. white-villous.)

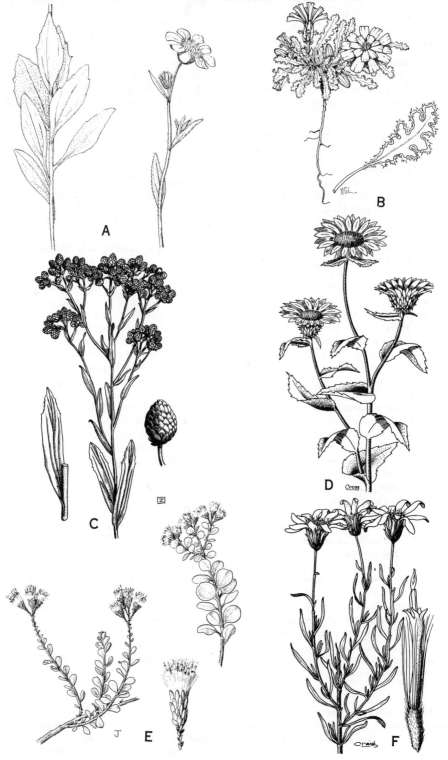

Plate 16. Fig. A, *Geraea canescens*. Fig. B, *Glyptopleura marginata*. Fig. C, *Gnaphalium californicum*. Fig. D, *Grindelia camporum*. Fig. E, *Haplopappus cuneatus*. Fig. F, *Haplopappus linearifolius*.

(Blake, S. F. A revision of Encelia and some related genera. Proc. Am. Acad. 49: 355–357. 1913.)

Heads radiate; phyllaries densely ciliate .... 1. *G. canescens*
Heads discoid; phyllaries densely glandular .... 2. *G. viscida*

1. **G. canéscens** T. & G. [*Simsia c.* Gray. *Encelia eriocephala* Gray.] Desert-Sunflower. Plate 16, Fig. A. Annual, often with several stems from the base, 2–6 dm. high, white-hirsute and glandular, asperous; lvs. lanceolate or oblanceolate to broadly ovate, acute, entire or few-toothed, narrowed to a margined petiole, 1–7 cm. long, the upper ones reduced to bracts; heads solitary or paniculate, peduncled; invol. 7–12 mm. high, the green lance-acuminate phyllaries prominently white-villous-ciliate; rays 10–21, golden, oblong, up to 20 mm. long; aks. 6–7 mm. long; pappus-awns 3 mm. long; $n=18$ (Powell & Turner, 1963).—Sandy desert floors, below 3000 ft.; Creosote Bush Scrub; mostly e. Mojave and Colo. deserts, Inyo Co. to L. Calif.; Utah, Ariz., Son. Feb.–May, Oct.–Nov.

2. **G. víscida** (Gray) Blake. [*Encelia v.* Gray.] Short-lived perennial with few coarse leafy stems 5–8 dm. high, hirsute and densely glandular-puberulent throughout; lvs. thin, ovate-oblong, acute to obtuse, clasping at base, irregularly dentate, 3–9 cm. long; heads corymbose-paniculate, discoid, 1.5–4 cm. across; phyllaries lance-oblong; aks. 7–10 mm. long; pappus-awns 3–4 mm. long, villous.—Dry hillsides, 2000–4000 ft.; Chaparral; se. San Diego Co.; n. L. Calif. May–June.

64. *Glyptopléura* D. C. Eat.

Depressed small winter annuals with compact rosette of pinnatifid lvs. having a toothed white crustaceous edge. Heads many, short-peduncled, white or pale yellow, often pinkish in drying; invol. of ca. 7–12 equal lanceolate scarious-margined phyllaries and a basal group of ± spatulate bractlets which are pinnatifid to toothed on their crustaceous margins. Aks. oblong or columnar, obtusely 5-angled, each face with 2 rows of tubercles, the ak.-apex abruptly beaked; pappus-bristles white, in several series, the outer falling separately. Two often recognized spp. (Greek, *glyptos*, carved, and *pleura*, side, referring to the sculptured aks.)

1. **G. marginàta** D. C. Eat. Plate 16, Fig. B. Plant 3–5 cm. high; lvs. obovate to spatulate, 2–4 cm. long, sinuately lobed with a conspicuous white margin cut into short teeth; invol. 10–12 mm. high; ligules white, turning pink, but little exserted; $2n=18$ (Stebbins et al., 1953).—Dry sandy flats, at 2000–4500 ft.; Creosote Bush Scrub; Mojave Desert (Kelso, Fenner, Goffs, Box "S" Springs, Keeler, Nelson Range, etc.); to Ore., Utah. April–June. Variable and for the most part giving way to a form whose lvs. have very narrow white margin, narrower than the acuminate teeth; ligules 10–15 mm. long, much exserted, mostly creamy to yellowish; more abundant w. Mojave Desert. This form is doubtfully distinct; it has been called *G. setulòsa* Gray and *G. marginata* var. *s.* Jeps.

65. *Gnaphàlium* L. Cudweed. Everlasting

Annual to perennial woolly herbs, with alternate entire lvs. Heads disciform, white, yellowish or tinted rose, arranged in panicles, corymbs or spikes. Invol. ovoid or campanulate; phyllaries slightly to evidently imbricated, scarious at tip or almost throughout. Receptacle naked. Numerous outer fls. slender, ♀, the few inner ones coarser and perfect. Pappus of capillary bristles, sometimes ± thickened at summit, sometimes united at base. Aks. small, nerveless. Over 100 spp., widely distributed. (Greek, *gnaphalon*, a lock of wool, these plants floccose-woolly.)

1. Pappus-bristles united at base, falling away in a ring; phyllaries but little imbricate.
    2. Lower lf.-surface loosely villous-lanate; invol. nearly buried in wool, 3–4 mm. high
    .... 9. *G. peregrinum*
    2. Lower lf.-surface closely white-pannose, the subappressed hairs tightly enmeshed; invol. densely woolly at base only, 4–6 mm. high .... 10. *G. purpureum*

1. Pappus-bristles not or only partly united at base, deciduous separately or in small groups.
    3. Heads small, the invol. mostly 2–4 mm. high; glomerules of heads leafy-bracted; plants less than 2 dm. high, usually much branched .... 8. *G. palustre*
    3. Heads larger, mostly 4–7 mm. high; glomerules of heads not leafy-bracted; plants mostly 2–10 dm. high, the stems often simple below the infl.
        4. Lvs. green in age, at least on the upper surface.
            5. Mature lvs. usually green on both surfaces; plants with rather strong balsamlike odor; annual to biennial.
                6. Infl. corymbose; invol. rounded, white; lvs. lanceolate to oblong. Common, widespread .... 3. *G. californicum*
                6. Infl. paniculate; invol. narrower, often pinkish; lvs. lance-linear. Occasional, Orange Co. n. .... 11. *G. ramosissimum*
            5. Mature lvs. green above, white beneath; plants not strongly scented, perennial.
                7. Lvs. oblong to broadly linear, broadly auriculate at base; stems tending to branch, 6–9 dm. high .... 2. *G. bicolor*
                7. Lvs. narrowly linear, with short-decurrent base; stems mostly simple, 3–6 dm. high
                    5. *G. leucocephalum*
        4. Lvs. permanently tomentose.
            8. The lvs. not or scarcely decurrent.
                9. Phyllary-tips pearly white or ± straw color. Native plants.
                    10. Middle phyllaries ca. twice as long as broad, rounded-obtuse at apex. Cismontane and montane .... 7. *G. microcephalum*
                    10. Middle phyllaries ca. 3 times as long as broad, acutish. Desert mts. or edge of desert .... 13. *G. wrightii*
                9. Phyllary-tips greenish or brownish. Introd. weed .... 6. *G. luteo-album*
            8. The lvs. strongly decurrent.
                11. The stems usually simple above; heads in dense terminal clusters; phyllaries all decidedly obtuse .... 4. *G. chilense*
                11. The stems usually branched above; heads in rather open corymbose panicles, at least the inner phyllaries abruptly pointed.
                    12. Lvs. of the young shoots oblanceolate, up to 5.5 cm. long; slender-stemmed plants of the pine belt .... 12. *G. thermale*
                    12. Lvs. of the young shoots linear, up to 10 cm. long; stout plants below the pine belt
                        1. *G. beneolens*

1. **G. beneòlens** Davids. Rather stout perennial, 4.5–10 dm. tall, persistently white-woolly throughout, sometimes greenish-yellow, sweet-scented; lvs. of the new shoots and the basal lvs. linear, narrowly acute, 4–10 cm. long, 2–3 mm. wide; stem lvs. but little reduced upward on the stems, evidently and often conspicuously decurrent; infl. usually a long narrow panicle, to 3 dm. long, with ascending branches; heads glomerulate on the ultimate branches; heads 44–55-fld., pale yellow; invol. campanulate, 5–6 mm. high, 4–5-seriate; phyllaries obtuse to acute, the inner more narrow, the exposed tips papery and opaque.—Dry places below 5000 ft.; Coastal Sage Scrub, Chaparral; s. Calif. to Ore. July–Nov. Santa Rosa, Santa Cruz ids.

2. **G. bicólor** Bioletti. Biennial or perennial, with several stout branched stems 4–9 dm. high, very leafy, white-tomentose; lvs. glabrate and becoming green above, white-tomentose beneath, lance-oblong, 2–7 cm. long, closely sessile by a broad auriculate base, ± crisped along margins; heads in rather loose corymbs; invol. campanulate, ca. 6 mm. high; phyllaries whitish, the outer ovate, the inner narrow-oblong.—Common in dry open places below 2500 ft.; Coastal Sage Scrub, Chaparral; cismontane, from L. Calif. to cent. Calif. San Clemente, Santa Catalina, Santa Rosa, Santa Cruz ids. Jan.–May.

3. **G. califórnicum** DC. [*G. decurrens* var. *c.* Gray.] Plate 16, Fig. C. Biennial, stoutish, 4–8 dm. high, corymbosely branched at summit, leafy, green, glandular, strongly scented; lower lvs. oblong-lanceolate, 4–10 cm. long, cauline gradually reduced and more lanceolate; infl. a large terminal paniculate corymb; invol. 6–7 mm. high, rounded, phyllaries mostly white, blunt to broadly rounded.—Dry hills, disturbed places, etc. below 4500 ft.; Chaparral, etc.; L. Calif. to Ore. Jan.–July. Santa Catalina Id.

4. **G. chilénse** Spreng. Annual or biennial, several-stemmed, 2–6 dm. high, usually quite leafy and with greenish-yellow tomentum; lvs. lanceolate to narrowly spatulate,

2–5 cm. long, gradually reduced upward; heads usually in a single rather close glomerule at end of each stem, sometimes paniculate; invol. 5–6 mm. high, greenish-yellow; phyllaries all obtuse.—Rather moist, often waste places, below 6000 ft.; many Plant Communities; cismontane Calif.; occasional on desert; to B.C., Mont., Tex. June–Oct. Most of our ids. An occasional plant lower in stature, densely leafy to the compact cluster of heads is var. *confertifòlium* Greene.

5. **G. leucocéphalum** Gray. Perennial, with few erect leafy permanently tomentose stems mostly 3–6 dm. high; lvs. narrowly linear, or lowermost oblance-linear, 2–8 cm. long, green above, white-tomentose beneath, attenuate; infl. a corymbose panicle; invol. pearly white, 6–7 mm. high; phyllaries papery, obtuse, ovate.—Occasional, dry disturbed hills and waste places, low elevs.; Coastal Sage Scrub, Chaparral; cismontane from Riverside and Orange cos. n. to Ventura Co.; e. to Tex. Aug.–Sept.

6. **G. lùteo-álbum** L. Permanently tomentose annual weed 2–6 dm. high, the stems erect from a decumbent base, leafy, loosely branched; lvs. linear-oblanceolate, 2–4 cm. long, 2–4 mm. wide, obtuse, the upper cauline becoming linear-lanceolate and acute to attenuate; heads in usually several cymosely arranged terminal clusters; invol. 4–5 mm. high, the hyaline-tipped phyllaries greenish or brownish-stramineous, the inner ones lance-oblong; $n=7$ (Shetty, 1967).—Occasional along roadsides, etc., Riverside Co., Los Angeles Co., Santa Barbara Co.; to cent. Calif.; introd. from Old World. On Santa Catalina, San Clemente, Santa Rosa, Santa Barbara ids. Flowering all year.

7. **G. microcéphalum** Nutt. [*G. albidum* Jtn.] Biennial or short-lived perennial, 5–10 (–15) dm. high, permanently densely white-tomentose throughout, loosely branched; lvs. oblanceolate to spatulate, 2–5 cm. long, 4–10 mm. wide, sessile; panicle corymbose, the heads in small clusters at the ends of the branches; invol. 5–6 mm. high; phyllaries whitish, the outer ovate, the inner oblong.—Dry slopes and open places, below ca. 4000 ft.; largely Chaparral; n. L. Calif. to cent. Calif. July–Oct. Santa Catalina and Santa Cruz ids.

8. **G. palústre** Nutt. Annual, commonly branched at base, 0.5–2 (–3) dm. high, floccose-tomentose, especially upwards and on the stems; lvs. spatulate, 1–3 cm. long, the uppermost oblong or lanceolate and subtending the glomerules of heads; invol. 3–3.5 mm. high, densely woolly below, the phyllaries not much imbricate, brown, usually with whitish tips; pappus-bristles distinct, falling separately.—Frequent on damp flats and banks, from sea-level to 9500 ft.; many Plant Communities; cismontane and montane Calif.; to B.C., Rocky Mts. May–Oct. San Clemente, Santa Catalina, and Santa Cruz ids.

9. **G. peregrìnum** Fern. [*G. spathulatum* auth.] Like *G. purpureum*, simple or loosely branched; lvs. spatulate, 2–4 cm. long, greenish, loosely villose-lanate; heads nearly buried in dense gray wool; invol. 3–4 mm. long.—Occasional weed, especially in Los Angeles Co.; to Atlantic Coast. March–May.

10. **G. purpùreum** L. [*G. ustulatum* Nutt.] Annual or biennial, simple or branched, 1–5 dm. high, closely woolly-canescent; lower lvs. spatulate to oblanceolate, petioled, 2–5 (–9) cm. long, ± bicolored, closely white-pannose beneath; cauline lvs. gradually reduced up the stem; spike terminal, dense or ± interrupted, leafy-bracted in lower part; invols. crowded, lanate at base only, 4–6 mm. long, brown to chestnut or purple; pappus-bristles united at base; $2n=28$ (Huynh, 1965).—Dry, often disturbed places, near the coast from Santa Barbara n.; e. U.S., S. Am. Santa Catalina, Santa Rosa, Santa Barbara ids. April–July.

11. **G. ramosíssimum** Nutt. Biennial, with 1–several slender erect stems 5–12 dm. high, glandular, sweet-scented, early greenish and glabrate; lvs. lance-linear, mostly 2–6 cm. long; infl. a large terminal ± oblong panicle; invol. 4–5 mm. high, turbinate; phyllaries often pinkish, ± rounded to pointed.—Occasional, dry slopes, below 1500 ft.; Coastal Strand, Coastal Sage Scrub, Chaparral; near the coast from San Diego Co. to cent. Calif. March–Sept.

12. **G. thermàle** E. Nels. Perennial, 2–4 dm. tall, loosely woolly throughout, stems erect and simple below the infl., ± densely leafy below; lower lvs. and those on new shoots

oblanceolate to spatulate, 2.5–5 cm. long; stem lvs. sessile, noticeably decurrent, 1–3 cm. long; infl. usually short, ± cymose; heads ca. 40–50-fld., the invol. woolly only at extreme base, 4–5 mm. high, usually 3-seriate; phyllaries thin, hyaline, shining, or with an opaque tip.—Mostly at 5000–7000 ft.; Yellow Pine F.; mts. of s. Calif. to B.C. and Rocky Mts. July–Sept.

13. **G. wrightii** Gray. Perennial with leafy stems; to ca. 3 dm. high; lvs. oblanceolate and not decurrent; heads ca. 35–40-fld.; invol. 2–3-seriate; phyllaries few, thin, hyaline and shining, the outer usually obtuse, the inner sharply acute.—Uncertainly present in Calif., possibly at Campo (San Diego Co.), Little San Bernardino and New York mts.; to Tex., Mex.

66. *Grindèlia* Willd. Gum-Plant

Annual, biennial or perennial herbs, mostly with a taproot, rarely suffrutescent at base, ± resinous, especially on the invol. Lvs. alternate, punctate, usually serrate and sessile, often clasping. Heads medium to large, yellow, usually radiate, solitary at branch-tips. Invol. multiseriate, imbricate, the phyllaries thickish, with pale appressed base and narrow often squarrose or revolute herbaceous tip. Receptacle flattish, foveolate. Ray-fls. 10–45, uniseriate, fertile. Disk-fls. usually fertile. Style-branches with slender hispidulous appendages. Aks. compressed to subquadrangular, few-angled, glabrous; pappus of 2–8 stiff often curved deciduous corneous or paleaceous awns. Ca. 50 spp. of w. N. and S. Am. (For Prof. D. H. *Grindel*, 1776–1836, botanist of Dorpat and Riga.)

(Steyermark, J. A. A monograph of the N. Am. spp. of the genus Grindelia. Ann. Mo. Bot. Gard. 21: 433–608. 1934.)

1. Tips of phyllaries erect or spreading, some gradually curved but not sharply reflexed. Ventura Co. .... 4. *G. hirsutula*
1. Tips of phyllaries (at least some of middle and outer ones) sharply reflexed or looped.
    2. Insular succulent plants .... 5. *G. latifolia*
    2. Interior non-succulent plants; stems ± erect.
        3. Lvs. callous-serrulate. Rare introds.
            4. Heads discoid. E. Mojave Desert .... 1. *G. aphanactis*
            4. Heads radiate. W. Mojave Desert .... 7. *G. squarrosa*
        3. Lvs. sharply toothed or entire, not callous-serrulate.
            5. Invols. mostly 1 cm. in diam.; ligules 5–8 mm. long. Cuyamaca Mts., San Diego Co.
               3. *G. hallii*
            5. Invols. 12–25 mm. in diam.; ligule (when present) mostly 8–15 mm. long.
                6. Tips of phyllaries spreading or reflexed, but rarely looped. N. Los Angeles Co.
                   2. *G. camporum*
                6. Tips of phyllaries looped back to form a tight ring .... 6. *G. robusta*

1. **G. aphanáctis** Rydb. Herbaceous biennial; stems slender, corymbosely much branched, glabrous, 3–7 dm. high, uniformly leafy throughout; lvs. ± resinous-punctate, the upper and middle cauline entire or crenulate-serrate, the lower and basal crenate to pinnatifid, the main cauline 2.5–7 cm. long, oblong or oblanceolate, the middle and upper amplexicaul; heads discoid; invol. 5–6-seriate, the upper third to half of the phyllaries loosely or moderately reflexed, 4–12 mm. long, with subulate tips; aks. 2.3–3 mm. long, deeply furrowed or ribbed; awns 2–3 slender, subentire to moderately setulose-serrulate. Apparently collected in Mountain Pass, Clark Mt., e. Mojave Desert at 5000 ft.; Sept. Native from cent. Ariz. to Colo., w. Tex.

2. **G. campòrum** Greene. [*G. robusta* var. *rigida* Gray.] Plate 16, Fig. D. Several erect herbaceous stems, 5–12 dm. high, simple or openly branched, subglabrous; lvs. subcoriaceous, very resinous, saliently dentate, narrowly oblong to broadly oblanceolate, the cauline 2–8 cm. long, 7–15 mm. wide; heads 2.5–4 cm. across, resinous, the green tips of the multiseriate elongate phyllaries strongly recurved to hooked; rays 18–35, 8–15 mm. long; $n = 12$ (Raven et al., 1960).—Dry places at 3000–4000 ft.; n. Los Angeles Co.; (Elizabeth Lake, Rock Creek of San Gabriel Mts.); to San Francisco Bay. May–Oct.

3. **G. hállii** Steyerm. Several glabrous herbaceous stems, 3–6 dm. high, corymbosely branched above; lvs. mostly sharply and regularly serrate, subcoriaceous, resinous-punc-

tate, the basal in often persistent rosettes, oblanceolate, 5–7 cm. long, the cauline oblong, smaller; heads many, 2–3 cm. across, strongly resinous, the green tips of the outer phyllaries strongly recurved or hooked; rays 13–21, 5–8 mm. long; $n=6$ (Raven et al., 1960). —Dry flats, largely Yellow Pine F., Cuyamaca Mts.; San Diego Co. July–Oct.

4. G. hirsùtula H. & A. Herbaceous perennial 3–8 dm. high; stems erect, slender, simple or commonly corymbosely branching above, ± villous with crisped hairs, especially on peduncles; lvs. chartaceous, gray-green, the basal oblanceolate or spatulate, obtuse, remotely serrate or lobed, sometimes crenate, 10–22 cm. long, the cauline reduced in size upwards, entire, the uppermost bractlike and clasping; heads 2.5–4 (–5.5) cm. across; invol. ± pubescent, the tips of the phyllaries erect or nearly so, not caudate; rays 20–35, 1–2 cm. long; $n=12$ (Raven et al., 1960).—Arid slopes, Coastal Sage Scrub; Ventura Co.; Monterey Co. to Napa Co. April–July.

5. G. latifòlia Kell. [*G. robusta* var. *l.* Jeps.] Very succulent and leafy herbaceous perennial 4–6 dm. high, the stout decumbent or ascending branches one-headed or topped by a cluster of 2–3 large heads, these often subtended by lvs.; lvs. principally cauline, thick, irregularly serrate to sharply dentate, scabro-ciliate, lance-ovate to oblong, amplexicaul to subcordate, rounded at apex, 3–8 cm. long, 1.5–4 cm. wide; heads 3–5 cm. across, milky-resinous, the outermost phyllaries foliaceous, their green tips and usually those of many inner ones squarrose; rays 30–45, 10–15 mm. long; $n=12$ (Solbrig et al., 1962).—Coastal Salt Marsh, Coastal Strand, Coastal Sage Scrub; Santa Rosa and Santa Cruz ids.; mainland coast northward. May–Sept.

6. G. robústa Nutt. Stems few, erect from a subligneous crown, stout, corymbosely branched above, 5–12 dm. long; lvs. sharply toothed to finely serrate or entire, the basal oblanceolate, to 18 cm. long, the cauline much reduced, lance-ovate to linear-oblong, clasping; heads 3–5 cm. across, often strongly and translucently resinous, the long green tips of the phyllaries rolled back in a loop; rays 25–45, 8–15 mm. long; $n=12$ (Raven et al., 1960).—Dry slopes and fields, below 4000 ft.; Coastal Sage Scrub, Chaparral; largely cismontane, n. L. Calif to Santa Barbara Co. March–Sept. Discoid plants occur and are the var. **bracteòsa** (J. T. Howell) Keck. [*G. b.* J. T. Howell.]

7. G. squarròsa (Pursh) Dunal. [*Donia s. Pursh.*] Erect biennial or short-lived perennial 2–10 dm. high, openly branched above and bearing many heads; lvs. regularly callous-serrulate, sometimes sharply toothed or even entire, mostly oblong, 2–5 cm. long, the upper clasping; heads 2–3 cm. across, strongly resinous, the green tips of the phyllaries strongly rolled back; rays 25–40, 7–15 mm. long.—Dry places, Joshua Tree Wd.; sparingly introd. in Antelope V., w. Mojave Desert. Native of the Great Plains. July–Sept.

67. *Guizòtia* Cass.

Annual herbs with opposite lvs. or the upper alternate. Heads peduncled, axillary and terminal; ray-fls. ♀, 1-seriate, fertile, yellow; disk-fls. perfect, fertile. Invol. campanulate; phyllaries 2-seriate, the outer subfoliaceous, the inner like the paleae. Receptacle conic or convex. Paleae flat, scarious. Disk-corollas pubescent without at the base. Aks. glabrous, dorsally compressed, rounded at tip; pappus none. A few spp. of trop. Afr.

1. G. abyssínica (Lf.) Cass. [*Polymnia a.* L. f.] RAMTILLA. Stout, erect, leafy, smooth or scabrid, 3–9 dm. high; lvs. sessile, 7–16 cm. long, linear or lance-oblong, partly amplexicaul, obtuse, serrate; heads 1–2 cm. in diam.; peduncles naked, 3–5 cm. long; outer phyllaries broadly elliptic or ovate, green; ligules few, broad; $n=15$ (Shetty, 1967).—Occasional weed in waste places, cismontane s. Calif. Native of Afr. and cult. in India, etc. for the oil.

68. *Gutierrèzia* Lag. MATCHWEED

Perennial herb or subshrub, glabrous to hirtellous. Lvs. alternate, entire, filiform to narrowly oblanceolate, usually punctate-glandular. Heads very small, radiate, yellow, numerous, scattered or crowded in cymes or panicles. Invol. cylindrical to turbinate-globose, the imbricated phyllaries coriaceous, appressed, whitish. Receptacle foveate, sometimes hairy. Ray-fls. ♀, fertile. Disk-fls. perfect, sometimes sterile. Aks. obovoid or oblong, pubescent; pappus of 10–12 oblong, unequal, free scales. Ca. 25 spp., native of w. N. and S. Am. (Named for P. *Gutierrez*, Spanish nobleman.)

(Solbrig, O. T. Cytotaxonomic and evolutionary studies in the N. Am. spp. of Gutierrezia. Contr. Gray Herb. 188: 1–63. 1960. The Calif. spp. of Gutierrezia Madroño 18: 75–84. 1965.)

Heads with only 2–3 fls.; invol. very narrow, to 1.5 mm. wide; aks. of disk fls. aborted. From desert fringes .... 2. *G. microcephala*
Heads with more than 4 fls.; invol. turbinate, 1.5 or more mm. wide; aks. of disk-fls. fertile.
    The heads mostly solitary at the ends of branchlets; fls. usually more than 10; invol. 2–5 mm. wide .... 1. *G. bracteata*
    The heads clustered at the ends of branchlets; fls. 5–10; invol. less than 5 mm. wide
                                                                                                                 3. *G. sarothrae*

1. **G. bracteàta** Abrams. [*G. californica* var. *b.* Hall.] Subshrub 3–6 dm. high, nearly glabrous to densely hirtellous; leaves spreading to deflexed, up to 5 cm. long, mostly 1 mm. wide; heads usually solitary at ends of branchlets; invol. 5–6.5 mm. high, conical to turbinate, the phyllaries in ca. 3 rows, narrow, elongate, carinate or strongly convex, to 4 mm. long and 2 mm. wide; ligulate fls. usually 5 (3–6), 6–8 mm. long; tubular fls. usually 4 (2–6), 4–6 mm. long; $n=8$ (Solbrig et al., 1964).—Dry hills and plains below 3500 ft., occasional higher; Chaparral, V. Grassland; cismontane L. Calif. n. to cent. Calif. May–Oct.

2. **G. microcéphala** (DC.) Gray. [*Brachyris m.* DC. *G. lucida* Greene.] Many-stemmed, 3–6 dm. high, strongly resinous, essentially glabrous, the slender stems striate-angled, much-branched above; lvs. 2–5 cm. long, or a few solitary; phyllaries yellowish-white, without green tips, with prominent hyaline margin, the inner row of only 2 members, as long as disk-fls.; ray- and disk-fls. 1–2 each, the ligule up to 2.5 mm. long; ray-aks. fertile, appressed-pilose, ca. 2 mm. long; ray-pappus of ca. 8 oblong paleae ca. 0.8 mm. long; disk-pappus of ca. 12 paleae 1 mm. long; $n=12$ (Solbrig et al., 1964).—Open desert up to 7500 ft.; Shadscale Scrub, Creosote Bush Scrub, Joshua Tree Wd.; Mojave Desert from the White Mts. to Antelope V. and Little San Bernardino Mts.; e. to Colo., Tex., n. Mex. July–Oct.

3. **G. saròthrae** (Pursh) Britt. & Rusby. [*Solidago s.* Pursh. *Xanthocephalum s.* Shinners.] Subshrub 1.5–6 dm. high, mostly hirtellous, the numerous slender stems cymosely paniculate above; lvs. punctate, 2–5 cm. long, 1–2 mm. wide; infl. flat-topped, the numerous resinous heads scattered or usually in small glomerules; invol. narrowly turbinate, 4–5 mm. high, 2–3 mm. thick, the phyllaries often with obscurely thickened herbaceous tips; ray- and disk-fls. 3–8 each, the ligule ca. 3 mm. long; aks. subsericeous-pilose; pappus of ray ca. 0.7 mm. long, of disk ca. 1.5 mm. long; $n=8$ (Solbrig et al., 1964).—Desert plains and slopes below 10,000 ft.; San Diego and San Bernardino cos. to Can., Kans., etc. May–Oct.

69. *Haplopáppus* Cass.

Herbs or shrubs, very varied in habit, often glandular. Lvs. alternate, entire to bipinnatifid, often thickish, sometimes glandular-punctate. Heads radiate or discoid, large to small, solitary to numerous and cymose or paniculate, yellow, rarely creamy white. Invol. cylindric to turbinate to hemispheric, the phyllaries numerous, subequal to strongly graduate, usually narrow and indurate or chartaceous, at least below. Receptacle ± alveolate. Ray-fls. ♀, rarely sterile; disk-fls. fertile, their style-branches ovate to subulate. Aks. terete or angled, linear-fusiform to turbinate, glabrous to silky-pilose. Ca. 150 spp., all Am., chiefly w. U.S., Mex., and Chile. (Greek, *haploos*, simple, and *pappos*, pappus.)

(Hall, H. M. The genus Haplopappus. Carnegie Inst. Wash. Publ. 389: 1–391. 1928.)

1. Lvs. dentate to bipinnatifid, the teeth spinulose- or bristle-tipped; aks. turbinate, 2–3 mm. long.
  2. Annual or biennial.
    2a. Plant 6–25 cm. high; ligules 7–12 mm. long. Native on the eastern deserts . 13. *H. gracilis*
    2b. Plant 5–15 dm. tall; ligules 12–18 mm. long. Introd. at Ventura River estuary
                                                                                                        6. *H. ciliatus*
  2. Perennials.

3. Tufted with several slender erect stems from a suffrutescent base; lvs. very narrow; heads radiate.
    4. Phyllaries prominently glandular-puberulent and scabrous; lvs. 2–5 cm. long. E. Mojave Desert .... 12. *H. gooddingii*
    4. Phyllaries beset with granular glands; lvs. 1–2 cm. long. Sw. San Diego Co.
                                                                                                 14. *H. junceus*
  3. Rigidly branched shrub; lvs. mostly oval, merely dentate; heads discoid. Inyo Co.
                                                                                                 4. *H. brickellioides*
1. Lvs. various, but if toothed, the teeth not as above; aks. nearly prismatic, subcylindric or fusiform, 3 mm. or more long.
  5. Perennial herbs with shoots of the season arising from prominent leafy rosettes surmounting a deep fusiform taproot.
    5a. Heads sessile or short-pedunculate; invol. 6–8 mm. high. E. Mojave Desert
                                                                                                    22. *H. racemosus*
    5a. Heads mostly long-pedunculate; invol. 10–13 mm. high. San Bernardino Mts
                                                                                                    24. *H. uniflorus*
  5. Shrubs or subshrubs.
    6. Stems scapiform, monocephalous, cespitose from a woody caudex. Inyo Mts. 1. *H. acaulis*
    6. Stems not scapiform; plants not cespitose.
      7. Low intricately branched shrub 2–3 dm. high.
      7a. Lvs. 2–3 mm. wide, 6–12 mm. long; ray-fls. 4–6 .... 11. *H. gilmanii*
      7a. Lvs. 3–6 mm. wide, 10–30 mm. long; ray-fls. 0 .... 17. *H. macronema*
      7. Usually more than 3 dm. high, or if not, the lvs. wider.
        8. Invol. hemispheric, 10–18 mm. wide; herbage glandular-punctate; lvs. entire
                                                                                   16. *H. linearifolius*
        8. Invol. turbinate or subcylindric.
          9. Disk-corollas with slender tube abruptly dilated to a much broader throat; heads discoid, in terminal cymes.
            10. Phyllaries with green but thin tips; stems brownish. Largely cismontane
                                                                                      25. *H. venetus*
            10. Phyllaries with conspicuous thickened gland or resin-pocket near tip; stems white and shining. Deserts .... 2. *H. acradenius*
          9. Disk-corollas only slightly ampliate upward.
            11. Heads large, the invols. 8–15 mm. high, tightly imbricate, mostly 6–8-seriate; herbage without distinct resin-pits.
              12. Herbage tomentose. Insular.
                13. Lvs. oblanceolate, thinnish, glabrate in age, closely and finely serrulate or entire. San Clemente Id. .... 5. *H. canus*
                13. Lvs. narrowly ovate or obovate, thickish, rarely glabrate, coarsely serrate. Anacapa, Santa Rosa, Santa Cruz ids. .... 9. *H. detonsus*
            12. Herbage not tomentose. Mainland .... 23. *H. squarrosus*
            11. Heads smaller, the invols 3–8 mm. high, loosely imbricate, 2–6-seriate; herbage with distinct resin-pits.
              14. Ray-fls. present; lvs. filiform; heads ± paniculate.
                15. Outer phyllaries ± caudate-tipped.
                  16. Lvs. 10–35 mm. long, with shorter fascicles in the axils; aks. pilose. Cismontane Los Angeles Co. s. .... 20. *H. pinifolius*
                  16. Lvs. 4–12 mm. long, scarcely exceeding the axillary fascicles; aks. glabrous or sericeous. Los Angeles Co. n. .... 10. *H. ericoides*
                15. Outer phyllaries obtuse .... 18. *H. palmeri*
            14. Ray-fls. reduced or wanting (up to 5 mm. long in *H. laricifolius*).
              17. Lvs. filiform to linear, less than 3 mm. wide.
                18. Heads solitary or racemose-paniculate, discoid. San Diego Co. s.
                                                               21. *H. propinquus*
                18. Heads cymose.
                  19. Lvs. 0.5–2 cm. long; broad rounded shrubs seldom more than 1 m. high. Deserts.
                    20. Herbage glabrous; lvs. subterete; without persistent fascicles in the old axils; ray-fls. 3–11 .... 15. *H. laricifolius*
                    20. Herbage hairy; lvs. flat; with persistent fascicles in the old axils; ray-fls. 0–2 .... 7. *H. cooperi*

19. Lvs. 3-6 cm. long; arborescent shrubs 1-3 m. high. Cismontane
.... 3. *H. arborescens*
17. Lvs. oblanceolate to obovate, 3-10 mm. wide.
21. Heads 9-12-fld., discoid; erect shrub 2-5 m. high; lvs. 2-6 cm. long, mostly tapering to base and apex .... 19. *H. parishii*
21. Heads 16-30-fld., sometimes radiate; spreading subshrub mostly less than 1 m. high; lvs. cuneate .... 8. *H. cuneatus*

1. **H. acáulis** (Nutt.) Gray. [*Chrysopsis a.* Nutt.] Stems scapiform, numerous, monocephalous, cespitose from a much-branched woody caudex, 5-10 cm. high, densely clothed at base with marcescent lvs., the whole mat up to 5 dm. across; lvs. linear-oblanceolate, to spatulate, mostly erect, entire, cuspidate-tipped, 1-6 cm. long, 2-7 mm. wide; invol. hemispheri,c 7-10 mm. high; phyllaries 2-3-seriate, pallid; ray-fls. 6 10, the ligules 6-10 mm. long; aks. densely sericeous to glabrous.—Dry often rocky places, 8000-10,500 ft.; Sagebrush Scrub, Bristle-cone Pine F.; Inyo Mts.; to Ore., Utah, Sask. May-Aug.

2. **H. acradènius** (Greene) Blake. ssp. **acradènius**. [*Bigelovia a.* Greene. *Isocoma a.* Greene.] Shrub 3-10 dm. high, the stems erect, quite woody, brittle, white-barked, striate, glabrous; lvs. linear-spatulate to oblong, 1-4 cm. long, 1-5 mm. wide, thick, entire, mostly mucronate and glabrous, minutely impressed-punctate, with axillary fascicles of shorter lvs.; heads in cymes, nearly sessile, 6-13-fld.; invol. 5-6.5 mm. high, phyllaries with conspicuous thick subepidermal resin-pocket near the rounded or blunt tip, the very narrow hyaline margin fimbrillate; fls. 6-13; aks. 3-4 mm. long, silvery-villous; pappus yellowish; $n=12$ (Raven et al., 1960).—Subalkaline or sandy flat, 1500-3500 ft.; Alkali Sink, Shadscale Scrub, Joshua Tree Wd.; sw. Mojave Desert to Nev., Ariz. Aug.-Nov.

Ssp. **eremóphilus** (Greene) Hall. [*Isocoma e.* Greene.] Lvs. denticulate or dentate or even saliently lobed, with some entire, 2-5 cm. long, to 7 mm. wide; heads 15-25-fld.; invol. 6-8 mm. high; $2n=12$ (DeJong & Mcntgomery, 1963).—Occasional to 7000 ft.; Creosote Bush Scrub, Alkali Sink; s. border of Mojave Desert, Colo. Desert; to L. Calif., Ariz.

3. **H. arboréscens** (Gray) Hall. [*Linosyris a.* Gray. *Ericameria a.* Greene.] GOLDEN FLEECE. Stout erect shrub 0.6-3 m. high, fastigiately branched, glabrous, resinous, prominently glandular-punctate; lvs. narrowly linear to filiform, 3-6 cm. long, up to 2 mm. wide, thick, crowded; heads discoid, 18-23-fld., many, in rounded terminal cymes or cymose panicles; invol. turbinate, 4-5 mm. high; phyllaries 4-seriate, graduate, lanceolate to linear, thin and chaffy except for the thickened costa; aks. turgid, obscurely 5-angled, less than 2 mm. long, finely sericeous; pappus very fragile; $n=9$ (Raven et al., 1960).—Dry foothills below 4000 ft.; Chaparral; Ventura Co.; to Ore. Aug.-Nov.

4. **H. brickellioìdes** Blake. Rigidly branched rounded shrub 2-8 dm. high, the older stems white-barked and ± pilose, the branches yellowish, some hairs thickened and tipped with yellow glands; lvs. oval, elliptic or obovate-cuneate, 1-3.5 cm. long, 5-25 mm. wide, spinescent-tipped, dentate with 1-4 pairs of spinescent teeth, triple-nerved; heads discoid, ca. 12-fld., rather small, sessile or subsessile in 1's-3's towards tips of leafy branchlets; invol. 6-7 mm. high; phyllaries 4-5-seriate, hispidulous and glandular; aks. oblong, hispidulous; pappus sparse.—Rare in rocky canyons, 2000-6500 ft.; Creosote Bush Scrub; Inyo Co., Last Chance Mts. and Saline V. to Death V. and Nev. April-Sept.

5. **H. cànus** (Gray) Blake. [*Diplostephium c.* Gray. *Hazardia c.* Greene.] Rounded shrubs to ca. 1 m. high; twigs stout, soft-woody, lanate-tomentose; lvs. oblanceolate, closely and finely serrulate or subentire, thinly tomentulose above and densely so beneath, glabrate in age, membranous, 4-10 cm. long, gradually reduced up the stems; infl. compound, thyrsiform; invol. turbinate 10-12 mm. high; phyllaries with tuft of loose woolly pubescence at apex, the inner being glabrous; heads discoid, sometimes purplish in age, with a ring of ca. 8-13 ♀ fls. around the margin, these having corollas 0.3-0.4 mm. wide near the base, the cent. perfect fls. ca. 1 mm. wide at base; aks. 4-5 mm. long; $n=5$ (Raven, 1963).—Dry bluffs, Coastal Sage Scrub; San Clemente Id.; Guadalupe Id. June-July.

6. **H. ciliàtus** (Nutt.) DC. [*Donia c.* Nutt.] Annual or biennial erect herb 5-15 dm. high; stems very leafy to top, glabrous; lvs. oval to oblong, dentate with spine-tipped teeth,

very obtuse, 3–8 cm. long; heads few, in open cymes; invol. 12–18 mm. high, the phyllaries in several series, the outer ± squarrose; rays many, the ligules 12–18 mm. long.—Introd. Ventura R. estuary; native Mo. to Tex., New Mex. Aug.–Oct.

7. **H. coóperi** (Gray) Hall. [*Bigelovia c.* Gray. *Haplopappus monactis* Gray. *Ericameria m.* McClat.] Low flat-topped fastigiately branched shrub 3–6 (–12) dm. high, glutinous, puberulent; lvs. linear-spatulate, 6–15 mm. long, to 1.5 mm. wide, with much smaller persistent ones in fascicles in lower axils; heads pediceled in small cymes; invol. 4–5 mm. high; phyllaries 2–3-seriate, 9–15, the outer ovate, the inner oblong, ± puberulent, the glandular thickening of the costa obscure; ray-fls. 0–2; disk-fls. 4–7 (–11), much exceeding the invol.; aks. silky-pilose; $n=9$ (Solbrig et al., 1964).—Common on dry mesas, 2000–6000 ft.; Creosote Bush Scrub, Joshua Tree Wd.; Mojave Desert from Antelope V. and Little San Bernardino Mts. to Mono Co.; Nev. Rare in cismontane s. Calif. March–June.

8. **H. cuneàtus** Gray. [*Ericameria c.* McClat. *H. c.* var. *spathulatus* (Gray) Blake.] Plate 16, Fig. E. Deep green spreading much-branched shrub 1–5 (–10) dm. high, glabrous, balsamic-resinous, glandular-punctate; lvs. crowded, cuneate to orbicular-obovate, entire, 5–20 mm. long, 3–10 mm. wide, thick; heads compactly cymose; invol. turbinate, 5–7 mm. high; phyllaries 4–6-seriate, imbricate, ± glandular-thickened along the costa, linear-oblong to lance-ovate, the outer passing into minute ovate peduncular bracts; ray-fls. usually wanting; disk-fls. 16–28; aks. densely appressed-pilose; $n=9$ (Solbrig et al., 1964).—Cliffs and slopes, especially in clefts of granitic rocks, 4000–7500 ft.; Pinyon-Juniper Wd., Yellow Pine F.; mts. n. L. Calif. to ranges surrounding Mojave Desert; to Sierra Nevada, Nev., Ariz. Sept.–Nov.

9. **H. detónsus** (Greene) Raven. [*Corethrogyne d.* Greene.] Resembling *H. canus*, but lvs. thick, nearly leathery, narrowly ovate to obovate, coarsely serrate with a few teeth or subentire, densely short-tomentulose; infl. subcorymbose, compound; disk-corollas 8.3–9.5 mm. long; $n=5$ (Raven, 1963).—Coastal Sage Scrub; Anacapa, Santa Rosa and Santa Cruz ids. June–Dec.

10. **H. ericoìdes** (Less.) H. & A. subsp. **ericoìdes**. [*Diplopappus e.* Less. *Ericameria e.* Jeps.] Broad compact fastigiately branched shrub 3–8 (–15) dm. high; herbage sparsely pilosulous, somewhat resinous; lvs. very numerous, nearly filiform, divaricate, 4–12 mm. long, subterete, with dense fascicles of scarcely shorter lvs. in the axils; heads cymose-paniculate, terminating leafy branches; invol. turbinate, 5–6 mm. high; phyllaries loosely 3–5-seriate, villous-ciliate, the outer lance-ovate, the inner oblong, the costa thickened above into a filiform gland; ray-fls. 2–6; disk-fls. 8–14; aks. glabrous.—Sand dunes on and near the coast; Coastal Strand; Los Angeles region to cent. Calif. Reported from San Miguel Id. Aug.–Nov.

Subsp. **blàkei** C. B. Wolf. Aks. moderately sericeous.—Sand hills away from the immediate coast and below 1500 ft.; Coastal Sage Scrub; Ventura Co. and n.

11. **H. gilmánii** Blake. Low rounded aromatic shrub 2–3 dm. high, with sticky twigs; lvs. numerous, often with axillary fascicles, spatulate, 6–12 mm. long, 2–3 mm. wide, often conduplicate; heads solitary or cymose, narrowly campanulate; invol. 7–9 mm. high, resinous; phyllaries 4–6-seriate, imbricate, the outer ovate-lanceolate, often with thickened green squarrose tip, the inner oblong, chartaceous, ciliolate, appressed; ray-fls. 4–6, white or pale yellow; disk-fls. 15–18; aks. silky.—Often in limestone, 7000–11,000 ft.; Pinyon-Juniper Wd., Bristle-cone Pine F.; Inyo and Panamint ranges, Inyo Co. Aug.–Sept.

12. **H. gooddíngii** (A. Nels.) M. & J. [*Sideranthus g.* A. Nels.] Taprooted perennial with stiffly erect or ascending slender stems 2–6 dm. high, ± glandular-puberulent, sometimes also canescent; lvs. scattered, ventrally lanate, scabrid dorsally, pinnatifid, the axis and remote lobes linear and bristle-tipped, 2–5 cm. long, the upper entire and reduced; heads terminating long branches; invol. ± hemispheric, 6–9 mm. high, 10–18 mm. wide; phyllaries linear-lanceolate, well imbricated, greenish, prominently glandular-puberulent and scabrous, with short apical bristle; ray-fls. 30–45, the ligules 6–10 mm. long; pappus of numerous tawny unequal bristles.—Rocky places, below 2000 ft.; Creosote Bush Scrub; e. San Bernardino Co.; Nev., Ariz. Feb.–May.

13. **H. grácilis** (Nutt.) Gray. [*Dieteria g.* Nutt. *Sideranthus g.* Nels.] Annual herb mostly 6–25 cm. high, usually divaricately branched, hirsute throughout, strigose; lvs. numerous, the lower oblanceolate, pinnatifid or bipinnatifid, 1.5–3 cm. long, 3–7 mm. wide, the upper linear, reduced, serrate-dentate to subentire, the lobes and teeth tipped with a prominent white bristle; heads cymose or solitary; invol. hemispheric, 6–7 mm. high, 8–12 mm. wide; phyllaries linear-lanceolate, imbricate, green with hyaline margin, pubescent, glandular; ray-fls. 16–28, the ligules 7–12 mm. long; pappus tawny, the bristles slightly dilated basally; $n=2$ (Jackson, 1957), (Turner & Flyr, 1966).—Sandy or rocky places, below 5000 ft.; Creosote Bush Scrub, Joshua Tree Wd.; e. Mojave Desert; to Colo., Tex., Mex. April–June.

14. **H. juncèus** Greene. [*Sideranthus j.* Davids. & Mox.] Stems tufted, 4–10 dm. tall, suffrutescent at base, slender, branching, sparingly strigose, slightly glandular near heads; lvs. chiefly linear, pinnatifid or serrate with bristle-tipped teeth, 1–2 cm. long, the upper reduced, entire; heads solitary on long scaly-bracted branches or in open cymes; invol. hemispheric, 5–8 mm. high, the phyllaries imbricate, linear, with granular glands, bristle-tipped; ray-fls. 15–25, the ligules 5–6 mm. long; pappus of many tawny unequal bristles.—Dry brushy hillsides, below 3000 ft.; Coastal Sage Scrub, Chaparral; s. San Diego Co.; n. L. Calif., Son. June–Oct.

15. **H. laricifòlius** Gray. [*Ericameria l.* Shinners.] TURPENTINE-BRUSH. Fastigiately branched broadly rounded shrub 3–10 dm. high, the herbage resinous, impressed-punctate, glabrous; lvs. linear, 1–2 cm. long, 1–2 mm. wide, usually subterete, mucronate, sometimes with smaller ones in axillary fascicles; heads in small leafy cymes; invol. broadly turbinate, 3–5 mm. high; phyllaries loosely 3–4-seriate, lance-acuminate, firm, the tip soft and ciliolate; the costa thickened into an olive-brown gland; ray-fls. 3–11; disk-fls. 10–16, much exceeding the invol.; aks pilose.—Rocky places, 3500–6500 ft.; Creosote Bush Scrub, Pinyon-Juniper Wd.; mts. of e. Mojave Desert; to Tex. and adjacent Mex. Sept.–Oct.

16. **H. linearifòlius** DC. [*Stenotus l.* T. & G. *H. interior* Cov.] Plate 16, Fig. F. Much-branched shrub 4–15 dm. high, essentially glabrous but usually puberulent below the heads, the fastigiate twigs resinous; lvs. crowded, sometimes fasciculate, glandular-punctate, nearly linear, narrowed toward base, entire, 1–4 cm. long, 1–2.5 mm. wide, flat to subterete; heads many, terminal on nearly leafless peduncles; invol. hemispheric, 8–14 mm. high, the phyllaries 2–3-seriate, scarcely graduate, lance-oblong to linear, acuminate, beset with granular glands, with greenish center and lacerate-ciliate scarious margin; ray-fls. 13–18, the ligules 8–15 mm. long; aks. silky-pilose; pappus white; $n=9$ (Raven et al., 1960), (Solbrig et al., 1964).—Dry slopes and banks, up to 6500 ft.; Chaparral, Creosote Bush Scrub to Pinyon-Juniper Wd.; interior cismontane s. Calif. across Mojave Desert to Inyo Co.; Ariz., L. Calif., cent. Calif. March–May. A ssp. *interior* Hall has been recognized based on short lvs. (1–2 cm.) and small invols. (8–10 mm. high).—Mojave Desert.

17. **H. macronèma** Gray. [*Macronema discoideum* Nutt.] Undershrub 1–4 dm. high, with numerous erect pannose-tomentose twigs of the season from a woody base; lvs. many, oblong to oblanceolate, sessile, entire or the margin crisped, 1–3 cm. long, 3–6 mm. wide, green, stipitate-glandular; heads discoid, yellow, solitary and terminal or few, subracemose, 10–26-fld.; invol. turbinate-campanulate, 11–15 mm. high, glandular-puberulent; phyllaries subequal, 2–3-seriate, the outer broader, more herbaceous, the inner acuminate to attenuate, thin, dry; aks. appressed-pilose; pappus brownish; $n=9$ (Anderson & Reveal, 1966).—Rocky mostly open slopes; Bristle-cone Pine F.; Inyo Mts., at 10,000 to 11,000 ft.; to Ore., Colo., Wyo. July–Sept.

18. **H. pálmeri** Gray subsp. **pachylèpis** Hall. Shrub 0.5–1.5 m. high, with numerous ascending puberulous stems, resinous, glandular-punctate, very leafy; lvs. mostly 8–12 mm. long, filiform, subterete, fasciculate; heads many, in thyrsoid panicles; invol. cylindro-turbinate, 6–7 mm. high; phyllaries 16–25, thickish, 4-seriate, broadly lanceolate to oblong, the bullate costal gland in apical half only; ray-fls. 1–4; disk-fls. 5–10; aks. densely sericeous; $n=9$ (Raven et al., 1960).—Rather common on dry plain below 2500 ft.;

Coastal Sage Scrub; s. Ventura Co. to desert borders, Riverside Co. Santa Catalina Id. Aug.–Dec.

Subsp. **pálmeri.** One to 4 m. high; lvs. 1.5–4 cm. long; invol. 5–6.5 mm. high; phyllaries 30–40, loosely 4–5-seriate, glabrous or the outer glandular-atomiferous, ciliolate at tip, the costa thickened most of its length into a linear-oblong gland; ray-fls. 4–10; disk-fls. 8–20.—Coastal Sage Scrub, s. San Diego Co.; L. Calif. Sept.–Nov.

19. **H. párishii** (Greene) Blake. [*Bigelovia p.* Greene. *Ericameria p.* Hall.] Erect shrub 2–5 m. high, arborescent, glabrous, resinous, glandular-punctate, the shoots densely leafy; lvs. linear-oblanceolate to lance-elliptic, 2–6 cm. long, 3–10 mm. wide, flat, coriaceous; heads discoid, 9–12-fld., in compact rounded cymes, the short pedicels with scalelike bracts; invol. ca. 5 mm. high; phyllaries 4-seriate, lanceolate to lance-oblong, acutish to acuminate, whitish, firm, carinate by the glandular-thickened costa; aks. appressed-pilosulous; pappus copious, fragile; $2n=18$ (DeJong & Montgomery, 1963).—Dry slopes, 1500–7000 ft.; Chaparral, s. slopes of San Gabriel and San Bernardino mts.; southward to L. Calif. July–Oct.

20. **H. pinifòlius** Gray. [*Ericameria p.* Hall.] PINE-BUSH. Plate 17, Fig. A. Stout shrub 6–25 dm. high, the main stems trunklike, the branches often with terminally arranged tufts of dense twigs and lvs.; lvs. filiform, 1.5–3.5 cm. long, resinous-dotted, less than 1 mm. wide, with axillary fascicles of shorter lvs.; autumnal heads cymose-paniculate, crowded or in elongate panicles, leafy-bracteate; heads solitary and few in spring flowering; invols. 5–7 mm. high, the phyllaries in 4 series, lance-acuminate to oblong, the outer often ± caudate-tipped, the inner short-tipped or merely acute, the tips green, margin scarious, ciliate, costa sometimes glandular-thickened above; ray-fls. 5–10 in autumn heads, 15–30 in spring; disk-fls. 12–18; aks. sparsely pilose, with buff or reddish pappus.—Common on dry fans and banks, away from the coast and below 5500 ft.; S. Oak Wd., Coastal Sage Scrub, Chaparral; n. Los. Angeles Co. to s. San Diego Co. April–July, Sept.–Jan.

21. **H. propínquus** Blake. [*Bigelovia brachylepis* Gray. *Ericameria b.* Hall. *H. b.* Hall, not Phil.] Shrub 1–2 m. high, rigidly branched, glabrous, ± resinous, glandular-punctate; lvs. crowded, linear-filiform, 1–2 cm. long, 0.5–1 mm. wide, flattish or subterete, often mucronate, sharply ascending, with axillary fascicles; heads discoid, 9–14-fld., yellow, terminal or racemose-paniculate; invol. 4.5–5.5 mm. high; phyllaries 3–4-seriate, strongly graduate, ovate to linear-oblong, the outer grading into the scaly bracts of the peduncles, the costa glandular-thickened throughout its length; aks. densely villous.—Dry brushy slopes below 4500 ft.; Chaparral; s. San Diego Co.; n. L. Calif. Sept.–Dec.

22. **H. racemòsus** (Nutt.) Torr. ssp. **glomeràtus** (Nutt.) Hall. [*Homopappus g.* Nutt.] Plate 17, Fig. B. Stems from a stout taproot, 2–7 dm. high, the herbage gray-green or glaucous, foliage stiff and thickish, the basal lvs. tufted; lvs. entire to sharply denticulate-serrate, the lateral veins obscure, cauline lvs. 1–3 cm. long, less than 1.5 cm. wide; heads sessile or short-pedunculate, racemose or spicate, sometimes ± glomerate; invol. 6–8 mm. high; phyllaries firm, thickish, lance-oblong to oblong, acute to obtuse, pale below the green tip, glabrous or ciliate, rarely more hairy; ray-fls. 12–30, the ligules 5–12 mm. long; aks. densely villous, sordid.—Damp alkaline places, below 6000 ft.; Alkali Sink; Tecopa region, Inyo Co.; San Joaquin V. and n. Inyo Co. to Ore., Utah. July–Oct.

Ssp. **sessiliflòrus** (Greene) Hall. [*Pyrrocoma s.* Greene.] Stems 2–5 dm. long; herbage blue-glaucous; lvs. mostly entire, ciliolate, the basal tufts often grasslike, 5–10 cm. long; heads often in clusters of 2's or 3's; phyllaries often squarrose at the tip.—Alkali Sink, Sagebrush Scrub; Cow Creek, Death V. and Deep Springs in Eureka V., Inyo Co. July–Oct.

23. **H. squarròsus** H. & A. ssp. **grindelioìdes** (DC.) Keck. [*Pyrrocoma g.* DC.] Many-stemmed shrub, 3–10 dm. tall, the stems often tomentulose near the heads; lvs. many, oblong to cuneate-obovate, obtuse, sharply serrate throughout, clasping at base, often ± hairy on upper surface, resinous, 1.5–4 cm. long, 1–2 cm. wide; heads discoid, 15–30-fld., spicate or racemosely paniculate; invol. 8–11 mm. high; phyllaries prominently cinereous on both surfaces of the green usually squarrose tip, but glandular only marginally or not at all; aks. glabrous or sparsely pilose; pappus red-brown.—Dry slopes below 4500

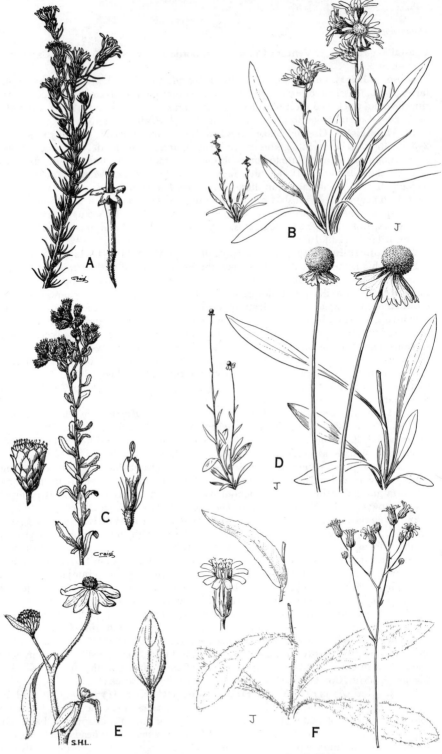

Plate 17. Fig. A, *Haplopappus pinifolius*. Fig. B, *Haplopappus racemosus*. Fig. C, *Haplopappus venetus* ssp. *vernonioides*. Fig. D, *Helenium bigelovii*. Fig. E, *Helianthus petiolaris*. Fig. F, *Hieracium albiflorum*.

ft.; Coastal Sage Scrub, Chaparral; Santa Barbara to n. L. Calif. Channel Ids. July–Oct.

Ssp. **obtùsus** (Greene) Hall. [*Hazardia o.* Greene.] Heads relatively large, 18–25-fld.; invol. broadly turbinate, 13–15 mm. high; phyllaries rather broad, very blunt, mucronate, the resinous-glandular pallid tips glabrous, appressed.—Canyons below 4000 ft.; mts. of Kern Co. w. of Tejon Pass to Ojai, Ventura Co. Sept.–Nov.

24. **H. uniflòrus** (Hook.) T. & G. ssp. **gossypinus** Hall. [*Pyrrocoma g.* Greene.] Stems 1–3 dm. high from a fusiform crown, decumbent, usually anthocyanous, tomentulose; lvs. mostly basal, lanceolate, tapering to both ends, the blade 2–8 cm. long, 6–18 mm. wide, entire to laciniate-dentate, the cauline much reduced and sessile; heads mostly solitary, on long peduncles; invol. hemispheric, 10–13 mm. high; phyllaries 2–3-seriate, obscurely graduate, typically herbaceous from apex to base at least medianly, the margin scarious, glabrous to tomentose; ray-fls. 18–32, the ligules 6–9 mm. long; aks. silky.—Moist alkaline meadows, 6000–7500 ft.; Yellow Pine F.; San Bernardino Mts. July–Sept.

25. **H. venètus** (HBK) Blake. ssp. **furfuràceus** (Greene) Hall. [*Bigelovia f.* Greene.] Shrubby; stems slender, decumbent or curved, 3–5 dm. long; herbage glabrous, the resinous exudate sometimes forming a scurf especially on the lvs.; lvs. crowded, narrow, 1–3 (–4) cm. long, linear to linear-spatulate, acute or obtuse, entire or with a few minute teeth near apex, with prominent fascicles in the axils, mostly glabrous, sometimes woolly; heads few, in close rounded terminal cymes, these simple or compound; invol. 6–8 mm. high; phyllaries mostly obtuse but mucronate, green-tipped, resinous, the thin margin erose; aks. silky-villous.—Dry sandy mesas; Coastal Sage Scrub; s. San Diego Co.; San Clemente and Santa Catalina ids.; n. L. Calif. April–Nov.

KEY TO SUBSPECIES

Lvs. mostly entire, acute or if obtuse then also small.
  Plant low, inclined to spread; heads few, the cymes nearly simple, compact, rounded; lvs. mostly 1–3 (–4) cm. long .... ssp. *furfuraceus*
  Plant tall, erect; heads many; cymes openly paniculate; lvs. mostly 3–5 cm. long
                                                       ssp. *oxyphyllus*
Lvs. mostly dentate, lobed or incised, obtuse to moderately acute or ending in an acute lobe.
  Shrub 4–12 dm. high; lvs. linear to spatulate-oblong; heads in compact cymes
                                                       ssp. *vernonioides*
  Shrub prostrate or decumbent; lvs. obovate; heads capitate .... var. *sedoides*

Ssp. **oxyphýllus** (Greene) Hall. [*Isocoma o.* Greene.] Robust shrub 1–2 m. high, loosely villous to glabrate; lvs. oblanceolate to narrowly spatulate, acute or acuminate, entire, 3–5 cm. long; cymes openly paniculate.—Common in San Diego area and L. Calif. June–Nov.

Var. **sedoìdes** (Greene) Munz. [*Bigelovia venetus* var. *s.* Greene.] Prostrate or decumbent, stout, almost glabrous; lvs. succulent, obovate, obtuse, toothed; heads large, in a capitate cluster. Santa Rosa, Santa Cruz, San Miguel, Anacapa ids. and adjacent coast from Point Mugu to Morro.

Ssp. **vernonioìdes** (Nutt.) Hall. [*Isocoma v.* Nutt.] Plate 17, Fig. C. Four–12 dm. high, the stems usually simple below the infl.; lvs. spatulate-oblong, 1–3 cm. long, 3–8 mm. wide, spinulose-dentate to almost lobed; heads in rounded terminal cymes; invol. 5–7 mm. high, the phyllaries with green acutish tips.—Common on dry slopes below 1200 ft.; Coastal Strand, Coastal Sage Scrub, Coastal Salt Marsh; San Diego n. to cent. Calif. Channel Ids. April–Dec.

70. *Hecastoclèis* Gray

Low rounded subshrub, glabrous, with rigid branches and rigid lvs., the cauline broadly linear, cuspidate and with spinose teeth especially in the primary ones, less spinose in the axillary fascicled lvs.; uppermost lvs. oval or ovate, and forming a loose envelope around the heads, reticulate, semitranslucent, sparsely beset with slender marginal prickles. Heads 1-fld., in a fascicle. Invol. cylindraceous, of several linear-lanceolate rather rigid cuspidate-acuminate phyllaries. Fls. perfect; corolla with linear spreading lobes. Aks. glabrous. Pappus coroniform, laciniate-dentate, corneous. One sp. (Greek, *ekastos*, each, and *kleio*, to shut up, each fl. in its own invol.)

1. **H. shóckleyi** Gray. Plants 4–6 dm. high and somewhat broader; cauline lvs. 1.5–3.5 cm. long, the primary longer and spinier than the secondary fascicled ones; floral envelope 1.5–2.5 cm. long, chartaceous; invol. ± woolly; fls. reddish-purple within before opening, greenish-white at anthesis.—Dry slopes, washes, etc.; 4000–7000 ft.; Creosote Bush Scrub, Shadscale Scrub; Death V. region and e. slope of Inyo Mts.; w. Nev. May–June.

71. *Hedýpnois* Schreb.

Low annual, simple or branched from base. Heads rather small, mostly peduncled, yellow-fld.; invol. with short outer phyllaries, the principal ones in one row, hardened in fr., enveloping the outer aks. Aks. cylindrical, ribbed, not beaked; pappus of outer aks. crownlike, of the inner usually in 2 series and consisting of free or basally connate scales ending in bristles and of short outer scales. A small genus of Medit. region. (Name given by Pliny to a kind of wild endive.)

1. **H. crètica** (L.) Willd. [*Rhagadiolus c.* Calif. refs.] Branches diffuse or spreading, 1–3 dm. long; basal lvs. petioled, often lobed, the cauline oblanceolate, 3–10 cm. long, serrate; peduncles rather long, mostly naked; invol. 8–10 mm. high; inner phyllaries strigose with short stiff bristles; ligules yellowish with purple tip; aks. cylindric, 6–7 mm. long; pappus-paleae 2 mm. long, the bristles 5 mm. long; $2n = 8$ (Stebbins et al., 1953).—Local but rather widespread weed, natur. from e. Medit. March–May.

72. *Helènium* L. SNEEZEWEED

Annual or perennial herbs, with simple or branched stems. Lvs. alternate, glandular-punctate, usually decurrent. Heads solitary or corymbose, pedunculate, usually radiate, golden-yellow, or the disk purple-brown. Invol. 1–3-seriate, the phyllaries subequal and usually soon deflexed. Receptacle naked, convex to subglobose. Ray-fls. ♀ or neutral. Disk-fls. very many, perfect. Aks. turbinate or obpyramidal, 4–5-angled with intermediate ribs. Pappus of 5–10 scarious, often awn-tipped scales. Ca. 40 spp., restricted to the New World. (A Greek name for some plant, said by Linnaeus to be named for Helen of Troy.)

Rays conspicuous, usually exceeding the disk, 10–20 mm. long. Pine belt . . . . 1. *H. bigelovii*
Rays inconspicuous, much shorter than the disk, 2–8 mm. long. Low elevs. . . . . . 2. *H. puberulum*

1. **H. bigelòvii** Gray. Plate 17, Fig. D. Stout, stems solitary or in clumps, 4–8 dm. tall, subglabrous, having rather pleasant odor; lvs. quite persistent, the lower petioled, the cauline sessile, decurrent, oblanceolate to linear-lanceolate, up to 22 cm. long, and 4 cm. wide, the upper rapidly reduced; heads mostly solitary on long naked peduncles; disk depressed-globose, 1.5–2 cm. wide; rays 13–30, 8–22 mm. long, reflexed; aks. 2 mm. long; pappus-paleae 6–8; $n = 16$ (Raven & Kyhos, 1961).—Moist meadowy places, 5000–8500 ft.; Yellow Pine F.; mts. of s. Calif.; to Ore. June–Aug.

2. **H. pubérulum** DC. Habit of *H. bigelovii*, herbage puberulent; basal rosette and lower cauline lvs. gone before flowering, the cauline lvs. lance-oblong to linear, 3–15 cm. long, sessile, prominently decurrent; heads with a globose disk; rays 5–10, 3–8 mm. long, reflexed and often nearly concealed; aks. 1.4–2 mm. long, sparsely hairy; pappus-paleae 5; $n = 29$ (Raven & Kyhos, 1961).—Moist places, mostly below 4000 ft.; many Plant Communities; cismontane s. Calif.; to Ore., n. L. Calif. Reaching edge of desert as at Victorville. Santa Catalina Id. June–Sept.

73. *Heliánthus* L. SUNFLOWER

Coarse annual or perennial herbs with simple lvs., at least the lowermost opposite. Heads radiate, large, solitary at branch-tips or in corymbs. Phyllaries imbricate or subequal, ± herbaceous. Receptacle plano-convex, the persistent paleae clasping the aks. Ray-fls. conspicuous in ours, yellow, neutral. Disk-fls. yellow or reddish, fertile. Aks. narrowly obovate, quadrangular but obcompressed, generally glabrous. Pappus of 2 principal awns paleaceous at base, rarely with additional short squamellae, all readily caducous. Some 50 spp. of temp. N. and S. Am. (Greek, *helios*, sun, and *anthos*, fl.)

(Heiser, C. B., Jr. et al. The N. Am. sunflowers. Mem. Torrey Bot. Club 22 (3): 1–218. 1969.)
Plants annual.
    Outer phyllaries ovate, abruptly caudate, strongly ciliate .... 1. *H. annuus*
    Outer phyllaries lanceolate, gradually attenuate, not obviously ciliate.
        Lvs. whitish to grayish; herbage densely canescent .... 5. *H. niveus*
        Lvs. green or bluish-green; herbage mostly strigulose to hispid .... 7. *H. petiolaris*
Plants perennial.
    Plants strongly glaucous and rhizomatous. Introd. weed .... 3. *H. ciliaris*
    Plants not strongly glaucous and rhizomatous.
        Foliage white with strigose pubescence. Subshrub .... 5. *H. niveus*
        Foliage greenish; perennial herbs.
            Outer phyllaries shorter than the disk. Dry places .... 4. *H. gracilentus*
            Outer phyllaries longer than the disk. Wet places.
                Phyllaries 3–4 mm. wide at base, considerably longer than the disk, reflexed at tip
                      2. *H. californicus*
                Phyllaries 2–3 mm. wide at base, slightly longer than the disk, erect ... 6. *H. nuttallii*

1. **H. ánnuus** L. ssp. **jaègeri** (Heiser) Heiser. [*H. j.* Heiser.] Resembling the next ssp., but with lvs. lance-ovate, cuneate to truncate at base, serrulate; heads with reddish disk 1.5–2.5 cm. across; phyllaries lance-ovate; $n=17$ (Heiser, 1948).—Wet alkaline areas; Alkali Sink; e. Mojave Desert (Tecopa Hot Springs, Soda Dry Lake, etc.); to Owens V., and Nev. July–Oct.

Ssp. **lenticulàris** (Dougl.) Ckll. [*H. l.* Dougl. ex Lindl.] COMMON SUNFLOWER. Stem usually stout, 3–20 or more dm. high, often openly branched, very hispid; herbage rough-hairy; lvs. petioled, the blade 6–15 cm. long, narrowly to broadly ovate, truncate or cordate at base, mostly serrate, the uppermost often entire; heads large, the low-convex often reddish disk 2–3.5 cm. wide; phyllaries narrowly to broadly ovate, abruptly and slenderly acuminate, densely scabrous and usually at least medianly hirsute-hispid dorsally, strongly hispid-ciliate; paleae often 3-toothed, not conspicuously hairy; $2n=34$ (Tahara, 1915).— Roadsides and waste places, frequent at lower elevs., occasional up to 5000 ft.; many Plant Communities; throughout our area except for the Mojave Desert; to Can., Mex. Feb.–Oct.

2. **H. califórnicus** DC. Herbaceous perennial with stout smooth erect stems up to 5 m. high from somewhat tuberous woody roots, branching above; lvs. lanceolate, tapering to apex and usually to base, short-petiolate, usually entire, up to 2 dm. long and 5 cm. wide, hispidulous and gland-dotted; heads in loose corymbs, the disk yellow, 18–25 mm. across; phyllaries lanceolate, the white-ciliolate basal part narrowed to a hispidulous spreading attenuate tip exceeding the disk; rays 2–3 cm. long; $n=51$ (Heiser, 1955).— Boggy meadows, stream banks and moist ground, below 6500 ft.; Freshwater Marsh, Chaparral, V. Grassland, etc.; scattered localities, n. L. Calif. and San Diego Co. to cent. Calif. June–Oct.

3. **H. ciliàris** DC. BLUEWEED. Perennial herbs spreading by strong rhizomes; stems 2–5 dm. high, striate; herbage blue-glaucous; glabrate or the margins strigose; lvs. linear-lanceolate to somewhat broader, 3–8 cm. long, 8–15 mm. wide; entire to crisped-undulate; heads on often lone naked peduncles, the disk 12–16 mm. across; phyllaries ovate to oblong, obtusish, white-ciliolate, otherwise glabrous, shorter than the disk; paleae canescent at tip; $n=51$ (Heiser, 1955).—A pernicious weed introd. here and there; native from Ariz. to Tex. June–Nov.

4. **H. gracilĕntus** Gray. Perennial from a branching caudex, the stems erect, 6–15 dm. tall, rough-hispidulous, slender, often purplish; lvs. opposite or the uppermost alternate, lanceolate to lance-ovate, 5–12 cm. long, with short margined petiole; infl. a few-

fld. panicle, the heads on long quite naked peduncles; disk 1–1.5 cm. broad; phyllaries lance-oblong to ovate, acuminate, apiculate, densely hispidulous-puberulent; rays 1.5–2 cm. long; $n=17$ (Heiser, 1955).—Dry slopes below 6000 ft.; Chaparral, Yellow Pine F.; n. L. Calif. and San Diego Co. through cismontane s. Calif. to cent. Calif. May–Oct.

5. **H. níveus** (Benth.) Bdg. ssp. **canéscens** (Gray) Heiser. [*H. petiolaris* var. *c.* Gray.] Mostly annual; stem usually erect, 3–12 dm. high, densely canescent; lvs. lanceolate to ovate, entire to serrate, cuneate to subcordate, petiolate, to 12 cm. long and 6 cm. wide, densely grayish or silvery canescent with fine appressed hairs; peduncles long; disk 0.8–2.5 cm. wide; phyllaries usually slightly exceeding disk, 2–3 mm. broad, canescent, usually drying dark green, erect or nearly so; disk-corollas red to deep purple; paleae 3-cuspidate; ak. mostly ca. 3 mm. long; pappus of 2 awns; $n=17$ (Heiser, 1948).—Open desert; Creosote Bush Scrub; w. Colo. Desert; to Tex., Mex. March–June.

Ssp. **tephròdes** (Gray) Heiser. [*Viguiera t.* Gray.] Perennial or sometimes annual from stout taproot, erect or decumbent, 5–15 dm. tall; stem appressed with white silky hairs; lvs. deltoid or deltoid-ovate, entire or serrulate, 3–7 cm. long, 2–4 cm. wide, appressed white-silky-villous; heads several, with reddish disk 1.5–2.5 cm. in diam.; phyllaries lanceolate, not exceeding the disk; disk-corollas very slender, puberulent at bulbous base of tube and on lobes; pappus of 2 long awns and several squamellae; $n=17$ (Heiser).—Sandy desert area; Creosote Bush Scrub; between Yuma and Imperial V.; to Son. March–May, Oct.–Jan.

6. **H. nuttállii** T. & G. ssp. **nuttállii**. Herbaceous perennial with short rhizomes and tuberous roots; stems to 4 m. high, glabrous, glaucescent; lvs. scabrous and glandular-dotted, linear-lanceolate, tapering to base and apex, up to 15 cm. long and 3 cm. wide, entire to serrate; heads loosely corymbose, the yellow disk 15–23 mm. across; phyllaries linear-lanceolate, 2–3 mm. wide at base, scabro-puberulent, the narrow tip erect, slightly exceeding the disk; rays 1.5–3 cm. long; $n=17$ (Long, 1966).—Moist meadows at ca. 4000 ft.; Chaparral; n. San Bernardino Mts.; at higher elevs. to Modoc Co., Rocky Mts. July–Sept.

Ssp. **párishii** (Gray) Heiser. [*H. p.* Gray. *H. oliveri* Gray.] Lvs. canescent, the soft, ± erect hairs scarcely pustulate; summit of peduncles and invols. densely white-villous.—Wet ground, 1000–1500 ft.; Los Angeles, San Bernardino and Orange cos. Aug.–Oct. Nearly if not entirely extinct.

7. **H. petiolàris** Nutt. Plate 17, Fig. E. Annual; stems 3–20 dm. high, subglabrous or sparingly hispid; lf.-blades 3–8 cm. long, lanceolate, entire or nearly so, the appressed hairs from pustulate bases but not harsh as in *H. annuus*; heads with a reddish disk ca. 2 cm. across; phyllaries lanceolate, gradually acuminate, hispidulous, not prominently hispid-ciliate; cent. paleae white-bearded at tip; $n=17$ (Heiser, 1947).—Ruderal of waste places, uncommon in s. Calif. (San Bernardino V. and Mts., Hemet V., Etiwanda, Guasti, etc.); common to Tex., Mo. Sask. May–Sept.

74. *Hemizònia* DC. TARWEED

Annual or perennial herbs or shrubs, usually very glandular and aromatic, mostly fall-flowering xerophytes, a very few spring-flowering. Basal lvs. variously lobed, rarely subentire; upper lvs. and bracts not terminated by open pit-glands. Phyllarie shalf-enclosing the ray-aks. Receptacular bracts in a single row (and often ± united) outside the disk-fls. or scattered. Fls. yellow or white. Ray-aks. beaked in our s. Calif. spp., triquetrous, the odd angle posterior, epappose, fertile. Disk-aks. usually bearing a paleaceous pappus. Self-sterile. Thirty-one spp., all in Calif. and L. Calif. (Greek, *hemi*, half, and *zone*, girdle, the phyllaries but half-enclosing the ray-aks.)

1. Plants shrubby.
    2. Flocs present in axils of older lvs.; phyllaries scarcely keeled; anthers black. Insular .... 3. *H. clementina*
    2. Flocs absent; phyllaries strongly keeled; anthers yellow. Santa Susanna Mts., Los Angeles Co. .... 10. *H. minthornii*

1. Plants annual herbs.
    3. Lvs. not spine-tipped; receptacular bracts confined to a row surrounding the outer disk-fls. and united into a cup; pappus of quadrate or oblong paleae.
        4. Disk-aks. sterile.
            5. Phyllaries not keeled; anthers yellow.
                6. Heads glomerate, small; ray-fls. 5; disk-fls. 6. Mojave Desert ..... 11. *H. mohavensis*
                6. Heads not glomerate.
                    7. Ray-fls. 5; disk-fls. 6 .... 8. *H. kelloggii*
                    7. Ray-fls. 8 or more; disk-fls. 10 or more.
                        8. Radical lvs. lobed; herbage villous below; ligules pale yellow; pappus obvious. Cismontane .... 12. *H. pallida*
                        8. Radical lvs. entire or obscurely toothed; herbage hispid-hirsute below; ligules deep yellow; pappus none or vestigial. Transmontane .... 1. *H. arida*
            5. Phyllaries keeled; anthers black.
                9. Ray-fls. 5; disk-fls. 6.
                    10. Heads paniculate; stems tall, scarcely hispid, flexuous and much-branched. San Luis Obispo Co. south .... 15. *H. ramosissima*
                    10. Heads in dense glomerules; stems lower, rather hispid above, with strict divaricate branches. Coastal plain .... 5. *H. fasciculata*
                9. Ray-fls. 8; disk-fls. 13–21. San Diego Co. .... 4. *H. conjugens*
        4. Disk-fls. mostly fertile
            10a. Radical lvs. pinnatifid or bipinnatifid; herbage grayish-hirsute; ray-fls. 8–13, the ligules as broad as long. Widespread .... 13. *H. paniculata*
            10a. Radical lvs. merely lobed; herbage bright green, soft-pubescent; ray-fls. 13–20, the ligules half as broad as long. S. San Diego Co. .... 7. *H. floribunda*
    3. Lvs. tipped with a rigid spine or apiculation; receptacular bract subtending each disk-fl. free, persistent; pappus none or of very narrow paleae.
        11. Pappus none; anthers yellow.
            12. Receptacular bracts pungent .... 14. *H. pungens*
            12. Receptacular bracts obtuse or ± acute, not cuspidate .... 9. *H. laevis*
        11. Pappus present; anthers black.
            13. Receptacular bracts fleshy at tip, not long-villous; pappus-paleae 3–5 . 2. *H. australis*
            13. Receptacular bracts long-villous at tip; pappus-paleae 8–12 .... 6. *H. fitchii*

1. **H. árida** Keck. Annual, 2–4 dm. high, intricately corymbosely branched throughout; herbage hispid-hirsute and hirsutulous, lightly glandular and mildly odorous; lvs. ca. as in *H. pallida*; heads many, cymose-paniculate; invol. 4 mm. high, 5 mm. wide, hispid-hirsute and stipitate-glandular; ligules 5–6 mm. long, 3–4 mm. wide; pappus vestigial or none; $n=12$ (Clausen et al., 1935).—Rare; Creosote Bush Scrub; Red Rock Canyon, Kern Co. at 2500 ft. May–Nov.

2. **H. austràlis** (Keck) Keck. [*H. parryi* ssp. *a*. Keck.] Stem erect, the central leader of medium length and with lax divaricate branches, the twiggery above dense; herbage dark green; lvs. and bracts densely glandular-puberulent and villous; peduncular bracts not exceeding the invol.; heads small; phyllaries 4–5.5 mm. long; ray-fls. often fading saffron; anthers black; $n=11$ (Venkatesh, 1956).—Lowlands near the coast; V. Grassland; Santa Barbara Co. to n. L. Calif. June–Sept.

3. **H. clementìna** Bdg. [*Zonanthemis c.* Davids. & Mox.] Shrub 3–8 dm. high, with many erect or ascending to decumbent branches from base, densely leafy above, the older lvs. deciduous; herbage sparsely hirsute, viscid; lower lvs. opposite, narrowly linear, 3–8 cm. long, 1.5–6 mm. wide, remotely dentate; upper lvs. alternate, entire, with fascicles or leafy shoots in their axils; heads in compound cymes; invol. 5–8 mm. high; ray-fls. 13–14, the ligules yellow, 4.5–6.5 mm. long; receptacular bracts in 2 series; disk-fls. 18–30, the cent. ones without bracts; ray-aks. 2–3 mm. long, transversely rugose, the recurved beak 1.5–3 times as long as thick; disk-aks. sterile; pappus of 7–10 (–15) paleae 1–3 mm. long, unequal; $n=12$ (Johansen, 1933).—Heavy soils; Coastal Scrub; Anacapa, S. Barbara, S. Nicolas, S. Catalina and S. Clemente ids. May–Aug.

4. **H. cónjugens** Keck. Stem 2–5 dm. high, branching as in *H. fasciculata*; foliage and invols. soft-hirsute and sometimes hispidulose, especially the invol. bearing large flat and small capitate sessile or subsessile glands; heads solitary on short peduncles or sub-

sessile in few-fld. glomerules; ray-fls. 8–10; disk-fls. 13–21.—Mesas, Coastal Sage Scrub; sw. San Diego Co. May–June.

5. **H. fasciculàta** (DC.) T. & G. [*Hartmannia f.* DC. *H. glomerata* Nutt.] Stem 1–10 dm. high, mostly branching from above the middle, the branches rigid, sharply ascending and with comparatively few twigs; heads subsessile in glomerules of 3 to many, the glomerules terminating short leafy branches and sometimes a few solitary heads below; otherwise similar to *H. ramosissima*; $n = 12$ (Johansen, 1933).—Dry coastal plains, below 1000 ft.; Coastal Sage Scrub, S. Oak Wd., V. Grassland; San Luis Obispo Co. to Riverside Co. and n. L. Calif. May–Sept.

6. **H. fítchii** Gray. [*Centromadia f.* Greene.] SPIKEWEED. Stem 2–8 dm. high, rigid, erect, diffusely branched above; herbage dark, densely villous and beset with prominent stalked glands, unpleasantly heavy-scented; lvs. crowded, at least in the rosette, very slender, the lower pinnatisect, up to 15 cm. long, the upper entire, rigid, linear-subulate, acerose-tipped, often with fascicles in their axils, the uppermost involucroid, spreading, radiating around the head which they overtop; invol. hemispheric, villous, with few prominent glands, the phyllaries rigid; receptacular bracts hairy, enfolding the disk-fls.; ray-fls. 10–20, the ligules short, bifid; pappus of 8–12 soft white oblong paleae long fimbriate above; $n = 13$ (Clausen et al., 1934).—Abundant on dry hills and plains below 3000 ft.; V. Grassland; Joshua Tree Wd.; n. base of San Bernardino Mts., S. Cruz Id.; to n. Calif., s. Ore. May–Nov.

7. **H. floribúnda** Gray. Stem 3–10 dm. high, corymbosely branched especially above with ascending branches; herbage pilose and densely stipitate-glandular, mildly odorous; lvs. mostly entire, the lower oblanceolate, the middle cauline linear-lanceolate, 1.5–3 cm. long, 2–3.5 mm. wide; heads short-peduncled, in compound cymes or panicles, not glomerate; invol. 5–6 mm. high, 6–7.5 mm. wide, soft-pubescent and very glandular; ray-fls. 13–20; disk-fls. ca. 28, their aks. all or mostly fertile; pappus of 6–9 oblong or elliptic closely fimbriate rufous and flecked paleae; $n = 13$ (Johansen, 1933).—Dry slopes and valleys; below 3500 ft.; Coastal Sage Scrub, Chaparral; s. San Diego Co., n. L. Calif. Aug.–Oct.

8. **H. kellóggii** Greene. [*H. wrightii* Gray.] Stem 2–10 dm. high, corymbosely branching above or intricately branching throughout, soft-villous at base, hispid-hirsute and densely glandular above; lower lvs. narrowly oblong, remotely sharp-toothed or pinnatifid, 3–8 cm. long, 3–10 mm. wide; upper lvs. entire; heads pedunculate, in an open panicle or thyrse; invol. 4–5.5 mm. high, 3.5–5 mm. wide, densely stipitate-glandular, usually hirsute; ligules 4–7 mm. long, 3.5–5.5 mm. wide; pappus of 6–12 white linear to oblong paleae; $n = 9$ (Johansen, 1933).—Abundant below 2500 ft.; V. Grassland, S. Oak Wd.; San Diego Co. n. to cent. Calif., found also in Red Rock Canyon and in Imperial V. April–July.

9. **H. laèvis** (Keck) Keck. [*H. pungens* ssp. *l.* Keck.] Similar to *H. pungens* but of somewhat lower habit; upper lvs. and bracts sparsely setose-ciliate, otherwise very glabrous; heads small, scattered, or approximate in loose glomerules; receptacular bracts obtuse or slightly acute, sometimes minutely but weakly mucronate, not cuspidate; $n = 9$ (Keck).— Grassy fields, low elevs.; V. Grassland; San Diego Co. to San Bernardino V. and Los Angeles region, away from the immediate coast; sw. Kern Co. April–Sept.

10. **H. minthórnii** Jeps. Shrub 6–10 dm. high, 10–30 dm. wide, with up to 500 stiff woody ascending stems from base; herbage short-hirsute, viscid, fragrantly resinous; lvs. alternate, somewhat thickened, like those of *H. clementina*; heads mostly solitary on long peduncles, paniculate; invol. 5.5–6 mm. high, 4–6 mm. broad; ray-fls. 8, the ligules yellow, 5.5–6.5 mm. long; receptacular bracts in ca. 3 series; disk-fls. 18–23, sterile, each subtended by a bract; ray-aks. 2.5–3 mm. long, smooth, the beak scarcely longer than thick; pappus of 8–12 linear ± fimbriate subequal paleae 1.2–2.6 mm. long; $n = 12$ (Clausen et al., 1934).—Rare, Chaparral; Santa Susanna Mts. July–Oct.

11. **H. mohavénsis** Keck. Stem 1.5–3 dm. high, subsimple or divaricately branched above; herbage soft-pubescent and viscid, pleasantly odorous; lower lvs. oblanceolate, subentire, the upper oblong-lanceolate, amplexicaul, much reduced toward the corym-

bose infl.; heads sessile in glomerules at the ends of the branches; invol. 5–6 mm. high, 5 mm. wide; ligules 5–6 mm. long; pappus of 6–8 quadrate ± connate paleae 0.5 mm. long; $n=11$ (Clausen et al., 1934).—Rare, 3000–4000 ft.; Chaparral, Joshua Tree Wd.; Mohave R. at Deep Creek; Mt. San Jacinto. July–Sept.

12. **H. pállida** Keck. Stem 2–8 dm. high, branching above or throughout with ascending or divergent branches, whitish or reddish; herbage villous-hirsute and hispidulous, lightly glandular and mildly odorous; lower lvs. linear to oblanceolate, remotely sharp-toothed or cleft, 5–10 cm. long, 3–6 (–10) mm. wide, the uppermost entire; heads many, cymose; invol. 4.5–6.5 mm. high, 5–8 mm. wide, hispid-hirsute and minutely stipitate-glandular; ligules 6–10 mm. long, 3–5 mm. wide; pappus of 4–8 narrow distinct paleae 0.8 mm. long or less; $n=9$ (Johansen, 1933).—Plains and hills below 2500 ft.; V. Grassland, etc.; Antelope V. to head of San Joaquin V. April–May.

13. **H. paniculàta** Gray ssp. **paniculàta**. Stems 2–8 dm. high, the cent. shaft replaced ca. midway by numerous ascending branches which ramify to form a twiggy infl.; herbage moderately hispid-hirsute (especially below) and stipitate-glandular (especially above), fragrant; lower lvs. persistent until anthesis, linear to oblanceolate, up to 25 mm. wide; heads obviously peduncled; invol. 5–7 mm. high, 5–7 mm. wide, the phyllaries densely glandular-pubescent; ray-fls. 8; disk-fls. 8–13; disk-aks. sparsely pubescent, some fertile and dark-colored, the majority sterile and pale; pappus of 6–12 white oblong fimbriate paleae; $n=12$ (Johansen, 1933).—Dry hills and mesas, low elevs.; V. Grassland; w. Riverside Co. to San Diego Co., n. L. Calif. May–Nov.

Ssp. **incréscens** Hall ex Keck. Plants erect or somewhat spreading, deep green, even the ultimate twigs rather rigid, never capillary; ray-fls. 8–13; disk-fls. 14–30; $n=12$ (Johansen, 1933).—Abundant in fields near the coast; Santa Rosa Id., Mainland, Santa Barbara Co. to Monterey Co. May–Nov.

14. **H. púngens** (H. & A.) T. & G. [*Hartmannia? p.* H. & A. *Centromadia p.* Greene. *Hemizonia p.* ssp. *interior* Keck.] SPIKEWEED. Stem 1–12 dm. high, divergently and rigidly branched above or throughout, leafy, hirsute, not glandular; herbage yellowish-green, inodorous; basal lvs. linear-lanceolate in outline, 5–15 cm. long, 1–4 cm. wide, bipinnately parted; upper cauline lvs. with fascicles in their axils, mostly entire, spine-tipped, stiffly ciliate and scabrous-puberulent; heads subsessile in upper axils and terminal; invol. hemispheric, 3–6 mm. high, usually overtopped by bracts; phyllaries keeled, pungent, scabrous, persistent; ray- and disk-fls. many, the ligules 2-toothed; ray-aks. angular, often rugose; disk-aks. in part sterile, the fertile ones ± fusiform; $n=9$ (Johansen, 1933). —Interior dry valleys and foothills below 1000 ft.; largely V. Grassland; San Joaquin V.; introd. Los Angeles, San Bernardino and San Diego cos. and s. Ore. April–Oct.

15. **H. ramosíssima** Benth. [*H. fasciculata* var. *r.* Gray.] Stem erect, 2–10 dm. high, simple below and corymbosely branched above or with ascending branches from the base upward; herbage moderately hirsute to glabrate, the glands (most frequent on invols.) sessile, yellow; lower lvs. mostly gone by anthesis, linear-oblanceolate, remotely dentate, 3–15 cm. long, 0.3–3 cm. wide; upper lvs. becoming entire, bractlike; heads pedunculate, many, ± remote and solitary, sometimes in 2's at ends of the branches; ray-fls. 5; disk-fls. 6; pappus of 6–8 lanceolate to oblong, entire to lobed paleae; $n=12$ (Johansen, 1933). —Dry fields and hillsides, below 1500 ft.; Santa Barbara Co., less common s. to n. L. Calif. Channel Ids. May–Aug.

75. *Heterothèca* Cass. TELEGRAPH WEED

Coarse erect herbs with yellow-fld. heads arranged in terminal corymbose panicles. Lvs. alternate. Invol. hemispheric; phyllaries narrow, in several series. Ray-fls. 1-seriate fertile, the ak. triangular-compressed, its pappus none or deciduous; disk-fls. many, fertile, the ak. cuneiform, the pappus double, the copious inner bristles capillary, the outer setose, short. Three or more spp. in s. U.S. and Mex. (Greek, *heteros*, different, and *theke*, case or ovary, from the unlike aks. of ray and disk.)

Upper lvs. narrowed to a sessile base; heads relatively large, invol. 7–10 mm. high, glandular-pubescent, but not also canescent .... 1. *H. grandiflora*
Upper lvs. subcordate-clasping at base; heads smaller; invol. 6–8 mm. high, glandular-puberulent and canescent .... 2. *H. subaxillaris*

1. **H. grandiflòra** Nutt. [*H. floribunda* Benth.] Annual or biennial, the stout stem simple below, 5–20 dm. tall, hirsute, the ample infl. glandular-pubescent and heavy-scented; lvs. thickish, densely villous, ovate to oblanceolate, 2–7 cm. long, obtuse, serrate, the lower petioled; invol. 7–9 mm. high; rays-fls. 25–35, the corolla 6–8 mm. long, 1 mm. wide, revolute from the tip, with hairy tube; disk-fls. 50–65, slender; pappus brick red, the outer series inconspicuous; $n=9$ (Heiser & Whitaker, 1948).—Sandy open coastal valleys below 3000 ft., behaving as a weed; Coastal Sage Scrub, Chaparral, S. Oak Wd.; cismontane s. Calif. to cent. Calif. Jan.–Dec.

2. **H. subaxillàris** (Lam.) Britt. & Rusby. [*Inula s.* Lam. *H. scabra* Nutt.] Annual or biennial, simple below or openly branching, 5–20 dm. high, hispid-hirsute, glandular above; lvs. rather coarsely hirsute, glandular, ovate to lance-oblong, subentire to serrate-dentate, the lower petioled, the upper subcordate-clasping; invol. 6–8 mm. high, glandular and canescent; ray-fls. ca. 20–28, disk-fls. 40–60; pappus rufous, the outer series usually conspicuous; $n=9$ (Smith, 1965).—Sandy places; Creosote Bush Scrub; e. Mojave and Colo. deserts to Atlantic Coast. Apparently introd. at San Gabriel, in Ventura Co., etc. Aug.–Oct.

76. *Hieràcium* L. HAWKWEED

Perennial herbs, often hairy, sometimes glandular, mostly caulescent. Lvs. entire or dentate, never deeply lobed. Heads variously panicled. Receptacle flat, usually naked. Invol. cylindric or campanulate, the phyllaries in 1–3 series, subequal or ± imbricate, not thickened on the back, with a few small calyculate ones at the base. Ligules white or yellow in our spp., truncate and 5-toothed at apex. Aks. oblong or columnar, truncate, 10–15-ribbed. Pappus of 1–2 series of brownish or sordid fragile capillary bristles. More than 700 spp., mostly of Eu. and S. Am. (Greek, *hierax*, a hawk; the ancients believed that the hawks used the sap to sharpen their eyesight.)

Ligules white or ochroleucous; stem (except at base) and invol. glabrous, or invol. with a few long hairs .... 1. *H. albiflorum*
Ligules yellow; stems and invols. pubescent or hirsute.
    Lvs. toothed; pappus almost white .... 2. *H. argutum*
    Lvs. entire; pappus brownish .... 3. *H. horridum*

1. **H. albiflòrum** Hook. Plate 17, Fig. F. Stems 1–several, slender, erect, 4–8 dm. high, leafy below, quite naked above, the lower parts densely hirsute with whitish or tawny hairs, the upper glabrous; lvs. mostly basal, oblong to oblanceolate, 8–15 cm. long, 2–4 cm. wide, the lower with winged petiole, the upper sessile, reduced; heads in a loose panicle, white-fld.; invol. 9–10 mm. high, glabrous or often glandular and with black hairs; phyllaries linear-subulate, the outer very short; ligules 3–4 mm. long; aks. red-brown, 2–3 mm. long; pappus dull white or tawny; $n=9$ (Kruckeberg, 1967).—Dry open wooded slopes 5000–9500 ft.; Montane Coniferous F.; mts. from San Diego Co. n.; to Alaska, Colo. June–Aug.

2. **H. argùtum** Nutt. ssp. **argùtum.** Stem simple, leafy below, slender, 3–10 dm. high, shaggy below with brown hairs, subglabrous above, ending in a long racemose panicle; lvs. oblong to oblanceolate, remotely dentate, the lower 8–16 cm. long, the upper reduced; invol. 7–10 mm. high, with dark stipitate glands, the inner phyllaries linear-attenuate, the outer shorter; ligules yellowish, ca. 8 mm. long; aks. dark brown, ca. 2.5 mm. long; pappus sordid.—Dry slopes; Closed-cone Pine F., etc.; Santa Cruz and Santa Rosa ids.; mainland Santa Barbara and San Luis Obispo cos. June–Aug.

Var. **párishii** (Gray) Jeps. [*H. p.* Gray. *H. grinnellii* Eastw.] Invol. with light-colored stipitate glands.—Below 6000 ft.; Chaparral, Yellow Pine F.; s. face of San Gabriel and San Bernardino mts. to Ventura Co. June–Oct.

3. **H. hórridum** Fries. Stems few to several, paniculately branched above, 1–3.5 dm. high; herbage densely shaggy and with whitish or brownish long soft hairs; lvs. spatulate to oblong, entire, 3–10 cm. long, not densely crowded; infl. open, with many small heads; invol. narrow, 6–9 mm. high, the phyllaries lanceolate with medial part ± glandular and long-hairy; ligules ca. 15, bright yellow, ca. 1 cm. long; aks. 10-ribbed, red-brown, ca. 3 mm. long; pappus brownish; $2n = 18$ (Stebbins et al., 1953).—Common in dry ± rocky places, 5000–9500 ft.; Montane Coniferous F.; Santa Rosa and San Jacinto mts. n. through Sierra Nevada to Ore. July–Aug.

77. *Holocárpha* (DC.) Greene. TARWEED

Annual, mostly very glandular and aromatic. Lvs. linear, the basal remotely serrulate to dentate with slender teeth, the upper reduced to entire bracts usually bearing fascicles or peduncles in their axils, truncated at their apex by a crateriform gland. Phyllaries half-enclosing the ray-aks., bearing on back and at apex stoutly stalked pit glands. Receptacular bracts subtending each disk-fl., enclosing the ak. and corolla-tube, free, persistent. Fls. yellow, the rays 3-lobed. Ray-aks. laterally beaked, obcompressed, triquetrous, the odd angle anterior. Disk-fls. sterile or outermost fertile. Pappus none. Self-sterile. Four spp., all in Calif. (Greek, *holos*, whole, and *karphos*, chaff, the entire receptacle chaffy.)

Anthers yellow; herbage densely puberulent above .... 1. *H. heermannii*
Anthers black; herbage not puberulent above .... 2. *H. virgata*

1. **H. heermánnii** (Greene) Keck. [*Hemizonia h.* Greene.] Stem 2–12 dm. high, simple below and virgately few-branched above to diffusely branched throughout, pilose or hispid below, cinereous above, with interspersed glandular pubescence throughout; lvs. canescent, the upper also stipitate-glandular, otherwise like *H. virgata*, the peduncular bracts usually overlapping each other but not the base of the invol., appressed to moderately recurved, flattened to apex, 2–5 mm. long; heads paniculate or openly racemose; invol. 5–6 mm. high, puberulent and viscid in addition to having gland-tipped processes; $n = 6$ (Clausen et al., 1934). In hard-baked soils, below 4000 ft.; V. Grassland, etc.; Ventura Co. (Ojai) and Tehachapi Pass; to cent. Calif. May–Oct.

2. **H. virgàta** (Gray) Keck. ssp. **elongàta** Keck. Stem slender, 5–12 dm. high, multibranched, the ultimate branchlets gracefully curving, elongated; herbage canescent or hispidulous above; basal lvs. 6–15 cm. long, 3–10 mm. wide, mostly lost before anthesis, the upper reduced to bracts, often fasciculate; heads scattered on leafy peduncles up to 15 cm. long; invol. 5–6 mm. high, the phyllaries bearing 5–20 stout terete ascending gland-tipped processes, otherwise glabrous; ray-aks. 2.4–3.5 mm. long.—V. Grassland, S. Oak Wd.; mesas w. of Murrieta, Riverside Co. and about San Diego. June–Nov.

78. *Húlsea* T. & G.

Annual, biennial or perennial rather fleshy viscid-pubescent aromatic herbs from a taproot. Lvs. alternate, the usually very numerous basal ones broadly petiolate, the cauline sessile, entire to pinnatifid. Heads many-fld., radiate, yellow to purplish-red. Invol. hemispheric, 2–3-seriate, the herbaceous subequal phyllaries persistent and at length reflexed. Receptacle convex, naked. Ray-fls. numerous, ♀, fertile. Disk-fls. perfect, fertile. Aks. linear, 2–3-angled, compressed, softly villous especially on the ribs. Pappus of 4 hyaline nerveless paleae united at base. Ca. 8 spp. of w. U.S. (G. W. *Hulse*, U.S. Army surgeon and collector.)

1. Heads mostly solitary at the ends of ± scapose stems except in reduced alpine plants.
  2. Plants strongly glandular-pubescent. Alpine in White Mts. .... 1. *H. algida*
  2. Plants ± white-tomentose .... 5. *H. vestita*
1. Heads several, ± corymbose.
  3. Basal and lower lvs. green and glandular to white-villous; cauline lvs. mostly 15–35 mm. wide.

4. Cauline lvs. saliently dentate.
      5. Rays reddish, ciliate; aks. 6–7 mm. long. San Jacinto Mts. to San Gabriel Mts., Panamint Mts., and to mts. of Santa Barbara Co. .... 3. *H. heterochroma*
      5. Rays yellow, not ciliate; aks. 4–5 mm. long. S. San Diego Co. .... 4. *H. mexicana*
   4. Cauline lvs. mostly entire; basal or lower lvs. white-villous, the upper ± glandular.
      6. Plants annual or biennial; heads mostly 2–3.5 cm. across; phyllaries hirsute. Laguna and Cuyamaca mts. .... 2. *H. californica*
      6. Plants perennial; heads mostly 1–2 cm. across; phyllaries subglabrate. Palomar Mts. to San Jacinto and Santa Rosa mts. .... 5. *H. vestita callicarpha*
3. Lvs. white-tomentose, mostly entire; rays often with some red or orange. San Bernardino and San Gabriel mts., Frazier Mt., Sierra Nevada .... 5. *H. vestita vestita*

1. **H. álgida** Gray. Rather succulent strongly glandular-pubescent disagreeably odorous perennial with few to many erect unbranched leafy monocephalous stems 1–4 dm. high, from the branching subterranean caudex; lvs. to 15 cm. long including the margined petiole, not woolly; invol. 13–20 mm. high, usually lanate; phyllaries narrow; rays 25–50, yellow; $n=19$ (Ornduff et al., 1967).—Alpine Fell-fields, 10,000–13,500 ft.; White Mts. to Sierra Nevada, Ore., Mont. July–Aug.

2. **H. califórnica** T. & G. Annual or biennial; stems branching, erect, few, 5–10 dm. high; herbage loosely woolly when young, becoming moderately hirsute and glandular; lower lvs. ample, 6–10 cm. long, 2–5 cm. wide, sinuate-dentate, the upper much reduced; heads several, racemose or corymbose; invol. 15 mm. high; phyllaries acuminate; rays 20–35, broadly linear, 10 mm. long, yellow; aks. 5 mm. long; pappus-paleae unequal, those over the angles 2–3 mm. long.—Open places up to 6000 ft.; burns in Chaparral, Yellow Pine F.; Laguna and Cuyamaca mts. to n. L. Calif. April–June.

3. **H. heterochròma** Gray. Plate 18, Fig. A. Biennial or short-lived perennial with several erect leafy stems, viscid-villous and heavy-scented, 4–12 dm. high; lvs. oblong, saliently dentate, up to 10 cm. long and 35 mm. wide, sessile or clasping; heads racemose or corymbose; invol. 12–15 mm. long; phyllaries long-acuminate; rays narrowly linear, purplish-red, hirsute and glandular; aks. 6–7 mm. long; pappus-paleae strongly unequal, lacerate; $n=19$ (Ornduff et al., 1967).—Infrequent in forest openings or chaparral, 3000–8000 ft.; San Jacinto, San Bernardino and San Gabriel mts. to Monterey Co.; Panamint Mts.; Nev. June–Aug.

4. **H. mexicàna** Rydb. Similar to *H. heterochroma*, but with larger heads, yellow glabrous corollas and aks. ca. 4–5 mm. long.—Near Table Mt., s. San Diego Co.; L. Calif. June–July.

5. **H. vestìta** Gray. ssp. **vestìta**. Plate 18, Fig. B. Scapose perennial with branching underground caudex; stems several, 0.4–2.5 dm. high, leafless or with few spatulate to linear-oblong reduced lvs.; basal lvs. persistent, densely clustered, 2–5 cm. long, 1.5–2 cm. broad, spatulate to obovate, narrowed to a petiolar base, ± entire, densely white-woolly-villous; phyllaries 8–10 mm. long, lanceolate, glandular and villous; rays 8–12 mm. long, golden-yellow on inner surface, often orange-red to red on outer; aks. 5–7 mm. long; pappus-paleae 1.5–2 mm. long, the pairs nearly equal, the margins lacerate; $n=19$ (Raven & Kyhos, 1961).—Gravelly or sandy places, 6000–9000 ft.; Montane Coniferous F.; occasional in San Bernardino and San Gabriel mts., Frazier Mt.; Sierra Nevada; w. Nev. June–Aug.

KEY TO SUBSPECIES

Flowering stems branched, 1.5–6 dm. tall; lower lvs. entire .... ssp. *callicarpha*
Flowering stems unbranched, mostly 0.4–1.5 dm. tall, ± scapose.
   Lvs. strongly dentate or lobed.
      Pairs of pappus-paleae subequal; rays yellow. Coso Mts. to Inyo Co. and w. Nev. at 5400–6700 ft. .... ssp. *inyoensis*
      Pairs of pappus-paleae ± unequal; rays with orange-red or red on outer side.
         Lvs. woolly beneath. At 7000–9000 ft., San Gabriel Mts. to Little San Bernardino Mts.
                                                                                                 ssp. *parryi*
         Lvs. not woolly beneath. At 10,000–11,500 ft. San Gorgonio Peak .... ssp. *pygmaea*

Plate 18. Fig. A, *Hulsea heterochroma*. Fig. B, *Hulsea vestita*. Fig. C, *Hymenopappus filifolius lugens*. Fig. D, *Iva axillaris*. Fig. E, *Jaumea carnosa*. Fig. F, *Lasthenia chrysostoma*.

Lvs. mostly entire; rays often with some red or orange on outer surface. San Bernardino and San Gabriel mts., Frazier Mt.; Sierra Nevada .... ssp. *vestita*

Ssp. **callicárpha** (Hall) Wilken. [*H. vestita* var. *c.* Hall.] Robust and well branched above the base; basal lvs. entire; invol. 10–13 mm. high.—Dry slopes, 4000–9000 ft.; Montane Coniferous F.; Palomar, San Jacinto, Santa Rosa mts.

Ssp. **inyoénsis** (Keck) Wilken. [*H. californica* ssp. *i.* Keck. *H. i.* Munz.] Plants low; lvs. green, glandular, toothed to entire; rays yellow.—Largely at 5000–7200 ft.; Pinyon-Juniper Wd.; Coso Mts. to n. Death V.; w. Nev.

Ssp. **párryi** (Gray) Wilken. [*H. p.* Gray.] Much like ssp. *vestita*, but with narrower dentate lvs., narrower phyllaries; ray corollas reddish.—At 7000–9000 ft.; Montane Coniferous F.; e. San Gabriel Mts. to Little San Bernardino Mts.

Ssp. **pygmaèa** (Gray) Wilken. [*H. v.* var. *p.* Gray.] Compact depressed dwarf mostly less than 6 cm. high, with 1-headed peduncles and no floccose pubescence on lower surface of lvs.—At 10,000–11,500 ft.; Subalpine; San Gorgonio Peak.

79. *Hymenoclèa* T. & G.

Low diffusely branched xerophytic shrubs. Lvs. alternate, filiform, entire or the lower pinnately parted into few lobes. Heads small, very numerous, scattered or glomerate-paniculate, both sexes usually found in the same lf.-axils, the ♂ above the ♀. Staminate heads several-fld.; invol. shallow, 4–6-lobed. Pistillate heads 1-fld.; invol. ovoid or fusiform, becoming indurate, pericarplike and beaked in fr., the phyllaries persistent as scarious horizontal wings near the middle. Two spp., sw. U.S. and Mex. (Greek, *humen*, membrane, and *kleio*, to enclose. The pollen of these plants is said to cause hay fever.)

Wings of the fruiting invol. cuneate or obovate, in a single whorl of 7–12, 1–2 mm. long
                                                                                                      1. *H. monogyra*
Wings of the fruiting invol. reniform-orbicular, spirally arranged, 3.5–8 mm. long . 2. *H. salsola*

1. **H. monogyra** T. & G. More strictly erect and more leafy than *H. salsola*; shrub 1–3 m. high; scales of the ♀ invol. 7–12, spreading in a single whorl around the middle of the considerably less indurate invol., obovate to cuneate, neither prominently veined, mucronate, not with pits in their axils; $n = 18$ (Payne et al., 1964).—Sandy washes, below 1800 ft.; Chaparral; Rialto, around San Diego; more common e. to Tex. and Coahuila, s. to L. Calif. Aug.–Nov.

2. **H. salsòla** T. & G. var. **pentalèpis** (Rydb.) L. Benson. [*H. p.* Rydb.] Very much like var. *salsola*, but the scales of the ♀ invol. in a single radiating whorl of usually 5 or 6 members.—E. Colo. Desert; to Son., L. Calif.

Var. **salsòla**. Erect or spreading twiggy bush 6–10 (–15) dm. high, the herbage yellowish-green, resinous, minutely canescent or glabrous; lvs. sparse, 2–5 cm. long; scales of the ♀ invol. spirally arranged from the base to near the middle, reniform, parallel-veined, mucronate-tipped, narrowed to a clawlike base, often with a deep pit in the axils, sometimes imbricated like a cone, sometimes all spreading; $n = 18$ (Payne et al., 1964).— Common in sandy washes, often in ± alkaline places; Creosote Bush Scrub, Shadscale Scrub, Pinyon-Juniper Wd.; n. Inyo Co. to Colo. Desert and head of San Joaquin V.; Nev., Utah, Ariz., L. Calif. March–June.

80. *Hymenopáppus* L'Hér.

Biennial and perennial, subscapose to leafy-stemmed herbs. Stems slender to stout, erect, angled and sulcate. Lvs. in a basal rosette, mostly bipinnately dissected, to entire, reduced up the stem, impressed-punctate. Infl. a cymose panicle. Heads discoid or radiate, subturbinate to broadly campanulate, on slender peduncles. Invol. of 6–14 subequal phyllaries in 2–3 series, membranous at apex or rarely throughout. Receptacle dome-shaped to flattish. Rays when present ♀ and fertile, with conspicuous white ligules.

Disk-fls. regular, yellow to white or red-purple, the tube mostly glandular, lobes reflexed after anthesis. Aks. 4-sided, obpyramidal to incurved, glabrous to pubescent. Pappus 0 or of 12–22 hyaline scales, the medial nerve completely included. Ca. 10 spp. of w. N. Am. (Greek, *humen*, membrane, and *pappos*, pappus, because of the hyaline paleae.)

(Turner, B. L. A cytotaxonomic study of the genus H. Rhodora 58: 163–186, 208–242, 250–269, 295–308. 1956.)

1. **H. filifòlius** Hook. var. **eriópodus** (Nels.) Turner. [*H. e.* Nels.] Perennial, 4–8 dm. high, ± tomentose below, glabrate above; rosette-lvs. 1–2 dm. long, 3–7 cm. wide, bipinnately dissected into filiform divisions, 1–2 cm. long; stem-lvs. 3–7, green, glabrate; heads discoid, 3–8 on a stem, 30–60-fld., on peduncles 9–16 cm. long; phyllaries 7–10 mm. long, yellow-membranous at tip; corollas ochroleucous, 4–5 mm. long, the tube glandular, ca. 2 mm. long, throat campanulate; aks. ca. 6 mm. long, pubescent; pappus of 12–16 linear-oblong scales 1.5–2 mm. long.—Dry limestone slopes, 4900–7000 ft.; Pinyon-Juniper Wd.; New York Mts., Clark Mt., e. Mojave Desert; to Utah. May–June, Oct.

KEY TO VARIETIES

Corolla 4–7 mm. long; anthers 3–4 mm. long; aks. 5–7 mm. long.
  Fls. whitish; peduncles 8–16 cm. long; stem-lvs. 2–7; aks. with hairs 0.5–1 mm. long
.... var. *eriopodus*
  Fls. yellow; peduncles 2–12 cm. long; stem-lvs. 0–6; ak.-hairs 1–2 mm. long.
    Stem-lvs. 2–7; tips of basal lvs. 5–30 mm. long; phyllaries 8–12 mm. long. Mid-Hills, Providence Mts. .... var. *megacephalus*
    Stem-lvs. 0–3; tips of basal lvs. 3–15 mm. long; phyllaries 6–10 mm. long. San Bernardino Mts. to Cuyamaca Mts. .... var. *lugens*
Corolla 3–4 mm. long; anthers 2–3 mm. long; aks. 4.5–5.5 mm. long. Inyo-White Mts.   var. *nanus*

Var. **lùgens** (Greene) Jeps. [*H. l.* Greene.] Plate 18, Fig. C. $n = 17, 34$ (Turner, 1956). —Dry slopes, 4000–7500 ft.; largely Yellow Pine F.; mts. from San Diego Co. to San Bernardino Mts.; to Utah, New Mex., L. Calif. June–Aug.

Var. **megacéphalus** Turner. At ca. 4900–5000 ft.; Pinyon-Juniper Wd.; Mid-Hills, Providence Mts., e. San Bernardino Co.; to Colo., Ariz.

Var. **nànus** (Rydb.) Turner. [*H. n.* Rydb.] Dry rocky places, 5500–10,000 ft.; Pinyon-Juniper Wd., Bristle-cone Pine F.; Inyo-White Mts.; to Utah, Ariz. July–Aug.

81. *Hymènothrix* Gray

Slender-stemmed annual or perennial herbs. Lvs. alternate, bi- or tri-ternately dissected into linear divisions. Heads few to many, cymose-panicled, radiate or discoid, yellow, white or purple. Disk-corollas zygomorphic, the lobes unequal, erect or merely spreading. Aks. narrowly obpyramidal, 4–5-angled; pappus-scales with medial nerve extending to the apex or excurrent into a distinct awn. Several spp. of sw. N. Am. (Greek, *humen*, membrane, and *thrix*, bristle.)

(Turner, B. L. Taxonomy of H. Brittonia 14: 101–119. 1962.)
Invol. 4–6 mm. high; pappus ca. 2 mm. long .... 1. *H. loomisii*
Invol. ca. 10 mm. high; pappus ca. 4 mm. long .... 2. *H. wrightii*

1. **H. loòmisii** Blake. Annual or biennial, 3–6 dm. high, finely incurved-puberulous throughout; lvs. alternate, the lobes deltoid, 1–8 cm. long; heads several to many; invol. 4–6 mm. high; corollas whitish, 5–6 mm. long; aks. 3–5 mm. long; pappus ca. 2 mm. long.—Taken at stations along Santa Fe Ry. tracks (as at San Dimas, Riverside); cent. Ariz. Nov.–Dec.

2. **H. wrìghtii** Gray. [*Hymenopappus w.* Hall. *H. w.* var. *viscidulus* Jeps.] Perennial; stems leafy, few to several, 3–8 dm. tall, glandular and hirsute; lf.-blades ovate to roundish in outline, 1.5–2.5 cm. long, the lower on equally long petioles, gradually somewhat

reduced up the stems; heads few, in open infl.; invol. ca. 1 cm. high, subglabrous; fls. white or purplish, ca. 7 mm. long.—Dry slopes ca. 5000 ft.; Yellow Pine F., Laguna and Cuyamaca mts.; L. Calif. to Son., New Mex. June–Sept.

82. *Hymenóxys* Cass.

Annual or perennial herbs with basal or alternate entire to pinnatifid or ternate lvs. Heads mostly radiate; rays ♀ and fertile, broad, yellow, mostly 5–35. Phyllaries partly or wholly herbaceous, in 2–3 similar or sharply differentiated series. Receptacle naked, ± rounded. Disk-fls. ± numerous, yellow, perfect, fertile. Aks. turbinate, hairy, mostly 5-angled. Pappus of a few, mostly 5, hyaline often aristate or awned scales. Ca. 20 spp. of w. N. Am. and S. Am. (Greek, *humen*, membrane, and *oxus*, sharp, because of the pointed pappus scales.)

Lvs. entire, mostly basal .... 1. *H. acaulis*
Lvs. ± dissected, mostly or partly cauline.
    Invol. ca. 12 mm. broad, hemispheric; outer phyllaries 12–14; rays 12–14, ca. 12–25 mm. long
                                                                                2. *H. cooperi*
    Invol. ca. 6–9 mm. broad; outer phyllaries ca. 8; rays 8–10, ca. 7–10 mm. long . 3. *H. odorata*

1. **H. acáulis** (Pursh) Parker var. **arizónica** (Greene) Parker. [*Tetraneuris a.* Greene. *Actinea a.* var. *a.* Blake.] Tufted perennial from a branched caudex; stems scapose, 1–3 dm. high, subglabrous; lvs. gray-silky, basal, linear-oblanceolate, 2–5 cm. long; heads solitary; invol. 6–8 mm. high, white-villous; rays bright yellow when young, ca. 1 cm. long; aks. silky-villous; pappus of 5–7 hyaline paleae ca. 2 mm. long, abruptly contracted into a very short awn; $n=30$ (Parker, 1946); 15, 30 (Strother, 1966).—Rocky largely limestone ridges, 4000–8000 ft.; Joshua Tree Wd., Pinyon-Juniper Wd.; Chuckwalla, Providence, Clark, New York mts.; to Colo., Ariz.

2. **H. coóperi** (Gray) Ckll. var. **canéscens** (D. C. Eat.) Parker. [*Actinella richardsonii* var. *c.* D. C. Eat.] Plants more canescent than in var. *cooperi*; stems 1–2.5 dm. high; heads larger, the disk often 15–25 mm. wide.—Rocky places, 9000–10,100 ft.; Pinyon-Juniper Wd., Bristle-cone Pine F.; White Mts., Inyo Co.; to Ida., Ariz. July–Aug.

Var. **coóperi.** Biennial or short-lived perennial, with 1–few stems 3–7 dm. high, puberulent and glandular-punctate, divided into 3–5 linear divisions, gradually reduced up the stem; heads ± open corymbose, few to many; disk to ca. 15 mm. wide; invol. 5–7 mm. high; outer phyllaries ca. 10–14, keeled, pointed, ± woolly, the inner more rounded, subglabrous; ligules yellow, 8–14 mm. long; pappus-paleae 5, oblong or ovate, obtuse, erose to ± acuminate; $n=15$ (Parker, 1947).—Dry slopes, 4000–7500 ft.; largely Pinyon Juniper Wd.; Little San Bernardino Mts., Providence, Clark and New York mts.; to Utah, Ariz. May–June, Sept.–Oct.

3. **H. odoráta** DC. [*H. chrysanthemoides* var. *excurrens* Ckll.] Slender-stemmed annual 3–6 dm. high, branched mostly above or at base, ± pubescent; lvs. 3–6 cm. long, pinnately dissected into linear divisions; heads many, corymbose; invol. ca. 6 mm. high; outer phyllaries ca. 8, lanceolate, united at the thickened base; pappus-paleae ovate, acuminate, awn-pointed; $n=11$ (Parker, 1947).—Lowlands along Colo. River.; Creosote Bush Scrub; Parker Dam to Yuma; to Kans., Mex. Feb.–May.

83. *Hypochoèris* L. Cat's Ear

Annual or perennial herbs with lvs. in radical rosette or cluster and naked stems bearing a solitary head or a somewhat corymbose cluster of long-peduncled heads. Fls. yellow. Invol. cylindric or campanulate, with rather few lanceolate erect imbricated phyllaries. Receptacle flat, the bracts scarious, chaffy, thin. Aks. glabrous, upwardly scabrous, the body 10-ribbed, narrow-oblong or fusiform, truncate or beaked. Pappus-bristles plumose or some of the outer shorter and simple. Ca. 70 spp., 12 in Eu., the others S. Am. (Greek name used by Theophrastus for this or some other genus.)

Annual, usually glabrous; outer aks. truncate, inner beaked .... 1. *H. glabra*
Perennial, pubescent; aks. all beaked .... 2. *H. radicata*

1. **H. glàbra** L. Lvs. usually not deeply lobed, spatulate-oblong, 2–10 cm. long; stems 1–4 dm. high; heads campanulate; invol. 12–16 mm. high; ligules scarcely exceeding invol.; aks. dark brown, the outer almost 4 mm. long; pappus ca. 9 mm. long, tinged yellowish or brownish; $n=5$ (Shetty, 1967).—Weed in fields and waste places; natur. from Eu. March–June.

2. **H. radicàta** L. Perennial; lvs. hispid, pinnatifid, 6–14 cm. long; stems 4–15 dm. long; ligules longer than the invol; aks. brown, the body ca. 3.5 mm. long; $2n=8$ (Stebbins et al., 1953).—Weed about lawns, waste places, etc.; scattered stations in most of cismontane Calif., more abundant northward; from Eu. May–Nov.

84. Iva. L.

Herbs or low shrubs. Lvs. alternate or opposite, entire in ours, with small nodding heads racemosely disposed in the axils. Phyllaries few, roundish, green, sometimes with a shorter inner scarious series. Paleae of the receptacle narrowly linear, membranaceous. Fls. greenish-white, inconspicuous, the marginal ones ♀, fertile, 1–5, their corollas tubular or obsolete, the disk-fls. perfect, sterile, their corollas funnelform, the anthers almost distinct. Aks. obovoid in ours, glabrous, epappose. Ca. 15 spp., all N. Am. (After the mint, *Ajuga iva*, because of the similar odor.)

Aks. densely long-villous; lvs. or their lobes linear-filiform .... 1. *I. acerosa*
Aks. not long-villous; lvs. or their lobes not linear-filiform.
    Phyllaries united into a cup .... 2. *I. axillaris*
    Phyllaries distinct .... 3. *I. hayesian*

1. **I. aceròsa** (Nutt.) Jackson. [*Oxytenia a.* Nutt.] Suffrutescent strong-scented perennial 5–20 dm. high; lvs. canescent, 5–10 cm. long, pinnatifid with linear-filiform subterete divisions 0.5 mm. wide, or the upper entire; heads hemispheric, 5 mm. wide, the fls. whitish, fragrant; aks. 2 mm. long, obovoid, with wide-beaked areola, densely villous, epappose; $n=18$ (Jackson, 1960).—Saline stream beds and plains, 500–2000 ft.; Alkali Sink; Amargosa Desert, Funeral Mts., Death V. region; to sw. Colo., New Mex. Aug.–Dec. Often nearly leafless and rushlike. Said to cause human dermatitis.

2. **I. axillàris** Pursh ssp. **robústior** (Hook.) Bassett. [*I. a.* var. *r.* Hook. *I. a.* var. *pubescens* Gray.] POVERTY WEED. Plate 18, Fig. D. Slightly woody or almost wholly herbaceous, spreading from slender rhizomes, the several stems erect from an often decumbent base, mostly 2–6 dm. high; herbage strigulose or hirsutulose, red-glandular-punctate; lvs. oblanceolate, obtuse, entire, subsessile, 0.5–4 cm. long, 0.6–1.6 cm. wide, upper lvs. distinctly smaller than lower; heads solitary in the axils, 4–6 (–8) mm. broad; peduncles 2–15 mm. long; ♂ fls. 9–25 (–50), ♀ fls. 3–10; aks. 2–2.5 mm. long; $2n=36$ (Bassett et al., 1962), $n=18, 27$ (Payne et al., 1964).—Alkaline plains or edges of saline marshes; sea-level to 6500 ft.; Coastal Salt Marsh, Alkali Sink and elsewhere; widespread weedy plant from w. desert to coast; n. to B.C., S. Dak., New Mex. May–Sept.

3. **I. hayesiàna** Gray. Frutescent perennial, several-stemmed, 5–10 dm. high, virgate or racemosely branching; lvs. oblong-oblanceolate, entire, obtuse, gradually reduced upward; invol. hemispheric, 5–6 mm. wide, the ovate to obovate phyllaries distinct.—Alkaline places below 1000 ft.; Alkali Sink; sw. San Diego Co., L. Calif. April–Sept.

85. Jaùmea Pers.

Perennial herbs or subshrubs. Lvs. opposite, entire, narrow. Heads radiate or rarely discoid, terminal or axillary, peduncled. Invol. of rather few broad herbaceous graduated phyllaries in 3–4 series. Receptacle flat or conic, naked. Ray-fls. (when present) ♀, fertile, yellow, deciduous. Disk fls. fertile, yellow, glabrous. Style-branches with very short blunt appendages. Aks. oblong, 10-nerved. Pappus of scales or bristles, or none. Ca. 6 spp., one in the Pacific states, the others from Mex. to S. Am. and S. Afr. (Honoring the French botanist, I. H. *Jaume* St. Hilaire.)

1. **J. carnòsa** (Less.) Gray. [*Coinogyne c.* Less.] Plate 18, Fig. E. Glabrous perennial with creeping branched rhizomes, and numerous mostly simple decumbent or ascending stems 1–3 dm. high; lvs. linear-oblanceolate, fleshy, 2–4 cm. long, 2–5 mm. wide; heads mostly solitary; invol. 8–12 mm. high, the phyllaries rounded and often pinkish above; rays 6–10, inconspicuous, 3-toothed; aks. glabrous, epappose, ca. 3 mm. long; $n=19$ (Raven & Kyhos, 1961).—Coastal salt marshes and tidal flats; Coastal Strand, Coastal Salt Marsh; along the coast from n. L. Calif. to B.C. Channel Ids. May–Oct.

86. *Lactùca* L. LETTUCE

Leafy-stemmed herbs with mostly panicled heads. Heads several- to many-fld. Invol. cylindrical, or in fruit, conical; phyllaries imbricated in 2 or more sets of unequal length. Rays 5-toothed at summit. Aks. contracted into a beak which bears copious early deciduous pappus of soft capillary bristles falling away separately. Perhaps 90 spp.; widely distributed. *L. sativa* L. is Garden Lettuce. (Ancient name from Latin, *lac*, milk, because of the milky sap.)

1. **L. serriòla** L. [*S. scariola* L.] WILD LETTUCE. Annual, ± prickly-bristly especially below, 6–15 dm. tall; lvs. spinose-denticulate, pinnatifid, sessile or sagittate-clasping, 4–16 cm. long, soft-prickly beneath along midrib; heads many, 9–14-fld.; invol. 10–12 mm. high, with phyllaries in ca. 4 lengths; fls. yellow, often drying purplish; aks. oblanceolate, with muriculate apex, 5–7-ribbed on each face, the beak filiform; pappus white; $2n=18$ (Thompson et al., 1941; Gadella & Kliphuis, 1966).—Common and widespread weed, even present on the deserts; from Eu. May–Sept. A form occurring with the pinnatifid one is *f. integrifòlia* Bogenh. [Var. *integrata* Gren. & Godr.]

L. ludoviciàna (Nutt.) DC. reported from San Bernardino by Ferris. Differs from *L. serriola* bv having aks. prominently 1-nerved on each side or face.

87. *Lagophýlla* Nutt.

Annuals, with basal lvs. serrate-dentate to subentire, cauline lvs. entire, readily caducous. Invol. turbinate to hemispheric, the phyllaries completely enfolding the obcompressed ray-aks. and caducous with them. Fls. yellow, the heads opening toward evening and closing in the morning. Receptacle penicillate-pubescent centrally, its bracts in a single row between ray and disk, slightly united. Ray-fls. 5, fertile. Disk-fls. 6, sterile. Pappus none. Five spp. of w. U.S. (Greek, *lagos*, a hare, *phyllon*, lf., alluding to the copious silky pubescence of the upper lvs. of the original sp.)

1. **L. ramosíssima** Nutt. [*L. minima* Kell.] Stiffly erect, mostly 2–10 dm. high, branched above, sometimes from the base, sometimes simple; herbage grayish or dull green, densely white-hirtellous or white-sericeous, the prominent yellow stipitate glands confined to upper heads and lvs.; lvs. linear-oblanceolate to spatulate, withering before anthesis; heads short-peduncled or subsessile, racemosely disposed along branchlets and in small clusters at their ends; invol. 4.4–6.7 mm. high, the phyllaries villous-ciliate; ray-aks. 2.5–4 mm. long; $n=7$ (Johansen, 1933).—Open places with dry hard soils, below 3500 ft.; V. Grassland, Coastal Sage Scrub, etc.; San Diego Co. through cismontane s. Calif. to Wash., Ida. May–Oct.

88. *Lasthènia* Cass. GOLDFIELD

Annual or perennial. Lvs. opposite, entire to pinnatifid. Heads terminal, peduncled. Invol. cylindrical to hemispheric, of free or partly or wholly united phyllaries. Ray-fls. ♀, whitish or greenish to yellow, inconspicuous to showy. Disk-fls. perfect, the corollas 4–5-lobed. Pappus, if present, monomorphic or dimorphic, of awns or scales or both. Receptacle hemispheric to conic or subulate, smooth to pitted or muriculate, glabrous or pubescent. Aks. linear, clavate to obovate, glabrous or pubescent. Anther tips subulate lanceolate or deltoid; style branches deltoid or rounded. A genus of 16 spp., 1 in Chile, the others largely Californian. (Named for a Greek girl who attended the lectures of Plato in the garb of a man.)

(Ornduff, R. A biosystematic survey of the goldfield genus Lasthenia. Univ. Calif. Publ. Bot. 40: 1–92. 1966.)

Phyllaries united into a cup over ⅔ their length .... 3. *L. glabrata*
Phyllaries free from each other.
    The phyllaries ca. 4; anther tips subulate or deltoid with wartlike glands .... 4. *L. microglossa*
    The phyllaries ca. 6–14; anther tips not as above.
        Lvs. all essentially entire; corollas turning a deep red in dilute aqueous alkali
                                                                                1. *L. chrysostoma*
        Lvs., especially the middle ones, usually pinnately lobed or cleft; corollas remaining yellow in dilute aqueous alkali.
            Herbage usually glandular; pappus, when present, ± monomorphic, or, if dimorphic, the longer members not awns .... 2. *L. coronaria*
            Herbage not glandular; pappus, when present, strongly dimorphic, the longer members awns .... 5. *L. minor*

1. **L. chrysóstoma** (F. & M.) Greene. [*Baeria c.* F. & M. *B. gracilis* DC. *B. hirsutula* Greene.] Plate 18, Fig. F. Low, simple or usually branched from base, to 4 dm. tall, with spreading or appressed white hairs; lvs. linear, hirsute or strigose; heads sometimes nodding in bud; invol. hemispheric, the free phyllaries up to 13, oblong to lance-ovate, to 10 mm. long; rays to 10 mm. long, entire or toothed at apex, bright yellow; aks. to 3 mm. long, glabrous or pubescent, or with yellow papillae, epappose or with pappus of 1–7 awns or scales; $n = 8, 16$ (Ornduff, 1966).—Abundant in dry open places, especially sand, below 3000 ft.; Coastal Sage Scrub, V. Grassland, Creosote Bush Scrub, Shadscale Scrub, etc.; cismontane and edge of deserts; L. Calif. to Ore.; Channel Ids. Feb.–June.

2. **L. coronària** (Nutt.) Ornduff. [*Ptilomeris c.* Nutt. *P. anthemoides* Nutt. *Baeria californica* (Hook.) Chamb. *B. parishii* Wats.] Erect annual, to 4 dm. high, pubescent with short glandular hairs and longer eglandular ones; lvs. to 6 cm. long, entire and linear or pinnately divided or laciniate into long segms.; invol. densely pubescent, turbinate to hemispheric, glandular-pubescent, the phyllaries free, to 7 mm. long; rays yellow, 6–15, to 10 mm. long; pappus exceedingly variable; aks. to 2.5 mm. long, linear; $n = 5$ or 4 (Ornduff, 1966).—Open places; Creosote Bush Scrub, Coastal Sage Scrub, etc.; from se. Ventura Co. and sw. San Bernardino Co. to L. Calif. March–May.

3. **L. glabràta** Lindl. ssp. **còulteri** (Gray) Ornduff. [*L. g.* var. *c.* Gray.] Plate 19, Fig. A. Annual to 6 dm. high, glabrous or puberulent; lvs. linear or subulate, obscurely toothed or entire, fleshy, glabrous, connate at base, to 15 cm. long; invol. to 10 mm. high, depressed-hemispheric; rays yellow, oblong, ca. 7–15, to 14 mm. long; aks. pubescent and with rusty to yellowish papillae.—Salt marshes, vernal pools and damp alkaline spots; Alkali Sink, Coastal Salt Marsh, etc.; s. San Diego Co. to Kern Co. and Twentynine Palms. Santa Rosa Id. April–May.

4. **L. microglóssa** (DC.) Greene. [*Baeria m.* (DC.) Greene. *Burrielia m.* DC.] Annual with very slender stems, weak, simple to few-branched, 5–12 cm. high; plants ± hirsute; lvs. linear, entire, 1–3 cm. long, 1–3 mm. wide; peduncles 1–3 cm. long; invol. 5–8 mm. high, the 3–4 phyllaries acute, oblong; pappus of 2–4 very narrow flattened squamellae attenuate into subulate awns; $n = 12$ (Ornduff, 1966).—Occasional in shade of overhanging rocks; Creosote Bush Scrub; Palm Springs, Victorville, Mohave, etc.; to middle Calif. March–May.

5. **L. minor** (DC.) Ornduff. [*Monolepis m.* DC. *Baeria m.* Ferris. *Dichaeta uliginosa* and *tenella* Nutt.] Rather stout and succulent, lax, 1–3.5 dm. high, the stem ± woolly-villous when young; lvs. ligulate, thin, 2–15 cm. long, with rachis 2–10 mm. wide and few or several salient linear lobes 2–15 mm. long, or lvs. linear and entire in slender plants; heads 10–13 mm. across; phyllaries 8–13, ovate, hairy on margin, midrib and within the tip; rays 4–8 mm. long, pale yellow; receptacle with pedicellike processes that bear the florets;

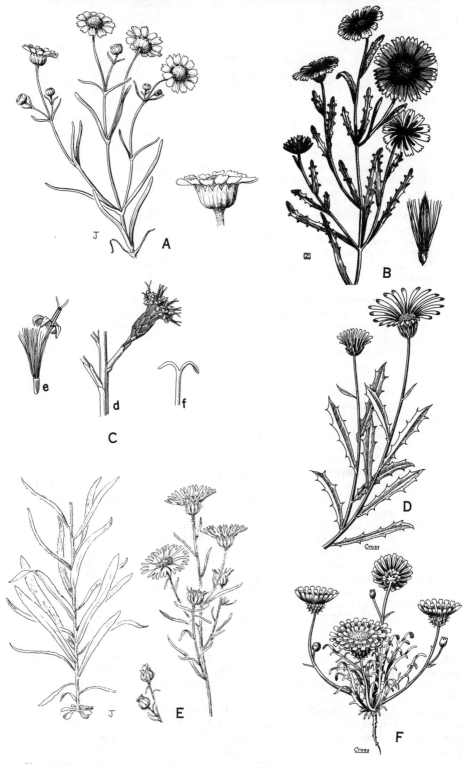

Plate 19. Fig. A, *Lasthenia glabrata*. Fig. B, *Layia platyglossa*. Fig. C, *Lepidospartum squamatum*. Fig. D, *Machaeranthera tortifolia*. Fig. E, *Madia elegans*. Fig. F, *Malacothrix glabrata*.

aks. 2–2.5 mm. long, ± strigose; pappus, when present, of 4–6 truncate fimbriate paleae with 2–4 awns; $n=4$ (Ornduff, 1966).—Touching our n. border in Kern Co., but probably not in s. Calif.

89. Làyia Hook. & Arn.

Spring annuals, mostly with alternate subentire or toothed to pinnatifid narrow lvs. Heads many-fld., on terminal usually naked peduncles, radiate in our spp. Both ray- and disk-fls. fertile. Invol. campanulate to broadly hemispheric, the thin lower margins of the phyllaries infolded and enclosing the ray-aks. Receptacle broad, ± chaffy. Ray-fls. 18–24; ligules 3-lobed or -toothed, white, yellow, or yellow with white tip. Disk-fls. many, yellow; anthers black or yellow. Ray-aks. obcompressed, commonly smooth and glabrous, epappose. Disk-aks. mostly pubescent and pappose. Pappus of numerous bristles, awns or paleae, the bristles often plumose below. A genus of 16 spp., all occurring in Calif. (Named for G. Tradescant *Lay*, botanist to Capt. Beechey's voyage in 1827.)

Pappus deciduous; stem prominently fistulous; ligules creamy white or pale yellow
.... 2. *L. heterotricha*
Pappus persistent; stem not prominently fistulous; ligules white or yellow or yellow with white tips.
    Anthers black; ligule yellow with white tip or yellow throughout; pappus mostly scabrid or, if plumose, also with inner wool .... 3. *L. platyglossa*
    Anthers yellow; ligule all white or all yellow; pappus plumose, lacking inner woolly hairs.
        Stems hispid; pappus bristles 10–12. Widespread .... 1. *L. glandulosa*
        Stems not hispid; pappus-bristles ca. 30. Local in San Jacinto Mts. .... 4. *L. ziegleri*

1. **L. glandulòsa** (Hook.) H. & A. subsp. **glandulosa**. [*Blepharipappus g.* H. & A.] Stems commonly corymbosely branched, 1–5 dm. high, often anthocyanous; lvs. hispid, often strigose on upper surface, the basal dentate or lobed, the cauline mostly entire; invol. 6–9 mm. high, short-hispid and usually stipitate-glandular; ligules white, often fading rose-purple, 6–15 mm. long, 5–10 mm. wide; disk-fls. 25–30 (–100), the corolla 4–6 mm. long; ray-aks. 3–4 mm. long; disk-aks. 3.5–6 mm. long; pappus of 10–12 glistening white flattened linear-attenuate paleae, plumose to above middle with capillary hairs outside and tangled woolly hairs inside; $n=8$ (Clausen et al., 1934).—Common in sandy places, below 8000 ft.; many Plant Communities; deserts and cismontane, L. Calif. to cent. Calif., Utah, New Mex. March–June.

Ssp. **lùtea** Keck. Ligules golden yellow; $n=8$.—Chaparral; mts. from Ventura Co. to cent. Calif.

2. **L. heterotrícha** (DC.) H. & A. [*Madaroglossa h.* DC.] Stem stout, 2–9 dm. high; herbage pallid, ± succulent, heavily odorous; lvs. short-hispid, strigose and glandular, crenate or short-dentate, the upper entire; invol. 8–12 mm. high, densely hispid and glandular; ligules creamy white to pale yellow, not tipped with white, often fading rose-purple, 6–15 (–22) mm. long, 4.5–12 (–15) mm. wide; disk-fls. 35–85, the corolla 4.5–7 mm. long; ray-aks. 4–5 mm. long; disk-aks. 4.3–6.3 mm. long; pappus of 14–19 very slender white glistening deciduous bristles 3.5–6 mm. long, plumose to above middle with straight capillary hairs nearly equaling the bristles; $n=8$ (Clausen et al., 1941).—Scattered, up to 5000 ft.; V. Grassland, Foothill Wd.; Mt. Pinos and Tehachapi Pass; to w. side of San Joaquin V. March–June.

3. **L. platyglóssa** (F. & M.) Gray ssp. **campéstris** Keck. [*Madaroglossa elegans* Nutt. *L. p.* var. *breviseta* Gray.] Stem erect, the side branches ascending to erect, to 5 dm. high; lvs. short-hirsute to pilose, the basal and lower with short narrow lobes; invol. 6–12 mm. high, the phyllary-tips rounded, not dilated; ligules 6–15 mm. long, 5–10 mm. wide, mostly yellow with white tip, sometimes yellow throughout; disk-fls. 2.8–5 mm. long; pappus of 18–32 scabrid bristles or these plumose and interlaced with woolly hairs within; $n=7$ (Johansen, 1933).—Grassy slopes and openings, mostly below 4000 ft.; Coastal Sage Scrub, Grassland, Chaparral, etc.; coastal plains, L. Calif. to n. Calif. March–May. Channel Ids.

Ssp. **platyglossa**. [*Callichroa p.* F. & M. *Blepharipappus p.* Greene.] TIDY TIPS. Plate 19, Fig. B. Stem prostrate or decumbent, stout, succulent, corymbosely branched, 1–3 dm.

long; lvs. with short round salient lobes; phyllary-tips round-dilated; pappus of scabrid bristles; $n=7$ (Johansen, 1933).—Coastal bluffs and grassy places; Santa Cruz Id. to Mendocino Co. May–June.

   4. **L. ziègleri** Munz. Divaricately branched from base or not, 1–3 dm. high, with spreading both stiff and soft hairs, as well as ± appressed ones; basal lvs. 2–4 cm. long, 3–4 mm. wide, winged-petiolate, dentate in distal part; cauline lvs. largely turned to one side, gradually reduced upward, entire, subsessile; heads hemispheric; phyllaries 13–14, 7–8 mm. long; ray-aks. ca. 3 mm. long; disk-fls. ca. 35, the corolla 4–5 mm. long; pappus-bristles ca. 30, scarcely plumose or flattened, without woolly hairs near the base; disk-aks. dark, ca. 3 mm. long, white-strigose.—An uncertain taxon, possibly belonging to *L. platyglossa* ssp. *campestris*; from grassy meadowlike slopes at 4750 ft., Hemet V., San Jacinto Mts., Riverside Co. May–June.

90. *Leóntodon* L. HAWKBIT

Low acaulescent perennials with toothed or pinnatifid basal lvs. and simple or forked scapes bearing 1 or more yellow many-fld. heads. Invol. oblong, scarcely imbricated, but with several bractlets at the base, the principal phyllaries subequal. Aks. finely striate, narrowed at summit or beaked; pappus persistent, of plumose bristles enlarged and flattened at the base or the outer sometimes paleaceous. Ca. 45 spp., of Old World. (Greek, *leon*, lion, and *odous*, tooth, because of the toothed lvs.)

   1. **L. leysseri** (Wallr.) G. Beck. [*Thrincia l.* Wallr.] Lvs. and scapes densely clustered, the former oblanceolate, 5–15 cm. long; scapes slender, 1–3 dm. high, simple; heads solitary, the invol. 7–9 mm. high; phyllaries 6–12, the outer very small; inner aks. roughened, with a plumose pappus with an outer ring of fine bristles; $2n=8$ (Stebbins et al., 1953).—Taken as a lawn weed, San Bernardino Valley; from Eu. May–Aug.

91. *Lepidospártum* Gray.

Rigid broomlike shrubs with alternate entire lvs. Heads numerous in small close terminal racemes arranged in large panicles. Phyllaries imbricated in 3–4 series, chartaceous, oblong, obtuse. Receptacle naked. Heads discoid. Fls. yellow, with long tube, short campanulate throat and rather long lobes. Anthers exserted. Aks. terete; pappus of copious minutely scabrous capillary bristles in several series. Two spp. (Greek, *lepis*, scale, and *sparton*, the broom-shrub.)

Branches permanently tomentose, plainly striate; heads 4–5-fld.; invol. 9–10 mm. high, usually tomentose .... 1. *L. latisquamum*
Branches glabrate in maturity, not much striate; heads 10–15-fld.; invol. 4–6 mm. high, subglabrous .... 2. *L. squamatum*

   1. **L. latisquàmum** Wats. [*L. striatum* Cov.] Rather narrow shrub 1–2.5 m. high; stems strongly striate; lvs. numerous, needlelike, 2–3 cm. long; invol. 8–10 mm. high, tomentose, subcylindric; phyllaries 2–2.5 mm. wide; aks. white-villous, ca. 5 mm. long; pappus whitish, 8–9 mm. long.—Dry slopes and benches, 5000–8000 ft., mostly on limestone; Pinyon-Juniper Wd., Shadscale Shrub, Sagebrush Scrub; n. slope of San Gabriel Mts., Clark Mts., Inyo-White Mts.; Nev. June–Oct.

   2. **L. squamàtum** (Gray) Gray. [*Linosyris s.* Gray. *Tetradymia s.* Gray.] SCALE-BROOM. Plate 19, Fig. C. A ± round-topped shrub 1–2 m. high, from creeping underground stems; the spring growth white-tomentose, densely leafy, with entire obovate lvs. ca. 1 cm. long; mature growth with many ascending virgate branches, glabrate, green, with discrete lf.-scales; heads many; invol. 4–6 mm. long, campanulate; phyllaries not extending down on to the peduncles or intergrading into scaly lvs., 1–1.5 mm. wide; aks. narrow-clavate, ca. 3.5 mm. long, glabrous; pappus whitish, ca. 5 mm. long; $n=30$, ca. 45, etc. (various workers).—Common in washes and gravelly places below 5000 ft.; Coastal Sage Scrub, Chaparral, Joshua Tree Wd., etc.; cismontane from n. L. Calif. n.; Mojave and Colo. deserts; to cent. Calif. Santa Cruz Id. Aug.–Oct.

A local form about the Whitewater and Palm Springs areas (Riverside Co.) has phyllaries extending 4–5 mm. down on to the peduncles and grading into the scaly lvs. It is called var. **pálmeri** (Gray) Wheeler. [*Linosyris s.* var. *p.* Gray. *L. s.* var. *obtectum* Jeps.]

92. Lessíngia Cham.

Annual herbs with tomentose or glabrate, generally glandular lvs. and stems. Lvs. alternate, sessile, entire or serrate or lobed, 3-nerved, the basal rosulate and petioled, the cauline sessile. Heads small, panicled or in spikes, 3–many-fld. Invols. turbinate to narrowly campanulate, the phyllaries imbricated in several series, the green tips appressed or recurved. Fls. all discoid, perfect, yellow, lavender, rose or white, the outer corollas enlarged, more deeply cleft on the inner side, the ligulelike limb palmately 5-lobed, ± reflexed. Style-branches tipped with a short obtuse appendage bearing a cup or subulate prolongation, among apical bristles. Aks. turbinate, silky-villous. Several spp., almost restricted to Calif. (Honoring the *Lessing* family of Germany, one of whom, Christian F. was the author of a work on Asteraceae.)

(Howell, J. T. A systematic study of the genus Lessingia Cham. Univ. Calif. Publ. Bot. 16: 1–44. 1929.)

Style-appendages subulate, bearing a long cusp, nearly or quite as long as the stigmatic portion. W. Mojave Desert and adjacent mts. .... 2. *L. lemmonii*
Style-appendages less than half as long as the stigmatic portion, the cusp if present usually short. Cismontane, except var. of *L. glandulifera*.
  Phyllaries closely imbricate, ca. 6-seriate. Often stout summer-flowering annuals
      1. *L. glandulifera*
  Phyllaries loosely imbricate, ca. 4-seriate. Slender spring-flowering annuals .... 3. *L. tenuis*

1. **L. glandulífera** Gray. var. **glandulífera**. [*L. germanorum* var. *g.* J. T. Howell.] Plants 1–5 dm. high, freely branched, very glandular with spheroidal, short-stipitate glands, with a strong sour odor, tomentose or glabrate; basal lvs. oblanceolate, 2–5 cm. long, pinnately lobed; cauline 0.5–2.5 cm. long, generally glabrate, ± glandular, acute, lobed or entire, uppermost bracteate; heads solitary, 20–30 fld.; invol. 5–8 mm. high; corollas yellow; style-branches 1–1.3 mm. long, with an appendage 0.3–0.6 mm. long; ak. 2–3 mm. long; pappus bristles 30 or more, free to the base; $n = 5$ (Raven et al., 1960). —Open ground below 5500 ft.; cismontane valleys from interior San Diego Co. and adjacent L. Calif. to interior cent. Calif. May–Nov.

Var. **tomentòsa** (Greene) Ferris. [*L. t.* Greene. *L. germanorum* var. *t.* J. T. Howell.] Plants depressed, 1–1.5 dm. high, densely leafy; lvs. persistently white-tomentose, glandular along the margin only, oblong, entire, less than 1 cm. long; phyllaries 5–6 mm. high, with large marginal glands.—W. edge of Colo. Desert near Warner's Ranch.

2. **L. lemmònii** Gray. var. **lemmònii**. [*L. germanorum* var. *l.* J. T. Howell.] Low and spreading, branching from the base or below the middle, 0.6–3 dm. high; spring-flowering plants tomentose almost throughout, by summer the tomentum restricted to base and plant finely hirtellous above, usually with glands only on bract- and phyllary-margins; corollas yellow without purplish band in throat; heads 12–22-fld., the phyllaries glandular-puberulent and with large marginal glands; appendages of style-branches 0.7–1 mm. long, slender-subulate; pappus-bristles abundant, mostly free.—In sandy places 1000–6000 ft.; Creosote Bush Scrub to Pinyon-Juniper Wd.; Mojave Desert, San Gabriel Mts. to Mt. Pinos and Inyo Co.; to Ariz. May–Oct.

Var. **peirsònii** (J. T. Howell) Ferris. [*L. germanorum* var. *p.* J. T. Howell.] Having a dense matted tomentum which is only partly deciduous in age and the invols. persistently woolly.—Open slopes, 3000–4000 ft.; Chaparral, Foothill Wd.; n. Los Angeles Co., s. and e. Kern Co. July.

Var. **ramulosíssima** (Nels.) Ferris. [*L. r.* Nels. *L. germanorum* var. *r.* J. T. Howell.] Very intricately branched; herbage and phyllaries very glandular; invols. narrow, the peduncles more bracteate.—Below 7000 ft.; Joshua Tree Wd., Pinyon Juniper Wd.; w. Mojave Desert to Owens V.

3. **L. ténuis** (Gray) Cov. [*L. ramulosa* var. *t.* Gray. *L. heterochroma* Hall.] Three to 15 cm. high, diffusely branched, the stems and slender divergent branches glandular, loosely tomentose below when young, the older plants glabrous; basal lvs. spatulate, mostly entire, 1–3.5 cm. long; invol. glandular-granuliferous, larger marginal glands essentially none; outer corollas pinkish or whitish; inner ones yellow; pappus bristles free or ± united. —Dry open slopes, 4000–7000 ft.; Yellow Pine F.; Mt. Pinos region, Ventura Co.; to S. Coast Ranges of cent. Calif. May–July.

93. *Leucelène* Greene

Low perennial herbs with numerous tufted stems from a buried cordlike rootstock. Lvs. alternate, linear or subulate, or the lower narrowly spatulate. Heads solitary at the ends of the branches, radiate. Invol. turbinate; phyllaries ca. 3-seriate, imbricate, herbaceous with narrow scarious margin. Ray-fls. ♀, fertile; ligules white, often turning reddish. Disk-fls. perfect, fertile, yellowish. Style-branches with obtuse appendages shorter than the stigmatic portion. Aks. linear. Pappus of long subequal bristles. Monotypic. (Etymology unknown.)

1. **L. ericoìdes** (Torr.) Greene. [*Inula*? *e.* Torr. *Aster ericaefolius* Rothr., not *A. ericoides* L.] Tufted, heathlike, the numerous slender very leafy crowded stems from the cespitose caudex 5–15 cm. high; herbage hoary-strigose, ± glandular; lvs. hispid-ciliate, bristle-tipped, 4–12 mm. long; invol. 5–6 mm. high; phyllaries strigulose, slightly glandular-puberulent, lance-oblong, acuminate, green, thickened, with white fimbrillate scarious margin; rays 12–21, 5 mm. long; $n = 8 + f$ (Turner & Flyr, 1966).—Dry rocky slopes, 3700–9400 ft.; Joshua Tree Wd., Pinyon-Juniper Wd., Bristle-cone Pine F.; mts. of e. San Bernardino Co. to Inyo and Mono cos.; to Nebr., Tex., Mex. April–Aug. Variable in pubescence.

94. *Lygodésmia* D. Don

Herbs or ± woody, with lvs. largely near the base. Heads terminal on the branchlets, erect, 3–12-fld. Receptacle naked. Invol. slender, of a few equal phyllaries and outer minute ones. Fls. pink or rose. Aks. slender, few-ribbed, truncate at apex; pappus bristles many, capillary, stiffish or soft, unequal, not plumose, deciduous separately. A small N. Am. genus. (Greek, *lugos*, a plant twig, and *desme*, bundle, the type sp. with fascicled stems.)

Plant annual, not spiny, not woolly at base .... 1. *L. exigua*
Plant perennial, the branches rigid, spine-tipped; plant with mats of wool at the base
2. *L. spinosa*

1. **L. exígua** Gray. Diffusely branched glabrous annual 1–4 dm. high; lvs. oblanceolate, mostly basal, entire or runcinate-dentate or -pinnatifid, 2–4 cm. long, the upper few and reduced; heads ca. 3–4-fld., terminal on very slender branchlets; invol. 4–5 mm. high, the inner phyllaries lanceolate, the outer few, minute; ligules inconspicuous; aks. 3–4 mm. long, the pappus bright white, conspicuous, 3–4 mm. long.—Well distributed but not common, open places, rocky hills, etc. up to 6200 ft.; Creosote Bush Scrub, Joshua Tree Wd., Pinyon-Juniper Wd.; both deserts; to Colo., Tex. March–May.

2. **L. spinòsa** Nutt. Perennial from a woody branched root-crown, with many slender, rigid, zigzag, glabrous to puberulent branches 3–4 dm. long, woolly-matted at base, almost leafless with spiny branchlets; lower lvs. linear, entire, 0.5–4 cm. long, the upper reduced to scales; heads 3–5-fld., subsessile or on very short branchlets; invol. 8–9 mm. high, the inner phyllaries lanceolate, the outer broad, calyculate; ligules well exserted; aks. 4 mm. long, ± 5-angled; pappus 4 mm. long.—Dry sandy or rocky places, 5000–11,000 ft.; Sagebrush Scrub, Pinyon-Juniper Wd.; Bristle-cone Pine F.; desert slopes of San Gabriel and San Bernardino mts., Panamint Mts., Inyo-White Range; to B.C., Mont., Ariz. July–Sept.

95. **Machaeránthera** Nees

Annual to perennial herbs or subshrubs, with a taproot system. Lvs. alternate, spinulose-tipped, entire or more often spinulose-dentate to pinnatifid or pinnately dissected. Heads solitary to many, small to large. Invol. several-seriate; phyllaries chartaceous or coriaceous toward base, greenish toward tip, appressed to squarrose. Ray-fls. usually present, ♀, fertile, bluish to lavender or whitish. Style-branches flattened, with hairy appendage. Aks. ± compressed, mostly several-nerved, hairy or glabrous; pappus of ± numerous unequal barbellate bristles. Ca. 25 spp. of temp. w. N. Am. (Greek, *machaira*, sword, and *anthera*, anther, from the lanciform anther-tips, which however do not distinguish the genus from its relatives.)

(Cronquist, A. and Keck, D. D. A reconstitution of the genus M. Brittonia 9: 231–239. 1957.)
The plants glandular annuals; invol. 4 mm. high. Mojave Desert .... 1. *M. arida*
The plants of longer duration; invol. longer.
    Plants herbaceous; heads usually several or many; biennials or short-lived perennials.
        Phyllaries narrowly linear, not glandular, the long narrow loose green tips longer than the indurated base .... 7. *M. tephrodes*
        Phyllaries not narrowly linear, the erect or squarrose tips shorter than the indurated base.
            Stems viscid-glandular to base; cauline lvs. ovate, mostly reduced to small bracts. Kelso, Mojave Desert .... 5. *M. leucanthemifolia*
            Stems often glandular above, but not to base; cauline lvs. usually larger and more elongate.
                Invol. 12–15 mm. high; phyllaries 6–10-seriate, their short broad canescent tips not squarrose; herbage and invol. not noticeably glandular .... 4. *M. lagunensis*
                Invol. 6–10 mm. (rarely 14) mm. high; phyllaries strongly squarrose; herbage often stipitate-glandular above .... 2. *M. canescens*
    Plants shrubby or woody at base; heads large, usually solitary at branch-tips, the disk mostly 2–4 cm. wide, the invol. 13–19 mm. high.
        Stem glandular and hispid, often also tomentose; lvs. chiefly lanceolate or lance-linear, pubescent or rarely merely ciliate; phyllaries hirtellous or pilose; scarcely shrubby. Widespread on deserts .... 8. *M. tortifolia*
        Stem glabrous or slightly glandular toward apex; lvs. chiefly oblong, glabrous or sparsely ciliate; phyllaries glabrous dorsally, stipitate-glandular on margin
            Outer phyllaries linear-attenuate; equaling or exceeding the inner; pedicels glandular above .... 3. *M. cognata*
            Outer phyllaries lanceolate, acuminate, much shorter than the inner; pedicels glabrous
            .... 6. *M. orcuttii*

1. **M. árida** Turner & Horne. [*Psilactis coulteri* Calif. refs., not Gray.] Divaricately branched annual, 5–30 cm. high, stipitate-glandular throughout and usually hispidulous or pilose; lvs. oblanceolate or oblong, mostly laciniate-pinnatifid, the teeth spinescent-tipped, the lower petiolate, 2–3 cm. long, the upper closely sessile and appressed, becoming much reduced, sometimes entire; invol. ca. 4 mm. high; phyllaries lance-oblong, acute, granular-glanduliferous except on the whitish basal chartaceous portion; rays 20–35, whitish to lavender, 5 mm. long; aks. pubescent; $n=5$ (Solbrig et al., 1964).—Uncommon, dry open sandy flats, ca. 300–3000 ft.; Creosote Bush Scrub, Alkali Sink; cent. and e. Mojave Desert from Inyo Co. s. to Whipple Mts.; to Nev., Ariz., Son. March–June.

2. **M. canéscens** (Pursh) Gray ssp. **canéscens**. [*Aster c.* Pursh.] Taprooted biennial or short-lived perennial; stems several, paniculately or racemosely branched, 2–7 dm. high; herbage cinereous-puberulent, often also stipitate-glandular above; lvs. spinulose-toothed, the basal 2–8 cm. long, the cauline smaller, often linear, the upper entire and often bractlike; heads numerous; invol. 6–10 (–12) mm. high, canescent or glandular or both; phyllaries 4–8-seriate, ± squarrose; rays mostly 8–25, pale bluish-purple, 5–10 mm. long; $n=4$ (Raven et al., 1960).—Dry gravelly slopes, meadow-borders, etc., 4500–9500 ft.; Sagebrush Scrub, Pinyon-Juniper Wd., Yellow Pine F., Joshua Tree Wd., etc.; San Jacinto Mts. n. to Mt. Pinos, mts. of Mojave Desert and Death V.; n. to B.C., Rocky Mts. July–Oct.

Ssp. **ziègleri** Munz. More stipitate-glandular in the upper parts; invol. 14–15 mm. high, all except the uppermost phyllaries strongly squarrose and very glandular; ligules mostly 20–22 mm. long.—At 5000–8100 ft., mostly Yellow Pine F.; Santa Rosa Mts., Riverside Co.

3. **M. cognàta** (Hall) Cronq. & Keck. [*Aster c.* Hall. *A. standleyi* Davids. *Xylorhiza c.* Davids.] Rounded openly branched shrub 5–15 dm. high, with white-barked stems; herbage with scattered stipitate glands; lvs. lance-ovate to elliptic, sessile, coarsely spinose-dentate, 2–4 cm. long, coriaceous, veiny, extending nearly to the solitary terminal heads; invol. 15–25 mm. high, hemispheric, glandular; phyllaries ca. 5-seriate, the outer green and often exceeding the inner which are broader and indurate at base; rays 21–35, pink-lavender or violet, 15–20 mm. long; aks. densely long-villous; pappus unequal, ± flattened, reddish.—Locally in gypsum clays, etc., below 500 ft.; Creosote Bush Scrub; region near and n. of Mecca, Riverside Co., Cucopa Mts.; n. L. Calif. Jan.–June.

4. **M. lagunénsis** Keck. Much like *M. canescens*, 4–7 dm. high, ± finely cinereous-puberulent throughout, not glandular; lvs. linear-oblong to linear, remotely spinose-dentate or entire, up to 8 cm. long; heads several to many, solitary at tips of branches; invol. 12–15 mm. high; phyllaries 6–10-seriate, the tips appressed or somewhat spreading, scarcely glandular; rays 13–30, deep lavender, 10–17 mm. long; aks. pubescent; pappus reddish.—Dry slopes at 5000–7000 ft.; Yellow Pine F.; Laguna Mts., San Diego Co., Bear V., San Bernardino Mts. July–Aug.

5. **M. leucanthèmifolia** (Greene) Greene. [*Aster l.* Greene.] Short-lived pallid perennial; stems few, erect, divaricately branched, 2–4 dm. high, cinereous-puberulent; lvs. canescent, ± glandular, spinulose-serrate-dentate, the basal rosulate, 2–3 cm. long, 8–12 mm. wide, the cauline amplexicaul smaller; heads inconspicuous, terminating the branchlets; invol. 6–7 mm. high, canescent and glandular-viscid; phyllaries ca. 5-seriate, mucronate-tipped; rays 8–15, pale purple, 6–8 mm. long, fertile; aks. pubescent, 3.2 mm. long; pappus brownish-white; $n=5$ (Raven et al., 1960).—Sand dunes 2200 ft.; Creosote Bush Scrub; Kelso, Mojave Desert. sw. Nev. May–June.

6. **M. orcúttii** (Vasey & Rose) Cronq. & Keck. [*Aster o.* Vasey & Rose. *Xylorhiza o.* Greene.] Base woody; stems many, 4–9 dm. high; leafy to the heads; lvs. spinulose-toothed, the lower oblanceolate, the middle lance-elliptic to oblong, sessile by a broad base, 2–5 cm. long, coriaceous; heads terminal, solitary; invol. 15–20 mm. high; phyllaries ca. 6-seriate, indurate, sparingly glandular-ciliolate; rays 30–45, lavender or purple, ca. 2 cm. long; aks. with a dense silky tomentum; pappus tawny.—Occasional in gypsum soils below 1000 ft.; Creosote Bush Scrub; sw. Colo. Desert, Imperial and San Diego cos.; n. L. Calif. March–April.

7. **M. tephròdes** (Gray) Greene. [*Aster canescens* var. *t.* Gray. *A. t.* Blake.] Erect biennial, 2–7 dm. high, cinereous-puberulent or canescent, with occasional gland-tipped hairs; lvs. entire to somewhat spinose-dentate, the lower 4–10 cm. long, oblanceolate, the middle and upper narrower, reduced, sessile; heads solitary at ends of branches; invol. 8–10 mm. high; phyllaries 5–7-seriate, with spreading or squarrose tip; rays 25–40, lavender to violet, 10–15 mm. long; aks. velutinous.—Low ground, below 3500 ft.; Coastal Sage Scrub, Creosote Bush Scrub; Mojave and Colo. deserts, about Hemet and Riverside; to Utah, Son., L. Calif. April–May, Sept.–Dec.

8. **M. tortifòlia** (Gray) Cronq. & Keck. [*Aplopappus t.* T. & G. *Aster t.* Gray, not Michx. *A. abatus* Blake. *Xylorhiza t.* Greene.] DESERT ASTER. Plate 19, Fig. D. Suffruticose, the leafy sparingly branched stems rather numerous, 3–7 dm. high; herbage tomentose at least when young, sometimes merely stipitate-glandular and sparingly hairy; lvs. linear to lanceolate or oblong, sessile, spinulose-toothed, 3–6 cm. long, coriaceous; heads 3.5–6.5 cm. wide, solitary at ends of rather naked peduncles; invol. 12–16 mm. high, usually canescent and glandular; phyllaries 4–5-seriate; rays ca. 40–60, blue-violet to lavender, to almost white, 12–25 mm. long; aks. silky villous, with tawny unequal pappus; $n=6$ (Raven et al., 1960).—Dry rocky places and washes, 2000–5500 ft.; largely Creosote Bush Scrub; Mojave and n. Colo. deserts from White Mts. to Riverside Co.; to Utah, Ariz. March–May; Oct.

96. Màdia Mol. TARWEED

Herbs, usually very glandular and heavy-scented. Lvs. linear or oblong, the basal entire to remotely denticulate. Invol. ± angled by its deeply sulcate phyllaries which completely enclose the ray-aks. Receptacle bearing between rays and disk a single row of chaffy bracts, these usually ± united into an often persistent and prismatic cup. Ray-fls. few to many, fertile. Disk-fls. fertile or sterile. Ray-aks. usually laterally compressed, with flat sides, narrow back, and sharp ventral angle (see also *M. minima*), longitudinally striate. Pappus usually none in ray-aks., sometimes present in disk-aks. Ca. 18 spp. of Pacific N. and S. Am. (*Madi*, the Chilean name of the original sp.)

Disk-fls. 1 (or 2), glabrous, phyllary embracing the ak. and sulcate to tip; tiny herbs.
    Ray-aks. laterally compressed, with sharp ventral angle, glabrous; invol. strongly stipitate-glandular, the glands yellow .... 2. *M. exigua*
    Ray-aks. obcompressed, not angled, pubescent; invol. minutely stipitate-glandular, the glands black .... 5. *M. minima*
Disk-fls. more than 1 (except rarely in *M. glomerata*), pubescent; phyllary with empty flat tip.
    Ray-fls. 0–3; ray-aks. broadest at middle, truncate at both ends with broad areola
                                                                                              3. *M. glomerata*
    Ray-fls. 8–16.
        Disk-aks. sterile; receptacle pubescent; ray-fls. conspicuous .... 1. *M. elegans*
        Disk-aks. fertile; receptacle glabrous; ray-fls. inconspicuous.
            Glandular above the middle, slender and flexuous; heads small, to 9 mm. high. Plants largely of interior .... 4. *M. gracilis*
            Glandular well down toward the base; stem stout; heads larger, the invol. 8–12 mm. high. Coastal, behaving as a ruderal .... 6. *M. sativa*

1. **M. élegans** D. Don ssp. **élegans**. [*Madaria e*. DC.] Plate 19, Fig. E. Stem commonly corymbosely branched above, 2–8 dm. high, villous below, ± glandular above; lvs. linear to broadly lanceolate, the basal few or in a small rosette, the lower cauline often crowded, the upper well spaced; invol. to 10 mm. high, mostly hirsute and stipitate-glandular, the attenuate tips of the phyllaries equaling the basal portion; ray-fls. 8–16, the ligules 6–15 mm. long, yellow or with maroon blotch at base; disk-fls. 25 or more, yellow or maroon; anthers purple-black; $n=8$ (Johansen, 1933).—Rather dry slopes, 3000–8000 ft.; Chaparral to Montane Coniferous F.; San Diego Co. to Ore. June–Aug.

Ssp. **densifòlia** (Greene) Keck. [*M. d.* Greene.] Basal lvs. forming a large rosette, the lower cauline closely imbricated.—Dry places below 3000 ft.; Chaparral, etc.; occasional in cismontane s. Calif., more abundant n. Aug.–Nov.

Ssp. **wheèleri** (Gray) Keck. [*Hemizonia w.* Gray. *M. w.* Keck.] Commonly divaricately branched from the middle, sparingly glandular; basal rosette small or none, cauline lvs. scattered; invol. 4.5–5.5 mm. high; ligules 4–5 mm. long, yellow; anthers yellow; $n=8$.— Dry places under conifers; largely Yellow Pine F.; San Jacinto Mts. to Mt. Pinos; Sierra Nevada and L. Calif. June–Aug.

2. **M. exígua** (Sm.) Gray. [*Sclerocarpus e.* Sm.] Corymbosely or paniculately branched, the stems very slender, 0.5–3 dm. high, prominently stipitate-glandular above; lvs. strigose, linear, 1–4 cm. long, 2 mm. or less wide; heads on filiform peduncles; invol. 2.5–4.8 mm. high; phyllaries early deciduous with the mature fr., lunate in outline laterally, with prominent stalked glands; ray-fls. 5–8, the ligule 1 mm. long; ray-aks. 1.8–2.8 mm. long, crescentic; disk-ak. fertile; $n=16$ (Johansen, 1933).—Dry open places at low or moderate elevs.; Chaparral, etc.; L. Calif. through cismontane Calif. to B.C. May–July.

3. **M. glomeràta** Hook. Rigid, very leafy, simple or with assurgent branches, 1.5–8 dm. high, villous to hispid, with yellow stipitate glands above, unpleasantly odorous; lvs. narrowly linear, the lower 3–9 cm. long; heads in dense terminal glomerules of 5–30, or in more open cymes; invol. narrowly ovoid, 5.5–9 mm. long; ligules inconspicuous, greenish-yellow to purplish, 1.5–2.5 mm. long; disk-fls. 1–10; aks. 4–6 mm. long; $n=14$ (Clausen et al., 1934).—Grassy places in Montane Coniferous F.; San Bernardino Mts.; to Alaska, Rocky Mts. July–Sept.

4. **M. grácilis** (Sm.) Keck. [*Sclerocarpus g.* Sm. *Madia dissitiflora* (Nutt.) T. & G.] GUMWEED. Stem usually slender, simple or flexuously branching from the middle, the branches not overtopping the main stem, 1–10 dm. high; not so viscid as in *M. sativa*; lvs. not very crowded even at base, mostly linear, sessile by a narrow base, to 10 cm. long and 5 mm. wide; heads paniculate to racemose, not congested, the leafy bracts rarely prominent; invol. ovoid to depressed-globose, 6–9 mm. high; phyllaries thickly beset with stout gland-topped hairs, the acuminate tips short; ray-fls. 8–12; ligules 3–8 mm. long; disk-fls. 15–35; anthers black, included; ray-aks. 3–5 mm. long, often mottled; disk-aks. similar but straighter; $n = 16, 24$ (Clausen et al., 1945).—Abundant on wooded hillsides and open places, below 8000 ft.; mostly Yellow Pine F.; mts. from San Diego Co. n.; to B.C., L. Calif., S. Am. April–Aug.

5. **M. mínima** (Gray) Keck. [*Hemizonia m.* Gray. *Hemizonella m.* Gray.] Stems 1 to several from near the base, divaricately branched to form often a hemispheric plant, villous below, glandular-puberulent above, 2–15 cm. high; lvs. often in little clusters at the nodes, otherwise usually scattered, linear-oblong, 1–2 cm. long; heads solitary or in small terminal glomerules, 2–3 mm. high; phyllaries loosely appressed, becoming arcuate with the ripening aks., rounded on the back, their glands tiny, on prominent pustulate processes; ray-aks. incurved, ± beaked; $2n = 32$ (Keck, 1949).—Gravelly slopes, 3500–8600 ft.; mostly Yellow Pine F.; mts. of San Diego Co. to Mt. Pinos; cent. Calif., B.C. May–July.

6. **M. sativa** Mol. CHILE TARWEED. Stem usually stout, often rigidly branched above, glandular well down toward base, 5–10 (–20) dm. high; herbage strongly odorous; lvs. rather crowded, sessile by a broad base, up to 18 cm. long and 12 mm. wide; heads paniculate, racemose or subspicate; often approximate along the branchlets, not foliose-bracted; ray-aks. falcate-oblanceolate; disk-aks. oblanceolate, the sides sometimes 1-nerved; $n = 16$ (Johansen, 1933).—Close to the coast, up to 1000 ft.; behaving as a ruderal, Los Angeles Co. n.; to Alaska; native of Chile, Argentina. May–Oct. San Clemente and Santa Catalina ids.

97. Malacòthrix DC.

Annual to perennial, sometimes suffrutescent. Lvs. usually basal and stems scapose, sometimes leafy. Lvs. toothed to pinnatisect. Heads small to medium in size, yellow or white, solitary or panicled, never sessile, commonly nodding in bud. Receptacle bristly or naked. Invol. of subequal inner phyllaries and much shorter outer ones, or strongly graduated, the phyllaries scarious-margined. Aks. columnar, truncate, ribbed. Pappus-bristles soft, scabrous, deciduous ± in a ring, usually 1–8 of them stiffer and persistent, the ak. also often crowned with a ring of minute teeth. Ca. a dozen spp. of w. N. Am. (Greek, *malakos*, soft, and *thrix*, hair.)

(Williams, Elizabeth W. The genus M. Am. Midl. Natur. 58: 499–512. 1957.)
1. Outer phyllaries orbicular, 4–5 mm. wide; with broad scarious margins; persistent pappus-bristles 1–4 .... 3. *M. coulteri*
1. Outer phyllaries lanceolate to linear, narrower; with narrow pale margins.
    2. Plants annual.
        3. Lf.-segms. linear-filiform, elongate; persistent pappus-bristles 2 or more.
            4. Stems simple, without leaves and ending in solitary heads; invol. 12–15 mm. high
                1. *M. californica*
            4. Stems few-branched, each branch ending in a head; stems with a few lvs.; invol. 5–12 mm. high .... 6. *M. glabrata*
        3. Lf.-segms. lanceolate to oblong, short, usually toothed; persistent pappus-bristles 1 or 0.
            5. Ligules ca. 1 cm. long; aks. 15-ribbed .... 10. *M. sonchoides*
            5. Ligules shorter or, if as long, the lvs. with tufts of white wool on the margins; aks. often only 5-ribbed.
                6. Stems leafy at base only.
                    7. Fls. white to pink; pappus all deciduous; aks. 5-ribbed .... 4. *M. floccifera*
                    7. Fls. yellow; 1 pappus-bristle and a crown of setulose teeth usually persistent.
                        8. Aks. less than 1.7 mm. long, fusiform, with 5 of the 18 ribs more prominent than the others; invol. less than 8 mm. high.

    9. Aks. brown or straw-colored; cauline lvs. often toothed; plants usually unbranched below. L. Calif. to Tehama Co. and Glenn Co. . 2. *M. clevelandii*
    9. Aks. dark purplish-brown, rarely paler; cauline lvs. entire; plants often well branched from base. Santa Cruz Id., Hueneme Beach, Ventura Co.
  9. *M. similis*
   8. Aks. more than 1.7 mm. long, subcylindrical, gray-brown to straw-colored, with 15 equally prominent ribs; invol. 7–10 mm. high. Deserts, Inyo Co. to e. San Diego Co. .... 11. *M. stebbinsii*
  6. Stems leafy throughout. Insular only .... 5. *M. foliosa*
 2. Plants perennial, suffrutescent; lvs. linear to lanceolate or oblanceolate.
   9a. Herbage densely white-tomentose when young; inner phyllaries obtuse .. 7. *M. incana*
   9a. Herbage essentially glabrous; inner phyllaries attenuate .... 8. *M. saxatilis*

1. **M. califórnica** DC. Annual with dense rosette of radical lvs. which are conspicuously woolly when young, 6–12 cm. long, pinnatifid into linear-filiform lobes; scapes several, ample, not leafy, 1-headed, 1.5–3.5 dm. high; invol. 12–15 mm. high, the wool often persistent, the phyllaries in ca. 3 ranks; ligules pale yellow, 12–16 mm. long; fls. fragrant, opening only in sunlight; aks. narrow, lightly striate; pappus white, of 2 persistent bristles and some intervening teeth; $2n=14$ (Stebbins et al., 1953).—Dry sandy places below 6000 ft.; V. Grassland, Coastal Sage Scrub, Joshua Tree Wd., etc.; cismontane s. Calif. to Sacramento V., to the coast in Monterey Co., etc. and extreme w. edge of desert; Santa Cruz and Santa Catalina ids. March–July.

2. **M. clevelándii** Gray. Annual, usually branched from base, 1–5 dm. high, glabrous, slender; lvs. mostly basal, many, 3–10 cm. long, pinnatifid into lobes 1–3 mm. wide; stems paniculately branched above, with many heads; heads cylindrical to narrow-campanulate, 4–8 mm. high, 20–60-fld.; ligules yellow; aks. truncate-fusiform, 1.4–1.8 mm. long, slightly curved, 5 of the 15 ribs more prominent than the others, apex of the ak. flared, bordered by a ring of 14–17 white-scarious teeth; persistent seta on ak. 1; $n=7$ (Davis & Raven, 1962). Burns and disturbed places, below 5000 ft.; Chaparral, Coastal Sage Scrub, etc.; cismontane, to cent. Calif. April–June.

3. **M. coúlteri** Gray. [*M. c.* var. *cognata* Jeps.] SNAKE'S HEAD. Erect annual, simple or branched above, glabrous and rather pale, leafy, 1–5 dm. high; lvs. oblong, the lower sinuate-pinnatifid to dentate, 3–6 (–10) cm. long, the upper shorter, usually less deeply toothed; heads short-peduncled, subglobose; invol. 12–16 mm. high, the phyllaries much imbricated, the outer orbicular; ligules light yellow, 5–8 mm. long; aks. 15-ribbed, 4–5-angled; 1–4 pappus-bristles persistent; $2n=14$ (Stebbins et al., 1953).—Occasional on dry slopes, about bushes, in washes, etc.; below 3500 ft.; Creosote Bush Scrub, V. Grassland, Coastal Sage Scrub, etc.; cismontane and desert s. Calif., upper San Joaquin V.; Santa Rosa and Santa Cruz ids.; to Utah, Ariz.; Argentina. March–May.

4. **M. floccífera** (DC.) Blake. [*Senecio f.* DC. *M. obtusa* Benth.] Annual, the stems 1–several from the base, 1–4 dm. tall, almost leafless; lvs. basal, oblong in outline, usually with tufts of wool in the axils and near the margin, 2–8 cm. long, dentate-pinnatifid; heads many; invol. 4–6 mm. high, the main phyllaries linear, subequal, acute, often purplish at tip; ligules 3–6 (–10) mm. long, white or pale-yellow, often tinged pinkish; aks. obovate-oblong, ca. 2 mm. long, entire at top, rather conspicuously 5-ribbed; persistent pappus-bristles 0; $2n=14$ (Stebbins et al., 1953).—Sandy and rocky places below 5000 ft.; Coastal Sage Scrub, Chaparral, etc.; Ventura Co. n. to n. Calif. April–June.

5. **M. foliòsa** Gray. Annual, 2–5 dm. high, glabrous, branched above, leafy nearly to the heads; lvs. lanceolate, mostly laciniate-pinnatifid, 3–10 cm. long, the uppermost linear-attenuate; heads many, short- and slender-peduncled; invols. 8–10 mm. high, the outer phyllaries in 1 or 2 series, lance-oblong, much shorter than the inner ones, all with narrow scarious margin and ± obtuse; fls. yellow, 4–6 mm. long; aks. subequally 15-ribbed; pappus wholly deciduous.—Occasional in dryish places, Coastal Sage Scrub. March–July. Two other spp. have been proposed and variously combined as vars., all of these occurring only on the islands off our coast and to me quite mixed up as to separating characters and distribution. One is *M. indecora* Greene, diffuse, depressed or erect, the lvs.

more succulent and more obtusely lobed; invol. largely 6–7 mm. high. The other is *M. squalida* Greene, with the phyllaries supposedly more imbricated (a character apparent in young heads but not in older ones on same plant), the outer in 3–4 series, round to ovate and with scarious margins. These converge also with *M. insularis* Greene of Coronado Ids.

6. **M. glabràta** Gray. [*M. californica* var. *g.* Gray ex D. C. Eat.] DESERT-DANDELION. Plate 19, Fig. F. General habit of *M. californica*, but the stems few-branched, bearing 2 or more heads each and a few lvs.; plants quite glabrous, 1–4 dm. high; lvs. pinnatifid into linear lobes; invol. 5–12 mm. high, mostly glabrous; $2n=14$ (Stebbins et al., 1953).—Dry sandy plains and washes below 6000 ft.; Creosote Bush Scrub, Joshua Tree Wd., Shadscale Scrub, etc.; both deserts, inner cismontane valleys from San Diego Co. to Santa Barbara Co.; Ida., Ariz. March–June.

7. **M. incàna** (Nutt.) T. & G. [*Malacomeris i.* Nutt.] Perennial with stout root and woody crown; herbage white-woolly, ± glabrate in age; stems several, 1.5–2.5 dm. high; lvs. 5–10 cm. long, pinnately lobed to subentire, the lobes short, broad, blunt; peduncles slightly exceeding the lvs., mostly 1–2-headed; invol. ca. 2 cm. long, much imbricated; ligules lemon-yellow, ca. 10–12 mm. long; aks. 15-striate, ca. 1.5 mm. long; pappusbristles all deciduous; $2n=14$ (Stebbins et al., 1953).—Coastal Strand; Ventura Co. to San Luis Obispo Co.; Santa Rosa and San Miguel ids. Most months.

8. **M. saxátilis** (Nutt.) T. & G. var. **saxátilis**. [*Leucoseris s.* Nutt.] Plate 20, Fig. A. Perennial, suffrutescent, with several stems 3–9 dm. long, diffuse or decumbent, branched, minutely tomentulose when young; lvs. many, well distributed, succulent, 3–10 cm. long, lanceolate to spatulate, obtuse, entire to dentate; infl. rather few-headed; invol. 10–16 mm. high, the phyllaries in 3–4 series; ligules ca. 15–18 mm. long, white with some rose; aks. 10–15-costate, ca. 5 of the striae riblike, the ak. crowned with a minute denticulate white border; none of the pappus persistent; $2n=18$ (Stebbins et al., 1953).—Sea bluffs, road-cuts, etc.; Coastal Sage Scrub; nw. of Santa Barbara to Gaviota. March–Sept.

Stems several, arising from the root at the surface of the ground, suffrutescent at base or herbaceous.
    Lvs. leathery, obtuse, oblanceolate, or broadly linear, entire or the lowest sometimes irregularly lobed; plants of ocean bluffs, Santa Barbara Co. .... var. *saxatilis*
    Lvs. not leathery, pinnatifid to pinnately or bipinnately divided into linear divisions, if entire, then linear-filiform.
        Lvs. bipinnately divided; stems woody, densely leafy to the infl.; Channel Ids. var. *implicata*
        Lvs. pinnately parted; stems not densely leafy to the infl.; mostly herbaceous; Coastal mts. of s. Calif. .... var. *tenuifolia*
Stems partly subterranean, arising singly from the branches of a deep-seated root, essentially herbaceous. Mts. from Tehachapi to Santa Monica mts. .... var. *altissima*

Var. **altíssima** (Greene) Ferris. [*M. a.* Greene.] Leafy-stemmed, 1–2 m. high, mostly glabrous; lvs. thin, 8–20 cm. long, deeply and irregularly laciniate-lobed and also denticulate.—Chaparral, etc., mts. of Santa Barbara and Kern cos. to Santa Monica Mts.

Var. **implicàta** (Eastw.) Hall. [*M. i.* Eastw.] Lvs. twice pinnatifid into linear or filiform segms.; fls. purplish-tinged.—Sandy and rocky places, Coastal Strand, Coastal Sage Scrub; San Nicolas, San Miguel, Santa Rosa and Santa Cruz ids.

Var. **tenuifòlia** (Nutt.) Gray. [*Leucosyris t.* Nutt. *M. s.* var. *tenuissima* Munz.] Lower lvs. acute, not succulent, often toothed or pinnately lobed with broad rachis, the upper linear-filiform, entire; infl. open, diffuse, slender-stemmed, many-headed.—Dry slopes and ridges, below 2000 ft.; Coastal Strand, Coastal Sage Scrub, etc.; Laguna Beach, Santa Ana Mts., Redlands to Kern and San Luis Obispo cos.; Santa Catalina Id. Most months.

9. **M. símilis** Davis & Raven. Annual, usually branched from the base; basal lvs. linear-lanceolate, entire to pinnatifid; cauline lvs. subentire; heads narrow-campanulate, 6–10 mm. high, 30–70-fld.; ligules yellow; aks. truncate-fusiform, 1.4–1.7 mm. long, slightly curved, 5 of the 15 ribs prominent; ak.-apex with 18 white-scarious teeth, persistent seta 1; $n=14$ (Davis & Raven).—Santa Cruz Id., Hueneme Beach in Ventura Co.; L. Calif. A segregate from the traditional *M. clevelandii*.

10. **M. sonchoìdes** (Nutt.) T. & G. [*Leptoseris s.* Nutt.] Freely branched annual, gla-

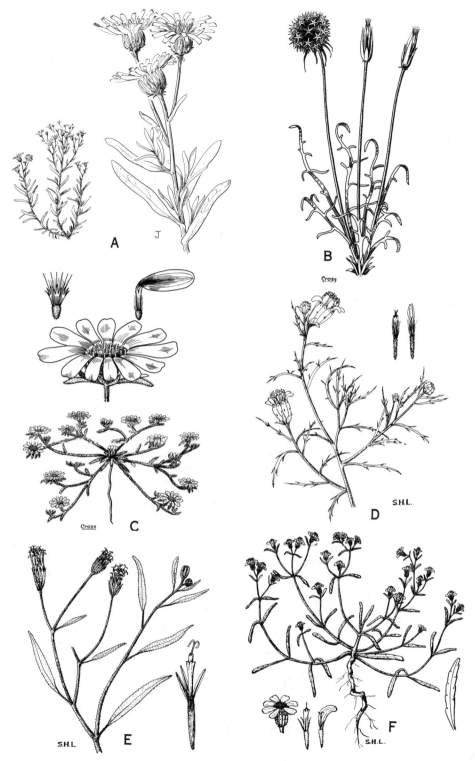

Plate 20. Fig. A, *Malacothrix saxatilis*. Fig. B, *Microseris linearifolia*. Fig. C, *Monoptilon bellioides*. Fig. D, *Nicolletia occidentalis*. Fig. E, *Palafoxia linearis*. Fig. F, *Pectis papposa*.

brous or early glabrate, the stems several from the base, 3–10 cm. long, oblong in outline, pinnatifid into short callus-toothed lobes; uppermost lvs. much reduced; heads short-peduncled; invol. 7–8 mm. high, the phyllaries linear-acuminate; ligules bright yellow, ca. 10–12 mm. long; aks. 15-striate, 5 of the ribs best developed; persistent bristles 0, but the ak. with a finely denticulate whitish crown.—Occasional in open ± sandy places, 2000–5000 ft.; Creosote Bush Scrub, Joshua Tree Wd.; w. and n. Mojave Desert to e. Modoc Co., Ida., Nebr. April–June.

11. **M. stébbinsii** Davis & Raven. A segregate from *M. clevelandii*; usually simple at the base; basal lvs. lanceolate to oblanceolate, dentate, rarely pinnatifid; heads campanulate, 7–10 mm. high, 20–70-fld.; aks. 1.7–2.3 mm. long, rarely curved, with 15 equally prominent ribs, apex bordered by a ring of 14–17 white-scarious teeth, persistent seta usually 1.—Deserts, Inyo Co. to e. San Diego Co.; Nev., Son.

98. *Malpèria* Wats.

Erect slender annual with narrow alternate mostly sessile lvs. Heads loosely cymose; invol. turbinate; phyllaries narrow, thin, unequal, several-nerved, scarious-margined. Receptacle flat, naked. Style-branches thickened, exserted only in age. Aks. slender, 5-angled. Pappus of 3 hispidulous setae as long as the corolla and alternating with minute truncate erose paleae. (Etymology obscure.)

1. **M. ténuis** Wats. Stems branching above, 1–3 dm. high, minutely scabrous; lvs. linear, acuminate, 1.5–5 cm. long, mostly entire; invol. 8–10 mm. high; phyllaries linear, strigulose; fls. brownish; aks. somewhat hispid on the angles; $2n = 20$ (Raven, 1969).—Sandy places, Creosote Bush Scrub; Colo. Desert (Signal Mt., Split Mt., Fish Mt.); n. L. Calif. March–April.

99. *Matricària* L.

Annual or perennial herbs with alternate pinnatifid or finely dissected lvs. Heads small, solitary or somewhat corymbed, radiate or discoid; rays white, ♀, usually fertile, or wanting. Phyllaries dry, of ca. 2–3 series, ± imbricated, with scarious margins. Receptacle naked, ± conic to hemispheric. Disk-corollas yellow, 4–6-toothed. Aks. 3–5-ribbed, wingless, generally nerved on the margin and ventrally, nerveless dorsally, glabrous or roughened. Pappus a short crown or none. Ca. 35 spp. of N. Hemis and S. Afr. (Latin, *matrix*, because used medicinally.)

1. **M. matricarioìdes** (Less.) Porter. [*Artemisia m.* Less. *M. suaveolens* (Pursh) Buch., not L.] Pineapple Weed. Branched erect aromatic annual 1–3 dm. high; lvs. 1–5 cm. long, 1–3-pinnatifid into linear or filiform segms.; heads several to many, rayless, the disk mostly 4–10 mm. wide; phyllaries with broad hyaline margin, the invol. 5–7 mm. high, much shorter than the conical disk; aks. with 2 marginal and 1 rather weak ventral nerves; pappus a short crown, minute, entire; $2n = 18$ (Rutland, 1941).—Common weed in waste places through most of Calif.; to Alaska, Rocky Mts., L. Calif. Santa Catalina and Santa Rosa ids. Apparently introd. into e. states. May–Aug.

M. occidentàlis Greene of the San Joaquin and Sacramento valleys has been reported from s. Calif. The pappus-crown is 2-lobed with elliptic or oblanceolate glands.

100. *Melampòdium* L.

Herbs with opposite ample mostly sessile lvs. Heads radiate, small or medium-sized, pedunculate, in leafy cymes. Invol. 2-seriate, dimorphous, the outer phyllaries few, loose, often foliaceous, the inner smaller, completely enfolding the marginal aks. and deciduous with them as a thickened covering. Rays white or yellow, ♀, fertile. Disks yellow, perfect, sterile. Pappus 0. Less than a dozen spp. of chiefly subtrop. Am. (Greek, *melas*, black, and *pous*, foot, meaninglessly applied to these plants.)

1. **M. perfoliàtum** HBK. [*Alcina p.* Cav.] Coarse widely branched annual, 2–12 dm. high; lvs. deltoid-ovate or hastate, to oval-oblong, membranaceous, 5–15 cm. long,

scabrid, dentate, narrowed to a winged petiole, these connate in pairs; outer phyllaries 4–5, ovate, green, scabro-ciliate, 9–13 mm. long, united at base; rays ca. 10, yellow, shorter than the invol. and scarcely longer than the disk; $n=12$ (Turner & King, 1962).—Rare weed of waste lands in the Los Angeles V.; adv. from Mex. May–Nov.

101. Micròpus L.

Rather small, floccose-woolly annuals. Lvs. entire, alternate. Heads terminal, rather clustered, discoid, several-fld. Receptacle usually broader than long, rather flat. Phyllaries open, scarious, surrounding the fl.-bearing bracts of the receptacle; these woolly, conduplicate, each enclosing a fertile ♀ fl. of which only the corolla-tube and style are exserted through a slit. Cent. perfect fls. sterile and mostly naked. Aks. with a lateral corolla and style. Pappus wanting sterile cent. aks.; ak. obovate, laterally compressed. Several spp. of Eu., N. Am. (Greek, *micros*, small, and *pous*, foot.)

1. **M. califórnicus** F. & M. Stem slender, erect, usually simple, 5–35 cm. high, leafy; lvs. linear-oblong, short-acuminate, 5–15 mm. long; phyllaries commonly ca. 5, rounded to ovate, scarious, with a cent. green spot; receptacular bracts loose-woolly, 4–6, indurate in age, ca. 3.5 mm. long; sterile fls. ca. 3.—Common in dry open places below 5000 ft.; many Plant Communities; cismontane Calif.; to Ore. Channel Ids. April–June.

102. Micróseris D. Don

Annual to perennial herbs, from a taproot or rhizomatous or vertical caudex. Lvs. principally basal and rosulate, sometimes cauline, entire to pinnatifid. Infl. of 1 to many 1-headed naked or bracteate scapes. Heads 5–300-fld., usually nodding in the bud. Invol. of inner subequal lanceolate phyllaries and sometimes outer imbricate shorter ones. Receptacle alveolate. Corollas ligulate, yellow to white, often striped red. Aks. truncate-columnar to slender-fusiform, not beaked, ca. 10-ribbed. Pappus of 2–60 silvery to sordid, mostly persistent paleae tapering into bristlelike awns. Ca. 15–20 spp. of w. N. Am., Chile, Australasia. (Greek, *micros*, small, and *seris*, a lettucelike plant.)

(Chambers, K. L. A biosystematic study of the annual spp. of M. Contr. Dudley Herb. 4: 207–312. 1955.)

Plants perennial with pale taproots; pappus of 5–10 parts paleaceous below and with a plumose awn above .... 5. *M. sylvatica*
Plants annual; pappus-parts 2–5.
  Heads always erect; pappus-paleae deciduous; awns filiform, white .... 4. *M. linearifolia*
  Heads nodding or semierect, at least in the bud; pappus-paleae persistent; awns stouter, brown or yellow.
    Paleae narrowly lanceolate, ± bifid at apex, one or more of outer phyllaries lanceolate after anthesis .... 3. *M. heterocarpa*
    Paleae not bifid at apex; outer phyllaries ovate or deltoid after anthesis.
      The paleae villous or scabrous, 2–5 in number, 5–13 mm. long .... 1. *M. douglasii*
      The paleae relatively smooth, 5 in no., 3.5–7 mm. long .... 2. *M. elegans*

1. **M. doúglasii** (DC.) Sch.-Bip. [*Calais d.* DC.] ssp. **doúglasii**. Acaulescent annual; lvs. 5–25 cm. long, glabrous or furfuraceous, entire to pinnatifid; scapes 1–6 dm. high; heads 5–100-fld., the inner phyllaries 5–18, subequal, lanceolate to lance-ovate, acute or acuminate, carinate, 5–15 mm. long; outer phyllaries 4–12, calyculate; aks. 4–10 mm. long, gray to strawcolor, brown or blackish, often dark-spotted; pappus parts 2–5, rarely 0, persistent, 5–13 mm. long, the basal paleaceous part 1–6 mm. long, round to ovate or lanceolate, tapering into a stout or slender barbellulate awn 3–8 mm. long; $2n=18$ (Chambers, 1955).—Grassy places, V. Grassland, Coastal Sage Scrub, etc.; much of cismontane s. Calif.; to s. Ore. San Nicolas Id. March–April.

Ssp. **platycárpha** (Gray) Chamb. [*Calais p.* Gray.] The paleae scabrous, 0.5–2 mm. shorter than the ak. (instead of 1–6 mm. as in ssp. *douglasii*); awn 1–4 mm. long.—Grassy places, Los Angeles Co. to L. Calif. Santa Catalina and San Clemente ids.

Ssp. **tenélla** (Gray) Chamb. [*Calais t.* Gray.] Paleae not more than 1 mm. long, as compared to more than 1 mm. in the 2 other ssp.; ak.-ribs linear, not flared out at the tip as in the other 2 ssp.—Malibu region; mostly n. of Santa Barbara.

2. **M. élegans** Greene ex Gray. Acaulescent annual; lvs. 5–20 cm. long, mostly toothed or pinnatifid; scapes 1–3.5 dm. high, mostly slender; heads 10–100-fld., inner phyllaries 5–15, lance-ovate, acute to acuminate, keeled on midrib, 4–8 mm. long, the outer calyculate, 6–12; aks. 1.5–3.5 mm. long, gray-brown to brown or almost black, not spotted; pappus-parts 5, persistent, 3.5–7 mm. long, the paleaceous base to 2 mm. long, tapering into a hairlike awn 3–5 mm. long; $2n = 18$ (Chambers, 1955).—Grassy flats and hills; largely V. Grassland; much of cismontane Calif.; to L. Calif. and n. Calif. Santa Cruz and San Clemente ids. April–May.

3. **M. heterocárpa** (Nutt.) Chamb. [*Uropappus h.* Nutt.] Annual, caulescent or acaulescent, glabrous or furfuraceous; lvs. 5–30 cm. long, linear to narrow-elliptic, entire to pinnatifid; scapes 1–6 dm. high; heads 10–125-fld.; invol. 1–3 cm. high, the inner phyllaries subequal, lanceolate or lance-ovate, often with thickened midrib, green or striped with red; outer phyllaries imbricate or very short; aks. 5–10 mm. long, brown to strawcolor, violet or gray, plain or dotted, the ribs lightly scabrous; pappus-parts 5, persistent, the broad paleaceous part 5–10 mm. long, apically toothed, ending in awn 4–8 mm. long; $n = 18$ (Chambers, 1955).—Grassy places, below 2000 ft.; V. Grassland, etc.; cismontane, n. to Butte Co.; Channel Ids.; L. Calif. March–May.

4. **M. linearifòlia** (DC.) Sch.-Bip. [*Calais l.* DC.] Plate 20, Fig. B. Subacaulescent glabrous or villous annual 1–6 dm. tall; lvs. basal and lower-cauline, 3–25 cm. long, linear to narrow-elliptic, acuminate, entire or with a few retrorse teeth or pinnatifid; heads terminal on scapose peduncles; invol. 15–30 mm. high, the inner phyllaries 3–18, somewhat unequal, lanceolate to lance-ovate, green or striped red, glabrous or with a few black hairs; outer phyllaries 2–8, imbricate; aks. 8–15 mm. long, black or dark brown, fusiform, narrowed upward; pappus of 5 narrow scales 6–20 mm. long, each 2-cleft at apex and slender-awned from between the lobes; $2n = 18$ (Stebbins et al., 1953).—Grassy and open places below 6000 ft.; many Plant Communities; cismontane and w. Mojave Desert; to Wash., Utah, New Mex., L. Calif. April–June.

5. **M. sylvática** (Benth.) Gray. [*Scorzonella s.* Benth.] Perennial with 1–few stems 3–6 dm. high and arising from a tuft of lvs. above the root-crown; glabrous to densely mealy-puberulent; lvs. erect, linear to lanceolate, mostly laciniate-pinnatifid; heads solitary; invol. 16–20 mm. high; ligules light yellow, with a median pinkish line; aks. ca. 10-ribbed, 6–9 mm. long; pappus brownish or sordid, 12–14 mm. long; $2n = 18$ (Stebbins et al., 1953).—Grassy places, Tehachapi Mts., n. through Central V. March–May.

103. *Monolòpia* DC.

Erect floccose-woolly annual herbs. Lvs. opposite below, the upper alternate, sessile. Heads rather large, pedunculate, terminating the branches. Invol. hemispheric or campanulate; phyllaries distinct to base or connate into a cup, the tips usually with some black hairs. Receptacle high-conic. Ray-fls. ♀, fertile, the corolla with a minute posterior lobe. Disk-fls. perfect, fertile, the tube glandular-hairy, the throat dilated. Aks. oblanceolate, 3- to 4-angled, epappose. Four spp., all in Calif. (Greek, *mono*, one, and *lopos*, covering, referring to the uniseriate invol.) (Crum, Ethel. A revision of the genus M. Madroño 5: 250–270. 1940.)

1. **M. lanceolàta** Nutt. Stems 1–6 dm. high, simple to branched, lanate but glabrate below; lower lvs. broadly lanceolate, 3–10 cm. long, the upper lanceolate to linear; heads mostly terminal, on peduncles 1–12 cm. long; invol. 6–10 mm. high; phyllaries ca. 8, free, lanceolate to lance-ovate; rays mostly 8, bright yellow, 10–20 mm. long; aks. gray-strigose, 2.2–3.8 mm. long; $n = 10$ (Carlquist, 1956).—Grassy places below 5000 ft.; V. Grassland, S. Oak Wd., Chaparral, etc.; Riverside and Orange cos. n.; to Fresno Co.; occasional on w. Mojave Desert. March–May.

104. Monóptilon T. & G. Desert Star

Small, depressed desert annuals, branched from base, with hispid herbage. Lvs. alternate, linear to spatulate, entire. Heads radiate, daisylike, terminal on the branchlets, subtended by the upper lvs. Invol. nearly uniseriate; phyllaries equal, linear, firm, ± keeled below, herbaceous. Receptacle flat, naked. Ray-fls. ♀, showy; lamina elliptic, white to pinkish. Disk-fls. fertile, the corolla yellow or sometimes purplish. Aks. oblong-obovate, compressed, pubescent. Pappus of unequal bristles alternating with short paleae, or of a short scarious cup and one apically plumose bristle. Two spp. of w. deserts. (Greek, *monos*, single, and *ptilon*, feather, referring to the pappus of the original sp.)

Pappus a cup of minute scales and a single apically plumose bristle; disk-corollas densely pilose below .... 1. *M. bellidiforme*
Pappus of 1–12 nonplumose bristles, alternating with shorter lacerate paleae; disk-corollas glabrate to sparsely pilose below .... 2. *M. bellioides*

1. **M. bellidifórme** T. & G. ex Gray. Stems several from the base, slender, decumbent or ascending, white-hirsute throughout, 1–7 cm. long; lvs. 5–20 mm. long, spatulate or narrower; invol. 5 mm. high; rays 12–20, ca. 5 mm. long; aks. 2 mm. long; pappus a minute crown with a single long plumose awn.—Uncommon, sandy desert, 2000–4000 ft.; Creosote Bush Scrub; n. border of Colo. Desert, Riverside Co. to Inyo Co., Mojave Desert; w. Nev., Utah, Ariz. March–June.

2. **M. bellioìdes** (Gray) Hall. [*Eremiastrum b.* Gray. *E. orcuttii* Wts.] Plate 20, Fig. C. Like the preceding sp., but often larger; rays to 7 mm. long; pappus of few to many unequal, tawny, merely scabrid bristles and as many deeply and finely lacerate white squamellae half as long; $n=8$ (Raven et al., 1960).—Abundant on sandy and stony desert plains, below 3000 ft.; Creosote Bush Scrub; both deserts; Ariz., Son. Feb.–May; Sept.

105. Munzothámnus Raven

Shrub 1–1.5 m. tall; lvs. in tufts at ends of branches, large, coarsely sinuate or lobulate. Invols. 9–12-fld., narrow. Florets all ligulate, purple. Pappus setae thick, few, plumose. One sp. from San Clemente Id. (*Munz*, and *thamnos*, shrub.)

1. **M. blaìrii** (M. & J.) Raven. [*Stephanomeria b.* M. & J. *Malacothrix b.* M. & J.] Coarse straggly shrub; flowering branches woody, 8–10 mm. thick; lvs. crowded, obovate to oblong-obovate, 5–13 cm. long, glabrate, coarsely sinuate or lobulate; infl. paniculate, leafless, 1–2 dm. long; heads many, 9–12-fld.; invol. ca. 6 mm. high; ligules light purple, ca. 10 mm. long; aks. 3–3.5 mm. long; pappus-bristles many, ca. 4 mm. long, plumose.—Rocky canyon-walls; Coastal Sage Scrub; San Clemente Id. July–Sept.

106. Nicollètia Gray

Perennial herbs with slender woody rootstocks atop a deep-seated root; herbage glabrous, glaucous, somewhat succulent. Lvs. alternate, pinnately parted into linear gland- and bristle-tipped lobes. Heads large, terminating the branches. Invol. turbinate; phyllaries oblong, abruptly acute, with a prominent gland near tip, all equal and uniseriate, or with a smaller and shorter calyculate series of subtending bracts. Receptacle convex. Ray-fls. few, fertile, purplish or pinkish. Disk-fls. many, fertile, yellow, aging pink. Aks. linear-rusty-pubescent. Pappus double, the outer of many capillary bristles, the inner of 5 attenuate hyaline-margined awn-tipped paleae. Five spp. of sw. U.S. and Mex. (Named for J. N. *Nicollet*, early Am. astronomer and explorer.)

1. **N. occidentàlis** Gray. Plate 20, Fig. D. Stout, ill-scented, 1–4 dm. high; lvs. 3–5 cm. long; heads nearly sessile among the upper lvs.; phyllaries 8–12, hyaline-margined, 11–15 mm. long; rays normally 8, narrow, lurid-purple, 5–7 mm. long, striped dorsally with pink; $n=10$ (Raven & Kyhos, 1961).—Sandy washes and flats, 2500–4500 ft.;

Creosote Bush Scrub, Joshua Tree Wd.; w. borders of the Colo. and Mojave deserts, San Diego Co. to ne. Kern Co. April–June.

107. *Palafóxia* Lag.

Annual or perennial. Lvs. mostly alternate, entire. Heads corymbose, narrow, discoid. Phyllaries subequal, mainly uniseriate, herbaceous. Receptacle small, flat. Disk-fls. all alike or the outer with very unequal lobes; corolla with short tube, long throat, narrow-lobed limb. Aks. narrow-tetragonal, nearly as long as the invol., pubescent. Pappus-paleae 4–8, slender, unequal, with an excurrent nerve, nearly as long as the ak., hence well exserted. Less than a dozen spp., natives of s. U.S. and adjacent Mex. (Named for J. *Palafox*, Spanish general, defender against the armies of Napoleon.)

1. **P. lineàris** (Cav.) Lag. var. **gigántea** Jones. [*P. l.* var. *arenicola* Nels.] Annual or becoming perennial and woody at the base, 1–2 m. high, the herbage green and scabrous; lvs. lanceolate, 3-nerved; invol. 15–20 mm. high, eglandular; fls. 25 or more; pappus-paleae 8, of 4 long and 4 short; $n=12$ (Raven & Kyhos, 1961).—Common in sand hills west of Yuma, Imperial Co.; Creosote Bush Scrub. March–May.

Var. **lineàris.** [*Ageratum l.* Cav. *Stevia l.* Cav.] SPANISH NEEDLES. Plate 20, Fig. E. Erect, branching, 2–7 dm. high, hispid or scabrous, glandular above; lvs. ± canescent, linear or lance-linear, tapering to base and apex, 2–6 cm. long; invol. 13–18 mm. long; phyllaries linear, acuminate; fls. 10–20, the corolla white, styles exserted, pink; aks. strigose, the cent. ones longer, equaling the invol.; pappus-paleae usually 4; $n=12$ (Raven & Kyhos, 1961).—Sandy places, below 2500 ft.; Creosote Bush Scrub, Alkali Sink; Mojave and Colo. deserts, but more abundant on the latter; to Utah, Ariz., Mex. March–May.

108. *Péctis* L.

Low, slender-stemmed, aromatic, annual or perennial herbs. Lvs. opposite, entire, dotted with oil-glands, usually ciliate with stiff setae toward the base. Heads small, radiate, yellow, solitary or cymose. Invol. 1-seriate. Ray-fls. ♀, fertile, the rays often tinged purplish. Style-branches short, obtuse, hispidulous. Aks. slender, terete or ± angled. Pappus various, of scales, awns, or bristles, or reduced to a mere crown. Ca. 70 spp., from sw. U.S. and Fla. through Mex. to S. Am. (Greek, *pecteo*, to comb, the lvs. of most spp. being pectinately ciliate.)

1. **P. pappòsa** Harv. & Gray ex Gray. CHINCH WEED. Plate 20, Fig. F. Slender dichotomously branched yellowish-green annual 1–2.5 dm. high, heavy-scented; lvs. narrowly linear, 1–6 cm. long, 1–2 mm. wide, with 2–5 pairs of marginal bristles near base and prominent marginal oil-glands; heads ± clustered in leafy cymes, yellow-fld.; invol. 5 mm. high; phyllaries 7–10, linear, strongly round-keeled and clasping the ray-aks., dotted with oil-glands; aks. sparsely hispidulous; pappus of disk-fls. of 15–20 short weak tawny plumose bristles, of ray-fls. coroniform from very short united bristles, or obsolete; $n=12$ (Raven & Kyhos, 1961).—Flats below 5000 ft.; Creosote Bush Scrub, Joshua Tree Wd.; Death V. region to Needles, Colo. Desert; to Utah, New Mex., n. Mex. Usually appearing after summer rains.

109. *Perèzia* Lag.

Perennial herbs, ours branched and with leafy stems. Lvs. alternate, sessile and often clasping; usually spinulose-toothed. Heads many, cymose-paniculate in ours. Receptacle flat, usually naked. Invol. imbricated. Fls. all perfect, bilabiate, the upper lip deeply 2-lobed, the lower broad, 3-toothed at apex. Aks. subcylindric or fusiform, densely glandular or glandular-hispidulous. Pappus of many scabrous bristles. Perhaps 70 spp. of warmer parts of Am. (L. *Perez*, 16th-century medical botanist of Toledo.)

1. **P. microcéphala** (DC.) Gray. [*Acourtia m.* DC.] Rather stout, ca. 1 m. high; lvs. scabrous-puberulent and minutely glandular, 1–2 dm. long, 3–8 cm. broad, sessile by broad clasping base, denticulate; panicle 3–5 dm. long, 1–3 dm. broad; invol. 7–9 mm.

high, the phyllaries oblong, abruptly acuminate or mucronate; corollas rose to white, ca. 1 cm. long; pappus white, soft, ca. 8 mm. long.—Common on dry slopes, below 4000 ft.; Chaparral; San Luis Obispo Co. to San Diego Co.; Channel Ids. June–Aug.

110. *Pericome* Gray

Tall branching perennial herbs with minutely puberulent and glandular-punctate herbage. Lvs. opposite, petiolate, hastate or deltoid-lanceolate, often caudate-acuminate. Heads in terminal corymbiform cymes, discoid, yellow, the fls. much exserted. Phyllaries thin, uniseriate, loosely united into a cup. Receptacle small, low-conic. Fls. perfect, the slender cylindric corolla ± glandular throughout, 4-toothed. Anthers much exserted. Style-branches slender, elongate. Aks. black, flattened, hirsute on the callus margin. Pappus a crown of lacerate-ciliate squamellae, sometimes with a pair of short marginal awns. Two spp., of sw. U.S. and Mex. (Greek, *peri*, around, and *come*, a tuft of hairs, referring to the ak.-margin.)

1. **P. caudàta** Gray. Widely spreading aromatic herb up to 2 m. high, suffrutescent at base; lvs. deltoid-lanceolate or subhastate, usually long-caudate, 3–5 cm. long, on petioles 1–5 cm. long; invol. broadly campanulate, 5–6 mm. high; phyllaries 20–25, linear, their tips tomentose within; corollas 4–4.5 mm. long; aks. narrow, 4–5 mm. long; pappus-paleae ca. 1 mm. long; $n=18$ (Turner & Flyr, 1966).—Dry stony canyons, 4000–7500 ft.; Pinyon-Juniper Wd.; Inyo-White Mts., mostly e. face Sierra Nevada to Tulare Co.; to Okla., Tex., Chihuahua. July–Oct.

111. *Perityle* Benth.

Annual herbs or low branched subshrubs, subglabrous to tomentose, often glandular and aromatic. Lvs. opposite at least below, the upper often alternate, petioled, small, toothed, lobed or parted, sometimes entire. Heads small or medium-sized, radiate or discoid, yellow or the rays white. Invol. hemispheric to turbinate, 1–2-seriate; phyllaries ± boat-shaped. Receptacle flat or low-convex, naked, alveolate. Ray-fls. ♀, fertile, yellow or white. Disk-fls. fertile, numerous, yellow, the tube glandular-puberulent, the limb 4-toothed. Aks. mostly flattish, sometimes ciliate on the margin which may be calloused. Pappus of a crown of squamellae which may be inconspicuous and vestigial and with or without 1–2 (–3) barbellate bristles. Genus (here including *Laphamia*) of ca. 50 spp., largely of sw. U.S. and n. Mex. (Greek, *peri*, around, and *tyle*, a callus, the aks. often callus-margined.)

Rays present; pappus of laciniate hyaline squamellae and often a slender bristle, the scarcely thickened margin of the ak. densely short-ciliate with stiff tawny hairs. Widespread below 3000 ft., deserts and along the coast .... 1. *P. emoryi*
Rays absent; pappus wanting; aks. with a thickened margin which is sparsely short-pubescent. Desert plants at 5000–9000 ft.
    Stems scabrous-puberulent .... 3. *P. megalocephala*
    Stems pilose-villous to loosely villous.
        Lvs. lance-ovate, 8–12 (–20) mm. long, serrate-lobed. Cerro Gordo Peak, Inyo Mts.
        .... 2. *P. inyoensis*
        Lvs. broadly ovate, 12–22 mm. long, entire or with 1–3 short pointed lobes on each margin. Hanaupah Canyon, Panamint Mts. .... 4. *P. villosa*

1. **P. émoryi** Torr. in Emory. Plate 21, Fig. A. Winter annual 1–4 dm. high with ascending brittle branches, glandular-puberulent and sparsely hirsute; lvs. mostly alternate, broadly cordate to ovate, deeply toothed to palmately lobed, the lobes laciniate-toothed, 1–4 cm. long with petioles as long; invol. 5–6 mm. high, biseriate, the oblanceolate carinate thin-margined glandular-puberulent phyllaries ciliate at tip; rays 10–13, white, rather inconspicuous; aks. arcuate-oblanceolate, flat, black, those of the ray puberulent on the faces, the scarcely thickened margin densely short-ciliate with stiff tawny hairs, 2.5–3 mm. long; pappus a crown of very small lanceolate squamellae and

a slender awn ca. as long as disk-corolla, or the awn wanting; $n = 51-54$ (Powell, 1968).—Crevices of cliffs and among boulders on dry desert slopes up to 3000 ft.; Creosote Bush Scrub; e. Mojave Desert and throughout Colo. Desert from s. Inyo Co. to Imperial and San Diego cos.; Coastal Sage Scrub; Santa Monica Mts. near the coast, Ventura and Los Angeles cos.; Channel Ids.; to s. Nev., sw. Ariz., Sinaloa, L. Calif. Feb.–June.

2. **P. inyoénsis** (Ferris) Powell. [*Laphamia i*. Ferris.] Plants 1.2–2.5 dm. high, in dense leafy clumps, pilose-villous; lvs. 8–12 (–20) mm. long, 6–12 (–15) mm. wide, pilose-villous and with short glandular hairs, ovate to roundish, serrate-lobed; petioles 5–20 (–50) mm. long; heads solitary or in small clusters of 2–3; peduncles usually 8–40 mm. long; invols. campanulate, the heads 7–8.5 (–9) mm. high; phyllaries 14–21, sublanceolate to linear-lanceolate, keeled; rays none; disk-fls. 35–60, yellow, the corollas 4–5 mm. long, with acute lobes; aks. mostly 3–3.3 mm. long, rounded or angled on one or both surfaces, with thin callous margins, pubescent on margins and both surfaces; pappus absent except for an inconspicuous callous crown; $n =$ ca. 18 (Powell).—At 5900–8500 ft.; Pinyon-Juniper Wd., etc.; about Cerro Gordo Peak, Inyo Mts. Spring–Fall.

3. **P. megalocéphala** (Wats.) Macbr. var. **megalocéphala**. [*Laphamia m*. Wats.] Plants 30–55 cm. high, in ± dense suffrutescent clumps, much branched, densely hirtellous; lvs. 7–17 mm. long, 4–9 (–12) mm. wide, ovate to suborbicular, the margins entire, seldom serrate or ± lobed, acute to rounded at apex; petioles 1–5 (–6) mm. long; heads 1–few, loosely aggregated; peduncles 1–5 (–8) cm. long; invols. campanulate; heads 6–9 (–10) mm. high; the phyllaries 14–19, 5–6 mm. long; rays none; corollas yellow, 3.5–4.2 mm. long; aks. 2.5–3 mm. long, rounded or sometimes angled on one or both surfaces, with thin callous margins, pubescent on margins and both surfaces; pappus none, of a callous crown or of vestigial hyaline squamellae especially evident above the ak.-margins; $n = 17$ (Powell).—Dry cliff-crevices at 5000–9000 ft.; Pinyon-Juniper Wd.; Inyo-White, Panamint and Cottonwood mts., Inyo Co; w. Nev.

Var. **oligophýlla** Powell. Plant 15–35. cm. tall, densely hirtellous, much branched; lvs. 7–17 mm. long, 1–4 (–6) mm. wide, lanceolate to linear; heads solitary or in 2's or 3's; peduncles 2–5 cm. long; $n = 34$ (Powell).—At 6000–8500 ft.; Pinyon-Juniper Wd.; Panamint, Cottonwood mts. Spring–Fall. Material referred here has largely passed as var. *intricàta*, a Nevada plant.

4. **P. villòsa** (Blake) Shinners. [*Laphamia v*. Blake.] Plants 13–20 cm. high, villous; lvs. 12–22 mm. long, 4–10 mm. wide, villous, ± ovate, entire or with 1–3 short pointed lobes on each margin; petioles 3–6 mm. long; heads solitary or in 2's or 3's; peduncles 10–25 mm. long; heads 7.5–9.5 mm. high; phyllaries 13–23, oblong-lanceolate to linear-lanceolate; rays none; disk-fls. 30–75, yellow, the corollas 4–5 mm. long, with acute lobes; aks. 3–3.5 mm. long, typically rounded on both surfaces, with callous margins, short-pubescent on the margins and both surfaces; pappus a conspicuous to inconspicuous callous crown, or in some heads of 1–2 bristles 1–2 mm. long; $n = 10-22$ II plus 3–19 I. (Powell).—Known only from Hanaupah Canyon, Panamint Mts.

112. *Petradòria* Greene

Suffrutescent herbs with woody caudex; stems several, annual, leafy to the infl., resinous, prominently striate; lvs. linear to lanceolate or oblanceolate, 3–5-parallel-veined, coriaceous, entire; infl. open, racemose to corymbose; invol. cylindric, the phyllaries in ± vertical ranks, keelless or nearly so, ovate-oblong, fls. 4–7, yellow, the rays present or absent; aks. somewhat compressed, glabrous; pappus brownish, the bristles somewhat unequal, capillary, finely twisted. Two spp. of sw. U.S. (Greek, *petra*, a rock, and *Doria*, an early name for the Goldenrod.)

Ray-fls. absent; invol. 11 or more mm. high .... 1. *P. discoidea*
Ray-fls. present; invol. 5–9 mm. high .... 2. *P. pumila*

1. **P. discoìdea** L. C. Anderson. [*Chrysothamnus gramineus* Hall.] Many-stemmed from a branched woody caudex, 2.5–6 dm. high, light green, essentially glabrous, the striate-

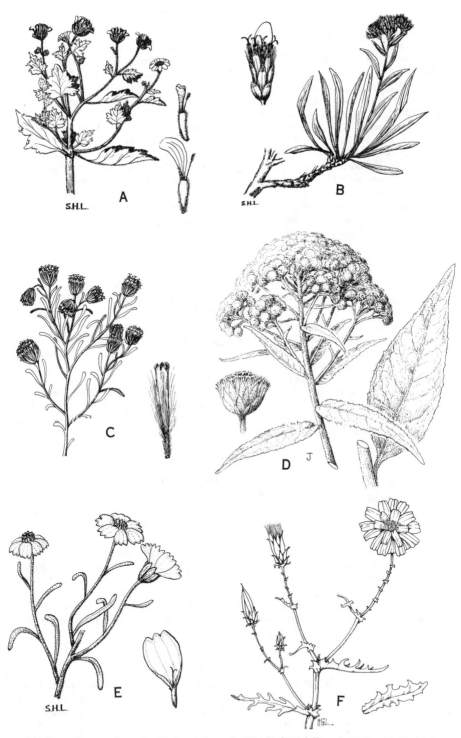

Plate 21. Fig. A, *Perityle emoryi*. Fig. B, *Petradoria pumila*. Fig. C, *Peucephyllum schottii*. Fig. D, *Pluchea purpurascens*. Fig. E, *Psilostrophe cooperi*. Fig. F, *Rafinesquia neomexicana*.

angled stems simple or erect-branched above; lvs. narrowly linear-lanceolate, 3–7 cm. long, 3–8 mm. wide, acuminate, 3–5-ribbed, coriaceous; heads mostly subsessile in small terminal clusters, pale yellow; invol. 10–13 mm. high; phyllaries 4–6-seriate, stramineous, oblong, abruptly cuspidate at the truncate or retuse ciliolate apex; corolla ca. 10 mm. long; aks. glabrous, ca. 6 mm. long; $2n = 18$ (L. Anderson, 1964).—Rocky wooded slopes, 7500–9500 ft.; Pinyon-Juniper Wd., Bristle-cone Pine F.; Inyo and Panamint Mts., Inyo Co.; Charleston Mts., Nev. July–Aug.

2. **P. pùmila** (Nutt.) Greene. [*Chrysoma p.* Nutt. *Solidago p.* T. & G., not Crantz.] Plate 21, Fig. B. Rather rigid flat-topped plant with numerous simple erect leafy stems from the caudex, 1–2.5 dm. high, glabrous, resinous, light green; lvs. crowded in basal rosettes, fewer above, obscurely punctate, 3–5-nerved, the basal mostly linear to oblanceolate, slenderly petiolate, 5–8 cm. long, 3–7 mm. wide, the cauline sessile, reduced to bracts upward; invol. 5.5–8 mm. high; phyllaries lance-oblong to lanceolate, apiculate, glabrous; ligules 2 mm. long; $2n = 36$ (Anderson, 1964).—Dry, stony, often limestone hillsides, 3500–7000 ft.; Pinyon-Juniper Wd.; Clark, New York and Providence mts., e. Mojave Desert; to Wyo., Tex. July–Oct.

113. *Peucephýllum* Gray. Pigmy-Cedar

Aromatic shrub, rather openly branched and with crowded alternate terete resinous-punctate lvs. near the ends of the branches. Heads solitary, discoid; invol. broadly campanulate; phyllaries in 2 series, linear-subulate, the outer subterete, the inner flattened. Receptacle flat, naked. Fls. perfect, fertile. Corolla-tube short, throat cylindric; lobes short. Aks. oblong-obconic, obscurely 10-striate, appressed-hirsute. Pappus of many scabrous bristles or some of them flattened and with hyaline margins. One sp. (Greek, *peuke*, the fir, and *phyllon*, lf.)

1. **P. schóttii** (Gray) Gray. [*Psathyrotes s.* Gray. *Inyonia dysodioides* Jones.] Plate 21, Fig. C. Plant 0.5–2.5 m. high with trunklike stem; lvs. bright green, 5–25 mm. long; heads scattered, subsessile; invol. ca. 8 mm. high, bright green; fls. yellow; aks. slender, 4 mm. long, dark brown; pappus 3–4 mm. long, tawny; $n = 20$ (Orndorf, 1963).—Rocky places, canyons, etc. below 3000 ft.; Creosote Bush Scrub; Death V. region through e. Mojave Desert to Colo. Desert; Nev., Ariz., L. Calif. Dec.–May.

114. *Pícris* L. Ox Tongue

Coarse rough-bristly annuals or biennials with numerous yellow fls. in the terminal heads. Stems leafy. Outer phyllaries loose or spreading. Aks. with 5–10 rugose ribs; pappus of 1–2 rows of plumose bristles. Perhaps 50 spp. of Old World. (Greek, *picros*, bitter, the name of some allied herb.)

1. **P. echioìdes** L. Three to 8 dm. high; lvs. oblong to oblanceolate, 5–20 cm. long, sessile, coarsely toothed; heads short-peduncled or along the branches; outer phyllaries ovate, subcordate, the inner long-acuminate, ± spinose-pinnatifid toward tip, becoming thickened below, 15–20 mm. long; aks. beaked, the body oblong, brownish, ca. 3 mm. long, plainly rugose; pappus densely plumose; $2n = 10$ (Schnack & Covas, 1947).— Well established weed in waste places, fields, etc., especially in heavy soil; from Medit. region. Catalina Id. June–Dec.

115. *Pleurocorònis* King & Robinson

Woody subshrub. Lvs. simple to compound, opposite in lower portion of plant, becoming alternate above. Peduncles not strongly differentiated, with alternate lvs. Infl. polycephalic. Heads many-fld. Phyllaries in multiple series, outer ones short-ovate; inner lanceolate. Corolla slender, tubular with glandular hairs on the outer surface. Anthers elongate, with large erect appendages. Style-branches elongate, slightly enlarged at tip, cells at surface slightly bulging; basal node of style glabrous. Pappus of 3–6 elongate scabrate setae with intervening short erose squamae, sometimes bearing many additional setae in a distinct inner series. Aks. 5-ribbed, lateral surfaces densely pubescent. One sp. (Greek, *pleuro*, rib or side, and Lat., *corona*, crown.)

1. **P. plurisèta** (Gray) King & Robinson. [*Hofmeisteria p.* Gray.] ARROW LEAF. Low rounded shrubs mostly 3–8 dm. high, intricately much branched; older stems with white shreddy bark; young growth glandular-puberulent; lf.-blades 4–10 mm. long, deltoid-lanceolate, entire or few-toothed; petioles 2–4 cm. long; heads few, in small terminal corymbs; invol. 7–9 mm. high; phyllaries 3-striate; corollas whitish; stigmas purplish at tips; aks. 2–3 mm. long; $2n = 18$ (Raven in King, 1967).—Common in rocky places below 4000 ft.; Creosote Bush Scrub; both deserts; to Utah, Ariz. March–May.

116. *Plùchea* Cass.

Tall leafy shrubs or herbs with alternate lvs. Heads many, in terminal corymbose infl., discoid, with numerous purplish fls. Marginal fls. ♀ and perfect, the corolla narrowly tubular, truncate, entire or 2–3-toothed; style 2-cleft. Cent. fls. few, perfect, but sometimes sterile, with 5-cleft corolla. Invol. imbricate. Receptacle flat, naked. Aks. grooved; pappus of 1 series of capillary bristles. A genus of some size, from warmer regions. (N. A. Pluche, Parisian 18th cent. naturalist.)

Plant a glandular-puberulent herb; lvs. narrowly ovate .... 1. *P. purpurascens*
Plant a sericeous shrub; lvs. lance-linear to elliptic .... 2. *P. sericea*

1. **P. purpuráscens** (Sw.) DC. [*Conyza p.* Sw. *P. camphorata* Calif. refs.] MARSH-FLEABANE. Plate 21, Fig. D. Annual to perennial, erect, branched above, 3–12 dm. high, green; lvs. ovate to lanceolate, glandular-dentate, 5–10 cm. long, 2–3 cm. wide, the lower petioled; corymbs large; invol. chartaceous, campanulate, ca. 5 mm. high; phyllaries lance-ovate; pappus of all the fls. similar, setaceous, without dilated tips; $n = 10$ (Turner et al., 1902).—Occasional in ± alkaline wet places, at low elevs.; Coastal Salt Marsh, Freshwater Marsh, etc.; cismontane s. Calif. n. to cent. Calif.; deserts from Inyo Co. s.; to Atlantic Coast. July–Nov. Santa Catalina and Santa Cruz ids.

2. **P. serícea** (Nutt.) Cov. [*P. borealis* Gray. *Polypappus s.* Nutt.] ARROWWEED. Slender-stemmed willowlike shrub 1–4 m. high; herbage silvery-silky; lvs. entire, lanceolate, 1–4 cm. long, 3–6 mm. wide, acute, sessile; invol. 5–6 mm. high; outer phyllaries ovate, coriaceous, the inner narrower, less firm; pappus of perfect fls. with clavellate tips.—Frequent in wet places like river-bottoms; Coastal Sage Scrub, Creosote Bush Scrub; cismontane s. Calif. from n. Santa Barbara Co. to L. Calif.; deserts, e. to Tex. March–July.

117. *Porophýllum* (Vaill.) Adans.

Herbs or subshrubs with glabrous foliage. Lvs. alternate or opposite, simple, with translucent oil-glands near the margin. Heads discoid. Invol. cylindric or campanulate, uniseriate, the phyllaries 5–9, equal, free or united only at base, usually with 2 rows of oil-glands, without accessory bractlets. Fls. perfect, fertile, purplish or yellow. Style-branches elongated, subulate, hirtellous. Aks. slender, striate, often tapering to apex. Pappus of numerous scabrous bristles. Ca. 30 spp., native from sw. U.S. and Mex. to Panama and S. Am. (Greek, *poros*, a passage or pore, and *phyllon*, lf., the translucent glands making the lvs. appear punctate.)

1. **P. grácile** Benth. Suffrutescent, much branched, 1.5–6 dm. high, the branches slender, sparingly leafy, ill-scented; lvs. bluish-green, filiform-linear, 1–4 cm. long; invol. 10–14 mm. high, often purplish, the phyllaries 5, oblong, purplish, obtuse, corolla 7–9 mm. long, purple or sometimes whitish with purple veins; aks. linear, hispidulous; pappus fuscous, of capillary bristles; $n = 24$ (Raven & Kyhos, 1961).—Dry rocky slopes, washes, mesas, sometimes alkaline, 200–5000 ft.; Creosote Bush Scrub, Coastal Sage Scrub; Clark Mt., e. San Bernardino Co. to Little San Bernardino Mts., s. through Colo. Desert, to the coast in San Diego and Orange cos.; to Ariz., Son. Oct.–June.

118. *Psathyròtes* Gray

Low herbs, aromatic, ours scurfy or tomentose, with numerous spreading-decumbent branches. Lvs. alternate, petioled, with broad rounded rather small blades. Heads solitary in the forks of the branches, peduncled, discoid; fls. all tubular, perfect, yellow, often turning purple in age. Phyllaries biseriate, the outer more herbaceous and often a little shorter than the inner. Receptacle flat, naked. Corolla-tube short; throat elongate, cylindric; lobes short, woolly or glandular. Aks. oblong-obconic, obscurely 10-striate, appressed-hirsute. Pappus of many scabrous capillary bristles, white to reddish. Four spp. of sw. N. Am. (Greek, *psathurotes*, brittleness, referring to the stems.)

Plant lanate-tomentose as well as scurfy; outer phyllaries much broader than the inner, the obtuse tips spreading or recurved .... 2. *P. ramosissima*
Plant scurfy-tomentose; outer phyllaries not very different from the inner, all with erect tips
1. *P. annua*

1. **P. ánnua** (Nutt.) Gray. [*Bulbostylis a.* Nutt.] Much like *P. ramosissima*, less lanate, more greenish; lvs. thinner; invol. 5–6 mm. high.—Dry sandy often ± alkaline places, below 6500 ft.; Shadscale Scrub, Alkali Sink, etc.; s. Mojave Desert to Mono Co.; to Utah, Ariz. June–Oct.

2. **P. ramosíssima** (Torr.) Gray. [*Tetradymia r.* Torr.] Compact round and rather flat plant 5–12 cm. high, 5–30 cm. broad, white-tomentose, largely annual, with strong turpentinelike odor; lvs. numerous, thick, coarsely and irregularly toothed, 6–20 mm. long, suborbicular to reniform; invol. 6–8 mm. high, the outer phyllaries 5 in number, the inner more membranous; corolla yellow to purplish, long-hairy above; aks. ca. 2 mm. long; pappus tawny, ca. 3 mm. long.—Common, hard dry soil of flats and ledges largely below 3000 ft.; Creosote Bush Scrub, both deserts, especially in e. part; to Utah, Ariz., nw. Mex. March–June, sometimes in winter.

119. *Psilocárphus* Nutt.

Small woolly annuals with opposite entire lvs. Heads disciform, small, round, terminal commonly subtended by a pair of branches. True invol. lacking although the heads are commonly subtended by foliage lvs. Receptacle subglobose to obpyriform. Pistillate fls. in several series, with short filiform-tubular corolla, each loosely enclosed in a saccate woolly bract which bears a hyaline appendage below the summit. Cent. fls. few, bractless, functionally ♂, sterile; pappus none. Aks. small, smooth, nerveless, turgid or compressed. Five spp. (1 in w. S. Am.). (Greek, *psilos*, bare, and *karphos*, chaff.)

(Cronquist, A. A review of the genus Psilocarphus. Res. Studies State Coll. Wash. 18: 71–89. 1950.)

Well developed receptacular bracts ca. 3 mm. long at maturity; lvs. lance-oblong to ± linear
1. *P. brevissimus*
Well developed receptacular bracts ca. 2 mm. long; lvs. spatulate or oblanceolate
2. *P. tenellus*

1. **P. brevíssimus** Nutt. [*P. globiferus* Nutt.] Usually prostrate and branching, the stems seldom up to 1 dm. long; herbage loosely and copiously woolly; lvs. 5–25 mm. long, generally broadest near their base and well surpassing heads; ♀ fls. 8–80, receptacular bracts mostly 2.5–4 mm. long; aks. obliquely oblanceolate, flattened, 1–2 mm. long.—Dried beds of vernal pools, below 3000 ft., or even much higher; many Plant Communities, especially V. Grassland; San Diego Co. to n. Calif.; Wash., Mont. April–June.

2. **P. tenéllus** Nutt. Slender annual, at first erect, later branched and prostrate; tomentum generally thin and rather loose; lvs. spatulate to oblong, 4–15 mm. long; ♀ fls. 25–46 in a head; receptacular bracts 2 mm. long; aks. 0.6–1.2 mm. long, moderately compressed, turgid, ± obovate.—Dried vernal pools, cismontane and montane, to 6000 ft.; n. L. Calif. to B.C. April–June. Channel Ids.

120. *Psilóstrophe* DC. PAPERFLOWER

Perennial herbs or low shrubs from a woody taproot, ± woolly. Lvs. alternate, linear or spatulate, entire, or the lower ones pinnatifid. Heads yellow, radiate, corymbosely arranged. Invol. cylindric to campanulate; phyllaries subequal, distinct, obscured by the thick tomentum. Receptacle small, naked. Ray-fls. ♀, the prominent persistent ligules becoming papery and paler in age. Disk-fls. perfect, somewhat hispidulous and glandular, the throat narrowly cylindric. Style-branches truncate-capitellate. Aks. slender, terete, obscurely angled. Pappus of 4–6 hyaline entire or laciniate paleae. Six spp. of sw. U.S. and n. Mex. (Greek, *psilos*, naked, and *strophe*, to turn.)

1. **P. coóperi** (Gray) Greene. [*Riddellia c.* Gray.] Plate 21, Fig. E. Shrubby, with many ascending stems from the woody caudex, 2–5 dm. high, white-lanose on younger parts; lvs. entire, linear, 3–7 cm. long; heads terminating the upper branches, pedunculate; invol. 5–8 mm. high; phyllaries 10–20, the outer much firmer than the inner; rays 4–8, broadly oval, 8–16 mm. long, shallowly 3-lobed at the broad apex; disk-fls. 5–20, often much exserted; aks. glabrous; pappus-paleae various, lance-oblong to ovate, fulvous, shorter than the aks.; $n = 16$ (Raven & Kyhos, 1961).—Rocky desert mesas and sandy fans, 2000–5000 ft.; Creosote Bush Scrub, Joshua Tree Wd.; e. Mojave Desert, San Bernardino Co. from Kingston and Clark mts. to Little San Bernardino Mts. and n. edge of Colo. Desert; to Utah, New Mex., n. Mex. April–June, Oct.–Dec.

121. *Pulicària* Gaertn.

Annual to perennial herbs, with alternate simple lvs. Heads many; invol. imbricated; phyllaries narrow, scarious, or the outermost broader and often herbaceous. Receptacle naked, flat or convex. Ray-fls. often with much reduced ligule. Disk-fls. tubular, perfect, yellow; anthers caudate, stigmatic branches flattened, slightly broader upward. Pappus 2-rowed, the outer row of short scales, free or joined into a cup, the inner of hairs. Aks. cylindrical, ribbed. An Old World genus of some size. (Latin, *pulicarius*, flea-like.)

1. **P. hispánica** (Boiss.) Boiss. Annual to perennial herb, to ca. 1 m. tall, branched, short-villous throughout; lvs. alternate, oblong to narrow-oblanceolate, clasping at base, 1–6 cm. long, entire; heads many, 7–10 mm. diam.; phyllaries linear, attenuate, 3–4.8 mm. long, villous; rays yellow, the ligules 1.5–2 mm. long; aks. cylindrical, 1 mm. long, brownish.—Adventive along streams, etc., from Orange and Riverside cos.; natur. from Medit. region.

122. *Rafinésquia* Nutt.

Glabrous or slightly puberulent ± fistulous or stoutish branching annuals with toothed or pinnatifid lvs. and ± cymosely branched panicles. Heads rather large, solitary at ends of branchlets, 15–30-fld.; invol. cylindro-conical of ca. 7–15 equal lanceolate acuminate scarious-margined phyllaries and of some shorter unequal outer ones. Receptacle flat, naked. Ligules white or slightly rose in tinge, unequal; fls. fragrant. Aks. subfusiform, tapering into a beak; pappus of 10–15 capillary bristles, long-plumose from base to tip. Two spp. (C. S. *Rafinesque*, early Am. naturalist.)

Rays white, 5–8 mm. long; ak.-beak slender, ca. as long as the body; pappus dull or brownish, the bristles plumose to the tip with straight hairs .... 1. *R. californica*
Rays white but veined outside with rose-purple, 10–15 mm. long; ak.-beak shorter than the body; pappus bright white, the bristles not plumose near the tips, with soft, subarachnoid hairs
.... 2. *R. neomexicana*

1. **R. califórnica** Nutt. Plants 2–15 dm. high, branched especially above; lvs. oblong in outline, 5–20 cm. long, subentire to pinnatifid, the lower petioled, the others auriculate-clasping, the uppermost much reduced; heads scattered, the invol. 15–18 mm. high; ligules white, 5–8 mm. long; $2n = 16$ (Stebbins et al., 1953).—Most frequent on burns and in disturbed places, at low elevs.; Coastal Sage Scrub, Chaparral, etc.; cismontane from n.

L. Calif. to n. Calif.; Channel Ids.; occasional in Creosote Bush Scrub, Joshua Tree Wd., deserts; to Utah, Ariz. April–July.

2. **R. neomexicàna** Gray. Plate 21, Fig. F. Stems rather weak; invol. 18–22 mm. high; ligules 15–18 mm. long; $2n=16$ (Stebbins et al., 1953).—Common in shade of shrubs, in canyons, etc. Creosote Bush Scrub, Joshua Tree Wd.; deserts to Utah, Tex. Feb.–May.

123. *Raillardélla* Gray in Benth. & Hook.

Perennial herbs with simple entire or subentire alternate or basal lvs. and solitary or few heads, these discoid or rayed. Phyllaries subequal, subherbaceous, uniseriate or nearly so. Receptacle naked, flat. Anthers subtruncate at base. Style-branches flattened, with introrsely marginal stigmatic lines and a slender hispidulous appendage. Aks. linear, subcompressed, several-nerved. Pappus of ± numerous plumose bristles somewhat flattened toward base. Ca. 5 spp. (Diminutive of *Raillardia*, a shrubby genus of Hawaii.)

1. **R. argéntea** (Gray) Gray. [*Raillardia a.* Gray.] Scapose perennial 1–10 cm. high, conspicuously silky-tomentose; lvs. mostly 1–6 cm. long; heads strictly discoid; invol. ca. 9–14 mm. high; phyllaries linear, acuminate; fls. yellow; pappus-bristles flattened.—Dry open rocky places, mostly 10,000–11,500 ft.; Subalpine F.; San Gorgonio Mt., San Bernardino Mts.; Sierra Nevada to Ore. July–Aug.

124. *Rigiopáppus* Gray

Slender wiry annuals with linear alternate entire lvs. Heads solitary at the ends of the simple stems or the filiform branches, radiate, many-fld., all fls. fertile. Invol. irregularly 2–3-seriate, the outer phyllaries concave and half-embracing the ray-aks., the inner narrower and shorter and sometimes encroaching upon the receptacle. Ray-fls. 1-seriate, very narrow, scarcely exceeding the disk; disk-corollas linear, 5-toothed. Aks. linear, transversely wrinkled, pubescent. Pappus of 3–5 rigid tapering awnlike scales, rarely wanting. One sp. (Greek, *rigios*, stiffened, and *pappos*, pappus.)

1. **R. leptoclàdus** Gray. Stems erect, 1–3 dm. high, sparsely pubescent, leafy, the branches naked below and overtopping the main stem; lvs. 1–3 cm. long, mostly less than 1 mm. wide; invol. turbinate, 5–8 mm. high; fls. yellow, usually suffused with reddish, the rays ca. 5–15, the disk-fls. ca. 5–35; aks. all alike, with short appressed hairs having swollen tips; pappus-scales ochroleucous, often with a purplish spot at the ± united bases; $n=9$ (Raven & Kyhos, 1961).—Dry places up to 7000 ft.; San Gabriel Mts. to Mt. Pinos, Tehachapi; then to Wash., Ida., Utah. April–June. Several Plant Communities.

125. *Santolìna* L.

Shrubs, sometimes herbs with aromatic herbage. Lvs. alternate, pinnately toothed or lobed or finely divided. Heads many-fld., discoid, yellow or rarely white, solitary on long peduncles. Invol. mostly campanulate; phyllaries appressed, imbricated. Receptacle with chaffy bracts. Aks. angled, without pappus. Ca. 8 spp., mostly of Medit. region. (Derivation of name uncertain.)

1. **S. chamaecyparíssus** L. LAVENDER-COTTON. Much-branched, evergreen, woody, 4–6 dm. high; herbage silvery-gray and tomentose; lvs. 1.5–3.5 cm. long, pinnately divided into small ovate-oblong segms.; heads yellow, subglobose, 12–15 mm. in diam.; phyllaries ± carinate, subtomentose.—Occasionally established, as near Santa Monica and in cent. Calif.; native of Medit. region. May–June.

126. *Sanvitàlia* Lam.

Low mostly branching annual herbs. Lvs. opposite, ± petioled, entire. Heads small, terminating the branches or subsessile in upper lf.-axils, radiate. Invol. 1–3-seriate, the phyllaries rather firm and dry. Receptacle strongly conical in fr. in ours, its chaffy bracts cuspidate-tipped. Ray fls. ♀, fertile, the ligules sessile and persistent on the ak. Disk-fls. perfect, fertile. Ray-aks. bearing apically 3–5 divaricate horny cusps or awns. Disk-aks. 4-angled, tuberculate, epappose or nearly so. Ca. 4 spp. of sw. U.S. and n. Mex. (Name honoring *Sanvitali*, a noble Italian family.)

    1. **S. abértii** Gray. Erect or rather diffuse annual 1–3 dm. high, the opposite ascending or divaricate branches slender, with rather long internodes; herbage scabrid-hispidulous with white curly appressed hairs to glabrate; lvs. linear-lanceolate, up to 7 cm. long and 10 mm. wide, short-petiolate; ray-fls. 6–11, the ligule yellow, green-veined dorsally; disk-fls. greenish-yellow; ray-aks. 3–4 mm. long, narrowly 4-sulcate, smooth to ± papillate, white, 3-cuspidate-awned; disk-aks. mostly 4-angled or -ribbed, warty, dark; $n = 11$ (Turner & Flyr, 1966).—Rocky slopes and dry washes, at 5400 ft.; Pinyon-Juniper Wd.; Clark Mts., e. San Bernardino Co.; to Tex., n. Mex. Aug.–Sept.

    S. procúmbens Lam. with semi-prostrate stems and orange ray-fls. has been found in an unused garden in Riverside. July. Natur. from Mex.

127. *Senècio* L. Groundsel. Ragwort

Herbs shrubs or vines, sometimes arborescent. Lvs. alternate or basal, entire to toothed or divided. Heads solitary to many, cylindric to campanulate or hemispheric, mostly many-fld., usually radiate; rays ♀ and fertile, or none. Phyllaries essentially equal, mostly uniseriate, often with smaller bractlets at their base. Receptacle flat or convex. Disk-fls. perfect, fertile. Anthers entire to minutely sagittate. Style-branches ± flattened, truncate. Aks. subterete, 5–10-nerved; pappus of many usually white soft bristles. Ca. 1000 spp., of very wide distrib. (Latin, *senex*, old man, because of the white pappus.)

1. Plants annual.
  2. Rays well developed, conspicuous. Native .... 5. *S. californicus*
  2. Rays lacking or very inconspicuous.
    3. Bracteoles at base of invol. well developed, black-tipped. Common introd. weed
                                                                                                    17. *S. vulgaris*
    3. Bracteoles at base of invol. none or poorly developed, not black-tipped. Natives.
      4. Phyllaries ca. 8. Near coast .... 1. *S. aphanactis*
      4. Phyllaries ca. 13. Deserts .... 12. *S. mohavensis*
1. Plants perennial.
  5. Stems climbing; lvs. ivylike. Introd. .... 11. *S. mikanioides*
  5. Stems not climbing; lvs. not ivylike.
    6. Lvs. pinnatifid into narrowly linear segms. or sometimes entire and linear; plants ± woody at base. Mostly from below Yellow Pine belt.
      7. Lf.-lobes obtuse, numerous; rays 6–8 mm. long; invol. 6–7 mm. high. Insular
                                                                                          10. *S. lyonii*
      7. Lf.-lobes acute, mostly few; rays 8–15 mm. long; invol. mostly 7–11 mm. high. Largely from mainland.
        8. Phyllaries mostly ca. 21; bractlets at base of invol. rather long and conspicuous
                                                                                           7. *S. douglasii*
        8. Phyllaries 8–13; bracteoles short, inconspicuous .... 15. *S. spartioides*
    6. Lvs. not narrowly linear or divided into linear segms.; plants largely herbaceous. Mostly from the Yellow Pine belt.
      9. Lvs. well distributed along the stems, the upper only slightly or gradually reduced.
        10. Stems 0.5–3 dm. long; lvs. mostly 1–2.5 cm. long. Subalpine ..... 8. *S. fremontii*
        10. Stems 5–15 dm. long; lvs. 4–20 cm. long.
          11. Lower lvs. with deltoid to subcordate base. At 6600–9500 ft. in s. Calif. mts.
                                                                                       16. *S. triangularis*
          11. Lower lvs. tapering at base. At 8000 ft., Inyo-White Mts. .... 14. *S. serra*
      9. Lvs. largely basal, the cauline rapidly reduced up the stem.
        12. Lvs. toothed, not lobed, crenate or pinnatifid; heads discoid ...... 2. *S. astephanus*
        12. Lvs. lobed, crenate or pinnatifid, sometimes entire; heads mostly radiate.
          13. Basal lvs. dissected or pinnatifid to lyrate.
            14. Stems solitary, mostly 3–8 dm. high; invol. ca. 8–11 mm. high. N. Los Angeles Co. to Tehachapi Mts., thence to Tulare Co. .... 4. *S. breweri*

14. Stems usually several, 1–3 dm. high; invol. mostly 5–8 mm. high. E. Mojave Desert to Death V., etc. .... 13. *S. multilobatus*
13. Basal lvs. entire or few-lobed, not usually long-pinnatifid or lyrate.
15. Heads several or rather many; lvs. nearly all tufted at base, the cauline few and much reduced; plants permanently white-woolly; invol. 4–8 mm. high.
16. Stems appearing subscapose, the lvs. nearly all tufted at base, the cauline lvs. few and abruptly much reduced. San Bernardino Mts. . 3. *S. bernardinus*
16. Stems not appearing subscapose; lvs. not all tufted at base, the cauline, though smaller than the basal, relatively better developed. Inyo Mountains 6. *S. canus*
15. Heads mostly 1–4; cauline lvs. progressively reduced; plant ephemerally white-tomentose, becoming subglabrous; invol. ca. 8–10 mm. high. Tehachapi, San Gabriel, San Bernardino mts. .... 9. *S. ionophyllus*

1. **S. aphanáctis** Greene. Slender annual, simple or branched, 1–2.5 dm. high, glabrous except for some woolly pubescence in infl.; lvs. small, sessile, 1–4 cm. long, coarsely toothed to pinnately lobed; heads several, narrow, especially near top; invol. ca. 5 mm. high, with ca. 8 2–4-nerved phyllaries not black-tipped; rays inconspicuous; aks. densely canescent; $n=20$ (Ornduff, 1963).—Dry open places, Coastal Sage Scrub, Chaparral; near the coast from San Diego Co. to cent. Calif. and n. L. Calif. Santa Cruz Id. Feb.–March.

2. **S. astéphanus** Greene. [*S. ilecetorum* A. Davids.] Erect coarse perennial herb 4–10 dm. tall, floccose-woolly; lvs. well distributed, sharply dentate, elliptic to broadly oblanceolate, thin, 1–3 dm. long, the lower petioled, the uppermost sessile and passing into leafy bracts; heads several to rather many, in compact or more often open infl., discoid, yellow; invol. 9–10 mm. high; phyllaries ca. 19–23, green-tipped; outer bracteoles few, short, narrow; aks. glabrous.—Steep rocky slopes, 2500–5000 ft.; Chaparral; s. face of San Gabriel and San Bernardino mts.; to San Luis Obispo Co. May–July.

3. **S. bernardínus** Greene. [*S. ionophyllus* var. *b.* Hall.] Perennial from a root-crown, the stems 1–several, 1–3 dm. tall; herbage mostly permanently white-tomentose; lvs. in a basal tuft, many, the blades round-ovate to spatulate-cuneate, 1–2 cm. long, subentire to ± toothed, on long slender petioles; cauline lvs. few, much reduced; heads ca. 3–20, in flat-topped rather open clusters; invol. campanulate, tomentose, 6–8 mm. long; rays yellow, 8–10, ca. 7–10 mm. long; aks. glabrous or hispidulous; $n=23$ (Ornduff, 1963).—Dry rocky slopes, 6400–7500 ft.; Yellow Pine F.; Bear and Holcomb valleys, San Bernardino Mts. May–July.

4. **S. bréweri** Davy. Perennial from a short caudex; stems largely solitary, 3–8 dm. high; herbage mostly glabrous or with tufts of tomentum in axils; lvs. thin, pinnatifid, mostly with 1–9 pairs of lateral segms., the terminal segm. enlarged, lobulate-dentate, 4–8 cm. long and almost as broad; petioles to as long as the blades; cauline lvs. ± reduced up the stem; heads several to many, in a flat-topped infl.; invol. ca. 8–11 mm. high; phyllaries ca. 13 or 21, rather wide; bracteoles small or 0; rays yellow, 8–15 mm. long; aks. glabrous; $n=23$ (Ornduff et al., 1963).—Open or brushy or ± wooded slopes, below 5500 ft.; Chaparral, Foothill Wd.; n. Los Angeles Co., Tehachapi Mts.; to Tulare Co. April–June.

5. **S. califórnicus** DC. Plate 22, Fig. A. Annual, the stems simple or branched from the base, glabrous or ± arachnoid-villous, 1–5 dm. high; lvs. narrowly oblong-lanceolate to -linear in outline, acute or obtuse, remotely toothed to pinnatifid, all but the lower with auriculate base, 1–6 cm. long; heads several in an open infl., rarely solitary; invol. 5–7 mm. high, quite naked at the base; phyllaries usually black-tipped; rays ca. 1 cm. long, usually several; aks. canescent.—Mostly open dry places below 3500 ft.; Coastal Strand, Coastal Sage Scrub, Chaparral, etc.; cismontane, n. L. Calif. to cent. Calif.; w. edge Colo. Desert. March–May.

6. **S. cànus** Hook. Perennial from a ± branching caudex; stems several, slender, 1–3 dm. high; herbage white-tomentose, sometimes ± glabrate in age, especially on upper surface of lvs.; lower lvs. ± tufted, round to elliptic, entire or rarely toothed, the blades mostly 1–4 cm. long, on petioles as long or shorter; middle and upper lvs. few, reduced;

Plate 22. Fig. A, *Senecio californicus*. Fig. B, *Senecio douglasii* var. *monoensis*. Fig. C, *Senecio multilobatus*. Fig. D, *Senecio triangularis*. Fig. E, *Solidago californica*. Fig. F, *Tetradymia comosa*.

heads several in flat terminal clusters; invol. 4–8 mm. long; phyllaries 13 (–17), green-tipped; bracteoles none or inconspicuous; rays mostly 5–8, yellow, 6–12 mm. long, rarely lacking; aks. glabrous, ca. 3 mm. long; $n = 23$, 46, 69 (Ornduff; Taylor et al.).—At ca. 9400 ft.; Bristle-cone Pine F.; Inyo Mts.; to B.C., Rocky Mts. May–Aug.

7. **S. doúglasii** DC. var. **doúglasii.** Shrubby, branched, bushy, ca. 1–1.6 m. high, leafy to the infl.; stems striate; herbage white-tomentose when young, the lvs. glabrate and gray-green above, whitish tomentose beneath; most lvs. 3–10 cm. long, pinnate usually into 5–9 linear lobes 2–6 cm. long and 1–2.5 mm. wide, or the upper lvs. 3-lobed or entire; principal lvs. often with axillary fascicles of smaller ones; infl. corymbiform, of several to many heads; invol. broadly turbinate, 8–10 mm. high, the longer phyllaries mostly 16–21, with quite well developed shorter bracteoles at their base; rays 10–13, yellow, showy, ca. 10–15 mm. long; aks. canescent, ca. 4 mm. long; $n = 20$ (Vasek).—Common in washes and dry stream beds, below 6000 ft.; Foothill Wd., Coastal Sage Scrub, Chaparral, V. Grassland, etc.; cismontane from n. L. Calif. to n. Calif.; w. parts of desert; Santa Cruz and Santa Catalina ids. June–Oct.

Var. **monoénsis** (Greene) Jeps. [*S. m.* Greene.] Plate 22, Fig. B. Less shrubby, often scarcely woody even at base, 3–10 dm. high, more yellowish-green, glabrous or nearly so; lvs. from scarcely dissected with long linear segms. to more so and with shorter ones; invol. ca. 7–8 mm. high, the outermost bracteoles quite conspicuous; aks. ca. 5 mm. long; $n = 20$ (Ornduff et al., 1961).—Dry slopes and washes, 2000–6500 ft.; Creosote Bush Scrub to Pinyon-Juniper Wd.; n. Colo. Desert, Mojave Desert n. to Mono Co.; to Utah, Ariz. Mostly March–May.

8. **S. fremóntii** T. & G. var. **occidentàlis** Gray. [*S. o.* Greene.] Doubtfully distinct from typical *fremontii*. Perennial from a branching caudex, the stems decumbent at base, rootstock-like, leafy, ca. 1–1.5 dm. long; lvs. ± succulent, glabrous, round-ovate to suborbicular, with blunt shallow teeth and ca. 1.5–2.5 cm. long; heads solitary at ends of branches, with short naked peduncles; invol. 8–12 mm. high, the principal phyllaries ca. 13, shorter bracteoles few; rays yellow, 6–10 mm. long; aks. mostly glabrous; $n = 20$ (Ornduff).—About rocks, 10,000–11,500 ft.; Subalpine F.; San Bernardino Mts.; Sierra Nevada to B.C., n. Rocky Mts. July–Aug.

9. **S. ionophýllus** Greene. Perennial from a root-crown, the stems several, 1.5–3.5 dm. tall, subglabrous to thinly tomentulose, leafy at base, sparingly so above; lower lvs. 2–5 cm. long, obovate to round or flabellate, few-toothed to lyrately pinnatifid; heads 1–few; invol. 8–10 mm. high; phyllaries ca. 15–21; rays yellow, 6–9 mm. long; aks. glabrous, 5–6 mm. long; $n = 23$ (Ornduff, 1963).—Dry places, 5000–9000 ft.; Montane Coniferous F.; Tehachapi and Piute mts., San Gabriel and San Bernardino mts. June–July. Plants with pinnatifid lvs. but growing with the typical few-toothed form have been called var. *sparsilobatus* (Parish) Hall. *S. s.* Parish.

10. **S. lyònii** Gray. Rather shrubby, 6–12 dm. high, branched, tomentose on younger parts, soon glabrate but with persistent tufts of wool in the axils; lvs. well spaced along the stems, 2–12 cm. long, 1–2-pinnate into linear or broadly linear segms. and lobes, green above, ± white-tomentose beneath, sessile and auriculate or petioled; infl. corymbiform, on a long bracteate peduncle-like stem; invol. 6–7 mm. high, broadly turbinate; phyllaries linear; rays yellow, 6–8 mm. long; aks. linear, canescent, almost 3 mm. long; $n = 20$ (Ornduff et al., 1961).—Coastal bluffs; Coastal Sage Scrub; Santa Catalina and San Clemente ids. March–June.

11. **S. mikanioìdes** Otto. GERMAN-IVY. Glabrous perennial with slender twining stems to 6 m. long; lvs. roundish-cordate, sharply 5–7-angled, 2–8 cm. long, ca. as wide; peduncle as long or longer; infl. condensed, pedunculate, axillary toward the summit of the stems; invol. 3–4 mm. long; principal phyllaries ca. 8; rays 0; aks. glabrous; $n = 10$ (Ornduff, 1963).—Natur. in canyons and gullies along coast of s. and cent. Calif.; native of S. Afr. Dec.–March.

12. **S. mohavénsis** Gray. Slender erect annual, mostly freely branched, 1.5–4 dm. tall, entirely glabrous; lvs. well distributed, broadly oblong to oblong-ovate, obtuse, 2–6 cm. long, obscurely dentate, mostly with broad clasping base; heads in loose corymbs;

invol. ca. 5–7 mm. high; phyllaries mostly ca. 13, linear; rays inconspicuous or wanting; aks. canescent; $n=20$ (Ornduff, 1963).—Occasional in washes and canyons below 2500 ft.; Creosote Bush Scrub; Colo. Desert, Death V. region; to Ariz., Son. March–May.

13. **S. multilobàtus** T. & G. [*S. stygius* Greene.] Plate 22, Fig. C. Perennial from a short taproot; stems mostly several, suberect, 1.5–3.5 dm. high, glabrous especially in age; lower lvs. lyrate-pinnatifid, 3–8 cm. long, including petiole, the terminal segm. much longer than the others, ± rounded, dentate; upper lvs. sessile, reduced; heads in a corymbose cyme; invol. campanulate, mostly 5–8 mm. high; bracteoles poorly developed; rays ca. 8, yellow, 6–10 mm. long; aks. glabrous to hirtellous; $n=23$ (Ornduff, 1963).— Frequent on dry slopes, 4000–6500 ft.; Joshua Tree Wd., Pinyon-Juniper Wd.; mts. of e. Mojave Desert; to Mono Co.; Colo., Ariz. May–July.

14. **S. sérra** Hook. Much like *S. triangularis* in stature, but the lvs. tapering or contracted toward the base, sharply toothed or subentire; heads often more numerous and in a larger more branched infl.; invol. 6–8 mm. high; phyllaries ca. 8–13, often black-tipped; bracteoles rather elongate; rays ca. 5–8 in no., 5–8 mm. long; aks. glabrous.—Moist banks, 8000 ft.; Wyman Creek, White Mts.; Sierra Nevada to Wash., Rocky Mts. July–Aug.

15. **S. spartioìdes** T. & G. [*S. serra* var. *sanctus* Hall.] Perennial with a heavy woody branched caudex, 2–6 dm. high, several-stemmed, leafy to the infl.; herbage usually glabrous; lvs. many, mostly linear and entire, the lower small, the others 3–10 cm. long, 1.5–5 mm. wide, often spreading or ± recurved; infl. paniculate-corymbiform, leafy-bracteate; invol. 6–11 mm. high, subcylindric, the principal phyllaries 8–13, the outer bracteoles inconspicuous; rays 8–15 mm. long, yellow; aks. strigulose to subglabrous, ca. 2.5–3 mm. long; $n=20$ (Ornduff et al., 1961).—Dry slopes and edge of meadows, 8000–10,500 ft.; Red Fir F., Bristle-cone Pine F.; San Bernardino Mts., Panamint Mts., Inyo-White Mts.; to Rocky Mts., L. Calif. July–Aug.

16. **S. triangulàris** Hook. Plate 22, Fig. D. Perennial herb from short stout rootstocks and fibrous roots; stems several, simple, often coarse, mostly 5–15 dm. tall, glabrous, rarely villous-puberulent; lvs. many, rather well distributed and only gradually reduced upward, the lower triangular-ovate to -lanceolate, truncate or cordate at base, long petioled, the upper short-petioled or sessile, often narrower; lf.-blades mostly ca. 4–20 cm. long, generally sharply or sinuately toothed; heads several to many, in a flat-topped infl.; invol. 7–10 mm. high; phyllaries ca. 9–13, often black-tipped, the outer bracteoles few, narrow, ± elongate; rays 6–12, ca. 7–13 mm. long; aks. glabrous, ca. 4 mm. long; $n=20$, 40 (Ornduff et al., 1963).—Common in wet meadows and on stream banks, 6600–9500 ft.; Montane Coniferous F.; San Jacinto, San Bernardino, San Gabriel mts., Piute Mts. (Kern Co.); to Alaska, Rocky Mts. July–Sept.

17. **S. vulgàris** L. COMMON GROUNDSEL. Annual, 1–5 dm. high, leafy throughout, simple or branched from base, sparsely strigose to glabrous; lvs. rather coarsely pinnatifid and then toothed or simply coarsely toothed, ca. 3–10 cm. long, the lower petioled, the upper clasping; heads discoid, the disk 5–10 mm. wide; invol. ca. 5–8 mm. high; phyllaries mostly ca. 21 and black-tipped; bracteoles well developed; aks. strigulose-hirtellous; $n=20$ (Ornduff, 1963).—Common weed in gardens and waste places; introd. from Eu. Most months.

128. *Silybum* Adans. MILK THISTLE

Stout annual or biennial herbs with large prickly sinuate-lobed or pinnatifid mottled lvs. Heads large, solitary, terminal, nodding. Fls. tubular, purple. Invol. broadly subglobose, the phyllaries in many rows, stiff, spiny-margined and -tipped. Receptacle fleshy, bristly. Aks. glabrous, shining flattened; pappus of several series of minutely barbed bristles and falling away from the ak. together because of the united base. Two spp. of Old World. (Greek name applied to thistlelike plants.)

1. **S. mariànum** (L.) Gaertn. [*Carduus m.* L.] Erect, branched, 1–2 m. tall; lvs. 3–7 dm. wide, with clasping bases and wavy or lobed margins bearing many yellow prickles; upper lf.-surface ± mottled with white blotches; heads 2.5–5 cm. broad; phyllaries

leathery, the spine 1–2 cm. long, or outer phyllaries mucronate; aks. glabrous, ca. 6 mm. long, spotted brown, shining white; $2n=34$ (Heiser & Whitaker, 1948) (Moore & Frankton, 1962).—Common weed in cismontane pasture and waste places, especially in heavy soil; natur. from Medit. region. Santa Catalina and Santa Cruz ids. May–July.

129. *Solidàgo* L. GOLDENROD

Perennial herbs with leafy usually simple stems arising from rhizomes or a caudex. Lvs. alternate, entire or toothed. Heads numerous, radiate, yellow in ours, small, campanulate to subcylindric, panicled racemose or cymose. Invol. few-seriate, graduate or subequal, the phyllaries usually with obscurely herbaceous tips. Receptacle usually alveolate. Ray-fls. small, fertile; disk-fls. perfect, fertile, their anthers subentire at base, their style-branches with mostly lanceolate appendages. Aks. short, pubescent in ours, usually few-nerved. Pappus copious, setose, whitish. A large genus. (Latin, *solidus*, whole, and the suffix *-ago*, from its reputed medicinal value.)

Plants with well developed creeping herbaceous rhizomes; stems rather equably leafy, the lowest lvs. not prominently different from the upper and at length deciduous.
    Lvs. glandular-punctate, lance-linear; infl. ample, copiously bracteate, the heads in terminal cymose clusters; ray-fls. 15–25 .... 4. *S. occidentalis*
    Lvs. not punctate, broader; infl. more compact, not interrupted, the heads not glomerate in cymes; ray-fls. 8–13.
        Lvs. densely puberulent on both faces, the middle and upper usually elliptic and entire. Widespread native .... 2. *S. californica*
        Lvs. close-pubescent beneath, scabrous above, the middle and upper lanceolate. Sparingly introd. .... 1. *S. altissima*
Plants with short rather woody rhizomes or a branched caudex; stems with lower and basal petiolate lvs. much larger than the upper reduced sessile ones; infl. without recurving branches.
    Phyllaries acute or acuminate; rays mostly 8 .... 3. *S. confinis*
    Phyllaries obtusish; rays mostly 13 .... 5. *S. spectabilis*

1. **S. altíssima** L. Stems grayish-puberulent or -pilose, 0.7–2 m. high; lvs. numerous, crowded, lanceolate, long-acuminate, subentire or remotely serrate, cinereous with close pubescence beneath, scabrous above, the median 5–13 cm. long, 1–2 cm. broad; panicle pyramidal, 0.5–3 dm. long, with recurved-spreading branches; heads 12–18-fld.; invol. 3–5 mm. high; phyllaries blunt; disk-corollas 3–4 mm. long.—Occasional as escape from gardens, as at Pasadena, Ventura; native of e. states. Summer.

2. **S. califórnica** Nutt. CALIFORNIA GOLDENROD. Plate 22, Fig. E. Stems 2–12 dm. high, they and the lvs. densely cinereous-puberulent; basal and lower cauline lvs. spatulate to obovate or oval, obtuse to acute, attenuate to base, firm, crenate or serrate, 5–12 cm. long, 1–3.5 cm. wide, the upper cauline much reduced; elliptic, entire, sessile; infl. mostly a narrow dense thryse, sometimes more open; invol. 3–4.5 mm. high; phyllaries narrowly oblong to lance-linear, sharply acute to obtuse, puberulent or glabrous; ray-fls. 8–13; disk-fls. 4–12; aks. hispidulous; $n=9$ (Raven et al., 1960).—Common in dry or moist fields below 8000 ft.; from Coastal Sage Scrub to Yellow Pine F.; cismontane and montane, n. L. Calif. to Ore.; rare in desert region like N.Y. Mts. July–Oct.

3. **S. confìnis** Gray. Stems usually stout, terminating short rhizomes or on a caudex, 3–14 dm. high, glabrous throughout; lvs. thick, pale green, entire, sometimes scabrid on the margins, the basal spatulate to oblanceolate or obovate, petioled, 15 cm. or less long, the cauline gradually reduced, lance-elliptic to linear, sessile; panicle usually oblong, very dense, to 2.5 dm. long, the branches erect, rarely spreading; invol. 3.5–4.5 mm. long; phyllaries only slightly imbricate, linear-lanceolate, acute or acuminate, rather firm; ray-fls. 6–10, disk-fls. 11–21; aks. sparsely hispidulous to canescent; $n=9$ (Raven et al., 1960).—Dry or moist banks, below 8200 ft.; Coastal Sage Scrub, Chaparral, Yellow Pine F.; n. L. Calif. through montane and cismontane s. Calif. to San Luis Obispo Co.; Death V. region; July–Oct.

4. **S. occidentàlis** (Nutt.) T. & G. [*Euthamia o.* Nutt.] WESTERN GOLDENROD. Stem

stout, from creeping rhizome, much-branched, 6–20 dm. high, glabrous, often glutinous above; lvs. lance-linear, acuminate, sessile, entire, 3–5-nerved, 4–10 cm. long, 3–9 mm. wide, glandular-punctate, the margin often scabrid; infl. ample, leafy-bracteate, interrupted-elongate or rounded, the heads in small cymose clusters; invol. 4 mm. high; phyllaries firm, lance-oblong to lance-linear, acute; ray-fls. 15–25, 1.5–2.5 mm. long; disk-fls. 7–14; aks. pilose; $n=9$ (Raven et al., 1960).—Wet meadows, stream-banks, etc., usually below 2000 ft., occasional to 5000 ft.; Coastal Salt Marsh, Freshwater Marsh, V. Grassland, Sagebrush Scrub, etc.; cismontane s. Calif., rare in desert, as at Victorville; L. Calif. to n. Calif., B.C. Rocky Mts., Nebr. July–Nov.

5. **S. spectábilis** (D. C. Eat.) Gray. [*S. guiradonis* var. *s.* D. C. Eat.] Stem rather stout, terminating a short stout rhizome or caudex, 4–13 dm. high, glabrous throughout or ± hispidulous in the infl.; lvs. usually entire, the basal oblanceolate, tapering to a long winged petiole, 9–28 cm. long, 1–4 cm. wide; the upper cauline linear-lanceolate and often much reduced, scabrid-ciliolate, otherwise glabrous; panicle usually oblong and very dense, mostly less than 10 cm. long; invol. ca. 4 mm. high; phyllaries linear-oblong, obtusish; ray-fls. 11–15, short; disk-fls. 15–22; aks. puberulent.—Alkaline meadows or bogs, sea-level to 7000 ft.; Alkali Sink, Shadscale Scrub, Sagebrush Scrub; Mojave Desert (N.Y. Mts., etc.), Tecopa (Inyo Co.) Furnace Creek, Deep Springs V.; to Ore., Utah. July–Sept.

130. *Solìva* R. & P.

Rather small, depressed annuals with short stiff leafy branches. Lvs. petioled, pinnately dissected. Fls. greenish, in small discoid, burlike heads sessile in the forks of the stems. Invol. of 5–12 subequal phyllaries in not more than 2 series. Receptacle flat. Outer series of florets ♀, apetalous; innermost fls. perfect, but sterile, the corolla 4-toothed. Aks. obcompressed, membranous, callus-margined or winged, pointed with the hardened persistent style. Pappus none. Several spp. of temp. N. and S. Am. (Dr. Salvador *Soliva*, of Spain.)

(Crampton, B. Observations on the genus S. in Calif. Leafl. W. Bot. 7: 196–8. 1954.)
Aks. with wings reduced to a hardened marginal callus, or if toothed, the teeth minute and little or not projecting above the ak.-body .... 1. *S. daucifolia*
Aks. with broad membranous entire wings, each wing projecting above the body of the ak. as a tooth 1–1.5 mm. long .... 2. *S. sessilis*

1. **S. daucifòlia** Nutt. Like *S. sessilis* in appearance; aks. wingless but sometimes with a minute apical tooth on each margin, little or not projecting above the 2–2.3 mm. long body of the ak.—Near the immediate coast on bluffs above the beaches, etc.; Santa Barbara Co. n. As a lawn weed in Escondido, Los Angeles, Riverside, Ventura. April–July.

2. **S. séssilis** R. & P. Plants 5–25 cm. across, pubescent to villous; lvs. ca. 1–3 cm. long, with 3–5 primary divisions, these parted into 3–5 narrowly lanceolate lobes; first heads sessile at very base of plant, the stems radiating from there; invol. ca. 2.5–3 mm. long; ♀ fls. 9–12; each wing of the ak. terminating above in an incurved tooth; ak.-body 3–3.5 mm. long; disk-fls. ca. 7–9, their styles stout, conspicuously exserted.—Moist open places at low elevs.; various Plant Communities; San Diego, Santa Barbara to Del Norte Co.; Chile. April–July.

131. *Sónchus*. L. Sow-Thistle

Leafy-stemmed mostly smooth and glaucous rather coarse herbs, with corymbed or umbellate heads. Invol. campanulate, the phyllaries few, thin, with many shorter ones at the base; these becoming callus-thickened. Fls. many, ligulate, yellow. Aks. obcompressed, ribbed or striate, not beaked; pappus copious, of very white exceedingly soft and fine bristles mainly falling away together. Ca. 45 spp. of Old World. (The ancient Greek name.)

Plant perennial with running rootstocks; heads large, 15–25 mm. high; aks. oblong, but little compressed .... 1. *S. arvensis*
Plant annual; heads smaller, not more than 13 mm. high.

Lvs. serrate or with broad lobes; plants stout; aks. broad, flat, thin-edged.
  Basal auricles of the lvs. rounded; aks. 3-nerved on each side, otherwise smooth . 2. *S. asper*
  Basal auricles of the lvs. acute; aks. striate, transversely wrinkled .... 3. *S. oleraceus*
Lvs. pinnately parted into narrow lobes; plant slender; aks. narrow, thickish . 4. *S. tenerrimus*

1. **S. arvénsis** L. Creeping by underground rootstocks; flowering stems 4–12 dm. high, leafy below; lvs. many, runcinate-pinnatifid to spinulose-dentate, short-petioled, 5–15 cm. long, the upper sessile, clasping, often not divided; heads few to many, on glandular-hispid peduncles; invol. 15–25 mm. high, the phyllaries linear-lanceolate, in ca. 3 series; aks. ca. 3 mm. long, transversely wrinkled on the ribs; $2n=54$ (Gadella & Kliphuis, 1967).—At one time natur. in peat lands near Wintersburg, Orange Co.; from Eu.

2. **S. ásper** (L.) Hill. Like *S. oleraceus* but the auricles of the clasping base of the lvs. rounded; aks. more plainly 3-nerved on each side, otherwise smooth; $2n=18$ (various workers).—Common weed; natur. from Eu. Most months. Occasional on deserts; Channel Ids.

3. **S. oleráceus** L. Rather simple, 5–25 dm. high; lower lvs. petioled, 1–2 dm. long, lyrate-pinnatifid, the lobes lanceolate, spinulose-dentate, the upper lobe large and deltoid; upper lvs. reduced; auricles acute; invol. 9–15 mm. high; aks. longitudinally striate on each face, transversely rugose, 2.5 mm. long; $2n=32$ (Ishikawa, 1916).—Common weed in waste places, gardens, etc.; natur. from Eu. San Clemente, Santa Catalina and Santa Barbara ids. Most months.

4. **S. tenérrimus** L. Much-branched annuals 3–10 dm. tall, leafy to the infl.; lvs. oblong, 1–2 dm. long, with linear to narrowly lanceolate lobes; aks. narrow, longitudinally striate and transversely rugose, 2 mm. long; $2n=14$ (Stebbins et al., 1953).—Reported as a weed from scattered locations; native of Eu. April–June.

132. *Stephanoméria* Nutt.

Annual or perennial herbs with slender strict or paniculately branched stems. Lvs. alternate, linear to oblong, entire to pinnatifid, the uppermost usually much reduced. Heads usually paniculate, small or middle-sized, rose- or flesh-colored. Invol. of several equal phyllaries, with some small outer bracts, sometimes more regularly graduated. Aks. columnar, 5-angled; pappus-bristles 1-seriate, plumose at least above, often paleaceous toward the base, sometimes connate into groups. A dozen or more spp. of w. N. Am. (Greek, *stephane*, wreath, and *meros*, division; of uncertain application.)

Plants annual.
  Pappus-bristles plumose only above, naked or only hirsutulous below the middle, the base thickened or paleaceous; plants mostly below 5 dm. high .... 2. *S. exigua*
  Pappus-bristles plumose almost throughout, the base scarcely thickened; plants mostly 5–30 dm. high .... 6. *S. virgata*
Plants perennial.
  Invol. strongly imbricated; receptacle pitted; young herbage woolly .... 1. *S. cichoriacea*
  Invol. not imbricated, but calyculate at the base with minute bractlets; receptacle naked; herbage mostly glabrous.
    Stems herbaceous; invol. 10–14-fld., 12–15 mm. high; lvs. callus-margined. Deserts
      4. *S. parryi*
    Stems ± woody at the base; invol. 3–9 fld.; less than 12 mm. high; lvs. not callus-margined.
      Pappus-bristles plumose essentially to the base; invol. 5–8 mm. high. Occasional
        3. *S. myrioclada*
      Pappus-bristles plumose only above, naked and merely scabrous toward the base; invol. 8–10 mm. high. Common .... 5. *S. pauciflora*

1. **S. cichoriàcea** Gray. [*Ptiloria c.* Greene.] Perennial, woody at the base, with erect simple or virgately branched stout stems 4–12 dm. high; herbage woolly when young, later ± glabrate; lvs. oblong to oblanceolate, 5–18 cm. long, sessile, acute, subentire to remotely and saliently toothed; heads on short bracteate peduncles along the branches, ca. 12-fld.; invol. 12–15 mm. high, the phyllaries imbricate; receptacle with hirsute pits

Stephanomeria

ligules pink, 15–20 mm. long; aks. faintly 5-angled, smooth, ca. 3 mm. long; pappus-bristles 18–22, plumose throughout, brownish; $2n=16$ (Stebbins et al., 1953).—Rocky slopes and canyons below 6000 ft.; Chaparral, Yellow Pine F., etc.; from San Bernardino and Santa Ana mts. near the coast to Monterey Co., inland to Tejon Pass; Santa Cruz Id. Aug.–Oct.

2. **S. exígua** Nutt. [*Ptiloria e.* Greene.] Annual, 2–6 dm. tall, diffusely branched, glabrous below, minutely glandular-pubescent above; lower lvs. pinnatifid with linear or lanceolate divisions, oblong in outline, auriculate-clasping, 2–5 cm. long; upper lvs. small, bractlike; heads many, scattered or paniculate; invol. 6–7 mm. high; fls. mostly 5; aks. 5-angled, with a double row of tubercles between the angles; pappus-bristles 10–18, naked on lower third, commonly united into 4–5 groups by the thickened bases.—Low dry places below 8500 ft.; many Plant Communities; cismontane and desert Calif. to Monterey and Mono cos.; to Rocky Mts. May–Sept. Santa Cruz Id. Variable, a number of forms have been recognized: var. *deanei* Macbr., a conspicuously glandular form with pappus deciduous above the paleaceous base so as to leave a crown of setose scales; and var. *pentachaeta* (D. C. Eat.) Hall, not glandular, the pappus-bristles ca. 5, all distinct and dilated at base. Occasional with the var. *exigua*.

3. **S. myrioclàda** D. C. Eat. in King. [*Ptiloria tenuifolia m.* Blake.] From a woody rootcrown, the branches very many, slender, often flexuous, 3–6 dm. high; lvs. linear, inconspicuous, occasionally 1–2-toothed; heads mostly at ends of branches, often only 3–4-fld.; invol. 5–8 mm. high, the phyllaries 3–5; ligules pink; aks. striate, not at all tuberculate, ca. 4–5 mm. long; pappus-bristles plumose essentially to the base and only slightly thickened there.—Occasional in dry places, 3000–6500 ft.; Creosote Bush Scrub to Pinyon-Juniper Wd.; w. Mojave Desert to Owens V. and Eureka V.; Nev. May–Aug.

4. **S. párryi** Gray. Perennial, with 1 to few, weak branching herbaceous stems 2–4 (–6) dm. high, glabrous; lvs. runcinate-pinnatifid, thickish, 2–8 cm. long, the lobes turned downward, somewhat spinulose-tipped; heads short-peduncled, 10–14-fld.; invol. ca. 15 mm. high; ligules whitish, 15–20 mm. long; aks. 3–4 mm. long, ribbed, not rugose; pappus-bristles sordid, naked at base, often united basally into 2's and 3's.—Occasional in dry open places, 2000–7000 ft.; Creosote Bush Scrub, Joshua Tree Wd., etc.; w. Mojave Desert to Mono Co.; to Utah, Ariz. May–June.

5. **S. pauciflòra** (Torr.) Nutt. [*Prenanthes p.?* Torr. *Ptiloria p.* Raf. *S. runcinata* Calif. refs.] Stems woody at base, rigid, intricately and divaricately branched, forming rounded bushes 3–6 dm. high; pale, glabrous; basal lvs. 3–7 cm. long, runcinate-pinnatifid with narrow divisions, the upper lvs. entire, often reduced to scales; heads solitary, short-peduncled, 3–5-fld.; invol. 8–10 mm. high, principal phyllaries 5; aks. 5-angled, striate, often ± rugulose; pappus-bristles tinged brownish, not plumose to very base; $2n=16$ (Stebbins et al., 1953).—Common in washes and open places below 6000 ft.; Creosote Bush Scrub to Pinyon-Juniper Wd.; both deserts and adjacent mts.; occasional in innermost cismontane valleys (Mint C., Claremont, Lytle Creek, etc.); to Kans., Tex., L. Calif. May–Sept.

Var. **párishii** (Jeps.) Munz. [*S. cinerea* Blake.] Plant densely white-tomentose.—Occasional on the deserts; to w. Nev.

6. **S. virgata** Benth. [*Ptiloria v.* Greene. *P. pleurocarpa* Greene.] Stiff erect annual 5–20 (–30) dm. tall, glabrous or puberulent, virgate or usually with virgate branches from ca. the middle; lower lvs., oblong or spatulate, 1–2 dm. long, often sinuate, dying before anthesis; upper lvs. small, linear, quite entire; heads subsessile along the elongated naked branches, 4–15-fld.; invol. 6–8 mm. high; phyllaries obtusely narrowed at tip; ligule purplish on back, pinkish to white above, ca. 7–9 mm. long; aks. ribbed, linear-clavate, to oblong, ± rugose, ca. 3–4 mm. long; pappus-bristles clear white, 4–5 mm. long, plumose almost throughout; $2n=16$ (Stebbins et al., 1953).—Common late summer annual in deserted fields and disturbed places below 6000 ft.; from Coastal Strand and Coastal Sage Scrub to Yellow Pine F.; widely distributed, cismontane; to Ore., Nev., L. Calif. July–Oct. Channel Ids. A plant doubtfully distinct may be referred to as var. **tomentòsa**

(Greene) Munz, from Santa Cruz Id. and Morro Bay; white-tomentose at least when young; heads on short bracted peduncles.

133. *Stylocline* Nutt.

Low woolly annuals resembling *Micropus*. Receptacle longer than broad, oblong to nearly linear, the membranous bracts attached spirally and the winged margins spirally imbricate in a subglobose slightly conic head. True invol. absent. Perfect cent. fls. usually subtended by plane bracts; sterile cent. aks. generally with a few pappus-bristles. The ♀ marginal fls. completely enclosed in woolly nonindurate phyllaries. A small genus of Eurasia and N. Am. (Greek, *stulos*, a column, and *kline*, bed.)

Bracts of the ♂ (sterile) fls. thick, with conspicuous terminal hooklike cusps; fertile fls. 5–9
.... 1. *S. filaginea*
Bracts of the ♂ (sterile) fls. thin, not ending in hooks; fertile fls. more.
   The ♀ fls. enclosed in the cent. portion of wholly hyaline bracts which have winglike margins. Cismontane .... 2. *S. gnaphalioides*
   The ♀ fls. enfolded by the basal part of the bracts which do not have winglike margins. Deserts
.... 3. *S. micropoides*

   1. **S. filagínea** (Gray) Gray. [*Ancistrocarphus f.* Gray.] Plants branched at or above the base, mostly 3–10 cm. high, appressed woolly-canescent; lvs. linear to oblance-linear, 5–12 mm. long, the uppermost broader; fertile fls. 5–9, their boat-shaped bracts ca. 3 mm. high, firm except at the hyaline tip; sterile fls. surrounded by 5 empty bracts, these 4–5 mm. long and each ending in a rigid incurved beak; sterile fls. without pappus.—Dry often brushy places, below 5500 ft.; Coastal Sage Scrub, Chaparral, etc.; w. Riverside Co. to Kern Co., Coast Ranges n. April–May.
   2. **S. gnaphalioìdes** Nutt. More or less branched from base, 5–18 cm. high; lvs. oblong- to spatulate-linear, 5–12 mm. long; fr.-bearing bracts many, barely 3 mm. long, ovate, nearly plane on outer surface, the basal part firm, forming a sac enclosing the ak., the rest hyaline; ak. without pappus; sterile fls. little shorter than their bracts and with scanty pappus.—Dry slopes, burns, etc. below 5500 ft.; Chaparral, Coastal Sage Scrub, Yellow Pine F., etc.; cismontane from San Diego Co. n., w. edge Colo. Desert. Channel Ids. March–May.
   3. **S. micropoìdes** Gray. Branched from base, 2–8 cm. high; lvs. linear, 5–8 cm. long, those below the heads lance-linear; ♀ fls. many, with oblong woolly bracts lacking hyaline margins and ending in a small hyaline tip, the whole body enveloping the ak.; sterile fls. naked or slightly subtended by glabrous oblong paleae and bearing a pappus of 3–4 slender deciduous bristles.—Occasional in sandy washes and on dry slopes below 4000 ft.; Creosote Bush Scrub, Joshua Tree Wd.; n. Colo. Desert, Mojave Desert; to w. Utah, New Mex. March–May.

134. *Syntrichopáppus* Gray

Low branched floccose-woolly annuals, with mostly alternate spatulate to linear lvs., often 3-lobed at apex. Heads many, small, radiate, solitary. Invol. narrowly campanulate, 1-seriate; phyllaries 5–8, oblong, concave and partly enfolding the ray-aks. Receptacle flat. Ray-fls. ♀, fertile, as many as the phyllaries. Disk-fls. fertile, yellow. Anther-appendages, very slender, acuminate. Aks. clavate, 5-angled. Pappus of numerous barbellate bristles united at base into a ring, falling off together, or none. Two spp. (Greek, *syn*, together, and *thrix*, hair, and *pappos*, pappus, from the united pappus bristles.)
   1. **S. fremóntii** Gray. Low floccose annual 2–10 cm. high, diffusely much-branched throughout; lvs. 5–20 mm. long, linear-spatulate to cuneate, entire or 3-toothed at apex; heads on short terminal or axillary peduncles; invol. narrowly campanulate to subcylindric, 5–7 mm. high, the phyllaries becoming boat-shaped, firm, scarious-margined; rays mostly 5, golden-yellow, 3–5 mm. long; aks. 2–3 mm. long, hoary; pappus bright white, of 30–40 barbellate bristles united at base, deciduous; $n = 6$ (Carlquist, 1956).—Frequent

sandy places, 2000–7500 ft.; Creosote Bush Scrub, Joshua Tree Wd.; Mojave Desert: Death V. region, sw. to Kern Co., Los Angeles and San Bernardino cos.; to Utah, Ariz. April–June.

2. **S. lemmònii** (Gray.) Gray. [*Actinolepis l.* Gray.] Thinly floccose to glabrate annual 2–8 cm. high, the thin reddish stem corymbosely branching above the middle or throughout; lvs. linear, rounded at apex, entire, 3–8 mm. long; invol. campanulate to narrow-turbinate, 4–5 mm. high, the 6–8 phyllaries scarious-margined; rays 6–8, white with cream-yellow base, ventrally, rose with bright red veins dorsally, 2–3 mm. long; aks. 2.5 mm. long, sparsely strigulose, epappose or sometimes with pappus; $n=7$ (Carlquist, 1956).—Sandy places, 3000–5000 ft.; Chaparral, Joshua Tree Wd.; sw. border of Mojave Desert and adjoining slopes of San Gabriel and San Bernardino mts.; Hemet V., s. slope of San Jacinto Mts., *Ziegler*. April–May.

135. *Tagètes* L. Marigold

Annual or sometimes perennial strong-scented herbs. Lvs. opposite, or the upper sometimes alternate, pinnate or pinnatifid, or sometimes simply serrate, conspicuously gland-dotted. Heads of various sizes, solitary at the ends of branches or in leafy cymes. Invol. of united phyllaries forming a cylindric or campanulate tube, naked at the base. Ray-fls. usually present, ♀, fertile. Disk-fls. fertile, orange or yellow like the rays. Aks. angled or flattened. Pappus of few entire mostly unequal, often ± united paleae, with 1–2 subulate bristles longer than the rest. Ca. 20 spp., from New Mex. to Argentina. (Perhaps from *Tages*, an Etruscan god.)

Fls. obscure, extending only 1–2 mm. out of the invol. .... 1. *T. minuta*
Fls. showy, extending several mm. above the invol. .... 2. *T. patula*

1. **T. minùta** L. Leafy branched glabrous annual 2–8 dm. high; lvs. pinnate, 5–15 cm. long, the lfts. 11–17, linear-lanceolate, sharply serrate, 2–4 cm. long; heads numerous, in congested cymes; invol. ca. 10 mm. long, 2.5–3 mm. broad; phyllaries 5, with rounded free lobes; rays usually 3, yellow, 1 mm. long.—Once reported from Riverside, later from Tulare Co.; native of S. Am.

2. **T. pátula** L. French Marigold. Glabrous annual 2–5 dm. high, with branching often purplish stem; lvs. pinnate, 5–10 cm. long, the lfts. linear-lanceolate, sharply toothed, 1–3 cm. long; heads cymose arranged on long peduncles; invol. campanulate, ca. 12 mm. long, 6–7 mm. broad; phyllaries 5–7, with deltoid-acuminate or acute free lobes; rays ca. 5, orange-yellow, 12–15 mm. long; $2n=48$ (Eyster, 1939).—Natur. in vacant lots and on dumps, Santa Barbara; native of Mex.

136. *Tanacètum* L.

Aromatic annual or perennial herbs, sometimes suffrutescent. Lvs. alternate, entire to more commonly pinnately dissected or ternate. Heads small or medium-sized, globular, discoid or short-radiate, in a capitate or corymbiform infl. Outer fls. ♀, with short tubular corolla which may have a short ray, or ♀ fls. sometimes wanting. Invol. of many imbricate phyllaries, the margins and tips of at least the inner usually ± scarious. Receptacle flat or low-conic, naked. Disk-fls. perfect, the tubular yellow corolla 5-toothed. Aks. mostly 5-ribbed or -angled, commonly glandular; pappus a short crown or none. Ca. 50 spp., mostly of N. Hemis. of Old World. Near to *Chrysanthemum*. (Name obscure.)

1. **T. cànum** D. C. Eat. Stems several from a woody root-crown, mostly 1–2 dm. high, slender; herbage closely tomentose-canescent; lvs. mostly 1–2 cm. long, well distributed, subsessile, linear-oblong and entire to narrowly obovate and with 3 linear-oblong segms.; corymbs compact, 2–several-headed; heads 3–6 mm. in diam.; phyllaries ovate, canescent, round-oblong; disk-fls. lemon-yellow.—Dry rocky places, 10,000–11,000 ft.; Bristlecone Pine F.; Panamint Mts. and Inyo Mts.; Sierra Nevada to Ore., Nev. July–Aug.

137. Taráxacum Zinn. DANDELION

Perennials or biennials with lvs. in a basal tuft and with heads solitary and terminal on naked hollow slender scapes. Lvs. pinnatifid to toothed. Fls. ligulate, many, yellow; invol. double, the outer phyllaries short, in several series, often reflexed or spreading, the inner in a single row, erect, linear. Receptacle naked. Aks. fusiform, to oblong-ovoid, 4–5-ribbed, the ribs roughened; apex prolonged into a very slender beak and with abundant soft pale capillary pappus. Inner invol. closing after anthesis and the slender beak elongating and exposing the pappus, the whole invol. finally reflexing and exposing the globular head of frs. to the wind. A rather large genus according to many workers, but many of the taxa representing apomictic races. (Name possibly of Arabian origin.)

Outer phyllaries erect, not reflexed, lance-ovate; lvs. toothed, usually not lobed to near the midrib. Native, San Bernardino Mts. . . . . 1. *T. californicum*
Outer phyllaries reflexed or spreading, ± linear; lvs. mostly lobed to near the midrib. Introd. weeds.
    All or some of the phyllaries corniculate-appendaged; aks. ± reddish at maturity; lvs. tending to be deeply cut their whole length . . . . 2. *T. laevigatum*
    All or nearly all of the phyllaries unappendaged; aks. pale-gray-brown to olive-brown when mature; lvs. tending to have a petiolar base . . . . 3. *T. officinale*

1. **T. califórnicum** M. & J. [*T. ceratophorum* var. *bernardinum* Jeps.] Usually glabrous, 0.5–2 dm. high; root rather thick; lvs. ascending to spreading, light green, oblanceolate, 5–12 cm. long, 1–2 (–3) cm. wide, subentire or sinuate-dentate, rarely incised; heads stout, broadly cylindrical, 10–15 mm. long, the outer phyllaries many, erect, lance-ovate, 5–7 mm. long, the inner lance-linear, 12–15 mm. long; fls. many; aks. pale brown, the body 3 mm. long, the beak 7–9 mm. long.—Moist meadows, 6500–8300 ft.; Yellow Pine F.; San Bernardino Mts.; May–July.

2. **T. laèvigatum** (Willd.) DC. [*Leontodon l.* Willd. *T. erythrospermum* auth.] RED-SEEDED DANDELION. Slender, the scapes 20–25 cm. high; lvs. narrowly oblanceolate, slender-petioled, 5–25 cm. long, 1–5 cm. wide, deeply cleft into remote divergent to reflexed narrow lobes with many smaller intermediate shorter slender lobes; invol. 13–25 mm. long, the inner phyllaries mostly appendaged near the tip; ligules sulphur-yellow above; aks. narrow-oblanceolate, red or red-purple, smooth below, sharply muricate upward, 2.8–3.5 mm. long, the beak 6–8 mm. long; pappus sordid; $2n=24$ (Poddubnaja & Dianowa, 1934).—Occasional as a weed at widely scattered stations; native of Old World. May–July.

3. **T. officinàle** Wiggers. [*T. vulgare* (Lam.) Schrank.] COMMON DANDELION. Lvs. slightly petioled, 5–30 cm. long, oblong or spatulate, sinuate-pinnatifid to subentire, the longer marginal lobes toothed and with intermediate small teeth; heads 2–5 cm. broad, orange-yellow; phyllaries green to fuscous, mostly not appendaged; aks. 2–4 mm. long, drab or olivaceous, tubercled at summit; pappus whitish; $2n=16, 24, 48$ (Woess, 1949). Damp often low places in scattered localities in much of cismontane and montane Calif.; natur. from Eu. Most months.

138. Tetradýmia DC.

Low rigid shrubs with dense matted or floccose wool which may be deciduous. Lvs. alternate, entire, solitary or fascicled, the primary ones often modified into spines. Heads discoid, 4–9-fld., usually in small corymbs or racemes. Invol. cylindric to oblong; phyllaries 4–6, firm, concave, overlapping, often enlarged and thickened at base. Receptacle flat, small. Fls. yellow, the lobes longer than the throat. Anthers strongly sagittate, almost caudate. Aks. terete, obscurely 5-nerved, glabrous to densely long-hairy; pappus of numerous white or whitish, soft, capillary bristles. Several spp., w. N. Am. (Greek, *tetra*, four, and *dymos*, together, the first sp. known had 4 fls.)

Aks. with numerous long ascending hairs which almost equal and conceal the pappus.
    Heads peduncled, scattered or loosely racemose; primary lvs. transformed into stiff slender spines.

Phyllaries subglabrous, with a broad longitudinal central green band and broad hyaline margins .... 2. *T. axillaris*
Phyllaries white, villose-tomentose especially along a longitudinal central band, the margins scarcely hyaline, ± tomentose .... 6. *T. spinosa*
Heads nearly sessile in a close terminal corymb; primary lvs. foliose, often spine-tipped
.... 4. *T. comosa*
Aks. glabrous or with much shorter hairs than the pappus which is evident and conspicuous.
Heads 5-fld., terminal; primary lvs. modified into spines.
Aks. canescent; spines 2–3 cm. long, perpendicular to the twigs; tomentum on young twigs complete, not separated into ridges .... 7. *T. stenolepis*
Aks. glabrous; spines 0.8–1.5 cm. long, ascending; tomentum on young twigs separated into ridges .... 1. *T. argyraea*
Heads 4-fld.; primary lvs. not changed into spines.
Primary lvs. linear to oblanceolate, 1–4 mm. wide, 20–30 mm. long .... 3. *T. canescens*
Primary lvs. erect, slender-subulate, softly spinescent, not over 1 mm. wide, 5–10 (–20) mm. long .... 5. *T. glabrata*

1. **T. argyraèa** Munz & Roos. Six–15 dm. tall, the younger branches with a densely matted silvery-white wool forming firm ridges on the internodes; primary lvs. linear, becoming acicular spines 8–15 mm. long, at first woolly, later glabrate and yellowish; fascicles of secondary lvs. 5–12 mm. long, sometimes present in the axils; heads 1–few, in close clusters at ends of branchlets; phyllaries 5, equal, 6–8 mm. long, rigid, thick, obtuse, silvery-white with dense matted wool; fls. 5, ca. 8 mm. long; aks. greenish, glabrous, ca. 4.5 mm. long; pappus copious, yellowish-white, 7–8 mm. long.—Dry rocky slopes, 5000–7000 ft., Pinyon-Juniper Wd.; Kingston and Clark mts., e. Mojave Desert. Aug.–Sept.

2. **T. axillàris** A. Nels. Closely resembling *T. spinosa* var. *longispina* in vegetative parts and differing from it particularly in its green phyllaries which are glabrous and green-banded longitudinally, with broad hyaline margins.—Uncommon, at 4500–7500 ft.; largely Pinyon-Juniper Wd.; Panamint Mts., Emigrant Spring, Chloride Cliff and Inyo Mts. April–May.

3. **T. canéscens** DC. Stems 1–3 (or more) dm. high, freely branched, sometimes curved downward toward the tips and with numerous short erect flowering branches; lvs. and heads white-tomentose; lvs. linear or lance-linear, numerous, rarely fascicled, 1–2 cm. long; heads in short-peduncled close corymbs; invol. 8 mm. high, 4-fld.; phyllaries 4–5, carinate, oblong; aks. subglabrous to villous, 7–8 mm. long; pappus yellowish or sordid, ca. 1 cm. long; $n=30$ (Ornduff et al., 1967).—Dry slopes, 4000–10,000 ft.; Sagebrush Scrub, Pinyon-Juniper Wd., Yellow Pine F., etc.; San Bernardino and San Gabriel mts. to Frazier Mt. (Ventura Co.), Panamint and Inyo mts.; n. to B.C. Mont., Utah. July–Aug.

4. **T. comòsa** Gray. Plate 22, Fig. F. Erect bush with many stems and virgate branches, 5–12 dm. high, permanently and densely white-tomentose; primary lvs. linear, 2.5–5 cm. long, white-tomentose, the early ones soft, the later rigid and spine-tipped; secondary fascicled lvs. when present short and narrow; heads in close terminal corymbs; invol. 8–10 mm. long, 6–9-fld.; phyllaries oblong, woolly, obtuse, 5–6 in number; aks. long- and soft-woolly.—Dry places, below 5000 ft.; Coastal Sage Scrub, Chaparral; interior cismontane from n. Los Angeles Co. s. to n. L. Calif.; occasional on Mojave Desert. June–Sept.

5. **T. glabràta** Gray. Branched, the stems ascending or arched, 3–9 dm. high, ± white-tomentose; lvs. early glabrate and greenish, the primary rigid, subulate, cuspidate, 5–10 mm. long, the secondary fascicled, soft, persistent; heads in small terminal clusters; invol. 7–8 mm. high, tomentose to green, 4-fld.; phyllaries 4–5, oblong, keeled; aks. villous, ca. 3 mm. long; $n=62$ (Ornduff et al., 1963).—Dry open places, 2000–6800 ft.; Joshua Tree Wd., Sagebrush Scrub, etc.; w. Mojave Desert from Antelope V. north; to Ore., Utah. May–July.

6. **T. spinòsa** H. & A. var. **longispina** Jones. Cotton-Thorn. Rigidly branched, densely white-tomentose, 6–12 dm. high, the foliage deciduous in the dry season; primary lvs. modified into slender rigid spines mostly 2–3 cm. long; secondary fascicled lvs. ca. 1

cm. long, linear, green; heads on short stout axillary peduncles; invol. short-woolly, 8–10 mm. high, even the margins ± tomentose; fls. usually 6–7; phyllaries 5–6, obtuse; aks. ca. 5 mm. long, with long soft white hairs nearly equaling the pappus.—Common on dry slopes and flats, 3000–7000 ft.; mostly Joshua Tree Wd., Pinyon-Juniper Wd.; Mojave Desert from Victorville region and Antelope V. to Panamint, Piute and White mts.; to Utah, Ariz. Apparently sometimes intergrading with *T. axillaris*.

7. **T. stenolèpis** Greene. Four–8 dm. high, much-branched, permanently white-tomentose; lower primary lvs. oblanceolate, ca. 2 cm. long, mucronate, most secondary fascicled lvs. ca. 1 cm. long; heads in close terminal corymbs; invol. 1–1.2 cm. high, 5-fld.; phyllaries 5, oblong, obtuse; aks. canescent, ca. 9 mm. long, glabrate; pappus rather stiff, whitish, ca. 1 cm. long.—Occasional in dry places, 2000–5000 ft.; Creosote Bush Scrub, Joshua Tree Wd.; Mojave Desert from Antelope V. to N.Y. Mts. May–Aug.

139. *Townséndia* Hook.

Low taprooted many-stemmed perennials or rarely of shorter duration. Lvs. alternate, entire, linear to spatulate. Heads medium to large, asterlike, with white to rosy or violet rays and a yellow disk. Invol. broad, few- to several-seriate, ± graduate; phyllaries appressed, mostly lanceolate, with green center and a narrow colored scarious usually ciliate margin. Aks. obovate or oblong, flattened, usually thick-margined, pubescent with 2-forked or glochidiate hairs. Pappus of rather numerous, rigid, narrow barbellate awns, often united at base, or of few awns and several squamellae or of squamellae only, often reduced on ray-aks. Ca. 20 spp., centering in the Rocky Mt. region; 1 in Mex. (For David *Townsend*, amateur botanist of West Chester, Pa.)

(Larsen, E. L. A revision of the genus T. Ann. Mo. Bot. Gard. 14: 1–46. 1927. Beaman, J. H. Chromosome numbers, apomixis, and interspecific hybridization in the genus T. Madroño 12: 169–180. 1954.)

1. **T. scapígera** D. C. Eat. GROUND-DAISY. Subscapose perennial from a taproot, densely cespitose, with numerous simple stems from the caudex, 2–8 cm. high, strigose-hirsute throughout; basal lvs. spatulate, 1–4 cm. long including the long slender petiole, 2–5 mm. wide; heads on scapiform 1–2-leaved stems; invol. 8–13 mm. high, broadly campanulate to hemispheric; phyllaries 4–5-seriate, graduate, lance-oblong, acuminate; rays 8–35, pink or lavender to almost white, 6–12 mm. long, densely stipitate-glandular dorsally; aks. and pappus (of awns) of ray and disk similar.—Rare, rocky ridges and flats, 4500–11,000 ft.; Sagebrush Scrub, Pinyon-Juniper Wd.; Bristle-cone Pine F.; Inyo-White Mts., Last Chance Mts.; to Modoc Co.; Nev. May–Aug.

140. *Tragopògon* L. GOAT'S BEARD

Stout nearly glabrous biennial or perennial herbs with entire grasslike clasping lvs. and large solitary heads of many yellowish or purple fls. Invol. of 8–13 (–17) lanceolate equal acuminate phyllaries. Aks. long-beaked, muricate; pappus of ± flattened plumose bristles. Ca. 45 spp. of Eurasia. (Greek, *tragos*, goat, and *pogon*, beard.)

1. **T. porrifòlius** L. SALSIFY. OYSTER PLANT. Stems from a stout taproot, leafy below, 6–12 dm. high; lvs. linear-lanceolate, long-acuminate, 1–3 dm. long, tapering uniformly from base to apex, not crisped on margins or curled backward at tip; peduncles strongly inflated toward apex; fls. averaging ca. 90 in a head; phyllaries mostly 8–9, not margined with purple; aks. 3–3.5 cm. long (including beak), tapering abruptly to the slender beak which exceeds the body; pappus brownish; $2n = 12$ (Poddubnaja et al., 1935).—Occasional weed in orchards, fields and waste places; natur. from Eu. April–June.

A sp. with phyllaries not longer than the chrome-yellow corollas and margined with purple and to be watched for in s. Calif. is *T. praténsis* L.

141. Trichocorònis Gray

Small annuals, the several stems creeping at base or spreading, branching, leafy. Lvs. opposite, sessile. Heads peduncled, terminal. Receptacle convex, naked. Phyllaries 12–18, greenish, subequal, nerveless. Corolla abruptly much dilated above the narrow tube, flesh-color to rose-purple. Aks. grayish-black, ± 5-angled, short-hispidulous toward the summit; pappus forming a sort of crown of unequal paleae and awns. Perhaps 3 spp., of sw. N. Am. (Greek, *trichos*, hair, and *koronis*, crown.)

1. **T. wrìghtii** (T. & G.) Gray. [*Ageratum? w.* T. & G.] Stems 1–2 dm. long, ascending, rooting at lower nodes; lvs. oblong-ovate to narrowly lanceolate, entire or serrate, auricled at base, 0.8–2 cm. long; heads 4–5 mm. broad; aks. ca. 1 mm. long, with barbellate bristles alternating with small fimbriate paleae; $n=15$ (Turner et al., 1962).—Mud-flats and shores; Mystic Lake near Moreno, Riverside Co., occasional in Cent. V. of Calif.; s. Tex., n. Mex. May–Sept.

142. Trichoptílium Gray

Low diffusely branched annual. Lvs. crowded, mostly alternate, oblanceolate, incised-dentate with spinescent teeth. Heads solitary on long slender peduncles, in the upper axils, yellow, discoid, the outer florets slightly enlarged. Invol. hemispheric, 2-seriate. Receptacle hemispheric, naked. Aks. turbinate, hairy. Pappus of 5 thin paleae almost as long as the corolla, palmately deeply fimbriate into slender bristles. Monotypic. (Greek, *trichos*, hair, and *ptilon*, feather, referring to the dissected pappus-paleae.)

1. **T. incìsum** (Gray) Gray. [*Psathyrotes i.* Gray.] Plate 23, Fig. A. Stems several to many from the taproot, dichotomously branching, forming plants 5–20 cm. high and as broad or much broader, white-woolly; lvs. mostly 1–3 cm. long, 3–9 mm. wide, petiolate, at last glabrate; peduncles not woolly but glandular-puberulent, much overtopping the lvs.; heads depressed-globose, 7–11 mm. high; phyllaries lanceolate, tomentose and glandular; florets 35–80; pappus bright white to fulvous; $n=13$ (Raven & Kyhos, 1961). —Common on desert pavements or sand, up to 2200 ft.; Creosote Bush Scrub; Colo. Desert and s. Mojave Desert; s. Nev., w. Ariz., L. Calif. Feb.–May; Oct.–Nov.

143. Tríxis P. Br.

Shrubs or herbs with alternate lvs. that are lanceolate, entire to denticulate, densely sessile-glandular beneath. Heads 9–12-fld., yellow, solitary or not, in ours in corymbosely branched leafy clusters. Invol. double, the outer series of phyllaries several, linear to elliptic, herbaceous; inner series ca. 8, linear, acuminate, corky-thickened at base. Corollas all 2-lipped, the outer lip 3-toothed, the inner 2-cleft. Aks. densely hispidulous, tapering or ± beaked, 5-ribbed. Pappus of many strawcolored soft bristles. Ca. 40 spp., ranging from U.S. to Chile. (Greek, *trixos*, 3-fold, because of the 3-cleft corolla-lobe.)

1. **T. califórnica** Kell. Plate 23, Fig. B. Erect, bushy, to ca. 1 m. high, leafy to the heads, green, minutely puberulent; lvs. lanceolate, narrow at base, acute, 2–5 cm. long, remotely shallowly denticulate, subsessile; invol. ca. 15 mm. high, subtended by leafy bracts; corolla ca. 13–17 mm. long; aks. brown, glandular, ca. 11–12 mm. long, narrowed and beadlike at summit; pappus sordid, ca. 1 cm. long; $n=27$ (Turner et al., 1962).— Frequent in canyons, washes, etc., below 3000 ft.; Creosote Bush Scrub; Colo. Desert, n. to Sheephole Mts., e. San Bernardino Co.; to w. Tex., n. Mex. Feb.–April.

144. Venegàsia DC.

Tall leafy branching perennial herbs. Lvs. alternate, deltoid-ovate, cordate or truncate at base, acuminate, crenate-dentate to subentire. Heads large, radiate, yellow-fld., in the upper lf.-axils and terminal. Invol. hemispheric, the broadly oval membranaceous phyllaries 3-seriate, the intermediate ones widest. Receptacle nearly flat, hairy. Ray-fls. ♀, numerous, showy, fertile. Corolla-tube of ray and disk glandular-hairy. Aks. clavate,

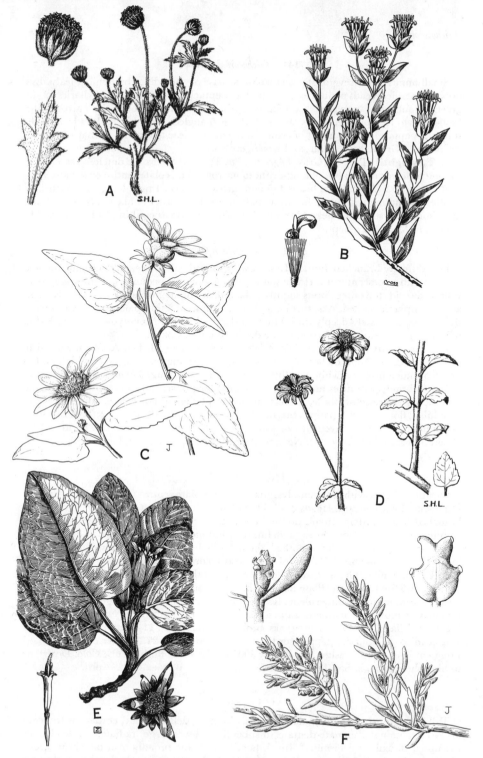

Plate 23. Fig. A, *Trichoptilium incisum*. Fig. B, *Trixis californica*. Fig. C, *Venegasia carpesioides*. Fig. D, *Viguiera deltoides* var. *parishii*. Fig. E, *Wyethia ovata*. Fig. F, *Batis maritima*, coalescent pistils.

muricate, ca. 10-striate, epappose. Monotypic. (For M. *Venegas*, early missionary writer upon California.)

1. **V. carpesioìdes** DC. [*Parthenopsis maritimus* Kell. *V. deltoides* Rydb.] Plate 23, Fig. C. Coarse widely branched, from a somewhat woody base, 1–2.5 m. high, glabrous or sparsely pubescent above; lvs. thin, 5–15 cm. long, gland-dotted beneath; heads usually solitary, on slender peduncles 2–6 cm. long; outer phyllaries foliaceous, loose; the erect inner ones nearly circular, thin, puberulent toward base, the innermost narrower; rays 13–21, 15–20 mm. long, obscurely toothed at the blunt tip; aks. ca. 6 mm. long; $n=19$ (Raven & Kyhos, 1961).—Rocky canyon walls, shaded slopes and moist stream banks, not far from the coast, below 2700 ft.; Coastal Sage Scrub, Chaparral, S. Oak Wd.; n. L. Calif. to cent. Calif.; Channel Ids. Reaching inland as far as Temecula region, San Dimas, etc. Feb.–Sept.

145. *Venídium* Less.

Annual or perennial herbs, ± tomentose, much like *Arctotis*, but pappus lacking or of 4 minute scales. Ca. 20 spp. of S. Afr. (Name of obscure application.)

1. **V. fastuòsum** (Jacq.) Stapf. [*Arctotis f.* Jacq.] MONARCH-OF-THE-VELD. Annual, 3–8 dm. high, branched from base, cobwebby when young; lvs. oblanceolate or obovate, 5–8 cm. long, irregularly lobed to sublyrate, the lower on petioles to 5 cm. long; heads 7–10 cm. across; rays bright orange with dark purplish-brown zone at base; disk yellowish-brown; $2n=18$ (Bliquez, 1951).—Occasional escape from cult. March–May.

146. *Verbesìna* L. CROWN-BEARD

Ours annual or perennial branching herbs. Lvs. opposite or alternate, usually toothed. Heads corymbose on long peduncles, medium-sized and radiate in ours, yellow. Invol. hemispheric to campanulate, its linear to ovate phyllaries 2–4-seriate. Receptacle conical, its bracts thin, clasping the aks. Disk-aks. flat, 2-winged, each wing running up into an awn in ours. Over 100 spp. of warm temp. to trop. N. and S. Am. (From *Verbena*, because of a foliage resemblance.)

Lvs. sessile, green on both faces; phyllaries broadly oblong .... 1. *V. dissita*
Lvs. petioled, grayish at least beneath; phyllaries lance-linear, densely canescent . 2. *V. encelioides*

1. **V. díssita** Gray. Suffrutescent, 5–10 dm. high, glabrate, scabrid; lvs. mostly opposite, rather remote, lance-ovate, acute, remotely serrate to entire, 4–8 cm. long, 1.5–4 cm. wide; heads several in a terminal corymb; invol. 10–12 mm. high, 15–20 mm. wide, graduate, the phyllaries spatulate and rounded (outer) to oblong and acute (inner), sparsely strigulose; rays 13–18 mm. long, neutral; aks. glabrate, the thin wings broadening above.—Once adv. in Mill Creek, San Bernardino Mts. and at Arch Beach, Orange Co.; n. L. Calif. May.

2. **V. encelioìdes** (Cav.) Benth. & Hook. var. **exauriculàta** Rob. & Greenm. [*Ximenesia exauriculata* Rydb.] Much-branched erect annual 3–12 dm. high with a taproot; stems canescent; lvs. narrowly lanceolate to deltoid-ovate, acute or acuminate, the blade 4–10 cm. long, saliently dentate, white-strigose, greener on upper surface, on slender petioles; invol. 7–12 mm. high, 10–15 mm. wide, scarcely graduate, densely strigose, the phyllaries slender; rays 10–15 mm. long, fertile; disk-aks. when fully mature turning from black to whitish-olive and the canescent corky wing of uniform width becoming more prominent; $n=17$ (Carlquist, 1954).—A weed of field-borders, etc.; V. Grassland, Coastal Sage Scrub, etc.; Riverside Co. to Ventura Co.; cent. valley of Calif.; Ariz. to Kans., Mex. May–Dec. The var. *encelioides* with lvs. auriculate at base, occurs from Miss. V. to Fla.

147. *Viguiéra* HBK.

Herbs or shrubs with leafy usually branching stems. Lvs. opposite at least below, linear to ovate. Heads medium-sized, on slender peduncles, solitary or cymose, yellow. Invol.

2-3-seriate, the phyllaries linear-lanceolate. Paleae of the convex receptacle clasping the aks., persistent. Ray-fls. neutral, the ligules dorsally pubescent. Disk-fls. numerous, fertile. Aks. laterally compressed, 4-angled. Pappus of 2 awns and several intervening scales, or none. Ca. 150 spp., distributed from s. U.S. to temp S. Am. (Honoring L. G. A. *Viguier*, physician and botanist of Montpelier.)

(Blake, S. F. A revision of the genus V. Contr. Gray Herb. n.s. 54: 1–205. 1918.)
Shrubs or subshrubs; phyllaries indurated at base; pappose; aks. pubescent.
    Lvs. ovate; invol. canescent.
        Lvs. tuberculate-hispidulous on both surfaces, the blade 1–2.5 cm. long ... 1. *V. deltoidea*
        Lvs. silvery appressed-canescent on upper surface, crisped-pubescent and prominently reticulate on lower, the blade 3–8 cm. long .... 4. *V. reticulata*
    Lvs. lanceolate; invol. green. San Diego Co. .... 2. *V. laciniata*
Herbs; phyllaries herbaceous; epappose; aks. glabrous .... 3. *V. multiflora*

1. **V. deltoìdea** Gray var. **párishii** (Greene) Vasey & Rose. [*V. p.* Greene.] Plate 23, Fig. D. Rounded subshrub 3–10 dm. high, multibranched, the stems harshly tuberculate-strigillose; lvs. deltoid-ovate, strongly toothed and reticulate, the fine spreading hairs from prominent tuberculate bases; heads on elongated peduncles, solitary at branch tips or in small cymes; invol. 5–9 mm. high, canescent, the phyllaries broadly lanceolate, acuminate; paleae much thinner, medianly hairy, hyaline-margined, several-nerved, abruptly acute; ray-fls. ca. 8, the narrow ligules 12–15 mm. long; aks. appressed-pilose; pappus-awns 2–3 mm. long, the fimbriate squamellae less than 1 mm. long; $n = 18$ (Heiser, 1960).—Sandy desert canyons and mesas below 5000 ft.; Creosote Bush Scrub; Colo. Desert and e. Mojave Desert, w. to coastal San Diego Co.; to L. Calif., Ariz., Nev. Feb.–June, Sept.–Oct.

2. **V. laciniàta** Gray. Multibranched, 6–12 dm. high, ± resinous and harshly hispidulous; lvs. many, lanceolate from a broad often hastate base, 2–5 cm. long, subentire to pinnatifid, acute, coriaceous, short-petioled, with axillary fascicles; heads in rather compact corymbs; invol. 5–8 mm. high, the outer phyllaries ovate, abruptly acuminate; ray-fls. 8–13, 10–15 mm. long; pappus paleae lanceolate, awn-tipped, deciduous, 2.5 mm. long.—Dry slopes below 2500 ft.; Coastal Sage Scrub, Chaparral; s. San Diego Co.; n. L. Calif. Feb.–June.

3. **V. multiflòra** (Nutt.) Blake var. **nevadénsis** (A. Nels.) Blake. [*Gymnolomia n.* A. Nels.] Perennial, sometimes suffrutescent at crown, the several slender erect stems 3–9 dm. long; herbage strigose to scabrous-puberulent and gland-dotted; lvs. linear to linear-lanceolate, entire, revolute, 2–5 cm. long, 2–5 mm. wide; heads few, loosely paniculate; invol. 6–7 mm. high, the narrowly lanceolate phyllaries exceeding the paleae; ray-fls. 8–13, the ligules 10–15 mm. long; aks. glabrous; pappus none.—Rocky places, 4000–7500 ft.; Sagebrush Scrub, Pinyon-Juniper Wd.; desert ranges of Inyo Co. and ne. San Bernardino Co.; to Utah, Ariz. May–Sept. Var. *multiflora* occurs n. and e. through the Great Basin.

4. **V. reticulàta** Wats. Suffrutescent, the several pallid pilose stems 5–15 dm. high; lvs. broadly ovate, often subcordate, entire or undulate, strongly reticulate, short-petiolate, densely white pilose with appressed hairs beneath; heads well overtopping the lvs. in a nearly naked cymose panicle; invol. 4–6 mm. high, densely canescent, the phyllaries lance-oblong, thickened; ray-fls. 10–15, the ligules 8–12 mm. long; pappus awns 1 or 2, 1.5–2.6 mm. long, stoutish, the squamellae coherent at base, less than 1 mm. long. Gravelly washes and rocky canyons below 5000 ft.; Creosote Bush Scrub; desert ranges of Inyo Co. Feb.–June.

148. *Wyèthia* Nutt.

Coarse perennial herbs with erect or ascending usually unbranched stems from a branching crown surmounting a thick fusiform taproot. Lvs. in a basal tuft, ample, the stems short with solitary or few heads. Invol. broad, the phyllaries in 2 or 3 series, the outer enlarged. Receptacle quite flat, with narrow rigid bracts. Fls. in ours yellow, fertile in both ray and disk. Pappus present, consisting of a persistent crown of unequal scales, these united at base and often awn-tipped, or lacking. Fourteen spp. of w. U.S. (Named for Capt. N. J. *Wyeth*, who found it in 1833.)

1. **W. ovàta** T. & G. [*W. coriacea* Gray.] Plate 23, Fig. E. Tomentose, the stems leafy, 5–10 cm. high; lvs. glabrate, firm-coriaceous, entire, ovate or oval, 5–18 cm. long, on somewhat shorter petioles; heads almost sessile to short-peduncled, surpassed by the lvs., 2.5–3 cm. broad; rays few, 1–1.5 cm. long; pappus coroniform, projected into short awns on the angles.—Grassy, openly wooded hillside, 1200–6000 ft.; Chaparral, V. Grassland, Yellow Pine F.; mts. from San Diego Co. to Los Angeles Co.; s. Sierra Nevada, n. L. Calif. May–Aug.

149. Xánthium L. Cocklebur

Coarse weedy annuals with stout branching stems. Lvs. alternate, usually lobed or toothed, petioled. Staminate heads borne above the ♀ in terminal and axillary clusters, many-fld.; invols. subglobose, the phyllaries free, in 1–3 series; fls. monadelphous; anthers free, very slender. Invol. of ♀ heads enclosing the 2 florets and becoming an indurated bur, 2-beaked, covered with rigid hooked prickles; corolla and pappus none; style-branches exserted through the beaks. Possibly only 3 spp., of which the 2 treated here are cosmopolitan weeds of doubtful origin. (Greek, *xanthos*, yellow, from the ancient name of some plant, the fr. of which was used to dye hair.)

Lvs. lanceolate, tapering at both ends, short-petiolate, with conspicuous 3-forked spines in their axils .... 1. *X. spinosum*
Lvs. broadly ovate or cordate, long-petiolate, without spines .... 2. *X. strumarium*

1. **X. spinòsum** L. Spiny Clotbur. Much-branched, 3–10 dm. high, to 15 or more dm. wide; lvs. lanceolate, entire or few-toothed or lobed, 3–8 cm. long, 6–25 mm. wide, subglabrous or strigulose above, silvery-tomentulose beneath, armed with 3-forked stout yellow spines 1–2.5 cm. long in their axils; ♂ fls. rusty-pubescent; burs many, but not very crowded, cylindric, ca. 1 cm. long, puberulent, the 2 beaks inconspicuous and unarmed, the body puberulent and beset with numerous uncrowded slender sharply hooked prickles; $n = 18$ (Heiser & Whitaker, 1948).—Common weed of old pastures and waste places at low elevs., cismontane; to the Atlantic Coast. Santa Catalina and San Miguel ids. July–Oct.

2. **X. strumàrium** L. var. **canadénse** (Mill.) T. & G. [*X. c.* Mill. *X. italicum* Moretti. *X. campestre, californicum, acutum,* and *palustre* Greene.] Coarse branching annual 5–15 dm. high; lf.-blades 3–12 cm. long and equally wide, on slender petioles as long, deltoid-ovate, ± cordate at base, dentate or serrate and somewhat lobed, thickish, scabridulous; burs narrowly to broadly ovoid, greenish- or yellowish-brown, mostly 2–3.5 cm. long, the ± crowded often very stout prickles hispid with stout gland-tipped hairs in lower half, the beaks short and very stout, with short incurved tips.—The common form of this cosmopolitan weed in Calif., often abundant in moist v. floors throughout. July–Oct. Var. *strumarium*, with a straight-beaked greenish, puberulent, small bur ca. 1.5 cm. long may occur in Calif. It is widespread in Am. tropics and in Eu.

11. Balsáminaceae. Jewel-Weed Family

Succulent herbs. Lvs. alternate, simple. Fls. irregular, showy, or the later fls. small and apetalous. Sepals 3, the 2 lateral small, greenish, the posterior sepal large, petaloid, saccate and spurred. Petals 3 or 5, with 2 cleft into unequal lobes. Stamens 5, short; fils. with scalelike appendages on the inner side and ± united; anthers connivent or coherent. Ovary oblong, 5-loculed; style short or obsolete; stigma 5-lobed. Ovules several in each locule. Caps. slender, elastically dehiscent into 5 coiled valves. Two genera and ca. 200 spp., largely of trop. Asia.

1. *Impàtiens* L.

Characters of the family. (The Latin name for *impatient*, because of the sudden bursting of the caps. when touched.)

1. **I. balfoúrii** Hook. f. Perennial, to 1 m. tall; lvs. 7–12 cm. long, with many sharp recurved teeth; infl. 6–8-fld.; lateral sepals ca. 6–7 mm. long; spur ca. 2 cm. long; anterior petal pink-purple.—Along streams near Mentone, San Bernardino Co., *Roos*, also in n. Calif. Garden introd.

12. Basellàceae. BASELLA FAMILY

Climbing fleshy perennial herbs with ca. 20 spp., mostly native in trop. Am. Rootstocks tuberous. Lvs. alternate, usually petioled, entire, mostly fleshy, broad, glabrous. Fls. bisexual, regular, racemose, small, with 2 bracts, 2 sepals, 5 persistent petals remaining closed. Stamens 5, opposite the petals. Ovary superior, 1-loculed, 1-ovuled; styles usually 3, with cleft or entire stigmas. Fr. indehiscent, fleshy, inclosed by the persistent corolla.

1. *Boussingáultia* HBK.

Stems much branched. Fls. in axillary and terminal spikelike racemes. Sepals nearly flat, not winged. Ca. 14 spp. (After J. B. *Boussingault*, 1802–1887, French agricultural chemist.)

1. **B. grácilis** Miers var. **pseudo-basellòides** Bailey. MADEIRA VINE. MIGNONETTE VINE. A twining vine 3–6 m. tall, producing little tubercles in the lf.-axils by means of which propagation occurs; lvs. ovate, 2.5–7.5 cm. long, subcordate at base, short-petioled; racemes to 3 dm. long, many-fld.; fls. white, aging black, fragrant.—Late summer. Occasional escape from cult., as at mouth of Ventura R.

13. Batidàceae. BATIS FAMILY

Low maritime bushy plant. Lvs. opposite, entire, fleshy, stipulate, with a small basal loose flange. Dioecious, the fls. crowded in bracteate catkinlike spikes which are sessile, axillary. Staminate fls. with 2-lobed calyx; stamens 4; petaloid staminodia 4. Pistillate fls. without calyx or corolla, consisting of a 4-loculed ovary with solitary ovule in each locule, and a sessile capitate stigma. Seed without endosperm. A single genus of warmer parts of the world.

1. *Bàtis* L. SALTWORT

Characters of the family. One sp. (Greek name of some seashore plant.)
1. **B. marítima** L. [*B. californica* Torr.] Plate 23, Fig. F. Stems prostrate or ascending, woody at base, 1–10 dm. long, glabrous; lvs. subterete, linear-oblanceolate, 1–2 cm. long; ♂ spikes sessile, ovoid-cylindric, 5–10 mm. long; bracts rounded, broader than long; calyx shorter than bracts; stamens exserted, longer than the white triangular staminodia; ♀ spikes short-peduncled, ca. 1 cm. long in fr.; pistils coalescent to form a fleshy fr.— Coastal Strand, Coastal Salt Marsh; Ventura Co. to L. Calif.; W.I., Atlantic Coast, S. Am. July–Oct.

14. Berberidàceae. BARBERRY FAMILY

Herbs or shrubs. Lvs. alternate or basal, simple or compound. Fls bisexual, regular, solitary or racemed. Sepals and petals hypogynous, usually imbricated in 2 or more series, usually similar and in sets of 3. Stamens as many as petals and opposite them, sometimes twice as many; anthers usually opening by 2 uplifting valves. Pistil of 1 carpel; style short or lacking; ovary 1-celled; ovules 2–many, basal or parietal. Fr. a berry or caps. Ca. 10 genera and 200 spp., of temp. regions, largely of the N. Hemis.

1. *Bérberis* L. BARBERRY

Deciduous, or in our spp. evergreen, spiny or unarmed shrubs, often with underground rootstocklike branches so as to form large patches. Wood and inner bark yellow. Lvs.

Berberis

simple, or in ours once-pinnate, alternate, with lfts. 3 to many, prickly, glabrous. Fls. perfect, yellow, in drooping racemes which are fascicled or solitary. Sepals 6, in 2 series, petallike, deciduous, subtended by 3 bractlets. Petals 6 in 2 series, usually with a pair of glands near the base. Stamens 6, opposite the petals. Ovary superior, 1-celled, consisting of 1 carpel; stigma sessile. Fr. a few-seeded berry, sometimes becoming dry. As here treated including *Mahonia* (unarmed and with pinnate lvs.). A genus of over 200 spp.; N. and S. Am., Eurasia, N. Afr. Many cult. as ornamentals. (*Berberis*, an Arabic name.)

(Ahrendt, L. W. A. Berberis and Mahonia. J. Linn. Soc. 57: 1–408. 1961.)
1. Racemes densely many-fld., the floral bracts deltoid-ovate, obtuse or acute; lfts. 2–3.5 cm. wide.
    2. Lfts. glossy-green above and beneath, the lower surface not microscopically papillate, the lfts. crowded on the rachis, often overlapping .... 7. *B. pinnata*
    2. Lfts. dull above and beneath, the lower surface microscopically papillate.
        3. Lfts. cordate at base, suborbicular, with 8–20 teeth on each edge .... 1. *B. amplectens*
        3. Lfts. not cordate at base, oblong-ovate, mostly with 3–10 teeth on each edge 2. *B. dictyota*
1. Racemes loosely 3–9-fld., the floral bracts mostly lanceolate, subacuminate; lfts. mostly 1–2 cm. wide.
    4. Lfts. stiff-rigid, somewhat folded along the midribs, the 2 sides elevated and strongly crisped with some of the stout marginal spines almost vertical to the lf.-surfaces.
        5. Berries dry, inflated, 10–15 or more mm. in diam.; seeds almost black, 4–5 mm. long; terminal lft. rarely more than twice as long as wide. Cushenbury Springs, Mojave Desert
        3. *B. fremontii*
        5. Berries juicy, not inflated, ca. 6–8 mm. in diam.; seeds red-brown, ca. 3 mm. long.
            6. Terminal lft. mostly 2–5 times as long as wide; berries plum-colored. E. Mojave Desert
            4. *B. haematocarpa*
            6. Terminal lft. mostly not more than twice as long as wide; berries yellowish-red. Se. San Diego Co. .... 5. *B. higginsae*
    4. Lfts. less rigid, almost plane, the bristlelike spines almost in the plane of the lf.-surfaces; berries red .... 6. *B. nevinii*

1. **B. ampléctens** (Eastw.) Wheeler. [*Mahonia a.* Eastw.] Stems 1.5–6 dm. high; lvs. 8–15 cm. long; lfts. 5–7 (3), suborbicular, ± closed-cordate at base, 3–8 cm. long, 2–8 cm. wide, veiny, stiff-coriaceous, dull green above, glaucous, paler and dull beneath, fairly plane, but with 8–20 spinose teeth on each side, the basal lfts. 2–4 cm. from base of petiole; racemes fascicled, rather lax, 3–6 cm. long; inner sepals ca. 5 mm. long; $2n=14$ pairs (Raven et al., 1965).—Rocky slopes, 3000–6100 ft.; Chaparral, Yellow Pine F.; Santa Rosa Mts., Riverside Co. to Laguna Mts., San Diego Co. April–May.

2. **B. dictyòta** Jeps. [*Mahonia d.* Fedde.] Stems erect, few-branched, 3–18 dm. high; lvs. 5–10 cm. long; lfts. mostly 5 (3 or 7), broadly oblong-ovate, blunt at apex, obtuse to subtruncate at base, 3–7 cm. long, 2–7 cm. wide, stiff-coriaceous, dull above, paler (glaucous) and dull beneath, prominently veined, the margins strongly crisped and with 3–8 stout spinose teeth on each side, the lowest lfts. 0.5–3 cm. from base of petioles; racemes fascicled, rather open, mostly 3–9 cm. long; inner sepals 5–7 mm. long; berries ovoid, blue-black, 6–7 mm. long.—Dry rocky places, 2000–6000 ft.; Chaparral, Foothill Wd., Yellow Pine F.; mts. of Orange and s. Riverside cos. to Kern Co. and Ventura Co., then n. April–May.

3. **B. fremóntii** Torr. Much like *B. haematocarpa*, but the mature frs. red, dry, inflated, spongy, 10–15 mm. in diam.—Rare, dry rocky places, 3000–5000 ft.; Joshua Tree Wd., etc.; s. Mojave Desert; to Utah, Colo.

4. **B. haematocárpa** Woot. [*B. nevinii* var. *h.* L. Benson.] Glaucous erect shrub, 1–2 m. high, with stiff erect branches; lvs. stiff-rigid, compact, crowded, strongly crisped and spine-toothed; terminal lft. ± lance-ovate, mostly 3–6 (–10) cm. long, 0.7–1.5 cm. wide, long-acuminate; lateral lfts. ovate to lance-ovate, mostly pointed, 1.5–3 cm. long, 1–1.5 cm. wide; fls. ca. 4.5–5 mm. long; berries juicy, purplish-red, with a bloom, ca. 7–8 mm. in diam.; seeds dark red-brown, plump, ca. 3 mm. long.—Dry rocky places, 4500–5500 ft., Pinyon-Juniper Wd.; New York Mts. and Old Dad-Granite Mts., e. Mojave Desert; to w. Tex., n. Mex. May–June.

Plate 24. Fig. A, *Berberis pinnata*, 6 stamens with broad fils., 6 petals. Fig. B, *Alnus rhombifolia*, ♂ and ♀ catkins. Fig. C, *Chilopsis linearis*. Fig. D, *Amsinckia tessellata*, seed like a mosaic. Fig. E, *Coldenia plicata*. Fig. F, *Cryptantha intermedia*, seed tubercled.

5. **B. hígginsae** Munz. [*Mahonia h.* Ahrendt.] With much the habit of the 2 preceding, stiff, 1–2.5 m. high, glaucous; lvs. rigid, strongly crisped, spine-toothed, with spines 2–3.5 mm. long; lfts. 1–7, the terminal squarish-quadrate, to lance-ovate, 1.5–3.5 cm. long, 1–2 cm. wide; lateral lfts. ± squarish-ovate, 1–2.5 (–3) cm. long, 1–2 cm. wide; racemes ca. 3.5–6 cm. long; fls. yellow, ca. 5 mm. long; berries yellowish-red, 6–7 mm. in diam., drying dark; seeds dark red-brown, plump, 2.7–3.2 mm. long.—Dry rocky points and slopes, ca. 2500–4000 ft.; Chaparral; Pinyon-Juniper Wd.; se. San Diego Co. (Boulevard, Dubber) and adjacent L. Calif. May.

6. **B. nevínii** Gray. [*Mahonia n.* Fedde.] Large rounded shrub, 1–4 m. tall, with stiff branched stems; lvs. 4–8 cm. long; lfts. mostly 3–5, lanceolate or the lateral lance-ovate, somewhat coriaceous, 2–4 cm. long, 0.6–1 cm. wide, blue-green and rather dull on both surfaces, paler beneath, rather plane, spinulose-serrate with 5–16 bristlelike teeth on each side ca. 1 mm. long, the basal lfts. 0.5–1 cm. from the base of the petiole; racemes loosely fld., 2.5–5 cm. long; inner sepals 3–4 mm. long; berries juicy, yellowish red to red, rounded, 6–8 mm. long; seeds plump, brownish, 3.5–4 mm. long.—Sandy and gravelly places below 2000 ft.; Coastal Sage Scrub, Chaparral; n. Los Angeles Co. (San Francisquito C.) through San Fernando V. and Arroyo Seco; San Timoteo C. near Redlands; Dripping Springs near Aguanga. March–April.

7. **B. pinnàta** Lag. ssp. **insulàris** Munz. Stems 2–6 (–8) m. high, leaning on trees for support; lfts. subentire or shallowly dentate, oblong, thin, 2–9 cm. long, 1.5–5 cm. wide.—Among trees; Closed-cone Pine F.; Santa Cruz and Santa Rosa ids. March.

B. pinnàta ssp. **p.** [*Mahonia p.* Fedde.] Plate 24, Fig. A. Stems branched, erect, 3–16 dm. high; lvs. 5–12 cm. long; lfts. 5–9 (–17), crowded on the rachis, usually overlapping, ovate to oblong, 2.5–5 cm. long, 2–3.5 cm. wide, glossy-green above, paler but glossy beneath, rather thin, rather deeply and sinuately spinulose-dentate with 10–20 teeth on each side, the lowest lfts. near base of petiole; racemes fascicled, 3–6 cm. long; inner sepa's to 6 mm. long; berries ca. 8 mm. long, blue-glaucous.—Rocky exposed places on exposed slopes below 4000 ft.; Chaparral; Santa Barbara, s. face of San Gabriel Mts., mts. of s. San Diego Co.; to Ore. March–May.

15. **Betulàceae.** BIRCH FAMILY

Deciduous trees and shrubs with alternate simple, usually serrate lvs. and deciduous stipules. Monoecious, both kinds of fls. in aments and appearing before the lvs. Staminate aments pendulous, elongate, and with 1–3 fls. in each bract-axil; calyx membranous, 2–4-parted or 0; stamens 1–10. Pistillate aments erect or drooping, relatively short; calyx 0 or adnate to the 1–2-celled ovary; style 2-cleft. Ovules 1–2 in each cell, pendulous. Fr. a small 1-seeded winged nutlet in conelike clusters or larger and enclosed in an invol. of enlarged bracts. Six genera of N. Hemis.

Pistillate catkins in short clusters, short-ellipsoid in fr.; fruiting scales woody, obscurely toothed, falling away as a cone .... 1. *Alnus*
Pistillate catkins solitary, cylindrical in fr.; fruiting scales thin and 3-lobed, falling away separately at maturity .... 2. *Betula*

1. **Álnus** Hill. ALDER

Trees and shrubs with scaly bark and few-scaled long lf.-buds. Lvs. dentate to serrate. Staminate aments clustered at ends of branchlets, pendulous. Staminate fls. 3–6 in each axil, the bract short-stalked, peltate; calyx 3–5-parted; stamens 3–5; bractlets 4–5. Pistillate fls. 2 in axils; calyx 0; bractlets 2–4, adnate to scales and woody in fr., forming an ovoid to roundish cone. Nutlet flat, with lateral wings or membranous border. Ca. 15–25 spp., mostly of N. Hemis., some in Andes. (Ancient Latin name.)

1. **A. rhombifòlia** Nutt. WHITE ALDER. Plate 24, Fig. B. Tree 10–35 m. high with whitish or gray-brown bark; lvs. oblong-ovate or -rhombic, apically rounded to acute, basally cuneate, finely or coarsely doubly serrate, dark green and glabrous or ± pubes-

cent above, lighter green and puberulent beneath, 5–11 cm. long, on petioles 1–2 cm. long; ♂ catkins 2–several in a cluster, 3–8 cm. long; ♀ cones 1–1.8 cm. long, the scales somewhat thickened and lobed at apex; nutlets with thin narrow margin.—Along streams below 6500 ft.; Chaparral, Yellow Pine F., etc.; most of cismontane s. Calif.; to B.C., Ida. Jan.–Apr. Variable, a pubescent form with cones 0.8–1 cm. long has been named var. *bernardina* M. &. J. and occurs in the Yellow Pine F.

2. *Bétula* L. BIRCH

Bark smooth, aromatic, with long lenticels and often separating in papery sheets. Lvs. variously toothed. Staminate aments solitary or clustered, elongate; fls. 3 in each bract-axil, the 2 lateral subtended by bractlets; calyx 2–4-lobed. Stamens 2, the fils. forked and ending in 2 pollen sacs. Pistillate fls. 3 in axils of bracts of erect catkins; sepals 0. Ovary sessile, style branched. Frs. small compressed nutlets, with winged margins. Cone scales deciduous with nutlets. Ca. 30 spp., of cooler regions. (Ancient Latin name.)

1. **B. fontinàlis** Sarg. [*B. occidentalis* auth., but a nom. confus.] WATER BIRCH. Tall shrub or tree to 8 m. tall; bark dark bronze, shining, not peeling as layers; twigs rough with large resinous glands; lvs. broadly ovate, sharply serrate except near the entire base, gland-dotted when young, bicolored; fruiting aments 2–3 cm. long, the bractlets glabrous or puberulent, ciliate, with short lateral lobes; nutlet-body not narrower than wings.— Moist places, mostly Montane Coniferous F., the s. limit apparently in Jail Canyon, Panamint Mts., also in White Mts., then n. to Can. and e. to New Mex. Apr.–May.

16. **Bignoniàceae**. BIGNONIA FAMILY

Trees to erect or scandent shrubs. Lvs. mostly opposite. Fls. large and showy, ± bilabiate. Functional stamens 4 in 2 sets, the 5th often rudimentary. Style 1; stigmas 2. Fr. a long woody 2-valved caps. resembling a silique, the valves pulling away from the broad partition which bears the numerous winged seeds. Ca. 100 genera and 600 spp., widely distributed in warmer regions.

1. *Chilópsis* G. Don. DESERT-WILLOW

A large shrub, with simple entire narrow, usually alternate lvs. Fls in short terminal racemes. Calyx inflated, deeply 2-lipped, the upper lip 3-, the lower 2-toothed. Corolla funnelform, 5-lobed, the lobes erose. Antheriferous stamens 4, the 5th rudimentary. Caps. terete, linear. Seeds with wings dissected into copious long hairs. One sp. (Greek, *cheilos*, lip, and *opsis*, resemblance.)

(Fosberg, F. R. Varieties of the desert willow, Chilopsis linearis. Madroño 3: 362-366. 1936.)

1. **C. lineàris** (Cav.) Sweet. [*Bignonia l.* Cav. *C. saligna* D. Don.] Two–6 m. high, few- to many-stemmed; twigs very slender, ± puberulent; lvs. deciduous, linear or lance-linear, 10–15 cm. long, ± arcuate, attenuate at ends; racemes ± pubescent to almost woolly especially on calyces, to ca. 1 dm. long; calyx ca. 1 cm. long; corolla 3–3.5 cm. long, lavender or pink or whitish, with purplish lines and markings; caps. 1.5–3 dm. long; seeds ca. 8 mm. long, 2–3 mm. wide, oblong, flat, thin, bearing a coma 1–1.5 cm. long at each end; $2n = 40$ (Bowden, 1945).—Common along washes and watercourses below 5000 ft.; Creosote Bush Scrub, Joshua Tree Wd.; Mojave Desert from Cushenbury Springs and Daggett to Kelso and Clark Mt.; Colo. Desert; Chaparral, San Jacinto V., Aguanga, etc.; to Tex., Mex. Variable, our form is referred to var. *arcuata* Fosb. by its author. May–Sept. (Plate 24, Fig. C.)

17. **Boráginaceae**. BORAGE FAMILY

Herbs, shrubs or trees; ours small, usually rough-hairy. Lvs. simple, alternate, or sometimes opposite or whorled, mostly entire. Fls. perfect, mostly regular, axillary or in 1-sided scorpioid cymes or racemes. Bracts usually between, to one side, or opposite the fls. Calyx usually 5-parted or -lobed, usually slightly irregular. Corolla 5-lobed, sometimes crested or appendaged in the throat. Stamens 5, alternate with the corolla-lobes, inserted mostly in the corolla-tube. Ovary superior, bicarpellate, 2- or 4-ovulate, becoming tough or even bony, globose and entire or divided, in ours breaking up to form uniovulate nutlets. Nutlets 1–4, smooth to variously roughened and even winged or appendaged. Style lobed or entire, seated in the pericarp at the apex of the fr. or borne between the nutlets directly on the receptacle or upon an upward prolongation of the receptacle called the gynobase. Endosperm usually absent. Almost 2000 spp., of worldwide distribution, but particularly in the w. U.S.

1. Style deeply cleft, the branches each with a capitate stigma .... 3. *Coldenia*
1. Style entire, terminated by a single simple or obscurely lobed stigma.
   2. The style borne on the summit of the fr. on the pericarp, falling away with the nutlets; stigma discoid, usually surmounted with a short sterile appendage .... 7. *Heliotropium*
   2. The style borne between the lobes of the fr. (nutlets) and attached directly and independently to the gynobase or receptacle; stigma capitate, lacking a sterile apical appendage.
      3. Mature calyx very irregular, burlike, two of the lobes united and enclosing the fr., becoming horned with 7–9 long glochidiate appendages; ovules 2 ...... 6. *Harpagonella*
      3. Mature calyx not conspicuously irregular, not armed with barbed hornlike appendages; ovules usually 4.
         4. Nutlets widely spreading in fr., flat, armed with hooked prickles ..... 11. *Pectocarya*
         4. Nutlets erect, often not armed with prickles.
            5. Corolla rotate; anthers connivent into a cone .... 2. *Borago*
            5. Corolla tubular, funnelform or salverform; anthers not forming a cone.
               6. Attachment of nutlet surrounded by an annular rim, strongly convex and leaving a pit upon the low receptacle .... 10. *Lycopsis*
               6. Attachment of nutlet not surrounded by a rim and not leaving a pit upon the receptacle.
                  7. Nutlets attached to the low receptacle by their base. Yellow-fld. perennial of e. Mojave Desert .... 9. *Lithospermum*
                  7. Nutlets attached ± laterally to an elongate or conical receptacle.
                     8. The nutlets conspicuously armed with armed prickles .... 8. *Lappula*
                     8. The nutlets not armed with conspicuous prickles.
                        9. Corolla bright blue. Locally adventive plants 4–9 dm. high .. 5. *Echium*
                        9. Corolla white or yellow or orange. Natives.
                           10. Fls. white.
                              11. Nutlets attached laterally along a usually partly opened and generally basally forked groove or slit in the pericarp .... 4. *Cryptantha*
                              11. Nutlets lacking a distinct ventral groove or slit in the pericarp, this replaced by an elevated keel and ± conspicuous attachment-scar
                                                    12. *Plagiobothrys*
                           10. Fls. yellow or orange.
                              12. Plants perennial .... 4. *Cryptantha*
                              12. Plants annual .... 1. *Amsinckia*

1. *Amsinckia* Lehm. FIDDLENECK

Annual herbs, usually pungent-bristly, with erect or spreading branched stems. Fls. in usually naked or sparsely bracted scorpioid spikes. Calyx 5-parted or with 1 to several lobes ± united, persistent, often tawny or even brown-hairy. Corolla yellow to orange, tubular to salverform, the tube frequently exserted, the limb narrow to broad, occasionally obscurely irregular, the throat unappendaged. Stamens inserted in throat or tube. Style filiform; stigma capitate, 2-lobed. Nutlets crustaceous, 1–4, smooth or rough, attached by a caruncular scar placed immediately at lower end of ventral keel or attached along lower part of groove or slit in pericarp on ventral angle of nutlet. Cotyledons each 2-parted by a dorsiventral longitudinal cleft. Gynobase pyramidal, ca. half as long as nutlets. Perhaps 20 spp., w. N. Am. and s. S. Am. (W. *Amsinck*, early 19th century patron of botanic garden in Hamburg.)

(Ray, P. M. and H. F. Chisaki, Am. Jour. Bot. 44: 529–554. 1957.)
1. Calyx 3-lobed by fusion of 4 of the 5 members in pairs.
   2. Nutlets smooth and shining; corolla 3–4 mm. broad, yellow .... 8. *A. vernicosa*
   2. Nutlets tessellate.
      3. Heterostylic; pollen tricolporate; corolla 6–10 mm. broad, orange .... 1. *A. douglasiana*
      3. Homostylic; pollen tetracolporate.
         4. Corolla 8–14 mm. broad, the tube expanding upward .... 3. *A. gloriosa*
         4. Corolla 3–5 mm. broad, the tube cylindrical .... 7. *A. tessellata*
1. Calyx 5-lobed, or the axial pair of lobes fused; nutlets muricate and variously rugose.
   5. Axial pair of calyx-lobes usually united; lvs. mostly erose-dentate; nutlets dark, 1.5–2 mm. long .... 6. *A. spectabilis*
   5. Axial pair of calyx-lobes not united.
      6. Corolla 2–3 times as long as calyx, 8–14 mm. broad, orange .... 2. *A. eastwoodae*
      6. Corolla 1–2 times as long as calyx, 5–10 mm. broad, orange to orange-yellow.
         7. Corolla 5–10 mm. broad, 1–2 times as long as calyx .... 4. *A. intermedia*
         7. Corolla 2–3 mm. broad, 1–1.5 times the calyx .... 5. *A. menziesii*

1. **A. douglasiàna** A. DC. [*A. lemmonii* Macbr. *A. macrantha* Suksd.] Erect, usually rather slender, 3–5 dm. high, simple up to infl. or branched below, thin bristly below, more abundantly so and the bristles more spreading above; lvs. lance-linear to lanceolate, appressed-hirsute and subcinereous, weakly pustulate; spikes becoming 1–1.5 dm. long; fls. heterostylic; calyx rusty-pubescent, 1 or 2 pairs of lobes united, 6–12 mm. long in fr.; corolla orange, 12–16 mm. long, 6–10 mm. broad, pollen tricolporate; nutlets broadly ovoid, 4 mm. long, flattish and tessellate on the back, sometimes denticulate on lateral angles; scar ovate, median; $n=6$ (Ray & Chisaki, 1957).—Dry open slopes and valleys, below 4500 ft.; V. Grassland, Foothill Wd.; Ventura Co. n. to cent. Calif. March–May.

2. **A. eastwoódae** Macbr. [*A. intermedia* var. *e.* Jeps.] Corolla deep orange, 14–18 mm. long; nutlet rather large; $n=12$ (Kamb, 1952).—Scattering in cismontane s. Calif. from Riverside Co. n.; to Shasta Co.

3. **A. gloriòsa** Eastw. ex Suksd. Resembling *A. douglasiana*; fls. homostylic; pollen tetracolporate; corolla 10–12 mm. long; $n=12$ (Ray & Chisaki, 1957).—Grassy places, Los Angeles Co. to Colusa Co.

4. **A. intermèdia** F. & M. Erect, slender and simple to widely branched, 2–8 dm. tall, sparsely bristly, otherwise subglabrous except for pubescence toward infl.; lvs. linear to lanceolate, 2–15 cm. long, the lower petioled, thinly hirsute on both sides with spreading often pustulate hairs; racemes leafy-bracteate at base, 5–20 cm. long in age; fls. homostylic, nearly radially symmetrical; calyx 5–10 mm. long in fr., the lobes separate, rusty-hispid on backs, white-hirsute on edges; corolla orange-yellow, 8–10 mm. long, the tube not much exserted, the limb 2–6 mm. broad; nutlets 2–3 mm. long, ovoid or angular-ovoid, tuberculate, usually medially keeled and with some oblique ridges, usually granulate; scar usually small; $n=15, 17, 19$ (Ray & Chisaki, 1957).—Common in grassy or open places, below 5000 (7500 ft.); many Plant Communities; throughout cismontane Calif., occasional on deserts; to Wash., Idaho, Ariz., L. Calif. March–June. Variable in lf.-shape, pubescence, nutlets; Suksdorf recognizing over 100 segregates.

5. **A. menzièsii** (Lehm.) Nels. & Macbr. [*Echium m.* Lehm. *A. retrorsa* Suksd. *A. parviflora* Heller.] Erect with few ascending branches above, mostly simple below, 1–6 dm. tall, bristly hirsute and also often fine-strigose; lvs. linear or upper lanceolate, hirsute on both sides with appressed or ascending hairs, the blades 3–12 cm. long, the lower long-petioled; racemes strict or ascending, bractless, 5–15 cm. long; calyx 5–11 mm. long in fr., the lobes distinct, white- to tawny-hispid; corolla pale yellow, 5–7 mm. long, the tube scarcely exserted, the limb ca. 2 mm. broad; nutlets 2.5–3.5 mm. long, triangular-ovoid to ovoid, usually incurved, 1–2 frequently aborted, tuberculate, usually with a narrow dorsal keel and often with some broken oblique ridges, the surface minutely granulate or spiculiferous; scar rather large, expanded over the pericarp; $n=8, 13, 17$ (Ray & Chisaki, 1957). —Occasional in dry grassy places, below 5000 ft.; many Plant Communities; largely valleys away from the coast, San Diego Co. to Wash., Utah.

6. **A. spectábilis** F. & M. [*A. maritima* Eastw.] Loosely branched, usually becoming

spreading or prostrate, the stems 1–6 dm. long, sparsely hispid, pustulate; lvs. lanceolate, acuminate, usually bright green, sparsely but pungently hispid, pustulate beneath; spikes becoming 8–10 cm. long; calyx 3–6 mm. long in fr., the 5 lobes linear-lanceolate (2 or 3 usually partly united), hispid and pilose, usually fulvous; corolla orange, 6–12 mm. long, 4–6 mm. wide; nutlets ovoid, usually somewhat incurved, 1.5–2 mm. long, dark but pale-tuberculate, usually also with narrow oblique ridges and scattered papillae; scar submedian, narrow; $n=5$ (Kamb in 1952).—Sandy places and edges of marshes; Coastal Strand, Coastal Salt Marsh; along the coast from San Diego Co. to Tillamook Bay, Ore., L. Calif. March–June.

Var. **nícolai** (Jeps.) Jtn. ex Munz. Spikes bracted throughout.—San Nicolas, San Miguel and San Clemente ids.

7. **A. tessellàta** Gray. [*A. pustulata* Heller.] Plate 24, Fig. D. Stems 2–6 dm. tall, hispid, pustulate; lvs. linear to lanceolate, or the upper ovate to subcordate, mostly pustulate especially above, with many spreading or appressed hairs; spikes becoming 5–12 cm. long; calyx-lobes fused into 3 or 4, hispid, white-hirsute on the margins, 8–12 mm. long in age; corolla orange, 8–12 mm. long, 2.5–5 mm. broad; nutlets ovoid, 3–3.5 mm. long, the back low, usually pale and crustose, angled on the margin, tessellate, often transversely rugose; $n=12$ (Kamb in 1952).—Common in dry mostly sandy or gravelly places, below 6000 ft.; Creosote Bush Scrub, Joshua Tree Wd., Sagebrush Scrub, etc.; deserts; interior s. parts of s. Calif.; inner side of Coast Ranges w. of San Joaquin V.; to e. Wash., Ariz., L. Calif. March–June.

8. **A. vernicòsa** H. & A. [*A. carnosa* Jones.] Erect, sparsely setose and pustulate, or nearly glabrous, 3–5 dm. tall; lvs. somewhat glaucous and fleshy, abundantly pustulate above, sparsely so beneath, middle and upper lvs. lanceolate to lance-ovate, acuminate, 4–8 cm. long, with somewhat clasping base; spikes 3–12 cm. long; mature calyx setose on back, pale ciliate on margins, 9–18 mm. long, sometimes 2 or more of the lobes partly united; corolla 10–12 mm. long, golden-yellow, 3–6 mm. broad; nutlets gray, smooth, shining, 4–6 mm. long, plane on back and lateral surfaces, the scar lineate; $2n=14$ (Ray, 1954).—Dry places below 4500 ft.; Joshua Tree Wd.; w. Mojave Desert; to cent. Calif. March–May.

2. Boràgo L. BORAGE

Erect, strigose-hispid herbs. Lvs. alternate. Fls. blue, in loose, leafy cymes and on long pedicels. Calyx with 5 linear segms. Corolla rotate, with 5 acute lobes, the throat with scales or hairy crests. Stigma subentire. Nutlets attached by the base. Old World. (The old name, probably from folk origin.)

1. **B. officinàlis** L. Coarse hairy annual, 4–6 dm. high; lvs. oblong or ovate, to 15 cm. long; corolla 2 cm. wide; stamens exserted, forming a cone 6 mm. high; nutlets verrucose. —Escape from cult., as at Santa Barbara and more n. cities. From Medit. region.

3. Coldènia L.

Low herbs or subshrubs, mostly depressed, canescent to hispid. Lvs. small, alternate, broad, usually with impressed veins. Fls. sessile, usually at forks of stems, but also axillary or terminal, clustered or solitary. Calyx 5-lobed. Corolla white, blue or pink; the tube short, naked or scaly within; throat open; limb spreading. Stamens 5, included, attached in the tube. Style 2-cleft or 2-parted. Ovary glabrous or pubescent, 2-celled, sometimes 4-celled by the septum-like placentae, entire and subglobose or lobed, breaking up at maturity into uniovulate nutlets, 1–3 of these frequently abortive. Ca. 20 spp., mostly of W. Hemis. (C. *Colden*, lieutenant-governor of N.Y. and correspondent of L.)

Fr. globose until completely mature, not lobed, bearing the style on its rounded summit, at last breaking into quarters to form the 4 nutlets; lvs. obscurely veined; fls. borne singly in lf.-axils or at forks of stems .... 1. *C. canescens*
Fr. early divided into distinct nutlets and bearing the style between the apices; lvs. with evidently impressed veins; fls. in clusters at forks.

Plant annual; corolla pink or white; sepals with short pungent hairs; style shorter than calyx
2. *C. nuttallii*
Plant perennial; corolla blue or bluish; sepals villous; style longer than calyx.
Lvs. with 2-3 pairs of shallowly impressed veins; calyx glabrous or short-pubescent within
3. *C. palmeri*
Lvs. with ca. 6 pairs of deeply impressed veins; calyx long-villous within ...... 4. *C. plicata*

1. **C. canéscens** DC. [*C. c.* var. *subnuda* Jtn.] Low, spreading, often matted, the older stems woody and gnarled, 5-25 cm. long, white-tomentose throughout, with intermixed pallid bristles; lf.-blades 6-10 mm. long, ovate to lance-oblong, with shorter petioles; calyx 4-6 mm. long; corolla white, 6-7 mm. long; style shortly exserted from calyx; fr. ca. 2 mm. in diam., glabrous or sparsely hairy near summit.—Rocky slopes and ridges, below 4000 ft.; Creosote Bush Scrub, Joshua Tree Wd.; e. Mojave and Colo. deserts; to Nev., Tex., Mex. March-May. A larger-fld. form (corolla 9-12 mm. long, blue or lavender) is the var. *pulchella* Jtn., e. Colo. Desert; adjacent Ariz.

2. **C. nuttállii** Hook. Prostrate cinereous annual, finely pubescent and short-hispid, somewhat glutinous; stems slender, 5-15 cm. long; lf.-blades ovate to roundish, 4-8 mm. long, with 2-3 pairs of veins; fls. in compact clusters in forks and at ends of branches; calyx 4-5 mm. long, finely hispidulous with scattered bristles; corolla pink or almost white, 3-4 mm. long, 2-2.5 mm. broad; nutlets oblong-ovoid, almost 1 mm. long, smooth, shining.—Sandy places in washes and on slopes or subalkaline flats below 7000 ft.; Creosote Bush Scrub, Joshua Tree Wd., Sagebrush Scrub; Mojave Desert n. to Lassen Co.; Wash., Utah, Wyo. May-Aug.

3. **C. pálmeri** Gray. [*C. brevicalyx* Wats.] Perennial, 1-3 dm. tall, 2-10 dm. broad, slightly glutinous, with suffrutescent trailing stems, thinly hirsutulous, exfoliating in age; lf.-blades obovate to ovate, 4-9 mm. long, usually longer than petioles, sparsely hispid and strigose, with 2-3 pairs of veins; calyx 2-3.5 mm. long; corolla 5-7 mm. long; nutlets ca. 1 mm. in diam. rounded, 1 or more frequently aborted.—Common in sandy places below 500 ft.; Creosote Bush Scrub; Colo. Desert and along Colo. R. to above Needles; w. Ariz., L. Calif. April-June.

4. **C. plicàta** (Torr.) Cov. [*Tiquilia brevifolia* var. *p.* Torr. *C. palmeri* of many auth.] Plate 24, Fig. E. Matted perennial from a deep woody root, the stems prostrate, 1-4 dm. long, the branchlets dichotomous, puberulent to subtomentose; lf.-blades obovate to round-ovate, 4-9 mm. long, densely canescent- or almost silvery-strigose, conspicuously plicate with 4-7 pairs of deeply impressed veins; fls. clustered; calyx 2-3 mm. long; corolla 4-6 mm. long, blue or lavender; nutlets ca. 1 mm. long, smooth, shining, ovoid to rounded, usually 1 or more aborted.—Sandy places below 3000 ft.; Creosote Bush Scrub; Colo. Desert and e. Mojave Desert; Nev., Ariz., L. Calif. April-July.

4. *Cryptántha* Lehm.

Annual or perennial herbaceous or suffruticose plants, usually setose or hispid. Earliest lvs. opposite, the others alternate. Fls. in bractless or bracteate usually scorpioid spikes or racemes, rarely somewhat cymose-paniculate. Calyx usually 5-parted, persistent or readily deciduous. Corolla white, rarely yellow, usually with a short included tube and spreading limb, the throat with intruded appendages or scales. Ovules 4, rarely 2. Nutlets 1-4, erect, affixed along a ventral slit in the pericarp to a columnar subulate or pyramidal gynobase, rough or smooth, the margin rounded or angled or winged. Ca. 65 spp. of New World, mostly w. N. Am., some in sw. S. Am. (Greek, *cryptos*, hidden, and *anthos*, fl., because of the minute corolla in the first known spp.)

(Johnston, I. M. The N. Am. species of Cryptantha. Contr. Gray Herb. 74: 1-114. 1925. Payson, E. B. A monograph of the sect. Oreocarya of Cryptantha. Ann. Mo. Bot. Gard. 14: 211-358. 1927.)

1. Plants annual, with slender stems (perennial in *C. racemosa*).
    2. Calyx circumscissile at maturity; low diffuse plants; infl. compact, each fl. in axil of leafy bract.
        3. Corolla 1-4 mm. broad; pollen grains 7-9 μ long, oblong .... 4. *C. circumscissa*

3. Corolla 4–6 mm. in diam.; pollen grains 5.5–6.5 μ long .... 30. *C. similis*
2. Calyx not circumscissile.
   4. Gynobase subulate, protruding beyond the nutlets, bearing a sessile stigma on its top; root and base of plant with a purple dye; each fl. in the axil of a leafy bract 20. *C. micrantha*
   4. Gynobase shorter than the nutlets; style developed; root and herbage usually with very little or no purple dye; fls. all or in part bractless except in *C. maritima*.
     5. Nutlets roughened or in *C. maritima* at least one of them so.
        6. Margin of nutlets decidedly winged or knifelike.
           7. Pedicels usually evident, slender, 1–4 mm. long; lateral angles of nutlets distinctly winged.
              8. Nutlets homomorphic, broadly winged .... 15. *C. holoptera*
              8. Nutlets heteromorphic, narrowly winged .... 27. *C. racemosa*
           7. Pedicels obscure or none, less than 1 mm. long.
              9. Nutlets 1 or 2; odd nutlet axial .... 34. *C. utahensis*
              9. Nutlets 4.
                10. The nutlets homomorphous, obscurely rugose, the back high-convex
                      8. *C. costata*
                10. The nutlets heteromorphous, verrucose or muricate.
                  11. Lateral margins of the nutlets distinctly winged; corolla 1 mm. broad
                      26. *C. pterocarya*
                  11. Lateral margins of the nutlets knifelike or acute.
                    12. Corolla 1 mm. broad; nutlets 1.3–1.7 mm. long ..... 16. *C. inaequata*
                    12. Corolla 1–2.5 mm. broad; nutlets ca. 1 mm. long ... 2. *C. angustifolia*
        6. Margins of the nutlets rounded or obtuse.
           13. Nutlets decidedly heteromorphous, 1–4, the large nutlet axial and sometimes less roughened.
              14. Mature calyces conspicuously recurved or deflexed, most hirsute on axial side; nutlet 1, bent; ovules 2 .... 28. *C. recurvata*
              14. Mature calyces spreading or erect, most hirsute on the abaxial side; nutlets straight.
                15. Calyx closely appressed to the flattened rachis, persistent, gibbous on the axial side due to the basal development of the roughened odd nutlet
                      10. *C. dumetorum*
                15. Calyx ascending or spreading, deciduous, not at all gibbous; odd nutlet somewhat smooth.
                  16. Nutlets 4, triangular-ovate, 0.7–0.9 mm. long; mature calyx subglobose, minute, the lobes scarcely longer than the nutlets ..... 21. *C. micromeres*
                  16. Nutlets 1–2, oblong-lanceolate, 1–2 mm. long; mature calyx oblong, the lobes surpassing the nutlets .... 19. *C. maritima*
           13. Nutlets homomorphic.
              17. Nutlets lanceolate.
                18. Nutlets only 1, rarely 2, in a normal fr.; style not more than half as long as nutlet.
                  19. Corolla less than 1 mm. broad. Deserts .... 9. *C. decipiens*
                  19. Corolla 2–3.5 mm. broad. Cismontane .... 7. *C. corollata*
                18. Nutlets 4 in normal fr.; style often more than half as long as nutlets.
                  20. Corolla conspicuous, 2–8 mm. broad; plant hirsute, mostly cismontane
                      17. *C. intermedia*
                  20. Corolla inconspicuous, 1–2 mm. broad; desert plants.
                    21. Fruiting calyx 6–12 mm. long.
                    22. Plant bristly with spreading hairs .... 3. *C. barbigera*
                    22. Plant strigose with appressed hairs .... 24. *C. nevadensis*
                  21. Fruiting calyx 3–4 mm. long .... 2. *C. angustifolia*
              17. Nutlets ovate to triangular-ovate.
                23. Dorsal side of nutlets obtuse and with at least a faint median ridge; style surpassing the nutlets .... 23. *C. muricata*
                23. Dorsal side of nutlets flat or low-convex, without median ridge; style not surpassing nutlets.
                  24. Nutlets minute, less than 1.5 mm. long; spikes with numerous bracts. Insular .... 32. *C. traski*
                  24. Nutlets larger, 2–2.5 mm. long; spikes naked or bracted at base only. Largely montane.

25. Plant closely strigose, pallid, usually 2-3 dm. tall; spikes commonly in 2's or 3's .... 31. *C. simulans*
25. Plant spreading-hispid, mostly lower; spikes usually solitary
11. *C. echinella*
5. Nutlets smooth and shining, not roughened.
26. Calyx armed with pale encrusted spreading hooked or curved hairs 11a. *C. flaccida*
26. Calyx armed with straight ± tawny unencrusted hairs.
27. Nutlets with excentric groove, one side of nutlet on lower surface appearing as if somewhat deformed. Montane .... 1. *C. affinis*
27. Nutlets with a centrally placed groove.
28. Dorsal side of nutlets low-convex or flat; nutlets 1-2, homomorphic
19. *C. maritima*
28. Dorsal side of nutlets rounded-convex, the lateral angles rounded or obtuse.
29. Calyx broadly conic at base, densely appressed-hispid-villous, mostly lacking conspicuous spreading bristles. Deserts .... 13. *C. gracilis*
29. Calyx rounded at base, with both appressed and spreading bristles. Cismontane, except *C. ganderi*.
30. Style less than half as long as nutlets; fruiting calyx 1.5-2 mm. long. Widely distributed .... 22. *C. microstachys*
30. Style almost as long as nutlets; calyx longer.
31. Fruiting calyx 2-4 mm. long, moderately white-bristly . 5. *C. clevelandii*
31. Fruiting calyx 6-10 mm. long, conspicuously tawny-bristly
12a. *C. ganderi*
1. Plants perennial or biennial, coarse.
32. Nutlets smooth and shining on the back.
33. Corolla pale yellow, 12-14 mm. long; stems erect, 1.5-5 dm. tall ..... 6. *C. confertiflora*
33. Corolla white, 3-4 mm. long; stems decumbent to prostrate, 0.5-1.5 dm. long
18. *C. jamesii*
32. Nutlets rough or wrinkled on back.
34. Corolla-tube distinctly longer than calyx; nutlet-scar open and with elevated margin
12. *C. flavoculata*
34. Corolla-tube not exceeding calyx; nutlet-scar various.
35. Inner surface of nutlets smooth or nearly so, the scar narrowly subulate but open at base; calyx 5-7 mm. long in fr. .... 25. *C. nubigena*
35. Inner surface of nutlets conspicuously rough or tubercled, the scar open and broadened toward base.
36. Nutlets 2.5-3 mm. long; calyx-lobes 5-7 mm. in fr.
37. Stems less than 1 dm. long, strigose and appressed-setose; nutlets irregularly ridged .... 29. *C. roosiorum*
37. Stems 1.5-3 dm. high, conspicuously hirsute and retrorse-pubescent; lvs. retrorse-hirsutulous and sparsely bristly; nutlets tuberculate .... 14. *C. hoffmannii*
36. Nutlets 4-4.5 mm. long; calyx-lobes 8-12 mm. in fr.
38. Stems 2-3, from a biennial root; infl. broad; calyx-lobes 10-12 mm. in fr.; nutlets prominently ridged dorsally .... 35. *C. virginensis*
38. Stems several to many, from a perennial root-crown; infl. narrow; calyx-lobes 8-10 mm. long in fr.; nutlets not prominently ridged dorsally ... 33. *C. tumulosa*

1. **C. affinis** (Gray) Greene. [*Krynitzkia a.* Gray. *C. geminata* Greene.] Erect sparsely branched annual 1-4 dm. high, pubescent with short upwardly curved hairs and scattered longer stiff ones; lvs. oblanceolate, obtusish, mostly 2-3.5 cm. long, sparsely short-hispid, pustulate; spikes in 1's or 2's, slender, 2-8 (-10) cm. long, becoming loosely fld., with some leafy bracts below; calyx in fr. 2.5-4 mm. long, ca. as thick, laterally compressed; corolla 1.5 mm. broad; nutlets ovoid or oblong-ovoid, 1.8-2.5 mm. long, smooth, grayish, frequently mottled, low-convex on back, rounded on sides, the ventral groove closed, simple or short-forked at base; style usually surpassed by nutlets.—Dry slopes, 3000-9500 ft.; Yellow Pine F., Red Fir. F., Lodgepole F.; Cuyamaca and San Bernardino mts.; Sierra Nevada to Wash., Wyo. June–Aug.

2. **C. angustifòlia** (Torr.) Greene. [*Eritrichium a.* Torr.] Annual, diffusely branched from base, cinereous with appressed and spreading stiffer hairs; stems 0.5-3 dm. long; lvs. linear, 1.5-3.5 cm. long, hispid and pustulate; spikes in 2's, ca. 3-5 cm. long, mostly

bractless, fairly dense; calyx in fr. ovoid-oblong, 3–4 mm. long, hirsute; corolla 1–2 mm. broad; nutlets usually 4, oblong-ovoid, usually brown with pale muriculations, the odd nutlet slightly larger, the others ca. 1 mm. long; style usually surpassing all.—Common, sandy and gravelly places, below 3500 ft.; Creosote Bush Scrub; deserts from Inyo Co. s.; to L. Calif., Son., Tex., Utah. March–May.

3. **C. barbígera** (Gray) Greene. [*Eritrichium b.* Gray.] Erectly branched annual 1–4 dm. tall, very bristly, sparsely if at all strigose except in the infl.; lvs. linear to broadly oblong, ± pilose and hirsute, pustulate; spikes mostly in 2's, bractless, 3–15 cm. long; fruiting calyx oblong-ovoid to -lanceolate, 5–10 mm. long, the lobes linear-lanceolate to lanceolate, connivent with recurved tips, the margins usually conspicuously long-white-villous, midribs thickened and hirsute; corolla 1–2 mm. broad; nutlets 1–4, lance-ovoid, 1.5–2.5 mm. long, strongly verrucose, the back convex, the edges obtusish, the groove open or closed, with a triangular areola below; style reaching to ca. the tip of the nutlets. —Sandy or gravelly places, below 5000 ft., rarely higher; Creosote Bush Scrub to Pinyon-Juniper Wd.; deserts n. Inyo Co. to n. L. Calif., Son., New Mex., Utah. Feb.–May.

4. **C. circumscíssa** (H. & A.) Jtn. [*Lithospermum? c.* H. & A. *Greenocharis c. C. c.* var. *hispida* (Macbr.) Jtn. *Greeneocharis h.* Macbr.] Low annual forming a dense hemispheric mass 2–10 cm. high, appressed-hispidulous to strigose, ± branched above; lower lvs. narrow-oblongish, the upper linear, to 1.5 cm. long; fls. in axils of foliaceous bracts; fruiting calyx 2.5–4 mm. long, united to middle, circumscissile just below the middle; corolla white, 1–2 mm. broad; nutlets 4, homomorphous or nearly so, smooth or obscurely muriculate, triangular-ovoid to oblong-lanceolate, 1.2–1.7 mm. long, with angled margins; style ca. as long as nutlets; $n=12$ (Mathew & Raven, 1962).—Sandy or gravelly places, mostly below 9500 ft.; Creosote Bush Scrub to Lodgepole F.; Mojave Desert and adjacent mt. ranges, n. to B.C., mts. along w. edge of Colo. Desert to L. Calif.; to Utah, Colo., Ariz.; S. Am. April–Aug.

5. **C. clevelándii** Greene. [*C. abramsii* Jtn. *C. brandegei* Jtn.] Annual, stems erect to decumbent, 1–5 dm. high, strigose to hirsute, usually branched; lvs. usually many at base, scattered upward, lance-linear to linear, 1–5 cm. long, thinly appressed-hispidulous with some setae on margins; spikes in 1's or 2's, slender, bractless, 4–10 cm. long; calyx 2–5 mm. long in fr.; ovoid to somewhat oblong, the lobes linear, connivent above with spreading tips, the 3 outer hispid on thick midribs, all densely appressed-hirsutulous; corolla 1–2 mm. broad; nutlets 1–4, oblong-ovoid to broadly lanceolate, 1.5–2 mm. long, convex on back, smooth, shining, the groove closed, broadly forked at base, rarely with a small areola; style ca. ⅔ as high as nutlets.—Slopes and rocky places, below 2500 ft.; Coastal Sage Scrub, Coastal Strand, Chaparral; coastal s. Calif. from Ventura Co. s.; Santa Barbara Ids., Santa Catalina Id., San Clemente Id.; L. Calif. April–June. A form usually stouter and with more pedunculate infl.; corolla 2–5 mm. broad; from San Diego Co. to Marin Co. is the var. *florosa* Jtn. [*C. hispidissima* Greene.]

6. **C. confertiflòra** (Greene) Pays. [*Oreocarya c.* Greene. *O. lutea* Greene ex Fedde.] Perennial from a branched root-crown and stout woody root; stems erect, 1.5–5 dm. tall, strigose and sparsely setose above, the bases white-hairy; lvs. linear-oblanceolate to oblanceolate, 3–10 cm. long; infl. a terminal cymose head 2.5–5 cm. across and with smaller pedunculate ones from upper lf.-axils; calyx-lobes lance-linear, 10–12 mm. long in fr., loosely strigose and weakly bristly; corolla-tube 9–13 mm. long, the limb 10 mm. broad, pale yellow or cream-colored; nutlets broadly ovoid, 3 mm. long, sharply 3-angled, glossy, smooth, the scar nearly closed.—Dry rocky places in limestone, 4000–9000 ft.; Sagebrush Scrub, Pinyon-Juniper Wd., to Lodgepole F.; n. base of San Bernardino Mts., scattering mts. in Mojave Desert (Clark, Kingston, Funeral, etc.) to Inyo-White Range and e. slope of Sierra Nevada; to Utah, Ariz. May–July.

7. **C. corollàta** (Jtn.) Jtn. [*C. decipiens* var. *c.* Jtn.] With much the same habit as *C. decipiens*; appressed-hispidulous throughout; calyx in fr. ascending, ca. 3 mm. long, white-strigose, the outer lobes also spreading-bristly, the tips erect; corolla 2–3.5 mm. broad; nutlet 1, lance-ovoid, acuminate, 2 mm. long, brownish, low convex on back with a low broad ridge toward base, granulate and papillate-muriculate on both sides, the groove

closed throughout and raised into a narrow keel, basally dilated into a small areola.—Dry slopes and ridges below 4500 ft.; Foothill Wd., V. Grassland, etc.; Ventura Co. to Monterey and San Benito cos.; March–June.

8. **C. costàta** Bdg. [*C. seorsa* Macbr.] Coarse stiff relatively few-branched annual 1–2 dm. tall, usually densely villous-strigose and ± hirsute; lvs. linear to lance-linear, 1–3 cm. long; spikes rigid, in 1's or 2's, sparsely bracted, 2–5 cm. long; mature calyx ovate-oblong, 4–6 mm. long, the lance-linear lobes with hirsute midribs, strigose margins; corolla ca. 2 mm. broad; nutlets 4, triangular, barely 2 mm. long, strongly convex on back, the margin sharp, the groove shallow, closed above, dilated toward base.—Uncommon, sandy and gravelly places, below 1500 ft.; Creosote Bush Scrub; Colo. Desert, Needles; w. Ariz. Feb.–May.

9. **C. decípiens** (Jones) Heller. [*Krynitzkia d.* Jones.] Slender loosely branched annual 1–4 dm. tall, strigulose and often short-hispid; lvs. few, linear, obtuse, 1–3 cm. long; spikes mostly in 2's, slender, 4–10 cm. long; calyx in fr. 3–4 mm. long, the lobes linear, connivent, with spreading tips; nutlets 1, sometimes 2, lance-ovoid, 1.5–2.4 mm. long, muriculate-granulate to tuberculate, usually brownish, convex on back, rounded on sides, the open or closed groove dilated basally into an areola; style well surpassed by nutlet.—Sandy or gravelly places, below 4000 ft.; Creosote Bush Scrub, Joshua Tree Wd.; deserts from Inyo Co. to Ariz., Nev. March–May.

10. **C. dumetòrum** (Greene ex Gray) Greene. [*Krynitzkia d.* Greene ex Gray. *C. intermedia* var. *d.* Jeps.] Laxly branched strigose bright green annual, at first erect, later commonly elongated and sprawling or scrambling up through bushes and 1–4 dm. high; lvs. remote, linear or wider, obtuse, 1–3 cm. long, pustulate, strigose or short-hispid; spikes in 1's or 2's, usually remotely fld., 5–10 cm. long, mostly bractless except sometimes at base, the 3 outer lobes reflexed-hispid; corolla ca. 1 mm. broad; nutlets 4, heteromorphous, granulate-muricate, the odd one axial, broadly lanceolate, 2–3 mm. long, the others 1.5–2 mm. long, lanceolate, earlier deciduous; style not longer than nutlets.—Sandy and gravelly places, below 4500 ft.; Creosote Bush Scrub; n. Colo. Desert, in scattered localities in the Mojave Desert to Inyo Co.; sw. Nev. April–May.

11. **C. echinélla** Greene. [*C. ambigua* var. *e.* Jeps. & Hoov.] Erect, sparsely branched, mostly 0.5–3 dm. high, short-hirsute; lvs. ± oblanceolate; spikes solitary or in 2's, often leafy-bracted below, 1–5 cm. long; calyx in fr. oblong-ovoid, spreading, 5–6 mm. long, the lobes lance-linear, connivent above, with ± recurved tips, the midribs pale tawny-hispid, the margins appressed-hirsutulous; corolla 1–1.6 mm. broad; nutlets 4, broadly ovoid, grayish, 2 mm. long, papillate-echinate on the rounded back, the groove closed or nearly so, widely forked at base; style not quite reaching to nutlet-tips.—Dry places, 2500–9000 ft.; Yellow Pine F. and above, Pinyon-Juniper Wd., Bristle-cone Pine F.; San Bernardino Mts., San Gabriel Mts., Panamint Mts., Inyo-White Mts.; to Sierra Nevada; w. Nev. June–Aug.

11a. **C. fláccida** (Dougl.) Greene. [*Myosotis f.* Dougl. ex Lehm.] Wiry annual, 1–5 dm. tall, strigose; lvs. linear or lance-linear, 2–5 cm. long; spikes in 1's to 5's, ebracteate; calyx in fr. 2–4 mm. long, usually appressed to rachis; corolla 1–4 mm. wide; nutlet mostly 1, lance-ovoid, smooth.—Cismontane, below 6000 ft.; to Wash. April–June.

12. **C. flavoculàta** (Nels.) Pays. [*Oreocarya f.* Nels.] Perennial with cespitose woody caudex; stems 1–many, strigose and hispid, 1–3 dm. long; lvs. oblanceolate, 3–8 cm. long, appressed silky-pubescent with scattered bristles; infl. ± cylindrical, with upper cymules congested, the lower more scattered; fl.-bracts lance-linear; calyx-lobes densely bristly with ± yellowish setae, lance-linear, ca. 10 mm. long in fr.; corolla-tube 7–10 mm. long, the white or pale yellow limb 7–10 mm. broad; nutlets ovate to lanceolate, 2.5–3.5 mm. long, obtusish, with acute-angled margins, somewhat glossy but muriculate on back, the scar open, conspicuously margined.—Dry rocky slopes, 4600–10,000 ft.; Pinyon-Juniper Wd., Sagebrush Scrub, Bristle-cone Pine F.; Panamint Mts., Inyo Mts., Last Chance Mts.; to Mono Co.; Wyo., Colo. May–July.

12a. **C. gánderi** Jtn. Bushy annual 1–4 dm. high; hirsute-hispid; lvs. linear; spike 1, 5–15 cm. long; calyx 6–10 mm. long in fr.; corolla inconspicuous; nutlet usually 1, lanceolate, 2–2.5 mm. long, smooth, rounded on edges, the groove with a narrow areola. —Creosote Bush Scrub, e. San Diego Co. Feb.–May.

13. **C. grácilis** Osterh. Erectly branched annual 1–2.5 dm. high, densely short-hispid; lvs. linear to oblance-linear, 1–3 cm. long, short-hispid, pustulate; spikes in 1's or 2's, usually dense and 1–2 cm. long; calyx in fr. ovoid, 2–2.8 mm. long, divaricately spreading, the lobes lanceolate, erect at apex, tawny hirsute-villous, with a few bristles on the midribs; corolla to 1 mm. broad; nutlets 1 (2–3), lanceolate, 1.5–2 mm. long, smooth, shining, the back nearly flat, the sides rounded at least toward apex, the groove usually open to above middle, scarcely forked below; style ca. ¾ as high as nutlets.—Dry slopes and mesas, 3000–7000 ft.; Joshua Tree Wd., Pinyon-Juniper Wd., Creosote Bush Scrub; scattered localities, Little San Bernardino Mts. and Providence Mts. to Inyo Co.; to Ida., Colo. Apr.–June.

14. **C. hoffmánnii** Jtn. [*Oreocarya h.* Abrams.] Root biennial or perennial; stems 1– several, 1.5–3 dm. high, hirsute and retrorsely pubescent; lower lvs. spatulate, 2–3 cm. long, on longer petioles, sparsely bristly and retrorsely pubescent, pustulate; infl. 5–15 cm. long, ± interrupted, with many cymules; calyx-lobes 5–7 mm. long in fr., linear-lanceolate, hirsutulous and bristly; corolla white, the tube 3 mm. long, the limb 5–6 mm. broad; nutlets ovoid, ca. 3 mm. long, tuberculate and slightly rugose, the ventral scar open, extending almost to apex.—Dry rocky places, 6400–9000 ft.; Pinyon-Juniper Wd., Bristle-cone Pine F.; Last Chance Mts., Inyo-White Mts. June–Aug.

15. **C. holóptera** (Gray) Macbr. [*Eritrichium h.* Gray.] Erect hirsute and strigose annuals, 1–5 dm. high, with rather numerous short ascending branches; lvs. linear-lanceolate, the upper sessile, the lower petioled, 2.5–5 cm. long, conspicuously pustulate and hispid; racemes in 1's or 2's, sparsely bracted, usually 4–5 cm. long; calyx oblong-ovate, 2.5–3.5 mm. long in fr., hirsute; corolla 1–1.5 mm. broad; nutlets 4, broadly ovoid or triangular-ovoid, 1.5–2.5 mm. long, dark with pale tuberculations, the margins ± winged; style clearly longer than nutlets.—Uncommon in gravelly or rocky places, below 2000 ft.; Creosote Bush Scrub; Colo. and e. Mojave deserts, but very scatteringly; Ariz. March–April.

16. **C. inaequàta** Jtn. Ascendingly branched annual 1–3 dm. tall, strigose and bristly; lvs. linear, 1.5–4 cm. long, hispid or strigose, pustulate; spikes in 1's or 2's, rather dense, minutely sparsely bracted; calyx in fr. 2.5–3.5 mm. long, ovoid-oblong, hirsute; corolla 1 mm. broad; nutlets 4, triangular-ovoid, with pale tuberculations, heteromorphous, the odd one 1.5–2 mm. long, the consimilar 3 slightly shorter; style much surpassing nutlets. —Gravelly and rocky places, below 4000 ft.; Creosote Bush Scrub, Joshua Tree Wd.; Death V. region to ne. Colo. Desert; w. Nev. March–May.

17. **C. intermèdia** (Gray) Greene. [*Eritrichium i.* Gray. *C. i.* var. *johnstonii* Macbr.] Plate 24, Fig. F. Erectly branched, commonly stiff annual 1.5–5 dm. tall, frequently somewhat strigose, usually very hirsute; lvs. lanceolate to linear, hispid or strigose, 1.5–5 cm. long, ± pustulate; spikes bractless, in 2's to 5's, 5–15 cm. long; calyx in fr. oblong-ovoid, ca. 4–6 mm. long, the lance-linear lobes connivent with ± spreading tips, the thick midribs pungently hirsute, the margins short-hispid or villous; corolla 3–7 mm. broad; nutlets mostly 4, lance-ovoid, 1.5–2.3 mm. long, tubercled or verrucose, the back convex, the groove narrow or closed, with small basal areole; style usually reaching ca. to tip of nutlets; $n = 12$ (Di Fulvio, 1965).—The common cismontane sp., mostly below 3000 ft., but reaching 7000 ft.; Coastal Sage Scrub, Chaparral, Foothill Wd., etc., to desert borders; n. to n. Calif.; L. Calif. March–July.

18. **C. jàmesii** (Torr.) Pays. var. **abortìva** (Greene) Pays. [*Oreocarya a.* Greene.] Perennial, the stems prostrate or laxly ascending, the root woody; plant leafy, pallid, 0.5–1.8 dm. high, tomentulose, sparsely hispidulous; lvs. oblanceolate, 4–10 cm. long, silky-strigulose; infl. open, not much elevated above lvs.; bracts conspicuous; calyx-lobes lance-ovate, becoming 4–5 mm. long, tomentulose, sparsely short-setose; corolla-tube 3–4 mm. long, the limb white, ca. 5 mm. broad; nutlets 1–4, lance-ovoid, 2–2.5 mm. long, smooth, glossy, the back strongly rounded from base toward apex, the center decidedly triangular in cross section, with the narrow groove along the crest of its median inner

angle.—Dry gravelly soil in open woods, 6000–12,000 ft.; Sagebrush Scrub, Yellow Pine F. to Subalpine F.; San Bernardino Mts., Inyo-White Range; w. Nev. May–Aug.

19. **C. marítima** (Greene) Greene. [*Krynitzkia m.* Greene.] A stiffish ascending loosely branched annual 1–3.5 dm. high, rather sparsely strigose or sparsely hispid, brown or reddish; lvs. linear to lance-linear, hispid and pustulate, 1–3.5 cm. long; spikes in 1's or 2's, 1–6 cm. long, completely or partially bracted; fruiting calyx asymmetrical, 2–3 mm. long, the lobes connivent, firm, with 3 hispid on midribs and ± villous; corolla 0.5–1 mm. broad; nutlets 1–2 heteromorphous, the axial one smooth, shining, lance-oblong, 1–2 mm. long, the abaxial grayish, minutely tubercled, slightly shorter, often abortive; style ca. as high as smaller nutlet.—Sandy and gravelly places below 4000 ft.; Creosote Bush Scrub, Joshua Tree Wd., Coastal Sage Scrub; deserts from Inyo Co. to Imperial Co., coastal mesas, Santa Barbara Ids., San Diego; L. Calif., Ariz. March–May. A form with long white hairs so as to be more conspicuously villous has been named var. *pilosa* Jtn. With the typical form.

20. **C. micrántha** (Torr.) Jtn. ssp. **lépida** (Gray) Mathew & Raven. [*Eritrichium m.* var. *l.* Gray.] Plants slightly taller, coarser than in ssp. *micrantha*; corollas 1–3.5 mm. broad; $n=12$ (Mathew & Raven, 1962).—Montane slopes and valleys, 2000–7000 ft.; mts. of s. Calif. to Mono Co.; L. Calif.

Ssp. **micrántha.** [*Eritrichium m.* Torr. *C. m.* (Torr.) Jtn.] Small slender strigose ascendingly branched dichotomous annual 5–15 cm. tall; lvs. oblong-oblanceolate, 3–7 mm. long; spikes many, solitary or geminate, dense, 1–4 cm. long; fruiting calyx ovate-oblong, ca. 2 mm. long; corolla 0.5–1.2 mm. broad, white; nutlets 4, homomorphous or nearly so, 1–1.3 mm. long, smooth or tubercled, the groove narrow, scarcely broadened at base; $n=12$ (Mathew & Raven, 1962).—Dry sandy places below 7500 ft.; Creosote Bush Scrub to Yellow Pine F., Coastal Sage Scrub; deserts and inner cismontane valleys; to L. Calif., Ore., Utah. March–June.

21. **C. micromères** (Gray) Greene. [*Eritrichium m.* Gray.] Slender dull green usually erect annual 1–5 dm. tall, short-hirsute throughout; lvs. linear or ± oblong, 1–4 cm. long, rather numerous, short-hispid; spikes commonly ternate, slender, bractless, 2–8 cm. long; calyx in fr. 1–2 mm. long, early deciduous, the lobes connivent, slender-hispid, the hairs on the midribs frequently hooked; corolla 0.5 mm. broad; nutlets 4, triangular-ovoid, the margins subangulate, odd nutlet ca. 1 mm. long, the others 0.7 mm. long and papillate, the groove open, gradually dilated into a small open areola.—Frequent in dry open places such as burns, below 2000 ft.; Chaparral; scattered places in cismontane s. Calif. to L. Calif. and cent. Calif. Santa Catalina, Santa Cruz and Santa Rosa ids. March–June.

22. **C. microstàchys** (Greene ex Gray) Greene. [*Krynitzkia m.* Greene ex Gray.] Erect annual, slender-branched from base or above, usually very hirsute, 1–5 dm. tall; lvs. linear to lance-linear, 1–4 cm. long, obtuse, hirsute-hispidulous, sometimes strigose; spikes commonly in 2's or 3's, slender and loosely fld. in age, 3–8 cm. long; calyx ovoid to somewhat oblong in fr., 1–2 mm. long, usually strict, hirsute, the lobes lance-linear; corolla 0.5–1 mm. broad; nutlet 1, lanceolate or attenuate-ovoid, 1.5 mm. long, smooth, rounded on back and sides, the groove closed, simple or forked at very base; style ca. half as tall as nutlets.—Dry openings, below 4000 ft.; Chaparral, Foothill Wd., etc.; San Diego Co. to cent. Calif. April–June.

23. **C. muricàta** (H. & A.) Nels. & Macbr. var. **jònesii** (Gray) Jtn. [*Krynitzkia j.* Gray. *C. j.* Greene.] Annual; stems commonly solitary and erect, or several and fastigiate, 2–10 dm. tall, bearing from tip to below middle many short floriferous branchlets forming an elongate leafy paniculate infl.; corolla 1–2.5 mm. broad; fr. as in var. *muricata*.— Below 5500 ft.; many Communities; L. Calif. and San Diego Co. to n. cent. Calif.; Nev., Ariz.

Var. **muricàta.** [*Myosotis m.* H. & A. *C. m.* Nels. & Macbr.] Erect hirsute annual 1–5 dm. tall, loosely branched, usually with several well developed ascending laterals; lvs. linear, cinereous, short-hirsute, 1.5–3 cm. long; spikes in 2's to 5's, bractless, terminating the well developed sparsely leafy pedunculate stem or branches; calyx in fr. ovoid, 2–4 mm. long, the lanceolate lobes connivent, tawny-hirsute; corolla 2–6 mm. broad; nutlets 4, glossy or dull, verrucose or tuberculate, grayish, the lateral angles usually acute, the ventral groove narrow or closed, broadly forked or opened into basal areola; style surpassing nutlets.—Gravelly or rocky open places, below 8000 ft.; many Plant Communities; Orange Co. to Tehachapi Mts., then cent. Calif. April–June. Minor variations that have been named are var. *clokeyi* (Jtn.) Jeps. [*C. c.* Jtn.] with calyx 6–9 mm. long; nutlets whitish; Mojave Desert n. of Barstow, and var. *denticulata* (Greene) Jtn. [*Krynitzkia d.* Greene] with corolla 1–2 mm. broad; plant loosely branched occurring in dry places, 4000–8000 ft. from Santa Rosa Mts. to cent. Sierra Nevada.

24. **C. nevadénsis** Nels. & Kenn. var. **nevadénsis**. [*Krynitzkia barbigera* var. *inops* Bdg.] Slender-stemmed annuals 1–5 dm. tall, the 1–several stems erect to flexuous, short-strigose; lvs. linear to linear-oblanceolate, 1–3.5 cm. long; spikes slender, in 2's, or 3's, sometimes bracteate, somewhat congested, or elongate up to 15 cm. long; fruiting calyx lanceolate, 8–12 mm. long, the linear lobes connivent above, with recurved tips, the margin somewhat villous, the midribs hirsute; corolla 1–2 mm. broad; nutlets 4, lanceolate, long-acuminate, ca. 2.5 mm. long, verrucose on back, muricate at tip, the groove open or closed, dilated below into small areola.—Sandy or gravelly places, below 6000 ft.; Creosote Bush Scrub to Pinyon-Juniper Wd.; deserts from Inyo Co. to L. Calif., Ariz., Utah. March–May.

Var. **rígida** Jtn. [*C. intermedia* var. *r.* Brand.] Erect, not flexuous, less slender-stemmed; calyx 5–10 mm. long; nutlets oblong-ovoid, merely acute, ca. 2 mm. long, verrucose but not noticeably muricate.—At 3000–6000 ft.; Creosote Bush Scrub, Joshua Tree Wd.; w. Mojave Desert; to s. cent. Calif.; Ariz.

25. **C. nubígena** (Greene) Pays. [*Oreocarya n.* Greene.] Stems slender, several, from a densely leafy and branched root-crown, erect, 0.5–1.5 dm. high, retrorsely pubescent and spreading-setose; lvs. linear-oblanceolate or spatulate, 2–3 cm. long, thinly hirsute-pubescent and setose, ± pustulate; infl. short-spicate, leafy-bracteate; calyx-lobes 6–7 mm. long in fr., slender-setose and with shorter finer hairs; corolla white, the tube ca. 3 mm. long, the limb ca. 4 mm. broad; nutlets lance-linear, 3 mm. long, narrowly wing-margined, tubercled or rugose on back, groove straight, narrow but open, extending almost entire length of nutlet.—Rocky and gravelly places, 8000–11,000 ft.; Bristle-cone Pine F.; Inyo Range; s. Sierra Nevada. July–Aug.

26. **C. pterocàrya** (Torr.) Greene. [*Eritrichium p.* Torr.] Erect ascendingly branched annual 1–5 dm. tall, strigulose or short-hirsute; lvs. linear or the upper somewhat wider and reduced; spikes mostly in 2's or 3's, 2–5 (–10) cm. long, usually bractless; mature calyx accrescent, ± ovoid, 3–5 mm. long, the connivent lobes ± hispid; corolla white, to 1(2) mm. broad; nutlets 4, heteromorphous, the lance-oblong or lanceolate body 2–2.8 mm. long, verrucose or muricate, the axial one unmargined, the other 3 with broad crenulate or lobulate wing-margins; style longer than nutlets.—Sandy and gravelly places below 6000 ft.; Creosote Bush Scrub to Pinyon-Juniper Wd.; deserts from L. Calif. to Lassen Co.; to Wash., Utah. March–June. A form with nutlets all wing-margined has been called var. *cycloptera* (Greene) Macbr. and another with one nutlet unmargined, the others with margins narrow, knifelike is var. *purpusii* Jeps.

27. **C. racemòsa** (Wats.) Greene. [*Eritrichium r.* Wats. *C. r.* var. *lignosa* Jtn.] Woody perennial but flowering the first year, mostly bushy, 2–8 dm. tall, the ultimate branches slender, strigose; lvs. subulate to linear-oblanceolate, 0.8–3.5 cm. long; racemes slender, lax, 2–4 cm. long; pedicels slender, often recurved, 1–4 mm. long; calyx-lobes hirsute and strigose, 2–3 mm. long; corolla white, 1 mm. broad; nutlets 4, triangular-ovate, with pallid tuberculations and knifelike margin; odd nutlet 1–2 mm. long, the others 0.8–1.5 mm. long; style much surpassing nutlets.—Rocky places below 4500 ft.; Creosote Bush Scrub, Joshua Tree Wd.; e. Mojave and the Colorado desert, Inyo Co. to Imperial Co.; Nev., Ariz. March–May.

28. **C. recurvàta** Cov. Branched erect rather slender annual 1–3 dm. high, strigose; the lower lvs. oblanceolate, 1.5–2 cm. long, the cauline remote, linear to lanceolate, 0.5–1 cm. long, strigose and pustulate; spikes bractless, lax, in 1's or 2's; fruiting calyx slender, bent and recurved, 3 mm. long, the thickened midribs hispid; corolla ca. 1.5 mm. broad; nutlets 2, one oblong-lanceolate, curved, brownish, granulate-muriculate, axial, 1.5–2 mm. long, the groove somewhat oblique, narrow or closed, the other abortive; style

shorter than perfect nutlet.—Uncommon in sandy places, 2500–6500 ft.; Creosote Bush Scrub, Joshua Tree Wd.; Kelso (Mojave Desert), Death V. region to White Mts.; Owens V.; to Ore., Colo. April–June.

29. **C. roosiòrum** Munz. Perennial near to *C. nana* var. *ovina* and doubtfully distinct in lower stems (0.1–0.3 dm. long) strigose and appressed-setose, not tomentose; corolla 5 mm. broad; nutlets irregularly ridged.—At 8400–10,600 ft.; Bristle-cone Pine F.; Inyo Mts.

30. **C. símilis** Mathew & Raven. Differing from *C. circumscissa* in having the corolla 4–6 mm. broad; pollen grains 5.5–6.5 µ long; $n=6$ (Mathew & Raven, 1962).—Occasional on Mojave Desert.

31. **C. símulans** Greene. Erect, slender-stemmed annual 1–4 dm. tall, with few strict branches, whitish-strigose; lvs. linear to oblance-linear, 1–3 cm. long, strigose; spikes in 1's to 3's, bractless, loosely fld., elongate; calyx in fr. 3–7 mm. long, the lance-linear lobes connivent above with spreading green tips, the midribs with retrorse or spreading bristles, the margins with ascending hairs; corolla 1–2 mm. broad; nutlets 4, broadly ovoid, 2–2.5 mm. long, tessellate-papillate, the angles rounded, the back low-convex, the groove mostly closed throughout, broadly forked below; style slightly shorter than nutlets.—Dry wooded slopes, 2500–7500 ft.; Yellow Pine F. and above; mts. from San Diego Co. to n. Calif. to Wash., Ida. May–July.

32. **C. tráskiae** Jtn. Loosely branched slender strigose annual 1–2 dm. high, lvs. linear 0.5–2 cm. long strigose or hispidulous; spikes in 1's or 2's, dense, 1–5 cm. long, the lower fls. bracteate; calyx in fr. ovoid, 2–3 mm. long, hirsute; corolla 1.5 mm. broad; nutlets 4, ovoid, finely granulate, the back convex, ± tubercled toward tip, the margin obtuse, the groove closed, slightly dilated to a minute basal areola; style not exceeding nutlets.—Sandy places, San Clemente and San Nicolas ids. April–June.

33. **C. tumulòsa** (Pays.) Pays. [*Oreocarya t.* Pays.] A strong perennial, 1–2.5 dm. tall, somewhat villous and slender-setose; lvs. oblanceolate to obovate-spatulate, 3–5 cm. long, appressed-setose, obscurely pustulate; infl. becoming loosely cylindrical, 2–2.5 cm. thick, with yellowish bristles; calyx-lobes 7–10 mm. long in fr., setose-spreading with retrorse bristles; corolla white, the tube 3.5–4 mm. long, the limb 6–7 mm. broad; nutlets ovoid, 4 mm. long, irregularly and coarsely tuberculate, sometimes obscurely rugulose, the scar open.—Dry places in limestone, 4500–6000 ft.; Pinyon-Juniper Wd.; Providence Mts., Mid Hills, New York Mts., Ivanpah Mts., e. San Bernardino Co.; s. Nev. April–June.

34. **C. utahénsis** (Gray) Greene. [*Krynitzkia u.* Gray.] Erectly branched slender annual 1–3 dm. high, strigose or ± appressed-hirsute; lvs. mostly linear, 3–5 cm. long, appressed short-hispid and pustulate especially beneath; spikes bractless, mostly in 2's; calyx ovoid or oblong-ovoid, 2–3 (–4) mm. long, the midrib only rarely setose; corolla 2–3 mm. broad; nutlets broadly lanceolate, usually 1–2 maturing, granulate to muricate-papillate, 1.7–2.5 mm. long, the back low or flat, the margin sharp, knifelike.—Sandy and gravelly places below 6500 ft.; Creosote Bush Scrub, Joshua Tree Wd.; deserts from Santa Rosa Mts., Riverside Co. to Inyo Co.; to Nev., Utah, Ariz. March–May.

35. **C. virginénsis** (Jones) Pays. [*Krynitzkia glomerata* var. *v.* Jones.] Biennial or short-lived perennial, the stems tufted, 1–4 dm. tall, hispidulous and hispid; lvs. oblanceolate to spatulate, 5–10 cm. long, the lower crowded, bristly, somewhat tomentulose, pustulate; infl. thyrsoid, loose, 3–6 cm. thick, fulvescent, hirsute; calyx-lobes 7–12 mm. long, in fr., lance-linear, setose; corolla white, the tube 3–4 mm. long, the limb 3–7 mm. broad; nutlets ovoid, short-acuminate, 3–4 mm. long, coarsely tuberculate, and irregularly rugose, carinate on back, the scar narrowly triangular.—Dry rocky slopes, 3500–10,200 ft.; Creosote Bush Scrub to Pinyon-Juniper Wd., Bristle-cone Pine F.; Panamint Mts., Inyo Mts. to Barstow, Baker, Kingston Mts., Clark Mt.; to Utah. April–June.

5. Èchium L. Viper's Bugloss

Herbs or shrubs, usually scabrous, hispid or canescent. Lvs. alternate. Fls. in scorpioid simple or forked cymes, with bracts small or leafy. Calyx with 5 narrow lobes. Corolla tubular to funnelform, with dilated oblique throat, without appendages, the corolla-lobes roundish, unequal. Stamens 5, inserted below the middle of the corolla-tube, unequal and exserted. Style filiform. Nutlets 4, erect, wrinkled. Ca. 35 spp. of the Old World. (Greek, *echis*, a viper, the nutlets supposed to represent a viper's head.)

1. **E. plantagíneum** L. Annual or biennial, erect or diffuse, 4–9 dm. high, villous-hirsute, with hairs pustulate at base; lvs. obtuse, strigose, the basal oval and petioled, the upper lance-oblong, subsessile; fls. mostly bracteate; calyx-lobes lanceolate, acuminate, ca. 10 mm. long; corolla blue, 15–20 mm. long, irregular with conspicuous throat and oblique limb; $2n=16$ (Britton, 1951).—Locally adventive as at De Luz, San Diego Co., Santa Barbara, etc.; native of s. Eu. May–June.

6. Harpagonélla Gray

Small annuals with stems branching from base. Lvs. narrow. Fls. in leafy-bracted false racemes. Pedicels twisted and laterally deflexed at maturity. Calyx becoming burlike, with 3 narrow distinct lobes, the 2 others fused, accrescent and indurate to form a galeate structure enclosing the upper part of 1 nutlet and armed dorsally with 5–9 soft hooked spines. Corolla minute, subbracteate. Style entire. Ovary 2-parted. Nutlets 2, dissimilar; the enclosed nutlets flattened on inner surface to form a margined strigose areole, otherwise round and glabrous; the free nutlet angled, completely strigose, bent so as to parallel the enclosed. Monotypic. (Diminutive of Latin *harpago*, grappling hook.)

1. **H. pálmeri** Gray. Plate 25, Fig. A. Loosely spreading, with strigose disarticulating stems 5–30 cm. long; lvs. strigose, ± linear, 1–3.5 cm. long; pedicels short, stout, recurved in fr.; bracts 2–8 mm. long; calyx-lobes 1–1.5 mm. long in fl., 2–3.5 mm. in fr.; corolla white, ca. 2 mm. long, enclosed in nutlet ca. 3 mm. long, minutely muriculate; free nutlet smooth.—Rare and local, on dry slopes and mesas below 1500 ft.; Chaparral; cismontane s. Calif. from Los Angeles Co. to L. Calif. Santa Catalina Id. March–April.

7. Heliotròpium L. HELIOTROPE

Herbs or shrubs with lvs. usually alternate, petioled, mostly entire. Fls. in spicate or racemose scorpioid cymes or borne along leafy stems, usually between or opposed to the lvs. Calyx 5-toothed or 5-lobed. Corolla white, blue or yellow, funnelform or with a spreading limb and short tube, rarely subtubular, the throat open. Stamens 5, attached to corolla-tube, included, the anthers obtuse to acuminate. Ovary glabrous or pubescent, with 4 fertile cells or fewer by abortion, entire or rounded with 2–4 lobes, at maturity breaking up into uniovulate or biovulate nutlets. Style apical, simple; stigma consisting of a fertile discoid base and a superimposed sterile apical frequently bifid appendage. Ca. 125 spp., of warmer regions. (Greek, *helios*, sun, and *trope*, turning, referring to the summer solstice when the spp. were supposed to come into fl.)

Corolla purplish. Garden escape .... 1. *H. amplexicaule*
Corolla white or pale. Native.
  Annual, rough hairy, not fleshy; fls. short-pedicellate .... 2. *H. convolvulaceum*
  Perennial, glabrous, succulent; fls. sessile .... 3. *H. curassavicum*

1. **H. amplexicáule** Vahl. Heavy-rooted perennial; stems decumbent, 3–6 dm. long, branched, rather soft-hirsute; lvs. deep green, lance-oblong, sinuate-dentate, 2–6 cm. long, subsessile, hirsute on veins beneath; cymes peduncled, 3–5-branched, dense at first, later loose, to ca. 1 dm. long; calyx stiff-pubescent and with short gland-tipped hairs; corolla lilac-purple, 5–6 mm. broad; fr. shallowly 2-lobed.—Occasionally established and persisting, as Yucaipa, n. of Claremont, and about Santa Barbara; native of S. Am. Much of year.

2. **H. convolvuláceum** (Nutt.) Gray var. **califórnicum** (Greene) Jtn. [*H. c.* Greene. *Euploca convolvulacea* ssp. *c.* Abrams.] Loosely branched annual, hispid with spreading and appressed yellowish-white hairs with pustulate bases; stems 5–30 cm. long; lvs. ovate to

Plate 25. Fig. A, *Harpagonella palmeri*, free nutlet. Fig. B, *Heliotropium curassavicum*. Fig. C, *Pectocarya penicillata*, 4 nutlets. Fig. D, *Plagiobothrys californicus*, nutlet. Fig. E, *Arabis glaucovalvula*. Fig. F, *Arabis holboellii* var. *retrofracta*.

oblong-ovate, rounded at base, 1.5–3 cm. long; fls. solitary in lf.-axils, fragrant; calyx-lobes linear, 4–5 mm. long; corolla white, 8–14 mm. broad; fr. 3–4 mm. broad, silky.—Occasional, in sandy places below 2000 ft.; Creosote Bush Scrub; e. Mojave Desert, Colo. Desert; to Ariz., Son. April–May.

3. **H. curassávicum** L. var. **oculàtum** (Heller) Jtn. [*H. o.* Heller.] Plate 25, Fig. B. Perennial with underground rootstocks which send up scattered shoots; stems diffuse, 1–5 dm. long, fleshy, glaucous, glabrous throughout; lvs. succulent, usually glaucous, oblanceolate to spatulate, cuneate at base, 1–4 cm. long, short-petioled; fls. in dense geminate or solitary scorpioid cymes 2–10 cm. long, bractless; calyx 2–3 mm. long; corolla 3–6 mm. broad, white with yellow spots in throat and usually becoming ± purple about the center; fr. glabrous, 1.5–2 mm. broad, subglobose, at last separating into 4 nutlets.—Common in saline or alkaline soils, below 7000 ft.; many Plant Communities; throughout California; to Ariz., Utah. March–Oct.

8. *Láppula* Moench. STICKSEED

Annual or biennial herbs with linear or oblong lvs. Fls. in paniculate leafy-bracted racemes, white or somewhat yellowish, usually on erect pedicels. Calyx 5-parted. Corolla with short tube and rounded ascending lobes, the throat usually appendaged. Stamens short, attached to corolla-tube. Style short, surmounting the columnar gynobase, usually surpassing the nutlets. Stigma subcapitate. Nutlets erect, smooth to roughened, attached along their ventral keel to the elongate gynobase, commonly with 1 or more marginal series of glochidiate appendages. Ca. 14 spp., mostly N. Temp. (Diminutive of Latin, *lappa*, a bur.)

(Johnston, I. M. Contr. Gray Herb. 70: 47–51. 1924.)

Corolla ca. 3 mm. broad; nutlets with 2 rows of slender marginal spines that are not confluent at base .... 1. *L. echinata*
Corolla ca. 2 mm. broad; nutlets with 1 row of marginal spines that are ± confluent at their base
2. *L. redowskii*

1. **L. echinàta** Gilib. Erect annual, 1.5–5 dm. high, villous-hirsute with hairs appressed upward; lower lvs. oblanceolate, others linear, sessile, ascending, 2.5–5 cm. long; racemes bracteate; pedicels 1–3 mm. long; calyx-lobes linear, ca. 3 mm. long; corolla blue; nutlets 3.5–4 mm. long, muricate-prickly on backs.—Occasional adventive; native of Eurasia. June–Aug.

2. **L. redówskii** (Hornem.) Greene. [*Myosotis r.* Hornem.] Erect annual 2–8 dm. tall, finely hispid or hispid-villous, somewhat cinereous; lvs. linear to lance-linear or oblong, the upper sessile, 1–2 cm. long, the lower petioled and longer; pedicels 1–2 mm. long; calyx-segms. lance-linear, ca. 3 mm. long; corolla blue to ochroleucous, 2–3.5 mm. long; nutlets 2.5 mm. long, tubercled or muricate and with a single row of barbed prickles 1–2 mm. long and nearly distinct.—Dry open slopes, 5000–8000 (–11,000) ft.; mostly Yellow Pine F., Pinyon-Juniper Wd., Sagebrush Scrub; San Bernardino Mts., Panamint Mts., White Mts.; to Modoc Co.; Wash., Eurasia, Argentina. April–July. Occasional plants tend to have marginal prickles confluent and ± cupulate and have been named var. *desertorum* (Greene) Jtn. They tend to occur at 2000–7000 ft.

9. *Lithospérmum* L. GROMWELL. PUCCOON

Annual to perennial hairy herbs, the roots usually with red or violet coloring matter. Stems few or several, simple below. Lvs. alternate, mostly rather narrow. Fls. small, yellow, white or blue, in leafy spikes or upper axils. Calyx 5-parted. Corolla funnelform or salverform, with rounded lobes imbricated in bud. Stamens 5, included, inserted in corolla-throat; fils. short; style slender; stigma truncate-capitate or 2-lobed. Nutlets 4 or fewer, erect, attached by their bases to the nearly flat receptacle, the attachment-scar flat. Ca. 50 spp., of all continents except Australia. (Greek, *lithos*, stone, and *sperma*, seed.)

1. **L. incìsum** Lehm. Perennial with branched root-crown; stems several, 1–2 dm. high, canescent-strigose; lvs. crowded, lance-linear, with attenuate apex, strigose, 3–6 cm. long; fls. in upper axils; calyx-lobes narrow; corolla bright yellow, the tube 12–30 mm. long, the limb 12–20 mm. wide, the lobes erose; nutlets smooth, shining.—Sandy slopes among rocks, 5650 ft.; Pinyon-Juniper Wd.; Keystone Canyon, New York Mts., e. San Bernardino Co.; to Mont. and B.C. May.

10. *Lycópsis* L. BUGLOSS

Coarse setose annuals with alternate lvs. Racemes leafy-bracted, spikelike, scorpioid. Calyx 5-parted. Corolla salverform with curved tube and slightly unequal limb, the throat closed by stiff hairs. Stamens and style included. Nutlets 4, rough-wrinkled. A few spp., of Old World. (Greek, *lycos*, wolf, and *opsis*, appearance.)

1. **L. arvénsis** L. Erect, 3–6 dm. high, rough-bristly; lvs. lanceolate, 3–5 cm. long; calyx 3–4 mm. long; corolla blue or purplish, 4–5 mm. broad.—Taken long ago as adventive at Upland; native of Eu.

11. *Pectocárya* DC. ex Meissn.

Low mostly spreading annuals, with slender stems and narrow lvs., strigulose-canescent. Fls. in leafy-bracted false racemes which constitute most of the plant. Pedicels decurved in fr. Calyx 5-parted, the lobes finally spreading. Corolla white, small, the lobes short, ascending, the throat with small intruded appendages. Stamens included. Style very short. Stigma capitate. Nutlets 4, linear to obovate, divaricate, commonly paired, usually margined, bearing hooked hairs, radially spreading or recurving. Gynobase low and broadly pyramidal. Ca. 10 spp. of w. N. Am. and w. S. Am. (Greek, *pectos*, combed, and *karua*, nut because of comb-like margins on some nutlets.)

1. Nutlet-margins entire, lacking fringelike teeth and uncinate hairs; nutlet-body cuneate-obovate; plant erect or nearly so .... 6. *P. setosa*
1. Nutlet-margins lacerate or undulate or uncinate-bristly; nutlet-body linear or oblong.
    2. Nutlets with margins entire or undulate along sides, armed only at distal end where densely uncinate-bristly.
        3. Nutlets all wing-margined .... 3. *P. penicillata*
        3. Nutlets heteromorphic, 1 of each divergent pair winged, the other wingless or merely margined .... 1. *P. heterocarpa*
    2. Nutlets with margins pectinately lacerate or toothed most of their length, also usually uncinate-bristly about the distal end.
        4. Nutlet-margins narrow and inconspicuous, the teeth nearly or quite distinct.
            5. Nutlets straight or somewhat falcate; marginal appendages stout ..... 2. *P. linearis*
            5. Nutlets conspicuously recurved; marginal appendages very elongate ... 5. *P. recurvata*
        4. Nutlet-margins broad and conspicuous, the teeth obviously confluent ... 4. *P. platycarpa*

1. **P. heterocárpa** (Jtn.) Jtn. [*P. penicillata* var. *h.* Jtn.] Habit of *P. penicillata*; fruiting nutlets on principal branches heteromorphic, with 2 evidently margined and like those of *P. penicillata* and 2 unmargined and somewhat reflexed.—Gravelly or sandy slopes below 3000 ft.; Creosote Bush Scrub, Joshua Tree Wd.; Mojave and Colo. deserts; n. to Stanislaus Co.; to Utah, Son., L. Calif. Feb.–May.

2. **P. lineàris** DC. var. **ferócula** Jtn. Stems very slender, prostrate to ascending, cinereous-strigose; lvs. filiform-linear to oblance-linear, 1–4 cm. long, strigose, numerous; calyx strigose, 1.5–2 mm. long; corolla 2 mm. long; nutlets homomorphous, the body linear- or somewhat spatulate-oblong, 2–3.5 mm. long, 0.5–1 mm. broad, divaricate or slightly falcate-recurved, the margin generally not conspicuous, divided into short crowded teeth, each of which is abruptly terminated by an uncinate bristle equal to or surpassing the total width of the cartilaginous margin beneath it.—Open grassy mesas and slopes below 2500 ft.; Coastal Sage Scrub, Chaparral, etc.; L. Calif. through cismontane s. Calif. to San Benito Co. Channel Ids. March–May.

3. **P. penicillàta** (H. & A.) A. DC. [*Cynoglossum p.* H. & A.] Plate 25, Fig. C. Stems many, slender, prostrate or widely ascending, cinereous-strigose, 5–20 cm. long; lvs. linear to narrow-spatulate, 1–3 cm. long, to 2 mm. wide; calyx almost as long as nutlets; corolla 2 mm. long; nutlets homomorphous, with body oblong, divaricate, straight, 1.6–2.4 mm. long, 0.5–0.8 mm. wide, with a distinct upturned or incurved margin broadest near base and apex, subentire and armed only at rounded distal end with crowded hooked bristles; $2n = 12$ pairs (Raven, Kyhos & Hill, 1965).—Dry sandy and gravelly places below 4500 ft.; many Plant Communities; cismontane away from the coast and to desert edge; to L. Calif., B.C., Wyo. March–May.

4. **P. platycárpa** (M. & J.) M. & J. [*P. gracilis* var. *p.* M. & J.] Stems slender, stiffish, prostrate or widely ascending, 8–16 cm. long, cinereous-strigose; lvs. linear or oblance-linear, 1–3.5 cm. long, to 1.2 mm. broad; calyx ca. as long as nutlets; corolla 2 mm. long; nutlets usually heteromorphous, the body linear- or spatulate-oblong, 2.5–3 mm. long, 0.6–1 mm. broad, at least 3 nutlets with conspicuous wide coarsely toothed pallid margin of which the irregular teeth are tipped with short hooked bristles that are much shorter than the cartilaginous margin beneath them, the odd nutlet may be narrower with more dissected wing and more pubescent body.—Dry gravelly places below 4000 ft.; Creosote Bush Scrub, Joshua Tree Wd.; Mojave and Colo. deserts from cent. San Bernardino Co. s.; to Utah, Nev., L. Calif. March–May.

5. **P. recurvàta** Jtn. With the habit of *P. linearis*; calyx surpassed by mature nutlets, these with a linear body, reflexed, distinctly falcate or even scorpioid, 3 mm. long, 0.8 mm. broad; nutlet-margin dissected, reduced to a series of quite distinct teeth which are crowned by elongate hooked hairs of equal or greater length, the slender subulate marginal appendages thus formed well spaced and surpassing the width of the fr.; $n = 12$ (Di Fulvio, 1965).—Sandy and gravelly slopes below 4000 ft.; Creosote Bush Scrub, Joshua Tree Wd.; Mojave and Colo. deserts; to Nev., Ariz., Son. L. Calif. March–May.

6. **P. setòsa** Gray. [*P. s.* vars. *aptera*, and *holoptera* Jtn.] Stems rather stiff and coarse, erect or ascending, scattered-setose and thinly strigose, 5–20 cm. high; lvs. oblong to spatulate or oblanceolate, 5–25 mm. long, 1–3.5 mm. broad, pustulate, stiff-hairy; calyx surpassing fr.; corolla 2 mm. long; nutlets divergent in pairs, ± heteromorphous, 3 usually with broad entire thin margin, rarely all 4 margined or all marginless; body of nutlets broadly cuneate-obovate; 2 mm. long, the margin and face with slender hooked bristles. —Dry gravelly or sandy places below 5000 (7000) ft.; Creosote Bush Scrub, to Pinyon-Juniper Wd.; Mojave and Colo. deserts; to Wash., Ida., Ariz., L. Calif. April–May.

12. *Plagiobóthrys* F. & M.

Annual or perennial herbs, usually with slender appressed hairs, at times bristly but not pungently so. Lower lvs. opposite or rosulate and crowded. Fls. generally in slender racemes or spikes, occasionally glomerate or axillary, frequently bracted. Calyx divided to middle or lower, usually persistent, frequently tawny or brown, often accrescent. Corolla white, the tube short and usually included in calyx. Stamens included. Ovules 4. Nutlets 1–4, erect or incurved, roughened or rarely smooth, the margin round or angled, the back occasionally appendaged, attached medianly or basally to the low convex gynobase through a mostly caruncular scar, the scar decurrent on lower part of keel, or projected on a short basal prolongation of keel, or more commonly located at lower end of keel and ± sunken beneath the level of its crest. Ca. 100 spp., w. N. and S. Am. (Greek, *plagios*, on the side, and *bothrys*, a pit, referring to the caruncular scar of *P. fulvus* the first known sp.)

(Johnston, I. M. A synopsis and redefinition of the genus Plagiobothrys. Contr. Gray Herb. 68: 57–80. 1923. The Allocarya section of the genus Plagiobothrys in the western U.S. Contr. Arnold Arbor. 3: 1–82. 1932.)

1. Lvs. all alternate.
    2. Caruncle of nutlet elongate, apparently extending along crest of ventral keel; nutlets trigonous; corolla 1–2.5 mm. broad; nutlets tessellate. Desert .... 7. *P. jonesii*

2. Caruncle round or nearly so, at or below end of ventral keel.
      3. Lowest lvs. not in a rosette; caruncle weakly developed, borne at tip of a short ventral stipe; nutlets lacking a broad transverse ventral groove. Montane and cismontane
         .... 4. *P. californicus*
      3. Lowest lvs. mostly in a rosette; caruncle well developed, sessile on nutlet; nutlet with a shallow transverse ventral groove.
         4. Calyx circumscissile in fr., less than 4 mm. long; lobes usually connivent over fr.; mature nutlets usually only 1 or 2.
            5. Infl. a long simple bracted raceme; nutlets highly arched in lateral outline, 1–2.5 mm. long; corolla 2–3 mm. broad. W. desert .... 2. *P. arizonicus*
            5. Infl. forked, bracted at base only if at all; nutlets low and flattened in lateral outline, 2–3 mm. long; corolla 3–9 mm. broad. Cismontane .... 9. *P. nothofulvus*
         4. Calyx not cirsumscissile, or if so, the strongly accrescent calyx over 4 mm. long; calyx-lobes erect or spreading; nutlets usually 4.
            6. Transverse dorsal crests of nutlets very narrow and sharp, with median keel enclosing polygonal granulate areoles; stems rather coarse .... 5. *P. canescens*
            6. Transverse dorsal crests of nutlets very low and broad, separated only by low lineate ridges.
               7. Nutlets ovoid, usually constricted at apex only, the base rounded or rarely weakly constricted, dark; plant dye-stained .... 13. *P. torreyi*
               7. Nutlets cruciform due to the abrupt constrictions at apex and base, glossy; plant rarely dye-stained .... 12. *P. tenellus*
1. Lvs. opposite at least below.
   8. Nutlets attached to gynobase by a ± well developed stipelike ventral projection
      .... 4. *P. californicus*
   8. Nutlets attached directly to gynobase, without a stipelike projection.
      9. Scar large, deeply excavated, ¼ to ½ the length of the nutlet; nutlets with glochidiate appendages .... 1. *P. acanthocarpus*
      9. Scar small, not much if at all excavated, ⅛ the length of the nutlet or less.
         10. Nutlet attachment basal, frequently substipitate; plants ± succulent; calyx-lobes strongly ribbed .... 8. *P. leptocladus*
         10. Nutlet attachment lateral to obliquely basal; calyx-ribs only rarely thickened.
            11. Scar linear or nearly so, usually borne on a knifelike attachment .. 15. *P. undulatus*
            11. Scar broad, the attachment not at all knifelike, or if so the nutlets rather asymmetric.
               12. Stems with distinctly spreading hairs. Deserts.
                  13. Nutlets less than 2 mm. long, transverse ridges on dorsal side prominent; scar lateral in a groove formed by ridges; racemes with 1 or 2 bracts near the base
                     10. *P. parishii*
                  13. Nutlets 2–2.5 mm. long, with low transverse ridges on dorsal side; scar suprabasal, oblique, not sunken in a groove; racemes well-bracted .. 11. *P. salsus*
               12. Stems strigose or appressed-hispidulous.
                  14. Scar near or at base of nutlet, oblique to the ventral keel .... 3. *P. bracteatus*
                  14. Scar distinctly lateral, parallel with the ventral keel.
                     15. Nutlets usually muriculate or minutely bristly; scar linear-oblong. Montane
                        6. *P. hispidulus*
                     15. Nutlets not muriculate or bristly; scar ovate or deltoid. Cismontane
                        14. *P. trachycarpus*

1. P. acanthocárpus (Piper) Jtn. [*Allocarya a.* Piper.] Strigose erect to spreading annual; the stems 1–4 dm. long; lower lvs. linear or spatulate-linear, 2–6 cm. long, upper linear to oblanceolate; racemes bracted, becoming loose and elongate; pedicels 1–2 mm. long; calyx broad and loose in fr., 3–6 mm. long, ± tawny; corolla 1–2.5 mm. broad; nutlets ovoid, 1.5–2.5 mm. long, the back strongly medially keeled, with lateral keels ± developed and with transverse ridges so as to be reticulate-rugose, tubercled in the interspaces, ± armed with stiff subulate apically glochidiate appendages; ventral side with large deeply excavated deltoid or ovate scar that is broadly flanged.—Moist flats or beds of winter pools, below 2000 ft.; Chaparral, V. Grassland; mesas near San Diego; L. Calif., San Joaquin V. and n. March–May.

2. P. arizónicus (Gray) Greene ex Gray. [*Eritrichium canescens* var. *a.* Gray.] Erect or loosely ascending annual, usually ± branched below the middle, 1–4 dm. tall, hirsute-hispid and somewhat villous; lvs. with stiff somewhat appressed hairs, ± pustulate, the lower in a rosette, linear-oblanceolate, 1.5–5 cm. long, the upper scattered on the slender stems, lanceolate, reduced; roots, petioles, etc. with purple dye; spikes lax in fr., bractless; calyx ca. 3 mm. long, lobed to ca. the middle, usually tawny-hirsute, connivent, circumscissile in age; corolla 2–2.5 mm. broad; nutlets 1–4, ovoid, abruptly acute, the back marked off into rectangular areoles by narrow keels and ridges and transverse rugae; scar median, in a sunken area at base of keel.—Dry places under desert influence, below 4000 (7000) ft.; Foothill Wd., Pinyon-Juniper Wd.; w. edge of Colo. Desert to Providence Mts., Tehachapi Mts., etc. then to cent. Calif.; New Mex., Son. March–May. Calyx-lobes less connivent over the nutlets characterize the var. *catalinensis* Gray of Santa Catalina Id.

3. **P. bracteàtus** (Howell) Jtn. [*Allocarya b.* Howell.] Sparsely strigose annual, the slender stems with long branches, usually ascending, 1–4 dm. long; lower lvs. 3–10 cm. long, linear, the upper linear to oblong; racemes slender, bracted toward base; mature calyx with linear to lanceolate lobes 2–4 mm. long; corolla 1–3 mm. broad; nutlets 1.2–2 mm. long, the back ± keeled above the middle, granulate, with irregular ± oblique transverse wrinkles or ridges, the lower somewhat broken into tubercles; ventral sides keeled and angulate to beyond the middle, the lower part forming a ridge surrounding the small narrowly ovate to elliptical scar.—Moist places or dried beds of ditches, etc. below 5000 ft.; Coastal Sage Scrub, Chaparral, etc.; much of cismontane s. Calif. especially in San Diego region; to Ore., L. Calif.

4. **P. califórnicus** (Gray) Greene var. **califórnicus**. [*Echidiocarya c.* Gray.] Plate 25, Fig. D. Light green annual with rather slender spreading or prostrate stems 1–4 dm. long, spreading-hirsute; lvs. often numerous below, oblanceolate or spatulate, 1–4 cm. long, thinly hirsute or appressed canescent, the upper cauline reduced, linear; racemes simple, elongate, leafy-bracted at least in lower half; calyx deeply lobed, 3 mm. long in fr., hirsute and sparingly hispid; corolla 4–7 mm. broad; nutlets usually 4, ovoid, 1.5–2 mm. long, the ridges and keels usually prominent, the ridges loosely reticulate or almost parallel, sometimes broken up into a series of tubercles; scar elevated on a stipe near the ak.-base.—Grassy places below 1500 ft.; V. Grassland, Coastal Sage Scrub; cismontane s. Calif. from San Bernardino and Los Angeles cos. to L. Calif. March–May.

Var. **fulvéscens** Jtn. Habit of var. *californicus*, but herbage rougher, short-hispid; corolla 2–3 mm. broad.—Dry places, 2000–6500 ft.; Chaparral, Yellow Pine F.; San Diego Co. to Santa Barbara Co. and San Luis Obispo Co. Anacapa and Santa Rosa ids. L. Calif.

Var. **grácilis** Jtn. Mature calyx 2–3 mm. long; corolla 1.5–2 mm. broad; nutlets 1–1.5 mm. long.—About La Jolla and San Diego, San Clemente, Santa Catalina and Santa Cruz ids.

Var. **ursìnus** (Gray) Jtn. [*Echidiocarya u.* Gray.] Of dense compact habit, 2–8 cm. high, stout, much-branched; racemes short; fls. obscured by lvs. and bracts; nutlets 1.7–2 mm. long.—Montane valleys, 4500–6800 ft.; Yellow Pine F.; San Bernardino and San Jacinto mts.; L. Calif.

5. **P. canéscens** Benth. Stems usually several from base, decumbent or prostrate, rarely erect, 1–6 dm. long, villous or finely hispid; lvs. linear or the basal linear-oblanceolate, 1.5–5 cm. long, the cauline well developed; roots, stems, etc. with some purple dye; spikes elongate and loosely fld. in age, leafy-bracteate; calyx 4–6 mm. long in fr., densely villous-hirsute, cleft over half way; corolla ca. 2–4 mm. broad; nutlets usually 4, round-ovoid, narrowed into a short beaklike apex, strongly incurved, obscurely tuberculate, 1–2 mm. long, usually with prominent transverse rugae forming rectangular papillate intervals; scar medial, annular, slightly raised.—Grassy slopes and flats below 4500 ft.; S. Oak Wd., V. Grassland, Coastal Sage Scrub, etc.; cismontane s. Calif. including islands, w. Mojave Desert; to n. Calif. March–May.

6. **P. hispídulus** (Greene) Jtn. [*Allocarya h.* Greene.] Grayish-green annual, usually branched at base, prostrate or loosely ascending, the strigose stems 0.5–4 dm. long; lvs. appressed-hispidulus, ± pustulate, especially beneath, the basal linear, 1–5 cm. long, the upper linear to oblanceolate; racemes usually elongate, loosely fld., single, bracted in

lower parts; mature calyx strigose, the narrow lobes 2–3 mm. long, ascending; corolla 1–2 mm. broad; nutlets ovoid to lance-ovoid, 1.5–2 mm. long, usually abruptly rounded at base, the back keeled above the middle, obliquely transverse-ridged, the ridges often broken, anastomosing, frequently dentate with papillae, or bearing branched hairs, the intervening spaces granular and tuberculate; ventral side keeled and angulate to below the middle, the lowermost part with a ridge encircling the linear-oblong scar.—Moist meadows or drying flats, 4000–11,000 ft.; Montane Coniferous F.; mts. of San Diego Co., San Bernardino Mts.; mts. of cent. and n. Calif.; to Wash., Wyo. June–Sept.

7. **P. jònesii** Gray. Annual, erect or ascending; stems 1–several, divergently branched, conspicuously bristly and fine-pubescent, pustulate; lower lvs. oblanceolate to linear, 2–6 cm. long, the cauline mostly lanceolate; racemes scorpioid, leafy-bracted, rather dense, terminal and in upper axils, but even lower lvs. may subtend some fls.; calyx-lobes linear-subulate, 6–10 mm. long; corolla 1–2 mm. broad; nutlets 2 or 3, incurved and 4-angled by the dorsal and ventral keels and lateral ridges, 2.5–3.5 mm. long, tuberculate on angles, tessellate between; scar narrow, merging into the keel above and with a diverging lateral ridge extending to either side.—Gravelly and rocky places, below 5700 ft.; Creosote Bush Scrub to Pinyon-Juniper Wd.; Whipple Mts. of e. San Bernardino Co. to White Mts., Argus Mts., etc.; to Ariz. April–May.

8. **P. leptoclàdus** (Greene) Jtn. [*Allocarya l.* Greene.] Somewhat succulent annual, the stems prostrate, strigose, 1–3 dm. long; lvs. subglabrous above, sparsely strigose and pustulate beneath, the lower linear or somewhat spatulate, 3–10 cm. long, the upper oblance-linear; racemes unilateral, rather lax in fr.; mature calyx usually accrescent, sparsely strigose, the lobes with ± indurated ribs, lanceolate to somewhat spatulate, 4–5 mm. long, keeled dorsally only above the middle, tuberculate somewhat rugulose, smooth to penicillate-hairy; ventral side angulate, keeled the entire length; scar basal, not surrounded by a ridge, frequently with downward directed dorsal flange.—Heavy usually alkaline soils that have been wet, below 2000 ft.; V. Grassland, Coastal Sage Scrub; coastal valleys of s. Calif.; to L. Calif.; Ore., Ida., Utah. March–May.

9. **P. nothofúlvus** (Gray) Gray. [*Eritrichium n.* Gray.] POPCORN FLOWER. Erect annual 2–5 dm. high, the stems branched mostly above, villous with short curly hairs or finely hispid; lvs. largely in a basal rosette, oblanceolate, 3–10 cm. long, sparsely villous, the cauline few, lance-linear, reduced upward; root, stem, petioles, etc. with copious purple dye; infl. once or twice forked, the spikes slender, mostly bractless, elongating and lax; calyx densely appressed-silky-villous, usually tawny, 2–3 mm. long in fr., the lobes erect, ca. as lnog as tube, this circumscissile in fr.; corolla 6–8 mm. broad; nutlets 1–4, round-ovoid, abruptly constricted into an acute apex, the back with ± rectangular granulate areas between narrow ridges and keels; scar annular, median at base of the narrow ventral keel; $n=12$ (Di Fulvio, 1966).—Common, grassy fields and hillsides, mostly below 2500 ft.; occasional to 5000 ft.; V. Grassland, S. Calif. Woodland, Coastal Sage Scrub, almost throughout cismontane s. Calif., occasional at desert edge; to L. Calif., Wash. March–May.

10. **P. párishii** Jtn. [*Eritrichium cooperi* Gray, not *P. c.* Gray. *Allocarya c.* Greene.] Diffusely branched annual, the stems prostrate, branched, 0.5–3 dm. long, with short stout spreading hairs; lvs. linear or the upper oblong, somewhat hispidulous and pustulate beneath, 1–5 cm. long; racemes slender, elongate, few-bracted in lower part; mature calyx hispidulous, tending to be deciduous, 2–3 mm. long, the lobes ligulate to lanceolate; corolla 3–5 mm. broad; nutlets 1–1.8 mm. long, ovoid or lance-ovoid, ± heteromorphic with the axial nutlet somewhat larger, plumper with a triangular-ovate scar; other nutlets with sublinear scar; back medianly keeled only near apex, transversely ridged, breaking up into tuberculations near base; scar in a groove.—Occasional in wet places, 2500–4500 ft.; largely Joshua Tree Wd.; w. Mojave Desert from Rabbit Springs and Lovejoy Springs to Owens V. and Mono Co. April–June.

11. **P. sálsus** (Bdg.) Jtn. [*Allocarya s.* Bdg.] Stems branched from or near the base, erect or ascending, 6–16 cm. long, sparsely hirsute; lvs. linear to narrowly oblong-oblanceolate, 3–6 cm. long, bristly-ciliate on the margins and sometimes sparsely hirsute on the upper surface; racemes leafy-bracted, loosely fld.; calyx subsessile, 4–5 mm. long, bristly

hirsute, strongly ridged at base; corolla 3-4 mm. broad; nutlets lanceolate, 2-2.5 mm. long, the dorsal side rugulose with transverse ridges above the middle, granulate below the middle but without ridges; keel on ventral side low, veinlike, extending to the small ovate-lanceolate basal scar, the sides with a few indistinct ascending lines.—Moist alkaline mud flats 3 miles se. of Death Valley Junction, se. Inyo Co.; at 2000 ft.; Alkaline Sink; Nev. to se. Ore. May-July.

12. **P. tenéllus** (Nutt.) Gray. [*Myosotis t.* Nutt. ex Hook.] Annual, with 1-several slender stems, erect, 1-3 dm. tall, soft-villous; lvs. of basal rosette lance-oblong, villous, 1.2-5 cm. long, the cauline few, ± ovate, shorter; racemes loosely-fld., elongate, slender, bracted only near base; calyx ca. 3 mm. long in fr., short-villous, whitish or reddish; corolla 2-3 mm. broad; nutlets thick-cruciform, 1.5-2 mm. long, usually light-colored, sharply ridged dorsally and on edges, commonly tubercled on edges.—Grassy and gravelly places, below 5000 ft.; Chaparral to Yellow Pine F.; much of cismontane Calif.; to L. Calif., Ariz., B.C., Ida. March-May.

13. **P. tórreyi** (Gray) Gray. [*Eritrichium t.* Gray.] Annual with 1 to several slender stems, erect and few-branched or more often diffuse, 0.5-1.5 dm. high, spreading-hirsute; purple dye unusually abundant; lvs. oblong to somewhat linear or the uppermost subovate, sessile, 0.5-2 cm. long; spikes lax in fr., leafy-bracted throughout; calyx 2.5 mm. long in fr., hirsute and somewhat hispid; corolla 1.5-2 mm. broad; nutlets mostly 4, round-ovoid, 1.5-2.2 mm. long, abruptly narrowed at apex; keeled on back especially in upper half, shining with ca. 7 transverse ridges with narrow intervening sinuses, smooth or with few whitish tubercles; scar small, borne in the transverse groove.—Meadows and flats, 4000-11,000 ft.; Yellow Pine F. to Subalpine F.; San Bernardino Mts.; Sierra Nevada. June-July.

14. **P. trachycárpus** (Gray) Jtn. [*Krynitzkia t.* Gray.] Laxly ascending or prostrate annual, strigose or appressed-hispidulous, branched at base, the stems 0.5-4 dm. long; lower lvs. linear, obtuse, 5-10 cm. long, the upper oblanceolate; racemes lax, leafy-bracted throughout; mature calyx strigose, the narrow lobes 3-5 mm. long, usually with tawny hairs near the tip; corolla 1-2 mm. broad; nutlets ovoid, rather angulate, ca. 2 mm. long, the back keeled to the middle or beyond as well as laterally and with transverse ridges, the interspaces tuberculate; ventral side obtusely angled with a strong keel; scar suprabasal, broad, ovate or deltoid, concave, closely surrounded by a strong ridge.—Flats and dried pools below 3200 ft.; Coastal Sage Scrub, V. Grassland, Chaparral; coastal Los Angeles Co. to Contra Costa and San Joaquin cos. April-June.

15. **P. undulàtus** (Piper) Jtn. [*Allocarya u.* Piper.] Slender-stemmed sparsely strigulose annual, at first erect, later often sprawling, the branches 1-3 dm. long; lvs. usually sparsely strigose or appressed-hispidulous beneath, subglabrous above, 2-6 cm. long, the lower linear, the upper wider; spikes loose, slender, with some leafy bracts below; calyx somewhat accrescent in age, appressed-villous-hispidulous, usually tawny toward tips, ca. 2 mm. long; corolla 1.5-2 mm. broad; nutlets ovoid to narrower, 1-1.6 mm. long, medially dorsally keeled toward apex, transversely rugose with low rounded undulating ridges, keeled laterally; ventral side keeled ca. $\frac{3}{5}$ the length then replaced by the linear scar which is slightly elevated and lying in an elongate depression.—Mud flats below 1200 ft.; several Plant Communities; San Diego Co. to Mendocino Co. March-June.

18. **Brassicàceae.** Mustard Family

Herbs or rarely shrubby plants, with a pungent watery juice. Lvs. alternate, entire, lobed, dissected or pinnate, without stipules. Fls. in terminal racemes or corymbs, rarely solitary and terminal. Sepals 4, erect or somewhat spreading, rather alike, deciduous. Petals 4, commonly clawed, the blades spreading in the form of a cross. Stamens 6, rarely 4 or 2, two of them inserted lower and shorter. Nectar glands commonly 4. Ovary superior, usually 2-celled by a thin partition stretched between the 2 marginal placentae, from which, when ripe, the valves usually separate. Style 1; stigma 2-lobed or entire. Fr. a 2-celled caps., long and narrow (silique) or short (silicle), sometimes indehiscent and

1-celled, sometimes separating across into 1-seeded joints. Seeds without endosperm, filled with large embryo which is curved and folded so that the cotyledons have the margins of one side against the radicle (accumbent) or the back of one cotyledon against the radicle (incumbent), or the cotyledons may be folded upon themselves (conduplicate). A large family, widely distributed, with many plants of economic or horticultural value (cabbage, mustard, radish, turnip, rutabaga, cauliflower, stock, etc.).

1. Fr. indehiscent, 1-celled, 1-seeded, thin and flat; pedicels recurved in fr., at least at their tips.
   2. Silicles not winged, with hooked hairs; pubescence of stems of branched hairs . 2. *Athysanus*
   2. Silicles wing-margined all around, suborbicular, not with hooked hairs; stems glabrous or with simple hairs .... 37. *Thysanocarpus*
1. Fr. dehiscent by valves or breaking transversely into joints, normally containing 2 or more seeds.
   3. Siliques at length breaking transversely into seed-bearing indehiscent joints.
     4. Frs. transversely 2-jointed, 2-seeded; petals 5–10 mm. long.
       5. Fls. purplish or whitish. Plants of sea-beaches .... 5. *Cakile*
       5. Fls. yellow with darker veins. Occasional waif .... 28. *Rapistrum*
     4. Fr. breaking irregularly into joints, several-seeded; petals 15–20 mm. long . 27. *Raphanus*
   3. Siliques or frs. not breaking transversely into seed-bearing joints, but dehiscent by valves.
     5a. Fr. a silique, elongate, several times longer than wide or thick.
       6. Siliques long-stipitate, the stipe 1–2 cm. long; anthers twisted; sepals spreading or recurved at anthesis .... 32. *Stanleya*
       6. Siliques sessile or with much shorter stipes; anthers usually not twisted; sepals usually erect.
         7. Siliques compressed contrary to the narrow partition; racemes leafy
                                                                               38. *Tropidocarpum*
         7. Siliques not compressed contrary to the partition; racemes leafless.
           8. Stigma-lobes situated over the valves; anthers sagittate at base; petals usually linear.
             9. Calyx urn-shaped or flask-shaped at anthesis; siliques sessile or subsessile.
                10. Siliques conspicuously beaked, pendent, sessile, strongly compressed; valves of mature fr. remaining attached at tip; seeds flat, winged. Annual
                                                                                   33. *Streptanthella*
                10. Siliques not or inconspicuously beaked, erect to deflexed, the valves separating completely. Annual to perennial.
                   11. The siliques flattened; seeds flat, usually winged .... 34. *Streptanthus*
                   11. The siliques terete or 4-sided; seeds wingless or nearly so . 10. *Caulanthus*
             9. Calyx open at anthesis, not urn-shaped; siliques sometimes short-stipitate, mostly terete or 4-sided when mature .... 35. *Thelypodium*
           8. Stigma-lobes situated over the placentae; anthers usually not sagittate at the base; petals frequently wider than linear.
             12. Basal lvs. forming basal rosettes, from which arise the flowering stems.
                13. Fls. yellow; stems angled .... 3. *Barbarea*
                13. Fls. white; stems terete.
                   14. Lvs. simple; undivided to sinuate or lyrate .... 1. *Arabis*
                   14. Lvs. deeply pinnatifid or pinnate.
                      15. The lvs. pinnate with definite lfts.; seeds wingless .... 8. *Cardamine*
                      15. The lvs. pinnatifid; seeds winged .... 30. *Sibara*
             12. Basal lvs. not in well-formed rosettes.
                16. Frs. with a long indehiscent conical to flattened beak.
                   17. Seeds globose, in 1 row in each locule .... 4. *Brassica*
                   17. Seeds ellipsoid to ovoid, in 2 rows in each locule.
                      18. Petals 8–10 mm. long; siliques terete, the beak slender-conical
                                                                            14. *Diplotaxis*
                      18. Petals 15–20 mm. long; siliques 4-angled, the beak flattened . 17. *Eruca*
                16. Frs. not with indehiscent beak, but dehiscent to tip.
                   19. Pubescence simple or wanting.
                      20. Siliques strongly flattened.
                         21. Valves of silique nerved; lvs. simple .... 1. *Arabis*
                         21. Valves of silique not nerved; lvs. mostly ± pinnate .. 8. *Cardamine*
                      20. Siliques terete or 4-angled.

22. Lvs. entire, simple, deeply auriculate-clasping ...... 11. *Conringia*
22. Lvs., at least the lower, pinnate to pinnatifid or deeply lobed.
    23. Plants with creeping base, perennial; seeds biseriate . 29. *Rorippa*
    23. Plants without creeping bases, mostly annual or biennial.
        24. Pods somewhat 4-angled; seeds uniseriate; biennial or perennial of wet places .... 3. *Barbarea*
        24. Pods terete; annuals.
            25. Siliques 2–10 cm. long (if shorter, closely appressed); seeds uniseriate. Weeds .... 31. *Sisymbrium*
            25. Siliques 0.4–1.2 cm. long, divaricate; seeds biseriate. Wet places .... 29. *Rorippa*
19. Pubescence of forked or stellate hairs.
    26. Lvs. pinnatifid to pinnate.
        27. Plants annual, with finely dissected to bipinnatifid lvs. 13. *Descurainia*
        27. Plants perennial from heavy caudex; lvs. subentire to coarsely dentate or lobed .... 19. *Halimolobos*
    26. Lvs. entire or nearly so.
        28. Fls. yellow to orange .... 18. *Erysimum*
        28. Fls. white to rose or purple.
            29. Petals 3–15 mm. long; stigmas not horned on back; epidermal cells of septum of silique not reticulate. Widespread native . 1. *Arabis*
            29. Petals ca. 20 mm. long; stigmas horned on back; epidermal cells of silique-septum reticulate. Natur. along seacoast . 25. *Matthiola*
5a. Fr. a silicle, 1–2 (–3) times longer than wide.
    30. Silicles evidently flattened.
        31. Silicles flattened parallel to the broad partition.
            32. Pubescence of appressed 2-branched hairs attached at middle; silicles orbicular with flat margins .... 23. *Lobularia*
            32. Pubescence stellate or forked, but not as above; silicles not margined
                                                                                    16. *Draba*
        31. Silicles flattened contrary to the narrow partition.
            33. Silicles twinlike, the 2 locules somewhat rounded with deep notch between both at summit and base.
                34. The silicles 6–18 mm. wide; pubescence stellate .... 15. *Dithyrea*
                34. The silicles ca. 2 mm. wide; pubescence not stellate .... 12. *Coronopus*
            33. Silicles not twinlike.
                35. The silicles inverted-triangular, broad at summit; annuals; fls. ca. 2 mm. long. Introd. weed .... 7. *Capsella*
                35. The silicles orbicular to oval or cuneate.
                    36. Petals 18–20 mm. long. Suffrutescent perennial of Colo. Desert
                                                                                    24. *Lyrocarpa*
                    36. Petals 1–5 mm. long.
                        37. Cells of the silicle 1-seeded .... 21. *Lepidium*
                        37. Cells of the silicles 2–many-seeded.
                            38. Silicles notched to subtruncate at apex, 4–18 mm. long; petals 3–5 mm. long .... 36. *Thlaspi*
                            38. Silicles rounded at apex, 2–4 mm. long; petals 1 mm. long
                                                                                    20. *Hutchinsia*
    30. Silicles turgid or inflated, not flattened.
        39. Plants in ± scapose tufts, the lvs. mostly basal.
            40. Silicles didymous, 10–15 mm. long; petals yellow, 10–12 mm. long. E. Mojave Desert .... 26. *Physaria*
            40. Silicles not didymous, 3–6 mm. long.
                41. Lvs. 2–12 mm. long; silicles ovoid, with mostly simple hairs ..... 16. *Draba*
                41. Lvs. 20–50 mm. long; silicles subglobose, ± stellate-pubescent
                                                                                    22. *Lesquerella*
        39. Plants with leafy stems.
            42. Petals white, 2–3 mm. long; long-lived perennials .... 9. *Cardaria*
            42. Petals yellow; annual to perennial.
                43. Valves of the silicles 1-nerved; cauline lvs. sagittate-clasping, not lobed
                                                                                    6. *Camelina*

43. Valves of the silicles not nerved; cauline lvs. not sagittate-clasping.
  44. Lvs. pinnatifid; Plants of wet places .... 29. *Rorippa*
  44. Lvs. entire. Plants of dry places .... 22. *Lesquerella*

1. *Árabis* L. Rock-Cress

Biennial or perennial herbs, glabrous or pubescent with simple, forked or stellate hairs. Stems leafy, simple or branched. Basal lvs. petioled, entire, dentate or sometimes dissected, the cauline sessile or sometimes petioled, often auricled. Infl. racemose, bractless, elongating. Sepals erect, oblong to subovate, uniform, or the outer pair sometimes saccate. Petals spatulate to oblong, white to purplish, rarely yellowish. Stamens 6, tetradynamous; anthers oblong. Siliques linear, straight or curved, erect to reflexed, sessile or rarely stipitate, flattened parallel to partition, 2-valved, the valves mostly 1-nerved. Style prominent to none. Stigma mostly entire. Seeds many, pendulous, orbicular to almost oblong, flat or plump, winged or not, uniseriate to biseriate, with accumbent cotyledons. Large genus of Am. and Eurasia. (Named for *Arabia*.)

(Rollins, R. C. Monographic study of Arabis in w. N. Am. Rhodora 43: 289–325, 348–411, 425–481. 1941 [Contr. Gray Herb. 138.])

1. Siliques 3–8 mm. wide; seed wing 1–3 mm. wide, or, if slightly less, then cauline lvs. petioled; seeds (including wings) 2.5–5 mm. long.
   2. Siliques and pedicels reflexed; plant hoary. Desert .... 4. *A. glaucovalvula*
   2. Siliques and pedicels erect or ascending.
      3. Lower stem lvs. with winged petioles; basal lvs. oblanceolate to broadly spatulate, 1–3 cm. wide; petals scarcely longer than sepals .... 15. *A. repanda*
      3. All stem lvs. sessile; basal lvs. linear to oblanceolate, less than 0.8 cm. wide; petals definitely exceeding sepals.
         4. Lvs. and lower stems hoary with a minute pubescence; pedicels pubescent. Mts. bordering Mojave Desert .... 2. *A. dispar*
         4. Lvs. and stems green, not hoary; pedicels glabrous; basal lvs. oblanceolate to spatulate. Montane .... 12. *A. platysperma*
1. Siliques usually less than 3 mm. wide; seed-wing less than 1 mm. wide or seeds wingless; seeds (including wings) less than 2 mm. long.
   5. Basal lvs. obovate to broadly oblanceolate, obtuse and rounded at apex, often forming a flat rosette at base of stems.
      6. Seeds biseriate; siliques semi-terete; fls. cream-yellow or rarely lilac; stem-lvs. ample, ± ovate, usually glaucous .... 3. *A. glabra*
      6. Seeds uniseriate; siliques flat; fls. white; stem-lvs. obovate to oblong, rarely glaucous
         .... 5. *A. hirsuta*
   5. Basal lvs. linear to linear-oblanceolate, acute to rarely obtuse (if broader, then minutely pubescent or with reflexed siliques or both).
      7. Plant hoary with a very minute pubescence. Mostly of the desert or bordering mts.
         8. Siliques reflexed; pedicels reflexed to pendulous .... 13. *A. pulchra*
         8. Siliques erect or ascending to widely spreading; pedicels erect to perpendicular to axis.
            9. Styles 1–8 mm. long; basal lvs. linear.
               10. The styles 4–8 mm. long; seeds with narrow wings; siliques 1–2 cm. long. San Bernardino Mts. .... 10. *A. parishii*
               10. The styles 1–3 mm. long; seeds broadly winged; siliques 3–5 cm. long. San Jacinto Mts. .... 9. *A. johnstonii*
            9. Styles less than 1 mm. long; basal lvs. narrowly oblanceolate.
               11. Seeds biseriate, essentially wingless, plump, ca. 1 mm. broad. N. slope of San Bernardino Mts. .... 16. *A. shockleyi*
               11. Seeds uniseriate, winged, flat, 1.5–2.5 mm. broad.
                  12. Seed-wing over 0.5 mm. wide; siliques 2.5–3.5 mm. wide, divaricately ascending. San Bernardino Mts. n. .... 2. *A. dispar*
                  12. Seed-wing less than 0.5 mm. wide; siliques ca. 2 mm. wide, spreading at right angles .... 8. *A. inyoensis*
      7. Plant greenish, not hoary.
         13. Mature fruiting pedicels erect to ascending; siliques erect or ascending, sometimes arcuate.

14. Lower fruiting pedicels 2–4 cm. long, glabrous; siliques somewhat curved. Santa Cruz Id. .... 6. *A. hoffmannii*
14. Lower fruiting pedicels less than 2 cm. long, pubescent or glabrous; siliques straight or arcuate.
   15. Plants 0.6–2 dm. high; basal lvs. broadly spatulate, pubescent with 3-forked hairs, 1–3 cm. long .... 1. *A. breweri*
   15. Plants 2–9 dm. high.
     16. Basal lvs. 3–10 cm. long; siliques 6–12 cm. long .... 17. *A. sparsiflora*
     16. Basal lvs. 2–3 cm. long; siliques 4–6 cm. long .... 8. *A. inyoensis*
13. Mature fruiting pedicels diverging at right angles to strictly reflexed; siliques straight to arcuate, diverging at right angles to stem or even deflexed.
   17. Basal lvs. always ciliate with large hairs, oblanceolate; siliques reflexed, appressed to rachis. San Bernardino Co. n. .... 14. *A. rectissima*
   17. Basal lvs. not ciliate.
     18. Seeds biseriate; cauline lvs. linear; petals 8–20 mm. long ...... 13. *A. pulchra*
     18. Seeds uniseriate; cauline lvs. wider, petals less than 12 mm. long.
       19. Mature fruiting pedicels descending to strictly deflexed, straight, not widely spreading; siliques mostly straight, pendulous to appressed .. 7. *A. holboellii*
       19. Mature fruiting pedicels spreading at right angles to stem, straight or arched downward; siliques straight and spreading at right angles or arcuate.
         20. Basal lvs. entire, finely pubescent; stems mostly strigose below
                                                                                         8. *A. inyoensis*
         20. Basal lvs., at least in part, dentate, coarsely pubescent; stems mostly hirsute below.
           21. Pedicels slender, glabrous, 10–20 mm. long; outer basal lvs. broadly oblanceolate, obtuse; petals 6–9 mm. long .... 11. *A. perennans*
           21. Pedicels stout, pubescent, 5–12 mm. long; outer basal lvs. narrowly oblanceolate, acute; petals 8–12 mm. long .... 17. *A. sparsiflora*

1. **A. bréweri** Wats. var. **bréweri**. Cespitose perennial from a woody branched caudex; stems several, simple, densely hirsute below with mostly simple hairs, often glabrous above, 6–20 cm. high; basal lvs. broadly spatulate, mostly entire, obtuse, short-petioled, pubescent with 3-forked hairs, 1–3 cm. long, 4–6 mm. wide; cauline lvs. sessile, auriculate, usually less than 2 cm. long; pedicels 5–9 mm. long, mostly pubescent; sepals pubescent, 4–5 mm. long; petals red-purple to pink, 6–9 mm. long; siliques divaricate, arcuate or almost straight, 3–7 cm. long, ca. 2 mm. wide, 1-nerved on lower third; style 0; seeds orbiculate, uniseriate, 1 mm. or wider, narrowly winged.—Dry rocky places, 1500–7000 ft.; largely Yellow Pine F.; s. Ore. s. apparently reaching its s. limit at Rose Lake in Ventura Co. March–July.

A. b. var. **pecuniària** Roll. Pedicels 3–4 mm. long; siliques 2–3 cm. long.—At 9000–10,500 ft.; Subalpine F.; San Bernardino Mts. July–Aug.

2. **A. díspar** Jones. Perennial from a cespitose base; stems several, simple or branched above, densely pubescent, especially below, 1–2.5 dm. high; basal lvs. many, entire, erect, spatulate to narrowly oblanceolate, hoary, with dense fine dendritic pubescence, 1.5–2.5 cm. long, 2–4 mm. wide; cauline lvs. sessile, broadly linear, hoary; pedicels suberect to divaricate, pubescent, 1–2 cm. long; sepals pubescent, ca. 4 mm. long; petals purplish, obovate, 5–6 mm. long; siliques divaricate to ascending, glabrous, 5–7 cm. long, 2.5–3.5 mm. wide, with midnerve in lower half; stigma subsessile; seeds suborbicular, imperfectly uniseriate, ca. 2 mm. wide, broadly winged.—Rare, gravelly slopes, 4000–8000 ft.; Creosote Bush Scrub, Joshua Tree Wd., Pinyon-Juniper Wd.; Panamint and Argus mts. and farther n., to San Bernardino and Little San Bernardino mts. April–May.

3. **A. glàbra** (L.) Bernh. [*Turritis g.* L. *A. perfoliata* Lam.] TOWER-MUSTARD. Biennial or occasionally perennial; stems 1–few, simple or rarely branched above, usually hirsute below with simple or forking spreading hairs, glabrous and glaucous above, 4–12 dm. high; lower lvs. oblanceolate to oblong, petioled, usually ± dentate or even repand, with coarse forked or dendritic hairs, 6–15 cm. long, 1–3 cm. wide; cauline lvs. lanceolate to ovate, sessile, auricled, mostly entire and glabrous; racemes long, strict; pedicels suberect, commonly 1–1.5 cm. long; sepals oblong, yellowish, 3–5 mm. long; petals yellowish-white

or rarely purplish; siliques erect, semiterete, glabrous, 4–10 cm. long, ca. 1 mm. wide, valves nerved at least in lower half; style short, stout; stigma expanded; seeds oblong to almost round, wingless or narrowly winged, mostly biseriate, ca. 1 mm. long.—Shaded canyons and mts., mostly below 7000 ft.; many Plant Communities; throughout cismontane s. Calif.; to B.C. and Atlantic Coast. March–July.

4. **A. glaucoválvula** Jones. Plate 25, Fig. E. Perennial from a woody branched caudex; stems 1–several, mostly simple, hoary throughout, 1.5–4 dm. high; basal lvs. narrowly oblanceolate, entire, obtuse, densely pubescent with dendritic hairs, hoary, 2–5 cm. long, 2–5 mm. wide; cauline lvs. lanceolate, or lance-linear, sessile, not auricled; pedicels stout, strongly recurved, pubescent, 5–10 mm. long; sepals pubescent, 4–5 mm. long; petals whitish to pink, 6–8 mm. long; siliques reflexed, oblong, obtuse, glabrous, glaucous, 2–4 cm. long, 5–7 mm. wide; style less than 1 mm. long; seeds biseriate, orbicular, 5–6 mm. wide, with broad wing.—Dry stony places, 2500–5300 ft.; Creosote Bush Scrub, Joshua Tree Wd.; desert from Bishop Creek, Inyo Co. to New York, Eagle and Little San Bernardino mts. March–May.

5. **A. hirsùta** (L.) Scop. var. **glabràta** T. & G. [*A. pycnocarpa* var. *g.* M. Hopk.] Biennial or perennial, the stems erect, 1–few, simple or branched above, hirsute below, glabrous above, 2–7 dm. high; basal lvs. obovate to oblanceolate, mostly entire, obtuse, petioled, 3–7 cm. long, 1–2.5 cm. wide, hirsute to subglabrous; cauline lvs. oblong to obovate, entire to rarely few-toothed, sessile, auricled; racemes rather long in fr.; pedicels erect to ascending, 5–15 mm. long; sepals mostly glabrous, 3–4.5 mm. long; petals white, spatulate, 5–9 mm. long; siliques erect to slightly divaricate, glabrous, flat, 3–6 cm. long, 1 mm. wide; style 0.5–1 mm. long; stigma subentire; seeds brown to dark, rounded to rectangular, winged at distal end to wingless, uniseriate, 1–1.5 mm. long.—Moist places in mts., 4000–8200 ft.; Montane Coniferous F.; San Bernardino Mts.; Sierra Nevada to B.C. and Wyo. May–July.

6. **A. hoffmánnii** (Munz) Roll. [*A. maxima* var. *h.* Munz.] Perennial with scaly caudex; stems 1–few, branched above, subglabrous, 5–7 dm. high; basal lvs. many, linear-lanceolate, sinuate-dentate, obtuse, dendritic-pubescent on lower surface, 5–10 cm. long, 6–10 mm. wide; cauline lvs. sessile, crowded, linear-oblong; pedicels divaricately spreading, glabrous, 1–4 cm. long; sepals 4–5 mm. long; petals white, 8–10 mm. long, narrow; siliques divaricate, straight or slightly arcuate, nerveless, glabrous, 6–10 cm. long, 2–3.5 mm. wide; style almost obsolete; seeds orbicular, biseriate, narrowly winged, ca. 1 mm. wide. —Rocky places; Chaparral; Santa Cruz Id. March–April.

7. **A. holboéllii** Hornem. var. **pinetòrum** (Tidestr.) Roll. [*A. p.* Tides.] Stems hirsutulous below, glabrous above, 3–9 dm. high; basal lvs. densely pubescent with coarse dendritic hairs; pedicels arched downward, rarely geniculate, usually glabrous; siliques slightly curved inward, sometimes straight, pendulous, 4–7 cm. long, 1.5–2 mm. wide; $2n=21$ (Rollins, 1966).—With the next var. from San Diego Co. to B.C. and Nebr. May–July.

Var. **retrofrácta** (Grah.) Rydb. [*A. r.* Grah.] Plate 25, Fig. F. Biennial or perennial from simple or branched caudex; stems 1–several, erect, simple or branched above, 2–8 dm. high, densely pubescent with fine appressed dendritic hairs to glabrous above; basal lvs. ± pannose, usually entire, oblanceolate to spatulate, 1–5 cm. long, 3–6 mm. wide; cauline lvs. revolute-margined, auriculate, the upper finely pubescent; pedicels pubescent, strongly reflexed, usually geniculate, 6–12 mm. long; sepals 2–4 mm. long; petals pinkish to whitish, 6–10 mm. long; siliques glabrous or sometimes finely pubescent, strongly reflexed, usually appressed to the stem, straight or nearly so, 3.5–8 cm. long, 1–1.5 mm. wide; seeds mostly uniseriate, orbicular, ca. 1 mm. wide, narrowly winged all around; $2n=14$ (Rollins, 1966).—Dry stony places below 8000 ft.; largely Montane Coniferous F.; San Bernardino Mts. n. to Alaska and to Colo. May–July.

8. **A. inyoénsis** Roll. Perennial from branched caudex; stems several, erect, densely pubescent below with fine dendritic hairs, glabrate above, 2–5 dm. tall; basal lvs. many, narrowly oblanceolate to spatulate, entire, acute, densely pubescent, 2–3 cm. long, 2–5 mm. wide; cauline lvs. sessile, oblong, auricled; fruiting pedicels usually horizontal, some-

times ascending, 6–12 mm. long, subglabrous; sepals pubescent, ca. 4 mm. long; petals pink to purplish, 7–9 mm. long; siliques wide-spreading, sometimes ± descending or ascending, nerved below, glabrous, 4–6 cm. long, ca. 2 mm. wide; seeds orbicular, uniseriate, ca. 1.5–2 mm. wide, narrowly winged.—Dry stony or rocky places, 5000–11,000 ft.; Pinyon-Juniper Wd. to Bristle-cone Pine F.; Argus and Panamint mts. n.; to Sierra Nevada. May–July.

9. **A. johnstònii** Munz. Perennial, densely pubescent with fine dendritic hairs; stems 1–few, ascending, simple, densely pubescent, 1–2 dm. high; basal lvs. narrowly spatulate to oblanceolate, entire, hoary with fine dense pubescence, 1–2 cm. long, 1.5–3.5 mm. wide; cauline lvs. few, sessile, not auricled; pedicels ascending-pubescent, 6–10 mm. long; sepals pubescent, 4.5–6 mm. long; petals purple, spatulate, 8–10 mm. long; siliques suberect, glabrous, 3–5 cm. long, 2–3 mm. wide, acuminate at apex; style slender, 1–2 mm. long; seeds suborbicular, uniseriate, ca. 1.5 mm. wide, winged.—Dry rocky knolls, 4500–5000 ft.; lower edge of Yellow Pine F.; s. end of San Jacinto Mts., Riverside Co. May–June.

10. **A. párishii** Wats. Plate 26, Fig. A. Tufted perennial from branched caudex; stems simple, slender, densely pubescent below with dendritic hairs, fewer above, 3–12 cm. high; basal lvs. many, linear-oblanceolate, entire, short-petioled, hoary with fine dendritic hairs, 5–15 mm. long, 1–2 mm. wide; cauline lvs. few, sessile, not auricled, linear; pedicels erect to slightly divergent, pubescent, 3–7 mm. long; sepals green or purplish, 3–4 mm. long; petals purple or lavender with white base, spatulate, 8–12 mm. long; siliques ascending, glabrous, 1–2 cm. long, 2–3 mm. wide, acuminate; styles filiform, 4–8 mm. long; seeds elliptical to suborbicular, imperfectly uniseriate, 1–1.5 mm. wide, narrowly winged.—Dry sunny slopes, 6500–9800 ft.; Yellow Pine F., Red Fir F.; Bear V. and Sugarloaf Peak, San Bernardino Mts. April–May.

11. **A. perénnans** Wats. [*A. arcuata* var. *p.* Jones.] Perennial with simple or branched woody caudex; stems few to many, simple or branched above, 1.5–6 dm. high, pubescent below with dendritic hairs; lower lvs. oblanceolate to wider, petioled, dentate to entire, 2–6 cm. long, 4–20 mm. wide, densely pubescent with dendritic hairs; cauline lvs. lanceolate, auricled, entire or sparsely toothed; pedicels spreading and arched down in fr., glabrous, 1–2 cm. long; siliques wide-spreading to pendulous, glabrous, curved inward, usually nerveless, 4–6 cm. long, 1.2–2 mm. wide; style 0; seeds orbicular, 1–1.5 mm. wide, winged, uniseriate.—Dry stony slopes, 1000–7000 ft.; Creosote Bush Scrub, Joshua Tree Wd., Pinyon-Juniper Wd., Yellow Pine F.; San Gabriel and San Bernardino mts. to Panamint Mts., Laguna Mts. and across the deserts to Colo., New Mex., L. Calif. April–July.

12. **A. platyspérma** Gray. Plate 26, Fig. B. Perennial, the lower stems, basal lvs., and sepals with dendritic pubescence; stems erect to subdecumbent, simple or branched, 1–4 dm. high; basal lvs. many, oblanceolate or narrower, entire, 2–5 cm. long, 3–8 mm. wide; cauline lvs. few, remote, oblong to linear-lanceolate, sessile, not auriculate; pedicels divaricately ascending, straight, 5–15 mm. long; sepals 3–4 mm. long; petals pink to white, spatulate, 4–6 mm. long; siliques erect to divaricately ascending, straight, flat, acuminate, 3–7 cm. long, 3–5 mm. wide; style 0 to almost 1 mm. long; seeds orbicular, uniseriate, 3–4 mm. wide, winged.—Dry stony flats and slopes, 5500–11,000 ft.; Red Fir F. to Subalpine F.; San Jacinto to San Gabriel mts.; n. to Ore., Nev. June–July.

13. **A. púlchra** Jones. Perennial, rather woody at the simple or branched caudex; stems 1–several, simple or branched, densely pubescent with minute appressed dendritic hairs or glabrous above, 2–6 dm. high; basal lvs. linear, entire, or slightly toothed, obtuse, densely pubescent, 4–8 cm. long, 3–6 mm. wide; cauline lvs. linear, sessile, not auricled; pedicels ascending at anthesis, geniculately reflexed in fr.; densely pubescent, 8–20 mm. long; sepals pubescent, 5–8 mm. long; petals purple, 10–20 mm. long; siliques strictly appressed to the stem, densely pubescent, straight, 4–7 cm. long, 2.5–3.5 mm. wide; stigma subsessile; seeds suborbicular, biseriate, 1.5–2 mm. wide, rather prominently winged. —Rocky slopes, 2000–7000 ft.; Creosote Bush Scrub, Joshua Tree Wd., Pinyon-Juniper Wd.; deserts from Inyo Co. to L. Calif.; Nev. April–May. A form growing with the

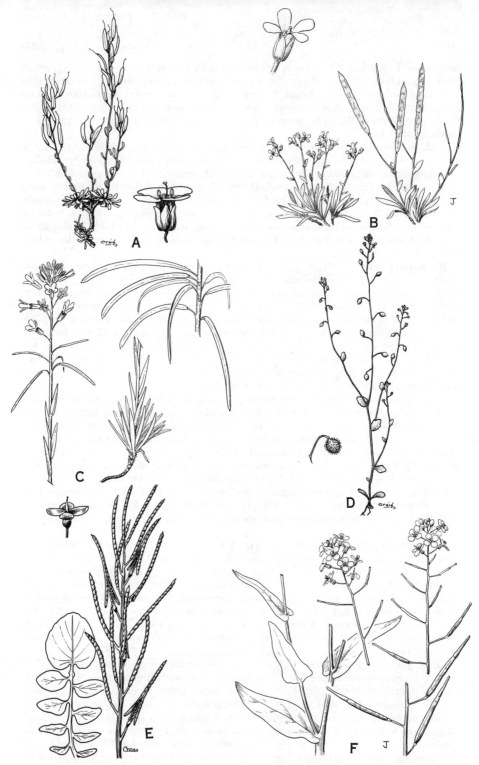

Plate 26. Fig. A, *Arabis parishii*. Fig. B, *Arabis platysperma*. Fig. C, *Arabis sparsiflora*. Fig. D, *Athysanus pusillus*. Fig. E, *Barbarea orthoceros*. Fig. F, *Brassica rapa* ssp. *sylvestris*.

typical one but with coarser and less dense pubescence and glabrous pedicels and these arched downward, not geniculately reflexed is the var. *gracilis* Jones. Another more local one in Calif. (about Darwin, Inyo Co.) and extending to Utah is var. *munciensis* Jones with pubescent pedicels gently spreading downward.

14. **A. rectíssima** Greene. Biennial with 1–several stems simple to branched above, glabrous to sparsely hirsute with simple hairs, 2–8 dm. high; basal lvs. spatulate to oblanceolate, petioled, entire, 1–3 cm. long, 4–10 mm. wide, hirsute with simple and coarse forked hairs; cauline lvs. crowded below, oblong to sublanceolate, ± auricled; pedicels glabrous, strictly reflexed, 4–12 mm. long; sepals oblong, sparsely hirsute at apex, 2–3 mm. long; petals white or pinkish, 4–6 mm. long; siliques many, straight, strictly reflexed, appressed to stem, glabrous, 1-nerved near base, 5–8 cm. long, 1.5–2.5 mm. wide; style almost none; seeds orbicular, ca. 1.5 mm. wide, uniseriate, winged all around.—Dry slopes and benches, 4000–9000 ft.; Yellow Pine F., Red Fir F.; San Bernardino Mts.; Sierra Nevada to Ore., Nev. June–July.

15. **A. repánda** Wats. Biennial or possibly perennial, with forked or dendritic hairs; stems mostly 1 or 2, ascending, densely pubescent below, sparsely so or glabrous above, 2–8 dm. tall; basal lvs. obovate to broadly oblanceolate, petioled, repand-toothed, 3–8 cm. long, 1–3 cm. wide; cauline lvs. few, oblanceolate; pedicels stout, straight, erect to ascending, usually pubescent, 3–8 mm. long; sepals pubescent, 4–5 mm. long; petals white to pinkish, linear, 4–6 mm. long; siliques divaricately ascending, straight or often falcate, pubescent to glabrous, 4–10 cm. long, 2–4 mm. wide; style slender, to 1 mm. long; seeds roundish, 2–4 mm. wide, broadly winged.—Dry slopes under pines, 4600–9000 ft.; Yellow Pine F. to Lodgepole F.; San Jacinto Mts. to Mt. Pinos; s. Sierra Nevada. June–Aug.

16. **A. shóckleyi** Munz. Perennial from thick simple caudex; stems 1–few, simple, hoary, 1–3 dm. high, densely pubescent with minute dendritic hairs; basal lvs. crowded, spatulate, entire, hoary, 1–2 cm. long; pedicels ascending, densely pubescent, 8–10 mm. long; sepals pubescent, 5–7 mm. long; petals pink, 8–11 mm. long; siliques divaricate, crowded, straight to slightly curved, subglabrous, 5–8 cm. long, ca. 2 mm. wide; stigma subsessile; seeds oblong, biseriate, essentially wingless, ca. 1 mm. wide.—Rare, dry rocky places; Pinyon-Juniper Wd.; n. slopes of San Bernardino Mts.; w. Nev. May–June.

17. **A. sparsiflòra** Nutt. in T. & G. var. **arcuàta** (Nutt.) Roll. [*Streptanthus a.* Nutt.] Perennial; stems rather slender, simple or branched, pubescent throughout, the lower part hirsute with large simple or branched hairs, 3–8 dm. high; lower long-petioled, entire, linear-oblanceolate, acute, coarsely pubescent, 3–10 cm. long; upper lvs. and stems hirsute; pedicels spreading, 5–15 mm. long; sepals 4–6 mm. long; petals pink or purple, 8–14 mm. long; siliques divaricately ascending, strongly arcuate, glabrous, nerved below the middle, 6–12 cm. long, 1.5–2 mm. wide; seeds orbicular, uniseriate, 1 mm. wide, narrowly winged.—Dry slopes, 2500–6000 ft.; Foothill Wd., Chaparral, Yellow Pine F.; rare, San Diego Co. n. to n. Calif. where more common. March–May.

Var. **califórnica** Roll. Plate 26, Fig. C. Stems coarse, pubescent throughout with fine dendritic hairs; basal lvs. large, coarsely toothed, densely pubescent; pedicels pubescent with appressed hairs; $2n=22$ (Raven, Kyhos & Hill, 1965).—Common on dry stony places, mostly below 5000 ft.; Chaparral, Coastal Sage Scrub; Los Angeles Co. to L. Calif. Feb.–May.

2. *Athýsanus* Greene

Diffuse slender-stemmed annual, branched from near base, pubescent with forked spreading hairs. Lvs. all near base, simple, few-toothed, ovate-oblong. Fls. minute, in loose unilateral elongate racemes. Pedicels slender, recurved. Sepals equal. Petals linear or 0. Stamens 6, almost equal. Ovary 1-celled, 2-ovuled. Silicles orbicular, indehiscent, 1-seeded, uncinate-hispid. Seeds round, glabrous, flat, light brown. One sp. of w. U.S. (Greek, *a*, without, and *thusanos*, fringe, the fr. wingless.)

1. **A. pusíllus** (Hook.) Greene. [*Thysanocarpus p.* Hook.] Plate 26, Fig. D. Plants 1–3 dm. high; lvs. few, 5–20 mm. long; fls. ca. 1.5 mm. long; silicles 1.5–2 mm. long.—Com-

mon but inconspicuous, in grassy and brushy places, mostly below 5000 ft.; many Plant Communities; almost throughout Calif. to desert edge; n. to B.C., Ida. March–June.

3. Barbarèa Scop. WINTER-CRESS

Glabrous biennial or perennial herbs with overwintering basal rosettes and angled stems. Lvs. pinnatifid, the cauline with clasping bases. Fls. racemose, yellow. Sepals erect, the 2 outer slightly saccate at base. Petals spatulate, clawed. Siliques linear, often somewhat 4-angled, with pointed style and somewhat bilobed stigma. Seeds in 1 row in each cell, marginless, flat, oblong, with cotyledons accumbent. Ca. 7 spp. of temp. zones. (Named for St. Barbara.)

1. **B. orthóceras** Ledeb. [*B. americana* Rydb.] Plate 26, Fig. E. Stems rather stout, strict, 2–4 dm. tall, usually few-branched above; basal lvs. 3–10 cm. long, petioled, elliptic to suborbicular, simple or with 2–4 small lfts. and large terminal one; middle and upper cauline lvs. lyrate-pinnatifid; racemes dense in anthesis, looser in fr.; pedicels thick, ascending, 3–6 mm. long; sepals yellow-green, ca. 3 mm. long; petals pale yellow, 4–6 mm. long; siliques erect, appressed, 2.5–3.5 cm. long, 1.5 mm. wide; seeds brown, minutely rugulose, 0.8–1 mm. long; $n=8$ (Rollins, 1966).—Banks of streams, springy places, meadows, etc., mostly 2500–11,000 ft.; usually Montane Coniferous F.; mts. from San Diego Co. n.; to Alaska, Asia and Atlantic Coast. A form with siliques spreading or ascending and 2.5–5 cm. long occurs with the typical form and is the var. *dolichocarpa* Fern., intergrading freely with var. *orthoceras*.

4. Brássica L. MUSTARD

Erect branched annual to perennial herbs. Basal lvs. pinnatifid, cauline dentate or subentire. Fls. showy, yellow, in elongated racemes. Lateral sepals ± gibbous at base. Petals with long claw and spreading limb. Siliques elongate, slender or thickish, subterete or 4-sided, with a stout indehiscent flat or conic beak and convex 1–3-nerved valves. Stigma truncate or 2-lobed. Seeds subglobose, marginless, in 1 row in each cell; cotyledons conduplicate. Ca. 100 spp. of Eu., Asia and N. Afr.; many, like cabbage, broccoli, cauliflower, turnip, etc., are important food plants. (Latin name for cabbage.)

1. Upper stem lvs. with clasping base.
    2. Lvs. thin, the basal with scattered hairs; petals 6–8 mm. long; siliques 3–6 cm. long
        8. *B. rapa*
    2. Lvs. thickish, mostly glabrous; petals 10–25 mm. long; siliques 5–10 cm. long.
        3. Infl. open and 10–25 cm. long at anthesis; petals 12–26 mm. long ...... 7. *B. oleracea*
        3. Infl. crowded, the blooming part 3–5 cm. long; petals 11–14 mm. long .... 5. *B. napus*
1. Upper stem-lvs. not clasping.
    4. Beak of silique terete or conic, often seedless.
        5. Pedicels mostly shorter than sepals; siliques appressed against the stem.
            6. Plant annual; petals ca. 8 mm. long .... 6. *B. nigra*
            6. Plant perennial; petals ca. 5 mm. long .... 1. *B. geniculata*
        5. Pedicels longer than sepals; siliques divergent.
            7. Petals ca. 3 mm. wide; beak of silique 6–10 mm. long, the apex narrower than the stigma
                3. *B. juncea*
            7. Petals ca. 1.5 mm. wide; beak of silique 10–16 mm. long, the apex as wide as the stigma
                9. *B. tournefortii*
    4. Beak of silique flat or conspicuously angled; beak usually 1-seeded.
        8. Lvs. petioled, pinnatifid; pedicels 5–15 mm. long; beak ensiform, equal to or longer than the bristly dehiscent part of the silique .... 2. *B. hirta*
        8. Lvs. subsessile, the upper merely toothed; pedicels 3–7 mm. long; beak 4-angled, 2-edged, ca. half as long as the smooth or sparsely hairy body of the silique .... 4. *B. kaber*

1. **B. geniculàta** (Desf.) J. Ball. [*Sinapis g.* Desf. *Sinapis incana* L. *Hirschfeldia adpressa* Moench.] Biennial or perennial, 4–8 dm. tall, ± canescent-hirsute; basal lvs. lyrate-pinnatifid, 4–10 cm. long, with large terminal lobe, the upper cauline smaller, dentate to

lobed; racemes many, terminal on the branches; pedicels appressed, 1–3 mm. long at anthesis; sepals ca. 3 mm. long; petals 5–6 mm. long, light yellow; siliques appressed, 8–12 mm. long, torulose, the beak flattened, frequently 3–4 mm. long and 1-seeded; seeds ca. 1 mm. long, ovoid or oblong-ovoid, red-brown, somewhat alveolate.—Common weed in waste places, along roadsides, etc. in much of cismontane Calif.; natur. from Eu. May–Oct.

2. **B. hírta** Moench. [*Sinapis alba* L. *B. a.* Rabenh., not Gilib.] WHITE MUSTARD. Annual, 3–7 dm. high, ± hirsute; lower lvs. broad, lyrately pinnate or pinnatifid, 1–2 dm. long, petioled, the terminal lobe or lft. large; upper lvs. short-petioled, lanceolate or oblong; pedicels spreading in fr., 5–15 mm. long; sepals 5–6 mm. long; petals yellow, 8–11 mm. long, 4–5 mm. wide; siliques white-bristly, 2–3 cm. long (including beak), the valves prominently 3-nerved, few-seeded, the beak flattened, broad, ca. as long as the rest of the silique; seeds pale yellow, subglobose, 1.5–2 mm. thick, minutely alveolate; $2n=24$ (Manton, 1932).—Cult. as a source of mustard and greens; natur. in scattered localities; from Eurasia. March–Aug.

3. **B. júncea** (L.) Coss. [*Sinapis j.* L.] Pale subglabrous annual, 3–12 dm. high; lower lvs. runcinate-pinnatifid and crenate, petioled, 3–12 cm. long, the upper nearly sessile, lanceolate or linear, entire or dentate; pedicels slender, divergent, 8–12 mm. long; sepals 4–5 mm. long; petals yellow, spatulate, 7–8 mm. long; siliques 3–4 cm. long, ascending, the beak 5–8 mm. long; seeds subglobose, ca. 1.5 mm. in diam., red-brown to yellowish, weakly reticulate; $n=18$ (Manton, 1932).—At scattered stations in cismontane s. Calif., growing as a weed in waste places and fields; natur. from Eu. June–Sept.

4. **B. káber** (DC.) Wheeler. [*Sinapis k.* DC.] CHARLOCK. Erect annual, ± hispid at base, 3–10 dm. tall; lower lvs. obovate, lyrate-pinnatifid, 5–15 cm. long, petioled, the upper oblong to lanceolate, toothed; pedicels short, thick, 2–3 mm. long, ascending; sepals 4–5 mm. long; petals 8 mm. long; siliques 2–2.5 cm. long, the beak 0.6–1.2 cm. long; seeds globose, 1–1.5 mm. thick, red-brown or darker, minutely alveolate; $n=9$ (Yarnell, 1956).—Scattered stations as a weed, natur. from Eurasia. Our material often referred to var. *pinnatifida* (Stokes) Wheeler. March–Oct.

5. **B. nápus** L. RAPE. Much like *B. rapa*, but more glabrous; terminal lobe of basal lvs. very large and obtuse; fls. paler; sepals 6–8 mm. long; petals 9–14 mm. long; siliques 5–9 cm. long, the beak 1–2 (–3) cm. long; seeds 1–1.5 mm. thick, obscurely purple-brown, minutely reticulate-alveolate; $n=18$ (Morinaga, 1929).—Occasional weed, as at Narod, San Bernardino Co.; natur. from Eu. April–June.

6. **B. nígra** (L.) Koch. [*Sinapis n.* L.] BLACK MUSTARD. Erect annual branched above, 5–25 dm. high, sparsely pubescent or subglabrous; lower lvs. 1–2 dm. long, deeply pinnatifid, with large terminal lobe and few small lateral ones; cauline lvs. gradually reduced but not clasping, the uppermost pendulous; sepals 3.5–4.5 mm. long; petals bright yellow, 7–8 mm. long; pedicels 2–3 mm. long, erect; siliques appressed, 1–2 cm. long, the beak subulate, empty, 1–3 mm. long; seeds ca. 1–1.3 mm. thick, dark red-brown, finely reticulate; $2n=16$ (Manton, 1932).—Common on dry grassy hills, in grain fields, etc., cismontane areas; to Atlantic Coast; natur. from Eu. April–July.

7. **B. olerácea** L. CABBAGE. Stout glaucous biennial or perennial, the stem 3–10 dm. high, or often decumbent; lower lvs. thick, fleshy, obovate to oblong, 1.5–3 dm. long; stem-lvs. narrow, some clasping; fls. in long racemes, whitish-yellow; sepals 6–12 mm. long; petals 12–26 mm. long; siliques 5–10 cm. long, including the long conical beak; seeds slightly compressed, 2–4 mm. thick, somewhat angled-striate; $n=9$ (Karpechenko, 1924), $2n=18$ (Sampson, 1966).—Occasional in Santa Barbara and Ventura cos. and n.; natur. from Eu. March–June. Persisting for only a short time.

8. **B. ràpa** L. ssp. **sylvéstris** (L.) Janchen. [*B. campestris* L.] FIELD MUSTARD. Plate 26, Fig. F. Erect annual 3–12 dm. high, with slender roots, glaucous and quite glabrous except for the scattered hairs on the lower lvs.; these petioled, ± pinnatifid or lobed, 1–2 dm. long; upper lvs. sessile, auriculate-clasping, lance-oblong, subentire, glabrous; pedicels spreading, 1–2 cm. long; sepals narrow-oblong, yellowish, 4–5 mm. long; petals yellow, spatulate, 6–8 mm. long; siliques not torulose, terete, 2–5 cm. long, stout with a

stout beak an additional 1–1.5 cm. long; seeds 1.5–2 mm. thick, dark, reticulate; $n=10$ (Karpechenko, 1924).—Common weed especially in orchards and waste places, widely distributed in Calif.; natur. from Eu. Jan.–May. The common TURNIP, with thickened roots is a cultivar of the sp.

9. **B. tournefórtii** Gouan. Annual, 1–7 dm. tall, branched at base, ± hirsute below; basal lvs. lyrate-pinnatifid, short-petioled, the upper reduced, sessile, oblong or linear; fls. crowded at anthesis; pedicels 3–10 mm. long, ascending; sepals 3 mm. long; petals pale yellow, 5–7 mm. long; siliques 3.5–6.5 cm. long, the beak 1–2 cm. long; seeds brown-purple, finely silvery-reticulate, ca. 1 mm. thick; $n=10$ (Sikka, 1940).—Fields and roadsides, Imperial Co. to Riverside Co. and w. San Bernardino Co.; natur. from N. Afr. Jan.–June.

5. *Cakile* Hill. SEA-ROCKET

Fleshy branched glabrous annuals. Lvs. deeply crenate to pinnatifid. Fls. purplish or whitish. Silicles short, transversely 2-jointed, fleshy, becoming dry and corky when ripe, sessile, flattened or ridged, the joints 1-celled and 1-seeded or the lower sometimes seedless. Seed erect in the upper, suspended in the lower joint. Cotyledons accumbent. Ca. 4 spp. of sea and lake shores of N. Am., Eu. and Afr. (Old Arabic name.)

Lvs. sinuate-dentate; petals ca. 6 mm. long; pods without hornlike processes at apex of lower joint .... 1. *C. edentula*
Lvs. pinnatifid; petals 9–10 mm. long; pods having 2 triangular protuberances at apex of lower joint .... 2. *C. maritima*

1. **C. edéntula** (Bigel.) Hook. ssp. **califórnica** (Heller) Hult. [*C. c.* Heller.] Branched from base, the branches often decumbent, up to 6 dm. long; lvs. oblanceolate to narrow-obovate, rounded at apex, sinuate-dentate, petioled, 4–8 cm. long; racemes dense; pedicels stout, 3–5 mm. long; sepals 3–4 mm. long; petals tinged purple, 6 mm. long; silicles 12–15 mm. long, the lower joint obovoid, 5–7 mm. long, the upper broadly ovoid, 8–10 mm. long, ribbed, flattened at apex; seeds somewhat ovate, compressed, brown, 5–6 mm. long; $n=9$ (Kruckeberg, 1948).—Beach sands at scattered points, from San Diego n.; to B.C. Channel Ids. May–Sept.

2. **C. marítima** Scop. Branching from base, procumbent or decumbent; lvs. 4–8 cm. long, deeply pinnatifid into oblong lobes with rounded tips; pedicels stout, ca. 2 mm. long; sepals 3 mm. long; petals pink to purplish, 8–10 mm. long; silicles ca. 15 mm. long, the upper joint flattened, ca. twice as long as lower which has 2 divergent protuberances at apex; seeds 4–5 mm. long; $n=9$ (Jaretzky, 1929).—Coastal Strand; Los Angeles Co. and Channel Ids. n. Natur. from Eu. June–Nov.

6. *Camelina* Crantz. FALSE-FLAX

Erect annual caulescent herbs. Lvs. entire to pinnatifid, sagittate-clasping. Fls. small, yellowish, in terminal racemes. Sepals equal. Stamens 6. Style slender. Silicle obovoid or pyriform, slightly parallel with the broad septum, margined, the valves convex, 1-nerved. Seeds several to many in each locule, biseriate, oblong, marginless, with incumbent cotyledons. Ca. 5 spp. of Eurasia. (Greek, *camai*, dwarf, and, *linon*, flax.)

Silicles ca. 5 mm. long, twice as long as style .... 1. *C. microcarpa*
Silicles 7–9 mm. long, 3–4 times as long as style .... 2. *C. sativa*

1. **C. microcárpa** Andrz. Stems with simple and branching rather elongate hairs; silicles ca. 5 mm. long, 4–5 mm. wide and twice as long as the style; $2n=40$ (Manton, 1932).—Reported long ago from San Gabriel Mts. (Swartout V. at 5900 ft.); more common in n. and e. states; native of Eu.

2. **C. satìva** (L.) Crantz. [*Myagrum s.* L.] Erect, 3–8 dm. high, usually branched; stems glabrous or with minute appressed stellate hairs; lower lvs. oblanceolate, petioled, 5–8 cm. long, toothed or entire, acutish; cauline lvs. largely clasping, entire, smaller;

racemes with many fls.; pedicels 1–2.5 cm. long; petals yellowish, 4–5 mm. long; silicles mostly 7–9 mm. long, 6–7 mm. broad; seeds yellow-brown, 3-angled, 1–2 mm. long, finely tubercled; $2n = 40$ (Manton, 1932).—Occasional weed in fields and roadsides; natur. from Eu. May–July.

7. Capsélla Medic. SHEPHERD'S-PURSE

Annual or biennial, with forked pubescence. Basal lvs. tufted. Fls. small, racemose, white or pink. Silicle obcordate-triangular, flattened contrary to the narrow partition, the valves boat-shaped, keeled. Style almost none. Seeds several, marginless, with cotyledons incumbent. Ca. 5 spp. of Eurasia. (Latin, *capsella*, a little box.)

1. **C. búrsa-pastòris** (L.) Medic. [*Thlaspi b.-p.* L.] Plate 27, Fig. A. Erect, 2–5 dm. tall, branched or almost simple, ± hirsute at base; basal lvs. in a rosette, runcinate-pinnatifid, petioled, 3–8 cm. long; cauline lvs. lanceolate, sessile with auricled base; pedicels slender, spreading or ascending, 10–15 mm. long; fls. white, ca. 2 mm. long; silicles obcordate, flattened, ca. 6 mm. wide, with straight or slightly convex sides; seed narrow-oblong, brownish, smooth, ca. 1 mm. long; $n = 8$ (Shull, 1937); $2n = 32$ (Löve & Löve, 1956).—A variable sp. growing as a common weed below 7000 ft. almost throughout Calif. and to Atlantic Coast and Alaska; natur. from Eu. Most of the year.

8. Cardámine L.

Annual to perennial herbs; glabrous or with simple hairs. Lvs. simple to pinnate. Inner sepals slightly saccate. Petals mostly white or purple. Fr. a strongly compressed silique, with valves coiling spirally from base at dehiscence, veinless or with indistinct median vein. Style short or distinct. Stigma slightly 2-lobed. Seeds in 1 row in each locule. Over 100 spp. of temp. regions. (Greek, *kardamon*, used for some cress.)

(Detling, L. R. E. The Pacific Coast spp. of Cardamine. Am. J. Bot. 24: 70–76. 1937. Detling, L. R. E. The genus Dentaria in the Pacific states. Am. J. Bot. 23: 570–576. 1936.)

Basal lvs. forming basal rosettes, from which arise the flowering stems.
    Plants from slender rootstocks; petals 4–6 mm. long.
        Some lower lvs. simple, others 3–7-foliolate; petals ca. 4 mm. long. High montane
            1. *C. breweri*
        Some lower lvs. 7-foliolate, but mostly more than that; petals 5–6 mm. long. Below 3000 ft.
            3. *C. gambelii*
    Plants from a slender taproot; petals ca. 2 mm. long . . . . 4. *C. oligosperma*
Basal lvs. from a thickened fleshy horizontal rhizome; petals mostly 9–14 mm. long 2. *C. californica*

1. **C. bréweri** Wats. Plate 27, Fig. B. Perennial from creeping rootstocks; stems often procumbent at base and rooting at nodes, usually glabrous, 2–6 dm. tall; lvs. mostly cauline, simple or 3–5-foliolate, the simple ones usually near the base, terminal lfts. ovate, sometimes with cordate base, 1–4 cm. long, lobed to toothed, lateral lfts. smaller, ovate to lanceolate; raceme lax; fruiting pedicels 7–15 mm. long; sepals 2 mm. long; petals white, ca. 5 mm. long; siliques ascending or erect, 1.5–2.5 cm. long, 1–1.5 mm. wide; style less than 1 mm. long; seeds 10–20, round-oblong, ca. 1 mm. long; $n = 42$–48 (Rollins, 1966).—Wet places in San Bernardino Mts. at ca. 7500 ft.; Sierra Nevada n. to B.C., Wyo. May–July.

2. **C. califórnica** (Nutt.) Greene. [*Dentaria c.* Nutt. *D. integrifolia* var. *c.* Jeps.] Plate 27, Fig. C. Glabrous perennial from deep-seated ovoid rhizomes 4–8 mm. thick; rhizomal lvs. mostly 3-foliolate, the lfts. broadly ovate, often cordate, 2–5 cm. broad, sinuate to dentate; stems slender, erect, 1–4 dm. high; cauline lvs. 2–3 (–5), 3–5-foliolate to -lobed, the lfts. lanceolate to ovate, toothed or sinuate to entire; racemes many-fld.; fruiting pedicels ascending, commonly 1–2.5 cm. long; petals pale rose to white, mostly 9–14 mm. long; siliques 2–5 cm. long, 1–2 mm. broad; style stout, 2–6 mm. long; $2n = 32$ (Manton, 1932), $n = 8$, 16 (Rollins, 1966).—Shaded banks and slopes, mostly below 2500 ft., but sometimes up to 6000; many Plant Communities; San Diego Co., Catalina Id., up through Santa Barbara Co. and on Channel Ids.; to Ore., L. Calif. Feb.–May.

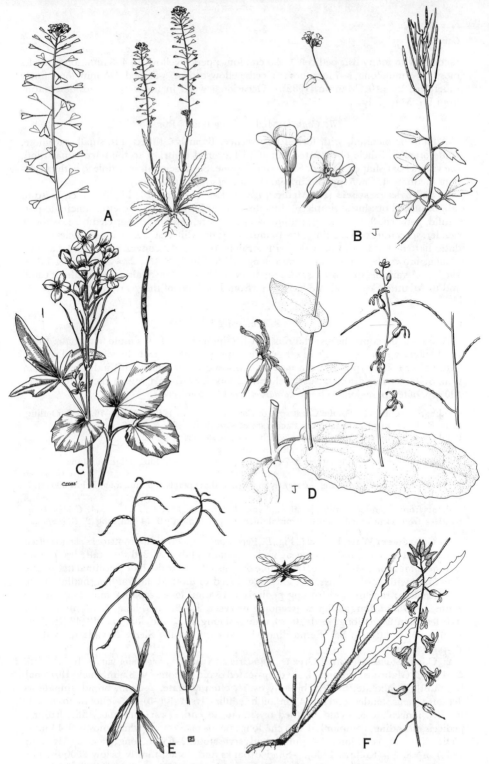

Plate 27. Fig. A, *Capsella bursa-pastoris*. Fig. B, *Cardamine breweri*. Fig. C, *Cardamine californica*. Fig. D, *Caulanthus amplexicaulis*. Fig. E, *Caulanthus cooperi*. Fig. F, *Caulanthus coulteri*.

3. **C. gambélii** Wats. [*Nasturtium g.* O. E. Schulz.] Perennial from slender horizontal rootstock; stems erect or procumbent and rooting at base, glabrous to somewhat villous, 5–9 dm. long; lvs. numerous, scattered along stem, 7–13-foliolate, lfts. sessile, oblanceolate to obovate, or roundish on lowest lvs., 5–20 mm. long, entire to few-toothed; racemes densely many-fld.; fruiting pedicels spreading to reflexed in age, 10–15 mm. long; sepals 3–4 mm. long; petals white, 6–7 mm. long; siliques erect, somewhat curved, 1.5–2.5 cm. long, ca. 1 mm. wide; style 1.5–2 mm. long; seeds 20–36.—Occasional in swampy places at low elevs. L. Calif. to Santa Barbara Co. April–June. Becoming much scarcer than in former years.

4. **C. oligospérma** Nutt. Annual or biennial from a slender taproot; stems glabrous to hispidulous, 1–3 dm. high; lvs. thin, frequently in a basal rosette and cauline, 5–11-foliolate, 2–9 cm. long; lfts. of rosette round to ovate, crenately 3–5-lobed, 3–9 mm. long, those of upper lvs. usually more elongate, oblanceolate to narrow-ovate, 4–15 mm. long; raceme 2–10-fld.; fruiting pedicels ascending to sub-erect, 3–7 mm. long; sepals ca. 1 mm. long; petals white, ca. 2 mm. long; siliques erect, 1.2–2 cm. long, 1–1.5 mm. wide, 8–28-seeded; style less than 1 mm. long; seeds winged.—Moist open woods or canyons, occasional in more open places, below 3000 ft.; Chaparral, S. Oak Wd.; Yellow Pine F., etc.; Los Angeles Co. n. to B.C. March–July.

9. *Cardària* Desv.

Perennial rhizomatous herbs. Stems branched, erect or decumbent, pubescent. Lvs. oblong, dentate, the lower petioled, the upper clasping. Fls. small, in corymbed racemes. Sepals alike at base, scarious-margined. Petals white. Stamens 6; fils. free, toothless. Silicles ovoid, subglobose or cordate, inflated, nearly or quite indehiscent, with thin fenestrate or entire septum; style slender. Seeds 2–4, pendulous, wingless. Four spp. of Old World. (Name from the *cordiform* fr. of first sp. described.)

(Rollins, R. C. On two weedy crucifers. Rhodora 42: 302–306. 1940.)

Silicles glabrous, broader than long, notched at base; style 1–2 mm. long . . . . 1. *C. draba*
Silicles pubescent, longer than broad, not notched at base; style 2–3 mm. long . . 2. *C. pubescens*

1. **C. dràba** (L.) Desv. [*Lepidium d.* L.] Hoary Cress. Pubescent or somewhat tomentulose, several-stemmed, 3–4 dm. high, leafy; lvs. 3–6 cm. long, the lower petioled, the upper narrowed to an auriculate base; racemes in a terminal paniculate cluster; pedicels spreading, slender, 6–12 mm. long; fls. 2–3 mm. long; silicles 3–4 mm. long, reniform or depressed-cordate; style slender, ca. 1 mm. long; seeds ovoid or ellipsoid, brown, almost smooth, compressed; $2n = 64$ (Manton, 1932).—Widespread as a pernicious weed in fields and waste places at low elevs.; to Wash. and Atlantic Coast; native of Eu. March–June.

2. **C. pubéscens** (C. A. Mey.) Roll. var. **elongàta** Roll. [*Hymenophysa p.* auth.] Whitetop. Minutely pubescent, 1–4 dm. high; lvs. 1–3.5 cm. long, clasping; pedicels slender, 3–8 mm. long; silicles pubescent, subglobose, 2.5–3.5 mm. long; style almost as long; seeds brown, pitted-reticulate, ca. 1.3 mm. long.—Occasional weed in waste places, alfalfa fields, etc.; to Wash., e. Coast; probably from Asia. April–July.

10. *Caulánthus* Wats.

Mostly annual, sometimes perennial herbs, glabrous or pubescent with simple hairs; stems simple or branched. Basal lvs. usually not forming a conspicuous rosette; cauline lvs. short-petioled or sessile. Fls. racemose, purple, yellow or white. Calyx somewhat flask-shaped, closed at anthesis or nearly so. Petals with narrow often crisped blades. Stamens 6, equal, distinct or sometimes united in pairs. Siliques divaricate, erect or deflexed, terete or only slightly flattened, sessile or nearly so; style usually short; stigma entire or 2-lobed. Seeds wingless or with narrow wings. Ca. 18 spp. of arid w. N. Am. (Greek, *kaulos*, stem, and *anthos*, fl., referring to cauliflower, since some spp. can be used like it.)

(Payson, E. B. A monographic study of Thelypodium and its immediate allies. Ann. Mo. Bot. Gard. 9: 233–324. 1922.)
1. Cauline lvs. sessile and auriculate at the base.
  2. Plants glabrous or inconspicuously pubescent.
    3. Stems conspicuously inflated; siliques erect; stigma deeply 2-lobed ..... 7. *C. inflatus*
    3. Stems not conspicuously inflated.
      4. Stigma entire or indistinctly 2-lobed.
        5. Fls. purplish; siliques erect or divaricate, 6–8 cm. long. Mostly s. montane
                                                                                                           1. *C. amplexicaulis*
        5. Fls. yellow; siliques reflexed, 2–4.5 cm. long. Deserts .... 2. *C. cooperi*
      4. Stigma distinctly 2-lobed; siliques 2–4 cm. long; fils. all distinct ...... 3. *C. coulteri*
  2. Plants evidently hirsute or pilose.
    6. Stigma distinctly 2-lobed.
      7. Calyx yellowish; stigma shallowly lobed. Riverside Co. s. .... 10. *C. simulans*
      7. Calyx purplish; stigma deeply lobed. Los Angeles Co. to Monterey and Madera cos.
                                                                                                                3. *C. coulteri*
    6. Stigma very small, entire. S. San Diego Co. .... 11. *C. stenocarpus*
1. Cauline lvs. sessile or petioled; but not auriculate.
  8. Calyx densely hispid-hirsute. W. edge of Colo. Desert .... 6. *C. hallii*
  8. Calyx glabrous to sparsely pilose.
    9. Plants pilose to hirsute, especially near the base. Inyo Co. north .... 9. *C. pilosus*
    9. Plants glabrous.
      10. Sepals hairy; stems ± inflated; perennial. E. Mojave Desert ..... 4. *C. crassicaulis*
      10. Sepals glabrous; stems mostly not inflated.
        11. Plants perennial; lvs. usually lyrate; petals purplish. San Gabriel and San Bernardino mts. eastward .... 8. *C. major*
        11. Plants annual or perennial; lvs. usually entire; petals greenish. Grapevine Mts. to White Mts. .... 5. *C. glaucus*

1. **C. amplexicáulis** Wats. [*Euclisia a.* Greene. *Streptanthus a.* Jeps.] Plate 27, Fig. D. Glabrous glaucous annual, simple or branched, 1–5 dm. high; basal lvs. oblong-oblanceolate, 3–10 cm. long, sinuate-dentate, with broadly winged petiole; cauline lvs. oblong, 1–7 cm. long, deeply amplexicaul, sinuate-dentate to entire, obtuse; racemes lax, few-fld., 5–20 cm. long; pedicels ascending, 8–15 mm. long; sepals purple, somewhat saccate, 5–7 mm. long, the tips recurved; petals purplish, linear, 8–10 mm. long, crisped above; siliques spreading, curved, terete, 6–8 cm. long, ca. 1 mm. thick, subsessile; style 1.5–2 mm. long; stigma entire; $n=14$ (Kruckeberg, 1957).—Loose dry slopes, 4500–8500 ft.; Yellow Pine F.; San Bernardino and San Gabriel mts., Mt. Pinos; less frequent in Joshua Tree Wd.; Antelope V., w. Mojave Desert. May–July.

2. **C. coóperi** (Wats.) Pays. [*Thelypodium c.* Wats.] Plate 27, Fig. E. Glabrous or subglabrous annual with simple or branched somewhat flexuous stems 2–6 dm. tall; radical lvs. oblanceolate, sinuate, 2–6 cm. long, with short winged petioles; cauline lvs. oblong to lanceolate, mostly entire, 1–3 cm. long, with clasping base; racemes lax; pedicels stout, recurved, 1–3 mm. long; sepals greenish, 5–6 mm. long; petals yellowish, 7–9 mm. long, somewhat crisped; siliques deflexed, terete, often arcuate, 2–4 cm. long; style 1–2 mm. long; stigma small, somewhat 2-lobed; seeds brown, oblong, not winged, ca. 1.5 mm. long, faintly cellular-reticulate; $2n=14$ pairs (Raven, Kyhos & Hill, 1965).—Common, often among shrubs, in washes and on slopes below 7000 ft.; Creosote Bush Scrub, Joshua Tree Wd.; both deserts; to Ariz., Nev. March–April.

3. **C. còulteri** Wats. [*Streptanthus c.* Greene.] Plate 27, Fig. F. Annual, hirsute-pubescent especially below, erect, the stem simple or branched, 3–8 dm. high; lower lvs. oblong-oblanceolate, sinuate-pinnatifid, 5–10 cm. long, petioled; cauline oblanceolate to oblong to lanceolate, 4–8 cm. long, subentire to sinuate-dentate, mostly amplexicaul; racemes lax, with terminal tufts of purplish buds; pedicels hirsute, reflexed, 3–10 mm. long; sepals purple, later yellowish or greenish, glabrous to hirsute, unequal, scarcely saccate, 8–14 mm. long; petals whitish with purple veins, widely spreading, crisped, 12–18 mm. long; longer fils. united; siliques divergent-ascending or pendent to ascending, glabrous, subterete to somewhat flattened, 5–10 cm. long, 2–3 mm. wide; style ca. 1 mm. long; stigma-

lobes ca. 1 mm. long; seeds dark brown, oblong, not winged, ca. 2 mm. long, faintly cellular-reticulate; $n=14$ (Rollins, 1966).—Dry slopes below 5000 ft.; V. Grassland, Chaparral, Foothill Wd.; foothills of Sierra Nevada from Madera Co. to Kern Co., s. to Eliz. Lake, Los Angeles Co. March–May.

4. **C. crassicáulis** (Torr.) Wats. [*Streptanthus c.* Torr.] Short-lived perennial; stems inflated, glaucous, glabrous, simple, 3–10 dm. high; lower lvs. oblanceolate, subentire or sinuate-dentate to runcinate, petioled, 5–15 cm. long; upper lvs. reduced, not clasping; racemes lax, to ca. 4 dm. long; pedicels stout, hirsute, 3–5 mm. long; sepals purplish, densely white-hairy, subequal, 10–15 mm. long; petals dark purple with white margins, narrow-oblong, 10–15 mm. long, not crisped; fils. distinct; siliques erect or ascending, rather stout, 10–13 cm. long; stigma subsessile, with 2 broad lobes; seeds narrowly oblong-obovate, brown, ca. 3 mm. long; $n=12$ (Rollins, 1939).—Uncommon, dry slopes and canyons, 4000–8000 ft.; Pinyon-Juniper Wd.; Clark Mt., Kingston Mts. (e. Mojave Desert); to Wyo. April–June.

5. **C. glàucus** Wats. [*Streptanthus g.* Wats.] Glabrous glaucous perennial, the stems several, 3–7 dm. high; lower lvs. suborbicular to oblong-ovate, obtuse, 5–12 cm. long, entire or lobed at base of blade only, petioled; cauline lvs. few, narrower, reduced, not amplexicaul; racemes lax, becoming 5–6 dm. long; pedicels erect or ascending, 7–16 mm. long; sepals greenish or purplish, glabrous, 8–10 mm. long; petals greenish, narrow, recurved at tip, 15–17 mm. long; siliques divaricate, frequently arcuate, 5–10 cm. long; stigma subsessile, deeply 2-lobed; seeds oblong-ovate, ca. 1.3 mm. long, not winged.— Dry rocky slopes, 5000–7500 ft.; Pinyon-Juniper Wd.; Grapevine Mts., Inyo Co. to White Mts.; Nev. May–June.

6. **C. hállii** Pays. [*Streptanthus h.* Jeps.] Simple or branched annual, sometimes with inflated stems, 2–5 (–9) dm. high, somewhat stiff-hairy on lower stems, lvs. and infl.; lower lvs. oblanceolate to oblong, coarsely sinuate-dentate to pinnatifid, 3–10 cm. long, short-petioled; cauline lvs. reduced, not auricled, mostly coarse-dentate; racemes very lax; pedicels divergent to recurved, 6–15 mm. long; sepals yellowish, hirsute, lanceolate, 6–8 mm. long; petals yellowish-white, 8–9 mm. long, spatulate, not crisped; fils. distinct; siliques divaricate, 6–10 cm. long; style 1.5–2 mm. long, stigma deeply 2-lobed; seeds dark brown, oblong, not winged.—Occasional in dry places and about washes, 2000–6000 ft.; Creosote Bush Scrub, Joshua Tree Wd., Pinyon-Juniper Wd.; Little San Bernardino Mts. along w. edge of Colo. Desert to e. San Diego Co. April–May.

7. **C. inflàtus** Wats. [*Streptanthus i.* Greene.] DESERT CANDLE. SQUAW-CABBAGE. Plate 28, Fig. A. Glabrous erect annual, sometimes slightly hirsute at base, usually simple-stemmed, conspicuously inflated, 2–7 dm. high; lvs. oblong to ovate, or the lower oblanceolate, all clasping at base, entire or denticulate, 2–7 cm. long; racemes lax below the terminal bud-tuft; pedicels ascending, glabrous to pilose, ca. 3 mm. long; sepals purple in bud, at anthesis white with purple tips, glabrous, subequal, acute, 8–10 mm. long, scarious on margins; petals white, linear, crisped near tip, slightly exceeding sepals; fils. coherent; siliques stout, erect or ascending, 5–10 cm. long; stigma subsessile, deeply 2-lobed; seeds oblique-oblong, dark brown, wingless, 2.5–3.5 mm. long; $n=10$ (Rollins, 1966).—Common on open flats and among shrubs, below 5000 ft.; Creosote Bush Scrub, Joshua Tree Wd.; w. Mojave Desert from region about Barstow, nw. to w. Fresno Co. March–May.

8. **C. màjor** (Jones) Pays. [*C. crassicaulis* var. *m.* Jones. *Streptanthus m.* Jeps.] Plate 28, Fig. B. Glabrous, somewhat glaucous perennial, with few erect stems 3–8 dm. high, often inflated, simple or few-branched; lower lvs. tufted, oblanceolate in outline, runcinate or lyrate, 5–18 cm. long, petioled; upper lvs. few, remote, linear to lanceolate, reduced, not auriculate; racemes rather few-fld., lax; pedicels stout, ascending, 3–5 mm. long; sepals glabrous, purple or yellowish with purple tips, 7–9 mm. long; petals purplish, broadly linear, somewhat crisped, ca. 12–13 mm. long; siliques erect or ascending, 8–12 cm. long; stigma subsessile, shallowly 2-lobed; seeds oblong-elliptic, brown, not winged, ca. 2.5 mm. long.—Occasional on dry loose and rocky slopes, 5500–7500 ft.; Joshua Tree Wd., Pinyon-Juniper Wd.; San Gabriel and San Bernardino mts., Grapevine Mts. (Death V.), New York and Providence mts.; to Utah. May–July.

Plate 28. Fig. A, *Caulanthus inflatus*. Fig. B, *Caulanthus major*. Fig. C, *Descurainia richardsonii*. Fig. D, *Dithyrea californica*. Fig. E, *Draba cuneifolia*. Fig. F, *Erysimum capitatum*.

9. **C. pilòsus** Wats. [*Streptanthus p.* Jeps.] Biennial or short-lived perennial, sparingly to densely hirsute about base, 3–10 dm. high; lower lvs. runcinate-pinnatifid, 5–12 cm. long, petioled, oblanceolate in outline; cauline lvs. reduced, less lobed, not auriculate; racemes lax, several dm. long; pedicels ascending, 5–8 mm. long; sepals green to purplish, 5–8 mm. long, ± pilose; petals whitish, 8–10 mm. long, crisped; siliques spreading, often arcuate, 6–12 cm. long; style ca. 1 mm. long; stigmas evidently 2-lobed; seeds narrow-oblong, not winged, almost 2 mm. long.—Dry slopes, washes, etc.; 4500–9000 ft.; Joshua Tree Wd., Pinyon-Juniper Wd., Shadscale Scrub; from Darwin, Inyo Co. to e. Ore., Ida., Utah. April–July.

10. **C. símulans** Pays. [*Streptanthus s.* Jeps.] Erect branching annual, hirsute near base, 3–4 dm. high; lower lvs. oblong to lanceolate, sparingly hirsute, deeply sinuate-dentate to subentire, 2–6 cm. long; upper cauline lvs. reduced, amplexicaul; racemes lax; pedicels recurved, hairy, 3–5 mm. long; sepals yellowish, sparsely hirsute, 5–6 mm. long; petals whitish, narrow, crisped, recurved at tip, 8–10 mm. long; siliques slender, straight, descending, terete, 4–6 cm. long; style ca. 1 mm. long; stigma 2-lobed; seeds oblong, not winged, ca. 1 mm. long.—Uncommon in rocky places, 2000–5500 ft.; Chaparral, Pinyon-Juniper Wd.; Santa Rosa Mts., Riverside Co. to interior San Diego Co. April–June.

11. **C. stenocárpus** Pays. Slender, erect, hirsute annual, simple or branched, 3–4 dm. high; cauline lvs. scattered, linear-lanceolate, subentire, sessile, clasping, 1–2 cm. long, the lowermost sinuate-dentate; racemes lax; pedicels hirsute, recurved, 1–3 mm. long; sepals purple, subglabrous, 4 mm. long; petals veined with purple, 6 mm. long, broadly linear; siliques descending, subterete, 2–4.5 cm. long, subglabrous; style 1–2 mm. long; stigma small, entire; seeds not winged.—Dry slopes, especially on burns; Chaparral, San Diego Co. (Harbison Canyon, near Dehesa, Bernardo); n. L. Calif. (Ensenada). April–May.

11. *Conríngia* Link. Hare's-Ear

Erect glabrous annuals. Lvs. sessile, clasping, elliptic, the lower sometimes narrowed at base. Fls. in elongate, terminal racemes. Sepals and petals narrow, the latter light yellow. Siliques long, slender, 4-angled, somewhat rigid, the dehiscent valves 1–3-nerved. Style short. Stigma entire or 2-lobed. Seeds uniseriate, oblong, marginless. Ca. 7 spp. of Eurasia. (Named for H. *Conring*, 1606–1681, professor at Helmstadt.)

1. **C. orientàlis** (L.) Dumort. [*Brassica o.* L.] Simple or slightly branched, 3–5 dm. tall; lvs. elliptic, 3–8 cm. long, deeply cordate-clasping; petals ca. 8 mm. long; siliques ascending, 6–10 cm. long, ca. 2 mm. thick; beak ca. 1.5 mm. long; seeds ovate, dark brown, 2–2.5 mm. long; $n=7$ (Jaretzky, 1929).—Occasional weed in waste places as w. of Yuma, at Upland, Independence; native of Eurasia. April–June.

12. *Corónopus* Trev. Wart-Cress

Strong-smelling diffuse or prostrate annuals or biennials, pubescent with simple hairs. Lvs. pinnately parted. Fls. minute, greenish-white, the capitate clusters elongating in fr. into short racemes. Sepals oval, spreading. Petals minute. Stamens 2 or 4. Silicles flattened contrary to the narrow partition, the 2 valves strongly wrinkled or tubercled, 1-seeded, indehiscent. Styles not evident. Seeds with narrow incumbent cotyledons. Ca. 6 spp. of wide distribution. (Greek, *korone*, crown, and *pous*, foot, from the deeply cleft lvs.)

1. **C. dídymus** (L.) Sm. [*Lepidium d.* L.] Stems 1.5–2 dm. long, leafy, somewhat hairy; lvs. 1–2 cm. long, with narrow divisions; pedicels 2–3 mm. long; fls. less than 1 mm. long; silicles notched, 1 mm. long, 2 mm. wide, rough-wrinkled; $2n=32$ (Manton, 1932).—Occasional weed in cismontane areas; to Atlantic Coast; natur. from Europe. March–July.

13. *Descuràinia* Webb & Berthel. Tansy-Mustard

Annual or biennial herbs, erect, branched, especially above. Lvs. ovate to obovate or oblanceolate in outline, 1–2–3-pinnate, ultimately finely or coarsely dissected, the basal in a rosette withering early. Pubescence stellate or forked. Fls. racemose, yellow or whitish, small. Siliques linear-cylindric to subclavate, somewhat torulose, the valves opening from below upward, 1-nerved. Style short or obsolete; stigma entire. Seeds in 1 or 2 series in each locule, elliptic, yellowish to brown. Ca. 20 spp. of temp. Eurasia and Am. (Named for F. *Descurain*, 1658–1740, French botanist.)

(Detling, L. R. E. A revision of the N. Am. spp. of Descurainia. Am. Midl. Nat. 22: 481–520. 1939.)

1. Upper lvs. 2–3-pinnate, mostly into linear segms.; siliques 10–30 (typically ca. 20) mm. long; silique-septum with 2–3 longitudinal nerves .... 5. *D. sophia*
1. Upper lvs. pinnate, the lfts. often deeply incised; siliques mostly less than 15 mm. long; silique-septum 1-nerved.
    2. Seeds in 1 row in each locule of silique; style conspicuous.
        3. Siliques 9–15 mm. long; stems glandular-pubescent, especially above .. 4. *D. richardsonii*
        3. Siliques 3–7 mm. long; stems without gland-tipped hairs .... 1. *D. californica*
    2. Seeds in 2 rows in each locule; style nearly or quite obsolete.
        4. Siliques linear, 12–20 mm. long .... 2. *D. obtusa*
        4. Siliques clavate to oblong-elliptic, mostly less than 12 mm. long .... 3. *D. pinnata*

1. **D. califórnica** (Gray) O. E. Schulz. [*Smelowskia c.* Gray.] Biennial, 3–8 dm. tall, moderately pubescent, not glandular, openly branched; lvs. 2–6 cm. long, the lower pinnate with 2–4 pairs of lanceolate entire to incised pinnae; sepals 1–1.5 mm. long, yellow or greenish; petals slightly longer, yellow; pedicels spreading, 3–7 mm. long; siliques 3–7 mm. long, narrowed toward both ends, ± erect; style prominent, 0.5–0.7 mm. long; seeds uniseriate, 1–3 per locule, elliptic, brownish, 1–1.5 mm. long.—Dry slopes, 7000–11,000 ft.; Montane Coniferous F.; and Pinyon-Juniper Wd.; Providence Mts., Mojave Desert, to White Mts.; Ore., Wyo., New Mex. May–Aug.

2. **D. obtùsa** (Greene) O. E. Schulz ssp. **adenóphora** (Woot. & Standl.) Detl. [*Sophia a.* Woot. & Standl.] Coarse strict biennials, 5–12 dm. tall, canescent, glandular especially in infl.; lvs. 1–6 cm. long, pinnate, with 2–5 pairs of linear to lanceolate obtuse entire to incised pinnae; pedicels divaricate-spreading, becoming 1–2 cm. long; sepals 2–2.5 mm. long; petals whitish to light yellow, 2–3 mm. long; siliques linear, 12–20 mm. long, 1–1.5 mm. wide, straight or subarcuate, very short-beaked; seeds 0.8–1 mm. long, obscurely biseriate, crowded, 24–32 in a locule; $2n = 14, 42$ (Baldwin & Campbell, 1940).—Dry slopes, 3000–7000 ft.; Joshua Tree Wd. to Yellow Pine F.; San Gabriel and San Bernardino mts. to Santa Rosa Mts.; to New Mex., L. Calif. May–June.

3. **D. pinnàta** (Walt.) Britton ssp. **menzièsii** (DC.) Detl. [*Cardamine m.* DC.] Pubescent annual, 1–6 dm. tall, simple to short-branched; lower lvs. 3–9 cm. long, bipinnate or again pinnatifid, the ultimate segms. mostly obovate and obtuse; upper lvs. pinnate to bipinnate, the segms. linear to oblanceolate; pedicels wide-spreading, 8–15 mm. long; sepals 1.5–2.5 mm. long; petals yellow, almost as long; siliques clavate, 5–12 (–15) mm. long, usually curved, 1.5–2 mm. wide; style minute; seeds biseriate, 6–10 in each locule, ca. 0.8–1 mm. long.—Dry sandy often waste places, below 8000 ft.; many Plant Communities; along the coast from Contra Costa Co. to San Diego and into interior Calif. and deserts. March–June.

D. pinnata ssp. *pinnata* occurs in the eastern states; the sp. is represented in the western states by a number of intergrading series which have been keyed out as follows for our area:

Segms. of upper lvs. narrowly oblong to linear; racemes mostly glandular-pubescent.
    Petals 2–2.5 mm. long, bright yellow; plants strict .... ssp. *menziesii*
    Petals 1–3 mm. long, whitish to yellow; plants branched below .... ssp. *halictorum*
Segms. of upper lvs. ovate to oblanceolate; racemes glabrous; petals scarcely 2 mm. long, yellow; seeds 8–12 in a locule .... ssp. *glabra*

Ssp. **glàbra** (Woot. & Standl.) Detl. [*Sophia g.* Woot. & Standl.] Largely in deserts, Santa Barbara Co. to New Mex., Chihuahua.

Ssp. **halictòrum** (Ckll.) Detl. [*Sophia h.* Ckll.] E. San Diego Co. to Siskiyou Co.

4. **D. richardsònii** (Sweet) O. E. Schulz ssp. **incìsa** (Engelm.) Detl. [*Sisymbrium i.*

Engelm.] Subglabrous to moderately pubescent, nonglandular; petals 1.5–2 mm. long; seeds ca. 0.8 mm. long; $n=21$ (Baldwin & Campbell, 1940).—Dry disturbed and rocky places, 4000–10,000 ft.; Montane Coniferous F.; mts. of L. and s. Calif. to n. Calif.; Son., Mont. June–Aug.

Ssp. **viscòsa** (Rydb.) Detl. [*Sophia v.* Rydb.] Plate 28, Fig. C. Slender biennials with mixed simple and stellate hairs, also glandular-pubescent, 3–12 dm. high; lvs. 2–10 cm. long, the lower pinnate or again pinnatifid, the ultimate segms. rather broad, obtuse, the upper pinnate with simple or toothed segms.; pedicels 6–10 mm. long, divaricate-spreading; sepals 1.5–2.5 mm. long; petals bright yellow, 2–3.5 mm. long; siliques 9–15 mm. long, linear, straight or curved upward, short-beaked; seeds uniseriate, 4–14 to a locule, red-brown, 1–1.2 mm. long; $n=7$ (Baldwin & Campbell, 1940).—Dry places, 5000–11,000 ft.; Montane Coniferous F., San Bernardino Mts.; Sierra Nevada to Wash., Rocky Mts. May–Aug.

5. **D. sophìa** (L.) Webb. [*Sisymbrium s.* L.] Leafy branched annual 2.5–6 dm. high, stellate-pubescent; lvs. 2–9 cm. long, 2–3-pinnate with fine linear to oblanceolate segms.; pedicels divaricate-ascending, 7–14 mm. long; sepals mostly equalling or exceeding the greenish-yellow petals, 2–2.5 mm. long; siliques linear, often curved, 1–3 cm. long, ca. 1 mm. thick, loosely ascending; style very short; seeds 10–20 in each locule, oblong-ellipsoid, ca. 0.8 mm. long; $2n=28$ (Manton, 1932).—Occasional weed in dry waste places below 8000 ft., from widely scattered localities in Calif.; to Atlantic Coast; natur. from Eu. May–Aug.

14. *Diplotáxis* DC.

Annual to perennial herbs much like *Brassica* in habit. Lvs. toothed to pinnatifid. Fls. yellow, white or purplish. Siliques linear, ± flattened parallel with the partition, short-beaked or beakless. Seeds ovoid, in 2 rows in each locule. Ca. 20 spp. of Medit. region and cent. Eu. (Greek, *diplous*, double, and *taxis*, row, because of the biseriate seeds.)

Annual; lvs. mostly basal, oblanceolate .... 1. *D. muralis*
Perennial; lvs. up to the infl. lanceolate .... 2. *D. tenuifolia*

1. **D. muràlis** (L.) DC. [*Sisymbrium m.* L.] Sand-Rocket. Annual, branched at base, glabrous to somewhat hispid; stems slender, 3–5 dm. high, leafy at base only; lvs. oblanceolate, sinuate-lobed, 3–10 cm. long, petioled; racemes elongate, lax in fr.; pedicels ascending, 1–2 cm. long; sepals 3–5 mm. long; petals yellow, 5–8 mm. long; siliques erect, flattish, 2–2.5 cm. long, 2 mm. wide, sessile; seeds brownish, smooth; ca. 1 mm. long; $2n=42$ (Manton, 1940).—Occasional weed along Santa Ana R. system and about Santa Barbara; native of Eu. Through the year.

2. **D. tenuifòlia** (L.) DC. [*Sisymbrium t.* L.] Wall-Rocket. Perennial, subglabrous, bushy, 3–6 dm. tall, leafy to the infl.; lvs. lanceolate; subentire to pinnatifid, 6–12 cm. long; racemes loose in fr.; pedicels 1–4 cm. long; sepals 5–8 mm. long; petals 8–10 mm. long; siliques 2–3 cm. long, suberect, stipitate; seeds brownish, smooth, ca. 1.2 mm. long; $2n=22$ (Manton, 1932).—Occasional weed especially in Los Angeles and Orange cos.; natur. from Eu. March–June.

15. *Dithyrèa* Harv. Spectacle-Pod

Annual or perennial, stellate-pubescent. Stems leafy, erect or decumbent. Lvs. sinuate-dentate to nearly entire. Fls. racemose. Sepals erect, stellate-tomentose, connivent above. Petals white to purplish or yellowish, broadly spatulate, slender-clawed. Stamens 6; anthers linear, sagittate. Silicles indehiscent or tardily dehiscent, strongly obcompressed, didymous, i.e. with 2 round locules side by side. Style almost none. Stigma large. Seeds 1 in each locule, with accumbent cotyledons. Three spp. of sw. N. Am. (Greek, *dis*, two, and *thureos*, shield, referring to the twin fr.)

Plants annual; upper lf.-blades oblong-ovate; silicles 8–10 mm. broad. Deserts . 1. *D. californica*
Plants perennial; upper lf.-blades roundish; silicles 14–15 mm. broad. Coastal Strand 2. *D. maritima*

1. **D. califórnica** Harv. [*Biscutella c.* Brew. & Wats.] Plate 28, Fig. D. Annual, with several decumbent stems from base, 1–3 dm. long; lvs. 2.5–7 cm. long, thickish, the lower obovate to oblong-ovate, coarsely sinuate-dentate, petioled, the upper oblong-ovate, subsessile; racemes dense at anthesis; pedicels ca. 2 mm. long; sepals 6–7 mm. long; petals white, 10–12 mm. long; silicles 8–10 mm. broad, notched above and below, with thickened tomentose margin; seeds flat, oblong, almost 3 mm. long; $n=10$ (Raven et al., 1965).—Common in sandy places, below 4000 ft.; Creosote Bush Scrub; both deserts from Inyo Co. to Nev., Ariz., L. Calif. March–May.

2. **D. marítima** A. Davids. Perennial from heavy cordlike underground rhizomes; lvs. to 1 dm. long, the blades rounded, 2–5 cm. in diam., fleshy, subentire; silicles 14–15 mm. broad; $n=$ca. 40 (Thompson orally for count by Miss Bartholomew).—Coastal Strand, Los Angeles Co. to San Luis Obispo Co. San Nicolas Id. Much of year.

16. *Dràba* L.

Mostly low, annual or perennial herbs, tufted, and often with stellate or branched hairs. Stems scapose or leafy. Lvs. simple. Fls. racemose, perfect, white or yellow. Sepals equal at base. Petals entire or bifid. Silicles elliptical, oblong or rarely linear, usually flat, with nerveless dehiscent valves compressed parallel with the partition. Stigma nearly entire. Seeds biseriate, numerous, usually marginless, with accumbent cotyledons. Ca. 250 spp. of the N. Hemis. (Greek, *draba*, acrid, applied by Dioscorides to some cress.)

(Hitchcock, C. L. A revision of the Drabas of w. N. Am. Univ. Wash. Publ. Biol. 11: 7–132. 1941.)
Plants annual; style scarcely if at all evident.
    Lvs. all in a basal rosette; petals bifid .... 5. *D. verna*
    Lvs. not all in a basal rosette; petals entire to emarginate.
        Pedicels mostly pubescent; lvs. usually dentate .... 2. *D. cuneifolia*
        Pedicels always glabrous; lvs. usually entire .... 4. *D. reptans*
Plants perennial; style usually well developed.
    Lvs. 5–12 mm. long; stems 1–3 cm. tall; silicles with short spreading simple hairs
        3. *D. douglasii*
    Lvs. 15–30 mm. long; stems 5–20 cm. high; silicles with minute branched hairs . 1. *D. corrugata*

1. **D. corrugàta** Wats, var. **corrugàta.** Perennial, tufted, with simple or branching crown; lvs. mostly in cushion-like rosettes, oblanceolate, 15–30 mm. long, 3–10 mm. wide, entire, grayish with dense coarse stiff simple and branched or stellate hairs; stems 1–several, mostly branched, leafy, hirsute, 5–20 cm. high; racemes usually many-fld.; pedicels pubescent, 2–10 mm. long; sepals ca. 2 mm. long, with branched hairs; petals pale yellow, fading white, 3–5 mm. long, emarginate; silicles flat or contorted, often purplish, with minute branched hairs, linear-lanceolate or somewhat wider, 7–15 mm. long, 2–3 mm. wide; styles 2–3.5 mm. long; seeds 20–30, ca. 1.2–1.5 mm. long.—Shaded slopes and rocky places, 7000–11,000 ft.; Montane Coniferous F.; San Bernardino Mts. July–Aug. Forma *vestita* (A. Davids.) C. L. Hitchc. [*D. v.* A. Davids.] has the silicles grayish with looser branched hairs; mostly San Gabriel Mts., but also on Mt. San Gorgonio.

Var. **saxòsa** (A. Davids.) M. & J. [*D. s.* A. Davids.] Stems mostly simple, leafless; silicles with some branched hairs.—Dry slopes 8500–11,500 ft.; Montane Coniferous F., Santa Rosa and San Jacinto mts. July–Aug.

2. **D. cuneifòlia** Nutt. ex T. & G. var. **cuneifòlia.** Plate 28, Fig. E. Annual, simple or branched from base, 5–25 cm. high; at least part of pubescence on lower stems not branched; lvs. ovate to obovate or oblanceolate, 1–5 cm. long, mostly with some teeth at least in lower half; infl. 10–70-fld., stellate-pubescent, usually less than half the height of the plant; pedicels 2–5 mm. long; sepals with branched hairs and 1.5–2.5 mm. long; petals white, 3.5–5 mm. long; silicles glabrous or hispidulous with simple hairs, linear to oblong-obovate, 5–15 mm. long, ca. 2–2.5 mm. broad; style nearly lacking; seeds 40–200, ca. 0.6 mm. long.—Moist sandy soil, 5000–6000 ft.; Pinyon-Juniper Wd.; New York Mts., Panamint Mts. (e. Mojave Desert); to Texas. May.

Var. **integrifòlia** Wats. [*D. i.* Greene. *D. sonorae* Greene.] Pubescence practically all branched; racemes half as long as whole stem; silicles with evident style, stellate or glabrous.—Rather common in shaded places below 5500 ft.; Chaparral, Coastal Sage Scrub, Creosote Bush Scrub to Pinyon-Juniper Wd.; Los Angeles Co. to L. Calif., Inyo Co. to Imperial Co.; to w. Fresno Co. and Utah and n. Mex. Feb.–May.

3. **D. doúglasii** Gray var. **cróckeri** (Lemmon) C. L. Hitchc. [*D. c.* Lemmon.] Tufted perennial; lvs. basal, thick and leathery, oblanceolate with prominent midrib, 5–12 mm. long, 1–2 mm. wide, ciliate with stiff simple or forked hairs and often with some hairs on surfaces; scapes 1–3 cm. tall, pubescent with slender straight unbranched hairs; racemes 2–10-fld.; pedicels ca. as long as silicles; sepals 2–2.5 mm. long, subglabrous; petals white, 4–5 mm. long; silicles ovoid, with leathery walls, little flattened, 4–7 mm. long, with short spreading simple hairs; styles 0.5–1.5 mm. long; seeds 1–2, ca. 2 mm. long.—Occasional on dry rocky ridges and slopes, 6500–7500 ft.; Yellow Pine F.; Bear V., San Bernardino Mts.; to n. Calif., w. Nev. May–June.

4. **D. réptans** (Lam.) Fern. var. **stellífera** (O. E. Schulz) C. L. Hitchc. Annual, much like *D. cuneifolia*, but with lvs. usually entire and at least some branched hairs on upper surface; stems with few stalked stellate hairs below and glabrous above; pedicels always glabrous; silicles usually nearly erect, linear, seldom as much as 2 mm. wide, with stiff appressed hairs.—Surprise Canyon, Panamint Mts. and Crooked Creek, White Mts.

5. **D. vérna** L. Annual, all the lvs. basal, spatulate to oblanceolate, 1–2.5 cm. long, entire or denticulate, with short branched hairs; scapes slender, 5–20 cm. long, glabrous or pubescent below; racemes 3–30-fld.; pedicels ascending, 15–25 mm. long; sepals ca. 1.5 mm. long; petals white, ca. 2.5 mm. long, bifid; silicles elliptic to elliptic-oblanceolate, 5–10 mm. long, 1.5–4 mm. broad, glabrous; styles ca. 0.1 mm. long; seeds 30–60; $n=7$, 15, 32 (Manton, 1932).—Rare weed in s. Calif. as at Palos Verdes Hills. Spring.

D. stenolòba Ledeb. var. **nàna** (O. E. Schulz) C. L. Hitchc. Annual; fls. yellow; silicles 8–12 mm. long.—Recently collected at 8100 ft. at Lily Springs, Mt. Hawkins, San Gabriel Mts. *Thorne.*

17. *Erùca* Mill. GARDEN-ROCKET

Annual or biennial, erect, branched. Lvs. pinnatifid to toothed. Fls. racemose, rather large, ochroleucous to yellowish or purplish, with violet veins. Sepals erect. Silique linear-oblong, thickish, somewhat 4-sided, long-beaked, the valves 3-nerved. Seeds many, in 2 rows in each cell, ellipsoid, slightly compressed; cotyledons conduplicate. Ca. 5 spp. of Medit. region. (Classical Latin name used by Pliny.)

1. **E. satìva** Hill. [*Brassica eruca* L.] Rather succulent, glabrous, 3–5 dm. high; lower lvs. 8–15 cm. long, pinnatifid or pinnately lobed, with oblong-spatulate lobes 1–3 cm. long; upper lvs. smaller, less deeply lobed; sepals 10–12 mm. long; petals 15–18 mm. long; siliques erect-appressed on stout pedicels, 1.5–2.5 cm. long and with valves keeled on back, the beak flat, almost as long as body; seeds 1.5–2 mm. long; $2n=22$ (Manton, 1932).—Waste places and fields at widely scattered localities; to Wash. and e. Coast. Natur. from Eu. May–July.

18. *Erýsimum* L. WALLFLOWER

Annual, biennial or perennial, leafy-stemmed herbs, stout, with appressed 2–3-forked hairs. Lvs. narrow, entire to dentate. Fls. in ours yellow to orange, in terminal racemes. Sepals narrow, erect. Petals clawed, with obovate spreading blades. Silique narrow, 4-sided or terete or compressed, the valves keeled by a strong midrib. Stigma broadly lobed. Seeds uniseriate, oblong, marginless or ± winged. Ca. 90 spp. of wide distribution in temp. zone. (Greek, *eryomai*, help or save, because of supposed medicinal value of some spp.)

(Rossbach, George B. New taxa and combinations in the genus Erysimum in N. Am. Aliso 4: 115–124. 1958.)

Plants not suffrutescent; caudex not noticeably elongate or long-branched above ground, erect, or at least not sprawling or widely spreading, mostly without or with only very short sterile branches; lowest lvs. mostly not marcescent.

Seeds distally winged; siliques tetragonal, fairly protrusively keeled on flatter surfaces. Largely away from the coast .... 2. *E. capitatum*
Seeds distally winged and ± along one side; siliques strongly compressed. Largely near coast.
 Siliques stiffly divaricate, usually upcurved; petals bright golden yellow, 15–20 mm. long. Native on Coastal Strand, San Diego Co. and Santa Rosa Id. .... 1. *E. ammophilum*
 Siliques erect; petals yellow, orange to purplish, 20–25 mm. long. Occasional escape as on Catalina Id. .... 3. *E. cheirii*
Plants suffrutescent; caudex elongate, long-branched above ground, sprawling or widely spreading, with elongate sterile branches; lower lvs. mostly marcescent.
 Branched base usually spreading-upcurved; lvs. commonly 2–3 mm. broad; siliques compressed parallel to septum or squarish in cross section. Coastal Strand .... 5. *E. suffrutescens*
 Branched base sprawling; siliques plump, squarish in cross section to compressed perpendicular to septum. Insular .... 4. *E. insulare*

1. **E. ammóphilum** Heller. Biennial or short-lived perennial from a long taproot; flowering stems usually simple, low; lvs. linear-oblanceolate to oblanceolate acute, narrowest toward base of stem where narrowly elongate, 1.5–3 (–6) mm. wide; stems commonly 1.5–5 dm. long to base of raceme; fls. bright golden yellow; pedicels ca. 6–10 mm. long.—Coastal Strand; San Diego Co. and Santa Rosa Id.; Monterey Bay. Feb.–May.

2. **E. capitàtum** (Dougl.) Greene. [*Cheiranthus c.* Dougl. in Hook. *E. asperum* Calif. auth.] Plate 28, Fig. F. Biennial, erect, relatively simple, coarse-stemmed, 2–8 dm. high, strigose; basal lvs. lanceolate, the lower 4–15 cm. long, 4–10 mm. wide, usually dentate or denticulate, acute to subacute, some or all of the upper foliar hairs 3-parted (2-parted hairs frequently also present or exclusively so in southern plants); pedicels stout in fr., 4–6 mm. long; sepals 8–12 mm. long; petals orange, yellow, brick-red, orange-brown to purplish maroon, 15–20 mm. long; siliques 5–10 cm. long, 1.5–2 mm. broad, 4-angled, fairly protrusively keeled on flatter surfaces; style thick, 1–2 mm. long; seeds oblong-elliptic, distally winged, ca. 1.5 mm. long; $2n = 36$ (Mulligan, 1966).—Frequent in dry stony places below 8000 ft.; many Plant Communities; largely away from the coast, through most of cismontane and montane Calif. and w. desert; to B.C., Ida. March–July. Exceedingly variable.

3. **E. cheìrii** (L.) Crantz. [*Cheiranthus c.* L.] WALLFLOWER. Erect perennial, 3–7 dm. high, strigose, grayish; lvs. lanceolate to narrower, mostly entire, acute, 4–8 cm. long, usually crowded beneath the fls. and at the ends of sterile shoots, with 2-parted hairs; fls. 2–2.5 cm. long, yellow, orange, brown-orange or purplish; siliques erect, 5–6 cm. long, rather thick, angled, the style short, bearing a bicornate stigma with long arching lobes.—Occasional garden escape as on Santa Catalina Id.; native of s. Eu.

4. **E. insulàre** Greene. Plants succulent, strongly suffrutescent, with a much branched sprawling base, tufted, densely leafy, ca. 2–3 dm. high, cinereous with minute appressed 2-forked hairs; lvs. crowded, lanceolate, entire, firm, 2–3 (–5) mm. wide; petals yellow, ca. 15 mm. long; siliques rather strict, quadrangular, 2–6 cm. long, 2.5–3.5 mm. broad; style stout, ca. 1 mm. long; seeds not winged, turgid, ca. 1.5 mm. long.—Coastal Strand; San Miguel and Santa Rosa ids.; Surf. March–May.

5. **E. suffrutéscens** (Abrams) G. Rossb. [*Cheiranthus s.* Abrams.] Suffrutescent, the branched base usually spreading-upcurved and with long vegetative stems; plants succulent; lvs. narrowly linear-lanceolate, 3–7 cm. long, 2–3 (–5) mm. wide, notably marcescent below; petals bright yellow, 15–20 mm. long; siliques coarse, squarish in cross section or compressed parallel to septum, divergent-spreading, 5–9 cm. long, 1.5–2 mm. wide; seeds ca. 1.5–2 mm. long, compressed, slightly winged distally.—Mostly Coastal Strand; s. Calif. to San Luis Obispo Co. Jan.–May.

19. *Halimolòbos* Tausch

Biennial or perennial herbs with terete mostly erect glabrous or pubescent stems. Lvs. simple, entire to deeply lobed, the basal often caducous, the cauline often differentiated. Racemes leafless. Sepals oblong, erect, pubescent. Petals white to yellow, spatulate, with narrow claw. Stamens 6; anthers small. Siliques terete on slender pedicels, not very strongly beaked. Stigma capitate. Seeds many, mostly biseriate, ellipsoid, wingless, crowded. Ca. 14 spp. of N. and S. Am. (Greek, *alimos*, of the sea, and *lobos*, pod, name used because of the resemblance to *Alyssum halimifolium*.)

1. **H. diffùsa** (Gray) O. E. Schulz var. **jaègeri** (Munz) Roll. [*Sisymbrium d.* var. *j.* Munz.] Diffusely branched perennial, suffrutescent, cinereous-tomentose; stems leafy, 3–5 dm. tall; lower lvs. 5–8 cm. long, sharply and deeply sinuate-toothed or -lobed, on short winged petioles; upper lvs. gradually reduced, sessile; racemes numerous, 5–10 cm. long in fr. with spreading pedicels 4–7 mm. long; sepals 2.5–3.5 mm. long; petals white, 3.5–4 mm. long; siliques widely spreading, pubescent, 1.5–2 cm. long, less than 1 mm. thick; seeds ca. 0.6 mm. long.—Dry rocky places, 5000–8200 ft.; Sagebrush Scrub, Pinyon-Juniper Wd.; mts. of e. San Bernardino Co. (New York, Clark, Kingston, etc.) to Inyo and White mts. and Alabama Hills.; w. Nev. May–Sept.

20. Hutchìnsia R. Br.

Low annuals or winter annuals, ± pubescent with forked hairs. Lvs. entire to pinnately lobed. Fls. racemose, small, white. Stamens 6. Silicles elongate-ovate to elliptic or lanceolate, entire at apex, compressed contrary to the narrow partition, the valves wingless, 1-nerved. Seeds 2–many in each locule. Ca. 8 spp. of N. Am. and Eurasia. (Named for Ellen *Hutchins*, Irish botanist, 1785–1815.)

1. **H. procúmbens** (L.) Desv. [*Lepidium p.* L. *Capsella p.* Fries and var. *davidsonii* Munz.] Slender annuals, the stems branching from base, erect to procumbent, glabrous to pubescent, 5–18 cm. high; lower lvs. entire to pinnately lobed, oblanceolate, short-petioled, 1–2 cm. long; cauline lvs. scattered, linear or oblanceolate, pedicels slender, spreading, becoming 7–12 mm. long; fls. ca. 1 mm. long; silicles 2–4 mm. long, obtuse, elliptical or oval; seeds light brown, smooth, oblong, 0.5–0.6 mm. long; $2n = 12$ (Manton, 1932).—Scattered, moist alkaline places, up to 8600 ft.; many Plant Communities; San Diego Co. n., from coast to desert edge, White Mts., Providence Mts., etc.; to B.C. and e. coast; Old World. Santa Barbara Ids. March–July.

21. *Lepídium* L. Peppergrass

Annual to suffrutescent perennials, glabrous to hirsute with simple hairs. Lvs. entire to bi- or tri-pinnate, sometimes clasping or perfoliate. Fls. minute, in racemes. Pedicels divaricate, terete to flattened. Sepals often pubescent. Petals white to yellow or absent. Stamens 2, 4 or 6. Silicles round, ovate, elliptic or obovate, reticulate to smooth, glabrous to hirsute, strongly obcompressed, usually notched or lobed at the ± winged apex. Style lacking or present. Seeds 2, flattened. Ca. 130 spp., widely distributed; one grown for salad. (Greek, *lepidion*, a little scale, from shape of pods.)

(Hitchcock, C. L. The genus Lepidium in the U.S. Madroño 3: 265–320. 1936.)
1. Cauline lvs. perfoliate; lower lvs. with linear segms. . . . . 10. *L. perfoliatum*
1. Cauline lvs. not perfoliate.
   2. Styles none, or less than 0.3 mm. long, usually shorter than notch of fr.
      3. Silicles notched at apex, but not or slightly winged.
         4. Pedicels terete or only slightly flattened.
            5. Sepals persisting until frs. are almost mature; pedicels somewhat narrowly wing-margined; cauline lvs. pinnatifid to lobed.
               6. Silicles ovate, prominently reticulate, with 2 acute winged divergent apical teeth; pedicels winged . . . . 13. *L. strictum*
               6. Silicles oval to obovate, not reticulate, with rounded apex; pedicels not winged
                                                                            9. *L. oblongum*
            5. Sepals deciduous along with petals and stamens or soon after; pedicels not wing-margined; cauline lvs. mostly entire.
               7. Petals wanting or shorter than sepals; cotyledons incumbent in the seed (with the back of 1 against the radicle).

            8. Lower lvs. coarsely toothed; silicles 2.5–3.5 mm. long.
                9. Branches bearing many short corymbiform racemes in the lf.-axils as well as
                    the longer terminal ones .... 12. *L. ramosissimum*
                9. Branches bearing simple long naked racemes .... 1. *L. densiflorum*
            8. Lower lvs. pinnatifid with incised-dentate divergent lobes; silicles 1.5 mm. long
                                                                11. *L. pinnatifidum*
        7. Petals equalling or exceeding sepals; cotyledons incumbent or accumbent (the
            edges against the radicle) .... 15. *L. virginicum*
    4. Pedicels strongly flattened.
        10. Plant freely branched from base, pubescent; stamens 2 or 4; petals sometimes lacking
                                                                5. *L. lasiocarpum*
        10. Plant usually simple below, glabrous; stamens 6; petals evident ... 8. *L. nitidum*
    3. Silicles plainly winged at apex with 2 prominent divergent lobes or teeth.
        11. Petals not more than 1.2 mm. long; silicles ca. 3.5 mm. long .... 2. *L. dictyotum*
        11. Petals 2–4 mm. long; silicles 5.5–7 mm. long .... 6. *L. latipes*
2. Styles well developed, at least 0.3 mm. long and exceeding the notch of the frs.
    12. Fls. white.
        13. Plants perennial, somewhat woody at base, glabrous or puberulent.
            14. Fls. 4 mm. long; silicles 5–6 mm. long .... 4. *L. fremontii*
            14. Fls. 2 mm. long; silicles 2 mm. long .... 7. *L. montanum*
        13. Plants annual to biennial, hirsute to villous .... 14. *L. thurberi*
    12. Fls. yellow; plants annual, prostrate or nearly so .... 3. *L. flavum*

1. **L. densiflòrum** Schrad. Diffuse annual 3–5 dm. tall, puberulent to pubescent; lvs. mostly oblanceolate, the basal 4–6 (–8) cm. long, coarsely toothed, the divisions also toothed, the cauline entire to somewhat toothed; racemes many, 6–15 cm. long; pedicels scarcely flattened, almost twice as long as thick; sepals ca. 1 mm. long; petals mostly lacking; silicles ca. 3.5 mm. long, glabrous, narrowly notched.—Reported from Barstow, Mojave Desert; to Wyo., New Mex.

2. **L. dictyòtum** Gray var. **dictyòtum.** Low pubescent annual, the branches decumbent to ascending, 0.2–2 dm. long; lower lvs. usually pinnatifid with 2–5 pairs of linear lobes, these sometimes cleft; cauline lvs. mostly entire, linear; racemes many-fld., usually rather compact; pedicels flattened, 1.5–3.5 mm. long, somewhat pubescent; sepals ca. 1 mm. long, pubescent; petals usually lacking; silicles ca. 3.5 mm. long, glabrous to hairy, reticulate, mostly ovate in outline, with winged apices less than 1 mm. long, usually rounded or obtuse, sometimes acute; style none; seeds ca. 2 mm. long, with incumbent cotyledons.—More or less alkaline places, below 2500 ft.; Alkali Sink; San Diego Co. to Wash., Utah. March–May.

Var. **acùtidens** Gray. Racemes often loosely fld.; silicles mostly over 4 mm. long, with acuminate divergent winged apices over 1 mm. long.—With the var. *dictyotum*; to Ore. April–May.

3. **L. flàvum** Torr. Prostrate or decumbent glabrous yellowish-green annual, the branches to 3 dm. long; basal lvs. spatulate to lanceolate, 2–5 cm. long, irregularly lobed to pinnatifid, the cauline more cuneate, somewhat smaller, entire to pinnatifid; racemes subcapitate to looser; pedicels 2–3 mm. long, terete or nearly so; sepals oblong, yellow-green, ca. 1–1.3 mm. long; petals sulphur-yellow, ca. 2 mm. long; silicles oval, 2–3 mm. long, glabrous, reticulate with distinct divergent apices; style 1.5–2 mm. long; seeds ca. 1 mm. long.—Common in semi-alkaline flats and washes below 4500 ft.; Creosote Bush Scrub, Joshua Tree Wd.; deserts from Inyo Co. s.; to Nev., L. Calif. March–May. A form occurs in Borrego V., e. San Diego Co. with roundish silicles 3–4.5 mm. long and style 1–1.5 mm. long; it is var. *felipense* C. L. Hitchc.

4. **L. fremóntii** Wats. Plate 29, Fig. A. Rounded suffrutescent perennial with many branching stems 2–5 dm. high, glabrous and glaucous; lvs. linear, acute, 2–5 (–10) cm. long, entire or pinnatifid into few salient elongate lobes; infl. much branched, somewhat leafy; pedicels 5–8 mm. long, slender; sepals 1.5–2 mm. long; petals white, ca. 3 mm. long; silicles broadly ovate to obovate, 4–7 mm. long, faintly nerved, with wide winged margins; styles 0.4–0.8 mm. long; seeds almost 2 mm. long.—Common in rocky and

Plate 29. Fig. A, *Lepidium fremontii*. Fig. B, *Lepidium lasiocarpum*. Fig. C, *Lesquerella kingii* ssp. *bernardina*. Fig. D, *Lesquerella palmeri*. Fig. E, *Lyrocarpa coulteri* var. *palmeri*. Fig. F, *Raphanus sativus*.

sandy places, below 5000 ft.; Creosote Bush Scrub, Joshua Tree Wd.; Inyo Co. to n. Riverside Co.; to Utah, Ariz. March–May.

5. **L. lasiocárpum** Nutt. Plate 29, Fig. B. Branched spreading annual, the branches 0.5–2.5 dm. long, hirsute-hispid; lvs. linear to oblanceolate, toothed to pinnatifid, 1–2 cm. long or the lower to 6 cm. long and petioled; racemes 3–8 cm. long; pedicels distinctly flattened on both surfaces, 2–5 mm. long, usually pubescent on lower side; sepals ca. 1 mm. long; petals narrow, usually shorter than sepals, sometimes none; silicles subglabrous to hispid, suborbicular to somewhat longer than wide, 3–4.5 mm. wide, finely reticulate, slightly winged at apex; style almost or quite lacking; seeds ca. 1.2 mm. long.— Common on grassy slopes and sandy flats, below 5000 ft.; Creosote Bush Scrub, Shadscale Scrub, deserts from Inyo Co. to Imperial Co., less frequent in cismontane Calif. in Coastal Sage Scrub, V. Grassland, etc. from Santa Barbara Co. s. and to Utah. Feb.–May. Intergrading with var. *georginum* (Rydb.) C. L. Hitchc. a less pubescent, fewer branched form, with pedicels glabrous on lower side.—Occasional on deserts; to Utah.

6. **L. látipes** Hook. Low spreading pubescent annual, branched from base, the stems from 0.3–1.5 dm. long; lvs. 5–10 (–14) cm. long, at least the lower pinnatifid into 3–10 pairs of linear entire or dissected divisions, the upper entire; racemes mostly 2–6 cm. long, very dense; pedicels 2–3 mm. long, wide and flat; sepals pubescent, ca. 1.3 mm. long; petals 2–4 mm. long, greenish; silicles coriaceous, glabrous to pubescent, oblongovate, 5.5–7 mm. long, with acute winged apices ca. 2 mm. long and very narrow sinus; seeds ca. 1.6 mm. long.—Alkaline flats and beds of winters pools, below 2000 ft.; largely V. Grassland; San Diego Co. to n. Calif. San Clemente and Santa Cruz ids. March–May.

7. **L. montànum** Nutt. ssp. **cinèreum** (Thell.) C. L. Hitchc. Biennial or perennial, 2–3 dm. tall, 1- to few-stemmed, densely cinereous with hairs ca. twice as long as thick; basal lvs. pinnately divided into entire segms. 1–2.5 mm. wide; cauline lvs. few, reduced, entire to lobed; racemes mostly 2–4 cm. long and many-fld.; pedicels terete, slender, 5–8 mm. long; sepals whitish, ca. 1 mm. long; petals white, ca. 2 mm. long; silicles 3–3.5 mm. long, glabrous, with minute apical notch; style 0.3–0.5 mm. long; seeds ca. 1.2 mm. long.—Dry places, 2600–5000 ft.; Creosote Bush Scrub to Pinyon-Juniper Wd.; New York Mts. and Mesquite V. near Kingston, e. Mojave Desert; to nw. Ariz. April–May.

8. **L. nítidum** Nutt. Annual, usually erect and simple at base, sometimes with spreading branches from base; stems glabrous to moderately pubescent, 0.5–4 dm. long; lower lvs. 3–10 cm. long, pinnately parted into narrow segms., the cauline smaller, pinnatifid to entire; racemes rather lax in fr.; pedicels densely pubescent, very much flattened; sepals ovate, ca. 1 mm. long; petals spatulate, 0.5–1.5 mm. long; silicles ovate to suborbicular, convex below, somewhat concave above, glabrous, 3.5–6 mm. long, without divergent apices, the margins upturned; stigma subsessile; seeds ca. 2 mm. long.—Common in open places below 3000 ft., throughout cismontane s. Calif.; Chaparral, V. Grassland, Coastal Sage Scrub; to L. Calif. and Wash. Channel Ids. Feb.–May. A form in the w. Mojave Desert is var. *howellii* C. L. Hitchc. with densely pubescent stems and silicles minutely pubescent on margins.

9. **L. oblóngum** Small. Much branched diffuse annual, 0.5–2 dm. tall, hirtellous to villous; lvs. pinnatifid to laciniately lobed or cleft, the basal lvs. ca. 3 cm. long, with lobed pinnae, the cauline smaller, laciniate with central rachis to 4 mm. wide; racemes many, 6–9 cm. long; pedicels somewhat flattened, scarcely wing-margined; sepals slightly over 1 mm. long, not long-persistent; petals 0 or minute; silicles glabrous or sparsely pectinate, flat, oval to oblong-obovate, 2.5–3.5 mm. long, indistinctly reticulate, narrow-winged, the apices rounded with a small v-shaped sinus; stigma subsessile; seeds ca. 1 mm. long.— Well distributed in s. Calif., including some of the ids.; apparently a native of S. Am. March–May.

10. **L. perfoliàtum** L. Shield-Cress. Erect annual, mostly glabrous, branched, 2–5 dm. high; lower lvs. bipinnatifid into linear segms., the middle cauline entire, auriculate, the upper perfoliate; pedicels spreading, slender, terete, 4–8 mm. long; sepals pilose, ca.

Lepidum

1 mm. long; petals yellow, narrow, slightly longer; silicles rhombic-ovate, ca. 4 mm. long, minutely notched; style ca. equal to sinus; seeds elliptic-oblong, almost 2 mm. long, narrow-winged; $n=8$ (Jaretzky, 1929).—Sparingly but widely natur. in Calif. below 7000 ft.; to Wash.; from Eu. March–June.

11. **L. pinnatífidum** Ledeb. Erect, annual or biennial, subglabrous, branched, leafy, 2–4 dm. high; basal lvs. broadly lanceolate in outline, 5–8 cm. long, pinnatifid with incised-dentate, divergent lobes; cauline lvs. smaller, less divided; racemes terminal and axillary; pedicels pubescent, slender, terete, divergent, 3–4 mm. long; sepals lance-ovate to sub-elliptic, white-margined, setulose, 1 mm. long; petals rudimentary; silicles broadly- to ovate-elliptic, minutely wing-margined, obtusely emarginate, ca. 1.5 mm. long; seeds ovoid-ellipsoid, somewhat tuberculate, brownish, less than 1 mm. long.—Reported from Oak View, Ventura Co. and Smeltzer, Orange Co.; native of Old World. May.

12. **L. ramosíssimum** A. Nels. Much-branched, puberulent, apparently biennial, 1.5–5 dm. tall; lower lvs. few-toothed, the upper linear, entire; infl. with many short corymbiform racemes in lf.-axils as well as the longer terminal racemes; pedicels somewhat wing-margined; sepals ca. 1 mm. long, pubescent; petals linear, shorter; silicles 2.5–3.5 mm. long, elliptic, shallowly notched and winged at apex; stigma sessile.—Reported from Chollas V., e. San Diego Co.; Rocky Mts.

13. **L. stríctum** (Wats.) Rattan. Pubescent annual, erect to spreading with branches 0.5–2 dm. long, glabrous or nearly so; racemes crowded, 2–5 cm. long; pedicels divergent, somewhat flattened and wing-margined, 1–2 mm. long; sepals scarcely 1 mm. long, pilose, persistent; petals minute; silicles ovate to rounded, 2.2–3 mm. long, reticulate, biconvex, with small wings at apex and open sinus; style 0; seeds ca. 1 mm. long; $2n=16$ (Manton, 1932).—Common in hard beaten soil along paths, etc.; Los Angeles Co. to Ore., Utah. Apparently introd. from S. Am. March–May.

14. **L. thúrberi** Woot. Annual, 1–6 dm. tall, branched, canescent-pubescent, with longer and shorter hairs; basal lvs. 3–6 cm. long, petioled, pinnatifid with 3–8 pairs of segms. which are usually lobed to parted; the cauline lvs. reduced, pinnatifid to entire; racemes many-fld.; sepals 1–1.5 mm. long; petals white, 2–3 mm. long, narrowly wing-margined near apex, with shallow notch; style 0.3–0.6 mm. long.—Reported from Barstow, San Bernardino Co.; New Mex. and Ariz.

15. **L. virgínicum** L. var. **pubéscens** (Greene) Thell. [*L. intermedium* var. *p.* Greene. *L. bernardinum* Abrams.] Freely branched annual, 1.5–6 dm. tall, pubescent; basal lvs. incised to pinnate, 5–15 cm. long; cauline lvs. incised to simple; racemes numerous, many-fld., elongate; pedicels slender, terete, usually somewhat longer than frs.; sepals ca. 1 mm. long; petals white, equal to or longer than sepals; silicles usually somewhat longer than broad, 2.5–4 mm. long, scarcely margined, shallowly notched at apex; style almost lacking; seeds 1–1.5 mm. long.—Widespread in waste places, along roadsides, etc., below 7000 ft.; Chaparral, Coastal Sage Scrub, etc.; cismontane from L. Calif. to n. Calif. Channel Ids. March–Aug.

Var. **robinsònii** (Thell.) C. L. Hitchc. [*L. r.* Thell.] Plants 1–2 dm. tall; cauline lvs. parted or with narrow lobes; Coastal Sage Scrub, Chaparral; from Los Angeles Co. s. to L. Calif. Channel Ids. Jan.–April.

22. *Lesquerélla* Wats. BLADDER-POD

Low annual to perennial herbs, densely stellate-pubescent. Lvs. simple. Fls. racemose, yellow in our spp. Sepals oblong to elliptical. Petals longer than sepals, obovate to spatulate. Stamens 6; anthers sagittate. Silicles generally inflated, subglobose to obovoid or ellipsoid; valves nerveless, dehiscent; locules 2–15-seeded. Style slender; stigma entire or nearly so. Seeds flattened, marginless or narrowly winged. Ca. 40 spp., mostly N. Am. (L. Lesquereux, 1805–1889, American botanist.)

(Payson, E. B. Monograph of the genus Lesquerella. Ann. Mo. Bot. Gard. 8: 103–236. 1921.)
Plants annual, without basal rosette; style shorter than fr. .... 1. *L. palmeri*
Plants perennial, with basal rosette.

Pods obtuse at the apex, subglobose .... 2. *L. kingii*
Pods acute at the apex, ovoid or ellipsoid .... 3. *L. wardii*

1. **L. pálmeri** Wats. [*L. gordonii* var. *sessilis* Wats.] Plate 29, Fig. D. Annual, with slender ascending stems 1–3 dm. long; basal lvs. entire to lyrate with few lobes, oblanceolate, 2–5 cm. long, petioled; cauline lvs. linear to oblanceolate, subsessile, entire; fruiting pedicels sigmoid, ascending to recurved, 1–1.5 cm. long; sepals ca. 3 mm. long; petals broadly spatulate, ca. 5–7 mm. long; silicles subglobose, 3–4 mm. thick, sparsely stellate-pubescent; seeds round-oblong, flat, not winged, almost 2 mm. long; $n=5$ (Rollins, 1966).
—In sandy places, below 3500 ft.; Creosote Bush Scrub; e. Mojave and ne. Colo. deserts; to Ariz., Utah. March–May.

2. **L. kíngii** Wats. [*Vesicaria k.* Wats.] ssp. **bernardína** (Munz) Munz. [*L. b.* Munz.] Plate 29, Fig. C. Resembling ssp. *kingii*, but petals 9–10 mm. long; styles 6–9 mm. long.—Dry flats, at 6600–6700 ft.; Yellow Pine F.; e. end of Bear V., San Bernardino Mts. May–June.

Ssp. **kíngii**. Silvery-stellate perennial; stems decumbent, 5–15 cm. long; basal lvs. ovate to suborbicular, entire, obtuse, 2–5 cm. long, petioled; cauline lvs. oblanceolate, 0.5–1.5 cm. long; pedicels spreading to recurved, sigmoid, 8–10 mm. long; sepals ca. 5 mm. long; petals narrowly spatulate, 6–7 mm. long; silicles subglobose, 3–5 mm. thick, stellate-pubescent; style 3–4 mm. long; seeds 2–4, not winged, brown, flattened, with thickened edge, ca. 2 mm. long.—Dry rocky slopes, 5000–9000 ft.; mostly Pinyon-Juniper Wd.; mts. of e. Mojave Desert (Clark, New York, Providence, Panamint, White); to Nev. March–June. A form in the Inyo-White Range at 10,000 ft. and above is var. *cordiformis* (Roll.) Maguire & Holmgren. [*Physaria c.* Roll.] Lvs. often narrow-spatulate; silicles 2–4 mm. long.

3. **L. wárdii** Wats. Perennial, the stems mostly prostrate, 4–10 cm. long; radical lvs. 1–4 cm. long, ovate or rounded, mostly entire, with a slender petiole; cauline lvs. broadly oblanceolate to sublinear, 8–18 mm. long; petals ca. 7 mm. long; pedicels 5–7 mm. long; pods ovoid or ellipsoid, 4–10 mm. long; style 3–5 mm. long.—At 5700–5800 ft.; Pinyon-Juniper Wd.; New York Mts., e. Mojave Desert; to Utah, Nev. May–July.

23. *Lobulària* Desv. Sweet-Alyssum

Low perennials with narrow entire lvs. and appressed 2-pointed hairs. Petals white, entire. Silicles orbicular as in *Alyssum*. Ca. 5 spp. of Medit. region. (Latin, *lobulus*, small lobe, possibly referring to 2-lobed hairs.)

1. **L. marítima** (L.) Desv. [*Clypeola m.* L. *Alyssum m.* Lam.] Much-branched decumbent strigose perennial, the stems 0.5–2.5 dm. long; lvs. linear to linear-lanceolate, 1–5 cm. long; fls. sweet-smelling, 3–4 mm. wide; pedicels 5–8 mm. long; silicles ca. 2.5 mm. long; seeds 2, wingless, brown, round-oblong, ca. 1.3 mm. long; $n=12$ (Jaretzky, 1928).
—Common escape from gardens and growing in waste places, roadsides, etc., at low elevs.; native of Eu. Flowering much of year.

24. *Lyrocárpa* Hook. & Harv.

Herbaceous annuals or perennials, densely pubescent with dendritic hairs. Stems branched, 2–8 dm. high. Lvs. petioled, repand to pinnatifid. Infl. loosely racemose. Calyx cylindrical, pubescent, the outer sepals subsaccate, longer than inner. Petals much longer than sepals, the blade often twisted. Stamens tetradynamous; 2 discoid nectar-glands at base of each stamen. Silicles obcordate to panduriform, pubescent, flattened contrary to the narrow partition; stigma large, bifid; style short or obsolete. Seeds 3–10 per locule, orbicular, brown, wingless. Two spp. of sw. N. Am. (Greek, *lyra*, a lyre, and *karpos*, fr.)

(Rollins, R. C. A revision of Lyrocarpa. Contr. Dudley Herb. 3: 169–173. 1941.)

1. **L. còulteri** Hook. & Harv. var. **pálmeri** (Wats.) Roll. [*L. p.* Wats.] Plate 29, Fig. E. Stems several, 3–5 dm. long, with flexuous branches, suffrutescent, lvs. mostly

canescent, lyrately pinnatifid, 3–5 cm. long, short-petioled, the segms. linear to oblong; fruiting pedicels 2–6 mm. long; sepals ca. 1 cm. long; petals tawny, 18–20 mm. long; silicles obcordate, 10–15 mm. long, nearly as wide, ± curled; seeds 3–5 in a locule, round, flat, ca. 3 mm. wide; $2n=20$ (Rollins, 1941).—Among rocks and in canyons, below 2000 ft.; Creosote Bush Scrub; w. edge of Colo. Desert, Borrego region to L. Calif. Dec.–April.

25. Matthiòla R. Br. STOCK

Biennial or perennial herbs with a close stellate tomentum, erect, branched at least above. Lvs. oblong to linear. Fls. purple to white, sweet-scented, racemose. Sepals erect, the lateral saccate at base. Petals with long claws and broad rounded blades. Siliques elongate, subterete, torulose; the valves dehiscent, keeled, 1-nerved. Stigmas thickened or horned at back. Seeds uniseriate, wing-margined, suborbicular. Ca. 50 spp. of w. Asia and Medit. region. (Named for P. A. *Matthioli*, 1500–1577, Italian botanist and physician.)

1. **M. incana** (L.) R. Br. [*Cheiranthus i.* L.] Plants 3–5 dm. tall; lvs. entire to sinuately dentate, 5–15 cm. long; sepals ca. 1 cm. long; petals ca. 2 cm. long; siliques 8–16 cm. long, 4–5 mm. thick; seeds almost 2 mm. wide; $n=7$ (Allen, 1924).—Sandy places and bluffs along the seacoast, occasional inland; natur. from Eu. March–May.

26. Physària Gray. DOUBLE BLADDERPOD

Low silvery-stellate cespitose perennials; stems simple. Basal lvs. usually many, petioled, oblanceolate to obovate or almost round, entire to dentate; cauline lvs. few, entire to dentate. Fls. racemose, yellow. Sepals linear-oblong, pubescent, often cucullate at apex. Petals usually spatulate. Stamens 6. Silicles didynamous, pubescent, dehiscent, inflated in ours, with apical sinus. Style slender, persistent. Seeds brown, wingless, several in each locule. Ca. 14 spp. of sw. N. Am. (Greek, *phusa*, bellows, because of the inflated pod.)

1. **P. chàmbersii** Roll. Stems 5–15 cm. long; radical lvs. round to obovate, entire or dentate, petioled, 3–6 cm. long; cauline lvs. entire, spatulate, 1–2 cm. long; sepals 6–8 mm. long; petals 10–12 mm. long; silicles 1–1.5 cm. long; style 6–8 mm. long; seeds round, flat, brown, 2–3 mm. broad; $2n=10$ (Rollins, 1966), $n=4, 8, 12$ (Mulligan, 1967).—Dry limestone slopes, 5000–8000 ft.; Pinyon-Juniper Wd.; Clark Mts., e. San Bernardino Co. and Grapevine Mts., Inyo Co.; to Utah, Nev. May.

27. Ráphanus L. RADISH

Annuals or biennials, erect, branched. Lvs. lyrate. Fls. showy, purple or yellow, fading white. Petals long-clawed. Siliques torulose or cylindric, coriaceous, indehiscent, several seeded, continuous and spongy within between the seeds, with no proper partitions, tapering above into the long, persistent slender style. Seeds globose. Ca. 4 spp. of Eurasia. (Greek, *raphanos*, quick appearing, because of rapid germination of seeds.)

Fr. markedly constricted between the seeds and breaking readily into 1-seeded joints; beak slender
    1. *R. raphanistrum*
Fr. not markedly constricted between the seeds and not breaking up into joints; beak long-conical
    2. *R. sativus*

1. **R. raphanístrum** L. JOINTED CHARLOCK. Like the next sp., but fls. yellow, aging white; siliques nearly cylindric when fresh, constricted between the seeds when dry, 4–6 mm. thick, slender-beaked, transversely divided into 1-seeded segms. when ripe; $n=9$ (Karpechenko, 1924).—Occasional weed in waste places; natur. from Eu. April–June.

2. **R. satívus** L. WILD RADISH. Plate 29, Fig. F. Freely branched, 5–12 dm. tall; subglabrous to scattered-hispid; lower lvs. pinnately parted, 1–2 dm. long, with large rounded terminal segm.; upper lvs. toothed; pedicels 1–2 cm. long, ascending; sepals narrow, ca. 9–10 mm. long; petals white with purplish or rose veins, or yellowish and purplish, 15–20 mm. long; seeds brownish, faintly reticulate, 2–3 mm. in diam.; $n=9$ (Karpechenko,

1924).—Common weed in waste places, fields, etc. through much of Calif.; natur. from Eu. Feb.–July. The Garden Radish is a cult. form.

28. Rapistrum Crantz

Annual to perennial, ± branched. Lvs. mostly lyrate-pinnatifid to bipinnate. Fls. rather small; sepals erect-divaricate, not or scarcely saccate. Petals yellow, rarely white, clawed. Siliques on thickened pedicels, transversely 2-jointed, appressed, the upper joint globose, 8-ribbed, abruptly slender-beaked and much thicker than the lower joint. Seeds ovoid or ellipsoid. Ca. 8 spp., largely Medit. (Greek, *rhapis*, rape, and *astrum*, appearance.)

1. **R. rugòsum** (L.) All. [*Myagrum r.* L.] Annual or biennial, 2.5–6 dm. high, ± stiff-hairy; lower lvs. 5–15 cm. long, petioled, the upper subentire; sepals 2.5–3.5 mm. long; petals yellow with darker veins, 5–7 mm. long; pedicels thickened; siliques with upper joint conspicuously beaked and 3–4 times as long as the lower joint; seeds 1–2 mm. long, smooth, yellow-brown; $2n=16$ (Mandon, 1932).—Occasional weed as at Vista, San Diego Co.; native of Eu. Summer.

29. Rorippa Scop.

Aquatic to terrestrial annual to perennial herbs. Lvs. usually glabrous, commonly pinnate to pinnatifid. Fls. white or yellow, in terminal racemes. Petals with nectariferous glands. Frs. short silicles or siliques, slender to subglobular, terete or nearly so, the valves usually convex, nerveless to weakly nerved. Seeds numerous, small, marginless, turgid usually in 2 ± irregular rows in each locule. A fairly large genus of temp. regions. (Name from an old Saxon word, *rorippen*.)

(Green, Peter S. Watercress in the New World. Rhodora 64: 32–43. 1962.)
1. Stems rooting at the nodes; fls. white .... 2. *R. nasturtium-aquaticum*
1. Stems from underground rhizomes (if perennial); fls. yellow.
   2. Plants annual; petals shorter than sepals.
      3. Pods strongly curved; lf.-segms. linear to oblong, usually acute; style to 0.5 mm. long, stout; pedicels 1–2 mm. long .... 1. *R. curvisiliqua*
      3. Pods straight or ± curved; lf.-segms. often rounded to obovate; style longer; pedicels longer.
         4. Pedicels 2–4 mm. long; style ca. 1 mm. long; stems branched from base .. 3. *R. obtusa*
         4. Pedicels 5–6 mm. long; style shorter; stems branched above .... 4. *R. palustris*
   2. Plants perennial with creeping rootstocks; petals longer than sepals .... 5. *R. sinuata*

1. **R. curvisilíqua** (Hook.) Bessey. [*Sisymbrium c.* Hook.] Annual or biennial, glabrous, the stems diffusely branched, 1–3 dm. long; lvs. pinnately divided into usually acute, toothed or entire lanceolate to oblong lobes; lower lf.-blades 2–8 cm. long, with somewhat shorter petioles; cauline lvs. gradually reduced and subsessile; pedicels mostly 1–2 mm. long; sepals 1–2 mm. long; petals somewhat shorter; siliques linear, terete, usually curved, 6–10 mm. long, 1–1.5 mm. thick; style less than 1 mm. long; seeds brown, ca. 0.5 mm. long, finely cellular-reticulate; $n=8$ (Rollins, 1966).—Frequent in wet or damp places, below 10,000 ft.; many Plant Communities; throughout cismontane Calif.; to L. Calif., Rocky Mts. Variable. April–Sept.

2. **R. nastúrtium-aquáticum** (L.) Schinz & Thell. [*Sisymbrium n.-a.* L. *Nasturtium officinale* R. Br.] WATER-CRESS. Plate 30, Fig. A. Aquatic perennial with prostrate or ascending stems 1–6 dm. long, rooting freely; lvs. pinnate, glabrous, 1–10 cm. long, with 3–11 ovate (or terminal rounded) lfts. 5–25 (–40) mm. long, subentire; pedicels divergent; fls. 3–4 mm. long; siliques ascending, curving; style scarcely evident; seeds ca. 1 mm. long, brown, reticulate; $2n=32$ (Howard & Manton, 1946).—Common in quiet water, on wet banks, etc., below 8000 ft.; many Plant Communities; cismontane and desert; natur. from Eu. March–Nov. The cult. Water-Cress.

3. **R. obtùsa** (Nutt.) Britton. [*Nasturtium o.* Britt.] Glabrous or subglabrous annual, the stems diffusely branched at the base, 1–3 dm. long; lvs. pinnatifid with rounded or obovate sinuately toothed divisions, the lower lvs. less divided, with blades 2–5 cm. long,

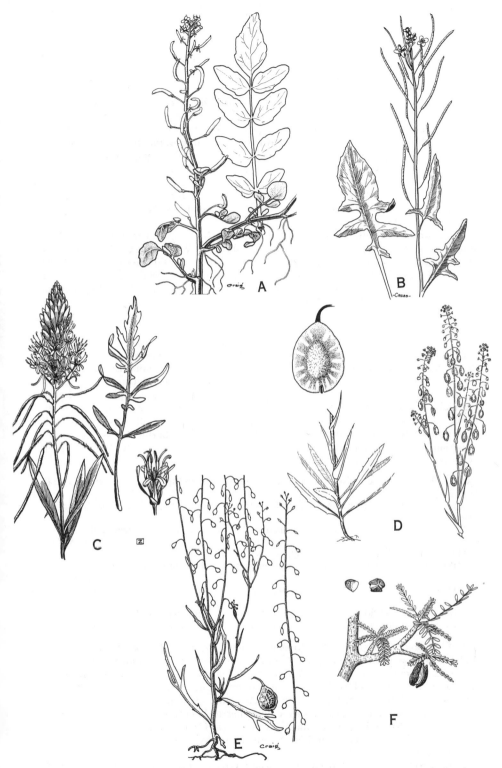

Plate 30. Fig. A, *Rorippa nasturtium-aquaticum*. Fig. B, *Sisymbrium irio*. Fig. C, *Stanleya pinnata*. Fig. D, *Thysanocarpus curvipes*. Fig. E, *Thysanocarpus laciniatus*. Fig. F, *Bursera microphylla*.

petioled, the upper more divided, 1–3 cm. long, subsessile; pedicels spreading, 2–4 mm. long; fls. 1 mm. long; siliques subglobose to oblong, 3–8 mm. long, 2–3 mm. thick; style ca. 1 mm. long; seeds light brown, ca. 0.5 mm. long, round-oblong.—Wet places, mostly 5000–8000 ft.; Montane Coniferous F.; San Bernardino and San Gabriel mts.; Sierra Nevada to B.C. and Atlantic states. June–Sept.

4. **R. palústris** L. ssp. **occidentàlis** (Wats.) Abrams. [*R. islandica* var. *o.* (Wats.) Butters & Abbe.] Annual or biennial, nearly or quite glabrous, the stems erect, branched above, 3–8 dm. high; lower lvs. lyrate-pinnatifid, 8–14 cm. long, wing-petioled, the upper nearly sessile, dentate or somewhat lobed; racemes long, lax in fr.; pedicels slender, widely divaricate, 5–6 mm. long; fls. 2 mm. long; siliques oblong-linear, straight or ± curved, 6–12 mm. long; style less than 1 mm. long; seeds brown, ca. 0.4 mm. long.—Occasional in wet places, often with roots immersed, below 6000 ft.; several Plant Communities; San Diego Co. to Alaska. April–July.

5. **R. sinuàta** (Nutt.) Hitchc. [*Nasturtium s.* Nutt.] Perennial from creeping deep-seated rootstocks; stems several, decumbent, glabrous, 1–3 dm. long; lvs. 4–8 cm. long, oblong to lance-elliptic, regularly sinuate- or pectinate-pinnatifid, with subentire, linear-oblong lobes; pedicels 6–10 mm. long; sepals 3–4 mm. long; petals ca. 6–7 mm. long; seeds ca. 1 mm. long; $n = 8$ (Rollins, 1966).—At 3000–5000 ft.; Little Lake, Inyo Co. and in n. Calif.; to Wash., Ill., Tex. May–Sept.

30. *Sìbara* Greene

Annual or biennial herbs. Stems single or several from the base, divaricately branched, glabrous or sparsely pubescent below with simple or branched hairs. Lvs. pectinate to runcinate-pinnatifid, the upper cauline sometimes almost entire, glaucous. Fls. small, in lax racemes. Sepals narrowly ovate to almost oblong. Petals white to pink or purplish, spatulate to almost oblong. Nectar-glands small, subtending or surrounding single stamens, otherwise absent or obsolete. Siliques linear, flattened parallel to septum or subterete, the valves nerveless to nerved below. Seeds oblong to suborbicular, winged or not, uniseriate, with accumbent or incumbent cotyledons. Ca. 11 spp. of N. Am. (Anagram of *Arabis*.)

(Rollins, R. C. Generic revisions in the Cruciferae: Sibara. Contr. Gray Herb. 165: 133–143. 1947.)

Lvs. pectinate; siliques ascending to reflexed, less than 1.5 mm. wide; seeds wingless. Insular and deserts.
    Mature siliques pendulous to reflexed, less than 2 cm. long; pedicels somewhat pubescent
        1. *S. deserti*
    Mature siliques divaricately ascending, mostly 2–5 cm. long; pedicels glabrous.
        Pedicels 6–12 mm. long; basal lvs. caducous; styles not expanded toward apex. Insular
        2. *S. filifolia*
        Pedicels 2–3 mm. long; basal lvs. persistent; styles expanded toward apex. Desert
        3. *S. rosulata*
Lvs. runcinate-pinnatifid, the lobes oblong to obovate; siliques divaricate, 1.5–2 mm. wide; seeds winged. Cismontane .... 4. *S. virginica*

1. **S. déserti** (Jones) Roll. [*Thelypodium d.* Jones. *Arabis d.* Abrams.] Stem 1, 1–3 dm. high, sparsely pubescent with minute branched hairs; basal lvs. caducous; cauline lvs. petiolate, 2–4 cm. long, sparsely pubescent, the lower pinnate with segms. 4–8 mm. long, the upper entire; pedicels widely spreading to descending, sparsely pubescent, 3–4 mm. long; sepals pubescent, ca. 2 mm. long; petals white, spatulate, 2–3 mm. long; siliques flattened, sparsely pubescent, slightly descending to loosely reflexed, 1–1.5 cm. long; style stout, 1–1.5 mm. long; seeds oblong, wingless, ca. 1 mm. long; $n = 13$ (Rollins, 1947). —Rare, 2000–4000 ft.; Creosote Bush Scrub; Death V., Panamint V., Saline V.; Nev. March–April.

2. **S. filifòlia** (Greene) Greene. [*Cardamine f.* Greene. *Arabis f.* Greene.] Slender, glabrous, glaucous annual, 1.5–3 dm. high; basal lvs. early caducous; cauline pinnate with

narrowly linear segms., petiolate, 2–4 cm. long, the segms. 5–10 mm. long; pedicels slender, divaricate, 6–12 mm. long; sepals glabrous; petals spatulate, pink to purplish, 3–5 mm. long; siliques slender, flattened, divaricate, 2.5–4 cm. long, less than 1 mm. wide; style ca. 1 mm. long; seeds oblong, wingless, ca. 0.8 mm. long.—Shaded n. slopes, Chaparral; Santa Cruz Id., formerly on Santa Catalina Id. April.

3. **S. rosulàta** Roll. Annual with few divaricately branched stems 1–3 dm. high, glabrous to sparsely pubescent; basal lvs. rosulate, deeply pinnately lobed, petiolate, 3–5 cm. long, the lobes 4–8 mm. long, 1–2 mm. wide; cauline lvs. somewhat lobed or the upper entire, linear; fruiting pedicels divaricately ascending, glabrous, 2–3 mm. long; sepals glabrous to sparsely pubescent, 1.5–2 mm. long; petals white, narrowly spatulate, 2.5–3 mm. long; siliques divaricately ascending, glabrous, flattened, 1.5–3 cm. long, 1–1.5 mm. wide; seeds wingless, oblong, ca. 1 mm. long.—Sandy and rocky places, ca. 2000–3000 ft.; Creosote Bush Scrub; Saline V., about Death V.; March–April.

4. **S. virgínica** (L.) Roll. [*Cardamine v.* L. *Arabis v.* Poir.] Annual or biennial, hirsute below, 1–4 dm. high; lvs. lyrate-pinnatifid, 3–8 cm. long, 7–12 mm. wide; fruiting pedicels glabrous, 3–7 mm. long; sepals 1–2 mm. long; petals white to faintly pink, 1.5–3 mm. long; siliques 2–2.5 cm. long, straight to slightly curved, erect to divaricate, glabrous; style to 0.5 mm. long; seeds round, uniseriate, ca. 1.5 mm. long, narrowly winged; $2n = 16$ (Rollins, 1966).—Rare about drying pools; V. Grassland, Coastal Sage Scrub; San Diego, Inglewood, Gardena, San Pedro Hills, etc.; to n. Calif., e. U.S. April–May.

31. Sisýmbrium L.

Ours annuals or biennials, erect, mostly branched. Lvs. dentate to pinnatifid, or finely dissected. Fls. yellow or white, small, in terminal racemes. Pubescence when present of simple hairs. Sepals spreading, oblong to linear. Stamens 6. Siliques cylindric and prismatic, or long-subulate, the valves 1–3-nerved, dehiscent. Stigmas 2-lobed. Seeds oblong, marginless, in 1 row in each cell, with incumbent cotyledons. A genus of some size as here recognized; native of the temp. parts of the world. Ours all introd. from Eu. (Greek name for some crucifer.)

Siliques closely appressed, 1–1.5 cm. long; lvs. pinnatifid .... 3. *S. officinale*
Siliques not appressed, 4–10 cm. long.
  Fruiting pedicels ca. as thick as the siliques; petals 6–9 mm. long.
    Upper lvs. pinnatifid with linear divisions; siliques ca. 1 mm. wide .... 1. *S. altissimum*
    Upper lvs. entire to hastate; siliques 1–2 mm. wide .... 4. *S. orientale*
  Fruiting pedicels more slender than the siliques; petals 3–4 mm. long .... 2. *S. irio*

1. **S. altíssimum** L. TUMBLE-MUSTARD. Erect annual, branched above, 5–10 dm. high, glabrous or nearly so; lower lvs. 10–15 cm. long, runcinate-pinnatifid with runcinate lateral lobes and a large subdeltoid terminal one, petioled; upper lvs. reduced, pinnatifid into linear divisions; pedicels spreading-ascending, 5–8 mm. long; sepals 4–5 mm. long; petals yellowish-white, 6–8 mm. long; siliques spreading, rigid, narrow-cylindric, 5–10 cm. long, scarcely 1 mm. thick; seeds 0.6–0.8 mm. long; $n=7$ (Rollins, 1966).—Common weed in waste places below 7500 ft., throughout the state, especially on the Mojave Desert; native of Eu. May–July.

2. **S. irio** L. LONDON-ROCKET. Plate 30, Fig. B. Erect annual, 2–8 dm. tall, glabrous, branched above; lower lvs. 10–15 cm. long, runcinate-pinnatifid, with lanceolate lateral lobes and a large subdeltoid terminal one; upper lvs. reduced; racemes long, many-fld.; pedicels filiform, ascending, 6–10 mm. long; sepals 2–2.5 mm. long; petals yellow, 3–4 mm. long; siliques ascending, 3–4.5 cm. long, less than 1 mm. thick; seeds light brown, shining, 0.6–0.8 mm. long; $n=7, 14, 21, 28$ (Khoshov, 1955).—Common weed in orchards, waste places, etc.; cismontane, occasional on desert; natur. from Eu. Jan.–April.

3. **S. officinàle** (L.) Scop. [*Erysimum o.* L.] HEDGE-MUSTARD. Annual, stiffly erect, with few widely divaricate branches, ± hirsute especially near base, 2–10 dm. tall; lvs. of basal rosette lyrate-pinnatifid, hirsute, 5–10 cm. long, the upper hastate with long narrow

sinuate-dentate terminal lobe; racemes long and narrow; pedicels erect in fr., ca. 2 mm. long; sepals ca. 2 mm. long; petals yellowish, 3 mm. long; siliques closely appressed, 10–15 mm. long, acuminate; seeds dark brown, ovoid, ca. 1 mm. long; $n=7$ (Wulff, 1937).—Common weed of waste places and waysides, mostly below 5000 ft.; natur. from Eu. April–July.

4. **S. orientàle** L. Annual or biennial, branched, 2–6 dm. high, ± hirsute-pubescent; lvs. pinnate or the upper pinnatifid, petioled, 3–12 cm. long, with terminal lobe hastate, lance-linear to ovate, the lateral lobes paired; racemes long, lax in fr.; pedicels stout, 3–10 mm. long, ascending; sepals 4 mm. long; petals yellow, 7–8 mm. long; siliques ascending-spreading, 4–9 cm. long, 1–1.5 mm. thick; seeds scarcely 1 mm. long; $n=7$ (Rollins, 1966).—Roadside and sidewalk weed, scattered cismontane localities; natur. from Eu. May.

32. *Stánleya* Nutt. PRINCE'S PLUME

Annual or perennial or even suffruticose, glabrous or pubescent with simple hairs. Stems simple or branched. Radical lvs. forming a basal rosette or not; cauline petioled to sessile and auriculate, pinnately dissected to entire. Infl. racemose. Sepals linear-oblong, spreading or reflexed at anthesis. Petals white to yellow. Fils. subequal, with glandular tissue surrounding the base of single stamens but obsolete or only on underside of paired fils. Siliques linear, flattened parallel to septum or subterete, long-stipitate; stigma subsessile. Seeds oblong, numerous, marginless. Several spp. of w. U.S. (Named for Lord Edward *Stanley*, English ornithologist of 19th century.)

(Rollins, R. C. The cruciferous genus Stanleya. Lloydia 2: 109–127. 1939.)
Lower lvs. entire or runcinate; inner surface of petal-claw glabrous . . . . 1. *S. elata*
Lower lvs. pinnate; inner surface of petal-claw villous . . . . 2. *S. pinnata*

1. **S. elàta** Jones. Perennial, erect, with 1–several stems simple or branched above, 6–15 dm. high, glabrous, glaucous; lvs. petioled, lance-ovate, 1–2 dm. long, entire or with a few small basal lobes; infl. becoming 1–5 dm. long; pedicels glabrous, 5–10 mm. long; sepals oblong-linear, yellowish, glabrous, 8–12 mm. long; petals yellow to whitish, 8–10 mm. long, the blade only ca. 1 mm. wide; stipes ca. 2 cm. long; siliques subterete, arcuate or spreading, 5–10 cm. long; seeds brown, oblong, wingless, 2 mm. long; $n=12$ (Rollins, 1939).—Desert washes and slopes, 4500–6500 ft.; Joshua Tree Wd., Pinyon-Juniper Wd.; White and Last Chance mts. to Panamint and Argus mts.; Nev., Ariz. May–July.

2. **S. pinnàta** (Pursh) Britton ssp. **pinnàta.** [*Cleome p.* Pursh. *S. pinnatifida* Nutt.] Plate 30, Fig. C. Suffrutescent, glabrous to pubescent, the stems simple or usually branched above, glaucous, 4–15 dm. high; truly radical lvs. lacking, the lower cauline usually pinnatifid into lanceolate segms., rarely bipinnate, broadly lanceolate in outline, 5–20 cm. long, petioled, the upper cauline oblanceolate, entire or divided, 3–6 cm. long; infl. becoming 2–6 dm. long in fr., many-fld.; pedicels 6–12 mm. long, glabrous to somewhat pilose; sepals glabrous, linear, dilated at base, 10–15 mm. long; petals yellow, 12–16 mm. long, the sharply differentiated blade oblong to almost oval, 3–6 mm. long, the claw villous, brownish; fils. pilose at base; siliques 3–8 cm. long, linear, subterete, spreading, arcuate to almost straight; stipe 1–2.5 cm. long; seeds brown, wingless, oblong, plump, ca. 2 mm. long, 1 mm. wide; $n=12$ (Rollins, 1939).—Seleniferous soil, desert slopes and washes, 1000–7500 ft.; Creosote Bush Scrub to Pinyon-Juniper Wd.; Santa Rosa mts. to Cuyama V. and Inyo Co.; e. to N. Dak., Kans., Tex. April–Sept.

Ssp. **inyoénsis** Munz & Roos. Plant a distinct shrub with a basal trunk 4–8 cm. thick; herbage yellow-green throughout; sepals 8–10 mm. long; petals 10 mm. long.—Sandy places at ca. 3000 ft.; Creosote Bush Scrub; se. end of Eureka V., e. base of Inyo Mts.

33. *Streptanthélla* Rydb.

Glabrous annuals with slender usually branched stems. Lvs. entire or shallowly dentate, linear-lanceolate, narrowed at base. Fls. small, in racemes. Sepals with lateral pair saccate at base. Petals narrow, crisped or channeled. Siliques pendent on recurved pedicels, not stipitate, strongly compressed, dehiscent at base but not separating at apex and provided with a conspicuous beak simulating a persistent style. Seeds oblong, flat, winged. One sp. (Diminutive of *Streptanthus*.)

1. **S. longiróstris** (Wats.) Rydb. [*Arabis l.* Wats.] Plants glaucous, 2–6 dm. tall; lvs. 2–5 cm. long, the upper reduced and entire; pedicels 2–5 mm. long; sepals greenish or tipped with purple, 4–6 mm. long; petals yellowish, linear-spatulate, 5–8 mm. long; siliques 3–6 cm. long; seeds almost 2 mm. long; $n=14$ (Rollins, 1966).—Common, especially about shrubs, sandy flats and washes, below 6000 ft.; Creosote Bush Scrub, Sagebrush Scrub, Joshua Tree Wd., Pinyon-Juniper Wd.; deserts from L. Calif. to e. Wash., Wyo., New Mex. and inner Coast Ranges to Monterey Co., Calif. March–May. A nonglaucous form with most lvs. pinnatifid into few, narrow divaricate lobes is the var. *derelicta* J. T. Howell. It occurs in the Colo. Desert.

34. Streptánthus Nutt.

Annual or perennial herbs, glabrous or pubescent with unbranched hairs. Lvs. entire to lyrate-pinnatifid, the upper usually clasping. Infl. racemose or paniculate. Calyx flask-shaped, closed or nearly so at anthesis, often brightly colored. Petals usually narrow, with crisped or channeled blades. Stamens often in 3 pairs according to length, the longest pair often connate and with reduced anthers. Siliques linear, flattened parallel to the septum, erect, divaricate or pendent, sessile or subsessile. Style usually short. Stigma entire or slightly 2-lobed. Seeds flat, usually winged. Ca. 25 spp. of sw. U.S. (Greek, *streptas*, twisted, and *anthos*, fl., because of the petals.)

Plants essentially glabrous.
    Petals white, 7–9 mm. long. San Gabriel Mts. to Laguna Mts. . . . . 1. *S. bernardinus*
    Petals purple, sometimes with white tips or edges, 10–14 mm. long.
        Siliques arcuate, divergent, 1.5–2.5 mm. wide; plants biennial. San Bernardino Mts. to L. Calif. . . . . 2. *S. campestris*
        Siliques almost straight, suberect, 2.5–4 mm. wide; plants perennial. New York and Panamint mts. n. . . . . 3. *S. cordatus*
Plants ± setose-hispid especially toward the base.
    Perennial; fils. distinct; petals 7–9 mm. long. Los Angeles Co. to San Diego Co.
                                                                                                                  1. *S. bernardinus*
    Annuals.
        Siliques pendent; fils. distinct; petals 12–14 mm. long. Los Angeles Co. to L. Calif.
                                                                                                            4. *S. heterophyllus*
        Siliques erect; fils. united; petals 9–12 mm. long. Ventura Co. to San Benito and Monterey cos. . . . . 5. *S. insignis*

1. **S. bernardínus** (Greene) Parish. [*Agianthus b.* Greene.] Glaucous, mostly perennial, with rather slender stems 3–6 dm. high; basal lvs. obovate to broadly oblanceolate, 3–6 cm. long, thickish, petioled, coarsely dentate, sometimes setose-ciliate; cauline lvs. subentire, lanceolate to lance-ovate, 1.5–4 cm. long, auriculate-clasping; racemes lax, becoming 1–2.5 dm. long; pedicels 3–6 mm. long, stout in fr.; sepals broad-oblong, whitish, 5 mm. long, somewhat crisped; petals white, 7–9 mm. long; siliques subsessile, 5–8 cm. long, 1–1.5 mm. wide; style 1–2 mm. long; stigma scarcely 2-lobed; seeds brown, ca. 2 mm. long, winged.—Dry slopes, 4000–7500 ft.; mostly Yellow Pine F.; San Gabriel Mts. to Laguna Mts. June–July.

2. **S. campéstris** Wats. Apparently largely biennial, glabrous, glaucous, the stems few, stout, 6–15 dm. tall, mostly rather simple; basal lvs. obovate or broadly oblanceolate, 5–10 cm. long, thickish, rounded at the apex, petioled, dentate with setosely tipped teeth and ± ciliate margins; cauline lvs. lance-oblong, auriculate-clasping; racemes lax, becoming 2–3.5 dm. long; pedicels 5–10 mm. long, divergent-ascending, stout in fr.; sepals purple, narrow-oblong, ca. 8 mm. long, bristle-tipped; petals purple, recurved, ca. 1 cm. long; fils. distinct; siliques ascending-divergent, arcuate, 7–14 cm. long, 1.5–2.5 mm.

wide; style 1–2 mm. long; stigma slightly 2-lobed; seeds brown, oblong, 2–3 mm. long, winged.—Occasional, dry rocky slopes, 3000–7000 ft.; Chaparral, Pinyon-Juniper Wd., Yellow Pine F.; San Bernardino and Little San Bernardino mts. to n. L. Calif. May–July.

3. **S. cordàtus** Nutt. Short-lived glabrous, glaucous perennial, the stems mostly simple, stoutish, 3–8 dm. high; basal lvs. spatulate-obovate, variously dentate, often setose-ciliate, petioled, 2.5–7 cm. long; cauline lvs. broadly oblong, 2–6 cm. long, sagittate at base, entire, rounded at apex, tending to be rather crowded on stem; racemes rather lax, 1–3 dm. long; pedicels ascending, 5–10 mm. long, stout in fr.; sepals greenish or purplish, broadly oblong, 5–8 mm. long, with terminal tuft of bristles; petals purple, white-margined, recurved, 10–14 mm. long; stamens distinct; siliques ascending or spreading, 5–8 cm. long, 3–4 mm. wide; style 2–3 mm. long; stigma distinctly 2-lobed; seeds broadly oblong, winged, brown, ca. 2.5 mm. long; $n=12$ (Rollins, 1939).—Dry stony slopes, 4000–10,500 ft.; Pinyon-Juniper Wd., Bristle-cone Pine F.; New York, Panamint and Inyo-White mts.; to Ore., Colo., Ariz. May–July.

4. **S. heterophýllus** Nutt. [*Caulanthus h.* Pays.] Annual, hirsute below, erect, simple or branched, 3–10 dm. high; lvs. linear to lanceolate, pinnatifid with divaricate lobes, 5–10 cm. long, all but the lowest amplexicaul; racemes lax, 2–4 dm. long; pedicels 4–8 mm. long, hirsute, recurved or reflexed, sepals purplish or greenish, 8–9 mm. long, linear-lanceolate, somewhat divergent at tips; petals pale with purple veining, linear, recurved, 12–14 mm. long; fils. distinct; siliques pendent, straight, subcompressed, glabrous, 5–8 cm. long, 1.5–2 mm. wide; style ca. 1 mm. long; stigma 2-lobed; seeds narrowly winged. —Occasional, disturbed places such as burns, below 3500 ft.; Coastal Sage Scrub, Chaparral; Los Angeles Co. to L. Calif. March–May.

5. **S. insígnis** Jeps. Annual, slender, sparsely hirsute-hispid, usually branched, 1–4.5 dm. high; lvs. deeply lobed, almost pinnatifid, ± lanceolate in outline, 2.5–6 cm. long, sessile, auriculate; racemes lax, with a terminal tuft of dark sterile fls.; pedicels spreading, 2–3 mm. long; sepals dark purple, sparsely short-hispid, 5–6 mm. long; petals narrow, white with dark midveins, crisped, 9–12 mm. long; siliques flattened, erect, straight, 5–6 cm. long, ca. 2 mm. wide, bristly-hispid; style ca. 1 mm. long; stigma somewhat 2-lobed; seeds round-oblong, winged, 1.5–2 mm. long; $n=14$ (Rollins, 1966).—Largely on serpentine, below 4500 ft.; Chaparral, Pinyon-Juniper Wd.; Santa Ynez Mts. of Ventura Co. n. April–May.

35. *Thelypòdium* Endl.

Annual to perennial herbs with mostly erect simple or branched stems; glabrous to hirsute with simple hairs. Basal lvs. mostly petioled, the cauline petioled to sessile. Infl. usually racemose, at least in fr. Sepals scarcely if at all saccate. Petals linear to oblanceolate or oblong, clawed, plane or crisped, white or cream to purple or lilac. Stamens 6, ± exserted. Siliques stipitate to subsessile, terete or somewhat flattened parallel to the partition, erect to pendent; style short; stigma small, entire to slightly 2-lobed. Seeds oblong, somewhat flattened, not winged. Ca. 16 spp. of w. N. Am. (Greek, *thelus*, female, and *pus*, foot, referring to presence of stipe.)

(Payson E. B. A monographic study of Thelypodium and its immediate allies. Ann. Mo. Bot. Gard. 9: 233–324. 1922.)

1. Cauline lvs. sagittate or auriculate-clasping at base; petals narrowly linear, 1½ times the sepals. San Bernardino Mts. .... 4. *T. stenopetalum*
1. Cauline lvs. not sagittate or amplexicaul.
    2. Plants deep-rooted perennials with woody base; petals purplish, 11–14 mm. long. Inyo Mts. .... 2. *T. jaegeri*
    2. Plants annual or biennial; petals whitish or faintly colored.
        3. Fruiting racemes dense, spikelike; plants biennial .... 1. *T. integrifolium*
        3. Fruiting racemes open; plants annual .... 3. *T. lasiophyllum*

1. **T. integrifòlium** (Nutt.) Endl. [*Pachypodium i.* Nutt. *T. affine* Greene.] Glabrous biennials, erect, branched above, 1–2 m. tall; basal lvs. oblong-elliptical, subentire, 1–3

dm. long, acute, gradually narrowed into the rather long petiole, the cauline linear to oblanceolate, 2–6 cm. long, subsessile; infl. paniculate, the racemes dense, 5–15 cm. long; pedicels 3–8 mm. long; sepals greenish with hyaline margins, 4–5 mm. long; petals white to bluish or pinkish, spatulate, 6–9 mm. long; fils. scarcely exserted; stipes ca. 1 mm. long; siliques horizontal-ascending, subterete, torulose, 1.5–2.5 cm. long; seeds ca. 1.5 mm. long, faintly cellular-pitted.—Mostly about alkaline seeps, 3000–7000 ft.; Joshua Tree Wd., Pinyon-Juniper Wd.; w. edge of Mojave Desert, n. base of San Bernardino Mts., Antelope V., Tehachapi Mts., White Mts.; to Wash., Nev. July–Oct. Variable as to fl.-color, no. of nectary-glands, etc.

2. **T. jaègeri** Roll. Cespitose, deep-rooted perennial, with woody base, many branching stems, glabrous, glaucous, 1–3 dm. high; cauline lvs. slender-petioled, repand to few-lobed, 2.5–6 cm. long, 1–2.5 cm. wide; infl. lax, 4–10 cm. long; pedicels 8–14 mm. long; sepals oblong, purplish, 5–7 mm. long, blunt; petals purple to pink-purple, spatulate, purple-veined, 11–14 mm. long; stamens not exserted; siliques subsessile, subterete, linear, 3–5 cm. long; seeds almost 1 mm. long.—Shaded rock-crevices, 6000–8000 ft.; Pinyon-Juniper Wd.; s. end of Inyo Mts. ca. 17–20 miles n. of Darwin, Inyo Co. May–June.

3. **T. lasiophýllum** (H. & A.) Greene var. **lasiophýllum**. [*Turritis l.* H. & A.] Annual, usually hirsute at base, erect, simple or branched above, 2–10 (–18) dm. high; lvs. all petioled, the lower oblong or oblanceolate, 3–12 cm. long, irregularly pinnatifid with divaricate segms. or only sinuate-dentate, the upper much reduced; infl. much elongate in fr.; pedicels 2–4 mm. long, usually reflexed in age; sepals greenish, oblong, 3 mm. long; petals yellowish-white, spatulate, ca. 6 mm. long; fils. included, the anthers not apiculate; siliques reflexed, terete, sessile or subsessile; 3–6 cm. long; seeds faintly reticulate.—Frequent on slopes, especially on burns, in washes, canyons, etc. below 5000 ft.; grasslands and open woodlands, most of cismontane s. Calif.; Channel Ids., San Clemente and Catalina ids.; to Wash., L. Calif. March–June.

Var. **utahénse** (Rydb.) Jeps. [*T. u.* Rydb.] Lvs. thin, glabrous, with rounded lobes; siliques reflexed, usually curved outward.—Common, washes and among bushes; Creosote Bush Scrub, Joshua Tree Wd., Pinyon-Juniper Wd.; Mojave and Colo. deserts; to Utah, Ariz. April–May.

4. **T. stenopétalum** Wats. Biennial, glabrous, glaucous, branched from base, with simple stems 3–5 dm. tall; radical lvs. soon withering, entire or repand, oblanceolate, the cauline oblong-lanceolate, erect, sagittate at base, 1–5 cm. long; infl. lax, 1–2 dm. long; pedicels ascending, 4–6 mm. long; sepals linear, greenish-purple, 6–9 mm. long; petals narrowly linear, whitish, crisped above, 10–14 mm. long; fils. 8–14 mm. long, the anthers coiled when dry, apiculate, ca. 5 mm. long; siliques sessile, ascending, 4–5 cm. long, slender; seeds oblong, ca. 1 mm. long.—Uncommon, stony slopes, 6500–7000 ft.; Yellow Pine F.; Bear V., San Bernardino Mts. June–July.

36. *Thláspi* L. PENNY-CRESS

Low erect annual or perennial glabrous herbs. Lvs. undivided, the cauline sagittate and clasping. Fls. small, white or purplish, in terminal racemes. Sepals short, oval, obtuse. Petals oblanceolate to obovate. Stamens 6; anthers short, oval. Silicles round or obovate or obcordate, flattened contrary to the narrow partition, the midrib of the valves extending into a wing. Seeds 2–8 per locule; cotyledons accumbent. Ca. 60 spp. of wide distribution. (Greek, *thlaein*, to crush, from the flattened silicle.)

1. **T. arvénse** L. Erect, 2–5 dm. high; lowest lvs. petioled, narrowly obovate, 3–5 cm. long, the cauline oblong, sessile, clasping, somewhat toothed, 2–4.5 cm. long; pedicels spreading, slender, commonly 6–8 mm. long; sepals white-margined, 1.5–2 mm. long; petals white, 3–3.5 mm. long; silicles suborbicular to round-oblong, 1–1.8 cm. long, broadly winged, deeply emarginate; style almost lacking; seeds compressed, oblong, blackish, concentrically ridged, 2–2.3 mm. long; $2n=14$ (Manton, 1932).—Occasional adv. in waste places; natur. from Eu. May–Aug.

37. *Thysanocárpus* Hook. LACE-POD. FRINGE-POD

Erect slender simple or branched annuals. Fls. minute, white to purplish, in slender racemes. Sepals ovate, spreading. Petals cuneate to spatulate. Stamens mostly 6, subequal. Ovary 1-celled, compressed, 1-ovuled. Silicle indehiscent, plano-convex or biconvex, strongly compressed, rounded in outline, with the wing ± crenate or even perforate and with ± highly developed radiating nerves. Ca. 4 or 5 spp. of w. N. Am. (Greek, *thusanos*, fringe and *karpos*, fr.)

Cauline lvs. auriculate, the basal lvs. rosulate, usually hirsute .... 1. *T. curvipes*
Cauline lvs. not auriculate, the basal not rosulate, usually glabrous .... 2. *T. laciniatus*

1. **T. cúrvipes** Hook. var. **cúrvipes**. Plate 30, Fig. D. Erect, 2–5 dm. tall, ± pubescent or hirsute at base, branched above; basal lvs. in rosette, sinuate-dentate to subentire, oblong in outline, 2–5 (–10) cm. long; cauline lvs. lanceolate, sagittate, clasping, the upper entire; pedicels slender, recurved, 3–7 mm. long; fls. ca. 1 mm. long; silicles round-obovate, often plano-convex, glabrous to pubescent, mostly 3–4 mm. wide including the wing, the latter entire or crenate, sometimes perforate, with broad rays; style ca. 0.5 mm. long; $2n = 28$ (Manton, 1932).—Common on grassy or brushy slopes, mostly below 5000 ft.; V. Grassland, Coastal Sage Scrub, Chaparral, Foothill Wd., etc.; most of cismontane Calif.; to B.C., Mont. March–May. Variable.

KEY TO VARIETIES OF T. CURVIPES

Wing of silicle rayed; plants mostly pubescent below. Cismontane.
    Silicles 3–4 mm. wide, the wing usually not perforate .... var. *curvipes*
    Silicles 5–6 mm. wide, the wing often perforate .... var. *elegans*
Wing of silicle mostly plane, without rays; plant glabrous and glaucous. Deserts .. var. *eradiatus*

Var. **élegans** (F. & M.) Rob. in Gray. [*T. e.* F. & M.] Silicles 5–7 mm. wide, the wings often perforated; style ca. 1 mm. long.—Grassy slopes, below 5000 ft.; especially in V. Grassland and Foothill Wd., less common in Chaparral, Yellow Pine F., etc.; cismontane Calif. March–May.

Var. **eradiàtus** Jeps. Plant glabrous and glaucous; wing of silicles membranous, without rays; silicles pubescent, ca. 4–5 mm. wide.—Washes and canyons below 5000 ft.; Creosote Bush Scrub to Pinyon-Juniper Wd.; deserts of Calif.; to Utah, Ariz. April–May.

2. **T. laciniàtus** Nutt. ex T. & G. var. **laciniàtus**. Plate 30, Fig. E. Stems slender, simple or branched, usually glabrous and glaucous, 1–4 dm. high; lvs. not rosulate at base, 1–5 cm. long, the lower narrow, subentire or pinnatifid into narrow segms., the upper narrowly linear, narrowed at base and not or scarcely auriculate; pedicels recurved in fr., capillary, 2–5 mm. long; silicles usually glabrous, elliptical to orbicular, 3–4 mm. wide, the wing continuous, subentire, lacking well defined rays.—Open places such as grassy slopes, below 3500 ft.; V. Grassland, Coastal Sage Scrub, Chaparral, etc., cismontane s. Calif. and from the w. edge of the desert s. of Santa Barbara and Inyo cos.; to L. Calif. March–May.

KEY TO VARIETIES OF T. LACINIATUS

Pedicels recurved in fr.; silicles mostly 3–5 mm. wide.
  Silicles essentially flat.
    Wings lacking well defined rays.
      Silicles usually glabrous. Mostly cismontane .... var. *laciniatus*
      Silicles with minute clavate hairs. Desert .... var. *hitchcockii*
    Wings with well-defined rays.
      Silicles 3–4 mm. wide. Widely distributed .... var. *crenatus*
      Silicles 4–4.5 mm. wide. Insular .... var. *ramosus*
  Silicles cup-shaped. Santa Cruz Id. .... var. *conchuliferus*
Pedicels spreading, almost straight; silicles 2.5 mm. wide. San Diego Co. .... var. *rigidus*

Var. **conchulíferus** (Greene) Jeps. [*T. c.* Greene.] Plant 4–10 cm. high; fls. pink; silicles boat-shaped, glabrous, the wings lavender, perforate or parted into spatulate lobes.—Santa Cruz Id. March–April.

Var. **crenàtus** (Nutt.) Brew. [*T. c.* Nutt.] Wing rayed and notched or perforate between the rays.—With var. *laciniatus* and n. to Contra Costa Co.

Var. **hitchcóckii** Munz. [probably *T. desertorum* Heller. *T. l.* ssp. *d.* Abrams.] Much branched from base, 1–1.5 (–2) dm. high, glabrous and glaucous; silicles 2.5–3 mm. wide, scabrellous with minute clavate hairs, the wing subentire and scarcely rayed.—Mostly from 2500–7500 ft.; Joshua Tree Wd., Pinyon-Juniper Wd.; w. Mojave Desert. April–May.

Var. **ramòsus** (Greene) Munz. [*T. r.* Greene.] Silicles glabrous, larger, 4–4.5 mm. wide, the wings strongly rayed, often perforate.—Santa Cruz and Santa Rosa ids.

Var. **rígidus** Munz. Plants compact, stiff, 3–12 cm. high, purplish, with pinnatifid lvs.; pedicels spreading, almost straight, not recurved; silicles glabrous, 2.5 mm. wide, subcrenate.—At 4500–5500 ft.; Yellow Pine F.; Laguna Mts., San Diego Co.; n. L. Calif. May.

38. *Tropidocárpum* Hook.

Slender, branched annuals, erect or more generally diffusely spreading, pubescent with simple and forked hairs. Lvs. pinnatifid, well distributed on stems. Fls. small, in loose leafy racemes. Sepals ovate-oblong, spreading. Petals yellow, spatulate-obovate. Stamens tetradynamous. Style slender; stigma obscurely lobed. Silique elongate, completely or partially 2-celled or 1-celled, flattened contrary to the narrow partition; valves 2 or 4, opening from above. Seeds in 2 or 4 rows, flattened not winged, with incumbent cotyledons. Spp. 2, Calif. (Greek, *tropis*, keel, and *karpos*, fr., referring to the keeled siliques.)

1. **T. grácile** Hook. Plant from simple and erect to (more usually) several-stemmed and decumbent or prostrate; stems 1–3 dm. long, pubescent throughout, the hairs sometimes branched; lvs. 1–5 cm. long, largely basal, pinnatifid into linear or oblong segms., these entire to cleft; upper lvs. reduced, pedicels axillary, 6–15 mm. long; sepals ca. 3 mm. long; petals ca. 4 mm. long; siliques narrow, strongly obcompressed throughout, with narrow partition most of the length, glabrous to more commonly strigose or pubescent, 2–5 cm. long, tardily dehiscent; style to ca. 1 mm. long; seeds in 2 rows, brown, narrow-oblong, 1.2–1.4 mm. long, cellular-reticulate; $n = 8$ (Rollins, 1966).—Common dry grassy slopes and in open places below 3500 ft.; V. Grassland, S. Oak Wd., Joshua Tree Wd., etc.; cismontane s. Calif. and w. Mojave Desert; to n. Calif. March–May. Some plants have the siliques 1-celled and not so strongly obcompressed below, twisted, then 2-celled and obcompressed in upper portion. They have been called var. *dubium* (A. Davids.) Jeps. [*T. d.* A. Davids.]

19. **Burseràceae.** Torchwood Family

Deciduous aromatic shrubs or trees. Lvs. mostly alternate, usually pinnately compound or decompound, the rachis often winged, estipulate. Fls. bisexual or unisexual, solitary or in panicles. Sepals 3–5, ± connate basally. Petals 3–5, sometimes 0. Disc present, annular to cup-shaped. Stamens in 1–2 whorls; fils. naked. Pistil 1, the ovary superior, 2–5-loculed; ovules usually 2 in each locule. Fr. drupelike, containing 1–5 stones. Seeds without endosperm. Ca. 20 genera and 500 spp., largely trop.

1. *Búrsera* Jacq. Elephant Tree

Petals inserted on the edge of the disk. Stamens 8–10. Ovary 3-loculed; stones covered by an aromatic pulp. Ca. 40 spp., of trop. Am. (J. *Burser*, 16th-century botanist.)

1. **B. microphýlla** Gray. [*Elaphrium m.* Rose.] Plate 30, Fig. F. Arborescent, 1–3 m. tall; older branches cherry-red; lvs. once pinnate, the pinnae 7–33, oblong-linear, 5–7 mm. long; fls. 5-merous; sepals ca. 1 mm. long; petals 4 mm. long; stamens 10; drupes hanging, the exocarp splitting into 3 valves; stone yellow, 6 mm. long.—Local, rocky

places; Creosote Bush Scrub; e. San Diego Co. and w. Imperial Co.; Ariz. to L. Calif. Early summer.

20. **Buxàceae.** BOX FAMILY

Trees, shrubs or perennial herbs, monoecious or dioecious. Lvs. mostly persistent, simple, alternate or opposite, exstipulate. Fls. regular, bracted, solitary or clustered. Calyx present or not. Petals none. Staminate fls. with 4–12 stamens. Pistillate fls. 2–4 (mostly 3)-loculed, with 1–2 anatropous ovules in each locule. Styles as many as locules, simple. Fr. a caps. or drupe; endosperm fleshy or scanty. Seven genera and ca. 30 spp.; temp and trop.

1. *Simmóndsia* Nutt. GOATNUT. JOJOBA

Dioecious shrubs, with opposite persistent leathery, entire lvs. Fls. on short axillary peduncles, the ♂ clustered, the ♀ solitary. Sepals 5 (4–6). Stamens 10–12. Ovary mostly 3-loculed; styles 3. Fr. a caps. with a firm wall, partly enclosed by the enlarged sepals. Seed 1. One sp. (F. W. *Simmonds*, English naturalist.)

1. **S. chinénsis** (Link) C. K. Schneid. [*Buxus c.* Link. *S. californica* Nutt.] Plate 31, Fig. A. Shrub commonly 1–2 m. tall, with stiff branches, pubescent on younger growth; lvs. oblong-ovate, subsessile, 2–4 cm. long, dull green, ± canescent-puberulent; peduncles 3–10 mm. long; sepals in ♂ fls. greenish, 3–4 mm. long, those of ♀ becoming 10–20 mm. long; caps. coriaceous, nutlike, ovoid, obtusely 3-angled, ca. 2 cm. long, filled by the large oily puberulent seed; $2n = 26$ pairs (Raven, Kyhos & Hill, 1965), ca. 100 (Stebbins & Major, 1965).—Locally common on dry barren slopes below 5000 ft.; Creosote Bush Scrub, Joshua Tree Wd.; Little San Bernardino Mts. and region of Twentynine Palms to Imperial Co.; also in Chaparral, interior cismontane s. Calif., w. Riverside Co. and San Diego Co. to Son., L. Calif. The foliage is in some areas quite glaucous. The nuts have a useful oil and there has been some interest in cultivating the sp. March–May.

21. **Cactàceae.** CACTUS FAMILY

Perennial, herbaceous or woody, ± succulent plants, with columnar globose terete or flattened stems, often jointed. Ours leafless, except for small subulate early caducous lvs. in *Opuntia*. Branches, spines, fls. and other parts developed from special structures called areoles which are situated in the lf.-axils when lvs. are present. Fls. usually perfect, solitary and sessile. Perianth with or without a tube and consisting of numerous outer segms. or sepals that commonly intergrade with the inner parts or petals, all imbricated in several rows. Stamens many, the fils. inserted on the perianth-throat. Style 1; stigmas 2 to many. Ovary inferior, 1-celled; placentae 3 or more, parietal, many-ovuled. Fr. a berry or dry, often spiny or glochidiate, usually many-seeded. An Am. family with perhaps 1500 spp.; found largely in dryer trop. and subtrop. regions; many with very showy fls.

(Benson, L. The native cacti of California. pp. 1–243. 1969.)
1. Stems jointed, flattened or cylindrical; areoles with numerous minute barbed bristles (glochids) as well as stiffer spines; lvs. small, subulate, shed early .... 9. *Opuntia*
1. Stems not jointed, leafless; areoles without glochids.
    2. Fl. on an area of the stem developed 1–several years earlier, not on the new growth of the current season, therefore below the apex of the stem or branch; fr. spiny.
        3. Stems ribbed, the spines borne in bundles on definite ribs; fl.-bud borne within the spiniferous areole or just above it.
            4. Stems to 15 dm. long; fls. not white.
                5. Stems 0.5–1.5 m. long, branching near the base; fls. yellow .... 1. *Bergerocactus*
                5. Stems 1–3 dm. long, cespitose; fls. red to purplish .... 5. *Echinocereus*
            4. Stems 4–15 m. tall, branching above; fls. whitish .... 2. *Carnegiea*
        3. Stems with separate tubercles that bear the spines; fl.-bud not associated with the spiniferous areole .... 7. *Mammillaria*
    2. Fl. borne on the new growth of the current season, hence near the apex of the stem or branch; fr. scaly but not spiny.

Plate 31. Fig. A, *Simmondsia chinensis*. Fig. B, *Echinocactus polycephalus*. Fig. C, *Ferocactus acanthodes*. Fig. D, *Mammillaria tetrancistra*. Fig. E, *Opuntia basilaris*. Fig. F, *Opuntia littoralis*.

6. Flower-bearing portion of the areole near to and merging into the spine-bearing portion, the fl. or fr. crowded against the edge of the spine cluster.
   7. Ribs continuous, not undulate-tuberculate; principal spines ± flattened, markedly annulate.
      8. Areoles of ovary not woolly; fr. fleshy for some time after maturing; stems mostly simple .... 6. *Ferocactus*
      8. Areoles of ovary woolly; fr. drying soon after maturity; stems in clumps
                                                                                                         4. *Echinocactus*
   7. Ribs distinctly undulate-tuberculate; spines not annulate .... 10. *Sclerocactus*
6. Flower-bearing portion of the areole distant from the spine-bearing portion, the fl. or fr. standing apart from the spine-cluster.
   9. Tubercles of mature stems coalescent basally into ribs; fr. dry at maturity, opening lengthwise .... 8. *Neolloydia*
   9. Tubercles of mature stems remaining separate, not forming ribs; fr. fleshy, green or red, indehiscent .... 3. *Coryphantha*

1. *Bergerocáctus* Britt. & Rose

Low, with branching cylindrical, erect, ascending and procumbent stems. Ribs low, many; spines numerous, interlaced, bright yellow, acicular. Fls. small, with widely spreading pale yellow petals; scales on the short tube and ovary small, with wool and slender spines in their axils; perianth-segms. obtuse. Fr. globular, dry, covered with slender straight spines. Seeds obovate, black, pitted; embryo curved. Monotypic. (Named for the German botanist *A. Berger*.)

   1. **B. émoryi** (Engelm.) Britt. & Rose. [*Cereus e.* Engelm. *Echinocereus e.* Rümpler.] Branches to 1.5 m. long, spreading then rooting, 2–4 cm. in diam., the yellow spines darkening in age; ribs 18–25, 2–3 mm. high; spines 10–30 at an areole; the central up to 5 cm. long; fls. 2–2.5 cm. broad, lemon-yellow tinged with green; fr. globular, 2–3.5 cm. in diam., dry, seeds 2.5–3 mm. long.—Dry bluffs and cliffs; Coastal Sage Scrub; along the coast, Orange Co. to L. Calif. Santa Catalina and San Clemente ids. May–June.

2. *Carnégiea* Britt. & Rose

Large, columnar, with erect stout stems and branches; ribs many, with crowded areoles bearing spines and tufts of brown felt. Fls. borne singly at uppermost areoles, funnelform-campanulate, nocturnal but remaining open the next day; scales on the tube felted in the axils; inner perianth-segms. white, waxy. Ovary oblong. Stamens many; stigma-lobes linear, slightly longer than stamens. Fr. ovoid to ellipsoid, sparingly spinose, fleshy, with a red edible pulp and small black seeds. Monotypic. (Named in honor of Andrew *Carnegie*.)

   1. **C. gigántea** (Engelm.) Britt. & Rose. [*Cereus g.* Engelm.] S AHUARO, GIANT CACTUS. Plants 4–10 m. tall, the branches and stems 3 or more dm. thick; ribs 12–24, 5–10 cm. apart and 1–3 cm. high; spines in clusters of 10–25, the cent. up to 5 cm. long; fls. in crownlike clusters, 10–12 cm. long; fr. 6–9 cm. long, green without, red within; seeds tessellate, shining; $n=11$ (Stockwell, 1935).—Local in Calif. on gravelly slopes and flats below 1500 ft.; Creosote Bush Scrub; near Colo. R. as in Whipple Mts. and near Laguna Dam; to Ariz., Son. May–June.

3. *Coryphántha* (Engelm.) Lemaire

Solitary or cespitose plants with globose to cylindric stems having conspicuous spirally arranged tubercles. Tubercles mammillate, narrowly grooved from apex to base. Fls. borne near top of plant in axils of young tubercles, showy; perianth withering-persistent. Ovary naked or sparsely scaly. Fr. ovoid to oblong, greenish to yellowish. Seeds mostly brown, smooth to finely reticulate; hilum subbasal. A fairly large genus ranging s. into Mex. (Name Greek, *koryphe*, summit, and *anthos*, fl.)

   1. **C. vivípara** (Nutt.) Britton & Rose var. **alversònii** (Coulter) L. Benson. [*Cactus*

radiosus var. *a*. Coult. *Mammillaria r.* var. *a*. K. Schum. *M. a.* Coult.] FOXTAIL CACTUS. Stems 1 or few, short-cylindrical, 1–2 dm. high; cent. spines 12–16, straight, stout, divergent, dark above the white base, 1–2 cm. long; radial spines ca. 30–35, more slender, nearly concealing the plant surface; fls. ca. 3 cm. long, magenta with deeper red midvein, the outer segms. ciliate; stigmas white.—Stony slopes, 2000–5000 ft.; Creosote Bush Scrub, Joshua Tree Wd.; Little San Bernardino Mts. to Eagle and Chuckwalla mts.; Ariz. May–June.

Var. **désertii** (Engelm.) Marshall. [*Mammillaria d.* Engelm. *Coryphantha d.* Britton & Brown. *M. arizonica* Calif. refs.] Stems mostly single, 7–20 cm. high; cent. spines 4–6, whitish with red-brown tips, 1–1.5 cm. long; radial spines 15–25, gray; fls. ca. 2.5 cm. long, yellowish or amber tipped with pink; outer perianth-segms. fimbriate; stigmas yellow or tinged pink.—Dry stony slopes, 1500–6000 ft.; mostly Joshua Tree Wd.; Clark & Ivanpah mts., e. Mojave Desert to Nev., Ariz. April–May.

Var. **ròsea** (Clokey) L. Benson. [*Coryphantha r.* Clokey.] Solitary or branching, 7–15 cm. long; cent. spines 10–12, white with red tips, 2–2.5 cm. long, the radial ca. 12–18, white; fls. ca. magenta to purplish.—Dry slopes, 4000–6000 ft.; Pinyon-Juniper Wd.; e. Mojave Desert (New York Mts., Cima, etc.); to s. Nev. May–June.

4. *Echinocáctus* Link & Otto

Single or cespitose, the globose or cylindrical stems clothed with a dense mass of wool or naked at apex. Ribs few to many. Areoles very spiny, large. Fls. borne on the plant-crown, yellow or pink; fl.-tube covered with imbricate, persistent, pungent scales. Ovary with narrower scales having axillary mats of wool. Fr. densely white-woolly, dry, thin-walled, oblong. Seeds black, smooth. Ca. a dozen spp. (Greek, *echinos*, hedgehog or spine, and *cactus*.)

1. **E. polycéphalus** Engelm. & Bigel. NIGGER-HEADS. Plate 31, Fig. B. Stems several, subglobose to short cylindrical, 2–3.5 dm. thick, in clumps of 10–30; ribs 10–20, scarcely tubercled; cent. spines 3–4, red or gray, 4–7 cm. long, spreading or curved; radial spines 6–8, like the centrals but shorter; fls. yellow, 3–6 cm. long, enveloped in abundant wool; fr. ovoid, densely woolly, 2–4 cm. long; seeds angled.—Rocky slopes, 2000–5000 ft.; Creosote Bush Scrub; n. Colo. Desert, Mojave Desert from Randsburg, San Bernardino Mts. and Panamint Mts. e.; to Utah, Ariz. March–May.

5. *Echinocèreus* Engelm.

Stems low, erect or prostrate, sometimes pendent, single or cespitose, globular to cylindric, usually 1-jointed, strongly ribbed. Areoles borne on the ribs. Spines of sterile and flowering areoles similar. Fls. usually large, diurnal, solitary at lateral areoles. Perianth campanulate to short-funnelform, scarlet, crimson, or purplish, the tube and ovary spiny, segms. few to many. Stigma-lobes green. Fr. fleshy, spiny, ± colored, thin-skinned. Seeds black, tuberculate. Perhaps 30 spp. of w. U.S. and Mex. (*Echinos*, hedgehog, and *cereus*, in reference to the spiny fr.)

Fls. scarlet; cent. spine(s) not flattened .... 2. *E. triglochidiatus*
Fls. lavender to purple; principal cent. spine flattened .... 1. *E. engelmannii*

1. **E. engelmánnii** (Parry) Lem. var. **engelmánnii**. [*Cereus e.* Parry ex Engelm.] HEDGEHOG CACTUS. Stems 5–15, in rather open mounds, erect, 20–25 cm. tall, 5 cm. in diam.; spines mostly yellowish, pink or gray, the lower cent. spine deflexed and flattened, 1.5–2 cm. long, stout, rigid, the other cent. spines straight, the largest nearly equaling the lower cent.; fls. crimson-magenta to paler, 4–8 cm. long; fr. red, ca. 3 cm. long; $n=22$ (Stockwell, 1935).—Gravelly slopes and flats, below 6000 ft.; Creosote Bush Scrub, to Pinyon-Juniper Wd.; deserts from s. San Bernardino Co. to w. Imperial Co.; Ariz., L. Calif. April–May.

Var. **aciculàris** L. Benson. Spines pinkish or yellowish, the lower cent. like the others,

2.5–3.5 cm. long, the other cent. weak and flexible.—At 1000–3000 ft.; e. Riverside Co.; to Ariz. and s.

Var. **armàtus** L. Benson. The lower deflexed and flattened cent. spine longer (4 cm. to 4.5 cm.), the others curving and twisting.—S. Mojave Desert in Victorville region and Argus Mts., at ca. 3000 ft.

Var. **chrysocéntrus** (Engelm. & Bigel.) Engelm. ex Rümpler. Spines reddish to reddish brown or yellow, dark or light, the lower deflexed and flattened spine slightly longer, stiff and swordlike.—At 3000 to 6500 ft.; Pinyon-Juniper Wd., etc.; White Mts. of Inyo Co. to mts. of e. Riverside Co.; Ariz.

Var. **múnzii** (Parish) Pierce & Fosberg. [*Cereus m.* Parish.] Spines at first pink, pale gray to tannish in age; lower deflexed flattened spine 2.5–5 cm. long, curving and twisted, other cent. spines half as long.—Dry stony places, 4500–7000 ft.; Pinyon-Juniper Wd., lower edge Yellow Pine F.; San Bernardino Mts. e. of Baldwin Lake, San Jacinto Mts. in Hemet V., se. of Julian, San Diego Co.; se. of Tecate, L. Calif.

2. **E. triglochidiàtus** Engelm. var. **mojavénsis** (Engelm. & Bigel.) L. Benson. [*Cereus m.* Engelm. & Bigel.] MOUND CACTUS. Cespitose, forming clumps or mounds; stems many, globose to oblong, 5–20 cm. long, pale green; ribs usually 10–12, strongly tuberculate at the areoles; spines gray, pink, or at first straw-colored, usually 4.5–9 cm. long, striate, smooth or angled; cent. spines 1 or 2, light, usually twisting; fls. slender, 4–5 cm. long, dull scarlet; fr. oblong, spiny, 2–3 cm. long.—Rocky places, 3500–9000 ft.; Creosote Bush Scrub to Pinyon-Juniper Wd.; Inyo-White Mts. to mts. of Riverside Co.; to Utah, Ariz. April–June.

Var. **melanacánthus** (Engelm.) L. Benson. [*Cereus coccineus* var. *m.* Engelm.] Occasional plants from Cushenberry Canyon and in Joshua Tree Nat. Mon. and in Clark Mts. are placed here by Dr. Benson. The fls. are somewhat shorter, the cent. spines 1–3, spreading or the longest deflexed. The range is to Colo. and Texas.

6. *Ferocáctus* Britton & Rose. BARREL CACTUS

Globular or cylindrical, often massive cacti. Ribs prominent, thick, often ± twisted spirally about the stem. Spines heavy, often hooked, the cent. ones usually flattened and annulate. Areoles large, having fls. just above the spine clusters and ± woolly-felted when young. Fls. conspicuous, broadly funnelform to campanulate, short-tubed, growing near the apex of the stem or branch. Fr. fleshy, usually dry at maturity and dehiscing by a large basal pore. Seeds black, pitted. Perhaps 35 spp. of U.S. and Mex. (Latin, *ferus* fierce, and *cactus*, referring to heavy spines.)

Plants cylindrical, 6–25 dm. high; fls. yellow. Deserts .... 1. *F. acanthodes*
Plants subglobose, 1–3 dm. high; fls. greenish-yellow. Coastal .... 2. *F. viridescens*

1. **F. acanthòdes** (Lem.) Britton & Rose var. **acanthòdes.** [*Echinocactus a.* Lem.] Plate 31, Fig. C. At first globular, later cylindric, to 2 m. or more tall, simple and erect, or 1–2-branched near the base, stout, 3–4 dm. thick; ribs usually 20–28, 1–2 cm. high; principal cent. spine 7.5–15 cm. long, the apex curving but not recurved; inner 6–8 radial spines 4–9 cm. long; petaloid perianth parts 4–6 mm. wide; fls. 3–6 cm. long; fr. greenish, 3 cm. long; seeds cellular-pitted.—Rocky slopes and walls, gravelly fans, etc. below 2000 ft.; Creosote Bush Scrub; New York and Whipple mts.; to Ariz. and Mex. April–May.

Var. **lecóntei** (Engelm.) Lindsay. [*Echinocactus l.* Engelm.] Principal lower cent. spine 5–7.5 cm. long, the apex curving a little; inner 6–8 radial spines 4–5 cm. long; petaloid perianth parts 6–9 mm. wide.—At 2500–5000 ft.; largely Joshua Tree Wd., Pinyon-Juniper Wd.; deserts from w. San Diego Co. to ne. San Bernardino Co.; Utah, Ariz., etc.

2. **F. viridéscens** (Nutt.) Britton & Rose. [*Echinocactus v.* Nutt.] Simple or rarely branched, globose or nearly so, 2–3 (–4) dm. high; ribs 10–21, tuberculately irregular; cent. spines 3–4, stout, brown, flattened, 2–5 cm. long; radial spines 10–20, acicular, unequal, 1–2 cm. long; fls. yellow-green, 3–4 cm. long, each petal with a cent. reddish

stripe; fr. greenish-red, 1.5-2 cm. long; seeds 1 mm. long.—Dry hills; Coastal Sage Scrub, V. Grassland; about San Diego, nw. L. Calif. May-June.

7. Mammillària Haw.

Stems simple or branching, ovoid to cylindric, globose or turbinate, with separated tubercles. Spines smooth; cent. none to several, straight, curved or hooked, acicular; radial spines usually smaller and lighter in color, 10-80 per areole, not hooked, acicular. Fls. and frs. on the growth of preceding seasons, below the apex of the stem or branch, between the tubercles and not obviously connected with them. Fr. fleshy, not splitting open, the floral tube deciduous. Seeds dark, rugose-reticulate, reticulate-pitted or smooth and shiny, or tuberculate. Perhaps 100 spp., sw. U.S. to Cent. Am. (Latin, *mammilla*, nipple.)

Hooked cent. spine 1; base of seed not corky; radial spines 12-28 per areole; stems of older plants usually branching.

Radial spines 12-18 per areole; stems with minor areoles between the tubercles, these bearing tufts of hairs and a few spines .... 1. *M. dioica*
Radial spines 18-28 per areole; stems lacking spine-bearing areoles between the tubercles
.... 2. *M. microcarpa*
Hooked cent. spines mostly 2-4; base of seed corky, the corky part half as large as the seed body or larger; stems unbranched .... 3. *M. tetrancistra*

1. **M. dioìca** K. Bdg. [*Neomammillaria d.* Britton & Rose. *M. incerta* Parish. *M. d.* var. *i.* Munz.] Stems frequently branched, especially below, 5-25 cm. high; areoles 4-6 mm. high, woolly when young; axils with 5-15 bristles; cent. spines 1-4, brown, acicular, the lowest spine stout, ca. 6-20 mm. long, hooked; radial spines 10-20, white, slender, 3-12 mm. long, concealing the plant surface; fls. incompletely dioecious, 1.5-2.5 cm. long, the petals creamy or whitish with colored midrib; fr. scarlet, 1-2 cm. long; seeds black, minutely pitted; $n-22$ (Lenz). Sandy places below 500 ft.; Coastal Sage Scrub and Chaparral; and dry places 500-5000 ft.; Creosote Bush Scrub and Pinyon-Juniper Wd.; w. edge Colo. Desert from Santa Rosa Mts. to n. L. Calif. Feb. April.

2. **M. microcárpa** Engelm. [*Neomammillaria m.* Britton & Rose. *M. grahamii* Engelm.] Low, the stems solitary or in clumps, 5-15 cm. high; tubercles cylindrical, 7 mm. high; axils naked; cent. spines 1-3, purplish-brown to light tan, 1-1.8 cm. long, one hooked, the others straight; radial spines 20-30, whitish or with brown tip, 6-12 mm. long; fls. 2-2.5 cm. long, the inner perianth-segms. pink or lavender with darker mid-stripe, the outer ciliate; frs. scarlet, 2-2.5 cm. long and cylindric-clavate or sometimes green, subglobose and ca. 1 cm. long; seeds black, glossy, ca. 1 mm. long.—Reported from Whipple Mts., e. San Bernardino Co.; to Tex., Chihuahua. April.

3. **M. tetrancístra** Engelm. [*Phellosperma t.* Fosb. *Neomammillaria t.* Britt. & Rose.] Plate 31, Fig. D. Body usually simple, oblong, 1-2.5 dm. high; tubercles 4-7 mm. high, the axils with some wool when young and a few long bristles, cent. spines 1-4, dark, acicular, one or more of them hooked, 1-1.5 cm. long; radial spines 30-60, mostly very slender and white, ca. 1 cm. long; fls. 2.5 cm. long, the inner perianth-segms. white with rose or lavender mid-stripe, outer segms. ciliate; stigma-lobes cream; fr. scarlet, subcylindric, 1.2-2 cm. long; seed black, rugose, 2-3 mm. long, the corky hilum ca. as large as the body.—Occasional on dry slopes below 2000 ft.; Creosote Bush Scrub; from Panamint Mts. s. to the border and to Utah., Ariz. April.

8. Neolloỳdia Britton & Rose

Stems branched or unbranched, ovoid to cylindroid, with 8-21 ribs and tubercles separate through half or less of their height. Spines smooth, the cent. 1-8 per areole, mostly straight, the radial 3-32 per areole, straight. Fls. on new growth of current season near apex of stem, each on the upper side of a tubercle in a felted area. Floral tube funnelform to obconic. Fr. dry, with few-20 membranous scales, ellipsoid to short-cylindroid, dehiscent basally by slits. Seeds reticulate or papillate, black. Twelve to 15 spp. of sw. U.S. and Mex. (Named for Prof. Francis E. *Lloyd,* Canadian botanist.)

1. **N. johnsònii** (Parry) L. Benson. [*Echinocactus j.* Parry ex Engelm. *Ferocactus j.* Britt. & Rose. *Echinomastus j.* Baxter.] Oblong, simple or sometimes few-branched, 1–2 dm. high, 8–12 cm. thick; ribs usually 18–20, narrow and somewhat tubercled; spine-clusters close, concealing the surface; spines reddish with gray, the centrals 4–8 or 9, nearly straight, widely divaricate, tapering, stout, rigid, 2.5–3.5 cm. long, the radials ca. 10–12, much shorter; fls. 5–7 cm. long, deep red to pink; fr. pale green, cylindrical, 2 cm. long; seeds black, pitted, ca. 3 mm. long.—Infrequent, dry rocky places and washes, 2000–4000 ft.; Creosote Bush Scrub and above; Death V. region and mts. to the n.; to Utah, Ariz. April–May.

9. *Opúntia* Mill. PRICKLY-PEAR. CHOLLA

Somewhat woody plants with short-jointed stems; joints cylindrical or flattened, often tubercled, but never ribbed. Lvs. small, fleshy, subulate, soon shed. Areoles at axils of lvs. and bearing many short easily detached bristles or glochids and usually longer and stouter spines. Fls. on joints of previous year and on same areoles with the spines. Floral tube bearing scales resembling the lvs., short, cup-shaped. Outer perianth-segms. or sepals thick, green or partly colored and grading into the longer colored inner segms. or petals. Stamens shorter than petals. Ovary spiny or spineless; stigma-lobes short. Fr. fleshy or dry; seeds bony, pale, ± discoid or angled. Probably ca. 300 spp., from s. Canada to Straits of Magellan. (Old Latin name used by Pliny, formerly belonging to some other plant.)

(Benson, L. The native cacti of California pp. 77–172. 1969.)
1. Joints of stem cylindrical or clavate, tuberculate; glochids usually small.
    2. Spines acicular, smooth, with a full-length thin and paperlike sheath.
        3. Stem nearly covered with low flat platelike tubercles; spines solitary or wanting; branches slender, 6–10 mm. thick, the old ones with solid woody axis . . . . . 18. *O. ramosissima*
        3. Stem either smooth or with simple tubercles; spines more than 1; ultimate branches stouter, the woody axis with a reticulate cylinder.
            4. Ultimate joints disarticulating readily; fr. fleshy when first mature.
                5. Frs. proliferous, persistent; fls. red to red-purple. On the coast . . . . 17. *O. prolifera*
                5. Frs. not proliferous or persistent after first season; fls. ± yellowish green. Deserts.
                    6. Plants mostly 1–2 m. tall, with a crown of usually shorter branches; terminal joints to 5 cm. in diam. . . . . 3. *O. bigelovii*
                    6. Plants mostly 2–5 m. high, with major branches elongate; terminal joints ca. 2.5 cm. in diam. . . . . 11. *O. munzii*
            4. Ultimate joints firmly attached; fr. dry at maturity.
                7. Terminal joints 7–9 mm. in diam.; cent. spine mostly 1, radial spines 6–8; spines of fr. weakly barbed. E. Colo. Desert . . . . 21. *O. wigginsii*
                7. Terminal joints mostly at least 15 mm. in diam.; cent. and radial spines indistinguishable; spines of fr. not barbed.
                    8. Spines of fr. few, 1–2 in the areoles of the 2 uppermost nearly horizontal series. Mostly cismontane . . . . 14. *O. parryi*
                    8. Spines of fr. many, several per areole in the upper nearly horizontal series. Deserts.
                        9. Tubercles 1–2 times as long as broad; longer terminal joints 10–15 cm. long.
                        6. *O. echinocarpa*
                        9. Tubercles 3–several times as long as broad; larger joints 15–40 cm. long
                        1. *O. acanthocarpa*
    2. Spines ± basally flattened, rugose or papillate, with a sheath at only the extreme apex. Desert . . . . 19. *O. stanlyi*
1. Joints of the stem flattened after the seedling stage, not tuberculate.
    10. Fr. dry on maturity; plants small or low, the areoles 0.5–1.5 cm. apart.
        11. Fr. lacking an apical rim of horizontally spreading, strongly barbed spines; joints mostly spineless; fls. orchid to rose or rarely white . . . . 2. *O. basilaris*
        11. Fr. with at least an apical rim of horizontally spreading, strongly barbed spines; joints usually spiny; fls. yellow or tinged with pink.
            12. Spines not flattened basally; joints mostly 5–10 cm. long, obovoid to roundish
            16. *O. polyacantha*

12. Spines ± flattened basally; joints mostly 10–15 cm. long, ± oblong . 7. *O. erinacea*
 10. Fr. fleshy and juicy; plants fairly large, the areoles mostly 15–50 mm. apart.
  13. Plant treelike, 3–7 m. tall. Cultivated spp. escaped about old homesites or in deep valley soil.
   14. Joints glabrous, green, 3–7 dm. long; fr. commonly yellowish, glabrous, 5–7 cm. long
    8. *O. ficus-indica*
   14. Joints densely short-woolly, green or whitish, mostly 1.2–2 dm. long; fr. red, tomentose .... 20. *O. tomentosa*
  13. Plant not treelike, rarely 2 m. tall.
   15. Spines present.
    16. Spines subcircular to elliptic in cross section, usually 1–3 per areole, not all yellow.
     17. Plants low and mat-forming, usually prostrate, the largest joints 5–8 (–10) cm. long .... 10. *O. macrorhiza*
     17. Plants commonly 3–6 dm. or taller; largest joints usually 10–18 cm. long or more
      9. *O. littoralis*
    16. Spines flattened at least basally, usually 3 or more per areole.
     18. Spines not all deflexed, spreading in all directions.
      19. The spines not all yellow, with at least reddish or brownish bases.
       20. Fr. purplish, obovoid to elongate; spines 1–6 (–11) per areole, straight or nearly so.
        21. Mature spines not white or pale gray, spreading.
         22. Spines brown or brownish-red or sometimes with lighter tips, the longest mostly 3.5–6 cm. long. Deserts .... 15. *O. phaeacantha major*
         22. Spines not necessarily all same color, gray, red, brown, yellow or mixtures of these; the longest 1.2–2 cm. long. Cismontane
          12. Hybrid population "*occidentalis*"
        21. Mature spines white or pale gray or younger sometimes brownish, 1 in each areole deflexed. Mostly desert margins .. 15. *O. phaeacantha discata*
       20. Fr. red, subglobose; spines mostly 8–16 per areole, the lower ones curving and twisting gently. Coastal .... 13. *O. oricola*
      19. Spines all yellow, becoming gray or dark in age
       5. Hybrid population "*demissa*"
     18. Spines all deflexed (1–7 per areole), clearly yellow, becoming black in age; fr. grayish, subglobose to ellipsoid, tinged purple; plant 1–2 m. tall.
      4. *O. chlorotica*
   15. Spines almost none; plant creeping or sprawling, up to 3–6 dm. high; joints 7–10 cm. long, glaucous; fr. reddish to red-purple .... 9. *O. littoralis austrocalifornica*

1. **O. acanthocárpa** Engelm. & Bigel. var. **coloradénsis** L. Benson. CHOLLA. Treelike or sometimes shrubby, 1–2 m. high, the relatively few joints forming acute angles; joints 1.5–3 dm. long, ca. 2.5 cm. in diam.; tubercles conspicuous, sharply raised and compressed from side to side, 3 or more times as long as broad; spines 2.5–3 cm. long, 10–15 per areole, tan to whitish, turning brown then black in age, with conspicuous papery sheaths; inner perianth-segms. usually purplish or red or yellow; fr. dry at maturity, with numerous spreading spines except at very base, obovoid-turbinate, 2.5–3 cm. long; seeds pale tan.—Sandy or gravelly places, 2000–4300 ft.; Creosote Bush Scrub, Joshua Tree Wd.; e. San Bernardino Co. to Imperial Co.; to Utah, Ariz. May–June.

Var. **gánderi** (Wolf) Benson. [*O. a.* ssp. *g.* Wolf.] Ca. 1 m. tall, the joints numerous, ascending, 2–5 dm. long, ca. 3 cm. thick; spines 15–25 per areole, 15–30 mm. long.—At 1000–3000 ft.; Creosote Bush Scrub; w. Colo. Desert.

Var. **májor** (Engelm. & Bigel.) L. Benson. [*O. echinocarpa* var. *m.* Engelm. & Bigel.] Sprawling or diffuse shrub, 1 m. or somewhat taller, the joints forming acute and obtuse angles, the joints 12–20 cm. long; spines 2.5 cm. long.—E. San Bernardino Co.; to Ariz., Son.

2. **O. basiláris** Engelm. & Bigel. var. **basiláris.** BEAVERTAIL CACTUS. Plate 31, Fig. E. Stems low and spreading, 1–3 dm. long, branched at base of joints, these flattened, obovate to roundish, canescent to glabrous, spineless, 7–15 (–20) cm. long, 5–12 cm. wide, glaucous, often purplish, frequently transversely wrinkled; areoles small, closely placed, with many short brown glochids; fls. orchid to rose, rarely white, clustered at upper edge

of joints; perianth-segms. to ca. 4 cm. long; frs. dry when ripe, spineless, obovoid, to ca. 3 cm. long; $2n=22$ (Takagi, 1938).—Frequent, dry benches and fans, up to 6000 or more ft.; Creosote Bush Scrub, Joshua Tree Wd.; Colo. and Mojave deserts; to Utah, Son. March–June.

Var. **brachyclàda** (Griffiths) Munz. [*O. brachyclada* Griffiths.] Joints small, reddish, thick, 3–6 cm. long, branched near base, varying from subterete to distinctly flat; frs. 1–2 cm. long, subglobose.—Dry slopes 4000–7500 ft.; Joshua Tree Wd.; desert slopes of San Gabriel and San Bernardino mts., Providence Mts. May–June.

Var. **ramòsa** Parish. Joints branched above, narrow-obovate, 2–3 dm. long; areoles spineless, 1–1.5 cm. apart.—Washes and slopes below 6000 ft.; Chaparral, S. Oak Wd., Coastal Sage Scrub; San Gorgonio Pass to Los Angeles Co.; Greenhorn Range, Kern Co. and Vulcan Mts., San Diego Co.

Var. **trelèasei** (Coulter) Toumey. [*O. t.* Coult.] Joints 5–15 cm. long; areoles many, commonly with 1–3 spines 6–25 mm. long.—Reported from Turtle Mts., se. San Bernardino Co. and from sw. Mojave Desert; San Joaquin V.; Ariz.

3. **O. bigelovii** Engelm. var. **bigelòvii.** JUMPING CHOLLA. Erect, usually with a single trunk 6–15 (–20) dm. tall, with short lateral branches forming a close group above, old basal part of plant black; younger branches cylindrical, 5–25 cm. long, 3.5–4.5 cm. thick, densely set with straw-colored spines; branches readily detached; tubercles pale green, somewhat 4-sided, ca. 1 cm. long, almost as wide; spines 6–8, 1.5–2.5 cm. long, barbed near apex, with persistent sheaths; petaloid perianth-segms. yellow to pale green or with some lavender, 3–4 cm. long; frs. green, obovoid, fleshy at maturity, 2–2.5 cm. long, spiny or almost spineless, with many glochids; seeds light brown when present, ca. 3–4 mm. long; sterile plants propagating by the joints falling to the ground.—Locally often very common on fans, benches and lower slopes mostly below 3000 ft.; Creosote Bush Scrub; Colo. Desert and se. Mojave Desert; to Nev., Ariz., L. Calif. April.

Var. **hoffmánnii** Fosb. [*O. fosbergii* Wolf.] Usually branching more diffusely, the main branches longer and again branched; tubercles ca. 1 cm. long, half as wide; spines reddish brown, 1.5–2 cm. long; fr. red-spined, sterile.—Creosote Bush Scrub; sw. part of Colo. Desert.

4. **O. chlorótica** Engelm. & Bigel. PANCAKE-PEAR. Erect, treelike, 1–2.5 m. tall and almost as broad, with stout trunk and ascending branches; joints circular to broadly obovate, yellow-green, 1–2 dm. long, flat; areoles 1–2 cm. apart, prominent, each with many yellow glochids ca. 5–6 mm. long and 3–6 slender unequal yellow mostly reflexed spines 2–3.5 cm. long; fls. yellow, the inner segms. 2–2.5 cm. long; frs. subglobose, tinged purple, fleshy, not spiny, ca. 3.5–5 cm. long; seeds thickish, 2.5–3 mm. long.—Usually on dry rocky walls, 3000–5500 ft.; Creosote Bush Scrub, Joshua Tree Wd., Pinyon-Juniper Wd.; e. Mojave Desert w. to Little San Bernardino Mts., along w. edge of Colo Desert to L. Calif.; Nev., New Mex., Son. May–June.

5. O. hybrid population "demissa". [*O. demissa* Griffiths.] A hybrid population related to *O. ficus-indica*, with joints elliptic, obovate, or rhombic, 2–5 dm. long and 12–20 cm. wide; spines yellow, 3 to 6 in each upper areole of the joint and up to 2–3.5 cm. long; petals yellow or with some reddish pigment basally.—Coastal Sage Scrub, V. Grassland, etc. below 1000 ft. mostly fairly near the coast from Ventura Co. s. Santa Catalina Id.

6. **O. echinocárpa** Engelm. & Bigel. var. **echinocárpa.** SILVER CHOLLA. GOLDEN CHOLLA. Intricately branched erect, 6–12 (–15) dm. tall, with a short woody trunk and dense crown of cylindrical branches; joints mostly 5–15 cm. long and 15–20 mm. thick, detachable, the tubercles conspicuous, 8–10 mm. long, 5–6 mm. high and wide, with ca. 3–10 silvery to golden spines 2–3 cm. long; glochids minute; fls. clustered at ends of branches; petals greenish yellow, the outer sometimes streaked with red, the inner ca. 2.5 cm. long; frs. dry, many-spined, often not maturing, 1.5–2 cm. long; seeds white, ca. 3 mm. wide.—Dry washes and mesas below 6000 ft.; Creosote Bush Scrub, Joshua Tree Wd., Pinyon-Juniper Wd.; deserts from Mono Co. to L. Calif., Utah, Ariz. April–May.

Var. **wólfii** L. Benson. [*O. e.* var. *parkeri* auth.] Plants more sparingly branched; joints

3–4 cm. thick; frs. more strongly tuberculate.—Dry places below 4000 ft.; Creosote Bush Scrub; w. edge Colo. Desert.

7. **O. erinàcea** Engelm. & Bigel. var. **erinàcea**. OLD MAN. MOJAVE PRICKLY-PEAR. Forming small low clumps, with erect or ascending branches; joints ± oblong to obovate, flat, 6–15 (–20) cm. long, 4–7 cm. wide, green but thickly beset with mostly grayish spines; areoles ca. 8–10 mm. apart, practically all spine-bearing; spines 4–7 (–9), white to pale gray, or brownish especially when young, slender but ± stiff, 0.5–5 (–6) cm. long, spreading-reflexed; glochids brownish, 1.5–3 (–6) mm. long; fls. yellow, sometimes reddish in age, the inner petals 2–2.5 cm. long; frs. dry, short-spiny, subcylindric, 2–3 cm. long; seed 5–6 mm. wide.—Dry gravelly and rocky places, 1500–8000 ft.; Creosote Bush Scrub to Pinyon-Juniper Wd.; Mono Co. to Santa Rosa Mts., Riverside Co.; to Utah, Ariz., Colo., etc. May–June.

Var. **ursìna** (A. Weber) Parish. [*O. u.* Weber.] GRIZZLY BEAR CACTUS. Spines, at least in part, threadlike and flexible, 3–12 cm. long; petals 3–3.5 cm. long.—Rocky places, mostly 4000–5500 ft.; White Mts. to n. Riverside Co. to Utah, Ariz.

Var. **utahénsis** (Engelm.) L. Benson. [*O. sphaerocarpa* var. *u.* Engelm. *O. e.* var. *xanthostemma* Benson. *O. rhodantha* K. Schum.] Areoles of lower half of joints spineless or nearly so; spines of upper areoles 1–4.—White Mts. to Colo., New Mex.

8. **O. ficus-índica** (L.) Miller. [*Cactus f.-i.* L.] INDIAN FIG. Tree 3–5 m. high, the trunk ca. 1 m. high, 2–3 dm. thick; joints green, obovate to oblong, 3–6 dm. long, 2–4 dm. wide; spines none to many, white or eventually some are tan, 1–6 per areole, some spreading, some deflexed, straight, to 4 cm. long, not barbed; fls. yellow or orange; fr. yellow or orange, fleshy and edible, 5–8 cm. long, persisting for some months.—Cult., rarely natur.; native to trop. Am. May–June. Hybridizing freely with native spp.

9. **O. littoràlis** (Engelm.) Ckll. var. **littoràlis**. [*O. engelmannii* var. *l.* Engelm.] Plate 31, Fig. F. Suberect or sprawling, commonly 3–6 dm. high and of greater diam., without a trunk; joints green, narrowly obovate or narrowly elliptic, 12–22 cm. long, 7–10 cm. wide; spines over the entire plant, some gray, some yellow, some mixtures of these and red, 3–4.5 cm. long, 5–11 per areole; petals yellow, 2.5–4.5 cm. long; style yellowish; fr. reddish to reddish-purple, fleshy, 3.5–4 cm. long; seed light tan or gray, nearly discoid. At low elevs., Coastal Sage Scrub, etc.; Santa Barbara Co. s. getting inland ca. 15–40 miles. May–June.

Joints green, not glaucous; spines 5–11 per areole. Near the coast from Santa Barbara Co. to L. Calif. .... var. *littoralis*
   Spines none or a few along the top of the joint; spines 6–12 (–20) mm. long; fls. magenta. Glendora to Riverside Co., mostly below 2000 ft. .... var. *austrocalifornica*
   Spines usually on most of the joint; spines 25–69 mm. long; fls. yellow or the center reddish.
Joints moderately glaucous; spines 1–4 (–6) per areole. Away from the coast.
   Spines brown or dark gray. Newhall (L.A. Co.) to San Bernardino and Riverside cos., at mostly below 2000 ft. .... var. *vaseyi*
   Spines reddish to gray.
     The spines reddish with yellow-white tips; fls. yellow. Largely at 3000–7000 ft., San Gabriel, San Bernardino and San Jacinto mts. .... var. *piercei*
     The spines red and yellow to gray; fls. often with a reddish center. Mts. of e. Mojave Desert
                                                                                            var. *martiniana*

Var. **austrocalifórnica** Benson & Walkington. Joints bluish, narrowly obovate to suborbicular, 7–15 cm. long, spineless or with a few along the top of the joint; spines white, gray or golden, 0 to 2 per areole; fls. red, flame or magenta.—Mostly below 2000 ft., shaded places, Chaparral, Coastal Sage Scrub; Glendora (L.A. Co.) to Riverside Co.

Var. **martiniàna** (Benson) Benson. [*O. macrocentra* var. *m.* Benson.] Moderately glaucous, the joints obovate to orbiculate, 12–18 cm. long; spines well distributed, 1–4 (–6) per areole; fls. yellow or the center red.—At 4500–6500 ft.; Pinyon-Juniper Wd.; e. Mojave Desert to Utah.

Var. **piércei** (Fosberg) Benson & Walkington. [*O. phaeacantha* var. *p.* Fosberg.] Moderately glaucous, the joints narrowly obovate to elliptic or roundish, 12–25 cm. long, spiny in upper half or more; spines 3–5 cm. long, 1–4 (–6) per areole; fls. yellow or the center reddish.—At 5000–7300 ft.; Yellow Pine F.; San Bernardino and San Jacinto mts.

Var. **vàseyi** (Coult.) Benson & Walkington. [*O. mesacantha* var. *V.* Coult.] Moderately glaucous, the joints obovate, 10–16 cm. long; spines on almost the whole joint, 1–4 (–6) per areole, ca. 2.5 cm. long; fls. yellow.—Below 2000 ft.; Chaparral, Coastal Sage Scrub, V. Grassland; Newhall (L.A. Co.) to Temecula, Riverside Co.

10. **O. macrorhìza** Engelm. Clump-forming, usually 7–12 cm. high and 3–18 dm. in diam.; main root usually tuberous; joints bluish green, round to obovate, 5–10 cm. long; spines from the upper areoles white or gray to brownish, 1–6 per areole, mostly deflexed, 4–7 cm. long; fls. yellow or tinged basally with red; frs. fleshy, purple or reddish-purple, 2.5–4 cm. long; seed tan or gray.—Reported from Clark Mt., e. San Bernardino Co.; to Mich., Tex., etc.

11. **O. múnzii** C. B. Wolf. Plant erect, arborescent, 2–4 m. tall, almost as wide, the main trunk 10–15 cm. thick; lower branches rather bare; ultimate joints near the ends of the branches somewhat pendulous, 6–25 cm. long, 3–5 cm. thick, readily detached; tubercles strongly raised, 10–16 mm. long, 5–6 mm. wide; areoles with short tan glochids and 10–12 (–15) yellowish subequal spines 1–2 cm. long; fls. few; petals yellowish green, 1.5–2 cm. long; frs. mostly sterile, subglobose, 2.5–3.5 cm. long; the occasional seeds subglobose.—Dry gravelly places; Creosote Bush Scrub; Chocolate and Chuckwalla mts., Colo. Desert. May.

12. O. hybrid population "occidentalis". [*O. o.* auth., not Engelm. & Bigel.] Suberect or sprawling, 1–2 m. high, 1–5 m. or more in diam.; trunk none; joints ± glaucous, narrowly elliptic or obovate to broader, 16–30 cm. long; spines in nearly all of the areoles, each brown or red on at least the lower part, sometimes all white or gray, 4–7 in the upper areoles, spreading or some deflexed, straight or rarely curved, 2–3 cm. long, flattened, not barbed; fl. large, the petals yellow to orange-yellow, becoming reddish in age; fr. red to purple, fleshy, obovoid, 3.5–7 cm. long; seeds nearly orbiculate.—Sandy, gravelly places to ca. 3000 ft. elev.; Chaparral, Coastal Sage Scrub, etc.; cismontane s. Calif. from e. Los Angeles Co. to San Bernardino and Riverside cos. and interior foothills to San Diego Co. A series of hybrids between cult. prickly pears and the native spp., hence in many different forms.

13. **O. orícola** Philbrick. Treelike with a trunk to ca. 3 dm. high, or more shrubby, whole plant 1–3 m. tall; joints green, ± rounded or broad, 1.5–2 dm. long; spines dense on the joint, yellow and at first translucent, darker in age, mostly 8–16 per areole, spreading, the lower ones larger and ± curved or twisted; petals yellow, 3–4 cm. long; fr. red, fleshy, subglobose, 2.5–3 cm. long.—Sandy places below 500 ft.; Coastal Sage Scrub and V. Grassland; Santa Barbara to n. L. Calif., extending inland to Santa Ana Canyon in Orange Co. April–May.

14. **O. párryi** Engelm. var. **párryi**. [*O. bernardina* Engelm.] VALLEY CHOLLA. Plate 32, Fig. A. Plants few-stemmed, erect, 6–24 dm. tall; branches ascending, cylindric, the branching open; joints to 3 dm. long, 1–3 cm. thick, detachable; tubercles prominent, 2–2.5 cm. long, compressed, 4–6 mm. wide, the areole at the upper end white-woolly, with long yellow glochids and 6–20 unequal yellow to brown slender spines 1–2.5 cm. long; fls. clustered at ends of branches, the outer perianth-segms. tinged red, the inner yellow or green, 2.5–3 cm. long; frs. globular to broadly obovoid, 1.5–2.5 cm. long, the areoles with glochids and sometimes the uppermost 1- to few-spined; sterile frs. drying early, the fertile green; seeds pale, ca. 5 mm. wide.—Common on dry gravelly fans and slopes below 6000 ft.; Coastal Sage Scrub, Chaparral, Yellow Pine F.; Santa Barbara Co. to San Diego Co. and to w. edge of Colo. Desert. May–June.

Var. **serpentína** (Engelm.) L. Benson. [*O. s.* Engelm.] Prostrate or suberect; joints 2–2.5 cm. in diam.; tubercles raised, but not usually compressed, ca. 3 times as long as broad.—Sandy places and dry slopes; Coastal Sage Scrub, Chaparral; canyons about San Diego; n. L. Calif. April–May.

15. **O. phaeacántha** Engelm. var. **discàta** (Griffiths) Benson & Walkington. [*O. d.* Griffiths.] Prostrate or sprawling, the joints broadly obovate or nearly orbiculate, usuall·

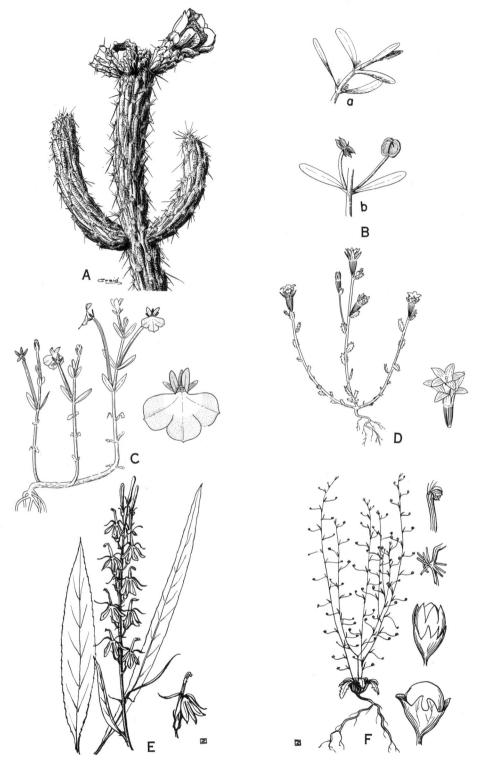

Plate 32. Fig. A, *Opuntia parryi*. Fig. B, *Callitriche marginata*. Fig. C, *Downingia cuspidata*. Fig. D, *Githopsis specularioides*. Fig. E, *Lobelia cardinalis*. Fig. F, *Nemacladus ramosissimus*.

12–25 cm. long, 10–20 cm. wide; spines in all or but a few basal areoles on the joint, 2.5–6 cm. long, 1–4 (–10) per areole; perianth-parts yellow or the bases red; fr. wine-colored or purplish, fleshy, 3–6 cm. long.—Growing mostly between 1500 and 4000 ft.; largely Creosote Bush and above; desert slopes of mts. from San Diego Co. to San Bernardino Co. and in mts. of e. San Bernardino Co.; to Tex.

Var. **màjor** Engelm. [*O. mojavensis* Engelm.] Plants open, the main stems subhorizontal, to ca. 1.5 m. long; joints suborbicular to broadly elliptic, 1.5–3 dm. long, the areoles remote; spines red-brown at base, lighter distally, the principal spine porrect, 3–6 cm. long, the secondary spreading and shorter; fls. pale yellow to orange, the inner petals 3.5–4.5 cm. long; fr. spineless, obovoid, red-purple, 3.5–7 cm. long; seeds 3–4 mm. broad. —Dry washes and slopes, 2000–5000 ft.; Joshua Tree Wd., Pinyon-Juniper Wd.; desert slopes, San Diego Co. n. to San Luis Obispo Co. and e. to Texas. May–June.

16. **O. polyacántha** Haw. var. **rufispìna** (Engelm. & Bigel.) L. Benson. [*O. missouriensis* var. *r.* Engelm. & Bigel.] Plants mat-forming; clumps 7–15 cm. high, and 3– rather many dm. in diam.; joints roundish or broadly obovate, 5–10 cm. long, not readily detached; spines in all the areoles of the joint, those of the lower rigid, straight, 12–30 mm. long, in the upper areoles the longest 20–30 mm. long, white, gray or sometimes reddish brown, mostly deflexed; fls. yellow or occasionally tinged with pink; fr. dry, spiny with barbed spines, obovoid, 2–3.5 cm. long.—Largely Pinyon-Juniper Wd., New York Mts. and mts. of Inyo Co.; to Colo., Tex.

17. **O. prolífera** Engelm. Plant erect, ± arborescent, 1–2 m. tall and of greater width, with many spreading branches; older stems to ca. 1 dm. thick, the terminal joints 3–15 cm. long, 2–3 (–4) cm. thick, fleshy, easily detached, the tubercles 1–2 cm. long, ca. half as broad, with 4–12 rusty-yellow spines 1–2 cm. long and many long glochids; fls. light rose to red purple, the inner petals ca. 1.5 cm. long; fr. subglobose, not tuberculate, fleshy, proliferous, usually sterile, 2–3 cm. long, spineless or nearly so, the glochids soon wearing off.—Forming vegetatively propagated thickets on arid slopes below 600 ft.; Coastal Sage Scrub; near and along the coast, Ventura Co. s. to L. Calif. Channel Ids. April–June.

18. **O. ramosíssima** Engelm. [*O. tessellata* Engelm.] PENCIL CACTUS. Freely branched, matted to erect, 3–15 (–20) dm. high; branches cylindrical, grayish, 6–10 mm. thick, with quite solid woody core, spiny to almost spineless, covered with low diamond-shaped or obovate plates (tubercles), each of these with areole at the apical notch; spines solitary, 2–5 cm. long, yellow-sheathed, each with a deflexed bristlelike spine at the base; glochids minute; fls. yellow-green, tinged red, the perianth ca. 1 cm. long; frs. at ends of short branches, dry, subcylindric, 2–2.5 cm. long, covered with tawny spines ca. 1 cm. long; seeds light tan, discoidal, 4–5 mm. wide.—Dry washes, slopes and mesas, below 4000 ft.; Creosote Bush Scrub, Joshua Tree Wd.; Calif. deserts; to Nev., Son. April–May. Variable as to color, statute and spininess.

19. **O. stánlyi** Engelm. var. **párishii** (Orcutt) L. Benson. [*O. p.* Orcutt.] DEVIL'S CACTUS. Stems low, creeping, forming dense patches 1–1.5 dm. high and 1–3 m. across; terminal joints clavate, ascending, 3–7 (–12) cm. long, 2.5–4 cm. thick, the tubercles 1–1.5 cm. long, high, almost rounded by the dense armature of stout flattened, rugulose spines, the areoles bearing a broad cent. spine up to 4 cm. long, then a ring of rounded or angled spines almost as long, then an outer set of slender spines 1–2 cm. long, surrounded by stiff glochids; young spines reddish, later whitish-gray; fls. yellow or with a reddish tinge, the inner petals ca. 2.5 cm. long; frs. clavate, 4–5 cm. long, half as thick, yellowish-green, strongly tubercled, spineless but covered with yellowish bristles; seeds light brown, ca. 4 mm. broad.—Local on dry flats, 3000 to 5000 ft.; Creosote Bush Scrub, Joshua Tree Wd.; Little San Bernardino Mts. to Clark Mt.; w. Nev. May–June.

20. **O. tomentòsa** Salm-Dyck. Trees to 8 m. tall, spreading; trunk 1–2 m. high; larger terminal joints velvety, dark green, 2–3 dm. long, 0.7–1.3 dm. broad; spines none or sometimes 1 or a few in some upper marginal areoles; petals red in the middle, orange on the margins; fr. red, fleshy, tomentose, obovoid, 3–6 cm. long.—Occasional escape from cult.; native of Mex.

21. **O. wígginsii** L. Benson. Shrub, 3–6 dm. high, the joints somewhat expanded upward, 5–10 cm. long; tubercles 1½ times as long as broad, up to 6 mm. long; spines red or pink but with straw-colored sheaths, 6–8 per areole, those toward end of joint much the longer, cent. spine 2–4 cm., straight; radial spines almost hairlike, to 6 mm. long; fr. dry at maturity, its spines barbed, flexible.—Sandy soils at up to 1000 ft.; Creosote Bush Scrub; e. San Diego Co. to Ariz.

10. *Sclerocáctus* Britton & Rose

Stems mostly solitary, cylindric to globose, with 12–17 ribs. Tubercles coalescent into the ribs through $\frac{1}{2}$–$\frac{4}{5}$ of their height. Spines smooth, not annulate; the cent. gray, white, yellow, red or brown, mostly 1–6 per areole, usually 1 or more hooked; radial mostly 6–11 per areole, straight, shorter than the cent. Fls. and frs. on the growth of the current season near the apex, each on the side of a tubercle in a felted area adjacent to and merging with the new spiniferous areole. Petals white, yellow, pink to reddish-purple. Fr. green, thin-walled, becoming dry and reddish, naked or with a few thin broad scales. Seeds black, papillate-reticulate, angled on 1 side of the hilum, rounded on the other, broader than long. (Greek, *skleros*, hard, and *cactus*.)

1. **S. polyancístrus** (Engelm. & Bigel.) Britton & Rose. [*Echinocactus p.* Engelm. & Bigel.] Simple, oblong to columnar, 2–3 (–4) dm. high, 0.7–1.2 dm. thick; ribs 13–17, somewhat tubercled; spines dense, hiding the surface; cent. spines ca. 8–10, almost perpendicular to the stem, the upper ones flattened, white, 4–12 cm. long, the others reddish, terete, often hooked; radial spines white, acicular, 1–2.5 cm. long; fls. magenta to rose-purple, 4–5 cm. long; ovary-scales few; fr. red-purple, 3–6 cm. long, obovoid; seeds ca. 4 mm. long.—Occasional, gravelly mesas and slopes, 2500–6500 ft.; Creosote Bush Scrub to Pinyon-Juniper Wd.; Mojave Desert from Red Rock Canyon and Inyo Co. e.; Nev. April–June.

22. Callitrichàceae. Water-Starwort Family

Mostly aquatic herbs, with very slender tufted stems. Lvs. opposite, entire, linear to obovate, estipulate. Fls. small, axillary, perfect or imperfect, with or without 2 saclike bracts. Perianth none. Stamen 1, the fil. elongate; anthers cordate, 4-loculed, dehiscent by longitudinal slits. Pistillate fl. a single 4-locular ovary of 2 carpels; styles 2, filiform, distinct. Fr. nutlike, compressed, 4-lobed, separating into 4 1-seeded mericarps; seeds pendulous; endosperm present. One genus, with ca. 20 spp., widely distributed.

1. *Callítriche* L.

Characters of the family. (Greek, *kallos*, beautiful, and *trichos*, hair, because of the slender stems.)

(Fassett, N. C. Callitriche in the New World. Rhodora 53: 138 ff. 1951.)
Frs. pedicelled.
    The frs. somewhat wider than high; lvs. spatulate, longer than wide, short-petioled
        3. *C. marginata*
    The frs. higher than wide; lf.-blades often wider than long, long-petioled . 2. *C. longipedunculata*
Frs. sessile or subsessile.
    Carpels broadly winged all around the margin .... 2. *C. longipedunculata*
    Carpels wingless or winged at summit, the wing narrower or lacking on the sides.
        Margins of carpels narrowly winged at summit.
            Fr. broader than high, 0.5–0.8 mm. wide, scarcely reticulate on sides. Greenhouse weed
                4. *C. peploides*
            Fr. as high as wide or a little higher, exceeding the width by 0.2 mm., plainly reticulate on the sides, the reticulations tending to be in vertical rows .... 5. *C. verna*
        Margins of carpels wingless or nearly so, the reticulations usually not in vertical rows
            1. *C. heterophylla*

1. **C. heterophýlla** Pursh ssp. **bolánderi** (Hegelm.) Calder & Taylor. [*C. b.* Hegelm.] Habit of *C. verna*; fr. 0.9–1.2 mm. high and wide, the carpels more broadly rounded at summit than at base, so that the outline of the fr. is ± heart-shaped, the margins wingless or nearly so, the reticulations on the face not in vertical rows; $2n=20$ (Taylor, 1968). —Freshwater Marsh at ca. 2200 ft.; Santa Rosa Plateau, w. Riverside Co.; vernal pools San Diego; cent. Calif. to B.C.

2. **C. longipedunculàta** Morong. [*C. marginata* var. *l.* Jeps.] Stems threadlike, forming mats; lvs. often crowded towards the ends of the branches, less than 1 cm. long, the blade often wider than long, abruptly narrowed to a long margined petiole; fr. oblong, with nearly parallel or slightly rounded sides, 0.8–1.2 mm. wide, 1–1.4 mm. long, almost black when ripe, distinctly pitted, the wing very narrow but equally wide all the way around, pale, slightly entering the lower sinus; commissural groove wide and flat-bottomed; pedicel 1–25 mm. long.—Water of vernal pools and later on mud, below 3000 ft.; Coastal Sage Scrub, Chaparral; San Diego Co., w. Riverside Co.; cent. Calif. March–April.

3. **C. marginàta** Torr. [*C. nuttallii* Jeps., not Torr.] Plate 32, Fig. B. Stems slender, 0.5–1 dm. long; lvs. nearly uniform, spatulate, the larger 3.5–8 mm. long, 0.8–2 mm. wide, 1–3-nerved; fr. on ± inflated pedicels, 0.8–1.2 mm. wide, not quite so high, the wing 0.1–0.2 mm. wide; stigmas to 1 mm. long, sharply deflexed.—Drying mud of vernal pools, below 2000 ft.; largely Coastal Sage Scrub, Chaparral; scattered stations from n. L. Calif., through cismontane s. Calif. to Ore. March–May.

4. **C. peploides** Nutt. var. **semialata** Fass. Stems apparently creeping at base with erect branches to 5 cm. long; lvs. 2–5 mm. long, 1–3-nerved or appearing nerveless; fr. 0.5–0.8 mm. wide, not quite so high, the margins of the carpels narrowly winged at summit.—A Mexican sp. recently reported from a greenhouse in Ontario.

5. **C. vérna** L. [*C. palustris* of many auth.] Slender, 0.5–2.5 dm. long; lvs. variable, the lower submerged, often linear, 1-nerved, to 1 mm. wide, the upper often dilated, the terminal in a floating rosette, obovate, ca. 4–6 mm. long, on petioles ca. as long; fr. 0.6–1.4 mm. wide, somewhat longer, sharply reticulate on the face, the reticulations in ± well defined vertical rows, scarious-winged at summit; $n=10$ (Sokolovskaja, 1932).— Shallow water or on mud, up to ca. 11,000 ft., various Plant Communities; Victorville and San Bernardino Mts.; cent. Calif. to Alaska, Atlantic Coast; Eurasia. May–Aug.

23. Campanulàceae. BELLFLOWER FAMILY

Plants herbaceous or rarely suffrutescent, usually with milky juice. Lvs. simple, alternate, estipulate. Fls. mostly perfect, usually 5-merous except as to carpels. Ovary inferior or partly so, a fl.-tube mostly not much developed above it. Sepals or calyx-lobes mostly 5, persistent. Corolla sympetalous, regular or irregular. Stamens distinct or united. Style 1; ovary 2–5-loculed, sometimes without internal septum. Fr. a caps. with many minute seeds. Ca. 60 genera and 1600 spp. of temp. and trop. regions; many of ornamental use.

1. Corolla regular; anthers and fils. distinct.
    2. Lvs. broad to roundish; caps. dehiscent on the side.
        3. Caps. opening by uplifting of small lids; ovary somewhat elongate ...... 7. *Triodanis*
        3. Caps. opening by irregular fissures; ovary short and broad .... 3. *Heterocodon*
    2. Lvs. narrow, oblong; caps. dehiscent at apex .... 2. *Githopsis*
1. Corolla irregular.
    4. Anthers all alike, distinct; fls. minute, inconspicuous.
        5. Lvs. mostly in a basal rosette; fls. racemose; caps. dehiscent by apical valves or irregularly
        .... 5. *Nemacladus*
        5. Lvs. basal and cauline; fls. capitate; caps. circumscissile .... 6. *Parishella*
    4. Anthers united into a tube, 2 shorter than the others; fls. mostly showier.
        6. Fls. sessile in the axils of foliaceous bracts, the ovary linear and simulating a pedicel
        .... 1. *Downingia*
        6. Fls. pedicelled; plants perennial; corolla 20–40 mm. long .... 4. *Lobelia*

1. Downíngia Torr.

Low glabrous soft-stemmed annuals, rather succulent and tender. Lvs. sessile, lanceolate to subulate, or the uppermost broader, mostly entire, the upper passing into leafy bracts. Fls. perfect, 5-merous, sessile in the axils of the bracts, but appearing pedicelled because of the long inferior twisted ovaries which "invert" the fls. Sepals rather narrow, ± unequal. Corolla mostly blue, bilabiate, blotched on lower lip, the 2 upper lobes usually smaller than the 3 fused lower. The 2 smaller anthers each with a terminal tuft of bristles and often with a terminal hornlike process. Caps. linear to fusiform, many times longer than thick. Seeds many, fusiform, light brown, mostly smooth. Ca. 13 spp., w. N. Am. and s. S. Am. (Named for A. J. *Downing*, Am. horticulturist.)

(McVaugh, R. A monograph of the genus Downingia. Mem. Torr. Bot. Club 19(4): 1–57. 1941.)

Anther-tube 2.5–3.5 mm. long, the fil.-tube long and prominently exposing the anthers; lower corolla-lip with 3 basal purple spots .... 3. *D. pulchella*

Anther-tube 1.5–2.5 mm. long, usually not prominently exserted, the fil.-tube shorter; lower lip lacking 3 basal purple spots.

Lower lip of corolla without any purple areas; seeds appearing twisted .... 2. *D. cuspidata*
Lower lip of corolla with 1 purple spot; seeds not appearing twisted .... 1. *D. concolor*

1. **D. concólor** Greene var. **brévior** McVaugh. Stems 5–20 cm. long; lvs. 5–18 mm. long; infl. 3–15 cm. long with bracts 5–16 mm. long; sepals ascending or rotately spreading, 3–8 mm. long, subequal; corolla 7–13 mm. long, blue, the lower lip with a quadrate or 2-lobed purple spot at base; tube 3–4 mm. long; fil.-tube 2.5–4 mm. long; anther-tube 1.8–2.3 mm. long; caps. 1.2–2.5 cm. long, the valves soon opening.—Vernal pools, etc. from San Diego mesas to Cuyamaca Lake; largely Chaparral. May–July.

2. **D. cuspidàta** (Greene) Greene. [*Bolelia c.* Greene. *D. immaculata* M. & J.] Plate 32, Fig. C. Plants 5–25 cm. high; lvs. 3–13 mm. long; infl. 3–12 cm. long; sepals 3–8 mm. long; corolla 7–15 mm. long, from bright or pale blue to lavender or almost white, lower lip with a cent. white area with a yellow spot or 2 ± confluent ones; the 2 upper lobes 3–6 mm. long, often darker than lower; tube 3–4 mm. long, the lower lobes ovate; fil.-tube 2.5–4 mm. long, all minutely white-tufted at tip, the 3 shorter anthers also with hornlike process; caps. 2–4 cm. long; seeds shining, twisted; $n=11$ (Wood, 1961).—Drying vernal pools, below 2000 ft.; w. Riverside and San Diego cos.; Chaparral, etc.; cent. and n. Calif. March–June.

3. **D. pulchélla** (Lindl.) Torr. [*Clintonia p.* Lindl.] Characterized by its long fil.-tube exposing the anthers and by its 3 purple spots on the lower lip of the corolla.—Reported from Ventura Co.; n. to Colusa Co.

2. Githópsis Nutt.

Small annuals, simple or diffusely branched, the stems commonly angled. Lvs. mostly narrow, inconspicuous, scattered, sessile or nearly so. Fls. in upper axils or terminal or both. Sepals usually conspicuous, lanceolate to acicular, firm. Corolla minute to exserted, salverform or campanulate. Ovary usually enlarged upward, elongate, 3-loculed, lineate or ribbed. Caps. clavate, opening by a terminal perforation. Seeds minute, smooth, shining. Ca. 5–6 spp., of w. N. Am. (Resembling *Githago*.)

(Ewan, J. A review of the genus Githopsis. Rhodora 41: 302–313. 1939.)

Stems elongate-filiform, the branches few, widely spreading, almost twining; fls. minute, 3–6 mm. long; ovary 3–5 mm. long. San Diego .... 2. *G. filicaulis*

Stems stout to slender but not filiform, few- to many-branched; fls. over 4 mm. long; ovary 4–12 mm. long.

The corolla 3–5 mm. long; caps. linear, slightly enlarged upward, not ribbed .. 1. *G. diffusa*
The corolla 4–7 mm. long; caps. inflated above the middle, ribbed .... 3. *G. specularioides*

1. **G. diffùsa** Gray. [*G. gilioides* Ewan.] Strict, slender plants, simple or few-branched, glabrous except for slightly ciliate bases of sepals, or bushier and setulose-pubescent

throughout; cauline lvs. obovate, ± serrulate, 4–7 mm. long; fls. few, scattered, in upper axils; sepals linear, erect, 4–6 mm. long in fr.; corolla subcylindrical, light purple, 3–5 mm. long; caps. 5-nerved, slightly enlarged upward, ca. 6–10 mm. long; seeds 0.6–0.8 mm. long, shining, subterete.—Banks and disturbed places, 1500–4700 ft.; Chaparral; hills, Santa Barbara to w. Riverside cos. April–June. Some plants are more tufted and pubescent and constitute the *G. gilioides* Ewan.

2. **G. filicáulis** Ewan. Stems filiform, 1–2.5 dm. long, hirsutulous on the angles; lvs. inconspicuous, ovate-oblong, serrulate, subglabrous, 1–3 mm. long; sepals minute, linear, acute; corolla shorter than the sepals; caps. 3–5 mm. long; seeds 0.6–0.9 mm. long, shining, red-brown.—Mission Canyon, San Diego. May.

3. **G. specularioìdes** Nutt. ssp. **specularioìdes**. Plate 32, Fig. D. Stems stiff, usually branched slightly above base, angled below, 5–15 cm. tall, gray-hispidulous; lowermost lvs. ovate to broader, small, withering early; middle cauline cuneate-obovate, few-toothed, ca. 3–6 mm. long; fls. in upper axils and terminal; sepals linear-subulate, 6–10 mm. long at anthesis, later 8–14 mm.; corolla ± salverform, 4–7 mm. long, bright blue, with acute lobes; caps. ribbed, mostly 5–9 mm. long, inflated above the middle; seed ± angled, light brown.—Open places, burns, etc., below 4000 ft.; Chaparral, etc.; Santa Barbara Co. to Wash. May–June. Santa Cruz Id.

Ssp. **cándida** Ewan. Plants 15–20 cm. tall; corolla white, 4–5 mm. long.—Burns and disturbed places; Chaparral; San Diego and w. Riverside cos.

3. *Heterocòdon* Nutt.

Delicate annual, simple or few-branched. Lvs. roundish, dentate, sessile. Fls. solitary, axillary, of 2 kinds: the lower cleistogamous with inconspicuous corollas and the uppermost with showy corollas; sepals round-ovate, few-toothed; corolla open-campanulate. Caps. short and broad, mostly 3-loculed, bursting by irregular fissures. One sp. (Greek, *heteros*, different, and *kodon*, bell, the campanulate fls. of 2 kinds.)

1. **H. rariflòrum** Nutt. [*Specularia r.* McVaugh.] Stems 5–30 cm. long, scattered bristly-pubescent; lvs. 5–8 mm. long; sepals of lower fls. ca. 1.5–2 mm. long, of upper 3 mm.; corolla of upper fls. blue, 3–4 mm. long; caps. ca. 2.5 mm. in diam.—Damp grassy or shaded places below 7500 ft.; Chaparral to Montane Coniferous F.; San Diego Co. to B.C. Ida., Mont. April–July.

4. *Lobèlia* L.

Annual or perennial herbs, or sometimes woody. Lvs. alternate, the cauline sometimes reduced to bracts. Fls. blue, red, or white, in terminal racemes, spikes or panicles, even solitary. Fls. inverted in anthesis, the pedicel twisted. Sepals 5, short. Corolla-tube often split to the base on one side, the limb 5-lobed, the upper 3 lobes forming the "upper" lip, the other 2 (1 on each side of the cleft) erect or recurved. Anthers united into a tube or ring around the style, 2 or all hairy at tips. Ovary largely inferior, 2-loculed, many-ovuled. Fr. a 2-valved caps. Ca. 250 spp.; a widely distributed genus. (Named for de *Lobel* or *L'Obel*, a 16th century Flemish botanist.)

Corolla bright red, 2.5–4 cm. long, the tube split to the base on the upper side .. 1. *L. cardinalis*
Corolla blue, less than 2 cm. long, the tube not cleft, at least at the upper end ..... 2. *L. dunnii*

1. **L. cardinàlis** L. ssp. **gramínea** (Lam.) McVaugh. [*L. g.* Lam. *L. splendens* Willd.] SCARLET LOBELIA. Plate 32, Fig. E. Perennial by means of slender offsets; stems simple, erect, 3–10 dm. high, glabrous to pubescent; lvs. mostly lanceolate or narrower, 5–12 cm. long, irregularly serrulate, the lowermost petioled; racemes 1–3 dm. long in fr.; pedicels 5–15 mm. long; sepals subulate, ca. 8–10 mm. long; anther-tube bluish-gray, 3–4 mm. long; caps. papery, ca. 4 mm. long, ca. 10-ribbed; seeds many, ± obpyramidal, pitted.—Occasional in boggy places below 6000 ft.; Coastal Sage Scrub, Chaparral, etc.; cismontane s. Calif. from Los Angeles Co. s., occasional on desert as in Panamint Mts.; to Tex., Panama. Aug.–Oct. Our plants sometimes referred to var. *pseudosplendens*

McVaugh when the lvs. are 8–14 times as long as wide and the infl. short, or to var. *multiflora* (Paxt.) McVaugh when with lvs. 5–8 times as long as wide and the infl. ample, often leafy.

2. **L. dúnnii** Greene var. **serràta** (Gray) McVaugh. [*Palmerella debilis* var. *s.* Gray.] Perennial, decumbent or erect, 3–5 dm. tall; lvs. linear-lanceolate to oblanceolate, glabrous, 3–7 cm. long, serrate, all except the lower sessile; infl. sometimes puberulent, densely fld.; pedicels 4–10 mm. long; bracts pubescent; anther-tube bluish-gray, 2–3 mm. long, pilose; caps. 6–10 mm. long; seeds ellipsoid-lenticular, smooth.—Occasional in moist canyons below 3000 ft.; Coastal Sage Scrub, Chaparral, etc.; cismontane from L. Calif. to Monterey Co. July–Sept.

5. *Nemácladus* Nutt.

Small annuals with capillary diffusely branched stems. Basal lvs. in compact rosette, cauline largely bracteate. Fls. minute, loosely racemose, borne on capillary pedicels from most axils. Fl.-tube ± evident above the ovary. Sepals entire. Corolla ± bilabiate, the lower lip 2-, the upper 3-lobed. Stamens monadelphous, the fils. separate below and sometimes above, the staminal tube curved near the tip and surmounted by the stellately spreading distinct anthers. Ovary with 3 flattened rounded glands near the base of the free part and opposite the lobes of the upper lip. Fils. between these glands with small stipelike appendages, each of which bears 1 or more rodlike transparent cells. Caps. dehiscent by 2 valves, and these splitting. Seeds many, cylindrical, longitudinally ridged and with transverse lines forming pits. Ca. a dozen spp. of sw. U.S. (Greek, *nemos*, thread, and *clados*, branch, because of the capillary stems.)

1. Corolla-tube mostly 2–5 mm. long, much exceeding the calyx; caps. superior, free from the calyx its entire length and exceeding it .... 4. *N. longiflorus*
1. Corolla-tube 1.5 mm. long or less, scarcely or not exceeding the calyx; caps. ca. half inferior, the free part not or scarcely exceeding the calyx.
    2. Corolla-lobes united at base, so that the tube is almost or ca. as long as the free lobes.
        3. Stems straight, not at all zigzag; seeds subglobose, with ca. 6 pits in each row
        6. *N. ramosissimus*
        3. Stems ± zigzag; seeds longer than thick, with 8–12 pits in a longitudinal row.
            4. Basal lvs. usually pinnatifid; bract at base of pedicel sublinear, not or scarcely enfolding the base of the pedicel; mature caps. 3–4 mm. long .... 5. *N. pinnatifidus*
            4. Basal lvs. entire or toothed; bract at base of pedicel ovate to lanceolate, enfolding and concealing the base of the pedicel; caps. 1.5–2.5 mm. long.
                5. Basal lvs. rhombic-ovate to elliptic, entire to crenate-dentate; flowering branches of large plants often repeatedly and intricately branched .... 9. *N. sigmoideus*
                5. Basal lvs. narrower (oblong-oblanceolate to ± spatulate), crenate-dentate to subpinnatifid; flowering branches simple or sparingly branched .... 3. *N. gracilis*
    2. Corolla-lobes scarcely united at base, the tube very short.
        6. Calyx and ovary much enlarged in fr., ca. 2½ times as long as at anthesis; plant compact, very robust .... 7. *N. rigidus*
        6. Calyx and ovary not notably enlarged in fr.; plant diffuse with very slender stems.
            7. Lobes of upper lip of corolla ciliate on margin; stamens well exserted.
                8. Anthers 0.4–0.8 mm. long; fils. 2–3 mm. long .... 8. *N. rubescens*
                8. Anthers 0.1–0.3 mm. long; fils. 1.3–2 mm. long .... 2. *N. glanduliferus*
            7. Lobes of upper lip of corolla not ciliate; stamens mostly not much exserted
            1. *N. capillaris*

1. **N. capillàris** Greene. [*N. rigidus* var. *c.* Munz.] Stems glabrous or minutely pubescent, ± lustrous, brownish or purplish, forked, 0.5–1.6 dm. long; rosette-lvs. few, ovate, entire or nearly so, 0.3–1.5 cm. long; pedicels mostly straight, spreading or ascending, 8–12 mm. long; fl.-bracts 1–3 mm. long, flat or nearly so; caps. 1.5–2.5 mm. long, half inferior; sepals 0.6–1.3 mm. long; anthers scarcely 0.2 mm. long; corolla white; fils. 0.8–1.2 mm. long; seeds ellipsoid, with 8–10 narrow ridges and rows of 9–12 pits between. —Dry slopes and burns, 2000–4500 ft.; Chaparral, Yellow Pine F.; Ventura Co. n. to cent. Calif. May–July.

2. **N. glandulíferus** Jeps. var. **glandulíferus.** Plants erect, pubescent below to glabrous, the stems brownish to purplish below, dull, 1–2.5 dm. tall, with ascending somewhat flexuous branches; basal lvs. mostly oblanceolate, toothed or pinnatifid, 1–2.5 cm. long; pedicels spreading, 7–13 mm. long, usually curved near the tip; bracts linear to lanceolate, 2–5 mm. long, ± flat; caps. ca. half inferior, 2–3 mm. long; sepals 1.5–3 mm. long; corolla-lobes ± purplish at tip, at least the 3 upper ciliate; fils. 1–2 mm. long; anthers 0.2–0.35 mm. long; seeds 0.5–0.6 mm. long, with 6–8 longitudinal ridges divided by 15–20 transverse lines.—Open sandy places, below 2000 ft.; Creosote Bush Scrub; w. Colo. Desert to s. Mojave Desert; n. L. Calif. March–May.

Var. **orientàlis** McVaugh. Pedicels stiffly spreading-ascending, not or scarcely curved at tip; sepals 0.8–1.8 mm. long.—More common than var. *glanduliferus* in sandy washes, etc. below 6000 ft.; Creosote Bush Scrub, Joshua Tree Wd.; deserts from Inyo Co. to L. Calif., Son., Utah.

3. **N. grácilis** Eastw. [*N. ramosissimus* var. *g.* Munz.] Stems ± zigzag, pubescent, to 1 dm. tall; rosette-lvs. spatulate, coarsely dentate, 5–8 mm. long; bracts of infl. involute, falcately recurved, entire, 2–4 mm. long; pedicels capillary, sigmoidally recurved-spreading, 8–12 mm. long; sepals ca. 1 mm. long; corolla 1.5 mm. long; anthers 0.3–0.4 mm. long; seeds with ca. 10 rows of ca. 13 pits.—V. Grassland, w. Kern Co. to Merced Co. March–April.

4. **N. longiflòrus** Gray. var. **longiflòrus.** Simple to diffuse, 0.3–2 dm. high, usually ± pubescent at least below; rosette-lvs. oblanceolate to obovate, the blades 0.5–1 cm. long, mostly entire; pedicels spreading or ascending, 1–2.5 cm. long; fl.-tube almost none; sepals 1–3 mm. long; corolla 5–8 mm. long, tubular for more than half its length; fils. 3.5–7.5 mm. long; caps. 3–4.5 mm. long; seeds ca. 0.5 mm. long, with 9–10 weak ridges and ca. 10–12 cells in each row.—Dry disturbed places, burns, etc. below 6000 ft.; Coastal Sage Scrub, Chaparral; Monterey Co. to n. L. Calif. April–June.

Var. **breviflòrus** McVaugh. A poorly marked taxon with corolla 3–3.5 mm. long.— Creosote Bush Scrub, etc.; desert from Inyo Mts. to e. San Diego Co.

5. **N. pinnatifidus** Greene. [*N. ramosissimus* var. *p.* Gray.] Stems slightly zigzag; basal lvs. mostly deeply pinnatifid with toothed lobes; pedicels 7–12 mm. long; sepals 1.2–2 mm. long; corolla 1.6–2 mm. long; caps. 3–4 mm. long, ca. half inferior; seeds ca. 0.6 mm. long, with 8–10 rows of 8–10 broad pits each.—Burns, disturbed places, etc., 1000–4000 ft.; Coastal Sage Scrub, Chaparral; s. base of San Gabriel and San Bernardino mts. to n. L. Calif. May–June.

6. **N. ramosíssimus** Nutt. [*N. tenuissimus* Greene.] Plate 32, Fig. F. Stems not at all zigzag, glabrous or puberulent, to 2.5 dm. long; rosette-lvs. oblanceolate, dentate, the blades 0.5–2 cm. long; pedicels spreading, 8–20 mm. long, glabrous; bracts 3–6 mm. long; ovary half inferior; sepals 0.6–1 mm. long; corolla 1.5–2.5 mm. long; stamens pubescent near the anthers; caps. exceeding calyx; seeds with ca. 10 rows of ca. 6 pits each.—Sandy places, burns, etc. below 5000 ft.; Chaparral, Coastal Sage Scrub; L. Calif. to Monterey Co. April–May.

7. **N. rígidus** Curran. Compact, coarse-stemmed, glabrous or finely pubescent, the stems strongly zigzag, to 1.5 dm. tall, somewhat shining, purple; rosette-lvs. elliptic, 5–10 mm. long; pedicels spreading, stiff, straight, 10–12 mm. long, subtended by bracts 2–3 mm. long; caps. ca. half inferior, 3–4 mm. long; sepals unequal, 1–2.5 mm. long; corolla deeply divided, ± purplish, 1–1.5 mm. long; seeds with 8–10 longitudinal flattened ridges and rows of ca. 15 narrow pits.—Dry caked adobe, etc., 4000–7500 ft.; Sagebrush Scrub, Pinyon-Juniper Wd., etc.; s. Inyo Mts. to e. Ore. May–June.

8. **N. rubéscens** Greene. [*N. rigidus* var. *r.* Munz. *N. adenophorus* Parish.] Plate 33, Fig. A. Mostly diffusely branched, 0.5–2 dm. tall, the stems slightly pubescent to glabrous, the lower parts silvery-gray; rosette-lvs. elliptic to oblanceolate, with winged petiole, entire to ± toothed, 1–3 cm. long, $\frac{1}{2}$–$\frac{1}{3}$ as wide; pedicels ± ascending, slender, subtended by bracts lanceolate to ovate, 1–2.5 mm. long, ± conduplicate; caps. 1.5–2 mm. long; sepals 0.7–1.3 mm. long; corolla yellow with purplish-brown markings, the lobes long spreading, the upper 3 ciliate on margins; fils. 2–3 mm. long; anthers 0.6 mm. long; seeds

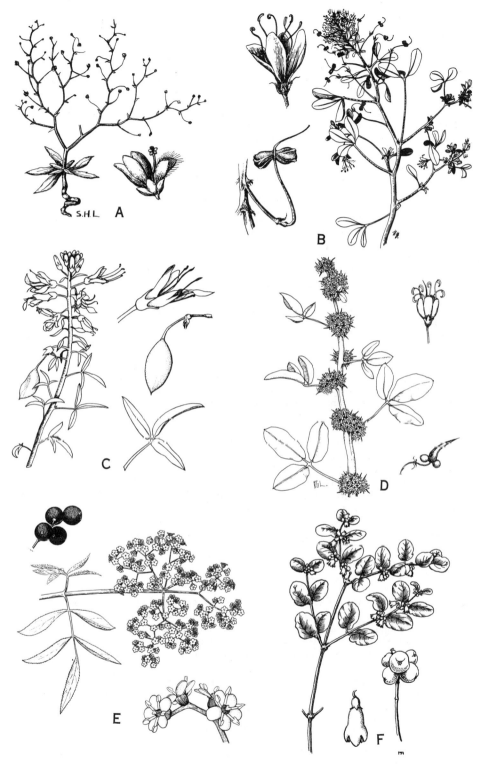

Plate 33. Fig. A, *Nemacladus rubescens*. Fig. B, *Cleomella obtusifolia*. Fig. C, *Isomeris arborea*. Fig. D, *Oxystylis lutea*. Fig. E, *Sambucus mexicana*. Fig. F, *Symphoricarpos mollis*.

with 8–10 ridges and obscure pits.—Dry sandy to rocky places, below 4000 ft.; Creosote Bush Scrub, Joshua Tree Wd.; deserts from Kern Co. to L. Calif., Ariz., Nev. April–May. A form with narrower lvs. and more finely capillary pedicels is the var. *tenuis* McVaugh, ranging from Inyo Co. to Colo. Desert.

9. **N. sigmoìdeus** Robbins. Like *N. gracilis*, but coarser; pedicels largely 12–14 mm. long, the subtending bracts more clasping; corolla 2–2.5 mm. long, the 3 lower lobes white with pinkish or yellowish tips and bases; anthers ca. 0.3 mm. long.—Common in sandy and gravelly places, 1500–7500 ft., mostly Joshua Tree Wd., Pinyon-Juniper Wd.; Inyo Co. to Mt. Pinos (Ventura Co.) and San Bernardino Co. and e. San Diego Co. April–June.

6. *Parishélla* Gray

Low spreading annual with lvs. and fls. in subcapitate tufts, the first at the base of the plant, the others at the ends of short naked proliferous branches. Ovary half inferior. Sepals spatulate. Corolla rotate, almost equally 5-parted. Stamens with appendages and ovary with glands as in *Nemacladus*. Caps. turbinate, circumscissile just above the sepals. One sp. (Named for S. B. and W. F. *Parish*, early s. Calif. botanical collectors.)

1. **P. califórnica** Gray. Stems 2–5 cm. long; lvs. oblanceolate to spatulate, 2–5 mm. long, the basal petioled; corolla white, 3–4 mm. long; seeds many, brown, oblong-cylindrical, ca. 0.4 mm. long, with ca. 10 rows of ca. 10 pits.—Local, gravelly places, 2500–5000 ft.; largely Joshua Tree Wd.; Victorville and Barstow regions and locally to San Luis Obispo Co. April–May.

7. *Triodànis* Raf.

Low annuals, simple or with ascending branches; stems leafy. Lvs. well distributed, gradually passing upward into leafy bracts. Fls. solitary, axillary, sessile, those from lower nodes normally cleistogamous, at least some of the upper open with well developed corollas. Fils. abruptly dilated and ciliated at the base. Caps. ovoid to clavate or linear, opening at apex or near middle, 3-loculed, dehiscing by small valvelike openings in ours. Eight spp., 1 Medit., the others Am. (Greek, *treis*, three, and *odons*, tooth.)

1. **T. biflòra** (R. & P.) McVaugh. [*Campanula b.* R. & P. *Specularia b.* F. & M.] Stems slender, simple or branched, to 4 dm. high, ± sparsely hirsute below, scabrous to retrorsely hispid above; lvs. glabrous to hispid beneath along the veins, ovate to elliptic, 1–3 cm. long, mostly sessile; axillary fls. mostly 1–4 at a node, the open ones blue to violet or lilac, 5–10 mm. long; caps. 4.5–7 mm. long, with oval or roundish pores; style 4–6 mm. long; sepals ± deltoid; seeds smooth, shining; $2n = 28$ (Löve and Solbrig, 1965).—Open disturbed and burned places, below 6500 ft.; many Plant Communities; cismontane Calif.; to Ore., Atlantic Coast, S. Am. Santa Cruz Id. April–June.

24. Capparàceae. Caper Family

Herbs or shrubs with ill-smelling foliage and watery sap. Lvs. alternate, usually palmately compound with 3–5 (–7) entire lfts. Fls. perfect, regular to irregular, solitary or in bracted terminal racemes. Sepals in ours 4, free or united, often persistent. Petals in ours 4, not much clawed. Stamens 6 to many, subequal, inserted on the receptacle which is often thickened or produced between the stamens and petals. Ovary superior, of 2 united carpels, usually 1-celled with 2 parietal placentae, few- to many-ovuled. Stipe usually well developed. Fr. a caps. or fleshy. Seeds ± reniform, without endosperm; cotyledons usually coiled. Ca. 35 genera and 450 spp., mostly of warmer regions, one furnishing the capers of commerce, a few used for ornamentals.

Shrubs with inflated caps.; petals over 1 cm. long .... 3. *Isomeris*
Herbs with caps. not or scarcely inflated; petals often less than 1 cm. long.
    Fls. in dense axillary glomerules; style spinescent in fr. .... 4. *Oxystylis*
    Fls. axillary or racemose, not in glomerules.
        Fr. didymous, 2-celled, each valve enclosing a seed and falling away with it .. 5. *Wislizenia*

Fr. not didymous, 1-celled, several- to many-seeded.
    Caps. 12–45 mm. long, linear to oblong; petals 6–12 mm. long, generally clawed
                                                                                             1. *Cleome*
    Caps. 3–6 mm. long, often somewhat wider; petals 1.5–7 mm. long, not clawed
                                                                                            2. *Cleomella*

1. *Cleòme* L.

Annual herbs to woody plants. Lvs. simple to 3–7-foliolate. Fls. solitary or racemose. Sepals distinct or united at base. Petals entire, cruciate. Stamens 6, rarely 4. Ovary stipitate, with a gland at base; style short or none. Caps. linear to oblong, erect or pendulous. Seeds round-reniform, several to many. Ca. 75 spp., mostly in trop. of Am. and Afr.; some grown for the fls. (Ancient name of some mustardlike plant.)

Sepals united at base; stamens long-exserted.
    Fls. pink or purplish, to white; lfts. 3. Rare waif .... 2. *C. serrulata*
    Fls. yellow; lfts. largely 5–7. Desert native .... 1. *C. lutea*
Sepals separate to base; stamens not exceeding petals .... 3. *C. sparsifolia*

    1. **C. lùtea** Hook. Erect annual, 3–8 dm. high; lfts. 3–5 (–7), oblong or lance-oblong, 1.5–5 cm. long; petals yellow, 6–8 mm. long; pod linear, 2–4 cm. long, on a stipe ca. 1 cm. long; seeds reniform-orbicular, yellowish, ca. 2 mm. long, scattered-tuberculate; $n = 17$ (Mulligan, 1966).—Sandy flats, 3600–7250 ft.; Creosote Bush Scrub, Sagebrush Scrub, Pinyon-Juniper Wd.; Mountain Pass, e. San Bernardino Co. and in Inyo Co.; to Wash., Nebr., Nex Mex. May–Aug.

    2. **C. serrulàta** Pursh. ROCKY MOUNTAIN BEEPLANT. Erect annual, to several dm. high; lower petioles to 5 cm. long; lfts. lanceolate to oblanceolate, 2–7 cm. long; racemes dense in fl., to 2.5 dm. long in fr.; calyx ca. 2 mm. long; petals 10–12 mm. long; caps. linear, long-stiped; $n = 17$ (Raven et al., 1965).—Rare, as a waif in s. Calif.; native n. Calif. to e. Wash., Great Plains. May–Aug.

    3. **C. sparsifòlia** Wats. Glabrous branched annual, 1–4 dm. high; lvs. few, the petioles 1–4 cm. long; lfts. 3, cuneate to obovate, 5–10 mm. long; racemes few-fld.; pedicels 5–8 mm. long; sepals 1–2 mm. long; petals greenish with white margin, 7–8 mm. long; stipe 3–5 mm. long; caps. 1.5–2.5 cm. long; seeds gray with dark mottling, cellular-punctate.—Sandy places, 3000–6500 ft.; Creosote Bush Scrub; Eureka V., Inyo Co.; to Mono Co. and w. Nev. May–Aug.

2. *Cleomélla* DC. STINKWEED

Glabrous to pubescent erect or diffuse annuals. Lvs. mostly trifoliolate, petioled. Fls. yellow, small, racemose or solitary. Stamens 6. Caps. rhombic, broader than long, small, few-seeded, the valves hemispheric or laterally produced into short horns, readily shed. Seeds reniform-orbicular, somewhat compressed. Ca. 12 spp. of N. Am. (Diminutive of *Cleome*.)

Stipes much longer than the caps.
    Lvs. and frs. pubescent; sepals ciliate .... 2. *C. obtusifolia*
    Lvs. and frs. glabrous; sepals not ciliate .... 4. *C. plocasperma*
Stipes shorter than the caps.
    Fls. in axils of ordinary lvs.; pedicels 2–3 mm. long .... 1. *C. brevipes*
    Fls. racemose; pedicels 10–20 mm. long .... 3. *C. parviflora*

    1. **C. brévipes** Wats. Diffusely branched from base, somewhat scaberulous, 5–30 cm tall; lvs. subsessile; lfts. linear-spatulate, 5–15 mm. long; fls. solitary, in axils of most lvs.; pedicels ca. 2–3 mm. long, recurved in fr.; calyx ca. 0.5 mm. long, the lobes long-acuminate; petals ca. 1.5 mm. long; caps. scarcely stipitate, ovoid, 3 mm. long; style ca. 0.4 mm. long; seeds stramineous, shining.—Alkaline seeps, 1400–4000 ft.; Alkali Sink; Mojave Desert (Camp Cady, Tecopa, Keeler, etc.); w. Nev. May–Oct.

    2. **C. obtusifòlia** Torr. & Frém. [*C. o.* vars. *florifera* and *jonesii* Crum.] Plate 33,

Fig. B. Diffuse, 8–15 cm. high, or with longer trailing stems; stems glabrous; petioles 1–3 cm. long; lfts. obovate to oblong, somewhat fleshy, 8–15 mm. wide, pubescent beneath; stipules fimbriate; fls. beginning near base of stems; pedicels 6–12 mm. long; calyx ciliate, ca. 1.5 mm. long; petals 5–6 mm. long, yellow; stipe ca. 6 mm. long, reflexed in fr.; caps. ca. 4 mm. long, 6–8 mm. broad, the valves pubescent, laterally conical to hornlike; seeds suborbicular, smoothish.—Alkaline flats below 4000 ft.; Creosote Bush Scrub, Joshua Tree Wd.; Colo. Desert to Inyo Co.; Ariz., Nev. April–Oct. Plants with pubescent stems have been called var. *pubescens* A. Nels.

3. **C. parviflòra** Gray. Mostly branched from base, glabrous, 1–3.5 dm. high; petioles 2–10 mm. long; lfts. linear, 10–25 mm. long; infl. lax, somewhat recurved; pedicels 10–25 mm. long; stipes scarcely 1 mm. long; calyx ca. 0.5 mm. long, the lobes acuminate; petals yellow, 1.5–2 mm. long; caps. ovoid-rhombic, ca. 4 mm. long; seeds several, smooth, stramineous, obovoid, compressed, ca. 1.2 mm. long.—Alkali flats, 2500–6700 ft.; Alkali Sink, Creosote Bush Scrub, Joshua Tree Wd., etc.; w. San Bernardino Co. to e. Lassen and Mono cos.; w. Nev. May–Aug.

4. **C. plocaspérma** Wats. var. **mojavénsis** (Pays.) Crum. [*C. m.* Pays. *C. p.* var. *stricta* Crum.] Glabrous and often glaucous, erect, widely branched, 3–6 dm. high; lfts. oblong-linear, 15–25 mm. high, on usually somewhat shorter petioles; racemes dense at anthesis; pedicels and stipes each ca. 1 cm. long; calyx ca. 1 mm. long; petals ca. 5 mm. long, pale yellow; seeds few, strawcolor, sometimes mottled.—Alkaline seeps and flats, 3000–5600 ft.; Alkali Sink and Creosote Bush Scrub; Mojave Desert, w. San Bernardino Co. to Inyo Co. May–Oct. Late-season collections have racemes in fr. to 6 dm. long and are the basis of var. *stricta*.

3. *Isómeris* Nutt. BLADDERPOD

Ill-scented widely branched glaucous puberulent shrubs. Lvs. alternate, trifoliate, petioled. Fls. yellow, in dense terminal bracted racemes. Calyx persistent. Petals oblong, sessile, the 2 lower more spreading than the 2 upper. Stamens 6, long-exserted. Torus hemispheric. Caps. large, inflated, pendulous, coriaceous, long-stipitate, tardily dehiscent in 2 valves. Seeds few, large, smooth, somewhat obovoid with pointed base. Monotypic. (Greek, *isos*, equal, and *meris*, part.)

1. **I. arbòrea** Nutt. var. **arbòrea.** [*Cleome Isomeris* Greene.] Plate 33, Fig. C. Rounded, erect, 6–15 (–25) dm. high; petioles 0.5–2 cm. long; lfts. oblong to oblong-elliptic, 1–3.5 cm. long, 0.3–0.8 cm. wide, entire, the uppermost sometimes solitary; racemes mostly becoming 3–15 cm. long, the green bracts simple, to ca. 1 cm. long; pedicels 6–12 mm. long; calyx 6–8 mm. long; petals 10–16 mm. long; stipe 1–2 cm. long in fr., stout, recurved; caps. 2.5–5 cm. long, 1–1.5 cm. thick, gradually narrowed at base; seeds smooth, 6–7 mm. long; $2n = 20$ pairs (Raven et al., 1965).—Frequent in subsaline places, such as coastal bluffs and stabilized dunes; Coastal Sage Scrub from n. L. Calif. to s. cent. Calif., and along desert washes; Creosote Bush Scrub, Joshua Tree Wd.; w. Mojave and Colo. deserts. Fls. most of the year. Variable, the following ill-defined extremes may be noted:

Var. **angustàta** Parish. [*I. a.* Parish.] Caps. narrow, 5–12 mm. thick, strongly attenuate at both ends.—Sandy washes below 4000 ft.; Mojave and Colo. deserts to the coast at San Diego.

Var. **globòsa** Cov. [*I. g.* Heller.] Caps. subglobose, 2–3 cm. long, abruptly narrowed at both ends.—With the typical form in s. Calif. and to w. Fresno and e. Monterey cos.

Var. **insulàris** Jeps. Caps. almost as thick as long, but gradually narrowed at ends, and with sharply pointed apex.—Santa Rosa, Santa Catalina and Coronado ids.

4. *Oxýstylis* Torr. & Frém.

Branching erect minutely scaberulous annual. Lvs. trifoliate, petioled. Fls. in headlike axillary glomerules. Sepals 4, oblong-linear. Petals 4, yellow, oblong-ovate. Stamens 4, exserted, borne on a somewhat fleshy elevated torus. Ovary didymous, borne on a short stout stipe; style elongate, subulate, forming a stiff spine. Fr. of 2 1-seeded rounded faintly reticulate nutlets. One sp. (Greek, *oxus*, sharp, and *stylis*, column or style.)

1. **O. lùtea** Torr. & Frém. Plate 33, Fig. D. Stems yellowish, stout, 3–9 dm. high, flowering from very base; petioles 3–6 cm. long; lfts. green, elliptic, 1–3 cm. long, mucronulate; petals ca. 1.5 mm. long; fr. pedicels recurved, 3–4 mm. long; stipe 1–2 mm. long; nutlets globose-ovoid, 1.5 mm. in diam.; style 4–7 mm. long; $2n = 20$ pairs (Raven et al., 1965).—Alkaline washes and flats, below 2000 ft.; Creosote Bush Scrub, Alkali Sink; Death V. region to Tecopa and adjacent Nev. March–Oct.

5. Wislizènia Engelm.

Erect branched ill-scented annuals. Lvs. mostly trifoliate, with small bristlelike deciduous stipules. Fls. yellow, racemose, inconspicuously bracteate. Sepals and petals 4. Stamens 6, much exserted. Stipe reflexed in fr. Fr. 2-seeded, didymous, each valve closely contracted upon its single seed and falling away with it like a nutlet. Style elongate, persistent. A small genus of sw. U.S. and nw. Mex. (Named for A. *Wislizenius*, early botanical collector in the Southwest.)

1. **W. refrácta** Engelm. [*W. californica* Greene.] JACKASS-CLOVER. Two to 15 dm. tall, subglabrous; lfts. oblanceolate or wider, 1–3 (–4) cm. long, scaberulous, mucronulate, entire; raceme dense at anthesis, later elongate; pedicels ca. 5–10 mm. long; sepals ca. 1.5 mm. long, united at base; petals 3 mm. long; stipe 5–10 mm. long; nutlets obovoid to almost round, 2–3 mm. long, reticulate-ridged, ± tubercled at summit; style 4–5 mm. long, hairlike.—Subalkaline soil; Alkaline Sink, and adjacent areas; L. Calif. and Sonora to Sacramento V. April–Nov. Plants from the s. desert areas tend to have narrower lfts. than the more n. ones and are the basis for var. *palmeri* (Gray) Jtn. [*W. p.* Gray.]

25. Caprifoliàceae. HONEYSUCKLE FAMILY

Shrubs or trees or vines, sometimes herbs. Lvs. opposite, mostly without stipules. Fls. bisexual, regular or irregular, 4–5-merous, with inferior ovaries. Corolla gamopetalous, rotate to tubular; stamens alternate with corolla-lobes, inserted on the tube. Ovary 1–5-loculed, each locule 1–many-ovuled; placentation axile; style 1 or obsolete; stigmas 1–5. Fr. an ak., berry, caps. or drupe. Seed-coat adherent to the fleshy endosperm; embryo small. Ca. 14 genera and 1400 spp., mostly of N. Temp. Zone; many of ornamental use.

Corolla rotate or nearly so, regular; style short, 3–5-lobed; lvs. pinnately compound . 2. *Sambucus*
Corolla tubular to funnelform, usually irregular; style elongate, mostly with capitate stigmas; lvs. entire to lobed.
    Berry red or black, few-seeded; corolla tubular .... 1. *Lonicera*
    Berry white, 1–2-seeded; corolla open-campanulate or tubular-funnelform .. 3. *Symphoricarpos*

1. Lonícera L. HONEYSUCKLE

Erect or twining shrubs, with simple entire lvs., the uppermost pair often connate-perfoliate. Fls. in terminal spikes or heads or in small axillary clusters. Sepals 5 or obsolete. Corolla campanulate to tubular, ± gibbous at base, the limb from somewhat irregular to strongly bilabiate with 4 lobes in upper lip and 1 in lower. Stamens 5, adnate to corolla-tube. Ovary 2–3-loculed. Fr. a fleshy few-seeded berry. Ca. 100 spp., mostly N. Temp. (Named for A. *Lonitzer*, latinized Lonicerus, a German herbalist.)

Fls. in peduncled axillary pairs.
    Stems twining; fls. 3–4 cm. long. Introd. plant .... 4. *L. japonica*
    Stems erect; fls. 1.2–1.6 cm. long. Native plant .... 3. *L. involucrata*
Fls sessile, in whorls; plants ± twining or trailing.
    Lvs. all distinct, the uppermost pair rarely slightly connate; infl. glandular-pubescent
                                                                                  5. *L. subspicata*
    Lvs. mostly with the uppermost connate.

The lvs. mostly with stipules; corolla pinkish to purplish; infl. glandular-pubescent
1. *L. hispidula*
The lvs. without stipules; corolla yellow; infl. glabrous . . . . 2. *L. interrupta*

1. **L. hispídula** Dougl. var. **vácillans** Gray. [*L. h.* var. *californica* Rehd. *L. catalinensis* Millsp.] Climbing shrubs 2–6 m. high, with long glabrous twigs; lvs. elliptic to oblong-ovate, obtusish, 3.5–8 cm. long, puberulent and glaucous beneath, green and mostly glabrous above, the several upper pairs mostly connate-perfoliate, the lower with petioles 2–3 mm. long; fls. in many whorls, forming spikes or loose panicles, the infl. glandular-pubescent; corolla purplish to pink, funnelform, 12–18 mm. long, bilabiate, glandular-pubescent; fr. red, subglobose, 5–6 mm. in diam.—Along streams and on wooded slopes, below 2500 ft.; Chaparral, etc.; Santa Catalina, San Clemente and Santa Cruz ids., Santa Ynez Mts.; to s. Ore. April–July.

2. **L. interrúpta** Benth. Bushy evergreen intricate shrub with the branches twining or leaning on other vegetation; twigs glabrous, glaucous; lvs. orbicular to elliptical, entire, 1.5–3.5 cm. long, glabrous or puberulent, glaucous beneath, green above, the uppermost pair usually connate; spikes interrupted, 3–16 cm. long, in an open panicle, subglabrous; corolla yellowish, funnelform, 10–14 mm. long, glabrous without; fils. pubescent; fr. red, subglobose, ca. 5 mm. long.—Dry slopes 1000–6000 ft.; Chaparral to Yellow Pine F.; N. Coast Ranges to San Gabriel, San Bernardino and San Jacinto mts. May–July.

3. **L. involucràta** (Richards.) Banks. [*Xylosteon i.* Richards.] TWINBERRY. Upright shrub, 6–30 dm. high, deciduous, glabrous to ± pilose and glandular-pubescent; lvs. obovate to ovate or oval, acutish, darker and more glabrous above, often ciliate, 3–12 cm. long, on petioles 5–12 mm. long; fls. in axillary pairs, the peduncle 1.5–2.5 cm. long, rather coarse, with 2 bracts at its summit, ovate to oblong, 1–1.3 cm. long, often turning reddish or purplish; bractlets united, broad, resembling the bracts; corolla yellow or yellowish, subcylindric, ca. 12–16 mm. long, viscid-pubescent, the lobes subequal; ovaries not united; fr. black, ca. 8 mm. in diam., almost enclosed in the bractlets; $2n=18$ (Jan Ammal & Saunders, 1952).—Moist places along the immediate shore in Santa Barbara Co.; ascending to high altitudes farther n.; to Alaska, Quebec, Mex. March–April with us. The coastal form is often referred to var. **ledeboùrii** (Esch.) Zabel and often has a reddish tinge to the corolla and the stamens included.

4. **L. japónica** Thunb. Half evergreen climber to 6 or 8 m., the twigs mostly hairy when young; lvs. ovate to oblong, 3–8 cm. long, pubescent especially beneath; fls. white, changing to yellow, sometimes purplish on outside, borne in axillary pairs on short peduncles with 2 ovate bracts; bractlets half as long as the ovary; corolla 3–4 cm. long, pubescent; fr. black; $2n=18$ (Jan Ammal & Saunders, 1952).—Occasional escape from cult.; native of Asia. Spring and summer.

5. **L. subspicàta** H. & A. var. **subspicàta**. Evergreen clambering shrub 1–2.5 m. high, with puberulent twigs; lvs. linear-oblong to oblong, 1–3 cm. long, 0.6–1 cm. wide, rounded at ends, ± revolute, entire, coriaceous, bicolored, pubescent especially beneath; petioles 1–5 mm. long; fl.-whorls several, compact, in short leafy spikes 2–12 cm. long; sepals broadly pubescent, 1 mm. long; corolla yellowish or somewhat cream-color, 8–10 mm. long, glandular-pubescent, with a 2-lipped often recurved limb; fils. pubescent at base; berry 5–7 mm. long, yellowish or red, ellipsoid.—Dry slopes below 3000 ft.; Chaparral; Santa Barbara region. June–July.

Var. **johnstònii** Keck. [*L. j.* McMinn.] Lvs. whitish and pubescent beneath, oblong-ovate to suborbicular less than twice as long as wide; corolla 1–1.4 cm. long.—Common on dry slopes below 5000 ft.; Chaparral; mts. from San Diego Co. n., occasional to cent. Calif. April–June.

Var. **denudàta** Rehd. [*L. d.* Davids. & Mox.] Lvs. of width of var. *johnstonii*, but yellowish and subglabrous beneath.—Chaparral, from about San Diego to Temecula, Riverside Co. April–June.

2. Sambùcus L. ELDERBERRY

Shrubs or small trees with opposite odd-pinnate lvs. and serrate lfts. Twigs with large pith. Fls. small, mostly whitish, in compound cymes, jointed with their pedicels. Sepals 5. Corolla regular, rotate, 5-lobed. Stamens 5, inserted at the base of the corolla. Ovary 3-5-loculed; style short; stigmas 3-5. Fr. a berrylike drupe with 3-5 cartilaginous nutlets. Ca. 20 spp. of temp. and subtrop. regions. (Greek, *sambuke*, a musical instrument made of Elder wood.)

Berry blue or white; cymes flat-topped, the axis not or seldom extended beyond the lowest branches.
    Lfts. mostly 5-9, oblong-lanceolate to lanceolate, 6-15 cm. long, gradually short-acuminate, sharply and often rather deeply serrate; infl. 0.5-2 dm. across. Mostly in the Pine belt
                                                                                              1. *S. caerulea*
    Lfts. mostly 3-5, roundish to ovate or oblong-lanceolate, rather abruptly acuminate, mostly 1.5-6 cm. long, finely serrate; infl. mostly 0.3-1 dm. across. Generally below Pine belt
                                                                                              2. *S. mexicana*
Berry bright red at maturity; cymes dome-shaped or thyrsoid, the axis extending beyond the lowest branches. High elevs. .... 3. *S. microbotrys*

1. **S. caerùlea** Raf. [*S. glauca* Nutt.] Two to 8 dm. tall; lvs. glabrous to pubescent or hispidulous beneath, the lfts. usually quite asymmetrical at base; fls. 5-6 mm. wide; berries nearly black but densely glaucous, thus appearing bluish, 5-6 mm. in diam.—Open places largely in Montane Coniferous F.; mts. from San Diego Co.; to B.C., Alta., Ida. June-Sept.

2. **S. mexicàna** Presl. Plate 33, Fig. E. Much like the preceding, the lfts. mostly fewer, smaller, more deciduous in the dry season; infl. smaller; berries either blue or white under the white bloom, often drier at maturity.—Open flats and cismontane valleys and canyons below 4500 ft ; largely Coastal Sage Scrub, Chaparral, S. Oak Wd.; n. L. Calif. n. to cent. Calif.; occasional in desert mts. and to Ariz., etc. March-Sept.

3. **S. microbótrys** Rydb. Low, 0.5-1 m. high, with rank odor; lvs. thin, glabrous or nearly so; lfts. 5-7, oval to elliptic, 3-8 cm. long, coarsely serrate; fls. cream-color; frs. red, 4-5 mm. in diam. Occasional at 9000-10,000 ft., Subalpine F.; San Bernardino Mts.; to n. Calif., Rocky Mts. June-Aug.

3. Symphoricárpos Duhamel. SNOWBERRY

Shrubs with simple, short-petioled, entire or ± toothed or lobed opposite deciduous lvs. Fls. white or pink, 2-bracteolate, in small terminal or axillary clusters, sometimes solitary, 4-5-merous. Corolla campanulate to long-funnelform, regular or nearly so. Stamens inserted on the corolla. Ovary inferior, 4-loculed, 2 of the locules with several abortive ovules, the other 2 each with 1 pendulous ovule. Fr. a berrylike drupe, in ours white, with 2 nutlets. Ca. 17 spp. of N. Am. and 1 in China. Some in cult. (Greek, *sumphoreo*, to bear together, and *karpos*, fr., because of the clustered frs.)

Corolla short-campanulate, the lobes ca. as long as tube; twigs closely pubescent; lvs. round-oval, 1-4 cm. long .... 2. *S. mollis*
Corolla elongate-campanulate to salverform, the lobes shorter than the tube; lvs. lanceolate to elliptical, 0.5-1.5 cm. long.
    The corolla 6-7 mm. long. Montane Coniferous F. .... 3. *S. parishii*
    The corolla 11-13 mm. long. Deserts .... 1. *S. longiflorus*

1. **S. longiflòrus** Gray. Low, spreading, the branches somewhat declined, 5-10 dm. long; young twigs glaucous, glabrous or sparsely pubescent; lvs. lanceolate to elliptical, entire, 6-15 mm. long, pale, glabrous to sparsely pubescent; fls. 1-2 in upper axils or in small racemes; sepals deltoid, ca. 1 mm. long; corolla salverform, pink, 11-13 mm. long; style 5-7 mm. long; fr. ellipsoid, 8-10 mm. long.—Dry slopes, often on limestone, 4500

to 10,000 ft.; mostly Pinyon-Juniper Wd., Bristle-cone Pine F.; mts. of e. San Bernardino and n. Inyo cos. to e. Ore., Colo., Tex. May-June.

2. **S. móllis** Nutt. in T. & G. Plate 33, Fig. F. Mostly low, trailing, diffusely branched, the stems 3-10 dm. long; sometimes erect or nearly so and twice that long; twigs closely pubescent; lvs. thin, mostly round-oval and entire, sometimes lobed, 1-4 cm. long, pubescent; fls. in pairs or small clusters; sepals 0.5-0.8 mm. long, deltoid, ciliate; corolla pink, 3-5 mm. long, the lobes 2-3 mm. long, sparsely pilose within near base; style 2 mm. long; frs. 4-6 mm. in diam.—Common on shaded slopes, mostly below 5000 ft.; Chaparral, S. Oak Wd.; n. L. Calif. to n. Calif. April-June. Catalina and Santa Cruz ids.

3. **S. párishii** Rydb. [*S. parvifolius* Eastw.] Low, spreading, the declined branches 5-10 dm. long and frequently rooting at the tips; young twigs short-pubescent, or internodes sometimes glabrous; lvs. glaucous, oval to narrow-elliptic, usually acutish, 1-2 cm. long, 0.5-1.3 cm. wide, grayish-green, short-pubescent, frequently lobed on young shoots; fls. 2 to several; sepals ca. 1 mm. long; corolla pink, 6-7 mm. long; fr. 6-8 mm. long.—Dry rocky places, 4000-11,000 ft.; Pinyon-Juniper Wd. to Subalpine F.; mts. from San Diego Co. to Santa Barbara Co. and s. Sierra Nevada; to Nev., Ariz. June-Aug.

26. Caryophyllàceae. Pink Family

Annual or perennial herbs. Lvs. mostly opposite, entire, simple, often connected at base by a transverse line, with or without stipules, these if present often scarious. Fls. symmetrical, 4-5-merous, with or without petals, solitary or in cymes, mostly perfect. Sepals free or united into a tube. Petals as many as the sepals, often small or none. Stamens up to 10, free from one another; anthers 2-celled, dehiscing longitudinally. Ovary superior, sessile or stipitate, 1-loculed (rarely 3-5-loculed), with free central placentation; styles 2-5, rarely united into 1. Fr. a dry caps., usually opening by valves or apical teeth, or sometimes a utricle. Seeds many or 1, with endosperm and a ± curved peripheral or excentric embryo. Ca. 1500 spp. in 75 genera, most abundant in temp. and cooler regions; many grown for their fls.

1. Fr. a 1-seeded indehiscent utricle; petals 0; fls. very small, greenish or whitish.
    2. Stipules present, scarious.
        3. Sepals united below into a short tube.
            4. Annual, prostrate or spreading; styles 2-cleft .... 1. *Achyronychia*
            4. Perennial, erect; styles 3-cleft .... 11. *Scopulophila*
        3. Sepals quite or almost distinct.
            5. Annual; lvs. elliptic-oblanceolate; stipules minute .... 5. *Herniaria*
            5. Perennial; lvs. subulate; stipules conspicuous .... 3. *Cardionema*
    2. Stipules lacking; annual with subulate lvs.; fls. clustered, greenish ...... 10. *Scleranthus*
1. Fr. a several- to many-seeded caps.; petals usually present; fls. small to large.
    6. Sepals distinct or nearly so; petals when present without claws and borne on a basal disc or at base of sessile ovary.
        7. Stipules wanting.
            8. Caps. ovoid or ellipsoid, dehiscent by as many or twice as many valves or teeth as there are carpels.
                9. Styles opposite sepals and usually 3.
                    10. Petals deeply notched or bifid, rarely wanting; valves of caps. bifid or 2-parted 15. *Stellaria*
                    10. Petals entire or nearly so; valves of caps. entire or bifid or 2-parted . 2. *Arenaria*
                9. Styles alternate with sepals, 4-5 .... 8. *Sagina*
            8. Caps. cylindrical, often bent near summit, dehiscent by twice as many teeth as there are carpels .... 4. *Cerastium*
        7. Stipules present, scarious.
            11. Styles 3-5, distinct; petals usually present.
                12. Styles and caps.-valves 5; lvs. appearing whorled .... 13. *Spergula*
                12. Styles and caps.-valves 3; lvs. opposite .... 14. *Spergularia*
            11. Style 1, 3-cleft or -toothed; petals minute or 0.
                13. Lvs. flat, oblong or obovate; stipules scarious .... 7. *Polycarpon*

13. Lvs. subulate; stipules setaceous .... 6. *Loeflingia*
 6. Sepals united into a tubular or cuplike calyx; petals clawed and borne on the carpophore (stipe of ovary).
   14. Styles 3, rarely 4; caps. dehiscent by 6, rarely 3, 4, or 8 apical teeth .... 12. *Silene*
   14. Styles 2; caps. mostly 4-valved.
      15. Calyx ovoid, 5-ribbed, wing-angled; petals not appendaged .... 16. *Vaccaria*
      15. Calyx tubular, 20-nerved, not wing-angled; petals appendaged at base of blades
                                                                      9. *Saponaria*

1. Achyrónychia T. & G.

Low glabrous annuals. Lvs. spatulate, those in each pair unequal; stipules hyaline. Fls. in dense, axillary, cymose clusters. Calyx 5-lobed, scarious. Petals 0. Stamens 10 to 15, only 1–5 fertile. Style 2-cleft. Fr. a utricle, included in calyx. Seeds minute, black, round-reniform. Two spp. of sw. U.S. and Mex. (Greek, *achuron*, chaff, and *onyx, onychos*, fingernail, referring to the silvery chaffy calyx.)

   1. **A. coóperi** T. & G. Plate 34, Fig. A. Stems several, prostrate, 3–15 cm. long; lvs. spatulate, 5–15 mm. long; fls. white, green at the base, 2 mm. long; calyx-lobes oval, white-scarious above the greenish base; seeds shining.—Common on sandy flats and in washes, below 3000 ft.; Creosote Bush Scrub; Colo. and e. Mojave deserts; Ariz., L. Calif. Jan.–May.

2. Arenària L. SANDWORT

Low branched annual to perennial herbs, commonly tufted or matted. Lvs. sessile, usually exstipulate, subulate to ovate. Fls. small, white, occasionally rose or cream, terminal, cymose or capitate, rarely solitary and axillary. Sepals 5. Petals 5, entire or somewhat emarginate, sometimes wanting. Stamens 10. Styles usually 3, opposite as many sepals. Caps. globose to oblong, splitting into as many or twice as many valves as there are styles. Seeds few to many, reniform to globose. Ca. 150 spp., the genus almost worldwide. (Latin, *arena*, sand, in which many spp. grow.)

(Maguire, B. Arenaria in America n. of Mexico. A conspectus. Amer. Midl. Nat. 46: 493–511, 1951.)

1. Plants annual.
   2. Petals shorter than sepals.
      3. Lvs. ovate, 3–5-nerved; stems 10–20 cm. long .... 13. *A. serpyllifolia*
      3. Lvs. narrowly lanceolate, 1-nerved; stems 2–5 cm. long .... 11. *A. pusilla*
   2. Petals longer than sepals, or at least as long.
      4. Lvs. somewhat lanceolate, 2–5 mm. long, obtuse .... 2. *A. californica*
      4. Lvs. filiform, mostly 0.5–1.5 cm. long, pointed .... 5. *A. douglasii*
1. Plants perennial.
   5. Lvs. linear-lanceolate to lance-oblong, 2–7 mm. wide, not pungent.
      6. Petals 2–4 mm. long; fls. in terminal cymes; stems puberulent, scarcely angled.
         7. Stems from running rootstocks; lvs. mostly 2–8 cm. long; fls. 1–few; seeds appendaged with a pale spongy strophiole at the hilum .... 8. *A. macrophylla*
         7. Stems from a branching root-crown; lvs. 0.5–2 cm. long; fls. numerous; seeds not so appendaged .... 3. *A. confusa*
      6. Petals 5–6 mm. long; fls. solitary, axillary; stems glabrous, angled .... 10. *A. paludicola*
   5. Lvs. filiform or subulate, ca. 1–1.5 mm. wide, often pungent.
      8. Sepals distinctly 3-nerved.
         9. The sepals glabrous; lvs. 1–2 cm. long .... 4. *A. congesta*
         9. The sepals glandular-pubescent; lvs. mostly less than 1 cm. long.
            10. Calyx ca. 3–4 mm. long; petals 2–3 mm. long .... 12. *A. rubella*
            10. Calyx 4.5–6.5 mm. long; petals ca. 5–6 mm. long .... 9. *A. nuttallii*
      8. Sepals 1-nerved, or if with lateral nerves, these indistinct.
         11. The entire plant glandular-pubescent; lvs. 0.3–1 cm. long.
            12. Lvs. 3-nerved; caps.-valves entire .... 9. *A. nuttallii*
            12. Lvs. 1-nerved; caps. valves 2-toothed .... 6. *A. kingii*
         11. The plant glabrous or at most slightly glandular-pubescent in infl.
            13. Lvs. 0.4–2 cm. long.

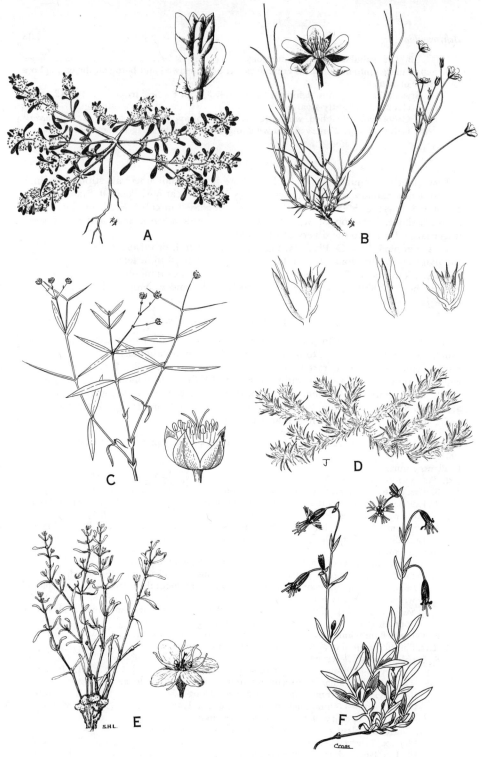

Plate 34. Fig. A, *Achyronychia cooperi*. Fig. B, *Arenaria macradenia*. Fig. C, *Arenaria macrophylla*. Fig. D, *Cardionema ramosissimum*. Fig. E, *Scopulophila rixfordii*. Fig. F, *Silene lemmonii*.

14. Pedicels 5-20 mm. long, the cymes open.
   15. Plants glaucous; lvs. rigid, straight .... 1. *A. aculeata*
   15. Plants not glaucous.
     16. Lvs. ± recurved. San Bernardino Mts. .... 14. *A. ursina*
     16. Lvs. rigid, straight. Inyo-White Mts. .... 6. *A. kingii*
14. Pedicels usually shorter, the cymes condensed .... 4. *A. congesta*
13. Lvs. 2-6 cm. long.
   17. Infl. capitate; sepals 3-4 mm. long .... 4. *A. congesta*
   17. Infl. an open cyme; sepals 4-6.5 mm. long. ..... 7. *A. macradenia parishiorum*

1. **A. aculeàta** Wats. Perennial from a woody caudex, much-branched and matted at base, the erect stems glabrous, or glandular-pubescent toward apex, 1-1.8 dm. high; lvs. many, mostly basal, glaucous, stiffly pungent, 1-2 cm. long; cyme open; pedicels 5-20 mm. long; sepals ovate, obtusish to acute, 4-5 mm. long, broadly scarious-margined; petals 6-7 mm. long; caps. slightly exceeding calyx to twice as long, the valves 2-toothed; seeds light brown, flat, broadly winged, 2.5-3 mm. long, with fine radiating reticulations.—Rocky slopes, 7500-9700 ft.; Sagebrush Scrub, Bristle-cone Pine F.; Inyo-White Mts.; to Ore., Mont. May-July.

2. **A. califórnica** (Gray) Brew. [*A. brevifolia* var. *c*. Gray.] Glabrous, erect, simple or branched annual 3-10 cm. high; lvs. lanceolate, obtuse, somewhat fleshy, 2-5 mm. long; fls. loosely cymose; pedicels 6-15 mm. long; sepals oblong-ovate, 2-3 mm. long, obtusish, 1-3-nerved, scarious-margined; petals oblong-ovate, 3-4 mm. long; caps. ovoid, slightly surpassing calyx; seeds minutely roughened.—Grassy slopes and disintegrated rock, below 2500 (4000) ft.; V. Grassland, S. Oak Wd.; Los Angeles Co. and Tehachapi Mts. n. to Ore. March-June.

3. **A. confùsa** Rydb. [*A. saxosa* auth.] Retrorsely puberulent perennial with branched root-crown and slender spreading stems 1-3 dm. long, leafy in lower half; lvs. lance-oblong, 1-2 cm. long, 2-3 (-5) mm. wide; cymes paniculate; pedicels 5-15 mm. long; sepals ovate, carinate, sharply acute, 3 mm. long; petals ca. as long; caps. ovoid, 3-4 mm. long; seeds black, shining, almost 1 mm. long.—Damp places near meadows, 6300-8500 ft.; Red Fir F., Lodgepole F., San Bernardino Mts.; San Pedro Martir Mts. in L. Calif. July-Aug.

4. **A. congésta** Nutt. ex T. & G. var. **charlestonénsis** Maguire. Perennial from a short branched woody caudex, with lvs. bunched at base and stems 4-10 cm. high; lvs. 1-1.5 cm. long, spreading-recurved, not pungent, subfiliform; fls. few, capitate; sepals ovate, acute, pungent, 4.5-5.5 mm. long; petals ca. 6 mm. long; caps. somewhat exserted; seeds brownish, ca. 2 mm. long, winged, radiately marked.—Sandy ridge at 7300 ft.; Pinyon-Juniper Wd.; New York Mts.; s. Nev. June.

Var. **subcongésta** (Wats.) Wats. [*A. fendleri* var. *s*. Wats.] Infl. slightly umbellate; sepals 4-5 mm. long, acute.—Dry slopes, 5500-8200 ft.; Pinyon-Juniper Wd.; Panamint Mts.; to Nev., Utah. June-July.

5. **A. dóuglasii** Fenzl ex T. & G. Delicate erect annual with slender freely branched stems 0.5-2.5 dm. high, subglabrous to somewhat glandular-pubescent, loosely cymose above; lvs. filiform, 0.5-2 cm. long; pedicels filiform, 6-30 mm. long; sepals ovate, 1-3-nerved, scarious-margined, 2-3 mm. long; petals white, conspicuous, obtuse, obovate, 3-6 mm. long; caps. subglobose, slightly exceeding calyx; seeds brown, margined, 1-1.5 mm. broad.—Locally common in dry sterile places below 7000 ft.; Coastal Sage Scrub to Yellow Pine F.; cismontane to desert edge; L. Calif. to Ore. April-June.

6. **A. kíngii** (Wats.) Jones var. **glabréscens** (Wats.) Maguire. [*A. fendleri* var. *g*. Wats.] Perennial with woody caudex; stems slender, 1-2 dm. high, glandular-pubescent, somewhat leafy; lvs. largely basal, acerose, strict or ascending, 1-2 cm. long; cymes open, few- to several-fld.; pedicels mostly 1-2 cm. long, glandular-pubescent; sepals narrow-ovate, subacuminate, glandular-pubescent, broadly hyaline-margined, 3.5-4.5 mm. long; petals oblong-obovate, exceeding sepals; caps. 4.5-6 mm. long; seeds 1.5-2 mm. long, almost smooth, not winged.—Frequent on dry rocky slopes, 8000-11,000 ft.; Bristle-cone Pine F.; Inyo-White Mts.; to Nev., Ore. June-Aug. Found in the same area, but at

somewhat higher elevs. is ssp. **compácta** (Cov.) Maguire with fewer stem-lvs., these 3–6 mm. long; sepals 2.5–3.5 mm. long.

7. **A. macradènia** Wats. var. **macradènia**. Plate 34, Fig. B. Perennial from a branched woody caudex, the stems ascending, mostly 2–4 dm. tall and with 5–8 pairs of lvs., usually glabrous; basal lvs. subulate, rather stout, ascending, straight, 2–5 cm. long, pungent, rigid, scabrous-ciliate; cauline lvs. ascending, 0.8–1.2 cm. long; infl. an open cyme; pedicels 1–3 cm. long; sepals ovate, broadly acute, 5.5–6.5 mm. long, hyaline-margined; petals conspicuously longer; caps. oblong, exceeding calyx; seeds dark brown, flat, ca. 1.5 mm. long, papillose.—Dry rocky slopes below 6000 ft.; Creosote Bush Scrub, Joshua Tree Wd., Pinyon-Juniper Wd.; deserts from Inyo and Los Angeles cos. to San Bernardino and Riverside cos.; to Utah, Ariz. May–June.

KEY TO VARIETIES

Sepals 5.5–6.5 mm. long; stems woody at base.
  Petals conspicuously exceeding sepals; cauline lvs. 5–12 pairs.
    Cauline lvs. ascending, 0.8–1.2 mm. broad .... var. *macradenia*
    Cauline lvs. strongly arcuate, 1.2–2 mm. broad.
      Infl. and sepals glabrous .... var. *arcuifolia*
      Infl. and sepals glandular .... var. *kuschei*
  Petals scarcely or not longer than sepals; cauline lvs. fewer .... var. *parishiorum*
Sepals 4–5 mm. long; stems scarcely woody at base .... ssp. *ferrisiae*

Var. **arcuifòlia** Maguire. Lvs. on stems 6–12 pairs, strongly arcuate; infl. glabrous; sepals 5.5–6 mm. long.—Dry slopes 2000–5000 ft.; largely Joshua Tree Wd.; from Kern co. to n. base of San Gabriel Mts. May–July.

Ssp. **férrisiae** Abrams. Infl. glabrous; sepals 3.5–5 mm. long; petals not longer than sepals or much longer.—Dry stony places to 10,000 ft.; Sagebrush Scrub, Pinyon-Juniper Wd.; Inyo Co. and possibly in San Jacinto Mts.; to Nev., Utah.

Var. **kúschei** (Eastw.) Maguire. Infl. glandular, congested, the pedicels 2–5 mm. long; sepals densely glandular, 6–7 mm. long.—Forest Camp, Mojave Desert.

Var. **parishiòrum** Rob. Cauline lvs. mostly fewer than 5 pairs; sepals 5.5–6.5 mm. long; petals not longer.—Our most common form, 3000–7000 ft.; mostly in Joshua Tree Wd., Pinyon-Juniper Wd.; Inyo Co.; Nev., Ariz. April–June.

8. **A. macrophýlla** Hook. Plate 34, Fig. C. Slender-stemmed perennial from running rootstocks, ascending to suberect, puberulent, 5–15 cm. high; lvs. lanceolate to oblanceolate, bright green, 1.5–6 cm. long, 4–8 mm. wide; fls. 1–5, in short cymes; pedicels 5–20 mm. long; sepals ovate, 3–4 mm. long; petals longer in ♂ fls., shorter in ♀; caps. shorter than calyx; seeds usually with a pale spongy strophiole at the hilum.—Occasional on shaded slopes; Yellow Pine F., Cuyamaca Mts., San Diego Co.; cent. and n. Calif. to B.C., Atlantic Coast. April–June.

9. **A. nuttállii** Pax ssp. **grácilis** (Gray) Maguire. Perennial with many loosely matted prostrate branching stems 5–15 cm. long, densely leafy, glandular-puberulent; lvs. subulate, pungent, overlapping, strict or somewhat recurved, 5–8 mm. long, with fascicles of secondary lvs.; fls. rather few, in open cymes; pedicels 4–10 cm. long; sepals lance-subulate, pungent, acute 4–5 mm. long, 1-nerved; petals shorter; caps. ovoid, shorter than calyx; seeds 1–1.2 mm. broad, low-papillate.—Dry granitic gravel, 6500–11,400 ft.; Montane Coniferous F.; Mt. Williamson, San Gabriel Mts. and San Bernardino Mts.; Sierra Nevada. July–Aug.

10. **A. paludícola** Rob. Glabrous flaccid perennial with several subsimple procumbent stems rooting at lower nodes, angled, leafy throughout, 3–7 dm. long; lvs. rather uniform, flat, 1-nerved, lance-linear, 1.5–4 cm. long, acute, somewhat connate; fls. solitary, axillary; pedicels 1.5–4 cm. long; sepals lance-ovate, 3–4 mm. long, nerveless; petals oblong- obovate, 5–6 mm. long; caps. ca. as long as calyx.—Occasional in swamps; Freshwater Marsh; San Bernardino, Los Angeles, Santa Barbara, etc. to Wash. May–Aug.

11. **A. pusílla** Wats. Slender-stemmed glabrous annual, simple or few-branched, 2–5

cm. high; lvs. lanceolate, 2–5 mm. long, the cauline 1–3 pairs; pedicels capillary, 2–10 mm. long; sepals narrow-lanceolate, acute, 2–3.5 mm. long; petals shorter or none; caps. shorter than calyx; seeds smooth, minute.—Occasional at ca. 5000 ft., Laguna Mts., San Diego Co. and 3600 ft., Rose Lake, Ventura Co.; n. to Wash. April–July. Plants with 3–6 pairs of cauline lvs. and sepals 1.8–2.5 mm. long are the var. *diffusa* Maguire.

12. **A. rubélla** (Wahl.) Sm. [*Alsine r.* Wahl.] Tufted perennial, densely glandular-puberulent, 2–5 cm. high; lvs. crowded at base, linear-subulate, 3-nerved, 3–10 mm. long, pungent; fls. 2–few, cymose; sepals lanceolate, 2.5–3 mm. long, 3-nerved; petals somewhat shorter; caps. narrow-ovoid, ca. as long as calyx.—Dry rocky places, 11,000–11,450 ft.; Mt. San Gorgonio; Sierra Nevada to Alaska, Atlantic Coast; Eurasia. July–Aug.

13. **A. serpyllifòlia** L. Puberulent annual, simple to branched, 5–20 cm. high; lvs. ovate, acuminate, ciliate, 3–5-nerved, 3–7 mm. long; cymes open, leafy-bracted; pedicels 3–10 mm. long, capillary; sepals lance-ovate, hispidulous, 3–3.5 mm. long; petals oblong, ca. 2 mm. long; caps. ovoid, exceeding calyx; seeds globose-reniform, rugose; $2n = 40, 20$ (Woess, 1941).—Weed in lawns and moist places, mostly below 5000 ft.; occasional on mainland and Santa Cruz Id.; natur. from Eu. April–June.

14. **A. ursìna** Rob. Perennial with cespitose caudex, the stems many, erect or ascending, 6–15 cm. high, glandular-puberulent above; lvs. subulate, straight or recurved, glaucous, ciliate, 4–12 mm. long, rigid; cymes few-fld.; pedicels 4–12 mm. long; sepals ovate, nerveless, 3–4 mm. long, scarious-margined; petals 4–5 mm. long; caps. 3–5 mm. long; seeds brown, faintly reticulate, ca. 1.2 mm. long.—Dry slopes, 6000–7000 ft.; Yellow Pine F.; San Bernardino Mts. June–July.

3. *Cardionèma* DC.

Low tufted perennial herbs with numerous short branched ± woolly stems, the upper internodes short and usually covered with persistent papery hyaline commonly bifid stipules. Lvs. subulate, pungent, with fascicles of secondary lvs. and of fls. in the main axils. Calyx 5-parted, the segms. unequal, hooded, woolly about base, with erect lower portion and with terminal portion (especially of the 3 outer) spreading and ending in a spine. Petals minute, scalelike. Stamens 3–5, inserted at base of calyx-segms. Style short, bifid. Fr. a utricle, inclosed in a rigid persistent calyx. Several spp. of w. N. and S. Am. (Greek, *cardia*, heart, and *nema*, thread, because of the obcordate stamens.)

1. **C. ramosíssimum** (Weinm.) Nels. & Macbr. [*Loeflingia r.* Weinm. *Pentacaena r.* H. & A.] Plate 34, Fig. D. Stems 5–20 cm. long, prostrate; lvs. 8–13 mm. long; stipules 4–6 mm. long; calyx 4–6 mm. long, pubescent below the spines.—Sandy places, Coastal Strand, Coastal Sage Scrub; mostly near the coast; L. Calif. to Wash.; w. Mex., Chile. April–Aug. Santa Cruz and Santa Rosa ids.

4. *Cerástium* L. Mouse-Ear Chickweed

Annual or perennial pubescent herbs. Fls. white, in terminal dichotomous cymes. Bracts green or scarious. Sepals 5, rarely 4. Petals as many, 2-lobed or -cleft, or wanting. Stamens 10 or 5. Styles mostly 5, opposite the sepals. Caps. slender, elongate, usually exceeding the calyx, often curved, dehiscent at apex by usually 10 teeth. Seeds many, rough. Ca. 50 spp., widely dispersed in cool and temp. regions. (Greek, *cerastes*, horned, referring to shape of caps.)

Annual, without basal persistent sterile offshoots; petals lacking or about as long as sepals
　　　　　　　　　　　　　　　　　　　　　　　　　　　　　　　　1. *C. glomeratum*
Perennial, with prostrate or creeping basal branches or offshoots; petals ca. as long as sepals
　　　　　　　　　　　　　　　　　　　　　　　　　　　　　　　　2. *C. vulgatum*

1. **C. glomerátum** Thuill. [*C. viscosum* auth.] Erect annual, viscid, simple to freely branched, 1–3 dm. high; lvs. elliptic to narrow-obovate, obtusish, hairy, 1–2.5 cm. long;

bracts small, green; infl. a glomerate cyme, lax in age; pedicels scarcely or not longer than calyx; sepals ovate-lanceolate, sharply acute, with scarious margins, 3.5–4.5 mm. long; petals 2-cleft, ca. as long or slightly shorter than calyx; caps. slender, 5–9 mm. long; seeds muriculate.—Common in waste places, in pastures, along roads, etc., at low elevs. through most of cismontane s. Calif.; natur. from Eu. Feb.–May.

2. **C. vulgàtum** L. Short-lived perennial, matted and with leafy basal offshoots; flowering stems glandular-pubescent, 1–4 dm. high, ascending or decumbent; lvs. oblong or the lower oblong-spatulate, 1–2.5 cm. long; bracts similar but smaller, the cymes loose; pedicels 2–4 times as long as calyx; sepals 4–6 mm. long, lance-ovate, acute, scarious-margined; petals 4–8 mm. long, 2-lobed; caps. twice as long as calyx; seeds tubercled.—Mostly in lawns in Calif.; occasional in meadows; natur. from Eu. March–Aug.

5. *Herniària* L.

Annual or perennial herbs with much branched prostrate or spreading stems. Lvs. opposite to alternate, small, entire; stipules minute, scarious. Fls. minute, crowded in sessile axillary clusters. Calyx 4–5-parted, the segms. unequal. Petals 0. Stamens 2–5. Style minute, 2-cleft. Fr. a utricle, membranaceous, included. Seed 1. Ca. 20 spp., of Eurasia and N. Afr. (Latin, *hernia*, rupture, one of the spp. being a supposed cure.)

1. **H. cinèrea** DC. Forming mats 5–20 cm. across, annual, leafy, hispidulous throughout; lvs. oblong-oblanceolate, 4–10 mm. long; fls. in all the axils; 1 mm. long, hispidulous. —Sparingly and fleetingly natur. in s. Calif., more frequent in cent. Calif.; s. Eu. April–Sept.

6. *Loeflíngia* L.

Low spreading rigid glandular-pubescent annuals, branching from base. Lvs. inconspicuous, subulate; stipules setaceous. Fls. small, sessile, axillary, solitary or fascicled. Sepals 5, carinate, narrow, with rigid setaceous straight or recurved tips. Petals 3–5, and small, or none. Stamens 3–5. Style 1 or none; stigmas 3. Caps. 3-valved. Seeds several, oblong. Ca. 5 spp. of Eu., Asia, w. N. Am. (Named for Peter *Loefling*, Swedish naturalist of 18th century.)

Outer sepals straight, entire .... 1. *L. pusilla*
Outer sepals recurved and with setaceous tooth on each side .... 2. *L. squarrosa*

1. **L. pusílla** Curran. Much like the next sp., but the sepals not rigid or squarrose; entire; petals none; caps. equaling calyx.—A local sp. from semi-alkaline places, Tehachapi Mts. April–May. Intergrading with *L. squarrosa*.

2. **L. squarròsa** Nutt. Stems 3–14 cm. long, glandular-pubescent; lvs. cuspidate, 4–6 mm. long; calyx ca. 4 mm. long, the sepals squarrose, rigid, toothed on each side; petals very minute; caps. shorter than calyx; seeds many, semitranslucent, ca. 0.4 mm. long.— Locally frequent in sandy places, mostly below 2000 ft.; Coastal Sage Scrub, Chaparral, V. Grassland; largely in interior valleys from L. Calif. and San Diego Co. to Sacramento V.; Ariz. April–May.

7. *Polycárpon* L.

Low much-branched annuals. Cauline lvs. numerous, flat; stipules and bracts scarious. Fls. small, cymose, numerous. Sepals 5, ± keeled and scarious-margined. Petals 5, hyaline, entire or emarginate, smaller than sepals. Stamens 3–5. Style 1, short, 3-cleft. Caps. 3-valved. Seeds ovoid, several. Ca. 6 spp. of wide distribution. (Greek, *polus*, many, and *karpos*, fr., because of the many caps.)

Sepals ca. 1 mm. long; lvs. spatulate, opposite .... 1. *P. depressum*
Sepals almost 2 mm. long; lvs. obovate to oblong, often appearing to be in 4's . 2. *P. tetraphyllum*

1. **P. depréssum** Nutt. Prostrate, with many slender stems 1–5 cm. high, much-branched, glabrous; lvs. spatulate, 3–6 mm. long, with slender petioles; sepals 1 mm. long,

inconspicuously keeled; petals linear; caps. spherical.—Occasional in sandy open places below 2000 ft.; Coastal Sage Scrub, Chaparral; L. Calif. to Monterey Co. April–June.

2. **P. tetraphýllum** (L.) L. [*Mollugo t.* L.] Glabrous, diffuse or prostrate, the stems 4–10 cm. long; lvs. oblong to obovate, 5–10 mm. long, short-petioled; cymes leafless, many-fld.; stipules and bracts lance-acuminate; sepals obovate or oblong, strongly keeled, almost 2 mm. long; petals thin, oblanceolate; caps. ovoid, almost as long as calyx; seeds somewhat angled, almost transparent, ca. 0.5–0.8 mm. long.—Occasional in beaten soil of paths and along roads, in waste places, etc., cismontane; natur. from Eu. May–July.

8. *Sagìna* L. Pearlwort

Low annual or perennial herbs, tufted or matted. Lvs. filiform or subulate, scarious-connate at base. Fls. small, whitish, terminal on stems or branches. Sepals 4 or 5. Petals 4 or 5 or 0, undivided. Stamens as many as or twice as many as sepals. Styles as many as sepals and alternate with them. Caps. many-seeded, 4–5-valved to base, the valves opposite the sepals. Seeds many, smooth or resinous-dotted. Ca. 10 spp. of cool or temp. regions. (Latin, *sagina*, fattening, once applied to Spergula, used in Eu. for forage.)

Plants annual, without sterile basal rosettes; pedicels straight; stems filiform.
    Lf.-bases ciliolate; fls. apetalous, usually 4-parted .... 1. *S. apetala*
    Lf.-bases not ciliate; fls. with petals, 5-parted .... 2. *S. occidentalis*
Plants perennial, with sterile basal rosettes; pedicels often curved at tip; stems ± fleshy
    3. *S. saginoides*

1. **S. apétala** Ard. var. **barbàta** Fenzl. Minutely glandular-pubescent annual 2–5 cm. high, with 1–few filiform stems; lvs. linear-subulate, 3–7 mm. long, long-ciliolate near base; pedicels capillary 3–12 mm. long; sepals 4, 1.5 mm. long; petals usually 0; caps. ovoid, ca. 2 mm. long.—Infrequent inconspicuous weed, apparently in Tehachapi region and perhaps also farther south; natur. from Eu. April–June.

2. **S. occidentàlis** Wats. Minute inconspicuous annual, 2–10 cm. high; lvs. filiform or the upper subulate, 6–10 mm. long, pedicels 6–15 mm. long; calyx hispidulous-glandular, 1.5 mm. long; petals 5, almost as long; caps. 2–2.5 mm. long; seeds smoothish, 0.3 mm. long.—Brushy and wooded places below 8000 ft.; many Plant Communities; San Diego Co. n. to B.C. Santa Cruz Id. March–June.

3. **S. saginoìdes** (L.) Karst. var. **hespéria** Fern. [*S. linnaei* auth.] Low matted or tufted perennial, glabrous, the stems 2–9 cm. long, numerous; lvs. thickish, linear, 5–10 mm. long; pedicels 5–12 mm. long, often curved at tip; sepals oblong-linear, obtuse, 1.5–2 mm. long; petals ca. 1 mm. long; caps. 2.5–3 mm. long; seeds papillose.—Frequent on moist banks, 4000–12,000 ft.; Yellow Pine F. to Alpine Fell-fields; mts. from Palomar Mts. n.; to B.C. and Rocky Mts. May–Sept.

9. *Saponària* L. Bouncing Bet

Annuals or perennials, erect or diffuse. Stems leafy. Lvs. mostly broad. Calyx-tube ovoid to cylindric, obscurely 15–25-nerved. Petals 5, long-clawed, entire or emarginate, crowned with an appendage or scale at base of blade. Stamens 10. Styles 2. Caps. 4-toothed. Seeds flat, reniform. Ca. 20 spp. of Eurasia and N. Afr. (Latin, *sapo*, soap, some spp. having saponaceous juice.)

1. **S. officinàlis** L. Stout perennial 4–7 dm. tall, forming quite large patches; lvs. oval-lanceolate, 5–7 cm. long; fls. pink, commonly double; calyx 15–22 mm. long, the tube cylindric, the lobes lanceolate, 2 mm. long; petals 3.5–4 cm. long; seeds black, 1.6–1.8 mm. in diam., pitted; $2n=28$ (Favarger, 1946).—Persisting from old gardens and occasionally natur., especially in cool damp places; from Eu. June–Sept.

10. *Scleránthus* L. KNAWEL

Low annual or perennial herbs, glabrous or pubescent, the stems rigid, forking. Lvs. connate at base, subulate, pungent. Stipules none. Fls. small, greenish, clustered in axils or terminal and cymose. Calyx 4–5-toothed or -lobed. Petals 0. Stamens 1–10. Styles 2, distinct. Utricle included; seed 1, lenticular. Ca. 10 spp. of Old World. (Greek, *scleros*, hard, and *anthos*, fl., from the hardened calyx-tube.)

    1. **S. ánnuus** L. Much-branched, spreading, somewhat pubescent, 2–10 cm. high; lvs. 5–8 mm. long; fls. sessile in the forks, 3–4 mm. long; calyx-lobes scarcely margined; $2n=22$ (Rohweder, 1939; Ehrenberg, 1945).—Sparingly natur. in dry places, below 5000 ft.; S. Oak Wd.; in region of Pala, San Diego Co.; to n. Calif.; native of Eu. March–June.

11. *Scopulóphila* Jones

Low perennial, arising from a dense woody root-crown. Stems glabrous, erect, few-branched. Lvs. oblance-linear; stipules scarious, lacerate. Fls. 1–3, sessile in axillary clusters. Sepals 5, united at very base, hyaline with cent. elongate green spot. Petals 0. Stamens 10, the 5 fertile ones alternating with 5 longer lanceolate staminodia. Style 3-cleft. Seed 1. Monotypic. (Greek, *skopla*, a high place, and *philos*, fond of.)

    1. **S. rixfórdii** (Bdg.) M. & J. [*Achyronychia r.* Bdg. *S. nitrophiloides* Jones.] Plate 34, Fig. E. Root-crown woolly; plant pallid; stems several, 6–18 cm. high; lvs. 5–15 mm. long; calyx-tube scarcely 0.5 mm. long, the lobes ca. 1.5 mm. long; staminodia ca. the same.—Dry rocky places, probably only on limestone, 3600–7500 ft.; Creosote Bush Scrub, Joshua Tree Wd.; Death V. region to Owens V.; sw. Nev. April–July.

12. *Siléne* L. CATCHFLY. CAMPION

Annual or perennial herbs, usually somewhat viscid. Lvs. opposite, exstipulate. Fls. solitary or more often cymose. Calyx cylindrical to ovoid or campanulate, 5-toothed, 10–many-nerved. Petals 5, clawed, usually with a scalelike appendage at the base of the blade which is usually cleft or toothed. Stamens 10. Styles 3, rarely 4. Ovary on a well developed stipe which with connate fil.-bases and petal-bases forms a carpophore. Caps. 1- or incompletely 2–4-locular, opening apically by 3 or 6 teeth. Seeds reniform to globose, striately muricate to tuberculate. Perhaps 250 spp., chiefly from temp. and cool regions of N. Hemis. (Name said to have come from *Silenus*, foster-father of Bacchus, supposedly covered with foam, many spp. having a viscid secretion.)

(Hitchcock, C. L. and B. Maguire. A revision of the N. Am. spp. of Silene. Univ. Wash. Pub. Biol. 13: 1–73. 1947.)

1. Annuals, mostly introd. weedy plants.
  2. Calyx with 20–30 well developed ribs.
    3. The calyx 8–12 mm. long; petals lacking appendages . . . . 10. *S. multinervia*
    3. The calyx 20–30 mm. long; petals with well developed appendages . . . . . . 3. *S. conoidea*
  2. Calyx usually with 10 ribs, if more, the ribs indistinct.
    4. Plant subglabrous or minutely puberulent, the upper internodes usually with sticky bands
                                                                                                        1. *S. antirrhina*
    4. Plant densely pubescent.
      5. Petals not deeply bilobed; appendages linear . . . . 6. *S. gallica*
      5. Petals deeply bilobed; appendages short, truncate . . . . 5. *S. dichotoma*
1. Perennials; native except *S. cucubalus*.
  6. Corolla bright red; lvs. largely lanceolate . . . . 7. *S. laciniata*
  6. Corolla white to pink or purplish or yellow.
    7. Petal-blades bilobed half their length; appendages lacking; carpophore 2–3 mm. long; calyx much inflated in fr. and ± papery. Introd. weed . . . . 4. *S. cucubalus*
    7. Petals, appendages and calyces not as above.
      8. Petals yellowish, scarcely exserted, 25–35 mm. long; calyx 15–30 mm. long
                                                                                                                    11. *S. parishii*

8. Petals shorter, not yellow.
  9. Corolla mostly less than 10 mm. long; calyx 5–8 mm. long; lower lvs. 2–8 mm. long
     .... 9. *S. menziesii*
  9. Corolla usually more than 12 mm. long; calyx 8–18 mm. long; lower lvs. longer.
    10. Fls. nodding at anthesis; stamens and style long-exserted .... 8. *S. lemmonii*
    10. Fls. erect or nearly so; stamens included or little exserted.
      11. Basal lvs. usually less than 2 cm. long and 1–4 mm. wide; plants tufted, usually not more than 2 dm. high .... 2. *S. bernardina*
      11. Basal lvs. mostly more than 3 cm. long, often more than 4 mm. wide; plants mostly more than 3 dm. tall.
        12. Blades of petals with 4 subequal linear lobes; calyx 13–16 mm. long. Inyo-White Mts. n. .... 2. *S. bernardina*
        12. Blades of petals bilobed, the appendages 2; calyx 10–12 mm. long. Widely distributed .... 12. *S. verecunda*

1. **S. antirrhìna** L. Mostly erect annual, simple or usually branched, 2–8 dm. tall, mostly retrorse-pubescent below, glabrous above, usually with glutinous bands on some of the internodes; basal lvs. oblanceolate to spatulate, the cauline oblanceolate to linear, 3–6 cm. long, 2–12 mm. wide; pedicels 1–4 cm. long; calyx 10-nerved, glabrous, somewhat contracted at orifice, 4–8 mm. long; petals white to pink, equaling or slightly exceeding calyx, the blades 2-lobed, appendages reduced; caps. 4–8 mm. long; seeds with 3–4 dorsal rows of papillae.—Common weed in open sandy places, on burns, etc., mostly below 5000 ft.; cismontane and desert s. Calif. to B.C. and Atlantic Coast. April–Aug. Variable, with some named forms.

2. **S. bernardína** Wats. [*S. montana* Wats. ssp. *b*. Hitchc. & Maguire.] Perennial from long woody taproot; stems many, 1.5–4.5 dm. long, glandular to base or nearly so; lvs. linear-oblanceolate to linear-lanceolate, 1.5–2 (–3) mm. broad, the cauline 2–4 pairs; infl. mostly several-fld.; pedicels 1–2 cm. long, glandular; calyx tubular, 13–16 mm. long, the ribs prominent; petals white, with some pink or purple, the blades 4–6 mm. long, cleft into 4 linear subequal lobes, the appendages 1–2.5 mm. long; carpophore 2–5 mm. long; caps. exceeding calyx; seeds with low radially elongate tubercles.—Pinyon-Juniper Wd. to Alpine Fell-fields; Inyo White Mts.; e. slope of Sierra Nevada. June–Aug.

3. **S. conoìdea** L. Erect annual 5–8 dm. high; lvs. 5–12 cm. long, 8–12 mm. wide; pedicels 1–3 cm. long, glandular and puberulent; calyx becoming conic-ovoid, 2–3 cm. long, with ca. 30 nerves; petals white to reddish, conspicuously exceeding the calyx.—Occasional weed; native of Eurasia.

4. **S. cucùbalus** Wibel. [*S. inflata* Sm. *S. latifolia* Britt. & Rend.] BLADDER CAMPION. Robust perennial from heavy creeping rootstocks; stems 2–8 dm. tall, glabrous; lvs. 3–8 cm. long, 1–3 cm. wide; pedicels 1–3 cm. long; calyx becoming much inflated, 1–2 cm. long, pale green to purplish, the nerves reticulate; petals white, the blades 3.5–6 mm. long, deeply cleft, the appendages reduced; caps. ovoid-globose on a carpophore 2–3 mm. long; seeds papillose; $2n=24$ (Blackburn, 1928).—Occasional weed; introd. from Eu. Summer, fall.

5. **S. dichótoma** Ehrh. Simple or branched annual 3–8 dm. high, strikingly hirsute; lvs. 3–8 cm. long, 0.5–3.5 cm. wide; pedicels 1–3 mm. long; calyx narrow-tubular, 10–15 mm. long; corolla white to reddish, the blades 5–9 mm. long, bilobed, with short truncate appendages; carpophore 2–4 mm. long.—Rare as a weed as at Ojai, Ventura Co.; from Eu.

6. **S. gállica** L. [*S. anglica* L.] Annual, usually erect, simple to branched, 1–4 dm. high, hirsute and strigulose, glandular-pubescent above; basal lvs. oblanceolate to spatulate, obtuse, the cauline somewhat narrower, 1–3.5 cm. long; infl. leafy-bracted, 1-sided, the pedicels 2–5 mm. long; calyx 10-nerved, 6–9 mm. long, inflated in age; petals whitish to pinkish, slightly exceeding calyx, blades entire or toothed, the linear appendages 1 mm. long; carpophore 1 mm. long; caps. 6–8 mm. long; seeds finely rugulose.—Common weed on vacant lots, in fields and waste places below Yellow Pine F.; cismontane and Channel Ids.; to B.C. and east; from Eu. Feb.–June.

7. **S. laciniàta** Cav. ssp. **màjor** Hitchc. & Maguire. Perennial from a deep fleshy

taproot, the stems weak, 2–7 dm. long, retrorsely puberulent and glandular-pubescent; lvs. narrowly linear to lance-oblong, mostly 5–10 cm. long, 2–14 mm. wide; infl. open, mostly 5–9-fld.; pedicels 2–4 cm. long, glandular-pubescent; calyx 15–20 mm. long, tubular-cylindrical; petals scarlet, conspicuous, the blades 8–15 mm. long, deeply cleft into 4 linear to lanceolate lobes, the 2 appendages broad, toothed 1–1.5 mm. long; carpophore 2–4 mm. long; caps. oblong-ovoid, 14–18 mm. long; seeds with prominent marginal papillae; $n=48$ (Kruckeberg, 1954).—Frequent on grassy or brushy slopes below 5000 ft.; Coastal Strand, Coastal Sage Scrub, Chaparral; L Calif. to cent. Calif. May–July. Two vars. in this ssp. have been proposed: *angustifolia* Hitchc. & Maguire for plants from near the coast, with lvs. linear to linear-lanceolate and calyces 18–20 mm. long; and *latifolia* Hitchc. & Maguire from the interior with lanceolate lvs. and calyces 16–18 mm. long.

8 **S. lemmònii** Wats. Plate 34, Fig. F. Perennial from a multicipital caudex, the stems slender, 1.5–4.5 dm. long, ± retrorsely fine-pubescent, glandular toward top; lvs. lance-elliptic to oblanceolate, 2–3 cm. long, 5–10 mm. wide, the cauline mostly 2 pairs; infl. open; pedicels 1–2 cm. long, slender; calyx narrow-campanulate, 8–10 mm. long, with prominent green ribs; petals yellowish-white to pinkish, the blades 4.5–8 mm. long, 2-lobed, the 2 appendages 0.5–1 mm. long; carpophore 2–2.5 mm. long; seeds papillose.— Open woods, 3000–8000 ft.; largely Montane Coniferous F.; Cuyamaca Mts. to Sierra Nevada and s. Ore. June–Aug.

9. **S. menzièsii** Hook ssp. **dórrii** (Kell.) Hitchc. & Maguire. [*S. d.* Kell.] Perennial from slender rootstocks; stems 0.5–2 dm. long, usually decumbent, short-pubescent below, the hairs 1–3-celled, gland-tipped; lvs. many, 2–6 cm. long, 0.3–1.8 cm. wide; infl. leafy; calyx tubular-campanulate, 5–8 mm. long; petals white, the blades 1.5–3 mm. long, cleft or lobed, the 2 appendages entire, very short; carpophore ca. 1.5 mm. long; seeds almost smooth, the sides obscurely reticulate.—At 6000–8000 ft.; Montane Coniferous F.; San Bernardino Mts.; White Mts., Sierra Nevada; w. Nev. June–July.

10. **S. multinérvia** Wats. Erect annual, simple or branched, 2–6 dm. high, villous and rather viscid; lvs. lance-linear, or the lowest oblanceolate, 2–8 cm. long; infl. ± open; pedicels 1–4 cm. long, glandular; calyx ovoid, 8–12 mm. long, ca. 25-nerved; petals white to pinkish, ca. as long as calyx, the blades 2-cleft, 1–3 mm. long, without appendages; caps. ovoid, 7–9 mm. long; seeds with 3 dorsal rows of raised papillae.—Mostly in disturbed places such as burns, below 3000 ft.; Chaparral, etc.; San Diego Co. to cent. Calif. Santa Cruz Id. April–May.

11. **S. párishii** Wats. var. **párishii.** Many-stemmed perennial with woody base; stems 2–4 dm. tall, glandular-pubescent, or eglandular below, the hairs mostly 1–3-celled; lvs. subelliptic to lanceolate, mostly 3–6 cm. long, the cauline 5–9 pairs and mostly less than 8 mm. wide; infl. 5–15-fld., rather compact; calyx tubular, 25–30 mm. long, very glandular, the lobes $\frac{1}{2}$–$\frac{1}{3}$ the length of the tube; petals whitish yellow, the blades 7–8 mm. long, deeply cleft into several linear lobes, the appendages oblong, ca. 2 mm. long; carpophore ca. 3 mm. long; seeds with tessellate margins; $n=24$ (Kruckeberg, 1954).—Dry rocky or gravelly places, 6000–11,000 ft.; Yellow Pine F. to Subalpine F.; San Bernardino Mts. July–Aug.

Var. **latifòlia** Hitchc. & Maguire. Plants mostly less than 2 dm. high, strigose and eglandular below; lvs. oblong-elliptic to obovate-oblanceolate, 2–3.5 cm. long, 0.8–1.5 cm. wide.—At 6800–9000 ft.; San Gabriel Mts. July–Aug.

Var. **víscida** Hitchc. & Maguire. Glandular-pubescent throughout, the hairs several-celled; larger cauline lvs. mostly over 1 cm. wide.—At 7000–10,500 ft.; San Jacinto and Santa Rosa mts. July–Aug.

12. **S. verecúnda** Wats. ssp. **andersònii** (Clokey) Hitchc. & Maguire. [*S. a.* Clokey.] Several-stemmed perennial 1–3 dm. high; lower stems scabrous-puberulent; basal lvs. linear-oblanceolate to obovate, 2–6 cm. long, 2–10 mm. broad, the cauline lanceolate to oblanceolate; infl. elongate, short-glandular-pubescent; calyx also glandular, tubular-campanulate, 10–12 mm. long; petals greenish white, the blades 4–6 mm. long, bilobed, the appendages usually less than 1 mm. long; seeds with concentric rows of papillae.—

Open slopes, 5000–9000 ft.; Sagebrush Scrub, Pinyon-Juniper Wd.; desert slopes from e. San Bernardino Co. to Mono Co.; to Nev., Utah. June–July.

Ssp. **platyòta** (Wats.) Hitchc. & Maguire. [*S. p.* Wats.] Lvs. 5–9 cm. long, 2–6 mm. wide; calyx glandular, usually greenish; petals white to greenish or rose, the appendages 1–2 mm. long; $n=24$ (Kruckeberg, 1954).—Dry benches and slopes, 5000–11,000 ft.; Yellow Pine F. to Subalpine F.; mts. of s. Calif. to cent. Calif.; L. Calif. June–Aug. Eglandular plants have been designated as var. *eglandulosa* Hitchc. and Maguire.

13. *Spérgula* L. SPURREY

Annuals, with narrowly linear, apparently whorled lvs. because of crowding in axils. Stipules small, scarious. Fls. in terminal cymose panicles. Sepals 5. Petals 5, white, entire. Stamens 5 or 10. Styles 5. Caps. 5-valved, the valves opposite the sepals. Seeds compressed, acutely margined. Two or 3 spp. of the Old World. (Latin, *spargere*, to scatter, because of sowing seeds to produce forage.)

1. **S. arvénsis** L. Erect, branched at base, 1–4 dm. high, glabrous or glandular-pubescent; lvs. linear, 1–3 cm. long; pedicels shorter, often reflexed in fr.; sepals 4–5 mm. long; petals as long or slightly longer; caps. ovoid, slightly exceeding calyx; seeds dark, round, plump, ca. 1 mm. in diam., minutely papillose, with very narrow wing-margin; $2n=18$ (Rohweder, 1939).—Occasionally natur. in vacant lots and in fields of cismontane s. Calif., especially toward the coast; native of Eu. Most months.

14. *Spergulària* J. & C. Presl. SAND-SPURREY

Low annual to perennial herbs, usually of saline areas. Lvs. narrowly linear or subfiliform, often fleshy, semiterete, with scarious stipules. Fls. whitish to pink, in terminal racemose cymes. Sepals 5. Petals 5, entire. Stamens 2–10. Styles 3. Caps.-valves 3, rarely 5. Seeds compressed to reniform-globose, smooth or roughened, often wing-margined. Ca. 40 spp., widely distributed. (Name derived from *Spergula*.)

(Rossbach, R. B. Spergularia in N. and S. Am. Rhodora 42: 57–83, 105–143, 158 193, 203–213 1940.)

1. Lvs. densely fascicled; stipules lance-acuminate; plants perennial (sometimes annual in *S. rubra*).
2. Sepals mostly 5–10 mm. long; seeds 0.7–0.9 mm. long; styles 0.6–3 mm. long
                                                        3. *S. macrotheca*
2. Sepals 3–5 mm. long; seeds 0.4–0.6 mm. long; styles 0.2–0.8 mm. long.
3. Stem below the infl. mostly glabrous; caps. equaling calyx; seeds reticulate, not winged
                                                         6. *S. rubra*
3. Stems below the infl. mostly glandular-pubescent; caps. longer than calyx; seeds papillate, usually winged .... 7. *S. villosa*
1. Lvs. not densely fascicled; stipules mostly deltoid, not long-acuminate; plants annual.
4. Seeds black, 0.6–0.8 mm. long, rounded, often iridescent .... 1. *S. atrosperma*
4. Seeds brown.
5. Stamens 6–10 .... 2. *S. bocconii*
5. Stamens 2–5.
6. Sepals 1.6–5 mm. long; cymes not much compound, leafy; seeds ca. 0.6–1.4 mm. long
                                                         4. *S. marina*
6. Sepals 0.8–1.6 mm. long; cymes much compounded, bracteate; seeds ca. 0.4 mm. long
                                                         5. *S. platensis*

1. **S. atrospérma** R. P. Rossb. Few-stemmed annual, 5–18 cm. high, glabrous or glandular-villous; lvs. fleshy, not fascicled, 1–2.5 cm. long; stipules broadly triangular, 2–2.8 mm. long; sepals ovate-lanceolate, 3–4 mm. long; petals ovate, white to pink, 2–2.6 mm. long; stamens 4–8; styles 0.5–0.8 mm. long; caps. 3–5 mm. long; seeds black, finely areolate, 0.6–0.8 mm. long, iridescent, not winged.—Alkaline places; Alkali Sink; w. Riverside Co.; Tulare Co. to Sierra Co.; w. Nev. April–May.

2. **S. boccònii** (Scheele) Foucaud. [*Alsine b.* Scheele.] Annual, 0.5–3 dm. high, glab-

Plate 35. Fig. A, *Spergularia macrotheca*. Fig. B, *Stellaria longipes*. Fig. C, *Stellaria media*. Fig. D, *Euonymus occidentalis* var. *parishii*. Fig. E, *Ceratophyllum demersum*. Fig. F, *Aphanisma blitoides*.

rous below, glandular-pubescent in infl.; lvs. not or but slightly fascicled, 1–2 cm. long; stipules deltoid, 2–4 mm. long, scarcely acuminate; bracts foliaceous; sepals ovate, glandular-pubescent, 2.5–5.4 mm. long; petals white to pink, shorter than calyx; stamens 6–10; styles 0.4–0.6 mm. long; caps. 3–5 mm. long; seeds plump with broad swollen rim, minutely papillate, not winged, 0.4–0.5 mm. long.—Along paths, beach dunes and alkaline places, cismontane; natur. from Eu. Santa Catalina, San Clemente and Santa Cruz ids. April–Sept.

3. **S. macrothèca** (Hornem.) Heynh. var. **macrothèca**. [*Arenaria m.* Hornem. *Tissa m.* var. *scariosa* Britton.] Plate 35, Fig. A. Perennial from a heavy branched caudex and stout fleshy root; stems 1–many, prostrate to ascending, stout, 1–4 dm. long, glabrous or glandular-pubescent, branched; lvs. linear, mostly fascicled, fleshy, 1–3.5 cm. long; stipules conspicuous, triangular-acuminate, 5–10 mm. long; infl. glandular-pubescent, lax or dense; sepals broadly lanceolate, 5–10 mm. long; petals mostly pink, ovate, 4–7 mm. long; stamens 10; styles 0.6–1.2 mm. long; caps. usually 6–8 mm. long; seeds mostly smoothish, 0.6–1 mm. long, usually with a narrow wing.—Near salt marshes along the coast and on sea bluffs, occasional inland; Coastal Strand, Coastal Sage Scrub, etc.; L. Calif. to B.C. Most of the year. Channel Ids.

Var. **leucántha** (Greene) Rob. [*Tissa l.* Greene. *T. l.* var. *glabra* A. Davids.] Sepals usually 5–6 mm. long; petals usually white; styles 1.2–1.8 mm. long.—Alkaline places; Alkali Sink, Freshwater Marsh; interior cismontane valleys and w. Mojave Desert, from San Diego Co. to cent. Calif.

4. **S. marìna** (L.) Griseb. [*Arenaria rubra* var. *m.* L. *Tissa m.* Britton.] More or less diffuse annual; stems 0.5–3 dm. long, fleshy, ± glandular-pubescent; lvs. scarcely fascicled, linear, 2–4 cm. long; stipules deltoid, 2–4 mm. long; infl. lax, usually glandular-pubescent; sepals ovate, 2.5–5 mm. long, blunt; petals white to pink, 2–4 mm. long; stamens 2–5; styles 0.4–0.6 mm. long; caps. 3.6–6 mm. long; seeds smooth or glandular-papillose, 0.6–0.8 mm. long, usually not winged; $2n=36$ (Löve & Löve, 1942).—Common along seashore and in alkaline places of interior and occasional on deserts; Coastal Strand, Alkali Sink, near Freshwater Marsh, etc.; to Wash., Atlantic Coast; Eurasia. March–Sept.

5. **S. platénsis** (Camb.) Fenzl. [*Balardia p.* Camb. in St. Hil.] Glabrous annual, the stems slender, 0.5–3 dm. long; lvs. linear-filiform, 1–3 cm. long; stipules deltoid, subacuminate, 1.5–2.5 mm. long; infl. much compounded, glabrous; sepals broadly lanceolate, 1–1.6 mm. long; petals white, narrowly ovate, 0.6–1 mm. long; stamens 5; styles 0.3–0.4 mm. long; caps. 1.5–2.5 mm. long; seeds tuberculate, not winged, 0.3–0.4 mm. long.—Drying mud flats of vernal pools; Chaparral, Coastal Sage Scrub; San Diego Co. to Riverside and Los Angeles cos.; to Tex., S. Am. April–June.

6. **S. rùbra** (L.) J. & C. Presl. [*Arenaria r.* L. *Tissa r.* Britton.] Annual or short-lived perennial, much branched at crown, ± matted; stems 0.6–3 dm. long, slender, glabrous below, glandular-pubescent toward infl.; lvs. mostly 6–12 mm. long, fascicled; stipules triangular-lanceolate, 3–5 mm. long; cymes many-fld., leafy; sepals lanceolate, 4–5 mm. long; petals pink, ovate, 2.5–3.8 mm. long; stamens 6–10; styles 0.6–0.8 mm. long; caps. 3.5–5 mm. long; seeds minutely papillate, 0.4–0.6 mm. long; $2n=36$ (Löve & Löve, 1942).—Rare in s. Calif.; more abundant weed n.; natur. from Eu.

7. **S. villòsa** (Pers.) Camb. [*Spergula v.* Pers. *Tissa clevelandii* Greene.] Perennial from a rather heavy taproot; stems usually viscid-glandular, 1–3 dm. long; lvs. fascicled, narrow, 1–4 cm. long; stipules broadly lanceolate, acuminate, 3–8 mm. long; cyme many-fld., lax, glandular-pubescent; sepals linear-lanceolate, 3–5 mm. long; petals white, 2.6–5 mm. long; stamens 7–10; styles 0.4–0.6 mm. long; caps. mostly 5–6 mm. long; seeds pyriform, black-papillate to smooth, winged or sometimes not, 0.4–0.6 mm. long.—Generally in sandy places and near the coast; near Coastal Salt Marsh and Coastal Strand; occasional in interior valleys, Coastal Scrub; L. Calif. to Ore.; introd. from S. Am. April–July.

15. Stellària L. CHICKWEED. STARWORT

Low diffuse annuals or perennials. Fls. white, cymose or solitary, terminal or seemingly lateral. Sepals 4–5. Petals 4–5, deeply 2-cleft, rarely 0. Stamens 8, 10 or fewer. Styles 3, rarely 4, opposite as many sepals. Caps. ovoid to globose, 1-celled, dehiscent by twice as many valves as there are styles. Seeds several to many, smooth or rough. (Latin, *stella*, a star, because of the star-shaped fls.)

Plants annual.
    Internodes with longitudinal line of hair; lvs. ovate .... 5. *S. media*
    Internodes lacking line of hair; upper lvs. lance-linear .... 6. *S. nitens*
Plants perennial.
    The plants glandular-puberulent; lvs. narrow-lanceolate, 5–10 cm. long ..... 3. *S. jamesiana*
    The plants not glandular.
        Petals equal to or longer than the sepals.
            Fls. many, the fruiting pedicels divergent; caps. pale .... 2. *S. graminea*
            Fls. solitary to few, the fruiting pedicels suberect; caps. dark .... 4. *S. longipes*
        Petals much shorter than the sepals or wanting .... 1. *S. crispa*

1. **S. críspa** Cham. & Schlecht. [*Alsine c.* Holz.] Perennial with slender leafy rootstocks; stems glabrous, weak, ± simple, 1–4 dm. long; lvs. ovate, subsessile, thin, 0.8–2 cm. long, usually crisped on margins; pedicels axillary, 6–20 mm. long; sepals 3–4 mm. long, lanceolate, scarious-margined; petals shorter than sepals or 0; caps. ovoid, pale, exceeding calyx; seeds ca. 0.5 mm. long, slightly reticulate.—Moist banks and meadows, 7000–9200 ft.; San Jacinto and San Bernardino mts.; largely Red Fir F., Lodgepole F.; Sierra Nevada; to Alaska. May–Aug.

2. **S. gramínea** L. [*Alsine g.* Britton.] Glabrous, weak, ascending from creeping rootstocks; stems 1.5–3 dm. long, 4-angled; lvs. sessile, lanceolate, 2–2.5 cm. long; cymes diffuse, terminal; pedicels slender, spreading; sepals lanceolate, 3-nerved, 3.5–5 mm. long; petals ca. as long, 2-cleft; caps. pale brown, oblong-ovoid; seeds finely roughened. —Occasional seed in lawns; native of Eu. May–July.

3. **S. jamesiàna** Torr. [*Alsine j.* Heller.] Glandular-pubescent perennial, diffusely branched; stems 1–3.5 dm. long, erect or ascending, from slender rootstocks which may have thickened tuberlike enlargements; lvs. sessile, lanceolate to lance-ovate, 4–10 cm. long; cymes loose, terminal and axillary; pedicels 5–20 mm. long; sepals 3–5 mm. long; petals broadly notched, 6–10 mm. long; caps. broadly ovoid, shorter than the calyx; seeds muriculate, ca. 2 mm. long.—At ca. 7000–8000 ft.; Montane Coniferous F.; San Bernardino Mts., Frazier Mt., Tehachapi Mts.; n. to Wash., Rocky Mts. May–July.

4. **S. lóngipes** Goldie. [*Alsine l.* Cov.] Plate 35, Fig. B. More or less tufted perennial from creeping rootstocks; 1–2.5 dm. high, erect or ascending, rather simple; lvs. lance-linear, 1–2.5 cm. long, rigid; fls. solitary or few, terminal on suberect pedicels; sepals lance-oblong, 3–5 mm. long; petals cleft, as long as sepals; caps. narrow-ovoid, dark, exceeding calyx; seeds nearly smooth, ca. 0.8 mm. long; $n=26$ (Bøcher & Larsen, 1950). —Frequent in moist places, 5000–9200 ft.; Montane Coniferous F.; San Bernardino Mts.; Sierra Nevada to Alaska, Atlantic Coast. May–Aug.

5. **S. mèdia** (L.) Vill. [*Alsine m.* L.] COMMON CHICKWEED. Plate 35, Fig. C. Annual with weak procumbent stems 1–4 dm. long and with a line of hair running down each internode; lvs. ovate, acute, 1–3 cm. long, short-petioled or the upper sessile; cymes leafy; sepals pubescent, ovate, 4–5 mm. long; petals white, 2-parted or wanting; caps. ovoid, slightly exceeding calyx; seeds minute, roughened; $2n=28$ (Pal, 1952), 40 (Negodi, 1935), 42, 44 (Peterson, 1936).—Common weed in shaded places, such as in orchards, etc., through most of cismontane Calif.; native of Eurasia. Feb.–Sept.

6. **S. nìtens** Nutt. [*Alsine n.* Greene.] Erect forked annual with filiform stems 1–2 dm. high, glabrous or slightly pubescent at base; lvs. near base linear to lance-linear, or even wider, 5–10 mm. long; pedicels erect, mostly 5–20 mm. long; sepals very acute, scarious-margined, subulate-lanceolate, 3–4 mm. long; petals half as long or wanting; caps. oblong, almost equaling calyx; seeds angled, ca. 0.5 mm. long, somewhat roughened.—Common

but inconspicuous in grassy places, below 5500 ft.; V. Grassland to Yellow Pine F.; most of cismontane Calif.; to B.C., Rocky Mts. March–June.

16. Vaccària Medic. Cow-Herb. Cockle

Glabrous annuals with erect forked stems. Lvs. clasping, ovate to lanceolate. Cymes open, terminal, with slender pedicels. Calyx tubular at anthesis, inflated, ovoid and 5-angled in fr. Petals red or pink, surpassing calyx, without appendages. Stamens 10. Styles 2. Caps. 4-toothed. Seeds rounded, laterally attached. Ca. 3 spp. of Eurasia. (Latin, *vacca*, cow, because of use as fodder.)

1. **V. pyramidàta** Medic. [*V. segetalis* Garcke. *Saponaria v.* L.] Plants 3–10 dm. high; lvs. 3–7 cm. long, connate at base; calyx 10–12 mm. long, the angles green, terminal lobes 2–3 mm. long; petals reddish, crenulate, exserted 5–10 mm.; seeds blackish, 2 mm. broad and long; $2n=24$ (Löve, 1942).—Largely a weed of grainfields, below 5000 ft.; adv. from Eu. May–Aug.

27. Celastràceae. Staff-Tree Family

Trees, shrubs or woody vines. Lvs. alternate or opposite, simple, deciduous or persistent. Stipules small and caducous, or none. Fls. small, regular, mostly bisexual, commonly on jointed pedicels. Sepals 4–6, imbricate, persistent. Petals 4–6, spreading. Disk broad, flat or lobed, fleshy. Stamens usually as many as the petals, sometimes more, inserted under or on the disk. Ovary sessile, the base free from or adherent to the disk, 3–5-loculed. Style 1, short; stigma entire or 3–5-lobed. Ovules mostly 2 in a locule. Fr. a caps., loculicidal or indehiscent. Seeds usually with an aril; embryo large. Ca. 500 spp. of temp. and trop. areas.

Lvs. opposite; plants 2–6 m. high; lvs. 3–9 cm. long .... 1. *Euonymus*
Lvs. alternate, smaller; plants lower.
  Twigs brownish or olive; lvs. spatulate to oblanceolate or elliptic; stamens often twice as many as petals .... 2. *Forsellesia*
  Twigs yellowish, hispidulous; lvs. very thick, elliptic; stamens as many as petals . 3. *Mortonia*

1. Euónymus L. Burning Bush

Shrubs with opposite petioled lvs.; evergreen or deciduous. Stipules minute or 0. Fls. greenish or purplish, small, in axillary few-fld. cymes, mostly 5- or 4-merous. Sepals spreading or recurved. Stamens inserted on the broad disk. Ovary 3–5-loculed, short; stigma 3–5-lobed. Caps. 3–5-loculed and -lobed, or rounded. Seeds enveloped in a red aril. Ca. 175 spp., of N. Temp. regions, mostly Eurasian. (Greek, *eu*, good, and *onoma*, a name.)

1. **E. occidentàlis** Nutt. ex Torr. var. **párishii** (Trel.) Jeps. [*E. p.* Trel.] Plate 35, Fig. D. Shrub 1–2 m. tall, with slender whitish branchlets; lvs. thin, glabrous, ovate, serrulate, 3–9 cm. long, usually obtusish; petioles 5–15 mm. long; peduncles slender, the cymes 3–5-fld.; petals 5, rounded, brown-purple, finely dotted, scarious at edges, 3–4 mm. long; caps. depressed, smooth, deeply 3-lobed; seeds ca. 6 mm. long.—Occasional in canyons, 5000–6500 ft.; Yellow Pine F.; San Jacinto Mts. to Cuyamaca and Palomar mts. May–June.

2. Forsellèsia Greene

Small deciduous intricately branched shrubs with slender greenish angled spinescent branches having decurrent lines from the nodes. Lvs. small, simple, entire, alternate, usually with minute stipules. Fls. usually solitary in axils, minute, mostly 5-merous. Petals white, narrow-oblanceolate, deciduous, inserted under the edge of the crenately lobed disk. Stamens equal or unequal, 4–10. Carpels 1–3, distinct, ovoid, sessile, attenuate to the stigma. Ovary superior, 1-loculed, 1–2-ovuled. Fr. a coriaceous follicle, striate, opening along the ventral suture. (Named for J. H. *Forselles*, 19th century Swedish botanical writer.)

(Ensign, M. A revision of the Celastraceous genus Forsellesia (Glossopetalon). Am. Midl. Nat. 27: 501–511. 1942.)

Low matted shrubs, not spiny, 5–20 cm. high; lvs. elliptical, spine-tipped; fls. terminal, 5-merous
2. *F. pungens*
Taller intricately branched shrubs, ± spiny, 2–30 dm. high; lvs. oblong to oblanceolate, acute to acuminate; fls. axillary, 4–5-merous.
Young branches ca. 1 mm. in diam.; petals 4–7 mm. long. Death V. to Mojave Desert
1. *F. nevadensis*
Young branches ca. 0.5 mm. in diam.; petals 6–9 mm. long. White Mts. ...... 3. *F. stipulifera*

1. **F. nevadénsis** (Gray) Greene. [*Glossopetalon n.* Gray.] Divaricately branched, the stems ribbed, pubescent; young branches more than 1 mm. thick, yellowish in age; lvs. oblong to oblanceolate, 5–12 mm. long, pubescent; petioles 1 mm. long; stipules subulate, less than 1 mm. long, adnate to a persistent often thickened glandular base; fls. axillary, 4–5-merous; pedicels 3–5 mm. long; with several bracts; sepals entire, 1–3 mm. long; petals 4–7 mm. long; stamens 6–10, unequal; carpels 1–2 mm., to 5 mm. in fr.—At 3500–7000 ft., often on limestone; Joshua Tree Wd., Pinyon-Juniper Wd.; Mojave Desert from n. base of San Bernardino Mts. to Inyo Mts., Death V. region; Nev., Ariz. April–May. Forma *glabra* Ensign, mostly glabrous; with the sp.

2. **F. púngens** (Bdg.) Heller. var **glàbra** Ensign. Stems and lvs. glabrous, the latter 6–10 mm. long, estipulate; pedicels 3–4 mm. long, with 3–4 scarious bracts at base; sepals 5, ovate, acuminate, 2- or 3-spinose-tipped, denticulate, hyaline-margined; petals 5, 6–8 mm. long; stamens 10, those opposite the sepals longer; carpels ovoid, less than 1 mm. long.—Limestone cliffs, ca. 5500–6500 ft.; Pinyon-Juniper Wd.; Clark Mts., e. San Bernardino Co. May–June.

3. **F. stipulífera** (St. John) Ensign. [*Glossopetalon s.* St. John.] Freely branched, 1–3 m. tall, occasionally spinescent; young branches 0.5 mm. thick; lvs. glabrous, oblanceolate, 6–17 mm. long; petioles 1 mm. long; stipules mostly over 1 mm. long; pedicels axillary, 2–5 mm. long; sepals 2 mm. long; petals 6–9 mm. long; stamens 5–8, equal; carpels solitary, 3–5 mm. long.—At 5700 ft.; White Mts., Inyo Co.; to Wash., Ida. May.

3. Mortònia Gray

Rather low evergreen much-branched shrubs. Lvs. alternate, crowded, subsessile, coriaceous, 1-nerved, revolute with minute glandlike caducous stipules. Fls. small, white, in narrow terminal thyrsoid cymes. Calyx tubular at base, 10-ribbed, obconic, 5-lobed. Petals 5. Stamens 5, with short fils. Ovary 5-loculed; style columnar; stigmas 5; ovules 2 in a locule, erect, basal. Fr. dry, indehiscent, 1-seeded. Seed not arillate, oblong. A small genus of se. N. Am. (Named for Dr. S. G. *Morton*, Am. naturalist of the last century.)

1. **M. utahénsis** (Cov.) A. Nels. [*M. scabrella* var. *u.* Cov. ex Gray.] Low, 9–12 dm. high, yellow-green, hispidulous, densely leafy; lvs. broadly oval to roundish, 8–12 mm. long, scabrellous, with thickened margin; calyx-lobes 2 mm. long, with scarious margins; petioles almost lacking; infl 3–6 cm. long, with lanceolate bracts; calyx-lobes 2 mm. long, with scarious margins; petals obovate, 3 mm. long; caps. oblong, 4 mm. long, glabrous.—On limestone, 3400–5500 ft.; mostly Joshua Tree Wd.; Pinyon-Juniper Wd.; Clark, Kingston and Panamint mts., e. Mojave Desert; to Utah, Ariz. March–May.

28. Ceratophyllàceae. Hornwort Family

Aquatic herbs, submerged, with slender branching stems. Lvs. whorled, dissected into filiform, stiffish, serrulate divisions. Plants monoecious, the ♂ and ♀ fls. usually on separate nodes, sessile, without perianth, but surrounded by a calyxlike 8–12-cleft invol. Stamens distinct, 8–18, with subsessile anthers Pistil usually 1; ovary 1-celled, 1-ovuled, with persistent style. Fr. a spinose or tuberculate ak. A single genus.

1. Ceratophýllum L. HORNWORT

Characters of the family. Ca. 3 widely distributed polymorphous spp. Sometimes used in aquaria. (Greek, *keras*, horn, and *phullon*, lf., because of the narrow rigid divisions.)

1. **C. demérsum** L. [*C. apiculatum* Cham.] Plate 35, Fig. E. Stems to 2.5 m. long; lvs. 5–12 in a whorl, 1–2.5 cm. long, forked 2–3 times; fr. oval, ca. 5 mm. long, smooth or tubercled, with or without a lateral spur on each side of the base of the style; $2n = 24$ (Langlet & Söderberg, 1927).—Ponds and slow streams to ca. 7000 ft.; many Plant Communities; widely distributed in cismontane Calif. and reaching the desert edge as at Victorville; all continents. June–Aug.

29. Chenopodiàceae. GOOSEFOOT FAMILY

Herbs or shrubs, often succulent or scurfy, often weedy and frequently of saline or subsaline places Lvs. simple, without stipules, mostly alternate, sometimes reduced to scales. Fls. perfect, polygamous, or imperfect, small, greenish, usually in small cymose glomerules which may be variously arranged. Calyx free, imbricated in bud, persistent, mostly including the fr., sometimes wanting in ♀ fls., green or membranous, of 5 or fewer sepals. Corolla none. Stamens as many as sepals or fewer, opposite them. Ovary 1-loculed, usually becoming a 1-seeded, thin-walled utricle, rarely an ak. Styles 2–3. Seed with or without endosperm; embryo coiled into a ring or conduplicate or spiral. Ca. 100 genera and 1400 spp., world-wide; many are weeds; some as beets and spinach, grown as vegetable.

1. Sepals strongly imbricate, scarcely united, strongly chartaceous; lvs. opposite, united at base
    14. *Nitrophila*
1. Sepals slightly or not imbricate, herbaceous when young; lvs. mostly alternate.
    2. Lvs. foliaceous, flattened, but sometimes linear, not particularly fleshy or scaly.
        3. Fls. perfect, sometimes also ♀, all with calyx and not enclosed in a pair of bracts.
            4. Calyx transversely winged in fr. .... 8. *Cycloloma*
            4. Calyx not transversely winged in fr.
                5. Calyx of 1 segm.; stamens 1–3.
                    6. Perianth-segm. and stamen 1; lvs. mostly triangular-lanceolate to oblanceolate or spatulate. Widespread .... 13. *Monolepis*
                    6. Perianth-segm. 1; stamens 1–3; lvs. linear. Eureka V. .... 7. *Corispermum*
                5. Calyx with 3–5 segms.; stamens 1–5.
                    7. Stamen 1; fls. axillary, solitary or clustered. Coastal .... 2. *Aphanisma*
                    7. Stamens 4–5; fls. in clusters.
                        8. Calyx indurate at base in age, with ovary partly sunk .... 5. *Beta*
                        8. Calyx not indurate, with ovary superior.
                            9. Calyx-lobes not with hooked spines; plants glabrous, mealy, or glandular-pubescent .... 6. *Chenopodium*
                            9. Calyx-lobes with stout hooked spines; plants pilose .... 4. *Bassia*
        3. Fls. imperfect, the ♀ enclosed in 2 accrescent bractlets.
            10. Frs. not hairy; lvs. plane, not revolute.
                11. Bracts compressed, the margins never wholly united; lvs. ± farinose .. 3. *Atriplex*
                11. Bracts obcompressed, wholly united into a sac; lvs. usually glabrate ... 10. *Grayia*
            10. Frs. conspicuously hairy; lvs. linear, revolute .... 9. *Eurotia*
    2. Lvs. not or scarcely flattened but fleshy and sublinear, or scaly and spiny.
        12. The lvs. scalelike; stems and branches fleshy.
            13. Branches opposite; fl.-clusters opposite .... 15. *Salicornia*
            13. Branches alternate; fl.-clusters alternate .... 1. *Allenrolfea*
        12. The lvs. not scalelike, but fleshy and sublinear, or spiny-tipped.
            14. Plants with staminate fls. in spikes and ♀ solitary and axillary .... 17. *Sarcobatus*
            14. Plants with perfect fls., or perfect and ♀.
                15. Lvs. not tipped with spine or bristlelike hair.
                    16. Fruiting calyx transversely winged .... 12. *Kochia*
                    16. Fruiting calyx not transversely winged .... 18. *Suaeda*

15. Lvs. tipped with spine or bristly hair.
  17. Lower lvs. mostly 3–5 cm. long; fruiting sepals winged on back .... 16. *Salsola*
  17. Lower lvs. mostly 0.6–2 cm. long; fruiting sepals ending in wings . 11. *Halogeton*

1. *Allenrólfea* Kuntze

Much-branched glabrous erect succulent shrub or half-shrub, with alternate articulate green branches. Lvs. short, scalelike. Fls. perfect, sessile, arranged spirally by 3's or 5's in the axils of fleshy peltate bracts and forming cylindrical spikes. Perianth small, angled, of 4 or 5 concave carinate lobes. Stamens 1 or 2, exserted. Stigmas 2 or rarely 3, short, usually distinct. Utricle ovoid, compressed, with free membranous pericarp. Seed erect, oblong, smooth; embryo partly enclosing the copious endosperm. One sp. (Named for *Allen Rolfe*, English botanist.)

1. **A. occidentàlis** (Wats.) Kuntze. [*Halostachys o.* Wats. *Spirostachys o.* Wats.] IODINE BUSH. Plate 103, Fig. C. Plants 5–15 dm. high, woody below or nearly throughout, somewhat glaucous, the stems jointed; lvs. very short, triangular, some deciduous; spikes many, 6–25 mm. long; calyx closely enveloping the fr.; seed brown, ca. 0.6 mm. long.— Moist alkaline places; Alkali Sink; deserts and to cent. Calif.; to Ore., Utah, Son., L. Calif. June–Aug.

2. *Aphanisma* Nutt. ex Moq. in DC.

Glabrous succulent annual with slender stems. Lvs. alternate, entire, mostly sessile or subsessile. Fls. perfect, green, solitary or in clusters, axillary, sessile. Calyx 3–5-cleft, the segms. concave, subequal, unchanged in fr. Stamen 1. Style 1. Stigmas 3. Fr. a depressed-globose utricle, indurate, finely costate. Seeds horizontal, lenticular, rugulose; embryo somewhat annular, surrounding the abundant endosperm. One sp. (Greek, *aphanes*, inconspicuous.)

1. **A. blitoides** Nutt. Plate 35, Fig. F. Branched from base, the stems decumbent or ascending, 1–5 dm. long; lvs. ovate, 1–2.5 cm. long, subsessile or the lower obovate to spatulate, longer and petioled; fr. 1–2 mm. across.—Bluffs; Coastal Sage Scrub, Coastal Strand; coastal Los Angeles Co. to L. Calif. Most Channel Ids. April–May.

3. *Atriplex* L. SALTBUSH

Herbs or shrubs, usually grayish or whitish, scurfy with inflated hairs. Lvs. alternate or opposite. Plants monoecious or dioecious, the fls. small, greenish, in axillary clusters or glomerules, or in panicled spikes. Staminate fls. with 3–5-parted calyx, bractless; stamens 3–5. Pistillate fls. consisting of a naked pistil enclosed between a pair of appressed foliaceous bracts which enlarge in fr. and may be partly united, ± expanded, and variously thickened and appendaged. Styles 2. Utricle with a usually free membranous pericarp; seed flattened, erect or inverted, rarely horizontal; embryo annular, surrounding the scanty endosperm. Over 100 spp., the genus essentially cosmopolitan. (The ancient Latin name.)

(Hall, H. M. and F. E. Clements. Carnegie Inst. of Wash. Pub. 326: 235–346. 1923.)
1. Plants herbaceous, slightly if at all woody at the base, mostly monoecious.
  2. Lvs. green or greenish on both surfaces, sparsely mealy and sometimes grayish when young.
    3. Fruiting bracts united only near the base; ♂ fls. mixed with ♀ or in very short spikes; lvs. mostly hastate.
      4. Bracts hastate to rounded or cuneate at base, the tips close together; radicle of embryo pointing downward .... 15. *A. patula*
      4. Bracts mostly with rounded earlike lobes near base, the tips far apart; radicle pointing upward .... 16. *A. phyllostegia*
    3. Fruiting bracts united to above the middle; ♂ glomerules in elongate terminal spikes or panicles; lvs. not hastate .... 20. *A. serenana*
  2. Lvs. gray or whitish with a fine scurf, at least on the lower surface.
    5. Bracts thickened, fleshy, ovate, strongly nerved, 3.5–5 mm. long. Introd. perennial
      19. *A. semibaccata*

Atriplex

    5. Bracts not thickened or fleshy. Mostly native annuals and perennials.
      6. Fruiting bracts broadest below their middle.
        7. Staminate glomerules in long naked terminal spikes; perennial, the plant forming tangled mats; lvs. mostly opposite. S. coast .... 25. *S. watsonii*
        7. Staminate glomerules in upper lf.-axils or in spikes not more than 1 cm. long (*except californica*).
          8. Lvs. coarsely toothed; bracts hard and indurated. Introd. .... 18. *A. rosea*
          8. Lvs. entire; bracts not becoming hard. Native.
            9. Plant prostrate with perennial fusiform root; bracts distinct .. 2. *A. californica*
            9. Plant with ascending intricately branched stems; bracts united half their length
              13. *A. parishii*
      6. Fruiting bracts broadest at or above their middle.
        10. The bracts truncate at summit and cuneate at base .... 23. *A. truncata*
        10. The bracts not truncate at summit or cuneate at base.
          11. Bracts becoming hard and almost bonelike. Introd. weeds .... 18. *A. rosea*
          11. Bracts not becoming particularly hard. Native.
            12. Staminate and ♀ fls. mostly mixed in the clusters.
              13. Bracts irregularly toothed at summit.
                14. Lvs. narrowly ovate to elliptic, narrowed at base, 3–10 mm. wide; bracts mostly 3–4 mm. long .... 5. *A. coronata*
                14. Lvs. cordate- to lance-ovate, broad and often subhastate at base, mostly 10–40 mm. wide; bracts 4–8 mm. long .... 1. *A. argentea*
              13. Bracts evenly toothed all around .... 7. *A. elegans*
            12. Staminate and ♀ fls. mostly in separate clusters.
              15. Bracts not compressed, elliptic-globose, 5–7 mm. long; seed 2.5–3 mm. long. Immediate coast .... 10. *A. leucophylla*
              15. Bracts usually ± compressed, 1–4 mm. long; seed 1.5 or less mm. long.
                16. Fruiting bracts orbicular, toothed all around .... 7. *A. elegans*
                16. Fruiting bracts obovate to round-cuneate, entire near base.
                  17. Plants annual.
                    18. Lvs. mostly dentate, broader in their lower half; bracts dentate above middle .... 20. *A. serenana*
                    18. Lvs. entire, broader in their upper half; bracts mostly entire
                      12. *A. pacifica*
                17. Plants perennial; stems 3–10 dm. long, slightly woody at base; lvs. entire
                  6. *A. coulteri*
1. Plants shrubby, definitely woody, mostly dioecious.
    19. Fruiting bracts with conspicuous extra wings or crests arising at middle of face.
      20. Lvs. narrowly spatulate to narrowly oblong, 1.5–5 cm. long; fruiting bracts stalked. Common .... 3. *A. canescens*
      20. Lvs. oblong to obovate, 1.2–2 cm. long; fruiting bracts hardly stalked. Rare weed
        24. *A. vesicaria*
    19. Fruiting bracts not with extra lateral wings.
      21. Plants not spiny.
        22. Lf.-margin entire, flat or crisped but not toothed.
          23. Bracts 2–4 mm. long. Desert plants.
            24. Lf.-blades 3–18 mm. long, oblong to spatulate .... 17. *A. polycarpa*
            24. Lf.-blades 15–50 mm. long, oblong to ovate-deltoid .... 9. *A. lentiformis*
          23. Bracts 4–10 mm. long. Cismontane.
            25. Lvs. oblong-ovate; bracts 4–7 mm. long. San Francisco to Orange Co.
              9. *A. lentiformis breweri*
            25. Lvs. round-spatulate; bracts 9–10 mm. long. Playa del Rey .. 11. *A. nummularia*
        22. Lf.-margin dentate.
          26. Bracts entire, reticulate-veined; lf.-blades 1–3 cm. long. Deserts . 8. *A. hymenelytra*
          26. Bracts sinuate-dentate, not reticulate; lf.-blades usually 3–5 cm. long. Coast at Playa del Rey .... 11. *A. nummularia*
      21. Plants spiny, the spines consisting of sharp-pointed twigs from which the lvs. and bracts have fallen.
        27. Bracts 2–4 mm. long.
          28. The bracts orbicular; lvs. truncate or cuneate at base, the blades 1.5–4 cm. long.
            29. Twigs sharply angled; mostly from above 2000 ft. .... 22. *A. torreyi*

29. Twigs terete, not sharply angled. Mostly from below 2000 ft. ... 9. *A. lentiformis*
28. The bracts with broad summit; lvs. cordate at base, the blades 0.5–1.5 cm. long
14. *A. parryi*
27. The bracts 6–15 mm. long.
30. Body of bract small, not contracted beneath the free terminal wings; lvs. all entire
4. *A. confertifolia*
30. Body of bract large, thick, contracted to a neck beneath the free terminal wings; lvs. sometimes somewhat hastate at base .... 21. *A. spinifera*

1. **A. argéntea** Nutt. ssp. **expánsa** (Wats.) Hall & Clem. [*A. e.* Wats. *A. e.* var. *mohavensis* Jones.] Erect much-branched bushy scurfy-gray annual 1.5–6 dm. high; lower lvs. alternate, ± petioled, the upper sessile or nearly so, the blades 2–5 cm. long, triangular- to rounded-ovate, often slightly hastate, entire to undulate; plants monoecious, the fls. in axillary glomerules and terminal interrupted spikes, with ♂ and ♀ usually mixed in clusters; fruiting bracts obovate, compressed, 4–8 mm. long, united to near summit, the free green margins subentire to variously laciniate and faces smooth to crested; seed 1.5–2 mm. long, brown.—Alkaline places, below 7000 ft.; several Plant Communities; coast from San Francisco through cismontane s. Calif. to L. Calif.; to New Mex., Tex. July–Nov.

2. **A. califórnica** Moq. in DC. Prostrate perennial from fusiform taproot, the many much-branched stems 2–5 dm. long, white-scurfy, later glabrate; lvs. many, alternate or the lowest opposite, crowded, sessile, lanceolate to oblanceolate, gray-scurfy, acute, 5–18 mm. long; fls. in mixed axillary clusters or the ♂ in terminal spikes; fruiting bracts ovate, acute, sessile, scarcely united, entire, ca. 3 mm. long; seed dark, ca. 2 mm. long—Sea bluffs and sandy coast; Coastal Strand and edge of Coastal Salt Marsh, Coastal Sage Scrub, etc.; Marin Co. to n. Calif. Most of Channel Ids. April–Nov.

3. **A. canéscens** (Pursh) Nutt. ssp. **canéscens.** [*Calligonum c.* Pursh.] Erect much-branched shrub 4–20 dm. high, grayish-scurfy, with spreading or ascending terete branches; lvs. linear-spatulate to linear-oblong, sessile, 1.5–5 cm. long, 2–8 mm. wide, revolute; plants dioecious, the ♂ glomerules in terminal panicles of dense spikes, the ♀ in dense leafy-bracted spikes and panicles; fruiting bracts stalked, the body hard, not compressed, the bracts 6–15 mm. long, 4–8 (–10) mm. wide, with a second pair of longitudinal wings from middle of each bract, the 4 wings entire to deeply and coarsely dentate, smooth or appendaged; seed brown, 1.5–2.5 mm. long.—Common on dry slopes, flats and washes, below 7000 ft.; Alkali Sink, Creosote Bush Scrub, to Pinyon-Juniper Wd.; both deserts; less frequent in subsaline places; Coastal Strand to V. Grassland; cismontane valleys from San Diego Co. n.; to e. Wash., S. Dak., Kans., Tex., Mex. June–Aug. A poorly marked form with laciniate bracts 7–8 mm. wide is var. *laciniata* Parish, from s. Mojave Desert to Salton Sea. Another is var. *macilenta* Jeps. with bracts 1.5–3 mm. wide and occurring about Holtville, Imperial Co.

Ssp. **lineáris** (Wats.) Hall & Clem. [*A. l.* Wats.] Lvs. linear, usually not more than 2 mm. wide; fruiting bracts 4–8 mm. long, irregularly dentate or laciniate, each bract with 2 thin wings 2–3 mm. wide.—Alkaline Sink; from Salton Sea through e. Colo. Desert to n. L. Calif., Son.

4. **A. confertifòlia** (Torr. & Frém.) Wats. [*Obione c.* Torr. & Frém.] Erect rigidly branched spiny rounded shrub 2–10 dm. high, with stout scurfy branchlets; lvs. crowded, deciduous, alternate, round-ovate or -obovate to elliptic, obtuse, subsessile or short-petioled, 1–2 cm. long, entire, gray-scurfy; plants dioecious, the ♂ glomerules in the upper axils, almost spicate, the ♀ 1–few, in each of upper lf.-axils; fruiting bracts sessile, convex over the seed, suborbicular, free-margined, entire, 6–12 mm. long; seeds red-brown, 1.5–2 mm. long.—Common, alkaline flats and slopes, below 7000 ft.; Shadscale Scrub, Creosote Bush Scrub; Mojave Desert to e. Ore., N. Dak., n. Mex. April–July.

5. **A. coronàta** Wats. var. **notàtior** Jeps. [*A. sordida* Standl.] CROWNSCALE. Erect or decumbent annual, 1–3 dm. tall, much-branched, furfuraceous; lvs. crowded, alternate, sessile, ovate-deltoid to ovate, 5–20 mm. long, whitish; fls. monoecious, in dense axillary glomerules; fruiting bracts sessile, cuneate-orbicular, 3.5–5 mm. long, united to the middle,

Atriplex

coarsely dentate and toothed-crested, subglobose.—Alkaline flats, San Jacinto V. May–Aug.

6. **A. còulteri** (Moq.) D. Dietr. [*Obione c.* Moq.] Spreading perennial, slightly woody at base, much branched, the stems 3–10 dm. long, sparsely scurfy; lvs. many, alternate, elliptic to lanceolate or ovate, acute, entire, 7–20 mm. long; ♂ glomerules in upper axils and short terminal spikes; ♀ below; fruiting bracts sessile, obovate, 2–3 mm. long, united half way, the free margins sharply dentate, the backs sometimes tubercled; seeds brown, 1.3–1.5 mm. long.—Somewhat alkaline low places; V. Grassland, Coastal Sage Scrub; Los Angeles Co. to w. San Bernardino Co. and L. Calif. March–Oct.

7. **A. elegáns** (Moq.) D. Dietr. ssp. **fasciculàta** (Wats.) Hall & Clem. [*A. f.* Wats. *A. saltonensis* Parish.] Erect to decumbent annual with many branches 5–20 cm. long, slender, scurfy; lvs. many, mostly alternate, subsessile, elliptic-spatulate to oblong, entire, 5–20 mm. long, densely scurfy; ♂ and ♀ fls. mixed in the same axillary clusters; fruiting bracts subsessile, strongly compressed, round, united throughout except at the margins, 2–4 mm. wide, minutely toothed to subentire; seed brown, 1–1.4 mm. long.—Rather saline places, Creosote Bush Scrub; deserts from Inyo Co. to Mex. border and w. Ariz. March–July.

8. **A. hymenelỹtra** (Torr.) Wats. [*Obione h.* Torr.] DESERT-HOLLY. Plate 36, Fig. A. Low rounded compact shrub, white-scurfy, 2–10 dm. tall; lvs. numerous, persistent, alternate, rounded to rhombic, obtuse at apex, rounded to subcordate at base, 15–35 mm. long, silvery, irregularly and sharply dentate, with petioles somewhat shorter; plants dioecious; ♂ fls. in short dense leafy panicles, the ♀ in short dense spikes; fruiting bracts strongly compressed, roundish, entire, 6–10 mm. long, reticulate-veined; seed brown, ca. 2 mm. long.—Dry alkaline slopes and washes; Creosote Bush Scrub; deserts to Utah, Son., L. Calif. Jan.–April.

9. **A. lentifórmis** (Torr.) Wats. ssp. **lentifórmis**. [*Obione l.* Torr. in Sitg.] Erect widely spreading shrub 1–3 m. high, often broader, the twigs terete, gray-scurfy when young, mostly not spinose; lvs. oblong to ovate-deltoid, sessile or short-petioled, 1.5–4 cm. long, mostly 1–2.5 cm. wide, entire, gray-scurfy; plants dioecious, the fl.-clusters in profuse terminal panicles; fruiting bracts flattish, round-ovate, 3–4 mm. long, united to ca. middle, entire to minutely crenulate; seed brown, ca. 1.4 mm. long.—Alkaline places, mostly below 2000 ft.; Alkali Sink; Mojave and Colo. deserts; to Utah, Son., L. Calif. Aug.–Oct.

Ssp. **bréweri** (Wats.) Hall & Clem. [*A. b.* Wats.] Plate 36, Fig. B. Lf.-blades mostly 3–5 cm. long, 1.5–4 cm. wide; plants monoecious or dioecious; fruiting bracts convex, 4–7 mm. long, entire or undulate.—Saline places; V. Grassland, Coastal Sage Scrub; edge of Coastal Salt Marsh; Orange Co. and w. Riverside Co. to cent. Calif. San Clemente Id. July–Oct.

10. **A. leucophýlla** (Moq.) D. Dietr. [*Obione l.* Moq. in DC.] Plate 36, Fig. C. Prostrate perennial somewhat woody at base, the branches many, coarse, 3–5 dm. long, white-scurfy; lvs. many, sessile, mostly alternate, crowded, round-ovate to -elliptic, or oblong, 1–4 cm. long, entire, obtuse or rounded; ♂ fls. in dense terminal spikes, the ♀ in few-fld. axillary clusters; fruiting bracts sessile, almost completely united, broadly ovate, 5–7 mm. long, spongy, not compressed, entire or dentate, usually with wartlike projections on the faces; seed dark red-brown, 2.5–3 mm. long.—Sea-beaches; largely Coastal Strand; L. Calif. n. to n. Calif.; occasionally inland as at Lake Elsinore. April–Oct. Most Channel Ids.

11. **A. nummulària** Lindl. [*A. johnstonii* C. B. Wolf.] Erect shrub 2–3 m. tall, gray-scurfy, with striated twigs; lvs. many, blue-green, 3–6.5 cm. long, ca. as wide, slightly crisped, entire to low-serrate, rounded to obtuse, short-petioled; plants semidioecious; ♂ fls. crowded in short spikes forming large panicles; ♀ in dense compound panicles; fruiting bracts sessile, thick and corky, united halfway, roundish, 5–12 mm. long, subentire to coarsely few-toothed; seeds brown, ca. 2 mm. in diam.—Sandy coastal bluffs, Playa del Rey, Los Angeles Co.; natur. from Australia. Dec.–June.

12. **A. pacífica** Nels. [*Obione microcarpa* Benth. *A. m.* D. Dietr. not Waldst. & Kit.] Prostrate annual forming tangled masses 3–10 dm. in diam., lightly furfuraceous on young

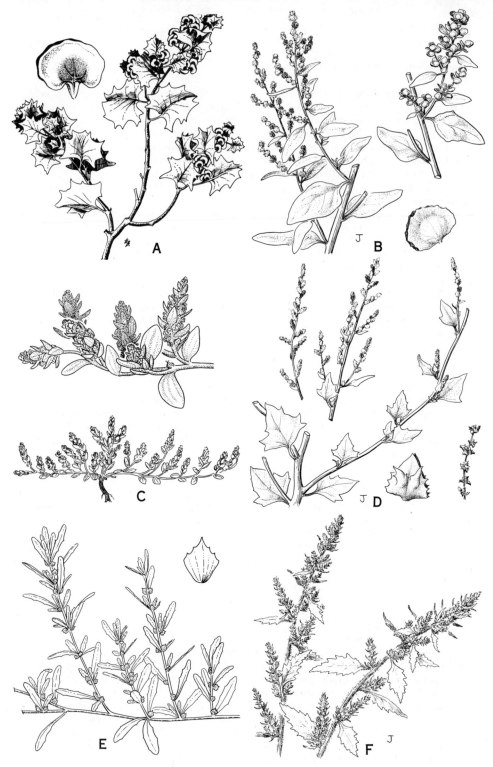

Plate 36. Fig. A, *Atriplex hymenelytra*. Fig. B, *Atriplex lentiformis* ssp. *breweri*. Fig. C, *Atriplex leucophylla*. Fig. D, *Atriplex patula* ssp. *hastata*. Fig. E, *Atriplex semibaccata*. Fig. F, *Chenopodium ambrosioides*.

parts; lvs. mostly alternate, numerous, sessile or the lower short-petioled, oblanceolate or spatulate-elliptic, 5–18 mm. long, entire, greenish above, more scurfy beneath; plants monoecious, the ♂ glomerules largely in the upper leafless axils, the ♀ in lower axils; fruiting bracts subsessile, suborbicular to obovate, 1–1.5 mm. long, united to middle, minutely 3–5-toothed at apex, otherwise entire; seed 0.8 mm. long.—Largely on seabluffs; Coastal Sage Scrub; Los Angeles Co. to L. Calif.; San Clemente, Santa Catalina and Santa Rosa ids. March–Oct.

13. **A. párishii** Wats. [*A. depressa* Jeps.] Spreading annual, the branches almost horizontal, fragile, white-scurfy, appearing almost pubescent, 5–20 cm. long; lvs. opposite or alternate, numerous, sessile, lanceolate to ovate, 4–10 mm. long, entire, rigid, gray to white, rounded at base; ♂ fls. mostly in upper axils and the ♀ in lower; fruiting bracts sessile, slightly compressed, ovate or rhombic, united half way, 3 mm. long, entire or with few teeth on each side, the faces smooth or tuberculate; seeds dark, ca. 1.2 mm. long.— Alkali flats, largely V. Grassland; cismontane s. Calif. to edge of deserts; to Central V. of Calif. June–Oct.

14. **A. párryi** Wats. Erect much-branched rounded shrub 2–4 dm. high, white-scurfy, with slender rigid spiny twigs; lvs. crowded, alternate, somewhat petiolate or the upper sessile, round-ovate to almost reniform, entire, 5–18 mm. long, dioecious; glomerules of fls. in interrupted leafy panicles of spikes; fruiting bracts compressed, thick, partly united, 3–4 mm. long, entire, with broad summit; seed brownish, 1–1.5 mm. long.—Alkaline places below 4000 ft.; Alkali Sink; Mojave Desert to Owens V.; Nev. May–Aug.

15. **A. pátula** L. ssp. **hastàta** (L.) Hall & Clem. [*A. h.* L.] Plate 36, Fig. D. Annual, simple to much-branched, glabrous or slightly mealy, the lower branches widely divergent, 3–10 dm. long; lvs. green, broadly triangular- to ovate-hastate, often dentate, 3–8 cm. long; infl. interrupted-spiciform, only the lower glomerules with leafy bracts; fruiting bracts rhombic-oval, green, truncate or broadly rounded at base, 2–5 mm. long, mostly smooth, united only at base; seed 1–2 mm. long; $2n=18$ (Löve, 1961).—Common in moist saline places along the coast and in interior valleys; Alkali Sink, Coastal Salt Marsh; to B.C., Atlantic Coast, Eurasia. June–Nov.

16. **A. phyllostègia** (Torr.) Wats. [*Obione p.* Torr.] ARROWSCALE. Much-branched erect annual 0.5–4 dm. high, rounded, bushy, sparsely mealy; lvs. mostly alternate, the blades 1–4 cm. long, rhombic-triangular to lanceolate, commonly hastate at base, on petioles 1–3 cm. long; plants monoecious or wholly ♀, the ♂ fls. in small axillary glomerules near the ends of the branches, the ♀ in axillary clusters; fruiting bracts 5–20 mm. long, sessile or stalked, united only near the base, lanceolate or lance-oblong, 3-ribbed, often hastate, the tips attenuate, widely separated; seed brown, 1.2 mm. long.—Alkaline places, Creosote Bush Scrub, Alkali Sink; w. Mojave Desert n. to cent. Calif., Ore., and to Utah, Ariz. April–Aug.

17. **A. polycárpa** (Torr.) Wats. [*Obione p.* Torr.] Erect intricately branched shrub 1–2 m. tall, gray-scurfy, with slender divaricate branches; lvs. crowded on young twigs, early deciduous, alternate, usually sessile, oblong to spatulate, 3–20 mm. long, gray-scurfy, entire; plants dioecious, ♂ fls. in dense or interrupted, simple or paniculate naked spikes, ♀ crowded in small sparsely leafy clusters arranged in paniculate spikes; fruiting bracts sessile, somewhat compressed, ± united, cuneate-orbicular, 2–4 mm. long, shallowly to deeply lobed, plane or tuberculate; seed pale brown, 1–1.5 mm. long.—Alkaline soils, below 5000 ft.; Creosote Bush Scrub, Shadscale Scrub, Sagebrush Scrub, Alkali Sink; Owens V. to L. Calif., Son., Utah. July–Oct.

18. **A. ròsea** L. REDSCALE. Erect annual, 1–10 dm. high, the branches arched-ascending, mealy or glabrate; lvs. mostly alternate, short-petioled to subsessile, ovate or rhombic-ovate, sinuate-dentate above the cuneate base, 2–5 cm. long, the upper reduced, entire; fls. in axillary glomerules or interrupted terminal spikes; ♂ calyx 5-cleft; fruiting bracts rhombic or ovate, 4–6 mm. long, united to ca. the middle, firm, often warty; seed 1.5–2 mm. long; $2n=18$ (Mulligan, 1957).—Alkaline places below 7000 ft.; many Plant Communities; widespread in Calif.; to L. Calif., Wash., Atlantic Coast; natur. from Eurasia. July–Oct.

19. **A. semibaccàta** R. Br. Australian Saltbush. Plate 36, Fig. E. Prostrate suffrutescent perennial, the stems much-branched, 2–12 dm. long, at first scurfy, then glabrate; lvs. many, alternate, short-petioled, elliptic-oblong, 1–3 cm. long, acute or obtuse, irregularly repand-dentate to subentire; ♂ fls. in small terminal glomerules, ♀ 1–few in the axils; fruiting bracts fleshy, becoming red, sessile, rhombic, 3.5–5 mm. long, compressed, united at base, entire to denticulate, the faces nerved, otherwise plane; seed dark, 1.5–2 mm. long.—Abundant in saline waste places, roadsides etc.; many Plant Communities; cismontane from Monterey Co. to n. L. Calif. and in Imperial V. Natur. from Australia. April–Dec.

20. **A. serenàna** A. Nels. Erect or decumbent annual, usually branched, often forming tangled mats 5–20 dm. across with ascending twigs, sparsely scurfy; lvs. many, alternate, lanceolate to oblong or oval, subsessile, 2–4 cm. long, sharply dentate to entire; plants monoecious, the ♂ glomerules in terminal spikes or panicles, the ♀ clusters small, axillary; fruiting bracts sessile or subsessile, somewhat compressed, united half way, cuneate-orbicular, 2–2.5 mm. long, smooth or tubercled, toothed above middle; seed brown, 1–1.3 mm. long.—Alkaline valleys at low elevs.; V. Grassland, Coastal Sage Scrub, etc.; Sacramento V. to L. Calif.; W. Nev. May–Oct. A form with lvs. 1–2 cm. long; terminal ♂ clusters short, subglobose, occurs from Los Angeles to Balboa and Laguna Beach and is var. **davidsònii** (Standl.) Munz. [*A. d.* Standl.]

21. **A. spinífera** Macbr. Much like *A. confertifolia*, taller than broad, the twigs becoming rigid divergent spines; lvs. deltoid-ovate to elliptic, 1–2 cm. long, short-petioled or subsessile, entire or subhastate, gray-scurfy; fruiting bracts strongly convex below, 7–15 mm. long, oblong to orbicular, entire or dentate, the faces plane or somewhat cristate; seed 2–2.8 mm. long.—Alkaline soils below 2500 ft.; largely Alkali Sink; from Kramer and Daggett through w. Mojave Desert to Fresno and San Luis Obispo cos. April–June.

22. **A. tórreyi** (Wats.) Wats. [*Obione t.* Wats.] Erect, much-branched gray-scurfy shrub 6–15 dm. tall, rather stiff, the twigs acutely angled by prominent striae; stiff and somewhat spiny by loss of lvs. and bracts; lvs. oblong to ovate-hastate, short-petioled, gray-scurfy, 1.5–3 cm. long, entire; plants dioecious; fls. in panicles of dense narrow spikes; fruiting bracts flattish, suborbicular, 2–4 mm. in diam., crenulate, distinct; seed brown, ca. 1.3 mm. in diam.—Occasional in alkaline places, 2000–5000 ft.; Alkali Sink, Shadscale Scrub; Mojave Desert to Owens V.; Utah, Ariz. June–Oct.

23. **A. truncàta** (Torr.) Gray. [*Obione t.* Torr. ex Wats.] Wedgescale. Erect to somewhat decumbent annual, the branches 2–5 dm. long, grayish-furfuraceous; lvs. alternate, subsessile or the lower petioled, round-ovate or deltoid-ovate, 1.5–4 cm. long; plants monoecious, the fls. in small axillary glomerules; fruiting bracts almost sessile, united to summit, broadly cuneate, truncate at summit, 2–3 mm. long, the faces smooth or obscurely tuberculate; seeds 1–1.5 mm. long, light brown to amber.—Alkaline places mostly below 7000 ft.; Creosote Bush Scrub to Lodgepole F.; w. Mojave Desert, then along e. slope of Sierra Nevada to B.C., Rocky Mts. June–Sept.

24. **A. vesicària** Heward in Hook. f. Bushy shrub with a scaly white tomentum; lvs. oblong to subobovate, 12–20 mm. long, short-petioled; ♂ fls. in small clusters forming dense leafless spikes 12–25 mm. long; ♀ fls. few together, in axillary clusters; fruiting bracts suborbicular, 6–10 mm. in diam., entire, flat, but each with a membranous inflated appendage on the disk nearly as large as the bract itself.—Weed in the Northridge area, Los Angeles Co., *Fuller*; native of Australia.

25. **A. watsònii** A. Nels. [*A. decumbens* Wats., not R. & S.] Prostrate suffrutescent perennial forming tangled mats 1–3 m. across, white-scurfy; lvs. numerous, mostly opposite, sessile, broadly elliptic to ovate, entire, thick, white-scurfy, 8–25 mm. long, acutish; plants dioecious, the ♂ glomerules in naked terminal spikes; ♀ clusters small, axillary; fruiting bracts sessile, ovate to rhombic, 5–8 mm. long, entire to erose, united to above the middle, the sides plane; seed light brown, ca. 1 mm. long.—Coastal bluffs and beaches; Coastal Strand, Coastal Salt Marsh, Coastal Sage Scrub; Santa Barbara Co. to L. Calif.; Santa Catalina, San Clemente ids. March–Oct.

4. Bássia All.

Annual to suffrutescent plants. Lvs. alternate, narrow, flat or subterete, entire. Fls. minute, solitary or glomerate in the axils, sessile, perfect and ♀. Calyx globose or depressed, the 5 lobes incurved, armed on the back with a spine, this usually hooked. Stamens mostly 5. Ovary ovoid; style short; stigmas 2–3. Utricle enclosed by the coriaceous calyx; seed free from the pericarp, orbicular, horizontal; embryo annular. Several spp. of Old World. (Named for F. *Bassi*, 1710–1774, Italian botanist.)

1. **B. hyssopifòlia** (Pall.) Kuntze. [*Salsola h.* Pall. *Echinopsilon h.* Moq.] Plant a grayish densely pilose annual with stems branched from base, prostrate, 3–5 dm. long and less hairy in maturity; lvs. narrowly linear-lanceolate, flat, 2–4 cm. long; fls. in small axillary glomerules; calyx-lobes broadly ovate, villous, ca. 1 mm. long, each with a spreading hooked spine; seed lenticular, ca. 1 mm. across.—Becoming a common weed in rather alkaline places through much of Calif. and in other w. states; from Eurasia. July–Oct.

5. Bèta L. BEET

Ours glabrous biennial herb with large fleshy roots. Lvs. alternate, the basal rosulate, large, long-petioled, the upper reduced and subsessile. Fls. perfect in glomerules of 3 or more, in panicled spikes. Calyx 5-parted, the segms. indurate and closed in fr. Stamens 5. Ovary sunk in the succulent base of the calyx; styles 2–3. Fr. ultimately opening by a lid. Seed horizontal, smooth, roundish; embryo annular. Five or more spp. of the Old World. (Perhaps Celtic, *bett*, red, because of the red roots.)

1. **B. vulgàris** L. GARDEN BEET. SUGAR BEET. Stems 3–12 dm. tall, paniculately branched above; lower lf.-blades 1–2 dm. long, ovate-oblong; calyx-segms. narrow-oblong, ca. 2 mm. long, carinate in fr.; fr. ca. 2.5 mm. long, somewhat wider; $2n=18$ (Levan, 1942).—Escaping from gardens and sometimes natur. in low damp places; native of Eu. July–Oct.

6. Chenopòdium L. GOOSEFOOT. PIGWEED

Annual or perennial herbs, usually mealy (farinose) or glandular. Lvs. alternate, entire to lobed. Fls. perfect, rarely unisexual, ebracteate, in small clusters or glomerules arranged in panicles or spikes. Calyx 5 (rarely 4)-parted or -lobed, the segms. persistent, flat or keeled, ± enveloping the fr. Stamens mostly 5. Styles 2, rarely 3. Fr. a utricle with membranous pericarp free from or adherent to the seed. Seed lenticular, horizontal or vertical, the embryo coiled partly or entirely around the albumen. A large genus, essentially cosmopolitan. (Greek, *chen*, goose, and *pous*, foot, referring to the shape of the lvs. in some spp.)

(Wahl, H. A. A preliminary study of the genus Chenopodium in N. Am. Bartonia 27: 1–46. 1954.)
1. Plants ± glandular-pubescent or resinous-glandular, especially about the calyx, not mealy or farinose.
   2. Calyx 3–5-toothed, becoming saccate and reticulate; lvs. pinnatifid ..... 17. *C. multifidum*
   2. Calyx 5 (rarely 4)-parted, not changing in fr. or merely becoming fleshy; lvs. various.
      3. Fls. in glomerules, these in capitate clusters or in small spikes.
         4. Glomerules small, headlike, all axillary; seed vertical .... 20. *C. pumilio*
         4. Glomerules in short spikes, the upper in panicles; seed mostly horizontal
            2. *C. ambrosioides*
      3. Fls. solitary in small cymes, these spreading-recurved and in elongating panicles
         5. *C. botrys*
1. Plants mostly mealy or farinose, not glandular-pubescent, but sometimes with nonglandular hairs.
   5. Seeds horizontal or almost entirely so.
      6. Lvs. shining on upper surface, ± rhombic; only younger parts mealy; plant branched from base .... 18. *C. murale*
      6. Lvs. dull on upper surface, various in shape.

7. Lf.-blades cordate or subcordate at base, bright green, thin, almost glabrous, acuminate and with large acuminate teeth .... 10. *C. gigantospermum*
7. Lf.-blades rounded to truncate or attenuate at base.
   8. Pericarp closely adherent to seed and removable with difficulty; at least lower lvs. conspicuously sinuate-dentate.
      9. Pericarp and seed surface essentially smooth.
         10. Seeds chiefly 1.2–1.5 mm. broad; sepals largely covering fr. .... 1. *C. album*
         10. Seeds 0.9–1.2 mm. broad.
            11. Sepals largely covering fr.; lower lvs. coarsely toothed; infl. branches densely fld., arching .... 16. *C. missouriense*
            11. Sepals exposing fr. at maturity; lower lvs. low-serrate; infl. of strict axillary spikes or terminal loose panicles .... 22. *C. strictum*
      9. Pericarp and seed-surface foveolate-reticulate; seed 1–1.3 mm. broad
                                                                                  4. *C. berlandieri*
   8. Pericarp free from or easily removed from the seed, or if adherent, the lower lvs. linear or entire or 3-lobed.
      12. Lf.-blades linear to narrow-lanceolate or narrow-oblong, short-petioled, the blades mostly 1–3-nerved.
         13. Lvs. linear, 1-nerved, mostly 2–3 mm. wide .... 14. *C. leptophyllum*
         13. Lvs. narrow-lanceolate or broader, the lower 4–18 mm. wide . 8. *C. desiccatum*
      12. Lf.-blades lance-ovate to ovate or broader, long-petioled, pinnately veined.
         14. Main lf.-blades definitely longer than broad.
            15. Pericarp separable; lvs. mostly entire; seed 1 mm. broad.
               16. Lvs. oblong or oval, 0.8–2 cm. long, mostly less than ⅓ as broad
                                                                                      8. *C. desiccatum*
               16. Lvs. ovate to triangular-oblong, 1.5–3 cm. long, mostly more than ⅓ as wide .... 3. *C. atrovirens*
            15. Pericarp attached; lvs. sometimes toothed.
               17. Seeds 1–1.5 mm. broad; plants mostly 3–12 dm. high, openly branched
                                                                                           13. *C. incognitum*
               17. Seeds 0.7–0.8 mm. wide; plants 2–3 dm. high, bushy-branched
                                                                                               19. *C. nevadense*
         14. Main lf.-blades scarcely if at all longer than wide.
            18. Lvs. thin, the blades deltoid, 1–4 cm. broad; plants upright, 3–10 dm. tall
                                                                                   9. *C. fremontii*
            18. Lvs. membranous to coriaceous, less than 1.5 cm. broad; plants bushy, 1–3 dm. high .... 12. *C. incanum*
5. Seeds vertical for the most part.
   19. Lvs. densely white-farinose beneath, at least when young.
      20. The lvs. rhombic to deltoid-rhombic, 1–4 cm. wide; calyx almost completely concealing the utricle .... 15. *C. macrospermum*
      20. The lvs. broadly lanceolate to oblong or subovate, 0.3–1.5 cm. wide; calyx concealing only a small part of the utricle .... 11. *C. glaucum*
   19. Lvs. sparsely or not at all farinose beneath.
      21. Plants annual, glabrous; stems up to 6 dm. long.
         22. Lf.-blades mostly 3–8 cm. long, sinuate-dentate; fls. in leafy spikes; calyx fleshy
                                                                                                    21. *C. rubrum*
         22. Lf.-blades 1–3 cm. long, entire to few-toothed; fls. in axillary glomerules; calyx not fleshy .... 7. *C. chenopodioides*
      21. Plants perennial, mealy on upper parts; stems 3–8 dm. long or more . 6. *C. californicum*

1. **C. álbum** L. PIGWEED. LAMB'S-QUARTERS. Erect annual, pale green, red-veined, 2–20 dm. high, branching, farinose; lvs. glaucous, farinose beneath, rhombic-ovate or the upper lanceolate, ± sinuate-dentate or -serrate, 1–5 cm. long; petioles slender, ca. half as long; fl.-glomerules thick, in rather dense heavy spikes in upper axils, forming panicles; calyx farinose, enclosing the fr., carinate; pericarp adherent; seed horizontal, black, nearly smooth, shining, ca. 1.3 mm. broad; $2n = 36$ (Cooper, 1935), 54 (Kjellmark, 1934). —Common weed in waste and fallow places below 6000 ft.; widely distributed over N. Am.; naturalized from Eu. Variable.

2. **C. ambrosioìdes** L. MEXICAN-TEA. Plate 36, Fig. F. Annual or perennial, erect or

ascending, coarse, strong-scented, the stems 4–10 dm. long, simple or branched, smoothish; lvs. short-petioled, oblong or lanceolate, subentire to repand-toothed, 2–10 cm. long, gradually reduced upward; infl. a slender pyramidal panicle of densely-fld. spikes, elongate, leafy or intermixed with lvs.; calyx ca. 1 mm. long, gland-dotted, enclosing the fr.; pericarp very thin, deciduous, gland-dotted; seed horizontal or vertical, ca. 0.7 mm. broad; $2n = 16, 32, 48$? (Kawatani & Ohno, 1950).—Weed in waste, especially damp places, particularly in s. Calif., but widespread on Pacific Coast and to New Eng.; natur. from trop. Am. June–Oct. A form more common in the San Joaquin V., but found also in s. Calif. is var. **anthelmínticum** (L.) Gray. [*C. a.* L.] with especially the lower lvs. more deeply divided, even laciniate-pinnatifid and with spikes almost or quite leafless.

3. **C. atròvirens** Rydb. Erect annual 1–5 dm. high, slender, almost simple or few-branched, green, subglabrous; lvs. ovate to triangular-oblong, mostly entire, 1.5–3 cm. long, obtuse; petioles slender, somewhat shorter; fl.-glomerules farinose, in ± paniculate spikes; calyx-lobes sharply carinate, obovate, enclosing the fr.; pericarp free; seed horizontal, 1 mm. broad.—Occasional in dryish places, 4000–11,000 ft.; Pinyon-Juniper Wd. to Bristle-cone Pine F.; Baldwin Lake, Cushenbury Springs, etc., w. Mojave Desert to e. Ore., Rocky Mts. July–Sept.

4. **C. berlándieri** Moq. var. **sinuàtum** (J. Murr.) Wahl. Like *C. album* but with thin membranous lvs. and a stronger unpleasant odor, slender leafy spikes, lvs. dentate but not lobed; calyx-lobes more sharply keeled; seeds mostly puncticulate and 1–1.3 mm. in diam. In waste places. Var. *zschackei* (Murr.) Murr. [*C. z.* Murr.] is difficult to distinguish, but has larger lvs.; seeds 1.2–1.5 mm. in diam. and seems to occur in alkaline spots from San Bernardino Co. to Lassen Co.

5. **C. bòtrys** L. JERUSALEM-OAK. Erect, densely glandular-villous aromatic annual 2–6 dm. high, viscid; lvs. sinuate-pinnatifid, 1–4 cm. long, oblong to oval, with obtuse angled lobes; petioles short; infl. a virgate elongate panicle of loosely spreading leafless cymes; fls. subsessile, pubescent, ca. 1 mm. long, the calyx-segms. oblong or ovate; pericarp adherent; seed vertical or horizontal, ca. 0.6 mm. broad.—Occasional weed, especially in sandy places below 7000 ft.; to Wash., Atlantic Coast; natur. from Eurasia. June–Oct.

6. **C. califórnicum** (Wats.) Wats. [*Blitum c.* Wats.] Perennial with stout fleshy root and several decumbent or ascending stems 3–8 dm. long, sparsely farinose on younger parts, stout, not much branched; lvs. deltoid, 3–10 cm. long, truncate or cordate at base, sharply and unequally sinuate-dentate; petioles slender, the lower as long as the blades; fl.-glomerules small, in long dense terminal spikes; calyx cleft to ca. the middle, green with broad lobes, shorter than fr.; pericarp adherent; seed vertical, compressed-globose, ca. 2 mm. broad.—Common on dryish plains and slopes, below 5000 ft.; many Plant Communities; much of cismontane Calif., to edge of deserts; L. Calif. March–June. Most Channel Ids.

7. **C. chenopodioìdes** (L.) Aellen. [*C. humile* Hook.] Low annual, widely branched at base, the branches 3–20 cm. long, ascending, slender, whitish; lvs. almost round to rhombic-ovate, 0.8–3 cm. long, entire to shallowly sinuate, green, glabrous; petioles shorter than blades; fls. green, in dense axillary glomerules; calyx 3–5-lobed, shorter than fr.; pericarp green; seed vertical, 0.8–1.0 mm. wide.—Moist alkaline places, mostly above 5000 ft.; San Bernardino Mts.; to n. Calif. and Atlantic Coast. Aug.–Oct.

8. **C. desiccàtum** Nels. var. **desiccàtum**. [*C. pratericola* var. *d.* Aellen.] Much like *C. leptophyllum*, low, diffusely branched from base, 1–3 dm. high; lvs. oblong or oval, entire, 0.8–2 cm. long, greenish above, farinose beneath; calyx-lobes enclosing the fr. obtuse, white-margined.—Dry places, 4000–10,000 ft.; Montane Coniferous F., Pinyon-Juniper Wd.; San Bernardino and Panamint mts.; to Ida., S. Dak., New Mex. July–Sept.

Var. **leptophylloìdes** (J. Murr.) H. A. Wahl. Erect, 2–8 dm. high; lvs. narrow-lanceolate to oblong, entire or slightly 3-lobed or hastate, the lower 3-nerved, 4–18 mm. wide. —Occasional in dry places below 9000 ft.; mostly Montane Coniferous F.; San Jacinto Mts. n. to Lassen Co.

9. **C. fremóntii** Wats. Plate 37, Fig. A. Erect annual 3–10 dm. high, slender, usually much-branched throughout with ascending branches, sparsely farinose or glabrous; lf.-

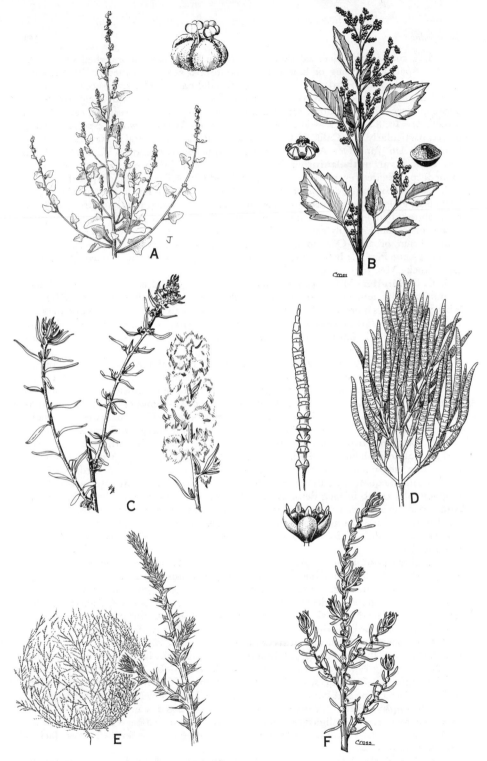

Plate 37. Fig. A, *Chenopodium fremontii*. Fig. B, *Chenopodium murale*. Fig. C, *Eurotia lanata*. Fig. D, *Salicornia subterminalis*. Fig. E, *Salsola iberica*. Fig. F, *Suaeda californica*.

blades broadly triangular-hastate, 1-3 cm. long, almost equally wide, bright green above, whitish beneath, obtuse; petioles slender, shorter than the blades; fl.-glomerules small, in rather slender paniculate spikes; calyx ca. 1 mm. broad, deeply cleft, rather sparsely farinose, completely enclosing the fr.; pericarp free; seed horizontal, 1 mm. broad, dark, smooth.—Rather frequent in dry places, mostly at 5000-8500 ft.; Pinyon-Juniper Wd., Yellow Pine F.; Santa Rosa Mts. and Cuyamaca Mts. to Mt. Pinos and Clark Mt., n. to Mono Co. and B.C., then to N. Dak. and Mex. June-Oct.

10. **C. gigantospérmum** Aellen. Glabrous erect bright green annual 2-14 dm. high, widely branched, the branches slender; lvs. thin, bright green, broadly to triangular-ovate, acuminate, 3-15 cm. long, with 1-few triangular-acuminate lobes; petioles slender, ca. half as long as lf.-blades; fl.-glomerules small, in loose terminal leafless panicles; calyx with thin segms., farinose, rounded on back, imperfectly enclosing the fr.; pericarp thin, adherent to seed; seed horizontal, lenticular, black, shiny, ridged, 1.5-2 mm. broad; $2n=38$ (Löve, 1954).—Occasional in moist somewhat shaded places, as at Westgaard Summit, Inyo Co.; transcontinental. July-Oct.

11. **C. glaúcum** L. ssp. **salìnum** (Standl.) Aellen. [*C. s.* Standl.] Freely branched prostrate to ascending annual, the branches subglabrous, 1-3 dm. long; lvs. many, well distributed, broadly lanceolate to oblong or subovate, the blades 1.5-3 cm. long, acute, sinuately toothed, somewhat hastate, pale green and somewhat farinose above, densely so beneath; petioles short; fls. in small axillary spikes shorter than the lvs., the infl. villous on the branches; calyx green, the lobes obovate, imperfectly enclosing the fr.; pericarp green, free; seed vertical or horizontal, ca. 0.9 mm. broad, dark red-brown, shining, finely tuberculate.—Occasional alkaline flats and shores as at Elsinore, Baldwin Lake, etc.; to Alta., New Mex. July-Oct.

12. **C. incànum** Heller. [*C. fremontii* var. *i.* Wats.] Plant 1-3 dm. tall, diffusely branched from base; lf.-blades with a basal lobe; seeds 0.9-1.1 mm. broad, covered by the calyx.—Much like *C. fremontii*. At 2500-5000 ft.; Joshua Tree Wd., Pinyon-Juniper Wd.; Mojave Desert; to Rocky Mts. April-Aug.

13. **C. incógnitum** H. A. Wahl. Plants erect, 3-12 dm. tall, branched from base; lvs. thin, ovate to deltoid ovate, 1.5-3.5 cm. long, entire or with small basal lobes, farinose beneath; fls. crowded in terminal and axillary spikes; sepals 5, farinose, narrowly keeled, ca. half covering the fr. at maturity; pericarp finely rugulose, attached to the seed; seed black, flattened, 1.2-1.5 mm. in diam., with rounded margin.—Dry places, 6000-8000 ft.; Montane Coniferous F.; from San Jacinto Mts. n. to Ore., Wyo., New Mex. July-Aug.

14. **C. leptophýllum** Nutt. [*C. inamoenum* Standl.] Densely farinose annual, erect, simple or branched below, usually much-branched above; lvs. short-petioled, linear, entire, 1-nerved, 1-4 cm. long, usually 2-3 mm. wide, roundish at apex; fls. in dense glomerules, these in interrupted or dense paniculate spikes; calyx completely enclosing the fr., the lobes keeled; pericarp free; seed horizontal, dark, smooth, shining, 1 mm. broad.—Dryish alkaline places, 5000-9000 ft.; Pinyon-Juniper Wd., Sagebrush Scrub, Yellow Pine F.; San Bernardino Mts., Panamint Mts.; to Ore., Wyo., Tex. July-Sept.

15. **C. macrospérmum** Hook. f. var. **farinòsum** (Wats.) J. T. Howell. [*C. murale* var. *f.* Wats.] Annual, branched from base, the stems stout, ascending to erect, glabrous, 1-5 dm. long; lvs. rhombic to deltoid-rhombic, 1.5-5 cm. long, obtuse, glabrous above, farinose beneath, sinuate-dentate; petioles shorter than to ca. as long as blades; glomerules in dense spikes, sessile in axils of reduced upper lvs.; calyx barely 1 mm. long, the segms. rounded; seed vertical, ca. 1 mm. long, dark.—Moist places, Coastal Strand; San Diego Co. to n. Calif. July-Oct.

16. **C. missouriénse** Aellen. Differing from *C. album* in having seeds 0.9-1.2 mm. broad (not 1.3 mm.). Collected in San Bernardino V.; native of cent. U.S.

17. **C. multífidum** L. [*Roubieva m.* Moq.] Glandular annual herb with taproot; strong-scented; stems branched, prostrate, 2-7 dm. long, villous in younger parts; lvs. oblong in outline, alternate, pinnatifid, 1-3 cm. long, the lobes mucronulate, narrow; fls. sessile, 1-3 in the axils; calyx urceolate, shallowly 5-dentate, puberulent, ca. 2 mm. long, becoming obovoid, coriaceous and reticulate in age; stamens 5, included; styles 3,

exserted; fr. membranaceous, compressed, gland-dotted, enclosed in perianth.—Uncommon wayside weed, below 5000 ft.; s. Calif. to cent. Calif.; natur. from S. Am. June–Nov.

18. **C. muràle** L. Plate 37, Fig. B. Rather stout annual, glabrous or sparsely mealy, ill-scented, branched from base, the branches ascending, 2–5 dm. long; lvs. dark green, rhombic-ovate, 2–6 cm. long, irregularly sinuate-dentate; petioles equal to or shorter than blades; fls. in small glomerules in lax or dense axillary and terminal short panicles; calyx 1.5 mm. broad, deeply cleft, the lobes oblong, obscurely keeled, incompletely enclosing the fr.; pericarp green, adherent; seed horizontal, ca. 1.5 mm. broad, puncticulate, with sharp edge; $2n = 18$ (Winge, 1917).—Common weed about orchards and gardens; widespread in N. Am. Most Channel Ids. Natur. from Eu. Most of the year, but especially in the spring.

19. **C. nevadénse** Standl. Erect much-branched annual 2–3 dm. high, obscurely farinose; lvs. rhombic-ovate to ovate-oblong, 0.8–2 cm. long, obtuse or rounded, subentire; petioles slender, usually shorter than the blades; fl.-glomerules very small, in diffuse cymose panicles; calyx farinose, enclosing the fr., the lobes acutish, slightly keeled; subovate; pericarp adherent; seed ca. 0.5 mm. long, dark brown.—Rare in valleys near Inyo-White Mts.; w. Nev. June–Aug.

20. **C. pumílio** R. Br. [*C. carinatum* auth., not R. Br.] Annual, branched from base, glandular-villous, depressed or ascending, the branches 2–4 dm. long; lvs. oblong, or oblong-ovate, the blades 1–2 cm. long, or uppermost reduced, coarsely sinuate-pinnatifid with obtuse lobes; petioles slender, from shorter than to longer than blades; fls. in short axillary clusters; stamen usually 1; calyx 0.6 mm. long, the lobes carinate, only partly enclosing the fr.; pericarp thin; seed vertical, 0.5 mm. broad.—Occasional weed, mostly below 5000 ft., through most of cismontane Calif.; natur. from Australia. June–Sept.

21. **C. rùbrum** L. Erect glabrous annual 2–6 dm. high, simple or branched with stout reddish stems; main lvs. deltoid- or rhombic-ovate, 3–8 cm. long, sinuate-dentate, cuneate at base; petioles ca. as long; fls. in short axillary leafy spikes, crowded; calyx 3–5-lobed, fleshy, not keeled; pericarp green; seed usually vertical, 0.6–0.8 mm. long, with rounded margin.—Occasional in low rather saline places, widely scattered in the state; to the Atlantic Coast; native of Eu. July–Oct.

22. **C. stríctum** Roth var. **glaucophýllum** (Aellen) H. A. Wahl. [*C. g.* Aellen.] Near *C. album*, the lower lvs. low-serrate, the median oblong, entire; fls. in strict axillary spikes or terminal loose panicles; sepals exposing fr. at maturity; seeds 0.9–1.2 mm. broad, smooth.—Occasional in low places, cismontane Calif.; to Atlantic Coast. Aug.–Oct.

7. *Corispérmum* L. BUGSEED

Low branching annuals, glabrous or sparsely pubescent. Lvs. alternate, entire, narrow, 1-nerved. Fls. perfect, in narrow loose terminal spikes, each fl. subtended by a leaflike scarious-margined bractlet broader than the foliage lvs. Calyx of a single delicate sepal on the inner side, deciduous. Stamens 1–3. Styles 2, persistent. Utricle oval, flat, with the outer face rather convex and the inner concave, sharp-margined; seed vertical, the embryo slender, coiled around a cent. albumen. A Eurasian genus. (Name from Greek, *coris*, a bedbug, and *sperma*, seed.)

1. **C. hyssopifòlium** L. Stiffish, much branched, to several dm. high, ± stellate-villous when young; lvs. narrowly linear, 2–7 cm. long; spikes dense, 4–8 mm. wide, the imbricated bracts lanceolate to ovate, scarious-margined, pointed, the lower 6–10 mm. long; fr. strongly winged, 3.5–4.5 mm. long, the wing ca. 0.5 mm. wide.—Sandy places, dunes, etc. at ca. 3400 ft.; mouth of Marble Canyon, w. side of Eureka V., Inyo Co.; introd. from Eurasia. May–June.

8. *Cyclolòma* Moq. WINGED PIGWEED

Erect or spreading branched annual. Lvs. alternate, petioled, oblong in outline, early deciduous, coarsely sinuate-dentate. Plants polygamo-monoecious, the fls. small, bracteate, solitary or in glomerules. Calyx 5-lobed, the lobes inflexed, strongly carinate, the basal tube in age developing a broad membranous wing. Stamens 5. Stigmas 3. Fr. depressed-globose, with membranaceous pericarp. Seed horizontal, smooth; embryo annular. One sp. (Greek, *cyclos*, a circle, and *loma*, referring to the calyx-wing.)

1. **C. atriplicifòlium** (Spreng.) Coult. Diffusely branched, 1–5 dm. tall, villous-tomentose in younger parts; lvs. 2–6 (–8) cm. long; infl. a broad panicle; calyx 3–4 mm. broad, villous, reddish in age; pericarp tomentulose; seed black, 1.5 mm. in diam.—Occasional weed in fields and groves, s. Calif.; to Man., Ind., Texas. May–Sept.

9. *Euròtia* Adans.

Low stellate-tomentose shrubs. Lvs. alternate or fascicled, slender, entire. Monoecious or dioecious. Staminate fls. bractless in dense axillary clusters forming spikes; calyx 4-parted, stamens 4. Pistillate fls. without calyx, but with a pair of conduplicate connate bracts which form a membranous silky sac that enlarges in fr., becoming 4-angled and beaked above with 2 short horns; styles 2, elongate; ovary ovoid. Fr. a utricle. Seeds vertical, obovoid. Two spp., one Asiatic. (Greek, *euros*, mould, because of hairy covering.)

1. **E. lanàta** (Pursh) Moq. [*Diotis l.* Pursh.] WINTER FAT. Plate 37, Fig. C. Erect shrubs 3–8 dm. high, white- or rusty-stellate, often with longer simple hairs interspersed; lvs. linear to lanceolate, 1.5–5 cm. long, 2–8 mm. wide, those of the fascicles shorter; fruiting bracts lanceolate, 5–7 mm. long, with dense spreading tufts of long silvery or rusty hairs; ovary hairy; seeds brown, ca. 2 mm. long.—Common on flats and rocky mesas, mostly above 2000 ft.; Creosote Bush Scrub to Pinyon-Juniper Wd.; Mojave Desert, to e. San Luis Obispo Co., along e. base of Sierra Nevada to Lassen Co.; Wash., Rocky Mts., Tex., Son. March–June. Important as a forage plant. Var. *lanàta* has longer hairs among the stellate and stems woody at base only; while those plants with short stellate hairs only and woodier stems are var. *subspinòsa* (Rydb.) Kearn. & Peebles. Both forms may grow together.

10. *Gràyia* H. & A.

Stiff much-branched shrubs, scurfy-pubescent. Lvs. alternate, entire, sessile, rather fleshy. Dioecious or rarely monoecious. Staminate fls. small, pedicellate, glomerate, bractless; calyx mostly 4-parted; stamens 4–5. Pistillate fls. racemose, with 2 orbicular flattened bracts which are dorsally winged in fr. and united to form a reticulate sac; calyx none; stigmas 2, filiform. Utricle compressed, included in the accrescent bracts; seed free, orbicular, with annular embryo. Two spp. of w. U.S. (Named for Asa *Gray*, 1810–1888, distinguished Am. botanist.)

1. **G. spinòsa** (Hook.) Moq. [*Chenopodium s.* Hook.] HOP-SAGE. Erect, 3–10 dm. high, gray-green, often spinose, with mealy younger parts; lvs. oblanceolate to oblong-lanceolate, 1–3 (–4) cm. long; ♂ clusters axillary and forming dense terminal spikes; ♀ in dense terminal racemose spikes; fruiting bracts membranous, reddish to whitish, glabrous, 6–12 mm. long, the wings thin, entire; seed flat, brown, round, cellular-reticulate, ca. 2 mm. across.—Common on mesas and flats, mostly 2500–7500 ft.; Creosote Bush Scrub to Pinyon-Juniper Wd.; Mojave Desert and rare in w. and extreme nw. Colo. Desert; to Siskiyou Co.; e. Wash., Wyo., Ariz. March–June.

11. *Halogèton* C. A. Mey.

Annual, glabrous or pubescent, fleshy. Lvs. alternate, terete, bristle-tipped. Polygamous, the fls. small, few, in glomerules in lf.-axils, bibracteate. Sepals 5, inconspicuous in some fls., becoming membranous and fanlike in some. Stamens 5 or 3. Stigmas 2. Utricle ovoid, thin-walled. Seeds vertical; embryo spiral. Three spp. of Eurasia. (Greek, *hals*, sea, salty, and *geiton*, a neighbor, from the habitat.)

1. **H. glomeràtus** (Bieb.) C. A. Mey. in Led. [*Anabasis g.* Bieb.] Branched from base, the stems divergent, then ascending, 0.5–5 dm. long; lvs. sessile, 6–20 mm. long, blunt,

then tipped with conspicuous bristlelike hair; fls. greenish-yellow, exceedingly numerous the sepals usually conspicuously membranous; seed ± compressed, ca. 1 mm. long.—Alkaline or disturbed places, etc.; Sagebrush Scrub; Joshua Tree Wd.; e. San Bernardino Co., e. Lassen Co.; Nev., Utah. Summer.

12. Kòchia Roth

Annual or perennial herbs or low shrubs. Lvs. alternate or opposite, narrow, often terete. Fls. perfect or ♀, small, sessile, solitary or glomerate in the axils of the lvs. Calyx 5-lobed, the lobes incurved, mostly coriaceous in age and then developing horizontal wings, these membranaceous or scarious, distinct or confluent. Stamens 5, exserted, the fils. compressed. Stigmas 2, rarely 3, the ovary subsessile. Utricle depressed-globose. Seeds horizontal; embryo annular, surrounding the scanty endosperm. Ca. 35 spp., mostly Old World. (Named for W. D. J. Koch, 1771-1849, German botanist.)

Plants perennial; native.
    Stems branched mostly at base, relatively simple above; lvs. subterete ...... 1. *K. americana*
    Stems paniculately branched above; lvs. flat, 1-3 mm. wide .... 2. *K. californica*
Plants annual; introd. weeds .... 3. *K. scoparia*

1. **K. americàna** Wats. GREEN-MOLLY. Perennial from a woody branching crown; stems many, erect, 1-3 dm. high, usually villose-tomentose when young, later glabrate; lvs. many, 6-25 mm. long, terete, fleshy, erect or ascending, ca. 1 mm. wide, acutish, sparsely silky to glabrous; fls. solitary or in 2's or 3's, white-tomentose; calyx 2 mm. broad in fr., the wings fan-shaped, 2 mm. long, membranous, striate, crenulate; utricle glabrate; seed brown, flat, rounded in outline, ca. 2 mm. in diam.—Occasional and local on alkaline flats, 4000-6000 ft.; Shadscale Scrub, Joshua Tree Wd.; Barstow and Inyo Co. to se. Ore., Sask., Kans. May-Aug. More densely and permanently pubescent plants have been, designated as var. *vestita* Wats.

2. **K. califórnica** Wats. Perennial, 1.5-5 dm. high, erect; stems simple below, paniculately branched above, grayish-rusty throughout with dense pubescence; lvs. spreading, linear-oblong, 5-15 mm. long, 1.5-3 mm. wide, flat, sericeous; fls. solitary or in clusters of 2-5, tomentulose; calyx 2 mm. in diam. in fr., with flabellate wings 2 mm. long, scarious, finely nerved; seeds 2 mm. in diam.—Alkaline flats; Creosote Bush Scrub, Joshua Tree Wd., V. Grassland, etc.; Mojave Desert from Victorville and Lancaster regions to Inyo Co.; and San Joaquin V.; Nev. May-Sept.

3. **K. scopària** (L.) Schrad. [*Salsola s.* L.] SUMMER-CYPRESS. Erect annual 4-15 dm. high, pyramidal to ovoid, densely leafy, pubescent; lvs. lanceolate to lance-linear, often ciliate at base, soft-strigose especially beneath, 1-4 cm. long; fls. in small axillary clusters; midrib of each sepal becoming prominently thickened.—A weed at scattered stations as in Los Angeles, Long Beach, etc.; e. U.S.; introd. from Eurasia. Aug.-Oct. A very hairy form which sometimes appears is the var. *subvillosa* Moq.

13. Monolèpis Schrad.

Low branched annual herbs. Lvs. alternate, sessile or petioled, entire or hastate. Plants polygamo-dioecious, the fls. sessile, ebracteate, densely clustered in axillary glomerules or sometimes solitary. Calyx of 1 persistent sepal, not changed in fr. Stamen 1 or 0. Styles 2, slender. Fr. ovoid, with thin pericarp, which may be slightly adherent to the seed. Seed erect, compressed, the embryo annular. Three spp. (Greek, *monos*, one, and *lepis*, scale, because of the single sepal.)

Lvs. 10-50 mm. long, frequently hastate; pericarp pitted .... 1. *M. nuttalliana*
Lvs. 5-15 mm. long; entire; pericarp papillose .... 2. *M. spathulata*

1. **M. nuttalliàna** (Schult.) Greene. [*Blitum n.* Schult.] Stems several from base, ascending, stout, succulent, mealy when young, 1-2 (-3) dm. long; lvs. triangular- lance-

olate or narrower, usually hastately lobed at base, the blades 1-4 cm. long, short-petioled or the lower with long petioles; fl.-clusters dense, sessile, often reddish; sepal spatulate or obovate, acutish, ca. 1 mm. long; pericarp 1 mm. broad, minutely pitted; seed 1 mm. broad, dark, with acute margin.—Rather common, dry or moist often saline places and on burns, mostly below 5000 ft., but up to 10,000; many Plant Communities, cismontane and desert; Channel Ids.; to Alaska, Mo., Tex., Son., S. Am.; Asia. April–Sept.

2. **M. spathulàta** Gray. Branched from base, the stems 3-15 cm. long, decumbent or ascending; lvs. narrowly spathulate to oblanceolate, 5-15 mm. long, fleshy, entire; fl.-clusters many-fld., sessile; sepal spathulate, obtuse, ca. 0.5 mm. long; pericarp papillose, free from the seed, ca. 0.5 mm. broad; seed brown, shining, ca. 0.4 mm. in diam.—Rare, moist subalkaline places, 5000-8000 ft.; Montane Coniferous F.; San Bernardino Mts.; Sierra Nevada to Ore., Ida. June–Sept.

14. *Nitróphila* Wats.

Low perennial herb. Lvs. opposite, linear to oblong, fleshy, amplexicaul, entire. Fls. perfect, axillary, 2-bracteolate, small, entire or in 3's. Calyx chartaceous, 5-parted, the segms. concave, carinate, 1-nerved. Stamens 5, united at base into a short perigynous disk. Style filiform; stigmas 2, subulate. Utricle ovoid, beaked by the persistent style, included by the connivent calyx-segms. Seed vertical, lenticular; embryo annular. Several spp., N. and S. Am. (Greek, *nitron*, carbonate of soda, and *philos*, fond of, i.e., alkali-loving.)

Stems to 1 dm. high; lvs. ovate, 0.2-0.3 cm. long .... 1. *N. mohavensis*
Stems 1-3 dm. high; lvs. linear, 1-2 cm. long .... 2. *N. occidentalis*

1. **N. mohavénsis** Munz & Roos. Stems 5-8 cm. high, erect from extensive heavy underground rootstocks; lvs. round-ovate, amplexicaul, concave above, 2-3 mm. long; calyx-segms. oblong-ovate, rose-colored when fresh; seed shining, black, 1.2 mm. long, 1 mm. broad.—Heavy alkaline mud, 2050 ft.; Alkali Sink; Amargosa Desert, Inyo Co. May–July.

2. **N. occidentàlis** (Nutt.) Moq. [*Banalia n Moq.*] Plants glabrous, the stems oppositely much-branched, 1-3 dm. long, decumbent from deep rootstocks; lvs. linear or the lowest oblong, sessile, 1-2 cm. long, mucronate, not much reduced up the stem; bracts like lvs. but shorter, 2-3 times the calyx; fls. quite sessile; calyx-segms. ca. 2 mm. long, broadly oblong, pinkish when fresh, straw-colored on drying; fr. brown; seed shining, black, 1 mm. broad.—Moist alkaline places, below 7000 ft.; several Plant Communities; cismontane s. Calif. and deserts; to Ore., Nev. May–Oct.

15. *Salicórnia* L. GLASSWORT. SAMPHIRE. PICKLEWEED

Annual to suffrutescent herbs with succulent leafless jointed stems, opposite branches and short internodes. Lvs. opposite, reduced and scalelike. Fls. mostly perfect, in cylindric fleshy spikes made up of very short internodes with fls. sunk in groups of 3-7 on opposite sides of the joints. Calyx fleshy, with a truncate or 3-4-toothed margin. Stamens 1 or 2. Styles 2, united at base. Utricle oblong or ovoid, included in spongy calyx. Seed vertical, mostly without endosperm; embryo thick, the cotyledons incumbent upon the radicle. Ca. 10 spp. of saline places; world-wide. (Greek, *sal*, salt, and *cornu*, a horn, being saline plants with hornlike branches.)

Plants annual.
    Joints of spike thicker than long, the scales acuminate .... 1. *S. bigelovii*
    Joints of spike longer than thick, the scales barely acute .... 2. *S. europaea*
Plants perennial from creeping rootstocks.
    Seeds glabrous; spikes slender, the upper part sterile .... 3. *S. subterminalis*
    Seeds pubescent; spikes stoutish, fertile to the top.
        Joints of the flowering spike 7-10; spikes 4 mm. thick .... 4. *S. utahensis*
        Joints of the flowering spikes 12-18; spikes 3 mm. or less thick .... 5. *S. virginica*

1. **S. bigelòvii** Torr. Erect annual, simple-stemmed or with strongly ascending branches mostly from the base, 1–5 dm. high, the joints 1–2.5 cm. long, 2–3 mm. thick, mostly thicker than long; spikes 2–10 cm. long, 4–6 mm. thick; scales of spike triangular-ovate, sharply mucronate; middle fl. half higher than the lateral and reaching nearly to summit of joint; seed pubescent, 1–1.5 mm. long.—Coastal Salt Marsh; Los Angeles Co. to San Diego; Atlantic Coast. July–Nov.

2. **S. europaèa** L. [*S. rubra* A. Nels. *S. depressa* Standl.] Low densely bushy annual, branched from base, 1–2 dm. high, with slender joints 1–2.5 mm. thick; spikes 2–6 cm. long, 2–3 mm. thick, with cent. fl. definitely higher than the lateral ones; seed 1–2 mm. long, pubescent.—Occasional in saline places; Coastal Salt Marsh, Alkali Sink; San Diego, Point Mugu, near Tehachapi; n. and to Atlantic Coast; Old World. July–Nov.

3. **S. subterminàlis** Parish. [*Arthrocnemum s.* Standl.] Plate 37, Fig. D. Perennial; stems widely spreading or erect and compact, 1.5–3 dm. high, the joints 2–20 mm. long; branchlets many, crowded, 2–3 mm. in diam., the joints 5–15 mm. long; spikes 2–3.5 cm. long, 2–2.5 mm. thick at base, with few to several lf.-bearing scales below and as many slender sterile ones above; fls. subequal; seeds glabrous, brown, ca. 1 mm. long.—Salt marshes and low alkaline places; Coastal Salt Marsh, Alkali Sink, etc.; San Francisco Bay and San Joaquin V. to Mex. April–Sept. Some Channel Ids. and inland to Perris, Riverside Co.

4. **S. utahénsis** Tides. Perennial, suffrutescent at base, 1.5–3 dm. high, the stems erect or decumbent, sparsely or much branched, branches erect or ascending, with joints 7–18 mm. long, 2–5 mm. thick, sheaths rounded or lobes acutish; spikes 10–20 mm. long, 4–5 mm. thick; fls. 3 in a cluster, subequal, of ca. the same height, extending nearly to the top of the joint.—Alkali Sink; Death V.; to Utah.

5. **S. virgínica** L. [*S. ambigua* Michx. *S. pacifica* Standl.] Forming extensive colonies, perennial, suffrutescent, decumbent, 2–6 dm. long, rooting along the trailing bases; branches stout, erect or ascending, the joints 6–20 mm. long, 2–4 mm. thick; spikes 1.5–5 cm. long, ca. 3 mm. thick, the fls. almost subequal; seed ca. 1 mm. long.—Coastal Salt Marsh and nearby alkaline flats; some Channel Ids.; L. Calif. to B.C.; Atlantic Coast. Aug.–Nov.

16. *Salsòla* L. Russian-Thistle

Annual or perennial herbs, usually much branched. Lvs. usually alternate, sessile or clasping, narrow, often with pungent apex. Fls. perfect, small, solitary or fascicled, axillary, bibracteolate; calyx 5 (4)-parted, with lanceolate to oblong segms., concave, usually transversely keeled and often winged in fr. Stamens 5, sometimes fewer. Stigmas 2 (3), subulate. Utricle flattened, included in the calyx. Seed horizontal, orbicular; embryo spirally coiled; endosperm lacking. Ca. 50 spp., widely dispersed. (Latin, *salsus*, salty.)

Fruiting calyx 3–6 mm. broad; lvs. 3–5 cm. long. Common .... 1. *S. iberica*
Fruiting calyx 8–9 mm. broad; lvs. 1.5–3 cm. long. Rare .... 2. *S. paulsenii*

1. **S. ibèrica** Sennen & Pau. [*S. pestifera* A. Nels. *S. kali* L. var. *tenuifolia* Tausch. *S. k.* ssp. *ruthnica* (Ilgin) Soo.] Plate 37, Fig. E. Annual 3–10 dm. high, densely and intricately branched, forming a round bushy clump, usually quite glabrous; lower lvs. terete, fleshy, linear, 3–5 cm. long, pungent-tipped; bracts ovate, short-acuminate, prickly-pointed, indurate in fr.; fruiting calyx 3–6 mm. broad, the wings membranous, conspicuously veined, often reddish; seed black, shining, 1.5–2 mm. broad.—Common as a tumbleweed in cult. fields and open areas; many Plant Communities through most of Calif. and w. N. Am. to Atlantic Coast; native of Eurasia. July–Oct.

2. **S. paulsènii** Litv. Plant 1–5 dm. tall, ca. 5–6 dm. across, glabrous or sparsely papillose; lvs. 1.5–3 cm. long, semicylindrical, mucronate, yellow; bracts ovate at base, with a linear spinose apex; bracts partly connate with the solitary fls.; perianth with a short tube and small stiffly erect spinose tips to the segms.; wings membranous, veined, the fruiting calyx 8–9 mm. wide.—Locally common in disturbed Creosote Bush Scrub

2 miles e. of Barstow and at 6000 ft., e. of Baldwin Lake, San Bernardino Mts.; occasional to Utah; introd. from Eurasia.

17. Sarcobàtus Nees. GREASEWOOD

Much-branched spinescent shrubs. Lvs. alternate or opposite, linear, fleshy, sessile. Monoecious or dioecious; ♂ fls. spirally arranged in terminal spikes, ebracteate, without perianth; stamens 2–3, covered by a peltate scarious scale. Pistillate fls. sessile, 1 or 2, each in axil of a lf.; calyx compressed, ovoid or oblong, adnate to base of the 2 recurved subulate stigmas. Fr. coriaceous, with broad crenulate scarious wing at middle, the lower part turbinate, the upper conic. Seed erect, flat, orbicular; embryo spirally coiled; endosperm none. (Greek, *sarx*, flesh, and *batos*, bramble, possibly because of the fleshy lvs. and spiny stems.)

1. **S. vermiculàtus** (Hook.) Torr. [*Batis v.* Hook. *Fremontia vermicularis* Torr.] Shrub 1–2 m. high, the younger branches yellowish-white; lvs. 0.5–3 cm. long; staminate spikes 1–3 cm. long, 3–4 mm. thick; the peltate scales roundish; fruiting calyx-wing 8–12 mm. in diam., the fr.-body glabrate, 4–5 mm. long.—Alkali places, 3000–7000 ft.; Rabbit Springs, s. Mojave Desert; Inyo Co. to Wash., N. Dak., Tex. May–Aug.

18. Suaèda Forsk. SEA-BLITE. SEEP-WEED

Annual or perennial herbs or shrubs. Lvs. alternate, entire, terete to spatulate. Fls. small, mostly perfect, solitary or glomerate in the lf.-axils, bracteate. Calyx 5-lobed, fleshy, globose, turbinate or urceolate, the lobes equal and unappendaged, or 1 or more larger and corniculate-appendaged. Stamens 5, with short fils. Ovary subglobose; stigmas usually 2, short, subulate, recurved. Utricle enclosed in the calyx, compressed or depressed. Seed horizontal or vertical; endosperm none or scanty; embryo coiled into a flat spiral. Ca. 50 spp. of wide distribution. (An Arabic name.)

Plants annual; calyx-lobes unequal, corniculately appendaged .... 2. *S. depressa*
Plants perennial, ± woody at the base; calyx lobes equal, not appendaged.
   Lvs. terete; branches of infl. stout.
      Fls. 2–3 mm. wide; lvs. of infl. crowded; calyx cleft halfway. Coastal ...... 1. *S. californica*
      Fls. 1–1.5 mm. wide; lvs. of infl. not crowded; calyx cleft more than halfway. Alkaline places mostly away from the coast .... 3. *S. fruticosa*
   Lvs. strongly flattened; branches of infl. very slender .... 4. *S. torreyana*

1. **S. califórnica** Wats. [*Dondia c.* Heller.] Plate 37, Fig. F. Glaucous, quite glabrous branching suffrutescent perennial, the stems ascending or decumbent, stout, 3–8 dm. long; lvs. many, rather crowded, spreading, subterete, 1.5–3.5 cm. long, not much reduced in infl.; fls. 2–3 mm. wide; calyx equally 5-cleft, not appendaged, glaucous; seed black, shining, 1.5–2 mm. broad.—Coastal Salt Marsh; San Francisco Bay to L. Calif. July–Oct. Two variants are (1) var. *pubescens* Jeps. [*Dondia brevifolia* Standl.] Lvs. 3–10 mm. long, densely pubescent; fls. 1–1.5 mm. broad; Coastal Salt Marsh, Ventura Co. to San Diego. (2) var. *taxifolia* (Standl.) Munz. [*Dondia t.* Standl.] Stems, lvs. and calyces pubescent, the lvs. 1.5–3 cm. long; fls. 2–3 mm. wide.—Coastal Salt Marsh, Santa Barbara to San Diego.

2. **S. depréssa** (Pursh) Wats. var. **depréssa.** [*Salsola d.* Pursh.] Mostly annual, branched from base, glabrous, with short decumbent branches 2–5 dm. long; lvs. linear, subterete, 1–2.5 cm. long, numerous, beoming shorter in the infl.; fls. crowded, barely 2 mm. wide; one or more of calyx-lobes longer than others and corniculate-appendaged; seed black, 1 mm. broad.—Alkaline places, at 6900 ft.; Pinyon-Juniper Wd.; Baldwin Lake, San Bernardino Mts.; Owens V. n. to Wash., Minn., Tex. July–Sept.

Var. **erécta** Wats. [*S. minutiflora* Wats.] Erect, the stem simple or branched above, 3–6 dm. high; lvs. 2–5 cm. long.—Alkaline flats and saline places, below 5000 ft.; Coastal Salt Marsh, Coastal Sage Scrub, etc.; largely cismontane s. Calif. to Modoc Co., Man., Nebr., Tex. July–Oct.

3. **S. fruticòsa** (L.) Forsk. [*Chenopodium f.* L.] Glaucous quite glabrous perennial, shrubby at base, 2–8 dm. high, much-branched, erect or ascending; lvs. many, narrow-linear, subterete, spreading, 1–2 cm. long, gradually reduced upward; branches of infl. slender, ascending; calyx deeply cleft, ca. 1 mm. broad; seed black, ca. 0.8 mm. broad; $2n = 36$ (Joshi, 1935).—Alkaline and saline places below 3500 ft.; Coastal Salt Marsh, Coastal Sage Scrub, Creosote Bush Scrub, etc.; coastal s. Calif. and through interior valleys n.; to Alta., also to W. Indies. Old World. July–Oct.

4. **S. torreyàna** Wats. [*Dondia t.* Standl.] Green glabrous erect suffrutescent perennial with usually slender ascending or spreading branches, sparsely leafy; lvs. linear, 1.5–3 cm. long, strongly flattened, much shortened in infl.; fls. 1–4 in the axils; calyx deeply cleft, green, with obtuse lobes, rounded on back; seeds 1–1.5 mm. broad, minutely tuberculate.—Occasional in alkaline places mostly away from the coast and below 5000 ft.; Coastal Sage Scrub, Creosote Bush Scrub, Alkali Sink, etc.; San Diego Co. to San Joaquin V., Lassen Co., etc. to Wyo., New Mex. May–Sept. Desert plants are more apt to be pubescent and constitute the var. *ramosissima* (Standl.) Munz.

30. Cistàceae. Rock-Rose Family

Shrubs or herbs. Lvs. opposite, sometimes alternate, simple, entire; stipules well developed to none. Fls. regular, usually bisexual, solitary or in racemes or panicles. Sepals 5, persistent, 2 wholly external, smaller and bractlike or lacking, the 3 inner persistent. Petals 5, rarely fewer or 0, usually ephemeral. Stamens many. Ovary superior, sessile, 1-celled or falsely 5–10-celled; ovules few, or many, on parietal placentae. Style simple; stigma entire or 3–10-lobed. Fr. a 3–5-valved caps.; seeds orthotropous, exalbuminous. Ca. 8 genera and 160 spp., mostly of N. Hemis., some grown as ornamentals.

Valves of caps. 5 or 10; lvs. ovate to oblong. Shrub escaped from cult. . . . . 1. *Cistus*
Valves of caps. 3; lvs. linear to lanceolate. Native herbs or sub-shrubs . . . . 2. *Helianthemum*

1. Cístus L. Rock-Rose

Shrubs. Lvs. opposite. Fls. showy, in terminal cymes. Sepals 3 or 5; petals 5. Stamens many. Ovary with 5 placentae; style with 5–10-lobed stigma. Caps. 5- or 10-valved. Ca. 20 spp. of Medit. region; several in cult. (Ancient Greek name.)

1. **C. villòsus** L. Erect shrub ca. 1 m. high; lvs. ovate to ovate-oblong, 3–7 cm. long; fls. 1–4; sepals 5, ovate, acuminate, 13–14 mm. long; petals 20–30 mm. long, mauve or purplish; caps. oblong-globose, ca. 1 cm. long; seeds brownish, shining. Native of Medit. region. Occurs in cult. in a number of forms, more than one of which has become natur. Var. *tauricus* Grosser is sparsely hairy, not glandular. Var. *undulatus* Grosser is glandular.

2. Heliánthemum Mill.

Herbs or undershrubs. Lvs. in ours largely alternate. Fls. yellow, the broad petals crumpled in the bud. Stigma 3-lobed. Caps. 3-valved. Ca.120 spp. of Medit. region and N. and S. Am. Our spp. particularly abundant after fires. (Greek, *helios*, the sun, and *anthemon*, fl., since the fls. open only in the sun.)

(Schreiber, B. O. The genus Helianthemum in Calif. Madroño 5: 81–85. 1939. Daoud, H. S. and R. L. Wilbur. A revision of the N. Am. species of Helianthemum (Cistaceae). Rhodora 67: 63–82, 201–216, 255–312. 1965.)

Infl. corymbose or short-paniculate, densely glandular. Insular . . . . 1. *H. greenei*
Infl. racemose or paniculate, not glandular. Insular and Mainland . . . . 2. *H. scoparium*

1. **H. greènei** Rob. [*H. occidentale* Greene, not Nym.] Low tufted perennial, 1–2 (–3) dm. high, from a woody base, stellate-villous; lvs. linear-lanceolate to oblanceolate, 1.5–3 cm. long, 3–8 mm. wide, the upper linear; infl. strongly glandular-pubescent; outer sepals lanceolate to linear, 3–5 mm. long, the inner thick, ovate, acuminate, 6–7 mm. long;

petals 5–9 mm. long; caps. sharply acute, 7–8 mm. long; seeds irregular, subpyramidal, black mottled white, cellular-pitted, ca. 1 mm. long.—Dry slopes and stony ridges; Chaparral; Santa Catalina, Santa Rosa and Santa Cruz ids. March–May.

2. **H. scopàrium** Nutt. Plate 38, Fig. A. Suffrutescent, with many ascending or spreading stems 2–3 dm. long, divaricate, ± mottled, but with ultimate twigs erect or not, greenish, stellate-pubescent to glabrate; lvs. narrowly linear, 1–2.5 cm. long; infl. a terminal panicle, leafy, few-fld.; outer sepals linear, 0.5–4 mm. long, inner 2.5–7 mm. long, ovate, acuminate; petals 3–11 mm. long, 3–8 mm. wide; stamens 12–47; caps. 2.5–4 mm. long; seeds irregular, ± angled, black, cellular-papillose, ca. 1 mm. long. Variable; largely Coastal Sage Scrub, Chaparral below 4000 ft.; Mendocino Co. to n. L. Calif. March–June.

In need of study. A tentative key based on Wilbur to what he annotates as "phases" is:

Plants low and divaricate, matted, with ultimate twigs erect; infl. usually leafy, few-fld.; inner sepals 4–5 mm. long; petals 5–7 mm. long. Coastal; Mendocino Co. to Santa Barbara Co. (Santa Rosa, Santa Cruz ids.) .... var. *scoparium*
Plants tall, rush-like; infl. many-fld., sparsely leafy; inner sepals 2–6 mm. long; petals 4–12 mm. long.
Panicle extremely open; petals 8–12 mm. long; inner sepals 5–6 mm. long. Sw. San Bernardino Co. to n. L. Calif. .... var. *aldersonii* (Greene) Munz
Panicle usually narrow; petals ca. 4–6 mm. long; inner sepals 2–3.5 mm. long. Catalina and Santa Cruz ids.; Lake Co. to L. Calif. .... var. *vulgare* Jeps.

31. **Convolvulàceae.** MORNING-GLORY FAMILY

Annual or perennial herbs, chiefly twining or trailing. Lvs. alternate or the plants leafless parasites. Fls. complete and perfect, regular, axillary, solitary or cymose. Calyx of mostly 5 ± united imbricated sepals, persistent. Corolla hypogynous, ± campanulate to funnelform, convolute in the bud, the limb entire or lobed. Stamens 5, borne on the corolla, included. Pistil of 2 ± united carpels, the ovary superior, 2- (sometimes 1-) celled, on a fleshy disk. Fr. a caps., 1–5-celled, with few large seeds. Ca. 50 genera and over 1100 species, of warmer regions.

Leafless twining parasites, with scalelike lvs. .... 4. *Cuscuta*
Leafy green plants.
  Styles 2, distinct or at least partly so; fls. less than 1 cm. long.
    Ovary deeply 2-lobed; creeping matted herb .... 5. *Dichondra*
    Ovary entire; erect herb .... 3. *Cressa*
  Style entire or cleft at apex only; fls. more than 1 cm. long.
    Stigma capitate .... 6. *Ipomoea*
    Stigma filiform or oblong-cylindric.
      Stigma oblong, ± cylindrical, the stigmatic area and style distinct; ovary with an incomplete septum .... 1. *Calystegia*
      Stigma linear, the stigmatic area and the style ± continuous; ovary with a complete septum .... 2. *Convulvulus*

1. *Calystègia* R. Br. MORNING-GLORY

Resembling *Convolvulus*, but with pollen sphaeroidal. Stigma oblong, ± cylindrical with blunt apices, the stigmatic area and style distinct. Caps. 1-locular with an incomplete septum. A fairly large genus of which the spp. have previously largely been referred to *Convolvulus*. (Greek, *kalux*, cup, and *stegos*, a covering.)

1. Calyx enclosed or closely subtended by a pair of large sepallike bracts.
  2. Corolla purple to rose.
    3. Lvs. ovate-hastate, thin, acute to acuminate, 6–10 cm. long; climbing swamp plants
                                                                                         7. *C. sepium*
    3. Lvs. reniform, obtuse, fleshy, 2–5 cm. broad; prostrate seaside herbs ... 8. *C. soldanella*
  2. Corolla white to cream, sometimes pinkish in age.

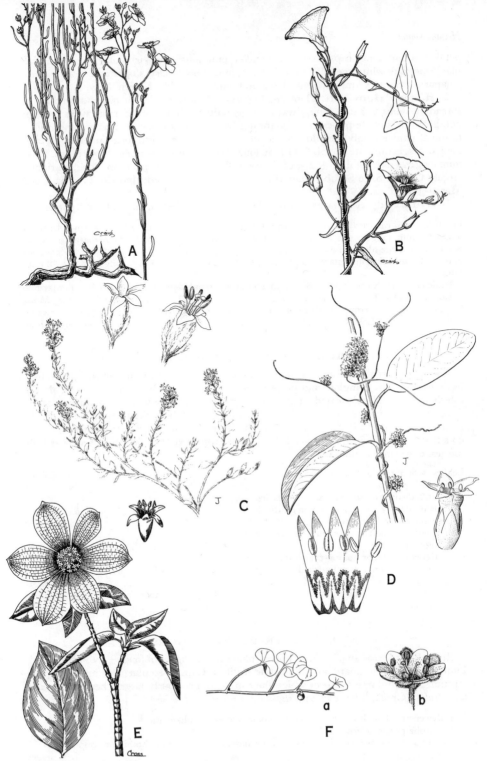

Plate 38. Fig. A, *Helianthemum scoparium*. Fig. B, *Calystegia macrostegia* ssp. *cyclostegia*. Fig. C, *Cressa truxillensis*. Fig. D, *Cuscuta ceanothi*, with fringed scales inside corolla. Fig. E, *Cornus nuttallii*. Fig. F, *Dichondra occidentalis*.

4. Stems mostly over 1 m. long, twining or trailing.
    5. Plants of dry places, the stems ± woody at base .... 3. *C. macrostegia*
    5. Plants of swamps and marshes, the stems entirely herbaceous .... 7. *C. sepium*
4. Stems usually 1–4 dm. long, ascending or trailing; plants variously pubescent
                                                                                          4. *C. malacophylla*
1. Calyx subtended by more remote bracts that are not much like sepals.
  6. Bracts hastately lobed, like the upper lvs. Interior Calif. from San Diego Co. n.
                                                                                                   1. *C. fulcrata*
  6. Bracts entire.
    7. Bracts broadly oblong to oval, attached near the base of the calyx. Desert slopes of the San Gabriel Mts. .... 5. *C. peirsonii*
    7. Bracts subulate to narrowly lanceolate, usually well below the calyx.
      8. Plants climbing; basal lobes of lvs. usually broad and toothed .... 6. *C. purpurata*
      8. Plants not climbing; basal lobes of lvs. linear, entire .... 2. *C. longipes*

1. **C. fulcràta** (Gray) Brummitt. [*Convolvulus f.* Gray. *C. luteolus* var. *f.* Gray.] Rootstocks slender; stems slender, trailing, 1–5 dm. long; herbage light green, minutely puberulent to subvillous; lvs. triangular-hastate, 1.5–5 cm. long, the basal lobes half as long to almost as long as the lanceolate to lance-ovate cent. lobe; petioles shorter than blades except on lowest lvs.; peduncles slender, 1-fld., mostly longer than lvs.; bracts well below the calyx, ± hastate; sepals unequal, oblong, blunt to truncate, pubescent, usually mucronate; corolla white to cream, 2.5–3.5 cm. long; stigmas linear; seeds dark, ca. 4 mm. long, minutely reticulate-ridged.—Dry slopes, 4000–8500 ft.; Montane Coniferous F.; Chaparral; Laguna Mts. to San Gabriel Mts.; Sierra Nevada. May–Aug.

2. **C. longìpes** (Wats.) Brummitt. [*Convolvulus l.* Wats.] Base woody, branched; stems erect to ascending, branched, glabrous, slender, 3–10 dm. high, sometimes twining; lvs. remote, linear to lance-hastate, 1–3 (–5) cm. long, the lower with well developed spreading basal lobes, the upper gradually reduced to linear bracts; petioles 1–2 cm. long; peduncles slender, 1–2-fld., 5–15 cm. long; bracts lance-linear, near to or remote from the calyx; sepals broadly oval, rounded and mucronate at the apex, unequal, 6–10 mm. long; corolla white to cream, often with lavender veins 2.5–3.5 cm. long; stigmas linear-oblong; seeds dark, smooth, ca. 4 mm. long.—Dry slopes, mostly 1500–4000 ft.; Creosote Bush Scrub, Joshua Tree Wd., Chaparral; w. edge of both deserts to Inyo Co., and to San Luis Obispo Co.; w. Nev. May–July.

3. **C. macrostègia** (Greene) Brummitt ssp. **árida** (Greene) Brummitt. [*Convolvulus a.* Greene.] Much like ssp. *macrostegia*, but the lvs. and stems cinereous with dense tomentulose puberulence; bracts mostly greenish, closely investing the calyx, lance-ovate, 12–15 mm. long, mostly acute; sepals like bracts but shorter; corolla 3–3.5 cm. long.—Dry slopes below 3000 ft.; Coastal Sage Scrub, Chaparral; interior, from San Gabriel and San Bernardino mts. to San Jacinto and Santa Ana mts. May–June.

KEY TO SUBSPECIES OF C. MACROSTEGIA

Bracts largely 2–3 cm. long; corolla 5–6 cm. long. Insular. .... ssp. *macrostegia*
Bracts 1–1.5 cm. long; corolla 2–4.5 cm. long. Mostly mainland.
  The bracts mostly subcordate at base, membranous and purplish. Near the coast, n. to Monterey Co. .... ssp. *cyclostegia*
  The bracts mostly rounded at the base, firm and greenish. Largely away from the coast.
    Lvs. and stems cinereous with dense tomentulose puberulence. Interior s. Calif. . ssp. *arida*
    Lvs. and stems glabrous or nearly so.
      Middle lobe of lvs. narrowly linear; corolla 2–2.5 cm. long. W. Riverside Co. to n. L. Calif.
                                                                                                       ssp. *tenuifolia*
      Middle lobe of lvs. narrowly to deltoid lanceolate; corolla 3–3.5 cm. long.
        Basal lobes of lvs. less than half as long as middle lobe, not strongly divergent. Ventura to Orange Co. Catalina Id. .... ssp. *intermedia*
        Basal lobes of lvs. at least half as long as middle lobe, strongly divergent. Dry hills about San Diego and to w. Riverside Co. and e. Orange Co. .... ssp. *longiloba*

Ssp. **cyclostègia** (House) Brummitt. [*Convolvulus c.* House.] Plate 38, Fig. B. With habit of ssp. *macrostegia*; lf.-blades 2–4.5 cm. long, the basal lobes usually toothed or angled; petioles much shorter than blades; bracts near to and like the calyx, 10–15 mm. long, ± pointed; corolla 2.5–4.5 cm. long; seeds ca. 3 mm. long, reticulate-ridged.—Dry slopes up to 1500 ft. elev.; Coastal Sage Scrub, Chaparral; mostly near the coast; n. Los Angeles Co. to Monterey Co. Catalina Id. March–Aug.

Ssp. **intermèdia** (Abrams) Brummitt. [*Convolvulus aridus* ssp. *i.* Abrams.] Lvs. and stems subglabrous; basal lobes of lvs. less than half as long as middle lobe, not strongly divergent; corolla 3–3.5 cm. long.—Coastal slopes, Topatopa, Santa Monica, and Santa Susanna mts., Ventura Co. to w. Orange Co. March–Aug.

Ssp. **longilòba** (Abrams) Brummitt. [*Convolvulus aridus* ssp. *l.* Abrams.] March–May.

Ssp. **macrostègia.** Stems wiry, trailing or twining, 1–4 m. long; lvs. rather fleshy, deltoid-hastate, broader than long to slightly less broad than long, 4–10 cm. long, with somewhat spreading coarsely 2–3-toothed basal lobes; petioles shorter than to longer than blades; fls. mostly solitary; bracts thinnish, 2–3 cm. long; corolla 5–6 cm. long, seeds minutely reticulate-papillose.—Rocky places, Coastal Sage Scrub, Chaparral; most Channel Ids. to Guadalupe Ids. April–July.

Ssp. **tenuifòlia** (Abrams) Brummitt. [*Convolvulus aridus* ssp. *t.* Abrams.] Dry places back from the immediate coast; San Diego Co. to n. L. Calif. March–May.

4. **C. malacophýlla** (Greene) Munz ssp. **pedicellàta** (Jeps.) Munz. [*Convolvulus villosus* var. *p.* Jeps. *Calystegia fulcrata* ssp. *p.* Brummitt.] Stems 1–3 dm. long; lvs. deltoid-hastate, the basal lobes commonly entire; outer sepals hairy-tomentose.—Dry slopes and ridges, mostly 1500–6000 ft.; Chaparral, Foothill Wd.; Mt. Pinos and vicinity to cent. Calif.

Ssp. **tomentélla** (Greene) Munz. [*Convolvulus t.* Greene. *Calystegia fulcrata* ssp. *t.* Brummitt.] With slender rootstocks; stems 2–4 dm. long, prostrate to ascending; herbage slightly tomentose; lvs. deltoid-hastate, 1.5–3 cm. long; bracts near the calyx, 8–10 mm. long; sepals tomentose; corolla 3–3.5 cm. long.—Dry slopes, 4000–6500 ft.; mts. of Kern Co.

Ssp. **tomentélla** var. **deltoìdea** (Greene) Munz. [*Calystegia fulcrata* ssp. *t.* var. *d.* Brummitt. *Convolvulus d.* Greene.] Lvs. 1.2–1.5 cm. long, deltoid, somewhat broader; herbage short-canescent.—At 3000–5000 ft.; Foothill Wd., Yellow Pine F.; Tehachapi Mts., Mt. Pinos.

5. **C. peirsònii** (Abrams) Brummitt. [*Convolvulus p.* Abrams.] Rhizomes slender; stems 2–4 dm. long; herbage glabrous and glaucous; lvs. lance-hastate, the cent. lobe 1.5–2 cm. long, the lateral lobes entire or with 1–2 teeth; bracts oval, closely subtending calyx, 5–8 mm. long; calyx 10–12 mm. long, rounded at apex; corolla 3–4 cm. long.—Dry slopes, 3000–4500 ft.; Creosote Bush Scrub, Joshua Tree Wd.; n. base of San Gabriel Mts. May–June.

6. **C. purpuràta** (Greene) Brummitt. [*Convolvulus luteolus* var. *p.* Greene.] Tall perennial, mostly climbing over plants, glabrous; lf.-blades 2.5–4 cm. long, the lobes sharply pointed, the lateral quite distinct from the cent. which is lanceolate; bracts entire, 5–15 mm. below the calyx, 5–10 mm. long; sepals unequal, 6–12 mm. long, mostly lance-ovate; corolla white to purplish, 4–4.5 cm. long.—Chaparral near the coast, Ventura Co. to n. Calif.

7. **C. sèpium** (L.) R. Br. [*Convolvulus s.* L. ssp. *americana* (Sims) Brummitt.] Perennial from creeping rootstocks, twining or creeping, the stems 1–3 m. long; lvs. broadly ovate, longer than broad, 4–6 cm. long, with sharply angulate lobes; fls. on angled peduncles, from many axils; bracts cordate, 1.5–2.5 cm. long; sepals enclosed in bracts; corolla whitish, ca. 5 cm. long.—Occasional in swampy places; Freshwater Marsh; San Bernardino, Riverside, etc.; e. U.S.

Ssp. **binghàmiae** (Greene) Brummitt. [*Convolvulus b.* Greene.] Glabrous, 1–2 m. high; lvs. mostly obtuse at apex; bracts ca. 8–10 mm. long, half as long as sepals.—Coastal Salt Marsh; Orange Co., Santa Barbara Co.

8. **C. soldanélla** (L.) R. Br. [*Convolvulus s.* L.] BEACH MORNING-GLORY. Prostrate

perennial with fleshy stems 1–5 dm. long from deep-seated rootstocks; herbage glabrous, rather fleshy; lf.-blades reniform, shining, sometimes somewhat angled, 2–5 cm. broad, scarcely as long, on petioles 4–6 cm. long; bracts subtending the calyx round-oval, 8–14 mm. long; sepals oval-ovate, mostly 15–20 mm. long in fr.; corolla rose to purplish, 4–6 cm. long; stigmas ovoid; caps. subglobose, 12–15 mm. long; seeds ca. 5 mm. in diam.—Common in sand; Coastal Strand; San Diego Co. to Wash.; S. Am. Old World. April–May.

2. Convólvulus L. MORNING-GLORY. BINDWEED

Twining or trailing herbs, usually perennial, sometimes suffrutescent. Lvs. alternate, usually petioled, commonly cordate or sagittate. Peduncles axillary, usually with a pair of bracts below the calyx. Fls. showy, white to purplish or rose. Sepals equal or unequal. Corolla funnelform to campanulate, plaited, usually 5-angled or -lobed. Stamens included, inserted on the corolla-tube. Pollen ± elongate. Stigmas 2, linear, ± applanate, acutate at apices, the stigmatic area and styles ± continuous; ovary 2-locular, the septum complete. Seeds ± ovoid, usually 4, glabrous. A fairly large genus of wide distribution. (Latin, convolvere, to entwine.)

Plants annual; corolla ca. 6 mm. long, deeply cleft .... 3. C. simulans
Plants perennial; corolla 20–60 mm. long, not cleft.
  Corolla purple to rose, 2.5–3 cm. long; stems climbing .... 1. C. althaeoides
  Corolla white or with some pink, 1.5–2 cm. long; stems largely prostrate ...... 2. C. arvensis

1. **C. althaeoìdes** L. Perennial; stems climbing, 1–3 m. long; herbage ± strigose; lower lvs. cordate-ovate, crenate, the upper ovate in outline, 3–6 cm. long, 3–7-lobed into irregularly lobed or toothed divisions; petioles mostly 1–2.5 cm. long; peduncles 1–2-fld., exceeding lvs.; bracts subulate, 1–1.5 cm. below the calyx; sepals oval, 7–8 mm. long; corolla purple to rose, 2.5–3 cm. long; $2n=40$ (Sa'ad, 1967).—Locally established as a weed; native of Medit. region. May–Oct.

2. **C. arvénsis** L. BINDWEED. Deep-rooted perennial with prostrate or somewhat twining stems 3–10 dm. long; herbage glabrous to pubescent; lf.-blades oblong sagittate to ovate, obtuse to rounded at apex, mostly 1.5–3.5 cm. long; petioles slender, shorter than blades; peduncles mostly 1-fld.; bracts well below the calyx, subulate to narrowly spatulate; sepals oblong, obtuse, ca. 3 mm. long; corolla white or with some pink, 1.5–2 cm. long; seeds dark, finely punctate, obovoid, 3–4 mm. long; $2n=50$ (Wolcott, 1937), $n=24$ (Khoshoo & Sachideva, 1961).—A very deep-rooted and troublesome weed in orchards, fields and waste places below 5000 ft.; to Atlantic Coast; natur. from Eurasia. May–Oct.

3. **C. símulans** L. M. Perry. [*Breweria minima* Gray, not Aubl. *C. pentapetaloides* auth., not L.] Diffuse minutely puberulent to pubescent annual 1–3 dm. high; lvs. oblong to linear-lanceolate, 1.5–4 cm. long, narrowed gradually at base to a shorter somewhat winged petiole; peduncles 1-fld., short; bracts spatulate, to subulate, 4–5 mm. below the calyx, 3–8 mm. long; sepals oblong-ovate, pubescent, 3–4 mm. long, scarious-margined; corolla pinkish, deeply 5-cleft, 6 mm. long; seeds dark, ca. 2 mm. long, minutely papillate.—Occasional, grassy and rocky places, below 1000 ft.; V. Grassland, Coastal Sage Scrub; cismontane s. Calif., Catalina Id.; to L. Calif. and n. to Contra Costa Co. March–May.

3. Créssa L. ALKALI WEED

Low much-branched perennial herbs with erect or diffuse stems. Lvs. alternate, small, canescent, entire. Fls. solitary in upper axils, small, perfect, 5-merous. Sepals 5. Corolla funnelform, white, persistent, 5-lobed. Stamens 5, exserted. Ovary 2-celled, 4-ovuled; styles 2, distinct; stigmas entire, capitate. Fr. a caps., often 1-seeded by abortion. A small genus of warm temp. and trop. regions around the world. (Greek, *Kressa*, a Cretan woman.)

1. **C. truxillénsis** HBK. var. **vallícola** (Heller) Munz. [*C. v.* Heller.] Plate 38, Fig. C. Low gray tufted woolly-villous plants 1–2 dm. tall; lvs. oblong-ovate, 5–10 mm. long,

sessile; pedicels mostly 2–5 mm. long; calyx canescent, 4–5 mm. long, the lobes ca. 2 mm. wide; corolla ca. 6 mm. long, the lobes ca. 2 mm. long, spreading or reflexed, ovate-lanceolate, acute, 1–1.5 mm. wide; ovary and caps. pubescent; $2n=28$ (Heiser & Whitaker, 1948).—Saline and alkaline places, mostly below 4000 ft.; many Plant Communities from the seashore to the interior of most of cismontane Calif.; to Ore., Tex., Mex. May–Oct.

Var. **mínima** (Heller) Munz. [*C. m.* Heller.] Plants appressed-silky; pedicels 1–4 mm. long; calyx-lobes 2.5–3 mm. wide; corolla-lobes round-ovate, obtuse, 2–2.5 mm. wide.— Alkali flats and swamps; Alkali Sink; Death V. region to Modoc Co.; w. Nev.

4. Cuscùta L. DODDER

Parasitic plants without chlorophyll; stems slender, twining, yellow to orange and fastened to their hosts by knobs or haustoria; lvs. reduced to minute scales; fls. small, perfect, mostly waxy-white, in lateral often compact cymose clusters. Calyx 4–5-cleft. Corolla 4–5-lobed, campanulate to urnshaped or cylindrical, the lobes imbricate in the bud, the tube usually with small crenulate or appendaged scales alternating with the corolla-lobes; stamens alternate with the lobes. Pistil 1, superior; ovary 2-celled; styles 2; stigmas capitate or linear. Caps. globose or ovoid, circumscissile or irregularly dehiscent or indehiscent. Seeds 1–4, with fleshy endosperm. Ca. 100 spp. of wide distribution. (Name supposed to be of Arabic derivation.)

(Yuncker, T. G. The genus Cuscuta. Mem. Torrey Bot. Club 18: 113–331. 1932.)
1. Caps. globose, ± depressed, sometimes thickened about the style-bases.
    2. Ovary without a thickened stylopodium at apex.
        3. Corolla with scalelike appendages attached to the tube below the stamens.
            4. Corolla-lobes obtuse, the tips not inflexed .... 7. *C. obtusiflora*
            4. Corolla-lobes acute, spreading but with the tips inflexed.
                5. Fls. as broad as or broader than long; corolla subglobose.
                    6. Calyx-lobes broadly overlapping at sinuses to form angles; fls. 1.5–2 mm. long
                        9. *C. pentagona*
                    6. Calyx-lobes not overlapping at sinuses to form angles; fls. 2–3 mm. long
                        2. *C. campestris*
                5. Fls. longer than broad; corolla campanulate to funnelform .... 11. *C. suaveolens*
        3. Corolla lacking scalelike appendages at base of stamens.
            7. Anthers oval; corolla-lobes spreading, starlike in fr.; fls. sessile ...... 8. *C. occidentalis*
            7. Anthers linear-oblong; corolla not star-shaped; fls. pedicelled .... 1. *C. californica*
    2. Ovary and caps. with a thickened stylopodium surrounding base of styles ... 5. *C. indecora*
1. Caps. ovoid, conic or beaked, commonly longer than thick.
    8. Calyx-lobes acute to acuminate, the edges entire.
        9. Calyx shorter than the corolla-tube; fls. ca. 5 mm. long; calyx-lobes overlapping
            3. *C. ceanothi*
        9. Calyx as long as or longer than corolla-tube; fls. mostly smaller; calyx-lobes not markedly overlapping.
            10. Corolla-lobes long-attenuate at tips; stems pale; seeds ca. 1 mm. long. On plants in mt. meadows .... 12. *C. suksdorfii*
            10. Corolla-lobes not markedly long-attenuate; stems often orange; seeds ca. 1.5 mm. long. At lower elevs.
                11. Fls. 2–3 mm. long; corolla-lobes ovate-lanceolate; scales attached to the corolla-tube most of their length; anthers oval, the fils. well developed ...... 10. *C. salina*
                11. Fls. 3–4 mm. long; corolla-lobes lanceolate; scales commonly free; anthers oval-oblong .... 6. *C. nevadensis*
    8. Calyx-lobes round, obtuse, denticulate .... 4. *C. denticulata*

1. **C. califórnica** H. & A. Stems medium, yellow; fls. 3–5 mm. long, on short pedicels, forming loose cymose-paniculate clusters; calyx somewhat shorter than to longer than corolla-tube, the lobes lanceolate to triangular, acuminate to acute, overlapping; corolla cylindric-campanulate, often saccate between the stamen-attachments, the lobes narrow; lanceolate, acute, reflexed, longer than the tube; scales lacking; stamens shorter

than lobes; styles much longer than ovary; caps. globose, not circumscissile; seeds slightly over 1 mm. long, light brown, rounded but flattened somewhat on 2 sides, the hilum short.—On many herbs and shrubs, sea-level to 8200 ft.; many Plant Communities, most of cismontane Calif., occasional on deserts; to Wash., L. Calif. May–Aug. Variants are: var. *apiculata* Engelm., with granulate corolla and pointed ovoid ovary; known from a single collection on the Colo. River. Var. *papillosa* Yunck., with densely papillate fls. and pedicels; uncommon with the typical *californica*.

2. **C. campéstris** Yunck. [*C. arvensis* auth., in part.] Stems medium, light yellow; fls. 2–3 mm. long, often glandular, mostly on shorter pedicels and in compact globular clusters; calyx ca. equal to corolla-tube, the lobes oval or round, mostly overlapping when young, but not forming angles; corolla-lobes broadly triangular, acute, the tips often inflexed, ca. as long as the campanulate tube; stamens shorter than corolla; scales ovate, fringed, exserted; styles slender, shorter than ovary; caps. depressed-globose; seeds dull pinkish- or grayish-tan, ovoid, ca. 1.5 mm. long; $2n=56$ (Fogelberg, 1938).—On many hosts, especially of the family *Asteraceae*; cismontane and transmontane s. Calif., at Los Angeles, Riverside; to Wash., Atlantic Coast; W. Indies, S. Am. July–Nov.

3. **C. ceanòthi** Behr. [*C. subinclusa* Dur. & Hilg.] Plate 38, Fig. D. Stems rather coarse, orange; fls. ca. 5–6 mm. long, sessile or on short pedicels, the fl.-clusters scattered or in continuous masses; calyx usually less than half as long as corolla-tube, the lobes lanceolate to ovate, acute, overlapping; corolla cylindrical, the lobes slightly overlapping, erect to spreading, shorter than tube; scales oblong, half as long as tube, short-fringed; styles slender, much longer than ovary; caps. ovoid, pointed, with a collarlike thickening about the intrastylar aperture; seeds ca. 1.8 mm. long, globose.—Mostly on woody plants (*Ceanothus, Rhus, Prunus, Salix, Quercus*, etc.) below 5000 ft.; many Plant Communities; cismontane Calif., occasional e. of Sierra Nevada; to Ore., L. Calif. June–Oct.

4. **C. denticulàta** Engelm. Stems very slender, pale yellow; fls. ca. 2 mm. long, subsessile, in shortened few-fld. infl., often with lance-ovate, acute bracts; calyx-lobes roundish, obtuse, denticulate, overlapping, enclosing the corolla-tube; corolla campanulate, the lobes oval to ovate, spreading, somewhat overlapping, ca. as long as tube; scales denticulate, oblong-ovate; styles shorter than ovary; caps. conic; seeds ca. 1 mm. long, round-ovoid; $2n=15$ pairs (Raven et al., 1965).—On desert shrubs like *Larrea, Haplopappus, Hymenoclea*, etc., mostly below 4000 ft.; Creosote Bush Scrub, Joshua Tree Wd.; Mojave Desert to Mono Co. and n. edge of Colo. Desert; to Utah, Ariz. May–Oct.

5. **C. indécora** Choisy. Stems rather coarse, yellow; fls. 3–4 mm. long, loosely or compactly clustered on pedicels as long or longer; calyx shorter than the corolla-tube, the lobes triangular-ovate, mostly obtusish, ± granular; corolla broadly campanulate, the lobes erect or somewhat spreading; fleshy and glandular toward the tips; stamens equaling or longer than the ovary; caps. globose; seeds ca. 1.7 mm. long, somewhat scurfy; $n=15$ (Raven et al., 1965).—On many hosts, particularly *Ambrosia, Asclepias, Aster, Artemisia, Medicago*, etc. and often in moist places like alfalfa fields; to Ill., Mex., se. U.S., W. I., S. Am. July–Oct.

6. **C. nevadénsis** Jtn. [*C. veatchii apoda* and *C. salina apoda* Yunck.] Fls. 3–4 mm. long, the lobes of the calyx and corolla more lanceolate than in *C. salina* and also somewhat longer; scales mostly broader and commonly free; anthers oval-oblong, subsessile.—On *Atriplex*, etc., Towne's Pass, Panamint Mts.; Nev.

7. **C. obtusiflòra** HBK. var. **glandulòsa** Engelm. Stems medium, light yellow; fls. glandular, subsessile, in compact glomerules; calyx-lobes broadly ovate, obtuse, ca. as long as corolla-tube; corolla campanulate, the tube short, lobes triangular-ovate, acutish, spreading or reflexed; scales large, ovate, prominently fringed; stamens shorter than corolla-tube; styles stoutish, subulate, ca. as long as globose ovary; caps. depressed-globose, bursting irregularly; seeds ca. 1.5 mm. long.—Often on *Polygonum*; reported from El Monte, San Bernardino; to Fla., W. I., Mex. May–Oct.

8. **C. occidentàlis** Millsp. Stems medium, yellow; fls. ca. 3 mm. long, often glandular, mostly sessile in compact clusters; calyx equaling or exceeding corolla-tube, the lobes

lance-ovate, acuminate, fleshy and thickened at base; corolla globose, saccate between the stamen-attachments, becoming globular about the developing caps.; lobes lanceolate, acuminate, spreading in age; stamens shorter than the lobes; scales lacking; styles longer than ovary; caps. globose; seeds ca. 1.5 mm. long.—On hosts like *Grindelia, Solanum*, etc.; many Plant Communities; from the coastal counties of Calif.; Catalina Id.; to Wash., Colo. June–Aug.

9. **C. pentagòna** Engelm. Stems slender, pale yellow; fls. mostly in small loose clusters, ± glandular, ca. 1.5 mm. long, on pedicels ca. as long as fls.; calyx mostly ca. as long as corolla-tube, the lobes broadly ovate, obtuse, broadly overlapping and calyx appearing angular; corolla-lobes spreading, lanceolate, the tips acute, inflexed, ca. as long as tube; stamens shorter than lobes, the fils. slender, equal to or longer than anthers; scales becoming exserted, ovate, fringed at top; styles slender, not longer than ovary; caps. globose or ± depressed, protruding from withered corolla; seeds depressed-globose, ca. 1 mm. long; $2n = $ ca. 56 (Fogelberg, 1938).—On many hosts, *Xanthium, Trifolium*, etc.; reported from San Diego and Santa Barbara cos.; to e. U.S.

10. **C. salìna** Engelm. Stems very slender, orange; fls. 2–3 mm. long, on mostly shorter pedicels, in umbellate-cymose clusters; calyx-lobes lance-ovate, ca. as long as corolla-tube; corolla-lobes ca. as long as subcampanulate tube, lance-ovate, erect or spreading, ± overlapping; scales oblong, narrow, shorter than tube, fringed; fils. not longer than anthers; styles not longer than ovary; seeds ca. 1.5 mm. long, round-ovoid.—On *Chenopodium, Cressa*, etc., in saline places at fairly low elevs.; to B.C., Utah, Ariz. May–Sept. Var. *major* Yunck. has fls. 3–4.5 mm. long, the corolla-lobes more ovate; largely on *Salicornia*; San Diego to B.C.

11. **C. suaveòlens** Ser. Stems slender, strawcolored; fls. 3–4 mm. long, ± glandular, membranous, on mostly shorter pedicels, in racemose clusters; calyx-lobes shorter than corolla-tube, triangular-ovate, acutish, not overlapping, the sinuses ± rounded; corolla campanulate to funnelform, globular in age, the lobes triangular-ovate, erect, with tips inflexed and ca. half or more the length of the tube; stamens shorter than lobes; scales mostly not reaching the stamens; fringed; styles slender, mostly ca. as long as ovary; caps. globose, bursting irregularly; seeds 1.5–2 mm. long.—Mostly on legumes, often a pest in alfalfa fields; introd. from S. Am. Aug.–Oct.

12. **C. suksdórfii** Yunck. var. **subpedicellàta** Yunck. Stems slender, pale; fls. 4–5-merous, 2–3 mm. long, sessile or subsessile, in few-fld. umbellate clusters; calyx enclosing the corolla, the lobes triangular, acute; corolla-tube campanulate, the lobes triangular-ovate, acuminate, erect; stamens shorter than lobes; scales oblong, represented by 2 shallowly dentate wings; styles shorter than ovary; caps. globose or depressed-globose, glandular; seeds ca. 1 mm. long, globose.—On *Calyptridium, Aster, Trifolium*, etc., at 5000–8000 ft.; Montane Coniferous F.; San Bernardino Mts.; Sierra Nevada. July–Sept.

5. *Dichóndra* Forst. & Forst. f.

Creeping perennial herbs with glabrous to silky foliage and slender stems forming extensive mats. Lvs. small, orbicular to reniform, petioled. Fls. small, solitary on axillary bractless peduncles. Calyx deeply 5-parted into subequal oblong to spatulate or obovate segms. Corolla deeply 5-parted, the lobes spreading. Stamens 5, short, with slender fils. Ovary deeply 2-parted. Caps. 2, utricular, 1–2-seeded. Seeds rather large, round-obovoid. Several spp. of warmer regions, some grown as ground covers. (Greek, *di*, two, and *chondra*, a grain, from the fr.)

(Tharp, B. C. and M. C. Johnston. Brittonia 13: 346–360. 1961.)
Fr. very deeply bilobed, the carpels usually 1-seeded.
    Calyx-lobes tapered and short-pointed; pedicels filiform. Escape from cult. .... 3. *D. repens*
    Calyx-lobes apically rounded; pedicels erect, but sharply recurved in upper part. Native in cent. Calif. .... 1. *C. donelliana*
Fr. slightly bilobed, the carpels often with 2 seeds in each carpel. Native of s. Calif.
                                                                2. *C. occidentalis*

1. **D. donelliàna** Tharp & Johnston. Internodes largely 1–2 mm. thick; lf.-blades mostly 10–15 mm. long; petioles 2–4 cm. long; calyx-lobes ca. twice as long as broad; corolla 2–3 mm. long; utricles 2.5–3.9 mm. long, pilose toward the summit; seed 2.4–3 mm. long.—This is the plant that has been treated in Calif. literature as *D. repens*, but is native from Monterey Co. to Ore.

2. **D. occidentàlis** House. Plate 38, Fig. F. Rather coarse, the internodes largely 1–2 cm. long, ± sparsely pubescent; lf.-blades broadly reniform, 2–5 cm. broad; petioles 2.5–6 cm. long; calyx-lobes lanceolate-oblong, to 2 mm. long; corolla 3–3.5 mm. long; utricles 2.8–3.8 mm. high, silky-pubescent; seeds 2.5–3 mm. long.—Mostly dry sandy banks in brush or under trees; Coastal Sage Scrub, Chaparral, S. Oak Wd.; coastal San Diego and Orange cos., Catalina, Santa Cruz and Santa Rosa ids.; n. L. Calif. March–May.

3. **D. rèpens** Forst. & Forst. f. Pedicels long, filiform, essentially straight; calyx-lobes narrow, tapered, and pointed, to 5 mm. long; corolla ca. 3.5 mm. long.—Native of New Zealand, etc.; cult. widely as a lawn plant in Calif. and becoming locally established.

6. *Ipomoèa* L. Morning-Glory

Ours twining or trailing herbs similar to *Calystegia* and *Convolvulus*. Calyx ebracteate, although the pedicels may bear alternate bracts below the calyx. Corolla salverform to campanulate, the limb entire or slightly lobed. Style undivided. Stigma capitate or of 2–3 globular lobes. Caps. globular, 2–4-valved. A large genus of warm regions, one sp. (*I. batatas*) the sweet-potato, others grown for their large fls. (Greek, *ips*, a worm, and *homoios*, like; from the twining habit.)

Sepals ca. 7–12 mm. long.
  Corolla 5–6 cm. long .... 3. *I. purpurea*
  Corolla 1.5–2 cm. long .... 4. *I. triloba*
Sepals ca. 20–30 mm. long.
  Corolla 2–4 cm. long, lvs. not canescent .... 2. *I. nil*
  Corolla 6–8 cm. long; lvs. silvery-canescent .... 1. *I. mutabilis*

1. **I. mutábilis** Ker-Gawl. Weedy pubescent perennial; lvs. to 1 dm. long, ca. as wide, cordate, whitish-tomentose beneath; calyx 2–3 cm. long; corolla vivid ultramarine blue, to ca. 8 cm. across.—Escaping as a weed in waste places, as in Santa Barbara and Ventura cos.; native of Mex.

2. **I. níl** (L.) Roth. [*I. hederacea* auth.] Annual, with slender twining ± retrorsely pubescent stems; lvs. cordate-ovate in outline, deeply 3-lobed, 3–8 cm. long, sparsely strigose; peduncles short, axillary, 1–3-fld.; calyx-lobes 12–20 mm. long, hirsute below, with long linear tips; corolla 2–4 cm. long, the limb light blue or purple; seeds glabrous.—Occasional garden escape; native of se. U.S. July–Nov.

3. **I. purpúrea** (L.) Roth. [*Convolvulus p.* L.] Common Morning-Glory. Annual with twining hairy stems; lvs. broadly cordate-ovate, 7–12 cm. long, entire, short-acuminate, pubescent; petioles shorter; peduncles 1–5-fld., with evident pedicels; sepals lanceolate to oblong, acute, 12–16 mm. long, pubescent; corolla funnelform, purple to blue, pink or white, 5–6 cm. long; seeds glabrous; $n=15$ (A. Jones in 1964).—Frequent escape from gardens and established in waste places, city dumps, etc.; native of trop. Am. June–Nov.

4. **I. trilòba** L. Climbing annual; stems glabrous; lvs. cordate, entire or 3-lobed, 3–6 cm. long, 2–5 cm. wide, glabrous, subacuminate; petioles 3–5 cm. long; infl. axillary, peduncled, 1–5-fld., umbellate; calyx 8 mm. long, the sepals ciliate, long-acuminate; corolla ca. 15 mm. long; $n=15$ (A. Jones in 1964).—Escape in Imperial and Riverside cos.; native of trop. Am.

32. **Cornàceae.** Dogwood Family

Trees, shrubs or herbs, usually with opposite, sometimes whorled or entire lvs. and perfect or imperfect fls. in capitate or cymose infls. Sepals 4–5, minute. Petals 4–5, rarely more, inserted at base of epigynous disk. Stamens as many as petals or more. Ovary

Plate 39. Fig. A, *Cornus occidentalis*. Fig. B, *Dudleya edulis*. Fig. C, *Dudleya hassei*. Fig. D, *Dudleya saxosa*. Fig. E, *Sedum niveum*. Fig. F, *Crossosoma californica*.

inferior, 1–2-loculed; style 1; stigma terminal. Ovule solitary in each locule, pendulous. Fr. a drupe; stone 1–2-celled; endosperm present. With 16 genera and ca. 80 spp., mostly of N. Hemis.

1. Córnus L. Dogwood

Fls. small, 4-merous. Ca. 25 spp., of n. temp. regions and s. to Peru. (Latin, *cornu*, horn, because of the hard wood.)

(Rickett, H. W. Cornaceae. N. Am. Fl. 28B: 299–316. 1945.)

Fls. in a head or umbel subtended by conspicuous persistent bracts, usually appearing before the lvs. .... 2. *C. nuttallii*
Fls. in a bractless cyme, appearing with or after the lvs.
    Veins 3–4 on each side of midrib; lf.-blades 3–5 cm. long .... 1. *C. glabrata*
    Veins 4–7 on each side of midrib; lf.-blades mostly 5–10 cm. long.
        Style less than 2.5 mm. long; petals 2–3 mm. long; cyme and lower surfaces of lvs. variously pubescent but not hirsute; endocarp smooth, usually at least as long as broad
                                                                                                    4. *C. stolonifera*
        Style 2.5 mm. long or more; petals 3–4 mm. long; cyme and lower surfaces of lvs. densely hirsute; endocarp ridged, mostly broader than long .... 3. *C. occidentalis*

1. C. glabràta Benth. [*Svida catalinensis* Millsp. *Svida g.* Heller.] Brown Dogwood. Shrubs to small trees 1.5–6 dm. tall, often forming thickets by means of underground shoots; twigs nearly or quite glabrous, brownish to reddish-purple, slender; lvs. lanceolate to elliptic, acute at both ends, subglabrous, gray-green, paler beneath, the blades mostly 2–5 cm. long, 1.5–2.5 cm. wide, on petioles 3–8 mm. long; infl. 2.5–4.5 cm. across, ± strigulose; pedicels 2–3 mm. long; sepals 0.8 mm. long; petals 4.5–5 mm. long; styles 3.5 mm. long; drupes white to bluish, subglabrous, 8–9 mm. in diam., the stone 5–6 mm. broad, almost smooth.—Damp places below 5000 ft.; many Plant Communities; cismontane from San Diego Co. to Ore. Santa Catalina Id. Uncommon in s. Calif. May–June.

2. C. nuttállii Aud. Mountain Dogwood. Plate 38, Fig. E. Arborescent or a tree, 4–25 m. tall; twigs at first green, later dark red to almost black, strigulose; lf.-blades commonly 6–12 cm. long, 3–7 cm. broad, elliptic to obovate, cuneate at base, shortacuminate, minutely strigulose above, paler beneath and pubescent with appressed and looser hairs; petioles 5–10 mm. long; infl. appearing in autumn, subtended by 2 lvs. and 2 bracts that persist until spring; at anthesis the head of fls. subtended by 4–7 white or ± pinkish petaloid bracts 4–6 cm. long, also by smaller bracts; calyx 2.5 mm. long; petals 4 mm. long; style 2 mm. long; drupes red, 1–1.5 cm. long, ± ellipsoid, the stone smooth.—Mountain woods below 6500 ft., largely Montane Coniferous F.; occasional in our s. mts. from San Diego Co. to Los Angeles Co.; more common n. of our area; to B.C., Ida. April–July.

3. C. occidentàlis (T. & G.) Cov. [*C. sericea* var. ? *o*. T. & G.] Plate 39, Fig. A. Much like *C. stolonifera*, but the twigs ± hirsute; lf.-blades mostly 6–10 cm. long, 3–5 cm. broad, sparsely strigulose above, paler and densely hirsute beneath; infl. ± hirsute; sepals 0.7 mm. long; petals 3–4.5 mm. long; style 2.5–3 mm. long; drupes white, ca. 8 mm. in diam., the stone with 3 often broad and low ridges on each face and furrowed laterally.—Moist places below 8000 ft.; many Plant Communities; San Diego Co. through cismontane Calif.; n. to B.C. May–July.

4. C. stolonífera Michx. American Dogwood. Spreading shrubs 2–5 dm. tall, often rooting at tips of branches; twigs strigulose, bright red-purple; lf.-blades commonly 5–9 cm. long, 1.5–5 cm. wide, lanceolate to ovate or elliptic, acute to acuminate, subglabrous above, strigulose beneath, villous-tufted in axils of veins, minutely papillose; petioles 5–8 (–12) mm. long; infl. 3–6 cm. across, strigulose to hirtellose; pedicels mostly 1–5 mm. long; sepals 0.5 mm. long; petals 2–3 mm. long; styles ca. 2 mm. long; drupes white, 7–9 mm. in diam., subglobose, the stone smooth on the faces, furrowed laterally, 4–5 mm. broad; $2n = 22$ (Taylor, 1967).—Moist places below 9000 ft.; mostly in Montane Coniferous F. Mostly n. of our area.

C. × califórnica C. A. Mey. A series of segregates even in our area that combine characters of *C. occidentalis* and *C. stolonifera* and may be hybrids of the two.

33. Crassulàceae. STONECROP FAMILY

Herbaceous or somewhat woody, mostly succulent and glabrous. Lvs. in ours entire, exstipulate. Fls. in cymes or panicles, rarely solitary, regular, usually perfect. Calyx free from the ovary, mostly persistent, 4–5-parted or -lobed. Petals 4–5, distinct or ± united, usually persistent. Stamens as many or twice as many as petals. Carpels as many as calyx-segms., distinct or united below, usually with a scale at base of each; styles filiform or subulate. Fr. of 1-loculed follicles, dehiscent along ventral suture. Seeds minute, mostly narrow, pointed at both ends. Ca. 25 genera and 900 spp., widely distributed, many of horticultural value.

1. Plants annual; stamens mostly 3–5 .... 2. *Crassula*
1. Plants perennial; stamens twice as many as petals.
    2. Flowering stems lateral, arising from axils of lvs. of basal rosette .... 3. *Dudleya*
    2. Flowering stems terminal, not arising in axils of basal rosette.
        3. Corolla with petals connate at base; lvs. 5–10 cm. long. Garden escape .... 1. *Cotyledon*
        3. Corolla with parts free or essentially so; lvs. mostly smaller. Native .... 4. *Sedum*

1. Cotylèdon L.

Plants ± woody at base. Lvs. sessile, opposite or alternate. Fls. erect or drooping, yellow red, or greenish, in terminal cymes. Calyx 5-parted. Corolla tubular, cylindrical, gamopetalous, usually much longer than calyx. Stamens 10, mostly included. Carpels 5, free, with narrow scale at base of each. Fr. of several-seeded follicles. Ca. 30 spp., largely of S. Afr. (Greek, *cotule*, a cavity, from the cup-like lvs. of some spp.)

1. **C. orbiculàta** L. Stout, branched, 6–12 dm. high; lvs. opposite, oblong to round-obovate or spatulate, 5–10 cm. long, flat, glaucous, powdery, often with red or purple margins; fls. red, drooping, the cyme on a peduncle ca. 5–6 dm. long; corolla-tube 2–2.5 cm. long, 4–5 times the calyx-length; $n = 9$ (Uhl, 1948).—Natur. on bluffs as at Newport Beach, Orange Co., Santa Barbara.

2. Cràssula L.

Our spp. much-branched glabrous annual herbs, with opposite entire minute lvs. Fls. very small, axillary, subsessile to pedicelled. Calyx 3–5-parted (in ours mostly 4-). Petals distinct or united at very base, 3–5 (mostly 4). Carpels 3–5, distinct, with short subulate styles and 1–12-seeded, becoming follicles. The whole genus including many perennials has perhaps 250 spp. and is largely of S. Afr. (Diminutive of Latin, *crassus*, thick.)

Fls. solitary; carpels several-seeded. Mud flats .... 1. *C. aquatica*
Fls. clustered; carpels 1–2-seeded. Dry flats .... 2. *C. erecta*

1. **C. aquática** (L.) Schoenl. [*Tillaea a.* L.] Stems spreading or decumbent, 1–6 cm. long; lvs. oblong-linear, entire, 4–6 mm. long, connate at base; fls. solitary, axillary, sessile or short-pedicelled; fl.-parts mostly 4; sepals 1 mm. long, ovate; petals longer; carpels several-seeded.—On dry mud-flats below 2000 ft.; many Plant Communities; widely scattered localities, cismontane s. Calif.; to Wash., e. U.S.; Mex., Eurasia. Catalina Id. Occasional at higher elevs., as Hidden Lake, San Jacinto Mts., at 8000 ft. March–July.

2. **C. erécta** (H. & A.) Berger. [*Tillaea e.* H. & A.] Tufted, erect, branched, 2–7 cm. high, becoming reddish; lvs. fleshy, ovate or oblong, connate at base, 1.5–3 mm. long; fls. in axillary bracted clusters; sepals mostly 4, ovate, 1 mm. long; petals lanceolate, slightly shorter; seeds mostly 1 in each carpel.—Common often in great masses in dry places, burns, etc. below 2500 ft.; many Plant Communities in s. Calif.; to Ore., Ariz., L. Calif.; Chile. Feb.–May.

3. Dúdleya Britt. & Rose. LIVE-FOREVER

Perennial herbs with simple or branched rootstocks or small globose to oblong corms; lvs. principally in basal rosettes, fleshy, flattened to subterete, ± ovate to linear. Flowering stems axillary, with cauline lvs. much reduced, and ± like fleshy bracts. Fls. in terminal paniculate or cymose clusters, the branches of which are here termed "cincinni". Calyx deeply 5-lobed into erect, lance-linear to ovate segms. Corolla white to yellow or red, cylindric or campanulate, the petals united near the base, erect or spreading from near the middle or near the tips. Stamens 10, borne on the corolla-tube. Carpels 5, ± united below, erect or divergent. Seeds many, minute, narrow, pointed. Perhaps 40 spp. of sw. N. Am., often very variable and apparently hybridizing freely. (Named for W. R. *Dudley*, early professor of Botany at Stanford Univ.)

1. Primary stem a ± elongating epigaeous caudex; rosette lvs. mostly persistent.
   2. Corolla tubular, the petals erect, with only the tips slightly spreading.
      3. Petals united for ⅓ or more of their length.
         4. Fls. spreading to pendent at anthesis, erect in fr. by a sharp bending of the pedicel; corolla red, the lobes ca. as long as the tube; plant densely white-pulverulent.
            5. Caudex 4–9 cm. thick; rosette-lvs. 8–25 cm. long; pedicels 5–30 mm. long. Coastal .... 17. *D. pulverulenta*
            5. Caudex 1–4 cm. thick; rosette -lvs. 5–15 cm. long; pedicels 5–15 mm. long. Deserts .... 2. *D. arizonica*
         4. Fls. erect or ± spreading, the pedicels not sharply bent.
            6. Rosettes large, 1–5 dm. in diam., usually of 30–80 lvs. or more; rosette-lvs. mostly 6–25 cm. long, 1.6–8 cm. wide. Insular .... 6. *D. candelabrum*
            6. Rosettes smaller, less than 1 dm. in diam., mostly of 10–25 lvs.; rosette-lvs. mostly 2–8 cm. long, 0.5–3 cm. wide. San Bernardino Mts. to L. Calif. .... 1. *D. abramsii*
      3. Petals united for less than ⅓ of their length.
         7. Plant branching by stolons; predehiscent carpels ascending; petals erect, yellow. San Joaquin Hills, Orange Co. .... 19. *D. stolonifera*
         7. Plant with dichotomous branching or none; predehiscent carpels mostly erect.
            8. Rosettes large, 1–6 dm. in diam., usually solitary, usually with 20–45 lvs.; petals white to pale yellow or pink. Insular .... 6. *D. candelabrum*
            8. Rosettes smaller, solitary to many; mostly less than 1.5 dm. in diam., or if larger, with not more than 25 rosette lvs.
               9. Plants leafless in summer; rosette-lvs. 5–12 (–15), 1.5–4 cm. long, 0.3–1.2 cm. wide; infl. mostly of 1–2 simple concinni. Santa Monica Mts.
                  10. Petals bright yellow, sharply acute, 10–14 mm. long; rosette-lvs. 5–12 mm. wide; pedicels 5–12 mm. long .... 7. *D. cymosa marcescens*
                  10. Petals pale yellow, acute, 8–12 mm. long; rosette-lvs. 3–6 mm. wide; pedicels 1–3 mm. long .... 16. *D. parva*
               9. Plants evergreen; rosette-lvs. mostly 10–25 or more, if fewer, then more than 12 mm. wide.
                  11. Caudex elongate, mostly 1–6 dm. long, often much branched; rosette-lvs. usually rather thick. Maritime.
                     12. Rosette-lvs. oblong-ovate, 2–6 cm. long, 1–2.5 cm. wide; petals pale yellow, the exposed margin of each petal commonly separated from each adjacent petal. Los Angeles Co. n. on sea bluffs .... 10. *D. farinosa*
                     12. Rosette-lvs. longer or narrower; petals pale yellow to bright yellow or red, the exposed margin of each petal usually appressed against the adjacent petal.
                        13. Petals bright yellow to red, 8–16 mm. long, erect; rosette-lvs. thick or thin. Monterey Co. to Los Angeles Co. .... 5. *D. caespitosa*
                        13. Petals pale yellow, 8–12 mm. long, sometimes curved outwardly at apex; rosette-lvs. thick. Insular .... 11. *D. greenei*
                  11. Caudex short, erect, rarely over 0.5 dm. long, simple or cespitosely few-branched; rosette-lvs. relatively thin. Coastal to inland.
                     14. Pedicels 0.5–6 mm. long.
                        15. Petals thin, often erose, 8–13 mm. long, 2–3 mm. wide, pale yellow and with red lines, connate for 1.5–4.5 mm.; rosette-lvs. 1.5–6 (–11) cm. long,

0.5–2 cm. wide. San Luis Obispo Co. and San Bernardino Mts. to L. Calif.
..... 1. *D. abramsii*
15. Petals thick, entire, 10–16 mm. long, 3.5–5 mm. wide, pale or bright yellow to red, connate for 1–2 mm.; rosette-lvs. 5–20 (–30) cm. long, 1–4 cm. wide. Santa Barbara Co. to L. Calif. .... 13. *D. lanceolata*
14. Pedicels 5–20 mm. long.
16. Rosette-lvs. oblanceolate to spatulate, acuminate to cuspidate, and sometimes tapering from base and acute; petals thin, sharply acute; fl.-stems 0.5–2.5 (–4) dm. tall; infl. rather dense, the cincinni mostly 1–5 cm. long. N. and cent. Calif. to Santa Monica and San Bernardino mts.
..... 7. *D. cymosa*
16. Rosette-lvs. oblong-lanceolate, acute to subacuminate; petals thick to thin, acute; cincinni often elongate. S. Calif.
17. Rosette-lvs. 5–30 cm. long, 1–4 cm. wide; fl.-stems 1.5–7.5 dm. tall; petals yellow to red, connate for 1–2 mm. Santa Barbara Co. to L. Calif.
..... 13. *D. lanceolata*
17. Rosette-lvs. 3–15 cm. long, 0.5–2.5 cm. wide; fl.-stems 0.5–4 dm. tall; petals yellow, connate for 1–4 mm. .... 18. *D. saxosa*
2. Corolla with segms. widely spreading from near the middle.
18. Lvs. flattened.
19. Petals bright yellow; lvs. 4–6 times as broad as thick. Santa Barbara Id.
..... 20. *D. traskae*
19. Petals white or tinged with red; lvs. often narrower.
20. Herbage viscid. Mainland .... 23. *D. viscida*
20. Herbage not viscid. Mostly insular.
21. Lvs. glaucous; petals white. Catalina Id. .... 12. *D. hassei*
21. Lvs. green; petals reddish. Catalina and San Clemente ids. .... 22. *D. virens*
18. Lvs. terete or nearly so, except at very base.
22. Branches of infl. usually 2 or 3, simple; caudex mostly less than 1 cm. thick; mature carpels ascending, slightly gibbous. S. San Diego Co. .... 3. *D. attenuata*
22. Branches of infl. several, once or twice bifurcate; caudex mostly more than 1 cm. thick; mature carpels abruptly divergent, strongly gibbous.
23. Herbage mealy; styles 2.5–3 mm. long; pedicels 2–4 mm. long. San Gabriel Mts.
..... 8. *D. densiflora*
23. Herbage slightly glaucous but not mealy; styles 1.5–2 mm. long; pedicels 0–2 mm. long. W. Riverside and Orange cos. south .... 9. *D. edulis*
1. Primary stems hypogaeous, usually cormlike; rosette-lvs. vernal.
24. Fls. white, with a sweet odor.
25. Petals erect to ascending, 7–14 mm. long; lf.-bases 4–12 mm. wide. Santa Cruz Id.
..... 15. *D. nesiotica*
25. Petals wide-spreading, 5–10 mm. long; lf.-bases 1–4 mm. wide. Mainland and Santa Rosa Id. .... 4. *D. blochmanae*
24. Fls. yellow, odorless.
26. Rosette-lvs. oblanceolate, strongly narrowed below, 1–7 cm. long; petals connate for 0.5–1 mm. San Diego region .... 21. *D. variegata*
26. Rosette-lvs. linear-lanceolate, not narrowed at base, 4–15 cm. long; petals connate for 1–2 mm. Los Angeles, Riverside and Orange cos. .... 14. *D. multicaulis*

1. **D. àbramsii** Rose. [*Echeveria a.* Berger. *D. tenuis* Rose.] Caudex short, 1–1.5 cm. thick; rosette-lvs. gray, ± glaucous, lanceolate, 2–6 cm. long, acuminate; fl.-stems slender, 0.5–1.6 dm. long; cauline lvs. mostly in upper parts, ovate, acute, ± cordate, the lower 2–5 mm. long; cicinni 2–3, rather few-fld., 2–12 cm. long; pedicels 2–5 mm. long; calyx 3–5 mm. long, greenish, the lobes lance-ovate; petals pale yellow with reddish stripes, 7–12 mm. long, connate for 2–4 mm., acute; $n=17$ (Uhl & Moran, 1953).—Dry rocky places, 2500–9000 ft.; Chaparral to Yellow Pine F.; desert edge from San Bernardino Mts. to n. L. Calif. April–June.

2. **D. arizónica** Rose. [*Echeveria lagunensis* Munz.] Caudex 1–4 cm. thick; lower lvs. obovate to spatulate, 5–15 cm. long, 2–5 cm. wide; fl.-stems 2–6 dm. high; cauline lvs. 1–3 cm. long, clasping, broadly ovate; pedicels rather slender, 5–15 mm. long; calyx 5–7 mm. long, the lobes lanceolate; petals 12–14 mm. long, brick-red; $n=17$ (Uhl & Moran,

1953).—Dry slopes 2000–4000 ft.; Creosote Bush Scrub, Joshua Tree Wd.; desert slopes of Laguna and Cuyamaca mts. and of mts. in e. Mojave Desert (Kingston, Old Dad, Old Woman); to Nev., Ariz., Son., L. Calif. May–July.

3. **D. attenuàta** (Wats.) Moran ssp. **orcúttii** (Rose) Moran. [*Stylophyllum o.* Rose. *S. parishii* Britton. *Echeveria palensis* Berger.] Caudex often elongate and branched, 0.5–1 cm. thick; rosette-lvs. 10–15, linear-oblanceolate, acute, 3–10 cm. long, glaucous when young; fl.-stems 1–2.5 dm. long; infl. usually with 1–3 simple branches; pedicels 1–3 mm. long; calyx-lobes deltoid-ovate to oblong-ovate, acute, connate 1–2 mm.; carpels 4–7 mm. long, ± spreading in age; seeds red-brown, 0.8–1 mm. long; $n = 17$ or 34 (Uhl & Moran, 1953).—Dry gravelly or rocky places below 1000 ft.; Coastal Sage Scrub, Chaparral; s. San Diego Co. from Pala s.; adjacent L. Calif. May–July.

4. **D. blóchmanae** (Eastw.) Moran ssp. **blóchmanae**. [*Sedum b.* Eastw. *Hasseanthus kessleri* Davids.] Corms globose to fusiform, mostly not more than twice as long as thick; rosette-lvs. 5–8, linear-oblanceolate to linear-spatulate, yellow-green, 2–7 cm. long; fl.-stems 4–15 cm. long; cauline lvs. deltoid-lanceolate to broadly ovate, 0.5–2.5 cm. long, not more than half as wide; infl. a cyme of 2–several branches; fls. subsessile; calyx 1.5–4 mm. long, the lobes ± oblong-ovate; petals white, marked with red or purple, 5–9 mm. long, lanceolate; follicles spreading, red, 4–6 mm. long; $n = 17$, 34, or 51 (Uhl & Moran, 1953).—Dry stony places below 1500 ft.; often on serpentine; Coastal Sage Scrub, near the coast from San Luis Obispo Co. to n. L. Calif. May–June.

Ssp. **brevifòlia** (Moran) Moran. Rosette-lvs. 0.7–1.5 cm. long, subglobular; fl.-stems 2–11 cm. high; cauline lvs. 0.2–1 cm. long, almost as wide; $n = 17$ (Uhl & Moran, 1953).—Dry sandstone bluffs; Chaparral; Del Mar to La Jolla, San Diego Co. April.

Ssp. **insulàris** (Moran) Moran. Rosette-lvs. 15–30; fl.-stems 3–4 cm. long, more glaucous than in ssp. *blochmanae*; $n = 17$ (Uhl & Moran, 1953).—Santa Rosa Id.; March–April.

5. **D. caespitòsa** (Haw.) Britt. & Rose. [*Cotyledon c.* Haw.] Much like *D. farinosa*, rosette-lvs. ± shining, only the inner glaucous, 5–20 cm. long, 1–5 cm. wide; cincinni becoming 3–11 cm. long, 3–14-fld.; calyx 4–8 mm. broad, 4–6 mm. high, subtruncate to tapering at base; petals bright yellow to red, 8–16 mm. long; $n = 51$ or 68 (Uhl & Moran, 1953).—Sea bluffs, Monterey Co. to Los Angeles Co. (Santa Monica Mts.); Santa Cruz Id., Anacapa Id. April–July.

6. **D. candelàbrum** Rose. [*Echeveria c.* Berger.] Caudex simple, 2–8 cm. thick, to 2 dm. long; rosette-lvs. green, oblong-lanceolate to obovate, 10–15 cm. long, 3–7 cm. wide; fl.-stems 1.5–3.5 dm. tall; infl. flat-topped, 0.5–2.5 dm. broad, mostly with ca. 3 cincinni, each with 5–15 fls.; calyx 6–9 mm. long; petals 8–12 mm. long; $n = 17$ (Uhl & Moran, 1953).—Rocky places, Santa Cruz and Santa Rosa ids. April–July.

7. **D. cymòsa** (Lem.) Britt. & Rose ssp. **marcéscens** Moran. Caudex 0.2–0.7 cm. thick; rosette-lvs. withering in summer, 1.5–3.5 cm. long, 0.5–1.2 cm. wide, 1–2 mm. thick; fl.-stems 0.4–1 dm. long; infl. of 1–2 simple cincinni each 3–5-fld.; pedicels 5–12 mm. long; calyx-lobes ± ovate, 3–4 mm. long; petals bright yellow, often marked with red, 10–14 mm. long, 2.5–3.5 mm. wide; $n = 17$ (Uhl & Moran, 1953).—Shaded rocky slopes at 1100 ft.; Chaparral; Little Sycamore Canyon, Santa Monica Mts., s. Ventura Co. May–June.

Ssp. **mìnor** (Rose) Moran. [*D. m.* Rose. *D. pumila* Rose. *D. bernardina* Britt.] Caudex 1–2 (–3.5) cm. thick; plant glaucous; rosette lvs. rhomboid-oblanceolate to spatulate, 2–6 (–10) cm. long, mostly 1–3 cm. wide; abruptly acuminate; fl.-stems slender, 0.5–1.5 dm. long; cauline lvs. ovate, cordate, 0.4–1.2 cm. long; cincinni mostly 1–3 cm. long, 3–6-fld.; pedicels slender, 5–14 mm. long; calyx 2.5–4 mm. wide, often truncate at base, 3–4 mm. long; petals orange to salmon-red to bright yellow, 10–12 mm. long, connate for ca. 2 mm., lanceolate; $n = 17$ (Uhl & Moran, 1953).—Dry rocky places, 2000–8500 ft.; Chaparral, Foothill Wd., Yellow Pine F., etc.; mts. from Monterey Co. to San Bernardino Co. April–July.

Ssp. **ovatifòlia** (Britt.) Moran. [*D. o.* Britt.] Caudex 1–1.5 cm. thick; rosette lvs. green, glabrous, ovate to elliptic, 6–10, shining, acute, 2–5 cm. long, 1.5–2.5 cm. wide;

fl.-stems 0.4–1.5 dm. high; cauline lvs. 0.5–1 cm. long, cordate; cincinni 1–3 cm. long, 3–5-fld.; calyx 2.5–3 mm. high; petals bright yellow, ca. 10 mm. long, lanceolate, 2–2.5 mm. wide; $n=17$ (Uhl & Moran, 1953).—Rocky places; Coastal Sage Scrub, Chaparral; Santa Monica Mts., Los Angeles Co. March–May.

8. **D. densiflòra** (Rose) Moran. [*Stylophyllum d.* Rose. *Cotyledon nudicaulis* Abrams, not Lam.] Caudex branched, 1–2 cm. thick; rosette-lvs. 20–25, linear or ± enlarged upward, acute, 5–10 cm. long, persistently glaucous; fl.-stems 1.5–3 dm. long; cauline lvs. turgid, acute, 1–4 cm. long; infl. dense, ± rounded, 3–several-branched; pedicels 2–5 mm. long; calyx-lobes deltoid-ovate, acutish, 1.5–2.5 mm. long; petals white or tinged pink, narrow-ovate, acute, 5–10 mm. long; carpels abruptly divergent, 5–10 mm. long; seeds 1 mm. long; $n=17$ (Uhl & Moran, 1953).—Rocky cliffs between 800 and 2000 ft.; Chaparral; s. base of San Gabriel Mts. in canyons near San Gabriel Canyon. June–July.

9. **D. édulis** (Nutt.) Moran. [*Sedum e.* Nutt.] Plate 39, Fig. B. Caudex usually short, 1–3 cm. thick; rosette lvs. 10–20, linear, subacuminate, 5–15 cm. long, slightly glaucous; fl.-stem 1.5–5 dm. long; cauline lvs. turgid, deltoid-lanceolate, acute, 1–5 cm. long; infl. elongate, rather open; pedicels to 2 mm. long; calyx-lobes oblong, acute, 2.5–4.5 mm. long; petals oblong-lanceolate, acutish, cream-white, 7–10 mm. long, connate for 1–1.5 mm.; carpels 6–8 mm. long, abruptly divergent; seeds 0.8–2 mm. long; $n=17$ (Uhl & Moran, 1953).—Rocky hillsides, below 3500 ft.; Coastal Sage Scrub, Chaparral; from Riverside and Orange cos. to n. L. Calif. May–June.

10. **D. farinòsa** (Lindl.) Britt. & Rose. [*Echeveria f.* Lindl.] Caudex stout, 1–3 cm. thick, usually with several rosettes, each with 15–30 lvs.; rosette-lvs. densely white-mealy to green, ovate-oblong, acute, flat on upper surface, lightly rounded beneath, 2.5–6 cm. long, 1–2.5 cm. wide; fl.-stems stout, mostly white-mealy, 1–3.5 dm. high; cauline lvs. many, triangular-ovate, 1–2.5 cm. long, concave; cincinni 1–3.5 cm. long, 3–11-fld.; pedicels stout, 1–5 mm. long; calyx mostly 5–6 mm. wide, rounded at base, 5–8 mm. long, the lobes deltoid-ovate; petals lemon-yellow, 10–14 mm. long, oblong, acute; carpels 5–8 mm. long; $n=17, 68, 85, 119$ (Uhl & Moran, 1953).—Sea bluffs; Coastal Sage Scrub; Los Angeles Co. from the Santa Monica Mts. n.; to Ore. May–Aug.

11. **D. greènei** Rose. [*D. hoffmannii, D. regalis,* and *D. echeverioides* Johansen.] Cespitose with branched caudex 2–5 cm. thick and several rosettes; rosette-lvs. densely white-mealy to green, oblong-oblanceolate to -obovate, 3–11 cm. long, 1.5–3 cm. wide, acute, with reddish tinge in age; fl.-stems 1.5–4 dm. tall, stout; cauline lvs. scattered, lance-ovate, cordate, mostly 1–2 cm. long; cincinni several, each 2–15-fld., the infl. ca. 1 dm. broad; pedicels mostly 1–4 mm. long; calyx-lobes triangular-lanceolate, 2–5 mm. long; petals pale yellow, 10–12 mm. long, connate for 2 mm.; $n=34$ or 51 (Uhl & Moran, 1953).—Rocky bluffs; Chaparral, Coastal Sage Scrub; Santa Cruz, Santa Rosa, San Miguel and Santa Catalina ids. May–July.

12. **D. hássei** (Rose) Moran. [*Stylophyllum h.* Rose.] Plate 39, Fig. C. Caudex 1–3 cm. thick, to 3 dm. long, much branched; rosette 6–10 cm. in diam.; lvs. 15–30, farinose, linear-lanceolate, obtuse, 3–10 cm. long, 5–15 mm. wide, 2–4 mm. thick; floral stems 1–3 dm. long, with 2–4 simple or forked branches; cymes 2–8 cm. long, 4–14-fld.; pedicels erect, 1–5 mm. long; sepals deltoid-ovate, acute, 2–3 mm. long; petals white, 8–10 mm. long, connate for 1.5–2 mm., spreading from middle; $n=34$.—Catalina Id.; L. Calif. May–June.

13. **D. lanceolàta** (Nutt.) Britt. & Rose. [*Echeveria l.* Nutt. *E. monicae* Berger. *D. congesta* and *D. robusta* Britt. *D. elongata, D. brauntonii, D. parishii, D. lurida* and *D. hallii* Rose.] Caudex short, simple; rosette-lvs. pale green, ± glaucous, lanceolate, long-acuminate, 5–20 (–30) cm. long, 1–3 cm. wide; fl.-stems 2–6 dm. tall, fairly stout, ± reddish; cauline lvs. lanceolate, cordate or sagittate at base, 0.5–3 cm. long; cincinni several, 5–12 cm. long, many-fld.; pedicels stout, 3–12 mm. long; calyx-lobes lance-ovate, 3–5 mm. long; petals orange or pale green with red tinge, 10–15 mm. long; $n=34$ (Uhl & Moran, 1953).—Common on dry slopes and banks below 3500 ft.; Coastal Sage Scrub, Chaparral; Santa Barbara and Kern cos. to n. L. Calif. May–July.

14. **D. multicáulis** (Rose) Moran. [*Hasseanthus m.* Rose. *H. elongatus* Rose. *Sedum oblongorhizum* and *sanctae-monicae* Berger.] Corms oblong, 1.5–5 cm. long, 0.3–1.8 cm.

thick; rosette-lvs. 6–15, linear, subterete, 3–15 cm. long; flowering stems 4–35 cm. long; cauline lvs. like the basal, but shorter; infl. cymose, 2–3-branched; fls. sessile or short-pedicelled; calyx-lobes deltoid-ovate to linear-oblong, 2–4 mm. long; petals yellow, elliptic-lanceolate, 5–9 mm. long, often flecked with red; follicles spreading, 5–8 mm. long; $n=17$ (Uhl & Moran, 1953).—Dry stony places, below 2000 ft.; Coastal Sage Scrub, Chaparral; Los Angeles Co. to w. San Bernardino, Riverside and Orange cos. and San Onofre Mt., San Diego Co., Beauchamp. May–June.

15. **D. nesiótica** (Moran) Moran. [*Hasseanthus n.* Moran.] Corms subglobose to irregular, 1–2 (–3) cm. long; rosette lvs. 8–16, oblanceolate to spatulate, 2.5–5 cm. long, 0.5–2 cm. wide; fl.-stems 3–10 cm. tall; cauline lvs. triangular-lanceolate to -ovate, turgid, 1.5–2 cm. long; infl. usually of 2 simple branches, each with 3–8 fls.; pedicels 1–2 mm. long; calyx 5–6 mm. broad, 4–6 mm. high, the lobes triangular-ovate, 3–4 mm. long; petals white above, greenish-yellow on the keel, elliptic, 7–14 mm. long, 3.5–5.5 mm. wide; carpels ascending; $n=34$ (Uhl & Moran, 1953).—Sea bluffs, Santa Cruz Id. March–April.

16. **D. párva** Rose & Davids. Caudex 3–5 cm. long, branched, purplish; rosette lvs. linear to oblanceolate, 1.5–4 cm. long, 0.3–0.6 cm. wide, acute, slightly glaucous, rather early fugacious; fl.-stems 0.5–1.5 dm. high; cauline lvs. well distributed, lanceolate to triangular-ovate, acute, 0.5–1.5 cm. long; infl. of 1–2 cincinni 3–4 (8) cm. long, each with 5–10 fls.; pedicels stout, 1–3 mm. long; calyx 3–5 mm. broad, ± rounded below, glaucous, 3–5 mm. high, the lobes triangular-ovate; petals pale yellow, sometimes with red flecks, 8–12 mm. long, connate for 1–2 mm., elliptic-oblong; $n=17$ (Moran, 1948).—Bare rocky slopes, 1000 ft.; Chaparral, Coastal Sage Scrub; Conejo Grade and Arroyo Santa Rosa, s. Ventura Co. May–June.

17. **D. pulverulénta** (Nutt.) Britt. & Rose. [*Echeveria p.* Nutt. in T. & G.] Plant covered with a white mealy powder throughout; caudex 4–9 cm. thick and up to 4 dm. long; rosette-lvs. 30–80, obovate-spatulate, 8–25 cm. long, 4–10 cm. wide, spreading; fl.-stems stout, 4–8 dm. high; cauline lvs. many, broadly ovate, cordate-clasping, acute, 1–2 (–4) cm. long; cincinni 2-several, ascending, 1–4 dm. long, 10–30-fld.; pedicels slender, 5–30 mm. long, spreading; calyx-segms. lanceolate, acute, red to glaucous, ca. 4–8 mm. long; petals deep red, 12–18 mm. long, connate nearly to the middle; carpels ± distinct, erect, ca. 7–8 mm. long; seeds brown, slender, ca. 0.5 mm. long; $n=17$ (Uhl & Moran, 1953).—Rocky cliffs and canyons, mostly below 3000 ft.; Coastal Sage Scrub, Chaparral, etc.; near the coast from San Luis Obispo Co. to L. Calif. May–July.

18. **D. saxòsa** (Jones) Britt. & Rose ssp. **saxòsa**. [*Cotyledon s.* Jones.] Plate 39, Fig. D. Pale green or ± glaucous; caudex 1–1.5 cm. thick; rosette-lvs. 10–25, narrow-lanceolate, semiterete, 3–10 cm. long, ca. 1–1.5 cm. wide; fl.-stems ± reddish, 0.5–2 dm. tall; cauline lvs. ovate-lanceolate, 0.5–2 cm. long, slightly clasping; infl. 0.5–1.5 dm. wide, the cincinni ascending, ± reddish; pedicels 1–2 cm. long; calyx-lobes red, lance-ovate, ca. 5 mm. long; petals yellow, ± reddish in age, 10–12 mm. long; $n=68, 85$ (Uhl & Moran, 1953).—Dry stony slopes, 3000–7000 ft.; Creosote Bush Scrub to Pinyon-Juniper Wd.; Panamint Mts., Inyo Co. May–June.

Ssp. **aloìdes** (Rose) Moran. [*D. a.* Rose. *D. grandiflora* and *D. delicata* Rose.] Scarcely to quite glaucous; fl.-stem 1–3.5 dm. high; fls. yellow; $n=17$ (Uhl & Moran, 1953).—Dry rocky places, 800–5500 ft.; Creosote Bush Scrub, Chaparral, Pinyon-Juniper Wd.; desert mts. of San Bernardino Co., desert slopes of San Jacinto and Laguna Mts. April–June.

19. **D. stolonífera** Moran. Caudex branching by slender stolons that become 1 dm. long; rosette-lvs. green, with purple tinge, not glaucous, oblong-obovate, 3–7 cm. long, 1.5–3 cm. wide; fl.-stems 0.8–2 dm. tall; cauline lvs. cordate-ovate, acute, 0.5–1.3 cm. long; infl. with cincinni 1–6 cm. long and 3–9-fld.; pedicels erect, 3–8 mm. long; calyx 6–7 mm. wide, 3–4 mm. high, truncate at base; petals yellow, 10–11 mm. long, 3–3.5 mm. wide, connate for 1–2 mm.; carpels 5–7 mm. high, separating before dehiscence; $n=17$ (Uhl & Moran, 1949).—Cliffs in Coastal Sage Scrub; canyons near Laguna Beach, Orange Co. May–July.

20. **D. tráskae** (Rose) Moran. [*Stylophyllum t.* Rose.] Caudex short, branched to form clumps with 20–100 heads; rosette-lvs. 25–35, strap-shaped to ± oblanceolate, subacuminate, 4–15 cm. long, 1–4 cm. wide, glaucous at least when young; fl.-stems 2–3 dm. long; cauline lvs. deltoid-ovate, acute, horizontal, mostly 1–3 cm. long; infl. flat-topped; pedicels stout, 1–3 mm. long; calyx-lobes deltoid to deltoid-ovate, acute, 2.5–4 mm. long; petals connate, 1–1.5 mm., narrow-ovate, 8–10 mm. long, bright yellow, later ± red-veined, curving outward in upper half; carpels ± spreading in age, 7–8 mm. long; $n = 34$ (Uhl & Moran, 1953).—Santa Barbara Id. April–May.

21. **D. variegàta** (Wats.) Moran. [*Sedum v.* Wats. *Hasseanthus v.* Rose.] Corms ovoid to rounded or oblong, 1–3 cm. long; rosette lvs. 5–8, oblanceolate to spatulate, blue-green, 1.5–5 cm. long; fl.-stems 5–25 cm. long; cauline lvs. lanceolate to narrow-deltoid, 0.5–2 cm. long, 0.3–0.8 cm. wide; infl. open, cymose, 2–3-branched; calyx-lobes deltoid-ovate, to ovate-oblong, erect, 2–4 mm. long; petals yellow with some red, elliptic-lanceolate, 5–7 mm. long; follicles spreading, red; $n = 17$ (Uhl & Moran, 1953).—Dry stony places, below 500 ft.; Coastal Sage Scrub, Chaparral; sw. San Diego Co.; n. L. Calif. May–June.

22. **D. vìrens** (Rose) Moran. [*Stylophyllum v.* Rose.] Caudex tending to be elongate and sprawling over rocks; rosette-lvs. bright green, strap-shaped, gradually tapering from base, 4–25 cm. long, acute; fl.-stems 1–5 dm. long, 2–8 mm. thick; cauline lvs. deltoid-lanceolate, 1–5 cm. long; infl. 5–10 cm. wide; calyx-lobes deltoid-ovate, 1.5–3 mm. long; petals elliptic-oblong, reddish, 7–10 mm. long; carpels spreading, when mature; $n = 17$ (Uhl & Moran, 1953).—Rocks and cliffs, San Clemente and Santa Catalina ids.; Point San Vicente. April–June.

23. **D. víscida** (Wats.) Moran. [*Cotyledon v.* Wats. *Stylophyllum v.* Britt. & Rose.] Caudex short; rosette lvs. 15–35, linear-deltoid, acute, 6–10 cm. long, dark green, viscid; fl.-stems 2–3 dm. long; cauline lvs. deltoid-lanceolate, subacute, 1–3 cm. long; calyx-lobes deltoid-ovate to oblong-ovate, acute, 1.5–4 mm. long; petals elliptic-oblong, acute, 6–9 mm. long, white strongly marked with red, spreading from near middle; carpels slender, 7–9 mm. long, not widely spreading; seeds caudate, 1.2–1.4 mm. long; $n = 17$ (Uhl & Moran, 1953).—Dry rocky places below 1200 ft.; Coastal Sage Scrub, from near San Juan Capistrano to near Oceanside. May–June.

4. Sèdum L. Stonecrop

Herbs or subshrubs, usually glabrous; lvs. mostly alternate, often small and imbricated, sometimes in terminal or basal rosettes. Fls. in terminal cymes, often with secund branches. Calyx 4–5-parted. Petals 4–5, distinct or partly united. Stamens mostly twice as many, sometimes of same number as petals. Pistils mostly 4–5, ± distinct, mostly many-ovuled. Seeds minute, slender, pointed at both ends, mostly brownish. Ca. 350 spp., N. Temp. Zone and mts. of tropics. (Latin name, to assuage, because of the healing properties of Houseleek, to which *Sedum* was applied by some writers.)

Rosette-lvs. larger and quite different from cauline lvs. .... 3. *S. spathulifolium*
Rosette-lvs. not much different from cauline lvs.
    Lvs. not narrowed toward base; petals ca. 4 mm. long. Escape from cult. .... 1. *S. album*
    Lvs. narrowed toward base; petals ca. 8 mm. long. Native of pine belt .... 2. *S. niveum*

1. **S. álbum** L. Glabrous, creeping, evergreen, forming large mats; fl.-stems 7–15 cm. high; lvs. alternate, linear-oblong to obovate or even globular, 3–15 mm. long, terete or flattened above, sessile; fls. ca. 9 mm. in diam.; petals white, obtuse, ca. as long as stamens; fr. erect, white, streaked red.—Becoming natur. from cult. Native of Eurasia. June.

2. **S. níveum** A. Davids. Plate 39, Fig. E. Prostrate glabrous branching perennial with fleshy rhizomatous stems 1–3 dm. long and short ascending branches; lvs. many, green, alternate, oblong-obovate, 5–6 mm. long, fleshy; fls. solitary or in small cymes; calyx-lobes lanceolate, 2.5–3 mm. long; petals white, with pinkish median stripe, lance-

olate, 6–7 mm. long; follicles erect.—Rocky shaded ledges, 7000–9500 ft.; Upper Montane Coniferous F.; San Bernardino Mts., Santa Rosa Mts. June–July.

3. **S. spathulifòlium** Hook. ssp. **anómalum** (Britt.) Clausen and Uhl. [*Gormania a.* Britt.] Perennial with slender rootstocks and sterile stems 1–8 cm. long; lvs. in prominent rosettes, green, minutely crenulate, 0.5–3 cm. long; sterile stems 1–1.5 mm. thick, usually naked except for the terminal rosette; fl.-stems 5–30 cm. long; cauline lvs. spatulate to elliptic-oblong, 6–20 mm. long; infl. a simple to compound 3-parted cyme, 12–50-fld.; sepals lanceolate to lance-ovate, 2–4 mm. long; petals yellow, rarely orange to white, lanceolate, 5–8 mm. long; follicles yellow-green, erect or divergent, 4–7 mm. long; $n=15$ (Clausen & Uhl, 1944).—Rocky places, 2500–7000 ft.; Chaparral, Yellow Pine F.; San Bernardino and San Gabriel mts. to cent. Calif. June–July.

34. Crossosomatàceae. Crossosoma Family

Glabrous shrubs, deciduous or with dry lvs. in dry season. Lvs. alternate, entire, simple. Fls. solitary, perfect, regular, borne at ends of naked or bracted peduncles. Fl.-tube turbinate, lined with a thin glandular disk. Sepals 5, persistent. Petals 5, white, deciduous. Stamens 15–50, inserted in several series on the disk lining the fl.-tube; fils. ± dilated at base; anthers basifixed. Pistils 2–9, unicarpellate, distinct, stipitate, with capitate stigmas. Fr. follicular. Seeds several, with a conspicuous fringed aril; endosperm present. Perhaps two spp.

1. *Crossosòma* Nutt.

Characters of the family. (Greek, *krossoi*, fringe, and *soma*, body, because of the aril.)

Lvs. 5–15 mm. long, fascicled; petals spatulate to oblong, distinctly clawed. Deserts
1. *C. bigelovii*
Lvs. 25–90 mm. long, scattered; petals rounded, scarcely clawed. Insular. .... 2. *C. californicum*

1. **C. bigelòvii** Wats. Spreading and subprostrate to 1–2 m. tall, much-branched with rigid subspinescent branchlets; lvs. elliptic to oblong-ovate, subsessile, apiculate, gray-green; sepals 4–5 mm. long, rounded; petals oblong, white to rose, 9–12 mm. long; follicles 1–3, 8–10 mm. long; seeds 2–5, ca. 2 mm. in diam.; $n=12$ (Raven et al., 1963).—Dry rocky canyons below 4000 ft.; Creosote Bush Scrub; w. Colo. Desert to Death V. region; w. Nev., Ariz., L. Calif. Feb.–April.

2. **C. califórnicum** Nutt. Plate 39, Fig. F. Shrub 1–2 m. tall, or arborescent and to 5 m., with stout gray-brown branches; lvs. oblong, rarely obovate, mucronate, subsessile, pale green; sepals ca. 1 cm. long, round-ovate, scarious on margins; petals 1.2–1.5 cm. long, rounded, white; follicles 2–7 (–9), cylindric, 1.5–2 cm. long, ± recurved; seeds ca. 20 or more, black, shining, round, somewhat flattened, ca. 2.5 mm. in diam., with yellowish fringed aril; $n=12$ (Raven et al., 1963).—Dry rocky slopes and canyons; Chaparral, Coastal Sage Scrub; San Clemente and Santa Catalina ids.; Guadalupe Ids. Feb.–May.

35. Cucurbitàceae. Gourd Family

Annual or perennial herbs, mostly with soft stems. Lvs. alternate, broad, mostly palmately lobed or veined, glabrous to scabrous or hairy, sometimes compound. Tendrils lateral, simple or branched, opposite the lvs. Fls. regular, unisexual, the plants mostly monoecious. Calyx 5-lobed. Corolla 5-lobed or almost polypetalous; stamens seemingly 3, but really 5 with 2 pairs united, mostly monadelphous by contorted anthers; fils. free or connate. Ovary inferior, with parietal placentation and many ovules; carpels commonly 3. Fr. mostly an indehiscent pepo (fleshy berrylike structure with rind and spongy interior), sometimes a papery bladdery ± spiny pod. Ca. 90 genera and 700 spp., largely trop.

Fr. fleshy, with hard or firm rind, indehiscent, often dry when mature; fls. yellow.
  Corolla campanulate, distinctly gamopetalous, 5-lobed, 5–12 cm. long; plants perennial from

Plate 40. Fig. A, *Brandegea bigelovii*, 1-seeded fr. Fig. B, *Cucurbita foetidissima*. Fig. C, *Marah fabaceus*. Fig. D, *Datisca glomerata*, ♂ fl. above, ♀ fl. with forked styles and 4 stamens. Fig. E, *Dipsacus sativus*. Fig. F, *Scabiosa atropurpurea*.

large fleshy root .... 4. *Cucurbita*
Corolla almost rotate, almost polypetalous, less than 2 cm. long; annuals.
    Anther-connective without terminal appendage; seeds with obtuse margin .... 2. *Citrullus*
    Anther-connective terminated by an appendage; seeds not margined .... 3. *Cucumis*
Fr. not fleshy, ± spinose, dry; fls. white or cream.
    Fr. less than 1 cm. long, indehiscent, 1-seeded; corolla 1.5 mm. in diam. Deserts . 1. *Brandegea*
    Fr. more than 2 cm. long, dehiscent, 2–many-seeded; corolla 5–30 mm. in diam. Cismontane
        5. *Marah*

1. *Brandègea* Cogn.

Perennial herbs with large thick roots. Lvs. 3–5-parted. Fls. small; corolla rotate, 5-parted. Ovary 1-loculed, oblique, 1-ovuled. Fr. narrowly obovoid, asymmetrical, indehiscent, sparsely echinate, less than 1 cm. long, thin-walled. Four or 5 spp. of sw. N. Am. (Named for *T. S. Brandegee*, pioneer western botanist.)

    1. **B. bigelovii** (Wats.) Cogn. [*Elaterium b.* Wats. *Echinocystis parviflora* Wats.] Plate 40, Fig. A. Stems slender, trailing or climbing over shrubs, 1–2 m. long, glabrous; lf.-blades ± round in outline, 1–5 cm. across, shallowly or deeply 3–5-lobed, the upper lobes lance-oblong to triangular; the upper surface closely covered with disklike pustules; petioles shorter than blades; ♂ fls. few, in small axillary clusters; corolla 1.5 mm. broad, ± cup-shaped; body of fr. 5–6 mm. long; styles 5 mm. long.—Locally common in washes and canyons below 2500 ft.; Creosote Bush Scrub; Colo. Desert to Sheep Hole Mts., s. Mojave Desert; Ariz., L. Calif. March–April.

2. *Citrúllus* Neck.

Monoecious annuals or perennials, ± hairy, with long trailing stems bearing branched tendrils. Lvs. deeply pinnatifid, the lobes again lobed. Fls. of medium size, light yellow, solitary in axils; corolla 5-parted nearly to base. Ovary with 3 placentae and many ovules; stigmas 3. Ca. 4 spp. of trop. Old World. (Latin diminutive of *citrus*, having a similar odor and flavor.)

Fr. hard, white-fleshed, not edible from the hand .... *C. lanatus citroides*
Fr. not so firm, with red, green or yellow flesh, edible .... *C. lanatus*

    1. **C. lanàtus** (Thunb.) var. **lanàtus.** Mansf. [*C. vulgaris* Schrad.] WATERMELON. Hairy annual; lf.-blades ovate in outline, 3–8 cm. long, pinnately divided into 3–4 pairs of lobes, these again lobed and toothed, the segms. broad at apex; corolla ca. 4 cm. in diam.; fr. hemispherical to ellipsoid, hard, smooth, green-striped; seeds white to black, in a sweet red to greenish or white flesh.—Occasional escape from cult.; Native of Afr.

    Var. **citroìdes** (Bailey) Mansf. CITRON. Fr. hemispherical, small, green with white spots, white-fleshed; seeds greenish or tan.—Occasional weed as in asparagus fields in Imperial V.

3. *Cucùmis* L.

Herbaceous scabrous plants; tendrils simple; lvs. 3–7-lobed, rounded-obtuse; pepo globose. (Greek, *kykyon*, cucumber.) An Afr. genus of ca. 40 spp.

Lvs. prominently lobed; fr. ca. 2 cm. in diam. .... 2. *C. myriocarpus*
Lvs. scarcely lobed; fr. 5–6 cm. in diam. .... 1. *C. melo*

    1. **C. mèlo** L. var. **dudaím** (L.) Dunal. [*C. d.* L.] Lvs. scarcely lobed; fls. relatively large; fr. medium, orange, smooth, 5–6 dm. in diam., very fragrant.—Weed in asparagus fields w. of El Centro, Imperial Co., *Fuller*.

    2. **C. myriocárpus** Naud. Green trailing annual with angulate-striate, rough-hairy branches; lvs. 4–5 cm. long, long-petioled, the prominent lobes and sinus rotundate, the middle lobe large; fls. very small; pepo ca. 2 cm. in diam., subglobose, marked with darker green bands and beset with weak deciduous prickles; $2n = 24$ (Shimotsuma, 1905).—Native of S. Afr. Reported from near Ballard, Santa Barbara Co., *Fuller*.

4. Cucúrbita L. Gourd. Melon. Pumpkin

Annual and perennial trailing or scandent herbs, with fibrous to tuberous roots; usually monoecious. Lvs. entire to lobed; tendrils opposite lvs. well developed. Fls. large, yellow to yellowish, campanulate, solitary in axils, the ♂ long-, the ♀ short-peduncled. Fils. distinct; anthers linear, contorted. Fr. morphologically 3-loculed with 3–5 placentae often in 1 cavity; stigmas 3–5, each 2-lobed. Seeds ovate or oblong-ovate, white to tawny or black, flat. Spp. ca. 25, of the warmer parts of Am.; many grown for food, ornaments, etc. (Latin name for the gourd.)

Lf.-blades triangular-ovate, 1–2.5 dm. long; fr. 3-celled, on a slender peduncle without thickened ridges .... 2. *C. foetidissima*
Lf.-blades palmately 5-lobed or -cleft, less than 1 dm. long; fr. seemingly 5-celled, with thickened ridges near the summit of the peduncle.
    Lvs. lobed to near the base, the lobes very narrow .... 1. *C. digitata*
    Lvs. lobed ca. half way, the lobes ± deltoid .... 3. *C. palmata*

1. **C. digitàta** Gray. Perennial with a deep fleshy fusiform root; stems trailing, slender, to ca. 1 m. long; lvs. 5-parted to the base, gray-green, with usually lighter midribs, the lobes linear-lanceolate, 5–10 cm. long, asperulate with whitish points and some short stiff setae; calyx-lobes lance-linear, ca. 1 cm. long; corollas 4–5 cm. long; fr. subglobose, ca. 8 cm. in diam., green with narrow whitish-green stripes; seeds white, 10–11 mm. long; $n=20$ (Groff & Bemis, 1967).—Occasional in sandy places, Creosote Bush Scrub at Whitewater; Chaparral at Buckman Springs, e. San Diego Co.; Coastal Sage Scrub, Orange Co.; to New Mex., Son. Aug.–Oct.

2. **C. foetidíssima** HBK. Calabazilla. Plate 40, Fig. B. Coarse rough strong-smelling perennial with an immense fusiform root; stems mostly trailing, 2–4 m. long; lvs. erect, triangular-ovate, somewhat cordate at base, 1–2.5 dm. long; ♂ fls. 10–12 cm. long, ribbed, veiny, with broad lobes; ♀ fls. 9–10 cm. long; calyx-lobes narrow, 8–10 mm. long; fr. slightly oblong-globose, 6–9 cm. long, dull light green with 5–6 main cream-white stripes and a few intermediate ones; seeds ca. 12 mm. long; $2n=40$ (McKay, 1931).—Common in sandy and gravelly places below 2000 ft.; Coastal Sage Scrub, Coastal Strand, V. Grassland, etc.; cismontane Calif.; to ca. 4000 ft. in Shadscale Scrub, etc. Mojave Desert to Tex., Nebr. June–Aug.

3. **C. palmàta** Wats. [*C. californica* Torr. ex Wats.] Stems 3–12 dm. long; lvs. grayish, palmately 5-lobed, ca. 3–9 cm. long and broad, with low bulbate prickles and some longer stiff setae, the lobes triangular to deltoid-lanceolate; corolla 4–6 cm. long; fr. globose, 8–9 cm. long, dull light green with broad ill-defined bands and splashes of greenish-white; seeds white, narrow-ovate, 10–14 mm. long; $2n=40$ (McKay, 1931).—Occasional in dry sandy places below 4000 ft.; Creosote Bush Scrub; Colo. and Mojave deserts; Coastal Sage Scrub, V. Grassland, interior cismontane s. Calif.; to L. Calif. & San Joaquin V. Apr.–Sept. Plants with triangular lf.-lobes rather than lanceolate are *C. californica*.

Plants of **C. pèpo** L. the Field Pumpkin with large orange frs. occasionally are spontaneous, as in Ventura Co.

5. Màrah Kell. Big-Root. Wild Cucumber

Climbing or trailing herbs with annual stems arising from large fusiform to subglobose perennial tubers. Lvs. suborbicular, ± cordate and 5–7-lobed or -cleft. Tendrils 1–3-fld.; petioles well developed. Staminate fls. in axillary racemes or panicles, tardily deciduous; sepals small, obsolete; corolla rotate to campanulate, mostly 5-merous. Stamens 3–4. Pistillate fls. solitary, from same axil as ♂ infl.; style short or obsolete; stigma discoid or subglobose. Ovary largely 4-loculed; ovules 1–4 in a locule. Fr. a turgid caps., subglobose to ellipsoid, pendent, with many large spines or smaller, sometimes almost none, irregularly dehiscent near the apex. Seeds large, brownish-gray to olive or tan. Ca. 7 spp. of Pacific Coast. (An aboriginal name.)

(Stocking, K. M. Some taxonomic and ecological considerations of the genus Marah. Madroño 13: 113–137. 1955.)
Mature corolla rotate or only slightly cup-shaped.
   Corollas rotate, cream; fr. globose; ovules 4 or fewer. Ventura and Kern cos. n.
                                                                                                                           1. *M. fabaceus*
   Corollas slightly cup-shaped, white; fr. oblong-cylindrical; ovules more than 4. Santa Barbara Co. s. .... 3. *M. macrocarpus*
Mature corollas campanulate, white .... 2. *M. horridus*

   1. **M. fabàceus** (Naud.) Greene. var. **agréstis** (Greene) Stocking. Plate 40, Fig. C. Stems 3–7 m. long, subglabrous to ± pubescent; lvs. roundish, 5–10 cm. broad, ± deeply 5–7-lobed, the lobes less than half the lf.-length, acute to obtuse at apices, the basal sulcus 1–3 cm. deep; ♂ fls. 8–20 in a raceme; pedicels 3–6 mm. long; ♂ corolla rotate, 7–11 mm. in diam., mostly cream; ♀ 10–15 mm. in diam., the lobes unequal; fr. globose below, tapering to a tip, 4–5 cm. in diam., the spines soft, less than 5 mm. long; seeds mostly 1–3, seldom flattened laterally.—Below 5000 ft.; from Ventura and Kern cos. n. to n. Calif. Feb.–Apr.

   2. **M. hórridus** (Congd.) S. T. Dunn. [*Echinocystis h.* Congd.] Stems 1–4 mm. long; lvs. rounded, 10–15 cm. in diam., usually rather deeply 5–7-lobed; petioles 3–8 cm. long; ♂ fls. 5–12 in a raceme; pedicels 5–15 mm. long; ♂ corolla campanulate, 10–12 mm. in diam., white; ♀ 13–17 mm. in diam.; fr. oblong-ellipsoid, 9–15 cm. long, 6–9 cm. in diam., very spinose, the spines 5–35 mm. long, 3–7 mm. wide at base; seeds lenticular, oblong-obovoid, 26–32 mm. long, light olive.—Dry slopes below 2500 ft.; Foothill Wd.; n. Los Angeles Co. and Tehachapi Mts. to Tuolumne Co. April.

   3. **M. macrocárpus** (Greene) Greene. [*Echinocystis m.* Greene.] Much like *M. fabaceus*; lf.-blades 5–10 cm. broad; ♂ fls. 8–13 mm. in diam., white, the lobes ovate, 3–12 mm. long; ♀ 15–20 mm. in diam.; fr. cylindrical, mostly 8–12 cm. long, beaked, densely spiny, the flattened spines 5–30 mm. long, 1–3 mm. wide at base; seeds oblong, 15–20 mm. long, 12–18 mm. wide, 11–14 mm. thick, brown to tan; $n=16$ (McKay, 1931), 32 (Whitaker, 1950).—Dry places, mostly below 3000 ft.; Coastal Sage Scrub, Chaparral, S. Oak Wd.; cismontane s. Calif., n. L. Calif. Jan.–April. Insular material tends to have lvs. 15–25 cm. broad; ♂ fls. 15–30 mm. broad; seeds 28–33 mm. long and is an uncertain taxon with the name var. **màjor** (S. T. Dunn) Stocking. [*M. m.* S. T. Dunn.]

36. Datiscàceae. Datisca Family

   Herbs or trees. Lvs. alternate, simple or pinnate, estipulate. Fls. dioecious or rarely bisexual, in spikes or racemes. Staminate fls. with 3–9 calyx-lobes; petals small, 8 or 0; stamens 4–25. Pistillate and perfect fls. with inferior ovary, the stamens reduced to staminodia or developed. Ovary 1-celled, open or closed at apex, the placentae parietal; styles free; ovules many. Caps. opening among the styles; seeds many, minute, with little endosperm and cylindric embryo. Small family of n. trop. and subtrop.

1. *Datísca* L.

   Stout perennial glabrous herbs. Lvs. pinnately incised. Corolla 0. Staminate calyx with 4–9 unequal lobes, 8–12 stamens. Pistillate calyx 3-toothed, sometimes with 2–4 stamens. A small genus of Eurasia and N. Am. (Meaning unknown.)
   1. **D. glomeràta** (Presl) Baill. [*Tricerastes g.* Presl.] Plate 40, Fig. D. Erect, 1 m. or more high, branched; lvs. ovate to lanceolate in outline, acuminate, 1–2 dm. long, the segms. lanceolate, sharply incised-serrate; petioles 2–3 cm. long; fls. several in each axil of a leafy raceme; ♂ calyces 2 mm., ♀ 5–8 mm. long; styles ca. 6 mm. long; caps. ca. 8 mm. long; seeds light brown, subcylindric, ca. 1 mm. long and with ca. 11–12 rows of small pits.—Dry stream beds and washes in the mts. and hills, below 6500 ft.; Yellow Pine F.; Foothill Wd., Chaparral, etc.; cismontane from L. Calif. to n. Calif., occasional in canyons on desert edge. May–July.

Plate 41. Fig. A, *Shepherdia argentea*. Fig. B, *Elatine californica*, fl. & seed. Fig. C, *Arctostaphylos insularis*, and separate stamen. Fig. D, *Chimaphila menziesii*. Fig. E, *Pyrola picta*. Fig. F, *Vaccinium occidentale*.

37. Dipsacàceae. Teasel Family

Annual to perennial herbs with lvs. opposite or whorled, entire, toothed or pinnatifid, exstipulate. Fls. small, in dense bracted and involucrate heads or interrupted spikes. Calyx cup-shaped, discoid or divided into spreading bristles. Corolla limb 4–5-lobed. Stamens 2–4, inserted on throat of corolla. Ovary inferior, 1-loculed; style filiform; stigma simple. Ovule 1. Fr. an ak. Ca. 11 genera and 160 spp., of Old World.

Scales of the elongate receptacle pointed; plant ± thistlelike .... 1. *Dipsacus*
Scales of the receptacle not prickly, but herbaceous, inconspicuous and concealed among the fls.; plant not thistlelike .... 2. *Scabiosa*

1. *Dipsacus* L. Teasel

Tall stout herbs with opposite lvs., the cauline united at base. Fls. pinkish, in dense terminal peduncled oblong heads surrounded by invol. of elongated bracts. Bracts subtending the fls. (i.e. the receptacular) shorter, rigid and spinelike in fr. Calyx and corolla 4-lobed. Stamens 4. Ca. 15 spp. of Eurasia and N. Afr. (Greek, *dipsa*, thirst, since the connate lf.-bases in some spp. hold water.)

1. **D. satìvus** (L.) Honckeny. Plate 40, Fig. E. Heads 5–8 cm. long; involucral bracts spreading ± horizontally; receptacle-bracts almost equaling the fls., spinose-ciliate, ending in a stiff recurved spine; corolla-tube 10–13 mm. long; $2n = 18$ (Kachidze, 1929).—Locally natur. throughout cismontane Calif., scarce in the so. part; grown in Eu. and used for textile mills. May–July.

2. *Scabiòsa* L. Scabiosa. Mourning-Bride

Fairly large herbs with opposite lvs., no prickles, and blue or white or purplish fls. in peduncled involucrate heads. Bracts of invol. herbaceous, separate. Scales on receptacle small or lacking, not sharp-pointed. Calyx of 5 setae united at base. Corolla oblique, 5-lobed, the marginal fls. larger with the upper lobes smaller than the lower. Stamens 4, rarely 2. Ca. 80 spp., mostly temp., Eurasia and Afr. (Latin name meaning scurfy, the plants having been used for skin diseases.)

1. **S. atropurpúrea** L. Plate 40, Fig. F. Branching annual, 5–6 dm. tall; basal lvs. oblong-spatulate, simple or lyrate, coarsely dentate; cauline lvs. pinnately parted, the lobes oblong, dentate or cut; fls. dark purple, rose or white, in heads to 5 cm. in diam. Natur. occasionally as an escape from gardens; native of Eu.

38. Elaeagnàceae. Oleaster Family

Shrubs or trees, with alternate or opposite, mostly silvery-scaly or stellate-pubescent simple entire lvs. Fls. perfect or imperfect, clustered or solitary, on nodes of previous year. Fl.-tube in ♂ fls. cup-shaped or saucer-shaped, in ♀ or perfect fls. tubular or urn-shaped and persistent. Sepals 4. Petals 0. Stamens 4 or 8. Disk lobed or annular. Ovary superior, 1-loculed, 1-ovuled. Style slender. Fr. drupe-like, the fleshy fl.-tube enclosing the ak. Ca. 3 genera and 20 spp., widely distributed.

1. *Elaeágnus* L.

Fls. small in lateral clusters on twigs of the current year. Fl.-tube tubular. Disk present or not. Stamens scarcely exserted. Ca. 25 spp., of N. Temp. Zone. (Greek, *elais*, olive, and *agnos*, chaste-tree.)

1. **E. angustifòlia** L. Oleaster. Russian-Olive. Shrub or tree, 3–7 m. tall, densely silvery; lvs. lanceolate, alternate, 3–10 cm. long; fls. 12–15 mm. long; fr. yellow with silvery scales.—Occasional escape in wet places, as at Victorville, San Bernardino Co., Inyo Co., etc. Native of Eu. May–June.

Shephérdia argéntea Nutt. BUFFALO-BERRY. Plate 41, Fig. A. With general grayish and scaly appearance of *Elaeagnus*, but with opposite lvs. and 8 stamens instead of 4. Occurs in Cuyama V., n. Ventura Co., in Inyo Co. and north and east.

39. Elatinàceae. WATERWORT FAMILY

Low herbs or shrubs. Lvs. opposite or whorled, simple, with paired membranaceous stipules. Fls. small, regular, axillary, perfect, solitary or cymose. Sepals 3–5, free, imbricate, persistent. Stamens as many or twice as many, free, hypogynous; anthers 2-celled, opening by longitudinal slits. Ovary superior, 3–5-loculed; placentation axile; styles free, 3–5, introrsely stigmatose, or stigmas sessile. Ovules many. Fr. a septicidal caps.; seeds oblong-cylindric, straight or curved, without endosperm; cotyledons thick and short. Cosmopolitan. Two genera.

Plants glandular-pubescent; fls. 5-merous; sepals pointed, with thickened midrib; caps. ovoid.
Terrestrial .... 1. *Bergia*
Plants glabrous; fls. 2–4-merous; sepals obtuse, without midrib; caps. globose or depressed.
Aquatic or semiaquatic .... 2. *Elatine*

1. *Bérgia* L.

Diffuse or ascending plants, glandular-pubescent. Fls. pedicelled, solitary or fascicled, 5-merous. Sepals cuspidate, with thickened midrib and scarious margin. Petals oblong. Stamens 5 or 10. Caps. firm. Ca. 20 spp., largely trop. (Named for P. J. *Bergius*, 1730–1790, Swedish botanist.)

1. **B. texàna** (Hook.) Seub. [*Merimea t.* Hook.] Diffusely branched annual, 5–40 cm. high; lvs. ovate to obovate, 1–2 cm. long, glandular-serrate; sepals 3–4 mm. long; petals white; seeds slightly curved, ca. 0.5 mm. long.—Occasional on mud flats; largely V. Grassland; Elsinore, Murrieta Hot Springs, Los Angeles, Santa Barbara, etc. n. to Wash., e. to Miss. V. July–Aug.

2. *Elátine* L. WATERWORT

Dwarf glabrous annuals or subperennials, often rooting at the nodes. Sepals 2–4; petals 2–4; stamens mostly 2–4; styles or sessile capitate stigmas 2–4. Caps. membranaceous, 2–4-celled, several–many-seeded, the partitions left attached to the axis, or evanescent. Ca. 10 spp. of temp. regions. (Classical name of some low creeping plant.)

(Fassett, N. C. Elatine and other aquatics. Rhodora 41: 367–377. 1939. Mason, H. L. New spp. of Elatine in Calif. Madroño 13: 239–240. 1956.)

Caps. with 4 carpels, pedicelled; seeds very strongly curved, differently rounded at the 2 ends
2. *E. californica*
Caps. with 2–3 carpels; seeds straight to slightly curved, almost alike at both ends.
  Seeds with pits in rows of 16–35.
    Pits of seeds much broader than long, the transverse ridges more prominent than the longitudinal.
      Seeds 6–10 in each locule; lvs. all linear-spatulate .... 4. *E. gracilis*
      Seeds 15–40 in each locule; lvs. linear-spatulate to round-obovate .... 3. *E. chilensis*
    Pits of seeds nearly as long as broad, the longitudinal and transverse ridges almost equally prominent .... 5. *E. rubella*
  Seeds with pits in rows of 9–15; lvs. linear to narrow-oblong .... 1. *E. brachysperma*

1. **E. brachyspérma** Gray. [*E. triandra* var. *b.* Fassett.] Plants densely matted; lvs. linear to narrow-oblong, 1.5–4 mm. long (or longer in submerged plants); fls. 2–3-merous; caps. depressed; seeds short-oblong, 0.4–0.6 mm. long, with 6–8 rows of 9–15 pits in each irregular row.—Many Plant Communities; mostly in cismontane s. Calif. below 9100 ft.; to Ore., Ill., Tex. April–Sept.

2. **E. califórnica** Gray. Plate 41, Fig. B. Matted, 2–14 cm. across; lvs. oblanceolate or wider, rounded or obtuse at apex, 2–4 mm. long; fls. pedicelled, 4-merous; seeds V- to

Elatine

U-shaped, rounded at one end, truncate and subapiculate at the other, with ca. 25 pits in each row.—Water borders and mud-flats below 5000 ft.; San Diego Co. to Modoc and Lake cos. Catalina Id. March–Aug.

3. **E. chilénsis** Gay. Plant to 1 dm. long, aquatic, or if on wet mud, creeping and rooting; lvs. obovate to broadly spatulate, 3–4 mm. long, 1–3 mm. broad, with entire stipules; fls. 1–2 at a node, sessile; sepals 2 or with a 3rd reduced; petals round; stamens 3, alternate with carpels; seeds slightly curved, 24–33 in a locule, erect, with 25–35 short broad pits and appearing transversely rugose.—Ponds and muddy shores; Santa Rosa Plateau, w. Riverside Co.; n. Calif.; S. Am.

4. **E. grácilis** Mason. [*E. triandra* Calif. refs.] Plants slender, erect, 2–4 cm. high; lvs. ca. equal to internodes, narrowed to petiolelike base with attenuate lacerate stipules; fls. solitary; sepals 2, a 3d sepal reduced or wanting; petals 3, thin-membranous, suborbicular; stamens 3–1 (–0), alternate with carpels; seeds nearly or quite straight, 7–8 to a locule, the areoles in 9–10 rows of 20–30 each, the horizontal ridges more conspicuous than the longitudinal.—Shallow water or mud-banks, below 9500 ft.; many Plant Communities; cismontane from Riverside Co. n. to Canada, Atlantic Coast. April–Sept.

5. **E. rubélla** Rydb. [*E. triandra* auth., not Schkuhr.] Prostrate to erect and 1.5 dm. high, often reddish; lvs. opposite, lanceolate to linear-spatulate, 2–15 mm. long; fls. sessile, 1–2 at a node, 3-merous; sepals 2–3, often unequal; stamens 3, alternate with carpels; seeds 12–30 in a locule, cylindrical or slightly curved, longitudinally ribbed and with rows of 16–25 hexagonal pits.—Ponds, vernal pools, ditches, etc. through much of Calif.; w. Am.

40. Ericàceae. Heath Family

Trees, shrubs or herbs. Lvs. simple, usually alternate, usually evergreen, coriaceous, sometimes reduced to bracts. Fls. regular and symmetrical, white to red, 4–5-lobed or -petaled. Stamens free from the corolla, as many or twice as many as the corolla-lobes; anthers 2-celled, often opening by terminal pores, often with 2 awnlike appendages. Style 1; stigma entire or lobed. Ovary 3–10-celled, usually with axile placentae. Fr. a caps., drupe or berry. Many spp. of ornamental use or of value for their edible frs.

1. Plants ± herbaceous, although sometimes with suffrutescent base.
    2. Plants with slender subterranean rootstocks, mostly with green lvs.; pollen in tetrads.
        3. Fls. in racemes; valves of caps. with cobwebby margins .... 7. *Pyrola*
        3. Fls. in corymbs; valves without threads on margin .... 3. *Chimaphila*
    2. Plants fleshy-stemmed, without green lvs.; pollen simple.
        4. Corolla 12–18 mm. long; anthers not horned .... 9. *Sarcodes*
        4. Corolla 7–8 mm. long; anthers horned .... 6. *Pterospora*
1. Shrubs or trees.
    5. Fr. a dry caps.
        6. Corolla 3–5 cm. long, white or yellowish; lvs. 2.5–10 cm. long .... 8. *Rhododendron*
        6. Corolla less than 1 cm. long, rose-color; lvs. 0.6–1.5 cm. long .... 5. *Phyllodoce*
    5. Fr. a berry or drupe.
        7. Ovary superior; fr. reddish.
            8. Berry many-seeded, granular on surface; lvs 7–15 cm. long. Tree ...... 1. *Arbutus*
            8. Berry drupe-like; lvs. less than 6 cm. long. Shrubs.
                9. Fr. granular on surface or warty .... 4. *Comarostaphylis*
                9. Fr. smooth, not granular or warty.
                    10. Lvs. revolute, not vertical; fils. slender; fr. with a solid 3–5-celled stone
                        .... 11. *Xylococcus*
                    10. Lvs. plane, vertical; fils. dilated at base; fr. usually with separable nutlets, rarely with a solid stone .... 2. *Arctostaphylos*
        7. Ovary inferior; fr. a blue berry .... 10. *Vaccinium*

1. *Arbùtus* L.

Evergreen trees or shrubs with bark of main trunks fissured or smooth and exfoliating. Lvs. alternate, entire or toothed, coriaceous, petioled. Fls. perfect, in terminal panicles.

Calyx 5-lobed, rather persistent. Corolla white or pale, urn-shaped, the 5-lobes shorter than the swollen tube. Stamens 10, included; fils. dilated below, usually pubescent; anthers broad, each with 2 slender awns. Ovary usually 5-celled. Fr. hard-stoned, berry-like, ± globose, with a rugose or granular surface. Ca. 20 spp. of Medit. region, s. Asia, N. and S. Am. (Latin name of *A. unedo*.)

1. **A. menzièsii** Pursh. MADRONE. Widely branched, 5 to 40 m. tall; bark freely exfoliating, leaving a polished reddish or brownish surface, fissured and darker on very old trunks; lvs. elliptic to subovate, entire or serrulate, glabrate when mature, 5–12 cm. long; panicles 6–15 cm. long; bracts ovate, whitish, 4–6 mm. long; calyx-lobes ovate, ciliate, ca. 1 mm. long; corolla white to pink, 6–8 mm. long; style 5 mm. long, columnar; berry red to orange, 8–10 mm. in diam.; seeds bony, ca. 2.5 mm. long; $2n = 26$ (Stebbins & Major, 1965).—Wooded slopes and canyons, below 5000 ft.; S. Oak Wd., etc. at scattered stations in s. Calif.; to L. Calif.; abundant n. to B.C. March–May.

2. *Arctostáphylos* Adans. MANZANITA

Woody evergreen shrubs, varying from low and prostrate to small trees, usually with crooked branches, these usually smooth with thin red bark that exfoliates freely; sometimes with fibrous more persistent bark. Lvs. simple, alternate, coriaceous, entire to serrulate, sessile to petioled, mostly vertical. Fls. in terminal simple racemes or in panicles and borne on pedicels with basal bract. Calyx persistent, 4–5-lobed, the lobes broad. Corolla small, urn-shaped, the lobes rounded, recurved, white to rose. Stamens 10 (8), included; anthers with 2 recurved appendages on the back and opening by small terminal pores. Ovary on a hypogynous disk, 4–10-celled, with 1 ovule in each cell. Fr. berrylike, with rather copious granular pulp, or with thin pericarp and dry, enclosing 4–10 nutlets, or these variously fused in 2's or 3's or even united into a solid stone. A genus of perhaps 50 spp., from Cent. Am. n.; 1 sp. circumpolar. Centering in Calif. Hybridizing freely. Some spp. are characterized by forming at an early stage a basal burl or enlarged root-crown from which sprouting takes place after fire; others lack this. (Greek, *arktos*, bear, and *staphyle*, a bunch of grapes, referring to the common name of the first known sp.)

(Adams, J. E. A systematic study of the genus Arctostaphylos. Jour. Elisha Mitchell Sci. Soc. 56: 1–62. 1940. Wells, P. V. New taxa, combinations, and chromosome numbers of Arctostaphylos. Madroño 19: 193–210. 1968.)

1. Branches prostrate or procumbent, usually less than 6 dm. high and rooting in contact with the ground.
   2. Hairs of branches and infl. not gland-tipped .... 7. *A. parryana*
   2. Hairs of branchlets and infl. gland-tipped .... 8. *A. patula*
1. Branches erect or ascending, taller, not usually rooting.
   3. Bracts of infl. shorter than pedicels, deltoid to almost subulate, even the lower bracts not or scarcely foliaceous.
      4. Pedicels glabrous; basal burl lacking and the plants not stump-sprouting.
         5. Lvs. with stomates on lower surface only. Santa Cruz Id. .... 5. *A. insularis*
         5. Lvs. with stomates ca. equally distributed on both surfaces. Mainland.
            6. Fr. depressed-globose, the nutlets separable .... 10. *A. pungens*
            6. Fr. somewhat ovoid, the nutlets forming a solid stone .... 8. *A. patula*
      4. Pedicels ± glandular-pubescent.
         7. Basal burl lacking; shrubs not stump-sprouting.
            8. Bracts of infl. 5–6 mm. long, lanceolate, deciduous; pedicels 10–15 mm. long; sweet-scented shrub from cismontane interior .... 9. *A. pringlei*
            8. Bracts of infl. 1–3 mm. long, deltoid, persistent.
               9. Stomates restricted to lower surface of lvs. Santa Cruz Id. .... 5. *A. insularis*
               9. Stomates on both surfaces of lvs. Mainland .... 4. *A. glauca*
         7. Basal burl present; crown-sprouting shrubs .... 3. *A. glandulosa*
   3. Bracts of infl. (at least the lower) foliaceous and equaling or exceeding pedicels.
      10. Basal burl present; crown-sprouting shrubs.
         11. Stomates absent from upper surface of lvs. Insular .... 12. *A. tomentosa*
         11. Stomates on both surfaces of lvs. Mainland .... 3. *A. glandulosa*

10. Basal burl absent; shrubs not crown-spreading.
    12. Ovary glabrous; lvs. cordate, clasping. Santa Ynez Mts. .... 11. *A. refugioensis*
    12. Ovary pubescent, often glandular.
        13. Lvs. auriculate-clasping, subsessile or very short-petioled; branchlets, petioles, rachis of infl. with long white hairs, not glandular. Santa Cruz Id. 13. *A. viridissima*
        13. Lvs. not auriculate-clasping.
            14. Branchlets not glandular.
                15. Branchlets glabrous to pubescent but not bristly.
                    16. Fr. with very little pulp, viscid, globose or slightly elongate, 12–15 mm. in diam., with a solid stone; corolla 8–9 mm. long .... 4. *A. glauca*
                    16. Fr. with well developed pulp, often depressed-globose, 6–10 mm. in diam., the nutlets generally separable; corolla 6–8 mm. long ..... 3. *A. glandulosa*
                15. Branchlets bristly with long hairs as well as variously pubescent
                                                                            3. *A. glandulosa*
            14. Branchlets with some hairs gland-tipped; stomates on both surfaces of lvs.
                17. Ovary and pedicels glandular-pubescent; lvs. elliptic to oblong. Mts. of s. San Diego Co. .... 6. *A. otayensis*
                17. Ovary and pedicels tomentose and sparsely glandular; lvs. cordate to truncate. Insular.
                    18. Bracts broadly ovate; infl. compact, crowded, with 3–4 branches from main rachis; sepals glandular-ciliate; corolla 5–6 mm. long; Santa Rosa Id.
                                                                                   2. *A. confertiflora*
                    18. Bracts narrowly deltoid to lanceolate; infl. more open, with 5–6 branches from main rachis; sepals not glandular-ciliate; corolla 7–10 mm. long. Catalina Id. .... 1. *A. catalinae*

1. **A. catalìnae** Wells. Arborescent shrub 2–6 m. high, without basal burl; bark smooth, red-brown; branchlets densely short hispid and with longer stiff hairs, largely at first gland-tipped; lvs. ovate or oblong, acute, subcordate at base to tapering, light green, with stomata above and below; petioles 2–6 mm. long; infl. openly paniculate with up to 5–6 branches from the main rachis; bracts leafy, longer than the 2–3 mm. pedicels; rachis and pedicels glandular-hispidulous; corolla 7–10 mm. long; ovary densely white-hairy and with sparsely gland-tipped hairs; fr. glabrescent; nutlets separable.—Chiefly volcanic rocks, Santa Catalina Id.

2. **A. confertiflòra** Eastw. Basal burl absent; lvs. with stomata on both surfaces; lvs. broadly round-ovate; infl. compact, crowded, with 3–4 branches from main rachis, densely glandular-hairy; bracts broadly ovate; pedicels and ovary tomentose and sparsely glandular; corolla 5–6 mm. long; fr. dull reddish brown; $n = 13$ (Wells, 1968).—Endemic to Santa Rosa Id.

3. **A. glandulòsa** Eastw. ssp. **glandulòsa**. With a basal burl, erect, spreading, 1.5–2.5 m. high, the stems smooth, reddish; branchlets coarse, conspicuously glandular-hairy; lvs. ovate to lance-ovate, dull green, with ca. the same number of stomates on each side; petioles glandular-hairy, 5–6 mm. long; infl. paniculate, short-spreading, with glandular-hairy rachises and peduncle; bracts leafy, lanceolate, glandular-hairy, 5–15 mm. long; pedicels ca. 5 mm. long; corolla white, 6–8 mm. long; ovary glandular; fr. depressed-globose, red-brown, ± viscid, ca. 8 mm. in diam., the nutlets separable, ridged on back; $n = 26$ (Wells, 1968).—Dry gravelly places, 1000–6000 ft.; Chaparral, Yellow Pine F., etc.; mts. of San Diego Co. to Ore. Jan.–Apr. Plants resembling this but eglandular are the forma *cushingiana* (Eastw.) Wells and are probably n. of our area.

KEY TO SUBSPECIES

Branchlets glandular-hairy .... Ssp. *glandulosa*
Branchlets not glandular-hairy.
    Lvs. dark green. Coast of San Diego Co. .... Ssp. *crassifolia*
    Lvs. light or yellow-green to pale.
        Branchlets puberulent. San Diego and Riverside cos. .... Ssp. *adamsii*
        Branchlets long white-hairy. Santa Barbara Co. to Los Angeles Co. .... Ssp. *mollis*

Ssp. **ádamsii** (Munz) Munz. [*A. glandulosa* var. *a.* Munz, Aliso 4: 95. 1958.] Branchlets puberulent, without longer hairs; infl. and ovary non-glandular; infl. small, with deltoid bracts.—Dry slopes, 800–6800 ft.; Chaparral, Yellow Pine F.; e. San Diego Co. to w. Riverside Co. hence largely along w. edge of Colo. Desert.

Ssp. **crassifòlia** (Jeps.) Wells. Plant 6–12 dm. high, spreading; branchlets tomentulose, sometimes with scattered longer hairs, not glandular; lvs. dark green, glabrate above, ± tomentulose beneath; fr. not glandular. Sandy mesas and bluffs, Chaparral; coast of San Diego Co.

Ssp. **móllis** (Adams) Wells. Branchlets dark, puberulent, with long soft hairs, lvs. light green, sparsely puberulent when young, glabrate in age, 2–4 cm. long; infl. small, compact, few-fld.; bracts hirsute-ciliate, the upper deltoid; ovary white-hairy, not glandular.—Dry slopes, 2500–6000 ft.; Chaparral, Yellow Pine F.; mts. from Riverside Co. to San Luis Obispo Co.

4. **A. gláuca** Lindl. [*A. g.* var. *eremicola* Jeps.] Without basal burl; large, erect, shrubby or arborescent, 2–4 (–6) m. high, with smooth red-brown bark; branchlets and lvs. glabrous and glaucous, the latter oblong, elliptic or ovate, with equal numbers of stomates on both surfaces, 2–4.5 cm. long, on petioles 7–10 mm. long; panicles dense, nodding, broad and short, the rachis glabrous to glandular; bracts 2–3 mm. long, spreading, broadly ovate; pedicels glandular-pubescent; corolla white or with some pink, 8–9 mm. long; ovary glandular; fr. globose, viscid, brownish, 12–15 mm. in diam., with thin leathery pericarp, the nutlets united into a solid smooth apiculate stone; $n=13$ (Wells, 1968).—Common on dry slopes below 4500 ft.; Chaparral; cismontane to Little San Bernardino Mts.; L. Calif., cent. Calif. Dec.–March. From Ventura and Santa Barbara cos. n. are plants with branchlets pubescent to glandular-puberulent and are known as var *puberula* J. T. Howell.

5. **A. insulàris** Greene. Plate 41, Fig. C. Erect, much branched, 1–2.5 m. or taller, without basal burl, the bark smooth, dark red-brown; branchlets smooth; lvs. bright green, without stomates on upper surface, shining, glabrous, ovate to elliptical, 2.5–4 cm. long, with petioles 3–6 mm. long; infl. an ample spreading panicle; bracts deltoid-acuminate, 1.5–2.5 mm. long; pedicels glandular-pubescent, 5–12 mm. long; calyx-lobes 1.5 mm. long; corolla white, ca. 6 mm. long; fr. bright orange-brown, 6–8 mm. broad; $n=13$ (Wells, 1968).—Confined to Santa Cruz Id. The more common form has glandular-pubescent branchlets and tomentose ovary and is f. *pubescens* (Eastw.) Wells.

6. **A. otayénsis** Wies. & Schreib. Without basal burl; erect, 1–2.5 m. high, with smooth dark red bark; branchlets glandular-pubescent; lvs. elliptic to oblong, 1.5–3.5 cm. long, gray-green to yellow-green, finely pubescent and glandular, with equal nos. of stomates above and beneath, acute or apiculate at apex, on petioles 4–7 mm. long; infl. of open panicles or racemes, glandular-pubescent; corollas white, 5–7 mm. long; fr. globose, 5–8 mm. in diam., pale brown, glabrate or microscopically glandular, rarely sparsely hairy, the nutlets separable or coalesced into a solid nut; $n=13$ (Wells, 1968).—Dry slopes, 1800–5000 ft.; Chaparral; mts. of San Diego Co. Jan.–March.

7. **A. parryàna** Lemmon. Without enlarged root-crown, diffuse widely spreading shrub 1–2 m. high with lateral branches commonly rooting and decumbent; bark smooth, dark red or red-brown; branchlets canescent to glabrous, not glandular; lvs. ovate to subovate, bright green, glabrous, somewhat shining, with equal nos. of stamens on both surfaces, acute to rounded at apex, 2–3 cm. long, on petioles 5–10 mm. long; infl. few-fld., simple or few-branched, with puberulent rachises; bracts short, deltoid-acuminate, ca. 2 mm. long; pedicels glabrous, 5–9 mm. long; calyx spreading, ca. 2 mm. long; corolla white, 6–7 mm. long; ovary glabrous; fr. dark red, round-ovoid, 8–12 mm. long, with a solid obscurely ridged stone; $n=26$ (Wells, 1969).—Dry stony slopes, 4000–7500 ft.; Chaparral, Yellow Pine F.; Tehachapi Mts., Mt. Pinos region to San Gabriel Mts. March–May.

8. **A. pátula** Greene ssp. **platyphýlla** (Gray) Wells. [*A. pungens* var. *p.* Gray. *A. pinetorum* Rollins. *A. parryana* var. *pinetorum* Wiesl. & Schreib.] Without basal burl and much like *A. parryana*, but branchlets, petioles and infl. glandular-pubescent.—Dry slopes,

5000–9000 ft., Montane Coniferous F.; San Gabriel Mts. to Santa Rosa Mts.; to Utah, Colo. May–June.

9. **A. prínglei** Parry var. **drupàcea** Parry. [*A. d.* Macbr.] Erect shrubs, sweet-scented, 2–4 m. high, without basal burl, with smooth dull red-brown bark; branchlets densely glandular-villous; lvs. oblong-ovate to elliptic, rarely obovate, gray-green, minutely glandular-pubescent or glabrate, with equal nos. of stomates on both surfaces, rounded to acute at apex, 2–4 cm. long, on petioles 3–7 mm. long; infl. ample, almost sessile, usually branched, glandular-pubescent to -hairy; bracts lanceolate, membranaceous, deciduous, pinkish, 5–6 mm. long; pedicels slender, pink, 10–15 mm. long, glandular-villous; calyx-lobes lanceolate, ca. 3 mm. long, glandular-villous, strongly ciliate; corolla rose, 7–8 mm. long; fr. round-ovoid, glandular-villous, red, 6–10 mm. in diam., with solid ribbed stone.—Dry slopes, 4000–7500 ft.; Chaparral, Yellow Pine F.; San Bernardino Mts. to L. Calif. Feb.–April.

10. **A. púngens** HBK. Without basal burl, erect, branched from base, 2–3 m. high, with smooth red-brown bark; branchlets canescent with a fine dense pubescence; lvs. bright green, glabrous or minutely puberulent, with stomates on both surfaces, elliptic to oblong or obovate or almost lanceolate, 1.5–3 cm. long, acute to obtuse, on petioles 3–4 mm. long; infl. small, dense, the rachises thickened upward, densely canescent; bracts short-deltoid, ca. 3 mm. long; pedicels glabrous, 5–7 mm. long; calyx-lobes ca. 1.5 mm. long, round-ovate, entire; corolla white, ca. 6 mm. long; ovary glabrous; fr. brownish-red, depressed-globose, 5–8 mm. broad, the nutlets separable or irregularly united; $2n = 26$ (Callan, 1941).—Dry slopes, 3000–7000 ft. in our area; Chaparral, Yellow Pine F.; common, San Diego Co. to Los Angeles and San Bernardino cos.; to Nev., Tex., Mex. Feb.–March.

11. **A. refugioénsis** Gankin. Erect, 2.5–4 m. tall, without basal burl; young growth with both short and long gland-tipped setose hairs; lvs. mostly sessile, cordate, clasping, entire to serrulate, imbricate, 2.5–4.5 cm. long, 2–3 cm. wide, with equal numbers of stomates on both surfaces; infl. branched, the rachis fine-pubescent and glandular-setose; bracts foliaceous, 5–10 mm. long; pedicels 6–9 mm. long; calyx-lobes 2 mm. long; corolla white to pinkish; ovary glabrous; fr. globose, 1–1.5 cm in diam., the nutlets coalesced; $n = 13$ (Wells, 1968).—At 2250 ft.; Chaparral; Refugio Pass region, Santa Ynez Mts.

12. **A. tomentòsa** (Pursh) Lindl. ssp. **insulícola** Wells. With a basal burl; erect, to more than 3 m. high; bark smooth, bright red-brown; branchlets tomentose, not setose or glandular; lvs. tomentose below, and with stomates on under side only, oblong to ovate or elliptic, 2.5–4 cm. long; infl. an ample open panicle; lower bracts leafy, 8–12 mm. long; pedicels stout, 2–4 mm. long; corolla white, 5–6 mm. long; ovary densely tomentose; fr. depressed-globose, brownish, 8–10 mm. in diam., the nutlets irregularly coalescent; $n = 26$ (Wells, 1968).—Chaparral; Santa Rosa and Santa Cruz ids.

Ssp. **subcordàta** (Eastw.) Wells. [*A. subcordata* Eastw.] Branchlets and infl. with glandular-setose hairs; lvs. 2–5 cm. long, rounded to subcordate at base; $n = 26$ (Wells, 1968). —Santa Cruz Id.

13. **A. viridíssima** (Eastw.) McMinn. Basal burl absent; branchlets, rachis, etc. densely hispidulous and setose with long white hairs, not glandular; lvs. with stomates on both surfaces, auriculate-clasping, dark green, glabrous, 2–4 cm. long; infl. compact; lower bracts lanceolate to linear; ovary and pedicels densely pubescent, not glandular; pedicels ca. 3 mm. long; corolla globose, 5–6 mm. long; fr. 10–15 mm. in diam., oblate-sphaeroidal; $n = 13$ (Wells, 1968).—Santa Cruz Id.

3. *Chimáphila* Pursh. PIPSISSEWA

Low evergreen suffrutescent perennials with running underground rootstocks and branching stems. Lvs. thick, shining, short-petioled. Fls. white to rose-pink, mostly in corymbs or racemes on end of a peduncle. Sepals 5. Petals 5, concave, round. Stamens 10; fils. dilated below, hairy in middle; anther-sacs prolonged into tubes at apex. Ovary 5-lobed; style short, straight; stigma disc-shaped, orbicular. Caps. 5-celled, splitting from apex downward. Seeds many, minute. Ca. 7–8 spp., N. Am., Asia. (Greek, *cheima*, winter, and *phelein*, to love, referring to one of the common names, *wintergreen*.)

Lvs. ovate; peduncle with 1–3 white fls. .... 1. *C. menziesii*
Lvs. oblanceolate; peduncle with 3–7 pink fls. .... 2. *C. umbellata*

1. **C. menzièsii** (R. Br. ex. D. Don) Spreng. [*Pyrola m.* R. Br. ex. D. Don.] Plate 41, Fig. D. Plants 1–1.5 dm. tall, few-branched; lvs. not distinctly whorled, ovate to lance-oblong, 1.5–3.5 cm. long, serrulate, sometimes mottled, dark green above, paler beneath; peduncles mostly 4–5 cm. long, with obsolete rather persistent bracts; sepals rounded, erose, ca. 5 mm. long; petals white (pinkish in age), round, ca. 6 mm. long; caps. 5–6 mm. in diam.—Shaded woods, 4900–8200 ft.; Montane Coniferous F.; San Diego Co. to Sierra Nevada, etc. to B.C., Ida. June–Aug.

2. **C. umbellàta** (L.) var. **occidentàlis** (Rydb.) Blake. [*C. o.* Rydb.] Stems stoutish, 1.5–3 dm. high; lvs. in whorls of 3–8, oblanceolate, 3–7 cm. long, serrate, mostly yellow-green beneath; peduncles mostly 6–8 cm. long; bracts linear-subulate, early deciduous; sepals ovate, fimbriate, ca. 3 mm. long; petals ciliolate, pink, 5–6 mm. long; caps. 6–7 mm. in diam.; $2n=26$ (Raven et al., 1965).—Dry shrubby slopes, 7000–9400 ft.; San Jacinto and San Bernardino mts.; Sierra Nevada; to Alaska, Mich. June–Aug.

4. *Comarostáphylis* Zucc.

Evergreen erect or spreading shrubs with exfoliating or persistent and shredded bark. Lvs. alternate, coriaceous, entire or toothed, petioled, often revolute. Fls. few to many, in terminal solitary or clustered racemes or panicles, 5 (rarely 4)-merous. Calyx persistent, the lobes exceeding the tube, reflexed or spreading at maturity. Corolla urn-shaped, the lobes broad and short, mostly recurved. Stamens 10 (8), included; fils. dilated below, pubescent; anthers broad, awned. Ovary 5 (4)-celled, glabrous or pubescent, rounded or ovoid; style columnar; stigma minute. Fr. fleshy, drupelike, warty or papillose, the nutlets united into a round stone. Ca. 20 spp., mostly Mexican. (Greek, *komaros*, arbutus, and *staphule*, grape, because of similarity to *Arbutus* and the clustered fr.)

1. **C. diversifòlia** (Parry) Greene. [*Arctostaphylos d.* Parry.] SUMMER-HOLLY. Erect, 2–5 m. high, with shredded bark and canescent-tomentulose twigs; lvs. oblong to elliptic, rounded to acute at apex, serrulate to green and shining above, tomentulose beneath, strongly revolute, 3–8 cm. long, on petioles 2–5 mm. long; racemes mostly solitary and terminal, 3–6 cm. long, tomentose; bracts lance-linear, 3 mm. long; pedicels recurved, 3–5 mm. long; calyx tomentulose, the lobes lanceolate, ca. 2 mm. long; corolla white, puberulent, 5–7 mm. long, the lobes very short; ovary pubescent; drupe globose, 5–6 mm. thick, red, granular-rugose.—Dry slopes, at low elevs.; Chaparral, near the coast, San Diego Co.; L. Calif. May–June.

Var. **planifòlia** Jeps. Lvs. not revolute; racemes 5–10 cm. long; bracts oblong-ovate, 3–7 mm. long; pedicels 8–12 mm. long.—Chaparral; Santa Monica Mts., Santa Ynez Mts.; Santa Rosa, Santa Cruz, Catalina ids. March–May.

5. *Phyllódoce* Salisb. MOUNTAIN-HEATHER

Low evergreen shrubs, much-branched, heathlike. Lvs. linear, needlelike, blunt or subacute, alternate, crowded, revolute. Fls. in umbellike infl., long-pedicelled from axils of persistent herbaceous bracts, nodding or erect. Calyx-segms. lance-ovate, acute, persistent, usually 5. Corolla urn-shaped to ovoid or open-campanulate. Stamens 7–10; fils. slender; anthers unappendaged, opening by oblique apical pores. Ovary usually 5-celled, globose; style filiform. Caps. ovoid to subglobose, septicidal. Seeds many, minute, narrowly winged. Ca. 8 spp., circumboreal. (Greek name of a sea nymph.)

1. **P. bréweri** (Gray) Heller. [*Bryanthus b.* Gray.] Semi-procumbent, rather lax, 1–3 dm. high; lvs. 6–15 mm. long, somewhat glandular, puberulent at margins, strongly revolute; pedicels 1–1.5 cm. long, glandular-pubescent; calyx-lobes oblong, obtuse, 3–4.5 mm.

long, glabrous except for the ciliolate margins; corolla rose-purple, to pinkish, open-campanulate, ca. 8-10 mm. long; stamens long-exserted; caps. round, 3-3.5 mm. in diam.—Rocky, sometimes rather moist places, 9000-11,400 ft.; Subalpine F.; San Bernardino Mts.; Tulare Co. to Mt. Lassen. July-Aug.

6. **Pteróspora** Nutt. PINEDROPS

Stout, simple, purple-brown, clammy-pubescent, herbaceous root-parasite. Lvs. scale-like, lanceolate, scattered. Fls. in long terminal racemes, nodding on recurved pedicels. Fls. white to red. Calyx 5-parted. Corolla urn-shaped, persistent, 5-lobed. Stamens 10, included; fils. slender; anthers with 2 dorsal appendages and opening by dorsal slits. Caps. 5-loculed, depressed-globose, loculicidal; style stout, short; stigma capitate-peltate, shallowly 5-lobed. Seeds many, reticulate-winged at apex. One sp. (Greek, *pteros*, wing, and *spora*, seed.)

1. **P. andromedèa** Nutt. Stems 3-10 dm. high; lvs. crowded below, 1.5-3.5 cm. long; sepals lance-linear, 4-5 mm. long; corolla 7-8 mm. long, the lobes rounded; caps. 8-12 mm. in diam.—Humus in forests, 5000-9000 ft.; Montane Coniferous F.; San Jacinto to San Gabriel mts.; to Sierra Nevada and Modoc Co.; to B.C., Atlantic Coast, Mex. June-Aug.

7. *Pýrola* L. SHINLEAF. WINTERGREEN.

Low perennial herbs with slender subterranean rootstocks. Lvs. in a dense basal cluster, rarely lacking, evergreen, petioled. Fls. in a simple raceme at end of leafless somewhat scaly scape, pedicelled, spreading or nodding, 5-merous. Calyx persistent, 5-parted. Petals 5, concave, ± converging. Stamens 10; fils. naked; anthers extrorse in bud, opening by pair of pores at base. Style elongate; stigma-lobes or rays 5. Caps. depressed-globose, 5-lobed. Seeds many, minute, with loose cellular-reticulate coat. Ca. a dozen spp. of N. Hemis. (Diminutive of *Pyrus*, the pear, from resemblance in the lvs.)

(Camp, W. H. Aphyllous forms in Pyrola. Bull. Torr. Club 67: 453-465. 1940.)
Style strongly deflexed at base; stigma subtended by a ring or collar.
    Calyx-lobes much longer than broad; petals red to purplish .... 1. *P. californica*
    Calyx-lobes scarcely longer than broad; petals greenish to whitish, rarely pinkish . 3. *P. picta*
Style straight, without a collar or ring below the stigma.
    Racemes spiral; corolla subglobose; style not exserted .... 2. *P. minor*
    Racemes 1-sided; corolla campanulate; style exserted .... 4. *P. secunda*

1. **P. califórnica** Krisa. [*P. asarifolia* Am. auth. *P. a.* var. *incarnata* Fern.] Extensively creeping; lvs. leathery, basal elliptic-ovate to rounded, the blades shining, 4-6 cm. wide, the petioles 7-9 cm. long; scapes slender, mostly 3-4 dm. long; bracts lance-ovate, 4.5-7.5 mm. long, usually as long as the pedicels, reddish; sepals 3-3.5 mm. long; petals pink to reddish purple; anthers 2.5-3 mm. long, short-pointed.—Moist shaded places, 7000-9300 ft.; San Bernardino Mts.; Montane Coniferous F.; to Alaska and Atlantic Coast. July-Sept.

2. **P. mìnor** L. Rootstock slender; basal bracts oblong, cuspidate; lvs. rounded, slightly crenulate, the blades dull, 1-2 (-3) cm. long, on petioles ca. as long; scapes slender, 6-15 cm. high; bracts at base of pedicels lance-oblong, 3-7 mm. long; calyx-lobes triangular, scarcely 2 mm. long; petals white to pink, 3-5 mm. long, with connivent tips; style included; caps. ca. 5 mm. broad; $2n=46$ (Hagerup, 1928).—Occasional in boggy shaded places, 8000-9200 ft.; Montane Coniferous F.; San Jacinto and San Bernardino mts.; Sierra Nevada to Alaska, Atlantic Coast; Eurasia. July-Aug.

3. **P. pícta** Sm. Plate 41, Fig. E. Rootstock branched; lvs. coriaceous, lance-ovate to broadly elliptic, mottled or veined with white, entire or serrulate, rounded at tip to acute, obtuse to subcuneate at base, the blades 2-7 cm. long, the petioles ca. as long; scapes 1-2 dm. high; bracts lance-deltoid, acuminate, 3-5 mm. long; calyx-lobes ovate, acutish, ca. 2 mm. long; petals oblong-obovate, greenish to cream, with hyaline margins, 7-8 mm.

long; anthers yellow; style 5 mm. long; caps. ca. 6–7 mm.. broad.—In humus in dry forests, below 8000 ft.; mostly Yellow Pine F.; mts. from San Diego Co. to B.C.; Rocky Mts. June–Aug.

Forma **aphýlla** (Sm.) Camp. [*P. a.* Sm.] Lvs. of flowering stems scalelike, of sterile stems sometimes green, 1–2 cm. long; calyx red-purple, ca. 2 mm. long.—Wooded slopes below 8000 ft.; Yellow Pine F.; mts. from San Diego Co. to B.C., Mont. June–Aug.

Ssp. **íntegra** (Gray) Piper. [*P. dentata* var. *i.* Gray.] Lvs. obovate to oblanceolate, mostly entire, the blades 1–4 cm. long, with little or no whitening along the veins.—Yellow Pine F.; San Jacinto Mts. n.; to Wash. June–July. Intergrading freely further north than our area with ssp. **dentata** (Sm.) Piper which tends to have dentate lvs.

4. **P. secúnda** L. Rootstocks long, creeping; caudex woody, branching; lower bracts lanceolate, ciliate; lvs. shining, elliptic to ovate, crenate-serrate, the blades 2–5 cm. long, on petioles usually somewhat shorter; scapes 5–15 cm. high; bracts of the 1-sided infl. ciliate; calyx-lobes ovate, obtuse, ca. 1 mm. long, ciliate-serrulate; petals oblong, erose, yellow-green, 4–5 mm. long; style exserted; caps. 4–5 mm. broad.—Rather dry shaded woods, mostly below 9000 ft.; Montane Coniferous F.; San Jacinto Mts., San Bernardino Mts.; Sierra Nevada to Alaska, Atlantic Coast, Eurasia. July–Sept.

8. *Rhododéndron* L. RHODODENDRON. AZALEA

Shrubs or small trees. Lvs. simple, alternate, deciduous or persistent, entire or toothed. Fls. showy, in terminal umbels or corymbs. Calyx small, saucer-shaped, with persistent lobes. Corolla funnelform to subcampanulate, regularly or irregularly 5-lobed. Stamens 5 or 10, the fils. slender, elongate, declined; anthers opening by terminal pores. Ovary 5-celled; style elongate, declined. Caps. septicidally 5-valved. Seeds many, wing-margined. Perhaps 200 spp., of wide distribution in N. Hemis., especially in Asia. (Greek, *rhodos*, rose, and *dendron*, tree.)

1. **R. occidentàle** (T. & G.) Gray. [*Azalea o.* T. & G.] WESTERN AZALEA. Loosely branched shrub, 1–3 m. tall, deciduous, with shredding bark; twigs stiff, divaricate, glabrous to pubescent, sometimes glandular; lvs. thin, light green, elliptic to obovate, 3–8 cm. long, ± pubescent, stiff-ciliate; petioles 4–8 mm. long; infl. terminal, ± glandular-pubescent; calyx-lobes oval to oblong-ovate, ciliate, 2–5 mm. long; corolla funnelform, 3.5–5 cm. long, slightly irregularly lobed, white or with pink tinge, the upper lobe with a yellowish blotch, the lobes lance-oblong; stamens exserted; caps. 1–2 cm. long; seeds tan, flat.—Stream banks, etc., 4500–7300 ft.; Cuyamaca Mts. to San Jacinto Mts.; cent. Calif. to Ore. April–Aug.

9. *Sarcòdes* Torr. SNOW PLANT

Red, fleshy, usually pubescent saprophyte. Stems simple, solitary or clustered. Lvs. scalelike, lanceolate, more crowded toward base of stem. Fls. many, in a stout spicate raceme, red, subtended by conspicuous red bracts. Sepals 5, glandular-pubescent. Corolla campanulate, 5-lobed, the lobes broad, red. Stamens 10, included; fils. slender. Caps. depressed-globose, 5-loculed; stigma subcapitate. Seeds small, pitted-reticulate, subovoid. One sp. (Greek, *sarx*, flesh, and *oeides*, like.)

1. **S. sangúinea** Torr. Plant mostly 1.5–3.5 dm. high; lvs. 2–8 cm. long, ciliate, the lower shorter and broader than the upper; sepals lance-ovate, 10–15 mm. long; corolla slightly longer; caps. 1–2 cm. in diam.—Thick humus of forests, 4000–8000 ft.; Montane Coniferous F.; Santa Rosa and San Jacinto mts. to Ore., Nev. May–July.

10. *Vaccínium* L. BLUEBERRY. HUCKLEBERRY

More or less woody plants, from slender and trailing to arborescent. Buds scaly. Lvs. simple, alternate, deciduous or persistent. Fls. perfect, solitary or clustered. Calyx-tube adnate to ovary which in fr. becomes a berry or drupe crowned with calyx-teeth. Corolla of 4–5 united petals, often urn-shaped to campanulate. Stamens twice as many as corolla-lobes; anthers erect, their locules partly separated or prolonged into a tubular appendage with apical pore; pollen in tetrads. Ovary mainly inferior, many-ovuled. Style filiform; stigma simple. Seeds berrylike, red to blue-black, sometimes with a bloom. Seeds compressed, with hard coat. Perhaps 150 spp.; quite cosmopolitan. (Ancient name of Bilberry.)

1. **V. ovàtum** Pursh. CALIFORNIA HUCKLEBERRY. Plate 41, Fig. F. Stout, erect, much-branched, to over 1 m. tall; branchlets pubescent; lvs. persistent, leathery, ovate to lance-oblong; 1.5–4 cm. long, serrate, glabrous and shining above, paler beneath, acutish apically; petioles 2–3 mm. long; fls. white to pink, bell-shaped, in few-fld., axillary racemes; bracts red, deciduous; calyx with 5 deltoid lobes; corolla 5–7 mm. long; anthers without awns; berry broadly ovoid, 6–9 mm. long, black without a bloom, sweet, edible; seeds brown, 1 mm. long; $2n=24$ (Darrow et al., 1944).—Dry canyon-slopes at ca. 2400 ft., Big Boulder Creek, Bottle Peak and El Cajon Mt., San Diego Co.; n. Santa Barbara Co. to B.C. Santa Rosa and Santa Cruz ids. March–May.

11. *Xylocóccus* Nutt.

Erect densely branched shrubs with shredding bark. Lvs. alternate or opposite, persistent, entire, with revolute margins. Fls. few, in terminal simple or branched panicles. Bracts small, scalelike. Calyx persistent, mostly 5-lobed, the broad lobes reflexed at maturity. Corolla oblong, urn-shaped, mostly 5-lobed. Stamens mostly 10, included; fils. dilated below; awned. Ovary mostly 5-celled, pubescent; style elongate; stigma minute. Fr. a dry drupe, with smooth surface and a thin pulp; nutlets united into a solid stone. One sp. (Greek, *xylon*, wood, and *kokkos*, berry.)

1. **X. bicólor** Nutt. [*Arctostaphylos b.* Gray.] Two to 3 m. high, with persistent shredding bark, and cinereous branchlets; lvs. ovate to oblong, acute at both ends; dark green and glabrous above, cinereous-tomentose beneath, 3–5 cm. long; short-petioled; panicles recurved, dense, tomentose, 1–2.5 cm. long; bracts ovate, tomentose, ca. 2 mm. long; pedicels 2–4 mm. long; calyx dark red, the lobes ovate, ca. 1.5 mm. long; corolla white or pink, 8–9 mm. long; fr. round, red to almost black, 5–8 mm. in diam.; stone, solid, smooth; $2n=26$ (Callan, 1941).—Dry slopes below 2000 ft.; Chaparral; scattered localities near the coast from Los Angeles Co. to L. Calif. Dec.–Feb. Santa Catalina Id.

41. **Euphorbiàceae.** SPURGE FAMILY

Herbs, shrubs or trees usually with a milky acrid sap; some succulent and cactuslike, others of very different habit. Lvs. simple, alternate, opposite or whorled, usually stipulate, sometimes with glands at the base. Plants monoecious or dioecious, the infl. variable; the fls. sometimes in a *cyathium*, i.e., the naked ♀ fl. surrounded by ♂ fls., each a single stamen jointed at union of pedicel to peduncle, and all surrounded by an invol. Calyx and corolla present or absent, frequently different in ♂ and ♀ fls., the parts free or rarely united. Disk present or reduced to glands. Stamens as many or twice as many as sepals or more numerous, or 1. Ovary superior, usually 3-celled, with 1 or sometimes 2 pendulous ovules in a cell; styles free or united. Fr. usually a 3-lobed caps., the lobes or carpels separating elastically from a persistent axis and 2-valved. Seeds with straight embryo and flat cotyledons. Ca. 280 genera and more than 8000 spp., widely distributed. Some furnish food, some valuable oils, some are of ornamental and medical use.

1. Fls. with a calyx, not in the base of an invol. (cyathium).
  2. Plants definitely woody.
    3. Lvs. palmately lobed, 1–5 dm. long .... 7. *Ricinus*
    3. Lvs. not palmately lobed, less than 0.5 dm. long.
      4. Fls. in umbellate axillary clusters; lvs. usually entire .... 9. *Tetracoccus*
      4. Fls. in racemes or spikes; lvs. dentate.
        5. Lvs. 6–10 mm. long, gray at least beneath, 1-veined. Deserts .... 2. *Bernardia*
        5. Lvs. 10–30 mm. long, often 3–5-veined from base. Cismontane .... 1. *Acalypha*
  2. Plants herbaceous, at most slightly woody at base.

6. The plants armed with stiff stinging hairs.
  7. Herbage gray-stellate as well as hispid. Cismontane annual .... 5. *Eremocarpus*
  7. Herbage green, hispid only, not stellate. Desert perennial .... 10. *Tragia*
6. The plants not armed with stinging hairs.
  8. Herbage densely stellate-scurfy .... 3. *Croton*
  8. Herbage not stellate-pubescent.
    9. Petals 5, straw-colored, most evident in ♂ fls.; styles 2-cleft .... 4. *Ditaxis*
    9. Petals lacking; styles entire .... 8. *Stillingia*
1. Fls. lacking a calyx, included in a cup-shaped and calyxlike invol. (cyathium) which surrounds several ♂ fls. and 1 ♀ fl. with a 3-lobed pistil .... 6. *Euphorbia*

1. Acalýpha L.

Herbs or shrubs. Lvs. alternate, simple, petioled. Monoecious. Fls. in spikes, staminate spikes amentlike, the fls. apetalous, calyx 4-parted; stamens usually 8, distinct, on a raised cent. receptacle. Pistillate fls. on short spikes, or 1–few at base of ♂ spikes; calyx 3-lobed; ovary 3-celled, 3-ovuled; styles 3, reddish, much dissected; caps. often surrounded by the enlarged bract. Ca. 250 spp. mostly of trop. of both hemis., often grown for their foliage. (Greek, *akalephes*, nettle.)

  1. **A. califórnica** Benth. CALIFORNIA COPPERLEAF. Slender-stemmed shrub; lvs. broadly ovate, crenate-dentate, glandular-pubescent, 1–2 cm. long; ♂ spikes 1–2 cm. long, peduncled; ♀ 4–5 mm. long, with conspicuous glandular-pubescent bracts; calyx ca. 1 mm. long; caps. ca. 3 mm. in diam.; seeds grayish-brown, minutely cellular-pitted; $2n = 10$ pairs (Raven, Kyhos, & Hill, 1965).—Locally frequent, dry granite slopes, 700–4000 ft.; Chaparral, S. Oak Wd., etc.; cismontane San Diego Co.; desert edge as at Borrego V., Santa Rosa Mts.; L. Calif. Jan.–June.

2. Bernárdia Houst. ex P. Br.

Ours shrubs with alternate stipulate serrate lvs. Fls. monoecious (in ours) or dioecious, in small spicate or racemose clusters, apetalous. Staminate calyx 3–5-parted; stamens 3–20, distinct, on a cent. receptacle. Pistillate calyx 3–6–9-parted; ovary 3-celled, 3-ovuled; styles 3; stigmas stout, 2-lobed. Ca. 20 spp., subtrop. or trop. Am. (P. F. *Bernard*, 1749–1825, French botanist.)

  1. **B. incàna** Mort. [*B. myricifolia* auth., not Wats.] Many-stemmed, 1–2 m. high, with close grayish stellate pubescence; lvs. oblong-ovate, 8–12 mm. long, thick, coarsely crenate; ♂ racemes axillary; calyx ca. 1 mm. long; stamens 5–7; ♀ fls. terminal; caps. ca. 1 cm. thick, stellate; seeds subglobose, mottled, light brown, 4–5 mm. long.—Occasional in dry rocky canyons, below 4000 ft.; Creosote Bush Scrub; Colo. Desert from s. Little San Bernardino Mts. and Eagle Mts. s.; Ariz. April–May. Oct.–Nov.

3. Cròton L. CROTON

Ours perennial, stellate-scurfy herbs, often suffrutescent. Lvs. alternate, entire. Plants usually monoecious. Staminate fls. in terminal racemes; calyx 5-parted, petals 5, small, alternating with the glands; stamens 5-many. Pistillate fls. mainly solitary, below the ♂ ones; calyx 5-parted; ovary 3-celled, the ovules solitary; styles twice cleft; caps. 3-lobed, globose. Perhaps 600 spp. of warm or hot regions; all continents but Eu. (Greek, *kroton*, a tick, the old name of Castor-bean, because of appearance of seeds.)

Caps. 10–11 mm. high; seeds 7–8 mm. long; lvs. 2–8 mm. wide. Sand dunes w. of Yuma
  .... 2. *C. wigginsii*
Caps. 4–7 mm. high; seeds 3–5.5 mm. long; lvs. 5–20 mm. wide.
  Lf.-blades mostly 2–4.5 cm. long; petioles 1–3.5 cm. long.
    Lvs. mostly 8–20 mm. wide; petioles 1–3.5 cm. long. Cismontane .... 1. *C. californicus*
    Lvs. mostly 5–10 mm. wide; petioles 0.5–1 cm. long. Coastal .... var. *tenuis*
  Lf.-blades mostly 0.5–2 cm. long; petioles 0.3–0.7 cm. long. Deserts .... var. *mohavensis*

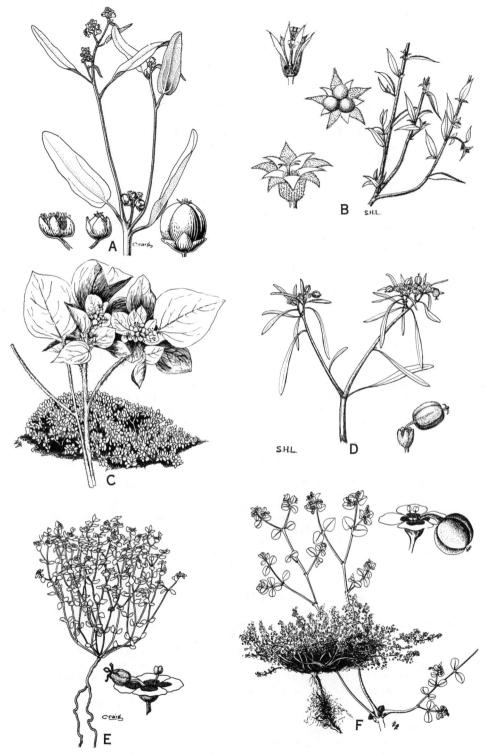

Plate 42. Fig. A, *Croton californicus*, ♂ & ♀ fls. Fig. B, *Ditaxis lanceolata*. Fig. C, *Eremocarpus setigerus*. Fig. D, *Euphorbia eriantha*, cyathium and exserted ♀ fl. Fig. E, *Euphorbia melanadenia*, cyathium with petaloid appendages and ♂ and ♀ fls. Fig. F, *Euphorbia polycarpa*.

1. **C. califórnicus** Muell.-Arg. var. **califórnicus.** Plate 42, Fig. A. Branching erect or spreading; stems 2–10 dm. long, ± hoary with stellate canescence; lf.-blades oblong-elliptic, 1.5–4 cm. long, 1–2 cm. wide, greener above, with petioles 1–3.5 cm. long; ♂ fls. corymbose in anthesis, the racemes becoming 10–15 mm. long, the fls. dropping off; calyx 1–2 mm. long; pistillate calyx 2–3 mm. long, the ♀ racemes few-fld.; styles 2–2.5 mm. long; caps. 5–7 mm. long; seeds 3–4 mm. long, black or mottled; $2n = 14$ (Szweykowski, 1965).—Locally frequent in dry sandy places, below 4000 ft.; Coastal Sage Scrub, Chaparral; Calif. to San Francisco region. March–Oct.

Var. **mohavénsis** Ferg. Lvs. pale olive-green, 1–2 cm. long; petioles less than 1 cm. long.—Sandy places, Creosote Bush Scrub; deserts; Ariz.

Var. **ténuis** (Wats.) Ferg. [*C. t.* Wats.] Stems very slender; lvs. generally less than 1 cm. wide; petioles less than 1 cm. long.—Sandy places, such as dunes and beaches; Coastal Strand; Ventura Co. to L. Calif.

2. **C. wígginsii** Wheeler. [*C. arenicola* Rose & Standl., not Small.] Woody, 5–8 dm. high, much-branched, densely silvery-stellate throughout; lvs. linear to lanceolate, 2–6.5 cm. long, 2–8 (–15) mm. wide; petioles 5–14 mm. long; dioecious; ♂ fls. in racemes 15–30 mm. long, the calyx-lobes ca. 2 mm. long; ♀ raceme ca. 2.5–3 cm. long; caps. 9–11 mm. high; seeds 7–8 mm. long, variegated with brown and gray.—Creosote Bush Scrub; sand dunes w. of Yuma, to Son. March–May.

4. *Ditáxis* Vahl.

Monoecious or rarely dioecious herbs, sometimes woody at base, mostly with 2-branched hairs (the branches appressed and extending in opposite directions). Lvs. simple, alternate, entire to serrulate. Fls. in short or reduced axillary, few-fld., usually bracted clusters. Sepals 5, nearly distinct. Petals 5, straw-color, reduced in ♀ fls. Stamens 10–15, united into a column, placed in sets of 5, the third or upper whorl sterile. Ovary 3-lobed, each cell with 1 ovule. Styles 3, 2-cleft. Caps. 3-lobed. Ca. 45 spp. of temp. and warmer parts of New World. (Greek, *dis*, two, and *taxis*, rank, the stamens in 2 whorls.)

Plants glabrous .... 2. *D. californica*
Plants strigose to pubescent.
    Gland-tipped teeth present on lvs. and bracts .... 1. *D. adenophora*
    Gland-tipped teeth mostly lacking.
        Pistillate sepals white-margined; plants herbaceous.
            Lvs. mostly broadly cuneate-spatulate with truncate coarsely toothed apices; depressions on seeds not with radiating ridges .... 5. *D. serrata*
            Lvs. oblanceolate, acute, serrulate or entire; depressions on seeds with minute radiating ridges .... 4. *D. neomexicana*
        Pistillate sepals not white-margined; lvs. linear-lanceolate, entire; plants woody at base
            3. *D. lanceolata*

1. **D. adenóphora** (Gray) Pax & K. Hoffm. [*Argythamnia a.* Gray. *A. clariana* Jeps.] Perennial, 2.5–4 dm. high, freely branched, rather thinly hairy; petioles 3–8 mm. long; stipules linear, 2 mm. long; lf.-blades lanceolate to ovate, 1.5–3.5 cm. long, 15–18 mm. wide, glandular-serrate; bracts lanceolate, 5 mm. long; racemes short, with 1 ♀ and 3–4 ♂ fls.; ♂ fl. with non-glandular obscurely margined sepals ca. 3 mm. long, the petals pinkish; stamens 10, in 2 series; ♀ fl. with glandular-serrulate sepals ca. 3 mm. long; caps. ca. 5 mm. thick; seeds papillose, pitted-rugose, ca. 3 mm. long.—Rare, sandy flats below 500 ft.; Creosote Bush Scrub; Coachella V.; Ariz., Son. Dec.–March.

2. **D. califórnica** (Bdg.) Pax & K. Hoffm. [*Argythamnia c.* Bdg.] Perennial becoming woody basally, glabrous, 2–4 dm. high; petioles 2–10 mm. long; stipules filiform; lf.-blades obovate to lance-oblong, green, minutely serrulate, 1–3.5 cm. long, conspicuously veined; racemes short, few-fld., the ovate hyaline bracts 1.5–2 mm. long; ♂ fls. 2–4, ♀ 1; sepals of ♂ fls. 2.5 mm. long, recurved, hyaline-margined; petals white, hyaline; stamens 5 and 5; ♀ fl. ca. 3.5 mm. long; style free, bifid; caps. ca. 3 mm. long; seeds brownish,

reticulate with low ridges.—Infrequent in sandy washes, etc. 400–3000 ft.; Creosote Bush Scrub; Santa Rosa Mts. to s. side of Eagle Mts. March–May, Oct.–Dec.

3. **D. lanceolàta** (Benth.) Pax & K. Hoffm. [*Serophyton l.* Benth. *Argythamnia l.* Muell.-Arg. *D. sericophylla* Heller.] Plate 42, Fig. B. Somewhat woody below, freely branched, brittle, silvery-strigose on younger twigs, 2–4.5 dm. high; petioles 1–3 mm. long; stipules 1 mm.; lf.-blades lance-linear to -ovate, 1–3 cm. long, 8–10 mm. wide, entire, greenish; racemes 6–8 mm. long, with 1 ♀ fl. and few ♂ fls.; bracts ovate, 1 mm. long; ♂ fls. 4 mm. long; ♀ 4–5 mm.; petals hairy; ovary hispid; style-branches hirsute; caps. ca. 5 mm. thick; seeds brownish-gray, with shallow circular pits having radiating lines.—Frequent, rocky places below 2000 ft.; Creosote Bush Scrub; Colo. Desert; Ariz., L. Calif. March–May.

4. **D. neomexicàna** (Muell.-Arg.) Heller. [*Argythamnia n.* Muell.-Arg.] Annual to perennial, not woody, 1–3.5 dm. high; lvs. lanceolate to oblanceolate, 1–2.5 cm. long, serrulate or entire, strigose with long setalike hairs; fls. few, congested in axils; ♂ fls. 1.5–2 mm. long, the petals longer than sepals; ♀ fls. 3.5–5 mm. long, somewhat hairy; styles united at base only, with narrow stigma-lobes; seeds brownish, shallowly foveolate, the depressions marked with minute radiating ridges.—Dry slopes, Creosote Bush Scrub; Mojave Desert to L. Calif., Tex., Son. March–Dec.

5. **D. serràta** (Torr.) Heller. [*Aphora s.* Torr. *Argythamnia s.* Muell.-Arg.] Bushy annual or perennial, 1–3.5 dm. high, densely strigose; petioles 2–5 mm. long; lvs. ovate to obovate, 1–2.5 cm. long, 7–12 mm. wide, usually serrulate, plainly veined; racemes with 1–2 ♀ and 2–4 ♂ fls.; bracts lanceolate, 1–1.5 mm. long; ♂ fls. 1–1.5 mm. long; ♀ ca. 3 mm.; styles 3, bifid; caps. 4 mm. thick; seeds grayish or brownish, reticulate.—Occasional in dry rocky and sandy places, below 2500 ft.; Creosote Bush Scrub; e. Mojave Desert (Newberry Springs e.) and Colo. Desert to Tex., Mex. April–Nov.

5. *Eremocárpus* Benth. TURKEY-MULLEIN. DOVE WEED

Low broad gray heavy-scented annual with alternate entire 3-nerved lvs., stellate-pubescent and with stinging longer hairs. Monoecious. Staminate fls. in terminal cymes; calyx 5–6-parted; petals 0; stamens 6–7. Pistillate fls. 1–3 in lower axils, without a calyx or corolla; ovary with 4 or 5 small glands at base, 1-celled, 1-ovuled; style undivided; caps. obovate-oblong, 2-valved. One sp. (Greek, *eremos*, solitary, and *karpos*, fr.)

1. **E. setígerus** (Hook.) Benth. Plate 42, Fig. C. Grayish, dichotomously branched, 3–20 cm. high, 5–80 cm. wide; lvs. ovate to suborbicular, 1–5 cm. long, on petioles as long; ♂ fls. pedicelled, calyx ca. 2 mm. long; pistil pubescent; caps. 4 mm. long; seeds dark, ± variegated, 3–4 mm. long; $2n = 20$ (Heiser & Whitaker, 1948).—Common in dry open places, mostly below 2500 ft.; Coastal Sage Scrub, V. Grassland, Oak Wd.; cismontane especially away from immediate coast, occasional at desert-edges; to Wash. May–Oct. Most growth during the dry season; seeds much eaten by quail.

6. *Euphórbia* L. SPURGE

Monoecious herbs or shrubs with milky acrid juice. Stems leafy and slender to almost leafless and fleshy, unarmed or armed. Lvs. simple, alternate or opposite, entire or dentate, stipulate or estipulate. Fls. in a cyathium (cupulate invol. resembling a calyx or corolla with united lobes) with 1–5 nectariferous glands on its margin alternating with the lobes and often with petaloid appendages from beneath the glands. Staminate fls. in 5 fascicles in the cyathium, 1–several per fascicle and each a single stamen jointed on its pedicel and usually subtended by a minute bract. Pistillate fl. solitary in center of cyathium, becoming exserted and consisting of a 3-celled, 3-ovuled ovary subtended by 3 small scales; styles 3, each usually 2-cleft. Caps. usually nodding, separating into 3 2-valved carpels. Seeds often carunculate. A highly diversified genus of over 1000 spp., mostly temp., many cactuslike or otherwise highly modified in habit. (*Euphorbus*, physician of Numidia.)

(Wheeler, L. C. Euphorbia, subgenus Chamaesyce in Canada and the U.S. exclusive of s. Florida. Rhodora 43: 97–154, 168–205, 223–286. 1941. Norton, J. B. S. Am. Spp. of Euphorbia section Tithymalus. Ann. Rep. Mo. Bot. Gard. 11: 85–144. 1900.)

1. Glands of the cyathium without petaloid appendages; lvs. essentially symmetrical.
    2. Lvs. linear; cyathia in terminal headlike clusters, their glands cup-shaped, concealed by the inflexed segms. of the invol. .... 5. *E. eriantha*
    2. Lvs. (at least in infl.) ovate to rounded; cyathia solitary or in cymes, their glands not cup-shaped.
        3. Glands elliptic or transversely oval; lvs. serrulate.
            4. Floral lvs. broad at base; caps. with fleshy ellipsoid-lenticular tubercles
                                                                                            28. *E. spathulata*
            4. Floral lvs. narrowed at base; caps. smooth; seeds subglobose-ovoid .. 8. *E. helioscopia*
        3. Glands crescent-shaped or 2-horned; lvs. largely entire.
            5. Stem-lvs. opposite, strongly decussate; caps. 8–12 mm. high .... 10. *E. lathyris*
            5. Stem-lvs. alternate; caps. 2–4 mm. high.
                6. Plants annual.
                    7. Caps. with thin keels; stem lvs. petioled; seeds with deep pits ..... 20. *E. peplus*
                    7. Caps. smooth; stem lvs. subsessile; seeds vermiculate-ridged to almost smooth
                                                                                            4. *E. crenulata*
                6. Plants perennial.
                    8. Glands hornless, lacerate; stem lvs. elliptic, tapering to acute tips .. 9. *E. incisa*
                    8. Glands horned, not lacerate; stem lvs. oblong to suborbicular .... 16. *E. palmeri*
1. Glands of the cyathium with petaloid appendages or, if these wanting, the lvs. all opposite and with ± inequilateral bases.
    9. Lvs. alternate or opposite, their bases quite symmetrical; stipules glandlike or none.
        10. Plant shrubby, 4–10 dm. tall; lvs. entire, glabrous .... 14. *E. misera*
        10. Plant a slender-stemmed annual, to ca. 1.5 dm. tall; lvs. serrulate to serrate, stiff-pubescent. Clark Mtn. .... 6. *E. exstipulata*
    9. Lvs. all opposite, their bases usually strongly inequilateral; plants herbaceous to suffrutescent, mostly prostrate or low.
        11. Ovary and caps. hairy.
            12. Perennials.
                13. Invols. or cyathia urceolate; ♂ fls. up to 12; appendages entire or crenate
                                                                                            3. *E. arizonica*
                13. Invols. or cyathia not urceolate; ♂ fls. 15–60.
                    14. Seeds scarcely angled, narrow-ovoid, encircled by 4–5 rounded ridges. Colo. Desert .... 19. *E. pediculifera*
                    14. Seeds quadrangular, smooth to slightly wrinkled.
                        15. Herbage with short straight spreading hairs .... 22. *E. polycarpa hirtella*
                        15. Herbage with appressed long and weak or matted hairs.
                            16. Appendages wider than the glands, with short spreading hairs beneath and on the margins. Desert of Inyo & Kern cos. .... 30. *E. vallis-mortae*
                            16. Appendages wider than to narrower than glands, glabrous or rarely with few short hairs beneath. Los Angeles Co. to San Diego and Colo. Desert
                                                                                            12. *E. melanadenia*
            12. Annuals.
                17. Invols. urceolate; appendages deeply 3–4-parted .... 27. *E. setiloba*
                17. Invols. obconic to campanulate.
                    18. Glands without appendages or with mere rudiments; seeds smooth; lvs. not more than 8 mm. long .... 13. *E. micromera*
                    18. Glands appendaged; seeds granular or ridged.
                        19. Caps. 1.5–1.9 mm. long; seeds 1–1.4 mm. long, not transversely ridged; appendages narrower than glands .... 26. *E. serpyllifolia hirtula*
                        19. Caps. 1–1.4 mm. long; seeds 0.9–1 mm. long, transversely ridged.
                            20. Seeds with low rounded transverse ridges not whitened on summit; caps. strigose; lvs. 4–17 mm. long; appendages narrower than glands
                                                                                            29. *E. supina*
                            20. Seeds with narrow sharp transverse ridges whitened on summit; caps. with spreading hairs; lvs. mostly 4–8 mm. long; appendages 1–2 times as wide as glands .... 23. *E. prostrata*
        11. Ovary and caps. glabrous.

21. Stipules united into a broad white membranous scale.
22. Plant perennial; ♂ fls. 12–30 .... 2. *E. albomarginata*
22. Plant annual; ♂ fls. 5–10 .... 25. *E. serpens*
21. Stipules not so united.
23. Plants perennial.
24. Glands discoid, without appendages; caps. 1.7 mm. long; lvs. 2–4 mm. long
17. *E. parishii*
24. Glands appendaged; lvs. mostly 3–10 mm. long.
25. Caps. 1.1–1.3 mm. long; ventral stipules united .... 22. *E. polycarpa*
25. Caps. 2.3–2.5 mm. long; stipules distinct .... 7. *E. fendleri*
23. Plants annual.
26. Lvs. linear, symmetric, glabrous; plants erect. E. Mojave Desert.
27. Caps. roundly 3-lobed, 2 mm. long; seeds 1.8 mm. long, smooth
18. *E. parryi*
27. Caps. sharply 3-angled, 1.3 mm. long; seeds 1–1.3 mm. long, with transverse rounded ridges .... 24. *E. revoluta*
26. Lvs. lanceolate to ovate or rounded, not symmetric at base.
28. Margins of lvs. entire or nearly so.
29. Glands radially elongate; seeds rounded on back, the face nearly flat
21. *E. platysperma*
29. Glands strictly discoid or transversely oblong; seeds ovoid or quadrangular, the face not flat.
30. Glands with wide appendages; seeds with 4–6 transverse rounded ridges. Colo. Desert .... 1. *E. abramsiana*
30. Glands not appendaged; seeds not thus transversely ridged.
31. Caps. 2–2.3 mm. long; seeds 1.1–1.4 mm. thick; ♂ fls. 40–60
15. *E. ocellata*
31. Caps. 1.3 mm. long; seeds 0.5 mm. thick; ♂ fls. 2–5 . 13. *E. micromera*
28. Margins of lvs. toothed, at least toward apex.
32. Stems mostly erect; lvs. oblong to oblong-lanceolate, 8–35 mm. long; seeds with finely rippled surface .... 11. *E. maculata*
32. Stems prostrate or nearly so.
33. Plants essentially glabrous; seeds not transversely ridged, 1–1.4 mm. long; appendages not as wide as glands .... 26. *E. serpyllifolia*
33. Plants pubescent; seeds with 4–6 transverse ridges and 0.6–0.7 mm. long
1. *E. abramsiana*

1. **E. abramsiàna** Wheeler. Prostrate annual, finely pubescent to subglabrous, the stems 5–25 cm. long; lvs. ovate- to elliptic-oblong, 2–12 mm. long, sometimes serrulate at apex; stipules distinct, 2–5-parted; cyathia mostly on congested lateral branches; invols. obconic, 0.6 mm. across; glands roundish, to transversely elliptic, ca. 0.2 mm. long, with wide white appendages entire or somewhat 2-lobed; ♂ fls. 3–5; caps. glabrous, round-oblong, ca. 1.5 mm. long; seeds sharply quadrangular, white, 1–1.4 mm. long, the facets with 4–6 irregular transverse rounded ridges.—Sandy flats; Creosote Bush Scrub; Imperial V. to Ariz., Sinaloa. Sept.–Nov.

2. **E. albomarginàta** T. & G. [*Chamaesyce a.* Small.] RATTLESNAKE WEED. Glabrous prostrate perennial, the stems 5–25 cm. long; lvs. rounded to oblong, 3–8 mm. long, with thin whitish entire margin; stipules united into a conspicuous whitish membrane entire or somewhat lacerate; cyathia solitary at the nodes, 1.5–2 mm. wide, campanulate to obconic; glands transversely oblong, concave, mostly maroon, 0.5–1 mm. long, the appendages conspicuous, white, entire or subcrenate; ♂ fls. 15–30; caps. glabrous, sharply 3-angled, 1.7–2.3 mm. long; seeds rounded-quadrangular, oblong, 1.3–1.7 mm. long, opaque-white, the facets smooth.—Common on dry slopes below 4000 ft.; Coastal Sage Scrub, Chaparral; Los Angeles Co. to San Diego Co.; and up to 7500 ft., Creosote Bush Scrub, to Pinyon-Juniper Wd.; deserts from Inyo Co. to Imperial Co.; to Utah, Tex., Mex. April–Nov.

3. **E. arizónica** Engelm. Perennial from a woody taproot, prostrate and matted to erect, the stems 1–3 dm. long, finely clavate-hairy; lvs. deltoid-ovate to ovate-oblong, entire, 2–10 mm. long, reddish; stipules minute; cyathia long-turbinate, constricted

above, ca. 1.5 mm. high; glands red, concave, almost twice as long as wide; the appendages oval, pinkish, to 1 mm. long; ♂ fls. mostly 6–7; caps. with spreading hairs, subglobose, ca. 1.5 mm. long; seeds quadrangular, whitish, 1–1.2 mm. long, the facets with low, often anastomosing ridges.—Rare; Creosote Bush Scrub; Colo. Desert from w. edge; to Tex., Mex. March–April.

4. **E. crenulàta** Engelm. [*Tithymalus c.* Heller.] Annual or of longer duration, glabrous; stems 1 or more, simple or branched below and above, 2–6 dm. high; cauline lvs. obovate to spatulate, obtuse, entire, 1.5–3.5 cm. long, subsessile; floral lvs. opposite or in 3's, subcordate, rhombic to deltoid, 5–15 mm. long; invol. ca. 2 mm. high, with denticulate lobes; glands crescent-shaped, the horns slender, sometimes cleft; caps. smooth, ca. 3 mm. long; seeds ash-colored, oblong-obovoid, irregularly vermiculate-ridged to almost smooth, 2–2.5 mm. long, the yellowish caruncle reniform.—Common and widespread in dry places below 5000 ft.; most Communities from Coastal Scrub to Yellow Pine F.; Santa Ana and San Gabriel mts. to Ore. March–Aug.

5. **E. eriántha** Benth. [*Poinsettia e.* Rose & Standl.] Plate 42, Fig. D. Erect green annual, 1.5–5 dm. tall, finely branched especially above the base, glabrous except at the strigose tips; lvs. narrow-linear, sparsely strigose, 2–5 cm. long, entire, short-petioled, the uppermost forming whorls near the fl.-clusters; invols. 1–4 at end of each branch, white-strigulose, ca. 2 mm. high; glands thin, 3–5, concealed by the inflexed segms. of the invol.; styles entire; caps. strigulose, 4–5 mm. long; seeds quadrate-oblong, whitish, compressed, coarsely wrinkled, 3.5–4 mm. long, with a stipitate reniform caruncle.—Infrequent, rocky canyons and mesas, below 3000 ft.; Creosote Bush Scrub; Colo. Desert from Eagle Mts. and Andreas Canyon to Tex., Mex. March–April.

6. **E. exstipulàta** Engelm. Erect, slender-stemmed annual, widely branched, to ca. 1.5 dm. high, stiff-pubescent; lvs. lanceolate to lance-linear, 2–5 cm. long, 1–2 mm. wide, opposite, sharply serrate, narrowed into short petioles; stipules glanduliform; ♂ fls. solitary; appendages of cyathia 2- or 4-lobed; styles equal to ovary, bifid; stigmas filiform; seeds 2 mm. long, subcubic, verrucose and transversely 2–3-costate.—S. side of Clark Mts., e. San Bernardino Co., at 6000 ft; to Tex. Sept.

7. **E. féndleri** T. & G. [*Chamaesyce f.* Small.] Glabrous perennial from a deep-set taproot, the stems decumbent to erect, 5–15 cm. long; lvs. entire, round-ovate to lance-ovate, 3–11 mm. long; stipules distinct, linear, mostly entire and glabrous; cyathia solitary at nodes, campanulate to turbinate, 1.3–1.7 mm. wide; glands reddish, 2–4 times as long as wide, up to 1 mm. long, the appendages white, crenate, ca. as wide as gland; ♂ fls. 25–35; caps. glabrous, rounded, ca. 2.4 mm. long; seeds quadrangular, 2–2.2 mm. long, white, with smooth front facets and slightly wrinkled back facets.—Occasional on dry slopes, 5000–7500 ft.; Pinyon-Juniper Wd.; n.e. San Bernardino Co. to White Mts.; to Nebr., Tex., Son. May–Oct.

8. **E. helioscòpia** L. [*Tithymalus h.* Hill.] WARTWEED. Annual with ascending stems 1–5 dm. high, rather stout, with few long scattered hairs above; lvs. all obovate, finely serrate, obtuse or retuse, subsessile, 1–2.5 cm. long; umbel usually 5-rayed, the branches then in 3's, ultimately 2's; floral lvs. narrowed to base; glands round to elliptical, yellow, stalked; caps. smooth, subglobose, 2.5–3 mm. high; seeds round-obovoid, 2 mm. long, dark, with coarse honeycomb-like reticulations and yellow reniform caruncle; $2n=42$ (Perry, 1943).—Natur. as weed, near Otay Mt., El Monte, etc.; from Eu. April–July.

9. **E. incisa** Engelm. [*E. schizoloba* Engelm. *Tithymalus s.* Nort.] With habit of *E. palmeri*, but stems more slender; cauline lvs. ovate to elliptic, 6–14 mm. long; floral lvs. ovate to broadly subcordate; invols. whitish, 2–3 mm. high; glands broad, irregularly toothed, hornless; caps. oblong-ovoid, ca. 4 mm. long; seeds oblong-cylindric, 2–2.5 mm. long, irregularly reticulate with low flat ridges, the caruncle small, conical.—Gravelly slopes, 3000–7000 ft.; Creosote Bush Scrub to Pinyon-Juniper Wd.; e. Mojave Desert; to Nev., Ariz. March–May.

10. **E. láthyris** L. [*Tithymalus l.* Hill.] CAPER SPURGE. Annual or biennial, coarse, 3–10 dm. high, glabrous, erect, usually simple below, glaucous; cauline lvs. opposite, strongly decussate, 5–14 cm. long, oblong-lanceolate with cordate clasping base; floral

lvs. ovate, cordate, 2–6 cm. long; cyathia ca. 4 mm. long; glands crescent-shaped, strongly 2-horned; caps. 8–12 mm. long; seeds oblong-ovoid, brown with darker spots, cellular-punctate and shallowly reticulate, 4–5 mm. long, the caruncle yellow, helmet-like.—Occasional weed in waste places, roadsides, etc.; San Diego Co. n. and to Atlantic Coast; from Eu. Most of year.

11. **E. maculàta** L. [*Chamaesyce m.* Small. *E. nutans* auth. *E. preslii* Guss.] Erect or ascending annual, 1–9 dm. high, simple or mostly branched, subglabrous except for the tips; lvs. oblong, 1–3 cm. long, serrulate, very short-petioled; stipules mostly united, broadly triangular-acuminate, ciliate; cyathia solitary or clustered, obconic, 0.7–1 mm. across; glands stipitate, round to transversely elliptical, with rudimentary appendages; ♂ fls. 5–11; caps. 2–2.3 mm. long, glabrous; seeds grayish to brown, ovoid-quadrangular, 1.1–1.6 mm. long, with finely rippled facets. $2n=28$ (Perry, 1943).—Occasional weed in cismontane Calif.; native, Atlantic Coast to Miss. V. and S. Am. April–Oct.

12. **E. melanadènia** Torr. [*Chamaesyce m.* Millsp.] Plate 42, Fig. E. Perennial with usually ascending forking stems 8–20 cm. high, hoary-tomentose throughout; lvs. ovate, 2–8 mm. long, entire; stipules on lower side of stem usually united, linear, hairy, the upper distinct; cyathia solitary, open-campanulate, 1.2–1.5 mm. in diam.; glands transversely oblong, dark red, usually with conspicuous white appendages; ♂ fls. 15–20; caps. tomentose, sharply angled, 1.5–1.7 mm. long; seeds white, quadrangular, 1.2–1.5 mm. long, the facets smooth or slightly wrinkled.—Dry stony slopes or flats, below 4000 ft.; Chaparral; Santa Monica Mts., San Gabriel Mts., e. San Diego Co.; Ariz., L. Calif. Dec.–May.

13. **E. micrómera** Boiss. [*Chamaesyce m.* Woot. & Standl.] Prostrate annual, glabrous or pubescent, the stems 1–2.5 dm. long; lvs. ovate or oblong, 2–7 mm. long, entire; stipules triangular, ciliate, those of lower side of stems distinct, of upper often united; invols. short-campanulate, ca. 0.9 mm. in diam.; glands discoid, pink or red, mostly unappendaged, 0.1–0.15 mm. wide; ♂ fls. 2–5; caps. glabrous to pubescent, subglobose, 1.3 mm. long; seeds sharply quadrangular, whitish, 1.1–1.3 mm. long, the facets smooth or faintly wrinkled, convex.—Sandy places, below 3000 ft.; Creosote Bush Scrub; deserts from Inyo Co. to San Diego Co.; to Utah, Tex., Mex.; Peru. Sept.–Dec. and April–June.

14. **E. mísera** Benth. Irregularly branched shrub 3–9 dm. tall, with grayish twigs and puberulent young growth; lvs. mostly fascicled, round-ovate, entire, 4–15 mm. long, on short slender petioles; stipules fimbriate; cyathia solitary, terminal, 2–3 mm. long, with short inflexed lobes; glands purple with whitish crenulate appendages; caps. 4–5 mm. long, rather smooth; seeds round-ovoid, 2.5 mm. long, slightly pitted or reticulate-wrinkled.—Occasional on sea-bluffs; Coastal Sage Scrub; Corona del Mar to San Diego Co., San Clemente and Santa Catalina ids. and Creosote Bush Scrub at Whitewater, Colo. Desert; L. Calif. Jan.–Aug.

15. **E. ocellàta** Dur. & Hilg. var. **ocellàta.** [*Chamaesyce o.* Millsp.] Glabrous prostrate annual, the stems 1–2 dm. long; median lvs. ovate-deltoid-falcate, 4–10 mm. long, blunt or mucronulate; stipules mostly distinct, filiform or wider, entire or parted; cyathia solitary at nodes, turbinate to campanulate, 1.5–2 mm. long; glands not appendaged, discoid or slightly elongate radially, ca. 0.6 mm. wide; ♂ fls. 40–60; caps. glabrous, smooth to rugose, cellular-punctate.—Dry sandy places, below 1500 ft.; V. Grassland, Coastal Sage Scrub, etc.; from Colton, San Bernardino Co. n. to Shasta Co. May–Oct.

Var. **arenícola** (Parish) Jeps. [*E. a.* Parish.] Glabrous; lvs. lance-ovate, scarcely or not falcate, 5–15 mm. long; seeds ovoid or slightly angled on back and sides, very smooth. —Occasional, sandy places, below 2500 ft.; Creosote Bush Scrub; Mojave Desert; to Utah, Ariz. May–Sept.

16. **E. pálmeri** Engelm. [*Tithymalus p.* Abrams.] Glabrous somewhat glaucous perennial with several ascending to erect stems from a heavy root-crown, 1–3.5 dm. high, these fairly simple below the few-rayed summit; cauline lvs. obovate, rounded or obtuse at apex, sessile, 0.5–2 cm. long, whorled and larger at the base of the umbel; floral lvs. round-ovate to subreniform; invols. ca. 3 mm. long, whitish, with rounded entire ciliate lobes; glands crenate, slightly 2-horned; caps. ovoid, 4 mm. long; seeds oblong-ovoid,

ashy, 2.5 mm. long, shallowly reticulate-ridged, the brownish caruncle subconic, easily removed.—Common in dry places, 4000–8500 ft.; Montane Coniferous F.; Mt. Pinos to Laguna Mts.; to Utah, Ariz.

17. **E. párishii** Greene. [*Chamaesyce p.* Millsp. *E. patellifera* J. T. Howell.] Glabrous perennial, forming prostrate mats; stems 1–2.5 dm. long; lvs. ovate, entire, 2–4 mm. long; stipules broadly linear, entire, ciliate, those on the lower surface of stem somewhat united; cyathia solitary, campanulate, 1–1.2 mm. wide; glands discoid, unappendaged, 0.5 mm. wide, yellow or reddish; ♂ fls. 40–50; caps. glabrous, sharply 3-angled, 1.7 mm. long; seeds white, quadrangular, 1.5 mm. long, the facets faintly wrinkled.—Infrequent, washes and fans, below 3000 ft.; Creosote Bush Scrub; both deserts from Death V. to e. San Diego Co.; Nev. April–Oct.

18. **E. párryi** Engelm. [*Chamaesyce p.* Rydb.] Glabrous annual, prostrate to ascending, branched from base, the branches 0.6–3.0 dm. long; lvs. linear, entire, 1–3 cm. long; stipules linear, distinct, entire or parted; cyathia on long peduncles, cupuliform, 1.5–1.7 mm. in diam.; glands transversely oval, ca. 0.4 mm. long, with narrow, white, entire appendages; caps. ca. 2 mm. long, with rounded lobes, glabrous; seeds triangular-ovoid, 1.8 mm. long, mottled brown and white.—Sand dunes, ca. 2300 ft.; Creosote Bush Scrub; Kelso, Mojave Desert; to Colo., Tex., Mex. May–June.

19. **E. pediculífera** Engelm. [*Chamaesyce p.* Rose & Standl.] Prostrate to ascending perennial from stout taproot, usually gray-strigulose throughout; stems many, forked, 1–4 dm. long; lvs. elliptic-ovate to oblong, 4–20 mm. long, entire; stipules minute, triangular, those on lower side of stem united; cyathia mostly solitary at the nodes, campanulate, 1.5–2 mm. long; glands transversely oblong, dark red-purple, the appendages from obsolete to 2 mm. wide; ♂ fls. 22–25; caps. strigulose, 2 mm. in diam., 2 mm. long, obtusely lobed; seeds white, narrow-ovoid, 1–1.3 mm. long, encircled by 4–5 rounded ridges.—Infrequent, dry slopes and washes, below 1500 ft.; Creosote Bush Scrub; Colo. Desert; to Ariz., L. Calif., Sinaloa. Jan.–April.

20. **E. péplus** L. [*Tithymalus p.* Hill.] PETTY SPURGE. Glabrous erect annual, 1–3.5 dm. high, simple or branched from below; cauline lvs. obovate to roundish, obtuse to retuse, 1–2.5 cm. long, distinctly petioled, entire; floral lvs. ovate, obtuse, 0.6–1.5 cm. long; cyathia 1.5 mm. high, campanulate; glands large, yellow, with spreading narrow horns; caps. 2 mm. long, each carpel with a broad thin dorsal keel; seeds oblong, 1.2 mm. long, cellular-punctate, ashy, subhexagonal, the 4 outer faces each with 3–4 rounded pits, the 2 inner with a longitudinal furrow, no caruncle; $2n=16$ (Perry, 1943).—Frequent in moist places as a garden weed, mostly about towns; to Atlantic Coast; from Eu. Feb.–Aug.

21. **E. platyspérma** Engelm. [*E. eremica* Jeps.] Glabrous annual, the stems prostrate, 1–2.5 dm. long; lvs. oblong to obovate, 5–10 mm. long; stipules 1.5–2 mm. long, mostly distinct, with 2–3 divisions; cyathia solitary at the nodes, shallow-campanulate, 1.5–2 mm. in diam.; glands slightly radially elongate, 1 mm. wide, the margins sometimes produced into 2 short rounded lobes, not appendaged; ♂ fls. ca. 50; caps. glabrous, round-ovoid, ca. 4 mm. long; seeds white, microreticulate, 2.5–3 mm. long, broadly oblong, the back rounded, smooth, the face with 2 smooth flat slightly concave facets separated by the raphe.—Rare, sandy soil; Creosote Bush Scrub; near Thousand Palms in Coachella V.; Yuma. May.

22. **E. polycárpa** Benth. var. **polycárpa**. [*Chamaesyce p.* Millsp.] Plate 42, Fig. F. Mostly glabrous, prostrate or ascending perennial from slender taproot, the stems 5–25 cm. long; lvs. roundish to oblong, 2–8 mm. long, entire; stipules lanceolate, mostly ciliate, those on ventral side of stem united; invols. solitary at nodes, campanulate, 1–1.5 mm. wide; glands transversely oblong, maroon, 0.5–0.7 mm. long, the appendages broader than the glands; ♂ fls. 15–32; caps. glabrous, rounded, sharply 3-angled, 1.1–1.3 mm. in diam.; seeds rather sharply quadrangular, whitish, 1–1.3 mm. long, the facets smoothish. —Dry slopes and washes, below 3000 ft.; Creosote Bush Scrub, Mojave and Colo. deserts and Coastal Sage Scrub and Chaparral, Ventura Co. to San Diego; to Nev., Ariz., Son., L. Calif. Most of the year.

Var. **hirtélla** Boiss. Herbage pubescent; appendages narrower than the glands.—Creosote Bush Scrub; s. Mojave Desert; to Nev., Son.

23. **E. prostràta** Ait. [*Chamaesyce p.* Small.] Prostrate to decumbent annual, crisped-hairy to subglabrous; stems 5–25 cm. long; lvs. broadly elliptic to elliptic-oblong, 3–11 mm. long, often serrulate; stipules triangular-subulate, short-hairy, often lacerate, those on ventral side of stem sometimes united; cyathia mostly on condensed lateral branchlets, 0.6–0.9 mm. in diam., obconic; glands transversely oval to oblong, 0.2–0.3 mm. long, the appendages white, 1–2 times as wide as gland; ♂ fls. 4; caps. glabrate on faces, hairy on angles, 1–1.4 mm. long; seeds sharply quadrangular, 1 mm. long, whitish, the facets with low transverse wrinkles.—Weed, as at San Bernardino, Ojai, etc.; natur. also in e. U.S.; from trop. Am. Aug.–Sept.

24. **E. revolùta** Engelm. [*Chamaesyce r.* Small.] Glabrous erect ± purplish annual, 0.5–2 dm. tall, diffusely branched from below; lvs. linear, entire, revolute, 1–2.5 cm. long; stipules entire, distinct; cyathia solitary in the forks, broadly obconic, ca. 1 mm. in diam.; glands round, 0.2–0.3 mm. wide, with appendages barely evident to radially elongate; caps. glabrous, 1.4 mm. long, sharply angled; seeds triangular-pyramidal to sharply 4-angled, 1–1.3 mm. long, whitish, with 2 transverse rounded ridges.—Rocky slope at 4000 ft.; Creosote Bush Scrub; n. slope of Clark Mt., e. Mojave Desert; to Colo., New Mex., Chihuahua. Aug.–Sept.

25. **E. sérpens** HBK. Like *E. albomarginata* in having stipules united, but is a prostrate annual with 5–10 ♂ fls. instead of 12 or more; lf.-blades 2–7 mm. long.—Hunter's Point, Salton Sea and Ojai; widely distributed in N. and S. Am.

26. **E. serpyllifòlia** Pers. [*Chamaesyce s.* Small.] Glabrous annual, prostrate to erect, the stems 5–35 cm. long; lvs. ovate, oblong or obovate, 3–14 mm. long, usually serrulate at least toward apex; stipules distinct, linear, entire or few-parted, mostly glabrous; cyathia solitary, 0.8–1.2 mm. in diam., campanulate; glands transversely oblong, 0.2–0.5 mm. long, with narrow white entire to subdentate appendages; ♂ fls. 5–18; caps. glabrous, sharply 3-angled, 1.5–1.9 mm. long; seeds acutely quadrangular, turgid-ovoid, 1–1.4 mm. long, the facets smooth to rugulose, white to brown.—Common in dry disturbed places, below 7000 ft.; many Plant Communities through most of Calif.; to B.C., Mich., Ia., Mex. Mostly Aug.–Oct.

Var. **hírtula** (Engelm.) Wheeler. Plants villous; lvs. 3–7 mm. long. San Diego n.

27. **E. setilòba** Engelm. [*Chamaesyce s.* Millsp.] Annual, mostly prostrate, soft-pubescent; stems 3–15 cm. long; lvs. oblong to oblong-ovate, 2–7 mm. long, entire; stipules not evident; invols. turbinate, constricted at summit, ca. 1.2 mm. high; glands red, mostly transversely oblong, 0.1–0.2 mm. long, the appendages white, ca. 1 mm. long and wide, parted into 3–5 narrow segms.; ♂ fls. 3–7; caps. hairy, globose, sharply angled, ca. 1.1 mm. in diam.; seeds brownish white, sharply quadrangular, 1 mm. long, the facets with low irregular wrinkles.—Sandy places, below 5000 ft.; Creosote Bush Scrub to Pinyon-Juniper Wd.; deserts from Inyo Co. to Imperial and San Diego cos.; to Nev., Tex., n. Mex. Most months.

28. **E. spathulàta** Lam. [*E. dictyosperma* F. & M. *Tithymalus d.* Heller.] Erect annual 2–4.5 dm. high, simple or branched, glabrous; lower lvs. alternate, oblong- or obovate-spatulate, serrulate, 1–3 cm. long, subsessile; floral lvs. opposite, broadly ovate, subcordate or truncate-based, 6–12 mm. long; infl. umbellike, the rays forked; invols. campanulate, ca. 1 mm. high; glands small, sessile, yellow, transversely oval; styles bifid; caps. subglobose, 2–3 mm. long, warty-tuberculate; seeds finely reticulate, ellipsoid-lenticular, 1.3–1.5 mm. long, yellowish brown.—Occasional, open and disturbed places below 4000 ft.; many Plant Communities; throughout cismontane s. Calif.; to Wash., Minn., Ala., S. Am.; March–June.

29. **E. supìna** Raf. [*Chamaesyce s.* Mold.] Annual, prostrate or ascending, ± villous; stems 1–4.5 dm. long; lvs. elliptic-ovate to oblong, serrulate to subentire, 4–17 mm. long; stipules subulate, sometimes lacerate; cyathia mostly on congested lateral branches, obconic, 0.8 mm. in diam.; glands transversely elongate, ca. 0.2 mm. long, the appendages narrow, white, crenulate; ♂ fls. mostly 4–5; caps. strigose, sharply angled, ca. 1.4

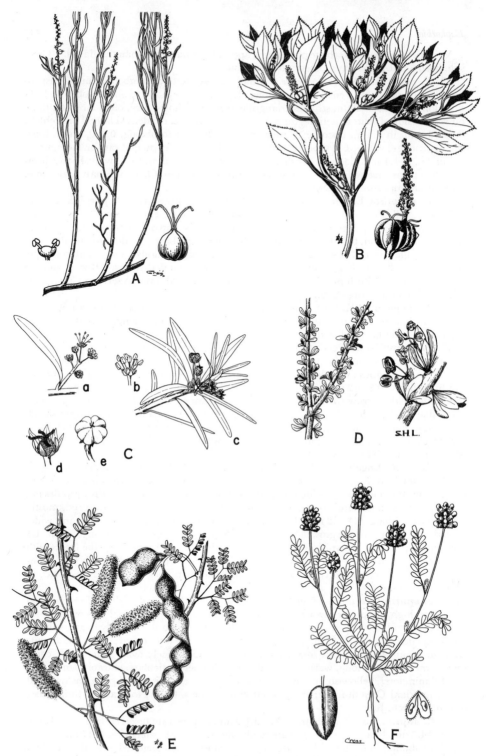

Plate 43. Fig. A, *Stillingia linearifolia*. Fig. B, *Stillingia spinulosa*. Fig. C, *Tetracoccus dioicus*. Fig. D, *Tetracoccus hallii*. Fig. E, *Acacia greggii*. Fig. F, *Astragalus didymocarpus* var. *dispermus*.

mm. long; seeds quadrangular, ca. 1 mm. long, the facets whitish-brown and with low transverse ridges.—Weed about waste places, especially near towns; throughout cismontane Calif.; introd. from e. U.S. May–Oct.

30. **E. vállis-mórtae** (Millsp.) J. T. Howell. [*Chamaesyce v.* Millsp.] Perennial, much branched, forming a dense rounded plant 5–15 cm. high, hairy-tomentose throughout; lvs. roundish to oblong-ovate, 4–8 mm. long, entire; stipules on lower side of stem united; the upper distinct, filiform, hairy; cyathia solitary, campanulate, ca. 2 mm. across; glands transversely oblong, mostly reddish, to 1 mm. long, the white appendages as wide as glands, entire or crenulate; ♂ fls. 17–22; caps. tomentose, 3-angled, 2 mm. long; seeds sharply 4-angled, 1.4–1.7 mm. long, white, the facets nearly smooth.—Dry sandy places below 4000 ft.; Creosote Bush Scrub; w. Mojave Desert from Red Rock Canyon to Owens Lake. May–Oct.

7. *Ricinus* L. CASTOR-BEAN

Monoecious arborescent shrub or in colder places grown as an annual; glabrous, glaucous. Lvs. large, alternate, peltate, palmately 5–11 lobed, the lobes serrate; petioles with conspicuous glands. Fls. in racemose or panicled clusters, ♀ above the ♂. Calyx 3–5-parted. Petals 0. Disk 0. Stamens many, the fils. much branched. Ovary 3-celled, 3-ovuled; styles 3, bifid, plumose, red. Caps. soft-spiny, with 3 2-valved carpels. Seeds glabrous, carunculate, variously marked and colored. One variable sp. with many horticultural forms. (Named for the Medit. sheep-tick, *Ricinus*, because the seed looks like a tick.)

1. **R. commùnis** L. With us a shrub 1–3 m. tall; lvs. 1–4 dm. broad; caps. 1–2 cm. thick; seeds compressed, oblong-ellipsoid, 10–14 mm. long, lustrous-silvery and brown, ± mottled, the caruncle 3–4 mm. wide, flattish; $2n=20$ (Hagerup, 1932).—Frequent as an escape in waste places; native of Old World. Flowers most of the year.

8. *Stillìngia* Garden ex L.

Glabrous monoecious herbs. Lvs. alternate, mostly 2-glandular at base. Fls. bracteolate, apetalous, in terminal or axillary spikes which are ♀ at base. Staminate fls. with 2-lobed calyx and 2 stamens. Pistillate fls. 1–6, the calyx 3-parted or lacking. Ovary 3-celled, 3-ovuled; styles 3, filiform, simple. Caps. 3-lobed, separating into 2-valved carpels. Seeds ovoid or ovoid-ellipsoid, small, the caruncle minute or lacking. Ca. 15 spp. of warm parts of New World and Pacific and Indian oceans. (Named for B. *Stillingfleet*, 1702–1771, English naturalist.)

(Rogers, D. J. A revision of Stillingia in the New World. Ann. Mo. Bot. Gard. 38: 207–259. 1951.)
Lvs. ovate, sharply serrate, 3-nerved. Mostly annual . . . . 3. *S. spinulosa*
Lvs. linear, entire or nearly so, 1-nerved. Perennial.
    Plants forming a rounded leafy bush; infl. not much higher than foliage; lvs. tending to have few slender teeth near base; caps. 3–3.5 mm. high . . . . 2. *S. paucidentata*
    Plants with fewer erect, sparsely leafy stems; infl. overtopping foliage; lvs. entire or serrulate at tip; caps. 2.5 mm. high . . . . 1. *S. linearifolia*

1. **S. linearifòlia** Wats. Plate 43, Fig. A. Perennial, 3–8 dm. tall, the stems slender, loosely branched; lvs. linear, usually entire, 1–4 cm. long, 1–2 mm. wide; spikes terminal, 1–2 mm. thick in ♂ part, lax, 2–5 cm. long; ♂ calyx 0.6 mm. long; stamens 1.5 mm. long; ♀ calyx almost obsolete; styles 1 mm. long; caps. 2.5 mm. high; seeds gray, obovoid, smoothish, 2 mm. long.—Occasional in washes and rocky places below 3500 ft.; Creosote Bush Scrub; Colo. and Mojave deserts; more frequent in interior cismontane valleys below 5000 ft.; Coastal Sage Scrub, Chaparral; San Bernardino Co. to San Diego Co.; to Ariz., Mex. March–May.

2. **S. paucidentàta** Wats. Perennial, 2–4.5 dm. tall, the stems freely branched above; lvs. linear or the lower lanceolate, 3–5 cm. long; with 2–3 setaceous teeth on each side near the base, short-petioled; spikes numerous, at tips of branches, 2–8 cm. long, 3–4

mm. thick, with subcrateriform glands; ♂ fls. many, the calyx ca. 1.5 mm. long and stamens ca. 3 mm. long; ♀ calyx ca. 2 mm. long; caps. 3–4 mm. long; styles 2 mm. long; seeds obovoid, gray, smooth, 2.5 mm. long.—Locally frequent, open places 2000–4000 ft.; Creosote Bush Scrub, w. half of Mojave Desert; below sea-level near Mecca; w. Ariz. April–June.

3. **S. spinulòsa** Torr. in Emory. [*Stillingia annua* Muell.-Arg.] Plate 43, Fig. B. Tufted annual or perennial 5–20 cm. high, much branched, densely lfy.; lvs. rhombic-ovate, spinulose-serrate, 3-nerved, 1–3 cm. long, acuminate, narrowed into short petiole; spikes mostly axillary, 5–12 mm. long, with trumpet-shaped glands at base of bracts; ♂ calyx ca. 0.5 mm. long; ♀ calyx ca. 1 mm. long; caps. 4–5 mm. long; styles ca. 3 mm. long; seeds cylindro-ovoid, grayish, with some brown mottling, ca. 3 mm. long, with numerous fine somewhat wavy longitudinal lines; caruncle obsolete.—Frequent in dry sandy places below 3000 ft.; Creosote Bush Scrub; Mojave and Colo. deserts; Nev., Ariz. March–May.

9. *Tetracóccus* Engelm.

Shrubs with alternate or subopposite lvs. Plants usually dioecious, the fls. small, apetalous, the ♂ in cymose racemes or umbellate clusters with 4–6–8-parted calyx and 4–6–9 stamens. Pistillate fls. solitary with 6–12-parted clusters. Ovary 3–4-celled; styles 3–4, distinct. Ovules 2 in each cell. Seeds usually solitary. Ca. 4 spp. of Calif., L. Calif. and Ariz. (Greek, *tetra*, and *kakkos*, fr., because of the 4-lobed caps. in the original sp.)

(Dressler, R. L. The genus Tetracoccus. Rhodora 56: 45–60. 1954.)
Plants not spiny; lvs. opposite or subopposite; caps. 4-celled.
  Lvs. linear, entire. Chaparral of cismontane San Diego Co. .... 1. *T. dioicus*
  Lvs. ovate to lance-ovate, toothed. Death V. region .... 3. *T. ilicifolius*
Plants spiny; lvs. alternate; caps. 3-celled. N. edge of Colo. Desert to L. Calif., Ariz.
                                                                          2. *T. hallii*

1. **T. dioìcus** Parry. Plate 43, Fig. C. Erect, spreading shrub to 1.5 m. high, the young branches reddish, glabrous, slender; lvs. linear, 2–3 cm. long, entire, on petioles 1–2 mm. long; ♂ fls. few, on short axillary pedicels, the calyx 6–10-parted, ca. 1 mm. long; fils. pubescent; ♀ fl. glabrous, the calyx 3–5 mm. long; caps. depressed-globose, 4-lobed, 8-grooved, ca. 8 mm. long, somewhat thicker; seeds red-brown, shining, asymmetrically oblanceolate in outline, somewhat flat, ca. 5 mm. long, the pale basal strophiole ca. 1 mm. long, somewhat 2-lobed.—Dry stony slopes below 2500 ft.; Chaparral; San Juan Camp on Ortega Highway, Orange Co. through cismontane San Diego Co. to Jacumba and L. Calif. April–May.

2. **T. hállii** Bdg. [*Halliophytum h.* Jtn.] Plate 43, Fig. D. Erect shrub 5–20 dm. high, with stiff divaricate spinescent branches; lvs. entire, pubescent when young, oblanceolate, 3–9 mm. long, borne in small fascicles on short spurlike branchlets; ♂ fls. in axillary umbels, the 4–6 sepals 1–2 mm. long; ♀ on short pedicels; caps. 3-celled, subglobose, pubescent, 3–12 mm. long, 3-seeded; seeds smooth, ± ovoid, 4–7 mm. long, wrinkled ventrally; $2n=24$ (Perry, 1943).—Dry rocky slopes below 3600 ft.; Creosote Bush Scrub; Eagle, Chuckwalla and Cottonwood mts. of Riverside Co. to near Needles and Ivanpah, San Bernardino Co.; sw. Ariz., L. Calif. March–May.

3. **T. ilicifòlius** Cov. & Gilman. Erect sparsely branched shrub, 3–15 dm. high; lvs. opposite, oblong-ovate to lance-ovate, 1–3 cm. long, thick, pinnately veined, 3–8-toothed on each side, villous when young, glabrate, short-petioled; ♂ fls. in branched pubescent clusters 2–3 cm. long, the calyx villous, ca. 1 mm. long; ♀ fls. villous, 4–5 mm. long; caps. 4-celled, oblong-orbicular, 5–6 mm. long, usually with 2 seeds per cell.—Dry slopes, 3000–5500 ft.; Falls Canyon, Grapevine Mts. and Tetracoccus Peak, Panamint Mts. May–June.

10. Tràgia L.

Perennial herbs, sometimes twining, with stinging hairs. Lvs. mostly alternate, stipulate, petiolate, simple or compound, serrate. Monoecious; fls. in bracteate racemes, these terminal or opposite the lvs.; ♂ fls. above, ♀ few and at the base, all apetalous and with small bracts. Staminate calyx 3(5)-parted; stamens mostly 3–5; fils. short. Pistillate calyx 3–8-parted; styles 3-cleft, the branches simple. Caps. 3-celled, bristly, 3-seeded, separating in 2-valved carpels. Seeds not carunculate. Ca. 100 spp., widely distributed in trop. and subtrop. of all continents but Eu. (*Tragus*, Latin name of Hieronymus Bock, 1498–1554, German herbalist.)

1. **T. ramòsa** Torr. [*T. stylaris* Muell.-Arg.] Perennial herb with woody caudex; stems several, slender, 1–3 dm. high; lvs. light green, lance-ovate to oblong-lanceolate, 1–2 cm. long, coarsely and sharply serrate; ♂ fls. 2–4, the 4–5-lobed calyx ca. 1 mm. long; stamens mostly 4–5; ♀ fls. solitary, the calyx 5-lobed, 1.5–2 mm. long; caps. depressed, 6–8 mm. across; seeds globose, brown, smooth, 2 mm. in diam.—Occasionally found, dry rocky slopes, 3000–5500 ft.; Shadscale Scrub, Pinyon-Juniper Wd.; Providence and Clark Mts., e. Mojave Desert to Colo., Tex. Apr.–May.

42. Fabàceae (Leguminòsae). PEA FAMILY

A very large family of herbs, shrubs or trees. Lvs. alternate, usually with stipules, mostly compound, the ultimate lfts. 1–many and usually entire. Fls. mostly perfect, regular or more generally irregular and papilionaceous. Calyx 5-toothed or -cleft. Petals usually 5, distinct or partly united, hypogynous, rarely 1 or none (if papilionaceous, the upper petal the "banner" or "standard", the lateral petals the "wings", and the 2 lower petals united to form the "keel"). Stamens few to many, often 10, distinct or more often the fils. variously united. Pistil 1, simple, free, superior, becoming a legume in fr. (a dehiscent 2-valved pod) or a loment (with several indehiscent segms.), or sometimes entire and indehiscent. Seeds mostly without endosperm. With 450–500 genera and many thousand spp., many of great economic importance for food, forage, dyes, wood, as ornamental plants, etc.

1. Corolla regular or nearly so, valvate in bud; lvs. bipinnate, the lfts. mostly small; fls. small, in many-fld. heads or spikes; petals inconspicuous; stamens conspicuous (Subfamily Mimusoideae).
    2. Stamens united below into a tube; fls. rose to reddish-purple .... 5. *Calliandra*
    2. Stamens distinct or nearly so; fls. yellow or whitish.
        3. The stamens more than 10; anthers not gland-tipped .... 1. *Acacia*
        3. The stamens 10; anthers gland-tipped .... 24. *Prosopis*
1. Corolla ± irregular, imbricate in the bud, the petals unlike in size or shape or both; fls. often larger and petals conspicuous; stamens mostly included.
    4. Upper petal internal in bud, enveloped by the lateral ones, therefore the corolla not papilionaceous (Subfamily Caesalpinoideae).
        5. Lvs. simple; fls. rose-purple .... 8. *Cercis*
        5. Lvs. compound; fls. ± yellowish.
            6. The lvs. once-pinnate; low subshrubs or herbs .... 6. *Cassia*
            6. The lvs. twice-pinnate.
                7. Plants low subshrubs or herbs without spines or thorns .... 12. *Hoffmannseggia*
                7. Plants large shrubs or trees with spines or thorns.
                    8. Rachis of the pinnae flattened, more than 8 cm. long; lfts. alternate; fls. in racemes up to 20 cm. long .... 20. *Parkinsonia*
                    8. Rachis of the pinnae terete, not more than 4 cm. long; lfts. opposite; fls. in short racemes or corymbs .... 7. *Cercidium*
    4. Upper petal external in bud, the corolla papilionaceous (Subfamily Papilionoideae).
        9. Stamens distinct or nearly so.
            10. Fls. yellow; plant an herb .... 29. *Thermopsis*
            10. Fls. purplish; plant a low shrub .... 22. *Pickeringia*
        9. Stamens all united or 9 or 5 of them united.

11. The lvs. simple; shrubs and mostly spiny.
  12. Corolla yellow, 15–25 mm. long.
    13. Calyx deeply 2-lipped; branches spinescent; lvs. acicular .... 31. *Ulex*
    13. Calyx cleft above, hence 1-lipped; branches not spinescent; lvs. ± oblong
        28. *Spartium*
  12. Corolla blue to red-purple, 8–10 mm. long.
    14. Herbage ± gland-dotted; stamens monadelphous; fls. in terminal spikes or racemes .... 10. *Dalea*
    14. Herbage not gland-dotted; stamens diadelphous; fls. in axillary racemes 2. *Alhagi*
11. The lvs. 3–many-foliolate.
  15. Lvs. palmate or pinnately 1–3-foliolate.
    16. Anthers strongly differentiated, some small and versatile, others larger and basifixed; stamens monadelphous and 10 in number; pods dehiscent.
      17. Lfts. 4–many. Native herbs or shrubs .... 15. *Lupinus*
      17. Lfts. 3 or fewer. Introd. shrubs.
        17a. Fls. solitary or racemose, terminal .... 9. *Cytisus*
        17a. Fls. solitary or umbellate, axillary .... 14. *Lotus*
    16. Anthers not much differentiated, or if they are, the plants with prickly pods; stamens 9 and 1, or 5 and monadelphous; pods largely indehiscent.
      18. Plants gland-dotted; lfts. 3–5 .... 25. *Psoralea*
      18. Plants not gland-dotted; lfts. mostly 3.
        19. Lvs. palmately 3-foliolate; fls. in ovoid to oblong heads; corolla persistent after flowering .... 30. *Trifolium*
        19. Lvs. pinnately 3-foliolate; fls. in spikes or racemes; corolla deciduous.
          20. Pods curved or spirally coiled; style subulate .... 16. *Medicago*
          20. Pods ovoid, straight; style filiform .... 17. *Melilotus*
  15. Lvs. pinnate, the lfts. more than 3, or sometimes partly replaced by tendrils.
    21. Rachis of lf. prolonged into a tendril or at least a short seta.
      22. Wings of corolla essentially free from keel; style bearded down the inner face
          13. *Lathyrus*
      22. Wings of corolla adherent to keel half or more its length.
        23. Style bearded in ring or tuft at apex .... 32. *Vicia*
        23. Style bearded down one side .... 23. *Pisum*
    21. Rachis of lf. not prolonged into a tendril.
      24. Plants gland-dotted; pods indehiscent, 1–2-seeded.
        25. Petal 1; lfts. numerous .... 3. *Amorpha*
        25. Petals 5.
          26. Pod prickly; fls. yellow-white; a perennial herb .... 11. *Glycyrrhiza*
          26. Pod not prickly; fls. purplish to rose or blue.
            27. Stamens 10, diadelphous, sometimes 9; shrubs or herbs .. 10. *Dalea*
            27. Stamens 5, monadelphous; herb .... 21. *Petalostemum*
      24. Plants not gland-dotted; pods mostly dehiscent.
        28. Fls. in umbels or solitary; lfts. mostly 3–7; fls. mostly yellow or whitish
            14. *Lotus*
        28. Fls. in racemes.
          29. Plants arboreous with odd-pinnate lvs. and usually spiny.
            30. Fls. whitish, 15–20 mm. long; pods linear, flat. Introd. ... 26. *Robinia*
            30. Fls. rose-purple, 10 mm. long; pods thick, torulose. Desert native
                18. *Olneya*
          29. Plants ± herbaceous.
            31. Lvs. even-pinnate; annual; lfts. linear-oblong; corolla yellowish, ca. 15 mm. long; pods 10–20 cm. long. Mostly a desert plant . 27. *Sesbania*
            31. Lvs. odd-pinnate.
              32. Keel petals not produced into a beak. Widespread .. 4. *Astragalus*
              32. Keel petals produced into a beak. High montane ... 19. *Oxytropis*

1. *Acàcia* Mill. ACACIA

Armed or unarmed shrubs or trees with bipinnate lvs. and small lfts. or with lvs. reduced to simple flat phyllodia. Fls. minute, usually yellow, in heads or spikes. Calyx usually 4–5-toothed or sepals sometimes distinct. Petals usually 4 or 5, united or distinct, sometimes 6. Stamens many, exserted; fils. filiform, distinct. Legume linear to oval, flat or ± turgid, often constricted between seeds. An immense genus, chiefly in subtrop. regions, but especially in Afr. and Australia. Many spp. of horticultural value. (Greek, *akakie*, from *ake*, a point, because of the prickles.)

Plants with spiny or thorny branches.
    Spines mostly 3–4 mm. long, recurved; lvs. with 2–3 pairs of pinnae, the pinnules scarcely 1 mm. wide. Deserts .... 3. *A. greggii*
    Spines mostly 8–30 mm. long, straight; lvs. with 2–8 pairs of pinnae, the pinnules scarcely 1 mm. wide. Cismontane .... 2. *A. farnesiana*
Plants not armed.
    Lvs. bipinnate, the pinnules exceedingly numerous .... 1. *A. decurrens*
    Lvs. simple and reduced to phyllodes, except sometimes in juvenile foliage as in young plants or on sprouts from stumps, pruned places, etc.
        Phyllodes 1-nerved; fls. yellow .... 6. *A. retinodes*
        Phyllodes mostly 2–5-nerved.
            Fls. in heads, cream-color; phyllodes somewhat sickle-shaped .... 5. *A. melanoxylon*
            Fls. in spikes, yellow; phyllodes curved alike on both edges .... 4. *A. longifolia*

1. **A. decúrrens** Willd. GREEN WATTLE. A tree to 15 m. tall, with angled unarmed branches; lvs. bipinnate, with 8–15 pairs of pinnae, each with 30–40 or more pairs of linear lfts. 4–7 mm. long; fls. yellow, in heads in axillary often compound racemes; legumes 5–10 cm. long. It and var. *mollis* Lindl. (*A. mollissima* Willd.) BLACK WATTLE (with lvs. dark green and shining above and fls. pale yellow), as well as var. *dealbata* F. Muell., SILVER WATTLE (with silvery-gray to light green lvs. and deep yellow fls.), are widely cult. in Calif. and occasionally establish themselves in the wild; native of Australia. Feb.–March.

2. **A. farnesiàna** (L.) Willd. [*Mimosa f.* L.] Arborescent shrub or small tree, deciduous; lfts. mostly 10–25 pairs on the pinnae, glabrous; fls. deep yellow, very fragrant, in heads on pubescent peduncles; pod cylindric, 4–8 cm. long, scarcely dehiscent, filled with a pith which separates the seeds.—Dry slopes, Chaparral near San Diego; probably introd. in mission times from trop. Am. Jan.–March. Widely cult. in warm countries as a source of perfume.

3. **A. gréggii** Gray. CATCLAW. Plate 43, Fig. E. Spreading straggling deciduous shrub, 1–2 m. high, or ± arborescent and taller; branches armed with short stout curved spines, the young growth ± pubescent; lvs. 2–5 cm. long, each pinna divided into 4–6 pairs of oblong to oblong-ovate pinnules 2–8 mm. long; fls. yellow, in cylindrical spikes 1–4 cm. long; pods compressed, ± constricted between the seeds, 2–12 cm. long; seeds dark brown, 7–9 mm. in diam.—Common in washes and canyons below 6000 ft.; Creosote Bush Scrub to Pinyon-Juniper Wd.; Colo. and s. Mojave deserts; to Tex., Son., L. Calif. Occasional in s. cismontane areas. April–June.

4. **A. longifòlia** Willd. GOLDEN WATTLE. Tree, 5–8 m. high, with light green foliage; phyllodia oblong-lanceolate, 5–15 cm. long, 2–4-veined, the 2 margins convex, the gland near the base; fls. bright yellow, in spikes 2–6 cm. long; legume terete until ripe, 4–12 cm. long; seeds black, the funicle fitting like a cap over 1 end.—Widely cult. Australian sp.; occasional escape. May.

5. **A. melanóxylon** R. Br. BLACKWOOD ACACIA. Tree, 7–20 m. high, with spreading thick head; lvs. dull dark green, lanceolate to oblanceolate, 5–10 cm. long, 1–2 cm. wide, 3–5-veined, with marginal gland near very base; fls. creamy-white, in heads on short racemes; legumes 7–12 cm. long; seeds encircled by the long red funicle.—Commonly cult. tree from Australia; occasional escape. May–June.

6. **A. retinòdes** Schlecht. Small tree, 5–8 m. tall or ± shrubby, with spreading branches drooping at the ends; phyllodia blue or yellow-green, 8–15 cm. long, 0.6–1.8 cm. wide, with a marginal gland slightly above the base; fls. yellow, in globose heads on lateral racemes; legumes 7–10 cm. long, 0.6 cm. wide.—Cult. Australian sp.; occasionally escape. Fls. much of year.

2. Alhàgi Desv. CAMEL THORN

Stiff spiny shrubs with small simple entire lvs. Stipules small. Fls. few, in axillary racemes. Calyx campanulate, with short subequal teeth. Corolla red, the banner short-clawed, obovate; wings oblong-falcate; keel incurved, obtuse. Stamens diadelphous. Pod linear, terete, incompletely 2-celled. Spp. 3, Medit. and s. Asia. (The Mauretanian name.)

1. **A. camelòrum** Fisch. Low, forming large impenetrable masses 3–8 dm. high; the branches striate, greenish, glabrous, with slender spines 1–2.5 cm. long; lvs. 8–15 mm. long, glabrous above, strigose beneath; fls. mostly 4–6, the calyx ca. 2 mm. long, with short broadly triangular teeth; corolla reddish purple, 8–9 mm. long; pod stipitate, torulose, few-seeded; $n=8$ (Baquar et al., 1965).—Locally a very troublesome weed in low hot places, as San Joaquin V., Colo. Desert, Afton Canyon, San Bernardino Co.; native of Asia Minor. June–July.

3. Amórpha L. FALSE INDIGO

Deciduous shrubs with gland-dotted and heavy-scented foliage. Lvs. odd-pinnate, with setaceous stipules and entire petiolulate lfts. Fls. many, small, ± purplish in long narrow terminal spikes. Calyx turbinate or obconic, 5-lobed, slightly oblique, persistent. Petals wanting except the banner, this erect, clawed. Stamens united at very base only. Style slender, bearded. Pod 1–2-seeded, exceeding the calyx, indehiscent. Seeds oblong or ± curved, rounded and broadened at apex. Ca. 15 spp. of N. Am. (Greek, amorphos, *deformed*, because of the corolla.)

Branchlets and lf.-rachises with pricklelike glands; calyx-teeth $\frac{1}{2}$ to $\frac{3}{4}$ as long as the tube
　　　　　　　　　　　　　　　　　　　　　　　　　　　　　　　　1. *A. californica*
Branchlets and lf.-rachises lacking pricklelike glands; calyx-teeth low-triangular, very short
　　　　　　　　　　　　　　　　　　　　　　　　　　　　　　　　2. *A. fruticosa*

1. **A. califórnica** Nutt. Shrub 1.5–3 m. high, the herbage pubescent; lvs. 1–2 dm. long, on short petioles; lfts. 11–27, oblong-elliptical, 1–3 cm. long; spikes 5–20 cm. long, the axis pilose; calyx pubescent, 4–5 mm. long, 10-nerved; banner red-purple, ca. 5 mm. long; pod 6–8 mm. long, turgid, puberulent, gland-dotted; seeds olive-green to brownish, 3–4 mm. long; $n=10$ (Kreuter, 1930).—Dry wooded or brushy slopes below 7500 ft.; Chaparral, Yellow Pine F., etc.; Santa Rosa and Santa Ana mts. to cent. Calif. May–July.

2. **A. fruticòsa** L. var. **occidentàlis** (Abrams) Kearn. & Peeb. [*A. o.* Abrams.] Minutely pubescent, 1–2.5 m. high; lvs. 1–2 dm. long; lfts. ovate to oblong, 2–5 cm. long; racemes 1 to few together, 1–2 dm. long, the axis ± pubescent; calyx canescent, especially at the margin, glandular, 2.5–3 mm. long; banner ca. 5 mm. long; pod 6 mm. long, glabrous, gland-dotted toward apex.—Stream banks and canyons below 5000 ft.; Coastal Sage Scrub, Chaparral; San Bernardino Co. to San Diego Co.; Ariz., New Mex., n. Mex. May–July.

4. Astrágalus L. MILKVETCH. LOCOWEED. RATTLEWEED

Ours annual to perennial, caulescent or acaulescent herbs. Pubescence simple and basifixed or dolabriform (hairs attached above the base) with ascending and descending arms. Stipules petiolar-cauline or connate opposite the petiole. Lvs. unequally pinnate, rarely 3-foliolate. Infl. axillary, racemose or spicate. Calyx 5-toothed. Corolla papilionaceous, the keel usually muticous. Stamens diadelphous, 9 and 1. Style glabrous. Pods sessile, stipitate or jointed to a stipelike gynophore, highly variable in texture, shape, pubescence; the valves often infolded dorsally to form a ± complete double-walled partition or false septum. A genus of nearly 2000 spp., of N. Hemis.; ca. 400 in N. Am. and 100 in S. Am. Many are poisonous to stock. (Greek, *astragalos*, ankle-bone, an early name for some leguminous plant.)

(Barneby, R. C. Atlas of N. Am. Astragalus. vols. 1–2; pp. 1–1188. 1964.)
　1. Pod strongly inflated.
　　2. The pod 1-celled.

Astragalus

3. The pod with true stipe or with an elongate gynophore, though these may not be longer than the calyx.
   4. Pod leathery, thick-walled; corolla 15–20 mm. long. E. Mojave Desert . 45. *A. preussii*
   4. Pod papery or membranous; corolla often shorter.
     5. The pod 2.5–4 cm. long, usually at least twice as long as thick; calyx without black hairs.
       6. Lfts. elliptical to oblong, mostly 25–39.
         7. Stipe 5–17 mm. long; stipules all free from each other opposite the petiole
            55. *A. trichopodus*
         7. Stipelike gynophore 2.5–6 mm. long; stipules, at least at the lowest nodes, ± united opposite the petiole .... 13. *A. curtipes*
       6. Lfts. orbicular, 9–23; stipe 4–8 mm. long. Ne. Mojave Desert .... 39. *A. oophorus*
     5. Pod 2–2.4 cm. long, less than twice as long as thick; calyx with black hairs.
       8. Pods glabrous, obovoid; corolla whitish, 10–12 mm. long. Mt. Pinos region
         56. *A. whitneyi*
       8. Pods strigulose, rounded-ellipsoid; corolla pink-purple, 8–10 mm. long. E. Mojave Desert .... 36. *A. nutans*
3. The pod sessile.
   9. Plants annual.
     10. Pods thick-walled, hard, 2–3 cm. long; corolla 21–28 mm. long. Colorado Desert
       12. *A. crotalariae*
     10. Pods thin-walled, membranous, or if firm, less than 2 cm. long; fls. smaller.
       11. Fls. purple to pink-purple, drying violet; pods 1 cm. or more thick.
         12. Hairs of herbage straight and appressed. From below 6500 ft.
           13. Lfts. 11–17; banner 5–7.5 mm. long; pod 7–14-ovulate, usually purple-speckled. Colo. Desert .... 22. *A. insularis*
           13. Lfts. 7–13; banner 8–10.5 mm. long; pod 19–24-ovulate, green, turning straw-colored. E. Mojave Desert .... 36. *A. nutans*
         12. Hairs of herbage incurved-ascending. Panamint Mts. at 6500–10,000 ft.
           20. *A. gilmanii*
       11. Fls. whitish.
         14. Herbage yellow-hirsutulous; pod stiffly papery, hirsutulous. Colorado Desert
           18. *A. sabulonum*
         14. Herbage hairs appressed: pod membranous.
           15. Pods ascending, silky-strigulose, mostly 3–6-ovulate. Colorado Desert
            3. *A. aridus*
           15. Pods horizontal or declined, strigulose, 7–28-ovulate. E. Mojave Desert
            57. *A. wootonii*
   9. Plants perennial.
     16. Pods shaggy to tomentose often with long white hairs.
       17. Pod short-tomentose, the longest hairs ca. 1 mm. long. San Bernardino and Santa Rosa mts. .... 28. *A. leucolobus*
       17. Pod shaggy, the hairs 2 mm. or more long. Desert mts. .... 46. *A. purshii*
     16. Pods not shaggy or tomentose with long white hairs.
       18. The pods hard, thick-walled, 2–3 cm. long; fls. 2 cm. long. Colorado Desert
         12. *A. crotalariae*
       18. The pods thin-walled, membranaceous, or if firm, less than 2 cm. long; fls. smaller.
         19. Fls. purplish-blue or with at least a purplish tinge.
           20. Banner 7–11 mm. long.
             21. Stems 2–5 dm. long. Colorado Desert .... 41. *A. palmeri*
             21. Stems less than 2 dm. long.
               22. The stems 6–15 cm. long. E. Mojave Desert and ne. Colo. Desert
                36. *A. nutans*
               22. The stems less than 2 cm. long. Subalpine, Inyo Mts. . 43. *A. platytropis*
           20. Banner 13–26 mm. long.
             23. Stems 1.5–4.5 dm. high; petals purple with white tips. Death V. region
               49. *A. serenoi*
             23. Stems less than 1 dm. high; petals purple to sordid.
               24. Pod crescentic, 2.5–5 cm. long Mts. of e. Mojave Desert 52. *A. tidestromii*
               24. Pod oblong-ovoid or ellipsoid, usually less than 2.5 cm. long, little curved. Needles .... 51. *A. tephrodes*

19. Fls. whitish or ochroleucous.
   25. Pod conspicuously hairy, ovoid, with narrow beaklike tip. W. Mojave Desert
                                                                                                         21. *A. hornii*
   25. Pod glabrous or inconspicuously pubescent when mature.
      26. Foliage densely white-hairy; pod ca. 2 cm. long. Insular 31. *A. miguelensis*
      26. Foliage not white, subglabrous or appressed-pubescent.
         27. Pod less than 2 cm. long, chartaceous and firm, turgid, ovoid, narrowed
            into acute tip; plant erect. Mts. of San Diego co. ..... 38. *A. oocarpus*
         27. Pod usually over 2 cm. long, membranous, abruptly narrowed at tip;
            stems decumbent or ascending.
            28. Petals strongly incurved and short-clawed; blades of the keel longer
               than the claws; fls. and frs. in loose racemes. Montane
                                                                 16. *A. douglasii*
            28. Petals not strongly incurved; blades of the keel not longer than their
               claws. Valleys.
               29. Peduncles 5.5–14 cm. long; racemes dense, 3.5–9 cm. long in fr.
                  Cismontane at low elevs. .... 44. *A. pomonensis*
               29. Peduncles 12–20 cm. long; racemes loose, 9–16 cm. long in fr. San
                  Diego region .... 14. *A. deanei*
2. The pod 2-celled.
  30. Pods shaggy-hairy or tomentose; plants nearly or quite acaulescent.
     31. Lfts. silvery-pilose with appressed hairs; herbage becoming hirsute late in the season.
        E. Mojave Desert .... 35. *A. newberryi*
     31. Lfts. silky or cottony-tomentose with fine entangled hairs.
        32. Pod densely villous-hirsute resembling a mass of wool, the vesture 2–3 mm. long,
           concealing the valves. Desert mts. .... 46. *A. purshii*
        32. Pod short villous-tomentulose, the vesture ca. 1 mm. long, scarcely concealing the
           valves. San Bernardino and Santa Rosa mts. .... 28. *A. leucolobus*
  30. Pods not as above, but strigulose to hoary, etc. with shorter hairs; plants caulescent.
     33. Terminal lfts. decurrent; herbage canescent or cinereous throughout. Sand dunes, s.
        Colo. Desert .... 29. *A. magdalenae*
     33. Terminal lfts. of Colo. Desert plants not decurrent; herbage variable. Widespread
        desert sp. .... 27. *A. lentiginosus*
1. Pods not strongly inflated.
  34. Plants annuals, perhaps occasionally biennials.
     35. Pods 3–4 mm. long, transversely wrinkled or ribbed.
        36. The pods deflexed, exserted from the calyx, conspicuously gray-hairy. Cismontane
                                                                        19. *A. gambelianus*
        36. The pods erect, little exserted from the calyx, usually minutely hairy. Cismontane and
           desert .... 15. *A. didymocarpus*
     35. Pods 12–18 mm. long, not transversely wrinkled or ribbed.
        37. Keel not beaked.
           38. Lfts. hoary; corolla purple, 8–10 mm. long.
              39. Pods linear, compressed; plant prostrate. Desert slopes of San Bernardino Mts.
                                                                              2. *A. albens*
              39. Pods lanceolate, thickened, somewhat obcompressed; plant erect. Widespread on
                 deserts .... 27. *A. lentiginosus*
           38. Lfts. greenish-canescent; corolla 4–8 mm. long, whitish to purplish.
              40. Calyx black-hairy; lfts. retuse to truncate, glabrate above; pods almost completely
                 2-celled. Coastal .... 50. *A. tener*
              40. Calyx white-hairy; lfts. retuse to acute, strigose on both surfaces; pods incom-
                 pletely 2-celled. Deserts .... 37. *A. nuttallianus*
        37. Keel produced into a porrect beak; lfts. canescent, retuse; calyx black-hairy. Deserts
                                                                       1. *A. acutirostris*
  34. Plants perennial.
     41. Dorsal and ventral sutures of the pod both prominent externally forming a thick cord-
        like ridge; pod coriaceous, rugulose.
        42. Pods not stipitate; plants hoary; lfts. ovate to elliptic. Mojave Desert 33. *A. mohavensis*
        42. Pods stipitate; plants not hoary.
           43. The pod compressed, essentially 2-celled, slightly curved; lfts. linear. Interior cis-
             montane .... 40. *A. pachypus*

43. The pod strongly obcompressed, 1-celled, strongly arcuate; lfts. oblong to lanceolate. San Gabriel and San Bernardino mts. .... 6. *A. bicristatus*
41. Dorsal and ventral sutures of the pod not both externally prominent, or with cordlike ridge (except dorsal suture in *A. casei*).
    44. Pods 1-celled.
        45. Midrib of lfts. stiff and running out as a callous mucro or spinule at apex; plants matted or prostrate. Inyo Co. .... 25. *A. kentrophyta*
        45. Midrib of lfts. not forming a spinule at apex.
            46. Pods not long-hairy.
                47. Pods stipitate; fls. yellowish to white.
                    48. Lfts. mostly 9–19; pods 3.5–6 mm. wide. Montane .... 17. *A. filipes*
                    48. Lfts. mostly 21–35; pods 5–20 mm. wide. Cismontane ... 55. *A. trichopodus*
                47. Pods sessile.
                  49. Banner purplish; herbage strigulose or canescent. N. deserts .. 9. *A. casei*
                  49. Banner greenish-yellow; herbage densely white silky-tomentose. Coastal
                              47. *A. pycnostachyus*
            46. Pods conspicuously hairy, sessile.
                50. Fls. bright red, 3–4 cm. long; pods curved, 2–3 cm. long. Deserts 11. *A. coccineus*
                50. Fls. white to bluish or pink-purple at tips, smaller.
                    51. Pods 3–5 cm. long; fls. 2 cm. long .... 18. *A. funereus*
                    51. Pods 0.5–2 cm. long; fls. shorter. Desert mts. .... 46. *A. purshii*
    44. Pods ± completely 2-celled.
        52. Pods strongly obcompressed.
            53. Pods sessile.
                54. Foliage grayish, almost hoary; lfts. mostly 3–8 mm. long; pods woolly, 2 cm. long. San Gabriel Mts. to Santa Rosa Mts. .... 28. *A. leucolobus*
                54. Foliage greener; lfts. 8–12 mm. long; pods villous, 30–60 mm. long. Mojave Desert .... 26. *A. layneae*
            53. Pods stipitate or with a stipelike gynophore.
                55. Banner with white tips, 12–15 mm. long; pod 15–35 mm. long. E. Mojave Desert .... 10. *A. cimae*
                55. Banner not white-tipped, 8–11 mm. long; pod 12–16 mm. long. Inyo Co.
                              23. *A. inyoensis*
        52. Pods not strongly obcompressed.
            56. Pods stipitate; fls. ± ochroleucous (see also *tricarinatus, malacus* and *bernardinus*).
                57. Plants villose-tomentose. Insular.
                    58. Lfts. 11–21; pods glabrous. San Clemente Id. .... 34. *A. nevinii*
                    58. Lfts. 21–29; pods short-woolly. San Nicolas Id., Santa Barbara Id.
                              53. *A. traskiae*
                57. Plants ± strigulose-cinereous; lfts. 9–15. Mojave Desert .... 24. *A. jaegerianus*
            56. Pods sessile; fls. mostly with some red or purple.
                59. Lvs. whitish beneath. (See also *A. panamintensis*.)
                    60. Lfts. not whitish above; hairs basifixed.
                        61. Lfts. 9–15, grasslike; stems 1–3 dm. tall. Inyo Co. .... 4. *A. atratus*
                        61. Lfts. 25–33, oblong; stems 6–12 dm. tall. Los Angeles Co. near the coast
                                7. *A. brauntonii*
                    60. Lfts. whitish above, with dolabriform hairs. Inyo Co. ...... 8. *A. calycosus*
                59. Lvs. greenish beneath, except in *A. panamintensis*.
                  62. Plants curly-villous to hirsute; lfts. 9–21.
                    63. Banner 12–17 mm. long; pods erect, nearly straight, not mottled. E. Mojave Desert .... 32. *A. minthorniae*
                    63. Banner 16–18 mm. long; pods deflexed, incurved, mottled. Ne. Mojave Desert .... 30. *A. malacus*
                62. Plants strigose.
                  64. Lvs. greenish beneath.
                    65. Lfts. 17–27, silvery above; banner whitish, 13–16 mm. long. N. Colo. Desert .... 54. *A. tricarinatus*
                    65. Lfts. 7–17, greenish; banner sordid-lilac, 7–10 mm. long. Mojave Desert ,,, 5. *A. bernardinus*
                64. Lvs. silvery-strigose; banner pink-purple; lfts. 5–9. Inyo Co.
                              42. *A. panamintensis*

1. **A. acutiróstris** Wats. Slender annual, cinereous or greenish, with incurved or contorted hairs; stems to 2.5 dm. long; stipules small; lvs. 1.5–4 cm. long, with 9–15 obovate to cuneate retuse lfts. 2–8 mm. long; peduncles erect, 1.5–6 cm. long; fls. 3–6, the racemes lax; calyx strigulose, the tube 1.5–2 mm. long, teeth 1–1.5 mm. long; banner purplish, 5–7 mm. long; keel incurved, 4.5–6 mm. long with a beaklike apex; pod linear-oblong, 1.2–2.2 cm. long, compressed-triquetrous, 2-locular.—Sandy or gravelly places, 2000–5200 ft.; Creosote Bush Scrub, Joshua Tree Wd.; both deserts especially in w. portions; to Nev. April–May.

2. **A. álbens** Greene. Short-lived perennial, silvery-white with appressed hairs, the stems several, flexuous, prostrate from base, 0.5–3 dm. long; stipules free; lfts. 5–9, elliptic-obovate, obtuse, 5–10 mm. long; racemes loosely 5–14-fld., the peduncles 2–5 cm. long; calyx white-haired, ca. 4 mm. long; petals pink-purple, the banner 7–9.5 mm. long; pod spreading, sessile, 13–18 mm. long, triquetrous, grooved dorsally, strigulose, 2-celled.—Gravelly places, 4000–5800 ft.; mostly in Pinyon-Juniper Wd.; n. slopes of San Bernardino Mts. March–May.

3. **A. áridus** Gray. Like *A. sabulonum*, but densely silvery-strigulose throughout; racemes 4–9-fld., the fls. ascending, the axis 1.5–5 cm. long in fr.; calyx-teeth 1–1.5 mm. long; petals whitish tinged with pink, the banner 3.5–6.5 mm. long; pod ascending, narrowly lunate-ellipsoid, cuneately acute at both ends, 10–17 mm. long, 4.5–7 mm. in diam., not grooved, the valves papery.—Sandy places like dunes, below 1200 ft.; Creosote Bush Scrub; Colo. Desert; n. L. Calif.; sw. Ariz.

4. **A. atrátus** Wats. var. **mensánus** Jones. Perennial, with several slender stems 1–3 dm. long, greenish to strigulose; stipules small, free; lfts. 9–15, oblanceolate, obtuse or retuse, 3–16 mm. long, glabrous above; peduncles 4–13 cm. long, loosely 7–16-fld.; calyx purplish, the tube 4.5–5.5 mm., the teeth 1–2.5 mm. long; petals lilac-purple, the banner 10–13 mm. long; pod pendulous, sessile, narrow-oblong, 16–22 mm. long, compressed, slightly arched downward, strigulose.—Uncommon, 5000–6000 ft.; Sagebrush Scrub, Pinyon-Juniper Wd.; Darwin Mesa and Nelson Range, Inyo Co. Apr.–June.

5. **A. bernardínus** Jones. Slender perennial, strigulose, the stems 1–5 dm. long, slender; stipules small, free; lvs. 4–12 cm. long, with 7–17 oblong to narrow-elliptic lfts. 5–25 mm. long; peduncles 3–10 cm. long, loosely 10–25-fld., the racemes ca. as long as peduncles; calyx-tube 2.5–4 mm., teeth 1–3 mm. long; petals sordid lilac, drying bluish, the banner 7–10 mm. long; pod arcuate-erect, on a gynophore 1–1.5 mm. long, slightly triquetrous, 2.5–3 cm. long, 2-locular, the valves papery.—Stony washes and mesas, 3000–7000 ft.; Joshua Tree Wd. and Pinyon-Juniper Wd.; from e. slope of San Bernardino Mts. to Little San Bernardino Mts., New York and Ivanpah mts. Apr.–June.

6. **A. bicristàtus** Gray. Perennial, strigulose, the lfts. sometimes silvery above; stems weakly ascending, 2–4.5 dm. long; lowest stipules connate into a scarious sheath, the others free, herbaceous; lvs. 3–13 cm. long, with 13–21 narrow lfts. 4–25 mm. long; peduncles 5–12 cm. long, bearing loosely 5–15-fld. racemes; calyx-tube 6–7.5 mm. long, the teeth 2–3.5 mm.; petals greenish-white or lilac-tinged, the banner 15–19 mm. long; pod stipitate, spreading or declined, the stipe stout, 6–12 mm. long; the body oblong-ellipsoid, narrowed at both ends, lunately incurved, 2–4 cm. long, roughly quadrangular when ripe, almost woody, reticulate.—Rocky or sandy places, 5800–9000 ft.; Montane Coniferous F.; San Gabriel and San Bernardino mts. May–Aug.

7. **A. brauntònii** Parish. Stout perennial, villous-tomentulose throughout, erect, 7–15 dm. high; stipules membranous, free, 3–10 mm. long; lvs. 4–14 cm. long, with 25–33 ovate or oblong lfts. 3–20 mm. long; peduncles 3–8 cm. long, with racemes 25–60-fld., 4–14 cm. long, the fls. subsessile; calyx-tube 3.5–4 mm. long, teeth 2.5–5 mm. long, filiform; petals dull lilac, banner 9–12 mm. long; pod deflexed, sessile, 6.5–9 mm. long, beaked, villous-tomentulose, the lower part with complete septum.—Brushy places, firebreaks, etc., below 1500 ft.; Chaparral; hills bordering the Los Angeles plain. Feb.–June.

8. **A. calycòsus** Torr. Low acaulescent tufted perennial, silvery-strigulose with dolabriform hairs; lvs. 2.5–7 cm. long, with 3–7 oblanceolate-obovate lfts. 5–19 mm. long; racemes shortly exserted, 2–6-fld.; calyx-tube 4–6.5 mm. long; petals whitish or pink-

tipped, fading bluish, the banner 10–17 mm. long; pod sessile, oblong, 1–1.5 cm. long, laterally compressed, 2-locular, the valves papery; $2n=22$ (Aedingham & Fahselt, 1964). —Mostly on limestone, 6000–10,000 ft.; Pinyon-Juniper Wd.; Panamint, Inyo-White mts. to Ida., Wyo., Ariz. May–June.

9. **A. càsei** Gray. Wiry, sparsely leafy perennial, strigulose, the lfts. often canescent above; stems 1.5–4 dm. long, erect or assurgent; stipules small, free; lvs. 3–10 cm. long, with 7–15 distant elliptic to linear-oblong lfts. 5–25 mm. long; peduncles 3–9 cm. long, bearing racemes with 6–26 fls.; petals purple with white wing-tips; banner 12–18 mm. long; pod deflexed, sessile, oblong to lance-ellipsoid, 2–5 cm. long, obcompressed in lower part, acuminate distally, leathery, reticulate; $2n=22$ (Trelease & Beath, 1949).— Sandy places, 4000–7200 ft.; Sagebrush Scrub, Pinyon-Juniper Wd.; Panamint Mts. to Mono Lake; w. Nev. Apr.–June.

10. **A. címae** Jones var. **címae.** Perennial, nearly glabrous except for the calyx; stems diffuse, 6–25 cm. long; stipules membranous, free; lvs. 5–11 cm. long, with 11–21 obovate to suborbicular lfts. 5–20 mm. long; peduncles 3–8 cm. long, with loosely 11–25-fld. racemes; calyx-tube 4.5–5.5 mm. long, the teeth 1.5–2.5 mm.; petals purple with white wing-tips, the banner 12–15 mm. long; pod stipitate, the thick stipe 6–8 mm. long, the body incurved-erect, oblong-ellipsoid, obcompressed, 1.5–2.5 cm. long, 8–12 mm. in diam., purple-dotted, becoming woody, with a complete septum.—Calcareous soils, 4700–6000 ft.; Sagebrush Scrub, Joshua Tree Wd.; e. Mojave Desert (New York Mts., Mid Hills). Apr.–May.

Var. **sufflàtus** Barneby. Pod larger, greatly inflated, the stipe 5–12 mm. long, the body 3–3.7 cm. long, 13–21 mm. in diam., the valves papery.—At 5000–6000 ft.; Pinyon-Juniper Wd.; e. slope of Inyo Mts. May.

11. **A. coccíneus** Bdg. Low acaulescent tufted perennial, with dense tangled, tomentose pubescence; stipules free; lvs. 5–10 cm. long, with 7–15 lfts.; peduncle 4–8 cm. long, with few crowded fls.; calyx-tube 10–11 mm. long, the teeth 2–5 mm.; petals scarlet, drying crimson, narrow, suberect, 3.5–4 cm. long; pod ovoid-acuminate or fusiform, 2.5–4 cm. long, somewhat obcompressed, 1-locular, silky-villous and tomentose with hairs 2–3 mm. long.—Gravelly ridges and canyons, 2100–7000 ft.; mostly Pinyon-Juniper Wd.; mts. bordering deserts from Inyo Co. to w. Colo. Desert; L. Calif.; Ariz. March–May.

12. **A. crotalàriae** (Benth.) Gray. [*Phaca c.* Benth.] Coarse malodorous annual or short-lived perennial, greenish, ± strigulose or pilosulous, the lfts. mostly glabrous above, 7–19, thick, oblong to subobovate, 5–35 mm. long; stems 1.5–6 dm. high; peduncles 7–17 cm. long; racemes 10–25-fld.; calyx reddish, the tube 7.5–12 mm. long, the teeth 1.5–2.5 mm. long; petals purple, drying violet; banner 21–28 mm. long; pod ascending, ovoid-ellipsoid, 2–3 cm. long, 10–14 mm. in diam., the valves stiffly papery; $2n=24$ (Trelease & Beath, 1949).—Sandy flats and fans, below 1000 ft.; Creosote Bush Scrub; Colo. Desert; to Ariz., L. Calif. Jan.–Apr.

13. **A. cúrtipes** Gray. Perennial, loosely strigose or subtomentulose, 2.5–4 dm. high, cinereous or green; stipules mostly connate; lvs. 5–16 cm. long, with 25–39 oblong to narrowly obovate lfts. 4–25 mm. long; peduncles 5–25 cm. long; racemes rather closely 15–35-fld., 2–11 cm. long in fr.; calyx-tube 4.5–5 mm. long, the teeth 1.5–3.5 mm.; petals ochroleucous, the banner 13–16 mm. long; pod ascending, to declined, on a stipelike gynophore 2.5–6 mm. long, the body obliquely ovoid, bladdery-inflated, 2.5–7 cm. long, the valves papery, strigulose.—Dry rocky and grassy places, at low elevs.; Coastal Sage Scrub, Coastal Strand; Santa Rosa and San Miguel ids. and mainland near Point Arguello; to San Luis Obispo Co. Jan.–June.

14. **A. dèanei** (Rydb.) Barneby. [*Phaca d.* Rydb.] Resembling *A. oocarpus* in growth-habit, nearly glabrous, the stems 3–6 dm. long; stipules free; lvs. 9–18 cm. long, with 19–29 lanceolate to ovate lfts. 1–2 cm. long; peduncles 1–2 dm. long; fls. 18–25; calyx-tube 3.5–4.5 mm. long, the teeth 2–3 mm. long; petals whitish, the banner 12–15 mm. long; pod loosely ascending, sessile, obliquely ovoid-ellipsoid, bladdery-inflated, 2–3 cm. long, 1-locular, the valves papery, strigulose.—Dry hillsides, burns, etc., at 800–1000 ft.; Chaparral, S. Oak Wd.; sw. San Diego Co. March–May.

15. **A. didymocárpus** H. & A. [*A. catalinensis* Nutt.] var. **didymocárpus**. Annual, the stems erect to prostrate, up to 3 dm. long; lfts. 7–17, mostly glabrous above; fls. subsessile in globose or ovoid heads, not deflexed; calyx-tube 1.5–2.5 mm. long, the teeth 0.8–1.4 mm. long; petals pale, the banner 3–6 mm. long; pods erect, sessile, invested by the calyx, ovoid to subglobose, 2.5–4 mm. long, divided dorsally by a narrow groove into 2 cross-wrinkled saclike 1-seeded lobes, the valves minutely strigulose; $n = 12$ (James, 1951).—Grassy hillsides and valleys, below 3100 ft.; Coastal Sage Scrub, V. Grassland; cismontane s. Calif. and ids.; occasional between 3500 and 4500 ft.; Creosote Bush Scrub and Joshua Tree Wd.; Mojave Desert to L. Calif., Nev. March–May.

Var. **dispérmus** (Gray) Jeps. [*A. d.* Gray.] Plate 43, Fig. F. Stems prostrate, radiating; herbage gray or silvery-pubescent; calyx-teeth 1.5–2.5 mm. long; $n = 13$ (James, 1951).—Sandy flats below 3000 ft.; Creosote Bush Scrub; deserts; Ariz., L. Calif. Feb.–May.

16. **A. doúglasii** (T. & G.) Gray. [*Phaca d.* T. & G. *A. tejonensis* Jones. *A. d.* var. *megalophysus* Munz & McBurney.] var. **doúglasii**. Perennial with several procumbent stems, forming mats 4–10 dm. in diam.; stipules free; lvs. 5–15 cm. long; lfts. 11–25, ovate or oblanceolate, 6–25 mm. long; peduncles incurved-ascending, 3–9 cm. long; fls. 10–35, spreading; calyx-tube 3–4 mm. long, the teeth 1.5–3 mm. long; petals ochroleucous, the banner 8–13 mm. long, strongly reflexed; pod spreading, sessile, obliquely ovoid-ellipsoid, 2.5–6 cm. long, the valves papery, strigulose, or glabrate in sage, not inflexed.—Dry places, below 7000 ft.; V. Grassland, Yellow Pine F., rarely Joshua Tree Wd.; from San Bernardino and San Gabriel mts. to cent. Calif. April–July.

Var. **párishii** (Gray) Jones. [*A. p.* Gray.] Calyx ± white-silky, the teeth deltoid, 1.2 mm. long or less, or subulate and up to 2 mm. long, the broad sinuses outlined with a band of short silvery hairs.—Brushy and wooded places, 4000–7700 ft.; S. Oak Wd., Yellow Pine F.; San Bernardino Mts. to n. L. Calif. May–Aug.

Var. **perstríctus** (Rydb.) Munz & McBurney. Calyx and fl. of var. *parishii*; stems stout, erect, 4–10 dm. high; lfts. 11–19; pod semiovoid, 4–6 cm. long.—Stony or sandy places, 3000–4000 ft.; S. Oak Wd., s. San Diego Co. May–June.

17. **A. fílipes** Torr. Perennial, thinly strigulose to glabrate, green or rarely cinereous; stems usually many, commonly erect, 3–7 dm. long; stipules dimorphic, the lowest connate into a sheath; lvs. 3–10 cm. long, lfts. 9–19, ± linear, 5–25 mm. long; peduncles 8–25 cm. long; fls. 7–25, nodding; calyx-tube 3–4 mm. long, the teeth usually deltoid, less than 1 mm. long; pod pendulous, stipitate, the stipe 8–15 mm. long, the body linear-oblong, straight or nearly so, 1.7–2.5 cm. long, 4–5 mm. wide, strongly compressed, 1-locular.—Dry hillsides and valleys, 3000–5000 ft.; Chaparral, Yellow Pine F.; mts. of interior s. Calif.; cent. Calif. to B.C. April–June.

18. **A. funèreus** Jones. Cespitose perennial with stems to ca. 1 dm. long, the plant densely villous with tangled hairs; lfts. 13–17, oval to obovate, obtuse, 5–8 mm. long, silky- or cottony-tomentose with tangled hairs; racemes subcapitate, 3–10-fld.; calyx-tube 7–8 mm. long, black-hairy, the subulate teeth ca. 3 mm. long; petals rose-purple, the keel 21–26 mm. long; pod 2.5–4 cm. long, densely hairy.—At 4000–5000 ft.; Death V. region. March–Apr.

19. **A. gambeliànus** Sheld. [*A. nigrescens* auth.] Annual, thinly pilosulus with ascending hairs, green or cinereous; stems erect and ascending, to 3 dm. long; stipules free; lvs. 1–4 cm. long; lfts. 7–13, narrowly cuneate-oblanceolate, 2–9 mm. long; peduncles 1.5–6 cm. long; fls. 4–15, the fls. at length recurved; calyx-tube 1–1.5 mm. long, the teeth less than 1 mm.; petals whitish with violet tinge; banner mostly 2.5–4 mm. long; pod deflexed, sessile, obcompressed, 3–4 mm. long, sulcate dorsally, the valves cross-wrinkled; $2n = 22$ (James, 1951).—Largely grassy places, below 3000 ft.; S. Oak Wd., Coastal Sage Scrub; L. Calif. and ids. and mainland; to Ore. March–June.

20. **A. gilmánii** Tidestr. [*A. triflorus* var. *morans* Crum.] Like *A. nutans*, but more loosely strigulose-villosulus; stems mostly 8–25 cm. long; peduncles 1.5–3.5 cm. long; racemes 3–9-fld., 1.5–4 cm. long in fr.; calyx-tube 2.5–3.5 mm. long, the teeth 1–2 mm.; banner 6–8 mm. long; pod sessile, 1.5–2.5 cm. long, 8–15 mm. in diam., green or minutely purple-dotted, strigulose with incumbent hairs.—Canyons and rocky hillsides, 6500–

10,000 ft.; Sagebrush Scrub, Pinyon-Juniper Wd.; Panamint Mts., Inyo Co. June–July.

21. **A. hórnii** Gray. Annual but of long duration; stems nearly glabrous, lvs. loosely strigulose; stems erect, later diffuse or procumbent, 2–12 dm. long, simple or later branched; stipules free, reflexed; lvs. 2.5–13 cm. long, the lfts. 15–33, 5–23 mm. long; peduncles 2–15 cm. long; racemes densely 10–35-fld., the fls. subsessile; calyx silky to pilose, the tube 3–4.5 mm., the teeth 2–2.5 mm. long; petals whitish, the banner 8–10 mm. long; pod sessile, spreading, ovoid, acuminate, bladdery-inflated, 9–18 mm. long, the valves papery, pubescent, not inflexed.—Alkaline places, Alkali Sink; at 2500–3700 ft.; w. Mojave Desert; at lower elevs. in San Joaquin V. and formerly in San Bernardino V.; nw. Nev. June–Oct.

22. **A. insulàris** Kell. var. **harwoódii** Munz & McBurney. Slender annual, simple or branched, appressed-cinereous; stems 4–35 cm. long; stipules free, small; lvs. 3–11 cm. long, the lfts. 11–19, narrow-oblong to oblanceolate, 3–18 mm. long; peduncles 1.5–7 cm. long; fls. 3–9, declined in age; calyx-tube 2–2.7 mm., the teeth 1.3–2.2 mm. long; petals reddish-purple, drying violet, banner 5.5–7.5 mm. long; pod declined, sessile, semiovoid, 14–23 mm. long, greatly inflated below, with a compressed beak, the valves papery, purple-specked.—Dunes and sandy places, below 1200 ft.; Creosote Bush Scrub, Colo. Desert.; Ariz., Son.

23. **A. inyoénsis** Shield. Perennial, strigulose-cinereous except on upper surface of lvs.; stems diffuse or prostrate, 1.5–6 dm. long; stipules small, free; lvs. 1.5–5 cm. long, the lfts. 11–21, crowded, 3–10 mm. long; peduncles 2.5–7 cm. long; fls. 6–15, calyx-tube 2.5–3.5 mm., the teeth 1–2.5 mm. long; petals dull pink-purple, the banner 8–11 mm. long; pod stipitate, the stipe 2–5 mm. long, the body incurved to erect, oblong-ellipsoid, obcompressed in lower part, 12–16 mm. long, semibilocular, the valves leathery, purplish, strigulose.—Open places, 5000–7800 ft.; Pinyon-Juniper Wd.; Darwin Mesa, Inyo Co., Inyo-White Mts. May–July.

24. **A. jaegeriànus** Munz. Perennial, cinereous-strigulose or greenish; stems weak, reclining or scrambling through bushes, 3–7 dm. long, branched; stipules small, free; lvs. 2–6 cm. long, with 9–15 linear-oblong lfts. 3–20 mm. long; peduncles 2.5–8 cm. long; fls. 5–15; calyx-tube 2.5–4 mm., teeth 1.5–2 mm. long; petals ochroleucous, veined with lavender, banner 7–10 mm. long; pod pendulous, stipitate, the stipe 3–5 mm. long, body narrow-oblong, 16–23 mm. long, straight, laterally compressed, 2-carinate by the sutures, 2-locular, valves fleshy, ± mottled, becoming brown and leathery.—Low granite hills, 3000–4000 ft.; cent. Mojave Desert. April–June.

25. **A. kentrophỳta** Gray var. **elàtus** Wats. Suffruticulose, with ± rigid erect to trailing stems 1–4 dm. long; lvs. stiff, prickly, early recurving, 1–2.5 cm. long, 5–7-foliolate strigulose, with mostly dolabriform hairs; lower stipules connate; lfts. linear-elliptic, 5–15 mm. long, the terminal spinule 1–2.5 mm. long; peduncles 1–6 mm. long; calyx 3.4–4.4 mm. long; petals whitish or faintly purple-veined, banner 5–6 mm. long; pod narrow ovoid-acuminate, 4–7 mm. long.—At ca. 10,000 ft.; Bristle-cone Pine F.; Inyo-White Mts.; to Colo., New Mex.

Var. **impléxus** (Canby) Barneby. Herbage with hairs basifixed, lvs. with 5–9 lfts. 1–9 mm. long, the terminal spinule 0.5–1 mm. long; fls. 2–3; petals commonly purplish, banner 4.5–8 mm. long.—Stony places, 9000–11,000 ft.; Sagebrush Scrub, Bristle-cone Pine F.; White Mts. to Mont. June–Sept.

26. **A. làyneae** Greene. Plate 44, Fig. A. Perennial from deepset horizontal rhizomes, villous-hirsute, greenish or canescent; stems 2–16 cm. long; stipules free; lvs. 6–16 cm. long, with 13–21 obovate lfts. 5–23 mm. long; peduncles stout, often reclinate in fr., 5–14 cm. long; fls. 15–45, the racemes becoming 5–25 cm. long in fr.; calyx black-hirsute, the tube 5–7 mm., the teeth 1–2 mm. long; petals ochroleucous tipped with dull purple, the banner 13–18 mm. long; pod arcuate-ascending or rarely reflexed, sessile, narrowly sickle-shaped, 3–6.5 cm. long, triquetrous with blunt angles, grooved dorsally, 2-celled, the valves hirsute, mottled.—Sandy places, 1500–5000 ft.; Creosote Bush Scrub; over much of Mojave Desert; to Death V., Nev., ne. Ariz. March–May.

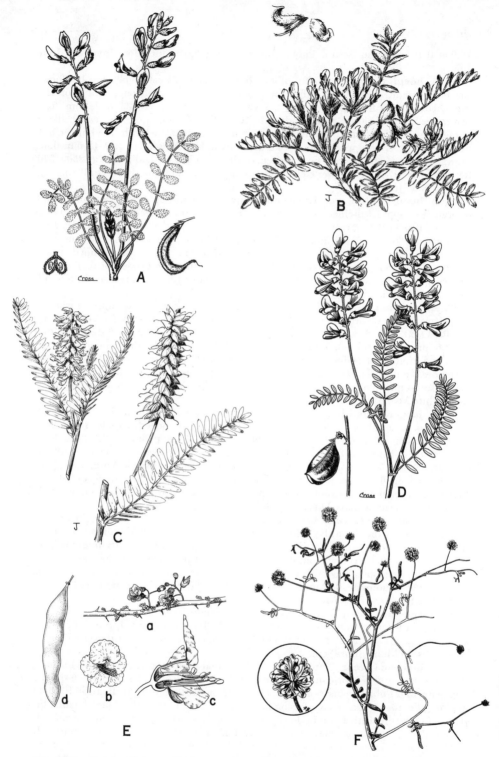

Plate 44. Fig. A, *Astragalus layneae*. Fig. B, *Astragalus purshii* var. *inctus*. Fig. C, *Astragalus pycnostachyus*. Fig. D, *Astragalus trichopodus* ssp. *leucopsis*. Fig. E, *Cercidium floridum*. Fig. F, *Dalea emoryi*.

27. **A. lentiginòsus** Dougl. Perennials and a few winter-annuals, erect or prostrate, glabrous, ± villous, strigulose or white-silky; fls. loosely or densely racemose, variable in size and color; calyx-tube deeply campanulate to cylindric; pod spreading, sessile, lance-ellipsoid to subglobose, mostly strongly inflated, abruptly deltoid-beaked, the leathery to papery valves inflexed as a complete septum.—A complex group of reticulately interrelated forms, most of them intergrading freely at the edge of their ranges.

KEY TO VARIETIES

1. Fls. small, the keel not more than 8 mm. long or, if longer, surpassing the wings and the banner whitish.
    2. Beak of the pod decurved; stems elongate, 3–10 dm. long; racemes cylindric, dense in fr.; lfts. canescent above .... var. *albifolius*
    2. Beak of the pod incurved or rarely erect.
        3. Racemes loose, 5 cm. or more in fr. Mostly below 5000 ft. .... var. *fremontii*
        3. Racemes compact, not over 4 cm. long in fr. Above 5000 ft.
            4. Calyx-teeth at least 1.5 mm. long. White Mts. .... var. *semotus*
            4. Calyx-teeth 1 mm. long or less. San Bernardino and San Gabriel mts.
                5. Lfts. silvery on both surfaces. San Gabriel Mts. .... var. *antonius*
                5. Lfts. glabrate above, ciliate. San Bernardino Mts. .... var. *sierrae*
1. Fls. larger, the keel 8.5 mm. long or more, exserted from the wings only in purple-fld. *variabilis*.
    6. Petals pale yellow; herbage villosulous. Kern & Santa Barbara cos. .... var. *nigricalycis*
    6. Petals purple; pubescence appressed or incumbent.
        7. Racemes compact, not over 4 cm. long in fr. Tehachapi Mts. and Mt. Pinos . var. *idriensis*
        7. Racemes loose, open, mostly more than 4 cm. long in fr. Deserts.
            8. Pod strongly inflated, the body ovoid or subglobose.
                9. Calyx-teeth 1–1.4 mm. long; pubescence variable, but very rarely densely white-silky; pod usually not more than strigulose. W. and s. Mojave Desert . var. *variabilis*
                9. Calyx-teeth 1.4–3 mm. long; plant white-silky.
                    10. Winter-annual or short-lived perennial; longest hairs of the herbage 0.5–1.2 mm. long. Coachella V., Riverside Co. .... var. *coachellae*
                    10. Strong perennial, indurate at base; longest hairs 1.2–2 mm. long. Eureka V., Inyo Co. .... var. *micans*
            8. Pod not strongly inflated, lance-acuminate in profile, silky-strigulose; racemes elongate, loose; herbage silky .... var. *borreganus*

Var. **albifòlius** Jones. Clay flats and seeps, prostrate or trailing, 2100–4700 ft.; Shadscale Scrub, Alkali Sink; Antelope V. to Inyo Co. April–July.

Var. **antònius** Barneby. Open slopes in Yellow Pine F., 5000–8500 ft.; e. division of San Gabriel Mts. (San Antonio Mts.).

Var. **borregànus** Jones. [*A. agninus* Jeps.] Dunes and sandy valleys, below 1000 ft.; Creosote Bush Scrub; e. Colo. and Mojave deserts; to Ariz., Son. Feb.–May.

Var. **coachéllae** Barneby. [Var. *coulteri* auth.] Sandy places below 1200 ft.; Creosote Bush Scrub; Coachella V., Riverside Co. Feb.–May.

Var. **fremóntii** (Gray) Wats. [*A. f.* Gray.] Sandy plains and fans, below 6500 ft.; Creosote Bush Scrub, to Pinyon-Juniper Wd.; e. Mojave Desert n. to White Mts. April–July.

Var. **idriénsis** Jones. [*A. tehatchapensis* Tidestr.] Grassy and brushy places, 2000–7000 ft.; Yellow Pine F. and below; Tehachapi Mts. and Mt. Pinos; to cent. Calif. April–July.

Var. **mìcans** Barneby. Dunes, 3000–3100 ft.; Eureka V., Inyo Co. April–June.

Var. **nigricàlycis** Jones. Plains and foothills, below 3200 ft.; V. Grassland; Kern Co. (region of Ft. Tejon, etc.) and Cuyama V.; to Fresno Co. March–May.

Var. **semòtus** Jeps. Sandy flats and ridges, 7000–11,000 ft.; Sagebrush Scrub and Bristle-cone Pine F.; Inyo-White Mts. June–Aug.

Var. **siérrae** Jones. Stony places, 6000–7000 ft.; Sagebrush Scrub, Yellow Pine F.; e. end of San Bernardino Mts. April–July.

Var. **variábilis** Barneby. Sandy places, below 4000 ft.; Creosote Bush Scrub; s. and sw. Mojave Desert. April–June.

28. **A. leucolòbus** Jones. [*A. purshii* var. *l.* Jeps.] Resembling the cismontane form of *A. purshii* var. *tinctus*, the petals poorly graduated, the banner 16–18 mm. long; pod lance-oblong in outline, 13–22 mm. long, ca. 7 mm. in diam., obcompressed, grooved dorsally, the leathery valves densely but quite shortly tomentulose with matted hairs ca. 1 mm. long, inflexed as a complete narrow septum.—Openings in sandy woods and stony shores in the mts. overlooking the deserts, 6000–8000 ft.; Yellow Pine F., Sagebrush Scrub; e. San Gabriel Mts., San Bernardino and Santa Rosa mts. May–July.

29. **A. magdalènae** Greene var. **pèirsonii** (Munz & McBurney) Barneby. [*A. p.* Munz & McBurney.] Stout perennial of short duration, flowering as a winter annual, at length woody below, with appressed hairs; stems erect, 2–7 dm. high, with short leafy spurs in the lower axils; stipules free; lvs. 5–15 cm. long, with broad flat rachis and 8–12 folded lateral lfts., the terminal one decurrent and continuous with the rachis; peduncles 7–10 cm. long; fls. 10–17; calyx-tube 3.5–4 mm., the teeth 1.5–5 mm. long; petals dull purple, the banner 10–14 mm. long; pod spreading, sessile, broadly ellipsoid with short deltoid beak, bladdery-inflated, 2–3.5 cm. long, the valves papery, strigulose.—Sand dunes below 800 ft.; Creosote Bush Scrub; s. Colo. Desert.

30. **A. málacus** Gray. Perennial, hirsute throughout with coarse spreading hairs; stems stout, erect or ascending, 3–23 cm. long; stipules free, conspicuous; lvs. 4–15 cm. long, with 9–21 lfts. 5–15 mm. long; peduncles stout, 4–10 cm. long; fls. many, rather crowded, declined; calyx-tube 6–8 mm. long, the teeth 2–4 mm. long; petals reddish-violet, the banner 16–18 mm. long; pod reflexed, short-stipitate, narrow-oblong, falcate, 1.5–3 cm. long, compressed-triquetrous, narrowly grooved dorsally, 2-locular, the valves mottled, papery, shaggy-hirsute.—Plains and stony hills, 4000–7500 ft.; largely Sagebrush Scrub, Pinyon-Juniper Wd.; Mono Co. to Ore., Ida. May–June.

31. **A. miguelénsis** Greene. Perennial, densely white-tomentulose throughout; stems decumbent and ascendent, 1.5–3 dm. long; stipules shortly connate, tomentulose; lvs. 4–12 cm. long, with 21–27 oblong-obovate lfts. 6–22 mm. long; peduncles 5–13 cm. long; fls. 10–30; calyx-tube 4–5 mm. long, teeth 2–3.5 mm.; petals yellowish, the banner 12–16 mm. long; pod spreading, sessile, bladdery-inflated, ovoid with deltoid beak, 1.5–3.5 cm. long, the valves papery, finely tomentulose, scarcely inflexed.—Dunes and sandy or talusy places; Coastal Strand; San Miguel, Santa Rosa, Santa Cruz, Anacapa and San Clemente ids. Spring.

32. **A. minthórniae** (Rydb.) Jeps. var. **villòsus** Barneby. Perennial, villous throughout; stems erect or ascending, 4–35 cm. long; stipules free, the lowest 8–12 mm. long; lvs. 5–18 cm. long, with 9–17 ovate-elliptic lfts. 5–26 mm. long; peduncles erect, 5–16 cm. long; fls. 5–35, loosely arranged in fr.; calyx-tube 4.5–6.5 mm. long, teeth 1.5–2.5 mm.; petals pink-purple with white wing-tips or livid, the banner 12–17 mm. long; pod erect or ascending, subsessile, narrow-oblong, 12–23 mm. long, subterete, becoming a little compressed, the valves leathery, long-villous, inflexed as a complete septum.—Open stony hillsides and in canyons, mostly limestone, 4400–7800 ft.; Pinyon-Juniper Wd.; e. Mojave Desert and n. slope of San Bernardino Mts.; Nev. April–June.

33. **A. mohavénisis** Wats. var. **mohavénsis.** Winter-annual or short-lived perennial, resembling *A. albens*, but coarser; lvs. 2–10 cm. long, with 5–11 obovate to elliptic lfts. 3–18 mm. long; fls. 3–18; calyx-tube 2.5–4.5 mm., teeth 2–3 mm. long; petals pink-purple, banner 9–12 mm. long; pod spreading or deflexed, sessile, oblong-obovate in outline, abruptly cuspidate-beaked, 1.5–3 cm. long, fleshy when young, the valves shrinking in age and leaving both sutures prominent, valves leathery, reticulate, ± pubescent, fully 2-locular.—Rocky places, 2500–7500 ft.; Creosote Bush Scrub, Joshua Tree Wd.; Little San Bernardino Mts. to Death V., Providence Mts., etc. April–June.

Var. **hemigỳrus** (Clokey) Barneby. [*A. h.* Clokey.] Pod longer and narrower; banner 7–9 mm. long.—Near Darwin, Inyo Co.; Charleston Mts., Nev.

34. **A. nevínii** Gray. Closely related to *A. traskae*, but lfts. 11–21; fls. smaller, the banner 10–12 mm. long; pod glabrous.—Dunes and sandy bluffs; Coastal Strand; San Clemente Id. March–July.

35. **A. newbérryi** Gray. Acaulescent tufted perennial, appressed-silky; stipules free,

Astragalus

5–9 mm. long; lvs. 2–13 cm. long, with 3–13 ovate-elliptic lfts. 4–15 mm. long; peduncles shorter than lvs.; fls. 3–8; calyx-tube 10–11 mm., teeth 2–5 mm. long; petals mostly pink-purple, the banner 22–27 mm. long; pod spreading, sessile, ovoid, incurved into a stiff deltoid laterally compressed beak, obcompressed below, 1.5–2.5 cm. long, 1-locular, densely hirsute-tomentose.—Dry stony places, mostly above 4000 ft.; Sagebrush Scrub, Pinyon-Juniper Wd.; e. Mojave Desert to Death V., Inyo-White Mts.; Ore., Ida. April–June.

36. **A. nùtans** Jones. [*A. deserticola* Jeps. *A. chuckwallae* Abrams.] Short-lived perennial flowering the first season, silvery-strigose or at length greenish; stems 6–15 cm. long; stipules small, free; lvs. 3–8 cm. long, with 7–13 oblanceolate-elliptic lfts.; peduncles 2–5 cm. long; fls. 6–10; calyx-tube 2.5–3 mm. long, the teeth 1.2–2 mm.; petals pink-purple, drying violet, the banner 8–10 mm. long; pod spreading, sessile or on a stipelike base to 2 mm. long, ovoid with conical beak, bladdery-inflated, 1-locular, valves papery, strigulose.—Uncommon, desert foothills and washes, 1500–6500 ft.; Creosote Bush Scrub, Joshua Tree Wd., Pinyon-Juniper Wd.; e. Mojave and n. Colo. deserts. March–June.

37. **A. nuttalliànus** DC. var. **imperféctus** (Rydb.) Barneby. Strigulose annual with stems to 3 dm. long; lfts. mostly oblong-elliptic, 9–15 in number, 2–8 mm. long; fls. few in a raceme; calyx 3.2–4.5 mm. long; petals purplish, the banner 4–6.5 mm. long; pod horizontally spreading on arched pedicels, sessile, linear-oblong, ± incurved, 12–21 mm. long, 2–3 mm. in diam., the valves papery, glabrous or strigulose, from almost 1-locular to fully 2-locular.—Sandy or stony places, washes, etc., below 5200 ft.; Creosote Bush Scrub, Joshua Tree Wd.; Death V. region to cent. and e. Mojave Desert; Utah, Ariz. March–June.

Var. **cedrosénsis** Jones. Stems filiform; lfts. of lower lvs. cuneate-obcordate, calyx 2.5–3 mm. long; pod 1.6–2 mm. in diam.—Desert flats and washes below 1000 ft.; Creosote Bush Scrub; w. edge of Colo. Desert; L. Calif., Son.

38. **A. oocárpus** Gray. Stout perennial from a woody base, subglabrous; stems erect or later supported by bushes, 6–13 dm. long; stipules free, early reflexed; lvs. 6–17 cm. long, with 19–35 thick, obovate-oblong lfts. 6–33 mm. long; peduncles stout, 1.5–7 cm. long; fls. 15–70; calyx-tube 3.5–4.5 mm. long, the deltoid teeth 0.8–1.8 mm. long, often strigulose on margins; petals ochroleucous, banner 10.5–12.5 mm. long; pod erect, sessile, persistent until dehiscence, ovoid, contracted at base into a thick neck, shortly deltoid-beaked, 1.5–2.5 cm. long, the leathery valves almost glabrous, inflated, 1-locular.—Dry brushy slopes, 2000–5000 ft.; S. Oak Wd.; interior San Diego Co. June–Aug.

39. **O. oóphorus** Wats. Perennial, subglabrous; stems 5–25 cm. long; stipules free, reflexed; lvs. 5–15 cm. long, with 9–23 ovate lfts. 4–20 mm. long; peduncles incurved, 4–10 cm. long; fls. 5–12; calyx glabrous, the tube 4.5–6.5 mm., the teeth 2–4 mm. long; petals purple with white wing-tips, banner 16–23 mm. long; pod pendulous, sessile on a slender gynophore 4–8 mm. long, the body broadly ellipsoid, bladdery-inflated, 3–5.5 cm. long, 1-locular, the valves papery, mottled, glabrous; $2n=24$ (Ledingham & Fahselt, 1964).—Bare canyons and hillsides, 5500–10,000 ft.; Sagebrush Scrub, Pinyon-Juniper Wd.; Panamint, Inyo and White mts.; to Ore., Nev.

40. **A. páchypus** Greene var. **páchypus**. Perennial from a woody caudex, cinereous-strigose, lfts. and stems mostly with whitish coat of short hairs; stems forming bushy plants 2–4.5 dm. high; stipules small, free; lvs. 6–15 cm. long, with 11–21 linear-oblong lfts. 8–25 mm. long; peduncles stiff, erect, 7–20 cm. long; fls. 8–26; calyx-tube 4.3–5.2 mm. long, the teeth 2.5–4.3 mm.; petals white, banner sometimes pinkish, drying yellowish, 17–22 mm. long; pod stipitate, the stipe 4–8 mm. long, body arcuate-erect, oblong, 15–25 mm. long, laterally compressed when ripe, 2-carinate by the cordlike sutures, 2-locular, the valves glabrous, transversely wrinkled; $2n=22$ (Head, 1957).—Dry clay hillsides, below 6500 ft.; Creosote Bush Scrub, V. Grassland; Mojave Desert from n. Los Angeles Co. and Kern Co. to S. Coast Ranges. March–July.

Var. **jàegeri** Munz & McBurney. Lfts. 17–27; fls. slightly smaller, yellow, the banner 15–27 mm. long.—Dry places below 2500 ft.; Chaparral; Banning to Aguanga and Temecula (Riverside Co.).

41. **A. pálmeri** Gray. [*A. vaseyi* Wats. *A. metanus* Jones. *A. v.* var. *johnstonii* Munz & McBurney.] Short-lived perennial flowering the first season, silvery-strigulose to greener; stems prostrate and ascending, forming loose mats 2–10 dm. across; stipules free; lvs. 3–15 cm. long, with 11–21 ovate-oblong to narrowly elliptic lfts. 5–25 mm. long; fls. 13–40; calyx white- or black-strigulose, the tube 2.5–3.5 mm. long, teeth 1–2.5 mm.; petals purplish to ochroleucous, banner 7–10 mm. long; pod spreading or ascending, sessile, obliquely ovoid-ellipsoid, 1–3 cm. long, turgid or moderately inflated in lower half, ± obcompressed, with a laterally flattened beak, valves papery, yellowish or purple-speckled, strigulose and often canescent.—Rocky places below 4500 ft.; Pinyon-Juniper Wd. to Creosote Bush Scrub; w. Colo. Desert e. to Eagle Mts.; L. Calif. April–May.

42. **A. panaminténsis** Sheld. Low perennial, silvery-strigulose; caudex matted and branched; stems 1–5 cm. long; stipules small, free; lvs. 2–12 cm. long, with filiform rachis and 5–9 distant linear-elliptic involute lfts. 2–12 mm. long; peduncles filiform, 1–6 cm. long; racemes 1–4-fld.; calyx-tube 3–4.5 mm. long, teeth 1.5–3 mm.; petals pink-purple, banner 9.5–14 mm.; peduncles filiform, 1–6 cm. long; pod spreading, sessile, oblong in outline, 8–18 mm. long, triquetrous, grooved dorsally, semibilocular, valves strigulose, purple-speckled.—Canyon walls, mostly on limestone, 4000–7000 ft.; Pinyon-Juniper Wd.; Panamint, Last Chance and Inyo mts. April–June.

43. **A. platytròpis** Gray. Small, tufted perennial, silvery-strigulose; stems almost none, clothed in imbricated stipules; lvs. 1–9 cm. long, with 5–15 obovate to elliptic, mostly obtuse lfts. 2–11 mm. long; peduncles scapiform, 2–6 cm. long; fls. 2–9; calyx-tube 2–3 mm., teeth 0.5–2 mm. long; petals ochroleucous, tinged purplish, 7–9 mm. long; pod spreading, sessile, bladdery-inflated, ovoid to broadly ellipsoid, with conic beak, 1.5–3.5 cm. long, the valves papery, strigulose, mottled, inflexed as a partition meeting a flange from the central suture.—Bare ridges above 8500 ft.; Inyo Mts.; n. to Mont. July–Aug.

44. **A. pomonénsis** Jones. Coarse procumbent perennial; stems stout, glabrous, 2.5–8 dm. long; herbage green; stipules free, although the lowest may be amplexicaul; lvs. 8–20 cm. long, with 25–41 oblong-oblanceolate lfts. 6–35 mm. long; peduncles stout, 5.5–14 cm. long; fls. 10–45 in dense racemes; calyx-tube 3.5–5 mm., teeth 1–3 mm. long; petals greenish-white or ochroleucous, banner 11–15 mm. long; pods densely racemose, spreading or ascending, obliquely ovoid, bladdery-inflated, 2.5–4.5 cm. long, 1–2 cm. in diam., valves papery, minutely strigulose, not inflexed.—Sandy valleys, grassy slopes, etc.; below 1000 ft.; Coastal Sage Scrub, S. Calif. Wd.; cismontane from n. L. Calif. to San Luis Obispo Co.

45. **A. préussii** Gray var. **préussii**. Resembling *A. crotalariae*, but of longer duration, nearly glabrous; fls. 4–16; calyx-tube 6–10 mm. long; petals pink- or lilac-purple, banner 15–21 mm. long; pod incurved-ascending or suberect, stipitate, the stipe 2–5.5 mm. long, the body ellipsoid-oblong or fusiform, cuspidate-beaked, 1.5–4 cm. long, 6–10 mm. in diam., subterete, valves papery, glabrous.—Barren clay flats; Shadscale Scrub, Alkali Sink; e. Mojave Desert at 2500–2600 ft.; to Utah. April–May.

Var. **laxiflòrus** Gray. [*Phaca davidsonii* Rydb.] Pod sessile or nearly so.—Alkaline flats, at ca. 2400 ft., sw. Mojave Desert; Utah, Ariz.

46. **A. púrshii** Dougl. var. **léctulus** Jones. Low acaulescent or short-caulescent tufted or matted perennials; pubescence arachnoid-villous or tomentose with fine tortuous hairs; stipules membranous, often attenuate; lfts. 5–9, hoary; racemes subcapitately 2–10-fld. on peduncles 4–8 cm. long; calyx 6–9 mm. long; corolla rose-lavender, or purplish, ca. 1 cm. long; pod moderately arched, ovoid to ovoid-acuminate, 7.5–15 mm. long, 4–8 mm. in diam., shaggy-villous with hairs 1.5–3.5 mm. long, the valves becoming leathery. —Dry rocky places, 6000–8000 ft.; Montane Coniferous F.; San Bernardino Mts.; Inyo Co. to Ore., Nev. April–July.

Var. **tinctus** Jones. [*A. p.* var. *gaviotus* Jeps. *A. p.* var. *longilobus* Jones. *A. p.* var. *ordensis* Jeps.] Plate 44, Fig. B. Lfts. 7–17; calyx ca. 10–15 mm. long; petals 14–18 mm. long.— Dry places, 3000–8000 ft.; Pinyon-Juniper Wd.; Sagebrush Scrub, Yellow Pine F.; Ord., Argus, Panamint mts., Mt. Pinos, Tehachapi Mts.; to Wash., Nev. May–June.

47. **A. pycnostàchyus** Gray var. **lanosíssimus** (Rydb.) Munz & McBurney. [*Phaca l.* Rydb.] Plate 44, Fig. C. Stout perennial, the herbage densely white silky-tomentose; stems several, erect, 4–9 dm. long; lowest stipules connate; lvs. 3–15 cm. long, with 25–41 narrowly oblong lfts. 5–30 mm. long; peduncles 2–4 cm. long; fls. numerous, dense, declined; calyx-tube 4–5 mm., the teeth 1.5 mm. long; petals ochroleucous, the banner 7–10 mm. long; pod 8–11 mm. long, thinly strigulose, the valves papery, not inflexed.—Coastal Marshes, Ventura and Los Angeles cos., if surviving at all.

48. **A. sabulònum** Gray. Winter annual, rarely apparently perennial, silky-hirsute or villosulous, stems decumbent and ascending, to 2.5 dm. long; stipules small, free; lvs. 1.5–6.5 cm. long, with 9–15 oblong to obovate-cuneate lfts. 4–13 mm. long; peduncles 1–3.5 cm. long; fls. 2–6, at length declined; calyx-tube 1.5–2.5 mm., teeth 2–3.5 mm. long; petals purplish, banner 5–7 mm. long; pod horizontal or declined, sessile, obliquely oblong-ovoid, 9–15 mm. long, with a short incurved beak, 1-locular, openly grooved dorsally.—Sandy valleys below 500 ft.; Creosote Bush Scrub; Colo. Desert; to Son., Utah. Feb.–May.

49. **A. serenoì** (Kuntze) Sheld. Sparsely leafy bushy-branched perennial, thinly strigulose; stems erect, 1.5–4.5 dm. long; stipules free, reflexed; lvs. 2–15 cm. long, with 5–11 remote linear lfts. 5–30 mm. long; peduncles stout, to 2.5 dm. long; fls. 5–25; calyx-tube 9–13 mm., teeth 2.5–4 mm. long; petals purple with white wing-tips, banner 17–26 mm. long; pod erect, sessile, broadly oblong, 17–30 mm. long, nearly straight, the valves glabrous, woody, with a partial septum; $2n=22$ (Trelease & Beath, 1949) 24? (Ledingham & Fahselt, 1964).—Bare places, 5000–7500 ft.; Pinyon-Juniper Wd.; Inyo, Grapevine, Cottonwood mts.; to Mono Co., w. Nev. May–July.

50. **A. téner** Gray var. **titì** (Eastw.) Barneby. [*A. t.* Eastw.] Annual, with stems to 2.5 dm. long; lfts. 7–17, broad, cuneate-obovate, 5–16 mm. long; fls. 3–12, in subcapitate racemes; petals purple, the banner 5–6 mm. long; pod 6–14 mm. long, spreading, sessile, flattened and grooved dorsally, the valves firmly papery, strigulose, inflexed as a complete septum.—Sandy places near the coast; Coastal Strand; near San Diego, Monterey Bay, Los Angeles. April–May.

51. **A. tephròdes** Gray. var. **remúlcus** (Gray) Barneby. [*A. r.* Jones.] Habit of *A. tidestromii*, but less densely pubescent, the hairs sometimes straight; calyx commonly strigulose; pod oblong, straight or nearly so, truncate-obtuse at base, obcompressed except for the deltoid laterally compressed beak, 2–2.5 cm. long, the valves leathery, strigulose.—Once collected supposedly at Needles; Ariz. and adjacent states. April–May.

52. **A. tidestròmii** (Rydb.) Clokey. [*A. marcus-jonesii* Munz, in part.] Subacaulescent tufted perennial, villous-tomentulose with curly hairs; stems rarely up to 7 cm. long; stipules free, 3–7 mm. long; lvs. 4–15 cm. long, with 7–19 roundish lfts. 4–13 mm. long; peduncles 3–12 cm. long, reclinate in fr.; fls. 5–13; calyx-tube 5–8 mm. long, teeth 1–2 mm.; petals sordid, the keel purple-tipped, banner ochroleucous or purplish, 13–16 mm. long; pod spreading, sessile, oblong-ellipsoid, incurved, obcompressed except at base and apex, 2.5–5 cm. long, 2-carinate with the prominent sutures, 1-locular, strigulose.—Sandy or gravelly washes and fans, 1000–3000 ft.; Creosote Bush Scrub; New York, Clark mts., n. base of San Bernardino Mts.; s. Nev. April–May.

53. **A. tráskiae** Eastw. Diffuse perennial, white villose-tomentose; stems leafy, 1.5–2 dm. long; stipules free; lvs. 5–10 cm. long, with 21–29 ovate-elliptic lfts. 5–15 mm. long; peduncles 4–14 cm. long; fls. 12–30, 2.5–3 cm. long in fr.; calyx-tube 5–6 mm., teeth 2.5–3.5 mm. long; petals ochroleucous, the banner 14–17 mm. long; pod pendulous, stipitate, the stipe 4–8 mm. long, the body obliquely oval-oblong in outline, gently incurved, abruptly acuminate into a narrow beak, 8–16 mm. long, triquetrously compressed, fully 2-locular, short-tomentose.—Dunes and sandy places below 700 ft.; Coastal Strand, Coastal Sage Scrub; San Nicolas and Santa Barbara ids. March–July.

54. **A. tricarinàtus** Gray. Perennial forming loose bushy plants, greenish except for the silvery upper surface of the lfts.; stems 1–2.5 dm. long; stipules free, reflexed; lvs. 7–20 cm. long, with 17–27 elliptic or obovate lfts. 3–12 mm. long, deciduous from the

persistent petioles; peduncles stout, 9–20 cm. long; fls. 6–15; calyx-tube 4–5 mm., teeth 2–3 mm. long; petals whitish, drying ochroleucous, banner 13–16 mm. long; pod like that of *A. bernardinus*, the gynophore 1–2 mm. long, the body 2.5–4.2 cm. long.—Gravelly places, 1400–4000 ft.; Creosote Bush Scrub, Joshua Tree Wd.; about the head of Coachella V. to Orocopia Mts. Feb.–May.

55. **A. trichopòdus** (Nutt.) Gray. [*Phaca t.* Nutt.] ssp. **trichopòdus**. Bushy-branched perennial, strigulose-villosulous, greenish cinereous to canescent, 2–6 dm. tall; stipules mostly free; lvs. 5–15 cm. long, with 21–39 ovate-oblong to oblanceolate lfts. 5–23 mm. long; peduncles 5–25 cm. long, 12–36-fld., the fls. nodding; calyx-tube 3.5–5.5 mm., teeth 1.5–3.5 mm. long; petals mostly greenish-white, the banner 11.5–15 mm. long; pod on a stipe 6–17 mm. long, the body turgidly and narrowly or sometimes plumply ellipsoid or half-ellipsoid, 1.5–3.5 cm. long, 6–10 mm. in diam., the sutures equally convex or the ventral one straight, the valves commonly glabrous, rarely strigulose.—Brushy hills and ocean-bluffs on the coast or inland hills below 1100 ft.; Coastal Strand, Coastal Sage Scrub; Ventura Co. and Santa Barbara Co. to Puente and Chino hills; Channel Ids. Feb.–June.

Ssp. **leucópsis** (T. & G.) Thorne. [*Phaca l.* T. & G. *A. t.* var. *lonchus* (Jones) Barneby.] Plate 44, Fig. D. Body of the pod 20–40 mm. long, bladdery-inflated, greatly oblique, usual'y openly sulcate ventrally, the ventral suture less convex than the dorsal one, pod usually pubescent.—Coastal Strand, Coastal Sage Scrub; Ventura Co. to n. L. Calif. Channel Ids. Feb.–June.

Ssp. **antisélli** (Gray) Thorne. [*A. a.* Gray. *A. a.* var. *phoxus* Jones. *A. hasseanus* Sheld. *A. gaviotus* Elmer.] Body of the pod laterally compressed, bicarinate by the sutures, the lateral face flat or nearly so; pod-body glabrous or rarely strigose, 1.5–3.6 cm. long, symmetric or oblique.—Dry inland slopes below 4000 ft.; Chaparral, Coastal Sage Scrub; Los Angeles region to Antelope V. (Mojave Desert) and cent. San Luis Obispo Co., reaching the coast above Santa Barbara. Feb.–June.

56. **A. whítneyi** Gray. [*A. hookerianus* var. *w.* Jones and ssp. *pinosus* Abrams.] Perennial, green and nearly glabrous to densely silvery-strigulose; stems decumbent or ascending, 3–20 cm. long; lower stipules connate; lvs. 3–8 cm. long, with 9–19 linear-elliptic to oblong-oblanceolate lfts. 2–15 mm. long; peduncles 2–10 cm. long, fls. 4–15; calyx-tube 3.5–4.5 mm., teeth 0.5–1.5 mm. long; petals purple with white wing-tips, banner 8.5–14 mm. long; pod pendulous, stipitate, the stipe 2–4.5 mm. long, the body bladdery-inflated, balloon-shaped, 1.5–3.5 cm. long, valves papery, translucent, mottled, glabrous, not inflexed; $2n = 22$ (Ledingham & Fahselt, 1964).—Dry places, 7000–12,000 ft.; Subalpine F., Alpine Fell-fields; Mt. Pinos, White Mts.; Sierra Nevada, w. Nev. May–Sept.

57. **A. wootònii** Sheld. Like *A. nutans* and *A. gilmanii*, but thinly strigulose, the 11–19 oblanceolate to linear-oblong lfts. glabrous above; peduncles 1.5–5 cm. long, shorter than the lvs.; racemes loosely 4–10-fld., 1–5 cm. long in fr.; calyx-tube ca. 2 mm., the teeth 2–3 mm. long; petals whitish, the banner 5–7 mm. long; pod spreading, sessile, ovoid-ellipsoid, symmetric or little oblique, obscurely deltoid-beaked, 1.5–3 cm. long, the valves papery, strigulose.—Sandy flats and plains, Creosote Bush Scrub; e. Mojave Desert (Goffs, 2600 ft.); to Tex., n. Mex. April.

5. *Calliándra* Benth.

Herbs, shrubs or trees, with slender branches. Lvs. bipinnate, with small lfts. Fls. in dense pedunculate heads arranged in racemes. Calyx 5-toothed or -lobed. Petals united to ca. middle into a funnelform or campanulate corolla. Stamens many, long-exserted; fils. united basally into a tube; anthers small. Style filiform. Legumes linear, flat, straight, elastically dehiscent from apex, the valves with raised margins. Seeds flattened. Perhaps 150 spp. of trop. and subtrop. Am., Afr., India. (Greek, *kallos*, beautiful, and *andra*, stamen.)

1. **C. eriophýlla** Benth. FAIRY DUSTER. Densely branched, ± spreading unarmed shrub ca. 2–3.5 dm. high, with gray pubescent twigs; pinnae 2–4 pairs with 5–10 pairs

of oblong lfts. 3–5 mm. long, strigose beneath; heads few-fld., sparingly strigose, rose to reddish-purple; calyx 1–1.5 mm. long; corolla 4–6 mm. long; stamens reddish, 15–20 mm. long; pod 5–7 cm. long, silvery-pubescent with dark red margins; seeds gray, narrow-obovate, 6–7 mm. long.—Sandy washes and gullies below 1000 ft.; Creosote Bush Scrub; Imperial Co. and e. San Diego Co.; to Tex., Puebla, L. Calif. Mostly Feb.–March.

6. Cássia L. SENNA

Herbs, shrubs or trees, with even-pinnate lvs. Fls. usually yellow, in racemes. Calyx deeply toothed or divided into subequal lobes. Corolla quite regular; petals 5, spreading, imbricated, clawed. Stamens 10 or 5, often unequal, the anthers opening by 2 pores at the summit, in ours 10 with 3 represented by short sterile fils. Ovary sessile or stipitate; pods flat or terete, often curved, many-seeded. A large genus of warm temp. and trop. regions. (Ancient Greek name.)

Lfts. in 1–4 pairs. Native plants, of the desert.
  Herbage nearly glabrous; lfts. 4–8 mm. long; racemes terminal .... 1. *C. armata*
  Herbage densely white-pubescent; lfts. 1–3 cm. long; racemes axillary .... 2. *C. covesii*
Lfts. in 6–8 pairs. Large shrub escaped from cult. .... 3. *C. tomentosa*

1. **C. armàta** Wats. Much-branched rounded shrub, 5–15 dm. high, with many yellow-green striate stems leafless much of year; lf.-rachis 5–15 cm. long, the 1–4 pairs of lfts. distant, oblong-ovate; racemes 5–15 cm. long, many-fld.; calyx ca. 6 mm. long; petals 8–10 mm. long, bright yellow; legume lance-cylindric, slightly compressed, spongy, 2–4 cm. long; seeds 7–9 mm. long; $2n = 14$ pairs (Raven, Kyhos & Hill, 1965).—Common in sandy washes and open places below 3700 ft.; Creosote Bush Scrub; Colo. and Mojave deserts; to Ariz., Nev. April–May.

2. **C. còvesii** Gray. Low suffrutescent plant 3–5 dm. high, densely and finely pubescent; lvs. 2–7 cm. long with 3 pairs of oblong-obovate lfts.; racemes few-fld.; calyx 6–7 mm. long; petals 10 mm. long, yellow, veiny; legume brownish, compressed, 2–5 cm. long.—Occasional, dry washes below 2000 ft.; Creosote Bush Scrub; Colo. Desert; Ariz., L. Calif. April–June.

3. **C. tomentòsa** L. f. Shrub 3–4 m. high; twigs and lower surface of lvs. tomentose; lfts. in 6–8 pairs, oblong, 2–4 cm. long, rounded and apiculate at the apex, each pair with a basal gland; fls. deep yellow, ca. 2.5 cm. across; pod compressed, pubescent, ca. 12–13 cm. long.—Occasionally escaped about dumps, etc.; Mex., S. Am.

7. Cercídium Tulasne. PALO VERDE

Shrubs or small trees with green bark and ± spinose twigs. Lvs. bipinnate, the pinnae with terete rachis and clearly borne below the spines. Infl. a short axillary corymb. Calyx-lobes short, valvate. Petals 5, yellow, clawed. Stamens 10, distinct, hairy at base of fils. Legume linear to oblong, flattened or cylindric, several-seeded. An Am. genus of ca. 10 spp. (Greek, *kerkidion*, a weaver's shuttle, referring to fr.)

Pinnae with 1–3 pairs of lfts., glabrous; pods flattish .... 1. *C. floridum*
Pinnae with 4–8 pairs of lfts., pubescent; pods cylindric .... 2. *C. microphyllum*

1. **C. flóridum** Benth. [*Parkinsonia f.* Wats. *P. torreyana* Wats. *C. t.* Sarg.] Plate 44, Fig. E. Tree up to 10 m. high with short trunk and smooth blue-green bark; leafless most of year; branchlets slender, glaucous, frequently spiny; lvs. few, scattered, 1–2 cm. long, with 1 pair of pinnae; fls. on slender pedicels; calyx ca. 6–7 mm. long; petals yellow, 8–10 mm. long; pod glabrous, 1–8-seeded, 4–8 cm. long, often somewhat constricted between seeds; seeds olive and brown, 8–10 mm. long; $2n = 14$ pairs (Raven et al., 1965). —Washes and low sandy places, below 1600 ft.; Creosote Bush; Colo. Desert; Ariz., Son., L. Calif. March–May. Occasionally hybridizing with next sp.

2. **C. microphýllum** (Torr.) Rose & Jtn. [*Parkinsonia m.* Torr.] Spiny shrub or tree

to 6 m. high with yellowish-green bark; lvs. pubescent, the pinnae with 4–8 pairs of minute lfts.; petals pale yellow, 5–6 mm. long; pod 4–8 cm. long, almost 1 cm. wide, turgid, puberulous, 1–4-seeded; seeds brown, oblong-obovoid, scarcely compressed, 7–8 mm. long.—Gravelly slopes and washes occasionally to 3500 ft.; Creosote Bush Scrub; Whipple Mts., e. San Bernardino Co.; Ariz., Son., L. Calif. April–May.

8. Cércis L. Redbud. Judas Tree

Large shrubs to small trees with simple broad lvs. Fls. red-purple in short lateral fascicles, appearing before the lvs. in the spring. Calyx broadly campanulate, 5-toothed. Corolla irregular; petals 5, the standard enclosed by the wings in the bud; keel larger than wings. Stamens 10, distinct. Ovary short-stipitate, many-ovuled. Legumes oblong or linear-oblong, flat, the upper suture margined. Seeds compressed, obovate to rounded. Ca. a half-dozen spp. of N. Am. and Eurasia. (The ancient Greek name.)

1. **C. occidentàlis** Torr. ex Gray. [*C. nephrophylla* Greene.] From 2–5 m. high, with clustered erect stems and glabrous twigs; lvs. round to reniform, cordate at base with open to almost closed sinus, palmately 7–9-veined, entire, ± coriaceous, 3–9 cm. broad; petioles 1–3 cm. long; fls. magenta-pink to reddish-purple, 8–12 mm. long; pedicels 7–11 mm. long; pods many, flat, 4–9 cm. long, 2–2.5 cm. wide; seeds 3–4 mm. in diam.; $n=7$ (Atchison, 1949); $2n=14$ (Taylor, 1967).—Dry slopes 400–5000 ft.; Chaparral, Yellow Pine F.; e. side of Laguna and Cuyamaca mts.; n. Los Angeles Co.; to n. Calif.; Utah, Ariz. The San Diego plants tend to be more coriaceous and later flowering than other Californian forms when grown in the Bot. Garden.

9. Cýtisus L. Broom

Evergreen or deciduous shrubs, mostly unarmed, sometimes nearly leafless. Lvs. 3-foliolate, sometimes unifoliolate, the lfts. entire. Fls. terminal, solitary or racemose, yellow, white or purple; calyx 2-lipped with short teeth; banner ovate to rounded; wings oblong or ovate; keel straight or curved. Stamens monadelphous. Pods compressed, several-seeded, with a callous appendage or strophiole near the base. Ca. 80 spp. Eurasia and N. Afr. (Greek, *kutisus*, a kind of clover.)

Shrub 2–8 dm. high; lfts. revolute, linear-oblong; fls. 12–14 mm. long .... 1. *C. linifolius*
Shrub 10–30 dm. high; lfts. flat, obovate; fls. 10–12 mm. long .... 2. *C. monspessulanus*

1. **C. linifòlius** (L.) Lam. [*Genista l.* L.] Low shrub, with erect appressed-silky branches; lfts. linear or linear-lanceolate, revolute at margin, nearly glabrous and shining above, silvery-pubescent beneath, 12–25 mm. long; fls. 12–14 mm. long; pod torulose.—Adventive on Santa Catalina Id. and in Santa Barbara Co. From Medit. region. March–April.

2. **C. monspessulànus** L. French Broom. Shrub 1–3 m. tall, with villous branchlets; petioles 3–5 mm. long; lfts. 3, ± obovate, 1–2 cm. long, subglabrous above, pubescent beneath; racemes subcapitate, terminating short lateral branches; calyx pubescent, 4–5 mm. long; petals bright yellow, 10 mm. long; pods densely villous, 2–2.5 cm. long; $n= \pm 23$ (Duarte de Castro, 1949).—Natur., often perniciously, near the coast from Ventura Co.; n. to Wash. Santa Catalina Id. Native of Canary Ids. March–May.

10. Dàlea Juss. Indigo Bush

Herbs, shrubs or small trees, usually gland-dotted. Lvs. odd-pinnate, rarely entire, with small subulate stipules. Fls. purple, white or yellow, in terminal spikes or racemes. Calyx persistent, ± equally 5-toothed. Petals clawed. Banner usually cordate or auriculate, mostly exceeded by the wings and keel. Stamens monadelphous. Pod ovoid, compressed, sessile or short-stipitate, generally 2 (rarely 4–6)-ovuled. Seeds reniform. Over 100 spp. of N. and S. Am. (T. *Dale*, an early English botanist.)

(Wiggins, I. L. Contr. Dudley Herb. 3: 41–64. 1940.)
1. Plants herbaceous; petals, except the banner, adnate to the stamens for about half their length.
    2. Fls. in a dense spike; calyx-lobes longer than the tube.
        3. Fls. 4–5 mm. long; calyx 3–4 mm. long; pubescence of whitish hairs, rarely turning rusty on drying .... 5. *D. mollis*
        3. Fls. 6–8 mm. long; calyx ca. as long as corolla; pubescence of infl. often turning rusty brown on drying .... 6. *D. mollissima*
    2. Fls. in loose spikelike racemes; calyx-lobes not longer than the tube .... 7. *D. parryi*
1. Plants shrubs or small trees; claws of petals free or adnate only to the very base of stamen-tube.
    4. Lvs. simple, sometimes wanting.
        5. Herbage silvery-silky, the lvs. often lacking except on young growth ..... 10. *D. spinosa*
        5. Herbage green, subglabrous, the lvs. mostly present during vegetative period 9. *D. schottii*
    4. Lvs. pinnate, present during vegetative periods.
        6. Fls. sessile in dense capitate spikes; corolla 5–6 mm. long.
            7. Lfts. 2–4 mm. long; glands many, conspicuous, flat .... 8. *D. polyadenia*
            7. Lfts. 5–20 mm. long; glands rather few, small, prickle-shaped .... 3. *D. emoryi*
        6. Fls. pedicelled, in loose spikes; corolla 8–12 mm. long.
            8. Plant densely villose-tomentose; fls. 10–12 mm. long .... 1. *D. arborescens*
            8. Plant strigose to subglabrous; fls. ca. 8 mm. long.
                9. Lfts. not decurrent on rachis or confluent .... 4. *D. fremontii*
                9. Lfts. decurrent on rachis or the upper ones confluent .... 2. *D. californica*

1. **D. arboréscens** Torr. ex Gray. [*Parosela a.* Heller. *P. neglecta* Parish.] Spinescent much-branched shrub 1–1.5 m. high, hoary-tomentose; lvs. 1.5–2.5 cm. long, the pinnae mostly 5–7, obovate, 5–10 mm. long, obscurely glandular; racemes 3–5 cm. long; calyx villous-tomentose, ca. 7 mm. long; corolla deep violet-blue, 10–12 mm. long; pod villous, gland-dotted, the body 1 cm., the beak 6 mm. long.—Low hills and mesas, 2000–2600 ft.; Creosote Bush Scrub; Barstow region, Mojave Desert. April–May.

2. **D. califórnica** Wats. [*Parosela c.* Vail. *P. c.* var. *simplifolia* Parish.] Much like *D. fremontii* var. *minutifolia*; lfts. 1–7, ± decurrent and the upper confluent, densely silky-pubescent; calyx-lobes not markedly different in width; pods subglabrous, conspicuously glandular. Dry places below 3500 ft.; Creosote Bush Scrub; Colo. Desert from Morongo V. to Twentynine Palms region, Shaver's Well, etc., San Jacinto V. April–May.

3. **D. émoryi** Gray. [*Parosela e.* Heller.] Plate 44, Fig. F. Densely branched shrub 3–15 dm. tall, white with feltlike tomentum and sprinkled with orange glands; lvs. 1–8 cm. long, the lfts. mostly 5–7, linear to oblong or obovate, 5–15 mm. long; calyx rusty-villous, 5–7 mm. long; corolla rose to purplish, 5–6 mm. long; pod obliquely obovoid, villous and glandular above.—Dry open places, below 1000 ft.; Creosote Bush Scrub; Colo. Desert; to L. Calif., Son., Ariz. March–May.

4. **D. fremóntii** Torr. var. **fremóntii.** [*Parosela f.* Vail.] Intricately branched shrub 5–20 dm. tall, younger growth strigose-canescent; lvs. 2–4 cm. long, closely strigose, the lfts. mostly 3–5, narrow-oblong, 6–8 mm. long, with small glands; racemes ca. 7–12 cm. long, rather lax; calyx ca. 6 mm. long; the upper 2 lobes triangular, ca. twice as wide as the lowest one; corolla dark purple-blue; pods ovoid, with brown glands, the body 8 mm. or more long.—Occasional, dry places 2500–4500 ft.; Sagebrush Scrub, Creosote Bush Scrub; about Owens V.; to Nev., Utah. April–May.

KEY TO VARIETIES

Pubescence closely appressed, the lvs. and young growth quite canescent.
    Lfts. oblong, mostly ca. 6 mm. long. Owens V. .... var. *fremontii*
    Lfts. linear to lanceolate, 6–20 mm. long. Death V. region to Colo. Desert .... var. *minutifolia*
Pubescence ± spreading, the lvs. and young growth greener. Mojave Desert ..... var. *saundersii*

Var. **minutifòlia** (Parish) L. Benson. [*Parosela johnsoni* var. *m.* Parish. *D. j.* Wats.] Plate 45, Fig. A. Lfts. 1–4 mm. broad; pods ca. 8 mm. long.—Common, below 5000 ft.; Creosote Bush Scrub, Joshua Tree Wd.; Mojave Desert from Death V. region to Riverside Co.; to Utah, Ariz.

Plate 45. Fig. A, *Dalea fremontii* var. *minutifolia*. Fig. B, *Dalea spinosa*. Fig. C, *Hoffmannseggia microphylla*. Fig. D, *Lathyrus laetiflorus* ssp. *alefeldii*. Fig. E, *Lotus argophyllus*. Fig. F, *Lotus crassifolius*.

Var. **sáundersii** (Parish) Munz. [*D. s.* Parish.] Lfts. 2.5–4 mm. broad, 6–13 mm. long; pubescence rather sparse, short, not closely appressed; body of pod 8–12 mm. long. —Dry places, 2000–6000 ft.; Creosote Bush Scrub, Sagebrush Scrub; Victorville region w. and n.; to White Mts.

5. **D. móllis** Benth. [*Parosela m.* Heller.] Short-lived perennial, flowering the first year, herbaceous; stems many, 1–3 dm. long, decumbent, soft-villous, dotted with small flat glands; lfts. 9–13, cuneate-oblong to obcordate, 3–8 mm. long, gland-dotted along the margins; spikes dense, 1.5–3.5 cm. long; calyx 3–4 mm. long, densely villous, the teeth filiform; corolla slightly longer, white with pinkish tinge; pods obovoid, 3 mm. long. —Open sandy places, below 3000 ft.; Creosote Bush Scrub; deserts s. and e. of Twentynine Palms; to L. Calif., Son. March–June.

6. **D. mollíssima** (Rydb.) Munz. [*Parosela m.* Rydb.] With habit and pubescence of *D. mollis*; calyx 6–8 mm. long, the filiform teeth plumose and longer than the tube; corolla included, the wing-petals rounded, without a gland near tip.—Creosote Bush Scrub; Colo. Desert n. to Inyo Co.; s. Nev., w. Ariz.

7. **D. párryi** T. & G. [*Parosela p.* Heller.] Diffuse, slender-stemmed, suffruticose perennial 3–5 dm. tall, the stems purplish, ashy-strigose, gland-dotted; lvs. pinnate, the lfts. 13–35, round to elliptical, 2–4 mm. long, ± strigose; spikes loose, 3–5 cm. long; calyx canescent, 3 mm. long, the teeth ovate; corolla blue-purple, with some white, ca. 6–8 mm. long; pod ca. 2 mm. long.—Dry gravelly or rocky places below 2500 ft.; Creosote Bush Scrub; se. Mojave Desert; to Nev., Ariz., L. Calif., Son. Feb.–June.

8. **D. polyadènia** Torr. ex Wats. [*Parosela p.* Heller.] Intricately branched shrub 3–20 dm. high, with canescent spine-tipped branches having amber glands; lvs. 1–2 cm. long, the pinnae 5–11, obovate, 2–4 mm. long; calyx 4 mm. long, canescent and gland-dotted, the teeth subulate; corolla rose to purplish, 5–6 mm. long; pods 3–5 mm. long.—Dry slopes and mesas, 2500–6000 ft.; Creosote Bush Scrub, Joshua Tree Wd.; Mojave Desert from Twentynine Palms region to Inyo and Mono cos. May–Sept.

9. **D. schóttii** Torr. [*Parosela s.* Heller.] Intricately branched shrub, spinescent, 1–3 m. high, the young growth bright green, subglabrous to minutely strigose, nearly glandless; lvs. usually simple, linear, 1–3 cm. long, 0.5–2 mm. wide; racemes 5–10 cm. long, lax; calyx 4–5 mm. long; corolla blue, 8–10 mm. long; pod gland-dotted, almost 1 cm. long.—Washes and benches, below 1000 ft.; Creosote Bush Scrub; Coachella V. to w. Ariz., L. Calif. March–May. A form with young branches, young lvs. and calyces canescent-puberulent is var. *pubérula* (Parish) Munz. [*Parosela s.* var. *p.* Parish.] $n = 10$ (Turner, 1963).—W. edge of Colo. Desert.

10. **D. spinòsa** Gray [*Parosela s.* Heller.] SMOKE TREE. Plate 45, Fig. B. Intricately branched shrub or small tree, spinose, 1–6 m. high, nearly leafless, ashy-gray throughout with close canescence, dotted with glands; lvs. early deciduous, simple, few, cuneate-oblong, 1–2 cm. long; spikes many, spinescent, 2–3 cm. long; calyx 3–4 mm. long, gland-dotted, the teeth ovate; corolla bright blue-purple, 8–10 mm. long, the keel not longer than other petals; pods obliquely ovoid, ca. 6 mm. long, with amber-colored glands; $2n = 10$ pairs (Raven et al., 1965).—Locally common in sandy washes below 1500 ft.; Creosote Bush Scrub; s. Mojave Desert from Daggett e., Colo. Desert; Ariz., Son., L. Calif. June–July.

11. *Glycyrrhìza* L. LIQUORICE

Perennial herbs with thick sweet rootstocks and often glandular odd-pinnate lvs. Fls. in axillary spikes or racemes, whitish to blue. Calyx 5-cleft, the 2 upper teeth somewhat shorter and ± united. Banner short-clawed, the banner oblong to ovate; wings oblong; keel acute to obtuse. Stamens mostly diadelphous, the alternate anthers shorter. Pods sessile, covered with glands or prickles, indehiscent, several-seeded. Ca. 15 spp. of N. Temp. Zone, S. Am., Australia. (Greek, *glukus*, sweet, and *rhiza*, a root.)

Pods glabrous; fls. bluish. Occasional weed .... 1. *G. glabra*
Pods burlike with hooked prickles; fls. yellow-white. Native .... 2. *G. lepidota*

1. **G. glàbra** L. Cultivated Licorice. Plants subglabrous; rootstocks sweet, woody, immensely developed; lfts. ovate, 2.5–5 cm. long; fls. pale blue; pods red-brown, 1.2–2.5 cm. long, glabrous; $2n=16$ (Tschechow, 1930).—Occasional escape as near Spadra (e. Los Angeles Co.); native of Eurasia. Spring.

2. **G. lepidòta** Pursh var. **glutinòsa** (Nutt.) Wats. Stems erect, 3–10 dm. high, ± viscid-pubescent; lfts. 11–19, oblong to ovate-lanceolate, 2–3 cm. long, mucronate; stipules linear-subulate; spikes broadly oblong, shorter than the lvs.; calyx very glandular, ca. 6 mm. long; corolla yellowish white, 8–12 mm. long; pod oblong, 12–15 mm. long, burlike with hooked prickles; seeds reniform, ca. 2.5 mm. long; $2n=16$ (Heiser & Whitaker, 1948).—Occasional as patches in low ground and moist waste places, below 4500 ft.; many Plant Communities; cismontane and occasional on Mojave Desert; to B.C., Ont., Mex. May–July.

12. *Hoffmannséggia* Cav.

Herbs or small shrubs. Lvs. bipinnate, glandular-dotted, with small stipules and lfts. Fls. and frs. in glandular racemes. Calyx 5-parted. Petals 5, subequal, orange or yellow. Stamens 10, distinct; fils. often glandular below. Legume flat, several-seeded, linear to ovate, 2-valved. Ca. 20 spp., Am. and Afr. (Count of *Hoffmannsegg*, coauthor of a flora of Portugal.)

Herb; lvs. with several pairs of pinnae; fls. orange .... 1. *H. densiflora*
Shrub; lvs. with 1 pair of lateral pinnae and the terminal pinna; fls. yellow .... 2. *H. microphylla*

1. **H. densiflòra** Benth. ex Gray. [*H. stricta* Benth. *H. falcata* Calif. auths.] Stems puberulent, 1–4 dm. tall, several, scattered, from underground rootstocks; lvs. several, often near the base, 6–12 cm. long, with 5–11 pinnae, each with 6–10 pairs of crowded oblong pubescent pinnules 2–6 mm. long; peduncle 3–7 cm. long, with equally long raceme, stipitate-glandular and pubescent; calyx 6–8 mm. long; corolla ca. 1 cm. long, the petals glandular on claws and lower margins; stamens puberulent, the alternate glandular; pod compressed, 2–3 cm. long, ± glandular.—Heavy alkaline soil, below 2600 ft.; Creosote Bush Scrub; Mojave and Colo. deserts; cismontane especially along railroads, probably as an adventive; to Tex., Mex. Apr.–June.

2. **H. microphýlla** Torr. Plate 45, Fig. C. Broad rounded subshrub usually 6–10 dm. tall and with many rushlike stems, sometimes to 2 m. high, pubescent; lvs. scattered, 2–3 cm. long, with 3 pinnae, the terminal with 10–14 pinnules, the 2 lateral each with 8–12; pinnules 3–5 mm. long; racemes many, slender, terminal and lateral, 8–20 or more cm. long; fls. 6–7 mm. long; stamens exserted, the fils. glandular-stipitate; pods lunate, 1.5–2.5 cm. long, with many stipitate glands; seeds brown, shining, ca. 3 mm. long.—Common about canyons and washes, below 4500 ft.; Creosote Bush Scrub; Colo. Desert; Ariz., Son., L. Calif. March–May.

13. *Láthyrus* L. Pea

Annual or mostly perennial herbs with rootstocks or sometimes taproots; stems erect to twining, angled to winged. Lvs. pinnate, the rachis usually produced into a tendril. Fls. sometimes solitary, mostly in axillary racemes. Calyx obliquely campanulate, the teeth subequal or the 2 upper much shorter. Corolla showy, the banner usually distinctly clawed; wings with small lunate ridge on inner face which fits into a fold in the keel. Stamens diadelphous. Style curved, flattened, hairy along the inner side. Pods flat or ± terete, 2-valved. Seeds several, ± spherical or compressed. Ca. 100 spp., of N. Hemis. and S. Am. (Ancient Greek name of some leguminous plant.)

(Hitchcock, C. L. A revision of the N. Am spp. of Lathyrus. Univ. Wash. Pub. Biol. 15: 1–104. 1952.)

1. Lfts. 2; introd. plants.
  2. Plants annual.
    3. Plants glabrous; lvs. linear to linear-lanceolate .... 7. *L. tingitanus*
    3. Plants rough-hairy; lfts. oval or oblong .... 4. *L. odoratus*

2. Plants perennial .... 3. *L. latifolius*
1. Lfts. at least 4 on some of the lvs.; native perennial plants.
    4. Corolla deep rich red, 3–4 cm. long, the banner reflexed and lying in nearly a straight line with wings and keel. Se. San Diego Co. .... 6. *L. splendens*
    4. Corolla mostly other than deep red, or if so, less than 3 cm. long; banner seldom reflexed more than 90°.
        5. Racemes mostly 2-fld.; calyx less than 5 mm. long .... 1. *L. hitchcockianus*
        5. Racemes more than 2-fld.
            6. Lateral calyx-lobes considerably broadened above their juncture with the tube, the lowest lobe usually as long as the tube .... 8. *L. vestitus*
            6. Lateral calyx-lobes usually broadest at their point of juncture with the tube, the lowest lobe often much shorter than the tube.
                7. Fls. white to flesh-color, often with pink or lavender veining and fading to yellow or tan .... 2. *L. laetiflorus*
                7. Fls. bluish or red, but sometimes paler and fading to brown.
                    8. Fls. over 17 mm. long, the plants widely distributed, pubescent to subglabrous
                    .... 2. *L. laetiflorus*
                    8. Fls. less than 17 mm. long; plants essentially glabrous, Tehachapi Mts. n.
                    .... 5. *L. pauciflorus*

1. **L. hitchcockiànus** Barneby & Reveal. Perennial, 1.5–2.5 dm. tall; lvs. with 1–4 lfts.; racemes mostly 2-fld.; calyx scarcely 5 mm. long; petals lilac-purple, drying bluish. —Coll. in 1891 in Wood Canyon, Grapevine Mts.; Nev.

2. **L. laètiflòrus** Greene ssp. **laètiflorus.** Pubescent to subglabrous perennial, scandent, the stems not winged, 1–3 m. high; stipules lanceolate to lance-ovate, usually dentate, ¼ to almost as long as lfts.; these mostly 8–12, linear to ovate, 2–5 cm. long; tendrils well developed; fls. 5–12, whitish to pale flesh-color, 16–22 mm. long; pedicels mostly 2–5 mm. long; calyx 10–14 mm. long, the upper teeth triangular, shorter than the narrower lower laterals; banner obcordate, usually reflexed at less than a right angle to the keel, the claw less than half as long as blade; pods 4–8 cm. long, 5–8 mm. wide, many-seeded; $2n = 7$ pairs (Raven et al., 1965).—Dry places below 5000 ft.; Coastal Sage Scrub, Chaparral; Santa Monica Mts. to San Jacinto and Santa Rosa mts. April–June.

Ssp. **bárbarae** (White) C. L. Hitchc. [*L. violaceus* var. *b.* White.] Corolla pink to lavender.—Near the coast from Santa Barbara to Orange and w. Riverside cos. Santa Cruz, Santa Rosa and San Clemente ids.

Ssp. **aleféldii** (White) Brads. [*L. a.* White. *L. strictus* Nutt., not Grauer.] Plate 45, Fig. D. Pedicels 3–10 mm. long; fls. deep red, the banner reflexed at more than a right angle to keel-axis; $n = 7$ (Hitchc., 1952).—Catalina Id. and w. Riverside and s. Orange cos.; to L. Calif.

3. **L. latifòlius** L. Everlasting Pea. Perennial from widespread rootstocks; stems broadly winged, 6–20 dm. tall; stipules 2–5 cm. long, mostly entire; lfts. 2.5–12 cm. long; fls. 5–15, purplish red to white, long-peduncled; calyx 8–12 mm. long; pods 6–10 cm. long; $n = 7$ (Senn, 1938).—Occasional garden escape; native of Eu. May–Sept.

4. **L. odoràtus** L. Common Sweet Pea. Crisp-hairy annuals, with winged stems; stipules 2-lobed, ca. half as long as the 2 lfts.; fls. 2–5, fragrant, 2–3 cm. long; calyx 12–15 mm. long; pods 3–6 cm. long; $n = 7$ (Senn, 1938; Bir & Sadhu, 1966).—Occasional garden escape in waste places, etc., especially near the coast; native of Medit. region. April–June.

5. **L. pauciflòrus** Fern. ssp. **brównii** (Eastw.) Piper. [*L. b.* Eastw.] Perennial, glabrous except for calyx-teeth; stems angled, erect, 2–6 dm. tall; lfts. thickish, subleathery, mostly 8–10, linear to ovate, 1.5–3.5 cm. long; fls. mostly 4–7, orchid or pinkish-lavender to violet-purple, aging bluish, 13–17 mm. long; calyx strongly 10-nerved; banner reflexed ca. 90°; pods 3–6 cm. long, glabrous.—Yellow Pine F.; Tehachapi Mts. to Ore. April–June.

6. **L. spléndens** Kell. Pride of California. Subglabrous to pubescent perennial, stems 5–30 dm. tall, not winged; stipules ca. ⅓ the length of the lfts., mostly few-toothed; lfts. mostly 6–10, linear to oval, 2–7 cm. long; tendrils well developed; fls. 4–12, deep

rich red or crimson, 3–4 cm. long; calyx pubescent, 8–12 mm. long; banner reflexed so as to lie almost in the same plane with the keel; pods 5–8 cm. long, glabrous; $n=7$ (Hitchc., 1952).—Dry slopes below 3500 ft.; climbing over shrubs in Chaparral; e. San Diego Co. near the Mex. border, adjacent L. Calif. April–June.

7. **L. tingitànus** L. TANGIER PEA. Glabrous annual, with narrowly winged stems 8–20 dm. tall; stipules lanceolate, entire or toothed; lfts. 2, linear to lance-elliptic, 4–10 cm. long; tendrils pinnate; fls. 1–3, rose-purple, 2.5–3 cm. long; calyx ca. 10 mm. long; pods glabrous, 7–10 cm. long; $n=7$ (Senn, 1938).—Occasionally natur. near the coast, Ventura Co. n.; Catalina Id.; native of Medit. region. May–July.

8. **L. vestìtus** Nutt. ex T. & G. ssp. **pubérulus** (White ex Greene) C. L. Hitchc. [*L. p.* White ex Greene.] Pubescent perennial usually over 4 dm. high, scandent; stipules undulate-dentate; lfts. mostly 10, oblong-lanceolate, 2–3.5 cm. long; tendrils well developed; fls. 5–20, pinkish to bluish-lavender, 15–20 mm. long; calyx 12–15 mm. long; pods 4–6 cm. long, pubescent.—Chaparral; Santa Monica Mts. n. to n. Calif. April–June.

14. *Lòtus* L. BIRD'S FOOT TREFOIL

Annual or perennial herbs or suffrutescent. Lvs. alternate, pinnately 1- to severalfoliolate. Stipules foliaceous, scarious, or glandlike. Fls. axillary, solitary to umbellate, sessile or on short to long, mostly leafy-bracteate peduncles. Calyx-teeth 5, subequal. Corolla white or yellow, often with some red or purple. Stamens diadelphous. Pod flattened or terete, straight to strongly arcuate, 1–many-seeded, dehiscent or indehiscent. Perhaps 150 spp., of all continents; but mostly of N. Hemis. (Ancient Greek name.)

(Ottley, A. M. A revision of the Calif. spp. of Lotus. Univ. Calif. Pub. Bot. 10: 189–305. 1923.)
1. Stipules expanded, not glandlike.
    2. The stipules green, resembling and almost as large as the 3 lfts. Introd. weed in lawns, etc.
        3. *L. corniculatus*
    2. The stipules membranous or hyaline, not like the several lfts.
        3. Pods 4–5 mm. wide. Plants of dry places .... 4. *L. crassifolius*
        3. Pods 1.5–2 mm. wide. Plants of wet places .... 14. *L. oblongifolius*
1. Stipules reduced to dotlike often dark or reddish glands.
    4. Pods dehiscent, straight or nearly so, abruptly short-beaked.
        5. Plants perennial.
            6. Lf.-rachis 2–5 cm. long; corolla 1.5–2.4 cm. long .... 6. *L. grandiflorus*
            6. Lf.-rachis not more than 1 cm. long.
                7. Plants erect, 3–9 dm. tall; corolla 1.3–2 cm. long; pods 3–4 cm. long; lvs. sparsely strigose .... 17. *L. rigidus*
                7. Plants prostrate to ascending, the stems 1–3 dm. long; corolla 8–12 mm. long; lvs. silvery-silky to appressed-canescent .... 2. *L. argyraeus*
        5. Plants annual.
            8. Fls. yellow, often aging red.
                9. Wings conspicuously longer than the keel; styles hairy at the base of the stigma; seeds not smooth.
                    10. Lfts. linear-oblong to elliptic, not noticeably succulent; herbage strigulose. Cismontane .... 20. *L. strigosus*
                    10. Lfts. mostly ± obovate, quite succulent; herbage canescently tomentulose. Deserts .... 22. *L. tomentellus*
                9. Wings ca. as long as or shorter than keel; styles glabrous; seeds smooth.
                    11. Fls. 1–5 on a peduncle; pods 1.5–2 mm. wide .... 18. *L. salsuginosus*
                    11. Fls. solitary, subsessile; pods ca. 3 mm. wide.
                        12. Plant subglabrous to strigose; calyx-teeth ca. as long as tube; pods 10–15 mm. long .... 21. *L. subpinnatus*
                        12. Plant densely villous-pubescent; calyx-teeth ca. twice as long as the tube; pods 5–10 mm. long .... 10. *L. humistratus*
            8. Fls. cream-white or red or pink.
                13. Calyx-teeth shorter than the tube; corolla pinkish or pale salmon, tinged or turning red; pods not deflexed; lfts. mostly 3–10 mm. long .... 11. *L. micranthus*

13. Calyx-teeth longer than the tube; corolla whitish, tinged with rose; pods deflexed; lfts. mostly 10–15 mm. long .... 16. *L. purshianus*
4. Pods indehiscent; often strongly curved, tapering to an elongate beak.
   14. Plants mostly annual.
      15. Umbels pedunculate; bract usually present, unifoliolate; corolla 5–7 mm. long; beak of pod glabrous .... 13. *L. nuttallianus*
      15. Umbels subsessile; bract lacking; corolla 3–4.5 mm. long; beak of pod strigose
                                                                                                   7. *L. hamatus*
   14. Plants perennial.
      16. Corolla mostly 3–9 mm. long.
         17. Pubescence spreading, ± woolly-villous .... 9. *L. heermannii*
         17. Pubescence appressed or almost lacking.
            18. Fls. on peduncles mostly more than 5 mm. long.
               19. Plant erect, sparsely strigose. Desert edge .... 8. *L. haydonii*
               19. Plants prostrate, densely strigose to silvery.
                  20. The plant ± canescent throughout, but scarcely silvery; calyx-teeth ca. half as long as the tube. Mts. of San Diego Co. and Inyo and Kern cos.; rare in Riverside and San Bernardino cos. .... 12. *L. nevadensis*
                  20. The plant more silvery.
                     21. Calyx-teeth almost as long as the tube; banner reflexed or at right angles to the wings. Mt. Pinos to Santa Rosa Mts. .... 5. *L. davidsonii*
                     21. Calyx-teeth ca. half as long as the tube; banner ascending, at an angle acute to the wings. San Gabriel and San Bernardino mts.
                                                                             1. *L. argophyllus* var. *decorus*
            18. Fls. sessile or subsessile, in lf.-axils.
               22. Plant plainly grayish or whitish with appressed pubescence.
                  23. Body of the pod scarcely exserted beyond the calyx, arcuate, the beak ca. as long as the body; herbage densely silvery-tomentose ... 1. *L. argophyllus*
                  23. Body of the pod well exceeding the calyx, somewhat curved, the beak shorter than the body; herbage with closely appressed pubescence
                                                                             15. *L. procumbens*
               22. Plant green, subglabrous except at growing tips; body of pod well exserted
                                                                          19. *L. scoparius*
      16. Corolla 10–12 mm. long.
         23a. Umbels on well developed peduncles; pods well exserted from the calyx. Insular.
            24. Plant decumbent to ascending, herbaceous; lfts. 3–7; fls. 12–20
                                                                        1. *L. argophyllus* var. *ornithopus*
            24. Plant erect, woody; lfts. 3; fls. 7–12 .... 19. *L. scoparius* var. *dendroideus*
         23a. Umbels sessile or nearly so; plants prostrate or decumbent.
            25. Body of the pod scarcely exserted from the calyx, ± arcuate; herbage silvery-silky, with ± spreading hairs. Santa Cruz Id. .... 1. *L. argophyllus* var. *niveus*
            25. Body of the pod well exserted from the calyx; plant silky-strigose. San Miguel Id.
                                                                                    19. *L. scoparius*

1. **L. argophýllus** (Gray) Greene ssp. **argophyllus**. [*Hosackia a.* Gray. *Syrmatium a.* Greene.] Plate 45, Fig. E. Prostrate or decumbent perennial, many-branched, densely silvery-tomentose, the stems 2–10 dm. long; lfts. 3–5, broadly oblanceolate to obovate, mostly 4–12 mm. long, obtusish; umbels sessile or nearly so, 3–8-fld.; bract unifoliolate; calyx ca. 4 mm. long, the teeth half as long as the tube; corolla yellow, 8–9 mm. long, the banner brown or purple in age; pods 1-seeded, the short arcuate body silky, scarcely exceeding the calyx, the beak ca. as long; seed solitary, scarcely 2 mm. long; $n=7$ (Grant & Sidhu, 1967).—Dry slopes mostly below 5000 ft.; Coastal Sage Scrub, Chaparral, Pinyon-Juniper Wd.; cismontane s. Calif. to desert edge and to cent. and L. Calif. April–July.

KEY TO SUBSPECIES

Fls. 7–9 mm. long; banner-claw shorter than the blade.
  Foliage rather loosely silky-canescent; calyx-teeth ca. half as long as tube. Below Yellow Pine F.
                                                                                                 ssp. *argophyllus*
  Foliage more closely silvery-strigose; calyx-teeth almost as long as tube. Largely from Yellow Pine F. .... ssp. *decorus*

Fls. 10–12 mm. long; banner-claw scarcely shorter than the blade.
    Plant suffrutescent, erect, with densely crowded lvs. San Clemente Id. .... ssp. *adsurgens*
    Plant herbaceous, the stems decumbent to ascending.
        Calyx-teeth ca. 1.5–2.5 mm. long; corolla well exceeding the calyx. San Clemente, Santa Catalina ids. .... ssp. *ornithopus*
        Calyx-teeth ca. 3.5 mm. long; corolla well exceeding calyx. Santa Cruz Id. ... ssp. *niveus*

Ssp. **adsúrgens** (Dunkle) Raven. [*L. argophyllus* var. *adsurgens* Dunkle.] A silvery suffrutescent plant. San Clemente Id.

Ssp. **décorus** (Jtn.) Munz, comb. nov. [*Hosackia a.* var. *d.* Jtn. Bull. S. Calif. Acad. Sci. 17: 63. 1918.] Glistening satiny-canescent, at least near the tips; peduncles to 1 cm. long.—Dry slopes 4000–7000 ft.; San Gabriel and San Bernardino mts.

Ssp. **níveus** (Greene) Munz comb. nov. [*Syrmatium niveum* Greene, Bull. Calif. Acad. 2: 148. 1886.] Pubescence dense, silky; umbels nearly or quite sessile; body of pod little or not exserted.—Santa Cruz Id.

Ssp. **ornithòpus** (Greene) Raven. [*Hosackia o.* Greene. *L. a.* vars. *argenteus* and *hancockii* Dunkle. *H. venusta* Eastw.] Lfts. acute; peduncles 0.5–4 cm. long; fls. 12–20, yellow; pods conspicuously exserted, 2-seeded; $2n = 7$ pairs (Raven et al., 1965).—Santa Catalina, San Clemente, and Santa Barbara ids.

2. **L. argyraèus** (Greene) Greene ssp. **argyraèus**. [*Hosackia a.* Greene.] Prostrate perennial forming silvery mats, the stems slender, branched, from a woody base, silky pubescent, 1–3 dm. long; lfts. 3–5, cuneate-oblanceolate to -obovate, mostly 4–10 mm. long, silvery, mostly obtuse, on a very short rachis; peduncles exceeding lvs., ebracteate, 1–3-fld.; calyx-tube 2.5–3 mm. long, the teeth subulate, almost as long; corolla 8–10 mm. long, yellow, aging red; pods 1–1.5 cm. long, 2–3 mm. wide, strigose; seeds 2-nerved, roundish-oblong, olive-brown, 1.5–1.8 mm. long.—Dry slopes and benches, 3500–8200 ft.; mostly Yellow Pine F.; San Bernardino, San Jacinto, Santa Rosa mts.; n. L. Calif. May–Aug.

Ssp. **multicáulis** (Ottley) Munz. [*L. wrightii* var. *m.* Ottley.] Stems 1–2 dm. long; herbage appressed-canescent, not silvery; lfts. linear-oblong to cuneate-obovate, 5–12 mm. long; fls. 1–2; calyx-tube 3–3.5 mm. long; corolla 8–12 mm. long.—Dry places, 4000–7200 ft.; mostly Pinyon-Juniper Wd.; New York and Providence mts.

3. **L. corniculàtus** L. Perennial, with ascending to procumbent slender stems 1–5 dm. long, glabrous or ± strigose; stipules foliar; lfts. 3, 5–15 mm. long, ± obovate; peduncles ca. 1–1.4 cm. long; bract 1–3-foliolate; fls. mostly 3–6; calyx 3–4 mm. long, the teeth ca. as long as the tube; corolla 8–12 mm. long, yellow or the banner reddish; pods linear, 2–2.5 cm. long; $n = 12$ (Grant & Sidhu, 1967), $2n = 32$ (Kodama, 1967).—Occasional in lawns, on roadsides, etc. as a waif; introd. from Eu. June–Sept.

4. **L. crassifòlius** (Benth.) Greene. [*Hosackia c.* Benth.] Plate 45, Fig. F. Perennial, glabrous or puberulent, the stems stout, 4–12 dm. high, erect; stipules triangular-lanceolate, membranous; lfts. 7–15, oval, rhombic or obovate, 1–3 cm. long; peduncles 3–8 cm. long; bract usually present, remote from umbel; fls. 8–15; pedicels 2–4 mm. long; calyx-teeth short, subulate-triangular, the tube ca. 3–4 mm. long; corolla 9–12 mm. long, greenish-yellow with some purplish-red; pods 3.5–6.5 cm. long, 4–5 mm. wide; seeds 7–12, dark; $2n = 14$ (Grant, 1965).—Dry banks and flats, 2000–8000 ft.; Yellow Pine F.; Red Fir F.; mts. of s. Calif. mts. n. to Wash. May–Aug.

5. **L. davidsònii** Greene. [*Syrmatium d.* Heller.] Much like *L. nevadensis*, but more silvery when young; fls. 4–8; bract unifoliolate; calyx 3 mm. long, the subulate teeth almost as long as the tube; corolla sulphur-yellow, aging red, 5–8 mm. long, the banner obovate, its blade at a right angle to the wings or reflexed; body of pod ca. as long as calyx, 1-seeded, with long curved beak.—Common on dry slopes, 4500–9000 ft.; Montane Coniferous F.; Santa Rosa Mts. to Mt. Pinos, largely replacing *L. nevadensis* in this area. May–July.

6. **L. grandiflòrus** (Benth.) Greene. [*Hosackia g.* Benth.] Perennial from a ± woody base, 2–6 dm. high, strigose throughout; stipules glandlike; lfts. mostly 7 or 9, obovate to elliptical, 0.7–2 cm. long; peduncles 4–8 cm. long; umbels 2–several-fld.; bract 1-, some-

times 3-foliolate, 1–2 cm. long; calyx-tube ca. 5 mm. long, the subulate teeth almost as long; corolla yellow, aging red, 1.5–2.4 cm. long, the claws not exceeding calyx; pods 3–4 cm. long; seeds many, ovoid, dark brown; $2n = 7$ pairs (Raven et al., 1965).—Dry slopes below 6000 ft.; Yellow Pine F., Chaparral, etc.; cismontane n. to n. Calif. April–July. Plants with more spreading short hairs are the var. *mutabilis* Ottley.

7. **L. hamàtus** Greene. Like *L. nuttallianus*, but umbels sessile or nearly so; bractless; fls. 3–4.5 mm. long; calyx-teeth subulate, shorter than tube; banner shorter than keel; pods strigose, even on the beak.—Dry slopes below 5000 ft.; Coastal Sage Scrub, Chaparral; n. L. Calif. to San Luis Obispo Co. March–June.

8. **L. haydònii** (Orcutt) Greene. [*Hosackia h.* Orcutt. *L. spencerae* Macbride.] Bushy erect perennial, 2–4 dm. high, with many slender wiry branches, sparsely strigose; lfts. 3, elliptic, 2–5 mm. long, the lvs. rather remote; fls. 1–2 on very short axillary peduncles without bracts; calyx strigose, 2.5 mm. long, the broadly linear teeth shorter than the tube; corolla yellow, 4–5 mm. long, the wings shorter than the keel; pods curved, ca. 5 mm. long; seed solitary, ca. 2.5 mm. long.—Dry rocky places, 2000–4000 ft.; Pinyon-Juniper Wd. to Creosote Bush Scrub; sw. Colo. Desert and adjacent L. Calif. March–June.

9. **L. heermánnii** (Dur. & Hilg.) Greene. [*Hosackia h.* Dur. & Hilg.] Perennial, with several to many prostrate stems 3–10 dm. long, much-branched and forming mats; herbage villous especially when young; lvs. 1–3 cm. long; lfts. 4–6, oblanceolate to obovate, 5–12 mm. long; umbels 4–10-fld., on peduncles up to 5 mm. long; bract unifoliolate; calyx ca. 3 mm. long, villous, the setaceous teeth ca. as long as tube; corolla 3–5 mm. long, yellow, aging red; pods usually 1-seeded, arcuate, villous, with a long incurved beak.—Moist banks and canyons, below 6000 ft.; Coastal Sage Scrub, Chaparral, etc. mostly in the interior and to desert edge; from San Diego Co. to cent. Calif. Plants from near the coast with more villous covering are the var. *eriophorus* (Greene) Ottley. [*L. e.* Greene.] March–Oct.

10. **L. humistràtus** Greene. Much like *L. subpinnatus*, but densely villous throughout; calyx-lobes ca. twice as long as the tube; pods 5–10 mm. long, densely villous; $n = 6$ (Grant & Sidhu, 1967).—Common below 6000 ft.; many Plant Communities; cismontane and occasional on deserts; to L. Calif., New Mex. March–June.

11. **L. micránthus** Benth. Diffuse slender-stemmed annuals 1–3 dm. high, glabrous or sparsely strigose; lvs. 1–1.5 cm. long; lfts. 3–5, oblong to oblanceolate or elliptical, 3–10 mm. long; peduncles shorter than to longer than lvs., 1-fld.; bract of 1–3 lfts.; calyx-tube turbinate-campanulate, ca. 1 mm. long, the subulate teeth shorter; corolla 4–5 mm. long, pinkish or pale salmon, turning red; pods 1.5–2 cm. long, ca. 2 mm. wide, constricted between seeds, glabrous; seeds less than 2 mm. long.—Open places below 5000 ft.; many Plant Communities; cismontane; to Wash. March–May.

12. **L. nevadénsis** Greene. [*Syrmatium n.* Greene.] Plate 46, Fig. A. Prostrate perennial with many slender wiry branches forming mats 3–8 dm. across; herbage villous-strigose, later glabrate; lfts. 3–5, obovate, 7–15 mm. long; lower peduncles 1–2.5 cm. long; umbels 1–several-fld.; bract 1–3-foliolate; calyx-tube ca. 2 mm. long, the teeth subulate, ca. 1 mm. long; corolla yellow, or tinged red, 5–7 mm. long, wings at least as long as keel; pods abruptly arcuate, ca. 6 mm. long, the slender curved beak longer than the body; seed 1, slender; $n = 7$ (Grant & Sidhu, 1967).—Dry slopes and benches, 3500–8500 ft.; Montane Coniferous F.; mts. from L. Calif. to Sierra Nevada, including Panamint, Argus, etc.; to w. Nev. May–Aug.

13. **L. nuttalliànus** Greene. Prostrate, annual or of longer duration, the branches slender, 3–8 dm. long, strigose when young; herbage thinly hirsutulose, later glabrate; lfts. 3–5 (–7), oblong-obovate, 4–12 mm. long; peduncles slender, 5–25 mm. long, usually with a unifoliolate bract; fls. 1–9; calyx 2–2.5 mm. long, the teeth triangular, 0.5 mm. long; corolla yellow, 5–7 mm. long, often red on banner and wing-tips; pods much longer than calyx, slender, arcuate, constricted between the 2 seeds, the slender beak shorter than the body; seeds ca. 2.5 mm. long.—Sandy places below 100 ft.; Coastal Strand, Coastal Sage Scrub; s. San Diego Co.; n. L. Calif. March–June.

Plate 46. Fig. A, *Lotus nevadensis*. Fig. B, *Lotus oblongifolius*. Fig. C, *Lotus rigidus*. Fig. D, *Lotus salsuginosus*. Fig. E, *Lupinus arizonicus*. Fig. F, *Lupinus densiflorus*.

14. **L. oblongifòlius** (Benth.) Greene. [*Hosackia o.* Benth.] Plate 46, Fig. B. Perennial with slender rootstocks, erect or ascending, 2–5 dm. high, ± strigose; stipules membranaceous, lanceolate; lfts. mostly 7–11, lance-linear to elliptical, acute, 5–20 mm. long; peduncles 5–12 cm. long; bract 1–3-foliolate, closely subtending the 1–5-fld. umbel; calyx-tube 2–3 mm. long, the teeth narrowly subulate, almost as long; corolla 9–14 mm. long, banner yellow, ± veined with purple; pods 2.5–4 mm. long, 1.5–2 mm. wide; seeds mottled, 1.5–2 mm. long.—Wet places below 8500 ft.; Chaparral, Montane Coniferous F.; cismontane and Panamint Mts. to cent. Calif. May–Sept.

15. **L. procúmbens** (Greene) Greene. [*Hosackia p.* Greene. *L. leucophyllus* Greene.] Silky-canescent, suffrutescent, much-branched, the stems procumbent to ascending, 3–8 dm. long; lfts. 3, narrowly elliptical, 5–12 mm. long; umbels sessile or nearly so, scattered, mostly 1–3-fld.; calyx-tube ca. 3 mm. long, the triangular-subulate teeth ca. 1 mm. long; corolla 6–8 mm. long, yellow, with exserted claws; pods reflexed, exceeding calyx, straight or curved, strigose; seeds 2 or more.—Dry slopes at 3000–7000 ft.; Creosote Bush Scrub to Pinyon-Juniper Wd.; desert slopes from Little San Bernardino Mts. to Kern Co., then to San Benito Co. April–June.

16. **L. purshiànus** (Benth.) Clem. & Clem. [*Hosackia p.* Benth. *H. americana* Benth.] Erect or ascending much-branched annual, 1.5–8 dm. tall, glabrate to villous; lvs. 1–2.5 cm. long, mostly 3-foliolate; lfts. 10–15 mm. long; peduncles 10–15 mm. long, 1-fld.; bract 1-foliolate; calyx-tube ca. 1.5 mm. long, the subulate teeth longer, corolla whitish, tinged pink, 4–7 mm. long; pods 1.5–2.5 cm. long, 2–3.5 mm. wide; seeds ca. 3 mm. long; $2n=14$ (Larsen, 1956).—Common in dry fields and disturbed places, below 7000 ft.; many Plant Communities; cismontane and sometimes on the desert; to B.C., Dak., Mex. May–Oct. A ± decumbent form, pilose to nearly glabrous and with the corolla 4–5 mm. long, occurs mostly in Yellow Pine F. and is known as var. *glaber* (Nutt.) Munz. [*Hosackia elata* var. *g.* Nutt.]

17. **L. rígidus** (Benth.) Greene. [*Hosackia r.* Benth. *L. argensis* Cov.] Plate 46, Fig. C. Perennial, erect, usually several-stemmed, woody at base, coarse, 3–9 dm. tall; lvs. remote, 1–2 cm. long, with 3–5 narrowly oblong lfts.; peduncles 5–12 cm. long, 2–3-fld.; bract 0 or 1-foliolate; calyx-tube 4–5 mm. long, strigose, the teeth subulate, nearly as long; corolla 1.3–2 cm. long, yellow, with some red or purple in age; pods subglabrous, 3–4 cm. long, ca. 3 mm. wide; seeds granulose, ca. 1.5 mm. in diam.; $2n=7$ pairs (Raven et al., 1965).—Common on dry slopes and in washes below 5000 ft.; Creosote Bush Scrub to Pinyon-Juniper Wd.; deserts from Inyo Co. s.; to L. Calif., Utah. March–May.

18. **L. salsuginòsus** Greene ssp. **salsuginòsus.** Plate 46, Fig. D. Decumbent or prostrate annual, the stems 1–3 dm. long; herbage subglabrous to ± strigose; lvs. slightly succulent, to 4 cm. long; lfts. 5–8, obovate to roundish, 4–14 mm. long; peduncles 1–4 cm. long; bract unifoliolate, broadly ovate; calyx-tube ca. 3 mm. long, the teeth broadly subulate, slightly shorter; corolla 6–9 mm. long, yellow; pod 1.5–3 cm. long, 1.5–2 mm. wide, glabrous; seeds smooth, 1–1.5 mm. long; $2n=7$ pairs (Raven et al., 1965).—Dry slopes and fields below 3500 ft.; Coastal Sage Scrub, Chaparral; San Diego Co. and n. L. Calif to cent. Calif.; Channel Ids. March–June.

Ssp. **brevivexíllus** (Ottley) Munz, comb. nov. [*L. s.* var. *brevivexillus* Ottley, Univ. Calif. Pub. Bot. 10: 217. 1923. *Hosackia humilis* Greene.] Corolla 3–5 mm. long, the keel exposed by the short wings and banner; pod more constricted between the seeds.—Sandy places below 6000 ft.; Creosote Bush Scrub; deserts from Panamint Mts. to L. Calif., Ariz.

19. **L. scopàrius** (Nutt. in T. & G.) Ottley ssp. **scopàrius.** [*Hosackia s.* Nutt. *Syrmatium glabrum* Vog. *L. s.* var. *dendroideus* (Greene) Ottley. *Syrmatium d.* Greene. *S. patens* Greene. *L. s.* var. *veatchii* (Greene) Ottley.] Bushy, suffruticose, ± suberect, 2–12 dm. high, with rather virgate green branches or ± silky; lfts. mostly 3, oblong to oblanceolate, 4–10 or more mm. long; umbels sessile in axils or short-peduncled, 1–5-fld.; calyx commonly 3 mm. long; corolla 7–10 (–12) mm. long, yellow or tinged red, the wings as long as keel; pods well exceeding calyx, to 1.5 cm. long, glabrous, tapering to a subulate beak; seeds 2–3, dark brown; $n=7$ (Grant & Sidhu, 1967).—Common on dry slopes and fans, especially as a successional stage after burns, below 5000 ft.; Coastal Sage

Scrub, Chaparral, Coastal Strand, etc.; cismontane from L. Calif. n. to Humboldt Co.; Channel Ids. March–Aug.

Ssp. **brevialàtus** (Ottley) Munz, comb. nov. [*L. s.* var. *b.* Ottley, Univ. Calif. Pub. Bot. 10: 229. 1923.] A slender-stemmed form, the wings shorter than the keel.—Chaparral, Joshua Tree Wd.; interior cismontane valleys, to w. edge of deserts, from San Diego Co. to Kern Co.

Ssp. **tráskae** (Eastw. ex Abrams) Raven. [*Syrmatium t.* Eastw. ex Abrams.] Pods 3–5 cm. long, with 4–8 seeds. San Clemente Id.

20. **L. strigòsus** (Nutt. in T. & G.) Greene. [*Hosackia s.* Nutt. *L. rubellus* (Nutt.) Greene.] Slender-stemmed, decumbent to ascending annual, the branches to 3 dm. long; herbage strigose; lvs. 1–2.5 cm. long, with flattened rachis; lfts. 6–10, linear-oblong to elliptic, 5–12 mm. long; lower peduncles 1-fld., upper 2–3-fld.; bract 0 or 1–3-foliolate; calyx-tube ca. 2 mm. long, the teeth subulate, almost as long; corolla mostly 6–10 mm. long, yellow turning reddish; pods 2–3 cm. long, 2–3 mm. wide, strigose; seeds quadrate, irregularly rugose, ca. 1 mm. long; $2n = 7$ pairs (Raven et al., 1965).—Common in dry disturbed places below 5000 ft.; many Plant Communities; cismontane from L. Calif. to cent. Calif. and inland to edge of deserts. March–June. A form with spreading pubescence occurs with the common form, but mostly in the interior and is var. *hirtellus* (Greene) Ottley.

21. **L. subpinnàtus** Lag. [*Hosackia s.* T. & G. *L. wrangelianus* F. & M.] Annual with decumbent or ascending stems 1–3 dm. long, subglabrous to strigose; lfts. 3–5, obovate, entire, 5–15 mm. long; fls. subsessile, solitary in axils; calyx-tube ca. 2 mm. long, the teeth ca. as long; corolla yellow, tinged red-purple in age, keel with attenuate beak; pod 10–15 mm. long, ca. 3 mm. wide, compressed, sparsely strigose; seeds 1.5–2 mm. long, notched at hilum; $n = 6$ (Grant & Sidhu, 1967).—Dry grassy slopes below 2500 ft.; many Plant Communities; most of cismontane Calif.; Chile. March–June.

22. **L. tomentéllus** Greene. [*Hosackia t.* Abrams.] Prostrate; pubescence canescently tomentulose; lvs. to ca. 2 cm. long, with 4–8 thick, cuneate-oblanceolate, obovate or oblong lfts. 3–10 mm. long; peduncles mostly shorter than subtending lvs.; umbels 1–2-fld.; bract 0 to 1–2-foliolate; calyx-tube 2–2.5 mm. long, the teeth lanceolate, somewhat shorter; corolla 5–7 mm. long, yellow, mostly lacking red on back, but aging red; pod 1–2 cm. long, 2 mm. wide; seeds globose to ovoid, rarely cubical, granulose, ± mottled.—Sandy places below 4500 ft.; Creosote Bush Scrub; deserts from Inyo Co. s. to L. Calif.; Ariz., Son., Nev. Feb.–May.

15. *Lupìnus* L. LUPINE

Annual or perennial herbs, or shrubs. Lvs. alternate, palmately compound, rarely unifoliolate. Stipules narrow. Fls. perfect, in terminal racemes. Calyx bilabiate, the lips entire or toothed, or the uppermost sometimes bifid, often with interstitial bracteoles. Corolla usually bluish, sometimes reddish, whitish, or yellow; banner commonly with reflexed sides, the back glabrous or ± pubescent; wings mostly glabrous; keel curved to almost straight, glabrous or ciliate on upper edges, sometimes on lower. Stamens 10, monadelphous, the anthers alternately of 2 forms, elongate and short. Pods flat, 2–12-ovuled. Seeds with a sunken hilum, which is thus often surrounded by a thickened ring. (Latin, from *lupus*, a wolf, because of an old idea that lupines rob the soil.)

1. Plants annual.
  2. Fls. definitely whorled or mostly so.
    3. Keel ciliate on upper margins near the claws.
      4. The keel ciliate on lower edges also, near the claws.
        5. Plants 0.5–1.5 dm. tall; peduncles 2–4 cm. long.
          6. Fls. ascending to suberect at anthesis, 10–11 mm. long. San Bernardino Co. n.
                                                                                                 28. *L. horizontalis*
          6. Fls. ± spreading, 6–7 mm. long. Riverside Co. s. . . . . 13. *L. brevior*
        5. Plants 2–10 dm. tall; fls. spreading in anthesis, 12–14 mm. long; peduncles mostly 3–15 cm. long.

Lupinus

```
        7. Fls. yellowish; pods ca. 1.5 cm. long ........................ 31. L. luteolus
        7. Fls. deep purple-blue, sometimes pink or white; pods 4–5 cm. long
                                                                          47. L. succulentus
     4. The keel not ciliate on lower edges.
        8. Banner 8–9 mm. long, 2–6 mm. wide; fls. suberect at anthesis ........ 42. L. ruber
        8. Banner 12–16 mm. long, 8–9 mm. wide; fls. spreading at anthesis.
           9. Fls. ± erect after anthesis, not secund, purple-red to rose-pink ... 46. L. subvexus
           9. Fls. spreading or secund after anthesis, white or yellowish, sometimes violet or rose
                                                                          20. L. densiflorus
  3. Keel ciliate on upper margins toward the apex or not ciliate at all.
    10. Pedicels slender, 3–9 mm. long; banner 8–15 mm. long, mostly broader than long
                                                                          33. L. nanus
    10. Pedicels 1–3 mm. long; banner largely 4–8 mm. long.
        11. Keel slender with long narrow ± upturned beak; reflexed portion of banner
            generally as long as the straight part ........................ 11. L. bicolor
        11. Keel short, broad, almost straight; reflexed portion of banner much shorter than
            the straight part ........................................ 39. L. polycarpus
2. Fls. not at all whorled.
  12. Keel not ciliate.
    13. Cotyledons petioled; ovules 2–4.
       14. Lfts. linear-spatulate, mostly subglabrous above; plants sparsely spreading-pilose;
           fls. bluish-purple ........................................ 2. L. agardhianus
       14. Lfts. spatulate-oblanceolate to obovate, amply pubescent above; plants abundantly
           spreading-pilose; fls. pinkish-white to lavender ................ 18. L. concinnus
    13. Cotyledons sessile and perfoliate; ovules usually 2 except in *odoratus*.
       15. Fls. mostly 5–8 mm. long; ovules 2.
           16. Pods constricted in middle; peduncles ca. 1 cm. long, the racemes not surpassing
               the foliage. Inyo Co. n. .................................. 41. L. pusillus
           16. Pods scarcely or not constricted; peduncles 3–10 cm. long, the racemes surpassing
               the lvs. Inyo Co. s.
               17. Racemes usually subcapitate, 1–2.5 cm. long; pods villous on sides; banner
                   longer than wide ...................................... 12. L. brevicaulis
               17. Racemes mostly 3–8 cm. long.
                   18. Pods loosely villous on sides; banner suborbicular ....... 23. L. flavoculatus
                   18. Pods scaly on sides; banner longer than wide ............ 43. L. shockleyi
       15. Fls. ca. 10 mm. long; ovules 2–6 ............................ 35. L. odoratus
  12. Keel ciliate at least on lower margin near the base.
    19. The plant covered with long stinging hairs; fls. reddish-violet, 13–15 mm. long; lfts.
        12–25 mm. broad ........................................... 26. L. hirsutissimus
    19. The plant not covered with stinging hairs; lfts. 2–12 mm. broad.
       20. Banner bright yellow, the wings rose to purple; keel whitish; racemes shorter than
           the peduncles ............................................ 45. L. stiversii
       20. Banner and other petals not so colored, or if so, with short peduncles.
          21. Keel stout, with short blunt acumen; plant subglabrous to sparsely strigulose
                                                                          48. L. truncatus
          21. Keel with slender acute acumen; plant usually ± villous.
             22. Petals blue, lilac or purple.
                23. Lfts. linear-filiform, 2–5 cm. long; bracts much exceeding the fl.-buds
                                                                          10. L. benthamii
                23. Lfts. linear to oblanceolate, 1–3 cm. long; bracts scarcely exceeding the fl.-
                    buds.
                   24. Fls. blue to lilac, 10–17 mm. long; lfts. mostly 1.5–3 mm. wide.
                       25. Pedicels mostly 3–5 mm. long; fls. 10–12 mm. long; pods 1–2 cm. long.
                           Widely distributed ............................. 44. L. sparsiflorus
                       25. Pedicels 6–12 mm. long; fls. 14–16 mm. long; pods 5–6 cm. long. San
                           Clemente Id. .................................. 25. L. guadalupensis
                   24. Fls. pale purplish-pink, 8–10 mm. long; lfts. mostly 4–12 mm. wide.
                       Deserts ........................................... 9. L. arizonicus
             22. Petals pale blue to whitish or yellowish, 6–7 mm. long. Deserts . 36. L. pallidus
1. Plants perennial.
  26. The plants shrubby or with definite woody stems of some length above the roots.
```

Fabaceae

27. The banner ± pubescent on the back; plants mostly appressed-silky; keel narrow with long slender acumen.
   28. Keel not ciliate on the edges or with only an occasional hair; petioles 2–3.5 cm. long. Coastal dunes and beaches .... 17. *L. chamissonis*
   28. Keel definitely ciliate on upper edge toward acumen; petioles 2–8 cm. long. Away from the immediate coast.
      29. Plants appressed-silky; fls. 10–13 mm. long, or if more, plant strigose; keel narrowed toward the base .... 3. *L. albifrons*
      29. Plants with appressed pubescence to tomentose; fls. 10–18 mm. long; keel not narrowed toward the base .... 22. *L. excubitus*
27. The banner glabrous on the back; keel mostly not markedly slender at the acumen.
   30. Herbage ± silvery; racemes mostly 1–2 dm. long above the peduncles.
      31. Petioles 2–6 cm. long; petals mostly yellow, sometimes lavender or blue. Immediate coast from Ventura Co. n. .... 6. *L. arboreus*
      31. Petioles 4–10 cm. long; petals blue to violet or orchid. Mostly away from the coast 22. *L. excubitus*
   30. Herbage greenish; racemes mostly 2–4 dm. long above the peduncles. Ventura Co. to L. Calif. .... 30. *L. longifolius*
26. The plants strictly herbaceous, i.e. stems not woody above the root-crown or caudex.
   32. Lvs. glabrous on upper surface or essentially so.
      33. Keel naked on upper edge; plants 5–15 dm. tall; corolla 12–14 mm. long. Plants of wet places .... 40. *L. polyphyllus*
      33. Keel ciliate on upper edges.
         34. Lfts. 1.5–4.5 cm. long, 2–5 mm. wide; stems 3–6 dm. tall; banner ± pubescent on back .... 8. *L. argenteus*
         34. Lfts. 3–10 cm. long, 5–30 mm. wide; stems 5–20 dm. tall; banner glabrous
                                                                              29. *L. latifolius*
   32. Lvs. ± hairy above.
      35. Banner glabrous on back.
         36. Lvs. mostly in dense basal clusters, the cauline few if any; plants cespitose.
            37. Racemes not exceeding lvs., but mostly down among them; fls. pale, 7–8 mm. long. White Mts. .... 15. *L. caespitosus*
            37. Racemes extending well above the lvs., 5–30 cm. long; fls. violet-purple, 10–14 mm. long. San Bernardino Mts. and Mt. Pinos .... 19. *L. confertus*
         36. Lvs. well distributed along the stems.
            38. Keel ciliate on upper edges at least toward tip.
               39. Bracts of infl. quite persistent, usually linear, mostly rather conspicuous, especially before anthesis .... 19. *L. confertus*
               39. Bracts of infl. early deciduous, mostly rather inconspicuous.
                  40. Lvs. densely woolly, with ± curly and interwoven hairs, the lfts. ca. 7–10 mm. wide. Inyo Co. .... 32. *L. magnificus*
                  40. Lvs. not woolly, but silky-strigose to strigulose.
                     41. Fls. 10–12 mm. long; plant long-villous .... 34. *L. nevadensis*
                     41. Fls. 13–18 mm. long; plant silky-strigose.
                        42. Lvs. silvery, crowded near base of plant; petioles 4–10 cm. long; fls. 14–18 mm. long. Kern Co. to L. Calif. . 22. *L. excubitus austromontanus*
                        42. Lvs. greenish, silky, not silvery, well distributed; petioles to 15 cm. long; fls. 13–15 mm. long. E. Inyo Co. .... 27. *L. holmgrenanus*
            38. Keel glabrous on upper edges or practically so.
               43. Plants less than 2 dm. high, matted, with ± prostrate stems from a woody base; fls. 4–9 mm. long. Montane .... 14. *L. breweri*
               43. Plants normally taller, erect or ascending; fls. mostly larger.
                  44. Fls. yellow, 1–1.2 cm. long, the banner obovate .... 1. *L. adsurgens*
                  44. Fls. blue or violet to yellowish-white.
                     45. Lvs. thinly strigose above, greenish.
                        46. Fls. 10–12 mm. long. San Bernardino Mts. and n. .... 5. *L. andersonii*
                        46. Fls. 13–16 mm. long. San Gabriel Mts. to San Jacinto Mts.
                                                24. *L. formosus* var. *hyacinthinus*
                     45. Lvs. ± silky-strigose above.
                        47. The lfts. silvery-silky; fls. 10–14 mm. long. Mt. Pinos and San Gabriel Mts. .... 21. *L. elatus*

47. The lfts. more loosely silky, less shiny.
   48. Fls. 10–14 mm. long, the banner obovate; racemes mostly 4–10 cm. long. San Diego Co. to n. Calif. .... 1. *L. adsurgens*
   48. Fls. 12–18 mm. long, the banner roundish; racemes mostly 10–25 cm. long. L. Calif. n. .... 24. *L. formosus*
35. Banner pubescent on the back, the pubescence sometimes almost concealed by the upper lip of the calyx.
  49. Calyx evidently spurred at the base just above the pedicel.
    50. Wing-petals with a patch of short dense pubescence on outer surface near upper distal corner .... 7. *L. arbustus*
    50. Wing-petals not pubescent or with a few scattered hairs.
      51. Pubescence appressed, the plants silky-strigose; fls. 10–13 mm. long
                                                      16. *L. caudatus*
      51. Pubescence spreading; fls. 8–10 mm. long .... 37. *L. palmeri*
  49. Calyx not spurred, sometimes ± gibbous at base.
    52. Plants matted, less than 2 dm. tall, with silvery-silky foliage; fls. 6–9 mm. long
                                                            14. *L. breweri*
    52. Plants not matted, taller; fls. mostly larger.
      53. Lfts. 10–30 mm. wide; fls. yellow, 10–12 mm. long. .... 38. *L. peirsonii*
      53. Lfts. 3–10 mm. wide.
        54. Fls. 14–18 mm. long. Mts. of s. Calif. ... 22. *L. excubitus* var. *austromontanus*
        54. Fls. 9–12 mm. long. Panamint Mts. n. .... 4. *L. alpestris*

1. **L. adsúrgens** E. Drew. Subappressed to appressed-pubescent, ± silky, 2–6 dm. tall, erect or nearly so; petioles 1–4 (–7) cm. long; lfts. 5–8, oblanceolate, 2–3.5 cm. long; peduncles 2–6 cm. long; racemes 4–10 cm. long; bracts ca. 5 mm. long; pedicels 2–4 mm. long, spreading-pubescent; upper calyx-lip notched or entire, lower entire; fls. 10–12 mm. long, pale yellow or blue to lilac, the banner obovate, glabrous, keel strongly curved, nonciliate; pods strigose, 2.5–3 cm. long; seeds 4–6, ca. 5 mm. long.—Dry slopes to ca. 5000 ft.; Chaparral, Yellow Pine F.; San Diego Co. to Ore. May–June.

2. **L. agardhiànus** Heller. [*L. concinnus* var. *a.* C. P. Sm.] Resembling *L. concinnus*, but lfts. linear-spatulate, mostly subglabrous above; plants sparsely spreading-pilose; fls. bluish-purple.—Sandy places; Coastal Sage Scrub to Yellow Pine F.; San Diego Co. to Santa Barbara Co. San Clemente, Santa Cruz and Santa Rosa ids.

3. **L. albifróns** Benth. var. **albifróns.** Rounded leafy shrub, appressed-silky, much-branched, 6–15 dm. high; petioles 2–4 cm. long; lfts. 7–10, oblanceolate to spatulate, 1–3 cm. long, 4–10 mm. wide, silvery-silky on both sides; peduncles 5–13 cm. long; racemes 8–30 cm. long, the fls. mostly whorled, 10–14 mm. long; bracts to ca. 5 mm. long, deciduous; pedicels 4–8 mm. long, with spreading hairs; upper calyx-lip cleft, lower entire, ca. 5–6 mm. long; petals blue to lavender or red-purple, the banner ± pubescent on back, with a light center; keel ciliate toward apex; pod 3–5 cm. long, villous-strigose; seeds 5–9; $2n = 24$ pairs (Raven et al., 1965).—Sandy places below 5000 ft.; many Plant Communities; Ventura Co. n. March–June.

Var. **doúglasii** (Agardh) C. P. Sm. Bracts mostly 10–15 mm. long, exceeding fl.-buds. Santa Cruz Id., San Nicolas Id.; n. Santa Barbara Co. to Marin Co.

Var. **éminens** (Greene) C. P. Sm. [*L. e.* Greene. *L. brittonii* Abrams.] Fls. mostly 14–16 mm. long; plants 10–25 dm. high.—Mostly below 3500 ft.; Chaparral, Coastal Sage Scrub, etc.; n. L. Calif. and San Diego Co. n.

4. **L. alpéstris** A. Nels. [*L. munzii* Eastw. *L. funstonianus* C. P. Sm.] Rather finely strigose, ± canescent perennial, with many leafy branched stems 5–7 dm. high; lowermost petioles 8–10 cm. long, upper 2–3 cm.; lfts. 5–8, unequal, linear-elliptic, 2–4 cm. long, 4–6 mm. wide; peduncles 2.5–3.5 cm. long; bracts 2.5–3 mm. long, deciduous; pedicels slender, 4–5 mm. long; upper calyx-lip ca. 4 mm. long, cleft, lower 4.5 mm., often gibbous at base, subentire; corolla 9–12 mm. long, light blue, the banner suborbicular, pubescent on back, keel somewhat curved, ciliate on upper edge toward slender apex.—Dry rocky places, 6500–10,800 ft.; Pinyon-Juniper Wd., Bristle-cone Pine F.; Panamint Mts.; to Mont., Colo., Ariz. June–July.

5. **L. andersònii** Wats. Herbaceous from a heavy root-crown, the stems slender, erect, branched above, 4–9 dm. high, leafy, thinly strigose; stipules linear; petioles 2–4 cm. long; lfts. 5–9, oblanceolate; racemes 6–18 cm. long, rather lax; fls. 10–12 mm. long; petals blue, purplish or yellowish-white; banner rounded, glabrous; wings covering most of the curved nonciliate keel; pods strigose, 3–4 cm. long; seeds 4–6.—Dry slopes under pines, 4000–6500 ft.; San Bernardino Mts., Mt. Pinos; to Ore., Nev. June–Sept.

6. **L. arbòreus** Sims. From suffrutescent to shrubby, mostly 1–2 m. high, with many short branches; herbage puberulent, or pubescent, to ± silky; petioles 2–6 cm. long; lfts. 5–12, oblanceolate, 2–6 cm. long, 5–10 mm. wide, strigose above and beneath or only beneath; peduncles 4–10 cm. long; racemes 10–30 cm. long, mostly lax, the fls. scattered or subverticillate, 14–17 mm. long; upper calyx-lip notched, the lower entire, 5–7 mm. long; petals broad, mostly yellow, sometimes lilac or blue or mixed, the banner roundish, glabrous, keel curved, ciliate on upper edges; pods brown, strigose, 4–7 cm. long; seeds 8–12; $2n = 40$ (Savchenko, 1935).—Sandy places below 100 ft.; Coastal Strand, Coastal Sage Scrub; Ventura Co. to Del Norte Co. and n. March–Aug.

7. **L. arbústus** Dougl. var. **montànus** (Howell) D. Dunn. [*L. laxiflorus* var. *m.* Howell.] Perennial with several erect stems 3–6 dm. high, strigose especially in upper parts; lvs. scattered, the lfts. 7–9, oblanceolate, 2–4 cm. long, strigose on both sides, but quite green; peduncles 2–4 cm. long; racemes 6–12 cm. long; bracts 3–4 mm. long; pedicels 2–3 mm. long; upper calyx-lip 2-toothed, lower 3-toothed; fls. 10–13 mm. long, the banner broad and largely hiding upper calyx-lip; keel ciliate at short pointed apex.—Reported from n. slope of San Gabriel Mts.; to Ore.

8. **L. argénteus** Pursh var. **tenéllus** (Dougl. ex D. Don) D. Dunn. [*L. t.* Dougl.] Short-strigose perennial, erect, branched above, 3–6 dm. tall; lvs. well distributed, lfts. 5–9, linear-oblanceolate, strigose to sericeous above, 1.5–4.5 cm. long; peduncles 2–3 cm. long; racemes 7–12 cm. long, ± lax; pedicels 3–6 mm. long; fls. 8–10 mm. long, blue or lilac; pods strigose-silky, 2–2.5 cm. long.—Dry rocky slopes, 9400–10,500 ft.; Bristle-cone Pine F.; Inyo-White Mts.; to Ore., Colo. July–Aug.

9. **L. arizónicus** (Wats.) Wats. [*L. concinnus* var. *a.* Wats. *L. sparsiflorus* var. *a.* C. P. Sm.] Plate 46, Fig. E. Branched fleshy annual, ± strigulose and with some spreading hair, ± fistulose, 3–6 dm. tall, light green; petioles 2–8 cm. long; lfts. 5–10, broadly oblanceolate, 1–3.5 cm. long, 4–12 mm. wide, thickish, rounded at apex, glabrous or nearly so above, ± villous beneath; peduncles 1–5 cm. long; racemes 5–30 cm. long, with fls. scattered, 8–10 mm. long, pale purplish-pink, often drying violet; pedicels ca. 2–3 mm. long; calyx 4–5 mm. long; banner roundish, keel ciliate near base on lower or both edges.—Common in sandy washes and open places, below 2000 ft.; Creosote Bush Scrub; deserts n. to e. Inyo Co.; Nev., Ariz. March–May.

10. **L. benthàmii** Heller. Erect villous annual, simple or branched, 3–6 dm. tall; petioles slender, 6–12 cm. long; lfts. 7–10, linear, 2–5 cm. long, villous; peduncles 5–10 cm. long; racemes 10–20 cm. long; bracts linear, longer than the buds; pedicels 3–6 mm. long; fls. 10–15 mm. long, light to deep blue; calyx mostly 4–5 mm. long; banner roundish, ca. 11–14 mm. long, with a yellow spot; keel curved, ciliate on lower edges near claws; pods 2–3.5 cm. long, ascending, 3–9-seeded.—Open slopes below 4000 ft.; V. Grassland; Foothill Wd.; n. Los Angeles and Kern cos. through interior valleys to Sacramento Co. March–May. Robust plants with hollow stems and fls. 14–15 mm. long are the var. *opimus*, C. P. Sm.

11. **L. bicólor** Lindl. ssp. **marginàtus** D. Dunn. Annual, the stems 1–several from the base, erect, 1–4 dm. tall, villous; petioles 2–7 cm. long; lfts. 5–7, oblanceolate to cuneate, 1–3 cm. long; peduncles 3–7 cm. long; racemes 1–7 cm. long, with 3–9 whorls; pedicels 1.5–3 mm. long; bracts subulate, deciduous, 4–6 mm. long; lower calyx-lip entire or nearly so; banner 4.5–7.5 mm. long, truncate or emarginate at apex, obovate, blue with cent. light spot; keel with a slender acumen, ciliate on upper edge near tip; pod strigose, 1.5–2 cm. long.—V. Grassland, Joshua Tree Wd., Yellow Pine F.; mt. valleys along w. edge of deserts, n. L. Calif. to Kern Co. March–June.

KEY TO SUBSPECIES OF L. BICOLOR

Lower calyx-lip deeply cleft, the teeth 1–5 mm. long; banner obovate, 6–9 mm. long; plants erect to decumbent. San Diego Co. n. .... ssp. *umbellatus* (Greene) Dunn
Lower calyx lip entire or with teeth not more than 1 mm. long.
    Banner oval, lemon-shaped when flattened, 3.6–6 mm. long, mucronate; keel with short blunt acumen. Open places below 5000 ft.; Coastal Sage Scrub to Yellow Pine F.; cismontane s. Calif. and on ids.; to Ore. .... ssp. *microphyllus* (Wats.) D. Dunn
    Banner obovate to oblong; keel with slender acumen.
        Banner 4.5–7.5 mm. long, truncate to emarginate at apex; whorls 3–9. Mt. valleys of s. Calif.
                                                                 ssp. *marginatus* D. Dunn
        Banner 5.7–8.6 mm. long, mostly rounded at apex; whorls mostly 1–2. San Diego Co. to Humboldt Co. .... ssp. *umbellatus* (Greene) D. Dunn

12. **L. brevicáulis** Wats. Densely villous annual, to 1 dm. tall, the main stem scarcely 1 cm. long; petioles 3–7 cm. long; lfts. 5–8, spatulate, 0.5–1.5 cm. long, glabrous above, villous beneath; peduncles 3–9 cm. long, spreading to subprostrate, racemes subcapitate, 1–2.5 cm. long; fls. 6–8 mm. long; bracts persistent, 2–3 mm. long; calyx villous, the upper lip truncate to bifid, 1–2 mm. long, the lower 4–6 mm. long, entire to tridentate; petals bright blue or yellowish-white on lower half, the banner ca. 6 mm. long, rounded or angled at apex, keel straight and nonciliate; pods ovate, ca. 1 cm. long.—Sandy places, 4000–7700 ft.; Pinyon-Juniper Wd., Sagebrush Scrub, Joshua Tree Wd.; Mojave Desert from e. San Bernardino Co. to White Mts.; to Ore., Colo. May–June.

13. **L. brévior** (Jeps.) Christian & Dunn. [*L. sparsiflorus* var. *b*. Jeps.] Close to *L. concinnus*; plants usually less than 15 cm. tall, glabrous or sparsely pilose; lfts. glabrous above, truncatish, 5–12 mm. long; fls. ca. 6 mm. long, off-white to bluish or purplish.—Creosote Bush Scrub; w. edge of Colo. Desert. March–April.

14. **L. bréweri** Gray. var. **bréweri.** Prostrate or decumbent branched matted perennial with woody base and silvery silky foliage and stems; lvs. crowded, the petioles 1–5 cm. long, the lfts. 7–10, oblanceolate to spatulate, 0.5–2 cm. long; peduncles 1–3 cm. long; racemes ca. 3–5 cm. long, mostly densely fld., pedicels 1–3 mm. long; bracts deciduous; upper calyx-lip cleft, lower entire or 3-toothed, ca. 3 mm. long; petals violet, the fls. 6–9 mm. long, the banner roundish to obovate, mostly glabrous on back, keel mostly glabrous; pods 12–16 mm. long, silky.—Dry stony slopes and benches, ca. 6000–7000 ft.; Yellow Pine F.; Mt. Pinos; to s. Ore.

Var. **bryoìdes** C. P. Sm. in Jeps. Infl. 2–3 cm. high, subcapitate; petioles ca. 1 cm. long, fls. 4–6 mm. long.—At somewhat higher elevs.; Ventura and Inyo cos. and s. Sierra Nevada.

Var. **grandiflòrus** C. P. Sm. [*L. b.* var. *clokeyanus* C. P. Sm. *L. campbellae* var. *bernardinus* Eastw.] Peduncles 3–8 cm. long; racemes 3–10 cm. long; fls. 9–11 mm. long, the banner mostly pubescent on back.—At 6500–10,000 ft.; Montane Coniferous F.; San Bernardino Mts.; to Tuolumne Co. June–Aug.

15. **L. caespitòsus** Nutt. ex T. & G. Subacaulescent, subappressed-silky, 5–12 cm. tall; petioles 5–10 cm. long, ± purplish; lfts. 5–7, oblanceolate, 1–2.5 cm. long, green and sparsely silky above, grayish-silky beneath; peduncles ca. 1 cm. long; racemes densely-fld., 4–8 cm. long, much exceeded by the lvs.; upper calyx-lip deeply cleft, ca. 3 mm. long; lower tridentate, ca. 4 mm. long; fls. 7–8 mm. long, pale blue or lilac to whitish, the banner almost twice as long as wide, wings narrow, keel short-ciliate on upper edges, almost straight; pods ca. 1.5 cm. long; seeds 3–4.—Rocky slopes and flats, 10,000–11,000 ft.; Sagebrush Scrub; White Mts., Inyo Co.; to Ore., Mont., Colo. June–July.

16. **L. caudàtus** Kell. Strigose, ± silky perennial, the stems 2–6 dm. tall, simple or branched; lower lvs. ± persistent until anthesis, with petioles to ca. 1 dm. long, upper 2–5 cm.; lfts. 5–9, oblanceolate, 2–5 cm. long, silky on both sides; peduncles 2–5 cm. long; racemes 5–15 cm. long; fls. scattered or subverticillate, 9–12 mm. long; bracts 4–5 mm. long, deciduous, subulate; upper calyx-lip bidentate, ca. 5 mm. long, the lower entire, ca. 7 mm.; spur ca. 1 mm. long; petals violet-blue to deep blue or almost white, banner

silky on back, keel ciliate on upper edges; pods 2.5–3 cm. long; seeds 5–6; $n=24$ (Phillips in 1957).—Dry open slopes, 4000–9500 ft.; Sagebrush Scrub, Pinyon-Juniper Wd.; n. Inyo Co. to Ore., Utah. May–Sept.

17. **L. chamissònis** Eschs. Erect branching shrub, 5–20 dm. tall, minutely silky-strigose, with many short leafy branches; petioles 2–3.5 cm. long; lfts. 6–9, oblanceolate, appressed-silky, 1–2.5 cm. long, 4–6 mm. wide; peduncles 2–5 cm. long; racemes 5–15 cm. long, the fls. ± verticillate, 12–16 mm. long; bracts 7–10 mm. long, early deciduous; upper calyx-lip cleft, 6–7 mm. long, lower entire, 7–8 mm. long; petals blue or lavender, the banner pubescent on back, broad, with a yellow spot, keel curved, nonciliate or nearly so, with slender upturned acumen; pods 2.5–3.5 cm. long, strigose; seeds 4–7.—Sandy beaches and dunes; largely Coastal Strand; Los Angeles Co. to Marin Co. March–July.

18. **L. concínnus** Agardh ssp. **concínnus.** Densely villous usually much branched annual, 0.5–2 dm. tall, with slender stems; lvs. well distributed, the petioles ca. 2–6 cm. long; lfts. 5–9, narrow-oblanceolate, 1–2 cm. long, 1.5–3 mm. wide, soft-hairy; peduncles 5–8 cm. long; the racemes 1–5 cm. long, villous; pedicels suberect soon after anthesis; fls. 7–9 mm. long, the upper calyx-lip cleft, the lower 3-toothed; petals mostly lilac, edged with red-purple, the banner 7–9 mm. long, 4–5 mm. wide, with yellow center; keel and wings not ciliate; pods 1–1.5 cm. long, 2–4-ovuled, hairy; seeds 2–3 mm. long.—Dry open and disturbed places below 5000 ft.; many Plant Communities; cismontane from San Diego to cent. Calif. March–May.

Ssp. **optàtus** (C. P. Sm.) D. Dunn. Lfts. mostly 4–7 mm. wide; stems thicker (1.5–2 mm.); fls. 9–11 mm. long.—Largely in Yellow Pine F.; mts. from San Diego Co. to Monterey Co.

Ssp. **orcúttii** (Wats.) D. Dunn. [*L. o.* Wats.] Lfts. 4–7 mm. wide; stems thickish; fls. 6–8 mm. long.—Dry sandy and open places; Creosote Bush Scrub, Joshua Tree Wd.; Inyo to Imperial cos.; Nev. to New Mex. March–May.

19. **L. confértus** Kell. Stout, densely white-silky perennial 1.5–3.5 dm. tall, with several stems from a stout root, densely leafy below, less so above; petioles 3–9 cm. long; lfts. mostly 7, elliptical-oblanceolate, 1.5–4 cm. long, 5–8 mm. wide, acutish; peduncles 2–7 cm. long; racemes dense, 5–30 cm. long; pedicels 1–2 mm. long; bracts persistent, subulate, 8–9 mm. long; upper calyx-lip ca. 5 mm. long, lower 7 mm.; fls. 10–14 mm. long, violet-purple, the banner elliptic-obovate, with pale cent. spot, keel short-acuminate, woolly-ciliate on upper edge; pods 10–18 mm. long, white silky-villous; seeds 2–5.—Meadows and dampish places, 3000–8500 ft.; Montane Coniferous F.; San Bernardino Mts. and Mt. Pinos; to cent. Calif. June–Aug.

20. **L. densiflòrus** Benth. var. **austrocóllium** C. P. Sm. Plate 46, Fig. F. Annual, pubescent with spreading or retrorse hairs 1–1.5 mm. long, 2–4 dm. tall, simple or branched especially above; petioles 4–10 cm. long; lfts. oblanceolate, 7–9, glabrous above, villous beneath; peduncles 0.5–2 dm. long, with racemes as long or longer; whorls 5–12, ± separate; calyx ca. 8 mm. long, bushy-villous near the base, the hairs 1.5–4 mm. long, drying tawny; fls. ca. 13 mm. long, white to pinkish, the banner acute at apex; keel and often the wings ciliate on the upper edges; pods ca. 1.5 cm. long; seeds 4–5 mm. long.—Dry open places below 2500 ft.; V. Grassland, Coastal Sage Scrub, Chaparral; s. Orange and w. Riverside cos. to w. San Diego Co. April–May.

Var. **glareòsus** (Elmer) C. P. Sm. [*L. g.* Elmer.] Pubescence of stems and peduncles very short (to 0.5 mm. long) and appressed; calyx 6.5–8 mm. long; petals light blue, the banner with a white center; lfts. blackening on drying.—At 4000–6000 ft.; Sagebrush Scrub, Pinyon-Juniper Wd.; Mt. Pinos, Ventura Co.

Var. **lácteus** (Kell.) C. P. Sm. [*L. l.* Kell.] Pubescence 1–1.5 mm. long; calyx rather inconspicuously pubescent, the hairs there barely 1 mm. long; fls. 11–16 mm. long, white or rose or purple.—Dry open places, below 6000 ft.; V. Grassland, S. Oak Wd., Yellow Pine F.; in the interior; San Diego Co. to Placer Co.

Var. **palústris** (Kell.) C. P. Sm. [*L. p.* Kell.] Stems, peduncles and lower part of calyx with hair over 1 mm. long; lower calyx-lip ± bent and inflated near the base;

banner mostly rounded at apex; fls. 12–17 mm. long.—V. Grassland, Chaparral, etc.; n. Los Angeles Co. to cent. Calif.

21. **L. elàtus** Jtn. Much like *L. andersonii*; erect, 5–9 dm. tall, branched above, densely short-silvery-hairy; petioles mostly 2–4 cm. long; lfts. 2–6 cm. long, silvery-silky above, duller beneath; fls. 10–14 mm. long, lavender to blue, the banner suborbicular, sometimes pubescent near middle of back.—Dry slopes among pines, 6000–8700 ft.; Montane Coniferous F.; Mt. Pinos and San Gabriel Mts. June–Aug.

22. **L. excùbitus** Jones var. **excùbitus**. Near *L. albifrons*; shrubby, 5–15 dm. tall, ± branched, densely silky-appressed-pubescent with short hairs; petioles 4–10 cm. long; lfts. 7–9, oblong-oblanceolate to spatulate, 2–4 cm. long, densely silvery above and beneath; peduncles 4–15 cm. long; racemes 10–25 cm. long; fls. 10–12 mm. long, in separate verticils, 2–5 cm. apart; bracts deciduous, 6–7 mm. long; pedicels 4–6 mm. long; upper calyx-lip notched, lower entire; petals blue to violet or orchid, the banner roundish, glabrous or somewhat pubescent on back, with yellowish center; wings and keel broad, the latter curved, ciliate on upper edge; pods 3–5 cm. long, silky; seeds 6–8.—Gravelly and rocky places, 4000–8700 ft.; Creosote Bush Scrub, Pinyon-Juniper Wd.; mts. of Inyo Co. s. and e. along desert margin to Little San Bernardino Mts. April–June.

KEY TO VARIETIES

Fls. 10–13 mm. long. Desert plants.
  Pubescence appressed; fl.-whorls mostly 2–5 cm. apart. Inyo Co. to San Bernardino Co.
                                                                            var. *excubitus*
  Pubescence tomentose; fl.-whorls mostly 1–2 cm. apart. Sw. Colo. Desert . . . . var. *medius*
Fls. 14–18 mm. long. Montane and cismontane.
  Stems scarcely if at all woody, but quite herbaceous. Mostly in pine belt . . var. *austromontanus*
  Stems woody at base.
    Plants 5–15 dm. high; foliage ± greenish, scarcely silvery. Below pine belt . . . . var. *hallii*
    Plants 1–2 dm. high; foliage silvery. Pine belt, San Gabriel Mts. . . . . var. *johnstonii*

Var. **austromontànus** (Heller) C. P. Sm. [*L. a.* Heller.] Stems mostly herbaceous, 2–5 dm. high; lvs. crowded near base, silvery; racemes mostly 5–10 cm. long, congested; fls. 14–18 mm. long, fragrant.—Dry slopes and rocky places, ca. 4000–8500 ft.; upper Chaparral, Yellow Pine F.; mts. from Tehachapi Mts. to n. L. Calif. May–July.

Var. **hállii** (Abrams) C. P. Sm. [*L. h.* Abrams. *L. paynei* Davids.] Plant rather coarse and green, the lvs. not very silky; whorls in infl. usually not more than 2.5 cm. apart; fls. 14–18 mm. long.—Gravelly and sandy places, below 4000 ft.; Chaparral, Coastal Sage Scrub; Ventura and San Bernardino cos. to n. L. Calif.

Var. **johnstònii** C. P. Sm. in Jeps. Basal stems woody, branched, to ca. 1–2 dm. long; racemes 6–12 cm. long, congested.—Dry slopes under pines, 5500–6600 ft.; San Gabriel Mts. May–July.

Var. **mèdius** (Jeps.) Munz. [*L. albifrons* var. *m.* Jeps.] Plants 3–7 dm. high, woody at base, finely white-tomentulose; fls. 11–13 mm. long, bluish-violet.—Washes; Creosote Bush Scrub; Mt. Springs Grade, w. Imperial Co. and e. San Diego Co. March–April.

23. **L. flavoculàtus** Heller. An almost acaulescent annual, 0.5–1.5 dm. high, villous; petioles mostly 2–5 cm. long; lfts. 7–9, broadly oblanceolate, 0.5–2 cm. long, glabrous above, sparsely villous beneath; peduncles wide-spreading, 3–10 cm. long; racemes dense, 2–8 cm. long; pedicels 1–4 mm. long; calyx sparsely villous, the upper lip 2–3 mm. long, bidentate, the lower 5–6 mm.; petals 7–8 mm. long, deep rich-violet-blue, the banner with a squarish yellow spot, keel glabrous; pods ovate, 8–11 mm. long, villous; seeds ca. 2 mm. in diam.—Dry open places, 2600–7000 ft.; Creosote-Bush Scrub to Pinyon-Juniper Wd.; Mojave Desert from e. San Bernardino Co. to White Mts.; w. Nev. April–June.

24. **L. formòsus** Greene var. **formòsus**. [*L. pasadenensis* Eastw.] Plate 47, Fig. A. Stems several, decumbent or ascending, 3–8 dm. long, 3–4 mm. thick, appressed-silky-pubescent, but somewhat greenish, leafy; petioles 3–7 cm. long; lfts. 7–9, silky on both sides, oblanceolate, 3–7 cm. long, mostly 5–15 mm. wide; peduncles 1–4 cm. long;

Plate 47. Fig. A, *Lupinus formosus*. Fig. B, *Lupinus longifolius*. Fig. C, *Lupinus nanus*. Fig. D, *Medicago polymorpha*. Fig. E, *Olneya tesota*. Fig. F, *Pickeringia montana*.

racemes 10–25 cm. long, mostly rather dense; pedicels spreading-pubescent, 3–4 mm. long; fls. mostly whorled, 12–14 mm. long, violet or blue to lilac or white, the banner roundish, glabrous, wings covering keel which is slender, curved, nonciliate; pods silky-hairy, ca. 3–3.5 cm. long; seeds 5–7.—Dry open fields and sandy places, mostly below 2500 ft.; Coastal Sage Scrub, Chaparral; occasional in pine belt; n. L. Calif. through cismontane Calif. to n. Calif. Variable; many plants have spreading pubescence and constitute the var. *bridgesii* (Wats.) Greene. April–Aug.

Var. **hyacinthinus** (Greene) C. P. Sm. [*L. h.* Greene.] Pubescence thin, plant decidedly green, erect; lfts. mostly sharp-pointed; fls. 13–16 mm. long, the blue or purplish banner with yellow center.—Dry slopes under pines, Montane Coniferous F.; San Jacinto Mts. and somewhat in Santa Rosa and e. San Gabriel mts. June–Aug.

25. **L. guadalupénsis** Greene. [*L. moranii* Dunkle. *L. aliclementinus* C. P. Sm.] Bushy annual, 3–9 dm. high, spreading-villous throughout; lvs. well distributed, the stipules 1–2 cm. long, linear; petioles 2–7 cm. long; lfts. 7–10, linear to narrow-oblanceolate, 1–4.5 cm. long; peduncles 5–8 cm. long; racemes 5–15 cm. long; fls. not in definite whorls, rather few, 14–16 mm. long; upper calyx bifid, 6–7 mm. long, lower tridentate; petals blue, the banner rounded, ca. 12 mm. long, wings 7–8 mm. wide, rounded at apex; keel arcuate, ciliate on upper edge; pods 5–6 cm. long, villous.—Banks and canyon-slopes; Coastal Sage Scrub; San Clemente Id.; Guadalupe Id. Feb.–April.

26. **L. hirsutíssimus** Benth. Robust annual, with stiff yellowish nettlelike hairs 3–5 mm. long, ± fistulous, 2–8 dm. high, usually few-branched from near base; lvs. well distributed, petioles 5–18 cm. long; lfts. 5–8, broadly cuneate-obovate, 2–5 cm. long; peduncles 5–8 cm. long; racemes 10–25 cm. long; fls. red-violet to magenta, scattered, 13–15 mm. long; calyx 8–10 mm. long; banner suborbicular, tending to have a yellow blotch, the keel densely ciliate on lower edges near the claws; pods 2.5–3.5 cm. long, hispid-bristly, several-ovuled.—Locally abundant in open wooded or brushy places, below 4000 ft.; Coastal Sage Scrub, Chaparral; cismontane from n. L. Calif. to cent. Calif. March–May. Santa Cruz Id., Catalina Id., San Clemente Id.

27. **L. holmgrenànus** C. P. Sm. Perennial, 4–7 dm. high, ± silky-pilose; lower petioles 8–15 cm. long; blades with 4–7 lfts., oblanceolate, 2–4 cm. long, silky-strigose; stipules 0.5–2 cm. long; peduncles ± fistulose, 6–10 cm. long, 3–4 mm. thick; racemes rather densely many-fld., 1–2 dm. long; bracts linear, 8–10 mm. long; pedicels 8–10 mm. long; fls. 13–15 mm. long, bluish-purple on drying; calyx slightly gibbous at base, appressed-villous; banner roundish, 10 mm. long, with yellowish center; wings ca. 9 mm. long, 6 mm. wide; keel ± ciliate on upper edge below the acumen, arcuate, narrow.— Dry slopes, 6000–7500 ft.; Pinyon-Juniper Wd.; Last Chance Range, e. Inyo Co.; w. Nev. May–June.

28. **L. horizontàlis** Heller var. **platypétalus** C. P. Sm. Diffuse short-villous annual, to 1.5 dm. high, with ascending branches; petioles 4–6 cm. long; lfts. 7–9, oblanceolate, 1–2 cm. long, 5–6 mm. wide, glabrous above; peduncles 3–4 cm. long; racemes 5–8 cm. long, equaling or slightly longer thar lvs.; bracts persistent; pedicels 1–1.5 mm. long; calyx villous, 6–8 mm. long; petals 13–15 mm. long, whitish to pale violet, the banner almost plane; keel straight, wings and keel ciliate on upper and lower edges near claws; pods ovate, villous, 12–15 mm. long, 2-ovuled.—Open sandy and gravelly places, below 3500 ft.; Creosote Bush Scrub, Joshua Tree Wd.; w. Mojave Desert to Barstow region and to Greenhorn Mts. in Kern Co. March–May.

29. **L. latifòlius** Agardh var. **latifòlius.** [*L. cystoides* Agardh. *L. rivularis* as used by Jeps.] Herbaceous perennial, erect, 3–12 dm. tall, subglabrous to minutely strigose, leafy; basal lvs. often dry by anthesis, the largest on mid-stem; petioles 5–20 cm. long; lfts. mostly 7–9, sometimes 5–12, broadly oblanceolate, 4–10 cm. long, 1–3 cm. wide; peduncles 8–20 cm. long; racemes 15–45 cm. long, rather lax, the fls. whorled or scattered, 10–14 mm. long; bracts deciduous, 8–12 mm. long; pedicels spreading-pubescent, 6–12 mm. long; petals blue or purplish or with some pink, fading brown, the banner roundish, 9–10 mm. wide; wings ± exposing the curved keel which has a slender acumen and is ciliate on upper margin; pods ca. 3 cm. long, hairy, 7–10-seeded; $n=24, 48$ (Phillips,

1957).—Common in open woods and thickets below 7000 ft.; many Plant Communities, from coast to Montane Coniferous F.; Santa Monica Mts. to Wash. April–July.

Var. **párishii** C. P. Sm. Fls. 14–18 mm. long, rose to lavender; stems to 2 m. tall, ± fistulous.—Moist places, 1000–9000 ft.; Chaparral to Montane Coniferous F.; mts. from San Diego Co. to Tulare Co. May–Aug.

30. **L. longifòlius** (Wats.) Abrams. [*L. chamissonis* var. *l.* Wats. *L. mollisifolius* A. Davids.] Plate 47, Fig. B. Erect, shrubby below, appressed-pubescent, greenish, 1–1.5 m. tall; petioles 4–7 cm. long; lfts. 6–9, elliptic- or oblong-oblanceolate, 2.5–6 cm. long, 5–15 mm. wide, subsilky on both sides; peduncles 6–12 cm. long; racemes 20–40 cm. long, the fls. scattered or subverticillate, 14–18 mm. long; bracts deciduous; pedicels 5–10 mm. long; petals blue to violet, the glabrous banner suborbicular, with cent. yellowish spot, wings broad, keel curved, ciliate along upper edges; pods 4–6 cm. long, 6–8-seeded.— Coastal bluffs and canyons inland to San Dimas and Santa Ana Mts.; Coastal Sage Scrub, Chaparral, S. Oak Wd.; n. L. Calif. to Ventura Co. April–June. *L. mollisifolius* has ± fistulose stems and only spring fls., and comes from Sierra Madre, La Cañada, etc., while coastal plants flower most of the year.

31. **L. luteòlus** Kell. [*L. bridgesii* Gray.] Much like *L. densiflorus*, but not fistulose, 3–8 dm. high, widely branched above, strigose; lfts. mostly 7–9, cuneate-oblanceolate, 2–3 cm. long, strigose on both surfaces or glabrous above; peduncles 5–15 cm. long; whorls crowded, few to many in racemes 5–20 cm. long; fls. light to pale yellow, sometimes pale lilac at first, 12–14 mm. long; calyx strigose, 10–11 mm. long; wings ciliate near base, keel ciliate on upper and lower edges near base; pods hirsute, ca. 1.5 cm. long.—Dry slopes and flats below 6000 ft.; Yellow Pine F., S. Oak Wd., Pinyon-Juniper Wd.; Ventura Co. to n. Calif., s. Ore. May–Aug.

32. **L. magníficus** Jones. Erect perennial 5–12 dm. tall, tomentose or tomentulose, with longer spreading hairs; lvs. near the base; petioles 1–2 dm. long; lfts. 5–9, elliptic-oblanceolate, 2–4.5 cm. long, 6–10 mm. wide; peduncles 1–3 dm. long; racemes 2–4 dm. long; bracts deciduous; pedicels 3–4 mm. long; fls. 16–18 mm. long, ± whorled; petals pinkish-purple, banner glabrous, roundish, the cent. yellow spot becoming dark purple; keel yellow, ciliate above on short upturned acumen; pods hairy, 3–7 cm. long, 5–8-seeded.—Dry gravelly banks, 5500–7500 ft.; Pinyon-Juniper Wd.; Panamint Mts., Inyo Co. May–June. Plants from the Coso Mts. have fls. 10–13 mm. long and are the var. **glarécola** Jones [*L. kerrii* Eastw.]

33. **L. nànus** Dougl. in Benth. var. **nànus**. Plate 47, Fig. C. Erect annual, ± villous and minutely pubescent or strigulose, 1–5 dm. tall; lvs. mostly well distributed; petioles 4–8 cm. long; lfts. 5–7, linear to spatulate, 1.5–3 cm. long, mostly less than 5 mm. wide; peduncles 5–12 cm. long, the racemes 6–24 cm. long, with well separated whorls; bracts 5–10 mm. long, deciduous; petals rich blue except for the yellowish or white spot on the banner; banner roundish, 8–11.5 mm. long, much reflexed, the wings mostly concealing the slender keel which is not much curved, ciliate on upper edges of slender acumen; pods strigose, 2–3.5 cm. long, 4–5.5 mm. wide, 4–8-ovuled; $2n = 48$ (Tuschnjakowa, 1935).— Local, grassy places and brushy slopes, below 3500 ft.; Santa Barbara Co. to Santa Cruz Co. April–May.

Ssp. **latifòlius** (Benth.) D. Dunn. Largest lfts. 5–15 mm. wide; pods 6–8.5 mm. wide. —Los Angeles Co. to Mendocino Co., largely in V. Grassland, etc.

34. **L. nevadénsis** Heller. Cespitose perennial with erect or ascending stems 2–4 dm. high, long-villous with shorter pubescence beneath; lvs. well distributed, the basal petioles to ca. 14 cm. long, the upper to 3 cm.; lfts. 7–10, elliptic-oblanceolate, 2.5–3.5 cm. long, shaggy-villous; peduncles 3–6 cm. long; racemes rather lax, whorled, 10–18 cm. long; pedicels 5–8 mm. long; bracts 5 mm. long, deciduous; fls. 10–12 mm. long, blue-violet, the banner broader than long, with a whitish cent. spot and purple dots; keel mostly ciliate on upper margins, strongly curved, with attenuate purple apex; pods 3–4 cm. long, hairy; seeds 3–4.—At ca. 4000–6000 ft.; Sagebrush Scrub; Grapevine Mts., Inyo Co. to Mono Co. and w. cent. Nev. April–June.

35. **L. odoràtus** Heller. Glabrous subcaulescent ± succulent annual, 1.5–3 dm.

high; lvs. basal, the petioles commonly 5–10 cm. long; lfts. cuneate to spatulate, 5–7, glabrous above, sometimes somewhat strigose beneath; peduncles 8–15 cm. long; racemes 5–10 cm. long, fairly dense; pedicels glabrous; bracts villous; calyx glabrous; fls. fragrant, mostly deep purple-violet with yellow spot on banner, the banner ca. 10 mm. long and wide, keel glabrous; pods 17–20 mm. long, villous on margins.—Locally common on dry flats, 2000–4000 ft.; Creosote Bush Scrub, Joshua Tree Wd.; Mojave Desert from Barstow region w. and n. to Lone Pine; w. Nev. April–May. Plants from around Nipton may have a villous pubescence and have been named var. *pilosellus* C. P. Sm.

36. **L. pállidus** Bdg. [*L. desertorum* Heller.] Near to *L. concinnus*; mostly rather coarse decumbent or prostrate plants, silky to strigose, usually less than 15 cm. tall; lvs. pubescent on both sides; fls. pale blue to whitish or yellowish, 5–7 mm. long; keel ciliate near the claws.—Occasional in sandy and gravelly places, at 800–5500 ft.; Creosote Bush Scrub; both deserts; to L. Calif. March–April.

37. **L. pálmeri** Wats. [*L. candidissimus* Eastw. *L. jaegerianus* C. P. Sm. *L. keckianus* C. P. Sm.] Plants 3–6 dm. tall, the several stems ± hoary with retrorsely spreading hairs 1–2 mm. long; petioles 4–10 cm. long; lfts. 6–9, elliptic-oblanceolate, 2.5–5 cm. long, densely sericeous; peduncles 4–7 cm. long; racemes 10–20 cm. long, lax, subverticillate; pedicels 2–7 mm. long; fls. blue, 8–10 mm. long, the banner suborbicular, reflexed above the midpoint; keel ciliate above toward the tip.—Dry stony places, 6500–7500 ft.; Pinyon-Juniper Wd.; s. White Mts. to Grapevine Mts., Inyo Co.; to w. Nev., Ariz., New Mex. May–June.

38. **L. peirsònii** Mason. Herbaceous silvery-silky perennial, 3–6 dm. tall, the stems many, ascending; lvs. subbasal; petioles 6–15 cm. long; lfts. 5–8, oblong-oblanceolate, 3–7 cm. long, 10–15 mm. wide, densely silky on both sides; peduncles 1–2.5 dm. long; racemes 1–1.5 dm. long; fls. subverticillate, 10–12 mm. long, the lower whorls distinct; bracts 6–7 mm. long, deciduous; calyx-lips subequal; petals yellow, the banner pubescent on the back, with a brownish spot above center, keel ciliate on upper straightish edges; pods silky, 3 or more cm. long, 3–5-seeded.—Rare, loose gravelly and rocky slopes, 4000–5000 ft.; Pinyon-Juniper Wd., Joshua Tree Wd.; desert slopes of San Gabriel and Tehachapi mts. April–May.

39. **L. polycárpus** Greene. [*L. micranthus* Dougl. in Lindl., not Gussone.] With habit and stature of *L. bicolor*, strigose and ± villous; lfts. 5–7, linear to oblanceolate, 1.5–4 cm. long, glabrous or sparsely hairy above, strigose beneath; peduncles 3–10 cm. long, racemes 1–8 cm. long, ± whorled; pedicels 1–2 mm. long; petals deep blue, the banner with a white spot having purple dots, cuneate-obovate to spatulate, 6–8 mm. long, ± emarginate; wings 5–6 mm. long; keel almost straight on upper edge, ciliate beyond the middle, the tip blunt; pods 2–3 cm. long, 6–7-seeded; $2n=48$ (Tuschnjakowa, 1935).—Rather heavy soils, below 5000 ft.; many Plant Communities; San Diego Co. to B.C. March–June.

40. **L. polyphýllus** Lindl. ssp. **bernardínus** (Abrams) Munz. [*L. b.* Abrams ex Eastw.] Erect, mostly unbranched, 5–15 dm. high, stout, fistulous, with minute spreading or appressed puberulence, few-lvd.; petioles 1.5–3 dm. long, stout; lfts. 5–9, glabrous or nearly so on both surfaces, 7–15 cm. long, ca. 1 cm. wide; peduncles 3–8 cm. long; racemes 15–60 cm. long, rather dense, fls. subverticillate, 9–11 mm. long; bracts to 1 cm. long, deciduous; pedicels 5–15 mm. long; calyx-lips entire; petals blue, purple, or reddish, glabrous, the banner ± oblong-obovate, keel curved, naked, with long pointed acumen; pods loosely hairy, 2.5–4 cm. long, 5–9-seeded.—Wet places, 6500–8500 ft.; Montane Coniferous F.; San Bernardino and San Jacinto mts. June–July.

Ssp. **supérbus** (Heller) Munz. [*L. s.* Heller.] Lfts. mostly 5–9, mostly 4–7 cm. long, 0.5–1.5 cm. wide, strigose beneath; fls. 11–14 mm. long, in lax racemes; keel minutely papillose on upper margin, but not ciliate.—Wet places, 4000–8500 ft.; Montane Coniferous F.; Piute Mts., Kern Co.; to n. Calif. May–July.

41. **L. pusíllus** Pursh ssp. **intermontànus** (Heller) D. Dunn. [*L. i.* Heller.] Loosely villous, 5–12 cm. high, with spreading branches; lvs. rather crowded, the petioles 3–7 cm. long; lfts. usually 5, oblong-oblanceolate, 1–3.5 cm. long, glabrous above, appressed-hairy beneath; peduncles barely 1 cm. long; racemes 3–5 cm. long; pedicels ± villous;

calyx ± villous; fls. 7–9 mm. long, bright violet-blue, fading white or pinkish, ca. 1.5 cm. long, constricted in middle.—Dry places, 4000–5000 ft.; Sagebrush Scrub; Deep Springs V., Inyo Co.; to e. Wash., Wyo., Colo. May–June.

42. **L. rùber** Heller. Villous annual, 1–4 dm. tall, branched from base; lvs. near the base, the petioles 4–8 cm. long; lfts. 5–8, oblanceolate to spatulate, 1–2 cm., glabrous above, villous beneath; peduncles 2–7 cm. long, the racemes 3–6 cm. long, of 2–5 whorls; bracts persistent, ca. 8 mm. long; upper calyx-lip ca. 2 mm., the lower 8 mm.; petals dull red to pink, the banner almost plane, lance-ovate, 2–4 mm. wide, 8–9 mm. long; wings narrow, scarcely ciliate, keel ciliate only above the claws; pods ovate, 12–17 mm. long, 2-ovuled.—Uncommon, dry open places below 5000 ft.; V. Grassland, Creosote Bush Scrub; L. Calif. to San Benito Co., inland to New York Mts., e. San Bernardino Co. March–June.

43. **L. shóckleyi** Wats. Acaulescent or with stems ca. 1 dm. long, densely pubescent with hairs to 1 mm. long; petioles 4–12 cm. long; lfts. 7–10, subappressed-silky beneath, glabrous above except near margins, ± spatulate, 1–2 cm. long; peduncles 3–10 cm. long; racemes 3–6 cm. long; fls. scattered, 5–6 mm. long; calyx densely white-pubescent; petals blue, purple or pink, the banner 5–6 mm. long, ± pointed; keel nonciliate; pods oblong-ovate, ca. 1.5 cm. long, ciliate, with scaly sides.—Dry sandy places, below 4000 ft.; Creosote Bush Scrub; Colo. and Mojave deserts; Ariz., Nev. April–May.

44. **L. sparsiflòrus** Benth. ssp. **sparsiflòrus**. [*L. subhirsutus* A. Davids.] Slender-stemmed annual, strigose and spreading-villous, mostly 2–4 dm. high, erect, branched; petioles 3–8 cm. long; lfts. 5–9, linear to oblanceolate, 1–3 cm. long; peduncles 2–8 cm. long; racemes 8–20 cm. long; fls. not crowded, light blue to lilac, 10–13 mm. long; calyx ca. 5 mm. long; banner suborbicular, with a yellow spot; keel curved, usually ciliate on lower edge near claws; pods 1–2 cm. long, 5–7-ovuled, strigose.—Open places, below 4000 ft.; Coastal Sage Scrub, Chaparral; Ventura to Riverside cos. March–May.

Ssp. **inopinàtus** (C. P. Sm.) Dziekanowski & Dunn. Fls. 7–11 mm. long; banner oblong-oval, longer than wide.—Cismontane, San Diego Co. and n. L. Calif.

Ssp. **mohavénsis** Dziekanowski & Dunn. Fls. 7–11 mm. long; banner orbicular, emarginate.—Creosote Bush Scrub; Mojave Desert to Ariz. & Son.

45. **L. stivérsii** Kell. Minutely pubescent annual, 1–4 dm. high, freely branched; lvs. scattered, the petioles 3–8 cm. long; lfts. 6–8, cuneate to obovate, 1–4 cm. long, strigose on both sides; peduncles 3–8 cm. long; racemes 1–3 cm. long; calyx pubescent; banner 13–15 mm. long, yellow, the wings rose-purple ca. 15–16 mm. long, keel whitish, ciliate above and below at base; pods 2–2.5 cm. long, glabrous, several-seeded.—Sandy or gravelly places, 1600–4600 ft.; locally in San Gabriel and San Bernardino mts.; Kern Co. n. April–July.

46. **L. subvéxus** C. P. Sm. Loosely villous annual, 1.5–4 dm. tall, simple or branched; lvs. basal or on lower and middle stem; lfts. mostly 5–9, oblanceolate, 1.5–3.5 cm. long, glabrous above, villous beneath; petioles 5–15 cm. long; peduncles 5–12 cm. long; racemes 5–15 cm. long, with 3–9 whorls; bracts persistent, soon reflexed; calyx 9–11 mm. long, ± inflated at base; petals dark violet-purple, lilac or rose-pink, the banner rounded at apex, 12–14 mm. long; wings usually nonciliate; keel straight, ciliate above near claws; pods 2-ovuled, ca. 1.5 cm. long.—Dry places below 2500 ft.; V. Grassland, Foothill Wd.; Santa Monica Mts., Ventura and Kern cos. n. to cent. Calif.; Creosote Bush Scrub, w. Mojave Desert. Santa Rosa Id.

47. **L. succuléntus** Dougl. ex Koch. [*L. affinis* of many Calif. auth., not Agardh.] Stout usually succulent or fistulous annual, mostly ± strigose, simple or branched, 2–8 dm. tall; lvs. well distributed, the petioles 6–12 cm. long; lfts. 7–9, cuneate to cuneate-obovate, 2–7 cm. long, rather dark green, glabrous above; peduncles 2–10 cm. long; racemes 6–30 cm. long, the fls. in whorls or groups; calyx substrigose, ca. 7–8 mm. long; fls. mostly deep purple-blue, the banner ca. 12–14 mm. long, usually with a yellow center; wings slightly ciliate at base; keel ± curved, ciliate near base on both edges; pods 4–5 cm. long, loosely pubescent, several-seeded.—Usually in heavy soil on grassy flats and slopes below 2000 ft.; many Plant Communities; n. L. Calif. to n. Calif. Feb.–May.

48. **L. truncàtus** Nutt. ex H. & A. Subglabrous to sparsely strigulose annual, branched, 3–7 dm. tall, rather deep green; petioles 5–10 cm. long; lfts. 5–7, linear, truncate to emarginate or toothed, 1–4 cm. long; peduncles 3–10 cm. long; racemes 5–15 cm. long; fls. not crowded, 10–12 mm. long, mostly purplish-blue, redder in age; calyx 5–6 mm. long; banner ca. 10 mm. long, 9 mm. wide, keel ± ciliate on lower edge toward claws, more so on upper; pods ca. 3 cm. long, villous, 6–7-ovuled.—Open grassy places, burns, etc.; Coastal Sage Scrub, Chaparral, n. L. Calif. to Monterey Co. Channel Ids.

16. *Medicágo* L. MEDICK

Annual or perennial herbs. Lvs. pinnately 3-foliolate, the lfts. mostly dentate; stipules adnate. Fls. in small heads, racemes or umbels. Calyx-teeth subequal, short. Banner obovate to oblong; wings oblong; keel obtuse. Stamens diadelphous, the upper 1 free. Style subulate. Pods small, 1–several-seeded, curved to spirally coiled, indehiscent, reticulate or spiny. Seeds small, ± beanlike. Ca. 50 spp. of Eurasia and Afr.; many important for hay and forage. (Greek, *medice*, name of alfalfa, since it came to Greece from Medea.)

Plant perennial; fls. blue .... 4. *M. sativa*
Plants usually annual; fls. yellow.
  Pods subreniform, 1-seeded; fls. many, in dense elongate spikelike racemes ... 1. *M. lupulina*
  Pods spirally coiled; fls. few, not in spikes.
    Stipules lanceolate, short-dentate at base; plant pilose .... 2. *M. minima*
    Stipules pectinate, with linear lateral segms.; plants subglabrous or nearly so
        3. *M. polymorpha*

1. **M. lupulìna** L. BLACK MEDICK. Annual or sometimes perennial, many-branched from base, prostrate to decumbent, the stems 2–6 dm. long, pubescent; petioles 2–15 mm. long; lfts. obovate to roundish, 5–15 mm. long; stipules lance-ovate, few-toothed; peduncles slender, 1–2.5 cm. long, with short terminal few-fld. spikes; fls. yellow, 1.5–2 mm. long, the calyx villous; fr. reniform, smooth, black when ripe, 1 seeded, unarmed.—Natur. in waste places; native of Eu. April–July. Var. **cupaniàna** (Guss.) Boiss. Perennial, forming dense mats that root at the nodes; fls. larger.—Weed in lawns, etc.; introd. from Eurasia.

2. **M. mínima** (L.) Desr. Resembling *M. polymorpha*, but softly pubescent; stipules short-dentate; lfts. 5–15 mm. long, pubescent on both sides; peduncles short; fls. 3–4 mm. long; pods subglobose, tightly coiled, pubescent between the prickles; $n=8$ (Ghimpu, 1929).—Occasional, as in Ventura and Riverside cos.; native of Eurasia. April–May.

3. **M. polymórpha** L. [*M. hispida* Gaertn.] BUR-CLOVER. Plate 47, Fig. D. Subglabrous annual, branching from base, the stems procumbent, 1–4 dm. long; petioles 1–4 cm. long; lfts. obovate or obcordate, 8–20 mm. long, sharply denticulate; stipules deeply divided with long acicular teeth; peduncles slender, 5–25 mm. long, 2–5-fld.; fls. 4–5 mm. long, the calyx sparsely villous; pods coiled 2–3 times, 4–6 mm. in diam., glabrous, with 2–3 rows of spines arising from a raised ridge and without any furrow between the rows, the spines usually hooked; $n=7$ (Fryer, 1930).—Common in grassy places of most of cismontane Calif. and valued for pasturage; native of s. Eu. March–June. Var. **brevispìna** (Benth.) Heyn. Pods with short nobs or no prickles.—With the sp., but less common.

4. **M. satìva** L. ALFALFA. Erect or ascending smooth perennial from an elongate taproot, much-branched, 4–9 dm. high; petioles 5–20 mm. long; stipules 5–8 mm. long, entire to toothed; lfts. oblanceolate to obovate, 1–2.5 cm. long, sharply serrate at the ± truncate tip; calyx 5–6 mm. long, sparsely villous; corolla 8–12 mm. long, blue-violet to purple; pods unarmed, coiled loosely 2–3 times, pubescent; $n=16$ (Fryer, 1930), 32, 64 (Tomé, 1947).—Commonly cult. and frequently established in waste places, along roadsides, etc.; native of Old World. Apr.–Oct.

17. Melilòtus Mill. Sweet-Clover

Annual or biennial herbs, with pinnately trifoliolate lvs. Fls. small, white or yellow, in spikelike racemes. Calyx subequally 5-toothed. Banner obovate; wings oblong; keel obtuse. Stamens diadelphous. Pod ovoid, coriaceous, wrinkled, longer than the calyx, scarcely dehiscent, 1–2-seeded. Ca. 20 spp. of Eurasia and Afr. (Greek, *meli*, honey, and *lotos*, some leguminous plant.)

Fls. white .... 1. *M. albus*
Fls. yellow.
    The fls. 2–3 mm. long, on pedicels less than 1 mm. long .... 2. *M. indicus*
    The fls. 5–7 mm. long, on pedicels 1.5–2 mm. long .... 3. *M. officinalis*

1. **M. álbus** Desr. Erect, 1–2 m. tall; petioles mostly 5–15 mm. long; stipules subulate, 5–7 mm. long; lfts. mostly lanceolate to oblanceolate, truncate, 1–2 cm. long, serrate; peduncles commonly 3–5 cm. long; racemes 5–10 cm. long; pedicels 1–2 mm. long; fls. 4–6 mm. long; pods ovoid, glabrous, ca. 3 mm. long; $n=8, 12$ (Atwood, 1936).—Abundantly natur. in waste places, especially in damp places; to B.C. and Atlantic Coast; native of Eurasia. May–Sept.

2. **M. índicus** (L.) All. Stems erect, 2–8 dm. high, glabrous or ± appressed-pubescent; petioles commonly 0.5–2 cm. long; stipules linear-subulate, 4–6 mm. long; lfts. cuneate-oblanceolate to -obovate, 1–3 cm. long, obtuse or truncate, denticulate; racemes 2–10 cm. long, including the peduncles; pods ovoid, reticulate, glabrous, 1.5–2 mm. long; $n=8$ (Tschechow, 1932).—Common in waste places or at low elevs., most of Calif.; native of Eurasia. Apr.–Oct.

3. **M. officinàlis** (L.) Lam. Like *M. indicus*, but taller; fls. larger; pods 2–3 mm. long; $n=8$ (Tschechow, 1932).—Less commonly natur. in Calif.; native of Eurasia. May–Aug.

18. Ólneya Gray. Desert-Ironwood

Spinose tree with thin scaly bark and pairs of spines below the lvs. Lvs. pinnate or odd-pinnate, densely canescent, the lfts. 8–24, oblong-cuneate to wider; stipules obsolete. Fls. few, in axillary racemes, appearing before the new growth of lvs. Calyx campanulate, 5-lobed, the 2 upper quite united. Petals short-clawed, the banner roundish, emarginate. Stamens 10, diadelphous. Ovary several-ovuled; style bearded above. Pod thick, torulose, puberulent, glandular-hispid. Seeds broadly ellipsoid. One sp. (Named in honor of S. T. *Olney*, New England botanist of the 19th century.)

1. **O. tesòta** Gray. Broad-crowned grayish tree 5–8 m. high; lvs. 3–10 cm. long; lfts. 5–20 mm. long; racemes 3–5 cm. long; calyx 4–5 mm. long; the lobes ovate, shorter than the tube; corolla pale rose-purple, 1 cm. long; pod 4–6 cm. long; seeds black, 8–9 mm. long.—Desert washes, below 2000 ft.; Creosote Bush Scrub; Colo. Desert; to Ariz., Son., L. Calif. April–May.

19. Oxýtropis DC. Locoweed.

Perennial caulescent or acaulescent herbs with the aspect of *Astragalus*. Lfts. asymmetrical at base. Keel-petals abruptly narrowed into a cuspidate or mucronate apex. Pods more deeply grooved ventrally than dorsally, the valves not infolded, but the seminiferous suture often intruded. A genus of perhaps 300 spp., of the N. Hemis, the N. Am. spp. chiefly of the Rocky Mts. and high latitudes. Several are notoriously toxic to cattle, sheep and horses. (Greek, *oxus*, sharp, and *tropis*, keel, in reference to the beaked lower petals.)

Plant silvery, the herbage silky-pilose, without glands; bracts pilose dorsally ... 1. *O. oreophila*
Plant green, the herbage glandular and resinous; bracts glabrous dorsally .... 2. *O. viscida*

1. **O. oreophìla** Gray. Densely cespitose, forming cushions to 15 cm. broad, the herbage silky-pilose; lvs. 1–3 cm. long, with 7–15 lfts. 3–8 mm. long; racemes shortly 2–8-fld.,

immersed in the foliage or slightly exserted; petals pink, the banner ca. 1 cm. long; pod ovoid to subglobose with a short conical-compressed beak, bladdery-inflated, 8–13 mm. long, the papery valves villous-hirsute.—Barren stony places, 11,000–11,500 ft., about the summit of Mt. San Gorgonio, San Bernardino Mts.; to Utah, n. Ariz. July–Aug.

2. **O. víscida** Nutt. Cespitose, the branches of the stout forking caudex beset with imbricated stipules; lvs. 6–10 cm. long, with 25–39 green, sparsely pilose lfts. less than 1 cm. long; racemes ca. 10-fld., slightly exserted on scapiform peduncles; petals whitish or red-purple, the banner ca. 12 mm. long; pod oblong-ellipsoid with acuminate beak.— Bare crests and talus-slopes, 11,500–12,200 ft.; Inyo White Mts., Sierra Nevada; to Alaska, Quebec, Colo. July–Aug.

20. *Parkinsònia* L. PALO VERDE

Low trees with green branches and twigs; spines present in pairs at each node and a spine terminating the primary lf.-rachis. Lvs. bipinnate; primary lfts. in 1–3 pairs, crowded, both the rachis and petiole almost obsolete; secondary lfts. many. Fls. in long racemes. Sepals reflexed. Banner yellow, spotted with red, turning red in age. Stamens 10, all functional; anthers versatile, splitting lengthwise. Legume bulging, strongly constricted between seeds. Two spp.; 1 Am., 1 Afr. (J. *Parkinson*, 1567–1650, English herbalist.)

1. **P. aculeàta** L. MEXICAN PALO VERDE. To 10 m. high; pinnae with flattened rachis to ca. 2 dm. long; lfts. scattered, 4–10 mm. long; petals 10–15 mm. long; pods 5–10 cm. long, short-stiped, sharp-pointed; $n = 14$ (Pantulu, 1942).—Common in cult., especially in hot arid places and becoming natur.; native Ariz. to Fla., W.I., S. Am.

21. *Petalóstemum* Michx. PRAIRIE-CLOVER

Herbs, mostly perennial, ± glandular-dotted. Lvs. odd-pinnate, the lfts. entire. Fls. perfect, in terminal spikes; bracts deciduous. Calyx campanulate, 10-nerved, the lobes triangular to lanceolate. Corolla indistinctly papilionaceous, pink to purple or yellowish; banner free, the wings and keel alternating with the free portion of the anthers and borne on the top of the filamentous sheath. Stamens 5, monadelphous below. Style filiform. Pod membranous, subglobose to obovoid. Ca. 50 spp. of N. Am. (Greek, words for *petal* and *stamen*, because of the peculiar union of these parts.)

1. **P. searlsiaè** Gray. Caudex branched, woody, the year's growth 30–50 cm. high, glabrous, glandular-dotted; lvs. 3–5 cm. long, with 3–5 oblong to oblanceolate lfts. 10–15 mm. long; peduncles 5–12 cm. long, the spikes ca. 1 cm. thick, 1–4 cm. long; calyx 4 mm. long, villous, the lobes lanceolate; corolla rose, ca. 4 mm. long, the petals narrow-clawed; pod 4 mm. long, pubescent; seed 1, kidney-shaped.—Dry gravelly and stony banks, 5000–7000 ft.; Pinyon-Juniper Wd.; Inyo Mts., New York and Providence mts.; to Utah, n. Ariz. May–June.

22. *Pickeringia* Nutt. CHAPARRAL-PEA

Spiny evergreen shrub with stiff branches. Lvs. small, palmately 1–3-foliolate, almost sessile, without stipules. Fls. large, purple, subsessile, solitary. Calyx campanulate, the border with 5 low broad teeth. Corolla with equal petals; standard roundish with reflexed sides, wings and keel petals oblong, the latter distinct and straight. Stamens distinct. Legume linear, flat, stipitate, straight, several-seeded. Monotypic. (Named for Charles *Pickering* of the Wilkes Exploring Expedition.)

1. **P. montàna** Nutt. [*Xylothermia m.* Nutt.] Plate 47, Fig. F. Densely and irregularly branched, 0.5–2 m. tall, the branchlets spinescent, greenish, subglabrous to pubescent; lvs. obovate, to oblanceolate, entire, firm, subglabrous, 4–12 mm. long; calyx ca. 6 mm. long; petals 16–19 mm. long; legume 3–5 cm. long, somewhat constricted between the seeds; seeds black; $2n = 14$ pairs (Raven et al., 1965).—Dry slopes and ridges below 5000 ft.; largely Chaparral; Santa Monica Mts. to Mendocino Co.; Santa Cruz Id. May–Aug.

Rarely fruiting, spreading largely by underground stems after fire. Plants from San Bernardino Mts. to n. L. Calif. tend to have the twigs and lvs. canescent and have been described as spp. **tomentòsa** (Abrams) Abrams. [*Xylothermia t.* Abrams.]

23. Pìsum L. PEA

Annual and perennial herbs with pinnate lvs., the pinnae 1–3 pairs, the rachis ending in a pinnate tendril. Stipules large and foliar. Fls. solitary or in small axillary racemes. Calyx oblique or gibbous near its base, with ± leafy lobes. Corolla white or colored, the wings somewhat adherent to the keel. Stamens 9 and 1. Style bearded down 1 side. Pod flattened, dehiscent, several-seeded. Ca. 6 spp., of Medit. region and w. Asia; grown for food and fodder. (*Pisum*, the classical name.)

    1. **P. satìvum** L. GARDEN PEA. Glabrous, glaucous annual, 1–2 m. tall, scandent; stipules mostly larger than lfts., the lower part denticulate; lfts. entire, oval to oblong, 2–5 cm. long; peduncles exceeding the stipules; fls. 1–3, white, 1.5 or more cm. long; pod 5–10 cm. long; $n=7$ (Sansome, 1933).—Occasional weed, escaping from cult., especially in waste places. Spring. The Field Pea, var. **arvénse** (L.) Poir., has short peduncles; fls. colored with violet banner and purplish wings. Occasional waif.

    Léns culinàris Medic. [*Ervum lens* L.] LENTIL, is much like *Pisum*, but the fls. small, inconspicuous, whitish; calyx-lobes very narrow; pod to 2 cm. long, almost as broad.—Occasional escape as in Ventura Co. From Eurasia.

24. Prosòpis L.

Deciduous shrubs and trees, armed with paired supra-axillary spines. Lvs. bipinnate with 1–2 pairs of pinnae and many small entire pinnules. Fls. small, greenish to yellow, regular, sessile in axillary spikes. Calyx campanulate, 5-toothed. Petals 5, united basally or free. Stamens 10, free and exserted; anthers tipped with a deciduous gland. Pods indehiscent, coriaceous, with numerous seeds separated by a spongy partition. Ca. 30–35 spp. of warm temp. and trop. regions. The pods of our spp. much eaten by cattle. (Greek, *prosopis*, ancient name for Burdock.)

(Johnston, M. C. The N. Am. mesquites. Prosopis and Algarobia. Brittonia 14: 72–90. 1962.)
  Fr. straight or curved, not spirally curved; lvs. bright green with 14–30 pinnules.
    Lfts. more than 5 times as long as broad. Native .... 1. *P. glandulosa*
    Lfts. less than 5 times as long as broad. Occasionally natur. .... 4. *P. velutina*
  Fr. tightly spirally coiled.
    Fls. in cylindrical spikes; lvs. canescent, with mostly 10–16 pinnules. Native .. 2. *P. pubescens*
    Fls. in round heads; lvs. glabrescent, with mostly 12–32 lfts. Escape about Bard, Imperial Co.
        3. *P. strombulifera*

    1. **P. glandulòsa** Torr. var. **torreyàna** (L. Benson) M. C. Jtn. [*P. julifera* and *P. chilensis* of Calif. refs. *P. odorata* Torr. & Frém.] MESQUITE. Plate 48, Fig. A. Low tree or large shrub with several trunks and crooked arched branches, 3–7 m. high; lfts. bright green, glabrous, or nearly so, 7–17 pairs, linear or nearly so, 1.5–2.3 cm. long, mostly 2–2.5 mm. wide; spikes slender, 4–6 cm. long; petals 2.5–3.5 mm. long; pods glabrous, flat, 1–several in a cluster, 5–15 cm. long, constricted between the seeds; seeds 6–7 mm. long, 4–5 mm. broad.—Common in washes and low places, below 4000 ft.; Creosote Bush Scrub, Alkali Sink; Colo. and Mojave deserts; interior cismontane s. Calif.; to Mex. and upper San Joaquin V. April–June.

    2. **P. pubéscens** Benth. SCREW-BEAN. TORNILLO. Plate 48, Fig. B. Shrub or small tree, to 10 m. high, with pale slender twigs and stout spines ca. 1 cm. long; lvs. canescently pubescent, 2–6 cm. long, the lfts. oblong, 3–8 mm. long; fls. yellow, in cylindrical spikes 4–7 cm. long; pods many, sessile, coiled with many turns into a tight springlike cylinder 2–4 cm. long; seeds 3 mm. long. Washes and canyons below 2500 ft.; Creosote Bush Scrub; Mojave Desert to San Joaquin V.; to Utah, Tex., n. Mex. May–July.

    3. **P. strombulífera** (Lam.) Benth. Near to *P. pubescens*, but glabrescent, the fls. in

Plate 48. Fig. A, *Prosopis glandulosa* var. *torreyana*. Fig. B, *Prosopis pubescens*. Fig. C, *Psoralea castorea*. Fig. D, *Psoralea orbicularis*. Fig. E, *Trifolium ciliolatum*. Fig. F, *Thermopsis macrophylla*.

globose heads; legumes twisted, 4–7 cm. long; $2n = 28$ (Schnack & Covas, 1947).—Native of Argentina and grown at the Experimentation Station at Bard, from which it is reported as having escaped.

4. **P. velutìna** Woot. Pinnules oblong, less than 5 times as long as wide; otherwise near to *P. glandulosa*. Native e. of Calif., occasionally natur. in this state, as in Orange Co., San Diego Co., Santa Barbara Co.

25. *Psoràlea* L.

Herbs or shrubs with heavy-scented foliage punctate with dots or glands. Lvs. alternate, 3–5-foliolate; stipules large. Fls. purple or whitish, mainly in pedunculate racemes or spikes. Calyx 5-cleft into subequal lobes or the lower longer, the upper sometimes united. Banner ovate to orbicular, clawed; wings oblong or falcate; keel obtuse, incurved. Stamens monadelphous or diadelphous. Ovary sessile or short-stipitate, 1-ovuled. Pod ovoid, indehiscent or circumscissile or bursting irregularly, seldom exceeding the calyx. Ca. 125 spp., widely distributed, especially in warmer regions. (Greek, *psoraleos*, scurfy or rough, from the glandular dots.)

Lvs. 4–5-foliolate, subpinnate; plants almost acaulescent, from deep-seated fusiform roots.
    Lowest calyx-lobe not much larger than the rest; seeds smooth.
        Pedicels 4–6 mm. long; cismontane; mature calyx villous .... 2. *P. californica*
        Pedicels 1–3 mm. long; mature calyx strigose. Deserts. San Bernardino Co. . 3. *P. castorea*
    Lowest calyx-lobe much larger than the rest; calyx shaggy-villous in maturity. E. Inyo Co.
        5. *P. mephitica*
Lvs. 3-foliolate; plants from rootstocks.
    Stems prostrate; petioles and peduncles erect, much elongated .... 6. *P. orbicularis*
    Stems erect.
        Corolla ochroleucous or whitish with purple-tipped keel; calyx-teeth subequal.
            Lfts. 1–1.5 times as long as wide, broadly ovate; calyx inflated in fr., the lobes 1–2 mm. long .... 7. *P. physodes*
            Lfts. 2–3 times as long as wide, lance-ovate; calyx not inflated in fr., the lobes 3–5 mm. long.
                Fls. racemose; calyx short-pubescent; plant 3–6 dm. high .... 8. *P. rigida*
                Fls. subcapitate or subumbellate; calyx conspicuously white-hairy; plant to 20 dm. high
                1. *P. bituminosa*
        Corolla purple, ca. 1 cm. long; the lower calyx-lobe much longer than others
            4. *P. macrostachya*

1. **P. bituminòsa** L. Half shrub to 2 m. tall; lvs. 3-foliolate; lfts. lanceolate, 2–5 cm. long; stipules lance-subulate, 5–7 mm. long; fls. lilac with an almost white keel and banner somewhat red in age.—Millard Canyon, San Gabriel Mts., *Griesel* at ca. 2000 ft. Possibly established here after being tested as a fire-resistant plant by forest personnel. From Old World.

2. **P. califórnica** Wats. Crown branched, the stems to 1 or 2 dm. high; herbage silvery-villous with appressed hairs; lower petioles 4–8 cm. long, the blades 5–6-foliolate; lfts. cuneate-obovate, 1–3 cm. long; upper lvs. reduced; racemes in upper axils, on peduncles 1–4 cm. long, dense, 1.5–3.5 cm. long; calyx densely white-villous, the campanulate tube ca. 4 mm. long, the lobes 5–6 mm. long, the lower somewhat broader than the other 4; corolla ca. 12 mm. long, the banner whitish, wings and keel purple; pod oblong-ovoid, 5–6 mm. long, the beak slightly more; seeds dark, ± mottled; $2n = 11$ pairs (Raven et al., 1965).—Dry slopes 1500–5500 ft.; Chaparral, Pinyon-Juniper Wd., etc.; scattered stations in n. L. Calif., San Jacinto and San Bernardino mts., Kern Co., Ventura Co. to cent. Calif. May–June.

3. **P. castòrea** Wats. Plate 48, Fig. C. Resembling *P. californica*; herbage more closely strigose; petioles 5–12 cm. long; lfts. 2–4 cm. long; pedicels stouter; calyx strigose, the lobes 8–10 mm. long, the lowest 1–2 mm. longer than the others; corolla bluish, 1 cm. long; body of pod ca. 8 mm. long, the beak 14–15 mm. long; seed ca. 7 mm. long; Creosote Bush Scrub, Joshua Tree Wd.; Mojave Desert from Victorville to Yermo; to Utah, Ariz. April–May.

4. **P. macrostàchya** DC. Stems 0.5–3 m. tall, subglabrous to finely pubescent; petioles 3–10 cm. long; lfts. 2–8 cm. long, lance-ovate to ovate-rhombic, obscurely gland-dotted; peduncles 4–10 cm. long; spikes 5–12 cm. long, silky-villous with black or white hairs; calyx villous, unequally lobed, 6–8 mm. long, the lowest lobe equaling or surpassing corolla, the latter purple, 8–10 mm. long; pod 6–8 mm. long, pubescent; $2n=11$ pairs (Raven et al., 1965).—Moist places below 5000 ft.; many Plant Communities, most of cismontane Calif.; to n. L. Calif. Variable as to pubescence, length of racemes, of lowest calyx-lobe, etc. May–Aug.

5. **P. mephítica** Wats. Resembling *P. castorea* and *P. californica* in acaulescent habit and grayish cast; petioles 4–6 cm. long; lfts. 1–3 cm. long; peduncles 2–5 cm. long, with spreading or reflexed white hairs; spikes ca. 2 cm. long, dense; calyx long-hairy; tube 4 mm. long, lobes 7–9 mm. long, the lowest one lanceolate, the other 4 subulate; corolla blue, ca. 12 mm. long; banner oblong.—Creosote Bush Scrub; Emigrant Canyon, Death Valley, e. Inyo Co.; to Utah, Ariz. May.

6. **P. orbiculàris** Lindl. Plate 48, Fig. D. Stems prostrate, creeping at nodes; petioles 10–50 cm. long; lfts. round-obovate, 3–8 cm. long, glabrous to short-pubescent; peduncles 20–70 cm. high; racemes dense, 5–30 cm. long; calyx densely villous, the tube 4–5 mm. long, the lobes linear-lanceolate, the lowest one ca. 15 mm. long, the others 10 mm.; corolla reddish-purple, ca. 15 mm. long, banner often with a white spot on each side; pod hirsute, ca. 8 mm. long, the beak small, straight.—Moist places, below 5000 ft.; many Plant Communities; from much of cismontane Calif. May–July.

7. **P. physòdes** Dougl. From a creeping rootstock, the stems erect, 3–7 dm. high, glabrous or ± black-hairy, grooved; petioles 2–5 cm. long; lfts. ovate, 2–6 cm. long; peduncles 3–10 cm. long; racemes dense, 1.5–2.5 cm. long, black-villous; calyx black- and white-hairy, the tube 4 mm. long, becoming 6–8 mm. in fr.; lobes deltoid, acuminate; corolla 10–12 mm. long; pod ca. 6 mm. long, hairy and punctate; $n=11$ (Raven et al., 1965).—Open places in brush or woods, below 7500 ft.; many Plant Communities; cismontane from Orange and San Bernardino co. to B.C. April–June.

8. **P. rígida** Parish. From a rootstock, erect, 3–6 dm. high, sparsely puberulent; petioles 2–4 cm. long, lfts. 3–10 cm. long; peduncles 3–7 cm. long; racemes dense, 2–3 cm. long; calyx short-pubescent, the tube 4–5 mm. long, the lobes 4–5 mm. long, the lowest slightly exceeding others; corolla 10–12 mm. long; pod 8–10 mm. long, strigose.— Dry slopes 1000–7000 ft.; Chaparral, Yellow Pine F.; Laguna Mts. to San Bernardino Mts. June–July.

26. *Robínia* L. LOCUST

Trees or shrubs, often with spiny stipules. Lvs. odd-pinnate, the lfts. ovate or oblong and with stipule-like appendage at base. Fls. showy, in drooping axillary racemes. Calyx with 5 short teeth, the 2 upper ± united. Banner large and rounded, not much exceeding wings and keel. Stamens 10, diadelphous. Pod linear, flat, several-seeded, tardily 2-valved. Ca. 8 spp. of e. N. Am. (Named for Jean and Vespasian *Robin*, 16th century, who first cultivated the Locust Tree in Eu.)

1. **R. pseùdo-acàcia** L. Tree with rough bark and subglabrous twigs and lvs.; lfts. 9–19, petiolulate, 2–5 cm. long; pedicels 6–12 mm. long; fls. whitish, fragrant, 15–20 mm. long; pods 5–10 cm. long; $n=10$ (Kreuter, 1930).—Natur. occasionally, especially where planted for erosion control, etc.; easily spreading by underground parts. May–June.

27. *Sesbània* Adans., corr. Scop.

Herbs or shrubs with abruptly pinnate lvs. and numerous lfts. Stipules small, scarious, caducous. Fls. solitary or in axillary racemes, commonly yellow. Calyx campanulate, the lobes shorter than the tube. Banner with a rounded reflexed blade; wings oblong; keel lunate, blunt. Stamens diadelphous. Pod slender, linear, terete, short-stipitate, many-seeded, partitioned between the seeds. Seeds many. Ca. 15 spp. of warm and trop. regions. (*Sesban*, Arabic name of 1 sp.)

1. **S. exaltàta** (Raf.) Cory. [*Darwinia e.* Raf. *S. macrocarpa* Muhl.] COLORADO-RIVER-HEMP. Glabrous annual 0.4–3 m. high, the stems striate; lfts. 20–80, linear-oblong, 1–2 cm. long, pale green; fls. few; pedicels 5–10 mm. long, slender; calyx 4–5 mm. long, the lobes triangular-subulate; corolla 15 mm. long, yellowish, the banner purple-dotted; pod 10–20 cm. long; $2n=12$ (Atchison, 1949).—Frequent in overflow lands, along ditches, etc.; Alkali Sink; Imperial Co. and e. Riverside Co.; occasional, San Diego and Los Angeles cos.; to Atlantic Coast and Cent. Am. April–Oct.

28. *Spártium* L. SPANISH BROOM

Mostly a rather tall virgately branched shrub with long slender leafless or few-lvd. rushlike branchlets. Lvs. alternate, simple, entire, small. Fls. yellow, in loose terminal racemes. Calyx split above, hence 1-lipped, with 5 minute teeth. Banner and keel longer than the wings, keel pubescent along its lower edge. Stamens monadelphous. Pod linear, compressed, many seeded. Seeds with basal strophiole. Monotypic. Medit. region. (Greek, *sparton*, broom.)

1. **S. juncèum** L. To ca. 3 m. high; lvs. oblance-oblong or narrower, 1–3 cm. long, ± strigose, short-petioled; fls. 2–2.5 cm. long, fragrant; pod 5–10 cm. long, ± strigose; $n=24$–26 (Tscheschow, 1931).—Natur. in dry and waste places, on roadcuts, etc. April–June.

29. *Thermópsis* R. Br. FALSE-LUPINE

Stout perennial herbs, with sheathing scales at base and mostly erect clustered stems. Lvs. palmately 3-foliolate, petioled, with free foliaceous stipules. Fls. yellow in ours, in a terminal raceme, with persistent bracts. Calyx campanulate, the teeth equal or the upper 2 united. Banner suborbicular, shorter than to ca. as long as the oblong wings; keel almost straight. Stamens 10, distinct. Legume narrow, flat, 2-valved, few- to many-seeded. Perhaps 20 spp., N. Am., Asia. (Greek, *thermos*, lupine, and *opsis*, likeness.)

1. **T. macrophýlla** H. & A. var. **macrophýlla**. [*T. californica* Wats.] Plate 48, Fig. F. Stems 3–8 dm. high, the whole plant ± short villous-tomentose with somewhat spreading hairs; stipules ovate-cordate, 2–3.5 cm. long; lfts. obovate-cuneate to oblanceolate, 4–7 cm. long, 3–4 cm. broad; racemes 1.5–2.5 dm. long; bracts 0.5–1 cm. long; pedicels 4–6 mm. long; calyx strigose, 5–7 mm. long; fls. 17–19 mm. long, the banner almost as long as wings and keel; pod densely strigose, erect, 6–8 cm. long, few-seeded.—Open places below 4500 ft.; Foothill Wd., etc.; Coast Ranges from Ventura Co. n. to Ore. April–June.

Var. **semòta** Jeps. Herbage densely velvety, almost shaggy.—Grassy places and among trees, 4000–5500 ft.; Yellow Pine F., mts. of e. San Diego Co. May–June.

30. *Trifòlium* L. CLOVER

Herbs with mostly palmately trifoliolate lvs. and adnate stipules. Lfts. sometimes 4–7. Heads white, yellow, pink to rose or purple, capitate to short-spicate, or subumbellate. Calyx 5-toothed, the lobes equal or unequal, entire to bifid or trifid. Petals usually persistent. Stamens diadelphous. Pods globose to elongate, 1–2–8 seeded, included within the persistent calyx, dehiscent or not. Ca. 300 spp., most abundant in N. Temp., but also in S. Am. and Afr. (Latin, *tres*, three, and *folium*, lf.)

1. Heads without an invol. at base of fls.
   2. Plants annual.
      3. Fls. pedicellate, reflexed in age; calyx mostly glabrous (except in *bifidum*).
         4. Fls. yellow; calyx 5-nerved. Introd. spp.
            5. Heads 8–15 mm. long; banner dilated, not folded over the pod, conspicuously veined
               .... 21. *T. procumbens*
            5. Heads 5–10 mm. long; banner not dilated, folded over the pod, not conspicuously veined .... 8. *T. dubium*
         4. Fls. purple; calyx 10-nerved. Native spp.

Trifolium

6. Plants ± villous on peduncles and calyx .... 4. *T. bifidum*
6. Plants nearly or quite glabrous.
   7. Calyx not ciliolate on the lobes.
      8. Stipules lance-ovate, 8–10 mm. long; lfts. obovate, the serrations not setaceous. Widely distributed .... 10. *T. gracilentum*
      8. Stipules lance-linear, 15–20 mm. long; lfts. almost linear, serrulate with setaceous teeth. Insular .... 19. *T. palmeri*
   7. Calyx ciliolate on the lobes with short flat appendages .... 5. *T. ciliolatum*
3. Fls. sessile, not reflexed in age; calyx densely villous.
   9. Calyx 20-nerved; stipules with a long bristle-tip; heads 1.5–2.5 cm. in diam. Introd. sp. 11. *T. hirtum*
   9. Calyx 10-nerved; stipules usually not bristle-tipped; heads often smaller. Native spp.
      10. Heads sessile, usually in pairs, subtended by the upper lvs. and their stipules 14. *T. macraei*
      10. Heads peduncled, solitary, not subtended by upper lvs. .... 1. *T. albopurpureum*
2. Plants perennial.
   11. Peduncles axillary; fls. on pedicels mostly 4–5 mm. long, recurved in fr.
      12. Stems erect or ascending, not rooting at nodes; stipules 10–25 mm. long, with long hairlike tip; corolla pink .... 12. *T. hybridum*
      12. Stems creeping, rooting at nodes; stipules 4–10 mm. long, lance-ovate; corolla mostly white .... 22. *T. repens*
   11. Peduncles terminal or subterminal; fls. sessile or on shorter pedicels.
      13. Heads with long peduncles. Native of mt. meadows .... 13. *T. longipes*
      13. Heads sessile, subtended by uppermost lvs. Introd. and escaped about fields, etc. 20. *T. pratense*
1. Heads with an invol. at the base of the fls., this reduced to a mere ring in *T. depauperatum*.
14. Corolla not inflated in age; involucral bracts united.
   15. Invol. campanulate to bowl-shaped.
      16. Lobes of the invol. entire; heads pubescent, small .... 15. *T. microcephalum*
      16. Lobes of the invol. toothed.
         17. Calyx-teeth 1–3 times trichotomously forked, glabrous .... 6. *T. cyathiferum*
         17. Calyx-teeth simple, or the upper one forked, all hairy or ciliate.
            18. The calyx-teeth plumose, the upper one forked; corolla purple . 3. *T. barbigerum*
            18. The calyx-teeth ciliate, all simple; corolla white to pink .... 16. *T. microdon*
   15. Invol. flat, rotate.
      19. Plants perennial.
         20. Corolla cream-color with purple-tipped keel; fls. 1–8 in a head . 17. *T. monanthum*
         20. Corolla white to light purple; fls. many .... 25. *T. wormskioldii*
      19. Plants annual.
         21. The plants viscid-pubescent and clammy; corolla 12–14 mm. long, pale with dark cent. spot .... 18. *T. obtusiflorum*
         21. The plants glabrous.
            22. Lfts. obovate to ± oblanceolate to oblong; invol. with 4–12 lobes, these 3–7-toothed; calyx-teeth simple or one bifid .... 24. *T. variegatum*
            22. Lfts. linear to lance-oblong; invol. unevenly laciniate but not lobed; calyx-teeth often with 2 small lateral teeth on shoulders or simple, entire .. 23. *T. tridentatum*
14. Corolla conspicuously inflated in age; involucral bracts distinct.
   23. The fls. 12–25 mm. long; involucral divisions 6–18 mm. long .... 9. *T. fucatum*
   23. The fls. 5–8 mm. long; involucral bracts to ca. 4 mm. long.
      24. Involucral lobes evident .... 2. *T. amplectens*
      24. Involucral lobes lacking, the invol. consisting of a mere ring ..... 7. *T. depauperatum*

1. **T. albopurpùreum** T. & G. [*T. traskae* Kennedy. *T. insularum* Kennedy.] Villous-pubescent decumbent to ascending annual, with slender stems 1–4 dm. long; stipules entire, ovate-lanceolate; lfts. obovate to cuneate-oblong, denticulate above the middle, 0.6–1.8 cm. long; peduncles 5–15 cm. long; heads ovoid, 8–15 mm. long; fls. sessile; calyx-teeth plumose-villose, subulate; corolla 6–7 mm. long, purple, slightly shorter than calyx; pods ± hairy at apex, 1-seeded; $2n=16$ (Wexelsen, 1928).—Open grassy places below 4700 ft.; V. Grassland, S. Oak Wd., etc.; cismontane Calif.; to B.C. Santa Catalina and Santa Cruz ids. March–June.

2. **T. ampléctens** T. & G. Light green annual, glabrous, with slender decumbent stems 1–2.5 dm. long; stipules broad, entire, with a subulate point; lfts. obovate to oblanceolate, emarginate to truncate, serrulate, 0.6–2 cm. long; peduncles mostly to ca. 4.5 cm. long; invol. deeply lobed, the lobes 3–4 mm. long, broad, scarious-margined, entire or 1–2-toothed; calyx ca. 4–5 mm. long, the lower teeth setose, longer than the tube; corolla ca. 6 mm. long, white to reddish to purple, much inflated in age; pods 4–6-seeded.—Grassy places below 2500 ft.; V. Grassland, Coastal Sage Scrub, S. Oak Wd., etc.; San Diego Co. to cent. Calif. where more common. April–June. Channel Ids. A form growing with the sp. and more common in s. Calif. is the var. *truncatum* (Greene) Jeps. with involucral bracts oblong, ca. as long as the calyx, without or with very narrow hyaline margins.

3. **T. barbígerum** Torr. Pubescent to subglabrous annual, with several rather stout procumbent stems 1–3 dm. long; stipules broadly ovate, ca. 10 mm. long, scarious, laciniate; lfts. obovate to deltoid, setate-serrulate, 4–12 mm. long; peduncles slender, 4–9 cm. long; invol. 10–14 mm. wide, shallowly bowl-shaped when young, flatter later, short-lobed and with many setaceous teeth; calyx-teeth awned, plumose, the upper one once or twice forked, slightly exceeding the dark purple corolla which is ca. 5–6 mm. long; pods 2-seeded.—Open low moist places; San Miguel and Santa Rosa ids.; to Del Norte Co. April–June.

4. **T. bífidum** Gray var. **decípiens** Greene. Pale green annual, with erect very slender stems 1.5–4 dm. high, usually with some long slender hairs on upper parts; petioles mostly 2–6 cm. long; stipules lance-ovate, entire, 8–15 mm. long; lvs. oblanceolate to obcordate, rounded to retuse at apex; peduncles slender, 3–10 cm. long; fls. 6–15; pedicels recurved, 1–3 mm. long; calyx-teeth subulate-setaceous, almost equaling the corolla; petals pale pink to purplish, 4–5 mm. long, the banner strongly veined; pod included, 1-seeded.—Grassy places near San Diego, in Santa Barbara Co.; n. to Ore. April–June.

5. **T. ciliolàtum** Benth. Plate 48, Fig. E. Glabrous pale green annual with erect ± fistulous stems 2–5 dm. high; petioles to ca. 10 cm. long; stipules lanceolate-acuminate, 15–30 mm. long; lfts. oblong to ± obovate, 1–3 cm. long, obtuse, entire to serrulate; peduncles 5–15 cm. long; heads ovoid, 10–20 mm. long; pedicels reflexed in age; calyx shorter than the corolla, the teeth lance-acuminate, ciliate with short flat appendages; corolla pinkish-purple, 6–7 mm. long, the banner inflated at the base; pods 1–2-seeded; $n=8$ (Gillett & Mosquin, 1967).—Frequent on open grassy slopes below 5000 ft.; many Plant Communities; most of cismontane Calif.; to Wash. March–June.

6. **T. cyathíferum** Lindl. Glabrous annual with erect or decumbent stems 1–3 dm. long; stipules lanceolate to ovate, laciniately toothed; lfts. obovate to elliptic-oblong, acute or obtuse, ± spinulose-denticulate, 1–2 cm. long; peduncles slender, mostly 2–9 cm. long; invol. 8–20 mm. broad, bowl-shaped, membranous, nerved and unequally toothed; calyx-teeth 1–3 times trichotomously forked, equaling the pink to paler corolla which is 7–10 mm. long; pods 2-seeded; $n=8$ (Gillett & Mosquin, 1967).—Moist places below 8000 ft.; Montane Coniferous F.; Mt. Pinos, Ventura Co.; to B.C., Ida. May–Aug.

7. **T. depauperàtum** Desv. Glabrous annual with ascending stems to ca. 12 cm. long; stipules entire, subulate-tipped; lfts. cuneate-oblong, denticulate, 0.6–2 cm. long; peduncles slender, 3–6 cm. long; invol. reduced to a small ring; calyx ca. 2–2.5 mm. long, the teeth unequal, triangular-subulate; corolla whitish to purple, 7–8 mm. long in age; pods 1–2-seeded; $2n=16$ (Mosquin & Gillett, 1965).—Santa Cruz Id.; V. Grassland, etc.; to Ore.; Chile. March–May.

8. **T. dùbium** Sibth. Much like *T. procumbens*; heads 5–10 mm. long; fls. 5–15, ca. 2.5–3.5 mm. long; banner much less conspicuously veined; $2n=14$, 16, 28 (several reports).—Waste places, lawns, etc.; Santa Barbara; cent. Calif. to B.C., Atlantic Coast; natur. from Eu. May–July.

9. **T. fucàtum** Lindl. var. **fucàtum.** Glabrous annual with stout fistulous diffuse stems 1–8 dm. long; stipules ovate, 1.5–2 cm. long, entire; lfts. rhombic-ovate to obovate, 6–30 mm. long, denticulate to almost setulose; peduncles 3–15 cm. long, stout; invol. with 5–9 entire, setulose-tipped scarious lobes; calyx scarious, 4–5 mm. long, the subulate

teeth simple, shorter than to longer than the tube; corolla cream, becoming pink and much inflated in age, then 20–25 mm. long, the keel dark purple; pods stipitate, 3–8-seeded; $n=8$ (Gillett & Mosquin, 1967).—Moist places, ± brackish, below 3000 ft.; many Plant Communities; cismontane Calif.; to Ore. April–June.

Var. **gambèlii** (Nutt.) Jeps. [*T. g.* Nutt.] Calyx 8–10 mm. long, the teeth 5, much elongate, bristlelike, 1 or more being 2–3-cleft; corolla 15–20 mm. long, white with yellow and purple tinge.—Grassy coastal bluffs; Los Angeles Co. to Contra Costa Co.; Channel Ids.

10. **T. graciléntum** T. & G. Practically glabrous annual, the stems slender, erect to procumbent, 1–4 dm. long; petioles commonly 2–7 cm. long; stipules lance-ovate, entire, 8–10 mm. long; lfts. obovate, 0.6–1.5 cm. long, serrulate, emarginate at apex; peduncles 2–6 cm. long; heads 6–10 mm. long; pedicels reflexed in age, 1–3 mm. long; calyx-teeth subulate-lanceolate, entire, shorter than the corolla; petals pink to reddish-purple, 5–6 mm. long; pods 1–2-seeded.—Common in open and grassy places; below 5000 ft.; V. Grassland, Chaparral, S. Oak Wd., etc.; cismontane and occasional along w. edges of deserts; to B.C., L. Calif. April–June. A poorly marked form is var. *inconspícuum* Fern., mostly less than 1.5 dm. high; fls. mostly pinkish, scarcely exceeding calyx.—Mostly toward s. part of range.

11. **T. bírtum** All. Densely long-villous annual, 1–4 dm. high; stipules narrow, with a long bristle-tip; lfts. narrowly cuneate-obovate, 0.8–2 cm. long; heads globose, sessile, subtended by paired lvs. or their stipules, 1.5–2.5 cm. in diam.; calyx 20-nerved, densely hirsute with ascending hairs, the setaceous lobes 4–6 mm. long, exceeding the tube; corolla purple, slender, longer than the calyx.—Goleta, Santa Ynez Mts., etc.; to cent. Calif.; natur. from Eu. April–May.

12. **T. hýbridum** L. ALSIKE CLOVER. Subglabrous perennial, with stems 3–6 dm. long, ± stout and succulent; petioles commonly 2–8 cm. long; stipules lance-ovate, 1–2.5 cm. long, with long hairlike tips; lfts. obovate to ovate, 1–3 cm. long, sharp-serrulate; peduncles to ca. 1 dm. long; heads many-fld., 1.5–2 cm. in diam.; pedicels 4–8 mm. long; calyx 3–4 mm. long, pubescent in the sinuses between the subulate teeth, corolla pink, 7–8 mm. long; pods 3–4-seeded; $2n-16$ (Kawakami, 1930).—Natur. in lawns, damp places, etc. in Santa Barbara region and n. to Wash., Atlantic Coast. Native of Eu. May–Oct.

13. **T. longipès** Nutt. ssp. **atrorùbens** (Greene) J. M. Gillett. [*T. rusbyi* var. *a.* Greene.] Plate 49, Fig. A. Perennial from slender taproot, the stems decumbent to erect, ± strigose, 0.5–4 dm. long; stipules green, lanceolate, entire, 0.5–2 cm. long; lfts. elliptic, 0.5–3 cm. long; peduncles mostly 3–8 cm. long; they and calyx densely white-pilose with tangled hairs; corolla ca. 10–12 mm. long, the banner usually purple, wings and keel ochroleucous; pods 2–4-seeded.—Moist places from 6500–7500 ft.; Montane Coniferous F.; San Jacinto and San Bernardino mts.; s. Sierra Nevada. June–Sept.

14. **T. macràei** H. & A. [*T. catalinae* Wats.] Decumbent or spreading, soft-pubescent, mostly strigose annual with several stems 1–3 dm. long; stipules ovate, often acuminate, green; lfts. oblanceolate to obovate, 5–15 mm. long, obtuse to emarginate, serrulate; heads sessile, often in pairs, in axils of uppermost sessile lvs.; calyx 6–7 mm. long, the subulate teeth densely plumose, linear-subulate; corolla purple, ca. as long as calyx; $2n=16$ (Gillett & Mosquin, 1967).—Open grassy places, various Plant Communities near the coast; Santa Barbara Co. n.; Channel Ids. Chile. March–May.

15. **T. microcéphalum** Pursh. Mostly slender-stemmed annual, ± villous, the stems procumbent to ascending, 2–4 dm. long; stipules ovate, acuminate, ± toothed; lfts. obcordate to oblanceolate, retuse, serrate, 0.5–1.5 cm. long; peduncles very slender, commonly 3–7 cm. long; invols. ca. 5–8 mm. broad, deeply campanulate, the lobes 7–10, lanceolate, entire, with scarious weblike margins; calyx ca. 4 mm. long, pubescent, the teeth subulate; corolla rose to white, ca. 6 mm. long; pods 1–2-seeded; $n=8$ (Gillett & Mosquin, 1967).—Open grassy places below 7500 ft.; many Plant Communities; cismontane Calif., Channel Ids.; to B.C., n. L. Calif., Nev. April–Aug.

16. **T. mìcrodon** H. & A. var. **pilòsum** Eastw. Rather slender branched, ± erect,

Plate 49. Fig. A, *Trifolium longipes*. Fig. B, *Trifolium monanthum*. Fig. C, *Trifolium tridentatum*. Fig. D, *Trifolium wormskioldii*. Fig. E, *Quercus agrifolia*, ♂ and ♀ fls. Fig. F, *Quercus dumosa*.

subglabrous annual, 2–several cm. high; stipules lance-ovate, abruptly acuminate, entire or somewhat toothed; lfts. obcordate to oblanceolate, 0.5–2 cm. long, ± setate-serrulate; peduncles slender, densely woolly-pubescent; invols. 5–10 mm. wide, deeply campanulate, densely woolly-pubescent; calyx 4–5 mm. long, the short triangular teeth abruptly apiculate, ciliate; corolla 5–6 mm. long, white to pale rose; pod 1-seeded.—San Nicolas and Santa Cruz ids.; March–May.

17. **T. monánthum** Gray. var. **monánthum.** Plate 49, Fig. B. Glabrous to sparingly villous perennial from a taproot and with slender decumbent to suberect stems 1–1.5 dm. long; stipules lanceolate, mostly subentire; lfts. obcordate to oblanceolate, retuse to truncate or rounded at the apex, 4–12 mm. long, 2–5 mm. wide, sparingly toothed; infl. 1–2 (–4)-fld.; peduncles straight, the fls. erect; invol. small, 2–4-lobed, the lobes 1–5 mm. long; calyx 2–4 mm. long, the teeth not longer than the tube; corolla cream-colored with purple-tipped keel, 8–12 mm. long; pod 1–3-seeded.—$2n=16$ (Gillett & Mosquin, 1967). —Wet places, 5000–7000 ft.; Yellow Pine F.; San Gabriel Mts.; to the Sierra Nevada; Nev. June–Aug.

Var. **grantiànum** (Heller) Parish. [*T. g.* Heller.] Glabrous to sparingly villous; stems 0.3–3 dm. long; infl. mostly 3–6-fld., usually on straight peduncles; involucral lobes mostly 3–5 mm. long.—At 5000–9000 ft.; Montane Coniferous F.; San Gabriel, San Bernardino, and San Jacinto mts.

18. **T. obtusiflòrum** Hook. Soft-pubescent annual, clammy, the stems ± fistulous, 3–5 dm. long, erect to decumbent-ascending; stipules rather broad, conspicuously setulose; lvs. lance-linear to narrow-obovate, 2–3 cm. long, setulose-serrate; peduncles 3–8 cm. long; invol. flat, 12–16 mm. wide, deeply laciniate with subulate divisions; calyx-tube mostly 20-nerved, the entire teeth dilated near the base; corolla 12–14 mm. long, pale with cent. dark spot; pods 2-seeded; $2n=16$ (Wexelsen, 1928).—Moist places, below 5000 ft.; many Plant Communities; cismontane Calif. to Ore. April–July.

19. **T. pálmeri** Wats. Glabrous, branched annual, the stems 1–3 dm. long; petioles mostly 2–5 cm. long; stipules 15–20 mm. long, lance-linear, with long setaceous tips; lfts. lance-linear, 1–3 cm. long, acute, serrulate with setaceous teeth; heads 10–15 mm. broad; calyx almost as long as purple corolla, the latter 6–7 mm. long; pods 2-seeded. Grassy places, San Clemente and Santa Catalina ids.; Guadalupe Id. March–May.

20. **T. praténse** L. Red Clover. Perennial, several-stemmed, 2–6 dm. high, pubescent; upper stipules ovate, membranous, conspicuously veined, subulate-tipped; lfts. oval to obovate, obtuse, pubescent, entire or crenulate, 2–5 cm. long, often with a pale blotch; heads subtended by 1–2 lvs., ovoid to obovoid, 2–3 cm. broad; fls. many, sessile; calyx 5–8 mm. long, villous, the teeth subulate, slightly longer than the tube; corolla red or pink, 10–20 mm. long; pod 2-seeded; $2n=14, 28, 56$ (Tatuno & Kodama, 1965).—A variable sp., widely cult. for hay and pasturage; natur. in many places in Calif.; native of Eurasia. April–Oct.

21. **T. procúmbens** L. Hop Clover. Erect annual, 2–5 dm. high, somewhat strigose; petioles 1–1.5 cm. long; stipules ovate, broadly rounded at base; terminal lft. petiolulate; heads 8–15 mm. long, globose to short-cylindric; fls. mostly 20–30, and 3.5–4.5 mm. long; style much shorter than the pod; $n=7$ (Larsen, 1955).—Reported from lawns, Santa Barbara, etc.; to B.C., Atlantic Coast; native of Eu. May–Sept.

22. **T. rèpens** L. White Clover. Glabrous perennial with creeping stems 1–3 dm. long and rooting at the nodes; stipules lance-ovate, 4–10 mm. long; lfts. 1–2 cm. long, obovate with cuneate base, serrulate, ± emarginate at apex; peduncles 5–30 cm. high; heads globose, 1.5–2.5 cm. broad; pedicels 2–4 mm. long; calyx 4–6 mm. long, the teeth subulate; corolla white to pale pinkish, 7–10 mm. long; $n=16$ (Atwood & Hill, 1940).— Planted in lawns and meadows; natur. in wet places, especially in the mts.; native of Eu. April–Dec.

23. **T. tridentàtum** Lindl. var. **tridentàtum.** Plate 49, Fig. C. Glabrous annual, with erect to ± decumbent stems 1–4 dm. long; lower stipules lanceolate, acuminate, entire, the upper round-ovate, laciniate; lfts. linear to lance-oblong, 1.5–3.5 cm. long, ± setate-denticulate; peduncles 5–9 cm. long; invol. flat, unevenly laciniate, but not

lobed, commonly 10–15 mm. broad; calyx-teeth shorter than the 10-nerved tube, dilated below, often with 2 small lateral teeth or shoulders; corolla 12–15 mm. long, red-purple, the banner pale toward tip, the wings dark; pods mostly 2-seeded; $2n = 16$ (Gillett & Mosquin, 1967).—Common in mostly grassy places below 5000 ft.; many Plant Communities; cismontane and to desert edges; Channel Ids.; to B.C. March–June.

Var. **aciculàre** (Nutt.) McDer. [*T. a.* Nutt.] Calyx-tube 20–25-nerved, the teeth mostly entire.—Commoner form in s. Calif. and cent. V. of Calif. Santa Catalina and San Clemente ids.

24. **T. variegàtum** Nutt. in T. & G. Glabrous annual, with usually several stems, decumbent or ascending, 1–6 dm. long; stipules ovate, laciniately toothed; lfts. obovate to ± oblanceolate or oblong, 5–15 mm. long, setose-serrulate; peduncles slender, 1–8 cm. long; fls. 1–many; invol. with 4–12 lobes, these 3–7-toothed; calyx 5–20-nerved, the teeth subulate-setaceous, simple or 1 bifid; corolla purple with white tip, 5–8 mm. long; pods 1–2-seeded; $2n = 16$ (Wexelsen, 1928), $n = 8$ (Gillett & Mosquin, 1967).—Moist places below 8000 ft.; many Plant Communities; common throughout cismontane Calif.; n. Channel Ids. A variable sp. April–July.

25. **T. wormskióldii** Lehm. [*T. involucratum* Ort., not Lam.] Plate 49, Fig. D. Glabrous perennial with creeping rootstocks and branched decumbent stems 1–3 dm. long; stipules lanceolate, laciniately toothed; petioles commonly 2–7 cm. long; lfts. oblanceolate to wider, 1–3 cm. long, acutish to mucronulate and obtuse, setulose-serrulate; peduncles 2–6 cm. long; invol. mostly 12–15 mm. broad, flattish, mostly ± lobed and then toothed; calyx 10-nerved, ca. 8–9 mm. long, the teeth subulate; corolla ca. 12 mm. long, the banner broad, white to light purple, the wings and keel dark; pods 2–6-seeded; $n = 16$ (Gillett & Mosquin, 1967), $2n = $ ca. 48 (Wexelsen, 1928).—In wet places mostly below 8000 ft.; Montane Coniferous F. and below; cismontane and montane Calif.; occasional in desert (Panamint, Lovejoy Buttes, etc.); to B.C. and Rocky Mts. Variable. May–Oct.

31. Ûlex L. Furze. Gorse

Densely branched shrubs with stiff spinescent branches. Lvs. simple, stiff, spinose. Fls. yellow, showy, solitary or racemose. Calyx yellow, deeply 2-lipped, the upper lip 2-toothed, the lower 3-toothed. Banner ovate, wings and keel oblong, obtuse. Stamens monadelphous. Pods short-oblong, several-seeded. Ca. 20 spp. of w. Eu. and N. Afr. (Old Latin name of some similar plant.)

1. **U. europaèus** L. Shrubs 1–2 m. tall, ± pubescent; lvs. acicular, 5–15 mm. long; calyx 10–15 mm. long, pubescent; corolla 15–18 mm. long; pods villous, dark, 15–18 mm. long; seeds dark, smooth, shining, ca. 2 mm. long.—Natur. sparingly in s. Calif. along the coast as at San Diego, Point Firmin in Los Angeles Co.; much more abundant and pernicious farther north; native of Eu. Feb.–July.

32. Vícia L. Vetch

Herbs, mostly vinelike, with pinnate, usually tendril-bearing lvs. and evident stipules. Fls. or peduncles axillary, the fls. 1–many, purple to yellowish or white. Calyx 5-cleft or -toothed, the 2 upper teeth often shorter, or the lowest longer. Petals clawed, the wings adhering to the middle of the keel. Stamens ± diadelphous (9 + 1), the orifice of the tube oblique. Style filiform, hairy all around or only on the back at the apex or beneath the stigma. Pod mostly laterally compressed, 2-valved, 2–several-seeded. Seeds globular to somewhat compressed. Ca. 130 spp. of N. Hemis. and temp. S. Am. (The classical Latin name.)

1. Fls. borne near the ends of evident peduncles.
  2. The fls. 4–5 mm. long; pods glabrous; slender-stemmed annual . . . . 5. *V. exigua*
  2. The fls. 10–18 mm. long.
    3. Lfts. 4–8 pairs; peduncles 3–9 (–10)-fld.
      4. Plants perennial, native; racemes mostly shorter than lvs. . . . . 1. *V. americana*

4. Plants annual, introd.; racemes equaling or exceeding lvs. .... 3. *V. benghalensis*
 3. Lfts. mostly 8-12 pairs; peduncles with dense 1-sided, many-fld. racemes; plants annual.
  5. Stems spreading-villous; lfts. to ca. 1 cm. long .... 9. *V. villosa*
  5. Stems glabrous or ± strigose; lfts. 1-2 cm. long .... 4. *V. dasycarpa*
1. Fls. 1-4 in lf.-axils, sessile or subsessile. Introd.
  6. Lvs. without tendrils; lfts. 1-3 pairs; stems mostly simple, erect .... 6. *V. faba*
  6. Lvs. with tendrils; lfts. more than 3 pairs.
   7. Banner pubescent on back; fls. 2-4, yellow-white, 20-25 mm. long ..... 7. *V. pannonica*
   7. Banner glabrous on back; fls. mostly 1-2.
    8. The fls. 18-25 mm. long, purple with violet wings; pod ± torulose; seeds 5 mm. broad
        8. *V. sativa*
    8. The fls. 10-18 mm. long, uniformly purple; pod plane; seeds 3 mm. broad
        2. *V. angustifolia*

1. **V. americàna** Muhl. ex Willd. [*V. oregana* Nutt. in T. & G. *V. truncata* Nutt. in T. & G. *V. a.* var. *linearis* Wats. *V. californica* Greene.] Trailing or climbing perennial, subglabrous to villous-pubescent, 2-12 dm. tall, often zigzag; stipules incisely toothed or laciniate; lfts. 4-8 pairs, ovate to elliptic or cuneate-obovate to narrow, 1-4 cm. long, obtuse to rounded or truncate at apex; peduncles shorter than lvs., 4-9-fld.; calyx-teeth deltoid-lanceolate, somewhat unequal; corolla 12-18 mm. long, purplish; pods 3-4 cm. long; seeds several.—Open places below 5000 (8000) ft.; many Plant Communities; most of cismontane Calif., occasional to Inyo and Mono cos.; to B.C., Ida. and far east. April-June.

2. **V. angustifòlia** Reichard. COMMON VETCH. Near to *V. sativa*, glabrous or glabrate; lfts. mostly 2-5 pairs, those of lower lvs. oblong, of upper linear, 1.5-3 cm. long; fls. 1-1.8 cm. long; calyx 7-11 mm. long; pods plane, 4-5 cm. long; seeds 3 mm. broad; $2n=12$ (Schwesnikova, 1927).—Natur. occasionally in waste places; native of Old World. April-June.

3. **V. benghalénsis** L. [*V. atropurpurea* Desf.] Soft-pubescent annual, 3-8 dm. high; stipules toothed; lfts. 5-8 pairs, elliptic to oblong, 1.5-2.5 cm. long; racemes equaling to exceeding lvs., 3-10-fld.; calyx villous, the teeth subulate, unequal, the lower exceeding the tube; corolla 12-14 mm. long, rose-purple, darker in age; pods 2.5-3.5 cm. long, 8-10 mm. wide, soft-pubescent, 4-6-seeded; $2n=12$ (Srivastaoa, 1963).—Occasionally natur. in waste places, as in Orange Co. and Santa Barbara region; native of Medit. region. May-June.

4. **V. dasycárpa** Ten. Much like *V. villosa*, but glabrous or sparingly strigose; lfts. mostly 1-2 cm. long; fls. fewer, 12-17 mm. long; calyx-teeth subglabrous, the lower shorter than the tube; corolla mostly purple-violet; pods 2-4 cm. long, 7-10 mm. wide; $2n=14$ (Schwesnikova, 1927).—On Santa Catalina Id., about Santa Barbara; native of Medit. region. Spring.

5. **V. exígua** Nutt. in T. & G. Slender-stemmed annual, 3-6 dm. high, climbing, ± strigose-villous; stipules entire, semisagittate; lfts. 2-6 pairs, linear, 1-2.5 cm. long, rounded to obtuse or emarginate, ± strigose, especially beneath; tendrils well developed; peduncles filiform, shorter than the lvs., 1-2-fld.; calyx ca. 2 mm. long, the teeth triangular-subulate; corolla white to purplish, ca. 5 mm. long; pods glabrous, 2-3 cm. long; 2-2.5 mm. wide.—Grassy, brushy or wooded slopes below 2000 ft., largely Coastal Sage Scrub, Chaparral, S. Oak Wd.; cismontane, L. Calif. n.; to Ore. Channel Ids. April-June.

6. **V. fàba** L. HORSE-BEAN. Erect annual, glabrous, 4-6 dm. high; stipules half-cordate, mostly entire; lfts. 2-3 pairs, ovate to elliptic, 4-10 cm. long, obtuse to acute; fls. 2-3 cm. long, subsessile, axillary, dull white with a large purple spot on the wings; pods 8-12 cm. long, 1-2 cm. wide, 2-5-seeded, the seeds flat.—Grown widely in Calif. and escaping; native of Old World.

7. **V. pannónica** Crantz. HUNGARIAN VETCH. Reclining or climbing annual, the stems 2-5 dm. long, weakly pubescent; stipules small; lfts. 7-9 pairs, linear to oblong, villous, 1-1.5 cm. long; fls. 2-4, short-peduncled, yellowish-white, 15-18 mm. long; pods nodding, 2.5-3 cm. long, 7-9 mm. broad, hairy; $2n=12$ (Heitz, 1931).—Sparingly natur. through Santa Barbara region; native of Eu.

8. **V. satìva** L. SPRING VETCH. Pubescent or subglabrous annual, 3–8 dm. high; stipules 5–15 mm. long, ± laciniate-toothed; lfts. 4–8 pairs, lance-oblong to obovate, pubescent when young, truncate to emarginate, 1–3 cm. long; fls. 1–2, subsessile, violet-purple, 1.8–2.5 cm. long; calyx 1–1.5 cm. long, the teeth as long as the tube; pods 4–8 cm. long, torulose; seeds 5 mm. broad; $n=6$ (Bir & Sadhu, 1966).—Natur. in waste places in much of cismontane Calif.; native to Eu. April–July.

9. **V. villòsa** Roth. WINTER VETCH. Annual or biennial, spreading-villous, the stems 6–12 dm. high; stipules narrow, entire, to ca. 1 cm. long; lfts. 8–12 pairs, linear to narrow-oblong, 1–2.5 cm. long; racemes 1-sided, densely many-fld.; calyx 5–6 mm. long, the unequal teeth linear-acicular, villous; corolla violet and white, 14–15 mm. long; pods oblong, 2–3 cm. long, glabrous; seeds 4–6, globular; $2n=14$ (Senn, 1948).—Natur. sparingly in waste places; native of Eu. April–July.

43. Fagàceae. BEECH FAMILY

Trees and shrubs with deciduous or persistent alternate petioled lvs. and small mostly deciduous stipules. Plants monoecious; fls. apetalous, the ♂ in catkins or capitate clusters, the ♀ solitary or in small clusters and subtended by an invol. of ± consolidated bracts which become indurated and form a cupule partly or completely enclosing the 1-locular and 1-seeded nut. Calyx of ♂ fls. 4–7-lobed; stamens 4–20. Calyx of ♀ fls. 4–8-lobed, adnate to the 3–7-locular ovary. Ovules 1–2 in each locule, usually only 1 ripening; styles 3. Endosperm none; cotyledons large, fleshy. Five genera and ca. 400 spp., mainly of N. Hemis. *Nothofagus* in the S. Hemis.

Fr. a spiny bur enclosing 1–3 nuts; lvs. persistent, yellowish-gray or rusty-tomentose beneath
.... 1. *Chrysolepis*
Fr. an acorn with cuplike invol.
    Staminate fls. in erect dense catkins with persistent bracts; ♀ fls. at base of ♂ catkins; lvs. with prominent parallel lateral veins .... 2. *Lithocarpus*
    Staminate fls. in drooping lax catkins with deciduous bracts; ♀ fls. in axillary clusters; lvs. usually not so veined .... 3. *Quercus*

1. Chrysólepis Hjelmquist. CHINQUAPIN

Evergreen trees or shrubs, the buds with imbricated scales. Lvs. simple. Catkins staminate or androgynous; fls. 3–7 (–11), fasciculate, staminate always bracteolate; calyx 5–6-parted; stamens several. Pistillate fls. at base of staminate, 3 in a cupule of 7 free valves (5 outer and 2 inner, the latter separating the 3 trigonous frs. from one another); styles 3. Fr. maturing in the second season, the spiny invol. enclosing the nuts, these angled. Two spp. of w. N. Am. (Greek, *chrysos*, gold, and *lepsis*, scale.) (Hjelmquist, Bot. Notiser, Suppl. 2: 117. 1948 and 113: 377. 1960.)

1. **C. sempervìrens** (Kell.) Hjelmquist. [*Castanea s.* Kell. *Castanopsis s.* Dudl.] BUSH CHINQUAPIN. Low spreading round-topped shrub 0.5–2.5 m. high, with smooth brown or gray bark; lvs. oblong or lance-oblong, mostly obtuse, 3–7.5 cm. long, subentire, yellowish gray-green above, golden or rusty-tomentose beneath, plane, on petioles 10–15 mm. long; ♂ fls. ill-smelling; burs 2–3 cm. thick, chestnutlike, 4-valved; seeds 8–12 mm. long, hard-shelled, with sweet kernel.—Dry rocky slopes and ridges, mostly above 6000 ft.; Montane Coniferous F.; San Jacinto Mts. to San Gabriel Mts.; Sierra Nevada and n. Coast Ranges to Ore. July–Aug.

2. Lithocárpus Blume. TANBARK OAK

Evergreen trees or shrubs with alternate lvs. and persistent stipules. Staminate catkins many, dense, elongate, erect, with clusters of 3 fls. in the axils of the round-ovate bracts with 2 lateral bractlets; stamens 10. Pistillate fls. at the base of the ♂ ament, solitary in the axils of the acute bracts and minute lateral bractlets; calyx 6-lobed; styles 3. Fr. an acorn, the cup with slender spreading scales. Ca. 100 spp. of se. Asia, 1 in w. N. Am. (Greek, *lithos*, rock, and *karpos*, fr., because of the hard acorn.)

1. **L. densiflòrus** (H. & A.) Rehd. [*Quercus d.* H. & A. *Pasania d.* Oerst.] Tree 20–45 m. tall, with narrow conical crown and tomentose young twigs; bark thick, fissured; lvs. oblong, acute, with prominent parallel lateral veins ending in sharp teeth, pale with whitish or yellowish stellate pubescence when young, ± glabrate later, especially on upper surface, 4–12 cm. long; petioles 1–2 cm. long; ♂ catkins 5–10 cm. long, ill-smelling; acorn maturing second year, short, thick-cylindric, with rounded base and gradually tapering toward apex, mostly 2.5–3.5 cm. long, tomentose when young; cup shallow, saucer-shaped, 1.5–2.5 cm. in diam., tomentose within, the scales narrow, spreading.—Wooded slopes below 4500 ft.; Ventura Co. and Santa Barbara Co.; to s. Ore. June–Oct.

3. Quércus L. Oak

Deciduous or evergreen trees or shrubs with hard wood, ± contorted branches and fairly slender twigs marked by lenticels. Staminate catkins slender, drooping or spreading, one or more from the lower axils of the current year's growth, the fls. solitary in the axils of caducous bracts; the calyx usually 6-lobed; stamens 5–12. Pistillate fls. solitary in many-bracted invols. in upper axils; calyx urn-shaped, adnate to the 3-celled inferior ovary; styles 3, short. Fr. an acorn, the nut set in the cuplike invol. Ca. 300 spp., widely distrib. in the N. Hemis. and into the mts. of the tropics. (Classical Latin name.)

1. Bark dark, not scaly but smooth, or on old trunks irregularly ruptured; stigmas on slender styles; involucral cups with thin closely imbricated scales; nut tomentose on inner surface. (Black Oaks.)
   2. Plant deciduous, the lvs. usually 1 dm. or more long, with large bristle-tipped lobes
      .... 8. *Q. kelloggii*
   2. Plant evergreen, the lvs. smaller, coriaceous.
      3. Acorns maturing the second autumn; lvs. plane, glabrous beneath .... 12. *Q. wislizenii*
      3. Acorns maturing the first autumn; lvs. mostly convex on upper surface, with hair beneath, especially in axils of lvs, ... 1. *Q. agrifolia*
1. Bark light in color, scaly, furrowed only on large trees; stigmas subsessile; involucral cups usually with tuberculate scales. (White Oaks.)
   4. Fr. maturing the first season; shell of the nut glabrous on the inner surface.
      5. Deciduous spp.
         6. Lvs. not blue-green, distinctly lobed.
            7. Acorn-cups deeply hemispheric; nut tapering at apex, 3–5 cm. long .. 9. *Q. lobata*
            7. Acorn-cups shallow; nut ± oblong, rounded at apex, 2–3 cm. long . 7. *Q. garryana*
         6. Lvs. blue-green, toothed or wavy-margined; cup shallow-cupulate; nut oval, acute
            .... 3. *Q. douglasii*
      5. Evergreen spp., or retaining at least some of the lvs. in winter.
         8. Tree; lvs. mostly 3–6 cm. long; cup enclosing nearly half the nut .. 6. *Q. engelmannii*
         8. Shrubs, occasionally somewhat arborescent; lvs. mostly 1–2.5 cm. long; cup more shallow.
            9. Lf.-blades shining and subglabrous above. Cismontane .... 4. *Q. dumosa*
            9. Lf.-blades stellate-pubescent above and dull. Largely desert ..... 11. *Q. turbinella*
   4. Fr. maturing the second season; shell of the nut tomentose on the inner surface.
      10. Trees; acorn-cup large, thick-walled, densely tomentose.
         11. Lvs. 4–8 cm. long, with prominent parallel veins. Insular .... 10. *Q. tomentella*
         11. Lvs. mostly 2–5 cm. long, the lateral veins not prominent or parallel. Widely distributed .... 2. *Q. chrysolepis*
      10. Shrubby; the cup with rather thin walls; branchlets rigid; lvs. spinose-toothed. Mostly desert-edged .... 5. *Q. dunnii*

1. **Q. agrifòlia** Neé. var. **agrifòlia**. Coast Live Oak. Encina. Plate 49, Fig. E. Broad-headed evergreen tree 10–25 m. high; trunk smooth or with broad checked ridges in old trees; lf.-blades oblong to oval or elliptic, 2–6 cm. long, harsh, strongly convex above, glabrous or somewhat stellate-pubescent beneath especially along the veins; petioles 5–15 mm. long; ♂ catkins ca. 3–6 cm. long; acorns slender, pointed, 2.5–3.5 cm.

long, 1–1.4 cm. thick toward base, glabrous; cup turbinate, 8–12 mm. deep, 10–16 mm. in diam., silky within, the scales brownish, thin, puberulent, ciliate; $n=12$ (Duffield, 1940).—Common in valleys and on not too dry slopes, below 3000 ft.; S. Oak Wd., Foothill Wd., etc.; n. L. Calif. and San Diego Co. to n. cent. Calif. Santa Catalina and n. Channel ids. March–April.

Var. **oxyadènia** (Torr.) J. T. Howell. [*Q. o.* Torr.] Lvs. densely stellate, especially beneath.—At 2000–4600 ft.; interior cismontane Riverside and San Diego cos.

A possible hybrid with *Q. kelloggii* has been named *Q.* × *ganderi* Wolf, with acorns maturing the first year and with deciduous lvs. somewhat lobed and having tufts of hair in vein axils. E. San Diego Co.

2. **Q. chrysólepis** Liebm. CANYON OAK. MAUL OAK. An evergreen ± roundish or spreading tree 6–20 m. tall, with pale gray rather smooth scaly bark and hoary-tomentose young twigs; lf.-blades coriaceous, grayish- or yellowish-tomentose beneath or later glabrate and glaucous, usually oblong, entire to spinulose-dentate, especially on young and sucker shoots, plane, 2–6 cm. long; petioles mostly 3–8 mm. long; acorns ovoid-oblong, 2.5–3 cm. long, 2–2.5 cm. thick, scanty-tomentose within; cup saucer-shaped, thick-walled, silky within, mostly 2–2.5 cm. in diam., the scales ± hidden by the feltlike tomentum; $n=12$ (Duffield, 1940).—Common in canyons and on moist slopes below 6500 ft.; many Plant Communities; cismontane Calif. and occasional on the desert at higher elevs.; to Ore., L. Calif., New Mex. Exceedingly variable, especially in cent. Calif.

3. **Q. dóuglasii** H. & A. BLUE OAK. Deciduous tree with ± rounded crown, 6–20 m. high; bark shallowly checked into small thin scales; twigs ± tomentose when young; lf.-blades oblong, 3–10 cm. long, entire to undulately shallowly few-lobed, blue-green above, minutely stellate-pubescent, paler and pubescent beneath; petioles 3–8 mm. long; acorns oval, acute, 2–3 cm. long, glabrous; cups shallow-cupulate, 10–12 mm. in diam., the scales with small warty processes; $n=12$ (Duffield, 1940).—Dry rocky slopes, mostly below 3500 ft.; Foothill Wd.; mostly on slopes bordering interior valleys from n. Los Angeles Co. to head of Sacramento V., hence reported from Castaic, Tehachapi, Santa Barbara, Santa Cruz Id. April–May.

4. **Q. dumòsa** Nutt. SCRUB OAK. Plate 49, Fig. F. Evergreen, mostly a shrub 1–3 m. high, sometimes arborescent, with stiff pubescent to glabrate and brownish young twigs; lf.-blades oblong to elliptic or roundish, mostly mucronate-dentate to entire or subspinose, coriaceous, 1.5–2.5 cm. long, green and shining above, paler and pubescent beneath; petioles mostly 2–3 mm. long; acorns ovoid, 1–3 cm. long, broad at base, rounded or acute at apex; cups hemispheric to ca. ⅔ spherical, 10–15 mm. in diam., the margins tapering inward, walls thick with scales mostly strongly tuberculate; $n=12$ (Duffield, 1940).—Common on dry slopes, mostly below 5000 ft.; Chaparral, Foothill Wd., etc.; n. L. Calif. through cismontane Calif. to n. part of state. March–May. Variable and hybridizing freely with *engelmannii*, with *garryana*, and with *lobata* (*Q. d.* var. *kinselae* C. H. Mull; *Q.* × *townei* Palmer).

5. **Q. dúnnii** Kell. [*Q. palmeri* Engelm.] Stiff evergreen shrub 2–5 m. high, with rigid, at first ± pubescent twigs; lf.-blades brittle, elliptic to round-ovate, wavy-spinose, crisped, 1–3 cm. long, gray-green above, ± tomentose beneath especially when young; petioles 2–5 mm. long; acorns ovoid, 2–3 cm. long, acute, the shell tomentose within; cup shallow, 12–30 mm. broad, 7–10 mm. deep, the walls thin, silky within, densely tomentose without. —Dry thickets and margins of Chaparral, mostly 3000–5000 ft.; Peachy Canyon Road, San Luis Obispo Co., Deep Creek at n. base of San Bernardino Mts., San Jacinto Mts. along w. edge of Colo. Desert to n. L. Calif., Ariz. April–May.

6. **Q. engelmánnii** Greene. ENGELMANN OAK. Spreading tree with rounded top, 5–18 m. high; bark covered with thin grayish scales; twigs tomentose; lf.-blades gray-green, obtuse, coriaceous, semipersistent, oblong to obovate, plane, 2–6 cm. long, subentire to sinuate-dentate, glabrate to somewhat pubescent; petioles 3–7 mm. long; acorns oblong-cylindric, 1.5–2.5 cm. long, 1.2–1.4 cm. thick, glabrous, ± acute; cup shallow to bowl-shaped, enclosing nearly half the nut, 1–1.5 cm. broad, 0.8–1 cm. deep, puberulent within, the scales light brown, tomentose, the lower tuberculate, the upper ciliate; $n=12$

(Duffield, 1940).—Dry fans and foothills, below 4000 ft.; S. Oak Wd.; Pasadena region inland to San Dimas and s. and e. to e. San Diego Co., but away from the coast; L. Calif. April–May. Santa Catalina Id.

Occasionally hybridizing in the Pasadena region with *Q. lobata*, forming large trees with lvs. up to 15 cm. long and ± lobed, also with *Q. dumosa* especially in the lower San Gabriel Mts., Santa Ana Mts. and e. San Diego Co., where the hybrid tends to be ± arborescent, with narrow oblong lvs. entire or often rather regularly and sharply serrate-lobed (*Q.* × *grandidentata* Ewan).

7. **Q. garryàna** Dougl. var. **semòta** Jeps. Rounded shrub, deciduous; twigs red-pubescent; lf.-blades mostly 5–9 cm. long, obovate to oblong in outline, coriaceous, coarsely pinnatifid into oblong-ovate lobes, these entire or 2–3-toothed; ♂ catkins commonly 3–5 cm. long; acorn ovoid to subglobose, rounded at tip, maturing the first autumn, 2–3 cm. long, subglabrous; cup shallow, puberulent within, 1.2–1.6 cm. in diam., with tubercled pubescent or tomentose scales.—Dry slopes, 2500–5000 ft.; Chaparral, Yellow Pine F.; Liebre Mts., n. Los Angeles Co. to Plumas Co.

8. **Q. kellóggii** Newb. CALIFORNIA BLACK OAK. Plate 50, Fig. A. A deciduous tree with broad rounded crown, 10–25 or more m. high; trunk thick with dark smooth bark that divides into ridges or checks deeply in age; young twigs subglabrous; lf.-blades broadly elliptic to obovate in outline, deeply and mostly sinuately lobed into ca. 3 main divisions on each side, each lobe with 1–4 coarse bristle-tipped teeth, bright green and mostly glabrous above, paler and often finely stellate-tomentose beneath when young, 7–20 cm. long; petioles 2.5–5 cm. long; ♂ catkins 3.5–7.5 cm. long; stamens 5–9; acorns maturing in the second year, oblong, 2.5–3 cm. long, 1.5–1.8 cm. thick, pubescent; cups 1.5–2.5 cm. deep, 2–2.8 cm. wide, puberulent within, the thin scales membranous, and ± ragged on the margins; $n=12$ (Duffield, 1940).—Common in hills and mts. mostly 4000–8000 ft.; Montane Coniferous F.; San Diego Co. n. through Sierra Nevada and Coast Ranges to Ore. April–May.

Q. × **morehus** Kell. ORACLE OAK. Evergreen tree 4–15 m. high; lf.-blades oblong to elliptic in outline, 4–12 cm. long, sinuately rather shallowly lobed, the lobes pointed forward and spinose-tipped, ± stellate-pubescent beneath; cups like in *Q. wislizenii*, thin-scaled; acorns cylindric, ca. 2.5 cm. long, pubescent.—At scattered stations and usually of 1 or few individuals, below 5000 ft.; usually near *Q. kelloggii* and *Q. wislizenii* and undoubtedly a hybrid, with some introgression shown. San Diego Co. to n. Calif.

9. **Q. lobàta** Neé. [*Q.hindsii* Benth.] VALLEY OAK. ROBLE. Stately graceful deciduous tree with open head, 12–35 m. high, with trunk to 4 m. thick; bark thick, cuboid-checked; twigs silky at first, glabrescent in 2nd year; lf.-blades oblong to obovate, 5–10 cm. long, deeply divided into 3–5 pairs of obtuse lobes which are mostly coarsely 2–3-toothed at apex, 5–12 mm. long, pubescent; petioles 5–12 mm. long; ♂ catkins 2.5–7.5 cm. long; acorns maturing the first autumn, long-conical, 3–5 cm. long, 1.2–2 cm. thick, glabrous; cup deeply hemispheric, 12–20 mm. deep, 12–25 mm. in diam., the lower scales conspicuously warty; $n=12$ (Duffield, 1940).—Rich loam, valleys and slopes, below 2000 ft.; S. Oak Wd., Foothill Wd.; San Marino and San Fernando V., Los Angeles Co.; to Central V. March–April.

Q. × **macdónaldii** Greene. Small deciduous tree, 5–15 m. tall with a compact rounded crown; bark scaly; young twigs densely tomentose; lf.-blades oblong to obovate in outline, 4–7 cm. long, with 2–4 blunt or sharp-pointed, bristle-tipped lobes on each side, glabrous above, pubescent beneath; petioles 3–8 mm. long; acorn oblong-conical, 2–3.5 cm. long, acute, glabrous; cup deeply hemispherical, 1.5–2.5 cm. in diam., with strongly tubercled pubescent scales.—Ravines and canyons; Chaparral, S. Oak Wd.; Santa Cruz, Santa Rosa, Santa Catalina ids. Possible hybrid between *Q. lobata* and *Q dumosa*.

10. **Q. tomentélla** Engelm. ISLAND OAK. Small, round-headed evergreen tree, 5–12 m. tall; bark red-brown, scaly; young twigs hoary-tomentose, later brown; lf.-blades dark green, oblong to oblong-lanceolate, acutish, often revolute, 4–8 cm. long, glabrate above, ± tomentose beneath, strongly pinnately veined, crenate-dentate; petioles 5–17

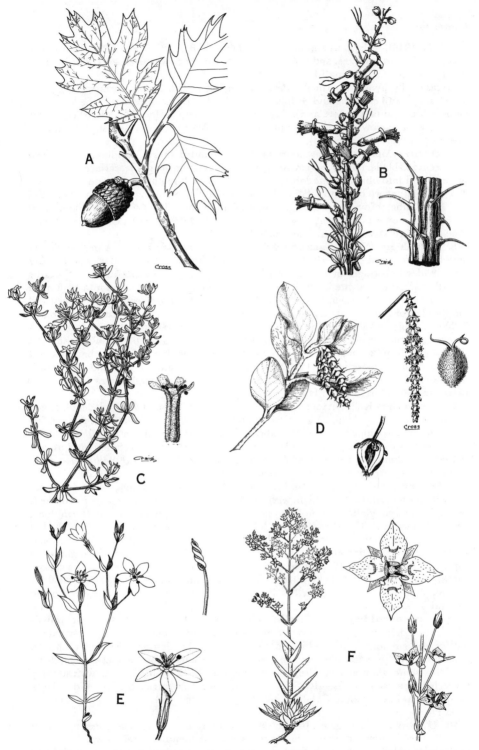

Plate 50. Fig. A, *Quercus kelloggii*. Fig. B, *Fouquieria splendens*. Fig. C, *Frankenia grandifolia*. Fig. D, *Garrya veatchii*, ♂ and ♀ fls. Fig. E, *Centaurium venustum*. Fig. F, *Frasera parryi*, gland on each petal.

mm. long; acorns subglobose, 2–2.5 cm. long, bluntish; cups shallow, 2–3.5 cm. in diam., the ovate-acute scales almost hidden in the dense tomentum.—Canyons and ravines; Chaparral, S. Oak Wd.; Santa Cruz, Santa Rosa, Santa Catalina and San Clemente ids.; Guadalupe Id. April–May.

11. **Q. turbinélla** Greene ssp. **turbinélla** [*Q. dumosa* var. *t.* Jeps.] Near to *Q. dumosa*, the young twigs densely gray-yellow tomentose; lf.-blades dull, gray to gray-green on upper surface, mostly oblong in outline, pointed apically, the margins spinose-dentate with definite short spines; fr. usually pedunculate, the acorns yellow-brown to buff, 1.2–2.3 cm. long, cylindric-ovoid to -ellipsoid, tapering abruptly toward apex; cups mostly ± hemispheric, the margins not turned inward.—At 4000–6000 ft.; Pinyon-Juniper Wd.; New York Mts., e. San Bernardino Co. to w. Nev., Tex., n. L. Calif.

Ssp. **califórnica** Tucker. Lf.-blades more irregularly dentate, elliptical to ovate or suborbicular in outline, often rounded at apex; fr. mostly sessile; acorns often brown, 2–3 cm. long; cup turbinate.—Dry slopes largely from 3000–6500 ft.; Joshua Tree Wd., Pinyon-Juniper Wd.; edge of w. Mojave Desert from Little San Bernardino Mts. w. and n. to San Benito Co. This ssp. hybridizes freely with *Q. dumosa* along the desert margins and with *Q. douglasii*, this latter hybrid and segregates form the *Q.* × *alvordiana* Eastw. with ± persistent lvs. of intermediate character, somewhat arborescent habit, etc. It occurs from hills w. of Salinas V. to Liebre Mts., n. Los Angeles Co. In the Joshua Tree National Monument is a large tree at Live Oak Tank growing with *Q. turbinella californica* and with lobed lvs. like those of the White Oak group. Seedlings from it vary tremendously. Tucker considers it a possible hybrid with *Q. lobata* and calls it *Q.* × *munzii*.

12. **Q. wislizènii** A. DC. var. **wislizènii**. INTERIOR LIVE OAK. Evergreen tree, mostly 10–22 m. high, with round top and smooth bark 5–6 cm. thick, broadly ridged below in old age; lf.-blades mostly oblong, varying to elliptic and lanceolate, rather firm, entire to spiny-toothed, plane, glabrous and shining above and beneath, but ± bicolored, 2–6 or more cm. long; petioles 0.5–1 cm. long; ♂ catkins ca. 3–6 cm. long; acorns maturing 2nd year, slender, ± conical, 2–4 cm. long, 0.7–1.3 cm. thick, often with longitudinal dark bands, glabrous; cup turbinate to cup-shaped, 12–20 mm. deep, 12–14 mm. in diam., with thin brownish pubescent and ciliate scales; $n = 12$ (Duffield, 1940).—Valleys and slopes, below 5000 ft., mostly Foothill Wd.; n. Calif. s. to Ventura Co. and in Hemet V., San Jacinto Mts. at 3000–4800 ft., *Ziegler*. March–May.

Var. **frutéscens** Engelm. [*Q. parvula* Greene.] A shrub 1–4 m. high, with very rigid twigs and lvs. 2–4 cm. long.—Chaparral; mts. of s. Calif. to n. Calif.; Santa Cruz Id.

44. Fouquieriàceae. OCOTILLO FAMILY

Resinous spiny shrubs with erect or spreading stems. Primary lvs. soon deciduous with dry weather, the petioles developing into heavy phyllodial thorns, in the axils of which later appear fascicles of secondary lvs. Fls. showy, bisexual, in terminal panicles. Sepals 5, imbricated, unequal. Petals 5, hypogynous, connate into a tube, imbricate. Stamens 10 or more, 1–2-seriate, hypogynous, with a basal toothed portion. Disk annular, small. Ovary 1-celled, with 3 parietal septiform placentae, each with ca. 6 ovules. Fr. a caps. Seeds oblong, compressed, winged, the wing breaking up into hairlike parts. Several spp., sw. N. Am.

1. *Fouquièria* HBK. OCOTILLO. CANDLEWOOD

The only genus. Several spp.; Mex. and sw. U.S. Sometimes planted as hedge-fences. (P. E. *Fouquier*, Parisian prof. of medicine.)

1. **F. spléndens** Engelm. Plate 50, Fig. B. Stems stout, several to many from the base, stiff, mostly simple and canelike, sometimes few-branched, 2–7 m. tall, gray with darker furrows and stout divaricate spines; lvs. fleshy, oblong-obovate, 1–2.5 cm. long, 1-nerved; panicles 1–2.5 dm. long, dense, many-fld.; sepals 4–7 mm. long, rounded; corolla tubular, scarlet, 2–2.5 cm. long, the lobes rounded, recurved; stamens 10–17, exserted; caps. 1.5 cm. long, 3-valved; seeds white, fringed; $2n = 12$ pairs (Raven et al., 1965).—Dry mostly

rocky places, below 2500 ft.; Creosote Bush Scrub; se. Mojave Desert and Colo. Desert; to Tex., Mex. March–July. Lvs. soon deciduous, but renewed with both spring and summer rains.

45. Frankeniàceae. FRANKENIA FAMILY

Low perennial herbs or undershrubs. Lvs. opposite, entire, subsessile, exstipulate. Fls. bisexual, regular, small, solitary or in cymes. Sepals 4–6, persistent, connate, tubular. Petals as many, clawed, imbricate, with scalelike appendage within. Stamens usually 6, hypogynous. Ovary superior, 1-celled, with 2–4 parietal placentae and few to many ovules. Fr. a caps., opening by valves. Seeds with endosperm and straight embryo. Four genera and ca. 65 spp., widely distributed in subtrop. and temp. regions.

1. Frankènia L.

Primary lvs. with axillary fascicles. Style 2–3-cleft. Caps. linear, angled. Ca. 50–60 spp. of maritime plants. (J. *Franke*, 1590–1661, the first writer on Swedish plants.)

Lvs. 5–15 mm. long, linear-oblanceolate to obovate, ± expanded; style 3-cleft . 1. *F. grandifolia*
Lvs. 2–4 mm. long, revolute and terete; style 2-cleft .... 2. *F. palmeri*

1. **F. grandifòlia** Cham. & Schlecht. Plate 50, Fig. C. Bushy, herbaceous or suffrutescent, 1.5–3 dm. high, glabrous to pubescent or subhirsute; lower lvs. obovate, somewhat revolute, 5–15 mm. long, subsessile, united in pairs by membranous base, upper narrower; calyx narrow-cylindric, 6–7 mm. long, furrowed, with acute teeth; petals pinkish, 2–4 mm. longer than calyx; stamens 4–7, commonly 5; caps. linear, ca. 5 mm. long; seeds brown, ca. 1 mm. long.—Salt marshes, beaches, alkali flats along or near the coast; Coastal Salt Marsh, Coastal Strand; L. Calif. n. to cent. Calif. Also on the islands. June–Oct. A slightly different form, more tufted; lvs. spatulate, more revolute; occurring on alkali flats of the interior below 5000 ft.; Alkali Sink, etc.; is var. *campestris* Gray. Riverside Co. and Orange Co. to Owens V. and Cent. V. of Calif.; Nev.

2. **F. pálmeri** Wats. Shrubby, 1–2 dm. high, densely leafy; lvs. oblong-linear, canescent, terete, 2–4 mm. long; calyx 3 mm. long; petals whitish; seeds few.—Coastal Salt Marsh; San Diego Co. to n. L. Calif., Son. May–July.

46. Garryàceae. SILK-TASSEL FAMILY

Evergreen dioecious shrubs or small trees with opposite simple rather leathery shortpetioled lvs. Fls. small, apetalous, imperfect, borne in pendulous catkinlike clusters. Staminate fls. pedicelled, in 3's in axil of each of the decussately connate bracts; calyx 4-parted, the tips often connate; stamens 4; fils. distinct. Pistillate fls. subsessile, borne 1 in each bract-axil; calyx with 2 lobes or obsolete; ovary inferior, 1-celled; styles 2, persistent, stigmatic on inner side. Fr. a berry, the bitter pulp surrounding the 1–2 seeds, dark purple to black, enclosed in and soon free from the dry brittle epicarp. Seeds with thin testa and horny endosperm; embryo minute.

1. Gárrya Dougl. SILK-TASSEL BUSH

One genus, with the characters of the family. Ca. 15 spp. of w. N. Am., many of which intergrade freely where they come in contact with each other. (Named for N. *Garry* of the Hudson Bay Co., friend of David Douglas.)

Lower surface of lvs. glabrous or nearly so; lvs. plane; fr. glabrous or pubescent when young
                                     3. *G. fremontii*
Lower surface of lvs. usually covered with hairs (glabrous in age in no. 2).
 Hairs of lower lf.-surface straight, upwardly appressed .... 2. *G. flavescens*
 Hairs of lower lf.-surface curly or wavy.

Lvs. oval or broadly elliptical, rounded or obtuse at apex, strongly undulate on margins
.... 1. *G. elliptica*
Lvs. lanceolate to ovate, ± acuminate, plane or nearly so .... 4. *G. veatchii*

1. **G. ellíptica** Dougl. Shrub or small tree to 8 m. high, the young twigs densely short-villous; lf.-blades elliptic to oval, 6–8 (–10) cm. long, green and subglabrous above, densely felty-woolly beneath with short ± curly hairs, mostly strongly crisped-undulate on margins; petioles 6–12 mm. long; ♂ catkins largely 8–15 cm. long; fruiting catkins 8–15 cm. long; fr. globose, 7–11 mm. in diam., white-tomentose, but ± glabrate in age.— $2n = 22$ (Van Horn, 1963).—Dry slopes and ridges, below 2000 ft. Chaparral, etc.; Ventura Co. n. to Ore. Jan.–March.

Plants from Santa Cruz Id. characterized by plane, not crisped-undulate lvs., have a strongly developed white tomentum on their under surface and seem to constitute a distinct taxon, but have not been described.

2. **G. flavéscens** Wats. var. **pállida** (Eastw.) Bacig. ex Ewan. [*G. p.* Eastw.] Erect shrub, 1.5–3 m. tall, the young twigs canescent-strigose, soon glabrate and red-brown; lvs. oblong-elliptic to oval, 3–6 cm. long, stiffly coriaceous, gray-green, silky beneath; petioles 3–10 mm. long; ♂ catkins 3–4 cm. long; fruiting catkins compact, densely silky, 3–5 cm. long; fr. broadly ovoid, 6–8 mm. broad, silky.—Dry slopes, 3000–8000 ft.; Chaparral, Yellow-Pine F., Pinyon-Juniper Wd.; San Diego to cismontane cent. Calif.; mts. of e. Mojave Desert where it approaches *G. flavescens* of the area to the east of Calif. in its more yellowish less glaucous appearance. Feb.–April.

3. **G. fremóntii** Torr. Erect shrub 1.5–3 m. tall, the young twigs strigose; lvs. oblong-elliptical to -ovate, 2–5 cm. long, usually tapered at ends, glabrous and shining above, and yellow-green in age, paler and glabrous to sparingly pilose beneath, plane on margins; petioles to ca. 1 cm. long; ♂ catkins solitary or clustered, simple, lax, 7–20 cm. long, yellowish; ♀ 4–5 cm. long or more in fr.; fr. ca. 6 mm. in diam., globose, almost black to purplish, subglabrous.—Dry brushy slopes mostly below 5500 ft.; Chaparral, Yellow Pine F.; mts. of w. Riverside Co., Orange Co., San Diego Co.; cent. Calif. to Wash. Jan.–April.

4. **G. veatchii** Kell. Plate 50, Fig. D. Shrub 1–2 m. tall, the young twigs densely white-tomentulose; lf.-blades lanceolate to ovate or ovate-elliptic, ± acuminate, 2.5–6 cm. long, green and glabrous above, felty-tomentose beneath, plane or slightly undulate on margins; petioles 6–14 mm. long; ♂ catkins 5–10 cm. long; fruiting 2.5–6 cm. long; fr. ovoid to rounded, 7–8 mm. in diam., buff to purple-brown, ± glabrescent in age.—Dry slopes below 7000 ft.; Chaparral, Foothill Wd., etc. L. Calif. to San Luis Obispo Co. Feb.–April.

47. Gentianàceae. GENTIAN FAMILY

Herbs with colorless bitter juice. Lvs. opposite or whorled, sometimes alternate, mostly sessile and simple, occasionally petioled, even trifoliolate. Fls. regular, perfect, axillary or in terminal infl. Calyx persistent, free from ovary, 4–12-lobed or toothed. Corolla sympetalous, rotate to funnelform, withering-persistent, the lobes mostly convolute in the bud. Stamens inserted on corolla-tube or -throat, alternate with the lobes. Ovary 1-locular, with 2 parietal placentae or nearly the whole inner face ovuliferous; style 1, simple or short-cleft; stigma entire or 2-lobed. Fr. a caps., usually dehiscent from above by 2 valves. Seeds often numerous, anatropous, with a minute embryo in fleshy albumen. A family of perhaps 65 genera and 600 spp., widely distributed, but most abundant in temp. regions.

Corolla rotate and with conspicuous glands on upper surface; fls. 4-merous; fls. in a terminal panicle .... 3. *Frasera*
Corolla campanulate to funnelform or salverform, without such glands.
  Anthers coiled or spirally twisted after anthesis; fls. mostly red or pink .... 1. *Centaurium*
  Anthers not coiled or twisted after anthesis.
    Style filiform; stamens inserted on corolla-throat. From below the yellow pine belt
                                                                                        2. *Eustoma*
    Style stout, short or lacking; stamens inserted on corolla-tube. From Yellow Pine belt or above
                                                                                          4. *Gentiana*

1. Centáurium Hill. CENTAURY

Glabrous branching erect annuals. Lvs. opposite, sessile or amplexicaul. Fls. 4–5-merous, mostly pink or rose, in terminal spikes or cymes. Calyx narrow, deeply parted into narrow keeled segms. Corolla salverform or funnelform, with slender tube, the lobes contorted, convolute in the bud. Stamens inserted on corolla-throat, alternate with the lobes; fils. slender; anthers commonly exserted, spirally twisting after shedding pollen. Ovary 1-celled, the parietal placentae sometimes intruded; style filiform, deciduous; stigmas oblong to fan-shaped. Caps. fusiform to oblong-ovoid, 2 valved. Seeds minute. Perhaps 30 spp., of N. Am., Eurasia and Afr. (Latin, *Centaurus*, Centaur, who is supposed to have discovered its medicinal properties.)

Corolla-lobes less than half as long as the tube; anthers 1.5–2.5 mm. long, oblong
    2. *C. exaltatum*
Corolla-lobes more than half as long as tube; anthers 3.5 mm. long, linear.
    Lvs. mostly oblanceolate to elliptic, tending to be wider in upper half. Se. desert
        1. *C. calycosum*
    Lvs. mostly oblong to ovate, wider in lower half. Largely cismontane ...... 3. *C. venustum*

1. **C. calycòsum** (Buckl.) Fern. [*Erythraea c.* Buckl.] Much like *C. venustum*, but with more open paniculate branching and often 4–5 dm. high; lvs. tending to be oblanceolate, especially those of lower half of plant; corolla-lobes apiculate.—Damp places, Colo. R. bottom; to Tex., Mex. April–May.

2. **C. exaltátum** (Griseb.) W. Wight. [*Cicendia e.* Griseb. in Hook.] Simple or usually branched, 1–3.5 dm. high; lvs. oblong-elliptic to -lanceolate, 1–3 cm. long, acute; pedicels 1–5 cm. long, in forks and at ends of branches; calyx-segms. subulate, 8–10 mm. long; corolla-tube 8–10 mm. long, the lobes pale pink to white, 3–4 mm. long, oblong, obtuse; anthers oblong, ca. 1 mm. long; stigma-lobes fan-shaped; seeds almost round, ca. 0.25 mm. in diam.—Damp somewhat alkaline places, below 5000 ft.; Coastal Sage Scrub, Chaparral, Creosote Bush Scrub, etc.; cismontane San Diego Co. to Los Angeles Co., w. edge of deserts n. through Inyo Co.; to e. Wash., Utah. May–Aug.

3. **C. venústum** (Gray) Rob. [*Erythraea v.* Gray.] CANCHALAGUA. Plate 50, Fig. E. Usually simple below, corymbosely branched above, 1–4 dm. high; lvs. ovate to oblong, 1–2.5 cm. long, obtusish; pedicels 2–25 mm. long; calyx-segms. 6–10 mm. long; corolla rose with red spots in the white throat, sometimes albino, the lobes mostly obtuse, 8–15 mm. long; anthers 4–6 mm. long; stigmas fan-shaped, divaricate; seeds dark roundish to oblong.—Dry slopes and flats, below 2500 ft., occasionally higher; Coastal Sage Scrub, Chaparral; w. San Diego Co. to Ventura Co., to desert edge. Santa Catalina Id. May–July.

2. Eustòma Salisb. CATCHFLY-GENTIAN

Annual or short-lived perennial herbs, glaucous. Stems erect or ascending, leafy. Lvs. opposite, sessile or clasping. Fls. blue or white to purplish, solitary or paniculate. Calyx deeply cleft, the lobes keeled, long-acuminate. Corolla campanulate-funnelform, 5–6-lobed, the lobes oblong to obovate, convolute in bud and often erose-denticulate. Stamens 5–6, inserted on throat of corolla; anthers oblong, versatile, straight or recurved. Ovary 1-celled; style filiform; stigma 2-lobed, the lobes broad, flattened. Caps. oblong to ellipsoid, 2-valved. Seeds many, small, honeycombed. Spp. 4; from s. U.S., Mex., W.I., n. S. Am. (Greek, *eu*, good, and *stoma*, mouth, the throat of the corolla large.)

1. **E. exaltàtum** (L.) Griseb. Stems solitary to few, erect or ascending, branched above, 4–7 dm. tall; basal lvs. obovate to broadly spatulate, with short broad petiole; cauline lvs. broadly oblong, sessile, 4–9 cm. long, subcordate-clasping, the uppermost bractlike; pedicels stout, 1–10 cm. long; calyx-lobes subulate, 1–1.5 cm. long; corolla blue or deep lavender, sometimes pale, the tube ca. 1 cm. long, the lobes 1.5–2 cm. long; style 4–5 mm. long; stigma-lobes ca. 2 mm. long; caps. 8–12 mm. long.—Occasional along streams below 1500 ft.; Coastal Sage Scrub, Creosote Bush Scrub; Santa Ana River of Orange and

Riverside and San Bernardino cos., wet spots on Colo. Desert; to Fla., W.I., Mex., L. Calif. Most months.

3. Fràsera Walt. GREEN GENTIAN

Biennial or perennial herbs with erect stems from bitter taproots. Lvs. opposite or whorled, entire, thickish. Fls. perfect, in a terminal panicle, 4-merous. Calyx-lobes deep, subulate, acute. Corolla rotate, parted nearly to base, the lobes convolute in bud and bearing 1–2 ± fringed glands. Stamens inserted at base of corolla. Ovary 1-celled, gradually attenuate into a filiform style; stigma 2-cleft; placentae parietal, many-ovuled. Caps. ovoid, compressed, 4–20-seeded. Ca. 15 spp. of N. Am. (Named for J. *Fraser*, an English collector.)

Stem-lvs. opposite, except occasionally in *F. puberulenta*.
  Gland lobed at apex or U-shaped; fls. in an open broad panicle.
    Herbage glabrous; plant 6–12 dm. high; corolla-lobes 10–15 mm. long. From Los Angeles Co. s. and e. .... 3. *F. parryi*
    Herbage puberulent; plant 1–3 dm. high; corolla-lobes 6–8 mm. long. Inyo-White Mts.
        4. *F. puberulenta*
  Gland entire at apex; fls. in a narrow, spikelike, often interrupted panicle; herbage glabrous. W. Mojave Desert .... 2. *F. neglecta*
Stem-lvs. whorled, the whorls of 3–4 narrow lvs. E. Mojave Desert .... 1. *F. albomarginata*

1. **F. albomarginàta** Wats. Taproot rather woody; stems 1–few, much-branched, 2–5 dm. high, glabrous; lvs. white-margined, pale, the lower narrowly oblanceolate, 4–8 cm. long; cauline lvs. in whorls of 3–4, lance-linear, sessile, 2–6 cm. long; panicle rather broad, diffuse, subcorymbose; pedicels 1–3 cm. long; calyx-lobes lance-linear, 5–6 mm. long; corolla-lobes oblong-obovate, acuminate, 8–10 mm. long, greenish-white with dark dots, the gland linear, fringed at margin, slightly 2-lobed above; caps. 10–15 mm. long, flattened-conic; seeds ca. 4 mm. long, spongy-pitted.—Dry rocky and gravelly places, 4500–7000 ft.; Pinyon-Juniper Wd.; New York and Providence mts., e. Mojave Desert; to Colo., Ariz. A form from Clark Mt. has glandular-puberulent stems and is var. *induta* Card.

2. **F. neglécta** Hall. Root-crown branched; stems slender, 2–4 dm. high; herbage glabrous, pale; lower lvs. linear to oblance-linear, 6–15 cm. long, inconspicuously white-margined, the cauline opposite, linear, sessile; panicle interrupted, the fls. in dense whorl-like clusters; pedicels 5–20 mm. long; calyx-lobes linear-lanceolate, white-margined, 5–7 mm. long; corolla-lobes oblong-obovate, 8–10 mm. long, acute, greenish-white with purple veins, the gland quadrate, fringed; crown much reduced, fimbriate; caps. ovoid, beaked, 6–7 mm. long; seeds ca. 5 mm. long, cellular-punctate.—Dry slopes, 4500–8000 ft.; largely Yellow Pine F.; desert slopes of San Bernardino and San Gabriel mts., Mt. Pinos region. May–July.

3. **F. párryi** Torr. Plate 50, Fig. F. Stems stout, usually 1, from a heavy taproot, 6–12 dm. high; plant glabrous; lvs. white-margined, the basal oblanceolate, 1–2 dm. long, narrowed into a short winged petiole; cauline lvs. lanceolate, sessile, opposite, 4–10 cm. long, the uppermost lance-ovate; panicle broad, 1.5–3 dm. long; pedicels 1–2.5 cm. long; calyx-lobes lanceolate, subacuminate, 10–15 mm. long; corolla-lobes ca. as long, greenish-white with black dots, the gland U-shaped, fringed all around; crown-scales wanting; caps. long-conic, 14–16 mm. long; seeds ca. 3.5–4 mm. long, spongy-pitted.—Frequent in rather dry places, 1500–6000 ft.; Coastal Sage Scrub, Chaparral, S. Oak Wd., Yellow Pine F.; San Gabriel and San Bernardino mts. s. to L. Calif.; to Ariz. April–July.

4. **F. puberulénta** A. Davids. Stems 1–2, stoutish, puberulent, 1–3 dm. high; lvs. with narrow white margin, oblong-lanceolate, mostly 5–12 cm. long, the cauline opposite, sessile; panicle broad, open, comprising half or more of the plant; pedicels 5–25 mm. long; calyx-lobes lanceolate, 6–8 mm. long; corolla-lobes obovate, short-acuminate, greenish-white with purple dots, ca. as long as calyx; gland oblong, almost covered by the pocket, the opening at top fringed; crown obsolete; caps. round-ovoid, ca. 1 cm. long including the 4 mm. beak.—Dry slopes at ca. 10,000 ft.; Inyo-White Mts.; to Sierra Nevada. June–Aug.

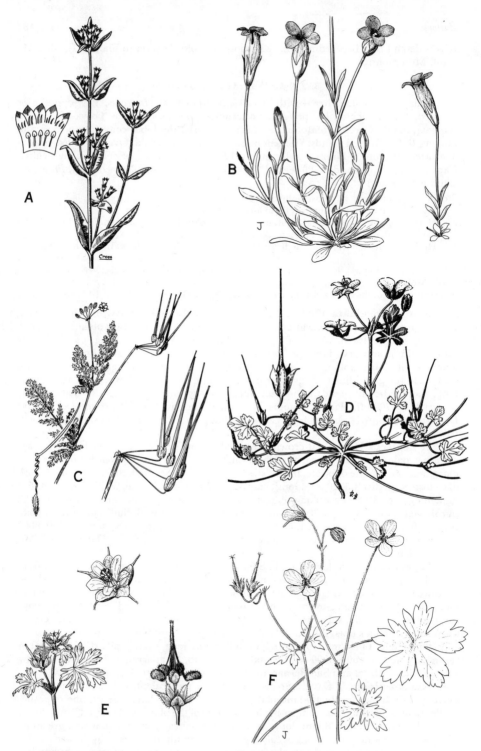

Plate 51. Fig. A, *Gentiana amarella*. Fig. B, *Gentiana holopetala*. Fig. C, *Erodium cicutarium*. Fig. D, *Erodium texanum*. Fig. E, *Geranium carolinianum*. Fig. F, *Geranium richardsonii*.

4. Gentiàna L. GENTIAN

Herbs with opposite mostly sessile lvs. Fls. solitary or clustered, axillary or terminal, frequently showy, blue, purple, white or even yellow, 4–5(–7) merous. Calyx tubular, lobed. Corolla tubular to funnelform or campanulate, lobed, often with intermediate plaited folds which bear appendages or teeth at the sinuses. Stamens inserted on corolla-tube; anthers versatile, straight or recurved in age. Style short or none; stigmas 2, persistent. Caps. 1-celled, 2-valved, sessile or stipitate. Seeds many, minute, winged or wingless, often with a loose cellular coat. Ca. 300 spp., of cool and temp. regions. (Named for King *Gentius* of Illyria, who was supposed to have discovered the medicinal virtues.)

Corolla 7–20 mm. long.
    The corolla with sinus-plaits; lvs. white-margined; corolla-lobes lacking a basal fimbriate crown, greenish-purple, the corolla 7–8 mm. long .... 2. *G. fremontii*
    The corolla lacking sinus-plaits; lvs. not white-margined; corolla-lobes with a basal fimbriate crown; the corolla 8–20 mm. long .... 1. *G. amarella*
Corolla 25–50 mm. long.
    Plants always with a simple stem, with mostly 3–6 pairs of lvs.; calyx-lobes not dark-ribbed; seeds subcylindric .... 4. *G. simplex*
    Plants mostly branched at base, the stems mostly with 1–3 pairs of lvs. above base; calyx-lobes usually dark-ribbed; seeds oval .... 3. *G. holopetala*

1. **G. amarélla** L. [*G. acuta* Michx. *Amarella californica* Greene.] FELWORT. Plate 51, Fig. A. Annual, mostly branched, 0.5–5 dm. tall, slender, erect; lvs. lanceolate to oblong, or the lower spatulate-obovate, 1.5–3.5 cm. long; fls. usually numerous, clustered in axils and terminal, on slender ascending branchlets, the pedicels 5–50 mm. long; calyx-tube 1–4 mm. long, the lobes unequal, 2–10 mm. long, lanceolate; corolla blue to bluish-lavender, tubular-campanulate, 8–20 mm. long, the lobes lanceolate, acute, each with a fimbriate crown at base; caps. sessile, fusiform-cylindric; seeds round-ovoid, ca. 0.6 mm. long; $2n = 36$ (D. Löve, 1953).—Moist places, 4500–11,000 ft.; Montane Coniferous F.; San Bernardino Mts.; Sierra Nevada to Alaska, Atlantic Coast, L. Calif.; Eurasia. June–Sept.

2. **G. fremóntii** Torr. Moss GENTIAN. Annual or biennial, mostly with several simple stems from base, 3–10 cm. high; lvs. oblong, or the lower broader, scarious-margined, 4–6 mm. long, erect; fls. solitary, terminal, subtended by bracts; calyx 6–7 mm. long; corolla tubular-funnelform, 7–8 mm. long, the lobes greenish-blue with whitish margins, the sinus-plaits rounded, minutely toothed; caps. ca. 5 mm. long; seeds ca. 1 mm. long.— Wet meadows, at ca. 8000 ft.; Montane Coniferous F.; San Bernardino Mts.; Rocky Mts. July–Aug.

3. **G. holopétala** (Gray) Holm. [*G. serrata* var. *h.* Gray.] SIERRA GENTIAN. Plate 51, Fig. B. Annual, 0.5–4 dm. high, usually branched at base, leafy below and terminating in long naked 1-fld. peduncles; lower lvs. spatulate-obovate, crowded, 1–4 cm. long; bracts lacking; calyx mostly 10–25 mm. long, the lobes dark-ribbed; corolla blue, infundibuliform, 3–5 cm. long, the lobes oblong, obtuse, entire or erose; caps. fusiform, 9–12 mm. long, on a stipe ca. as long; seeds to 1 mm. long.—Wet meadows, 6000–7000 ft.; Yellow Pine F.; San Bernardino Mts.; Sierra Nevada. July–Sept.

4. **G. símplex** Gray. Simple, erect 5–20 cm. high; lvs. mostly 3–6 pairs, the lowest clasping, the upper sessile, 6–25 mm. long; stem ending in a single fl. without subtending bracts; calyx lobed to ca. the middle, 1.5–2 cm. high, the 4 lobes lanceolate, acute; corolla blue, 2.5–4 cm. long, the lobes irregularly erose; caps. ca. 15 mm. long, on an equally long stipe; seeds ca. 1 mm. long.—Meadows, 7000–8000 ft.; Montane Coniferous F.; San Bernardino Mts.; Sierra Nevada to Ore., Ida., Nev. July–Sept.

48. Geraniàceae. GERANIUM FAMILY

Annual or perennial herbs or somewhat woody. Lvs. opposite or alternate, simple or compound, ours with stipules. Fls. bisexual, 5-merous, hypogynous. Sepals imbricate in bud, persistent. Petals usually imbricate. Disk often with 5 glands alternating with the petals. Stamens as many, or 2 or 3 times as many, as the petals, some of them frequently sterile; fils. tending to be somewhat connate basally. Ovary deeply lobed, the carpels 2-ovuled, 1-seeded, with axile placentation. Styles long, adnate to the elongate axis, separating and coiling when mature. Eleven genera and ca. 650 spp. of temp. and subtrop. regions.

Fls. without nectar-spur, and with glands alternating with the petals; petals alike.
    Lvs. palmately veined or divided; stamens all bearing anthers .... 2. *Geranium*
    Lvs. pinnately veined or divided; stamens having outer fils. without anthers ..... 1. *Erodium*
Fls. with nectar-spur at base of calyx (discovered by sectioning pedicel) and without glands; upper petals usually larger .... 3. *Pelargonium*

1. Eròdium L'Hér. FILAREE. STORKSBILL

Ours annual herbs. Lvs. usually at first forming close rosette on ground, pinnately veined, simple to pinnate, opposite, with 1 interpetiolar stipule on one side and 2 on the other. Pedicels usually retrocurved in fr. Sepals 5. Petals 5. Stamens 5, alternating with 5 scalelike staminodia. Style-column very elongate, the styles bearded inside, spirally coiled when freed from the cent. axis. Carpel-bodies narrow, spindle-shaped, indehiscent. Ca. 60 spp., widespread in temp. and semitrop. regions, some of importance for forage. (Greek, *erodios*, a heron, because of the long beak on the fr.)

1. Lvs. ± cordate at base, usually 3–more-lobed.
    2. Petals unequal, much exceeding sepals; pedicels strigulose, not glandular. Deserts
                                                                                                      7. *E. texanum*
    2. Petals equal, rarely longer than sepals. Cismontane.
        3. Lvs. deeply 3–5-parted; pedicels retrorse-pubescent, not glandular; carpel-bodies 5–6 mm. long .... 3. *E. cygnorum*
        3. Lvs. shallowly 5–7-lobed; pedicels glandular-pilose; carpel-bodies 8–10 mm. long
                                                                                                                  4. *E. macrophyllum*
1. Lvs. not cordate at base.
    4. The lvs. simple, often lobed or divided; style-column 5–12 cm. long.
        5. Style-column 5–9 cm. long; concavities at top of fr. subtended by a single fold, the upper part of the carpel-body ± pubescent; sepals with a short green tip ... 6. *E. obtusiplicatum*
        5. Style-column 9–12 cm. long; concavities at top of fr. subtended by 2 folds, the upper part of the carpel-body glabrous; sepals with a prominent reddish mucro ..... 1. *E. botrys*
    4. The lvs. pinnate; style-column 2.5–4 cm. long.
        6. Lvs. broad, coarsely toothed or serrate; sepals-tips not setose; claws of petals glabrous
                                                                                 5. *E. moschatum*
        6. Lvs. pinnately lobed or divided; sepal-tips setose; claws of petals ciliate . 2. *E. cicutarium*

1. **E. bòtrys** (Cav.) Bertol. [*Geranium b.* Cav.] Stems semiprostrate to suberect, 1–9 dm. long, retrocurved-hirsute; lf.-blades ovate to oblong-ovate, 3–8 cm. long, shallowly to deeply lobed or pinnatifid, setose-pilose on veins and margins; stipules ovate; peduncles 2–20 cm. long; pedicels 1–4, glandular-pubescent; sepals 7–8 mm. long, 13–15 mm. in fr., with prominent reddish mucro; petals ca. 15 mm. long, lavender; style-column 9.5–12.5 cm. long; carpel-bodies 8–10 mm. long, with short stiff spreading hairs, the apex subglabrous and its concavities surrounded by 2 folds; $n=20$ (Heiser & Whitaker, 1948).—Grassy places at low elevs., almost throughout Calif., except on the desert; native of Medit. region. March–May.

2. **E. cicutàrium** (L.) L'Hér. [*Geranium c.* L.] Plate 51, Fig. C. Stems slender, decumbent, 1–5 dm. long, strigulose and glandular-pubescent; lvs. commonly 3–10 cm. long, pinnate, the lfts. incisely pinnatifid; stipules lanceolate; peduncles 5–15 cm. long, glandular-pubescent; pedicels 2–10, glandular-pubescent 8–18 mm. long; sepals 3–5 mm. long, short-mucronate and with 1–2 white bristles; petals rose-lavender, 5–7 mm. long, ciliate at the base, 2-spotted; style column 2–4 cm. long; carpel-bodies 4–5 mm. long, stiff-pubescent, the apical concavities glabrous, without a subtending fold; $2n=40$ (Andreas, 1947).—Common everywhere in Calif. in open dry places below 6000 ft.; natur. very early from the Medit. region. Feb.–May.

3. **E. cygnòrum** Nees. Stems 1–3 dm. long, decumbent to ascending, pilose; lf.-blades ovate in outline, 2–4 cm. long, 3–5-parted into cuneate lobed divisions; stipules ovate; peduncles commonly 3–5-fld.; pedicels filiform, 1–2.5 cm. long, retrorse-pubescent; sepals 6–7 mm. long; petals 7–8 mm. long, blue; style-column 5–6 cm. long; carpel-bodies stiff-pubescent, 5–6 mm. long, the apical concavities subtended by a single fold.—Occasionally collected, as at San Diego, Corona, etc.; native of Australia.

4. **E. macrophýllum** H. & A. Stem proper very short, the lvs. and peduncles subbasal, the latter 1–3 dm. high; plant puberulent and ± glandular-pubescent; lf.-blades reniform-cordate, 2–5 cm. wide, crenate, usually shallowly lobed, the petioles 3–12 cm. long; stipules ovate; peduncles 2–3 (–6)-fld.; sepals 8–10 mm. long; petals white or sometimes rose-red to purple, 10–16 mm. long; style-column 4–5 cm. long; carpel-bodies truncate, 8–10 mm. long, pubescent.—Open places below 3500 ft.; Los Angeles Co. n.; Santa Cruz Id.; it has been found near San Diego. March–May.

5. **E. moschàtum** (L.) L'Hér. [*Geranium m.* L.] Stems rather fleshy, decumbent to ascending, 1–6 dm. long, glandular-pubescent; lvs. 6–40 cm. long (including petioles), pinnate, the lfts. ovate to oblong-ovate, 1–3 cm. long, serrate to cleft, the terminal 3–5-parted; stipules rounded-ovate; peduncles 6–20 cm. long; pedicels 6–13; sepals 6–7 mm. long, mostly without terminal setae; petals rose-violet, not spotted, somewhat longer; style-column 2–4 cm. long; carpel-bodies 4–5 mm. long, stiff-pubescent, the apical concavities glabrous, subtended by a concentric fold; $2n = 20$ (Gauger, 1937).—Common at low elevs., especially in heavy soils; throughout cismontane Calif.; natur. from Medit. region. Feb.–May. Channel Ids.

6. **E. obtusiplicàtum** (Maire, Weiller & Wilcz.) J. T. Howell. Stems ascending to erect, 1–4 dm. long, hirsute and somewhat glandular; lf.-blades oblong-ovate, 2–8 cm. long, the lower with 3–7 rounded shallow lobes, the upper more deeply pinnatifid; stipules ovate; pedicels 1–5; sepals 7–8 mm. long, to 11 mm. in fr., scarcely mucronate; petals lavender, 8–11 mm. long; style-column 5.5–8.5 cm. long; carpel-bodies 6–8 mm. long, the apex pubescent and its concavities surrounded by 1 fold; $2n = 40$ (Baker, 1954). —Local in open grassy places at low elevs.; San Diego Co. to Ore.; natur. from N. Afr. April–Aug.

7. **E. texànum** Gray. Plate 51, Fig. D. Stems few to several, decumbent, 1–4 dm. long, puberulent, not glandular; the peduncles and pedicels canescent-strigulose; lf.-blades ovate to rounded in outline, 1–3 cm. long, cordate, with 3 rounded lobes; stipules lanceolate to deltoid; peduncles 1–3-fld.; sepals 5–8 mm. long, silvery with purple veins; petals twice as long as sepals; style-column 4–6.5 cm. long; $2n = 20$ (Baker, 1954).—Dry sandy or gravelly places below 3500 ft.; Creosote Bush Scrub; Colo. and e. Mojave deserts, Jurupa Hills near Riverside; to Tex., L. Calif. March–May.

2. *Geràmium* L. CRANESBILL

Herbs with forking stems and swollen nodes. Lvs. palmately lobed or parted. Peduncles axillary, 1–3 (–7)-fld., often in subcymose arrangement. Fls. regular, bisexual, 5-merous. Sepals imbricated. Stamens 10, rarely 5; fils. sometimes somewhat connate at base; anthers all well formed, the 5 longer alternate with the petals and with basal glands. Style-column usually beaked, the styles in fr. nearly glabrous inside. Carpels turgid, permanently attached to the styles. Seeds smooth or pitted. Ca. 250 spp. of temp. regions around the world, some weedy annuals, a few of the perennials cult. (Greek, *geranos*, crane, from the beaklike fr.)

(Jones, G. N. and F. F. A revision of the perennial spp. of Geranium of the U.S. and Canada. Rhodora 45: 5–26, 32–53. 1943.)

Petals mostly 3–8 mm. long; mature style-column less than 2 cm. long; annuals or perennials, mostly introd.
    Sepals prominently awned or subulate-tipped; seeds reticulate.
        Plants perennial from a carrotlike root; no hairs with gland-tips .... 5. *G. retrorsum*
        Plants annual; some hairs of pedicels gland-tipped.

Carpel-bodies with short stiff subequal spreading hairs; lobes of upper lvs. acute; pits of seed square or rounded .... 3. *G. dissectum*
Carpel-bodies with long ± unequal ascending hairs; lobes of lvs. obtuse; pits of seed ± elongate .... 2. *G. carolinianum*
Sepals awnless, at most with callous tips; seeds smooth .... 4. *G. molle*
Petals 10–20 mm. long; mature style-column 2–5 cm. long; native perennials.
Style-branches 3–4 mm. long; petals white or whitish; pedicels usually with purplish glands. Sierra Nevada to s. Calif. .... 6. *G. richardsonii*
Style-branches 4.5–9 mm. long; petals pink or lavender; pedicels with yellowish glands. S. montane .... 1. *G. californicum*

1. **G. califórnicum** Jones & Jones. [*G. leucanthum* Small, not Griseb. *G. concinnum* Jones & Jones.] Perennial; stems 1-few, slender, 1–5 dm. tall, sparse-villous to retrorse-pubescent; lower petioles 5–25 cm. long, sparsely villous to retrorse-pubescent; blades 2–8 cm. broad, usually 5-parted, the divisions rhombic to cuneate, strigulose above and beneath or ± villous beneath; peduncles 2–10 cm. long; glandular-pubescent to glandular-villous and usually with longer eglandular hairs; sepals 6–9 mm. long, awned; petals rose-pink to lavender, 10–16 mm. long; style-column 2–2.5 cm. long, including the 3–5 mm. beak; carpel-bodies 4–5 mm. long, pubescent to glandular; seeds faintly reticulate, ca. 3 mm. long.—Damp places, 4000–7500 ft.; Montane Coniferous F.; mts. from San Diego Co. to cent. Sierra Nevada. June–July.

2. **G. caroliniànum** L. Plate 51, Fig. E. Annual, 2–4 dm. high, densely retrorse-hirsute; lvs. 3–6 cm. broad, 5–7-parted, the cuneate divisions ± cleft into linear or oblong, mostly obtuse lobes; lower petioles very long; peduncles mostly 2-fld., 1–3 cm. long, solitary or loosely aggregated; pedicels 2–7 mm. long; sepals 5–7 mm. long, with some gland-tipped hairs among the others, awn-tipped; petals pink or paler, ca. as long as sepals; style-column 10–14 mm. long, the beak 2 mm.; carpel-bodies 2–3 mm. long; seeds oblong, reticulate, with 25–35 rows of somewhat elongate areolae; $n=26$ (Shaw, 1952).—Frequent in grassy and shaded places below 5000 ft.; many Plant Communities; cismontane; to B.C., Atlantic Coast. April–June.

3. **G. disséctum** L. Usually freely branched annual, 3–6 dm. high, retrorsely pubescent; lvs. 3–6 cm. broad, deeply 5-parted, the divisions again divided into broadly linear acute segms.; peduncles mostly 2-fld.; sepals awn-tipped, 5–7 mm. long; petals rose-purple, equaling sepals; style-column ca. 12 mm. long including the 2 mm. beak; carpel-bodies 2–2.5 mm. long, with stiff spreading hairs; seeds strongly reticulate, with unequally thick-walled pits; $2n=22$ (Löve & Löve, 1944).—Waste and open places, mostly below 3000 ft.; widely distributed but not common in s. Calif.; to Atlantic Coast; natur. from Eu. March–May.

4. **G. mólle** L. Annual or biennial, and with thickened root; stems decumbent, 1–4 dm. long, villous-pubescent; lvs. 2–6 cm. broad, mostly 5–7-cleft, the broad cuneate segms. toothed or lobed; petioles pilose; peduncles 2-fld., 1–3 cm. long; pedicels glandular-pubescent, filiform; sepals 3–4 mm. long, awnless; petals rose-purple, 3–5 mm. long; style-column 8–12 mm. long including the filiform 1–2 mm. beak; carpel-bodies glabrous, usually cross-wrinkled, ca. 2 mm. long; seeds smooth, brown; $n=13$ (Warburg, 1938).—Occasional as lawn-weed, etc.; to B.C. and Atlantic Coast; native of Eu. May–July.

5. **G. retrórsum** L'Hér. Perennial with a taproot, retrorse-pubescent throughout with dull whitish hairs; plant 3–5 dm. high; lvs. incisely 3–5-parted, the cuneate divisions again obtusely 3–5-lobed; peduncles usually 2-fld., 1–4 cm. long; sepals awn-tipped, 3–4 mm. long; petals 5–6 mm. long, reddish-lilac; style-column ca. 1.5 cm. long; carpel-bodies ca. 2.5 mm. long, strigose; seeds finely reticulate.—Occasionally natur. in grassy places or edge of brush; Palomar Mts., Ventura Co., etc.; to n. and e.; native of Australasia. May–Sept.

6. **G. richardsònii** Fisch & Trautv. Plate 51, Fig. F. Perennial; stems 1-few, 3–9 dm. high, glabrous or sparsely pubescent; blades 3–15 cm. broad, 5–7-parted, the segms. rhombic, sparsely strigose above and on veins beneath, they and the pedicels glandular-

pilose, the hairs tipped with purplish glands; pedicels paired, 1-2 cm. long; sepals 6-12 mm. long, awn-tipped; petals 10-18 mm. long, white or pinkish, with purple veins; style-column 2-2.5 cm. long, including the 2 mm. beak; style-branches 3-5 mm. long; carpel-bodies 3-4 mm. long, sparingly pubescent and with some long stiff hairs; seed coarsely reticulate; $n=26$ (Shaw, 1952).—Meadows and moist places, 4000-9000 ft.; Montane Coniferous F.; San Jacinto Mts. n. through the mts. to B.C., Rocky Mts. July-Aug.

3. *Pelargònium* L'Hér. GERANIUM

Annual or perennial herbs, sometimes woody, often succulent, often strong-smelling. Lvs. alternate or opposite, palmately or pinnately veined, lobed or dissected. Peduncles axillary, umbellately 2-many-fld. Fls. of many colors, irregular. Sepals 5, imbricate, the posterior with a nectar-spur that is joined to the pedicel for much of the latter's length. Petals 5, the 2 upper mostly larger and more prominently colored. Stamens 10, part of them without anthers. Fr. of 5 carpels, the style-column elongate, the styles pubescent within and spirally coiled when free from the cent. axis. Nearly 250 spp., mostly from S. Afr., many grown for ornament and for their contained oils. (Greek, *pelargos*, stork, from the bill-shaped fr.)

Several spp. have been reported as natur. in Calif., but apparently most do not survive long. They are to be sought in waste places especially near the coast. The most likely to be found in s. Calif. are:

Stems fleshy at least when young; lvs. scarcely if at all lobed.
    Stems usually solitary, 2-4 dm. high; stipules pointed, longer than broad ..... 3. *P. zonale*
    Stems several, usually much taller; stipules rounded, broader than long .... 2. *P. hortorum*
Stems not fleshy, shallowly several-lobed .... 1. *P. grossularioides*

1. **P. grossularioìdes** (L.) Ait. [*Geranium g.* L.] Subglabrous annual or perennial, suberect, 3-6 dm. high, with pungent turpentinelike odor; lvs. 2-5 cm. wide, shallowly several-lobed, crenate-dentate; fls. 3-10; pedicels 2-5 mm. long; sepals ca. 4 mm. long; petals red, 6 mm. long.—Occasional in waste ground, Fallbrook, Santa Barbara, etc.; S. Afr. April-July.

2. **P. × hortòrum** Bailey. FISH GERANIUM. Freely branched, with fishy odor; lvs. round to reniform, somewhat scalloped, crenate-toothed, mostly with a color band; stipules broad, rounded; calyx-spur 2-3 cm. long; fls. of many colors.—The common bedding geranium, of horticultural origin; occasional escape near cult. areas.

3. **P. zonàle** (L.) Ait. [*Geranium z.* L.] HORSESHOE GERANIUM. Sparsely branched, ± pilose; lvs. round-cordate, crenate-dentate, obscurely many-lobed, mostly with a dark horseshoe band; stipules ovate, pointed; calyx-spur 2.5-3.5 cm. long; fls. red to pink. — Reported from Oceanside, etc.; from S. Afr.

49. **Haloragàceae.** WATER-MILFOIL FAMILY

Mostly aquatic perennial herbs with alternate or verticillate lvs., the blades of those submerged often pectinate-pinnatifid or pinnately divided into capillary divisions. Fls. perfect or imperfect, axillary, solitary or groups or in interrupted spikes. Fl.-tube adherent to ovary, prolonged little or none beyond it. Sepals 2-4 or obsolete. Petals small, 2-4 or 0. Stamens 1-8. Ovary inferior, 1-4-loculed; styles 1-4. Fr. indehiscent, angular, ribbed or winged, with 2-4 1-seeded carpels. Endosperm fleshy; cotyledons minute. Ca. 100 spp., widely distributed.

1. *Myriophýllum* L. WATER-MILFOIL

Perennial aquatics; lvs. often whorled, the submersed pinnately parted into capillary divisions. Fls. sessile, chiefly in the axils of the reduced upper lvs., usually above water in summer, the uppermost ♂. Calyx of the ♂ fls. of 4 sepals, of the ♀ of 4 teeth. Fr. nutlike, 4-loculed, deeply 4-lobed; stigmas 4, recurved; dehiscence into 4 indehiscent 1-seeded

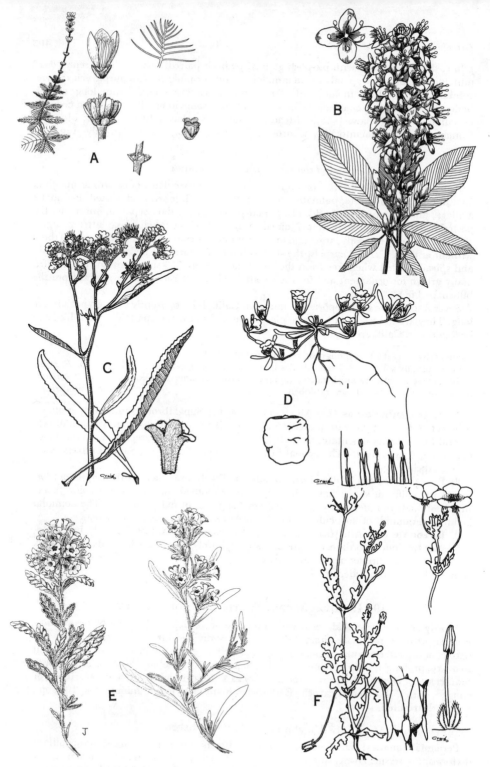

Plate 52. Fig. A, *Myriophyllum exalbescens*. Fig. B, *Aesculus californica*. Fig. C, *Eriodictyon trichocalyx*. Fig. D, *Nama demissum*. Fig. E, *Nama rothrockii*. Fig. F, *Nemophila menziesii*, with bract between calyx-lobes.

Myriophyllum

bony carpels. Ca. 20 spp. (Greek, *myrios*, numberless, and *phyllon*, leaf, alluding to the many divisions of the lvs.)

1. **M. exalbéscens** Fern. Plate 52, Fig. A. Stems to 1 m. long, simple or forked, purple, whitish on drying; lvs. in 3's or 4's, 1–3 cm. long, with 6–11 pairs of capillary segms.; fls. in emersed almost naked interrupted spikes, the lower ♀, the upper ♂; bracts rarely equaling the frs., the lower serrate, the upper entire; petals fugacious, 2.5 mm. long; stamens 8; fr. subglobose, 2.5–3 mm. long; $2n=14$ (Löve, 1954).—In quiet water below 8000 ft.; Montane Coniferous F.; San Bernardino Mts.; at a lower elev. in Murray Lake, San Diego Co.; to Alaska, Atlantic Coast. June–Sept.

50. **Hippocastanàceae.** BUCKEYE FAMILY

Trees or shrubs with opposite estipulate palmate lvs. Fls. polygamous, irregular, showy, borne on jointed pedicels in a terminal panicle or thyrse. Calyx tubular or campanulate, 5-parted, the segms. unequal. Petals 4–5, clawed, unequal. Stamens 5–8; fils. long, slender. Ovary 3-loculed; ovules 2 in each cell; style slender. Fr. a leathery caps., globose or slightly 3-lobed, smooth or spiny, 3-celled or by abortion 1–2-celled and -seeded. Seeds large, shining; endosperm none; cotyledons large and thick. Three genera and 18 spp. of N. Hemis.

1. *Aésculus* L. BUCKEYE. HORSE-CHESTNUT

Characters of the family. (Latin name of some oak.)

1. **A. califórnica** (Spach) Nutt. [*Calothyrsus c.* Spach.] Plate 52, Fig. B. Large bush or tree to ca. 10 m. tall, with broad round top; lfts. 5–7, oblong-lanceolate, serrulate, acute to acuminate, 5–15 cm. long, subglabrous to finely pubescent; thyrse erect, 1–2 dm. long; fls. white or pale rose; calyx ca. 7–8 mm. long, 2-lobed; petals ca. 13–15 mm. long; stamens 5–7, with orange anthers; fr. pear-shaped, smooth; seeds mostly 1, glossy brown, 2–3 cm. long; $n=20$ (Ornduff & Lloyd, 1965).—Common on dry slopes and in canyons below 4000 ft.; Foothill Wd.; n. Los Angeles Co. to n. Calif. May–June.

51. **Hippuridàceae.** MARE'S-TAIL FAMILY

Perennial aquatics with whorled entire lvs. Fls. perfect or polygamous, minute, sessile in the axils. Calyx entire. Style filiform, stigmatic down one side, lying in the groove between the lobes of the large anther. Fr. nutlike, 1-locular. One genus.

1. *Hippùris* L. MARE'S-TAIL

Flaccid or fleshy plants of cool regions. (Greek, meaning horse-tailed, from *hippos*, horse, and *oura*, a tail.)

1. **H. vulgàris** L. Stems erect, from creeping rhizomes, hollow, simple, cylindrical, 1.5–5 dm. long; lvs. 6–12 in a whorl, linear-attenuate, firm, 5–30 mm. long, or flaccid and longer; calyx tubular or barrel-shaped, subtruncate at summit, adherent to ovary, 1.5–2.5 mm. long in maturity.—Shallow water; collected at Lake Fulmor, San Jacinto Mts.; San Bernardino Mts.; Sierra Nevada to Alaska, Atlantic Coast; Patagonia, Eurasia. July–Sept.

52. **Hydrophyllàceae.** WATERLEAF FAMILY

Herbs or shrubs with opposite or alternate lvs. Fls. usually 5-merous, cymose or solitary. Calyx deeply lobed, the divisions alike or unlike, with or without basal auricles between the lobes. Corolla gamopetalous, rotate to campanulate or tubular, usually with a pair of scalelike appendages at base of each fil. Stamens hypogynous, inserted near base of corolla, equal or unequal, exserted or included. Pistil 1, consisting of 2 carpels; style subentire to deeply bifid; stigmas 2, mostly capitate. Fr. a caps., loculicidal with 2 valves or both loculicidal and septicidal and dehiscent by 4-valves, or irregularly dehiscent, 1-

Hydrophyllaceae

loculed or partially 2-loculed by intrusion of the parietal placentae. Ovules few to many. Seeds with endosperm; cotyledons entire. Ca. 25 genera and 300 spp., largely w. Am.

1. Styles ± united.
    2. Perennial herbs.
        3. Calyx-lobes similar.
            4. Lvs. entire to toothed, but not deeply divided or lobed.
                5. Plants caulescent; fls. in cymes or clustered, sometimes solitary and axillary.
                    6. Stamens unequally inserted, unequal in length .... 6. *Nama*
                    6. Stamens equally inserted, subequal in length .... 8. *Phacelia*
                5. Plants acaulescent; fls. solitary in lf.-axils of basal rosette .... 4. *Hesperochiron*
            4. Lvs. deeply divided or lobed, largely cauline .... 8. *Phacelia*
        3. Calyx-lobes of 2 kinds, the 3 outer cordate, enlarged and veiny in fr., the 2 inner narrower .... 10. *Tricardia*
    2. Annual herbs.
        7. Herbage viscid and scented; ovules borne on both faces of the placentae; calyx lacking auricles between the lobes .... 3. *Eucrypta*
        7. Herbage not viscid or scented; ovules borne only on the front of the placentae.
            8. Ovary 1-celled, with broad placentae; infl. not a clearly marked cyme; calyx usually with ± evident auricles between lobes.
                9. Stems recurved-prickly; caps. prickly or bristly; seeds without a cucullus
                                                                                                        9. *Pholistoma*
                9. Stems usually not recurved-prickly; caps. pubescent; seeds with a ± terminal cucullus
                                                                                                        7. *Nemophila*
            8. Ovary falsely 2-celled by intrusion of narrow placentae; infl. usually an evident coiled cyme.
                10. Stamens equally inserted, and if unequal in length, the seeds corrugate.
                    11. Corolla mostly deciduous, blue to purplish or white, if persistent, the fls. erect
                                                                                                        8. *Phacelia*
                    11. Corolla persistent, yellowish or reddish, pendulous .... 1. *Emmenanthe*
                10. Stamens unequally inserted and unequal in length; seeds ovoid, reticulate, often pitted .... 6. *Nama*
1. Styles separate to base.
    12. Corolla constricted at insertion of stamens, the fils. coherent laterally by their dilated bases; low annual .... 5. *Lemmonia*
    12. Corolla not constricted at point of stamen-insertion, the fils. distinct.
        13. Low annuals; seeds pitted or alveolate .... 6. *Nama*
        13. Perennial or shrubby.
            14. Herbaceous, the fls. in an elongate thyrsoid panicle of cymes; seeds longitudinally striate .... 11. *Turricula*
            14. Shrubby, the fls. not paniculate; seeds transversely corrugate or winged.
                15. Lvs. to 1.5 dm. long. Natives .... 2. *Eriodictyon*
                15. Lvs. to 4.5 dm. long. Escape from gardens .... 12. *Wigandia*

1. *Emmenánthe* Benth.

Annuals, resembling *Phacelia*, glandular-viscid, rather agreeably scented, erect. Fls. soon pendulous, the corolla cream-colored, persistent, without inner appendages. Stamens included. Style inclined, somewhat 2-cleft, deciduous. Caps. unilocular, partly divided by the intrusion of the placentae, loculicidal. Ovules many. Seed oval, flat, pitted-reticulate. One sp. (Greek, *emmeno*, to abide, and *anthos*, fl., because of the persistent corolla.)

1. **E. penduliflòra** Benth. WHISPERING BELLS. Erect, usually much branched, villous-pubescent and minutely viscid-glandular, 1–5 dm. tall; lvs. linear-oblong, 3–10 cm. long, the lower short-petioled, the upper sessile, pinnatifid into many oblong entire or toothed lobes; cymes in terminal paniclelike infl.; pedicels recurved, filiform, ca. 1 cm. long; calyx-lobes lance-ovate, 6–10 mm. long; corolla yellowish, tubular-campanulate, 8–12 mm. long, the lobes rounded; style ca. 2 mm. long; caps. ca. 1 cm. long; seeds ca. 15, light brown, ca. 2 mm. long; $n=18$ (Cave & Constance, 1942).—Common in dry places par-

ticularly after burns or disturbance, below 6000 ft.; Chaparral, Coastal Sage Scrub, Creosote Bush Scrub, etc.; cismontane from L. Calif. to Tehama Co., across the deserts from Inyo Co. to Utah, Ariz. April–July. A form with brownish herbage and pink corolla drying white is var. *rosea* Brand; from Ventura Co. to Santa Clara Co., away from the coast.

2. *Eriodictyon* Benth. YERBA SANTA

Aromatic shrubs, evergreen, with shredding bark and open weedy growth from running woody underground rootstocks. Lvs. alternate, somewhat coriaceous, toothed or entire, sessile to petioled, mostly crowded towards ends of branches. Fls. many, in terminal branched scorpioid open to subcapitate cymes. Calyx deeply divided. Corolla white to purple, deciduous, funnelform to campanulate, without appendages. Stamens included, equally inserted, the fils. often adnate. Style divided to base. Caps. cartilaginous, falsely bilocular by intrusion and union of narrow placentae, dehiscing by 4 valves. Seeds usually 2–6, ovoid, angled or flattened, finely longitudinally ridged. Ca. 8 spp., of sw. U.S. and Mex. (Greek, *erion*, wool, and *diktuon*, net, referring to under-surface of lvs.)

Upper surface of lvs. and the young stems glabrous or subglabrous, glutinous.
   Lvs. usually entire or subentire, subsessile, mostly 2–10 mm. wide.
      Calyx 2–4 mm. long; cyme open; corolla sparsely pubescent, 4–7 mm. long. E. Mojave Desert .... 1. *E. angustifolium*
      Calyx 6–8 mm. long; cyme subcapitate; corolla villous, 7–11 mm. long. Coastal Santa Barbara Co. .... 2. *E. capitatum*
   Lvs. usually toothed, petioled, 5–30 mm. wide; Los Angeles Co. s. .... 5. *E. trichocalyx*
Upper surfaces of lvs. and young stems pubescent to tomentose, except in a form of *E. crassifolium* of Santa Barbara and Ventura cos.
   Calyx and corolla glandular; corolla constricted at throat, 4–7 mm. long. .... 4. *E. traskiae*
   Calyx and corolla not glandular; corolla not constricted at throat, 10–17 mm. long
                                                                         3. *E. crassifolium*

1. **E. angustifòlium** Nutt. To ca. 1 m. tall, the branches glabrous and glutinous, lvs. linear to lance-linear, 4–10 cm. long, glabrous and glutinous above, sparsely pubescent to tomentose beneath, reticulate, short-petioled, entire to crenulate, ± revolute; cymes glutinous and sparsely pubescent; calyx-lobes 3–4 mm. long, sparsely hirsutulous; corolla narrow-campanulate, white, 5–6 mm. long, pubescent; caps. sparsely pubescent in upper half, ca. 2 mm. long; seeds almost 1 mm. long, transversely reticulate; $n=14$ (Cave & Constance, 1950).—Dry slopes 5000–5500 ft.; Pinyon-Juniper Wd.; New York Mts., e. Mojave Desert; to Utah, Ariz., L. Calif. June–July.

2. **E. capitàtum** Eastw. To almost 2 m. tall, the branches glabrous and glutinous; lvs. linear, glabrous and glutinous above, white-tomentose beneath, 4–9 cm. long, 2–4 mm. broad, revolute; cymes glabrous, capitate; calyx-lobes linear, 6–8 mm. long, villous; corolla lavender, tubular-funnelform, 8–12 mm. long; style 3–4 mm. long; caps. densely pubescent on entire surface, 3 mm. long; seeds ca. 1 mm. long; $n=14$ (Cave & Constance, 1950).—Brushy slopes below 1000 ft.; Closed-cone Pine F. and Chaparral; Santa Ynez Mts. and n. of Lompoc, Santa Barbara Co. May–June.

3. **E. crassifòlium** Benth. Shrub 1–3 m. tall, leafy above; twigs and both lf.-surfaces and calyx hoary-tomentose; lvs. lance-ovate to oval, 5–15 cm. long, crenate to coarsely dentate, plane, short-petioled; cymes tomentose, lax in fr.; calyx-lobes lance-linear, 3–5 mm. long; corolla lavender, broadly funnelform, 10–16 mm. long, the round lobes 2–3 mm. long; style 4–5 mm. long; caps. 2–3 mm. long, hirsute; seeds ca. 8–12, ca. 1 mm. long; $2n=14$ (Cave & Constance, 1947).—Dry gravelly and rocky places below 6000 ft.; Chaparral, Pinyon-Juniper Wd., etc.; cismontane, mostly from Santa Monica and San Gabriel mts. to w. edge of Colo. Desert and to San Diego. April–June. Variable, the following poorly defined: (1) var. *denudatum* Abrams. Lvs. greenish and glabrate above; corolla 8–10 mm. long.—Fillmore, Ojai, etc. to Santa Ynez Mts.; (2) var. *nigrescens* Brand. Lvs. dull green and short-tomentose above; corolla 6–7 mm. long.—Ventura and

Los Angeles cos. in mts. bordering w. end of Mojave Desert and apparently Thomas Mt., San Jacinto Mts.

4. **E. tráskiae** Eastw. ssp. **tráskiae**. White-woolly, especially on lvs.; lf.-blades mostly narrow-elliptic, ca. 1.5–2.5 cm. wide, 6–7 cm. long; calyx 6–6.5 mm. long; corolla 7.5–8 mm. long; caps. hairy on upper half, 4-seeded.—Chaparral; Santa Catalina Id. May–June. Ssp. **smithii** Munz, with blades 3–7 cm. wide, to 10 cm. long; calyx 4.5–5 mm. long; corolla 6–7 mm. long; in very n. part of our range from Santa Ynez Mts. to San Luis Obispo Co.

5. **E. trichocàlyx** Heller ssp. **trichocàlyx**. Plate 52, Fig. C. Shrubs 5–20 dm. tall, the lvs. glabrous and glutinous above, veiny and tomentulose beneath, lanceolate to oblong, 5–15 cm. long; cymes hirsutulous; calyx-lobes linear to lanceolate, 3–4 mm. long, mostly densely hirsute, not glandular; corolla pale purplish to white, 5–8 mm. long, the round lobes ca. 2 mm. long; style 3–4 mm. long; caps. 2–3 mm. long, densely hispidulous; seeds dark brown, 1–1.5 mm. long; $n=14$ (Cave and Constance, 1947).—Dry rocky slopes and fans, below 8000 ft.; Chaparral, Yellow Pine F., Pinyon-Juniper Wd., Joshua Tree Wd.; Ventura Co. near and through the San Gabriel and San Bernardino mts. May–Aug.

Ssp. **lanàtum** (Brand) Munz. [*E. californicum* Greene ssp. *australe* var. *lanatum* Brand (Pflanzenreich 4, Fam. 251: 142. 1913.)] Twigs pubescent; lvs. white-tomentose, obscurely reticulate beneath; calyx-lobes 2–3 mm. long, densely hirsute and somewhat glutinous; corolla-lobes ovate.—Dry slopes and ridges, 1000–6000 ft.; Chaparral, Yellow Pine F., Pinyon-Juniper Wd.; w. edge of Colo. Desert from Santa Rosa Mts. to L. Calif. April–June.

3. *Eucrýpta* Nutt.

Annual, erect or diffuse, viscid, scented, hispid. Lower lvs. opposite, the others alternate, pinnately divided, petioled or the upper sessile or clasping. Fls. several, in open, terminal or axillary cymes, with filiform pedicels. Calyx divided half way or more, the sinuses naked. Corolla whitish to yellowish or blue, campanulate, deciduous, shallowly lobed, longer than calyx. Stamens included, equally inserted on corolla, the appendages minute or none and with a V-shaped transverse fold between each pair of fils. near the throat. Style shortly bifid. Mature caps. 1-loculed, round to ovoid, surpassed by the calyx. Seeds usually 5–15, the inner lens-shaped, smooth, the outer terete, corrugated or all alike. Two spp. (Greek, *eu*, well or true, and *crypta*, secret, referring to the extra, hidden seeds.)

Lvs. 2–3-pinnatifid, mostly opposite; mature calyx stellate-rotate beneath the caps.
.... 1. *E. chrysanthemifolia*
Lvs. 1-pinnate, alternate, the lobes mostly entire or few-toothed; mature calyx erect and enclosing caps. .... 2. *E. micrantha*

1. **E. chrysanthemifòlia** (Benth.) Greene var. **chrysanthemifòlia**. [*Ellisia c.* Benth.] Erect, branched, 2–5 dm. tall, somewhat hirsute and glandular, with characteristic rather pleasant odor; lvs. 3–10 cm. long, broadly ovate to oblong in outline, pinnatifid, the 9–13 lance-oblong lobes again 1–2-pinnatifid; lvs. short-petioled or subsessile, auriculate at base; fls. loosely clustered; calyx pilose and finely pubescent, the lobes 1–2 mm. long, obtuse; corolla yellowish-white, open-campanulate, 6–8 mm. broad; caps. 2–4 mm. in diam., hirsute, the inner seeds 1–1.5 mm. long, the outer 0.8–1 mm.; $n=10$ (Cave & Constance, 1944).—Common on burns and in partly shaded places, below 3000 ft.; Coastal Sage Scrub, Chaparral, Oak Wd.; L. Calif. through cismontane Calif. to Marin Co., Channel Ids. March–June.

Var. **bipinnatífida** (Torr.) Const. [*Ellisia torreyi* Gray.] Weaker and more diffuse, glandular-pubescent only in infl., hairy below; lvs. 2–4.5 cm. long, 7–9-lobed, the lobes shallowly pinnatifid; fls. mostly 4–8, corolla 2–3 mm. broad, not exceeding calyx.—Occasional in shelter of rocks, below 7500 ft.; Creosote Bush Scrub, Joshua Tree Wd.; deserts from Inyo Co. to Nev., Ariz., L. Calif. March–May.

2. **E. micrántha** (Torr.) Heller. [*Phacelia m.* Torr. *Ellisia m.* Brand.] Erect but rather

weak and diffuse, the stems slender, often stipitate-glandular, 1–2.5 dm. tall; lower lvs. oblong to oval, 1.5–3 cm. long, pinnately parted into 7–9 oblong to spatulate, entire or few-toothed divisions; upper lvs. auriculate-clasping; fls. 4–12 on each branch of infl., the fruiting pedicels mostly ascending; calyx-lobes 1.5–2 mm. long, spatulate or oblong; corolla 2–3 mm. in diam.; seeds 7–15, homomorphic; $n=6$ (Cave & Constance, 1950).— Usually in partly shaded places, below 7000 ft.; Creosote Bush Scrub to Pinyon-Juniper Wd.; deserts from Inyo Co. to Imperial Co.; to Nev., Tex., L. Calif. March–June.

4. Hesperochiron Wats.

Acaulescent perennial herbs from a short vertical thick root. Lvs. in a rosette, entire, ovate to spatulate, petioled. Fls. solitary in the lf.-axils, on long slender erect or spreading pedicels. Calyx 5-parted, the lobes ± unequal. Corolla white to bluish, funnelform to rotate, deciduous, exceeding the calyx. Stamens often unequal, included, inserted on corolla-tube, the fils. dilated at base. Style included, shortly 2-cleft at apex. Caps. unilocular, ± ovoid. Seeds many, dark brown, ovoid, angular, somewhat pitted. Two spp. (Greek, *hesperos*, western, and *Chiron*, a centaur skilled in medicine.)

1. **H. califórnicus** (Benth.) Wats. [*Ourisia c.* Benth. *Capnorea leporina* Greene.] Plants somewhat pubescent or subglabrous save for the ciliate lvs. and calyx-lobes; lvs. many, narrow-oblong to oval, 1–5 cm. long, on somewhat shorter petioles; pedicels usually many, 2–8 cm. long; calyx-lobes oblong to subovate, 4–7 mm. long, ± glabrous on backs; corolla funnelform to narrow-campanulate, 1–2.5 cm. broad, the lobes ± oblong, 3–6 mm. long; caps. 5–10 mm. long; seeds ca. 2 mm. long; $n=8$ (Cave & Constance, 1950).— Moist places, ca. 4000–7000 ft.; largely Montane Coniferous F.; San Bernardino Mts., Mt. Pinos region; to Siskiyou Co. and Rocky Mts. A more pubescent form largely n. of s. Calif. has been called var. *watsonianus* (Greene) Brand.

5. Lemmònia Gray.

Small depressed annual, dichotomously branched. Lvs. alternate, entire, clustered in a basal rosette and near tips of branches. Fls. sessile, solitary in upper axils and forks and in terminal capitate cymes. Calyx deeply divided into 5 linear lobes. Corolla deciduous, campanulate, without appendages, shorter than calyx. Stamens unequal, included, the fils. coherent laterally at the base. Style deeply divided. Caps. falsely bilocular by intrusion of placentae, membranaceous, loculicidal. Placentae narrow, each with 2–3 ovules. Seeds ca. 4, oblong-ovoid, irregularly corrugated. One sp. (J. G. *Lemmon*, early Calif. botanist.)

1. **L. califórnica** Gray. Branches slender, pubescent, 2–10 cm. long; lvs. oblanceolate, 5–15 mm. long; calyx 3 mm. long; corolla white, 2.5 mm. long; seeds dark, ca. 1 mm. long; $n=7$ (Cave & Constance, 1950).—Dry sandy places, mostly 3000–8000 ft.; many Plant Communities; away from the coast, Lake Co. to Kern Co., thence in mts. along w. edge of Mojave Desert se. to San Bernardino Mts.; Nev. April–June.

6. Nàma L.

Low branching annual to somewhat woody perennials. Lvs. alternate, entire to sinuate-dentate, well distributed on the stems. Fls. in reduced terminal nonscorpioid cymes and axillary or in angles of branches, subsessile. Calyx deeply divided. Corolla deciduous, funnelform to ± tubular, longer than calyx. Stamens included, unequal or unequally inserted, appendaged at base or appendages obsolete. Style included, ± bifid. Caps. falsely bilocular by intrusion of the placentae, loculicidal and sometimes septicidal. Ovules many. Seeds ovoid, usually reticulate, sometimes shallowly pitted. Perhaps 45 spp. of sw. U.S., Mex., S. Am. and 1 in Hawaii. (Greek, *nama*, a spring.)

(Hitchcock, C. L. Am. J. Bot. 20: 415–430, 518–534. 1932–1933.)

Plants perennial, with ± woody base; fls. many, in terminal cymes . . . . 6. *N. rothrockii*
Plants annual; fls. solitary or few, in reduced cymes.

Calyx divided ca. ¾ its length, the tubular base adnate to the somewhat inferior ovary; calyx-
lobes indurate and recurved in fr. .... 7. *N. stenocarpum*
Calyx divided nearly to base, the ovary superior; calyx-lobes mostly erect and not indurate.
    Style shallowly 2-lobed at very apex; corolla 7–15 mm. long. .... 1. *N. aretioides*
    Style divided almost to base.
        Corolla campanulate, 8–15 mm. long, with expanded limb.
            Stems quite erect; adnate portion of fils. not winged; seeds many, dark brown
                                                                           4. *N. hispidum*
            Stems prostrate; adnate portion of fils. with narrow margin; seeds 15–25, yellowish
            brown .... 2. *N. demissum*
        Corolla tubular or nearly salverform, 3–5 mm. long.
            Upper lvs. obovate, abruptly narrowed into a petiole; calyx densely hirsute 5. *N. pusillum*
            Upper lvs. oblanceolate, gradually narrowed into petiole; calyx sparsely pubescent
                                                                       3. *N. depressum*

1. **N. aretioìdes** (H. & A.) Brand. [*Eutoca a.* H. & A.] Prostrate hirsute annual, few-branched from base, the branches densely leafy at tips, 2–8 cm. long; lvs. narrow-spatulate, 8–15 mm. long; fls. solitary, sessile; calyx-lobes 4–7 mm. long; corolla purple to rose-red, tubular-funnelform, 7–15 mm. long; styles 2–6 mm. long; caps. 10–35-seeded; seeds ovoid, 0.5–1.5 mm. long, deeply corrugate, $n=7$ (Cave & Constance, 1950).—Sandy to rocky places, 3500–6500 ft.; Pinyon-Juniper Wd.; Argus Mts., Inyo Co.; to Wash., Ida., Nev. May–June.

2. **N. demíssum** Gray. var. **demíssum**. Plate 52, Fig. D. Prostrate annual, few-branched, the stems slender, strigose, villous-hirsute, 3–15 cm. long, with long internodes; lvs. largely in terminal compact clusters, narrow-spatulate to obovate, 1–2 cm. long, 2–5 mm. wide, green; fls. solitary in axils of branches and numerous at leafy tips; calyx-lobes lance-linear, gray-hirsute, 5–8 mm. long; corolla purplish-red, tubular-campanulate, 9–12 mm. long; styles 3–5 mm. long; caps. ca. 4 mm. long, 8–18-seeded; seeds brown, 0.5 mm. long, shallowly pitted and minutely reticulate; $n=7$ (Cave & Constance, 1950). —Occasional on dry flats and slopes, 3000–5500 ft.; Creosote Bush Scrub to Pinyon-Juniper Wd.; e. Mojave Desert; to Utah, Ariz. April–May. Two forms are ± different: var. *deserti* Brand, with pubescence and aspect of var. *demissum*, but lvs. 1–4 cm. long and mostly 1–2 mm. wide; Creosote Bush Scrub; Inyo Co. to Imperial Co.; Ariz. The other is var. *covillei* Brand, gray-villous throughout; lvs. rhombic-obovate, 1–2 cm. long, 3–6 mm. wide; Creosote Bush Scrub; Death V. region.

3. **N. depréssum** Lemmon ex Gray. With general habit and size of *N. pusillum*, softly appressed-pubescent throughout; lvs. oblanceolate, 0.5–1 cm. long, short-petioled; calyx-lobes 3–4 mm. long; corolla 4 mm. long, pubescent without; seeds rhomboid-ovoid, with ca. 20 irregular pits; $n=7$ (Cave & Constance, 1959).—Sandy places below 3000 ft.; Creosote Bush Scrub; desert from Inyo Co. to San Bernardino and Imperial cos.; w. Nev. April–May.

4. **N. híspidum** Gray var. **spathulàtum** (Torr.) C. L. Hitchc. [*N. biflorum* var. *s.* Torr.] Hirsute annual, branched from base, suberect; branches 5–30 cm. long, short-hispid throughout; lvs. narrow-oblanceolate, 1–2 cm. long, plane or somewhat revolute, subsessile or short-petioled; fls. in terminal crowded rather leafy cymes; calyx-lobes linear-spatulate, 4–6 mm. long; corolla bright purple-red, narrow-campanulate, 10–12 mm. long; style 2–5 mm. long; caps. linear-oblong, 5–6 mm. long; seeds many, yellow-brown, reticulate, ca. 0.4 mm. long.—Occasional, flats and washes; Creosote Bush Scrub; s. Mojave Desert, Colo. Desert; to Tex., Mex. March–May.

Var. **revolùtum** Jeps. Lvs. grayish, soft-hirsute as well as hispid, strongly revolute.—Colo. Desert; to Ariz., Son., L. Calif.

5. **N. pusíllum** Lemmon ex Gray. Diffuse to matted annual, short-hirsute throughout, the branches 2–10 cm. long; lvs. ovate to spatulate-ovate, 5–8 mm. long; fls. nearly or quite sessile in upper forks; calyx-lobes linear, 3–4 mm. long; corolla whitish, cylindrical, 4–5 mm. long; style 1–1.5 mm. long; caps. 2.5–3 mm. long, 20–40-seeded; seeds angular, reticulate; $n=7$ (Cave & Constance, 1950).—Occasional in open and sandy places below 4000 ft.; Creosote Bush Scrub; Death V. region to s. Mojave Desert. March–May.

6. **N. rothróckii** Gray. Plate 52, Fig. E. Plants perennial from running underground rootstocks; stems 1.5–3 dm. tall, hispid and glandular, simple or few-branched; lvs. lanceolate to oblong, 2–5 cm. long, coarsely sinuate-dentate, hispid and glandular, veiny beneath; fls. many, in subcapitate terminal cymes; calyx-lobes linear, 10–15 mm. long, hirsute; corolla purplish-lavender, funnelform, 10–15 mm. long; stamens unequal; style 8–10 mm. long, divided to base; caps. 4–5 mm. long, loculicidal; seeds ca. 15, ca. 1.5 mm. long; $n = 7$ (Cave & Constance, 1959).—At ca. 7500 ft.; Yellow Pine F.; Holcomb V., San Bernardino Mts.; Fresno and Inyo cos. to Tulare Co.; w. Nev. July–Aug.

7. **N. stenocárpum** Gray. Annual, sometimes probably of longer duration; stems many from base, decumbent, 1–3 dm. long, hirsute throughout; lvs. oblanceolate, 1–2.5 cm. long, the upper sessile and clasping; fls. in terminal leafy clusters; pedicels 1–3 mm. long; calyx-lobes 4–5 mm. long, subspatulate, erect to recurved, indurate in fr.; corolla funnelform, 6–7 mm. long, pale violet; style 1.5–2 mm. long; caps. half-inferior; seeds many, straw-colored, ca. 0.3 mm. long, finely alveolate; $n = 7$ (Cave & Constance, 1950). —Occasional, muddy places below 1000 ft.; Los Angeles Co. to San Diego Co., across Colo. Desert; to Tex., n. Mex., L. Calif. March–May.

7. *Nemóphila* Nutt. ex Barton

Annual, usually branched and diffuse, sometimes prostrate, often weak, hispid to glabrous. Lvs. mostly opposite, variously toothed, lobed or pinnately divided, petioled. Fls. pedicelled, solitary in upper axils or opposite the lvs. Calyx deeply divided, the sinuses usually armed with sepallike spreading or reflexed auricles. Corolla white to blue, sometimes spotted, deciduous, rotate to campanulate, usually with a pair of appendages within at base of each fil. Style ± deeply bifid. Caps. uniloculed, round to ovoid, loculicidal. Ovules 2–several on each of the large parietal placentae. Seeds 1–20, ovoid, smooth to corrugate-tubercled, regularly to irregularly pitted or not, with a ± evident papillalike group of colorless cells (cucullus). Ca. 13 spp. of N. Am., some used as garden annuals. (Greek, *nemos*, grove, and *phileo*, to love.)

(Constance, L. Univ. Calif. Pub. Bot. 19: 341–398. 1941.)
Corolla 0.7–3.5 cm. broad, mostly deep blue .... 1. *N. menziesii*
Corolla 0.3–0.6 cm. broad, mostly paler.
    Lvs. oblong or oval, deeply divided, weakly cuneate to truncate at base .... 2. *N. pedunculata*
    Lvs. spatulate, shallowly lobed, strongly cuneate at base .... 3. *N. spatulata*

1. **N. menzièsii** H. & A. ssp. **menzièsii**. [*N. insignis* Dougl. ex Benth.] BABY BLUE-EYES. Plate 52, Fig. F. Diffuse, ± succulent, the stems obscurely winged or angled, pubescent, 1–3 dm. long, often growing up among other plants; lvs. opposite, oval to oblong in outline, mostly 2–5 cm. long, pinnately divided usually into 9–11 rounded to oblong divisions, these mostly again toothed, sparingly appressed-hispid; pedicels shorter to as long; pedicels slender, longer than lvs.; calyx-lobes lanceolate, 4–6 mm. long, the auricles narrow, 1.5–2.5 mm. long; corolla semirotate to bowl-shaped, 1.5–4 cm. broad, typically bright blue with a light center, the lobes longer than the tube; fil.-appendages broad or narrow, partly free or adherent along one edge or reduced to hairy lines; style 3–5 mm. long; caps. 5–12 mm. in diam.; seeds usually 10–20, oblong to ovoid, corrugate-tuberculate; $n = 9$ (Cave & Constance, 1942).—Flats and slopes below 5000 ft.; Coastal Sage Scrub, Chaparral, V. Grassland, S. Oak Wd., etc.; cismontane San Diego Co. to n. Calif. Feb.–June.

Ssp. **integrifòlia** (Parish) Munz. [*N. i.* Parish.] Lower lvs. with ca. 5–7 mostly entire lobes; upper lvs. entire or 3-lobed; corolla mostly pale blue, 7–10 mm. long, mostly with narrow scales with very long hairs; $n = 9$ (Cave & Constance, 1947).—Shaded slopes, 2500–6000 ft.; Chaparral, Yellow Pine F.; e. San Gabriel and San Bernardino mts. to e. San Diego Co. Running into and poorly separated from the following tendencies: Var. *incana* Brand. Plant quite densely villous; upper lvs. less different from lower; corolla 6–10 mm. long, pale blue, the appendages linear, with short hairs.—At 500–5000 ft.; San

Bernardino Co. to Ventura Co. Var. *rotata* (Eastw.) Chandl. Lvs. commonly 3–5-lobed; corolla light blue, 3–6 mm. long, the appendages triangular, attached only near the base, with hairs longer than scales.—Santa Ana Mts. to San Diego. Var. *annulata* Chandl. Fls. deep blue, 3–6 (–10) mm. long; appendages oblong, attached along one side, with very short hairs.—Between 3000–5000 ft.; Mojave Desert from Victorville region to Providence Mts.

2. **N. pedunculàta** Dougl. ex Benth. [*N. sepulta* Parish.] Stems weak, low, often prostrate, sparingly pubescent, 1–3 dm. long; lvs. all opposite, 1–3 cm. long, deeply pinnately divided into 5–9 short oblong to obovate divisions, these appressed-hispid, entire to 1–2-toothed; petioles to as long as the blades, expanded upward into the cuneate lf.-base; pedicels short, axillary; calyx-lobes lanceolate, 1–3 mm. long, the reflexed auricles ca. as long; corolla campanulate, 3–6 mm. broad, white to pale blue, mostly dark-spotted or -veined, each lobe with a purple blotch; appendages linear or reduced to linear lines; style barely 1 mm. long; caps. exceeding calyx; seeds mostly 2–8, 1–4 mm. long; $n=9$ (Cave & Constance, 1942).—Moist open or shaded places below 7000 ft.; Foothill Wd., Yellow Pine F.; n. Los Angeles Co. and Frazier Mt. region in Ventura Co.; to B.C. April–Aug. Santa Cruz Id.

3. **N. spatulàta** Cov. Stems few, weak, 1–2 dm. long, ± hispidulous; lvs. all opposite, thin, 1–3 cm. long, cuneate and passing into the equally long petioles, 3–5-lobed or -toothed, appressed-hairy; calyx-lobes 2.5–5 mm. long, the reflexed auricles 1–2 mm. long; corolla white to bluish, shallowly bowl-shaped, often dotted, sometimes with purple blotches on the lobes; fils. shorter than tube; appendages broad to narrow to lineate; caps. longer than calyx; seeds mostly 5–6, shallowly pitted; $n=9$ (Cave & Constance, 1947).—Shaded damp places, 4000–10,000 ft.; Montane Coniferous F.; mts. from Riverside Co. n.; to Plumas and Tehama cos.; w. Nev. May–July.

8. *Phacèlia* Juss.

Herbs, varying from annual to perennial, usually ± pubescent and often glandular. Lvs. mostly alternate, ranging from entire to dentate to pinnatifid to pinnate or bipinnate. Fls. few to many, in dense to lax (especially in fr.) simple or branched cymes. Calyx lobed almost to base. Corolla rotate to campanulate to tubular, ± lobed, purple to blue to white or yellow, deciduous or withering-persistent. Corolla-scales attached in pairs to the corolla-tube at base of each fil. and ± adnate to tube, sometimes absent. Stamens equally inserted at base of corolla-tube, ± equal, included or exserted. Style bifid at apex to divided almost to base. Caps. unilocular or nearly bilocular, round to ovoid or oblong, with 2–many ovules on the 2 linear placentae. Seeds round to oblong or ovoid, reticulate, pitted or transversely corrugate, sometimes excavated on each side of a ridge. Perhaps 200 spp. of New World, especially in w. N. Am., some of horticultural value. (Greek, *phakelos*, a cluster, because of crowded fls.)

1. Plants perennial or biennial.
  2. Ovules several to many on each placenta; lf.-blades round or nearly so, 1–2 cm. long; stems white-woolly below. Deserts .... 49. *P. perityloides*
  2. Ovules 2 to each placenta.
    3. Lvs. pinnate into toothed or pinnatifid lobes; calyx-lobes mostly spatulate; corolla 6–8 mm. long .... 51. *P. ramosissima*
    3. Lvs. entire to pinnate into entire lobes or lfts.; calyx-lobes mostly not widened upward.
      4. Infl. mostly elongate, narrow, dense, consisting of many short lateral cymes along a cent. axis; stems mostly simple.
        5. Lower lvs. mostly with 2–several pairs of lateral lfts.
          6. Plants greenish, biennial; calyx-lobes linear to linear-oblong .. 27. *P. heterophylla*
          6. Plants grayish, perennial; calyx-lobes oblanceolate to ovate
                                             29. *P. imbricata bernardina*
        5. Lower lvs. mostly entire, or with 1 or 2 pairs of basal pinnae; perennial. N. of our range .... *P. nemoralis*
      4. Infl. usually shorter, more open, widely branched; stems often branched except in reduced and alpine forms.

Phacelia

8. Lvs. mostly entire, silvery to gray-green.
   9. Stems less than 2 dm. tall .... 23. *P. frigida*
   9. Stems mostly over 2 dm. tall .... 38. *P. mutabilis*
8. Lvs. mostly lobed, green to gray or white but not silvery.
   10. Corolla with incurved lobes.
      10a. Corolla lavender to pale pink. San Gabriel and San Bernardino mts. above 5000 ft. .... 42. *P. oreopola*
      10a. Corolla white; calyx-lobes often overlapping in fr. Los Angeles Co. n. mostly below 5500 ft. .... 29. *P. imbricata*
   10. Corolla campanulate.
      11. Lvs. mostly with 1–2 pairs of lfts. .... 38. *P. mutabilis*
      11. Lvs. mostly with 3 or more pairs of lfts. .... 20. *P. egena*
1. Plants annual.
   12. Seeds not transversely corrugated, or if so, the ovules only 2 to a placenta.
      13. Lvs. pinnately toothed to compound, their divisions further toothed or divided.
         14. Seeds 10–40, less than 1 mm. long. Insular .... 33. *P. lyonii*
         14. Seeds 1–8, 1.5–3 mm. long.
            15. Calyx-lobes pinnatifid into 3–5 segms. San Clemente Id. .... 21. *P. floribunda*
            15. Calyx-lobes entire or at the most crenate.
               16. Calyx much enlarged and ± chartaceous and veiny in fr., the lvs. ovate to lance-oblong; corolla broadly campanulate, 8–10 mm. long .... 11. *P. ciliata*
               16. Calyx not much enlarged in fr., or if so, not coriaceous or veiny, the lobes then linear to spatulate.
                   17. Seeds terete or angled, usually pitted, but not excavated on either side of a salient ridge; plants not markedly viscid or ill-scented.
                      18. Corolla lavender or white; calyx-lobes linear or nearly so, up to 1 cm. long in fr., often conspicuously clawed and loosely enveloping the caps.
                         19. Corolla open-campanulate, 8–12 mm. long, much exceeding calyx.
                            20. Fls. many, crowded; caps. with pustulate bristles; stems mostly not strongly purple. Mostly cismontane .... 10. *P. cicutaria*
                            20. Fls. few, remote; caps. hirsutulous, stems mostly purple. Largely desert 55. *P. vallis-mortae*
                         19. Corolla tubular-campanulate, 4–7 mm. long, not much exceeding calyx. Deserts .... 14. *P. cryptantha*
                      18. Corolla blue or bluish; calyx-lobes linear to obovate, less than 1 cm. long and closely investing the caps.
                        21. Calyx-lobes 4–5 mm. long, mostly oblanceolate to obovate, corolla promptly deciduous; caps. hairy in whole upper half; stamens mostly not long-exserted .... 17. *P. distans*
                        21. Calyx-lobes linear to lance-linear, 6–8 mm. long; corolla tardily deciduous; caps. hairy only at tip; stamens long-exserted .... 54. *P. tanacetifolia*
                 17. Seeds flattened, excavated on each side of a salient ridge; plants very viscid, ill-scented.
                    22. Stamens and style plainly exserted.
                        23. Pedicels shorter than calyx; lobes of lf.-blades ± oblong.
                          24. Corolla 6–10 mm. long, open-campanulate.
                            25. Corolla purplish; seeds thick, transversely corrugated. Deserts 13. *P. crenulata*
                            25 Corolla white; seeds thin, not corrugated. Saline V., Inyo Co. 2. *P. amabilis*
                        24. Corolla 4–5 mm. long, tubular-campanulate .... 35. *P. minutiflora*
                    23. Pedicels at least as long as calyx; lobes of lf.-blades almost round 46. *P. pedicellata*
                    22. Stamens and style included.
                      26. Stems greenish; calyx-lobes oblanceolate, 3–4 mm. long; corolla 6 mm. long .... 3. *P. anelsonii*
                      26. Stems purplish; calyx-lobes lanceolate, 2–3 mm. long; corolla 3–4 mm. long .... 12. *P. coerulea*
13. Lvs. entire or pinnate, but then with entire divisions.
   27. Corolla-scales present, attached to tube at base of each fil.
      28. Corolla rotate to campanulate.

29. Calyx-lobes obovate to narrow-spatulate; lvs. oblong to ovate ... 25. *P. grisea*
29. Calyx-lobes linear to linear-oblanceolate.
  30. Plants glandular.
    31. Lvs. linear to oblanceolate, mostly entire; corolla 5–8 mm. long. Los Angeles Co. to Tulare Co. .... 36. *P. mohavensis*
    31. Lvs. broader, the lower 1–6-pinnate; corolla 3–4 mm. long. Sw. Inyo Co.
        41. *P. novenmillensis*
  30. Plants not glandular.
    32. Pedicels elongating noticeably in fr., spreading or recurving; seeds not over 1 mm. long.
      33. Lvs. mostly entire, or if pinnate, plants tending to be montane; seeds ca. 1 mm. long .... 15. *P. curvipes*
      33. Lvs. mostly pinnate to pinnatifid.
        34. Corolla 9–15 mm. long. Yellow Pine F. .... 16. *P. davidsonii*
        34. Corolla 10–12 mm. long. Below Yellow Pine F. .... 19. *P. douglasii*
    32. Pedicels not noticeably elongate in fr., but erect or ascending.
      35. Corolla 6–15 mm. long; lvs. elliptic to narrow-ovate .... 18. *P. divaricata*
      35. Corolla 3–6 mm. long.
        36. Fls. white to lavender or light blue .... 4. *P. austromontana*
        36. Fls. violet or purple .... 28. *P. humilis*
28. Corolla tubular to tubular-campanulate.
  37. Corolla 6–14 mm. long.
    38. Stamens glabrous; seeds 1–1.5 mm. long. Cismontane ...... 53. *P. suaveolens*
    38. Stamens ± pubescent; seeds 0.5–0.7 mm. long. Deserts.
      39. Cymes not pedunculate. Alkaline flats, near Kingston, e. Mojave Desert
        50. *P. pulchella*
      39. Cymes pedunculate. Rocky places, Death V. region ...... 37. *P. mustelina*
  37. Corolla 3–6 mm. long.
    40. Petioles ca. as long as or longer than blades; fls. in lax cymes.
      41. Lf.-blades roundish.
        42. Lf.-blades plainly toothed, 0.5–2 cm. wide; stems hirsutulous and glandular; stamens glabrous. Inyo Co. to Riverside Co. .... 52. *P. rotundifolia*
        42. Lf.-blades less toothed, 1.5–3.5 cm. wide; stems scarcely if at all hirsutulous; stamens hairy at base. Westgard Pass to Mono Co. 48. *P. peirsoniana*
      41. Lf.-blades oblong-oval to ovate; stamens glabrous. Inyo Mts. to Clark Mts.
        5. *P. barnebyana*
    40. Petioles shorter than lf.-blades; fls. in compact cymes.
      43. Cymes pedunculate; seeds ca. 25 and 1–1.5 mm. long. ..... 44. *P. parishii*
      43. Cymes subsessile; seeds 50–100, ca. 0.5 mm. long .... 31. *P. lemmonii*
27. Corolla-scales lacking, the filament-bases sometimes dilated or winged.
  44. Fils. with a dilation at base; style parted above the middle.
    45. Corolla open-campanulate, the tube ca. as long as the limb.
      46. Fls. purple to white.
        47. Corolla purple to violet, with paler center, 10–20 mm. long; fil.-dilations pubescent. Mostly below 3000 ft. .... 45. *P. parryi*
        47. Corolla white, sometimes bluish, 7–12 mm. long; fil.-dilations glabrous. From 3000–8000 ft. .... 32. *P. longipes*
      46. Fls. deep blue. W. Mojave Desert .... 39. *P. nashiana*
    45. Corolla tubular-campanulate or campanulate-funnelform, the tube ca. twice as long as the limb.
      48. The corolla purple, slightly constricted at throat. Cismontane . 34. *P. minor*
      48. The corolla deep blue, not constricted at throat. Deserts ... 9. *P. campanularia*
  44. Fils. without a basal dilation; style parted to below the middle.
    X. Cymes lax in fr., the pedicels becoming 1–2 cm. long; corolla 8–18 mm. long, blue with purplish or whitish center; fils. pubescent .... 56. *P. viscida*
    XX. Cymes dense, the pedicels less than 1 cm. long; corolla 12–30 mm. long, the center colored like the periphery; fils. glabrous .... 24. *P. grandiflora*
12. Seeds transversely corrugated; ovules many.
  49. Lvs. entire to crenate-dentate, not deeply lobed. Deserts.
    50. Corolla 5–7 mm. long; style 2–3 mm. long.
      51. Plant with black-headed glands throughout; racemes subpedunculate
        43. *P. pachyphylla*

51. Plant grayish-pubescent, without black-headed glands; racemes ± concealed by
foliage .... 40. *P. neglecta*
50. Corolla 6–10 mm. long; style 2.5–6 mm. long.
   50a. Corolla creamy white; seeds ca. 50 .... 8. *P. calthifolia*
   50a. Corolla blue to lavender; seeds 5–8 .... 26. *P. gymnoclada*
49. Lvs. pinnate to bipinnatifid.
  52. Corolla 2.5–6.5 mm. long, not longer than calyx.
    53. Calyx-lobes linear to narrow-oblanceolate; corolla white with yellowish tube
                                                                                                     30. *P. ivesiana*
      53a. Lvs. divided into narrow-oblong segms.; seeds with 8–12 transverse corrugations.
E. San Bernardino Co. .... 30. *P. ivesiana*
      53a. Lvs. divided into deltoid segms.; seeds with 5–7 transverse corrugations. Inyo Co.
to Imperial Co. .... 47. *P. pediculoides*
    53. Calyx-lobes spatulate; corolla lavender with yellowish tube .... 1. *P. affinis*
  52. Corolla mostly 7–15 mm. long, at least twice as long as calyx.
    54. Lvs. bipinnatifid to deeply pinnately lobed; seeds 12–20.
      55. The lvs. bipinnatifid; calyx-lobes linear to narrow-oblanceolate; stamens usually pubescent near the base. San Bernardino Co. to Modoc Co. ..... 6. *P. bicolor*
      55. The lvs. pinnate or pinnatifid; calyx-lobes spatulate; fils. glabrous.
        56. Corolla bright blue to deep blue. Deserts .... 22. *P. fremontii*
        56. Corolla white to pink. Cismontane .... 7. *P. brachyloba*
    54. Lvs. shallowly pinnately lobed; seeds 5–8. Inyo and Mono cos . 26. *P. gymnoclada*

1. **P. affìnis** Gray. Annual with 1–several ± erect stems to 3 dm. high, hirsutulous and cinereous, with retrorse hairs below, more glandular-puberulent upward; lvs. narrow-oblong, 2–6 cm. long, pinnatifid to pinnate into oblong or ovate segms., these entire to lobed, not noticeably glandular; petioles shorter; cymes simple or few-branched, few- to many-fld.; lvs. subsessile or on pedicels 1–2 mm. long; calyx-lobes spatulate, 4–5 mm. long, accrescent and 6–10 mm. in fr.; corolla pale lavender or whitish with pale yellow tube, narrow-campanulate, 3–5 mm. long, deciduous; scales inconspicuous; caps. ca. 5 mm. long, hirsutulose near apex; seeds plump, ca. 1 mm. long; $n = 11$ (Cave & Constance, 1947).—Occasional in sandy and gravelly places mostly below 4000 ft.; Creosote Bush Scrub, Joshua Tree Wd.; Pinyon-Juniper Wd.; deserts from Ventura Co. and Kern Co. to San Bernardino and e. San Diego cos.; to n. L. Calif., New Mex., Utah. March–June.

2. **P. amábilis** Const. Like *P. crenulata*, the corolla white, 7–8 mm. long; stamens 9–15 mm. long; scales with a broad free portion; seeds 3–4 mm. long, pale, with a thin broad margin.—At 1800 ft.; Saline V., Inyo Co. April–May.

3. **P. anelsònii** Macbr. Viscid-pubescent annual 2–4 dm. high, with simple erect green stems; lf.-blades oblong to oblanceolate, 2–8 cm. long, pinnately lobed or divided into crenate lobes; petioles to almost as long as blades; fls. many, short-pedicelled, in dense elongate cluster of cymes; calyx-lobes oblanceolate, 3–4 mm. long, 1–2 being wider than the others; corolla blue or violet, funnelform-campanulate, 6 mm. long, deciduous; scales lunate, narrowed upward; stamens included; styles included; caps. ovoid, glandular, puberulent at apex, 2.5–3 mm. long; seeds 4, ca. 3 mm. long, with median ridge on ventral surface, favose-pitted.—Rare in dry places, 4000–5500 ft.; Pinyon-Juniper Wd., Joshua Tree Wd.; e. Inyo and San Bernardino cos.; to Nev., Utah. April–May.

4. **P. austromontàna** J. T. Howell. [*P. humilis* var. *lobata* A. Davids. *P. l.* A. Davids.] Much like *P. mohavensis*, the lvs. often pinnately lobed; pedicels to 2–3 or more mm. in fr.; calyx-lobes 4–5 mm. in fr.; corolla 3–5 mm. long; stamens 2–4 mm. long; style 1–2 mm. long, hairy near base; seeds 2–4.—Dry loose slopes, 6000–9000 ft.; Montane Coniferous F.; San Jacinto, San Bernardino and San Gabriel mts., to Mt. Pinos, Panamint Mts., s. Sierra Nevada; sw. Utah. May–July.

5. **P. barnebyàna** J. T. Howell. Like *P. rotundifolia*; stems very slender, hirsutulous and glandular, mostly 3–9 cm. high; lvs. oblong-oval to ovate, 0.5–1.5 cm. long; fls. few, on pedicels 4–10 mm. long; calyx-lobes 2–2.5 mm. long in fl., 3–4 mm. in fr.; corolla pale lavender, 4.5–5 mm. long; stamens glabrous; style glabrous; seeds ca. 1 mm. long.—Rare, 5000–8000 ft.; Pinyon-Juniper Wd.; Clark Mt. to Inyo Mts.; w. Nev. May–June.

6. **P. bicólor** Torr. ex Wats. Annual; stem erect or spreading, 0.5–4 dm. long, com-

monly branched, hirsutulous and capitate-glandular, leafy; lower lvs. few, oblong, 2–4 cm. long, bipinnatifid into linear divisions; petioles shorter; cauline lvs. gradually reduced upward; cymes few- to many-fld., not elevated above leafy part of pl.; pedicels 1–4 mm. long; calyx-lobes linear to narrow-oblanceolate, 3–6 mm. long, somewhat more in fr.; corolla ± purplish with yellow tube, funnelform to campanulate, 9–16 mm. long; scales adnate; stamens included, 5–8 mm. long, the fils. usually pubescent below; style included, 4–5 mm. long, pubescent; caps. oblong-ovoid, ca. 4 mm. long, obtuse, hirsutulous at apex; seeds brown, ca. 1.5 mm. long, with 8–10 rounded or sharp corrugations; $n=13$ (Cave & Constance, 1950).—Sandy places, mostly 3000–10,000 ft.; Joshua Tree Wd., Pinyon-Juniper Wd., Sagebrush Scrub; deserts from w. San Bernardino Co. to Modoc Co.; Ore., Nev. May–Aug.

7. **P. brachylòba** (Benth.) Gray. [*Eutoca b.* Benth.] Annual, 1–6 dm. high, simple or branched from base and above, capitate-glandular and hirsutulous; lvs. oblanceolate, to narrow-elliptical, 3–7 cm. long, mostly near base of plant, pinnately lobed or pinnatifid into entire or few-toothed segms.; petioles shorter; upper cauline lvs. scattered, reduced; cymes paniculately clustered, forming upper half of plant, many-fld.; pedicels 1–4 mm. long; calyx-lobes linear-oblanceolate, 4–5 mm. long, glandular and hirsutulous, much enlarged in fr.; corolla white to pink with yellow tube, broadly funnelform to campanulate, 7–10 mm. long, deciduous; scales usually obsolete; stamens 4–5 mm. long, glabrous; style 3–4 mm. long, pubescent; caps. 4–5 mm. long, hirsutulous above; seeds ca. 20, plump, ca. 0.5 mm. long, pitted and with 5–7 transverse corrugations; $n=12$ (Cave & Constance, 1944).—Frequent in dry sandy ± disturbed places below 7000 ft.; Coastal Sage Scrub, Chaparral, n. L. Calif. to Monterey Co. May–June.

8. **P. calthifòlia** Brand. Much like *P. pachyphylla* Gray. Often 1.5–3 dm. high; corolla purple, broadly campanulate, ca. 1 cm. long; stamens 5–6 mm. long; style 5–6 mm. long; seeds ca. 50, ca. 1 mm. long; $n=11$ (Cave & Constance, 1950).—Below 3000 ft.; Creosote Bush Scrub; Inyo and n. San Bernardino cos.; w. Nev. March–May.

9. **P. campanulària** Gray ssp. **campanulària.** Much like *P. minor*, the corolla deep blue, rarely white, not constricted at the throat, 1.5–3 cm. long; dilation at base of stamens usually glabrous; style 2–3.5 cm. long, cleft $\frac{1}{3}$–$\frac{1}{2}$ its length; seeds 1–1.5 mm. long; $n=11$ (Cave & Constance, 1947).—Dry sandy and gravelly places, below 4000 ft.; Creosote Bush Scrub, w. Colo. Desert from Whitewater R. to Collins V. Feb.–April.

Ssp. **vasifórmis** Gillett. Corolla mostly 25–40 mm. long; style 3–4.5 cm. long, cleft $\frac{1}{4}$–$\frac{1}{2}$ its length.—Deserts from Victorville region and Morongo V. to Providence Mts. and Cottonwood Springs. March–May.

10. **P. cicutària** Greene var. **cicutària.** Annual 2–6 dm. high, setose-hispid and pubescent throughout and somewhat glandular-pubescent; stems erect or ascending or weaker in shade; lvs. ± ovate in outline, 4–15 cm. long, pinnate into oblong or lanceolate toothed to pinnatifid divisions; petioles mostly shorter than blades; fls. many, short-pedicelled, in dense few-branched cymes which become lax and up to 2 dm. in fr.; calyx-lobes linear to narrow-spatulate, densely ± yellowish and bristly-hispid and glandular, 6–8 mm. in fl., to 10 or 12 mm. in fr.; corolla dirty yellowish-white, broadly campanulate, 8–12 mm. long, deciduous; scales adnate to corolla on one side and to fil. on other, with free oval part; stamens 8–12 mm. long; style cleft to middle; caps. 3–4 mm. long, sparsely hispid, with dark spot at base of each hair; seeds 2–4, ca. 3 mm. long, reticulate-pitted; $n=11$ (Cave & Constance, 1944).—Dry rocky slopes below 4000 ft.; V. Grassland, Oak Wd.; Tehachapi Mts. n. March–May.

Var. **híspida** (Gray) J. T. Howell. [*P. hispida* Gray.] Plants more widely branched; sepal-lobes grayish-hispid; corolla lavender; scales more adnate; seeds usually 4, ca. 1.5 mm. long.—Common in dry rocky canyons, etc., below 4000 ft.; Coastal Sage Scrub, Chaparral, S. Oak Wd.; San Luis Obispo Co. to cismontane s. Calif. to edge of desert, Santa Barbara Ids.; L. Calif. A stout grayish-shaggy form with dense cymes has been named var. *hubbyi* (Macbr.) J. T. Howell and occurs from Santa Barbara Co. to Los Angeles Co.

11. **P. ciliàta** Benth. Annual, branched from base, 1–5 dm. high, glandular-puberulent throughout, also hispidulous; lvs. ovate to oblong in outline, 3–10 cm. long, pinnate

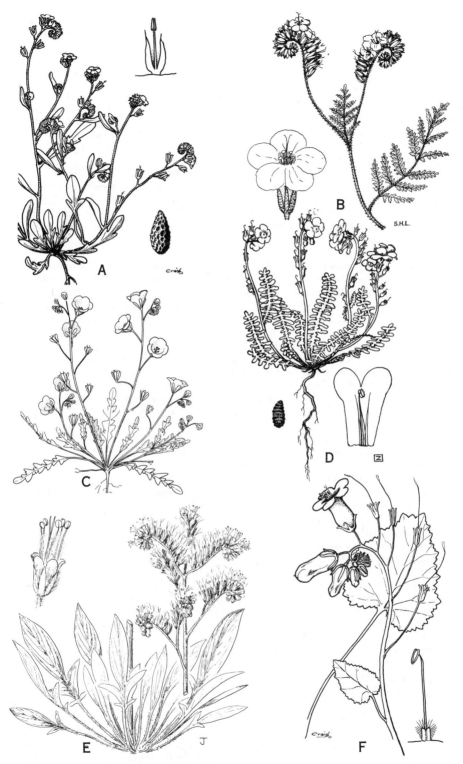

Plate 53. Fig. A, *Phacelia davidsonii*. Fig. B, *Phacelia distans*. Fig. C, *Phacelia douglasii*. Fig. D, *Phacelia fremontii*. Fig. E, *Phacelia imbricata*. Fig. F, *Phacelia minor*.

or pinnatifid into toothed or incised divisions; petiole rather short except in lower lvs.; fls. subsessile, crowded in few cymes; calyx-lobes broadly lanceolate, stiff-ciliate, 4–6 mm. long in fl., venulose and subcoriaceous, semitransparent and to 1 cm. in fr.; corolla blue with paler center, broadly campanulate, 8–10 mm. long, deciduous; scales suborbicular, partly free; stamens 9–13 mm. long; style 6–8 mm. long, pubescent below; caps. round-ovoid, 4–5 mm. long, hispidulous; seeds mostly 4, dark brown, 2.5–3 mm. long, deeply pitted; $n=11$ (Cave & Constance, 1942).—Grassy slopes, gravelly places, cultivated fields, mostly below 5000 ft.; V. Grassland, Foothill Wd., Coastal Sage Scrub, Glenn Co. to Kern Co., uncommon s. to L. Calif. March–May.

12. **P. coerùlea** Greene. Near to *P. anelsonii* but stems purplish, more slender; fls. subsessile; calyx-lobes lanceolate, 2–3 mm. long; corolla blue to white, broadly campanulate, 3–4 mm. long; stamens included, 2–3 mm. long; style as long; caps. roundish, 2–3 mm. in diam., somewhat pubescent; seeds 4, dark to light brown, 2–3 mm. long, with salient ventral ridge, favose-pitted dorsally; $n=11$ (Cave & Constance, 1950).—Dry rocky places at 5500 ft. Pinyon-Juniper Wd.; Clark Mt., e. Mojave Desert; to Texas, Chihuahua. April–May.

13. **P. crenulàta** Torr. var. **crenulàta.** Annual, stems simple or usually few-branched, 1–4 (–6) dm. high, hispid-pubescent but green and very glandular and strongly scented; lf.-blades mostly oblong in outline, 3–12 cm. long, the lower petioled, pinnately divided or undulately lobed, the uppermost subsessile, mostly coarsely crenate-dentate; cymes dense, in flattish terminal panicles; fls. many, short-pedicelled; calyx-lobes oblong, 4–5 mm. long, obtuse, not much longer in fr., densely glandular and hispid; corolla deep violet or bluish-purple, with white throat, open-campanulate, 6–10 mm. long; scales rectangular to lunate, acute, wholly adnate; stamens exserted, 10–14 mm. long, glabrous; style exserted, 12–15 mm. long; caps. ca. as long as calyx, sparsely hirsutulous; seeds mostly 2, oblong-ovoid, excavated on ventral side, shallowly pitted, ca. 3 mm. long; $n=11$ (Cave & Constance, 1963).—Common in gravelly washes and open places, mostly below 6000 ft.; Creosote Bush Scrub to Pinyon-Juniper Wd.; Mojave Desert from Barstow region to Twentynine Palms and to Inyo Co.; cent. Nev. and Utah. March–May.

Var. **ambígua** (Jones) Macbr. [*P. a.* Jones.] Stems often purplish, grayish hirsute, less glandular; infl. flat-topped; $n=11$ (Cave & Constance, 1944).—Mojave Desert e. of Twentynine Palms, Colo. Desert; s. Nev. March–May.

Var. **funèrea** J. Voss ex Munz. Stems purplish, mostly simple, slender, glandular-pubescent, mostly 4–8 dm. long; cymes scattered along upper stem, not in flat-topped cluster.—At ca. 4000–9500 ft.; mostly Joshua Tree Wd., Pinyon-Juniper Wd.; Death V. region to White Mts.; adjacent Nev. April–June.

14. **P. cryptántha** Greene. Resembling *P. cicutaria*, but mostly lower; calyx-lobes linear-oblanceolate, 4–7 mm. long (8–10 mm. in fr.); corolla tubular-campanulate, 4–7 mm. long; scales elongate, narrow, ending in a sharp point; stamens included; style 3–4 mm. long, pubescent below; caps. 2 mm. long, hispid; seeds 4, 1.5–2.5 mm. long, pitted; $n=11$ (Cave & Constance, 1944).—Dry rocky slopes below 5700 ft.; Shadscale Scrub, Joshua Tree Wd., Creosote Bush Scrub; w. Colo. Desert, Mojave Desert n. to Mono Co.; Nev., Ariz. March–May. A form occurring throughout the range and with calyx-lobes broader above (1.5–2 mm.) and corolla-scales without a sharp tip has been named var. *derivata* J. Voss.

15. **P. cúrvipes** Torr. ex Wats. Annual, branched, diffuse or ascending, soft-pubescent, somewhat glandular above, 4–15 cm. high; lvs. largely basal, oblong to oblong-obovate, mostly entire, 1–3 cm. long, the petiole ca. as long; upper lvs. few, reduced; cymes lax, few-fld., with pedicels 2–7 mm. long, curved, spreading or deflexed; calyx-lobes linear-spatulate, 3–6 mm. long, hirsute, 7–10 mm. long in fr., somewhat unequal; corolla bluish to violet with white throat, open-campanulate, 5–8 mm. long, deciduous; scales oblong, adnate; stamens 2–6 mm. long, sparsely hairy; style pubescent; caps. ovoid, 4–5 mm. long, appressed-hirsute; seeds 6–16, irregularly pitted, ca. 1 mm. long.—Occasional, dry rocky slopes 3500–8000 ft.; Chaparral, Yellow Pine F., Sagebrush Scrub, Foothill Wd., etc.; mts. of San Diego Co. to e. slope of Sierra Nevada; to Nev. April–June.

16. **P. davidsònii** Gray. [*P. curvipes* var. *d.* Brand.] Plate 53, Fig. A. Lower lvs. com-

monly pinnate; corolla 9-15 mm. long; $n=10$ (Cave & Constance, 1944).—Chaparral, Foothill Wd., Yellow Pine F.; mts. from San Diego Co. to s. Sierra Nevada.

17. **P. dístans** Benth. WILD-HELIOTROPE. Plate 53, Fig. B. Annual 2-8 dm. high, erect and simple or branched and decumbent, finely pubescent and scatteringly stiff-hairy, glandular mostly in upper parts; lvs. ovate to ovate-oblong in outline, 2-10 cm. long, short-petioled, 1-2-pinnate into oblong to lanceolate toothed to pinnatifid divisions; upper lvs. reduced, subsessile; cymes few, simple to somewhat branched; fls. many, subsessile; calyx-lobes mostly unequal, oblanceolate to linear, 4-5 mm. long, glandular-pubescent and hirsute; corolla blue or bluish, broadly campanulate, 6-8 mm. long, deciduous; scales half-ovate, free at tips; stamens included to short-exserted, the fils. glabrous to somewhat pubescent; style 7-12 mm. long, very deeply cleft, glabrous to somewhat pubescent below; caps. 2-3 mm. long, somewhat hairy; seeds 2-4, coarsely pitted; $n=11$ (Cave & Constance, 1942).—Common, cismontane fields and slopes, mostly below 5000 ft.; many Plant Communities; deserts n. to Mono Co. Santa Barbara Ids.; to Nev., Ariz., L. Calif. March-June.

18. **P. divaricàta** (Benth.) Gray var. **insulàris** (Munz) Munz. [*P. i.* Munz.] Annual, erect and simple or branched, 0.5-3 dm. high, puberulent and somewhat hirsute; basal lvs. mostly elliptic to narrow-ovate, 1.5-5 cm. long, entire or rarely few-lobed, on petioles to 6 cm. long; cauline lvs. scattered, reduced upward; cymes rather few-fld., pedunculate, the lower pedicels longer than the fruiting calyx; calyx-lobes often spatulate; corolla rotate-campanulate, lavender to violet, mostly ca. 10-12 mm. long; stamens 6-9 mm. long; scales ovate, adnate; caps. obovoid; seeds several, 1-1.3 mm. long.—Sand dunes; Coastal Strand; Santa Rosa and San Miguel ids. March-April.

19. **P. doúglasii** (Benth.) Torr. var. **doúglasii**. [*Eutoca d.* Torr.] Plate 53, Fig. C. Annual, mostly several-stemmed from base, the branches prostrate or ascending, simple or branched, 0.5-4 dm. long, softly hirsutulous to stiffer-haired, glandular in infl.; lvs. mostly basal, oblong in outline, pinnate to pinnatifid into unequal oblong or subovate lobes, the blades 3-8 cm. long, petioles ca. as long; cauline lvs. few, reduced; cymes lax, few-fld., mostly solitary at ends of branches; pedicels longer than fls., often recurved; calyx-lobes spatulate to oblanceolate, 4-7 mm. long, hirsute; corolla light blue to purplish, open-campanulate, 10-12 mm. long; scales lanceolate, adnate; stamens 3-7 mm. long, the fils. hairy or not; style 2-7 mm. long, hairy; caps. ovoid, pointed, 5-7 mm. long, hairy; seeds mostly 10-20, pitted, to 1 mm. long; $n=11$ (Cave & Constance, 1950).—Sandy places below 5000 ft.; cismontane Riverside Co. to San Francisco Bay; Cent. V. March-May.

Var. **cryptántha** Brand. Calyx 2.5-3 mm. long in fl., 5-6 mm. in fr.; style 1-2 mm. long.—Sandy places; Coastal Sage Scrub; Downey, Los Angeles Co., to San Diego Co.; L. Calif. March-April.

20. **P. egèna** (Greene ex Brand) Const. [*P. magellanica* ssp. *barbata* f. *e.* Brand.] Stems several, ascending to erect, rather slender, 2-6 dm. high; lvs. usually 3-7-pinnatifid into acute divisions; calyx-lobes oblanceolate to lanceolate-linear, 4-6 mm. long in fr. and venulose; corolla white, 7-9 mm. long.—Dry places, mostly below 5000 ft.; Yellow Pine F.; Chaparral, etc. San Gabriel Mts. and Sespe Creek in Ventura Co.; to n. Calif. May-June.

21. **P. floribúnda** Greene. Annual 3-5 dm. high, puberulent and hirsute throughout, somewhat glandular above; lvs. ovate in outline, 5-12 cm. long, short-petioled, pinnate, then pinnatifid into crenate-dentate lobes; fls. many, subsessile, in cymes crowded into dense panicles; calyx-lobes short-hirsute, obovate in outline, 4-5 mm. long, pinnatifid into 3 or 5 oblong lobes or 1 or 2 entire, narrow-spatulate, corolla pale blue, campanulate, 5-6 mm. long, deciduous; scales adnate, oblong-lanceolate; stamens 5-6 mm. long, deciduous; scales adnate, oblong-lanceolate; stamens 5-6 mm. long, slightly exserted; style included, deeply cleft; caps. 2-3 mm. long, pilose; seeds 1-4, oblong, ca. 1.5 mm. long, pitted.—Sheltered places; Coastal Sage Scrub; San Clemente, Santa Barbara, and Guadalupe ids. March-May.

22. **P. fremóntii** Torr. [*P. hallii* Brand.] Plate 53, Fig. D. Annual, the few to many stems ± spreading, 0.5-3 dm. long, retrorsely hirsutulous below, glandular above; lvs. mostly near base, 2-5 cm. long, pinnately divided into oblong to roundish coarse entire

or few-toothed divisions; petioles often longer; cauline lvs. scattered, reduced upward; cymes many-fld. dense, well projected above the lvs.; pedicels 1–4 mm. long; calyx-lobes spatulate, 4–6 mm. long, sparsely hirsutulose, densely glandular, slightly enlarged in fr.; corolla bright blue to deep lavender, funnelform to subcampanulate, 8–15 mm. long, deciduous; scales linear-lanceolate, adnate; stamens 3–8 mm. long; style 3–5 mm. long; seeds ca. 12, 1–1.5 mm. long, with ca. 6–8 corrugations; $n=13$ (Cave & Constance, 1942).
—Sandy or clayey slopes and flats, below 7000 ft.; Creosote Bush Scrub, to Juniper-Pinyon Wd.; Mojave Desert from Inyo Co. to Riverside Co. and Colo. Desert at n. base of Santa Rosa Mts.; to Cent. V.; Utah, Ariz. March–May.

23. **P. frígida** ssp. **dasyphýlla** (Macbr.) Heckard. Perennial from branched heavy caudex, gray-pubescent and hispid throughout; stems several, slender, decumbent to ascending, 0.5–2.5 dm. long; lower lf.-blades grayish, densely rosulate, mostly lanceolate, entire, 2–4 cm. long, prominently veined; petioles ca. as long; cauline lvs. few, reduced; fls. many, in dense short cymes in rather compact mostly elongate panicles; calyx-lobes linear, 3–4 mm. long, often purplish, hispid-ciliate, 6–8 mm. long in fr.; corolla lavender to white, 4–6 mm. long; stamens 6–8 mm. long; style ca. 8 mm. long; caps. ca. 4 mm. long, short-bristly; seeds 1–2, regularly pitted; $n=22, 33$ (Heckard, 1960), (Cave & Constance, 1947).—Gravelly to rocky places, mostly 7000–11,000 ft.; Sagebrush Scrub, Montane Coniferous F.; White Mts., Sierra Nevada. July–Sept.

24. **P. grandiflòra** (Benth.) Gray. [*Eutoca g.* Benth.] Like *P. viscida* but coarser, 5–10 dm. high; lf.-blades 5–15 cm. long; cymes very dense, with pedicels remaining less than 1 cm. long; corolla violet to bluish, 12–30 mm. long; seeds 0.8–1 mm. long; $n=11$ (Cave & Constance, 1944).—Dry slopes and disturbed places, especially burns, below 2500 ft.; Coastal Sage Scrub, Chaparral; along coast from n. L. Calif. to Santa Barbara Co., inland to Claremont. April–June.

25. **P. grísea** Gray. Annual, erect, branched, 2–6 dm. high, glandular-hirsute and hirsutulous; lvs. mostly basal, lanceolate to broadly ovate, 1–3 cm. long, entire or dentate-lobed, on petioles to 3 cm. long; cymes densely fld.; pedicels 1–2 mm. long; calyx-lobes narrowly spatulate to obovate, 3–4 mm. long, unequal and 6–8 mm. in fr.; corolla white to pale lavender, open-campanulate, 5–6 mm. long; scales oblong, adnate; stamens 7–8 mm. long, with papillate fils.; style 5–9 mm. long, hairy; caps. ovoid, 4–5 mm. long, bristly hairy; seeds 5–10, ca. 1.5 mm. long, pitted; $n=9$ (Cave & Constance, 1959).—Dry slopes below 3500 ft.; burns and disturbed places in Chaparral; Santa Ynez Mts. to Monterey Co. May–June.

26. **P. gymnoclàda** Torr. ex Wats. Annual, the stems branched from base, spreading, 5–20 cm. long, hirsutulous, somewhat glandular; lower lvs. oblong to oval, 1.5–2.5 cm. long, shallowly pinnately lobed to subentire, on petioles as long or longer; cauline lvs. few, subtending the infl.; cymes few-fld., the pedicels ca. 1–8 mm. long; calyx-lobes linear to linear-spatulate, 3–5 mm. long, hirsutulous, 5–8 mm. long in fr.; corolla blue to lavender with yellow tube, campanulate-funnelform, 6–10 mm. long; scales inconspicuous; stamens included, 3–6 mm. long, the fils. pubescent; style 2.5–4 mm. long, pubescent; caps. oblong-ovoid, 3–4 mm. long, pilose; seeds 5–8, oblong-ovoid, 3–4 mm. long, with 7–9 rounded corrugations; $n=13$ (Cave & Constance, 1950).—Uncommon, dry places, 5000–7000 ft.; Pinyon-Juniper Wd.; White Mts.; Nev., Ore. May–June.

27. **P. heterophýlla** Pursh. ssp. **virgàta** (Greene) Heckard. Biennial stems mostly solitary, simple, coarse, erect; lower stem and rosette lvs. lacking gland-tipped hairs and mostly with 5–7 pinnae; infl. virgate, of many short cymes; infl. and calyx-lobes usually with dull whitish hairs; corolla yellowish or greenish-white, campanulate, 4–5 mm. long; scales adnate; stamens with pubescent fils.; seeds 1–2, pitted; $n=11$ (Heckard, 1960).—Largely in Montane Coniferous F., from just n. of our area to Wyo.

28. **P. hùmilis** T. & G. var. **hùmilis**. [*P. irritans* Brand.] Annual, simple and erect or branched and wide-spreading, hirsute and hirsutulous, scarcely glandular, 5–20 cm. high; lowest lvs. opposite, others alternate, linear-oblong to ovate, 1–6 cm. long, entire, short-petioled or the upper subsessile; racemes 1–3, the fls. many, short-pedicelled; calyx-lobes linear or linear-oblong, 3–5 mm. long, white-hirsute; corolla violet, open-campanulate,

4-6 mm. long, deciduous; stamens 4-6 mm. long, the fils. pubescent; style 4-8 mm. long, subglabrous; caps. 2-3.5 mm. long, apiculate, hirsutulous; seeds 1-2, lance-ovoid, 1.5-2.5 mm. long, finely pitted; $n=11$ (Cave & Constance, 1943).—Flats and borders of meadows, 5000-9400 ft.; Sagebrush Scrub, Montane Coniferous F.; Last Chance Mts., e. Inyo Co., Sierra Nevada mostly along the e. side, from Inyo Co. to Wash. Nev. May-July.

Var. **dúdleyi** J. T. Howell. Stamens 6-8 mm. long; calyx 8-12 mm. long in fr.; seeds 2.5-3 mm. long.—Tehachapi Mts.

29. **P. imbricàta** Greene. ssp. **imbricàta**. [*P. californica* var. *calycosa* Dundas.] Plate 53, Fig. E. Perennial from a branched woody caudex, mostly 2-5 dm. high; herbage green to yellowish or occasionally grayish green; upper part of stem sparsely hispid, sometimes also glandular-puberulent; lvs. mostly basal, 5-25 cm. long, with 1-7 pairs of lateral segms.; upper lvs. reduced, even simple; fls. subsessile or nearly so, numerous, in dense cymes, in open panicles; calyx-lobes heteromorphic, the outer lobe lanceolate to ovate, 3-6 mm. long in fl., 5-10 mm. in fr.; corolla white, 4-7 mm. long, cylindrical with incurved or erect lobes or somewhat campanulate with slightly spreading lobes; seed usually 1, 2-2.5 mm. long, coarsely pitted in vertical rows; $n=11$ (Cave & Constance, 1942).—Common in rocky places, below 6500 ft.; mostly Chaparral, Yellow Pine F.; San Bernardino and San Gabriel mts., Mt. Pinos, n. through the Coast Ranges and Sierra Nevada.

Ssp. **bernardína** (Greene) Heckard. [*P. virgata* var. *b.* Greene.] Stout perennial to 10 dm. tall; herbage green to yellowish green, densely hispid and glandular-puberulent throughout; cyme-bearing branches often many and overlapping to form a virgate stem pattern; corolla tubular-campanulate with lobes slightly divergent after anthesis; calyx-lobes lanceolate, tawny- or white-hispid; $n=11$ (Heckard, 1960).—Dry canyons and mesas at ca. 1000 ft.; Coastal Sage Scrub, Chaparral; base of Santa Ana, San Bernardino and San Gabriel mts.

Ssp. **pátula** (Brand) Heckard. [*P. magellanica* f. *p.* Brand.] Stems ascending, to 8 dm. high, with fewer loosely arranged floral branches; foliage grayish or whitish; petioles and calyces lacking glandular hairs; basal lvs. with 1-3 pairs of lobes; $n=11$, 22 (Heckard, 1960).—At elevs. of 2500-7500 ft.; San Bernardino Mts. to Sierra Juarez. June-July.

30. **P. ivesiàna** Torr. Annual, with several to many erect or spreading stems 0.5-2.5 dm. long, ± hirsutulous and glandular-puberulent; lvs. scattered, the lower oblong to oblanceolate, 1-3.5 cm. long, usually divided to midrib into narrow-oblong segms., these entire or toothed; petioles ca. as long; upper lvs. reduced; cymes sessile, the pedicels becoming 2-6 mm. long; calyx-lobes linear to linear-oblanceolate, unequal, 3-5 mm. long, somewhat more in fr., hirsutulous on margins; corolla white with yellowish tube, tubular-funnelform, 2.5-4 mm. long; scales minute or none; stamens 1-2 mm. long; style ca. 1 mm. long, subglabrous; caps. ± oblong, ca. 4 mm. long, hirsutulous; seeds 10-15, with 8-12 transverse corrugations; $n=11$ (Cave & Constance, 1959).—Occasional, dry sandy places, ca. 2500 ft.; Creosote Bush Scrub; near Kelso, e. San Bernardino Co.; to Utah, Colo. March-June.

31. **P. lemmònii** Gray. Glandular-puberulent annual, erect, 7-20 cm. high, the stems ± reddish; lvs. well distributed, oblong-oval to broadly ovate, 1-2.5 cm. long, subentire to repand or dentately lobed; petioles mostly less than 1 cm. long; cymes many-fld., with short pedicels; calyx-lobes oblong-oblanceolate, slightly unequal, 2.5-4 mm. long, glandular-hirsutulous, 5-7 mm. in fr.; corolla whitish to pale purple, narrow-campanulate, 4-6 mm. long, deciduous; scales free from fils.; stamens 2-3 mm. long; style 2-2.5 mm. long, glandular-pubescent at base; caps. 3-4 mm. long; seeds numerous, 0.5 mm. long, coarsely pitted; $n=22$ (Cave & Constance, 1944, 24 in 1947).—Damp places, 1400-7000 ft.; Creosote Bush Scrub to Pinyon-Juniper Wd.; deserts from e. Inyo Co. to Riverside Co.; Nev., Ariz. March-June.

32. **P. lóngipes** Torr. ex Gray. Annual, freely branched from base, 1-4 dm. high, hirsute and glandular-puberulent throughout; lvs. ovate to rounded, crenate, the blades 1.5-4 cm. long; petioles longer; cymes lax, several-fld., with pedicels 5-10 mm. long; calyx-lobes linear-oblong, 3-6 mm. long, somewhat more in fr.; corolla white, sometimes

bluish, rotate-campanulate, 7–12 mm. long; stamens 10–15 mm. long, pubescent; style 8–15 mm. long, pubescent, divided halfway; caps. 5–6 mm. long, glandular-hispidulous; seeds 8–15, pitted, 1–1.5 mm. long; $n=11$ (Cave & Constance, 1944).—Dry loose slopes, mostly 3000–7000 ft.; Chaparral, Montane Coniferous F.; mts. from Santa Barbara Co. to San Gabriel Mts. April–July.

33. **P. lyònii** Gray. Annual, densely glandular and heavy-scented, sparsely hispid and strigulose throughout; stems simple or branched, 3–10 dm. high; lvs. ovate, 5–10 cm. long, short-petioled, pinnate, then pinnatifid into crenate-dentate divisions; fls. many, in a short dense panicle of cymes; fls. subsessile; calyx-lobes spatulate, glandular and hirsute, ca. 5 mm. long in fl., 8 mm. in fr.; corolla pale blue, campanulate, 5–7 mm. long, deciduous; scales broadly ovate, attached to fil. at base; stamens 4–5 mm. long; style ca. 3 mm. long, deeply cleft; caps. 5–7 mm. long; seeds many, 0.5–0.7 mm. long, pitted; $n=11$ (Cave & Constance, 1963).—Rocky places; Coastal Sage Scrub, Chaparral; San Clemente and Santa Catalina ids. April–June.

34. **P. minor** (Harv.) Thell. [*Whitlavia m.* Harv.] WILD CANTERBURY-BELL. Plate 53, Fig. F. Erect annual, simple or branched from base, 2–6 dm. high, hirsute and glandular-pubescent throughout; lvs. well distributed, the lower ovate or somewhat oblong, coarsely serrate, 2–7 cm. long, on somewhat longer petioles, the upper gradually reduced; cymes many-fld., lax, the pedicels becoming 1–1.5 cm. long; calyx-lobes linear-oblong, 6–8 mm. long; corolla purple, tubular-campanulate, slightly constricted at throat, 1.5–4 cm. long; stamens somewhat exserted, 2–4.5 cm. long, the fils. with an oblong, usually hairy basal dilation; style 1.5–3.5 cm. long; caps. 8–12 mm. long, glandular-hirsutulous and almost hispid; seeds ca. 100, ca. 1 mm. long, coarsely pitted; $n=11$ (Cave & Constance, 1942).— Common in dry disturbed places like burns, below 5000 ft.; Coastal Sage Scrub, Chaparral; cismontane from Santa Monica Mts. to edge of desert and to n. L. Calif. March–June.

35. **P. minutiflòra** J. Voss. Resembling *P. crenulata*, branched from base, 1–3 dm. high, glandular-pubescent, puberulent, and short-hirsute; lf.-lobes especially the terminal almost round, often not much divided; calyx-lobes unequal, 3–4 mm. long, 1 or 2 wider than others; corolla lavender to violet purple, tubular-campanulate, 4–5 mm. long; stamens 3–4 mm. long; caps. ca. 3 mm. long, scattered-pubescent; seeds ca. 4, ca. 2 mm. long, shallowly reticulate-pitted; $n=11$ (Cave & Constance, 1963).—Uncommon, sandy and rocky places, below 1500 ft.; Creosote Bush Scrub; Whipple Mts., se. San Bernardino Co. through Colo. Desert; to Ariz., Son. March–April.

36. **P. mohavénsis** Gray. Hirsute annual, pubescent and somewhat glandular throughout, 5–25 cm. high, simple or branched; lvs. linear to oblanceolate, 1–3 cm. long, mostly entire, short-petioled; fls. many, in 1 to few dense cymes; pedicels very short; calyx-lobes linear to linear-oblanceolate, 3–5 mm. long, unequal, glandular-hirsutulous, 5–15 mm. in fr.; corolla lavender, open-campanulate, 5–8 mm. long, deciduous; scales lanceolate; stamens 5–8 mm. long; style 5–8 mm. long, pubescent; caps. 3–5 mm. long, hirsutulous, glandular; seeds 4–8, pitted, 1.5 mm. long; $n=9$ (Cave & Constance, 1942). —Dry slopes, 4000–7300 ft.; Yellow Pine F.; San Bernardino and San Gabriel mts.; mts. of ne. San Bernardino Co. and e. Inyo Co. to s. Sierra Nevada. June–July.

37. **P. mustelìna** Cov. Like *P. rotundifolia*, but coarser; lf.-blades 0.5–3.5 cm. long; corolla violet, 6–10 mm. long; fils. sparsely short-hairy; style 3–5 mm. long; seeds 0.5–0.7 mm. long.—Rocky places, often in limestone, 3000–6000 ft.; Creosote Bush Scrub; mts. about Death V. March–June.

38. **P. mutábilis** Greene. [*P. californica* var. *jacintensis* Dundas.] Mostly short-lived perennial with branched caudex and 1–several stems, ascending to erect, 2–6 dm. high, slender, with ca. 5–12 cauline lvs.; rosette lvs. 5–12 cm. long, pinnate with 1–3 discrete lfts., frequently withered by anthesis; fls. many, in mostly few cymes; calyx-lobes green, usually glistening-hispid, slightly heteromorphic, the outer lobe wider than the inner linear ones, 3–5 mm. long; corolla white to yellow-white to deep lavender, campanulate to ± tubular, 3–5 mm. long; caps. 2–3 mm. long; seeds 1–4, irregularly pitted; $n=11, 22$ (Cave & Constance, 1958, 1942).—Rather dry places, 4000–8000 ft.; Montane Coniferous F.; San Jacinto Mts., Sierra Nevada from Kern Co. n.; to Wash. June–Aug.

39. **P. nashiàna** Jeps. Plant 1–2 dm. high, glandular-pubescent and with some longer non-glandular hairs; lvs. crowded in lower part of stem, roundish with cordate base or oval with obtuse base, crenate, thickish, mostly 1–2 cm. long, on petioles 2–3 times as long, or upper somewhat scattered, smaller; stems branched above with a few cymes elongating in fr.; calyx-lobes linear, ca. 5 mm. long in fl., 8 mm. in fr.; corolla deep bright blue with 5 pale basal spots, rotate-campanulate, 1.5–2.2 cm. broad, the lobes broad, rounded; stamens ca. as long as or shorter than corolla; fils. pilose, dilated and with a spreading tooth on each side near base; caps. many-seeded, 8–10 mm. long; seeds reticulate-favose, ca. 1 mm. long; $n=11$ (Cave & Constance, 1963).—Fine loose sand or gravel on steep slopes below 6500 ft.; Pinyon-Juniper Wd., Joshua Tree Wd.; w. Mojave Desert from Red Rock Canyon, Kern Co. to Nine-Mile Canyon, e. slope of s. Sierra Nevada, Inyo Co. May–June.

40. **P. negléct a** Jones. Like *P. pachyphylla* but more grayish-pubescent and lacking black-headed glands; racemes ± concealed by the foliage; corolla creamy-white, 5 mm. long; stamens 3 mm. long; style 2 mm. long; caps. tending to be deflexed, 4 mm. long; seeds ca. 100, less than 1 mm. long; $n=11$ (Cave & Constance, 1959).—Heavy alkaline soil below 3000 ft.; Creosote Bush Scrub; deserts from San Bernardino Co. to Imperial Co.; Nev., Ariz. March–May.

P. nemoràlis Greene. Short-lived perennial 5–20 dm. high, with many cauline lvs.; rosette-lvs. with 1–2 pairs of discrete lfts.; fls. white. S. Coast Ranges just n. of our area.

41. **P. novenmillénsis** Munz. Several-branched from base, 5–10 cm. high, setulose and finely pubescent with some hairs gland-tipped; basal lvs. 2–7 cm. long, mostly with 1–5 ± decurrent lanceolate to lance-elliptic lateral pinnae 0.8–1.2 cm. long and with a terminal broadly oblanceolate lft. 1.5–2.5 cm. long, greenish but setulose on veins underneath; upper lvs. reduced, entire; cymes compact, ca. 2 cm. long in fr., 8–14-fld.; pedicels 2–5 mm. long; calyx cleft to base, the lobes subequal, linear, 2–3 mm. long in fl., 9–10 mm. in fr.; corolla lavender, broadly tubular-campanulate, 3–4 mm. long, the lobes rounded, ca. 1.3 mm. in diam.; scales lanceolate, adnate to tube on one edge, connivent with adjacent scale near base; stamens glabrous; style glabrous, ca. 5 mm. long; caps. 2.5–3 mm. long, pubescent and finely hirsute, seeds narrow-ovoid, 1 6–8 mm. long, finely pitted.—Dry disturbed bank, 6500 ft.; Pinyon-Juniper Wd.; Nine-Mile Canyon, e. slope of Sierra Nevada, s. Inyo Co. May.

42. **P. oreópola** Heckard ssp. **oreópola.** Perennial; stems few, decumbent, to 3.5 dm. long; herbage grayish to white with dense tomentose indument, sparsely hispid; lvs. of mature rosette pinnately dissected with 3–5 (–7) pairs of lobes; cauline lvs. 2–4, reduced; floral branches 3–6; calyx-lobes lanceolate, slightly heteromorphic, 3–6 mm. long at anthesis, 6–12 mm. in fr.; corolla white to pale pink, the tube cylindrical to divergent with incurved lobes 4–6 mm. long; seeds 1–3, 2–2.5 mm. long; $n=22$ (Heckard, 1960).—Open woods at 5400 to 8500 ft.; Montane Coniferous F.; San Gabriel Mts. June.

Ssp. **símulans** Heckard. Stems several, ascending; rosette lvs. entire or with 1–2 pairs of lobes; corolla urceolate, lavender; $n=22$ (Heckard, 1960).—At 6500–9000 ft.; Montane Coniferous F.; San Bernardino Mts.

43. **P. pachyphýlla** Gray. Fleshy annual, simple or few-branched, erect, 1–1.5 dm. high, hirsutulous and with stalked black-headed glands throughout; stems and lvs. brittle; lvs. roundish, thick, entire to crenulate, 2–2.5 cm. in diam., largely near base of plant, with petioles longer than blades; upper cauline lvs. reduced; cymes dense, short, subpaniculate, the pedicels to 1 or 2 mm. long; calyx-lobes oblong, 2–3 mm. long, densely glandular-hirtellous; corolla purplish or violet, funnelform-campanulate, 5–7 mm. long, deciduous; scales linear, 1–2 mm. long; stamens 2.5–4 mm. long; style 2–3 mm. long; caps. 5–7 mm. long, glandular-puberulent; seeds over 100, 1–1.2 mm. long, transversely corrugated; $n=11$ (Cave & Constance, 1959).—Occasional on alkaline flats, mostly below 3000 ft.; Creosote Bush Scrub, Joshua Tree Wd.; deserts, Kern Co. to Imperial Co.; n. L. Calif. April–May.

44. **P. párishii** Gray. Annual diffusely branched from base, the stems decumbent then ascending, 5–15 cm. long, glandular-puberulent; lvs. largely basal, oblong to obo-

vate, 1–3 cm. long, entire to shallowly dentate, the lower on petioles 1–2 cm. long, the upper subsessile; cymes many-fld., with short pedicels; calyx-lobes very unequal, 4 of them linear-oblong, ca. 4 mm. long, 1.5–2.5 mm. wide, the 5th spatulate-obovate, 2.5–4 mm. wide; corolla lavender with the base of the tube yellowish, tubular-campanulate, 5–6 mm. long; stamens 2.5–3.5 mm. long, the fils. sparsely hairy at base; style 1.5–2 mm. long; caps. ca. 4 mm., hirsutulous; seeds ca. 25, finely pitted, 1–1.3 mm. long.—Uncommon, rather alkaline places, 2000–6000 ft.; Creosote Bush Scrub, Joshua Tree Wd.; Mojave Desert from e. of Victorville; s. Nev. April–June.

45. **P. párryi** Torr. Plate 54, Fig. A. Annual, mostly 1–4 dm. high, hirsute and glandular-pubescent throughout, simple or few-branched; lvs. ovate to somewhat oblong, 1.5–5 cm. long, dentate, the lower petioles longer than the blades, cauline lvs. somewhat reduced upward; cymes open, many-fld., pedicels 1–2 cm. long; calyx-lobes linear-spatulate, 5–8 mm. long, enlarging somewhat in fr.; corolla dark purple to violet, with paler center, open-campanulate to almost rotate, 1–2 cm. long, deciduous; stamens 1–2 cm. long, with basal wings on pubescent fils.; style 1–2 cm. long, cleft in upper third; caps. 6–10 mm. long, glandular-hirsutulous; seeds many, pitted, ca. 0.7 mm. long; $n=11$ (Cave & Constance, 1947).—Dry slopes and disturbed places like burns, below 2500 (4000) ft.; Chaparral, Coastal Sage Scrub, Creosote Bush Scrub; cismontane from Monterey Co. to L. Calif., w. edge of Colo. Desert. March–May.

46. **P. pedicellàta** Gray. Robust openly branched annual 2–5 dm. high, densely glandular-pubescent and short-hirsute throughout; lf.-blades broadly ovate in outline, 4–12 cm. long, pinnate with 3–7 distinct rounded lfts., these lobed to serrate, often with crinkled margins, 4–12 cm. long; petioles shorter than blades; upper lvs. less deeply divided, but still petioled; fls. many, in short dense paniculately arranged cymes; pedicels 3–6 mm. long; calyx-lobes linear-oblanceolate, 3–4 mm. long, glandular and finely hirsute; corolla pinkish to bluish or paler, open-campanulate, 5–7 mm. long, deciduous; scales short-rounded; stamens 6–8 mm. long, glabrous; style 6–8 mm. long; caps. 3–3.5 mm. long, pubescent; seeds 4, shallowly pitted, 2.5–3 mm. long.—Desert washes and canyons, below 4500 ft.; Creosote Bush Scrub, Joshua Tree Wd.; Panamint Mts. to e. Mojave Desert, Colo. Desert; w. Nev., Ariz., L. Calif. March–May.

47. **P. pediculoìdes** (J. T. Howell.) Constance. [*P. ivesiana* var. *p.* J. T. Howell.] Lvs. divided or lobed into deltoid segms.; seeds with 5–7 transverse corrugations; $n=23$ (Cave & Constance, 1959). Occasional in sandy places, below 4000 ft.; Creosote Bush Scrub, Joshua Tree Wd.; Colo. Desert through Mojave Desert to Inyo Co.; Ariz., Nev. March–May.

48. **P. peirsoniàna** J. T. Howell. Like *P. rotundifolia*, the lvs. less prominently toothed, roundish, 1.5–3.5 cm. wide; stems densely glandular-pubescent but scarcely if at all hirsutulous; calyx-lobes more oblong, 3–4 mm. long in fl., 7–8 mm. in fr.; corolla dull purple to white, 5 mm. long; stamens sparsely hairy at base; seeds 1–1.5 mm. long; $n=12$ (Cave & Constance, 1959).—Desert canyons, 4500–8000 ft.; Pinyon-Juniper Wd.; Mono Co. to Westgard Pass, Inyo Co.; adjacent Nev. May–Aug.

49. **P. perityloìdes** Cov. var. **perityloides.** Perennial, with branched woody rootcrown, often with many diffusely branched stems, white-woolly below, glandular-pubescent above, 1–4 dm. long; lvs. well distributed, round to ovate in outline, 1–2 cm. in diam., coarsely dentate to shallowly lobed, on longer slender petioles; cymes lax, few-fld., with pedicels ca. 1 cm. long; calyx-lobes oblong-spatulate, 5–6 mm. long; corolla white, with yellowish or purplish throat, campanulate-funnelform, 10–12 mm. long, deciduous; scales linear; stamens 3–6 mm. long; style 4–5 mm. long; seeds many, ca. 0.5 mm. long, shallowly pitted; $n=11$ (Cave & Constance, 1959).—Dry crevices in limestone cliffs, 3000–7500 ft.; Creosote Bush Scrub and higher; Titus Canyon and Panamint Mts. to White Mts., Inyo Co. April–June.

Var. **jàegeri** Munz. [*P. geraniifolia* Brand.] Calyx-lobes narrower, to 1 mm. wide; corolla 12–15 mm. long, the tube white to pale yellow.—Limestone crevices, 6000–7700 ft.; Pinyon-Juniper Wd.; Clark Mt., e. San Bernardino Co.; s. Nev. May–July.

50. **P. pulchélla** Gray var. **gooddíngii** (Brand) J. T. Howell. [*P. g.* Brand.] Widely

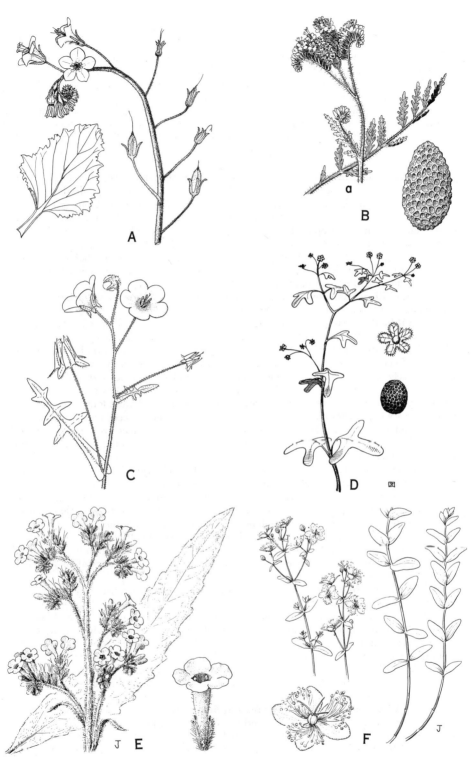

Plate 54. Fig. A, *Phacelia parryi*. Fig. B, *Phacelia tanacetifolia*. Fig. C, *Pholistoma auritum*. Fig. D, *Pholistoma membranaceum*. Fig. E, *Turricula parryi*. Fig. F, *Hypericum formosum*.

and openly branched annual 5–20 cm. high, finely glandular-puberulent and hirsutulous; lvs. oblong-ovate to roundish, 0.5–2 cm. long, crenate or dentate; lower petioles ca. as long; fls. many, short-pedicellate in lax cymes; calyx-lobes oblanceolate, 4–5 mm. long, glandular and hirsutulous, 6–9 mm. in fr.; corolla violet or mauve, with yellow tube, funnelform-campanulate, 8–12 mm. long; stamens 3–5 mm. long, the fils. sparsely pubescent at base; style 4–5 mm. long, pubescent below; caps. oblong, 4–5 mm. long, hirsutulous; seeds many, 0.5–0.7 mm. long, coarsely pitted; Rather alkaline flats, at 2600 ft.; Creosote Bush Scrub; Mesquite V. n. of Kingston, e. Inyo Co.; s. Nev., nw. Ariz. April–June.

51. **P. ramosíssima** Dougl. ex Lehm. var. **suffrutéscens** Parry. Perennial, with few to several branching stems 5–10 or more dm. long, ± suffrutescent, glandular-pubescent and hispid, with basal pustules; lvs. 5–15 cm. long, pinnate with oblong, crenate to serrate divisions up to 2.5 cm. wide; calyx-lobes oblanceolate, 6–7 mm. long; corolla dirty-white to bluish, campanulate, 6–8 mm. long; stamens glabrous, exserted; style glabrous, exserted; caps. 3–4 mm. long, pubescent; seeds usually 2 or 4, deeply pitted, 2–3 mm. long; $n=10$ (Cave & Constance, 1947).—Frequent in canyons, below 8000 ft.; mostly Chaparral and Yellow Pine F.; interior s. Calif. from L. Calif. to Tehachapi Mts. and Panamint Mts. and to Santa Clara Co. May–Aug.

Var. **austrolitorális** Munz. The stems ± pubescent and setose-hispid lacking gland-tipped hairs.—Coastal Strand; Santa Barbara Co. to L. Calif.

Var. **válida** Peck. Stems mostly 2–4 dm. long, slender, puberulent, with occasional longer fine hairs.—Dry slopes above 7000 ft.; Piute Mts., n. through Sierra Nevada.

52. **P. rotundifòlia** Torr. ex Wats. Annual, glandular-hirsutulous and hirsute, 0.5–3 dm. high, freely branched, the stems slender and fragile; lvs. roundish, mostly 0.5–2 cm. wide, coarsely toothed, on somewhat longer slender petioles; cymes few-fld., with pedicels 1–5 mm. long; calyx-lobes linear-oblanceolate, 2–4 mm. long, 5–6 mm. in fr.; corolla white to pinkish or pale violet, pale yellow below, tubular, ca. 5 mm. long; stamens 2–3 mm. long; style 1.5–2 mm. long; caps. 4 mm. long, puberulent; seeds many, pitted, less than 0.5 mm. long; $n=12$ (Cave & Constance, 1959).—Rocky places, crevices, etc., 3000–6000 ft.; Creosote Bush Scrub to Pinyon-Juniper Wd.; deserts from Inyo Co. to Riverside Co.; Utah, Ariz. April–June.

53. **P. suaveòlens** Greene var. **kéckii** (M. & J.) J. T. Howell. [*P. k.* M. & J.] Annual, hirsutulous and glandular throughout, mephitic, erect or spreading, 0.5–4 dm. high, simple or branched; lvs. well distributed, oblong to ovate or elliptic, 1–6 cm. long, serrate to shallowly lobed, on petioles to ca. 3 cm. long; cymes sessile or short-peduncled, many-fld., the pedicels 1–2 mm. long; calyx-lobes oblanceolate, 4–5 mm. long, 6–8 mm. in fr., glandular; corolla mostly 10–14 mm. long, lavender to purple with yellow tube, tubular-campanulate, deciduous; stamens 3–5 mm. long; style 3–4 mm. long, glandular-hirsutulous; caps. 3–5 mm. long; ovules 8–10; seeds coarsely pitted.—Dry slopes, 4000–5000 ft.; Chaparral, Closed-cone Pine F.; Santa Ana Mts. May–June.

54. **P. tanacetifòlia** Benth. Plate 54, Fig. B. Resembling *P. distans*, glandular-puberulent and hirsute throughout, often stouter; fls. many, short-pedicelled in prominent compact cymes; calyx-lobes mostly linear, 6–8 mm. long, densely pubescent and hispid; corolla blue, broadly campanulate, 6–9 mm. long, quite persistent; scales usually completely adnate; stamens 1.5–2 times as long as corolla; style glabrous, deeply cleft; caps. 3–4 mm. long, pubescent; seeds usually 2, coarsely pitted in transverse rows; $n=11$ (Cave & Constance, 1942).—Open flats and slopes below 4000 ft.; many Plant Communities, n. L. Calif. to Lake and Butte cos.; up to 6000 ft. in Mojave Desert; Santa Cruz Id.; to Nev., Ariz. March–May.

55. **P. vállis-mórtae** J. Voss var. **vállis-mórtae**. Resembling *P. cicutaria*, but stems slender, purplish, tending to be zigzag and leaning on other plants for support, the stiffer hairs rather slender, somewhat retrorse, finer hairs gland-tipped or not; lvs. rather few; calyx-lobes linear-oblanceolate, 4–6 mm. long, to 1 cm. in fr.; corolla lavender, open-campanulate, 8–10 mm. long, deciduous; scales 1.5–2 mm. long, the transverse portion rather inconspicuous; stamens 4–8 mm. long; style 6–10 mm. long; caps. 3–3.5 mm. long,

hirsutulous; seeds usually 4, ca. 3 mm. long, evenly pitted; $n=11$ (Cave & Constance, 1950).—Common, dry gravelly and rocky places, 2000–7500 ft.; Creosote Bush Scrub, Joshua Tree Wd., Pinyon-Juniper Wd.; Mojave Desert from Twentynine Palms to Death V. and Owens V.; Nev., Ariz. April–May.

Var. **helióphila** (Macbr.) J. Voss. [*P. hispida* var. *h*. Macbr.] Corolla pale with darker veins, 10–12 mm. long; stamens exserted; $n=11$ (Cave & Constance, 1942).—V. Grassland, w. edge of San Joaquin V. s. to Kern Co. April–May.

56. **P. víscida** (Benth.) Torr. [*Eutoca v.* Benth.] Annual, 1–7 dm. tall, erect, simple or few-branched, hirsute and glandular throughout; lvs. oblong-ovate to rounded, 4–9 cm. long, doubly serrate or irregularly doubly dentate, with somewhat shorter petioles; cymes terminating the branches, many-fld., dense in fl., lax in fr., with pedicels becoming 1–2 cm. long; calyx-lobes linear-spatulate, 5–6 mm. long, somewhat enlarged in fr.; corolla blue with purplish or whitish center, rotate-campanulate, 8–18 mm. long, deciduous; stamens included, the fils. pilose; style 5–12 mm. long; caps. 8–12 mm. long, hirsute and glandular; seeds many, ca. 0.6 mm. long, finely pitted; $n=11$ (Cave & Constance, 1950). —Open sandy and disturbed places, below 2000 ft.; largely Coastal Sage Scrub, Chaparral; near the coast from San Diego Co. to Monterey Co. Channel Ids. March–June.

9. Pholístoma Lilja ex Lindbl.

Succulent annuals, prostrate or reclining, with brittle stems often having prickly angles. Lower lvs. opposite, the upper alternate, pinnate; petioles often winged and clasping. Fl. solitary, axillary and also several in open terminal cymes. Calyx lobed nearly to base, the sinuses naked or with a sepaloid auricle. Corolla subrotate, deciduous, white to violet or blue, lobed to about the middle. Stamens included, inserted on the corolla; appendages broad and triangular to minute or reduced to glands. Style cleft less than halfway. Caps. globose, 1-loculed, exceeding the enveloping or spreading calyx, loculicidal. Ovules 2–several on front of each of 2 large placentae. Seeds mostly 1–6, globose, reticulate or pitted. Three spp. (Greek, *pholis*, scale, and *stoma*, mouth.)

Calyx with auricles between the lobes, enveloping the caps.; stems with retrorse prickles on the angles.
Petioles broadly winged and auriculate-clasping; corolla 1–3 cm. broad ..... 1. *P. auritum*
Petioles narrowly winged, scarcely auriculate-clasping; corolla 0.6–1 cm. broad
3. *P. racemosum*
Calyx without auricles, stellate-rotate under the caps.; stems slightly scabrous, without retrorse prickles .... 2. *P. membranaceum*

1. **P. aurìtum** (Lindl.) Lilja. [*Nemophila a.* Lindl. *Ellisia a.* Jeps.] Plate 54, Fig. C. Straggling, with coarse loosely branched stems 3–10 dm. long, with retrorse prickles as well as pubescent; lower lvs. oblong to lance-ovate, 6–15 cm. long, with 7–13 divisions, these oblong or lanceolate, ± retrorse, the petioles broadly winged and auriculate-clasping; fls. cymose and terminal or solitary and axillary; pedicels 2–3 cm. long; calyx 5–10 mm. long, the lobes 4–7 mm. long, the auricles 1.5–3 mm. long; corolla lavender to purple with darker markings, 1.5–3 cm. broad, the tube pale; appendages purple, 2–3 mm. long; style ca. 5 mm. long; caps. 5–10 mm. long; seeds 1–4, pitted-reticulate, 2–3 mm. in diam.; $n=9$ (Cave & Constance, 1942).—Shaded slopes and deep canyons, below 4500 ft.; Coastal Sage Scrub, Chaparral; S. Oak Wd.; L. Calif. to cent. Calif. Channel Ids. March–May.

2. **P. membranàceum** (Benth.) Const. [*Ellisia m.* Benth.] Plate 54, Fig. D. Stems weak, procumbent, 2–5 dm. long, glaucous and glabrous except for a few scattered stiff hairs; lvs. 2–6 or more cm. long, the divisions 5–11, linear-oblong to ovate, ± hispidulous, the pedicels narrowly winged but not auriculate; fls. mostly cymose; calyx-lobes oval, 2–3 mm. long, the auricles lacking; corolla 0.4–1 cm. broad, white, often with narrow purple spot on each oval lobe; appendages triangular, minute; style 1.5–2 mm. long; caps. 2–4 mm. in diam.; seeds 1–2, reticulate, 2–3 mm. in diam.; $n=9$ (Cave & Con-

stance, 1942).—Occasional in shady places, below 3500 ft.; Foothill Wd., S. Oak Wd., Chaparral; San Diego Co. and n. L. Calif. to cent. Calif.; inland to deserts from Inyo Co. to Imperial Co.; Nev. March–May.

3. **P. racemòsum** (Nutt.) Const. [*Nemophila r.* Nutt. ex Gray. *Ellisia r.* Jeps.] Straggling, loosely branched, retrorse-scabrous and pubescent, the stems 2–6 dm. long; lvs. thin, sparsely pubescent, 3–10 cm. long, rather regularly pinnatifid into 5–9 ovate entire or round-toothed lobes and narrowed into petioles scarcely winged or auriculate; fls. solitary and cymose; calyx 5–7 mm. long, the lobes 2–3 mm. long, the auricles 1–2 mm. long; corolla 0.6–1 cm. broad, white or sometimes bluish; appendages small; style 1.5–2 mm. long; caps. 5–8 mm. in diam.; seeds usually 4–6, pitted, 1–2 mm. in diam.; $n=9$ (Cave & Constance, 1947).—Shaded places below 1000 ft.; Coastal Sage Scrub, Chaparral, S. Oak Wd.; Channel Ids., coastal San Diego Co., n. L. Calif. March–May.

10. Tricárdia Torr. ex Wats.

Perennial caulescent herbs, branched from heavy base. Lvs. alternate, largely in a basal rosette, petioled, entire. Fls. purplish, rather few, in loose short racemelike cymes, pedicelled. Calyx parted into 5 unequal lobes, the 3 outer large and cordate, becoming scarious and reticulate-veiny in fr., the 2 inner linear. Corolla broadly campanulate, deciduous, with 10 narrow appendages near base of stamens. Stamens included, equally inserted but unequal in length. Style included, 2-cleft. Caps. uniloculed, the walls scarious in maturity. Seeds 4–8, oblong, minutely alveolate. Monotypic. (Greek, *tri*, three, and *cardia*, heart, because of calyx.)

1. **T. watsònii** Torr. ex Wats. Stems few, 1–3 dm. tall, finely hairy; lvs. oblong- to lance-elliptic, 3–5 cm. long, the lower petioled; cymes usually simple; calyx 5–6 mm. long in fl., 15–25 mm. in fr.; corolla ca. 4 mm. long, 6–8 mm. broad, with a purplish throat and white limb with purple veining, the roundish lobes ca. 2 mm. long; caps. ca. 8 mm. long; seeds 3–4 mm. long; $n=8$ (Cave and Constance, 1959).—Occasional on dry slopes, below 7500 ft.; Creosote Bush Scrub, Joshua Tree Wd., Pinyon-Juniper Wd.; about Palm Springs, Riverside Co. across Mojave Desert to e. slope of Sierra Nevada and White Mts.; to Utah, Nev., Ariz. April–June.

11. Turrícula Macbr.

Very glandular stout erect ill-scented suffruticose perennial, branched from base. Lvs. alternate, entire or toothed, sessile, lanceolate, numerous. Fls. many, in terminal thyrsoid panicle of subsessile scorpioid cymes. Calyx deeply divided. Corolla purple, deciduous, funnelform, shallowly lobed, exceeding calyx. Stamens unequal, included; appendages obsolete. Style deeply divided. Caps. falsely bilocular, 4-valved at maturity. Ovules 6–8 in each locule. Seeds 6–10, longitudinally finely ridged and minutely transversely reticulate. One sp. Causes severe dermatitis to many persons, as do many of the glandular spp. of *Phacelia*. (Latin, little tower.)

1. **T. párryi** (Gray) Macbr. [*Nama p.* Gray. *Eriodictyon p.* Greene.] Plate 54, Fig. E. Coarse, 1–2.5 m. tall, viscid-villous; lvs. crowded, 5–12 cm. long; calyx-lobes 3–4 mm. long; corolla pubescent, 13–18 mm. long; fils. adnate; style 4 mm. long; caps. ovoid, ca. 3 mm. long; seeds 1–1.5 mm. long; $n=13$ (Cave & Constance, 1947).—Occasional in dry places, particularly on burns, 1000–8000 ft.; Chaparral, Yellow Pine F.; L. Calif. away from the immediate coast to Fresno and Kern cos. and San Luis Obispo Co., Panamint Mts., Tehachapi Mts. June–Aug.

12. Wigándia HBK.

Large perennial herbs or subshrubs, covered with stinging glistening hairs; lvs. very large, simple and dentate. Fls. sessile, in terminal scorpioid cymes; calyx deeply parted; corolla broadly campanulate, without scales inside, the large spreading limb 5-lobed. Stamens usually exserted and barbed. Ovary 1-celled or falsely 2-celled, with numerous ovules; the styles 2, filiform, with club-shaped capitate stigmas. Caps. usually loculicidally 2-valved. Ca. 5 spp. of trop. Am. (Named after Johannes *Wigand*, 1523–1587, Prussian writer on plants.)

1. **W. caracasàna** HBK. var. **macrophýlla** Brand. Erect robust subshrub to 3 m. or more tall, the pubescence yellow or silky; lvs. ovate, to 4.5 dm. long and 3 dm. wide, long-petioled, subcordate at base; infl. white-pubescent; fls. violet with white tube, 12–18 mm. long and ca. 2.5 cm. across. Occasionally escaped from cult. especially in the Santa Barbara region where it spreads widely underground, new shoots breaking through pavements and hard surfaces; native from s. Mex. to n. S. Am. Summer.

53. Hypericàceae. (*Clusiaceae.*) St. John's Wort Family

Herbs or shrubs with opposite entire, glandular-punctate, mostly sessile lvs. Fls. perfect, regular, hypogynous. Sepals 4 or 5, green, persistent. Petals 4 or 5, mostly oblique and convolute in the bud. Stamens usually many, distinct or ± united into 3–5 clusters. Ovary superior, 1- or 3–7-locular; ours with 3 styles. Caps. 1-loculed, with 2–5 parietal placentae or 3–7-loculed, mostly septicidal. Seeds many, small, anatropous, without endosperm. Ca. 10 genera and 300 spp. of temp. and warm regions.

1. *Hypèricum* L. St. John's Wort

Plants glabrous. Lvs. several-nerved from base, sessile. Fls. often yellow, solitary or in terminal cymes; petals deciduous. Ca. 200 spp. of N. Hemis., some in cult. (Ancient Greek name.)

Stems mostly prostrate; lvs. 4–12 mm. long; petals 2–3 mm. long .... 1. *H. anagalloides*
Stems not prostrate; lvs. larger; petals longer.
  Sepals linear-deltoid, acuminate; stems with sterile shoots in lower lf.-axils ... 4. *H. perforatum*
  Sepals ovate to obovate, acute to obtuse, stems lacking sterile shoots in lower axils.
    Lvs. 1–4 cm. long; herbs to 7 dm. tall .... 3. *H. formosum*
    Lvs. 5–7 cm. long; shrubs to 5 m. tall .... 2. *H. canariense*

1. **H. anagalloides** Cham. & Schlecht. Tinker's Penny. Annual or perennial, procumbent, often in mats, the stems rooting at nodes, 5–15 (–20) cm. long; lvs. elliptic to ovate or orbicular, obtuse, 5–7-nerved, 4–12 mm. long; fls. in few-fld. cymes, golden to salmon color; sepals 2.5–3 mm. long; petals ca. the same; stamens 10–25; caps. ca. 3 mm. long.—Wet places, 700–9500 ft.; many Plant Communities; mts. of San Diego Co., San Jacinto and San Bernardino mts.; cent. Calif. to B.C., Mont. June–Aug.

2. **H. canariénse** L. Shrub to 5 m. high; lvs. oblong-lanceolate, narrowed at the base, 5–7 cm. long; fls. 2–3 cm. in diam., in panicles; sepals ovate, acute, ciliate.—Established at Montecito and n. of Santa Barbara in the hills; native of Old World.

3. **H. formòsum** HBK. var. **scoùleri** (Hook.) Coult. [*H. s.* Hook.] Perennial with erect stems from running rootstocks, slender, 2–7 dm. high, simple or branched above; lvs. oblong-ovate, obtuse, 1–2.5 cm. long, several-nerved, black-dotted along margin; cymes panicled; sepals ovate, obtuse, 3 mm. long, black-dotted along margin; petals yellow, 7–10 mm. long, obovate; stamens many, in 3 groups; caps. 3-lobed, 6–7 mm. long; seeds brownish, ca. 0.6 mm. long, reticulate-pitted; $n=8$ (Kyhos, 1967).—Frequent, wet meadows and banks, 2500–7500 ft.; Yellow Pine F.; Chaparral; San Diego Co. n.; to B.C., Mont. June–Aug. (Plate 54, Fig. F.)

4. **H. perforàtum** L. Klamath Weed. Perennial with leafy basal offshoots, the stems rather simple, tough, 3–10 dm. high, branched above; lvs. linear- to elliptic-oblong, revolute, 1.5–2.5 cm. long, subtending short leafy branchlets; sepals linear-acuminate, 4–5 mm. long; petals orange-yellow, 8–12 mm. long; stamens many, in 3–5 groups; caps. 7–8 mm. long; seeds reticulate; $n=16$ etc. (Robson & Adams, 1968).—Established near Wynola, San Diego Co.; widely so in n. Calif.; native of Eu. Somewhat poisonous to livestock.

Plate 55. Fig. A, *Juglans californica*, ♂ and ♀ fls. below. Fig. B, *Krameria parvifolia*, spine of fr. barbed down the sides. Fig. C, *Agastache urticifolia*. Fig. D, *Hyptis emoryi*. Fig. E, *Monardella lanceolata*. Fig. F, *Monardella linoides stricta*.

54. Juglandàceae. WALNUT FAMILY

Aromatic deciduous trees and shrubs. Lvs. alternate, pinnate, without stipules. Monoecious, the fls. opening after the lvs. unfold; ♂ in bracteate aments on twigs of the previous year and consisting of a 3–6-lobed calyx, with 3–many stamens with short distinct fils. and oblong anthers of longitudinal dehiscence. Pistillate fls. 1–several on peduncles at the ends of the shoots of the season, bracted, then usually with 2 bracteoles and 3–5-lobed calyx. Petals sometimes present. Ovary inferior, 1- or incompletely 2–4-loculed; style short; stigmas 2, plumose. Fr. drupaceous, with a fibrous somewhat fleshy exocarp, indehiscent or 4-valved; endocarp bony, rugose or sculptured, forming a nut. Seeds solitary, 2-lobed, with large fleshy oily cotyledons. Six genera and ca. 40 spp. of N. Temp. Zone and n. Andes.

1. Jùglans L. WALNUT

Staminate fls. with 4–40 stamens, in 2 or more series. Pistillate fls. with a 4-lobed calyx and 4 petals. Ca. 10 spp. (Classical name of the Walnut.)

1. **J. califórnica** Wats. Plate 55, Fig. A. Low trees, usually with several trunks, 3–10 m. tall; young twigs brownish-tomentose; old bark dark, with broad irregular ridges; lvs. 1.5–2 dm. long, the petioles glandular-pubescent; lfts. finely serrate, cuneate or rounded at base, glabrous above with hairy tufts beneath in the vein-axils; ♂ aments 5–8 cm. long, brownish-pubescent; ♀ fls. in small spikes, with yellow recurved stigmas; fr. roundish, 2.5–3 cm. in diam., with dark brown pubescent husk; nut globose, thick-shelled, longitudinally grooved. Locally common below 4500 ft.; S. Oak Wd.; Orange Co. and w. cismontane San Bernardino Co. to Ventura Co. April–May.

Occasionally hybridizing with *J. regia* L., the English Walnut, when the latter is in cult. and forming taller more open trees with broader lfts.

54A. Koeberlíniaceae. JUNCO FAMILY

Almost leafless trees or shrubs with pale green, spine-tipped, stiff branchlets. Lvs. alternate, minute, scalelike, and soon deciduous. Fls. small, in axillary racemes. Sepals 4–5, minute. Petals 4–5, short-clawed, on a hypogynous disk lining the calyx, deciduous. Stamens as many as petals, hypogynous. Ovary of 2–5 united carpels, superior, borne on a short gynophore. Ovules 2–8 in each cell. Style subulate, persistent. Fr. in ours a berry; seeds wingless, hard.

1. Koeberlinia Zucc. ALL-THORN

Shrubs or small trees with rigid interlocking spinose branches. Racemes few-fld. Sepals 4. Petals 4, greenish-white. Stamens 8. Ovary 2-celled. Berry globose, 4–5 mm. in diam., green or with some red.

1. **K. spinòsa** Zucc. var. **spinòsa.** Branches with green or yellowish bark; petals 4–4.5 mm. long, rounded at apex; stamens equaling petals.—Desert washes, Creosote Bush Scrub. Reported from Chocolate Mts., Imperial Co. May–July. Ranging to Texas, L. Calif.

Var. **tenuispìna** Kearney & Peebles. With longer more slender branches having bluish-green bark; petals slightly narrower; stamens longer.—Reported from Chocolate Mts.; to Sonora, L. Calif.

55. Krameriàceae. KRAMERIA FAMILY

Shrubs or perennial herbs, mostly pubescent. Lvs. alternate, simple and entire, rarely 3-foliolate. Fls. irregular, axillary or racemose, ours purplish. Peduncles with paired opposite leafy bracts. Sepals 4–5, unequal. Petals 5, the 3 upper long-clawed, the other 2 sessile and smaller. Stamens 4 in ours, free or borne on the united claws of the upper petals; anthers 2-celled, dehiscent by a pore. Ovary 1-celled; ovules 2; style cylindric. Fr. globose, indehiscent, spiny, 1-seeded. A single genus, with ca. 20 spp.

1. Kramèria L.

Characters of the family. (Named for J. G. H. *Kramer*, 18th-century Austrian botanist.)

Barbs on spines of frs. in an umbrellalike group at apex; upper petals separate to base . 1. *K. grayi*
Barbs on spines of frs. scattered along whole upper portion; upper petals partly united at base of claws .... 2. *K. parvifolia*

1. **K. gràyi** Rose & Painter. [*K. canescens* Gray, not Willd.] Densely and intricately branched, thorny, 3–7 dm. high, of wider spread, silky-tomentose on young growth; lvs. linear to lanceolate, 8–18 mm. long; peduncles commonly 15–25 mm. long, with 1 pair of bracts; calyx 7–10 mm. long, red-purple within; upper petals spatulate, 4–5 mm. long, the lower cuneate-obovate, shorter; fr. densely canescent, 7–8 mm. in diam., armed with many slender spines each with a few barbs at very tip; $n=6$ (Turner, 1959).—Dry sandy and rocky places, below 4000 ft.; Creosote Bush Scrub; Colo. Desert and s. Mojave Desert; to Tex., n. Mex. April–May.

2. **K. parvifòlia** Benth. var. **imparàta** Macbr. [*K. i.* Britt.] Plate 55, Fig. B. Like the preceding, but lvs. more narrowly linear, 2–12 mm. long; peduncles sericeous, 10–15 mm. long, with 2–3 pairs of leafy bracts; sepals sericeous; spines on fr. with weak barbs along upper half or third.—Rather common, cent. Mojave Desert e. and s.; to Nev., Son., L. Calif. March–May.

Var. **glandulòsa** (Rose & Painter) Macbr. Peduncles and sepals with stalked glands. At 2000–4000 ft.; Creosote Bush Scrub; Death V. region, Kingston Mts., New York Mts., Santa Rosa Mts.; to Nev., Tex.

56. Lamiàceae. (*Labiatae.*) MINT FAMILY

Aromatic herbs or shrubs, rarely trees, mostly with 4-angled stems and opposite or whorled simple lvs. Fls. mostly bilabiate, variously clustered in ± cymose and bracteate infls. Calyx persistent, 2-lipped or regular, mostly 5-toothed or -lobed. Corolla usually 2-lipped with the upper lip 2-lobed to entire, the lower lip 3-lobed, sometimes almost regular. Stamens on the corolla-tube, mostly 4 and didynamous, or one of the pairs abortive; anthers 2-celled, or sometimes one of the cells abortive. Ovary 4-lobed or -parted; style single, arising between the lobes, divided at its summit. Fr. of 4 1-seeded nutlets, included in the persistent calyx. A family of over 150 genera and 3000 spp., widely distributed in temp. and trop. regions.

1. Ovary of 4 united nutlets; style not basal; nutlets laterally attached.
    2. Corolla strongly irregular, the upper lip very small; stamens moderately exserted. Desert annual .... 24. *Teucrium*
    2. Corolla almost equally 5-lobed, the lobes declined; stamens long-exserted. Montane or cismontane .... 25. *Trichostema*
1. Ovary of 4 distinct or nearly distinct nutlets; style basal; nutlets basally attached.
    3. Calyx 2-lipped, the lips entire.
        4. Shrub; calyx inflated in fr., not crested or gibbous on the back .... 19. *Salazaria*
        4. Herbs; calyx with a gibbous or helmetlike crest on the back .... 22. *Scutellaria*
    3. Calyx regular or 2-lipped, the lips not entire.
        5. Corolla strongly 2-lipped.
            6. Stamens included in the corolla-tube; calyx with 10 spinescent hooked teeth at the tip
            10. *Marrubium*
            6. Stamens exserted from the corolla or included in throat; calyx-teeth not hooked at tip.
                7. The stamens ascending, not declined and enveloped by the lower lip.
                    8. Upper lip of corolla concave.
                        9. Fertile stamens 4.
                            10. Upper pair of stamens longer than lower pair.
                                11. Anther-sacs parallel or nearly so; upper stamens declined; fls. rose-purple. Native .... 2. *Agastache*
                                11. Anther-sacs divergent. Introd. in waste places.
                                    12. Plants erect; corolla whitish, 10–12 mm. long .... 14. *Nepeta*

                12. Plants creeping; corolla purplish-blue, 16–22 mm. long .. 3. *Glecoma*
            10. Upper pair of stamens shorter than or equal to lower pair.
                13. Calyx 2-lipped, closed in fr., the upper lip truncate; fls. in dense spike,
                    purplish; low herb .... 17. *Prunella*
                13. Calyx 5-toothed, not closed in fr.; if 2-lipped, the upper lip not truncate.
                    14. Plants shrubby; fls. 2.5–4 cm. long; lvs. coarsely serrate .. 7. *Leonotis*
                    14. Plants herbaceous; fls. shorter.
                        15. Calyx-teeth not spine-tipped; corolla-tube without a hairy ring
                            within; nutlets sharply 3-sided; plants annual .... 6. *Lamium*
                        15. Calyx-teeth spine-tipped; corolla-tube with hairy ring within; nut-
                            lets ovoid; plants mostly perennial .... 23. *Stachys*
        9. Fertile stamens 2.
            16. Calyx 2-lipped, or in 1 sp. of *Salvia* entire and oblique at the orifice.
                17. Pollen-sacs in each anther 2, with very short connective between; plants
                    annual with spines on bracts and calyx-teeth .... 1. *Acanthomintha*
                17. Pollen-sacs in each anther 1 or 2, at the ends of elongate arms of connective
                    jointed to the fils.; plants shrubs to annuals, usually not spiny on bracts
                    or calyx-teeth .... 20. *Salvia*
            16. Calyx equally 5-cleft; plant annual. Desert .... 12. *Monarda*
        8. Upper lip of corolla plane.
            18. Fertile stamens 2; low purple-fld. plant of e. Mojave Desert .... 4. *Hedeoma*
            18. Fertile stamens 4, or if 2, on cismontane plants.
                19. Fls. solitary, or few in the lf.-axils.
                    20. Plants quite woody.
                        21. Corolla 2.5–4 cm. long; lvs. ± ovate or deltoid. Cismontane
                                                                                8. *Lepechinia*
                        21. Corolla 1–1.3 cm. long.
                            22. Lvs. round-ovate; corolla white. S. San Diego Co. .... 21. *Satureja*
                            22. Lvs. linear to linear-oblong; corolla pale purplish-blue. Mojave
                                Desert .... 16. *Poliomintha*
                    20. Plants strictly herbaceous; corolla with straight tube and 6–8 or 30–40
                        mm. long .... 21. *Satureja*
                19. Fls. in dense heads.
                    23. Plants annual, with round to spatulate lvs. not more than 2 cm. long
                                                                                15. *Pogogyne*
                    23. Plants perennial, with ovate to lance-ovate lvs. mostly 2–8 cm. long; calyx
                        naked in the throat; bracts green .... 18. *Pycnanthemum*
    7. The stamens declined and enveloped by the lower lip of the corolla; scurfy-tomentose
        desert shrub .... 5. *Hyptis*
 5. Corolla regular or nearly so, the lobes subequal.
    24. Fls. in terminal heads; stamens 4. Plants of dry places .... 13. *Monardella*
    24. Fls. in axillary whorls which may be aggregated in terminal spikes. Plants of wet
        places.
        25. Fertile stamens 2; fls. white .... 9. *Lycopus*
        25. Fertile stamens 4; fls. usually lavender to purplish .... 11. *Mentha*

1. *Acanthomintha* Gray. THORNMINT

Glabrous or pubescent, aromatic annuals, mostly branched from base, sometimes simple, with entire to denticulate or serrulate lvs. Fls. in distinct or at length remote whorls, each whorl subtended by a pair of lvs. and several conspicuous broad bracts armed with needlelike spines. Calyx 2-lipped; upper lip with 3 aristate teeth, lower with 2 oblong acute or spine-tipped lobes. Corolla 2-lipped, white or tinged with rose, the palate cream; upper lip entire or retuse, ± hooded; lower lip 3-lobed; tube exserted, with funnelform throat. Stamens inserted high on the throat; the lower pair antheriferous, the upper shorter and the anthers smaller or imperfect. Style slender, 2-lobed, the lower longer. Nutlets ovoid, smooth. Three spp. (Greek, *acantha*, thorn, and *mentha*, mint.)

Anthers glabrous, 2 fertile, 2 obsolete. San Diego Co. .... 1. *A. ilicifolia*
Anthers woolly or pubescent, all 4 polleniferous. From Ventura Co. n. .... 2. *A. obovata*

1. **A. ilicifòlia** (Gray) Gray. [*Calamintha i.* Gray.] Stems 0.5–1.5 dm. long, glabrous to ± puberulent; lf.-blades 0.5–1.5 cm. long, ovate, serrate-denticulate, on petioles ca. as long; bracts roundish, 7–9 mm. long, the spines 4–6 mm. long; calyx ca. 5 mm. long; corolla 12 mm. long, white with some rose, upper lip 3–4 mm. long, lower 5–6 mm.; antheriferous stamens 2, glabrous or papillate, the other pair abortive.—Clay depressions on mesas and slopes; Coastal Sage Scrub, Chaparral; sw. San Diego Co.; adjacent L. Calif. April–May.

2. **A. obovàta** Jeps. Stems 1–2 dm. high, minutely canescent; lvs. lance-oblong to obovate, denticulate or the upper spinose-serrate; blades 8–12 mm. long, the petioles often longer; bracts broadly ovate to almost round, sparingly puberulent, the spines 5–8 mm. long; calyx puberulent to pubescent, 8–9 mm. long; corolla white or with some lavender tinge near the tips, 1.5 cm. long, the upper lip entire, 4 mm. long, the lower ca. as long; anthers woolly-pubescent, those of upper stamens smaller.—Dry slopes, 1000–5000 ft.; Chaparral, Foothill Wd.; inner Coast Ranges from Ventura Co. (base of Mt. Pinos, 10 mi. w. of Lockwood V.) to San Benito Co. April–June.

2. Agástache Clayt.

Tall perennial herbs. Lvs. ovate, serrate, petioled. Fls. in dense sessile whorls which are mostly in continuous or interrupted cylindrical or tapering spikes. Calyx tubular to turbinate, 15-veined or more, 5-dentate, the teeth equal or the 3 posterior ± united, deltoid to subulate, often thin and colored. Corolla rose, violet or white, with oblique orifice, the upper lip erect, 2-lobed, the lower spreading, 3-lobed. Stamens 4, the lower pair shorter. Nutlets ovoid, smooth, with hispidulous apex. Ca. 20 spp., mostly of N. Am. (Greek, *agan*, much, and *stachus*, ear of grain, because of many spikes.)

1. **A. urticifòlia** (Benth.) Kuntze. [*Lophanthus u.* Benth.] Plate 55, Fig. C. Stems several, branched above, 1–2 m. tall, subglabrous; lf.-blades ovate or deltoid-ovate, the median 3.5–8 cm. long, obtuse or acute, with truncate or subcordate base, mostly coarsely serrate; petioles 1–2.5 cm. long; infl. 4–15 cm. long; calyx green or rose, hirtellous or glabrous, the tube 4–7 mm. long, the teeth deltoid-lanceolate, 2.5–5 mm. long; corolla rose or violet, 10–15 mm. long; nutlets brown, dull, ± striate, 1.5–2 mm. long.—Moist places below 9000 ft.; many Plant Communities but mostly in Montane Coniferous F.; San Jacinto Mts. northward; Kern Co. to B.C. and Rocky Mts. June–Aug.

3. Glecòma L.

Low mostly creeping herbs with petioled to suborbicular lvs. Fls. blue or blue-purple in verticillate clusters. Calyx oblong-tubular, 15-nerved, oblique at throat, unequally 5-toothed. Corolla-tube exserted, enlarged above, the limb 2-lipped, the middle lower lobe enlarged. Stamens 4, didynamous. Nutlets smooth, ovoid. Ca. 6 spp., of Eurasia. (Greek, *glechon*, old name for pennyroyal.)

1. **G. hederàcea** L. GILL-OVER-THE-GROUND. Stems 1–5 dm. long, retrorsely puberulent; lvs. round-reniform, crenate, 1–2.5 cm. broad, long-petioled; fls. axillary, short-pedicelled; calyx 5–6 mm. long, puberulent; corolla 16–22 mm. long, purplish-blue; $2n = 18, 24, 36$ (various auth., 1941–1947). Natur. in moist shaded places; occasional in s. Calif., more common northward. Native of Eu. March–May.

4. Hedeòma Pers. MOCK PENNYROYAL

Strongly aromatic herbs; the small entire or toothed lvs. sessile or short-petioled. Fls. in small cymules or solitary in upper axils. Calyx tubular, 13-ribbed, gibbous at base, pubescent, 5-toothed. Corolla blue or purple, 2-lipped, the upper lip erect, entire or 2-lobed and spreading. Stamens 4, usually only 1 pair fertile. Style 2-cleft at apex, glabrous. Nutlets ovoid, smooth. Ca. 25 spp., all Am. (Greek, *hedus*, sweet, and *osme*, odor.)

1. **H. nànum** (Torr.) Briq. ssp. **califórnicum** W. S. Stewart. [*H. thymoides*, Calif.

auth.] Perennial, cinereous-puberulent, 1–2 dm. tall, diffusely branched from base; lvs. ovate, 6–10 mm. long, petioled, the upper gradually reduced into bracts; upper axils with 1–2 short, 1–2-fld. peduncles; calyx tubular, 4–5 mm. long, the teeth subulate-aristate; corolla 7–8 mm. long, light purple, the lower lip with a white blotch and purple-lined.—Dry rocky slopes, 2800–6000 ft.; Joshua Tree Wd., Pinyon-Juniper Wd.; Providence, N.Y., Clark and Kingston mts. of e. Mojave Desert; to Nev., Ariz. May–June.

5. *Hýptis* Jacq.

Herbs or shrubs with opposite commonly toothed lvs. Fls. bilabiate, usually in dense axillary clusters. Calyx mostly equally 5-toothed. Corolla 2-lipped, declined, the lower lip saccate, abruptly deflexed at the contracted and callous base, the lobes of the upper lip and the lateral ones of the lower similar, flat, equal. Fertile stamens 4, the upper pair shorter; all included in the sac of the middle lower corolla-lobe. Nutlets smooth or slightly roughened. Ca. 350 spp. of the New World, widely distributed in Latin Am. (Greek, *huptios*, turned back, referring to the lower lip.)

1. **H. émoryi** Torr. DESERT-LAVENDER. Plate 55, Fig. D. Erect aromatic shrub, with numerous slender fairly straight branches, 1–3 m. tall, whitish-scurfy-tomentose; lvs. ovate, crenulate, 1–2 cm. long, on petioles less than half as long; fls. in axillary short-peduncled cymes which are in a ± paniculate arrangement at ends of branchlets; pedicels 1–4 mm. long; calyx stellate-woolly, 4–6 mm. long, with setaceous teeth; corolla violet, 4–6 mm. long; nutlets light brown, flattish-oblong, ca. 1.5 mm. long, smooth.—Common in washes and canyons below 3000 ft.; Creosote Bush Scrub; Colo. Desert and s. Mojave Desert; to Ariz., Son., L. Calif. Jan.–May.

6. *Làmium* L. HENBIT

Low herbs, with lvs mostly toothed, petiolate. Fls. small, verticillate in axillary and terminal clusters. Calyx tubular-campanulate, usually 5-nerved, 5-toothed, the teeth sharp-pointed, equal or the upper longer. Corolla 2-lipped, the tube somewhat longer than the calyx, upper lip ascending and concave, the lower spreading, 3-lobed, the middle lobe emarginate, contracted at the base. Stamens 4, didynamous. Nutlets smooth or tubercled. Ca. 40 spp. of the Old World. (Greek, *laimos*, throat, because of the gaping corolla.)

1. **L. amplexicaùle** L. Mostly annual, sparsely pubescent, the stems branched from the base, ± decumbent, 1–4 dm. long; lvs. broadly ovate to roundish, truncate or cordate at base, coarsely crenate, the lower petioled, the upper not, 2–2.5 cm. wide; fls. few, in axillary and terminal clusters; calyx pubescent, ca. 4–5 mm. long, the teeth erect; corolla purple-red, 12–16 mm. long, the tube very slender, the upper lip pubescent; nutlets ca. 2 mm. long, brownish, mottled; $2n = 18$ (Bernström, 1944).—In waste places below 4500 ft., occasionally natur., especially northward; native of Eu. April–Sept.

7. *Leonòtis* R. Br. LION'S-EAR

Herbs or shrubs with lanceolate to ovate, sessile to petioled lvs. Fls. white to orange, in dense axillary whorls. Calyx 8–10-ribbed, the tube funnel-shaped, arched, teeth often acerose-tipped. Corolla-tube as long as calyx or longer, the upper lip long and concave, hairy outside, the lower lip shorter, deflexed, the 3 lobes subequal. Stamens 4, arched. Ca. 12 spp. of S. Afr. and farther n. (Greek, *leo*, lion, and *otis*, ear.)

1. **L. leonùrus** (L.) R. Br. [*Phlomis l.* L.] Shrubby, 1–2 m. tall; lvs. lanceolate, 4–6 cm. long, coarsely serrate; calyx ca. 12 mm. long; corolla orange, 3.5–4 cm. long, densely pilose; stamens not exserted.—Occasional escape from cult.; native of S. Afr. July–Sept.

8. **Lepechínia** Willd. PITCHER SAGE

Shrubby or suffrutescent, aromatic. Fls. showy, solitary in axils of reduced upper lvs. and forming short racemes, or 2–6 in a verticil and forming spikes. Calyx campanulate, subequally 5-toothed, often enlarged in fr. Corolla with a broad tube, pilose-annulate at base within, 5-lobed, the lobes broad, rounded and plane, ± erect, the upper bifid, the lateral smaller, entire, the middle lower emarginate. Stamens 4, subequal or didynamous; fils. glabrous; anther-sacs divergent. Nutlets ovoid, black, stony, smooth. Ca. 38 spp. of w. S. Am., Calif., Hawaii. (Named for *Lepechin*, a Russian botanist and traveller.)

Calyx-teeth deltoid, mostly shorter than the mature tube.
    Lvs. narrowed toward the base, narrowly ovate; pedicels seldom 1.5 cm. long; nutlets puberulent. Ventura Co. n. and Santa Ana Mts. .... 1. *L. calycina*
    Lvs. strongly cordate, broadly ovate; pedicels 1–2.5 cm. long; nutlets glabrous. Santa Ana Mts.
        2. *L. cardiophylla*
Calyx-teeth lanceolate or acicular, equal to or longer than the mature tube; nutlets glabrous.
    Lvs. deltoid or subhastate, less often subcuneate, viscid-villous; calyx-teeth lanceolate, membranous. Los Angeles Co. and nearby ids. .... 3. *L. fragrans*
    Lvs. narrowed toward base, glabrate and green; calyx-teeth acicular, rigid. S. San Diego Co.
        4. *L. ganderi*

1. **L. calycìna** (Benth.) Epling in Munz. [*Sphacele c.* Benth.] Erect, 3–12 dm. high, pubescent to somewhat woolly; lf.-blades oblong-ovate, obtuse, veiny, punctate, 4–12 cm. long, the lower on petioles 5–20 (–40) mm. long, the upper sessile; pedicels mostly 5–15 mm. long; calyx at anthesis campanulate, 10–15 mm. long, the lobes variable, deltoid to lance-deltoid, the calyx enlarging, reticulate, membranous, 2.5–3 cm. long in fr.; corolla 2.5–3 cm. long, white or pink with purplish blotches and veins; nutlets broadly ellipsoid, ca. 3.5 mm. long.—Open slopes below 3000 ft.; Chaparral, Foothill Wd.; Coast Ranges from Ventura Co. to cent. Calif. April–June.

2. **L. cardiophýlla** Epl. Near to *L. calycina*, but with broader lvs. with cordate base; pedicels 1.5–3.5 cm. long; corolla 3–3.5 cm. long; nutlets glabrous, 3–4 mm. long.—Dry slopes, 2000–4000 ft.; Chaparral; Santa Ana Mts.

3. **L. fràgrans** (Greene) Epl. [*Sphacele f.* Greene.] Lvs. deltoid or subhastate, spreading-villous-tomentose; calyx-teeth lanceolate, villous, longer than the tube; corolla purplish, 2.5–3.5 cm. long; nutlets glabrous, 3.5 mm. long; $n=16$ (Epling, 1948).—Occasional in canyons, below 3000 ft.; Chaparral; San Gabriel and Santa Monica mts.; Santa Cruz, Santa Rosa and Catalina ids. March–May.

4. **L. gánderi** Epl. Stems hirtellous; lvs. oblong-ovate, cuneate at base on petioles to 1 cm. long; pedicels 1–3 cm. long; calyx-teeth rigid, acicular, 6–11 mm. long, the tube 8–10 mm. long; corolla lilac; nutlets glabrous, 2–3.5 mm. long; $n=16$ (Epling, 1948).—Dry slopes, 2500–3500 ft.; Chaparral; Otay Mt. and San Miguel Mt., San Diego Co. June–July.

9. **Lycòpus** L. WATER-HOREHOUND

Perennial herbs with slender rootstocks and erect or diffuse stems. Lvs. opposite, punctate, toothed to pinnatifid, usually petioled. Fls. small, white or whitish, in sessile densely capitate glomerules in axils of upper scarcely reduced lvs. Calyx campanulate, 10-ribbed, regular or nearly so, 4–5-lobed. Corolla funnelform or campanulate, nearly regular, the upper lip slightly broader, entire or notched, the lower 3-lobed. Fertile stamens 2, anterior, the posterior pair wanting or rudimentary; anther-sacs parallel. Nutlets smooth, truncate at top, narrowed below. Ca. 15 spp. of N. Temp. regions. (Greek, *lukos*, wolf, and *pous*, foot.)

Lower and main lf.-blades tapering to petioles or petiolelike bases; lvs. unevenly incised
        1. *L. americanus*
Lower and main lf.-blades sessile, evenly serrate .... 2. *L. lucidus*

1. **L. americànus** Muhl. Plant with stout nontuberous stolons; stems erect, slender, 2–9 dm. high, acutely 4-angled, glabrous or nearly so; lvs. lanceolate or oblong, acuminate, 3–10 cm. long, irregularly incised or pinnatifid, short-petioled, the upper narrow and merely sinuate; calyx-teeth with long subulate tips; corolla 2 mm. long; nutlets 1–1.5 mm. long, with entire or slightly undulate angles, the dorsal angular face soft, dark, summit entire; $2n = 22$ (Ruttle, 1932).—Wet places, below 2000 ft.; many Plant Communities; from Orange and Riverside cos. n.; to B.C. Atlantic Coast. Aug.–Sept.

2. **L. lùcidus** Turcz. ex Benth. Perennial with stolons; stems stout, 4–8 dm. high; herbage mostly puberulent; lvs. lance-oblong to lanceolate, 4–8 cm. long, acute to acuminate, subsessile, coarsely but evenly toothed; calyx-lobes lance-subulate, acuminate, hispidulous on margins; corolla scarcely longer than calyx; nutlets with inner angle granulose to base.—Swampy places at low elevs.; many Plant Communities; Inyo Mts. n. to Wash.; e. Asia. June–Oct.

10. Marrùbium L. HOREHOUND

Perennial mostly woolly herbs, caulescent, mostly with bitter sap. Lvs. rugose, toothed petioled. Fls. small, white or purple, in dense whorls. Calyx cylindric, 5–10-nerved, regularly 5–10-toothed, the teeth aristate to acute, spreading or recurved, subequal or the alternate smaller. Corolla bilabiate, the upper lip erect, entire or 2-cleft, the lower spreading, 3-cleft, the middle lobe often emarginate. Stamens 4, included, didynamous, the posterior pair shorter; anthers 2-celled. Nutlets smooth, rounded. Ca. 30 spp. of Old World. (From Hebrew word meaning bitter.)

1. **M. vulgàre** L. Branched from base, erect, white-woolly, 2–6 dm. tall; lvs. roundish, ovate, crenate, 1.5–4 cm. long, canescent above, tomentose beneath, on petioles ca. as long as blades; calyx 4–5 mm. long, the teeth subulate, recurved; corolla white, 5–6 mm. long; $2n = 34$ (Rutland, 1941).—Common pestiferous weed in waste places and old fields, the dried calyx forming a bur; natur. from Eu. Spring and summer.

11. Méntha L. MINT

Aromatic caulescent perennial herbs from rootstocks. Stems erect or diffusely branched. Lvs. opposite, punctate, toothed, usually petioled. Fls. in dense axillary clusters or in terminal spikes. Calyx campanulate to cylindric, 10-nerved, regular or slightly bilabiate, 5-toothed. Corolla funnelform or campanulate, bilabiate, the tube shorter than the calyx, upper lip entire or emarginate, lower 3-lobed. Stamens 4, equal, included or exserted; fils. glabrous; anther-sacs parallel. Nutlets ovoid, smooth. Ca. 30 spp. of N. Temp. regions. (*Mentha*, a Greek nymph supposed to have been changed into Mint.)

Fl.-whorls distant and in lf.-axils, mostly exceeded by the subtending lvs.
    Lvs. 1–2 cm. long, grayish; calyx-teeth dissimilar, the 2 lower lanceolate-subulate
        5. *M. pulegium*
    Lvs. 2–7 cm. long, light green; calyx-teeth quite alike, subequal .... 1. *M. arvensis*
Fl.-whorls crowded, usually in terminal spikes, or some of the lower more distant.
    Plants glabrous or nearly so.
        Lvs. sessile or subsessile; spikes narrow, mostly interrupted .... 7. *M. spicata*
        Lvs. petioled; spikes thick, dense.
            Calyx hirsute; lvs. lanceolate, acute .... 4. *M. piperita*
            Calyx glabrous; lvs. ovate, obtuse .... 2. *M. citrata*
    Plants villous or canescent.
        Lvs. 1–2 cm. long, pubescent, elliptic to oblong-ovate .... 5. *M. pulegium*
        Lvs. more than 2 cm. long.
            The lvs. round-ovate, 2.5–5 cm. long, woolly .... 6. *M. rotundifolia*
            The lvs. lanceolate to lance-ovate, 5–12 cm. long, pubescent to tomentose . 3. *M. longifolia*

1. **M. arvénsis** L. var. **arvénsis**. Stems usually branched, 1–8 dm. high, the angles in region of the first-flowering infl. always more pubescent than the sides; stem, petioles and lower surface of the lvs. slightly to very pubescent; lvs. lanceolate to oblong or ovate,

serrate, petioled, 2–5 cm. long, the upper ovate to elliptic, ± rounded at base; fl.-whorls all axillary; calyx ca. 3 mm. long, pubescent, the teeth triangular-subulate, ca. as long as tube; corolla lilac-pink to purplish, 5–6 mm. long; $2n = 12, 54, 60, 72, 92$ (various auth.). —Moist places below 7500 ft.; many Plant Communities; much of Calif., in several forms which may or may not be native here; to Atlantic Coast, Eurasia. July–Oct.

Var. **villòsa** (Benth.) S. R. Stewart. [*M. canadensis* L. and var. *villosa* Benth.] Lvs. in region of infl. lanceolate, with ± cuneate bases; in region of first flowering infl. the angles of stem more pubescent than sides.—More common than var. *arvensis*; to Alaska, Atlantic Coast. Variable, with a number of named forms.

2. **M. citràta** Ehrh. BERGAMOT MINT. Stems rather weak, glabrous, 3–5 dm. long; lvs. glabrous, ovate to round-ovate, obtuse, 2–4 cm. long, slender-petioled; whorls terminal and in upper axils, the mature spikes 2–2.5 cm. long; calyx-teeth subulate, glabrous, shorter than the tube; corolla rose-purple, 6–7 mm. long.—Occasional escape; natur. from Eu. July–Oct.

3. **M. longifòlia** Huds. [*M. silvestris* L.] Stoloniferous; stems erect, puberulent to tomentose, 4–12 dm. high; lvs. nearly sessile, lanceolate to lance-ovate, 5–12 cm. long, sharply serrate, pubescent to tomentose above, white-tomentose beneath; spikes thickish, ± dense, especially above; corolla purplish, 3 mm. long.—Escaped as at Santa Barbara, Ojai Valley; from Eurasia.

4. **M. piperìta** L. PEPPERMINT. Much like *M. spicata*, but with petioled lvs.; spikes 2–12 cm. long; calyx with hirsute teeth; corolla rose-purple to white; $2n = 36, 64, 66, 68, 70$ (various auth.).—Natur. occasionally in wet places at low elevs.; native of Eu. July–Oct.

5. **M. pulègium** L. PENNYROYAL. Stems erect and simple to ± decumbent, slender, 2–5 dm. long, white-pubescent; lvs. elliptic to oblong-ovate, 1–2 cm. long, canescent; fl.-whorls often many, rather remote, with reduced subtending lvs.; calyx 3–4 mm. long, short-hirsute on nerves and teeth; corolla lavender, ca. twice as long as calyx, the lobes villous; $2n = 10, 20, 30, 40$ (Morton, 1956).—Low moist places below 3000 ft.; many Plant Communities; San Diego Co. and in cent. and n. Calif.; natur. from Eu. June–Sept.

6. **M. rotundifòlia** (L.) Huds. [*M. spicata* var. *r.* L.] Herbage ± tomentose, and viscid; stems 5–8 dm. tall, simple or branched, mostly erect; lvs. round-ovate to broadly elliptical, cordate, subsessile, round-tipped, rugose-reticulated, 2.5–5 cm. long; spikes slender, 5–6 cm. long; calyx campanulate, puberulent, almost 2 mm. long, the subulate teeth ca. as long as the tube; corolla white, puberulent, ca. 4 mm. long; $2n = 18, 24, 54$ (several auth.).—Sparingly natur. in moist places; from Eu. July–Oct.

7. **M. spicàta** L. SPEARMINT. Nearly or quite glabrous, 3–12 dm. high, often purplish; lvs. oblong- or ovate-lanceolate, ± rounded at base, pubescent to villous, serrate, 2–6 cm. long, subsessile or nearly so; spikes slender, to 4 or 6 cm. long in fr., the bracts lance-subulate, ciliate, 4–8 mm. long; calyx with glabrous base, ca. 1.5 mm. long, the subulate teeth ciliate; corolla pale lavender, glabrous, 2.5–3 mm. long, $2n = 36, 48$ (Löve & Löve, 1942).—Moist fields and marshes below 5000 ft.; many Plant Communities; natur. from Eu. July–Oct.

12. *Monárda* L. HORSEMINT

Aromatic herbs with erect stems and dentate or serrate petioled lvs. Fls. in dense capitate bracteate axillary or terminal clusters. Calyx tubular, 15-ribbed, nearly regularly 5-lobed. Corolla 2-lipped, the throat dilated, upper lip narrow, erect or arched, entire or notched; lower lip spreading, 3-lobed, the middle lobe much the larger. Anther-bearing stamens 2, ascending close under the upper lip, the upper pair reduced or wanting; anthers narrow, 2-celled, versatile, with divergent sacs. Styles 2-cleft at apex. Nutlets ovoid, smooth. Ca. 12 spp. of N. Am. (Dedicated to N. *Monardes*, a 16th-century Spanish botanist.)

1. **M. pectinàta** Nutt. Stout annual, 2–4 dm. high, retrorsely puberulent; lvs. narrow-lanceolate to oblanceolate, distinctly serrulate, ciliate, punctate, 2–4 cm. long; bracts pale, lanceolate, 10–14 mm. long, mucronate-aristate; calyx puberulent; stamens scarcely

equaling the upper lip.—Dry slopes, 4000-5000 ft.; Pinyon-Juniper Wd.; New York Mts., e. San Bernardino Co.; to Nebr. and Tex. July–Sept.

13. Monardélla Benth.

Annual or perennial herbs, some ± woody at base, of pleasant odor. Lvs. rather small, entire or serrate. Fls. borne in terminal heads subtended by broad involucral bracts, which are frequently colored. Calyx tubular, narrow, ca. equally 5-toothed, 10–15-nerved, the teeth mostly erect, ± triangular. Corolla rose to purplish, lavender or white, the upper lip erect, 2-lobed, the lower 3-lobed, horizontal or declined; lobes linear-oblong. Stamens 4, all fertile, anther-cells divergent. Style unequally 2-cleft at apex. Nutlets broadly oblong, smooth. Ca. 20 spp., w. N. Am. (Diminutive of *Monarda*, which it resembles.)

(Epling, C. Monograph of the genus Monardella. Ann. Mo. Bot. Gard. 12: 1–106. 1925.)
1. Fls. in rather loose heads; limb of the corolla $\frac{1}{3}$–$\frac{1}{2}$ the length of the tube; calyx 10–25 mm. long.
   2. Corolla 35–45 mm. long, scarlet or yellowish; calyx 20–25 mm. long ..... 7. *M. macrantha*
   2. Corolla 25–30 mm. long, pale to pinkish; calyx 12–15 mm. long .... 8. *M. nana*
1. Fls. in dense heads; limb of corolla $\frac{1}{2}$–$\frac{2}{3}$ the length of the tube; calyx 5–10 mm. long.
   3. Plants perennial.
      4. The lvs. with well developed tomentum on under surface .... 4. *M. hypoleuca*
      4. The lvs. lacking a feltlike tomentum beneath.
         5. Lvs. 0.5–1.2 cm. long, round-ovate to lance-ovate, ± toothed; plants low cespitose dwarfs, mostly less than 1 dm. high. San Gabriel Mts. .... 2. *M. cinerea*
         5. Lvs. 1–several cm. long, mostly narrower, entire; plants taller.
            6. Stems subglabrous to pubescent, but not silvery-puberulent.
               7. Bracts firm, not membranaceous or chaffy; corolla-lobes blunt ... 11. *M. viridis*
               7. Bracts membranaceous; corolla-lobes rounded to a point.
                  8. Stems and lvs. glabrous or puberulent. Widely distributed .. 9. *M. odoratissima*
                  8. Stems and lvs. villous with spreading hairs. Little San Bernardino Mts.
                      12. *M. robisonii*
            6. Stems silvery with a dense minute puberulence. Cismontane to desert . 6. *M. linoides*
   3. Plants annual.
      9. Bracts not white even on tips or margins; calyx-teeth not white-tipped; corolla mostly colored.
         10. The bracts puberulent.
            11. Bracts acute, with green secondary veins forming a prominent network. From San Diego Co. n. .... 5. *M. lanceolata*
            11. Bracts abruptly acuminate, with white tips; secondary veins lacking or not conspicuous; calyx 15-nerved. Los Angeles Co. n. .... 1. *M. breweri*
         10. The bracts villous throughout. Near Colton, San Bernardino Co. .... 10. *M. pringlei*
      9. Bracts white or with white margin; calyx-teeth white or with white tips; corolla mostly white. W. Mojave Desert .... 3. *M. exilis*

1. **M. bréweri** Gray. Erect annual, 1.5–3 dm. tall, branched, puberulent above, usually purplish; lvs. lanceolate, 1.5–4.5 cm. long, puberulent, short-petioled; heads 2–3 cm. in diam.; bracts broadly ovate, 1–1.5 cm. long, abruptly acuminate, with 5–9 subparallel hispidulous veins, purplish above; calyx 6–8 mm. long, ± hirsute, the teeth lanceolate-deltoid; corolla rose, 12–14 mm. long; stamens well exserted.—Sandy flats below 4500 ft.; largely Foothill Wd.; Inner Coast Ranges from n. Los Angeles Co. to Alameda Co. May–Aug.

2. **M. cinèrea** Abrams. Dwarf perennial, woody at base, to 1 dm. high, cinereous; lvs. narrowly to broadly ovate, 5–12 mm. long, denticulate, soft subvillous, sessile; heads 1.5–2 cm. broad, purplish, the bracts broadly ovate, ca. 8 mm. long; calyx 7 mm. long, the teeth subulate, 2 mm. long; corolla rose-purple, 13–14 mm. long, the tube not exserted, the lobes 4 mm. long; stamens exserted.—Dry slopes 6000–10,000 ft.; Montane Coniferous F.; e. half of San Gabriel Mts. July–Aug.

3. **M. éxilis** (Gray) Greene. [*M. candicans* var. *e.* Gray.] Annual, erect, mostly simple below, purplish, retrorsely puberulent, 2–4 dm. high; lvs. oblong-lanceolate to lance-

ovate, puberulent to subglabrous especially above, short-petioled; heads on obvious peduncles, 1.5–2.5 cm. broad; bracts often purplish, abruptly acuminate, with few or no secondary veins; corolla limb 2.5–3 mm. long, white.—At 2000–6200 ft.; w. Mojave Desert; Joshua Tree Wd., Pinyon-Juniper Wd. May–June.

4. **M. hypoleùca** Gray ssp. **hypoleùca.** Suffrutescent perennial from creeping rootstocks, 2–5 dm. tall, simple or branched, pubescent; lvs. narrowly to broadly rhomboid-lanceolate, 2–4 cm. long, entire, obtuse, green and mostly glabrous above, white beneath with feltlike tomentum, slightly revolute, on petioles 3–10 mm. long; heads 3–4 cm. broad; bracts ovate, tomentose, 8–12 mm. long; calyx 6–7 mm. long, with triangular-ovate teeth; corolla white to pale lavender, 15–16 mm. long, the limb 5–6 mm. long; stamens well exserted.—Occasional on dry slopes, below 4500 ft.; Chaparral; mts. from Santa Barbara Co. to Orange Co. July–Sept.

Ssp. **lanàta** (Abrams) Munz. [*M. l.* Abrams.] Branchlets villous to lanate; lvs. short-pubescent to lanate above, white-tomentose beneath.—Chaparral; e. San Diego Co. June–July.

5. **M. lanceolàta** Gray. Plate 55, Fig. E. Erect annual 2–5 dm. tall, simple or branched, glabrous below, puberulent above; lf.-blades lanceolate to lance-oblong, entire, obtuse, 1.5–5 cm. long, on petioles 5–15 mm. long; heads 1.5–3 cm. broad; bracts lance-ovate, acute, scabrous, membranous but green, often purplish toward tips, 5–15 mm. long, pinnately veined; calyx 6–8 mm. long, pubescent, the teeth ciliate; corolla rose-purple or paler, 12–15 mm. long, the tube puberulent; stamens well exserted; nutlets brownish, ± mottled.—Locally common in dry places, especially disturbed areas, below 8000 ft.; many Plant Communities; San Diego Co. to n. Calif. Variable: a form from the San Gabriel Mts. with stalked glands in upper parts is var. *glandulifera* Jtn.; another with heads 1–2 cm. broad, from the Laguna Mts. to n. L. Calif. is var. *microcephala* Gray.

6. **M. linoìdes** Gray ssp. **linoìdes.** Stems several from a woody base, rather dense, erect, 4–6 dm. high, silvery with rather fine retrorse appressed pubescence; lvs. linear to linear-oblong, entire, thick, minutely silvery-pubescent, 1.5–2 cm. long, subsessile or short-petioled; heads mostly 2–3 cm. broad; bracts broadly ovate, membranous, whitish, puberulent, soft-ciliate, 8–15 mm. long; calyx ca. 8 mm. long, short-pubescent, the teeth with longer hairs; corolla pale rose-lavender, 12–15 mm. long; stamens exserted.—Dry slopes, 3000–9500 ft.; Pinyon-Juniper Wd. to Yellow Pine F.; desert slopes from Laguna Mts. to mts. of Mojave Desert and Mono Co. June–Aug.

Lvs. silvery-gray.
    The lvs. linear to linear-lanceolate, 1.5–2.5 cm. long. Desert slopes .... ssp. *linoides*
    The lvs. lance-oblong, 1–1.5 cm. long. Ventura and Kern cos. .... ssp. *oblonga*
Lvs. greenish.
    The lvs. 2–4 cm. long. W. San Diego Co. .... ssp. *viminea*
    The lvs. 1–1.6 cm. long. San Gabriel and San Bernardino mts. .... ssp. *stricta*

Ssp. **oblónga** (Greene) Abrams. [*M. o.* Greene.] Bracts ovate, pale purple, 1.5–2 cm. long.—Dry slopes 3000–7000 ft.; Yellow Pine F.; Foothill Wd.; mts. of Ventura and Kern cos.

Ssp. **strícta** (Parish) Epling. [*M. l.* var. *s.* Parish. *M. epilobioides* Greene.] Plate 55, Fig. F. Lvs. mainly linear-lanceolate, greenish; bracts lanceolate, purplish, acuminate, ca. 1 cm. long.—Dry banks and slopes, 5000–9000 ft.; Yellow Pine F.; San Gabriel and San Bernardino mts.

Ssp. **viminèa** (Greene) Abrams. [*M. v.* Greene.] Plants scarcely silvery; stems puberulent; lvs. narrowly linear-lanceolate; bracts lanceolate, greenish-white with tips often rose, to 14 mm. long.—Rocky washes below 1000 ft.; Coastal Sage Scrub; Chaparral; sw. San Diego Co.

7. **M. macrántha** Gray ssp. **macrántha.** Perennial from slender woody rootstocks; stems branched or simple, 1–3 (–5) dm. long, pubescent with ± recurved hairs; lf.-blades ovate to elliptic, obtuse, 1–3 cm. long, on petioles half as long, entire or nearly so, subglabrous above, pubescent beneath; heads 10–20-fld., 3–4 cm. broad, the bracts purplish, ciliate-villous, 1–1.5 cm. long; calyx 2–2.5 cm. long, glandular-pubescent, the teeth subu-

late, 4 mm. long; corolla scarlet to yellowish, 3–4.5 cm. long, slender-funnelform, sparsely pubescent, the limb 5–6 mm. long; stamens well exserted, anthers 1.3–1.5 mm. long; nutlets ca. 3 mm. long.—Dry slopes and ridges, 2500–6000 ft.; Chaparral, Yellow Pine F.; mts. of San Diego Co. and n. L. Calif., to San Gabriel Mts.; Santa Lucia Mts. June–Aug.

Ssp. **hállii** Abrams. Herbage densely villous-pubescent throughout; corolla frequently yellowish, the limb 6–10 mm. long.—San Gabriel and San Bernardino mts. to Cuyamaca and Santa Ana mts.

8. **M. nàna** Gray ssp. **nàna.** Stems from creeping woody rootstocks, simple or few-branched, 1–2 (–3) dm. long, pubescent; lvs. mostly 0.5–1.5 cm. long, ovate, almost glabrous above, ± cinereous beneath, with shorter petioles; glomerules 2–3.5 cm. broad, the bracts narrow-ovate, 1.5–2 cm. long, membranaceous, pale with purple tinge; calyx 12–15 mm. long, teeth ca. 3 mm. long; corolla pinkish or paler, 2–3 cm. long, pubescent, the lobes 5–6 mm. long; stamens ca. as long; nutlets ca. 2 mm. long.—Dry slopes, 4000–6000 ft.; Chaparral, Yellow Pine F.; Laguna and Cuyamaca mts.; n. L. Calif. June–July.

Stems with a short ± retrorse pubescence.
    Lf.-blades subglabrous above. Laguna and Cuyamaca mts. .... ssp. *nana*
    Lf.-blades cinereous above.
        Corolla-lobes acute, narrower at base than toward the middle, ca. 6–8 mm. long. Yellow Pine F., San Jacinto and Santa Rosa mts. .... ssp. *tenuiflora*
        Corolla-lobes often obtuse, oblong, 4–6 mm. long. Desert slopes, San Jacinto and Santa Rosa mts. .... ssp. *arida*
Stems villous with spreading hairs. Palomar Mts. .... ssp. *leptosiphon*

Ssp. **árida** (Hall) Abrams. [*M. macrantha* var. *a.* Hall.] Lf.-blades 5–10 mm. long, the petioles ca. ¾ as long; corolla-tube mostly 15–18 mm. long.—Canyons at 4000–5000 ft.; Pinyon-Juniper Wd.

Ssp. **leptosìphon** (Torr.) Abrams. [*M. villosa* var. *l.* Torr.] Stems and lvs. villous with spreading hairs; corolla-tube 20–25 mm. long, the limb 10 mm. long.—Chaparral and Yellow Pine F.; Palomar Mts.

Ssp. **tenuiflòra** (Wats.) Abrams. [*M. t.* Wats. ex Gray.] Lf.-blades ca. 1 cm. long, the petioles half as long; corolla-tube ca. 20 mm. long.—At 4500–8000 ft.; Yellow Pine F.

9. **M. odoratíssima** Benth. ssp. **austràlis** (Abrams) Epling. [*M. a.* Abrams.] Perennial from a branching woody base, the stems decumbent or ascending, subvillous; lvs. lanceolate to oblong, green or cinereous, 1–2.5 cm. long, on petioles 1–3 mm. long; bracts lanceolate, longer than the calyces; calyx ca. 8 mm. long; corolla rose-purple, 1.5 cm. long.—Dry slopes 4500–9500 ft.; Montane Coniferous F.; San Gabriel Mts. to San Jacinto Mts. June–Aug.

10. **M. prínglei** Gray. Cinereous-puberulent annual, branched, 2–4 dm. high; lvs. oblong to lance-ovate, 1.5–3.5 cm. long, ± finely pubescent, short-petioled, entire; heads 2–2.5 cm. broad; bracts ovate, 8–10 mm. long abruptly acuminate, with 5–7 subparallel finely villous veins; calyx 6–7 mm. long, pubescent with slender hirsute teeth; corolla rose, 11–13 mm. long; stamens exserted.—Sandy places; Coastal Sage Scrub; near Colton, San Bernardino Co. May–June.

11. **M. víridis** Jeps. ssp. **saxícola** (Jtn.) Ewan. [*M. s.* Jtn.] Perennial from creeping rootstocks, suffrutescent; stems branched below, 1–3 dm. long, puberulent; lvs. rhombic-lanceolate, 1.5–3 cm. long, glabrous above or nearly so; heads 2.5–3 cm. broad; bracts membranous to subfoliar; calyx 8–10 mm. long; corolla 14–16 mm. long, rose-purple, the tube scarcely exserted; stamens well exserted.—Dry rocky places, 1700–6000 ft.; Chaparral, Yellow Pine F.; San Gabriel Mts. June–Sept.

12. **M. robisònii** Epling in Munz. Woody at base, 3–5 dm. high, spreading-villous; lvs. 6–15 mm. long; heads 1–2 cm. wide; bracts pallid; calyx 7–8 mm. long; corolla pale.—About rocks, Joshua Tree Nat. Monument. June.

14. *Népeta* L. CATNIP

Perennial herbs with toothed lvs. Fls. usually white or blue, in crowded whorls. Calyx tubular, often incurved, 15-nerved, 5-toothed, obscurely 2-lipped, the upper teeth usually longer. Corolla dilated in throat, the upper lip erect, notched or 2-cleft, the lower 3-cleft with middle lobe the largest. Stamens 4, didynamous. Nutlets ovoid, compressed, smooth. Ca. 150 spp., of Eurasia. (Latin name of Catnip.)

1. **N. catària** L. Erect, 3–8 dm. high, with ascending branches; herbage densely canescent; lvs. ovate to oblong, acute, usually cordate at base, coarsely crenate-serrate, the blades 3–9 cm. long, the petioles 1–4 cm. long; calyx urceolate, very pubescent, ca. 6 mm. long; the teeth subulate; corolla whitish, dotted with purple, 10–12 mm. long; $2n=36$ (Sugiura, 1940).—Occasional as natur. weed; native of Eu. July–Sept.

15. Pogógyne Benth.

Small aromatic annuals with suborbicular to spatulate lvs. Fls. congested in cymules in axils of bracts, forming dense terminal spikes or the lower whorls distant. Bracts and calyces glabrate to hirsute-ciliate. Calyx deeply 5-cleft, 15-nerved, the 2 lower teeth longer. Corolla blue or purple, tube exserted, the upper lip entire, lower spreading, 3-lobed. Stamens 4, all antheriferous or the upper 2 sterile, ascending under the upper lip; fils. pubescent. Style ± hairy below the branches. Nutlets ± obovoid, concolorous or mottled. Ca. 5 spp. (Greek, *pogon*, beard, and *gune*, female, because of the bearded style.)

(Howell, J. T. The genus Pogogyne. Proc. Calif. Acad. Sci. (IV) 20: 105–128. 1931.)
Fl.-bracts and calyx-lobes conspicuously hirsute, bristly-ciliate .... 1. *P. abramsii*
Fl.-bracts and calyx-lobes subglabrous .... 2. *P. nudiuscula*

1. **P. àbramsii** J. T. Howell. Like *P. nudiuscula*, but fl.-bracts and calyx-lobes conspicuously hirsute and bristly-ciliate; calyx 6 mm. long; corolla 10–12 mm. long, rose-purple; nutlets 1–1.5 mm. long.—Beds of dried pools; Chaparral, Coastal Sage Scrub; mesas from San Diego to Miramar. April–June.

2. **P. nudiúscula** Gray. Erect and simple, or more commonly with ascending branches from the base, 1–3 dm. tall, minutely strigose; lvs. spatulate or narrower, obtuse, subglabrous, 1–2 cm. long, short-petioled; fls. whorls distant or crowded into a short subcapitate spike; bracts linear-oblanceolate, 10–14 mm. long; calyx ca. 8 mm. long, the lobes linear-subulate, the lower 3–5 mm. long; corolla 11–14 mm. long, lavender, sparsely pubescent; upper stamens pubescent, 2–3 mm. long; lower glabrous, 5–6 mm. long; style pubescent, 1–4 mm. below the stigma-lobes; nutlets 1.5 mm. long.—Moist flats; Chaparral, Coastal Sage Scrub; mesas from San Diego and Loma Alta, San Diego Co. May–June.

16. Poliomíntha Gray

Hoary-canescent shrub with linear or linear-oblong lvs. Fls. in small axillary clusters towards the end of the branches. Calyx 15-veined, the tube shaggy-pilose, the teeth subequal, ± connivent, throat strongly annulate. Corolla pale purplish-blue, ± annulate in the tube with coarse ascending hairs. Fertile stamens 2, ascending against the upper lip. Nutlets oblong, smooth. One sp. (Greek, *polios*, hoary, and *mentha*, mint.)

1. **P. incàna** (Torr.) Gray. [*Hedeoma i.* Torr.] Six to 10 dm. high; lvs. 1–2 cm. long, nearly sessile; calyx 6–7 mm. long; corolla ca. 12 mm. long.—Collected at 5400 ft., in a wet place above Cushenbury Springs, n. base of San Bernardino Mts.; Ariz. to Utah, Tex. June–July.

17. Prunélla L. SELFHEAL

Low perennial herbs with nearly simple stems from slender rootstocks. Lvs. petioled. Fls. rather small, in 3's in axils of round bractlike membranaceous lvs. that are imbricated in a dense spike or head. Calyx tubular-campanulate, usually 10-nerved, deeply 2-lipped, the upper lip truncate or with 3 short teeth, the lower cleft into 2 lanceolate teeth. Corolla slightly contracted at throat, 2-lipped, the upper lip erect, arched, entire, the lower reflexed-spreading, 3-cleft with lateral oblong lobes and rounded middle denticulate lobe. Stamens 4, didynamous. Nutlets smooth, obovoid. Ca. 5 spp., world-wide. (Origin of name doubtful, often written *Brunella*.)

1. **P. vulgàris** L. ssp. **vulgàris.** Stems mostly tufted or loosely ascending, 1–5 dm. high, simple or branched; lvs. in basal tufts and cauline, the main ones ovate to ovate-oblong, rounded at base, mostly obtuse, the blades 1–3 cm. long, on somewhat shorter petioles; calyx purplish, 4–5 mm. long; corolla bluish, violet or lavender, ca. 8–10 mm. long; $n=16$ (Hruby, 1932), $2n=28$ (Kurosawa, 1966).—Weed in lawns, etc.; native of Eu. Summer.

Ssp. **lanceolàta** (Barton) Hult. [*P. pennsylvanica* var. *l.* Barton.] Stems and lvs. commonly pilose; lvs. lance-ovate to -oblong, 2.5–5 cm. long, mostly acutish, narrowed to short petioles; bracts ± ciliate, often tinged purple; calyx 5–10 mm. long, often with ciliate teeth; corolla 10–20 mm. long, mostly violet; nutlets ca. 2 mm. long; $n=14$ (Nelson, 1966).—Moist woods, about ditches, etc. below 7500 ft.; many Plant Communities, but mostly Montane Coniferous F.; most of cismontane and montane Calif.; to Alaska, Atlantic States. May–Sept.

18. *Pycnánthemum* Michx. MOUNTAIN-MINT

Aromatic perennial herbs, mostly with branching erect stems. Whorls remote, many-fld., leafy-bracted. Calyx ovoid to cylindric, with equal teeth and 10–13-nerved. Corolla bilabiate, short, white or purple-dotted, the upper lip entire to emarginate, the lower 3-cleft with obtuse lobes. Stamens 4, didynamous; anther-sacs parallel. Nutlets smooth, pubescent or roughened. Ca. 17 spp. of N. Am., mostly eastern. (Greek, *pychnos*, dense, and *anthemon*, fl.)

1. **P. califórnicum** Torr. Stems from creeping rootstocks, canescent, simple or few-branched, 6–10 dm. tall; lvs. ovate to lance-ovate, finely pubescent, the upper canescent, serrulate, 3–8 cm. long, sessile or subsessile; heads 1–4, compact; calyx 4–5 mm. long, pubescent, the teeth woolly-villous; corolla 6–7 mm. long, white, the throat hairy, lobes ca. 2 mm. long; nutlets smooth, ca. 1.3 mm. long; $2n=40$ (Chambers, 1961).—Moist places in canyons, etc., below 5500 ft.; Chaparral, Yellow Pine F., Foothill Wd.; San Diego Co. to n. Calif. June–Sept.

19. *Salazària* Torr. BLADDER-SAGE

Low dense and divaricately branched shrub with spinescent branchlets. Lvs. small, entire or rarely toothed, short-petioled. Fls. in loose spicate racemes. Calyx equally 2-lobed, the lips entire, becoming inflated and globular in fr. Corolla purple, bilabiate, the upper lip arched, the lower with recurved sides and small lateral lobes. Anthers ciliate, those of upper pair of stamens 2-celled, of lower 1-celled. Nutlets tubercled. Monotypic. (Don Jose *Salazar*, Mexican commissioner on the Boundary Survey.)

1. **S. mexicàna** Torr. Plate 56, Fig. A. Six to 10 dm. tall, intricately branched, with canescent twigs; lvs. green, subglabrous or minutely hispidulous, oblong or broadly lance-olate, 1–1.5 cm. long, short-petioled; racemes 5–10 cm. long; calyx 8 mm. in fl., 2 cm. and papery in fr.; corolla 15–18 mm. long, pubescent without, the throat pale, lips darker; nutlets olive-brown, peltate on a raised gynobase, tessellate-tubercled.—Common in dry washes and canyons, below 5000 ft.; Creosote Bush Scrub, Joshua Tree Wd.; deserts from Inyo Co. to Riverside Co.; to Utah, Tex., n. Mex. March–June.

20. *Sálvia* L. SAGE

Herbs or shrubs, usually strongly aromatic. Lvs. opposite, sometimes mostly basal. Fls. usually in whorls, these in ± interrupted spikes or panicles or racemes. Calyx tubular to campanulate or ovoid, 2-lipped, the upper lip usually concave or arched, entire to 2-lobed, the lower 2-toothed or -cleft, sometimes teeth and lips suppressed. Corolla strongly 2-lipped, the upper usually erect, concave or arched, entire to 2-lobed, the lower lip spreading or drooping, 3-lobed. Stamens inserted in throat, the upper sterile or rudimentary, the lower pair fertile with anther-cells widely separate on a long filamentlike connec-

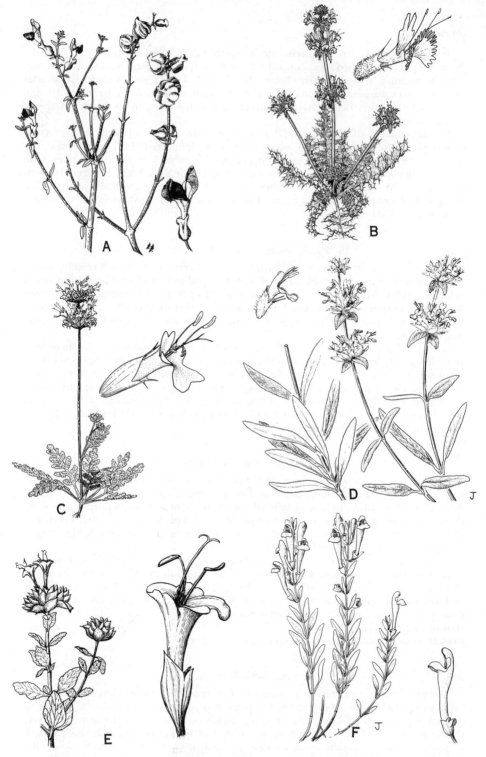

Plate 56. Fig. A, *Salazaria mexicana*. Fig. B, *Salvia carduacea*. Fig. C, *Salvia columbariae*. Fig. D, *Salvia mellifera*. Fig. E, *Salvia mohavensis*. Fig. F, *Scutellaria austinae*.

Salvia

tive which exceeds the fil. itself and is jointed to it, sometimes the lower arm rudimentary or suppressed and the anther deformed or obsolete. Style 2-cleft at summit; ovary deeply 4-parted. Nutlets smooth. A genus of over 500 spp., widely distributed in temp. and warmer regions; some spp. cult. as ornamentals or for flavoring. (Latin, *salveo*, the verb to *save*, because of medicinal use.)

(Epling, C. The Californian Salvias. Ann. Mo. Bot. Gard. 25: 95–188. 1938.)
1. Both branches of the stamen-connective evident and bearing fertile anthers; lvs. pinnatifid or spinose-sinuate or -lobed (sometimes entire in *S. funerea*).
   2. Plants herbs with lvs. largely basal.
      3. Corolla lavender, 25–35 mm. long, the lower lip fimbriate; lvs. spiny, tomentose
                                                                                                         4. *S. carduacea*
      3. Corolla blue, 12–16 mm. long, the lower lip not fimbriate; lvs. not spiny or tomentose
                                                                                                         6. *S. columbariae*
   2. Plants shrubby with lvs. all along the current year's growth.
      4. Calyx subglobose with subequal teeth; lvs. tomentose with long terminal spine and often lateral ones. Death Valley region .... 9. *S. funerea*
      4. Calyx elongate with unequal teeth; lvs. with greenish spinulose teeth. Colo. Desert
                                                                                                               10. *S. greatai*
1. Lower end of stamen-connective obsolete or forming only a tooth, rarely bearing an anther.
   5. Plants perennial herbs.
      6. Fls. 30 mm. long, purplish-red; lvs. 10–20 cm. long .... 19. *S. spathacea*
      6. Fls. 15–18 mm. long, bluish; lvs. 2–5 cm. long .... 18. *S. sonomensis*
      6. Fls. 25 mm. long, mostly bright blue; rare introduction .... 17. *S. pratensis*
   5. Plants shrubby.
      7. Infl. with dense glomerules in a ± interrupted spike, the spikes occasionally branched; corolla-tube usually longer than lower lip; upper lip well developed.
         8. Lf.-blades entire, mostly broader in upper half, not rugose; fl.-bracts membranous, usually colored.
            9. Corolla 13–15 mm. long, the tube pubescent within, but the hair not forming a definite band; upper lip of corolla at least half as long as tube .... 7. *S. dorrii*
            9. Corolla 15–22 mm. long, the tube with a definite band of inner hairs; upper lip of corolla $\frac{1}{4}$–$\frac{1}{3}$ as long as tube .... 16. *S. pachyphylla*
         8. Lf.-blades crenulate to crenate, broader in their lower half, rugose; bracts often not colored.
            10. Stamens not or scarcely exceeding the upper lip of the corolla, lying close under it; corolla 8–12 mm. long.
               11. Lvs. not revolute; corolla-tube with narrow band of hair within.
                  12. Lf.-blades oblong-elliptic, mostly 3–6 cm. long, ashy beneath; corolla mostly pale blue to whitish. Widely distributed .... 12. *S. mellifera*
                  12. Lf.-blades oblong-obovate, mostly 1–2 cm. long, ashy on both surfaces; corolla clear blue. S. San Diego Co. .... 14. *S. munzii*
               11. Lvs. revolute; corolla-tube with inner hair ± scattered. Santa Rosa Id.
                                                                                  3. *S. brandegei*
            10. Stamens well exserted from upper corolla-lip and not lying close under it; corolla 15–25 mm. long.
               13. Calyx-teeth not tipped with long spines.
                  14. Pubescence of simple hairs; calyx-teeth of lower lip evident.
                     15. Lvs. green and rugose on upper surface, not ashy beneath.
                        16. Fl.-bracts purplish-green, spinulose-tipped; corolla 20–25 mm. long. Colo. Desert .... 8. *S. eremostachya*
                        16. Fl.-bracts pale, mucronulate; corolla 15–16 mm. long. Mojave Desert
                                                                       13. *S. mohavensis*
                     15. Lvs. rugose, but tomentose beneath; corolla ca. 2 cm. long. San Diego Co.
                                                                                5. *S. clevelandii*
                14. Pubescence of short much-branched hairs; calyx-teeth and lips completely united. Orange Co. n. .... 11. *S. leucophylla*
               13. Calyx-teeth tipped with long spines; lvs. whitish; corolla white, 2 cm. long. Colo. Desert .... 20. *S. vaseyi*
      7. Infl. thyrsoid-paniculate with lax few-fld. glomerules; corolla-tube shorter than lower lip; upper lip very short. Cismontane and to desert edge .... 1. *S. apiana*

1. **S. apiàna** Jeps. WHITE SAGE. Shrubby below, 1–2 (–3) m. tall, white with minute appressed hairs, the growth of the current year of long erect branches with lvs. crowded at base and flowering above; lf.-blades lance-oblong, 3–9 cm. long, obtuse, crenulate, on petioles 0.5–2 cm. long; fls. few, in lax glomerules in open thyrsoid panicles 5–15 dm. long; bracts ovate-lanceolate, 5–8 mm. long; calyx 5–7 mm. long, upper lip entire or retuse, 1.5–2 mm. long, the lower 8–18 mm. long, abruptly bent at base, closing the orifice and with rounded erose, cupped middle lobe; stamens exserted; nutlets 2.5–3 mm. long; $n=16$ (Stewart, 1939).—Common on dry benches and slopes, mostly below 5000 ft.; Coastal Sage Scrub, Chaparral, Yellow Pine F.; cismontane from n. L. Calif. to Santa Barbara Co. April–July. A form along the desert edge with condensed spicate panicle is var. *compacta* Munz. *S. apiana* hybridizes with most other spp. with which it comes in contact, especially *S. mellifera*.

2. **S. × bernardína** Parish ex Greene. [*S. columbariae* var. *b*. Jeps.] Annual to somewhat woody; lvs. with texture of *S. mellifera*, but pinnately lobed, strongly bicolored; upper lip of calyx with 3 spinose teeth.—Apparent hybrid between *S. columbariae* and *S. mellifera* and found only where both spp. occur.

3. **S. brandegèi** Munz. [*Audibertia stachyoides* var. *revoluta* Bdg.] Near to *S. mellifera*; the twigs with branching hairs; lf.-blades linear to linear-oblong, 2–6 cm. long, 2–5 mm. wide, obtuse, subsessile, revolute, green and rugose above, white-tomentose beneath; glomerules few-fld., 1.5–2 cm. in diam. with ovate bracts with branching hairs; calyx 7–8 mm. long; corolla lavender with broader throat, the tube 7–8 mm. long, the lips 3–4 mm. long; stamens included; $n=15$ (Epling, Lewis & Raven, 1962).—Dry places; Coastal Sage Scrub; Santa Rosa Id. Reported from Santo Tomas n. L. Calif. April–May.

4. **S. carduàcea** Benth. THISTLE SAGE. Plate 56, Fig. B. Annual, simple or few-branched at base, the stems erect, scapelike from a rosette of basal lvs., 1–5 dm. high, ± white-woolly, bearing 1–4 capitate whorls of fls.; lvs. oblong in outline, sinuate-pinnatifid, spinulose-toothed, 3–10 (–15) cm. long, on shorter winged petioles; fls. 1–4 in a whorl, equaled or surpassed by the foliaceous lanceolate spinescent bracts; calyx woolly, 10–15 mm. long, the lobes spine-tipped; corolla lavender, 2–2.5 cm. long, the upper lip 2-cleft, with laciniate or denticulate segms., the lower with small, erose, lateral lobes and a large fan-shaped middle fimbriate lobe; fils. short, the connectives 12–14 mm. long; anthers vermilion or brick-red; nutlets tan-gray, ± mottled, dorsally flattened, 2.5–3.5 mm. long; $n=12$ (Stewart, 1939).—Sandy and gravelly places below 4500 ft.; Creosote Bush Scrub, Coastal Sage Scrub, V. Grassland, etc.; from interior Contra Costa Co. through the Cent. V. and inner Coast Ranges to Kern Co., then through interior cismontane s. Calif. and w. Mojave Desert; to n. L. Calif. March–June.

5. **S. clevelándii** (Gray) Greene. [*Audibertia c.* Gray.] A fragrant rounded shrub to 1 m. tall, ashy with retrorse hairs; lf.-blades elliptic-oblong, 1–3 cm. long, obtuse, crenulate, rugose and grayish-tomentulose, on petioles 3–6 mm. long; glomerules 1–3, separated, many-fld., with ovate bracts 7–8 mm. long; calyx 8–10 mm. long, only the lower teeth free; corolla dark blue-violet, ca. 2 cm. long, the tube narrow, well exserted, the middle lobe of the lower lip oblong, 3–4 mm. long, plane, retuse; stamens exserted; nutlets 1.5–2 mm. long, light yellow with small brown dots; $2n=30$ (Epling, Lewis & Raven, 1962).—Dry slopes below 3000 ft.; Chaparral, Coastal Sage Scrub; cismontane San Diego Co.; n. L. Calif. With characteristic wafted odor. A good substitute for *S. officinalis* in cookery.

6. **S. columbàriae** Benth. var. **columbàriae**. CHIA. Plate 56, Fig. C. Annual, simple or branching below, 1–5 dm. tall, ± cinereous with mostly retrorse hairs; lvs. mostly basal, finely pubescent, oblong-ovate in outline, the blades 2–10 cm. long, 1–2-pinnatifid into toothed or incised divisions, the petioles ca. as long; upper lvs. reduced; fls. in capitate glomerules, these 1–3, subtended by rounded, glabrous or hispidulous, colored awn-tipped bracts; calyx 8–10 mm. long, purplish, arcuate, the middle spinose tooth of the upper lip suppressed; corolla blue, 12–16 mm. long, upper lip small, emarginate, lower with small lateral lobes and larger middle lobe ± 2-lobed; nutlets tan-gray, ± mottled, and dorsally flattened, ca. 2 mm. long; $n=13$ (Epling, Lewis & Raven, 1962).—Common in dry

open disturbed places, below 4000 ft., occasional to 7000 ft.; Coastal Sage Scrub, Chaparral, Creosote Bush Scrub, etc.; inner Coast Ranges, Mendocino Co. s., throughout s. Calif.; to Utah, Ariz., Son., L. Calif. March–June.

7. **S. dórrii** (Kell.) Abrams. ssp. **dórrii**. [*Audibertia d.* Kell. *S. carnosa* var. *pilosa* Jeps.] Low broad much-branched shrub 3–8 dm. tall; branches densely scurfy-canescent, punctate-glandular, with short axillary fascicles; lvs. round-obovate to spatulate, ca. 7–15 mm. in diam., rounded to emarginate at apex, ± abruptly narrowed at base to a petiole 5–8 mm. long; infl. hispidulous, and ± pilose between the 3–4 glomerules which appear ± contiguous in age, 1.5–2.5 cm. in diam.; bracts purplish or greenish, oblong-elliptic to roundish, pilose on backs, conspicuously ciliate; calyx ca. 6 mm. long, the lower lip deeply 2-toothed, the upper entire; corolla blue, ca. 10–12 mm. long, the lower 3-lobed with middle lobe erose; stamens long-exserted, the upper pair short, sterile; nutlets ± mottled, ca. 3 mm. long; $2n = 30$ (Epling, Lewis & Raven, 1962).—Dry flats and slopes, 2500–8800 ft.; Joshua Tree Wd., Pinyon-Juniper Wd., Sagebrush Scrub; Mojave Desert (Los Angeles and San Bernardino cos.) to Lassen Co.; w. Nev. May–July.

KEY TO SUBSPECIES

Lf.-blades mostly round to spatulate, abruptly narrowed to petioles, mostly 4–15 mm. long.
    Bracts pilose on outer surface; glomerules mostly 3–4, usually approximate. San Bernardino Co. to Lassen Co. .... ssp. *dorrii*
    Bracts usually ± glabrous except for the ciliate margin; glomerules mostly 2–3.
        Glomerules mostly ca. 1.5 cm. in diam., usually 0.5–1.5 cm. apart on the slender rachis. Death V. region .... ssp. *gilmanii*
        Glomerules mostly 1.5–2.5 cm. in diam., generally crowded. E. Mono and Inyo cos.
                                                                                                                                ssp. *argentea*
Lf.-blades usually oval or elliptical, rarely obovate, mostly 15–30 mm. long, gradually narrowed at base. N. Calif. .... ssp. *carnosa*

Ssp. **argéntea** (Rydb.) Munz. [*Audibertiella a.* Rydb.] Lvs. 8–20 mm. in diam.; bracts mostly glabrous, purple or blue, shining.—At 5000–10,000 ft.; mostly Pinyon-Juniper Wd.; Panamint, Argus and Inyo mts.; w. Nev.

Ssp. **gilmánii** (Epl.) Abrams. [*S. carnosa* ssp. *g.* Epl.] Lvs. very scurfy-hoary, 4–7 mm. in diam., on petioles ca. 2.5 mm. long; bracts mostly opaque, glabrous or hispidulous, usually rose; $2n = 30$ (Epling, Lewis & Raven, 1962).—Dry places, 3000–7000 ft.; mostly Pinyon-Juniper Wd.; Panamint, Argus and Inyo mts.; w. Nev.

8. **S. eremostàchya** Jeps. Intricately branched shrub, 6–8 dm. high, the branchlets ashy with spreading glandular hairs; lf.-blades lance-oblong, 1.5–3.5 cm. long, 4–10 mm. wide, obtuse, subtruncate at base, crenulate, green and rugose above, hispidulous beneath, on petioles 3–8 mm. long; glomerules 2–3 in interrupted spikes, with thin round-ovate bracts; calyx ca. 11 mm. long, lower lobes free, weakly spinose, upper united; corolla pale blue to rose, 2–2.5 cm. long, the lower lip half as long as tube and with large notched middle lobe; stamens exserted; nutlets 3 mm. long; $2n = 30$ (Epl., Lewis & Raven, 1962).—Dry rocky and gravelly places, 1200–4500 ft.; Creosote Bush Scrub; w. edge of Colo. Desert. March–May.

9. **S. funèrea** Jones. Compact densely branched shrub, 5–8 (–12) dm. tall, white-woolly, densely leafy; lvs. ovate to lance-ovate, 1.5–2 cm. long, subsessile, acuminate-spinose at apex, mostly entire, sometimes with a lateral pair of spiny teeth; fls. mostly 3 in axils of iliciform lvs. and crowded into leafy spikes 3–8 cm. long; calyx white-woolly, 4.5–6 mm. long, with subequal deltoid teeth; corolla violet, 12–16 mm. long, the lower lip 5–6 mm. long, with large erosulate middle lobe; $n =$ ca. 32 (Epl., Lewis & Raven, 1962).—Dry washes and rocky places, 1000–3000 ft.; Creosote Bush Scrub; w. slope of Amargosa Range, in Grapevine Mts. and n. part of Panamint Mts. March–May.

10. **S. greàtai** Bdg. Shrubby, much like *S. funerea*, but with lvs. less crowded, less tomentose, the blades 2–3 cm. long and mostly with more than 1 pair of lateral spinulose teeth; whorls of infl. more remote; calyx 9–11 mm. long; corolla lavender, 14 mm. long,

the lower lip 3 mm. long; $2n=$ ca. 30 (Epl., Lewis & Raven, 1962).—Dry washes and fans below 600 ft.; Creosote Bush Scrub; Orocopia Mts., Riverside Co. to Chocolate Mts., Riverside Co. March–April.

11. **S. leucophýlla** Greene. [*Audibertia nivea* Benth., not *S. n.* Thunb.] Much-branched grayish-white tomentose shrub, 1–1.5 m. high; lf.-blades lance-oblong, or the lower somewhat wider, obtuse, crenulate, rugose, paler beneath, 2–6 cm. long, on petioles 3–8 mm. long; whorls capitate, 3–5, crowded or scattered, with oval to oblong bracts ca. 9–11 mm. long; calyx 10–12 mm. long, the teeth wholly united into 1 lip; corolla rose-lavender, ca. 2 cm. long, the lips subequal, the lower middle lobe 4–5 mm. long, oblong; stamens exserted; nutlets ± mottled, 3–3.5 mm. long; $2n=30$ (Epl., Lewis & Raven, 1962).— Common on dry barren slopes, mostly below 2000 ft.; Coastal Sage Scrub; Orange Co. to San Luis Obispo and Kern cos. May–July.

12. **S. mellífera** Greene. [*Audibertia stachyoides* Benth.] BLACK SAGE. Plate 56, Fig. D. Openly branched shrub 1–2 m. tall, with herbaceous leafy strigulose twigs, ± glandular; lf.-blades oblong-elliptical, 2–6 cm. long, obtuse, green and somewhat rugulose above, cinereous-tomentulose beneath, crenulate, subsessile or on petioles to 12 mm. long; fls. many in compact glomerules, 2–4 cm. in diam. with ovate greenish cuspidate bracts 5–10 mm. long, intervals between glomerules mostly 2–6 cm. long; calyx 6–8 cm. long, villous, ± glandular, the lower teeth free, 1.5–2 mm. long, the upper connate; corolla pale blue or whitish, sometimes lavender, ca. 12 mm. long, the lower lip almost equal to tube, with notched large middle lobe; stamens exserted; nutlets ca. 2 mm. long; $2n=30$ (Epl., Lewis & Raven, 1962).—Common on dry slopes and benches, below 2000 ft.; Coastal Sage Scrub, Chaparral; n. L. Calif. to cent. Calif. April–July. It hybridizes freely with *S. apiana*, *S. columbariae* (*S.* × *bernardina*), and *S. leucophylla*, sometimes with *S. carduacea* and *S. clevelandii*.

13. **S. mohavénsis** Greene. [*Audibertia capitata* Gray, not Schlecht.] Plate 56, Fig. E. Compact many-branched shrub, 2–6 (–9) dm. tall, the branchlets hispidulous; lf.-blades lance- to ovate-oblong, 1–2 cm. long, obtuse, crenulate, rugose and glandular above, hispidulous beneath, on petioles 5–10 mm. long; capitate whorls mostly solitary, with outer whitish bracts membranaceous and ovate, inner narrower; calyx 8–10 mm. long, hirtellous, the upper teeth joined; corolla pale blue or lavender, 15–16 mm. long, the lobes oblong, entire; stamens exserted; nutlets 2.5–3 mm. long; $n=15$ (Epl., Lewis & Raven, 1962).—Dry rocky washes and canyons, 1000–5000 ft.; Creosote Bush Scrub, Joshua Tree Wd.; deserts from Little San Bernardino Mts. and Sheephole Mts. to Clark and Turtle mts.; sw. Nev., nw. Son. April–June.

14. **S. múnzii** Epl. [*S. mellifera* var. *jonesii* Munz.] Like *S. mellifera*, but young stems more slender; lf.-blades oblong-obovate, mostly 1–2 cm. long, ashy on both surfaces, the upper hirtellous, the lower with minute appressed hairs; fl.-glomerules 1–1.5 cm. in diam., with oblong-elliptic bracts; calyx ca. 5 mm. long, hirtellous; corolla clear blue, ca. 10 mm. long; $2n=30$ (Epl., Lewis & Raven, 1962).—Coastal Sage Scrub; San Miguel Mt., San Diego Co. and adjacent L. Calif. Feb.–April.

15. **S. pachyphýlla** Epl. ex Munz. [*A. pachystachya* Parish not Trautv. *S. compacta* Munz not Kuntze.] Resembling *S. dorrii*, but more sprawling; lf.-blades oblanceolate to obovate, 2–4 cm. long, on petioles 5–15 mm. long; whorls in infl. crowded, forming a dense spike to 1.5 dm. long; bracts mostly purplish, 1–2.5 cm. long, obovate to oblong, rounded-truncate at apex, ± ciliate, glabrous to hirtellous on back; calyx 9–12 mm. long; corolla dark- to violet-blue, 15–22 mm. long, the tube cylindrical, the lower lip ca. ⅓ as long as tube and with erose middle lobe; stamens well exserted; nutlets ca. 3 mm. long; $2n=30$ (Epl., Lewis & Raven, 1962).—Dry rocky slopes, 5000–10,000 ft.; Pinyon-Juniper Wd., Yellow Pine F.; San Bernardino Mts. to n. L. Calif.; Panamint, Kingston, Clark, New York mts., July–Sept. The sp. hybridizes easily in the botanical garden with *S. clevelandii*, making plants with stature and odor of the latter and with large beautiful blue fls.

16. **S.** × **pálmeri** (Gray) Greene. Apparent hybrid between *clevelandii* and *apiana* with general aspect of former, but with paler foliage and more freely branched infl. having

more numerous heads; odor less aromatic than in *clevelandii*; upper lip of corolla short. Occasional in San Diego Co.

17. **S. praténsis** L. Perennial, to 7 dm. high, erect, pubescent; lvs. largely basal, oblong-ovate, those of infl. cordate-ovate; racemes glandular, subsimple, the whorls remote, 6-fld.; corolla bright blue, sometimes red or white, 2.5 cm. long.—Found as an escape near Lake Arrowhead, San Bernardino Mts.; native of Eu.

18. **S. sonoménsis** Greene. [*Audibertia humilis* Benth., not *S. h.* Benth.] Herbaceous perennial with creeping leafy somewhat matted stems and erect quite leafless flowering stems 1–4 dm. tall; lf.-blades elliptical-obovate, mostly 3–6 cm. long, rounded at tip, attenuate at base into subequal petioles, crenulate, green and rugulose above, closely whitish-tomentulose beneath; infl. a spike of 4–6 remote whorls, hispid, glandular; bracts 5–8 mm. long; calyx ca. 1 cm. long, lobes of upper lip wholly united, trimucronate; corolla blue-violet, ca. 15 mm. long, the lower lip 7–8 mm. long, with lateral lobes obsolete; nutlets ca. 2.5 mm. long; $2n = 30$ (Epl., Lewis & Raven, 1962).—Dry slopes below 6000 ft.; Chaparral, Yellow Pine F.; mts. of San Diego Co.; cent. Calif. May–June.

19. **S. spathàcea** Greene. [*Audibertia grandiflora* Benth., not *S. g.* Epl.] Pitcher Sage. Coarse perennial herb with creeping rhizomes, the annual stem stout, glandular-villous, 3–7 (–10) dm. high; lvs. numerous below, oblong-hastate, 8–20 cm. long, oblong, rugose, green above, ± ashy-tomentose beneath, mostly on petioles 3–8 cm. long; stem lvs. reduced upward, subsessile, with truncate base; infl. spicate, with several close or remote whorls, coarse, viscid, 1.5–3 dm. long; bracts ovate to lance-ovate, acuminate, 1.5–4 cm. long, mostly purplish; calyx 1.5–2 (–3) cm. long; $2n = 30$ (Epl., Lewis & Raven, 1962).— Grassy and shaded slopes below 2000 ft.; Chaparral, S. Oak Wd., etc.; Orange Co. n. to cent. Calif. March–May.

20. **S. vàseyi** (Porter) Parish. [*Audibertia v.* Porter.] Whitish rounded shrub to 1.5 m. tall; lf.-blades oblong-ovate, mostly 2–6 cm. long, crowded at base of wandlike branchlets, obtuse, crenulate, on petioles 5–10 mm. long; glomerules 4–8 (–14), in long interrupted long-peduncled spikes, with white bracts tipped with a stout bristle; calyx white with minute hairs, 8–10 mm. long, not counting the terminal bristles, these 3–7 mm. long; corolla white, ca. 2 cm. long, the tube 11–14 mm., the upper lip roundish, retuse, 3–4 mm. wide, the lower lip 7–12 mm. long, with erose subreniform middle lobe; stamens exserted; nutlets 2.5–3 mm. long; $n = 15$ (Epl., Lewis & Raven, 1962).—Dry rocky slopes and canyons, below 2500 ft.; Creosote Bush Scrub, w. edge of Colo. Desert from Morongo V. to n. L. Calif. April–June.

21. *Saturèja* L.

Perennial herbs or suffrutescent, with entire or toothed lvs. Fls. large to small, solitary, clustered in the lf.-axils. Calyx narrow-campanulate or tubular, 10–13–15-nerved, 5-toothed. Corolla small and little-exserted to larger and well exserted; upper lip erect, lower spreading. Stamens 4, all perfect. Style glabrous or hairy. Nutlets ovoid, smooth. Perhaps 150 spp., widely distributed. (Ancient Latin name used by Pliny.)

Trailing herbs; calyx 4 mm. long .... 2. *S. douglasii*
Erect subshrub; calyx 6–7 mm. long .... 1. *S. chandleri*

1. **S. chándleri** (Bdg.) Druce. [*Calamintha c.* Bdg.] Shrubby, branched, 2–5 dm. high, upper parts pubescent; lvs. round-ovate, 1–2 cm. long, short-petioled, entire to ± crenate, densely pubescent beneath, puberulent above; fls. 1–5 in axillary clusters; pedicels 1–2 mm. long; calyx tubular-campanulate, 6–7 mm. long; corolla white, pubescent, 10–13 mm. long; nutlets cellular-reticulate, ca. 1.2 mm. long.—Rocky canyons, below 2500 ft.; Chaparral; Santa Ana Mts. near Murietta (Riverside Co.) and San Miguel Mts. and Jamul Mt., San Diego Co. March–May.

2. **S. doùglasii** (Benth.) Briq. [*Thymus d.* Benth. *T. chamissonis* Benth.] Yerba Buena. Trailing evergreen perennial herb, the stems slender, rooting, 2–6 dm. long; lvs. round-ovate, 1.5–2.5 cm. long, short-petioled, crenate, ± pubescent, obtuse at apex, obtuse to subcordate at base; fls. solitary in axils on pedicels 1–1.5 cm. long; calyx tubular; corolla

white or ± purplish, 6–8 mm. long, pubescent; nutlets ca. 1 mm. long.—Shaded woods, below 3000 ft.; Chaparral, Closed-cone Pine F., etc.; Santa Monica Mts., Santa Catalina Id., Santa Ynez Mts.; to B.C. April–Sept.

22. *Scutellària* L. SKULLCAP

Herbs with opposite entire to toothed lvs. Fls. 1–3 in the axils or in bracted racemes or spikes. Calyx campanulate, gibbous, 2-lipped, the lips entire, the upper with a crestlike projection on the back. Corolla blue, violet or white, well exserted, bilabiate, dilated above the throat, the upper lip arched, entire or emarginate, the lower spreading or deflexed, with lateral lobes small and middle large. Stamens 4, in 2 pairs; anthers ciliate-pilose, the upper pair 2-, the lower 1-celled. Style unequally 2-cleft at apex. Nutlets 4, subglobose or depressed, papillose or tubercled. Ca. 100 spp. of wide distribution. (Latin, a tray or platter, referring to fruiting calyx.)

(Epling, C. The American species of Scutellaria. Univ. Calif. Pub. Bot. 20: 1–146. 1942.)
Corolla 5–7 mm. long; fls. in slender, axillary bracteate racemes .... 3. *S. lateriflora*
Corolla 12–33 mm. long; fls. solitary in axils of lvs.
    Petioles of all but lowest lvs. 1–3 mm. long; blades truncate-cordate at base ... 2. *S. bolanderi*
    Petioles 5–30 mm. long; blades of all but lowest lvs. not truncate-cordate.
        Stems with short ± curved hairs; corolla 24–29 mm. long .... 1. *S. austinae*
        Stems loosely villous with spreading hairs; corolla 15–20 mm. long .... 4. *S. tuberosa*

1. **S. áustinae** Eastw. Plate 56, Fig. F. Rhizomes slender; stems 1–3 dm. tall, glabrous to puberulent, with short ascending curved hairs; lvs. oblong-lanceolate to -linear, 15–25 mm. long, rounded apically, entire, subsessile; fls. axillary in upper half of plant; pedicels 3–5 mm. long; calyx puberulent, 4.5–6 mm. long; corolla deep blue-violet, 24–29 mm. long, the slender tube curved upward; lower stamens seated above the middle of the tube; nutlets black, rugose.—Gravelly or rocky places, below 6000 ft.; Chaparral, Foothill Wd., S. Oak Wd., Yellow Pine F.; San Diego Co. and Santa Rosa Mts. n. to n. Calif. May–July.

2. **S. bolánderi** Gray ssp. **austromontàna** Epl. Spreading by slender rhizomes; stems 2–4 dm. high, simple or loosely branched; lvs. thin, spreading, deltoid-ovate, 1–2 cm. long, on petioles 5–10 mm. long, the upper ± sessile; fls. few, solitary in axils; calyx at anthesis 4.5 mm. long, pubescent, ± glandular, the lower lip becoming 5–6 mm. long; corolla white, 12–15 mm. long, with spreading ± gland-tipped hairs; tube and galea 16–18 mm. long; lower fls. 9–11.5 mm. long, attached 6–7 mm. above the base of the tube; nutlets rugose, 1 mm. long.—Uncommon, in damp places at 3000–4500 ft.; Chaparral, S. Oak Wd.; interior s. Calif. (San Diego Co. to Victorville). June–July.

3. **S. lateriflòra** L. Glabrous or puberulent above; stems 2–6 dm. tall, slender, branched above; lvs. 3–7 cm. long, deltoid-ovate, acutish, crenate-serrate, on petioles 0.5–2 cm. long; fls. in lateral axillary racemes; calyx 1.5–2.5 mm. long at anthesis, hirtellous and glandular; corolla blue, 5–7 mm. long; nutlets smooth, subcompressed.—Collected in a wet place, Saline V., Inyo Co.; to Atlantic Coast and B.C. May–July.

4. **S. tuberòsa** Benth. Perennial with tuber-bearing rhizomes; stems 0.5–2 dm. tall, usually branched at base, ± viscid with rather long spreading hairs; lvs. ovate, 1–2 cm. long, mostly coarsely dentate, the lower on petioles 5–15 mm. long; fls. solitary in axils, the pedicels 2–3 mm. long; calyx pilose, the lower lip 4–6.5 mm. long; corolla 15–20 mm. long, the palate pilose, the tube glabrous within below the middle; lower stamens 8–11 mm. long; nutlets black, papillate.—Borders of brush and open woods, below 2000 ft.; Foothill Wd., Chaparral, etc.; cismontane from n. L. Calif. to cent. Calif. April–May.

23. *Stàchys* L. HEDGE-NETTLE

Commonly pubescent or hispid herbs. Fls. few to many, borne in whorls in dense or interrupted terminal spikes, or also in upper axils. Calyx usually ± campanulate, 5–10-

nerved, 5-toothed, the teeth subequal, erect or spreading. Corolla with cylindrical tube, little or not dilated at throat; upper lip erect, concave, entire or emarginate; lower longer, spreading, 3-lobed, the middle lobe larger, the lateral often deflexed. Stamens 4, didynamous, ascending under the upper lip. Nutlets ovoid or oblong. Ca. 150 spp., mostly of N. Temp. Zone, some of S. Temp. (Greek, *stachus*, ear of grain, hence a spike.)

(Epling, C. Preliminary revision of the Am. Stachys. Fedde Rep. Sp. Nov. Regni Veg. 80: 1–75. 1934.)

Ring of hairs on inner surface of corolla tube horizontal and not indicated on the outside by a constriction .... 3. *S. bullata*
Ring of hairs on inner surface of corolla-tube oblique and indicated on the outside by a constriction of the tube, this most noticeable on the anterior side.
    Pubescence of plant of very soft slender hairs that become tangled and cobwebby in age; corolla whitish .... 2. *S. albens*
    Pubescence of straight usually stiffish hairs.
        Lf.-blades oblong, narrowed at base, usually silky-strigose; corolla whitish .. 1. *S. ajugoides*
        Lf.-blades oval or ovate to oblong, scarcely narrowed at base, never silky; corolla pale rose-purple .... 4. *S. rigida*

1. **S. ajugoìdes** Benth. Stems simple and erect, or ± branched below and decumbent, 1–6 dm. long, villous and glandular; lvs. oblong, 1.5–7 cm. long, rounded at apex, narrowed at base, the lower on petioles to 3 cm. long; fls. in whorls of 6, the spikes bracted, approximate or interrupted, 1–2 dm. long in age; calyx 6–8 mm. long, villous, the lanceolate or ovate-deltoid teeth cuspidate; corolla white, pale rose or with purple veins, 10–15 mm. long, the upper lip 5–6 mm., the lower 6–7.5 mm. long.—Moist places below 2500 ft.; Coastal Sage Scrub; Orange and Los Angeles cos. n. to cent. Calif. May–Sept.

2. **S. álbens** Gray. Stems stout, 3–10 dm. high, usually branched, ± densely white-woolly; lvs. 3–12 cm. long, narrowly to broadly ovate, ± cordate at base, villous-tomentose beneath, crenate-serrate, on petioles to 5 cm. long; spikes 1–2 dm. long, somewhat interrupted in age; calyx 7 mm. long, woolly, the deltoid-ovate teeth cuspidate; corolla white or pinkish with purple veins, the tube 6–8 mm. long, upper lip 3.5–5.5 mm., lower 6–8 mm. long; fils. woolly.—Moist places below 8100 ft.; V. Grassland and Coastal Sage Scrub to Montane Coniferous F.; mostly away from the coast; Riverside Co. to Inyo-White Mts. and Lake Co. May–Oct.

3. **S. bullàta** Benth. [*S. californica* Benth.] Plate 57, Fig. A. Stems procumbent at base, slender, simple or branched, 4–8 dm. long, sparsely retrorse-hispid on angles, otherwise ± glandular and pubescent; lvs. ovate to oblong-ovate, 3–18 cm. long, mostly obtuse, crenate-serrate, the base subcordate, on petioles up to 6 cm. long; whorls 6-fld., rather distinct; calyx 6–7.5 mm. long, pilose, the teeth triangular, with apical spines; corolla purple, the tube 7–10 mm. long, upper lip 3.5–5 mm., lower 6–10 mm. long; stamens exserted 2–4 mm.; nutlets ca. 1.5 mm. long.—Dryish slopes and canyons, below 4000 ft.; Coastal Sage Scrub, Chaparral, etc.; Orange Co. to San Francisco. Santa Cruz and Santa Rosa ids. April–Sept.

4. **S. rígida** Nutt. ex. Benth. ssp. **rigida.** Plate 57, Fig. B. Stems erect or ± decumbent, simple or branched above, mostly 6–10 dm. long, villous-hirsute; lvs. deltoid-oblong to lance-oblong, acute, crenate-serrate, rounded or subcordate at base, 5–9 cm. long, the lower petioles 2.5–4 cm. long; spikes interrupted in age, often 1–2 dm. long; fls. 1–3 in axil of each bract; calyx 5–8 mm. long, the teeth narrow-deltoid, ca. as long as tube, villous-hirsute, mucronulate; corolla rose-purple or veined with purple, 12–16 mm. long, the upper lip 3–5.6 mm., the lower 7–9 mm. long; stamens exserted 2–3 mm.—Mostly low moist places, below 8000 ft.; Montane Coniferous F.; San Diego Co. n.; to Wash. July–Aug.

Ssp. **quercetòrum** (Heller) Epl. [*S. q.* Heller.] Plants 1.5–4 dm. high; lvs. ± ovate, the blades 4–8 cm. long.—Mostly below 5000 ft.; Coastal Sage Scrub, etc.; n. L. Calif. n. to s. Ore.

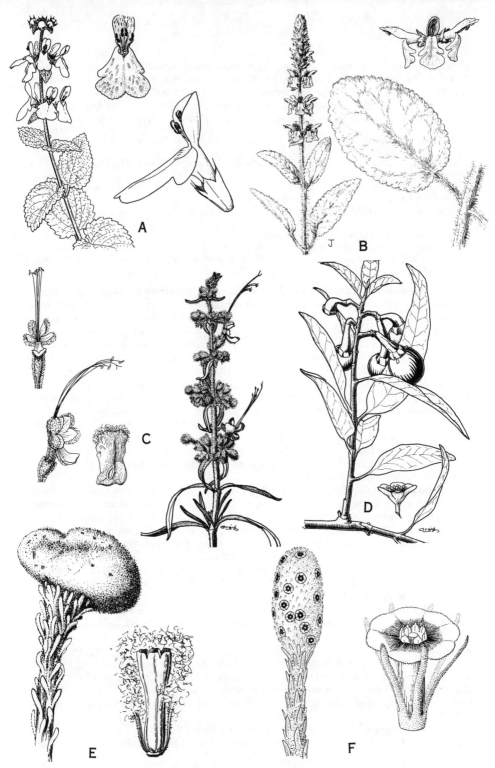

Plate 57. Fig. A, *Stachys bullata*. Fig. B, *Stachys rigida*. Fig. C, *Trichostema lanatum*. Fig. D, *Umbellularia californica*. Fig. E, *Ammobroma sonorae*. Fig. F, *Pholisma arenarium*.

24. Teùcrium L. Germander

Herbs or subshrubs with entire to laciniate lvs. Fls. solitary in the axils of ± modified fl. lvs. or crowded into linear bracteate terminal spikes, subsessile or pedicelled. Calyx campanulate, 10-veined, with deltoid or lance-deltoid teeth. Corolla with short or a longer funnelform tube, the lower lip prominent with oblong to obovate middle lobe and smaller lateral lobes, the upper lip small, 2-cleft. Stamens 4, paired. Style exserted with stamens, 2-parted at apex. Nutlets oval, smooth or sculptured. Ca. 100 spp., world wide. (*Teucer*, a Trojan king.)

Pedicels 4–12 mm. long; annual herb .... 1. *T. cubense*
Pedicels 15–40 mm. long; low shrub .... 2. *T. glandulosum*

1. **T. cubénse** Jacq. ssp. **depréssum** (Small) McClint. & Epl. [*T. d.* Small.] Mostly much-branched from base, 1–3 dm. long, glabrous below, pubescent above; median lvs. 1.5–2.5 cm. long, 3–5-lobed nearly to midvein, obovate-cuneate in outline, petioled; fl.-lvs. 3-parted, sessile; calyx 5–6 mm. long, the teeth ca. 4 mm. long, spreading, pubescent; corolla pale blue, 7–8 mm. long; nutlets corky-thickened, reticulate.—Rare, semialkaline flats; Creosote Bush Scrub; Colo. Desert (Hayfields, Palo Verde V.); to Tex., nw. Mex. March–May.

2. **T. glandulòsum** Kell. Branching, to ca. 1 m. high, glabrous; lvs. oblong-lanceolate, entire or some of the lower obscurely 3-lobed, 2–5 cm. long; fls. in elongate racemes; calyx campanulate, 7–11 mm. long, gland-dotted, the teeth lanceolate, 4–8 mm. long; corolla white, with violet streaks, pubescent, 15–20 mm. long, the lower lip 10–17 mm. long; nutlets 2.5–3 mm. long, grooved, apically hairy.—Rocky places at 1450 ft.; Creosote Bush Scrub; Whipple Mts., e. Colo. Desert; L. Calif. April–May.

25. Trichóstema L. Bluecurls

Strong-scented annual or perennial herbs or subshrubs with entire lvs. Fls. 1–many in axillary cymes or racemose. Calyx usually deeply 5-lobed, to unequally so. Corolla blue or lavender or pink to whitish, 5-lobed, the slender tube often exceeding the calyx, the limb oblique, deeply 5-cleft into oblong, ± declined segms. Stamens 4, usually ascending from near the throat, arched or nearly straight, anther-sacs divergent at maturity. Nutlets joined ca. ⅓ their length, alveolate to rugose-reticulate, the ridges ± prominent. Sixteen spp., N. Am. (Greek, *trichos*, hair, and *stemon*, stamen.)

(Lewis, H. Chromosome numbers and phylogeny of Trichostema. Brittonia 12: 93–97. 1960.)
Plants annual, with pale pubescence.
  Stamens 13–20 mm. long; corolla-tube 5–10 mm. long .... 3. *T. lanceolatum*
  Stamens 2–7 mm. long.
    Calyx-lobes ca. twice as long as the tube, approximately equal to each other in width
      1. *T. austromontanum*
    Calyx-lobes 1–2 times as long as the tube, the uppermost more slender than the other 4
      4. *T. micranthum*
Plants low shrubs, with ± purplish wool in the infl.
  Corolla-tube 9–14 mm. long; calyx ca. 8 mm. long, with lanceolate lobes; fils. exserted 2.5–3 cm.
    2. *T. lanatum*
  Corolla-tube 4–7 mm. long; calyx ca. 5 mm. long, with triangular lobes; fils. exserted 1–1.5 cm.
    5. *T. parishii*

1. **T. austromontànum** Lewis. Soft-villous annual 1–4 dm. high, the lvs. linear- to oblong-lanceolate, 1.5–3.5 cm. long, acute, 2–8 mm. wide; cymes axillary, few–many-fld., horizontal and spreading; calyx-lobes lance-subulate, the teeth longer than the tube; corolla 3 mm. long; stamens 5–6 mm. long; nutlets 1.2–2 mm. long, irregularly reticulate, puberulent; $n=14$ (Lewis, 1945).—Drying edges of meadows, 3500–7500 ft.; Montane Coniferous F.; mts. from San Diego Co. to Los Angeles Co. and Mono Co. A compact plant ca. 1 dm. high, from Hidden Lake, San Jacinto Mts. has been named ssp. *compactum* Lewis.

2. **T. lanàtum** Benth. WOOLLY BLUE-CURLS. ROMERO. Plate 57, Fig. C. Rounded shrub 5–15 dm. tall, many-branched, hirtellous; lvs. lance-linear, sessile, 3.5–7.5 cm. long, revolute, green above, lanate beneath, usually with axillary fascicles; infl. an interrupted spike of dense sessile cymes, woolly with blue, pink or whitish hairs 2–3 mm. long and concealing the pedicels; calyx 5–8 mm. long, the teeth subequal, ovate, acuminate to acute, equal to or longer than the tube; corolla floccose without, blue, the tube 9–14 mm. long, the lower lip 7–12 mm. long; stamens arched; nutlets 2–4 mm. long, irregularly reticulate; $n=10$ (Lewis, 1945).—Dry slopes, mostly below 4500 ft.; Chaparral, Coast Ranges near the coast from San Diego and Orange cos. to Monterey Co. May–Aug.

3. **T. lanceolàtum** Benth. VINEGAR WEED. Annual, mostly branching from base, 1–6 dm. high, mostly bushy, glandular-villous, sour-smelling; lvs. lanceolate to lance-ovate, mostly 2–5 cm. long, 4–14 mm. wide, sessile or nearly so; cymes in mostly simple, ± secund racemes; calyx 2.5–4 mm. long; corolla light blue, the tube 5–8 mm. long, the lower lip 4–8 mm. long, strongly deflexed; stamens 13–20 mm. long; nutlets alveolate, 1.6–3 mm. long; $n=7$ (Lewis, 1945).—Dry fields and open places, mostly below 3500 ft.; Coastal Sage Scrub, Chaparral, V. Grassland, etc.; n. L. Calif. to B.C. Aug.–Oct.

4. **T. micránthum** Gray. Erect annual 1–3 dm. tall, branched below, hirtellous and glandular-pubescent; lvs. lance-linear, 2–2.5 cm. long, 2–5 mm. wide, acute; cymules 2–10-fld.; calyx at anthesis 1.5–2.5 mm. long, the lobes scarcely longer than the tube; corolla 2–3 mm. long; stamens exserted ca. 1.5 mm.; nutlets 1.2–1.7 mm. long, reticulate; $n=7$ (Lewis, 1960).—Drying margins of wet places, 6500–7500 ft.; Yellow Pine F.; San Bernardino Mts. and n. L. Calif. July–Sept.

5. **T. párishii** Vasey. [*T. lanatum* var. *denudatum* Gray.] Resembling *T. lanatum*, the lvs. pubescent beneath; infl. open, the tomentum short and not concealing the pedicels; calyx 3–5 mm. long at anthesis, the teeth lanceolate to ovate; corolla-tube 4–7 mm. long, the lower lip 5–9 mm. long; nutlets 1.8–3.5 mm. long, irregularly reticulate; $n=10$ (Lewis, 1945).—Dry slopes, 2000–6000 ft.; Chaparral, Yellow Pine F., Joshua Tree Wd.; away from the coast, Acton (n. Los Angeles Co.) to n. L. Calif. May–Aug.

57. Lauràceae. LAUREL FAMILY

Aromatic trees and shrubs, usually evergreen. Lvs. alternate, rarely opposite, simple, exstipulate, mostly leathery, punctate. Fls. perfect or imperfect, yellow or greenish, regular. Calyx usually 6-parted, the segms. in 2 series. Petals 0. Stamens in 3–4 whorls of 3 each, some frequently reduced to staminodia; anthers 2–4-celled, opening by uplifting valves. Ovary usually superior and free, 1-celled, 1-ovuled, with a single style. Fr. a berry or drupe, indehiscent. Ca. 40 genera and 1000 spp., widely distributed in trop., less so in temp. regions. Some grown for ornament, some (avocado, cinnamon, and camphor) of economic importance.

1. *Umbellulària* Nutt. CALIFORNIA-BAY. CALIFORNIA-LAUREL.

Evergreen, pungently aromatic trees. Lvs. alternate, entire, coriaceous. Fls. bisexual, in simple peduncled umbels in upper axils. Sepals 6, deciduous. Stamens 9, the 3 inner with 2 stalked orange glands at base. Anthers 4-celled. Fr. a drupe. One sp. (Latin, *umbellula*, a small umbel.)

1. **U. califórnica** (H. & A.) Nutt. [*Tetranthera c.* H. & A.] Plate 57, Fig. D. Tree with broad crown and to 30 or 40 m. high, or an erect shrub in dryer places; bark greenish- to reddish-brown; lvs. oblong to oblong-lanceolate, 3–8 (–10) cm. long, 1.5–3 cm. wide, short-petioled, obtusely acuminate, shining, glabrous to somewhat pubescent, yellowish-green; fls. 6–10 on a peduncle, yellow-green; sepals 6–8 mm. long, oblong-ovate; drupe usually solitary, rounded-ovoid, 2–2.5 cm. long, greenish, becoming dark purple when ripe; stone light brown, ellipsoid, smooth; $2n=24$ (Bambacioni, 1941).—Common, canyons and valleys, below 5000 ft.; Chaparral, Foothill Wd., lower Yellow Pine F., etc.;

San Diego Co. n. to sw. Ore., where it is known as Myrtle. The wood is hard, strong, and takes a high polish; used for turning. Dec.–May.

58. Lennoàceae

Fleshy herbs parasitic on roots and without chlorophyll, ± brown when dry. Lvs. reduced to bractlike scales. Fls. perfect, in spikes or heads. Calyx deeply parted into almost distinct lobes. Corolla tubular with a ± flaring 5–8-lobed limb. Stamens as many as the corolla-lobes and inserted on the throat; anthers 2-celled, dehiscent by longitudinal slits. Pistil of 6–14 completely united carpels; style 1, simple; stigma ± peltate, crenulate or obscurely lobed. Fr. a caps., concealed in persistent perianth, finally breaking up into 12–28 nutlets. Seeds with endosperm and a rounded rather undifferentiated embryo. Three genera and 5–6 spp., w. N. Am.

Fls. on a dilated concave receptacle; sepals plumose .... 1. *Ammobroma*
Fls. in a dense terminal spicate or somewhat paniculate infl.; sepals glandular-puberulent
2. *Pholisma*

1. *Ammobròma* Torr. ex Gray. SAND FOOD

Stems simple, mostly buried in sand, fleshy, scaly, expanded at summit into a hollow saucer-shaped receptacle which is thickly covered with the fls. Sepals 6–10, filiform, plumose. Corolla regular, purple, tubular, 6–9-lobed, the lobes erect, plicate. Stamens 6–9. Carpels 6–10, divided by false partitions, the ovary 12–20-celled; stigma subcapitate, crenate on margin. Fr. globose, with fleshy exocarp and hard endocarp; nutlets 12–20. One sp. (Greek, *ammos*, sand, and *broma*, food.)

1. **A. sonòrae** Torr. ex Gray. Plate 57, Fig. E. Stems to 1 m. or so long, 1–2 cm. thick; lvs. oblong-linear, 1–3 cm. long, the lower glabrous, brown-purple, those below the head woolly-villous; disk of the infl. 3–10 cm. across, densely covered with fls. and the color of light sand because of the calyx-hairs; calyx-lobes filiform; corolla purple, ca. 8 mm. long, tubular, slightly longer than calyx; $2n=18$ (Moore, 1962).—Sand hills below 1000 ft.; Creosote Bush Scrub; Colo. Desert e. of Imperial V.; to Ariz., Son., L. Calif. April–May. Once an important food for the local Indians.

2. *Pholísma* Nutt. ex Hook.

Spike simple or compactly branched. Calyx-lobes 5–7, linear. Corolla narrow-funnelform, the limb expanded, obscurely lobed, purplish with white border, undulate-plicate. Ovary subglobose, 6–10-celled, each cell divided into two by a false partition. Stigma peltate, crenately 6–10-lobed. One sp. (Greek, *pholis*, scale, because of the scalelike lvs.)

1. **P. arenàrium** Nutt. ex Hook. [*P. paniculatum* Templeton.] Plate 57, Fig. F. Part above ground 1–2 dm. tall, the stem whitish, drying brown, fleshy, with brownish bractlike lvs. 8–14 mm. long; spikes oblong to branched, 2–8 cm. across; calyx shorter than corolla, glandular-puberulent; corolla 3–4 mm. broad; $n=18$ (Carlquist, 1953).—Occasional in sandy places, on roots of shrubs such as *Eriodictyon*, *Haplopappus*, *Chrysothamnus*, *Hymenoclea*, *Ambrosia*, etc., below 5000 ft.; Coastal Strand, Creosote Bush Scrub, Joshua Tree Wd.; coast from San Luis Obispo Co. to n. L. Calif. Mojave and Colo. deserts. April–July, Oct.

59. Lentibulariàceae. BLADDERWORT FAMILY

Rather small herbs, many insectivorous, and in water or wet places. Calyx often bilabiate. Corolla 2-lipped, the lower lip in ours with a narrow basal spur. Stamens 2, inserted on the corolla tube near its base, the anther-sacs of each stamen confluent. Ovary 1-locular with a free cent. placenta. Caps. 2–4-valved or often bursting irregularly. Seeds several, anatropous. Five genera and ca. 300 spp., rather cosmopolitan.

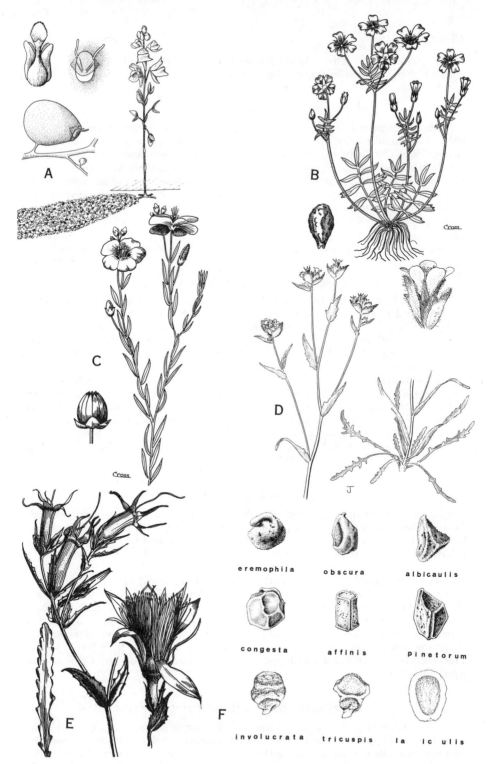

Plate 58. Fig. A, *Utricularia* sp. Fig. B, *Limnanthes gracilis*. Fig. C, *Linum lewisii*. Fig. D, *Mentzelia congesta*. Fig. E, *Mentzelia laevicaulis*. Fig. F, *Mentzelia* seeds.

1. Utriculària L. BLADDERWORT

Stems mostly submerged. Lvs. simple to much dissected and bearing small urn-shaped bladders which possess a kind of valvelike opening and trap insects and minute crustacea. Scapes emergent, with small auricled scales, 1–several-fld. Fls. perfect, racemose, yellow in ours. Calyx with 2 entire lips. Corolla with a projecting palate on lower lip, the upper lip erect. Stamens 2, twisted, flattened. Caps. irregularly dehiscent, many-seeded. A ± cosmopolitan genus. (Latin, *utriculus*, a little bladder.)

1. **U. vulgàris** L. Plate 58, Fig. A. Immersed stems 3–10 dm. long, coarse, few-branched; lvs. mostly tripinnately divided, 1.5–4.5 cm. long, with numerous traps ca. 2–2.5 mm. in diam.; scapes 1–3 dm. long; calyx 3.5–4 mm. long; lower lip of corolla 12–15 mm. long, yellow with brown stripes on the large palate; fruiting pedicels mostly arched-recurving, 1–1.8 cm. long; $2n = 36$–40 (Reese, 1952).—Ponds and quiet water of many Plant Communities in several localities from Victorville region n.; to Alaska, Atlantic Coast, Eurasia. July–Sept.

60. Limnanthàceae. FALSE MERMAID FAMILY

Low tender glabrous annual herbs of moist places. Lvs. alternate, pinnately dissected, exstipulate. Fls. bisexual, regular, solitary, 3–6-merous, subperigynous. Sepals valvate, persistent. Petals contorted. Stamens 6 or 10, free, some with a gland at the base. Carpels 3–5, nearly distinct, connected by a common gynobasic style; ovules solitary in each carpel. Fr. of semidrupaceous free ± tuberculate nutlets. Seeds erect, without endosperm; embryo straight. Two genera, N. Am.

1. *Limnánthes* R. Br. MEADOW-FOAM

Lvs. usually pinnate, then pinnatifid. Sepals and petals usually 5, sometimes 4 or 6. Petals with a U-shaped band of hairs on the claw. Stigmas 5, capitate. Seven spp., largely Çalifornian. (Greek, *limne*, marsh, and *anthos*, fl., because of the habitat.)

(Mason, C. T. A systematic study of the genus Limnanthes. Univ. Calif. Pub. Bot. 25: 455–512. 1952.)

1. **L. grácilis** Howell var. **párishii** (Jeps.) C. T. Mason. [*L. versicolor* var. *p.* Jeps.] Plate 58, Fig. B. Stems mostly with widely divaricate branches from base, 1–2 dm. long, glabrous; lvs. 2–6 cm. long, with ca. 5 pinnae, these pinnatifid into 3–5 oblanceolate segms., glabrous; pedicels 2–8 cm. long; sepals lance-ovate, glabrous or with a few long hairs, acuminate, 5–8 mm. long; fls. bowl-shaped; petals white, aging pink, obovate, emarginate, 8–10 mm. long, glabrous; stamens 3–3.5 mm. long; nutlets 4 mm. long, with low tubercles; $n = 5$ (Mason, 1952).—Moist lake shores and wet places, 4500–5000 ft.; Yellow Pine F.; Cuyamaca and Laguna mts. April–May.

61. Linàceae. FLAX FAMILY

Herbs or shrubs. Lvs. simple, alternate or opposite; stipules small or lacking. Fls. bisexual, regular. Sepals 5 (rarely 4), free or partially united, imbricate. Petals 5 (or 4) and alternate with sepals, convolute. Stamens 5 (or 4) and alternate with petals; fils. connate at base, often alternating with small staminodia; anthers introrse, 2-celled. Ovary superior, 2–5-celled or seemingly 4–10-celled by false septa; ovules 2 in each cell; styles as many as ovary-cells, ± free, filiform, with simple subcapitate stigmas. Fr. a caps., often septicidal; seeds compressed, shining, with straight embryo and flat cotyledons. Ca. 14 genera and 150 spp., widely distributed, a few important as ornamentals or for flax, linseed oil, etc.

Petals blue, 10–20 mm. long; styles 5; caps. 5–10 mm. long .... 2. *Linum*
Petals white to pink or yellow, 2–8 mm. long or 15 mm. in *puberulum*; caps. 1–4 mm. long.

Styles united almost to summit; fls. yellow .... 2. *Linum*
Styles quite free; fls. white to pinkish .... 1. *Hesperolinon*

1. Hesperolinon Small

Slender-stemmed annuals, ± glaucous, essentially glabrous; lvs. in whorls of 4 at basal nodes, irregularly whorled at upper nodes and opposite or alternate on ultimate branches, usually early caducous, entire, linear or lanceolate to oblong, sessile, fleshy, the uppermost being small bracts. Fls. in cymes; pedicels filiform, pseudocleistogamous. Sepals 5, united below. Petals 5, caducous, clawed. Stamens 5; fils. filiform. Ovary of 2 or 3 carpels; ovules 4 or 6. Styles as many as carpels, with minute stigmas. Seeds plump, shining. (Greek, *hesperos*, western, and *linos*, flax.) Ca. a dozen spp. of sw. U.S.

1. **H. micránthum** (Gray) Small. [*Linum m.* Gray.] Annual, freely branched at or above the base, 1–4 dm. high, the branches very slender, minutely pubescent above the nodes; lvs. linear, 1–2.5 cm. long; pedicels 5–16 mm. long, filiform; sepals lanceolate or somewhat oblong, 1–2 mm. long with some glandular teeth; petals pale pink to white, mostly 2–3 mm. long, the cent. scale ciliate, the lateral often almost obsolete; styles 3, scarcely 1 mm. long; caps. 1–2 mm. long; seeds brown, with dark spots.—Open slopes and ridges, often on burns, 1000–5500 ft.; Chaparral, S. Oak Wd., Yellow Pine F.; mts. of San Diego Co. to Piute Mt., Santa Monica and Santa Ynez mts., to Kern Co. and to Ore.; w. edge of Colo. Desert. May–July. Catalina Id.

2. Linum L. FLAX

Annual or perennial herbs, with tough fibrous cortex. Lvs. sessile, entire, without stipules or with stipular glands. Fls. in terminal or axillary racemes, corymbs or cymes, red, yellow, blue or white. Sepals, petals, stamens 5, the petals fugacious. Ovary 5- or 10-celled. Caps. dehiscent or indehiscent. Seeds mucilaginous. Ninety or more spp. of temp. and warm regions. (*Linum*, Latin for flax.)

Fls. blue or red, the petals 1–2 cm. long.
  Stigmas nearly as broad as long; petals mostly 1.5–2 cm. long. Native perennial . 2. *L. lewisii*
  Stigmas elongate. Introd. plants.
    Sepals not longer than caps.; fls. blue, the petals ca. 1.5 mm. long. ...... 4. *L. usitatissimum*
    Sepals longer than caps.; petals red, 1.5–2 cm. long .... 1. *L. grandiflorum*
Fls. yellow, the petals coppery-orange to yellow, 10–15 mm. long. Native of e. Mojave Desert
        3. *L. puberulum*

1. **L. grandiflòrum** Desf. Annual, 2–5 dm. high, much branched; main lvs. lanceolate, glaucous, 1–2 cm. long; pedicels 2–7 cm. long, slender; sepals 7–8 mm. long, lanceolate, ciliated with rather stiff hairs; petals rich red, 1.5–2 cm. long; styles 5, with linear stigmas; caps. shorter than sepals.—Grown in fl.-gardens and occasionally escaping as about Santa Barbara; native of n. Afr. April–June.

2. **L. lewísii** Pursh. [*L. perenne* L. ssp. *l.* Hult.] Plate 58, Fig. C. Glabrous perennial, usually with several densely leafy stems from a branched caudex, 1.5–7.5 dm. high; lvs. linear to lance-linear, acute, 1–2 cm. long; fls. in leafy 1-sided racemes, blue; pedicels mostly 1–3 cm. long; sepals ovate, 4–6 mm. long, obtuse to short-mucronate, with smooth hyaline margins; petals 1–1.5 cm. long; seeds shining, 3.5–4.5 mm. long.—Dry slopes and ridges between 4000 and 9000 ft.; Montane Coniferous F., Pinyon-Juniper Wd., Bristlecone Pine F.; mts. from n. L. Calif. through San Diego Co. and s. Calif. mts. to Mt. Pinos, mts. of e. Mojave Desert; to Alaska, James Bay, Tex., n. Mex. Resembling the European *L. perenne*, but with strong breeding barriers between them.

3. **L. pubérulum** (Engelm.) Heller. [*L. rigidum* var. *p.* Engelm.] Short-lived pale green perennial, 1–3 dm. tall, branched from base, puberulent throughout, the stems striate-angled; lvs. crowded below, linear to subulate, 5–10 mm. long, the upper lvs. and bracts gland-toothed; pedicels in open bracteate cymes; sepals unequal, lanceolate to lance-ovate, gland-toothed, 4–6 mm. long; petals coppery-orange to yellow, with reddish

base, 10–15 mm. long; styles 5, united to near apex; caps. ca. 4 mm. long.—Dry slopes, 4500–6000 ft.; Pinyon-Juniper Wd.; e. Mojave Desert (New York Mts., Clark Mt.); to Colo., Tex. May–July.

4. **L. usitatíssimum** L. COMMON FLAX. Slender-branched annual, 3–9 dm. high, glabrous, with stems somewhat angled above; lvs. ± lanceolate, erect or ascending, 3-nerved, 1–3.5 cm. long; pedicels erect, 2–2.5 cm. long; sepals acuminate, unequal, 6–9 mm. long; petals blue, occasionally white, 1–1.5 cm. long; styles nearly distinct, with clavate stigmas; caps. 7–10 mm. high; $2n = 30$ (Ray, 1944), 32 (Kostoff, 1940). Cult. as a field crop and escaping as a casual weed in waste places and along roads; introd. from Eu. Feb.–May.

62. Loasàceae. STICK-LEAF FAMILY
(Henry J. Thompson and Joyce Roberts)

Annual or perennial herbs, woody in one species (Mexico). Pubescent with barbed or sometimes stinging hairs. Leaves alternate or rarely opposite, entire or lobed. Flowers regular, bisexual, epigenous. Sepals persistent. Corolla apopetalous to sympetalous, sometimes with petaloid staminodia alternate with the petals. Stamens 5 to many, sometimes fascicled, filaments narrow to petaloid. Style 1, filiform, stigma not enlarged, usually persistent. Ovary inferior, 1-locular with 1–5 placentae. Ovules 1 to many. Fruit dry, dehiscent, or indehiscent and 1-seeded. Seeds variable. Embryo straight or slightly hooked. Ca. 14 genera and 200 species in temperate and tropical America, and 1 genus of 2 species in the Old World.

A. Stamens 5, capsules indehiscent, 1-seeded .... 3. *Petalonyx*
AA. Stamens many, capsules dehiscent by apical valves, few to many seeded.
   B. Placentae 5, seeds minute, flowers green or white, if white, the plant with stinging hairs
      1. *Eucnide*
  BB. Placentae 3, seeds 1–4 mm. long; flowers white to golden yellow; never with stinging hairs
      2. *Mentzelia*

1. *Eucnìde* Zucc. ROCK NETTLE

Plants herbaceous, pubescent, annual or perennial, sometimes woody at the base. Leaves alternate, petiolate, broadly ovate to suborbicular. Hairs smooth, needlelike or reflexly barbed, often stinging. Inflorescence racemose, usually bracteate. Flowers pedicellate. Petals united at the base to the short filament tube, or corolla sympetalous. Ovary spherical to clavate, with 5 intruded parietal placentae. Fruit dehiscent apically by 5 triangular valves interior to and opposite the persistent sepals. Seeds minute, very numerous, oblong, longitudinally ribbed. Eleven spp., Guatemala, Mexico and sw. U.S. (Greek, *eu*, pretty, and *knide*, nettle, "Schonnessel").

(Thompson, H. J. & W. R. Ernst, Floral Biology and Systematics of *Eucnide* (Loasaceae). Journ. Arn. Arbor. 48: 56–76. 1967.)
A. Corolla green, 1–1.5 cm. long, united for more than ½ its length into a tube with a ring of hairs below the sessile anthers (Sect. Sympetaleia) .... 1. *E. rupestris*
AA. Corolla white to cream, 3–5 cm. long, united only at the very base into a short, inconspicuous tube; ring of hairs lacking; all anthers on long filaments (Sect. Mentzeliopsis)
      2. *E. urens*

1. **E. rupéstris** (Baill.) Thompson & Ernst. [*Loasella r.* Baill.; *Sympetaleia r.* (Baill.) S. Wats.] Annual, up to 3 dm. tall; leaf blades suborbicular, 5–8 cm. wide, cordate at base, upper surface shiny green; earliest flowers often solitary and axillary, later ones in crowded inflorescences; corolla green, 1–1.5 cm. long, cylindrical, the lobes 3–5 mm. long, erect, inconspicuous on the plants; stamens epipetalous, included, the anthers sessile or subsessile in very short, stout filaments, with a ring of upwardly curved hairs below the stamens; style short and thick, not exceeding the lower anthers; fruit cylindric, 1–1.5 cm. long, pedicels short at anthesis, elongating up to 2 cm.; seeds numerous, minute, less than

1 mm. long; $n=21$.—Steep rocky slopes, occasional in washes below 3000 ft.; Creosote Bush Scrub, in Calif. known only from Painted Gorge, Imperial Co.; to sw. Ariz., and Mex. on the islands and shores of the Gulf of Calif. Jan.–March.

2. **E. ùrens** Parry. [*Mentzelia u.* Parry ex Gray.] Herbaceous perennial, woody at base, up to 1 m. tall, pubescent and with stinging hairs; leaf blades 2–6 cm. long, 1–5 cm. wide, ovate, coarsely toothed, grayish-green; inflorescences of several flowers, terminal, the flowers sometimes subtended by small leaves; corolla pale yellowish white, 3–5 cm. long, the petals rigid, spreading, fused with the filament bases to form a very short, inconspicuous tube about 2 mm. long; stamens 1–2 cm. long, included, the filaments thick and slanted inward around the stout style; stigma exceeding the anthers; fruit clavate, 1–2 cm. long, 8–12 mm. wide, pedicels to about 1.5 cm. long, neither reflexed nor elongated; seeds numerous, minute, less than 1 mm. long; $n=21$.—Cliffs or rocky slopes, occasional in washes, below 4000 ft.; Creosote Bush Scrub; Mojave Desert from Red Rock Canyon e.; to sw. Utah, Ariz., s. along Colorado R. to n. Baja Calif. March–May.

2. Mentzèlia L. BLAZING STAR

Annual or perennial herbs. Hairs barbed but not stinging. Leaves alternate, ovate to linear, lobed or entire, petiolate or sessile. Flowers solitary and axillary or in terminal bracteate more or less cymose inflorescences. Perianth 5-merous, usually 2-seriate. Stamens few to many, filaments filiform to petaloid, staminodes present or absent. Ovary with 3 (in Calif.) –5 placentae. Ovules few to many. Fruits dehiscent, cup-shaped, turbinate to cylindrical, sessile to short-pedicillate. Seeds irregularly angled, flattened and winged, or oblong, surface smooth, tuberculate or wrinkled, sometimes striate. Ca. 60 species in temperate and tropical America, mostly in sw. U.S. and Mex. (C. *Mentzel,* 1622–1701, German botanist.)

A. Petals 5 and clearly distinct from the outer stamens.
  B. Outer stamens with narrow filaments.
    C. Perennial; petals lanceolate, longer than 4 cm. .... 12. *M. laevicaulis*
    CC. Annual; petals shorter than 2.5 cm., broadly oval or obovate (Sect. Trachyphytum in part).
      D. Seeds (and maturing ovules) angular, prismatic, regularly grooved along 3 sides; capsules not narrowed at base.
        E. Petals ovate, apex acute, leaves deeply and irregularly lobed; below 5000 ft.
          1. *M. affinis*
        EE. Petals broadly oval with rounded tips, or obovate with retuse tips; leaves entire or dentate; above 4500 ft.
          F. Leaves, at least upper, orbicular; capsules less than 1 cm.; above 7500 ft.; habit compact; San Antonio Mts. .... 6. *M. dispersa* var. *obtusa*
          FF. Leaves entire or dentate, linear to lanceolate; capsules longer than 1 cm.; between 4500–8000 ft.; habit erect, open .... 20. *M. pinetorum*
      DD. Seeds not regularly grooved, irregularly angled or round; capsules narrowed toward base. (*Note*: the following species form a polyploid complex which can only be distinguished with difficulty; up to 5 species may grow together.)
        E. Bracts lobed; all capsules erect.
          F. Bracts without white area at base .... 2. *M. albicaulis*
          FF. Bracts with white area at base.
            G. Petals 0.7–1.7 cm. long; styles 0.6–1.1 cm. long; rare, Pine Mt.
              8. *M. gracilenta*
            GG. Petals mostly 2–7 mm. long; styles 2–7 mm. long.
              H. Flowers terminal, not solitary in axils; bracts concealing fruit
                4. *M. congesta*
              HH. Flowers and bracts not as above.
                I. First formed capsules 0.9–1.7 cm. long; cauline leaves often entire, linear; above 4500 ft. .... 16. *M. montana*
                II. First formed capsules 1.4–3 cm. long; leaves never entire; below 5000 ft.
                  J. Petals yellow to deep orange; branching strict; bracts longer than wide
                    27. *M. veatchiana*

JJ. Petals yellow; branching spreading; bracts as wide or wider than long
.... 23. *M. ravenii*
EE. Bracts usually entire; oldest capsules recurved.
F. At least the oldest capsules recurved more than 90°.
G. Seeds uniformly light in color, not tessellate.
H. Seeds rounded or sharply but irregularly angled, surface smooth; leaves with short rounded lobes; on fine sandy soils .... 5. *M. desertorum*
HH. Seeds rounded, surface with slight, pointed papillae; leaves with long, pointed lobes .... 18. *M. obscura*
GG. Seeds moderately to heavily tessellate.
H. Seeds rounded.
I. Calyx lobes 1.1–1.5 cm. long; styles 1–1.4 cm. long; leaves deeply lobed
.... 7. *M. eremophila*
II. Calyx lobes less than 5 mm. long; styles less than 7 mm. long; upper leaves often entire .... 17. *M. nitens*
HH. Seeds irregularly angled.
I. Leaves deeply and sharply lobed; styles 5–6 mm. long   3. *M. californica*
II. Leaves with rounded, short to medium lobes; style 0.6–1 cm. long
.... 11. *M. jonesii*
FF. None of the capsules recurved more than 90°.
G. Rosette leaves with short to medium lobes, upper leaves broad-based, somewhat clasping; petals obovate or broadly ovate .... 15. *M. mojavensis*
GG. Rosette leaves entire or with long, irregularly toothed lobes, upper leaves not clasping, often entire; petals ovate .... 2. *M. albicaulis* (in part)
BB. Outer stamens with broad filaments.
C. Desert annuals; bracts laciniate; seeds round or obovate with 1 or 2 grooves on both faces, surface tuberculate (Sect. Bicuspidaria in part).
D. Bracts scarious with conspicuous white area and narrow green margin
.... 10. *M. involucrata*
DD. Bracts green without white area.
E. Leaves sessile; petals orange .... 9. *M. hirsutissima*
EE. Leaves, at least upper, petiolate; petals cream-colored.
F. Lateral cusps of stamens longer than central anther-bearing cusp; seeds ovate, constricted at middle .... 25. *M. tricuspis*
FF. Lateral cusps of stamens shorter than central cusp; seeds rounded, constricted above and below middle .... 26. *M. tridentata*
CC. Coastal annual; bracts ovate, entire, cupped; seeds prismatic, surface smooth
.... 14. *M. micrantha*
AA. Petals more than 5, or if appearing to be 5, then these grading through a whorl of petallike staminodia to stamens with broad filaments.
B. Petals 8, less than 8 mm., about as long as calyx lobes; seeds obovate, grooved at middle on both faces .... 24. *M. reflexa*
BB. Petals 10, or 5 with an inner series of 5 petallike staminodia, longer than 8 mm.; seeds flattened, winged (Sect. Bartonia in part).
C. Leaves, at least the upper, cordate-clasping at base .... 19. *M. oreophila*
CC. Leaves often sessile, but never cordate-clasping.
D. Petals 10, ovate-lanceolate, tip acute; on gypseous clay slopes, Valley Wells Station, San Bern. Co. .... 21. *M. pterosperma*
DD. Petals 10, or 5 with an inner series of 5 petallike staminodia, obovate or oblanceolate, tip rounded; not on gypseous clay soils.
E. Petals 10; leaves lanceolate, irregularly lobed; seeds obovate, longer than 2 mm., wing broad; on sandy soils .... 13. *M. longiloba*
EE. Outer petals 5 with inner whorl of 5 petallike staminodia; leaves oblanceolate or broadly oval, shallowly lobed or coarsely dentate; seeds oblong, less than 2 mm.; on cliffs or rocky gravelly slopes .... 22. *M. puberula*

1. **M. affinis** Greene. [*Acrolasia viridescens* Heller.] Plants annual; 2–4.5 cm. tall, stems erect, several branched from near base; rosette leaves broad with short, wide, blunt lobes, cauline leaves broad-based with longer, more pointed lobes, leaves often entire in desert populations; bracts ovate-lanceolate, entire or somewhat 2 lobed; flowers opening in the morning; calyx lobes 1–6 mm. long; petals yellow, ovate, apex acute,

4–12 mm. long; capsules cylindrical, erect or recurved, not narrowed at base, 0.7–3.2 cm. long; seeds prismatic, deeply grooved along 3 sides, ends truncate, surface smooth; $n=9$.—Open areas in V. Grassland, Foothill W., Creosote Bush Scrub, below 4000 ft.; throughout s. Calif.; nw. of Sierras to San Joaquin Co., e into Ariz., s. into Baja Calif. March–May.

2. **M. albicáulis** (Hook.) T. & G. [*Bartonia a.* Hook.] Plants annual; 1–4 dm. tall, stems erect or decumbent, sparingly or many branched; rosette leaves with slender, pointed lobes, upper leaves linear and entire, linear and few-lobed, or ovate-lanceolate and entire or few-lobed; bracts usually linear and entire or ovate and entire, or in some populations at higher altitudes angular to slightly 3-lobed; flowers opening in the morning; calyx lobes 2–4 mm. long; petals 3–10 mm. long, yellow with an orange spot at base, usually ovate with an acute apex, sometimes slightly obovate; stamens 3–5 mm. long; style 3–5 mm. long; capsules mostly erect, narrowed at base, the first-formed capsules sometimes recurved to about 90°, in some populations to 180°, 1.5–3.5 cm. long, about 2 mm. wide; seeds irregularly angled, surface with large, pointed papillae; $n=36$.—Desert slopes between 1500 and 5500 ft.; Creosote Bush Scrub, Joshua Tree Wd., Sagebrush Scrub, Pinyon-Juniper Wd., throughout Calif.; s. into Mex., n. into Wash., e. to Colo. and Tex. Late March–June.

3. **M. califórnica** Thompson & Roberts. Plants annual, erect, branching habit spreading, 2–4 dm. tall; rosette leaves irregularly lobed with long pointed lobes, upper leaves ovate-lanceolate, fewer-lobed or rarely entire; bracts entire, ovate-lanceolate, not white at base; flowers opening in the morning; calyx lobes 4–6 mm. long; petals 6–11 mm. long, yellow with an orange spot at base, apex rounded or acute; stamens 5–6 mm. long; style 5–6 mm. long; capsules recurved to 180°, 1.5–3.5 cm. long, about 1.5 mm. wide, narrowed at base; seeds irregularly angled, slightly tessellate, surface with medium, rounded to slightly pointed papillae; $n=27$.—Desert plains and roadside embankments, usually below 3000 ft.; Creosote Bush Scrub; throughout n. Mojave Desert of Calif.; e. into Nev. March–April.

4. **M. congésta** T. & G. Plate 58, Fig. D. Plants annual; 2–5 dm. tall, erect, several to many branched from base; leaves linear or linear-lanceolate, the rosette leaves entire or with short rounded lobes, upper leaves usually entire, sometimes ovate-lanceolate; outer floral bracts enclosing or concealing the mature fruit, shallowly and bluntly 5-lobed with a conspicuous white area and a narrow green margin, inner bracts narrower, usually 3-lobed, with a white basal area; flowers opening in the morning; calyx lobes 1–3 mm. long; petals 2–10 mm. long, yellow with an orange spot at base, obovate, apex retuse; stamens 3–7 mm. long; style 3–7 mm. long; capsules erect, narrowed at base, 7–10 mm. long; seeds irregularly angled and slightly rounded, lightly tessellate, the surface with rounded papillae; $n=9$.—In isolated, discrete colonies on disturbed slopes or open areas; Yellow Pine F. above 5500 ft. in mountain ranges of s. Calif.; Sagebrush Scrub ne. of crest of Sierra Nevada to El Dorado Co., in mountains of Nev., locally near Fort Hall, Idaho, between 5500 and 8000 ft. Late May–July.

5. **M. desertòrum** (Davids.) Thompson & Roberts. [*Acrolasia d.* Davids.] Plants annual; stems erect, compact and much branched from base, the stems rather slender; rosette leaves with very short, rounded lobes, upper leaves more ovate-lanceolate, few lobed as in the rosette leaves, or sometimes entire, but lobes always short; bracts entire, ovate or ovate-lanceolate, not white at base; flowers opening in the morning; calyx lobes 2–3 mm. long; petals 2–5 mm. long, yellow with an orange spot at base, ovate or ovate-lanceolate, not white at base; calyx lobes 2–3 mm. long; petals 2–5 mm. long, yellow with an orange spot at base, ovate, apex acute or rounded; stamens 2–4 mm. long; style 2–4 mm. long; capsules recurved to 180°, narrowed at base, 1.3–2.5 cm. long, narrow, about 1 mm.; seeds more or less rounded, or sharply but irregularly angled, surface smooth, not tessellate; $n=9$.—On fine sandy desert flats below 2000 ft.; Creosote Bush Scrub; common throughout the Sonoran Desert n. sporadically to extreme s. Inyo Co.; s. into Baja Calif. and Son., Mex. Late Feb.–March.

6. **M. dispérsa** Wats. var. **obtùsa** Jeps. Plants annual; erect, 1–3 dm. tall; rosette

leaves absent or loosely formed, shallowly and bluntly lobed, at least the upper cauline leaves orbicular, entire; bracts orbicular, entire; flowers opening in the morning; calyx lobes 1-3 mm. long; petals yellow, broadly ovate, tips rounded, 2-4 mm. long; stamens 2-3 mm. long; style 2-3 mm. long; capsules cylindrical, not narrowed at base, usually less than 1 cm. long, about 1 mm. wide; seeds prismatic, grooved along 3 sides, ends truncate, about as long as wide, less than 1.2 mm.; $n=9$.—In open areas in montane forests above 7000 ft.; San Gabriel Mts., Mt. Baldy, San Antonio Mts., Icehouse Canyon. July-Aug.

7. **M. eremóphila** (Jeps.) Thompson & Roberts. [*M. lindleyi* T. & G. var. *eremophila* Jeps.] Plants annual; erect, stems stout, 3-6 dm. tall, several branched; rosette and upper leaves with very long, slender pointed lobes; bracts entire, lanceolate, not white at base; flowers opening in the morning; calyx lobes 1-1.5 cm. long; petals 1.5-2.5 cm. long, clear yellow without an orange basal spot, obovate, apex acute or mucronate; stamens 6-10 mm. long; style 1-1.4 cm. long; capsules narrowed at base, recurved to 180°, strongly fibrous, often splitting longitudinally into 3 pieces, 1.7-3.7 cm. long; seeds rounded, slightly tessellate, hilum enlarged, slightly tessellate, surface with large, rounded papillae; $n=9$.—On canyon slopes below 4000 ft.; Creosote Bush Scrub; e. margins of Kern Co. and the nw. corner of San Bernardino Co. Late March-May.

8. **M. gracilénta** T. & G. Plants annual; 2-7 dm. tall, stems stout, several branched from lower portion of stem; rosette leaves linear-lanceolate with short, rounded lobes, upper leaves more ovate-lanceolate with broad blades and short blunt lobes, the lobes sometimes more pointed in uppermost leaves; bracts pinnately 7-10-lobed with a white basal area, closely appressed to capsule, usually longer than broad; flowers opening in the morning; calyx lobes 4-7 mm. long; petals 0.7-1.7 cm. long, yellow with an orange spot at base, broadly obovate with a retuse apex; stamens 6-8 mm. long; style 6-11 mm. long; capsules erect, narrowed at base, 1.3-2.3 cm. long, about 3 mm. wide; seeds irregularly angled, large, about 1.5 mm. long, heavily tessellate, the surface with slightly rounded papillae; $n=18$.—On steep, coarse, talus slopes in Foothill Wd.; rare in s. Calif. at Pine Mt.; throughout Monterey, San Benito and Fresno cos. April-May.

9. **M. hirsutíssima** Wats. var. **stenophýlla** (U. & G.) Jtn. [*M. s.* U. & G.; *M. peirsonii* Jeps.] Plants annual; erect, branching at base, 1.5-4 dm. tall; rosette leaves oblanceolate, crenate, cauline leaves lanceolate to oblanceolate, sharply lobed; bracts laciniate, at base of capsule; flowers opening in the morning; calyx lobes 1.5-2 cm. long; petals orange, obovate, tip mucronate, 1.5-2.5 cm. long; stamens reflexed toward style at anthesis, about 1 cm. long, outermost filaments with a band of orange near top; style 1-1.5 cm. long; capsules cylindrical, tapering toward base, 2-2.5 cm. long; seeds obovate, somewhat flattened, grooved through the middle on both faces, surface tuberculate.— On coarse rubble and talus slopes, below 2000 ft.; Creosote Bush Scrub; locally in s. Calif. in Imperial Co. at Mountain Springs Grade, Box Canyon and Palm Canyon, San Diego Co.; s. into Baja Calif. March-April.

10. **M. involucràta** Wats. Plants annual; stems erect, usually much branched, 1-4.5 dm. tall; rosette leaves lanceolate, dentate or lobed, lower leaves medium to broad in width, irregularly and sharply lobed, or dentate, upper leaves sessile, broad at base, lanceolate to ovate; bracts laciniate, white with a green margin; flowers opening in the morning; calyx lobes 0.7-2 cm. long; petals satiny, cream-colored with reddish veins and reddish at base, obovate with mucronate tip, 1.3-6.5 cm. long; stamens 1-2 cm. long, filaments flattened, outermost with band of orange near tip of filament, reflexed toward style at anthesis; style 1.3-2.5 cm. long; capsules broadly cylindrical, 0.8-2.5 cm. long, 4-10 mm. wide; seeds irregular, somewhat flattened, more or less rounded with broadest portion at the middle, grooved above and below the middle on both faces, surface tuberculate; $n=9$.—Roadside embankments and desert slopes below 4500 ft.; Creosote Bush Scrub; Mojave and Colorado deserts; s. into Baja Calif., e. into Ariz. Feb.-April.

A. Petals 1.3-2.6 cm. long; style 1.3-1.6 cm. long, usually inbreeding . . . . . ssp. *involucrata*
AA. Petals 3-6.5 mm. long; style 1.8-2.5 cm. long; usually outcrossing . . . . ssp. *megalantha*

Ssp. **megalántha** Jtn. Plants up to 4.5 dm. tall; generally larger than the typical ssp.; usually w. of Corn Springs, San Bernardino Co.

11. **M. jònesii** (U. & G.) Thompson & Roberts. [*M. albicaulis* (Hook.) T. & G. var. *j.* U. & G.] Plants annual; stems erect, or often spreading and viney, slender; rosette leaves linear-lanceolate with short to medium-long, rounded lobes, upper leaves ovate-lanceolate with short, pointed lobes, or sometimes ovate and entire; bracts entire, ovate-lanceolate, not white at base; flowers opening in the morning; calyx lobes 6–8 mm. long; petals 0.8–2 cm. long, yellow with or without an orange spot at base, ovate to somewhat obovate, apex acute or rounded; stamens 5–8 mm. long; style 6–10 mm. long; capsules narrowed at base, recurved to 180°, 1.3–3 cm. long, 2–2.5 mm. wide; seeds irregularly angled, slightly to moderately tessellate, the surface with large, more or less rounded papillae; $n = 18$.—In coarse soils on desert plains and slopes, often growing up through shrubs, below 4000 ft.; Creosote Bush Scrub, Joshua Tree Wd.; s. cent. Inyo Co. s. throughout the Mojave Desert; e. to s. Nev. and w. Ariz. along the Colorado River. March–April.

12. **M. laevicáulis** (Doug. ex Hook.). T. & G. [*Bartonia l.* Doug. ex Hook.] Plate 58, Fig. E. Plants perennial; erect, branching from the upper portion of the stem, 2–7.7 dm. tall; rosette leaves lanceolate, shallowly to somewhat deeply lobed, lower cauline leaves lanceolate, medium to broad, margins sinuate, toothed, or shallowly lobed, upper leaves usually toothed; bracts at base of capsule, linear to lanceolate, entire, toothed only at base, or laciniate with 4–5 lobes; flowers opening in early evening; calyx lobes 2–4 cm. long; petals 5, yellow, lanceolate, 3–7 cm. long, 0.7–2 cm. wide; stamens 1.5–4.5 cm. long, more or less equal in length, the outer 5 filaments broadened to about 2 mm. in width; style 2.4–5.8 cm. long; capsules cylindrical, 1.5–4 cm. long; seeds obovate, about 2.6 cm. long, wing to 0.5 mm. wide, light brown, surface granulate; $n = 11$.—On rocky talus, sandy flats and washes, often on road cuts, in a variety of plant communities, inner cismontane s. Calif., throughout Calif.; n. to B.C., e. through Utah to w. Wyo. and Mont. June–Oct.

13. **M. longilòba** Darl. Plants perennial, sometimes flowering first year; erect, few to many branched from base, 2–7 dm. tall, stems often glabrous; rosette leaves linear-lanceolate to linear-oblanceolate, blades medium to broad, irregularly short-lobed, cauline leaves similar; bracts at base of capsule, entire; flowers opening in late afternoon; calyx lobes 5–10 mm. long; petals 5 or 10, golden yellow fading to light yellow, 1–2.5 cm. long, 4–10 mm. wide, broadly obovate or oblanceolate, tip obtuse or rounded; stamens grading in length, inner 4 mm., outer 1.2 cm. long, filaments grading from narrow inner ones to outer ones to 5 mm. wide; style 0.6–1.2 cm. long; capsules short-cylindrical, to 1.8 cm. long; seeds flattened, obovate, wing very broad to 1.3 mm. wide, surface colliculate; $n = 9$.—On fine sandy soils; Creosote Bush Scrub; cent. Mojave Desert, e. to Ariz., s. into Baja Calif. and n. cent. Son., Mex. May–Aug.

14. **M. micrántha** (Hook. & Arn.) T. & G. [*Bartonia m.* Hook. & Arn.] Plant annual; erect; leaves linear-lanceolate, lanceolate, or ovate, the rosette leaves irregularly toothed or dentate, or with sinuate margins, upper leaves ovate-lanceolate or somewhat orbicular, the margins sinuate or dentate, rarely entire; bracts ovate, entire or with sinuate margins, the apex blunt or acute, often concealing the inflorescence; flowers opening in the morning; calyx lobes 1–3 mm. long; petals 2–5 mm. long, yellow, ovate with an acute apex; stamens with the outer 5 with dilated filaments and bifid apices, inner stamens with narrow filaments, 2–4 mm. long; style 2–5 mm. long; capsules short-cylindrical, not narrowed at the base, usually erect, 5–12 mm. long, about 2 mm. wide; seeds few, more or less prismatic, the surface smooth, not tessellate; $n = 9$.—Disturbed areas below 5000 ft.; Chaparral, Foothill Wd.; coastal Calif. from Baja Calif. n. to Trinity Co., and on the Channel Islands. April–May.

15. **M. mojavénsis** Thompson & Roberts. Plants annual; erect, branching pattern moderately spreading, 2–4 dm. tall; rosette leaves medium in width with short to medium, rounded lobes; upper leaves broadly ovate-lanceolate and lobed, sometimes slightly

clasping at base; bracts broadly ovate, entire or 3–5-lobed, rarely with a faint white area at base; flowers opening in morning; calyx lobes 2–5 mm. long; petals 6–8 mm. long, yellow with an orange spot at base, obovate or broadly ovate, apex acute or rounded, rarely retuse; stamens 4–5 mm. long; style 4–5 mm. long; capsules narrowed at base, erect or the earliest capsules recurved to about 90°, 1.2–2.5 mm. long, about 2–3 mm. wide; seeds irregularly angled, slightly or moderately tessellate, surface with somewhat pointed papillae; $n = 27$.—On desert plains and roadside embankments below 3500 ft.; Creosote Bush Scrub; along the w. margins of the Mojave Desert in Los Angeles and Kern cos. Late March–April.

16. **M. montána** (Davids.) Davids. [*Acrolasia m.* Davids.] Plants annual; stems slender, 2–4 dm. tall, usually sparingly branched, erect; rosette leaves linear, entire, with short rounded lobes, or with medium-long, somewhat pointed lobes; upper leaves linear and entire or narrow and few lobed; bracts usually 3-lobed, sometimes angular, with a white area at base, not concealing the mature fruit; flowers opening in the morning; calyx lobes 2–4 mm. long; petals 2–8 mm. long, yellow with an orange basal spot, obovate, apex rounded or retuse; stamens 2–6 mm. long; style 2–6 mm. long; capsules narrowed at base, erect, or occasionally slightly curved, but never more than 45°, 0.5–2 cm. long; seeds irregularly angled, heavily tessellate, surface with small pointed papillae; $n = 18$.— On talus slopes and open areas in montane forests throughout Calif. between 4500 and 6000 ft.; Sagebrush Scrub and Yellow Pine F.; Oreg., Colo., Tex., Ariz., N. Mex., Baja Calif. Early May–July.

17. **M. nìtens** Greene. Plate 59, Fig. A. Plants annual; erect, 2–3 dm. tall, branching habit spreading; rosette leaves narrow to medium in width with short rounded lobes, upper leaves ovate-lanceolate, entire or with a few sharp lobes; bracts entire, lanceolate, not white at base; flowers opening in the morning; calyx lobes 3–5 mm. long; petals 6–16 mm. long, yellow, with an orange spot at base, ovate or obovate, the apex rounded or acute; stamens 3–6 mm. long; style 4–7 mm. long; capsules narrowed at base, recurved to 180°, 1.4–2.5 cm. long; seeds rounded, about 1 mm. long, 1 mm. wide, slightly tessellate, the surface with large, rounded papillae; $n = 9$.—Roadside embankments and desert plains, often growing up through desert shrubs; Creosote Bush Scrub, Sagebrush Scrub; 4000–5000 ft., s. Inyo Co.; n. into Mono Co., w. Nev. April–May.

18. **M. obscùra** Thompson & Roberts. Plants annual; erect or spreading, many branched from base, often compact and rounded; rosette leaves with long, pointed, irregular lobes, upper leaves ovate-lanceolate, usually entire or with a few long pointed lobes; bracts entire, ovate or ovate-lanceolate, not white at base; flowers opening in the morning; calyx lobes 2–5 mm. long; petals 4–8 mm. long, yellow with an orange spot at base, ovate or occasionally obovate, apex rounded or acute; stamens 3–6 mm. long; style 3–6 mm. long; capsules narrowed at base, recurved to 180°, 1.3–3 cm. long; seeds more or less rounded, not tessellate, the surface with slightly pointed papillae, about 1 mm. long and 1 mm. wide; $n = 18$.—On disturbed sites along roadside embankments and desert slopes; Creosote Bush Scrub, Joshua Tree Wd.; throughout the Mojave and Sonoran Deserts from n. cent. Inyo Co. s. into Baja Calif.; e. into w. Ariz. and Nev., locally in Utah. Feb.–March.

19. **M. oréophila** Darl. Plants perennial, sometimes flowering first year; erect, much branched from base, 1.5–4 dm. tall; rosette leaves oblanceolate, crenate or shallowly lobed, the lowermost leaves petiolate, upper leaves ovate or oblong, sessile, cordate-clasping at base, entire, sinuate, crenate or toothed on the margins; bracts at or just below the base of capsule, linear or lanceolate, entire; flowers opening in late afternoon; calyx lobes 6–10 mm. long; petals 5, golden yellow, 0.8–1.5 cm. long, 3–5 mm. wide, somewhat obovate to narrowly oblanceolate, tip rounded to obtuse; stamens grading in length, inner 3 mm., outer 10 mm. long, filaments graded in width from inner ones narrow to outer 5 to 2.5 mm. wide; style 5–9 mm. long; capsules cup-shaped, 4–10 mm. long; seeds obovate to somewhat oblong, 2–2.5 mm. long, wing 0.4–0.8 mm. wide, surface white, strongly colliculate; $n = 11$.—Between 2000 and 7200 ft.; Creosote Bush Scrub,

Plate 59. Fig. A, *Mentzelia nitens*. Fig. B, *Petalonyx thurberi*. Fig. C, *Buddleja utahensis*. Fig. D, *Lythrum californicum*. Fig. E, *Eremalche rotundifolia*. Fig. F, *Horsfordia newberryi*.

Joshua Tree Wd., Pinyon-Juniper Wd.; canyons of Inyo Co., locally in the Clark Mts. in San Bernardino Co. May-Aug.

20. **M. pinetòrum** Heller. Plants annual; erect, single or many branched from base, 1-4 dm. tall; rosette leaves lacking or loosely formed, lower leaves linear, entire or dentate, widely varying in width, sometimes with short blunt lobes, upper cauline leaves linear or lanceolate, entire, sinuate or dentate; bracts ovate, entire, at base of capsule; flowers opening in the morning; calyx lobes 1-3 mm. long; petals yellow, broadly oval with rounded tips, 3-5.5 mm. long; stamens 2-3 mm. long; style 2-3 mm. long; capsules cylindrical, not narrowed at base, not recurved, 1-3 cm. long, about 1 mm. wide; seeds prismatic, grooved along 3 sides, ends truncate, less than 1.2 mm. long, longer than wide; $n=18$.—Open areas and roadside embankments, usually below 8000 ft.; montane forests of s. Calif.; n. throughout the Sierras. June-July.

21. **M. pterospérma** Eastw. Plants perennial, sometimes flowering the first year; 0.8-3 dm. tall, white, pubescent; rosette leaves oblanceolate, crenate or with short rounded lobes, lower leaves oblanceolate, ovate or lanceolate, crenate or with short, more or less pointed to rounded lobes, lowermost leaves petiolate, upper leaves more ovate, sessile, margins crenate or toothed; bracts at or just below base of capsule, linear and entire; flowers opening in late afternoon; calyx lobes 7-12 mm. long; petals 10, yellow, 0.8-2.2 cm. long, 3-8 mm. wide, ovate-lanceolate, tip acute; stamens grading in length, inner 3 mm., outer 1.4 cm. long, in graded series from narrow inner filaments to outer filaments up to 2.5 mm. wide; style 5-15 mm. long; capsules more or less cup-shaped to subcylindrical, 1-2 cm. long; seeds flattened, obovate to almost rounded, 2-2.3 mm. long, wing about 0.5 (Calif. only) mm. wide, surface white, granulate to slightly colliculate; $n=11$.— On gypseous clay slopes locally in Calif. near Valley Wells, San Bernardino Co.; Creosote Bush Scrub; nw. Ariz., s. Nev., Colo., Utah. April-June.

22. **M. pubérula** Darl. Plants perennial, sometimes flowering first year; erect, 1-3 dm. tall, few to many branched at base; rosette leaves not known, lower leaves oblanceolate to broadly oval, coarsely dentate or shallowly lobed, lowermost leaves petiolate, upper leaves similar but smaller, sessile but not clasping at base; bracts at or just below base of capsule, linear and entire; flowers opening in late afternoon; calyx lobes 4-7 mm. long; petals 5, golden yellow, 5-12 mm. long, 3-6 mm. wide, obovate to broadly obovate, tip rounded to obtuse; stamens grading in length, inner 3 mm., outer 8 mm. long, inner stamens with narrow filaments, grading to outer 5 filaments to 3 mm. wide; style 5-7 mm. long; capsules cup-shaped, 7-10 mm. long; seeds flattened, more or less oblong, 1.8-2 mm. long, wing to 0.6 mm. wide, surface light tan, granulate; $n=11$.—On cliffs or rocky, gravelly slopes; Creosote Bush Scrub; in mountains in se. Calif. along the Colorado River from Valley of Fire, Nev., s. to the Yuma area; into n. cent. Baja Calif. Late Feb.- May.

23. **M. ràvenii** Thompson & Roberts. Plants annual; erect but spreading, much branched from base, stout, 2-4 dm. tall; rosette leaves lanceolate, broad, with short, rounded lobes, upper leaves more ovate-lanceolate, broad, with fewer, sharply pointed lobes; bracts broadly 3-5-lobed with a white area at base; flowers opening in the morning; calyx lobes 3-4 mm. long; petals 5-10 mm. long, yellow with an orange spot at base, obovate, apex retuse; stamens 3-7 mm. long; style 4-7 mm. long; capsules narrowed at base, erect, 0.9-2.3 cm. long, about 3 mm. wide; seeds irregularly angled, slightly to moderately tessellate, the surface with rounded papillae; $n=18$.—On roadside embankments and canyon slopes below 4000 ft.; Creosote Bush Scrub; desert margin areas in Los Angeles Co. and w. Riverside Co. March-April.

24. **M. refléxa** Cov. Plants annual; rounded, compact, much branched at base, to 2 dm. tall; rosette leaves lanceolate to oblanceolate, crenate, cauline leaves petiolate, broader; bracts lanceolate; flowers opening in morning; calyx lobes 6-8 mm. long; petals 8, in 2 series, shorter or equal to sepals; stamens in irregular series, outermost about as long as petals with broadened filaments, inner shorter, 2-3 mm. long; style 2-3 mm. long; capsules cylindrical, tapering slightly at both ends, first formed capsules reflexed; seeds obovate, grooved at middle on both faces, surface tuberculate; $n=10$.—On coarse

rubble, below 3000 ft.; Creosote Bush Scrub; Panamint Valley and Death Valley, occasionally as far south as Kelso, San Bernardino Co. March–May.

25. **M. tricúspis** G. [*Bicuspidaria t.* Rydb.] Plants annual; erect, 0.5–3 dm. tall, several branched from base; rosette leaves lanceolate, crenate, cauline leaves lanceolate to oblanceolate, short-petiolate, short lobed or dentate; bracts laciniate, green; flowers opening in morning; calyx lobes 6–8 mm. long; petals cream colored, broadly obovate, tip mucronate, 1–2.5 cm. long; stamens 5–8 mm. long, reflexed towards style, lateral cusps longer than central anther-bearing cusp; style 5–8 mm. long; capsules cylindrical, tapering towards base, 1–1.6 cm. long, the first formed capsules reflexed; seeds ovate, broadest at top, constricted at the middle on both faces, surface tuberculate; $n=10$.— On sandy slopes and roadside embankments; Creosote Bush Scrub; cent. and w. Mojave Desert, s. along the Colorado River to the Mex. border. March–April.

26. **M. tridentàta** (David.) Thompson & Roberts. [*Acrolasia t.* Davids. *M. tricuspis* G. var. *brevicornuta* Jtn.] Plants annual; erect, 0.5–1.5 dm. tall, several branched from base; rosette leaves lanceolate, crenate, cauline leaves lanceolate to oblanceolate, short-petiolate, short lobed or dentate; bracts laciniate, green; flowers opening in morning; calyx lobes 6–7 mm. long; petals cream-colored, broadly obovate, tip mucronate, 1–2 cm. long; stamens 5–8 mm. long, reflexed towards style, lateral cusps shorter than central anther-bearing cusp; style 5–8 mm. long; capsules cylindrical, tapering towards base, 1–1.5 cm. long, first formed capsules reflexed; seeds rounded, constricted above and below the middle on both faces, surface tuberculate; $n=10$.—On talus slopes of mesas around Barstow, San Bernardino Co., sporadically in canyons of Inyo Co., Creosote Bush Scrub. March–April.

27. **M. veatchiàna** Kell. Plants annual; erect, 2–5 dm. tall, several to many branched from base, branching pattern strict; rosette leaves broad or narrow, with short, rounded lobes, or long, irregularly toothed lobes, the upper leaves similar, or ovate-lanceolate, sometimes clasping at base; bracts 3–5 lobed, broad or occasionally angular and narrow, with white spot at base; flowers opening in the morning; calyx lobes 2–5 mm. long; petals 3–8 mm. long, yellow with a deep orange spot at base, or petals deep orange, broadly obovate, apex retuse; stamens 3–5 mm. long; style 3–5 mm. long; capsules narrowed at base, erect, or sometimes the first formed capsules slightly recurved to 45°, 1.2–3 cm. long, about 3 mm. wide; seeds irregularly angled, somewhat cubicle, moderately tessellate, the surface with large, rounded or pointed papillae; $n=27$.—Open slopes and disturbed areas along roadsides, usually below 3500 ft., Creosote Bush Scrub, Joshua Tree Wd.; in Grassland and Foothill Wd. below 5000 ft.; throughout s. Calif. away from the coast; n. into s. Ore., s. into Baja Calif., e. into Nev. and Ariz. March–April.

3. *Petalónyx* G. Sandpaper Plant

Shrubs or sub-shrubs. Leaves alternate, petiolate or sessile, entire, toothed, or crenate. Inflorescence racemose. Calyx lobes linear to lance-linear. Petals 5, cream to white, clawed, inner surface glabrous, outer hispid, claws free or united to form a tube, limbs usually reflexed. Stamens 5, originating between the petals but outside the petals at anthesis. Ovary uniloculate, mature fruit cylindrical, constricted near apex then flared above, 3–5 nerved, one-seeded, indehiscent. Seed pointed at hilum, plain or with fine striations on surface. (Greek, *petalon*, petal, and *onyx*, claw.) Five species of southwestern U.S. and Mex.

(Davis, W. S. & H. J. Thompson, A Revision of *Petalonyx* (Loasaceae) with a Consideration of Affinities in Subfamily Gronovioidae. Madroño 19: 1–18. 1967.)

A. Petals distinct .... 1. *P. linearis*
AA. Petals connivent, claws forming a tube.
    B. Petals longer than 5 mm.; leaves petiolate, all similar in size .... 2. *P. nitidus*
    BB. Petals shorter than 5 mm.; leaves sessile, reduced upward along the stem .. 3. *P. thurberi*

Petalonyx

1. **P. lineàris** Greene. Suffrutescent, rounded, with numerous, erect branches, 1.5–10 dm. tall; leaves sessile to very short-petiolate, all similar in size, linear to oblanceolate, acute or occasionally obtuse, narrowed basally becoming cuneate to obtuse, entire or rarely dentate, green, somewhat shiny; inflorescence a short, capitate terminal spike, 4–10 cm. long, in fruit elongating to 21 cm.; bracts ovate to round-ovate, acute to obtuse, rarely retuse, cordate, entire, sinuate or crenulate at base; calyx lobes 1–2 mm. long; petals white, 2–6 mm. long, claw 1–3 mm. long, linear, not connivent, limb 2–3 mm. long, ovate, acute; stamens 3–7 mm. long, filaments without epidermal papillae; style 3–6 mm. long; fruit 5-nerved, 3-ribbed, 2–4 mm. long; $n = 23$.—Sandy soils, occasionally in rocky places in canyons, below 3000 ft.; Creosote Bush Scrub; Colorado Desert from Coachella Valley to Baja Calif. March–May.

2. **P. nítidus** Wats. Suffructicose, erect, many branched, 1–5 dm. tall; leaves petiolate, all similar in size, ovate to broadly ovate, acute, obtuse to rarely truncate, serrate to coarsely few-toothed, dark green, vernicose; inflorescence a terminal raceme, 3–4.5 cm. long, not much elongated in fruit; bracts lance-ovate, acuminate, crenate; calyx lobes 1–3 mm. long; petals cream colored, 5–11 mm. long, claws linear, 4–7 mm. long, upper ¼ connivent, limbs 1–3 mm. long, ovate, acute; stamens 7–14 mm. long, filaments with scattered epidermal papillae at base; style 0.8–1.5 cm. long; fruit 5-nerved, 5-ribbed, 1–3 mm. long; $n = 23$.—Rocky canyons, washes, roadside banks above 3500 ft. in Mojave Desert; Creosote Bush Scrub, Joshua Tree Wd., Pinyon-Juniper Wd.; Inyo and San Bernardino cos. of Calif.; s. Nev., sw. Utah, nw. Ariz. May–July.

3. **P. thúrberi** G. ssp. **thúrberi**. Plate 59, Fig. B. Suffrutescent, often broader than tall, to 10 dm. tall; leaves sessile, reduced in size along the branches, variable in shape, acute to acuminate, cuneate to cordate-clasping, entire to few toothed, gray-green, dull to somewhat shiny; inflorescence a short, dense, naked, terminal spicate raceme, 1–4 cm. long, slightly elongated in fruit; bracts ovate, acute to acuminate, obtuse to sub-cordate, entire or crenate at base; calyx lobes 1–2 mm. long; petals cream-colored, 3–7 mm. long, claw linear, 2–4 mm. long, upper ⅕ irregularly connivent, limb 1–3 mm. long, ovate, acute; stamens 5–10 mm. long, filaments without epidermal papillae; style 4–11 mm. long; fruit 2–3 mm. long, obscurely 5-ribbed or smooth; $n = 23$.—Dry sandy or gravelly places below 4000 ft.; Mojave and Colo. Deserts, Inyo Co. to Nev.; Ariz., Son. and Baja Calif. May–June.

A. Leaf and stem pubescence very soft, perpendicular to stem; stamens less than 6 mm. long
 ssp. *gilmanii*
AA. Leaf and stem pubescence harsh, retrorse on stems; stamens longer than 6 mm.
 ssp. *thurberi*

Ssp. **gilmánii** (Munz) Davis & Thompson. [*P. g.* Munz.] Plants up to 10 dm. tall, pubescence perpendicular to stems; leaves deltoid-ovate, acute to abruptly acuminate, truncate to cordate-clasping, dull, never shiny; petals pale cream, 3–4 mm. long; stamens 5–8 mm. long; style 4–6 mm. long. Rare in washes and canyons, restricted to Inyo Co. May–June.

63. **Loganiàceae.** LOGANIA FAMILY

Herbs to vines or trees. Lvs. simple, usually opposite, stipulate. Infl. of leafy interrupted spikes to cymose. Fls. regular, usually perfect, 4–5-merous. Calyx-lobes imbricate. Corolla sympetalous, the lobes valvate, imbricate or contorted. Stamens as many as the corolla-lobes, alternate with them. Ovary superior, 2-loculed. Style usually simple; stigma capitate or 2-lobed. Fr. in ours a caps. A family of ca. 600 spp., largely tropical.

Fls. in headlike clusters in an interrupted spike; lvs. densely woolly; plant 2–3 dm. high. E. Mojave Desert .... 1. *Buddleja*
Fls. in paniculate cymes; lvs. glabrous above; plant 4–5 m. tall. Escape from cult., Santa Monica Mts. .... 2. *Chilianthus*

1. Búddleja L.

Shrubs to trees; lvs. simple, entire to dentate. Fls. mostly 4-merous. Calyx campanulate. Corolla salverform or rotate-campanulate, the lobes ovate or rounded. Anthers subsessile on throat or tube of corolla. Fr. a septicidal caps.; valves 2-cleft at apex. Seeds many. Ca. 70 spp. of warm N. & S. Am., Asia, S. Afr. (After Adam *Buddle*, English botanist, 1660–1715.)

1. **B. utahénsis** Cov. Plate 59, Fig. C. Much-branched shrub 2–3 dm. high, densely lanate-tomentose; lvs. subsessile, linear, with revolute margins, 1–3 cm. long; axils usually with fascicles of very small lvs.; fls. in glomerules forming 2–4 heads in an interrupted spike ca. 10–15 mm. thick; corolla creamy-yellow to purple, 4–5 mm. long, the lobes rounded, 1 mm. long, the tube tomentulose.—Dry rocky limestone slopes, 3500–5500 ft.; Joshua Tree Wd., Pinyon-Juniper Wd.; Kingston, Panamint, Clark, Last Chance and Nopah mts.; to Utah. May–Oct.

2. Chiliánthus Burchell

Arborescent shrubs. Lvs. entire to toothed or lobed. Fls. in terminal pyramidal or subspherical cymes. Fls. small, 4-merous. Corolla-tube short, scarcely exserted. Stamens elongate, exserted. Caps. 2-loculed. Seeds small. A S. African genus of ca. 4 spp.

1. **C. oleàceus** Burchell. To 4 or 6 m. tall; stems 4-angled; lvs. lanceolate, 7–10 cm. long, rusty-scurfy beneath, but soon glabrate, ± revolute; fls. fragrant, creamy-white, 2.5 mm. long, in panicles to 1 dm. across.—Escaping from cult. as at Saddle Peak, Santa Monica Mts. at 2400 ft.

64. Lythràceae. Loosestrife Family

Herbaceous or woody, mostly with opposite lvs., no stipules. Fls. perfect, solitary or clustered, regular, mostly minute in ours. Fl.-tube persistent, enclosing but free from the 1–4-loculed ovary, with 4–6 minute sepals and sometimes accessory teeth in the sinuses. Petals when present as many as the primary calyx-teeth, inserted with the 4–17 stamens on the fl.-tube. Anthers versatile, longitudinally dehiscent. Style 1; stigma 2-lobed; ovules many or few, anatropous. Caps. 1–several-celled, variously dehiscent or indehiscent. Seeds without endosperm.

Fl.-tube short, campanulate to globular; lvs. opposite; petals 4 .... 1. *Ammannia*
Fl.-tube cylindric; petals usually 6; lvs. mainly alternate .... 2. *Lythrum*

1. Ammánnia L.

Subglabrous annuals with angled stems and opposite sessile narrow lvs. which are auricled at the base. Fl.-tube 4-angled; sepals 4, with small accessory teeth in the sinuses. Petals 4, deciduous. Stamens 4–8. Ovary 2–4-loculed. Seeds many, angled, minutely pitted. Ca. 20 spp., mostly in warmer regions. (Named for J. *Ammann*, 18th-century German botanist.)

1. **A. coccínea** Rottb. Ascending or depressed, 1–4 dm. high; lvs. 2–5 cm. long, linear-lanceolate, cordate-auriculate, with acute or acuminate apex; fls. subsessile in lf.-axils; petals purple, 1–2 mm. long, deciduous; caps. 4 mm. long; style persistent.—Occasional in wet places at low elevs.; cismontane s. Calif.; to Wash.; Catalina Id.; to Atlantic Coast, Brazil. May–Oct.

2. Lýthrum L.

Slender herbs with angled stems and opposite, alternate or whorled lvs. Fls. solitary in axils to spicate or subpaniculate. Fl.-tube cylindric, 8–12-ribbed, the sepals 4–6, with as many intervening appendages. Petals 4–6, rarely none. Stamens 4–12. Ovary oblong,

Lythrum

2-loculed; style filiform. Caps. membranous, included, 2-valved or bursting irregularly. Seeds many, minute, flat or angled. Ca. 30 spp., widely distributed. (Greek, *lytron*, a name used by Dioscorides for *L. salicaria*.)

Fls. sessile or nearly so .... 2. *L. hyssopifolia*
Fls. short-pedicelled .... 1. *L. californicum*

1. **L. califórnicum** T. & G. Plate 59, Fig. D. Erect, somewhat woody at base, 5–18 dm. high, pale green, glabrous; lvs. linear to linear-oblong, 1–3 cm. long, entire; fl.-tube 5–6 mm. long; petals purple, 4–6 mm. long; seeds linear-lanceolate, ca. 1 mm. long, 0.5 mm. broad.—Moist places, below 6000 ft.; many Plant Communities, cismontane, n. to cent. Calif.; occasional on deserts and s. to L. Calif. April–Oct.

2. **L. hyssopifòlia** L. Simple or branched, pale green, glabrous annual or perennial (apparently sometimes called *L. adsurgens* Greene if annual), the stems 1–5 dm. long, sometimes rooting at nodes; lvs. linear to oblong, 6–15 mm. long, sessile, obtuse; fls. solitary and sessile, whitish to pale purple, the petals 1.5–2 mm. long; stamens included; fl.-tube ca. 4 mm. long in fr.; seeds obliquely ovoid, scarcely 1 mm. long, almost as wide; $2n=20$? (Tischler, 1929).—Moist places below 5000 ft.; cismontane Calif.; to Wash., Atlantic Coast; Eu. April–Oct.

65. **Malvàceae.** Mallow Family

Herbs or soft-woody shrubs or even trees, usually with stellate or branched pubescence. Lvs. alternate, palmately ribbed and usually lobed, with small deciduous stipules. Fls. regular and bisexual, occasionally unisexual, mostly 5-merous, white, yellow, pink or red to purple. Calyx 5-lobed, often subtended by calyxlike bracts forming an involucel. Petals 5, convolute. Stamens many, hypogynous, cohering into a tube or column around the styles and often adnate to the petals. Pistil of several united carpels, the ovary superior, rarely 1-celled, mostly with as many cells as styles or stigmas; styles mostly united below. Fr. a loculicidal caps. or sometimes berrylike or more often separating into carpels which are also or follicles. Ca. 50 genera and 1000 spp. of temp. and trop. regions, many important economically for their fibers (cotton), food products, and as ornamentals.

1. Fr. a caps., the carpels not separating from each other or the axis; fls. solitary in upper axils; ovules several in each carpel .... 4. *Hibiscus*
1. Fr. of several carpels, usually separating from each other and the axis when mature.
    2. Style-branches terminating in capitate or truncate stigmas.
        3. Involucel of bractlets below calyx wanting.
            4. Carpels not sharply differentiated into a reticulate basal and a wider smooth apical portion; mostly herbaceous; petals yellow or orange; carpels 2–9-seeded . 1. *Abutilon*
            4. Carpels sharply differentiated into a reticulate basal and a wider winged smooth apical portion; rather woody .... 5. *Horsfordia*
        3. Involucel of 1 to several bractlets below calyx present.
            5. Carpels sharply differentiated into a reticulate basal and a wider winged smooth apical portion; plants herbaceous or suffruticose .... 12. *Sphaeralcea*
            5. Carpels not so differentiated.
                6. Ovules 2 or more per carpel; carpels hirsute, 4 mm. high, reniform, 2-celled
                9. *Modiola*
                6. Ovules solitary; carpels not hirsute.
                    7. Plants not lepidote (beset with small scurfy scales); fls. white, pink, lavender, or purplish.
                        8. Carpels indehiscent, rugose on the sides; annuals .... 3. *Eremalche*
                        8. Carpels splitting into halves at maturity, not rugose; shrubby . 7. *Malacothamnus*
                    7. Plants lepidote; petals yellowish .... 10. *Sida*
    2. Style-branches filiform, longitudinally stigmatic on the inner side.
        9. Involucel of 3–9 connate bractlets.
            10. Axis of fr. surpassing the carpels, forming a cone or projection in the center
            6. *Lavatera*
            10. Axis of the fr. not extending above the ring of carpels .... 2. *Althaea*

9. Involucel of 1-3 distinct bractlets or lacking.
    11. Bractlets of involucel 3; weedy introd. annuals .... 8. *Malva*
    11. Bractlets 0 or 1; native annuals or perennials .... 11. *Sidalcea*

1. Abùtilon Mill. INDIAN-MALLOW. FLOWERING-MAPLE

Perennial or annual, herbaceous or shrubby, stellate-canescent or -tomentose, or hirsute with simple hairs. Lvs. alternate, petioled, crenate or dentate, cordate at base, not or obscurely lobed. Fls. solitary and axillary or in leafy panicles, often drooping, without involucel, the calyx 5-cleft and sometimes bright-colored. Corolla trumpet-shaped or bell-shaped, commonly orange or yellow; stamineal column anther-bearing at apex. Fr. truncate-cylindric or subglobose, the carpels smooth-sided, dehiscent nearly to base when ripe; ovules 2 or more per carpel. Ca. 100 spp. in many warm regions, some cult. as ornamentals. (Name of Arabic origin.)

Plants annual; lvs. often 10-20 cm. wide; carpels more than 10, with long divergent awns
                                                                    3. *A. theophrasti*
Plants perennial; lvs. 1-5 cm. wide; carpels usually fewer.
    Carpels usually 7 or more; petals orange, 15-20 mm. long .... 1. *A. palmeri*
    Carpels seldom more than 5; petals pink or red, 4-6 mm. long .... 2. *A. parvulum*

1. **A. pálmeri** Gray. Woody at base, 7-8 dm. high, stellate-villous; lvs. velvety, round-cordate, 3-5 cm. wide, obscurely 3-lobed; petioles 3-5 cm. long; calyx 9-12 mm. long, the lobes round-ovate, acuminate; petals rich orange, ca. 2 cm. long; fr. very villous, ca. 10 mm. long.—Dry slopes, below 2500 ft.; Creosote Bush Scrub; base of Laguna Mts., e. San Diego Co., Chuckwalla Mts., Riverside Co.; to Ariz., Son., L. Calif. April–May.

2. **A. párvulum** Gray. Cespitose perennial; stems slender, 1-4 dm. tall, cinereous-stellate, diffusely branched; lvs. ovate-cordate, dentate, 1-2.5 cm. wide, the petioles 5-10 mm. long; calyx 3-4 mm. long, reflexed in fr.; petals 4-6 mm. long; fr. 7-8 mm. long, stellate-puberulent.—Arid rocky slopes, 3000-4000 ft.; Shadscale Scrub; Providence Mts.; to Colo., Ariz., Mex. April–May.

3. **A. theophrásti** Medic. VELVET LEAF. Plant 1-2 m. tall; lvs. ovate-orbicular, cordate, velvety; fls. yellow, 6-8 mm. long; $2n = 42$ (Smith, 1965).—Occasional escape (Riverside, Santa Ana River Canyon, San Diego, Santa Barbara); native of s. Asia. July–Aug.

2. Althaèa L. HOLLYHOCK

Biennial to perennial herbs, tall and with leafy stems. Lvs. lobed or parted. Fls. solitary or racemose, axillary, usually toward the top of the stem. Involucel of 6-9 bractlets. Otherwise like *Malva*. A genus of ca. 15 spp. of Old World temp. region. (Greek, *althaino*, to cure.)

1. **A. ròsea** (L.) Cav. [*Alcaea r*. L.] Mostly biennial, 2-3 m. tall, the stems strict, hairy; lvs. round-cordate, 5-7-lobed or -angled, commonly 1-2 dm. wide, crenate, long-petioled; fls. 6-10 cm. wide, subsessile in elongate terminal spicate raceme, of many colors.—Occasional escape from cult.; native of China. Summer.

3. Eremálche Greene

Low annual herbs. Lvs. orbicular or palmately parted, stellate-pubescent. Fls. solitary or in pairs in the upper lf.-axils. Involucellate bractlets 3, distinct, persistent. Sepals somewhat united at base. Petals white to rose-purple, hairy along the margins of the claws. Stamineal column simple, glabrous. Style-branches from one and one-half to two times as long as stamineal column, filiform, as many as the carpels. Stigmas capitate. Carpels 10-40, indehiscent, 1-ovulate, glabrous, reticulate or transversely ridged on the back and angles. Embryo of the solitary seed forming an incomplete circle; endosperm scanty. (Greek, referring to the desert habitat.) Four spp. of Calif. and sw. U.S.

Lvs. crenate; petals 2-3 cm. long, each with a dark spot .... 3. *E. rotundifolia*
Lvs. 5-7-lobed; petals not spotted.
    Petals 5-6 mm. long; calyx 3-5 mm. long .... 1. *E. exilis*
    Petals 12-30 mm. long; calyx 10-14 mm. long .... 2. *E. parryi*

1. **E. éxilis** (Gray) Greene. [*Malvastrum e.* Gray.] Annual, decumbent or prostrate, with several stems from base, these 1-4 dm. long, stellate-pubescent; lvs. suborbicular, 8-18 mm. wide, palmately 3-5-cleft with rounded lobes; petioles slender, 1-3.5 cm. long; bractlets slender, 3-4 mm. long; calyx 3-5 mm. long; petals whitish to pink or lavender, 5-6 mm. long; carpels not more than 15, rather turgid, transversely wrinkled, grayish, 1-1.5 mm. long; $2n=20$, 40 (Krapovickas, 1957).—Common in dry open places, below 5000 ft.; Creosote Bush Scrub; Mojave and Colo. deserts; to Utah, Ariz.; has been reported from near Riverside. March-May.

2. **E. párryi** (Greene) Greene. Stems several from near base, or ± simple, erect or ascending, subglabrous to somewhat stellate-pubescent; lvs. twice-lobed or -cleft, 2-4 cm. wide; petioles 1-5 cm. long, slender; fls. pedicelled, in loose terminal corymbs; bractlets linear, 9-11 mm. long; calyx loosely stellate-pubescent, 10-14 mm. long, the lobes ovate, 4-5 mm. wide, abruptly acuminate; petals pinkish lavender to purplish, darker on drying, mostly 1.5-3 cm. long; carpels turgid, ca. 1.5 mm. long, rugose-reticulate, somewhat brownish.—Dry flats and hills, below 4500 ft.; Pinyon-Juniper Wd., V. Grassland; largely about the San Joaquin V., s. to Tehachapi Mts. and Mt. Pinos area. March-May.

3. **E. rotundifòlia** (Gray) Greene. [*Malvastrum r.* Gray.] DESERT FIVESPOT. Plate 59, Fig. E. Erect annual, 1-4 (-6) dm. high, simple or branched, the stems and petioles hispid with long mostly simple hairs; lvs. remote, suborbicular, cordate, coarsely crenate, 2-5 cm. long, on petioles 2-10 cm. long; fls. in racemose corymbs; bractlets filiform, 6-10 mm. long; calyx 10-15 mm. long, stellate-hispid, the lobes ovate, acuminate; petals 2-3 cm. long, rose-pink to lilac, drying purplish, each with a conspicuous darker spot, the corolla subglobose when fresh; carpels many, thin, flat, black at maturity, reticulate near the edges, 2.5-3 mm. high.—Frequent in washes and on mesas, below 3800 ft.; Creosote Bush Scrub; Mojave and Colo. deserts; Ariz., Nev. March-May.

4. *Hibíscus* L. ROSE-MALLOW

Herbs or shrubs. Lvs. palmately veined, lobed or parted. Fls. bisexual, 5-merous, mostly bell-shaped, axillary or paniculate, often large and showy. Involucel of few to several slender to broad bractlets. Calyx 5-toothed or -cleft. Corolla of 5 petals. Stamineal column anther-bearing below the naked truncate 5-toothed summit. Style-branches 5. Ovary 5-celled, with 2-many ovules in each cell; fr. a 5-valved loculicidal caps. Seeds several in each locule, long-hairy. A genus of ca. 200 spp. of warmer regions around the world; many used for ornament, some for food and fibers. (Ancient Greek and Latin name for some large mallow.)

Plants with toothed, not parted, lvs.; perennial with a dense yellowish stellate tomentum; petals pinkish-lavender to whitish .... 1. *H. denudatus*
Plants with 3-5-parted lvs.; annual; petals yellow with dark basal spot .... 2. *H. trionum*

1. **H. denudàtus** Benth. Tufted suffrutescent perennial 3-7 dm. tall, yellowish-tomentose; lvs. ovate, 1-2.5 cm. long, serrulate; fls. in upper axils; bractlets of involucel 2-5 mm. long, setaceous; calyx 10-14 mm. long; petals whitish to pinkish-lavender, 15-22 mm. long; caps. 5-7 mm. long, dehiscent to base; seeds ca. 2.5 mm. long.—Rocky slopes and canyons below 2500 ft.; Creosote Bush Scrub; Colo. Desert; to Tex., n. Mex. Feb.-May.

2. **H. triònum** L. FLOWER-OF-AN-HOUR. Hairy, 3-6 dm. high; lvs. 2-3 cm. wide, 3-5-parted, the divisions coarsely toothed; fls. axillary; solitary; bractlets of involucel linear; calyx inflated, papery, 5-winged, with many dark nerves, becoming 2.5 cm. long in fr.; petals ca. 2 cm. long, sulphur-yellow with blackish basal spot; caps. 15-18 mm.

long; seeds reniform, ca. 2 mm. long.—Occasional weed in waste places and about gardens, Imperial Co., Orange Co., Riverside Co., etc.; native of cent. Afr. Aug.–Sept.

5. Horsfórdia Gray

Erect, sparingly branched, rather woody, densely stellate-canescent or -tomentose, somewhat yellowish. Lvs. cordate to lanceolate, thickish, somewhat denticulate or crenulate, truncate or subcordate at base. Fls. axillary, 1 to few, peduncled, often in leafy panicles; bractlets none. Calyx 5-lobed. Carpels 8–12, disjoined at maturity, the apical part scarious, winged, the basal part firm and reticulate. Seeds of the 2 parts of the carpel unlike. Four spp., sw. U.S. and adjacent Mex. (F. H. Horsford, a New England botanist.)

Petals pink, drying bluish, 10–15 mm. long; plant dull green .... 1. *H. alata*
Petals yellow, ca. 8 mm. long; plant vivid yellow-green .... 2. *H. newberryi*

1. **H. alàta** (Wats.) Gray. [*Sida a.* Wats.] Stems 1–3 dm. tall; lvs. ovate-lanceolate, subcordate, 2–9 cm. long; pedicels 8–18 mm. long; calyx-lobes ovate, acuminate; petals obovate, hairy at base, 15–18 mm. long; carpels 10–12, becoming 8 mm. long, the upper part 2-winged and 3 times the length of the lower part; seeds solitary. Rocky canyons and sandy washes, at ca. 500 ft.; Creosote Bush Scrub; Coral Reef Ranch, s. side of Coachella V.; Riverside Co.; to Ariz., Son., L. Calif. March–April. Nov.–Dec.

2. **H. newbérryi** (Wats.) Gray. [*Abutilon n.* Wats.] Plate 59, Fig. F. Stems 1–2.5 m. tall, becoming woody but flowering the first year; lvs. cordate, ± ovate, 3–9 cm. long; pedicels 5–12 mm. long; calyx-lobes subdeltoid, acute; petals rounded, 7–8 mm. long; carpels 8–9, ca. 6 mm. long in fr., the wings not much longer than lower part; seeds 2–3 per carpel.—Dry rocky places, below 2500 ft.; Creosote Bush Scrub; w. Colo. Desert; to Ariz., Son., n. L. Calif. March–April, Nov.–Dec.

6. Lavatèra L. Tree-Mallow

Ours erect to arborescent herbs or shrubs. Lvs. angled or lobed, maplelike in ours, long-petioled. Fls. showy, axillary or in terminal racemes, each with a 3-lobed involucel in ours. Calyx 5-parted. Petals reflexed after anthesis, clawed, emarginate or truncate. Fr. depressed, the carpels in ours 5–8, smooth, 1-seeded. Ca. 25 spp. of Medit. region to Asia, Australia, Canary Ids., ids. off Calif. and L. Calif. (*Lavater*, two Swiss brothers, of the time of Tournefort.)

Involucral lobes lanceolate; petals 2.5–4.5 cm. long .... 2. *L. assurgentiflora*
Involucel-lobes rounded-ovate; petals 1–2 cm. long.
    Involucel surpassing calyx; lvs. densely soft-downy .... 1. *L. arborea*
    Involucel shorter than calyx; lvs. sparsely pubescent .... 3. *L. cretica*

1. **L. arbòrea** L. Shrubby, becoming 1–3 m. tall; lvs. 5–20 cm. broad, softly stellate-pubescent on both sides, shallowly and unequally 5–9-lobed, crenate; petioles 3–8 cm. long; fls. many, in short leafy racemes or axillary clusters, short-pedicelled; involucel exceeding calyx, with broad rounded lobes; calyx ca. 4 mm. long at anthesis; petals pale purple-red with dark purple veins at base; 1.5–2 cm. long; carpels 7–9, subglabrous, ± reticulate; $n=22$ (Skovsted, 1935).—Natur. on bluffs and dunes near the coast, Ventura and Santa Barbara n.; introd. from Eu. June–July.

2. **L. assurgentiflòra** Kell. Malva Rosa. Plate 60, Fig. A. Erect, shrubby, 1–4 m. tall, with thick, glabrous to pubescent twigs; lvs. 5–15 cm. wide, 5–7-lobed, the lobes ovate-triangular, coarsely and irregularly toothed, pale beneath; petioles 5–15 cm. long; bractlets ca. 3, lance-ovate, 5–9 mm. long; calyx densely stellate-pubescent, 12–15 mm. long; petals rose with darker veins, 2.5–4.5 cm. long, with long narrow glabrous claws and a pair of dense hairy tufts at base; carpels woody, ca. 6 mm. high, rounded on back, triangular in cross-section, glabrous to pubescent.—Variable, native of sandy flats and

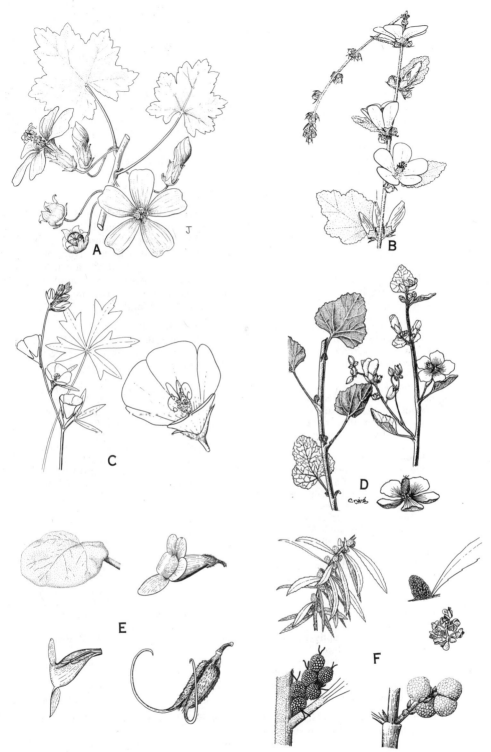

Plate 60. Fig. A, *Lavatera assurgentiflora*. Fig. B, *Malacothamnus densiflorus*. Fig. C, *Sidalcea malvaeflora*. Fig. D, *Sphaeralcea ambigua*. Fig. E, *Proboscidea* sp. Fig. F, *Myrica californica*.

rocky places; Coastal Sage Scrub; s. Santa Barbara Ids., cult. and escaped along mainland coast. March–Nov.

3. **L. crètica** L. Winter annual, 1–3 m. tall, sparsely pubescent; lower lvs. suborbicular, the upper shallowly and broadly 5-lobed, crenate, truncate to subcordate at base, 4–10 cm. wide, on longer petioles; involucel shorter than calyx, with broad rounded lobes; calyx ca. 4 mm. long at anthesis, later much enlarged and surrounding the fr.; petals pinkish to lilac, 10–16 mm. long; fr. depressed, of 7–10 glabrous or puberulent relatively smooth carpels; $n =$ ca. 56 (Skovsted, 1935).—Occasionally natur. about Santa Barbara and in Ventura Co.; n. along the coast; native of Medit. region. May–July.

7. Malacothámnus Greene

Shrubs, sometimes merely suffrutescent, with long usually flexuous branches, ± densely stellate-tomentose or -pubescent. Lvs. petioled, ± evidently and palmately 3–5-lobed. Fls. pink or lavender, 1 to several at a node, and in axillary ± capitate clusters forming elongated interrupted spikes or in open panicles. Calyx subtended by an involucel of bractlets. Sepals 5, united at base. Petals 5, distinct, but often joined at base to the stamen-tube. Stamens many, the fils. united into a tube. Carpels several, completely dehiscent (splitting into halves at maturity), muticous, unappendaged, not rugose, 1-ovuled. A genus of 20 spp., of sw. N. Am. (Greek *malakos*, soft, and *thamnos*, shrub.)

(Kearney, T. H. The genus Malacothamnus. Leafl. W. Bot. 6: 113–140. 1951.)
1. Calyx-lobes mostly 4–8 mm. wide at base; bractlets subtending the calyx ovate to broadly lanceolate; lvs. ± cordate at base .... 1. *M. aboriginum*
1. Calyx-lobes mostly not more than 3 mm. wide at base; bractlets narrowly lanceolate to filiform.
    2. Calyx conspicuously and densely white-lanate, the hairs ± concealing the calyx-lobes; stems loosely pubescent. S. edge of Mojave Desert .... 7. *M. orbiculatus*
    2. Calyx not conspicuously and densely lanate, the pubescence not concealing the calyx-lobes, or if the calyx sometimes rather woolly, then the infl. very narrow and few-fld.
        3. Hairs of calyx long, simple to few-branched (the arms to 2 mm. or more long).
            4. Stems conspicuously shaggy-tomentose with grayish hairs; lvs. angulately 3–5-lobed, bicolored, soft-tomentose beneath. San Clemente Id. .... 2. *M. clementinus*
            4. Stems closely pubescent with yellowish hairs; lvs. simple or with rounded lobes, not strongly bicolored or tomentose beneath. Mainland .... 4. *M. densiflorus*
        3. Hairs of calyx short (to 1 mm. long) and many-armed.
            5. Stems loosely pubescent.
                6. Lvs. deeply 3–5-lobed; petioles and lf.-veins very stout; bractlets not more than ⅓ the calyx. San Fernando V. .... 3. *M. davidsonii*
                6. Lvs. simple or shallowly lobed; petioles and lf.-veins slender; bractlets usually at least half as long as calyx.
                    7. The lvs. cordate at base; calyx-lobes ovate, 3–7 mm. long. Desert borders, Inyo Co. to San Bernardino Co. .... 7. *M. orbiculatus*
                    7. The lvs. mostly truncate at base; calyx-lobes 6–8 mm. long. San Gabriel Mts. to Fresno Co. .... 6. *M. marrubioides*
            5. Stems appressed-pubescent, or if pubescence loose, then the calyx-lobes abruptly acuminate.
                8. Calyx-lobes abruptly acuminate .... 4. *M. densiflorus*
                8. Calyx-lobes acutish.
                    9. Lvs. thickish, truncate or cuneate at base, not or scarcely lobed; infl. an open panicle. Near San Bernardino .... 8. *M. parishii*
                    9. Lvs. thinnish, mostly cordate at base, distinctly but shallowly lobed; infl. paniculate. Widely distributed .... 5. *M. fasciculatus*

1. **M. aboríginum** (Rob.) Greene. [*Malvastrum a.* Rob.] Coarse, woody below, 1–3 m. tall, densely felted-tomentose on stems and twigs; lf.-blades gray-green, broadly ovate, 3–6 cm. long, 3–5-lobed, crenate-dentate, cordate at base, stellate-pubescent; petioles 1–3.5 cm. long; fls. sessile in headlike clusters in upper axils, forming an elongate ± naked infl.; bractlets 3, ovate, 6–8 mm. long; calyx strongly angled in bud, 8–10 mm. long, the

lobes 4–6 mm. long and somewhat wider; petals rose, 12–14 mm. long; carpels ca. 3 mm. high.—Dry rocky summit of Monument Peak, San Diego Co. at 5600 ft.; Yellow Pine F.; Foothill Wd. of S. Coast Ranges of Monterey, San Benito and Fresno cos.

2. **M. clementìnus** (M. & J.) Kearn. [*Malvastrum c.* M. & J.] Rounded subshrub to ca. 1 m. tall, with numerous shaggy, stellate-tomentose branches; lf.-blades 3–5-lobed, deeply cordate at base, broadly ovate, 3–5 cm. wide, greenish above, whiter and soft-stellate beneath; petioles 1–1.5 cm. long, fls. many, subsessile and densely glomerate in uppermost axils forming interrupted spikes 1–2 dm. long; bractlets filiform, almost as long as calyx; calyx 7–8 mm. long, loosely stellate-tomentose, the lobes broadly lanceolate, acute, 4 mm. long; petals pink, ca. 13 mm. long, oblong-obovate; carpels stellate-tomentose at summit, 2.5–3 mm. high; seeds almost 2 mm. long.—Rocky canyon-walls; Coastal Sage Scrub; San Clemente Id. April.

3. **M. davidsònii** (Rob.) Greene. [*Malvastrum d.* Rob.] Erect coarse shrub, 2–5 m. high, the branches stout, densely shaggy with stellate tomentum; lf.-blades thick, round-cordate, 2.5–10 cm. wide, 5-angled or shallowly 3–5-lobed, densely stellate-tomentose; petioles 1.5–3 cm. long; fls. numerous, in short racemes forming panicles 2.5–4.5 dm. long; bractlets 3 mm. long calyx densely stellate-tomentose, 5–8 mm. long, the lobes ovate, acute, 2–4 mm. long, obscured by the tomentum; petals pink or rose, 12–15 mm. long; carpels tomentulose at summit, ca. 3 mm. high, incised; seeds dark; $n=17$ (Webber, 1936).—Sandy washes and flats; Coastal Sage Scrub; San Fernando V., Los Angeles Co. June–Sept.

4. **M. densiflòrus** (Wats.) Greene var. **densiflòrus** [*Malvastrum d.* Wats.] Plate 60, Fig. B. Erect, 1–2 m. tall, with yellowish close scurfy pubescence; lvs. rounded-ovate, simple or shallowly rounded-lobed, truncate or cordate at base, 2–5 cm. wide, green, sparsely pubescent beneath; petioles 5–20 mm. long; infl. an interrupted spike of dense sessile heads, ± glandular; bractlets filiform, viscid, 7–16 mm. long; calyx 10–14 mm. long, loosely hirsute with long subsimple or stellate hairs, the lobes lance-ovate, 5–12 mm. long; petals rose-pink, 10–15 mm. long; carpels 2.2–2.8 mm. high, shallowly incised.—Dry slopes, mostly 1000–4000 ft.; Chaparral; Santa Ana Mts. to Palm Springs and to Cuyamaca Mts. April–July.

Var. **víscidus** (Abrams) Kearn. [*Malvastrum v.* Abrams.] Bractlets 4–6 mm. long; calyx-lobes 3–7 mm. long, broadly ovate; $n=17$ (Webber, 1936).—San Diego Co., n. L. Calif. March–June.

5. **M. fasciculàtus** (Nutt.) Greene var. **fasciculàtus**. [*Malva f.* Nutt. ex T. & G.] Shrub 1–5 m. tall, with long slender wandlike branches covered with soft short tomentum; lf.-blades round-ovate, mostly shallowly lobed, 2–4 (–6) cm. wide, densely canescent beneath, less so and somewhat greener above, crenate or crenate-dentate; petioles 0.5–1 cm. long; fls. in sessile or nearly sessile headlike clusters, forming an interrupted spicate infl.; bractlets linear, 2–4 mm. long; calyx 6–8 mm. long, stellate-pubescent, the lobes deltoid, acute to obtuse; petals pink, 12–18 mm. long; carpels stellate-pubescent at summit, 2.5–3 mm. long; seeds ca. 1.5 mm. long; $n=17$ (Webber, 1936).—Common on dry fans and slopes, below 2500 ft.; Coastal Sage Scrub, Chaparral; s. Riverside Co. to L. Calif. April–July.

KEY TO VARIETIES

Infl. of interrupted spike of headlike clusters, or with lower branches sometimes to 6 cm. long; calyx loosely pubescent with long hairs.
    Lvs. small, to 4 cm. long, not conspicuously bicolored. S. Riverside Co. to n. L. Calif.
                                                              var. *fasciculatus*
    Lvs. larger, to 7 cm. long, quite conspicuously bicolored. Catalina Id. . . . . . . var. *catalinensis*
Infl. open-paniculate, the lower branches often long and loosely fld.; calyx closely pubescent.
    Lvs. essentially concolored, 2–4 cm. wide. Coastal Santa Barbara and Ventura cos.
                                                              var. *nuttallii*
    Lvs. bicolored, paler beneath.
        Lf.-blades 3–7 cm. wide; panicle many-branched, the ultimate divisions rigid, not racemose. Santa Cruz Id. . . . . var. *nesioticus*

Lf.-blades 2–4 cm. wide; panicle few-branched, the ultimate divisions elongate, commonly racemose. San Bernardino Co. to Riverside, Orange and Ventura cos. .... var. *laxiflorus*

Var. **catalinénsis** (Eastw.) Kearn. [*Malvastrum c.* Eastw.] Lvs. thin, subglabrous above, deeply lobed, 5–7 cm. long; fls. in congested clusters in short-branched panicle; calyx stellate-tomentose; carpels 3.2–3.8 mm. long.—Catalina Id. Closely matched by plants from Santa Monica Mts. May–Aug.

Var. **laxiflòrus** (Gray) Kearn. [*Malvastrum thurberi* var. *l.* Gray.] Lvs. somewhat bicolored; infl. branched and paniculate; carpels 2–3 mm. high; $n=17$ (Webber, 1936). —Below 5500 ft., Orange and n. Riverside cos. to Cajon Pass to s. Ventura Co. May–Aug.

Var. **nesióticus** (Rob.) Kearn. [*Malvastrum n.* Rob.] Lvs. thinnish, 5–7 cm. long, cordate, subglabrate above, deeply lobed; panicles open, terminal, quite leafless; calyx finely canescent; carpels ca. 4 mm. high.—Steep canyons, Santa Cruz Id. June–July.

Var. **nuttállii** (Abrams) Kearn. [*M. n.* Abrams.] Lvs. closely hairy-stellate on both surfaces; panicle elongate, few-branched; carpels 3–5 mm. high; $n=17$ (Webber, 1936). —Chaparral; Ventura and Santa Barbara cos.

6. **M. marrubioìdes** (Dur. & Hilg.) Greene. [*Malvastrum m.* Dur. & Hilg. *M. gabrielense* M. & J.] Shrubby below, 6–20 dm. high, with erect slender branches and close stellate pubescence; lvs. rounded, obscurely lobed, pale green, 1.5–4.5 cm. wide; petioles 1–2.5 cm. long; fls. few, in glomerules at upper axils, forming a spicate or subpaniculate scarcely glandular infl.; bractlets subulate, 7–12 mm. long; calyx 10–12 mm. long, the lobes ovate-lanceolate, 6–8 mm. long, densely stellate-tomentose; petals pink, 16–18 mm. long; carpels stellate-pubescent, at apex, 2.5–3.5 mm. high; shallowly incised.—Dry stony slopes, 1500–7000 ft.; Chaparral, Foothill Wd.; n. slopes of San Gabriel Mts. to Mt. Pinos and n. June–Aug.

7. **M. orbiculàtus** (Greene) Greene. [*Malvastrum o.* Greene.] Suffruticose, 5–20 dm. high, with mostly stout simple branches densely soft stellate-tomentose; lvs. rounded, thinnish, 3–7 cm. wide, coarsely crenate, sometimes 3–5-lobed, greener above than beneath; petioles mostly 1–2 cm. long; fls. rose, few in sessile axillary clusters forming long interrupted spicate panicles; bractlets linear, shorter than calyx; calyx ± hispid-stellate to densely lanate, 7–10 mm. long, the segms. ovate, 3–7 mm. long, acute to acuminate; petals 12–16 mm. long; carpels pubescent on upper half, 2–2.5 mm. high; seeds 1.5 mm. long; $n=17$ (Webber, 1936).—Dry stony slopes, 3000–8500 ft.; Pinyon-Juniper Wd., Yellow Pine F.; borders of Mojave Desert from e. slope of Sierra Nevada, Inyo Co. to n. slope of San Bernardino Mts. June–Oct.

8. **M. párishii** (Eastw.) Kearn. [*Malvastrum p.* Eastw.] Erect shrub, white-tomentose; lvs. thin, rhombic-ovate, distinctly but shallowly 3-lobed, cuneate at base, crenate, ca. 5 cm. long, 4 cm. wide, green above, densely white-tomentose beneath; petioles ca. 1 cm. long; infl. open-paniculate with elongate lower branches and usually 3 or more fls. per node; bracteoles filiform, ca. 2 mm. long; calyx 8–9 mm. long, the lobes ovate, acute, nerved, 5 mm. long; petals rose, ca. 12 mm. long; carpels 3 mm. high, conspicuously stalked.—Near San Bernardino, at 1000–1500 ft. June–July.

8. *Málva* L. MALLOW. CHEESES

Mostly annual or biennial herbs, sparsely pubescent or glabrate. Lvs. alternate, orbicular or reniform, ± lobed or dissected. Fls. solitary or clustered in lf.-axils. Involucel of 3 bractlets like an outer calyx. Calyx 5-cleft into broad lobes. Petals 5, emarginate, white to pink or purplish. Styles numerous, stigmatic down the inner side. Fr. depressed, disklike, separating into maturity into the many 1-seeded compressed reniform indehiscent carpels. Ca. 30 spp. of the Old World; ours natur. (Latin name from Greek, *malache*, referring to the emollient lvs.)

Bractlets of involucel ovate to oblong.
    Petals 2–4 times the length of the calyx; corolla 2–3 cm. wide .... 3. *M. sylvestris*

Petals 1–2 times the length of the calyx; corolla 1–1.5 cm. wide . . . . 1. *M. nicaeensis*
Bractlets linear to nearly so.
Petals twice the length of the calyx; carpels quite smooth on the back, rounded at margins
 . . . . 4. *M. verticillata*
Petals scarcely longer than calyx; carpels rugose-reticulate on back, sharp-margined; calyx
strongly accrescent . . . . 2. *M. parviflora*

1. **M. nicaeénsis** All. Lvs. semicircular in outline, scarcely cordate, shallowly 5–7-lobed, crenate, 3–10 cm. wide, long-petioled; bractlets lance-ovate, 4–5 mm. long; calyx 4–5 mm. long at anthesis, enlarged, veiny and closed over fr.; petals pinkish to blue-violet, 10–12 mm. long; carpels 7–9, rugose-reticulate, sharp-edged, glabrous to pubescent; $n=21$ (Skovsted, 1935).—Occasional weed in waste places, as about towns; introd. from Eurasia. March–Sept.

2. **M. parviflòra** L. CHEESEWEED. Erect subglabrous to pubescent annual, 2–5 dm. high, with ascending branches; lvs. roundish, somewhat angular-lobed, 2–8 cm. wide, on somewhat longer petioles; fls. axillary, short-pedicelled; bractlets linear, 1–2 mm. long; calyx ca. 3 mm. long at anthesis, enlarged, reticulate and accrescent in fr.; petals whitish to pinkish, 4–5 mm. long; carpels ca. 11, rugose-reticulate, thin-margined, denticulate on angles, pubescent on back in our form; $2n=42$ (Skovsted, 1941).—Common weed in orchards, waste places, almost throughout Calif.; native of Eurasia. Can be found in bloom most of year.

3. **M. sylvéstris** L. [*M. mauritiana* L.] Biennial or perennial, ascending, 3–10 dm. tall, branched, rough-hairy; lvs. 3–7 cm. wide, round-cordate or reniform, with 5–7 triangular sharply pointed lobes, long-petioled; fls. 2–6 in the axils; involucels 4–5 mm. long, the lobes short and broad; petals rose-purple with deeper veins, 2–2.5 cm. long, deeply notched; carpels ca. 10, wrinkled-veiny on back, sharp-edged, glabrous or nearly so; $2n=42$ (Skovsted, 1935).—Reported as garden escape in scattered localities.

4. **M. verticillàta** L. [*M. crispa* L.] Erect smoothish annual 6–18 dm. high; lvs. crisped, 5–7-lobed, crenate, long-petioled; fls. sessile or nearly so, white to purple, crowded in axils; bractlets narrow; calyx ca. 4–5 mm. long, the lobes acuminate; petals 8–10 mm. long; carpels glabrous, smoothish or obscurely reticulate, the edges rounded; $n=$ ca. 56 (Skovsted, 1935).—Occasional escape, as at Santa Barbara, etc.; native of Old World.

9. *Modìola* Moench

Low diffuse perennial herb. Lvs. rounded, coarsely crenate to palmately incised. Fls. small, purplish to orange, solitary in axils, each with involucel of 3 bractlets. Calyx in ours enlarging in fr., the 5 lobes ovate, hirsute, acute. Petals obovate. Fr. depressed, of 14–25 thin-coriaceous reniform carpels, each septate between the 2 seeds. Small genus of Am. and Afr. (Latin, *modiolus*, the nave of a wheel, because of the fr.)

1. **M. caroliniàna** (L.) G. Don. [*Malva c.* L.] Stems spreading, pubescent, 3–6 dm. long; lvs. 2–5 cm. wide, on slender petioles; calyx 5–6 mm. long at anthesis, slightly enlarged in fr.; petals dull red, 5–8 mm. long; carpels hirsute, ca. 4 mm. high, beaked at outer edge of top.—Rather widely natur. in cismontane Calif. at lower elevs.; native of trop. Am. April–Sept.

10. *Sìda* L.

Ours low canescent stellate or cinereous-puberulent herbs. Lvs. crenate, serrate or lobed. Fls. axillary, solitary or in small cymules. Involucel usually lacking, in ours mostly of 1–3 linear deciduous bractlets. Calyx 5-lobed. Petals 5. Carpels 5–10, 1-seeded, indehiscent or dehiscent only part way from apex, ± rugose and often reticulate on the sides. Ca. 150 spp. of warmer parts of world. (Unexplained Greek name of some plant.)

1. **S. lepròsa** (Ort.) K. Schum. var. **hederàcea** K. Schum. [*S. h.* (Dougl.) Torr.] ALKALI-MALLOW. Stems from elongate rootstocks, decumbent or prostrate, whitish-stellate, 1–4 dm. long; lvs. round-reniform to broadly deltoid, dentate, rounded at apex, 1.5–4.5 cm. wide, on petioles 1–3 cm. long; calyx 6–7 mm. long; petals yellowish, 10–12

mm. long; carpels 6–10, indehiscent, reticulate on sides; $2n=22$ (Heiser & Whitaker, 1948).—Moist ± saline places, below 6000 ft.; many Plant Communities; widely distributed in Calif.; to Wash., Okla., Mex. May–Oct.

11. Sidálcea Gray. CHECKER

Annual or perennial herbs. Lvs. rounded, frequently palmately or pedately parted or lobed, with small stipules, petioled. Fls. in terminal racemes or spikes, perfect or with abortive anthers; these ♀ fls. small. Involucel lacking or of 1 bractlet. Calyx 5-lobed. Petals 5, emarginate or truncate, purple to rose-pink or white. Stamen-tube pubescent, with double series of fils., the outer series of phalanges (sets of united fils.) often distinctly below the inner. Carpels 5–9, beakless or beaked, 1-seeded, dehiscent. Ca. 22 spp. of w. N. Am., used somewhat in horticulture. (*Sida*, and *Alcea*, two malvaceous genera.)

(Hitchcock, C. L. A study of the perennial spp. of Sidalcea. Univ. Wash. Pub. Biol. 18: 1–79. 1957.)

Lvs. not deeply parted or divided; stamineal column not conspicuously double . . 1. *S. hickmannii*
Lvs. (at least the upper) usually deeply parted or divided; stamineal column conspicuously double, the outer phalanges narrow and cleft.
    Plants with enlarged rather fleshy taproots or fascicled roots, not at all rhizomatous; stems usually hirsute toward base; calyx mostly hirsute (at least in part with pustulose hairs; racemes mostly elongate and loosely fld.; pedicels slender and at least as long as the calyx; carpels smooth to lightly reticulate on the sides .... 3. *S. neomexicana*
    Plants usually with tough and fibrous rather than fleshy roots, often rhizomatous; stems often stellate-pubescent at base; calyx seldom with pustulate hairs; racemes often spicate and very closely fld.; carpels frequently conspicuously reticulate-alveolate.
        Carpels smooth; stems leafless or nearly so except at their softly hirsute bases; racemes spikelike, but usually rather open and loosely fld.; pedicels 1–3 mm. long; lvs. 3–5-lobate, the cauline ternately dissected into linear segms. San Bernardino Mts. .... 4. *S. pedata*
        Carpels usually roughened or plant not otherwise as above. Widely distributed. Stems often trailing and rooting .... 2. *S. malvæflora*

    1. **S. hickmánnii** Greene ssp. **párishii** (Rob.) C. L. Hitchc. Perennial from thick woody root-crown, the stems 4–7 dm. high, coarsely grayish-stellate; lvs. 1–6 cm. broad, roundish, coarsely dentate; fls. numerous, in many racemose spikes with lance-ovate bracts; calyx 6–8 mm. long; corolla pale pink, 9–15 mm. long; carpels glabrous, with a few transverse wrinkles.—Local on dry slopes at 5000–6000 ft.; Chaparral, Yellow Pine F.; San Bernardino Mts., Mission Pines, Santa Barbara Co.

    2. **S. malvæflòra** (DC.) Gray ex Benth. ssp. **malvæflòra**. [*Sida m.* DC.] Plate 60, Fig. C. Perennial with rather widely spreading rootstocks from heavy root; stems 1.5–6 dm. tall, coarsely pubescent below with simple to 4-rayed spreading to appressed hairs, more stellate upward; lvs. mostly 2–6 cm. broad, often markedly fleshy, usually simply to bifurcately hirsute beneath and cruciately stellate above, long-petioled, the basal from roundish to ± reniform, shallowly 7–9-lobed and coarsely crenate, the cauline similar or more deeply lobed, the floral often divided to base; racemes mostly simple, congested to elongate and open; pedicels 3–20 mm. long; calyx 8–12 mm. long; finely appressed-stellate and with longer coarse 2–4-rayed hairs; petals 1–2.5 cm. long, usually white-veined; carpels ca. 4 mm. long, sparsely glandular-pubescent, coarsely reticulate-faveolate, short-beaked; $2n=40$ (Krapovickas, 1967).—Frequent on open grassy slopes and mesas at low elevs. near the coast, Los Angeles Co. to Mendocino Co.

KEY TO SUBSPECIES

Lvs. chiefly basal, the cauline few, not dissected into linear segms.; infl. open, elongate; rootstocks usually poorly developed; carpels 2.5–3 mm. long, lightly reticulate-alveolate. Kern and e. Santa Barbara cos. to L. Calif. .... ssp. *sparsifolia* C. L. Hitchc.
Lvs. usually also on the stem; infl. typically few-fld., more congested; rootstocks usually well developed; carpels often over 4 mm. long, usually prominently reticulate-alveolate.

Stems hairy throughout with 4 (or more)-rayed spreading soft-stellate hairs; lvs. stellate on both surfaces or with a mixture of forked hairs ventrally; calyx grayish with uniform coarse many-rayed stellae. Coastal mts. of Santa Barbara and Ventura cos.
.... ssp. *californica* (Nutt.) C. L. Hitchc.
Stems usually partially hirsute below, or the lvs. not stellate ventrally, or the calyx finely stellate.
  Lvs. finely stellate on dorsal surfaces; stems mostly hirsute with slender simple hairs; carpels ca. 3 mm. long; calyx rather sparsely finely stellate and somewhat hirsute with longer simple or 4-rayed hairs. San Bernardino Mts. .... ssp. *dolosa* C. L. Hitchc.
  Lvs. coarsely stellate dorsally; stems usually coarsely hirsute (to stellate); carpels 3.5–4 mm. long; calyx coarsely stellate, often ± lanate at base. Near coast, Mendocino Co. to Los Angeles Co. .... ssp. *malviflora*

3. **S. neomexicàna** Gray ssp. **thúrberi** (Rob. ex Gray) C. L. Hitchc. Stems 2–9 dm. long from fleshy fusiform roots, hirsute to stellate below to glabrous, frequently ± stellate above; lvs. 1.5–4.5 cm. wide, the lower 5–9-lobed or -parted into lobed or toothed divisions, the upper into linear entire or lobed divisions; infl. slender-racemose, many-fld., stellate-pubescent and with some longer hairs; petals rose, mostly 6–12 mm. long; carpels lightly and coarsely reticulate on the sides, usually smooth on back, ca. 2 mm. long; $2n = 20$ (Kruckeberg, 1957).—Alkaline usually wet places; Coastal Sage Scrub, Chaparral, Creosote Bush Scrub; Los Angeles, Orange, Riverside and San Bernardino cos.; to New Mex. April–June.

4. **S. pedàta** Gray. Stems few to several from the root-crown, ascending, simple, subscapose, 1–3 dm. tall, glabrous to hirsute, reddish; lvs. many, 2–4 cm. wide, pedately 5–7-parted or -divided, the divisions narrowly cuneate, 3-lobed, the ultimate segms. linear to oblong; infl. a many-fld., at length elongate spicate, stellate-pubescent raceme; bracts 3–5 mm. long, simple to bifid; calyx 4–5 mm. long, the lobes lanceolate, widening in fr.; petals rose-purple, 8–12 mm. long; carpels 5, rounded, smooth, ca. 3 mm. high; $2n = 20$ (Kruckeberg, 1957).—Wet meadows, 6500–7500 ft.; Montane Coniferous F., San Bernardino Mts. May–July.

12. *Sphaerálcea* St. Hil. Globemallow

Perennial herbs or suffruticose plants, usually closely stellate-pubescent. Lvs. fairly thick, shallowly dentate to pedately dissected. Infl. racemose or paniculate. Involucel of 2 or 3 bractlets. Petals 5, red (grenadine), sometimes pink or lavender. Calyx 5-lobed. Stamens in one series. Stigmas capitate. Carpels 5 or more, 1–3-ovuled, 1-celled, relatively thick, less than 8 mm. high, differentiated into a dehiscent unreticulate apical portion and an indehiscent reticulate basal portion. Seeds reniform, usually pubescent. Perhaps 60 spp. of warmer parts of the New World, a few S. African. (Greek, *sphaera*, globe, and *alkea*, mallow, because of the spherical fr.)

(Kearney, T. H. The North American species of Sphaeralcea, subgenus Eusphaeralcea. Univ. Calif. Pub. Bot. 19: 1–128. 1935.)

1. Plants annual or biennial; petals orange; the reticulate basal part of the carpel forming $\frac{2}{3}$ or more of the whole and conspicuously wider than the unreticulate apical part.
    2. Plant yellowish-canescent; lvs. thick and firm, crenulate; carpels deeply notched
        .... 5. *S. orcuttii*
    2. Plant grayish-pubescent; lvs. thin and soft, coarsely crenate; carpels shallowly notched
        .... 3. *S. coulteri*
1. Plants perennial; petals red (grenadine) to white, pink or lavender; the reticulate part of the carpel forming less than $\frac{2}{3}$ of the whole and not conspicuously wider than the unreticulate part.
    3. Lvs. not more than $\frac{1}{3}$ as wide as long; pubescence yellowish .... 2. *S. angustifolia*
    3. Lvs. at least half as wide as long.
        4. Calyx at anthesis 11–20 mm. long.
            5. Reticulate part of carpel rugose or muricate on back, the reticulations prominent and coarse. Widely distributed on deserts .... 1. *S. ambigua*
            5. Reticulate part of carpel smooth or nearly so on the back, the reticulations less prominent and finer. Panamint Mts. .... 7. *S. rusbyi*

4. Calyx at anthesis 4–10 mm. long.
   6. Reticulate part of carpel rugose or muricate on back.
      7. Fr. truncate-conical; lvs. ca. as wide as long; infl. usually an open panicle
         1. *S. ambigua*
      7. Fr. hemispherical or nearly so; lvs. definitely longer than wide; infl. narrow
         4. *S. emoryi*
   6. Reticulate part of carpel smooth or nearly so on back; pubescence of plant dense, grayish; lf.-blades finely crenate; carpels broadly ovate, acutish. Inyo Co.
         6. *S. parvifolia*

1. **S. ambígua** Gray ssp. **ambígua**. [*S. a.* var. *keckii* Munz.] DESERT-HOLLYHOCK. DESERT-MALLOW. Plate 60, Fig. D. Perennial with thick woody crown and woody lower stems, whitish- or yellowish-canescent; stems few to many, 5–10 dm. high; lvs. thickish, ovate to suborbicular, ± cordate at base, blunt at apex, crenulate to crenate, 3-lobed near middle, 2–6 cm. long and ca. as wide; infl. usually open-paniculate with few to many fls.; bractlets 3–5 mm. long; calyx at anthesis 6–20 mm. high, the lobes lanceolate, acuminate; petals grenadine, 15–35 mm. long; fr. hemispherical; carpels 12–16, with thickish chartaceous walls, 3.5–6 mm. high, helmet-shaped, rather deeply notched, usually rugose and muricate dorsally, the indehiscent part ca. ⅓ of the carpel, coarse-reticulate; seeds usually 2, pubescent; $n = 10$ or 15 (Webber, 1936).—Dry rocky slopes and canyons, mostly below 4000 ft.; Creosote Bush Scrub, Joshua Tree Wd.; throughout the Calif. deserts; to Utah, Son., L. Calif. March–June. Exceedingly variable.

KEY TO SUBSPECIES

Stems usually woody above base and more than 5 dm. high; infl. usually an open long-branched panicle.
  Petals grenadine .... ssp. *ambigua*
  Petals pink or lavender, often drying violet .... ssp. *rosacea*
Stems usually herbaceous above the crown and less than 5 dm. high; infl. racemiform or narrow-thyrsoid.
  Pubescence of stems usually whitish; carpels coarsely reticulate; lvs. thin, not conspicuously rugose-veined .... ssp. *monticola*
  Pubescence of stems usually yellowish; carpels finely reticulate; lvs. thickish, conspicuously rugose-veined .... ssp. *rugosa*

Ssp. **montícola** Kearn. [*S. pulchella* Jeps.] Stems herbaceous, whitish-pubescent; lvs. 1.5–3 cm. long; infl. narrow, few-fld.; petals 15–20 mm. long; coarsely reticulate portions of the slightly rugose carpels small; $n = 10$ (Webber, 1936), 15 (Bates, 1967).—Dry rocky slopes, mostly 4000–7000 ft.; Pinyon-Juniper Wd.; Mono Co. to San Bernardino Co.; Nev. April–June.

Ssp. **rosàcea** (M. & J.) Kearn. [*S. r.* M. & J.] Stems woody at base, white-pubescent; petals pale purplish-pink, drying rose-violet; anthers usually purple; $n = 15$ (Webber, 1936).—Below 3000 ft.; Creosote Bush Scrub; e. base of Inyo Mts. to w. Colo. Desert; Ariz., n. L. Calif. March–May.

Ssp. **rugòsa** Kearn. Stems herbaceous, yellowish-pubescent; infl. narrow, thyrsoid, interrupted; fls. many, 12–15 mm. long; carpels finely reticulate; $n = 10$ (Bates, 1967).—Dry slopes, 3000–6000 ft.; Joshua Tree Wd., Pinyon-Juniper Wd., Yellow Pine F.; s. Mojave Desert and San Jacinto Mts. to n. L. Calif.; occasional at San Jacinto, etc.

2. **S. angustifòlia** (Cav.) G. Don ssp. **cuspidàta** (Gray) Kearn. [*S. c.* Britton.] Perennial with thick woody crown, densely canescent, erect, 8–18 dm. high; lvs. thickish, prominently veined, often revolute, linear-lanceolate, 2–7 cm. long, usually ca. ⅕ as wide, subhastately lobed basally; petioles 0.5–2.5 cm. long; infl. long, narrow, interrupted, many-fld., leafy; calyx at anthesis 5–9 mm. long, the lobes lanceolate; petals grenadine to almost pink, 7–12 mm. long; fr. truncate-conical, carpels 10–16, with chartaceous walls, 4–6 mm. high, shallowly notched, the indehiscent portion prominently reticulate; seeds 1–3, usually glabrous; $n = 5$ or 10 (Webber, 1936).—Reported from near Indio and

Hayfields (Riverside Co.), Tecopa, Blythe, etc. on the desert, Los Angeles; Kans. to Mex., Ariz. Summer.

3. **S. còulteri** (Wats.) Gray. [*Malvastrum c.* Wats.] Annual, with slender taproot, rather sparsely pubescent, erect, 2–10 dm. tall, rather slender; lvs. thin, soft, broadly ovate to suborbicular, truncate to cordate at base, scarcely lobed to deeply 3–5-lobed, coarsely crenate, 1–3 cm. long; petioles 1–4 cm. long; infl. thyrsoid, few- to many-fld.; bractlets linear, 2–3 mm. long; calyx at anthesis 5–7 mm. long, the lobes lanceolate, acuminate; petals ± orange, 8–15 mm. long; fr. hemispheric; carpels 14–22, thin-walled, 2–2.5 mm. high, reniform, shallowly notched, the reticulate part forming $\frac{2}{3}-\frac{3}{4}$ of the carpel, fenestrate; seed 1, glabrous or sparsely pubescent; $n=5$ (Webber, 1936).—Rare, dry sandy places, low elevs.; Creosote Bush Scrub; Imperial Co.; to Ariz., Sinaloa. March–April.

4. **S. émoryi** Torr. in Gray ssp. **émoryi.** Perennial with a stout woody crown, several stems 3–10 (–12) dm. tall, grayish-canescent; lvs. thickish, ovate-oblong, usually subcordate, and somewhat rounded-angulate at base, crenulate to dentate, 2–9 cm. long, $\frac{1}{2}-\frac{3}{4}$ as wide; petioles 1–6 cm. long; infl. a many-fld. usually narrow thyrse, leafy; bractlets linear, 1–2 mm. long; calyx at anthesis 5–10 mm. long, the lobes deltoid-ovate to lanceolate, acute to acuminate; petals grenadine, sometimes pink or lavender, 10–20 mm. long; fr. truncate-conical; carpels 11–16, with chartaceous walls, 3.5–6 mm. high, rather deeply notched, with prominent ventral beak, the reticulate part forming $\frac{1}{3}-\frac{1}{2}$ of carpel, usually rugose-tuberculate on back; seeds usually 2, usually pubescent; $n=15, 25$ (Webber, 1936). —Sandy or loamy places along roads and in fields, below 2000 ft.; Creosote Bush Scrub; desert parts of Riverside and Imperial cos.; Nev., Ariz., L. Calif. March–May. Exceedingly variable.

KEY TO SUBSPECIES

Lf.-blades $\frac{1}{3}-\frac{3}{4}$ as wide as long, cleft near base, 3–9 cm. long.
   Lvs. $\frac{1}{2}-\frac{3}{4}$ as wide as long, mostly cordate at base; carpels rather thick-walled and coarsely reticulate.
      Stems and lvs. densely pubescent, grayish-canescent; lvs. thickish, scarcely lobed
                                                                                 ssp. *emoryi*
      Stems and lvs. less pubescent, greener; lvs. thinnish, definitely lobed . . . . . . ssp. *variabilis*
   Lvs. $\frac{1}{3}-\frac{1}{2}$ as wide as long, at most subcordate; carpels thin-walled, finely reticulate
                                                                                   ssp. *nevadensis*
Lf.-blades ca. as wide as long, cleft far above base, 2–5 cm. long . . . . ssp. *arida*

Ssp. **árida** Kearn. [*S. arida* Rose.] Lvs. as wide as long, not more than 5 cm. long; carpels 8–12, thinly walled, finely reticulate; $n=10$ (Webber, 1936).—Creosote Bush Scrub; e. Imperial Co.; to Nev., Son., Sinaloa. April.

Ssp. **nevadénsis** Kearn. Lvs. $\frac{1}{3}-\frac{1}{2}$ as wide as long; carpels thin-walled, finely reticulate; Creosote Bush Scrub; e. Riverside Co. to e. Inyo Co.; Nev., Ariz. March–May.

Ssp. **variábilis** (Ckll.) Kearn. [*S. variabilis* Ckll.] Herbage greenish, sparsely pubescent; lvs. thinnish, 3-cleft, with broad to narrow lateral lobes and longer ± pinnatifid terminal lobe; $n=10$ or 15 (Webber, 1936).—Dry places in fields and along roads, below 3000 ft.; Coastal Sage Scrub, Creosote Bush Scrub; Upland, Colton, Redlands to s. Mojave and n. Colo. deserts; Ariz. March–May.

5. **S. orcúttii** Rose. Annual or biennial, with large taproot, erect, densely yellowish canescent, 6–10 dm. tall, stout; lvs. thick, firm, deltoid-ovate, shallowly 3-lobed near base, usually subcordate, crenulate, 3–5 cm. long, $\frac{3}{5}-\frac{4}{5}$ as wide; petioles 1–2.5 cm. long; infl. long, narrow, many-fld.; bractlets linear, 2–3 mm. long; calyx 4–7 mm. high at anthesis, the lobes lance-ovate, acuminate; petals ± orange, 8–12 mm. long; fr. hemispherical; carpels 12–17, with thin scarious walls, 2.5–3 mm. high, deeply notched, the indehiscent part forming $\frac{3}{4}-\frac{4}{5}$ of the carpel and much wider than the dehiscent part, reticulate, fenestrate; seeds 1, sometimes 2; $n=5$ (Webber, 1936).—Dry semialkaline places at low elevs.; Imperial Co. to L. Calif., Ariz.

6. **S. parvifòlia** A. Nels. Perennial, grayish- or whitish-canescent; stems several, 6–10

dm. high; lvs. thickish, prominently veined beneath, broadly ovate to suborbicular, 1.5–4 cm. long, not lobed to shallowly 3-lobed near middle; petioles 1–3 cm. long; infl. many-fld., narrowly thyrsoid-glomerate; calyx densely pubescent, 4–8 mm. long at anthesis, the lobes lance-ovate, short-acuminate; petals grenadine, 8–18 mm. long; carpels 9–12, with chartaceous walls, 3–5 mm. high, broadly notched, the indehiscent part ca. ¼ of the carpel, finely rather faintly reticulate; seeds usually 2; $n = 5$ or 10 (Webber, 1936).—Dry slopes, 5000–7000 ft.; Pinyon-Juniper Wd.; Inyo Co. to Colo., New Mex. June–July.

7. **S. rúsbyi** Gray ssp. **eremícola** (Jeps.) Kearn. [*S. e.* Jeps.] Perennial with thick woody crown, green and sparsely pubescent, 3–6 dm. high; lvs. thin, round-cordate, 4–20 mm. long, 3–5-parted, the lobes then cleft and toothed; fls. few, in a loose panicle; calyx woolly, 11–14 mm. long; petals apricot, 14–16 mm. long; carpels 10–12, with chartaceous walls, 4–5 mm. high, shallowly and broadly notched, the indehiscent part faintly reticulate; seeds 2; $n = 5$ (Webber, 1936).—At 4000–4400 ft.; Creosote Bush Scrub; Emigrant Canyon, Panamint Mts., Inyo Co. May–June.

66. **Martyniàceae.** MARTYNIA FAMILY

Viscid-pubescent herbs with simple opposite or alternate lvs. and terminal racemose infl. Calyx 4–5-lobed. Corolla campanulate, obscurely 2-lipped. Stamens 4, didynamous, or 2, with the other pair forming staminodia. Ovary superior, of 2 carpels, but 1-celled, with 2 parietal placentae. Style 1; stigma with 2 flat lobes. Caps. horned, with fleshy deciduous exocarp and woody endocarp, often crested on median line above and sometimes below. Seeds few to many, black, nearly oblong, sculptured, somewhat compressed. A small family of the warmer parts of the New World.

Calyx composed of 5 free sepals; body of fr. echinate .... 1. *Ibicella*
Calyx ± spathaceous, dentate or lobed above, cleft to base below; body of fr. roughly sculptured
2. *Proboscidea*

1. *Ibicélla* Van Es.

Lvs. ovate or suborbicular, entire. Infl. dense, terminal. Calyx of 5 sepals, the 3 upper linear-lanceolate to obovate, the 2 lower much broader. Corolla yellow, oblique-campanulate. Fertile stamens 4. Fr. including beak 1.5–2 dm. long, the body 8–10 cm. long, cylindrical-ovoid, echinate. Two spp., native of S. Am. (Diminutive of *ibex*, the wild goat or chamois, because of the curved horns of the fr.)

1. **I. lùtea** (Lindl.) Van Es. [*Martynia l.* Lindl.] Glandular-pubescent, spreading, branched, the stems 3–5 dm. long; lvs. opposite or alternate, the blades rounded, dentate, ± angularly incised or subcordate, 8–14 cm. wide, on petioles 10–15 cm. long; racemes few-fld., dense, scarcely surpassing lvs.; calyx ca. 2 cm. long; corolla deeper colored within, often dotted with red, ca. 3 cm. long; fls. thick, with purple spots; caps. with a slender horn longer than the short-spined body; seed 6–12 mm. long, compressed, rugose; $2n = 32$ (Covas & Schnack, 1947).—Occasional weed, in Sacramento and San Joaquin valleys and s.; natur. from S. Am. Late summer.

2. *Proboscídea* Keller in Schmid. UNICORN-PLANT (Plate 60, Fig. E)

Coarse viscid-pubescent plants. Lvs. petioled, the lower mostly opposite, large, entire to shallowly lobed. Fls. few, large, showy, purplish, pinkish or yellowish. Calyx 5-lobed, split ventrally to base, viscid-pubescent. Corolla limb flaring. Stamens 4, didynamous. Ovary 1-celled; style long, cylindrical; stigmas with 2 flat lobes. Caps. with an ovoid or cymbiform sculptured body and ending in a long incurved dehiscent horn. Seeds many, large. Ca. 9 spp. (Greek, *proboscis*, beak.)

Lvs. averaging 5 cm. or less wide; fls. yellowish or copper-colored .... 1. *P. althaeifolia*
Lvs. averaging 10 cm. or more across; fls. deep reddish-purple to pinkish, or at least with purplish blotches .... 2. *P. parviflora*

1. **P. althaeifòlia** (Benth.) Dcne. [*Martynia a.* Benth.] Coarse spreading perennial 3–4 dm. tall; lvs. reniform to suborbicular, usually broader than long, 3–5 cm. wide, on petioles 4–8 cm. long; raceme 3–7-fld. on an axis 5–8 cm. long; pedicels often as long as axis of infl.; calyx ca. 1 cm. long; body of caps. 5–6 cm. long, the horns ca. twice as long, crested on both dorsal and ventral edges.—Occasional, sandy places; Creosote Bush Scrub; Imperial Co.; to Ariz., Mex. Summer.

2. **P. parviflòra** (Woot.) Woot. & Standl. [*Martynia p.* Woot.] Annual, spreading or matted, the stems up to 8 dm. long; lvs. ± opposite, broadly triangular, to round-ovate, subentire to obtusely 5–7-lobed, cordate, obtuse, 8–12 (–25) cm. wide; corolla red-purple to almost white, often dotted or blotched with red-purple and streaked with yellow; body of fr. 5–7 cm. long, the horns ca. twice as long; seeds black, 6–8 mm. long.—Occasional, as at Whittier, La Verne, Panamint Mts., Hunter Canyon in Saline V. e. of the Inyo Mts.; s. Nev. to Mex., Tex. Summer.

67. Moràceae. Mulberry Family

Herbs to trees, often with milky juice. Lvs. mostly alternate and simple. Fls. small, inconspicuous, regular, mostly imperfect, usually in heads or spikes, sometimes the ♂ racemose. Perianth single, entire or parted; stamens of same no. as perianth-parts and opposite them. Styles or stigmas 2; ovary 1-celled, ovule pendulous. Fr. an ak. or drupe, often imbedded in fleshy perianth or axis.

Plant an erect herb; lvs. digitate, alternate .... 1. *Cannabis*
Plant a twining vine; lvs. lobed, opposite .... 3. *Humulus*
Plants trees, with lobed lvs.
   Fls. lining a fleshy receptacle which forms a fig .... 2. *Ficus*
   Fls. in small catkinlike spikes, the ♂ soon falling away, the ♀ ripening into a juicy blackberry-like cluster .... 4. *Morus*

1. Cánnabis L. Hemp. Marijuana

Stout erect rough puberulent annual herbs, mostly dioecious. Lvs. 5–11-divided; fls. greenish, axillary, the ♂ panicled, the ♀ spicate. Stamens 5, drooping; sepals free. Pistillate fl. in axil of small enclosing bract, the calyx inconspicuous, adherent to the long ovary. Fr. an ak., held in the persistent bract. (*Cannabis*, the ancient classical name.)

1. **C. satìva** L. Plants 4–12 or more dm. high; lvs. linear-lanceolate, 5–10 cm. long, coarsely toothed; $2n=20$ (Melvedeva, 1935).—Occasionally adventive, but now not grown without a permit; native of Old World and of use for fiber. Producing a resinous oil, the source of hashish and marijuana.

2. Fìcus L. Fig

Woody plants with milky sap and mostly alternate lvs. The fr. a synconium, the minute fls. being on the inside of a closed receptacle or branch that arises from the axil of a lf. or is supernumerary, the receptacle ripening into a ± fleshy body, with the many aks., succulent pedicels and interfloral scales on the inner walls; ♂ perianth 2–6-parted, with 1 or more erect stamens; ♀ perianth sometimes wanting; receptacles unisexual or bisexual, usually the latter; pollination by special insects. An enormous genus of warm regions. (*Ficus*, Latin name for fig.)

Lvs. wavy-margined or lobed; fr. ± pear-shaped .... 1. *F. carica*
Lvs. deeply lobed; fr. roughened .... 2. *F. pseudo-carica*

1. **F. carìca** L. Common Fig. Tree 4–8 m. high; lvs. 3–5-lobed, the lobes ± wavy-margined or lobed, and with palmate veins; fr. single, axillary, pear-shaped. Occasional escape from cult. Catalina Id., Tehachapi, etc. Native of Asia Minor.

2. **F. pseudo-carìca** Miq. Lvs. more deeply cut, 3- or sometimes 5-nerved; fr. axillary,

round, roughened.—Reported as occasional escape in Santa Barbara region; native of Abyssinia.

3. *Hùmulus* L. Hops

Twining annuals and perennials, dioecious. Lvs. opposite. Staminate fls. in panicled, tassellike racemes, with 5-parted calyx and 5 stamens. Pistillate fls. 2 together under large imbricate, persistent bracts in a spike that becomes a conelike structure or hop, each fl. with an entire calyx that surrounds the ovary. Fr. an ak. in the tight enlarged calyx. Three spp. of N. Temp. (Latin name of uncertain derivation.)

1. **H. lùpulus** L. Perennial with rough stems; lvs. mostly 3-lobed to ca. the middle, the terminal lobe broadest; margins coarsely serrate, upper lf.-surface rough, lower less so and sparsely glandular; petioles ca. as long as blade; fr.-spikes 3–5 cm. long at maturity with thin light-colored resinous-dotted scales.—Occasional escape from cult.; grown for brewing purposes; native of Eurasia.

4. *Mòrus* L. Mulberry

Trees with milky juice; lvs. alternate, lobed. Fls. monoecious or dioecious, in small cylindrical catkinlike spikes, the ♂ soon falling, the ♀ ripening into a blackberry-like juicy cluster; perianth 4-parted; stamens 4; each ovary becoming a drupelet inclosed in the large fleshy perianth. Ca. a dozen spp. (Morus, the classical name.)

1. **M. álba** L. White Mulberry. Tree to 15 m.; lvs. broad-ovate, the blade 5–15 cm. long, largely scallop-toothed, often irregularly lobed; fr. 2–5 cm. long, whitish to purple, sweet.—Occasionally natur.; originally from China.

68. **Myoporàceae.** Myoporum Family

Woody plants, glandular or woolly; lvs. alternate or rarely opposite, simple, entire or rarely toothed; fls. bisexual, regular or irregular, axillary, solitary or fascicled, subsessile or pedicellate; calyx 5-parted or -cleft, persistent; corolla gamopetalous, the limb 5–6-lobed, sometimes bilabiate; stamens 4, didynamous, the 5th represented by a staminode, epipetalous; ovary superior, 2-celled or falsely 3–10-celled; ovules usually 1 or 2, rarely 8 in a cell; style 1. Stigmas 1–2; fr. drupaceous. A family of 5 genera and ca. 90 spp. of Australia and other warm places to e. Asia, Hawaii, W. Indies.

1. *Myóporum* Banks

More or less heathlike shrubs, with mostly alternate lvs., entire or toothed, with pellucid glands; fls. small or medium-sized, mostly white; corolla-tube short and campanulate or longer and funnelform; stamens 4; fr. a drupe. Species 25–30. (Greek, *myein*, to close, and *poros*, a pore, from the translucent dots on the lvs.)

1. **M. laètum** Forst. f. To ca. 6 m. tall; lvs. lanceolate, 5–10 cm. long, finely serrate above the middle, bright green, shining, almost fleshy; fls. 2–6 in a fascicle, white with purple spots, 8–16 mm. across, the lobes rounded and hairy inside; drupe reddish-purple. —Natur. near Ventura; native of New Zealand.

69. **Myricàceae.** Wax-Myrtle Family

Monoecious or dioecious shrubs or trees with simple alternate deciduous or persistent resinous-dotted often fragrant lvs. without stipules. Both kinds of fls. in short scaly aments. Differing from *Betulaceae* mostly in the 1-locular ovary with a single erect ovule and the drupelike nut. Perianth and invol. none. Two genera and ca. 40 spp., of wide distribution.

1. Myrica L.

Lvs. entire, dentate or lobed. Staminate aments oblong or thick-cylindrical; stamens 2–20, the fils. somewhat united below; anthers 2-locular. Pistillate aments ovoid or subglobose, the ovary subtended by 2–4 short bractlets. Fr. globose or ovoid. Ca. 30 spp. of temp. and trop. regions. (Greek, *myrike*, an old name of a fragrant shrub.)

1. **M. califórnica** Cham. & Schlecht. WAX-MYRTLE. Plate 60, Fig. F. Shrub 2–4 m. tall or even arborescent; branches slender, ascending; bark gray or light brown; lvs. oblong to oblanceolate, dark green, glossy, persistent, 5–11 cm. long, subentire to remotely serrate, with a petiole 3–10 mm. long; ♂ catkins 1–2 cm. long, borne in lower axils; stamens 7–16; ♀ catkins in upper axils, 8–12 mm. long; ovary ovoid, bractlets minute; fr. brown-purple, 6–8 mm. in diam., covered with a whitish wax.—Canyons and moist slopes below 500 ft.; Chaparral, Coastal Sage Scrub, etc.; Santa Monica Mts. n. to Wash. March–April.

70. Myrtàceae. MYRTLE FAMILY

Aromatic trees and shrubs, mostly with opposite persistent thickish entire short-petioled, exstipulate lvs., ± pellucid-punctate. Fls. bisexual, regular, solitary in axils or in racemes or corymbs, usually bracted. Fl.-tube ± adnate to the ovary, sometimes elongate. Sepals 4–5, usually persistent. Petals 4–5, imbricate, rarely lacking. Stamens usually numerous, often in fascicles opposite the petals, the fils. distinct or partly united. Ovary inferior, 1- to many-loculed, with 1–many ovules in each locule, the placentation mostly axile; style simple. Fr. a berry caps. or nut. Ca. 75 genera and 3000 spp., mostly trop. or subtrop., particularly in Australia and S. Am.

1. *Eucalýptus* L. Hér. GUM TREE

Lvs. mostly vertical, alternate, stiff, pinnate-veined, those of the young shoots often opposite, sessile, horizontal. Fls. in umbels or heads. Fl. tube turbinate or campanulate, adnate to ovary at base, the free part entire or 4-toothed. Sepals and petals united to form lid or cap which is present on the bud and drops off at anthesis. Stamens many, in several series, white to highly colored. Ovary 3–6-loculed; ovules many. Fr. a caps., loculicidally dehiscent at top by 3–6 valves. Seeds minute. A large genus, largely Australasian, many spp. important for wood, oil, ornament. Many are grown in Calif. and some occasionally establish themselves outside of cult. (Greek, *eu*, well, and *kalyptos*, covered.)

Fls. 6–12 mm. across .... 1. *E. camaldulensis*
Fls. ca. 4 cm. across .... 2. *E. globulus*

1. **E. camaldulénsis** Dehnhardt. Shrub or small tree, finely pubescent; lvs. narrowly lanceolate, sharply pointed; fls. white, solitary; stamens 4–8, together, in a stalked umbel; lid conical, not beaked; fr. subglobular. Occasionally natur. as about Santa Barbara.

2 **E. glóbulus** Labill. BLUE GUM. Tree to 80 m. tall, the bark deciduous; lvs. of older branches 1.5–2 dm. long; fls. white, solitary in axils, ca. 4–5 cm. wide; lid of bud warty; fr. angular, 2–2.5 cm. across. Dec.–May.

71. Nyctaginàceae. FOUR-O'CLOCK FAMILY

Largely herbs or shrubs, with ± swollen nodes and fragile stems. Lvs. subentire, petiolate, exstipulate, mostly opposite. Fls. perfect, regular, subtended by bracts which are frequently united into calyxlike invols. Perianth corolla-like, usually campanulate or funnelform or salverform, 4–5-lobed or -toothed, constricted above the ovary. Petals none. Stamens hypogynous; fils. filiform; anthers 2-celled. Ovary superior, enclosed by the persistent perianth-tube, 1-celled, 1-ovuled; style short or long; stigma capitate to linear. Fr. an anthocarp, closely invested by the hardened base of the perianth, usually striate, angled or winged and enclosing the free ak. Seed erect, the embryo curved in ours, the cotyledons enclosing the mealy or fleshy endosperm. Ca. 20 genera and 150 spp. of warmer parts of the world.

Stigma linear; fr. usually winged; perianth salverform .... 1. *Abronia*
Stigma subglobose; fr. rarely winged; perianth usually funnelform or campanulate.
    Fr. lenticular, the margins inrolled, dentate, the dorsal surface with 2 rows of stipitate glands; invols. 3-parted .... 3. *Allionia*
    Fr. not lenticular or with dentate margins or glands.
        Invol. of distinct bracts, 1 at base of each fl.
            Perianth 10 or more cm. long, tubular, white tinged purple, usually solitary
                                                                                             2. *Acleisanthes*
            Perianth not over 4 cm. long.
                The perianth not more than 1 cm. long; fr. 5-angled or -ribbed........ 4. *Boerhaavia*
                The perianth 2–4 cm. long.
                      Fr. ellipsoid, smooth, with vertical lines; each pedicel partially attached to the leafy subtending bract; perianth purplish-red .... 5. *Hermidium*
                      Fr. with 3–5 hyaline wings; pedicels not attached to subtending bracts; perianth greenish .... 8. *Selinocarpus*
        Invol. of united bracts, resembling a gamosepalous calyx.
            Fr. 5-ribbed; invol. enlarged and papery in fr. .... 7. *Oxybaphus*
            Fr. smooth or nearly so; invol. but little changed in fr. .... 6. *Mirabilis*

1. *Abrònia* Juss. SAND-VERBENA

Annual or perennial, often viscid, branching herbs, usually prostrate or decumbent. Lvs. opposite, unequal, slightly to quite fleshy, petioled, the base often unequal. Fls. perfect, capitate, the heads on axillary or terminal peduncles and subtended by an invol. of 5–8 scarious bracts. Perianth fragrant, often showy, salverform with slender tube, enlarged at throat, the limb with 4–5 emarginate lobes and withering-persistent. Stamens 4–5, unequal, included in the tube and inserted upon it. Stigma fusiform, included, on a filiform style. Fr. an anthocarp, turbinate or biturbinate, deeply lobed or winged. Seed adherent to the pericarp, dark brown, elliptic-oblong or narrower, one cotyledon broad, the other abortive. Perhaps 25 spp., of w. N. Am. (Greek, *abros*, graceful.)

Perianth yellow; lvs. as wide as long. Seashore .... 1. *A. latifolia*
Perianth white to rose or purple-red.
    Frs. with 2–5 evident wings.
        Fls. deep dark red; fr. 12–20 mm. broad. Seashore .... 2. *A. maritima*
        Fls. white to rose; fr. often narrower.
            Plants with a perennial branched caudex; lvs. all basal; fls. scapose. Montane . 4. *A. nana*
            Plants annual or perennial, but not with a branched caudex; lvs. largely on elongate stems; fls. not scapose.
                Wings of fr. translucent, extended around the body (above and below it); fls. mostly 4-merous .... 3. *A. micrantha*
                Wings of fr. opaque, interrupted above and beneath the body; fls. mostly 5-merous.
                      The wings of the fr. usually 2; bracts broadly ovate; plants annual. Deserts
                                                                           5. *A. pogonantha*
                    The wings of the fr. usually 3–5; bracts commonly lanceolate.
                        Stems and perianth densely long-villous; body of fr. rugose-veined, the veins coarse and extending into the wings. Annual, away from the coast .. 8. *A. villosa*
                      Stems and perianth glabrous to short-villous; body of fr. not rugose-veined. Mostly annual. Seashore .... 7. *A. umbellata*
    Frs. not winged, merely angled; fls. 15–30 in a head; plants annual. W. edge of Mojave Desert
                                                                                6. *A. turbinata*

    1. A. latifòlia Eschs. Much-branched perennial with stout fleshy roots and prostrate stems 3–10 dm. long, densely glandular-pubescent; lf.-blades round to oval, 1–4 (–6) cm. long, thick and succulent, glandular-puberulent to glabrate; petioles stout, 1–6 cm. long; peduncles 2–6 cm. long, viscid; bracts 5–6, ovate to lance-ovate, 6–8 mm. long; perianth

13–18 mm. long, the tube slender, glandular-villous, yellow-green, the limb 5–8 mm. broad, yellow, with shallowly emarginate lobes; fr. 8–15 mm. long, turbinate, with 5 winglike reticulate-veined lobes attenuate upward on the ak.-body; seed elliptic-oblong, brown, 4–5 mm. long.—Coastal Strand; San Miguel Id.; Surf n. to B.C. May–Oct. Hybridizes freely with *A. maritima* and *A. umbellata*, various introgressants appearing.

2. **A. marítima** Nutt. ex Wats. Perennial with fleshy roots and succulent much-branched prostrate glandular-puberulent and somewhat villous stems 2–10 dm. long; lf.-blades oval to oval-oblong, thick, 2–6 cm. long, viscid-puberulent; petioles 1–2 cm. long; peduncles 3–8 cm. long; bracts lanceolate to somewhat wider, 6–8 mm. long; perianth 12–14 mm. long, dark crimson to red-purple, the limb 3–5 mm. broad; fr. 10–14 mm. long, puberulent above, turbinate, with mostly 5 winglike lobes truncate above, irregular, attenuate below, coarsely reticulate-veined; seed oblong, dark, 5 mm. long.—Coastal Strand; San Luis Obispo Co. to L. Calif. Feb.–Oct. Channel Ids. Hybridizes with *A. umbellata* and *A. latifolia*.

3. **A. micrántha** Torr. Erect or procumbent branched annuals with viscid-puberulent stems; peduncles 1–2 cm. long; bracts lanceolate to lance-ovate, 6–10 mm. long, attenuate; perianth ca. 1.5 cm. long, viscid-puberulent, the limb greenish-white, 3–4 mm. broad; fr. 1.5–3 cm. long, the body puberulent to subglabrous, the wings 2–3, finely veined; seed brown, slightly angled, ca. 6–8 mm. long.—Sand dunes; Creosote Bush Scrub; Kelso, Mojave Desert; to N. Dak., Kans., New Mex. April–May.

4. **A. nàna** ssp. **covíllei** (Heimerl) Munz. [*A. c.* Heimerl.] Densely cespitose perennial from a thick woody root and branched caudex, the branches 2–6 cm. long; lvs. densely clustered, the blades glaucous, oblong-ovate, 0.5–2 cm. long, obtuse or rounded at apex, minutely puberulent; petioles 1–4 cm. long, puberulent and sometimes with longer hairs; peduncles scapelike, slender, reddish, puberulent, sometimes pubescent, 2–10 cm. long; bracts lanceolate, scarious, 2–3 mm. wide; perianth white or pinkish, 11–15 mm. long, the limb 6–8 mm. wide; fr. turbinate, 7–8 mm. long, obcordate with 5 thin-walled regular winglike lobes; seed dark, narrow, 2.5 mm. long, longitudinally veined.—Dry sandy places, 5000–9400 ft.; Pinyon-Juniper Wd.; Montane Coniferous F., Sagebrush Scrub; San Bernardino Mts., New York Mts., Inyo Mts. to Mono Co.; sw. Nev. June–Aug.

5. **A. pogonántha** Heimerl. [*A. angulata* Jones.] Much-branched annual with ascending to decumbent stems 1–4 dm. long, villous and viscid-puberulent; lf.-blades oblong-ovate to suborbicular, rounded to obtuse at apex, villous on midveins, 1–4 cm. long; petioles 1–3 cm. long, villous and glandular; peduncles slender, 2–5 cm. long; bracts scarious, oval to rounded, 6–9 mm. long; perianth glandular, white to rose-pink, 12–18 mm. long, the tube very slender, the limb 6–8 mm. broad; fr. 3–6 mm. long, finely reticulate-veined, orbicular-obcordate with 2 (3) thin wings faintly veined; seed obovate, 2 mm. long.—Sandy places, mostly 2000–5000 ft.; Creosote Bush Scrub, Pinyon-Juniper Wd.; Mojave Desert; to w. Nev. and e. slope of Sierra Nevada. April–July.

6. **A. turbinàta** Torr. Much branched annual, the stems 1.5–5 dm. long, viscid-puberulent when young, later glabrate; lf.-blades round to oblong-ovate, 1–4 cm. long, rounded to obtuse at apex, irregular at base; petioles slender, 1–5 cm. long; peduncles 3–9 cm. long, slender, viscid-puberulent; bracts lanceolate, 5–8 mm. long; perianth white or pinkish, 15–20 mm. long, the limb 5–6 mm. broad; frs. short-villous, 4–7 mm. long, often beaked, the inner broadly turbinate, deeply lobed, the lobes compressed, and winglike, wrinkled; seed 2 mm. long.—Occasional, sandy places, mostly at 4000–8000 ft.; Creosote Bush Scrub to Pinyon-Juniper Wd.; Mt. Pinos region to Ore., Nev. A form occurring with the typical *turbinata* and having ovate obtusish bracts and frs. wingless or with 2 incurved wings, has been described as *A. exalata* Standl.

7. **A. umbellàta** Lam. Perennial with slender prostrate stems 2–10 dm. long, sparsely to much-branched, succulent, often reddish, viscid-puberulent to glabrous; lf.-blades oval to oval-rhombic, to lance-oblong, irregular in outline, 1.5–6 cm. long; petioles slender, 1–5 cm. long; peduncles slender, 2–12 cm. long, viscid-puberulent; bracts lanceolate, 4–6 mm. long; perianth rose, with central whitish eye-spot, the tube glandular-pubescent, 12–15 mm. long, the limb 8–10 mm. broad, with emarginate lobes; fr.-body

rather hard, 7–12 mm. long, glandular-villous above, short-beaked, with 2–5, usually 4, wings faintly net-veined, widened then truncate or narrowed above, not or scarcely prolonged above the body; seed elliptic-oblong, ca. 4 mm. long.—Coastal Strand; Los Angeles Co. to Sonoma Co. Most of the year. Channel Ids. Hybridizes with *A. latifolia*, *A. maritima* and *A. villosa*, some of the retrogressants having been named (*A. alba*, *A. platyphylla* and *A. variabilis*).

8. **A. villòsa** Wats. var. **villòsa**. Plate 61, Fig. A. Much-branched annual with stems 1–5 dm. long, procumbent to ascending, villous, usually viscid; lf.-blades 1–3.5 cm. long, rhombic-ovate to almost round, often very unequal at base, sparingly glandular; petioles 1.5–3 cm. long; peduncles slender, 2–7 cm. long, viscid-villous; bracts lanceolate to lance-linear, 6–8 mm. long, viscid-viscous, scarious, attenuate; perianth 12–16 mm. long, purplish-rose, the limb ca. 1 cm. broad; fr. 5–8 mm. long, rugose-veined so as to appear somewhat pitted, villous above, with 3–4 wings truncate or prolonged slightly above the body; seed narrow-oblong, 2.5 mm. long.—Common in open sandy places, mostly below 3000 ft.; Creosote Bush Scrub; Mojave and Colo. deserts; to Nev., Ariz., Son., L. Calif. Feb.–July.

Var. **aurìta** (Abrams) Jeps. [*A. a.* Abrams. *A. pinetorum* Abrams.] Perianth 16–26 mm. long; body of fr. with almost no transverse veins, hence not pitted; wings thin, broad, prolonged above the body; $n = $ ca. 45 (Snow).—Sandy places below 5000 ft.; Coastal Sage Scrub, Chaparral; from head of Coachella V. to interior Riverside, Orange and San Diego cos. March–Aug.

2. Acleisánthes Gray

Perennial herbs or low shrubs. Lvs. thick, opposite, the blades unequal, entire, petioled. Fls. axillary or terminal, each subtended by 1–3 small narrow bracts, solitary or in 2–3-fld. cymes. Perianth funnelform, with long tube and shallowly 5-lobed limb, constricted above the ovary. Stamens 2–5, unequal, exceeding perianth-tube. Ovary ovoid or oblong; style exserted; stigma capitate. Fr. narrow-ellipsoid, 5-ribbed or -angled, the ribs smooth, sometimes ending in a gland. Seed with testa adherent to pericarp; cotyledons broad, enclosing the copious endosperm. Several spp. of sw. N. Am. (Greek, *a*, without, *cleis*, something which closes, and *anthos*, fl., i.e. without an invol.)

1. **A. longiflòra** Gray. YERBA DE LA RABIA. Stems slender, decumbent or ascending, scabrous-puberulent, 1–3 dm. long; lvs. deltoid to lanceolate, 1.5–4.5 cm. long, on petioles 3–8 mm. long; fls. mostly solitary, sessile, opening at night; bracts subulate, 2–3 mm. long; perianth white, tinged purple, 10–16 cm. long, the limb 1.5–2 cm. broad; fr. 5–6 mm. long, 5-angled, striate between the ribs.—Dry stony places; Creosote Bush Scrub; Maria Mts., e. Riverside Co.; to Tex., Chihuahua. May.

3. Alliònia L.

Annual or perennial herbs, usually glandular-pubescent, with dichotomous prostrate branches. Lvs. petioled, very unequal in the pair, oblong to broadly ovate. Fls. perfect, in axillary peduncled clusters of 3, each subtended by a bract that encloses the fr. Perianth campanulate-rotate, 4–5-lobed. Stamens 4–7, exserted. Stigma capitate. Fr. coriaceous, flattened, the dorsal (seemingly inner) face bearing 2 rows of stipitate glands. Embryo curved, the broad cotyledons enclosing the endosperm. Ca. 3 spp., extending into S. Am. (C. *Allioni*, 1725–1804, Italian botanist.)

1. **A. incarnàta** L. [*Wedelia i.* Kuntze. *Wedeliella i.* Ckll.] WINDMILLS. Plate 61, Fig. B. Perennial or winter annual with slender trailing simple or branched villous to pubescent glandular stems; lf.-blades ovate or broader, usually rounded at base, 2–3 cm. long (or upper reduced); petioles shorter to as long; peduncles 5–25 mm. long; bracts round-ovate, 5–9 mm. long; perianth rose-magenta, rarely white, opening in the morning, 6–15 mm. long; frs. 3–4.5 mm. long, straw-colored, the inner side 3-nerved, the margins usually toothed, incurved.—Dry stony benches and slopes, below 5000 ft.; Creosote Bush Scrub; Mojave and Colo. deserts; to Colo., Chihuahua, S. Am. April–Sept.

Plate 61. Fig. A, *Abronia villosa*. Fig. B, *Allionia incarnata*. Fig. C, *Boerhaavia annulata*. Fig. D, *Mirabilis bigelovii retrorsa*. Fig. E, *Mirabilis californica*. Fig. F, *Mirabilis froebelii*.

4. Boerhaávia L.

Annual or perennial herbs, usually branched from base, ± pubescent, and usually with viscous areas on the internodes. Lvs. opposite, petioled, frequently unequal. Fls. perfect, small, variously arranged, bracteate, the bracts usually small. Perianth funnelform, campanulate or subrotate, corollalike, shallowly 5-lobed. Stamens 1–5; fils. unequal, filiform, united near base. Ovary stipitate; style filiform; stigma peltate. Anthocarp obovoid or obpyramidal, 3–5–10-nerved, -angled, or -winged. Seed with curved embryo. About 30 spp., in the warm parts of both hemis. (Hermann *Boerhaave*, 1668–1738, botanist of Leiden.)

Fr. 10-nerved; perianth funnelform .... 1. *B. annulata*
Fr. 3–5-nerved or -angled; perianth campanulate.
  Plant perennial; fr. pubescent .... 2. *B. coccinea*
  Plant annual; fr. glabrous
    Ultimate branches of infl. with fls. in racemose spikes.
      Bracts persistent, as long as the mature fr., which is 4-angled; stamens 3–4 . 6. *B. wrightii*
      Bracts deciduous, shorter; fr. 5-angled; stamens 1–2 .... 3. *B. coulteri*
    Ultimate branches of infl. with fls. in umbels or solitary on slender pedicels.
      Fls. in umbels; fr. ca. 3 times as long as thick .... 4. *B. erecta*
      Fls. mostly solitary, forming cymes; fr. ca. twice as long as thick .... 5. *B. triquetra*

1. **B. annulàta** Cov. Plate 61, Fig. C. Coarse perennial with 1–few suberect stems 3–10 dm. tall, these subglabrous, with reddish viscid semioblique band on each internode; lvs. oblong-ovate to suborbicular, bright green above, paler beneath, reddish-veined, hirsute, the hairs with small dark pulvini at base; blades 2–8 cm. long, rounded to acute at apex, irregularly repand-dentate, often cordate at base; petioles 1–6 cm. long; infl. paniculate, leafless, 3–8 dm. long, the fls. in dense headlike clusters at the ends of the branches; bracts short, persistent, hairy; perianth 6–8 mm. long, pinkish or greenish; stamens 3, exserted; fr. glabrous, 4–5 mm. long, thick-fusiform.—Sandy washes and gravelly slopes, below 3000 ft.; Creosote Bush Scrub; Death V. region of Calif. and Nev. April–May.

2. **B. coccínea** Mill. [*B. caribaea* Jacq. *B. hirsuta* Willd.] Perennial with branching decumbent or prostrate ± glandular-pubescent stems 2–9 (–14) dm. long; lvs. round-ovate to oblong-ovate, mostly obtuse, irregularly and shallowly undulate, green above, paler beneath, glabrous to hirsute, even viscid, the blades mostly 1–3 cm. long, the petioles 0.5–2 cm. long; infl. cymose, with many slender glandular-pubescent branches, 1–4 dm. long, the fls. in heads on slender peduncles; perianth purplish-red, 2 mm. long; stamens 1–3, short-exserted; fr. clavate, glandular-pubescent, 5-ribbed, the sulci not wrinkled, broader than the ribs, 2.5–3.5 mm. long.—Dry disturbed places, such as road-banks, railroads, washes, etc.; mostly below 3000 ft.; Creosote Bush Scrub, Colo. Desert; V. Grassland, etc., San Jacinto V., Santa Ana R.; to trop. Am.

3. **B. còulteri** (Hook. f.) Wats. [*Senkenbergia c.* Hook. f.] Erect or decumbent annual 3–8 dm. high, much-branched, puberulent; lvs. lanceolate to rhombic-ovate, acutish, entire to sinuate, crisped, green above, paler beneath, glabrous or nearly so, the blades 1.5–5 cm. long, the petioles 0.5–2 cm. long; infl. much-branched, with slender branches ending in very slender spicate racemes; bracts ca. 1 mm. long, lanceolate or wider, deciduous; obscurely ciliolate; perianth pink or white, glabrous, 1–1.5 mm. long; stamens 1–2, included; fr. 2.5 mm. long, narrow-obovoid, 5-angulate, the angles broad, smooth, almost touching each other, the sulci linear, scarcely rugose.—Rare, washes and flats below 4000 ft.; Creosote Bush Scrub; Colo. Desert (Hayfields, Chuckwalla Mts., Valley Wells, San Felipe V.); to Ariz., Mex. Sept.–Nov.

4. **B. erécta** L. var. **intermèdia** (Jones) Kearn. & Peebles. [*B. i.* Jones.] Erect or ascending annual 2–5 dm. high, freely branched, puberulent; lvs. oblong-lanceolate or broader, obtuse to acute, glabrous or nearly so, entire or somewhat sinuate, green above, paler beneath, frequently brown-punctate, the blades 1.5–4 cm. long, the petioles 0.3–2 cm. long; infl. cymose, with small umbels at tips of very slender branches; bracts lanceolate, minute, persistent; perianth pink, 1.5–2 mm. long; stamens 2–3, scarcely or

slightly exserted; fr. narrowly obpyramidal, 2–2.5 mm. long, truncate, 5-angled, the sulci narrow, transverse-rugulose.—Infrequent, dry washes and flats, below 4000 ft.; Creosote Bush Scrub, Joshua Tree Wd.; s. San Bernardino Co. (Little San Bernardino Mts., Clark Mt.) s. and e. to Imperial Co.; to Tex., Mex. Aug.–Oct.

5. **B. triquètra** Wats. Slender-stemmed, ascending to spreading annual, branched from base, 1.5–6 dm. high, minutely puberulent; lvs. oblong to narrowly lanceolate, bicolored, undulate-crisped, ± brownish-punctate, the blades 1–3 cm. long; infl. open-cymose, with slender glabrous branches; fls. pink, mostly solitary, short-pedicellate; bracts minute; perianth 1–1.3 mm. long; stamens 2–3, ca. as long as perianth; fr. obpyramidal, truncate, 3–5-angled, 2–2.5 mm. long, more than half as thick, the angles broad, smooth, acute, with open rugulose furrows.—Sandy washes and open gravelly flats below 5500 ft.; Creosote Bush Scrub, Joshua Tree Wd.; Little San Bernardino Mts. s. and e.; Ariz., L. Calif. Sept.–Dec.

6. **B. wrightii** Gray. Erect annual, 2–6 dm. high, branched from base, with stems slender, glandular-pubescent; lvs. toward base of plant oblong-ovate to lanceolate, green above, paler and glandular-punctate beneath, undulate, the blades 1–4 cm. long, the petioles 0.5–2 cm. long; infl. glabrous to glandular-pubescent, open and branched, the fls. in loose spikes; bracts 2–3 mm. long, ovate or almost orbicular, often reddish, villous, persistent; perianth pink, 1–2 mm. long; stamens 3–4, included; fr. broadly clavate, 2 mm. long, mostly angled, the angles broad, acute, with broad rugulose sulci.—Sandy and gravelly flats and washes, below 4500 ft.; Creosote Bush Scrub, Joshua Tree Wd.; Mojave and Colo. deserts; to Nev., Tex., Mex. Aug.–Dec.

5. *Hermídium* Wats.

Perennial nearly glabrous herbs with erect dichotomous stems. Lvs. opposite, thick, entire, short-petioled. Fls. in axillary or terminal peduncled headlike clusters, each fl. subtended by a broad foliaceous bract of which the midvein is partially attached to the pedicel of the fl. Perianth campanulate-funnelform; stamens 5–7; fls. unequal, united at base. Ovary globose; style capillary; stigma capitate. Fr. ellipsoid, smooth, usually with 10 vertical lines. Seeds adherent to pericarp; embryo uncinate, the cotyledons orbicular. Monotypic. (Diminutive of *Hermes*, Greek god.)

1. **H. alipès** Wats. Plants 2–4 dm. high, the stems sparsely branched, glaucous, glabrous or obscurely puberulent upward; lf.-blades 2–7 cm. long, round to broadly ovate, on petioles 1–10 mm. long; peduncles 3–12 mm. long; heads 4–6-fld.; bracts oblong to ovate, 1–2 cm. long, green; perianth purplish-red, 2–2.5 cm. long; fr. 6–7 mm. long, almost globose.—Dry slopes and flats, 4000–6500 ft.; Sagebrush Scrub, Pinyon-Juniper Wd.; Panamint Mts. to White Mts.; to Nev., Utah. May–June.

6. *Mirábilis* L. Four-O'clock

Perennial often suffrutescent herbs with repeatedly dichotomous stems. Lvs. opposite, petiolate to sessile, entire. Fls. 1–several in a 5-lobed calyxlike usually campanulate invol.; invols. axillary, often clustered near ends of branches. Perianth surpassing invol., funnelform to campanulate, white to red or red-purple. Stamens 5 (–3), unequal; fils. capillary. Fr. subglobose to elongate, smooth or obscurely angled. Ca. 20 spp. of warmer parts of Am. (Latin, *mirabilis*, wonderful.)

Invol. 1-fld.; perianth campanulate.
  Perianth 3–5 cm. long; fr. 5-angled. Garden escape .... 4. *M. jalapa*
  Perianth 1–1.5 cm. long; fr. smooth. Native.
    Involucral lobes lanceolate, longer than the tube; invol. 11–13 mm. long; lf.-blades 3–4 cm. long. W. Colo. Desert .... 5. *M. tenuiloba*
    Involucral lobes subovate, not exceeding tube; invol. 5–9 mm. long; lf.-blades mostly less than 2.5 cm. long.
      Fls. white to pale pink; lvs. viscid-villous. Deserts .... 1. *M. bigelovii*

Fls. rose to red; lvs. subglabrous to rough-pubescent. Cismontane ...... 2. *M. californica*
Invol. 3–10-fld., 2.5–4 cm. long; perianth funnelform .... 3. *M. froebelii*

1. **M. bigelòvii** Gray var. **bigelòvii**. Erect or ascending, much-branched, villous-viscid, 3–5 dm. high, the stems slender; lf.-blades ovate to reniform-ovate, 1–3 (–4) cm. long, obtuse, viscid-villous; petioles 3–10 mm. long; invols. clustered at ends of branches, short-peduncled, 5–6 mm. long, the lobes ovate, shorter than the tube; perianth white to pale pink or pale lavender, 8–12 mm. long; fr. dark, ovoid, elongate, often mottled, smooth or slightly rugulose, ca. 3 mm. long.—Rocky places, especially in canyons, below 7000 ft.; Creosote Bush Scrub to Pinyon-Juniper Wd.; mostly e. Colo. and e. Mojave deserts from Inyo Co. s.; Nev., Ariz. March–June, Oct.–Nov.

Var. **áspera** (Greene) Munz. *M. a.* Greene. Stems viscid-villous; fr. subglobose, with 10 pale vertical lines.—Above 3000 ft., with the sp. and also in the w. part of the deserts.

Var. **retròrsa** (Heller) Munz. [*M. r.* Heller.] Plate 61, Fig. D. Stems subglabrous below, scabrous-puberulent above with short retrorse hairs, also somewhat glandular; fr. subglobose, sometimes striate.—Common, mostly below 4000 ft.; Creosote Bush Scrub; both deserts, n. to Mono Co.; to Utah, Ariz. March–June.

2. **M. califórnica** Gray var. **califórnica**. Wishbone Bush. Plate 61, Fig. E. Somewhat woody at base with many ascending to decumbent stems, slender, subglabrous or commonly viscid-puberulous, repeatedly forked, 1.5–8 dm. long; petioles 2–15 mm. long; blades ovate, mostly acute, 1–2.5 (–3.5) cm. long; invols. clustered near ends of branches, short-peduncled, campanulate, 5–8 mm. long, the lobes subovate, shorter than tube; perianth rose to purplish-red, 10–14 mm. long; fr. broadly ovoid, 5 mm. long, dark, sometimes mottled or pale-striate, smooth.—Common in dry stony washes and on slopes below 2500 ft.; Coastal Sage Scrub, Chaparral, Foothill Wd.; cismontane cent. Calif. to San Diego Co. and to edge of desert. Channel Ids. Some plants on San Clemente Id. with more scabrous pubescence seem to be

Var. **cedrosénsis** (Standl.) Macbr. [*Hesperonia c.* Standl.], a taxon variable and of uncertain worth. Other plants from the same id. with stems viscid-pubescent and with perianth 5–8 mm. long have been named var. *cordifolia* Dunkle.

3. **M. froebélii** (Behr) Greene var. **froebélii** *Oxybaphus f.* Behr. Plate 61, Fig. F. From a thick woody tuberous root, erect or ascending or procumbent, 3–8 dm. long, much-branched, densely villous-viscid throughout; petioles stout, 3–20 mm. long; lf.-blades broadly ovate, 3–8 cm. long, often subcordate at base, acute to rounded at apex; invols. peduncled, solitary in lower axils, or clustered at tips of branches, campanulate, 2.5–4 cm. long, the lobes ca. $\frac{1}{3}$ the whole; perianth rose-purple to deep pinkish, 3.5–4.5 cm. long, funnelform, the limb 2–2.5 cm. broad; fr. elliptic-ovoid, smooth, 8 mm. long, with 10 vertical light-colored lines.—Dry stony places, below 6500 ft.; Creosote Bush Scrub to Pinyon-Juniper Wd.; deserts from L. Calif. n. to interior San Luis Obispo Co. and Mono Co.; Nev. April–Aug.

Var. **glabràta** (Standl.) Jeps. *Quamoclidium f.* ssp. *g.* Standl. Herbage glabrous.—Occasional, Providence Mts., New York Mts. and w. edge of Colo. Desert; Nev. to L. Calif.

4. **M. jalápa** L. Cult. Four-O'Clock. Erect, 3–9 dm. high, glabrous or nearly so; lvs. ovate, 5–10 cm. long, truncate or cordate at base, acuminate at apex, the petioles half as long as blades; fls. ca. 2.5 cm. across, of many different colors, opening in late afternoon; $2n = 58$ (Showalter, 1935).—The cult. Four-O'Clock occasionally reported as a garden escape; native of trop. Am. Summer.

5. **M. tenuilòba** Wats. Woody at base, erect to decumbent, 3–5 dm. high, glandular-pubescent to -villous; lf.-blades ovate to broadly deltoid, 2.5–5 cm. long, rounded to cordate at base, usually acute, glandular-pubescent; petioles 2–8 mm. long; invols. almost sessile, 11–13 mm. long, narrow-campanulate, the lobes slightly unequal, slightly longer than tube, narrow-lanceolate; perianth whitish, pubescent, 12–15 mm. long; fr. 5 mm. long, ovoid, smooth, brown.—Occasional, rocky slopes, below 1500 ft.; Creosote Bush Scrub; w. part Colo. Desert; L. Calif. March–May.

7. Oxybaphus L'Hér. ex Willd.

Perennial herbs, the stems tall and erect to low and decumbent, mostly from a woody root. Lvs. opposite, usually thickish, entire, sessile or petiolate. Fls. perfect, 1-5 in a calyxlike invol. which is gamophyllous, 5-lobed, enlarged and papery in fr., reticulate-veined. Perianth campanulate to short-funnelform, slightly oblique, the limb 5-lobed. Stamens 3-5, unequal, slightly united below. Stigma depressed-capitate. Fr. ± obovoid, constricted at base, 5-angled or -ribbed, mucilaginous when wet. Seed adherent to pericarp; embryo curved, the cotyledons enclosing the abundant endosperm. Ca. 25 spp., of the warmer parts of Am.; 1 Himalayan.

Lf.-blades sessile or subsessile, at least 5 times as long as wide.
    Perianth bright red, 3-4 times as long as invol. E. Mojave Desert .... 1. *O. coccineus*
    Perianth pink or purplish-red, ca. twice as long as invol. Cismontane .... 3. *O. linearis*
Lf.-blades plainly petioled, usually not more than 3 times as long as wide.
    Invols. usually glabrous at anthesis except at very base, 15-20 mm. broad in fr.
        4. *O. nyctagineus*
    Invols. viscid-pilose at anthesis, not over 15 mm. wide in fr.
        Stems much-branched, viscid and pilose or villous to base; lvs. scarcely longer than wide
            5. *O. pumilus*
        Stems branched only in infl., glabrous below or puberulent in lines; lvs. longer than wide
            2. *O. comatus*

1. **O. coccíneus** Torr. Stems 1-many from an elongate woody root, erect or ascending, glabrous, 3-6 dm. long; lf.-blades linear or filiform, 2-12 cm. long, 1-5 mm. wide; infl. loosely cymose; peduncles slender; invols. short-pilose, 4-5 mm. long at anthesis, deeply lobed, 5-6 mm. at maturity; perianth opening at night, deep red, 12-15 mm. long, the tube 3-4 mm. in diam., the limb 11-15 mm. broad; stamens exserted; fr. 5-ribbed, somewhat rugose, 5 mm. long; seed 2.5 mm. long.—Dry rocky slopes and along washes; 4000-5300 ft.; Pinyon-Juniper Wd.; Providence, New York and Ivanpah mts.; to New Mex., Son. May-July.

2. **O. comàtus** (Small) Weath. [*Allionia c.* Small.] Stems 1-few, 1-5 dm. long, ascending, simple at least below the infl., glabrous below or with lines of hair; lf.-blades narrow-deltoid, 1.5-4 cm. long, ± puberulent, on petioles 5-15 mm. long; invols. few to many, 3-5 mm. long at anthesis, ca. 8 mm. long in fr., viscid-pilose; fr. obovoid, 3-4 mm. long, the angles broad, tuberculate; seeds 2-3 mm. long.—Dry slopes; Pinyon-Juniper Wd.; Ivanpah Mts., e. San Bernardino Co.; Nev. to Tex., Mex. May-June.

3. **O. lineàris** (Pursh) Rob. [*Allionia l.* Pursh.] Much like *O. coccineus* in habit, the stems glabrous or puberulent below, more viscid-pubescent above; lvs. linear to lance-linear, 3-10 cm. long, mostly 1-5 mm. wide; invol. ca. 4 mm. long at anthesis, 6-10 mm. in fr., viscid-villous; perianth pale pink to purple-red, ca. 10 mm. long; fr. ca. 5 mm. long, the angles smooth, the sides transversely rugose.—Occasional escape along Santa Ana R. system (Riverside, Rancho Santa An.); native, Ariz. to S. Dak., Mo., Mex. July-Oct.

4. **O. nyctagíneus** (Michx.) Sweet. [*Allionia n.* Michx.] Stems many, erect to decumbent, 3-10 dm. tall, subglabrous; lf.-blades lance-ovate to subdeltoid, 2-8 cm. long, on petioles 1-3 cm. long; infl. cymose; invol. 5-6 mm. long in anthesis, 10-15 mm. in fr., puberulent or short-pilose near the base, the glabrous lobes ciliate; perianth white or pale pink, ca. 10 mm. long, the limb 12-15 mm. broad; fr. 5 mm. long, short-pilose, the angles broad, the sides rugulose; $2n=58$ (Bowden, 1945).—Adv., as at Upland; native from Wis. and Mont. to Mex. June-Aug.

5. **O. pùmilus** (Standl.) Standl. [*Allionia p.* Standl.] Stems few to many from a woody root, ascending to procumbent, 1-5 dm. long, short-pilose throughout; lvs. deltoid to ovate-deltoid, 2-6 cm. long, succulent, puberulent or short-pilose, on petioles 1-2 cm. long; infl. axillary or in narrow cymes, viscid-pilose, plainly bracteate; invols. 3-4 mm. long at anthesis, 7-8 mm. in fr.; perianth 8-10 mm. long, pale pink, pilose; stamens exserted; fr. 5 mm. long, short-pilose, rugose; seed 2.5-3 mm. long, obovoid.—Dry rocky

and gravelly places, 4500–7800 ft.; Pinyon-Juniper Wd., Yellow Pine F.; about the Mojave Desert, as San Bernardino Mts., Ivanpah Mts., Clark Mt., Kingston Mts., Inyo Mts.; to Nev., New Mex. June–Aug.

8. Selinocárpus Gray

Perennial herbs with dichotomous branching. Lvs. opposite, petioled, the blades thick. Fls. in clusters, sessile or pedicelled, each subtended by 2–3 narrow bracts, axillary or terminal. Perianth tubular-funnelform. Stamens 3–5. Ovary oblong. Fr. with 3–5 hyaline wings; seed with testa adherent to pericarp; embryo conduplicate. Several spp. of sw. N. Am. (*Selinum*, a genus of *Apiaceae*, and *karpos*, fr.)

1. **S. diffùsus** Gray. Diffusely branched, 1–3 dm. high, with short appressed white inflated hairs and sparsely glandular; lf.-blades oval to roundish, 1–2.5 cm. long; petioles 3–20 mm. long; fls. solitary or in 2's, often cleistogamous, short-pedicellate, the bracts linear-subulate, 3–6 mm. long; perianth greenish, 3–4 cm. long; fr. 6–7 mm. long, 4–5-winged, truncate, the body puberulent.—Dry often rocky places; Creosote Bush Scrub, Joshua Tree Wd.; San Bernardino Co.–Nev. border. June–Sept.

72. Oleàceae. OLIVE FAMILY

Trees, shrubs or more rarely herbs. Lvs. usually opposite, simple to pinnate, deciduous in ours, exstipulate. Fls. regular, perfect or more commonly unisexual, small, in compact clusters. Calyx free from the ovary, small, usually 4-lobed, sometimes wanting. Corolla sympetalous or polypetalous or none, 2–4-merous when present. Stamens 2, rarely 3 or 4, hypogynous or inserted on the corolla-tube. Ovary superior, 2-locular, with few ovules in each locule; style 1 or the stigma sessile. Fr. a caps., samara, berry, or drupe. Seeds erect or pendulous; endosperm present or lacking; embryo straight. Ca. 25 genera and 300 spp., widely distributed in warmer regions.

Fr. a samara; lvs. largely pinnately compound .... 2. *Fraxinus*
Fr. not a samara; lvs. simple.
    Fr. a drupe; lvs. opposite, large shrubs .... 1. *Forestiera*
    Fr. a caps.; lvs. mostly alternate; small shrubs .... 3. *Menodora*

1. Forestiéra Poir.

Deciduous shrubs with simple opposite lvs. Fls. small, inconspicuous, polygamo-dioecious, appearing before the lvs. Calyx minute, unequally 5–6-cleft, sometimes lacking. Corolla usually wanting, rarely 2–3-petaled. Stamens 2 or 4. Ovary 2-celled, with 2 ovules in each cell; style slender. Fr. a drupe, round to ovoid, usually 1-seeded, glabrous or pubescent. Ca. 15 spp., of N. and S. Am. (Named for M. *Forestier*, French physician.)

1. **F. neomexicàna** Gray. [*Adelia n.* Kuntze. *A. parvifolia* Cov.] DESERT-OLIVE. Glabrous shrub, 1.5–3 m. high, with smooth gray bark and spiny branchlets; lvs. spatulate-oblong, obtuse to acuminate, entire or serrulate, 1–5 cm. long, often fascicled; fls. in sessile fascicles; ♂ fls. sessile, ♀ on slender pedicels; drupe blue-black, ellipsoid, 5–7 mm. long; seed longitudinally ridged; $2n = 46$ (Taylor, 1945).—Occasional on dry slopes and ridges, below 6700 ft.; Creosote Bush Scrub, Chaparral, Coastal Sage Scrub, Foothill Wd., etc.; inner Coast Ranges from Contra Costa Co. s. to interior cismontane s. Calif. and on Mojave Desert and bordering ranges, Inyo Co. to Santa Rosa Mts., Riverside Co.; to Colo., Tex. March–April.

2. Fráxinus L. ASH

Trees or arborescent shrubs with deciduous usually pinnately compound opposite lvs. Largely dioecious or polygamous, the fls. small, in crowded panicles which appear mostly before the lvs. Calyx 4-cleft or -toothed. Petals in ours 2 or 0. Stamens usually 2,

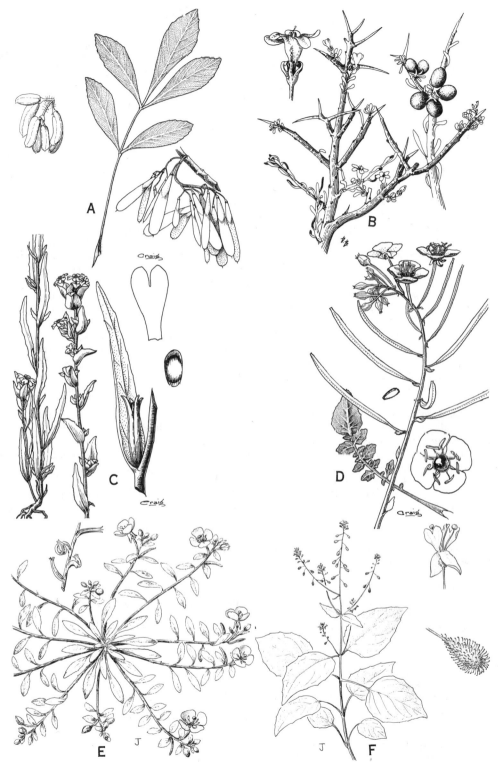

Plate 62. Fig. A, *Fraxinus velutina* var. *coriacea*. Fig. B, *Menodora spinescens*. Fig. C, *Boisduvalia densiflora*. Fig. D, *Camissonia brevipes*. Fig. E, *Camissonia cheiranthifolia*. Fig. F, *Circaea alpina* ssp. *pacifica*.

rarely 3 or 4. Fr. a 1-seeded samara, with terminal wing; seed usually 1, oblong. Ca. 40 spp. of N. Hemis. (Latin name of the ash.)

(Miller, Gertrude N. Fraxinus in N. Am. Cornell Univ. Agric. Exp. Sta. Mem. 335. 1955.)
Corolla present, consisting of 2 white petals; fls. bisexual; style obscurely lobed; samara flat, broadly wing-margined to base .... 2. *F. dipetala*
Corolla absent; style conspicuously 2-lobed.
    Lfts. 1 (3); branchlets of the season 4-sided; body of the samara flattened and broadly wing-margined to base .... 1. *F. anomala*
    Lfts. 5-7; branchlets terete; body of samara subterete and narrowly winged along the side for ½ the way. Los Angeles and Inyo cos s. .... 3. *F. velutina*

    1. **F. anómala** Torr. ex Wats. Shrub or small tree 2-5 m. high, bushy; branchlets glabrous or somewhat pubescent, 4-sided; lvs. mostly simple, round-ovate, entire or crenulate, 2-5 cm. long, sometimes trifoliolate; petioles slender, ca. as long as blades; fls. perfect or ♀; calyx 1.5 mm. long; petals 0; samaras 1.5-2.5 cm. long, 6-8 mm. wide, the wing decurrent down the sides almost to the base.—Dry canyons and gulches, 3000-11,000 ft.; mostly Pinyon-Juniper Wd.; e. Mojave Desert (Panamint, New York, Providence, Clark mts.) to Last Chance Mts.; to Colo., Tex. April-May.

    2. **F. dipétala** H. & A. FLOWERING ASH. Arborescent shrub or small tree 2-7 m. high, erect, glabrous or the young parts somewhat pubescent; branchlets slender, the youngest somewhat 4-angled; lvs. 4-12 cm. long, the lfts. 3-7 (-9), thin, ovate to obovate, serrate, 2-4 cm. long, on short petiolules; panicles 3-12 cm. long, many-fld.; the fls. perfect or polygamous; calyx ca. 1.5 mm. long; petals 2, white, 5 mm. long; samaras 2-3 cm. long, 7-9 mm. wide, the body flattened, winged along the sides, often retuse at apex.—Dry slopes, mostly below 3500 ft.; Chaparral, Foothill Wd., etc. from n. Calif. s.; in s. Calif. largely in Chaparral and mostly from Santa Ana Mts. n., rare in San Diego Co. April-May.

    3. **F. velùtina** Torr. var. **velùtina.** ARIZONA ASH. Tree 5-10 m. tall; branchlets terete, ± puberulent; lfts. 3-7, lanceolate to ovate or obovate, 2-8 cm. long, thickish, ± pubescent underneath especially when young, the lateral commonly on petiolules 3-10 mm. long; plant dioecious; calyx ca. 1.5 mm. long; petals 0; samaras 1.5-2.5 (-3.5) cm. long, 4-7 mm. wide, the wing usually decurrent barely to the middle of the body.—Canyons and along streams, below 5000 ft.; deserts from Inyo Co. s.; to Nev., Ariz., L. Calif.

    Var. **coriàcea** (Wats.) Rehd. [*F. c.* Wats.] Plate 62, Fig. A. Cismontane plants and some desert plants with more strongly veined and subglabrous lfts. are this var.

3. Menodòra Humb. & Bonpl.

    Low subshrubs or suffruticose herbs. Lvs. alternate or the lower opposite, simple, sessile or subsessile. Fls. perfect, solitary and terminal or dichotomously corymbose or paniculate. Calyx deeply 5-15-lobed. Corolla subrotate to salverform, 5-6-lobed. Stamens 2, inserted near base of corolla or in throat; fils. long or short. Ovary superior, bilocular; style slender; stigma depressed-capitate. Ovules 2 or 4 in each locule. Fr. a caps., membranaceous, indehiscent or circumscissile. Seeds 4 (2) in each cell, attached laterally, the outer coat spongy and reticulate; endosperm absent. Ca. 17 spp. of N. and S. Am. and Afr. (Greek, *menos*, force, and *doron*, gift.)

Plants unarmed; corolla-lobes longer than the tube, yellow.
    Lvs. all foliose, with well developed blades; stems 0.5-3 dm. high .... 1. *M. scabra*
    Lvs., at least the upper, reduced to bracts; stems 3-8 dm. high .... 2. *M. scoparia*
Plants spinose; corolla-lobes shorter than the tube, white .... 3. *M. spinescens*

    1. **M. scàbra** Gray. [*M. s.* var. *laevis* Steyerm.] Stems erect, almost herbaceous, 0.5-3 dm. high, usually ± scabrous; lvs. 5-15 mm. long, 2-5 mm. wide, mucronate, glabrous or sparsely scaberulous, the lower ovate to oblong, the upper lanceolate; calyx glabrous to slightly scabrous, the lobes 7-11, linear, 4-5 mm. long; corolla bright yellow, subrotate, the lobes 7-8 mm. long, the tube ca. 4 mm. long; caps. 5-7 mm. high, 8-12 mm. broad, thin-walled; seeds 4 in each locule, 4-5 mm. long, ca. 3 mm. broad, flat, obovate,

greenish or brownish with a yellowish narrow wing; $2n = 44$ (Taylor, 1945).—Dry rocky places, 3500–5500 ft.; Joshua Tree Wd.; Eagle Mts., New York Mts., Lanfair V., e. Mojave Desert; to Tex.

2. **M. scopària** Engelm. ex Gray. Paniculately branched erect suffruticose perennial 3–8 dm. high, with many slender, mostly glabrous stems; lvs. rather few, oblong-obovate to oblanceolate, the upper remote and rudimentary, the lower 1–2.5 cm. long; fls. few, cymose; calyx-lobes 5–7, subulate, 3–5 mm. long; corolla yellow, subrotate, 10–12 mm. long, the lobes ovate, ca. 7 mm. long; caps. 4–6 mm. long; seeds 4 in each locule, brown, angled on ventral side, rounded on back, 4–5 mm. long, ca. 3 mm. broad; $2n = 22$ (Bowden, 1945).—Dry slopes, 2000–6000 ft.; Joshua Tree Wd., Pinyon-Juniper Wd., Shadscale Scrub; w. edge of Colo. Desert, e. Mojave Desert; to Ariz., Mex. May–July.

3. **M. spinéscens** Gray. [*M. s.* var. *mohavensis* Steyerm.] Plate 62, Fig. B. Low shrub 1.5–8 dm. high, with irregular divergent branches with short stout spiny puberulent branchlets; lvs. 4–12 mm. long, linear to spatulate-oblong, entire, fleshy, appressed-puberulent, alternate; fls. solitary or clustered, short-pedicelled; calyx-lobes 5–7, linear-subulate, ca. 4 mm. long; corolla white, tinged brownish-purple without, 9–15 mm. long, funnelform, the lobes oblong-ovate, spreading, 3.5–4.5 mm. long; caps. 5–7 mm. long, 8–9 mm. broad; seeds 2 in each cell, 5–6 mm. long, 3–4 mm. broad, somewhat rounded, pitted, dark brown.—Dry mesas and slopes, mostly 3500–6500 ft.; Shadscale Scrub, Joshua Tree Wd.; e. Mojave Desert to Owens V.; Nev., Ariz. April–May.

73. Onagràceae. EVENING-PRIMROSE FAMILY

Herbs, shrubs or trees with simple alternate or opposite lvs. Stipules 0 or ± glandular. Fls. perfect, mostly symmetrical, axillary or in terminal racemes, the parts mostly in 2's or 4's. Fl.-tube adnate to ovary and usually prolonged beyond. Sepals 4, sometimes 2, 5, 6, inserted at summit of fl.-tube. Stamens as many or twice as many as sepals and petals. Style 1, slender; stigma 2–4-lobed or discoid, capitate or elongate. Ovary inferior, mostly 2–4-loculed. Fr. a caps., berry, or nutlike and indehiscent. Pollen grains often connected by cobwebby threads. Seeds 1–many, exalbuminous. Ca. 600 spp., of wide distribution, especially in w. N. Am.

1. Sepals persistent, petals 5 or 0; stamens 10 or 3–6 .... 9. *Ludwigia*
1. Sepals deciduous after anthesis.
  2. Fls. 2-merous; fr. indehiscent, obovoid, bristly with hooked hairs .... 3. *Circaea*
  2. Fls. normally 4-merous.
    3. Seeds with tuft of hairs (coma) at one end.
      4. Fl.-tube lacking to less than 1 cm. long, without a ring of internal scales; fls. not scarlet
      5. *Epilobium*
      4. Fl.-tube 2–3 cm. long, funnelform, with a row of 8 scales within at ca. ½ its length; fls. scarlet .... 11. *Zauschneria*
    3. Seeds without a coma.
      5. Fr. a dehiscent caps. (although sometimes tardily so).
        6. Ovary 4-loculed; fl.-tube prolonged beyond the ovary.
          7. Stigma capitate, entire; sporogenous tissue continuous in the anther-locules
          2. *Camissonia*
          7. Stigma 4-lobed.
            8. Petals 2-lobed, small; pollen in tetrads .... 1. *Boisduvalia*
            8. Petals entire, usually larger; pollen not in tetrads.
              9. Anthers erect, attached near the base, innate; sporogenous tissue in the anther-locules divided into packets by sterile tissue .... 4. *Clarkia*
              9. Anthers versatile, attached near their middle; sporogenous tissue continuous
              10. *Oenothera*
        6. Ovary 2-loculed; fl.-tube not prolonged beyond the ovary; fls. minute; stem branches capillary .... 7. *Gayophytum*
      5. Fr. indehiscent, nutlike and hard.
        10. Stigma discoid, shallowly lobed; fls. opening in morning; anthers basifixed
        8. *Heterogaura*
        10. Stigma with 4 linear lobes; fls. opening near sunset; anthers versatile .. 6. *Gaura*

1. Boisduvália Spach

Caulescent erect or decumbent annuals. Lvs. generally alternate, sometimes opposite, simple, sessile. Fls. small in ours, in leafy spikes or solitary in lf.-axils. Fl.-tube produced above the ovary, short in ours, funnelform. Sepals 4, erect. Petals 4, deeply emarginate, purplish to pinkish to white. Stamens 8, the epipetalous with shorter fils.; anthers versatile, attached near base in small-fld. spp.; pollen shed in tetrads. Stigma 4-lobed. Caps. 4-loculed, sessile, loculicidal or in 1 sp., the partitions between the locules forming a 4-winged central column. Seeds without a coma, very finely papillose, with a pellucid margin at one end. Six spp. of w. N. and S. Am. (Named for J. A. Boisduval, French naturalist and physician.)

(Raven, P. H. and D. M. Moore. A revision of Boisduvalia. Brittonia 17: 238–254. 1965.)
Caps. with a tough wall, the seeds in 2 rows in each locule; plant often subglabrous . 2. *B. glabella*
Caps. with a pliable wall, the seeds in a single row in each locule; plants not subglabrous.
  The caps. not conspicuously beaked, the central axis holding together at maturity, 4-winged
  1. *B. densiflora*
  The caps. ± conspicuously beaked, the central axis disintegrating at maturity, the septa adhering to the valves at dehiscence .... 3. *B. stricta*

1. **B. densiflòra** (Lindl.) Wats. [*Oenothera d.* Lindl.] Plate 62, Fig. C. Simple or branched (especially above), erect or nearly so, mostly 3–10 dm. tall, villous or strigulose; lvs. well distributed, mostly lanceolate, 1–8 cm. long, passing into fl.-bracts ovate and sometimes overlapping in the dense long-spicate infl.; sepals 2–4 mm. long; petals rose-purple, mostly 3–8 mm. long; caps. stout, straight, 4–10 mm. long, the central column persisting; seeds 4–6 in each locule; $n=10$ (Lewis et al., 1958).—Moist places, below 8500 ft.; many Plant Communities, cismontane and montane, from San Diego Co. to B.C., Ida., Nev. May–Aug.

2. **B. glabélla** (Nutt.) Walp. [*Oenothera glabella* Nutt. in T. & G.] Mostly freely and decumbently branched from base, the stems 1–3 dm. long, subglabrous or pubescent on lf.-veins or throughout, rather uniformly leafy; lvs. lance-ovate to -oblong, serrulate, 1–1.5 cm. long; fls. axillary, sometimes even in lowest axils; sepals 2 mm. long; petals purplish, 2–4 mm. long; caps. straight, 6–8 mm. long, pointed at tip; seeds many, grayish-brown, narrowly subfusiform, angled, 1 mm. long; $n=15$ (Raven & Moore, 1965).—Dry mud flats and vernal pools, below 5000 ft.; many Plant Communities; cismontane and montane from San Diego Co. n.; to w. Can.; n. L. Calif., Argentina. May–Aug.

3. **B. strícta** (Gray) Greene. [*Gayophytum s.* Gray.] Stems 1–5 dm. tall, simple and erect or virgately branched from near base, pilose and ± canescent throughout; lvs. linear to lance-linear, 1–4 cm. long, 2–4 mm. wide, acute, entire to sharply denticulate, the upper narrower than the lower; fls. axillary, often from near the base of the plant; sepals 1 mm. long; petals rose-purple or violet, 1.5–2 mm. long; caps. 8–10 mm. long, membranous, slender, tardily loculicidal, usually curved outward and attenuate; seeds ca. 1 mm. long, cellular-pitted, ca. 6–8 in each locule; $n=9$ (Lewis et al., 1958).—Moist places below 8500 ft.; many Plant Communities.—Victorville, San Bernardino Co.; Kern Co. n. to Wash., Ida., Nev. May–July.

2. Camissònia Link

Annual or perennial herbs, rarely subshrubs, from a taproot which in some spp. branches to give rise to new plants. Lvs. basal or cauline, alternate, largest near the base, reduced upward. Fls. opening near sunrise, or in some spp. toward evening, the stigma receptive and anthers shedding pollen simultaneously and immediately and the fls. lasting less than a day, mostly 4-merous. Petals yellow, rose-purple or white. Stamens and style yellowish, the stigma greenish yellow. Stamens 8, the episepalous longer, or rarely epipetalous stamens lacking; anthers versatile or basifixed, the sporogenous tissue con-

Camissonia

tinuous in the anther-locules. Pollen shed singly or in tetrads. Stigma capitate or hemispherical. Caps. loculicidal, straight or contorted, the seeds in 1 or 2 rows in each of the 4 locules. (Raven, P. H. A revision of the genus Camissonia. Contr. U.S. Nat. Herb. 37: 161–396. 1969.) Following key to species after Raven.

1. Ovary with a long slender sterile projection below the fl.-tube, the plants acaulescent or nearly so, annual.
   2. Petals 5–18 mm. long; style 3–5.5 mm. long; plants pilose .... 12. *C. graciliflora*
   2. Petals 2–3.5 mm. long; style 1–2.2 mm. long; plants strigose .... 23. *C. palmeri*
1. Ovary lacking a sterile projection, the plants only occasionally appearing acaulescent, and then when immature.
   3. Seeds with a thick papillate wing surrounding the concave face; petals white, yellow near the base .... 24. *C. pterosperma*
   3. Seeds not winged; petals yellow, white or purple.
      4. Caps. on well-defined pedicel, not coiled or contorted; seeds in 2 rows in each locule.
         5. Fl.-tube 0.4–0.8 mm. long; lvs. basal or cauline, often pinnately divided or lobed; pollen shed singly.
            6. Petals lavender in anthesis; fl.-tube 2–5 mm. long; petals 2–6 mm. long
                                                                        14. *C. heterochroma*
            6. Petals yellow or white at anthesis.
               7. Caps. distinctly clavate, more than 2 mm. thick.
                  8. Mature caps. and pedicels sharply deflexed; corolla bright yellow . 21. *C. munzii*
                  8. Mature caps. and pedicels spreading; or ascending; corollas white or yellow
                                                                        10. *C. claviformis*
               7. Caps. elongate, often linear, usually less than 2 mm. thick; pedicels often inconspicuous.
                  9. Stigma surrounded by anthers at maturity; petals less than 6 mm. long; infl. erect in bud .... 30. *C. walkeri*
                  9. Stigma elevated above anthers at maturity; petals usually more than 6 mm. long .... 4. *C. brevipes*
         5. Fl.-tube 4.5–40 mm. long; lvs. all cauline, simple, usually cordate-orbicular; pollen shed in tetrads.
            10. Fl.-tube 18–40 mm. long; style 30–58 mm. long .... 1. *C. arenaria*
            10. Fl.-tube 4.5–14 mm. long; style 8–23 mm. long .... 7. *C. cardiophylla*
      4. Caps. sessile (see also *kernensis*), often coiled or contorted; seeds in 1 row in each locule.
         11. Petals white, the fls. vespertine.
            12. Caps. not thickened near the base; plants lacking fls. near lower nodes.
               13. Petals 3.5–7 mm. long; stigma held well above the anthers at anthesis
                                                                        27. *C. refracta*
               13. Petals 1.8–3 mm. long; stigma surrounded by the anthers at anthesis
                                                                        8. *C. chamaenerioides*
            12. Caps. thickened near the base; plants with or without fls. at lower nodes. Mojave Desert and n. and w. .... 3. *C. boothii*
         11. Petals yellow; the fls. matinal.
            14. Lvs. pinnately lobed; petals usually with a pattern of minute maroon flecks near the base; seeds often with purple dots; caps. often sharply reflexed at maturity
                                                                        5. *C. californica*
            14. Lvs. subentire; petals unspotted or with 1 or 2 bright red spots near the base; seeds unspotted; caps. not reflexed.
               15. Caps. quadrangular, at least when dry; plants usually with fls. at the basal nodes; lvs. mostly lanceolate to ovate; seeds dull, flattened, usually over 1 mm. long.
                  16. Plants perennial, often somewhat woody near the base .... 9. *C. cheiranthifolia*
                  16. Plants annual, except occasionally in *bistorta*.
                     17. Stigma evidently held above the anthers at anthesis, the petals mostly 7–15 mm. long .... 2. *C. bistorta*
                     17. Stigma surrounded by at least the anthers of the longer set of stamens at anthesis, the fls. often smaller.
                        18. Caps. 2.8–3.5 mm. thick near the base, straight or nearly so, deeply grooved along the lines of dehiscence. San Clemente and Guadalupe ids.
                                                                        13. *C. guadalupensis*

18. Caps. 0.8–2.2 mm. thick near the base, straight or curved into 1 or more spirals, not deeply grooved.
19. Pollen 25–100 percent 4- or 5-pored .... 28. *C. robusta*
19. Pollen up to 5 percent 4-pored.
20. Caps. markedly quadrangular in transection. Los Angeles Co. to L. Calif. .... 19. *C. lewisii*
20. Caps. not markedly quadrangular in transection, at least in living material.
21. Upper lvs. evidently petiolate, attenuate at the base; caps. usually ± contorted; plants subglabrous to strigulose, often reddish. Cismontane .... 16. *C. ignota*
21. Upper lvs. usually not evidently petiolate, rounded or truncate at the base; caps. straight to coiled; plants strigulose to villous.
22. Plants conspicuously gray-strigose, of the deserts, the branches usually decumbent .... 22. *C. pallida*
22. Plants not conspicuously gray-strigose, barely reaching the desert-edge.
23. Caps. 0.75–0.9 mm. thick; upper lvs. ± ovate; plant erect
15. *C. hirtella*
23. Caps. 0.9–1.8 mm. thick; upper lvs. lanceolate to lance-ovate; plants erect or decumbent.
24. Plants with decumbent branches from the basal rosette; upper lvs. narrowly lanceolate; infl. usually without short glandular hairs .... 20. *C. micrantha*
24. Plants erect; upper lvs. lanceolate to lance-ovate; infl. with short glandular hairs.
25. Petals 5–10.5 mm. long; style 4.5–7.5 mm. long; plants gray-villous .... 11. *C. confusa*
25. Petals 1.5–4 mm. long; style 2–3.5 mm. long; plant usually not gray-villous .... 17. *C. intermedia*
15. Caps. subterete; plants lacking fls. in the basal nodes; lvs. mostly linear or very narrowly elliptic; seeds shining, often under 1 mm. long.
26. Sepals reflexed singly; plants often with conspicuous spreading pubescence.
27. Stigma elevated well above the anthers at anthesis; petals 8–16 mm. long
18. *C. kernensis*
27. Stigma surrounded by the anthers at anthesis; petals 1.8–4 mm. long.
28. Plants 0.4–1.8 mm. wide; fl.-tube 0.8–1.6 mm. long; style 1.6–3.2 mm. long .... 26. *C. pusilla*
28. Plants 2–6 mm. wide; fl.-tube 1.3–3 mm. long; style 3.2–4.1 mm. long
25. *C. pubens*
26. Sepals usually reflexed in pairs, remaining attached.
29. Stigma held above the anthers at anthesis; petals 4–15 mm. long
6. *C. campestris*
29. Stigma surrounded by the anthers at anthesis; petals 2–4 mm. long
29. *C. strigulosa*

1. **C. arenària** (A. Nels.) Raven. [*Chylisma a.* Nels. *Oenothera cardiophylla* var. *splendens* M. & J. Var. *longituba* Jeps.] Plants bushy, perennial, to 1.8 m. tall, villous, sometimes with some glandular hairs in the infl.; lvs. reduced upward, the petioles to 6 cm. long, the blades to 6 cm. long, ovate, cordate at base, villous; infl. less compact than in *C. cardiophylla*: fl.-tube 18–40 mm. long; sepals 8–15 mm. long; petals 8–20 mm. long, 7–28 mm. wide; fils. 5–9 mm. long; anthers 5–8 mm. long; style 30–58 mm. long; stigma above the anthers at anthesis; caps. 3–4.4 cm. long, 2.5–3.5 mm. thick; seeds 0.5–0.7 mm. long; $n=7$ (Raven, 1962).—Sandy and rocky places, below 3000 ft.; Creosote Bush Scrub; Mecca to Needles and s. to Son. March–May.

2. **C. bistórta** (Nutt. ex T. & G.) Raven. [*Oenothera b.* Nutt. *Oe. heterophylla* Nutt. ex H. & A.] Annual or of longer duration, occasionally simple, usually with several prostrate to ascending stems, these often with reddish tinge, the older epidermis exfoliating; stems rather slender, 0.5–8 dm. long; lvs. green, pubescent to pilose, narrowly elliptic in the basal rosette, to lanceolate on the stems, rarely linear, 1–12 cm. long, to 1.5 cm. wide,

± denticulate, acute, the base usually ± cuneate; petioles to 4 cm. long in the rosette, ± subsessile above; infl. with short-erect and long-villous hairs; fl.-tube 2–6 mm. long; sepals 3–10 mm. long; petals 5–15 mm., usually with a bright red spot near the base; stamens unequal, the epipetalous shorter; anthers 1–2.5 mm. long; style commonly 7–12 mm. long; stigma held well above the anthers at anthesis; caps. 1.2–2 cm. long, sharply quadrangular, 2–2.5 mm. thick, blunt or with a beak to 5 mm. long in the typical form of Orange and San Diego cos. on the immediate coast, caps. longer, 20–40 mm. long, 1.5–2 mm. thick, with a beak 3–10 mm. long away from the coast (*Oenothera b.* var. *veitchiana* Hook. not recognized by Raven); $n=7$ (Lewis et al., 1958).—Coastal Strand, Coastal Sage Scrub, Chaparral, etc. in disturbed places; Ventura and Kern cos. to n. L. Calif. March–June.

3. **C. bóothii** (Dougl. in Hook.) ssp. **decórticans** (H. & A.) Raven. [*Gaura d.* H. & A. *Oenothera d.* Greene. *Oe. b.* ssp. *d.* Munz.] Erect annual, simple or branched below, subglabrous below, finely pubescent and glandular above, the shining epidermis exfoliating readily; lvs. largely near the base, bright green or tinged red, glabrous or finely pubescent, subentire, 2–8 cm. long, the petioles almost as long, upper lvs. reduced; infl. a fairly compact spike, to 3 dm. long in fr.; fl.-tube 4–6 mm. long; sepals 4–5 mm. long; petals white, 5 mm. long, not so wide, reddish only in age; stamens unequal; caps. subfusiform, thickest in lower half, 2 mm. thick, 1.5–2.5 cm. long, attenuate into a slender beak, simply curved so that the beak spreads away from the stem; seeds ash-color, linear-obovoid, ± angled, 1 mm. long; $n=7$ (Lewis et al., 1958).—Loose slopes and disturbed places, mostly below 3000 ft.; Foothill Wd., V. Grassland, etc., away from the coast; n. Los Angeles Co. and Tehachapi Mts. to Monterey and San Benito cos. March–June.

KEY TO SUBSPECIES

Mature caps. merely curved or bent, not distinctly curved or contorted, linear or subfusiform in shape.
    Lvs. well distributed, glandular-pubescent to -villous; stem epidermis exfoliating tardily if at all; caps. 10–15 mm. long .... ssp. *intermedia*
    Lvs. largely near the base of the plant, subglabrous to strigulose, lance-ovate to oblanceolate; stem epidermis exfoliating promptly; caps. 15–25 mm. long.
        The caps. not more than 2 mm. thick at base, not conspicuously quadrangular or indurated at the angles, scarcely woody; plant slender, 2–5 dm. tall.
            Caps. with simple curve ca. ⅓ the way from the base, so that the tip spreads away from the stem axis.
                Base of mature caps. ca. 2 mm. thick, the body curved and with spreading tip.
                    Exfoliating epidermis of stem straw- or flesh-color; petals 4.5–5 mm. long, white except possibly when aging. Monterey Co. to Kern and Los Angeles cos.
                                                                                                         ssp. *decorticans*
                    Exfoliating epidermis of stems white to reddish; petals 3.5–4 mm. long, red. Mts. about the w. end of the Mojave Desert .... ssp. *rutila*
                Base of mature caps. ca. 1 mm. thick, the body straight or curved; petals 3–3.5 mm. long. Inyo Mts. .... ssp. *inyoensis*
            Caps. often contorted so that the tip points down; base of caps. 1–1.5 mm. thick; epidermis of stems white. W. Mojave Desert .... ssp. *desertorum*
        The caps. 2.5–3 mm. thick at base, conspicuously quadrangular and much thickened and indurated at the angles, quite woody; plants low and coarse, rarely more than 2 dm. high. Deserts from e. Inyo Co. to Imperial and e. San Diego cos. .... ssp. *condensata*
Mature caps. usually distinctly coiled or contorted, not merely bent and curved, not subfusiform in shape; lateral stems often prominent; plants finely pubescent to short-villous. San Bernardino Co. n. .... ssp. *alyssoides*

Ssp. **alyssoìdes** (H. & A.) Raven. [*Oenothera a.* H. & A. *Oe. boothii* ssp. *a.* Munz.] Usually branching from the base, the cent. stem erect, the stems rather slender, 5–30 cm. long, finely pubescent throughout or short-villous especially below; lowest lvs. lance-ovate to oblanceolate, 0.5–1.5 cm. wide, with slender petioles; upper lvs. gradually reduced, subsessile; infl. of subsecund spikes; fl.-tube 3–8 mm. long; sepals 4–5 mm.; petals white, 4–5 mm.; style 7–12 mm.; caps. 14–23 mm. long, thickened at base (1.5

mm.) gradually attenuate toward the beaklike tip, much coiled or only curved; seeds pale; $n=7$ (Lewis et al., 1958).—Rare in s. Calif., as in the Victorville region, s. Mojave Desert; to Ore., Utah. May–Aug. Intergrading with ssp. *desertorum*.

Ssp. **condensàta** (Munz) Raven. [*Oenothera decorticans* var. *c.* Munz.] Similar to ssp. *desertorum*, but shorter, mostly 5–20 cm., the stems thick, the infl. more crowded; caps. 2–3.8 mm. thick near the base, ± quadrangular, tapering abruptly, the midribs of the valves very prominent; $n=7$ (Lewis et al., 1958).—Open plains below 3000 ft.; Creosote Bush Scrub; from Death V. region to Imperial and e. San Diego cos.; Nev., Ariz. Old fruiting plants persisting in a dead woody condition and shedding their seeds tardily.

Ssp. **desertòrum** (Munz) Raven. [*Oenothera decorticans* var. *d.* Munz.] Similar to ssp. *decorticans* but usually less than 3.5 dm. high; caps. flexuous-contorted, the beak often directed downward, 1–1.6 mm. thick near the base; $n=7$ (Lewis et al., 1958).—Open places below 6000 ft.; Creosote Bush Scrub, Joshua Tree Wd.; deserts from Kern and Inyo cos. to n. Colo. Desert.; Nev.

Ssp. **intermèdia** (Munz) Raven. [*Oenothera boothii* ssp. *i.* Munz.] Similar to *alyssoides* but densely villous with a dense mixture of glandular hairs in the infl., the plants usually 5–20 cm. tall, the lvs. usually less than 2.5 cm. long; seeds dimorphic; $n=7$ (Raven, 1969).—Sandy places at 5000–7000 ft.; Sagebrush Scrub; ne. San Bernardino Co. to Inyo Co.; w. Nev.

Ssp. **inyoénsis** (Munz) Munz, comb. nov. [*Oenothera boothii* ssp. *i.* Munz, N. Am. Fl. ser. II, part 5: 153. 1965.] Resembling ssp. *desertorum* and included in that taxon by Raven, but with a more slender caps. (1 mm. thick at base), the body straight or curved; petals 3–3.5 mm. long. Growing along the e. base of the Inyo Mts. and with quite a different habit and appearance than in *desertorum*. April–May.

Ssp. **rùtila** (Davidson) Munz, comb. nov. [*Oenothera rutila* Davidson, Erythea 2: 62. 1894.] Included in ssp. *decorticans* by Raven and very near to that, but with a tendency to red stems, shorter red petals (3.5–4 mm. instead of 4.5–5 mm.) and growing at 1400–2250 m. (instead of usually below 1000 m.) on the desert slopes of the San Gabriel Mts. and nw. to Mt. Pinos.

4. **C. brévipes** (Gray) Raven ssp. **brevipes.** [*Oenothera b.* Gray.] Plate 62, Fig. D. Annual, usually 1–few-stemmed from base, 5–70 cm. tall, the stems unbranched above, villous especially below; lvs. mostly in a basal rosette, the uppermost reduced to ovate or lanceolate bracts, the lower 5–15 cm. long, subglabrous to villous, frequently noticeably bicolored, with conspicuous reddish veins beneath, simple to pinnately compound, terminal lfts. ovate; infl. a terminal, somewhat pedunculate nodding raceme, becoming 2–4 dm. long in fr., usually glandular; fr.-tube 4–8 mm. long, villous; sepals usually villous, with free subapical tips; petals bright yellow, 6–18 mm. long; stamens subequal; style 10–18 mm. long, the stigma above the anthers at anthesis; caps. terete, elongate, 2–9 cm. long, spreading or ascending, on a pedicel 2–20 mm. long; seeds 1–1.5 mm. long; $n=7$ (Lewis et al., 1958).—Dry slopes and washes, below 5000 ft.; Creosote Bush Scrub, Joshua Tree Wd.; deserts from Imperial Co. to w. San Bernardino Co. and Inyo Co.; Nev., Ariz. March–May.

Ssp. **pallídula** (Munz) Raven. [*Oenothera brevipes* var. *p.* Munz.] Slender, usually branched above, the stems mostly strigose; fl.-tube 4–5 mm. long; sepals strigose, without projections or these poorly developed; petals 7–12 mm. long; caps 2–4 cm. long; $n=7$. Dry flats, Inyo and Riverside cos. to Utah, Ariz.

5. **C. califórnica** (Nutt. ex T. & G.) Raven. [*Eulobus c.* Nutt. ex T. & G. *Oenothera leptocarpa* Greene.] Erect, fairly coarse-stemmed annual, simple or with a few short branches; stems subglabrous; lvs. few, mostly in a basal rosette, these lanceolate in outline, pinnatifid, 5–15 cm. long, dying early; cauline smaller, remote, the upper pendulous; fls. not crowded; fl.-tube obconic, 1 mm. long, orange and pubescent within; sepals 5–8 mm. long, glabrous to pubescent; petals yellow or orange, drying pink, often with reddish spots near the base, 6–14 mm. long, rhombic-obovate; stamens of 2 lengths; style 4–10 mm. long; stigma held at the level of the anthers of the longer stamens; caps. straight or slightly curved, 5–11 cm. long, sharply reflexed at maturity; seeds 1.3–1.6

mm. long, often flecked with purple dots; $n = 7$, 14 (Lewis et al., 1958).—Dry and disturbed places, such as burns, below 5000 ft.; Coastal Sage Scrub, Chaparral, etc. cismontane s. Calif. to cent. Calif., deserts to Ariz., L. Calif., Son. April–May.

6. **C. campéstris** (Greene) Raven. [*Oenothera c.* Greene. *Oe. dentata* var. *c.* Jeps. *Oe. d.* var. *johnstonii* Munz.] Slender-stemmed annual, usually erect, well branched, glabrous, strigulose or glandular-pubescent, the stems 0.5–2.5 dm. long, with white exfoliating epidermis; lvs. linear to narrowly elliptic, serrulate, to 3 cm. long, cauline; infl. often ± glandular and with non-glandular hairs; fl.-tube 1.5–4 mm. long; sepals 3–8 mm. long, reflexed in pairs; petals yellow, 5–15 mm. long, usually with 1 or 2 red dots at base of each one; episepalous stamens longer than epipetalous; stigma held well above the anthers at anthesis; caps. 2–4 cm. long, 0.7–2 mm. thick, subsessile, linear; seeds 0.8–1.6 mm. long; $n = 7$ (Lewis et al., 1958).—Open sandy flats, below 3000 ft.; Creosote Bush Scrub; V. Grassland, etc.; interior cismontane and desert s. from Inyo Co. interior Contra Costa Co. to interior San Diego Co. March–May. Variable as to pubescence and fl.-size, but these variants do not seem to be tenable taxa.

7. **C. cardiophýlla** (Torr.) Raven. ssp. **cardiophylla.** [*Oenothera c.* Torr.] Suffrutescent perennials flowering the first year, caulescent, not flowering at basal nodes; infl. nodding; plants viscous or glandular, 1–7 dm. tall, erect, usually branched; lvs. round-cordate to ovate, the blades to 5.5 cm. long, erose-dentate, petioles to 7 cm. long; fl.-tube 4–12 mm. long, villous to glandular-pubescent; sepals 3–9 mm. long; petals yellow or cream, 3–12 mm. long; stamens unequal; style 8–23 mm. long; stigma often held above the anthers at anthesis; caps. 2–5 cm. long; pedicel 1–18 mm.; $n = 7$ (Lewis et al., 1958). —Desert mesas and canyons, below 5000 ft.; Creosote Bush Scrub; s. San Bernardino Co. to e. San Diego Co.; L. Calif., Ariz. March–May.

Ssp. **robústa** (Raven) Raven. [*Oenothera cardiophylla* ssp. *robusta* Raven.] Plant glandular-pubescent and with scattered long, nonglandular hairs; lvs. broadly ovate, often cordate at base; fl.-tube 9–14 mm. long; petals 7–11 mm. long; style 14–20 mm. long, the stigma usually held above the anthers at anthesis; $n = 7$ (Raven).—Rocky canyons, 2000–4000 ft., Creosote Bush Scrub; mts. along w. and s. margins of Death V., Inyo Co.

8. **C. chamaenerioìdes** (Gray) Raven. [*Oenothera c.* Gray. *Sphaerostigma c.* Small.] Erect, usually branched annual, 1–5 dm. tall, the stems slender, often reddish, glandular-puberulent below, the same above and strigulose; lf.-blades subglabrous, lance-ovate to lanceolate, 4–8 cm. long, with petioles 1–3 cm. long; infl. a corymbose raceme with linear bracts and becoming 2 dm. long in fr.; fl.-tube 2.5–3 mm. long; sepals 2.5 mm. long; petals white, often reddish in age, ca. 3 mm. long; stamens subequal; stigma surrounded by the stamens at anthesis; caps. terete, linear, divaricately spreading, 2.5–5 cm. long, scarcely beaked; seeds pale, linear, 1 mm. long; $n = 7$ (Lewis et al., 1958).—Open desert below 7500 ft.; Creosote Bush Scrub to Pinyon-Juniper Wd.; from White Mts. to Imperial Co.; Utah, Tex. March–June.

9. **C. cheiranthifòlia** (Hornem. ex Spreng.) Raimann in Engl. & Prantl. ssp. **cheiranthifolia.** [*Oenothera c.* Hornem. ex Spreng. *Sphaerostigma c.* F. & M. *Oe. nitida* Greene.] Plate 62, Fig. E. Perennial, flowering the first year, with several prostrate to decumbent stems radiating from a cent. rosette, these 1–6 dm. long; plants grayish-green to rarely green and glabrous (*Oe. nitida*) but usually with strigose pubescence; lvs. thick, those of the rosette oblanceolate, 1–7 cm. long, with petioles 1–2 cm. long; lower cauline lvs. lance-oblong, subsessile to short-petioled, 2–4 cm. long, the upper shorter and wider; fls. in axils, mostly above the base of the stems; fl.-tube 2.5–5 mm. long; sepals 4–10 mm. long; petals bright yellow, 6–11 mm. long, rarely with 1 or 2 red spots near base; stamens unequal; stigma surrounded by both sets of stamens at anthesis; caps. coiled, quadrangular, pubescent, short-beaked or not, 12–22 mm. long; seeds 1 mm. long, obovoid; $n = 7$ (Kurabayashi et al., 1962).—Coastal Strand; Point Conception, Santa Barbara Co. to Ore. Channel Ids. April–Aug.

Subsp. **suffruticòsa** (Wats.) Raven. [*Oenothera c.* var. *s.* Wats.] Plant usually suffrutescent, often with dense silvery pubescence; fl.-tube 5–8.5 mm. long; sepals 6–11.5 mm. long; petals 10–22 mm. long, usually with one or two red dots near base; stigma held well above the anthers at anthesis; $n=7$ (Raven).—Coastal Strand from Point Conception to L. Calif.

10. **C. claviförmis** (Torr. & Frém.) Raven ssp. **claviförmis.** [*Oenothera c.* Torr. & Frém.] Annual, 6–50 cm. tall, glabrous or strigose below and occasionally glandular, the lvs. sometimes purple-dotted, mostly in a basal rosette, simple and irregularly dentate with ovate blades commonly 2–5 cm. long and petioles ca. as long, rarely pinnatifid; cauline lvs. much reduced; infl. racemose, somewhat pedunceled, the fls. quite crowded at anthesis, pedicels 8–25 mm. long; fl.-tube and sepals glabrous, each ca. 5 mm. long; petals white, often drying reddish and often red-brown at base, 4–6 mm. long; stamens subequal, the anthers with white spreading hairs; stigma held well above the stamens at anthesis; caps. 1.2–3 cm. long, subglabrous, on a pedicel 10–25 mm. long; $n=7$ (Lewis et al., 1958).—Sandy plains and washes, below 4000 ft.; Creosote Bush Scrub, Joshua Tree Wd.; from Inyo and Kern cos. to s. San Bernardino Co.; w. Nev. March–May.

KEY TO SUBSPECIES

Lower parts of plant villous with spreading hairs; sepals with free caudate projections arising below the apices .... ssp. *peirsonii*
Lower parts of plant strigulose or glabrous; sepals entire or with caudate projections.
    Petals white, the fl.-tube orange-brown.
        Infl. and buds usually glabrous; lateral lfts. usually well developed .... ssp. *claviformis*
        Infl. and buds variously pubescent; lateral lfts. often reduced.
            Lateral lfts. reduced in number, the lvs. often nearly simple; basal rosette compact. Mono and Inyo cos. .... ssp. *integrior*
            Lateral lfts. generally well developed and numerous, the basal rosette not compact.
                Sepals often with projections arising below the apices; terminal lfts. often large and nearly cordate; buds and infl. often silky-strigose. Death V. region ...... ssp. *funerea*
                Sepals usually entire; terminal lfts. usually inconspicuous; buds and lfts. not silvery
                                                                       ssp. *aurantiaca*
    Petals yellow, the fl.-tube yellow or orange-brown.
        Infl. strigose; sepals sometimes with free projections .... ssp. *yumae*
        Infl. usually glabrous; sepals entire; fl.-tube dark .... ssp. *lancifolia*

Ssp. **aurantiaca** (Wats.) Raven. [*Oenothera scapoidea* var. *a.* Wats. *Chylismia s.* var. *a.* Davids. & Moxley. *Oe. clavaeformis* var. *a.* Munz.] Plants strigose, especially below; lateral lfts. up to 25 on each side of rachis, terminal lft. to 3 cm. long; fl.-tube strigose without; sepals strigose; petals 2.5–8 mm. long, white, often fading purple; caps. 1.3–3 cm. long on pedicels 8–25 mm. long; $n=7$ (Lewis et al., 1958).—Creosote Bush Scrub; Inyo Co. to n. L. Calif., Ariz., Nev.

Ssp. **funèrea** (Raven) Raven. [*Oenothera c.* var. *f.* Raven.] Plants strigose below; lateral lfts. not or well developed; terminal lft. to 8 cm. long; fl.-tube 3–5.5 cm. long, strigose; sepals strigose, with conspicuous free caudate projections; petals white, often fading purple; caps. strigose, on pedicels 8–22 mm. long; $n=7$ (Raven).—Creosote Bush Scrub; Eureka and Saline valleys to Death V. region.

Ssp. **intégrior** (Raven) Raven. [*Oenothera c.* ssp. *i.* Raven. *Oe. scapoidea* var. *purpurascens* Wats.] Plants strigose below, usually glandular-pubescent above; lateral lfts. few or none; terminal lft. to 7 cm. long; fl.-tube 3–6 mm. long, strigose, often glandular; sepals strigose to subglabrous, rarely with caudate portions; caps. 13–25 mm. long; pedicel 15–40 mm.; $n=7$ (Raven).—At 4000–6000 ft.; Mono and ne. Inyo cos.; w. Nev. to e. Ore.

Ssp **lancifòlia** (Heller) Raven. [*Chylismia l.* Heller.] Plants strigose below, glabrous and often glaucous above; rosette not strictly basal; lateral lfts. few or none; terminal lft. lanceolate, to 5 cm. long; fl.-tube and sepals glabrous or the latter sparsely strigose near the apices; petals bright yellow, usually red-dotted near the base; caps. 1–2.8 cm. long, glabrous or nearly so; $n=7$ (Raven).—Sandy places, 4000–5500 ft.; s. Mono and Inyo cos.

Ssp. **peirsònii** (Munz) Raven. [*Oenothera c.* var. *p.* Munz.] Plants spreading-villous below and often above; lateral lfts. usually well developed; sepals villous or strigose, with conspicuous free caudate segms.; petals mostly yellow; caps. spreading pubescent.—Creosote Bush Scrub; Salton Sea to Imperial and e. San Diego cos.; n. L. Calif.

Ssp. **yùmae** (Raven) Raven. [*Oenothera cl.* var. *y.* Raven.] Plants strigose below, sometimes glandular-pubescent above; lateral lfts. well developed or reduced; sepals strigose, with short free caudate portions; petals pale yellow, fading reddish or not changing color; caps. 1.2–3.2 cm. long, strigose; $n=7$ (Raven).—Dunes and sandy flats, se. Imperial Co. to Ariz., Son., L. Calif.

11. **C. confùsa** Raven. Robust annual, densely cinereous-villous, rarely strigose, similar to *C. micrantha*, up to 6 dm. tall, with erect branches radiating from a central rosette; upper lvs. lanceolate or lance-ovate, 1–5 cm. long, sparsely denticulate, undulate, acuminate; infl. glandular, rarely with other hairs too; fl.-tube 2–3.8 mm. long; sepals 3.2–8 mm. long; petals 5–10 mm. long; episepalous stamens longer than epipetalous; up to 5 percent of pollen 4-pored; caps. suberect to once or twice contorted; $n=14$ (Raven). —Dry slopes away from the coast, cent. San Luis Obispo Co. to the San Bernardino Mts. and s. San Diego Co., largely in Chaparral; Ariz.

12. **C. graciliflòra** (H. & A.) Raven. [*Oenothera g.* H. & A. *Taraxia g.* Raimann in Engl. & Prantl.] Cespitose annual forming a single acaulescent tuft or with several short horizontal branches, finely pubescent to hirsute throughout; lvs. linear to linear-oblanceolate, entire or nearly so, 2–10 cm. long; upper sterile portion of ovary filiform, 1.5–4 dm. long; fl.-tube proper 2 mm. long; sepals 6–8 mm. long; petals yellow, aging red, 8–14 mm. long, shallowly notched with an apical tooth; stamens unequal, the longer anthers surrounding the stigma at anthesis; caps. ovoid-oblong, 8–12 mm. long, coriaceous, 4-angled near base, each angle expanding upward into a broad wing; seed strawcolor with grayish blotches, obovoid, 1.5–2 mm. long; $n=7$ (Raven).—Grassy places, below 3500 ft.; V. Grassland, Foothill Wd., Joshua Tree Wd.; n. Los Angeles Co. to s. Ore. March–May.

13. **C. guadalupénsis** (Wats.) ssp. **clementìna** (Raven) Raven. [*Oenothera g.* ssp. *c.* Raven.] Erect annual with short branches from lower parts, with exfoliating epidermis densely villous, mostly 2–25 cm. high; stems leafy, the lvs. narrowly elliptic in basal rosette to narrow-ovate on stems and in infl.; lvs. subsessile; infl. somewhat glandular; fl.-tube 1.6–2.4 mm. long; sepals 2–3 mm. long; petals 3–4 mm. long, sometimes with basal red dot; stigma surrounded by anthers at anthesis; caps. 10–18 mm. long, 3–3.5 mm. thick at sessile base, quadrangular, ca. 10–15 mm. long; seeds 0.8–0.9 mm. long; $n=7$ (Raven).—Dunes, San Clemente Id. April–June.

14. **C. heterochròma** (Wats.) Raven. [*Oenothera h.* Wats. *Oe. h.* ssp. *monensis* Munz. *Chylismia h.* Small.] Annual, simple or branched at base and above, glandular-pubescent throughout or stems subglabrous, 2–5 dm. tall; lvs. in lower portion only, irregularly serrate, ovate, conspicuously veined beneath, villous, 2–6 cm. long, on petioles ca. as long; upper lvs. reduced; pedicels capillary, 2–5 mm. long; petals 3–5 mm. long, purplish; caps. clavate, 8–13 mm. long, 2 mm. thick; stigma surrounded by anthers at anthesis; seeds obovoid, brown; $n=7$ (Lewis et al., 1958).—Dry loose soil at 2200–7000 ft.; Creosote Bush Scrub to Pinyon-Juniper Wd.; Death V. to Mono Co.; Nev. Early summer.

15. **C hirtélla** (Greene) Raven. [*Oenothera h.* Greene. *Sphaerostigma h.* Small. *Oe. micrantha* var. *jonesii* Munz.] Erect annual, with one or more ascending branches from the basal rosette; upper lvs. mostly ovate, with cordate to truncate base; infl. almost always with short glandular hairs as well as longer nonglandular ones; fl.-tube 1–3 mm. long; sepals 2.5–6 mm. long; petals yellow, 2–9 mm. long, sometimes red-dotted near base, sometimes with salient tooth at emarginate apex; stamens unequal; stigma surrounded by anthers at anthesis; caps. 0.7–0.9 mm. thick, terete, once or twice contorted; $n=7$ (Raven).—Brushy places, often on burns; Coastal Sage Scrub, Chaparral; L. Calif. to cent. Calif. March–May.

16. **C. ignòta** (Jeps.) Raven. [*Oenothera micrantha* var. *i.* Jeps. *Oe. i.* Munz.] Erect, often reddish-tinged, similar to *C. micrantha*, with one or more ascending branches from the basal rosette; plants to 5 dm. tall, finely strigulose with some longer hairs; upper lvs.

narrowly elliptic, to 6 cm. long, acute, mostly petiolate; infl. subglabrous or glandular-pubescent and with some longer hairs or strigose; fl.-tube 1.8–3 mm. long; sepals 2.6–5.5 mm. long; petals 4–8 mm. long, yellow, sometimes red-dotted near base; stamens unequal; stigma surrounded by anthers at anthesis; caps. 2–3 cm. long, terete in life, slender and usually contorted; seeds 1.2–1.3 mm. long; $n=7$ (Raven).—Coastal Sage Scrub, Chaparral, interior valleys and mts.; San Diego Co. to cent. Calif.

17. **C. intermèdia** Raven. Densely villous erect herb, similar to *C. micrantha*, to 6 dm. tall, with several erect branches from a cent. rosette and upper lvs. broader and with glandular hairs in infl. as well as nonglandular; some pollen grains 4-pored; $n=14$ (Raven).—Disturbed places in Chaparral; n. L. Calif. to cent. Calif.

18. **C. kernénsis** (Munz) Raven ssp. **kernénsis**. [*Oenothera k.* Munz.] Erect annual with few spreading branches from near base or simple and erect, with short spreading nonglandular hairs in lower parts; lvs. well distributed, the basal oblanceolate, 1–3.5 cm. long, 2–5 mm. wide, subsessile, ascending, plane; fls. few, solitary in upper axils in a loose raceme; fl.-tube 2–3 mm. long, pubescent; petals bright yellow, 10–14 mm. long, often with 2 reddish spots near the base; stamens unequal; stigma held well above the anthers at anthesis; caps. pedicelled, 1.5 mm. thick, not beaked; seeds ca. 1 mm. long; $n=7$ (Raven).—Desert washes and in canyons, 2500–6000 ft.; Joshua Tree Wd., Pinyon-Juniper Wd.; w. Kern Co. May.

Ssp. **gilmánii** (Munz) Raven. [*Oenothera dentata* var. *g.* Munz. *Oe. k.* ssp. *g.* Munz and ssp. *mojavensis* Munz.] Plants to 3 dm. tall, covered with a short glandular pubescence, usually with a few scattered hairs, caps. subsessile or on a very short pedicel; $n=7$ (Raven). —At 2500–6000 ft.; widely distributed over Mojave Desert; to Nev.

19. **C. lewísii** Raven. Villous annual similar to *C. micrantha*, but with short (13–20 mm.) usually one-coiled caps. that are square in cross section in fresh material; $n=7$ (Raven).—Open grassland and sandy places; Coastal Strand, V. Grassland, etc.; below 1000 ft.; Los Angeles Basin to L. Calif. April–May.

20. **C. micrántha** (Hornem. ex Spreng.) Raven. [*Oenothera hirta* Link, not L. *Oe. micrantha* Hornem. *Sphaerostigma hirtum* F. & M.] Annual, mostly with sprawling branches up to 6 dm. long, ± densely villous all over; lvs. very narrowly elliptic in the basal rosette, cauline tending to be broader near base, 1–12 cm. long, denticulate; lower petioles to 2 cm. long, upper reduced; infl. more densely villous than rest of plant, usually grayish; fl.-tube 1.2–2 mm. long; sepals 1–1.5 mm. long; petals yellow 1.5–4 mm. long, occasionally red-dotted near base; stamens unequal, the anthers surrounding the stigma at anthesis; caps. subterete in living material, straight or somewhat curved; $n=7$ (Raven). —Coastal dunes or beaches inland to Elsinore, Upland, etc.; Channel Ids. March–May.

21. **C. múnzii** (Raven) Raven. [*Oenothera m.* Raven.] Annual, 8–50 cm. tall, with many branches at base and above, the lvs. mostly in a subbasal rosette; stems and lvs. strigose; lvs. pinnate, to 2 dm. long, with well developed lateral lfts.; infl. nodding, elongating in mature bud; fl.-tube 2–3 mm. long, strigose; sepals 4–7 mm. long, strigose; petals bright yellow, red-dotted near base, 3–10 mm. long; stamens subequal, the stigma held well above the anthers in anthesis; caps. 0.8–2.4 cm. long, 1.5–2 mm. thick, clavate; pedicel 8–28 mm. long, sharply deflexed at maturity; seeds pale brown, 0.8–1.6 mm. long; $n=7$ (Raven).—Dry slopes and washes, 2000–5000 ft.; Creosote Bush Scrub; ne. Mojave Desert to Inyo Co.; Nev.

22. **C. pállida** (Abrams) Raven subsp. **pállida**. [*Sphaerostigma pallidum* Abrams.] Similar to *C. micrantha*, but covered with dense appressed pubescence; lvs. entire or nearly so, mostly 1–3 cm. long, with petiole ca. 2 mm. long; infl. occasionally with some gland-tipped hairs, gray-pubescent; fl.-tube 1.8–4 mm. long; sepals 3.5–6 mm. long; petals yellow, 3.5–6 mm. long, rarely red-dotted at base; stamens unequal, surrounding the stigma at time of anthesis; caps. ± quadrangular, 1–1.2 mm. thick, straight to once or more coiled; $n=7$ (Raven).—Desert flats and slopes below 6000 ft.; Creosote Bush Scrub to Pinyon-Juniper Wd.; deserts from n. L. Calif. to Inyo Co. March–May.

Ssp. **hállii** (Davids.) Raven. [*Sphaerostigma h.* Davids. *Oenothera h.* Munz.] Fl.-tube 3.8–4.2 mm. long; sepals 4.7–8 mm. long; petals 6.5–13 mm. long, each with 1–3 red dots near the base; longer stamens just equal to stigma; $n=7$ (Lewis et al., 1958).—Dry sandy places below 6000 ft.; Creosote Bush Scrub, Joshua Tree Wd.; deserts from Riverside and San Bernardino cos.

23. **C. pálmeri** (Wats.) Raven. [*Oenothera p.* Wats. *Taraxia p.* Small.] Dwarf cespitose annual, finely strigulose throughout, forming acaulescent tufts 2–6 cm. long or with a few short horizontal branches; stems pubescent, exfoliating, tough and almost woody in age; lvs. linear-lanceolate to -oblanceolate, subentire or minutely denticulate, 2–6 cm. long; sterile part of ovary filiform, 8–15 mm. long; fl.-tube proper 1–2 mm. long; sepals 2–3 mm. long; petals yellow, 3–5 mm. long, the fls. diurnal; stigma surrounded by anthers; stamens unequal; caps. ovoid, 5–7 mm. long, coriaceous, 4-angled below, each angle becoming truncate wing above and dehiscing along the wing's edge; seeds few, 1.5 mm. long; $n=7$ (Lewis et al., 1958).—Dry open places, 2000–4000 ft.; Creosote Bush Scrub, Joshua Tree Wd.; Barstow region to Temblor Range in w. Kern Co.; e. Ore., Nev. April–May.

24. **C. pterospérma** (Wats.) Raven. [*Oenothera p.* Wats.] Low annuals, 5–12 cm. tall, simple or few-branched, pilose below, finely glandular above; lvs. oblong to lance-ovate, often with a shoulder on each side near the tip, entire, subsessile, 5–20 mm. long; fls. axillary, pinkish-white; pedicels 5–8 mm. long; fl.-tube 1–2 mm. long; sepals 1.5–2.5 mm. long; petals white, obcordate, ca. as long; stigma surrounded by the anthers at anthesis; caps. cylindric-clavate, slightly curved, 10–16 mm. long; seeds oblong, brownish, bordered with a revolute winglike tubercled margin; $n=7$ (Raven).—Rare, dry places 4500–8000 ft.; Sagebrush Scrub, Pinyon-Juniper Wd.; Panamint and Inyo mts.; to Utah. May–June.

25. **C. pùbens** (Wats.) Raven. [*Oenothera strigulose* var. *p.* Wats. *Sphaerostigma p.* Rydb.] Similar to *C. pusilla* but differing in the more robust habit, the stouter stems to 38 cm. long; lvs. narrowly lanceolate, 1.5–4.5 cm. long; fl.-tube 1.3–3 mm. long; sepals 2.2–3.8 mm. long; petals yellow, 3–4 mm. long, often red-dotted near base; caps. 2.6–5 cm. long, 0.8–1.2 mm. thick, cylindrical; $n=14$ (Raven).—Sandy or sagebrush-covered places, 3000–9000 ft.; Sagebrush Scrub, Pinyon-Juniper Wd.; s. Inyo to Lassen cos.; Nev. May–June.

26. **C. pusílla** Raven. [*Oenothera contorta* var. *flexuosa* Munz, in part.] Slender annual, 2–22 cm. tall, usually branched from base; glandular-pubescent and often with longer nonglandular hairs; lvs. linear, 1–3 cm. long, 0.4–1.6 mm. wide, serrulate; fl.-tube 0.8–1.6 mm. long; sepals 1.2–2 mm. long; petals 1.8–3 mm. long, yellow; stigma surrounded by anthers at anthesis; caps. 1.8–3.2 cm. long, 0.6–0.9 mm. thick, subsessile or very short-petioled; not beaked; $n=7$ (Raven).—Largely Sagebrush Scrub, 400–9500 ft.; San Bernardino Mts. to se. Wash.; Utah.

27. **C. refrácta** (Wats.) Raven. [*Oenothera r.* Wats.] Annual, to 4 dm. high, erect, with open divaricate branching, usually glandular-puberulent and ± strigose, the stems slender, mostly with some red; lvs. well distributed, but the lower largest, oblong-linear to oblanceolate, 2–5 cm. long, entire to denticulate, 2–5 cm. long, sessile to short-petioled; infl. racemose, sometimes paniculate; 5–15 cm. long; fl.-tube 5–6 mm. long; sepals lance-oblong, 5–6 mm. long; petals white, roundish, 4–7 mm. long; stamens unequal, the stigma held well above the anthers at anthesis; caps. linear, spreading or reflexed, 3–5 cm. long, straight or coiled, mostly not beaked; seeds 1 mm. long; $n=7$ (Lewis et al., 1958).—Open sandy places, below 5500 ft.; Creosote Bush Scrub, Joshua Tree Wd.; deserts from Inyo to Imperial cos.; Ariz., to Utah. March–May.

28. **C. robústa** Raven. Resembling *C. micrantha*, but with an erect robust habit; upper lvs. wider, to 12 cm. long; pollen mostly 4-porate; stigma surrounded by anthers at anthesis; caps. ± quadrangular, stout, 1.5–2 mm. thick; $n=21$ (Raven).—San Clemente, Santa Cruz, and Santa Catalina ids.; Coastal Sage Scrub, San Diego Co.; L. Calif.

29. **C. strigulòsa** (Fisch. & Meyer) Raven. [*Sphaerostigma s.* F. & M. *Oenothera contorta* var. *s.* Munz.] Mostly erect annual, to 5 dm. tall, usually well branched, with wiry stems and no well defined basal rosette, the stems with exfoliating epidermis; strigose and often also glandular-pubescent especially in the infl.; lvs. linear to narrowly elliptic, to

3.5 cm. long, sparsely serrulate; fl.-tube 1.6–2.7 mm. long; sepals 1.6–4 mm. long, reflexed in pairs; petals yellow, 2–4 mm. long, sometimes with 2 red dots at base of each one; stigma surrounded by anthers at anthesis; usually less than 10 percent of pollen with 4 pores; caps. straight or somewhat flexuous, to 4.5 cm. long, 1.4–3 mm. thick, subsessile, with a sterile beak 1–3 mm. long; $n=14$ (Raven).—Open sandy grassland; from Sonoma Co. to n. L. Calif.; V. Grassland, etc. Santa Rosa Id.

30. **C. wálkeri** (Nels.) Raven ssp. **tórtilis** (Jeps.) Raven. [*Oe. scapoidea* var. *t.* Jeps.] Annual or short-lived perennial, villous below, nearly leafless above and with well-developed basal rosette; lvs. pinnate or even bipinnate, the lateral lfts. to 3 cm. long; terminal lft. 1–5 cm. long; infl. branching, erect; fl.-tube 1–1.5 mm. long, glandular-pubescent to hispid; sepals 1.5–4 mm. long, often purple-dotted; petals 3–6 mm. long, bright yellow; stamens unequal, the stigma surrounded by the anthers at anthesis; caps. 1–4.5 cm. long, on a pedicel 5–30 mm. long, often twisted at maturity; $n=7$ (Raven).— Rocky debris, 2000–6000 ft.; Creosote Bush Scrub to Pinyon-Juniper Wd.; ne. San Bernardino Co. and Inyo Co. to Nev., Utah. May–June.

3. *Circaèa* L. Enchanter's Nightshade

Low slender perennial herbs with subterranean rootstocks. Lvs. opposite, thin, petioled. Fls. small, white to pinkish, paniculately arranged in racemes. Pedicels capillary. Fl.-tube short, deciduous and with a ringlike disk within. Sepals 2, reflexed. Petals 2, notched. Stamens 2, alternate with petals. Ovary 1–2-loculed, each locule 1-ovuled. Fr. nutlike, 1–2-seeded, obovoid, indehiscent, usually with hooked hairs. Ca. 6 spp. of N. Hemis. (*Circe*, the enchantress.)

1. **C. alpìna** L. ssp. **pacífica** (Asch. & Magnus) Raven. [*C. p.* Asch. & Magnus.] Plate 62, Fig. F. Rootstock tuberous-thickened; stem simple, 2–4 dm. tall, usually ± strigulose below the infl.; lf.-blades ovate to roundish, usually rounded at base, sometimes ± cordate, entire or minutely denticulate or obscurely repand-denticulate, 2–6 cm. long, acuminate; petioles 2–3 cm. long; racemes bractless; sepals and petals ca. 1 mm. long; caps. narrow-obovoid, 1-loculed, 1.5–2 mm. long, with hooked hairs; $n=11$ (Lewis et al., 1958).—Deep woods, below 8000 ft.; Montane Coniferous F.; San Bernardino Mts.; Sierra Nevada etc. to B.C., Rocky Mts. June–Aug.

4. *Clárkia* Pursh.

Annual, with slender to stoutish stems, simple or branched, usually with epidermis exfoliating below. Lvs. simple, pinnately veined, linear to ovate, entire to denticulate, sessile or short-petioled. Infl. a leafy spike or raceme, the axis straight or recurved in bud; buds erect, deflexed or pendulous; fls. mostly showy, the fl.-tube obconic, campanulate to funnelform to slender and elongate, mostly with a ring of hairs within. Sepals 4, often colored, reflexed individually, in pairs, or united and deflexed to one side at anthesis. Petals 4, cuneate to obovate to oblanceolate, sometimes lobed, sessile to clawed, lavender to pink or purple or white, often variously spotted or flecked or blotched. Stamens 8 and in 2 series or 4 and in 1 series; anthers basifixed, the sporogenous tissue of the anthers divided into packets. Stigma 4-lobed. Caps. clavate to cylindrical, terete to quadrangular, sessile to pedicelled, beakless or beaked, often 4- or 8-grooved; seeds brown to gray, minutely tubercled to scaly, usually angular, cubical or elongate, sometimes spindle-shaped, mostly crested. Ca. 33 spp. of temp. N. Am. & Chile. (Capt. William *Clark*, of Lewis and Clark Expedition.)

(Lewis, H. and M. E. Lewis. The genus Clarkia. Univ. Calif. Pub. Bot. 20 (4): 241–392. 1955.)
1. Petals with 3 lobes, the middle one a prominent tooth in the emarginate sinus of the petal
   .... 12. *C. xantiana*
1. Petals not lobed.
    2. The petal constricted at base into a definite claw which is expanded into a pair of lateral teeth .... 8. *C. rhomboidea*

2. The petals not clawed, or if so, the claw entire.
   3. Petals white to cream, fading pink, not flecked with purple, mostly more than 10 mm. long
                                                                 5. *C. epilobioides*
   3. Petals colored, of if nearly white, flecked with purple, mostly more than 10 mm. long.
      5. Buds, at least the older ones, pendulous or deflexed.
         6. Claw of the petals slender, equal to or longer than the limb . . . . . . 11. *C. unguiculata*
         6. Claw of the petals short or lacking.
            7. Ring of hairs within the fl.-tube at or near the summit or lacking; fl.-tube 1–3, rarely 4 mm. long.
               8. Sepals 6–10 mm. long; petals usually not more than 12 mm. long.
                  9. Rachis of infl. reflexed at tip, becoming straight as buds mature; petals pink, often with reddish-purple flecks near the base . . . . 9. *C. similis*
                  9. Rachis of infl. straight, the buds deflexed; petals pink to rose-lavender
                                                                          3. *C. delicata*
               8. Sepals mostly 10–20 mm. long; petals mostly 10–30 mm. long.
                  9a. Rachis of infl. deflexed at tip, becoming straight as the buds mature. Riverside Co. to Tuolumne Co. . . . . 4. *C. dudleyana*
                  9a. Rachis of infl. straight, only the buds deflexed. Orange Co. to Monterey Co.
                                                                            2. *C. deflexa*
            7. Ring of hairs toward the middle of the fl.-tube; fl.-tube 3–5 mm. long; petals 10–35 mm. long, usually with bright purple-red base . . . . 1. *C. cylindrica*
      5. Buds erect.
         10. Plants of sea-coast, usually prostrate or decumbent; lf.-blades usually obtuse; petals 10–15 mm. long, usually with deltoid red spot. Santa Rosa Id.
                                                                                       6. *C. prostrata*
         10. Plants away from the immediate coast, usually erect; lf.-blades acute.
            11. Ovary densely spreading-pubescent.
               12. Petals 5–15 mm. long, often with a wedge- or shield-shaped dark spot at or above the middle; stamens equaling the stigma
                                                                    7. *C. purpurea* ssp. *quadrivulnera*
               12. Petals 15–25 mm. long, often with a dark blotch in center; stamens shorter than stigma . . . . 7. *C. p.* ssp. *viminea*
            11. Ovary strigulose puberulent with upwardly curved hairs . . . . . 10. *C. speciosa*

1. **C. cylíndrica** (Jeps.) Lewis & Lewis. [*Godetia bottae* var. *c.* Jeps. *G. c.* Hitchc.] Erect, usually branched, 2–5 dm. tall, puberulent above; lvs. lanceolate to lance-linear, 2–5 cm. long, 3–6 mm. broad; infl.-axis recurved in bud, erect at anthesis; buds pendent; fl.-tube 3–5 mm. long, the ring of hairs near the middle; sepals lanceolate, 1–2.5 cm. long; petals 1–3.5 cm. long, purple to pinkish-lavender, shading to white near the middle, usually flecked with red-purple and with a bright red-purple base; outer anthers with bluish, the inner with yellow or bluish-gray pollen; caps. 2–5 cm. long, 1–2 mm. broad, usually with a beak to 1 cm. long, 4-grooved when immature, often enlarged upward; seeds brown, 1–1.5 mm. long, the crest to 0.1 mm. long; $n=9$ (Håkansson, 1943).—Dry slopes and flats below 4000 ft.; V. Grassland, Foothill Wd., Chaparral; n. Los Angeles Co. n. to cent. Calif. April–July.

2. **C. defléxa** (Jeps.) Lewis & Lewis. [*Godetia d.* Jeps.] Plate 63, Fig. A. Erect, 3–9 dm. tall, stout, mostly glabrous; lvs. lanceolate, 3–8 cm. long, 5–18 mm. wide, sparsely puberulent, short-petioled; infl.-axis erect; buds deflexed; fl.-tube 2–3 mm. long, the ring of hairs near summit; sepals 10–20 mm. long, united and deflexed to one side at anthesis; petals fan-shaped, 1–3 cm. long, pale lavender to pinkish-lavender, mostly white toward base, red-flecked, scarcely clawed; outer anthers lavender, inner yellowish, ± flecked with red; caps. subterete to quadrangular, 3–4 cm. long, scarcely beaked, from sessile to having a pedicel 3 cm. long, obscurely 4-grooved when young; seeds brown or gray, 1.2–1.8 mm. long, the crest ± developed; $n=9$ (Håkansson, 1941).—Dry openings below 3000 ft.; Coastal Sage Scrub, Chaparral; S. Oak Wd.; Orange and w. Riverside cos.; to Monterey Co. April–June.

3. **C. delicàta** (Abrams) Nels. & Macbr. [*Godetia d.* Abrams.] Erect, 2–7 dm. tall, the stems slender, simple or branched, subglabrous and glaucous above; lvs. lanceolate to ovate, 1.5–4 cm. long, 4–15 mm. wide, serrate or denticulate or upper entire, sub-

Plate 63. Plate A, *Clarkia deflexa*. Fig. B, *Clarkia purpurea* ssp. *viminea*. Fig. C, *Clarkia unguiculata*. Fig. D, *Epilobium adenocaulon* var. *parishii*. Fig. E, *Epilobium glaberrimum*. Fig. F, *Gaura coccinea*.

glabrous, with petioles 5–10 mm. long; infl.-axis erect; buds deflexed; fl.-tube 2 mm. long, the inner ring of hairs near summit; sepals 6–9 mm. long, united and deflexed to one side at anthesis; petals spatulate to obovate, 8–12 mm. long, 4–8 mm. broad, pale pink to bright rose-lavender, lighter in basal half, short-clawed; outer stamens ± orange-red, inner cream; caps. quadrangular, 1.5–3.5 cm. long, ca. 2 mm. broad, subsessile, 8-grooved when young; seeds brown, 1–1.5 mm. long; $n = 18$ (Lewis & Lewis, 1955).— Dry slopes, below 4000 ft.; mostly Chaparral, S. Oak Wd.; San Diego Co.; n. L. Calif. May–June.

4. **C. dudleyàna** (Abrams) Macbr. [*Godetia d.* Abrams.] Erect, 3–7 dm. tall, simple or branched, puberulent above; lvs. lanceolate, subentire or denticulate, 1.5–7 cm. long, 3–15 mm. broad, with petioles 3–10 mm. long; infl.-axis recurved in bud, erect in anthesis; buds pendulous; fl.-tube slender, 1–3 mm. long, the inner hair-ring near summit; sepals 10–22 mm. long, united and deflexed to one side at anthesis; petals fan-shaped, 10–28 mm. long, lavender-pink, often flecked with red, lighter toward base; outer stamens with bluish or whitish pollen, inner with cream pollen; caps. 4- or 8-sided, ribbed, 1–3 cm. long, 2–2.5 mm. broad, sessile or nearly so; seeds brown, ca. 1 mm. long, short-crested; $n = 9$ (Lewis & Lewis, 1955).—Dry slopes below 6000 ft.; Coastal Sage Scrub, Chaparral, Yellow Pine F.; Los Angeles Co. to w. Riverside Co.; Kern Co. to Tuolumne Co. May–July.

5. **C. epilobioìdes** (Nutt.) Nels. & Macbr. [*Oenothera e.* Nutt. in T. & G. *Godetia e.* Wats.] Erect, 2–7 dm. tall, simple or branched, sparsely puberulent above; lvs. linear to narrowly lanceolate or oblanceolate, 15–25 mm. long, 2–4 mm. broad, on short petioles; infl.-axis reflexed in bud, erect in anthesis; buds pendulous; fl.-tube 0.5–1.5 mm. long, the ring of hairs in upper part or lacking; sepals 6–10 mm. long, united or in pairs at anthesis; petals white, ± pinkish in age, obovate, 5–10 mm. long, scarcely clawed; anthers white or cream; caps. slender, 4-sided, 1–3 cm. long, 1–1.5 mm. wide, on a pedicel to 1 cm. long; very short-beaked; seeds brown, 0.5–1 mm. long, short-crested; $n = 9$ (Lewis & Lewis, 1955).—Shaded places below 2500 ft.; Chaparral, Coastal Sage Scrub, S. Oak Wd., etc.; San Diego Co. to San Francisco; Santa Cruz, Santa Catalina ids.; n. L. Calif. March–May.

6. **C. prostràta** Lewis & Lewis. Prostrate or decumbent, the stems simple or branched, 2–5 dm. long; lvs. elliptic to oblanceolate, 1–2.5 cm. long, 4–8 mm. broad, subsessile; infl.-axis straight; buds erect; fl.-tube 4–7 mm. long, the ring of hairs in lower third; sepals lanceolate, 6–8 mm. long, usually reflexed in 2's; petals truncate-obovate or fan-shaped, 10–15 mm. long, lavender-pink, ± cream or yellowish toward base, usually with a deltoid purple blotch near middle; fils. greenish, anthers cream; caps. quadrangular, 2–3 cm. long, 2.5–3 mm. broad; seeds brown or grayish, 1–1.5 mm. long, short-crested; $n = 26$ (Lewis & Lewis, 1955).—Coastal bluffs, etc.; Santa Rosa Id.; San Luis Obispo Co. n. April–July.

7. **C. purpùrea** (Curt.) Nels. & Macbr. ssp. **quadrivúlnera** (Dougl.) Lewis & Lewis. [*Oenothera q.* Dougl. *Godetia q.* Spach.] Stems erect; lvs. linear to lanceolate, 1.5–5 cm. long, less than ⅛ as wide; infl. lax or congested; petals 5–15 mm. long, lavender or purple, often deep red, the paler fls. often with a wedge-shaped or shield-shaped purple spot; caps. quadrangular, 1–3.5 cm. long, 2–3 mm. thick, not noticeably enlarged at middle, terete and 8-nerved when fresh, or square and conspicuously ribbed when dry, sessile or nearly so, short-beaked; $n = 26$ (Håkansson, 1941).—Common in open spots, below 6000 ft.; many Plant Communities; L. Calif. through cismontane s. Calif. to Wash., Ariz. April–July.

Ssp. **vimínea** (Dougl.) Lewis & Lewis. [*Oenothera v.* Dougl. *Godetia v.* Spach.] Plate 63, Fig. B. Stems erect or decumbent; lvs. linear to lance-linear, 3–7 cm. long, less than ⅛ as broad; infl. lax; petals 1.5–2.5 cm. long, lavender to purple, usually with dark spot in upper part; $n = 26$ (Lewis & Lewis, 1955).—Dry open places below 5500 ft.; Chaparral, Foothill Wd., Yellow Pine F.; in much of cismontane Calif.; to Ore. May–July.

8. **C. rhomboìdea** Dougl. [*Oenothera r.* Lév.] Erect, simple or few-branched, 2–11 dm. tall, puberulent above; lvs. few, subopposite, lance-ovate to elliptic, 2–7 cm. long, 0.4–2.3

cm. wide, acute, entire or nearly so, with petioles 1–2.5 cm. long; infl.-rachis recurved, erect at anthesis; buds nodding; fl.-tube 1–3 mm. long, with white hairs at summit; sepals green, usually distinct at anthesis, 6–8 mm. long; petals 6–12 mm. long, pinkish-lavender, with or without darker flecks, often red at base, the claw with a pair of projections near the base; stamens subtended by ciliated scales to 1.5 mm. long; anthers lavender to purple; caps. quadrangular, straight to curved, 1–2.5 cm. long, 2–2.5 mm. broad, with a beak to 3 mm. long, sessile or nearly so; seeds 1–1.5 mm. long, sometimes mottled, short-crested; $n=12$ (Lewis & Lewis, 1955).—Rather dry slopes, below 8000 ft.; Chaparral to Montane Coniferous F.; San Diego Co. to Wash., Mont., Utah, Ariz. May–July.

9. **C. símilis** Lewis & Ernst. Resembling *C. epilobioides*, the lvs. 2–4 cm. long, 3–8 mm. broad; fl.-tube 1.5–2 mm. long; petals 6–12 mm. long, oblanceolate to rhombic or obovate, whitish to light pink, flecked with purple in lower half, with a basal claw ca. 1 mm. long; caps. 1.5–3 cm. long, 1–1.5 mm. broad, sessile or nearly so; seeds brown, ca. 1 mm. long, obscurely or short-crested; $n=17$ (Lewis & Lewis, 1955).—Shaded places below 3500 ft.; Chaparral, Foothill Wd., S. Oak Wd.; San Diego and w. Riverside cos. to San Benito Co. April–May.

10. **C. speciòsa** Lewis & Lewis ssp. **polyántha** Lewis & Lewis. Erect, with many-fld., wandlike branches; lvs. linear to linear-lanceolate, 1–6 cm. long, 2–6 mm. wide, sessile or short-petioled; infl.-axis erect; buds erect; fl.-tube mostly 8–15 mm. long; petals 10–27 mm. long, purple or lavender or yellowish, usually lighter near base, conspicuously spotted purple-red near the center; $n=9$ (Lewis & Lewis, 1955).—At 2000–5000 ft.; Tehachapi and Piute mts., Kern Co. to Fresno Co. May–July.

11. **C. unguiculàta** Lindl. [*C. elegans* Dougl., not Poir.] Plate 63, Fig. C. Erect, 3–10 dm. high, simple or branched, glabrous and glaucous; lvs. lanceolate to elliptic or ovate, 1–6 cm. long, 0.5–2 cm. wide, with petioles 3–10 mm. long; infl.-axis erect; buds deflexed; fl.-tube obconic to campanulate, 2–5 mm. long, with broad band of inner hairs in upper half and other hairs at margin; sepals pilose, 10–16 mm. long, united and deflexed to one side at anthesis, subglabrous to densely pilose; petals deltoid to rhombic, 1–2 cm. long, lavender-pink to salmon or purplish or dark red-purple, with a slender claw almost half the total length; outer anthers red-orange to dull red, the inner lighter to whitish; caps. straight or curved, subterete to subquadrangular, 1.5–3 cm. long, 2–3 mm. broad, scarcely beaked, sessile or nearly so, 8-ribbed; seeds brown, 1–1.5 mm. long, minutely crested; $n=9$ (Lewis, 1951).—Dry often shaded places, below 5000 ft.; Chaparral, V. Grassland, Coastal Strand, etc.; San Diego Co. to n. Calif. May–June.

12. **C. xantiàna** Gray. Erect, 2–8 dm. tall, simple or branched, glabrous and glaucous; lvs. linear to lance-linear, subentire, 2–6 cm. long, 1.5–8 mm. wide, sessile or nearly so; infl.-axis erect; buds deflexed; fl.-tube 2–5 mm. long, obconic to campanulate, with broad band of white hairs at upper margin; petals 6–20 mm. long, lavender to red-purple, 2-lobed with a subulate tooth in the sinus, conspicuously clawed, often with a red-purple spot; anthers 8, lavender to purple; caps. 1.5–2.5 cm. long, 1.5–2.5 mm. wide, 4-sided, usually short-beaked; seeds brown, 1.3–1.5 mm. long, minutely crested; $n=9$ (Lewis & Lewis, 1955).—Dry slopes, 800–5700 ft.; Chaparral, etc.; n. slope of San Gabriel Mts. to Tulare and Inyo cos. May–June.

5. *Epilòbium* L. WILLOW-HERB

Mostly herbs, sometimes suffrutescent; annual or usually perennial and wintering over by *turions* (tuberlike buds with swollen scales which persist about the base of the next year's stem) or *rosettes* (at first ± fleshy) or slender *stolons*. Lvs. opposite or alternate, nearly or quite sessile, entire or denticulate. Fls. axillary or in terminal racemes or panicles, perfect. Fl.-tube short or not prolonged beyond the ovary. Sepals 4. Petals 4, usually notched, purplish, pink or white, sometimes yellow. Stamens 8, the alternate shorter. Stigma oblong to 4-lobed. Caps. elongate, subcylindric to fusiform or clavate, 4-loculed, loculicidal. Seeds with *coma* (tuft of silky hairs) at upper end. Over 100 spp., cosmopolitan except in trop. (Greek, *epi-*, upon, and *lobon*, a caps.)

1. Fl.-tube not prolonged above the ovary; fls. slightly irregular, the petals 1–2 cm. long, entire, spreading .... 2. *E. angustifolium*
1. Fl.-tube prolonged above the ovary; fls. regular, mostly smaller.
    2. Plants annual; stems with exfoliating epidermis. Of dry situations ...... 9. *E. paniculatum*
    2. Plants perennial, sometimes blooming the first year; epidermis not exfoliating from the stems; mostly of wet situations.
        3. Rootstocks bearing globose or ovoid turions with fleshy overlapping scales.
            4. Petals 5–10 mm. long; stems simple to divaricately branched at summit; lvs. 5–12 cm. long .... 5. *E. exaltatum*
            4. Petals 3–5 mm. long; lvs. ovate to lanceolate, 2–5 cm. long .... 3. *E. brevistylum*
        3. Rootstocks not turioniferous.
            5. Plant pallid, glaucous and almost glabrous throughout; rootstocks branched, scaly and rather tough .... 6. *E. glaberrimum*
            5. Plant not glaucous, but green, or canescent with pubescence.
                6. Stems usually 1–3 dm. tall, simple above, with few pairs of opposite lvs. High montane.
                    7. Lvs. sessile, oblong or linear, suberect; stem slender, perennial with subfiliform stolons .... 8. *E. oregonense*
                    7. Lvs. ± distinctly petioled and spreading; plant with subterranean scaly branches
                    7. *E. lactiflorum*
                6. Stems mostly 3–10 dm. tall and freely branched especially above, if shorter, the upper lvs. alternate and more numerous. Mostly midmontane to low elevs.
                    8. Lvs. firm, sessile or with short broad petioles; coma mostly persistent; papillae of seeds ± rounded. Widespread .... 1. *E. adenocaulon*
                    8. Lvs. thin, flaccid, pale green, tapered to definite slender petioles; coma caducous; papillae of seeds ± conical. E. Calif. .... 4. *E. ciliatum*

1. **E. adenocáulon** Hausskn. var. **adenocáulon.** Perennial, flowering the first year, innovations by fleshy rosettes; stem erect, 3–10 dm. tall, glabrous below with some hair on decurrent lines below the nodes, glandular-pubescent in infl., simple or weakly branched below, freely branched above; lvs. glabrous or nearly so, ovate to elliptic-lanceolate, 3–6 cm. long, acute to obtuse, serrulate, rounded into very short flat petioles, upper lvs. gradually reduced, ± pubescent; sepals 2 mm. long; petals white to pale or ± reddish, 3–4 mm. long; fruiting pedicels 3–8 mm. long; caps. slender, ± reddish, 4–6 cm. long. with gland-tipped hairs, glabrate in age; seeds obovoid, abruptly short-beaked, ca. 1 mm, long, longitudinally ridged and rounded-papillose.—Moist places below 10,000 ft.; many Plant Communities; most of cismontane and montane Calif.; to Alaska and Atlantic states. July–Sept.

KEY TO VARIETIES

Infl. glandular-pubescent .... var. *adenocaulon*
Hairs of infl. not gland-tipped.
  Herbage green, the stems and lvs. scarcely pubescent, the infl. strigulose ...... var. *parishii*
  Herbage grayish, the stems and lvs. densely soft-pubescent .... var. *holosericeum*

Var. **holosericeum** (Trel.) Munz. [*E. h.* Trel.] Canescent throughout with soft appressed hairs; petals pink to purple, 4–5 mm. long.—Occasional in moist places, below 2500 ft.; many Plant Communities; much of cismontane Calif.

Var. **párishii** (Trel.) Munz. [*E. p.* Trel. *E. californicum* Hausskn. *E. c.* var. *p.* Jeps.] Plate 63, Fig. D. Infl. with a whitish, ± appressed pubescence; petals white to pink, 2–4 mm. long; $n = 18$ (Lewis et al., 1958).—Wet places mostly below 6000 ft.; many Plant Communities; common in cismontane s. Calif.; to n. Calif. and B.C.; occasional on deserts.

2. **E. angustifòlium** L. ssp. **circumvàgum** Mosquin. FIREWEED. Perennial from underground stocks; stems mostly simple, few, 6–25 dm. tall, glabrous below, commonly puberulent above, rather densely leafy; lvs. alternate, lanceolate, 60–200 mm. long,

10–40 mm. wide, glabrous to pubescent on abaxial midribs, nearly entire, sessile or nearly so; fls. many, in long terminal racemes with small almost linear bracts; pedicels 5–12 mm. long; fl.-tube not evident above the ovary; sepals lance-linear, 8–12 mm. long; petals lilac-purple, rose, rarely white, clawed, obovate, 8–18 mm. long; stamens 8, in a single series, often unequal, shorter than petals; fils. dilated below; style exceeding stamens; stigma-lobes slender and elongate; caps. 5–8 cm. long, canescent; seeds oblong, 1–1.4 mm. long, with long dingy coma; $n = 36$ (Mosquin, 1966).—In disturbed areas, such as burns, 5000–9500 ft.; Montane Coniferous F.; San Jacinto, San Bernardino and San Gabriel mts.; to s. Canada, Eurasia and Atlantic Coast. July–Sept.

3. **E. brevístylum** Barb. Perennial with well-formed compact turions, the dried scales of which persist at the base of the stem of the following season; stems erect, simple or subsimple, slender, 2–6 dm. tall, glabrous below, crisp-pubescent or ± glandular in infl.; lvs. ovate to lanceolate, denticulate, with rounded sessile base, mostly opposite, 2–4 cm. long, mostly shorter than internodes; fls. several, fl.-tube to ca. 1 mm. long; sepals 2–3 mm. long; petals purplish or paler, emarginate, 3–5 mm. long; fruiting pedicels 5–15 mm. long; caps. 4–6 cm. long; seeds ca. 1.5 mm. long, papillate, with whitish coma.— Wet places, 5500–9300 ft.; Montane Coniferous F.; San Jacinto and San Bernardino mts.; to Wash., Rocky Mts. June–Aug. A form with spreading white hairs on lvs. and lower stems and growing with the typical form has been named var. *ursinum* (Parish) Jeps. [*E. u.* Parish], and seems scarcely worth taxonomic recognition.

4. **E. ciliàtum** Raf. [*E. adenocaulon* var. *perplexans* Trel. *E. americanum* Hausskn.] Like *E. adenocaulon*, but more slender, mostly 1–3 dm. high, scarcely glandular, but with crisped hairs in the infl., mostly unbranched; lvs. thin, tapering at the base into narrow petioles 2–10 mm. long; fls. whitish, 3–4 mm. long.—Occasional in wet places e. of Sierra Nevada and at 7000 ft. in White Mts.; to Rocky Mts. and Atlantic Coast. July–Oct.

5. **E. exaltàtum** E. Drew. Perennial with large turions; stems 3–9 dm. tall, slender, ± pubescent, often freely branched above with very slender branches; lvs. lance-ovate, serrulate, subsessile, 5–12 cm. long, the uppermost much reduced; fls. near ends of glandular-pubescent branches; pedicels 5–10 mm. long in fr.; fl.-tube 2–3 mm. long, almost as broad; sepals suberect, 3–4 mm. long; petals pink to rose-purple, 5–10 mm. long, spreading; caps. 3–5 cm. long; seeds beaked, rugose, 1 mm. long, with white coma; $n = 18$ (Lewis et al., 1958).—Wet places, at 6000–7500 ft.; Montane Coniferous F.; San Bernardino Mts.; Sierra Nevada n.; to Ore., Mont. June–Aug.

6. **E. glabérrimum** Barb. Plate 63, Fig. E. Perennial with several stems from branching scaly tough rootstocks; stems simple or nearly so, slender, erect, from ± decumbent base, glabrous and glaucous, sometimes slightly glandular-puberulent above, often purplish, mostly 3–6 dm. tall; lvs. glabrous, glaucous, ascending, oblong-lanceolate, obtuse, entire or minutely denticulate, sessile, 2–5 cm. long, gradually reduced up the stem; fls. erect or ± drooping; fl.-tube 1–2 mm. long; sepals 1–2 mm. long; petals 4–7 mm. long, purplish to almost white; fruiting pedicels 1–2 cm. long; caps. slender, suberect, 4–7 cm. long; seeds papillate, ca. 1 mm. long, with whitish coma; $n = 18$ (Lewis et al., 1958).—Stream banks and wet places, 4000–9300 ft.; Montane Coniferous F.; San Jacinto Mts., San Bernardino Mts., e. San Gabriel Mts.; n. through Sierra Nevada to Wash., Nev. July–Aug. Growing with the above described plant but ranging farther n. and e. is one of questionable status, mostly 1–3 dm. high; lvs. ovate, 1.5–2.5 cm. long, more crowded; petals 2–5 mm. long; $n = 18$ (Lewis et al.). It is var. *fastigiatum* (Nutt.) Trel. [*E. platyphyllum* Rydb.]

7. **E. lactiflòrum** Hausskn. Perennial with subterranean and epigaeous scaly branches; stems slender, simple, lvs. delicate, pale, subentire or obscurely denticulate, elliptic to oblong-ovate, obtuse, 2–5 cm. long; fls. few; petals white or pink-tipped, ca. 3–4 mm. long; caps. slender, erect, linear, less than 1 mm. thick, 4–5 cm. long; seeds smooth, ca. 1 mm. long, beaked, with dingy coma; $2n = 36$ (Löve & Löve, 1956).—At 7000–9500 ft.; San Jacinto and San Bernardino mts.; to Alaska, Eurasia, e. U.S.

8. **E. oregonénse** Hausskn. Perennial with subfiliform stolons; stems simple, slender,

erect, 1–3 dm. high, glabrous except in the sparsely glandular-pubescent infl., often purplish above; lvs. oblong-linear to -ovate, suberect, entire to ± denticulate, obtuse, sessile, somewhat crowded on lower stem, remote above, 1–2.5 cm. long; fls. 1–few; sepals often purplish, 1–2 mm. long; petals cream to purplish, 4–7 mm. long, deeply emarginate; pedicels 1–3.5 cm. long in fr.; caps. erect, slender, 2–5 cm. long, often purplish; seeds smooth, scarcely 1 mm. long, with white coma.—Boggy places, 7000–9000 ft.; Montane Coniferous F.; San Jacinto and San Bernardino mts.; Sierra Nevada to B.C., Ida., Nev. July–Aug.

9. **E. paniculàtum** Nutt. ex T. & G. Erect annual with stem simple and shreddy below, paniculately branched above, 3–20 dm. tall, glabrous except for the tips of the infl. which are slightly glandular-puberulent; lvs. linear-lanceolate to linear, 2–5 cm. long, usually alternate, short-petioled, remotely denticulate, with thickened acute tip and teeth, quite early caducous and with fascicles of smaller lvs. in the axils; fls. in lax racemes on filiform branches of the panicle; bracts subulate; pedicels 5–15 mm. long, usually slightly glandular-puberulent; fl.-tube 2–3 mm. long, subglabrous; sepals 2–3 mm. long; petals pink to almost white, 3–6 mm. long; caps. 2–2.5 cm. long, 4-angled, linear-clavate, beaked, usually slightly glandular-puberulent; seeds obovoid, 2 mm. long, with tawny coma.—Open usually dry disturbed ground, below 7500 ft.; many Plant Communities, most of cismontane and montane Calif.; to B.C., S. Dak., New Mex. June–Sept. Variable; plants with pedicels densely glandular-pubescent have been named f. *adenocladon* Hausskn.; $n=12$ (Lewis et al., 1958); those with pedicels and caps. quite glabrous, f. *subulatum* Hausskn. A larger fld. plant, also of questionable worth as a taxon is var. *laevicaule* (Rydb.) Munz. [*E. l.* Rydb.] Pedicels glabrous or nearly so; fl.-tube 4–6 mm. long; sepals ca. 3 mm. long; petals 5–8 mm. long, rose to pink; caps. subglabrous.—Occasional with the typical plant.

6. Gaúra L.

Annual to perennial caulescent herbs with alternate lvs. Fls. white or pink, rarely yellow, in terminal spikes or racemes. Fl.-tube narrow, short; sepals 4, deciduous. Petals 4, clawed. Stamens 8, often with scalelike appendage at base of each fil. Ovary 4-loculed, usually with 1 ovule in each locule. Stigma 4-lobed, with cuplike border near base. Caps. nutlike, obovoid, nearly or quite indehiscent, 1–4-seeded. Ca. 18 spp. temp. N. A., Argentina. (Greek, *guaros*, proud, some spp. being showy.)

Caps. sessile or nearly so, not narrowed into pedicellike base; petals 1.5–2 mm. long
 3. *G. parviflora*
Caps. narrowed into a distinct thick or slender stipelike base; petals 5–10 mm. long.
  Fl.-tube 2.5–4 mm. long; fils. 7–10 mm. long; fr. 5–9 mm. long, 1–1.5 mm. wide . 4. *G. sinuata*
  Fl.-tube 5–12 mm. long; fils. 3–6 mm. long; fr. 5–11 mm. long, 2–3 mm. wide.
    Sepals 10–13 mm. long; petals 7–8 mm. long; fl.-bracts lance-ovate, caducous; main cauline lvs. 5–25 mm. wide .... 2. *G. odorata*
    Sepals 5–9 mm. long; petals 3–6 mm. long; fl.-bracts lanceolate to linear, mostly persistent; main cauline lvs. mostly narrower .... 1. *G. coccinea*

1. **G. coccínea** (Nutt.) Pursh. [*Malva c.* Nutt. ex Fraser.] Plate 63, Fig. F. Perennial, the stems several to many, branched so as to form a bushy plant 1–5 dm. tall, strigose-canescent; lvs. many, sessile, oblong-lanceolate to linear, entire to repand-dentate, acute to obtuse, 1–3 cm. long; fl.-bracts 3–6 mm. long; spikes short, 1–2 dm. long in fr.; fl.-tube 6–10 mm. long; sepals 6–9 mm. long; petals 5–8 mm. long; frs. canescent, short-obovoid, 4-angled in upper part, 5–7 mm. long, with stout terete base; $n=7$ (Johansen, 1929).—Dry slopes, mostly near limestone, 3000–5000 ft.; Joshua Tree Wd., Pinyon-Juniper Wd.; mts. of e. Mojave Desert; to S. Dak., Tex. April–June. Natur. as a weed in Los Angeles and Orange cos.

2. **G. odoràta** Ses. ex Lag. Apparently biennial or winter annual, mostly branched at base, the stems ascending, 2–5 dm. long, slender, grayish-pubescent; lower lvs. oblanceolate, sinuate-dentate, 2–6 cm. long, upper narrower, shorter; bracts lance-ovate, 3–6 mm. long; fl.-tube 7–8 mm. long; petals ca. 8 mm. long; caps. glabrous to strigulose,

8–11 mm. long, the lower third terete, gradually enlarged upward into an ovoid-pyramidal 4-angled part, each face 2.5 mm. wide with median nerve; $n = 7$, 14 (Lewis et al., 1958).—Weed from San Diego to Santa Barbara cos. and n.; native from Tex. to cent. Mex. May–Sept.

3. **G. parviflòra** Dougl. ex Hook. Biennial or winter annual, mostly 5–20 dm. tall, erect, rather simple below, freely and widely branched above, glandular-pubescent; lvs. simple, those of basal rosette broadly oblanceolate, 5–15 cm. long, the cauline reduced upward to sessile lanceolate lvs. 3–10 cm. long; bracts lance-linear; spikes slender, nodding at tips, becoming 1–3 dm. long; fl.-tube 1.5–2.5 mm. long; sepals ca. as long; petals 1.5–2 mm. long; caps. sessile, 6–10 mm. long, ca. 2 mm. thick, subfusiform, 4-nerved, puberulent; seeds 1–2, brown.—Occasional weed as in Orange and Santa Barbara cos.; native from Mo. and Utah to Mex., Argentine. June–Aug. Variable as to pubescence, some plants with spreading hairs on the fr. have been called var. *lachnocarpa* Weatherby, but seem to appear in any region with the normal form and seem unworthy of taxonomic recognition.

4. **G. sinuàta** Nutt. ex Ser. in DC. Perennial, 3–8 dm. tall, simple or branched above base, subglabrous; basal lvs. oblanceolate to oblong-lanceolate, 3–8 cm. long, sinuate-dentate, short-petioled, cauline crowded, lanceolate to spatulate, 1–5 cm. long; fl.-bracts 1–3 mm. long, caducous; fl.-tube 2.5–3 mm. long, subglabrous to sparsely strigulose; sepals strigulose, 7–10 mm. long; petals 8–10 mm. long; caps. subglabrous, the body fusiform, obtusely 4-angled, 5–9 mm. long, 1–1.5 mm. wide, gradually tapered into the thick pedicellike base 2–5 mm. long; $n = 14$ (Lewis et al., 1958).—Occasional weed in s. and cent. Calif.; native from Okla. to n. Mex. June–Sept.

7. *Gayophỳtum* Juss.

Slender-stemmed caulescent annuals, simple or more usually branched, often with the older epidermis exfoliating and often with a tinge of red on stems and calyces. Lvs. alternate, entire, linear or lanceolate and largely subsessile, or the lower opposite, usually narrowly oblanceolate and petioled. Uppermost lvs. usually passing into narrow bracts of the leafy racemes. Fls. small. Fl.-tube not prolonged beyond the ovary. Sepals 4, usually reflexed in anthesis. Petals 4, small, rhombic-spatulate to -obovate, white, frequently drying pink or red. Stamens 8, the alternate set much reduced and usually sterile except in the larger-fld. spp.; anthers versatile, attached near the base, sporogenous tissue continuous. Stigma capitate, hemispherical to subglobose. Caps. ± pedicellate, 2-loculed, 4-valved, linear to cylindric-clavate. Seeds few to many, in one row in each locule, narrowly obovoid, not comose. A small genus of w. U.S. and s. S. Am. (C. *Gay*, author of Flora of Chile, and Greek word for plant.)

(Lewis, H. and J. Szweykowski, The genus Gayophytum. Brittonia 16: 343–391. 1964.)
1. Plants flowering from near the base, the first fls. at 1–4 nodes above the cotyledons.
    2. The plants branched only in their lower portion, secondary branches few or none, the branching not dichotomous.
        3. Caps. with dorsal and ventral valves remaining attached to the septum at maturity, the lateral valves free; seeds usually obliquely placed in the caps. and often ca. 15–18 in a locule .... 4. *G. humile*
        3. Caps. with all 4 valves free from the septum at maturity, the seeds usually vertically placed in the caps. and 5–10 in a locule .... 6. *G. racemosum*
    2. The plants branched throughout, secondary branches evident.
        4. Branches mostly separated by 2–8 nodes; seeds in even rows, not staggered, those in one row not alternate with those in the other; caps. not torulose .... 1. *G. decipiens*
        4. Branches mostly at every node or every other node; seeds often plainly staggered, those in one row alternate with those in the other; caps. often plainly torulose
                                                                2. *G. diffusum parviflorum*
1. Plants not flowering near the base, but only several fo many nodes above the cotyledons.
    5. Caps. irregularly lumpy by failure of part of the ovules to develop; a high percentage of pollen grains empty .... 3. *G. heterozygum*

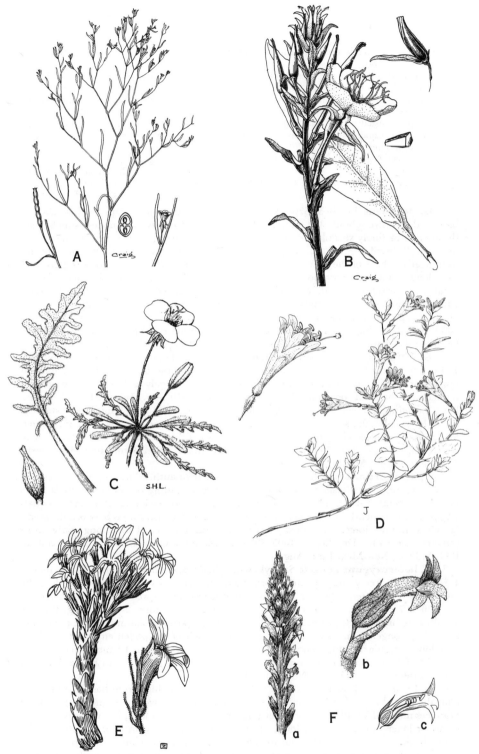

Plate 64. Fig. A, *Gayophytum diffusum* ssp. *parviflorum*. Fig. B, *Oenothera hookeri* T. & G. Fig. C, *Oenothera primiveris*. Fig. D, *Zauschneria californica* ssp. *mexicana*. Fig. E, *Orobranche californica* ssp. *feudgei*. Fig. F, *Orobranche cooperi*.

5. Caps. ± regular in outline, entire or torulose, all of the ovules developing; pollen grains almost all good.
  6. Petals 0.5–1 mm. long; caps. 2–5 mm. long, mostly shorter than the pedicels. Last Chance and White mts. north .... 7. *G. ramosissimum*
  6. Petals mostly 1.5–7 mm. long; caps. 3–12 mm. long, equaling or longer than the pedicels.
    7. Seeds 1–5 (–6) in a caps., each ca. 1.5 mm. long; caps. ca. twice as long as pedicels; petals usually 1.5–2 mm. long .... 5. *G. oligospermum*
    7. Seeds usually 6–10 or more in a caps., each ca. 1–1.2 mm. long; petals mostly 1.5–7 mm. long.
      8. Petals generally 1.5–2.5 mm. long; style not surpassing stamens.
        2. *G. diffusum parviflorum*
      8. Petals 3–5 mm. long; style surpassing stamens; sepals 2–3 mm. long
        2. *G. diffusum diffusum*

1. **G. decípiens** Lewis & Szweykowski. Plants 1–3 dm. tall, simple or branched at base and ± throughout, the branches mostly separated by 2–8 nodes, subglabrous throughout or finely strigulose especially above or with some short spreading hairs; lvs. linear, well distributed; the lower petioled, 2–3 cm. long, the uppermost bracteate; fls. from near the base of the plant or beginning higher; pedicels 0–3 cm. long; sepals ca. 1 mm. long, commonly strigulose; petals slightly shorter; pistil slightly shorter than the petals; caps. erect or curved, often subclavate, not strongly contracted between the seeds, with same pubescence as sepals; seeds evenly spaced, not staggered, usually more than 10 in a cap., vertical, ca. 1 mm. long, glabrous or pubescent; $n=7$.—Dry places at 7000–10,000 ft.; Sagebrush Scrub, Montane Coniferous F.; San Bernardino and San Gabriel mts., Panamint and Inyo mts.; to Wash., Rocky Mts. June–Aug.

2. **G. diffùsum** T. & G. ssp. **diffùsum.** Erect, simple or ± successively branched, 1–5 or more dm. tall, subglabrous below, usually strigulose or with spreading hairs in upper parts; lower lvs. lance-linear to linear, petioled, largely 2–4 cm. long, gradually reduced up the stem, becoming linear bracts above 3–10 mm. long; pedicels 3–8 mm. long, ascending to erect; sepals 2–3 mm. long; petals 3–4 (–5) mm. long; longer stamens ca. as long as petals; style somewhat surpassing longer stamens; caps. 5–12 mm. long, longer than pedicels, glabrous or pubescent; seeds 1–1.2 mm. long; $n=14$.—Occasional on dry slopes at ca. 6500–7000 ft.; Montane Coniferous F.; San Bernardino Mts.; Sierra Nevada to Wash., Mont., Nev. June–Aug.

Ssp. **parviflòrum** Lewis & Szweykowski. [*G. nuttallii* auth.] Plate 64, Fig. A. Pedicels 1.5–5 mm. long, mostly shorter than caps., erect to spreading or deflexed; sepals 1–1.5 mm. long; petals 1–2.5 mm. long; longer stamens from somewhat shorter than to equalling petals; style nearly as long as petals; caps. often with constrictions between the seeds, largely 4–12 mm. long; seeds mostly 3–10, often staggered, ca. 1 mm. long, glabrous or pubescent; $n=14$.—Dry slopes, 4000–8000 ft.; Montane Coniferous F.; n. L. Calif. to B.C., S. Dak., New Mex. June–Aug.

3. **G. heterozỳgum** Lewis & Szweykowski. [*G. diffusum villosum* Munz.] Plants mostly 1.5–5 dm. tall, freely and repeatedly branched, the cent. axis more prominent than the lateral branches, often ± zigzag, subglabrous to strigulose or spreading-pubescent; lower lvs. to 6 or 7 cm. long, early deciduous, cauline largely subsessile, 2–4 cm. long; fls. only on ultimate branchlets; pedicels ca. 2–7 mm. long; sepals 1.5–2.5 mm. long; petals 2–3 mm. long; longer stamens ca. as long as petals, much of the pollen empty; caps. 5–10 mm. long, irregularly lumpy since part of the ovules abort; seeds 1–1.5 mm. long; $n=7$.—At 4000–7000 ft.; Chaparral, Montane Coniferous F.; San Jacinto, San Bernardino and San Gabriel mts. n.; to Wash., Nev. June–Aug.

4. **G. hùmile** A. Juss. Low, mostly 5–15 cm. long, branching from base and in lower half, the branches relatively simple, the internodes of the lower stem and of the lateral branches much alike; plant glabrous or minutely glandular in infl.; lvs. well distributed, linear or lance-linear, 1–3 cm. long, short-petioled; flowers from near the base of the plant, beginning at 1–4 nodes from the cotyledons; sepals ca. 1 mm. long; petals ca. 1 mm. long; caps. suberect, not torulose, flattened, 6–12 mm. long; seeds mostly oblique in the locules, usually more than 10 in a locule, 0.5–0.7 mm. long; $n=7$. Mostly in damp or

drying places, largely coniferous forest; San Bernardino Mts.; to Wash., Ida., Mont.; Chile. July–Aug.

5. **G. oligospérmum** Lewis & Szweykowski. Plants 2–8 dm. tall, repeatedly branched above, erect, largely glabrous, reddish; lower lvs. early deciduous, the cauline linear, short-pedicelled to subsessile, 2–5 cm. long, passing upward into linear bracts 5–12 mm. long; pedicels 3–6 mm. long; sepals 1–1.5 mm. long, often strigulose; petals 1.5–2 mm. long; style slightly surpassing stamens; caps. ca. as long as pedicels, mostly strigulose, constricted between the seeds; seeds 1–5 in a caps., those in one locule alternating with those in the other; $n=7$.—Dry places, 4000–8000 ft.; largely Montane Coniferous F.; Cuyamaca Mts. to San Bernardino and San Gabriel mts.

6. **G. racemòsum** T. & G. [*G. caesium* T. & G. *G. helleri* Rydb.] Low, 5–25 cm. tall, subsimple or branched (mostly in lower half), the ultimate branchlets rather simple, subglabrous to pubescent; lvs. linear to linear-oblanceolate, 1–3 cm. long, sessile or short-petioled, only gradually reduced up the stem; fls. beginning near the base of the plant; pedicels almost lacking to 2 mm. long; sepals 0.5–1 mm. long; petals ca. 1 mm. long; caps. terete, narrowly linear, 6–14 mm. long, not very torulose; seeds usually more than 10 in each caps., suberect, glabrous or pubescent, 1 mm. long; $n=14$.—In places that have been moist; Montane Coniferous F.; Ventura Co. to Wash., Alberta, S. Dak. Summer.

7. **G. ramosíssimum** T. & G. Plant 1–4 dm. high, diffusely branched, mostly glabrous; lvs. lance-linear, 1–3 cm. long, short-petioled; pedicels capillary, mostly 3–5 mm. long, frequently spreading-deflexed; sepals 0.5–0.7 mm. long; petals 0.5–1 mm. long; caps. plump, 2–5 mm. long, terete, not very torulose; seeds crowded, overlapping, usually with 2 well formed rows; $n=7$.—At 6500–10,000 ft.; Inyo Co. (Last Chance Mts., White Mts.); to Wash., Mont., Colo., Ariz. Summer.

8. *Heterogáura* Rothr.

Caulescent annual with alternate lvs. Fls. pink, in terminal spicate racemes. Fl.-tube short, obconic. Sepals 4, deciduous. Petals 4, clawed. Stamens 8, erect, the 4 epipetalous sterile; fils. not appendaged. Stigma discoid, entire, without a basal cuplike border. Ovary 4-loculed, with 1 ovule in each locule. Fr. 2–4-celled, 1–2-seeded, indehiscent. One sp. (Greek, *heter*, different, and *Gaura*.)

1. **H. heterándra** (Torr.) Cov. [*Gaura h.* Torr. *H. californica* Rothr.] Erect, stem simple or paniculately few-branched, 1–4 dm. tall, minutely puberulent throughout; lvs. oblong-ovate to lanceolate, entire to remotely and shallowly denticulate, the blades 2–5 cm. long; ca. half as wide, on petioles 5–10 mm. long; pedicels 1–1.5 mm. long; fl.-tube 2–3 mm. long; sepals ca. the same; petals pink, aging lavender, spatulate, 3–5 mm. long; alternate stamens fertile, 2 mm. long, opposite sterile, 1 mm. long; caps. ridged, often triquetrous, 3 mm. long; seeds slender, 2 mm. long; $n=9$ (Lewis et al., 1958).—Dry shaded places, 2000–5000 ft.; Chaparral, Yellow Pine F., Foothill Wd.; Los Angeles and San Bernardino cos. to n. Calif. May–June.

9. *Ludwígia* L.

Our spp. herbaceous, mostly of wet places, sometimes floating in open water, sometimes with basal vegetative shoots creeping and rooting at the nodes. Underwater parts often swollen and spongy. Lvs. alternate or opposite, mostly simple, membranaceous or rarely coriaceous. Stipules present, at least in upper part of plant. Fls. yellow or white, solitary and axillary, or in terminal spikes or heads. Fl.-tube not prolonged beyond the ovary, usually with 2 bracteoles at the base of the ovary or summit of the pedicel. Fls. diurnal, regular, 3–7-merous, but mostly 4–5-merous. Sepals persistent after anthesis. Petals 0–7, caducous. Stamens in 1–2 series, each series usually as many as the sepals; anthers usually versatile, basifixed or nearly so in very small fls. Ovary with as many locules as sepals; style simple, ± produced above the disc; stigma capitate or hemispheric, often slightly lobed. Ovary cylindric to obconic, many-ovuled, the ovules pluriseriate to uniseriate in each locule. Pluriseriate seeds naked and with evident raphe, the uniseriate surrounded by an endocarp. Caps. dehiscing by a terminal pore or by flaps separating from the valvelike top or more irregularly.

Stamens in 2 series, mostly 8 or 10 in number; petals 10–20 mm. long.
    Flowering stems usually floating or creeping; lvs. oblong, 1–10 cm. long; bracteoles at base of ovary deltoid; caps. mostly 2–3 mm. thick .... 2. *L. peploides*
    Flowering stems usually erect; lvs. ± lanceolate, mostly 5–10 cm. long; bracteoles lanceolate; caps. 3–4 mm. thick .... 4. *L. uruguayensis*
Stamens in 1 series, 4–5 in number; petals none or small and quickly shed.
    Ovary with 4 evident longitudinal bands; basal bracteoles from not evident to ca. 1 mm. long; petals none .... 1. *L. palustris*
    Ovary lacking green bands; bracteoles above the base and 1–5 mm. long; petals present, but easily shed .... 3. *L. repens*

1. **L. palústris** (L.) Ell. var. **pacífica** Fern. & Griscom. Glabrous, with floating or creeping stems bearing erect branches 1–3 dm. tall; lvs. mostly short-petiolate, the blades mostly 1–3 cm. long, usually not half as wide; sepals acuminate, 1–2 mm. long; petals none; caps. with 4 evident longitudinal green bands, 2–2.8 mm. thick, ± 4-angled; seeds yellowish, 0.5 mm. long.—Ponds and muddy places below 3000 ft.; several Plant Communities; Coast Ranges from San Diego and Ventura cos. n.; to B.C. June–Sept.

2. **L. peploìdes** (H.B.K.) Raven. [*Jussiaea p.* HBK. *J. repens* var. *californica* Wats.] Mostly glabrous perennial herb with oblong, obtuse to acute, subentire, plainly and evenly pinnate-veined, the blades 1–4 cm. long, on petioles 1–2.5 cm. long, or longer especially on floating lvs.; pedicels 1–4 cm. long in fr.; fls. mostly 5-merous, pubescent about the base of the stamens and style; sepals lanceolate, 4–7 mm. long; petals yellow, obovate, pinnately veined, 10–14 mm. long; stamens ca. ⅓ as long; pistil as long as stamens; caps. hard, quite cylindric, ca. 2 cm. long, at length reflexed; seeds in 1 row in each locule, included in the endocarp, somewhat triangular in cross-section, 1–1.5 mm. long.—Pools and slow streams, below 2000 ft.; s. Calif. to Ore., S. Am. May–Oct.

3. **L. rèpens** Forster var. **stipitàta** (Fern. & Griscom) Munz. [*L. natans* Ell. var. *s.* Fern. & Gris.] Habit of *L. palustris*; lf.-blades 1–4.5 cm. long, rhombic-ovate, rather long-petiolate; fls. short-pedicelled; sepals 4, triangular-acuminate; petals shorter than sepals; pedicel of caps. 2–4 mm. long; caps. 6–8 mm. long, 3–3.5 mm. broad, light brown, ± 4-angled, without longitudinal green bands; seeds ca. 0.5 mm. long.—Ponds and wet places below 5000 ft.; San Bernardino, Victorville, Lake Arrowhead, Lone Pine Canyon near Cajon Pass. July–Sept.

4. **L. uruguayénsis** (Camb.) Hara. [*Jussiaea u.* Camb. in St. Hil. *J. grandiflora* Michx.] Perennial herb from creeping rhizome, rooting freely at the nodes; stems slender and floating, or erect to ascending and ± soft-hirsute, often freely branched, creeping and forming mats (depending on whether in deeper water, emergent from shallow water, or growing on sand or mud banks); lvs. on erect flowering branches linear-lanceolate to oblanceolate, short-petioled; fls. on pedicels 1–2 cm. long; bracteoles lanceolate, 5–13 mm. long; sepals 5 (6), lanceolate, 6–13 mm. long; petals yellow, 12–20 mm. long; stamens unequal; caps. cylindric, usually hairy, 13–25 mm. long, 3–4 mm. thick; seeds enclosed in the hard endocarp, 1.5 mm. long.—Introd. in wet places, San Diego, etc.; se. U.S. to S. Am. June–Aug.

10. *Oenothèra* L. EVENING-PRIMROSE

Annual to perennial, caulescent or acaulescent, sometimes suffruticose; lvs. alternate or basal. Fls. in ours opening near sunset, white to yellow or rose, often aging reddish to purplish. Fl.-tube prolonged beyond the ovary, deciduous after anthesis. Sepals 4, reflexed in anthesis. Petals 4. Stamens 8, equal, or if unequal, the epipetalous shorter; anthers versatile, the sporogenous tissue continuous. Stigma with 4 linear lobes, the lobes receptive all around. Caps. many-seeded, sometimes short and ± indehiscent. Seeds many, naked. A large genus of the New World, mostly of temp. regions. (Greek, meaning wine-scenting, a name given to some unknown plant once used for that purpose.)

1. Caps. terete or round-angled (winged above in *Oe. speciosa*); stems well developed and bearing lvs. and fls.
   2. Fls. yellow, the buds erect.
      3. The caps. angled, tapering from base toward apex; seeds prismatic-angled.
         4. Fl.-tube 3–5 cm. long. Largely cismontane and montane .... 5. *Oe. hookeri*
         4. Fl.-tube 8–12 cm. long. E. Mojave Desert .... 7. *Oe. longissima*
      3. The caps. terete, not narrowed in upper part; seeds not angled.
         5. Plants erect, leafy, biennial or perennial, 2–6 dm. tall; lvs. denticulate; sepals 12–20 mm. long .... 11. *Oe. stricta*
         5. Plants mostly annual, semiprostrate, 1–7 dm. tall; lvs. sinuate-pinnatifid to subentire; sepals 6–12 mm. long .... 6. *Oe. laciniata*
   2. Fls. white to rose, the buds often nodding.
      6. Caps. sterile and slender in lower part, thicker, fertile and ± winged in upper; seeds in more than 2 rows in each locule; fl.-tube 0.4–2 cm. long.
         7. Petals white to pink, 2.5–4 cm. long; plants with running underground rootstocks; buds nodding .... 10. *Oe. speciosa*
         7. Petals rose to red-violet, 0.5–1 cm. long; plants from a ± woody caudex; buds erect .... 9. *Oe. rosea*
      6. Caps. cylindric, sessile, not sterile in lower part; seeds in 1 row in a locule; fl.-tube 2–4 cm. long.
         8. Plants annual or surviving longer from a deep taproot; basal lvs. ± rhombic, the blades 2–10 cm. long; caps. usually woody; plant with exfoliating epidermis .... 4. *Oe. deltoides*
         8. Plants perennial, largely from running underground rootstocks; basal lvs. tending to be smaller and more narrow; caps. usually not woody.
            9. Plants greenish and subglabrous to strigose; caps. 2.5 mm. thick at base. S. Calif. mostly west of the deserts .... 3. *Oe. californica*
            9. Plants canescent to hoary, usually with some spreading hairs especially in the upper parts. Desert plants .... 1. *Oe. avita*
1. Caps. crested or tubercled; plants acaulescent or nearly so, cespitose.
   10. Fls. white .... 2. *Oe. caespitosa*
   10. Fls. yellow .... 8. *Oe. primiveris*

1. **Oe. ávita** (W. Klein) W. Klein ssp. **ávita**. [*Oe. californica* ssp. *a.* W. Klein.] Resembling *Oe. californica* but the lvs. oblanceolate or spatulate, the cauline more deeply incised or lobed; buds 17–22 mm. long at maturity; petals 25–35 mm. long; caps. narrower, 22–50 mm. long, 2 mm. in diam.; $n=7$ (Klein, 1962).—Dry plains below 5700 ft.; Joshua Tree Wd., Pinyon-Juniper Wd., etc.; deserts from Mojave Desert to Utah and Mono Co. April–June.

Ssp. **eurekénsis** (Munz & Roos) W. Klein. [*Oe. deltoides eurekensis* Munz. & Roos.] With deep-seated fleshy perennial rootstocks, bushy, 3–6 dm. tall, densely white-strigose and ± spreading-villous; lvs. crowded, deltoid-ovate, 1–3.5 cm. long; petals 1.5–2.5 cm. long; old fruiting shoots becoming buried in sand and resuming growth at tips.— Sand dunes, Creosote Bush Scrub, s. end Eureka V., Inyo Co.

2. **Oe. caespitòsa** Nutt. var. **marginàta** (Nutt.) Munz. [*Oe. m.* Nutt. ex H. & A. *Pachylophis m.* Rydb.] Cespitose perennial with thick caudex, acaulescent or with short stem, villous-hirsute throughout; lvs. linear-lanceolate, sinuate-pinnatifid, the blades 3–10 cm. long, on winged petioles almost as long; fls. fragrant, vespertine, white, aging pink; fl.-tube 5–8 cm. long; sepals 2.5–3.5 cm. long; petals 2.5–4 cm. long; stamens subequal, glabrous; caps. 3–4 cm. long, pedicelled, linear-cylindric, scarcely ridged, with low tubercles; seeds dark brown, ca. 3 mm. long, obovoid, minutely cellular-roughened, conspicuously furrowed along the raphe; $n=7$ (Kurabayashi, 1962).—Occasional on dry stony slopes, 3000–10,000 ft.; mostly Pinyon-Juniper Wd., Shadscale Scrub, Yellow Pine F.; desert slopes from Santa Rosa Mts., Riverside Co. to White Mts., Inyo Co. and to Utah, e. Wash. April–Aug.

Subsp. **crinìta** (Rydb.) Munz. [*Pachylophis c.* Rydb.] Caespitose from a woody caudex;

stems 3–8 cm. long; lvs. densely hirsute or pilose, crowded, the lanceolate blades obtuse, sinuate, 10–15 mm. long, 3–5 mm. wide; fl.-tube 2–4 cm. long; caps. 1–1.4 cm. long, 3–5 mm. thick.—Dry rock-crevices, 7500–10,000 ft.; Pinyon-Juniper Wd.; Bristle-cone Pine F.; Panamint Mts., Clark Mt.; to Utah. June–Sept.

3. **Oe. califórnica** Wats. [*Anogra c.* Small.] Perennial from underground rootstocks, with mostly branched stems 1–5 dm. tall, frequently decumbent, ashy with short appressed hairs; epidermis exfoliating; lvs. variable, the lower blades oblanceolate to spatulate in outline, cauline oblong to lanceolate, all subentire to deeply and regularly sinuate-dentate, 1–6 cm. long, sessile or short-petioled; fls. several; buds nodding; fl.-tube slender, 2–4 cm. long, strigulose; sepals 1.5–2 cm. long, free tips wanting or very short; petals orbicular-obovate, 2–3 cm. long, white, aging pink; stamens subequal; stigma lobes 4–6 mm. long; caps. terete, mostly divaricate, somewhat curved upward, 2–5 cm. long, ca. 3 mm. thick at base; seeds plump, obovoid, 1.5 mm. long, brown with dark spots; $n=14$ (Klein, 1962).—Mostly sandy dry places, below 6000 ft.; Coastal Sage Scrub, Chaparral, S. Oak Wd., Ventura Co. to n. L. Calif. April–June.

4. **Oe. deltoìdes** Torr. and Frém. [*Oe. d.* var. *cineracea* (Jeps.) Munz.] Coarse spring or winter annual, simple or more often with cent. erect stem 0.5–3 dm. high and few to several decumbent branches naked at the base and 1–10 dm. long; stems pale, with exfoliating epidermis, glabrous below, spreading-pubescent or strigose above; lower lvs. in loose rosette, the blades rhombic-obovate to -lanceolate or oblanceolate, subentire to denticulate, 2–8 cm. long, with petioles ca. as long; cauline lvs. gradually somewhat reduced, the upper sessile and dentate; fls. solitary in upper axils, the buds nodding; fl.-tube slender, 2–4 cm. long; sepals 1.5–3.5 cm. long; petals white, aging pink, 2–4 cm. long; stigma-lobes 3–6 mm. long; caps. spreading to ± reflexed, woody, with exfoliating epidermis, prismatic-cylindric, 4–6 cm. long, 2–3 mm. thick at base; seeds narrowly obovoid, 1.5–2 mm. long, light brown, usually with purple spots; $n=7$ (Lewis et al., 1958).—Common in sandy places below 3500 ft.; Creosote Bush Scrub, Joshua Tree Wd.; Mojave and Colo. deserts from Adelanto region e. and s.; to Ariz., L. Calif.

5. **Oe. hoókeri** T. & G. Plate 64, Fig. B. ssp. **angustifòlia** (Gates) Munz. [*Oe. h.* var. *a.* Gates.] Biennial, ± branched below, mostly 3–10 dm. high, erect to ascending, simple-stemmed, muricate-hirsute, strigose and glandular-pubescent (in upper parts); fl.-tube mostly red, 2.5–4.5 cm. long; sepals mostly red, 2.5–3.5 cm. long, the tips 3–5 mm. long; petals 2–4 cm. long; caps. mostly 2.5–4.4 cm. long, ca. 5 mm. thick at base; seeds 1–1.5 mm. long; $n=7$ (Cleland in Munz, 1949).—Moist places, 5000–9000 ft.; Montane Coniferous F.; San Jacinto Mts. n. through Sierra Nevada to Ore., Utah, Colo. Summer.

Ssp. **grísea** (Bartlett) Munz. [*Oe. venusta* var. *g.* Bartlett.] Plants 1–1.5 m. tall, freely branched, appressed-pubescent with fine white and occasional longer hairs; stems usually reddish; fl.-tube 2.5–4.2 cm. long; sepals 2–4.5 cm. long, not papillose or glandular, greenish, the tips 3–6 mm. long; petals 2.5–4.5 cm. long; $n=7$ (Cleland in Munz, 1949). —Moist places below 4000 ft.; Coastal Sage Scrub, Chaparral; Ventura to San Bernardino and San Diego cos.; L. Calif.

Ssp. **venústa** (Bartlett) Munz. [*Oe. v.* Bartlett.] Mostly 1.5–2.5 m. tall, freely branched, gray-pubescent with long hairs with basal red papillae, also finely pubescent or strigose, upper parts also glandular-pubescent; stems mostly reddish; fl.-tube green, 4.5 cm. long; sepals 3–4.5 cm. long, the slender tips 3–5 mm. long; petals 3–4.5 cm. long; $n=7$ (Cleland in Munz, 1949).—Moist places below 5000 ft.; Chaparral, Coastal Sage Scrub, etc.; Sierra Nevada foothills s., cismontane s. Calif.; to L. Calif.

6. **Oe. laciniàta** Hill. [*Raimannia l.* Rose. *Oe. sinuata* L.] Annual or perennial, simple and erect or more commonly branched, semiprostrate to ascending, 1–6 dm. high, strigose and villous or hirsute; lvs. siniate-pinnatifid to subentire, lance-oblong to oblanceolate, 2–10 cm. long, the lower petioled, the upper sessile or nearly so; fls. solitary in upper axils; fl.-tube 1.5–3.5 cm. long, with free tips 1–2 mm. long; petals yellowish, 5–18 mm. long; caps. cylindrical, ± arcuate, 1–3.5 cm. long; seeds brownish, evenly pitted, ca. 1 mm. long; $n=7$.—Occasionally natur. in waste places; native of se. U.S. May–July.

7. **Oe. longíssima** Rydb. ssp. **clùtei** (Nels.) Munz. [*Oe. clutei* Nels.] Biennial to

short-lived perennial, simple to branched, erect, 1–3 m. tall, ± hirsute, especially above, somewhat muricate on stems, hair mostly appressed, upper parts also glandular-pubescent; lvs. of rosette oblanceolate, the blades 1–2 dm. long, 1.5–3 cm. wide, with winged petioles; cauline lvs. linear-lanceolate, plane, stiffly spreading-ascending, gradually reduced upward to sessile lanceolate bracts soon exceeded by buds; fl.-tube 8–12 cm. long, ± reddish; sepals 3.5–4.5 cm. long, the tips 3–5 mm. long; petals obovate, ca. 4 cm. long, yellow; anthers 14–18 mm. long; caps. subquadrangular, 3.5–4.5 cm. long, 4.5–5.5 mm. thick; seeds 1–1.5 mm. long; $n=7$ (Cleland in Munz, 1949).—Moist places, 3400–5500 ft.; Creosote Bush Scrub to Pinyon-Juniper Wd.; Providence and New York mts., e. San Bernardino Co.; to Utah, Ariz. July–Sept.

8. **Oe. primivèris** Gray ssp. **primivèris**. [*Lavauxia p.* Small.] Plate 64, Fig. C. Annual or winter annual with long taproot system, acaulescent or nearly so, villous or pilose-pubescent throughout; lf.-blades oblanceolate in outline, 1–12 cm. long, 0.3–3 cm. wide, mostly deeply and rather regularly deeply pinnatifid into lanceolate to ovate lobes which are in turn toothed or lobed, blades gradually narrowed into petioles of ca. same or less length; fl.-tube 3–5 cm. long, villous; sepals 10–20 mm. long; petals yellow, aging orange-red, 10–22 mm. long; anthers mostly 8–10 mm. long; caps. 17–30 mm. long, square in cross section, with heavy rib down middle of each face, reticulate, not winged or reticulate, 6–8 mm. thick at base, gradually tapering to an attenuate apex; seeds in 2 rows in each locule, somewhat roughened-tuberculate, 2.5–3 mm. long, with narrow raphal groove.—Dry plains and slopes, below 5000 ft.; Creosote Bush Scrub, Joshua Tree Wd., Pinyon-Juniper Wd.; deserts from e. San Bernardino Co. to Utah, Tex., Son. March–May.

Ssp. **bufònis** (Jones) Munz. [*Oe. bufonis* Jones.] Acaulescent or nearly so; fl.-tube 3–5 cm. long; sepals mostly 20–25 mm. long; petals largely 25–35 (–40) mm. long; caps. 25–40 (–50) mm. long.—Mostly at 1600–5000 ft.; Creosote Bush Scrub to Pinyon-Juniper Wd.; Inyo Co. to Riverside Co.; w. Nev.

Ssp. **cauléscens** (Munz) Munz. Caulescent, the stems 1–4 dm. long, erect or ascending, ± red purple, strongly muricate with red papillae at base of bristly hairs of stems and frs.; fl.-tube 4–6 cm. long; sepals 20–28 mm. long, densely hirsute; petals mostly 3–4 cm. long; caps. 2–3.5 (–4) cm. long, muricate-bristly.—From below 900 ft.; Creosote Bush Scrub; Imperial Co.; sw. Ariz.

9. **Oe. ròsea** L'Herit. ex Ait. [*Hartmannia r.* G. Don.] Perennial, blooming the first year, from a somewhat woody caudex, freely branched, to a m. or more tall, ± strigulose throughout; lvs. not crowded, the lower oblanceolate or wider, subentire to pinnatifid, 2–5 cm. long, the cauline gradually reduced up the stem; fls. in simple, slender, erect, bracteate racemes; fl.-tube 4–8 mm. long; sepals 5–8 mm. long; petals rose to red-violet, 5–10 mm. long; caps. proper obovoid, 8–10 mm. long, 3–4 mm. thick, with 4 wings to ca. 1 mm. wide, caps. passing at base into a hollow ribbed pedicel 5–20 mm. long; $n=7$.—Tex. to S. Am.; occasionally natur. in Calif.

10. **Oe. speciòsa** Nutt. [*Hartmannia s.* Small.] Perennial from a running rootstock, the stems 1–5 dm. high, erect to almost prostrate, mostly strigose, ± villous; lvs. oblanceolate to obovate, the cauline 3–8 cm. long, oblong, the lower deeply pinnatifid, the upper sinuate-dentate to subentire; fls. in uppermost axils; buds nodding; fl.-tube 1–2 cm. long; sepals lance-linear, acuminate, 1.5–3 cm. long; petals white to pinkish, 2.5–4 cm. long; caps. stout, 10–15 mm. long, the basal part cylindrical, 1.5–2 mm. thick, sterile, the upper part 2–5 mm. thick, fertile, ± winged; seeds brown, asymmetrically obovoid, ca. 1 mm. long; $n=7$.—Occasional escape from gardens; native of s. cent. states. May–June.

A form closely related to *Oe. speciosa*, with rose fls., the petals 2–2.5 cm. long is cult. as "Mexican Primrose" and may be called **Oe. delessertiana** Steud. (*Oe. berlandieri, Oe. speciosa* var. *childsii*); $n=14$. Se. Tex. to adjacent Mex.

11. **Oe. strícta** Ledeb. ex Link. [*Oe. arguta* Greene.] Decumbent biennial or perennial, simple or few-branched, 2–6 dm. tall; stems reddish, finely pubescent below, often villous above; foliage green, the lower lvs. oblanceolate to narrower, 5–8 cm. long, ± denticulate, the cauline rather remote, lanceolate, plane, sessile, 2–7 cm. long; uppermost becom-

ing lance-ovate bracts, each with an axillary fl.; fl.-tube 1.5–2.5 cm. long; sepals 12–20 mm. long; petals yellow, aging reddish, 12–25 mm. long; caps. short-villous, enlarged in upper half, 2–2.5 cm. long; seeds brown, narrow-obovoid, smooth, 1.5 mm. long.—Occasional reported as natur.; native of Chile. April–Sept.

11. Zauschnèria Presl. CALIFORNIA-FUCHSIA

Erect or decumbent perennials, sometimes woody at the base, with shredding epidermis on lower stems, usually much branched and leafy. Lvs. sessile or nearly so, opposite or alternate, ± fascicled. Infl. spicate, the fls. large, horizontal, fuchsia-like. Fl.-tube scarlet, globose at base, narrowed into a long tube bearing within its narrow part 8 lobe-like appendages, 4 erect and 4 deflexed. Sepals 4, red. Petals 4, red. Stamens 8, the alternate shorter; anthers versatile; pollen grains shed singly. Ovary 4-loculed, elongate; style elongate; stigma somewhat 4-lobed, peltate to capitate. Caps. imperfectly 4-loculed, many-seeded, loculicidally dehiscent. Seeds oblong, narrowed at base, comose at apex. Four spp. (Named for Dr. M. Zauschner, professor of natural history at Univ. of Prague.)

Plants herbaceous throughout; lvs. broadly lanceolate to ovate, the principal ones more than 6 mm. wide.
  Lvs. rather coriaceous, ± sharply denticulate, the lateral veins evident, surface subglabrous to pilose, obscurely if at all glandular, broadly ovate. Mojave Desert .... 3. Z. garrettii
  Lvs. not coriaceous, the lateral veins obscure, subentire to denticulate, variously pubescent, often tomentose, often glandular, broadly lanceolate to ovate. Montane
                                                                              1. Z. californica latifolia
Plants suffrutescent at base; lvs. linear to lanceolate, mostly less than 6 mm. wide.
  Stems slightly woody; lvs. mostly 3–5 mm. wide, ± pilose, not canescent
                                                                              1. Z. californica mexicana
  Stems obviously woody at the base; lvs. 2–3.5 mm. wide, densely tomentose-canescent.
    Lvs. linear, 2.5–3.5 mm. wide, moderately fasciculate; fls. 3–4 cm. long
                                                                              1. Z. californica californica
    Lvs. subfiliform, not more than 2 mm. wide, very densely fasciculate; fls. 2.5–3.5 cm. long
                                                                              2. Z. cana

1. Z. califórnica Presl. ssp. **califórnica.** [*Z. c.* ssp. *angustifolia* Keck.] Suffrutescent at base, the stems 3–9 dm. long; lvs. linear, densely tomentose-canescent, 0.5–4 cm. long, 2–3.5 mm. wide, the lower often opposite or subopposite, the upper usually alternate, lateral veins usually not evident; fls. largely 3–4 cm. long; fl.-tube largely 2–3 cm. long; sepals erect, lanceolate, 8–10 mm. long, scarlet; petals 2-cleft, scarlet, 8–15 mm. long; stamens well exserted; style and stigma surpassing stamens; caps. sessile or short-pedicelled, linear, 4-angled, 8-nerved, with a short beak, often curved, variously pubescent, 1.5–2 cm. long; seeds many, 1.5 mm. long; $n=30$.—Dry slopes below 2000 ft.; Coastal Sage Scrub, Chaparral; San Diego Co. to Monterey Co. Catalina Id. Aug.–Oct.

Ssp. **latifòlia** (Hook.) Keck. [*Z. californica* var. *l.* Hook. *Z. l.* Greene.] Plants herbaceous, the stems slender, 1–5 dm. long; lvs. mostly opposite, ovate to lance-ovate, tapering to both ends or rounded at base, 7–17 mm. wide; plant green to grayish-green, villous to tomentose, often silky, often ± glandular; $n=30$ (Clausen et al., 1940).—Dry slopes and ridges below 10,000 ft.; mostly Montane Coniferous F.; mts. from San Diego Co. to Ore. Aug.–Sept. Variable, plants with broadly ovate to elliptic lvs. more crowded on stem and stems 1–3 dm. long, herbage more viscid, San Jacinto Mts. to s. Sierra Nevada, have been called *Z. latifolia* var. *viscosa* (Moxley) Jeps. They occur mostly above 7000 ft., while those at 3500–6500 ft., 2.5–5 dm. tall, villous, ± glandular, with lvs. lance-ovate to elliptical and growing in mts. from L. Calif., Palomar Mts. to San Gabriel Mts. Little San Bernardino Mts., Eagle Mts. have been called var. *Johnstonii* Hilend.

Ssp. **mexicàna** (Presl.) Raven. [*Z. mexicana* Presl. *Z. villosa* Greene.] Plate 64, Fig. D. Suffrutescent at base, the stem to 9 dm. long; lvs. lanceolate to linear-lanceolate, mostly 3–5 mm. wide, green to gray-pilose; fls. 3–4 cm. long; $n=30$.—Dry slopes below 2000 ft.; Coastal Sage Scrub, Chaparral; L. Calif. to cent. Calif.

2. **Z. càna** Greene. Suffrutescent at base, the stems 3–6 dm. long; herbage tomentose-canescent, entirely gray, mostly not very glandular; lvs. narrowly linear-lanceolate to subfiliform, not over 2 mm. wide, entire or nearly so, much fascicled; fl.-tube 2–3 cm. long; sepals 8–10 mm. long; petals 8–12 mm. long; caps. 1.5–2 cm. long, curved or almost straight, beaked or not; $n = 15$.—Dry slopes below 2000 ft.; Coastal Sage Scrub, Chaparral; Los Angeles Co. to Monterey Co. Santa Cruz, Anacapa and Catalina ids. Aug.–Oct.

3. **Z. garréttii** Nels. Herbaceous, the stems 1.5–3 dm. long; lvs. quite coriaceous, subglabrous to pilose, sharply denticulate, broadly ovate, 1–3.5 cm. long, 0.7–1.5 cm. wide; fls. 2.8–3.2 cm. long; $n = 15$ (Clausen et al., 1940).—Dry rocky places, at 5500 ft.; Pinyon-Juniper Wd.; Kingston Mts., e. Mojave Desert; to Utah, Wyo. June–July.

74. Orobanchàceae. Broom-Rape Family

Root-parasitic rather fleshy herbs, without chlorophyll, having alternate scales in place of lvs. Fls. complete, with persistent calyx. Corolla tubular, ± 2-lipped, the upper lip mostly 2-lobed, the lower 3. Stamens 4, in 2 pairs. Ovary 1-celled, ovoid, pointed with a long style. Caps. 2–4-valved, each valve with 1 or 2 placentae. Seeds many, very small.

Fils. hairy at base; caps. 4-valved; anther-cells closely parallel, blunt at base .... 1. *Boschniakia*
Fils. not hairy; caps. 2-valved; anther-cells deeply separated from below, mucronulate or aristulate at base .... 2. *Orobanche*

1. *Boschniákia* C. A. Mey.

Stems simple, glabrous, brown, thick, from a cormlike basal thickening and junction with root of host. Lvs. scalelike, mostly imbricated. Fls. nearly or quite sessile, in a dense spike, ± concealed by the subtending bracts. Calyx cup-shaped, 5-toothed, somewhat truncate. Corolla with throat broader than tube, ± curved, bilabiate, the upper lip entire or emarginate, the lower lip 3-lobed. Stamens 4, from almost as long as corolla to somewhat exserted. Stigma generally lobed (2–4) according to no. of carpels. Caps. 2–4-valved. Seeds many, favose. Few spp., widely distributed. (*Boschniaki*, a Russian botanist.)

1. **B. strobilàcea** Gray. [*Kopsiopsis s.* G. Beck.] Plants 1–2.5 dm. tall, the spike 3.5–6 cm. thick through flowering parts, dark reddish-brown; bracts 1.5–2 cm. long; fl. 1–1.5 cm. or more long; calyx truncate or 1–3-toothed, the teeth mostly shorter than the tube; corolla noticeably bent at middle, the lower lip as long as upper.—Reported from spp. of *Arctostaphylos* and from *Arbutus*, below 10,000 ft.; many Plant Communities; from San Jacinto Mts n.; to s. Ore. May–July.

2. *Orobánche* L. Broom-Rape

Usually viscid-pubescent plants with purplish to yellowish fls. Calyx 5-cleft into acute or acuminate lobes. Corolla curved, the upper lip erect or arched, the lower spreading or erect. Stamens mostly included. Stigma 2-lobed or peltate. Placentae 4. Ca. 100 spp., widely distributed. (Greek, *orobos*, vetch, and *anchone*, choke, because of the parasitic habit.)

1. Fls. solitary on long naked scapose pedicels that are longer than the corolla, without bracts at base of calyx; corolla-lobes orbicular.
    2. Caudex somewhat woody, with 5–10 firm bracts and several axillary pedicels that are rarely longer than the caudices; calyx-lobes equal to or shorter than the calyx-tube
                                                                                          5. *O. fasciculata*
    2. Caudex soft, herbaceous, with 1–several greatly elongated pedicels from the axils of the few basal bracts; calyx-lobes longer than the calyx-tube .... 8. *O. uniflora*
1. Fls. densely spicate, sessile or on pedicels shorter than the corolla, with 2 bracts at the base of the calyx; corolla-lobes tapered to the pointed or rounded apex.
    3. Stem arising from an enlarged tuberlike thickening; infl. parts glandular-puberulent
                                                                                                    1. *O. bulbosa*

3. Stem not having an enlarged tuberlike base.
    4. Corolla-lips, especially margins, villous with predominantly nonglandular hairs; calyx-lobes often dark purplish.
        5. The corolla-lobes triangular, acute, the upper erect or reflexed, the lower reflexed.
            6. Corolla 20–35 mm. long, the lips 4–8 mm. long; anthers often hairy. Deserts
            .... 3. *O. cooperi*
            6. Corolla 12–15 mm. long, the lips 3–4 mm. long; anthers glabrous. San Gabriel Mts.
            .... 9. *O. valida*
        5. Corolla-lobes rounded, obtuse, the upper erect, the lower lobes extended or only slightly reflexed. Deserts .... 6. *O. multiflora*
    4. Corolla-lips with scattered short, glandular hairs; calyx-lobes pale to brownish, sometimes with the tips suffused with lavender or red.
        7. Corolla 25–50 mm. long, the lips 10–14 mm. long and widely spreading
            .... 2. *O. californica*
        7. Corolla 15–25 mm. long, or more in *O. corymbosa*, the lips 4–9 mm. long, nearly erect to moderately spreading.
            8. Infl. subcorymbose; corolla-lobes purplish to brownish red on inner surface, grayish to brownish on exterior surface; anthers densely villous; northern desert ranges
            .... 4. *O. corymbosa*
            8. Infl. racemose to spicate; corolla-lobes whitish to pinkish or buff, often with lavender or purplish veins; anthers glabrous or hairy mainly along the dehisced margins.
                9. Lobes of the lower corolla-lip lanceolate and pointed; stigma-lobes strongly recurved, rounded apically and laterally; calyx-lobes and bracts in dried specimens dark brown with the venation not apparent .... 10. *O. vallicola*
                9. Lobes of the lower corolla-lip oblong to ovate with blunt apices; stigma usually funnelform with the lobes spreading; calyx-lobes and bracts thin, usually light brown or tan with conspicuous venation in dried specimens ...... 7. *O. parishii*

1. **O. bulbòsa** (Gray) G. Beck. [*O. tuberosa* Heller, not Vell.] Stout, dark, 1–3 dm. tall, pruinose-puberulent throughout, from thickened tuberlike base; cauline bracts lanceolate to acuminate, closely placed, ca. 1 cm. long; infl. densely pyramidal, thyrsoid-paniculate, the fls. nearly or quite sessile; calyx ca. 1 cm. long, somewhat exceeding the subtending bracts, unequally divided into lanceolate segms. equal to or exceeding the tube; corolla 12–15 mm. long, yellowish to purplish or brownish, the lips erect, 2.5–3.5 mm. long, usually with acute and sublanceolate lobes and without palatal folds in throat; anthers white, subglabrous, somewhat apiculate; stigma discoid; caps. ca. as long as calyx.—On *Adenostoma* and associates; below 5000 ft.; Chaparral; San Diego Co. to cent. Calif.; Channel Ids., Catalina Id. Mostly April–July.

2. **O. califórnica** C. & S. ssp. **feùdgei** (Munz) Heckard. [*C. grayana* var. *f.* Munz.] Plate 64, Fig. E. Plants 1–3 dm. long; stems mostly 10–18 cm. long below the infl., up to 2 cm. thick, usually simple below; fls. borne in a subcapitate or subcorymbose cluster, or sometimes in a raceme to 12 cm. long; calyx-lobes largely 14–20 mm. long; corolla 25–35 mm. long, pale or brownish externally, the inner surface of the lips commonly reddish to occasionally brownish, the entire corolla pale or yellowish, throat 8–10 mm. wide, the upper tube broad and gradually tapering to the sinus; corolla-lips 10–12 mm. long, the upper corolla-lobes broadly oblong and rounded with the apex sometimes emarginate or shallowly retuse, slightly spreading and sometimes reflexed; lower corolla-lobes narrowly oblong, subacute to obtuse, usually spreading; stigma bilamellate with triangular lobes.—Parasitic on *Artemisia tridentata*, perhaps on *Eriogonum* and *Eriodictyon* spp. From ca. 2500–8000 ft.; n. L. Calif. to Piute Mts., Kern Co. and to Mono Co. May–July.

Ssp. **grándis** Heckard. Plants 8–30 cm. long; calyx-lobes 15–20 mm. long; corolla 35–50 mm. long, the tube buff or yellowish with the lips pinkish or pale brownish red with darker veins, the throat 9–10 mm. broad; corolla-lips widely spreading, 12–14 mm. long, the upper lobes broad and obtuse; lower corolla-lobes lanceolate to lance-ovate, acute with blunt apices; stigma bilamellate with triangular lobes.—Reported on *Adenostoma*, *Artemisia*, *Haplopappus*; sandy places near the coast, from Los Angeles Co. to Luis Obispo Co. Santa Rosa Id. April–Sept.

3. **O. coóperi** (Gray) Heller. [*Aphyllon c.* Gray. *O. ludoviciana* var. *c.* G. Beck. *O. l.* var.

latiloba Munz.] Plate 64, Fig. F. Viscid-pubescent, 1–3 dm. tall, quite simple and spicate; cauline scales obtuse, 5–10 mm. long; pedicels to 1.5 cm. long; calyx 5–8 mm. long, the segms. ± unequal; corolla 15–28 mm. long, the lobes purplish, lanceolate, acute, the lips 3–9 mm. long, erect; anthers glabrous or pubescent; caps. equaling or slightly exceeding calyx.—Largely below 4000 ft., parasitizing *Larrea*, *Ambrosia*, *Hymenoclea*, and in Imperial V., tomato plants; to New Mex., Son. Jan.–May.

4. **O. corymbòsa** (Rydb.) Ferris. [*Myzorrhiza c.* Rydb.] Plant mostly 0.5–1.2 dm. high, light colored, glandular-pubescent, subcorymbose; cauline bracts lance-ovate, 8–10 mm. long; calyx 12–18 mm. long, the lobes lanceolate-subulate, ca. 10 mm. long; corolla purple, 25–30 mm. long, the upper lip darker than the lower, the lips 4–7 mm. long, corolla-throat curved toward outer part of the infl., the palatal folds evident; upper lip shallowly lobed or cleft, the lobes rounded or retuse; lower lip cleft to the base into lanceolate acutish divisions; anthers woolly; stigma peltate.—At 6000–8500 ft., mostly on *Artemisia tridentata*; Panamint, Inyo and White mts.; to B.C., Ida., Utah. June–Aug.

5. **O. fasciculàta** Nutt. Plant purplish, the caudex 3–10 mm. thick, 3–12 cm. long with 5–12 bracts and several erect axillary pedicels 3–10 cm. long; calyx 6–8 mm. long, the triangular lobes not longer than the tube; corolla usually purple, 15–22 mm. long, constricted at base, 2–5 mm. wide at throat, the lobes suborbicular; anthers glabrous to pubescent.—Occasional, mostly at 4000–10,000 ft.; on *Artemisia*, *Eriogonum*, *Eriodictyon*, etc., through much of Calif.; to B.C., Mich., New Mex. April–July.

6. **O. multiflòra** Nutt. var. **arenòsa** (Suksd.) Munz. [*Aphyllon a.* Suksd.] Slender, 5–15 cm. high, pale, viscid-pubescent; fls. subsessile; cauline bracts lance-ovate, 5–10 mm. long; calyx 9–12 mm. long, with lance-linear lobes; corolla purplish or yellowish, 15–20 mm. long, the upper lip ca. 5 mm. long, the lower ca. 4 mm., the lobes rounded apically; anthers glabrous.—On *Artemisia*, etc., 3500–5500 ft., Joshua Tree Wd., Pinyon-Juniper Wd.; mts. of e. Mojave Desert; to e. Wash., Colo. May–June.

7. **O. párishii** (Jeps.) Heckard ssp. **parishii**. [*O. californica* var. *p.* Jeps.] Stem simple, 8–26 cm. tall; infl. narrow, subspicate, 5–14 cm. long, the lower fls. on pedicels up to 5 cm. long or subsessile, the upper mostly sessile; cauline bracts of infl. and calyx often thin and submembranous in pressed specimens, conspicuously veined; calyx-lobes 10–16 mm. long, pinkish-tinged within, tips often recurved; upper fls. slightly curved, the middle and lower sharply curved at right angles to axis; corolla 18–25 mm. long, yellow or straw-colored, the lobes with 3 conspicuous pink or purplish veins on inner surfaces or occasionally ± suffused with purple; corolla-lips 6–8 mm. long; upper lip slightly spreading or retroflexed, cleft 2–3 mm. into 2 oblong or oblong-ovate lobes with rounded to truncate and often retuse or erosulate apices; lower corolla-lobes narrower, rounded or obtuse at tip; spreading; anthers glabrous or hairy; stigma bilobed.—On *Eriodictyon*, *Adenostoma*, *Arctostaphylos*, etc., mostly between 2000 and 6000 ft.; s. Tehachapi and Topatopa mts. to cent. San Diego Co. May–July.

Ssp. **brachylòba** Heckard. Stems simple or few-branched, 5–20 cm. tall; infl. subspicate, moderately dense, 3–12 cm. long; fls. sessile or the lower with pedicels to 2 cm. long; calyx-lobes 7–11 mm. long, purplish; corollas 15–22 mm. long, directed upward or sometimes curving horizontally, sparsely glandular, buff or yellowish, purple-tinged on lips and inner surfaces; corolla-lips 4–6 mm. long, the upper lobes 2–3 mm. long, broadly ovate, rounded to pointed apically, erect or slightly spreading; lower lobes erect to slightly spreading, ovate to lance-oblong, rounded apically to pointed; $n=48$ (Raven & Thompson).—On *Haplopappus*, sandy soil near beaches; San Diego, Channel Ids. April–Aug.

8. **O. uniflòra** L. ssp. **occidentàlis** Ferris. Stems slender, 2–4 mm. thick, 0.5–5 cm. long, largely subterranean with 2–6 crowded bracts; pedicels 1–3, slender, 3–10 cm. long; calyx-lobes narrow-subulate, from a broad base; corolla usually deep purple, 15–20 mm. long, the tube 4–5 mm. wide at the throat, the lobes rounded; anthers glabrous.—Largely on Saxifragaceae, below 7000 ft.; San Bernardino and Los Angeles cos.; to B.C. and Rocky Mts. March–July.

9. **O. válida** Jeps. [*O. ludoviciana* var. *v.* Munz.] Plant dark, brownish-purple, spicate,

1-2 dm. high, slender, simple or few-branched; cauline bracts lanceolate, ca. 1 cm. long; pedicels short, the longest to 1 cm.; calyx 7-9 mm. long, with lance-acuminate segms.; corolla 12-14 mm. long, brownish-purple, the upper lip 3-4 mm. long, purple, the 2 lobes acute, the lower lip yellowish, 3-4 mm. long, each lobe with dark purple midvein; anthers glabrous.—On *Eriodictyon*, etc., 4000-7000 ft.; Yellow Pine F.; San Gabriel Mts., Los Angeles Co. June-July.

10. **O. vallícola** (Jeps.) Heckard. [*O. grayana* var. *v.* Jeps. *O. californica* var. *claremontensis* Munz. *O. californica* var. *californica* auth.] Spicate to paniculate, 1-3 dm. tall, glandular-pubescent; bracts on stems few, ovate to lanceolate or suborbicular; lower pedicels 5-20 mm. long; fls. crowded, many; calyx 1-2 cm. long, the segms. narrow and acuminate; corolla pale with darker veins, 20-25 mm. long, the upper lip ca. 8 mm. long, its lobes 2.5-3 mm. long, the lower lip 6-8 mm. long with acute lobes; anthers subglabrous. —On *Sambucus*, *Quercus*, etc.; below 5000 ft.; Los Angeles Co. to n. Calif. July-Feb.

75. Oxalidàceae. Oxalis or Wood-Sorrel Family

Annual or perennial herbs to shrubs or trees, with acid sap. Lvs. usually compound, either digitate or pinnate, alternate or the basal opposite. Fls. bisexual, regular; sepals and petals 5, sometimes somewhat united at the base. Stamens 10, hypogynous, the outer 5 opposite the petals; fils. joined near base and at least some with basal glandular appendages. Styles 5, separate; stigmas capitate or somewhat bifid. Ovary superior, 5-celled, 2- to many-ovuled. Fr. a caps. or berrylike. Ca. 10 genera and over 500 spp., widely spread in temp. and trop. regions.

1. Óxalis L. Wood-Sorrel

Caulescent or acaulescent, annual or perennial, often with bulbous or tuberous underground parts or creeping rootstocks. Lfts. 3 or more, sensitive to light, folding at night. Fls. 1-several in umbellike cymes on axillary peduncles, yellow, white, pink to red, or violet. Stamens 5 long and 5 short. Styles 5. Fr. a loculicidal caps. Nearly 500 spp., many of horticultural value, some persistent weeds, a few grown for the edible tuberous parts. (Greek, *oxus*, sour.)

Plants acaulescent, with underground bulblets; petals 20 mm. long .... 3. *O. pes-caprae*
Plants caulescent, without bulblets; petals 8-16 mm. long.
   Stems creeping from a slender taproot and rooting at the nodes. Garden weed . 2. *O. corniculata*
   Stems erect or decumbent, not rooting at nodes. Native plants.
      Stems glabrous or appressed-puberulent with very short crisped hairs 0.3-0.5 mm. long; pedicels to 4 times longer than their mature caps.; caps. strigulose .. 1. *O. albicans californica*
      Stems with straight hairs appressed upward, to 1.2 mm. long; pedicels to 1.6 times longer than mature caps. .... *O. albicans pilosa*

1. **O. albicáns** ssp. **califórnica** (Abrams) Eiten. [*Xanthoxalis c.* Abrams.] Perennial with stout woody taproot; plants loosely cespitose, the stems slender, decumbent, 1-4 dm. long, with appressed pubescence; stipules narrow or almost obsolete; petioles 2-7 cm. long; lfts. 5-15 mm. long, ca. as wide, obcordate; pedicels 1.5-3 cm. long; sepals 4.5-6 mm. long; petals yellow, 8-12 mm. long; styles 2.5-3.5 mm. long; caps. cylindrical, 12-18 mm. long, closely puberulent; seeds 1.5-1.6 mm. long.—Common on dry brushy and stony slopes, below 2000 ft.; Coastal Sage Scrub, Chaparral; Santa Barbara to L. Calif. Santa Cruz. and S. Catalina ids. March-May.

Ssp. **pilòsa** (Nutt.) Eiten. [*O. pilosa* Nutt. in T. & G.] Much like ssp. *californica* but the stems with hairs to 1.2 mm. long; pedicels to 1.6 times longer than mature caps.; caps. with retrorse non-septate hairs and often also with intermixed spreading septate hairs; styles mostly 1-2 mm. long; seeds 1.2-1.5 mm. long.—Occasional, open hills and brushy hillsides mostly near the immediate coast; Coastal Scrub, Chaparral; San Diego Co. to Mendocino Co.

2. **O. corniculàta** L. Stems from slender taproot, creeping, 0.5-3 dm. long, often

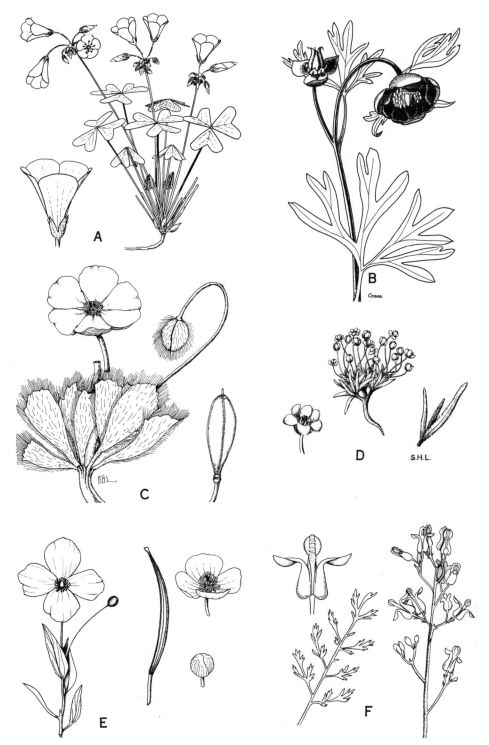

Plate 65. Fig. A, *Oxalis pes-caprae*. Fig. B, *Paeonia californica*. Fig. C, *Arctomecon merriamii*. Fig. D, *Canbya candida*. Fig. E, *Dendromecon rigida*. Fig. F, *Dicentra chrysantha*.

rooting at nodes, ± pubescent; petioles slender, pubescent, 1-5 cm. long; stipules broad, brown or purple; lfts. glabrous or subglabrous above, hairy beneath, 6-10 mm. long; peduncles few-fld., pubescent, ca. as long as petioles; pedicels 4-10 mm. long, becoming deflexed; sepals 2.5-4.5 mm. long; petals yellow, 4-8 mm. long; caps. prismatic-cylindric, 1-2.5 cm. long, evenly and closely puberulent; seeds brown, 1.2-1.5 mm. long; $2n = 24$, 36, 42, 48 (Eiten, 1963).—Common weed in lawns, gardens through most of the state; native of Eu. Most months of the year.

3. **O. pés-cáprae** L. [*O. cernua* Thunb.] BERMUDA-BUTTERCUP. Plate 65, Fig. A. Perennial from a deep rootstock with scaly bulbs, acaulescent; lvs. on glabrous petioles 1-2 dm. long; lfts. obcordate-bilobed, 1-2.5 cm. long, green, glabrous; pedicels 3-10 or more, at first cernuous, then erect, pubescent, 1-2 cm. long; sepals lanceolate, 5-7 mm. long; petals deep yellow, 1.5-2.5 cm. long; caps. 5-7 mm. long.—Becoming abundantly natur. in orchards, fields and waste places at low elevs.; Fla., W. I.; native of S. Afr. Nov.–March.

76. **Paeoniàceae.** PEONY FAMILY

Large herbs or shrubs with spirally arranged estipulate lvs. Fls. large, terminal, usually solitary, perfect, actinomorphic, hypogynous. Calyx mostly of 5 free sepals. Corolla of 5-10 (-13) large free petals. Stamens many, maturing in centrifugal sequence, often united below into a ring. Carpels 2-5, free, with fleshy walls and mounted on a fleshy disk. Ovules several, each with 2 integuments of which the outer projects beyond the inner. Fr. of 2-5 large follicles each with several seeds. A single genus, differing from *Ranunculaceae*, in which often included, in having persistent sepals and stamens maturing in centrifugal sequence.

1. *Paeònia* L. PEONY

Ours perennial herbs with fascicled fleshy roots; stems one to several, many 2-4 dm. high, branched or simple; lvs. basal and cauline, alternate, glaucous, biternate, usually somewhat fleshy. Fls. large, solitary, terminal. Sepals 5 or 6, persistent, green. Petals mostly 5 or 6, with the numerous stamens borne on a fleshy disk adnate to the base of the calyx. Pistils 2-5, free, becoming large fleshy, dehiscent follicles, with several large seeds. Ca. 30 spp. of the N. Hemis., largely Asian, many of horticultural value. (Greek, *Paeon*, the physician of the gods.)

1. **P. califórnica** Nutt. ex T. & G. [*P. brownii* var. *c.* Lynch.] Plate 65, Fig. B. Stems 5-30, mostly branched, 3.5-7 dm. high; lvs. 7-12 per stem, green, thin, easily wilting, the primary segms. cuneate at base, sessile or with short winged petiolules, mostly 3-7 cm. long, 1-4 cm. wide, the ultimate lobes lanceolate or narrow-elliptic, usually acute; petals elliptic, 15-25 mm. long, 11-18 mm. wide, deep blackish-red at center, pink at margins; fils. 5-8 mm. long; anthers 3-7 mm. long; follicles 3-4 cm. long; seeds blackish, subterete, slightly curved, 1.3-2 cm. long, 5-8 mm. thick; $n=5$ (Stebbins, 1938).—Brushy places below 4000 ft.; Chaparral, Coastal Sage Scrub, S. Oak Wd.; San Diego Co. to Monterey Co. Jan.–March.

77. **Papaveràceae.** POPPY FAMILY

Herbs or shrubs, with milky yellow or colorless sap. Lvs. alternate or sometimes opposite, without stipules. Fls. regular or irregular, complete, usually solitary and pedunculed. Sepals 2, rarely 3 or 4, sometimes united, often caducous. Petals mostly twice as many as sepals. Stamens mostly numerous, hypogynous, distinct. Pistil 1, of 2–several usually united carpels; ovary superior, usually 1-celled, with parietal placentae; style short; stigma simple or lobed. Fr. a caps., usually dehiscent by apical pores or valves; seeds usually many. Ca. 25 genera and 400 spp., widely distributed, but most abundant in w. N. Am.

Fls. regular; stamens mostly numerous.
    Lvs. mainly opposite or whorled.
        Stamens many; carpels 6–25, torulose and separating when mature; petals rather persistent
                11. *Platystemon*
        Stamens 6–12; carpels mostly 3, united; petals falling early . . . . 9. *Meconella*
    Lvs. mainly alternate or basal.
        Plants shrubby, at least at base, mostly 1 m. or more tall.
            Fls. white; lvs. lobed . . . . 12. *Romneya*
            Fls. yellow; lvs. entire . . . . 4. *Dendromecon*
        Plants herbaceous, usually less than 1 m. tall.
            Plant spiny or conspicuously long-hairy.
                Herbage long-hairy; lvs. bunched at base of plant . . . . 1. *Arctomecon*
                Herbage spiny; lvs. well distributed . . . . 2. *Argemone*
            Plant not spiny or conspicuously long-hairy.
                Lvs. multifid into linear segms. . . . . 6. *Eschscholzia*
                Lvs. not multifid into linear segms.
                    Caps. long-linear; stigma 2-lobed; upper lvs. clasping. Rare garden escape
                            8. *Glaucium*
                  Caps. ovoid to turbinate, obovoid or globose.
                    Plants 2–3 cm. high; stamens 6–9. Deserts . . . . 3. *Canbya*
                    Plants usually 20–120 cm. high; stamens many. Cismontane.
                        Style slender, short; stigma capitate . . . . 13. *Stylomecon*
                        Style lacking; stigma discoid . . . . 10. *Papaver*
Fls. irregular; stamens 4 or 6.
    Outer petals both saccate or spurred at base . . . . 5. *Dicentra*
    Outer petals with one spurred, the other not . . . . 7. *Fumaria*

1. *Arctomècon* Torr. & Frém.

Rather low perennial herbs with stout taproot. Lvs. many, long-hirsute, largely basal, cuneate-obovate, mostly toothed at apex. Fls. large, 1–several at ends of long naked stems, nodding in the bud. Sepals 2 or 3. Petals 4 or 6. Stamens many. Styles united, short; stigmas united, 3–6, cordate-bilobed. Caps. ovoid to oblong-linear, 3–6-valved. Seeds rather few, oblong-obovoid. Three spp. of sw. U.S. (Greek, *arctos*, a bear, and *mecon*, poppy, because of the hairiness.)

    1. **A. merriàmii** Cov. BEAR POPPY. Plate 65, Fig. C. Plants 2–4 (–5) dm. high, branched at base, glaucous; lvs. 2.5–7.5 cm. long, gradually narrowed at base into petioles ca. as long; peduncles ca. 2–3.5 dm. long, naked, glabrous, one-fld.; sepals 3, villous, 1.5–2 cm. long; petals 6, white, obcordate, 2.5–4 cm. long; caps. narrow-obovoid, 2.5–3.5 cm. long; seeds black, shining, 1.5–2 mm. long, rather coarsely reticulate-ridged; $n=12$ (Ernst, 1958).—Rare, rather loose rocky slopes, 3000–4500 ft.; Creosote Bush Scrub; Death V. region; to Clark Co., Nev. April–May.

2. *Argémone* L. PRICKLY POPPY

Caulescent annual or perennial herbs with yellow or orange latex. Stems glaucous, 1–several, erect or ascending, 2–20 dm. tall, cymosely branched, smooth to closely prickly, oblanceolate to obovate below, becoming elliptical to ovate up the stem, the uppermost often bracteate, the margin dentate and prickly, the lamina unlobed to variously lobed or pinnatifid, the surfaces smooth to closely prickly; buds oblong, obovoid to hemispherical, the sepals normally 3 (2–6), imbricated, smooth to hispid, each horned, the horn tipped with an indurated prickle. Fls. 3–15 cm. in diam. Petals yellow, lavender or white. Stamens 20–250 or more; fils. filiform; anthers linear. Stigma 3–7-lobed. Style to 3 mm.; stigma 3–7-lobed. Caps. 3–7-carpellate, unilocular, the valves apically splitting away, caps. surface commonly spinescent. Seeds many. Ca. 30 spp. in N. Am., S. Am., Hawaii. (Greek, *argema*, cataract of the eye, for which the juice of a poppylike plant was a supposed remedy.)

Stamens ca. 100–120; latex orange when fresh; caps. ca. 25–30 mm. long, armed with rather large even-sized spines; lvs. smooth to sparingly prickly on the veins beneath; sepal-horns smooth .... 1. *A. corymbosa*
Stamens ca. 150–250; latex yellow when fresh; caps. ca. 35–55 mm. long, with widely spaced spines to closely prickly; lvs. smooth or very sparingly prickly to closely prickly throughout; sepal-horns smooth to closely prickly .... 2. *A. munita*

1. **A. corymbòsa** Greene. Perennial with orange latex; stems 5–8 dm. tall, armed with scattered stout spines; lvs. coriaceous, the basal shallowly lobed, the middle and upper shallowly to unlobed; buds sparingly prickly, the sepal-horns smooth; fls. 5–9 cm. in diam.; petals white; caps. 25–30 mm. long; seeds ca. 1.5 mm. long; $n=28$ (Ownbey, 1957), 14 (Ernst, 1959).—Dry slopes and flats, 1400–3500 ft.; Creosote Bush Scrub; Mojave Desert. April–May.

2. **A. munìta** Dur. & Hilg. ssp. **munìta**. Annual to perennial, with pale yellow latex; stems 6–15 dm. long, sparingly spinescent to closely prickly; lvs. lobed ca. halfway to midrib below, more shallowly upward, upper surfaces sparingly prickly on larger veins and also often between veins; sepal-horns scatteringly prickly at base; fls. 5–13 cm. in diam.; petals white; caps. 35–55 mm. long, moderately spinescent with spreading and a few lesser spines and prickles; seeds ca. 1.8–2.6 mm. long; $n=14$ (Ownbey, 1957).—Occasional, below 6000 ft.; Chaparral, etc.; San Luis Obispo Co. to L. Calif. Aug.

KEY TO SUBSPECIES

A. Stems with ca. 0–80 prickles per sq. cm. at 5 cm. below the oldest caps.; lf.-surfaces smooth or sparingly prickly.
  Stems with ca. 0–10 prickles per sq. cm. at 5 cm. below the oldest caps.; lf.-surfaces usually totally smooth above and with a few prickles on main veins beneath; caps. armed with large widely spaced spines. Santa Ana Mts .... ssp. *robusta* G. Ownbey
  Stems more copiously prickly; lf.-surfaces prickly on the veins and usually also sparingly prickly between the main veins; caps. armed with large spines interspersed with unevenly sized smaller ones.
    The stems with ca. 10–20 prickles per sq. cm. at 5 cm. below the oldest caps.; lf.-surfaces sparingly prickly on and usually between the veins .... ssp. *munita*
    The stems with ca. 50–80 prickles per sq. cm. at 5 cm. below the oldest caps.; lf.-surfaces more prickly on and between the veins, but not closely so .... ssp. *munita* × ssp. *rotundata*
A. Stems with ca. 80–500 prickles per sq. cm. at 5 cm. below the oldest caps.; lf.-surfaces moderately to closely prickly.
  Stems usually purplish, usually closely armed with uneven-sized prickles of moderate length; buds usually with numerous larger compound prickles interspersed with smaller uneven-sized ones.
    The stems with ca. 80–180 prickles per sq. cm. at 5 cm. below the oldest caps.; bud-prickles all simple; less copiously prickly throughout than the following ssp. ssp.*munita* × ssp. *rotundata*
    The stems with ca. 120–500 prickles per sq. cm. at 5 cm. below the oldest caps.; largest bud-prickles usually compound; usually very copiously prickly throughout ..... ssp. *rotundata*
  Stems usually greenish-white, usually closely armed with very long uneven-sized prickles, i.e., with ca. 100–300 prickles per sq. cm. at 5 cm. below the oldest caps.; buds usually with numerous simple more even-sized prickles .... ssp. *argentea*

Ssp. **argéntea** G. Ownbey. Perennial; stems 4–8 dm. high, greenish-white, mostly closely armed with long slender perpendicular prickles; lf.-surfaces usually copiously prickly; buds copiously armed with simple ascending prickles; sepal-horns prickly; caps. armature consisting of numerous uneven-sized spines and prickles, the surface partially obscure; $n=14$ (Ownbey, 1957).—At 1500–3000 ft.; Creosote Bush Scrub; dry desert mt.-ranges, Inyo Co. to San Diego and Imperial cos.; s. Nev., w. Ariz. March–May.

Ssp. **rotundàta** (Rydb.) G. Ownbey. [*A. r.* Rydb.] Stems 4–10 dm. tall, usually closely prickly throughout; lf.-surfaces usually closely prickly above and below; buds usually copiously prickly, the larger prickles often compound; sepal-horns very prickly; caps. closely armed with usually simple spines interspersed with smaller spines and prickly; $n=14$ (Ownbey, 1957).—Usually at 4000–8500 ft.; Chaparral, Pinyon-Juniper Wd.,

Montane Coniferous F.; mts. of Mojave Desert, San Gabriel and San Bernardino mts.; to n. Calif., Utah, Ariz. Apparently hybridizing with other sspp. in Coast Ranges to San Diego Co.

3. Cánbya Parry

Minute almost acaulescent annuals, glabrous, tufted. Lvs. alternate, linear-oblong, fleshy, entire. Scapes filiform, 1-fld. Sepals 3, caducous. Petals 6, persistent, obovate. Stamens 6–9. Ovary 1-celled, with 3 placentae; style 0; stigmas 3, linear. Caps. ovoid, 3-valved from apex down. Seeds many, elongate-oblong, slightly arcuate. Spp. 2. Ore. to Calif. (W. M. *Canby*, Delaware botanist of 19th century.)

1. **C. cándida** Parry. Plate 65, Fig. D. Plants 2–3 cm. high; lvs. 5–7 mm. long; petals white, 3 mm. long, closing over the caps. after anthesis; caps. 2–2.5 mm. long; seeds shining, brown, 0.6 mm. long; $n=8$ (Ernst, 1958).—Sandy flats, 2000–4000 ft.; Creosote Bush Scrub, Joshua Tree Wd.; w. Mojave Desert from Walker Pass to Victorville region. April–May.

4. Dendromècon Benth. TREE POPPY

Glabrous openly branched evergreen shrubs. Lvs. alternate, simple, entire, coriaceous, yellowish to somewhat grayish green. Fls. solitary, terminal on the short branchlets. Sepals 2. Petals 4, yellow. Stamens numerous, with short fils. Stigmas oblong, 2. Caps. linear, 1-celled, elastically 2-valved from the base upward. Seeds obovoid to rounded, finely pitted, carunculate at the hilum. Here treated as 1 sp., although 20 have been proposed. (Greek, *dendron*, tree, and *mekon*, poppy.)

1. **D. rígida** Benth. ssp. **rígida.** Plate 65, Fig. E. Stiff rounded shrub 1–3 (–6) m. high, the main stems with grayish or whitish shredding bark; lvs. lance-linear to lance-oblong, 2.5–10 cm. long, 0.7–2.5 cm. wide, coriaceous, reticulate, subacuminate, scabrous-denticulate under a lens, vertical; petioles usually twisted, 2–8 mm. long; peduncles exceeding lvs., sepals falling early, round, 8–10 mm. long; petals yellow, obovate to rounded, 2–3 cm. long; caps. linear, arcuate, 5–10 cm. long; seeds brownish olive, 2–2.5 mm. long, finely reticulate, the caruncle pale, subpeltate, an additional 0.5–1 mm. long, $n=28$ (Ernst, 1958).—Rather common on dry slopes and in stony washes, especially after disturbance such as burns; Chaparral; cismontane ranges from n. L. Calif. to cent. Calif. and to Shasta Co. April–July. Variable.

Ssp. **harfórdii** (Kell.) Raven. [*D. h.* Kell.] More or less rounded erect shrub or tree 2–6 m. high, but with some branches spreading or even drooping, glaucous-green; lvs. crowded on branches, elliptic to oblong-ovate, usually not more than twice as long as wide, the main ones 3–8 cm. long, usually rounded at tip; peduncles not exceeding lvs.; petals 2–4 cm. long; caps. arcuate, 7–10 cm. long.—Brushy slopes; Chaparral; Santa Cruz and Santa Rosa ids. April–July, but with some fls. most of the year.

Ssp. **rhamnoides** (Greene) Thorne. [*D. r.* Greene. *D. harfordii* var. *r.* Munz.] Lvs. pale green, less crowded, mostly 2–3 times as long as wide, the main ones 6–13 cm. long, acute-mucronulate at tip; $n=28$ (Lenz, 1950).—Brushy slopes, Santa Catalina and San Clemente ids.

5. Dicéntra Bernh.

Perennial herbs. Lvs. basal or basal and cauline, dissected. Fls. solitary or in racemes or panicles, irregular, usually flattened and heart-shaped. Sepals small. Outer 2 petals saccate at base, spreading at apex; inner 2 narrow, clawed, usually cohering above and crested on back. Stamens 6, in 2 sets opposite the outer petals. Caps. several-seeded, elongated, 2-valved. Seeds crested or not. A genus of ca. 16 spp. of N. Am. and Asia; of some ornamental value. (Greek, *dis*, twice, and *centron*, spur.)

(Stern, K. R. Revision of Dicentra. Brittonia 13: 1–57. 1961.)

Fls. golden yellow, 12–15 mm. long, the outer petals spreading or recurved to middle
  1. *D. chrysantha*
Fls. off-white to cream, 20–25 mm. long, the outer petals spreading only at tips . 2. *D. ochroleuca*

1. **D. chrysántha** (H. & A.) Walp. [*Diclytra c.* H. & A.] GOLDEN EAR-DROPS. Plate 65, Fig. F. Glaucous, erect, with several coarse stems from stout roots, 5–15 dm. high; lvs. basal and on lower stems, bipinnate, rather stiff, 15–30 cm. long (including petiole), the pinnules 1–3 cm. long and further divided into usually narrow lobes; panicle loose, narrow, 2–5 dm. long, terminal, many-fld.; sepals suborbicular, 3–4 mm. long, caducous; corolla oblong, bright yellow, slightly cordate, 12–15 mm. long, the outer petals saccate below the tip, spreading in upper half; crest of inner petals narrow, crisped; caps. mostly 15–25 mm. long, lance-ovoid; seeds black, subreniform to almost round, densely papillate, 1.5–2 mm. long; $n = 12$ (Ernst, 1965).—Frequent on burns and in disturbed places, dry slopes mostly below 5000 ft.; Chaparral, Yellow Pine F., S. Oak Wd.; n. L. Calif. through cismontane s. Calif.; to Mendocino Co. April–Sept.

2. **D. ochroleùca** Engelm. [*Diclytra o.* Greene.] Habit of *D. chrysantha,* but lvs. 3-pinnate, the pinnules dissected into narrow lobes; sepals ca. 5 mm. long; corolla off-white to cream, 20–25 mm. long, only the purplish tips of the outer petals spreading; inner petals purple-tipped, broad-crested; seeds ca. 1.2–1.4 mm. long, densely papillate; $n = 16$ (Ernst, 1965).—Occasional in dry disturbed places below 3000 ft.; Chaparral; Santa Ana Mts. to Santa Ynez Mts. and Santa Lucia Mts. May–July.

6. *Eschschólzia* Cham. in Nees. CALIFORNIA POPPY

Annual or perennial herbs with colorless juice. Lvs. alternate, mostly glabrous, ternately finely dissected. Fls. yellow to red-orange, or in cult. forms with other colors, peduncled. Torus dilated to form a funnel-shaped base for the pistil. Sepals 2, united to form cap. (calyptra) pushed off by the expanding petals. Petals 4, rarely 6 or 8. Stamens many, to as few as 16; fils. short; anthers linear. Ovary cylindric, 1-celled, with 2 placentae; style short; stigma with 4–6 linear divergent lobes. Caps. elongated, 10-nerved, 2-valved from base toward apex. Seeds subglobose to slightly elongate, reticulate to rough-tubercled or pitted. Of much horticultural interest, many color-forms having been developed. Ca. a dozen spp., from Columbia R. to n. Mex. (Named for Dr. J. F. *Eschscholtz,* 1793–1831, surgeon and naturalist with Russian expeditions to Pacific Coast in 1816 and 1824.)

1. Torus with 2 rims, the inner erect and hyaline, the outer spreading; perennials or annuals with 2-cleft cotyledons .... 2. *E. californica*
1. Torus with only an erect hyaline rim, the outer rim absent or rudimentary; annuals with entire cotyledons.
   2. Stems scapose, the lvs. mostly crowded in a basal tuft.
      3. Lf.-blades mostly 4-times ternate into very short sharp-pointed lobes; seeds weakly reticulate. S. Mojave Desert and Colo. Desert .... 6. *E. parishii*
      3. Lf.-blades mostly 3-times ternate into blunt lobes.
         4. Calyptra long-apiculate; seeds reticulate to almost burlike .... 1. *E. caespitosa*
         4. Calyptra acute but not long-apiculate; seeds with deep remote pits. Mojave Desert
            4. *E. glyptosperma*
   2. Stems leafy.
      5. Petals 3–8 mm. long. Deserts.
         6. The petals 3–6 mm. long. Widely distributed .... 5. *E. minutiflora*
         6. The petals 8 mm. long. N. Mojave Desert .... 3. *E. covillei*
      5. Petals 10 or more mm. long.
         7. Lvs. mostly 3 times ternate into blunt lobes .... 1. *E. caespitosa*
         7. Lvs. mostly 4 times ternate.
            8. Ultimate lobes of lvs. sharp-pointed; calyptra long-apiculate. Deserts . 6. *E. parishii*
            8. Ultimate lobes of lvs. blunt; calyptra short-apiculate. Insular ..... 7. *E. ramosa*

1. **E. caespitòsa** Benth. ssp. **caespitòsa.** Annual, usually with several stems from a tuft of basal lvs., glabrous or with patches of short stiff hairs, sometimes glaucous, 1–3 (–4) dm. high; stems scapose to leafy; lvs. dissected into many narrow divisions mostly under 1 cm. long; buds erect; torus turbinate, without a spreading outer rim; calyptra ovoid-elliptic, apiculate, 10–18 mm. long; petals bright yellow, 10–25 mm. long; caps.

5–8 cm. long, 1.5–2 cm. thick; seeds subglobose to somewhat longer than thick, 1–1.3 mm. long, gray-brown with a network of thin low roughened ridges forming rows of ca. 5–6 irregular meshes; $n=6$ (Smith, 1937).—Dry flats and brushy slopes below 3500 ft.; V. Grassland, Foothill Wd., Chaparral; Orange and San Bernardino cos. to Central V. March–June.

Ssp. **kernénsis** Munz. Stems coarse, leafy; petals deep orange, 25–40 mm. long; caps. 2.5–4 mm. thick; seed 1.3–1.5 mm. long, rounded, the edges fluted, prominent, some seed almost burlike.—Heavy soil, 1000–2000 ft., V. Grassland; Tejon Pass region, Kern Co. March–April.

In the e. part of the Mojave Desert grow plants very much like those of the Tejon Pass region and identified by W. Ernst as *E. mexicana* Greene, but apparently with entire cotyledons, not the bifid ones of the *caespitosa* group.

2. **E. califórnica** Cham. Annual to perennial, flowering the first year, freely branched, ± glaucous, generally nearly glabrous, the stems becoming 2–6 dm. long and falling over in age; lvs. ternately several times dissected into narrow segms., the blades 2–6 cm. long, the basal lvs. petioled; peduncles 3–15 cm. long; torus with 2 rims, the inner erect and hyaline, the outer spreading and 2–4 mm. wide; calyptra 1–4 cm. long, variable in shape; petals deep orange to light yellow, 2–6 cm. long; caps. 3–8 (–10) cm. long; seeds gray-brown, round-oblong, 1.2–1.5 mm. long, reticulate with ca. 8–10 meshes per row; $n=6$ (Lawrence, 1930).—Common in grassy and open places, after burns, etc., up to 6500 ft.; many Plant Communities; most of cismontane Calif. and w. part of Mojave Desert. Feb.–Sept. Exceedingly variable with over 50 spp. proposed for the state; much in need of study; the following tendencies seem most evident:

(1) The typical form from dunes and bluffs along the coast, a heavy-rooted glaucous perennial, with smooth broad compact lvs. and yellow fls.; Channel Ids. n. to Mendocino Co.

(2) An inland form or at least away from the immediate coast; perennial, glaucous, smooth-lvd., less compact, marked by great seasonal variation in fl.-size and color (large and deep orange in spring, smaller and light yellow in late season). Columbia R. to s. Calif. This is the var. *crocea* (Benth.) Jeps., but this may well not be the earliest name available, which may be var. *douglasii* (Benth.) Gray. [*E. d.* Benth.]

(3) No. 2 is largely replaced in the San Joaquin V. and in s. Calif. by an annual form, var. *peninsularis* (Greene) Munz. [*E. p.* Greene, for which the earliest varietal name may be *E. arvensis* var. *dilatata* Greene.]

(4) A perennial with very gray roughish puberulent lvs. (appearing pitted when dry and under a lens) and prostrate stems; sand dunes, San Miguel Id., Surf to Monterey. This is var. *maritima* (Greene) Jeps. [*E. m.* Greene.]

3. **E. covíllei** Greene. [*E. minutiflora* var. *darwinensis* Jones.] Like *E. minutiflora* Wats. but with petals 8 mm. long; $n=12$ (Mosquin, 1961) and of more northern distribution.

4. **E. glyptospérma** Greene. Glaucous glabrous tufted annual with many slender erect scapose stems 1–3 dm. high; lvs. basal, of equal length, the blades 1–2 cm. long, very finely dissected; buds erect, torus turbinate, without spreading outer rim; calyptra ovoid to lance-ovoid, 8–12 mm. long, acute but not long-apiculate; petals yellow 1–2.5 cm. long; caps. 4–7 cm. long; seeds gray, globose, 1–1.3 mm. in diam., with deep rather remote pits; $2n=14$ (Lewis & Snow, 1951).—Open flats and slopes up to 5000 ft.; Creosote Bush Scrub, Joshua Tree Wd.; Mojave Desert, Inyo Co. to n. Riverside Co.; to Utah, Ariz. March–May.

5. **E. minutiflòra** Wats. Mostly glabrous somewhat glaucous annual, with a number of ascending stems 1–4.5 dm. long, branched, leafy throughout; lvs. thickish, the blades mostly 1–3 cm. long, finely to coarsely dissected; fls. on short peduncles scattered along the stems; torus short-turbinate, the outer rim erect, with hyaline inner edge; calyptra 4–6 mm. long, short-apiculate; petals yellow, 3–6 mm. long; stamens ca. 12; caps. 3–5 (–6) cm. long; seeds round-oblong, ca. 1 mm. long, dark brown, the ridges of the reticulum rather firm and forming rows of ca. 8 cells; $n=18$ (Ernst, 1959).—Common in sandy and gravelly places, mostly below 5000 ft.; Creosote Bush Scrub to Piñyon-Juniper Wd.; deserts of Calif. from L. Calif. to Mojave Desert. March–May.

6. **E. párishii** Greene. Annual, erect, several-stemmed, glabrous, somewhat glaucous, 2–3.5 dm. high; lvs. largely basal, dissected into linear lobes, the lower blades mostly 2–4 cm. long, the cauline scattered and reduced; peduncles very slender, short; torus turbinate, the outer margin mostly erect with an inner evident hyaline edge; calyptra ca. 1.5 cm. long, conspicuously apiculate; petals yellow, 1.5–3 cm. long; stamens ca. 24, caps. mostly 5–7 cm. long; seeds dark brown, 1–1.5 mm. long, round to oblong, the ridges of the reticulum very low and indefinite; $2n = 12$ (Lewis & Snow, 1951).—Dry rocky slopes below 4000 ft.; mostly Creosote Bush Scrub; s. Mojave Desert from Barstow, Sheephole Mts. and Cottonwood Springs to Colo. Desert. March–April.

7. **E. ramòsa** Greene. [*E. elegans* of Calif. refs. *E. wrigleyana* Millsp. & Nutt.] Stout, leafy-stemmed glabrous glaucous annual, 1.5–4 dm. high; lvs. conspicuously glaucous, the blades 1–3 cm. long, very finely dissected; peduncles 1–4 cm. long, erect in bud; torus broadly turbinate, without a spreading outer rim; calyptra ovoid, short-apiculate, 4–10 mm. long; petals yellow, often orange at base, 5–15 mm. long; caps. 4–7 cm. long; seeds round, dark, 0.6–0.7 mm. thick, reticulate with low rough ridges; $n = 17$ (Ernst, 1958).—Grassy and rocky places; Channel Ids., Guadalupe Id. March–May.

7. *Fumària* L. FUMITORY

Annuals with branching leafy stems. Lvs. decompound and finely dissected. Fls. small, in dense racemes or spikes. Sepals 2, scalelike. Petals 4, one of the outer pair spurred at base, the inner pair narrow, coherent at apex, keeled or crested on back. Stamens in 2 bundles. Style deciduous; ovary subglobose, 1-ovuled. Fr. 1-seeded, indehiscent. Seeds crestless. Ca. 15 spp. of Old World. (Latin, *fumus*, smoke, presumably because of color of fresh roots.)

Fls. purple, 5–7 mm. long; nutlet depressed-globose .... 1. *F. officinalis*
Fls. cream with inner petals purple at tip, 3–4 mm. long; nutlet apiculate .... 2. *F. parviflora*

1. **F. officinàlis** L. Glabrous, diffusely branched, the stems 2–6 dm. long; lvs. 2–6 cm. long (including petiole), finely dissected into segms. ca. 2 mm. wide; racemes narrow, 3–7 cm. long; pedicels 2–4 mm. long; sepals 2–3 mm. long; nutlet ca. 2 mm. in diam.; $n = 16$ (Mulligan, 1967).—Occasional in waste places as a weed; native of Eu. April–July.

2. **F. parviflòra** Lam. Much like the preceding, but lf.-segms. narrower, channeled; fls. paler, smaller; sepals 0.5–1 mm. long; fr. not depressed at apex; $2n = 28$ (Negodi, 1936).—Occasional as ruderal; native of Eu. March–May.

8. *Glaùcium* Mill. SEA POPPY

Glaucous herbs with saffron-colored sap. Lvs. alternate, pinnatifid or dissected, the basal petioled, the cauline clasping. Fls. solitary, axillary and terminal. Sepals 2. Petals 4, yellow or in some spp. red. Stamens many. Stigma nearly sessile, 2-lobed, the lobes convex, dilated. Caps. long-linear, dehiscent to the base. Seeds ovoid-reniform, with numerous shallow depressions. Ca. 6 spp. mostly Medit. (Greek name referring to glaucousness.)

1. **G. flàvum** Crantz. [*Chelidonium g.* L.] Somewhat papillose-hairy biennial, stout, branched, 6–9 dm. high; lvs. ovate to oblong in outline, 5–15 cm. long, pinnatifid to sinuate-lobed; fls. golden to orange; sepals acute, pilose; petals 2–3 cm. long; caps. 15–20 cm. long; seeds 1.5 mm. broad; $n = 6$ (Sugiura, 1936); $2n = 14$ (Larsen, 1954).—Occasional in waste places, as Elsinore; native of Eu. June–Sept.

9. *Meconélla* Nutt. in T. & G.

Low slender-stemmed annuals, glabrous or with a few short hairs on sepals. Basal lvs. spatulate, distinctly narrowed at base, the blades ± deltoid to orbicular; upper lvs. ± linear. Stamens 4–6 in 1 series or ca. 12 in 2 series. Carpels 3, the ovary 1-loculed. Fr.

frequently spirally twisted. Seeds lustrous black, free. Three spp. (Greek, *mekon*, poppy, and *ella*, diminutive.)

1. **M. denticulàta** Greene. [*M. californica* var. *d.* Jeps.] Glabrous, 5–30 cm. high; basal lvs. ovate to obovate, 2.5–4 cm. long, petioled; cauline lvs. spatulate to linear, 1–3 cm. long, sometimes denticulate, subsessile; petals 2–4 mm. long, white; anthers 6, linear, 1 mm. long; almost equal to or longer than fils.; caps. linear, 2–3 cm. long, twisted; seeds reniform-obovoid, black, shining, ca. 0.5 mm. long; $n=8$ (Ernst, 1958).—Frequent but inconspicuous, shaded canyons below 3300 ft.; Coastal Sage Scrub, Chaparral; San Diego Co. to Monterey Co. Santa Cruz Id. March–May.

10. *Papáver* L. Poppy

Annual or perennial herbs with yellowish or milky juice. Lvs. pinnately lobed or divided. Fls. solitary on long peduncles, drooping in the bud. Sepals 2. Petals 4. Stamens many. Ovary and caps. obovoid to subglobose, with 4–20 intruded placentae; caps. dehiscent by transverse pores under the edge of the sessile round flat or subconical disk formed by the united radiating stigmas. Seeds many, minute, reniform, variously striate and pitted. Ca. 50 spp., mostly of Old World, some well known as garden fls. (Classical name of the *Poppy*.)

Cauline lvs. with broad spreading base; caps. 3–5 cm. long .... 3. *P. somniferum*
Cauline lvs. not clasping, but with narrowed base; caps. 1–2 cm. long.
    Peduncles with weak spreading hairs or glabrous; caps.-disk 5–18-rayed; lvs. pinnately lobed into toothed or lobed segms. .... 1. *P. californicum*
    Peduncles stiff-strigose; caps.-disk with fewer rays; lvs. 2–3 times pinnately lobed
        2. *P. hybridum*

1. **P. califórnicum** Gray. Plate 66, Fig. A. Slender glabrous or pilose annual, 3–6 dm. tall; lvs. 3–9 cm. long, pinnately divided into oblong or rounded toothed or lobed segms.; sepals 8–10 mm. long, pilose; petals 1–2 cm. long, brick red with basal greenish spot; caps. clavate-turbinate, 1–1.6 cm. long, the disk flat; seeds dark, coarsely rugose-reticulate, 0.4–0.5 mm. long; $2n=28$ (Ernst, 1958).—Burns and disturbed places below 2500 ft.; Chaparral and Oak Wd.; San Diego Co. to cent. Calif. April–May.

2. **P. hýbridum** L. Annual, 2–5 dm. tall, stiff-strigose; lvs. 2–3 times pinnately lobed into bristle-pointed narrow segms.; fls. 2–4 cm. in diam.; sepals bristly; petals round-obovate, crimson with a blackish basal spot; caps. ± globose, 1–1.5 cm. in diam.—Reported from vineyards, etc. from w. Kern and e. San Luis Obispo cos. north; native of Eurasia.

3. **P. somníferum** L. Opium Poppy. Glaucous annual, glabrous or slightly hairy, 6–12 dm. high, with rather coarse stems and peduncles; lower lvs. petioled, the upper oblong, clasping, cordate, unequally coarse-toothed to lobed, 1–2 dm. long; sepals 1.5–2 cm. long; petals round, entire, white, pink, red or purple, 4–6 cm. long; caps. globose, 3–5 cm. long, the disk plane; seeds dark, reticulate-favose, 0.6–0.7 mm. long; $n=11$ (Furusato, 1940).—Common Garden Poppy, reported as occasional escape in waste places; native of Old World. Summer.

11. *Platystèmon* Benth. Cream Cups

Low villous annuals with lvs. not narrowed at base, ± alternate below, opposite or whorled above. Fls. terminal on long peduncles. Sepals 3. Petals 6, white to yellowish. Stamens many, hypogynous. Carpels more than 3, each forming a locule around a central chamber; carpels at first united, separate in fr., each several-ovuled, breaking transversely into indehiscent 1-seeded joints. Stigmas linear, free. Seeds dark, quadrate, variously sculptured and some tissue of the caps. wall adhering. A genus of Calif. and adjacent areas with apparently 1 variable sp., for which many segregates have been proposed. (Greek, *platus*, broad, and *stemon*, stamen.)

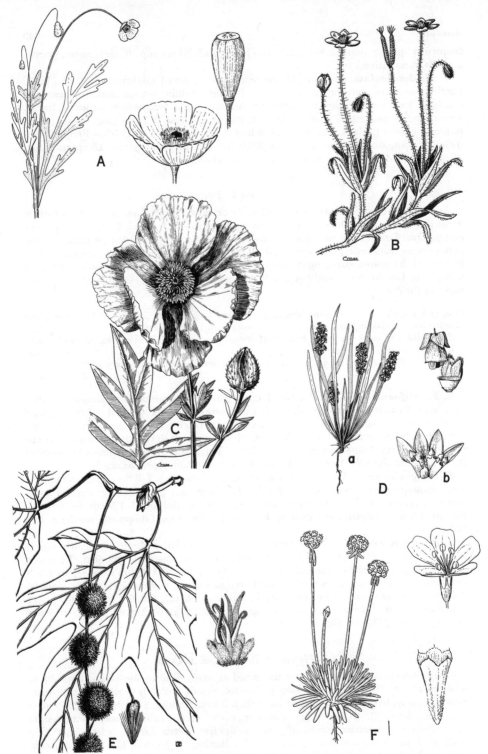

Plate 66. Fig. A, *Papaver californicum*. Fig. B, *Platystemon californicus*. Fig. C, *Romneya trichocalyx*. Fig. D, *Plantago erecta*, upper right a circumscissile caps. Fig. E, *Platanus racemosa* with ♀ fl. to right. Fig. F, *Armeria maritima* var. *californica*.

1. **P. califórnicus** Benth. Plate 66, Fig. B. Plant 1-3 dm. high, with many stems from the base, soft-pilose, the branches ± decumbent; lvs. largely on lower part of plant, lance-linear, subsessile, 2-5 (-8) cm. long; fls. subscapose, solitary on peduncles 1-2 dm. long; sepals villous, forming rounded to oblong-ovoid long-hairy buds 6-10 mm. long; petals usually cream, 8-16 (-20) mm. long; carpels forming an erect glabrous or setose oblong-cylindric head 10-16 mm. long beaked by the persistent styles (an added 4-8 mm.); seeds ca. 1 mm. long; $n=6$ (Sugiura, 1937).—Common in open grassy clay or sandy places, also on burns, usually below 3000 ft.; V. Grassland, Chaparral, Oak Wd., etc., most of cismontane Calif., w. edge of deserts; to Utah, Ariz., L. Calif. March-May.

Exceedingly variable, the extreme variations may be keyed:

1. Mature fr. nodding. W. San Diego Co., Santa Cruz and Santa Rosa ids. . . . . var. *nutans*
1. Mature fr. erect.
    2. Plants low, often less than 1 dm. tall; fr. scarcely torulose.
        3. The plants glabrous. San Miguel, San Nicolas, Santa Rosa ids.
                                                    var. *ornithopus* (Greene) Munz
        3. The plants sparingly short-pilose. Santa Barbara Id. . . . . var. *ciliatus* Dunkle
    2. Plants over 1 dm. tall; fr. usually torulose.
        4. Petals spreading nearly rotately from the base; young carpels white with stiffish hairs. W. foothills of Sierra Nevada . . . . var. *horridulus* (Greene) Jeps. [*P. h.* Greene]
        4. Petals cupped-ascending; young carpels glabrous to hairy.
            5. The petals cream; plant moderately hairy. General cismontane . . . . . var. *californicus*
            5. The petals yellow; plant excessively long-hairy. Cuyama V. and Tehachapi Mts. along desert slopes to L. Calif. . . . . var. *crinitus* Greene

12. *Rómneya* Harv. MATILIJA POPPY

Large glabrous bushy suffruticose somewhat glaucous perennial, with colorless bitter juice and creeping underground rootstocks giving rise to large patches from an original plant. Lvs. gray-green, alternate, pinnatifid. Fls. few at summit of stems. Sepals 3. Petals 6, white, very large. Stamens many. Style 0; stigmas 7-12, partly cohering in a ring. Ovary and coriaceous caps. oblong to ovoid, strigose-hispid, with 7-12 placentae, some or most of which meet in the axis and so form partitions. Seeds ovoid, roughened, dull, slightly incurved. Spp. 2, s. Calif. and n. L. Calif.; cult. for the handsome fls. (T. *Romney* Robinson, Irish astronomer, friend of T. Coulter, the discoverer of the genus.)

Sepals and summit of peduncle glabrous; peduncles not conspicuously leafy near summit; lobes of lvs. mostly more than 12 mm. wide . . . . 1. *R. coulteri*
Sepals and summit of peduncles setose; peduncles leafy to top; lobes of lvs. mostly less than 10 mm. wide . . . . 2. *R. trichocalyx*

1. **R. còulteri** Harv. Stems heavy, 1-2.5 m. tall, branching, leafy; lvs. firm, round-ovate in outline, petioled, 5-20 cm. long, pinnately parted or divided into 3-5 main lanceolate to ovate divisions, these in turn sparingly dentate to 2-3-cleft and mostly 10-20 mm. wide, sparsely spinulose-ciliate; fls. ca. 5-8 per stem; buds glabrous, round-ovoid, somewhat beaked, 2.5-3 cm. long; petals crinkled, white, 6-10 cm. long; stamens yellow; caps. 3-4 cm. long; seeds dark brown, 1.3-1.5 mm. long, microscopically papillose and with larger roughenings or subreticulate ridges; $2n=38$ (Bilquez, 1951).—Dry washes and canyons below 4000 ft.; Chaparral, Coastal Sage Scrub; away from the immediate coast, Santa Ana Mts. to San Diego Co. May-July.

2. **R. trichocàlyx** Eastw. [*R. coulteri* var. *t.* Jeps.] Plate 66, Fig. C. Stems more slender; lvs. mostly 3-10 cm. long, the lobes 3-10 mm. wide; peduncles leafy, spreading-bristly below receptacle; buds round, not beaked, 2-2.5 cm. long, appressed-setose; petals 4-8 cm. long; seeds strawcolor to brown, rather smooth; $n=19$ (Ernst, 1959).—Below 3600 ft.; Chaparral, Coastal Sage Scrub, Ventura and San Diego cos. to L. Calif. May-July.

Hybrids between the 2 spp. are in the nursery trade.

13. Stylomècon G. Tayl.

Erect annual, the stem simple or branched, with yellow sap. Lvs. alternate, pinnatisect or pinnatifid, the lobes entire or in turn dissected or lobed. Fls. axillary, on elongate peduncles. Sepals 2, deciduous. Petals 4, orange-red with basal purplish spot above the green claw. Stamens many. Caps. turbinate, glabrous, with 4–11 parietal placentae and dehiscing near their apex. Style slender, short, persistent; stigma capitate. Seeds many, reniform, rugose-reticulate. One sp. (Greek, *stylus*, style, and *mekon*, poppy.)

1. **S. heterophýlla** (Benth.) G. Taylor. [*Meconopsis h.* Benth.] Plants 3–6 dm. high, glabrous, or somewhat pilose below; lvs. 2–12 cm. long, including the petioles; peduncles 5–20 cm. long; sepals 4–10 mm. long; petals 1–2 cm. long; caps. clavate-obovoid, 8–15 mm. long, with yellowish ribs; seeds dark, ca. 0.4 mm. long; $2n=56$ (Ernst, 1958).—Occasional on grassy and brushy slopes, below 4000 ft.; Chaparral, V. Grassland; S. Oak Wd.; L. Calif. to cent. Calif. Channel Ids. April–May.

78. Phytolaccàceae. POKEWEED FAMILY

Herbs to trees. Lvs. alternate, entire, usually exstipulate. Fls. bisexual or unisexual, mostly in axillary or terminal racemes, regular. Calyx 4–5-parted, persistent. Petals usually none. Stamens as many as calyx-segms. and alternate with them, or more numerous, hypogynous, distinct or basally united. Ovary usually superior with 1–many distinct or united pistils; styles as many as pistils or none; stigmas linear or filiform. Ovules solitary. Fr. in ours of several locules and forming a berry. Ca. 17 genera and 100 spp., of warmer parts of Am. and S. Afr.

1. Phytolácca L. POKEWEED. POKEBERRY

Tall, stout, perennial herbs. Lvs. large, petioled in ours, ovate to oblong-lanceolate. Fls. small. Sepals 5, petaloid. Stamens 5–30. Ovary of 5–12 carpels united in a ring. Fr. a 5–12-locular berry, with a single vertical seed in each locule. Ca. 35 spp. of trop. or subtrop. regions, especially Am. (Greek, *phyton*, plant, and Latin, *lacca*, crimson lake, because of the color in the berries.)

1. **P. americàna** L. [*P. decandra* L.] Strong-smelling, 1–3 m. high; lvs. 1–3 dm. long; fls. bisexual, white or pinkish, ca. 5–6 mm. across, in peduncled racemes 5–20 cm. long; fr. dark purple, ca. 10–12 mm. across; $2n=36$ (Suzuka, 1950).—Occasional weed in cult. areas, San Diego Co. to n. Calif.; native of e. U.S. Aug.–Oct.

79. Pittosporàceae. PITTOSPORUM FAMILY

Trees and shrubs with alternate lvs. simple, usually leathery, exstipulate. Fls. mostly bisexual, regular. Sepals 5, distinct or nearly so, imbricated. Petals 5, hypogynous, the claws often connivent or coherent. Stamens 5, hypogynous, free, alternating with petals. Pistil 1; ovary superior, of 2, rarely 3–5 carpels, with few or many ovules on parietal placentae. Style simple, with terminal stigma entire or minutely lobed. Fr. a loculicidally dehiscent caps. or an indehiscent berrylike body. Nine genera and ca. 200 spp., mostly Australian.

Fls. white or greenish; fr. dehiscent .... 1. *Pittosporum*
Fls. blue; fr. indehiscent .... 2. *Sollya*

1. Pittósporum Banks.

Evergreen trees and shrubs; lvs. entire or sinuate-dentate, often seemingly whorled on young growth; fls. in terminal panicles or corymbose clusters or solitary. Sepals, petals and stamens 5. Ovary mostly incompletely 2-celled; style short; ovules few to many. Fr. a caps., the 2–4 valves leathery or woody. (Greek, *pitta*, resin, and *sporos*, seed.)

Lf. mostly obtuse, thick and leathery; revolute on margin .... 1. *P. tobira*
Lf. acuminate, rather thin, the margin undulate .... 2. *P. undulatum*

1. **P. tobira** Ait. A winter-flowering shrub 3–5 m. high with lvs. thick, leathery, obovate, obtuse, 5–10 cm. long, glabrous, revolute-margined; fls. in terminal umbels, white or greenish, ca. 12 mm. long; fr. ovoid, 6–12 mm. long, angled, short-hairy.—Occasionally reported as established from cult.; native of e. Asia.

2. **P. undulàtum** Vent. Tree to 10 m.; lvs. oblong-lanceolate, 7–15 cm. long, acuminate, entire, shining, the margins somewhat crisped; fls. white, ca. 10–12 mm. long, very fragrant; fr. subglobose, smooth, ca. 12 mm. long.—Occasional escape from cult.; native of Australia.

2. *Sóllya* Lindl.

1. **S. fusifórmis** Briq. An evergreen climbing subshrub with narrow lvs. 2.5–5 cm. long and cymes of 4–8 blue fls. 8–12 mm. long; caps. 12–15 mm. long, pubescent. Occasional escape from gardens. Native of W. Australia.

80. **Plantaginàceae.** PLANTAIN FAMILY

Herbs, annual or perennial, mostly with basal longitudinally ribbed lvs. and scapose bracteate spikes, occasionally with branched stems and opposite linear lvs. Fls. regular, 4-merous, sometimes imperfect. Calyx of 4 imbricated persistent sepals, mostly with scarious margins. Corolla salverform or tubular, with 4 erect or spreading scarious persistent veinless lobes. Stamens 4 or 2, alternate with corolla-lobes and inserted on the tube. Anthers 2-loculed. Ovary superior, 2–4-celled, with 1–several ovules in each cell. Style 1, with long hairy stigma. Caps. 2–several-seeded, circumscissile. Seeds mostly with flattened or concave surfaces, and straight embryo in fleshy endosperm. Ca. 250 spp. and 3 genera, of which the largest and most cosmopolitan is *Plantago*.

1. *Plantàgo* L. PLANTAIN

Characters as given for the family. Seeds with mucilaginous coat, hence of some use as laxative (psyllium). (Latin, from *planta*, footprint.)

(Pilger, R. Das Pflanzenreich, IV, 269: 39–432. 1937.)
1. Lvs. in pairs along slender stems; peduncles axillary .... 6. *P. indica*
1. Lvs. in basal rosettes; infl. scapose.
  2. Corolla-lobes spreading or reflexed; stamens 4.
    3. Corolla-tube pubescent externally. Strictly coastal.
      4. Plants mostly annual; lvs. acutely salient-toothed except in very depauperate individuals .... 3. *P. coronopus*
      4. Plants perennial; lvs. entire to remotely and sparsely denticulate ..... 10. *P. maritima*
    3. Corolla-tube glabrous externally. Generally distributed.
      5. Anterior sepals connate; stamens conspicuously exserted .... 8. *P. lanceolata*
      5. Anterior sepals separate; stamens not conspicuous.
        6. Plants heavy-rooted perennials; bracts not longer than calyx; lvs. broadly elliptic to cordate-ovate .... 9. *P. major*
        6. Plants annual, or if perennial, with lower bracts much exceeding calyx.
          7. Plants dark green; lower bracts 4–10 times as long as sepals. Introd. weed
            1. *P. aristata*
          7. Plants light green to white-woolly; lower bracts 1–3 times as long as sepals. Mostly common natives.
            8. Lowest bracts lanceolate to subulate, scarcely membranous on sides, 1–3 times as long as sepals; seeds brown, dull .... 11. *P. purshii*
            8. Lowest bracts ovate with broad membranous margins and not exceeding calyx.
              9. Seeds brown, dull; bracts not like sepals and ca. half their length 4. *P. erecta*
              9. Seeds reddish-yellow, shining; bracts almost exactly like sepals
                7. *P. insularis*

2. Corolla-lobes often erect in fertile fls. and forming a beak over the caps.
   10. Lvs. linear to subfiliform; stamens 2.
      11. Corolla-lobes mostly erect in age, forming a beak; seeds 4, ca. 0.8–1.8 mm. long .... 12. **P. pusilla**
      11. Corolla-lobes spreading or reflexed in age, not forming a beak; seeds 4–9, 1.5–2 mm. long .... 2. **P. bigelovii**
   10. Lvs. linear-lanceolate to ovate; stamens 4.
      12. Plants annual, rarely biennial; spikes less than 1 dm. long; introd. weeds.
         13. Sepals rounded at apex; seeds pale brown .... 14. **P. virginica**
         13. Sepals long-attenuate; seeds red .... 13. **P. rhodosperma**
      12. Plants perennial; spikes 1–2.5 dm. long. Native .... 5. **P. hirtella**

1. **P. aristàta** Michx. Annual or perennial, dark green, mostly loosely villous; lvs. linear, 5–15 cm. long, narrowed to margined semiclasping petioles; scapes 1–3 dm. high, slender; spikes 2–8 cm. long; bracts linear, 5–25 mm. long, divergent; sepals spatulate-oblong, ca. 2.5 mm. long; corolla-lobes round-ovate, ca. 2 mm. broad; seeds 2, oblong, finely pitted, brown, 2–2.5 mm. long; $2n=20$ (Heitz, 1927).—Occasional weed, as at Loma Linda, San Marcos Pass, etc.; from cent. U.S. June–July.

2. **P. bigelòvii** Gray ssp. **bigelòvii**. Plant ± pubescent, decumbent to semierect, 3–5 cm. high; lvs. entire, linear to subfiliform, 1–7 cm. long; bracts ovate, 2 mm. long, hyaline-margined; sepals broadly obovate, ca. 2 mm. long; corolla-lobes spreading to sharply reflexed in fr., 0.5–1 mm. long; caps. oblong-ovoid, 2–3 mm. long, circumscissile just below the middle; seeds mostly 4–5, dark brown to black, oblong, slightly angled, irregularly and coarsely pitted, 1.5–2 mm. long; $2n=20$ (Bassett, 1966).—Well distributed in cismontane Calif., mostly near the coast in saline and alkaline places, Coastal Strand, V. Grassland; to B.C. April–June.

Ssp. **califórnica** (Greene) Bassett. [*P. c.* Greene.] Plants 4–8 cm. tall; lvs. often with a few teeth; seeds mostly 6–9.—Mostly inland (Elsinore, Inglewood); cent. Calif. to L. Calif., Son.

3. **P. coronòpus** L. [*P. parishii* Macbr.] Mostly annual, coarsely pubescent; lvs. lance-linear in outline, hairy, sharply and acutely salient-toothed or -incised or subentire in depauperate individuals, the blades 2–14 cm. long, with winged petioles; scapes 1–4 dm. high, the spikes nodding before anthesis, becoming 2–12 cm. long, dense, with closely appressed fls.; bracts ovate, long-acuminate from a rounded body, green or purplish-red, ciliate, ca. 2 mm. long; sepals ovate, ciliate, ca. 2 mm. long; corolla-tube pubescent; corolla-lobes lanceolate, 1 mm. long; caps. ca. 2 mm. long; seeds commonly 3, elliptic, glaucous, ca. 1 mm. long.—Sea cliffs, about salt marshes, etc.; Coastal Strand, Coastal Salt Marsh, Closed-cone Pine F.; Coastal Sage Scrub; Catalina Id., Santa Barbara; to n. Calif.; native of Eu. April–July.

4. **P. erécta** Morris. ssp. **erécta**. [*P. hookeriana* var. *californica* (Greene) Poe.] Plate 66, Fig. D. Villous annual; lvs. filiform to linear-lanceolate, entire or with small remote denticulations, 3–12 cm. long; scapes 5–25 cm. tall, erect to arcuate-ascending; spikes capitate to short-cylindric, 5–25 mm. long, dense; bracts ovate, broad at base, scarious-margined at least half their length, not exceeding calyx; sepals scarious-margined, oblong, 3 mm. long, villous; corolla-lobes spreading, 1–2 mm. long; caps. ellipsoid, ca. 3 mm. high; seeds 2, dull, brown, 2–2.5 mm. long, finely pitted; $2n=20$ (Moore, 1965).—Common, dry open places, below 2500 ft.; Coastal Sage Scrub, V. Grassland, Chaparral, etc.; cismontane Calif.; to L. Calif., Ore. Channel Ids. March–May.

Ssp. **rigídior** Pilger. With more rigid spreading hairs; $2n=42$ (Moore, 1965). Mesas near San Diego.

5. **P. hirtélla** HBK. var. **galeottiàna** (Dcne.) Pilger. [*P. g.* Dcne.] Perennial, ± hirsute-pubescent on scapes, lf.-veins, etc., occasionally subglabrous; lvs. narrowly elliptic to somewhat oblanceolate, entire to denticulate, the blades 1–2 dm. long, tapered gradually into broad shorter petioles; scapes arcuate-ascending, 1.5–4 dm. high; spikes dense, cylindrical, 1–2.5 dm. long; bracts elliptic-lanceolate, ca. 3 mm. long, ciliate and stiff-pubescent on the sharp keel; sepals unequal, elliptic to broadly lanceolate, 2.5–3 mm.

long; corolla-lobes erect, ca. 2 mm. long; stamens 4; seeds 3, ca. 1.6 mm. long.—Occasional on moist banks, low elevs.; Coastal Salt Marsh, Coastal Sage Scrub, etc.; along the coast and inland to San Bernardino; San Diego to n. Calif. and cent. Mex. May–Sept.

6. **P. índica** L. Caulescent annual, branched, ± pilose, 2–6 dm. high; lvs. opposite, linear, attenuate, 2–4.5 cm. long; peduncles axillary, 3–7 cm. long; spikes dense, 1–2 cm. long; lower bracts concave, 4–5 mm. long, round-ovate at base, with narrow tip; sepals obtuse, 4 mm. long; corolla-lobes narrow-ovate, 2 mm. long; caps. 2 mm. long; seeds red-brown, shining; $2n=12$ (Rahn, 1966).—Reported from sandy and waste places; it is often used as a constituent of commercial bird-seed, poultry-feed, etc. and escapes. Native of Old World. July–Nov.

7. **P. insulàris** Eastw. Habit and stature of *P. erecta*; plant villous; bracts usually with brown midribs, ovate to roundish, 2.5–3 mm. long; sepals elliptic to obovate-elliptic, ca. 2.5 mm. long; seeds reddish-yellow when mature, shining, 2–2.5 mm. long.—Occasional on Coastal Strand, Coastal Sage Scrub, along the coast from Santa Barbara Co. to n. L. Calif. Channel Ids. Feb.–April. Inland the sp. becomes more white-woolly and the bracts often have green midribs; this form has been called var. *fastigiata* (Morris) Jeps. Extending through Calif. deserts to Utah, Ariz.

8. **P. lanceolàta** L. RIBGRASS. ENGLISH PLANTAIN. Usually perennial with strong caudex, somewhat short-villous; lvs. lanceolate to lance-oblong, erect or spreading, the blades 5–20 cm. long, attenuate at apex and gradually narrowed into rather slender somewhat shorter petioles; scapes 2–8 dm. high, arched-ascending, rather slender; spikes dense, ovoid-conic at first, cylindric and 2–8 cm. long in fr.; bracts broadly ovate, somewhat pubescent on back, scarious-margined, ca. 2 mm. long; front sepals connate, ca. 3 mm. long; corolla almost rotate; anthers well exserted; caps. oblong-ovoid, dehiscing below middle; seeds 1–2, brown, shining, deeply hollowed on one face, ca. 3 mm. long; $2n=12$ (Nakajima, 1930), 24, 96 (MacCullagh, 1934).—Weed in lawns and moist wet places; introd. from Eu. April–Aug.

9. **P. màjor** L. COMMON PLANTAIN. Mostly perennial, acaulescent; lvs. thick, ascending, usually roughish, with minute hairs, broadly elliptic to somewhat cordate-ovate, the blades 5–15 cm. long, obtuse, with several conspicuous nerves converging at base and apex; petioles winged, mostly shorter than and rather abruptly expanded into blades; spikes linear-cylindric, dense, 0.5–4 (–7) dm. high, curved-ascending to erect; bracts broadly ovate, scarious-margined, mostly shorter than calyx; sepals 1.5–2 mm. long, broad; corolla-lobes pointed, 0.5 mm. long; caps. broadly conic, brown or purplish, mostly 6–10-seeded; seeds brown, reticulate, papillate, scarcely 1 mm. long; $2n=12$ (Gadella & Kliphuis, 1967).—Weed of damp waste places; natur. from Eu. April–Sept. Variable, with a number of named forms.

10. **P. marítima** L. var. **califórnica** (Fern.) Pilger. [*P. juncoides* var. *c.* Fern.] Perennial with thickish root; lvs. linear-oblanceolate to subspatulate, obtuse, very fleshy, spreading to depressed, usually 3–6 cm. long; scapes depressed to arched, usually much exceeding lvs.; bracts usually ovate, ca. 1.5 mm. long; sepals broadly oblong, ciliolate, almost 2 mm. long; corolla-tube pubescent; corolla-lobes ca. 1 mm. long; caps. thick-ellipsoid, ca. 3 mm. long; seeds 2–3, dark brown, ca. 1.5 mm. long.—Coastal Salt Marsh, etc.; Santa Rosa Id. n. along the coast. May–Sept.

11. **P. púrshii** R. & S. var. **oblónga** (Morris) Shinners. [*P. oblonga* Morris. *P. p.* var. *picta* Pilg.] Erect tufted annual, 5–20 cm. tall; lvs. entire, acute, 3–10 cm. long, 1–4 mm. wide, villous; scapes erect or ascending, 2–15 cm. high, mostly villous; spikes cylindrical, dense, 1–8 cm. long; basal bracts 2–3 times as long as calyx, scarcely if at all membranous on sides at base; sepals 2–3 mm. long; corolla-lobes 1–2 mm. long; caps. ellipsoid; seeds dark brown, finely pitted, 1.5–2 mm. long.—Dry places, mostly 2500–7000 ft.; Pinyon-Juniper Wd., Shadscale Scrub, Joshua Tree Wd., etc.; interior s. Calif. to w. edge of Colo. Desert and mts. of Mojave Desert; L. Calif., Ariz. April–June.

12. **P. pusílla** Nutt. Plant pubescent to glabrous, erect or strongly ascending, 2–10 cm. high; lvs. $\frac{1}{2}$–$\frac{2}{3}$ as high as scape, all basal; scapes several to many, spikes 1.5–6 cm.

long; bracts triangular-ovate, scarious-margined, slightly shorter than to equaling the calyx, 1.5–2 mm. long; sepals obovate; corolla-lobes 0.5 mm. long, mostly erect and forming a beak over the caps.; caps. ovoid, circumscissile below the middle, ca. 2 mm. long; seeds 4, dark brown, pitted, ca. 1.8–0.8 mm. long; $2n=12$ (Bassett, 1966).—Near San Diego; introd. from e. U.S.

13. **P. rhodospérma** Dcne. Like *P. virginica*, but sepals bristly and attenuate, 2.5–3 mm. long; seeds red, 2.5–3 mm. long.—Occasional weed, as at San Diego, Corona del Mar; Mo. and Kans. to Mex. May.

14. **P. virgínica** L. Villous annual or biennial; lvs. narrow-obovate, the blades 1–12 cm. long, soft-villous, entire or nearly so, narrowed gradually into somewhat shorter petioles; scapes long-villous, suberect, 5–30 cm. high; spikes dense, 2–12 cm. long; bracts lanceolate to narrow-elliptic, with fleshy green keel, ca. 2 mm. long; sepals elliptic to oblance-elliptic, 2–2.5 mm. long, rounded at apex, the midrib scarcely projected; corollas of fertile fls. closed over maturing caps., the lobes 2–3 mm. long; caps. ovoid, ca. 3 mm. long; seeds 2, pale, 1.5–2 mm. long; $2n=24$ (Fujiwara, 1956).—Collected near Loma Linda, San Bernardino, etc.; to Ore., Atlantic Coast. May.

81. **Platanàceae.** Sycamore Family

Trees with thin exfoliating bark. Lvs. large, alternate, deciduous, palmately lobed. Stipules thin, sheathing, entire or toothed. Petioles dilated at the base and largely covering the buds. Infl. of spherical unisexual heads, in our sp. these racemose on long slender peduncles. Fls. imperfect, minute. Calyx of 3–8 minute scalelike sepals. Petals minute, 3–6 in the ♂ fls., cuneiform-sulcate, scarious-pointed and longer than the sepals, in the ♀ fls. acute and longer than the rounded sepals. Stamens as many as the sepals and opposite them. Carpels as many as the sepals, 1-loculed, surrounded by persistent hairs and a few staminodia. Style terminal with the stigmatic surface ventral. Ovule 1 (2). Fr. a dense globose head of aks. with intermingled hairs and staminodis. Seeds elongate-oblong, pendulous, with fleshy endosperm.

1. *Plátanus* L. Sycamore. Plane-Tree

Characters of the family. Nine spp., of the N. Temp.

1. **P. racemòsa** Nutt. [*P. californica* Benth.] Plate 66, Fig. E. Large tree, 10–25 m. tall, with smooth pale bark; young growth rusty-tomentose; lvs. 1.5–2.5 dm. broad, ca. as long, deeply 5-lobed, the lobes subentire, tomentose on both surfaces when young, glabrescent above; petioles 3–8 cm. long; stipules 2–3 cm. long; ♂ heads several, 8–10 mm. in diam.; ♀ heads 3–5, sessile, 2–2.5 cm. in diam. in fr. Along stream beds and watercourses below 4000 ft.; many Plant Communities; cismontane s. Calif. to L. Calif. and cent. Calif. Feb.–April.

82. **Plumbaginàceae.** Leadwort Family

Mostly acaulescent perennial herbs with basal lvs. and scapose panicles, spikes or heads, sometimes with elongate branched stems. Fls. perfect, regular, 5-merous. Calyx bracted at base, tubular or funnelform, plaited, 5–15-ribbed, often scarious and colored. Corolla usually of nearly distinct petals, convolute or imbricate in bud. Stamens opposite the petals, often adnate to the base of claw. Ovary superior, 1-celled with 1 ovule; styles 5, separate or united. Fr. a utricle or ak., usually inclosed by calyx. Seeds 1 with membranous testa; endosperm mealy or lacking; embryo straight. Ca. 10 genera and 300 spp., widely distributed, usually of saline or calcareous places; some grown as ornamentals.

Infl. racemose, corymbose or paniculate; lvs. broad .... 2. *Limonium*
Infl. capitate; lvs. linear .... 1. *Armeria*

1. Armèria Willd. THRIFT

Tufted acaulescent perennial herbs. Lvs. narrowly linear, basal, persistent. Fls. in a globose head at the end of a naked scape, the heads subtended by 2 or more whorls of usually scarious bracts producing a sort of invol., the 2 outer bracts sheathlike and reflexed. Calyx funnelform, 5-toothed, 10-ribbed, scarious. Petals 5, united somewhat or distinct. Stamens 5. Styles united below. Fr. membranous, rarely dehiscent, 5-pointed apically. Ca. 50 spp., mostly of cooler climates. Several spp. cult. (Name said to be Celtic.) (Lawrence, G. H. M. The genus Armeria in N. Am. Am. Midl. Nat. 37: 757-779. 1947.)

1. **A. marítima** (Mill.) Willd. var. **califórnica** (Boiss.) Lawrence. Plate 66, Fig. F. Taproot long, tough; lvs. linear, 4-15 cm. long, 2-2.5 mm. wide, glabrous; scapes 1-several, 5-45 cm. long; outer sheathing bracts 1-3 cm. long, the inner invol. bracts narrow-deltoid to lance-oblong, acutish, 8-15 mm. long; calyx ca. 6-7 mm. long, with pubescent ribs, the lobes broadly triangular, ca. 1 mm. long, scarious; petals rose-pink, exceeding calyx.—Coastal bluffs and sandy places below 700 ft.; Santa Rosa Id.; San Luis Obispo Co. to B.C. April-Aug.

2. Limònium Mill. SEA-LAVENDER. MARSH-ROSEMARY

Perennial herbs with broad flat lvs. in a basal tuft. Fls. secund in loose spikes or clusters at ends of much-branched scape. Calyx campanulate or tubular, usually 10-ribbed, the limb 5-toothed, scarious. Petals 5, distinct, clawed. Stamens 5. Styles mostly 5, separate. Fr. membranous, indehiscent. Almost 200 spp., of wide distribution, often along the seaside. (Ancient Greek name, *Leimonion*, supposedly from *leimon*, a marsh.)

Lvs. not pinnatifid; branches of infl. not winged.
    Fls. ca. 5-6 mm. long, and 2 mm. wide. Native along coast .... 1. *L. californicum*
    Fls. ca. 9 mm. long and 4-5 mm. wide. Occasional escape on coast .... 2. *L. perezii*
Lvs. pinnatifid; branches of infl. winged. Occasional escape .... 3. *L. sinuatum*

1. **L. califórnicum** (Boiss.) Heller. [*Statice c.* Boiss. in DC.] Plate 67, Fig. A. Caudex heavy, woody, reddish; lvs. oblong to oblong-obovate, mostly obtuse, the blades 5-20 cm. long, tapering into petioles ca. as long; scape stout, 2-5 dm. high, loosely paniculate, with branches densely fld. at tips with small secund stipes 1-3 cm. long; calyx obconic, the ribs pubescent to above the middle, lobes whitish, deltoid-ovate, ca. 0.6 mm. long, acutish; petals pale violet, oblong, slightly exceeding calyx; $2n=18$ (Baker, 1954).—Coastal Salt Marsh, Coastal Strand; San Diego Co. to n. Calif. July-Dec. A form with glabrous calyx and extending from n. L. Calif. to cent. Calif. has been called var. *mexicanum* (Blake) Munz.

2. **L. perèzii** F. T. Hubb. [*Statice p.* Webb.] Woody at base; lvs. rhombic-ovate to deltoid, basal, the blades 8-15 cm. long, subtruncate at base, on petioles exceeding the blades; scapes branched, 4-6 dm. high; calyx purplish-blue, pubescent on ribs;·petals pale yellow.—Cult. and occasionally established as at beach at Ventura; native of Canary Ids. March-Sept.

3. **L. sinuàtum** (L.) Mill. [*Statice s.* L.] Rough-hairy perennial or biennial; lvs. basal, lyrate-pinnatifid with rounded lobes and sinuses, the blades 3-12 cm. long, short-petioled; scapes corymbosely panicled, 1.5-4 dm. high, winged, the wings ending in foliose lance-linear appendages 1-4 cm. long; calyx blue to white, funnelform, ca. 1 cm. long; petals yellowish-white.—Beaches and coastal marshes, San Diego and Los Angeles cos. and farther n.; native in Medit. region. June-Oct.

83. Polemoniàceae. PHLOX FAMILY

Annual or perennial herbs or shrubs or vines. Lvs. alternate or opposite, simple or dissected or compound. Infl. usually a paniculate glomerate or flat-topped cyme, or the fls. in heads or even solitary. Fls. complete, regular, mostly 5-, rarely 4- or 6-merous as to

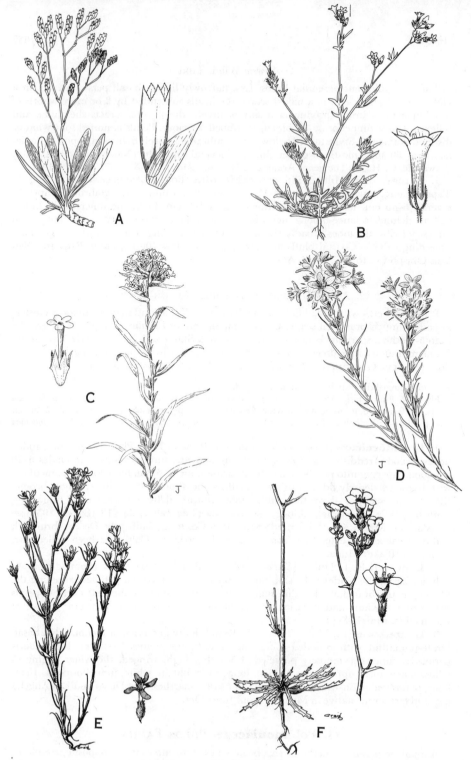

Plate 67. Fig. A, *Limonium californicum*. Fig. B, *Allophyllum gilioides*. Fig. C, *Collomia linearis*. Fig. D, *Eriastrum densifolium*. Fig. E, *Eriastrum sapphirinum* ssp. *dasyanthum*. Fig. F, *Gilia latiflora* ssp. *davyi*.

perianth and stamens. Calyx herbaceous to variously membranous or chartaceous, accrescent after anthesis or distended or ruptured by the caps., variously cleft, the lobes often with hyaline membrane between. Corolla gamopetalous, campanulate to funnelform to salverform, usually regular. Stamens equally or unequally inserted, the fils. equal or unequal. Ovary mostly 3-loculed; style simple; stigma-lobes usually 3. Fr. a caps., usually regularly dehiscent. Seeds 1 to many. Ca. 18 genera and 350 spp.; most numerous in w. N. Am.

1. Calyx growing with the caps. and not ruptured by it, becoming chartaceous in age.
    2. Calyx wholly green-herbaceous, the sinuses not distended; lvs. pinnately compound; mostly perennial .... 12. *Polemonium*
    2. Calyx ± scarious between the lobes, the sinuses distended or replicate; lvs. entire to bipinnately dissected; mostly annual .... 2. *Collomia*
1. Calyx at length ruptured by the maturing caps., commonly with a membranous pseudotube formed by the coalescence of the membranes bordering the sepals.
    3. Lvs. usually opposite, at least near base of plant, the upper sometimes alternate (if the lower are alternate, plant usually woody and lvs. prickly).
        4. Stamens unequally inserted; lvs. entire.
            5. Plants perennial; fls. showy .... 11. *Phlox*
            5. Plants annual; fls. inconspicuous .... 9. *Microsteris*
        4. Stamens equally inserted; lvs. mostly palmately cleft.
            6. Plants annual or perennial herbs, usually not very woody below; lvs. not spinose
                8. *Linanthus*
            6. Plants subshrubby; lf.-lobes mostly ending in ± spinose tips .... 7. *Leptodactylon*
    3. Lvs. mostly alternate, entire to pinnately dissected.
        7. Calyx-lobes subequal; fls. solitary, cymose or in heads.
            8. Corolla-lobes essentially alike; lf.-lobes mostly not setose- or spine-tipped.
                9. Upper cauline lvs. usually much reduced (see also *G. capillaris, G. filiformis*); tips of lf.-lobes not conspicuously mucronate; bracts mostly subtending groups of fls.; seeds small, rounded, brown .... 4. *Gilia*
                9. Upper cauline lvs. quite well developed.
                    10. Tips of lf.-lobes not conspicuously mucronate; bracts mostly subtending groups of 2–8 fls.; upper lvs. digitately lobed; seeds round-ovoid, mostly dark brown to black .... 1. *Allophyllum*
                    10. Tips of lf.-lobes bearing horny mucros; bracts subtending individual fls.; upper lvs. pinnatifid to entire; seeds long and slender to shorter and angled, ± curved, pale brown to straw-colored .... 5. *Ipomopsis*
            8. Corolla-lobes ± unlike; lf.-lobes setose- or spine-tipped .... 6. *Langloisia*
        7. Calyx-lobes unequal; fls. in dense bracteate heads.
            11. Plants cobwebby-pubescent, at least the infl. with a feltlike mass of interlaced hairs; caps. dehiscent from top; lvs. and bracts rarely with rigid spinose lobes . 3. *Eriastrum*
            11. Plants not having a feltlike mat of interlaced hairs; caps. dehiscent mostly from below or indehiscent; lvs. and bracts usually with rigid spinose lobes ..... 10. *Navarretia*

1. *Allophyllum* (Nutt.) A. & V. Grant

Erect annual herbs 1–6 dm. tall with slender or stouter stems, leafy, usually well branched cymosely, sometimes simple, lightly pubescent to villous or glutinous. Lvs. alternate, simple and entire to irregularly toothed, or pinnately to bipinnately lobed, the uppermost digitately 3–5-lobed; segms. linear-lanceolate to oblong, sessile or petiolulate. Fls. in loose to somewhat congested 2–8-fld. cymose glomerules, each glomerule subtended by a lf.; pedicels usually much elongate in fr. Calyx glandular-pubescent, accrescent, the lobes joined by the sinus-membrane to ca. the middle. Corolla funnelform, regular to ± bilabiate, light to dark blue-violet, red-violet, or white. Stamens unequally or equally inserted in upper half of tube, subequal to very unequal in length, included or exserted. Caps. subspherical, 3-loculed, with 1–3 seeds in each locule. Seeds black when ripe, 1.3–2.8 mm. long, plump, rounded on one side, or angular on 1 end when 2–3 seeds occur in same locule. Five spp. of sw. N. Am. (Greek, *allos*, other, and *phullon*, lf.)

(Grant, Alva and V. Grant. The genus Allophyllum. Aliso 3: 93–110. 1955.)
Segms. of cauline lvs. lanceolate to oblong, 3–15 mm. wide, the terminal segms. 3–8 times longer than wide; fls. pale to medium blue-violet, white or red-violet to purple with pink lobes.
  Corolla regular, white to red-violet to purple with pink lobes; at least some of stamens included; lvs. usually not in a basal tuft.
    Corolla-tube 6–16 mm. long, red-violet to purple with pink lobes; infl. usually of 4–8 fls. subtended by a lf. At 1000–9600 ft. .... 1. *A. divaricatum*
    Corolla-tube 5–8 mm. long, it and the limb white; infl. usually of 2 fls. subtended by a lf. At 4500–9000 ft. .... 4. *A. integrifolium*
  Corolla slightly bilabiate, pale to medium blue-violet; stamens all usually exserted; basal tuft of lvs. usually well developed .... 3. *A. glutinosum*
Segms. of lvs. linear to linear-lanceolate, not over 3 mm. wide, the terminal segms. of upper lvs. 8–14 times longer than broad; fls. dark blue-violet.
  Infl. glomerate with 4–8 fls. subtended by a lf.; many lvs. pinnately lobed; infl.-branches usually arising above the middle of a plant. Below 6000 ft. .... 2. *A. gilioides*
  Infl. loose with 2–3 fls. subtended by a lf.; few lvs. pinnately lobed; infl.-branches arising throughout the plant. At 4000–8500 ft. .... 5. *A. violaceum*

    1. **A. divaricàtum** (Nutt.) A. & V. Grant. [*Gilia d.* Nutt. and var. *volcanica* Brand.] Robust, to 6 dm. high, with stout stems lightly pubescent to villous and viscid and with skunklike odor; lower part of plant not densely leafy; lower lvs. entire to pinnately lobed with 1–6 pairs of lobes; infl. congested in early stages, but with pedicels elongating in fr.; calyx in fr. 6–7 mm. long; corolla regular, 7.5–22 mm. long, the tube red-violet and lobes pink, or tube purple and lobes pink-violet, the tube 2.5–4.5 times as long as lobes; stamens unequally inserted, unequally long; stigma included or exserted; locules of caps. 2–3-seeded; $n=8, 9$ (Grant and Grant, 1955).—In open ± disturbed places, burns, etc., 1000–6000 ft.; Chaparral, Foothill Wd., etc.; San Bernardino and San Gabriel mts., Mt. Pinos, to n. Calif. April–June.
    2. **A. gilioìdes** (Benth.) A. & V. Grant. [*Collomia g.* Benth. *Gilia g.* Greene.] Plate 67, Fig. B. Main stems 1–several, 1–4 dm. long, usually arising from a leafy base, and with many pinnately lobed lvs. up to the infl.; herbage puberulent, ± glandular in infl.; lower lvs. with 2–5 pairs of linear lobes 5–15 mm. long; upper reduced; infl. glomerate, each fl.-branch subtended by a lf. and terminated by mostly 4 short-pedicellate fls.; calyx in fr. ca. 5 mm. long; corolla dark blue-violet, 6.5–9.5 mm. long; stamens subequal, mostly included; locules of caps. 1-seeded; $n=9$ (Grant & Grant, 1955).—Open rather dry slopes and flats, below 6000 ft.; mostly Foothill Wd., Yellow Pine F.; mts. San Diego Co. to s. Sierra Nevada; Pinyon-Juniper Wd., Panamint Mts.; Ariz. April–June.
    3. **A. glutinòsum** (Benth.) A. & V. Grant. [*Collomia g.* Benth. *Gilia traskiae* Eastw.] With 1–several rather stout stems from near base, 1–6 dm. long, villous and viscid; with skunklike odor; lower lvs. in a basal tuft, with 3–10 pairs of lobes which are often irregularly toothed or lobed, the segms. 1–2 mm. wide, those of upper lvs. 3–11 mm. wide; infl. loose, 2–8-fld., the pedicels subequal, 5–20 mm. long in fr.; calyx 4.7–7.5 mm. long in fr.; corolla pale to medium violet-blue, 6–11 mm. long, the lobes 2.5–5 mm. long, spreading widely in bilabiate manner; style and stamens lying along lower lip and curving upward at tips; stamens ± unequally inserted and usually all exserted; caps.-locules mostly 2–3-seeded; $n=9$ (Grant & Grant, 1955).—Shaded brushy slopes and ravines, below 5000 ft.; mostly Coastal Sage Scrub and Chaparral; n. L. Calif. to cent. Calif.; Catalina Id. April–June.
    4. **A. integrifòlium** (Brand) A. & V. Grant. [*Gilia gilioides* var. *i.* Brand.] Plant 1–2.5 dm. high, the stems villous, with 1 main leader; lower lvs. few, not forming a basal tuft, oblong to lanceolate, entire or irregularly toothed, the segms. of the cauline lvs. oblong, 5–15 mm. wide, the terminal broadest; fls. mostly in loose pairs; pedicels 6–18 mm. long in fr.; calyx in fr. 5–6 mm. long; corolla white or pale blue, 6.5–11 mm. long, the tube 5.5–8 mm. long; stamens unequal, included; caps.-locules mostly 1-seeded; $n=9, 8$ (Grant & Grant, 1955).—Open places, 4500–9000 ft.; Yellow Pine F. to Lodgepole F.; San Gabriel and San Bernardino mts.; to Shasta Co. June–Aug.

5. **A. violàceum** (Heller) A. & V. Grant. [*Gilia v.* Heller.] Plants 0.7–4 dm. high with 1–several slender stems ± dichotomously branched to form a diffuse infl.; puberulent throughout, glandular in infl.; basal and lower stem-lvs. few, 1–5 cm. long, entire or with 1–2 pairs of lobes 1.5–8 mm. long; lvs. of infl. shorter; infl. dense, the fls. typically in 2's on a fl.-branch; fruiting calyx 3–5 mm. long; corolla dark blue-violet, 5–8 mm. long; stamens subequally inserted, mostly included; caps.-locules 1-seeded; $n=9$ (Grant & Grant, 1955).—Sandy and gravelly places, 4000–8500 ft.; Pinyon-Juniper Wd., Yellow Pine F. and above; mts. from San Diego Co. to Santa Barbara Co., Death V. region; n. to Plumas Co.; w. Nev. May–July.

2. Collòmia Nutt.

Annual or rhizomatous perennial herbs, erect to prostrate. Stems simple or branched. Lvs. alternate, linear to subovate, entire to pinnately dissected. Fls. in cymes or capitately congested or solitary and axillary. Calyx herbaceous or rarely with a chartaceous membranelike area below the sinuses. Corolla narrow-funnelform to trumpet-shaped, red, purplish, blue, white or salmon; the lobes spatulate to lanceolate. Stamens equally or unequally inserted on the throat; fils. equal or unequal. Pistil included or exserted. Caps. ellipsoid or obovoid, the locules mostly 1–2-seeded. Seeds oblong, often ± mucilaginous when wetted. Ca. 15 spp., w. N. Am. and s. S. Am. (Greek, *kolla*, glue, because of mucilaginous wetted seeds of most spp.)

(Wherry, E. T. Am. Midl. Nat. 31: 216–231. 1944.)
Corolla 15–30 mm. long, the tube much longer than the calyx, the limb salmon-yellow
                                  1. *C. grandiflora*
Corolla 5–15 mm. long, the tube little exceeding the calyx, the limb pink or white . 2. *C. linearis*

1. **C. grandiflòra** Dougl. ex. Lindl. [*Gilia g.* Gray.] Erect, usually simple annual 1–10 dm. high, leafy throughout, glabrous or puberulent or glandular; lvs. lanceolate to linear, entire, sessile, 3–5 cm. long, passing upward into ovate leafy bracts; fls. sessile in dense terminal, sometimes axillary heads; calyx obconic, 7–10 mm. long, becoming chartaceous, the lobes lanceolate; corolla yellow-salmon to almost white, 15–30 mm. long; stigma included; caps. obovoid, ca. 5 mm. long; seeds 1 in each locule, brown, ca. 3 mm. long; $n=8$ (Flory, 1937).—Dry open and wooded slopes below 8000 ft.; Chaparral, Yellow Pine F., etc.; mts. from San Diego Co.; to B.C., Rocky Mts. April–July.

2. **C. lineàris** Nutt. [*Gilia l.* Gray.] Plate 67, Fig. C. Annual, usually simple and erect, 1–6 dm. high; lvs. lanceolate to linear, entire, sessile, 1.6–5 cm. long, mostly acute; fls. sessile in bracteate heads, terminal on branches and sometimes in upper axils; bracts leafy; calyx campanulate, 4–7 mm. long; corolla pink to ± purplish, 8–15 mm. long; stamens included; stigma included; caps. ellipsoid, ca. 5 mm. long; seed 1 in a locule, ca. 2 mm. long; $n=8$ (Flory, 1937).—Dry places, 5000–7500 ft.; mostly Montane Coniferous F.; San Bernardino Mts.; to Alaska, Quebec, Ariz. May–Aug.

3. *Eriástrum* Woot. & Standl.

Annual or perennial, herbaceous or suffrutescent, ± erect to spreading, simple to branched, puberulent to arachnoid or woolly. Lvs. alternate, entire and linear or pinnately divided into linear divisions. Fls. mostly sessile in bracteate heads, rarely solitary and pedicelled. Calyx deeply cleft into subulate pungent lobes, the sinuses ± filled with hyaline membrane. Corolla blue to white or yellow, even pink, funnelform to subsalverform. Stamens inserted at base of corolla-throat or just below the sinuses, exserted to included. Anthers often sagittate, sometimes cordate or elliptic. Caps. ellipsoid or obovoid, often splitting into valves. Seeds 1–several in a locule, usually mucilaginous when wetted. A genus of ca. 14 spp., of w. N. Am. (Greek, *erion*, wool, and *astrum*, star, the plants woolly and with star-like fls.)

(Mason, H. L. The genus Eriastrum, etc. Madroño 8: 65–91. 1945.)
Plants perennial, with woody base; corolla 15–30 mm. long, the tube 2–3 times as long as calyx
                                        1. *E. densifolium*
Plants annual, not at all woody; corolla 7–18 mm. long, the tube rarely much longer than calyx.
 Stamens inserted in the sinuses of the corolla; anthers 2–2.5 mm. long. W. Mojave Desert
                                        5. *E. pluriflorum*
 Stamens inserted well below the sinuses.
  Corolla 15–16 mm. long, irregular, the lobes ca. 6 mm. long; stamens very unequal; bracts 5–9-lobed. Deserts .... 3. *E. eremicum*
  Corolla smaller, or if as long, regular and with stamens subequal and bracts 3-lobed.
   Lobes of corolla ca. as long as tube. Cismontane .... 6. *E. sapphirinum*
   Lobes of corolla conspicuously shorter than tube.
    Fils. cf stamens long-exserted. Cismontane .... 4. *E. filifolium*
    Fils. of stamens not long-exserted, the anthers sometimes showing.
     Corolla 9–12 mm. long, the lobes ca. 4 mm. long. Ne. Mojave Desert . 8. *E. wilcoxii*
     Corolla 7–8 mm. long, the lobes 2–3 mm. long.
      Stems spreading from base, thinly pilose, later glabrate. Mojave and Colo. deserts
                                           2. *E. diffusum*
      Stems erect, mostly floccose. Little San Bernardino Mts. and north
                                         7. *E. sparsiflorum*

  1. **E. densifólium** (Benth.) Mason. Plate 67, Fig. D. [*Huegelia d.* Benth. *Gilia d.* Benth.] subsp. **austromontànum** (Craig) Mason. [*Gilia d.* var. *a.* Craig.] Erect perennial, much branched from the woody base, 1–4 dm. high, glabrate; lvs. well distributed along the stems, irregularly pinnatifid into linear ± pungent segms. 2–5 cm. long; fls. up to 20 in a cluster, sessile in terminal bracteate heads; bracts 3–5-lobed, the terminal lobe much the longest, acerose; calyx 6–7 mm. long, the lobes subequal, pungent, united below by a white-lanate membrane; corolla salverform, deep blue, 15–19 mm. long; stamens inserted just below the sinuses, the anthers white, sagittate, exserted; caps. ellipsoid, 3–4 mm. long, the locules several-seeded; seeds cellular-reticulate, ca. 2 mm. long; $n=7$ (Grant, 1958).—Dry slopes, 4000–8000 ft.; Montane Coniferous F.; L. Calif. to Santa Barbara and Inyo cos. May–Sept.
  Ssp. **mohavénse** (Craig) Mason. [*Gilia d.* var. *m.* Craig.] Plant 1–3 dm. high, canescent-lanate; lvs. with broad rachis and short teeth; corolla pale blue to lavender, ca. 15 mm. long.—Dry slopes, 2500–8500 ft.; Joshua Tree Wd., Pinyon-Juniper Wd.; Mojave Desert from Kelso and Little San Bernardino Mts. to Kern and Inyo cos. June–Oct.
  Ssp. **sanctòrum** (Mlkn.) Mason. [*Gilia d.* var. *s.* Milliken.] Entire plant lanate even in age, 25–75 cm. high; fls. blue, 25–32 mm. long.—Below 1500 ft.; Coastal Sage Scrub; along Santa Ana River. June–Aug.
  2. **E. diffùsum** (Gray) Mason. [*Gilia filifolia* var. *d.* Gray.] Annual, diffusely branched, 3–15 cm. high, thinly pilose, then glabrate; lvs. simple, linear to pinnatifid into 3–5 linear lobes; fl.-clusters 3–20-fld., woolly; bracts 3–7-lobed, ± arched; calyx 5–7 mm. long, the lobes unequal; corolla pale blue to white, 7–8 mm. long, almost regular, the lobes ca. half as long as the tube; stamens 1–2 mm. long, inserted in middle of throat; caps. 3–4 mm. long, the locules 2–3-seeded.—Dry sandy places, mostly below 8000 ft.; Creosote Bush Scrub, Joshua Tree Wd., Pinyon-Juniper Wd.; deserts from Inyo Co. to L. Calif., Tex. March–May.
  3. **E. erèmicum** (Jeps.) Mason. [*Huegelia e.* Jeps. *Navarretia densifolia* var. *jacumbana* Brand.] Erect or spreading annual, 3–25 cm. high, floccose to glabrate, usually much-branched from base; lvs. 1–4 cm. long, pinnatifid into 5–9 linear lobes; fl.-clusters lanate, 2–10-fld.; bracts recurved at tips; calyx segms. linear, subequal; corolla ± bilabiate, violet, ca. 15–16 mm. long, the lobes ca. 6 mm. long; stamens inserted at base of throat, unequal, the longest exserted; stigma exserted; caps. 4–6 mm. long, the locules several-seeded; seeds ca. 2 mm. long.—Common in sandy places below 7500 ft.; Creosote Bush Scrub, Joshua Tree Wd., Pinyon-Juniper Wd.; deserts from Imperial Co. to Inyo Co.; Nev. April–June.

4. **E. filifòlium** (Nutt.) Woot. & Standl. [*Gilia f.* Nutt.] Annual, pilose to subglabrous, 0.4–4 dm. high, simple and virgate or branched; lvs. linear, filiform, entire to pinnately 3–5-lobed; heads 3–15-fld., lanate; bracts 1–2 cm. long, lanate; calyx 5–7 mm. long; corolla-funnelform, ± regular, 8–9 mm. long, blue, the lanceolate lobes 2–2.5 mm. long, the tube yellow; stamens 2.5 mm. long, attached at base of throat, exserted; caps. 4–5 mm. long, cylindric; seeds several in a locule, ca. 1 mm. long.—Occasional, dry slopes below 5000 ft.; Coastal Sage Scrub, Chaparral, etc.; cismontane, L. Calif. to Santa Barbara Co., San Clemente and Catalina ids. May–June.

5. **E. pluriflòrum** (Heller) Mason. [*Gilia p.* Heller.] ssp. **pluriflòrum**. Erect annual, 1–4 dm. high, simple to branched, floccose to glabrate; lvs. sessile, 1.5–5 cm. long, with 3–9 filiform lobes; infl. densely arachnoid, dense, 1.5–4 cm. broad, 8–50-fld.; calyx 8–11 mm. long; cleft into linear unequal lobes, densely arachnoid; corolla funnelform, nearly or quite irregular, 1–2 cm. long, deep blue-violet or with some yellow in throat, the lobes ⅔ as long as tube; stamens inserted in sinuses, 4–5 mm. long; stigma exserted; caps. 4 mm. long, the locules 2–3-seeded.—Dry plains and slopes below 6000 ft.; n.w. Mojave Desert to San Joaquin V.; Chaparral, Yellow Pine F., etc. May–July.

Ssp. **sherman-hoỳtae** (Craig) Mason. [*Gilia s.* Craig.] Erect, 6–12 cm. high; infl. 2–8-fld.; corolla pale blue to lavender, the lobes not more than ½ as long as tube; stamens 3–4 mm. long.—Sandy flats, ca. 2500 ft. Joshua Tree Wd.; Antelope V. region, w. Mojave Desert as at Muroc Lake. May.

6. **E. sapphirìnum** (Eastw.) Mason. [*Gilia s.* Eastw.] ssp. **sapphirìnum**. Erect annual 1–3 dm. high, loosely paniculately branched, sparsely leaved, commonly viscid-glandular, sometimes lightly floccose in infl.; lvs. linear, commonly entire, 1–3 cm. long, the upper sometimes with 2 short lateral lobes; fls. sessile in few-fld. cymes, the bracts broadly ovate, 3-lobed, often hyaline-margined; calyx ca. 8 mm. long, with broad hyaline membrane between the lobes; corolla funnelform, 8–10 mm. long, sapphire-blue with yellow tube and throat, the tube 1–2 times the throat; stamens inserted at base of throat, 7–8 mm. long; stigma exserted; caps. ca. 4 mm. long, the locules several-seeded; seeds 1 mm. long; $n=7$ (Grant, 1958).—Dry places, mostly 4000–8000 ft.; Chaparral, Yellow Pine F.; San Bernardino Mts. to L. Calif. June–Aug.

KEY TO SUBSPECIES

Flower-heads 1–3 (–4)-fld.; corolla usually less than 12 mm. long, the lobes as long as the tube. Calyx rarely at all lanate, frequently glandular; bracts 1–3-lobed; corolla usually 10–12 mm. long. Montane.
    Fls. sessile, mostly more than one .... ssp. *sapphirinum*
    Fls. pedicelled, solitary. S. San Diego Co. .... ssp. *gymnocephalum*
Calyx always lanate; bracts 2–5-lobed; corolla usually less than 10 mm. long. Deserts and bordering mts. .... ssp. *ambiguum*
Flower-heads 3–10-fld.; corolla ca. 14 mm. long. Cismontane .... ssp. *dasyanthum*

Ssp. **ambíguum** (Jones) Mason. [*Gilia floccosa* var. *a.* Jones.] Bracts broad, 3–5-lobed; corolla pale to deep blue, 8–10 mm. long.—Dry places, 2500–7000 ft.; Joshua Tree Wd., Pinyon-Juniper Wd.; w. Mojave Desert and bordering mts.; Santa Ana Mts. May–Sept.

Ssp. **dasyánthum** (Brand) Mason. [*Navarretia virgata* var. *d.* Brand.] Plate 67, Fig. E. Heads dense, mostly 5–10-fld., white-woolly; corolla blue, 14–15 mm. long.—Common in dry places below 2500 ft.; Coastal Sage Scrub, Chaparral; cismontane L. Calif. to Ventura Co. and n. May–Sept.

Ssp. **gymnocéphalum** (Brand) Mason. [*Navarretia virgata* ssp. *g.* Brand.] Fls. solitary and pedicelled, rarely in pairs.—Chaparral and Yellow Pine F.; s. San Diego Co.; n. L. Calif. May–Sept.

7. **E. sparsiflòrum** (Eastw.) Mason. [*Gilia s.* Eastw.] Erect annual, 1–4 dm. high, mostly floccose, usually branching from below; lvs. linear or with a pair of short lobes near the base; bracts 3–5-lobed or simple; fl.-clusters 2–5-fld., arachnoid; calyx 5 mm. long, deeply cleft into equal lobes; corolla 7–8 mm. long, pale blue to whitish or pinkish,

the lobes ca. 3 mm. long; stamens 2–2.5 mm. long, inserted at base of throat, scarcely evident externally; caps. 3–5 mm. long; seeds light brown, ca. 1 mm. long, slightly wing-angled.—Dry slopes below 8000 ft.; Joshua Tree Wd., Sagebrush Scrub, Pinyon-Juniper Wd., Yellow Pine F.; Little San Bernardino Mts. to Panamint Mts., Tehachapi Mts.; to s. Sierra Nevada and to Ore., Nev. June–July.

8. **E. wilcóxii** (A. Nels.) Mason. [*Gilia w.* A. Nels.] Branched annual, floccose, 1–2 dm. high, the lowest branches longest, making a flat-topped crown; lvs. with 5 (3) linear lobes; bracts 3–5-lobed; heads 3–5-fld., often several heads crowded together; calyx unequally lobed; corolla ± funnelform, 9–11 mm. long, blue to pale blue, the lobes ca. 4 mm. long; stamens slightly exserted; caps. ca. 5 mm. long, ellipsoid; seeds several to a locule, ca. 1 mm. long, ± wing-angled.—Dry places below 9000 ft.; Sagebrush Scrub, Pinyon-Juniper Wd.; Panamint Mts. along e. face of Sierra Nevada and White Mts., to Wash., Ida., Utah. June–Aug.

4. *Gília* R. & P. GILIA

Annual to rarely perennial herbs. Lvs. alternate, entire or more commonly pinnately lobed or dissected, mostly well developed below, reduced upward, often largely in a basal rosette. Fls. solitary in the lf.-axils or in paniculately branched or thyrsoid infls. or crowded into compact heads. Calyx-lobes nearly equal, cleft deeply and often margined by membranes which may unite them. Corolla tubular-funnelform to salverform, mostly regular, blue, lavender, pink to yellow or white. Stamens equally inserted in the corolla-tube or the throat, often in the sinuses of the corolla-lobes, equal or unequal in length. Caps.-valves remaining joined below, spreading above at dehiscence. Seeds few to many in a locule. A genus of ca. 50 spp., Calif. and S. Am. (Felipe *Gil*, Spanish botanist.)

(Grant, V. or A. and V. A series of papers, genetic and taxonomic, on Gilia, in Aliso mostly in vols. 2 and 3, 1950–1956.)
1. Pollen cream or yellow; calyx-lobes broadest at base, narrowing to a point, joined by sinus-membrane at base only, or in *G. latifolia* and *G. ripleyi*, ca. to middle; corolla campanulate, rotate or funnelform. Largely deserts.
    2. Corolla funnelform, pink.
        3. Plant annual or perennial; corolla pink within, buff or white without; calyx 6–8 mm. long
           .... 19. *G. latifolia*
        3. Plant perennial; corolla pink within and without; calyx 4–5 mm. long .. 29. *G. ripleyi*
    2. Corolla campanulate golden yellow .... 14. *G. filiformis*
1. Pollen blue (sometimes white); calyx-lobes narrow or broad, not wider at base than middle, joined by sinus-membrane to above middle; corolla salverform or funnelform.
    4. Plants arachnoid-woolly on lower stems and lvs. Mainly deserts and mts.
        5. Stems glabrous at base (except sometimes in *G. latiflora*), rarely glandular-pubescent at base.
            6. Corolla-throat and tube dark purple, the former narrow and flaring; lower lvs. pinnate with linear lobes. Santa Rosa Id. .... 34. *G. tenuiflora*
            6. Corolla-throat yellow and white or purple in lower part and yellow and white above, usually full and broadly expanded; cauline lvs. shallowly dentate or entire with broad clasping base tapering to a narrow apex.
                7. Fls. 7–12 mm. long, with lobes 1.5–4 mm. long; stamens and styles short; pollen white .... 31. *G. sinuata*
                7. Fls. 8.2–35 mm. long, the lobes 3.5–11 mm. long; stamens unequal, the longest well exserted; style exceeding longest stamens; pollen blue or white.
                    8. Corolla-tube and -throat combined 1½–2 times as long as calyx; corolla 8.2–12 mm. long. Mostly s. of the San Gabriel Mts. .... 13. *G. diegensis*
                    8. Corolla-tube and -throat combined 2½–6 times as long as calyx; corolla 9.5–35 mm. long. Mainly n. of San Gabriel Mts. .... 18. *G. latiflora*
        5. Stems cobwebby-pubescent at base (glabrous in *G. ochroleuca* and sometimes in *G. mohavensis*).
            9. Upper corolla-throat violet or white, sometimes yellow; throat lacking dark purple except sometimes streaked with purple in *G. aliquanta*; fls. 4–32 mm. long; calyx

Gilia

glabrous to glandular-dotted or cobwebby-pubescent; caps. mostly globular to broadly ovoid.

10. Fls. 4–13.5 mm. long; calyx glabrous to lightly pubescent with small glandular dots or cobwebby hairs.

11. Calyx glabrous to sparsely cobwebby-pubescent; pedicels divaricate.

12. Corolla-throat broadly expanded and with 2 bands of color (yellow at base, blue-violet above); pedicels and calyces glabrous or pedicels glandular immediately beneath the fls. and then the calyces cobwebby-pubescent .... 27. *G. ochroleuca*

12. Corolla-throat narrowly flaring and without a clear banding of color; pedicels glandular beneath the glabrous or sometimes cobwebby calyx . 2. *G. aliquanta*

11. Calyx sparsely to moderately glandular; pedicels strict or at least not strongly divaricate.

13. Calyx-lobes acuminate, prolonged ca. 1–1.5 mm. beyond the bordering sinus-membrane; calyx very sparsely pubescent; basal lvs. bipinnately dissected.

14. Corolla 7–12 mm. long, the tube well exserted from the calyx

28. *G. ophthalmoides*

14. Corolla 3.5–5 mm. long, the tube included in the calyx ...... 12. *G. clokeyi*

13. Calyx-lobes acute, terminating less than 0.5 mm. beyond the bordering sinus-membrane; calyx moderately glandular; basal lvs. once pinnate.

13a. Corolla-lobes pointed; corolla-tube violet to white .... 36. *G. transmontana*

13a. Corolla-lobes rounded at apex; corolla-tube deep purple . 22. *G. malior*

10. Fls. 10–32 mm. long; calyx moderately pubescent with large glandular dots or cobwebby hairs.

15. Upper corolla-throat light violet; stamens short, barely exserted or to ⅔ as long as corolla-lobes.

16. Corolla-tube long-exserted from calyx, or, if included, then infl. full, divaricately branched, with shortest pedicels at least half as long as longest

7. *G. cana*

16. Corolla-tube included or but slightly exserted from calyx ... 20. *G. leptantha*

15. Upper corolla throat white or yellow; stamens long, subequal to or exceeding corolla-tube .... 20. *G. leptantha*

9. Upper corolla-throat yellow or purple or partly purple; fls 4–20 mm. long; calyx moderately to heavily glandular or cobwebby; caps. ± ovoid.

17. Infl. ± congested to glomerate; calyx glandular and at maturity as long as or longer than caps.

18. Stamens and style long-exserted.

19. Corolla with broad throat, white to violet. Deserts and e. of Sierra Nevada

6. *G. brecciarum*

19. Corolla slender in form, deep violet with purple tube. Inner S. Coast Ranges from Ventura and Santa Barbara and Kern cos. .... 17. *G. jacens*

18. Stamens and style appearing at orifice of corolla or slightly exserted.

20. Cauline lvs. deeply lobed, the lateral lobes usually much longer than width of rachis; corolla-veins beneath the sinuses violet .... 6. *G. brecciarum*

20. Cauline lvs. coarsely dentate, or deeply lobed and then the lateral lobes longer than the width of the rachis; corolla-veins beneath the sinuses white, or rarely violet and then cauline lvs. shallowly dentate

25. *G. modocensis* (see also 35. *G. tetrabreccia*)

17. Infl. diffuse; calyx moderately glandular, at maturity shorter than to longer than the caps.

21. Stamens very short, maturing at orifice; corolla-throat purple or if yellow, then purple-spotted or -veined; lf.-parts very narrow, 0.5–1.2 mm. wide; calyx shorter than caps. at maturity .... 24. *G. minor*

21. Stamens unequal, the longest exserted; corolla-throat yellow or with some purple; lf.-parts very narrow to broader, 0.5–3 mm. wide; calyx ca. as long as or longer than caps. at maturity.

22. Calyx 2.6–4.6 mm. long; corolla-lobes 2.0–4.5 mm. long. Piute Range to s. Sierra Nevada .... 16. *G. interior*

22. Calyx 2–2.5 mm. long; corolla-lobes 1.4–2.8 mm. long. Cuyama V. to Boron in w. San Bernardino Co. .... 5. *G. austrooccidentalis*

4. Plants not arachnoid-woolly on lower stems or lvs.

23. Infl. with loose to headlike glomerules mostly of 3–many fls. subtended by a single lf.; upper cauline lvs. well developed and pinnately dissected. Cismontane.
 24. Fls. in dense spherical heads; calyx composed of relatively narrow herbaceous bands and narrow to broad hyaline sinuses; stamens as long as or longer than corolla-lobes
   9. *G. capitata*
 24. Fls. solitary, in loose cymes, or glomerate to loosely capitate with 12–25 fls. in a head; calyx composed of broad herbaceous bands and narrow hyaline sinuses; stamens shorter than to subequal to corolla-lobes.
  25. Corolla blue-violet, rarely white, the tube sometimes darker than the lobes.
   26. Corolla-throat full and equal to or exceeding tube; lf.-segms. broad. Mainland
    1. *G. achilleaefolia*
   26. Corolla-throat narrow and exceeded by tube; lf.-segms. very narrow and linear. Insular .... 26. *G. nevinii*
  25. Corolla bi- or tri-colored with yellow tube and purple-spotted or yellow throat.
   27. Corolla campanulate, 5–12 mm. wide when pressed; style exserted, the stigmas beyond the anthers; corolla deciduous after anthesis .... 3. *G. angelensis*
   27. Corolla funnelform, 3–5 mm. wide when pressed; style included, the stigmas scarcely beyond the anthers .... 11. *G. clivorum*
23. Infl. loose, consisting of 1–8 fls. subtended by a single lf.; upper cauline lvs. much reduced, or, if well developed, then linear to linear-oblong.
 28. Plant body leafy throughout, the basal rosette absent or very little developed; lvs. linear to linear-oblong, usually entire .... 8. *G. capillaris*
 28. Plant scapose with reduced lvs. above and basal rosette well developed; lvs. sinuate to much dissected or the cauline linear and entire.
  29. Basal lvs. 1–3-times pinnate with narrow rachis; cauline lvs. pinnately lobed except sometimes the uppermost bracts entire.
   30. Caps. oblong-ovoid; lvs. finely divided; basal lvs. pubescent with multicellular hairs, uppermost lvs. usually entire.
    31. Stamens inserted in corolla-sinuses, shorter than corolla-lobes; fls. pink to whitish or pale violet.
     32. Corolla 10–36 mm. long, usually pink; style exserted. San Jacinto Mts. and north .... 32. *G. splendens*
     32. Corolla 5–9 mm. long, whitish or pale violet; style included. Little San Bernardino Mts. and San Bernardino V. to San Diego Co. . 4. *G. australis*
    31. Stamens inserted in middle of corolla-throat, exserted beyond corolla-lobes; fls. blue-violet. San Diego Co. .... 10. *G. caruifolia*
   30. Caps. broadly ovoid; lvs. finely to coarsely divided; basal lvs. pubescent with multicellular or geniculate hairs; uppermost lvs. toothed or lobed.
    33. Corolla-tube included in the calyx; lower lvs. with white geniculate hairs
     33. *G. stellata*
    33. Corolla-tube well exserted; lower lvs. with straight translucent hairs
     30. *G. scopulorum*
  29. Basal lvs. strap-shaped and sinuate to bipinnatifid; cauline lvs. narrowly linear and entire, or the lowermost sometimes like the basal. Deserts.
   34. Corolla 2.5–7.5 mm. long; basal lvs. pinnately toothed or lobed and the lobes not exceeding twice the width of the lf. between sinuses; lf.-rachis or -blade 2–10 mm. wide.
    35. Corolla-throat narrow, the tube and throat threadlike; sinus-membrane of calyx U-shaped between lobes in fr. .... 21. *G. leptomeria*
    35. Corolla-throat full, the tube and throat not threadlike; sinus-membrane of calyx V-shaped between lobes in fr. .... 23. *G. micromeria*
   34. Corolla 8–14 mm. long; basal lvs. bipinnately lobed and the lobes longer than twice the rachis-width; lf.-rachis 1–2 mm. wide. .... 15. *G. hutchinsifolia*

1. **G. achilleæfòlia** Benth. ssp. **achilleæfòlia**. Erect annual, mostly ± branched above, 1–7 dm. high, often ± floccose below and glandular above; lower lvs. largely bipinnate, 4–10 cm. long, the ultimate segms. ± falcate, 1–25 mm. long, 0.5–2 mm. wide; lf.-axils floccose; infl. a fairly dense fan-shaped head or heads on naked peduncles 1–15 cm. long; fls. 8–25, subsessile on pedicels 1–2 mm. long; calyx 4.5–7 mm. long, floccose or with some glands, the lobes acute with hyaline blue-violet wings; corolla funnelform, mostly 10–20 mm. long, blue-violet, the lobes oval; stamens included; styles 1–2 mm. longer than

mature corolla; caps. ovoid, dehiscent, 10–18-seeded; seeds 1–2 mm. long; $n=9$ (Grant, 1954).—Loose soils of open places, below 4000 ft.; Coastal Strand, Chaparral, etc.; Santa Cruz Id.; n. Santa Barbara Co. to cent. Calif. May–June.

Ssp. **multicáulis** (Benth.) V. & A. Grant. [*G. m.* Benth.] Infl. loosely cymose, with groups of 2–7 fls. on pedicels 1–60 mm. long; corolla 5–10 mm. long; $n=9$ (Grant, 1954). —Shaded woods, with the sp., from Ojai and Santa Ynez Mts. n.

2. **G. aliquánta** A. and V. Grant. ssp. **aliquánta.** Small erect plant with a main stem and basal rosette of lvs., later secondary stems 1–1.5 dm. long, glabrous to cobwebby or glandular below, lightly glandular above; basal lvs. 1–3 cm. long, with opposite linear lobes 2–3 mm. long; middle cauline lvs. somewhat reduced; infl. loose; pedicels unequal, 1–11 mm. long; calyx glabrous to sparsely cobwebby, 3.5–5 mm. long at anthesis; corolla 6–12 mm. long, 2–3 times the calyx, bright yellow with some yellow in throat, the lobes ca. as long as tube and throat combined; stamens long exserted; $n=9$ (Grant et al., 1956).—Rocky and gravelly slopes, 2500–4100 ft.; Joshua Tree Wd., Pinyon-Juniper Wd.; deserts base of San Gabriel and San Bernardino mts. to n. Kern Co. April–May.

Ssp. **brevilòba** A. & V. Grant. Corolla-lobes 1.3–2.5 mm. long, the tube included to well exserted from the calyx; stamens scarcely exserted; $n=9$ (Grant, 1958).—Rocky and gravelly slopes, 2800–6200 ft.; Joshua Tree Wd., Pinyon-Juniper Wd.; Providence Mts., e. San Bernardino Co. to White Mts.; to Utah. March–May.

3. **G. angelénsis** V. Grant. Erect annual 0.7–7 dm. high, usually with spreading branches; stems glabrous or floccose below, sometimes glandular above; lower lvs. unipinnately or bipinnately dissected, 2–5 cm. long, the ultimate segms. ± falcate, 5–15 mm. long, 0.5–1.5 mm. wide, the axils of the lvs. floccose; infl. cymose, the heads usually ca. 5-fld., several, on naked peduncles 1–5 cm. long; pedicels 2–3 mm. long; calyx 3–4 mm. long, mostly floccose, the lobes acute, the cent. part green, the margins hyaline, 0.1–0.3 mm. wide at sinus; corolla campanulate, 7–8 mm. long, the limb blue-violet to white, the tube pale yellow, the lobes ovate, 2–3 mm. wide; fils. ca. 1 mm. long; caps. ovoid, dehiscent, 20–30-seeded; seeds 0.5–1 mm. long; $n=9$ (Grant, 1952).—Loose often sandy or gravelly soil, below 6300 ft.; Coastal Sage Scrub, Chaparral, Yellow Pine F.; cismontane to Santa Barbara Co. and n., and s. to L. Calif. March–May. Catalina Id.

4. **G. austràlis** (Mason & A. Grant) V. Grant. [*G. splendens* ssp. *a.* Mason & A. Grant.] Branched scapose annual 1–3 dm. high, with basal rosette of villous lvs.; stems glabrous; infl. glandular, sometimes also sparsely pubescent throughout; lvs. 1-, 2-, 3-pinnate, 2–7 cm. long, the lateral lobes 3–6 on each side, narrow, 5–15 mm. long; infl. cymose; fls. in 2's on unequal pedicels 4–26 mm. long; calyx 2–4 mm. long; corolla funnelform, 5–10 mm. long when pressed; tube 2–3 mm. long, limb 4–9 mm. in diam., pale violet or whitish, with yellow spots in throat, the lobes commonly streaked purple without; stamens subequal; stigmas 1 mm. long, included; caps. ellipsoid, 4–6 mm. long, 20–30-seeded; seeds ovoid, 0.5–1 mm. long; $n=9$ (Grant & Grant, 1954).—Sandy places, below 4000 ft.; Coastal Sage Scrub, Chaparral, San Bernardino V. to e. San Diego Co.; Creosote Bush Scrub and Joshua Tree Wd., Little San Bernardino Mts. to Palm Springs region; L. Calif. March–June.

5. **G. austrooccidentàlis** (A. & V. Grant) A. & V. Grant. Resembling *G. interior*. Lower stem and lvs. rather lightly pubescent; calyces mostly glandular.—W. Mojave Desert from Boron to Cuyama V.

6. **G. brecciárum** Jones. [*G. inconspicua* ssp. *sinuata* var. *deserti* Brand.] ssp. **brecciárum.** Erect or decumbent, 1–3 dm. high, leafy, cobwebby-pubescent below, glandular above; basal rosette semi-erect, loose, the lower lvs. bipinnately to tripinnately lobed, the rachis 1.5–3 mm. wide, the lobes ca. as wide; calyx glandular, 2.5–4 mm. long at anthesis, the broad herbaceous lobes connected by narrower sinus-membranes; corolla 7–11 mm. long; throat and orifice narrow to broad, the veins beneath sinuses violet; lobes 2–3.5 mm. long, violet to white; stamens and style not or slightly exserted; caps. 4–6.8 mm. long, broadly oval; $n=9$ (Grant et al., 1956).—Sandy slopes, below 7500 ft.; Creosote Bush Scrub to Pinyon-Juniper Wd., Yellow Pine F., etc.; w. Mojave Desert to Cuyama V. and e. of Sierra Nevada to Ore., Nev. April–June.

Ssp. **argusàna** A. & V. Grant. Corolla 9–20 mm. long, the form long and slender, the tube well exserted; stamens and style long exserted.—At 2100–5800 ft.; Creosote Bush Scrub and above; w. borders of Mojave Desert from Red Rock Canyon to Darwin, etc. March–May.

Ssp. **neglécta** A. & V. Grant. Corolla 9–20 mm. long, corolla-form full and stout, the tube not or scarcely exserted; stamens and style long exserted.—Sandy places, 2400–7000 ft.; Creosote Bush Scrub to Pinyon-Juniper Wd.; w. borders of Mojave Desert in Inyo and Kern cos.

7. **G. càna** (Jones) Heller ssp. **bernardína** A. & V. Grant. Erect annuals with 1 to several rather stout stems 1–3 dm. high; lvs. in a basal rosette, densely cobwebby, bipinnately lobed or toothed, the rachis 0.5–3 mm. wide, the lobes broader; infl. somewhat crowded, with short internodes; shortest pedicels usually much less than half as long as the longest; infl. strict, the branches not divaricate; corolla 17–23 mm. long, the tube purple, the throat narrow, yellow below, violet above; stamens not or slightly exserted; caps. 5–9 mm. long; $n=9$ (Grant et al., 1956).—Sandy flats and washes, 2700–4800 ft.; Joshua Tree Wd.; n. base of San Bernardino Mts. April–May.

Ssp. **specifórmis** A. & V. Grant. Basal lvs. moderately cobwebby-pubescent; corolla 15–29 mm. long, the tube stout, 2–4 times the calyx, gradually widened into the slender throat; lobes 3–8 mm. wide.—Sandy places, 2800–3800 ft.; largely Joshua Tree Wd., Mojave Desert from Ord Mts. to Amargosa and Black mts. April–May.

Ssp. **speciòsa** (Jeps.) A. & V. Grant. [*G. tenuiflora* var. *s*. Jeps.] Densely cobwebby-pubescent below; corolla 20–32 mm. long, with narrow orifice, the lobes broadly oval, 4–8 mm. wide; $n=9$ (Grant et al., 1956).—Canyons and fans, 2500–4000 ft.; Joshua Tree Wd., Creosote Bush Scrub, Red Rock Canyon to Nine-mile Canyon, w. Mojave Desert. April–May.

Ssp. **tricéps** (Brand) A. & V. Grant. [*G. tenuiflora* var. *t*. Brand.] Basal lvs. moderately cobwebby-pubescent; corolla 8–23 mm. long, the slender tube expanding abruptly into the throat; lobes 2–6 mm. wide; $n=9$ (Grant et al., 1956).—Dry places, 2800–5200 ft.; Creosote Bush Scrub to Pinyon-Juniper Wd.; Barstow and Kelso to Panamint and White mts.; Nev. April–May.

8. **G. capillàris** Kell. Erect annuals, 0.2–3 dm. high, well branched to subsimple, stipitate-glandular throughout; lvs. simple or rarely with a pair of basal lateral lobes, linear, 0.5–5 mm. wide, the lower and middle lvs. 1–4 cm. long; infl. cymose, the first pedicels shorter; calyx slender, 3–4 mm. long, the lobes acuminate, glandular; corolla funnelform, 3–6 mm. long, pale violet to pink or white, often streaked purple, the tube sometimes yellow, the throat sometimes with purple spots; fils. and stigmas included; caps. ovoid, 3–4 mm. long; seeds 6–12, 1–1.6 mm. long; $n=9$ (Grant, 1958).—Sandy places, 4500–7500 ft.; mostly Montane Coniferous F.; mts. from San Diego Co. to Mt. Pinos; at other elevs. farther n.; to Wash., Ida. June–Aug.

9. **G. capitàta** Sims ssp. **abrotanifòlia** (Nutt. ex Greene) V. Grant. [*G. a.* Nutt. ex Greene.] Commonly branched from base, 2–8 dm. high, floccose or glabrous below and glandular or glabrous above; lower lvs. unipinnately or bipinnately dissected, the ultimate segms. 0.5–2 mm. wide, the axils of lvs. usually floccose; heads several, on naked peduncles 4–25 cm. long, 1.5–3 cm. in diam.; pedicels 1–2 mm. long; calyx greenish to purplish-brown, lightly floccose, the acuminate lobes recurved at tip; corolla 8–12 mm. long, light blue-violet to whitish, the oval lobes 1.5–3 mm. wide; fils. 2–4 mm. long; style 1–3 mm. longer than corolla; seeds 9–13, ovoid, 1–1.5 mm. long; $n=9$ (Grant, 1952).—Loose sandy places below 6000 ft.; Coastal Sage Scrub, Chaparral, Yellow Pine F., etc.; near the coast from Santa Barbara Co. s., inland to Tehachapi Mts., Santa Ana Mts., etc. to L. Calif. Santa Catalina, Santa Cruz ids. April–May.

10. **G. caruifòlia** Abrams. [*G. latiflora* var. *c.* Jeps. *G. tenuiflora* var. *c.* Munz.] Scapose annual, 3–6 or more dm. high, with erect cent. axis and many branches spreading from base; base of plant with many large bipinnate or tripinnate lvs. 3–7 (–30) cm. long, with narrow rachis and deeply cut segms.; upper stems with reduced linear bracts; herbage glabrous except for the glandular upper branches of the cymose infl.; fls. in pairs on

unequal pedicels 1–10 mm. long; calyx 3–4 mm. long; corolla funnelform; tube 3–5 mm. long, pale blue-violet; throat 3 mm. long, whitish with yellow spots in upper part; limb 15–20 mm. in diam., pale blue-violet, often with a pair of purple spots at base of each lobe; stamens unequal, the longer fils. 6–7 mm. long; stigmas 2–3 mm. long; caps. 4 mm. long, 12–20-seeded; seeds 1 mm. long; $n=9$ (Grant & Grant, 1954).—Dry hills and openings in brush and woods, 4500–7500 ft.; Chaparral, Yellow Pine F.; mts. of San Diego Co.; n. L. Calif. May–Aug.

11. **G. clivòrum** (Jeps.) V. Grant. [*G. multicaulis* var. *c.* Jeps.] Erect or divaricately branched annual 0.6–3.5 dm. high, ± floccose below, glandular above; lvs. unipinnate or bipinnate, 1–6 cm. long, the ultimate segms. frequently elliptical, sometimes filiform, 0.5–2 mm. wide; fls. 2–5 in glomerules on peduncles 0.5–3 cm. long; pedicels 1–10 mm. long; calyx floccose or glandular, 4–5 mm. long in fl., 5–6 mm. in fr.; corolla funnelform, 6–8 mm. long, the limb 3–5 mm. wide when pressed, the tube yellow, the throat with 5 pairs of purple spots, the limb blue-violet with oval spreading lobes 1–2 mm. wide; stamens included; caps. ovoid, 24–36-seeded; seeds ovoid, 1–1.5 mm. long; $n=18$ (Grant, 1954).—Open fields and slopes; V. Grassland, Foothill Wd., etc.; cent. Calif. to Santa Barbara and Ventura cos., n. San Diego Co. Santa Barbara Ids. March–May.

12. **G. clòkeyi** Mason. Differing from *G. ophthalmoides* Brand in having corollas 3.5–5 mm. long, the tube included in the calyx, the throat pale yellow below, white above; $n=9$. Inyo and e. San Bernardino cos. (as at 4000 ft., Kingston Mts.); to New Mex.

13. **G. diegénsis** (Munz) A. & V. Grant. [*G. inconspicua* var. *d.* Munz.] Erect, 1–4 dm. high, with stout cent. leader; lvs. in dense basal cluster, glabrous to cobwebby beneath, 1–7 cm. long, strap-shaped, sinuately toothed to pinnately lobed, the rachis 1–6 mm. wide; cauline lvs. much shorter; infl. glandular-pubescent, congested or loose, with unequal pedicels; calyx glandular-pubescent, 3–5 mm. long in fl.; corolla 8–12 mm. long, the tube largely included, purple, the throat yellow above; lobes light violet, 2.5–4.4 mm. long; longest stamens slightly exserted; caps. 4–6.7 mm. long; $n=9$ (Grant et al., 1956).— Sandy openings, 1800–7200 ft.; Coastal Sage Scrub, Chaparral, Yellow Pine F.; interior s. Calif. from Los Angeles Co. to Santa Rosa Mts., Anza; to n. L. Calif. April–June.

14. **G. filifórmis** Parry ex Gray. Erect annual, mostly simple below, branched above, glabrous, the stems very slender; lvs. mostly alternate, simple, entire, linear, 1–5 cm. long; fls. few, each solitary and opposite a lf. on a slender pedicel 6–12 mm. long; calyx deeply cut, ca. 3 mm. long; corolla campanulate, yellow, 5–6 mm. long, the tube obsolete, the lobes 3–4 mm. long; stamens extending beyond the throat; pistil exserted; caps. ca. as long as calyx; $n=9$ (Grant, 1958).—Sandy or gravelly places below 6000 ft.; Creosote Bush Scrub to Pinyon-Juniper Wd.; Mojave Desert; to Utah, Ariz. March–May.

15. **G. hutchinsifòlia** Rydb. [*G. leptomeria* ssp. *rubella* Mason & A. Grant.] Near to *G. leptomeria*, the basal lvs. more deeply cut and mostly bipinnatifid; middle cauline lvs. well developed, 0.5–3.5 cm. long, narrowly linear and entire or with a few linear lobes; corolla 8–14 mm. long, the tube 3.5–7 mm. long, violet, lower throat yellow, upper white, the lobes white with pale violet streaks on outsides; $n=9$ (Grant, 1958).—Sandy places, largely 3000–5000 ft.; Creosote Bush Scrub, Joshua Tree Wd., etc.; w. Mojave Desert (Red Rock Canyon), Death V. region, to e. of White Mts.; to Utah, Ariz. April–May.

16. **G. intèrior** (Mason & A. Grant) A. Grant. Annual, 8–25 cm. high, cobwebby-pubescent below, ± glandular above; basal and lower lvs. 1- to 2-pinnately lobed with rachis 0.5–1.8 mm. wide; pedicels slender, unequal; calyx 2.6–4.6 mm. long; corolla-lobes 2.0–4.5 mm. long, pinkish violet, the throat yellow with purple markings in upper part; stamens unequal, the longest exserted; caps. broadly ovoid; $n=9$ (Grant et al., 1956).—At ca. 3500–4500 ft.; Piute Mts., Kern Co.; to Kern R. drainage, and s. Sierra Nevada.

17. **G. jàcens** A. & V. Grant. Resembling *G. brecciarum* in habit of branching and fl.-size and *G. leptantha* in lf.-dissection. It differs in its slender form and deep violet to purple corollas which are 5–7 mm. long.—At ca. 4500 ft.; Sagebrush Scrub; Mt. Pinos region; toward the n.

18. **G. latiflòra** (Gray) Gray ssp. **latiflòra.** [*G. tenuiflora* var. *l.* Gray.] Erect plant 1–3 dm. high, with 1–many stems, glabrous and glaucous at the base, ± glandular upward; basal lvs. 2–7 cm. long, strap-shaped, sinuately toothed to pinnately lobed with serrate divisions, lightly cobwebby; cauline lvs. much smaller; infl. loose, with unequal, gland-dotted pedicels; calyx 4–5 mm. long at anthesis; corolla 16–22 mm. long, with a slender purple tube 1–1.5 mm. wide, slightly exserted; throat full, abruptly expanded, with orifice 6–9 mm. wide; lobes white at base, pale violet at tips, 6–8 mm. long; stamens unequal, the longest ± exserted; caps. 6–9 mm. long; $n=9$ (Grant et al., 1956).—Sandy washes and flats, 2500–3600 ft.; Creosote Bush Scrub, Joshua Tree Wd.; sw. Mojave Desert. April–May.

KEY TO SUBSPECIES

Stems glabrous and glaucous at base; corolla 10–23 mm. long.
  Corolla-throat yellow and white or more than half yellow and white.
    Calyx 2.3–3.3 mm. long; corolla 10–15 mm. long. S. Coast Ranges ...... ssp. *cuyamensis*
    Calyx 4–5 mm. long; corolla 15–20 mm. long. Mojave Desert .... ssp. *latiflora*
  Corolla-throat purple in its lower half or three-fourths .... ssp. *davyi*
Stems cobwebby-pubescent near base.
  Corolla-tube and -throat together 14–20 mm. long; erect plants. Nw. Mojave Desert.
    Corolla 2½ to 3½ times as long as the lobes, the form stout and full .... ssp. *excellens*
    Corolla 3–6 times as long as the lobes, the form slender. .... ssp. *elongata*
  Corolla-tube and -throat together 7.5–8.5 mm. long; low spreading plants. Coso Mts.
    ssp. *cosana*

Ssp. **cosàna** A. & V. Grant. Basal lvs. 2–3.5 cm. long; calyx 3–3.5 mm. long; corolla 12–13 mm. long, tube twice as long as calyx.—Coso Mts., Inyo Co. June.

Ssp. **dàvyi** (Mlkn.) A. & V. Grant. [*G. d.* Mlkn.] Plate 67, Fig. F. Basal lvs. 2–7 cm. long; calyx 4–6.6 mm. long; corolla richly colored, 18–24 mm. long, the tube and throat combined 2–3 times the calyx; $n=9$ (Grant et al., 1956).—Open flats, 2500–4000 ft.; Joshua Tree Wd., Creosote Bush Scrub; w. Mojave Desert to Cuyama V. and s. San Luis Obispo Co. March–May.

Ssp. **cuyaménsis** A. & V. Grant. Basal lvs. moderately cobwebby, 1.5–3.5 cm. long; calyx 2.3–3.3 mm. long at anthesis; corolla pale, 10–15 mm. long, the tube and throat 2½–3 times as long as calyx; $n=9$ (Grant et al., 1956).—Sandy places, 2000–5000 ft.; V. Grassland, Pinyon-Juniper Wd.; Liebre Mts., Mt. Pinos region, etc. to Ventura and Kern cos.

Ssp. **elongàta** A. & V. Grant. Basal lvs. pinnately to bipinnately lobed; calyx 4–6 mm. long; corolla 21–34 mm. long, the combined tube and throat 3–6 times the calyx.—Sandy places, 2300–3300 ft.; mostly Creosote Bush Scrub; Mojave Desert from n. of Barstow to Rand and El Paso mts. April–May.

Ssp. **excelléns** (Brand) A. & V. Grant. [*G. tenuiflora* ssp. *latiflora* var. *e.* Brand.] Basal lvs. bipinnately lobed; calyx 4.7–7 mm. long; corolla 21–30 mm. long, the tube and throat combined 2–4 times the calyx; $n=9$ (Grant et al., 1956).—Open sandy soil, 2500–3000 ft.; Creosote Bush Scrub, Joshua Tree Wd.; w. Mojave Desert from near Mohave to Kramer Hills and beyond El Paso Mts.

19. **G. latifòlia** Wats. [*Navarretia l.* Kuntze.] Erect annual or perennial with rank odor, simple or branched, 1–3 dm. high, glandular-pilose throughout; lvs. mostly on lower part, sometimes quite basal, ovate to roundish, 2–8 cm. long, coarsely serrate to laciniate with mucronate teeth, sessile or the lower lvs. short-petioled; fls. many in corymb-like panicles, sessile or with slender pedicels 1–2 cm. long; calyx campanulate, 6–8 mm. long, the lobes subulate, 3–4 mm. long; corolla narrow-funnelform, 6–11 mm. long, bright red-pink within, pale or buff outside, the lobes 2–3 mm. long; stamens unequal, barely exserted; pistil included; caps. 5–7 mm. long; seeds many, deep red-brown; $n=18$ (Grant, 1958).—Common on desert mosaics and in washes, below 2000 ft.; Creosote Bush Scrub; both deserts n. to Inyo Co.; to Utah. March–May.

20. **G. leptántha** Parish ssp. **leptántha.** Erect plants 1.5–4.5 dm. high, cobwebby-

pubescent below, glandular above; basal lvs. 2.5–6.5 cm. long, bipinnately to almost tripinnately lobed; infl. diffuse, with 2–3 fls. above a bract; pedicels very unequal, 1–27 mm. long; calyx lightly gland-dotted, 3.5–4 mm. long at anthesis; corolla 12–23 mm. long, the tube very slender, expanding into a narrow throat, the lobes 1.4–2.8 mm. wide, 4–6 mm. long, bright pink, tube and throat ± yellow; stamens unequal, the longest well exserted; caps. 3–4 mm. long; $n = 9$ (Grant et al., 1956).—Sandy and gravelly places, 5000–7700 ft.; largely Yellow Pine F.; upper Santa Ana R. system, San Bernardino Mts. June–Aug.

Ssp. **pinetòrum** A. & V. Grant. Plants low, glandular in infl.; calyx 2.8–4.2 mm. long at anthesis; corolla 11–13 mm. long, the tube 4.2–6.4 mm. long, pale violet, the lobes pale violet, 3–4 mm. long. Among pines, 5100–9000 ft.; Mt. Pinos, Ventura Co. May–June.

Ssp. **purpùsii** (Mlkn.) A. & V. Grant. Cent. stem to 6 dm. high; infl. diffuse, glandular-pubescent; basal lvs. 2–6 cm. long; pedicels very unequal; corolla 13–29 mm. long, widening abruptly into the short throat; lobes 2.5–4.5 mm. wide, pinkish-violet; longest stamens long-exserted; $n = 9$ (Grant et al., 1956).—Open stream beds and slopes, 2500–8500 ft.; nw. of Inyokern and Nine-mile Canyon, nw. Mojave Desert.

21. **G. leptomèria** Gray. Erect annual, 0.5–2 dm. high, 1–several-stemmed, ± glandular-puberulent throughout; basal lvs. in rosette, many, broadly strap-shaped, coarsely dentate to pinnatifid with broad rachis, mostly 2–5 cm. long; infl. corymbose-paniculate; fls. pedicelled; calyx 1.6–2.7 mm. long at anthesis; corolla tubular, 4.7–6.5 mm. long, the tube well exserted from calyx, 2.5–3 mm. long, throat narrow and lobes frequently tridentate, 1–1.5 mm. long, usually each with a diffused purple streak; stamens shorter than corolla-lobes; caps. 3–4 mm. long.—Open sandy places, 2600–6750 ft.; Creosote Bush Scrub to Pinyon-Juniper Wd.; Mojave Desert and Mono Co.; to e. Wash., Rocky Mts. April–June.

22. **G. màlior** A. Day & V. Grant. Close to *G. transmontana*, but with corolla-lobes broad with rounded apex and violet in color; corolla-tube deep purple; pollen bright blue; secondary branches of stem often strongly decumbent.—Arid foothills and plains, 1800–5000 ft.; w. Mojave Desert from Mohave, Inyokern, Death V., etc. to ne. Calif., nw. Nev.

23. **G. micromèria** Gray. Much like *G. leptomeria*, 1–3 dm. high; basal lvs. strap-shaped, coarsely dentate to pinnatifid with narrow to broad rachis; calyx 1.4–2.8 mm. long at anthesis, becoming much exceeded by caps.; corolla tubular, 2.5–7.5 mm. long, the tube included or slightly exserted; lobes usually oval, acute, 1.5–3 mm. long; caps. 3–5.5 mm. long; $n = 8, 17, 25$ (Grant, 1958).—Sandy places, 2100–9000 ft.; Creosote Bush Scrub to Pinyon-Juniper Wd.; Mojave Desert to Ore., Ida., Colo. April–June.

24. **G. mìnor** A. & V. Grant. Mostly 1–2 dm. high, with 3–8 stems from a cluster of basal lvs., cobwebby below, lightly glandular above; basal lvs. 1.5–4 cm. long, pinnately lobed with 4–6 pairs of lobes 1.5–4 mm. long, rarely bipinnatifid; cauline lvs. reduced upward; infl. strict; pedicels unequal; calyx 2–3 mm. long in fl.; corolla 4–7.5 mm. long, the upper tube and throat dark violet or with some yellow, the lobes pale violet; stamens barely exserted; caps. 5–7 mm. long; $n = 9$ (Grant et al., 1956).—Sandy flats and washes, 1000–3500 ft.; Creosote Bush Scrub to Pinyon-Juniper Wd., V. Grassland; w. Mojave Desert to inner Coast Ranges. March–April.

25. **G. modocénsis** Eastw. Small to large often spreading plant branching from the base, the ± decurrent secondary branches 1.5–4.5 dm. long, ± cobwebby-pubescent near base, ± glandular above; basal rosette usually well developed, the lvs. 2–7 cm. long, pinnately lobed with rachis 1–5 mm. wide; cauline lvs. ± clasping at base, dentate to deeply lobed; infl. ± congested in fl., looser in fr., with 1–5 fls. borne above a bract; corolla 6–11 mm. long, tube and usually lower throat purple, middle and upper throat with yellow spots; lobes violet or white; stamens short; caps. 4.4–6.5 mm. long; $n = 18$ (Grant et al., 1956).—Sandy places, 1000–7000 ft.; Yellow Pine F., San Bernardino Mts.; San Gabriel Mts. and also Joshua Tree Wd., Mojave Desert to Mt. Pinos region, and Mono Co. and Ore.

26. **G. nevínii** Gray. Erect, simple or branched annual, finely glandular-pubescent throughout, leafy, 1–4 dm. high; lvs. 1–7 cm. long, bipinnate or tripinnate into fine and numerous linear segms.; fls. in capitate clusters; calyx 5–6 mm. long in fl.; corolla narrowly funnelform, 10–14 mm. long, the tube exceeding the calyx, the lobes 2 mm. long, with limb and tube blue, throat yellowish; caps. 6.5–8 mm. long; $n=18$ (Grant, 1958).—Open slopes, Catalina and San Clemente ids. March–May.

27. **G. ochroleùca** Jones ssp. **ochroleùca**. Delicate annual, with short cent. leader 6–15 cm. high and many secondary ± spreading branches to 30 cm. long, glaucous and glabrous except for the glandular infl.; lvs. mostly basal, sparsely cobwebby-pubescent, 2–3 cm. long, pinnately lobed with narrow rachis and lobes 1 mm. wide and 3–10 mm. long; pedicels 3–10 mm. long, in open cymose panicles; calyx 2.5–3 mm. long, glabrous; corolla tubular-funnelform, yellowish, with some pale violet in upper throat; stamens not exserted; caps. 3–4.5 mm. long; $n=9$ (Grant et al., 1956).—Sandy slopes and plains, 2500–5000 ft.; Creosote Bush Scrub, Joshua Tree Wd.; w. half of Mojave Desert. April–June.

Ssp. **bizonàta** A. & V. Grant. Corolla 8–13.5 mm. long, the tube usually included in calyx and shorter than the throat, the lobes light pink; lower stems usually glabrous; pedicels glabrous just below fls.; $n=9$ (Grant et al., 1956).—Sandy flats, 3000–6700 ft.; Foothill Wd., Yellow Pine F., etc.; Santa Ynez Mts., Mt. Pinos; to San Luis Obispo Co.

Ssp. **éxilis** (Gray) A. & V. Grant. [*G. latiflora* var. *e.* Gray.] Lower stems usually cobwebby-pubescent; pedicels glandular just below the fls.; corolla-tube slightly exserted from calyx; corolla 8–10.5 mm. long, shallow-throated.—Sandy or gravelly places, 800–7000 ft.; Coastal Sage Scrub, Chaparral, Yellow Pine F.; largely interior cismontane, from San Bernardino Co. s.

Ssp. **vívida** (A. & V. Grant) A. & V. Grant. [*G. leptantha* ssp. *v.* A. & V. Grant.] Plant compact, lower stems glabrous to cobwebby; basal rosette dense, the lvs. densely cobwebby-pubescent; calyx 2.8–3.7 mm. long, on a glandular pedicel; corolla 9–14 mm. long, the tube equalling or slightly longer than calyx, violet, the throat yellow below the middle, deep violet above, lobes deep pinkish violet.—At 6000–8000 ft., mostly Yellow Pine F. San Gabriel Mts., Tehachapi Mts.

28. **G. ophthalmoìdes** Brand. Much-branched annual with many stems from near base, 1.5–3 dm. high, cobwebby-pubescent below, glandular above; basal lvs. in dense rosette, bipinnately lobed with primary lobes 2–3 times as long as width of rachis; infl. diffuse; calyx 2.5–4.5 mm. long at anthesis, glabrous or sparsely glandular; corolla 7–12 mm. long, the tube well exserted, filiform, light violet, the lower throat ± yellow, pale violet above, the lobes pinkish-violet or pink, 1.4–2.8 mm. long; $n=18$ (Grant et al., 1956).—Sandy places, 3800–8500 ft.; Joshua Tree Wd., Pinyon-Juniper Wd.; desert mts. from e. San Bernardino Co. to Mono Co.; to Utah. May–July.

29. **G. rípleyi** Barneby. [*G. gilmanii* Jeps.] Suffrutescent perennial branching from base, glutinous with capitate hairs; lvs. hollylike, 2–6 cm. long, 1–4 cm. wide, mostly simple with a broad obovate blade and long aristate teeth; petioles 0.5–2 cm. long; infl. cymose-paniculate with divaricate pedicels; calyx 4–5 mm. long, the lobes aristiform; corolla vespertine, 5–7 mm. long, the lobes 2–3 mm. long, pink, the tube white, throat narrow; stamens unequal; style included; caps. 4–5 mm. long, included.—Limestone crevices, 3000–4500 ft.; Creosote Bush Scrub; Panamint Mts. to sw. Nev. May–June.

30. **G. scopulòrum** Jones. Glandular-villous annual, erect and simple or with ascending branches 1–3 dm. long, the stems paniculately branched; lower lvs. 3–9 cm. long, viscid, thin, broadly oblanceolate in outline, 1–2-pinnatifid into ovate coarsely toothed segms.; infl. paniculate, the fls. solitary at the ends of stipitate-glandular pedicels; calyx 3–4 mm. long, enlarging in fr.; corolla funnelform, rose-lavender, 10–14 mm. long, the tube and throat paler or yellowish; stamens subequal, shorter than corolla-lobes; caps. 4.5–5.5 mm. long; $n=9$, 18 (Grant, 1958).—Dry washes and slopes mostly below 4000 ft.; Creosote Bush Scrub; Chuckawalla Mts. to Death V. region and Owens V.; to sw. Utah. April–May.

31. **G. sinuàta** Dougl ex Benth. [*G. inconspicua* var. *s.* Gray.] Stiffly erect plant with

1 main stem or several, from basal rosette, glabrous and glaucous near base, moderately glandular-pubescent above, 1–3 dm. high; basal lvs. cobwebby-pubescent, strap-shaped, the rachis 2–6 mm. wide, the lobes toothed or short-lobed; cauline lvs. much reduced; infl. ± loose; calyx 2.5–3.6 mm. long in fl.; corolla 7–11 mm. long, the tube well exserted; lobes 1.5–3 mm. long, pale violet to pinkish or whitish; stamens very short; caps. 4–7 mm. long; $n=18$ (Grant et al., 1956).—Sandy deserts below 6000 ft.; Creosote Bush Scrub, Joshua Tree Wd., Sagebrush Scrub, etc.; Mojave Desert to Plumas Co.; to Wash., Colo., Ariz. April–June.

32. **G. spléndens** Dougl. ex Lindl. ssp. **spléndens**. [*G. tenuiflora* var. *altissima* Parish.] Robust subscapose branched annual 1–6 or more dm. high; stems glabrous to glandular especially below; basal lvs. in a rosette, villous, 2–14 cm. long, with narrow rachis, bipinnatifid or tripinnatifid with deeply cut segms.; infl. cymose, the fls. in pairs on unequal pedicels 3–20 mm. long; calyx 3–4 mm. long, with hyaline sinuses and erect acute lobes 1 mm. long; corolla funnelform, the limb usually pink, 12–20 mm. in diam., the lobes broadly oval, 6–10 mm. long, throat 4–7 mm. long, white to pale violet, tube 3–6 mm. long, slender, red to purple; stamens unequal; caps. 4–8 mm. long; $n=9$ (Grant & Grant, 1954).—Openings in brush or woods, 1000–7000 ft.; Chaparral, Foothill Wd., Yellow Pine F.; San Jacinto Mts. to Monterey Co. April–July.

Ssp. **grántii** (Brand) V. Grant. [*G. collina* var. *g.* Brand.] Corolla-limb tube 10–18 mm. long; longest fils. 2–3 mm. long; $n=9$ (Grant & Grant, 1954).—Open slopes and woods, 2700–7000 ft.; mostly Yellow Pine F.; San Bernardino and San Gabriel mts. May–July.

33. **G. stellàta** Heller. Erect annual 1–4 dm. high; stems stout, paniculately branched, with bent hairs below and stipitate-glandular above; lower lvs. 2–3-dissected, 2–10 cm. long, the ultimate segms. callous-tipped, pubescent; upper lvs. reduced; infl. paniculate, the fls. solitary on short stout pedicels; calyx 3–4 mm. long, larger in fr.; corolla funnelform, 6–10 mm. long, pale blue or lavender to white, with row of purple spots in throat; stamens subequal, shorter than corolla-lobes; caps. 5–7 mm. long; $n=9$ (Grant, 1958).—Sandy and gravelly places below 4000 ft.; Creosote Bush Scrub; Colo. and Mojave deserts; to Utah, nw. Mex. March–May.

34. **G. tenuiflòra** Benth. ssp. **hoffmánnii** (Eastw.) A. & V. Grant. [*G. h.* Eastw.] Erect annual 6–12 cm. high, with a cent. leader, rather densely glandular-pubescent; basal lvs. ± strap-shaped, with 4–6 pairs of short lobes; infl. somewhat congested, with 1–2 fls. borne above a bract; corolla-tube and -throat together 13–14 mm. long, the limb 6–7 mm. long; calyx 4.6–5.7 mm. long; style equaling stamens.—Sandy soil, Coastal Sage Scrub; Santa Rosa Id. April.

35. **G. tetrabréccia** A. & V. Grant. Near *G. modocensis*, stems 2–4 dm. long; lower lvs. densely matted with cobwebby pubescence; cauline lvs. deeply lobed; corolla 6.5–7.8 mm. long, the tube dark purple, throat and base of lobes white with yellow spots, lobes violet; $n=18$ (Grant and Grant, 1956).—Sandy slopes, Yellow Pine F.; Mt. Pinos region. May–June.

36. **G. transmontàna** (Mason & A. Grant) A. & V. Grant. [*G. ochroleuca* ssp. *t.* Mason & A. Grant.] Erect to ascending, 1–3 dm. high, cobwebby-pubescent near base; leafy below middle; lower lvs. cobwebby, pinnately lobed with linear lobes 3–9 times as long as rachis width; infl. strictly branched; calyx moderately gland-dotted; corolla 4.8–7 mm. long with tube and often part of throat included in calyx, the tube violet to white, the throat yellow and white, lobes pale violet or white; stamens slightly exserted; $n=18$ (Grant et al., 1956).—Sandy places, 2000–6500 ft.; Creosote Bush Scrub and above; Mojave Desert to Cuyama V. and Nev. March–May.

5. *Ipomópsis* Michx.

Annual to perennial herbs, often with a basal rosette of lvs. and with leafy stems; one sp. a shrub. Herbage frequently villous, sometimes with stipitate glands, sometimes glabrous. Lvs. pinnatifid, the lateral segms. well developed, sometimes reduced, the segm.-

tips with horny mucros. Infl. cymose, varying from loosely racemose to tightly congested, bracteate, each fl. usually subtended by a bract. Calyx of equal herbaceous mucronate lobes with broad hyaline sinuses. Corolla salverform or tubular, violet, red, white or shades of yellow or pink. Stamens inserted in the tube or sinuses of corolla, often unequal in length and point of insertion; pollen blue, white or yellow. Caps. dehiscent, the locules 1–2-several-seeded. Seeds long and slender to rounded or angular, smooth or corrugated. Ca. 23 spp. from w. N. Am., Fla.; 1 sp. in S. Am. (Greek, *ipo*, to strike and *opsis*, appearance.) (Grant, V. A synopsis of Ipomopsis. Aliso 3: 351–362. 1956.)

Plants annual.
    Lvs. entire or irregularly toothed; lower fls. solitary in lf.-axils . . . . 3. *I. depressa*
    Lvs. pinnatifid; fls. all in terminal clusters . . . . 4. *I. polycladon*
Plants perennial or biennial.
    Corolla 0.6–1 cm. long, white; infl. subcapitate . . . . 2. *I. congesta*
    Corolla 2–5 cm. long, red or pink to yellowish; infl. not subcapitate.
        Plant scarcely woody at base, the stems largely simple; corolla-lobes subequal, lanceolate; plants usually with basal rosettes. San Bernardino Co. n. . . . . 1. *I. aggregata*
        Plant with a ± woody base, the stems much branched; corolla-lobes unequal and unequally cleft; mature plants lacking a basal rosette of lvs. S. San Diego Co. . . . . 5. *I. tenuifolia*

1. **I. aggregàta** (Pursh) V. Grant ssp. **arizónica** (Greene) V. & A. Grant. [*Callisteris a.* Greene.] Plant biennial to perennial, 1–3 dm. high, often lanate-pubescent, strict; lvs. mostly 1–2 cm. long, pinnately dissected; infl. a ± thyrsoid elongate panicle, the fls. short-pedicelled; calyx united ½–¾ its total length of 4–6 mm.; corolla red, 2–2.5 cm. long; tubular-funnelform, the lobes lanceolate, rotately spreading; stamens mostly included; caps. ca. as long as calyx.—Dry washes and rocky places, mostly 4500–10,500 ft.; Pinyon-Juniper Wd., Bristle-cone Pine F.; Providence, New York, Clark, Panamint and Inyo mts.; to Utah, Ariz.

2. **I. congésta** (Hook.) V. Grant ssp. **montàna** (Nels. & Kennedy) V. Grant. [*Gilia m.* Nels. & Kennedy.] Densely matted plants forming cushionlike rosettes; basal lvs. palmately divided into short broad segms.; fls. in single or clustered heads; calyx cylindric, the lobes united below by a membrane, densely arachnoid, ca. 4.5 mm. long; corolla white, salverform, 4–6 mm. long, the yellow tube 3–4 mm. long; stamens exserted; caps. obovoid, seeds 1–2 in a locule.—Dry slopes, 7000–12,000 ft.; Bristle-cone Pine F., Alpine Fell-fields; White Mts., Sierra Nevada; to Ore. July–Sept.

3. **I. depréssa** (Jones) V. Grant. [*Gilia d.* Jones.] Divaricately branched annual 2–10 cm. high, glandular-pubescent throughout, even canescent; lvs. well distributed, linear to elliptic-lanceolate, 1–2 cm. long, short-petioled, the upper entire or with 2 teeth; fls. crowded in leafy subcapitate racemes; calyx scarious, 5–6 mm. long, with subulate teeth; corolla white, tubular-funnelform, sometimes ± irregular, ca. 5 mm. long; caps. subglobose; seeds 4–5 in a locule; $n = 7$ (Grant, 1956).—Occasional in dry disturbed places, 3500–5000 ft.; Creosote Bush Scrub, Joshua Tree Wd., Sagebrush Scrub; w. Mojave Desert from Rabbit Springs to Owens V.; to Utah, Nev. April–June.

4. **I. polyclàdon** (Torr.) V. Grant. [*Gilia p.* Torr.] Annual with several naked spreading branches from base, these 0.5–1.5 dm. long, puberulent throughout, ± glandular-villous above; lvs. few, 1–2 cm. long, mostly basal and crowded about the subcapitate fl.-clusters, pinnatifid into short mucronulate segms.; calyx 3–4 mm. long; corolla white, tubular, 4–6 mm. long, the lobes ca. 1 mm.; stamens inserted in the sinuses; seeds 2 in a locule; $n = 7$ (Grant, 1956).—Sandy, gravelly or rocky slopes below 5200 ft.; Creosote Bush Scrub, Pinyon-Juniper Wd., etc.; Mojave Desert to Ore., Colo., Tex. April–June.

5. **I. tenuifòlia** (Gray) V. Grant. [*Loeselia t.* Gray.] Suffruticose perennial, 1–4 dm. high, much branched from base, subglabrous below, minutely puberulent above, leafy throughout; lvs. narrowly linear, entire, or some pinnately dissected into few linear acerose lobes; fls. solitary in axils of upper lvs., forming subcorymbose clusters; calyx narrow-campanulate, ca. 6 mm. long, membranous between ribs, the lobes ca. ¼ the tube; corolla scarlet, 1.5–2.5 cm. long, tubular-funnelform, the limb ± irregular, the lobes 5–6 mm. long, retuse with middle triangular tooth; stamens and style long-exserted;

Ipomopsis

$n = 7$ (Grant, 1956).—Dry rocky slopes, 1500–3500 ft.; Creosote Bush Scrub, Pinyon-Juniper Wd., Chaparral; Mt. Springs Grade to Jacumba and Campo, se. San Diego Co.; n. L. Calif. March–May.

6. Langloìsia Greene

Diffuse low rigid annuals. Lvs. alternate, cuneate to linear, pinnately toothed, the lower divisions reduced to slender bristles, the upper bristle-tipped. Fls. in few-fld. terminal bracteate heads, the bracts leafy, with bristly teeth. Calyx lobed, the lobes equal, spinescent-tipped, the tube scarious between the lobes and splitting to base. Corolla tubular-funnelform, almost regular to 2-lipped, with 3 lobes in upper lip. Stamens 5, inserted in corolla-throat. Caps. 3-sided. Seeds 2–9 in each caps., slightly angled, mucilaginous when wet. A small genus of our deserts. (Named for Father *Langlois* of Louisiana.)

Corolla distinctly 2-lipped; lvs. pinnatifid with ligulate or spatulate rachis and single marginal bristles.
 Calyx 5–6 mm. long; corolla-lobes rounded or truncate at apex, almost as long as tube
  1. *L. matthewsii*
 Calyx 3–4 mm. long; corolla-lobes pointed at apex, $\frac{1}{3}$ to $\frac{1}{2}$ as long as tube ..... 3. *L. schottii*
Corolla almost regular; lvs. abruptly dilated toward apex, the marginal bristles mostly in pairs.
 Calyx 8–9 mm. long; corolla-lobes purple-dotted, almost as long as tube ..... 2. *L. punctata*
 Calyx 6 mm. long; corolla-lobes not dotted, ca. $\frac{1}{3}$ as long as tube .... 4. *L. setosissima*

1. **L. matthèwsii** (Gray) Greene. [*Loeselia m.* Gray.] Plate 68, Fig. A. Branched from base or above, tufted, 3–15 cm. high, somewhat broader; stems whitish, thinly tomentulose; lower lvs. broadly linear, 1.5–4 cm. long, pinnately toothed, the narrow lobelike teeth bristle-tipped; bracts somewhat shorter and wider; calyx 5–6 mm. long; corolla 2-lipped, whitish to pink, the upper lobes with an elaborate red and white pattern, the tube 8–12 mm. long, the lobes nearly as long, oblong, sometimes shallowly toothed, rounded or retuse at tips; stamens and style well exserted; caps. ca. 3 mm. long; seeds irregular, plump, ± wing angled, ca. 1.2–1.5 mm. long.—Common in great masses, sandy and gravelly areas, below 5000 ft.; Creosote Bush Scrub, Joshua Tree Wd.; deserts from Inyo Co. to Imperial Co.; Nev., Son. April–June.

2. **L. punctàta** (Cov.) Goodd. [*Gilia setosissima* var. *p.* Cov.] Tufted or simple, 3–15 cm. high, or flat-topped and 15–20 cm. wide, thinly tomentulose to glabrate; lower lvs. linear, 2–3.5 cm. long, finely bristle-toothed, the upper wide, with deltoid 3–5-toothed apex; calyx 8–9 mm. long; corolla almost regular, 15–20 mm. long, lilac, the lobes entire, almost as long as tube, obtuse, purple-dotted; stamens exserted; caps. narrow-oblong, 7–8 mm. long; seeds angled, plump, ca. 1 mm. long.—Dry gravelly places, below 5000 (8000) ft.; Creosote Bush Scrub to Pinyon-Juniper Wd.; White Mts. of Inyo Co. s. through Mojave Desert to San Bernardino Mts.; Nev., Ariz. April–June.

3. **L. schóttii** (Torr.) Greene. [*Navarretia s.* Torr. *L. flaviflora* A. Davids.] Tufted, 2–10 cm. high, scattered villous-tomentulose; lvs. linear to linear-oblanceolate, pectinately pinnatifid, 1–3 cm. long, the teeth mostly bristle-tipped; bracts cuneate-oblanceolate, 1–1.5 cm. long, 3–5-toothed; calyx 3–4 mm. long; corolla pale lavender to white to yellowish or pinkish, 8–12 mm. long, the upper lip 3-lobed with purple spots, the lower lip 2-lobed, $\frac{1}{3}$ to $\frac{1}{2}$ as long as tube, usually spotted near base; stamens well exserted; caps. 3–4 mm. long, 2–6-seeded; seeds ± angled; $n = 7$ (Grant, 1958).—Dry gravelly or sandy places below 5000 ft.; Creosote Bush Scrub, Joshua Tree Wd., etc.; Imperial Co. to Inyo Co., Fresno Co., etc.; to Utah, Son., L. Calif. March–June.

4. **L. setosíssima** (T. & G.) Greene. [*Navarretia s.* T. & G.] Plate 68, Fig. B. Tufted, 2–7 cm. high, or with prostrate branches to 1 dm. long, tomentulose to glabrate; lvs. 1–2 cm. long, cuneate, with 3 large apical teeth and one pair of lateral, or the basal lvs. linear-subulate, not apically dilated; bracts like upper lvs.; calyx ca. 6 mm. long; corolla almost regular, light violet, not spotted, 12–16 mm. long, the lobes oval-oblong, ca. $\frac{1}{3}$ as long as tube; stamens shorter than corolla-lobes; caps. ca. 6 mm. long; $n = 7$ (Grant, 1958).—Dry sandy places, below 3500 ft.; Creosote Bush Scrub; Mojave and Colo. deserts; to Ore., Ida., Utah, Son., L. Calif. April–June.

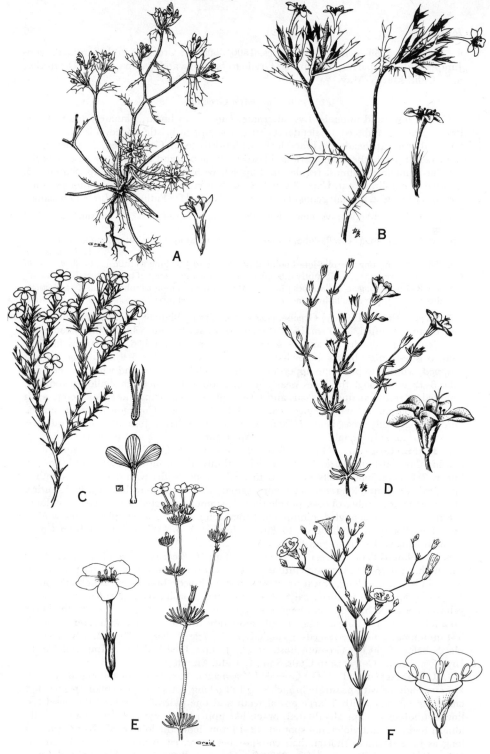

Plate 68. Fig. A, *Langloisia matthewsii*. Fig. B, *Langloisia setosissima*. Fig. C, *Leptodactylon pungens* ssp. *hallii*. Fig. D, *Linanthus aureus*. Fig. E, *Linanthus breviculus*. Fig. F, *Linanthus liniflorus pharnaceoides*.

7. **Leptodáctylon** H. & A.

Subshrubs or perennial herbs, straggly or compact, leafy, often tufted. Lvs. opposite or alternate, palmately or subpinnately parted into linear pungent lobes, usually glandular and with axillary fascicles. Fls. mostly in terminal congested cymes or glomerules, rarely solitary in axils, ± subsessile. Calyx-lobes entire, pungent, the sinuses ca. ⅔ filled with membrane. Corolla narrow-funnelform or salverform, usually showy, white or cream to rose or lilac. Stamens included, inserted in tube or throat; fils. ca. as long as anthers. Style included. Caps. ± cylindric; locules several- to many-seeded. A small genus of w. N. Am. (Greek, *leptus*, narrow, and *dactylon*, finger, because of lf.-lobing.)

(Wherry, E. T. Two Linanthoid genera. Am. Midl. Nat. 34: 381–387. 1945.)

Lvs. opposite, 3-cleft; plants less than 1 dm. tall. San Jacinto Mts. .... 2. *L. jaegeri*
Lvs. mostly alternate; plants mostly 3–10 dm. high. More widely distributed.
  Plants mostly 1–3 dm. high; corolla tubular-funnelform, mostly pale or whitish; corolla-lobes 7–10 mm. long .... 3. *L. pungens*
  Plants usually 3–10 dm. high; corolla salverform, rose to lilac; corolla-lobes 10–15 mm. long
                                                                          1. *L. californicum*

1. **L. califórnicum** H. & A. ssp. **califórnicum**. [*Gilia c.* Benth.] PRICKLY-PHLOX. Erect widely branched shrub 3–10 dm. high, the stems tomentose, not glandular, thickly set with fascicles of prickly lvs. in axils of current and old main lvs.; the latter palmately 5–9-lobed into acerose segms. 3–12 mm. long, unequal, linear, acerose, glabrous or subglabrous; fls. sessile, in terminal congested clusters with ± villous leafy bracts; calyx ca. 1 cm. long, deeply cleft into acerose lobes, the sinuses ca. half or ⅔ filled with hyaline membrane; corolla salverform, bright rose or rose-lavender, sometimes almost white, 2–2.5 cm. long, the continuous tube and throat 10–15 mm. long, the lobes 10–15 mm. long; stamens inserted near middle of tube, the fils. ca. 1 mm. long; stigma included; caps. elongate, the locules several-seeded; $n=9$ (Flory, 1937).—Dry slopes and banks below 5000 ft.; Chaparral, Foothill Wd.; San Luis Obispo Co. to w. San Gabriel Mts. and Santa Ana Mts. March–June.

Ssp. **glandulòsum** (Eastw.) Mason. [*Gilia c.* var. *g.* Eastw.] Hairs of upper stems and infl. gland-tipped.—Below 3000 ft.; Chaparral; San Gabriel Mts. to San Bernardino and Santa Ana Mts.

2. **L. jàegeri** (Munz) Wherry. [*Gilia j.* Munz.] Low, cespitose, with woody base, the branches many, 0.2–1 dm. high, glandular-pubescent, densely leafy; lvs. mostly opposite, 1–1.5 cm. long, mostly 3-cleft into flat linear prickly lobes, the terminal 6–15 mm. long, the lateral 4–10 mm. long; fls. crowded, solitary in upper axils; calyx 8–9 mm. long, membranous between the unequal spiny lobes; corolla whitish, funnelform, 25–30 mm. long, the tube exserted, the 5–6 broadly oblanceolate lobes 7–9 mm. long; stamens 5–6, inserted at base of throat; fils. 2–3 mm. long; caps. 4-loculed.—Dry rocky places, ca. 8800 ft.; upper Montane Coniferous F.; San Jacinto Mts. July–Aug.

3. **L. púngens** (Torr.) Rydb. ssp. **hállii** (Parish) Mason. [*Gilia hallii* Parish.] Plate 68, Fig. C. Low branching shrub 1–4 dm. high, with several to many stems, grayish-pubescent, somewhat glandular; lvs. many, mostly alternate, palmately cleft into 3–7 lobes, the middle one 2–4 times as long as the next outer one; fls. sessile or nearly so, tending to open in evening, in upper axils and in terminal or subterminal glomerules; calyx 8–10 mm. long, with unequal acerose linear lobes; corolla mostly cream, sometimes purplish, 15–20 mm. long, the tube very slender, long-exserted, the lobes 7–10 mm. long; stamens included; pistil ca. ¼ as long as corolla-tube; the locules many-seeded.—Dry slopes, 4000–9000 ft.; Pinyon-Juniper Wd. and higher; mts. in and on edge of desert; Inyo Co. to e. San Diego Co.; L. Calif., Nev. April–July.

Ssp. **pulchriflòrum** (Brand) Mason. [*Gilia p.* ssp. *p.* Brand.] Plant more viscid; middle lf.-segm. less than twice as long as next outer; corolla white with some pinkish or purplish, 15–25 mm. long, the lobes narrow-obovate.—Dry rocky and sandy places, 5000–12,000 ft.; mostly Montane Coniferous F.; San Bernardino Mts. and Sierra Nevada and N. Coast Ranges to Ore. May–Aug.

8. Linánthus Benth.

Annuals or perennials, with simple or divaricately or dichotomously branched stems. Lvs. opposite or alternate, palmately or somewhat pinnately parted into linear segms., rarely simple. Infl. from open cymose to subcapitate at ends of branches, sometimes also with solitary fls. in upper axils. Lvs. sessile or nearly so, or on long pedicels. Calyx deeply cleft, with or without a pseudotube of hyaline membrane in the sinuses, the free margins of the lobes with or without such membrane. Corolla campanulate to funnelform or salverform. Stamens equally inserted in the throat or more rarely in the tube, the fils. equal or sometimes unequal, included or inserted. Pistil from short and included to long and exserted; stigma 3–4-lobed. Caps. ellipsoid to cylindrical; the locules adhering at base during dehiscence, 1–several-seeded. Seeds unaffected by water or producing spiracles or mucilage when wetted. A genus of ca. 40 spp., mostly of w. N. Am. and Chile. (Greek, *linon*, flax, and *anthos*, flower.)

1. Plants annual.
   2. Calyx with a conspicuous hyaline membrane in the sinuses, either forming a pseudotube or present on the margins of the lobes.
      3. Fls. on pedicels at least 5 mm. long.
         4. Corolla barely exceeding calyx, glabrous within; fils. glabrous; stems 2–10 cm. long; lvs. with segms. 2–6 mm. long .... 23. *L. pygmaeus*
         4. Corolla mostly 2–5 times as long as calyx, with a hairy ring within or fils. hairy at base.
            5. Fils. pubescent at base, inserted on lower part of corolla-throat; corolla 1–3 cm. long; stems 1–5 dm. long .... 16. *L. liniflorus*
            5. Fils. glabrous, inserted on upper half of throat; corolla 0.6–1.3 cm. long; stems 0.5–1.6 dm. long .... 3. *L. aureus*
      3. Fls. sessile or on pedicels less than 5 mm. long.
         6. Calyx-lobes free to base, membrane-margined but not with an obvious hyaline membrane below the sinuses; corolla campanulate. Small desert plants.
            7. Lvs. 3–5-cleft or entire, 6–10 mm. long with 2 dark lines below base of each lobe. Inyo Co. .... 10. *L. demissus*
            7. Lvs. entire, 2–5 mm. long; corolla 4–5 mm. long with vermilion spot on each lobe. Little San Bernardino Mts. .... 17. *L. maculatus*
         6. Calyx-lobes united by their bordering membranes to form a pseudotube which may be quite short; corolla funnelform to salverform.
            8. Corolla bearing ± reniform arches in throat below each lobe, blue, white or yellow; internodes usually short, concealed by lvs. Tufted desert plants ..... 22. *L. parryae*
            8. Corolla without reniform arches below the lobes.
               9. Fls. yellow or cream.
                  10. Calyx 4–5 mm. long; corolla 5–8 mm. long.
                     11. Stamens inserted just below the corolla-sinuses; plants pubescent throughout. Cismontane and desert .... 15. *L. lemmonii*
                     11. Stamens inserted at base of throat; plant glabrous or minutely puberulent. Rare, Mojave Desert .... 2. *L. arenicola*
                  10. Calyx 6–8 mm. long.
                     12. Lvs. entire; pedicels and calyces with stalked glands. Deserts . 13. *L. jonesii*
                     12. Lvs. 5–9-parted; pedicels and calyces without stalked glands. Mostly cismontane .... 1. *L. androsaceus*
               9. Fls. white to pink or lilac.
                  13. The corolla-tube proper (below the throat) included in calyx.
                     14. Calyx 3–4 mm. long; lvs. 3-cleft, 2–4 mm. long. Se. San Diego Co. 4. *L. bellus*
                     14. Calyx 6–14 mm. long.
                        15. Calyx-tube almost wholly membranous, the teeth villous.
                           16. Calyx ca. 10 mm. long; corolla-tube 1–2 mm. long, the lobes with 2 dark linear basal spots. San Gabriel Mts. .... 9. *L. concinnus*
                           16. Calyx ca. 6 mm. long; corolla-tube 4–5 mm. long, the lobes with a single elongate basal spot. San Bernardino Mts. .... 14. *L. killipii*
                        15. Calyx-tube membranous only below sinuses.
                           17. Fls. white to cream or with some purplish tinge; mostly vespertine, the corolla-lobes entire.

18. Calyx glabrous.
  19. Lvs. divided, 3-7-parted, each stamen with a basal hairy pad; corolla 1.5-3 cm. long .... 12. *L. dichotomus*
  19. Lvs. mostly entire, rarely 2-3-cleft; stamens without basal pads; corolla 1-1.5 cm. long .... 6. *L. bigelovii*
18. Calyx with stalked glands; corolla 9-10 mm. long ..... 13. *L. jonesii*
17. Fls. pink to lilac, rarely white, diurnal, the corolla-lobes dentate.
  19a. Lvs. entire, 10-20 mm. long; calyx 10-16 mm. long
  11. *L. dianthiflorus*
  19a. Lvs. mostly 3-lobed, 2-4 mm. long; calyx 3-4 mm. long . 4. *L. bellus*
13. The corolla-tube proper well exserted from the calyx.
  20. Corolla funnelform, the tube stout, 1-1½ times as long as calyx. San Diego Co.
  21. *L. orcuttii*
  20. Corolla salviform, the tube slender, subfiliform, 2-4 times as long as calyx.
    21. Bracts of infl. pilose to short-hispid, mostly not conspicuously coarsely ciliate
      22. Calyx 5 mm. long; corolla 10-20 mm. long .... 19. *L. nudatus*
      22. Calyx 7-8 mm. long; corolla 18-35 mm. long.
        23. Corolla-tube glabrous; sinus-membrane of calyx extending more than half way; seeds rugose .... 7. *L. breviculus*
        23. Corolla-tube ± puberulent; sinus-membrane of calyx very short; seeds tuberculate .... 1. *L. androsaceus*
    21. Bracts of infl. conspicuously and coarsely ciliate; calyx 7-10 mm. long; corolla 12-25 mm. long .... 8. *L. ciliatus*
2. Calyx not membranous in sinuses or on margins of lobes, or very inconspicuously so.
  24. Corolla-lobes 5-8 mm. long. From mainland .... 1. *L. androsaceus*
  24. Corolla-lobes 3-5 mm. long. San Clemente and Catalina ids. .... 5. *L. bicolor*
1. Plants suffrutescent perennials, 1-4 dm. high; lvs. 3-9-cleft; calyx not membranous below sinuses.
  25. Stems 1-4 dm. high; fls. 12-15 mm. long .... 20. *L. nuttallii*
  25. Stems shorter, almost matted; fls. ca. 8-10 mm. long .... 18. *L. melingii*

1. **L. androsàceus** (Benth.) Greene ssp. **lutèolus** (Greene) Mason. [*L. l.* Greene.] Annual, decumbent to ascending, the stems mostly several-branched, 0.5-2.5 dm. long, pubescent; lvs. remote, 5-9-parted into oblanceolate or narrower segms. 1-2.5 cm. long; fls. sessile in terminal dense bracteate heads; calyx 6-8 mm. long, deeply cleft into subulate lobes, the sinuses with a short narrow membrane; corolla salverform, the limb mostly 6-10 (-12) mm. wide, the tube subfiliform, much exceeding calyx; stamens glabrous, inserted near middle of throat; styles often exserted; seeds somewhat wing-angled and tuberculate.—Dry slopes below 5000 ft.; Chaparral, Foothill Wd., Yellow Pine F. Pinyon-Juniper Wd.; Monterey Co. to San Diego Co. and w. edge of Mojave Desert. April-May.

Ssp. **lùteus** (Benth.) Mason. [*Leptosiphon l.* Benth.] Stems erect, simple or few-branched from base; styles scarcely if at all exserted; corolla-limb 6-10 mm. broad, lilac to pink or yellow.—At ca. 4500 ft.; upper Sespe Creek, Ventura Co. to cent. Calif. April-May.

Ssp. **micránthus** (Steud.) Mason. [*Gilia m.* Steud.] Stems erect; styles conspicuously long-exserted; fls. yellow to pink or lilac.—Below 4500 ft.; V. Grassland, Coastal Sage Scrub, Chaparral; L. Calif. to cent. Calif. April-May.

2. **L. arenícola** (Jones) Jeps. & Bail. [*Gilia a.* Jones.] Small erect annual 1-8 cm. high, compactly branched, glabrous to minutely puberulent; lvs. mostly 3-5-cleft above their base, pilose above, glabrous beneath, 3-12 mm. long; fls. vespertine, solitary and sessile in forks of cymes or at tips of branches; calyx 4-5 mm. long, the sinuses ca. ⅔ filled with membrane; corolla yellow, sometimes with purple in throat, 5-7 mm. long; stamens inserted in base of throat, glabrous; seeds brownish, with bordered whitish angles.— Rare gypsophilous plant, 2500-4000 ft.; Joshua Tree Wd.; Mojave Desert (Barstow, Kelso, Daggett, Searles Lake, Needles, Trona, etc.) and Nipton, Nev. March-April.

3. **L. aúreus** (Nutt.) Greene. [*Gilia a.* Nutt.] Plate 68, Fig. D. Annual, subglabrous

or puberulent or ± glandular, 1-several-stemmed from base, the stems very slender, simple or few-branched, 5–16 cm. long, erect or ascending; lvs. remote, 3–7-cleft into linear-oblong lobes 3–6 mm. long; cyme open, the pedicels filiform, 5–15 mm. long; calyx narrow-campanulate, 4–6 mm. long, deeply cleft into lance-ovate lobes, the hyaline membrane between the sinuses extended ca. ⅔ their length; corolla funnelform, 6–13 mm. long, deep to pale yellow, with orange to brownish-purple throat; tube included in calyx, shorter than throat and with hairy ring at summit; stamens glabrous; stigma exserted; caps.-locules several-seeded; $n = 9$ (Flory, 1937).—Locally common over large areas, in sandy places, below 6000 ft.; Creosote Bush Scrub, Joshua Tree Wd.; Mojave and Colo. deserts, occasional in interior cismontane valleys from L. Calif. to Ventura Co.; Nev., New Mex. March–June. A form with whitish corolla and often with a dark throat, the corolla 7–15 mm. long, occurs with the yellow form, but often in separate areas of some extent. It is var. *decorus* (Gray) Jeps.

4. **L. béllus** (Gray) Greene. [*Gilia b.* Gray.] With general aspect of *L. dianthiflorus*; stems capillary, subglabrous; lvs. mostly tripartite, 2–4 mm. long; calyx 3–4 mm. long, sessile, the lobes with purple basal band and margined to tip; corolla 1–1.5 cm. long, lilac to pink, with yellow throat with purple spots; stamens glabrous; caps. almost as long as calyx; seeds ca. 0.3 mm. long, light brown, sharply white-angled.—Dry slopes and flats, ca. 3000–4000 ft.; Chaparral; se. San Diego Co. (Jacumba, Tecate, etc.); adjacent L. Calif. April–May.

5. **L. bìcolor** (Nutt.) Greene. [*Leptosiphon b.* Nutt.] Annual, simple or few-stemmed from base, erect or ascending, 3–15 cm. high, puberulent; lvs. remote, 3–7-cleft into linear segms. 3–10 mm. long, hispid-ciliate, the middle segm. broader than others; fls. sessile in leafy-bracteate heads; calyx hispid, 6–8 mm. long, deeply cleft into subulate lobes, not or scarcely scarious in the sinuses; corolla salverform, 15–30 mm. long, bicolored, the lobes red to pink or white, 2–3 mm. long, round-obovate, the throat yellow, 1–2 mm. long, the tube yellow, stout, 12–22 mm. long, puberulent; stamens exserted; caps. 3–4 mm. long, the locules 2–4-seeded; seeds tuberculate.—Grassy and open wooded places, mostly below 5000 ft.; V. Grassland, Foothill Wd., Chaparral; Santa Catalina and San Clemente ids. and Piute Mts., Kern Co. n. to n. Calif. and Wash. March–June.

6. **L. bigelòvii** (Gray) Greene. [*Gilia b.* Gray.] Habit of *L. dichotomus*; lvs. mostly entire, rarely 2–3-cleft, 1–3 cm. long; calyx 8–12 mm. long; corolla 10–15 mm. long, the lobes half as long as tube; stamens 2–3 mm. long, on upper half of tube, glabrous, included; seeds red-brown, angular-cylindric, 1 mm. long, with a bladdery close-fitting testa which makes whitish angles.—Occasional, dry slopes and valleys, mostly below 5000 ft.; Creosote Bush Scrub, Joshua Tree Wd.; w. Colo. Desert and Mojave Desert to Inyo Co.; V. Grassland, etc. from Ventura Co. to Monterey and Stanislaus cos.; to Utah, Tex. March–May.

7. **L. brevículus** (Gray) Greene. [*Gilia b.* Gray.] Plate 68, Fig. E. Annual, erect, simple or branched from base, 1–2.5 dm. high, the stems slender, ± cymosely branched above, puberulent or subglabrous; lvs. 3–5-parted, 3–10 mm. long, minutely pilose; fls. sessile in small bracteate compact clusters; calyx 7–8 mm. long, glandular, stiff-pilose, deeply cleft, the sinus-membrane extending more than halfway; corolla salverform, 15–25 mm. long, the tube glabrous, slender, 2–4 times the calyx, the lobes 4–6 mm. long, white to pink or bluish, the throat often purple; stamens 2 mm. long, glabrous, inserted on the throat; anthers at the orifice; caps.-locules several-seeded; seeds light brown, angled, ± rugose.—Dry open slopes, below 7000 ft.; Yellow Pine F., Pinyon-Juniper Wd., Joshua Tree Wd.; San Gabriel and San Bernardino mts., to Liebre and Ord mts. and adjacent Mojave Desert. May–Aug.

8. **L. ciliàtus** (Benth.) Greene. [*Gilia c.* Benth.] Erect annual, simple or branched from base, rather stiff, 1–3 dm. high, rather stiff-pubescent; lvs. rather remote, 5–11-cleft into linear, hispid ciliate lobes 5–20 mm. long; fls. sessile in leafy-bracteate heads; bracts hispid-ciliate; calyx 7–10 mm. long, deeply cleft into ciliate hispid acerose segms., the hyaline membrane of the sinuses extending ca. halfway; corolla rose to white, salverform, 12–25 mm. long, the tube slender, long-exserted, pubescent without, the throat yellow,

short, the lobes obovate, 2-4 mm. long; stamens glabrous, slightly exserted; caps. ca. 6 mm. long, few-seeded; seeds ± rugose, with narrow whitish wings on the angles.—Dry open places, 5000-7000 ft.; Yellow Pine F.; mts. from San Diego Co. to Ventura Co., then n. to Ore., Nev. April-July.

9. **L. concínnus** Mlkn. [*Gilia c.* Munz.] Tufted annual 1-2 cm. high or loosely branched and 5-12 cm. high, glandular-puberulent; lvs. opposite or alternate, 3-5-lobed, 8-15 mm. long, with some weak curly white hairs; fls. subsessile, in 3-7-fld. crowded cymes; calyx ca. 1 cm. long, deeply cleft into linear lobes, the broad hyaline membrane extending over halfway; corolla white, funnelform, 10-15 mm. long, the tube 1-2 mm., the narrow throat 6-8 mm. long, the lobes entire or denticulate, with 2 linear dark basal spots; stamens pubescent; caps. included; seeds several, ellipsoid, red-brown.—Dry rocky slopes, 5000-8500 ft.; Montane Coniferous F.; San Gabriel Mts. May-July.

10. **L. demíssus** (Gray) Greene. [*Gilia d.* Gray.] Annual, diffusely branched from base, 2-10 cm. high, sparingly glandular-puberulent; lvs. 3-5-cleft or entire, the acicular lobes 6-10 mm. long; fls. sessile or subsessile, in leafy terminal clusters; calyx ca. 3 mm. long, deeply cleft into lanceolate scarious-margined lobes, the membranes sometimes united below; corolla campanulate, white, 6-8 mm. long, the tube very short, the throat ca. 4 times the tube, the oblong lobes with 2 dark streaks below base; stamens glabrous, included; caps. ca. half as long as calyx; seeds several in each locule, slightly irregular in shape but not tubercled.—Occasional on dry plains and in washes below 4000 ft.; Creosote Bush Scrub, Joshua Tree Wd.; Mojave Desert from Victorville to Inyo Co.; to Utah, Ariz. March-May.

11. **L. dianthiflòrus** (Benth.) Greene ssp. **dianthiflòrus**. [*Fenzlia d.* Benth. *Gilia dianthoides* Endl.] GROUND-PINK. Annual, usually with several very slender at first spreading, then erect stems 5-12 cm. long, puberulent; lvs. entire, filiform, mostly opposite, 1-2 cm. long, shorter than internodes; fls. solitary or in few-fld. leafy cymes, subsessile or with short pedicels; calyx 10-16 mm. long, deeply cleft into linear lobes with hyaline membrane filling lower half of sinus and for short distance on the lobes; corolla short-funnelform, pink to lilac or white, 10-25 mm. long, with dark basal spots and yellow throat, the lobes dentate, longer than tube and throat; stamens inserted at base of throat; seeds many, round-oblong with whitish angled membrane; $n=9$ (Flory, 1937).—Common, open sandy places, below 4000 ft.; Coastal Sage Scrub, Chaparral, V. Grassland; Santa Barbara to San Diego and w. edge of Colo. Desert; n. L. Calif. Feb.-April.

Ssp. **farinòsus** (Brand) Mason. [*Gilia dianthoides* var. *f.* Brand.] Lvs. oblong-linear, 5-10 mm. long; calyx-lobes spatulate, farinose-pubescent.—Occasional, 1500-4500 ft.; San Bernardino, Hemet V. (San Jacinto Mts.) to Oak Grove (n. San Diego Co.).

12. **L. dichótomus** Benth. [*Gilia d.* Benth.] EVENING SNOW. Erect annual, simple or dichotomously branched, 5-20 cm. high, the stems slender, usually glabrous and ± glaucous; lvs. opposite, usually 3-7-parted into linear-filiform lobes 1-2 cm. long; fls. on short pedicels in the forks or terminal, vespertine; calyx glabrous, 8-14 mm. long, the green linear subulate divaricate tips 3-5 mm. long above the hyaline membrane; corolla funnelform, 1.5-3 cm. long, white or with some purple at the throat and on lobes when closing; stamens included, each with a dilated basal hairy pad; caps. several-seeded; seeds pale, ± angled-cubic, ca. 0.7 mm. long; $n=9$ (Flory, 1937).—Dry often sandy or gravelly places, mostly below 5000 ft.; Creosote Bush Scrub, Joshua Tree Wd.; Mojave Desert and w. edge of Colo. Desert. A form near Victorville with undivided lvs. has been named *Gilia dichotoma* var. *integra* Jones.

13. **L. jònesii** (Gray) Greene. [*Gilia j.* Gray.] Much like *L. bigelovii*, 3-15 cm. high; lvs. filiform, 1-2 cm. long; pedicels and calyx with stalked glands; calyx 7-8 mm. long, the membrane filling ca. ¾ of the sinus; corolla tubular-funnelform, 9-10 mm. long, whitish or yellowish, the obovate lobes ca. 4 mm. long; stamens glabrous, inserted in tube, ca. 2 mm. long; seeds reniform or nearly so, red-brown, with rounded broad irregularities, but not angled or with loose testa.—Uncommon, sandy slopes and washes, below 2500 ft.; Creosote Bush Scrub; from Death V. region through e. Mojave and Colo. deserts to Ariz., L. Calif. March-May.

14. **L. kíllipii** Mason. With aspect of *L. concinnus*; lvs. 3–10 mm. long, 5–7-cleft; fls. sessile; calyx ca. 6 mm. long, the membrane extending on to the lobes; corolla 10–15 mm. long, the tube 4–5 mm. long, each lobe with an elongate spot near the base; fils. glabrous. —Dry slopes, 5000–7000 ft.; Pinyon-Juniper Wd.; Cactus Flat to Baldwin Lake, San Bernardino Mts. May–July.

15. **L. lemmònii** (Gray) Greene. [*Gilia l.* Gray.] Annual, with 1–several very slender stems from base, 5–15 cm. long, short-pubescent throughout, glandular above; lvs. remote, 3–5-cleft into linear lobes 2–5 mm. long; fls. in subcapitate terminal and subterminal clusters; calyx 4–5 mm. long, deeply cleft into linear-hispid lobes, the sinuses with a white membrane ca. halfway; corolla short-funnelform, 5–8 mm. long, cream to yellowish, the lobes 2–3 mm. long; the tube with an inner hairy ring at top; stamens inserted just below the corolla-sinuses, the fils. glabrous; seeds many.—Sandy open places below 5500 ft.; Chaparral, Coastal Sage Scrub, Yellow Pine F.; interior cismontane from San Bernardino Co. to L. Calif., reaching w. edge of Colo. Desert. April–June.

16. **L. liniflòrus** (Benth.) Greene ssp. **pharnaceoìdes** (Benth.) Mason. [*Gilia p.* Benth.] Plate 68, Fig. F. Annual, 1–4 dm. high, with ± dichotomous branching, glabrous to puberulent; lvs. 3–9-cleft into linear segms. 1–3 cm. long; pedicels slender, in cymose panicles and 1–2.5 cm. long; calyx 3–4 mm. long, the lobes linear, the sinus-membrane ca. ⅔ as long as lobes; corolla short-funnelform, 1–3 cm. long, the throat 1–2 times as long as the tube, the fl. white or with some pink or lilac; fils. pubescent at base, inserted at bottom of throat; seeds 1–2 in each locule.—Occasional on dry slopes below 5000 ft.; many Plant Communities, cismontane and montane s. Calif., w. Mojave Desert, n. to Wash., Ida. April–July.

17. **L. maculàtus** (Parish) Mlkn. [*Gilia m.* Parish.] Minute annual 1–3 cm. high, with few spreading branches from base and ± pilose; lvs. alternate, oblong, entire, cuspidate, thick, 2–5 mm. long, puberulent above, glabrous beneath; fls. crowded, in leafy clusters at ends of slender branches; calyx 2–3 mm. long, parted almost to base; corolla campanulate, 4–5 mm. long, white with vermilion spot on each spreading oblong-obovate lobe; stamens included; caps. round-ovoid, ca. 1.5 mm. long; seeds brownish, unequally low-tubercled.—Rare, sandy places, 500–4000 ft.; Creosote Bush Scrub, Joshua Tree Wd.; Little San Bernardino Mts. April–May.

18. **L. melíngii** (Wiggins) V. Grant. [*Leptodactylon m.* Wiggins.] Like *Linanthus nuttallii* ssp. *floribundus*, but compact almost matted and with fls. 8–10 mm. long.—At 8500–9500 ft.; Inyo-White Mts., w. Nev. and n. L. Calif.

19. **L. nudàtus** Greene. [*Gilia n.* Brand.] Like *L. breviculus*; lvs. 5–11-cleft into linear lobes 3–12 mm. long, the upper lvs. and bracts densely hirsute-ciliate, ± glandular, the divisions joined by a hyaline membrane in their lower half or third; fls. sessile in compact heads; calyx ca. 5 mm. long; corolla 1–2 cm. long, the tube pubescent; stamens exserted; caps.-locules 1-seeded.—Open slopes, 2000–7000 ft.; Jeffrey Pine F., Foothill Wd., Chaparral; Tehachapi Mts., to Sierra Nevada of Tulare Co. May–July.

20. **L. nuttállii** (Gray) Greene ex Mlkn. ssp. **nuttállii**. [*Gilia n.* Gray. *Leptodactylon n.* Rydb. *Linanthastrum n.* Ewan.] Bushy perennial from a woody base, the erect stems thickly branched, mostly 1–2 dm. high, ± villous-hispid; lvs. opposite, 5–9-cleft into linear-oblanceolate lobes 1–1.5 cm. long and with fascicled smaller lvs. in axils; fls. sessile or subsessile in upper axils, forming subcapitate infl.; calyx narrow-campanulate, 8–9 mm. long, the tube ca. 2 mm. long, the lobes lance-subulate, puberulent, with scant hyaline membrane in the sinuses; corolla funnelform to salverform, 12–15 mm. long, the usually yellowish tube ca. 8 mm. long, pubescent, the throat short, lobes white or nearly so, 4–5 mm. long, oblanceolate; stamens inserted at base of throat, barely exserted; caps. ca. 5 mm. long, with 1–4 seeds in a locule; seeds shallowly rugose-reticulate.—Dry rocky or brushy places, 4000–12,000 ft.; Sagebrush Scrub to Subalpine F.; San Bernardino Mts., e. slope of Sierra Nevada and in Inyo-White Mts.; to Wash., Rocky Mts. May–Aug.

Ssp. **floribúndus** (Gray) Munz. [*Gilia f.* Gray.] Stems 1–4 dm. high; lvs. 3–5-cleft into almost filiform lobes 8–20 mm. long, often without axillary fascicles; fls. often in more open cymose clusters.—Dry places below 7000 ft.; Chaparral, Yellow Pine F., S. Oak Wd.; San Jacinto, Santa Rosa, Santa Ana mts. s. to L. Calif.; to Colo., Chihuahua, etc. May–July. A form from s. slope of Santa Rosa Mts. has been called *L. floribundus* ssp. *hallii* Mason.

21. **L. orcúttii** (Parry & Gray) Jeps. ssp. **pacíficus** (Mlkn.) Mason. [*L. p.* Mlkn.] Several-stemmed, suberect, puberulent, 5–10 cm. high; lvs. remote, 3–7-parted into linear lobes 5–12 mm. long, sparsely hairy; fls. solitary or few in small heads; calyx 6–10 mm. long, deeply cleft, the linear segms. membrane-margined and with sinus membrane below; corolla funnelform, 15–25 mm. long, pink to blue, the stout tube 5–15 mm. long, the short throat yellow, the lobes 5–8 mm. long and each with a purple basal reniform spot; stamens inserted at base of throat; caps. 6–12-seeded.—Uncommon, 4000–5000 ft.; Chaparral; Palomar Mts. and Laguna Mts., San Diego Co. May–June.

22. **L. párryae** (Gray) Greene. [*Gilia p.* Gray.] Tufted annual, 1–6 (–9) cm. high, mostly compactly branched, glandular-villous; lvs. crowded, 3–7 parted, the linear lobes 5–15 mm. long; fls. crowded, sessile to subsessile, in leafy cymes; calyx deeply cleft, 6–8 mm. long, the segms. scarious-margined, the membranes united at base only; corolla funnelform, white, less frequently blue-purple or even cream, 10–15 mm. long, the tube included in calyx, the lobes 6–12 mm. long, entire or erose, the throat bearing below each lobe ± reniform arches; stamens glabrous; seeds with a whitish somewhat angled outer membrane.—Common over large areas of dry sandy flats, mostly below 3500 ft.; Creosote Bush Scrub, Joshua Tree Wd.; w. Mojave Desert to Inyo and Mono cos., arid inner Coast Ranges to e. Monterey Co. March–May.

23. **L. pygmaèus** (Brand) J. T. Howell. [*Gilia p.* Brand. *G. pusilla* auth., not Benth.] Simple or with several very slender branches 2–10 cm. long, erect or diffuse, minutely hispid-puberulent; lvs. remote, 3–5-cleft into linear segms. 2–6 mm. long; fls. in open leafy dichotomous cymes, the capillary pedicels 4–15 mm. long; calyx narrow-cylindric, 3–5 mm. long, the sinuses half filled with membrane; corolla narrow-funnelform, 3–5 mm. long, white to pale blue, the lobes definitely shorter than tube plus throat; stamens glabrous; locules of caps. several-seeded.—Occasional on dry slopes below 5000 ft.; Chaparral, Yellow Pine F., etc.; mts. of interior San Diego Co. and w. Riverside Co., Tehachapi Mts., Coast Ranges to cent. Calif.; L. Calif. April–June.

9. *Micrósteris* Greene

Small branching annuals. Lvs. entire, the lower opposite, spatulate to oblanceolate, the upper alternate, lanceolate. Fls. small, pedicelled, usually in pairs in upper axils, white to purplish-pink. Calyx cylindrical, 5-lobed, the herbaceous lobes acute, subequal, ± united by a hyaline membrane. Corolla salverform with a slender tube, the lobes ovate to obcordate. Stamens included, unequally inserted in the corolla-tube. Ovary globose, 3-loculed; style slender; stigma 3-lobed. Caps. ovoid, 3-valved, rupturing the calyx in age. Seeds solitary in locules, lenticular, mucilaginous when wet. One polymorphous sp. of N. and S. Am. (Greek, *mikros*, small, and *aster*, star.)

1. **M. grácilis** (Hook.) Greene. [*Gilia g.* Hook. *Phlox g.* Greene.] ssp. **grácilis**. Usually with erect slender stems 1–2 dm. high, generally simple and glabrous to pilose below, branched and glandular-pubescent above; lvs. 1–3 cm. long, short-petioled, the upper sessile, somewhat reduced; infl. cymose, glandular, usually dense; calyx 5–8 (–10) mm. long, the free lobes ca. as long as tube; corolla 8–12 mm. long, well exserted, the tube ± yellowish, the lobes rose to white or lavender, 1–2 mm. long; stamens included; caps. ca. 5 mm. long; seeds ca. 3 mm. long; $n=7$ (Kinch, 1956).—Frequent in open grassy places and woods, below 10,000 ft.; many Plant Communities; cismontane and montane Calif.; to B.C., Mont. March–Aug. Intergrading with

Ssp. **hùmilis** (Greene) V. Grant. [*M. h.* Greene.] Diffuse from base, usually broader than high, ± glandular-canescent; corolla largely 5–8 mm. long, barely exserted.—Mts. of s. Calif. from San Diego Co. to ne. Mojave Desert and Mt. Pinos; to e. Wash., Rocky Mts., Ariz.

10. Navarrètia R. & P.

Annuals, mostly with rigid stems, erect and simple to divaricately branched from base or above, the branches often spreading or prostrate. Lvs. alternate, entire to once or twice pinnate, sometimes palmately lobed, the upper bracteate, acerose or spine-tipped; lobes of bracts or lvs. sometimes proliferating on back with extra segms. from rachis. Fls. 5- or 4-merous, sessile or nearly so, in spiny densely bracted heads, the bracts usually with broad coriaceous rachis. Calyx cleft to base into ± unequal entire to toothed segms. which are usually acerose and united by a scarious membrane in their lower fourth to two-thirds. Corolla funnelform to salverform, white or yellow to pink, blue or purple, the lobes with 1 or 3 main veins, which may be simple or branched. Stamens equally or unequally inserted in throat or in sinuses of corolla-lobes, exserted or included; fils. glabrous. Stigma entire to 2-3-lobed, included or exserted. Caps. ovoid to obovoid, 1-3 loculed, 3-8-valved, with irregular or regular dehiscence, membranous to chartaceous, usually circumscissile near base or above and then splitting into valves. Seeds mostly brown, often minutely pitted, 1-many in a cell. Ca. 30 spp., mostly w. N. Am., one in s. S. Am. (Named for Dr. F. *Navarrete*, a Spanish physician.)

1. Main stems finely retrorse-pubescent with crisped white hairs; mature caps. thin-membranous, the walls not splitting into distinct valves.
   2. Stigma minutely 2-lobed; lobes of the foliaceous bracts soft-herbaceous when fresh
   6. *N. prostrata*
   2. Stigma deeply 2-3-lobed; lobes of the foliaceous bracts at base of the head rigidly acerose, the rachis expanded at base.
      3. Stigma 2-cleft.
         4. Each corolla-lobe with 1 main vein; the corolla 5-9 mm. long .... 3. *N. intertexta*
         4. Each corolla-lobe with 3 main veins; corolla 10-14 mm. long ...... 7. *N. pubescens*
      3. Stigma 3-cleft, each corolla-lobe with 3 main veins .... 9. *N. tagetina*
1. Main stems with spreading hairs to almost glabrous, or if the hairs introrse, not crisped and appressed against the stems; mature caps. with thicker wall, with definite scheme of dehiscence.
   5. Seeds 1-2 (if more, fls. yellow); caps. chartaceous.
      6. Stems erect, simple or branched.
         7. Bracts constricted in middle, broad above and below; caps. circumscissile at middle. Piute Mts. .... 8. *N. setiloba*
         7. Bracts not as above, but laciniately linear-lobed from a slightly expanded base; caps. circumscissile at base .... 7. *N. pubescens*
      6. Stems prostrate or spreading, several from base; bract-rachis broadest above the middle
      4. *N. mitracarpa*
   5. Seeds few to many; caps. coriaceous.
      8. Lvs. with filiform rachis, the lateral linear lobes much exceeded by the elongate terminal segm.; plants not heavily glandular .... 5. *N. peninsularis*
      8. Lvs. with broader rachis, the terminal lobe not greatly elongate; plants heavily glandular.
         9. Terminal segm. of bract with 3 diverging acerose teeth; corolla-throat narrow; stamens included, unequally inserted .... 1. *N. atractyloides*
         9. Terminal segm. of bract as above; stamens exserted, equally inserted ... 2. *N. hamata*

1. N. atractyloìdes (Benth.) H. & A. [*Aegochloa a.* Benth.] Plate 69, Fig. A. Erect, usually freely branched, glandular-hirsute, 5-20 cm. high; lvs. well distributed, sessile, mostly 2-5 cm. long, pinnately dissected from a broad rachis, with pungent lobes or teeth and usually 3 terminal lobes; bracts ovate, coriaceous, with divergent spines; calyx 5-7 mm. long, ± membranous in lower part, the lobes entire or toothed; corolla narrow-funnelform, ca. 1 cm. long, the lobes 2.5 mm. long, white or blue or yellow; stamens included, 1-1.5 mm. long, unequally inserted; stigma included; caps. coriaceous, dehiscent from base up; seeds pitted, irregularly angled; $n=9$ (Grant, 1958).—Mostly in dry places, below 2000 ft.; many Plant Communities, but especially Coastal Sage Scrub, Chaparral; L. Calif. to n. Calif. May-July.

2. N. hamàta Greene. Close to *N. atractyloides* but tending to have a skunky odor; corolla with an ample throat and lobes usually purple, to 4.5 mm. long; stamens exserted,

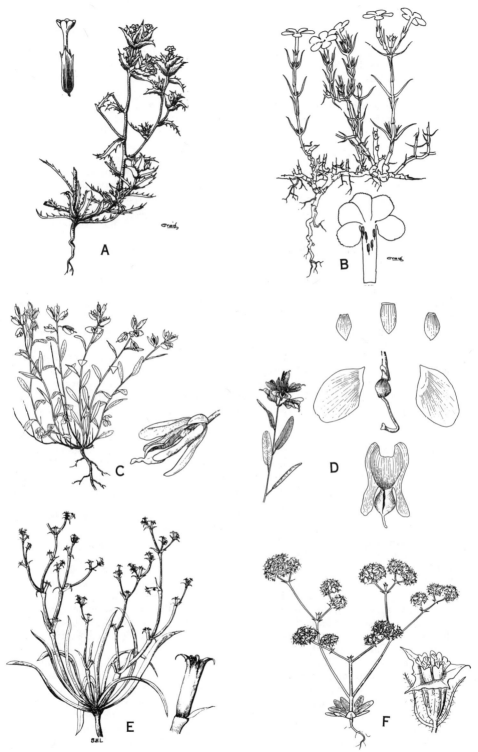

Plate 69. Fig. A, *Navarretia atractyloides*. Fig. B, *Phlox austromontana*. Fig. C, *Polygala californica*. Fig. D, *Polygala cornuta*, sepals 5 (the 2 inner being the enlarged lateral wings), 3 united petals below (2 lateral and 1 keel-shaped). Fig. E, *Chorizanthe brevicornu*. Fig. F, *Chorizanthe staticoides*, invol. with spreading teeth.

unequal in length, but equally inserted.—Dry often rocky slopes mostly below 3000 ft.; Coastal Sage Scrub, Chaparral; L. Calif. to San Luis Obispo Co. Santa Catalina and San Clemente ids. April–June. Some plants with softer lvs. and whitish fls. have been called var. *foliacea* (Greene) Thorne. A form with long-exserted corolla-tube, ranging from the San Diego region south is ssp. *leptantha* (Greene) Mason.

3. **N. intertéxta** (Benth.) Hook. [*Aegochloa i.* Benth.] Simple and erect, or branching from base, the stems brown, 5–20 cm. high, retrorse-pubescent below the heads; lvs. 1–5 cm. long, 1–2-pinnate, the rachis linear, glabrous or pubescent on upper surface; bracts foliaceous, pinnate, 1–2 cm. long, ± villous on upper surface, broadly membranous-margined; calyx 8–10 mm. long, the entire or toothed lobes narrow, the sinus-membrane conspicuously pubescent; corolla pale blue to white, slender-funnelform, 5–9 mm. long, the lobes 4–6 mm. long, 1-veined; stamens unequally inserted; stigma deeply 2-lobed; seeds irregularly angled.—Vernal pools and moist places, below 6000 ft.; many Plant Communities; cismontane and montane from San Diego Co. n.; to B.C. May–July.

4. **N. mitracárpa** Greene. Erect to spreading, 2–15 cm. high, the stems slender, brown, pubescent, often glandular; lvs. 1–2-pinnate, 1–3 cm. long, with acicular spines-cent lobes and flattened rachis having marginal teeth above the middle; bracts 5–15 mm. long, more rigid, less dissected; infl. 1.5–2.5 cm. broad; calyx 5–9 mm. long, unequally lobed, the segms. entire to 3–5-lobed, sinus-membrane narrow; corolla funnelform, 8–11 mm. long, the lobes purple to pink, 2–3 mm. long, 3-veined; stamens 5–8 mm. long; stigma exserted, 2-lobed; caps. chartaceous, circumscissile at base; seed 4-angled, ca. 2 mm. long.—Dry gravelly and clayey slopes, below 3500 ft.; Foothill Wd., Chaparral; Ojai, Ventura Co. to cent. Calif. April–June.

5. **N. peninsuláris** Greene. Sparsely glandular-pubescent, to ca. 1.5 dm. high; lvs. with a rachis ca. 1 mm. wide; bracts pinnate; calyx 4–5 mm. long, the segms. unequal; corolla 6 mm. long, the lobes ca. 2 mm. long, pale-purplish; stamens unequal; caps. many-seeded, chartaceous, dehiscent from top down; seeds obscurely pitted.—Rare, damp disturbed places, at 7000–7500 ft.; Holcomb V., San Bernardino Mts.; Tehachapi Mts.; to L. Calif. June–Aug.

6. **N. prostràta** (Gray) Greene. [*Gilia p.* Gray.] Cent. head nearly or quite sessile, usually with prostrate branches from beneath, these 4–10 cm. long, retrorse-pubescent; lvs. 3–7 cm. long, glabrous, pinnate to bipinnate, the rachis and lobes linear; bracts leafy, 1–4 cm. long, ciliate; calyx 4–5 mm. long, the lobes entire to trifid, pubescent, the sinus-membrane ciliate; corolla broadly funnelform, ca. 8 mm. long, white to violet, the lobes 1.5–2 mm. long; stamens 2.5 mm. long; stigma minutely 2-lobed; caps. indehiscent, 2-celled.—Vernal pools and moist places, below 2000 ft.; Los Angeles and w. San Bernardino cos.; to Monterey Co. April–May.

7. **N. pubéscens** (Benth.) H. & A. [*Aegochloa p.* Benth.] Erect, simple or branched mostly above, retrorse-canescent and somewhat glandular; lvs. 1.5–5 cm. long, 1–2-pinnate into linear divisions, the rachis sometimes expanded toward tip especially in upper lvs.; bracts with prominent rachis and pungent lobes; calyx ca. 1 cm. long, the segms. unequal, the longer often toothed, the sinus-membrane densely pilose; corolla funnelform, blue with darker veins, 10–14 mm. long, the lobes 3-veined, 2.5 mm. long; stamens unequal, ± exserted; stigma 2-cleft; caps. chartaceous, circumscissile at base.—Dried depressions and dry slopes below 3500 ft.; Chaparral, Foothill Wd., etc.; Santa Monica Mts. n. to n. Calif. May–July.

8. **N. setilòba** Cov. Erect, racemosely branched, 1–2 dm. high, puberulent; lvs. bipinnate, 1–3 cm. long, the terminal lobe broad, often purple, with irregularly and finely laciniate margin; bracts with lanceolate terminal lobe; calyx 7–10 mm. long, puberulent, with coarser hairs in middle, the sinus-membrane extending over halfway, the lobes entire or with small teeth; corolla purple, funnelform, ca. 10 mm. long, the lobes 2.5–3 mm. long; stamens well exserted; stigma 2-lobed; caps. circumscissile at middle, 1-seeded; seed 4-angled.—Slopes and flats, 5000–7000 ft.; Yellow Pine F.; Piute Mts., Kern Co.; to Tehachapi Mts. April–June.

9. **N. tagetìna** Greene. Erect, 5–25 cm. high, simple or racemosely branched, ± retror-

sely pubescent, especially up the stems; lower lvs. 1–2-pinnate, 2–5 cm. long, the lobes narrow, cuspidate; upper lvs. 2–3-pinnate, passing into the bracts, the rachis rather broad; bracts rigid, cup-shaped at base, the lobes proliferating on under surface, densely pubescent; calyx 8–10 mm. long, the lobes unequal, acerose, 3–5-toothed, pubescent on outer surface; corolla pale blue, funnelform, ca. 10 mm. long, the lobes 2–2.5 mm. long, 3-veined; stamens inserted in lower throat, 2–2.5 mm. long; caps. not regularly dehiscent; seeds pitted.—Moist and grassy places, below 5000 ft.; Cuyamaca Lake; cent. Calif. to Wash. April–June.

11. Phlóx L.

Perennial or annual, herbaceous to suffrutescent, erect, diffuse or cespitose. Lvs. largely opposite, entire. Fls. in terminal corymbiform or paniculate cymes or sometimes solitary, often showy, red or purplish or blue to white. Calyx tubular or narrow-campanulate, 5-cleft and ribbed, the lobes acute or acuminate, usually scarious on the margins and with a scarious membrane between the lobes. Corolla salverform, the tube slender and throat constricted. Stamens short, included, unequally inserted. Ovary 3-celled, ovoid to oblong; style mostly slender. Seeds 1–few in each cell, oblong, usually not emitting spiral threads when wet. Ca. 45 spp. of N. Am. and n. Asia, some of garden use. (Greek, *phlox*, flame, ancient name of Lychnis.)

(Wherry, E. T. The genus Phlox. Morris Arb. Mon. III, 1–174. 1955.)
1. Infl. without gland-tipped hairs.
    2. Calyx-membranes manifestly carinate; lvs. mostly 12.5–25 mm. long, not ciliate. San Gabriel to San Bernardino and Cuyamaca mts. . . . . 1. *P. austromontana*
    2. Calyx-membranes flat; lvs. 8–15 mm. long, sparsely finely ciliate. San Gabriel Mts.
    3. *P. diffusa*
1. Infl. with some gland-tipped hairs.
    3. Corolla-tube 7–10 mm. long; lvs. 3–7 mm. long, concave, coarse-ciliate, mostly pilose. San Bernardino and Inyo mts . . . . 2. *P. covillei*
    3. Corolla-tube 13–45 mm. long; lvs. 12–60 mm. long.
        4. Corolla-tube 13–17 mm. long; lvs. 2–9 mm. wide, 30–40 mm. long. Mts. of c. San Bernardino and of Inyo cos. . . . . 6. *P. viridis*
        4. Corolla-tube 26–45 mm. long.
            5. Corolla-tube 35–45 mm. long; lvs. 3–8 mm. wide. Bear V., San Bernardino Mts.
            4. *P. dolichantha*
            5. Corolla-tube 26–33 mm. long; lvs. 2–5 mm. wide. Inyo-White Mts. . . . . 5. *P. superba*

1. **P. austromontána** Cov. Plate 69, Fig. B. Cespitose perennial from a woody caudex, the stems 5–10 (–15) cm. long; herbage gray-green, not glandular, but ± white-pubescent; lvs. mostly 1–1.5 cm. long, acerose to pungent, pubescent above, often glabrous beneath; fls. solitary at the ends of branchlets, short-pedicelled; calyx plicate in sinuses, 6–10 mm. long, the acerose teeth slightly longer than the angled tube; corolla white to pink or lavender, the tube 11–14 mm. long, lobes obovate, rounded, 5–7 mm. long; styles 4–5 mm. long; caps. 4–5 mm. long.—Dry rocky places, 4500–8000 ft.; Pinyon-Juniper Wd., Yellow Pine F.; San Gabriel Mts. to Santa Rosa Mts., Cuyamaca Mts.; to Ariz., Utah, L. Calif. May–July.

2. **P. covíllei** E. Nels. Cushionlike from a woody densely cespitose perennial base; branchlets 1–3 cm. long, subglabrous; lvs. crowded, linear-oblong, appressed or ascending, thick, mostly 3–5 (–7) mm. long, abruptly narrowed to an apiculate tip, the margins thick, riblike, ciliate, the surfaces glandular-puberulent, the lower surface grooved; fls. mostly solitary, sessile, terminal; calyx 5–6 mm. long, glandular-puberulent, the lobes like the lvs., shorter than or equaling the tube; corolla white or pale pink, the tube 8–10 mm. long, the lobes round-obovate, ca. 4–6 mm. long; style ca. 2 mm. long; caps. ca. 4 mm. long.—Dry slopes and benches, especially on travertine and limestone, 6000–10,000 ft.; Pinyon Juniper Wd., Yellow Pine F. and above; San Bernardino Mts., Inyo-White Mts.; to Sierra Nevada; Nev. June–Aug.

3. **P. diffùsa** Benth. Suffrutescent perennial from stout branching base; stems 5–15

cm. high, ± decumbent; herbage subglabrous to thinly tomentose, not glandular; lvs. subulate, 10–15 mm. long, acerose, 1–2 mm. wide, sparsely ciliate with fine hairs; fls. short-pedicelled; sepals 8–11 mm. long, somewhat pilose, the membrane flat; corolla pink to lilac or white, the tube 9–13 mm. long, the lobes obovate, 6–7 mm. long, rounded at apex; style 3–5 mm. long; caps. 4–5 mm. long; $2n = 28$ (Flory, 1937).—Dry slopes and flats, 3300–10,000 ft.; e. end of San Gabriel Mts.; to Ore. May–Aug.

Ssp. **subcarinàta** Wherry. Lvs. tending to be flatter and 1–1.5 mm. wide, intercostal membrane of calyx somewhat carinate.—Rare, dry slopes, 4000–11,000 ft.; Montane Coniferous F.; San Bernardino Mts., Mt. Pinos; to Wash., w. Nev.

4. **P. dolichántha** Gray. [*P. bernardina* M. & J.] Perennial herb from underground rootstocks, scarcely if at all suffrutescent; stems 1–3 dm. high, branched, subglabrous to somewhat glandular in upper parts; lvs. sessile, lance-linear, 2–4.5 cm. long, 2–6 mm. wide, acute to acuminate; fls. few, in loose cymose clusters, white or rose to pinkish lavender; pedicels 3–20 mm. long; calyx 10–12 mm. long, glandular-puberulent; corolla-tube 35–45 mm. long, the limb 1.5–2 cm. broad, the oblong lobes 8–10 mm. long, sub-entire; stamens somewhat exserted; style 25–30 mm. long; caps. 5–8 mm. long, 1-seeded. —Open places in Montane Coniferous F.; Bear V., San Bernardino Mts. June–July.

5. **P. supérba** Brand. [*P. stansburyi* ssp. *superba* Wherry.] Woody-based perennial 1–3 dm. high; lvs. 2–6 cm. long, 2–5 mm. wide; infl. 3–18-fld., glandular-pubescent; pedicels 7–40 mm. long; sepals 9–14 mm. long, the membrane plicate; corolla-tube 26–33 mm. long, petal-blades purple or pink, 9 mm. long, obtuse; styles 18–28 mm. long.—Dry gravelly slopes and washes, 5000–9000 ft.; Sagebrush Scrub, Pinyon-Juniper Wd.; Inyo-White Mts.; to Nev. and n. April–June.

6. **P. víridis** E. Nels. ssp. **compácta** (Brand) Wherry. [*P. longifolia compacta* Wherry.] Woody-based perennial 5–20 cm. tall; lvs. linear to lanceolate, acuminate, sparsely ciliate, 3–4 cm. long, 2–3 mm. wide; infl. compact, 6–15-fld., glandular-pubescent; pedicels 8–20 mm. long; sepals 10–14 mm. long, the membrane strongly plicate-carinate; corolla light purple to pink or white, the tube 13–17 mm. long, blades ca. 9 mm. long; styles 10–14 (–18) mm. long.—Dry rocky slopes to 9000 ft.; Creosote Bush Scrub to Pinyon-Juniper Wd.; Argus, Panamint, Providence, and Clark mts. to Death V. region and n. and to Rocky Mts.

12. *Polemònium* L.

Erect, spreading or decumbent annuals, or cespitose or rhizomatous perennials, with simple or branched stems. Lvs. pinnate, the lfts. entire or divided. Fls. in terminal or axillary cymes, congested or lax. Calyx green, ± campanulate, the lobes deltoid to acuminate. Corolla rotate-campanulate to narrow-funnelform, the tube and throat not sharply distinct, the lobes spatulate to roundish, blue or purple to pink, white or yellowish. Stamens usually equally inserted on the tube; fils. and style included to exserted. Caps. ovoid, trilocular, each locule 1–10-seeded. Seeds elongate, oblong or pointed at ends, in some spp. mucilaginous when wet. Ca. 20 spp. of cooler parts of N. Hemis. and of S. Am. (Ancient name, possibly for *Polemon*, early Athenian philosopher.)

(Davidson, J. F. The genus Polemonium. Univ. Calif. Pub. Bot. 23: 209–282. 1950.)
Plants annual; corolla 2–6 mm. long .... 2. *P. micranthum*
Plants perennial; corolla ca. 13–15 mm. long .... 1. *P. caeruleum*

1. **P. caerùleum** L. ssp. **amygdalìnum** (Wherry) Munz. [*P. occidentale* Greene.] Perennial from a rootstock; stems solitary, erect, glandular-pubescent, 2–9 dm. high; lvs. pinnate, the lfts. mostly 19–27, 1–3.5 cm. long; infl. strict; bracts pinnatifid to entire; calyx campanulate, 4–10 mm. long, the lobes ca. as long as tube, glandular-pubescent; corolla blue, rotate-campanulate, 13–15 mm. long; stamens ca. as long as corolla; style exceeding corolla; caps. 3 mm. long; seeds 2 mm. long.—Wet places, S. Fork of San Bernardino Mts. at 7000–9000 ft.; Montane Coniferous F.; to Alaska and Rocky Mts. June–Aug.

2. **P. micránthum** Benth. Annual, the stems 5–25 cm. tall, solitary to diffusely

branched, slender, subglabrous to glandular-pilose, leafy; lfts. 7–15, 1–5 mm. long, glandular-pubescent; infl. cymose, each fl. on a branch with 1 lf.; pedicels 2–15 mm. long; calyx campanulate, 3–7 mm. long; corolla white to bluish, 2–6 mm. long, open-campanulate, the rounded lobes ca. as long as the tube; stamens included; caps. 3 mm. in diam.; $n=9$ (Grant, 1958).—Dry open places, 2000–6000 ft.; Santa Ynez Mts., etc.; to B.C., Rocky Mts. March–May.

84. Polygalàceae. Milkwort Family

Herbs, shrubs or trees. Lvs. alternate, rarely opposite or whorled, simple; stipules 0 or small glands; petioles short. Fls. bisexual, irregular, usually racemose or spicate, each subtended by a bract and 2 bractlets. Sepals 5, free, imbricate, the 2 inner (wings) larger and often petaloid. Petals 3, rarely 5, the upper 2 free or united with the lowermost or anterior (keel) which is boat-shaped, often with a terminal beak or fimbriate crest, the 2 lateral rarely present. Stamens usually 8, with united fils. forming a sheath split above, often adnate to the petals; anthers usually confluently 1-celled, opening by a subterminal pore. Disc present or reduced to gland at base of ovary or lacking. Pistil of 1 or 2, rarely 3–5 united carpels; style 1; stigma 2-lobed. Ovules usually solitary, pendulous. Fr. a caps., drupe or samara, dehiscent or not. Seeds usually pubescent, with conspicuous aril. Ca. 10 genera and 1000 spp. from around the world in trop. and temp. regions; some of hort. importance.

1. Polýgala L. Milkwort

A widespread genus of 500–600 spp. (Greek, *polus*, much, and *gala*, milk; some spp. supposed to increase the flow of milk.)

Branches spreading-puberulous or -pubescent, spine-tipped.
    Fls. 4–5 mm. long; yellowish; plants 12–90 cm. high .... 1. *P. acanthoclada*
    Fls. 10–11 mm. long; pink-purple and yellow; plants 5–15 cm. high .... 4. *P. subspinosa*
Branches glabrous or strigulose, not spine-tipped.
    Fls. of 2 kinds, those in racemes at base of plant lacking petals; others with petals; sepals glabrous except for the ciliate margins; caps. thin-walled, reticulate .... 2. *P. californica*
    Fls. all having petals; sepals often pubescent; caps. thick-walled, scarcely reticulate 3. *P. cornuta*

1. **P. acanthoclàda** Gray. Spiny cinereous much-branched shrub, 4–7 dm. high, the pubescence short, white, spreading; lvs. oblanceolate to linear-spatulate, 8–15 mm. long, short-petioled, puberulous; fls. yellowish; outer sepals 2–3 mm. long, the wings 4–5 mm. long; petals ca. 3 mm. long, slightly purplish at apex; caps. orbicular-truncate, notched, ca. 4 mm. long; seed straw-colored, somewhat pubescent, cylindro-obovoid, carunculate, ca. 2–2.5 mm. long.—Dry stony slopes and mesas, 2500–6000 ft.; Shadscale Scrub, Joshua Tree Wd., Pinyon-Juniper Wd.; e. Mojave Desert (Eagle, New York and Shadow mts.); to Ariz., Colo. May–Aug.

2. **P. califórnica** Nutt. Plate 69, Fig. C. Stems slender, 3–30 cm. high, with short racemes of cleistogamous fls. near base; lvs. 1–4 cm. long, ± elliptic, short-petioled; terminal racemes loosely several-fld.; bracts linear; fls. bright rose to paler; outer sepals ciliolate, 5–6 mm. long, the wings 10–11 mm. long; keel ca. 12 mm.; caps. roundish, ca. 7 mm. long.—Chaparral; Santa Cruz Id.; to s. Ore. April–June.

3. **P. cornùta** Kell. ssp. **físhiae** (Parry) Munz, comb. nov. [*P. fishiae* Parry. Proc. Davenport Acad. 4: 39. 1884.] Plate 69, Fig. D. Stems many, slender, 2–5 dm. high; lvs. ovate to sublinear, strigulose on veins, 2–4 cm. long; fls. 4–20, red-purple at least in age; pedicels 2–4 mm. long; outer sepals ca. 4 mm. long, puberulent, the wings 7–8 mm. long, glabrous except on ciliate margins; keel 8–9 mm. long including the 1 mm. beak; caps. ciliolate, ca. 8 mm. long; $n=9$ (Raven in herb.).—Shaded rocky places in canyons below 3000 ft.; S. Oak Wd., Chaparral; Santa Monica Mts., Mt. Wilson, Santa Ana Mts., etc.; L. Calif. June–Aug.

Ssp. **pollárdii** (Munz) Munz, comb. nov. [*P. cornuta* var. *p.* Munz, Aliso 4: 93. 1958.] Fls. greenish-white; wings ca. 7–8 mm. long, densely puberulent; keel ca. 9–10 mm. long, including the 1 mm. beak.—Shaded canyons, below 2000 ft.; Chaparral, S. Oak Wd.; Santa Barbara and Ventura cos. May–July.

4. **P. subspinòsa** Wats. var. **heterorhýncha** Barneby. Much-branched tufted undershrub 5–15 cm. high, pale green, densely puberulous, the branches spine-tipped; lvs. short-sessile, elliptic to elliptic-oblanceolate, 1–2 cm. long, sparsely ciliolate, ± pubescent; fls. 6–9, pink-purple and yellow; sepals elliptic, hyaline, the outer puberulent along their middle, 4–6 mm. long, the wings ca. 10 mm. long, glabrous; petals ca. 8 mm. long, the keel ca. 13 mm. long with a blunt entire straight beak 1.5–3 mm. long; caps. oval, reticulate, pubescent especially on the margin, 7–8 mm. long; seed ellipsoid, pubescent, 4 mm. long.—Alkaline calcareous hills, 3000–4000 ft.; Shadscale Scrub; e. Inyo Co., Calif. and Nye Co., Nev. April–May.

85. **Polygonàceae.** BUCKWHEAT FAMILY

Herbs, shrubs or climbers, rarely trees. Lvs. alternate or rarely opposite or whorled, simple, with stipules sometimes dilated into a membranous sheath (ocrea) above the swollen node. Fls. mostly perfect, small, rarely solitary. Calyx ± persistent, 2–6-cleft or -parted, in 1 or 2 series, often petaloid. Petals 0. Stamens usually 6–9, inserted near the base of the calyx. Pistil solitary, superior, mostly 1-celled; styles 2–4, usually free. Ovule 1, basal. Fr. a lenticular or angled ak., sometimes enclosed in the ripening perianth which may become fleshy. Ca. 30–40 genera and 800 spp., widely distributed in cold and temp. regions. A few spp. are used for ornament and food (*Rheum*, rhubarb; *Fagopyrum*, buckwheat; etc.).

1. Lvs. without stipules.
    2. Fls. subtended by bracts or not, but not enclosed in an invol.
        3. Fls. without subtending bracts; stamens 9 .... 4. *Gilmania*
        3. Fls. with 1 or more subtending bracts; stamens 3–9.
            4. Bracts woolly; fls. in headlike clusters; calyx glabrous .... 5. *Nemacaulis*
            4. Bracts not woolly; fls. not in heads.
                5. Bracts solitary, 2-lobed, enlarged, reticulate and saccate in fr.; lvs. rounded
                    .... 9. *Pterostegia*
                5. Bracts whorled, hooked at tip; lvs. linear .... 1. *Chorizanthe*
    2. Fls. enclosed in a tubular to campanulate invol.
        6. Invol. with spine- or bristle-tipped teeth.
            7. Invol. commonly 1-fld. and mostly 4–5-toothed, with the tube generally cylindric or prismatic and the teeth often hooked at their tips .... 1. *Chorizanthe*
            7. Invol. commonly 2-several-fld. and mostly 4–5-lobed, with the tube turbinate and the lobes ending in straight spines or bristles .... 7. *Oxytheca*
        6. Invol. with 3–8 teeth or lobes, these not bristle- or spine-tipped .... 3. *Eriogonum*
1. Lvs. with evident stipular sheaths.
    8. Calyx 4- or 6-parted or urn-shaped and 6-lobed.
        9. Calyx urn-shaped, becoming hard and burlike in fr.; outer lobes spine-tipped . 2. *Emex*
        9. Calyx 4- or 6-parted, the outer lobes not spine-tipped.
            10. Lvs. reniform; sepals 4; stigmas 2 .... 6. *Oxyria*
            10. Lvs. not reniform; sepals 6; stigmas 3 .... 10. *Rumex*
    8. Calyx 5-parted; aks. often enclosed by the somewhat enlarged fruiting calyx . 8. *Polygonum*

1. *Chorizánthe* R. Br. ex. Benth.

Low annuals (or perennial in S. Am.) dichotomously or trichotomously branched, erect to prostrate. Lvs. basal or cauline and alternate, entire, the upper commonly reduced to opposite or whorled bracts. Infl. capitate or cymose. Invols. sessile, cylindric, urn-shaped or funnelform, sometimes lacking, 1-fld. or sometimes 2–3-fld., 3–6-angled, or -ribbed, 3–6-toothed or cleft, the teeth divaricate, awned or uncinate. Fls. pedicellate or subsessile, included or partly exserted; bractlets lacking. Calyx 6-parted or -cleft (rarely 5), colored. Stamens 9, or 6 or 3. Styles 3. Aks. glabrous, 3-angled. An Am. genus of 50–60 spp. (Greek, *chorizo*, to divide, and *anthos*, fl., because of the divided calyx.)

(Goodman, G. J. A revision of the N. Am. spp. of the genus Chorizanthe. Ann. Mo. Bot. Gard. 21: 1–102. 1934.)
1. Invols. lacking; lvs. linear, the cauline in whorls of 4–5 .... 4. *C. coriacea*
1. Invols. present.
   2. Bracts entire.
      3. Invols. 6-toothed and -ribbed (see also 12. *C. polygonoides*).
         4. Teeth of the invol. without a scarious membrane along their margin.
            5. Involucral teeth equal or the outer 3 subequal, the anterior not greatly enlarged.
               6. Calyx-lobes entire or nearly so.
                  7. Outer and inner calyx-lobes alike or nearly so.
                     8. Plants erect; stems easily breaking up into joints when mature. Deserts
                                                                                       1. *C. brevicornu*
                     8. Plants prostrate to ascending; stems not very fragile. Cismontane
                                                                                       14. *C. procumbens*
                  7. Outer and inner calyx-lobes definitely unlike, the inner much smaller.
                     9. Involucral teeth straight, not hooked at tip; calyx-lobes erose; plants decumbent .... 11. *C. parryi fernandina*
                     9. Involucral teeth uncinate.
                        10. Bracts not leaflike, or those of the first node rarely so, but soon deciduous.
                           11. Lf.-blades 1–3 cm. long; invols. mostly in subcapitate clusters; calyx not half-exserted .... 17. *C. staticoides*
                           11. Lf.-blades 0.5–1.5 cm. long; invols. scattered, not crowded; calyx almost half-exserted .... 8. *C. leptotheca*
                        10. Bracts leaflike on main stems and branches.
                           12. Calyx 5–6 mm. long. Plants from e. of the coast ranges. . 21. *C. xanti*
                           12. Calyx 3–4 mm. long. Plants from near the coast.
                              13. Heads of invols. 1–1.5 cm. in diam.; plants gray-pubescent, suberect. Santa Monica Mts. to Santa Barbara .... 20. *C. wheeleri*
                              13. Heads of invols. less than 1 cm. in diam.; plants greenish-yellow above, prostrate or decumbent. Los Angeles Co. to Riverside Co.
                                                                                       11. *C. parryi*
               6. Calyx-lobes, at least the inner 3, fimbriate .... 6. *C. fimbriata*
            5. Involucral teeth very unequal, the anterior one usually longer than the involucral tube, the others relatively short .... 3. *C. clevelandii*
         4. Teeth of the invol. with a scarious membrane along their margins; invols. woolly, the membranes whitish; lvs. sublinear, basal and on lower stems ...... 9. *C. membranacea*
      3. Invols. not both 6-ribbed and 6-toothed.
         14. Involucral tube angled or ribbed, not transversely corrugated.
            15. Involucral teeth and ribs 4–5, one tooth much exceeding others; plants prostrate, with leafy bracts spinose-tipped. W. Mojave Desert .... 16. *C. spinosa*
            15. Involucral teeth and ribs 3, no one tooth greatly exceeding others.
               16. Involucral teeth 6, with alternate shorter. Cismontane ...... 13. *C. polygonoides*
               16. Involucral teeth 3.
                  17. Plants prostrate; involucral teeth uncinate, equal. Point Loma
                                                                                       10. *C. orcuttiana*
                  17. Plants erect; involucral teeth straight, unequal. Deserts ..... 15. *C. rigida*
         14. Involucral tube cylindric, corrugated but not ribbed or angled.
            18. Involucral teeth 5, one leaflike and much enlarged. W. Mojave Desert
                                                                                       19. *C. watsonii*
            18. Involucral teeth 3, equal. Death V. and s. .... 5. *C. corrugata*
   2. Bracts 3-lobed.
      19. Invols. with 3–6 divaricate spurs at base; bracts small.
         20. Spurs 3, saccate, rather broad; terminal involucral teeth not hooked . 18. *C. thurberi*
         20. Spurs 6, spinelike; terminal involucral teeth hooked .... 7. *C. leptoceras*
      19. Invols. not spurred at base.
         21. Bracts unilateral; invol. cylindrical, 2–3 in the axils. .... 2. *C. californica*
         21. Bracts orbicular-perfoliate; invols. 4-angled, mostly solitary ...... 12. *C. perfoliata*

1. **C. brevicórnu** Torr. ssp. **brevicórnu.** Plate 69, Fig. E. Erect, yellowish, glabrous to strigulose or pubescent, the stems several, ascending, 0.5–2 dm. long, breaking at nodes when dry; lvs. oblanceolate, largely basal, the blades 1.5–3 cm. long, petioles ca. as long;

lower bracts foliaceous, lanceolate to oblanceolate, upper reduced, all opposite, apiculate; invols. solitary in axils of branches of cymose infl., narrow-subcylindric, conspicuously ridged, slightly curved, ca. 4 mm. long, with 6 subequal short uncinate teeth; fls. short-pedicelled, 3-4 mm. long, glabrous, the calyx-lobes whitish, similar, scarcely exserted, linear-oblong, ca. 1 mm. long; stamens 3; aks. narrow, ca. 2 mm. long.—Dry gravelly and stony slopes, below 5000 ft.; Creosote Bush Scrub, Joshua Tree Wd.; deserts from Mono Co. to Nev., Utah, L. Calif. March-June.

Ssp. **spathulàta** (Small) Munz. [*C. s.* Small.] Stems and invols. with more red; lvs. broadly spathulate; invols. more broadly cylindric, less conspicuously ridged.—Dry sandy and gravelly places, 5000-7500 (-10,000) ft.; Sagebrush Scrub, Pinyon-Juniper Wd.; Panamint Mts. to Mono Co.; Ida., Nev. June-July.

2. **C. califórnica** (Benth.) Gray. [*Mucronea c.* Benth.] Stem or stems erect or ascending, branching near base, glandular-hirsute, 1-2.5 dm. long; lvs. basal, spatulate to obovate, short-petioled, 1-3 cm. long; bracts deeply 3-lobed, mostly 5-10 mm. long, somewhat broader, sessile, clasping at base, the lobes lance-ovate, spine-tipped; invols. 1-3 at each node, the tube cylindric, 2.5-3 mm. long, obscurely ribbed, the teeth usually 3, unequal, curved outward, ending in a straight spine; calyx white, somewhat exserted, the lobes oblong, obtuse, entire, pubescent; aks. narrow, papillose upward.—Occasional, dry sandy places, below 3000 ft.; Coastal Sage Scrub, San Luis Obispo Co. to San Bernardino and San Diego. April-July. A form growing on sand dunes at Surf, Playa del Rey, Ballona Harbor, with bracts 1-2 cm. long, oblong and more obtuse; invols. urceolate is var. *suksdorfii* Macbr.

3. **C. cleveléndii** Parry. Stems several from base, decumbent, tomentose, 5-25 cm. long; lvs. basal, oblanceolate, petioled, grayish-pubescent, 1-3 cm. long; lower bracts leafy, upper acerose; invols. in scattered headlike clusters, the tubes urn-shaped, 3-angled, hoary-pubescent, 3-3.5 mm. long; teeth all uncinate, the anterior much longer than the others; calyx 3.5 mm. long, strigose on veins, the outer lobes ovate, ca. 1 mm. long, minutely erose, the inner shorter and conspicuously erose; stamens 3; aks. grayish, almost wing-angled above, 2.5 mm. long.—Dry rocky and sandy places, below 5000 ft.; Chaparral, Foothill Wd.; Ventura and Kern cos. n. to Mendocino Co. June-July.

4. **C. coriàcea** Goodm. [*C. lastarriaea* Parry, in part.] Brittle slender-stemmed prostrate to ascending plants, the stems diffusely branched, 0.5-2.5 dm. long, pubescent; lvs. linear, basal and whorled, ciliate, 2-2.5 cm. long, scarcely 1 mm. wide; passing above into whorled green bracts with hooked awns, these connate at base, 4-8 mm. long; fls. solitary in the axils, without invols.; calyx coriaceous, tubular, mostly 5-cleft to ca. the middle, 4 mm. long, lobes divergent, long-spinose, hooked, 2 lobes shorter than others; stamens 3; ak. triangular.—In dry sandy places and openings, below 2500 ft.; largely Coastal Sage Scrub, Chaparral; cismontane from L. Calif. to cent. Calif., inland to San Bernardino, Palm Springs. Catalina and Santa Rosa ids. April-June.

5. **C. corrugàta** (Torr.) T. & G. [*Acanthogonum c.* Torr.] Erect, tufted, the stems several from base, much branched above, ± tomentose, 5-13 cm. high; basal lvs. round-ovate, 1-2 cm. broad, petioled, woody especially beneath, the lower cauline narrower, spatulate, the upper more bractlike, acicular; invols. solitary in axils, numerous, the upper congested, the tube cylindrical, transversely corrugated, 2-3 mm. long, glabrate, the teeth 3, lanceolate, squarrose, as long as tube or longer; calyx subcylindric, 2-2.5 mm. long, white, the segms. oblong, pubescent, subequal; stamens 6; aks. slightly exserted, subcylindric, ca. 1 mm. long, papillose at apex.—Dry rocky places, below 3000 ft.; Creosote Bush Scrub; deserts from Death V. region to L. Calif., Ariz. Feb.-May.

6. **C. fimbriàta** Nutt. Stems 1 to several from base, erect or ascending, pubescent, 5-30 cm. high; lvs. basal, sparsely tomentose beneath, pubescent above, petioled, obovate-spatulate, 2-5 cm. long; bracts rarely leafy, spreading-acerose; invols. in small dense terminal clusters, cylindric, mostly thinly pubescent, 4-6 mm. long, the teeth widely divergent, uncinate, the 3 inner noticeably smaller than the outer; calyx white or pink, 6-7 mm. long, the segms. with fimbriate margins and linear entire terminal portion; stamens 9; aks. slightly 3-angled.—Dry slopes, below 3000 ft.; Coastal Sage Scrub, Chap-

arral; w. San Diego to e. Orange and w. Riverside cos.; n. L. Calif. Feb.–June. Var. **laciniàta** (Torr.) Jeps. [*C. l.* Torr.] Calyx 7–9 mm. long, the segms. finely laciniate so that the terminal lobe is scarcely broader than the others.—Below 5000 ft.; Chaparral, Pinyon-Juniper Wd.; s. San Jacinto Mts. to e. San Diego Co. May–June.

7. **C. leptóceras** (Gray) Wats. [*Centrostegia l.* Gray. *Eriogonella l.* Goodm.] Stems few, slender, prostrate or decumbent, glabrous, forked, 5–15 cm. long; lvs. basal, oblanceolate, glabrous, 1.5–5 cm. long, short-petioled; bracts 3-lobed, pubescent, 3–6 mm. long, the lobes spine-tipped; invols. 1–3 at an axil, glandular-pubescent, 4–6 mm. long, the tube cylindric, with 6 terminal long-awned unequal ciliate teeth and a ring of 6 hooked spine-like spurs near base; calyx partly exserted, pubescent, the lobes spatulate, subequal; stamens 6; aks. dark gray, ovoid, ca. 1.3 mm. long, the beak pyramidal.—Occasional, sandy places, Coastal Sage Scrub, San Fernando V. to San Bernardino V. and Elsinore. April–June.

8. **C. leptothèca** Goodm. Resembling *C. staticoides* in habit, but with more numerous stems 1–2 dm. long, scantily pubescent; invols. 4–5 mm. long, not crowded, the tube very slender, curly-pubescent, the 6 teeth spreading-uncinate, the 3 inner smaller than the outer; calyx well exserted, 4.5–5 mm. long, the outer segms. linear, obtuse, the inner smaller; stamens 9; aks with a stout beak.—Dry slopes, 3000–6000 ft.; Chaparral; San Jacinto Mts. to e. San Diego Co.; adjacent L. Calif. May–July.

9. **C. membranàcea** Benth. [*Eriogonella m.* Goodm.] Erect, mostly simple below, few-branched above, woolly, floccose in age, 1–4.5 dm. high; lvs. basal and alternate on lower stem, linear to narrowly oblanceolate, 1.5–6 cm. long, glabrate above, white-woolly beneath; bracts opposite or whorled, like the lvs., but reduced; invols. except for the very lowest, in dense heads at upper nodes and terminal, woolly, urn-shaped, 4–5 mm. long, 3-angled, the 6 teeth reddish, slender, hooked, subequal, united by a broad pale membrane; calyx woolly, the outer segms. obovate, the inner spatulate, scarcely exserted; stamens 9; aks. somewhat 3-angled, beaked.—Dry rocky slopes, below 4500 ft.; Chaparral V. Grassland; Ventura Co. and Kern Co. to n. Calif. April–July.

10. **C. orcuttiàna** Parry. Branched from base, somewhat pubescent, the stems prostrate, 3–10 cm. long; lvs. basal, narrowly oblanceolate, petioled, 1–3 cm. long, lower bracts opposite, foliaceous, the upper reduced, acerose; invols. mostly solitary in axils of the cymes, 3-angled, the tube ca. 2 mm. long, the 3 teeth almost as long, squarrose, hooked at tip; calyx cylindric, 2–2.5 mm. long, with erect lance-linear pubescent lobes; stamens 9; aks. 3-angled.—Sandy places; Coastal Sage Scrub; Point Loma and Kearney Mesa, San Diego Co. March–April.

11. **C. párryi** Wats. Prostrate or decumbent, the stems several from base, 0.4–2.5 dm. long, repeatedly forked, strigose; lvs. basal, oblong-lanceolate, to -oblanceolate, narrowed to the petiole, strigose, 2–7 cm. long; lower bracts leafy but more mucronate, the upper acicular; invols. rather openly distributed in small clusters, urn-shaped, 6-ribbed, the tube ca. 2.5–3 mm. long, appressed-canescent, the teeth long, spreading, uncinate, the 3 outer much longer than the inner; calyx white, 2.5–3 mm. long, the outer lobes obovate-oblong, to oblong, erosulate, the inner linear-lanceolate; stamens 9; aks. lance-ovoid, 3-angled upward, ca. 2 mm. long.—Dry sandy places below 2500 ft.; mostly Coastal Sage Scrub; e. Los Angeles Co. to San Gorgonio Pass, w. Riverside Co. April–June. Var. **fernandìna** (Wats.) Jeps. [*C. f.* Wats.] Involucral teeth not hooked; calyx lobes subequal.—San Fernando V. to Orange Co. and San Diego Co.

12. **C. perfoliàta** Gray [*Mucronea p.* Heller.] Stems branching mostly just above base, diffuse, 1.5–3 dm. high, sparingly glandular-puberulent; lvs. basal, spatulate, 2–5 cm. long; bracts perfoliate, orbicular or 3-lobed, spine-tipped, the larger bracts to 2 cm. across; invols. mostly solitary at each node, 4–8 mm. long, the tube 4-angled, corrugated between the ribs, the teeth 4, divergent, spine-tipped; calyx puberulent, white, exserted, the lobes fimbriate or erose at summit; stamens 6.—Dry flats and slopes at low elevs.; Chaparral, Coastal Sage Scrub, Joshua Tree Wd.; w. Mojave Desert to Mt. Hamilton (Stanislaus Co.). April–June.

13. **C. polygonoìdes** T. & G. ssp. **longispìna** (Goodman) Munz, comb. nov. [*Acan-*

thogonum p. var. *l.* Goodman., Leaflets W. Bot. 7: 236. 1955.] Villous, the stems prostrate, 3-10 cm. long; basal lvs. oblanceolate to elliptic, petioled, 2-5 cm. long; bracts opposite, the lower like the lvs., but smaller, the upper acicular; invols. 3-4 mm. long, glabrate, solitary in axils or in small clusters, the tube obpyramidal, 3-angled, transversely corrugated, the 3 outer teeth conspicuously spinose-hooked, the alternate shorter and inconspicuous; calyx whitish, scarcely exserted, the lobes oblong, erect; stamens 9; aks. rather long-beaked.—Dry places below 5000 ft.; Chaparral; w. Riverside and San Diego cos.

14. **C. procúmbens** Nutt. [*C. uncinata* Nutt.] Stems several, procumbent, few-branched, brittle, 5-25 cm. long, strigose-tomentulose; lvs. mostly basal, oblong-oblanceolate, 2-7 cm. long, curly-strigulose; lower bracts like the lvs., upper acicular; invols. in small terminal clusters and axillary, yellowish-green, 2-2.5 mm. long, the tube cylindric, 6-ribbed, pubescent to glabrate, the teeth 6, spreading, with uncinate or curved tips, mostly somewhat unequal; calyx yellow, 1.5-2 mm. long, the lobes mostly oblong, obtuse, entire, scantily strigose; stamens 9; aks. lance-ovoid, somewhat 3-angled upward.—Mostly sandy places below 2500 ft.; Coastal Sage Scrub, Chaparral; San Fernando and San Bernardino to San Diego and L. Calif. April-June. Var. **albiflora** Goodm. Calyx white.—Pala to Fallbrook, San Diego Co.

15. **C. rígida** (Torr.) T. & G. [*Acanthogonum r.* Torr.] Stem erect, usually simple below, often short-branched above, 2-10 cm. high, ± woolly; principal lvs. broadly elliptic, to obovate, 2-5 cm. long, petioled, woolly beneath, well distributed along the stem; secondary lvs. bractlike, lanceolate or narrower, spine-tipped, becoming hard and thornlike; infl. very dense, with clusters of invols. in axils of bracts; involucral tube 2 mm. long, 3-angled, reticulate between angles, with 3 lanceolate spreading unequal straight or spine-tipped teeth 4-15 mm. long; calyx yellowish, scarcely exserted, the lobes oblong, hairy on back; stamens 9; aks. prominently beaked.—Common in dry open stony places, such as desert mosaics, below 3000 ft.; Creosote Bush Scrub; deserts from Inyo Co. to Utah, Ariz., Son., L. Calif. March-May. The old dried plants persist as dense spiny masses for some time.

16. **C. spinòsa** Wats. [*Eriogonella s.* Goodm.] Prostrate, loosely branched, the stems 5-25 cm. long, puberulent; lvs. basal, broadly oblong, petioled, 2-3 cm. long, woolly beneath, villous above; bracts in whorls of 3, lanceolate, stiff, 5-15 mm. long, spine-tipped, the floral subulate-acicular; invols. in small axillary clusters, canescent, 3-4 mm. long, 4-5-toothed, the teeth straight, unequal, 1 usually much exceeding the others; calyx white, well-exserted, the 3 outer lobes broader than long, the 3 inner narrower and smaller; stamens 9; aks. subovoid, with a stout, 3-angled papillose beak.—Occasional, dry sandy and gravelly places, 2500-3500 ft.; Creosote Bush Scrub, Joshua Tree Wd.; w. Mojave Desert e. to Rabbit Springs. April-July.

17. **C. staticoìdes** Benth. ssp. **staticoìdes**. [*C. nudicaulis* Nutt. *C. discolor* Nutt.] TURKISH RUGGING. Plate 69, Fig. F. Stems erect or ascending, 1-2 dm. high, 1-few from base, mostly simple below, branched above, usually reddish purple, ± pubescent; lvs. basal, oblong to oblong-ovate, green above, tomentose beneath, petioled, 2-6 cm. long; bracts subulate, often recurved; invols. solitary in the forks or congested at the ends of the branchlets, cylindric, 6-ribbed, 3-4 mm. long, 6-toothed, the teeth spreading, not membraned on margins, hooked, 1-2 mm. long, 3 larger than the alternate; calyx mostly rose, sometimes paler, 4-5 mm. long, hairy externally, the lobes narrow-oblong, subentire, obtuse; stamens 9; aks. narrow, somewhat 3-angled toward apex.—Dry slopes and flats below 4000 ft.; Coastal Sage Scrub, Chaparral; San Diego and San Bernardino cos. to Monterey Co. Variable, with some forms named: var. *brevispina* Goodm. Involucral teeth less than 1 mm. long.—Below 5000 ft., s. base of San Gabriel Mts. Var. *elata* Goodm. Plants 3-6 dm. high; lvs. 7-10 cm. long.—San Gabriel to San Bernardino and Santa Ana mts., often on chaparral-burns and may be ecological. Var. *latiloba* Goodm. Outer calyx lobes obovate, truncate.—Desert slopes of San Gabriel Mts.

Ssp. **chrysacántha** (Goodm.) Munz. [*C. c.* Goodm.] Infl. with few large dense clusters; invols. 5-7 mm. long, the tube 4-5 mm. long; calyx 5-5.5 mm. long, the outer

Plate 70. Fig. A, *Chorizanthe thurberi*. Fig. B, *Eriogonum cinereum*. Fig. C, *Eriogonum deflexum*. Fig. D, *Eriogonum elongatum*. Fig. E, *Eriogonum fasciculatum*. Fig. F, *Eriogonum gracile*.

lobes broadly ovate-oblong, ca. 2 mm. long, the inner shorter, 3 mm. long.—Ocean bluffs, Coastal Sage Scrub; Orange Co. April–May.

18. **C. thúrberi** (Gray)Wats. [*Centrostegia t.* Gray ex Benth.] Plate 70, Fig. A. Erect, branched at or near the base, forked above with spreading branches, glandular-hispidulous, 5–20 cm. high; lvs. in basal rosette, oblong to broadly spatulate, almost sessile, 1–3 cm. long, subglabrous; bracts 3-lobed, spine-tipped, 2–4 mm. long; invols. solitary, scattered but numerous, 4–6 mm. long, 5-toothed at apex, with erect spiny, straight tips, and 3-angled with 3 basal spreading saccate horns; calyx included, pubescent, deeply parted into narrow segms.; stamens 6 or 9; aks. lance-ovoid, ca. 1.5 mm. long.—Common, dry sandy places, below 7500 ft.; Creosote Bush Scrub, Pinyon-Juniper Wd.; deserts and occasional in inner S. Coast Ranges to the north; to Utah, Ariz. April–June.

19. **C. watsònii** T. & G. Erect to ascending, canescent-strigose, the stems branched, 4–15 cm. long, sometimes reddish; basal lvs. oblanceolate, petioled, 2–3 cm. long, the lower cauline somewhat reduced, the upper bractlike; lower invols. solitary; upper in small clusters, canescent, cylindric, inconspicuously ribbed, ca. 4 mm. long, 5-toothed, the teeth short, recurved to hooked except the anterior which is foliaceous, lanceolate, 3–12 mm. long, terminally hooked; calyx 3–4 mm. long, yellow, scantily pubescent; stamens 9; ak. beaked.—Dry sandy or gravelly places, 2500–7500 ft.; Joshua Tree Wd., Sagebrush Scrub, Pinyon-Juniper Wd.; w. Mojave Desert n. to e. Wash., Ida., Utah. April–July.

20. **C. wheèleri** Wats. [*C. insularis* R. Hoffm.] Erect to spreading, simple to branched at base, 6–25 cm. high, gray-pubescent; lvs. basal, 2–3 cm. long, oblong to ovate-spatulate, tomentose beneath, villous above, lower bracts leaflike, floral bracts acerose; invols. in crowded clusters, cylindric, 2.5–3 mm. long, pubescent to glabrate, the teeth short, widely divergent, uncinate; calyx cylindric, 3–3.5 mm. long, the lobes entire or nearly so, the outer elliptic-oblong to lanceolate, 1.3–1.5 mm. long, obtuse; the inner shorter and narrower; stamens 6; aks. linear, 3-angled upward, ca. 3 mm. long.—Dry slopes, Chaparral; Santa Cruz and Santa Rosa ids.; mainland coast from near Santa Barbara to Santa Monica Mts. April–July.

21. **C. xánti** Wats. ssp. **xanti**. Erect plants, simple or branched from base, 5–25 mm. high, with gray appressed pubescence; lvs. basal, petioled, oblong-ovate to ovate, 2–5 cm. long; lower bracts leaflike, upper acerose; invols. not in subcapitate clusters but in more open cymes, cylindric, 4.5–6 mm. long, the tube ca. 4 mm. long, canescent, the teeth 6, divergent, hooked, the 3 outer larger than inner; calyx white to pink, subcylindric, 5–6 mm. long, the outer lobes oblong or elliptic, ca. 2 mm. long, the inner somewhat shorter and narrower; stamens 9; aks. very slender, with a stout beak.—Dry slopes and washes, 1000–6000 ft.; Pinyon-Juniper Wd., Foothill Wd., Joshua Tree Wd.; San Bernardino Mts. to Inyo and Kern cos. April–June.

Ssp. **leucothèca** (Goodm.) Munz, comb. nov. [*C. x.* var. *leucotheca* Goodm., Ann. Mo. Bot. Gard. 21: 60. 1934.] Invols. white-woolly, 4 mm. long; calyx scarcely evident.—Below 3500 ft., Creosote Bush Scrub; edge of Colo. Desert from n. base of Santa Rosa Mts., to foot of San Bernardino Mts. April–May.

2. *Eméx* Neck.

Glabrous monoecious annuals. Lvs. petioled, alternate, with membranaceous or scarious sheathing stipules. Fls. small, in axillary fascicles or in racemelike terminal infl.; ♂ fls. sessile and below; ♀ on jointed filiform pedicels and above. Staminate fls. with 5–6-parted calyx with narrow segms.; stamens 4–6. Pistillate calyx urn-shaped, with ovoid tube and 6 lobes in 2 series. Ovary 3-angled; styles 3, short. Fruiting calyx hard, 3- or 6-angled, burlike, the 3 outer segms. spine-tipped. Aks. 3-angled, inclosed by the spiny perianth. Two spp. of Old World. (Latin, *ex*, out of, and *Rumex*, having been transferred from that genus.)

1. **E. spinòsa** (L.) Campd. [*Rumex s.* L.] Plants decumbent to ascending, the stems 3–8 dm. long; lvs. oblong-ovate to somewhat triangular, 5–12 cm. long, on somewhat shorter to longer petioles; outer segms. of fruiting calyx ca. 5–6 mm. long, tipped with

divergent spines, the inner erect, 6–7 mm. long; aks. shining, light brown, ca. 4 mm. long; $n=10$ (Sugiura, 1937).—Weed in orchards, etc., Ventura Co. to San Diego Co.; native of Medit. region. July–Nov.

3. Erióngonum Michx. WILD BUCKWHEAT

Annual or perennial herbs or shrubs with basal or cauline, alternate lvs. and usually whorled scalelike or foliaceous bracts, entire and estipulate. Fls. perfect or sometimes also imperfect, borne in invols. Invols. campanulate to turbinate or cylindric, 4–10-lobed or -toothed, awnless, few- to many-fld., sessile or peduncled. Pedicels ± exserted, intermixed with setaceous bractlets and jointed at summit with the base of the perianth or with a slender and stalklike stipitate base. Perianth commonly called "calyx", 6-parted or -cleft, petaloid, with 2 series of 3 segms. each. Stamens 9, the fls. filiform, often pilose at the base, inserted at base of perianth. Ovary 1-celled, 3-angled or -winged; styles 3; stigmas capitate. Aks. mostly 3-angled, sometimes lenticular. A N. Am. genus of over 200 spp., mostly w.; some of importance as bee-plants, others with some horticultural possibility. (Greek, *erion*, wool, and *gonu*, joint or knee, the type of the genus *E. tomentosum* Michx. being hairy at the nodes.)

1. Calyx stipelike at the attenuated base (see also *E. saxatile* and *E. crocatum*); bracts leafy, indefinite in number (2–several).
  2. Calyx pubescent externally; invols. solitary, terminal. Inyo Co. n. ..... 9. *E. caespitosum*
  2. Calyx glabrous externally; invols. clustered and subtended by 2 to several bracts
        65. *E. umbellatum*
1. Calyx not stipelike at the base; bracts not leafy, regularly 3 in number.
  A. Stems not jointed internally.
    3. Invols. campanulate to turbinate, not angled or ribbed, 4- or 5-toothed or -lobed, rarely obscurely nerved at the base, mostly peduncled.
      4. Lvs. basal and also on the lower nodes, tomentose or floccose except in *E. spergulinum*.
        5. Basal lvs. linear, revolute, pilose; invols. 0.5–1 mm. long ..... 61. *E. spergulinum*
        5. Basal lvs. oblanceolate to oblong-obovate, not revolute for most part, tomentose or floccose.
          6. Invols. ± glandular-puberulent to pubescent externally.
            7. Calyx-segms. dissimilar, the outer segms. ovate, elliptic or roundish, the inner narrowly lanceolate or oblong and longer.
              8. Outer segms. obovate to elliptic, not obviously inflated, or if so, only near the base, the inner segms. spatulate; stamens conspicuously exserted
                1. *E. angulosum*
              8. Outer segms. elliptic to roundish or obovate, obviously inflated at maturity, the inner segms. narrowly lanceolate; stamens included.
                9. Outer segms. inflated at the base and middle, the sides of the segms. incurved below, the inner segms. obtuse to acute; peduncles and invols. glandular-puberulent, with non-capitate hairs .... 40. *E. maculatum*
                9. Outer segms. inflated above the middle, the apex curved inward, the inner segms. acute to acuminate; peduncles and invols. with capitate-glandular hairs .... 66. *E. viridescens*
            7. Calyx-segms. similar or nearly so, the outer segms. oblong, not inflated; invols. 2 mm. long .... 28. *E. gracillimum*
          6. Invols. glabrous externally; calyx hispid; plants to 7 dm. tall ..... 45. *E. ordii*
      4. Lvs. strictly basal, pilose or tomentose at least beneath.
        10. Invols. 4-lobed or -toothed.
          11. Calyx pubescent with hooked hairs externally.
            12. Invols. 2-fld.; aks. exserted; calyx ca. 1 mm. long .... 31. *E. hirtiflorum*
            12. Invols. 4–6-fld.; aks. not exserted; calyx ca. 1.5 mm. long ..... 34. *E. inerme*
          11. Calyx puberulent or hispidulous externally, the hairs not hooked.
            13. Calyx white, 1.5–2 mm. long, apex notched or apiculate; lvs. spatulate, ciliate or pilose .... 2. *E. apiculatum*
            13. Calyx pink or yellow, 0.5–1 mm. long, the segms. not apiculate.
              14. Calyx pink, 0.5–0.7 mm. long; lvs. spatulate, hirsute; invols. 0.5–0.7 mm. long. Montane .... 49. *E. parishii*

14. Calyx yellow, 1–2 mm. long; invols. 0.7–2 mm. long. Deserts.
15. Lvs. short-hirsute, suborbicular, 1–2.5 cm. long; invols. 0.7–1 mm. long
.... 63. *E. trichopes*
15. Lvs. floccose or glabrous, obovate to oblanceolate, 2–8 cm. long; invols.
1–2 mm. long .... 45. *E. ordii*
10. Invols. 5-lobed or -toothed.
16. Calyx pubescent or puberulent externally.
17. Invols. glabrous externally.
18. Outer calyx-segms. not saccate-dilated.
19. Lf.-blades short-hirsute; invols. turbinate.
20. Calyx pink or whitish with reddish midveins; lvs. hirsutulous and
slightly glandular .... 25. *E. glandulosum*
20. Calyx yellow.
21. Plants strictly annual; branchlets numerous and whorled at each
node; invols. often 4-lobed, occasionally 5-lobed ... 63. *E. trichopes*
21. Plants perennial but flowering the first year; branchlets few at each
node, not in whorls .... 35. *E. inflatum*
19. Lf.-blades woolly; calyx 1 mm. long; invols. broadly campanulate
54. *E. reniforme*
18. Outer calyx-segms. saccate-dilated at each side of the cordate base at
maturity, sparsely puberulent at the base of the perianth-tube, yellow
maturing reddish or reddish with white lobes .... 62. *E. thomasii*
17. Invols. glandular-puberulent externally.
22. Calyx white to red, the outer segms. rounded and narrowed abruptly to a
clawlike base, slightly glandular externally at the base and with a tuft of
white hairs within .... 64. *E. thurberi*
22. Calyx yellow, the outer segms. obovate, smooth, glandular on entire outer
surface, glabrous within .... 52. *E. pusillum*
16. Calyx glabrous externally.
23. Lvs. pilose-hispid, not woolly. Inyo and Mono cos. .... 20. *E. esmeraldense*
23. Lvs. tomentose beneath.
24. Outer calyx-segms. panduriform, crisped, not cordate; peduncles slender,
5–25 mm. long .... 10. *E. cernuum*
24. Outer calyx-segms. not panduriform or crisped.
25. Outer calyx-segms. cordate at the bases; peduncles stoutish.
26. Peduncles deflexed.
27. Stems glabrous.
28. Invols. narrowly turbinate to turbinate; peduncles 0–15 mm. long;
calyx white, oblong.
29. Invols. 1.5–3 mm. long; plants variously branched; calyx 1–2.5
mm. long, not gibbous at maturity .... 15. *E. deflexum*
29. Invols. 1–1.5 mm. long; plants branching in a series of flat-
topped layers, pagoda-like; calyx 1.5 mm. long, gibbous at the
base at maturity .... 55. *E. rixfordii*
28. Invols. hemispheric, sessile; calyx yellow to reddish-yellow, suborbicular .... 33. *E. hookeri*
27. Stems glandular, short and stout; crowns flat-topped 7. *E. brachypodum*
26. Invols. erect on peduncles, sessile or peduncles less than 5 mm. long.
26a. Stems short, less than 3 cm. long, often hidden by the basal lvs.;
plants 1–4 dm. high and 3–15 dm. across, the crowns often spreading and flat-topped.—Pahrump and Steward valleys, Inyo Co.,
Calif. and Nye Co., Nev. .... 5. *E. bifurcatum*
26b. Stems long, 2–20 cm. long, the mature plants with stems not hidden
by the basal lvs.; plants 2–12 dm. high and 1–5 dm. across, the
crowns strict and erect.—Deserts, Imperial and e. San Diego cos.
to Inyo Co.; to Utah .... 36. *E. insigne*
25. Outer calyx-segms. obtuse at the base.
30. Plants glandular; invols. deflexed, sessile or subsessile above, peduncled
to 10 mm. below; calyx white becoming reddish .... 19. *E. eremicola*
30. Plants glabrous; peduncles erect, less than 1 mm. long; calyx white to
reddish .... 32. *E. hoffmannii*

3. Invols. cylindric to cylindric-turbinate to turbinate or prismatic, often 5-6-nerved, angled or ribbed, mostly sessile, solitary or congested into heads, the teeth usually short.
  31. Plants annual; lvs. mostly in basal rosettes.
    32. Flowering branches with short branchlets, usually of a single internode; invols. axillary and terminal; calyx 1-1.5 mm. long, yellow. Mojave Desert 42. *E. mohavense*
    32. Flowering branches elongate, virgate, and bearing invols. at the nodes, the lateral ones appressed.
      33. Invols. 2-5 mm. long.
        34. Lvs. oblong-obovate to oblanceolate; stems tomentose.
          35. Invols. 4-5 mm. long, cylindric, with minute teeth; outer calyx-segms. narrowly obovate; aks. 2 mm. long .... 57. *E. roseum*
          35. Invols. 2-3 mm. long, turbinate, with prominent teeth; outer calyx-segms. broadly obovate; aks. ca. 1 mm. long.
            36. Stems and infls. tomentose to floccose; petioles slender, not winged; inner Coast Ranges from the Sacramento V. s. into s. Calif., mainly in the mts.
              26. *E. gracile*
            36. Stems and infls. glabrous; petioles often winged. Near the coast from Ventura Co. n. .... 12. *E. citharæforme*
        34. Lvs. rounded or nearly so; stems usually glabrous or only sparsely floccose; outer calyx-segms. more than twice as long as wide. Mts. . . . 14. *E. davidsonii*
      33. Invols. 1-1.5 mm. long.
        37. Outer calyx-segms. fan-shaped, their sides incurved below the broad truncate apices, yellowish to reddish; branches usually incurved at the summit in age; invols. few-fld. Mojave Desert .... 43. *E. nidularium*
        37. Outer calyx-segms. not as above.
          38. Outer calyx-segms. not hastate, glabrous or glandular, 1-2 mm. long, white or yellow; lvs. strictly basal.
            39. Calyx-segms. 1.5-2 mm. long.
              40. Stems densely tomentose; fls. 3-10 per invol.; plants densely branched and spreading. San Bernardino Co. to Inyo Co. . . 47. *E. palmerianum*
              40. Stems glabrous to sparsely floccose; fls. more than 10 per invol.; plants usually sparsely branched and erect. San Bernardino Co. n. to e. side of Sierra Nevada ,................................... 4. *E. baileyi*
            39. Calyx-segms. 1-1.5 mm. long; calyx yellow. Mojave Desert
              6. *E. brachyanthum*
          38. Calyx-segms. hastate at the base when mature, 0.8-1.2 mm. long, white to pink; lvs. basal or also cauline; stems tomentose. San Bernardino Mts. south
            22. *E. foliosum*
  31. Plants perennial; lvs. often cauline as well as basal.
    41. Invols. solitary at the nodes, the lateral ones appressed to the branches.
      42. Calyx stipitate with long, winged attenuate bases, 5-7 mm. long.
        43. Infl. 1-1.5 dm. across, open and lax; calyx narrowed to a 3-angled base, pinkish to white or yellowish; aks. ca. 2 mm. long. Mts. of s. Calif.
          59. *E. saxatile*
        43. Infl. 0.3-0.8 dm. across, dense; calyx narrowed to a tubular base, sulphur-yellow; aks. ca. 3 mm. long. N. base of Santa Monica Mts. . . . 13. *E. crocatum*
      42. Calyx astipitate, 2-5 mm. long.
        44. Invols. 6-7 mm. long; loosely branched white tomentulose herbs 8-18 dm. high. Cismontane .... 18. *E. elongatum*
        44. Invols. 2-6 mm. long.
          45. Infl. with invols. in cymes or panicles.
            46. Lf.-blades more than 2 cm. long .... 44. *E. nudum*
            46. Lf.-blades less than 2 cm. long.
              47. Calyx glabrous.
                48. Infl. a compact terminal cyme; invols. tomentose or glabrous; outer calyx-segms. subcordate at the base .... 41. *E. microthecum*
                48. Infl. a divaricately branched panicle; invols. glabrous.
                  49. Outer calyx-segms. round, subcordate at the base; branches green, dichotomous, ascending .... 30. *E. heermannii*
                  49. Outer calyx-segms. obovate, narrowed at the base; branches grayish, mostly horizontal, tiered .... 51. *E. plumatella*

47. Calyx silky-villous, yellowish; plants shrubby, 6–15 dm. high. Colo. Desert .... 16. *E. deserticola*
45. Infl. with invols. placed racemosely along the branches; invols. tomentose.
  50. Plants shrubby, up to 15 dm. high; lvs. elliptic to oblong, 2–3 cm. long; invols. racemosely arranged on the ends of the fragile branches
    38. *E. kearneyi*
  50. Plants low subshrubs less than 3 dm. high.
    51. Plants suffrutescent and much branched at the base or densely cespitose; lvs. many, lance-elliptic to oblanceolate, ± revolute; invols. 2–3 mm. long; calyx 2–4 mm. long .... 67. *E. wrightii*
    51. Plants not suffrutescent at the base or densely cespitose; basal lvs. few, rounded to broadly ovate.
      52. Basal lvs. roundish to broadly ovate, 1.5–4 cm. long, on petioles 1.5–4 cm. long; plants from a highly branched, woody, spreading caudex, suberect; invols. and calyx 3–5 mm. long
        48. *E. panamintense*
      52. Basal lvs. oblong to elliptic, 2.5–3.5 cm. long, on petioles 3–7 cm. long; plants arising from short-branched, compact caudices; invols. 3–4 mm. long. .... 53. *E. racemosum* and 58. *E. rupinum*
41. Invols. mostly clustered or in heads.
  53. Calyx-segms. dissimilar, the outer segms. often twice as wide as the inner; lobes dividing the calyx to the swollen basal joint.
    54. Outer calyx-segms. plane, not inflated; lvs. 6–20 (–60) mm. long. Montane
      46. *E. ovalifolium*
    54. Outer calyx-segms. rounded and inflated; lvs. 2–4 mm. long. Panamint Mts.
      24. *E. gilmanii*
  53. Calyx-segms. similar or nearly so.
    55. Plants cespitose, matted, herbaceous, the caudex much branched and woody; flowering stems scapelike.
      56. Calyx villous externally and internally; ovary sparsely pilose; lvs. lanate-tomentose. Inyo Co. .... 60. *E. shockleyi*
      56. Calyx and ovary not as above.
        57. Invols. membranaceous and indistinctly forming a tube, 2–3 mm. long; calyx rose to red. White Mts. .... 27. *E. gracilipes*
        57. Invols. a distinct rigid tube, 3–4 mm. long.
          58. Infl. compactly cymose-umbellate, the rays to 5 mm. long; calyx whitish becoming reddish, finely glandular-hairy. Kern Co. .. 8. *E. breedlovei*
          58. Infl. tightly capitate.
            59. Calyx yellow; lf.-blades 5–35 mm. long, greenish-tomentose above
              56. *E. rosense*
            59. Calyx white; lf.-blades mostly less than 5 mm. long, densely white-tomentose on both surfaces .... 39. *E. kennedyi*
    55. Plants shrubby, or if herbaceous, then not cespitose.
      60. Plants essentially herbaceous, only the base woody.
        61. Lvs. spreading, oblong-ovate to ovate, obtuse, mostly 2–6 cm. long.
          62. Invols. 2–5 mm. long. Mainland .... 44. *E. nudum*
          62. Invols. 5–6 mm. long. Insular .... 29. *E. grande*
        61. Lvs. erect, ± lanceolate, 4–15 cm. long. Panamint Mts. and northward
          17. *E. elatum*
      60. Plants definitely shrubby.
        63. Lvs. narrowly linear or nearly so, the blades less than 2 cm. long, strongly fascicled; shrubs with terminal cymose or subumbellate infl.
          21. *E. fasciculatum*
        63. Lvs. linear-oblong to orbicular, or if linear, more than 2 cm. long.
          64. Heads in dense compound cymes. Insular.
            65. Lvs. linear or narrowly oblong, ± revolute, 2–3 cm. long
              3. *E. arborescens*
            65. Lvs. oblong-ovate, plane, 3–10 cm. long .... 23. *E. giganteum*
          64. Heads terminal on 2-forked peduncles or scattered along the stems. Mostly mainland.
            66. Calyx glabrous externally; lf.-blades 5–15 mm. long 50. *E. parvifolium*

66. Calyx white-villous; lf.-blades 15–30 mm. long .... 11. *E. cinereum*
A. Stems internally jointed. Death Valley .... 37. *E. intrafractum*

1. **E. angulòsum** Benth. Erect annuals, with spreading dichotomous ± angled stems 1–4 (–9) dm. long, whitish-tomentose to glabrate; basal lvs. oblanceolate to oblance-oblong, the blades 1–3 cm. long, short-petioled, glabrate above, tomentose beneath, revolute and crisped on margins; stem-lvs. well distributed, 0.5–2 cm. long; peduncles arising from most axils, slender, 1–2 cm. long, glabrous or sparsely tomentose; invols. open-turbinate, 1.5–2.5 mm. long, ± puberulent, with broad rounded lobes; calyx rose, tipped with white, ca. 1.5 mm. long, minutely glandular-puberulent, the outer calyx-segms. obovate to elliptical, deeply concave, the inner narrow-spatulate, longer; stamens 2–3 mm. long, usually well exserted; aks. ca. 1 mm. long, with a sharply angled triangular beak.—Dry open places, mostly below 4000 ft.; V. Grassland, Joshua Tree Wd., Pinyon-Juniper Wd., etc.; San Diego Co. to w. Mojave Desert and to Contra Costa Co. May–Nov.

2. **E. apiculàtum** Wats. Erect annuals, usually simple at base, dichotomously or trichotomously branched above, spreading, 2–9 dm. high, somewhat glandular-pubescent in lower portions of internodes and peduncles, with ultimate very slender branchlets; lvs. strictly basal, oblanceolate to obovate, the blades 1.5–4 cm. long, pilose, glandular, the petioles ca. as long; bracts 1–2 mm. long; peduncles in forks and scattered along the branchlets, filiform, 2–35 mm. long, often deflexed; invols. 1.5 mm. long, glabrous, 4-lobed ca. half their length; calyx white, 1.5–2 mm. long, puberulent, the segms. oblong-obovate, notched to apiculate; aks. ca. 1.5 mm. long, not distinctly beaked.—Dry open places in disintegrated granite, 3600–8000 ft.; Joshua Tree Wd., Pinyon-Juniper Wd., Yellow Pine F.; Joshua Tree National Monument, San Jacinto, Santa Rosa, Palomar and Cuyamaca mts. July–Aug.

3. **E. arboréscens** Greene. Loosely branched shrubs 6–15 (–20) dm. tall, the stems to 1 dm. thick, with shreddy bark; branchlets tomentose when young, later glabrate, purplish and glaucous; lvs. in crowded terminal tufts, linear to oblong, revolute, 2–3 cm. long, densely white-tomentose below, glabrate above; infl. dense terminal leafy-bracted cymes, 5–15 cm. across; invols. tomentose, 3 mm. long, turbinate, with obtuse oval teeth; calyx whitish to pink, 2 mm. long, villous at the base; fls. glabrous; aks. lance-ovoid, shining, angled, ca. 2.5 mm. long; $n=20$ (Reveal, 1967).—Rocky slopes and canyon walls; Coastal Sage Scrub, Chaparral; Santa Cruz, Santa Rosa and Anacapa ids.

4. **E. bàileyi** Wats. Erect annuals, diffusely branched from base, 1–4 dm. high, forming broad round-topped crowns, glabrous except at the white-woolly base of the stems; lvs. basal, suborbicular, 5–20 mm. wide, densely white-tomentose on both surfaces, on somewhat longer petioles; invols. tubular-campanulate, 1–1.5 mm. long, glabrous except for the ciliated margins, sessile at the nodes of the subvirgate branchlets, several-fld.; calyx white to pink, ca. 1.5 mm. long, the outer segms. oblong or oblong-obovate, somewhat constricted near the middle and flaring above, the inner narrower, glabrous or glandular; aks. dark brown, ca. 1 mm. long, the base globose, the beak stout, muriculate, 3-angled.—Dry sandy or gravelly flats and banks, mostly 2500–7500 ft.; Creosote Bush Scrub, Joshua Tree Wd. to Yellow Pine F.; e. end of San Bernardino Mts. across the Mojave Desert and along the e. side of the Sierra Nevada to e. Ore., w. Nev. May–Sept.

5. **E. bifurcàtum** Reveal. Low spreading herbaceous annual, 1–4 dm. high, 3–15 dm. across, nearly glabrous throughout, arising from a slender taproot; lvs. strictly basal, round-cordate, 1–3 cm. long and wide, on petioles 1–4 cm. long, densely white-tomentose beneath, less so to floccose and greenish above, rounded at apex, ± cordate at base; flowering stems often concealed by the lvs., the branches openly branched, forming infls. 3–15 dm. across; a peduncled invol. in each of main forks, and other invols. racemosely arranged on branches, subsessile or short-peduncled, turbinate, 2–2.5 mm. long, the 5 teeth obtuse, glabrous, with thin hyaline margin; pedicel 2–2.5 mm. long, 10–20-fld.; fls. white with greenish to reddish midribs and bases, 1.5–2 mm. long, the segms. dis-

similar, the outer obovate, the inner lanceolate; stamens exserted.—Creosote Bush Scrub; Pahrump V. and Stewart V., e. Inyo Co.; Nye Co., Nev. May–June.

6. **E. brachyánthum** Cov. [*E. baileyi* var. *b.* Jeps.] Erect annuals, diffusely branched from the base, 1–3 dm. high, forming broad round-topped crowns, glabrous except at the white-woolly base of the stems; lvs. basal, rounded to broadly ovate, 5–20 mm. wide, densely white-tomentose on both surfaces or ± glabrate above, on petioles somewhat longer; invols. turbinate, 1 mm. long, glabrous, closely appressed to the branchlets, few-fld.; calyx yellow, 0.6–0.8 mm. long, the outer segms. oblong to oblong-obovate, the inner slightly narrower, glabrous; aks. brown, ca. 1 mm. long, the globose bases narrowed into long exserted 3-angled beaks.—Dry sandy to gravelly slopes, mostly 2500–7500 ft.; Creosote Bush Scrub to Pinyon-Juniper Wd.; Mojave Desert n. to Mono Co.; w. Nev. May–Aug.

7. **E. brachypòdum** T. & G. [*E. parryi* Gray. *E. deflexum* var. *b.* Munz.] Annuals with 1–several stems, 5–40 cm. high; lvs. basal, 2–4 cm. long, 2–5 cm. wide, orbicular to cordate, densely white-tomentose below, less so to subglabrous above; stems stout, erect, 2–7 cm. long, glandular; branches horizontal in a low flat-topped crown or forming a subglobose one, trichotomous at the first node, usually dichotomous above, the ultimate branches often with alternating secondary branches which too have alternating branches of various lengths, becoming shortest in the last of the secondary branches and toward the tips of the main ones, glandular throughout; invols. peduncled up to lengths of 15 mm., deflexed; invols. turbinate to campanulate, 1–2.5 mm. long, 1.5–2.5 mm. wide, 5-lobed; calyx white becoming reddish at maturity, 1–2.5 mm. long, glabrous, the outer segms. ovate to oblong, the base cordate to auriculate, the inner segms. narrower and shorter; aks. brown to nearly black, 1.5–2 mm. long, base subglobose, beak 3-angled; $n=20$ (Reveal, 1967).—Below 7500 ft.; Creosote Bush Scrub to Pinyon-Juniper Wd.; Kern and San Bernardino cos. across Inyo Co. to Nev., s. Utah, nw. Ariz. March–Oct.

8. **E. breedlóvei** (J. T. Howell) Reveal. [*E. ochrocephalum* var. *b.* J. T. Howell.] Caudex cespitose, branched, low, the plants densely olive-green to whitish tomentose with glandular hairs nearly throughout; lvs. crowded, broadly elliptic, plane or with margins slightly revolute, lf.-blades 2–8 (–10) mm. long, 2–4 (–6) mm. wide, densely tomentose below, less so and green above, petioles to 1 cm. long, petiole-bases broadened; flowering stems scapose, erect, 1.5–6 cm. tall, densely glandular-hairy, the 3 triangular bracts ca. 2 mm. long; infl. compactly cymose-umbellate, 3.5–4 mm. long and wide, glandular-puberulent, 7–9-lobed, the triangular to acute lobes 1–1.5 mm. long, forming a distinct tube; calyx whitish, becoming reddish, 2.5–3.5 mm. long, the segms. cuneate, somewhat dissimilar, the outer wider than the inner, finely glandular-hairy within and without; fils. ± pilose basally; aks. lance-ovoid, 2–3 mm. long, sparsely and minutely hairy above the middle.—Dry rocky outcrops, 7800–8200 ft.; Red Fir F.; Piute Mts., Kern Co.; Baker Point, s. Tulare Co. June–Aug.

9. **E. caespitòsum** Nutt. Low compact matted perennial from much-branched woody caudex; lvs. elliptic to oblong-spatulate, densely white-tomentose, 5–15 mm. long, short-petioled, 1–3 mm. wide, crowded on the tips of the short branches, ± revolute; flowering stems scapelike, bractless, slender, 3–8 cm. high, somewhat loosely tomentose; invols. solitary, terminal, the tube turbinate, ca. 3 mm. long, with somewhat longer linear lobes that become reflexed; calyx yellow, 2.5–4 mm. long in anthesis, later reddish and 4–6 (–10) mm. long in fr., pubescent especially toward the stipelike base, the segms. similar, oblance-oblong; fils. pilose basally; aks. lanceolate in outline, somewhat 3-angled, ca. 3 mm. long, often pubescent apically.—Dry gravelly slopes and flats, 5000–8600 ft.; Sagebrush Scrub, N. Juniper Wd., Yellow Pine F.; e. side, White Mts., Inyo Co. n.; to Ida., Mont., Colo. May–July.

10. **E. cérnuum** Nutt. [*E. c.* var. *tenue* T. & G.] Annuals, glabrous and glaucous, diffusely branched from or above the base, 1–3 dm. high, less so at higher elevs.; lvs. basal, rounded, 1–2 cm. long, densely white-woolly below, subglabrate above; the petioles to 4 cm. long; peduncles capillary, deflexed, straight or curved, 5–25 mm. long; invols. turbinate, 1.5–2 mm. long, 5-lobed; calyx white, often becoming rose, glabrous,

1–1.5 mm. long, attenuate at base; segms. oblong-obovate, crisped and often emarginate; aks. slender, ca. 1.5 mm. long.—At 7000–10,000 ft.; Panamint and Inyo-White mts.; e. slope of Sierra Nevada, to Ore., Rocky Mts., Great Plains. June–Sept.

11. **E. cinèreum** Benth. Plate 70, Fig. B. Freely branched shrubs, 6–15 dm. high, tomentulose, leafy below the infl.; lvs. ovate, 1.5–3 cm. long, obtuse, cuneate at the base, greenish-cinereous above, white-tomentulose below, crisped-undulate, short-petioled; flowering stems elongate, dichotomous, with scattered heads; invols. cylindric-turbinate, tomentulose, 3–4 mm. long, somewhat angled, 5-toothed; calyx densely white-villous, ca. 3 mm. long, the segms. narrow-obovate, whitish to pinkish; fils. subglabrous; aks. brown, deltoid-ovoid, sharply angled, somewhat roughened, ca. 2 mm. long; $n=40$ (Reveal, 1967).—Beaches and bluffs near the coast; Coastal Strand, Coastal Sage Scrub; Santa Barbara to San Pedro, Santa Rosa Id. June–Dec.

12. **E. citharaefórme** Wats. [*E. vimineum* var. *c.* Stokes. *E. gracile* var. *c.* Munz.] Erect annuals 2–3 dm. high; lvs. basal and cauline, the basal leaf-blades oblong-lanceolate, 1–2 cm. long, tomentose beneath, floccose to glabrate above, the petioles 1–5 cm. long, winged, the cauline elliptic, on lower nodes only; flowering stems slender, 0.5–1 dm. long, glabrous; infl. spreading and open, 5–25 cm. long, the branches often curved; invols. turbinate, 2.5–3 mm. long, glabrous, 5-lobed; calyx white to rose, 1.5–2 mm. long, glabrous, the lobes oblong-obovate; stamens 1–1.5 mm. long; aks. with prominent beak. —Dry slopes and waste places; Chaparral, Coastal Sage Scrub; Ventura Co. to San Luis Obispo Co. May–Aug. A form with lvs. elliptic to ovate or nearly rounded and petioles not winged is the var. *agninum* (Greene) Reveal. [*E. a.* Greene. *E. vimineum* ssp. *polygonoides* Stokes.] Santa Barbara and San Luis Obispo cos.

13. **E. crocàtum** A. Davids. Perennial subshrubs, the caudices loosely branched and clothed with old lvs.; lvs. sheathing up the stems, the blades broadly ovate, 1.5–3.5 cm. long, the petioles shorter to ca. as long; flowering stems terminal, 1–2 (–3) dm. long, 1–2-forked at right angles, forming rather dense cymes 3–8 cm. across; invols. broadly campanulate, 3–4 mm. long, white-woolly; calyx sulphur-yellow, 5–6 mm. long, narrowed into a tubular stipelike base, the outer calyx-segms. oblance-oblong, the inner wider than the outer, glabrous; aks. brownish, lance-ovoid, ca. 3 mm. long, somewhat 3-angled; $2n=40$ (Stokes and Stebbins, 1955).—Rocky slopes at ca. 500 ft.; Coastal Sage Scrub, Conejo Grade, n. base of Santa Monica Mts., Ventura Co. April–July.

14. **E. davidsònii** Greene. [*E. vimineum* var. *d.* Stokes. *E. molestum* of auth., not Wats.] Erect annual, few-branched or usually simple at the base, glabrous and glaucous except on the lvs., 1–5 dm. high; lvs. basal, rounded to reniform, 1–2 (–4) cm. long and wide, densely white-tomentose especially below, crisped or undulate, on longer petioles; invols. cylindric-turbinate, 3–5 mm. long, glabrous, scarious between the ribs, sessile and remote, few on a branch; calyx white or tinged pink to rose-red, 1.5–2 mm. long, the outer segms. oblong-obovate, the inner slightly narrower, glabrous; fils. pilose basally; aks. narrow-ovoid, ca. 2 mm. long, narrowed slightly to muriculate, 3-angled beaks.—Occasional in dry places, 3000–7000 ft.; Chaparral, Pinyon-Juniper Wd., Joshua Tree Wd., Yellow Pine F.; L. Calif. and San Diego Co. to Monterey and Tulare cos. June–Sept.

15. **E. defléxum** Torr. ssp. **defléxum**. Plate 70, Fig. C. Annuals with 1–several stems from the base, glabrous and glaucous nearly throughout, up to 7 dm. high; lvs. basal, cordate, reniform to nearly orbicular, 1–4 cm. long, 2–5 cm. wide, densely white-tomentose below, less so to subglabrous and green above; infl. mostly spreading, open to diffuse; peduncles deflexed, up to 3 mm. long, usually less; invols. 1.5–2 mm. long, turbinate, 5-lobed; calyx white to pinkish, 1–2 mm. long, the outer segms. ovate to ovate-elliptic, or oblong, usually cordate at bases, the inner segms. lanceolate to ovate; aks. brown, 2–3 mm. long, with stout 3-angled beaks; $2n=20$ (Reveal, 1964).—Common in sandy to gravelly washes and slopes, mostly below 6000 ft.; Creosote Bush Scrub to Pinyon-Juniper Wd.; Mojave and Colo. deserts, Inyo Co. to San Diego Co. and Imperial; to Utah, Ariz., L. Calif. May–Oct.

Ssp. **baràtum** (Elmer) Munz, comb. nov. [*E. baratum* Elmer, Bot. Gaz. 39: 52. 1905. *E. d.* var. *b.* Reveal.] Plants up to 10 dm. high, slender or stout, stems and branches often inflated, forming erect strict crowns, the branches few, often elongated, and whiplike; peduncles mostly 5–15 mm. long; invols. 2.5–3 mm. long, narrowly turbinate; calyx 2–2.5 mm. long; $n=20$ (Reveal, 1968).—Common on talus, gravelly slopes up to 9500 ft.; Montane Coniferous F.; s. Mono Co. and Inyo Co. s. to Ventura and Los Angeles cos.; w. Nev. June–Oct.

16. **E. desertícola** Wats. Erect shrubs 6–12 (–15) dm. tall, much branched, the ultimate branchlets white-tomentose and leafy when young, becoming glabrous, green and leafless in age; lvs. oblong-ovate to round-oblong, 5–15 mm. long, sometimes wider, white-woolly, on petioles 5–12 mm. long; invols. solitary, terminal, woolly, 1.5 mm. long, with 4 rounded teeth, short-peduncled; calyx yellow with green or reddish mid-ribs, silky-villous, the segms. oblong-obovate, 2–3 mm. long; fils. pubescent basally; aks. dark, lance-ovoid, strigose, 3 mm. long.—Locally common along sandy washes, dunes, etc.; Creosote Bush Scrub; Imperial Co. from Salton Sea to dunes w. of Yuma. Sept.–Dec.

17. **E. elàtum** Dougl. ex Benth. ssp. **glabréscens** Stokes. [*E. e.* var. *villosum* Jeps.] Caudex woody, branched or simple; lvs. basal, erect, the blades lanceolate to lance-ovate, 4–15 cm. long, acutish, green and glabrate above, somewhat tomentose below, the petioles ca. as long, villous; flowering stems 4–8 dm. long, villous-pubescent, repeatedly trichotomous above, sometimes inflated; invols. in terminal clusters of 2–4, sometimes solitary, turbinate, ca. 4 mm. long, 5-toothed; calyx white or pinkish, 2.5 mm. long, pubescent without, the segms. obovate; fils. glabrous except at very base; aks. ca. 4 mm. long, with 3-angled beak.—Dry rocky slopes; Sagebrush Scrub, N. Juniper Wd., etc.; Panamint Mts. to Siskiyou Co.; w. Nev. June–Sept.

18. **E. elongàtum** Benth. Plate 70, Fig. D. Perennial herbs, usually loosely branched at base, whitish-tomentose nearly throughout, leafy in lower portion, passing into elongate leafless paniculately forked infls. above, 6–12 (–18) dm. high; lvs. lance-oblong to narrowly ovate, crisped-undulate, somewhat glabrate above, white-tomentose beneath, 3–5 cm. long, cuneate at base, short-petioled; invols. remotely scattered, oblong-cylindric, 6–7 mm. long, tomentose, truncate, obscurely 5-toothed; calyx white or pinkish, glabrous, 2.5–3 mm. long, the segms. obovate, the inner slightly longer than the outer, somewhat pubescent within; fils. glabrous; aks. dark, narrow, 2–2.5 mm. long, glabrous, somewhat 3-angled; $2n=34$ (Stokes & Stebbins, 1955).—Dry rocky places below 6000 ft.; Coastal Sage Scrub, Chaparral, etc.; n. L. Calif. to Monterey Co. Aug.–Nov.

19. **E. eremícola** Howell & Reveal. Annuals with one or occasionally several stems from the caudex, 8–25 cm. high; lvs. basal, 1–2.5 cm. long and wide, rounded, subcordate at the base, densely white-tomentose below, less so to glabrous above, on petioles up to 3 cm. long; stems slender, 3–10 cm. long, capitate-glandular; branches spreading so as to form a subglobose crown, ± open, trichotomous at the first node, dichotomous or trichotomous above, glandular; peduncles slender, deflexed, sessile or subsessile above, 1–1.5 mm. wide, 5-lobed, sparsely glandular; calyx whitish, becoming reddish at maturity, 2–2.5 mm. long, glabrous, the segms. ovate-oblong, the bases obtuse to subcordate; aks. 2 mm. long, with a stout, 3-angled beak.—Sandy to gravelly soils from 7500–10,000 ft.; Pinyon-Juniper Wd., Yellow Pine F., Bristle-cone Pine F.; Panamint Mts. June–Sept.

20. **E. esmeraldénse** Wats. Glabrous annuals, 1- to few-branched, then repeatedly branched, 1–3 dm. high, the ultimate branches very slender; lvs. basal, somewhat round-obovate, 6–15 mm. long, pilose-hispid, on petioles ca. as long; peduncles filiform, 5–15 mm. long; invols. narrow-turbinate, 1 mm. long, 5-lobed, few-fld.; calyx white to pink. glabrous, the segms. ± oblong, obtuse or retuse; aks. narrow-ovoid, shining, ca. 1 mm-long, gradually narrowed to a stout beak.—Dry gravelly places, 6000–9800 ft.; Pinyon, Juniper Wd.; foothills and low mts. of Inyo and Mono Co. (New York Butte, Waucoba Peak); w. Nev. July–Sept.

21. **E. fasciculàtum** Benth. ssp. **fasciculàtum.** CALIFORNIA BUCKWHEAT. Plate 70, Fig. E. Low spreading shrubs, the stems ± decumbent, 6–12 dm. long, branched, leafy; branchlets loosely pubescent to subglabrous, ending in leafless peduncles 3–10 (–15) cm. long, bearing ± open cymose infl. with many capitate clusters at the tips; lvs. numerous, fascicled, oblong-linear to linear-oblanceolate, green and glabrate above, white-woolly

beneath, 6–15 mm. long, strongly revolute; invols. prismatic, 3–4 mm. high, glabrous, with 5 short acute teeth; calyx white or pinkish, ca. 3 mm. long, nearly or quite glabrous without, the outer segms. broadly elliptic, the inner obovate; fils. subglabrous basally; aks. lance-ovoid, shining, ca. 2 mm. long; $2n=20$ (Reveal, 1967).—Dry slopes and canyons near the immediate coast; Coastal Sage Scrub; Santa Barbara to n. L. Calif. Much of year.

Ssp. **flavovíride** (M. & J.) Stokes. Low, 2–3 dm. tall; lvs. yellow-green, subglabrous above, strongly revolute; peduncles glabrous; invols. and calyx subglabrous, the latter quite reddish.—Rocky places, below 4000 ft.; Creosote Bush Scrub; Eagle Mts., e. Riverside Co. to Little San Bernardino and Sheephole mts., San Bernardino Co. March–May.

Ssp. **foliolòsum** (Nutt.) Stokes. Upper surface of lvs., outer surface of calyx, invols., etc. pubescent; peduncles 1–2 dm. long; $2n=80$ (Stebbins, 1942).—Common on interior cismontane slopes and mesas, below 3000 ft.; Coastal Sage Scrub, Chaparral; n. L. Calif. to Monterey Co. Very variable. March–Oct.

Ssp. **polifòlium** (Benth.) Stokes. [*E. p.* Benth.] Plants commonly 2–5 (–8) dm. tall; lvs. densely canescent to hoary above, commonly less revolute than other sspp.; invols. and calyx pubescent; heads solitary or in reduced cymes; $2n=40$ (Stebbins, 1942).—Common on dry slopes below 7000 ft.; Sagebrush Scrub to Pinyon-Juniper Wd.; both deserts to Inyo Co. and San Joaquin V. and interior cismontane s. Calif.; to Utah, Ariz., L. Calif. April–Nov.

22. **E. foliòsum** Wats. [*E. baileyi* var. *tomentosum*, in part.] Spreading annuals with imperfectly dichotomously branched tomentose branches 1–3 dm. long, these often becoming glabrate toward their tips at maturity; basal lvs. ovate to oblong, 5–10 mm. long, 3–10 mm. wide, densely white-tomentose beneath, less so to glabrous above, on tomentose petioles up to 2 cm. long; cauline lvs. when present in pairs at the nodes, narrowly ovate, sessile or subsessile; invols. turbinate, 0.8–1.2 mm. long, solitary, sessile, 5-lobed; calyx white with pink midribs, becoming pink or rose at maturity, 0.8–1.2 mm. long, the outer segms. broadly hastate at the base when mature, obtuse at the apices, the margins ± crispate, the inner segms. narrower, oblong, slightly longer than the outer; fils. minutely pubescent basally; aks. 1–1.2 mm. long, with long sharp 3-angled beaks.—Rare in sandy to gravelly places below 4000 ft.; Pinyon-Juniper Wd.; San Bernardino Mts. s. to n. L. Calif. March–Aug.

23. **E. gigantèum** Wats. ssp. **gigantèum**. St. Catherine's Lace. Coarse rounded branching shrubs, open, 3–20 (–35) dm. high, the cent. trunk to 1 dm. thick, the younger branches tomentose, then glabrate and dark, with lvs. toward the tips; lf.-blades leathery, oblong-ovate, to ovate, 3–7 (–10) cm. long, closely white-tomentose below, cinereous and somewhat glabrate above, on stout petioles 1–3 cm. long; peduncles stout, 1–3 dm. long, tomentose, later glabrate, bearing large, 2–3-forked horizontal cymes often several dm. across and with leafy bracts at the forks; invols. crowded, campanulate, 3–4 mm. long, tomentose, with short obtuse teeth, subsessile or on short slender peduncles; calyx white, becoming rusty in age, ca. 2 mm. long, white-hairy, the segms. obovate; fils. hairy; aks. brown, narrow-ovoid, angled above, 2 mm. long; $n=20$ (Reveal, 1967).—Dry slopes, Chaparral, Coastal Sage Scrub; Santa Catalina Id. May–Aug.

Ssp. **compáctum** (Dunkle) Munz, comb. nov. [*E. formosum* var. *c.* Dunkle, Bull. So. Calif. Acad. Sci. 41: 130. 1943.] Lower in stature; lvs. oblong; tomentum on young growth looser; involucral peduncles very stout.—Santa Barbara Id.

Ssp. **formòsum** (K. Bdg.) Raven. [*E. g.* var. *f.* K. Bdg.] Lvs. oblong-lanceolate.—San Clemente Id.

24. **E. gilmánii** Stokes. Low compact perennial from an elongated woody root, the caudex covered with old lvs.; lvs. crowded in a basal rosette, few to ca. 14, sub-erect, densely white-tomentose, the blades 2–4 mm. long, elliptic, with petioles margined and ca. as long; scapes solitary and 1–2 cm. high, with 2–3 cymosely arranged invols.; invols. turbinate, 1.5 mm. long; calyx reddish, glabrous, the outer segms. inflated, rounded on the back and orbicular in shape so as to form a globose segm. 3–4 mm. in diam., the

inner segms. narrower, longer, slightly exserted; aks. 2.5 mm. long, acutely 3-angled.—At 6200 ft., Pinyon Mesa, Panamint Mts., Inyo Co. Aug.–Sept.

25. **E. glandulòsum** Nutt. [*E. carneum* (J.T. Howell) Reveal. *E. g.* var. *c.* J. T. Howell.] Annuals, usually with 1 stem from the caudex, erect, 1–2.5 dm. high, glandular with small tack-shaped glands, numerous on the lower nodes, less so above; lvs. basal, broadly elliptic, 6–12 mm. long, 5–10 mm. wide, pilose-hirsutulous and slightly glandular, on petioles 3–12 mm. long, stiffly hispid; bracts sparsely hirsutulous within and apically, glandular along upper margins, 1 mm. long; branches numerous, spreading, trichotomous at first node with 1–several branchlets at this node, dichotomous above; peduncles slender, deflexed or nearly so, straight, 2–3 mm. long, glandular; invols. narrowly turbinate, 0.8–1.2 mm. long, glabrous, 5-lobed, few-fld.; calyx white, maturing pinkish, with red midribs, 1–1.8 mm. long, externally pilose; stamens included; aks. with globose base and short beak.—Dry sandy soil, 3500–4500 ft.; Sagebrush Scrub, Pinyon-Juniper Wd.; extreme e. Inyo and ne. San Bernardino cos.; sw. Nev. June–Aug.

26. **E. grácile** Benth. [*E. vimineum* ssp. *g.* Stokes.] Plate 70, Fig. F. Erect annuals, strictly or rather diffusely branched from base, 2–5 dm. high, thinly to floccose-tomentose throughout, the branchlets slender, ascending; lvs. mostly basal, oblanceolate to oblong, the lf.-blades 1–3 (–4) cm. long, tomentose, especially beneath, on petioles ca. as long; cauline lvs. becoming strongly reduced above; invols. 1.8–2 (–3) mm. long, turbinate, subglabrous, the 5 teeth conspicuous, rigid; calyx white, pinkish or yellowish, 1.5–2 mm. long, the outer segms. broadly obovate, the inner oblong, glabrous; aks. ca. 1 mm. long, ovoid, with 3-angled beaks; $2n = 22$ (Stokes & Stebbins, 1955).—Common in dry cismontane washes, on mesas, below 5000 ft.; Coastal Sage Scrub, Chaparral, S. Oak Wd., etc.; L. Calif. to cent. Calif. July–Oct.

27. **E. gracílipes** Wats. [*E. kennedyi* ssp. *g.* Stokes.] Caudex cespitose, branched, low, the plants with a dense whitish tomentum; lvs. crowded, oblanceolate to elliptic, plane, the blades 1–2 cm. long, with petioles ca. as long, tomentose and glandular below, less so above; flowering stems scapelike, 3–8 cm. long, slender, glandular; infl. capitate, of 5–7 invols.; invols. campanulate, membranaceous, not rigid, 2–3 mm. long, flaring at the throat, the 5 lobes deeply divided to near the base; calyx white with reddish becoming rose at maturity, glabrous, 2–3 mm. long, the segms. obovate, similar; fils. pilose basally; aks. lance-ovoid, ca. 2 mm. long.—Dry rocky slopes and ridges, 10,000–13,000 ft.; Bristlecone Pine F.; Inyo-White Mts.; w. Nev. July–Sept.

28. **E. gracíllimum** Wats. [*E. angulosum* var. *g.* Jones.] Annuals, freely branched at or just above the base, 1–4 dm. high, thinly woolly nearly throughout; basal lvs. oblong to oblanceolate, 2–4 cm. long, short-petioled, densely tomentose beneath, less so above, with revolute crisped edges; cauline lvs. lance-oblong, well distributed, similar to the basal lvs.; peduncles filiform, 8–25 mm. long, glabrous; invols. subcampanulate, angled, 2 mm. long, glandular-puberulent, shallowly 5-lobed; calyx rose, tipped with white, 2–2.5 mm. long, the segms. similar, oblong to elliptic, frequently crenulate; aks. black, ca. 1 mm. long, shortly beaked; $n = 20$ (Reveal, 1967).—Common on sandy plains below 3500 ft.; Grassland, Foothill Wd., Joshua Tree Wd., Coastal Sage Scrub; interior cismontane s. Calif. and w. Mojave Desert to Monterey Co. April–Sept.

29. **E. gránde** Greene ssp. **gránde**. Caudex woody at the base, few-branched with elongated leaf-bearing areas 2–3 dm. long at the base; lf.-blades oblong-ovate, 3–10 cm. long, greenish above, closely white-woolly beneath, strongly undulate-crisped, the petioles much longer; flowering stems 8–15 dm. long, glabrous, glaucous, forking above; invols. in clusters of 1–3, turbinate, 5–6 mm. long, subglabrous without; calyx whitish, ca. 3 mm. long, the segms. oblong-obovate, spreading; fils. pilose basally; aks. 2.5–3 mm. long; $2n = 40$ (Stokes & Stebbins, 1955).—Bluffs and cliffs; Coastal Sage Scrub, Chaparral; Santa Cruz, Santa Catalina, Anacapa and San Clemente ids. June–Oct.

Ssp. **rubéscens** (Greene) Munz, comb. nov. [*E. rubescens* Greene, Pittonia 1: 39. 1887. *E. g.* var. *r.* Munz.] Lower and more decumbent with a tendency toward subcapitate cymes with red fls.—San Miguel Id. and w. end of Santa Cruz Id.

Ssp. **timòrum** (Reveal) Munz, comb. nov. [*E. grande* var. *t.* Reveal, Aliso 7: 229.

1970.] Plants 1–2 dm. high; lf.-blades 2–3.5 cm. long; invols. 4–5 mm. long; fls. white.—San Nicolas Id.

30. **E. heermánnii** Dur. & Hilg. ssp. **heermánnii**. Woody and branched, the stems erect, woody and floccose in lower portions, glabrous and light green above, 3–7 (–15) dm. high; lvs. lance-oblong to oblanceolate, 1–2 (–2.5) cm. long, green above, floccose beneath, somewhat undulate, short-petioled; infl. a cymose panicle of dichotomously branched rigid branchlets, almost or quite smooth, subterete, the lower internodes 2–4 (–6) cm. long; invols. solitary in the forks or terminal, broadly turbinate, 2 mm. long, rather deeply lobed, glabrous; calyx yellowish white, 3–4 mm. long, the outer segms. orbicular, subcordate at the bases, the inner oblong, glabrous; aks. narrow, 3-angled, 2–2.5 mm. long.—Dry slopes and ridges, 2000–7000 ft.; Foothill Wd., Joshua Tree Wd., Pinyon-Juniper Wd.; San Bernardino and San Gabriel mts. through w. Mojave Desert; to San Benito Co.

KEY TO SUBSPECIES

Branchlets of infl. floccose-tomentose; fls. 2–3 mm. long. Little San Bernardino Mts. to Providence, New York, Clark mts. .... ssp. *floccosum*
Branchlets of infl. not tomentose.
    Fls. 3–4 mm. long, the branchlets not scabrellous, ca. 1 mm. in diam.
        Branchlets with smooth internodes.
            Invols. racemosely arranged on tips of stoutish rigid branchlets.
            San Bernardino and San Gabriel mts. to San Benito Co. .... ssp. *heermannii*
            Invols. scattered along slender branchlets. Inyo-Mono cos. to w. Nevada . ssp. *humilius*
        Branchlets strongly angled and grooved. Panamint and Kingston mts. ..... ssp. *sulcatum*
    Fls. 2 mm. long, the branchlets very slender, ca. 0.5 mm. in diam., scabrellous under a lens
        ssp. *argense*

Ssp. **argénse** (Jones) Munz, comb. nov. [*E. sulcatum* var. *a*. Jones, Contrib. W. Bot. 11: 15. 1903.] Low subshrubs 1–2 dm. high; stems numerous, slender and rather delicate; infl. very compact and intricate, the lower internodes 0.5–1.2 cm. long, the branchlets glabrous and scabrellous; calyx ca. 2 mm. long.—Dry rocky places, 5000–8000 ft.; Pinyon-Juniper Wd.; Inyo and Mono cos.

Ssp. **floccósum** (Munz) Munz. [*E. h.* var. *floccosum* Munz, Man. So. Calif. Bot. 121, 597. 1935.] Lower internodes of infl. 2–3 cm. long, the branchlets floccose-tomentose; fls. 2–3 mm. long.—Largely in Pinyon-Juniper Wd.; mts. of e. San Bernardino Co.

Ssp. **humílius** Stokes. [*E. h.* var. *h.* Reveal.] Like ssp. *heermannii*, but with invols. scattered along the slender branches.—Inyo and Mono Co. s. to w. Nev.

Ssp. **sulcàtum** (Wats.) [*E. s.* Wats.] Intricately branched, 1.5–4 dm. high, the branchlets strongly angled and grooved; fls. 3–4 mm. long.—At 6300–7000 ft.; Pinyon-Juniper Wd.; Panamint and Kingston mts., e. Mojave Desert.

31. **E. hirtiflòrum** Gray ex Wats. Annuals, 5–15 cm. high, repeatedly dichotomously branched, glandular-puberulent; lvs. basal and at the lower nodes, obovate to spatulate, 1–2.5 cm. long, ciliate, narrowed into winged petioles; invols. sessile in the forks and along the branches, or short-peduncled, narrow, 2-fld., ca. 1 mm. long; calyx reddish, ca. 1 mm. long, hirsutulous with hooked hairs, the segms. oblong; aks. narrow, ca. 1 mm. long, exceeding the calyx, with broad angled beaks.—Occasional, dry gravelly places, below 6000 ft.; Chaparral, Foothill Wd., Yellow Pine F.; w. San Gabriel Mts., San Bernardino Mts., Ventura Co. to n. Calif. June–Oct.

32. **E. hoffmánnii** Stokes ssp. **hoffmánnii**. Annual; scapes usually one, rarely more, 1–5 dm. high; lvs. basal, 1–4 cm. long, 2–4 cm. wide, suborbicular to subcordate, densely white-tomentose beneath, less so to glabrous and green above; stems less than 5 cm. long, glabrous; with spreading glabrous branches; peduncles lacking; invols. erect, turbinate, 1–2 mm. long, 5-lobed; calyx 1.5 mm. long, white with greenish midribs, glabrous, the segms. spatulate; aks. 2 mm. long, with 3-angled beaks; $n=20$ (Reveal, 1968).—Dry talus slopes, 4000–5000 ft.; Pinyon-Juniper Wd.; Panamint Mts. July–Sept.

Ssp. **robústius** (Stokes) Munz, stat. nov. [*E. h.* var. *robustius* Stokes, Leafl. W. Bot.

3: 16. 1941.] Plants to 10 dm. high, stems to 4 dm. long; lvs. basal or sheathing up the stems 1–5 cm., the blades 2–5 cm. long, 3–8 cm. wide, crisped; infl. erect and strict, with long whiplike branches; calyx white or reddish with reddish midribs, the segms. ovate.—Dry sandy washes, 1000–2000 ft.; Creosote Bush Scrub; Black and Funeral mts., Inyo Co. Aug.–Nov.

33. **E. hoókeri** Wats. [*E. deflexum* ssp. *h.* Stokes.] Annual with only a single stem, rarely more, from the caudex, 1–4 (–6) dm. high; lvs. basal, 2–5 cm. long, 2–6 cm. wide, cordate to subreniform, margins often upward rolled on the edges, densely white feltytomentose below, white-tomentose above; stems glabrous, erect; branches spreading, the crown subglobose to flat-topped with the outer edge lower than the center so as to appear umbrella-like, glabrous; peduncles none; invols. deflexed, hemispheric, 1–2 mm. long, 1.5–3 mm. wide, 5-lobed; calyx yellow, becoming reddish-yellow at maturity, 1.5–2 mm. long, glabrous, the segms. dissimilar, the outer orbicular, cordate or hastate at base, the inner oblong-ovate and shorter, the outer expanding in fr., becoming nearly isodiametric, 2–2.5 mm. long and wide; aks. light brown, 2–2.5 mm. long, with 3-angled beaks; $n=20$ (Reveal, 1968).—Rare in Pinyon-Juniper Wd., at 5000–6000 ft.; n. Inyo Co.; to Wyo., Colo., Ariz. July–Oct.

34. **E. inérme** (Wats.) Jeps. ssp. **hispídulum** (Goodm.) Munz, stat. Nov. [*E. inerme* var. *hispidulum* Goodm., Am. Midl. Nat. 39: 508. 1948.] Annual, dichotomously or trichotomously forked just above or at the base, then repeatedly dichotomous, 5–30 cm. high, sparingly stipitate-glandular; lvs. basal, spatulate, 1–2 cm. long, sessile, ciliate, otherwise glabrous; bracts 3-lobed; invols. on pedicellike peduncles, 4-lobed nearly to the base, hispidulous, 1.5 mm. long; calyx rose, 1.5 mm. long, hispid with hooked hairs, the segms. oblong, the inner retuse and smaller than the outer; aks. scarcely longer than calyx. —Uncommon, 3000–6000 ft.; Yellow Pine F.; San Bernardino Mts., Kern Co. to Tuolumne Co. June–Aug.

35. **E. inflátum** Torr. & Frém. DESERT TRUMPET. Plate 71, Fig. A. Perennial, but flowering the first year, 2–10 dm. high, glabrous and glaucous, sometimes hirsute near base, with 1–several stems that are simple below and dichotomous above to form diffuse panicles, the stems usually conspicuously inflated in upper part of lower internodes; lvs. oblong-ovate to rounded, usually cordate or truncate at base, 1–3 cm. long, short-hirsute, green, somewhat crisped, on petioles 2–5 cm. long; peduncles capillary, 5–20 mm. long; invols. glabrous, 1–1.5 mm. long, turbinate, 5-lobed, the lobes occasionally with stipitate glands, several-fld.; calyx yellow, often with red-brown midribs, densely pubescent, 2–2.5 mm. long, the segms. lance-ovate; aks. brown, ca. 2 mm. long, lance-ovoid; $n=16$ (Reveal, 1965).—Common in washes and along mesas, below 6000 ft.; Creosote Bush Scrub, Joshua Tree Wd., Sagebrush Scrub, Pinyon-Juniper Wd.; Mojave and Colo. deserts, n. to Mono Co.; to Colo., Ariz., L. Calif. March–Oct. In the Colo. Desert and L. Calif. occur plants with stems not inflated; these are var. **deflàtum** Jtn. [*E. glaucum* Small.] In Death V. region others not inflated, annual, are var. **contíguum** Reveal.

36. **E. insigne** Wats. [*E. deflexum* var. *i.* Jones.] Annuals, usually with only a single stem from the caudex, up to 10 dm. high; lvs. basal, up to 8 cm. long and wide, subcordate to orbicular, the base subcordate, densely white-tomentose beneath, less so to subglabrous above; stems erect, stout, up to 2 dm. high, glabrous; branches erect and strict, dichotomous or trichotomous, often 4–5 times taller than wide, the long whiplike branches with alternating right-angled secondary ones which are also with alternating branches, the tips ending in racemes of 2–8 invols.; invols. erect, on peduncles to 3 mm. long; invols. turbinate, 2–2.5 (–3) mm. long, 1.5–2.5 mm. wide, 5-lobed; calyx 1.5–2.5 mm. long, white, glabrous, the outer segms. oblong, the base cordate, the inner narrower and shorter, rarely over 1 mm. wide; aks. 2–2.5 mm. long, tapering to long 3-angled beaks; $n=20$ (Reveal, 1968).—Sandy soils mostly below 4000 ft.; Creosote Bush Scrub to Sagebrush Scrub; San Diego Co. to s. Inyo Co.; sw. Utah, Ariz. May–Nov.

37. **E. intrafráctum** Cov. & Mort. Perennial, woody at the base, from distinct taproots; lvs. basal, oblong-ovate, somewhat whitish-pilose, the blades 2.5–7 cm. long, the petioles somewhat longer; flowering stems 6–12 dm. high, usually solitary, simple below,

Plate 71. Fig. A, *Eriogonum inflatum*. Fig. B, *Eriogonum nudum*. Fig. C, *Eriogonum parishii*, showing 1 fl. and invol. Fig. D, *Eriogonum parvifolium*. Fig. E, *Eriogonum saxatile*. Fig. F, *Oxytheca perfoliata*, invol. lobes awn-tipped.

sometimes branched in infl., rather stout, glabrous, glaucous, transversely jointed into hollow, ringlike segms., each segm. 3–10 mm. long, becoming easily fractured; infl. usually of 2–3 virgate branches, 2–4 dm. long and sometimes with shorter branches; invols. usually in whorls of 3 at each node, usually 1 in the axil of each of 3 bracts, 5-parted into oblong lobes which become more divided with expanding fls.; calyx yellow, tinged with red, pubescent, ca. 2 mm. long, the lobes oblong-lanceolate, subequal; aks. flask-shaped, almost 2 mm. long, 3-ridged in lower part, then narrowed into triangular beaks; $n = 20$.—Local and rare, limestone crevices, 2000–5000 ft.; Creosote Bush Scrub; Grapevine and Panamint mts., Inyo Co. May–Oct.

38. **E. keárneyi** Tidestr. ssp. **monoénse** Stokes. Subshrubs with few to several straggly fragile stems from a woody taproot, usually densely tomentose; stems 8–15 dm. long; lvs. on lower third of plant 2–3 cm. long, broadly oblanceolate to elliptic, densely tomentose; petioles to 1 cm. long; infl. making up more than half the height of the plant; invols. racemosely disposed, 2.5–3 mm. long, clustering at the ends of the branches; calyx white with reddish midribs, glabrous, 2.5–3 mm. long; aks. 2 mm. long, with 3-angled beaks.—Dry gravelly places, 6000–8500 ft.; Sagebrush Scrub, Pinyon-Juniper Wd.; Mono and Inyo cos. (Inyo Mts., Last Chance Mts.). July–Aug.

39. **E. kénnedyi** Porter ex Wats. ssp. **kénnedyi.** Caudex branched, woody, forming dense leafy mats with numerous lvs.; lf.-blades elliptic to oblong, tomentose, 2–4 (–5) mm. long, 0.5–2 mm. wide, ± revolute in some, subsessile; flowering stems scapelike, wiry, glabrous, 4–12 cm. long; infl. capitate, 4–8 mm. across; invols. few, sparsely tomentose to glabrous, turbinate, angled, 1.5–2.5 mm. long; calyx glabrous, white with reddish midribs, 1.5–2.5 mm. long, the segms. broadly elliptical, somewhat rounded at the base; fils. subglabrous; aks. ca. 2 mm. long, 3-angled.—Dry stony to gravelly slopes and ridges, 5000–7000 ft.; San Bernardino Mts. and Mt. Pinos. April–June.

KEY TO SUBSPECIES

Stems 1–2 cm. high; infl. capitate, 4–8 mm. wide. Mostly 9000–11,000 ft. ...... ssp. *alpigenum*
Stems 4–15 cm. high. Mostly 6000–8000 ft.
  Lvs. 2–6 mm. long; flowering from April–June.
    Calyx with reddish midribs; infl. 4–8 mm. across. San Bernardino Mts., Mt. Pinos
                                                    ssp. *kennedyi*
    Calyx with greenish midribs; infl. 8–15 mm. wide. Argus & Coso mts. n. ... ssp. *purpusii*
  Lvs. 6–12 mm. long; flowering from July–Aug. Bear V., San Bernardino Mts.
                                                    ssp. *austromontanum*

Ssp. **alpígenum** (M. & J.) Stokes. Mats very dense and woody; lvs. 2–4 mm. long; stems 1–2 cm. long, densely white-tomentose; infl. capitate, 4–8 mm. wide; invols. 1.5–2 mm. long, tomentose; calyx white to reddish with reddish brown midribs, 1.5–2.5 mm. long.—Dry granitic slopes and ridges, 10,000–11,500 ft.; San Bernardino Mts., lower to 8750 ft. in San Gabriel Mts. and on Mt. Pinos.—July–Aug.

Ssp. **austromontànum** (M. & J.) Stokes. Loosely matted; lvs. oblanceolate, the blades 6–10 mm. long; invols. tomentose, 2.5–4 mm. long; calyx white with reddish-brown midribs, 2–3 mm. long, oblong-obovate.—Dry stony slopes, 6300–6500 ft.; Yellow Pine F.; Bear V., San Bernardino Mts. July–Aug.

Ssp. **purpùsii** (Bdg.) Munz, stat. & comb. nov. [*E. purpusii* Bdg., Bot. Gaz. 27: 457. 1899. *E. k.* var. *p.* Reveal.] Lvs. white-tomentose, 3–6 mm. long; stems mostly glabrous, 4–10 cm. high; infl. 8–15 mm. high; invols. turbinate-campanulate, subglabrous, 1.5–2 mm. long; calyx white with greenish midribs, 2–2.5 mm. long.—Dry granite, 5000–8000 ft.; Sagebrush Scrub, Pinyon-Juniper Wd.; Argus and Coso mts., Inyo Co. to e. slope of Sierra Nevada. May–June.

40. **E. maculàtum** Heller. [*E. angulosum* ssp. *m.* Stokes.] Annual, with 1–several branches from the base, 1–2 (–3) dm. high, tomentose almost throughout; basal lvs. lanceolate to obovate, 1–3 (–4) cm. long, short-petioled, tomentose beneath, glabrate or nearly so above, occasionally revolute and crisped on margins; cauline lvs. sessile, 0.5–2

Eriogonum

cm. long, lanceolate to oblanceolate, subtended by scalelike bracts; branches widely spreading; peduncles filiform, 1-3 cm. long, axillary, often glandular-puberulent; invols. campanulate, 1-1.5 (-2) mm. long, glandular-puberulent externally, glabrous to woolly within, with 5 rounded lobes; calyx white to yellow, pink or red with rose-colored spots on outer inflated segms., 1-2.5 mm. long, glandular-puberulent, the outer segms. elliptic to roundish or obovate, with an inflated area at base and middle with the sides of the blades incurved below, inner segms. narrower, longer; aks. 1-1.5 mm. long, with 3-angled beaks; $n = 20$ (Reveal, 1965).—Dry, often sandy or gravelly places, below 7000 ft.; Creosote Bush Scrub, Joshua Tree Wd., Sagebrush Scrub, Pinyon-Juniper Wd.; deserts L. Calif. to e. Wash., Utah, Ariz. April-Nov.

41. **E. microthècum** Nutt. var. **ambíguum** (Jones) Reveal. [*E. aureum* Jones var. *ambiguum* Jones.] Low half-bushy shrubs 1-4 dm. high; stems usually floccose-tomentose above; lvs. narrow-elliptic, the blades 10-25 mm. long, 4-8 mm. wide, usually not revolute, whitish- or red-brown-tomentose beneath; infl. commonly 3-6 cm. across; invols. turbinate, 2-2.5 mm. long; calyx yellow.—Dry rocky and gravelly places, 5000-10,000 ft., Pinyon-Juniper Wd., Yellow Pine F.; Inyo-White range n. and to w. Nev. July-Sept.

KEY TO VARS. OF ERIOGONUM MICROTHECUM

1. Calyx white or with some pink, not yellowish.
   2. Tomentum whitish.
      3. Lvs. strongly revolute, almost linear, mostly 1-2 mm. wide. E. Mojave Desert
         var. *foliosum*
      3. Lvs. usually plane, ± elliptical, 2.5-6 mm. wide. E. Mojave Desert and n.
         var. *laxiflorum*
   2. Tomentum brownish to reddish.
      4. Plants shrubby, 3-6 dm. high.
         5. Stems and infl. tomentose when young, becoming floccose at maturity. Panamint and Inyo mts. .... var. *panamintense*
         5. Stems and infl. lanate, even at maturity. San Gabriel and San Bernardino mts.
            var. *corymbosoides*
      4. Plants subshrubs, 0.5-1.5 dm. high.
         6. Lvs. elliptic to ovate, 3-5 mm. wide; invol. 2.5-3 mm. long. San Gabriel Mts.
            var. *johnstonii*
         6. Lvs. elliptic, 1-4 mm. wide; invol. 3-3.5 mm. long. Inyo Mts. .... var. *lapidicola*
1. Fls. yellowish to yellow, not white. Inyo-White range n. .... var. *ambiguum*

Var. **corymbosoìdes** Reveal. Stems often very slender, ± semiprostrate, white-tomentose, 1-3 dm. long; lvs. mostly not revolute, white-tomentose beneath, 5-10 mm. wide; fls. deep red to whitish-brown.—Dry rocky limestone slopes, 6400-9500 ft.; Pinyon-Juniper Wd.; n. slopes of San Gabriel Mts. to Sugarloaf Mt. and Cactus Flats, San Bernardino Mts. July to Sept.

Var. **foliòsum** (Torr.) Reveal. [*E. effusum* var. *foliosum* T. & G.] Suberect, 1-12 dm. high; lvs. sublinear in appearance, because strongly revolute, mostly 10-25 mm. long; fls. white.—At 4500-6800 ft., dry limestone areas, Pinyon-Juniper Wd.; mts. of e. Mojave Desert (N.Y. Mts., Clark Mt., Kingston Mts., Last Chance Mts.). July-Oct.

Var. **johnstònii** Reveal. Lower, more branched, less woolly; lf.-blades mostly less than 1 cm. long, 3-5 mm. wide, not strongly revolute, white-woolly beneath; fls. whitish brown with red-brown midribs.—Dry rocky places, 8500-9500 ft., Montane Coniferous F.; Cucamonga Mt. region and Mt. San Antonio, San Gabriel Mts. July-Aug.

Var. **lapidícola** Reveal. Low, 0.5-1.5 dm. high; lvs. 3-7 mm. long, 1-4 mm. wide, densely red-brown tomentose beneath, not revolute; fls. whitish-red with red midribs and bases. Rocky places at 6000-8500 ft., Pinyon-Juniper Wd.; Inyo Mts. to Nev. July-Sept.

Var. **laxiflòrum** Hook. Plants 2-4 dm. high; lvs. not or only slightly revolute, narrow, 2.5-5 mm. wide, 1-2 cm. long; fls. whitish with some pink.—Mostly dry slopes, 4500-10,000 ft., Pinyon-Juniper Wd; Panamint and Kingston mts. to Lassen Co. and Nev. and Wash. June-Oct.

Var. **panamínténse** Stokes. Large, round to flat-topped, 3–6 dm. high; lvs. mostly broadly elliptical, 3–8 mm. wide, 6–18 mm. long, ± brownish-tomentose beneath, plane; invol. 2–2.5 mm. long; fls. whitish-brown with large red-brown midribs and base.—Dry slopes, Sagebrush Scrub and Pinyon-Juniper Wd., 6000–9000 ft., Panamint and Inyo mts. July–Oct.

42. **E. mohavénse** Wats. Erect annuals, diffusely and repeatedly dichotomously or trichotomously branched at or above the base, 1–3 dm. high, glabrous and green throughout except at the nodes and lvs., the ultimate branchlets capillary; lvs. basal, rounded or broadly oblong, closely white-woolly, 6–20 mm. long, on petioles up to twice as long; invols. sessile in the forks and often terminal on the branchlets, hence in subcymose infls., glabrous except at the throat, turbinate, 1.7–2 mm. long; calyx yellow, glabrous, ca. 1 mm. long, the outer segms. oblance-oblong to subelliptic, the inner narrower; aks. dark, ca. 1 mm. long, with stout muriculate beaks.—Dry sandy and gravelly places, mostly ca. 2000–4000 ft.; Creosote Bush Scrub, Joshua Tree Wd.; Mojave Desert from e. base of San Bernardino Mts. to Owens V. May–Aug.

43. **E. nidulàrium** Cov. Erect annuals repeatedly forked from near the base, 5–20 cm. long, floccose-tomentose almost throughout, forming dense masses of numerous branches with short internodes and at maturity the tips of the branches becoming curved inward; lvs. basal, rounded, sometimes with cordate base, 1–2 cm. broad, on much longer petioles; invols. cylindrical-turbinate, 1 mm. long, sessile in all the forks and along the branches; calyx red, white or yellow, 1.5–2 (–3) mm. long, the outer segms. obovate, dilated and truncate at the apices, fan-shaped, the inner narrower, glabrous; fils. glabrous; aks. ca. 1 mm. long, with long scaberulous beaks.—Common in dry gravelly and rocky places, mostly below 7000 ft.; Creosote Bush Scrub, Joshua Tree Wd., Pinyon-Juniper Wd.; deserts from San Bernardino Co. n.; to Ore., Utah, Ariz. April–Oct.

44. **E. nùdum** Dougl. ex Benth. var. **dedúctum** (Greene) Jeps. [*E. d.* Greene.] Plate 71, Fig. B. Perennial, the caudex short, mostly few-branched, the stems glabrous, glaucous, 2–3 dm. high, slender, with basal lvs., the blades 1–2 cm. long, oblong, well petioled; infl. branched, with largely solitary invols. subcylindric, 3–4 mm. long; calyx 2–2.5 mm. long, mostly white with some pink, usually glabrous; aks. 1.5–2.5 mm. long.— Dry rocky places, at ca. 9000 ft., Pinyon-Juniper Wd.; Inyo Mts.; to Sierra Nevada.

Ssp. **pauciflòrum** (Wats.) Munz, comb. nov. [*E. nudum* var. *pauciflorum* Wats. Proc. Am. Acad. Arts & Sci. 12: 264. 1877. *E. latifolium* ssp. *p.* Stokes.] Caudex rather simple with lvs. crowded; lf.-blades oblong-ovate, 1.5 cm. long, green or glabrate above, white-woolly beneath, on petioles as long or longer; flowering stems 3–8 dm. high, slender, glabrous, forked several times; invols. 1, rarely 2, at a place, rather few on a branch, 5–7 mm. long; calyx whitish, pubescent, 2 mm. long.—Dry slopes, 5000–9000 ft.; Montane Coniferous F.; Cuyamaca Mts. to Santa Rosa and San Bernardino mts. Aug.–Oct.

Ssp. **saxícola** (Heller) Munz, stat. nov. [*E. saxicola* Heller, Muhlenhercia 2: 191. 1901. *E. nudum* var. *publiflorum.* Benth. in DC.] Flowering stems glabrous, glaucous, 3–6 dm. high, cymose above; invols. clustered, rarely solitary, subcampanulate, calyx yellow to white, pubescent without; $n = 20$ (Reveal, 1965).—Dry hot places, below 6000 ft.; largely Foothill Wd., Joshua Tree Wd., Pinyon-Juniper Wd., Yellow Pine F.; Santa Ana Mts., San Gabriel Mts. through w. Mojave Desert to n. Calif. June–Sept.

45. **E. órdii** Wats. Annuals, diffusely paniculate, with many capillary ultimate branches, floccose near the base, glabrous above, 4–7 dm. high; basal lvs. oblong-obovate to oblong-oblanceolate, the blades 2–8 cm. long, on equally long petioles, floccose-tomentose to ± woolly, especially below; bracts of the lower nodes often in foliaceous whorls and like the basal lvs., the upper bracts reduced; peduncles capillary, 1–2 cm. long; invols. solitary, turbinate, 1 mm. long, 4-toothed; calyx white, tinged with pink or pale yellowish, 1.5 mm. long, densely pubescent, 1–3 fls. in an invol.; calyx-segms. oblong-ovate; aks. ca. 1.5 mm. long, stout-beaked.—Dry disturbed and barren places, below 3000 ft.; Foothill Wd. from n. Los Angeles Co. n.; also Creosote Bush Scrub and Pinyon-Juniper Wd. Colo. Desert, w. Ariz. March–June.

46. **E. ovalifòlium** Nutt. ssp. **ovalifòlium.** Cespitose perennial with closely branched

woody caudices thickly beset with lvs., densely white-tomentose; lvs. basal, round to obovate, the blades 5–15 mm. long, short-petioled; flowering stems scapose, slender, white-woolly, 1–2 dm. high; infl. capitate, 1.5–2.5 cm. in diam.; invols. several, white-woolly, commonly 4–5 mm. long, narrow-cylindric; calyx whitish or rarely yellow to ochroleucous, glabrous, 4–5 mm. long, the outer lobes elliptic, subcordate at base, the inner segms. spatulate, exserted; aks. 2–2.5 mm. long, 3-angled; $n = 20$ (Reveal, 1965).—Dry slopes and flats, mostly 5000–7000 ft.; Sagebrush Scrub, Pinyon-Juniper Wd.; Inyo Mts., Last Chance Mts., e. slope of Sierra Nevada to Alta., Rocky Mts.

Ssp. **vinèum** (Small) Stokes. [*E. v.* Small.] Lf.-blades round-ovate, 7–12 mm. long, white, felty-tomentose; flowering stems ca. 1 dm. high; infl. capitate, up to 4 cm. in diam.; calyx 5–7 mm. long, pale.—Limestone slopes at ca. 5000–5500 ft.; Joshua Tree Wd.; Cushenbury region, n. slope of San Bernardino Mts. May–June.

47. **E. palmeriànum** Reveal. [*E. plumatella* var. *palmeri* T. & G. *E. baileyi* var. *tomentosum* Wats.] Densely branched and spreading annuals, 1–3 dm. high, forming broad, often flat-topped crowns up to 3 dm. across, densely tomentose nearly throughout; lvs. basal, suborbicular or cordate, 0.5–1.5 cm. long, densely woolly beneath, less so and often green above, on petioles 1–3 cm. long; invols. campanulate, 1–1.5 mm. long, sparsely tomentose with ciliated margins, sessile at the nodes and along the subvirgate branchlets, few-fld.; calyx white to pink with reddish-brown midribs, 1.5–2 mm. long, the outer segms. oblong to fan-shaped, somewhat constricted near the middle and flaring above, the inner narrower; aks. with scabrous 3-angled beaks.—Dry sandy or gravelly flats mostly below 7000 ft.; Creosote Bush Scrub, Joshua Tree Wd., Pinyon-Juniper Wd.; e. San Bernardino Co. to n. Inyo Co.; s. Nev. to Colo., Ariz. June–Oct.

48. **E. panaminténse** Morton ssp. **panaminténse.** Caudex low and matted, branched, woody, often up to 4 dm. across; lvs. basal and along the lower nodes, not crowded, densely white-tomentose below, less so above, elliptic, ovate or obovate, 15–40 mm. long, on tomentose petioles 1–5 cm. long; flowering stems several to many, white-tomentose, 1.5–3 dm. long, 1–3 times dichotomous; cauline lvs. whorled, ± orbicular, smaller than basal; invols. solitary, sessile, in the forks and racemosely along the branches, 3–5 mm. long, glabrous, the segms. ± similar, oblanceolate; calyx white to whitish-brown, 3.5–5 mm. long, the segms. ± alike, oblanceolate; aks. narrow, 3-angled.—Dry rocky slopes, 5000–9400 ft.; Pinyon-Juniper Wd., Bristlecone Pine F.; Inyo-White Mts. to Providence and Clark mts.; sw. Nev. May–Oct.

Ssp. **mensícola** (Stokes) Munz, comb. nov. [*E. mensicola* Stokes, Leafl. W. Bot. 3: 16. 1941. *E. p.* var. *m.* Reveal.] Lvs. strictly basal, the blades 1–1.5 cm. long, roundish, densely tomentose on both surfaces; bracts not leafy, mostly 2–6 mm. long; invols. few, scattered along the stems, 2–4 mm. long; calyx 3–4 mm. long.—Dry rocky slopes, 5000–8000 ft.; Pinyon-Juniper Wd.; Inyo-White and Panamint mts.; s. Nev.

49. **E. paríshii** Wats. Plate 71, Fig. C. Annuals with 1–3 erect stems 1–3 dm. high, diffusely branched so as to form a dense rounded mass of very slender ultimate branchlets, glaucous, glabrous except for short-stipitate glands above the nodes; lvs. basal, spatulate, 2–6 cm. long, hirsute; peduncles capillary but rigid, 4–12 mm. long; invols. solitary, 5-lobed, ca. 0.8 mm. long; calyx pinkish, 1–2 per invol., minutely puberulent, ca. 0.6 mm. long, the outer segms. ovate, the inner oblong-spatulate; aks. ca. 1 mm. long, dark brown, with a stout somewhat angled beak; $n = 20$ (Reveal, 1967).—Dry gravelly places, 4000–9000 ft.; Pinyon-Juniper Wd., Yellow Pine F., Lodgepole F.; n. L. Calif. to s. Sierra Nevada. July–Sept.

50. **E. parvifòlium** Sm. in Rees. Plate 71, Fig. D. Shrubs with loosely branched decumbent or prostrate stems 3–10 dm. long, thinly floccose, densely leafy to the summit; lvs. fascicled, round-ovate to lance-oblong, thickish, revolute, sometimes cordate at the base, 5–15 mm. long, on shorter petioles, the blades green and glabrate above, densely white-tomentose beneath; flowering stems few, mostly 2–5 cm. long, simple or forked, bearing compact heads 1–2 cm. in diam.; invols. glabrate or somewhat woolly, turbinate-campanulate, 3–4 mm. long; calyx white or tinged rose, glabrous, ca. 3 mm. long, the segms. obovate; fils. pilose basally; aks. ovoid-deltoid, 2.5 mm. long; $2n = 40$ (Stokes &

Stebbins, 1955).—Common on bluffs and dunes along the coast; Coastal Strand, Coastal Sage Scrub; San Diego Co. to Monterey Co. Mostly summer, but with some fls. most of the year. An ill-defined form is var. **pàynei** (Wolf ex Munz) Reveal, with lanceolate lvs. 1.5–3 cm. long; white fls. in heads scarcely 1 cm. in diam. and broad infl. 1–2 dm. in diam.; from Santa Paula Canyon, Ventura Co.

51. **E. plumatélla** Dur. & Hilg. Rather woody at the base, with several erect stems 3–6 dm. high, leafy on lower portions, white-tomentose almost throughout, forked above; lvs. oblanceolate to oblong-lanceolate, 8–15 mm. long, revolute, acute, hoary-tomentose, with short slender petioles; invols. borne on mostly horizontal branches which spread in tiers to one side of the main axis and form an intricate mass ending in short-noded branchlets; invols. solitary but close together, glabrous, sessile, turbinate-cylindric, 2.5 mm. long; calyx glabrous, white, 2 mm. long, the outer segms. obovate, the inner narrower; fils. pilose basally; aks. slightly angled, scaberulous above.—Dry stony places below 4500 ft.; Creosote Bush Scrub to Pinyon-Juniper Wd.; Mojave Desert from n. base of San Gabriel and San Bernardino mts. to Kern R. region and e. to Ariz. Aug.–Oct. Var. **jàegeri** (M. & J.) Stokes ex Munz [*E. nodosum* var. *j.* M. & J.] Upper half of plant green and glabrous. Victorville region e. to Ariz.

52. **E. pusíllum** T. & G. Annual, erect, simple or branched at base, trichotomously branched above, with glabrous glaucous stems 1–3 dm. high; lvs. basal, the blades rounded to oblong-ovate, 1–3 cm. long, densely white-woolly on both surfaces, sometimes greenish above, on petioles ca. as long; peduncles very slender, 1–3 cm. long; invols. broadly turbinate to campanulate, glandular-puberulent, yellow, later with reddish midribs, the outer segms. obovate, the inner oblong, both elongating to ca. 3 mm. in fr.; aks. dark, 0.6–0.8 mm. long, with short stout beaks; $n=16$ (Reveal, 1965).—Common on plains and mesas, mostly 2500–6500 ft.; Creosote Bush Scrub to Pinyon-Juniper Wd.; deserts from Palm Springs region n. to Mono Co.; Nev., Utah. March–July.

53. **E. racemòsum** Nutt. Caudex woody, few-branched, compact; lvs. basal, oblong to oblong-ovate, the blades 2.5–3.5 cm. long, glabrate above and closely white-tomentose beneath, on petioles as long or longer; flowering stems slender, mostly 1.5–3 dm. high, tomentose, trichotomous once or twice, usually not leafy-bracted; invols. solitary and arranged racemosely along the upper branches, tubular-campanulate, 3–4 mm. long, tomentose; calyx cream-white, 2.5–3 mm. long, the segms. oblong-oblanceolate; fils. pilose basally; aks. lance-ovoid, ca. 2 mm. long, 3-angled; $n=18$ (Reveal, 1967).—At ca. 6000–7000 ft.; Pinyon-Juniper Wd.; Inyo-White Mts.; to Nev. July–Aug.

54. **E. renifórme** Torr. & Frém. Annuals with 1–several stems, ± floccose below, glabrous above, divergently trichotomously branched, 0.5–2.5 dm. high; lvs. basal, ± rounded, mostly 1–2 cm. wide and white-woolly on both surfaces, or more glabrous above, with somewhat crisped margins; peduncles 4–15 mm. long, capillary; invols. subcampanulate, glabrous, ca. 2 mm. long, broadly 5-lobed, several-fld.; calyx pale yellow, 1–1.5 mm. long, glandular-puberulent, the outer segms. elliptic-ovate, the inner narrower; aks. scarcely beaked; $n=16$ (Reveal, 1965).—Common in sandy places below 4500 ft.; Creosote Bush Scrub, Joshua Tree Wd.; Mojave and Colo. deserts, L. Calif. to Inyo Co.; s. Nev. March–June.

55. **E. rixfórdii** Stokes. Annuals with several stems 2–4 dm. high; lvs. basal, 1–3 cm. long and wide, cordate to orbicular, margins often crisped, densely white-tomentose beneath, less so to green above; stems erect, the cent. main one stout and 3–20 cm. long, dividing repeatedly at the first node, the outer stems slender and 4–12 cm. long, glabrous; branches diffuse, of varying lengths, the main center the longest, dichotomous, with angles above the 2nd to 4th nodes near to 90°, intricately and divaricately branched so as to form a nearly globose crown with secondary flat-topped branches; invols. turbinate-campanulate, 1–1.5 mm. long, 5-lobed, glabrous; calyx white with greenish midribs, 1.5 mm. long, glabrous, the outer segms. oblong-oval, to bases cordate and gibbous at maturity, the inner narrowly lanceolate and longer; aks. lenticular, 1.5 mm. long; $n=20$ (Reveal, 1968).—Dry sandy to gravelly soils below 5000 ft.; Creosote Bush Scrub to Pinyon-Juniper Wd.; Panamint Mts., Last Chance Mts., Death V. July–Oct.

56. **E. rosénse** Nels. & Kenn. [*E. anemophilum* auth. *E. ochrocephalum* var. *agnellum* Jeps.*]* Caudex densely cespitose, branched, low, the foliage dense olive-green to white-tomentose; lvs. crowded, oblanceolate, 4–15 mm. long, white-tomentose beneath, green-tomentose above, the petioles not exceeding blades; flowering stems scapelike, 1–9 cm. tall, slender, glandular; infl. capitate, 6–15 mm. across; invols. few, turbinate, 3–3.5 mm. long, with 6–8 teeth; calyx yellow, becoming reddish, often glandular, 2–3 mm. long, the segms. obovate, alike; aks. lance-ovoid, 1.5 mm. long; $n=20$ (Reveal, 1967).—Dry granitic and volcanic soils, 9000–12,000 ft.; Montane Coniferous F.; Inyo Co. n.; Nev. July–Aug.

57. **E. ròseum** Dur. & Hilg. [*E. virgatum* Benth.] Erect annuals, simple or with few ascending virgate branches, floccose-tomentose nearly throughout, 1–8 dm. high; lvs. at the base and lower nodes, the basal oblong-oblanceolate, 1–3 cm. long, with equal petioles; invols. cylindric, 4–5 mm. long, 5-toothed, sessile, rather remote, tomentose; calyx yellow to pink or white, 2 mm. long, the outer segms. narrowly obovate, the inner oblong, glabrous; aks. ca. 2 mm. long, with broad beaks; $2n=18$ (Stokes & Stebbins, 1955).—Dry, sandy to rocky places, below 5000 ft.; Chaparral, Foothill Wd., Yellow Pine F., etc.; n. Los Angeles and n. Ventura cos. to s. Ore. June–Oct.

58. **E. rupìnum** Reveal. Like *E. panamintense*, but stems 3–7 from base; plants 3–5 dm. high, lvs. 2.5–4 cm. long, 1.5–2.5 cm. wide; tomentum less dense and with a greenish hue rather than whitish; from *E. racemosum* it differs in 3–5 invols. per branch, not congested, the tubes 3–4 mm. long and wide.—Known from 10 mi. up Wyman Creek, White Mts., Inyo Co.; w. Nev. July–Oct.

59. **E. saxátile** Wats. [*E. bloomeri* Parish.] Plate 71, Fig. E. Perennial herbs with few-branched caudices clothed with the crowded closely white-felted lvs.; lvs. basal, rounded or broadly obovate, the lf.-blades 1–2.5 cm. long, on petioles at least as long; flowering stems ascending, rather slender, 1–3 (–4.5) dm. high, closely tomentose or floccose, forking above with ascending or spreading branches; invols. turbinate, 3–4 mm. long, solitary at the nodes, scattered along the branches, tomentulose, many-fld.; calyx white, pinkish or yellowish, 5–7 mm. long, glabrous, narrowed to a sharply triangular base, the outer segms. oblanceolate, the inner obovate and larger; aks. ca. 2 mm. long, 1 winged and 3-angled, not beaked; $n=20$ (Reveal, 1967).—Dry rocky slopes and ridges, mostly 4000–11,000 ft.; Joshua Tree Wd. to Subalpine F.; San Jacinto and Little San Bernardino mts. w. and n. to San Gabriel, Argus, Panamint mts.; Sierra Nevada to Fresno Co., Santa Lucia Mts.; Nev. May–July.

60. **E. shóckleyi** Wats. Caudex pulvinate, branched, low, the plants with a dense brownish-white tomentum; lvs. crowded, oblanceolate to spatulate, plane, the blades 3–8 mm. long, with petioles at least as long, densely tomentose; flowering stems scapelike, 0–2 cm. long, slender, tomentose; infl. capitate, with 4–10 invols.; invols. campanulate, ± membranaceous but forming distinct tubes, 2–2.5 mm. long, the 5–7 lobes dividing the tube to near the middle; calyx brownish with tan to rusty midribs, densely pubescent without, slightly so within, 2.5–3 mm. long, the segms. obovate, similar; fils. pilose basally; aks. lance-ovoid, ca. 2.5 mm. long, sparsely pubescent.—Dry rocky slopes and clay hills, 6000–8000 ft.; Sagebrush Scrub, Pinyon-Juniper Wd.; Last Chance, Inyo Co.; e. to Utah, w. Colo., n. Ariz. May–July.

61. **E. spergulìnum** ssp. **reddingiànum** (Jones) Munz, stat. nov. [*Oxytheca r.* Jones. *E. s.* var. *r.* (Jones) J. T. Howell.] Erect, slender-stemmed annuals, forking freely above with widely spreading branches 1–5 dm. high, the internodes generally stipitate-glandular; basal lvs. linear, 2–3 cm. long, short-petioled; cauline lvs. linear, whorled, sessile, becoming bracteate above; peduncles filiform, 5–12 mm. long; invols. solitary, glabrous, 0.5–1 mm. long, deeply 4-lobed; calyx white with rose midribs, 1.5–2.5 mm. long; the segms. oblong; anthers 0.2–0.3 mm. long, roundish; aks. ca. 1.5 mm. long, narrowed gradually into narrow beaks.—At 4000–11,000 ft.; Montane Coniferous F.; Mt. Pinos, Ventura Co. and Piute Mts., Kern Co.; n. to Ore., Ida., Nev. June–Sept.

62. **E. thómasii** Torr. Annuals, 1–several-stemmed at the base, 1–2.5 dm. high, glabrous and glaucous nearly throughout, repeatedly trichotomous; lvs. basal, round to

round-reniform, 1–2 cm. long, often glabrate above, densely white-woolly beneath, on petioles 2–5 cm. long; peduncles filiform, 5–15 mm. long; invols. ca. 1 mm. long, deeply 5-lobed, several-fld.; calyx at first yellow, later white to rose, ca. 1 mm. long, externally hispidulous, the outer segms. ovate, with a saclike dilation on each side of the cordate base when mature, the inner segms. spatulate; aks. ca. 1 mm. long, with 3-angled beaks; $n=20$ (Raven et al., 1965).—Common in dry sandy places, below 5000 ft.; Creosote Bush Scrub, Joshua Tree Wd.; Mojave and Colo. deserts n. to Inyo Co.; to Utah, Ariz., L. Calif. March–June.

63. **E. trichopès** Torr. [*E. trichopodum* Torr. ex Benth.] Annuals with 1–several stems from base, trichotomous at the first node and di- or tri-chotomous at upper nodes, with few to many whorled branchlets at each node especially the lower ones, glabrous and glaucous nearly throughout, infrequently inflated at the lower internodes, lvs. in a basal rosette, round to somewhat round-oblong, ± cordate at the base, hirsute, often somewhat crinkled, the lf.-blades 1–2 cm. long, on petioles as long to twice as long; peduncles capillary, 8–15 mm. long; invols. turbinate, scarcely 1 mm. long, glabrous, 2–few-fld., especially 4-lobed; calyx yellow to green, 1 mm. long in anthesis, 1.5 mm. in fr., white-strigulose, the segms. ovate; aks. 1.5 mm. long, narrow-ovoid, 3-angled with stout beaks; $n=16$ (Reveal, 1965).—Common in washes and on mesas, below 5500 ft.; mostly Creosote Bush Scrub, Joshua Tree Wd.; Mojave and Colo. deserts, inner S. Coast Ranges; to Utah, New Mex., Son., L. Calif. April–Aug.

64. **E. thurberi** Torr. Annuals, simple or several-stemmed from the base, diffusely and trichotomously branched, 1–3 dm. high, floccose at least in the lower parts; lvs. basal, oblong-ovate, 1–3 cm. long, densely white-woolly below, glabrate above, on petioles 1–3 cm. long; peduncles capillary, 5–25 mm. long, ± glandular-puberulent, invols. broadly turbinate, glandular-puberulent, 2 mm. long, 5-lobed to near the middle; calyx rose to whitish, 1–1.5 mm. long, glandular-puberulent near the base, the outer segms. roundish or broadly ovate, with a clawlike base, and a tuft of white-cottony tomentum within, the inner segms. narrowly lanceolate; aks. dark, sharp-beaked.—Sandy places below 5000 ft.; Coastal Sage Scrub, Chaparral; Los Angeles region to San Diego Co.; Creosote Bush Scrub, Joshua Tree Wd.; Colo. Desert and occasional on Mojave Desert; to Ariz., L. Calif. April–July.

65. **E. umbellàtum** Torr. ssp. **umbellàtum.** Caudex open or depressed, the plants cespitose to subshrubby, 2–6 dm. high; lvs. spatulate to suborbicular, tomentose below, less so above, mostly less than 2 cm. long and not more than 3–4 times as long as broad, tomentose to glabrate, the petioles short to long; flowering stems erect, 1–3 dm. high, tomentose to glabrate, erect; umbels few-rayed, bractless; invol.-tubes 2–3.5 mm. long, with reflexed lobes as long; calyx yellow, 4–7 mm. long; $n=40$ (Reveal, 1965).—Dry slopes, 8000–9000 ft.; Yellow Pine F.; Piute Mts.; Sierra Nevada, etc. to Wash. and Colo. June–Sept.

KEY TO SUBSPECIES

1. Primary rays of umbels simple, not branched or bracteate in the middle.
    3. Calyx bright yellow, 4–7 mm. long. Piute Mts. .... ssp. *umbellatum*
    3. Calyx whitish to red, or pale yellow.
        4. Lvs. greenish, ± glabrate above; calyx mostly whitish to yellow, or if yellow, then plants distinctly from above 9000 ft.; scapes 0.4–1.2 dm. high.
            5. Calyx cream-color to pale yellow with a tannish midrib. San Bernardino and San Gabriel mts. to e. slope of Sierra Nevada and Last Chance Mts. ..... ssp. *aridum*
            5. Calyx red-brown to pink with a large reddish or purplish midrib. Mts. of ne. San Bernardino Co. and Inyo Mts. .... ssp. *versicolor*
        4. Lvs. densely white-woolly on both surfaces; calyx deep red. San Gabriel Mts.
                                                                                                        ssp. *minus*
1. Primary rays of umbels usually branched, or if not, then with bracts near the middle.
    6. Lvs. sparsely tomentose above and below, the pubescence even on both surfaces. Argus, Inyo-White, Panamint mts. to Utah, sw. Colo. .... ssp. *ferrissii*
    6. Lvs. densely tomentose below, less so above. San Jacinto and San Bernardino mts. to Mt. Pinos region .... ssp. *munzii*

Ssp. **áridum** Stokes. [*E. a.* Greene. *E. u.* var. *dicrocephalum* Gand.] Lf.-blades tomentose on both surfaces, or subglabrous to glabrous above, 1.5–2 cm. long; scapes 1–2.5 dm. high; rays few, 1–3 cm. long; calyx whitish, cream-colored or pale yellow, often with a large tannish midrib, 4–8 mm. long, including the stipe.—Occasional, below 10,000 ft.; Pinyon-Juniper Wd. to Lodgepole F.; San Gabriel Mts., to Inyo-White Mts., Sierra Nevada, n. Nev., w. Utah. June–Aug.

Ssp. **mìnus** (Jtn.) Munz, comb. nov. [*E. u.* var. *minus* Jtn., Bull. Torrey Bot. Club 64: 515. 1937.] Low and densely matted; lf.-blades round-ovate, 4–10 mm. long, densely white-woolly on both surfaces; scapes 3–12 cm. long; rays of umbels 1–3, 5–20 mm. long; calyx 4–5 mm. long, mostly deep red.—Dry stony slopes, 8000–10,000 ft.; Lodgepole F.; San Gabriel Mts. July–Sept.

Ssp. **múnzii** (Reveal) Thorne ex Munz, comb. nov. [*E. umbellatum* var. *munzii* Reveal, Aliso 7: 218. 1970.] Plants spreading and forming mats up to 4 dm. across; lvs. in rather compact rosettes, the blades elliptic, 1–2 cm. long, densely white-tomentose beneath, floccose above, sometimes glabrous when old; rays of umbels compoundly branched with foliaceous bracts at the base of each division, 3–10 cm. long; peduncles slender, 0.5–2 cm. long, tomentose; invols. with tubes 2–3 mm. long and lobes 1–2 mm. long; calyx bright yellow, 3–8 mm. long.—Dry places, 4000–8000 ft.; Montane Coniferous F.; San Jacinto and San Bernardino mts. to Mt. Pinos region.

Ssp. **férrissii** (A. Nels.) Stokes [ssp. *subaridum* Munz.] Lvs., very finely and closely tomentose on both surfaces, not densely tomentose beneath; lf.-blades 1–1.5 cm. long; infl. compound; calyx mostly yellow, occasionally rather pale, 6–7 mm. long.—Dry rocky slopes, 5000–9000 ft.; Pinyon-Juniper Wd.; Inyo-White, Argus and Panamint mts., across the Mojave Desert to s. Nev., Utah, sw. Colo. July–Aug.

Ssp. **versicólor** (Stokes) Munz, comb. nov. [*E. u.* ssp. *aridum* var. *v.* Stokes, Leafl. W. Bot. 31: 17. 1941.] Low matted perennials less than 1.5 dm. high; lvs. lightly tomentose on both surfaces, the blades elliptic, 0.5–1.5 cm. long, the petioles ca. as long; scapes slender, less than 1.5 dm. long, the umbel rays few, less than 1.5 cm. long; calyx reddish-brown to rose or pink, with a large reddish or purplish midrib, 3–6 mm. long.—Occasional, below 9000 ft.; Pinyon-Juniper Wd.; Panamint, Inyo and Grapevine mts.; s. Nev.

66. **E. viridéscens** Heller. [*E. angulosum* var. *v.* Jones.] Annuals with 1–several stems from the base, 1–2 (–3) dm. tall, tomentose almost throughout; basal lvs. lanceolate to obovate, 2–4 cm. long, with slightly shorter petioles, tomentose beneath, sparsely pubescent to glabrate above; cauline lvs. sessile, 0.5–2 cm. long, more reduced above; branches widely spreading; peduncles filiform, 1–2 cm. long, axillary, sparsely finely pubescent with capitate hairs; invols. campanulate, 2–3 mm. long, 2–4 mm. wide, externally pubescent, internally tomentose, with 5 rounded lobes; calyx white, pink or red, often with reddish spots on the outer segms., 1–2.5 mm. long, calyx-segms. dissimilar, the outer broader and blunter; aks. 1–1.5 mm. long, with long 3-angled beaks.—Dry plains and hills; Creosote Bush Scrub, Joshua Tree Wd.; w. Mojave Desert; to inner Monterey Co. May–Oct.

67. **E. wrìghtii** Torr. ex Benth. ssp. **wrìghtii.** Low, highly branched subshrubs from branched woody caudices, 1.5–4 dm. high; lvs. on lower half of plants, crowded, oblanceolate to elliptic, 0.5–1.5 cm. long, mostly entire, densely white-tomentose above and beneath, or rarely greenish above, the petioles 2–5 mm. long; flowering stems several, tomentose or rarely glabrous, up to 2.5 dm. long; infl. racemose, 5–30 cm. long, once or twice dichotomous or trichotomous; invols. solitary, turbinate, 2–2.5 mm. long, ± tomentose; calyx whitish or pink, 2.5–3.5 mm. long, the segms. obovate; aks. 3-angled, somewhat scaberulous above, 2.5–3 mm. long; $2n=34$ (Stokes & Stebbins, 1955).—Gravelly and rocky places, mostly below 5000 ft.; Pinyon-Juniper Wd.; Mojave and Colo. deserts; to Tex. Aug.–Oct.

Ssp. **membranàceum** Stokes. Woody, 2–4 dm. high; petiole bases dilated into glabrate brownish sheaths which clasp the stems; lvs. strongly involute, 3–6 (–10) mm. long; invols. 2–3 mm. long, often glabrous; calyx white to pink, 3–4 mm. long.—Dry stony places below 6000 ft.; Chaparral to Pinyon-Juniper Wd.; Little San Bernardino and Santa Monica mts. to San Diego Co.; L. Calif. Aug.–Oct.

Ssp. **nodòsum** (Small) Munz, comb. nov. [*E. nodosum* Small, Bull. Torrey Bot. Club 25: 49. 1898. *E. w.* var. *n.* Reveal.] Shrubby, densely white-lanate, to 6 dm. high; petiole bases dilated but not extending completely around the stems; lvs. 8–12 mm. long, tomentose, not revolute; invols. 1.5–2.5 mm. long; calyx white, 3–4 mm. long.—Dry stony places below 3000 ft.; Creosote Bush Scrub, Pinyon-Juniper Wd.; w. edge of Colo. Desert e. to Twentynine Palms; L. Calif. Aug.–Nov.

Ssp. **subscapòsum** (Wats.) Stokes. Low and loosely matted, 1–2.5 dm. high; lvs. crowded, 5–12 mm. long, grayish- or brownish-white; flowering stems slender, to 1.5 dm. long, glabrous to floccose; infl. 5–15 cm. long; invols. 1.5–4 mm. long.—Rocky and gravelly places, 5000–11,000 ft.; Montane Coniferous F.; San Jacinto and San Bernardino mts. to Sierra Nevada. July–Oct.

Ssp. **trachygònum** (Torr. ex Benth.) Stokes. [*E. t.* Torr. ex Benth.] Subshrubby, 2–4 dm. high; lvs. elliptic, 1.5–3 cm. long, densely tomentose beneath, usually less so above, not revolute; invols. 2–3 mm. long, ± tomentose; calyx pink or white, 3–4 mm. long.— Exposed and open rocky places, mostly below 6500 ft.; Foothill Wd., Yellow Pine F.; n. Los. Angeles Co. to n. Calif. Aug.–Oct.

4. *Gilmánia* Cov.

Prostrate annual with divergent branches and yellowish herbage. Basal lvs. in a rosette, petioled, 3-nerved, ± obovate. Cauline lvs. 3 at the nodes, the uppermost sessile. Fls. yellow, pedicelled, fascicled at the nodes, without invols. or bracts. Pedicel jointed at base. Calyx 6-parted. Stamens 9. Styles 3. Ak. ovoid-triangular; cotyledons rounded. One sp. (M. French *Gilman*, 1871–1944, California naturalist.)

1. **G. luteòla** (Cov.) Cov. [*Phyllogonum l.* Cov. *Eriogonum l.* Jones.] Stems several, 3–15 cm. long, glabrous except for the thinly pilose infl.; lf.-blades 10–15 mm. long, entire, obtuse; pedicels 2–5 mm. long; sepals ca. 1.5 mm. long, linear-oblong; ak. buff-color, slightly longer than the calyx, smooth, shining.—Below 1200 ft., barren alkaline slopes; Alkali Sink; Death V. March–April. Local and developing in wet years only.

5. *Nemacáulis* Nutt.

Slender-stemmed diffuse but sparingly branched annual. Lvs. mostly radical, spatulate, white-woolly, without stipules. Fls. small, in crowded sessile subglobose heads, perfect, each fl. with a free herbaceous bract. Calyx 6-cleft, enclosing the ak. Stamens 3. Styles 3. Ak. short-ovoid, obscurely 3-angled. One sp. (Greek, *nema*, thread, and *kaulos*, stem.)

1. **N. denudàta** Nutt. [*N. nuttallii* auth.] var. **denudàta**. Stems several from the base, prostrate or usually ascending, reddish, 1–3.5 dm. long, glabrate; lvs. oblance-spatulate, 2–5 cm. long, woolly, ± crisped; cauline lvs. bractlike; bracts of infl. oblong-obovate, 2 mm. long, whorled, glabrous without, woolly within; fls. yellowish or pinkish, 1 mm. long, short-petioled; ak. brown, shining, ovoid, ca. 0.6 mm. long, acute.—Sandy places such as sea beaches, dunes, etc.; Coastal Strand; Los Angeles Co. to L. Calif.; rare about Palm Springs. April–Sept.

Var. **grácilis** Goodman & Benson. Wool of infl. very long and copious, almost or quite concealing fls.—Creosote Bush Scrub, w. Colo. Desert. March–May.

6. *Oxýria* Hill. MOUNTAIN SORREL

Low glabrous perennial, with acid juice. Caudex thick, covered with persistent lf.-bases. Stems erect, slender. Lvs. alternate, but mostly basal, long-petioled, round or reniform, with cylindric stipule-sheaths. Fls. perfect, small, whorled, arranged in compact panicles. Sepals 4, the outer larger than the inner. Stamens 6, included, the fils. short-subulate. Ovary 1-celled, 1-ovuled. Styles 2, short; stigmas fimbriate. Aks. lenticular, nearly flat, broadly winged, the body ovate. Two spp. (Greek, *oxus*, sour.)

1. **O. dígyna** (L.) Hill. [*Rumex d.* L.] Caudex branched; stems scapiform, simple or few-branched, 6–25 cm. high; basal lvs. round-reniform, 1.5–3 cm. wide; petioles 5–12 cm. long; pedicels slender, recurved; sepals red or greenish, 1.5–2 mm. long, the inner erect in fr. and becoming 4–6 mm. long, the outer reflexed; aks. ca. 3 mm. long; $n=7$ (Taylor & Brockman, 1966).—Rocky places, 9400–10,700 ft.; San Jacinto and San Bernardino mts.; Sierra Nevada and White Mts. n.; to Arctic Am., Atlantic Coast, Eurasia. July–Sept.

7. Oxythèca Nutt.

Slender-stemmed dichotomously branched annuals with small stipitate glands at the nodes. Lvs. rosulate at the base. Bracts ± connate, often in 3's, foliaceous. Invols. few-fld., ± pedicellate, campanulate to turbinate, 4–5-cleft, each lobe terminated by a bristle or awn, or 7–20-ribbed and each rib ending in a long awn. Fls. pedicelled; calyx 6-parted, glabrous or pubescent. Stamens 9, inserted at base of calyx. Ak. ± ovoid, the cotyledons accumbent, orbicular. Ca. 8–9 spp. of w. N. and S. Am. (Greek, *oxus*, sharp, and *theke*, a case, referring to the spiny invol.)

Invol. not lobed, its tube short, 7–30 ribbed, each rib ending in a long bristle ... 5. *O. parishii*
Invol. lobed.
  Involucral divisions unequal, 5; plant prostrate; calyx woolly, yellow .... 4. *O. luteola*
  Involucral divisions subequal; plants erect or ascending; fls. not yellow.
    Bracts of upper nodes connate-perfoliate into a conspicuous disk 1–2 cm. broad; invols. sessile or nearly so .... 6. *O. perfoliata*
    Bracts of upper nodes not so united; invols. peduncled.
      The invols. 4-lobed.
        Lvs. revolute, linear to linear-oblanceolate, acute, hirsutulous. Inyo Co. n.
          2. *O. dendroidea*
        Lvs. plane, spatulate, rounded at apex, hispidulous. San Bernardino Co. . 8. *O. watsonii*
      The invols. 5-lobed.
        Invols. deeply lobed.
          Calyx-lobes entire, greenish or reddish, very short .... 1. *O. caryophylloides*
          Calyx-lobes deeply cleft, white, almost as long as tube .... 7. *O. trilobata*
        Invols. very shallowly lobed, reddish, forming a round disk with scarious margins
          3. *O. emarginata*

1. **O. caryophylloìdes** Parry. Erect, 1- or several-stemmed at base, diffusely branched above, 1–4 dm. high, minutely glandular; lvs. oblong-spatulate, or wider, 2–5 cm. long, subglabrous, short-petioled; bracts 3-parted; invols. in forks and terminal, peduncled, deeply 5-parted, glabrous, the divisions oblanceolate to narrow-oblong, 3 mm. long, with awns an additional 1 mm. long; fls. 2–3, calyx greenish, strigose, barely 1 mm. long, the lobes short, entire; aks. ca. 1 mm. long.—Occasional, 4000–7000 ft.; mostly Yellow Pine F.; Ventura Co., San Gabriel Mts. to San Jacinto Mts. July–Sept.

2. **O. dendroìdea** Nutt. Stems erect, then dichotomously or trichotomously branched, 1.5–4 dm. high, the ultimate branchlets very fine, somewhat stipitate-glandular; lvs. basal, linear to linear-oblanceolate, 1.5–3 cm. long, acute, short-petioled, thinly hirsute; bracts entire; invols. solitary, narrowly turbinate, 2–4 mm. long (including spines), 4-lobed, the spines 1–2 mm. long; fls. 2–3; calyx pinkish to whitish, 1.5 mm. long, hispidulous.—Dry sandy and gravelly flats and slopes, 4000–7000 ft.; Sagebrush Scrub, Pinyon-Juniper Wd., Yellow Pine F.; Inyo-White Mts. n. to Wash., Wyo., Nev. June–Sept.

3. **O. emarginàta** Hall. Stem or stems erect, then dichotomous or trichotomous, with widely spreading branches 5–15 cm. long, reddish, glandular-pubescent; lvs. oblanceolate, pubescent, the blades 1.5–4.5 cm. long, on petioles ca. as long or shorter; bracts deeply 3-lobed; invols. mostly peduncled, broadly funnelform, shallowly 5-lobed, reddish, 4–6 mm. long, scarious-margined, the awns ca. 1 mm. long; fls. 3–4; calyx whitish,

6-parted, pubescent, with fimbriate lobes.—Dry slopes, 4000–8000 ft.; Pinyon-Juniper Wd., Yellow Pine F.; San Jacinto Mts. and Santa Rosa Mts., Riverside Co. July–Aug.

4. **O. luteòla** Parry. Prostrate with several stems from base, 0.3–1 dm. long, yellowish-green, pubescent; lvs. at base and in pairs at lower nodes, the blades green above, woolly beneath, rounded, 2–5 mm. long, on petioles 10–20 mm. long; bracts linear, acerose, 4–5 mm. long; invols. in forks and at the nodes, 5-parted, 3–5 mm. long, the divisions unequal, lance-linear, long-awned; fls. several; calyx 1 mm. long, the tube globose, woolly, the lobes glabrous, yellow; aks. ca. 1 mm. long, stout-beaked.—Alkaline flats, dry lakes, etc.; Alkali Sink, V. Grassland, Creosote Bush Scrub, etc.; w. Mojave Desert to Mono Co. and Cent. V. May–Aug.

5. **O. paríshii** Parry ssp. **paríshii.** Stems erect with ascending branches, 2–6 dm. high, glaucous, glabrous except for the remote internodes; lvs. broadly spatulate, 2–4 cm. long, short-ciliate; bracts 3-lobed, 1–2 mm. long; invols. solitary on slender pedicels 1.5–5 cm. long, broadly turbinate, the basal part scarcely lobed, ca. 2 mm. long, and 14–30-ribbed, each rib ending in a bristle 3–5 mm. long; calyx whitish, 6-cleft, the segms. oblong, entire, strigulose; aks. ca. 2 mm. long, the base and apex quite contracted.—Dry granitic slopes and flats, 4000–8300 ft.; mostly Yellow Pine F.; Topatopa Mts., San Gabriel and San Bernardino mts. June–Sept.

Ssp. **abrámsii** (McGreg.) Munz, comb. nov. [*Acanthoscyphus a.* McGreg., Bull. Torrey Bot. Club 36: 605. 1909.] Invol. plainly lobed and with 4–12 bristles, these mostly 2–3 mm. long.—Similar situations, Big Pine Lookout, Santa Barbara Co., Mt. Pinos to San Bernardino Mts. June–Sept.

6. **O. perfoliàta** T. & G. Plate 71, Fig. F. Stem or stems erect, then dichotomous or trichotomous with horizontal branches 1–3 dm. long, green then reddish, glandular in lower part of each internode; basal lvs. 1.5–4 cm. long, oblong-oblanceolate; bracts of upper nodes mostly 3 and connate-perfoliate into a slightly angled cupulate disk 1–2 cm. broad, with short spinose tips; invols. solitary, narrowly turbinate, 3–4 mm. long, 4-lobed to middle, the lobes ending in spines ca. 3 mm. long; fls. several; calyx whitish, 1.5 mm. long, minutely scaly; aks. ca. 2 mm. long, with stout stipe and beak.—Sandy or gravelly places, 2400–6000 ft.; Creosote Bush Scrub to Pinyon-Juniper Wd.; Mojave Desert to Lassen Co.; Nev., Ariz. April–July.

7. **O. trilobàta** Gray. Stem or stems erect, trichotomous, with spreading forked branches 1–6 dm. long, sparingly glandular; lvs. basal, spatulate, 1–6 cm. long, somewhat hairy; bracts deeply 3-lobed, the lobes lanceolate, 2–3 mm. long, spine-tipped; invols. mostly on slender peduncles, broadly turbinate, deeply 5-lobed, the lobes lanceolate to subovate, ca. 3 mm. long, with terminal bristly awns ca. as long; fls. few; calyx white with reddish base, the segms. 3-cleft into lanceolate lobes with erosulate sides; aks. ca. 1.5 mm. long, beaked.—Fairly frequent, dry slopes, mostly 3000–7000 ft.; Chaparral, Joshua Tree Wd., Yellow Pine F.; San Gabriel Mts. to Little San Bernardino Mts. and n. L. Calif. July–Sept.

8. **O. watsònii** T. & G. Stems erect, dichotomously branched, glaucous, 1–2 dm. high, sparingly stipitate-glandular; lvs. spatulate, 1–2.5 cm. long, hispidulous; bracts lance-ovate, awned; invols. turbinate, 4-lobed, ca. 4 mm. long, including the awns; fls. several; calyx white, puberulent, 1–1.5 mm. long.—Joshua Tree Wd.; Cushenberry Springs, sw. Mojave Desert; Nev. May–July.

8. *Polýgonum* L. KNOTWEED. SMARTWEED

Annual or perennial herbs, sometimes shrubs, with fibrous roots or thick rootstocks and often swollen joints. Lvs. entire, alternate; stipules usually scarious, sheathing and forming an ocrea, with which the petiole may be jointed. Fls. on jointed pedicels, variously disposed, mostly perfect. Calyx 4–6 (mostly 5)-parted, the divisions often petaloid and brightly colored, enlarging, persistent and surrounding the lenticular or 3-angled ak. Stamens 3–9. Style and stigmas 2 or 3, deciduous. Embryo curved halfway around the albumen and placed in a groove on the outside of it. Ca. 200 spp. of wide geographical

Polygonum

range and falling in several natural groups often recognized as genera. (Greek, *poly*, many, and *gonu*, knee or joint, because of the thickened joints on the stem.)

1. Fls. in axillary fascicles, or in spikes with leafy bracts; ocreae hyaline and finally mostly 2-lobed and lacerate.
    2. Lvs. jointed with the ocreae and 1-nerved; fls. mostly 2 or more in the axils, short-pedicelled.
        3. Stems terete or nearly so, not sharply angled, often spreading to prostrate.
            4. Aks. dull or scarcely shining; fls. borne in axils of ± reduced lvs.; calyx with pinkish to purplish margins.
                5. Branch-lvs. much smaller than stem-lvs.; perianth persistent, divided almost to base; fr. trigonous with concave sides .... 4. *P. aviculare*
                5. Branch-lvs. not much smaller than stem-lvs.; perianth divided for half its length; fr. with 2 sides convex, 1 side concave .... 2. *P. arenastrum*
            4. Aks. smooth and shining.
                6. Upper lvs. bracteate; calyx with pinkish margins .... 3. *P. argyrocoleon*
                6. Upper lvs. not reduced to bracts; calyx with yellowish margins 17. *P. ramosissimum*
        3. Stems and branches strongly angled, usually suberect.
            7. Pedicels recurved at least in fr. .... 8. *P. douglasii*
            7. Pedicels erect.
                8. Fls. in small axillary clusters, not in spicate infl.; plants 1–4 dm. high; calyx 3–4 mm. long .... 8. *P. douglasii*
                8. Fls. in terminal leafy-bracteate spikes; plants 3–8 cm. long; calyx ca. 2 mm. long
                    11. *P. kelloggii*
    2. Lvs. not jointed with the ocreae, 3-nerved; fls. solitary in the axils, sessile, the calyx 1.5 mm. long. Cuyamaca Mts. .... 13. *P. parryi*
1. Fls. fascicled in terminal dense to open spikelike racemes or in panicles without leafy bracts; ocreae hyaline or lacerate.
    9. Stems not twining; infl. elongate, spiciform, not in ample panicles or lvs. not cordate-sagittate.
        10a. Stems simple, from fleshy rootstocks; lvs. basal and long-petioled or cauline and short-petioled or sessile, infl. a single terminal spike. Mountain meadows, etc.
            6. *P. bistortoides*
        10a. Stems branched; lvs. all cauline and similar; spikes terminating main stems and branches.
            11. Ocreae eciliate or nearly so, at least in maturity.
                12. Perennial plants with creeping rhizomes and stolons; in and about water
                    1. *P. amphibium*
                12. Annuals; not aquatic.
                    13. Peduncles with stalked glands; spikes erect, oblong-cylindric; aks. suborbicular, 2.2–2.5 mm. in diam. .... 14. *P. pensylvanicum*
                    13. Peduncles with or without sessile glands; spikes often arching, slender-cylindric; aks. subovate, 1.8–2.2 mm. long .... 12. *P. lapathifolium*
            11. Ocreae fringed with bristly cilia.
                14. Plants annual.
                    15. Calyx punctate with glands; spikes slender, arching, white; aks. shining
                        16. *P. punctatum*
                    15. Calyx not glandular-punctate; spikes dense, erect .... 15. *P. persicaria*
                14. Plants perennial.
                    16. Internodes fusiform-thickened; ocreae somewhat ciliate when young. Colorado River bottoms .... 9. *P. fusiforme*
                    16. Internodes cylindric; ocreae bristly-ciliate .... 10. *P. hydropiperoides*
    9. Stems twining; infl. of heads or panicles and the lvs. broad, or the infl. of small axillary clusters and terminal spike, with the lvs. cordate-sagittate.
        10a. Plants annual; lvs. ovate-sagittate; calyx 3.5–4 mm. long .... 7. *P. convolvulus*
        10a. Plants perennial; lvs. not sagittate; calyx 4–5 mm. long. .... 5. *P. baldschuanicum*

1. **P. amphíbium** L. var. **stipulàceum** Coleman. [*P. natans* East. *P. hartwrightii* Gray.] WATER SMARTWEED Plate 72, Fig. A. Perennial with slender rhizomes; stems elongate, leafy, glabrous, simple; lvs. lanceolate to lance-oblong, or even broader in the floating form, 5–10 cm. long, petioled, obtusish; peduncles glabrous; fls. in thick terminal short-cylindric to -ovoid spikes 3–4 cm. long, rose; calyx 4–5 mm. long; stamens 5; aks.

Plate 72. Fig. A, *Polygonum amphibium* var. *stipulaceum*. Fig. B, *Polygonum bistortoides*. Fig. C, *Polygonum douglasii* var. *johnstonii*. Fig. D, *Rumex crassus*, ak. with 1 callosity. Fig. E, *Calandrinia ciliata* var. *menziesii*. Fig. F, *Calandrinia maritima*.

lenticular, 2.5–3 mm. in diam.—Ponds and lakes below 10,000 ft.; many Plant Communities; San Diego Co. n. (as Laguna Mts., San Bernardino Mts., etc.), occurring both as strand and floating forms; to Alaska and Atlantic Coast. July–Sept.

Var. **emérsum** Michx. [*P. coccineum* Muhl. *P. e.* Britton.] Plants like var. *stipulaceum* but rather coarse, glabrous to strigose; lvs. to 2 dm. long, lanceolate to lance-ovate, somewhat acuminate; spikes mostly 4–10 cm. long, on somewhat glandular peduncles; calyx ca. 3 mm. long.—Terrestrial or with floating forms; lakes and ponds, San Diego Co. to San Bernardino Co. as at Victorville, San Bernardino Mts., etc., Ventura Co.; to B.C. and Atlantic Coast.

2. **P. arenástrum** Bor. Forming dense prostrate mats 1–16 dm. across; lvs. to 20 mm. long, 5 mm. wide; infl. 2–3-fld.; calyx greenish white or pink; fr. 1.5–2.5 mm. long, brown to black.—San Clemente and Santa Catalina ids.; near Murrieta, Riverside Co., Santa Cruz Id. Introd. from Eu.

3. **P. argyrocòleon** Steud. ex Kunze. Pale green annual, erect with ascending striate branches, rather lax; lvs. linear, 1–3 cm. long, narrowed at both ends; ocreae silvery above and lacerate, brown toward base; upper lvs. bracteate, so that infl. seems racemose, although fls. are several at a node; calyx green with roseate margins, 1.5 mm. long; stamens 6; aks. 3-angled, ovoid, smooth and shining, ca. 1.5 mm. long; $2n=40$ (Heiser & Whitaker, 1948).—Becoming rather widespread in Calif. as a weed about orchards, alfalfa fields, etc.; native of Asia. June–Oct.

4. **P. aviculàre** L. Annual with prostrate or ascending bluish-green slender stems 1–12 dm. long; lvs. lanceolate to almost oblong, 5–20 mm. long, blue-green, scattered to approximate, not much reduced upward; stipule-sheaths silvery, soon torn; fls. 1–5 in axillary clusters; calyx 2–3 mm. long, greenish with pinkish to purplish margins; stamens 8, rarely 5; aks. dull or slightly shiny, 2–2.5 mm. long, somewhat roughened; $2n=40$, 60 (Löve & Löve, 1942).—Common weed in dooryards, waste places, etc.; natur. from Eurasia. May–Nov. Variable.

5. **P. baldschuánicum** Regel. Tall vigorous hardy perennial vine, twining, glabrous, 6 m. or taller; lvs. ovate to oblong-ovate, the blade 3.5–5 cm. long, acute to obtuse; fls. ca. 5 mm. long, in large dense drooping panicles, bright rose, the 3 outer segms. winged; growing vigorously in Mission Creek Canyon, Santa Barbara; natur. from USSR.

6. **P. bistortoìdes** Pursh. Plate 72, Fig. B. Perennial with a thick horizontal rootstock and several erect slender simple glabrous stems 2–6 dm. high; lvs. mostly near base, the lower oblong to oblanceolate, 1–2.5 dm. long, on petioles ca. as long and with broad midrib; upper lvs. sessile, lanceolate, reduced; ocreae narrow-cylindric, 3–7 cm. long with oblique summit; spikes terminal, thick-cylindric, 1–6 cm. long; calyx pink or white, 4–5 mm. long; stamens exserted; aks. shining, angled, 3.5–4 mm. long.—Wet meadows and along streams, 5000–9000 ft.; Montane Coniferous F.; San Jacinto and San Bernardino mts.; Sierra Nevada to Alaska and Atlantic Coast. June–Aug.

7. **P. convólvulus** L. BLACK BINDWEED. Slightly roughish annual with twining stems 2–10 dm. long; lvs. ovate-sagittate, 2–5 cm. long, on shorter petioles; ocreae short, with entire margins; fls. few, in axillary clusters or in subracemose infl. at ends of branches; calyx green, 3.5–4 mm. long; aks. dull, minutely roughened, 3.5–4 mm. long; $2n=20, 40$ (Löve & Löve, 1942).—Occasional weed at elevations below 5000 ft.; natur. from Eu. May–Sept.

8. **P. doúglasii** Greene var. **doúglasii**. Slender erect annual, rather pale green, loosely few-branched, 1–4 (–6) dm. high, subglabrous; lvs. lance-oblong to linear, rather remote, 1–4 cm. long, subsessile; fls. 1–3 per axil, drooping, reddish; pedicels 2–3 mm. long, deflexed; calyx 3–4 mm. long; stamens 8, aks. 3-angled, smooth, shining, ca. 3.5 mm. long.—Fairly dry areas as at edge of meadows, 5000–7500 ft.; Montane Coniferous F.; San Diego Co. n.; to B.C. and Quebec. June–Sept.

Var. **johnstònii** Munz. [*P. sawatchense* Small.] Plate 72, Fig. C. Pedicels erect, ca. 1 mm. long; calyx green, 2–3 mm. long, pale on margins.—Dryish places, 5000–9600 ft.; Montane Coniferous F.; Santa Rosa Mts. to Mt. Pinos; Sierra Nevada to Wash., Rocky Mts. July–Sept.

9. **P. fusifórme** Greene. Perennial, the stems dark red, ascending, 6–10 dm. high, the internodes fusiform-thickened above the nodes; lvs. lance-linear, 7–13 cm. long, subsessile, glabrous or strigulose along the midribs; ocreae strigulose, 1–2 cm. long; racemes dense, narrow-cylindric, mostly 3–4 cm. long, not more than 6 mm. thick; calyx red in bud, white at anthesis, 2.5 mm. long; stamens 4–5; aks. shining, ca. 2 mm. long.—Moist places, Colo. River Valley; Ariz. July–Oct.

10. **P. hydropiperoìdes** Michx. var. **asperifòlium** Stanf. Perennial, the stems glabrous, erect or ascending, 3–10 dm. high; lvs. lanceolate, 5–15 cm. long, somewhat scabrous-strigulose beneath, scabrous-ciliate; ocreae cylindric, truncate, strigose, bristly ciliate; racemes slender, somewhat interrupted, 2–7 cm. long, erect, in terminal panicles; calyx white to rose, 2–3 mm. long; stamens 8, included; aks. 2.5–3 mm. long, shining.—Moist places, mostly below 4000 ft.; many Plant Communities; cismontane; to Cent. Valley. June–Oct.

11. **P. kellóggii** Greene. Glabrous often tufted annuals 3–8 cm. tall, with very short internodes; lvs. linear or lance-linear, 0.5–1 cm. long, spreading, not imbricate, acute, sessile; bracts green like small lvs., mostly 3–5 mm. long; infl. spicate, 0.5–3 cm. long; pedicels scarcely 1 mm. long; calyx green, almost 2 mm. long, with white margins; stamens 3; aks. ovoid, 1.5 mm. long, dull.—Usually in damp, silty or gravelly places, 4500–9500 ft.; Montane Coniferous F.; San Jacinto and San Bernardino mts., Mt. Pinos; Sierra Nevada to Wash., Colo. June–Sept.

12. **P. lapathifòlium** L. Erect or ascending annual, the stems simple or branched, usually swollen at nodes and 5–15 dm. high; ocreae cylindric, peduncles often with subsessile glands; spikes slender, cylindric, 1–6 cm. long, often nodding; calyx white to pink or purplish, ca. 2 mm. long; $2n = 22$ (Löve & Löve, 1948).—Common in moist places below 5000 ft.; several Plant Communities; throughout temperate N. Am. Eurasia. June–Oct.

13. **P. párryi** Greene. Compact tufted annual 2–5 cm. high, sometimes more lax and taller, leafy; lvs. narrowly linear, 0.5–2 cm. long, cuspidate, 3-nerved; ocreae much lacerate and concealing fls.; spikes dense, the fls. solitary at each axil, sessile; calyx 1.5 mm. long; stamens 8; aks. oblong-ovoid, 1.5 mm. long, brown, angled, smooth, shining. —Reported from Cuyamaca Peak; Sierra Nevada to Wash. June.

14. **P. pensylvánicum** L. PINKWEED. Ascending to erect annual to ca. 1 m. high, the upper stems and peduncles stipitate-glandular; lvs. lanceolate, 0.4–2 dm. long, acuminate, ciliate, otherwise mostly glabrous; ocreae cylindric, without cilia; fls. in panicles of racemes, these erect, compact, not interrupted, oblong-cylindric; bracts of infl. subglabrous; calyx reddish to purplish, 3–4 mm. long; stamens included; aks. lenticular, 2.2–2.5 mm. in diam., somewhat shining.—Occasional introd. from the e. U.S. July–Sept.

15. **P. persicària** L. LADY'S THUMB. Almost glabrous annual, erect or ascending, 2–6 dm. high, simple or branched; lvs. lanceolate, 3–9 cm. long, acuminate, short-petioled; ocreae cylindric, truncate, with cilia 1–2 mm. long; spikes erect, densely fld., paniculate; calyx pink to purplish, not glandular-punctate, 2.5–4 mm. long; aks. shining, 2–3 mm. long; $2n = 44$ (Jaretzky, 1928).—Moist waste places, below 5000 ft.; to Wash., Atlantic Coast; native of Eu. June–Nov.

16. **P. punctàtum** Ell. Commonly perennial with slender branching rootstocks; stems erect or ascending, 3–10 dm. high, simple or branched, subglabrous; lvs. lanceolate to lance-elliptic, 5–10 cm. long, punctate, acutish to acuminate; ocreae scarious, ciliate; racemes in a naked or leafy panicle, linear-cylindric, 1.5–5 cm. long, suberect, loosely fld. at base; calyx greenish, conspicuously glandular-punctate, 2–3 mm. long; aks. shining, 2.5–3.5 mm. long.—Common in low moist places; cismontane and at desert edge as at Victorville; to Wash., Atlantic Coast; S. Am. July–Oct.

17. **P. ramosíssimum** Michx. Rather bushy erect yellowish-green annual 2–10 dm. high; lvs. narrowly lanceolate to linear, 2–5 cm. long, the upper not bracteate; ocreae silvery, deeply cut; fls. several in each upper axil; calyx yellowish or with yellowish margins, ca. 3 mm. long; aks. ovoid, 3-angled, black, 2.5–3.5 mm. long.—Dry open places,

where occasional as a weed (Blythe, Baldwin Lake, San Diego Co.); to Wash. and Atlantic Coast. July–Sept.

9. Pterostègia F. & M.

Annual with dichotomous diffusely branched slender stems. Lvs. opposite, fan-shaped to broadly elliptical, entire or 2-lobed. Bracts foliaceous, often reddish, each subtending a fl., 2-lobed and enlarged in fr., scarious and reticulate, loosely enclosing the ak., 2-gibbous on the back. Calyx reddish, mostly 6-parted into oblong-lanceolate segms. Stamens 3 or 6. Style 3-cleft. Aks. ovoid-triangular, glabrous. One sp. (Greek, *pteron*, a wing, and *stege*, covering, referring to the bract.)

1. **P. drymarioìdes** F. & M. Stems pubescent, mostly prostrate, 1–4 dm. long; lvs. short-petioled, 0.5–2 cm. long, often broader; bracts 1.5–2 mm. long in fr.; calyx ca. 1 mm. long; ak. ca. 1 mm. long; $2n = 28$ (Sugiura, 1936).—Common, especially in shade of shrubs, rocks, etc., below 5000 ft.; Many Plant Communities; cismontane to Ore., L. Calif.; occasional in desert. March–July.

10. Ruméx L. Dock. Sorrel

Annual or mostly perennial herbs with simple or branched grooved stems. Fls. numerous, mostly greenish, small, usually crowded and commonly whorled in panicled racemes. Lvs. often mostly basal, the cauline alternate, the petioles sheathing at base. Calyx of 6 sepals, the 3 outer herbaceous, sometimes united at base, usually appressed to the margins of the inner; the 3 inner larger, somewhat colored and convergent over the 3-angled ak., veiny, called *valves* in fr., often bearing a grainlike tubercle (callosity) on the back. Stamens 6. Styles 3; stigmas tufted. Fls. perfect or imperfect, varying with the spp. Ca. 140 spp., widely distributed. (The ancient Latin name.)

(Rechinger, K. H., Jr. The North American species of Rumex. Field Mus. Pub. Bot. 17 (1): 1–151, 1937.)

1. Plant dioecious, low, slender, acid to the taste; valves small, without callosities.
  2. Valves not connate with the nutlet or ak. .... 1. *R. acetosella*
  2. Valves connate with the nutlet or ak. .... 2. *R. angiocarpus*
1. Plant monoecious, usually tall and weedy, not acid to the taste.
  3. Stems ascending to decumbent, or axillary from a creeping rhizome, with axillary shoots or lf.-tufts; without basal lvs. even when young.
    4. Valves without callosities; mostly montane .... 3. *R. californicus*
    4. Valves, at least one or two often callosity-bearing.
      5. Callosity occupying nearly the whole breadth of the valve; plants drying brownish.
        6. Valves relatively large, 4–5 mm. long; lvs. 2–3 times longer than broad
          5. *R. crassus*
        6. Valves much smaller; lvs. narrower .... 14. *R. salicifolius*
      5. Callosity much narrower than the breadth of the valve; plants drying yellowish green.
        7. Low subaquatic plant with flexuous branches; lvs. usually elliptic or ovate, often papillose; valves 2.1–2.5 mm. long, with small callosities .... 10. *R. lacustris*
        7. Ascendent or suberect; lvs. usually lanceolate, always glabrous; valves usually larger .... 15. *R. triangulivalvis*
  3. Stems erect, without axillary shoots or tufts; with basal lvs. when young.
    8. Valves 8–14 mm. long, without callosities; nutlets 3–6 mm. long. .. 8. *R. hymenosepalus*
    8. Valves smaller, at least one valve with a distinct callosity.
      9. Valves entire.
        10. Lvs. small, flat, truncate; valves very small, scarcely broader than the thick callosities; whorls remote, nearly all foliate .... 4. *R. conglomeratus*
        10. Lvs. large, somewhat crisped or undulate, often narrowed at base, seldom truncate; valves large, much broader than the callosities; only the lower whorls foliate and occasionally remote.
          11. Lvs. rather narrow, broadest at middle, mostly much undulate, gradually narrowed to base; petiole somewhat canaliculate on upper side; valves 4–6 mm. long .... 6. *R. crispus*

11. Lvs. broader, often broadest below the middle, abruptly narrowed toward base, truncate or slightly cordate, less undulate; petiole flat on upper side; valves larger; callosities smaller in proportion to valves. Rare .... 9. *R. kerneri*
9. Valves denticulate.
    12. Plants perennial.
        13. Lvs. small; pedicels short, not longer than fr., articulate at middle . 13. *R. pulcher*
        13. Lvs. larger; pedicels long, slender, nearly twice as long as fr., articulate toward base .... 11. *R. obtusifolius*
    12. Plants annual.
        14. Lvs. up to 3 times longer than broad; plants tall .... 16. *R. violascens*
        14. Lvs. more than 3 times as long as broad.
            15. Valves triangular; callosity fusiform, narrowed .... 7. *R. fueginus*
            15. Valves ovate; callosity thickish, rounded .... 12. *R. persicarioides*

1. **R. acetosélla** L. SHEEP SORREL. Plant very similar to the next; valves in fr. free, not connate with the ak.; $2n=42$ (Ono, 1930).—Weed, less common than the next, mostly at 3000–7000 ft.; Chaparral, Yellow Pine F.; Cuyamaca Mts. to San Gabriel Mts.; to Alaska, Greenland; introd. from Eurasia. March–Aug.

2. **R. angiocárpus** Murbeck. Perennial with running slender rootstocks; stems tufted, slender, erect, or with decumbent base, 1–4 dm. high, glabrous; lvs. lanceolate or linear, the blades 2–6 cm. long, at least the lower with hastate base, the upper usually entire; lower petioles often longer than blades; fls. nodding, in naked terminal panicles, yellowish, not reddish in age; pedicels jointed at summit; calyx ca. 1 mm. long, green, the sepals scarcely enlarged in fr., united with the nutlet into a single body; $n=7$ (Löve, 1941).—Weed, especially in damp and cooler places, and largely from Santa Barbara Co. n.; widely natur. from Eu. March–Aug.

3. **R. califórnicus** Rech. f. [*R. salicifolius* f. *ecallosus* J. T. Howell.] Perennial from stout taproot; stems 2–6 dm. high, ascending or suberect, slender but firm, often branching, leafy; lvs. plane, glabrous, lance-linear to lanceolate, to 10 cm. long; petioles as long as the width of the blades; panicle-branches slender, divergent, the lower multiflowered glomerules ± remote, the upper approximate, or all approximate; pedicels jointed near base; outer sepals ovate-lanceolate, ca. 1.8 mm. long, the valves in fr. 3 mm. long, 2.5 mm. wide, triangular, truncate at base, somewhat denticulate, veiny, brown-red.—Widespread in moist places below 9000 ft.; many Plant Communities; mts. from San Diego Co. to Ventura Co.; to L. Calif., Nev., Ariz., n. Calif. May–Sept.

4. **R. conglomerátus** Murr. Perennial with taproot; stems smoothish, rather slender, 8–15 dm. high; lower lvs. oblong to lance-oblong, cordate at base, 1–2 dm. long, long-petioled, slightly crisped, obtuse, the upper reduced; panicle leafy, lax, 1–5 dm. long, the branches interrupted-spicate, subsimple, 1–3 dm. long; pedicels ca. 4 mm. long, stoutish, geniculate and jointed near base; valves 2.5–3 mm. long, oblong, obtuse, entire, each with a large smooth callosity; aks. ca. 2 mm. long; $2n=18$ (Sugiura, 1936).—Low moist places, cismontane valleys at low elevs.; natur. from Eu. Catalina Id. April–Oct.

5. **R. crássus** Rech. f. Plate 72, Fig. D. Stems procumbent or flexuous-ascending, 2–5 dm. long; lvs. fleshy, drying coriaceous, lance-oblong to lance-ovate, ca. 2.5–3.5 times as long as wide, broadly cuneate at base; panicle-branches short, simple, compact, the multiflowered glomerules contiguous; pedicels jointed near base; outer sepals ca. 2 mm. long, the valves in fr. 4–5 mm. long, ovate, the margin minutely and irregularly crenate-denticulate, one valve with a prominent callosity which almost covers the surface of the valve; $2n=20$ (Sarkar, 1958).—Coastal dunes and rocky ocean bluffs; Coastal Strand, Coastal Sage Scrub; Los Angeles Co. n. to Wash. May–Sept.

6. **R. críspus** L. CURLY DOCK. Perennial with taproot; stem smooth, rather slender, 5–12 dm. high; lower lvs. lanceolate to oblong-lanceolate, 1–3 dm. long, with long petioles, strongly crisped marginally, acute, the upper reduced; panicle strict, narrow, 1–5 dm. long; pedicels 5–10 mm. long, with swollen joints near the base; outer sepals barely 1 mm. long; the valves in fr. 4–6 mm. long, round-ovate, subcordate, entire to minutely erose, usually with 3 equal or unequal callosities, rarely with only 1; aks. 2 mm. long; $2n=40$ (Löve, 1967), 60 (Jensen, 1937).—Common weed in low places, through much of

N. Am.; native of Eurasia. On most of our islands and occasional on desert. Most of year.

7. **R. fueginus** Phil. Annual or biennial, scabrid-pubescent, the stem simple or diffusely branched, 1–6 dm. high; lower lvs. lanceolate, with ± crisped margin and cordate to truncate base, 3–15 cm. long, on shorter petioles, the uppermost lvs. narrower; fl.-whorls crowded in leafy compact or interrupted spikes; pedicels jointed toward base; valves in fr. 1.7–2.5 mm. long, rhombic-oblong, lance-pointed, each with 2–3 awnlike bristles on each side and a prominent oblong callosity ca. 1 mm. long; $2n=40$ (Löve & Löve, 1967).—Wet often brackish places, below 7000 ft.; many Plant Communities; cismontane to n. Calif.; B.C., Atlantic Coast, S. Am. May–Sept.

8. **R. hymenosèpalus** Torr. CANAIGRE. WILD-RHUBARB. Perennial with a cluster of tuberous roots; stems smooth, somewhat reddish, 6–12 dm. high, stout; lvs. very fleshy, oblong or oblong-elliptic, 0.6–3 dm. long, on shorter petioles, acute, crisped on margins, the sheaths 1–3 cm. long; panicle compact, 1–3 dm. long, pinkish; pedicels 8–12 mm. long, jointed near middle; outer sepals ca. 2 mm. long, the valves in fr. 8–14 mm. long, cordate-ovate, reticulate; nutlets 4–6 mm. long; $2n=40$ (Löve & Patil, 1967), 100 (Kihara, 1927).—Common in dry sandy places, mostly below 5000 ft.; Coastal Sage Scrub, V. Grassland, Chaparral, Joshua Tree Wd., Creosote Bush Scrub; L. Calif. to Kern and San Luis Obispo cos.; Wyo., Tex. Jan.–May.

9. **R. kérneri** Borb. Perennial; stem strict or subflexuous, often branching above, 6–15 dm. high; lower lvs. oblong-elliptic, 10–22 cm. long, 3–7 cm. wide, with long petioles; panicle open, the lower glomerules remote; the upper contiguous at maturity; pedicels jointed in lower third; outer sepals 2–2.5 mm. long, the valves in fr. 6–8 mm. long, rotund-cordate, subentire to minutely denticulate, the callosities ovate-globose, 2 mm. long, 1.5 mm. wide; aks. 3 mm. long.—Known from Hope Ranch, Santa Barbara Co.; introd. from Eu.

10. **R. lacústris** Greene. Aquatic or terrestrial, the former with erect fistulous stems 5–9 dm. high emerging from the water, the terrestrial state with decumbent or ascending stems 2–4 dm. long, the oblong-lanceolate lvs. pubescent, ± obtuse; panicle reduced, comparatively few-fld., the branches short; pedicels jointed below the middle; outer sepals lanceolate, 1.3 mm. long, the valves in fr. 2–2.5 mm. long, narrowly ovate, entire, all subequally callose, the callosities fusiform, 1.5–2 mm. long.—Marshes, alkali sinks, ephemeral ponds, 4500–8000 ft.; Montane Coniferous F.; San Jacinto Mts.; n. Calif., s. Ore. July–Sept.

11. **R. obtusifòlius** L. ssp. **agréstis** (Fries) Danser. Perennial with stout taproot; stem stout, 6–12 dm. high, simple or few-branched; lower lvs. broadly lance-oblong, cordate at base, flat, 1 3.5 dm. long, long-petioled, the upper with rounded base and smaller; panicle open, with divergent branches; valves in fr. ca. 6 mm. long, triangular-ovate, the margin with few spreading spinose teeth, usually only 1 with a callosity; nutlets ca. 2 mm. long.—Low moist places at low elevs., s. base of San Gabriel Mts.; cent. Calif. n.; common weed in N. Am.; native of Eu. June–Dec.

12. **R. persicarioìdes** L. Similar in habit to *R. fueginus*; stems simple or branched, strict or angular-flexuous, 1–5 dm. high; lvs. mostly linear-oblong, obtuse at base and apex, on rather short slender petioles; infl. branched, the fl.-whorls crowded in leafy spikes; valves in fr. 2–2.5 mm. long, ca. 1 mm. wide, lance-ovate, each with 1–3 awnlike bristles exceeding the width of the valve on each side and bearing a thick swollen obtuse callosity nearly hiding the face of the valve.—Coastal Salt Marsh, Freshwater Marsh; Ventura Co., cent. Calif. n. to Ore., Atlantic Coast. July–Sept.

13. **R. púlcher** L. FIDDLE DOCK. Dark green perennial, the stem slender, 3–6 dm. high, with spreading branches; lower lvs. oblong, 5–12 cm. long, long-petioled, obtuse-cordate at base, usually contracted above the base, the upper short-petioled; panicle with divergent branches, lax, the clusters remote; valves 4.5–6 mm. long, the margin long-toothed toward the base, the apex linguiform, usually all bearing a callosity, but these often unequal; nutlets 3–4 mm. long.—At low elevs., as a weed in waste places; native of Medit. region.

14. **R. salicifòlius** Weinm. In habit and lf.-shape like *R. californicus*; stems 3–9 dm. high; lvs. to 13 cm. long; petioles equalling or exceeding width of lvs.; panicle-branches short, simple, ascending or often elongated; pedicels jointed near base; outer sepals narrowly lanceolate, 1.2–1.5 mm. long, the valves in fr. 2.2–3 mm. long, deltoid, acute, subentire or minutely denticulate, one valve with a large ovate callosity.—Moist places below 9000 ft.; many Plant Communities; L. Calif. to n. Calif. May–Sept.

15. **R. trianguliválvis** (Danser) Rech. f. [*R. salicifolius* ssp. *t.* Danser.] Stouter and with broader lvs. than in *R. californicus*; stems erect, 4–10 dm. high; lower lvs. 12–15 cm. long; panicle ample, the branches subelongate; pedicels jointed near base; outer sepals 1.6–1.8 mm. long, the valves in fr. mostly 3 mm. long and wide, triangular, entire or irregularly crenulate, all subequally calliferous, the callosities narrowly fusiform, 1.8–2.5 mm. long.—Polymorphic and widespread, 3000–9000 ft.; many Plant Communities; San Bernardino Mts.; Sierra Nevada to B.C., Atlantic Coast, Mex. July–Sept.

16. **R. violáscens** Rech. f. Annual or biennial with taproot; stem stout, 3–7 dm. high; lower lvs. narrow-oblong to spatulate, 6–10 cm. long, rather flat, long-petioled, the upper reduced and lance-linear; panicle narrow, leafy-bracted, 5–12 cm. long, with ascending branches; pedicels about as long as fr., jointed below middle; valves in fr. 2.5–3 mm. long, nearly as wide, triangular-ovate, erose or toothed near the base, all with an ovate-oblong callosity, these unequal, the largest 1.5–2 mm. long; aks. 1.7 mm. long.—Alkali Sink, Creosote Bush Scrub; Colo. Desert, Lake Elsinore region, Palmdale; to Tex., Mex. March–Aug.

86. **Portulacàceae** Purslane Family

Annual or perennial herbs, ± succulent. Lvs. entire, alternate or opposite, or mostly basal. Fls. perfect, nearly or quite regular. Sepals 2, rarely more. Petals 3–16, commonly 5, sometimes 0, generally hypogynous, ephemeral. Stamens few–many, sometimes only 1, opposite the petals if of the same number and usually adnate to their bases. Ovary commonly superior, 1-locular, with few to many ovules on free cent. or basal placenta; styles 1–8, united below or distinct, stigmatic along the inside. Caps. mostly 1-locular, dehiscing by 2–3 valves or circumscissile. Seeds 1 to many, mostly round-reniform to orbicular; embryo curved. Ca. 20 genera and over 200 spp. of wide temperate distribution. Some, like garden *Portulaca*, of hort. value.

1. Calyx adnate to lower part of ovary, its lobes coming off the summit of the circumscissile caps.
       8. *Portulaca*
1. Calyx and ovary free.
   2. Caps. circumscissile; sepals 2–8; stamens 5 to many .... 5. *Lewisia*
   2. Caps. opening by 2–3 valves; sepals 2; stamens 1–5, except in *Calandrinia*.
      3. Style 1, with 2 stigmas; caps. 2-valved; infl. secund .... 2. *Calyptridium*
      3. Style-branches 3; caps. 3-valved; infl. mostly not markedly secund.
         4. Fls. in leafy racemes or panicles; petals mostly red; stamens 5 to many; seeds 18–25
                1. *Calandrinia*
         4. Fls. in naked or bracteate racemes; petals white to pink; stamens 1–5; seeds 1–6.
            5. Perennial or annual, with a basal rosette of lvs. surmounting a taproot, rhizome or deep-seated bulbous root; pollen tricolpate .... 3. *Claytonia*
            5. Perennial or annual branched above the base and lacking a basal rosette except in 2 spp. where it is early deciduous; pollen often dodecacolpate, sometimes polyrugate with each section of the wall with a hemispheric spine-covered saccus.
               6. Annuals.
                  7. Cauline lvs. opposite in ours .... 6. *Montia*
                  7. Cauline lvs. alternate .... 7. *Montiastrum*
               6. Perennial with vegetative propagules at the ends of runners having opposite scalelike lvs. .... 4. *Crunocallis*

1. *Calandrínia* HBK.

Annual or perennial herbs. Lvs. alternate, entire. Fls. red, rarely white, ephemeral, in rather leafy racemes or panicles. Sepals 2, persistent. Petals 3–7, usually 5. Stamens 3–14, shorter than the petals, seldom of the same no. as the petals. Style-branches 3. Caps. 3-valved from apex. Seeds many, somewhat flattened, rounded, very dark, usually minutely pitted but shining. Ca. 75 spp. of W. Am. and Australia. (Named for J. L. Calandrini, Swiss botanist of the 18th century.)

Fls. in panicles of umbels; seeds naked at hilum; petals white. Desert .... 1. *C. ambigua*
Fls. ± racemose; seeds with a strophiole; petals usually red. Cismontane.
    Plants green; seeds black and shining.
        Caps. ca. twice as long as the calyx; lvs. mostly basal; pedicels often reflexed in fr.
                2. *C. breweri*
        Caps. scarcely exceeding calyx; lvs. well distributed; pedicels ascending to erect
                3. *C. ciliata*
    Plants glaucous; seeds dull and gray .... 4. *C. maritima*

1. **C. ambígua** (Wats.) Howell. [*Claytonia a.* Wats.] Very fleshy annual with several stems from base, erect or spreading, 3–15 cm. long; lvs. linear-spatulate, 2–5 cm. long, well distributed, obtuse, subterete, channeled beneath; fls. in rather compact lateral and terminal umbellate clusters; pedicels 2–7 mm. long; sepals ovate, white-scarious at margin, 3–5 mm. long, obtuse; petals white, 3–5, obovate, ca. as long as or shorter than sepals; caps. equal to calyx; seeds shining, black, not strophiolate, 0.6–0.8 mm. long.—Somewhat alkaline washes and slopes, below 2000 ft.; Creosote Bush Scrub, Colo. Desert and drainage of Mohave and Amargosa rivers to Death V.; Ariz. Jan.–April.

2. **C. bréweri** Wats. Glabrous annual, the stems several from the base, 1–4 dm. long, prostrate or nearly so; lvs. lance-ovate or spatulate, the basal 2–8 cm. long, petioled, the cauline reduced and nearly or quite sessile; fls. in elongate leafy-bracted racemes, not numerous; pedicels 7–20 mm. long, generally reflexed in fr.; sepals glabrous or sometimes ciliate, 5–6 mm. in fr., scarious-margined towards base; petals rose-red, 4–5 mm. long; caps. 10–12 mm. long, well-exserted; seeds many, black, shining, minutely tuberculate, ca. 1 mm. long.—Mostly on burns and in disturbed places, gravelly slopes, below 3500 ft.; Chaparral, Coastal Sage Scrub; L. Calif. to cent. Calif. March–June.

3. **C. ciliàta** (R. & P.) DC. var. **menzièsii** (Hook.) Macbr. [*Talinum m.* Hook.] RED MAIDS. Plate 72, Fig. E. Annual, simple or more usually with several spreading stems from base, these 1–4 dm. long, subglabrous; lvs. well distributed, petioled, narrowly oblanceolate to linear, 2–8 cm. long, somewhat fleshy; fls. in leafy racemes, the pedicels suberect, 0.4–2 cm. long; sepals ovate, short-acuminate, glabrous or hispidulous on margin and midrib, 3–8 mm. long; petals 5, rose-red, rarely white, 4–10 (–14) mm. long; caps. ovoid, 4–7 mm. long; seeds many, black, shining, minutely tuberculate, 0.8–1 mm. in diam.; $2n = 24$ (Sugiura, 1936). Common, especially in open grassy places and cult. fields, below 6000 ft.; mostly V. Grassland, Foothill Wd., to a lesser extent in other communities; cismontane Calif., occasional at desert edge; to L. Calif., B.C., Ariz., Son. Mostly Feb.–May. Variable.

4. **C. marítima** Nutt. Plate 72, Fig. F. Glaucous annuals branched at base, the stems spreading or ascending, 7–25 cm. long; lvs. mostly at base, spatulate-obovate, 2–6 cm. long, petioled, fleshy; fls. in a lax naked terminal panicle of cymes; pedicels 5–15 mm. long; sepals round-ovate, dark-veined, short-acute, 4–5 mm. long; petals red, ca. 4 mm. long; caps. ovoid, 5–6 mm. long; seeds dull, grayish, minutely roughened, ca. 0.6 mm. long.—Sandy places, sea bluffs; Coastal Sage Scrub; Santa Barbara Co. to L. Calif. Channel Ids. March–May.

2. *Calyptridium* Nutt. in T. & G.

Annual or perennial herbs with alternate or basal spatulate lvs. Fls. small, perfect, in scorpioid spikes or groups of spikes. Sepals 2, rounded or round-reniform, scarious or scarious-margined. Petals 2 or 4. Stamens 1–3. Style simple, with 2 stigmas. Caps. membranous, 2-valved, few–many-seeded. Ca. 6 spp. of w. N. Am. (Greek, *kaluptra*, a cap or covering, because of the way the petals close over the caps.)

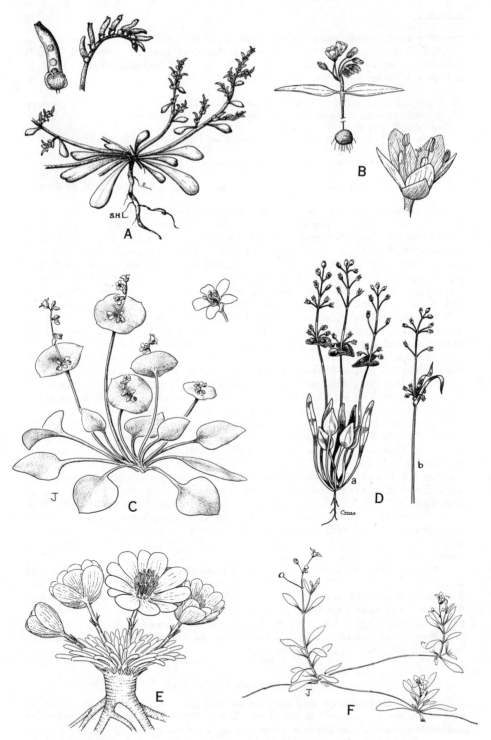

Plate 73. Fig. A, *Calyptridium monandrum*, elongate caps. Fig. B, *Claytonia lanceolata*. Fig. C, *Claytonia perfoliata*. Fig. D, *Claytonia spathulata*, and to right, var. *tenuifolia*. Fig. E, *Lewisia rediviva*. Fig. F, *Crunocallis chamissoi*.

Calyptridium

Style short; petals in age folding as a cap over the caps.; caps. elongate, linear-oblong or ovoid, 6–many-seeded.
    Caps. linear-oblong to elliptic, the valves widest near middle and with many nerves near base; fls. sessile or subsessile, deciduous.
        Caps. 2–4 times as long as sepals; sepals 1–2 mm. long .... 1. *C. monandrum*
        Caps. less than 2 times as long as sepals; sepals 2–3.5 mm. long .... 2. *C. parryi*
    Caps. ovoid, the valves widest near base, with 5–6 main nerves; lowest fls. distinctly pedicelled, persistent .... 3. *C. pygmaeum*
Style long-filiform; petals in age twisting about the style; caps. round or nearly so 4. *C. umbellatum*

    1. **C. monándrum** Nutt. in T. & G. Plate 73, Fig. A. Annual, branched at base; stems frequently flat on ground, 5–18 cm. long; lvs. mostly in a basal rosette, linear-spatulate, 2–6 cm. long; cauline lvs. scattered, somewhat reduced; fls. subsessile in short spikes in a terminal panicle, the branchlets scorpioid when young, secund in age; sepals reniform-ovate, or deltoid, white-margined, 1–2 mm. long; petals commonly 3, white, ovate, ca. 1 mm. long; caps. compressed, linear-oblong, striate, 4–8 mm. long; seeds 5–10, black, shining, ca. 0.5 mm. long.—Common in sandy and open places, usually below 6000 ft.; Creosote Bush Scrub, Joshua Tree Wd.; deserts from Mono Co. s.; also on burns, etc.; Chaparral, lower Yellow Pine F., Coastal Sage Scrub, etc.; L. Calif. to Monterey Co.; Nev., Son. March–June.
    2. **C. párryi** Gray. ssp. **párryi**. Habit of *C. monandrum*; Lvs. 1–3 cm. long; sepals 2–3.5 mm. long; petals largely 4, shorter than sepals; caps. oblong, ca. 1.5–2 times the length of the calyx; seeds dull, muriculate throughout, 0.6–0.7 mm. long.—Disturbed places, 5000–10,000 ft.; Montane Coniferous F.; San Jacinto Mts. to Mt. Pinos; to Sierra Nevada. June–July.
    Ssp. **nevadénse** (J. T. Howell) comb. nov. [*C. parryi* var. *n*. J. T. Howell, Leafl. W. Bot. 4: 216. 1945.] Fls. more easily deciduous; abaxial sepal in fr. reniform, with wide scarious margin; seeds shining, almost smooth.—Dry slopes, 7000–10,000 ft.; Pinyon-Juniper Wd.; Panamint Mts.; w. Nev. June–July.
    3. **C. pygmaèum** Parish ex Rydb. Diffuse annual; stems 1–2.5 cm. high; lvs. mostly basal, spatulate, 5–10 mm. long; fls. few, in 1-sided racemes; bracts ovate, scarious, early deciduous; pedicels 1–3 mm. long; sepals ovate, acutish, ca. 2 mm. long, scarcely scarious-margined; petals 4, ca. 3 mm. long; caps. ovoid, pointed, ca. 4 mm. long; seeds black, smooth, ca. 0.4 mm. long.—Rare, dry to moist sandy or gravelly places, 7000–11,500 ft.; Lodgepole to Subalpine F.; San Bernardino Mts.; se. Sierra Nevada. June–July.
    4. **C. umbellàtum** (Torr.) Greene. [*Spraguea u.* Torr.] Pussy Paws. Annual to perennial; stems several, spreading to suberect, 5–15 cm. long, ± scapelike; lvs. largely basal, in dense rosette, spatulate, 2–7 cm. long; cauline lvs. almost none or several, reduced; infl. umbellate-cymose, rather lax to subcapitate, consisting of scorpioid spikes; fls. deciduous, imbricate-crowded, very short-pedicellate; sepals pink or white, round-reniform, mostly 5–8 mm. long, with broad scarious margins; petals 4, oblong or ovate, 3–6 mm. long; caps. ovoid, 3–4 mm. long; seeds compressed, round-reniform, smooth, shining, 0.8–1 mm. long.—Rather common locally, in sandy or gravelly places, 4500–10,800 ft.; mostly Yellow Pine F. to Subalpine F.; San Jacinto Mts. to Mt. Pinos; Sierra Nevada n. to B.C., Rocky Mts., L. Calif. May–Aug. Variable.

3. *Claytònia* L.

    Glabrous perennial herbs with deep-seated corms or fleshy roots to annuals with well developed basal rosettes, which are generally late deciduous. Scapose flowering branches upright from axils of basal lvs., with a single pair of free or united opposite lvs. subtending a generally simple infl. Bracts subtending each fl. or only the lowest fl. with a subtending bract. Pollen mostly tricolpate, some grains 6-rugate. Sepals 2, herbaceous, persistent. Petals mostly 5, rose to white or yellowish. Seeds few, often shining. Largely N. Hemis. (Named for John *Clayton*, 18th century Am. botanist.)

(Swanson, J. R. A synopsis of relationships in Montioideae. Brittonia 18: 229–240. 1966.)
Plant with a globose corm; basal lvs. 0–2 .... 1. *C. lanceolata*
Plants annual; basal lvs. several.
   Stem-lvs. usually united on both sides, forming a flattish rounded somewhat angled disk
                                                                            2. *C. perfoliata*
   Stem-lvs. united on part of 1 side, not forming a flattish disk .... 3. *C. spathulata*

1. **C. lanceolàta** Pursh. var. **peirsònii** M. & J. Plate 73, Fig. B. Corm globose, 1–2 cm. in diam.; stems largely subterranean, 5–7 cm. long; basal lf. thick, the blade ca. 1 cm. long, cauline lvs. short-petiolate, widest below the middle, 7–12 mm. wide; fls. few, subumbellate, the pedicels 1–2 cm. long; sepals 3.5–5 mm. long; petals pinkish, 8–12 mm. long, ± emarginate at apex; caps. ovoid, ca. 4 mm. long; seeds shining, black, ca. 2 mm. long.—Dry ridges, e. end of San Gabriel Mts., at ca. 8000 ft.; Lodgepole F. May–June.

2. **C. perfoliàta** Donn. var. **perfoliàta**. [*Montia p.* Howell.] MINER's-LETTUCE. Plate 73, Fig. C. Glabrous green annual, branched from base, 1–3 dm. high; basal lvs. rhombic-ovate to elliptic-obovate, long-petioled, 5–20 cm. long; cauline lvs. 2, opposite, connate into an oblique orbicular disk 1–8 cm. broad; racemes ± elongate, sessile or peduncled, the fls. usually ± whorled; pedicels commonly 2–8 (–10) mm. long, often recurved in fr.; sepals rounded, ca. 3–5 mm. long; petals white, clawed, obovate, 4–6 mm. long; seeds black, shining, rounded, minutely punctate, 1–2 mm. in diam.; $n = 6, 12, 18$ (Raven, 1962).—Common in ± shaded and vernally moist places, below 5000 ft.; Coastal Sage Scrub, Chaparral, S. Oak Wd., etc.; L. Calif. to B.C., inland to desert edge. Feb.–May. Running into innumerable forms, of which the following are most notable:

KEY TO VARIETIES

Basal lvs. largely ovate to deltoid.
   Stems suberect, 1–3 dm. high; racemes elongate, often peduncled; sepals 3–5 mm. long.
     Cismontane, mostly below Yellow Pine F. .... var. *perfoliata*
   Stems spreading, mostly less than 1 dm. long; racemes short, compact, sessile; sepals ca. 2 mm.
     long. Yellow Pine F. .... var. *depressa*
Basal lvs. largely linear to oblanceolate or spatulate.
   Stems mostly 1–3 dm. high. Cismontane and montane .... var. *parviflora*
   Stems mostly not over 1 dm. high. Deserts .... var. *utahensis*

Var. **depréssa** (Gray) v. Poellnitz. [*C. parviflora* var. *d.* Gray. *Montia perfoliata* var. *d.* Jeps.] Stems spreading, mostly 5–10 cm. long; plant ± reddish; basal lvs. rhombic-ovate to subdeltoid; infl. short, sessile; sepals 2 mm. long.—Rather dry shaded places, largely 4000–7500 ft.; Yellow Pine F.; mts. of s. Calif.; n. to B.C., Mont. April–June.

Var. **parviflòra** (Dougl. ex Hook.) Torr. [*Claytonia p.* Dougl. *Montia p.* Howell.] Earlier basal lvs. linear, the later oblanceolate; infl. tending to be elongate and open; sepals ca. 2 mm. long.—With var. *perfoliata*.

Var. **utahénsis** Poelln. [*Montia perfoliata* var. *u.* Munz.] Basal lvs. linear to spatulate, otherwise much as in var. *depressa*.—Semishade, 3500–7500 ft.; largely in Pinyon-Juniper Wd., Joshua Tree Wd.; mts. of Mojave Desert from Inyo Co. to Ariz., Utah. April–June.

3. **C. spathulàta** Dougl. ex Hook. var. **spathulàta**. [*Montia s.* Howell.] Densely tufted glaucous annual, the stems 2–6 cm. high; basal lvs. numerous, linear to linear-spatulate, 3–9 cm. long, mostly surpassing the stems; cauline lvs. 2, opposite, linear to lance-ovate, usually joined on one side, 1–2 cm. long; racemes mostly 3–6-fld., not over 2 cm. long; sepals ovate, acute, ca. 1.5 mm. long; petals 5, white or pinkish, 2.5–3 mm. long; caps. ca. 2 mm. high; seeds 2–3, black, shining, minutely tuberculate, 0.6–1 mm. long; $2n = 48$ (Nilsson, 1966).—Rather open dry slopes, below 2500 ft.; Chaparral, V. Grassland, etc. S. Coast Ranges n. of our area; to B.C. Feb.–April.

Cauline lvs. 1–2 cm. long, not exceeding infl.; petals ca. 4 mm. long .... var. *exigua*
Cauline lvs. 2–6 cm. long, surpassing the infl.

Plant glaucous; basal lvs. linear or nearly so .... var. *tenuifolia*
Plant green; basal lvs. oblanceolate .... var. *viridis*

Var. **exígua** (T. & G.) Piper. [*C. e.* T. & G.] Plate 73, Fig. D. Plant 5–15 cm. high, the stems much surpassing the basal lvs.; stem lvs. mostly 1–2 cm. long, 1–2 mm. wide; petals ca. 4 mm. long; seeds 1.5–2 mm. long.—Dry places, often on serpentine, below 5500 ft.; Chaparral, etc.; mts. of San Diego Co.; to Wash., Nev. April–July.

Var. **tenuifòlia** Gray. [*C. t.* T. & G. *Montia t.* Howell.] Glaucous, 5–10 cm. high; basal lvs. linear; cauline lvs. linear to oblance-linear, 2–4 cm. long; petals 3–4 mm. long; seeds 1–1.2 mm. long.—Dry somewhat shaded places, below 6500 ft.; Chaparral, Yellow Pine F., etc.; mts. of San Diego Co.; to n. Calif. April–June.

Var. **víridis** (A. Davids.) Munz, comb. nov. [*Montia spathulata* var. *v.* A. Davids., Bull. S. Calif. Acad. Sci. 5: 61. 1907.] Green, not glaucous, 5–15 cm. high; basal lvs. narrow-oblanceolate; stem lvs. lanceolate, 2–5 cm. long, surpassing infl.; petals ca. 3 mm. long.—Shaded somewhat damp places, 4000–7200 ft.; largely Yellow Pine F.; San Diego Co. to Ventura Co. May–July.

4. *Crunocállis* Rydb.

Glabrous, perennial, stoloniferous herbs, with terete aerial stems and axillary stolons from stem base with small opposite scales; lvs. simple, oblanceolate-obovate or spatulate; stomata on both sides of lf.; infls. 1–4, with few fls.; pedicels evenly recurved after anthesis; petals and stamens 5; style-branches 3; valves 3, dehiscing to the grooves; ovules 3; pollen spherical, punctate and spinulate or smooth. Two spp.

(Nilsson, O. The genus Crunocallis Rydb. Bot. Notiser 123: 119–143. 1970.)

1. **C. chamissoì** (Ledeb.) Rydb. [*Montia c.* Dur. & Jackson, *Claytonia c.* Eschs. in Ledeb.] Plate 73, Fig. F. Aerial stem 2–30 cm. long; stolons spreading shallowly in the ground, ending in small bulblets; lvs. 1–7 cm. long; fls. few; petals 4–10 mm. long; caps. 2–3.5 mm. long; $n = 11$ (Lewis et al.).—Wet places, meadows, along streams, etc. 4500–8500 ft.; Montane Coniferous F.; mts. from San Diego Co. to Ventura Co.; Sierra Nevada, etc. to Alaska, Minn., New Mex. June–Aug.

5. *Lewísia* Pursh

Fleshy perennials from thick fleshy root and short caudex, or from a globose corm. Lvs. largely in a basal rosette, entire, with wide mostly hyaline petiole-base; cauline lvs. few, often bracteate. Infl. bracteate, sometimes disjointing in age. Sepals 2–6 (–8), persistent. Petals 4–18, often unequal. Stamens 5–many. Styles 3–8, united at base. Caps. circumscissile near base, globose or ovoid. Seeds 6–many, dark, shining, smooth or finely tuberculate. Ca. 18 spp. of w. N. Am. (Named for Capt. Meriwether *Lewis*, of the Lewis and Clark Expedition of 1806/07.)

Bracts remote from sepals and not resembling them.
  Sepals herbaceous or hyaline, not petaloid.
    Sepals glandular-toothed and conspicuously veined in age .... 3. *L. pygmaea*
    Sepals entire or with a few nonglandular teeth, not so conspicuously veined . 2. *L. nevadensis*
  Sepals petaloid, becoming scarious in age; infl. readily disjointing in fr.; the sepals several, 10–25 mm. long .... 4. *L. rediviva*
Bracts just below the sepals and resembling them .... 1. *L. brachycalyx*

1. **L. brachycàlyx** Engelm. ex Gray. Caudex short, thick; stems many, 2–6 cm. long; lvs. many, fleshy, 3–6 cm. long, oblanceolate; fls. usually solitary, sessile in pair of calyx-like bracts; sepals lance-ovate, acute, 8–10 mm. long, gland-dentate on margin; petals creamy-white, 8–15 mm. long; caps. ovoid, ca. 8 mm. long; seeds black, shining, 1.5 mm. long.—Wet meadows, 4500–7500 ft.; Yellow Pine F.; Cuyamaca and San Bernardino mts.; to Utah, Ariz., New Mex. May–June.

2. **L. nevadénsis** (Gray) Robins. in Gray. [*Calandrinia n.* Gray.] Habit and foliage

much as in *L. pygmaea*; sepals broadly ovate, acute, 5–10 mm. long, the margin quite entire or if slightly dentate, then not gland-toothed; petals 9–15 (–18) mm. long, white; seeds ca. 1.3 mm. long.—Wet banks and meadows, 4500–11,000 ft. Yellow Pine F. to Subalpine F.; San Bernardino Mts., Mt. Pinos; Sierra Nevada to Wash., Rocky Mts. May–July.

3. **L. pygmaèa** (Gray) Rob. in Gray. [*Talinum p.* Gray.] Root fleshy, fusiform; caudex short; stems several to many, 2–6 cm. long, partly underground; lvs. many or fewer, basal, linear to narrowly oblanceolate, 3–8 cm. long; bracts hyaline, lanceolate, 6–10 mm. long, opposite, and near middle of stem; fls. 1–3 per stem; sepals rounded, 4–5 mm. long, glandular-dentate (glands light in color, not stipitate); petals white to pink, 7–10 mm. long; caps. 4–6 mm. long; seeds 15–20, dark, shining, 1–1.2 mm. long; $n=$ ca. 33 (Wiens & Halleck, 1962).—Damp gravel, 5000–9000 ft.; Montane Coniferous F.; San Bernardino Mts.; Sierra Nevada to Wash. Rocky Mts. July–Sept.

4. **L. redivìva** Pursh var. **mìnor** (Rydb.) Munz. [*L. m.* Rydb.] BITTERROOT. Plate 73, Fig. E. Caudex short; taproot fleshy, much branched; stems many, 1–3 cm. long, 1-fld.; lvs. basal, clavate to narrow-oblanceolate, 2–4 cm. long; bracts 5–8, whorled, subulate, scarious, at base of pedicel; pedicels mostly 3–8 mm. long, readily disjointing in age; sepals several, mostly ca. 1 cm. long; petals ca. 1.5 cm. long, rose or white, many; caps. ellipsoid, 5–6 mm. long; seeds dark, shining, round-reniform, 2–2.5 mm. long.—Loose gravel and rocky places, 6500–9000 ft.; Pinyon-Juniper Wd., Yellow Pine F.; San Jacinto Mts., San Bernardino Mts. to Mt. Pinos, Panamint Mts., White Mts.; w. Nev. May–June.

6. *Móntia* L.

Perennials or annuals of varying habit, but branched above the base and lacking a basal rosette in our sp. Cauline lvs. largely opposite; infls. from the lf.-axils or terminating the main branches, or both. Fl. much as in *Claytonia*. Pollen mostly dodecacolpate or with a number of rugae, or in some spp. polyrugate and bearing a spiniferous saccus on each segm. of the pollen wall. A genus of both hemis. (Named for Giuseppe *Monti*, 1682–1760, Italian botanist.)

1. **M. fontàna** L. ssp. **amporitàna** Sennen. [*M. hallii* (Gray) Greene.] Annual, with stems slender, spreading, branched, 3–10 or more cm. long, often rooting at the nodes; lvs. opposite, the lower sublinear to oblanceolate, 8–20 mm. long, mostly 5–10 (–14) mm. wide; fls. axillary and terminal, nodding, solitary or in small terminal clusters; pedicels mostly 3–15 mm. long; sepals ca. 1 mm. long; petals 2–5 in no.; seeds 0.6–1.2 mm. long, shining, with 7–11 rows of slender tubercles around the keel.—Rainpools, etc., below 6000 ft.; L. Calif. to B.C. March–June.

Ssp. **variábilis** Walters [*M. funstonii* Rydb.] Like ssp. *amporitana*, but with tubercles on seed low, flat or even lacking.—Wet places above 6000 ft.; Montane Coniferous F.; L. Calif. to B.C. March–June.

7. *Montiástrum* (Gray) Rydb.

Small perennial or annual herbs. Lvs. alternate, ± linear and sessile, sheathing at the base. Infl. terminal or pseudo-lateral, pedunculate, common axis elongate, bract normally one. Invol. longer than the caps., usually distinctly parallel-veined. Petals 5, almost free. Stamens 5 or 3. Caps. dehiscing almost to base; seeds normally 3, usually distinctly keeled, with smooth testa; pollen grains pantocolpate, spherical, tholate. A genus of 4 spp. (*Monti*, an Italian botanist, and *astron*, star.)

(Nilsson, Ö. The genus Montiastrum. Bot. Notiser 124: 87–121. 1971.)

1. **M. lineáre** (Dougl.) Rydb. [*Claytonia l.* Dougl. in Hook. *Montia l.* Greene.] Erect branched annual 5–20 cm. high; lvs. alternate, linear, 2–3.5 cm. long, with enlarged scarious bases; racemes terminal, secund, lax, mostly 2–7-fld.; pedicels 6–15 mm. long, ± recurved; sepals reniform-orbicular, ca. 4 mm. long, white-margined; petals unequal,

obovate, white, ca. 5 mm. long; caps. ovoid, 4 mm. long; seeds 3, lenticular, microscopically muricate, especially on the margin, 1.5–2 mm. in diam.—Moist banks, meadows, etc., below 7500 ft.; Montane Coniferous F.; mts. in San Diego Co., San Bernardino Mts.; Sierra Nevada, then to Modoc Co.; to B.C., Mont. April–June.

8. Portuláca L.

Fleshy herbs, with mostly scattered alternate or partly opposite lvs. Fls. terminal, sessile, mostly opening only in sunshine. Sepals 2, united below, the tube cohering with the ovary. Petals 5, rarely 4 or 6–many, inserted on the calyx. Stamens 7–20. Style deeply 3–8-parted. Caps. 1-locular, globular, many-seeded, circumscissile near the middle. Seeds many, round-reniform. Ca. 100 spp. of warmer regions. (Old Latin name, of uncertain application.)

Petals pink or purplish; lvs. subterete; plant conspicuously hairy in axils. Desert native
  1. *P. mundula*
Petals yellow; lvs. flat; plant nearly or quite glabrous. Introd. weed .... 2. *P. oleracea*

1. **P. múndula** Jtn. Annual, branched from base, the stems ascending, 5–10 cm. long, copiously long-hairy in the axils; lvs. linear-subulate, 6–12 mm. long; petals pink or purplish, at least in age, 3–4 mm. long; caps. 4–5 mm. long; seeds granular, ca. 0.3 mm. long.—Sandy washes, 3500–4000 ft., Joshua Tree Wd., Pinyon-Juniper Wd.; Little San Bernardino Mts., Valley Wells, e. San Bernardino Co.; to Mo., Tex.

2. **P. olerácea** L. PURSLANE. Prostrate annual, the stems 1–2 dm. long; lvs. obovate or cuneate, 5–25 mm. long; fls. sessile, 3–6 mm. broad; sepals keeled, acute, 3–4 mm. long; petals pale yellow; style deeply 5–6-parted; caps. 4–8 mm. long; seeds black, finely rugulose-pitted, ca. 0.6 mm. long; $2n = 54$ (Steiner, 1944).—Widely scattered weed in gardens, waste places, cismontane and desert, natur. from Eu. May–Sept.

87. Primuláceae. PRIMROSE FAMILY

Scapose or caulescent annual to perennial herbs. Lvs. simple, exstipulate, opposite, whorled or alternate. Fls. perfect, regular, mostly 5-merous. Calyx deeply lobed, often persistent. Corolla deeply parted to lobed, the lobes spreading or reflexed. Stamens of same number as corolla-lobes and inserted opposite them on the corolla-tube. Ovary 1-celled, usually superior, sometimes half-inferior; ovules on a free cent. or basal placenta. Style 1; stigma capitate. Fr. a caps. commonly 2–6-valved, sometimes circumscissile at apex. Seeds few to many, with endosperm. Ca. 25 genera and 600 spp., widely distributed but most common in N. Hemis.

Ovary partly inferior, the lower portion included in and adnate to the calyx-tube; plant with leafy stems; fls. small, racemose .... 4. *Samolus*
Ovary superior and free.
  Plants scapose, the lvs. all basal; fls. in bracteate umbels.
    Corolla-lobes spreading or erect, 1–2 mm. long .... 2. *Androsace*
    Corolla-lobes reflexed, much larger .... 3. *Dodecatheon*
  Plants with leafy stems; fls. not in umbels; the lvs. alternate or opposite .... 1. *Anagallis*

1. Anágallis L.

Low annual or perennial herbs, diffusely branched or tufted. Lvs. alternate, opposite or whorled, mostly entire, sessile or nearly so. Fls. small or minute, solitary in axils. Calyx 4–5-parted, persistent. Corolla rotate. Stamens 4–5; fils. distinct or united at base. Caps. globose, circumscissile. Seeds many, angled. Ca. 20 spp., mostly Eurasian. (Greek, *ana*, again, and *agallein*, to delight in, since the fls. open when the sun strikes them.)

Lvs. 5–20 mm. long; pedicels 10–30 mm. long; petals to ca. 6 mm. broad .... 1. *A. arvensis*
Lvs. 2–5 mm. long; pedicels shorter; petals smaller .... 2. *A. minimus*

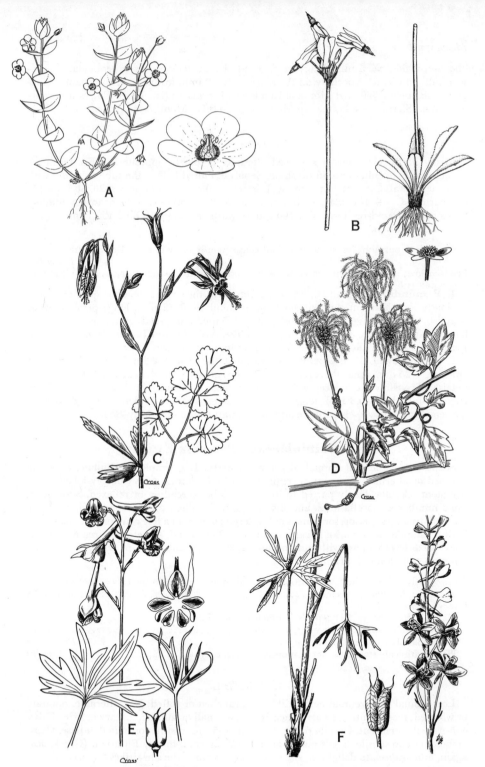

Plate 74. Fig. A, *Anagallis arvensis*. Fig. B, *Dodecatheon clevelandii*. Fig. C, *Aquilegia formosa* var. *truncata*, petals spurred. Fig. D, *Clematis lasiantha*, pistils tailed. Fig. E, *Delphinium cardinale*, fl. spaced to show sepals (uppermost spurred) and petals. Fig. F, *Delphinium parishii*.

1. **A. arvénsis** L. PIMPERNEL. Plate 74, Fig. A. Glabrous diffusely branched annual, the stems 1–2.5 dm. long; lf.-blades ovate to oval, sessile, 0.5–2 cm. long; pedicels slender, 1–3 cm. long; calyx-lobes lanceolate, 3–5 mm. long; corolla mostly salmon, sometimes blue (var. *caerulea*, Lodi) 8–10 mm. across; caps. 3–4 mm. long; seed 1 mm. long, triangular, dark, finely pitted; $2n=40$ (Wulff, 1937).—Common weed at low elevs.; to Atlantic Coast; natur. from Eu. Mostly March–July.

2. **A. mínimus** (L.) Krause. [*Centunculus m.* L.] Annual, stems 3–10 cm. long; lvs. spatulate to broadly obovate, 2–5 mm. long; fls. almost sessile, mostly 4-merous; calyx ca. 3 mm. long; corolla shorter, pink; seeds red-brown, minutely pitted, ca. 0.5 mm. long; $2n=22$ (Hagerup, 1941).—Occasional in vernal pools and moist places, low elevs.; Coastal Sage Scrub, Chaparral, etc.; San Diego Co. to n. Calif. Santa Rosa Id.; to B.C., Atlantic Coast, Eu., S. Am. April–May.

2. Andrósace L.

Small annual or perennial herbs. Lvs. basal, rosulate. Scapes 1–several, with involucrate umbels of small fls. Calyx 5-cleft, the tube becoming scarious. Corolla salverform, with short tube, constricted throat, and emarginate or obcordate lobes. Stamens 5, included; fils. short; anthers oblong. Ovary globose; style short. Caps. 5-valved. Seeds few to many, ovoid to triquetrous, minutely pitted. Ca. 60 spp. of N. Hemis., especially in colder parts. (Old Latin name.)

(Robbins, G. Thomas. N. Am. species of Androsace. Am. Midl. Nat. 32: 137–163. 1944.)
Corolla included in calyx. From below 4000 ft. .... 1. *A. elongata*
Corolla not included in calyx. From ca. 11,000 ft. .... 2. *A. septentrionalis*

1. **A. elongàta** L. ssp. **acùta** (Greene) Robbins. [*A. a.* Greene.] Annual, puberulent, 2–8 cm. high, the scapes mostly 1–6, erect to divergent; lvs. linear-lanceolate, subacuminate. 5–20 mm. long, hispidulous on margins; involucral bracts narrow-ovate, 3–5 mm. long; pedicels 2–6, unequal, 1–4 cm. long; calyx-tube obpyramidal, 2 mm. high, whitish, the lobes ca. as long, subulate to acerose, apiculate; corolla included; seeds brown, minute.—Dry grassy places, below 4000 ft.; Coastal Sage Scrub, Chaparral, S. Oak Wd., etc.; scattered stations in cismontane s. Calif.; to L. Calif., s. Ore. Occasional at desert edge as near Victorville. March–May.

2. **A. septentrionàlis** L. ssp. **subumbellàta** (A. Nels.) Robbins. Annual or weak perennial; lvs. linear-lanceolate, 5–20 mm. long, entire or weakly denticulate, puberulent; scapes 1–3 cm. high, erect or spreading, unequal, 1–3 cm. long; calyx subcampanulate, angled, the base scarious between the ridges, ca. 2 mm. long, the lobes 1–1.5 mm. long, lanceolate; corolla equaling or surpassing calyx; caps. included; seeds dark brown, ca. 1 mm. long.—Dry rocky places at ca. 11,000 ft.; Subalpine F.; San Bernardino Mts.; White Mts., Sierra Nevada; to B.C. Rocky Mts. July–Aug.

3. Dodecátheon L. SHOOTING STAR

Rather low perennial herbs, glabrous or glandular-puberulent, with short rootstocks and fleshy-fibrous roots. Lvs. basal. Stems scapose, naked, ending in an umbel of few to many fls. Fls. 4–5 merous, nodding, on slender pedicels. Calyx deeply 5-cleft, persistent, the segms. reflexed at anthesis, later erect. Corolla 5-parted, with short tube, dilated throat and reflexed lobes. Stamens opposite the corolla-lobes, exserted, the fils. short, broad, often united; anthers basifixed, mostly erect, approximate. Style filiform, exserted. Caps. partially 5-valved or circumscissile. Seeds many, small, punctate. Ca. 14 spp. of N. Am., mostly western. (Greek, *dodeca*, twelve, and *theos*, god, a name given by Pliny to the Primrose as being under the care of the 12 leading gods.)

(Thompson, H. J. The biosystematics of Dodecatheon. Contr. Dudley Herb. 4 (5): 73–154. 1953.)
Stigma enlarged, more than twice the diam. of the style; connective rugose, dark maroon to black; fils. 1 mm. long, free or joined by a thin membrane, but not united into a tube.

Anthers 4, subulate, obtuse to truncate; corolla-tube not covering base of anthers. San Jacinto Mts. and San Bernardino Mts. .... 1. *D. alpinum*
Anthers 5, lanceolate, acute; corolla-tube covering base of anthers. General montane
4. *D. redolens*
Stigma not enlarged; connective rugose to smooth, dark or yellow; fils. united into a tube 2–4 mm. long.
    Rice-grain bulblets produced at flowering time; connective dark maroon to black, never yellow. San Bernardino Mts. .... 3. *D. hendersonii*
    Rice-grain bulblets not produced; connective usually yellow or dark with a yellow spot; fils. with a dark tube. Low elevs. .... 2. *D. clevelandii*

1. **D. alpìnum** (Gray) Greene ssp. **màjus** H. J. Thomps. Plants glabrous; roots white, without bulblets; lvs. linear to linear-oblanceolate, 6–16 cm. long, 0.9–1.6 cm. wide, acute, entire; scape 1.4–3 dm. high, ± glandular above; umbels 4–10-fld.; pedicels 1–3 cm. long; calyx-tube 1.5–2.5 mm. long, the lobes 4–5 mm. long; corolla-tube maroon, yellow above, the lobes 9–16 mm. long, magenta to lavender; fils. free, black, 0.5 mm. long; anthers 5–6 mm. long; caps. 6–8 mm. long, ovoid; seeds ca. 1.5 mm. long, wing-angled.—Wet places, 4000–11,000 ft.; Montane Coniferous F.; San Jacinto and San Bernardino mts.; Sierra Nevada to Ore., Utah. May–Aug.

2. **D. clevelándii** Greene ssp. **clevelándii**. Plate 74, Fig. B. Glandular-pubescent; roots white, without bulblets; lvs. including petioles 5–11 cm. long, 0.8–3 cm. wide, oblanceolate to spatulate, crisped, dentate, rarely entire; scape 1.8–4 dm. high; umbel 5–16-fld.; pedicels 2–5 cm. long at anthesis; calyx-tube 1.5–2.5 mm. long, the lobes 3–5 mm. long; corolla-tube dark maroon with yellow band above, the lobes 10–20 mm. long, magenta varying into white; fil.-tube 2.5–4 mm. long, 3–4 mm. wide, dark maroon to black; anthers 3–5 mm. long, lanceolate, the connective yellow, the anther-sacs yellow, each with dark line; caps. 8–13 mm. long, operculate; seeds brown, angled, rough, ca. 1.3 mm. long; $n=22$ (Thompson, 1951).—Cismontane grassy slopes and flats, below 2000 ft.; V. Grassland, Coastal Sage Scrub; Los Angeles Co. to n. L. Calif. Jan–Apr.

KEY TO SUBSPECIES

Connective yellow; fil.-tube without a yellow spot below each anther .... ssp. *clevelandii*
Connective maroon to black.
    Yellow spot on fil.-tube lacking .... ssp. *insulare*
    Yellow spot present below each anther on fil.-tube .... ssp. *sanctarum*

Ssp. **insulàre** H. J. Thomps. Lvs. 6–18 cm. long, 1–6 cm. wide; scape 1.2–4.5 dm. tall; umbels 5–9-fld.; pedicels 2–10 cm. long; calyx-lobes 4–7 mm. long; corolla-lobes 10–25 mm. long; anther-connective dark, the anther-sacs usually yellow; $n=22$ (Thompson, 1951).—Below 2500 ft.; Chaparral, V. Grassland, etc.; Channel Ids., mainland, Santa Barbara Co. n.; Guadalupe Ids. March–April.

Ssp. **sanctárum** (Greene) Abrams. [*D. s.* Greene.] Lvs. 4–6 cm. long; scape 1.5–3 dm. high; umbels 3–7-fld.; calyx-lobes 3–4 mm. long; corolla-lobes 10–20 mm. long; fil.-tube dark with a light spot at base of each anther; $n=22, 33, 44$ (Thompson, 1953).—Below 2000 ft.; Foothill Wd.; nw. Los Angeles Co. to San Francisco.

3. **D. hendersònii** Gray. Roots white, with bulblets at flowering time; lvs. including petioles 5–15 cm. long, 2–6 cm. wide, spatulate to elliptic; scape 1.2–4.8 dm. high, glandular to glabrous, the umbel 3–17-fld.; pedicels 2–7 cm. long at anthesis; fls. 5-merous; calyx-tube 2 mm. long, the lobes 3–5 mm.; corolla-tube maroon, yellow above, the lobes 6–23 mm. long, magenta to deep lavender or white; fil.-tube 1.5–3 mm. long, 1.5–2.5 mm. wide, as broad as anther whorl, dark; anthers 4–5 mm. long, lanceolate, the pollen-sacs dark red to black; caps. 8–15 mm. long, operculate; seeds pitted, angled, 1.3 mm. long; $n=22, 33, 66$ (Thompson, 1951).—Mostly in shaded places, at ca. 6500 ft.; Yellow Pine F.; Bear V., San Bernardino Mts.; Sierra Nevada, etc. to B.C. Feb.–May.

4. **D. rédolens** (Hall) H. J. Thomps. Plant heavily glandular-pubescent; roots white, without bulblets; lvs. including petiole 2–4 dm. long, 2.5–6 cm. wide, oblanceolate, en-

tire; scape 2.5–6 dm. high; umbel 5–10-fld.; pedicels 3–5 cm. long in fl.; fls. 5-merous; calyx-tube 3–4 mm. long, the lobes 5–8 mm. long; corolla-tube yellow, never maroon, covering the anther-bases, the lobes 15–25 mm. long, magenta to lavender; fils. less than 1 mm. long, free, black; anthers 7–11 mm. long, dark maroon to black; caps. 8–14 mm. long, valvate.—Moist places, 8000–11,500 ft.; Montane Coniferous F.; San Jacinto Mts. to s. Sierra Nevada; Bristlecone Pine F.; Panamint Mts., White Mts.; to Utah, Nev. July–Aug.

4. Sámolus L. Water-Pimpernel

Glabrous perennial herbs. Lvs. alternate, entire. Fls. small, white, in terminal racemes. Calyx-tube adnate to ovary below, the limb 5-cleft. Corolla somewhat campanulate, perigynous, 5-lobed or -cleft. Stamens 5, inserted on corolla-tube opposite the lobes and alternating with 5 staminodia. Caps. globose, 5-valved at summit. Seeds many, minute, triangular. Ca. 10 spp., semicosmopolitan. (Celtic name, supposed to refer to curative properties.)

1. **S. parviflòrus** Raf. [*S. floribundus* HBK.] Stems usually solitary, erect, simple or branched above, 1.5–4 dm. high; basal lvs. often in a rosette, the blades round-obovate to oblong-spatulate, obtuse to rounded at apex, 2–5 cm. long, narrowed gradually into short winged perennials; cauline lvs. similar, shorter; fls. in loose elongate often panicled racemes; pedicels slender, divaricate, 1–2 cm. long; calyx 1–2 mm. long, with short triangular teeth; corolla white, ca. 1.5 mm. broad; caps. ca. 2.5 mm. broad; seeds ca. 0.3 mm. long.—Frequent in moist places, below 4000 ft.; Coastal Sage Scrub, Chaparral; cismontane to B.C.; Santa Cruz Id.; Atlantic Coast, Mex., S. Am. June–Aug.

88. Rafflesiàceae. Rafflesia Family

Dioecious or monoecious herbs, parasitic on roots or branches, the vegetative body reduced to mycelium-like tissues or thalloid. Lvs. usually scalelike. Fls. minute to large, unisexual. Calyx of 4–10 ± distinct segms. Corolla 0. Stamens indefinite; anthers sessile, usually 2-celled. Pistillate fls. with single pistil, the ovary ⊥ inferior, unilocular, of 4–6–8 carpels with parietal placentation and many ovules. Style 1 or 0. Stigma capitate or discoid or lobed. Fr. a berry or caps.; seed with endosperm. Seven genera and ca. 30 spp., mostly of Old World.

1. Pilóstyles Guill.

Minute stem-parasite; lvs. reduced to fl.-bracts. Fls. brown, minute; bracts roundish, imbricated; sepals 4–5, distinct, similar to the subtending bracts. Fls. with a thick cent. column expanded at apex into a fleshy disk. Anthers in ♂ fls. many, borne under margin of disk. Stigmas in ♀ fls. ring-shaped. Ovary inferior; fr. a many-seeded caps. Several spp., mostly of trop. Am. (Latin, *pilus*, hair, and *stylus*, a pillar or stylus.)

1. **P. thúrberi** Gray. Only the small brown fls. and their imbricated bracts visible on the outside of the host, the whole 1.5–2 mm. long.—On *Dalea*, especially *D. emoryi*; Creosote Bush Scrub; Imperial Co.; to Ariz. Jan.

89. Ranunculàceae. Crowfoot Family

Herbs, sometimes small shrubs or woody climbers. Lvs. alternate or opposite or basal, the blades simple to compound; petioles with dilated base, exstipulate. Fls. with the parts usually present, free and distinct, but exceptions are many; petals may be lacking and sepals petaloid; sepals and petals may be spurred; pistils may be somewhat united. Sepals mostly 3–15. Petals usually the same. Stamens usually many, hypogynous, maturing in centripetal sequence. Pistils few or many; ovary superior, 1-celled, with 1–many ovules and basal or parietal placentae. Fr. an ak. or follicle, sometimes a berry or caps. Ca. 35 genera and perhaps 1500 spp., mostly of N. Temp. and Arctic regions.

Pistils few- to several-ovuled, becoming follicles or berries when ripe.
  Fls. irregular; upper sepal spurred; petals 4, the upper pair spurred and included in the spurred sepal .... 5. *Delphinium*
  Fls. regular.
    Petals produced backwards into long spurs .... 3. *Aquilegia*
    Petals not spurred.
      Fls. solitary, cymose or panicled, 4-9 mm. long .... 6. *Isopyrum*
      Fls. racemose, 2-3 mm. long; fr. a berry .... 1. *Actaea*
Pistils 1-ovuled, becoming aks.
  Stem-lvs. opposite or whorled; petals none.
    The stem-lvs. forming a single whorl; herbs .... 2. *Anemone*
    The stem-lvs. many, opposite; woody vines .... 4. *Clematis*
  Stem-lvs. alternate or none.
    Petals present, rarely none in *Ranunculus*.
      Sepals spurred; small annuals with basal linear lvs. .... 7. *Myosurus*
      Sepals not spurred; lvs. usually cauline as well as basal .... 8. *Ranunculus*
    Petals absent; sepals greenish or petaloid; lvs. decompound; fls. usually unisexual
                                                                               9. *Thalictrum*

1. *Actaèa* L. BANEBERRY

Erect perennial herbs from short branching rootstocks. Stems tall, with 1-3 large 2-3-times ternate lvs. Fls. small, white, in terminal racemes. Sepals 4-5, petaloid, caducous. Petals 4-5, small, flat, spatulate, with slender claws. Stamens many. Pistil 1, with a broad sessile 2-lobed stigma; ovary 1-celled with 2 parietal placentae and several to many ovules. Fr. a berry; seeds smooth, lens-shaped, in 2 horizontal rows. Ca. 6 spp. of rich woods of the N. Temp. Zone. (Greek name of the elder.)

1. **A. rùbra** (Ait.) Willd. ssp. **argùta** (Nutt.) Hult. [*A. a.* Nutt. ex T. & G.] Stems 1-several, 2-6 dm. tall, branching above, sparsely pubescent; lvs. all cauline, the lowest 1-2 dm. wide, with thin mostly ovate, incised and serrate lfts. 2.5-6 cm. long and petioles to ca. 1 dm. long; racemes 1-5, dense in fl., 6-10 cm. long in fr.; pedicels 4-8 mm. long; sepals and petals 2-3 mm. long, falling away; stamens 5 mm. long; berries mostly red, sometimes white, oblong-ovoid to rounded, 6-8 mm. long; seeds dark, compressed, pitted, ca. 3 mm. long; $n=8$ (Rodriguez, 1949).—Rich moist woods, 5000-8000 ft.; Montane Coniferous F.; San Bernardino Mts.; Sierra Nevada to Alaska, Rocky Mts. May-June.

2. *Anemòne* L. ANÉMONE. WINDFLOWER

Perennial herbs, low to fairly tall. Lvs. mostly radical, lobed, divided or dissected, sometimes compound; stems lvs. 2 or 3 together forming an invol. subtending the fl. or remote from it. Fls. mostly showy, apetalous, the sepals colored or petaloid, 4-20. Stamens many, shorter than the sepals. Pistils many, ovary 1-ovuled. Fr. tailed or appendaged; aks. ribless, flattened, forming a head. Styles short. Ca. 100 spp., mostly of the N. Temp. Zone; a few of hort. importance. (Ancient Greek name, from *anemos*, wind.)

1. **A. tuberòsa** Rydb. Plate 75, Fig. B. Root tuberous; stems 1-3 dm. tall, subglabrous; lvs. few, 3-5 cm. wide, twice ternate, subglabrous, the divisions cuneate, ternately cleft; involucral bracts similar but short-petioled; peduncles 1-2, strigose; sepals rose, linear-oblong, 10-14 mm. long; aks. densely woolly, in an ellipsoid head; styles filiform, 1.5 mm. long; $2n=16$ (Joseph & Heimberger, 1966).—Dry rocky slopes, 3000-5000 ft.; Joshua Tree Wd., Pinyon-Juniper Wd.; w. edge Colo. Desert, e. Mojave Desert; to Utah, New Mex. April-May.

3. *Aquilègia* L. COLUMBINE

Perennial herbs from a thick caudex. Stems usually several, erect, branched. Lvs. largely basal, 2-3-ternate; the cauline gradually reduced up the stem. Fls. showy, terminating the branches, pendent or erect. Sepals 5, alike, colored. Petals 5, usually with a broad lamina projected to the front and a long hollow nectariferous spur projecting backward. Stamens many, with filiform fils.; innermost stamens represented by a sheath of flattened staminodia. Pistils usually 5, free, sessile, with terminal styles. Follicles separate, erect, reticulate-veined. Seeds black, shining, many, narrow-obovoid. Ca. 65 spp. of the N. Temp. Zone; many grown for their fls. (Derivation uncertain, possibly from Latin, *aquila*, eagle, because the spurs suggest the claws, or *aquilegus*, water-drawer, since many grow in moist places.)

(Munz, P. A. The cultivated and wild columbines. Gentes Herbarum 7: 1–150. 1946.)
Basal lvs. biternate; lfts. 2–5 cm. long; laminae of petals 1–4 mm. long .... 2. *A. formosa*
Basal lvs. mostly triternate.
    Lfts. grayish, 0.5–3 cm. long, with rounded lobes; laminae 3–5 mm. long. Deserts
                                                                       3. *A. shockleyi*
    Lfts. green, 1–4 cm. long, with pointed tips to lobes; laminae of petals lacking. Mt. Pinos region n. .... 1. *A. eximia*

1. **A. exímia** Van Houtte ex Planch. Plant densely glandular-pubescent throughout, 5–10 dm. high; basal lvs. triternate; petioles 1–2 dm. long; lfts. broadly cuneate-obovate to suborbicular, 1–4 cm. long, greenish above, glaucous beneath, cleft to near the middle, then lobed and with pointed teeth; fls. nodding, viscid-puberulent; sepals spreading or reflexed, reddish, ovate-lanceolate, 15–25 mm. long; laminae of petals obsolete; spurs scarlet with yellow near the orifice, 18–30 mm. long, 8–10 mm. wide; stamens 15–25 mm. long; follicles 15–25 mm. long, the styles almost as long; seeds ca. 2 mm. long.—Springy places, often on serpentine, below 6000 ft.; Chaparral, etc.; Mt. Pinos region, Ventura Co. to Mendocino Co. May–Aug.

2. **A. formòsa** Fisch. in DC. var. **truncàta** (F. & M.) Baker. [*A. t.* F. & M.] Plate 74, Fig. C. Stems glabrous or sparingly pubescent, 5–10 dm. high; basal lvs. ternate to biternate, pale green above, sometimes glaucous beneath; petioles 5–15 cm. long; lfts. rounded-obovate, deeply cleft, 1–2 (–3) cm. wide, the divisions few-toothed or -lobed; stem-lvs. few, reduced; pedicels short; fls. glandular-puberulent; sepals spreading, red, 10–20 mm. long; petal-laminae yellow, 1–2 mm. long; spurs red, 10–20 mm. long; stamens long; follicles glandular-puberulent, 15–25 mm. long, the styles an additional 15–25 mm.; seeds ca. 2 mm. long; $2n = 14$ (Skalinska, 1931).—Common in moist places below 8000 ft.; many Plant Communities; Riverside Co. to s. Ore., w. Nev. April–Aug. A form with viscid-pubescent stems and sepals 10–20 mm. long is the var. **hypolasia** (Greene) Munz. [*A. h.* Greene.] occurring below 7000 ft.; Los Angeles Co. to L. Calif. A more well marked one is var. **pauciflora** (Greene) Boothman [*A. p.* Greene.] with lvs. mostly basal; stems simple, scapelike, 1.5–3 dm. high and occurring from 5000 to 10,000 ft.; mts. in San Bernardino Co. to Shasta Co. June–Aug.

3. **A. shóckleyi** Eastw. [*A. mohavensis* Munz.] Stems 4–8 dm. high, glabrous and glaucous except for the branched glandular-pubescent infl.; basal lvs. mostly triternate, pale green above, more glaucous beneath, glabrous or somewhat pilose; petioles 5–20 cm. long; lfts. cuneate-obovate to suborbicular, 5–30 mm. long, cleft to middle or deeper, the principal segms. 2–3-lobed; cauline lvs. gradually reduced upward; fls. nodding, glandular-pubescent; sepals red or with some green or yellow, spreading or reflexed, lanceolate or elliptic, 10–20 mm. long, acutish; laminae yellow, 3–5 mm. long; spurs yellow-red to red, 12–25 mm. long; stamens 10–16 mm. longer than laminae; follicles glandular-puberulent, 17–25 mm. long, the styles ca. 12 mm. more.—Springy places, 5000–7700 ft.; Pinyon-Juniper Wd.; Panamint, Argus, Clark and New York mts., Mojave Desert; Nev. June–July.

4. *Clématis* L. CLEMATIS. VIRGIN'S-BOWER

Half-woody vines which climb by clasping or twining petioles, sometimes erect perennial herbs. Lvs. opposite, entire to pinnately compound. Fls. small to large, solitary to paniculate, sometimes individually showy, sometimes as a mass, on axillary peduncles, perfect to imperfect (plants dioecious or polygamo-dioecious). Sepals 4–5, valvate in bud. Petals 0, or sometimes represented by staminoid structures. Stamens many. Pistils many, 1-ovuled, forming 1-seeded, mostly tailed aks. in a head. Over 200 spp., mostly of temp. regions; many of hort. importance. (Ancient Greek name for some climbing plant.)

Fls. many in cymose panicles; lfts. 5–7 .... 2. *C. ligusticifolia*
Fls. 1–3 on a peduncle; lfts. 3–5.
    Sepals 1.5–2.5 cm. long, pubescent above; aks. pubescent .... 1. *C. lasiantha*
    Sepals mostly 0.8–1.2 cm. long, glabrous above; aks. glabrous .... 3. *C. pauciflora*

1. **C. lasiántha** Nutt. in T. & G. Plate 74, Fig. D. Woody climber to 4–5 m. high; branchlets woolly-pubescent; lfts. mostly 3, usually broad-ovate, 2–5 cm. long, rounded to subcordate at base, coarsely toothed to 3-lobed, with rounded teeth, strigose-pubescent especially beneath; petioles commonly 2–5 cm. long, the petiolules 0.3–1 cm. long; peduncles mostly 1-fld., 4–12 cm. long; sepals broadly oblong, silky-woolly without and within, 1.4–2.5 cm. long; aks. pubescent, the styles in fr. plumose, 2.5–3 cm. long; $n=8$ (Meurman & Th., 1939).—Clambering over shrubs and in low trees, in canyons and near streams, below 6000 ft.; largely Chaparral, sometimes Yellow Pine F., etc.; L. Calif. to Trinity and Shasta cos. March–June.

2. **C. ligusticifòlia** Nutt. in T. & G. Climbing, 4–6 or more m. high; branchlets almost glabrous except in infl.; lvs. pinnately 5–7-foliolate; lfts. lanceolate to lance-ovate (or ovate), rounded or subcuneate at base, mostly acuminate at apex, subentire to 3-lobed or coarsely toothed from ca. the middle, 2–8 cm. long, somewhat strigose especially beneath; petioles commonly 3–7 cm. long, the petiolules 1–3 cm.; peduncles densely strigose, 3–10 cm. long; fls. few to many; sepals white, oblong-oblanceolate, densely woolly-silky within and without, ca. 1 cm. long; aks. pubescent; the styles plumose in fr., 2–4 cm. long; $n=8$ (Meurman & Th., 1939).—Climbing in bushes and in trees, largely along streams and in moist places, below 7000 ft.; cismontane and montane, many Plant Communities; n. L. Calif. to B.C. and Rocky Mts. Santa Catalina and Santa Cruz ids. March–Aug. A s. Calif. form with ± silky-canescent lvs. is var. *californica* Wats. and one from Inyo Mts. n. with lfts. broad, subglabrous and cordate at base, is var. *brevifolia* Nutt.

3. **C. pauciflòra** Nutt. in T. G. Woody climber, 2–4 m. high; branchlets rather sparsely hairy; lfts. 3–5, round-ovate, toothed or lobed, mostly rounded at base and acute at apex, 1–2 cm. long, glabrous to sparsely strigose; petioles mostly 1–3 cm. long, the petiolules to 1 cm.; peduncles 1–3-fld., mostly 2–3 cm. long; sepals white, oblong, 8–12 (–15) mm. long, thin, woolly-pubescent without, glabrous within; aks. glabrous; styles in fr. plumose, 2–4 cm. long.—Clambering over shrubs, in canyon rocky slopes and in canyons, below 4000 ft.; Chaparral; Los. Angeles Co. s., inland to Little San Bernardino Mts.; to L. Calif. Santa Cruz Id. Jan.–April.

5. *Delphínium* L.

Erect branching herbs, ours perennial from fleshy or fibrous roots. Lvs. palmately lobed or divided. Fls. irregular, in showy spikes or racemes which may be paniculate. Sepals 5, blue, white, yellow or red, the posterior sepal prolonged into a spur into which project the nectary-bearing spurs of the 2 upper petals. Petals 4, in unequal pairs, the 2 lateral small and short-clawed to obsolete. Stamens many, usually included. Pistils 1–5, sessile, maturing into many-seeded dehiscent follicles. Seeds obpyramidal, the angles winged or wingless, sometimes inclosed in a loose papery membrane. Ca. 250 spp., of the N. Temp. Zone, some important in hort. and some as stock-poisoning plants especially for cattle and horses. (Latin, *delphinus*, dolphin, because of fl.-shape in some spp.)

(Lewis, H. & C. Epling. A taxonomic study of California delphiniums. Brittonia 8: 1–22. 1954.)
1. Fls. usually scarlet, sometimes yellowish .... 1. *D. cardinale*
1. Fls. blue to purple or bluish-white.
  2. Stem attenuate at base and very easily separable from the tuberous roots, glabrous
                                                                                                                            8. *D. patens*
  2. Stems not attenuate at base and rather firmly attached to woody or fibrous roots.
    3. Plants mostly 1–2 m. tall, leafy to the dense many-fld. raceme. Mt. meadows
                                                                                                                       2. *D. glaucum*
    3. Plants usually much less than 1 m. tall, the uppermost lvs. reduced to leafy bracts.

4. Fls. deep blue to purple; sepals 8–24 mm. long .... 6. *D. parishii*
4. Fls. violet or pale blue to whitish.
   5. Stems essentially glabrous.
      6. Fls. mostly bluish; sepals 8–12 mm. long; follicles 8–14 mm. long . 6. *D. parishii*
      6. Fls. white; sepals 6–7 mm. long; follicles 7–9 mm. long. Mt. Pinos region
                                                                                          4. *D. inopinum*
   5. Stems puberulent to hairy.
     5a. Sepals deep blue to purple, 10–15 mm. long .... 7. *D. parryi*
     5a. Sepals paler.
        7. Sepals pale purple, 6–8 mm. long; follicles glabrous. Mts. of Riverside and San Diego cos. .... 3. *D. hesperium cuyamacae*
        7. Sepals violet, 12–18 mm. long; follicles pubescent. San Clemente Id.
           8. Stem 4–5 dm. tall, ± spreading-pilose above; sepals 16–18 mm. long, oblong to subovate, the lf.-lobes subfalcate .... 5. *D. kinkiense*
           8. Stem 1–2.5 dm. tall, spreading- or deflexed-pilose; sepals ca. 12 mm. long, elliptic; lf.-lobes not subfalcate .... 9. *D. variegatum* ssp. *thornei*

1. **D. cardinàle** Hook. SCARLET LARKSPUR. Plate 74, Fig. E. Roots deep, thickened, woody; stems erect, simple or branched above, 1–2 m. tall, hollow, mostly puberulent throughout; basal lvs. withered at anthesis, 5–20 cm. wide, 5-parted, the primary divisions cuneate, shallowly to deeply 3-lobed, petioles 1–2.5 dm. long; cauline lvs. 5–7-parted, the primary divisions deeply divided into divergent linear to oblanceolate segms., subglabrous or thinly hairy on the veins; fls. in open racemes or panicles; bracts linear; pedicels mostly 2–6 cm. long; sepals ovate, obtuse, scarlet, rarely yellow, ovate, 10–16 mm. long; upper petals exserted, yellow with scarlet tips; lower petals narrow-oblong, the sinus closed, 1 mm. deep; follicles glabrous, erect, 1–1.5 cm. long; $2n = 16$ (Lewis et al., 1951).—Dry openings in brush and woods, below 5000 ft.; Coastal Sage Shrub, Chaparral, etc.; n. L. Calif. to cent. Calif. May–July.

2. **D. gláucum** Wats. Caudex stout, woody; stems coarse, glabrous, glaucous, fistulous, leafy, 1–2 m. high; lvs. 8–15 (–20) cm. broad, palmatifid into broad cuneate acutely incised and toothed segms., glabrous or pubescent beneath; petioles 4–10 cm. long; racemes 4–10 dm. long, many-fld., glabrous to somewhat pubescent, sometimes branched at base; pedicels ascending, 1–4 cm. long; lower bracts leafy, upper subulate; sepals light to dark violet-purple, rhombic-ovate, abruptly acute, 8–12 mm. long, puberulent on back; spur rather stout, 8–10 mm. long; upper petals narrow, notched, purple-tipped; lower petals rhombic-ovate, the sinus nearly closed, 2–2.5 mm. deep; follicles erect, glabrous or puberulent, 10–18 mm. long; seeds ovoid, 3 mm. long, wing-angled; $2n = 16$ (Lewis et al., 1951).—Wet meadows, 8000–9000 ft.; Montane Coniferous F.; San Bernardino and San Gabriel mts.; to Alaska, Rocky Mts. July–Sept.

3. **D. hespérium** Gray ssp. **cuyamácae** (Abrams) Lewis & Epling. Rootstock stout, elongate and with cluster of fibrous roots; stems erect, simple, rather slender, puberulent, 3–6 (–9) dm. tall; lvs. tending to wither early, well distributed along the stem, the blades 2–4 cm. wide, 3–5-palmatifid, the primary divisions lobed with apiculate segms., pubescent on both surfaces; petioles 3–6 cm. long; racemes strict, densely fld., 5–20 cm. long; pedicels ca. 5–15 mm. long; sepals pale blue to pale violet, 6–8 mm. long; upper petals clavate-oblique, the lower rounded, bearded; follicles glabrous.—Grassy meadows, 4000–5000 ft.; lower edge of Yellow Pine F.; San Jacinto Mts. to Palomar and Cuyamaca mts. May–July.

4. **D. inopìnum** (Jeps.) Lewis & Epling. [*D. parishii* var. *i.* Jeps. *D. p.* var. *pallidum* Munz. *D. amabile* ssp. *pallidum* Ewan.] Much like *D. parishii*; lvs. mostly basal, generally glabrous, ± fleshy; fls. small, cupped, white; sepals 6–7 mm. long, somewhat recurved at tips; spur 7–10 mm. long; follicles 7–9 mm. long, glabrous; $2n = 16$ (Lewis et al., 1951).—Open places, 5000–8000 ft.; Yellow Pine F.; Mt. Pinos region, upper Kern River Canyon. June–July.

5. **D. kinkiénse** Munz. Perennial, 4–5 dm. tall, from a woody branching root; stems 2.5–3 mm. thick, pubescent, with ca. 7–9 internodes; lower lvs. trifid almost to base, the main divisions ± palmately lobed into ultimate segms. 5–10 mm. wide, rounded at apex

to mucronulate, other lvs. only gradually reduced up the stem, the blades 5–8 cm. wide, the main division subfalcate; petioles 2–14 cm. long; infl. a raceme or few-branched panicle, pilose, the main axis 8–10-fld.; bracts lance-linear, 2–3.5 cm. long; pedicels 1.5–3 cm. long; sepals whitish to pale violet, 16–18 mm. long, oblong to subovate; upper petals whitish, the blades dolabriform, glabrous, those of lower petals deeply cleft, bearded; anthers dark; follicles 3, hairy.—Grassy places, San Clemente Id. March–April.

6. **D. paríshii** Gray ssp. **paríshii**. [*D. amabile* Tidestrom.] Plate 74, Fig. F. Perennial with several woody roots; stems erect, 1.5–6 dm. tall, often hollow, glabrous to pubescent, ± glaucous; lvs. largely basal or well distributed, often withered at anthesis, deltoid in outline, 2–8 cm. wide, the primary divisions with linear to oblong lobes 2–5 mm. wide, subglabrous to pubescent; petioles 3–15 cm. long; racemes compact to open, ca. 5–25-fld.; sepals ovate, lavender to light- or azure-blue, obtusish, 8–12 mm. long, mostly pubescent; spur 8–13 mm. long; upper petals whitish, notched; lower bluish or violet, mostly hairy; follicles glabrous to puberulent, erect, 8–14 mm. long; seeds with copious loose white pellicle; $2n = 16$ (Lewis et al., 1951).—Gravelly benches and washes, below 7500 ft.; Creosote Bush Scrub, Joshua Tree Wd.; Pinyon-Juniper Wd.; Mojave Desert from Mono Co. to w. end of Coachella V., Riverside Co. April–June.

Ssp. **purpùreum** Lewis & Epling. Stems largely fistulous, puberulent, glaucous; lvs. dissected into narrow lobes which are conspicuously pubescent with straight or curled hairs; sepals pinkish-purple.—Pinyon-Juniper Wd.; Mt. Pinos region, Cuyama V., Kern and Ventura cos. May–June.

Ssp. **subglobòsum** (Wiggins) Lewis & Epling. [*D. s.* Wiggins.] Stems slender, not at all fistulous; fls. dark blue-purple; sepals 8–10 mm. long; spur 10–13 mm. long, the tip curved downward; upper petals bluish to whitish, notched; lower rounded, villous, bright blue, the sinus ca. 2.5 mm. deep; follicles sparsely hairy.—Dry stony fans and slopes, below 5000 ft.; Creosote Bush Scrub, Chaparral, Pinyon-Juniper Wd.; w. edge of Colo. Desert, from Santa Rosa Mts. to L. Calif. March–May.

7. **D. párryi** Gray. Caudex with woody deep-seated roots; stems slender, usually simple, usually puberulent, 3–9 dm. tall; lvs. chiefly cauline, the lower withering early, main lf.-blades 2–8 cm. wide, 3–5-parted, then further divided into linear lobes, puberulent; petioles to ca. 1 dm. long; racemes 6–18 cm. long, usually loose, few- to many-fld.; pedicels 1.5–3 cm. long; bracts filiform; sepals deep purplish-blue, ovate, rounded to acutish, 10–15 mm. long, puberulent; spur ca. as long as sepals; upper petals exserted, oblique, whitish; lower rounded, purplish-blue, floccose; follicles puberulent, 10–15 mm. long; seeds nearly wingless to almost invested; $2n = 16$ (Lewis et al., 1951).—Common, open slopes and mesas below 6500 ft.; V. Grassland, Coastal Sage Scrub, Chaparral, S. Oak Wd., Yellow Pine F.; Santa Barbara Co. to San Diego Co., L. Calif. Santa Barbara Ids. Varying in width of lf.-segms.; if 5–7 mm. wide are f. *maritimum* (Davids.) Ewan, coastal bluffs, etc.

8. **D. pàtens** Benth. ssp. **hepaticoìdeum** Ewan. Roots tuberiform, in a globose cluster; stem slender, procumbent, 2–5 (–9) dm. long, glabrous; lvs. near base, long-petioled, 4–7 (–9) cm. wide, with 3 broadly obovate primary divisions, the cent. shallowly 3-lobed, the lateral 2-lobed; racemes lax, with long pedicels; sepals 13–15 mm. long, glabrous, dark blue; petals whitish or cream, with blue lines; follicles 13–14 mm. long, glabrous.—Shaded canyons, mostly below 5000 ft.; Chaparral, S. Oak Wd.; Santa Barbara Co. to n. L. Calif. April–May.

Ssp. **montànum** (Munz) Ewan. [*D. parryi* var. *m.* Munz.] Rather stout, 2.5–3.5 dm. high, glabrous; lvs. mostly basal and with main divisions shallowly lobed, a few lvs. cauline and deeply palmatisect; racemes compact, 6–12-fld.—At 5000–7500 ft.; Yellow Pine F.; Ventura Co. to San Bernardino Co. May–June.

9. **D. variegàtum** T. & G. ssp. **thórnei** Munz. Erect perennial from a slender simple subfusiform woody root, the stem slender, ± spreading- or deflexed-pilose with white hairs to ca. 0.5 mm. long, the stem 1–2.5 dm. tall, varying from simple and 4–5-fld. to 7–10-fld., sometimes forked near summit; cauline lvs. mostly green at anthesis, basal withered, cauline internodes ca. 4–8, the midcauline blades 1–2 cm. wide, deeply pin-

natisect, then pinnatifid with 1–3 almost linear divisions 5–12 mm. long, 1.5–2 mm. wide, densely pubescent; bracts subfoliose; pedicels divergent, 1–3.5 cm. long; sepals elliptic, violet, ca. 12 mm. long, 6–7 mm. wide, glabrous except near the acutish apex and ciliate margin; upper petals whitish, 9–10 mm. long, glabrous; lower whitish violet, somewhat hairy; follicles 3, pubescent.—Grassland at ca. 1600 ft.; San Clemente Id. April–May.

6. Isopỳrum L.

Low slender glabrous perennial herbs, ours with fleshy fibrous roots. Lvs. 2–3-ternate. Fls. mostly whitish, solitary or in cymes or panicles. Sepals 5–6, petaloid. Petals 0 in ours. Stamens many; fils. enlarged upward. Pistils few to many, sessile or stipitate, several-ovuled, becoming follicles. Seeds smooth. Ca. 25 spp. of N. Temp. Zone. (*Isopyron*, the Greek name for a sp. of *Fumaria*.)

1. **I. occidentàle** H. & A. Stems 1–several, slender, 10–25 cm. high, branched above; basal lf. long-petioled (5–10 cm.) with cuneate 2–3-lobed lfts. 1–2 cm. long, glaucous beneath; cauline lvs. few, reduced; sepals white, occasionally pink or purplish, oblong-obovate, 7–9 mm. long; stamens ca. 5 mm. long; follicles 8–10 mm. long, compressed, sessile; seeds several.—Infrequent and local, shaded places as among shrubs, below 5000 ft.; Chaparral, Foothill Wd., Yellow Pine F.; from Bouquet Canyon, n. Los Angeles Co. to Mt. Pinos and Tehachapi Mts.; to n. Calif. March–April.

7. Myosùrus L. MOUSE-TAIL

Small tufted annuals with fibrous roots. Lvs. basal, entire, linear or linear-spatulate. Fls. minute, greenish-yellow to whitish, solitary on naked scapes. Sepals 5, sometimes 6–7, spurred at base. Petals of same number if present, greenish-yellow, each with a nectar-bearing pit at summit of claw. Stamens 5–25. Pistils many, on a cylindrical axis which is long and spikelike in fr. Aks. apiculate or aristate. Ca. 5 spp., largely of moist flats, all continents. (Greek, *mus*, mouse, and *oura*, tail.)

(After Donald E. Stone in Mason, H. L. A flora of the marshes of Calif., 499. 1957.)
Mature fruiting spikes with divergent beaks, i.e., with the beak of each fr. spreading from the plane of the axis of the spike.
    Aks. thick, square, with horseshoe-shaped basal and lateral margin surrounding a large beak. Desert .... 2. *M. cupulatus*
    Aks. longer than broad, lacking a horseshoe-shaped margin. Montane ..... 1. *M. aristatus*
Mature fruiting spikes with the beaks of the fr. absent or closely appressed to the spike, i.e., directed in a plane parallel to the axis of the spike, the spike thus appearing smooth.
    Fruiting spikes extending beyond the lvs. .... 3. *M. minimus filiformis*
    Fruiting spikes not extending beyond the lvs. .... 3. *M. minimus apus*

1. **M. aristàtus** Benth. ex Hook. Tufted annual, 3–8 cm. tall; lvs. linear to linear-spatulate, 1–5 cm. long; scapes very slender, 2–5 cm. tall; sepals 5–6, oblong, the spur variable, from ca. ⅓ as long to as long as the blade; petals linear, often lacking at maturity; fruiting spikes 5–10 mm. long; aks. somewhat quadrate in outline, the back sharply keeled, extending to the elongate, divergent, sometimes falcate beak.—Moist places, 6000–7500 ft.; Yellow Pine F.; San Bernardino Mts.; n. Calif. to B.C., Rocky Mts. May–July.

2. **M. cupulàtus** Wats. Tufted, 3–8 cm. tall; lvs. linear to linear-spatulate, 1–5 cm. long; sepals 5–6, oblong, the spur to ⅓ the length of the blade; petals linear-filiform; scapes 3–8 cm. high; fruiting spikes slender, 2–5 cm. long, aks. with a thickened margin producing a cuplike depression on back from which protrudes the divergent beak; seed short-oblong, flattened.—Dry limestone slopes, 3500–5500 ft.; Joshua Tree Wd., Pinyon-Juniper Wd.; Little San Bernardino Mts., Providence, New York and Kingston mts.; to New Mex. April–May.

3. **M. mínimus** L. var. **àpus** Greene. Plate 75, Fig. A. Tufted annual, 2–6 cm. tall; lvs. linear, terete to subspatulate, 2–6 cm. long, extending beyond the spike; spike sessile

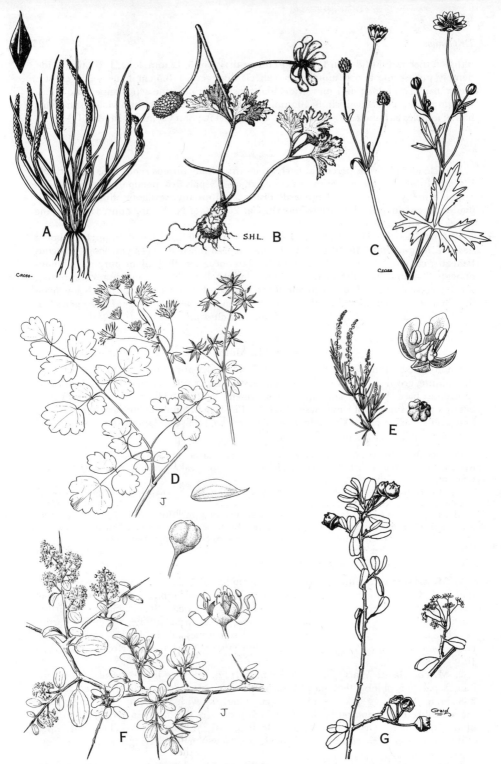

Plate 75. Fig. A, *Myosurus minimus* var. *apus*, with separate ak. Fig. B, *Anemone tuberosa*. Fig. C, *Ranunculus californicus*. Fig. D, *Thalictrum fendleri* ♂ and ♀ fls. and ak. Fig. E, *Oligomeris linifolia*, petals 2, caps. open above. Fig. F, *Ceanothus cordulatus*. Fig. G, *Ceanothus megacarpus*.

to short-stalked, cylindrical to subconical, 1–2 cm. long; stalk when present very stout; sepals 5–10, the spur blunt; beak of ak. closely appressed to spike, approximately ½ as long as the body.—Vernal pools and alkaline marshes, below 1500 ft.; San Diego to w. Riverside Co. April–May.

Var. **filifórmis** Greene. Scape and spike very slender; beak of the ak. very short to inconspicuous.—Vernal pools below 3000 ft.; mesas near San Diego; Cent. V. and Mendocino Co.; n. L. Calif. March–May.

8. *Ranúnculus* L. BUTTERCUP. CROWFOOT

Annual and perennial herbs. Root systems from fibrous to fleshy to tuberous. Lvs. alternate, entire to compound, often largely basal. Fls. solitary on ends of stem or the many branches, or sometimes ± panicled or corymbose, yellow or white. Sepals usually 5; petals mostly 5, sometimes fewer or more, each with a nectary at the base. Stamens mostly rather many, short. Pistils 5 to many, the ovary 1-loculed and 1-ovuled, ripening into a small hard ak. with a persistent beak (style). Frs. mostly clustered on the receptacle to form a headlike group. Perhaps 250 spp., widely dispersed in temp. and cold countries, often in damp or wet places; some of hort. value. (Latin, little frog, because of the moist habitat of many spp.)

(Benson, L. A treatise on the N. Am. Ranunculi. Am. Midl. Natur. 40: 1–261. 1948.)
1. Petals white; aquatics with submersed lvs. dissected; aks. transversely ridged . 3. *R. aquatilis*
1. Petals yellow or red, rarely white in age; plants mostly terrestrial; aks. not transversely ridged.
   2. The petals red, ca. 1.5 cm. long; aks. 7–14 mm. long .... 2. *R. andersonii*
   2. The petals yellow, usually shorter; aks. mostly 1–5 mm. long.
      3. Plants annual; petals 1–3 mm. long.
         4. Ak.-beak ca. 0.5 mm. long; basal lvs. 2–2.5 cm. broad .... 9. *R. hebecarpus*
         4. Ak.-beak 2 mm. long; basal lvs. 4–14 cm. broad .... 11. *R. uncinatus*
      3. Plants perennial.
         5. The plants, at least the stems, hirsute or evidently pubescent.
            6. Petals 2.5–6 mm. long .... 11. *R. uncinatus*
            6. Petals 7–18 mm. long.
               7. Receptacle pubescent or hispid; sepals reflexed from base; petals 9–16
                                                                           4. *R. californicus*
               7. Receptacle glabrous.
                  8. Body of ak. mostly 4–5 mm. long, the beak with broad thin base 0.6–1 mm. wide; petals 5–10 .... 5. *R. canus*
                  8. Body of ak. mostly 2–3 mm. long, the beak with base less than 0.5 mm. wide.
                       9. Petals mostly 5–6, mostly 1–2 times as long as broad ..... 10. *R. occidentalis*
                     9. Petals mostly 8–15, mostly 2–3 times as long as broad ..... 4. *R. californicus*
        5. The plants glabrous or nearly so.
           10. Lower lvs. 3–7-parted or -foliate or even dissected.
               11. Stem 0.4–1.5 dm. high; ak.-beaks straight. High montane .... 7. *R. eschscholtzii*
               11. Stems 1.5–9 dm. long; ak.-beaks mostly curved.
                   12. Petals 6–15 mm. long.
                       13. Petals mostly 5–6 .... 10. *R. occidentalis*
                       13. Petals mostly 7–16 .... 4. *R. californicus*
                   12. Petals 2.5–4 mm. long .... 11. *R. uncinatus*
           10. Lower lvs. entire or shallowly lobed.
               14. Stems not rooting at the nodes .... 1. *R. alismaefolius*
               14. Stems rooting at the nodes.
                   15. The stem not differentiated into stolons and scapes; receptacle glabrous, 1–2 mm. long in fr. .... 8. *R. flammula*
                   15. The stems differentiated into stolons and scapes; receptacle hairy, cylindroid, 4–7 mm. long in fr. .... 6. *R. cymbalaria*

1. **R. alismaefòlius** Geyer ex Benth. var. **alisméllus** Gray. [*R. alismellus* Greene.] Glabrous perennial with gradually tapering roots; stems ascending, 1–3 dm. high, mostly 1–1.5 mm. thick; blades of basal lvs. lance-ovate, 2–4 cm. long, 4–10 mm. wide, abruptly

narrowed at base; sepals spreading, yellowish-green; petals ca. 6 mm. long; receptacle glabrous, ca. 1 mm. long; aks. in a rounded head, glabrous, 1.5–2.5 mm. long, the beak not recurved.—Meadows and wet banks, 7000–9000 ft.; Montane Coniferous F.; San Jacinto and San Bernardino mts.; Sierra Nevada to Wash., Mont.; L. Calif. June–July.

2. **R. andersònii** Gray. [*Beckwithia a.* Jeps.] Glabrous perennial with short caudex and slender roots; scapes erect, 1–2 dm. high, 1–2-fld.; lvs. cordate or rounded in outline, 1–3 cm. long, 2–4 cm. broad, 3-foliolate, the lfts. dissected into ovate to linear segms. 7–8 mm. long, 1–3 mm. wide; petioles 3–6 cm. long; sepals reddish, thin, spreading, 7–8 mm. long, persistent; petals 5, pink to red, cuneate to round, 1.5 cm. long; stamens many; aks. 15–25 in a round cluster, each utricular, obovoid, 7–14 mm. long, with very short beak.— Dry rocky slopes, 3000–7500 ft.; Sagebrush Scrub, Pinyon-Juniper Wd.; Panamint Mts., Grapevine Mts., Inyo Mts.; to Modoc Co.; Ore., Ida., Nev. June.

3. **R. aquátilis** L. var. **capillàceus** (Thuill.) DC. Aquatic perennial with submersed stems 2–several dm. long, 1–2.5 mm. in diam.; lvs. all submersed and dissected into filiform divisions; sepals light green, 2–3 mm. long; petals 5, white, 4–8 mm. long; aks. glabrous or nearly so, 1–1.5 mm. long, with minute beaks; receptacle with tufted hairs.— Ponds and slow streams, below 7500 ft.; many Plant Communities; widely distributed in Calif.; to Alaska, Rocky Mts., Atlantic Coast; Old World. April–Aug.

Var. **hispídulus** E. Drew. Submersed lvs. dissected into filiform divisions; floating lvs. 1–10, simple, 1–2 cm. wide, 3-lobed or -parted, the lobes again forked or parted; pedicels 1–3 cm. long; petals 4–6 mm. long.— Ponds, ditches and slow streams. At 4000–5000 ft.; Cuyamaca Mts., San Diego Co.; cent. Calif. to Alaska, Utah, Wyo. April–July.

4. **R. califórnicus** Benth. var. **califórnicus**. Plate 75, Fig. C. Perennial with slender roots; stems erect, 3–7 dm. high, 1.5–5 mm. in diam., hirsute to subglabrous, branched above; basal lvs. 2–7 cm. broad, long-ovate or orbicular, 3–5-foliolate or 3-parted, the lfts. or lobes cuneate, lobed, mostly hirsute, the ultimate lobes acute; petioles 5–25 cm. long; sepals greenish-yellow, reflexed, 4–8 mm. long; petals mostly 9–16, yellow, 8–15 mm. long, 3–5 mm. broad; stamens 30–60; aks. 5–35, in a rounded head, each obovoid-discoid, 2–2.5 mm. long, the beaks stout, recurved, 0.4–0.8 mm. long; $2n=28$ (Coonen, 1939).—Vernally moist slopes and meadows, below 3000 ft.; V. Grassland, S. Oak Wd., etc.; L. Calif. n. to s. Ore. Santa Rosa, Santa Cruz, San Miguel ids. Feb.–May.

Var. **cuneàtus** Greene. Stems hirsute to pubescent, prostrate, 1–2.5 dm. long; lvs. with ultimate lobes cuneate, rounded or obtuse.—Coastal bluffs and hills; Coastal Sage Scrub; Santa Cruz and San Miguel ids.; to s. Ore. Feb.–May.

Var. **austromontànus** L. Benson. Densely pilose-hirsute; lvs. fanshaped; petals 6–8 mm. long.—Meadows and stream banks, 4500–7500 ft.; Montane Coniferous F.; San Bernardino Mts. to San Diego Co. May–July.

5. **R. cànus** Benth. var. **ludoviciànus** (Greene) L. Benson. [*R. l.* Greene.] Perennial with slender roots; stems erect, very hairy, 2.5–4.5 dm. high, 3–5 mm. thick, branched above; lvs. 2–5 cm. wide, 3-parted into cuneate, appressed-hirsute lobes; sepals greenish-yellow, reflexed at middle, deciduous, 5–7 mm. long; petals yellow, 10–23, 10–13 mm. long, 3–5 mm. broad; stamens 40–70; aks. 12–30, in a rounded head, each discoid, 4–5 mm. long, the beak deltoid, 0.5–1 mm. long, recurved at tip.—Vernally moist ground, 3300–7500 ft.; mostly Yellow Pine F.; mts. from Kern Co., San Bernardino Mts., San Diego Co. March–May.

6. **R. cymbalària** Pursh. var. **saximontànus** Fern. Glabrous or sparingly hairy perennials with slender roots, stoloniferous; stems ca. 1 mm. thick, several dm. long; basal lvs. simple, cordate, ovate or even reniform, 1–4 cm. long, 1–3 cm. broad, mostly crenate, on petioles 2–5 cm. long; scapes erect, 5–30 cm. high, 1–several-fld.; sepals greenish-yellow, spreading; petals 5–12, yellow, obovate, 4–8 mm. long; stamens 20–35; aks. many, in an elongate head, each cuneate-oblong, 1.5 mm. long, the beak 0.3 mm. long, not curved; receptacle hairy.—Muddy places, often subalkaline; below 10,500 ft.; many Plant Communities; Santa Ana River system, San Bernardino Mts., Mojave Desert, to Mono Co., Sierra Nevada to n. Calif., B.C., Nebr., Tex., L. Calif. June–Aug.

7. **R. eschschóltzii** Schlecht. var. **oxynòtus** (Gray) Jeps. [*R. o.* Gray.] Perennial; caudex 1.5–5 cm. long, 5–12 mm. thick, usually branched; lf.-sinuses and -lobes rounded or obtuse, the middle lobe of the basal lvs. usually entire; stipular lf.-bases persistent, thickened; stems erect or decumbent, 4–15 cm. long, scapose; sepals yellow, tinged lavender, glabrous or sparsely hairy, 4–8 mm. long; petals 5, yellow, cuneate-obovate, 7–11 mm. long, 5–10 mm. broad; stamens 20–40; aks. in ovoid head, each oblong-obovoid, glabrous, ca. 1.5 mm. long, the beak slender, rather straight, ca. 1 mm. long; receptacle glabrous.—At 8750–11,400 ft., in rock-crevices, etc.; Subalpine F.; San Jacinto and San Bernardino mts., White Mts.; to Modoc Co. July–Aug.

8. **R. flámmula** L. var. **ovàlis** (Bigel.) L. Benson. Almost glabrous perennials with filiform roots; stems reclining, usually rooting at nodes, slender, 1–4 dm. long; lvs. often fascicled at rooting nodes, simple, entire, 1–5 cm. long, linear-spatulate to oblanceolate; petioles 1–6 cm. long; pedicels from any node, 1–8 cm. long; sepals greenish-yellow, spreading or somewhat reflexed, 3–5 mm. long, glabrous to strigose; petals 5 or 10, yellow, obovate, 3–6 mm. long; stamens 20–30; aks. 10–25 in a rounded head, each 1–1.5 mm. long, with very short recurved beak; receptacle glabrous.—Muddy and marshy places, Montane Coniferous F.; reported from San Bernardino Mts.; to Alaska, Atlantic Coast. July–Aug.

9. **R. hebecárpus** H. & A. Annuals with filiform roots; stems erect, slender, pubescent, 1–3 dm. high, freely branched throughout; basal lvs. cordate-reniform in outline, 1–2 cm. wide, 3-parted and then lobed, the ultimate segms. oblong-ovate, acute, strigose; petioles 3–9 cm. long; pedicels 1–16 mm. long; sepals yellow, 1.5 mm. long, scarious-margined, pubescent; petals 0–5, yellow, 1.5 mm. long; stamens 8–15; aks. 4–10 in a round head, each 2 mm. long, papillate with hooked bristles, the beak stout, 0.6 mm. long, hooked at tip; receptacle glabrous.—Shaded places, below 3000 ft.; Coastal Sage Scrub, Chaparral, S. Oak Wd.; L. Calif. through cismontane Calif. to Wash., Ida. March–May.

10. **R. occidentàlis** Nutt. var. **eisènii** (Kell.) Gray. [*R. e.* Kell.] Perennial with slender roots; stems erect, 3–7 dm. high, 2–5 mm. thick, branched above, mostly hirsute; basal lvs. fan-shaped or semicircular in outline, 2–5 cm. wide, 3-parted, the lobes cuneate, again lobed, the ultimate teeth acutish; petioles 3–12 cm. long, hirsute; pedicels 2–10 cm. long; sepals 5, greenish-yellow, reflexed at middle, 4–8 mm. long; petals usually 5, yellow, elliptic, 7–12 mm. long, 4–8 mm. wide; stamens 25–50; aks. mostly 8–15, in hemispheric cluster, each discoid to discoid-obovoid, 2–3.5 mm. long, glabrous to strigose, the beak 0.5–1 mm. long, usually recurved.—Vernally moist places, below 6000 ft.; Tehachapi Mts. to Ore. March–May.

11. **R. uncinàtus** D. Don in G. Don. Annual or perennial, glabrous to sparsely hirsute, 2–4.5 dm. high; basal lvs. cordate-reniform in outline, 2–9 cm. broad, 3-parted, the parts lobed, glabrous, with ultimate segms. mostly obtuse; petioles 5–20 cm. long; cauline lvs. often larger than basal; pedicels to 2 cm. long in fl., twice that in fr.; sepals reflexed, 3 mm. long; petals 5, yellow, 2.5–3 mm. long; stamens 10–15; aks. 5–30, in a round head, the body discoid, 2–2.5 mm. long, the beak 1–1.5 mm. long, recurved, hooked at tip.— Occasional, moist shaded places below 7500 ft.; Montane Coniferous F.; San Bernardino Mts.; to Alaska, Rocky Mts. Plants with reddish-brown instead of white hairs, with ultimate lf.-segms. acute instead of obtuse and with ak.-beaks 2 mm. long may be the var. *parviflorus* (Torr.) L. Benson and may occur in San Bernardino Mts. May–July.

9. *Thalíctrum* L. MEADOW-RUE

Erect perennial herbs from a short caudex. Lvs. alternate, ternately decompound, ample, the amplexicaul petioles dilated at the base. Fls. often unisexual, paniculate, subcorymbose or rarely racemose, numerous. Sepals 4 (–7), fugacious, greenish or petaloid. Petals none. Stamens free, numerous, exceeding the sepals at anthesis, the fils. often colored, erect or pendent. Carpels several on a small receptacle, with a single pendulous ovule, the aks. often compressed, sometimes inflated, longitudinally ribbed or veined.

A genus of ca. 120 spp., most abundant in temp. N. Am. and Eurasia. (Greek, *thalictron*, name given to the Meadow-Rue by Dioscorides.)

(Boivin, B. American Thalictra and their Old-World allies. Contr. Gray Herb. 151. 1944.)

Fls. perfect, paniculate; anthers obtuse .... 3. *T. sparsiflorum*
Fls. unisexual, paniculate; anthers apiculate.
    Carpels and under surface of upper lvs. glandular-puberulent, the herbage green; aks. ovate-lanceolate, strongly 3–5-nerved, not obviously reflexed or inflated. Pine belt. . 1. *T. fendleri*
    Carpels and lvs. entirely glabrous, the herbage glaucous; aks. obovate, 1-nerved, or the several veins ± reticulate, ± reflexed and when fresh turgid. Below the pine belt .. 2. *T. polycarpum*

1. **T. féndleri** Engelm. ex Gray. Plate 75, Fig. D. Habit of *T. sparsiflorum*, 6–15 dm. high, minutely glandular-puberulent above, especially on the carpels, pedicels and under surface of the lvs. not glaucous; lvs. 2–4-ternate, the lfts. roughly 3-lobed and crenate; panicle open, foliose; sepals narrowly to broadly elliptic, 2–3 mm. long, scarious; aks. oblanceolate to obovate, ± ventricose and compressed, subsessile, 5–6 mm. long, 3–4-ribbed on each side, the nerves rarely branched but not anastomosing; $n=14$ (Langlet, 1927, Clausen, 1940).—Moist soil near streams and meadows, often in shade and thickets, occasional in dryer places, 4000–10,000 ft.; Montane Coniferous F.; n. L. Calif. and Laguna Mts., San Diego Co. to cent. Ore., Wyo., Tex. May–Aug.

2. **T. polycárpum** (Torr.) Wats. More robust, 6–18 dm. high, glabrous throughout and ± glaucous; aks. many, in globular heads, vesicular-inflated when fresh, subglobose to obovoid, obliquely ventricose, subsessile, 4–7 mm. long, often with 1, sometimes 2, longitudinal wavy nerves on each side with branched anastomosing nervelets; otherwise as in *fendleri*.—Frequent, mostly below 2500 ft.; S. Oak Wd., Chaparral, etc.; San Diego Co. and cismontane s. Calif. n. to Columbia R. March–June.

3. **T. sparsiflòrum** Turcz. [*T. s.* var. *nevadense* Boivin.] Erect, slender, branching above, 3–12 dm. high, minutely glandular-puberulent throughout, especially on the carpels and under surface of lvs.; lvs. 2–4-ternate, the upper subsessile, the lfts. 1–2 cm. long, rotund or cordate at base; panicle foliose; sepals 3–4 mm. long; aks. 4–12, semi-obovate (scimitar-shaped), strongly compressed, the dorsal angle straight, beaked and stipitate, the 4–5 lateral nerves curving upward; $n=21$ (Kuhn, 1928).—Moist stream banks and bogs, 5000–10,000 ft.; Montane Coniferous F.; San Jacinto Mts., San Bernardino Mts.; White Mts., etc., n. to Alaska, Siberia. Not common with us. July–Aug.

90. Resedàceae. MIGNONETTE FAMILY

Annual or perennial herbs with watery juice, rarely woody. Lvs. alternate, simple or pinnately divided, with small glandlike stipules. Fls. small, perfect, irregular, in racemes or spikes. Calyx 4–7-parted, persistent. Petals small and inconspicuous, 2–7 or 0, often laciniate, clawed. Stamens 3–40, usually in a ± one-sided hypogynous disk; fils. free or united at base; anthers 2-celled, introrse. Ovary of 2–6 free or connate carpels, 1-celled, closed or opening at top, each carpel with its own stigma. Fr. a gaping caps. or a berry; seeds many, subreniform, without endosperm. Ca. 6 genera and 65 spp., mostly of Medit. region.

Petals 2; disk lacking; lvs. entire .... 1. *Oligomeris*
Petals 4–7; disk cup-shaped; lvs. often pinnatifid .... 2. *Reseda*

1. Oligómeris Camb.

Low branching herbs. Lvs. linear, entire. Fls. small, greenish, in terminal spikes. Sepals 4. Petals 2, entire or lobed, persistent. Stamens 3–10. Caps. 4-lobed, each lobe sulcate on back. Seeds many. Ca. 5 spp., sw. U.S., Mex., Afr., Asia. (Greek, *oligos*, little, and *meris*, parts.)

1. **O. linifòlia** (Vahl) Macbr. Plate 75, Fig. E. Annual, erect, glabrous, rather fleshy, 1–3 dm. high; lvs. often fascicled, 1–2.5 cm. long; spikes bracteate, densely fld., 2–10 cm. long; fls. ca. 1.5 mm. long; petals oblong, whitish, obscurely lobed; stamens 3; caps. 1.5

mm. long, 2.5 mm. thick, 4-toothed; seeds ca. 0.5 mm. long.—Common in open often subsaline places, below 3000 ft.; Creosote Bush Scrub, Alkaline Sink; deserts; occasional on sea bluffs, alkaline places, Santa Rosa and Catalina and Santa Barbara and San Nicolas ids.; to L. Calif., Tex., Mex. Feb.–July.

2. Resèda L. MIGNONETTE

Erect or decumbent herbs, sometimes almost woody at base. Lvs. entire, lobed or pinnatifid. Fls. small, in terminal spikes or racemes. Petals 4–7, cleft, unequal. Disk cupshaped. Stamens 8–40. Caps. 3–6-horned or -angled, opening only at the top at maturity. Ca. 50 spp., mostly of Medit. region and about the Red Sea. (Latin, *resedare*, to assuage or calm, because of supposed sedative properties.)

Fls. greenish-white; stamens 12–15; all petals cleft .... 1. *R. alba*
Fls. pale yellow; stamens 15–20; lowest petal not cleft .... 2. *R. lutea*

1. **R. álba** L. Biennial to perennial, glabrous, erect, 3–8 dm. high; lvs. 3–12 cm. long, deeply pinnatifid into lance-oblong segms.; fls. white in dense spikelike racemes; petals 6 (5), 4–5 mm. long, 3-cleft; caps. ovoid-oblong, 1–1.2 cm. long, usually 4-toothed; seeds subreniform, dark, dull, minutely papillose, ca. 1 mm. long; $2n=20$ (Eigsti, 1936).— Established occasionally in waste places, roadsides, etc.; native of Eu. May–Nov.

2. **R. lùtea.** L. Biennial to perennial, pubescent to glabrous; stems slender, ascending to suberect, 3–8 dm. long; lvs. bipinnatifid to pinnate-parted, 5–10 cm. long, the segms. oblong to linear; racemes narrow, 1–2 dm. long; sepals mostly 6; petals 6, all but the lowest cleft, 3–4 mm. long; stamens 15–20; caps. ovoid-oblong, 7–8 mm. long, usually apically toothed; seeds obovoid, shining, smooth, 1.4–1.8 mm. long; $2n=48$ (Eigsti, 1936).—Sparingly established in waste places; native of Eu. May–Sept.

91. **Rhamnàceae.** BUCKTHORN FAMILY

Shrubs or small trees, sometimes climbers, often thorny. Lvs. simple, alternate, sometimes opposite, with small deciduous stipules or these sometimes thick, corky, persistent. Fls. small, regular, perfect or imperfect, usually in terminal or axillary cymes, corymbs or panicles made up of small umbels. Calyx ± tubular at base, 4–5-lobed, lined with a disk on edge of which are inserted the petals and stamens. Petals 4–5, usually clawed, hooded, sometimes lacking. Stamens 4–5, opposite the petals; anthers short. Ovary 2–3-loculed, partly immersed in the disk; ovules 1–2 in each locule. Styles and stigmas ± united. Fr. a caps. or drupe. Ca. 45 genera and 600 spp. of temp. and trop. regions.

Fr. fleshy, drupelike.
    Drupe with 1 nutlet; petals clawed or none; spinose desert shrubs.
        Petals lacking; fr. 3–6 mm. long .... 4. *Condalia*
        Petals present; fr. 6–15 mm. long .... 5. *Condaliopsis*
    Drupe with 2–3 nutlets; petals clawless or none; plants not usually spinose ..... 6. *Rhamnus*
Fr. dry, capsular.
    Calyx-tube free from lower part of ovary, the sepals persistent; style jointed. Sw. San Diego Co.
                                                                                                             1. *Adolphia*
    Calyx-tube joined to base of ovary, the sepals deciduous in fr.; style not jointed.
        Pedicels and calyx glabrous; calyx-lobes or sepals petaloid; petals showy, hooded and longclawed, often spreading away from the stamens. Common, widespread ...... 2. *Ceanothus*
        Pedicels and calyx tomentose; calyx-lobes not petaloid; petals small, sessile, surrounding the stamens. Rare, desert plant .... 3. *Colubrina*

1. *Adólphia* Meissn.

Shrubs with opposite divaricate spinose twigs which are articulated with the stems. Lvs. opposite, small, mostly caducous, with stipules. Fls. inconspicuous, solitary or in few-fld. axillary clusters. Sepals 5. Petals 5, hooded. Caps. 3-loculed, 3-lobed, the lower third surrounded by but mostly free from the cuplike calyx-base. Style 3-cleft, articulate near the base. Seeds 1 in each locule, smooth, bony. Ca. 2 spp. of sw. N. Am. (*Adolphe Brongniart*, French botanist and student of Rhamnaceae.)

1. **A. califórnica** Wats. Intricately branched, ca. 1 m. high, the branches short-pubescent, green, stiff, divaricate, striate, the ultimate spinescent; lvs. oblong or obovate, entire or nearly so, 3–12 mm. long, short-petioled, entire, puberulent; fls. solitary or few, short-pedicelled; calyx pubescent, greenish-white; petals white, ca. 2 mm. long; caps. globose, 4–5 mm. in diam., 3-grooved; seed dark, ca. 3.5 mm. long, rounded, oblong, plump.—Dry canyons and washes; Chaparral; sw. San Diego Co.; n. L. Calif. Dec.–April.

2. *Ceanòthus* L. CALIFORNIA-LILAC

Shrubs or small trees, often with divaricate, sometimes spiny twigs. Lvs. alternate or opposite, deciduous or in most of ours persistent, frequently serrate, 3-nerved from the base or strictly pinnately veined, ± petioled. Stipules present. Fls. small but showy, white to blue or purplish, sometimes lavender or pinkish, borne in terminal or lateral panicles or umbellike cymes. Sepals 5, somewhat petallike, united at the base with the urn-shaped receptacle that is filled with a glandular disk in which the ovary is immersed. Petals 5, distinct, hooded and clawed. Stamens 5, opposite the petals, with elongate fils. Ovary 3-loculed, 3-lobed, with a short 3-cleft style. Fr. a 3-lobed caps., separating at maturity into 3 parts. Seeds smooth, convex on 1 side. Between 50 and 60 spp. of temp. N. Am., many of considerable hort. value. Hybridizing freely. (Greek, *keanothus*, name used by Dioscorides for some spiny plant.)

(McMinn, H. E. A systematic study of the genus Ceanothus. Part II of Ceanothus. Santa Barbara Bot. Gard., 1942.)

1. Stipules thin and early deciduous; lvs. alternate, with stomata not sunken; caps. without horns, but often with ridges or crests; fls. in simple or compound panicles.
  2. Ultimate branchlets flexible, not rigidly divaricate and spinose.
    3. Lvs. with 3 distinct veins from the base, the lateral sometimes obscure.
      4. The lvs. entire, ± deciduous, mostly more than 2 cm. long; fls. mostly white.
                                                                                        8. *C. integerrimus*
      4. The lvs. serrate, serrulate or glandular-denticulate; fls. blue or purple.
        5. Branchlets angled and with scattered brownish glands. San Diego Co. . 5. *C. cyaneus*
        5. Branchlets terete, not angled or striate.
          6. Lvs. tomentose beneath.
            7. Lf.-blades 3–8 cm. long. Insular .... 1. *C. arboreus*
            7. Lf.-blades 1–2.5 cm. long. Mainland .... 16. *C. tomentosus*
          6. Lvs. ± pubescent beneath, chiefly on the veins.
            8. Branchlets distinctly hispid or villous; lvs. usually pubescent above. San Diego Co. north .... 11. *C. oliganthus*
            8. Branchlets subglabrous or slightly pubescent; lvs. usually glabrous above. Orange and Riverside cos. north .... 14. *C. sorediatus*
    3. Lvs. with 1 main vein from the base, even though the basal lateral pair may be longer than the other pairs.
      9. Lvs. entire, mostly rather glabrous.
        10. Shrubs evergreen; lvs. rather firm and thick.
          11. Fls. white; caps. 5–7 mm. broad, conspicuously crested. At elevs. greater than 3000 ft. .... 12. *C. palmeri*
          11. Fls. blue to almost white; caps. 4–5 mm. broad, scarcely crested. From below 3000 ft. .... 15. *C. spinosus*
        10. Shrubs deciduous; lvs. thinnish; fls. mostly white .... 8. *C. integerrimus*
      9. Lvs. dentate to serrulate, usually pubescent at least beneath.
        12. Lf.-margins ± revolute, the upper lf.-surface glandular-papillate. Orange Co. north .... 13. *C. papillosus*
        12. Lf.-margins not revolute, the upper lf.-surface glossy. San Diego Co. north
                                                                                                   6. *C. foliosus*
  2. Ultimate branchlets rigidly divaricate and spinose; lvs. persistent, plane.
    2a. Lvs. with 1 main vein from base, mostly entire and glossy on both surfaces   15. *C. spinosus*

2a. Lvs. with 3 main veins from the base, the 2 lateral sometimes obscure, gray-glaucous on both surfaces.
    2b. Shrubs 2–4 m. tall; fls. whitish to blue, the clusters 3–8 cm. long. Mostly below Yellow-Pine belt .... 9. *C. leucodermis*
    2b. Shrubs less than 2 m. high; fls. white, the clusters 1.5–3 cm. long. Mostly above Yellow-Pine belt .... 2. *C. cordulatus*
1. Stipules with thick corky persistent bases; lvs. alternate or opposite, the stomata in sunken pits; caps. usually with 3 horns; fls. in lateral umbels; lvs. firm-coriaceous, persistent.
  13. Lvs. alternate at least in part.
    14. The lvs. all alternate, broadly obovate to almost round; caps. 4–6 mm. in diam. W. San Diego Co. .... 17. *C. verrucosus*
    14. The lvs. both alternate and opposite, oblanceolate to broadly elliptic; caps. 8–12 mm. in diam. Insular .... 10. *C. megacarpus*
  13. Lvs. all opposite.
    15. Lvs. densely white-tomentose beneath, mostly revolute .... 3. *C. crassifolius*
    15. Lvs. not densely white-tomentose beneath, mostly plane and minutely canescent.
      16. Lvs. mostly not concave on upper surface; horns of caps. apical or subapical, conspicuous. Cismontane .... 4. *C. cuneatus*
      16. Lvs. mostly concave on upper surface; horns of caps. lateral and spreading, usually small. Mojave Desert and inner cismontane Calif. .... 7. *C. greggii*

1. **C. arbòreus** Greene. Tall ± arborescent shrub, evergreen, 3–7 m. high, with smooth gray bark and soft-pubescent twigs; lvs. broadly ovate to elliptical, rounded at base, acute to obtuse at tip, 3–8 cm. long, 3-veined, dull green and puberulent above, paler and canescent-tomentose beneath, serrulate to serrate; petioles 8–25 mm. long; fl.-clusters compound, 5–12 cm. long, short-peduncled; fls. pale blue; caps. triangular, roughened all over, 6–8 mm. broad, ± crested on the backs of the lobes, almost black when mature; seeds round-obovoid, olive-brown, ca. 2 mm. long, shining; $n=12$ (Nobs, 1942).—Brushy slopes, Chaparral; Santa Cruz, Santa Rosa, Santa Catalina ids. Feb.–May. Some subglabrous plants from the n. islands have been designated as var. *glaber* Jeps.

2. **C. cordulatus** Kell. SNOW BUSH. Plate 76, Fig. F. Intricately branched, spreading, 1–2 m. high, with smooth whitish bark, the whole plant grayish-glaucous, the ultimate twigs stiff, divaricate, spinose; lvs. persistent, ovate to elliptic, mostly 1–2 cm. long, acute to obtuse, 3-veined, glabrous or ± puberulent, green and glaucous above, grayer beneath, mostly entire; petioles 3–6 mm. long; fls. white, the clusters dense, 1.5–3 cm. long; caps. triangular, mostly lobed, slightly crested, 4–5 mm. broad, ± viscid when young; seeds olive-brown, round-obovoid, almost 2 mm. long; $n=12$ (Nobs, 1942).—Dry open flats and slopes, 5000–10,000 ft.; Montane Coniferous F. mostly above Yellow Pine F.; San Jacinto Mts., San Bernardino Mts., San Gabriel Mts., Panamint Mts., Mt. Pinos; to Ore., L. Calif. May–July.

3. **C. crassifòlius** Torr. Rather stiffly and openly branched, 2–3.5 m. high, with grayish, brownish, or whitish branches and ± stout tomentose twigs; lvs. opposite, persistent, broadly elliptic to ± elliptic-obovate, rounded to obtuse, 1.5–3 cm. long, revolute, thick and leathery, olive-green and glabrous above, white-tomentose beneath, coarsely and pungently dentate, sometimes almost entire; petioles 4–7 mm. long; fl.-clusters umbellike, almost sessile, short; fls. white; caps. globose, 7–8 mm. broad, viscid, with short subdorsal horns, roughened at summit; seeds black, shining, 3.5–4 mm. long.—Common on dry slopes and fans, below 3500 ft.; Coastal Sage Scrub, Chaparral; Santa Barbara Co. to L. Calif. Jan.–April. Plants intergrading with typical form but with the lvs. not or scarcely revolute, less tomentose beneath are the var. *planus* Abrams.

4. **C. cuneàtus** (Hook.) Nutt. BUCK BRUSH. Rigid shrub 1–3.5 m. tall; lvs. on spurlike branchlets, cuneate-obovate to spatulate, mostly obtuse, entire, finely tomentulose beneath, gray-green and glabrous above, firm, 0.5–1.5 cm. long, plane; fls. white, umbellate; caps. subglobose, 5–6 mm. broad with short erect horns near the top; seeds shining, black, round-oblong, ca. 4 mm. long; $n=12$ (Nobs, 1942).—Common on dry slopes and fans below 6000 ft.; Chaparral, Pinyon-Juniper Wd., Yellow Pine F.; cismon-

tane, L. Calif. to Ore. March–May. Variable and hybridizing with spp. with which it comes in contact; some forms have been named.

5. **C. cyàneus** Eastw. Arborescent shrub 1–5 m. high, with gray-green branches and glabrous to sparsely puberulent branchlets with scattered glands; lvs. persistent, ovate-elliptical, rounded at base, acute or obtuse at apex, 2–4.5 cm. long, 3-veined, green and glabrous above, lighter and thinly puberulent beneath, glandular-serrulate to subentire; petioles 3–6 mm. long; fl.-clusters compound, 5–18 cm. long, terminal; fls. bright blue; frs. subglobose, shallowly 3-lobed, ca. 4 mm. broad, smooth, with small or no crests; $n=12$ (Nobs, 1942).—Chaparral, San Diego Co. (Lakeside, Ramona, Alpine, El Capitan). April–June.

6. **C. foliòsus** Parry. [*C. austromontanus* Abrams.] Low shrub to ca. 1 m. high, evergreen, with flexuous pubescent glandular branchlets; lvs. oblong-elliptic, to broadly elliptic, 0.5–1.5 cm. long, glandular-denticulate, 1-veined or faintly 3-veined from base, obtuse at both ends, dark green and somewhat strigose above, paler and with somewhat longer hairs beneath on the veins; petioles 1–3 mm. long; fl.-clusters simple, subglobose or somewhat elongate, 0.5–2.5 cm. long, sometimes compound and longer; peduncles as long or longer; fls. blue; caps. subglobose, ca. 4 mm. broad, smooth, faintly lobed and crested; $n=12$ (Nobs, 1942).—Dry slopes, below 5000 ft.; Chaparral, Yellow Pine F.; region of Cuyamaca Mts., San Diego Co.; n. Calif. March–May.

7. **C. gréggii** Gray var. **vestìtus** (Greene) McMinn. Erect, rigidly branched, evergreen, 1–2 m. high with tomentulose branchlets; lvs. opposite, elliptic-ovate, to oblanceolate, commonly concave above, entire to dentate, grayish-green above, gray beneath with a close tomentum (later ± glabrescent), rigid and firm, 0.7–1.5 cm. long; petioles mostly 1–2 mm. long; fls. cream-white, umbellate, on short axillary peduncles; caps. ca. 5 mm. broad, with spreading dorsal horns scarcely 1 mm. long.—Dry slopes, 3500–7500 ft.; Joshua Tree Wd., Pinyon-Juniper Wd., Sagebrush Scrub; desert slopes from Mono Co. to Kern, Los Angeles and San Bernardino cos. and to San Luis Obispo Co.; Utah, Ariz. May–June.

Var. **perpléxans** (Trel.) Jeps. [*C. p.* Trel.] Lvs. roundish to broadly elliptical, or broadly obovate, 1–2 cm. long, mostly conspicuously toothed, yellowish-green and glabrous above, finely white-tomentose beneath in the areoles.—Dry slopes below 7000 ft.; Chaparral, Pinyon-Juniper Wd., Yellow Pine F.; s. face of San Bernardino Mts. to n. L. Calif. March–April.

8. **C. integérrimus** H. & A. DEER BRUSH. Loosely branched, ± deciduous, 1–4 m. tall, with green or yellowish branches, the twigs flexible, glabrous to strigose; lvs. broadly elliptical to ovate or suboblong, rounded at base, acute to obtuse at apex, 2.5–7 cm. long, light green and subglabrous to puberulent above, paler and mostly pubescent beneath, entire; petioles 6–12 mm. long; fl.-clusters mostly compound, 4–15 cm. long; peduncles ca. as long; fls. white to dark blue or pink; caps. globose to triangular, slightly depressed at summit, viscid, 4–5 mm. wide, often with small lateral crests; seeds dull brown, obovoid, ca. 2.5 mm. long.—Dry slopes and ridges between 1000 and 5000 ft.; Yellow Pine F., mostly in mts. from Ventura Co. to cent. Calif. Variable, with a number of named vars., such as var. *puberulus* (Green) Abrams [*C. p.* Greene.] Lvs. ovate-oval, puberulent above, pubescent beneath, more definitely 3-veined from base.—Yellow Pine F. from San Diego Co. n. to Kern Co. May–June. The sp. hybridizes with *C. cordulatus* and *tomentosus*.

9. **C. leucodérmis** Greene. [*C. divaricatus* auth., not Nutt.] Evergreen shrub 2–4 m. high, with rigid divaricate branchlets with pale green smooth bark and short spreading subglabrous spinescent branchlets; lvs. elliptical-oblong to ovate, rounded or subcordate at base, obtuse to acute at apex, entire to serrulate, usually glabrous and glaucous on both surfaces, sometimes pubescent, 1–2.5 cm. long; petioles 2–3 mm. long; fl.-clusters mostly simple, 3–8 cm. long, the fls. white to blue; caps. globose, slightly depressed at top, 4.5–6 mm. wide, viscid, not crested; seeds shining, dark olive-brown, flattened-obovoid, 2.5–3 mm. long; $n=12$ (Nobs, 1942).—Dry rocky slopes, below 6000 ft.; Chaparral, S. Oak Wd.; mts. of cismontane s. Calif., n. to Alameda Co. and Eldorado Co.; n. L. Calif. April–June.

10. **C. megacárpus** Nutt. ssp. **megacárpus**. [*C. macrocarpus* Nutt., not Cav.] Plate 75, Fig. G. Rather compact large shrub 1–4 m. tall, the branches grayish-brown or reddish, the young branchlets canescent-strigose; lvs. persistent, alternate, spatulate to obovate, cuneate at base, glabrous above, canescent beneath, ± revolute; petioles 2–3 mm. long; fl.-clusters on lateral branchlets, few-fld.; fls. white; caps. 8–12 mm. broad, scarcely lobed, laterally or dorsally horned; seeds olive, rather dull, 4.5–5 mm. long.— Dry slopes, below 2000 ft.; Chaparral, near the coast from Santa Barbara Co. to San Diego Co. Jan.–April. On San Clemente and Santa Catalina ids. intergrading with and hybridizing with

Ssp. **insulàris** (Eastw.) Raven. [*C. i.* Eastw.] Lvs. mostly opposite, 1.5–3.5 cm. long; fr. scarcely or not at all horned.—Chaparral; Santa Cruz and Santa Rosa ids. and in introgressing forms on Santa Catalina and San Clemente ids. Jan.–March.

11. **C. oligánthus** Nutt. in T. & G. Shrub, often arborescent, 1–3 m. tall, younger branches ± reddish, hairy, not spinescent; lvs. persistent, ovate or oblong-ovate to elliptical, obtuse, rounded or acute, 1.5–4 cm. long, mostly 3-veined, dark green and sparingly pubescent above, paler and ± hirsute-pubescent beneath, especially on veins, glandular-denticulate; petioles 3–8 mm. long; fl.-clusters mostly simple, rather loose, 1.5–5 cm. long; fls. mostly deep blue or purplish; frs. ± triangular, ca. 4 mm. broad, smooth or rough, crested, usually viscid; seeds shining, olive-brown, rounded, ca. 3 mm. long.—Dry slopes, below 4500 ft.; Chaparral; Los Angeles and w. Riverside cos. to San Luis Obispo Co. Feb.–April.

12. **C. pálmeri** Trel. Large spreading shrub 1–3.5 m. tall with gray-green bark and glabrous flexible twigs; lvs. oblong to oblong-ovate, mostly persistent, 1.5–3.5 cm. long, rounded to retuse at apex, rather firm and leathery, glabrous and light green above, pale and sometimes ± pubescent beneath on the midrib, 1-nerved, entire; petioles ca. 3–4 mm. long; fl.-clusters compound, 7–12 cm. long, on somewhat shorter peduncles; fls. white; caps. ± triangular; 5–7 mm. broad, 3-lobed, with glandular crests; seeds dark olive-brown, round-obovoid, ca. 3 mm. long.—Dry slopes, 3200–6000 ft.; Chaparral, Yellow Pine F.; mts. of Riverside, Orange and San Diego cos.; n. L. Calif.; also in Sierra Nevada. May–June.

13. **C. papillòsus** T. & G. ssp. **roweànus** (McMinn) Munz, n. comb. [*C. p.* var. *roweanus* McMinn, Madroño 5: 13. 1939.] Evergreen, loosely branched, 1–5 m. tall, with densely tomentose-pubescent young branchlets; lvs. rather crowded, linear to oblong, less than 1 cm. broad, ± retuse to truncate at tip, 1-veined from base, dark green, ± villous and glandular-papillose above, paler and hirsutulose to tomentose beneath, glandular-denticulate and somewhat revolute; petioles 4–6 mm. long; fl.-clusters mostly simple, narrow, dense, 2–5 cm. long, on naked peduncles ca. as long; fls. deep blue; fr. triangular, lobed, ca. 3 mm. wide, with low narrow subdorsal crests; seeds plump, dark, ca. 2 mm. long.—Dry slopes, 2000–4000 ft.; largely Chaparral; Santa Ana Mts. to Monterey and San Benito cos. Feb.–June.

14. **C. sorediàtus** H. & A. Densely and rigidly branched, mostly 1–3 m. tall, with smooth gray-green bark and rigid ± villous stiff nearly spinescent branchlets, evergreen; lvs. elliptic to ovate, mostly acute, 1–4 cm. long, finely strigose beneath especially along the veins; petioles 2–5 mm. long; fl.-clusters mostly simple, dense, 1–4 cm. long, on somewhat shorter peduncles; fls. light to deep blue; frs. globose, smooth, 4 mm. broad, scarcely lobed or crested, viscid; seeds dark, plump, ca. 2 mm. long; $n=12$ (Nobs, 1942). —Dry slopes, below 3500 ft.; Chaparral, from Orange and w. Riverside cos. n. to n. Calif. Feb.–May. Hybridizing with *C. oliganthus*, *C. papillosus* and *C. tomentosus*.

15. **C. spinòsus** Nutt. in T. & G. Large or arborescent, 2–6 m. tall, with smooth olive-green bark, the main branchlets flexible, ascending, glabrous, the ultimate short, stiff, divergent, spinescent; lvs. persistent, elliptic to oblong, obtuse to emarginate at apex, mostly entire, 1.2–3 cm. long, glabrous or somewhat strigose on midrib beneath, bright green above, paler beneath; petioles 4–8 mm. long; fl.-clusters mostly large, compound, 4–15 cm. long on somewhat leafy peduncles; fls. pale blue to almost white; caps. globose, 4–5 mm. broad, viscid, scarcely lobed or crested; seeds ca. 3 mm. long, dark

Plate 76. Fig. A, *Ceanothus tomentosus* ssp. *olivaceus*. Fig. B, *Condaliopsis parryi*. Fig. C, *Rhamnus crocea*. Fig. D, *Adenostoma fasciculatum*. Fig. E, *Amelanchier pallida*. Fig. F, *Cercocarpus betuloides*, plumose pistil.

olive-brown; $n = 12$ (Nobs, 1942).—Dry slopes mostly below 3000 ft.; Chaparral, Coastal Sage Scrub; mts. mostly near the coast; n. L. Calif. to San Luis Obispo Co. Feb.–May. Hybrids with *C. sorediatus* have been reported.

16. **C. tomentòsus** Parry. ssp. **olivàceus** (Jeps.) Munz, comb. nov. [*C. t.* var. *olivaceus* Jeps., Man. Fl. Pl. Calif., 621. 1925.] Plate 76, Fig. A. Evergreen shrub 1–3 m. high, with grayish-brown or reddish bark, the branches long and slender, young growth rusty-tomentose; lvs. ovate to elliptic, glandular-denticulate, gray-green beneath, 1–2.5 cm. long, 3- or 1-veined; petioles 1–5 mm. long; fl.-clusters compound, lateral or terminal, 2–5 cm. long, on shorter, 1–2-lvd. peduncles; fls. blue to almost white; frs. subglobose, slightly lobed at top, ca. 4 mm. broad, somewhat laterally crested, very sticky when developing and dark in age.—Dry brush-covered slopes, below 3500 ft.; Chaparral; mts. from Redlands and Santa Ana Mts. and San Diego Co. to L. Calif. March–May.

17. **C. verrucòsus** Nutt. in T. & G. Erect, stiff-branched, rounded evergreen shrub 1–3 m. tall with tomentulose young branchlets; lvs. alternate, crowded, round to deltoid-obovate, tapering to rounded at base, obtuse to retuse and apex, plane, firm, entire or nearly so, 0.5–1.4 cm. long, dark green and glabrous above, minutely canescent beneath; petioles mostly 1–3 mm. long; fl.-clusters dense, axillary, few-fld., 1–2 cm. long; fls. white; caps. globose, 5 mm. broad, shallowly lobed, laterally short-horned; seeds black, shining, flattened, 2.5–3 mm. long; $n = 12$ (Nobs, 1942).—Dry hills and mesas; Chaparral; w. San Diego Co.; adjacent, L. Calif. Jan.–April.

3. Colubrìna Rich. in Brongn.

Unarmed shrubs or trees, or with rigid, ± spiny twigs. Lvs. alternate, deciduous or persistent, entire or denticulate, pinnate or 3-nerved, usually with small stipules. Fls. inconspicuous, in sessile or peduncled axillary umbels. Sepals tardily deciduous, united at the base with the urn-shaped receptacle which is lined with the disk and adherent to the base of the ovary. Petals minute, hooded and partly enclosing the anthers, sessile or clawed. Style short, 3-lobed nearly to the base. Caps. 3-loculed, ± 3-lobed, with 1 seed in each locule. Ca. 18 spp. of sw. N. Am. and of S. Am.; 1 in Old World. (Latin, *coluber*, a serpent, of uncertain application.)

1. **C. califórnica** Jtn. [*C. texensis* var. *c.* L. Benson.] Intricately branched shrub 1–2.5 m. tall, the branches ± spinescent, usually divaricate, finely grayish-tomentose; lvs. oblong-obovate, 0.8–2 cm. long, entire, rounded to obtuse, dull gray-green, ± tomentose; fls. in small axillary clusters; sepals deltoid, tomentose, ca. 1 mm. long; petals yellowish, ca. 1 mm. long; caps. globose, 6 mm. in diam.—Occasional, dry canyons below 3000 ft.; Creosote Bush Scrub, Joshua Tree Wd.; Joshua Tree National Monument, n. Riverside Co. to Eagle Mts. and Chuckwalla Mts.; to Ariz., L. Calif. April–May.

4. Condàlia Cav.

Rigidly branched shrub or small tree, with spine-tipped twigs. Lvs. alternate, subsessile, coriaceous, entire, deciduous. Stipules minute, deciduous. Fls. solitary or fascicled in axils, small, short-pedicellate. Calyx-tube broadly obconic, shallow, with 5 ovate to deltoid lobes. Petals none. Disk flat, fleshy, 5-angled on free margin. Stamens 4–5; fils. subulate. Ovary 1–2-celled, 2-ovuled, free from disk; style short, 2–3-lobed or stigmas 3 and distinct. Fr. ovoid to globose, drupaceous, usually 1-locular. Seeds subglobose or compressed. A small genus of warm parts of New World. (A. *Condal*, Spanish physician.)

1. **C. globòsa** Jtn. var. **pubéscens** Jtn. Intricately branched spinose shrub with spreading twigs, puberulent, thick; to 4 m. tall; lvs. narrow-spatulate to oblanceolate, 5–15 mm. long, puberulent, thick; umbels 1–2-fld.; pedicels 3–4 mm. long; sepals deciduous, 1 mm. long; drupe obliquely ovoid, 4–5 mm. long, black and juicy.—Uncommon at low elevs.; Creosote Bush Scrub; e. Colo. Desert; to Ariz., n. L. Calif. March–April.

5. Condaliópsis Suesseng. in Engler & Prantl.

Shrubs with rigid branches and some spiny twigs. Lvs. alternate, entire, pinnately veined. Fls. solitary or fascicled in axils on short spur branchlets; pedicels short. Sepals 5. Petals 5. Disk lining calyx-tube subentire. Stamens 5. Ovary free, 2-celled by intrusion and overlapping of 2 opposite placentae. Fr. ellipsoid or obovoid, with dry-fleshy or juicy berries and bony or woody kernel, 2-celled. Seeds mostly 2. (*Condalia* and *opsis*, resemblance.) A small genus of sw. N. Am.

Plants canescent; infl. peduncled; drupe beakless, 6–10 mm. long .... 1. *C. lycioides*
Plants glabrous; infl. sessile; drupe beaked, ca. 15 mm. long .... 2. *C. parryi*

1. **C. lycioìdes** (Gray) Suesseng. var. **canéscens** (Gray) Suesseng. Rigid much-branched shrub 1–4 m. tall, with ± zigzag canescent-puberulent spiny twigs; lvs. elliptical to oblong-ovate, 8–15 mm. long, ± canescent, short-petioled, soon deciduous; umbel short-peduncled, 2–6-fld.; fls. minute; drupe ellipsoid, 6–10 mm. long, deep blue to black, with ± bloom.—Uncommon in sandy places below 1500 ft.; Creosote Bush Scrub; Colo. Desert; to Nev., Ariz., Son., L. Calif. April–June.

2. **C. párryi** (Torr.) Suesseng. in Engler & Prantl. [*Condalia p.* Weberb.] Plate 76, Fig. B. Large rounded shrubs 1–4 m. high, glabrous, the ultimate branchlets spiny, flexuous; lvs. mostly fascicled, obovate to oblong-elliptical, 8–20 mm. long, coriaceous, bright green, on short slender petioles; fls. 1–2 mm. long; drupe ellipsoid-ovoid, 1–2 cm. long, usually distinctly beaked, with dry thin flesh, the pedicels 1–1.5 cm. long.—Local on dry slopes and in canyons, 1200–3500 ft.; Joshua Tree Wd., Pinyon-Juniper Wd., upper Creosote Bush Scrub; w. edge of Colo. Desert from Morongo Pass to L. Calif. Feb.–April.

6. Rhámnus L. BUCKTHORN. CASCARA

Shrubs or small trees, evergreen or deciduous. Lvs. alternate, pinnately veined. Fls. small, greenish, bisexual or unisexual, in axillary clusters. Calyx 4–5-lobed or -toothed, the tube circumscissile after anthesis. Petals when present 4–5. Stamens usually 4–5, with short fils. Pistil 1, free from the disk; ovary 2–4-loculed. Fr. a berrylike drupe with 2–4 separate nutlets. Ca. 100 spp., of almost worldwide distribution, some of considerable medicinal value. (The ancient Greek name.)

(Wolf, C. B. The N. Am. species of Rhamnus. Rancho Santa Ana Bot. Gard. Mon. 1. 1938.)
1. Petals present; winter-buds without bud-scales .... 1. *R. californica*
1. Petals lacking; winter-buds with bud-scales.
  2. Branchlets somewhat divaricate, ending in weak spines; lf.-blades 10–17 mm. long 2. *R. crocea*
  2. Branchlets not divaricate, seldom ending in weak spines; lf.-blades 18–60 mm. long.
    3. Shrubs usually with many branches from the base; lvs. mostly serrate to dentate.
      4. Lvs. glabrous to pubescent, rarely slightly revolute. Widespread ..... 3. *R. ilicifolia*
      4. Lvs. pilose on both surfaces, revolute. San Diego Co. .... 4. *R. pilosa*
    3. Small trees with distinct trunks; lvs. mostly crenate to entire, occasionally serrate. Insular
                                                                     5. *R. pirifolia*

1. **R. califórnica** Esch. ssp. **califórnica.** COFFEEBERRY. Upright rounded shrub to low and spreading, 1–4 m. tall; bark of young twigs usually reddish; lvs. persistent, oblong to elliptic, plane or ± revolute, entire or serrate, acute to obtuse, 3–8 cm. long, usually shining and glabrous above, glabrous or with a few hairs beneath; fls. perfect, 5-, rarely 4-merous; umbels on peduncles 4–18 mm. long, 6–50-fld.; fls. 2–3 mm. long; berries green, black or red when ripe, somewhat elongate or globose, 10–12 mm. long; seeds mostly 2, green-brown, 7–9 mm. long, smooth; $2n=24$ (Bowden, 1945).—Sandy and rocky places along the coast, hillsides and ravines farther back, below 3500 ft.; Coastal Strand, Coastal Sage Scrub.

Ssp. **cuspidàta** (Greene) C. B. Wolf. [*R. c.* Greene.] Lvs. dentate, green above, white-tomentose beneath.—Creosote Bush Scrub to Pinyon-Juniper Wd. and Yellow Pine F.; Orange and Riverside cos. to Sierra Nevada.

Ssp. **tomentélla** (Benth.) C. B. Wolf. [*R. t* Benth.] Lvs. tomentose beneath, narrowly elliptical, entire or with blunt teeth.—Chaparral; L. Calif. and Los Angeles Co. to n. Calif.

Ssp. **ursìna** (Greene) C. B. Wolf. [*R. u.* Greene.] Lvs. green above, elliptical, serrate to almost entire, white-tomentose beneath.—Canyons, 4000–7000 ft.; Joshua Tree Wd., Pinyon-Juniper Wd.; Providence, N.Y. and Clark mts., e. Mojave Desert; to Ariz., New Mex. May–July.

2. **R. cròcea** Nutt. in T. & G. BUCKTHORN. REDBERRY. Plate 76, Fig. C. Spreading gray-green much-branched shrub mostly 1–2 m. high, with rigid often spinescent branchlets, evergreen; lvs. often fascicled, glabrous or slightly puberulent, shining, coriaceous, elliptic to obovate, 5–15 mm. long, usually glandular-serrulate; petioles 1–4 mm. long; fls. unisexual, 4 (–5)-merous; pedicels 1–4 mm. long; petals 0; berry red, 5–6 mm. long, obovoid, 2-seeded; seeds 4 mm. long, finely reticulate, rounded at apex, the outer side deeply grooved.—Dry washes and canyons, below 3000 ft.; Coastal Sage Scrub, Chaparral, S. Oak Wd.; L. Calif. to Lake Co. March–April.

3. **R. ilicifòlia** Kell. [*R. crocea* ssp. *i.* C. B. Wolf.] Branched from base or almost arborescent, 1.5–4 m. tall; petioles 2–8 mm. long; lf.-blades 18–40 mm. long, oval to roundish, glabrous or slightly pubescent beneath, subentire to spinosely serrate; fr. to 8 mm. long; seeds to 6 mm. long. $2n = 12$ pairs (Raven et al., 1965).—Dry slopes mostly below 5000 ft.; Chaparral, Yellow Pine F., etc.; L. Calif. and San Diego Co. to n. Calif.; Providence Mts., e. Mojave Desert; Ariz. March–June.

4. **R. pilòsa** (Trel.) Abrams. [*R. crocea* var. *p.* Trel.] Much like *R. ilicifòlia*; upright, sparsely branched, 1.5–2 m. high; young branchlets, lvs., floral parts densely pilose; lf.-blades 15–28 mm. long, orbicular to broadly ovate, mostly revolute.—Dry slopes at ca. 1500 ft.; Chaparral; San Diego Co.

5. **R. pirifòlia** Greene. [*R. crocea* ssp. *p.* C. B. Wolf. *R. c.* var. *insularis* Sarg.] Treelike, to 10 m. high; petioles 5–10 mm. long; lf.-blades elliptical to almost round, to 4 cm. long, mostly crenate to entire. Coastal Sage Scrub, Chaparral; Channel Ids. March–June.

92. **Rosàceae.** ROSE FAMILY

Herbs, shrubs and small trees, spiny or unarmed, evergreen or deciduous. Lvs. usually alternate and with stipules. Fls. mostly bisexual and regular, solitary or clustered. Sepals and petals at edge of a fl.-tube (hence perigynous) which is lined or rimmed with a glandular disc. Sepals 5 (4), often with alternating bractlets; petals 5 (4 or 0). Stamens mostly in whorls or cycles of 5, sometimes numerous and indefinite. Pistils 1–many, simple, distinct and free from fl.-tube or united into a 2–5-celled ovary which may be ± inferior. Fr. a follicle, ak., drupe, pome or cluster of drupelets. Seeds usually without endosperm. A family of over 100 genera and over 3000 spp.; many of great horticultural value.

1. Ovary or ovaries superior; fr. not a pome.
   2. Fr. of 1–5 dehiscent follicles.
      3. Lvs. opposite, 1–2 dm. long; fls. bisexual, in large corymbose panicles; sepals deciduous. Insular .... 18. *Lyonothamnus*
      3. Lvs. alternate, usually smaller; fls. not as above, or if so, unisexual; sepals persistent. Mainland.
         4. Foliage stellate-pubescent or -tomentose; petals 3–5 mm. long.
            5. Lvs. simple, generally 3–7-lobed; carpels inflated in maturity ..... 21. *Physocarpus*
            5. Lvs. twice pinnate; carpels coriaceous at maturity .... 7. *Chamaebatiaria*
         4. Foliage not stellate-pubescent; petals usually 1.5–2 mm. long.
            6. Fr. 1-seeded, tardily dehiscent; stamens scarcely exserted; plants bushy
               15. *Holodiscus*
            6. Fr. several-seeded, soon dehiscent; stamens well exserted; plants forming woody mats
               20. *Petrophytum*
   2. Fr. of indehiscent aks., or of drupelets, or of large drupes.
      7. The fr. of dry aks., which may be embedded in the surface of a fleshy receptacle (strawberry) or enclosed in a fleshy cup-shaped structure (rose).

    8. Lvs. simple, sometimes 3-lobed; plants woody.
        9. Petals present.
            10. Fls. solitary; lvs. cuneate, 3-lobed above .... 25. *Purshia*
            10. Fls. paniculate; lvs. not lobed.
                11. The lvs. sublinear; ovule and seed 1 .... 1. *Adenostoma*
                11. The lvs. obovate to ovate; ovules 2; seeds mostly 1 .... 15. *Holodiscus*
        9. Petals absent.
            12. Fl.-tube salver-shaped, the limb deciduous, the tubular base persistent; ak. with
                long feathery tail .... 5. *Cercocarpus*
            12. Fl.-tube campanulate; ak. not long-tailed .... 8. *Coleogyne*
    8. Lvs. compound or pinnately dissected.
        13. Plants shrubby.
            14. Pistil usually 1.
                15. Lvs. 3-lobed; fls. solitary .... 25. *Purshia*
                15. Lvs. pinnately dissected into minute segms.; fls. cymose-paniculate
                                                                                6. *Chamaebatia*
            14. Pistils 4-many.
                16. Carpels not enclosed in a fleshy fl.-tube; plants not prickly.
                    17. Lvs. pinnate; fls. ± yellow .... 22. *Potentilla*
                    17. Lvs. mostly 5-lobed; fls. white or cream.
                        18. Sepals alternating with bractlets; petals 13-14 mm. long .. 11. *Fallugia*
                        18. Sepals present; bractlets absent; petals 6-8 mm. long .... 10. *Cowania*
                16. Carpels enclosed in the fl.-tube which is fleshy in fr.; stems mostly prickly
                                                                                27. *Rosa*
        13. Plants herbaceous.
            19. Styles jointed with the ovary and deciduous.
                20. Bractlets lacking between the sepals; fls. without a stalked receptacle; fl.-tube
                    funnelform .... 24. *Purpusia*
                20. Bractlets present, alternating with sepals; fls. with a stalked receptacle; fl.-tube
                    not funnelform.
                    21. Stamens 5.
                        22. Lvs. pinnate, with more than 3 lfts. .... 17. *Ivesia*
                        22. Lvs. with 3 lfts. .... 30. *Sibbaldia*
                    21. Stamens 10, 15, 20 or many.
                        23. Fils. dilated, 10 in number; fl.-tube usually deep .... 16. *Horkelia*
                        23. Fils. filiform or narrow, mostly 20 or more; fl.-tube shallow or none.
                            24. Receptacle not enlarged in fr.
                                25. Uppermost lfts. confluent; petals usually clawed; carpels usually
                                    few; stamens inserted some distance above the receptacle and with-
                                    out an annular thickening at the base of the fils. ...... 17. *Ivesia*
                                25. Uppermost lfts. not confluent; petals usually sessile; carpels usually
                                    many; stamens inserted near the base of the receptacle on a ± evi-
                                    dent annular thickening .... 22. *Potentilla*
                            24. Receptacle enlarged in fr., fleshy .... 12. *Fragaria*
            19. Styles not jointed with the ovary, persistent.
                26. Pistils many; petals 4-10 mm. long .... 13. *Geum*
                26. Pistils 1; fls. smaller.
                    27. Fl.-tube not prickly; petals 0.
                        28. Lvs. palmately lobed; plants not over 1 dm. high .... 3. *Alchemilla*
                        28. Lvs. pinnate; plants 1-6 dm. high .... 29. *Sanguisorba*
                    27. Fl.-tube armed with prickles, these hooked; petals yellow; sepals 5
                                                                                2. *Agrimonia*
7. The fr. of drupelets or drupes.
    29. Carpels becoming drupelets, which are ± coherent and form a fleshy "berry"; plants
        ± woody, often prickly; lvs. mostly pinnate .... 28. *Rubus*
    29. Carpels becoming larger solitary drupes; trees or shrubs with simple lvs.; pistil 1
                                                                                23. *Prunus*
1. Ovary inferior; fr. a pome.
    30. Plant with deciduous lvs.
        31. Petals spreading; styles 2; fls. 1-3 in a sessile umbel.
            32. Low native shrub. Inyo Mts. .... 19. *Peraphyllum*

32. Natur. tree, the common pear. Cismontane .... 26. *Pyrus*
31. Petals erect; styles 2–5; fls. racemose .... 4. *Amelanchier*
30. Plants with persistent evergreen lvs.
    33. Carpels with leathery walls at maturity; styles 2–3; fls. in large corymbose panicles; lvs. sharply toothed. Common native .... 14. *Heteromeles*
    33. Carpels bony at maturity with 2-seeded nutlets; lvs. entire or crenate. Rare escape from cult. .... 9. *Cotoneaster*

1. Adenóstoma H. & A.

Unarmed evergreen shrubs with ± resinous herbage. Lvs. small, entire, alternate or fascicled, linear, rigid, numerous. Fls. small, white, crowded, in terminal panicled racemes. Fl.-tube obconical, 10-striate. Sepals 5. Petals 5, roundish. Stamens 10–15, inserted 2–3 together, alternating with petals. Pistil 1; style lateral. Ovary 1-celled, 1–2-ovuled. Aks. enclosed by the indurated fl.-tube. Two spp. (Greek, *aden*, gland, and *stoma*, mouth, because of glands at mouth of fl.-tube.)

Lvs. fascicled; bracts lance-linear, not scarious; fls. sessile .... 1. *A. fasciculatum*
Lvs. scattered; bracts broadly lanceolate, scarious-margined; fls. pedicelled .. 2. *A. sparsifolium*

1. **A. fasciculàtum** H. & A. CHAMISE. GREASEWOOD. Plate 76, Fig. D. Diffuse shrub, 0.5–3.5 m. high, with well developed basal burl; bark reddish, subglabrous on the twigs, becoming shreddy with age; stipules small, acute; lvs. narrow-clavate to linear, 4–10 mm. long; panicles 4–12 cm. long; bracts ca. 1 mm. long; fl.-tube green, almost 2 mm. long; sepals barely 1 mm. long; petals ca. 1.5 mm. $2n=9$ pairs (Raven et al., 1965).—Common dominant on dry slopes and ridges below 5000 ft.; Chaparral; L. Calif. to n. Calif. and Sierran foothills. May–June. Plants on dry mesas about San Diego tend to have pubescent twigs and obtuse lvs. 4–6 mm. long and are the var. **obtusifòlium** Wats.

2. **A. sparsifòlium** Torr. RIBBON BUSH, RED SHANK. Erect, arborescent, 2–6 m. high, the trunks red-brown and freely exfoliating; twigs green, resinous-glandular; lvs. filiform, 6–15 mm. long, resinous-glandular; fls. in open showy panicles less than 1 dm. long; fl.-tube ca. 1.5 mm. long; sepals ca. 2 mm. long; petals elliptic; ca. 2 mm. long; $2n=9$ pairs (Raven et al., 1965).—Dry slopes and mesas, below 6000 ft.; Chaparral; common from n. L. Calif. to San Gorgonio Pass; local in San Luis Obispo and Santa Barbara cos. July–Aug.

2. Agrimònia L. AGRIMONY

Perennial herbs with rootstocks. Lvs. pinnate, with crenate-serrate lfts. Fls. small, spicate-racemose. Fl.-tube turbinate or hemispherical, constricted at the throat, where beset with hooked bristles, indurate in fr. and enclosing the 2 aks. Sepals 5, connivent after flowering. Petals 5, small, yellow. Stamens 5–15. Styles terminal. Perhaps 12–15 spp. of N. Hemis. (Possibly from Greek, *argema*, an eye-disease, because of supposed medicinal value.)

1. **A. gryposèpala** Wallr. Stems 3–15 dm. high, hirsute and glandular-puberulent, lvs. remote, extending to base of infl.; larger lfts. 5–9, mostly lanceolate or oblanceolate, 4–12 cm. long, resinous-glandular beneath, sparingly hirsute on veins; racemes 2–4 dm. long; pedicels 2–10 mm. long; fruiting fl.-tube 4–5 mm. long, ca. as broad, the hooked bristles spreading from a thin horizontal flange, the short outer bristles often reflexed; petals obovate, ca. 3 mm. long.—Borders of woods, 3500–5500 ft.; mostly Yellow Pine F.; Palomar, Cuyamaca and San Bernardino mts.; scattered stations in n. Calif.; to Atlantic Coast. July–Aug.

3. Alchemilla L.

Annual to perennial herbs. Lvs. alternate, palmately lobed, with sheathing stipules. Fls. small, cymose in ours in small axillary clusters. Fl.-tube campanulate to urn-shaped. Bractlets 4–5 or sometimes obsolete. Sepals 4–5. Petals 0. Stamens 1–4. Pistils 1–8, free from fl.-tube; styles nearly basal. Widely distributed genus of perhaps 50 spp. (Valued in *alchemy*.)

1. **A. occidentàlis** Nutt. [*A. cuneifolia* Nutt.] Small slender-branched annual, to 1 dm. high, floriferous throughout; lvs. petioled, cuneate-flabelliform, 5–8 mm. long, ± short-hirsute, deeply parted into 3–5-cleft divisions; fl.-tube urn-shaped, ca. 1 mm. long; bractlets ± developed; sepals ca. 1–1.5 mm. long, ovate; stamen 1; pistils 1–2; aks. glabrous.—Open grassy or wooded places below 2000 ft.; V. Grassland, etc.; cismontane; L. Calif. to Wash. Santa Catalina, Santa Rosa and Santa Cruz ids. March–June.

4. *Amelánchier* Medic. SERVICE-BERRY

Deciduous shrubs to small trees, with unarmed branches and slender terete branchlets. Lvs. simple, alternate, pinnately veined, entire to serrate. Fls. perfect, regular, mostly in racemes, terminating short leafy branches of the season. Fl.-tube campanulate or urceolate, ± adnate to carpels, becoming globose or ellipsoid in fr. Sepals 5, persistent. Petals 5, white, oblanceolate to narrowly oval. Stamens 10–20, short. Carpels 2–5, ± united to form an inferior, 2–5-loculed ovary, each locule divided by a false partition, the cells 1-seeded. Fr. a pome, with firm carpel walls and fleshy outer tissue. Seeds small, smooth, dark brown. Perhaps 20 spp. of N. Temp. Zone. (The Savoy name of the Medlar.)

(Jones, G. N. Am. spp. of Amelanchier. Univ. Ill. Press. pp. 1–126. 1946.)

Lvs. glabrous at maturity or nearly so; top of ovary glabrous .... 1. *A. covillei*
Lvs. permanently puberulent to tomentulose, especially beneath; top of ovary pubescent.
    Petals oval to obovate, 8–11 mm. long; styles free nearly to the base; lvs. with 7–9 pairs of lateral veins. Cuyamaca Mts. to n. Calif. .... 2. *A. pallida*
    Petals mostly 6–8 mm. long, oblanceolate; styles united below; lvs. with 9–13 pairs of lateral veins. Laguna Mts., San Bernardino Mts. and mts. of e. Mojave Desert .... 3. *A. utahensis*

1. **A. covíllei** Standl. Shrub 1–2 m. tall, with stout glabrous branches; lvs. oblong-oval to obovate-orbicular, 1–2 cm. long, rather finely serrate to near the base, glabrous, pale green and ± glaucous; racemes short, rather densely few-fld.; fl.-tube glabrous; sepals 2.5–3 mm. long, reflexed; petals 5–6 mm. long.—Pinyon-Juniper Wd.; Panamint Mts. to Clark and Old Dad mts.; s. Nev. Doubtfully specifically distinct from *A. utahensis*.

2. **A. pállida** Greene. [*A. gracilis* Heller. *A. alnifolia* var. *cuyamacensis* Munz.] Plate 76, Fig. E. Shrubs 1–3 or more m. tall, with rigid erect to spreading branches; bark red-brown to gray; lvs. oval to elliptical or rounded, tomentulose or puberulent on both surfaces, the lower usually paler than upper, the blades 2–4 cm. long, 1.5–2.5 cm. wide, acute to roundish at apex, with 7–9 pairs of lateral veins, entire to toothed to or below the middle; racemes corymbose, 2–4 cm. long, 4–6-fld.; sepals lanceolate, 2–3 mm. long; petals 8–11 mm. long, oval or obovate; styles mostly 3–4; fr. subglobose, 4–6 mm. in diam.—Dry slopes below 11,000 ft.; largely Montane Coniferous F.; Palomar Mts., Cuyamaca Mts., San Gabriel Mts., Kern and Ventura cos. to Ore., w. Nev. Variable.

3. **A. utahénsis** Koehne. Shrubby, 1–5 m. high, white-pubescent on young growth; lvs. largely roundish to oval, grayish green, tomentulose or cinereous on both sides, the blades 1–3 cm. long, with 11–13 pairs of lateral veins, usually coarsely crenate-serrate from apex to near the base; racemes sublanate, 3–6-fld.; sepals narrow, 3 mm. long; petals 6–8 mm. long; styles mostly 3–4; fr. 6–10 mm. in diam.—Dry slopes 5000–7000 ft.; Yellow Pine F., Pinyon-Juniper Wd.; mts. of San Diego Co. to San Bernardino and San Gabriel mts., Kingston Mts. to Sweetwater Mts. to Mont., New Mex. April–May.

5. *Cercocárpus* HBK. MOUNTAIN-MAHOGANY

Evergreen shrubs or low trees with alternate simple ± coriaceous, straight-veined lvs. borne on spurlike branchlets. Fls. solitary or fascicled, small, axillary or terminal. Fl.-tube with a lower persistent subcylindric portion and an upper deciduous bowl-shaped deciduous part. Sepals 5, broad to narrow. Petals 0. Stamens 10–45, inserted in 2–3 rows. Pistil 1; style terminal; ovule 1. Fr. a cylindric-fusiform ak. with terminal elongate silky-plumose style. Seed cylindric. Ca. 8–10 spp. of w. and sw. N. Am. (Greek, *kerkos*, tail, and *karpos*, fr.)

Lvs. ± toothed, not strongly inrolled or resinous; anthers hairy.
    Upper surface of the lvs. with impressed veins, the lower conspicuously white-tomentose. Catalina Id. .... 5. *C. traskae*
    Upper surface of the lvs. with the veins not impressed, the lower surface subglabrous to subtomentose.
        Fls. 6–8 mm. broad; lvs. whitish, ± hairy beneath. Widely distributed .... 1. *C. betuloides*
        Fls. 2–4 mm. broad; lvs. greenish-yellow and glabrous beneath. S. San Diego Co.
4. *C. minutiflorus*
Lvs. entire, strongly revolute, resinous; anthers glabrous.
    The lvs. mostly less than 1 cm. long, linear, the margins revolute almost to the midrib. E. Mojave Desert .... 2. *C. intricatus*
    The lvs. mostly 1.5–3 cm. long, elliptic; the margins only slightly revolute. Montane
3. *C. ledifolius*

    1. **C. betuloìdes** Nutt. ex T. & G. Plate 76, Fig. F. Erect open shrub to small tree, 2–7 m. high, with stiff erect or graceful spreading terminal branches; bark smooth, gray; twigs subglabrous; lf.-blades obovate to oval or broadly elliptical, mostly 1–2.5 cm. long, cuneate and entire below the middle, serrate above, glabrous on upper surface, paler and somewhat pubescent beneath; petioles 3–6 mm. long; fls. mostly in clusters of 2–3, with short pedicels; fl.-tube silky-tomentose, 5–6 mm. broad at summit, later glabrescent, brownish and split part way down one side, 8–10 mm. long; sepals broad-triangular; styles 4–9 cm. long in fr.; $2n = 18$ (Morley, 1949).—Common on dry slopes and in washes below 6000 ft.; Chaparral; n. L. Calif. through cismontane s. Calif. to Ore. Variable, the most marked deviation being var. **bláncheae** (C. K. Schneid.) Little. [*C. alnifolius* Rydb.] Petioles 8–10 mm. long; lf.-blades 4–6 cm. long, dark green above, subglabrous or strigose beneath; fl.-tube strigose; tail of mature fr. 5–7 cm. long.—Chaparral, Santa Catalina, Santa Rosa and Santa Cruz ids., Santa Monica Mts. March–April.

    2. **C. intricàtus** Wats. Intricately branched, 1–3 m. tall, the young growth pubescent; lvs. 3–15 mm. long, 1–2 mm. wide, linear, inrolled to midrib, gray beneath, with petioles to 1 mm. long; fl.-tube 3–5 mm. long at anthesis and in fr.; fls. 1–2 mm. wide; ak.-tails 1–2 cm. long.—Dry slopes, 4000–9000 ft.; mostly Pinyon-Juniper Wd.; Providence and Clark mts. to White Mts.; to Sierra Nevada, Utah. May.

    3. **C. ledifòlius** Nutt. Shrub or tree to 9 m. high, with red-brown furrowed bark; twigs canescent when young; lvs. lance-elliptic, lanceolate to oblanceolate; 1–3 cm. long, 3–10 mm. wide, revolute, resinous and tomentulose beneath; fls. 1–3, 4–5 mm. wide; fl.-tube 4–6 mm. long at anthesis, 6–10 mm. in fr.; tail of fr. 4–7 cm. long.—Dry rocky slopes, 4000–10,000 ft.; Pinyon-Juniper Wd., Sagebrush Scrub; from Santa Rosa and San Jacinto mts. through mts. of Mojave Desert to Wash., Rocky Mts., L. Calif. April–May.

    4. **C. minutiflòrus** Abrams. Shrub 2–5 m. high, glabrous throughout; lvs. obovate to almost round, 1–2 cm. long, cuneate at base, serrate above, light green on upper surface, yellow-green on lower, with 3–5 lateral veins on each side of midrib; petioles 2–5 mm. long; fl.-tube 5–8 mm. long at anthesis, 11 mm. in fr.; tail of mature fr. 3–7 cm. long.—Dry slopes below 3000 ft.; Chaparral; Roblar Grade, Santa Margarita Mts., San Diego Co.; n. L. Calif. March–May.

    5. **C. tráskae** Eastw. [*C. betuloides* var. *t.* Dunkle.] A small tree, the coriaceous lvs. 3–6 cm. long, with impressed veins on upper surface, densely gray-tomentose beneath.—Salte Verde Canyon, Santa Catalina Id. March.

6. Chamaebàtia Benth.

Glandular-pubescent shrubs, evergreen, heavy-scented. Lvs. twice or thrice-pinnate with numerous minute segms. Fls. white, cymose-paniculate. Fl.-tube persistent, turbinate-campanulate. Sepals 5. Petals 5. Stamens many, in several series. Pistils solitary; style terminal, villous at base; ovule 1. Aks. coriaceous. Two spp. (Greek, *chamae*, low, and *batos*, bramble.)

    1. **C. austràlis** (Bdg.) Abrams. Shrub 6–20 dm. high, the twigs somewhat glandular-

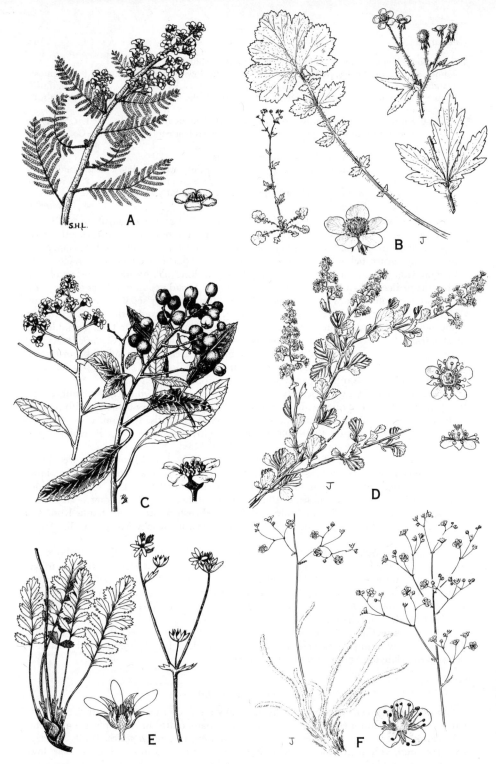

Plate 77. Fig. A, *Chamaebatiaria millefolium*. Fig. B, *Geum macrophyllum*. Fig. C, *Heteromeles arbutifolia*. Fig. D, *Holodiscus boursieri*. Fig. E, *Horkelia cuneata*. Fig. F, *Ivesia santolinoides*.

pubescent; lvs. 3.5–8 cm. long, twice- or weakly thrice-pinnate, the oval ultimate segms. tipped with sessile glands; fl.-tube mostly 3 mm. long; sepals ca. 3 mm. long; petals 4–5 mm. long.—Dry slopes, below 2200 ft.; Chaparral; s. San Diego Co.; n. L. Calif. Nov.–May.

7. *Chamaebatiària* (Porter) Maxim. FERN BUSH. DESERT SWEET

Low aromatic shrubs, ± stellate-pubescent. Lvs. twice-pinnate, with entire stipules. Fls. white, fairly large, in panicles. Fl.-tube turbinate. Sepals 5. Petals 5, ± united below; ovules pendulous. Follicles coriaceous, few-seeded, dehiscent at apex and down ventral suture. Seeds terete, with endosperm. Monotypic. (Resembling *Chamaebatia*.)

1. **C. millefòlium** (Torr.) Maxim. [*Spiraea m.* Torr.] Plate 77, Fig. A. Stout, densely branched, 0.6–2 m. tall, the young growth and herbage ± glandular and stellate-pubescent; lvs. 2–4 cm. long, oblong in outline, with 15–20 pairs of primary pinnae, these divided into 10–17 pairs less than 1 mm. long; panicle 3–10 cm. long, heavily glandular; sepals lanceolate, acute, 3–5 mm. long; petals roundish, 5 mm. long; follicles 5 mm. long; seeds yellowish, ca. 2.5 mm. long.—Dry rocky slopes, 3400–10,000 ft.; Sagebrush Scrub to Bristle-cone Pine F.; Panamint, Inyo-White Mts.; to Modoc and Siskiyou cos.; Ore., Wyo. June–Aug.

8. *Coleógyne* Torr. BLACKBUSH

Intricately branched shrub with opposite spinescent branches. Lvs. in opposite fascicles, linear-oblanceolate, entire, deciduous. Fls. solitary, terminating short branchlets, subtended by trifid bracts. Fl.-tube coriaceous, short, persistent. Sepals 4, persistent. Petals 0. Disk at mouth of fl.-tube tubular, separating stamens from pistils. Stamens 30–40; fils. filiform. Pistil 1; ovary 1-celled; ovule 1; style lateral, villous at base, bent and twisted, persistent. Ak. glabrous. Monotypic. (Greek, *koleos*, sheath, and *gune*, ovary.)

1. **C. ramosíssima** Torr. Three–20 dm. tall, ashy gray, with divergent branches; lvs. 5–15 mm. long, thickish, strigose; stipules persistent after lf.-fall; sepals 7–8 mm. long, strigose, the inner with scarious margins; sheath between stamens and pistil 4–5 mm. long; ak. 3–4 mm. long, brown.—Dry slopes, below 5000 ft.; Creosote Bush Scrub to Pinyon-Juniper Wd.; sparingly in w. Colo. Desert, more common in Mojave Desert; to Colo., Ariz. April–June.

9. *Cotoneáster* Medic.

Shrubs, sometimes arborescent, evergreen or deciduous, not thorny. Lvs. many, alternate, short-petioled, simple and entire, stipulate. Fls. white or pink, small but many, solitary or in cymose clusters terminating lateral spurs, appearing after the new lvs.; fl.-tube adnate to the ovary. Sepals 5, persistent. Petals 5. Stamens ca. 20. Pistil 1; ovary 2–5-celled; styles 2–5, distinct. Fr. a red or dark pome. Ca. 50 spp. of the Old World. (Latin, *quince-like*, from the lvs. of some spp.)

1. **C. pannòsa** Franch. Arching evergreen shrub to 3 m. tall; lvs. elliptic- to ovate-oblong, ± glabrous above, white-tomentose beneath, 1.5–4 cm. long; infl. corymbose; petals white, spreading, roundish; calyx tomentose; fr. red, globose-ovoid, 8 mm. in diam.—Escape from cult., as near mouth of Ventura River, etc. From China.

10. *Cowània* D. Don

Shrubs or small trees with alternate pinnatifid coriaceous, gland-dotted lvs. Fls. solitary, terminal on short branches. Fl.-tube funnelform, persistent. Sepals 5, imbricate. Petals 5, obovate, spreading. Stamens many, in 2 series. Pistils 4–12, villous-hirsute, the style terminal, plumose, elongate in fr.; ovules solitary. Ca. 5–6 ssp. of sw. N. Am. (Named for J. *Cowan*, British amateur botanist.)

1. **C. mexicàna** D. Don var. **stansburiàna** (Torr.) Jeps. Freely branched shrub to 3 m. tall, with dark shreddy bark, and red-brown glandular twigs; lvs. obovate in outline,

6–15 mm. long, glandular-punctate and green above, white-tomentose beneath, pinnately 3–5-divided into narrow revolute segms.; pedicels 3–8 mm. long, glandular-pubescent; fl.-tube ca. 5 mm. long; sepals 4–6 mm. long; petals cream, 6–8 mm. long, broadly obovate; pistils 5–10; styles 3–5 cm. long in fr.; $n=9$ (Baldwin, 1951).—Dry slopes and canyons, 4000–8000 ft.; Joshua Tree Wd., Pinyon-Juniper Wd.; mts. of e. Mojave Desert; to Colo., New Mex., Mex. April–July. Occasional plants occur with some fls. ♂ and with perfect fls. having only 2–3 pistils and short fruiting styles; they constitute the var. *dubia*. Bdg.

11. *Fallùgia* Endl. APACHE-PLUME

Low deciduous shrubs with flaky bark. Lvs. pinnately dissected into linear revolute divisions. Fls. terminal, showy, peduncled, solitary or few. Fl.-tube hemispheric, persistent, villous within. Sepals 5, with alternate linear bractlets. Petals 5, white, rounded, spreading. Stamens many, in 3 series. Pistils many, villous, on a conic receptacle; style terminal; ovules solitary, basal. Aks. villous, tipped by the plumose style. One sp. (Named for V. *Fallugi*, Italian abbot.)

1. **F. paradóxa** (D. Don) Endl. [*Sieversia p.* D. Don.] Much branched, 3–15 dm. tall, the young growth and lvs. ± rusty-lepidote and pubescent; lfts. mostly 5–10 mm. long, pinnatifid with 3–5 linear, obtuse segms.; fl.-tube 3–4 mm. long; sepals 5–7 mm. long; petals 13–14 mm. long; fruiting styles 25–30 mm. long; $n=9$ (Baldwin, 1951).—Dry slopes, 4000–5500 ft.; Joshua Tree Wd., Pinyon-Juniper Wd.; mts. of e. Mojave Desert; to Nev., Tex., Mex. May–June.

12. *Fragària* L. STRAWBERRY

Perennial herbs with scaly rootstock and runners which root at the nodes. Lvs. and fls. in basal tufts; lfts. 3; stipules membranaceous. Fls. white to pinkish, borne in cymes on a naked stalk. Fl.-tube almost flat. Bractlets and sepals 5. Petals 5, round to elliptical. Stamens 20, in 3 series, sometimes abortive; fils. short. Pistils many, borne on an elevated conical receptacle which enlarges and becomes fleshy in fr. Styles lateral. Aks. small, turgid, borne on the surface of the pulpy fr. Perhaps 30 spp., of N. Temp. Zone and Andes. (Latin, *fragum*, fragrant.) (Staudt, G. Taxonomic studies in the genus Fragaria. Can. J. Bot. 40: 869–996. 1962.)

1. **F. vésca** L. ssp. **califórnica** (C. & S.) Staudt. [*F. c.* C. & S.] Rootstock short; lvs. rather few, with brownish stipules to 1 cm. long; petioles 3–13 cm. long; terminal lfts. 2–5 cm. long, rounded-obovate, coarsely serrate, subsessile; lateral similar, shorter, very oblique; peduncles usually several, few-fld., villous; petals 5–8 mm. long; fr. to 1 or 1.5 cm. thick; $n=7$ (Ichijima, 1930).—Shaded fairly dampish places, 4000–7500 ft.; mostly Montane Coniferous F.; mts. of n. L. Calif. to San Bernardino Mts.; to n. Calif. March–June.

13. *Gèum* L. AVENS

Perennial herbs with rootstocks. Lvs. pinnate, stipulate. Stipules adnate to the sheathing petioles. Fls. rather large, solitary or cymose, yellow, white to purple. Fl.-tube persistent, turbinate or hemispheric, usually 5-bracteolate. Sepals 5. Petals 5. Stamens many. Pistils many, on a short clavate receptacle. Styles filiform, elongate in fr. Fr. an ak. tipped with an elongate style. Perhaps 50 spp. of temp. and cooler regions. (The ancient Latin name.)

1. **G. macrophýllum** Willd. Plate 77, Fig. B. Stems stout, erect, bristly-pubescent, 3–10 dm. high; stipules broad, foliaceous; basal lvs. lyrate-pinnate, 1–4 dm. long, including petiole, with large rounded terminal lft. ± 3-cleft and with 2–6 smaller principal lateral lfts. and lesser ones interspersed; middle cauline lvs. reduced; fls. few, the cyme open; bractlets small, linear; sepals 3–5 mm. long; petals yellow, 4–8 mm. long; receptacle oblong, short-pubescent; persistent part of style hooked, ca. 6–8 mm. long; $n=21$

(Raynor, 1952).—Moist places like meadows 5500–8100 ft.; Montane Coniferous F.; San Bernardino Mts.; mts. of cent. and n. Calif.; to Alaska, e. Asia, Labrador. May–Aug.

14. *Heterómeles* M. Roem. TOYON. CHRISTMAS-BERRY

Evergreen arboreous shrub, unarmed, with simple coriaceous toothed lvs. Fls. small, white, many, in large terminal corymbose panicles. Fl.-tube turbinate, partly adnate to ovary. Sepals 5, persistent. Petals 5, spreading, rounded, concave. Stamens 10, in pairs opposite the sepals; fils. dilated at base and ± connate. Ovary 2–3-celled, with 2 ovules in each cell; styles 2–3, distinct. Fr. a berrylike pome, red or yellow, ovoid, the thickened persistent sepals incurved over the carpels. One sp. (Greek, *heter*, different, and *malus*, apple, unlike related genera.)

1. **H. arbutifòlia** M. Roem. [*Photinia a*. Lindl.] Plate 77 Fig. C. Plant 2–10 m. high, freely branched, with gray bark and tomentulose young branchlets; lvs. elliptical to oblong, or lance-oblong, 5–10 cm. long, rather sharply toothed, paler beneath, glabrous or sparsely tomentulose; petioles 1–2 cm. long; fl.-tube ca. 3 mm. high; sepals triangular 1–1.5 mm. long; petals ca. 4 mm. long; fr. red, 5–6 mm. long, quite persistent through, the winter months; seeds 2.5–3 mm. long.—Common on semidry brushy slopes and in canyons, below 4000 ft.; Chaparral; mts. of s. Calif. to n. L. Calif. and n. Calif. Occasional plants growing with the typical ones bear yellow frs. and have been called *Photinia a*. var. *cerina* Jeps. There is a tendency for insular plants to have lvs. subentire, the frs. red, 8–10 mm. long, and these constitute a possible var. **macrocárpa** (Munz) Munz. [*Photinia a*. var. *m*. Munz.] Judging from their behavior in the botanic garden, they are not only more showy, but less readily eaten by birds.

15. *Holodíscus* Maxim.

Small to arborescent shrubs, often spreading. Lvs. alternate, simple, toothed, without stipules. Infl. terminal, racemose or paniculate, villous. Fls. whitish to pinkish, small, perfect. Sepals 5, 3-nerved, erect. Petals 5, rounded or short-clawed. Fl.-tube saucer-shaped, free from ovaries. Stamens usually 20, inserted on disk, with 3 opposite each petal, 1 opposite each sepal. Pistils 5, distinct, villous; styles terminal; ovules 2, pendulous. Frs. indehiscent, 1-seeded, laterally flattened, villous. Seeds broadly oblong. Ca. 8 spp. of w. N. Am. (Greek, *holo*, whole, and *diskos*, a disk, the disk not lobed.)

(Ley, Arline. A taxonomic revision of the genus Holodiscus. Bull. Torrey Bot. Club 70: 275–288. 1943.)

Lvs. toothed along the sides to below the middle, elliptic to orbicular in shape, not obovate.
  The lvs. broadly ovate to roundish, scarcely longer than broad, with 3–4 teeth on each side, not deeply toothed, pubescent to villous above. Montane .... 1. *H. boursieri*
  The lvs. elliptic to elliptic-ovate, or ovate, longer than broad, with 3–6 teeth on each side, usually deeply toothed, glabrous or with a scattered pubescence above. Coastal . 2. *H. discolor*
Lvs. toothed at top, rarely to the middle, obovate or spatulate. Montane ..... 3. *H. microphyllus*

1. **H. boursiéri** (Carr.) Rehd. in Bailey. [*Spiraea b*. Carr. *Sericotheca b*. Rydb.] Plate 77, Fig. D. Compact, to 1 m. high; young twigs often angled, pubescent to almost villous; lf.-blades broadly obovate to orbicular, the base sometimes cuneate, apex broad, rounded, many-toothed, the teeth broad, rounded, shallow, blades ± villous-pubescent above, villous to villous-tomentose beneath, often glandular, 1–3 cm. long; petioles 2–3 mm. long; infl. villous, 2.5–8 cm. long and wide; pedicels 3-bracted; sepals 1.5–2 mm. long; petals 2 mm. long, oval; carpels 1.5 mm. long.—Dry rocky slopes, 1500–5700 ft.; mostly Montane Coniferous F.; Santa Ana Mts.; to n. Calif., w. Nev. June–Aug.

2. **H. discólor** (Pursh) Maxim. var. **discólor**. [*Spiraea d*. Pursh.] OCEAN SPRAY. Spreading, 1.5–6 m. tall; older bark exfoliating; twigs straw-colored, pubescent or villous; lvs. broadly ovate, with a truncate base to ovate-elliptic with cuneate base, 5–9 cm. long, villous to tomentose beneath, deeply 3–7-toothed on each side, each tooth again divided; petioles 7–20 mm. long; infl. very compound, dense, 7–20 cm. long; sepals 1.5–2

mm. long; petals 2 mm. long, with several hairs at outer base; pistils villous, to 1.5 mm. long.—Woods and rocky places below 4500 ft.; Chaparral; Catalina and Santa Cruz ids.; Los Angeles Co. to B.C., Mont. May–Aug.

Var. **franciscànus** (Rydb.) Jeps. [*Sericotheca f.* Rydb.] Lf.-blades 3–4.5 cm. long, villous beneath; truncate at base; petioles 5–8 mm. long.—Brushy slopes below 4000 ft.; Chaparral; Orange Co. to s. Ore. May–July.

3. **H. microphýllus** Rydb. [*Sericotheca m.* Rydb.] var. **microphyllus.** Spreading, bushy, 0.2–2 m. high; twigs glabrescent to villous-pubescent, with gland droplets; lvs. obovate to spatulate, with cuneate base, many-toothed, the teeth usually small, broad, rounded and only above the middle of the blade which is pubescent above, villous beneath, 0.5–2 cm. long, on very short petioles; infl. villous, 2.5–3.5 cm. long; sepals 1–1.5 mm. long; petals 1.5–2 mm. long; carpels villous, 1 mm. long.—Dry rocky slopes, 5500–11,000 ft.; Pinyon-Juniper Wd., Montane Coniferous F.; San Bernardino Mts., mts. of e. Mojave Desert; to Mono Co.; Colo., Ariz. June–Aug.

Var. **seríceus** Ley. Lf.-blades villous above, white-silky beneath.—At ca. 7500 ft.; Montane Coniferous F.; San Jacinto Mts.; s. Nev., L. Calif.

16. *Horkèlia* Cham. & Schlecht.

Perennial herbs with a thick woody caudex or rootstock. Lvs. pinnate with several pairs of toothed, cleft or divided lfts., the uppermost confluent. Fls. white, rarely cream or pink, cymose, ± crowded. Fl.-tube deeply cup-shaped to hemispheric or saucer-shaped. Bractlets, sepals and petals 5. Petals round to spatulate. Stamens 10; fils. usually dilated. Carpels many, sometimes few, on a dry or fleshy conical or hemispherical, mostly hairy receptacle; styles subterminal, glandular-thickened at base, deciduous. Aks. smooth or rugulose, mostly brown. Ca. 17 spp. of w. N. Am. (Named for J. *Horkel*, a German physiologist.)

(Keck, D. D. Revision of Horkelia and Ivesia. Lloydia 1: 75–142. 1938.)
Bractlets ovate, sepaloid; pistils more than 50.
    The bractlets equaling or exceeding sepals, often toothed; fl.-tube cylindrically cup-shaped, 3–5.5 mm. deep. Yellow Pine F., San Bernardino Mts. .... 3. *H. elata*
    The bractlets shorter than the sepals, entire; fl.-tube deeply saucer-shaped, 1.5–2 mm. deep. Cismontane.
        Lfts. 5–10 pairs; the uppermost confluent; petals oblanceolate to narrowly obovate
                2. *H. cuneata*
        Lfts. 1–3 pairs, terminal lft. petiolulate; petals orbicular .... 5. *H. truncata*
Bractlets lanceolate to narrowly linear, much smaller than the sepals; pistils fewer.
    Pedicels recurving in fr.; cyme diffuse .... 6. *H. wilderae*
    Pedicels never recurving, erect.
        Lfts. serrate to shallowly lobed. Mts. of San Diego Co. and San Jacinto Mts. to Mt. Pinos
                1. *H. bolanderi*
        Lfts. deeply palmatifid. White Mts. .... 4. *H. hispidula*

1. **H. bolánderi** Gray ssp. **cleveládii** (Greene) Keck. [*Potentilla c.* Greene.] Cespitose clumps; stems few, 1.5–5 dm. long; rosette-lvs. 6–18 cm. long, light green, sparsely glandular, the lfts. 5–15 mm. long, densely silky-pubescent; infl. with short erect branches; fl.-tube finely pubescent, somewhat glandular; sepals lance-ovate, 4 mm. long; petals spatulate-oblong, a little longer than sepals.—Dampish places, 4000–7500 ft.; Yellow Pine F.; San Jacinto Mts. to n. L. Calif. June–Aug.

Ssp. **párryi** (Wats.) Keck. [*H. bolanderi* var. *p.* Wats.] Stems 1.5–7 dm. long; basal lvs. 4–18 cm. long; lfts. 5–15 mm. long.—Mats in moist places, below 9500 ft.; Yellow Pine F.; San Bernardino Mts. to Mt. Pinos region and to Monterey Co. June–Aug.

2. **H. cuneàta** Lindl. [*Potentilla lindleyi* Greene.] Plate 77, Fig. E. Stems erect, 2–5 dm. high; herbage glandular-villous to subglabrous; lower lvs. 1–2 dm. long; lfts. 5–10 pairs, the uppermost confluent, 1–3 cm. long, sharply serrate above the base; infl. mostly congested; fl.-tube deeply saucer-shaped, 4.5–7 mm. wide, 1.5–2 mm. deep, densely

pilose within; bractlets shorter than sepals, entire; sepals 4.5–6.5 mm. long; petals narrowly oblanceolate; pistils 40–80.—Open sandy fields and in woods; S. Oak Wd., Coastal Sage Scrub, etc.; San Diego Co. to cent. Calif. April–Sept.

Ssp. **pubérula** (Greene) Keck. [*Potentilla p.* Greene.] Fls. in open cymes; herbage moderately glandular-pubescent; fl.-tube obscurely pilose to glabrous within.—Away from the coast, San Diego Co. to San Luis Obispo Co.

3. **H. elàta** (Greene) Rydb. [*Potentilla e.* Greene.] Three to 8 dm. high, pilose, glandular; lower lvs. 7–25 cm. long; lfts. 5–10 pairs, cuneate-oblanceolate to broadly flabelliform, 1–3-cleft and incised into laciniate teeth; cyme forked; fl.-tube 4–6 mm. wide, 3–3.5 mm. deep; bractlets 1–2 mm. long; sepals 3–5 mm. long, ca. as long as petals; pistils 50–100.—Moist places, 4500–5500 ft.; Yellow Pine F.; San Bernardino Mts.; cent. Calif. May–Aug.

4. **H. hispídula** Rydb. [*Potentilla h.* Jeps.] Caudex simple or few-branched; stems slender, ± villous, scarcely leafy, 1–2.5 dm. long; basal lvs. 0.5–1 dm. long, densely hispidulous, setose-ciliate, obscurely glandular; lfts. 6–12 pairs, 3.5–5 mm. long, flabelliform; cyme few-fld.; fl.-tube hemispheric, 3.5–4 mm. wide; bractlets linear, 2–3 mm. long; sepals 3.5–4.2 mm. long; petals obovate; stamens 1-rowed; pistils 12–18.—Dry slopes at ca. 10,000 ft.; Bristle-cone Pine F.; White Mts. June–Aug.

5. **H. truncàta** Rydb. [*Potentilla t.* Munz & Jtn.] Erect, sparsely leafy, 2–5 dm. high, glandular-pubescent; lower lvs. to 12 cm. long, with 1–3 pairs of lateral lfts. 1–3 cm. long and a ± petiolulate terminal lft.; cyme forked; fls. usually solitary; fl.-tube saucer-shaped, 5 mm. wide, glabrous within; bractlets ovate, entire, ciliate; sepals 4–5.5 mm. long; pistils 50–80.—Dry slopes, 2000–4000 ft.; Chaparral, S. Oak Wd.; San Diego Co.; n. L. Calif. May–June.

6. **H. wìlderae** Parish. [*Potentilla w.* M. & J.] Caudex from a deep taproot, stems spreading widely, slender, leafy, diffusely branched, 1–3 dm. long, finely glandular-pubescent; basal lvs. to 1 dm. long, the lfts. 4–7 pairs, 5–10 mm. long, deeply flabellate-dissected into linear-oblong lobes; cyme diffuse; fl.-tube cupulate; bractlets narrow-oblong, 0.7 1 mm. long; sepals 1.7–2.5 mm. long, ± purplish, glandular and pilose; petals longer; pistils 3–4.—Dry benches in Yellow Pine F. at 6000–8000 ft.; San Bernardino Mts. May–Aug.

17. *Ivèsia* T. & G.

Perennial with mostly heavy caudex. Lvs. pinnate, mostly basal; lfts. mostly parted or divided into narrow lobes, usually small, many and imbricated. Fls. yellow or white or purple, cymose, usually crowded. Fl.-tube turbinate or campanulate to saucer-shaped or disciform. Bractlets, sepals and petals usually 5. Petals linear to obovate. Stamens 5 or 20, rarely 10 or 15; fils. filiform, usually inserted on rim of fl.-tube. Carpels usually few (1–15); styles subterminal. Twenty-two spp. of w. Am. (Named for Lt. E. *Ives*, leader of a Pacific Ry. Survey.)

(Keck, D. D. Revision of Horkelia and Ivesia. Lloydia 1: 75–142. 1938.)
Inner row of stamens inserted at margin of receptacle; stamens 20; cespitose dwarfs with open few-fld. cymes; anthers 0.3 mm. long; aks. caruncled.
 Petals white, obovate; lfts. 2-lobed or entire. San Jacinto Mts. . . . . 2. *I. callida*
 Petals yellow, linear; lfts. 2–5-lobed. Clark Mt. . . . . 3. *I. jaegeri*
Inner row of stamens inserted well away from the margin of the receptacle.
 Pistils 4–8; sepals 2.5–3.5 mm. long . . . . 1. *I. argyrocoma*
 Pistil 1; sepals less than 2 mm. long . . . . 4. *I. santolinoides*

1. **I. argyrocòma** (Rydb.) Rydb. [*Horkelia a.* Rydb. *Potentilla a.* M. & J.] Stems slender, 1–2 dm. long; herbage densely silvery-silky; basal lvs. many, vermiculariform, 3–10 cm. long; lfts. 20–35 pairs, tightly imbricate, 1–3 mm. long; cyme often paniculately branched; fl.-tube turbinate to campanulate, 2.5–3.5 mm. wide, glabrous within; bractlets linear, half as long as sepals; sepals lanceolate, 2.5–3.5 mm. long; petals white, obovate, longer; pistils 4–8.—Dry meadows, 6500–7500 ft.; Montane Coniferous F.; San Bernardino Mts. June–Aug.

2. **I. cállida** (Hall) Rydb. [*Potentilla c.* Hall.] Stems 2–5 cm. long, spreading; herbage

hirsute and finely glandular throughout; lvs. 2–3 cm. long; lfts. 6–8 pairs, 2–3 mm. long; cyme open, 1–6-fld.; fl.-tube ca. 2 mm. wide; bractlets lance-linear, sepals lanceolate, 2.5–3.5 mm. long; petals equalling sepals; pistils 4–6.—Rock crevices at ca. 8000 ft.; Montane Coniferous F.; Tahquitz Peak, San Jacinto Mts. July–Aug.

3. **I. jaègeri** M. & J. [*Potentilla j.* Wheeler.] Stems subdecumbent, 5–12 cm. long, subscapose; herbage puberulent and finely glandular; lvs. many, thin, 3–8 cm. long; lfts. 4–8 pairs, not crowded, 3–6 mm. long; cyme open, few-fld., with filiform pedicels; fl.-tube disciform, ca. 2.5 mm. wide; bractlets ovate, minute; sepals 2–3 mm. long; petals yellow, linear, 1.5–2 mm. long.—Limestone crevices, at ca. 7500 ft.; Pinyon-Juniper Wd.; Clark Mts., e. San Bernardino Co.; sw. Nev. June–July.

4. **I. santolinoìdes** Gray. [*Potentilla s.* Greene.] Plate 77, Fig. F. Stems suberect, slender, scarcely leafy, 1–4 dm. high, diffusely branched above, subglabrous except at base and axils; lvs. mostly crowded in a basal rosette, vermiform, densely silvery-silky, terete, 3–10 cm. long; lfts. very numerous, tightly imbricate, minute, 1–1.5 mm. long; cyme many-fld., diffuse; fl.-tube broadly funnelform to disciform, ca. 2.5 mm. wide; bractlets roundish to oblong, minute; sepals less than 2 mm. long; petals white, 2–2.5 mm. long; fils. filiform; pistil 1.—Dry gravelly slopes and ridges, 6500–9000 ft.; Montane Coniferous F.; San Jacinto Mts. to Mt. Pinos; Sierra Nevada. June–Aug.

18. *Lyonothámnus* Gray. CATALINA IRONWOOD

Evergreen tree with bark exfoliating in narrow strips. Lvs. opposite, petioled, thickish, simple and entire to pinnately compound. Stipules deciduous. Fls. many, perfect, in large terminal corymbose compound panicles, with short pedicels. Fl.-tube campanulate, free from ovary, subtended by 1–3 bractlets. Sepals 5, persistent. Petals 5, clawless. Stamens ca. 15, inserted on a woolly disk lining the fl.-tube. Pistils 2, distinct, 1-loculed. Style stout, with subcapitate stigma. Ovules 4 in each ovary. Fr. a pair of small woody follicles. Seeds flat. One sp. (Named for W. S. *Lyon*, early resident of Los Angeles, and *thamnos*, shrub.)

1. **L. floribúndus** Gray ssp. **floribúndus**. Slender, 5–15 m. tall, with red-brown to grayish bark and narrow crown; young growth and twigs pubescent; lvs. lance-oblong, 10–16 cm. long, bicolored, glossy green above, ± pubescent beneath, entire to crenate-serrate with tendency in some lvs. to some dissection; petioles 1–2 cm. long; infl. 1–2 dm. broad; petals white, rounded, 4–5 mm. long; follicles glandular-pubescent, 3–4 mm. high; seeds ca. 2 mm. long.—Dry slopes and canyons, Santa Catalina Id. May–June.

Subsp. **aspleniòlius** (Greene) Raven. Lvs. broadly ovate in outline, pinnate into 2–7 parts then pinnatifid into numerous oblique lobes.—San Clemente, Santa Rosa and Santa Cruz ids.

19. *Peraphýllum* Nutt. SQUAW-APPLE

Low shrub with grayish bark and simple alternate lvs. crowded at ends of spurlike branchlets or scattered along new growth, oblanceolate, entire or serrulate. Fls. appearing with lvs., 1–3, perfect. Fl.-tube subglobose, adnate to ovary. Sepals 5, persistent, reflexed. Petals 5, orbicular. Stamens ca. 20, distinct. Pistil 1, with 2 carpels, but ovary 4-celled by false partitions; styles 2. Fr. a pome, globose, fleshy, bitter, the carpels cartilaginous. Monotypic. (Greek, *pera*, excessively, and *phullon*, leafy, i.e., very leafy.)

1. **P. ramosíssimum** Nutt. Intricately branched, 1–2 m. high; lvs. 2–4 cm. long, acute, strigose; sepals ca. 3 mm. long, woolly at least on inner surface; petals pale pink, 7–8 mm. long; fr. yellowish, 8–10 mm. thick.—Dry washes and slopes, 7000–8000 ft.; Pinyon-Juniper Wd.; Inyo Mts. as at Westgard Pass; to Ore., Colo. April–May.

20. *Petrophýtum* Rydb. ROCK-SPIRAEA

Woody cespitose plants forming mats on rocks. Lvs. persistent, crowded, oblanceolate to spatulate, entire. Fls. perfect, in spicate racemes. Sepals 5. Petals 5, white. Stamens ca.

Plate 78. Fig. A, *Petrophytum caespitosum*. Fig. B, *Potentilla egedei grandis*. Fig. C, *Potentilla glandulosa*, style lateral on ak. Fig. D, *Potentilla gracilis* ssp. *nuttallii*. Fig. E, *Potentilla saxosa*. Fig. F, *Prunus emarginata*.

20. Pistils 3–5; ovary pubescent; style filiform. Follicles dehiscent on both sutures. Ca. 4 spp. of w. N. Am. (Greek, *petra*, a rock, and *phyton*, plant.)

1. **P. caespitòsum** (Nutt.) Rydb. [*Spiraea c.* Nutt.] Plate 78, Fig. A. Depressed undershrub with prostrate branches forming dense mats 3–8 dm. across, silky-pubescent and clothed with rosulate tufts of spatulate lvs. 5–12 mm. long and 1-nerved; peduncles bracted, 3–10 cm. high, bearing a spike 2–4 cm. long and usually simple; sepals 1.5 mm. long; petals 1.5 mm. long; follicles 2 mm. long.—Limestone ledges and rocks, 6000–9000 ft.; largely Pinyon-Juniper Wd.; Providence, Clark, Panamint and Inyo mts.; to Rocky Mts. May–Sept.

21. *Physocárpus* Maxim. NINEBARK

Deciduous shrubs with exfoliating bark. Lvs. simple, alternate, petioled, palmately lobed. Fls. in terminal corymbs. Fl.-tube campanulate, stellate-pubescent. Sepals 5, persistent. Petals white, 5, rounded, spreading. Stamens 20–40, inserted on a disk in the throat of the fl.-tube. Pistils 1–5, ± united at base; styles filiform; stigmas capitate. Follicles inflated, 2–4-seeded. Ca. 10 spp., N. Am. and Asia. (Greek, *phusa*, bladder, and *karpos*, fr.)

1. **P. alternáns** (Jones) J. T. Howell. [*Neillia monogyna* var. *a.* Jones.] Densely branched, 0.5–1.5 m. high, with tawny or grayish-white bark, the young twigs stellate-pubescent; lvs. 5–18 mm. long, 3–7-lobed, the lobes doubly crenate, ± stellate-pubescent on both surfaces; petioles 5–10 mm. long; fls. 3–12; fl.-tube stellate without, glabrous within, 3–4 mm. wide; sepals 3 mm. long; stamens ca. 20; follicle ca. 5 mm. long.—Dry rocky slopes, 7000–10,000 ft.; Pinyon-Juniper Wd.; White Mts., Inyo Co. Plants from the Panamint Mts. with more densely stellate lvs. have been called ssp. *panamintensis* J. T. Howell and some from White Mts. with an inner ring of hair in fl.-tube, ssp. *annulatus* J. T. Howell.

22. *Potentílla* L. CINQUEFOIL. FIVE-FINGER

Perennial or sometimes annual; herbs, rarely shrubs. Lvs. pinnately or digitately compound. Fls. perfect, solitary or cymose. Fl.-tube persistent, flat to hemispherical, 5-bracteolate. Sepals 5. Petals 5. Stamens few to many; fils. not flattened. Pistils many, inserted on a hemispherical or conical receptacle; style terminal, lateral or basal, deciduous. Fr. aks. Perhaps 250 spp., of N. Hemis. (Diminutive of *potens*, powerful, because of supposed medicinal powers.)

(Rydberg, P. A. in N. Am. Fl. 22 (4): 293–355. 1908. Clausen, Keck & Heisey, Carnegie Inst. of Wash. 520: 26–195. 1940.)
1. Fls. solitary, axillary, long-pedicelled; stems slender, prostrate, often rooting at nodes.
  2. Stems and petioles pubescent; basal lvs. spreading; aks. with deep dorsal groove. Montane
                                                                                          1. *P. anserina*
  2. Stems and petioles subglabrous; basal lvs. erect; aks. without groove. Coastal   3. *P. egedei*
1. Fls. cymose, short-pedicelled; stems not stoloniferous.
  3. Styles terminal or nearly so.
    4. The styles fusiform and glandular; plants annual or biennial; lfts. not deeply cleft; petals shorter than sepals.
      4a. Basal lvs. with 5 or more lfts. .... 10. *P. rivalis*
      4a. Basal lvs. with 3 lfts.
        5. Sepals 2.5–3 mm. long; stamens ca. 10 .... 2. *P. biennis*
        5. Sepals 4–5 mm. long; stamens 15–20 .... 7. *P. norvegica*
    4. The styles filiform, not glandular; plants perennial.
      5a. Basal lvs. definitely pinnate into 5 or more lfts.
        6. Petals acutish or acuminate at apex. Plants from deserts.
          7. Stamens 15–35; anthers ca. 0.2 mm. long .... 11. *P. saxosa*
          7. Stamens 5–9; anthers ca. 0.7–1 mm. long .... 8. *P. patellifera*
        6. Petals emarginate or rounded at apex. Coastal Los Angeles Co. ... 6. *P. multijuga*
      5a. Basal lvs. palmate, 5–9-foliolate.

8. Stems ± prostrate; lvs. silky-villous; petals 4–5 mm. long. .... 12. *P. wheeleri*
8. Stems erect or ascending.
   8a. Lfts. divided ⅔ or more of the way to midrib and with linear segms.
                                                             9. *P. pectinisecta*
   8a. Lfts. less deeply divided or if ⅔ of way with broader segms. ..... 5. *P. gracilis*
3. Styles lateral or nearly basal, fusiform; lvs. pinnate .... 4. *P. glandulosa*

1. **P. anserìna** L. [*P. argentina* Huds.] SILVERWEED. Low stoloniferous perennial, with rosettes of horizontal pinnate lvs. 1–2 dm. long; lfts. ca. 9–31, with smaller ones interspersed, 1–4 cm. long, oblong to lance-oblong, deeply and sharply serrate, green and subglabrous above, white silky-tomentose beneath; petioles 1–5 cm. long; lvs. on stolons much reduced; peduncles axillary, solitary, 1-fld.; fl.-tube saucer-shaped; bractlets mostly slightly exceeding sepals; these 3–5 mm. long; petals yellow, oval, 7–10 mm. long; stamens 20–25; aks. corky, thick, deeply dorsally grooved; $2n = 28$ (Taylor, 1967).—Moist ± alkaline places, 6000–7500 ft.; Yellow Pine F.; San Bernardino Mts.; Sierra Nevada to Alaska, Nfld.; Eurasia. May–Oct. Lvs. white-silky on both surfaces are the var. *sericea* Hayne.

2. **P. biénnis** Greene. Stems 1–several, mostly erect, pubescent and glandular, 2–5 dm. tall; lvs. 3-foliolate, ± pubescent; lfts. roundish to broadly obovate, 2–4 cm. long, coarsely crenate; petioles 1–7 cm. long; cymes often appearing racemose; pedicels 4–8 mm. long; fl.-tube glandular-pubescent; bractlets shorter than sepals; sepals 2.5–3 mm. long; petals yellow, ca. 2 mm. long; stamens ca. 10; aks. whitish.—Moist mostly sandy places, 6000–10,000 ft.; Montane Coniferous; San Bernardino Mts., to Mt. Pinos, Panamint Mts.; Sierra Nevada to B.C., Rocky Mts. L. Calif. May–Aug.

3. **P. egèdei** Wormsk. var. **grándis** (Rydb.) J. T. Howell. [*P. anserina* var. *g.* T. & G.] Plate 78, Fig. B. Habit of *P. anserina*, but with stolons, petioles etc. subglabrous; lvs. suberect, 2–5 dm. long; lfts. 7–31, oblong to obovate, 3–6 cm. long, green above, subglabrous to white-tomentose beneath, the pubescence opaque or dull, not shining; bractlets 6–8 mm. long; sepals 5–6 mm. long; petals 10–12 mm. long.—Coastal Strand and Salt Marsh, Orange Co. n.; San Miguel Id. to Alaska, Asia. April–Aug.

4. **P. glandulòsa** Lindl. [*Drymocallis g.* Rydb. *P. wrangelliana* Fisch. & Avé-Lall.] ssp. **glandulòsa**. Plate 78, Fig. C. Perennial from a woody caudex; stems erect, 3–8 dm. high, leafy, glandular- or viscid-villous, branching above, often reddish; basal lvs. pinnate, glandular, dark green above, lighter beneath; lfts. 5–9, obovate, 1–4 cm. long, serrate; stem-lvs. reduced; cyme open, many-fld.; fl.-tube glandular-hirsute, 4–8 mm. broad; bractlets linear, 4–6 mm. long; sepals lance-ovate, 6–8 mm. long, more in fr.; petals pale yellow to creamy white, ca. as long as sepals; stamens ca. 25; pistils many; aks. brownish, veiny; $n = 7$ (Clausen et al., 1937).—Dryish to damp open places, mostly at low elevs., but up to 7000 ft.; many Plant Communities; along the coast from n. L. Calif. to B.C.; Santa Catalina Id.; Sierran foothills; to Ida. May–July. Variable.

KEY TO SUBSPECIES

Petals much longer than sepals; cyme open; herbage subglabrous to moderately pilose. At 5000–7000 ft. .... ssp. *nevadensis*
Petals slightly or not longer than sepals.
   Lfts. averaging more than 15 mm. long; stems mostly more than 1 dm. long; sepals mostly 7 or more mm. long.
      Petals broad, equaling or slightly exceeding sepals.
         Branches divaricate, leafy-bracted above, anthocyanous, prominently glandular; sepals ovate, to 12 mm. long .... ssp. *glandulosa*
         Branches erect, not leafy above, scarcely anthocyanous or glandular; sepals broadly lanceolate, to 10 mm. long. Piute Mts. .... ssp. *hansenii*
      Petals narrow, shorter than the sepals. Montane .... ssp. *reflexa*
   Lfts. up to 6 mm. long, extending most of the length of the rachis; herbage ± silky-pubescent; stems 8–12 cm. long, from slender stolons; sepals 5 mm. long .... ssp. *ewanii*

Ssp. **èwanii** Keck. Stems 8–12 cm. high; basal lvs. viscidulous, hirsute, 2.5–5 cm. long; lfts. 7–11, round-flabelliform, deeply incised-serrate, 4–7 mm. long; fls. few; sepals 4–5 mm. long.—Seeps at ca. 6500–7500 ft. Yellow Pine F.; Mt. Islip, San Gabriel Mts. June.

Ssp. **hansènii** (Greene) Keck. [*P. h.* Greene.] Stems 5–8 dm. high, villous; lfts. 9–11, coarsely serrate, 1–4 cm. long; sepals 5–7 mm. long.—Meadows, 4000–6000 ft.; Piute Mts., Kern Co.; Sierra Nevada.

Ssp. **nevadénsis** (Wats.) Keck. [*P. g.* var. *n.* Wats.] Stems slender, 2–4 dm. high; cyme open.—Moist places, 5000–7500 ft.; Montane Coniferous F.; San Jacinto Mts. n.; to Wash. June–Aug.

Ssp. **refléxa** (Greene) Keck. [*P. g.* var. *r.* Greene.] Stems 3–6 dm. high, villous, slightly glandular; lfts. 7, densely pubescent, 1–3 cm. long; fls. few; bractlets shorter than sepals, these 5–6 mm. long; petals ca. 4 mm. long.—Mostly dryish slopes, 4500–7000 ft.; Montane Coniferous F.; L. Calif.; to Ore. May–July.

5. **P. grácilis** Dougl. ex Hook. ssp. **nuttállii** (Lehm.) Keck. [*P. n.* Lehm.] Plate 78, Fig. D. Perennial with a short caudex; stems slender, 4–7 dm. high, ± villous; basal lvs. digitate; lfts. 5–7, somewhat paler beneath than above, hirsute to tomentulose to subglabrous, variously cut, 1–5 cm. long; cymes loose; fl.-tube hairy, 4–6 mm. broad; sepals acuminate; petals yellow, 5–7 mm. long; stamens usually 20; pistils many.—Meadows, etc. below 11,000 ft.; Montane Coniferous F., mts. of San Diego Co., n. to Alaska, S. Dak. June–Aug.

6. **P. multijùga** Lehm. Perennial with a taproot and almost no caudex, the stems to 6.5 dm. long, slightly silky, somewhat leafy; lvs. pinnate, the lower numerous, 1–3 dm. long; lfts. 11–27, 1–4 cm. long, with few coarse teeth above middle; fls. in strict cymes; fl.-tube 4–6 mm. wide; sepals 5–6 mm. long; petals yellow, ca. 7 mm. long; pistils many. —Originally in brackish meadows, Coastal Sage Scrub near Ballona, Los Angeles Co. Probably now extinct.

7. **P. norvègica** L. ssp. **monspeliénsis** (L.) Asch. & Graebn. [*P. m.* L.] Annual to short-lived perennial; stems erect or ascending, stout, branched above, 2–7 dm. tall, hirsute with mostly stiff spreading hairs; lvs. 3-foliolate; lfts. oblanceolate to obovate, coarsely serrate, ± hirsute, 3–5 cm. long; cyme leafy; fl.-tube hirsute, 7–8 mm. broad; bractlets as long as sepals; sepals 4–5 mm. long; petals mostly slightly shorter; stamens 15–20; $n=28$ (Löve, 1954).—Moist places, 4500–7500 ft.; Montane Coniferous F.; Cuyamaca Mts., scattered stations northward; to Atlantic Coast; Eurasia. June–Sept.

8. **P. patellífera** J. T. Howell. Much like *P. saxosa*, villous, slightly glandular, the stems 5–20 cm. long; fl.-tube 1.5–2 mm. in diam.; bractlets not over 0.5 mm. long; petals 0.5–1 mm. wide, 2–3 mm. long; stamens 5–9; anthers 0.8–1 mm. long.—Rock-crevices, 5500–6500 ft.; Pinyon-Juniper Wd.; Kingston Mts., ne. San Bernardino Co. June & Oct.

9. **P. pectinisécta** Rydb. Mostly 3–4 dm. tall; basal lvs. digitate, lfts. 5–9, appressed-silky on both sides, sometimes ± tomentose beneath, obovate, the linear segms. evenly and pectinately arranged; sepals 5–6 mm. long; petals 7–8 mm. long; $n=21$ (Clausen et al., 1940).—Moist places, 5000–7000 ft.; Pinyon-Juniper Wd., Yellow Pine F.; mts. about w. end of Mojave Desert (San Bernardino, San Gabriel, Mt. Pinos, Tehachapi) to White Mts. and e. slope of Sierra Nevada; to Ida., Mont., Colo. May–July.

10. **P. rivàlis** Nutt. var. **millegràna** (Engelm.) Wats. [*P. m.* Engelm.] Annual or biennial, stems diffusely branched from base; lvs. 3-foliolate; lfts. 2–5 cm. long; infl. leafy, paniculate-cymose; fl.-tube ca. 5 mm. broad, hirsute; sepals 3–4 mm. long; petals yellow, shorter than sepals; stamens 5–10; pistils many.—Bottom lands, as along Colo. River, scattered stations in Calif. to B.C., New Mex. April–Oct.

11. **P. saxòsa** Lemmon ex Greene. Plate 78, Fig. E. Low cespitose perennial, with thick woody root and caudex; stems slender, glandular-pubescent, 3–25 cm. long, leafy; basal lvs. pinnate, villous and somewhat glandular; lfts. 5–15, flabelliform, 5–15 mm. long, strongly toothed to dissected; cymes few-fld.; fl.-tube 2–4 mm. broad; sepals spreading, 2–3 mm. long; petals yellow, 2–3 mm. long; stamens 28–40; pistils 10 or more.— Rock-crevices and dry places, 3000–6000 ft.; Joshua Tree Wd., Pinyon-Juniper Wd.; w.

edges of Colo. Desert from n. L. Calif.; Little San Bernardino Mts. to Kingston Mts., Inyo Mts., etc. April–June.

12. **P. wheèleri** Wats. var. **wheèleri**. Low perennial with several to many spreading or prostrate stems 5–20 cm. long, freely branching, pubescent to silky; lvs. subpalmate, the basal with 5 lfts. which are cuneate to obovate, silky, 5–20 mm. long, with few large terminal teeth; petioles 1–6 cm. long; upper lvs. reduced; cymes loosely branched in age, with pedicels 4–16 mm. long; fl.-tube saucer-shaped; sepals ovate; petals yellow, obcordate, 4–5 mm. long; stamens 20; pistils many.—Edge of meadows, 6500–11,000 ft.; Montane Coniferous F.; San Bernardino Mts.; Sierra Nevada. June–Aug.

Var. **rimícola** M. & J. Lvs. not conspicuously silky, rather green and more glandular; branches of cyme and pedicels very slender.—Rock crevices, 7500–9000 ft.; San Jacinto Mts.; San Pedro Martir Mts., L. Calif. July–Aug.

23. *Prùnus* L. STONE-FRUITS

Trees and shrubs with simple, deciduous or persistent, mostly serrate lvs. often bearing glands on the petioles and base of blade. Stipules small, caducous. Fls. umbellate, corymbose or racemose, appearing before or with the lvs. Fl.-tube hemispheric or cup-shaped. Sepals and petals 5. Stamens many. Pistil 1; style terminal; ovules 2, pendulous. Fr. a drupe developed from a fleshy pericarp and a bony endocarp enclosing usually a single seed. Perhaps 150 spp. of temp. climates mostly in N. Hemis. Many valuable in horticulture. (*Prunus*, ancient Latin name of plum.)

1. Plants evergreen, with ± coriaceous or leathery lvs.; fls. racemose.
   2. Lvs. coarsely spinose-toothed, crisped, 2–5 cm. long; fr. red. Native . . . . . . 6. *P. ilicifolia*
   2. Lvs. not as above, mostly longer; fr. dark.
      3. Base of lf. acute, the blade ca. ⅓ as wide as long; racemes 2–3 cm. long. Scarce, introd. tree
         2. *P. caroliniana*
      3. Base of lf. rounded, the blade ca. ½ as wide as long; racemes ca. 5–12 cm. long. Native, insular . . . . 7. *P. lyonii*
1. Plants with deciduous lvs.
   4. Plants shrubby, many of branchlets spinose. Natives.
      5. Lvs. broadly ovate to rounded, the blades 1–2 cm. long; drupe puberulent. W. edge of Colo. Desert . . . . 5. *P. fremontii*
      5. Lvs. ± oblanceolate, obscurely if at all serrulate, mostly fascicled.
         6. Petals 5–6 mm. long, pink; pedicels 6–8 mm. long. Inyo Co. n. . . . . . . 1. *P. andersonii*
         6. Petals 2–3 mm. long, white; pedicels ca. 1 mm. long. Inyo Co. to Imperial Co.
            4. *P. fasciculata*
   4. Plants shrubby to arboreous, not having spinose branchlets.
      7. Fls. 12 to many, in elongate racemes. Native Choke Cherry . . . . 10. *P. virginiana*
      7. Fls. 1–10, not in well developed racemes.
         8. The fls. 1–2, subsessile; sepals not reflexed. Escape from cult. . . . . 9. *P. persica*
         8. The fls. several, on slender pedicels; sepals reflexed at anthesis.
            9. Corymbs short, subumbellate; drupe red, bitter; lvs. much longer than wide. Common native . . . . 3. *P. emarginata*
            9. Corymbs elongate; drupe black; lvs. almost as wide as long. Escape from cult.
               8. *P. mahaleb*

1. **P. andersònii** Gray. DESERT PEACH. Diffusely branched spreading deciduous shrub 1–2 m. tall, with short stiff spinescent branches; lvs. fascicled, oblanceolate, 1–2 cm. long, obscurely serrulate, on very short pedicels; fls. usually 1; pedicels 6–8 mm. long; fl.-tube 2.5 mm. long; sepals ca. 3 mm. long; petals broadly obovate, ± rose, 5–6 mm. long; fr. roundish, ca. 12 mm. long, brownish tomentulose; the pulp dryish.—Dry slopes and mesas, 3500–6500 ft.; Sagebrush Scrub, Yellow Pine F.; from Kern and Inyo cos. along e. slope of Sierra, Inyo Mts.; w. Nev. March–April.

2. **P. caroliniàna** Ait. CHERRY-LAUREL. Evergreen tree, 6–12 m. tall; lvs. lance-oblong, acuminate, 5–10 cm. long, entire, slightly revolute, glossy; racemes dense, 2–3 cm. long; fls. ca. 3–4 mm. across; fr. short-ovoid, black, dryish, 8–10 mm. long.—Occasional escape from cult., as in Ventura Co. Native of se. U.S.

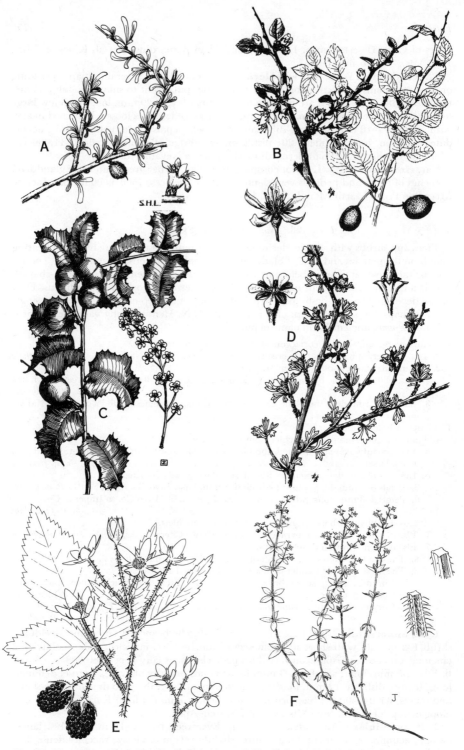

Plate 79. Fig. A, *Prunus fasciculata*. Fig. B, *Prunus fremontii*. Fig. C, *Prunus ilicifolia*. Fig. D, *Purshia glandulosa*. Fig. E, *Rubus ursinus*. Fig. F, *Galium californicum*.

3. **P. emarginàta** (Dougl.) Walp. [*Cerasus e.* Dougl.] BITTER CHERRY. Plate 78, Fig. F. Erect deciduous shrub or small tree, 1–6 m. high, with glabrous red shining twigs, older bark smooth; lvs. oblong-obovate to -elliptic, finely serrulate, subglabrous or slightly pubescent beneath, 2–5 cm. long, short-petioled; corymbs short, 3–10-fld.; fl.-tube campanulate, glabrous, ca. 3 mm. long; sepals 1.5–2 mm. long, oblong; petals 5–7 mm. long, obovate; drupe red, bitter, 6–8 mm. in diam.—Rocky ridges and slopes and canyons, 2000–8000 ft.; Chaparral, Montane Coniferous F.; mts. from San Diego Co. n.; to B.C., Ida. April–May.

4. **P. fasciculàta** (Torr.) Gray. [*Emplectocladus f.* Torr.] DESERT ALMOND. Plate 79, Fig. A. Divaricately much branched deciduous shrub 1–2 (–3) m. tall, with short stiff ± thornlike twigs, minutely pubescent when young; lvs. fascicled on short stubby branchlets, oblance-spatulate, mostly entire, 6–15 mm. long, pale green, minutely pubescent; fls. perfect or imperfect, subsessile; fl.-tube ca. 2 mm. long; petals oblanceolate, 2–3 mm. long; fr. ovoid, dry, 8–12 mm. long, pubescent; stone smooth.—Dry slopes and washes, 2500–6000 ft.; Creosote Bush Scrub to Pinyon-Juniper Wd.; Mojave and Colo. deserts; to Ariz., Utah. March–May.

5. **P. fremóntii** Wats. [*P. eriogyna* S. C. Mason.] DESERT APRICOT. Plate 79, Fig. B. Rigidly branched deciduous shrub or small tree, 1.5–4 m. high, the twigs often spine-tipped, glabrous, red-brown; lvs. roundish to broadly ovate, serrate, 1–2 cm. long, on petioles 3–4 mm. long; fls. 1-few; pedicels slender, 8–12 mm. long; fl.-tube campanulate, ca. 2.5 mm. long; sepals ciliate, ca. 1.5 mm. long; petals white, 4–6 mm. long; fr. elliptic-ovoid, yellowish, puberulent, 8–14 mm. long, dry; stone with a thick ridge on ventral side.—Canyons below 4000 ft.; Creosote Bush Scrub to Pinyon-Juniper Wd.; w. edge of Colo. Desert from Palm Springs region to L. Calif. Feb.–March.

6. **P. ilicifòlia** (Nutt.) Walp. [*Cerasus i.* Nutt. ex H. & A.] HOLLY-LEAVED CHERRY. Plate 79, Fig. C. Dense evergreen shrub or small tree 1–8 m. tall, glabrous, the twigs soon gray or reddish brown; lvs. coriaceous, ovate to roundish, 2–5 cm. long, coarsely spinose-toothed; crisped, the petioles 8–12 mm. long; fls. few to many, racemose; racemes mostly 3–6 cm. long; sepals ± deltoid, to ca. 1 mm. long; petals white, 2–3 mm. long; fr. red, rarely yellow, ovoid-ellipsoid, 12–15 mm. long, with thin sweetish pulp; stone smooth, apiculate.—Common on dry slopes and fans below 5000 ft.; Chaparral, etc.; n. L. Calif. to cent. Calif. San Clemente and Santa Catalina ids. April–May.

7. **P. lyònii** (Eastw.) Sarg. [*Cerasus l.* Eastw. *P. ilicifolia* ssp. *l.* Raven.] CATALINA CHERRY. Like *P. ilicifolia* but more arboreous, to 15 m. high; lvs. darker green, more narrowly ovate, mostly entire, plane, 4–10 cm. long; racemes many-fld., 5–12 cm. long; fr. almost black, 12–24 mm. long.—Chaparral; canyons of Santa Catalina, San Clemente, Santa Cruz and Santa Rosa ids. March–May. Occasionally escaping from cult. on mainland. Hybridizes easily with *P. ilicifolia*.

8. **P. mahaléb** L. Small tree with slender green twigs; lvs. light green, broadly ovate, to roundish, 3.5–6 cm. long, abruptly short-pointed, closely crenate-serrate, rather short-petioled; fls. several in racemose corymbs, white, fragrant, ca. 1–1.5 cm. broad; fr. ovoid, black, with little flesh, ca. 6 mm. long, on much longer pedicels. Cult. especially as budding stock, sometimes escaping; Eu. April.

9. **P. pérsica** Batsch. PEACH. Small deciduous tree with glabrous twigs; lvs. lanceolate, 5–20 cm. long, acuminate, coarsely crenate-serrate, glabrous; fls. solitary, sessile, in advance of lvs., pink, 1–5 cm. across; fr. 3–8 cm. in diam., pubescent, yellow or red, with white or yellow flesh and deeply pitted stone; $2n=16$ (Darlington, 1930).—Occasional escape from cult.; Asia. April–May.

10. **P. virginiàna** L. var. **demíssa** (Nutt.) Sarg. [*Cerasus d.* Nutt.] WESTERN CHOKE CHERRY. Erect deciduous shrub or small tree, 1–5 m. high, with smooth gray-brown bark; young twigs pubescent; lvs. oblong-ovate to obovate, 3–8 cm. long, finely serrate, ± pubescent beneath, abruptly pointed; ± subcordate at base; petioles ca. 1 cm. long; racemes many-fld., 5–10 cm. long, at ends of short lfy. branches; sepals barely 1 mm. long, glandular-pectinate; petals white, 5–6 mm. broad; fr. round, 5–6 mm. thick, dark red, bitter but edible, especially late in the season; stone smooth, globose.—Dampish places in

woods below 8000 ft.; Yellow Pine F.; mostly in mts. from San Diego Co. n.; to Wash., Ida. May–June.

24. Purpùsia Bdg.

Cespitose glandular perennials, with short caudex and thick roots. Lvs. mostly basal, odd-pinnate. Lfts. 5–11, roundish to elliptical or oblanceolate, crenate to palmately lobed or divided, glandular-pilose. Infl. cymose, few-fld., leafy-bracted. Fl.-tube funnelform. Bractlets mostly lacking. Sepals 5. Petals 5, white to yellowish. Stamens 5, opposite the sepals; fils. filiform. Pistils 6–13, on a stalked receptacle; styles subterminal. One sp. (Named for J. A. *Purpus*, 1860–1932, botanical collector.)

1. **P. saxòsa** Bdg. Stems 5–20 cm. high; lfts. 5–15 mm. long; pedicels 1–2 cm. long; fl.-tube glabrous within, glandular and pilose without, ca. 2 mm. deep; sepals 2.5–3 mm. long, lance-ovate, acuminate; petals ± oblanceolate, acuminate, 3–4 mm. long; receptacle 1.5–2 mm. long.—Limestone rocks and crevices, 4000–10,000 ft.; Pinyon-Juniper Wd. to Bristle-cone Pine F.; Inyo Mts., Grapevine Mts. and Funeral Mts., Inyo Co.; to Nev., Ariz. May–Aug.

25. Púrshia DC. ex Poir. ANTELOPE BUSH

Shrubs or small trees. Lvs. alternate, crowded, apparently fascicled, deeply 3-cleft with revolute margins. Fls. solitary at ends of short branches. Fl.-tube turbinate to funnelform, persistent. Sepals 5. Petals 5, cream to yellow. Stamens ca. 25, in one series. Pistil mostly 1, short-styled, 1-ovuled. Fr. an ak., tipped with the rather short persistent style. Two spp. (Named for F. *Pursh*, author of an early flora of N. Am.)

1. **P. glandulòsa** Curran. Plate 79, Fig. D. Greenish shrub 1–2 (–5) m. tall, the glabrous twigs prominently glandular; lvs. 5–10 mm. long, slightly tomentose beneath, glabrous and conspicuously punctate above with sunken glands; fl.-tube ca. 3–4 mm. long, tomentulose; sepals 3 mm. long; petals 6–8 mm. long, spatulate; ak. canescent, almost 2 cm. long including style; seed black.—Dry slopes, 2500–8000 ft.; Joshua Tree Wd., Pinyon-Juniper Wd.; w. edge of Colo. Desert, Cajon Pass, Mojave Desert to Mono Co.; Nev., Ariz., L. Calif. April–June.

26. Pỳrus L.

Woody plants with lvs. alternate, serrate or entire. Fls. white or pinkish. Sepals 5. Petals 5. Fl.-tube urn-shaped, the sepals reflexed or spreading. Fr. a fleshy pome, in our sp. pear-shaped and with a thick zone of grit-cells beneath the epidermis. An Old World genus. (Classical name of pear.)

1. **P. commùnis** L. PEAR. Lvs. oval to oblong-ovate, 5–10 cm. long.—Occasional escape from cult., as near Ventura; native of Old World.

27. Ròsa L. ROSE

Erect, sprawling to climbing shrubs, usually prickly. Lvs. alternate, deciduous to persistent, mostly odd-pinnate, with stipules adnate to petiole. Fls. solitary or in corymbs or panicles, rather large, ours mostly rose-pink. Fl.-tube fleshy, cup-shaped to urceolate. Sepals 5. Petals 5, rounded, spreading Stamens many, inserted on the disk at the edge of the fl.-tube Pistils few to many, free and distinct but included in the fl.-tube. Fr. a fleshy hip (ripened fl.-tube) containing the hairy aks. More than 100 spp., mostly from n. temp. regions. (Ancient Latin name.)

(Cole, D. A revision of the Rosa californica complex. Am. Midl. Nat. 55: 211–224. 1956.)
Prickles mostly slender, straight.
Sepals and styles deciduous in fr.; pistils few; pedicels 1–3 cm. long, ± reflexed in fr.; fr. 4–8 mm. thick at maturity . . . . 2. *R. gymnocarpa*

Sepals and styles persistent in fr.; pedicels usually less than 2 cm. long, not reflexed in fr.;
fr. more than 7 mm. thick at maturity .... 3. *R. woodsii*
Prickles stout, flattened, recurved .... 1. *R. californica*

1. **R. califórnica** C. & S. Erect, branched, 1–3 m. tall, armed with stout flattened usually recurved prickles; petiole and rachis pubescent, prickly, non-glandular to glandular; lfts. 5–7, oval, 1–3.5 cm. long, simply or doubly serrate, puberulent above, pubescent and often glandular beneath; stipules narrow; fls. in corymbs; pedicels glabrous to vilous, sometimes glandular; fl.-tube glabrous or pilose when young; sepals lanceolate, caudate-attenuate, sometimes glandular; petals 1–2.5 cm. long; fr. globose to ovoid, 8–16 mm. long, 10–15 mm. thick, usually with distinct neck; $n=14$ (E. Erlanson, 1932).
—Fairly moist places, canyons, near streams, etc., below 6000 ft.; many Plant Communities; cismontane; n. L. Calif. to s. Ore. May–Aug. Variable.

2. **R. gymnocárpa** Nutt. in T. & G. Wood Rose. Slender-stemmed, mostly ca. 1 m. high, armed with slender straight prickles, sometimes with stouter infrastipular prickles; petiole and rachis glandular-hispid, sometimes unarmed; lfts. 5–7, oval to roundish, 1–4 cm. long, doubly serrate, with gland-tipped teeth, glabrous on both surfaces; fls. usually solitary; fl.-tube smooth; sepals ovate, acuminate, glabrous on back, to ca. 1 cm. long; petals 8–12 mm. long; fr. ellipsoid, to globose, glabrous, 5–10 mm. long, red, without sepals at maturity; $n=7$ (Erlanson, 1932).—Shaded woods at ca. 5000 ft. or less; Palomar Mts., San Gabriel Mts.; cent. Calif. to B.C., Mont. May–July.

3. **R. woódsii** Lindl. var. **ultramontàna** (Wats.) Jeps. Erect, 1–3 m. tall, stout, armed with slender straight prickles, or floral branches ± unarmed; lfts. 5–7, oval, 1–4 cm. long, coarsely serrate, glabrous above, puberulent and ± pruinose beneath; stipules pubescent below and on margins, glandular-ciliate or -dentate; fls. in corymbs; pedicels glabrous; fl.-tube glabrous; sepals glabrous on back, caudate-attenuate, rarely glandular; petals 1.5–2 cm. long; fr. ellipsoid to globose, 7–10 mm. in diam. at maturity.—Dampish places, Montane Coniferous F.; Mt. Pinos region; to Sierra Nevada and then n. to B.C. June–Aug.

Var. **glabràta** (Parish) Cole. [*R. mohavensis* Parish.] Stems to ca. 1 m. high, well armed; lvs. glabrous; sepals glabrous.—Moist places about springs, 3000–4000 ft., Joshua Tree Wd.; n. base of San Bernardino Mts. May–July.

Var. **gratíssima** (Greene) Cole. [*R. g.* Greene.] Stems rather densely armed, even the floral branches usually heavily spiny; lfts. 1–2 cm. long; fls. 1–few; petals 1.5 cm. long.—Dry slopes Pinyon-Juniper Wd.; Panamint and San Bernardino mts. to Sierra Nevada. April–Aug.

28. *Rùbus* L. BLACKBERRY. RASPBERRY, etc.

Shrubs to trailing vines, prickly or unarmed. Stems in their first year shooting up and sterile (primocanes), in second year usually flowering and with different foliage (floricanes). Stipules adnate to petioles. Lvs. alternate, simple to pinnately compound or subpalmate, petioled. Fls. ± perfect, racemose or paniculate, sometimes few or solitary. Fl.-tube persistent, rotate to campanulate. Bractlets 0. Sepals mostly 5. Petals mostly 5. Stamens many. Carpels many, crowded on an elevated receptacle, becoming drupelets which coalesce and form an aggregate fr. Ovules 2, 1 abortive; style terminal, slender. Spp. estimated as some hundreds, of n. temp. regions and the Andes. Many of great horticultural value. (Latin name for bramble, related to *ruber*, red.)

(Bailey, L. H. Rubus in N. Am. Gentes Herb. 5: 1–932. 1941–1945.)
Lvs. simple, palmately lobed; stems unarmed, erect, with peeling or flaky bark; styles club-shaped
5. *R. parviflorus*
Lvs. mostly 3–5-foliolate; stems mostly prickly; styles filiform.
  Drupelets adhering to the fleshy receptacle and falling with it or separately. (Blackberries, etc.)
    Fls. borne in small clusters; prickles slender, ± setose.
      Primocanes pruinose or with a "bloom": fls. mostly functionally unisexual. Cismontane native .... 7. *R. ursinus*
      Primocanes not pruinose; fls. bisexual. Escape from cult. .... 1. *R. almus*

Fls. borne in large terminal panicles; prickles stout, broad-based. Escapes from cult.
   Lvs. deeply cut or dissected, not whitish-tomentose or canescent beneath . 3. *R. laciniatus*
   Lvs. not deeply cut, whitish-tomentose or gray-canescent beneath, unequally serrate or
      toothed .... 6. *R. procerus*
Drupelets forming a hollow cone which separates from the dry receptacle as a single aggregate
fr. (Raspberries)
   Sepals deflexed at anthesis; sepals and pedicels glandless or nearly so ..... 4. *R. leucodermis*
   Sepals not deflexed; pedicels and sepals with stipitate glands .... 2. *R. glaucifolius*

1. **R. álmus** (Bailey) Bailey. GARDENA DEWBERRY. Prostrate, dense, the primocanes to 2 m. long, glabrous except near the tips; prickles scattered, 3–4 mm. long; lfts. 5–7, soft-pubescent beneath, 7–8 mm. long, elliptic-ovate, narrowly cut-toothed; fls. few on each fl. shoot, on stout pilose pedicels; sepals heavily pubescent, prominently pointed; corolla 2–3 cm. broad; fr. oblong, to 3 cm. long, juicy.—Occasional escape from cult., as at Hawthorne; native of Texas. May–June.

2. **R. glaucifòlius** Kell. var. **gánderi** (Bailey) Munz. [*R. g.* Bailey.] Plant wholly prostrate; prickles few, 1–2 mm. long; lvs. green and with 3 lfts., acute, sharply dentate, ± tomentose beneath, lfts. 3–6 cm. long; fls. few, in umbellate clusters surpassed by lvs., the pedicels scarcely if at all glanduliferous; sepals lanceolate, not reflexed, 6–8 mm. long; petals ca. as long, white; fr. hemispheric to conic, red or purplish, with few drupelets.— Shaded woods, 4500–5500 ft.; Montane Coniferous F.; Palomar and Cuyamaca mts., San Diego Co. June.

3. **R. laciniàtus** Willd. CUT-LEAF BLACKBERRY. Much-branched, diffuse shrub to 3 m. high, stoutly armed; primocane lvs. 5-foliolate, the lfts. 3–8 cm. long, cut and parted into laciniate or lobed sublfts.; fls. mostly pinkish to rose, ca. 2 cm. across; sepals ± foliaceous, reflexed, with very narrow apices; fr. large, rounded, of few succulent drupelets. Occasional garden escape as in Palomar Mts.; European garden plant. May–July.

4. **R. leucodérmis** Dougl. ex T. & G. var. **bernardìnus** (Greene) Jeps. WESTERN RASPBERRY. Stems to ca. 2 m. long, arched and branched, rooting at tip, with heavy whitish bloom at least when young, the prickles many; primocane lvs. 3–5-foliolate, the lfts. ovate to almost lanceolate, the terminal 7–9 cm. long, irregularly doubly sharp-serrate; fls. mostly 3–10, 7–10 mm. across, the pedicels, fl.-tube and sepals ± stipitate-glandular; fr. firm, dark purple to yellow red, depressed-globose, to ca. 1.5 cm. in diam.— Dry flats and slopes, 4700–7500 ft.; Yellow Pine F.; Palomar Mts., San Bernardino and San Gabriel mts., Mt. Pinos. June–July.

5. **R. parviflòrus** Nutt. [*R. nutkanus* Moc.] THIMBLEBERRY. Deciduous, mostly 1–2 m. high, without prickles, the bark shreddy in age; lvs. palmately 5-lobed, unequally serrate, 1–1.5 dm. broad, ± cordate at base; petioles and peduncles hirsute-glandular; fls. few, in terminal corymbs, white to pink, 2–5 cm. broad; sepals pubescent, evidently glandular, terminated by a long appendage; petals elliptic, 15–20 mm. long; fr. scarlet, hemispheric, 1–1.6 cm. broad; $n=7$ (Darrow & Longley, 1933).—Open woods and in canyons, below 8000 ft.; Montane Coniferous F.; mts. from San Diego Co. n.; to Alaska. March–Aug.

6. **R. prócerus** P. J. Muell. HIMALAYA-BERRY. Robust, sprawling, ± evergreen, glandless, to 3 m. tall; primocanes pilose-pubescent, glabrescent, angled, furrowed, with broad-based prickles 6–10 mm. long; primocane lvs. 5-foliolate, with hooked prickles on petioles; lfts. broad, the terminal roundish to broad-oblong, 10–12 cm. long, coarsely serrate-dentate; floricane lvs. smaller; infl. large, terminal, branched, the fls. 2–2.5 cm. across; sepals soon reflexed, 7–8 mm. long; fr. roundish, shiny black, to 2 cm. long, with large succulent drupelets; $2n=28$ (Crane, 1936).—Occasionally reported as natur., as on Catalina Id., near Santa Barbara, etc.; native of Eu.

7. **R. ursìnus** C. & S. CALIFORNIA BLACKBERRY. Plate 79, Fig. E. Grayish or ± tomentose mound-builder with running or semiscandent stems; primocanes pruinose, with bristlelike prickles; lvs. of primocanes mostly 3-foliolate, on bristly glandless petioles, the terminal lfts. 5–12 cm. long; fls. 2–15, at or near summit of lateral leafy shoots, perfect or imperfect, on prickly mostly glandless pedicels; petals of ♂ fls. to 15 mm. long, of ♀ smaller; sepals tomentose, pointed, bristly; fr. oblong or conical, black, to 2 cm. long;

$n = 21, 28, 42$ (Gustafsson, 1943).—Waste places, canyons, etc., below 3000 ft.; many Plant Communities, L. Calif. through cismontane Calif. to Ore.

29. Sanguisórba L. BURNET

Annual or perennial herbs with unequally pinnate lvs. and stipules adherent to the petiole. Lfts. toothed or divided. Fls. small, often imperfect, each bracteate and bibracteolate. Fl.-tube urn-shaped, angled, usually winged. Sepals 4. Petals 0. Stamens 4–12 or more. Pistils 1–3, with terminal slender styles tipped by brushlike or tufted stigma. Aks. mostly solitary, enclosed in the angled dry thickish fl.-tube. N. Temp. Zone. (Named for *sanguis*, blood, and *sorbere*, to absorb, some spp. supposed to be styptic.)

Plants annual or biennial; lfts. incisely pinnatifid .... 1. *S. annua*
Plants perennial; lfts. toothed .... 2. *S. minor*

1. **S. ánnua** (Nutt. ex Hook.) T. & G. [*S. occidentalis* Nutt. *Poteridium o.* Rydb.] Glabrous, the stems branching, leafy, 1–4 dm. high; stipules of cauline lvs. foliaceous, pectinately divided; lower lvs. with 11–15 lfts., these obovate in outline, 7–20 mm. long, pectinately pinnatifid into linear segms.; spikes roundish to oblong, 5–25 mm. long; fls. perfect; bracts and bractlets ovate, green on midrib, with broad scarious margins; sepals oval, white-margined, ca. 2 mm. long; stamens 2, opposite inner sepals; fruiting fl.-tube 4-angled, with narrow wings and reticulate faces.—Dryish grassy places, 4500–4700 ft.; Cuyamaca Lake, e. San Diego Co.; to B.C., Mont. May–July.

2. **S. mìnor** Scop. Perennial, branched, 2–5 dm. high, leafy; stipules of cauline lvs. lunate, coarsely toothed; lfts. 7–21; lower fls. ♂, upper perfect or ♀; bracts and bractlets green, ciliate; sepals purple-tinged, 3.5–4 mm. long; stamens and stigmas purple-tinged; pistils 2; fruiting fl.-tube 4 mm. long.—Occasional natur. weed, as at Seminole Hot Springs, Santa Monica Mts.; native of Eurasia. May–July.

30. Sibbáldia L.

Low tufted perennial herbs with short cespitose rootstocks. Lvs. ternate. Fls. cymose, on scapelike peduncles. Fl.-tube small, saucer-shaped or cup-shaped. Bractlets, sepals and petals 5. Petals yellow. Stamens 5; fils. short, filiform. Pistils 5–20. Styles lateral. Ca. 5 spp. of n. regions. (Named for R. *Sibbald*, a Scotch botanist.)

1. **S. procúmbens** L. [*Potentilla sibbaldii* Haller f.] Rootstocks creeping or cespitose; petioles slender, strigose, mostly 1–4 cm. long; lfts. of basal lvs. obovate to oblanceolate, 1–2 cm. long, cuneate at base, 2–5-toothed at apex; stem-lvs. smaller; flowering stems to ca. 8 cm. high; cymes dense; fl.-tube 3–4 mm. in diam.; bractlets narrow, 1–1.5 mm. long; sepals ± ovate, 2.5–3 mm. long; petals 1.5 mm. long, narrow; $n = 7$ (Böcher, 1938). —Dry stony places, 9000–10,000 ft.; Montane Coniferous F.; San Bernardino Mts.; Sierra Nevada to Alaska, Greenland, Eurasia. June–Aug.

93. Rubiàceae. MADDER FAMILY

Herbs or woody plants with opposite entire lvs. connected by interposed stipules or in whorls without apparent stipules. Fls. perfect or imperfect, mostly 4-, sometimes 3- or 5-merous. Calyx sometimes obsolete. Corolla regular, the stamens as many as the corolla-lobes and inserted on the tube. Ovary in ours 2-loculed, splitting when dry into 2 or 4 indehiscent 1-seeded carpels; styles 1 or 2. Embryo in fleshy or horny endosperm. A large family, mostly trop., many of great economic importance (*Coffea, Cinchona*), others ornamental (*Gardenia, Bouvardia*).

Lvs. in whorls mostly of 4 or more; plants herbs or low shrubs.
    Fls. solitary or in cymes, pedicelled; corolla rotate .... 1. *Galium*
    Fls. in involucrate heads; corolla funnelform .... 3. *Sherardia*
Lvs. opposite; low perennial herb of montane pine belt .... 2. *Kelloggia*

1. *Gàlium* L. BEDSTRAW. CLEAVERS

Annual or perennial herbs, sometimes shrubby, with 4-angled rather slender stems and branches. Lvs. in apparent whorls because of the large leaflike stipules. Fls. small, perfect or imperfect, in cymes or these arranged in panicles. Sepals obsolete. Corolla rotate, 4- rarely 3-parted. Stamens 4 or 3, short. Ovary 2-lobed, 2-loculed, 2-ovuled; styles 2; stigmas capitellate. Fr. didymous, of 2 indehiscent carpels, dry or fleshy, separating when ripe. Ca. 300 spp., widely distributed. (Greek, *gala*, milk, certain spp. being used to curdle milk.)

(Ehrendorder, F. Evolution of the G. multiflorum complex in w. N. Am. I. Madroño 16: 109–122. 1961. Dempster, L. T. New names and combs. in the genus Galium. Brittonia 10: 181–192. 1958. A re-evaluation of G. multiflorum and related taxa. Brittonia 11: 105–122. 1959. Dempster, L. T. and G. L. Stebbins. The fleshy-fruited Galium spp. of Calif. I. Madroño 18: 105–112. 1965. Dempster & Stebbins. Fleshy-fruited Galium species of the Californias and s. Oregon. Univ. Calif. Publ. Bot. 46: 1–51. 1968. Dempster & Stebbins, The Galium angustifolium complex. Madroño 21: 70–95. 1971.)

1. Lvs. 5–8 in a whorl.
    2. Ovary and fr. glabrous or at most granulate, not bristly-hairy; stems ± retrorsely scabrous, at least when young; plants perennial .... 24. *G. trifidum*
    2. Ovary and fr. bristly-hairy.
        3. Fls. several on each side-branch, the whole upper part of the plant forming an open panicle; plant annual; fr. ca. 1 mm. in diam. .... 21. *G. parisiense*
        3. Fls. mostly 2–3 in axillary cymules; fr. larger.
            4. Plant perennial; stems not readily clinging to other vegetation; lvs. ± ovate
            25. *G. triflorum*
            4. Plants annual; stems easily clinging to other vegetation; lvs. linear to linear-oblong.
                5. Fr. 4–5 mm. long; corolla 2 mm. broad, white .... 3. *G. aparine*
                5. Fr. 1.5–3 mm. long; corolla 1 mm. broad, green .... 22. *G. spurium*
1. Lvs. mostly 4 in a whorl, occasionally 2 or 5.
    6. Plants not at all woody at the base.
        7. Annuals; frs. with hooked hairs; lvs. very unequal in the whorl, often only 2 in the upper whorls .... 5. *G. bifolium*
        7. Perennials; frs. glabrous; lvs. subequal .... 24. *G. trifidum*
    6. Plants woody at the base, at least below the growth of the current year.
        8. Mature fr. glabrous or slightly pubescent, the hairs if present much shorter than the body of the fr.
            9. Lvs. linear-subulate; plants often forming low dense mats, the stems densely leafy
            1. *G. andrewsii*
            9. Lvs. lance-linear or broader; plants taller, more open.
                10. Stems mostly retrorse-scabrous on the angles, woody; lvs. 2–6 (–10) mm. long, oval to linear-oblong .... 19. *G. nuttallii*
                10. Stems not retrorse-scabrous, scarcely woody; lvs. mostly longer.
                    11. Mature fr. dry; main cauline lvs. mostly 15–30 mm. long; fls. solitary or in few-fld. cymes, terminal or in uppermost axils of very leafy compound branch-lets; lvs. 1-nerved. Insular .... 8. *G. catalinense*
                    11. Mature fr. fleshy; main cauline lvs. mostly 6–20 mm. long.
                        12. Lf.-surfaces glabrous or nearly so.
                            13. Fls. mostly deep reddish-purple; lvs. mostly 3-nerved, acute to acuminate; woody stems tending to climb among shrubbery. Piute and Tehachapi mts. north .... 6. *G. bolanderi*
                            13. Fls. greenish or pale yellow; lvs. pungent, 1-nerved. Santa Ynez and Santa Monica mts. .... 9. *G. cliftonsmithii*
                        12. Lf.-surfaces and stems hispid or canescent throughout.
                            14. Woody stems wiry and very slender; coastal and island plants
                            7. *G. californicum*
                            14. Woody stems stouter, stiffer. San Gabriel Mts. at 2000–4000 ft.
                            10. *G. grande*
        8. Mature frs. densely hispid with hairs usually almost as long as the body of the fr.
            15. Corolla-lobes glabrous or minutely pubescent.

16. Hairs of fr. ascending and subappressed; corollas cleft ca. halfway. At 7000 ft. or higher. San Gabriel and San Bernardino mts. .... 14. *G. jepsonii*
16. Hairs of fr. spreading.
   17. Corolla reddish, with lance-acuminate lobes; plants often polygamous. E. Mojave Desert .... 26. *G. wrightii*
   17. Corolla mostly greenish-white, the lobes mostly acute.
      18. Lvs. linear to linear-oblong, broadest at or above the middle, the margins with short hairs directed exclusively forward (in ± hirsute forms with spreading tomentum); some fls. ± sessile; infl. large, pyramidal, not nodding
                                                                                 2. *G. angustifolium*
         19. Fruiting pedicels 1–4 times as long as frs.; upper nodes much longer than the lower, the lvs. often congested toward base of the plant; corolla ± cupped or campanulate .... 15. *G. johnstonii*
         19. Fruiting pedicels usually shorter than frs.; nodes ± equal; lvs. not congested toward base of plant; corolla rotate .... 2. *G. angustifolium*
      18. Lvs. ovate to orbicular.
         20. Herbage glabrous and shiny; lvs. acuminate, ± arcuate; infl. divaricately branched. n. Inyo Co. north .... 17. *G. multiflorum*
         20. Herbage canescent with many long fine hairs; lvs. obtuse or acute; at least some pedicels decurved in fr. Telescope Peak, Panamint Mts.
                                                                                     13. *G. hypotrichium*
15. Corolla-lobes mostly hairy to hispid.
   21. Lvs. acerose-acute or -acuminate, the midveins prominent, lateral veins mostly lacking, margin often strongly revolute. Deserts, mostly below 4500 ft.
                                                                                 23. *G. stellatum*
   21. Lvs. acute, but not acerose-acuminate, the lateral veins usually present, sometimes as prominent as the midvein.
      22. Fl.-clusters drooping; mature frs., including hairs, 4–5 mm. in diam.; corolla-lobes silky-hairy without. Mts. about the w. end of the Mojave Desert
                                                                                        11. *G. hallii*
      22. Fl.-clusters not drooping; frs. mostly smaller; corolla-lobes mostly bristly hairy (sometimes subglabrous in *G. argense*).
         23. Lvs. linear-oblong to oblong, 1–2 mm. wide; plants tufted, 0.5–2 dm. tall, cinereous-pubescent. San Gabriel Mts. to San Jacinto Mts.
                                                                            2. *G. angustifolium gabrielense*
         23. Lvs. lance-ovate to roundish, mostly wider.
            24. Herbage glabrous, except sometimes the uppermost bracts; lvs. mostly arcuate.
               25. Corolla campanulate. Death V. and north .... 17. *G. multiflorum*
               25. Corolla rotate.
                  26. Plants 1–3 dm. high, the stem lvs. much reduced; lvs. arcuate; corolla mostly hairy. Argus Mts. to Panamint, Kingston and Inyo mts.
                                                                         16. *G. matthewsii*
                  26. Plants 2–6 dm. high, more leafy; corolla usually almost glabrous. Argus and Panamint mts. .... 4. *G. argense*
            24. Herbage hispid.
               27. Lvs. of a given whorl unequal, 3–7 mm. long; fls. ca. 1.5 mm. in diam., borne in sessile glomerules on a virgate or sparsely branched infl. San Gabriel Mts. to Santa Rosa and Kingston mts.
                                                                                  20. *G. parishii*
               27. Lvs. of a given whorl subequal, 5–12 mm. long, plane.
                  28. Corolla campanulate; lvs. ovate to suborbicular. N. base of San Bernardino Mts. to Death V. and White Mts. .... 12. *G. hilendiae*
                  28. Corolla rotate; lvs. elliptical or ovate, tapering at both ends. Providence and New York mts. .... 18. *G. munzii*

1. **G. andréwsii** Gray ssp. **andréwsii.** Depressed-cespitose or matted prickly plants, with branching underground creeping stems from which arise tufted leafy slender erect flowering stems 3–8 cm. tall, glabrous or slightly scabrous; internodes generally shorter than lvs.; lvs. in 4's, crowded, subulate, rigid, cuspidate, 4–10 mm. long; ♀ fls. solitary, the ♂ in few-fld. terminal cymes; corolla greenish-white, 1.5 mm. wide, glabrous, the

lobes acute; fr. glabrous, 2–3 mm. wide, dark, baccate; $2n=22$ (Dempster, 1958).—Dry benches and ridges, 1000–6500 ft.; Chaparral, Foothill Wd., Yellow Pine F., Pinyon-Juniper Wd., etc.; L. Calif. to Ventura Co.; cent. Calif. April–June.

Ssp. **gaténse** (Demp.) Demp. & Steb. [*G. a.* var. *g.* Demp.] Plants often with internodes longer than the lvs., the stems and lvs. sometimes hispid, the lvs. often flat and linear or lanceolate; $2n=88$ (Demp. & Steb., 1968).—Sawmill Mt., San Gabriel Mts. at ca. 4500 ft.; to Contra Costa Co.

Ssp. **intermèdium** Demp. & Steb. Stems and lvs. not hispid, the lvs. usually narrower, but not subulate; $2n=44$ (Demp. & Steb., 1968).—Mostly low elevs., Los Angeles Co. to Monterey Co.

2. **G. angustifòlium** Nutt. ssp. **angustifòlium.** Plants 15–100 cm. high, the fertile branches arising singly or in tufts from the nodes of slender, woody scaffold stems; stems glabrous to hispid, the internodes commonly 2–7 cm. long; lvs. 5–27 mm. long, filiform to strapshaped, the margins usually with hairs; infl. usually diffuse, with abundant fls. and frs.; corolla cream to greenish-yellow; $2n=22$, 44 (Demp. & Steb., 1971).—Dry often shaded places below 8000 ft.; many Plant Communities; n. L. Calif. to Tehachapi Mts. and Santa Ynez Mts.; Santa Lucia Mts. Catalina Id. Spring.

KEY TO SUBSPECIES

1. Corollas usually hispid (not always in ssp. *jacinticum*).
   2. Stems glabrous or nearly so.
      3. Infl. narrow, relatively few-fld., the branching little compounded.
         4. Plants 6–16 cm. high; lvs. mostly 2–10 mm. long. San Gabriel and San Bernardino mts.
            7. ssp. *nudicaule*
         4. Plants 17–35 cm. high; lvs. mostly 11–26 mm. long. San Jacinto Mts. . 6. ssp. *jacinticum*
      3. Infl. pyramidal, many-fld., compoundly branched; plants 35–50 cm. high. Borrego Desert
         2. ssp. *borregoense*
   2. Stems not glabrous, but hispid, the hairs usually abundant and long like those on the lvs. E. Los Angeles and w. San Bernardino cos. .... 4. ssp. *gabrielense*
1. Corollas usually glabrous or not more hispid than stems and lvs.
   5. Internodes of scaffold stems short, often shorter than lvs.; plants usually glabrous, the lvs. very slender, often crowded. Santa Cruz, Santa Rosa, Anacapa ids. ...... 3. ssp. *foliosum*
   5. Internodes of scaffold stems long, generally much longer than the lvs., hence the lvs. not crowded.
      6. Plants tall and slender, essentially glabrous; lvs. early deciduous. Deserts, San Bernardino and Riverside cos. .... 5. ssp. *gracillimum*
      6. Plants stout, tall or lower, glabrous to canescent. Cismontane and montane
         1. ssp. *angustifolium*

Ssp. **borregoénse** Demp. & Steb. Similar to ssp. *gracillimum*, but corollas hispid with long hairs; lvs. larger and more glabrous.—Creosote Bush Scrub; Borrego V., e. San Diego Co.

Ssp. **foliòsum** (Hilend & Howell) Demp. & Steb. [*G. a.* var. *f.* H. & H.] Congested subshrubs with internodes mostly 2–15 mm. long; stems glabrous; lvs. narrowly linear, 3–16 mm. long; fls. and frs. congested; $2n=22$ (Demp. & Steb., 1971).—Rocky slopes, Santa Cruz, Santa Rosa and Anacapa ids.

Ssp. **gabrielénse** (M. & J.) Demp. & Steb. [*G. g.* M. & J.] Plants low, 6–35 cm. high, tufted from woody base; stems hispid; infl. narrow, relatively few-fld., little compounded; lvs. 2–14 mm. long, ± hispid; corollas yellowish to red, ± hispid externally; fr.-hairs ca. as long as or longer than the body; $2n=44$ (Demp. & Steb., 1971).—Dry rocky places, 4000–8700 ft.; largely Montane Coniferous F.; e. San Gabriel Mts.

Ssp. **gracíllimum** Demp. & Steb. [*G. a.* var. *diffusum* Hilend & Howell, in large part.] Plants very slender throughout, commonly 4 dm. long from the persistent woody glabrous stems; internodes 2–3 times as long as lvs.; lvs. 4–15 (–18) mm. long, scabrous; infl. open-paniculate, many-fld.; corollas mostly glabrous, yellowish or with pink tinge; frs.

less than 2 mm. long; $2n=22$ (Demp. & Steb., 1971). Rocky places, 400–4800 ft.; Creosote Bush Scrub, Joshua Tree Wd.; San Jacinto Mts. to Little San Bernardino Mts. to Providence Mts.

Ssp. **jacínticum** Demp. & Steb. Plants 17–35 cm. tall; stems glabrous; lvs. 11–26 mm. long, glabrous or very short-hairy; infl. narrow, few-fld.; corollas quite hispid; $2n=66$ (Demp. & Steb., 1971).—Largely Yellow Pine F., 4200–6500 ft.; w. side of San Jacinto Mts.

Ssp. **nudicáule** Demp. & Steb. Plants 6–16 cm. tall; stems glabrous, ± papillose; lvs. 2–10 (–15) mm. long, ± hispid with short hairs; infl. narrow, few-fld.; corollas usually red, sometimes yellow, externally hispid; fr. hairy; $2n=22$ (Demp. & Steb., 1971).—Dry slopes, 6700–8200 ft.; Montane Coniferous F.; cent. San Gabriel & e. San Bernardino mts.

3. **G. aparine** L. Annual, the stems weak, reclining or scrambling over other plants, retrorsely hispid at the angles, hairy at the nodes, 1–10 dm. long, rather slender; lvs. 6–8 in a whorl, mostly linear-oblanceolate, and 1.5–7 cm. long, bristle-tipped, with coarse divergent or reflexed marginal setae and ± hispidulous on surfaces; fls. 2 mm. in diam., whitish, 2–5 in cymes in the upper axils, the peduncles with a whorl of leaflike bracts; frs. bristly, 3–5 mm. in diam., the bristles with tuberculate bases; $2n=44$ (Kliphuis, 1962), ca. 66 (Löve & Löve, 1956).—Common on shaded banks to ca. 7500 ft.; many Plant Communities; cismontane Calif., Channel Ids., San Clemente and Catalina ids., occasional on deserts; to Alaska, e. Coast; said to be introd. from Eu. March–July.

4. **G. argénse** Dempster & Ehrend. Suffrutescent, mostly 2–6 dm. tall, lax; middle internodes 2.4–3.8 cm. long; herbage glabrous; lvs. coriaceous, lanceolate to ovate, plane or often arcuate, tapering gradually to the pungent apex, tapering or somewhat rounded at base; longest lvs. 8–20 mm. long; corollas white, ± rotate, glabrous or with a few slender hairs; frs. ca. 6 mm. broad; $2n=44$ (Dempster, 1965).—Dry stony loose slopes, Pinyon-Juniper Wd.; Argus and Nelson ranges, Inyo Co.

5. **G. bifòlium** Wats. Slender erect glabrous annuals, 5–15 cm. high, simple or few-branched; lvs. 2 or 4, often quite unequal, 1–2.5 cm. long, lanceolate to sublinear, mostly acute; pedicels axillary and terminal, 1-fld., recurved and equalling lvs. in fr.; fls. whitish, ca. 1 mm. wide; fr. uncinate-hispid, ca. 2 mm. in diam.—Moist partly shaded places, 6000–8500 ft.; Montane Coniferous F.; San Bernardino Mts., Mt. Pinos region; to Wash., Rocky Mts. June–Sept.

6. **G. bolánderi** Gray. Herbaceous shoots arising in tufts from a persistent woody base, 1.5–3.5 dm. high, or the woody stems longer and climbing among shrubs; stems sharply 4-angled, commonly glabrous; plants branching freely above; lvs. 5–20 (–27) mm. long, often unequal in size, narrowly to broadly elliptical, acute or apiculate, glabrous or scabrous-puberulent, commonly plane, usually obscurely 3-nerved; ♀ fls. often solitary in leaf axils, on pedicels recurved in fr.; fr. 4–6 mm. wide, mostly glabrous; $2n=66$ (Dempster & Stebbins, 1968).—At 4000–5500 ft.; Piute and Tehachapi mts.; to Ore.

7. **G. califórnicum** H. & A. (Plate 79, Fig. F.) ssp. **fláccidum** (Greene) Dempster & Steb. [*G. f.* Greene.] Plants entirely herbaceous above ground, or sometimes a few strictly prostrate stems wintering over and producing new axillary stems; lvs. 6–25 mm. long, usually broadly elliptical or scarcely ovate, obtuse or round at apex, apiculate, usually plane, but sometimes laterally inrolled, or the margins slightly revolute; stems, lvs. and corollas ± pubescent, rarely glabrous; hairs generally long, spreading, straight or slightly curved; ovaries and frs. usually pubescent, sometimes glabrous; $2n=88$ (Demp. & Steb., 1968).—Woodlands in sw. San Bernardino Co., and near Santa Barbara, Santa Cruz Id.; to Monterey Co.

Ssp. **miguelénse** (Greene) Demp. & Steb. [*G. m.* Greene.] Plants climbing or clambering on persistent woody stems; congested, the stems stout; lvs. 2–8 mm. long, broadly ovate to suborbicular, the apices round-apiculate.—San Miguel and Santa Rosa ids.

Ssp. **prìmum** Demp. & Steb. Plants tufted or somewhat decumbent, usually not at all woody; lvs. 6–25 mm. long; hairs generally spreading, fine.—W. San Jacinto Mts. at 4600 to 5500 ft.

8. **G. catalinénse** Gray ssp. **catalinénse.** Stems woody, stout, erect, 5–12 dm. high, quadrangular, mostly ± scabrous, with a swollen ring at the nodes and bearing above numerous leafy branches with short internodes and thus ± tufted; main lvs. mostly in 4's, narrow-elliptical, 1.5–2.5 cm. long, ca. $\frac{1}{5}$–$\frac{1}{6}$ as wide, ± retrorse-scabrous on under surface, the lateral veins inconspicuous; infl. tufted, leafy, cymosely branched; pedicels short, glabrous or pubescent; corolla whitish, 2–3 mm. across; fr. dry, ca. 2 mm. in diam., mostly with slender spreading hairs 0.5–1 mm. long, sometimes subglabrous.—Dry rocky canyons; Coastal Sage Scrub, Chaparral; Santa Catalina and San Clemente ids. April–July.

Var. **buxifòlium** (Greene) Demp. [*G. b.* Greene.] Nodes less conspicuously swollen; main cauline lvs. ca. $\frac{1}{3}$ to $\frac{1}{2}$ as wide as long, the lateral veins conspicuous; fr. with short appressed hairs less than 0.5 mm. long; $2n = 22, 44$ (Fagerlind, 1937).—Dry rocky places, Chaparral, Closed-cone Pine F.; San Miguel and Santa Cruz ids. March–July.

9. **G. cliftonsmíthii** (Demp.) Demp. & Steb. [*G. nuttallii* var. *c.* Demp.] Plants rather stout, 3–6 dm. high, often climbing a little; herbage shiny, coarse, pungent; stems sharply angular, coarsely scabrous; lvs. 7–15 mm. long, 1-nerved, ovate or elliptical, acuminate, tipped with a pungent hair, the margins slightly revolute; lf.-surfaces, with few or no hairs, shining, the margins usually with stout hairs; fls. few, the corollas greenish or pale yellow; ovaries and frs. glabrous; chromosome no. very high.—Light shade of *Quercus agrifolia*, etc., 700–4000 ft.; Santa Ynez Mts. and Santa Monica Mts.

10. **G. gránde** McClat. [*G. pubens* var. *g.* Jeps.] Plants large, to 6 dm. high and 12 dm. across; stems woody, weak, usually clambering over shrubs and hidden by the tufts of cinereous growth; stems and lvs. grayish with soft spreading hairs, the lvs. usually turning dark when dried; lvs. commonly 5–12 mm. long, somewhat fleshy; elliptical, short-petioled, acute at apex, usually tipped with a persistent hair; corolla 4 mm. across, pale yellow, hispid externally; ovaries and frs. pubescent; $2n$ perhaps 220 (Dempster, 1968).—Chaparral and Oak Wd.; 2200–3300 ft.; s. face of San Gabriel Mts.

11. **G. hállii** M. & J. Plate 80, Fig. A. Dioecious; stems many from a branched rootcrown, woody below with white exfoliating bark, erect and 2–6 dm. tall, or ± spreading, cinereous-pubescent throughout with white spreading hairs; lvs. oblong-elliptic to broadly ovate, 5–15 mm. long, ± revolute, with strong midvein and weakish lateral veins; panicles loose, leafy, the branches recurved at tips; corolla yellowish, 2–2.5 mm. in diam., pilose without; body of fr. ca. 3 mm. in diam., the dense white hairs 2 mm. long.—Dry rocky slopes and canyons, 4000–7000 ft.; Pinyon-Juniper Wd., Yellow Pine F.; desert slopes from San Bernardino Mts. to Mt. Pinos and Santa Ynez Mts. and Piute Mts. May–Aug.

12. **G. hiléndiae** Demp. & Ehrend. ssp. **hiléndiae.** Plants arising from a low woody base, 1–3 dm. tall; middle internodes 1.5–4 cm. long; plants hispid throughout with long hairs; lvs. membranaceous to subcoriaceous, ovate to nearly orbicular, 5–18 mm. long, plane, usually acute; infls. rather ample with divaricate long branches; corollas markedly campanulate, hispid externally; frs. small, ca. 5.5 mm. in diam.; $2n = 44$ (Demp. & Ehrend., 1965).—Dry rocky places, 5000–6000 ft.; n. base of San Bernardino Mts. to White Mts. and Death V.

Ssp. **cárneum** (Hilend & Howell) Dempst. & Ehrend. [*G. munzii* var. *c.* Hilend & Howell.] Plants 13–35 cm. tall; middle internodes 2–4.5 cm. long; herbage scabrous to hispid; lvs. lanceolate to ovate, usually tapering gradually toward the acute or sometimes ± pungent apex; longest lvs. 7–10 mm. long; corolla pinkish, glabrous or with a few short stout hairs; frs. ca. 5 mm. across; $2n = 44$ (Dempster & Ehrend., 1965).—At 5000–10,500 ft., s. half of Panamint Range; Pinyon-Juniper Wd., Bristle-cone Pine F.

Ssp. **kingstonénse** (Demp.) Demp. & Ehrend. [*G. munzii* var. *k.* Demp.] Slender, weak, the stems to 3.5 dm. long; middle internodes 1.8–4.8 cm. long, entire plant hispid with usually long scattered hairs; lvs. plane, lanceolate to elliptical or ovate, obtuse or acute, the longest 8–16 mm. long; corolla pink, hispid externally or glabrous; $2n = 44$ (Demp. & Ehrend., 1965).—Rocky places, 5500–6500 ft.; Pinyon-Juniper Wd., Kingston Mts., e. San Bernardino Co.

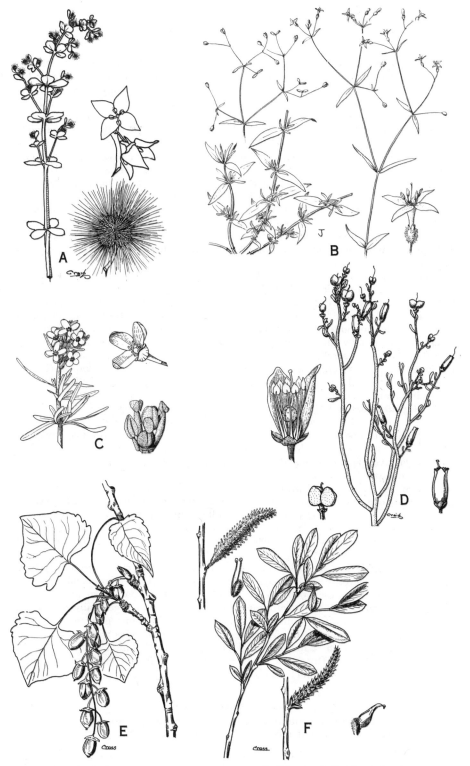

Plate 80. Fig. A, *Galium hallii*, ♀ branch, separate ♂ fls. Fig. B, *Kelloggia galioides*. Fig. C, *Cneoridium dumosum*, stamens 4 long, 4 short. Fig. D, *Thamnosma montana*, ovary stipitate. Fig. E, *Populus fremontii*, with capsulse. Fig. F, *Salix lasiolepis*, ♂ and ♀ aments.

13. **G. hypotríchium** Gray ssp. **tomentéllum** Ehrend. Plants 3–14 cm. tall, the middle internodes 7–20 mm. long; herbage canescent with many long fine hairs; lvs. somewhat fleshy, ovate below the usually gradual, obtuse or acute apex; longest lf. 5–9 mm. long; corolla externally pubescent, slightly campanulate, reddish, very small; fr. 5–8 mm. in diam., with long fine brownish tomentum; $2n=22$ (Demp. & Ehrend., 1965). —On talus, 10,500–11,000 ft.; Telescope Peak, Panamint Mts. in Bristle-cone Pine F.

14. **G. jepsònii** Hilend & Howell. [*G. angustifolium* var. *subglabrum* Jeps.] Perennial, the stems several and tufted from the ends of underground stems, erect, 1–2 dm. tall, glabrous, smooth; lvs. linear-oblong, or the lower broader, 5–15 mm. long, bristly-ciliate but glabrous on surfaces; plants dioecious; the fls. 1–3 in a cyme, the cymes few in a narrow infl.; corolla glabrous, ca. 2 mm. in diam.; body of fr. 2.5 mm. in diam., subpilose with appressed hairs 1–1.5 mm. long.—Rare, in dry rocky and gravelly places, 7000–8000 ft.; upper Montane Coniferous F.; San Gabriel and San Bernardino mts. July–Aug.

15. **G. johnstònii** Demp. & Steb. [*G. angustifolium* var. *pinetorum* M. & J.] Herbaceous, 1–4 dm. tall; upper internodes much exceeding lvs.; lvs. narrow, 1–3 cm. long; panicle lax, few-fld., 3–8 cm. long; corolla rotate to flaring campanulate, lobes longer than tube; $2n=66$ (Demp. & Steb.).—Occasional on dry slopes, 5000–7500 ft.; Yellow Pine F.; San Gabriel, San Bernardino and San Jacinto mts. June–July.

16. **G. matthèwsii** Gray. Suffrutescent, 13–30 cm. high; stems many, slender, stiff, with very small, remote lvs.; herbage glabrous, except for a few bristles on uppermost bracts; middle internodes 2–3.5 cm. long; lvs. leathery, often arcuate, ovate below the ± tapered and usually pungent apex; largest lvs. 4–8 mm. long; infl. much branched, the branches divaricate; corollas rotate, yellow or somewhat pinkish, very small, externally hispid; fr. 3–4 mm. across, the body usually not as long as the white hairs; $2n=22$ (Demp. & Ehrend., 1965).—Dry rocky places, 3400–8900 ft.; Pinyon-Juniper Wd.; Inyo, Argus, Panamint, Kingston mts.

17. **G. multiflòrum** Kell. Stems 1.5–4.5 dm. tall from the woody base; middle internodes 2–4 cm. long; herbage glabrous; lvs. usually shiny, the longest 5–12 mm. long; corollas mostly campanulate; fr. 5–9 mm. broad, ± densely clothed with long tawny or white hairs which are longer than the body.—Dry rocky places, 5000–8600 ft.; Pinyon-Juniper Wd.; Grapevine, Inyo and Last Chance mts.; to w. Nev.

18. **G. múnzii** Hilend & Howell. Stems mostly 1–3 dm. long, from a suffrutescent base; herbage coarsely hispid with usually long hairs; middle internodes 1–4 cm. long; lvs. generally ovate, sometimes lanceolate, tapered toward the often pungent apex, the longest 7–19 mm. long, usually abundant; corollas generally hispid, or merely puberulent, strictly rotate; frs. 3–5 mm. wide, the hairs white or yellowish; $2n=44, 45$ (Demp. & Ehrend., 1965).—Rocky places, 3500–7000 ft.; Pinyon-Juniper Wd.; Providence and New York mts.; to Utah.

19. **G. nuttállii** Gray. Slender-stemmed woody plants, mostly clambering or climbing over other plants; herbage early tinged with red, or becoming quite maroon, often densely matted at the nodes of the woody stems; internodes of the secondary stems 2–30 mm. long; stems merely rough or set with minute aculeolate hairs; lvs. generally coriaceous, linear to triangular-ovate, 3–8 mm. long, narrowed gradually to a pungent apex with stout and usually persistent hair; lf.-surfaces glabrous, the margins with relatively few hairs; fls. yellowish-green, glabrous; frs. glabrous, baccate, 2–3 mm. in diam.; $2n=22$ (Demp. & Steb., 1968).—Dry slopes, mostly below 1500 ft.; Coastal Sage Scrub, Chaparral; n. L. Calif. to Riverside and Santa Barbara cos. March–June.

KEY TO SUBSPECIES

Lvs. gradually narrowed to an acute apex, the terminal hair usually persistent.
    Stems and lvs. ± armed with retrorse aculeolate hairs. L. Calif. to Santa Barbara Co.
        ssp. *nuttallii*
    Stems and lvs. mostly glabrous. Catalina, Santa Rosa and Santa Cruz ids. .... ssp. *insulare*
Lvs. generally rounded or apiculate, the terminal hair ephemeral; lvs. elliptical or very broadly ovate. w. San Diego Co. to n. Calif. .... ssp. *ovalifolium*

Ssp. **insulàre** Ferris. Chaparral and S. Oak Wd. Insular.

Ssp. **ovalifòlium** (Demp.) Demp. & Steb. [*G. n.* var. *o.* Demp.] Herbage variously green to reddish; lvs. thin to coriaceous, never pungent.—Dry places below 5000 ft.; Chaparral to Yellow Pine F.; San Diego Co. to n. Calif. Catalina and Santa Cruz ids.; L. Calif.

20. **G. paríshii** Hilend & Howell. [*G. multiflorum* var. *parvifolium* Parish. *G. p.* Jeps. not Gaud.] Stems many, slender, scabrous-puberulent, 1–3 dm. long, from a branched woody root-crown, erect or spreading, tufted; lvs. in 4's, the lowermost scalelike, the main lvs. ovate, acute, abruptly mucronate, 3–7 (–10) mm. long, scabrous-puberulent; infl. spicate-paniculate, slender, with congested cymules; plants dioecious; fls. mostly reddish, ca. 1.5 mm. in diam., short-bristly externally; fr., including hairs, ca. 3 mm. in diam.; $2n = 22$ (Ehrend., 1961).—Common in dry rocky places, 6000–10,000 ft.; Montane Coniferous F.; Pinyon-Juniper Wd. to Lodgepole F.; San Gabriel Mts. to Santa Rosa Mts., New York Mts., Kingston Mts., etc.; s. Nev. June–Aug.

21. **G. parisiénse** L. Slender-stemmed diffusely branched annual, mostly 1.5–4 dm. high, minutely scabrous; lvs. chiefly 5–7 at a node, linear to oblanceolate, 5–12 mm. long, spinulose-ciliate, bristle-tipped; infl. a lax panicle of small several-fld. cymules, the upper bracts reduced or wanting; fls. ca. 1 mm. broad, greenish-white, glabrous; fr. 0.7–1 mm. wide, hispid-pubescent; $2n = 22, 44, 66$ (Fagerlind, 1937).—Waste places from Santa Barbara Co. to n. Calif.; introd. from Eu.

22. **G. spùrium** L. var. **echinospérmum** (Wallr.) Hay. Much like *G. aparine* but with narrower lanceolate long-mucronate lvs. 5–20 mm. long; fls. greenish, 1 mm. across, in axillary cymes of 3–9 fls. with only 2 leaflike bracts; frs. 1.5–3 mm. in diam., the hairs nontuberculate at the base.—Occasional in grassy and wooded places, where introd.; cismontane Calif. Native of Eu. Commonly mistaken for *G. aparine.*

23. **G. stellàtum** ssp. **erèmicum** (Hilend & Howell) Ehrendf. [*G. s.* var. *e.* Hilend & Howell.] Bushy, much branched above, the woody base, 2–7 dm. high; stems and lvs. ± scabrous-pubescent, the younger branches with shining white exfoliating epidermis; old lvs. whitish, persistent; lvs. of the current year lanceolate to almost linear, ± revolute, 4–10 mm. long, acuminate-cuspidate, pale green; corolla greenish-yellow, 2–2.5 mm. across, bristly-hairy outside; ♂ fls. in crowded panicles, ♀ on short pedicels and solitary at ends of leafy-bracted branchlets; fr. including the soft hairs ca. 4 mm. in diam., the body ca. 2 mm.; $2n = 22$ (Ehrendorfer, 1961).—Common on dry rocky slopes below 5000 ft.; Creosote Bush Scrub, Joshua Tree Wd.; deserts from Inyo to Imperial cos., to Nev., Ariz. March–April.

24. **G. trífidum** L. var. **pacíficum** Wieg. [*G. t.* var. *subbiflorum* Wieg.] Slender-stemmed perennial from slender rootstocks, ascending, branched and intertangled, 1.5–4 dm. long, the angles ± retrorse-scabrous; lvs. elliptic-oblong to linear-spatulate, obtuse, not bristle-pointed, in 4's, 5's or 6's, thin, obscurely scabrous on margins and midribs, 5–20 mm. long; fls. mostly 1.5–2 mm. broad, whitish, 3–4-merous, 1, 2, or 3 on axillary peduncles, the pedicels strongly arcuate in age, ± indistinctly scabrous, slender, mostly 5–18 mm. long; fr. glabrous, dry, each carpel 1–1.5 mm. in diam.—Wet places, below 8000 ft.; many Plant Communities; most of cismontane and montane Calif.; to Alaska, Atlantic Coast. June–Aug.

Var. **pusíllum** Gray. More glabrous throughout, ± matted, the weak stems mostly 15–18 cm. long, simple or forked; lvs. in 4's, oblanceolate, 4–8 mm. long, fleshy; fls. 1–2, on short stocky glabrous pedicels that thicken in fr.; fr. 1–2 mm. in diam.—Wet places, 5000–10,500 ft.; San Jacinto & San Bernardino mts.; to Wash.

25. **G. triflòrum** Michx. SWEET-SCENTED BEDSTRAW. Perennial, with weak simple or remotely forking smooth or ± scabrous herbaceous stems to 8 dm. long; lvs. mostly in 6's, oblong-ovate to ± obovate, thin, cuspidate, 1.5–6 cm. long, the surfaces mostly glabrous, margins and midribs beneath minutely scabrous; axillary peduncles 3-fld. or 3-forked, the fls. all pedicelled; corolla greenish-white; fr. densely bristled, scarcely 2 mm. broad.—Moist shaded places, below 8000 ft.; Montane Coniferous F.; San Bernardino Mts.; to Alaska, Atlantic Coast; Eurasia. May–July.

26. **G. wrightii** Gray. Suffrutescent, many-stemmed, glabrous, not at all scabrous, the stems slender, 1.5–6 dm. tall; lvs. in 4's, narrowly linear, rigid, 8–10 mm. long, the uppermost bracteate; plant polygamo-monoecious, the fls. in open paniculate infls. of small cymes; corolla usually red, 1–2 mm. in diam., the lobes long-acuminate; fr. sparsely hairy, including the bristles ca. 2 mm. broad, the bristles straight, scarcely as long as the body of the fr.—Dry rocky places, 5000–8000 ft.; Pinyon-Juniper Wd.; Clark Mtn. of e. Mojave Desert; to New Mex. May–June; Aug.–Sept.

2. *Kellóggia* Torr.

Slender perennial herbs with few stems from slender underground parts. Stems slender, simple or branched, subglabrous. Lvs. opposite, lanceolate, with interposed stipules. Fls. in loose forking terminal cymes, pinkish-white, small. Corolla funnelform, with exserted stamens. Ovary inferior, 2-celled; style filiform; stigmas 2, linear-clavate. Fr. dry, uncinate-hispid, separating into 2 closed carpels. One sp. (Named for Dr. A. *Kellogg*, pioneer Calif. botanist.)

1. **K. galioides** Torr. Plate 80, Fig. B. Stems 1–2.5 dm. tall; lvs. 1–2.5 cm. long, glabrous; pedicels filiform, 1–12 cm. long; corolla 3–4 mm. long, mostly 4-lobed; fr. broadly clavate, 3–4 mm. long.—Dry benches and slopes, 4500–9000 ft.; Montane Coniferous F.; Palomar Mts., San Jacinto and San Bernardino mts.; Sierra Nevada; to Wash., Ida., Utah, Ariz. May–Aug.

3. *Sherárdia* L. Field Madder

Slender, many-stemmed annual, procumbent, the stems 4-angled, subglabrous to hispidulous-roughened, 1–2 dm. long. Lvs. in whorls of 4–6, lance-ovate, pungent, mostly retrorse-hispid on margins and midrib beneath, 5–10 mm. long. Fls. surrounded by a gamophyllous invol., small, blue or pinkish, 2–3 in a head. Sepals persistent. Corolla funnelform, 4–5-lobed. Stamens 4–5. Style filiform, slightly 2-cleft. Fr. dry, didymous, forming 2 indehiscent 1-seeded carpels. Monotypic. (Named for Dr. W. *Sherard*, a patron of Dillenius.)

1. **S. arvénsis** L. Stems 1–2 dm. long, tufted; invol. deeply 6–8-lobed.—Natur. widely but usually not abundantly, in fields, lawns, meadows, etc. from San Diego Co. n.; native of Eu. Jan.–July.

94. Rutàceae. Rue Family

Aromatic herbaceous or mostly woody plants with punctate glands. Lvs. alternate or opposite, often compound, estipular. Fls. bisexual or unisexual, in axillary or terminal infl. Sepals and petals usually 3–5. Stamens as many or in ours twice as many, inserted on or at base of the hypogynous disk. Carpels 1–5, distinct or united; styles distinct or connate; stigma simple or lobed. Ovary superior; ovules usually 2 in each locule. Fr. a 2–5-celled caps. or drupelike. Seeds with or without endosperm. Ca. 110 genera and 900 spp., most abundant in trop. of S. Am. and Australia.

Lvs. simple.
    The lvs. opposite; ovary 1-loculed, forming a round berry. Cismontane San Diego Co.
        1. *Cneoridium*
    The lvs. alternate; ovary 2-loculed, forming a deeply 2-lobed caps. Deserts ... 3. *Thamnosma*
Lvs. pinnate; fr. a 4–5-lobed caps. .... 2. *Ruta*

1. *Cneorídium* Hook. f. in Benth. & Hook.

Low evergreen shrub with grayish bark and slender twigs. Lvs. opposite or fascicled, narrow, pellucid-punctate, rather heavily odorous. Fls. 1–3, usually on short axillary peduncles. Calyx 4-lobed. Petals 4, white. Stamens 8, of 2 lengths, with dilated fils. Ovary 1-celled, sessile; ovules 2; style short, flattened, arising from near base of ovary on 1 side;

stigma capitate. Fr. globose, fleshy, punctate. Seeds 1–2, subglobose, dark brown. One sp. (Resembling *Cneorum*, an Old World genus.)

1. **C. dumòsum** (Nutt.) Hook. f. [*Pitavia d.* Nutt. in T. & G.] BUSHRUE. Plate 80, Fig. C. Intricately branched, 1–2 m. tall; lvs. linear to oblong, glabrous, 1–2.5 cm. long, 1–3 mm. wide, gland-dotted, entire; calyx 1–1.5 mm. long; petals obovate, 5–6 mm. long; fr. 5–6 mm. thick, greenish to reddish, drying red-brown; seeds 5–6 mm. long; $2n=18$ pairs (Raven et al., 1968).—Frequent on mesas and bluffs, below 2500 ft.; Chaparral, Coastal Sage Scrub; Orange and San Diego cos., San Clemente Id.; n. L. Calif. Nov.–March.

2. *Rùta* L. RUE

Perennial herbs and subshrubs, strong-scented. Lvs. alternate, pinnate, gland-dotted. Fls. small, in erect terminal clusters; sepals and petals 4–5. Calyx persistent. Petals toothed or fringed. Stamens 8 or 10. Ovary 4–5-loculed; style central; ovules several in each locule. Fr. a caps., 4–5-lobed, dehiscent or not. Seeds angled. Ca. 60 spp.; of Eurasia. (*Ruta*, the classical name.)

1. **R. chalepénsis** L. Glaucous perennial, the stems 4–8 dm. tall, puberulent; lvs. bipinnate or tripinnate, oblong in outline, to 1 or more dm. long, the segms. entire, narrow-elliptic, ca. 1–1.5 cm. long; infl. corymbose; petals yellow, 6–8 mm. long; $2n=36$ (Negodi, 1939).—Occasional escape; native of Medit. region. Feb.–June.

3. *Thamnósma* Torr. & Frém. in Frém.

Strong-scented undershrubs covered with glands. Lvs. narrow, alternate, early deciduous. Fls. in racemose cymes, bisexual. Calyx persistent, 4-lobed. Petals 4, erect. Stamens 8, inserted on the disk. Ovary usually 2-celled, 2-lobed, stipitate; style filiform; stigma capitate. Fr. a leathery 2-lobed caps. opening at the apex. Seeds 4–6, reniform. Two spp. of sw. N. Am. (Greek, *thamnos*, bush, and *osme*, odor.)

1. **T. montàna** Torr. & Frém., TURPENTINE-BROOM. Plate 80, Fig. D. Stems branching, broomlike, yellowish-green, 3–6 dm. tall; lvs. oblance-linear, 5–15 mm. long; sepals ovate to round, 3–4 mm. long; greenish; petals ovate to oblong, 8–12 mm. long, purplish; stamens 4 long and 4 short with filiform fils.; style somewhat exceeding petals; ovules several; caps. with 2 subglobose glandular lobes, each 5 mm. thick; seeds whitish, ca. 4 mm. long.—Frequent on dry slopes, below 5500 ft.; Creosote Bush Scrub to Pinyon-Juniper Wd.; deserts from Inyo Co. to Imperial and e. San Diego cos.; to Utah, New Mex., Son., L. Calif. March–May.

95. Salicàceae. WILLOW FAMILY

Dioecious deciduous trees or shrubs with simple alternate lvs. mostly with stipules. Wood soft, light, mostly pale. Winter buds scaly. Roots often producing new shoots. Fls. early in spring, often before the lvs., in unisexual aments, the entire catkin falling away at once. Perianth none; each fl. subtended by a scalelike bract. Stamens 2–many, sometimes 1 by fusion. Ovary 1-celled, ovoid or globose; stigmas 2–4. Fr. a 2–4-valved caps., with many minute comose seeds.

Scales of the catkins laciniate; fls. surrounded by a cup-shaped often oblique disc; stamens many; buds with numerous scales .... 1. *Populus*
Scales of the catkins entire; disc a minute glandlike structure; stamens 1, 2, or more; buds with a single scale .... 2. *Salix*

1. *Pópulus* L. ASPEN. COTTONWOOD. POPLAR

Trees with pale furrowed bark and soft usually whitish wood. Winter-buds resinous. Lvs. alternate, deciduous, stipulate, broad to narrow, petioled. Stipules minute, deciduous. Bracts of the usually pendulous catkins laciniate, narrowed at base. Disk cup-shaped, lobed or entire, symmetrical or lengthened in front. Stamens 6–80, with distinct fils. Ovary sessile; style short; stigmas 2–4, entire or lobed. Caps. 2–4-valved, the valves narrowed at maturity. Ca. 30 spp., of wide distribution in N. Hemis. (Classical Latin name.)

Lf.-blades conspicuously white-woolly beneath, angulate-lobed. Introd. .... 1. *P. alba*
Lf.-blades not white-woolly beneath. Native spp.
    Stigma-lobes filiform; stamens 6–12; petioles flattened or compressed toward the summit; caps. oblong-conical, thin-walled; winter-buds 3–10 mm. long. Fish Creek, San Bernardino Mts. .... 4. *P. tremuloides*
    Stigma-lobes dilated; stamens 40–80; petioles terete or flattened; caps. ± globose, thick-walled; winter-buds 10–20 mm. long.
        Petioles terete; lvs. dark green above, rusty or silvery beneath .... 5. *P. trichocarpa*
        Petioles laterally compressed; lvs. yellowish-green, alike on both surfaces.
            Disk flattish, 2–3 mm. broad; lvs. bluish-green; petioles and young shoots densely pilose
                                                                                3. *P. macdougallii*
            Disk cup-shaped, 5–6 mm. broad; lvs. yellowish-green; petioles and young twigs glabrous or finely puberulent .... 2. *P. fremontii*

1. **P. álba** L. SILVER POPLAR. Tree with mostly grayish-white smooth bark; young twigs and winter-buds and young lvs. white-tomentose; older lvs. 3–5-angulate-lobed with few teeth, white-tomentose beneath; caps. woolly, 3–5 mm. long.—Occasional escape near old dwellings; introd. from Eu.

2. **P. fremóntii** Wats. var. **fremóntii**. Plate 80, Fig. E. Tree, 12–30 m. tall with broad open crown and whitish roughly cracked bark; twigs stout, quite glabrous; lvs. quite glabrous, deltoid, 4–7 cm. long, 4–9 cm. wide, slightly cordate or ± cuneate at the entire base, abruptly sharp-pointed at the apex, coarsely and irregularly serrate-dentate, bright green, lustrous; petioles laterally flattened, 3–6 cm. long; catkins 4–5 cm. long; stamens 60 or more; ovary glabrous; caps. 8–12 mm. long.—Moist places below 6500 ft.; several Plant Communities; cismontane and Mojave Desert to cent. Calif., Nev., Ariz. Catalina and San Nicolas ids. March–April.

Var. **pubéscens** Sarg. An ill-defined form with pubescent twigs and lvs., from San Bernardino, Orange and San Diego cos. Some examples, as at Snow Canyon have been called var. *arizonica* (Sarg.) Jeps.

3. **P. macdoùgallii** Rose. [*P. fremontii* var. *m.* Jeps.] Twigs coarser, pubescent; petioles pilose; lvs. bluish-green, mostly truncate at base and 4–8 cm. wide; disc ca. 3 mm. broad; caps. 10–12 mm. long.—Creosote Bush Scrub, but along the Colorado River bottom and ditches; Ariz., Nev., etc.

4. **P. tremuloìdes** Michx. QUAKING ASPEN. Slender tree, mostly 3–30 m. tall in our area, with extensive lateral roots sending up sucker shoots; bark smooth, greenish-white; twigs slender, often drooping; lvs. round-ovate or wider, crenulate or serrulate with many teeth or subentire, with a short sharp point at the apex, broadly rounded or subcordate at base, quite glabrous, pale beneath, the blades 2–4 (–6) cm. long; petioles slender, ca. as long; catkins 3–6 cm. long; bracts 3–5-lobed, fringed with long hairs; $2n=38$ (Smith, 1943).—Stream borders and ± damp slopes, 7000–7500 ft.; Montane Coniferous F.; Fish Creek, San Bernardino Mts.; Sierra Nevada to Alaska, Labrador, n. Mex. April–June.

5. **P. trichocárpa** T. & G. var. **trichocárpa**. BLACK COTTONWOOD. Tree 30–60 m. tall with broad open crown, the grayish bark furrowed in age; lvs. ovate, finely serrate, truncate or cordate at the base, acute to subacuminate at apex, dark green above, pale and somewhat glaucous beneath, 3–7 cm. long, on terete petioles 2–4 cm. long; catkins 4–8 cm. long; stamens 40–60; caps. subglobose, 4 mm. thick, pubescent; $2n=38$ (Smith, 1943).—Along streams, below 6000 ft.; many Plant Communities, cismontane to Alaska, w. Nev. Feb.–April.

Var. **ingràta** (Jeps.) Parish. Lvs. lance-ovate to lanceolate, less than 2 cm. wide.—Along streams, 6500–7500 ft.; San Bernardino and San Gabriel mts.

2. Sàlix L. Willow

Trees or shrubs with rather slender twigs and simple mostly narrow pinnately veined lvs. Winter-buds single-scaled. Fls. unisexual, the ♂ and ♀ borne in separate catkins that appear before, with, or after the lvs. Each fl. subtended by a small scale; sepals and petals none; 1–2 small glands at base. Stamens 1–10, mostly 2–5; pistil 1, forming a caps.; style evident or obsolete; stigmas entire or divided. A genus of 300 or more spp., mostly temp. and colder.

1. Lf.-blades entire or nearly so.
  2. Lvs. glabrous beneath, except possibly when very young.
    3. Lf.-blades 1.5–4 cm. wide, mostly 3–4 times longer.
      4. Fils. and caps. glabrous or nearly so; scales of catkins tawny; styles less than 0.5 mm. long .... 11. *S. lutea*
      4. Fils. and caps. hairy; catkin scales dark .... 10. *S. lemmonii*
    3. Lf.-blades 1–2 cm. wide, mostly ca. 5 times as long.
      5. Catkin-scales yellow; stamens 3–9. Trees.
        6. Petioles glandular above .... 8. *S. lasiandra abramsii*
        6. Petioles not glandular .... 5. *S. gooddingii*
      5. Catkin-scales dark; stamens 2; tree or shrub .... 9. *S. lasiolepis*
  2. Lvs. definitely pubescent or hairy beneath, even when quite mature.
    7. The lvs. linear or lance-linear, less than 1 cm. wide.
      8. Stamen-fils. pubescent near the base; catkin-scales yellow.
        9. Stigmas ca. 1 mm. long, on styles ca. 0.5 mm. long; caps. hairy .... 6. *S. hindsiana*
        9. Stigmas ca. 0.5 mm. long, sessile; caps. glabrous or nearly so, at least in maturity
          3. *S. exigua*
      8. Stamen-fils. glabrous; catkin-scales dark .... 9. *S. lasiolepis bracelinae*
    7. The lvs. oblanceolate or obovate, mostly more than 1 cm. wide.
      9a. Caps. glabrous; catkins subsessile .... 9. *S. lasiolepis*
      9a. Caps. gray-tomentose or silky.
        10. Catkin-bracts dark or almost black; caps. gray-tomentose .... 13. *S. scouleriana*
        10. Catkin-bracts yellowish with reddish tips; caps. silky .... 4. *S. geyeriana*
1. Lf.-blades distinctly serrulate.
  11. Lf.-blades glabrous beneath except sometimes when very young.
    12. Caps. pilose to silky.
      13. Catkin-scales yellow; tree 6–10 m. tall; lvs. gray-green on both surfaces; stamens 4–5
        5. *S. gooddingii*
      13. Catkin-scales dark; shrub 1–5 m. high; lvs. deep green above; stamens 2 10. *S. lemmonii*
    12. Caps. essentially glabrous.
      14. Lf.-blades 0.6–1.5 cm. wide; caps. 5.5–7 mm. long; stamens 4–5; lvs. gray-green
        5. *S. gooddingii variabilis*
      14. Lf.-blades 1.5–4 cm. wide.
        15. Stamens 3–9, the fils. hairy near base; twigs mostly red-brown.
          16. Shrub 1–4 m. high; lvs. not glaucous beneath. At 6000–8500 ft. . 2. *S. caudata*
          16. Tree 5–15 m. high; lvs. glaucous beneath.
            17. Petioles glandular above .... 8. *S. lasiandra*
            17. Petioles not glandular .... 7. *S. laevigata*
        15. Stamens 2, the fils. glabrous; twigs yellowish or brownish .... 11. *S. lutea*
  11. Lvs. rather permanently pubescent beneath.
    18. Trees or large shrubs mostly 3–15 m. high.
      19. Plants with "weeping habit" having long pendulous branches; lvs. largely 10–15 cm. long, very long-acuminate. Escape from cult. .... 1. *S. babylonica*
      19. Plants not conspicuous for pendulous branches. Native.
        20. Stamens 4–6; lvs. 1.5–3 cm. wide, 5–12 cm. long; style and stigmas minute
          7. *S. laevigata araquipa*
        20. Stamens 2; lvs. 0.2–1.4 cm. wide, 4–8 cm. long; style to ca. 0.5 mm. long; catkin-scales yellow.
          21. Stigma-lobes ca. 1 mm. long; style ca. 0.5 mm. long .... 6. *S. hindsiana*
          21. Stigma-lobes very short; style obsolete .... 12. *S. melanopsis bolanderiana*
    18. Shrubs mostly not over 3 m. tall; catkin-scales yellowish; lvs. linear .... 3. *S. exigua*

1. **S. babylónica** L. WEEPING WILLOW. Broad-headed large tree with long flexible hanging branches; lvs. 8–15 cm. long, long-acuminate, finely serrulate; stipules rarely developed.—Native of China. Escape from cult., as in San Bernardino, Santa Barbara cos.

2. **S. caudàta** (Nutt.) Heller var. **bryantiàna** Ball & Bracelin. Erect shrub 1–4 m. tall, several-stemmed; twigs and branches brownish-red, glabrous, lustrous; lf.-blades lanceolate, 6–12 cm. long, 1.5–3 cm. wide, dark green and glabrous above, paler but not glaucous beneath, glandular-serrulate, acuminate; petioles 3–6 mm. long, glabrous, glandular in upper part; stipules small or none; catkins appearing with the lvs.; scales yellow, subglabrous; ♂ catkins 2–4 cm. long; stamens 3–9; fils. hairy below; ♀ 2.5–5 cm. long; caps. glabrous; styles evident.—Along streams and in meadows, 6000–8500 ft.; upper Montane Coniferous F.; San Bernardino Mts.; Sierra Nevada to Alta., S. Dak., New Mex.

3. **S. exígua** Nutt. NARROW-LEAF W. Shrub 2–4 m. tall; lvs. linear, entire to remotely denticulate, canescent on both surfaces, sometimes silvery-tomentose beneath, 5–12 cm. long, 0.2–0.8 cm. wide, subsessile; catkins coming after the lvs. on leafy pedunculate branchlets; scales yellowish, lanceolate, ± hairy; stamens 2; fils. hairy below; caps. ca. 5 mm. long, glabrous or nearly so; stigmas sessile, ca. 0.5 mm. long; $2n = 38$ (Suda & Argus, 1968).—Wet places, below 8000 ft.; Sagebrush Scrub, Creosote Bush Scrub, etc.; deserts, Imperial Co. to Modoc Co. and B.C., Tex. March–May. Some plants with silvery-pubescent lvs. 2–5 mm. wide and silvery-pubescent caps. when young constitute the var. *stenophylla* (Rydb.) C. K. Schneid.

4. **S. geyeriàna** Anderss. Shrub to 4 or 5 m. tall with slender leafy olive-green to brown (later darker) mostly glabrous twigs; lvs. narrowly oblanceolate to elliptic-oblong, 2–7 cm. long, 0.6–1.5 cm. wide, acute at both ends, entire, dark green above, ± glaucous beneath, thinly silky on both sides; stipules none; petioles 4–10 mm. long; catkins coming with the lvs., short-pedunculate; scales yellowish with reddish tips; stamens 2; caps. 5–7 mm. long, silky.—At ca. 10,000 ft.; White Mts., Mono Co.; to B.C., S. Dak., Ariz. May–June.

5. **S. gooddíngii** Ball. var. **gooddingii**. Tree 6–10 (–20) m. tall with rough dark bark, sometimes shrubby; twigs yellowish; lvs. narrowly lanceolate, acute or acuminate, finely glandular-serrulate, grayish-green, 6–10 cm. long, often pubescent when young; petioles 6–10 mm. long; stipules usually ± glandular; catkins 4–8 cm. long, on lateral leafy branchlets 2–4 cm. long; stamens 4–5, pubescent, subtended by yellow bracts; caps. 5.5–7 mm. long, pilose; pedicels 2–3 mm. long; styles and stigmas very short; $2n = 19$ pairs (Raven et al., 1965).—Stream banks and wet places, mostly below 2000 ft.; mostly Creosote Bush Scrub, in drainage of Colo. River, less common in cismontane s. Calif.; to cent. Calif., Utah, Son., L. Calif. Largely replaced in cismontane Calif. by var. **variábilis** Ball.; caps. glabrous.—Coastal Sage Scrub, etc.

6. **S. hindsiàna** Benth. var. **hindsiàna**. SANDBAR W. Erect shrub or small tree, 2–7 m. high, with gray furrowed bark; young twigs gray-tomentose; lvs. linear to lance-linear, tapering at both ends, 4–8 cm. long, to scarcely 1 cm. wide; mostly entire and gray silky-villous to subtomentose; petioles 1–3 mm. long, not glandular; stipules wanting or small, to arcuate-lanceolate and 8 mm. long on sucker shoots; catkins appearing after the lvs., 2–4 cm. long, on leafy peduncles; stamens 2, with pubescent fils.; scales villous; caps. subsessile, 5–6 mm. long, villous-tomentose; style ca. 0.5 mm. long; stigmas 1 mm. long.—Common along ditches, sand bars, etc., below 3000 ft.; many cismontane Plant Communities; Santa Cruz Id., L. Calif. to Ore. March–May.

Var. **leucodendroìdes** (Rowlee) Ball. [*S. macrostachya* var. *l.* Rowlee.] Lvs. remotely denticulate, more pointed, mostly narrower; caps. usually less densely pilose; styles 0.5–0.7 mm. long.—San Diego to Ventura cos., and to cent. Calif.

Var. **parishiàna** (Rowlee) Ball. [*S. p.* Rowlee.] Lvs. gray to silvery, 2–3.5 mm. wide, ± revolute; style 0.1–0.2 mm. long.—Occasional at low elevs. from San Diego Co. to San Benito Co.

7. **S. laevigàta** Bebb var. **laevigàta**. Tree 5–15 m. high, with rough bark; twigs glabrous, red-brown to yellowish; lvs. lanceolate to oblong-lanceolate, closely serrulate,

light green above, paler and more glaucous beneath, 5-12 cm. long, 1.5-3 cm. wide; petioles 2-10 mm. long, not glandular; stipules small, glandular-toothed; catkins 3-10 cm. long, on short leafy lateral peduncles; stamens 4-6, the fils. pilose below; scales yellow; caps. glabrous, 3.5-5 mm. long.—Along streams below 5000 ft.; cismontane in many Plant Communities; Catalina Id.; to Utah, Ariz. March-May.

Var. **araquípa** (Jeps.) Ball. Young twigs, petioles, etc. pilose.—With var. *laevigata*, but more abundant southward.

8. **S. lasiándra** Benth. var. **lasiándra**. Tree 6-15 m. tall; bark rough; twigs reddish, shining, glabrous; lvs. lanceolate, acuminate, closely glandular-serrulate, dark green and shining above, glaucous beneath, 6-10 cm. long, 1.5-3.5 cm. wide; stipules small, rounded, acute, glandular; petioles stout, 5-15 mm. long, glandular at upper end; ♂ catkins 2-6 cm. long; ♀ 3-10 cm. long; scales yellow, lanceolate to ovate, usually toothed; stamens 4-5; caps. glabrous; $2n = 76$ (Wilkinson, 1944).—Stream banks below 8000 ft.; Coastal Sage Scrub to Montane Coniferous F.; most of cismontane s. Calif.; Santa Cruz Id.; to Alaska, Rocky Mts. March-May.

Var. **abrámsii** Ball. Twigs glabrous; lvs. 1-1.7 cm. wide, subentire to shallowly serrulate.—An indefinite form at 4000-8000 ft.; San Bernardino Mts.; Sierra Nevada to Ore., Ida.

Var. **lancifòlia** (Anderss.) Bebb. [*S. l.* Anderss.] Young twigs pilose-pubescent. With the typical var.

9. **S. lasiólepis** Benth. var. **lasiólepis**. ARROYO W. Plate 80, Fig. F. Erect shrub or small tree 2-10 m. high, with smooth bark and yellowish to dark brown twigs, usually pubescent; lvs. oblanceolate, 6-10 cm. long, 1-2 cm. wide, acute to obtuse; nearly entire, subrevolute, dark green and glabrous above, pubescent to glabrate and glaucous beneath; petioles 3-12 mm. long; stipules none or roundish; catkins mostly before the lvs., subsessile, 3-7 cm. long; scales dark, round-obovate, 1 mm. long, densely hairy; stamens 2, the fils. glabrous, united below; caps. 4.5-5 mm. long, glabrous; pedicels 0.5-1.2 mm. long; styles 0.5 mm. long; stigmas short; $2n = 38$ pairs (Raven et al., 1965).—Common along streams, below 7000 ft.; cismontane Plant Communities to Montane Coniferous F., occasional in the desert; Santa Rosa, Santa Cruz ids.; to Wash., Rocky Mts.

Var. **bracelinae** Ball. Twigs stoutish, pubescent to puberulent; lvs. linear to linear-elliptic or -lanceolate; petioles 5-15 mm. long.—At 2500-7500 ft.; San Diego Co. to Inyo and Sonoma and Alpine cos.; to Utah, Tex., Chihuahua, L. Calif.

10. **S. lemmònii** Bebb. Shrub 1-5 m. high, the younger twigs glabrous to thinly pubescent, yellowish and the older darker; lvs. oblanceolate to almost lanceolate, 3-10 cm. long, 0.8-2 cm. wide, acute to acuminate, entire or ± glandular-serrulate; petioles 3-10 mm. long; stipules minute to 8 mm. long; catkins mostly appearing with the lvs., 1-6 cm. long; scales dark; stamens 2; caps. slender, 6-9 mm. long, silky.—Moist places, 6700-9500 ft.; Montane Coniferous F.; San Bernardino Mts.; Sierra Nevada to Ore., Wyo. May-June.

11. **S. lùtea** Nutt. var. **watsònii** (Bebb) Jeps. Clustered shrub 2-5 m. high, with yellowish-white ± divaricate twigs; lvs. 4-6 cm. long, lanceolate, pointed, entire to serrulate, yellowish-green above, paler and more glaucous beneath, 4-6 cm. long, 1.5-4 cm. wide, on petioles 3-10 mm. long; stipules entire to serrulate; catkins appearing before and with the lvs., subsessile, 2-3 or 4 cm. long; scales tawny; stamens 2; fils. glabrous; caps. 4-5 mm. long, glabrous.—Wet places along streams, etc.; at 7200-8500 ft.; Montane Coniferous F.; San Jacinto and San Bernardino mts.; at 4700 ft., Deep Springs region, Inyo Co. and at 7300 ft., White Mts. May-June.

12. **S. melanópsis** Nutt. var. **bolanderiàna** (Rowlee) C. K. Schneid. Shrub or small tree to 3 or 5 m. tall; young growth and lvs. pubescent to grayish-pubescent, the lvs. oblanceolate to elliptic, 4-7 cm. long, 6-14 mm. wide, rather closely denticulate, dark and lustrous above, paler beneath; catkins 3-4 cm. long; scales nearly glabrous; stamens 2, distinct; caps. 4-5 mm. long, glabrous, subsessile.—Stream banks at ca. 7500 ft.; occasional in San Bernardino and San Gabriel mts.; Sierra Nevada to B.C., Rocky Mts. March-May.

13. S. scouleriàna Barr. Shrub or small tree 1–10 m. tall, pubescent; twigs stoutish, the younger yellowish, the older darker; lvs. oblanceolate to obovate, 3–10 cm. long, 1.5–3 cm. wide, dull green above, glaucous beneath; petioles 5–15 mm. long; stipules subreniform to semiovate; catkins before the lvs., dense, subsessile; scales obovate, 1.5–2.5 mm. long, black; stamens 2; caps. 7–9 mm. long, gray-tomentose; stigmas 0.5–1 mm. long; $2n=76$ (Suda & Argus, 1968).—At ca. 6000–9000 ft.; Montane Coniferous F.; San Jacinto, San Bernardino and San Gabriel mts. April–June.

96. Santalàceae. SANDALWOOD FAMILY

Herbs, shrubs or trees. Lvs. entire, estipulate. Fls. perfect or imperfect, axillary or terminal. Fl.-tube adnate to the base of the ovary. Calyx 4–5-merous, valvate in bud. Corolla none. Stamens as many as the sepals, opposite them and inserted on the fleshy disk. Style 1; ovary 1-loculed; stigma capitate. Ovules 2–4, suspended from the top of a free central placenta. Fr. a drupe or nut. Seed 1, round or ovoid, without testa; embryo small, axile at one end of the abundant endosperm. Ca. 26 genera and 250 spp.; mostly trop.

1. *Comándra* Nutt. BASTARD TOAD-FLAX

Smooth perennial semiparasitic herbs from rootstocks; stems striate. Lvs. alternate, entire, subsessile. Fls. greenish, in small terminal or axillary clusters, perfect. Fl.-tube campanulate or urn-shaped. Sepals 5. Anthers attached to sepals by a tuft of hairs. Fr. drupelike, crowned by the persistent calyx; seed globular. A polymorphic sp., with 1 ssp. in Eu. and 3 in N. Am. (Greek, *kome*, hair, and *ander*, man, referring to the hairy attachment of the stamens.)

(Piehl, M. A. The natural history and taxonomy of Comandra. Mem. Torrey Bot. Club 22 (1): 1–97. 1965.)

1. C. umbellàta (L.) Nutt. ssp. **califórnica** (Eastw. ex Rydb.) Piehl. [*C. pallida* Calif. refs.] Plate 81, Fig. A. Stems many, ± glaucous, the fertile shoots 1.5–4 dm. tall, often much branched and many-fld.; lvs. 1.7–5 cm. long, broadly elliptic and ± ovate to linear and lanceolate, attenuate at base, acute at apex; infl. corymbose to paniculate; fls. infundibuliform to rotate; sepals 2–3.5 mm. long, lanceolate to subovate; fr. 5–7.5 mm. in diam.; $n=14$ (Piehl).—On mostly dry slopes, Sagebrush Scrub and higher; Inyo and Kern cos. n.; to B.C. and Ariz. April–Aug.

97. Saururàceae. LIZARD-TAIL FAMILY

Perennial herbs, ours with creeping rootstocks. Lvs. alternate, simple, petioled, the stipules adnate to the petiole. Stems nodose, scapelike. Fls. perfect, in dense spikes or racemes; bracts conspicuous. Perianth absent. Stamens mostly 6 or 8, free or adnate to ovary at base or epigynous; anthers 2-celled. Pistil of 3–4 free or connate carpels; ovules 2–4 in each free carpel, or 6–8 on each parietal placenta, if carpels united. Fr. a caps. or berry. Seeds globose or ovoid, with copious perisperm. Three genera and 4 spp. of N. Am. and Asia.

1. *Anemópsis* Hook. YERBA MANSA

Stems nodose, scapelike, stoloniferous from thick creeping rootstocks. Lvs. mostly basal, minutely punctate. Spike conical, with basal, persistent, white invol. of several bracts; each fl., except the lowest, subtended by a small white bract. Ovary sunk in the rachis of the spike; stigmas 3–4. Fr. a caps. One sp. formerly used medicinally for diseases of skin and blood. (Greek, *anemone*, and *opsis*, anemonelike.)

1. A. califórnica Hook. Plate 81, Fig. B. Stems 1–5 dm. high, woolly-pubescent, with a broadly ovate clasping lf. above the middle and 1–3 small lvs. in this axil; basal lvs. elliptic-oblong, 4–18 cm. long, on equally long petioles, entire, cordate at base; spikes

Plate 81. Fig. A, *Comandra umbellata* ssp. *californica*, ovary inferior. Fig. B, *Anemopsis californica*, white petaloid invol. at base of flowering spikes, single fl. with bract. Fig. C, *Boykinia elata*. Fig. D, *Heuchera rubescens*, sepals 5, petals linear. Fig. E, *Jamesia americana* var. *californica*, styles distinct. Fig. F, *Lithophragma affine*.

1–4 cm. long; involucral bracts white or reddish beneath, 1–3 cm. long; floral bracts 5–6 mm. long; $2n=22$ pairs (Raven et al., 1965).—Common in wet, especially subalkaline places, below 6500 ft.; Alkali Sink and Coastal Salt Marsh to Yellow Pine F.; L. Calif. n. through cismontane and desert areas; to Nev., Tex., Mex. March–Sept.

98. Saxifragàceae. SAXIFRAGE FAMILY

Herbs or shrubs with opposite or alternate lvs., usually without stipules. Fls. perfect, perigynous, solitary or in racemes, cymes or paniculate clusters. Fl.-tube free or partly united to the ovary, usually persistent. Sepals 5 or 4. Petals 5 or 4, sometimes more. Stamens usually 5 or 10, sometimes fewer or more. Carpels mostly fewer than sepals, ± separate or combined into 1 pistil, with ovary superior to inferior. Fr. a caps., follicle or berry. Seed with endosperm. Ca. 700 spp., mostly of N. Temp. and Arctic regions, a few Andean.

1. Plants herbaceous, not woody.
  2. Fls. appearing before the lvs.; petals spatulate; stems from cormlike root .... 5. *Jepsonia*
  2. Fls. normally appearing with or after the lvs.
    3. Fertile stamens 10.
      4. Styles 3; petals clawed, usually laciniate or toothed; rootstocks usually slender, tuberous
        6. *Lithophragma*
      4. Styles 2; petals clawed or sessile, entire; rootstocks various. .... 10. *Saxifraga*
    3. Fertile stamens 5 or 3.
      5. Fertile stamens alternating with clusters of gland-tipped sterile staminodia; fls. solitary, scapose .... 7. *Parnassia*
      5. Fertile stamens not alternating with sterile staminodia.
        6. Ovary 2-celled with placentae axile; seeds minutely punctate ...... 1. *Boykinia*
        6. Ovary 1-celled with placentae parietal; seeds echinate with minute spines
          3. *Heuchera*
1. Plants definitely shrubby.
  8. Lvs. opposite; fr. a caps., partly inferior.
    9. Petals white, 7–10 mm. long; stamens 30–50 .... 8. *Philadelphus*
    9. Petals rose-pink to white, 3–7 mm. long; stamens 10.
      10. Lvs. crenate-serrate, 1–2.5 cm. long; petals 6–7 mm. long, rose-pink .... 4. *Jamesia*
      10. Lvs. entire, 0.6–1.4 cm. long; petals 3–4 mm. long, white .... 2. *Fendlerella*
  8. Lvs. alternate; fr. a berry, wholly inferior .... 9. *Ribes*

1. Boykínia Nutt.

Perennial herbs with scaly rootstocks. Lvs. several, alternate on the stem and in a basal tuft, reniform, toothed to lobed, usually stipulate. Infl. a leafy bracted paniculate or corymbose cyme. Fl.-tube campanulate, adnate to lower half of ovary. Sepals 5, lanceolate to lance-ovate. Petals white, 5, usually clawed and early deciduous. Stamens 5, opposite the sepals; fils. short. Ovary and caps. 2-loculed, styles 2, forming divergent beaks. Seeds many, ovoid, minutely punctate. Ca. 8 spp., N. Am. and e. Asia. (Dr. *Boykin*, resident in Georgia in early 19th century.)

Petals well exserted; stipules evident; lvs. cleft or incised .... 1. *B. elata*
Petals scarcely exceeding calyx; stipules not evident; lvs. shallowly round-lobed
    2. *B. rotundifolia*

1. **B. elàta** (Nutt.) Greene. [*Saxifraga e.* Nutt. in T. & G.] Plate 81, Fig. C. Stems slender, erect, 2–6 dm. high, with minute brown gland-tipped hairs; lower lvs. thin, 2–8 cm. wide, reniform to cordate-ovate, somewhat incised into 5–7 lobes with bristle-pointed teeth; petioles 5–15 cm. long; upper lvs. reduced; infl. a many-fld. panicle, densely glandular-puberulent; fl.-tube 2–3 mm. long; sepals lance-acuminate, 1.5–2 mm. long; petals ± oblanceolate; $n=7$ (Hamel, 1953).—Shaded springy places below 5000 ft.; Chaparral; Santa Monica Mts., to n. Calif. and Wash. June–July.

2. **B. rotundifòlia** Parry. Rather stout, 3–8 dm. high, densely glandular-villous;

lower petioles 8–16 cm. long, upper shorter; lf.-blades round-cordate, mostly 5–12 cm. wide, shallowly round-lobed with crenate-dentate teeth; panicle 2–15 cm. long; fl.-tube glandular-pubescent, striate, 3–4 mm. long; sepals 4–5 mm. long, lance-ovate; petals obovate-spatulate, clawed; $n=13$ (Hamel, 1953).—Wet places in canyons, below 6000 ft.; largely Chaparral; Elsinore and San Jacinto mts. to s. face of San Bernardino and San Gabriel mts.; Cuyama V. in Santa Barbara Co. June–July.

2. Fendlerélla Heller

Low shrub with shreddy bark. Lvs. opposite, entire. Fls. bisexual, small, in small dense terminal cymes. Fl.-tube turbinate-campanulate, adnate to lower half of ovary. Sepals 5, lanceolate. Petals 5, white. Stamens 10, alternately longer and shorter; fils. dilated toward base. Ovary 3-celled; styles 3, distinct. Caps. narrow, 3-valved, septicidal. Seeds solitary in each cell. Three spp., sw. N. Am. (Diminutive of *Fendlera* a related genus named for A. *Fendler*, early Texan collector.)

1. **F. utahénsis** (Wats.) Heller. [*Whipplea u.* Wats.] Much-branched, 3–6 dm. high, the young growth strigose; lvs. oblong-elliptic to -spatulate, subsessile, 6–14 mm. long, strigose; sepals ca. 1.5 mm. long; petals 3–4 mm. long; caps. ca. 4 mm. long.—Dry limestone slopes, at ca. 5500–6500 ft.; Pinyon-Juniper Wd.; Last Chance Mts. (Inyo Co.), Clark and Providence mts. (e. San Bernardino Co.); to Utah, Tex. June–Aug.

3. Héuchera L. ALUM-ROOT

Perennial herbs mostly with stout caudices or rhizomes. Lvs. mostly basal, long-petioled, roundish-cordate, lobed and notched or toothed; stipules united with petiole at base. Flowering stems lateral, scapose or ± leafy, with a terminal paniculate infl. Fl.-tube half epigynous, urceolate to cylindric to turbinate or saucer-shaped. Sepals 5, often unequal. Petals 5, entire, small, spatulate to linear or obovate, sometimes 0. Stamens 5, opposite the sepals. Ovary half inferior, 1-celled with 2 parietal placentae; styles filamentous to short and cylindrical. Caps. opening between the 2 ± divergent beaks. Seeds small, ovoid, mostly echinate with minute spines. Ca. 50 spp. of N. Am. (Named for J. H. von *Heucher*, 1677–1747, German medical botanist.)

(Rosendahl, Butters and Lakela. A monograph on the genus Heuchera. Minn. Studies in Plant Science 2: 1–180. 1936.)

1. Mature styles long and slender, the closed distal portion not less than 1.5 mm. long; fl.-tube urceolate or cylindric to deeply campanulate.
  2. Free portion of the fl.-tube shorter than the part adnate to the ovary; fls. mostly regular. Insular .... 8. *H. maxima*
  2. Free portion of the fl.-tube equal to or longer than the adnate portion; fls. ± irregular.
    3. Stamens long exserted, surpassing the sepals and ca. equal to petals; free portion of fl.-tube less than 1 mm. long; petals very narrow, filament-like; styles long-exserted.
      4. Fls. almost regular; petals equal, up to 0.3 mm. wide.
        5. Inferior part of ovary narrow-obconic, distinctly longer than broad. San Diego Co. 7. *H. leptomeria*
        5. Inferior part of ovary turbinate to broadly obconic, usually broader than long; fls. open-campanulate, 3–5 mm. long, 2–3 mm. wide; sepals villous. San Bernardino Co. n. .... 10. *H. rubescens*
      4. Fls. irregular; petals unequal, to 0.5 mm. wide. San Bernardino Mts.
        6. Fls. campanulate; sepals crisp-hairy .... 9. *H. parishii*
        6. Fls. narrow-campanulate, soon urceolate; sepals glandular-hirtellous . 2 *H. alpestris*
    3. Stamens mostly included, shorter than petals; free portion of fl.-tube more than 1 mm. long on its shorter side; petals distinctly wider than fils.; styles scarcely exserted.
      7. Fls. glandular-puberulent with a few longer hairs at sepal tips; petioles not hirsute. San Gabriel Mts. .... 1. *H. abramsii*
      7. Fls villous-hirsute; petioles ± hirsute.
        8. Stamens reaching about to the apex of the sepals; free portion of the fl. tube over 1 mm. long on its shorter side.

9. All sepals with oblong or ovate free parts. San Gabriel Mts. ...... 5. *H. elegans*
9. Sepals on long side of fl. with only very short semicircular tips free. San Jacinto Mts. .... 6. *H. hirsutissima*
8. Stamens distinctly shorter than sepals; free portion of fl.-tube ca. 0.7 mm. long. Laguna Mts., San Diego Co. .... 3. *H. brevistaminea*
1. Mature styles (true styles, not hollow carpel beaks) cylindrical, not more than 1 mm. long; fl.-tube open-campanulate to saucer-shaped. Mono Co. .... 4. *H. duranii*

1. **H. abrámsii** Rydb. Lvs. 0.5–1.8 cm. wide; deeply lobed; near to *H. alpestris* in general appearance, paucity of long pubescence on fls. and very short oblique fl.-tube, but fls. 4.5–6 mm. long; fl.-tube subcylindric; petals ca. 2.5 mm. long, oblanceolate; stamens included; seeds 0.6 mm. long.—At 9000–10,000 ft.; upper Montane Coniferous F.; e. part of San Gabriel Mts. July–Aug.

2. **H. alpéstris** Rosend., Butt. & Lakela. Near to *H. parishii*; lvs. 0.5–2 cm. wide; fl.-stems 0.5–2 dm. long; fls. irregular, narrow-campanulate, soon urceolate, ca. 4 mm. long; fl.-tube glandular-puberulent; sepals with some longer hairs near the tips; petals oblanceolate, 2.5–3 mm. long; stamens well exserted; seeds ca. 0.6 mm. long.—At 8500–11,400 ft.; San Bernardino Mts. July–Aug.

3. **H. brevistamínea** Wiggins. Caudex short, woody; flowering stems 1–3 dm. high, scapose, glandular-puberulent and with some longer hairs; basal lvs. roundish, 1.5–3 cm. wide, shallowly 5-lobed, the lobes shallowly toothed, white-ciliate; petioles white-hirsute, 2–4 cm. long; infl. narrow, rather lax; fls. 4–5 mm. long, the fl.-tube campanulate, reddish-purple, glandular, 1.7 mm. long; sepals oblong with rounded tips; petals ca. 3.5 mm. long, oblance-spatulate, conspicuously veiny; caps. ca. 5 mm. long; seeds ca. 0.7 mm. long.—Dry rocky places, 5000–6000 ft.; Yellow Pine F.; Laguna Mts., San Diego Co. May–July.

4. **H. duránii** Bacig. From a heavy caudex; lvs. basal, reniform to almost round, 1–2 cm. wide, shallowly crenately 5–9-lobed, glandular-puberulent above and beneath with some short hairs along the veins beneath; petioles 2–5 cm. long; flowering branches 1.5–3.5 dm. high, slender, glandular-puberulent; panicle narrow, 4–8 cm. long, glandular-puberulent; fls. yellowish with some pink, 2–2.5 mm. long, the fl.-tube broadly turbinate, minutely glandular-puberulent; sepals scarcely 1 mm. long.—Rocky hillsides, 9000–11,700 ft.; Bristle-cone Pine F.; Subalpine F.; White Mts., etc., Mono Co.; w. Nev. July–Aug.

5. **H. elegáns** Abrams. Lvs. 1.5–3.5 cm. wide, shallowly lobed; petioles more hirsute; fls. 4–7 mm. long, narrow-campanulate; fl.-tube round-turbinate at base, it and sepals hirsute; petals ca. 5 mm. long; stamens scarcely exserted; seeds 0.7 mm. long.—At 4000–8500 ft.; San Gabriel Mts. May–June.

6. **H. hirsutíssima** Rosend., Butt. & Lak. Near to *H. elegans*, but sepals of long side of fl.-tube united almost to top; anthers ± exserted; fl.-tube very oblique.—At 7000–10,800 ft.; upper Montane Coniferous F., Subalpine F.; San Jacinto and Santa Rosa mts. July–Aug.

7. **H. leptomèria** Greene var. **peninsulàris** Rosend., Butt. & Lak. Near to *H. rubescens*, but with lower part of fl.-tube narrow-obconical, distinctly longer than broad, glandular-puberulent, sparingly hirsute; petals 3–5 mm. long, mostly 1-nerved; stamens ca. as long as petals; seeds 0.5 mm. long.—At ca. 5000–6000 ft.; Yellow Pine F.; Hot Springs Mts., Cuyamaca Mts., San Diego Co.; L. Calif. May–June.

8. **H. máxima** Greene. With long heavy caudex; flowering branches stout, often 3–4-lvd., mostly hirsute; basal lvs. round-cordate, 6–18 cm. in diam., with deep basal sinus, 7–9 lobes and with aristate teeth, upper surface subglabrous, lower strigose-villous especially along the veins; petioles 8–20 cm. long, villous-hirsute; panicle narrowly cylindrical, 2–3 dm. long, glandular-puberulent and with some longer hairs; fls. 3.5–4.5 mm. long, the fl.-tube short-campanulate; petals whitish, ca. 2 mm. long; caps. 5–8 mm. long. —Canyon walls and cliffs; Chaparral, Coastal Sage Scrub; Santa Cruz, Santa Rosa and Anacapa ids. Feb.–April.

9. **H. paríshii** Rydb. Much like *H. rubescens* in habit; lvs. 1.5–4 cm. wide; flowering

stems 1–3 dm. high, glandular-puberulent and ± hirsute, the panicle narrow, 5–10 cm. long; fls. quite zygomorphic, 4–5 mm. long; fl.-tube with a turbinate base, glandular-puberulent; sepals oblong; petals oblanceolate, 2–3 mm. long; stamens exserted; styles glabrous except at very base; seeds ca. 0.9 mm. long.—Rocky places, 5000–8900 ft.; Montane Coniferous F.; San Bernardino Mts. July–Aug.

10. **H. rubéscens** Torr. var. **pachypòda** (Greene) Rosend., Butt. & Lak. Plate 81, Fig. D. Caudex mostly multicipital, cespitose; lvs. basal, coriaceous, 1–2.4 cm. wide, roundish, subcordate to rounded at base, ± deeply 3–5-lobed, rather sharply dentate, ciliate, glandular-puberulent to glabrate in age; petioles slender, 1–5 cm. long, glandular-puberulent and short-hirsute; flowering stems 1–3 dm. high, scapose; panicles narrow, 3–11 cm. long, ± secund; fls. 4–5.5 mm. long, only slightly irregular, the fl.-tube turbinate, oblique; sepals oblong; petals 3–4 mm. long, oblanceolate; stamens slightly exceeding petals; caps. 4.5–5.5 mm. long; seeds 0.8 mm. long.—Dry rocky places, 6000–11,000 ft.; Pinyon-Juniper Wd., Bristle-cone Pine F.; New York Mts., Panamint Mts., Clark Mt., Inyo-White Mts.; s. Sierra Nevada, w. Nev. May–July.

4. *Jàmesia* T. & G.

Low deciduous shrubs with opposite toothed lvs. Fls. in terminal cymose clusters. Fl.-tube hemispheric to turbinate, adnate to lower part of ovary. Sepals 5, lanceolate to subovate. Petals 5, spreading, oblong-obovate, pubescent within. Stamens 10, the alternate shorter; fils. dilated. Ovary imperfectly 3–5-celled; styles 3–5, distinct; stigmas terminal. Caps. conic, beaked. Seeds striate-reticulate. One sp. (Named for Dr. E. *James*, a member of the Long Expedition to the Rocky Mts. in 1829.)

1. **J. americàna** T. & G. var. **califórnica** (Small) Jeps. [*Edwinia c.* Small.] Plate 81, Fig. E. Low, 3–10 dm. high, with shreddy bark, pubescent on young growth; lvs. oblong to roundish, 1–2.5 cm. long, coarsely crenate-serrate, green and pubescent above, gray-strigose beneath, prominently veined; petioles mostly 2–6 mm. long; pedicels, fl.-tube and calyx strigose; sepals lance-ovate, 2–2.5 mm. long; petals rose-pink, narrowly oblong-ovate, 6–7 mm. long; styles 6–7 mm. long in fr.; caps. ca. 4 mm. long; n = 16 (Hamel, 1953).—Rocky places, 7000–9000 ft.; Pinyon-Juniper Wd. and higher; Panamint and Inyo mts.; s. Sierra Nevada; sw. Utah, Nev. July–Aug.

5. *Jepsònia* Small

Perennial herbs with cormlike rootstocks, few basal mostly vernal lvs., and slender autumnal scapes. Lvs. round-cordate, petioled, shallowly lobed. Fls. in terminal cymes. Fl.-tube ± campanulate, scarcely adnate to the ovary, the upper part free, veiny, mostly purplish-striate. Sepals 5. Petals 5, white, spatulate, clawed, withering-persistent. Stamens 10, shorter than the sepals; fils. filiform. Carpels 2, united to ca. the middle. Follicles veiny, beaked. Seeds small, 4-ridged. A small California genus. (W. L. *Jepson*, longtime professor of Botany at the Univ. of Calif. and authority on Calif. flora.)

Sepals usually shorter than the fl.-tube. Mainland .... 2. *J. parryi*
Sepals not shorter than fl.-tube. Insular .... 1. *J. malvaefolia*

1. **J. malvaefòlia** (Greene) Small. [*Saxifraga m.* Greene. *J. neonuttalliana* Millsp.] Lvs. appearing with the scapes after the first rains, 2–3.5 cm. wide; petioles 2.5–4 cm. long; scapes slender, 8–20 cm. long; fl.-tube mostly truncate at base; petals ca. 3.5 mm. long, the claw short and broad; carpels at anthesis united for ⅔ their length.—Canyons and mesas, Chaparral, etc.; Catalina, San Clemente, San Nicolas, Santa Cruz, Santa Rosa and Guadalupe ids. Nov.–Dec.

2. **J. párryi** (Torr.) Small. [*Saxifraga p.* Torr.] Lvs. few, 2–6 cm. wide, pubescent; scapes 1–3 dm. tall, glandular-puberulent to glabrate in age; petals 4–6 mm. long.—Moist shaded banks below 2000 ft.; Coastal Sage Scrub, Chaparral; sw. San Bernardino Co., Orange and w. Riverside cos., San Diego Co.; n. L. Calif. Oct.–Dec.

6. **Lithophrágma** Nutt. WOODLAND-STAR

Perennial herbs often with slender bulbet-bearing rootstocks and simple slender scapose flowering shoots. Lvs. largely basal, petioled, roundish to reniform in outline, mostly 3–5-lobed or -parted; petioles with stipulelike dilated bases. Fls. in simple racemes. Fl.-tube hemispheric to campanulate or obconic. Sepals 5, rather short. Petals 5, white to pinkish, clawed, entire or toothed or cleft, usually ± unequal. Stamens 10, included, the fils. short. Ovary 1-celled with 3 many-seeded parietal placentae; styles 3, unequal. Fr. a caps.; seeds ovoid, horizontal. Ca. 10 spp. of w. N. Am. (Greek, *lithos*, rock, and *phragma*, fence or hedge.)

(Taylor, R. L. The genus Lithophragma. Univ. Calif. Publ. Bot. 37: 1–122. 1965.)
1. Basal lvs. truly compound, trifoliolate; stems stout, 4–6 dm. high. San Clemente Id.
5. *L. maximum*
1. Basal lvs. not truly compound, but merely lobed; stems slender.
  2. The basal lvs. not lobed to near their bases.
    3. Petals entire or shallowly toothed.
      4. Fl.-tube with a ± rounded, not definitely acute base; pedicels 1–3 mm. long; stem-lvs. alternate.
        5. Base of petal-blade not involute or toothed; fl.-tube campanulate with a truncate base
4. *L. heterophyllum*
        5. Base of petal-blade somewhat involute, minutely toothed or laciniate; fl.-tube obtuse or rounded at base .... 2. *L. bolanderi*
      4. Fl.-tube with an acute base; pedicels 5–10 mm. long; stem lvs. mostly opposite
3. *L. cymbalaria*
    3. Petals deeply parted.
      6. Pedicels 1–2 mm. long; petals 4–7 mm. long; fl.-tube truncate at base
4. *L. heterophyllum*
      6. Pedicels 2–8 mm. long; petals 6–10 mm. long; fl.-tube obconic at base . 1. *L. affine*
  2. The basal lvs. lobed almost to their base.
    7. Petals mostly 3–5 mm. long; fl.-tube ± rounded at base .... 7. *L. tenellum*
    7. Petals mostly 6–10 mm. long; fl.-tube ± acute at base .... 6. *L. parviflorum*

1. **L. affìne** Gray ssp. **affìne.** Plate 81, Fig. F. Robust, 2–5 dm. tall, variously pubescent; basal lvs. palmately 3–5-lobed, each lobe sub-lobed into cuspidate lobules; cauline lvs. 1–3, more dissected; fls. 5–9 (–15); fl.-tube widely inflated, obconic below, the pedicels shorter than or equalling it; petals 6–13 mm. long, always 3-lobed; seeds not tuberculate; $2n = 14$, 21, or 28 (Taylor).—Grassy banks below 3500 ft.; many Plant Communities; Santa Barbara Co. to n. Calif.

Ssp. **míxtum** R. L. Taylor. [*L. tripartita* Greene.] Basal lvs. 3-lobed, each lobe often subdivided into spreading lobules; petals 4–9 mm. long, 3-lobed; $2n = 28$, 35 (Taylor).—Below 6500 ft.; n. L. Calif. to San Luis Obispo Co. Catalina Id.

2. **L. bolánderi** Gray. [*L. scabrella* Greene and var. *peirsonii* Jeps.] Flowering stalks 2.5–8.5 dm. tall, several; herbage pubescent; basal lvs. orbicular, 3–5-lobed; cauline lvs. 2–3, much reduced; fls. 3–5-many; pedicels not longer than fl.-tube which is campanulate, obtuse at base; corolla wide-spreading, the petals ovate-elliptic, 4–7 mm. long, mostly entire or with small serrations near base; seeds tuberculate; $2n = 14$, 28, 35, 42 (Taylor).—Foothill Wd., Yellow Pine F., etc.; Los Angeles and Kern cos. to n. Calif. May–June.

3. **L. cymbalària** T. & G. Slender, 2–4 dm. tall, sparingly pubescent; basal lvs. reniform, weakly 3-lobed; fls. 2–5 (–8), the pedicels to twice the turbinate fl.-tube; corolla wide-spreading, bowl-shaped, the petals 4–5 mm. long, entire, spatulate; seeds tuberculate; $2n = 14 + 1$ (Taylor).—Shaded woods, etc., below 3000 ft.; S. Oak Wd.; Palomar Mts., Ventura, Santa Barbara cos.; Santa Cruz and Santa Rosa ids.; to cent. Calif. March–May.

4. **L. heterophýllum** (H. & A.) T. & G. [*Tellima h.* H. & A.] Slender perennial 2–4 dm. tall, the flowering stalks 1–several; herbage glandular-pubescent or ± hirsutulose; basal lvs. round-reniform, 1.5–4 cm. wide, crenately shallowly lobed;

cauline lvs. alternate, usually much reduced, mostly deeply 3-cleft; fls. mostly 3–9; pedicels shorter than fl.-tube which is campanulate with a truncate base; corolla widely spreading; petals 5–12 mm. long, usually 3-, sometimes 5- or 7-lobed; seeds tuberculate; $2n=14$ (Taylor).—Mostly in S. Oak and Foothill Wd., below 4500 ft.; Los Angeles Co. n.

5. **L. máximum** Bacig. Stout perennial, the basal lvs. truly compound, the 3 lfts. rhombic, cuneate; petioles to 15 cm. long; fl.-stems stout, 4–6 dm. long; fl.-tube campanulate, ca. 6 mm. long at anthesis including the sepals, 8 mm. in fr.; petals 4 mm. long, digitately incised.—San Clemente Id.

6. **L. parviflòrum** (Hook.) T. & G. [*Tellima p.* Hook.] Slender, 2–5 dm. tall, nearly glabrous to densely pubescent; basal lvs. orbicular, 3-parted or digitately trifoliolate; cauline lvs. 2–3, much like the basal; fls. 4–14, the pedicels not longer than the fl.-tubes, which are oblong-obconic; petals white or pink, obovate-rhombic, 7–16 mm. long, always 3-cleft; seeds smooth or wrinkled; $2n=14, 21, 28, 35$ (Taylor).—Open slopes 2000–6000 ft.; Foothill Wd., Yellow Pine F.; Tehachapi Mts., San Bernardino Mts., San Diego Co.; to B.C. and Rocky Mts. March–June.

7. **L. tenéllum** Nutt. in T. & G. Stems 1.5–3 dm. tall; herbage light green and sparsely pubescent; basal lvs. round, simple and irregularly 3–5-lobed or digitately compound; cauline lvs. 2, pinnatifid; fls. 3–12; pedicels not longer than fl.-tube which is campanulate or hemispheric; petals mostly pink, 3–7 mm. long, palmately 5-parted; seeds smooth or wrinkled; $2n=14, 35$ (Taylor).—Occasional, 6000–8000 ft.; Montane Coniferous F.; San Gabriel and San Bernardino mts.; n. Calif. to Rocky Mts.

7. *Parnássia* L. Grass-of-Parnassus

Glabrous perennial scapose herbs, with short rootstocks. Lvs. entire, in basal tuft, or sometimes also with a small sessile lf. on scape. Fls. solitary, terminal, white or pale yellow. Fl.-tube short, poorly developed. Calyx deeply 5-lobed. Petals 5, conspicuously veined, each bearing at its base a group of gland-tipped sterile fils. (staminodia). Stamens 5, alternating with the sepals. Ovary superior to half inferior, 1 celled, with 3–4 projecting parietal placentae. Caps. 3–4-valved; seeds many, winged; endosperm none. Ca. 25 spp. of N. Temp. and Subarctic regions. (Named for Mt. *Parnassus*.)

Petals fimbriate on the sides of basal half, mostly 10–12 mm. long .... 1. *P. cirrata*
Petals entire, 10–18 mm. long .... 2. *P. palustris*

1. **P. cirràta** Piper. Basal lvs. 1–2 cm. long, on petioles 2–6 cm. long; scape 2–4 dm. high with lanceolate to ovate bract 5–10 mm. long; sepals lanceolate to lance-oblong, 5–7 mm. long; petals oblong-obovate, ca. 1 cm. long; staminodia with filiform gland-tipped fils.; fertile fils. 3–5 mm. long; $2n=36$ (Taylor & Brockman, 1966).—Wet places at ca. 7000–8000 ft.; upper Montane Coniferous F.; San Bernardino and San Gabriel mts.; n. Calif. Aug.–Sept.

2. **P. palústris** L. var. **califórnica** Gray. [*P. c.* Greene.] Basal lvs. ovate, 2.5–4 cm. long, with 5–7 principal veins, cuneate at base, on petioles 2–10 cm. long; scape 2.5–5 dm. high, with an ovate bract 5–10 mm. long, sessile above the middle or sometimes wanting; sepals lance-ovate, 4–6 mm. long; petals oblong-ovate to roundish, 3–7-veined; sterile fils. 15–24, capillary, gland-tipped, 3–4 mm. long, united in lower half; stamens ca. 7–8 mm. long.—Wet meadows, at 7000–8000 ft.; Montane Coniferous F.; San Bernardino Mts.; to Ore. July–Oct.

8. *Philadélphus* L. Mock-Orange. Syringa

Deciduous shrubs with opposite entire or toothed mostly petioled lvs. Fls. bisexual, regular, white, solitary or in few-fld. cymes in terminal clusters. Sepals 4–5, persistent. Petals 4–5, distinct. Stamens usually many; fils. subulate, free or united below; anthers short. Ovary almost completely inferior, mostly 4-celled; styles ± united. Fr. a many-seeded caps., loculicidal. Ca. 40 spp., of N. Temp. regions; some of hort. value. (Named for Ptolemy *Philadelphus* of ancient Egypt.)

(Hitchcock, C. L. The xerophyllous species of Philadelphus in sw. N. Am. Madroño 7: 35–56. 1943.)

1. **P. microphýllus** Gray ssp. **stramíneus** (Rydb.) C. L. Hitchc. [*P. s.* Rydb.] Much-branched rounded shrub 1–2 m. tall; young branches densely strigose; lf.-blades lance-ovate, 1–2.5 cm. long, strigose above and beneath, entire, on petioles 1–3 mm. long; fl.-tube 2–3 mm. long at anthesis, 3–5 mm. in fr., ± strigose; sepals 3–5 mm. long, lanate on inner surface, with stiff straight hairs without; petals 7–10 mm. long; stamens 30–50; styles 1–2 mm. long, free above or united to tips.—Dry rocky places, 5000–9000 ft.; Pinyon-Juniper Wd.; New York and Clark mts. to White Mts.; to Nev., L. Calif. May–July.

Ssp. **pùmilus** (Rydb.) C. L. Hitchc. [*P. p.* Rydb.] Lvs. mostly 8–12 mm. long, hirsute on upper surface.—At 7000–8500 ft.; Montane Coniferous F.; San Jacinto and Santa Rosa mts. July.

9. *Rìbes* L. CURRANT. GOOSEBERRY

Shrubs, either unarmed or with nodal spines and sometimes internodal bristles. Lvs. simple, alternate, usually palmately lobed, with stipules adnate to the petiole or lacking. Fls. usually in racemes, sometimes solitary, on 1–2-lvd. axillary shoots; pedicels subtended by bracts and usually bearing 2 bractlets. Fl.-tube adnate to the globose ovary and ± produced above it. Sepals 5, rarely 4; petals as many and usually shorter. Stamens alternate with the petals. Ovary inferior, 1-celled; style 2-lobed or -divided. Fr. a 1-celled pulpy berry, several-seeded; endosperm fleshy; embryo minute, terete. Ca. 120 spp. of the N. Temp. Zone and of the Andes. The Gooseberries are sometimes recognized as a separate genus, *Grossularia*. (*Ribes*, an ancient Arabic name.)

1. Nodal spines lacking; berry spineless; fls. usually several to many in a raceme.
    2. Lvs. not lobed, persistent, leathery; fl.-tube subrotate beyond the ovary. Insular
        18. *R. viburnifolium*
    2. Lvs. lobed, mostly deciduous; fl.-tube beyond the ovary subrotate to campanulate or cylindrical.
        3. Fls. yellow, sometimes tinged with red; lvs. glabrous or nearly so, almost alike on both surfaces .... 2. *R. aureum*
        3. Fls. not yellow; lvs. ± pubescent or glandular, the 2 surfaces unlike.
            4. The fls. white or greenish-white.
                5. Free portion of fl.-tube 2.5–3 times as long as broad; berry red ..... 5. *R. cereum*
                5. Free portion of fl.-tube 1–2 times as long as broad; berry blue or black
                    7. *R. indecorum*
            4. The fls. pink or red or purplish.
                6. Sepals erect; free part of fl.-tube short, bowl-shaped. Montane Coniferous F.
                    12. *R. nevadense*
                6. Sepals spreading or recurved; free part of fl.-tube cylindric to urceolate. Chaparral, etc.
                    7. Fls. bright rose; lvs. rugose; sepals 3–4 mm. long. From Riverside Co. n.
                        9. *R. malvaceum*
                    7. Fls. rose-purple; lvs. scarcely rugose; sepals 1.5 mm. long. S. San Diego Co.
                        4. *R. canthariforme*
1. Nodal spines present; berry often spiny; fls. mostly 1–few.
    8. Free part of fl.-tube inconspicuous, saucer-shaped; pedicels jointed beneath the ovary
        11. *R. montigenum*
    8. Free part of fl.-tube evident, campanulate to cylindric; pedicels not jointed beneath the ovary.
        9. Sepals 4, red, erect, ca. 1 cm. long; petals as long; stamens 2–3 times as long. Coastal canyons .... 15. *R. speciosum*
        9. Sepals 5, mostly not erect; petals shorter than sepals.
            10. Ovary and berry without spines or prickles.
                11. The ovary and berry glabrous.

12. Sepals greenish or purplish, 5–8 mm. long .... 6. *R. divaricatum*
12. Sepals yellowish, 2–3 mm. long.
  13. Free part of fl.-tube ca. 4 mm. long; berry red .... 8. *R. lasianthum*
  13. Free part of fl.-tube ca. 2.5 mm. long; berry black .... 13. *R. quercetorum*
11. The ovary and berry soft-pubescent or glandular .... 17. *R. velutinum*
10. Ovary and berry with spines or bristles.
  14. Free part of fl.-tube ca. as broad as long.
    15. Young twigs usually bristly on the internodes; lvs. glandular beneath. Santa Cruz Id. .... 10. *R. menziesii*
    15. Young twigs not bristly on internodes; lvs. not glandular. Mainland
                                                                3. *R. californicum*
  14. Free part of fl.-tube much longer than broad; young twigs usually not markedly bristly on internodes.
    16. Ovary and berry usually with all or nearly all of the bristles gland-tipped; lvs. glandular beneath .... 1. *R. amarum*
    16. Ovary and berry with eglandular spines; lvs. not glandular beneath.
      17. Sepals dull purplish-red; berry 14–16 mm. in diam. Mainland . 14. *R. roezlii*
      17. Sepals pinkish; berry 7–8 mm. in diam. Santa Cruz Id. ... 16. *R. thacherianum*

1. **R. amàrum** McClat. BITTER GOOSEBERRY. Erect, deciduous, 1–2 m. high, pubescent and glandular on young twigs, but not bristly; nodal spines brown, to 1 cm. long; lvs. roundish, with cordate base, 2–3 cm. wide, pubescent and ± glandular-puberulent above and beneath, 3–5-lobed, crenate; fls. 1–3, purplish, on glandular-pubescent peduncles; bracts broadly ovate; ovary densely glandular-bristly, free part of fl.-tube 5–6 mm. long, ca. half as wide, glandular-pubescent; sepals 7–8 mm. long; petals pinkish white, almost as long as fils.; stamens ca. 1 cm. long; berry 1.5–2 cm. in diam., with gland-tipped bristles 1–2 mm. long.—Wooded canyons; Chaparral, etc.; San Diego Co. to cent. Calif. March–April. A form with unequal spines 1–3.5 mm. long on the berry and from the Santa Barbara region is the var. *hoffmannii* Munz.

2. **R. aúreum** Pursh var. **aúreum.** GOLDEN CURRANT. Plate 82, Fig. A. Erect, 1–2 m. tall, almost glabrous; lvs. cuneate to subcordate at base, obovate to round-reniform in outline, 1.5–5 cm. wide, mostly 3-lobed, the lobes rounded, entire to toothed; racemes 5–15-fld., 3–7 cm. long; bracts 5–12 mm. long, ± glandular-ciliate; fls. with spicy odor; fl.-tube 6–10 mm. long (including ovary), slender, yellow; sepals spreading, 5–8 mm. long, erect after anthesis; petals orange in age, oblong, erose, 2–3 mm. long; berry round, mostly red or black, 6–8 mm. in diam.; $n = 8$ (Darlington, 1929).—Moist banks at ca. 8000 ft.; White Mts.; to Wash., S. Dak., New Mex.

Var. **gracíllimum** (Cov. & Britt.) Jeps. [*R. g.* Cov. & Britton.] Fls. odorless, soon deep red; fl.-tube more slender; sepals 3–4 mm. long; berry orange to yellowish.—Brushy often alluvial places, S. Oak Wd.; w. Riverside Co. to cent. Calif. Feb.–Apr.

3. **R. califórnicum** H. & A. var. **hespérium** (McClat.) Jeps. [*R. h.* McClat.] Compact, 6–14 dm. high, intricately branched, the young twigs usually without bristles, glabrous; nodal spines stoutish, usually 3; lvs. roundish, thin, pubescent, 1–3 cm. wide, 3–5-cleft, nonglandular; fls. mostly 1, greenish, dull white or purplish, the sepals reddish, mostly strigose, ca. 6–7 mm. long; petals white, ca. as long as fils.; berry ca. 9–10 mm. in diam., bristly.—Canyons below 2000 ft.; Chaparral, etc.; Santa Ana and San Gabriel mts. to Santa Barbara Co. Jan.–March.

4. **R. cantharifórme** Wiggins. Erect open shrub 1–2 m. tall, the young growth finely pubescent and stipitate-glandular; lvs. 4–6 cm. wide, bright green, cordate at base, 3-lobed, villous and slightly rugose above, more densely villous beneath; petioles 2.5–3.5 cm. long; racemes many-fld., 3–6 cm. long, the fls. rose-purple; fl.-tube villous-pubescent and with some gland-tipped hairs, broadly urceolate, ca. 2 mm. long, much broader; sepals spreading, 1.5 mm. long; petals less than 1 mm.; berry round-ovoid, 5–6 mm. in diam., dark, sparsely white-villous and with some stalked glands.—Chaparral in vicinity of Moreno Dam, San Diego Co. Feb.–April.

5. **R. cèreum** Dougl. SQUAW CURRANT. Compact, erect, much branched, 1–12 dm. high, fragrant, with glandular-pubescent young growth; lvs. round-reniform, 1–4 cm.

Plate 82. Fig. A, *Ribes aureum*, inferior ovary. Fig. B, *Ribes indecorum*. Fig. C, *Ribes malvaceum*. Fig. D, *Saxifraga californica*, broad fl.-tube and petals. Fig. E, *Saxifraga odontoloma*. Fig. F, *Antirrhinum coulterianum*, the lower corolla-lip a palate.

wide, puberulent and glandular or upper surface subglabrous and ± shining, 3–5-lobed, obtuse, crenulate; petioles 5–15 mm. long; racemes mostly 3–7-fld., glandular-puberulent; bracts toothed at subtruncate apex; fl.-tube tubular, 6–8 mm. long, greenish-white to pink; sepals round-ovate, 1–1.5 mm. long; petals minute, rounded; berry red, slightly glandular-hairy, ca. 6 mm. in diam.; $n=8$ (Tischler, 1927).—Dry rocky places, 5000–11,500 ft.; Pinyon-Juniper Wd. to Alpine Fell-fields; Santa Rosa and San Jacinto mts. n., mts. of e. Mojave Desert; Sierra Nevada to B.C., Rocky Mts.

6. **R. divaricàtum** Dougl. var. **paríshii** (Heller) Jeps. [*R. p.* Heller.] Widely branched, deciduous, 1–3 m. high; spines 1–3 at a node, young growth pubescent and often also bristly; lvs. thin, 2–5 cm. broad, 3–5-lobed, densely pubescent beneath; petioles 1–3 cm. long; racemes 2–5-fld., pubescent; ovary glabrous; free part of fl.-tube 4 mm. long, purple-red, pubescent; sepals 6–8 mm. long, oblong, greenish to purplish, spreading-reflexed; petals red, ca. 2 mm. long; berry round, smooth, dark.—Willow-thickets, swamps, etc.; Coastal Sage Scrub; San Bernardino region, Los Angeles Co. March–April.

7. **R. indécorum** Eastw. WHITE-FLOWERED CURRANT. Plate 82, Fig. B. Erect open shrub, fragrant, largely deciduous, 1.5–2.5 m. tall, new growth densely pubescent and glandular; lvs. 3–5-lobed, 1–4 cm. wide, thickish, dark green, stipitate-glandular and ± rugose above, whitish-tomentose beneath, the lobes rounded, crenulate, obtuse; petioles 5–25 mm. long; racemes 2–5 cm. long, rather densely fld., glandular-pubescent; pedicels 1–2 mm. long; fl.-tube cylindrical-urceolate, whitish, 4–5 mm. long; sepals white, obtuse, recurved, 1.5 mm. long; petals ca. 1 mm. long; berry viscid-pubescent and with some stipitate glands, globose, 6–7 mm. in diam.—Interior washes and canyons, usually below 2000 ft.; Chaparral, Coastal Sage Scrub; n. L. Calif. to Santa Barbara Co. Nov.–March.

8. **R. lasiánthum** Greene. [*Grossularia l.* Cov. & Britt.] Low spreading shrub to 1 m. high; spines 1–3 at each node; young shoots puberulous; lvs. roundish, 1–2 cm. wide, mostly glandular-puberulent on both surfaces, cleft into 3–5 blunt, toothed lobes; petioles pubescent, 3–6 mm. long; fls. lemon-yellow, 2–4 in a raceme; bracts broad, pubescent; fl.-tube 5–6 mm. long, cylindrical and pubescent in free portion; sepals early reflexed, then erect, ca. 2 mm. long; petals shorter; berry round, smooth, ca. 6–7 mm. in diam.—Rocky places, 6500–8000 ft.; Montane Coniferous F.; San Gabriel Mts.; Sierra Nevada. June–July.

9. **R. malvàceum** Sm. var. **malvàceum**. [*R. m.* var. *clementinum* Dunkle.] CHAPARRAL CURRANT. Plate 82, Fig. C. Erect, deciduous, 1–2 m. high, the young growth tomentose and with gland-tipped bristly hairs; lvs. mostly roundish in outline, rather thick, rugose, 2–6 cm. wide, dull green and rough above with stalked glands, glandular and gray pubescent beneath, 3–5-lobed, the lobes obtuse and doubly toothed; petioles 1–5 cm. long; racemes drooping, exceeding lvs., 10–25-fld., pubescent and glandular-hairy; bracts 6–9 mm. long; fls. bright rose, the fl.-tube 5–8 mm. long; sepals spreading, obovate, 3–4 mm. long; petals erect, roundish, shorter; berry purple-black, with a bloom, subglobose, ca. 6 mm. in diam., ± hairy and glandular.—Dry wooded or open hills, below 2500 ft.; Chaparral, Foothill Wd., Closed-Cone Pine F.; w. Los Angeles Co. n., Santa Cruz Id.; to n. Calif.

Var. **viridifòlium** Abrams. Lvs. greener beneath and ± scabrous with a coarser pubescence, more glandular; fl.-tube 8–12 mm. long.—Dry gullies and canyons below 5000 ft.; Chaparral; Santa Monica Mts. to San Jacinto and Santa Ana mts.

10. **R. menzièsii** Pursh. Loosely branched shrub 1–2 m. high, the young twigs densely bristly and pubescent; nodal spines mostly 3; lvs. rather firm, ± rounded in outline, cordate or truncate at base, 1.5–4 cm. long, subglabrous to roughish with glandular hairs above, velvety-pubescent and with gland-tipped hairs beneath, 3–5-lobed; fls. purplish, 1–3; pedicels glandular, 3–6 mm. long; ovary glandular-bristly and pubescent; free part of fl.-tube 2–3 mm. long, ca. as broad, ± glandular-hairy; sepals reflexed, oblong, 7–11 mm. long; petals whitish, ca. half as long as fils.; anthers apiculate; berry globular, ca. 1 cm. in diam., densely clothed with gland-tipped and nonglandular bristles. Canyons and flats below 1000 ft.; Santa Cruz Id.; San Luis Obispo Co. to Ore. March–April.

11. **R. montígenum** McClat. Straggling many-branched shrub 3–6 dm. tall, the twigs mostly bristly-prickly, glandular-pubescent; nodal spines 3–5; lvs. 5–25 mm. wide, 5-cleft almost to base, the lobes incised-serrate, glandular-pubescent on both sides; racemes few-fld.; pedicels 2–5 mm. long; fl.-tube saucer-shaped above the ovary, glandular-bristly; sepals 3–4 mm. long; petals purplish; berries red, ca. 5 mm. thick, glandular-bristly.—Dry rocky places, 7000–11,500 ft.; Montane Coniferous F., Subalpine F.; San Jacinto Mt. to Mt. Pinos, Panamint Mts.; to interior B.C. and Rocky Mts. June–July.

12. **R. nevadénse** Kell. SIERRA CURRANT. Slender-stemmed open deciduous shrub 1–2 m. tall, glabrous to puberulent on young growth; lvs. roundish in outline, rather thin, 3–7 cm. wide, 3–5-lobed, the lobes obtuse, bluntly toothed, resinous-dotted, glabrous above, ± pubescent and paler beneath; petioles 1–4 cm. long; racemes 8–12-fld., spreading or drooping; bracts 4–8 mm. long; pedicels 3–5 mm. long; fls. rose to rather deep red, the entire tube ca. 5 mm. long, the broad free part above the ovary 2 mm. long; sepals erect, pink to reddish, 4–5 mm. long; petals white, shorter; berry round, blue-black, with a bloom, ± glandular, ca. 8 mm. in diam.—Moist places and along streams, 4500–7500 ft.; Montane Coniferous F.; Palomar Mts., San Jacinto, San Bernardino and San Gabriel mts.; to Ore. A form from San Jacinto Mts., with glabrous lvs. has been called var. *jaegeri* Berger.

13. **R. quercetòrum** Greene. Rounded shrub 6–15 dm. tall, with arcuate-spreading branches, the young twigs puberulent, without bristles; nodal spines usually 1; lvs. roundish, glandular-puberulent, light green, 1–2 cm. wide, 3–5-cleft into toothed lobes; petioles ca. as long as blades; peduncles 2–3-fld.; fl.-tube 6–7 mm. long, the ovary subglabrous, the free portion tubular, puberulent, yellow; sepals yellow, ca. 3 mm. long; petals white, ca. 1 mm. long; berry glabrous, black.—Dry slopes, below 5000 ft.; S. Oak Wd., Foothill Wd.; n. L. Calif. and w. edge of Colo. Desert to Riverside region, Liebre Mts., Tehachapi and Piute mts.; to Tuolumne Co. March–May.

14. **R. roézlii** Regel. SIERRA GOOSEBERRY. Stout, 5–12 dm. high, with spreading branches; young twigs pubescent but not bristly; nodal spines 1–3, straight; lvs. roundish, 1.2–2.5 cm. wide, paler and short-pubescent beneath, 3–5-cleft into toothed lobes; petioles 6–18 mm. long; fls. 1–2; ovary white-hairy and bristly; free portion of fl.-tube ca. 6 mm. long, purplish, pubescent; sepals dull purplish-red, 7–10 mm. long, pubescent, lanceolate; petals whitish, 3–5 mm. long; fls. ca. as long as petals, anthers apiculate; berry purple or lighter, 14–16 mm. in diam., pubescent and with long pubescent spines.—Dry open slopes mostly 3500–8000 ft.; Montane Coniferous F.; mts. of San Diego Co. to Mt. Pinos and Tehachapi Mts.; to n. Calif. May–June.

15. **R. speciòsum** Pursh. FUCHSIA-FLOWERED GOOSEBERRY. Evergreen, 1–2 m. tall, with long simple spreading bristly branches; nodal spines 3, stout, 1–2 cm. long; lvs. roundish, 1–3.5 cm. wide, shining and dark green above, paler beneath, subglabrous to somewhat glandular-pilose, somewhat 3-lobed; peduncles drooping, 1–4-fld.; ovary glandular-bristly; free fl.-tube red, broadly campanulate, 2–3 mm. long, glandular-hairy; sepals 4, red, ligulate, glandular, erect, ca. 1 cm. long; petals ca. as long; stamens 2–3 times as long as sepals; berry glandular-bristly, ovoid, ca. 1 cm. long.—Common in shaded canyons below 1500 ft.; Coastal Sage Scrub, Chaparral, coastal and inland to San Dimas, etc.; n. L. Calif. and San Diego; to Santa Clara Co. Jan.–May.

16. **R. thacheriànum** (Jeps.) Munz. [*R. menziesii* var. *t.* Jeps.] Shrub 1–2.5 m. high, the young growth densely soft-pubescent, not glandular, sometimes somewhat bristly; nodal spines 0–3; lvs. mostly 2–3 cm. wide, greenish and subglabrous above, paler and soft-pubescent beneath, rather shallowly 5-lobed, the lobes coarsely crenate-dentate; petioles 1–2 cm. long; peduncles white-pilose, drooping, 2–4 cm. long, 1–2-fld.; ovary pilose; free fl.-tube subcylindric, 4–5 mm. long, white-pilose; sepals 9–10 mm. long, pinkish, not reflexed, soft-pilose; stamens and petals shorter than sepals; berry dark, ca. 7 mm. in diam., pilose-hairy and with nonglandular spines.—Ravines and stream-bottoms, Santa Cruz Id. March–April.

17. **R. velutìnum** Greene [*Grossularia v.* Cov. & Britt.] Stout, rigidly branched, to 2 m. tall, the recurved branches soft-pubescent, without bristles or glands; nodal spines 1–3,

stout; lvs. roundish, 1-2 cm. diam., deeply 5-cleft, soft-pubescent; fl.-tube cylindric; sepals ca. 3 mm. long; petals 2-2.5 mm. long; berry dark, 4-6 mm. in diam., velvety-pubescent but not usually glandular.—Dry slopes, 5000-9000 ft.; Sagebrush Scrub, Pinyon-Juniper Wd., Yellow Pine F., etc.; San Gabriel Mts. and mts. of e. San Bernardino Co. to Mt. Pinos and White Mts.; to Ore., Utah, Ariz. Plants more glandular-puberulent and with berries ± bristly are the var. *glanduliferum* (Heller) Jeps.

18. **R. viburnifòlium** Gray. Straggling, evergreen, to ca. 1 m. high, the branches tending to spread ± horizontally, young growth resinous-glandular; lvs. coriaceous, oval or ± obovate, dark green and glabrous above, pale and resinous-dotted beneath, not lobed, rounded-obtuse, 1.5-3.5 cm. long, on petioles 3-10 mm. long; racemes lax, pubescent, few-fld.; pedicels 2-5 mm. long; fl.-tube turbinate, pubescent; sepals spreading, purplish-brown, 2.5 mm. long; petals minute, greenish; berry red, ca. 6 mm. in diam., glabrous; n = 8 (Hamel, 1953).—Among shrubs and in canyons, largely Chaparral; Santa Catalina Id.; All Saints Bay, L. Calif. Feb.–April.

10. *Saxifraga* L. SAXIFRAGE

Herbs, largely perennial. Lvs. alternate or opposite or basal, entire to pinnatifid. Fls. perfect, solitary to cymose-paniculate. Fl.-tube ± developed, adnate to at least the base of the ovary. Sepals 5. Petals 5, deciduous, entire. Stamens 10, the fils. subulate to subpetaloid. Ovary from almost superior to partly inferior, 2-loculed, dehiscent between the beaks, sometimes of 2 almost separate follicles. Seeds many, small, smooth to roughish. Ca. 250 spp., mostly of cooler parts of N. Temp. Zone. (Latin, *saxum*, rock, and *frango*, to break; many spp. rooting in crevices.)

Lvs. mostly more than twice as long as wide; petals strap-shaped, obovate to elliptic. Plants from below 5000 ft. .... 1. *S. califórnica*
Lvs. roundish; petals rounded. Plants from 7000-8000 ft. .... 2. *S. odontoloma*

1. **S. califórnica** Greene. Plate 82, Fig. D. Caudex erect, short; lvs. basal, ovate to oblong-elliptic, denticulate to serrate, 1-5 cm. long, on somewhat shorter petioles; flowering stems 1-3 dm. tall, ± glandular-pubescent, loosely branched above; fl.-tube broad, shallow; sepals ovate, 1.5-2 mm. long, soon reflexed, often purplish; petals obovate to broadly elliptic, 3.5-5 mm. long; caps. 2.5-3.5 mm. long.—Common on shaded often grassy banks, mostly below 4000 ft.; Coastal Sage Scrub, Chaparral, etc.; cismontane from s. San Diego Co. n.; to Ore. Santa Cruz Id. Feb.–June.

2. **S. odontolòma** Piper. [*S. arguta* auth., not D. Don.] Plate 82, Fig. E. Perennial from a horizontal rhizome; lvs. basal, orbicular to orbicular-reniform, 1.5-7 cm. wide, glabrous, coarsely dentate, on petioles 2-20 cm. long; scapes 2-4 dm. high, slender, glandular-pubescent upward; infl. open, with spreading rather few-fld. cymules; fl.-tube scarcely evident; sepals lance-oblong, 1.5-2 mm. long; petals white with 2 yellow dots, rounded, 2.5-5 mm. long; fils. broadened upward.—Moist stream-banks, 7000-8200 ft.; Montane Coniferous F.; San Bernardino Mts.; Sierra Nevada to Wash., Rocky Mts. July–Aug.

99. **Scrophulàriaceae.** FIGWORT FAMILY

Herbs, occasionally shrubs or trees, with rounded stems. Lvs. simple, alternate to rarely whorled, mostly entire, sometimes parted or pinnatifid, estipulate. Fls. perfect, racemose or paniculate. Calyx usually 2-lipped, sometimes nearly regular, with 4-5 lobes, sometimes these grown into a leaflike structure split on one side. Corolla 4-5-lobed, usually bilabiate, sometimes almost regular. Stamens 5, sometimes 4 and didynamous, or only 2. Carpels 2, united into a superior 2-loculed ovary with styles usually united. Fr. a 2-celled, mostly 2-valved caps. Seeds few to many, wingless or winged, with fleshy endosperm. A family of ca. 200 genera and 3000 spp., widely distributed.

1. Fertile stamens 5; lvs. alternate; corolla nearly regular, rotate. Tall introd. weedy plants
19. *Verbascum*

1. Fertile stamens 4 or 2; lvs. opposite or alternate; corolla usually ± labiate.
    2. Plants acaulescent; corolla nearly rotate, 1.5 mm. long; anther-cells wholly confluent
    .... 9. *Limosella*
    2. Plants mostly caulescent; corolla usually not rotate, larger; anther-cells distinct.
        3. Stigmas distinct, flattened or platelike.
            4. Corolla campanulate, not distinctly 2-lipped, 40–50 mm. long, mostly purple; lvs. oblong-lanceolate .... 5. *Digitalis*
            4. Corolla nearly 2-lipped.
                5. Sepals nearly or quite distinct; fls. blue-violet .... 18. *Stemodia*
                5. Calyx of united sepals, usually angled; corolla usually purple or red to yellow or buff
                .... 12. *Mimulus*
        3. Stigmas wholly united, capitate or punctiform.
            6. Corolla spurred or saccate on lower side of base, usually also with prominent palate; caps. opening by pores or irregularly; lvs. usually alternate.
                7. Fertile stamens 2; corolla 15–30 mm. long. Desert annuals.......... 13. *Mohavea*
                7. Fertile stamens 4; corolla smaller if from desert, and plants annual.
                    8. Lvs. mostly in 3's; plant shrubby, with scarlet corollas 20–25 mm. long. Insular
                    .... 6. *Galvezia*
                    8. Lvs. alternate; plants not both shrubby and insular and with scarlet corollas.
                        9. Corolla gibbous or saccate at base.
                            10. Lvs. triangular or hastate or 5-lobed, or round-ovate and irregularly dentate with long bristly teeth. Deserts .... 11. *Maurandya*
                            10. Lvs. entire, ovate to lanceolate or linear. Cismontane and desert
                            .... 1. *Antirrhinum*
                        9. Corolla with narrow spur at base.
                            11. Fls. in terminal spikes or racemes; stems ± erect at least in upper part; lvs. linear to lanceolate or ovate .... 10. *Linaria*
                            11. Fls. in axils of broad lvs.; stems trailing or twining; fls. white or yellowish-white .... 8. *Kicksia*
            6. Corolla not spurred or saccate on lower side at base, 2-lipped or nearly regular, without prominent palate; caps. valvate.
                12. Corolla with upper lip flattened or widely arched, not forming a galea.
                    13. Upper lip of corolla 2-lobed.
                        14. Plants annual; corolla gibbous on upper side at base, the lower lip appearing 2-lobed, the middle lobe keel-shaped and concealed by the lateral lobes
                        .... 3. *Collinsia*
                        14. Plants perennial; corolla not gibbous on upper side.
                            15. Corolla ± brownish or darker, mostly less than 1 cm. long, inflated, with 4 erect lobes and 1 reflexed; sterile stamens reduced to a scale or wanting
                            .... 17. *Scrophularia*
                            15. Corolla mostly blue, lavender, rose, white, rarely brownish or yellow, 10–40 mm. long, tubular or funnelform, ± bilabiate; sterile stamen usually ca. as long as the fertile ones.
                                16. Nectary a hypogynous disc; plant a definite shrub; infl. cymose
                                .... 7. *Keckiella*
                                16. Nectary of glandular epistaminal hairs; plant mostly ± herbaceous; infl. racemose to cymose .... 16. *Penstemon*
                    13. Upper lip of corolla appearing 1-lobed by fusion of 2 fls.; small, mostly bluish
                    .... 20. *Veronica*
                12. Corolla with upper lip narrowly arched, forming a galea which encloses the anthers.
                    17. Anther-cells equal in size and position; seed-coat not loose and reticulate; perennials .... 15. *Pedicularis*
                    17. Anther-cells unequally placed, the upper attached by its middle, the lower normally attached or abortive; seed-coat mostly loose and reticulate.
                        18. Calyx-tube surrounding the lower portion of the corolla, its 4 (or 2 by fusion) lobes placed laterally.
                            19. Lower lip of corolla ca. as long as and larger than galea; plants annual
                            .... 14. *Orthocarpus*
                            19. Lower lip of corolla definitely shorter and smaller than galea; plants mostly perennial .... 2. *Castilleja*

18. Calyx-tube surrounding only base of corolla, or mostly entirely to the dorsal side and consisting of a bractlike structure, entire or minutely bidentate at apex
 4. *Cordylanthus*

1. Antirrhinum L. SNAPDRAGON

Erect or diffuse annual or perennial herbs. Lvs. alternate, or the lower opposite or whorled, entire. Fls. axillary to foliage lvs. or in terminal racemes. Calyx 5-parted. Corolla 2-lipped, gibbous or saccate at base; palate usually closing the throat. Fertile stamens 4, didynamous, the fils. often dilated toward apex. Caps. opening by 2 or 3 pores below base of style or bursting somewhat irregularly. Perhaps 40 spp., mostly in Medit. region and in sw. U.S. (Greek, *anti*, like, and *rhinon*, nose, because of snoutlike fls.)

Caps. ± oblique, dehiscing by fairly definite terminal or subterminal pores; stems self-supporting or supporting themselves by tortile branchlets.
 Corolla 3–5 cm. long; short-lived perennial escaping from gardens . . . . 5. *A. majus*
 Corolla mostly not over 2 cm. long; usually annual, native.
  Stems self-supporting, lacking filiform tortile branchlets; plant densely glandular-hirsute; lvs. lanceolate . . . . 6. *A. multiflorum*
  Stems in mature plants largely supported by tortile branchlets, or at least possessing them.
   Plant simple below, erect, glabrous below the glandular-villous minutely bracted spicate raceme; fls. whitish with the lower lip much enlarged . . . . 1. *A. coulterianum*
   Plant usually branched below and ± pubescent along the stem; infl. lax or fairly dense but not set off sharply by its pubescence and leaflessness from the upper stem.
    Palate and corolla-tube with 2 bands of hairs, the tips of which are conspicuously enlarged; pedicels 5–20 mm. long, exceeding calyx; fls. violet; corolla-tube merely gibbous at base, ca. as long as lower lip . . . . 7. *A. nuttallianum*
    Palate and corolla-tube minutely and uniformly puberulent or glandular-puberulent; pedicels mostly shorter than calyx; fls. whitish . . . . 4. *A. kingii*
Caps. not oblique, dehiscing by irregular bursting; stems twining and supported by the long capillary pedicels.
 Fls. yellow, 11–13 mm. long; stems very slender. Deserts . . . . 2. *A. filipes*
 Fls. blue, 13–15 mm. long; stems fairly stout at base. Cismontane . . . . 3. *A. kelloggii*

1. **A. coulteriànum** Benth. in DC. Plate 82, Fig. F. Erect annual 3–12 dm. high, glabrous except in infl., with rather coarse main stem, simple below and with many simple tortile branchlets above; lvs. lance-ovate to ovate, the lower opposite, with blades 1–3 cm. long and petioles 1–2 cm. long, the main cauline lanceolate, 2–9 cm. long, subsessile or short-petioled; raceme dense, subsecund, glandular-villous, with linear to lance-linear green bracts; pedicels 2–3 mm. long; calyx glandular-villous, the segms. subequal, 3–4 mm. long; corolla white, 10–14 mm. long, pubescent without, the spur saccate, broad, 1 mm. long, the palate yellowish; caps. 6–8 mm. high, glandular-pubescent.—Common in dry ± disturbed places like burns, below 5000 ft.; Coastal Sage Scrub, Chaparral, n. L. Calif. to Santa Barbara Co. and inland to Joshua Tree Nat. Monument. Plants with bluish fls. 8–9 mm. long, largely from San Diego Co. have been called the forma *orcuttianum* (Gray) Munz.

2. **A. fílipes** Gray [*Asarina f.* Penn. *Antirrhinum cooperi* Gray.] Plate 83, Fig. A. Filiform-stemmed, 3–8 dm. high, bright green, branched annual with twining stems and pedicels; lowermost lvs. ovate, the upper narrower; pedicels capillary, 3–8 cm. long; calyx obscurely glandular-puberulent, the lanceolate segms. subequal, ca. 4 mm. long; corolla bright yellow, 11–13 mm. long, the lips 5–6 mm. long, the hairy palate with dark spots; stamens 6–8 mm. long.—Twining among low shrubs in sandy places, below 5000 ft.; Creosote Bush Scrub, Joshua Tree Wd.; Calif. deserts n. to Inyo Co.; to Utah. Feb.– May.

3. **A. kellóggii** Greene. [*A. strictum* (H. & A.) Gray not Sibth. & Sm. *A. hookerianum* Penn.] Annual, glabrous except for the slight woolliness at very base, 3–10 dm. high, strict or branched, upper parts usually vinelike and climbing by pedicels; lower lvs. crowded, ovate, obtuse, 5–20 mm. long, petioled, the upper narrower, sessile, longer; pedicels 4–8 cm. long, solitary in upper axils; calyx-segms. 5–6 mm. long, lance-linear;

Plate 83. Fig. A, *Antirrhinum filipes*. Fig. B, *Castilleja affinis*, tubular calyx, elongate galea. Fig. C, *Collinsia torreyi* var. *wrightii*, lower corolla-lip with keel-shaped middle lobe. Fig. D, *Cordylanthus nevinii*, corolla 2-lipped. Fig. E, *Keckiella antirrhinoides*, bearded staminode, anther horseshoe-shaped. Fig. F, *Mimulus cardinalis*, angled calyx.

corolla blue, 13–15 mm. long, with reticulated pubescent palate; caps. globose. Dry slopes, especially burns, below 2000 ft.; Chaparral; L. Calif. to Marin Co. March–May.

4. **A. kíngii** Wats. Annual, 1–4 dm. high, erect, simple or branched at base, glabrous except for woolly base and glandular-puberulent infl.; lvs. lanceolate to linear, 5–30 mm. long, sessile to petioled; fls. short-pedicelled, beginning at lowest axils, the upper in a secund raceme; calyx 4 mm. long in fl., the upper segms. 5–7 mm. in fr.; corolla white, with purple veins, 7–8 mm. long.—Dry gravelly places, 5000–8000 ft.; Pinyon-Juniper Wd.; Clark Mts., e. San Bernardino Co., Inyo-White Mts.; to Utah. May–June.

5. **A. màjus** L. COMMON SNAPDRAGON. Much-branched, 4–8 dm. high, glandular-pubescent above; lvs. lanceolate; fls. of many colors; calyx-lobes broadly ovate, 3–5 mm. long; corolla 3–5 cm. long; caps. ca. 1 cm. long; $n=8$ (Propach, 1935).—Occasional as escape from cult. about dumps and waste places; native of Medit. region. Most of year.

6. **A. multiflòrum** Penn. [*A. glandulosum* Lindl., not Lejeune.] Stout widely branched annual or short-lived perennial, viscid, glandular-hirsute throughout, 6–15 dm. tall; branchlets spreading, non-tortile; lvs. many, entire, lanceolate, sessile, 1–6 cm. long, gradually passing upward into leafy bracts; infl. subsecund, 0.5–5 dm. long; pedicels appressed, 5–7 mm. long; calyx oblique, herbaceous, the upper segm. 10–13 mm. long, the others shorter; corolla rose-red with white or cream palate, 17–19 mm. long, saccate at base, the upper lip reflexed, 6–7 mm. long, the lower erect; caps. 8–9 mm. long; $2n=16$ pairs (Raven et al., 1965).—Dry slopes below 4000 ft.; Chaparral, Closed-cone Pine F.; s. face of San Bernardino Mts. to the coast and n. to cent. Calif. May–July.

7. **A. nuttalliànum** Benth. in DC. Annual or biennial, simple and erect or commonly diffusely branched, leafy, softly viscid, glandular-pubescent, 1–10 dm. high, with tortile horizontal branchlets; lvs. ovate, 0.5–4 cm. long; raceme lax, leafy-bracted; pedicels 5–20 mm. long; calyx 3–5 mm. long; corolla violet, 10–12 mm. long, the basal pouch pale, the palate large, with violet reticulations; $2n=16$ pairs (Raven et al., 1965).—Common in dry, especially disturbed places, below 2500 ft.; Coastal Sage Scrub, Chaparral; L. Calif. to Santa Barbara Co., San Clemente, Catalina, Santa Rosa, Santa Cruz ids. A small form with cleistogamous fls. from Point Loma, San Diego Co. and islands off the L. Calif. coast has been called forma *pusillum* (Bdg.) Munz.

2. *Castillèja* Mutis. PAINT-BRUSH

Herbs or suffrutescent plants, partially root-parasites. Lvs. alternate, sessile, entire or laciniate, passing above into usually more incised and colored conspicuous bracts of the terminal spikelike raceme. Bracteoles none. Calyx tubular, 4-lobed or seemingly 2-lobed by fusion of those on each side. Corolla tubular, ± compressed and often greenish, 2-lipped, the upper lip (galea) elongate, entire, enclosing the style and stamens; lower lip shorter, often rudimentary, 3-toothed, 3-carinate or -saccate below the teeth. Stamens 4, didynamous, all with unequal 2-celled anthers. Caps. many-seeded, glabrous, loculicidal, ovoid or cylindric-ovoid. Seeds with a loose reticulate coat. Ca. 200 spp. of the New World, with 1 going into Asia. (D. *Castilleja*, A Spanish botanist.)

(Holmgren, N. H. A taxonomic revision of the Castilleja viscidula group Mem. N.Y. Bot. Gard. 7 (4): 1–63. 1971.)
1. Plants annual; lvs. and bracts entire, linear-lanceolate. Wet places.
    2. Corolla scarcely exserted, 2.5–3 cm. long; galea ca. as long as corolla-tube. Cismontane and desert edge .... 18. *C. stenantha*
    2. Corolla well exserted, 12–20 mm. long; galea ca. half as long as corolla-tube. Rather alkaline places, e. Inyo Co. .... 14. *C. minor*
1. Plants perennial; lvs. often and upper bracts usually divided.
    3. Calyx-lobes united much further dorsally than ventrally and their tips usually curved upward; corolla usually curved forward through lower calyx-sinus.
        4. Floral bracts oblong to linear-oblong, either entire or with 3 relatively short, very blunt lobes toward their tips; foliage pubescent, the lvs. not involute; axillary shoots well developed; upper surface of galea ± densely and shaggily pubescent. Cismontane 10. *C. jepsonii*

4. Floral bracts cuneate in outline, deeply and digitately 3-cleft or -parted, the spreading lobes narrow and acute, green to rose-pink; foliage and stems glabrous and glaucous except in infl. and at very base, the narrowly linear lvs. involute; axillary shoots mostly early developed or lacking; dorsal surface of galea subglabrous to sparsely and finely pubescent. Transmontane .... 11. *C. linariaefolia*
3. Calyx-lobes not united much further dorsally than ventrally, the tips straight.
   5. Lower corolla-lip ca. half to almost as long as galea and with thin often whitish usually not incurved lobes.
      6. The lower corolla-lip thin and whitish, at least $\frac{2}{3}$ as long as galea. From Inyo-White Mts. north .... 16. *C. nana*
      6. The lower corolla-lip green and thickened at apex, ca. half as long as galea. San Bernardino Co. .... 4. *C. cinerea*
   5. Lower corolla-lip less than half as long as galea, usually green or dark and with minute and incurved lobes.
      7. Herbage with simple (not branched hairs), green to grayish-green.
         8. Plant evidently glandular-pubescent below the infl.
            9. Stems mostly less than 2 dm. long; galea 6–10 mm. long, the lower lip mostly ca. half as long. Piute Mts. north. .... 2. *C. breweri*
            9. Stems mostly more than 2 dm. long; galea 12–20 mm. long, the lower lip mostly less than $\frac{1}{5}$ as long.
               10. Calyx only $\frac{1}{2}$ the length of the corolla, 12–17 (–20) mm. long; corolla the most conspicuous part of the infl. coloration; lvs. almost linear. Piute Mts. north
                  5. *C. disticha*
               10. Calyx more than $\frac{1}{2}$ the length of the corolla, 16–25 (–28) mm. long; bracts the most conspicuous part of the infl. coloration. Widely distributed.
                  11. Galea usually the same length as the tube, conspicuous, 11–23 mm. long; adaxial calyx-cleft 5–10 mm. deep.
                     12. Primary calyx-lobes $\frac{1}{5}$–$\frac{2}{5}$ the total length of the calyx, the segms. broadly lanceolate to ovate, mostly obtuse to rounded at tips; red coloration of the infl. striking due to broad bract lobes and conspicuous galea. Coast Ranges and mts. of cent. Kern Co. .... 12. *C. martinii*
                     12. Primary calyx-lobes proportionately longer, $\frac{2}{5}$–$\frac{1}{2}$ the total length of the calyx, the segms. lanceolate to broadly lanceolate, usually acute; red coloration of infl. less conspicuous, being borne on narrower bract-lobes and usually restricted to margin of galea. Panamint and Inyo mts.
                        12. *C. martinii* var. *clokeyi*
                  11. Galea shorter than the tube, $\frac{2}{3}$ to almost as long, 10–12 (–15) mm. long; adaxial calyx cleft shallower, 4–5 (–6) mm. deep. San Bernardino Mts.
                     12. *C. martinii* var. *ewanii*
         8. Plant not evidently glandular-pubescent below the infl.
            13. The plant whitened by an arachnoid-lanose coat of long flexuous hairs; lvs. linear; sepals of each side united to a roundish tip. Channel Ids. .... 9. *C. hololeuca*
            13. The plant not grayish-woolly.
               14. Calyx-lobes mostly less than 2 mm. long, blunt; lvs. rather blunt, lance-oblong. Immediate coast .... 19. *C. wightii*
               14. Calyx-lobes 3–6 mm. long, pointed; lvs. narrower.
                  15. Main cauline lvs. entire.
                     16. Herbage glabrous or inconspicuously pubescent; lvs. mostly 2–5 cm. long
                        13. *C. miniata*
                     16. Herbage stiff-pubescent; lvs. 3–9 cm. long. .... 1. *C. affinis*
                  15. Main cauline lvs. mostly with 1 or 2 pairs of lobes.
                     17. Galea 16–23 mm. long. Cismontane .... 1. *C. affinis*
                     17. Galea 12–15 mm. long. Deserts .... 3. *C. chromosa*
      7. Herbage with branched hairs.
         18. Calyx cleft laterally more deeply than medianly, the upper lip 4–7 mm. long, the lower 6–9 mm. long; corolla 15–20 mm. long; infl. yellow. Desert slopes, San Gabriel and San Bernardino mts. .... 17. *C. plagiotoma*
         18. Calyx cleft medianly more deeply than laterally; corolla usually longer; infl. almost reddish.
            19. Plants ± grayish-tomentose to white-woolly.
               20. Calyx-lobes acute; stems villous-hirsute; lvs. oblong; corolla 17–18 mm. long.

Santa Rosa Id. .... 15. *C. mollis*
20. Calyx-lobes rounded; stems soft-tomentose; lvs. linear or oblong-linear.
21. Corolla ca. 15 mm. long; galea 7 mm. San Clemente Id. ...... 8. *C. grisea*
21. Corolla 18–26 mm. long; galea 7–18 mm. long. Catalina Id. and mainland
6. *C. foliolosa*
19. Plants pubescent, the infl. finely lanulose-villose; lvs. linear-lanceolate. San Gabriel Mts. .... 7. *C. gleasonii*

1. **C. affìnis** H. & A. var. **affìnis**. [*C. douglasii* Benth. *C. californica* Abrams.] Plate 83, Fig. B. Stems rather few from a ± woody base, rather stiff-pubescent, slender, with glandless hairs, 3–5 dm. tall; infl. hirsute; lvs. rather scabrous-pubescent, lanceolate, 3–9 cm. long, entire or with 1–3 pairs of slender lobes; bracts and calyces distally scarlet, the former with 2–3 pairs of lobes; calyx 18–25 mm. long, cleft medianly ca. halfway, laterally into linear-oblong to lance-ovate lobes 3–6 mm. long; corolla 2.5–3.5 cm. long, the galea 16–23 mm. long, finely pubescent dorsally, the lower lip 1.5–2 mm. long, dark green to brownish; caps. ca. 12–13 mm. long.—Dry wooded or brushy slopes; Coastal Sage Scrub, Chaparral, etc.; n. L. Calif. to cent. Calif. March–May.

Var. **contentiòsa** (Macbr.) Bacig. With vitreous cellular trichomes on the lf.-margins. —Santa Monica Mts. to Monterey Co. Anacapa and Santa Rosa ids.

Ssp. **insulàris** (Eastw.) Munz. [*C. latifolia* ssp. *i.* Eastw.] Corolla 1.5–2 cm. long.— Chaparral, Closed-cone Pine F.; Santa Cruz Id.

2. **C. brèweri** Fern. Stems many, 1–2.5 dm. tall, glandular-pubescent and usually with longer non-glandular hairs also; lvs. lanceolate or narrower, entire or the upper broader and with a pair of lobes; bracts and calyces distally red, the former 3-lobed; calyx 13–15 mm. long, medianly cleft more than ⅓ its length, laterally cleft 3–4 mm.; corolla 16–22 mm. long, the galea 7–10 mm. long, puberulent and greenish dorsally with thin red ventral margins.—Dry stony places and about meadows; upper Montane Coniferous F.; Piute Mts., Kern Co. to Mono and Eldorado cos. June–Aug.

3. **C. chromòsa** A. Nels. [*C. angustifolia* Calif. refs.] Herbaceous perennial from woody root-crown, the stems 1–4 dm. tall, simple or few-branched, subcinereous with rather stiff glandless hairs; lvs. lanceolate, 1–4 cm. long, entire or with 1–2 pairs of spreading narrow lobes; bracts and calyx distally scarlet, the former with 2 pairs of lobes; calyx 15–20 mm. long, cleft medianly ca. ⅓ its length, laterally into ovate or oblong lobes 4–5 mm. long; corolla 25–30 mm. long, the galea 12–15 mm. long, sparsely puberulent dorsally, with wide reddish thin margins, the lower lip dark green, 2–3 mm. long, included in calyx; $n=12, 24$ (Heckard, 1968).—Dry brushy slopes, 2000–9000 ft.; mostly Sagebrush Scrub, Shadscale Scrub, Joshua Tree Wd.; deserts from Pinto Mts., Riverside Co. to ne. Calif.; S. Ore. Rocky Mts. April–Aug.

4. **C. cinèrea** Gray. Stems several from the perennial root-crown, densely cinereous-pubescent, 1–2 dm. long; lvs. many, overlapping, lance-linear, cinereous, entire or the upper sometimes with 3 linear lobes; bracts and calyces purple to greenish yellow; calyx 15–20 mm. long, equally cleft into linear lobes; corolla 16–20 mm. long, yellowish, the galea 4–5 mm. long, with narrow purplish thin margins.—Local on dry benches and slopes, 5000–9800 ft.; Montane Coniferous F.; San Bernardino Mts. May–Aug.

5. **C. dísticha** Eastw. Perennial with slender stems mostly 3–7 dm. tall, often branched, sometimes suffrutescent at base, glandular-pubescent and hispid to villous; lvs. 2–5 cm. long, glandular-pubescent and sometimes with longer glandless hairs, lvs. mostly linear, wavy-margined, entire, rarely divided; infl. elongate, the fls. remote and often distichous, the coloration from the corolla except distally where bracts and calyx are also colored; bracts rather inconspicuous, entire, sometimes the upper with 2 lobes, tipped red to red-orange; calyx mostly 13–17 mm. long, subequally cleft for ⅔ to ½ its length into 2 primary lobes, these divided into segms. 1.5–4.5 mm. long; corolla mostly 28–35 mm. long, usually exserted beyond the calyx-tube; galea ± as long as tube, colored; lower lip often darker; caps. 9–12 mm. long; $2n=24$ (Holmgren, 1971).— Montane Coniferous F.; Piute Mts.; Sierra Nevada.

6. **C. foliolòsa** H. & A. Suffrutescent bushy perennial, white-woolly throughout, 3–6

dm. tall; lvs. linear or oblong-linear, 1–2.5 cm. long, entire or upper 3-lobed, obtuse, fairly crowded and with fascicles in lower axils; bracts and calyces distally scarlet, the former with 1–2 pairs of lobes, the latter ± yellow in middle; calyx 15–20 cm. long, cleft medianly ca. $\frac{2}{5}$ its length, laterally barely 0.5 mm. into rounded truncate lobes; corolla 1.8–2.2 cm. long, the galea 7–15 mm. long, minutely pubescent and greenish dorsally, with pale thin reddish margins, the lower lip 2 mm. long; $n = 12$ (Heckard, 1968).—Dry open ± rocky places, below 5000 ft.; Coastal Sage Scrub, Chaparral, etc.; n. L. Calif. to n. Calif. Catalina Id. March–June.

7. **C. gleasònii** Elmer. Perennial, ± grayish with mostly branched hairs, the stems simple or branched, 3–7 dm. tall; lvs. linear-lanceolate, 2–6 cm. long, entire or the upper pair with a pair of lobes; bracts and calyces scarlet, the former with a pair of lobes; calyx-lobes ± ovate, 1–3 mm. long; corolla 2.5–3 cm. long, the galea 15–20 mm. long, yellowish and pubescent dorsally, with wide red thin margins, the lower lip dark green, ca. 2 mm. long, included in calyx.—Rocky places, 5000–7100 ft.; Yellow Pine F. about Mt. Gleason, San Gabriel Mts. May–June.

8. **C. grísea** Dunkle. Perennial, grayish-cinereous throughout, with short arachnoid tomentum; stems woody and much branched, the herbaceous distal portion short, the plants altogether 5–6 dm. high; lvs. linear, obtuse, entire, foliolose in most axils; bracts and calyces green or brownish green, the former with 1 pair of lobes; calyx 13 mm. long, cleft medianly $\frac{2}{3}$ its length, laterally 5 mm. or less; corolla ca. 15 mm. long, the galea 7 mm. long, with pale narrow thin margins, the lower lip 2 mm. long, dark green.—Bluffs, Coastal Sage Scrub; San Clemente Id.

9. **C. hololeùca** Greene. Covered with a dense white-woolly felt, shrubby, 3–6 dm. high, much branched; lvs. linear, dense, 1–3 cm. long, entire, obtuse, mostly with fascicles in axils; bracts and calyces distally red, the former with a pair of lobes; calyx 15–18 mm. long, cleft medianly $\frac{2}{3}$ its length, not at all laterally, the 2 sides rounded at their tips; corolla 2–2.5 cm. long, the yellowish galea 12–13 mm. long, puberulent dorsally, with pale thin margins, the dark green lower lip 2–3 mm. long, included.—Canyon walls and rocky slopes; Coastal Sage Scrub, Chaparral; San Miguel, Anacapa, Santa Cruz, Santa Rosa ids. March–Aug.

10. **C. jepsònii** Bacig. & Heckard. Pilose and puberulose, leafy, gray-green perennial, the usually many strict stems 6–12 dm. tall, striate, with sterile axillary, quite leafy shoots; cauline lvs. mostly linear-oblong, mostly entire, with long and short hairs, the lvs. 3–8 cm. long, 2–6 mm. wide; infl. eventually as long as 4 dm., with some gland-tipped hairs often among the septate pilose ones; upper bracts often tridentate, distally rose-pink; calyx narrow, striate, 2.5–5 cm. long, the galea ca. 2 cm. long, the upper surface deep green, densely short-hairy, the margins narrow, pink, glabrous; lower lip 1–2.5 mm. long, dark red-purple; $n = 12$ (Heckard).—At 1500–7500 ft., se. San Diego Co., Los Angeles Co., Piute Mts. to San Benito Co.; L. Calif.

11. **C. linariaefòlia** Benth. Stems several to many, from a ± woody root-crown, 6–8 or more dm. long, simple or little branched, subglabrous or finely pubescent, ± hirsute in infl.; lvs. linear, mostly 1.5–8 cm. long, entire or with a pair of slender lobes; bracts and calyces distally red, the former with 1–2 pairs of slender lobes; calyx 2.5–3.5 cm. long, moderately cleft more deeply dorsally than ventrally, laterally cleft 5–7 mm. into lanceolate lobes which upcurve altogether; corolla 3–4 cm. long, decurved, the galea ca. as long as the tube, dorsally puberulent, greenish with thin red margins, the lower lip 3 mm. long.—Moist places, 2500–8000 ft.; Joshua Tree Wd., Pinyon-Juniper Wd., Sagebrush Scrub, etc.; s. edge of Mojave Desert n. to Ore., Mont., and to Nev. and New Mex. June–Sept.

12. **C. martínii** Abrams var. **martínii**. [*C. roseana* Eastw. *C. hoffmannii* Eastw. *C. gyroloba* Penn.] Perennial, the stems relatively tall, 2–5 dm. high, often branched, sometimes suffrutescent at base; lvs. 2–4.5 cm. long, lanceolate to ovate, rounded or obtuse, rarely acute; infl. relatively broad; bracts hispid to hirsute-ciliate; calyx 18–22 mm. long, the abaxial cleft 4.5–9 mm. deep, the adaxial cleft 6–10 mm. deep, the primary lobes broadly lanceolate to ovate; corolla 28–38 mm. long; galea $\frac{2}{5}$–$\frac{1}{2}$ the corolla length, red-

dish or orange-red; lower lip often projected forward by a broad throat; $2n=24, 48, 72$ (Holmgren, 1971).—Dry slopes and talus, 1000–8000 ft.; Chaparral, Yellow Pine F.; n. L. Calif. to cent. Kern Co. and Humboldt Co. March–July.

Var. **clòkeyi** (Penn.) N. Holmgren. [*C. c.* Penn.] Stems wholly herbaceous; lvs. 2–4 cm. long; calyx-lobes $\frac{2}{5}-\frac{1}{2}$ the total length of calyx; $2n=48$ (Holmgren, 1971).—Dry places, 7000–10,000 ft.; Pinyon-Juniper Wd.; Panamint and Inyo mts.; to Nev. May–July.

Var. **èwanii** (Eastw.) N. Holmgren. [*C. e.* Eastw.] Galea shorter than corolla-tube, 10–12 mm. long; adaxial cleft of calyx 4–5 mm. deep.—Pinyon-Juniper Wd., Yellow Pine F.; at 6000–9000 ft.; San Bernardino Mts. June–Aug.

13. **C. miniàta** Dougl. ex Hook. Erect herbaceous few-stemmed glabrous to somewhat pubescent perennial; stems 4–8 dm. tall, simple or nearly so; lvs. lanceolate, mostly 2–5 cm. long, acute, entire or the upper occasionally lobed; infl. villous-pubescent; bracts and calyx mostly distally scarlet, at least the upper bracts with a pair of slender lobes; calyx mostly 2–2.7 cm. long, medianly cleft $\frac{1}{2}-\frac{2}{3}$ its length, laterally cleft 3–7 mm. into lance-linear acuminate lobes; corolla 2.5–3.5 cm. long, the galea as long as the tube, with thin red margins; lower lip 1–2 mm. long; $n=12, 24$ (Heckard, 1968).—Along streams and wet places below 11,000 ft.; Montane Coniferous F.; mts. of San Diego Co.; to B.C., Rocky Mts. May–Sept.

14. **C. mìnor** (Gray) Gray. [*C. affinis* var. *m.* Gray.] Annual with lvs. and bracts entire, linear-lanceolate; corolla well exserted, 2.5–3 cm. long; galea as long as corolla-tube; otherwise much like *C. stenantha*.—Subalkaline wet places, e. of Sierra Nevada as at Deep Springs V., Inyo Co. to Rocky Mts.

15. **C. móllis** Penn. Suffrutescent, the stems diffusely branched, to 3 or 4 dm. long, with short axillary leafy shoots; lvs. oblong, mostly entire, 3–7 cm. long, distally rounded, soft-tomentose with branched hairs; axis of infl. villous-hirsute, and with some gland-tipped hairs; bracts pale green or yellowish, obovate, distally rounded or 3-lobed; calyx 16–18 mm. long, cleft medianly almost half its length and laterally into ovate acute lobes ca. 2 mm. long; corolla 17–18 mm. long, the galea 7–8 mm. long, truncate, strongly pubescent dorsally, with wide pale thin margins, the lower lip 2 mm. long, appressed, green.—Sand dunes; Coastal Strand, etc.; Santa Rosa Id.; mainland of San Luis Obispo Co. April–Aug.

16. **C. nàna** Eastw. Low villous-pubescent perennial; lvs. mostly 3–5-lobed; bracts spreading, they and calyces distally dull yellow to dull purplish-red; calyx 15 mm. long, deeply cleft into linear lobes; corolla 13–16 mm. long, the galea ca. 6 mm., dorsally greenish and puberulent, the thin margins white distally, purple proximally.—Dry rocky places, 8000–10,000 ft.; Inyo-White Mts.; Sierra Nevada. July–Aug.

17. **C. plagiótoma** Gray. Stems growing up through low shrubs, sparsely pubescent, 3–6 dm. tall; infl. more pubescent; lvs. linear, 1–3 cm. long, the upper lvs. and bracts divided into 3 linear lobes; bracts green, white-pubescent, the middle segms. broad, rounded; calyx 10–15 mm. long, white-woolly, cleft laterally more deeply than medianly, the upper lip 4–7 mm. long, the lower 6–9 mm., with lobes 2–3 mm. long; corolla 1.5–2 cm. long, the galea ca. as long as tube, straight, yellowish, pubescent dorsally, with wide thin pale margins, the lower lip 1.5 mm. long, included in calyx.—Dry flats and ridges, 2500–7500 ft.; Sagebrush Scrub, Joshua Tree Wd.; n. base of San Bernardino and San Gabriel mts. to Piute Mts. and San Luis Obispo Co.

18. **C. stenántha** Gray. Annual, erect, simple or branched, pubescent with some hairs gland-tipped, the stems slender, 3–12 dm. high; lvs. lanceolate to ascending, entire, lance-linear, 2–8 cm. long; bracts leaf-like, the lower green, the upper red-tipped, attenuate-acute; calyx green, 15–25 mm. long, cleft medianly $\frac{2}{3}$ its length, laterally 1–3 mm. into narrow lobes; corolla 2.5–3.5 cm. long, the galea 15–20 mm. long, dull reddish-yellow dorsally and ± puberulent; lower lip 2–3 mm. long, yellow, well exserted; $n=12$ (Heckard).—In moist places, below 7000 ft.; many Plant Communities; cismontane to the desert edge; to cent. Calif. May–Sept.

19. **C. wìghtii** Elmer. [*C. anacapensis* Dunkle.] Pubescent or pilose perennial with

gland-tipped hairs as well, the stems densely pubescent and villous in the infl., 3–8 dm. long, ± woody at the base; lvs. oblong, entire or the upper 3-lobed, these obtusish to rounded; main lvs. crowded, mostly 2–6 cm. long and with short fascicles in their axils; bracts distally dull red or yellowish, with a pair of rounded lobes; calyx 18–20 mm. long, the ± obtuse lobes 2–3 mm. long; corolla 21–25 mm. long, the galea 13–15 mm., bluntish, dorsally pubescent, the lower lip ca. 2 mm., dark green; $n = 12, 24$ (Heckard, 1968).—Dry slopes and banks, Coastal Strand; near the coast from Anacapa, Santa Rosa and Santa Cruz ids.; to n. Calif. March–July.

3. Collínsia Nutt.

Annuals with simple mostly opposite entire to crenulate lvs., the upper sessile or clasping and passing into ± foliose bracts. Fls. solitary to fascicled, in axils of uppermost lvs. Calyx 5-parted. Corolla 5-lobed, gibbous or saccate at base on upper side, bilabiate, the tube short, the throat well developed; upper lip 2-lobed, the lower 3-lobed, the middle lobe keel-shaped and enclosing the 4 declined stamens and style. Lower pair of fils. inserted higher on the corolla-tube than the upper pair. Caps. dehiscent, the valves 2-cleft. Seeds flattened, convex dorsally, concave ventrally, smooth to reticulate, ± winged. A genus of ca. 17 spp., all except 2 found on the Pacific Coast. (Zaccheus Collins, 1764–1831, Philadelphia botanist.)

(Newsom, V. A revision of the genus Collinsia. Bot. Gaz. 87: 260–301. 1929.)
Fls. congested in whorls, the pedicels mostly shorter than calyces in the lower whorls.
  Upper pair of fils. with distinct basal appendages 1–2 mm. long; upper lip of corolla usually distinctly paler than lower .... 5. *C. heterophylla*
  Upper pair of fils. without or with very rudimentary basal appendages.
    Keel sparsely bearded without; corolla bluish-lavender, the upper lip almost as long as lower and not evidently veined. From San Bernardino Mts. south .... 4. *C. concolor*
    Keel glabrous; corolla paler at least on upper lip, the lobes strongly veined. From Los Angeles Co. north .... 1. *C. bartsiaefolia*
Fls. pedicelled, solitary or in whorls, the lower pedicels as long as or longer than calyces.
  Infl. not glandular-pubescent, or if so, the glands minute and scarcely thicker than the supporting hairs; calyx-lobes exceeding caps.; seeds usually 3 or more to a locule.
    Calyx-lobes obtuse or obtusish; corolla 7–10 mm. long; upper fils. bearded. Santa Monica Mts. to San Bernardino Mts. .... 6. *C. parryi*
    Calyx-lobes acute to attenuate; corolla 4–7 mm. long; fils. glabrous. From San Diego Co. north .... 7. *C. parviflora*
  Infl. glandular-pubescent; calyx-lobes scarcely or not longer than caps.; seeds mostly 1–2 (–3) to a locule.
    Fruiting pedicels ascending or ascending-spreading; upper bracts of infl. at least 2 mm. long.
      Calyx membranous, rounded and ca. 3 mm. wide at base in fr., the lobes exceeding the caps.; seeds 2, ca. 3 mm. long; upper lvs. narrowed to base. From San Diego Co. north
                                                                                3. *C. childii*
      Calyx thickened, subtruncate and swollen, ca. 5 mm. wide at base in fr., the lobes ca. as long as caps.; seeds 6–8, ca. 2 mm. long; upper lvs. rounded to base. From San Bernardino Co. n. .... 2. *C. callosa*
    Fruiting pedicels deflexed-spreading; upper bracts of infl. obsolete, less than 2 mm. long; upper corolla-lip white, the lower blue. Mts. from San Bernardino Co. n. ..... 8. *C. torreyi*

1. **C. bartsiaefòlia** Benth. in DC. var. **davidsònii** (Parish) Newsom. [*C. d.* Parish.] Plants 0.5–2 dm. tall, simple to diffuse; stems canescent-puberulent, not at all glandular-puberulent; lf.-blades oblong-ovate, crenate-serrate, 1.5–4 cm. long, rounded at base; fls. several in a whorl, sessile or nearly so, the bracts shorter than fls.; calyx-tube glabrous to villous, 2 mm. long, the lobes ca. 3 mm. long; corolla rose-purple to almost white, often dark-veined and spotted, 8–20 mm. long; the lower lip with whitish to purple lateral lobes and glabrous purple-dotted keel.—At 2000–4000 ft.; Joshua Tree Wd., etc.; w. Mojave Desert to Greenhorn Mts. and San Benito Co. April–June.

2. **C. callòsa** Parish. Stems relatively stout, glandular-pubescent above, 0.5–2 dm. high; lf.-blade : thickened, glabrous, ± oblong, obtuse, entire, 1–3 cm. long, sessile or

nearly so; infl. lax, the bracts linear, pedicels 5–15 mm. long; calyx conspicuously broad in fr., 4–6 mm. long, the lobes half as long, lance-ovate; corolla 7–9 mm. long, rose-lavender, the upper lip purple-dotted on the whitish base, ca. as long as lower; caps. 4–6 mm. long.—Dry places, 3000–7500 ft.; Chaparral, Sagebrush Scrub, Yellow Pine F., Pinyon-Juniper Wd.; San Bernardino and San Gabriel mts. to Tehachapi, Greenhorn, Panamint mts. and e. slope of Sierra Nevada in Inyo Co. April–June.

3. **C. childii** Parry ex Gray. Erect, puberulent below, glandular above, 1–4 dm. tall; lf.-blades oblong-lanceolate, subentire to serrulate, 1–4 cm. long, short-petioled; infl. lax, the lower bracts foliose, the upper reduced; pedicels 5–25 mm. long; calyx 5–7 mm. long, glandular-puberulent, the lobes lanceolate, 3–4 mm. long; corolla 6–8 mm. long, the lips subequal, pale violet to whitish, the lobes rather narrow; caps. 3–4 mm. long; $n=7$ (Garber, 1956).—Dry shaded places, 3000–7000 ft.; Yellow Pine F., S. Oak Wd.; mts. from San Diego Co. to cent. Calif. April–June.

4. **C. concólor** Greene. Erect, occasionally diffusely branched, puberulent, sometimes minutely glandular in infl., 1.5–4.5 dm. tall; lf.-blades thin, lance-oblong, obtuse, crenate-serrulate to entire, 2–5 cm. long, with rounded to subsessile bases or the lowermost petioled; fls. sessile or on pedicels to 4 mm. long; bracts 5–10 mm. long; calyx-tube villous, 2–3 mm. long, the lobes minutely pubescent, 3–4 mm. long, oblong; corolla declined, 10–14 mm. long, bluish-lavender, the upper lip 6–10 mm. long, the lower slightly larger, keel sparsely bearded without; caps. 4 mm. long; $n=7$ (Garber, 1956).—Shade of bushes, etc. below 5500 ft.; Chaparral, Yellow Pine F.; interior s. Calif. from San Bernardino Co. to n. L. Calif. Apr.–June.

5. **C. heterophýlla** Buist ex Grah. var. **heterophylla.** [*C. bicolor* Benth. not Raf.] CHINESE HOUSES. INNOCENCE. Stem simple or diffusely branched, subglabrous to pubescent, sometimes glandular above, 2–5 dm. high; lf.-blades lanceolate to lance-oblong, subentire to serrulate, 1–7 cm. long, ± obtuse; fls. 2–7 in sessile or short-pediceled whorls; bracts 5–20 mm. long; calyx green to red-purple, the tube ca. 2 mm. long, the lanceolate pubescent lobes 5–6 mm., ± acute; corolla mostly 15–20 mm. long, pubescent, usually the white to lilac upper lip paler than the violet or rose-purple lower lip, keel glabrous; upper fils. with appendage 2 mm. long projecting into the nectar-pouch; caps. 5 mm. long; $n=7$ (Sugiura, 1940).—Common in ± shaded places, below 2500 ft.; many Plant Communities; cismontane from L. Calif. to n. Calif. March–June.

Var. **austromontàna** (Newsom) Munz. [*C. bicolor* var. *a.* Newsom.] Stems and under side of lvs. pubescent; upper lip of corolla mostly ca. half as long as lower.—Dry slopes, mostly 2000–5000 ft.; Chaparral, Yellow Pine F.; San Gabriel and San Bernardino mts. May–July.

6. **C. párryi** Gray. Stems simple to diffusely branched, minutely puberulent, 1–5 dm. tall; lf.-blades lanceolate, obtusish, entire to inconspicuously serrulate, glabrous, 1.5–4 cm. long, with rounded to subsessile bases; lowermost with petioles to 2 cm. long; infl. lax, the foliose bracts subtending 1–2 fls., the pedicels 1–4 cm. long; calyx puberulent, 4–7 mm. long, the lobes broadly lanceolate, 2–3 mm. long, ± ciliate; corolla 7–10 mm. long, glabrous, the lips violet-blue, the upper somewhat shorter than the lower and purple-dotted near base; caps. 4–5 mm. long.—Disturbed places, such as burns, below 5000 ft.; Chaparral; Santa Monica Mts. to San Bernardino Mts. March–June.

7. **C. parviflòra** Dougl. ex Lindl. Branched, ascending to erect, puberulent, 0.5–4 dm. tall; lf.-blades lance-oblong, obtuse, mostly entire; 2–4 cm. long, only the lower petioled; fls. 1–2 in bract-axils; pedicels glandular to puberulent, 3–15 mm. long; calyx 5–7 mm. long, glabrous, the lobes acuminate, ca. 2–3 mm. long; corolla 4–7 mm. long, the upper lip white, to violet-blue at tips, the lower longer, violet-blue, the lateral lobes longer than keel; caps. 3–4 mm. long; $n=7$ (Garber, 1956).—Moist ± shaded places, 5000–11,000 ft.; Sagebrush Scrub to Subalpine F.; mts. from San Diego Co. to B.C., Ontario, Colo. April–July.

8. **C. tórreyi** Gray var. **wrìghtii** (Wats.) Jtn. [*C. w.* Wats.] Plate 83, Fig. C. Erect, widely branched, 0.5–3 dm. tall, glandular-pilose on stems and in infl.; lf. blades broadly linear, 1.5–4 cm. long, sessile or short-petioled; infl. lax, with foliose lower bracts and

reduced upper; pedicels 5–10 mm. long; calyx 3–4 mm. long, the lobes linear-obtuse; corolla 4–5 mm. long, with broadly rounded basal pouch, the upper lip pale, the lower deeper blue.—Granitic sand, etc., mostly 7000–11,000 ft.; upper Montane Coniferous F.; San Bernardino and San Gabriel mts.; to n. Calif. June–Aug.

4. Cordylánthus Nutt. ex Benth. in DC. BIRD'S-BEAK

Branched annuals with yellow roots. Lvs. alternate, entire or pinnatifid into narrow divisions. Fls. dull yellow or purple, in spikes and subtended by outer bracts that are seldom colored. Inner flowering bract mostly foliose or modified into a calyx-like structure. Calyx forming a single piece, split almost to base ventrally, extending dorsally into a tongue-like structure, entire or bifid apically. Corolla 2-lipped, the upper lip galeate and hooding the stamens, the lower as long or shorter, ± inflated and minutely lobed. Stamens 4 or 2, the anther-cells placed unequally with the lower often smaller or abortive. Caps. turgid, glabrous, loculicidal. Seeds many, wingless, with loose reticulate testa. Ca. 35 spp. of w. N. Am. (Greek, *cordule*, club, and *anthos*, fl.)

1. Lvs. oblong to lanceolate, at least the lower entire; fls. in an elongate spike; calyx spathelike, enclosing proximal part of corolla.
    3. Corolla shorter than calyx, purple on lower lobes and thin margin of galea; bracts usually with a pair of short distal lobes. Coastal .... 7. *C. maritimus*
    3. Corolla longer than calyx. Transmontane.
        4. Corolla yellowish on lower lobes and edges of galea; bracts entire. E. of Sierra Nevada and north .... 2. *C. canescens*
        4. Corolla lavender; bracts lobed. E. Mojave Desert .... 12. *C. tecopensis*
1. Lvs. or their segms. filiform to linear; fls. in heads or scattered; calyx proper narrow, enclosing corolla at base only (the bract below the fl. often confused with the calyx).
    4a. Galea longer than and curved up away from lower lip, rather bright purple; fls. solitary, their outer bracts 3-lobed. Dry places, e. Mojave Desert .... 9. *C. parviflorus*
    4a. Galea not or scarcely longer than lower lip, mostly pale or dull in color.
        5. Corolla less than to ca. twice as long as wide, inverted with ventral side up; flowering bracts setose; anthers 1-celled. Mts. .... 8. *C. nevinii*
        5. Corolla more than twice as long as wide, the dorsal side up; flowering bracts with hairs like those of other lvs. of infl.; anthers 2-celled.
            6. Flowering (inner) bract with 1–3 pairs of lobes; calyx teeth ca. 1.5 mm. long. E. of Sierra Nevada .... 5. *C. helleri*
            6. Flowering (inner) bract entire or toothed at apex; calyx entire or with teeth to 0.5 mm. long.
                7. Outer bracts palmately 3–7-lobed, shorter than flowering bract and corolla; corolla-throat ventrally pubescent within.
                    8. Infl. purplish, contrasted with whitish foliage; main lvs. 2–2.5 cm. long. Panamint Mts. .... 10. *C. ramosus* ssp. *eremicus*
                    8. Infl. yellowish, only slightly contrasted with foliage; main lvs. 1–1.5 cm. long. San Bernardino Mts. and Nelson Range .... 1. *C. bernardinus*
                7. Outer bracts entire or 3-lobed; corolla-throat glabrous within ventrally or nearly so.
                    9. Outer bracts and their lobes not noticeably widened at their tips; spikes not markedly stiff setose-ciliate; plants widely branched.
                        10. Corolla 13–16 mm. long, the lower lip sparsely pubescent without; plants finely pubescent, the bracts ciliate. Piute Mts. n. .... 3. *C. ferrisianus*
                        10. Corolla 17–21 mm. long, the lower lip pubescent externally; plants soft-pubescent throughout. Santa Barbara Co. n. .... 6. *C. littoralis*
                    9. Outer bracts and their lobes distinctly widened at the tips; spikes rather harshly setose-pilose; plants with ascending branches.
                        11. Throat of corolla longer than wide, scarcely distinguishable from tube; width of bract-divisions more than half the length of the spreading setae; corolla 12–14 mm. long, ± yellowish .... 11. *C. rigidus*
                        11. Throat of corolla wider than long, strongly contrasted with tube, width of bract-divisions less than half the length of the spreading setae; corolla white with purple lines .... 4. *C. filifolius*

1. **C. bernardínus** Munz. Plants 2–4.5 dm. high, diffusely branched, yellow-green, finely and scabrous-pubescent throughout; lvs. many, linear, entire, involute, 0.5–2 cm. long; outer bracts olive green with purple tips, 3–5-parted, 5–17 mm. long, with linear lobes; fls. mostly 2–5 in heads; fl.-bract ca. 18 mm. long, bulbous-scabrous, glandular-pubescent and ciliate-villous, ± blunt at apex; calyx 15–16 mm. long, entire or obscurely bidentate; corolla 14–16 mm. long, yellowish, the galea minutely pubescent with white wide margin, the lower lip pubescent; stamens 4, the fils. bearded; caps. ca. 8 mm. long. —Alkaline seeps at ca. 3000 ft.; Joshua Tree Wd.; n. base of San Bernardino Mts. and Nelson Range, Inyo Co. Oct.

2. **C. canéscens** Gray. [*Adenostegia c.* Greene.] Plants 2–4 dm. tall, corymbosely much branched, pubescent with fine hairs; lvs. and bracts glaucous-green, the former lanceolate, 1–2.5 cm. long, the latter somewhat wider; calyx 13–16 mm. long, broadly lanceolate, canescent, the teeth ca. 0.5 mm. long; corolla 15–17 mm. long, the minutely pubescent galea pale yellow with wide pale thin margin, the lower lip with distally pubescent pouch and minute glabrous lobes; stamens 4.—Alkaline flats and marshes, 3700–5500 ft.; Sagebrush Scrub, Shadscale Scrub; Inyo Co. (Little Lake) to Modoc Co.; Ore., Utah. June–Sept.

3. **C. ferrisiànus** Penn. Plants 3–6 dm. tall, branched, finely pubescent with recurved glandless hairs; lvs. 1–3 cm. long, linear or with a pair of linear lobes; fls. 3–5 in heads subtended by several 3-lobed dorsally ± glabrous, but ciliate outer bracts; flowering bract 13–15 mm. long, lance-oblong, rounded at tip, ± setose-pilose apically; calyx 14–16 mm. long, lanceolate, entire; corolla 13–16 mm. long, white, the galea minutely pubescent, apically glabrous, the lower lip sparsely pubescent; stamens 4.—At 4500–7000 ft.; Montane Coniferous F.; Piute Mts. to s. Sierra Nevada. July–Sept.

4. **C. filifòlius** Nutt. ex Benth. in DC. Plants 3–10 dm. tall, much-branched, pubescent with some longer hairs, not glandular; lvs. 1–3 cm. long, with 3 sub-filiform lobes widened toward tips, ± setose-pilose; heads 5–15-fld.; flowering bract 15–17 mm. long, lance-oblong, obtusish, pubescent and ± setose-pilose; calyx 14–16 mm. long, minutely denticulate; corolla 13–16 mm. long, white with 2 wide purplish lines, the galea minutely pubescent, greenish-yellow with purplish margins, the lower lip inflated, the pouch pubescent; stamens 4.—Dry slopes and open places, below 6500 ft.; Coastal Sage Scrub, Chaparral, S. Oak Wd.; Los Angeles Co. s. to n. L. Calif. May–Aug.

5. **C. hélleri** (Ferris) Macbr. [*Adenostegia h.* Ferris.] Plant much branched, 1–3 dm. tall, soft grayish-pubescent, some of hairs gland-tipped; lvs. 0.5–1.5 cm. long, linear or with pair of linear lobes; bracts usually with 2 pairs of lobes; heads 1–4-fld.; calyx 2–2.2 cm. long, the 2 acute teeth 1–2.5 mm. long; corolla ca. 2 cm. long, the galea purplish, hairy-striate, glabrous on margin and tip, the lower lip with a purple-striped pouch externally hairy.—Rocky slopes, 5000–8000 ft.; Sagebrush Scrub, Pinyon-Juniper Wd.; Last Chance Mts., Deep Springs and Inyo-White Mts. to Mono Co. and w. Nev. July–Sept.

6. **C. littoràlis** (Ferris) Macbr. ssp. **platycéphalus** (Penn.) Munz. [*C. p.* Penn.] Plants 4–12 dm. high, finely pubescent with recurved glandless hairs; lvs. linear, 13–25 mm. long, entire or with a pair of linear lobes; heads 5–10-fld., subtended by several setose-pilose and finely ciliate bracts; fl.-bract 18–19 mm. long; calyx 17–19 mm. long, lance-oblong, minutely bidentate; corolla 17–20 mm. long, white, the throat with 2 dull purplish lines, the galea finely pubescent, glabrous distally and on margin, ± yellowish; lower lip equal to upper, with pubescent pouch and yellow apex; stamens 4.—Chaparral, S. Oak Wd.; hills back of Santa Barbara.

7. **C. marítimus** Nutt. ex Benth. in DC. [*Adenostegia m.* Greene.] Loosely and corymbosely much branched, often decumbent, the stems 2–4 dm. long; herbage pubescent, some of the hairs gland-tipped; lvs. and bracts glaucous-green, the former 0.5–2.5 cm. long, lance-oblong, the latter ± oblong and often with a pair of teeth near the summit; calyx 15–22 mm. long, oblong-lanceolate, the terminal sharp teeth less than 0.5 mm. long; corolla 15–20 mm. long, the galea finely pubescent dorsally, with wide purplish thin margin, the lower lip pilose-pubescent, stamens 4.—Coastal Salt Marsh, L. Calif. to Ore. May–Oct.

8. **C. nevínii** Gray. Plate 83, Fig. D. Slender, freely paniculately branched, 2–5 dm. tall, pubescent to ± hirsute with spreading glandless hairs; lower lvs. crowded, 2–2.5 cm. long, mostly with 3 linear divisions, the upper more remote, linear, shorter; fls. in 1–3-fld. heads, subtended by several 3-lobed outer bracts; inner bract 13–15 mm. long, oblong-lanceolate, stiff hairy; calyx 10–13 mm. long, ± purplish, linear-lanceolate, slightly 2-toothed; corolla 12–16 mm. long, purplish, glabrous, the galea dark with white lateral margins; stamens 4.—Dry slopes, 5000–8000 ft.; Yellow Pine F. and above; San Gabriel Mts. to Little San Bernardino Mts., Santa Rosa Mts. and mts. of San Diego Co. July–Sept.

9. **C. parviflòrus** (Ferris) Wiggins. Plants 2–4 dm. tall, branched, loosely pubescent with some hairs gland-tipped; lvs. linear, 1–1.5 cm. long; outer bracts 3-lobed, the flowering bracts 11–12 mm. long, purplish; calyx 10–14 mm. long, lanceolate, slightly 2-toothed; corolla purplish, 14–18 mm. long, inverted, the galea with thin margin, the lower lip pubescent; stamens 4.—Dry rocky limestone slopes, 4000–5700 ft.; Pinyon-Juniper Wd.; New York Mts., e. San Bernardino Co.; to Utah, Ariz. Aug.–Oct.

10. **C. ramòsus** Nutt. ex Benth. in DC. ssp. **erèmicus** (Cov. & Mort.) Munz. [*Adenostegia eremicus* Cov. & Mort.] Plants 1.5–3 dm. high, with many slender branches, gray-puberulent with minute recurved-spreading glandless hairs; lvs. filiform, 1–3.5 cm. long, entire or with 1 pair of filiform lobes; fls. 3–5 in dense clusters; bracts several, the outer much shorter than flowering bract and corolla; calyx ca. as long; corolla purplish, ca. 15 mm. long, the galea pubescent dorsally; stamens 4.—Dry rocky places, at ca. 7000 ft.; Pinyon-Juniper Wd.; Panamint Mts., Inyo Co. Aug.–Oct.

11. **C. rígidus** (Benth.) Jeps. ssp. **brevibracteàtus** (Gray) Munz. [*C. filifolius* var. *b*. Greene.] Plants 3–7 dm. tall, the branches strict and suberect, not glandular, finely pubescent and with some longer hairs on the stems; lvs. mostly 1–2 cm. long, linear and entire or with a pair of linear lobes, obtuse or ± truncate at tips; heads mostly 5–6-fld., the outer bracts setose-ciliate, otherwise but slightly setose; flowering bract 15–17 mm. long, lance-oblong, ± purplish; calyx 12–15 mm. long, lanceolate, subentire; corolla 12–15 mm. long, light yellow, the galea finely pubescent, distally and marginally glabrous; stamens 4.—Dry granitic slopes, 5000–6000 ft.; Chaparral, Yellow Pine F.; Liebre and Tehachapi mts. to Mariposa Co. July–Aug.

12. **C. tecopénsis** Munz. & Roos. Grayish, glandular, erect, 3–6 dm. high, diffusely branched; lvs. glaucous, lance-linear to narrowly lance-oblong, 5–15 mm. long; bracts 10–12 mm. long, narrow-ovate, with 1 pair of linear lobes ca. 2 mm. long; calyx 10–13 mm. long, entire; corolla pale lavender, ca. 1 cm. long, the galea puberulent, with glabrous thin margin, the lower lip ± saccate, pubescent, with very short teeth; stamens 4.—Alkaline meadows, below 2500 ft.; Creosote Bush Scrub; Tecopa Hot Springs, Saratoga Springs, etc., s. Inyo Co.; Nye Co., Nev. Aug.–Oct.

5. *Digitàlis* L. FOXGLOVE

Erect biennial or perennial herbs, with alternate lvs. and showy racemose fls. Calyx 5-parted. Corolla declined, with a somewhat inflated tube and short scarcely spreading limb, the orifice open. Stamens 4, didynamous, included. Stigmas distinct, lamellate. Caps. ovoid, loculicidal. Seeds many, reticulate, not winged. Ca. 30 spp. of Old World. (*Digitalis*, pertaining to the finger, as fingers of a glove, because of tubular corolla.)

1. **D. purpùrea** L. Stoutish pubescent biennial 5–18 dm. high, distally glandular; lower lvs. oblong-lanceolate, acute, dentate, 1–3 dm. long; somewhat petioled; fls. many, from linear-lanceolate bracts; pedicels 1.5–2.5 cm. long; calyx-lobes ovate, becoming 15–18 mm. long; corolla purple to white, spurred on lower paler side, 4–5 cm. long; caps. 12 mm. long; seeds 0.5 mm. long; $2n=56$ (Buxton & Newton, 1928).—Natur. in ± shaded places in canyons etc. near the coast, Santa Barbara region n. May–Sept.

6. Galvèzia Domb. ex Juss.

Shrubs or herbs, with opposite or whorled lvs. Fls. in a terminal raceme. Calyx 5-parted. Corolla saccate at base, tubular, red, strongly 2-lipped, with rather prominent palate. Stamens 4, didynamous, the fils. with 2 rows of tack-shaped glands. Caps. globose-ovoid, with irregular subterminal dehiscence. Seeds cylindric, with thin irregular winglike plates. Ca. 4 spp., from Calif. to Peru. (Jose Galvez, a Spanish administrator.)

1. **G. speciòsa** (Nutt.) Gray. [*Gambelia s.* Nutt.] Glabrous or pubescent spreading bright green shrub to 1 m. tall and 2 m. broad; lvs. in 3's, thickish, elliptic-ovate, entire, 2–4.5 cm. long, on petioles ca. 5–8 mm. long; fls. in terminal leafy-bracted rather dense soft-pubescent racemes; pedicels slender, 1–2 cm. long; calyx 7–10 mm. long, the lobes lance-attenuate, divided almost to base; corolla 2–2.5 cm. long, the lips ca. 5 mm. long; fils. dilated; caps. 6–7 mm. long; seeds dark, 1 mm. long; $2n = 15$ pairs (Raven et al., 1965).—Rocky canyons, Santa Catalina and San Clemente ids.; Guadalupe Id. Feb.–May.

7. Keckiélla Straw

A genus close to *Penstemon*, but differing in its shrubbiness, in having a nectariferous hypogynous disc and in lacking glandular hairs in the corolla. Upper lip of corolla subgaleate; anthers small and glabrous. All 5 fils. are fused to the corolla to ca. the height of the ovary and bear on their basal portions rows of broad, flat, simple hairs. (Named for David D. *Keck*, 1903——, American botanist and student of *Penstemon*.)

Corolla whitish, yellowish or fulvous, not distinctly tubular.
    Infl. spicate-racemose; pedicels shorter than calyces; fls. solitary or in 2's. San Jacinto Mts. and Panamint Mts. to s. Sierra, etc. .... 4. *K. rothrockii*
    Infl. paniculate or thyrsoid; pedicels longer than calyces; fls. usually in 2's or several.
        Staminode glabrous; corolla white, tinged with pink, long-hairy externally. From Los Angeles Co. n. .... 2. *K. breviflora*
        Staminode densely bearded; corolla short-pubescent externally, yellow. Cismontane and deserts .... 1. *K. antirrhinoides*
Corolla red, distinctly tubular.
    Lvs. opposite; stems not glaucous .... 3. *K. cordifolia*
    Lvs. ternate; stems glaucous .... 5. *K. ternata*

1. **K. antirrhinoides** (Benth.) Straw. ssp. **antirrhinoides**. [*Keckia a.* Straw. *Penstemon a.* Benth.] Plate 83, Fig. E. Shrub 1–2.5 m. high with spreading much-branched stems, ± puberulent throughout, only the fls. viscid; lvs. entire, linear to ovate-elliptic, 1–2 cm. long, 0.2–0.7 cm. wide, firm, crowded; panicle broad, leafy; calyx 3–6 mm. long, the lobes ovate to rounded; corolla yellow, tinged with brownish-red, 16–20 mm. long, ca. 8–10 mm. broad at throat, this abruptly much dilated, the upper lip broad, arching, the lower reflexed; staminode densely bearded with long yellow hairs, exserted.—Dry often rocky slopes, below 4000 ft.; Chaparral, interior cismontane s. Calif. from San Bernardino Co. (occasional in Los Angeles Co.) to L. Calif. April–May.

Subsp. **microphýlla** (Gray) Straw. [*Penstemon m.* Gray.] Herbage yellowish gray-green, canescent throughout, the twigs cinereous; calyx 5.5–8 (–10) mm. long, canescent, viscid, the lobes lance-oblong, acuminate.—Rocky places, below 5000 ft. Creosote Bush Scrub to Pinyon-Juniper Wd.; w. edge of Colo. Desert, s. and e. Mojave Desert; Ariz. April–June.

2. **K. breviflòra** (Lindl.) Straw. [*Penstemon b.* Lindl.] Shrub, 5–20 dm. high, the stems many, virgate, glabrous, glaucous, rather lax in age; lvs. 1–5 (–7) cm. long, 0.3–1.2 cm. wide, lanceolate, subsessile, entire to serrulate; thyrsus pyramidal, 1–5 dm. long, many-fld.; pedicels and calyx glandular-pubescent; calyx 6–10 mm. high, the lobes ovate-lanceolate to ovate; corolla white, flushed with rose, and with purplish guide lines, 15–18 mm. long, the upper lip arched, galeate, over half the entire length, the lower lip reflexed, lobed almost to its base, glandular-pubescent externally, ± hirsute apically; staminode glabrous; $n = 8$ (Keck, 1951).—Dry rocky slopes, below 8000 ft.; Los Angeles Co. to cent. Calif. May–July.

3. **K. cordifòlia** (Benth.) Straw. [*Penstemon c.* Benth.] Loosely branched scandent shrub 1–3 m. high; herbage dark green, glabrous to puberulent, more densely hairy and moderately glandular in infl.; lvs. 2–5 cm. long, 1–3 cm. wide, lance-ovate to cordate, remotely serrulate to sharply dentate, shiny, strongly veined; panicle compact, subsecund, drooping, hence the fls. resupinate and peduncles often reflexed, leafy; calyx 7–10 mm. long, the lobes lanceolate; corolla dull scarlet, 3–4 cm. long, tubular, the upper lip galeate, the lower widely spreading; staminode densely bearded with long yellow-brown hairs; $n=8$ (Keck, 1951).—Dry slopes and canyons below 5000 ft.; Chaparral; n. L. Calif. to San Luis Obispo Co. Channel Ids.

4. **K. rothróckii** (Gray) Straw. ssp. **rothróckii**. [*Penstemon r.* Gray.] Low rounded shrub 3–6 dm. high, with many slender strict stems, puberulent throughout; lvs. sparsely scabridulous, 0.5–1.5 cm. long, 0.2–0.7 cm. wide, subsessile, lance-oblong to ovate, entire or undulate-denticulate; raceme spiciform, ± glandular, the lower fls. geminate, the upper often alternate; calyx 4–6 mm. long, with lanceolate lobes; corolla dull yellow with purplish guide lines, 10–12 mm. long, the upper lip erect, the lower reflexed; staminode glabrous.—Dry rocky slopes, 7000–10,000 ft.; Pinyon-Juniper Wd. to Lodgepole F.; Panamint Mts. to Mono Co.; w. Nev. June–Aug.

Ssp. **jacinténsis** (Abrams) Straw. [*Penstemon j.* Abrams.] Corolla 13–16 mm. long. —Dry slopes, 7000–9500 ft.; Montane Coniferous F.; San Jacinto Mts.

5. **K. ternàta** (Torr. ex Gray) Straw ssp. **ternàta**. [*Penstemon ternatus* Torr. ex Gray.] Straggly shrub 5–15 dm. high, the wandlike branches glaucous, glabrous, erect or sometimes scandent; lvs. in whorls of 3, or lower opposite, 2–5 cm. long, 0.2–0.9 cm. wide, lanceolate, remotely serrate-dentate, thickish, often folded along middle; panicle elongate, many-fld.; calyx 3–5 mm. high, the lobes lance-ovate, acuminate; corolla scarlet, 23–30 mm. long, narrowly tubular, glandular-puberulent, the upper lip galeate, the lower spreading; staminode densely yellow-bearded; $n=8$ (Keck, 1951).—Dry slopes and canyons below 6000 ft.; Chaparral, Yellow Pine F.; San Gabriel and San Bernardino mts. to San Diego Co.; L. Calif. June–Sept.

Ssp. **septentrionàlis** (M. & J.) Straw. [*Penstemon ternatus* var. *s.* M. & J.] Pedicels and calyces glandular-pubescent.—Liebre Mts., n. Los Angeles Co. to Mt. Pinos and Tehachapi Mts.

8. *Kícksia* Dumort. FLUELLIN

Diffuse repent perennial herbs, hairy, with short-petioled scattered lvs. Fls. scattered, axillary, white or yellowish-white. Calyx 5-parted. Corolla ventrally spurred at base, 2-lipped with prominent palate closing throat. Stamens 4, didynamous, the anthers ± coherent, ciliate. Caps. globose, circumscissile by 2 pores. Seeds rounded, brown, alveolate with thin winglike irregular convolutions. Ca. 30 spp., mostly from Medit. region. (J. *Kickx*, Belgian professor in early 19th century.)

Sepals lanceolate; lvs. hastate-lobed at base; corolla 5 mm. long .... 1. *K. elatine*
Sepals ovate; lvs. rounded or cordate at base; corolla (without the spur) 6–8 mm. long
2. *K. spuria*

1. **K. elátine** (L.) Dumort. [*Antirrhinum e.* L. *Linaria e.* Mill.] Like the next sp., but lvs. ovate, 1–3 cm. long, 1–2.5 cm. wide; petioles 2–3 mm. long; pedicels 10–25 mm. long; corolla 5 mm. long, the spur an additional 4 mm.; caps. 3–4 mm. long; $n=9$ (Braun, 1932).—Occasionally natur. in old fields, river bottoms, etc., Orange Co., Santa Barbara Co. etc.; introd. from Eu. June–Sept.

2. **K. spùria** (L.) Dumort. [*Antirrhinum s.* L. *Linaria s.* Mill.] Much-branched, glandular and soft-hairy, forming mats a meter or so across; lvs. ovate to rounded, 2.5–4 cm. wide, on petioles 2–5 mm. long; pedicels 10–13 mm. long; calyx-lobes accrescent; upper lip of corolla violet, the lower yellow; spur 5 mm. long; caps. 3–4 mm. long; $2n=$ ca. 14 (Heitz, 1927).—Occasionally natur. as near Ventura; n. Calif. Introd. from Eu. June–Sept.

9. Limosélla L. MUDWORT

Stoloniferous or tufted glabrous perennial or annual herbs without ascending stems. Lvs. entire, petioled, palmately veined, in tufts. Fls. solitary on long pedicels from the tufts. Bracteoles none. Calyx 5-lobed. Corolla subrotate, white or pinkish to violet-tinged, 5-lobed, the lobes acute. Stamens 4, subequal; anther-cells wholly confluent. Stigma. united and capitate. Caps. septicidal, distally 1-celled, the septum not complete. Seeds many, minute. Ca. 12 spp., widely distributed. (Latin, *limus*, mud and *sella*, seat, the spps growing in wet places.)

1. **L. acáulis** Ses. & Moç. Lf.-blades linear-oblanceolate, 0.6–1.2 cm. long, attenuate to petioles several times as long; pedicels mostly ca. as long as petioles.—Muddy shores, mostly 4000–8000 ft.; largely Yellow Pine F.; mts. of s. Calif. to cent. Calif.; L. Calif., Mex. May–Oct.

10. Linària Mill. TOADFLAX

Mostly annual or perennial herbs with opposite or whorled lvs. or upper alternate, entire, dentate or lobed. Fls. in terminal spikes or racemes. Calyx 5-parted. Corolla with long tube, long-spurred at base, the throat often nearly closed by the palate. Fertile stamens 4, the fils. and styles filiform. Caps. dehiscent by 4 or more mostly 3-toothed pores or slits below the summit. Ca. 100 spp., many in cult. (Latin, *linum*, flax, as the lvs. of some spp. are flaxlike.)

Fls. yellow.
    Lvs. linear; corolla (including the spur) 2–3 cm. long .... 6. *L. vulgaris*
    Lvs. ovate to lanceolate; corolla (including the spur) 3.5–4 cm. long .... 3. *L. dalmatica*
Fls. not yellow.
    Plants perennial; lvs. linear; corolla including spur 12–18 mm. long; violet-purple with dark palate .... 5. *L. purpurea*
    Plants annual or biennial; lvs. linear or lance-linear.
        Spur strongly curved, placed transversely or obliquely; corolla including spur 15–20 mm. long.
            Corolla violet with whitish ridges instead of well formed palate. Native . 2. *L. canadensis*
            Corolla violet-purple with well-formed orange palate. Occasional waif .. 1. *L. bipartita*
        Spur nearly straight, vertical; corolla violet-purple with small yellowish patch on palate
                                                                                                      4. *L. maroccana*

1. **L. bipartìta** Willd. Annual ca. 3 dm. high, glabrous, glaucous; lvs. linear, 3–5 cm. long; corolla violet-purple with orange palate, 18–20 mm. long including spur; caps. ca. 2 mm. long; seeds minute, rugose.—Occasional garden escape, as at San Diego, Avalon, Claremont, Santa Barbara; native of Medit. region. Early summer.

2. **L. canadénsis** (L.) Dum-Cours. var. **texàna** (Scheele) Penn. [*L. t.* Scheele.] Slender annual or biennial, 1–6 dm. high, with short-trailing basal offshoots; cauline lvs. narrowly linear, sessile, 0.5–2.5 cm. long, those of the basal sterile stems ovate to linear; racemes spicate, slender, the pedicels 2–10 mm. long; calyx ± glandular-puberulent, 2–3 mm. long; corolla blue-violet, without the spur 9–12 mm. long, the spur 5–9 mm. long; seeds minutely tubercled; $2n=12$ (Lewis, 1958).—Dry slopes and waste places, often on burns, below 2000 ft.; Coastal Sage Scrub, Chaparral; L. Calif. to B.C., Atlantic Coast; S. Am. March–May.

3. **L. dalmática** (L.) Mill. [*Antirrhinum d.* L.] Rather coarse perennial to 1 m. high; lvs. ovate to oblong, clasping, 1–4 cm. long; calyx 5–7 mm. long; corolla yellow, 3.5–4 cm. long; caps. 5–6 mm. long; seeds black, rhomboid, angled, ca. 1 mm. long, pitted; $2n=12$ (Matsuura & Soto, 1935).—Occasional waif, San Diego Co., San Bernardino Co., Kern Co., Ventura Co., etc.; native of e. Medit. region. Summer.

4. **L. maroccàna** Hook. f. Annual to 5 dm. tall, glabrous below, viscid-pubescent above; lvs. remote, linear, 2–4 cm. long; corolla 35–38 mm. long including spur, violet-purple with whitish patch on palate; seeds obconic, with 4–6 ringlike wings; $n=6$ (Heitz, 1927).—Occasional escape from cult., as at Santa Barbara; native of N. Afr.

5. **L. purpùrea** (L.) Mill. [*Antirrhinum p.* L.] Perennial 3–7 dm. high, glabrous; lvs.

many, linear, 2–5 cm. long; racemes dense; calyx ca. 2.5 mm. long; corolla 15–18 mm. long, violet-purple with dark palate; caps. ca. 3 mm. long; seeds angled, rugose-reticulate. —Occasional escape, as in Ventura Co.; native of s. Eu. May–June.

6. **L. vulgàris** Hill. BUTTER-AND-EGGS. Perennial with several ascending stems to ca. 1 m. long; lvs. pale, numerous, narrow; raceme dense; corolla bright yellow, 2–3 cm. long, with rounded orange palate; caps. 9–12 mm. long; seeds flattened, 1.5 mm. long; $2n = 12$ (Gadella & Kliphuis, 1966).—Occasional weed or escape in much of cismontane Calif.; widely distributed in N. Am.; native of Eu. Summer.

11. Mauràndya Ort.

Perennial herbs with twining or prostrate stems. Lvs. alternate, petioled, coarsely toothed or hastately lobed. Calyx 5-parted. Corolla gibbous or saccate at base, bilabiate, usually with internal plaits, occasionally almost closed by true palate. Stamens 4, didynamous; fils. with 2 rows of tack-shaped glands. Caps. scarcely oblique, irregularly dehiscent near apex. Seeds oblong, with irregular corky ridges and tubercles. Ca. 10 spp., sw. N. Am. (Dr. *Maurandy*, a teacher of botany at Cartagena.)

Lvs. hastately lobed, triangular-ovate; stems twining or climbing, to ca. 1 m. long; fls. rose to purple. Kelso, Providence Mts. .... 1. *M. antirrhiniflora*
Lvs. rounded, irregularly dentate; stems ± pendent, ca. 1 dm. long; fls. yellow. Death V.
2. *M. petrophila*

1. **M. antirrhiniflòra** Humb. & Bonpl. ex Willd. [*Antirrhinum maurandioides* Gray. *Asarina a.* Penn.] Stems slender, glabrous, much-branched; lvs. glabrous, triangular, hastate to 5-lobed, 1–2.5 cm. long, on equally long somewhat flexuous petioles; fls. solitary, axillary, on slender pedicels 1–2 cm. long; calyx 10–12 mm. long; corolla 2.5–3 cm. long, the tube pale, the lobes rose to purple, the palate yellowish-white with dark lines, pubescent; stamens 17–19 mm. long; caps. subglobose, thin-walled; seeds with short corky tuberculate ridges; $n = 12$ (Heitz, 1927).—Limestone, 2500–4000 ft.; Joshua Tree Wd., Shadscale Scrub; Providence Mts.; to Tex., Oaxaca. April–May.

2. **M. petróphila** Cov. & Mort. [*Asarina p.* Penn.] Stems slender, branched, forming dense ± pendent tufts from a woody caudex; plants soft-hairy; lf.-blades round-ovate, 2–3 cm. long, irregularly dentate with long callose bristly teeth; petioles 1–2 cm. long; pedicels 1–3 mm. long; calyx 8–12 mm. long, the lobes lanceolate, irregularly dentate with slender teeth; corolla pale yellow, 15–30 mm. long, the throat cylindric, open with 2 ventral pilose ridges, the lobes spreading; seeds with lines of spongy tubercles.—Limestone crevices, 3500–5800 ft.; Creosote Bush Scrub; Grapevine Mts., Inyo Co. April–June.

12. Mímulus L. MONKEY-FLOWER

Annual or perennial herbs or low shrubs, often ± viscid or glandular-pubescent or pilose to slimy-viscid. Lvs. opposite, entire to dentate or even laciniate. Fls. axillary to foliage lvs. or bracts, often in open racemes. Bracteoles none. Sepals 5, united into a tubular or campanulate calyx, of which the tube is usually inflated and plicate-angled, the lobes or teeth equal to unequal. Corolla 2-lipped to subregular, purple or red or almost violet to yellow or buff, with a pair of bearded or naked ridges running down the lower side of the throat and forming a palate which may almost close the throat. Stamens 4, didynamous, usually all antheriferous. Stigmas 2, distinct and lamelliform, or adhering to form a funnellike structure. Caps. cylindrical, loculicidal, the septum unruptured or splitting. Seeds many, small, oblong to oval or fusiform, mostly yellowish, reticulate or almost smooth, wingless. Ca. 150 spp., mostly of w. N. Am., but also in S. Am., Asia, Australia, etc. (Diminutive of Latin, *mimus*, a comic actor, because of the grinning corolla.)

(Grant, A. L. A monograph of the genus Mimulus. Ann. Mo. Bot. Gard. 11: 99–388, 1924. Beeks, R. M. Variation and hybridization in southern California populations of Diplacus. Aliso 5: 83–122. 1962.)

1. Plants perennial.
    2. Pedicels longer than the calyces; plants not at all woody. Wet places.
        3. Corolla scarlet, the upper lobes arched; stamens exserted .... 8. *M. cardinalis*
        3. Corolla yellow.
            4. Mature calyx little or not much inflated, the lobes practically straight; corolla-throat open, the limb not conspicuously 2-lipped.
                5. Plants ± slimy; lvs. pinnately veined; stems mostly 1–3 dm. long . 23. *M. moschatus*
                5. Plants not slimy; lvs. palmately veined; stems mostly less than 1 dm. long
                    28. *M. primuloides*
            4. Mature calyx strongly inflated, the lower lobes curved against the others; corolla-throat ± closed by the palate formed by the large ventral ridges, the limb strongly 2-lipped.
                6. Stems mostly with definite racemes; pedicels usually shorter than corollas; rootstocks rarely fleshy or yellow .... 17. *M. guttatus*
                6. Stems mostly 1- to 3-, rarely 5-fld.; pedicels usually longer than corollas; rootstocks fleshy, yellow .... 34. *M. tilingii*
    2. Pedicels shorter than calyces; plants somewhat woody at base to definitely shrubby. Dry places.
        7. Plants woody at base only; lvs. finely glandular-pubescent on both surfaces; calyx usually swollen at base; caps. 9–12 mm. long. Santa Ana Mts. and e. San Diego Co.
            9. *M. clevelandii*
        7. Plants shrubby; lvs. not glandular-pubescent on both surfaces; calyx not swollen at base; caps. 12–25 mm. long.
            8. Corollas orange, copper, yellow, buff, cream or almost white, never true red.
                9. Calyx, lower lf.-surfaces, and upper branchlets glabrous or microscopically glandular-puberulent.
                    10. Lvs. slightly to distinctly impressed-veiny above; corolla-tube not usually exserted from calyx .... 3. *M. aurantiacus*
                    10. Lvs. not impressed-veiny above; corolla-tube usually exserted from calyx, the corolla light lemon-yellow. Se. San Diego Co. .... 2. *M. aridus*
                9. Calyx, lower lf.-surfaces, and upper branchlets obviously pubescent to almost woolly .... 20. *M. longiflorus*
            8. Corollas red or reddish.
                11. Calyx glandular-hairy; pedicels mostly less than 10 mm. long
                    20. *M. longiflorus rutilus*
                11. Calyx glabrous or nearly so; pedicels mostly 10–25 mm. long.
                    12. Lvs. mostly less than 3–4 times as long as wide; corolla not over 1.8 cm. broad, the lobes subequal, only slightly if at all notched. Insular ..... 14. *M. flemingii*
                    12. Lvs. mostly 3–4 times as long as wide; corolla 1.8 or more cm. broad, the lobes unequal and notched. Mainland mostly .... 29. *M. puniceus*
1. Plants annual.
    13. Pedicels conspicuous, normally longer than calyx; corolla mostly deciduous, dropping before shriveling and leaving the styles exposed (± persistent in *M. breweri*).
        14. Mature calyx strongly inflated, the lower lobes curving up; fls. yellow.
            15. Uppermost calyx-lobe rarely more than twice the length of the others, the lower teeth in maturity only partly closing the orifice .... 17. *M. guttatus*
            15. Uppermost calyx-lobe almost 3 times as long as others, the lower teeth in maturity folding up so as almost to close the orifice .... 24. *M. nasutus*
        14. Mature calyx not strongly inflated or if somewhat so, cylindric, the lobes permanently straight or nearly so.
            16. Fls. predominantly yellow, or sometimes partly white, sometimes with reddish spots.
                17. Calyx not ridged or low-winged along the sepal midribs; caps. glandular-puberulent .... 27. *M. pilosus*
                17. Calyx ridged or low-winged along the sepal midribs; caps. glabrous.
                    18. Calyx-lobes 1.5–3 mm. long; corolla 10–12 mm. long .... 19. *M. latidens*
                    18. Calyx-lobes 0.5–1 mm. long; corolla 4–15 (–20) mm. long.
                        19. Calyx-lobes triangular-acute to -acuminate; stem and lvs. strongly pubescent, the lvs. rounded to cordate at base .... 15. *M. floribundus*
                        19. Calyx-lobes rounded, either bluntly so or mucronate.
                            20. Corolla 5–9 mm. long.

21. Calyx-lobes ciliate; plants rather open in growth; pedicels 7–20 mm. long. Mts. along desert edge to White Mts. .... 31. *M. rubellus*
21. Calyx-lobes not ciliate; plants compact, bushy; pedicels 2–7 mm. long. San Jacinto Mts. to n. Calif. .... 33. *M. suksdorfii*
20. Corolla 10–17 mm. long, the tube little or not surpassing the calyx. Inyo and Mono cos. .... 22. *M. montioides*
16. Fls. predominantly purple, red or magenta.
22. Calyx 2 mm. long; corolla 2–2.5 mm. long; anthers glabrous. San Bernardino Mts. .... 13. *M. exiguus*
22. Calyx 4–12 mm. long; corolla 6–20 mm. long.
23. Calyx-lobes mostly 1–2 mm. long, acute to subulate-tipped (shorter in *M. palmeri*).
24. Anthers pubescent; lvs. sessile or nearly so by narrow base; corolla 15–20 mm. long. Mojave Desert .... 25. *M. palmeri*
24. Anthers glabrous; lvs. sessile by rounded base; corolla 10–12 mm. long.
25. Plant minutely glandular-puberulent; lvs. 1–2 cm. long; calyx inflated in fr. Cismontane .... 19. *M. latidens*
25. Plant densely glandular-villous; lvs. 1.5–4 cm. long; calyx not much inflated in fr. Mojave Desert .... 26. *M. parishii*
23. Calyx-lobes mostly ca. 0.5 mm. long, broader than long, obtuse to rounded.
26. Stigmas not fringed; basal tube of corolla scarcely or not exceeding calyx; corollas 6–10 mm. long.
27. Pedicels 5–20 mm. long, less than twice the length of the subtending bracts; anthers and stigmas included; lvs. entire, 3-veined from narrow base.
28. Plants glandular-puberulent; calyx-teeth ciliate, rounded. Mts. bordering deserts .... 31. *M. rubellus*
28. Plants glandular-pubescent; calyx-teeth not ciliate, deltoid. San Jacinto Mts. n. through Sierra Nevada .... 7. *M. breweri*
27. Pedicels 15–27 mm. long, 3–5 times the length of the bracts; anthers and stigmas exposed; lvs. crenate-dentate, scarcely veined, round-clasping at base. Tehachapi Mts. n. .... 1. *M. androsaceus*
26. Stigmas ciliate-fringed; basal tube of corolla exceeding calyx; corollas 10–15 mm. long.
29. Corolla-lobes distinctly unequal. San Bernardino Mts. .. 30. *M. purpureus*
29. Corolla-lobes subequal. San Jacinto and Santa Ana mts. .. 12. *M. diffusus*
13. Pedicels normally shorter than calyces; corolla mostly shriveling and persistent on the developing caps.
30. Fls. yellow.
30A. Corolla 1.5–2 cm. long; calyx-lobes 2.5–3.5 mm. long. Inyo Co. north 11. *M. densus*
30A. Corolla 20–50 mm. long; calyx-lobes 3–12 mm. long. Cismontane . 6. *M. brevipes*
30. Fls. purple to rose or magenta.
31. Caps. symmetrical, membranous, soon dehiscent, not gibbous or oblique at base.
32. Corolla salverform, the cylindrical tube expanding directly into the rotate lobes which are purple with pale margins. Mojave Desert .... 21. *M. mohavensis*
32. Corolla tubular-campanulate, with definite throat wider than the tube below.
33. Stigmas and longer stamens usually exserted from the corolla-throat.
34. Calyx-lobes subequal, the uppermost less than twice as long as the lowermost; mature calyx little or not at all inflated; lvs. mostly sessile. Inyo Co.
11. *M. densus*
34. Calyx-lobes strongly unequal, the uppermost ca. twice as long as lowermost; mature calyx distinctly inflated. Cismontane .... 18. *M. johnstonii*
33. Stigmas and anthers included in corolla-throat.
35. Inner face of corolla-lobes glabrous; lvs. ± lanceolate; corolla 2–2.5 cm. long; the lobes abruptly widely spreading .... 16. *M. fremontii*
35. Inner face of corolla-lobes pilose; lvs. ± elliptic.
36. Lowest pair of calyx-lobes more than half as long as uppermost lobe; corolla 12–30 mm. long; bracts acuminate to cuspidate. Deserts .. 4. *M. bigelovii*
36. Lowest pair of calyx-lobes not more than half as long as uppermost lobe; corolla 13–20 mm. long; bracts obtuse to acute. Largely cismontane
18. *M. johnstonii*
31. Caps. asymmetrical, cartilaginous, tardily or not dehiscent, gibbous or oblique at base.

Mimulus

37. Lower lip of corolla less than ⅓ as long as upper lip.
   38. Mature calyx 10–12 mm. long; corolla purple; plants 1–6 cm. high. Santa Cruz Id. .... 5. *M. brandegei*
   38. Mature calyx 17–20 mm. long; corolla purple with white upper lip; plants 10–14 cm. high. Catalina Id. .... 35. *M. traskiae*
37. Lower lip of corolla more than ⅓ as long as the upper.
   39. The lower lip of the corolla at least as long as the upper; stigmas subequal. Death V. .... 32. *M. rupicola*
   39. The lower lip of the corolla shorter than the upper; stigmas unequal, the lower longer. Cismontane .... 10. *M. congdonii*

1. **M. andròsàceus** Curran ex Greene. [*M. palmeri* var. *a*. Gray.] Minutely glandular-puberulent annual, erect, simple or branched, 2–8 cm. high; lvs. few, sessile, oblong-ovate to -lanceolate, entire or distally obscurely toothed, 0.3–0.7 cm. long; pedicels 2–8 mm. long; calyx reddish, 6–7 mm. long, lightly ridged, the lobes almost 1 mm. long; corolla red-purple, ca. 8 mm. long, the narrow throat yellowish or purplish, slightly purple-spotted on the yellowish ridges, the lobes rounded, spreading; anthers glabrous.—Rare, gravelly and stony places, below 4000 ft.; Chaparral, Foothill Wd.; Tehachapi Mts. to Lake Co. March–June.

2. **M. áridus** (Abrams) Grant. [*Diplacus a*. Abrams.] Glabrous glutinous decumbent subshrub 2–4 dm. tall, 6–12 dm. across; lvs. 2–4 (–6) cm. long, elliptic to ± lanceolate, yellow-green, slightly dentate, ± revolute; pedicels 3–6 mm. long; calyx 2.5–3 cm. long, funnelform, glabrous, flared above, the uppermost lobe 8–10 mm. long, the others ca. 4–6 mm.; corolla light lemon-yellow to golden-yellow, 3.5–5 cm. long, the tube exserted, the limb rotate; anthers glabrous; stigmas unequal, fimbriate; caps. splitting along upper suture; $n=10$ (Stebbins, 1951).—Local in dry rocky places, 2500–3500 ft.; Chaparral; se. San Diego Co.; adjacent L. Calif. April–July.

3. **M. aurantìàcus** Curt. ssp. **austràlis** (McMinn) Munz. [*Diplacus a*. McMinn.] Erect profusely branched shrub 6–12 or more dm. high, puberulent to subglabrous, but glandular and glutinous; lvs. 2.5–6 cm. long, slightly impressed-veiny, subglabrous above, with gland-tipped and nonglandular hairs beneath, often revolute; pedicels 5–10 mm. long; calyx tubular, scarcely enlarged upward, 2–2.5 cm. long; corolla orange-yellow, to light apricot or buff or white, 3.5–4.5 cm. long, the tube included.—Rocky and disturbed places, below 3500 ft.; Chaparral; Santa Ana Mts. through interior San Diego Co. to n. L. Calif. Hybridizing with *M. puniceus*. March–July.

4. **M. bigelòvii** (Gray) Gray. [*Eunanus b*. Gray.] var. **bigelòvii**. Densely glandular-pubescent annual, simple or branched, 5–25 cm. tall, erect; lvs. scattered, elliptic to obovate, ± acute, subentire, 1–3 cm. long, sessile or the lowest short-petioled; fls. mostly clustered near tip of stems; pedicels 1.5–4 mm. long; calyx broadly oblong, 8–11 mm. long, wing-ridged, space between the glandular ridges pale; lobes attenuate, unequal, 3–4 mm. long; corolla red-purple, 2–3 cm. long, externally finely pubescent, internally pubescent on the subrotate limb, the throat pale yellow below with purple dots; anthers ± pubescent; stigmas ciliolate, subequal; caps. dehiscing throughout.—Common in dry sandy washes and canyons below 7500 ft.; Creosote Bush Scrub, Joshua Tree Wd., Pinyon-Juniper Wd., Sagebrush Scrub, etc.; Mojave and Colo. deserts and bordering mts.; n. to Mono Co.; Nev., w. Ariz. March–June.

Var. **cuspidàtus** Grant. Stout, densely viscid-villous; upper lvs. almost round, abruptly cuspidate; pedicels 3–8 mm. long; calyx 8–10 mm. long; corolla 1.2–2 cm. long, the ventral ridges strongly constricting the orifice.—Dry rocky places below 7000 ft.; Barstow and Baker, Mojave Desert to Death V. region and White Mts. March–May, Sept.–Oct.

Var. **panaminténsis** Munz. Lvs. as in the sp.; calyx 6–8 mm. long; corolla 15–18 mm. long; caps. slender, curved, 10–12 mm. long.—Dry rocky places, 8000–10,000 ft.; Bristle cone Pine F., etc.; Panamint Mts. July–Aug.

5. **M. brandègei** Penn. Glandular-pubescent annual 2–3 cm. high; lvs. narrow-elliptic to spatulate-oval, obtuse, entire, short-petioled; pedicels 1–2 mm. long; calyx

10–12 mm. long, scarcely ridged, the lobes ± unequal, obtuse; corolla purple, 13–15 mm. long, the tube equal to calyx, the throat funnelform, the lobes rounded, the 2 upper 3–4 mm. long, the lower ca. 1 mm.—Santa Cruz Id.

6. **M. brévipes** Benth. Viscid-pubescent annual, erect, simple or branched, 1–8 dm. high; basal lvs. obovate, 5–8 cm. long, on slender petioles, the cauline remote, linear to lanceolate, 2–6 cm. long, entire to denticulate, sessile or nearly so, ± acuminate; pedicels 2–6 (–10) mm. long; calyx 2–2.5 cm. long, broadly tubular-campanulate, glandular-pubescent, the ridges green in contrast with white intervening spaces; corolla yellow, bilabiate, 2–5 cm. long, glabrous without, pubescent within on the ridges and base of lower lip, the lobes rounded; anthers glabrous; stigmas ciliolate, the lobes unequal; caps. 9–13 mm. long, dehiscing throughout; $n=8$ (Mukherjee & Vickery, 1962).—Dry exposed places, below 5000 ft.; Chaparral, Coastal Sage Scrub; L. Calif. to Santa Barbara Co. Santa Cruz Id. April–June.

7. **M. brèweri** (Greene) Cov. [*Eunanus b.* Greene.] Glandular-pubescent annual, simple or branched, 0.3–1.8 dm. tall; lvs. narrowly oblong-lanceolate to -oblanceolate, entire or nearly so, 0.5–3 cm. long, narrowed to sessile bases; pedicels 3–10 mm. long; calyx cylindrical, 5–7 mm. long, with 5 glandular-pubescent ridges, the lobes ca. 1 mm. long; corolla funnelform, pale pink to purplish, 6–10 mm. long, persistent, the throat with pale yellow ventral ridges, the lobes subequal; anthers glabrous.—Damp sandy places, 4000–10,000 ft.; Montane Coniferous F.; San Jacinto and San Bernardino mts., Mt. Pinos; n. to B.C., Mont., Nev. June–Aug.

8. **M. cardinàlis** Dougl. ex Benth. Plate 83, Fig. F. Freely branched viscid-villous perennial, erect or decumbent, the stems 2.5–8 dm. long, from a running rootstock; lvs. obovate to oblong, 2–8 cm. long, sessile, serrate, longitudinally 3–5-veined, at least the upper with broad clasping bases; pedicels 5–8 cm. long; calyx tubular or oblong-prismatic, 2–3 cm. long, angulate-winged, the teeth subequal, 4–5 mm. long, ovate; corolla strongly bilabiate, 4–5 cm. long, scarlet, sometimes yellowish, the throat narrow, yellowish, with yellow hairy ridges, the upper lip arched-ascending; anthers ciliate; stigmas fimbriolate; caps. 16–18 mm. long; $n=8$ (Brozek, 1932).—Stream banks, seeps, etc., below 8000 ft.; Coastal Sage Scrub to Montane Coniferous F.; most of cismontane and montane Calif.; to Ore., Nev., Ariz. Catalina, Santa Cruz ids. April–Oct.

9. **M. clevelándii** Bdg. [*Diplacus c.* Greene.] Erect freely branched suffrutescent perennial 3–9 dm. tall, glandular-villous; lvs. of early season lanceolate to oblong, 2.5–10 cm. long, with broad sessile base, serrate in upper half or entire, impressed-veiny and finely hairy above, distinctly hairy and somewhat glutinous beneath; lvs. of late season smaller, usually fascicled in main axils; pedicels 3–4 mm. long; calyx campanulate, 20–25 mm. long, ridge-angled, the tube narrow, constricted above the ovary, the lobes unequal, 6–9 mm. long; corolla golden-yellow, 3.5–4 cm. long, glandular externally, ventrally 2-ridged and with rows of orange dots on lower side; anthers glabrous; stigmas unequal, ciliate; caps. 10–12 mm. long; $n=10$ (Stebbins, 1951).—Dry disturbed places, 3000–6000 ft.; Chaparral, Yellow Pine F.; Santa Ana Mts. to s. San Diego Co.; n. L. Calif. May–July.

10. **M. congdònii** Rob. Annual, acaulescent to 6 or 8 cm. tall, coarsely pubescent, some of the hairs gland-tipped; lvs. obovate, 1.5–3.5 cm. long, subentire to bluntly dentate; pedicels 2–4 mm. long; calyx 9–12 mm. long, straight, the tube pale between the green-pubescent ridges, the lobes somewhat unequal, to 1 or 2 mm. long; corolla rose-pink or -purple, 1.5–2.5 cm. long, sparsely hairy without, the tube slender, 1.5–2 times the length of the calyx, the lobes rounded; anthers glabrous; stigmas unequal.—Damp grassy and disturbed places, below 3000 ft.; Pratt Canyon, Ventura Co. and Greenhorn Mts. north. March–May.

11. **M. dénsus** Grant. Glandular-pubescent annual, densely branched, 0.5–1.5 dm. tall; lvs. lanceolate to oblanceolate, 1–1.7 cm. long, entire, sessile; pedicels 1–4 mm. long; calyx 7–9 mm. long in fr., slightly angled, the ± unequal lobes 2.5–3.5 mm. long; corolla 1.5–2 cm. long, yellow and streaked with red, to red-purple, pubescent within and ± spotted; anthers pubescent, scarcely exserted; stigmas ciliate; caps. dehiscing throughout.

—Dry gravelly or sandy places, 4000–9000 ft.; Sagebrush Scrub, Pinyon-Juniper Wd., Yellow Pine F., etc.; White Mts. and e. slope of Sierra Nevada; w. Nev. May–July.

12. **M. diffùsus** Grant. Glandular-puberulent diffusely branched annual 0.5–2 dm. tall; lvs. oblong-ovate to -lanceolate, entire to deeply dentate-lobed, 1–2 cm. long, sessile or the lowermost petioled; pedicels 2.5–4.5 cm. long; calyx oblong-campanulate, 5–7 mm. long, weakly angled, the teeth obtuse, 0.5–1 mm. long; corolla 12–15 mm. long, purple to rose-violet, the lobes subequal, emarginate, irregularly marked with yellow and purple, the throat with 2 prominent yellow ridges; anthers glabrous; caps. 6 mm. long, dehiscing through the apex; seeds reticulate, oblong.—Damp sandy or gravelly places, 4000–6000 ft.; Chaparral, Yellow Pine F.; San Jacinto and Santa Ana mts.; to n. L. Calif. April–June.

13. **M. exíguus** Gray. Minutely glandular-puberulent, usually diffusely openly branched, 3–7 cm. tall, ± reddish throughout; lvs. few, linear to spatulate or elliptic, 3–6 mm. long, subentire, sessile; pedicels 15–20 mm. long; calyx campanulate, 2 mm. long, the lobes acutish, not ciliate; corolla funnelform, reddish-purple, 2–2.5 mm. long, the minute lobes scarcely opening; anthers glabrous; caps. tardily dehiscent apically.—Rare, moist disturbed places, as in Holcomb V., San Bernardino Mts.; n. L. Calif. June–July.

14. **M. flemíngii** Munz. [*Diplacus parviflorus* Greene, not *M. p.* Lindl.] Subglabrous semierect or spreading subshrub 1–6 dm. tall, with greater spread; lvs. 2–4.5 cm. long, obovate to oval-elliptic, dark green above, paler beneath, entire to glandular-denticulate, usually revolute; pedicels 10–20 mm. long; calyx tubular, 2–2.3 cm. long, the lobes linear, the uppermost 6–8 mm. long, the lower 3–4 mm.; corolla brick-red or orange-red, 2.5–4 cm. long, the tube included in calyx, the upper lobes subentire, the lower entire or slightly toothed; anthers glabrous.—Rocky places; Coastal Sage Scrub, Chaparral; Santa Cruz, Santa Rosa, Anacapa and San Clemente ids. March–July.

15. **M. floribúndus** Dougl. ex Lindl. Viscid-villous somewhat slimy annual, much branched, erect to decumbent, the stems 1–5 dm. long; lvs. thin, ovate to lance-ovate, scattered, 1.5–4 cm. long, mostly sharply dentate, subpalmately veined, rounded or cordate at base, with petioles almost as long as blades; pedicels 0.5–2.5 cm. long; calyx plicate-carinate, 6–8 mm. long, or more in fr., the lobes 1–1.5 mm. long; corolla yellow, 0.7–1.5 cm. long, the ventral ridges in the throat pubescent with clavate hairs and with reddish spots; anthers glabrous; $n=16$ (Mukherjee & Vickery, 1961).—Moist places below 8000 ft., most of cismontane Calif.; Panamint Mts., Catalina Id.; to B.C., Rocky Mts. April–Aug.

16. **M. fremóntii** (Benth.) Gray. Glandular-pubescent annual, ± reddish throughout, usually freely branched, 4–20 cm. high; lvs. oblanceolate to oblong, 1–3 cm. long, entire or nearly so, sessile; pedicels 2–4 mm. long; calyx campanulate, 8–10 mm. long, strongly angled, the throat oblique, the lobes subequal, ca. 2 mm. long; corolla broadly funnelform, 2–2.5 cm. long, rose-purple, pubescent externally, glabrous within except on the ventral yellow ridges, the lobes ± unequal; anthers included, glabrous; stigmas equal, ciliolate; caps. ca. as long as calyx.—Dry disturbed places, such as burns, below 7000 ft.; Chaparral, Foothill Wd., Joshua Tree Wd., Yellow Pine F.; edge of deserts and through cismontane s. Calif. to L. Calif. and San Benito Co. April–June.

17. **M. guttàtus** Fisch. ex DC. ssp. **guttàtus.** Plate 84, Fig. A. Usually perennial, glabrous below the ± glandular-puberulent infl., rooting at nodes, typically with stolons or rootstocks; stems ± fistulous, 0.5–10 dm. tall, mostly simple; lvs. oval, rounded to pinnatifid-dentate at base, 1–8 cm. long, the upper sessile, ± connate, the lower long-petioled; infl. ± racemose; pedicels 2–6 cm. long; calyx campanulate, glabrous to puberulent, often tinged or dotted with red, inflated and 1.5–2.5 cm. long in fr., strongly plicate-angled, the lobes acute and the lower infolding fr. so as partly to close the orifice, the upper tooth mostly less than 3 times the length of the others; corolla yellow, usually spotted red, mostly 1.5–4 cm. long, the throat nearly closed by the hairy ridges; anthers glabrous; stigmas fimbriolate; caps. stipitate, 7–9 mm. long; $n=14$ (Campbell, 1950).—Exceedingly common in wet places, below 10,000 ft.; many Plant Communities, most of Calif.; to Alaska, Rocky Mts., Mex. March–Aug. Variable.

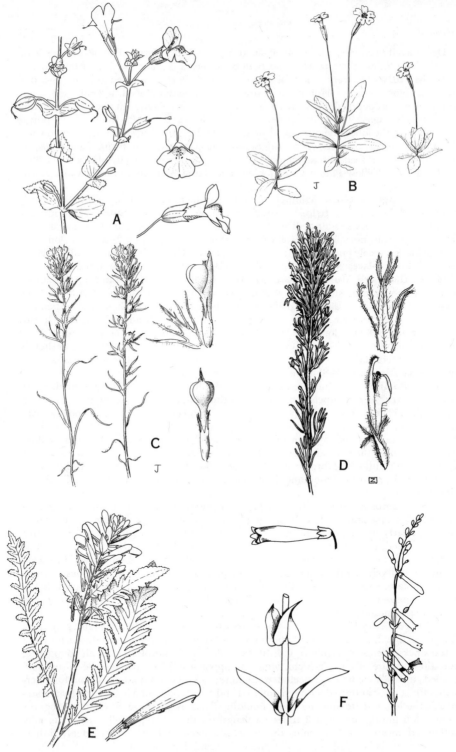

Plate 84. Fig. A, *Mimulus guttatus*. Fig. B, *Mimulus primuloides*. Fig. C, *Orthocarpus hispidus*, fl. and subtending bract. Fig. D, *Orthocarpus purpurascens*. Fig. E, *Pedicularis densiflora*, corolla tubular, arched. Fig. F, *Penstemon centranthifolius*, upper lvs. clasping, corolla tubular.

Ssp. **litoràlis** Penn. [*M. g.* var. *grandis* Greene.] Plants usually stout, softly pubescent in infl.; lvs. pubescent; calyx pubescent, 17–30 mm. long; corolla 3–4.5 cm. long.—Wet places, Coastal Strand; coast from Santa Barbara Co. to Wash. April to Oct.

18. **M. johnstònii** Grant. Glandular-pubescent annual, openly branched, 1–2.5 dm. tall; lvs. obovate to oblanceolate, entire, 1–2 cm. long, the lower petioled; pedicels 1–3 mm. long; calyx 8–9 mm. long in fr., angled, the acuminate lobes unequal, the upper 2–4 mm. long; corolla rose-purple, 1.3–2 cm. long, greenish-yellow and purple-spotted within, the lobes rounded; anthers pubescent; stigmas ciliolate; caps. dehiscing throughout.—Dry gravelly slopes, 4000–7000 ft.; mostly Yellow Pine F.; San Gabriel Mts., occasional at lower levels to Kern and Santa Barbara cos., San Diego Co. May–July.

19. **M. látidens** (Gray) Greene. [*M. inconspicuus* var. *l.* Gray.] Minutely glandular-puberulent annual, the stem simple or branched, 1–2.5 dm. high; lvs. broadly ovate, forming a rosette of petioled dentate lvs.; the cauline remote, sessile, 1–2 cm. long; pedicels 2–3 cm. long; calyx 7–8 mm. long in anthesis, 10–12 mm. in fr., plicate, the ovate lobes 1–2 mm. long; corolla white or yellowish, often flushed with pink and pink-dotted, 10–11 mm. long, the throat cylindric, the lobes short, spreading; anthers glabrous.—Drying mud-flats in heavy soil, V. Grassland; Menifee V., Riverside Co.; possibly also in San Diego Co.; n. and L. Calif. April–June.

20. **M. longiflòrus** (Nutt.) Grant ssp. **longiflòrus** [*Diplacus l.* Nutt.] Much-branched shrub, 3–12 dm. high, with upper stems, branches, under surface of lvs., pedicels and calyces densely pubescent and glandular-hairy; lvs. lanceolate to oblong, 2.5–8 cm. long, subglabrous, glandular and impressed-veiny above, often revolute, sessile; pedicels 3–7 mm. long; calyx 2.5–3.7 cm. long, the tube gradually expanded upward into a slightly wider throat that is contracted toward apex, the lobes linear-lanceolate, the uppermost 7–9 mm. long, the lower ca. half as long; corolla orange-yellow, to deep orange or buff to almost white, 5–6 cm. long, the tube included and expanding gradually into a throat ca. 1.8 cm. long, the lower surface with 2 deep orange bands; anthers included, glabrous; stigmas subequal, ciliolate; caps. dehiscing along upper suture.—Common on dry rocky slopes, to ca. 5000 ft.; Coastal Sage Scrub, Chaparral; cent. San Diego Co. to San Luis Obispo Co. inland to San Jacinto Mts. and Kern R. region; s. L. Calif. Hybridizing freely with other spp.

Ssp. **calycìnus** (Eastw.) Munz. [*Diplacus c.* Eastw.] Lvs. oblong-elliptic to ovate or lanceolate, yellow-green and glabrous above, glandular-pubescent beneath; calyx-tube with a more inflated throat; corolla light lemon-yellow, the tube usually well exserted; $n=10$ (Stebbins, 1951).—Rocky places below 4500 ft.; mts. of ne. San Diego Co. to Little San Bernardino Mts. and e. San Gabriel Mts. Apr.–July.

M. longiflòrus var. **rùtilus** Grant. Lvs. dark green, villous toward base and beneath; corolla deep velvety red. With ssp. *longiflorus*, particularly in interior Los Angeles Co., less so in Ventura and Riverside cos.

21. **M. mohavénsis** Lemmon. Puberulent, ± glandular, leafy, simple to many-branched annual, reddish-purple, 3–15 cm. high; lvs. elliptical to oblong or obovate, 1–1.8 cm. long, acute, entire, sessile or the lower petioled; pedicels 2–3 mm. long; calyx 10–12 mm. long in fr., angled, broadly campanulate; corolla salverform, 13–15 mm. long, red-purple in the tube and center of limb, the margin pale, ciliate or erose, with rounded-truncate lobes; stigmas rounded, the lower slightly larger.—Uncommon, sandy and gravelly places, 2000–3000 ft.; Creosote Bush Scrub, Joshua Tree Wd.; Mojave Desert in the Barstow-Victorville-Ord Mts. region. April–June.

22. **M. montioìdes** Gray. Glandular-puberulent annual, usually densely branched, 2–6 cm. tall; lvs. linear-oblanceolate to elliptic, entire, 6–13 mm. long, 1-veined, sessile or the lower subovate, petioled; pedicels 5–25 mm. long; calyx 5–7 mm. long, reddish, the teeth equal, broadly ovate, acute to rounded, ca. 0.6 mm. long; corolla yellow, 1–1.7 cm. long, the throat campanulate, minutely pubescent on the ventral ridges, the lobes spreading, rounded; caps. dehiscing through apex.—Moist places, 3000–6000 ft.; Pinyon-Juniper Wd.; Death V. region, Inyo Mts., s. Sierra Nevada; w. Nev. April–June.

23. **M. moschàtus** Dougl. ex Lindl. Perennial from slender creeping rootstocks,

glandular-pilose to -villous, ± slimy; stems 0.3–3 dm. long, creeping or decumbent; lvs. ovate to somewhat oblong, pinnately veined, 1–5 cm. long, subentire to denticulate, the upper sessile; pedicels slender, 1–5 cm. long; calyx campanulate, 9–12 mm. long, ± plicate-angled, the teeth lanceolate, 2–3 mm. long; corolla yellow, 18–26 mm. long, funnelform, the cylindrical tube exserted, ventrally 2-ridged; anthers subglabrous to pubescent.—Wet places, below 7500 ft.; Montane Coniferous F.; mts. from San Diego Co. to B.C., Rocky Mts., Atlantic Coast. June–Aug. Variable, plants with fls. 22–26 mm. long (var. *longiflorus* Gray) and with more sessile upper lvs. (var. *sessiliflorus* Gray) occurring throughout the range.

24. **M. nasùtus** Greene. Much like *M. guttatus*, the calyx 1–2 cm. long, usually dark-dotted, plicate-angled, appressed-puberulent, almost closed by the very acute lower lobes being eventually upcurved against the middle upper which is ca. 3 times as long and projects forward like a long index-finger.—Wet sandy and gravelly places, below 7500 ft.; many Plant Communities; through most of Calif.; to B.C., Rocky Mts., n. Mex. March–Aug.

25. **M. pálmeri** Gray. Glandular-pubescent annual, the stem erect, usually branched, 0.6–1.5 dm. tall; lvs. few, linear to oblanceolate, 1.2–1.7 cm. long, subentire, all but the lowest sessile; pedicels 1–2.5 cm. long; calyx cylindrical, 7–8 mm. long, slightly angled; corolla red-purple, funnelform, 1.5–2 cm. long, the ventral ridges yellow, pubescent, the lobes subequal, rounded, emarginate; anthers finely pubescent; caps. dehiscing slightly through the septum-apex.—Mostly rather damp places, below 7000 ft.; Chaparral, Yellow Pine F., Joshua Tree Wd.; San Bernardino Mts. and edge of Mojave Desert to Sierra Nevada. April–June.

26. **M. paríshii** Greene. Stout erect, densely glandular-villous annual 1–5 dm. high, simple or branched from base; lvs. ovate to lance-oblong, 1.5–4 cm. long, longitudinally 3-veined, remotely denticulate to dentate, sessile with broad base; pedicels ascending-erect, 15–18 mm. long; calyx oblong, 8–9 mm. long in fl., 10–12 mm. in fr., angled, the lobes acute, 1 mm. long; corolla pinkish to lilac or white, 10–12 mm. long, with subequal erect lobes; anthers glabrous; caps. 8–9 mm. long, dehiscing through the apex.—Occasional in wet sandy places, below 7000 ft.; mostly Joshua Tree Wd., Pinyon-Juniper Wd.; desert slopes of San Gabriel and San Bernardino mts. to Little San Bernardino Mts., New York Mts. and Death V. May–Aug.

27. **M. pilòsus** (Benth.) Wats. [*Herpestis p.* Benth. *Mimetanthe p.* Greene.] White-villous annual, erect or ascending, slightly viscid, simple or branched, 0.5–3.5 dm. high, flowering from near base; lvs. lanceolate to oblong, entire, 1–3 cm. long, sessile; pedicels 10–15 (–30) mm. long; calyx 6–7 mm. long, oblique at orifice, short-campanulate, plane, the lobes unequal, lance-ovate, the lowermost shortest and ca. as long as calyx-tube; corolla yellow, obscurely 2-lipped, 7–8 mm. long, the lower lip usually with maroon spots; anthers glabrous; caps. 4–7 mm. long, loculicidal along whole upper side and on lower side toward apex.—Common in moist sandy and gravelly places below 8500 ft.; many Plant Communities; cismontane and montane Calif.; to Wash., Utah, Ariz., L. Calif. April–Sept.

28. **M. primuloìdes** Benth. var. **piloséllus** (Greene) Smiley. [*M. p.* Greene.] Plate 84, Fig. B. Rosulate or short-stemmed perennial, rhizomatous, often stoloniferous, the stems villose to subglabrous, 1–5 cm. high; lvs. obovate, to oblong, with long soft white hairs above, which catch dew and glisten in the sun, longitudinally 3-veined, dentate to entire, 1–4 cm. long, somewhat cuneate at the subsessile base, usually in several pairs and closely set; pedicels erect, slender, 3–12 cm. long; calyx tubular, ca. 5 mm. long; corolla yellow, ca. 8–15 mm. long, funnelform with narrow-campanulate throat, the palate deeper yellow, densely hairy, the throat with red-brown spots below the lower lip; caps. 6–7 mm. long, dehiscing through the apex.—Moist places like meadows, below 11,000 ft.; Montane Coniferous F.; San Jacinto, San Bernardino mts., Mt. Pinos region; to Ore. June–Aug. In the San Bernardino Mts. some plants have some larger corollas and less villous lvs. and may represent var. **primuloides,** which is mostly more northern.

29. **M. puníceus** (Nutt.) Steud. [*Diplacus p.* Nutt.] Freely branched erect shrub 5–15

dm. tall, glabrous to puberulent, glutinous; lvs. 2.5–7 cm. long, linear-lanceolate to elliptic, dark green and glabrous above, paler and glabrous to puberulent beneath, entire to finely toothed, usually revolute; pedicels 1–3 cm. long; calyx 20–25 mm. long, tubular, usually reddish, the lobes linear, the uppermost 6–8 mm. long, the lower 3–4 mm.; corolla brick-red to orange-red, 3–5 cm. long, the tube included in the calyx, the upper lobes with a lateral notch and irregularly toothed, the 3 lower subentire to finely toothed at tip; anthers glabrous; caps. 15–20 mm. long, splitting along upper suture; $n=10$ (Stebbins, 1951).—Dry slopes and mesas, below 2500 ft.; Chaparral, Coastal Sage Scrub; Laguna Beach and Santa Ana Mts. to n. L. Calif. Santa Catalina Id. March–July.

30. **M. purpùreus** Grant. Glandular-pubescent annual 0.3–1 dm. tall, simple or with few basal branches; lvs. lance-oblong, obtuse, obscurely 3–5-veined, subentire, 1–1.5 cm. long; pedicels 3–4.5 cm. long; calyx 6–8 mm. long, slightly angled, the lobes round-ovate, mucronulate, not ciliate, 0.5–1 mm. long; corolla red-purple, 12–15 mm. long, the narrow throat scarcely ridged, glabrous, the lobes round-emarginate, anthers glabrous, red-purple; caps. ca. 6 mm. long, dehiscing slightly through the apex.—Moist sandy places, 6000–7300 ft.; Yellow Pine F.; San Bernardino Mts. May–July.

31. **M. rubéllus** Gray in Torr. Glandular-puberulent annual, simple or branched from base, ± reddish, 2–15 cm. tall; lvs. oblong to lanceolate, 3-veined, entire or nearly so, 0.5–2 cm. long, sessile or lowest subpetiolate; pedicels 7–20 mm. long; calyx 5–8 mm. long, tubular, the teeth broadly ovate; corolla yellow or reddish, 7–9 mm. long, the throat narrow, puberulent on ventral ridges, the lobes rounded, subequal, ± emarginate; anthers glabrous; caps. 4 mm. long, not dehiscent through apex.—Moist sandy places, 3000–8000 ft.; Joshua Tree Wd., Pinyon-Juniper Wd., along w. edge of Colo. Desert and of Mojave Desert, n. to White Mts.; w. Nev. to Colo., New Mex., L. Calif. April–June.

32. **M. rupícola** Cov. & Grant. Finely pubescent annual with some gland-tipped hairs, branched from base, tufted, 3–15 cm. tall; lvs. rhombic-ovate to elliptic-oblanceolate, acute at both ends, entire, sessile, or short-petioled, 2–7 cm. long, pinnately veined; pedicels 3–4 mm. long; calyx 14–16 mm. long in fr., distally plicate-ridged, the lobes lance-attenuate, glandular-pilose; corolla rose, 2.5–3 cm. long, puberulent externally, the tube ca. as long as calyx, glandular-puberulent within, each lobe with a purple basal spot; anthers glabrous; caps. 3–4 mm. long, slightly decurved, tardily dehiscent.—Crevices in limestone, below 5000 ft.; Creosote Bush Scrub; mts. e. side of Death V. March–May.

33. **M. suksdórfii** Gray. Bushy compact glandular-puberulent annual, 1–6 cm. tall, ± reddish; lvs. oblong to oblanceolate, entire, 5–12 mm. long, subsessile; pedicels 2–7 mm. long; calyx 4–6 mm. long, cylindrical, the lobes rounded-mucronate; corolla yellow, funnelform, 5–6 mm. long, the puberulent ventral ridges faintly brown-spotted, the lobes subequal, emarginate; anthers glabrous; caps. ca. 4 mm. long, not apically dehiscent.—Moist sandy places, mostly 5000–8000 ft.; Montane Coniferous F.; San Jacinto and San Bernardino mts., Grapevine Mts., White Mts.; to Wash., Rocky Mts., Ariz. May–Aug.

34. **M. tilíngii** Regel. var. **tilíngii.** Perennial with mass of yellowish slender rootstocks, the stems subglabrous, ± branched, 2–4 dm. tall, commonly stoloniferous; lvs. few, light green, often slimy, broadly ovate to subelliptic, usually saliently dentate, 1–3 cm. long, the lower short-petioled; the upper sessile; fls. few; pedicels 25–50 mm. long; calyx broadly campanulate, 1.5–2 cm. long in fr., often dotted red, strongly plicate-angled, the lowermost pair of teeth eventually curved against the median upper which is 3–5 mm. long; corolla yellow, 2.5–3.5 cm. long, the orifice almost closed by the brown-spotted ridges; anthers glabrous; stigmas slightly fimbriate; caps. short-stipitate, 7–8 mm. long, not dehiscing through the septum-apex; n mostly 14 (Mukherjee & Vickery, 1962).—Wet banks, 5000–11,000 ft.; Montane Coniferous F. to Alpine Fell-fields; San Jacinto and San Bernardino mts.; through Sierra Nevada to B.C., Rocky Mts. July–Sept.

Var. **corallinus** (Greene) Grant. A form with pubescent usually less clammy lvs. With the var. *tilingii*; reported to have $n=24$ (Mukherjee & Vickery, 1962).

35. **M. tráskiae** Grant. Glandular-pubescent annual 1–1.4 dm. tall; lvs. few, broadly

ovate, subglabrous, 3–4 cm. long, the lower short-petioled; pedicels 3–4 mm. long; calyx 17–20 mm. long, glandular-pubescent, the glandular-pubescent ridges green, the intervening space pale, the lobes lanceolate, very unequal, the uppermost 3–3.5 mm. long; corolla red-purple and white, 2–3.5 cm. long, glabrous, the tube ca. as long as the calyx, the 2 upper lobes white, 5–6 mm. long, glabrous, the lower purple, scarcely 1 mm. long; caps. ca. 7–8 mm. long, tardily or not dehiscent.—Santa Catalina Id. March. Not collected in recent years.

13. Mohávea Gray.

Annual herbs, erect, with lanceolate or slightly broader alternate lvs. Fls. in dense leafy spikes; bracteoles none. Calyx 5-parted. Corolla with a short tube, merely gibbous at base, and with ample subcampanulate bilabiate limb, the lips fan-shaped, the lower with large hairy palate. Fertile stamens 2, connivent, the other 3 abortive. Caps. ovoid, thin, bursting irregularly. Seeds ovate-discoid, flattened, each surrounded by its incurved wing. Two spp. (Name of stream where first collected by Frémont.)

Fls. pale yellow, 2.5–3.5 cm. long; palate purple-dotted; lower lip lobed to 6 or 8 mm. above the palate .... 2. *M. confertiflora*
Fls. lemon yellow, 1.5–2 cm. long; palate not conspicuously dotted; lower lip lobed to within 2 or 3 mm. of palate .... 1. *M. breviflora*

1. **M. breviflòra** Cov. With much the same habit as that of *M. confertiflora*, 0.5–2 dm. tall, densely glandular; lvs. ovate-lanceolate, 1–4 cm. long; pedicels 2–5 mm. long; calyx 10–11 mm. long; corolla 15–18 mm. long; stamens glabrous; caps. 8–10 mm. long; seeds 2–2.5 mm. long.—Dry sandy and gravelly places, below 2500 ft.; Creosote Bush Scrub; Death V. region, El Paso Range and s. to Kelso and Bagdad; w. Nev., nw. Ariz. March–April.

2. **M. confertiflòra** (Benth. in DC.) Heller. [*Antirrhinum c.* Benth. in DC.] GHOST FLOWER. Simple or few-branched, 1–4 dm. tall; lvs. 1–6 cm. long, linear to ovate-lanceolate, short-petioled; pedicels 5–10 mm. long; calyx 9–12 mm. long, enlarging in fr., the segms. narrow; corolla closed at throat, the tube and throat ca. ⅓ its entire length; stamens 9–10 mm. long, somewhat pubescent at base; caps. subglobose, 10–12 mm. long; seeds dark, barely 2 mm. long.—Common in sandy washes and on dry gravelly slopes, below 3000 ft.; Creosote Bush Scrub; Colo. and Mojave deserts w. to Daggett and Ord Mts.; to L. Calif., Ariz., Nev. March–April.

14. Orthocárpus Nutt.

Erect or diffuse annuals, with sessile alternate narrow entire to pinnate or parted lvs. Infl. spicate, prominently bracted. Calyx tubular-campanulate, 4-cleft, or cleft before and behind with the divisions lobed. Corolla narrow-tubular, strongly 2-lipped, valvate, the upper lip (galea) erect, entire, beaklike, not much longer than the inflated saccate lower lip which is tipped with 3 small teeth. Stamens 4, inserted near summit of tube; anthers 1- or 2-celled, thin, usually explanate and ciliate. Caps. loculicidal. Seeds few to many, with reticulate or alveolate often loose coat. Ca. 25 spp. of w. N. Am. and 1 in the Andes. (Greek, *orthos*, straight, and *karpos*, fr.)

(Keck, D. D. A revision of the genus Orthocarpus. Proc. Calif. Acad. Sci. IV, 16: 517–571. 1927.)
Anthers 1-celled; seed-coat tight-fitting or ridged; sacs of lower lip of corolla 3–4 mm. deep, yellow, the galea purple .... 3. *O. erianthus*
Anthers 2-celled; lower lip of corolla rather shallow, the sacs mostly 1–2 mm. deep; seed-coat loose-fitting.
　Bracts green throughout; galea equalling or barely exceeding lower lip.
　　Galea finely puberulent or pubescent; upper lvs. 3–5-cleft .... 4. *O. hispidus*
　　Galea densely white-villous; upper lvs. with 2 small lateral lobes ...... 5. *O. lasiorhynchus*
　Bracts tipped with purple or yellow; galea exceeding lower lip.
　　Corolla widened upward; spike usually broad and conspicuous.
　　　Galea nearly straight, pubescent; lower lvs. entire, the upper with a pair of lateral lobes
　　　　　　　　　　　　　　　　　　　　　　　　　　　　　　　　　2. *O. densiflorus*

Galea hooked at tip, densely bearded; lvs. deeply pinnatifid into several filiform divisions
6. *O. purpurascens*
Corolla linear, the lower lip shallow; spike narrow, pale and rather inconspicuous
1. *O. attenuatus*

1. **O. attenuàtus** Gray. Erect, simple or few-branched, 1-4 dm. high, canescent; lvs. lance-linear, 2-6 cm. long, the upper with 2 filiform lobes; spike narrow; bracts 1.5-2 cm. long, 3-lobed, white-tipped, sometimes purplish; calyx 2-2.3 cm., the subequal lobes white-tipped; corolla 1-2.5 cm. long, whitish or sometimes purplish-tinged, narrow, the shallow lower lobe purple-dotted; caps. ellipsoid, 6-10 mm. long; seeds brown, smooth, the loose reticulate coat 1 mm. long.—Grassy places below 5000 ft.; V. Grassland, Foothill Wd., Chaparral, Yellow Pine F., etc.; cismontane from n. L. Calif. and San Diego Co. n. to B.C. Chile. March-May.

2. **O. densiflòrus** Benth. var. **densiflòrus**. OWL'S-CLOVER. Stem erect, slender, often corymbosely branched above, 1-3.5 dm. high, puberulent, yellowish; lvs. linear or linear-lanceolate, 2-8 cm. long, the lower entire, the upper with a pair of lateral lobes; spike dense; bracts usually 3-lobed, 1-2 cm. long, the upper purple-tinged, finely pubescent; calyx 1-2.5 cm. long, the galea subulate, puberulent, the lower lip longer than deep, often yellowish with 3 purple spots; caps. ovoid, 7-10 mm. long; seeds black, in a loose reticulate coat.—Grassy places below 2500 ft.; Foothill Wd., Chaparral, etc.; Los Angeles Co. n. to. Calif. Santa Rosa, Santa Cruz, San Miguel ids. March-May.

Var. **grácilis** (Benth.) Keck. [*O. g.* Benth.] Bracts mostly 5-10 mm. long; corolla more exserted, the lower lip much inflated, as deep as long, often white.—Chaparral, S. Oak Wd.; Santa Lucia Mts. to L. Calif.

3. **O. eriánthus** Benth. BUTTER-AND-EGGS. Stems corymbosely branched from below upward, rarely simple, finely canescent and glandular, 0.5-3.5 dm. high; lvs. 1-5 cm. long, narrow-linear, pinnately divided into several filiform divisions, ± purplish; spike lax below, more congested above; bracts 0.5-1.8 cm. long, with 4-10 linear divisions; calyx 5-8 mm. long; corolla 1-2.5 cm. long, with long slender, exserted tube, the lower lip abrupt, purplish at base, the sacs inflated, light yellow, 3-4 mm. deep, teeth purplish, inconspicuous; galea dark purple, subulate.—Rare in San Diego Co.; common in grassy places, cent. Calif. to n. Calif. March-May.

4. **O. híspidus** Benth. Plate 84, Fig. C. Stems slender, erect, simple or with few erect branches, 1-4 dm. high, pubescent below, ± glandular and hirsute above; lvs. narrowly linear-lanceolate, 1-4 cm. long, the upper 3-5-cleft; spike slender; bracts ovate, 1-2.5 cm. long, palmately 3-7-cleft; calyx 8-10 mm. long; corolla yellow or white, 1.2-2 cm. long, pubescent, the lower lip small, galea usually straight, narrow; caps. oblong, 5-8 mm. long; seeds in a loose reticulate coat.—Meadows, 5000-7500 ft.; Yellow Pine F.; Palomar Mts., San Bernardino Mts.; Sierra Nevada; to Alaska, Ida., Nev. May-Aug.

5. **O. lasiorhýnchus** Gray. Stem slender, erect, often with erect branches, 1-3 dm. high, pilose; lvs. lance-linear, at least the upper with 2 small lateral lobes; spike lax; bracts 3-5-parted, 6-12 mm. long; calyx cleft halfway into 2 lobes each with 2 teeth; corolla yellow with 2 minute blackish dots at base of lower lip, 12-25 mm. long, the tube slender, gradually inflated into the obovoid lower lip; caps. ellipsoid, 6-9 mm. long, the seeds in loose alveolate coats.—Meadows, 4600-7400 ft.; Yellow Pine F., San Bernardino Mts. to Cuyamaca Mts. June-July.

6. **O. purpuráscens** Benth. var. **purpuráscens**. OWL'S-CLOVER. Plate 84, Fig. D. Stems erect, ± slender, often corymbosely branched from base upward, 1-4 dm. high, ± reddish, hirsute; lvs. 1-5 cm. long, deeply pinnatifid into several filiform thickened divisions; spikes dense, pale to deep purple; bracts 1-2 cm. long, palmately 5-7-lobed, typically with greenish hirsute base, greenish-purple in middle and velvety and rose-purple at tips; corolla 1.2-3 cm. long, crimson or purplish, scarcely exserted, the lower lip mostly purplish often with white or yellow toward the tip and with purple dots; galea slender, ± hooked at tip, bearded; caps. 10-15 mm. long; seeds in a loose alveolate coat. —Open fields, grassy slopes, etc.; below 3000 ft.; V. Grassland, Coastal Sage Scrub, etc.;

n. L. Calif. to n. Calif.; Ariz. March–May. Santa Rosa, S. Cruz, S. Catalina and San Miguel ids.

Var. **ornàtus** Jeps. Spike deep purple; corolla deep velvet-red, the outer third of the lower lip orange-yellow.—Open flats, 2000–3000 ft.; Creosote Bush Scrub, Joshua Tree Wd.; w. Mojave Desert.

Var. **pállidus** Keck. Lower of corolla white or pale yellow.—Sandy and disturbed places below 4000 ft.; Coastal Sage Scrub, Chaparral, V. Grassland; San Diego Co. to San Luis Obispo Co.

15. *Pediculàris* L. LOUSEWORT

Perennial herbs. Lvs. in ours alternate, or basal, pinnatifid or pinnate, rarely simple. Fls. in a usually spikelike raceme, white to yellow, red or purple. Calyx 2–5-cleft or -toothed. Corolla strongly 2-lipped, the upper lip galeate, strongly arched; lower lip shorter, with spreading or appressed lobes. Stamens 4, didynamous; anthers glabrous, the equal cells obtuse to subulate at tips. Caps. flattened, glabrous, loculicidal. Seeds several, turgid, often slightly winged. Ca. 500 spp., mostly of N. Temp. Zone. (Latin, *pediculus*, a louse; application uncertain.)

Upper lip of corolla produced above the galea into a slender curving proboscislike beak; raceme much exceeding lvs. White Mts. .... 1. *P. attolens*
Upper lip of corolla blunt and beakless; raceme often little exceeding the lvs.
    Corolla mostly purple-red, ca. 25 mm. long .... 2. *P. densiflora*
    Corolla pale yellow, 15–20 mm. long .... 3. *P. semibarbata*

1. **P. attòlens** Gray. LITTLE ELEPHANT'S HEAD. Glabrous below the white woolly infl., 1.5–4 dm. high; lvs. basal and on lower stem, the blades 3–9 cm. long, the pinnules narrow, serrate-dentate; petioles shorter than blades; bracts few-lobed; pedicels 1–2 mm. long; calyx ca. 5 mm. long; corolla body ca. 7 mm. long, lavender or pink, with purplish markings, the tube recurved through the ventral cleft of the calyx, then recurved and narrowed into an upward beak; caps. ca. 9 mm. long.—Meadows and moist places, high elevs.; White Mts.; Sierra Nevada to Ore. June–Sept.

2. **P. densiflòra** Benth. ex Hook. INDIAN WARRIOR. Plate 84, Fig. E. Stems several, simple, 1–5 dm. tall, pubescent, exceeding lvs.; lvs. largely in a basal rosette or well distributed on stems, pubescent to glabrate, the blades 5–15 cm. long, on shorter petioles, twice pinnatifid into many laciniate-dentate lobes; spikes dense, oblong, the bracts ca. as long as fls., oblong-lanceolate, with sharp salient distal teeth; calyx 8–12 mm. long, deep red; corolla 2–2.5 cm. long, deep purple-red, cylindric, glabrous, strongly arched; the galea 12–16 mm. long; caps. ca. 7 mm. long, dehiscing dorsally and ventrally.—Dry slopes below 6000 ft.; Chaparral, Foothill Wd., Yellow Pine F.; San Diego Co. to n. Calif.; Ore., L. Calif. Jan.–June.

3. **P. semibarbàta** Gray. Stems few, mostly underground, from the root-crown, not over 1 dm. long; lvs. in a rosette, bipinnatifid, 5–15 cm. long, the pinnules ovate and deeply cut into irregularly toothed segms.; petioles as long as blades; pedicels 4–5 mm. long; calyx ca. 10 mm. long, the lobes linear, entire or dentate; corolla 15–20 mm. long, yellowish with purplish tips, white-tomentulose without, the lower lip ca. 2 mm. shorter than the upper; anther-cells sharply acuminate; caps. ca. 9 mm. long, dehiscent dorsally.—Dry slopes, 5000–10,000 ft.; Yellow Pine F. to Subalpine F.; Cuyamaca Mts. to n. Calif.; Ore., Nev. May–June.

16. *Pénstemon* Mitch. BEARD-TONGUE

Perennial herbs or shrubs. Lvs. opposite, rarely whorled or the upper alternate, the lower usually petioled, the upper sessile. Fls. usually showy, in racemose, cymose, or thyrsoid panicles. Calyx 5-parted. Corolla tubular, the throat often inflated, the limb ± 2-lipped, the upper lip 2-lobed, the lower 3-cleft. Fertile stamens 4, paired, the fils. arched, a 5th stamen represented by a long sterile fil. (staminode) often dorsally bearded;

anthers 2-celled, the cells often confluent. Caps. septicidal, cartilaginous. Seeds many, with irregularly angled cellular coat. Over 200 spp., mostly of w. N. Am., 1 of e. Asia; several of horticultural importance. (Greek, *pente*, five, and *stemon*, stamen.)

1. Anther-sacs opening from free tips throughout or partially, almost always divaricate after dehiscence.
    2. Corolla scarlet.
        3. Herbage glaucous; anther-sacs dehiscent by a continuous slit extending across their contiguous parts; corolla subtubular. Cismontane .... 6. *P. centranthifolius*
        3. Herbage not glaucous; anther-sacs opening at their distal ends, the slits not confluent.
            4. Corolla obscurely 2-lipped, the short round lobes scarcely spreading. Desert mts. and borders below the Yellow Pine belt .... 9. *P. eatonii*
            4. Corolla strongly 2-lipped, the upper lip galeate, erect, the lower with sharply reflexed linear lobes. Yellow Pine belt and above .... 15. *P. labrosus*
    2. Corolla not scarlet, although it may be rose or carmine to blue, purple, etc.
        5. Lvs. of lower and middle stems entire or obscurely denticulate.
            6. Infl. not glandular-puberulent.
                7. Corolla 18–35 mm. long.
                    8. Fls. blue-purple, 25–35 mm. long; infl. secund; lvs. not glaucous. Borders of Mojave Desert n. along e. base of Sierra Nevada .... 22. *P. speciosus*
                    8. Fls. red-purple to carmine or paler, 18–27 mm. long; infl. not secund; lvs. ± glaucous.
                        9. Corolla white to flesh-color with lavender limb; lvs. 3–6 mm. wide. N. and e. Mojave Desert .... 11. *P. fruticiformis*
                        9. Corolla red to red-purple; lvs. wider.
                            10. Lvs. lance-ovate, 1–2.5 cm. wide; fls. red-purple. Cismontane
                                19. *P.* × *parishii*
                            10. Lvs. linear to lance-linear, 5–15 mm. broad; fls. rose-lavender or carmine. Desert.
                                11. Thyrsus racemiform; corolla glandular-pubescent, carmine; anther-sacs peltately explanate. San Bernardino Co. .... 27. *P. utahensis*
                                11. Thyrsus open, ± decompound; corolla glabrous, rose-lavender; anther-sacs boat-shaped. Owens V. region ....,.......... 8. *P. confusus*
                7. Corolla 7–18 mm. long, pink to lavender-rose.
                    12. Cauline lvs. cuneate-oblong to spatulate or obovate, 5–18 mm. wide.
                        13. Lvs. white-margined, rhombic-obovate. San Bernardino Co.
                            1. *P. albomarginatus*
                        13. Lvs. not white-margined, lance- or linear-oblong. Inyo Co. .... 8. *P. confusus*
                    12. Cauline lvs. linear, 1–3 mm. wide; plant 30–60 cm. high. Deserts 26. *P. thurberi*
            6. Infl. glandular-puberulent or -pubescent.
                14. Lvs. glabrous, the cauline lance-linear, 2–5 mm. wide; corolla violet with blue limb. Inyo Co. to San Bernardino Co. .... 14. *P. incertus*
                14. Lvs. puberulent.
                    15. The lvs. ashy-puberulent, mostly not much more than 1 cm. long, crowded on the stems.
                        16. Lvs. 3–5 mm. wide; corolla 2-ridged on floor within throat, the upper lip erect, the lower spreading. E. Mojave Desert .... 25. *P. thompsoniae*
                        16. Lvs. 1.5–3 mm. wide; corolla not plicate in throat, the 2 lips spreading. San Jacinto Mts. to L. Calif. .... 5. *P. californicus*
                    15. The lvs. usually not ashy, mostly 2 or more cm. long, not crowded on stems.
                        17. Corolla 14–20 mm. long, ± ventricose; anther-sacs not explanate, 1.3 mm. long. Inyo Co. mts. .... 17. *P. monoensis*
                        17. Corolla 11–14 mm. long, narrowly tubular; anther-sacs explanate, 0.5 mm. long. Grapevine and Providence mts. .... 4. *P. calcareus*
        5. Lvs. of lower and middle stems definitely toothed, usually coarsely so.
            18. Corolla abruptly inflated from a short tube ca. equaling the calyx, strongly 2-lipped, white or tinged pink, lavender or purple; staminode long-bearded, uncinate.
                19. Upper lvs. connate-perfoliate; ovary glandular-puberulent; thyrsus virgate. Mojave Desert .... 18. *P. palmeri*
                19. Upper lvs. distinct; ovary glabrous; thyrsus lax. Montane and to edge of San Joaquin V. .... 12. *P. grinnellii*

18. Corolla nearly tubular to inflated from tube twice as long as calyx, not whitish; staminode short-bearded or glabrous, not uncinate.
  20. Anther-sacs peltately explanate and glabrous; thyrsus virgate, secund; corolla pink to rose-purple.
    21. Corolla obviously 2-lipped and strongly inflated (except in ssp. *austinii*) more than 10 mm. wide (except in ssp.); lvs. glaucous .... 10. *P. floridus*
    21. Corolla nearly regular and tubular-funnelform, 4–8 mm. wide; lvs. green or glaucous.
      22. Calyx 4–6 mm. long; corolla moderately ampliate, 5–8 mm. wide.
        23. Lvs. thin, glaucous, dentate; corolla rose-purple, with dark guide-lines; stems to 10 dm. tall. E. deserts .... 20. *P. pseudospectabilis*
        23. Lvs. thick, green or glaucous, entire or dentate; corolla crimson or red-purple, without guide-lines; stems to 7 dm. tall. W. border of Colo. Desert
        7. *P. clevelandii*
      22. Calyx 3–4.5 mm. long; corolla essentially tubular, 4–5 mm. wide. E. Mojave Desert .... 24. *P. stephensii*
  20. Anther-sacs not explanate, ± scabro-ciliate at suture; thyrsus lax; corolla purplish with blue limb, obviously 2-lipped. Cismontane .... 23. *P. spectabilis*
1. Anther-sacs opening across their continuous apices, the free tips remaining saccate, parallel even after dehiscence, glabrous or somewhat hairy.
  24. Corolla scarlet, tubular; limb relatively long, the upper lip erect, the lower sharply reflexed. Montane .... 2. *P. bridgesii*
  24. Corolla bluish or purplish, usually ampliate; limb relatively short, with both lips ± spreading.
    25. Staminode bearded; infl. glandular-pubescent; corolla blue; lvs. mostly basal. Inyo-White Mts. .... 21. *P. scapoides*
    25. Staminode glabrous.
      26. Lvs. mostly basal; orifice of corolla bearded ventrally. San Bernardino Mts.
      3. *P. caesius*
      26. Lvs. well distributed; orifice of corolla glabrous.
        27. Infl. glandular-pubescent; peduncles divergent; corolla 20–30 mm. long
        16. *P. laetus*
        27. Infl. not glandular; peduncles appressed; corolla 25–35 mm. long
        13. *P. heterophyllus*

1. **P. albomarginàtus** Jones. Stems 1.5–3 dm. high, many, from an elongate fleshy deeply buried root, the whole plant pallid, glaucescent, glabrous; lvs. entire, spatulate to rhombic-ovate, 1–3 cm. long, on shorter petioles; lvs. and calyces narrowly bordered with a scarious white ± scabrid margin; thyrsus leafy, 5–12 cm. long; calyx-lobes linear-lanceolate to broadly ovate-oblong, 4–5 mm. long; corolla lavender-pink, whitish ventrally, with purple guide-lines within, 13–18 mm. long, up to 5 mm. wide at throat, the 2 palatal ridges bearded with flat yellow hairs; anther-sacs explanate; staminode glabrous; $2n=8$ pairs (Raven et al., 1965).—Deep sand at ca. 1800 ft.; Creosote Bush Scrub; near Lavic, s. Mojave Desert; to s. Nev., w. Ariz. March–May.

2. **P. brídgesii** Gray. Woody at base, 3–10 dm. high from a branched caudex; herbage yellow-green, glabrous or puberulent; lower lvs. 2–6 cm. long on somewhat shorter petioles, 2–12 mm. wide, linear-lanceolate to spatulate, the cauline linear to elliptic, sessile; thyrsus subsecund, glandular-pubescent, mostly rather narrow; calyx 4–8 mm. long; corolla scarlet to vermilion, 22–35 mm. long, 4–6 mm. wide pressed, tubular, sparingly glandular; anther-sacs dehiscent $\frac{1}{4}$–$\frac{1}{3}$ their length, the suture spinulose-ciliate; staminode glabrous; $n=8$ (Keck, 1951), $2n=42$ (Zhukova, 1967).—Dry slopes, 5000–10,700 ft.; Montane Coniferous F.; mts. from San Diego Co. to Alpine Co., Panamint and Tehachapi mts. to Last Chance and White mts.; to L. Calif., Colo., Ariz. June–Aug.

3. **P. caèsius** Gray. Loosely cespitose from a matty wooded caudex, the herbage ± glaucous, glabrous below, the stems erect, 1.5–4.5 dm. high; lvs. mostly basal, the lower roundish, 1–2 cm. long, on petioles as long, the cauline remote, reduced, the upper linear-oblanceolate; thyrsus glandular-pubescent, lax, rather few-fld.; calyx 4–7 mm. high; corolla purplish-blue, 17–23 mm. long, gradually ampliate, the lips equal, small,

glabrous within; anther-sacs short-toothed on the suture; staminode glabrous.—Dry rocky slopes 6700–11,200 ft.; Montane Coniferous F.; San Bernardino and San Gabriel mts.; San Emigdio Range and Greenhorn Mts. Kern Co.; s. Sierra Nevada. June–Aug.

4. **P. calcàreus** Bdg. [*P. desertorum* Jones.] Stems 0.5–2.5 dm. tall, densely pruinose-puberulent, purplish; lvs. firm, mostly entire, largely basal, ovate to elliptic, 1–3.5 cm. long, on petioles as long, the cauline narrower, smaller; thyrsus densely glandular-pubescent; calyx 6 mm. long in fl., to 11 mm. in fr.; corolla light rose to rose-purple, 12–16 mm. long, 2.5–4 mm. wide pressed, the palate sparingly pilose; anther-sacs widely divaricate, peltately explanate, 0.5 mm. long; staminode strongly bearded with coarse golden hairs.—Dry crevices in limestone, 3500–6000 ft.; Creosote Bush Scrub to Pinyon-Juniper Wd.; Providence Mts., Grapevine Mts., Last Chance Mts. April–May.

5. **P. califórnicus** (M. & J.) Keck. [*P. linarioides* var. *c.* M. & J.] Stems 0.5–1.5 dm. high, forming matted tufts, densely leafy below, the herbage cinereous-puberulent with retrorse appressed hairs; lvs. linear-oblanceolate, entire, thickish, 5–9 mm. long, mucronate, short-petioled; thyrse racemiform, minutely glandular; calyx-lobes ovate, the margin scarious; corolla blue with purplish cast, 14–18 mm. long, tubular-funnelform, strongly bilabiate; anther-sacs ovate-oblong; staminode yellow-bearded.—Local, stony slopes, 3500–7000 ft.; Chaparral, Yellow Pine F.; Hemet V., San Jacinto Mts., Santa Rosa Mts., Aguanga, n. L. Calif. May–June.

6. **P. centranthifòlius** Benth. SCARLET BUGLER. Plate 84, Fig. F. Glabrous, glaucous; stems few to several, virgate, 3–12 dm. high; lvs. entire, the basal spatulate, 3–7 cm. long, on much shorter petioles; the cauline lanceolate, the upper auriculate-clasping; thyrse virgate, half as tall as plant, secund; calyx 3–6 mm. long; corolla scarlet, tubular, 25–33 mm. long, 4.5–6 mm. wide pressed, the lobes scarcely spreading; anthers-sacs peltately explanate; staminode glabrous; $n=8$ (Keck, 1951).—Dry disturbed places below 6500 ft.; Chaparral; L. Calif. to cent. Calif. April–July.

7. **P. clevelándii** Gray ssp. **clevelándii**. Stems few to several, 3–7 dm. high; lvs. glaucescent to deep green, entire to moderately serrate, the basal ovate, 2–5 cm. long, petioled, the upper cauline triangular-lanceolate to cordate, distinct; thyrsus narrowly racemose, 1–3 dm. long, ± glandular-pubescent; calyx-lobes ovate to roundish, purplish, 4–5 mm. long; corolla crimson to red-purple, 17–24 mm. long, tubular-funnelform, the tube proper shorter than the gradually ampliate throat, the limb rotately spreading; anther-sacs explanate; staminode 9–11 mm. long, glabrous or feebly bearded; $n=8$ (Keck, 1951).—Dry open slopes, 2500–4500 ft.; Chaparral; e. San Diego Co. to L. Calif. March–May.

Ssp. **connàtus** (M. & J.) Keck. Glaucous; upper lvs. connate-perfoliate; infl. glabrous; corolla-limb not glandular; anthers not explanate; staminode bearded.—Below 5500 ft.; canyons bordering Colo. Desert, Riverside Co. to e. San Diego Co.

Ssp. **mohavénsis** Keck. Lvs. bright green, coarsely serrate; upper lvs. not connate-perfoliate; corolla contracted at orifice, narrow in throat; staminode 6–8 mm. long, bearded.—Dry rocky places, 3500–4500 ft.; Creosote Bush Scrub, Joshua Tree Wd.; Little San Bernardino Mts., Sheephole Mts., s. Mojave Desert. March–May.

8. **P. confùsus** Jones ssp. **pàtens** (Jones) Keck. Like *P. utahensis*, but lower, more leafy; thyrsus more open, compound; corolla rose-lavender to purplish, 14–20 mm. long, slightly ampliate, glabrous; anther-sacs not explanate, scabrid-ciliate at suture; staminode uncinate, papillose-bearded at apex; $n=8$ (Keck, 1951).—Dry loose soil, washes and slopes, 6000–7500 ft.; Pinyon-Juniper Wd., Sagebrush Scrub; Inyo Mts. May–June.

9. **P. eatònii** Gray. Stems few to several, glabrous, virgate, 3–10 dm. high; lvs. coriaceous, green or glaucescent, glabrous, the basal oblanceolate, 3–10 cm. long, on petioles almost as long, the cauline lance-oblong with clasping base, 4–10 cm. long; thyrsus strict, secund, glabrous, half the length of the stem; calyx 4–6 mm. long, the lobes elliptic to ovate, with narrow scarious margin; corolla scarlet, 25–30 mm. long, sub-tubular, obscurely 2-lipped, glabrous; anther-sacs parallel or divergent, puberulent; staminode glabrous or ± bearded; $n=8$ (Keck, 1951).—Dry gravelly slopes below 8000 ft.;

Pinyon-Juniper Wd. to Creosote Bush Scrub; desert slopes from San Bernardino Mts. to Clark Mt.; to Utah. March–July.

Var. **undòsus** Jones. [*P. munzii* Jtn.] Stems and lvs. puberulent; anthers often ± exserted.—With the spp., but more frequent.

10. **P. flóridus** Bdg. ssp. **flóridus.** Stems several, erect, virgate, 6–12 dm. high, the herbage blue-glaucous, glabrous below the infl.; lvs. irregularly spinose-dentate or the uppermost subentire, the basal oblong-ovate, 2–6 cm. long, on fairly long petioles, the lower cauline lance-ovate, 5–10 cm. long, sessile; thyrsus glandular-pubescent; calyx-lobes ± ovate, 4–7 mm. long; corolla rose-pink, with dark guide-lines within, 22–30 mm. long, abruptly inflated, the orifice oblique; staminode glabrous; $n = 8$ (Keck, 1951).—At 5500–8500 ft.; Pinyon-Juniper Wd.; in and near Inyo-White Mts.; w. Nev. May–July.

Ssp. **austínii** (Eastw.) Keck. [*P. a.* Eastw.] Corolla gradually ampliate, the orifice perpendicular.—At 3500–6500 ft.; Panamint Mts., Grapevine Mts., Inyo Mts. May–June.

11. **P. fruticifórmis** Cov. ssp. **fruticifórmis.** Glabrous glaucous shrub, much branched, 3–6 dm. high; lvs. ± entire, narrowly linear-lanceolate, 2–6 cm. long, 3–6 mm. wide; thyrsus lax, few-fld.; calyx 5–7 mm. long, the lobes ovate to roundish; corolla 20–27 mm. long, white or flesh-color, with pale lavender limb, the guide-lines purple, glabrous externally; $n = 8$ (Keck, 1951).—Dry rocky places, 3600–7500 ft.; Creosote Bush Scrub to Pinyon-Juniper Wd.; Panamint, Argus and Inyo mts. May–June.

Ssp. **amargòsae** Keck. Calyx-lobes lance-ovate to broadly ovate; corolla glandular puberulent externally, scarcely at all within.—Kingston Mts., e. San Bernardino Co.; w. Nev.

12. **P. grinnéllii** Eastw. ssp. **grinnéllii.** Stems decumbent at base, ± branched, 1–4 dm. high, forming low spreading plants, glabrous; lvs. light green, finely to coarsely spinulose-dentate, or uppermost entire, like those of *P. palmeri*, but not connate; thyrse more lax and open, glandular-pubescent, 1–2 dm. long, 4–8 cm. wide; calyx 4–6 mm. long; corolla whitish with flesh-pink or lavender tinge, 20–30 mm. long, the guide-lines prominent, otherwise like that of *P. palmeri*; $n = 8$ (Keck, 1951).—Dry gravelly mostly granitic slopes, 4500–9500 ft.; mostly Yellow Pine F. and above; San Gabriel Mts. to Santa Rosa Mts. May–Aug.

Ssp. **scrophularioìdes** (Jones) Munz. [*P. s.* Jones.] Plants larger, 3–6 dm. tall, glaucous; lvs. blue-green; corolla with violet to blue-violet limb.—Dry slopes, 1800–8000 ft.; Foothill Wd., Pinyon-Juniper Wd.; Mt. Pinos and Frazier Mt. region, Lebec, Tehachapi, etc.; to s. cent. Calif. April–July.

13. **P. heterophýllus** Lindl. var. **heterophýllus.** Plate 85, Fig. A. Woody at base, forming clumps 3–5 dm. high, glabrous throughout or sometimes minutely puberulent below, green or glaucous; lvs. linear, usually fasciculate, 2–4.5 cm. long, 2–4 mm. wide; thyrsus strict, subracemose, glabrous; calyx 4–6 mm. long, the lobes mostly oblanceolate to obovate; corolla rose-violet, with blue or lilac lobes, glabrous, 25–35 mm. long, gaping; anther-sacs ca. 2.5 mm. long, sagittate, arcuate, spinose on suture margin; staminode glabrous; $n = 8, 16$ (Keck, 1951).—Dry hillsides below 5500 ft.; Chaparral, Foothill Wd., Yellow Pine F.; San Diego Co. to n. Calif. April–July.

Var. **austràlis** M. & J. Puberulent almost throughout; lvs. narrow, fasciculate; calyx usually puberulent.—Below 5000 ft.; Chaparral; San Diego Co. to Monterey Co.

14. **P. incértus** Bdg. [*P. fruticiformis* var. *i.* M. & J.] Habit of *P. fruticiformis*; lvs. 2–3 mm. wide; thyrsus lax, moderately glandular; calyx 5–7 mm. high, the lobes lance-ovate to roundish; corolla 25–28 mm. long, violet with a reddish cast or purple, the limb deep blue without guide-lines, strongly 2-lipped; anther-sacs divaricate but not explanate, minutely denticulate-ciliate at suture; staminode densely bearded.—Dry gravelly slopes, 3500–5500 ft.; Joshua Tree Wd., Sagebrush Scrub, from n. Los Angeles Co. and n. base of San Bernardino Mts. to Argus Mts., Tehachapi, etc. May–June.

15. **P. labròsus** (Gray) Hook. [*P. barbatus* var. *l.* Gray.] Mostly bright green, glabrous throughout, the stems 1–few, erect, virgate, 3–7 dm. high; lvs. coriaceous, the basal linear-oblanceolate, obtuse, short-petioled, the blades 3–7 cm. long, on petioles almost as long; the cauline linear, reduced up the stem; thyrsus strict, slender, somewhat secund; calyx 4–5.5 mm. long, the lobes ovate, the margin scarious, erose to entire; corolla scarlet,

Plate 85. Fig. A, *Penstemon heterophyllus*, sagittate anther-sacs, staminode glabrous. Fig. B, *Penstemon spectabilis*, upper lvs. connate, corolla-throat ample. Fig. C, *Scrophularia californica* var. *floribunda*, corolla-tube subglobose, anthers 4. Fig. D, *Castela emoryi*, ♂ and ♀ fls. and diverging drupes. Fig. E, *Datura meteloides*, nodding spiny caps. Fig. F, *Lycium brevipes*.

32–40 mm. long, tubular, the limb ca. ⅔ the length of the whole, the upper lip erect, the lower more deeply divided, with reflexed linear lobes; anther-sacs divergent, opening by slits for ⅔ their length; staminode glabrous; $n=8$ (Keck, 1951).—Fairly dry slopes and benches, 5000–10,000 ft.; Montane Coniferous F.; mts. from Ventura Co. to n. L. Calif. July–Aug.

16. **P. laètus** Gray. Woody at base, 2–8 dm. high, the herbage gray- or yellow-green, mostly puberulent or canescent, the stems often purplish; lvs. linear to oblanceolate, the upper lanceolate, 1–6 cm. long, the lower petioled; thyrsus narrow, glandular-pubescent, somewhat lax; calyx 4–8 mm. high, the lobes lanceolate to narrow-ovate or oblong, acute to acuminate; corolla 20–30 mm. long, blue-lavender to blue-violet, tubular-campanulate, the limb bright blue, widely gaping, glabrous within; anther-sacs tinged purple, mostly white-hairy, the suture spinose; staminode glabrous; $n=8$ (Keck, 1951).—Dry rocky and disturbed slopes, 1200–8500 ft.; Foothill Wd. to Montane Coniferous F.; mts. from Ventura Co. to Tehachapi, n. to Yuba Co. May–July.

17. **P. monoénsis** Heller. Stems 1.5–3.5 dm. high, densely cinereous-puberulent; lvs. entire, the margins often crisped, densely scurfy-puberulent, oblong-ovate to lance-oblong, the basal 4–12 cm. long, with petioles 2–4 cm. long, the upper broadly clasping; thyrsus densely glandular-pubescent, of 4–8 dense clusters; calyx 7–8 mm. long, in fl., the lobes linear-lanceolate; corolla rose-purple to wine-red, 14–20 mm. long, tubular-funnelform, mostly nearly glabrous within; anther-sacs divergent, dehiscent quite to the proximal apices, not explanate; staminode strongly bearded; caps. 6–8 mm. long.—Dry stony places, 3800–6000 ft.; Joshua Tree Wd., Sagebrush Scrub; Last Chance Mts., Argus Mts., Inyo-White Mts. April–May.

18. **P. pálmeri** Gray. Gray-glaucous and glabrous below the infl., the stems 5–12 dm. tall; lvs. irregularly spinose-dentate or the uppermost subentire, the basal oblong-ovate, 2–8 cm. long, on petioles almost as long, the cauline lance-ovate, auriculate-clasping; the upper pairs connate-perfoliate; thyrsus virgate, glandular-pubescent, 2–6 dm. long; calyx 4–6 mm. long, the lobes broadly ovate; corolla whitish suffused with pink or lilac, with prominent guide-lines extending into throat from lower lip, 22–35 mm. long, the short tube expanding abruptly into the inflated throat; anther-sacs peltately explanate; staminode shaggy-bearded; caps. 10–14 mm. long; $n=8$ (Keck, 1951).—Dry rocky gullies, 4000–6000 ft.; Joshua Tree Wd., Pinyon-Juniper Wd.; mts. w. of Death V., Kingston, Clark, Providence and N.Y. mts.; to Utah, Ariz. May–June.

19. **P. × paríshii** Gray. Habit of *P. centranthifolius*, 3–8 dm. high; cauline lvs. entire to shallowly serrate, the uppermost clasping but not connate; thyrsus virgate; corolla red-purple, the gradually ampliate throat 6–9 mm. wide.—Occasional in areas where *P. centranthifolius* and *P. spectabilis* come together and apparently a hybrid between them.

20. **P. pseudospectàbilis** Jones. Habit of *P. floridus*; lvs. glaucous, prominently serrate, the basal lance-ovate to broadly ovate, the upper connate-perfoliate, forming disks up to 15 cm. long and 8 cm. broad; thyrsus sparingly glandular; calyx-lobes mostly ovate, short-acuminate, 4–6 mm. long; corolla rose-purple, 20–26 mm. long, 6–9 mm. wide pressed, moderately ampliate, viscid-puberulent at orifice; staminode glabrous; $n=8$ (Keck, 1951).—Desert washes and canyons, below 4000 ft.; Creosote Bush Scrub; Sheephole Mts. to Chuckwalla Mts. and Imperial Co.; Ariz. March–May.

21. **P. scapoìdes** Keck. Stems few from a matted branching caudex, erect, slender, 2–4 dm. high, glabrous, glaucous; lvs. mostly basal, ovate to roundish, often folded, 0.7–1.5 cm. long, on petioles as long, the cauline few, much reduced; thyrsus few-fld., glandular-pubescent, lax, the solitary fls. on long pedicels; calyx 3–4.5 mm. long, the lobes oblong to ovate; corolla 26–34 mm. long, blue, the throat pale beneath and within, only slightly ampliate, the lower lip somewhat exceeding the upper, the palate 2-ridged, yellow-hairy; anther-sacs pale, dehiscent less than half their length, toothed along the suture; staminode yellow-pilose below the tip.—Dry stony ridges and canyons, 7000–10,000 ft.; Pinyon-Juniper Wd., Bristle-cone Pine F.; White and Inyo mts. June–July.

22. **P. speciòsus** Dougl. ex Lindl. [*P. glaber* and var. *utahensis* of Jeps.] Herbage glabrous to pruinose-puberulent, sometimes glaucescent; stems in erect clumps, 2–8 dm.

high; lvs. entire, thickish, the basal lanceolate to oblanceolate, sessile, gradually reduced upward; thyrsus elongate, of many obscurely interrupted showy clusters, ± secund; calyx 4–6 (–8) mm. high, the lobes narrow-ovate to suborbicular; corolla bright blue-purple, 26–35 mm. long, glabrous, the tube rather long, abruptly flaring into the ample throat, the limb large, strongly 2-lipped; anther-sacs divaricate, opening from the apex $\frac{2}{3}$ their length; staminode glabrous or rarely bearded; caps. 6–12 mm. tall; $n=8$ (Keck). —Dry plains and slopes, 3500–8000 ft.; Sagebrush Scrub to upper Montane Coniferous F.; borders of Mojave Desert from San Bernardino Mts. to San Gabriel Mts. and Tehachapi, Mt. Pinos, Inyo-White, Grapevine and Last Chance mts.; to Wash., Ida., Utah. May–July.

23. **P. spectábilis** Thurb. ex Gray. Plate 85, Fig. B. Stems several from the base, erect, 8–12 dm. high, green or glaucescent, glabrous throughout; lvs. coarsely serrate, the lower broadly oblanceolate to ovate, 2–10 cm. long, the upper connate-perfoliate; thyrsus lax, often half as tall as the plant; calyx 4–7 mm. high, the lobes lance-ovate to roundish; corolla lavender-purple with blue lobes, whitish within, 25–33 mm. long, the tube rather abruptly expanded into the ample throat, the limb strongly bilabiate; anther-sacs not explanate, twice as long as wide; staminode glabrous toward tip; caps. 10–14 mm. long; $n=8$ (Keck).–Dry washes and recently disturbed places, below 6000 ft.; Coastal Sage Scrub, Chaparral; cismontane slopes from e. Los Angeles Co. to n. L. Calif. April–June.

Ssp. **subviscòsus** Keck. Pedicels and calyces glandular-puberulent.—San Gorgonio Pass to Santa Monica Mts. and Liebre Mts. (n. Los Angeles Co.).

24. **P. stèphensii** Bdg. Habit of *P. clevelandii*; lvs. mostly finely and sharply denticulate, the upper pairs connate-perfoliate; thyrsus sparingly glandular; calyx 3–4.5 mm. long, the roundish to broadly ovate lobes abruptly acute; corolla rose to magenta, 17–22 mm. long, without prominent guide-lines, the throat slightly dilated; anther-sacs glabrous, peltately explanate, at least as broad as long; staminode glabrous. Rocky slopes, 5000–6000 ft.; Sagebrush Scrub, Pinyon-Juniper Wd.; Kingston and Providence mts., e. San Bernardino Co. April–June.

25. **P. thompsòniae** (Gray) Rydb. [*P. pumilus* var. *t.* Gray.] Stems prostrate or ascending, arising from a woody caudex, forming tufts or mats, 0.2–0.5 dm. high and 1–2.5 dm. across; lvs. entire, oblanceolate to spatulate-oblong, 5–12 mm. long, cinereous-whitened with closely appressed hairs, narrowed to short petioles; thyrse racemiform, leafy, obscurely viscid; calyx-lobes acuminate to attenuate, 4–5 mm. long; corolla blue-violet, 13–18 mm. long, subtubular, the palate bearded; anther-sacs ovate-oblong; staminode golden-bearded.—Dry limestone slopes, at ca. 6000 ft.; Pinyon-Juniper Wd.; Clark Mt., e. San Bernardino Co.; to se. Utah, n. Ariz. May–June.

26. **P. thúrberi** Torr. [*P. ambiguus* var. *t.* Gray.] Intricately branched subshrub 3–6 dm. high, with numerous slender erect mostly simple stems, glabrous throughout; lvs. bright green, equally distributed, entire, ± scabrid on margin, mostly narrowly linear, involute, 1–3 cm. long; thyrsus racemose, the short divergent peduncles mostly 1-fld.; calyx 2–3 mm. long, with broad entire lobes; corolla lavender-rose or bluish, 12–15 mm. long, obliquely salverform, the limb prominent; anther-sacs explanate; staminode glabrous.—Local, in dry gravelly places, 800–5000 ft.; Creosote Bush Scrub, Joshua Tree Wd., Pinyon-Juniper Wd.; San Felipe V. in e. San Diego Co., Little San Bernardino Mts., Providence Mts.; to New Mex., L. Calif. May–June, sometimes also in fall.

27. **P. utahénsis** Eastw. Much like *P. centranthifolius*, 3–6 dm. high; cauline lvs. lance-oblong, 3–8 cm. long, broadest at clasping base; corolla carmine, 18–24 mm. long, the lobes glandular-pubescent without, glandular about orifice; staminode uncinate at apex, glabrous or slightly bearded.—Occasional in rocky places, 4000–5500 ft.; Shadscale Scrub to Pinyon-Juniper Wd.; New York and Kingston mts., e. Mojave Desert; to s. Utah, n. Ariz. April–May.

17. *Scrophulària* L. FIGWORT

Coarse perennial herbs with 4-angled stems. Lvs. opposite, petioled, with ± irregularly toothed to divided blades. Fls. small, in loose cymes forming terminal panicles. Calyx

5-parted, with rather broad lobes. Corolla greenish-purple to maroon or darker, the tube globular to ellipsoid, ventricose, the upper lip with its 2 lobes projected forward, the lower with the lateral lobes vertical and the cent. deflexed. Stamens 4, declined, usually included, the anther-cells divergent; sterile stamen scalelike to lacking. Caps. septicidal. Seeds many, rugose, plump. Ca. 100 spp. of N. Hemis. (From *scrofula*, since some spp. have rhizomal knobs that were supposed to cure this disease.)

(Shaw, R. J. The biosystematics of Scrophularia in western North America. Aliso 5: 147-178. 1962.)

Infl. villose, the hairs with small gland-tips; sepals acuminate to acute; sterile fil. nearly or quite lacking. Santa Catalina & San Clemente ids. .... 3. *S. villosa*
Infl. puberulent or short-pubescent, the hairs with large gland-tips; sepals often rounded; sterile fil. developed. Mainland.
    Lf.-bases truncate to cordate; corolla not distinctly bicolored .... 1. *S. californica*
    Lf.-bases cuneate; corolla distinctly bicolored .... 2. *S. desertorum*

1. **S. califórnica** C. & S. var. **floribúnda** Greene. Plate 85, Fig. C. Stems coarse, 10-18 dm. tall, finely pubescent; lf.-blades triangular-ovate to ovate, rather dark green, doubly dentate to ± incised, 5-15 cm. long, on petioles 2-6 cm. long; panicle rather lax, 2-4 dm. long; calyx-lobes mostly ca. 3 mm. long; corolla mostly 6-10 mm. long, red-brown to maroon; sterile fil. brown to purplish, clavate to obovate; caps. conic-ovoid, 6-8 mm. long; $n=$ ca. 45-46, 47-48 (Shaw, 1962).—Dryish to moist places below 6000 ft.; Coastal Sage Scrub, Chaparral, etc.; cismontane s. Calif. to n. L. Calif. and from Tehachapi Mts. etc. to n. Calif. away from the coast. March-May.

2. **S. desertòrum** (Munz) Shaw. [*S. californica* var. *d*. Munz.] Glabrescent perennial 7-12 dm. tall; lf.-blades lanceolate, narrowly acute, doubly serrate with cuspidate teeth, puberulent; larger lvs. with blades 10-13 cm. long, on petioles 7-10 cm. long; panicle elongate, glandular-puberulent; sepals 2-3 mm. long, slightly erose; corolla 7-9 mm. long, conspicuously bicolored, dorsal half maroon, ventral half cream with pinkish edges; sterile fil. clavate, light maroon, longer than broad, 1 mm. wide; caps. 6-8 mm. long; $n=48$ (Shaw, 1962).—Dry slopes below 10,000 ft.; Pinyon-Juniper Wd., etc.; Panamint Mts. to Inyo-White Mts.; w. Nev. May-Aug.

3. **S. villòsa** Penn. Lvs. ovate, acute, dentate with sharply acute teeth, the blades 10-15 cm. long, on petioles 3-5 cm. long; infl. glandular-villous with conspicuous hairs; panicle with widely spreading branches; calyx-lobes triangular-ovate, acute to acuminate; corolla 8-9 mm. long, deep maroon, the upper lobes dark, the lower slightly paler and deflexed; uppermost fil. a minute rudiment or lacking; $n=$ ca. 47-48 (Shaw).—Rocky canyons, Coastal Sage Scrub; San Clemente and Santa Catalina ids. April-Aug.

18. Stemòdia L.

Perennial herbs with opposite or alternate serrate clasping lvs. Fls. solitary in the axils of bracts of a loose raceme. Bracteoles 2 just beneath the calyx. Calyx 5-parted into linear-lanceolate segms. Corolla blue-violet, with cylindrical tube, 2-lobed arched upper lip and 3-lobed deflexed-spreading lower. Stamens 4, didynamous. Anther-cells separate and stipitate. Stigmas distinct, lamelliform. Caps. cylindric-ovoid, septicidal and loculicidal. Seeds many, obscurely reticulate. Ca. 20 spp., mostly of trop. Am. (Name abbreviated from *Stemodiacra*, meaning stamen with 2 tips.)

1. **S. durantifòlia** (L.) Sw. Stems several to many, ascending, 1-5 dm. tall, glandular-pubescent; lvs. lanceolate, subsessile with narrowed then dilated base, 2-4 cm. long, smaller upward; infl. spiciform, glandular, of 6-12 remote fascicles; lower pedicels 2-12 mm. long; calyx 5-6 mm. long; corolla 7-8 mm. long, purplish; caps. 5 mm. long.—Occasional in wet places, as near Palm Springs, in San Diego region; to Ariz., trop. Am. Much of year.

19. **Verbáscum** L. MULLEIN

Biennial or perennial herbs, erect, simple or virgately branched. Cauline lvs. simple, alternate, sessile, clasping or somewhat decurrent. Fls. in racemes or crowded spikes, ephemeral. Bracteoles 0. Calyx 5-parted. Corolla rotate, slightly irregular, commonly yellow. Stamens 5, the fils. villous-pubescent, alike or the lower pair different from the others. Stigmas united, capitate. Caps. ellipsoid to subglobose, septicidally 2-valved. Seeds many, wingless, pitted or roughened. Ca. 250 spp. of Eurasia. (Corrupted from *Barbascum*, the old Latin name.)

Plants with green herbage, with simple gland-tipped hairs; lvs. dentate.
    Pedicels 10–15 mm. long; lvs. sinuate-dentate, glabrous .... 1. *V. blattaria*
    Pedicels 3–5 mm. long; lvs. sinuately denticulate, pubescent .... 3. *V. virgatum*
Plants very woolly; lvs. entire; pedicels less than 10 mm. long .... 2. *V. thapsus*

1. **V. blattària** L. MOTH MULLEIN. Stems 4–12 dm. high, glandular-pubescent above; cauline lvs. 2–12 cm. long, elliptic to ovate, doubly serrate-crenate, not decurrent; calyx 5–8 mm. long; corolla yellow or white, 25–30 mm. broad; caps. 6–8 mm. long; $2n=30$, 32 (Håkansson, 1926).—Occasional weed in waste places, through most of cismontane Calif.; natur. from Eurasia. May–Sept.

2. **V. thápsus** L. COMMON MULLEIN. Stems stout, 5–18 dm. tall; basal lvs. in rosettes, oblong-obovate, to obovate-lanceolate, 15–40 cm. long including petioles; cauline lvs. elliptic-lanceolate, gradually reduced up the stem, decurrent; pedicels less than 2 mm. long; calyx 7–9 mm. long; corolla 20–25 mm. broad, yellow; caps. 5 mm. long, stellate-pubescent; $2n=36$ (Gadella & Kliphuis, 1966).—Common weed in waste places in n. Calif., occasional in s. Calif. especially above 4000 ft.; natur. from Eurasia. June–Sept.

3. **V. virgàtum** Stokes ex With. Similar to *V. blattaria*, but somewhat more pubescent and glandular; lvs. crenate; pedicels shorter than caps.; $2n=32$ (Håkansson, 1926).—Occasional weed; natur. from Eurasia.

20. **Verónica** L. SPEEDWELL

Annual or perennial, erect or prostrate herbs, with opposite lvs. or the upper bractlike, alternate. Calyx 4- or 5-parted. Corolla subrotate, white to blue, 4-lobed (the upper broad lobe formed by fusion of the normal upper pair). Stamens 2. Stigma entire. Caps. flattened, loculicidal. Seeds flattened, smooth or rarely roughened. (Possibly named for St. *Veronica*.)

(Pennell, F. W. Veronica in N. and S. Am. Rhodora 23: 1–22, 29–41. 1921.)
1. Main stem ending in a single racemelike infl.
  2. Plants perennial from underground rootstocks; fls. in upper axils only; caps. wider than long, deeply notched .... 8. *V. serpyllifolia*
  2. Plants annual; fls. from most axils.
    3. Corolla 2–2.5 mm. wide; pedicels 1–2 mm. long.
      4. Lvs. linear-oblong to spatulate; corolla white .... 5. *V. peregrina*
      4. Lvs. rounded to oval; corolla bright blue .... 3. *V. arvensis*
    3. Corolla 7–11 mm. wide; pedicels 10–25 mm. long .... 6. *V. persica*
1. Main stem with lateral racemes below the tip.
  5. Lvs. of main stems definitely short-petioled, much longer than wide; corolla 7–10 mm. wide
                                                                                                 1. *V. americana*
  5. Lvs. of main stems sessile, ± cordate-clasping; corolla 3–7 mm. wide.
    6. Caps. ovoid to roundish, scarcely notched; lvs. lanceolate to broader.
      7. Pedicels in fr. ascending, 6–8 mm. long; calyx mostly exceeding the caps.
                                                                                      2. *V. anagallis-aquatica*
      7. Pedicels in fr. ± horizontal, 4–6 mm. long; calyx shorter than caps. .... 4. *V. comosa*
    6. Caps. deeply obcordate, much broader than long; lvs. linear to linear-lanceolate
                                                                                                7. *V. scutellata*

1. **V. americàna** (Raf.) Schw. BROOKLIME. Glabrous ± succulent perennial, rhizomatose and creeping at base, the main stems 1–10 dm. long; principal lvs. lance-ovate to lanceolate, acutish, serrate or denticulate, 1–9 cm. long, short-petioled, the lowermost ± rounded; racemes lax, few–many-fld., the lower pedicels to ca. 12 mm. long; calyx 3 mm. long; corolla 7–10 mm. wide, violet-blue to lilac; caps. turgid, suborbicular, 3–4 mm. long; $2n=36$ (Schlenker, 1936).—Wet places along streams, sea-level to 10,000 ft.; many Plant Communities; cismontane and montane and occasional on deserts; to Alaska, Atlantic Coast, Mex., Asia. May–Aug.

2. **V. anágallis-aquática** L. Habit of *V. americana*; lvs. of flowering stems ± cordate-clasping, lance-oblong, acute; racemes mostly many-fld.; the fruiting pedicels ascending, 6–8 mm. long; calyx 4–4.5 mm. long; corolla 4–5 mm. broad, pale lavender, with violet lines; caps. scarcely notched; $2n=36$ (Ehrenberg, 1945).—Occasionally natur. in wet places; native of Eu. May–Sept.

3. **V. arvénsis** L. Erect or ascending annual, the slender stems 1–3 dm. long, pilose, nonglandular; lower lvs. rounded to oval, crenate-dentate, 5–15 mm. long, petioled, the upper ovate to lanceolate; smaller, sessile, subtending the fls.; pedicels 1–2 mm. long; calyx 3.5–4 mm. long; corolla bright blue, 2–2.5 mm. wide; caps. 2–2.5 mm. long, deeply emarginate; $2n=14, 16$ (Yamashita, 1937).—Occasional weed in lawns, waste places, etc.; native of Eu. April–July.

4. **V. comòsa** Richt. Similar to *V. anagallis-aquatica*; lvs. lance-oblong; pedicels 4–6 mm. long, horizontally divergent; calyx shorter than caps.; corolla whitish or pale rose, 3–4 mm. wide; caps. 2.5–4 mm. long, obcordately notched.—Occasional in wet places, Otay Lake (San Diego Co.), in San Bernardino Co., etc.; to Canada, Penn. July–Sept.

5. **V. peregrìna** L. ssp. **xalapénsis** (HBK) Penn. [*V. x.* HBK.] Erect annual, simple to few-stemmed, 1.5–3 dm. tall, glandular-pubescent on stems and caps.; lvs. linear-oblong to spatulate, obtuse, entire to ± dentate, sessile or the lower petioled, 1–2.5 cm. long; raceme lax, leafy-bracted; pedicels 1–2 mm. long; calyx-segms. subequal, 3 mm. long; corolla white, 2–2.5 mm. wide; caps. 3–3.5 mm. long, slightly emarginate; seeds 0.5 mm. long.—Moist places, sea-level to 10,000 ft.; many cismontane Plant Communities; to w. Canada, Mex., S. Am. April–Aug.

6. **V. pérsica** Poir. Annual, usually with several procumbent stems 1–4 dm. long, spreading-pubescent; lf.-blades roundish-ovate, crenate-serrate or coarsely dentate, 6–20 mm. long, on shorter petioles; racemes lax, pedicels 10–25 mm. long; calyx 5 mm. long; corolla deep blue, 7–11 mm. wide; caps. 2.5–3 mm. long, widely notched, with strongly divergent lobes.—Occasional weed in lawns, waste places; natur. from Eu. Feb.–May.

7. **V. scutellàta** L. Weak glabrous perennial, rhizomatose at base; stems slender, 1–6 dm. tall; lvs. sessile, linear to narrow-lanceolate, entire or minutely denticulate, 1.5–8 cm. long; racemes divergent, lax, small-bracted, rather few-fld.; pedicels 6–16 mm. long; calyx 3 mm. long; corolla lilac or blue-lavender, 5–7 mm. wide; caps. 3–4 mm. long, deeply obcordate, broader than long.—Wet places and swales; many Plant Communities in cent. and n. Calif.; possibly also at Bluff Lake, San Bernardino Mts.; to B.C., New England; Eurasia. May–Aug.

8. **V. serpyllifòlia** L. Much-branched perennial, with creeping base; stems to ca. 1.5 dm. high, closely and finely strigulose, not glandular; lvs. ovate or oblong, obscurely crenate, 0.5–1.5 cm. long, the lower petioled and rounded, the uppermost becoming bracteate; racemes terminal, small, with pedicels 4–5 mm. long; calyx 3–4 mm. long; corolla whitish or pale blue, with darker lines, 6–7 mm. broad; caps. ca. 3 mm. long, 4 mm. broad, obtusely notched.—Occasional weed in lawns, etc.; native of Eu. April–June.

100. Simaroubàceae. QUASSIA FAMILY

Shrubs or trees resembling *Rutaceae*, but the foliage without gland-dots; bark usually very bitter with contained oil-sacs. Lvs. in ours alternate. Fr. in ours a samara or drupe. Ca. 30 genera and 125 spp. of warmer regions.

Fr. a samara; plant a tree; lvs. large, odd-pinnate .... 1. *Ailanthus*
Fr. drupaceous; plant a shrub, thorny; lvs. scalelike .... 2. *Castela*

1. Ailánthus Desf. TREE OF HEAVEN

Trees, polygamo-dioecious; with large odd-pinnate lvs. Fls. small, greenish, in large terminal panicles, the ♂ very strong-smelling. Calyx of 5 imbricate segms. Petals 5, spreading. Stamens 10, inserted at the base of the disk. Ovary in ♀ fls. 2–5-cleft, with flat 1-celled divisions. Ovule solitary in each cell. Fr. a linear or oblong samara. Three spp. of se. Asia. (*Ailanto*, a Moluccan name.)

1. **A. altíssima** (Mill.) Swingle. [*A. glandulosa* Desf.] Tree 5–20 m. tall and spreading freely underground; lvs. 3–6 dm. long; lfts. lanceolate to oblong, acuminate, 7–18 cm. long, with 2–4 teeth near the base; samaras 3–5 cm. long, often ± reddish.—Rather widely natur. in cismontane Calif., especially about old dwelling sites; native of Asia. June. Introd. in early mining days by the Chinese.

2. Castèla Turp.

More or less thorny or spiny shrubs or small trees with small alternate simple lvs. or essentially leafless; fls. solitary or few together in the axils, dioecious, 4–8-merous, the sepals united at base; petals distinct; stamens ca. twice as many as the petals, reduced in the ♀ fls. and non functional; pistils 5–10 in ours, the bodies lightly cohering, tipped by divergent styles. Fr. of several dry diverging drupes. Seeds ovoid. Several spp., largely sw. N. Am.

1. **C. émoryi** (Gray) Moran & Felger. [*Holacantha e.* Gray.] CRUCIFIXION THORN. Plate 85, Fig. D. Low and spreading and ca. 1 m. high, or taller and more erect, appressed-canescent on younger branches; lvs. on mature plants scalelike, ovate to subulate, those of seedling ca. 1 cm. long; fls. densely pubescent without, the ♂ ca. 3 mm. long, ♀ smaller; drupes 6–8 mm. long, quite persistent; $2n = 16$ (Raven, 1967).—Occasional in dry gravelly places, Creosote Bush Scrub; Mojave Desert (Daggett, Ludlow, Amboy, Goffs) and Colo. Desert (Hayfields, Coyote Well); to Ariz., Son. June–July.

101. Solanàceae. NIGHTSHADE FAMILY

Herbs or shrubs with alternate lvs. Fls. perfect, regular or nearly so, terminal or axillary, solitary, umbellate, cymose or paniculate. Calyx 5-lobed or -toothed. Corolla 5-lobed, rotate to tubular, the lobes valvate or mostly plicate in the bud. Stamens 5, alternate with the corolla-lobes and inserted on the tube. Ovary entire, superior, mostly 2-celled; style 1; stigma terminal. Fr. a berry or caps. Seeds many. A family of over 3000 spp. in almost 100 genera, widely distributed. Many plants like tomato, potato, peppers, petunia, tobacco, etc. of economic importance.

1. Shrubs with spiny branches; fr. fleshy or dry and bony; corolla funnelform ...... 3. *Lycium*
1. Shrubs without spines or more commonly herbs.
    2. Corolla rotate or nearly so.
        3. Anthers connivent, longer than the fils.; calyx remaining small.
            4. Anthers dehiscing by terminal pores; plants prickly if with yellow fls. .... 10. *Solanum*
            4. Anthers dehiscing lengthwise; plants with yellow fls., but not prickly .. 4. *Lycopersicon*
        3. Anthers not connivent, mostly shorter than the fils.
            5. Calyx herbaceous, not inflated, closely investing the berry but open above; corolla with tomentose pads alternating with the stamens .... 1. *Chamaesaracha*
            5. Calyx large and bladdery in fr., nearly closed at top; corolla lacking tomentose pads.
                6. Ovary 2-loculed; fruiting calyx 5-toothed .... 8. *Physalis*
                6. Ovary 3–5-loculed; fruiting calyx 5-parted .... 5. *Nicandra*
    2. Corolla not rotate.
        7. Fr. a berry; corolla urceolate. Introd. woody perennial .... 9. *Salpichroa*
        7. Fr. a caps.; corolla tubular to funnelform.

8. Corolla 5–6 mm. long; plant a low annual, prostrate, with purplish-red fls. solitary in the axils .... 7. *Petunia*
8. Corolla 15–200 mm. long; plants mostly larger, more erect.
9. Caps. prickly, 25–50 mm. long; fls. solitary, axillary .... 2. *Datura*
9. Caps. not prickly, 5–13 mm. long; fls. in terminal racemes or panicles
6. *Nicotiana*

1. Chamaesarácha Gray

Low perennial herbs with leafy decumbent to prostrate branched stems. Lvs. entire to pinnatifid, with margined petioles. Fls. 1–5 in axillary fascicles, pedicelled. Calyx campanulate, 5-toothed or -lobed, slightly accrescent in fr. Corolla rotate, white to cream, with some purple tinge, with pubescent cushionlike appendages in the throat. Stamens inserted near base of corolla. Stigma obscurely 2-lobed. Berries globose, on recurved pedicels. Seeds flattened, reniform, finely rugose-favose. Ca. 9 spp., w. N. and S. Am. (Greek, *chamea*, low, and *Saracha*, a Solanaceous genus of S. Am.)

1. **C. coronòpus** (Dunal) Gray. [*Solanum c.* Dunal.] Diffusely branched, 0.5–1.5 dm. high, from slender underground rootstocks, green but scurfy-puberulent; lvs. many, lanceolate or linear, 2–6 cm. long, with cuneate-attenuate base, entire to laciniate-pinnatifid, short-petioled; pedicels slender, 1–18 mm. long; calyx copiously stellate-puberulent, ca. 5 mm. long in fr., the lobes narrow-triangular; corolla greenish white, the large appendages nearly contiguous, almost filling the throat; berry almost hidden in calyx; $n=24$ (Powell & Averett, 1967).—Dry clay soil, 4000–5000 ft.; Pinyon-Juniper Wd.; Barnwell, New York Mts.; to Utah, Kans., Mex. May–July.

2. Datùra L. Thorn-Apple. Jimsonweed

Ours erect or spreading coarse rank-smelling herbs. Lvs. alternate, short-petioled, entire to sinuate-dentate or lobed. Fls. solitary, large, erect in ours, on short peduncles in the forks of branching stems, whitish-purple to violet, opening in evening at least in some spp. Calyx tubular, 5-toothed, circumscissile near base in ours, with the lower part persistent as a collar below the caps. Corolla funnelform, convolute-plicate in bud. Stamens included; fils. filiform. Ovary 2- or falsely 4-celled. Caps. 2–4-valved from the top, sometimes splitting irregularly, prickly or spiny. Seeds large. Ca. 25 spp. in warmer parts of all continents. Plants ± poisonous and with narcotic properties; some spp. used by Calif. Indians in certain tribal rites. (The Hindoo name, *dhatura*.)

Corolla 15–20 cm. long; calyx tubular, not prismatic; caps. nodding .... 2. *D. meteloides*
Corolla 6–12 cm. long; calyx prismatic.
Caps. nodding, pubescent as well as prickly; stems cinereous. Desert native .... 1. *D. discolor*
Caps. erect, glabrous to puberulent; stems glabrous to puberulent, not cinereous. Introd. weed
3. *D. stramonium*

1. **D. discólor** Benth. Erect annual, 2–5 dm. tall, cinereous-pubescent; lvs. broadly ovate, sinuate-dentate, 5–14 cm. long, on somewhat shorter petioles; calyx prismatic, 4–6 cm. long, the persistent basal part mostly rotate; corolla white with purplish tinge in throat, 10–14 cm. long, the limb 4–8 cm. broad, the teeth 4–7 mm. long; anthers white, ca. 8 mm. long; caps. nodding, globose, 3–4 cm. thick, the spines 1–2 cm. long; seeds finely pitted; $n=12$ (Buchholz et al., 1935).—Low places, washes, etc., below 1500 ft.; Creosote Bush Scrub; Colo. Desert; Ariz., Mex., W. I. April–Oct.

2. **D. meteloìdes** A. DC. Plate 85, Fig. E. Perennial, erect, widely branched, 5–10 (–15) dm. tall, minutely grayish-pubescent; lvs. unequally ovate, 4–12 cm. long, irregularly repand to subentire, on somewhat shorter petioles; calyx 7–10 cm. long; corolla white, suffused with violet, 15–18 cm. long, 10–20 cm. broad, with 5 subulate teeth ca. 1–2 cm. long; anthers ca. 1.5 cm. long, white; caps. nodding, 2.5–3 cm. long, densely prickly and puberulent; the spines 5–12 mm. long; seeds smooth; $n=12$ (Buchholz et al., 1935).—Sandy and gravelly dry open slopes, below 4000 ft.; Coastal Sage Scrub, V.

Grassland, Joshua Tree Wd., Creosote Bush Scrub, etc.; cismontane and desert, cent. Calif. to Tex., Mex., n. S. Am. April–Oct. I am using the traditional species name, since there seems some uncertainty about the name *D. wrightii* Regel (1881) or the *D. inoxia* (Miller, 1786) being applicable to this concept.

3. **D. stramònium** L. Annual, erect, few-branched, 5–12 dm. tall, sparsely puberulent to glabrate; lvs. elliptic to ovate, 3.5–4.5 cm. long, on petioles ca. half as long; calyx 3.5–4.5 cm. long, the teeth 5–10 mm. long, unequal; corolla white, 6–8 cm. long, 3–5 cm. broad, the teeth 5–8 mm. long, unequal; caps. erect, ovoid, 4–5 cm. long, puberulent to glabrate and often with spines 3–10 mm. long; seeds pitted and rugulose; $n=12$ (Blakeslee, 1928).—Locally adventive, especially in waste places; native of trop. Am. Summer. A form with purplish stems and violet corolla and more prickly caps. is also occasionally adventive especially northward and is the var. *tátula* (L.) Torr. [*D. t.* L.]

3. *Lýcium* L. Box-Thorn

Armed shrubs, erect or spreading, sometimes scrambling over supports. Lvs. often fasciculate, entire to minutely dentate, glabrous and glaucous to glandular or pubescent. Fls. solitary or in 2's or 4's in the axils. Calyx campanulate to tubular, commonly ruptured by fr. Corolla whitish to purplish or greenish-purple, regular, tubular to funnelform, 4–6-lobed. Anthers affixed near their middle. Fr. 2-celled, from dry and bony to fleshy and juicy, 2–many-seeded. Embryo coiled. Ca. 100 spp., from ± arid regions in all continents. (*Lycia*, ancient country in Asia Minor.)

(Hitchcock, C. L. A monographic study of the genus Lycium of the W. Hemis. Ann. Mo. Bot. Gard. 19: 179–374. 1932.)

1. Fr. 2-seeded; lvs. fleshy-turgid, subterete in cross section; corolla-tube ca. as long as calyx. Coastal .... 4. *L. californicum*
1. Fr. several-seeded; lvs. ± flattened; corolla-tube longer than calyx.
  2. Calyx-lobes ⅔ as long as calyx-tube, or at least 2 mm. long.
    3. Fr. hard, greenish or purplish.
      4. Plant glaucous; fr. not transversely grooved near middle .... 7. *L. pallidum*
      4. Plant not glaucous, densely pubescent; fr. with 2 transverse grooves above the middle.
                                                                                5. *L. cooperi*
    3. Fr. soft, fleshy, red.
      5. Calyx and rest of plant densely pubescent; fls. mostly borne singly ... 8. *L. parishii*
      5. Calyx and rest of plant glabrous or sparsely pubescent; fls. mostly in 2's and 3's.
        6. Fils. glabrous, attached near summit of corolla. San Nicolas Id. .. 10. *L. verrucosum*
        6. Fils. pubescent at base, attached near middle of corolla-tube. Widely distributed
                                                                                3. *L. brevipes*
  2. Calyx-lobes less than ⅔ as long as the tube, or less than 2 mm. long.
    7. Scrambling shrubs, sometimes adventive near habitations of cismontane Calif. as a plant introduction .... 2. *L. barbarum*
    7. Stiff erect native shrubs of the deserts.
      8. Corolla-lobes ⅓ as long to as long as the tube .... 3. *L. brevipes*
      8. Corolla-lobes less than ⅓ as long as the tube.
        9. Plant densely pubescent.
          10. Calyx-tube 3–6 mm. long, the lobes shorter. From alkaline soil .. 6. *L. fremontii*
          10. Calyx-tube not more than 3 mm. long, the lobes usually as long. Not from alkaline soil .... 8. *L. parishii*
        9. Plant not densely pubescent.
          11. Corolla-lobes lanate-ciliate; some lvs. usually over 5 mm. broad ... 9. *L. torreyi*
          11. Corolla-lobes glabrous to ciliolate; no lvs. over 5 mm. broad .. 1. *L. andersonii*

1. **L. andersònii** Gray. Subglabrous or sparsely pubescent rounded shrub, much-branched, 1–2 (–3) m. high, the spines needlelike; lvs. spatulate, subterete, ± pear-shaped, 0.3–1.5 cm. long, 1–3 mm. broad; fls. mostly 1–2 in an axil, on pedicels 3–9 mm. long; calyx cup-shaped, 1.5–3 mm. long, glabrous; corolla whitish-lavender, tubular-funnelform, the tube 10–16 mm. long, the lobes 1.5–2.5 mm. long, entire to ciliolate; stamens exserted 2–3 mm.; fr. ellipsoid or ovoid, red, fleshy, 4–8 mm. long.—Dry stony hills

and mesas, below 6000 ft.; Creosote Bush Scrub to Pinyon-Juniper Wd., Sagebrush Scrub, Chaparral, Coastal Sage Scrub; common, Mojave and Colo. deserts n. to Mono Co.; occasional, cismontane s. Calif.; to Utah, New Mex., Mex. March–May. Occasional plants have lvs. 2–3.5 cm. long and flattish. They occur with the same distribution as the more normal form and have been called var. *deserticola* (C. L. Hitchc.) Jeps. [*L. a.* f. *d.* C. L. Hitchc.]

2. **L. barbàrum** L. [*L. halimifolium* Mill.] MATRIMONY VINE. Upright or spreading, with arching scrambling branches to 5 m. high; lvs. ovate to spatulate, 2–6 cm. long, glabrous, short-petioled; calyx mostly cupulate, the triangular lobes obtusish, from ca. half as long to as long as the tube; corolla rotate-campanulate, dull lilac-purple, 8–10 mm. broad; fr. ovoid, salmon to red, 10–15 mm. in diam.; $n=12$ (Delay, 1967).—Occasional escape from gardens; native of Eurasia. July–Oct.

3. **L. brévipes** Benth. var. **brévipes.** Plate 85, Fig. F. Thorny, 1–3 m. tall, subglabrous to pubescent, divaricately branched; lvs. oblanceolate to spatulate, 0.5–3 cm. long, 3–10 mm. broad, short-petioled; fls. few to many, on pedicels 1–10 mm. long; calyx campanulate, the tube 2–6 mm. long, the unequal lobes linear and as long as tube to triangular and much shorter; corolla lavender to whitish, funnelform, the tube 6–10 mm. long, the lobes 3–5 mm. long, glabrous or sparsely ciliolate; stamens slightly exserted; fr. ovoid, bright red, 4–9 mm. in diam., many-seeded.—Washes and hillsides, below 1500 ft.; Creosote Bush Scrub, edge of Alkali Sink; w. Colo. Desert and San Clemente Id.; L. Calif., Son. March–April.

Var. **hássei** (Greene) C. L. Hitchc. [*L. h.* Greene.] Calyx-lobes equal, ± spatulate, 1–3 times as long as tube.—Coastal Sage Scrub; coastal bluffs; Santa Catalina and San Clemente ids., mainland at San Diego and Santa Barbara. May now be extinct.

4. **L. califórnicum** Nutt. Dense intricately branched shrub 1–2 m. tall, the branches with very spinose tips; lvs. glabrous, fleshy, subterete, 3–10 mm. long, subsessile; fls. 1–2 in the axils, on pedicels 1–5 mm. long; calyx campanulate, ca. 2.5 mm. long, 2–4 (–5)-lobed, the lobes minute, broadly triangular; corolla white with a purplish tinge, the tube 2–3 mm. long, the lobes ca. as long, rotate to slightly reflexed; fils. pubescent below; fr. ovoid, reddish, firm, 2-seeded, 3–6 mm. long.—Dry bluffs and slopes, near the coast; Coastal Sage Scrub; Los Angeles Co. to L. Calif.; Channel Ids. and Santa Catalina and San Clemente ids. March–July.

5. **L. coóperi** Gray. Plate 86, Fig. A. Densely leafy compact thorny shrub 1–1.5 (–2) m. high, glandular-puberulent; lvs. oblanceolate to spatulate, 1–3 cm. long, 0.4–1 cm. broad; fls. pendent, the pedicels 1–3 in an axil, ca. as long as the calyx; calyx bowl-shaped, 8–15 mm. long, the lobes half as long as tube; corolla greenish-white, funnelform, 8–12 mm. long, the limb ca. as broad; stamens ca. as long as tube; fr. ovoid, dry, greenish, constricted above, 6–10 mm. long, several-seeded.—Dry mesas and slopes, below 5000 ft.; Creosote Bush Scrub to Pinyon-Juniper Wd.; Mojave and Colo. deserts; upper San Joaquin V.; to Utah, Ariz. March–May.

6. **L. fremóntii** Gray. Densely glandular-pubescent, much-branched, erect, 1–3 m. tall, spinose; lvs. oblanceolate-spatulate, 1–2.5 cm. long, 3–10 mm. broad; pedicels slender, 4–25 mm. long; calyx tubular, the tube 4–6 mm. long, the lobes mostly triangular, 1–2 mm. long; corolla lavender to pale violet, tubular to narrow-funnelform, glabrous, 10–15 mm. long, the spreading lobes 1.5–3.5 mm. long; stamens mostly shorter than corolla-lobes; fr. ovoid, red, juicy, 6–8 mm. long, many-seeded.—Rather alkaline places below 1500 ft.; Creosote Bush Scrub; Colo. Desert from near Mecca south; to Ariz., Son., L. Calif. March–April.

7. **L. pállidum** Miers var. **oligospérmum** C. L. Hitchc. Glabrous somewhat glaucous intricately branched very thorny shrub 1–2 m. tall; lvs. oblong-oblanceolate, 1–5 cm. long; fls. pendent on pedicels 8–12 mm. long; calyx campanulate, 5–8 mm. long, the ovate lobes ca. as long as tube; corolla white to lavender, narrow-campanulate, 15–20 mm. long, the limb 1.4–1.8 cm. broad; stamens exserted; fr. depressed-globose, somewhat fleshy but hard on outside, greenish-purple to -white, 8–10 mm. in diam.—Dry rocky hills and mesas, below 2500 ft.; Creosote Bush Scrub; Mojave Desert from near Barstow to Panamint and Death valleys. March–May.

Plate 86. Fig. A, *Lycium cooperi*, fr. with transverse grooves. Fig. B, *Nicotiana bigelovii*, fils. unequally inserted. Fig. C, *Nicotiana clevelandii*, corolla shorter. Fig. D, *Nicotiana glauca*, corolla-limb reduced. Fig. E, *Nicotiana trigonophylla*, lvs. with clasping base. Fig. F, *Physalis philadelphica*, calyx enlarging in fr.

8. **L. paríshii** Gray. Erect, intricately branched, spiny, 1–3 m. tall, pubescent, slightly glandular; lvs. 0.5–3 cm. long, oblanceolate to subelliptic, the longer short-petioled; fls. usually solitary; calyx campanulate, the tube 1.5–2.5 mm. long, the oblong-oval lobes from half as long to longer than tube; corolla purplish, tubular-funnelform, 8–12 mm. long, the tube 6–10 mm., the rounded lobes ± spreading; stamens ca. as long as corolla-lobes; fr. ovoid, red, several-seeded, 4–6 mm. long.—Rare, dry places, below 2000 ft.; Coastal Sage Scrub?, Creosote Bush Scrub; San Bernardino V. and w. Colo. Desert (Mountain Palm Springs, Vallecitos); Ariz., Son. March–April.

9. **L. tórreyi** Gray. Spreading, intricately branched, subglabrous, 1–3 m. tall, heavy-spined; lvs. broadly spatulate, 1–5 cm. long, 3–10 mm. broad; fls. mostly in small fascicles, the pedicels 5–20 mm. long; calyx cupulate to short-cylindric, 2.5–4 mm. long, the lobes 0.5–2 mm. long, ciliolate; corolla lavender-purple, tubular-clavate, 10–15 mm. long, the lanceolate to ovate lobes lanate-ciliate, 3–4 mm. long; stamens ca. equal to corolla-lobes; fr. red, juicy, many-seeded, 7–10 mm. long.—Along washes and benchlands, below 2000 ft.; Creosote Bush Scrub; Mojave R., Colo. R. and near streams of w. edge of Colo. Desert; to Utah, Tex., Mex. March–May.

10. **L. verrucòsum** Eastw. Much like *L. brevipes*, but calyx-lobes shorter than tube, pubescent; corolla 8–10 mm. long, the lobes ca. ¼ as long as tube; fils. adnate to bases of sinuses between corolla-lobes, glabrous; fr. ovoid, reddish.—San Nicolas Id.

4. *Lycopérsicon* Mill. TOMATO

Near to *Solanum*, but always unarmed; lvs. always pinnate to pinnatifid; fls. yellow; anthers projected into narrow or sharp sterile tips and dehiscing from top to bottom; fr. a red or yellow pulpy berry with 2–few cells that multiply under domestication. Ca. 6 spp. of S. Am. (Greek, *lykos*, wolf, and *persicon*, peach, because of supposed poisonous properties.)

1. **L. esculéntum** Mill. [*Solanum lycopersicum* L.] Annual or short-lived perennial, hairy-pubescent and ± strong-smelling and glandular, 1–2 m. tall, much branched; lvs. 1–3 dm. long, odd-pinnate with small lfts. interspersed between, to 5–9 main lfts.; fls. nodding, 2–2.5 cm. broad; fr. of various colors and sizes.—The cultivated tomato escapes from cult. and persists for some time in waste places. Summer–fall.

5. *Nicándra* Adans.

Annual herbs differing from *Physalis* in the 3–5-loculed ovary and deeply parted calyx. (Named for *Nicander*, poet of Colophon, who wrote on plants ca. 100 B.C.)

1. **N. physalòdes** Gaertn. To 1 m., much-branched; lvs. ovate to oblong, 10–15 mm. long, sinuate-toothed; fls. blue, 2.5–4 cm. long and broad, enclosed in the enlarged green calyx 2.5–4 cm. long. Escape as near Ventura, Santa Barbara; native of Peru.

6. *Nicotiàna* L. TOBACCO

Narcotic-poisonous heavy-scented annual or perennial herbs or shrubs, usually ± viscid-pubescent. Lvs. large, alternate, entire or repand. Fls. in terminal panicles or racemes. Calyx tubular-campanulate to ovoid, 5-cleft, persistent. Corolla funnelform, salverform, or nearly tubular, the limb usually shallowly 5-lobed and spreading. Fils. filiform. Ovary 2-, rarely 4-celled. Caps. 2- or 4-valved at summit. Seeds many, small, ovoid to reniform, minutely reticulate-punctate. Ca. 60 spp., mostly of N. and S. Am.; some important in commerce and some as ornamentals. (J. *Nicot*, French ambassador to Portugal, who introduced tobacco into France ca. 1560.)

(Goodspeed, T. H. The genus Nicotiana. Chron. Bot. 16: 1–536. 1954.)
Shrub; fls. yellow .... 4. *N. glauca*
Herbs; fls. whitish or cream, sometimes tinged violet.
 Lvs. auriculate-clasping at base; fls. open throughout day. Desert perennial or biennial
5. *N. trigonophylla*

Lvs. not auriculate-clasping; fls. closing in sun. Annuals.
  Cauline lvs. petioled; calyx-teeth deltoid; corolla 2.5–3 cm. long .... 1. *N. attenuata*
  Cauline lvs. sessile or nearly so, except near base of stem.
    Corolla 4–7 cm. long; calyx-teeth linear-lanceolate .... 2. *N. bigelovii*
    Corolla 1.5–2 cm. long; calyx-teeth linear .... 3. *N. clevelandii*

1. **N. attenuàta** Torr. Erect, simple or branched, glandular-pubescent to glabrate, 3–19 dm. high; lvs. 5–15 cm. long, ovate to lance-ovate, mostly petioled; infl. racemose or the racemes in a panicle; calyx ovoid-campanulate, 6–8 mm. long, the teeth deltoid; corolla white, 2.5–3 cm. long, ca. 1 cm. broad; caps. ca. 8–12 mm. long; seeds 0.6–0.7 mm. long; $n=12$ (Clausen, 1928).—Disturbed places below 9000 ft.; many Plant Communities; interior cismontane, occasional on deserts; to Wash., Ida., Utah, Tex. May–Oct.

2. **N. bigelòvii** (Torr.) Wats. var. **bigelòvii**. Plate 86, Fig. B. Annual, glandular-pubescent, ill-smelling, with ascending branches, mostly 4–12 dm. high; lvs. sessile or the lower petioled, ovate-oblong to lanceolate, 5–20 cm. long; fls. mostly racemose; calyx 1.5–2 cm. long, the linear-lanceolate teeth ± unequal; corolla white, tinged with green, 4–7 cm. long, the limb 3–5 cm. wide; caps. ovoid, ca. 1.5 cm. long; $n=24$ (Goodspeed, 1923).—Dryish plains, mesas, etc., below 3500 ft.; many Plant Communities, cismontane Calif. from Los Angeles Co. n.; to Ore. May–Oct. Intergrading with
  Var. **wallàcei** Gray. Corolla-limb 2–3 cm. wide.—Chaparral, Coastal Sage Scrub; San Diego Co. to Santa Barbara Co., especially in hot interior valleys.

3. **N. clevelándii** Gray. Plate 86, Fig. C. Viscid-pubescent annual 2–6 dm. tall, mostly branched; lvs. ovate or the upper lanceolate, sessile or short-petioled, 3–6 (–8) cm. long; calyx 9–12 mm. long, the linear lobes distinctly unequal, the longest much exceeding the tube; corolla whitish, tinged with violet, 1.5–2 cm. long, the limb 8–10 mm. broad; caps. 4-valved, 8–10 mm. long; $n=24$ (Clausen, 1928).—Occasional in sandy places below 1500 ft.; Coastal Strand, Coastal Sage Scrub; near the coast from Santa Barbara Co. to n. L. Calif.; Colo. Desert (Shaver's Well, Corn Springs, etc.); Santa Cruz Id.; Ariz. March–June.

4. **N. gláuca** Grah. TREE TOBACCO. Plate 86, Fig. D. Erect glabrous glaucous shrub or small tree, 2–6 m. tall, with loose branching and open panicles; lvs. ovate, entire to repand, 3–8 (–15) cm. long, on long petioles; fls. greenish-yellow, loosely paniculate; calyx ca. 1 cm. long, unequally 5-toothed; corolla tubular, 3–4 cm. long, somewhat contracted at throat, minutely pubescent, with narrow limb; caps. ovoid, 10–12 mm. long, 4-valved above; $n=12$ (Goodspeed, 1923).—Common, natur. in waste places, on burns, etc., below 3000 ft.; native of S. Am. Spring and summer.

5. **N. trigonophýlla** Dunal in A. DC. Plate 86, Fig. E. Viscid-pubescent, 2–8 dm. tall, erect, simple or few-branched, perennial or biennial; lowest lvs. petiolate, others oblong-ovate to -lanceolate, sessile and auricled, 2–8 cm. long; infl. loosely paniculate-racemose; pedicels 5–10 mm. long; calyx campanulate, 6–12 mm. long, with lance-subulate lobes ca. as long as tube; corolla greenish-white, 18–22 mm. long, constricted at orifice, the limb 8–10 mm. broad; caps. 2-valved, 8–10 mm. long; $n=12$ (Clausen, 1928).—Mostly about rocks, below 4000 ft.; Creosote Bush Scrub, Joshua Tree Wd.; Mojave and Colo. deserts from Mono Co. s.; to Tex., Mex. Mostly March–June.

7. Petùnia Juss.

Viscid annual or perennial herbs. Lvs. entire, the upper often subopposite. Fls. solitary, axillary. Calyx 5-parted. Corolla funnelform. Stamens 5, unequal, four being didynamous, the fifth shortest. Hypogynous disk fleshy. Caps. ovoid, with 2 undivided valves. Seeds minute, spherical. Perhaps 40 spp., largely S. Am.; some of hort. value. (*Petun*, an Indian name of tobacco.)

1. **P. parviflòra** Juss. Prostrate diffusely branched annual, glandular-pubescent; the stems 1–4 dm. long; lvs. oblong-linear to -spatulate, 0.4–1.2 cm. long, rather fleshy; calyx subsessile, 3–4 mm. long, with linear lobes; corolla 4–6 mm. long, purplish with

whitish tube, the short lobes somewhat unequal; caps. 2–3 mm. long; seeds ca. 0.6 mm. long, favose-reticulate; $n=9$ (Ferguson & Coolidge, 1932).—Sandy arroyos and dried beds of pools, below 4000 ft.; many Plant Communities; cismontane to cent. Calif.; Ariz., Tex., Fla., trop. Am. April–Aug.

P. violàcea Lindl. Viscid, the stems 1–1.5 dm. long, slender, branching; lvs. ovate; fls. variously colored, 2.5–3.5 cm. long.—Cult. and sometimes escaping; native of Argentina.

8. Phýsalis L. GROUND-CHERRY

Annual or perennial herbs. Lvs. entire to sinuate-dentate. Fls. solitary in axils, less often in clusters of 2–5. Pedicels slender. Calyx campanulate to tubular-campanulate, 5-toothed, in fr. enlarged and bladder-inflated, membranous, the teeth mostly connivent. Corolla open-campanulate, yellowish or whitish or purplish, obscurely 5-lobed, plicate. Stamens 5, inserted near base of corolla-tube. Style slender; stigma faintly 2-lobed. Fr. a berry, completely and loosely enclosed in the calyx. Seeds many, flattened, finely pitted. Perhaps 100 spp., mostly of New World, some from Eurasia and Australia. (Greek, *physalis*, a bladder, because of inflated calyx.)

(Waterfall, U. T. Rhodora 60: 106–114, 128–141, 152–173. 1958.)
Lvs. pubescent beneath with at least some stellate or forked hairs; plants perennial.
    Corolla 1.5–2 cm. in diam.; pubescence dense, mostly stellate. Introd. weeds ... 8. *P. viscosa*
    Corolla 0.8–1 cm. in diam.; pubescence largely of forked or simple hairs. Native of e. desert areas .... 5. *P. hederæfolia*
Lvs. lacking stellate or forked hairs beneath.
    Plants perennial; anthers yellow, rarely purplish. Desert plants.
        Pedicels shorter than fls.; foliage somewhat canescent; calyx-lobes more than ⅔ the length of tube at anthesis .... 5. *P. hederæfolia*
        Pedicels longer than fls.; foliage green; calyx-lobes not more than half as long as tube at anthesis .... 3. *P. crassifolia*
    Plants annual; anthers purple, green or blue, rarely yellow (as in *P. greenei*).
        Plant distinctly pubescent.
            Calyx-lobes deltoid, shorter than the tube at anthesis; pedicels longer, exceeding the fruiting calyx .... 4. *P. greenei*
            Calyx-lobes ± lanceolate, at anthesis often as long as tube; pedicels short, not exceeding the fruiting calyx.
                Lvs. subentire, acute; stems sharply angled; fruiting calyx membranaceous
                                                                                            7. *P. pubescens*
        Plants subglabrous.
            Pedicels exceeding fruiting calyx; corolla without dark center.
                Corolla whitish, with yellow center and 10–20 mm. broad; lvs. sinuate-dentate; fruiting calyx-lobes acuminate .... 1. *P. acutifolia*
                Corolla yellow, 3–8 mm. broad; lvs. shallowly sinuate; fruiting calyx-lobes acute
                                                                            2. *P. angulata*
            Pedicels shorter than fruiting calyx; corolla 10–15 mm. broad, yellow, with dark center
                                                                                 6. *P. philadelphica*

1. **P. acutifòlia** (Miers) Sandwith. [*P. wrightii* Gray.] Erect or ascending annual, branched, 2–10 dm. high, glabrous or sparsely strigose above; lf.-blades lanceolate, deeply sinuate-toothed, cuneate at base, 2–7 cm. long, on somewhat shorter petioles; pedicels 3–4 cm. long in fr.; calyx at anthesis campanulate, 3–5 mm. long, the narrow-deltoid lobes ca. as long as tube; corolla rotate, whitish with yellow center, 10–16 mm. broad; calyx in fr. 2–3 cm. long, ovoid, with acuminate lobes.—Orchards, roadsides and waste places as a weed; cismontane from Kern Co. to Los Angeles Co. and south; Colo. Desert; to Tex., Mex. July–Oct.

2. **P. angulàta** L. var. **lanceifòlia** (Nees) Waterfall. [*P. l.* Nees.] Like *P. acutifolia*, but lvs. entire to slightly toothed; corolla more yellow, 4–6 mm. broad; calyx-lobes in fr. broadly deltoid; $n=12$ (Gottschalk, 1954).—Occasional weed in cult. fields, as in Los Angeles, San Diego and Imperial cos.; to Tex., cent. Mex. July–Oct.

3. **P. crassifòlia** Benth. var. **crassifòlia**. Diffusely and intricately branched perennial, viscid-puberulent, 2–5 dm. tall; lf.-blades ovate, deltoid or cordate, thickish, 1–3 cm. long, often entire, on petioles equally long; pedicels slender, from little longer than calyces to 6–7 times as long; calyx campanulate, 3–5 mm. long at anthesis, the lobes broadly deltoid, 1–1.5 mm. long; corolla remaining yellow when dried, 10–15 mm. broad; calyx in fr. ovoid, 1.5–2.5 cm. long, obscurely angled; berry greenish.—Sandy and rocky places, below 4000 ft.; Creosote Bush Scrub; Colo. Desert and e. Mojave Desert; to Utah, Tex., L. Calif. March–May.

Var. **versicólor** (Rydb.) Waterfall. [*P. v.* Rydb.] Lvs. thin, often toothed; corolla often drying with a blue tinge; flowering calyces usually 3–4 mm. long, on pedicels 5–10 times as long.—Creosote Bush Scrub; Imperial and Riverside cos.

4. **P. greènei** Vasey & Rose. Erect-spreading annual 1–4 dm. tall, with puberulent slender, somewhat flexuous branches; lf.-blades ovate, acute, sinuate-dentate, 1.5–3 cm. long, puberulent, with equally long petioles; pedicels 2–3 cm. long, slender; calyx in fl. broadly tubular-campanulate, 5–6 mm. long, the lobes deltoid, shorter than the tube; corolla greenish-yellow, 12–15 mm. broad; calyx in fr. 1.5–2 cm. long, hispidulous on the angles.—Uncommon, Coastal Sage Scrub; coastal Orange and San Diego cos.; L. Calif. March–June.

5. **P. hederaefòlia** Gray var. **cordifòlia** (Gray) Waterfall. [*P. fendleri* var. *c.* Gray. *P. fendleri* Gray.] Perennial from underground rootstocks, mostly erect, ± canescent or cinereous, 1–2.5 dm. high; lf.-blades ovate to ± cordate, 1.5–4 cm. long, coarsely sinuate-dentate, on petioles ca. as long; pedicels 8–20 mm. long at anthesis; calyx tubular-campanulate, 5–8 mm. long at anthesis; corolla yellow, usually darker in the center, 12–14 mm. broad; fruiting calyx ovoid, 2–3 cm. long; berry yellow.—Dry rocky and gravelly places, 3000–6000 ft.; Creosote Bush Scrub to Pinyon-Juniper Wd.; mts. of e. San Bernardino Co.; to Tex., Mex. May–July.

6. **P. philadélphica** Lam. [*P. ixocarpa* Brot. ex Hornem.] TOMATILLO. Plate 86, Fig. F. Much-branched erect to spreading annual 3–9 dm. high, glabrous or the younger parts puberulent; lf.-blades ovate, cordate or cuneate at base, entire to sinuate dentate, 1.5–6 cm. long, on equally long petioles; pedicels 4–8 mm. long; calyx in fl. 3.5–4.5 mm. long, the deltoid lobes shorter than the tube; corolla yellow with dark center, 8–15 mm. broad; calyx in fr. ovoid, 1.5–2 cm. long; berry purple; $2n=24$ (Menzel, 1951).—Orchard weed and escape from gardens; cismontane n. to cent. Calif.; Atlantic Coast.

7. **P. pubéscens** L. var. **grísea** Waterfall. Diffusely branched annual, the stems with long jointed hairs, or with long and short, or densely short viscid-hairy, 1.5–4 dm. high; lvs. broadly ovate, usually short-hairy, the surfaces grayish, coarsely and irregularly 6–9-dentate; pedicels 5–9 mm. long; calyx in fl. 4–7 mm. long, in fr. ca. 2 cm. long.—Rare, as in Saline V. (Inyo Co.), San Gabriel, Elsinore; mostly in eastern states. More abundant is

Var. **integrifòlia** (Dunal) Waterfall. [*P. hirsuta* Dunal var. *i.* Dunal.] Plants ± villous; lf.-blades often entire, sometimes with 3–4 prominent teeth on a side, translucent or semitransparent; pedicels 5–9 mm. long; fruiting calyx 2–3 cm. long.—Weed at San Diego, Needles, etc.; to Penna. and Fla.

8. **P. viscòsa** L. var. **cineráscens** (Dunal) Waterfall. [*P. pensylvanica* L. var. *c.* Dunal.] Perennial with prostrate or spreading stems 4–8 dm. long; pubescence of upper parts cinereous, of minute branched or stellate hairs; lvs. ovate to elliptic, 3–10 cm. long, entire, short-cuneate at base, petioled; pedicels 1–2 cm. long; lobes of calyx at anthesis triangular, shorter than tube; calyx ovoid in fr., 2–3 cm. long; corolla greenish-yellow, with darker center.—Reported as occasional weed (Ventura, Yorba Linda); to Atlantic Coast.

9. *Salpichròa* Miers.

Herbs or shrubs with entire lvs. Fls. perfect, white or yellow, solitary, axillary. Calyx 5-parted. Corolla tubular or urceolate, 5 lobed. Stamens 5, inserted ca. the middle of the corolla-tube; fils. slender; anthers converging about the style, oblong. Ovary 2-celled, many-ovuled; style filiform; stigma entire to bilobed. Fr. an oblong or ovoid berry. Seeds round, compressed, with strongly curved embryo and little endosperm. Ca. 20 spp. of S. Am. (Greek, *salpe*, trumpet, and *chroa*, color or complexion because of form and color of fls.)

1. **S. origanifòlia** (Lam.) Thell. [*S. rhomboidea* (Gill. & Hook.) Miers, Calif. refs.] LILY-OF-THE-VALLEY VINE. Heavy-rooted perennial with running rootstocks; stems branched, spreading or trailing, 5–15 dm. long, sparsely hirsutulous; lf.-blades ovate to broadly elliptic, 1–3 cm. long, on somewhat shorter petioles; pedicels 4–6 mm. long; calyx-lobes lance-linear, 2–3 mm. long; corolla white, urceolate, 5–7 mm. long, the spreading round lobes 1–1.5 mm. long; berry yellowish, 10–12 mm. long; $n=12$ (Vilmorin & Simonet, 1928).—Established as a difficult weed in fields and orchards in much of cismontane Calif.; native of S. Am. July–Oct.

10. *Solànum* L. NIGHTSHADE

Herbs or shrubs, glabrous to pubescent or tomentose, often glandular, sometimes climbing, armed or unarmed. Lvs. simple and entire to lobed or parted. Fls. commonly in umbels or cymes, white or yellow to blue or purple. Calyx 5-cleft or -toothed, rotate to campanulate. Corolla rotate, 5-angled or -lobed, plaited in bud. Stamens 5, inserted on corolla-tube; fils. short; anthers connivent around style, dehiscent by a terminal pore or short introrse slit. Ovary 2-celled; stigma small, capitate or ± bilobed. Fr. a subglobose berry with several to many seeds, these ± flattened, with annular embryo. Over 1000 spp., of all continents, but especially trop. Am. (Latin, *solamen*, quieting, because of narcotic properties of some spp.)

1. Plants not prickly.
    2. Corolla deeply 5-cleft, whitish; peduncles mostly longer than pedicels.
        3. Lvs. entire or shallowly toothed.
            4. Stems conspicuously and persistently villous or hirsute; ripe berries greenish, yellow or reddish .... 10. *S. sarrachoides*
            4. Stems sparsely pubescent or glabrous, at least in maturity; ripe berries black.
                5. Corolla 6–11 mm. long; anthers 2.6–4 mm. long. Native half-shrub
                    .... 2. *S. douglasii*
                5. Corolla 4–6 mm. long; anthers 1.2–2.6 mm. long. Introd. weeds.
                    6. Infl. umbelliform; calyx-lobes distinct, reflexed at maturity; surface of berry glossy; anthers 1–1.2 mm. long. Frequent .... 7. *S. nodiflorum*
                    6. Infl. subracemose; calyx-lobes partly fused, not reflexed at maturity; surface of berry dull; anthers 1.8–2.4 mm. long. Occasional .... 6. *S. nigrum*
        3. Lvs. pinnatifid or deeply lobed; corolla 6–10 mm. broad, white ...... 12. *S. triflorum*
    2. Corolla angulately shallowly lobed, often purple; peduncles mostly shorter than pedicels.
        7. The lvs. mostly hastate with a pair of basal linear lobes. Se. San Diego Co.
            .... 11. *S. tenuilobatum*
        7. The lvs. seldom lobed.
            8. Pubescence of upper stems with at least some forked or branched hairs
                .... 13. *S. umbelliferum*
            8. Pubescence on stems lacking or of simple hairs.
                9. Stems distinctly pubescent.
                    10. Lvs. densely tawny-viscid-villous, oblong-ovate; berry 10–25 mm. in diam., purple or yellow. Insular.
                        11. Corolla 2–4 cm. broad; plant quite viscid; berry dark purple. Santa Catalina Id. .... 14. *S. wallacei*
                        11. Corolla 1.5–2 cm. broad; plant less viscid; berry yellow. Santa Cruz and Santa Rosa ids. .... 1. *S. clokeyi*
                    10. Lvs. glabrous to pubescent, not both tawny and viscid-villous, lanceolate to ovate; berry mostly 6–8 mm. in diam., greenish. Widely distributed
                        .... 15. *S. xanti*
                9. Stems glabrous or nearly so.
                    12. Lvs. acute or tapering at base .... 8. *S. parishii*
                    12. Lvs. obtuse or subcordate at base .... 15. *S. xanti*

1. Plants prickly.
    13. Plants annual; lvs. pinnatifid or bipinnatifid; berry partly or wholly invested by spiny calyx; corolla yellow .... 9. *S. rostratum*
    13. Plants perennial or shrubby; lvs. entire, sinuate or lobed, but not pinnatifid; berry not enclosed by the spiny or unarmed calyx.
        14. Lvs. entire to repand-dentate, permanently gray-canescent on both surfaces, linear-lanceolate to narrowly oblong .... 3. *S. elaeagnifolium*
        14. Lvs. lobed or deeply dentate, not permanently gray-canescent at least on upper surface.
            15. Berry less than 1 cm. in diam., orange; corolla light purplish-blue, ca. 1.5 cm. in diam., lvs. tomentose, but becoming greenish in age .... 4. *S. lanceolatum*
            15. Berry yellowish, 3–4 cm. in diam.; corolla whitish with some purple, 2–3.5 cm. in diam., lvs. with a whitish marginal band of tomentum above .... 5. *S. marginatum*

1. **S. clòkeyi** Munz. [*S. arborescens* Clokey, not Moench.] Like *S. wallacei* in foliage, but less viscid-villous; calyx 3–4 mm. long; corolla 1.5–2 cm. in diam.; berry yellow, 1–1.5 cm. thick. Santa Cruz and S. Rosa ids. March–July.

2. **S. douglásii** Dunal in DC. Perennial, ± woody, 1–2 m. tall; herbage puberulent to subglabrous, the stem-angles rough-pubescent; lf.-blades ovate, 2–10 cm. long, sinuate-dentate, subacuminate; petioles 1–2.5 cm. long; peduncles 1–3 cm. long; pedicels 0.5–1.2 cm. long; calyx 2–3 mm. long at anthesis; corolla whitish with greenish basal spots, 1–2 cm. broad, the lobes lance-oblong; anthers ca. 3–4 mm. long; berry black, 6–9 mm. in diam.; $n=12$ (Stebbins & Paddock, 1949).—Partly shaded slopes, in canyons, etc., below 3500 ft.; Coastal Sage Scrub, Chaparral, Coastal Strand, etc.; cismontane mainland, San Clemente and Catalina ids. and Channel Ids.; to cent. Calif., L. Calif., Mex. Most of year.

3. **S. elaeagnifòlium** Cav. SILVERLEAF-NETTLE. Short-lived perennial with dense fine stellate silvery canescence, 3–8 dm. high; stems unarmed or prickly; lvs. lance-oblong to lance-linear, 3–9 cm. long, ± cuneate at base, on shorter petioles; fls. cymose; peduncles, pedicels and calyces usually yellow-spinose; calyx-lobes narrow-lanceolate, to 1 cm. long; corolla violet or blue, 2–3 cm. in diam.; anthers 7–9 mm. long; berry 10–14 mm. in diam., yellow or brownish; $n=12$ (*Heiser* & Whitaker, 1048). Weed at stations as far apart as Needles and Catalina Id.; to cent. Calif.; native of cent. U.S. to Mex. May–Sept.

4. **S. lanceolàtum** Cav. Subshrubby, 1–2 dm. high, covered with white tomentum, the lvs. ± glabrescent above in age; main stems spiny; lower lvs. bluntly lobed, 1–2 dm. long; upper entire, lanceolate; fls. in long-peduncled cymose panicles in upper lf.-axils; corolla light purplish-blue, ca. 15 mm. in diam.; berry orange, 6–8 mm. in diam.—Occasional weed at Los Angeles, Saticoy (Ventura Co.); from trop. Am.

5. **S. marginàtum** L. f. Shrub 1–1.5 m. tall, with stout yellow spines; stems and lower lf.-surfaces white-stellate-tomentose; lvs. 1–2 dm. long, sinuate-lobate; upper lf.-surfaces with tomentum near edges; peduncles 1–2 cm. long, few-fld.; calyx 8–12 mm. long; corolla 2.5–3.5 cm. in diam., white with central purple star; berry yellow, 3–4 cm. in diam.; $2n=24$ (Vilmorin & Simonet, 1928).—Weed near Pala, San Diego Co., and n. Calif. coast; native of Afr. May–Aug.

6. **S. nìgrum** L. Annual 3–8 dm. high, branched, glabrous or nearly so; lvs. ovate, 3–7 cm. long, entire to ± serrate, with cuneate base; peduncles 1–2.5 cm. long; pedicels 3–8 mm. long at anthesis; calyx 1.5–2 mm. long; corolla white, 5–6 mm. in diam.; berry dull, dark, 6–9 mm. in diam.; $n=36$ (Stebbins & Paddock, 1949).—Sparingly established in waste places and about fields, cismontane and rare on desert (Darwin Falls, Inyo Co.); native of Eu. March–Oct.

7. **S. nodiflòrum** Jacq. [*S. nigrum* of auth., not L.] Straggling annual or perennial; stems 3–6 dm. long, glabrous to ± scabrous on angles; lvs. entire or sinuate-dentate, ovate to elliptic, 4–10 cm. long; peduncles slender, 2–3 cm. long; pedicels 5–12 mm. long; calyx-lobes quite distinct to base, 1–2.5 mm. long; corolla white or faint purple, 4–6 mm. in diam.; berry 5–6 mm. in diam., shining, black; $n=12$ (Stebbins & Paddock, 1949).—Common weed in damp fields and waste places, cismontane; native of Old World. April–Nov.

8. **S. paríshii** Heller [*S. xanti* var. *glabrescens* Parish in part.] Suffrutescent, 5–10 dm. high, the branches slender, glabrous or nearly so; lvs. elliptic to lance-oblong-ovate, 2–5 cm. long, on petioles to ca. 1 cm. long; peduncles short, few-fld.; calyx 4–5 mm. long; corolla lavender, 15–18 mm. broad; berries 7–9 mm. in diam.—Dry grassy and brushy slopes, below 6000 ft.; Chaparral, Yellow Pine F.; interior San Diego Co., w. Riverside Co.; interior n. Calif. April–July.

9. **S. rostràtum** Dunal. Buffalo-Bur. Annual, 4–8 dm. tall, with yellowish prickles, also stellate-pubescent; lvs. 1–2-pinnatifid, 5–10 cm. long; calyx nearly hidden by prickles; corolla yellow ca. 2 cm. broad; anthers dimorphic, some 6 mm., others 9–10 mm. long; $n=12$ (De Lisle, 1965).—Occasional weed; native in cent. Calif. May–Sept.

10. **S. sarrachoìdes** Sendt. ex Mart. Much-branched annual 1–5 dm. tall, short viscid-villous; lf.-blades subentire to sinuate-toothed, 2.5–6 cm. long, ovate; petioles 1–1.5 cm. long; peduncles 5–15 mm. long; pedicels somewhat shorter; calyx 2–2.5 mm. long, accrescent in fr.; corolla white, 3–5 mm. in diam., the lobes villous outside; berry ± yellowish; $n=12$ (Stebbins & Paddock, 1949).—Occasional weed in fields and waste places; native of Brazil. May–Oct.

11. **S. tenuilobàtum** Parish. Suffrutescent, hirsutulous above, 3–10 dm. high; lvs. linear to narrow-oblong, 2–3 cm. long, all but the upper hastate with a pair of linear basal lobes; petioles 3–5 mm. long; fls. few; peduncles ca. 1 cm. long; pedicels 1–1.5 cm. long; calyx 3–5 mm. wide; corolla blue-purple, 15–18 mm. broad, berry ca. 6–7 mm. in diam.—Dry open places, 1000–2700 ft.; Chaparral; s. San Diego Co.; L. Calif. March–April.

12. **S. triflòrum** Nutt. Annual, ± pubescent, branched, the stems 1–4 dm. long, decumbent; lvs. oblong, 2.5–4 cm. long, deeply pinnatifid with rounded sinuses; peduncles 1–3-fld., 5–15 mm. long; pedicels shorter; calyx 2.5–3 mm. long; corolla white, 7–9 mm. broad; berry 8–12 mm. in diam., green.—Dry places, mostly 5000–7000 ft.; Sagebrush Scrub, Pinyon-Juniper Wd.; Clark Mt. (E. San Bernardino Co.), Last Chance Mt. and Saline V. (Inyo Co.); adventive at Aguanga, Riverside Co. and Claremont; to B.C., Minn., Okla. June–Sept.

13. **S. umbellíferum** Eschs. var. **incànum** Torr. [*S. californicum* Dunal.] Rounded to spreading subshrub 6–10 dm. high, the stems white-tomentose with several-branched hairs; lf.-blades narrow-oblong, 1–2 cm. long, ± tomentose, ± entire; peduncles 5–15 mm. long; pedicels 1–2 cm. long; calyx broadly campanulate, 4–5 mm. long; corolla mostly blue, 1.5–2 cm. broad, with 2 greenish glands at base of each lobe; berry 10–15 mm. in diam.—Dry valleys and slopes below 4500 ft.; Chaparral, V. Grassland, etc.; interior valleys from Ventura Co. to cent. Calif.

Var. **glabréscens** Torr. Branches sparingly pubescent, the hairs partly simple and partly few-branched.—Dry slopes below 5000 ft.; Chaparral, Coastal Sage Scrub; cismontane from San Diego Co. to Santa Barbara Co.; Ariz.

14. **S. wallàcei** (Gray) Parish. [*S. xanti* var. *w.* Gray.] Suffrutescent, erect-spreading, 1–2 m. tall, densely tawny-villous, with many hairs gland-tipped; lvs. many, thickish, oblong-ovate, 3–14 cm. long, acute, with rounded or subcordate base; petioles 1–2 cm. long; peduncles 1–3 cm. long; with cymes of several fls.; calyx campanulate, 5–7 mm. long; corolla purplish-blue, 2–4 cm. broad, berry dark purple, 1.5–2.5 cm. in diam.—Dry rocky slopes and in canyons; Chaparral; Santa Catalina Id.; Guadalupe Ids. March–Aug.

15. **S. xánti** Gray var. **xánti**. Suffrutescent, 4–9 dm. tall, short-villous with white mostly nonglandular hairs, especially on the stems and veins; lvs. ovate, subentire, sometimes lobed at base, 2–4 (–7) cm. long, cuneate to subcordate at base, on petioles 5–18 mm. long; fls. mostly 6–10 in lateral subumbellate cymes; calyx 5–6 mm. long; corolla deep violet to dark lavender, 1.5–2.5 cm. in diam.; berry greenish, 6–8 mm. in diam.—Dry places about brush or woods, below 4000 ft.; Chaparral, Foothill Wd.; cismontane from Ventura and western Kern counties to San Luis Obispo Co. Santa Rosa and Santa Cruz ids. Feb.–June.

Var. **hoffmánnii** Munz. Stems glabrous; lvs. 3–6 cm. long, truncate to subcordate at

Solanum

base.—Chaparral; Gaviota Pass (Santa Barbara Co.), Warner's Ranch (San Diego Co.).

Var. **intermèdium** Parish. Plate 87, Fig. A. Subshrub; short-pubescent with gland-tipped hairs, the lvs. often ± cordate at base.—Dry slopes and canyons below 5000 ft.; Chaparral, Foothill Wd., S. Oak Wd.; cismontane.

Var. **montànum** Munz. Strictly herbaceous, the stems 1–4 dm. long, often in prostrate mats, densely grayish-pubescent; berry 9–10 mm. in diam.—Dry places, 5000–9000 ft.; Montane Coniferous F.; Santa Rosa and San Jacinto mts. to Sierra Nevada. May–Sept.

102. Sterculiàceae. Cacao Family

Trees, shrubs or herbs, chiefly with stellate pubescence. Lvs. alternate, simple or rarely compound, stipulate. Fls. small or large, perfect, regular or nearly so. Calyx usually 5-lobed. Petals 5 or 0, free or united with stamen tube. Fertile stamens 5, the fils. ± united below, with staminodia sometimes also present. Fr. a 1–5-celled caps. A family of ca. 60 genera and 700 spp., usually of warmer regions.

Fls. brownish, 0.4–0.7 mm. across; shrub 1–3 dm. high; lvs. not lobed .... 1. *Ayenia*
Fls. yellow, 3–8 cm. across; shrub 1–6 m. tall; lvs. usually lobed .... 2. *Fremontodendron*

1. *Ayènia* Loefl.

Herbs or ours a subshrub. Lvs. serrate-dentate. Fls. small, pedicellate, in axillary clusters. Calyx 5-lobed. Petals 5, long-clawed, hooded, the hoods inflexed, adnate to the calyx-tube and covering the anthers. Anthers 3-celled, solitary in sinuses of stamen-tube and alternating with truncate staminodia. Caps. stipitate, 5-celled, splitting into 5 1-seeded bivalvate carpels. Seeds in ours dark, pitted-rugose. Ca. 15 spp. of warmer parts of New World. (Named for the Duc d'*Ayen*.)

1. **A. compácta** Rose. [*A. californica* Jeps. *A. pusilla* auth., not L.] Stems several, slender, 1–2 (–3) dm. tall, strigulose; lvs. ovate to oblong-ovate, 6–10 mm. long, on petioles half as long; fls. brownish, 2 mm. long; caps. ca. 3 mm. long, finely pubescent and with short tubercles; seed almost 2 mm. long.—Occasional in dry rocky canyons, below 1500 ft.; Creosote Bush Scrub; w. edge of Colo. Desert and Eagle Mts.; L. Calif., Son. March–April.

2. *Fremontodéndron* Cov. Fremontia. Flannel Bush

Large shrubs or small trees, evergreen but losing many lvs. in dry season, stellate-pubescent and with mucilaginous inner bark. Lvs. simple, alternate, subentire to 3-, 5-, or 7-lobed. Fls. mostly bisexual, large, showy, regular, solitary, opposite the lvs. on the younger branches or on short lateral spurs. Calyx subtended by involucel of ca. 3 lance-subulate bractlets, petaloid, open-campanulate, 5-lobed to below the middle, externally stellate-pubescent, internally with a pit at the base of each lobe. Petals 0. Stamens 5, alternate with the sepals, joined by their fils. for ca. half their length. Ovary superior, 4- or 5-celled, surrounded by the base of the fil.-tube; style filiform, exserted beyond the stamens. Fr. a densely bristly-hairy caps., 4- or 5-valved, dehiscent from apex, persisting for months. Seeds 2 or 3 in a locule. (John C. *Frémont*, western explorer, and *dendron*, tree.)

Lvs. 1–3-veined at base; glands at base of sepals hairy; fls. 3–6 cm. wide, borne on lateral branchlets; seeds dull, with large terminal caruncle .... 1. *F. californicum*
Lvs. palmately 5–7-veined from base; glands at base of sepals glabrous; fls. 6–9 cm. across, borne on main twigs; seeds shining, without caruncle .... 2. *F. mexicanum*

1. **F. califórnicum** (Torr.) Cov. [*Fremontia c.* Torr.] Plate 87, Fig. B. Open spreading shrub 1.5–4 (–7) m. tall; lvs. and fls. mostly on short lateral spurlike branchlets; lvs. round to elliptic-ovate, dull green and sparsely stellate-pubescent above, densely tawny-stellate beneath, 1–3 (–7) cm. long; petioles 1–4 cm. long; calyx mostly clear yellow, flat, 3.5–6 cm. in diam., the basal glands usually with long hairs; caps. ovoid, acute, 2.5–3.5

Plate 87. Fig. A, *Solanum xanti intermedium*, fr. a berry. Fig. B, *Fremontodendron californicum*, 5 petaloid sepals subtended by smaller bracts. Fig. C, *Styrax officinalis* var. *fulvescens*, calyx quite gamosepalous. Fig. D, *Tamarix parviflora*, lvs. scalelike, fls. bisexual, seeds tufted with hairs. Fig. E, *Celtis reticulata*, *a*, drupe, *b*, bisexual fl., although ♂ and ♀ often separate. Fig. F, *Urtica holosericea*, with separate ♂ and ♀ fls.

cm. long; seeds 3–4 mm. long; $n=20$ (Lenz, 1950).—Dry mostly granitic slopes, 3000–7000 ft.; Chaparral, Yellow Pine F., Pinyon-Juniper Wd.; mts. in San Diego Co., San Bernardino and San Gabriel mts., region of Mt. Pinos, Tehachapi Mts.; to Shasta Co.; Ariz. May–June. In e. San Diego Co. some plants have subentire lvs. and petioles more than half as long as blade and have been called var. *diegensis* M. Harvey.

2. **F. mexicànum** Davidson. [*Fremontia m.* (A. Davidson) Macbr.] Tall, stiff and ca. 2–6 m. across; branches with dense stellate tomentum at first yellowish, later dark; lvs. roundish, 2.5–7 cm. across, cordate at base, with 5–7 main veins from base, shallowly lobed; petioles mostly 2–4 cm. long; calyx somewhat shallowly campanulate, 6–9 cm. across, orange becoming reddish at base on outside, somewhat puberulent within basal pits; caps. conical, mostly 3–4 cm. long.—Local in dry canyons, at ca. 1500 ft.; Chaparral, S. Oak Wd.; Otay Mt., Jamul, San Diego Co. and adjacent L. Calif. March–June.

103. **Styracàceae.** Storax Family

Shrubs or trees. Lvs. simple, alternate, exstipulate, entire or dentate. Fls. bisexual, regular, in axillary or terminal racemes or fascicles, rarely solitary. Calyx campanulate or tubular, 4–8-lobed, the petals often united only at base. Stamens twice as many as corolla-lobes, sometimes more, in 1 series, the fils. united at base. Ovary partly inferior, 2–5-loculed. Ovules 1–few in each locule. Stigma simple or 2–5-lobed. Fr. a berry or drupe, or dry and dehiscent by 3 valves. Seeds usually 1 in each locule; endosperm copious. Ca. 6 genera and 120 spp., warmer parts of N. and S. Am., e. Asia, Medit. region.

1. *Stýrax* L. Storax. Snowdrop Bush

Ours deciduous shrubs with axillary or leafy-racemose showy fragrant fls. on drooping peduncles. Calyx truncate. Corolla white. Stamens 10–16, united at base. Fr. globular, nearly dry, 1-loculed, commonly 3-valved. Ca. 100 spp., some grown for the attractive fls. (Ancient Greek name used for the sp. producing the gum storax.)

1. **S. officinàlis** L. var. **fulvéscens** (Eastw.) M. & J. [*S. californica* var. *f.* Eastw.] Plate 87, Fig. C. Erect, 1–4 m. high, with grayish twigs; lf.-blades round-ovate to obovate, obtuse to subcordate at base, obtuse to rounded at apex, 2–7 cm. long, pubescent above and stellate-tomentose beneath, the pubescence sometimes tawny; petioles 3–10 mm. long; fls. few, in terminal clusters on the branchlets; peduncles 6–12 mm. long; calyx unequally toothed, persistent; corolla 4–10-lobed, 12–18 mm. long; fr. ca. 12–14 mm. long; seed globose-ovoid, smooth.—On slopes and in canyons, below 5000 ft.; Chaparral, S. Oak Wd.; San Luis Obispo to San Bernardino and San Diego cos. April–May.

104. **Tamaricàceae.** Tamarisk Family

Shrubs or small trees. Lvs. alternate, entire, thickish, small, scalelike, exstipulate. Fls. regular, bisexual, small, 4–5-merous. Sepals free or ± united. Petals imbricated, sometimes withering and persistent. Stamens as many or twice as many as petals, free or united, inserted on a fleshy disk. Ovary 1-celled; styles 3–5; placentae 3–5, parietal. Caps. 3–5-valved; seeds many, usually with terminal tuft of hairs. Genera 4; spp. ca. 100.

1. *Tamaríx* L. Tamarisk

Branchlets slender, with minute appressed scaly lvs. Ca. 75 spp., Medit. region to E. Indies and Japan. (*Tamaris*, Spanish river.)

(Baum, B. R. Introduced and naturalized tamarisks in the United States and Canada. Baileya 15: 19–25. 1967.)

Fls. 4-merous; petals oblong, scarcely 2 mm. long; bracts diaphanous . . . . 6. *T. parviflora*
Fls. 5-merous.
    Lvs. vaginate; bracts not vaginate, but somewhat clasping . . . . 2. *T. aphylla*

Lvs. sometimes auriculate, but not vaginate or amplexicaul.
> Staminate fls. with fils. inserted below the disk.
>> Petals obovate, widened distally; bracts ovate, subobtuse. In saline places .... 7. *T. ramosissima*
>> Petals oblong-ovate, narrowed distally. Non-saline habitats .... 4. *T. chinensis*
> Staminate fls. with fils. inserted around the disk.
>> Petals more than 2.25 mm. long, subpersistent, narrowed slightly toward the apex; racemes 5–10 mm. broad .... 1. *T. africana*
>> Petals not more than 2 mm. long, soon deciduous; racemes 4–5 mm. broad.
>>> The petals elliptical, equally wide in both halves; disk with 5 rounded lobes, each between 2 stamens .... 3. *T. aralensis*
>>> The petals elliptic-ovate, definitely wider in lower half; disk not lobed, the fils. confluent at base. Rare .... 5. *T. gallica*

1. **T. africàna** Poir. Bark dark; lvs. sessile; racemes 3–7 cm. long, 6–9 mm. broad, those occurring on green branches of the current year somewhat smaller; bracts longer than pedicels; fls. 5-merous; sepals subentire, the outer 2 slightly keeled and longer than the inner more obtuse ones; petals ± ovate, 2.5–3 mm. long in vernal fls., 3 mm. or more in aestival; staminal fils. inserted on gradually tapering lobes of disk.—Mouth of Ventura R. and in Tex.; introd. from Medit. region.

2. **T. aphýlla** (L.) Karsten. Bark reddish-brown to gray; lvs. vaginate; racemes 3–6 cm. long, 4–5 mm. broad; bracts longer than pedicels; fls. 5-merous; sepals entire, the inner somewhat larger; petals elliptic-oblong to ovate-elliptic, 2–2.5 mm. long, caducous, sometimes 1 or 2 persisting; fils. inserted between the ± retuse disk-lobes.—Occasional escape in Calif., where used for windbreaks; native of Eurasia.

3. **T. aralénsis** Bunge. Bark reddish-brown to brown; lvs. sessile, racemes densely fld., 2–6 cm. long, ca. 3.5 cm. broad, with minutely papillose rachis; bracts longer than pedicels; fls. 5-merous; sepals deeply and irregularly denticulate, obtuse; petals elliptical, ca. 1.75 mm. long, deciduous; fils. inserted between the ± retuse lobes of the disk.—Rare; native of sw. Asia.

4. **T. chinénsis** Loureiro. Bark brown to black-purple; lvs. sessile; vernal infls. of many dense racemes, aestival loose, of slender racemes; racemes 2–6 cm. long, 5–7 mm. broad; bracts equaling or slightly exceeding pedicels; fls. 5-merous; sepals subentire, acute; petals elliptic to ovate, persistent, 1.5–2.25 mm. long; fils. inserted between lobes of disk but from its lower part near the margin; in aestival fls. 1–2 fils. are inserted in sinuses between the lobes and other 3–4 under the disk near the margin.—Frequent on the deserts; introd. from e. Asia.

5. **T. gállica** L. Bark blackish-brown to deep purple; lvs. sessile; racemes 2–5 cm. long, 4–5 mm. broad; bracts longer than pedicels, not exceeding calyx; fls. 5, 5-merous; sepals acute, entire or subentire; petals caducous, elliptical to ± ovate, 1.5–1.75 mm. long; fils. inserted on apices of the gradually attenuating lobes of the disk.—Rare in Calif.; introd. from Medit. region.

6. **T. parviflòra** DC. Plate 87, Fig. D. Bark brown to deep purple; lvs. sessile; racemes more often on last year's branches, 1.5–4 cm. long, 3–5 mm. broad; bracts diaphanous, longer than pedicels; fls. 4-merous; sepals eroded-denticulate, the outer 2 (shaped like a bricklayer's trowel) trullate-ovate, acute and keeled, the inner 2 ovate, obtuse; petals 2 mm. long; fils. emerging gradually from the disk-lobes.—Desert and cismontane; introd. from e. Mediterranean.

7. **T. ramosíssima** Ledeb. Bark reddish-brown; lvs. sessile; racemes 1.5–7 cm. long, 3–4 mm. broad; bracts longer than pedicels; fls. 5-merous; sepals ± acute, erose to irregularly denticulate, the inner 3 broader than the outer; petals 1–1.75 mm. long, obovate; fils. inserted under the disk near the margin, between the usually emarginate lobes. —Deserts of Calif.; introd from Asia.

105. Tropaeolàceae. Tropaeolum Family

Succulent, prostrate or vinelike herbs, often climbing by coiling petioles; annual or perennial. Roots sometimes tuberous. Lvs. alternate, digitately angled or peltate, sometimes lobed. Fls. solitary, axillary, very irregular, usually showy. Sepals 5, one produced into a spur. Petals 5, rarely fewer, clawed, the 2 upper unlike the others. Stamens 8, free, unequal. Ovary superior, 3-lobed, 3-celled; cells 1-ovuled; placentation axile. Style 1, apical; stigmas 3, linear. Fr. of 3 indehiscent 1-seeded carpels which separate from cent. axis when mature. Single genus with ca. 50 spp. of Latin Am.

1. *Tropaèolum* L. Nasturtium

Characters of family. (Greek, *trophy*, from the shieldlike lvs.)

1. **T. màjus** L. Garden Nasturtium. Annual or of longer duration, dwarf or climbing; lvs. peltate, 4–15 cm. wide, ca. 9-nerved; fls. yellow, red, etc., 3–6 cm. wide; spur 2–4 cm. long; $2n=28$ (Warburg, 1938).—Occasionally escaping and establishing itself in gullies, on bluffs, etc., along the immediate coast; native of S. Am. Fls. most of year.

106. Ulmàceae. Elm Family

Trees or shrubs with watery juice, terete branchlets, and alternate simple serrate pinnately veined mostly deciduous lvs. with stipules. Perfect or polygamous, the fls. clustered or the ♀ solitary. Calyx 4–9-parted or -lobed, small. Corolla none. Stamens 4–6. Styles 2; ovary 1-celled, with 1 suspended ovule. Fr. a samara, nut or drupe.

Fr. drupaceous; fls. imperfect .... 1. *Celtis*
Fr. a dry samara; fls. perfect .... 2. *Ulmus*

1. *Céltis* L. Hackberry

Deciduous, sometimes evergreen, bark mostly smooth, gray. Lvs. 3-nerved at base, serrate or entire. Fls. greenish, appearing with the lvs. Calyx 5–6-parted, the ♂ fls. in cymose clusters, the ♀ solitary or in few-fld. clusters in upper lf.-axils. Ovary ovoid; fr. a drupe with thick-walled nutlet. Perhaps 70 spp. of N. Hemis. and trop. (Greek name of a tree with sweet fr.)

1. **C. reticulàta** Torr. [*C. douglasii* Planch.] Plate 87, Fig. E. Small spreading tree, 2–6 m. tall, with slender glabrous or puberulent twigs; lvs. lance-ovate, acute or acuminate, serrate to almost entire; 3–6 cm. long, on petioles 2–6 mm. long, scabrous above, strongly reticulate-veined beneath; pedicels 1–1.5 cm. long; fr. globose, 6–8 mm. thick, orange-brown.—Occasional about widely scattered damp places, 2800–5000 ft.; Creosote Bush Scrub, Joshua Tree Wd., Pinyon-Juniper Wd., etc.; Laguna Mts. (e. San Diego Co.), Banning, Clark Mt., Providence Mts., etc. to n. Inyo Co., Kern Co., etc.; to e. Wash., Utah, Ariz. April–May.

2. *Úlmus* L. Elm

Trees or shrubs, the buds with many ovate rounded brown scales imbricated in 2 ranks. Lvs. simple or doubly serrate; stipules linear to obovate. Fls. in fascicles or cymes, appearing in spring before the lvs. Calyx persistent; stamens 5–6, fils. filiform or slightly flattened; fr. surrounded at base by remnants of calyx and winged most of the way around with a terminal notch and persistent style. Ca. 18–20 spp. of N. Temp. and colder regions. (Classical Latin name for Elm.)

Lvs. doubly serrate, unequal at base.
    Lvs. with scattered pubescence, scabrous above; branchlets pubescent until the second year
                                                                 3. *U. procera*

Lvs. with axillary tufts beneath, smooth above; branchlets subglabrous .... 1. *U. minor*
Lvs. usually simple serrate, subequal at base.
    Branchlets pubescent; fls. in late summer; samara ca. 8 mm. long .... 2. *U. parvifolia*
    Branchlets soon glabrous; fls. in spring before the lvs.; samara ca. 12 mm. long . 4. *U. pumila*

1. **U. mìnor** Mill. [*U. carpinifolia* Gleditsch.] Smooth-Lvd. Elm. Tree to 30 m. high, suckering; young twigs subglabrous; lvs. 5–8 cm. long, on petioles 6–12 mm. long; samara cuneate at base; $2n=28$ (Sax, 1933).—Occasional escape; native of Eurasia.

2. **U. parvifòlia** Jacq. [*U. chinensis* Pers.] Chinese Elm. Partially evergreen in mild climate; branchlets pubescent; lvs. elliptic to ovate, 2–3.5 cm. long, acute; fls. in late summer.—Seeding itself as at Pasadena, Santa Barbara; from e. Asia.

3. **U. prócera** Salis. English Elm. Tree to 35 m. tall, with oval head; usually suckering abundantly; bark deeply fissured; young twigs pubescent; lvs. 5–7 cm. long, ovate to broadly oval, short-acuminate, very unequal at base, scabrous above; samara to 2.5 cm. long; $2n=28$ (Sax, 1933).—Occasionally reported as weed tree; native of Eu.

4. **U. pùmila** L. Siberian Elm. Small tree; branchlets soon glabrous; lvs. elliptic to oblong-lanceolate, 2–3.5 cm. long, short-acuminate; fls. in the spring.—Cult. in interior and desert areas; native of Siberia.

107. Urticàceae. Nettle Family

Ours herbaceous; trop. forms may be woody. Stinging hairs often present. Lvs. simple, alternate or opposite, mostly stipulate. Fls. small, greenish, imperfect or perfect, in cymose racemes or fascicles. Corolla none. Calyx 2–5-cleft or of separate sepals. Stamens as many as the calyx-divisions and opposite them. Ovary superior, 1-celled, with a solitary orthotropous ovule. Style and stigma 1. Fr. an ak., with oily endosperm.

Plants with stinging hairs and opposite lvs.
    Pistillate calyx saccate, 2–4-toothed at apex .... 1. *Hesperocnide*
    Pistillate calyx 4-parted, the segms. almost distinct, the inner much the longer and inclosing the ak. .... 4. *Urtica*
Plants without stinging hairs and with alternate lvs.
    Fls. in axillary glomerate clusters .... 2. *Parietaria*
    Fls. solitary .... 3. *Soleirolia*

1. Hesperócnide. Torr.

Annual herbs similar to *Urtica urens* in appearance. Stipules minute; ♂ fls. in clustered axillary heads, mixed with pistillate. Calyx of ♂ fls. 4-parted, the lobes equal; that of ♀ fls. tubular, 2–4-toothed, with uncinate hairs. Stamens 4. Stigma tufted. Ak. enclosed by the membranous calyx. Two spp., one in Hawaii. (Greek, *hespera*, west, and *knide*, nettle.)

1. **H. tenélla** Torr. Stems slender, rather weak, 2–5 dm. long, simple or branched, with stinging hairs; lvs. opposite, ovate, crenate-serrate, the blades 0.5–2.5 cm. long; calyces 1–1.5 mm. long in fr.—Occasional on shaded slopes at low elevs.; Coastal Sage Scrub, Chaparral; cismontane San Diego Co. to Napa Co., desert edge. April–June. San Clemente, Santa Catalina, Santa Cruz ids.

2. Parietària L. Pellitory

Herbs to shrubs, without stinging hairs. Lvs. alternate, entire, 3-nerved, exstipulate. Plants polygamous, the fls. in axillary glomerate clusters, involucrate by small leafy bracts. Staminate calyx 4-parted. Pistillate calyx tubular-ventricose, 4-lobed. Stigma tufted. Ak. ovoid, enclosed in the persistent calyx. Ca. 7 spp. of wide distribution. (Ancient Latin name of the Italian sp.)

1. **P. floridàna** Nutt. [*P. debilis* Calif. refs.] Loosely branched annual with slender weak procumbent pilose stems 1–3 dm. long; lvs. ovate to lance-ovate, 0.5–3 cm. long,

entire, obtuse, short-petioled; clusters few-fld., the bracts narrow, 3 mm. long; sepals ca. 2 mm. long; ak. 1 mm. long, shining.—Moist shaded slopes about rocks, etc. at low elevs.; Coastal Sage Scrub, Chaparral, Creosote Bush Scrub, etc.; cismontane and desert s. Calif.; to cent. Calif.; to se. U.S. Channel Ids.

3. Soleiròlia Gaud.-Beaup.

Like *Parietaria* but creeping and rooting at the nodes; fls. solitary, surrounded by an invol. of 1 bract and 2 bracteoles; ak. enclosed by the perianth and invol.

1. **S. soleiròlii** (Req.) Dandy. [*Helxine s.* Req.] Pubescent very slender perennial 5–20 cm. long; lvs. 2–6 mm., suborbicular, 3-veined; fls. unisexual, the lower ♀, the upper ♂; $2n = 20$.—Cult. in shade gardens and escaping in damp shaded places near the coast, as at Santa Barbara; native of Eu.

4. Úrtica. L. Nettle

Annual or perennial herbs with stinging hairs and opposite petioled lvs. with 3–7 nerves, stipulate. Our spp. dioecious or monoecious, the fls. clustered in geminate racemes or heads. Staminate calyx deeply 4-parted and with 4 stamens. Pistillate calyx with unequal sepals, the longer inner ones enclosing the flattened ak. Stigma sessile, tufted. Perhaps 30 spp., of wide distribution. (The ancient Latin name from *urere*, to burn.)

Plant perennial, ♂ and ♀ fls. in separate clusters .... 1. *U. holosericea*
Plant annual; ♂ and ♀ fls. mixed in the same cluster .... 2. *U. urens*

1. **U. holosericea** Nutt. Plate 87, Fig. F. Perennial from underground rootstocks, the stems rather stout and simple, 1–2.5 m. tall, bristly and also densely fine-pubescent; lvs. lanceolate to narrow-ovate, 5–12 cm. long, coarsely serrate, densely soft-pubescent and grayish beneath, greener above, attenuate at apex, on petioles 1–4.5 cm. long; stipules narrow-oblong, 5–10 mm. long, ♂ clusters rather loose, almost as long as the lvs.; ♀ denser and shorter; fls. ca. 1 mm. long; ak. broadly ovate, smooth.—Low damp places, below 9000 ft., many Plant Communities; cismontane Calif.; to Wash., Ida.; occasional on desert edge. Santa Catalina and Santa Cruz ids.

2. **U. ùrens** L. Dwarf Nettle. Plate 88, Fig. A. Erect annual, simple or branched from the base, glabrous except for the stinging hairs, 1–5 dm. high; lvs. ovate, glabrous, coarsely laciniate-serrate, the blades 1.5–3 cm. long; petioles 1–2 cm. long; stipules 1 mm. long; fl.-clusters scarcely 1 cm. long; calyx almost 2 mm. long; ak. ca. 2 mm. long, ± yellow; $2n = 24, 26, 52$ (Löve & Löve, 1942).—Garden and orchard weed; natur. from Eu. Jan.–April.

108. Valerianàceae. Valerian Family

Herbs, sometimes shrubs, with opposite exstipulate lvs. Fls. small, bisexual or unisexual, in cymose or capitate infl. Calyx annular or variously toothed, often inrolled in fl. and forming a feathery pappus in fr. Corolla funnelform to almost salverform, the base often saccate or spurred on one side. Stamens 1–4, inserted near the base of the corolla-tube. Ovary inferior, mostly 3-loculed, one locule fertile, the other 2 sterile or almost lacking. Ovule 1, pendulous. Fr. dry, indehiscent. Ca. 9 genera and 300 spp., mostly of N. Hemis.

Perennial; corolla with a tubular spur; stamen 1. Natur. .... 1. *Kentranthus*
Annual; stamens 3. Native .... 2. *Plectritis*

1. Kentránthus Neck.

Annual or perennial herbs, sometimes suffruticose. Lvs. entire, dentate or pinnatisect. Fls. small, red to white, in dense terminal clusters. Calyx cut into 5–15 narrow divisions

Plate 88. Fig. A, *Urtica urens*, with ♂ fl. Fig. B, *Plectritis macrocera*, spurred ♀ fl., fr. with inrolled wings. Fig. C, *Verbena gooddingii*, *b* and *c*, fl., *d*, seed. Fig. D, *Verbena menthifolia*. Fig. E, *Viola adunca*, caulescent. Fig. F, *Viola lobata*, lobed and entire-lvd. vars.

infolded at anthesis, enlarging and spreading in fr. Corolla with a slender tube, the limb 5-parted; spur basal. Stamens 1. Fr. 1-loculed, narrow, crowned with a pappuslike calyx, 1-seeded. Ca. 12 spp., Old World. (Greek, *kentron*, a prickle or spur, and *anthos*, fl.)

K. rùber (L.) DC. [*Valeriana r.* L.] RED VALERIAN. Smooth, ± glaucous, 3–8 dm. high; lvs. lance-ovate to ovate, sessile and broad at base, or short-petioled, mostly entire, 3–10 cm. long; fls. many, red to pink and occasionally white, rather crowded; corolla-tube ca. 10 mm. long, the limb spreading, spur slender; stamens exserted; fr. an elongate narrow nut, 3–4 mm. long; $2n = 14$ (Poucques, 1949).—Occasionally established as wild plant; escape from cult. April–Aug.

2. Plectrìtis DC.

Subglabrous annuals, with erect to sprawling simple or few-branched stems, having tufts of hairs at nodes. Lvs. entire or with a few teeth, obovate-oblong, the lower short-petioled, the upper sessile and oblong to linear. Infl. capitate or interruptedly spicate; bracts linear-subulate, the base of several being fused to give a palmate appearance. Calyx obsolete. Corolla small, 5-lobed, sometimes strongly bilabiate, funnelform, usually spurred near the base. Stamens 3. Stigma-lobes 2, flat-reniform. Ovary inferior, 1-loculed, the fr. a wingless or winged ak., the wings being an outgrowth from the dorsal angles of the fertile locule. Ca. 4 spp., w. N. Am. and Chile. (Latin, *plecto*, to plait, because of the complex infl.)

(Morey, D. H. Changes in nomenclature in the genus Plectritis. Contr. Dudley Herb. 5: 119–121. 1959.)

Corolla bilabiate, the spur obsolete to usually less than ⅓ the length of the corolla; keel of the fr. rarely grooved .... 2. *P. congesta*
Corolla regular to bilabiate, the spur usually more than ⅓ the length of the corolla; keel of the fr. often with a dorsal groove.
    Corolla strongly bilabiate, pink to light red, usually with 2 red spots at the base of the ventral lip, slender and with a slender spur; keel of the fr. often with 2 brushlike rows of hairs
                                                                                                                                1. *P. ciliosa*
    Corolla essentially regular, white or light pink, usually lacking red spots at the base of the ventral lip, stout and with a clavate spur; keel of the fr. without 2 brushlike rows of hairs
                                                                                                                                3. *P. macrocera*

1. **P. ciliòsa** (Greene) Jeps. ssp. **insígnis** (Suksd.) Morey. [*Aligera i.* Suksd.] Slender, 1–5 dm. tall; lvs. obovate to oblong-fusiform; corolla deep pink with dark spots at base of ventral lobes, 1.5–3.5 mm. long, slender, bilabiate, the spur slender, usually exceeding ovary; fr. 3–4 mm. long, ± pubescent; keel rounded, obtusely angled, with a deep dorsal groove; wings variously expanded, bounded with a usually grooved marginal thickening. —Yellow Pine F. and below, inland cismontane valleys, San Diego Co. to Napa Co.; to Wash., n. L. Calif. March–May.

2. **P. congésta** (Lindl.) DC. ssp. **brachystèmon** (F. & M.) Morey. [*P. b.* F. & M.] Corolla bilabiate, often spurless, 1–3.5 mm. long, tubular-funnelform.—Brushy montane areas; Yellow Pine F.; Cuyamaca Lake (San Diego Co.); Monterey Co. n.

3. **P. macrócera** T. & G. Plate 88, Fig. B. Plants slender, 1–6 dm. tall; lvs. obovate to lance-ovate; corolla white to pale pink, 2–3.5 mm. long, stout, subcampanulate to broadly tubular, regular to weakly bilabiate, the spur stout, ca. twice as long as broad, 1–1½ times as long as tube; fr. pale straw-yellow to red-brown, 2–4 mm. long, ± pubescent; keel rounded to rather angular, obtusely angled, often with a dorsal groove; wings thick, expanded to obsolete, with a usually grooved marginal thickening; $n = 18$ (Raven, Kyhos & Hill, 1965).—Usually shaded places below 4000 ft.; Foothill Wd., S. Oak Wd., V. Grassland, etc.; cismontane; to s. Wash. April–May.

Ssp. **gràyii** (Suksd.) Morey. [*Aligera g.* Suksd.] Expanded wings thin, with a narrow marginal thickening scarcely grooved; fr. with a median ridge on ventral surface usually bearing a multiseriate row of bristles.—Open or shaded places, s. Calif. to Wash., Mont., Utah.

109. Verbenàceae. VERVAIN FAMILY

Herbs, shrubs or trees with opposite or whorled lvs. Fls. perfect, mostly ± irregular, in terminal or axillary spikes, racemes or panicles. Calyx persistent, 2-, 4-, or 5-toothed or -lobed. Corolla gamopetalous, the tube cylindrical, the limb 4–5-lobed. Stamens 4 and didynamous, or rarely 2 or 5, inserted on corolla-tube. Ovary 2–4-celled, ovules usually 1 in a cell; style simple; stigmas 1–2. Fr. dry, forming 2 or 4 bony nutlets or a drupe with 2–4 nutlets. Ca. 80 genera and 800 spp., mostly of warmer regions.

Nutlets 2; fls. in dense spikes or heads.
   Spikes flat-topped; fr. drupaceous .... 1. *Lantana*
   Spikes cylindrical or globose; fr. dry .... 2. *Lippia*
Nutlets 4; fls. in terminal spikes .... 3. *Verbena*

1. Lantána L.

Shrubs or herbs, scabrous-hirsute, pubescent or tomentose. Lvs. opposite, dentate, often rugose. Fls. small, sessile in the axils of bracts, forming dense spikes or heads, which are terminal or axillary. Calyx very small. Corolla somewhat irregular, but not bilabiate; tube slender; lobes 4–5. Stamens 4, included. Ovary 2-loculed, forming a fleshy drupe with 2 bony nutlets. (Old Latin name.)

   1. **L. montevidénsis** Briq. Stems weak, vinelike, ca. 1 m. long; lvs. ovate, 2–5 cm. long; fls. rose-lilac, in heads 2.5–3 cm. across.—Occasionally spontaneous, as at Claremont, Santa Barbara, etc.

   L. camára L. Taller, with fls. opening yellow or orange, later red to scarlet, has also been reported as an escape.

2. Lippia L.

Herbs or shrubs with opposite or whorled lvs. Fls. in spikes or heads on slender axillary peduncles. Calyx small, 2–4-cleft. Corolla-tube cylindrical, the limb bilabiate. Nutlets 2. Rather a large genus of over 100 spp., often divided into several. (Dr. A. *Lippi*, European naturalist.)

Erect shrub; fls. in narrow spikes; calyx 4-lobed (*Aloysia*) .... 4. *L. wrightii*
Low herbs; fls. capitate or nearly so; calyx 2-lobed (*Phyla*).
   Lvs. broadest below the middle, the blades 2–4 cm. long, serrate from below the middle to the apex; calyx-lobes ca. as long as tube .... 2. *L. lanceolata*
   Lvs. broadest above the middle, the blades mostly 0.5–2 cm. long, serrate only above the middle; calyx-lobes shorter than tube.
      Lf.-margins with spreading teeth; lvs. apically obtuse, tapering gradually to the ± sessile base .... 1. *L. incisa*
      Lf.-margins with teeth pointing forward; lvs. mostly acute at apex, abruptly narrowed to a petioled base .... 3. *L. nodiflora*

   1. **L. incìsa** (Small) Tides. [*Phyla i.* Small.] Spreading or creeping, rooting at nodes, much-branched, sparingly strigose; lvs. linear-cuneate, 1–3 cm. long; peduncles 3–7 cm. long; heads becoming 1–2 cm. long in age; bracts rhomboidal; calyx almost 2 mm. long; corolla white or bluish, 2.5–3 mm. long; fr. broadly obovoid, 1.5–2 mm. long.—Low wet places; Imperial and San Diego cos.; San Joaquin V.; New Mex., Tex. April–Oct.

   2. **L. lanceolàta** Michx. [*Phyla l.* Greene.] Perennial with procumbent stems rooting at base, 2.5–4 dm. long, strigulose-canescent; lvs. opposite, lanceolate to ovate, acute, 3–6 cm. long, cuneately narrowed at base to short petioles; peduncles in upper axils, exceeding lvs.; spikes 1–1.5 cm. long; bracts broadly ovate; calyx 2 mm. long; corolla pale blue to lavender, 2.5–3 mm. long, ± strigose toward summit; fr. subglobose, ca. 2 mm. thick; $2n=32$ (E. B. Smith, 1966).—Occasional in low wet places, cismontane and desert; to Atlantic Coast. May–Sept.

   3. **L. nodiflòra** Michx. var. **ròsea** (D. Don) Munz. [*Zappania n.* var. *r.* D. Don.] GARDEN LIPPIA. Plants matted, more or less woody near base, cinereous-strigulose; lvs.

Lippia 851

pale green, narrow-oblanceolate to -obovate, entire to toothed, 1–2 cm. long, ± acutish; peduncles 1–3 cm. long; spikes ovoid, mostly 5–8 (–10) mm. long; bracts ovate; calyx ca. 1 mm. long; corolla rose to white, 4–5 mm. long.—Used as a groundcover and well established in ± moist waste places in cent. and cismontane s. Calif.; from S. Am. Santa Catalina Id. May–Oct. Var. **canéscens** (HBK) Kuntze with spikes cylindric in age, 4–5 mm. thick; calyx 2 mm. long; corolla 3 mm. long, has been recorded from Imperial V.; S. Am.

4. **L. wrìghtii** Gray. [*Aloysia w.* Heller.] Aromatic shrub 6–15 dm. tall with slender spreading tomentose branchlets; lvs. canescent-tomentose, 5–10 mm. long, ± ovate, crenate; spikes slender, 2–2.5 cm. long; bracts lanceolate; calyx 2.5 mm. long; corolla white, 3 mm. long, 2 mm. wide; nutlets oblong, almost 2 mm. long.—Dry rocky places, 3000–5000 ft.; Joshua Tree Wd., Pinyon-Juniper Wd.; Little San Bernardino Mts., Clark and Providence mts.; to Tex., n. Mex. Spring and fall.

3. *Verbèna* L. Verbena. Vervain

Annual or perennial herbs or subshrubs. Lvs. mostly opposite. Fls. bracteate, in terminal corymbose or paniculate spikes. Calyx tubular, 5-ribbed, 5-toothed. Corolla with straight or incurved tube, the limb spreading, 5-lobed, regular or slightly 2-lipped. Stamens 4 in 2 pairs, rarely only 2. Ovary entire or slightly 4-lobed at apex, 4-celled, the cells 1-ovuled. Fr. dry, inclosed in calyx, separating into 4 narrow nutlets. Ca. 100 spp., chiefly in warmer parts of Am. (Ancient Latin name of the common European Vervain.)

1. Fls. in ± slender spikes; corolla 3–6 mm. long.
    2. Bracts longer than or equalling fls., conspicuous .... 1. *V. bracteata*
    2. Bracts shorter than fls., inconspicuous.
        3. Spikes panicled, much elongate, slender.
            4. Petioles scarcely winged, expanding abruptly into the blade .... 6. *V. scabra*
            4. Petioles winged, expanding gradually into the blade .... 4. *V. menthæfolia*
        3. Spikes in 3's at the ends of the branches, often congested at anthesis.
            5. The spikes 1–3 dm. long in fr.; lvs. canescent, not scabrous above; inner face of nutlets without whitish papillae .... 3. *V. lasiostachys*
            5. The spikes 0.3–1 dm. long in fr.; lvs. bright green, scabrous above, inner face of nutlets with whitish papillae .... 5. *V. robusta*
1. Fls. mostly in headlike clusters; corolla 10–20 mm. long.
    6. The ultimate lf.-divisions oblong, flat. Native of e. Mojave Desert .... 2. *V. gooddingii*
    6. The ultimate lf.-divisions linear, revolute. Escape from gardens .... 7. *V. tenuisecta*

1. **V. bracteàta** Lag. & Rodr. [*V. bracteosa* Michx.] Annual to perennial, diffuse to spreading, the stems 1–5 dm. long, ± hirsute; lvs. oblong to obovate, pinnately lobed or incised, 1–4 cm. long, narrowed at base into a winged petiole 0.5–1.5 cm. long; stems leafy to the spikes which are 3–6 (–10) cm. long in fr.; bracts 5–10 mm. long; calyx 3–4 mm. long; corolla blue, the tube 3–4 mm. long, the limb 2.5–3 mm. broad; nutlets ca. 2 mm. long; $n=7$ (Dermen, 1936).—Occasional in waste places, below 5000 ft.; many Plant Communities; deserts and cismontane; to B.C. and Atlantic Coast. May–Oct.

2. **V. gooddíngii** Briq. Plate 88, Fig. C. Perennial with several ascending stems 2–4.5 dm. high, hairy, ± glandular; lvs. rounded in outline, 1–2 cm. long, palmately 3-parted, then pinnately cleft, on petioles 0.5–1.5 cm. long; spikes capitate; bracts lance-linear, ca. 8 mm. long; calyx 7–8 mm. long; corolla purplish, the tube ca. 1 cm. long; the limb 8–10 mm. broad with retuse lobes; nutlets 3–3.5 mm. long, reticulate to near the striate base.—Dry canyons and slopes, 4000–6500 ft.; Joshua Tree Wd., Pinyon-Juniper Wd.; Providence, N.Y., and Ivanpah mts. to Clark Mts., e. Mojave Desert; to Utah, Son. April–June.

3. **V. lasiostàchys** Link. [*V. prostrata* R. Br., not Savi.] Perennial, much branched and ultimately diffuse and ± procumbent, the stems 3–8 dm. long, villous; lvs. oblong- to broadly-ovate, coarsely serrate to laciniately lobed, 2–6 cm. long, the cuneate base narrowed into a short petiole; spikes 1–3, 5–20 cm. long and lax after anthesis; calyx

hairy, 4–5 mm. long; corolla mostly purple, the tube 4–5 mm. long, the limb 3–4 mm. wide; nutlets oblong-trigonous, striate below, reticulate above on backs; $2n=7$ pairs (Raven et al., 1965).—Dry to moist places, below 8000 ft.; many Plant Communities; n. L. Calif. n.; to Ore. May–Sept. A form with calyx 2–3 mm. long (from San Diego Co. n.) is var. *abramsii* (Mold.) Jeps.

4. **V. menthaefòlia** Benth. Plate 88, Fig. D. Annual or short-lived perennial; stems 3–6 dm. long, ± strigose; lvs. obovate to oblong in outline, 1–5 cm. long, ± canescent-strigose, irregularly pinnatifid or serrate, the petioles to ca. 1 cm. long, winged; spikes becoming 5–20 cm. long in fr., slender; calyx 2.5–3 mm. long; corolla purple, the tube 3.5–4 mm. long, the limb 5–6 mm. broad; nutlets 2–2.5 mm. long, striate, reticulate above.—Dry places; Coastal Sage Scrub; s. Los Angeles Co. to w. Riverside and San Diego cos.; Ariz., L. Calif. to s. Mex. April–June.

5. **V. robústa** Greene. Much like *V. lasiostachys*, but lvs. greener and scabrous; spikes dense, usually 3–10 cm. long in fr.; calyx 3–4 mm. long; corolla-limb 2–3 mm. broad; nutlets muriculate and gray on commissural face.—Moist places, largely Coastal Sage Scrub, Chaparral, etc.; cismontane; Santa Catalina and Santa Cruz ids. to n. Calif., n. L. Calif. May–Oct.

6. **V. scàbra** Vahl. Perennial with underground rootstocks, 4–12 dm. long, hispidulous, branched above; lvs. oblong to lance-ovate, serrate, 2–8 cm. long, scabrous, short-petioled; spikes very slender, 6–20 cm. long; calyx 2 mm. long, hispidulous; corolla white or purplish, the tube ca. 2 mm. long, the limb 2 mm. broad; nutlets 1.5 mm. long, striate.—Moist places; Coastal Sage Scrub, V. Grassland; Los Angeles and Orange cos. to San Bernardino and Riverside cos., Santa Barbara Co.; to Atlantic Coast, W. I. Sept.–Oct.

7. **V. tenuisécta** Briq. Cespitose perennial, the stems branched, slender; lvs. 2–4 cm. long, the segms. mostly ca. 1 mm. wide, revolute; corolla-limb ca. 10 mm. wide.—Garden escape as in Santa Monica Mts., McAndrew Road, Ventura Co., in Kern Co.; native of S. Am.

110. **Violàceae.** Violet Family

Herbs (with us), shrubs or rarely trees. Lvs. alternate or basal, simple, entire to laciniate, with stipules. Fls., in ours, irregular, axillary, nodding, bisexual, the peduncles usually 2-bracted. Sepals 5, free or slightly connate, nearly alike, usually persistent. Petals 5, hypogynous, the lower one usually longer than the others and saccate or spurred. Stamens 5, alternate with petals, the fils. short, broad, continued beyond the anther-locules, often connate into a ring around the ovary. Ovary free, sessile, 1-celled, with 3 (2–5) parietal placentae bearing 1–many ovules; style usually clavate, with the simple stigma turned to one side. Fr. a caps., dehiscent by valves. Seeds large, with hard seed-coat and straight embryo. Ca. 15 genera and 400 spp. of wide distribution.

1. *Viòla* L. Violet

Annual to perennial herbs. Fls. in most spp. of 2 kinds, those of early season with showy petals and mostly fertile, of later season cleistogamous, apetalous, and often producing seeds. Lower petal spurred, the other 4 an upper usually larger pair and a lateral pair. Stamens 5, enclosing stamineal cavity, the 2 lower with nectarlike appendages projecting into the spur. Caps. ovoid, the 3 valves boat-shaped and with thick rigid keels and thin sides; the latter dry and contract when mature and cause the seeds to be discharged forcibly. Over 300 spp., mostly of temp. zones; many, like the Common Pansy and Garden Violet, of great beauty. (*Viola*, the classical name.)

(Baker, M. S. A series of papers in Madroño and Leafl. W. Bot.)
1. Stipules almost as large as lvs., leaflike with large terminal lobe and pinnatisect at base; annual. Introd.
    2. Petals scarcely longer than sepals; spur ca. equal to appendages .... 2. *V. arvensis*
    2. Petals longer than sepals; spur rather longer than appendages .... 12. *V. tricolor*

1. Stipules much smaller and very different from lvs.; perennials.
    3. Fls. with some yellow, at least at bases of petals and on spur; access to the spur-cavity blocked by the much-expanded retrorsely bearded thickened head of the style; bearding of lateral petals short and clavate; spur ca. as broad as long. Plants of dry habitats.
        4. Lvs. not cut or lobed, but with margins entire or serrate-crenate.
            5. Plants strictly erect with cauline lvs. and fls. crowded at ends of stems which are naked below; caps. acute. E. San Diego Co. mts. .... 5. *V. lobata*
            5. Plants erect to prostrate, but with lvs. and fls. scattered along the stem; caps. obtuse.
                6. Crown of rootstock near the surface; fls. 2 cm. or less across; cleistogamous fls. present in upper axils.
                    7. Blades of basal lvs. as long as or longer than wide.
                        8. Petals 6–10 mm. long; lvs. rather spreading, greenish to grayish, usually tinted purple; upper lvs. subentire or irregularly or few (2–4)-toothed. Yellow Pine F. or higher .... 9. *V. purpurea*
                        8. Petals 10–12 mm. long; lvs. erect, grayish, without purple tinting; upper lvs. rather regularly 5–6-dentate on each side. Below Yellow Pine F.
                        10. *V. quercetorum*
                    7. Blades of basal lvs. commonly wider than long.
                        9. Teeth of basal lvs. shallow and very irregular .... 9. *V. purpurea*
                        9. Teeth of basal lvs. deep (2–3 mm.) and regular. Mts. at w. edge of Mojave Desert .... 3. *V. aurea*
                6. Rootstocks deep, short (1–2 cm. long), soft; fls. more than 2 cm. across; cleistogamous fls. lacking. Below Yellow Pine F. .... 8. *V. pedunculata*
        4. Lvs. deeply cut or lobed.
            10. Lf.-blades palmately 3–7-lobed, these divisions broad and not dissected; stems naked below. Mts. of San Diego Co. .... 5. *V. lobata*
            10. Lf.-blades dissected into narrow ultimate segms.
                11. Lvs. twice palmately divided; petals pale yellow; cleistogamous fls. present. Plants of wooded or brushy slopes .... 11. *V. sheltonii*
                11. Lvs. 2–3 times pinnatifid; cleistogamous fls. lacking; plants of open grassy places
                4. *V. douglasii*
    3. Fls. white, blue or purple, without any yellow; access to spur cavity not blocked by the style (the stigma held above the floor of the lower petal); bearding of the lateral petals longer, not clavate; spur longer than broad. Plants of damp places.
        12. Plants with evident leaf-bearing stems; petals violet, 8–12 mm. long. Montane
        1. *V. adunca*
        12. Plants lacking leaf-bearing stems (although they may have stolons), the lvs. and peduncles arising from the base or crown.
            13. Petals white, 6–10 mm. long; lvs. rounded to ovate-cordate, mostly less than 2 cm. long. Widespread in mts. .... 6. *V. macloskeyi*
            13. Petals bright violet; lvs. mostly 2.5–5 cm. wide, or if less, not rounded. Montane
            7. *V. nephrophylla*

1. **V. adúnca** Sm. Plate 88, Fig. E. Usually puberulent, perennial, from a slender rootstock; stems 4–20 cm. long; lvs. thickish, round-ovate, obtuse, somewhat cordate at base, obscurely crenate, 1–4 cm. long; petioles 1–6 cm. long; stipules linear-lanceolate; sepals ca. 5 mm. long; petals deep to pale violet, 8–13 mm. long, the 3 lower white at base and veined purple, the lateral white-bearded; spur 5–12 mm. long, rather broad, often hooked at tip; cleistogamous fls. present; caps. 6–8 mm. long; $n=10$ (Gershoy, 1934).—Damp banks and edge of meadows at 5000–8000 ft.; Montane Coniferous F.; mts. from San Diego Co. n.; to Alaska, Quebec, Colo. March–July.

2. **V. arvénsis** Murr. Freely branched annual, spreading or decumbent, 1–3 dm. long, harsh-puberulent; lvs. oblong-lanceolate to -ovate, coarsely crenate-serrate, 1.5–3 cm. long, short-petioled; stipules green, large, cut into linear segms. and with a large terminal portion; peduncles 5–8 cm. long; sepals lance-linear, acuminate, 6–7 mm. long; petals pale yellow, 7–10 mm. long.—To be expected as an occasional escape especially near the coast; native of Eu.

3. **V. aúrea** Kell. ssp. **mohavénsis** Baker & Clausen. Grayish with a rather dense short pubescence, perennial; stems mostly 5–15 cm. long; cauline lvs. subacute, with rather pointed teeth, the blades rounded or subreniform to lance-ovate, 1–3 cm. long,

petioles 2–7 cm. long; stipules 5–10 mm. long; sepals 5–6 mm. long; petals deep yellow, 8–12 mm. long, the 2 upper purple-brown on back; spur saccate, 1–2 mm. deep; seeds ca. 1.7 mm. long; $n=6$ (Clausen, 1949).—Dry rocky or gravelly places, 3000–7500 ft.; Sagebrush Scrub, Chaparral, Yellow Pine F.; about the w. Mojave Desert from Mono Co. and Tehachapi Mts. to Mt. Pinos and San Bernardino Mts. April–June.

4. **V. douglásii** Steud. More or less pubescent perennial with a deep short rootstock; stems 5–10 cm. high, partly underground; lvs. ovate in outline, 2–5 cm. long, bipinnately 3–5-parted into 3–5 cleft linear or oblong segms.; petioles 2–6 cm. long; stipules entire to incised; peduncles 5–12 cm. long; sepals 6–12 mm. long, ciliate; petals mostly 8–16 mm. long, light golden-yellow with dark brown veins, the 2 upper brown to very dark on back, the lateral yellow-bearded; spur saccate; cleistogamous fls. absent; caps. green, acute, 6–9 mm. long.—Grassy slopes and flats, 3500–7500 ft.; S. Oak Wd., Yellow Pine F.; mts. from San Diego Co. n. to Tehachapi; to Ore. March–May.

5. **V. lobàta** Benth. Plate 88, Fig. F. Mostly pubescent perennials from a short thickish rootstock; stem erect, 1–3 dm. high; lvs. mostly at summit of stem, sometimes 1 from rootstock, ovate to wedge-shaped, or almost reniform in outline, 2.5–10 cm. wide, usually shorter, palmately 3–7 (–9)-cleft or -divided into lance-oblong lobes; stipules 10–15 mm. long, green; peduncles ca. as long as lvs.; sepals 6–10 mm. long; petals deep yellow, 8–15 mm. long, the 2 upper purple on back, all or lower 3 with purple-brown veins, the lateral yellow-bearded; spur saccate; caps. 8–10 mm. long; $n=6$ (Gershoy, 1934).— Rather dry slopes in open woods, etc., 4500–5500 ft.; mts. of San Diego Co.; Santa Lucia Mts. n. to Ore. April–July. Occasional s. Calif. plants have undivided lvs. and are in the var. *integrifolia* Wats.

6. **V. maclóskeyi** Lloyd. [*V. blanda* auth., not Willd.] Subglabrous stemless perennial from a slender creeping rootstock, sending out leafy stolons in late season; lvs. round-reniform to ovate-cordate, thin, 1–2 (–3) cm. long; petioles slender 1–10 cm. long; stipules greenish, subovate; peduncles 2–15 cm. long; sepals ovate-lanceolate, 3–4 mm. long, glabrous; petals 6–9 mm. long, white, the 3 lower with purple veins; cleistogamous fls. present; caps. ca. 6 mm. long; $n=12$ (Clausen, 1936).—Wet banks and meadows, 6000–8000 ft.; upper Montane Coniferous F.; San Jacinto Mts. to San Gabriel Mts.; n. Calif. to B.C. May–Aug.

7. **V. nephrophýlla** Greene. Nearly glabrous acaulescent perennial from a short rootstock with long fibrous roots; lvs. broadly cordate-ovate to subreniform, crenate, 2–7 cm. wide, the brownish petioles 5–15 cm. long; stipules somewhat gland-toothed; peduncles 10–25 cm. long; sepals 5–6 mm. long; petals 12–18 mm. long, deep blue-violet with white basal bearding and the 3 lower with dark veins; spur ca. 3 mm. long; cleistogamous fls. on shorter peduncles; caps. glabrous, 6–9 mm. long.—Occasional in shaded cool, wet places, 3500–6000 ft.; largely Yellow Pine F.; mts. of San Diego Co. to San Gabriel Mts.; to B.C. Atlantic Coast. May–June.

8. **V. pedunculàta** T. & G. JOHNNY-JUMP-UP. Perennial with several slender stems from a short thick deep rhizome; plants 1–3.5 dm. high; lvs. bright green, deltoid-ovate to subcordate, crenate, obtuse, 1.5–4 cm. long, the petioles 3–6 cm. long; stipules deeply toothed; peduncles much exceeding lvs.; sepals 6–8 mm. long; petals orange-yellow, 10–16 mm. long, the 2 upper red-brown on the back, the lateral bearded, the 3 lower veined with dark brown; spur 1–2 mm. long; no cleistogamous fls.; $n=6$ (Clausen, 1949). —Common in grassy places below 2500 ft.; V. Grassland, Coastal Sage Scrub; cismontane from L. Calif. to cent. Calif. Channel Ids.

9. **V. purpùrea** Kell. not Stev. ssp. **purpùrea.** Plate 89, Fig. A. Perennial from a strong woody taproot; stems 6–20 cm. high, depressed in sun, suberect in shade, retrorse-pubescent; radical lvs. 1–5, rounded, with subcuneate base, glabrate above, purple-tinted, 3.5 cm. long, on longer petioles; stipules scarious, adnate to petioles; cauline ovate, crenate-serrate, with greenish unequal stipules; peduncles exceeding lvs.; sepals 4–6 mm. long; petals deep lemon-yellow, 8–10 mm. long, the 2 upper purplish on back, the 3 lower with purple-brown veins and the 2 lateral bearded; spur saccate; cleistogamous fls. in upper axils; $n=6$ (Clausen, 1949).—Dry rocky places, 4000–6500 ft.;

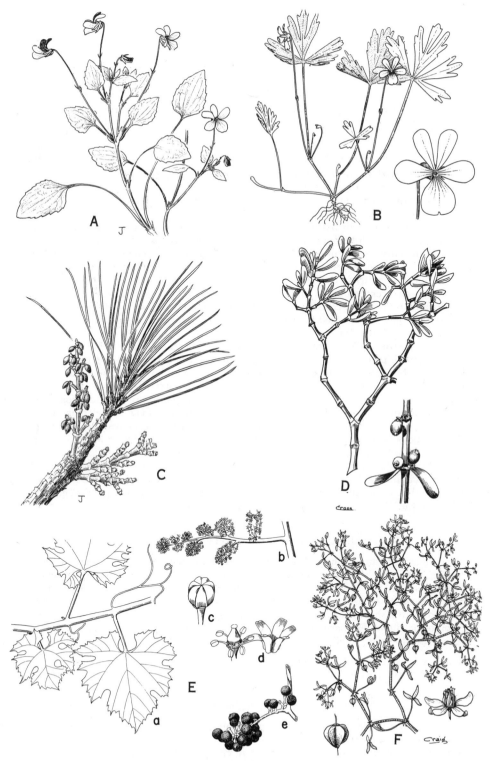

Plate 89. Fig. A, *Viola purpurea*. Fig. B, *Viola sheltonii*. Fig. C, *Arceuthobium* sp., with ♀ and ♂ infls. Fig. D, *Phoradendron bolleanum* ssp. *densum*. Fig. E, *Vitis girdiana*, *d*, opened fl., *e*, berries. Fig. F, *Fagonia laevis*, fr. 5-angled.

Yellow Pine F.; mts. of San Diego Co. to San Gabriel Mts. and Mt. Pinos; to Ore. April–June.

KEY TO SUBSPECIES

Foliage green; stems well developed above ground; petals 8–10 mm. long. Forest plants.
  Uppermost lvs. ovate, toothed; radical lvs. rounded to obtuse. Yellow Pine F. . . ssp. *purpurea*
  Uppermost lvs. ± lanceolate, often entire; radical lvs. acute. Mostly above Yellow Pine F.
    ssp. *mesophyta*
Foliage grayish; stems mostly buried; petals 6–8 mm. long. Open rocky places . . ssp. *xerophyta*

Ssp. **mesophyta** Baker & Clausen. Dry slopes mostly above Yellow Pine F. San Jacinto Mts. to Sierra Nevada. May–June.

Ssp. **xerophyta** Baker & Clausen. At 5000–10,000 ft.; San Jacinto and San Bernardino mts. to Mt. Pinos; to Sierra Co. May–July.

10. **V. quercetòrum** Baker & Clausen. Perennial from a woody taproot, the foliage grayish-green, with little purple, puberulent throughout; stems 5–15 cm. tall; lowest lvs. erect, rounded to ovate, obtuse, 2–5 cm. long, irregularly sinuate-dentate, on petioles 3–9 cm. long; upper lvs. smaller, more acute; stipules of lower lvs. subscarious, of upper green, unequal, entire to toothed; peduncles 4–13 cm. long; sepals 6–8 mm. long; petals yellow, ± darkened on back, 10–12 mm. long, the lateral clavate-bearded; spur short; cleistogamous fls. in upper axils; caps. puberulent; $n=12$ (Clausen, 1949).—Dry grassy or brushy slopes, below 7000 ft.; Chaparral, S. Oak Wd., etc.; cismontane, San Diego Co. to s. Ore. March–June.

11. **V. sheltònii** Torr. Plate 89, Fig. B. Glabrous perennial from a short stout rootstock; stems rising only a little above the ground; lvs. 2–6 (–10) cm. wide, broader than long, dark blue-green, palmately 3-divided, the cuneate-obovate divisions again palmately 3-parted and cleft into linear-oblong lobes, ± ciliolate; petioles 3–12 cm. long; stipules lance-ovate, membranous, the upper somewhat lacerate; peduncles somewhat overtopping the lvs.; sepals 6–8 mm. long; petals 10–15 mm. long, deep lemon-yellow, the 3 lower veined brown-purple, the 2 upper brown-purple on the back; spur saccate; cleistogamous fls. in upper axils; shaded places, 4200–5500 ft.; Chaparral, Yellow Pine F.; Santa Ana Mts., San Gabriel Mts.; to Wash. April–July.

12. **V. tricólor** L. WILD PANSY. Annual; corolla 15–25 mm. long, blue-violet or yellow or combination of these; spur up to twice as long as appendages.—Occasionally established n. of our area and to be sought along our coast as an escape from gardens. Native of Eu.

111. Viscàceae. MISTLETOE FAMILY

Hemiparasitic shrubs on the branches of trees; haustorial attachment single, without runners. Lvs. mostly opposite, occasionally alternate or absent. Fls. minute, unisexual. Perianth-segms. 2–4, valvate. Stamens opposite the perianth-segms., adnate to them or free; anthers 1- to many-celled, opening by pores; pollen spherical. Ovary inferior, 1-celled, with a short placental column containing the sporogenous cells; ovules absent; embryo sac single. Fr. baccate, the viscous layer within the vascular bundles. Ca. 11 genera and 450 spp.; in all continents.

Anthers 1-celled; ♀ sepals 2; berry on a recurved pedicel, compressed . . . . 1. *Arceuthobium*
Anthers 2-celled; ♀ sepals 3; berry sessile, globose . . . . 2. *Phoradendron*

1. Arceuthòbium Bieb.

Yellowish or brownish plants with fragile jointed stems, the segms. glabrous, often ± 4-angled. Lvs. decussate, reduced to connate scales. Fls. solitary or several from the same axil, unisexual; the ♂ usually 3-merous, compressed, with anthers a single round locule opening by a circular slit. Pistillate fls. ovoid, compressed, 2-toothed. Berry fleshy, compressed, dehiscing on a short recurved pedicel. Parasitic on conifers. When mature

Arceuthobium

the frs. discharge the viscid seeds with great force at slightest touch and seeds adhere to adjacent branches. Spp. several, of N. Hemis. (Greek, *arkeuthos*, juniper, and *bios*, life.) Plate 89, Fig. C.

KEY TO THE SOUTHERN CALIFORNIA SPECIES OF ARCEUTHOBIUM

BY DELBERT WIENS

1. Internodes ca. 1.5 mm. wide at or near the third internode above the base; mature fr. ca. 3.5 mm. high, ca. 2.0 mm. wide; parasite on *Pinus flexilis* or *P. monophylla*.
   2. Staminate plants less than 2 cm. high, often forming mat-like clusters over the host branches; parasite on *P. flexilis* .... 4. *A. cyanocarpum*
   2. Staminate plants usually over 8 cm. high, never forming mat-like clusters; parasite on *P. monophylla* and possibly on *P. quadrifolia* .... 5. *A. divaricatum*
1. Internodes ca. 2–4 mm. wide at or near the third internode above the base; mature fr. ca. 4.0 mm. or more high and 2.5 or more mm. wide; never parasitic on *Pinus flexilis* or *P. monophylla*.
   3. Parasite on *Abies concolor*; at least some of the shoots usually olive-green or yellowish-green
      1. *A. abietinum*
   3. Parasite on *Pinus*; shoots usually yellow to orange, but occasionally approaching greenish-yellow.
      4. Mature fr. ca. 4 mm. long and ca. 2.5 mm. wide; internodes not glaucous; anthesis primarily in July; parasite on *P. lambertiana* .... 2. *A. californicum*
      4. Mature fr. ca. 4.5–5.0 mm. long and ca. 3.0 mm. wide; some internodes at least usually slightly glaucous; anthesis primarily from late August to late November.
         5. Lateral staminate spikes less than 10 mm. long in summer; anthesis primarily in late August to mid September; seeds usually maturing in late September; parasite primarily on *Pinus ponderosa*, occasionally also on *P. attenuata*, *P. coulteri* and *P. jeffreyi*
            3. *A. campylopodum*
         5. Lateral staminate spikes more than 10 mm. long in summer; anthesis primarily in late October to early November; seeds usually maturing in November; parasite on *Pinus sabiniana* .... 6. *A. occidentale*

1. **A. abietinum** (Engelm.) Hawksworth & Wiens. [*A. douglasii* var. *a* Engelm. in Wats.] Main shoot height ca. 8 cm., but up to 22 cm.; shoots yellow green to yellow; third internode 4–23 mm. long (mean 14 mm.); mature fr. ca. 4 mm. long, 2 mm. wide.—On *Abies concolor*; to Wash. & Ariz.

2. **A. califórnicum** Hawksworth & Wiens. [*A. campylopodum* var. *cryptopodum* (Engelm.) Jeps.] Mean shoot height ca. 8 cm. but up to 12 cm.; shoots greenish to bright yellow, turning brown at base of older shoots; third internode 6–16 mm. (mean 10.5 mm.) long; mature fr. 4 mm. long, 2.5 mm. wide.—Principally on *Pinus lambertiana*.

3. **A. campylopòdum** Engelm. in Gray. Main shoot ca. 8 cm., but up to 13 cm.; shoot olive-green to yellow; ♂ plants brownish, ♀ plants greenish; third internode 7–22 mm. (mean 11.4 ± 3.8 mm.) long; mature fr. 5.0 mm. long, 3.0 mm. wide.—Parasite principally on *Pinus ponderosa* and *P. jeffreyi*, sometimes on *P. murrayana*.

4. **A. cyanocárpum** Coulter & Nels. Mean shoot height ca. 3 cm., but up to 5 cm.; shoots yellow green, densely clustered; third internode 2–14 mm. (mean 5.2 ± 2.0 mm.) long, 1–1.5 mm. wide; mature fr. 3.5 mm. long, 2.0 mm. wide.—Principally on *Pinus flexilis*.

5. **A. divaricàtum** Engelm. Main shoot height ca. 8 cm., but up to 13 cm.; shoots olive green to brown; third internode 6–15 mm. (mean 9.8 ± 2.4 mm.) long and 1–2 mm. wide; mature fr. 3.5 mm. long, 2.0 mm. wide.—In Calif., mostly on *Pinus monophylla*, rare possibly on *P. quadrifolia*.

6. **A. occidentàle** Engelm. Mean height ca. 8 cm., but up to 17 cm.; shoots yellowish, glaucous; third internode 7–18 mm. (mean 12.7 ± 2.0 mm.) long and mean (1.8 mm.) wide; mature fr. 4.5 mm. long, 3.0 mm. wide.—In s. Calif., the principal host is *Pinus sabiniana*, possible also on *P. muricata*.

2. Phoradéndron Nutt. MISTLETOE

Woody, with much branched rather brittle stems. Lvs. foliaceous, entire and faintly nerved, or reduced to connate scales. Fls. sunk in the jointed rachis, usually several in the axil of each bract. Staminate fls. mostly globose, the sepals 3 and with the sessile transversely 2-celled anthers at their base. Pistillate fls. with the fl.-tube adnate to the ovary. Berry sessile, ovoid to globose. A large genus of the Americas, largely trop. Largely spread from host to host by birds as they feed. (Greek, *phor*, a thief, and *dendron*, tree.)

Pistillate spikes with 2 fls. at each joint.
    Lvs. reduced to connate scales.
        Plants canescent; ♀ spikes 3–4-jointed; berries usually red .... 2. *P. californicum*
        Plants glabrous; ♀ spikes 1-jointed; berries white .... 3. *P. juniperinum*
    Lvs. green, oblanceolate .... 1. *P. bolleanum*
Pistillate spikes with 6 or more fls. at each joint; lvs. green.
    Internodes minutely canescent to canescent-tomentose; anthesis largely from Oct. through March; fr. glabrous .... 4. *P. tomentosum*
    Internodes densely short-villous; anthesis largely July through Sept.; fr. pubescent
                                                                           5. *P. villosum*

1. **P. bolleànum** (Seem.) Eichler ssp. **dénsum** (Torr.) Wiens. Plate 89, Fig. D. Much branched, dense, 1–2 dm. high, glabrous; lvs. oblanceolate, sessile, usually 1–1.5 cm. long, 3–4 mm. wide; spikes 3 mm. long, 1–2-jointed, the ♂ ca. 12-fld.; fr. 4 mm. thick, straw-colored, subglobose.—On *Juniperus*.

Ssp. **pauciflòrum** (Torr.) Wiens. Plant 2–3 dm. high; lvs. subpetiolate, 1.5–3 cm. long, 5–10 mm. wide.—On *Abies* and *Cupressus*.

2. **P. califórnicum** Nutt. Stems slender, terete, with ± pendulous clustered branches 1–5 dm. long and internodes 1–3 (–4) cm. long, often reddish, puberulent when young; lvs. scalelike; spikes axillary, the ♂ of 2–5 joints, the ♀ of 3–4 joints; fr. globose, reddish (rarely white), 3–4 mm. thick.—Mostly on *Prosopis*, occasional on *Acacia*.; to Son., L. Calif.

3. **P. juniperìnum** Engelm. ex Gray ssp. **juniperìnum**. [*P. ligatum* Trel.] Stems glabrous, often bluntly squarish, rather stout, the plant usually remaining erect; internodes rather short, to ca. 1 cm. long, microscopically granular; lvs. scalelike, spreading, 1–2 mm. long, not constricted at base; spikes solitary, glabrous, ca. 3 mm. long, with a single joint; ♂ fls. 6 or 8; ♀ 2; fr. straw- or wine-colored, ca. 3 mm. in diam.—On *Juniperus*; to Colo., Tex., Chihuahua.

Ssp. **libocèdri** (Engelm.) Wiens. Internodes usually over 1 cm. long; plant often pendulous in age.—On *Calocedrus*; to L. Calif. and Ore.

4. **P. tomentòsum** (Engelm. ex Gray) ssp. **macrophýllum** (Engelm.) Wiens. Plants stout, dioecious, 3–6 dm. high, the internodes 3–5 cm. long, minutely canescent to canescent-tomentose, later ± glabrate; lvs. thickish, elliptic-obovate to oblong-oblanceolate, obtuse, 1–5 cm. long, cuneately petioled; anthesis mostly from Oct. through March; spikes 1.5–5 cm. long, the ♂ 20–30-fld., the ♀ ca. 6- or 12-fld.; fr. glabrous, 4–5 mm. in diam.; $n = 14$ (Wiens, 1964).—On *Platanus, Populus, Salix, Fraxinus, Juglans*, etc.; to w. Tex.

5. **P. villòsum** (Nutt. in T. & G.) Nutt. Internodes and lvs. densely short-villous; lvs. elliptic-obovate; anthesis mostly from July through Sept.; fr. pubescent.—Mostly on *Quercus*, sometimes on other broad-lvd. trees; to Ore., Tex., Mex., L. Calif.

112. Vitàceae. GRAPE FAMILY

Woody vines climbing by tendrils opposite the lvs. Lvs. alternate, petioled. Fls. small, paniculate-cymose, bisexual or unisexual. Calyx entire or 4–5-toothed. Petals 4–5, separate or coherent, valvate, deciduous without expanding. Stamens as many as the petals, and opposite them. Ovary generally immersed in the disk, 2–6-loculed; ovules 1–2 in each locule. Fr. a berry, commonly 2-celled. Seeds with a bony testa and hard endosperm; embryo short. Ca. 500 spp., widely distributed in trop. and temp. regions.

1. Vìtis L. GRAPE

Lvs. palmately lobed or dentate, with small caducous stipules. Calyx minute, with entire limb. Fls. fragrant; petals hypogynous or perigynous, coherent in a cap. Style short, conic; ovules 2 in each locule. Berry round to ovoid, pulpy. Ca. 50 spp. of N. Hemis. (Classical Latin name.)

1. **V. girdiàna** Munson. Plate 89, Fig. E. Stems 1–6 m. long; lf.-blades broadly cordate-ovate, 5–16 cm. wide, mostly with triangular apex and with rather deep and narrow sinus, obscurely or distinctly lobed as well as irregularly sharply dentate, ashy tomentose beneath; petioles mostly 3–8 cm. long; petals and stamens 6; pedicels of fr. smooth; berries globose, 3–6 mm. in diam., black, with little bloom.—Canyon bottoms, along streams, etc.; below 4000 ft.; S. Oak Wd., Coastal Sage Scrub, etc.; Santa Barbara Co. to n. L. Calif.; occasional on desert edge from Inyo Co. s. Santa Catalina Id. May–June.

113. Zygophyllàceae. CALTROP FAMILY

Herbs or shrubs, often with branches jointed at nodes. Lvs. opposite or alternate, pinnate or 2–3-foliolate, not gland-dotted; stipules paired, persistent, often spinescent. Fls. bisexual, regular, 1 or 2 in the axils of the stipules. Sepals 5, rarely 4, usually free and imbricate. Petals 5, rarely 4 or 0, free, usually imbricate or contorted. Disk usually present. Stamens free, essentially hypogynous, usually twice as many as petals; anthers 2-celled, opening lengthwise. Pistil usually of 4 or 5 united carpels, or twice as many; ovary mostly superior, sessile; style simple; stigmas 1 or 5; ovules 2 or more in each cell. Fr. various; forming a caps. or splitting into 5–12 indehiscent nutlets. Seeds mostly with some endosperm; embryo straight with flat cotyledons.

Stipules spiny; lvs. 3-foliolate; fls. purplish .... 1. *Fagonia*
Stipules not spiny; lvs. 2–many-pinnate; fls. yellow or orange.
  Lfts. 1 pair; stamens with scalelike appendages.
    Plant a well developed shrub, fr. hairy, globose. Common desert native ...... 3. *Larrea*
    Plant suffrutescent; fr. glabrous, elongate. Rare waif .... 5. *Zygophyllum*
  Lfts. 4 or more pairs; stamens not appendaged.
    Fr. hemispheric or higher, not radiate, breaking up into 8–12 merely tubercled nutlets
        2. *Kallstroemia*
    Fr. flat, radiate, breaking up into 5 very spiny nutlets .... 4. *Tribulus*

1. Fagònia L.

Ours low suffrutescent branching spreading plants. Lvs. in ours digitately 3-foliolate; stipules spiny. Fls. solitary, purplish, small, 5-merous. Petals clawed, deciduous. Ovary with 5 2-ovuled cells. Fr. pyramidal-ovoid, deeply 5-angled, each carpel 1-seeded and ventrally dehiscent. Seeds erect, compressed, broadly oblong. Ca. 18 spp. of Medit. region, sw. Africa, Chile, sw. N. Am. (G. C. *Fagon*, French botanist of 17th century.)

(Porter, Duncan. Contr. Gray Herb. 192: 99–135. 1963.)
Fls. ca. 1 cm. in diam.; stipules 1–6 mm. long .... 1. *F. laevis*
Fls. ca. 1.5 cm. in diam.; stipules 3–16 mm. long .... 2. *F. pachyacantha*

1. **F. laèvis** Standl. [*F. californica* ssp. *l.* Wiggins.] Plate 89, Fig. F. Compact, intricately branched, 1–4 dm. high, the stems somewhat angled, smooth or minutely scabrous above; stipules subulate, 2–3 (–6) mm. long, spreading; lfts. lanceolate, 3–10 mm. long, 1–3 mm. wide; fls. in open cymes; sepals ca. 2 mm. long; petals 5–8 mm. long; fr. deflexed, 4–5 mm. long, reticulate, pubescent, with beak 1–2 mm. long; seeds subovate, brown, mottled, ca. 2 mm. long.—Common on dry often rocky slopes, mostly below 2000 ft.; Creosote Bush Scrub, through most of Colo. Desert, occasional on s. Mojave Desert; sw. San Diego Co., as at Otay Dam; to L. Calif., Son., Ariz. March–May, Nov.–Jan.

2. **F. pachyacántha** Rydb. in Vail & Rydb. [*F. californica* var. *glutinosa* Vail.] With ± the habit of the preceding; stipules 3–16 mm. long; lfts. 1–2 cm. long, 3–9 mm. wide; upper stems glandular; fls. to 1.5 cm. in diam.—Common, Indio to Orocopia Mts., Riverside Co., less so s.; to Son., L. Calif. Same months as sp. no. 1.

2. *Kallstroèmia* Scop.

Annual herbs much like *Tribulus*, but with fr. dehiscing into twice as many indehiscent, 1-seeded tuberculate, not spiny, nutlets and leaving a ± persistent cent. axis. Perhaps a dozen spp. of New World and Australia. (In honor of *Kallstroem*, obscure contemporary of Scopoli.)

Petals 4–12 mm. long, orange-yellow, fading whitish; beak of fr. 2–6 mm. long, strongly conic at base.
    Beak of fr. 2–3 mm. long, shorter than nutlets; petals 3–5 mm. long . . . . 1. *K. californica*
    Beak of fr. 4–6 mm. long, mostly longer than nutlets; petals 6–12 mm. long . . 3. *K. parviflora*
Petals 15–30 mm. long, orange; beak of fr. 8–11 mm. long, slender, scarcely enlarged at base
    2. *K. grandiflora*

1. **K. califórnica** (Wats.) Vail. Plate 90, Fig. A. Stems decumbent, branched, 1–6.5 dm. long, whitish-pubescent; stipules ca. 2 mm. long; lfts. 5–7 pairs, elliptic, 4–8 mm. long; pedicels 1–3 cm. long in fr.; sepals deciduous, 3–4 mm. long; petals 3–5 mm. long; fr. strigulose, ovoid-globose, ca. 3 mm. long, with sharp tubercles on back of carpels, the beak glabrous, shorter than carpels.—Occasional or common in some seasons, in sandy places, below 2500 ft.; Creosote Bush Scrub; Mojave Desert (Cronise V., 29 Palms) to Imperial Co.; to Ariz., Mex. Aug.–Oct.

2. **K. grandiflòra** Torr. in Gray. Stems suberect, rather diffusely branched, 2–7 dm. long, pubescent, somewhat hirsute; stipules 5–7 mm. long; lfts. 5–9 pairs, oblong, 10–25 mm. long; pedicels 2–6 cm. long in fr., thickened upward; sepals linear, hirsute, 8–15 mm. long, persistent; petals 12–30 mm. long; fr. 4–5 mm. long, tuberculate, the beak mostly 8–10 mm. long.—Sandy places, at 900 ft.; Creosote Bush Scrub; near Desert Center, n. Colo. Desert; Ariz. to Tex. and Mex. Sept.–Oct.

3. **K. parviflòra** Nort. Much like *K. californica*, but stipules 6–7 mm. long; lfts. 3–5 pairs; pedicels more conspicuously enlarged upward; sepals more persistent, 5 mm. long; petals 6–12 mm. long; fr. 3–4 mm. long, with rounded tubercles, the beak 4–6 mm. long. —Sandy slopes at 5000–6000 ft.; Joshua Tree Wd., Pinyon-Juniper Wd.; Clark Mts., e. Mojave Desert; Warner's Hot Springs, e. San Diego Co.; to Miss., Mex. Aug.–Oct.

3. *Lárrea* Cav. Creosote Bush

Evergreen strong-scented resinous shrubs. Lvs. opposite, with 2 divaricate sessile asymmetrical olive-green lfts. Fls. solitary, yellow. Sepals 5, unequal, imbricate, deciduous. Petals 5, clawed, oblong-spatulate. Stamens 10, on small 10-lobed disk. Pistil of 5 united carpels; style slender, with 5 stigmas, the ovary-cells ca. 6-ovuled. Fr. globose, hairy, separating at maturity into 5 indehiscent 1-seeded carpels. Perhaps 3–4 spp. of the warmer dry parts of the New World. (Named for J. A. de *Larrea*, Spaniard interested in science.)

1. **L. tridentàta** (Sessé & Moç. ex DC.) Cov. [*L. divaricata* of Calif. refs., not of Cav.] Plate 90, Fig. B. Branches 1–3 (–4) m. tall, brittle, grayish, irregular, with dark glandular bands at the nodes, densely leafy toward tips; stipules ca. 1 mm. long; lfts. obliquely lance-ovate, 5–10 mm. long, indistinctly 3-veined, entire; sepals round-ovate, strigose, 5–7 mm. long; petals 5–8 mm. long, partly twisted like vanes of a windmill; fr. 4–5 mm. long, rusty long-villous, the style slender, persistent, 5 mm. long.—The dominant shrub over great areas of desert from Inyo Co. s. on dry slopes and plains up to ca. 5000 ft.; occasional in interior cismontane valleys in Kern Co., Riverside and San Diego cos.; to Utah, Mex., L. Calif. Here considered as of a different sp. than S. Am. plants.

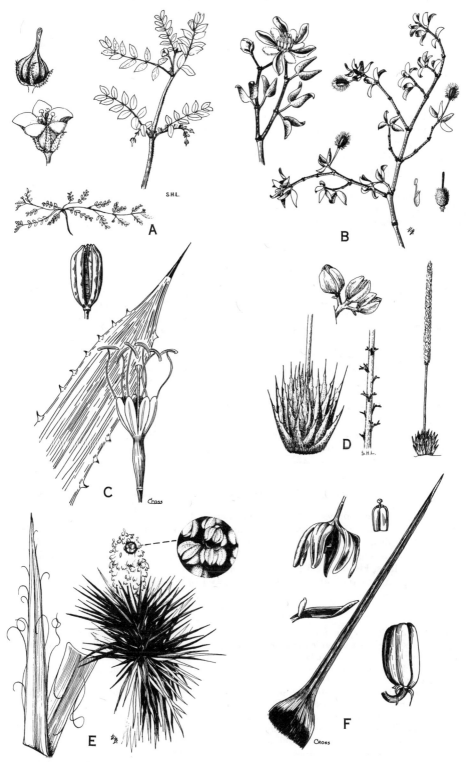

Plate 90. Fig. A, *Kallstroemia californica*, fl. and beaked fr. Fig. B, *Larrea tridentata*, with separate stamen and pistil. Fig. C, *Agave deserti*. Fig. D, *Agave utahensis* var. *nevadensis*. Fig. E, *Yucca schidigera*. Fig. F, *Yucca whipplei* var. *parishii*, with pistil above and caps. below.

4. Tríbulus L. CALTROP. PUNCTURE VINE

Pubescent herbs with weak often prostrate stems. Lvs. even-pinnate; stipules membranaceous. Fls. solitary, axillary, (4-) or 5-merous. Fils. 10 (8), slender, without appendages. Carpels as many as petals, surrounded at base by an annular 10-lobed disk; styles united into stout short column; ovules 3–5 per cell. Fr. depressed, spinose, separating into 5 bony indehiscent 3–5-seeded carpels. Ca. 12 spp., widely distributed, especially in trop. (Latin name of *caltrop*, the shape of which is suggested by the 3-pronged fr.)

1. **T. terréstris** L. Annual, trailing, branched from base, the stems 2–10 dm. long; lfts. 4–7 pairs, oblong or elliptic, 5–12 mm. long; petals yellow, obovate, 3–4 mm. long; carpels crested and armed with 2–4 spreading spines 4–6 mm. long; $2n = 24$ (Negodi, 1939); 48 (Schnack & Covas, 1947); $n = 6$ (Baquar et al., 1965).—Natur. in waste places, along roadsides, etc. through much of Calif. below 5000 ft., even on the deserts; to Atlantic Coast; from Old World. April–Oct. A serious pest.

5. Zygophýllum L. BEAN-CAPER

Much branched herbs or subshrubs with fleshy branches. Lvs. stipulate, fleshy, opposite, composed in ours of 1 pair of lfts. Fls. axillary, solitary. Sepals 4–5. Petals 4–5, clawed. Disk fleshy. Stamens 8–10, inserted at base of disk, each with a scale at base. Ovary sessile, angular, 4–5-celled, tapering into a style. Fr. angular or winged, indehiscent or dehiscent; each cell 1-seeded. Genus of ca. 70 spp. of Medit. region, S. Afr., Asia, Australasia. (Greek, *zygon*, yoke, and *phyllon*, lf., because of the paired lfts.)

1. **Z. fabàgo** L. var. **brachycárpum** Boiss. Herbaceous or suffrutescent, with deep strong root and erect branching tops, 3–6 dm. high, glabrous; lfts. obovate, 1–4 cm. long; fls. copper and yellow; sepals 5, oblong, 6–7 mm. long, rounded at apex; petals 5, 7–8 mm. long; pedicel reflexed in fr.; caps. 5-angled, ovate-oblong, 10–12 mm. long; seeds obovate, 2–3 mm. long.—Established near Rosamond, Muroc, etc., w. Mojave Desert and reported from near Blythe, e. Riverside Co., as well as Stanislaus and Tulare cos., introd. from Old World. Summer.

Subclass II. MONOCOTYLEDONES

Stems mostly with the vascular bundles closed and scattered irregularly throughout the pith, sometimes the whole stem woody, but not exhibiting xylem, then cambium, then phloem and cortex (bark). Lvs. usually parallel-veined, mostly alternate and entire, commonly sheathing the stem at the base and often with little or no distinction into petiole and blade. Fls. mostly 3-merous or 6-merous. Embryo with 1 cotyledon; the first lvs. of the germinating plantlet alternate. Perhaps ¼ of the angiosperms fall here, many occurring in the tropics.

KEY TO FAMILIES

1. Plants small, floating on water or stranded on mud, not differentiated into stem and lf., sometimes with 1 to few simple roots .... 12. *Lemnaceae* p. 918
1. Plants with stem and lvs., the latter sometimes scalelike.
    2. Perianth wanting or reduced, its parts often bristles or mere scales, not petallike in color and texture.
        3. Fls. in the axils of chaffy or husklike scales and ± concealed by them, or stamens and style protruding; fls. in spikes, spikelets or heads.
            4. Lf.-sheaths split lengthwise on the side opposite the blade; lvs. usually 2-ranked; stems mostly hollow and terete; anthers versatile .... 16. *Poaceae* p. 934
            4. Lf.-sheaths continuous around the stem or ruptured in age; lvs. mostly 3-ranked; stems often triangular in cross-section, usually with a pith; anthers basifixed
                                                                                                7. *Cyperaceae* p. 883
        3. Fls. not in the axils of chaffy bracts, or, if subtended by bracts, exceeding or equaling them and not concealed.

5. Plants floating or submerged; fls. floating or submerged, or barely raised above the surface of the water.
  6. Marine plants; fls. on 1 side of a flattened axis .... 22. *Zosteraceae* p. 1013
  6. Freshwater plants or of brackish water.
    7. Fls. in spikes or heads.
      8. Lvs. in dense rosettes; fls. on a spadix adnate to and shorter than the inclosing spathe .... 4. *Araceae* (*Pistia*) p. 882
      8. Lvs. not in rosettes; fls. not on a spadix.
        9. Fls. perfect, in peduncled axillary spikes; sepals 4 18. *Potamogetonaceae* p. 1007
        9. Fls. unisexual, in globose heads, the lower ♀, the upper ♂
            19. *Sparganiaceae* p. 1011
    7. Fls. axillary and solitary or very few in groups.
      10. Lvs. alternate; frs. in umbelliform clusters .... 18. *Potamogetonaceae* p. 1007
      10. Lvs. opposite.
        11. Carpels 2 or more; lvs. mostly 3–10 cm. long .. 21. *Zannichelliaceae* p. 1013
        11. Carpel 1; lvs. mostly 1–3 cm. long .... 14. *Najadaceae* p. 931
5. Plants of land or shallow water; usually lvs. and fls. well emersed.
  12. Fls. on a spadix (fleshy, spikelike) surrounded by a yellow or white spathe
            4. *Araceae* p. 882
  12. Fls. not on a spadix.
    13. Infl. a dense elongate spike.
      14. Plants 1–4 dm. high; ♂ and ♀ fls. intermingled, or fls. perfect
            11. *Juncaginaceae* p. 916
      14. Plants 10–20 dm. high; stamens usually 3, the ♂ fls. in a separate upper portion of the spike, the ♀ below .... 20. *Typhaceae* p. 1012
    13. Infl. of subglobose heads, racemes or open clusters.
      15. Fls. unisexual, the lower heads ♀ .... 19. *Sparganiaceae* p. 1011
      15. Fls. bisexual; perianth-segms. distinguishable.
        15a. Ovaries 3 or 6, separating at maturity .... 11. *Juncaginaceae* p. 916
        15b. Ovary 1, forming a caps. .... 10. *Juncaceae* p. 908
2. Perianth well developed, at least the inner segms. petaloid in color and texture.
  16. Carpels ± free, 1-loculed, mostly 1-ovuled, maturing into a bunch or whorl of aks.
            2. *Alismataceae* p. 868
  16. Carpels united into a mostly 3–12-loculed ovary maturing into a caps. or berry.
    17. Lvs. large, plicately folded in bud; plant arborescent, unbranched; fls. in a large panicle with a spathaceous bract .... 5. *Arecaceae* p. 882
    17. Lvs. not as above and the plant branched if arborescent.
      18. Plants quite woody, with long stiff ± swordlike lvs. .... 1. *Agavaceae* p. 863
      18. Plants herbaceous, or if woody, climbing vines.
        19. Ovary superior.
          20. Infl. with a spathe; submersed or floating aquatics with perianth-segms. partly connate .... 17. *Pontederiaceae* p. 1007
          20. Infl. without a spathe; plants not floating.
            21. Sepals green; petals colored; fls. in umbels; prostrate introd. plant
                6. *Commelinaceae* p. 883
            21. Sepals and petals concolored, or if unlike, the fls. not in umbels.
              22. Fls. in a scapose umbel subtended by ± membranous spathelike bracts
                3. *Amaryllidaceae* p. 871
              22. Fls. not in a scapose umbel with spathelike bracts ... 13. *Liliaceae* p. 919
        19. Ovary inferior (see also *Narcissus* in *Amaryllidaceae*).
          23. Aquatic plants with mostly submersed lvs. .... 8. *Hydrocharitaceae* p. 905
          23. Land plants, the lvs. not submersed.
            24. Fls. regular; ovary 3-loculed; stamens 3; lvs. equitant .. 9. *Iridaceae* p. 905
            24. Fls. irregular; ovary 1-loculed; stamens 1–2; lvs. not equitant
                15. *Orchidaceae* p. 931

1. Agaváceae. AGAVE FAMILY

Mostly xerophytic plants, with a rhizome. Stem short to arborescent. Lvs. fibrous, narrow, often thick or fleshy, usually crowded on or near the base of the stem, entire or prickly on the margin. Fls. bisexual or unisexual, regular or somewhat irregular, racemose or paniculate, they and the branches subtended by bracts. Perianth segms. usually partially united into a short or long tube, segms. subequal to unequal, petaloid, often fleshy. Stamens 6, on the tube or segm.-bases; fils. filiform to thickened, free; anthers linear, usually dorsifixed. Ovary superior or inferior, 3-loculed; style slender, often very short; stigma ± three-lobed. Fr. a berry or loculicidal caps. A family of ca. 19 genera and 500 spp., formerly partly put in *Liliaceae* and partly in *Amaryllidaceae*.

Ovary inferior; fls. yellow to greenish .... 1. *Agave*
Ovary superior; fls. white to purplish.
    Perianth less than 1 cm. long; fls. both perfect and imperfect .... 2. *Nolina*
    Perianth 3–13 cm. long; fls. all perfect .... 3. *Yucca*

1. Agáve L. Century Plant. Maguey

Stemless or sometimes short-stemmed perennials, from a thick fibrous-rooted crown. Lvs. fleshy, persistent, mostly in basal rosettes and spine-edged and -tipped. Fl.-stems tall, arising from the center of the rosette and forming panicles or spikes. Fls. numerous, relatively small, fleshy; perianth ± funnelform, with short tube, the 6 segms. narrow, subequal. Fils. filiform, exserted; anthers versatile. Ovary inferior, 3-loculed; style 1, awl-shaped; stigma capitate, but 3-lobed. Caps. oblong, loculicidal; seeds many, superposed in 2 rows, obovate, black, thin, flat. A genus of perhaps 300 spp., from warmer parts of W. Hemis.; many of horticultural importance, producing various fibers such as sisal and henequen. The young fl.-stalks yield a liquid which is fermented for drinks like pulque and distilled for tequila. (Greek, *agave*, noble or admirable.)

(Berger, A. Die Agaven, 1–288. 1915.)
Fls. 4 in a cluster, the panicle subracemose and narrow .... 4. *A. utahensis*
Fls. many, in terminal clusters on the branches of an open panicle.
    Lvs. mostly less than 6 dm. long, less than 10 cm. broad.
        Trunk subterranean; lvs. glaucous; prickles pale. Deserts .... 2. *A. deserti*
        Trunk rising above ground; lvs. deep green; prickles red. Coastal .... 3. *A. shawii*
    Lvs. 15–25 dm. long, 10–25 cm. wide .... 1. *A. americana*

1. **A. americàna** L. Large, often producing offsets from near the base and from spreading rootstocks; lvs. 1.5–2.5 m. long, gray-green and somewhat glaucous; terminal spine 2–3.5 cm. long; lateral teeth 5–8 mm. long, 1.5–6 cm. apart; flowering stalks to 12 m. tall, broadly paniculate-branched in upper part; fls. yellow, tinged green, 5–6 cm. long, numerous; caps. 4.5–6 cm. long.—Planted as an ornamental and establishing itself on dumps and in waste places.

2. **A. déserti** Engelm. [*A. consociata* Trel.] Plate 90, Fig. C. Densely cespitose, acaulescent, forming large colonies; lvs. gray-green, glaucous, triangular-lanceolate, 1.5–4 dm. long, edged with straight or curved pale prickles 3–7 mm. long and with dark terminal spine 1–3 cm. long; scape 2–5 m. high, the infl. slender; fls. yellow, 3.5–5 cm. long, tube 6–8 mm. long, segms. 12–18 mm. long; fils. 2–3 cm. long; anthers 10–15 mm. long; caps. 3–4.5 cm. long, short-pointed at apex; seeds ca. 5 mm. long; $n=30$ (5 long, 25 short) (McKelvey & Sax, 1933).—Washes and dry rocky slopes below 5000 ft.; Creosote Bush Scrub, Shadscale Scrub; w. edge of Colo. Desert to Providence, Old Dad, Granite and Whipple mts. of s. Mojave Desert; n. L. Calif. May–July.

3. **A. sháwii** Engelm. Cespitose, the trunks 1–3 dm. long, leafy; lvs. ovate to lanceovate, green, glossy, openly concave, 2–5 dm. long, with red hooked prickles 3–10 mm. long and terminal spine 2–2.5 cm. long; fl.-stems 1–3.5 m. high, the panicle congested; bracts and buds purplish-brown; fls. greenish-yellow, 7–10 cm. long; perianth tube 14–18 mm. long, the segms. 15–20 mm. long; fils. dilated in middle and toward base, 6–7 cm. long; anthers 2–3 cm. long; caps. 4–6 cm. long; seeds 6–8 mm. long; $n=30$ (5 long, 25 short) (Lenz, 1950).—Dry coastal bluffs; Coastal Sage Scrub; near Mexican boundary where now quite extinct; n. L. Calif. Sept.–May.

4. **A. utahénsis** Engelm. var. **nevadénsis** Engelm. ex Greenm. & Roush. [*A. n.*

Hester.] Plate 90, Fig. D. Acaulescent with subterranean trunk, usually cespitose; lvs. glaucous, curved inward at tips, 10–25 cm. long, 2–3 cm. wide, margined by ca. 8–10 lateral white teeth on each side and with a slender terminal spine 3–8.5 cm. long; fl.-stalk 1.5–2.5 m. high, with narrow subracemose panicle; fls. yellow, 2.5–3.5 cm. long; perianth abruptly expanded above the tube, this 1–3 mm. long, the segms. 12–15 mm. long; fils. 2–3 cm. long; anthers ca. 8 mm. long; caps. fusiform, 9–12 mm. thick, 18–20 mm. long; seeds 3.5–4 mm. long; $n=30$ (Lenz, 1951).—Dry stony limestone slopes, 3000–5000 ft.; Shadscale Scrub, Joshua Tree Wd.; Ivanpah, Potosi, Clark and Kingston mts., e. Mojave Desert. May–July.

2. Nolìna Michx. NOLINA

Perennial with a yuccalike aspect, the stem forming a thick woody trunk or underground. Lvs. many, narrowly linear with greatly distended rigid bases. Fls. many, imperfect, borne on a nearly naked stem in an elongate congested, compound panicle; the main branches subtended by long white bracts. Pedicels jointed, subtended by minute scarious bracts. Perianth small, usually whitish, persistent, the 6 segms. distinct, 1-nerved. Stamens 6, usually reduced in the ♀ fls.; fils. short and slender. Ovary sessile, 3-lobed, with 2 ovules per locule; style short; stigmas 3. Caps. membranous, broadly 3-winged, bursting irregularly; seeds subglobose to ovoid, light-colored, often solitary. Ca. 25 spp. of sw. N. Am. (C. P. *Nolin*, joint author of an agricultural essay, 1755.)

Lvs. mostly 15–30 mm. wide above the expanded base.
    The lvs. smooth on margins and with shredding fibers; perianth 2–2.5 mm. long
        1. *N. bigelovii*
    The lvs. hispid-serrulate on the margins, not fibrous; perianth 5 mm. long .... 4. *N. wolfii*
Lvs. mostly 8–15 mm. wide above the expanded base.
    Stems forming underground platforms bearing many rosettes; lvs. flat .... 2. *N. interrata*
    Stems erect or ascending, each ending in a rosette; lvs. concave above .... 3. *N. parryi*

1. **N. bigelòvii** (Torr.) Wats. [*Dasylirion b.* Torr.] Caudex many branched, 6–10 dm. high; lvs. flat, 8–12 dm. long, 1.5–3 cm. wide, glaucous, not or scarcely serrulate on the margins, but with shredding brown fibers; fl.-stalks 1–3 m. high, the panicle dense, 6–15 dm. long; perianth 2–2.5 mm. long; caps. 8–10 mm. wide, somewhat shorter; seeds grayish-white, oblong-ovoid, 3 mm. long; $n=19$ (Cave, 1964).—Dry slopes below 3000 ft.; Creosote Bush Scrub, w. and n. edges of Colo. Desert, Old Woman Mts., Sheephole Mts., Eagle Mts. (Mojave Desert); to Ariz., L. Calif., Son. May–June.

2. **N. interràta** Gentry. Plant with subterranean caudex 2–3 m. long and forming a platform bearing many aerial rosettes of 10–20 or more scabrous glaucous lvs. 7–9 dm. long, 8–15 mm. wide, the margins with minute teeth of 2 sizes; fl.-stalk 1.5–2 m. tall; perianth-segms. 3 mm. long, puberulent at the swollen tips; caps. 12–15 mm. wide; seeds yellowish, wrinkled, 5 mm. long, 4 mm. thick.—Dry slope; Chaparral; w. of Dehesa School, 8 miles e. of El Cajon, San Diego Co. June–July.

3. **N. párryi** Wats. [*N. bigelovii* var. *p.* L. Benson.] Caudex several-branched, erect, 3–10 dm. tall, ca. 3–4 dm. thick; lvs. in dense crown, thickish, ± gray-green, concave, scarcely rigid, 6–10 dm. long, 8–16 (–20) mm. wide, with serrulate margins, not becoming fibrous; fl.-stalk 1–1.5 m. high, the bracts 1–2 dm. long; infl. very dense, 5–6 dm. long; branches 1–3 dm. long; fls. 5 mm. long; caps. 12–15 mm. wide, scarcely as long; seeds light brown, wrinkled, ca. 4 mm. long, 3 mm. thick; $n=19$ (Lenz, 1950).—Dry slopes below 3000 ft.; Chaparral, Coastal Sage Scrub; San Diego, Orange, Riverside, Ventura cos.; 3500–5500 ft. Pinyon-Juniper Wd., Chaparral, desert slopes of Santa Rosa and San Jacinto mts. April–June.

4. **N. wólfii** (Munz) Munz, comb. nov. [*N. parryi* ssp. *w.* Munz, Aliso 2: 221–227. 1950.] Flowering plant 4–5 m. tall, the caudex simple or few-branched, 8–20 dm. high, 6–9 dm. in diam.; lvs. green, flat, stiff, conspicuously hispid-serrulate on margins, 12–15 dm. long, 25–35 mm. wide; fl.-stalk 3–4 m. high, the base bare for 1–2 m. and 12–15 cm.

thick; infl. proper 2–2.5 m. long, 0.8–1 m. thick; lower bracts 3–3.5 dm. long; perianth 5 mm. long, the segms. oblong; caps. 12–14 mm. wide, 10–11 mm. long; seeds light brown, oblong-spherical, 3–3.5 mm. long, slightly wrinkled. Dry slopes 3500–6000 ft.; Pinyon-Juniper Wd., Joshua Tree Wd.; Kingston, Eagle and Little San Bernardino mts. (Mojave Desert), se. San Jacinto Mts.; Kern Plateau, Kern Co. May–June.

3. Yúcca L. SPANISH BAYONET. YUCCA

Trees or shrubs, simple or branched. Lvs. narrow, linear, usually stiff, with expanded base and terminal spinose tip, ± persistent. Fls. large, pendent, numerous, in dense, often elongate, terminal panicles, the perianth-segms. subequal, fleshy, withering-persistent, lanceolate to ovate, the tips tending to curve inward. Stamens much shorter than the perianth; fils. fleshy. Ovary superior. Style short with 3 stigmatic lobes or with a subcapitate stigma. Fr. a caps. or fleshy. Seeds obovoid or compressed, black, in 2 rows in each locule. Pollination by moths. Ca. 40 spp., largely from the arid parts of N. Am. (Haitian name for manihot.)

(McKelvey, Susan D. Yuccas of the southwestern U.S. 1: 1–150, 1938; 2: 1–192. 1947. Webber, J. M. Yuccas of the southwest. U.S.D.A. Agriculture Mon. 17: 1097. 1953.)

Lvs. with free marginal fibers; fr. fleshy, eventually pendent. (Sect. Sarcocarpa.)
  Pistil 6–8 cm. long; perianth 5–13 cm. long; plant acaulescent or nearly so. E. San Bernardino Co. .... 1. *Y. baccata*
  Pistil 2–3 cm. long; perianth 3–5 cm. long; plants with a definite trunk. Widespread
      3. *Y. schidigera*
Lvs. without free marginal fibers; fr. spongy or dry, spreading to erect.
  Plant with definite trunk, arborescent; style stout with 3 stigmas. (Sect. Clistocarpa)
      2. *Y. brevifolia*
  Plant practically stemless; style slender with capitate stigma. (Sect. Hesperoyucca)
      4. *Y. whipplei*

1. **Y. baccàta** Torr. var. **baccàta**. Mostly simple, stemless, sometimes in clumps with 2–6 short procumbent stems; fl.-stem 3–10 dm. tall; lf.-cluster 6–7.5 dm. high, ca. twice as wide; lf.-base ca. 1 dm. long, ca. half as wide at the middle, reddish, the blade 5–7.5 dm. long, 5–7 cm. wide if flattened, often glaucous, with broad coarse recurved marginal fibers and thickened acute to acuminate apex ending in a spine 1.5–7 mm. long; infl. ca. 6–7.5 dm. high, the fl.-cluster fleshy, heavy, with red-purple tinge and ca. 15 branches; bracts subtending pedicels 2.5–5 cm. long; perianth campanulate, red-brown outside, cream-white inside, 6–13 cm. long; fr. ellipsoid, fleshy, 15–17 cm. long; seeds somewhat ridged, 8–12 mm. long.—Uncommon on dry slopes, 3000–4000 ft.; Joshua Tree Wd.; Clark, New York and Providence mts., e. San Bernardino Co.; to Utah, Tex. May–June.

Var. **vespertìna** McKelvey. Dense, cespitose, in larger colonies, acaulescent, blue-green; infl. short, 3.5–4 dm. long, with 10–12 branches, red-purple.—More common, at 4000–5000 ft., mts. of e. San Bernardino Co.; to Utah, Ariz. April–May.

2. **Y. brevifòlia** Engelm. in Wats. var. **brevifòlia**. [*Y. draconis* var. *arborescens* Torr. *Y. a.* Trel. *Clistoyucca a.* Trel. *C. b.* Rydb.] JOSHUA TREE. Tree 5–12 (–15) m. high, branched mostly at 1–3 m. above ground; stem stout, red-brown or gray, the bark checked into small squarish plates; lf.-clusters near ends of branches, 3–15 dm. long; lf.-bases whitish, 4–5 cm. wide, the blades 2–3.5 dm. long, rigid, denticulate, not fibrous on margin, the apical spine 7–12 mm. long; perianth 4–7 cm. long, fleshy, waxen, cream to greenish white; pistil 2.5–3 cm. long; fr. 6–10 cm. long, plump, but drying in age; seeds thin, smooth, 7–10 mm. long.—Dry mesas and slopes, 2000–6000 ft.; Joshua Tree Wd.; Mojave Desert to Owens V.; to Utah, w. Ariz. March–May.

Var. **jaegeriàna** McKelvey. [*Y. b.* var. *wolfei* Jones.] Plants 3–3.5 m. tall, branching at 7–10 dm. from the ground; lvs. mostly 1 dm. long; infl. ca. 3 dm. long.—Clark Mt. to New York Mts., e. Mojave Desert; s. Nev., sw. Utah.

Var. **herbértii** (J. M. Webber) Munz. [*Y. b. f. h.* J. M. Webber.] Plants with many stems, 1–5 dm. tall, arising from scaly underground rootstocks, forming clumps up to 10

m. in diam.—W. end of Mojave Desert, Los Angeles Co. to Monolith and Walker Pass, Kern Co.

In cult. in the Botanical Garden var. *brevifolia* tends to send out many underground offsets, more than it does in the desert, and may tend to approach var. *herbertii* in maturity, but grows taller.

3. **Y. schidígera** Roezl ex Ortgies. [*Y. mohavensis* Sarg.] MOHAVE YUCCA. Plate 90, Fig. E. Trunk 1–4.5 m. tall, simple or branched; head of lvs. 6–13 dm. long, 1–2 m. wide; lf.-base 2.5–7.5 cm. long, 4–11 cm. wide at insertion; lvs. 3–15 dm. long, 2.5–4 cm. wide, yellow-green, with few coarse marginal fibers, the spine 7–12 mm. long; infl. 6–12 dm. high, the scape 1.5–5 dm. long; fl.-cluster with 15–25 branches; perianth globose, 3–4.5 (–6) cm. long, cream or with purplish tinge; pistil 2–3 cm. long; capsule 5–7 (–10) cm. long; seeds thick, obovoid to compressed, 5–8 mm. long.—Dry rocky slopes and mesas, below 5000 (7800) ft.; Chaparral, Creosote Bush Scrub; coastal and desert areas in San Diego Co., deserts of Riverside and San Bernardino cos.; Nev., Ariz., L. Calif. April–May. A form in the region of Morongo V., San Bernardino Co. has blue-green lvs., a heritable character.

4. **Y. whípplei** Torr. ssp. **whípplei.** [?*Y. californica* Groenl. *Y. ortgesiana* Roezl. *Hesperoyucca w.* Trel.] OUR LORD'S CANDLE. Plant quite acaulescent, simple, with single fl.-stalk, the whole plant dying after fruiting; lf.-bases subrectangular, whitish to greenish, 4–7 cm. wide; lvs. in dense basal rosette, gray-green, 3–8 (–10) dm. long, mostly 8–10 mm. wide, ± serrulate, flattened, rigid, the slender terminal spine 1–2 cm. long; fl.-stalk 1.5–2.5 or more dm. long, the panicle compact, 5–12 (–18) dm. long, averaging 3 dm. thick; branches numerous; fls. pendent, 2.5–3.5 cm. long, often with a purple tinge; pistil 1–1.3 cm. long; caps. oblong, 3–4 cm. long; seeds compressed, thin, smooth, 6–8 mm. long; $n = 30$ (5 long, 25 short) (McKelvey & Sax, 1933).—Dry often stony slopes, 1000–4000 ft.; mostly Chaparral, sometimes Coastal Sage Scrub, Creosote Bush Scrub; San Jacinto and Santa Ana mts. to n. L. Calif. April–May.

KEY TO SUBSPECIES

Crown simple, the whole plant dying after fruiting.
    Fl.-stalk averaging 2–2.5 m. tall; lvs. 8–10 mm. wide. Riverside and Orange cos. south
                                                                                                      Ssp. *whipplei*
    Fl.-stalk averaging 4 m. tall; lvs. 12–20 mm. wide. Coastal slopes of San Gabriel and San Bernardino mts. .... Ssp. *parishii*
Crown branched, only the flowering branch dying after fruiting; both dead and living rosettes usually present.
    Stem branching by means of rhizomes 6–18 dm. long, forming colonies. Monterey Co. to Santa Barbara Co. .... Ssp. *percursa*
    Stem not forming rhizomes.
        The stem branching to form a clump of rosettes prior to flowering, with several infls. possible in one season. Desert edge .... Ssp. *caespitosa*
        The stem branching to form new rosettes only after the first fl.-stalk is formed, with one infl. in a season. Coastal Ventura and Los Angeles cos. .... Ssp. *intermedia*

Ssp. **caespitòsa** (Jones) Haines. [*Y. w.* var. *c.* Jones.] Stems branching above ground to form crowded cespitose clump; fl.-stalk averaging 2.5 m.; lvs. triquetrous, averaging 8 dm. long; fls. creamy white.—Dry slopes, mostly 2000–4000 ft.; Joshua Tree Wd., Pinyon-Juniper Wd.; Chaparral; edge of Mojave Desert from San Bernardino Mts. to Walker Pass; w. base of s. Sierra Nevada. May–June.

Ssp. **intermèdia** Haines. Stem branching by axillary buds only after it is well matured; fl.-stalk averaging 3 m.; lvs. flat, averaging 9 dm. long.—Below 2000 ft.; Coastal Sage Scrub, Chaparral; Santa Monica and Santa Susanna mts.

Ssp. **parishii** (Jones) Haines. [*Y. w.* var. *p.* Jones. *Y. graminifolia* Wood. *Y. w.* var. *g.* Trel.] Plate 90, Fig. F. Stem solitary; fl.-stalk averaging 4 m.; lvs. flat, averaging 1 m. long. On slopes and fans, 1000–8000 ft.; Chaparral, Coastal Sage Scrub, Yellow Pine F.; s. front of San Gabriel and San Bernardino mts.

Ssp. **percúrsa** Haines. Stem 3–6 dm. long, with rhizomes 6–18 dm. long; fl.-stalk averaging 3 m.; lvs. averaging 4.8 dm. long; fls. creamy-white.—Below 2000 ft.; Chaparral, Coastal Sage Scrub; Santa Ynez Mts., Santa Barbara Co. to Monterey Co.

2. Alismatàceae. WATER-PLANTAIN FAMILY

Aquatic or marsh plants with scapose stems. Lvs. basal, long-petioled, sheathing; the blades linear to ovate-rounded, longitudinally nerved. Fls. whorled, subtended by a whorl of bracts, perfect or unisexual, in racemes or panicles. Receptacle flat to globose. Sepals 3, green, imbricate, persistent. Petals white to pinkish, deciduous. Stamens hypogynous, 6 or more, free; anthers 2-celled, extrorse. Carpels ± free, 1-celled and mostly 1-ovuled, with persistent style. Fr. a whorl or bunch of aks. A family of ca. 10 genera, mostly of N. Hemis.

Carpels in fr. spreading stellately, long-beaked; petals toothed .... 2. *Damasonium*
Carpels not spreading stellately or long-beaked in fr.; petals entire.
    Lowest fls. with carpels only; ♂ fls. above; stamens many .... 4. *Sagittaria*
    Lowest fls. perfect.
        Upper fls. ♂; frs. winged; stamens 9–15 .... 4. *Sagittaria*
        Upper fls. bisexual; frs. not winged.
            Frs. in a ring; stamens 6 .... 1. *Alisma*
            Frs. in a dense head; stamens 6–30 .... 3. *Echinodorus*

1. Alísma L. WATER-PLANTAIN

Ours perennial herbs. Lvs. erect or floating, several-ribbed and with transverse veinlets. Fls. many, small, perfect, in pyramidal panicles of whorled branches, each of which has a simple or compound umbel. Sepals broad, usually ribbed. Petals entire, deciduous, white or pinkish, spreading, deciduous. Stamens 6, 2 opposite each petal; fils. slender; anthers short. Carpels few to many, in 1 whorl; aks. 2–3-ribbed. A small genus widely distributed in temp. and trop. regions. (*Alisma*, the Greek name.)

(Hendricks, A. J. A revision of the genus Alisma. Am. Midl. Natur. 58: 470–493. 1957.)

1. **A. triviàle** Pursh. [*A. Plantago-aquatica* auth. not L. *A. P.-a.* var. *t.* Farw. var. *michalettii* Buch. and var. *americanum* Schultes & Schult.] Scapes 6–12 dm. high, exceeding lvs.; lf.-blades oblong to ovate, cuneate to rounded at base, 3–15 cm. long, long-petioled; pedicels slender, erect-spreading, 2–4 cm. long; sepals scarious-margined, 3–4 mm. long in anthesis; petals white, 3–5 mm. long; stamens twice as long as the ovaries, the anthers 0.6–0.8 mm. long; styles shorter than the ovaries, often slightly curved; stigma papillae large; aks. 2.2–3 mm. long; $2n=28$ (Taylor & Brockman, 1966).—Mirror Lake, Ventura Co.; n. to B.C. and e.

2. Damasònium Juss.

Perennial, from a short erect cormlike rootstock. Lvs. erect, ascending or floating; blades ovate to linear-oblong, 3–5-veined, long-petioled. Fls. perfect, in a simple panicle of several whorls. Sepals 3, persistent, broad, ribbed. Petals 3, white or pink, sharply and unevenly toothed, deciduous. Stamens 6, 2 opposite each sepal; fils. flattened; anthers elongate. Carpels few, in 1 whorl; aks. ribbed on back; beak erect. (*Damasonion*, classical name.)

1. **D. califórnicum** Torr. in Benth. [*Machaerocarpus c.* Small.] Plate 91, Fig. A. Lf.-blades 2.5–8 cm. long; scapes usually 2 or more, 2–4 dm. high; pedicels spreading to recurved in fr., 2–5 cm. long; sepals 4–5 mm. long; petals suborbicular, 8–10 mm. long; aks. 7–12 mm. long.—Shallow water or mud usually below 5000 ft.; Bear V., Kern Co.; to Ore.

3. Echinodòrus Rich. BUR HEAD

Annual or perennial herbs, often with runners. Lvs. long-petioled, elliptic-ovate to lanceolate, often cordate, 3–9-ribbed and punctate with dots or lines. Scapes often exceed-

Plate 91. Fig. A, *Damasonium californicum*. Fig. B, *Echinodorus berteroi*. Fig. C, *Sagittaria latifolia*, ♂ fls. above, ♀ below. Fig. D, *Allium campanulatum*, ovary with 6 low crests. Fig. E, *Allium fimbriatum*, ovary with fimbriate crests. Fig. F, *Dichelostemma pulchella*, appendaged stamens.

ing lvs., bearing a paniculate or racemose infl. Fls. perfect, in umbellate whorls. Petals white. Stamens 6–30. Pistils many; style obliquely apical, persistent; aks. turgid, ribbed, beaked, forming spinose heads. Ca. 14 spp., widely distributed in Am., Eu., Afr. (Greek, *echinos*, hedgehog, and *doros*, utricle, leather bottle, referring to the fr.)

1. **E. berteròi** (Spreng.) Fassett. [*Alisma b.* Spreng. *E. cordifolius* Calif. references. *E. rostratus* Engelm.] Plate 91, Fig. B. Lvs. basal, broadly lanceolate to ovate, with cordate or truncate base, obtuse, the blade 2–15 cm. long; scape 1–5 dm. high; umbels remote, 3–12-fld., proliferous; sepals ovate, 4–5 mm. long; petals 6–10 mm. long; fruiting heads subglobose, 4–10 mm. diam.; aks. 2.5–3 mm. long, the beak 1.2–2 mm. long; $n=11$ (Heiser & Whitaker, 1948).—Occasional in wet places at low elevs.; Freshwater Marsh; San Diego Co. to central Calif., e. U.S., S. Am. Summer.

4. *Sagittària* L. ARROWHEAD

Mostly perennial aquatic or marsh herbs, stoloniferous or tuber-bearing, with milky juice. Lvs. with sheathing bases, at least the outermost without distinct blades, the lf.-blades arrow-shaped or lanceolate. Scapes erect or lax. Fls. conspicuous, pedicelled, in whorls of 3, the plants usually monoecious; ♂ fls. above, ♀ below. Sepals spreading or reflexed in fr.; petals white, exceeding sepals. Stamens many. Pistils many, crowded on a large globose receptacle, forming flat aks. in fr. Ca. 20 spp. of temp. and trop. regions, but largely Am. (Latin, *sagitta*, arrow, because of the lf.-shape.)

(Bogin, Clifford. Revision of the genus Sagittaria. Mem. N.Y. Bot. Gard. 9: 179–233. 1955.)

Sepals of the ♀ fls. erect and accrescent .... 1. *S. calycina*
Sepals of ♀ fls. reflexed or spreading.
    Pedicels of ♀ fls. thick, recurved in fr.; lvs. entire .... 5. *S. sanfordii*
    Pedicels of ♀ fls. slender, ascending; lvs. with basal lobes.
        Beak of ak. laterally inserted, horizontal, well developed .... 3. *S. latifolia*
        Beak of ak. apically inserted, erect, minute.
            Basal lobes of lvs. not longer than terminal lobe; ak. not winged .... 2. *S. cuneata*
            Basal lobes of lvs. longer than terminal lobe; ak. wing-margined .... 4. *S. longiloba*

1. **S. calycìna** Engelm. [*S. montevidensis* ssp. *c.* Bogin. *Lophotocarpus c.* J. G. Sm. *L. californicus* J. G. Sm.] Plants to 2 m. tall, glabrous; lf.-blades hastate or sagittate, 0.5–2 dm. long, with basal lobes long, acute, sharply divergent; scapes stout; sepals erect, obtuse, 1–1.5 cm. long in fr.; petals white, ca. 1.5 cm. long; fls. glandular-pubescent; heads 1.5–3 cm. in diam.; aks. 2–3 mm. long, winged; $2n=20$ (Taylor, 1925).—Sloughs and slowly moving water; Freshwater Marsh; reported from s. Calif.; central Calif.; Miss. V. July–Aug.

2. **S. cuneàta** Sheld. [*S. arifolia* Nutt. ex J. G. Sm.] Mostly emersed and 2–4 dm. high or submerged and longer; lvs. hastate-sagittate, the blades ovate, acute, 3–17 cm. long; scapes simple or branched; fertile pedicels ascending; sepals ovate, 4–8 mm. long; petals 6–10 mm. long; fils. glabrous, not dilated; heads 1–1.5 cm. in diam. in fr.; aks. obovate, 2–2.5 mm. long, with thickened margins; $n=11$ (Brown, 1946).—Shallow ponds and swampy places below 7500 ft.; Montane Coniferous F.; San Bernardino Mts.; lower Communities further n.; n. and e. to Atlantic Coast. June–Aug.

3. **S. latifòlia** Willd. [*S. variabilis* Engelm.] WAPATO. TULE-POTATO. Plate 91, Fig. C. Plants 2–12 dm. tall, glabrous; lvs. variable, 1.5–6 dm. long, the blades 1–4 dm. long, the basal lobes acuminate, divaricate, the terminal lanceolate to broadly ovate, obtuse to acute; scape simple or branched; sepals ovate, 5–11 mm. long; petals ca. 10–20 mm. long; fils. glabrous, not dilated; fruiting heads 1.5–3 cm. in diam.; aks. obovate, ca. 3 mm. long, broadly winged; $n=11$ (Oleson, 1941).—Edge of ponds or slow streams, meadows, below 7000 ft.; largely Freshwater Marsh; most of Calif.; to B.C., Atlantic states, Mex. July–Aug.

4. **S. longilòba** Engelm. [*S. greggii* J. G. Sm.] Plants 5–14 dm. high, stout; lf.-blades 1–3 dm. long, the terminal lobe ovate to lanceolate, acute to subacuminate, the basal lobes equalling it or longer, lanceolate, widely divergent; scape paniculately branched;

Sagittaria

fertile pedicels ascending; sepals ovate, 4–7 mm. long; petals 8–12 mm. long; fils. glabrous, not dilated; fruiting heads 8–15 mm. in diam.; aks. 1.2–2.3 mm. long, thin-margined.—Sloughs and sluggish water; Freshwater Marsh; Kern Co.; Cent. V.; Ariz. to Nebr., Tex., Mex. May–June.

5. **S. sanfórdii** Greene. Plants 9–15 dm. high, glabrous; lvs. entire, 5–9 dm. long, the blades 5–15 cm. long, lanceolate or wider; scapes shorter than the lvs., simple; fertile pedicels 15–20 mm. long, recurved in fr.; sepals spreading, ovate, 4–6 mm. long; petals somewhat longer; fils. glabrous, dilated; fruiting head 12–14 mm. in diam.; aks. 2 mm. long, winged on margins.—Mirror Lake, Ventura Co.; Cent. V. May–June.

3. **Amaryllidàceae.** AMARYLLIS FAMILY

Perennial herbs, with bulbs, corms, or rhizomes. Stems well developed or not. Lvs. radical or cauline, alternate, mostly linear, entire. Fls. in scapose umbels (solitary in *Ipheion*) subtended by ± spathaceous bracts. Fls. usually perfect, the perianth of 6 segms. or lobes, usually petaloid. Stamens usually 6 (sometimes 3 sterile or lacking) inserted in the throat or on the base of the segms.; fils. filiform or thickened; anthers introrse. Ovary superior or inferior, 3-loculed; ovules usually many, on axillary placentation; style 1; stigma usually 3-lobed. Fr. a caps. or a berry, usually loculicidal. Ca. 90 genera and 1200 spp. of temp. and warm regions. Many are valuable horticulturally and economically.

Ovary superior.
  Perianth-segms. distinct or nearly so; anthers versatile.
    Fils. not appendaged; pedicels not jointed.
      Pedicels not subtended by bractlets within the main spathelike bracts; plants with onion-like odor and taste .... 1. *Allium*
      Pedicels subtended by bractlets above the main spathelike bracts; plants lacking onion-like odor and taste .... 7. *Muilla*
    Fils. surrounded at base by cup-shaped winged appendages; pedicels jointed . 3. *Bloomeria*
  Perianth-segms. united into definite basal tube; anthers often basifixed.
    Fils. separate, not united to form a corona. Common and widely distributed.
      Anthers erect and appressed to style, attached near their base, 3 (6 in *Dichelostemma pulchella*); inner corm-coats brown; stigma ± 3-lobed.
        Lvs. narrow, rounded, not keeled on lower side; stigma 3-parted, with elongate recurved-spreading lobes; seeds obtusely angled .... 4. *Brodiaea*
        Lvs. flat, keeled on lower surface; stigma with 3 small lobes which run into wings on upper style; seeds sharply angled .... 5. *Dichelostemma*
      Anthers versatile, 6; corm-coats straw-color; stigma not evidently lobed.
        Spathe-valves 2; fls. solitary .... 6. *Ipheion*
        Spathe-valves 3 or more; fls. in umbels .... 9. *Triteleia*
    Fils. united to form a tubular corona with erect bifid segms. between the anthers. Possibly in e. deserts .... 2. *Androstephium*
Ovary inferior; perianth with a crown or corona .... 8. *Narcissus*

1. *Állium* L. WILD ONION

Scapose plants from tunicated bulbs; lvs. mostly basal, linear, flat or terete, sometimes hollow, sometimes convolute-filiform. Herbage with taste and smell of onions. Fls. variously colored, small, few to many in a terminal umbel which is subtended by a scarious, sometimes colored, 1-, 2-, 3- or more-lvd. sheath of ± connate bracts. Pedicels rather slender, not jointed. Perianth persistent, its segms. nearly or quite distinct, 1-nerved, erect or spreading. Stamens 6, usually attached to base of perianth; fils. often dilated at base. Ovary superior; stigma entire to 3-lobed. Fr. a loculicidal caps., tending to be obovoid-globose, obtusely 3-lobed, often with terminal crests. Seeds obovoid, wrinkled, black, finely cellular-punctate under a lens. A genus of perhaps 500 spp., widely spread in the N. Hemis. Many from the Old World grown, as onions, garlic, chives, shallots, leeks, etc., others as ornamentals. Some of our Am. spp. are well flavored. The so-called reticulations on the bulb coats characteristic of some spp. are the cells of the persistent inner epidermis and become evident only if the outer layers of the bulb coat have been removed by decay or otherwise.

1. Lvs. mostly 2–several, if solitary, strongly flattened, or the ovary not prominently crested; crests various.
    2. The lvs. concave-convex to subterete, never falcate, if flattened, more than 2 in no.; scape persistent in maturity.
        3. Bulbs oblong, long-necked, clustered on a ± well developed rhizome; scapes 1–4 dm. high; stamens included .... 9. *A. haematochiton*
        3. Bulbs round or ovoid, mostly separated or enclosed only by bulb-coats in common.
            4. Ovary prominently 6-crested; bulb-coat reticulations with thin sinuous walls.
                5. Lvs. 3–10 mm. broad, usually green at anthesis; perianth-segms. papery in fr., tips with neither strongly involute margins nor a pronounced keel ... 3. *A. bisceptrum*
                5. Lvs. 1–5 mm. wide, usually withering at anthesis; perianth-segms. becoming rigid, the tips strongly involute and with a pronounced keel .... 5. *A. campanulatum*
            4. Ovary crestless or with less prominent crests; reticulations on bulb-coats thick-walled, prominent under a lens.
                6. Reticulations reticulate-polygonal, ± isodiametric.
                    7. Perianth-segms. 8–12 mm. long, rounded at apex, white; bracts 15–20 mm. long. Garden escape .... 14. *A. neapolitanum*
                    7. Perianth-segms. 5–8 mm. long; bracts 8–12 mm. long. Native.
                        8. Pedicels 5–10 mm. long; scape 10–15 (–20) cm. tall. Santa Barbara Co. to Marin Co. .... 12. *A. lacunosum*
                        8. Pedicels 10–20 mm. long; scape 15–30 cm. tall. Kern Co. to Riverside Co. 7. *A. davisiae*
                6. Reticulations transversely elongate, commonly in regular serrate rows.
                    9. Ovary without crests or with very obscure crests; reticulations of bulb-coats rather irregularly arranged.
                        10. Perianth 5–9 mm. long, pale, becoming thin and membranous. Tehachapi Mts. to Sierra Nevada .... 11. *A. hyalinum*
                        10. Perianth 9–12 mm. long, rose to rose-purple, less papery. S. Calif. to L. Calif. 18. *A. praecox*
                    9. Ovary with ± evident, mostly cent. crests; reticulations of bulb-coats usually very regularly serrate and in vertical rows.
                        11. Perianth 6–7 mm. long, becoming papery; fls. many, in subglobose umbel; ovary with 6 lateral crests .... 1. *A. amplectens*
                        11. Perianth usually 8–14 mm. long, not papery in age; fls. in open umbels; ovary with 3 cent. crests.
                            12. Inner perianth-segms. undulate-crisped or serrulate ...... 6. *A. crispum*
                            12. Inner perianth-segms. plane and entire .... 17. *A. peninsulare*
    2. The lvs. flat, ± falcate, usually much longer than the short scape, breaking off with the scape at the surface of the ground at maturity.
        2a. Lf. 1 .... 4. *A. burlewii*
        2a. Lvs. 2 .... 19. *A. tribracteatum*
1. Lvs. solitary and terete; ovary prominently 6-crested.
    13. Stigma distinctly trifid, the divisions often slender and recurved (see also *A. parishii*).
        14. Stamens well included .... 8. *A. fimbriatum*
        14. Stamens ca. as long as perianth or exserted .... 10. *A. howellii*
    13. Stigma entire, capitate, at most 3-lobed, not distinctly trifid.
        15. Perianth-segms. 8–12 mm. long; pedicels slender, longer than the fls.
            16. Outer bulb-coats usually with ± distinct contorted reticulations; stamens usually less than ⅔ as long as the perianth. E. San Bernardino Co. .... 15. *A. nevadense*
            16. Outer bulb-coats without reticulations; stamens relatively longer. N. San Bernardino Co. to Mono Co. .... 2. *A. atrorubens*
        15. Perianth-segms. 12–20 mm. long; pedicels stout, much shorter than the fls.
            17. The perianth-segms. elliptic-lanceolate, acute, becoming papery and abruptly spreading at tips in fr. .... 16. *A. parishii*
            17. The perianth-segms. linear-lanceolate, attenuate, becoming rigid and widely divergent apically in fr. .... 13. *A. monticola*

1. **A. ampléctens** Torr. [*A. attenuifolium* Kell. *A. occidentale* Gray. *A. monospermum* Jeps. *A. attenuifolium* var. *m.* Jeps.] Bulb ovoid, 1–1.5 cm. long, the outer coats brownish,

Allium

with transverse broadly V-shaped regular reticulations in vertical rows, the inner coats reddish or whitish; scape 2–5 dm. high; lvs. 2–4, shorter than the scape, narrow, flattened, but convolute-filiform in age; bracts 2–3, broadly ovate, abruptly acuminate, 7–10 (–19) mm. long; perianth-segms. white to pinkish, lance-ovate, acute, 6–7 mm. long, papery after anthesis; stamens equal to or shorter than perianth; fils. slightly dilated at base; anthers yellow or purplish; stigma capitate; caps. ca. 3 mm. long, with 6 very low crests; seeds ca. 2 mm. long; $2n=14?$, 21, 28 (Levan, 1940).—Dry slopes and fields, mostly below 6000 ft.; Yellow Pine F., Foothill Wd.; Cuyamaca Mts.; cent. Calif. to B.C. March–June.

2. **A. atrorùbens** Wats. Bulb ovoid, ca. 15 mm. long, with sessile bulblets, or these on slender rhizomes, the coats red-brown with indistinct quadrate reticulations; scape 6–15 cm. high, terete; lf. 1, exceeding scape, terete above the tubular sheath, coiled above; bracts usually ovate, 12–18 mm. long, long-pointed; fls. 20–25; pedicels 10–15 mm. long; perianth-segms. red-purple, rarely white, stiff, spreading, lanceolate, attenuate, 9–12 mm. long; stamens and style included; fils. slender, dilated at base; anthers yellow or purple; stigma entire; caps. 3–4 mm. high, the 6 crests 1 mm. high, entire or toothed; seeds 2 mm. long; $n=7$ (Aase).—Dry hillsides, 5000–7000 ft.; Sagebrush Scrub; Death V. region to Mono Co. and w. Nev. May–June.

Var. **inyònis** (Jones) Traub. [*A. i.* Jones. *A. decipiens* Jones, not Fisch.] Perianth pale with dark midveins, broader, not attenuate, 8–12 mm. long.—At 4000–8000 ft.; Sagebrush Scrub, Pinyon-Juniper Wd.; Inyo and Mono cos.; adjacent Nev. May–June.

3. **A. biscéptrum** Wats. Bulb round-ovoid, 1–1.5 cm. long, with supplementary bulblets in a tight basal cluster, outer coats grayish, inner white, with very irregular reticulations, the vertical walls very sinuous; scapes 1–3, terete, 1–3 dm. high; lvs. 2–3, flat, 3–10 mm. wide, ca. as long as scape; bracts 2, lance-ovate, acuminate, 5–15 mm. long; perianth-segms. rose-purple, lance-ovate, acuminate, 6–10 mm. long, divaricate, greenish at base; stamens ¾ as long as perianth; fils. dilated at base; anthers dark; stigma subcapitate; caps. 3–4 mm. long, the 6 crests triangular, away from the style; seeds ca. 1.5 mm. long; $n=7$ (Aase).—Meadows and aspen groves or moist banks, 6500–10,000 ft.; Montane Coniferous F.; White Mts. and e. Sierra Nevada, Inyo Co.; to Ida., Utah. May–July.

4. **A. burlèwii** A. Davids. Bulb ovoid, 1.5–2 cm. long, the outer coats grayish or brownish, usually without reticulations, the inner white, obscurely cellular; scape 2–8 cm. long, somewhat flattened; lf. 1, flat, 1–10 mm. wide, exceeding scape; bracts 2–3, ovate, obtuse or abruptly pointed; fls. 8–18; pedicels stout, 1–2 cm. long; perianth-segms. pinkish-purple with dark midveins, 7–10 mm. long; stamens ca. equal to perianth or exserted; fils. dilated basally; anthers purple or yellowish; caps. 4 mm. long, with 6 low crests or crestless; seeds obovoid, 2.5–3 mm. long; $n=7$.—Dry granitic slopes and ridges, 6000–9000 ft.; Montane Coniferous F.; San Jacinto Mts. w. and n.; to s. Sierra Nevada. May–July.

5. **A. campanulàtum** Wats. [*A. bidwelliae* Wats. *A. acuminatum* var. *b.* Jeps. *A. austinae* Jones. *A. tenellum* Davids. in part. *A. bullardii* Davids.] Plate 91, Fig. D. Bulb ovoid, 1–2 cm. long, with supplementary bulblets either in a tight basal cluster or on filiform rhizomes up to 1 dm. long; outer bulb-coats brown, inner reddish to white, with minute quadrate reticulation or the horizontal walls faint and the vertical prominent and sinuous; scapes 1–2 in no., 1–3 dm. high; lvs. 2–3, flat, 1–5 mm. wide, shorter than to longer than scape, tending to wither by anthesis; bracts 2, abruptly acuminate, 7–15 mm. long; fls. 15–40, loosely arranged; pedicels slender, 1–2 cm. long; perianth-segms. pale rose, spreading, lance-ovate, acute, 5–8 mm. long; stamens ca. ¾ as long; fils. slender, basally dilated; anthers reddish; stigma subcapitate; caps. ca. 3 mm. high, with 6 low cent. triangular crests; seeds ca. 2 mm. long; $n=7$, 14 (Aase).—Dry slopes in woods, 2000–8500 ft.; Pinyon-Juniper Wd., Chaparral, Montane Coniferous F.; mts. of San Diego Co. n. to Ore., Nev. May–July.

6. **A. críspum** Greene. [*A. peninsulare* var. *c.* Jeps.] Bulb round-ovoid, 8–12 mm. long, outer coats gray-brown, inner whitish, all closely and regularly horizontal-serrate; scape

1–3 dm. high; lvs. 2, ca. 1.5 mm. wide, mostly shorter than scape; bracts 2, ovate, 1–2 cm. long, abruptly short-acuminate; fls. 10–40, in open umbels; pedicels stout, spreading, 1–3 cm. long; perianth-segms. bright red-purple to orchid, 9–12 mm. long, the outer oblong-ovate, acute, spreading at apex, the inner narrower, undulate-crisped; stamens shorter, broadly dilated at base; anthers yellowish; stigma slightly lobed to distinctly trifid; caps. ca. 4 mm. high, the crests cent., usually evident; seeds 3 mm. long; $n=7$ (Aase).—Heavy soil on rolling hills, below 2500 ft.; V. Grassland, Foothill Wd.; Figueroa Mt., Santa Barbara Co. and Sierran foothills, Kern Co.; n. to cent. Calif. March–May.

7. **A. davísiae** Jones. [*A. pseudobulbiferum* Davids.] Mostly 2–4 dm. high; lvs. 2–3, ca. as long as the scape; bracts 2, broadly ovate, 8–12 mm. long, abruptly acuminate; pedicels 10–20 (–25) mm. long, stoutish; perianth-segms. pale with reddish midveins, lance-ovate, acuminate, 6–8 mm. long; stamens ¾ as long; fils. dilated at base; caps. ca. 4 mm. high, 3-lobed, scarcely crested; $n=7$ (Aase).—Dry flats, 2000–5300 ft.; Creosote Bush Scrub, Joshua Tree Wd.; Kenworthy, San Jacinto Mts., n. slopes San Bernardino and San Gabriel mts., to Kern Co. April–May. Specimens from the Greenhorn Mts., Kern Co. provisionally placed here.

8. **A. fimbriàtum** Wats. Plate 91, Fig. E. Bulb globose-ovoid, 12–16 mm. long, the outer coats dark to grayish, the inner brownish to pinkish or whitish, obscurely cellular; scapes 7–30 cm. high, slender; lf. 1, one to 2 times as long, terete above the tubular sheath; bracts 2–3, setaceous at apex, 6–20 mm. long; pedicels 4–20 mm. long, stoutish; perianth-segms. pale to white with pink midveins to rose-purple, lanceolate to lance-oblong or oblong-ovate, 6–15 mm. long; stamens shorter; anthers yellow; stigma with 3 linear lobes; caps. 3–4 mm. high, 3-lobed, the 6 erect crests fimbriate to subentire; seeds ca. 2 mm. long; $n=7$ (Aase).—Much of Calif.

KEY TO VARIETIES
Perianth purple to rose.
  Perianth-segms. not conspicuously recurved-spreading at tips; ovary-crests usually fimbriate or toothed .... var. *fimbriatum*
  Perianth-segms. conspicuously recurved-spreading at tips; ovary-crests entire or nearly so
                                                                                       var. *denticulatum*
Perianth mostly pale pink to white.
  Ovary-crests coarsely dentate-laciniate; scape 7–25 cm. high. N. Ventura Co. northward
                                                                                       var. *diabolense*
  Ovary-crests entire to ± toothed.
    Scape 1–2.5 dm. long; from below 4000 ft.
      Perianth-segms. oblong-lanceolate, acuminate. W. Mojave Desert ...... var. *mohavense*
      Perianth-segms. oblong-ovate, ± rounded at apex. W. Riverside Co. ..... var. *munzii*
    Scape mostly 0.6–1.5 dm. long. From 4000–7000 ft., San Bernardino Mts. to n. L. Calif.
                                                                                                        var. *parryi*

Var. **denticulàtum** Ownbey & Aase. Inner perianth-segms. minutely denticulate; crests entire or toothed.—At 3000–5000 ft.; largely Joshua Tree Wd.; Piute Mts., Walker Pass, Kern Co. to Little San Bernardino Mts. April–May.

Var. **diabolénse** Ownbey & Aase. Scape 7–25 cm. high, rather slender; lf. 1–2 times as long; perianth-segms. 7–9 mm. long, pale with deep pink midribs, elliptic-lanceolate, obtuse or acute, erect, not recurved at tip; crests coarsely dentate-laciniate.—Serpentine, 1700–4000 ft.; Chaparral, etc.; inner Coast Ranges, Ventura Co. to Stanislaus Co. April–June.

Var. **fimbriàtum.** Scape 3–15 cm. long, slender; pedicels 4–15 mm. long; perianth-segms. purple to rose, not conspicuously recurved-spreading at tips; ovary-crests usually fimbriate or toothed.—At 2000–8000 ft.; Creosote Bush Scrub, Joshua Tree Wd.; Pinyon-Juniper Wd.; w. edge of Colo. Desert, w. Mojave Desert, inner Coast Ranges to Lake and Napa cos., n. L. Calif. March–July.

Var. **mohavénse** Jeps. [*A. m.* Tides.] Bracts often 3; perianth-segms. pale pink to white with pink midveins, lance-oblong, 7–10 mm. long, not recurved-spreading at tip;

ovary-crests toothed to subentire.—At 2500–4000 ft.; Creosote Bush Scrub; w. Mojave Desert. April–May.

Var. **múnzii** Ownbey & Aase. Scape slender, 1–2.5 dm. high; bracts 7–10 mm. long; pedicels 6–12 mm. long; perianth-segms. pink, 7–9 mm. long, oblong-ovate, usually rounded at apex, erect, not recurved at tip; crests ca. 1 mm. high, triangular-lanceolate, entire or coarsely toothed.—At 1000–2000 ft.; w. end of Riverside Co. April–May.

Var. **párryi** (Wats.) Ownbey & Aase. [*A. p.* Wats. *A. tenellum* Davids. in part. *A. kessleri* Davids.] Scape slender, 6–16 cm. high; bracts 8–10 mm. long; pedicels 10–15 mm. long; perianth-segms. 6–8 mm. long, pink or with pink midveins, lanceolate, acuminate; crests erect, 1–1.5 mm. high, entire or erose.—Dry slopes, 4000–7000 ft.; Yellow Pine F.; San Bernardino Mts. to n. L. Calif. June–July.

9. **A. haematochìton** Wats. [*A. marvinii* Davids.] Bulbs oblong, 2–3 cm. long, usually clustered, from a short rootstock, the coats membranous, deep red to white, with fine vertical striations; scape 1–4 dm. high, slightly compressed; lvs. several, flat, 1–2 dm. long, 1–4 mm. wide, with much wider sheath; bracts 2–4, connate, obtuse; fls. 10–30; pedicels 1–2 cm. long; perianth-segms. white to rose, with darker midvein, broadly ovate to lance-ovate, acute, 6–8 mm. long; stamens and style included; fils. dilated at base; anthers yellow; caps. 4 mm. high, obcordate, with 6 short rounded crests; seeds broadly obovoid, 2 mm. long; $n=7$ (Aase).—Dry slopes and ridges of clay or stony soil, below 2500 ft.; Chaparral, Coastal Sage Scrub, V. Grassland; L. Calif. to San Luis Obispo Co. March–May.

10. **A. howéllii** Eastw. var. **clòkeyi** Ownbey & Aase. Bulb 7–12 mm. long, the outer coats red-brown, the inner paler, obscurely cellular; scape 1.5–3 dm. high, stout; lf. 1, terete above the tubular sheath, ca. as long as scape; bracts usually 3, reddish, broad-ovate, 4–15 mm. long; fls. many, white, sometimes with reddish midveins; pedicels 5–15 mm. long; stamens and style usually exserted; stamens ca. as long as perianth; anthers yellow; stigma trifid; caps. 3 mm. high, with 6 red acuminate dentate-laciniate to entire crests; $n=7$ (Aase).—Heavy soil, 4500–6000 ft.; Sagebrush Scrub, etc.; region of Mt. Pinos, Ventura Co. April–June.

Var. **howéllii** has been reported from between Tehachapi and Keene, Kern Co., *Twisselmann*. In it the perianth is pale lilac-violet to deep rose and the stamens are usually exserted. It ranges northward.

11. **A. hyalìnum** Curran. Bulb ovoid, 6–10 mm. long, the coats gray, thin, with horizontal undulate reticulation in vertical rows; scape slender, 1.5–3 dm. high; lvs. 2, sometimes 1 or 3, 1–5 mm. wide, ca. as long as scape, often convolute; bracts 2, lanceolate, 1–2 cm. long, acuminate; fls. 6–15, in open umbel; pedicels slender, spreading, 2–2.5 cm. long; perianth-segms. white or pinkish, lance-ovate, acute, 5–9 mm. long, thin and membranous in age; stamens ⅔ as long, dilated at base; anthers pale; stigma subentire; caps. ca. 3 mm. high, crestless; seeds 1.5–2 mm. long; $n=7$ (Aase).—Shallow moist soil in grassy and rocky places, 500–1500 ft.; V. Grassland, Chaparral, Foothill Wd.; Tehachapi Mts. to Eldorado Co. March–June.

12. **A. lacunòsum** Wats. Bulb ovoid, often with remnants of older bulbs in a row above, outer coats gray-brown, with quadrate heavy-walled reticulations, inner thin, pale, the reticulations with thinner walls; scape 1–2 dm. high; lvs. 1–2, ca. as long as scape, terete, slender; bracts 2, ovate, short-acuminate, 8–12 mm. long; fls. 8–25; pedicels 5–10 mm. long; perianth-segms. white to pinkish, with green or red midveins, lance-oblong, acute or subacuminate, 5–7 mm. long; stamens almost as long, fils. dilated at base; anthers yellow; stigma capitate; caps. 3–4 mm. long, with an obtuse thickened ridge at summit of each side; seeds 2 mm. long; $n=7$ (Aase).—Open slopes and flats, often on serpentine, below 3000 ft.; Foothill Wd., Mixed Evergreen F.; Santa Barbara Co. to cent. Calif.; Santa Rosa Id. April–May.

13. **A. montícola** Davids. [*A. peirsonii* Jeps.] Bulb ovoid, 1–2 cm. long, sometimes with basal bulblets, the outer coats dark gray-brown, the inner whitish or pinkish, obscurely quadrate-reticulate; scape 5–15 cm. high, largely underground, stout; lf. 1, terete above the tubular sheath, exceeding scape, glaucous; bracts 3 or 2; broadly ovate,

1–1.8 cm. long, subacuminate; fls. 6–many; pedicels 5–12 mm. long, stoutish; perianth-segms. pale to deep pink, 12–20 mm. long, linear-lanceolate, attenuate, with spreading tips; fils. shorter, scarcely dilated at base; anthers purple; stigma entire; caps. ca. 5 mm. high, the 6 crests deltoid-lanceolate, 1–2 mm. high, entire or irregularly toothed; seeds 3 mm. long; $n=7$ (Aase).—Loose rock or talus, 4000–10,000 ft.; Montane Coniferous F.; San Gabriel and San Bernardino mts. May–July.

Var. **kéckii** (Munz) Ownbey & Aase. [*A. parishii* var. *k.* Munz.] Crests linear, 2–3 mm. high, papery; fls. deep purple.—Summits, 4000–5500 ft.; Chaparral, Yellow Pine F.; Santa Ana Mts., Topatopa Mts., Santa Ynez Mts. June.

14. **A. neapolitànum** Cyr. Bulb coats with quadratic reticulations having very heavy thick walls; scape 3–6 dm. high, subterete; lvs. 2–3 or more, lance-linear, loose-spreading, shorter than the scape, 1–3 cm. broad; pedicels 3–4 cm. long; fls. pure white, the perianth-segms. ovate, obtuse, 1–12 mm. high; stamens included; ovary not crested; stigma subentire.—Garden escape as at Fullerton, Orange Co., Carpenteria. From the Medit. region.

15. **A. nevadénse** Wats. var. **cristàtum** (Wats.) Ownbey. [*A. c.* Wats.] Bulb ovoid, 12–15 mm. long, with basal bulblets, outer coats grayish or brownish, the inner pinkish, obscurely or imperceptibly marked; scape 8–14 cm. long, mostly underground; lf. 1, terete above the tubular sheath, exceeding scape; bracts 2–3, lance-ovate, acuminate, 8–14 mm. long; perianth-segms. pink with red midveins, ovate-lanceolate, acute to acuminate, 8–12 mm. long; stamens included; fils. dilated at base; anthers yellow or purple; stigma subentire; caps. ca. 3 mm. long, with 6 erect, ± toothed crests ca. 1 mm. high; seeds ca. 2 mm. long.—Dry stony slopes, 5000–7000 ft.; Pinyon-Juniper Wd.; Providence, New York, Clark and Kingston mts. to Last Chance Mts. and e. to Utah, Ariz. Apr.–May.

16. **A. paríshii** Wats. Bulb round-ovoid, 10–15 mm. long, with pinkish nonreticulate coats, sometimes with basal bulblets; scape stout, 1–2 dm. long, largely underground; lf. 1, terete above the tubular sheath, much longer than the scape; bracts 2–3, ovate, 1–1.5 cm. long, acuminate; fls. 10–many; pedicels 6–12 mm. long; perianth-segms. pale pink, 12–15 mm. long, elliptic-lanceolate, acute, with spreading tips; fils. half as long, lance-olate; anthers yellow; stigma entire or slightly lobed, rarely trifid; caps. 2–3 mm. high, with 6 triangular entire or toothed crests ca. 1 mm. high; $n=7$ (Aase).—Open rocky slopes mostly at 3000–4000 ft.; Joshua Tree Wd.; n. base of San Bernardino Mts., Little San Bernardino Mts.; w. Ariz. April–May.

17. **A. peninsulàre** Lemmon [*A. montigenum* Davids.] Bulb round-ovoid, 1–1.5 cm. long, the outer coats gray-brown, the inner whitish, all with close regular horizontal-serrate reticulation; scape 2–4 dm. high; lvs. 2–4, almost as long as scape, 1–6 mm. wide; bracts mostly 2, lance-ovate, acuminate, 1–2 cm. long; fls. 6–25, in open umbels; pedicels slender, 15–30 mm. long, spreading; perianth-segms. deep rose-purple, shining, 10–13 mm. long, the outer ovate-lanceolate, acuminate and with spreading apex, the inner narrower, shorter, erect or spreading; stamens shorter; fils. broad at base; anthers yellowish; stigma ± distinctly 3-lobed; caps. ca. 4 mm. high, the crests very minute; seeds ca. 2 mm. long; $n=7$ (Aase).—Dry open or wooded slopes below 3000 ft.; V. Grassland, Foothill Wd.; n. L. Calif. to n. Calif. March–June.

18. **A. praecóx** Bdg. [*A. hyalinum* var. *p.* Jeps.] Bulbs round-ovoid, 7–15 mm. long, the outer coats gray-brown, inner paler, all with horizontal somewhat irregular and somewhat undulating reticulation; scape stout, 2–5 dm. high; lvs. 2–4, ca. as long as scape, flat, 1–5 mm. wide; bracts 2, lance-ovate, acuminate, 1–2.5 cm. long; fls. 6–30, in open umbel; pedicels stoutish, spreading, 15–30 mm. long; perianth-segms. ± papery, purple or lighter with dark midveins, lance-ovate, 9–12 mm. long, acuminate, the inner somewhat narrower, but ca. as long as the outer; stamens included, ± dilated at base; anthers yellowish or reddish; stigma slightly lobed; caps. 4–5 mm. high, minutely crested; seeds 3–4 mm. long; $n=14$ (Lenz, 1966).—Shaded slopes and canyons, below 2500 ft.; Chaparral, S. Oak Wd.; L. Calif. to San Bernardino Co. and e. Los Angeles Co.; San Miguel, Santa Cruz, Santa Rosa, Santa Catalina and San Clemente ids. March–April.

19. **A. tribracteàtum** Torr. [*A. parvum* var. *jacintense* Munz.] Bulbs ovoid, 1–2 cm.

long, with thin white coats with faint mostly oblong reticulations; scape 3–12 cm. long; lvs. 2, longer than the scape, 2–6 mm. wide; bracts 2–3, abruptly acuminate; fls. 10–20; pedicels 4–16 mm. long; perianth-segms. pale rose with dark purplish midveins, narrowly oblong to lanceolate, 7–11 mm. long, obtuse or acute; stamens and stigma included; fils. dilated at base; anthers yellow or purple; caps. 4 mm. high, 3-lobed, crestless or obscurely 3-crested; seeds 2 mm. long; $n=7$ (Aase).—Dry rocky ridges and slopes, 4000–8000 ft.; Montane Coniferous F.; San Jacinto Mts.; to n. Calif. April–July.

2. Androstèphium Torr.

Corm fibrous-coated. Lvs. linear, channeled. Bracts several, scarious. Fls. few; pedicels not jointed. Perianth funnelform. Stamens 6; fils. partly united into a tube. Caps. obtusely 3-angled. Three spp. of sw. U.S. (Greek, *andros*, stamen, and *stephanos*, crown, referring to united fils.)

1. **A. brevifòlium** Wats. [*Bessera b.* Jeps.] Scape 1–3 dm. high; lvs. several; fls. whitish to light violet-purple, drying brownish-yellow, 15–20 mm. long.—Known from deserts immediately east of Inyo Co. and to be sought in Calif.

3. Bloómeria Kell. GOLDEN STARS

Corm fibrous-coated; stem scapose. Lvs. basal, few, linear, carinate. Fls. yellow, many, in a loose umbel subtended by membranous bracts; pedicels jointed at summit. Perianth persistent, rotate, of 6 subequal oblong-linear segms. Stamens 6; fils. filiform, margined at the base by winglike or cupshaped appendages; anthers versatile, attached near the base. Style 1, persistent, splitting with the subglobose loculicidal caps. Seeds 1–several in each locule, black, subovoid, angular and wrinkled. A small California genus. (H. G. *Bloomer*, early San Francisco botanist.)

(Ingram, J. A monograph of the genera Bloomeria and Muilla. Madroño 12: 10–27. 1953.)

1. **B. crocèa** (Torr.) Cov. var. **crocea.** [*Allium c.* Torr. ?*B. gracilis* Borzi.] Corms ca. 15 mm. thick; scape 1.5–6 dm. high; lf. ca. half as long; bracts lanceolate; pedicels many, 2–6 cm. long; perianth-segms. orange-yellow, with median dark lines, 8–12 mm. long; fils. 6 mm. long, the appendages shallowly bicuspidate; style 5 mm. long; caps. 5–6 mm. high; seeds 2 mm. long; $n=9$ (Burbanck, 1944), (Lenz, 1966).—Common, dry flats and hillsides, often in heavy soil, up to 5000 ft.; Coastal Sage Scrub, Chaparral, V. Grassland, S. Oak Wd.; L. Calif. to Santa Barbara and w. Kern cos.; Channel Ids. April–June.

Var. **montàna** (Greene) Ingram. [*B. m.* Greene.] Perianth-segms. yellow, 11–13 mm. long; fil.-appendages with linear-attenuate cusps ca. as long as the body.—Largely Chaparral, Yellow Pine F.; Mt. Pinos, Tehachapi Mts.; to Temblor and San Emigdio ranges.

4. Brodiaèa Sm. BRODIAEA

Perennial herbs with underground corms bearing dark brown tunicated bulb-coats and few grasslike, rounded not carinate lvs. Scapes erect, bearing umbels subtended by usually scarious bracts. Pedicels jointed beneath the perianth, the latter ± tubular at the base. Perianth-segms. 6; stamens 3, the anthers erect, attached to the fils. on their back and immediately above the distinctly sagittate base. Stigma 3-lobed, the lobes spreading and recurving. Seeds slightly longer than thick, obtusely angled, longitudinally striate, with ridge on one side moderately developed. A small genus of Pacific N. Am. (J. J. *Brodie*, Scottish cryptogamic botanist.)

(Hoover, R. F. A definition of the genus Brodiaea. Bull. Torrey Bot. Club 66: 161–166. 1939; A revision of the genus B. Am. Midl. Nat. 22: 551–574. 1939.)

Perianth-tube in fr. firm and not splitting; staminodia obtuse or acute.
 Fils. 2–5 mm. long, dilated at base only if at all; staminodia not incurved at apex
  5. *B. terrestris*

Fils. 1 mm. long.
　　The fils. broadly triangular; staminodia with incurved apex .... 2. *B. jolonensis*
　　The fils. linear; staminodia erect. San Clemente Id. .... 3. *B. kinkiensis*
Perianth-tube thin-membranous and splitting in fr.; staminodia acute or lacking.
　　The fils. 1 mm. long; staminodia present .... 1. *B. filifolia*
　　The fils. 4–6 mm. long; staminodia lacking .... 4. *B. orcuttii*

　　1. **B. filifòlia** Wats. [*Hookera f.* Greene.] Scape 2–4 dm. high; lvs. several, shorter than or nearly as long as scapes, 1–2 mm. wide; pedicels 2–5 cm. long; perianth violet, the tube greenish, narrow-campanulate, 6–7 mm. long, membranous and splitting in fr.; segms. spreading, 9–12 mm. long; staminodia plane, linear, 6–7 mm. long when fresh, much shorter in dry fls., curved outward above; fils. 1 mm. long, triangular; anthers 4 mm. long, broad, notched; caps. short-ovoid; seeds 2–2.5 mm. long; $2n=32$ (Niehaus, 1965).—Local in heavy clay soil below 2000 ft.; Coastal Sage Scrub, Chaparral; Glendora, San Bernardino V., Perris, Vista. May–June.
　　2. **B. jolonénsis** Eastw. Scape 3–20 cm. high; lvs. usually longer; pedicels 2–12 cm. long; perianth violet, 18–27 mm. long, the tube rounded at base, 8–13 mm. long; segms. spreading, 10–14 mm. long; staminodia violet, involute, 4–6 mm. long, incurved at apex; fils. 1 mm. long, broader than anthers; these 3–5 mm. long, emarginate at apex; caps. ovoid; $2n=12$ (Niehaus, 1965).—Clay depressions on mesas and gentle slopes, below 4500 ft.; V. Grassland, Coastal Sage Scrub, Chaparral; Monterey Co. to n. L. Calif., Santa Cruz and Santa Catalina ids. April–June.
　　3. **B. kinkiénsis** Niehaus. Corm with heavy fibrous outer coat; lvs. 2–4 dm. long; scape 2–3 dm. tall; pedicels 3–8 cm. long; perianth-tube whitish with brown-purple midribs, rounded at base, 12 mm. long, 4–5 mm. wide, spreading, the outer oblong, the inner obovate; staminodia erect, 7 mm. long, 3 mm. wide, cuspidate; fils. 1 mm. long; anthers 4–5 mm. long, retuse; $2n=32$ (Niehaus, 1966).—Local at ca. 1400–1600 ft.; San Clemente Id.
　　4. **B. orcúttii** (Greene) Baker. [*Hookera o.* Greene. *B. filifolia* var. *o.* Jeps. *H. multipedunculata* Abrams.] Scape 1–4 dm. high; pedicels 2–8 cm. long; perianth violet, the tube 5–8 mm. long, narrow-campanulate, thin-membranous and brittle in fr.; segms. 10–18 mm. long, ascending; staminodia lacking; fils. slender, 4–6 mm. long; anthers 5–6 mm. long, subentire at apex; $2n=24$ (Lenz, 1966).—Near streams and about vernal pools and seeps, up to 5500 ft.; Chaparral, Yellow Pine F.; San Diego Co. April–July.
　　5. **B. terréstris** Kell. ssp. **kernénsis** (Hoov.) Niehaus. [*B. coronaria* var. *k.* Hoov., Am. Midl. Nat. 22: 561. 1929.] Scapes largely 1–3 dm. high; lvs. to ca. as long; pedicels rather few, 2–9 cm. long; perianth lilac or violet, the tube 10–14 mm. long, firm; segms. oblong-oval, 12–20 mm. long; stamens 3; fils. mostly 2–4 mm. long, dilated at base; staminodia erect, not bent toward stamens, exceeding the anthers; anthers 5–7 mm. long; seeds 1.5–2 mm. long; $2n=36$ (Niehaus, 1965).—Slopes and flats below 5000 ft.; V. Grassland, Foothill Wd., Yellow Pine F.; Kern Co. to San Diego Co. April–June.

5. *Dichelostémma* Kunth

　　Perennial herbs with scape and lvs. arising from a subglobose corm with brown fibrous coat. Lvs. 2–5, flattened, keeled on lower side. Fls. in an umbel, each subtended by a bract, the outer bracts enlarged, forming an invol. Pedicels jointed to the perianth; perianth-segms. similar, from an inflated or campanulate tube. Stamens opposite the outer segms. with reduced or sterile anthers; those opposite the inner with fils. entirely adnate to the perianth and anthers attached to the sagittate base and closely approximate around the style. Stigma small, 3-lobed. Caps. firm; seeds black, elongated, striate-reticulate, winged on the angles. A small genus of Pacific N. Am. (Greek, *dicha*, bifid, and *stemma*, garland, referring to the stamen-appendages.)
　　1. **D. pulchélla** (Salisb.) Heller. [*Hookera p.* Salisb. *Hookera p.* Greene. *Dipterostemon p.* Rydb. *B. capitata* Benth. *D. c.* Wood. *Milla c.* Baker. *Hookera c.* Kuntze. *Dipterostemon c.* Rydb. *B. parviflora* Torr. *Hookera p.* Ktze. *B. insularis* Greene. *Dichelostemma i.* Greene.

Dipterostemon i. Greene.] BLUE DICKS. WILD-HYACINTH. Plate 91, Fig. F. Scapes 3–6 (–9) dm. high, smooth; lvs. 1.5–4 dm. long, 5–12 mm. wide; bracts purple, usually ovate; pedicels 2–15 mm. long; perianth-tube pale, 4–8 mm. long, cylindro-campanulate; segms. violet, rarely white, ascending, 7–11 mm. long; fils. opposite the outer segms. dilated, 2 mm. long, bearing anthers 2–3 mm. long, those opposite the inner segms. adnate but extending beyond the anthers (which are 3.5–4.5 mm. long) as 2 lanceolate appendages; style 4–6 mm. long; caps. ovoid, 4–6 mm. long, sessile; seeds 2.5–4 mm. long; $2n = 18, 36, 45, 54, 72$ (Lenz,1966).—Common on plains and hillsides below Yellow Pine F., in most parts of Calif. w. of the Sierra Nevada; to Ore., L. Calif.; more uncommon e. of the Sierra, in Pinyon-Juniper Wd. and Yellow Pine F. to s. Utah and n. Ariz. Mostly March–May.

Var. **pauciflòra** (Torr.) Mort. [*Brodiaea capitata* var. *p.* Torr. *Milla c.* var. *p.* Baker. *Dichelostemma p.* Standl. *Dipterostemon p.* Rydb. *Hookera p.* Tides.] An uncertain form intergrading freely with the typical var. and characterized by lanceolate bracts, not or scarcely colored; pedicels 6–35 mm. long; pale blue, spreading perianth segms.—Dry open places; Creosote Bush Scrub and higher; e. Mojave Desert, w. edge Colo. Desert; to New Mex.

6. *Iphèion* Raf.

Scapose bulbose perennial with onion like odor when bruised. Bulbs with membranous tunic. Lvs. linear, not keeled. Scape 1-fld. the spathe split into 2 valves united basally. Fl. erect, funnelform; perianth-segms. ± equal, basally connate into a tube. Stamens 6, inserted on the tube; anthers versatile. Ovary superior. Style filiform, erect, the stigma obscurely 3-lobed. Ca. 10 spp. of s. S. Am. (Greek name of obscure origin.)

1. **I. uniflòrum** Raf. [*Triteleia u.* Lindl.] STAR FLOWER. Bulbs small, deep-seated; lvs. slightly glaucous, nearly flat, broad-linear; scape 1.5–2 dm. high, bracted ca. midway by the 2-valved dry spathe; fl. salverform, white with bluish tinge, 3–3.5 cm. in diam., the oblong segms. with darker blue and under sides ribbed bluish-brown. Native of Argentine. Widely cult. and occasionally becoming natur. as at Santa Barbara, *Pollard*.

7. *Muílla* Wats. MUILLA

Scape from a corm with fibro-membranous coats. Lvs. few, narrow, subterete. Umbel subtended by scarious acuminate bracts; pedicels slender, not jointed. Perianth subrotate, persistent, of 6 subequal slightly united, lance-oblong segms., whitish or greenish, with dark 2-nerved midribs. Stamens 6, inserted near the base; fils. filiform to dilated; anthers versatile. Ovules 8–10 in a locule; style short, clavate, persistent and finally splitting. Caps. globose, 3-angled, loculicidal. Seeds compressed, irregularly angled, black. A small alliumlike genus from the sw. U.S. and Mex., lacking odor and taste of onions. (Anagram of *Allium*.)

(Ingram, J. A monograph of the genera Bloomeria and Muilla. Madroño 12: 19–27. 1953.)
Perianth bright yellow .... 1. *M. clevelandii*
Perianth not bright yellow.
    Fils. greatly dilated. Transmontane .... 2. *M. coronata*
    Fils. filiform or subulate. Cismontane .... 3. *M. maritima*

1. **M. clevelándii** (Wats.) Hoov. [*Muilla c.* Hoov.] Scape 8–30 cm. tall; lvs. 6–15 cm. long; pedicels 2–3.5 cm. long; perianth-segms. 6–10 mm. long, yellow with greenish stripe; fils. 3–5 mm. long, with oblong appendages 2–3 mm. long; style 1.5 mm. long; caps. 4–5 mm. long; $2n = 14$ (Lenz, ined.).—Dry mesas and hillsides; Chaparral, Coastal Sage Scrub; sw. San Diego Co. May.

2. **M. coronàta** Greene. Scapes 3.5–15 cm. high; lvs. longer, scabrous; fls. 3–10; perianth-segms. 3–4 mm. long, bluish or whitish within, green without; fils. hyaline, conspicuously dilated, retuse at summit, their broad margins overlapping to form a cylindrical crown; anthers yellow.—Infrequent, in heavy soil, 3000–5000 ft.; Shadscale

Plate 92. Fig. A, *Muilla maritima*. Fig. B, *Carex alma*, with infl., ♀ scale above and perigynium below (right). Fig. C, *Carex nebraskensis*, lower left, ♀ scale, center broad ribbed perigynium. Fig. D, *Carex pansa*. Fig. E, *Cyperus odoratus*, with separate spikelet. Fig. F, *Eleocharis montevidensis*, scale and (above) ak. with bristles.

Scrub, Joshua Tree Wd.; Antelope V., El Paso Mts., w. side Indian Wells V., near Independence. March–April.

3. **M. marítima** (Torr.) Wats. [*Hesperocordium* (?) *m*. Torr. *Allium m*. Benth.; *Nothoscordum m*. Hook. f. *Bloomeria m*. Macbr. *M. tenuis* Congd. *M. serotina* Greene.] Plate 92, Fig. A. Corm 1–2 cm. thick; scapes 1–5 dm. high; lvs. equalling or exceeding scape, ± scabrous; bracts 3–6, lanceolate, mostly 3-nerved; fls. few to many; pedicels 1–5 cm. long; perianth-segms. greenish white with brownish midnerve, 3–6 mm. long, the inner generally wider; fils. subulate, dilated toward base; anthers usually purplish; caps. 5–8 mm. high; seeds 2–3 mm. long; $2n=20$ (Lenz, 1966).—Subalkaline flats and granitic or serpentine slopes below 7500 ft.; many Communities; L. Calif. through cismontane Calif., especially the Coast Ranges to Glenn Co. March–June.

8. *Narcissus* L.

Bulbous plants with linear lvs. and simple scapes of solitary to several fls. Fls. yellow to white, from a 1-valved thin tubular spathe; perianth salverform, with a short tube and subequal segms.; crown (corona) short to long. Stamens inserted in tube; fils. mostly included in crown; anthers erect and basifixed. Ovary inferior. A rather large genus, including daffodils, jonquils, etc. (Classical Latin name.)

1. **N. tazétta** L. POLYANTHUS N. The Paper-White with segms. white reported from behind beach dunes, mouth of Ventura R., *Pollard*, apparently natur. It and the Chinese-Lily with sulfur-yellow segms. to be expected as escape from gardens. Old World.

9. *Tritelèia* Dougl. ex Lindl.

Perennials from corms with reticulate-fibrous straw-colored coats. Lvs. 1–2, linear, flattened, keeled on lower side. Scape slender, bearing an umbel subtended by an invol. of distinct bracts. Pedicels jointed to the fls. Perianth-segms. 6, from a tube at base. Stamens 6, all fertile; anthers versatile; fils. alike or unequal. Style slender, the stigma small, 3-lobed. Caps. stipitate, loculicidal. Seeds black, rounded, ridged on one side, pitted, granulate or granulate-reticulate. Several spp., Pacific N. Am. (Greek, *tri*, three, and *treleios*, referring to ternary arrangement of floral parts.)

(Hoover, R. F. A systematic study of Triteleia. Am. Midl. Nat. 25: 73–100. 1941.)
Stamens alternate, attached at 2 different levels.
    Anthers 1.5 mm. long; fils. dilated toward base. San Clemente Id. . . . . 1. *T. clementina*
    Anthers 2–5 mm. long; fils. not or scarcely dilated. Mainland . . . . 3. *T. laxa*
Stamens all attached at one level.
    Fils. not forked at apex; perianth-tube ca. as long as perianth-segms. . . . . 2. *T. dudleyi*
    Fils. forked at apex; perianth-segms. shorter than the tube . . . . 4. *T. scabra*

1. **T. clementìna** Hoov. [*Brodiaea c*. Munz.] Scapes 3–4 dm. high, smooth; lvs. 3–5 dm. long, 4–10 mm. wide; pedicels 3–8 cm. long; perianth light blue, 17–25 mm. long, the tube acute at base, 7–10 mm. long; segms. erect, 10–15 mm. long; fils. 2 mm. long, triangular; anthers 1.5 mm. long; stipe of ovary ca. as long as body at anthesis; seeds 3–4 mm. long; $2n=16$ (Niehaus, 1965).—Damp clefts on rocky walls, Coastal Sage Scrub; San Clemente Id. March–Apr.

2. **T. dúdleyi** Hoov. [*Brodiaea d*. Munz.] Corm deep-seated; lvs. 4–8 mm. wide, at least as long as scape; bracts lance-acuminate; umbel with 3–25 fls.; pedicels slender, ascending, 15–25 mm. long; perianth pale yellow with dark midveins, purplish on drying, the tube slender-funnelform, 9–12 mm. long, the segms. lanceolate, spreading, 10–12 mm. long; fils. narrow-triangular, acuminate, inserted at the same level, the alternate long ones ca. 3.5 mm. long, the shorter ca. 2 mm. long; anthers oval, 1 mm. long, lavender; stipe of ovary at anthesis ca. as long as the body; $n=8$ (Niehaus).—Rare, at streamsides, 5000–6000 ft., Cloudburst Summit, Mt. Waterman in San Gabriel Mts., Piute Mts.; 6000–10,000 ft., Sierra Nevada n. to Mono Co. June.

3. **T. láxa** Benth. [*Brodiaea l.* Wats. *Milla l.* Baker. *Hookera l.* Ktze. *T. candida* Greene. *B. c.* Baker.] GRASS NUT. ITHURIEL'S SPEAR. Scapes 1–7 dm. high, smooth to scabrous or retrorse-pubescent below; lvs. 2–4 dm. long, 4–25 mm. wide; pedicels 2–9 cm. long; perianth blue to white, 20–45 mm. long, with pistil on lower side and fils. curved upward; tube attenuate at base, 12–25 mm. long, with narrow membranous appendages within from the adnate fil.-bases; segms. divergent, 8–20 mm. long; fils. of both rows 3–6 mm. long; anthers 2–5 mm. long; ovary-stipe 2–3 times the length of the body at anthesis; seeds 1.5–2 mm. long; $n=14, 21, 24$ (Burbanck, 1941); $2n=16, 32, 48$ (Lenz, 1966).—Heavy soil below 3000 ft., Coastal Sage Scrub, Loma Linda, San Bernardino Co., *Roos*; at higher elevs., Yellow Pine F., Tehachapi Mts., Kern Co.; to Ore. Apr.–June.

4. **T. scàbra** (Greene) Hoov. [*Calliprora s.* Greene. *Brodiaea s.* Baker. *B. lutea* var. *s.* Munz.] Scape mostly retrorsely hairy near base; lvs. often 8–12 mm. wide; perianth cream-color to deep golden yellow, the tube 4–6 mm. long, less than half as long as perianth-segms., the segms. 10–17 mm. long, at anthesis rotate or slightly deflexed, the inner usually broadly rounded at apex; fils. ± narrowed upward from base to below the terminal fork, tending to be contiguous and thus form a closed tube, the longer 5–7 mm. long, the shorter 4–5 mm. long; anthers 1–2 mm. long; stalk of anther usually black.—Yellow Pine F. to V. Grassland; Tehachapi Mts., *Twisselmann*; to Butte Co. March–May.

4. Aràceae. ARUM FAMILY

Herbs, but sometimes hard and woody, erect, prostrate or climbing. Lvs. large, mostly basal, simple or compound. Fls. reduced, often without a perianth, crowded on a ± fleshy spadix which is usually surrounded by a large, often colored bract or spathe Fls. usually imperfect, the ♂ above, the ♀ below, or perfect, or plants dioecious. Perianth of 4–6 scalelike segms. or lacking. Stamens 4–10; fils. short; anthers 2-celled. Ovules 1–several in a locule; style short or none; stigma terminal, usually minute and sessile. Fr. a berry or utricle. Ca. 100 genera and 1500 spp., mostly trop.

Plant floating; lvs. in dense rosettes .... 1. *Pistia*
Plant not floating; lvs. not in rosettes .... 2. *Zantedeschia*

1. *Pístia* L. WATER-LETTUCE

Mostly free-floating. Lvs. rounded, entire, with several ribs, sessile. Spathes small, axillary, white. Fls. without perianth; stamens 2. One trop. sp. (Greek, *pistos*, liquid, referring to aquatic habitat.)

1. **P. stratiòtes** L. Stoloniferous; lvs. 5–15 cm. long, plaited; spathe ca. 1 cm. long; $2n=28$ (Blackburn, 1933).—Drainage canal on the Calif. side of the Colo. R., near Yuma; Tex. to tropics.

2. *Zantedéschia* Spreng. CALLA-LILY

Perennial herbs from thick rhizomes. Lvs. basal, long-petioled, the blade hastate to cordate-ovate. Spathe showy, on long peduncles. Perianth lacking. Stamens 2–3. Fr. berrylike. Ca. 8 spp., of S. Afr. (F. *Zantedeschi*, Italian botanist.)

1. **Z. aethiòpica** (L.) Spreng. [*Calla a.* L.] Robust plant; spathe creamy or white; spadix yellow; $2n=32$ (Ito, 1942).—Much grown in Calif. as a garden plant and occasionally becoming natur. Nov.–May.

5. Arecàceae. (Palmae) PALM FAMILY

Ours trees with fibrous roots and cylindrical unbranched trunks. Lvs. in ours fan-shaped, long-petioled, plaited in early stages, borne in a large apical tuft. Plants commonly monoecious, the fls. small, on large compound axillary spadices. Perianth-segms. 6, in 2 series, usually firm in texture. Stamens 6; fils. dilated, ± united at base. Carpels 3,

± united, each 1-ovuled. Fr. a drupe or berry. Ca. 140 genera and some thousands of spp. of the warmer parts of the world, many of great economic and horticultural value.

1. Washingtònia Wendl. FAN PALM

Tall, columnar, the trunks naked or clothed with the persistent lf. bases. Lvs. large and heavy; petioles stout, flattish, ± spiniferous on the margins and projected into the blade with an "arrowhead" point or hastula. Spadix long, projecting from a flattened spathe 1 m. or more long. Fls. whitish, ca. 8 mm. long; calyx persistent, tubular; corolla-lobes reflexed, later deciduous. Fr. hard, short-oblong, 8–10 mm. long, blackish. Two spp., of sw. N. Am. (Named in honor of George *Washington*.)

1. **W. filífera** (Lindl.) Wendl. [*Pritchardia f.* Lindl. ex André. *P. filamentosa* Fenzi. *W. f.* var. *robusta* Parish, not *W. r.* Wendl.] Tree to 25 m. tall, the trunk 6–10 dm. thick; lvs. gray-green; petioles 1–2 m. long; blades 1–2 m. long, with 40–60 folds, segmented to ca. the middle, copiously fibrous; spadices to 3 or 4 m. long; $2n=36$ (Sato, 1952).—In groves in moist alkaline places about seeps, springs and streams, below 3500 ft.; Creosote Bush Scrub; w. and n. edge of Colo. Desert, region of 29 Palms; w. Ariz., n. L. Calif. Natur. in Kern Co. June. Much planted in subtrop. regions.

6. Commelinàceae. SPIDERWORT FAMILY

Annual or perennial herbs, with knotty and leafy stems. Lvs. alternate, entire, lance-ovate to linear, amplexicaul or vaginate. Fls. bisexual, usually almost regular. Sepals 3, sometimes slightly connate, green. Petals 3, free or united into a tube, alternate with sepals, colored. Stamens usually 6, in 2 series; fils. often hairy. Style 1; stigma simple or obscurely lobed; ovary 2–3-celled, few–many-ovuled. Fr. a caps. A family of ca. 500 spp., of wide circulation in warmer regions.

1. Tradescántia L.

Plants prostrate or erect. Lvs. variable. Peduncles solitary, fascicled or paniculate. Fls. mostly umbellate. More than 30 spp., temp. to trop. regions. (J. *Tradescant*, English gardener of the 17th century.)

1. **T. fluminénsis** Vell. WANDERING JEW. Prostrate, rooting at the nodes; lvs. oblong to oblong-ovate, 3–7 cm. long, acute, glabrous; sheath hairy at summit; umbels many-fld., subtended by 2 lance-ovate bracts; petals white, 6 mm. long; fils. hairy; $2n=60$ (Anderson & Sax, 1935).—Common in cult. and becoming natur. in damp places. From S. Am.

7. Cyperàceae. SEDGE FAMILY

Grasslike or rushlike herbs, perennial with rhizomes or annual with fibrous roots. Culms (stems) solid or rarely hollow, terete to variously angled. Lvs. mainly basal, alternate, commonly 3-ranked, the blades narrow and the sheaths closed. Infl. commonly subtended by 1 or more involucral lvs. and consisting of a terminal spikelet or a cluster of spikelets. Each spikelet of a series of scales (bracts) subtending individual fls. Scales spirally imbricated or 2-ranked, persistent or deciduous. Fls. perfect or unisexual. Perianth represented by several bristles or by an inner hyaline member, or absent. Stamens 1–3. Pistil 1, the ovary superior, 1-loculed, with 1 ovule and a single 2- or 3-fid style. Fr. a triangular or lenticular ak. Embryo minute. Endosperm mealy. A large cosmopolitan family of ca. 60 genera and 2600 spp.

1. Fls. perfect, or perfect and ♂; ak. naked.
  2. All the scales of the spikelet bearing aks., or empty basal scales not more than 2; fls. all perfect.
    3. Scales of the spikelet spirally imbricated.
      4. Perianth bristles present, but included within the scales.

5. Style-base deciduous from the summit of the ak., tubercle none; 1 or more involucral lvs. present .... 8. *Scirpus*
  5. Style-base persistent as a tubercle on the summit of the ak.; involucral lvs. none
                                                                                 4. *Eleocharis*
 4. Perianth-bristles absent.
  6. Hyaline perianth-member present between fl. and rachis of spikelet . 6. *Hemicarpha*
  6. Hyaline perianth-member absent.
   7. Style-base not swollen; plants 4–20 cm. high .... 8. *Scirpus cernuus*
   7. Style-base swollen; plant 20–70 cm. high .... 5. *Fimbristylis*
 3. Scales of the spikelet 2-ranked .... 3. *Cyperus*
 2. Basal 3–several scales of the spikelet empty; spikelet consisting of both ♂ and perfect fls.
  8. Bristles present; infl. a dense head .... 7. *Schoenus*
  8. Bristles absent; infl. a loose compound umbel .... 2. *Cladium*
1. Fls. all unisexual; ak. surrounded by a saclike bractlet (perigynium) .... 1. *Carex*

1. Càrex L. Sedge

Grasslike plants, perennial, with shorter or longer rootstocks. Culms mostly solid, sharply triangular to nearly terete, usually leafy mostly at the base. Lvs. 3-ranked, sheath closed, a small inconspicuous ligule usually present on the ventral side of the blade at the junction of the sheath. Uppermost lvs. (bracts) subtending the spikelets either scalelike or foliaceous. Fls. unisexual, the plants monoecious or dioecious, borne in the axils of bracts and ± congested into short or long, few- to many-fld. spikelets (spikes of Mackenzie), the spikelets solitary or many and arranged in spikes, heads, racemes or panicles, the spikelets unisexual or bisexual and either androgynous (♂ fls. above and ♀ fls. below) or gynaecandrous (♀ fls. above and ♂ fls. below). Perianth lacking; each ♀ fl. of a 2- or 3-stigmatic pistil enclosed in an urn- or flask-shaped bractlet (perigynium) through the small terminal opening of which protrudes the style. Each ♂ fl. of 3 (or 2) stamens. Ak. lenticular or triangular, falling off with the perigynium when mature. A genus of over 1000 spp. The classical Latin name. (Mackenzie, K. K. in N. Am. Fl. 18: 9–478. 1931–1935 and N. Am. Cariceae, 2 vols., plates. 1940.)

1. Perigynia pubescent; styles 3.
 2. Spike solitary; perigynia beakless .... 15. *C. exserta*
 2. Spikes more than one; perigynia beaked.
  3. Spikes with few to ca. 25 perigynia.
   4. Perigynia strongly 2-keeled with subglobose body; bracts of the non-basal ♀ spikes sheathless or nearly so.
    5. Perigynia finely many-ribbed as well as strongly 2-keeled; ♀ scales usually at least as long as perigynia .... 19. *C. globosa*
    5. Perigynia nerveless except for the 2 keels; ♀ scales mostly shorter than perigynia.
     6. The perigynia 2.5–3 mm. long; beak ca. 0.2–0.8 mm. long, shallowly bidentate
                                                                              7. *C. brevipes*
     6. The perigynia 3–4.5 mm. long; beak longer, strongly bidentate .. 41. *C. rossii*
   4. Perigynia sharply triangular, obovoid; bracts of the non-basal ♀ spikes sheathing
                                                                              54. *C. triquetra*
  3. Spikes with 25–50 perigynia.
   7. Perigynium-beak strongly bidentate; ♀ scales glabrous or ciliate. Widespread
                                                                              28. *C. lanuginosa*
   7. Perigynium-beak shallowly bidentate, hyaline-tipped; ♀ scales pubescent. San Jacinto Mts. and Kern Co. n. .... 43. *C. sartwelliana*
1. Perigynia glabrous.
 8. Styles 3; aks. triangular.
  9. Spike 1 on each culm.
   10. Perigynia 5–7 mm. long; culms 2–6 dm. tall. Widespread sp. ... 34. *C. multicaulis*
   10. Perigynia 3.5–4 mm. long; culms 1.5–2 dm. tall. White Mts. .... 51. *C. subnigricans*
  9. Spikes more than 1.
   11. Beak of perigynium strongly bidentate, with teeth from almost 1 to 2 mm. long.
    12. Perigynium 5–7 mm. long; teeth of beak 1.5–2 mm. long .... 9. *C. comosa*
    12. Perigynium 4–5 mm. long; teeth of beak not more than 1 mm. long . 42. *C. rostrata*

11. Beak of perigynium weakly if at all bidentate, the teeth very minute.
    13. Lf.-blades 2–6 mm. wide; scales ovate.
        14. Lowest bract strongly sheathing .... 29. *C. lemmonii*
        14. Lowest bract sheathless.
            15. Scales hyaline with green center; beak 0.5 mm. long .... 46. *C. serratodens*
            15. Scales reddish brown with lighter center, beak 1 mm. long .. 23. *C. heteroneura*
    13. Lf.-blades 7–14 mm. wide; scales narrowly ovate, serratulate-awned . 48. *C. spissa*
8. Styles 2; aks. lenticular.
    16. Terminal spikes ♂ and lateral spikes elongated, or if terminal spike gynaecandrous (♀ fls. above ♂), the lateral spikes peduncled.
        17. Lowest bract long-sheathing; ♀ spikes rather few-fld.; perigynia beakless.
            18. Perigynia granular, whitish-puberulent at maturity, not fleshy ... 21. *C. hassei*
            18. Perigynia not granular or puberulent, fleshy and golden-yellow or brownish at maturity .... 4. *C. aurea*
        17. Lowest bract sheathless, or if sheathing, the ♀ spikes with very numerous fls.; perigynia short-beaked.
            19. Lf.-sheaths not becoming filamentous when breaking; beak of perigynium 0.5–1 mm. long .... 36. *C. nebrascensis*
            19. Lf.-sheaths breaking and becoming filamentose.
                20. Perigynium-beak 0.5 mm. long, bidentate, hispidulous between the teeth
                    5. *C. barbarae*
                20. Perigynium-beak ca. 0.25 mm. long, emarginate or entire.
                    21. Lf.-sheaths strongly carinate dorsally; blades 6–12 mm. broad; culms 10–15 dm. tall .... 44. *C. schottii*
                    21. Lf.-sheaths slightly carinate or rounded dorsally; blades 3–5 mm. broad; culms 3–10 dm. tall .... 45. *C. senta*
    16. Terminal spike partly ♀, or if ♂, the lateral spikes short and sessile.
        22. Rootstocks long-creeping, the culms arising singly or few together.
            23. Perigynium-beak from half as long to nearly as long as the body.
                24. Culms generally smooth and less than 2 dm. long; lvs. folded-involute near the apex; perigynia finely nerved ventrally .... 13. *C. douglasii*
                24. Culms generally roughened above and more than 2 dm. long; lvs. flat or caniculate; perigynia nerveless ventrally .... 40. *C. praegracilis*
            23. Perigynium-beak one-third as long as body or less.
                25. Spikelets in dense globose heads; beak smooth .... 56. *C. vernacula*
                25. Spikelets in elongate heads; beak serrulate.
                    26. Perigynia 3.5–4.5 mm. long. Insular .... 39. *C. pansa*
                    26. Perigynia 1.8–3.5 mm. long.
                        27. Lvs. folded-involute near the apex, 1–1.5 mm. wide; perigynia 2.5–3 mm. long. White Mts. .... 14. *C. eleocharis*
                        27. Lvs. flattish, 1.5–4 mm. wide.
                            28. Perigynia 1.8–2.2 mm. long. Piute Mts. .... 47. *C. simulata*
                            28. Perigynia 2.5–3.5 mm. long. San Bernardino Mts. ... 37. *C. occidentalis*
        22. Rootstocks short-creeping; plants cespitose.
            29. Perigynia narrowly to broadly wing-margined.
                30. Lower bracts well developed, conspicuous or at least quite evident.
                    31. The lowest bracts much exceeding the head. San Bernardino Mts. and Kern Co. .... 3. *C. athrostachya*
                    31. The lowest bracts scarcely if at all exceeding the head.
                        32. Perigynia strongly nerved on outer face, faintly if at all on the inner.
                            33. The perigynia 3 mm. long, green. Montane .... 52. *C. teneraeformis*
                            33. The perigynia 4 mm. long, soon brownish. Santa Barbara Co.
                                34. Lvs. 1–2 mm. wide; spikes 3–6 .... 20. *C. gracilior*
                                34. Lvs. 2–4 mm. wide; spikes 5–10 .... 49. *C. subbracteata*
                30. Lower bracts inconspicuous or wanting.
                    35. Sheaths hyaline only at the mouth, green-striate below opposite the blades
                        17. *C. feta*
                    35. Sheaths hyaline below opposite the blades.
                        30. The sheaths strongly prolonged upward at the mouth opposite the blades
                            18. *C. fracta*
                        36. The upper sheaths concave or truncate at mouth opposite the blades.

37. Perigynia thin except where distended by the aks.
   38. The perigynia lanceolate to lance-ovate, the margin very narrow. White Mts. .... 31. *C. microptera*
   38. The perigynia usually ovate, the margin broad to the base.
      39. Culms erect, 3–10 dm. tall; scales dark brown. A widespread sp. 16. *C. festivella*
      39. Culms mostly decumbent or spreading, 1–3 dm. tall; scales blackish. White Mts. .... 22. *C. haydeniana*
37. Perigynia plano-convex, thickish.
   40. Beak of perigynium flat and serrulate to the tip.
      41. Lvs. 3–6 mm. wide; spikes ca. 10; perigynia nerved on both faces 35. *C. multicostata*
      41. Lvs. 1.5–3 mm. wide; spikes 4–8; perigynia nerved on inner face 50. *C. subfusca*
         41a. Perigynia strongly nerved on the ventral side. Santa Cruz Id. 33. *C. montereyensis*
         41a. Perigynia smooth and nearly or quite nerveless on the ventral side. Mainland .... 50. *C. subfusca*
   40. Beak of perigynium terete, smooth in upper part.
      42. Perigynia 3.5–5 mm. long.
         43. Spikes densely aggregated; perigynia brownish, hyaline-tipped 1. *C. abrupta*
         43. Spikes somewhat separate; perigynia greenish or straw-color, reddish-tipped .... 30. *C. mariposana*
      42. Perigynia less than 3 mm. long . ... 26. *C. integra*
29. Perigynia not evidently wing-margined.
   44. The spikes with ♂ fls. uppermost (androgynous).
      45. Perigynia corky, ⅓–½ the length of the body, the margin and the beak smooth. Natur. .... 53. *C. texensis*
      45. Perigynia not conspicuously corky at base, the beak mostly serrulate.
         46. Heads round to ovoid, 8–20 mm. long.
            47. Scales light brown with greenish midrib, to very pale.
               48. Scales chestnut brown with lighter keel; perigynia obsoletely nerved on inner face, 4–5 mm. long, with sharply bidentate beak. Montane 24. *C. hoodii*
               48. Scales very pale; perigynia often nerveless, 2.5 mm. long. Occasional as a lawn weed .... 8. *C. cephalophora*
            47. Scales dark brown with inconspicuous midvein; perigynia nerved on both faces, 3–4 mm. long, with shallowly bidentate beak ..... 27. *C. jonesii*
         46. Heads spicate, 25–50 or more mm. long.
            49. Lvs. 3–6 mm. wide.
               50. Perigynia 3.5 mm. long, lightly nerved on both faces .... 2. *C. alma*
               50. Perigynia 4–4.5 mm. long, strongly nerved on both faces . 10. *C. densa*
            49. Lvs. 1–2.5 mm. wide.
               51. Perigynia 2–2.5 mm. long, nerved on outer face. Mainland 11. *C. diandra*
               51. Perigynia 3.5–5 mm. long, nerved on both faces or only dorsally. Insular .... 55. *C. tumulicola*
   44. The spikes with ♀ fls. uppermost (gynaecandrous).
      52. Perigynia not puncticulate (cellular-dotted).
         53. Head dense, 8–14 mm. long; perigynia ovate, very dark .... 25. *C. illota*
         53. Head lax, 20–60 mm. long; perigynia lanceolate to lance-ovate, lighter in color.
            54. Spikes 3–4, widely separate; lf.-blades 1.5–2 mm. wide; perigynia brownish-yellow .... 38. *C. ormantha*
            54. Spikes 4–8, the lower separate.
               55. Lf.-blades 2.5–5 mm. wide; beak strongly bidentate. Montane native 6. *C. bolanderi*
               55. Lf.-blades 2–3 mm. wide; beak shallowly bidentate. Introd. at Santa Barbara .... 32. *C. molesta*
      52. Perigynia puncticulate (cellular-dotted). White Mts. .... 12. *C. disperma*

1. **C. abrúpta** Mkze. Plants densely cespitose; culms 4–6 dm. high, erect, much longer than the lvs.; lvs. 1.5–2.5 mm. wide, flat; infl. capitate, roundish, 9–17 mm. long, the spikelets gynaecandrous, with ♀ scales ovate, narrower and shorter than the perigynia; perigynia broadly lanceolate to subovate, 3.7–4 mm. long, planoconvex, membranaceous, nerved dorsally and ventrally, wing-margined, serrulate on margins above, slender-beaked, the beak ¼–⅓ the length of the body, terete, smooth, white hyaline-tipped.—Meadows and open slopes, below 9500 ft., Montane Coniferous F.; San Jacinto, San Bernardino and San Gabriel mts., Mt. Pinos, Panamint Mts.; Santa Cruz Id., Sierra Nevada to Ore. Summer.

2. **C. álma** Bailey. [*C. vitrea* Holm.] Plate 92, Fig. B. Lvs. and culms cespitose, the latter 3–12 dm. tall, erect; blades flat or ± folded along the middle, 3–6 mm. wide, strongly scabrous on the margins; infl. paniculate, 2.5–20 cm. long, the spikelets androgynous, congested in clusters along the branches; ♀ scales white-hyaline, mostly covering the perigynia; perigynia ovate or oblong-ovate, 3.5–4 mm. long, slightly nerved on both sides, serrulate on margin above the middle, narrowed to a serrulate beak, the beak bidentate, ca. ⅓ the length of the body.—Frequent along streams, in springy places, etc., below 8000 ft.; many Plant Communities; San Diego Co. to cent. Calif., Nev., Ariz. Summer.

3. **C. athrostàchya** Olney. Densely cespitose in small or large clumps; culms 0.5–6 dm. tall, longer or shorter than the lvs., smooth or a little roughened; blades flat, 1.5–4 mm. wide; infl. capitate or the lowest spikelets a little separate, the head mostly ovoid or rounded, 1–2 cm. long, the spikelets gynaecandrous, the lowest subtended by foliaceous bracts that usually exceed the infl.; ♀ scales oblong-ovate, a little narrower and shorter than the perigynia; perigynia flattened, lance-ovate, 3–4.5 mm. long, nerved dorsally, slightly nerved or nerveless ventrally, margined, serrulate above, tapering into a subterete or little-flattened beak ca. ⅓ as long as the body.—Moist places, 6500–7500 ft.; Montane Coniferous F.; San Bernardino Mts.; Kern Co. to Alaska and Rocky Mts.

4. **C. aúrea** Nutt. Rather loosely cespitose, sometimes more tufted; culms to 5 dm. tall, erect or spreading, shorter to longer than the lvs.; blades 2–4 mm. wide, flat, or channeled at the base; spikelets 4–6, the terminal ♂ or with a few ♀ fls., 3–10 mm. long, sessile or short-stalked, the lateral spikelets ♀, oblong or linear-oblong, 5–20 mm. long, the upper approximate, the lower ± widely separate, the lowest frequently basal, their stalks exserted from the sheaths of leaflike bracts; ♀ scales ovate to roundish, obtuse to cuspidate, spreading at maturity; perigynia roundish-obovate, 2–3 mm. long, coarsely ribbed, golden yellow or brownish and fleshy at maturity, rounded and beakless at apex.—Wet places below 9500 ft.; Montane Coniferous F.; San Bernardino Mts. to Mt. Pinos, White Mts.; to B.C. and the Atlantic.

5. **C. bárbarae** Dewey. [*C. wilkesii* Olney. *C. lacunarum* Holm.] Plants cespitose, with long stout rhizomes; culms 3–10 dm. tall, usually longer than the lvs.; blades flat above, channeled near the base, 3.5–9 mm. wide, the lf.-sheaths breaking and becoming filamentose; ♂ spikelets 1 or 2, linear, up to 6 cm. long, ♀ spikelets 2–5, the upper often androgynous, usually separate, linear to oblong, 2.5–8 cm. long, the lowest bract lf.-like, shorter or longer than the infl.; ♀ scales lance-ovate to ovate, narrower than the perigynia, and from shorter to longer; perigynia oblong-obovate, to roundish, the beak 0.5 mm. long, bidentate, the teeth hispidulous.—Open or brushy slopes that are moist in the spring, sea level to 3000 ft.; V. Grassland, Foothill Wd., etc.; coastal s. Calif. n. to Ore. Santa Cruz Id.

6. **C. bolánderi** Olney. Loosely cespitose, the clustered culms and lvs. from slender rootstocks; culms 1.5–9 dm. high, slender, lax and frequently spreading, longer than the lvs.; blades 2–5 mm. wide, flat; infl. ± spicate-capitate, the lower spikelets sometimes discrete, the upper approximate and capitate-congested, spikelets gynaecandrous; ♀ scales ovate or somewhat narrower, covering the perigynia-bodies; perigynia lanceolate, 4–4.5 mm. long, strongly nerved dorsally, lightly nerved ventrally, serrulate on upper margin, tapering above into a serrulate beak, the beak more than half as long as the perigynium-body, strongly bidentate.—Edge of streams, in meadows, etc., below 7500

ft., Montane Coniferous F.; San Bernardino and San Gabriel mts., Sierra Nevada and Coast Ranges n. to B.C., e. to Rocky Mts.

7. **C. brévipes** W. Boott. [*C. deflexa* var. *boottii* Bailey.] Densely cespitose, the clustered lvs. and culms arising from short rootstocks; culms to 18 cm. long, longer or shorter than the lvs.; blades 1.5–2.5 mm. wide, flat or channeled at the base; ♂ spikelets terminal, sessile or short-stalked, 4–12 mm. long; ♀ spikelets 3–5, the upper 1 or 2 approximate, sessile or stalked, the others widely separate, nearly basal; ♀ scales ovate, acute to cuspidate, shorter than the perigynia but ca. as wide; perigynia 2.5–3 mm. long, pubescent, the body roundish, 2-keeled, stipitate, contracted into a bidentulate beak 0.2–0.8 mm. long.—Dry meadows and slopes, at ca. 7000–8000 ft.; Montane Coniferous F.; San Bernardino and San Gabriel mts.; n. to Wash., e. to Nev. and Ida.

8. **C. cephalóphora** Muhl. Plants densely cespitose; culms 2–5 dm. tall, ca. as long as the lvs.; blades flat, 2–4.5 mm. wide; spikelets androgynous, few to several, in a compact or roundish head 1–2 cm. long; ♀ scales narrower and shorter than the perigynia; perigynia ovate, narrowed at the base, 2.5 mm. long, the margins raised ventrally, contracted to a serrulate beak, the beak bidentate, $\frac{1}{4}$–$\frac{1}{2}$ the length of the body.—Lawn weed as in Altadena, Arcadia, Santa Barbara; native of e. states.

9. **C. comòsa** Boott. Densely cespitose, forming large clumps; culms 5–15 dm. tall, usually shorter than the lvs.; blades flat, 6–16 mm. wide; spikelets 4–7, the terminal ♂, linear, 3–7 cm. long, the ♀ approximate or a little separate, the lower stalked and ± nodding, oblong, 1.5–7.5 cm. long, the lower bracts leaflike and longer than the infl.; ♀ scales long-awned, the base much shorter and narrower than the perigynia; perigynia lanceolate, 5–7 mm. long, shining, strongly ribbed, tapering into a slender beak, 1.5–2 mm. long, the beak bidentate with spreading awnlike teeth 1–2 mm. long; ak. continuous with the indurate style.—Swampy places, San Bernardino V.; cent. Calif. to Wash., e. to the Atlantic.

10. **C. dénsa** (Bailey) Bailey. [*C. brongniartii* var. *d.* Bailey. *C. chrysoleuca* Holm.] Densely cespitose; culms 3–7 dm. tall, erect, usually exceeding the lvs.; blades flat, 3–6 mm. wide; ligule conspicuous, as long as wide; infl. paniculate, the branches closely placed to form a head 2–5 cm. long; spikelets androgynous; ♀ scales acute to cuspidate, ca. as wide as the perigynia but shorter; perigynia ovate, 3.5–4.5 mm. long, ± nerved dorsally and ventrally, margin serrulate above the middle, narrowed to a serrulate beak half as long as the body or longer, bidentate.—Wet places, below 4600 ft.; Foothill Wd., V. Grassland; San Diego Co., Riverside Co., Ventura Co.; to Ore.

11. **C. diándra** Schrank. [*C. bernardina* Parish.] Loosely cespitose in large clumps, the culms and lvs. from somewhat prolonged rootstocks; culms 3–7 dm. tall, erect, usually longer than the lvs.; blades flat or folded along the middle, 1–2.5 mm. wide; infl. 2.5–5 cm. long, slender, the numerous androgynous spikelets arranged in an elongate, ± interrupted panicle; ♀ scales scarcely concealing the perigynia, acute or cuspidate; perigynia ovate, 2–2.7 mm. long, somewhat biconvex, turgid, nerved dorsally, nerved only at the base ventrally, serrulate on the margin above the middle, spongy at the rounded base, narrowed above into the serrulate beak, the rather long beak bidentulate; $2n = 60$ (Löve & Ritchie, 1966).—Swampy places, 1000–9400 ft.; in scattered localities, San Bernardino V.; cent. Calif. to Alaska, the Atlantic; Eurasia.

12. **C. dispérma** Dewey. Loosely cespitose, the culms and lvs. arising from elongate slender rootstocks; culms lax and spreading, 1.5–6 dm. long, slender, longer than the lvs.; blades 0.7–2 mm. wide, flat; infl. spicate-capitate, 1.5–2.5 cm. long, the spikelets androgynous, the lower separate, the upper congested; ♀ scales broadly ovate, hyaline, narrower and shorter than the perigynia; perigynia thick and ± biconvex, puncticulate, narrowly ovate, 2.5–3 mm. long, nerved dorsally and ventrally, entire-margined, abruptly contracted into a smooth beak, the beak entire, ca. $\frac{1}{10}$ as long as the body; $2n = 70$ (Löve & Löve, 1965).—At high elevs., White Mts.; Sierra Nevada to Yukon, e. to the Atlantic; Asia.

13. **C. douglásii** Boott. Rootstocks rather slender, tough, brown; lvs. and culms in small scattered tufts; culms 0.5–3 dm. long, erect, mostly longer than the lvs.; blades flattened below, folded along the middle above, erect or spreading, 1–2.5 mm. wide; spikelets few–many, densely aggregated into heads 1.5–5 cm. long, usually unisexual and

the plants dioecious, the ♂ heads narrowly oblong, the ♀ oblong to suborbicular; ♀ scales larger than the perigynia; perigynia lance-ovate, 3.5–4 mm. long, buff, ± nerved on both faces, stipitate, the beak nearly as long as body, serrulate, bidentulate, hyaline at apex.— Occasional in dry somewhat alkaline places below 9500 ft.; several Communities; Santa Rosa Mts., San Bernardino and San Gabriel mts., Little San Bernardino, White and Inyo mts., Mt. Pinos; to B.C., Neb. New Mex.

14. **C. eleócharis** Bailey. Rootstocks slender, long-creeping; lvs. and culms in small tufts; culms 2–20 cm. tall, erect, shorter or longer than lvs.; blades 1–1.5 mm. wide, flat or channeled; spikelets few–several, androgynous to ± unisexual, closely aggregated in heads 1–2.0 cm. long, the heads ± unisexual, the ♂ more slender, the ♀ broader; ♀ scales covering the perigynia; perigynia broadly ovate, 2.5–3 mm. long, becoming blackish at maturity, short-stipitate, frequently subtruncate at base, contracted above into a very short beak, upper body and beak serrulate.—Dry places at 11,500 ft.; White Mts.; to Yukon, Iowa, New Mex.

15. **C. exsérta** Mkze. [*C. filifolia* var. *erostrata* Kük.] Densely cespitose with very short rootstocks, the plants tufted, the slender wiry culms 5–25 cm. long, erect; blades filiform, channeled, 0.2–0.5 mm. wide; spikelet solitary, androgynous, narrowly lance-oblong, 7–20 mm. long; ♀ scales ovate to roundish, obtuse, with hyaline margins, shorter than the perigynia; perigynia obovate or roundish, 2.5 mm. long, faintly 2-ribbed, short-puberulent, almost beakless.—Long ago reported from Bear V., San Bernardino Mts.; Sierra Nevada to Ore., Nev.

16. **C. festivélla** Mkze. Plants densely cespitose, the culms and lvs. in rather large clumps from very short rootstocks; culms 3–10 dm. tall, erect, longer than the lvs.; blades 2–6 mm. wide, flat; infl. densely capitate, the head ovoid or suboblong, 1–2.5 cm. long; spikelets gynaecandrous; ♀ scales ovate, dark brown, narrower and shorter than the perigynia; perigynia thin, distended by the ak., ovate, 3.8–5 mm. long, ± lightly nerved dorsally and ventrally, strongly margined to base, serrulate above middle, tapering into a terete-tipped beak which is not serrulate above and ca. ⅓ as long as the body.—Reported as from mts. of s. Calif. and desert; n. to B.C., e. to Rocky Mts.

17. **C. féta** Bailey. [*C. f.* var. *multa* Bailey. *C. straminea* var. *mixta* Bailey.] Densely cespitose, culms 5–10 dm. high, obtusely triangular, smooth or nearly so; lvs. shorter, the blades flat, 2.5–4 mm. wide, sheath green-striate ventrally; infl. rather openly spicate to subcapitate, suboblong to rounded, 2–8 cm. long; spikelets gynaecandrous; ♀ scales ovate, narrower and shorter than the perigynia; perigynia ovate, 3–3.5 mm. long, ± nerved dorsally and ventrally, margined to the base, serrulate above the middle, narrowed above into a flat serrulate beak ½ the length of the body, the margined tip reddish.— Occasional below 6000 ft.; Foothill Wd., Yellow Pine F.; San Juan Capistrano, San Bernardino Mts.; Sierra Nevada and Coast Ranges to B.C.

18. **C. frácta** Mkze. Loosely cespitose; culms 5–12 dm. high, much longer than the lvs.; blades 3–6 mm. wide, flat or the margins a little revolute, sheaths hyaline ventrally and breaking easily, prolonged at the mouth into a fragile, scarious appendage; infl. slender, oblong, spicate or the spikelets ± approximate, 2.5–7.5 cm. long; spikelets gynaecandrous; ♀ scales lanceolate, long-acuminate or aristate, a little shorter and narrower than the perigynia, hyaline; perigynia lance-ovate, 3–4.5 mm. long, plano-convex but somewhat distended by the ak., nerved on each side, narrowly winged, serrulate above, contracted into a serrulate beak ca. ⅓ the length of the body, with terete, tawny, smooth tip.—Frequent moist places, 2600–9000 ft.; mostly Montane Coniferous F.; Cuyamaca Mts. n.; to Wash.

19. **C. globòsa** Boott. Culms and lvs. clustered, arising from somewhat extended rootstocks; culms 1.5–4 dm. long, slender, spreading, longer or shorter than the lvs.; blades flat, or channeled below, 1.5–2.5 mm. wide; ♂ spikelet terminal, solitary, short-stalked, 7–20 mm. long; ♀ spikelets 4–6, the uppermost 2 or 3 approximate, sessile or nearly so, the lowest nearly basal and widely separated on filiform stalks, the spikelets broadly oblong to suborbicular, 5–10 mm. long; ♀ scales ovate, ca. the size of the perigynia; perigynia 4–5 mm. long, globose or nearly so, with prominent stipe and beak,

short-pubescent, 2-keeled and finely several-ribbed, the beak 0.7–1.3 mm. long, bidentate.—In matted clumps on dry slopes below 5000 ft.; several Plant Communities; San Diego and San Bernardino cos. n.; to Humboldt Co. Santa Cruz and Santa Rosa ids.

20. **C. gracílior** Mkze. Loosely cespitose, culms 3–6 dm. high, erect, longer than the lvs.; blades 1–2 mm. wide, flat; infl. narrowly capitate-spicate, 1–2 cm. long, the gynaecandrous spikelets aggregated above but generally somewhat separate below; ♀ scales ovate, ca. as wide as the perigynia, but shorter, with white-hyaline margin; perigynia narrowly ovate, 3.5–4.5 mm. long, plano-convex, coriaceous, nerveless ventrally, few-nerved dorsally, narrowly wing-margined, serrulate above, contracted into a slender beak half as long as the body, smooth and hyaline at the apex.—Below 2000 ft., S. Oak Wd.; Santa Barbara, and Santa Cruz and Santa Rosa ids.; to n. Calif.

21. **C. hássei** Bailey. [*C. aurea* var. *celsa* Bailey.] Habit of *C. aurea*; scales ovate, acute, reddish-brown; perigynia roundish in cross-section, straw-color to white, pulverulent, obscurely nerved.—Wet meadows and banks, 3500–9500 ft.; Montane Coniferous F.; Palomar and Santa Rosa mts. n., Panamint Mts.; to Yukon, e. to Utah, Ariz., s. to L. Calif.

22. **C. haydeniàna** Olney. [*C. nubicola* Mkze.] Densely cespitose; culms 1–4 dm. long, equaling or exceeding the lvs., erect or frequently recurving; blades 1.5–4 mm. wide, flat; infl. capitate, ovoid or roundish, 1–2 cm. long, the spikelets gynaecandrous; ♀ scales ovate, blackish, shorter and narrower than the perigynia; perigynia thin, distended by the ak., lanceolate to broadly ovate, 4.5–6 mm. long, obscurely nerved or nerveless, wing-margined to the base, serrulate upward from below the middle, contracted into a terete-tipped beak, this ca. half as long as the body.—Rare, rocky slopes at 13,500 ft., White Mts.; Sierra Nevada to Ore. and Rocky Mts.

23. **C. heteroneúra** W. Boott. [*C. quadrifida* and vars. *caeca* and *lenis* Bailey.] Densely or loosely cespitose; culms 2.5–10 dm. tall, longer than the lvs.; blades flat, 2–5 mm. wide; spikelets 3–6, approximate, the terminal gynaecandrous or rarely ♂, the lateral ♀, oblong or somewhat narrower, 7–25 mm. long, sessile or short-stalked, the lowest bract leaflike; ♀ scales lance-ovate or wider, narrower and shorter than the perigynia or ca. as long; perigynia oval or obovate to roundish, 2.5–3.5 mm. long, strongly flattened, nerveless or nearly so, with a bidentulate beak ca. $\frac{1}{10}$ as long as the body.—Moist banks 7000–9500 ft.; Montane Coniferous F.; San Jacinto and San Bernardino mts., White Mts.; Sierra Nevada to Nev., Wyo.

24. **C. hoódii** Boott. [*C. h.* var. *nervosa* Bailey.] Densely cespitose, the culms slender, 2.5–8 dm. tall, longer than the lvs.; blades 1.5–3.5 mm. wide; spikelets few to several, androgynous, forming a compact oblong or roundish-ovoid head 1–2 cm. long, the lower spikelets rarely discrete; ♀ scales ca. the size of the perigynia and nearly covering them; perigynia somewhat spreading, ovate-elliptic, 3.5–5 mm. long, greenish or brownish, nerveless or nearly so, serrulate above the middle, with a serrulate bidentate beak ca. $\frac{1}{3}$ the length of the body.—Meadow borders and gravelly slopes, 7200–9100 ft.; Montane Coniferous F.; San Jacinto and San Bernardino mts.; Sierra Nevada to B.C. and Rocky Mts.

25. **C. illòta** Bailey. Cespitose from short rootstocks; culms erect to spreading, 1–3.5 dm. long, longer than the lvs.; blades 1.5–3 mm. wide, flat or channeled; infl. capitate, nearly round, 6–15 mm. long, the spikelets closely aggregated, gynaecandrous; ♀ scales broadly ovate, obtuse, blackish, wider than the perigynia and ca. as long as the body; perigynia lance-ovate, 3 mm. long, plano-convex, not conspicuously nerved, very narrowly margined or only sharp-edged, not serrulate, with a terete-tipped beak $\frac{1}{3}$ the length of the body and smooth or slightly serrulate.—Moist places, 7000–8000 ft.; Montane Coniferous F.; San Bernardino Mts.; Sierra Nevada to B.C. and Rocky Mts.

26. **C. íntegra** Mkze. Loosely cespitose; culms 1.5–5 dm. long, erect to spreading, longer than the lvs.; blades 1–3 mm. wide, flat; infl. spicate-capitate, the spikelets approximate or crowded into an oblong to ovate-oblong head 1–3 cm. long; spikelets gynaecandrous; ♀ scales ovate, nearly as wide as the perigynia but a little shorter; perigynia lanceolate, 2.2–2.8 mm. long, plano-convex, nerveless ventrally, lightly nerved dorsally, narrowly wing-margined, scarcely or not serrulate, with a slender beak $\frac{1}{2}$–$\frac{3}{4}$ the length of

the body.—Edges of meadows and forests at ca. 7000 ft.; Montane Coniferous F.; San Bernardino Mts.; Sierra Nevada to n. Calif., Ore., Nev.

27. **C. jònesii** Bailey. Cespitose, from ± elongate rootstocks; culms 2–6 dm. long, erect to ± spreading, longer than lvs.; blades flat, 1.5–2.5 mm. wide, sheaths ventrally truncate and hyaline at the mouth; infl. densely capitate, the head oblongish to nearly round, 8–18 mm. long, spikelets androgynous; ♀ scales broadly subovate, as wide as the perigynia but a little shorter; perigynia lance-ovate, 3–4 mm. long, nerved dorsally and ventrally, with entire margin, tapering into a smooth or slightly serrulate bidentate beak ca. $\frac{1}{5}$ as long as the body.—Wet banks, 7000–8000 ft.; Montane Coniferous F.; San Bernardino Mts.; Greenhorn Range n. to Wash. and to Rocky Mts.

28. **C. lanuginòsa** Michx. With long rootstocks; culms stout, 3–10 dm. tall, frequently exceeded by the lvs.; blades 1.5–5 mm. wide, flat; spikelets usually 4–6, the upper 2 usually ♂, linear or nearly so, 2–6 cm. long, the lower ♀, distant, sessile or short-stalked, oblong, 1.5–5 cm. long; bracts sheathless or nearly so, the lowest leaflike and exceeding the culm; ♀ scales lanceolate, longer or shorter than the perigynia; perigynia roundish-ovate or obovate, 2.5–3.5 mm. long, hairy, ribbed, contracted to a bidentate beak $\frac{1}{4}-\frac{1}{3}$ as long as the body.—Moist places below 7000 ft.; many Plant Communities; from San Diego Co., Orange Co., Los Angeles Co., San Bernardino Co. (Victorville to Big Meadows), Ventura Co., White Mts.; to n. Calif., B.C., e. to the Atlantic.

29. **C. lemmònii** W. Boott. [*C. abramsii* Mkze.] Rootstocks slender, ± elongated; culms 2–8 dm. tall, slender and somewhat spreading, exceeding lvs.; blades flat, 1.5–4 mm. wide; spikelets 3–5, the terminal ♂, linear, 5–25 mm. long, exceeding the uppermost ♀ spikelet, the lateral spikelets ♀, linear-oblong, 5–20 mm. long, the lower separate and on stalks long-exserted from the bract-sheaths; ♀ scales broadly ovate, a little narrower and shorter than the narrowly ovate perigynia, these 3–4 mm. long, 2-ribbed and finely nerved, narrowed into a bidentulate, hyaline-tipped beak $\frac{1}{5}$ the length of the perigynium-body.—Meadows at 6500–9500 ft.; Montane Coniferous F.; San Jacinto and San Bernardino mts.; Greenhorn Range n. to n. Calif.

30. **C. mariposàna** Bailey. Habit of *C. abrupta*, the spikelets gynaecandrous, 4–12, the lower slightly separate, the infl. 2–3.5 cm. long; ♀ scales ovate, much narrower and shorter than the perigynia; perigynia ovate, 3.5–5 mm. long, plano-convex, membranaceous, conspicuously nerved on both surfaces, narrowly winged, serrulate above the middle, tapering into a terete-tipped smooth beak ca. $\frac{1}{4}$ the length of the body, the beak scarcely hyaline at the tip.—Drier parts of meadows and on slopes, 6000–10,800 ft.; San Jacinto and San Bernardino mts.; Sierra Nevada to n. Calif., w. Nev.

31. **C. micróptera** Mkze. Densely cespitose; culms 3–10 dm. tall, erect, much exceeding lvs.; blades 2–4.5 mm. wide, flat; infl. capitate, rounded or ovoid, 1–2 cm. long, the spikelets gynaecandrous; ♀ scales lance-ovate, brownish, narrower and a little shorter than the perigynia; perigynia thin, distended by the ak., lanceolate to lance-ovate, 3.5–4.5 mm. long, lightly nerved on both faces, minutely margined to the base, serrulate above the middle, tapering into a terete-tipped beak, the beak not serrulate above, $\frac{1}{2}-\frac{1}{3}$ the length of the body.—At high altitudes, White Mts.; Sierra Nevada to B.C., Rocky Mts.

32. **C. molésta** Mkze. Cespitose, the culms 3–10 dm. high, roughened above, brownish black at the base; blades 1–3 dm. long, 2–3 mm. wide; spikes 4–8, gynaecandrous, in a head 2–3 cm. long, the spikelets subglobose, 6–9 mm. long; scales ovate, yellowish brown with 3-nerved green center and hyaline margins; perigynia ovate, 4.5 mm. long, rounded at base, ± nerved, tapering abruptly into a beak almost half the length of the body; beak flat, serrulate, shallowly bidentate, brownish-tipped.—Santa Barbara, *Pollard*; natur. from e. U.S.

33. **C. montereyénsis** Mkze. Loosely to densely cespitose; culms 8–10 dm. long, erect or rather lax, longer than the lvs.; blades flat, 2.5–3 mm. wide; infl. capitate, 1.5–2.5 cm. long, ovoid or oblongish, the spikelets closely aggregated, gynaecandrous; ♀ scales lance-ovate, nearly as long as the perigynia-bodies but somewhat narrower, cuspidate or short-awned; perigynia ovate, 3.25 mm. long, plano-convex, membranaceous, nerved dorsally and ventrally, narrowly wing-margined, serrulate above the middle, contracted

into a terete-tipped beak half as long as the body and slightly hyaline-tipped.—Occasional, open and brushy places below 2000 ft.; Santa Cruz Id.; Santa Barbara Co. to Sonoma Co.

34. **C. multicáulis** Bailey. Loosely to densely cespitose; culms terete or obtusely triangular, 2-6 dm. tall, green, much longer than the short lvs.; blades 1-1.5 mm. wide, rough above; spikelet solitary, androgynous, elongate, the ♂ part linear and 1-2.5 cm. long, the ♀ part consisting of 1-6 perigynia; ♂ scales oblong-obovate, with broad white-hyaline margin, the ♀ scales bractlike or leaflike below, the lower prolonged into a blade 1-4 cm. long above a widened hyaline-margined base, the upper lanceolate, cuspidate or awned; perigynia 5-7 mm. long, oblong-ovate, triangular, glabrous, 2-ribbed and obscurely nerved, minutely beaked.—Dry slopes at 4000-7200 ft.; largely Yellow Pine F.; San Diego Co. n.; to Ore. and Nev.

35. **C. multicostàta** Mkze. [*C. adusta* var. *congesta* W. Boott. *C. pachycarpa* Mkze.] Densely cespitose, the culms 3-9 dm. tall, exceeding the lvs.; blades flat, 2.5-6 mm. wide; infl. usually compact, the head oblong to roundish, 1.5-4 cm. long, the spikelets gynaecandrous; ♀ scales ovate, with hyaline margins, a little shorter and narrower than the perigynia; perigynia ovate, 3.5-5.5 mm. long, nerved dorsally, finely nerved or smooth ventrally, wing-margined, serrulate from below the middle, abruptly contracted into a beak $\frac{1}{4}$-$\frac{1}{3}$ the length of the body, the beak broad and flat, winged to the tip or nearly so.—Rare, meadows, 5000-11,000 ft., Montane Coniferous F.; San Jacinto and San Bernardino mts.; Sierra Nevada to Ore., Ida., Nev.

36. **C. nebrascénsis** Dewey. [*C. jacintoensis* Parish.] Plate 92, Fig. C. Cespitose, the clustered culms and lvs. from elongate rootstocks; culms 2.5-12 dm. long, shorter or longer than the lvs.; blades flat or somewhat channeled below, 3-8 mm. wide, frequently ± glaucous; ♂ spikelets 1-2, 1.5-4 cm. long, the ♀ 2-5, approximate to separate, 1.5-6 cm. long, the lowest bract leaflike, longer or shorter than the infl.; ♀ scales lanceolate, longer or shorter than the perigynia and narrower; perigynia oblong-obovate, 3-3.5 mm. long, granular and coriaceous, many-ribbed, sessile or subsessile, abruptly beaked, the beak 0.5-1 mm. long, bidentate, the teeth somewhat ciliate within.—Wet places below 9500 ft.; mostly Montane Coniferous F.; San Diego Co. n., Panamint Mts.; to B.C., Kans.

37. **C. occidentàlis** Bailey. Loosely cespitose; culms slender, erect or spreading, 2.5-7 dm. high, longer than the lvs.; blades flat, 1.5-2.5 mm. wide; spikelets androgynous, several to many, crowded into an elongate-oblong head 1.5-3 cm. long, the lower spikelets ± separate; ♀ scales ca. the size of the perigynia; perigynia elliptic-oblong, 2.5-3.5 mm. long, slightly nerved on both faces, serrulate only at the base of the beak, the beak serrulate, bidentate, ca. $\frac{1}{3}$ the length of the body.—Dryish places at 6300 ft.; Montane Coniferous F.; San Bernardino Mts.; Ariz., New Mex. to Wyo.

38. **C. ormántha** (Fern.) Mkze. [*C. echinata* var. *o.* Fern.] Cespitose; culms 1.5-4 dm. tall, slender, rather lax, longer than the lvs.; blades 1.5-2 mm. wide, flat or channeled; infl. spicate, 2-6 cm. long, the spikelets several, separate, the terminal gynaecandrous with a conspicuous elongate ♂ base, the lower either ♀ or gynaecandrous; ♀ scales ovate, obtuse, ca. as long as the bodies of the perigynia; perigynia lanceolate, 3.5-4 mm. long, many-nerved dorsally and ventrally, smooth on the margin except at the base of the beak, this serrulate and over half the length of the body.—Wet meadows at 5000-7500 ft.; Montane Coniferous F.; San Bernardino Mts.; N. Coast Ranges and Sierra Nevada to Wash.

39. **C. pánsa** Bailey. [*C. arenicola* ssp. *p.* Koyama & Calder.] Plate 92, Fig. D. Rhizomes long-creeping; culms 1.5-3 dm. tall, stiff, suberect or curving, longer than the lvs.; blades 1-4 mm. wide, flattish or canaliculate; spikelets few to several, androgynous, clustered in rather a dense ovoid head 1.5-2.5 cm. long; ♀ scales larger than the perigynia; perigynia ovate-lanceolate or elliptic, 3.5-4.5 mm. long, shining and dark brown at maturity, scarcely nerved ventrally, many-nerved dorsally, serrulate above the middle, stipitate, with a serrulate, bidentulate beak scarcely hyaline at the apex and ca. $\frac{1}{3}$ the length of the body.—Coastal Strand, Santa Rosa and San Miguel ids.; San Luis Obispo Co. to Wash.

40. **C. praegrácilis** W. Boott. [*C. douglasii* var. *brunnea* Olney. *C. usta* Bailey.] Rootstocks long-creeping; culms 2–7 dm. high, erect or laxly spreading, usually longer than the lvs.; blades 1.5–3 mm. wide, flat or canaliculate; spikelets few to many, androgynous or ± unisexual, closely aggregated in an oblong or ovoid head 1–5 cm. long, or the lower spikelets ± separate; ♀ scales mostly covering the perigynia; perigynia lance-ovate to ovate, 3–4 mm. long, dull and dark brown at maturity, nerveless ventrally, finely nerved dorsally, serrulate on the margin above the middle, the beak serrulate, bidentulate, hyaline at the apex, half as long as the body.—Meadows and wet places below 7000 ft.; many Plant Communities; desert and cismontane; Santa Catalina, Santa Cruz and Santa Rosa ids.; to Yukon, Mich., S. Am.

41. **C. róssii** Boott. Loosely to densely cespitose; culms 5–30 cm. long, mostly equaling or exceeding lvs.; blades 1–2.5 mm. wide, flat or folded; ♂ spikelet terminal, sessile or short-stalked, 3–15 mm. long; ♀ spikelets 3–5, the uppermost 1 or 2 approximate or a little apart, sessile or short-stalked, the others widely separated, nearly basal, the spikelets 3–5 mm. long; ♀ scales ovate, acute to acuminate or awned, wider but shorter than the perigynia; perigynia 3–4.5 mm. long, pubescent, the body nearly round, 2-keeled, with a bidentate beak 0.7–1.5 mm. long.—Tufts under pines and oaks, 7000–11,400 ft.; Montane Coniferous F.; San Jacinto, San Bernardino and San Gabriel mts.; White Mts.; to Yukon, Mich., Colo.

42. **C. rostràta** Stokes. [*C. utriculata* var. *globosa* Olney.] Loosely or scarcely cespitose, the culms erect, 3–12 dm. tall, shorter than the uppermost lvs.; blades flat or ± channeled at base, 3–12 mm. wide; ♂ spikelets 2–4, linear, 1–6 cm. long; ♀ spikelets 2–5, separate, erect, oblong, 1–15 cm. long; perigynia spreading-squarrose at maturity, the bracts leaflike, longer than the infl.; ♀ scales lanceolate to ovate, narrower than the perigynia, and longer or shorter; perigynia ± ovate, 3.5–8 mm. long, ± inflated, shining, strongly nerved, contracted into a beak 1–2 mm. long, the teeth erect, 0.5–0.8 mm. long; aks. continuous with the indurate style; $2n = 76$ (Löve & Löve, 1956).—Wet meadows, 6500–7800 ft.; Montane Coniferous F.; San Bernardino Mts.; Sierra Nevada, Coast Ranges n. to Alaska, to Atlantic, Greenland; Eurasia.

43. **C. sartwellìàna** Olney. [*C. yosemitana* Bailey.] Densely cespitose, frequently forming large clumps; culms 3–9 dm. long, much exceeding the lvs.; blades flat, 3–7 mm. wide, softly pubescent; spikelets 4 or 5, the terminal ♂ or with a few perigynia, linear, 1–3 cm. long, the ♀ spikelets thicker, 1–4 cm. long, approximate or separate, the bracts sheathless, the lowest leaflike and ca. equaling the infl.; ♀ scales ovate or lance-ovate, hairy, longer or shorter than the perigynia; perigynia ± obovate, 2.5–3.5 mm. long, pilose, 2-ribbed and obscurely nerved, with a bidentate beak $\frac{1}{3}-\frac{1}{2}$ the length of the body and hyaline-tipped.—Moist banks, 5000–7500 ft.; Yellow Pine F.; San Jacinto Mts.; Greenhorn Range to cent. Sierra Nevada.

44. **C. schóttii** Dewey. Forming large tussocks, the culms a meter or taller, sharply triangular, longer than the lvs.; blades flat, 6–12 mm. wide, the lf.-sheaths breaking and becoming filamentose; ♂ spikelets frequently 3, linear, 8–14 cm. long; ♀ spikelets 3, usually androgynous, separate, linear, 5–20 cm. long, the lowest bract leaflike, usually longer than the infl.; ♀ scales linear-lanceolate to oblong, narrower but usually longer than the perigynia; perigynia oval to obovate, 3–3.5 mm. long, few-nerved both dorsally and ventrally, short-beaked, the beak 0.25 mm. long, ± emarginate.—Wet places below 2500 ft.; Coastal Sage Scrub, S. Oak Wd.; San Diego Co. to Santa Clara Co.

45. **C. sénta** Boott. [*C. auriculata* Bailey. *C. austromontana* Parish. *C. bishallii* Clarke.] Loosely cespitose and stoloniferous; culms 3–10 dm. tall, longer than the lvs.; blades flat, 3–5 mm. wide, the leaf-sheaths breaking and becoming filamentose; ♂ spikelets 2–3, linear, 3–4.5 cm. long, often with a few perigynia near the base, the ♀ spikelets 1 or 2, rarely androgynous, 2.5–5 cm. long, the lowest bract leaflike; ♀ scales ± oblong, shorter and narrower than the perigynia; perigynia broadly ovate to obovate, 3–3.5 mm. long, granular, few-nerved on both faces, abruptly beaked, the beak 0.25 mm. long, entire.—Swampy and moist places, below 8500 ft.; several Plant Communities; cismontane and montane s. Calif.; Santa Cruz Id.; to n. Calif., Ariz., L. Calif.

46. **C. serrátodens** W. Boott. [*C. bifida* Boott., not Roth. *C. aequa* Clarke.] Loosely cespitose; culms 3–13 dm. tall, longer than the lvs.; blades flat, 1.5–4 mm. wide, glaucous-green; spikelets 3–6, the terminal ♂ or with some perigynia, narrow, 1.5–3 cm. long, the ♀ oblong or broader, 6–18 mm. long, ± separate, the lowest rather distant, sessile or short-stalked, the lowest bract leaflike; ♀ scales ovate, somewhat shorter and narrower than the perigynia; perigynia ± ovate, 3–5 mm. long, ca. 10-nerved, puncticulate, somewhat contracted into a roughened bidentate beak 0.5–1 mm. long, with hispidulous teeth.—Moist places, frequently on serpentine, to 6000 ft.; Pinyon-Juniper Wd., Montane Coniferous F.; Ventura and Kern cos.; to Ore.

47. **C. simulàta** Mkze. Loosely cespitose or in scattered tufts from long-creeping rootstocks; culms 3–5 dm. tall, erect or spreading, usually longer than the lvs.; blades 2–4 mm. wide, flat or canaliculate; spikelets few to many, androgynous or ± unisexual, in a narrowly oblong to broadly ovoid head 1–2.5 cm. long, the heads tending to be unisexual, the ♂ more slender, the ♀ broader; ♀ scales covering the perigynia; perigynia broadly ovate, 1.7–2.2 mm. long, brown and shining at maturity, few-nerved dorsally, almost nerveless ventrally, short-stipitate and frequently subtruncate at base, with a short terminal beak less than ⅓ the length of the body, the upper body and beak serrulate, hyaline at apex.—Moist places, Chaparral, Rancheria Creek, Piute Mountains, Kern Co.; to Wash., Rocky Mts.

48. **C. spíssa** Bailey. Loosely cespitose; culms 1–2 m. tall, longer than the lvs.; blades flat above, 7–14 mm. wide; ♂ spikelets 3 or 4, linear, 4–10 cm. long, the ♀ spikelets 3–7, androgynous, approximate or the lower somewhat separate, 6–14 cm. long, the bracts leaflike, the longest lower than the infl.; ♀ scales lanceolate to lance-ovate, narrower than but longer than the perigynia; perigynia broadly obovate, 3–4.5 mm. long, smooth and only obscurely nerved, abruptly beaked, the beak 0.5 mm. long, frequently bent, emarginate; style indurate, persistent, continuous with the ak.—Stream banks below 2000 ft.; Coastal Sage Scrub, Chaparral, Foothill Wd.; L. Calif. to San Luis Obispo Co.

49. **C. subbracteàta** Mkze. Loosely cespitose; culms 3–12 dm. tall, longer than the lvs.; blades 2.5–4 mm. wide, flat; infl. capitate, ovoid or roundish, 1.5–2.5 cm. long, the spikelets gynaecandrous; ♀ scales ovate, narrower and shorter than the perigynia, with white-hyaline margins; perigynia narrowly ovate, 3.5–4.5 mm. long, plano-convex, nerveless ventrally, finely nerved dorsally, narrowly margined, serrulate at apex.—Moist open places, Chaparral, etc.; Santa Rosa Id.; Santa Barbara to Humboldt Co.

50. **C. subfúsca** W. Boott. [*C. stenoptera* Mkze.] Densely cespitose; culms 2–6 dm. high, erect, smooth; blades 2–3.5 mm. wide, flat; infl. elongate-capitate, 1–3.5 cm. long, the spikelets gynaecandrous; ♀ scales ovate, much narrower and shorter than the perigynia; perigynia appressed in the spikelets, narrowly to broadly ovate, 3–3.5 mm. long, plano-convex, nearly nerveless ventrally, lightly nerved dorsally, the margin winged, serrulate, with a serrulate beak half as long as the body, the beak flattened and somewhat serrulate at the tip.—Dry meadows and forest borders, 4000–11,500 ft.; Montane Coniferous F., S. Oak Wd.; mts. San Diego Co. to Kern Co., Inyo Co.; to B.C., Nev., Ariz.

51. **C. subnìgricans** Stacey. Loosely cespitose; culms to 2 dm. tall, usually much longer than the lvs.; blades erect, subfiliform, folded along the middle; spikelet solitary, androgynous, lanceolate, 8–12 mm. long; ♀ scales 1-nerved, mostly covering the perigynia even at maturity; perigynia ascending, subappressed, little inflated, but frequently distended by the ak., lanceolate to lance-ovate, 3.5–4 mm. long, nerveless, brownish, stipitate, tapering into a short beak, this hyaline and obliquely cut.—Moist spots at high elevs.; White Mts.; to Ore., Utah, Ida.

52. **C. teneraefórmis** Mkze. Cespitose; culms 3–4.5 dm. tall, erect, slender, longer than the lvs.; blades flat, 1–2.5 mm. wide; infl. 1.5–2.5 cm. long, slender, spicate with the gynaecandrous spikelets discrete or approximate; ♀ scales ovate, narrower and shorter than the perigynia; perigynia loosely appressed or somewhat spreading in the spikelets, 2.7–3.2 mm. long, plano-convex, nerveless ventrally, nerved dorsally, the margin winged, serrulate above the middle, tapering above into a slender beak ca. ½ the length of the

body, the tip of the beak terete and nearly smooth.—Meadows and wet forests, 4000–9000 ft.; largely Montane Coniferous F.; mts. from San Diego Co. to Ore., Nev.

53. **C. texénsis** (Torr.) Bailey. [*C. rosea* var. *t.* Torr.] Cespitose, forming a loose turf; culms erect, slender, 1.5–3 dm. tall, longer than the lvs.; blades flat, 0.8–1.5 mm. wide; spikelets androgynous, few to several in a narrow head or interrupted spike 1–3 cm. long; ♀ scales ca. as long as the perigynium body, early deciduous; perigynia spreading to reflexed in age, lanceolate to subovate, 3 mm. long, nerveless, the margins a little raised, smooth, beak smooth, bidentate, ca. $\frac{1}{3}$ as long as the body.—Reported from gardens and waste ground in Los Angeles and Santa Barbara cos.; native in cent. and e. U.S.

54. **C. triquètra** Boott. [*C. monticola* Dewey.] Densely cespitose; culms 3–6 dm. tall, longer than the lvs.; blades 2.5–6 mm. wide, flat but with revolute margins; spikelets usually 3 or 4, the terminal ♂, 1–3 cm. long, the lateral ♀ or with a few terminal ♂ fls., the uppermost approximate, the others more remote, 1–4.5 cm. long; ♀ scales broadly ovate, ca. as wide as the perigynia but shorter; perigynia ovate to obovate, 4–4.5 mm. long, greenish, pubescent, obscurely nerved, very short-beaked, the beak bidentulate and less than $\frac{1}{10}$ as long as the body.—Dry places below 3000 ft.; Coastal Sage Scrub and Chaparral; L. Calif. to San Luis Obispo Co.; Santa Catalina, Santa Cruz ids.

55. **C. tumulícola** Mkze. Rootstocks somewhat elongate; culms erect or spreading, 2–8 dm. long, generally exceeding the lvs.; blades 1.5–2.5 mm. wide, flat, or folded along the middle; spikelets several to many, the lower ± separate, the upper approximate, the spike slender, 2–5 cm. long; ♀ scales ca. as large as perigynia and usually concealing them; perigynia elliptic to ovate, 3.5–5 mm. long, nerved dorsally and ventrally or nerveless ventrally, serrulate on the margin above the middle, stipitate, and with a serrulate beak $\frac{1}{3}$–$\frac{1}{2}$ the length of the body and bidentate.—Below 2000 ft.; Coastal Sage Scrub, Chaparral; San Clemente Id., Santa Rosa Id., Santa Cruz Id.; to Wash.

56. **C. vernácula** Bailey. Loosely cespitose; culms to 3 dm. tall, erect, slender, usually longer than the lvs.; blades flat or nearly so, 2–4 mm. wide; spikelets many, androgynous, crowded into a roundish or ovoid head ca. 1 cm. long; ♀ scales ca. as large as perigynia; perigynia flattened, lance-ovate, 3.5–4.5 mm. long, membranous, finely or obscurely nerved, stipitate, with a conspicuous smooth beak, the beak obliquely cut, becoming bidentulate.—Moist meadowy places at high elevs.; White Mts.; to Wash., Rocky Mts.

2. *Clàdium* R. Br. SAW-GRASS

Perennial with hollow leafy culms. Infl. a compound umbel of numerous spikelets. Spikelets small, few-fld., the lower scales empty or imperfect. Scales spirally arranged. Perfect fl. 1. Bristles none. Stamens 2. Style 2–3-fid. A genus of ca. 40 spp., trop. and temp. regions, especially in Australasia. (Greek, *klados*, branch, referring to the branched infl.)

1. **C. califórnicum** (Wats.) O'Neill in Tides. & Kittell. [*C. mariscus* var. *c.* Wats. *Mariscus c.* Fern.] Culms stout, 1–2 m. tall; lvs. 1–2 m. long, flat, 7–10 mm. wide, with serrate margins; umbels axillary, compound, the spikelets in glomerules of 3–6, oblong, acute, 3 mm. long, reddish brown, bearing 1 perfect floret and several ♂; ak. ovoid, smooth, ca. 2 mm. long, without a tubercle; $n=18$ (Pfeiffer, 1942).—Uncommon, Freshwater Marsh, Alkali Sink; deserts from Riverside to Inyo cos., cismontane San Bernardino Co. to San Luis Obispo Co.; Nev., Ariz., Mex.

3. *Cypèrus* L. UMBRELLA-SEDGE

Annual or perennial herbs. Culms in ours simple, triangular, leafy at base, striate. Involucral lvs. 1 to several, much exceeding the infl. Infl. umbellate or capitate, the rays commonly bearing divaricate clusters of spikelets. Spikelets flat or subterete, many-fld., falling away from the head, or persistent and then the scales deciduous. Rachis straight, offset or zigzag, unwinged or frequently bearing a pair of wings at each node, these being the decurrent bases of the next distal scale. Scales 2-ranked, keeled, all fertile or the lower ones empty. Fls. perfect. Bristles none. Stamens 1–3. Styles 2–3-fid.; the style-base deciduous from the summit of the ak. Ak. triangular or lenticular, naked or clasped by the rachis-wings. A genus of ca. 600 spp., of trop. and temp. regions. (Greek, *cypeiros*, the classical name.)

A. Spikelets elongated, many-fld., in ± loose umbels or heads.
  1. Style 2-fid; ak. lenticular.
    2. Ak. laterally flattened.
      3. Spikelets 2–2.5 mm. wide .... 10. *C. niger*
      3. Spikelets ca. 5 mm. wide .... 14. *C. unioloides*
    2. Ak. dorsally flattened .... 9. *C. laevigatus*
  1. Style 3-fid; aks. triangular.
    4. Spikelets persistent on the spike; scales deciduous.
      5. Rachis not winged.
        6. Perennial with short rhizomes.
          7. Spikelets 10–20 mm. long .... 6. *C. eragrostis*
          7. Spikelets 4–7 mm. long .... 2. *C. alternifolius*
        6. Annual, with fibrous roots.
          8. Scales acuminate or awned, the tip recurved.
            9. Low plants with celery-scented herbage; scales awned, several-nerved
                3. *C. aristatus*
            9. Taller plants without a marked odor; scales sharply acute, 3-nerved
                1. *C. acuminatus*
          8. Scales obtuse .... 5. *C. difformis*
      5. Rachis winged with a pair of inner hyaline appendages at each node.
        10. Plants perennial; aks. obtuse or mucronulate.
          11. Stoloniferous with tubers; lvs. about as long as culms.
            12. Scales dull brown or yellow-brown, scarcely keeled; stolons weak
                8. *C. esculentus*
            12. Scales shining red-brown, keeled; stolons ligneous .... 13. *C. rotundus*
          11. Short-rhizomatous; lvs. much shorter than culms .... 12. *C. parishii*
        10. Plants robust annuals; aks. distinctly mucronate .... 7. *C. erythrorhizos*
    4. Spikelets disarticulating above the basal pair of scales, or breaking up into 1-fruited joints; scales persistent.
      13. Spikelets disarticulating above the sterile basal pair of scales; rachis-wings thin and hyaline .... 12. *C. parishii*
      13. Spikelets breaking up into 1-fruited joints; rachis-wings firm and brown
          11. *C. odoratus*
A. Spikelets minute, with 1 perfect fl., in dense glomerate heads .... 4. *C. brevifolius*

    1. **C. acuminàtus** Torr. & Hook. Annual, with fibrous roots and slender cespitose culms 7–40 cm. tall; lvs. 2–4 on a culm, flat, ca. as long as culm, 0.5–2.5 mm. wide, with scaberulous margins; involucral lvs. 3–4, unequal, to 18 cm. long; infl. umbellate with 2–5 unequal rays, bearing globose heads of spikelets; spikelets ovate-oblong, obtuse, compressed, 4–10 mm. long; rachis straight, unwinged; scales ovate, acuminate, with a recurved tip, 3-nerved, pale green to light brown, reticulate on the surface; stamen 1; ak. triangular, oblong, 0.6–0.9 mm. long, short-stipitate, short-mucronate.—Uncommon at low elevs., in wet ground, from Ventura Co. n.; to Wash., e. states. June–Oct.

    2. **C. alternifòlius** L. UMBRELLA-PLANT. Cespitose perennial with short thick rhizomes and tall stout angled culms to 1.5 m. tall, clothed below with leafless brown sheaths; umbel terminal in an invol. of many long firm spreading leaflike bracts 1–3 dm. long; rays as many as the bracts, 2–10 cm. long; spikelets oblong, 5–10 mm. long, pale, 10–30-fld.; glumes mostly deciduous, 1.6–2 mm. long; stamens 3; $2n = 32$ (Tanaka, 1937).—Common in cult. and occasionally natur. as in Ventura and Santa Barbara cos.; Santa Cruz Id.; native of Afr. Most of year.

    3. **C. aristàtus** Rottb. [*C. inflexus* Muhl. *C. a.* var. *i.* Kükenth.] Small annual with fibrous roots and celery-scented foliage; culms cespitose, slender, 1–20 cm. long; lvs. 2–3 on a culm, flat, longer or shorter than culm, 0.5–3 mm. wide; involucral lvs. much longer than infl.; infl. umbellate, the rays lacking or to ca. 2 cm. long, bearing capitate clusters

of spikelets; these linear-oblong, 4–10 mm. long, compressed; rachis straight, deciduous after the scales; scales lanceolate, awned, the awn green to light brown, deciduous; stamen 1; style 3-fid; ak. triangular, obtuse, mucronulate, 0.7–1 mm. long, brown, puncticulate.—Wet ground below 8500 ft.; many Plant Communities; almost throughout Calif.; to B.C., Atlantic Coast, S. Am.; Asia, Afr. June–Nov.

4. **C. brevifòlius** (Rottb.) Hassk. [*Kyllinga b.* Rottb.] Perennial, with rhizomes; culms leafy, 2.5–7.5 dm. tall; involucral lvs. 2–4, 1–4 cm. long; head nearly globose, 5–7 mm. long; spikelets 3 mm. long; empty outer 2 scales ovate, acuminate, strongly several-nerved, enclosing the fertile scale and ak. at maturity; ak. lenticular, much flattened, obtuse; $2n=120$ (Tanaka, 1941).—Introd. weed in San Diego, Pasadena, etc.; native Am. trop.; Old World. July–Sept.

5. **C. diffórmis** L. [*C. lateriflorus* Torr.] Fibrous-rooted annual; culms smooth, cespitose, 1.5–5 dm. tall; lvs. 2–4 on a culm, 1–4 mm. wide, ca. as long as culms, minutely scaberulous on upper edges; involucral lvs. unequal, 2–3; infl. umbellate, the rays bearing globose heads of linear obtuse subcompressed spikelets 4–8 mm. long; rachis straight, unwinged; scales roundish, obtuse, 0.6–0.8 mm. long, membranous, green with brown sides, readily deciduous; stamen 1; ak. triangular, obovate, minutely mucronulate, 0.5 mm. long, pale greenish-brown.—Occasional weed in low wet spots, San Diego; common in cent. Calif.; native of Asia. July–Nov.

6. **C. eragróstis** Lam. [*C. vegetus* Willd. *C. serrulatus* Wats.] Perennial with short thick rhizomes; stems stout, erect, 2–8 dm. tall, triangular, smooth; lvs. basal, 6–9, ca. as long as culm, ± scaberulous; involucral lvs. 5–8, unequal, to 5 dm. long, scaberulous on margins and midrib; infl. a compact globose head, with rays to 1 dm. long; spikelets flat, 10–20 mm. long, 3–3.5 mm. wide, rachis straight; scales ovate, acute, keeled, 3-nerved, falling off with the ak. enclosed; stamen 1; style 3-fid.; aks. sharply triangular, obovoid, stipitate, mucronate, brown, puncticulate.—Moist places below 5000 ft.; Coastal Sage Scrub, Chaparral, etc., much of cismontane s. Calif.; to Ore., Mex., temp. S. Am. May–Nov.

7. **C. erythrorhìzos** Muhl. [*C. occidentalis* Torr.] Cespitose annual; culms smooth, bluntly triangular, 1–10 dm. tall; lvs. several, flat, ca. as long as culm, 2–10 mm. wide, scaberulous on margins and midrib; involucral lvs. 4–10, unequal, with scaberulous margins; infl. a compound or sometimes simple umbel; rays 1–30 cm. long, with numerous divaricate linear spikelets 3–10 mm. long, 1–1.5 mm. wide; rachis winged at each node; scales oblong-obovate, mucronulate, keeled, with green midrib and golden-brown sides; stamens 3; style 3-fid; ak. sharply triangular, sessile, mucronate, grayish white.—Marshy places below 5000 ft.; several Communities; Imperial Co. and cismontane s. Calif.; to Wash., Atlantic Coast. July–Oct.

8. **C. esculéntus** L. [*C. e.* var. *hermannii* Britton.] NUT-GRASS. Perennial with scaly stolons ending in edible tubers; culms stout, smooth, 1.5–5 dm. tall; lvs. many, flat, ca. as long as culms, 3–10 mm. wide, smooth; involucral lvs. 2–6; infl. umbellate, with 5–10 rays 0–12 cm. long and numerous remotely divaricate spikelets; spikelets linear, 6–30 mm. long, 2–3 mm. wide, flat; rachis winged with narrow hyaline persistent members; scales ovate, obscurely mucronulate, 7–9-nerved, light brown; stamens 3; style 3-fid; ak. triangular, oblong, obtuse, 1.3–2 mm. long, puncticulate; $n=$ ca. 54 (Hicks, 1929).—Noxious weed of gardens and low places, at low elevs.; s. Calif.; to Alaska, Atlantic Coast, trop. Am.; Old World. June–Oct.

9. **C. laevigàtus** L. Perennial with wiry rootstocks; culms smooth, 1.5–5 dm. tall; lvs. 2–5 on a culm, shorter than culm, sometimes reduced to basal sheaths; involucral lvs. 1–2, very unequal; infl. a capitate cluster of 1 to several spikelets appearing lateral; spikelets linear-oblong, compressed, 6–12 mm. long, 2–3 mm. wide; scales ovate, obtuse, ca. 1.5 mm. long and 1 mm. wide, whitish on base and keel and chocolate-brown on sides; stamens 2; style 2-fid; ak. lenticular, elliptic, compressed parallel to rachis, stipitate.—Moist places at low elevs.; Alkali Sink, Creosote Bush Scrub, Coastal Sage Scrub, etc.; desert and cismontane s. Calif. to Kern Co.; trop. Am.; Old World trop. July–Dec.

10. **C. nìger** R. & P. var. **capitàtus** (Britton) O'Neill. [*C. diandrus* var. *c.* Britton.

C. melanostachyus HBK.] Perennial with short rhizomes and cespitose culms; culms slender, smooth, erect, 1.5–5 dm. tall; lvs. 2 on a culm, narrow, shorter than culm, smooth, with reddish-brown sheaths; involucral lvs. 3, linear, sometimes very long; infl. a capitate cluster of spikelets; spikelets lanceolate, acutish, 5–12 mm. long, 2–2.5 mm. wide, with zigzag rachis; scales ovate, acutish to obtuse, keeled, 3–5-nerved, ochre-brown, deciduous; stamens 2; ak. lenticular, 1–1.3 mm. long, oblong, short-stipitate, brown or gray, puncticulate.—Wet places, below 5000 ft.; Chaparral, etc.; San Diego Co. n., Inyo Co.; to cent. Calif.; Tex., Mex. July–Nov.

Var. **castàneus** (Pursh) Kükenth. [*C. n.* var. *rivularis* V. Grant.] Annual; infl. umbellate; scales deep reddish-brown.—In Orange and San Bernardino cos.; to n. Calif.; Atlantic Coast.

11. **C. odoràtus** L. [*C. ferax* Rich. *C. californicus* Wats.] Plate 92, Fig. E. Annual, with 1–several culms 2–5 dm. tall, stout, smooth; lvs. several to many, equal to or longer than culms, the blades channeled above, 3–8 mm. wide, with stout prickles on the margin; involucral lvs. several, 5–30 cm. long, very unequal; infl. simulating a compound spike, the rays arising from the top of the culm and umbellately spreading, branching into secondary rays each in the axil of an involucel-lf.; spikelets linear, subterete, 10–25 mm. long, ca. 2 mm. wide, disarticulating above the basal pair of scales when ripe; fl. scales ovate, obtuse, strongly 7–9-nerved, yellowish-brown, sheathing the rachis and almost enclosed by the winglike margins of the rachis joint; ak. triangular, obovoid, 1–1.2 mm. long, brown.—Rather common in moist sandy places at low elevs.; several Plant Communities; Imperial V., w. Mojave Desert; cismontane s. Calif. to Cent. V.; transcontinental; trop. Am.; Asia. July–Oct.

12. **C. paríshii** Britton. Fibrous-rooted perennial; culms 1–2.5 dm. tall; lvs. several, much shorter than the culm, 3–5 mm. wide, minutely scaberulous; involucral lvs. 3–4; infl. umbellate, the rays to 5 cm. long; spikelets linear, acute, 12–20 mm. long, ca. 2 mm. wide; rachis winged with a pair of hyaline members at each node; scales ovate, acute, 2–3 mm. long, strongly several-nerved, with green keel and reddish-brown sides; stamens 3; style trifid; ak. triangular, obovoid-ellipsoid, 1–1.2 mm. long, mucronulate, almost black.—Coastal Sage Scrub, San Bernardino and Riverside cos.; to New Mex.

13. **C. rotúndus** L. PURPLE NUT-GRASS. Like *C. esculentus*, but scales shining reddish-brown; $n=54$ (Tanaka, 1937).—Noxious garden weed in s. Calif. and San Joaquin V.; se. U.S.; trop. Am.; Old World. July–Nov.

14. **C. unioloìdes** R. Br. [*C. bromoides* auth., not Link.] Perennial with slender rootstocks and slender culms 5–8 dm. high; lvs. mostly shorter, 2–4 mm. wide; involucral lvs. 3–5, one much longer than the infl.; infl. a capitate or loose cluster of spikelets; spikelets lance-ovate, flat, acute, 12–18 mm. long, many-fld.; scales yellow, acute; ak. obovate, dark.—Reported long ago from a swamp near Los Angeles; trop Am.; Old World.

4. *Eleócharis* R. Br. SPIKE-RUSH

Annual or perennial herbs with rhizomes, stolons or fibrous roots. Culms simple, terete or subterete, usually striate. Lvs. reduced to basal lf.-sheaths, sometimes with an apiculate tip. Spikelets solitary, terminal, erect, several- to many-fld., not subtended by an involucral lf. Scales ovate to lanceolate, spirally imbricated. Bristles 1–8 or wanting, downwardly barbed. Fls. perfect. Stamens 2–3. Style 2- or 3-fid. Aks. lenticular or triangular. Style-base persistent, forming a tubercle on the apex of the ak. A cosmopolitan genus of ca. 150 spp., inhabiting wet places, mostly in warm regions. (Greek, *helios*, marsh, and *charis*, grace, many spp. found in marshes.)

(Svenson, H. K. Monographic studies in Rhodora vols. 31–41. 1929–1939.)
1. Aks. plump or triangular; style trifid.
  2. Aks. with several longitudinal ridges and many fine horizontal lines.
    3. Dwarf annual with fibrous roots; montane marshes .... 3. *E. bella*
    3. Perennials with rhizomes or stolons; lowland marshes.
      4. Culms capillary, furrowed; stamens 3; scales brown .... 1. *E. acicularis*

4. Culms spongy, striate; stamens 2; scales straw-color .... 11. *E. radicans*
2. Aks. not longitudinally ribbed.
  5. Spikelets 2–7-fld.
    6. Culms 2–7 cm. tall; aks. pitted; bristles 3, rudimentary or wanting. Below 7000 ft.
                                                                              5. *E. coloradoensis*
    6. Culms 7–14 cm. long; aks. reticulate; bristles 2–6, well developed. From above 6500 ft.
      7. Culms green, erect; rhizomes not forming a close indurated turf . . 10. *E. quinqueflora*
      7. Culms grayish, arching; rhizomes forming a close indurated turf . . 4. *E. bernardina*
  5. Spikelets 10–many-fld.
    8. Tubercle conic or pyramidal, constricted at base.
      9. Spikelets ovoid, blunt; bristles 2–6; aks. pitted-reticulate under magnification
                                                                              8. *E. montevidensis*
      9. Spikelets linear-lanceolate, acute; bristles 6–7; aks. smooth or finely pitted under
        magnification .... 9. *E. parishii*
    8. Tubercle long-subulate, continuous with apex of ak. .... 12. *E. rostellata*
1. Aks. lenticular; style bifid.
  10. Ak. yellowish-brown; plants perennial with long-creeping rhizomes ....7. *F. macrostachya*
  10. Ak. shining black; plants annual with cespitose culms.
    11. Aks. 0.5 mm. long; bristles 2–4 or more .... 2. *E. atropurpurea*
    11. Aks. 1.0 mm. long; bristles 6–8 .... 6. *E. geniculata*

1. **E. aciculàris** (L.) R. & S. [*Scirpus a.* L. *E. a.* var. *occidentalis*. Svens.] Perennial with filiform stolons; culms matted, capillary, furrowed, 2–20 cm. tall; basal lf.-sheaths truncate, inconspicuous; spikelets ovate to linear, acute, 2–7 mm. long, 5–10-fld.; scales ovate-lanceolate, with brown sides, green midrib and hyaline margins; bristles 3–4 or wanting, equaling ak.; stamens 3; style trifid; ak. obovoid-oblong, with ca. 40 close transverse lines; tubercle compressed-conical; $n = 10$, 15–19, 25–29 (Tanaka, 1937; Hicks, 1929).—Muddy banks, meadows, vernal pools and marshes, sea level to 8000 ft., many Plant Communities; throughout the state except in the deserts; throughout N. Am. and Eurasia. May–Aug.

2. **E. atropurpùrea** (Retz.) Kunth. [*Scirpus a.* Retz.] Annual, with fibrous roots and cespitose culms; culms filiform, striate, 3–12 cm. tall; basal lf.-sheaths loose, obliquely truncate, with an attenuated tooth; spikelets ovoid, many-fld.; scales ovate, obtuse, with purple-brown sides, a green midrib and very narrow scarious margin; bristles 2–4 or none, slender, shorter than ak.; stamens 2–3; style bifid; ak. lenticular, obovoid, 0.5 mm. long, shining black, the surface smooth; tubercle minute, depressed, constricted at base, conic, white.—Weed in low wet places, Mentone near San Bernardino; cent. Calif. to Wash., cent. and s. U.S., trop. S. Am.; Old World. Aug.

3. **E. bélla** (Piper) Svens. [*E. acicularis* var. *b.* Piper.] Dwarf annual with fibrous roots and cespitose culms, often forming dense round tufts 5–10 cm. in diam.; culms capillary, furrowed, 2–6 cm. tall, light green; basal lf.-sheaths loose, obliquely truncated; spikelets 1–3 mm. long, mostly 8–10-fld.; scales with purplish-brown sides and green midrib; bristles 0; stamens 2; ak. with ca. 30 fine transverse lines; tubercle compressed-conical.— Marshes, from ca. 3000 to 8000 ft.; largely Montane Coniferous F.; San Diego Co. to San Bernardino Co.; Kern Co. to Wash., Ida., New Mex. May–Aug.

4. **E. bernardìna** Munz & Jtn. [*E. pauciflora* var. *b.* Svens.] Perennial, forming an almost impenetrable turf; stems gray-green, 5–14 cm. long, spreading and recurved; spikelets 4–7 mm. long, ovate, 2–7-fld.; scales lanceolate, acuminate; bristles 2–6; style 3-fid; ak. triangular, light brown, 1.8 mm. long.—Forming large patches in wet meadows, 7500–9000 ft.; Montane Coniferous F.; San Bernardino Mts. and Mt. Pinos. July–Sept.

5. **E. coloradoénsis** (Britt.) Gilly. [*Scirpus c.* Britton. *E. parvula* var. *c.* Beetle. *Heliocharis leptos* Svens. *H. l.* var. *johnstonii* Svens.] Dwarf sedge with somewhat arched stems 2–3 cm. tall; spikelets 4–6-fld.; scales with brown or purple sides and green midrib and hyaline margins; bristles 3 or 0, short; aks. subtriangular, the convex surface with an obscure keel, the surface papillose; tubercle pyramidal.—Muddy places up to 7000 ft.; several Plant Communities; widely scattered localities, Imperial, San Bernardino and Los Angeles cos.; to Ore., S. Dak., New Mex. June–Sept.

6. **E. geniculàta** (L.) R. & S. [*Scirpus g.* L. *E. caribaea* Blake.] Similar to *E. atropurpurea*; culms subfiliform, 5–25 (–40) cm. long; scales pale brown with a scarious margin; bristles 6–8, downwardly barbed, equaling ak.; ak. 1 mm. long; tubercle spongy.— Freshwater Marsh at low elevs.; Imperial, Orange, Riverside and San Bernardino cos.; to Ind., Fla., trop. Am. Old World. March–Dec.

7. **E. macrostàchya** Britton in Small. [*E. palustris* Am. auth. *E. mamillata* Lindb. f.] Perennial with long creeping rhizomes; culms loosely or densely cespitose, terete or oval in cross section, stout or slender, striated, erect, 3–10 dm. tall, pale to dark green; lvs. reduced to basal sheaths, obliquely truncated or sometimes mucronate; terminal spike 5–25 mm. long, subtended by 2–3 empty scales or 1 sterile clasping scale; fertile scales lanceolate, brown to purple or green, with a green midrib and scarious margin; bristles 4, sometimes 0, ca. as long as ak.; style mostly 2-fid; ak. lenticular, yellowish-brown, 1–1.5 mm. long; tubercle pyramidal or cone-shaped, constricted at base; $n = 5, 8, 16, 18, 19, 21, 23$, probably many reports not for this sp.—Common and widespread in marshes and wet places, below 8000 ft.; many Plant Communities, desert and cismontane s. Calif.; San Clemente and Catalina ids.; throughout temp. N. Am. April–Nov.

8. **E. montevidénsis** Kunth. [*E. montana* w. Am. auth. *E. arenicola* Torr.] Plate 92, Fig. F. Perennial with slender creeping reddish rhizomes; culms slender, striate, 1–3 dm. tall, in fascicles; lf.-sheaths reddish-brown, usually straw-color toward the apex, apex truncate, often with a minute tooth; spikelets narrowly ovoid to oblong, obtuse, 4–15 mm. long, many-fld.; scales ovate, obtuse, brownish or yellowish with a hyaline margin; bristles 4–6, equaling or shorter than the ak.; style trifid; ak. trigonous, obovoid, yellowish-brown to brown, minutely pitted or reticulate; tubercle conic, short.—Moist ground up to 6500 ft.; many Plant Communities; coastal s. Calif. to the desert edges; to Lake Co., Fla., Mex., S. Am. May–Sept.

9. **E. paríshii** Britton. [*E. montevidensis* var. *p.* V. Grant.] Like *E. montevidensis*; spikelets linear-lanceolate, acute, 10–15 mm. long; scales chestnut or dark brown with a short hyaline tip; bristles 6–7, sometimes exceeding the ak.; tubercle short-subulate to conic; $n = 5$ (Raven et al., 1965).—Moist places below 7000 ft.; many Plant Communities; Colo. and Mojave deserts n. to Mono Co.; Coast Ranges to mostly from Ventura Co., n. Calif.; Ariz., New Mex.

E. discifórmis Parish, an annual sp., otherwise like *E. parishii*, was collected in 1901 at the e. base of San Jacinto Mts. It remains an uncertain entity; perhaps it was an annual that had begun its development early and flowered in its first season.

10. **E. quinqueflòra** (Hartm.) O. Schwarz. [*Scirpus q.* Hartm. *E. pauciflora* (Lightf.) Link. *S. p.* Lightf.] Perennial with filiform rhizomes bearing small leafy tubers; culms capillary, grooved, erect, 7–14 (–40) cm. tall; basal lf.-sheaths 2–3 cm. long, truncate; spikelets 4–7 mm. long, ovate, 2–7-fld.; scales lanceolate, acuminate, purplish-brown; bristles 2–6, shorter than to exceeding ak.; style 3-fid; ak. triangular, the surface finely reticulate, yellowish-brown, 2 mm. long; tubercle a subulate beak merging into the dark base of the style.—High mt. meadows, 5000–9000 ft.; San Jacinto and San Bernardino mts.; White Mts.; Sierra Nevada n. to B.C., Atlantic Coast; Old World. July–Aug.

11. **E. radicáns** (Poir.) Kunth. [*E. acicularis* var. *r.* Britton.] Perennial with longcreeping rhizomes; culms soft and spongy, striate, 3–8 cm. tall; basal lf.-sheaths closely investing the culm; spikelets 2–4 mm. long; scales with scarious straw-colored sides; bristles 4, longer than ak.; stamens 2; tubercle narrowly conical, not compressed.— Moist places below 4500 ft.; Coastal Sage Scrub, etc.; cismontane Los Angeles and San Bernardino cos.; to Fresno Co. and Sonoma Co.; through U.S. to S. Am.

12. **E. rostellàta** (Torr.) Torr. [*Scirpus r.* Torr. *E. r.* var. *occidentalis* Wats.] Perennial from a short caudex; culms wiry, coarse, grooved, erect or somewhat arched, 2.5–15 dm. tall, or some procumbent and rooting at the tips; basal lf.-sheaths 2–7 cm. long; spikelets oblong, 6–12 mm. long, 10–20-fld.; scales lance-ovate; bristles 4–8, ca. as long as ak.; style 3-fid; ak. obtusely triangular, plump, 1.2–2 mm. long, finely reticulate; tubercle a subulate beak continuous with the apex of the ak.—Rare, alkaline marshes at low elevs.;

Coastal Sage Scrub, Creosote Bush Scrub; San Bernardino, Chino, Death V.; Los Angeles Co., Orange Co., Ventura Co.; throughout N. Am., Andes. May–Aug.

5. Fimbrístylis Vahl

Annual or perennial herbs. Culms erect, leafy at base. Involucral lvs. 1–several. Infl. a loose umbel or capitate cluster of spikelets. Scales spirally imbricated. Fls. all perfect. Bristles none. Stamens 1–3. Style 2–3-fid, pubescent or glabrous, the style-base swollen. Ak. triangular or lenticular. A genus of ca. 200 spp., trop. and temp. regions. (Greek, *fimbria*, fringe, and *stylus*, style, some spp. having a fringed style.)

1. **F. thermàlis** Wats. Perennial with rhizomes; culms slender, grooved, erect, 2–7 dm. tall; lvs. basal, the blades flat, 1–2.5 mm. wide, somewhat pubescent near the sheath; involucral lvs. few, short, narrow; infl. a loose umbel, the rays to 6 cm. long; spikelets oblong, 8–15 mm. long, many-fld.; scales oblong, mucronate, light brown; style 2-fid; pubescent; ak. lenticular, obovate, ca. 1.5 mm. long, slate-colored, with numerous rows of very fine pits.—Freshwater Marsh; Arrowhead Hot Springs, San Bernardino Co. and Death V.; to Nev.

6. Hemicárpha Nees & Arn.

Dwarf annuals with fibrous roots and cespitose culms. Culms erect, filiform, grooved. Lvs. basal, the blades convolute, shorter than the culms. Involucral lvs. 1–3, one much exceeding the infl. Infl. a solitary spikelet or capitate cluster of 2–3 spikelets. Scales spirally imbricated. Fls. all perfect. Bristles 0. A single hyaline perianth-member developed between the fl. and the rachis of the spikelet, or sometimes wanting. Stamens 1–3. Style 2-fid, short, not swollen at the base. Ak. oblong, finely pitted. A few spp., of warm regions. (Greek, *hemi*, half, and *carphos*, chaff, referring to the perianth member.)

(Friedland, S. The Am. spp. of Hemicarpha. Am. Jour. Bot. 28: 855–861. 1941.)

Scales short-awned, the awns not more than 0.5 mm. long .... 1. *H. micrantha*
Scales longer-awned, the awns ca. 1 mm. long .... 2. *H. occidentalis*

1. **H. micrántha** (Vahl) Pax var. **minor** (Schrad.) Friedl. [*Isolepis subsquarrosa* var. *m.* Schrad.] Culms mostly 1–3 cm. tall; lvs. basal, the sheaths weakly united below, open above, with loose hyaline margins; blades convolute, filiform; infl. a solitary spikelet or of 2–3 spikelets, these ovoid, ca. 2 mm. long, many-fld.; scales obovate, acuminate, the green midrib produced into a short recurved awn; stamen 1; ak. 0.5–0.8 mm. long, finely pitted.—Collected at Mirror Lake, Ventura Co., *Pollard*; to Wash., Atlantic Coast, S. Am. Aug.–Oct.

2. **H. occidentàlis** Gray. Similar to *H. micrantha*, but the scales with a longer awn.—Moist places at middle elevs., San Bernardino and San Jacinto mts.; cent. Calif. to Wash. June–Aug.

Kyllinga brevifolia Rottb., perennial with rhizomes; to 7 dm. tall, leafy; invol. lvs. 2–4, 1–4 cm. long; head nearly globose; spikelets 3 mm. long, with 2 empty outer acuminate scales which enclose the fertile scale and ak. at maturity; ak. flat, obtuse.—Introd. weed, reported from San Diego, Pasadena, Santa Barbara; native of trop. Am.; Old World. July–Sept.

7. Schoènus L.

Perennial with stiff quill-like culms and lvs. Lvs. basal, the blades convolute. Involucral lvs. 2, stiff, sharp-pointed, one of them to 8 cm. long. Infl. a capitate cluster of spikelets. Spikelets flattened, the scales in 2 series, the 3–5 basal scales empty. Fls. 5–8, perfect, ♀, and ♂. Bristles 6. Stamens 3. Style 3-fid. Ak. triangular, without a tubercle. Ca. 60 spp., trop. and warm temp. regions, largely Australasian. (Greek, *schoinos*, a rush.)

1. **S. nigricáns** L. BLACK SEDGE. Culms 2–7 dm. high; some of the lvs. nearly as long as the culms; basal lf.-sheaths shining chestnut-brown; bristles 6, plumose, exceeding the ak.; ak. triangular, white, ca. 2 mm. long; $n=22$ (Davies, 1956).—Marshes and hot

springs to 5000 ft.; San Bernardino and San Gabriel mts., Death V.; Texas, Fla., W. I.; Eu. Aug.–Sept.

8. Scírpus L. BULRUSH. TULE

Perennial, rarely annual herbs. Culms erect, triangular to terete, leafy or the lvs. reduced to basal sheaths. Invol. of several blade-bearing lvs., or reduced to a solitary lf., or rarely absent. Infl. an open umbel or close cluster of numerous spikelets, or a solitary terminal spikelet. Scales spirally imbricated. Fls. perfect. Bristles (0–) 1–6, barbed or ciliate, usually included within the scales, in 1 sp. exserted. Stamens 2–3. Style 2- or 3-fid. Ak. lenticular or triangular. A cosmopolitan genus of ca. 200 spp. (Latin, *scirpus*, the classical name.)

(Beetle, A. A. Scirpus, in N. Am. Fl. 18 (8): 481–504. 1947.)
2. Involucral lvs. 2–5, usually exceeding the infl.; culms leafy.
  3. Spikelets 3–6 mm. long; infl. a loose spreading compound umbel .... 6. *S. microcarpus*
  3. Spikelets 10–40 mm. long; infl. capitate or with 1–several elongated rays . 8. *S. robustus*
2. Involucral lf. solitary, often appearing as a continuation of the culm.
  4. Culm leafy, triangular or subterete; spikelets 1–12.
    5. Plants perennial with rhizomes; culms not filiform; bristles present.
      6. Lf.-blades convolute; involucral lf. 3–10 cm. long; plant to ca. 1 m. tall
      2. *S. americanus*
      6. Lf.-blades flat; involucral lf. 1–3 cm. long; plant to over 2 m. tall .... 7. *S. olneyi*
    5. Plants annual with fibrous roots; culms filiform; bristles absent.
      7. Scales sharply keeled, acute or acuminate; involucral lf. to 2.5 cm. long
      5. *S. koilolepis*
      7. Scales only slightly keeled at tip, obtuse; involucral lf. less than 0.5 cm. long
      4. *S. cernuus*
  4. Culm with the lvs. reduced to basal sheaths with short blades and not more than 8 cm. long; culm stout and terete; spikelets numerous, in umbels.
    8. Bristles 2–4, broad, ciliate or plumose, but never barbed .... 3. *S. californicus*
    8. Bristles usually 6, filiform, downwardly barbed.
      9. Scales viscid-pubescent, copiously flecked with red or brown, the midrib not prominent, projecting as a short mucro; culm firm in texture .... 1. *S. acutus*
      9. Scales glabrous or nearly so, minutely if at all spotted, the midrib prominent and excurrent as an awn; culm soft, easily compressed .... 9. *S. validus*

1. S. acùtus Muhl. [*S. occidentalis* (Wats.) Chase. *S. a.* var. *o.* Beetle.] HARD-STEM BULRUSH. Rhizome often drab or brown, stout; culms stout, erect, terete throughout, to 5 m. tall and 2 cm. thick, rather hard and firm, olive-green; lvs. reduced to basal sheaths with blades to 8 cm. long, the sheath-margins becoming fibrillose; involucral lf. solitary, terete, shorter than infl., appearing as a continuation of the culm; infl. capitate to umbellate; spikelets ovate to cylindric, 8–18 mm. long; scales oblong-ovate, red-dotted, viscid-villous above, short-awned; bristles slender, downwardly barbed; style 2-fid or 3-fid; aks. 2–3 mm. long, lenticular or sometimes triangular; $n = 20$ (Hicks, 1928).—Occasional, wet places below 5000 ft.; several Plant Communities; s. Calif.; to B.C., Atlantic Coast. May–Aug.

2. S. americànus Pers. [*S. a.* var. *polyphyllus* Beetle.] Rootstocks elongate; culms sharply triangular, erect or arched, to ca. 1 m. tall; lf.-blades to ca. 2 dm. long, keeled, convolute, 2–3 mm. wide; involucral lf. solitary, 3–10 cm. long; infl. a capitate cluster of 1–7 spikelets; spikelets oblong, acuminate, 8–12 mm. long; scales pale to chocolate brown, cleft at apex, short-awned; bristles 2–6, unequal, half as long as to longer than ak.; style 2–3-cleft; ak. lenticular or obtusely triangular, mucronate, 3 mm. long; $n = 38$, ca. 40 (Hicks, 1928, Wulff, 1937).—Occasional in wet places, usually at low elevs., s. Calif. to B.C., Atlantic Coast, S. Am.; Old World. May–Aug.

3. S. califórnicus (C. A. Mey.) Steudel. [*Elytrospermum c.* C. A. Mey.] Stout perennial; culms subterete to triangular, to 4 m. tall; lvs. reduced to basal sheaths; involucral lf. solitary, erect, shorter than infl.; infl. loosely umbellate; spikelets narrow, acute, 5–10 mm. long; scales ovate, reddish-brown; bristles 2–4, dark red or sometimes paler, broad and ciliate or plumose, not barbed; style 2-fid; ak. lenticular, 2 mm. long.—Freshwater Marsh and wet places in various Plant Communities, at low elevs., cismontane and occasionally desert; to cent. Calif., Atlantic Coast, temp. S. Am.; Hawaii. June–Sept.

4. **S. cérnuus** Vahl var. **califórnicus** (Torr.) Beetle. [*Isolepis pygmaea* var. *c*. Torr.] Annual with fibrous roots; culms cespitose, nearly filiform, 4–20 cm. high; one basal lf.-sheath bearing a convolute blade 2–5 cm. long, the others without blades or with short blades to 3 mm. long; involucral lf. solitary, appearing as if a continuation of the culm, 2–5 mm. long; spikelet solitary, ovoid, 2–5 mm. long; scales broad, obtuse, keeled at tip, pale to deep brown with green midrib; bristles 0; stamens 2–3; style 3-fid; ak. triangular, punctate, white becoming brownish at maturity, 1 mm. long.—Coastal Salt Marsh, Freshwater Marsh; low elevs. L. Calif. to Wash. Santa Rosa and San Miguel ids. May–Aug.

5. **S. koilolèpis** (Steud.) Gleason. [*Isolepis k.* Steud. *I. carinatus* H. & A.] Annual with fibrous roots; culms cespitose, 4–20 cm. tall; lf.-blades to 4 cm. long, obtuse; involucral lf. solitary, appearing as a continuation of the culm, obtuse, to 2.5 cm. long; spikelets 1–3 on a culm, ovate, 2–5 mm. long; scales strongly carinate, greenish-brown, acute to acuminate, with broad hyaline margins and a broad green midrib ending as a short mucro; bristles 0; style 3-fid; ak. sharply triangular, punctate, light to dark brown, 1.5 mm. long.—Reported from marshes along the coast, but doubtful in s. Calif.; known from cent. and n. Calif.; s. U.S. April–June.

6. **S. microcárpus** Presl. Plate 93, Fig. A. Rootstocks creeping, stout; culms stout, erect, leafy, subterete, 6–16 dm. tall; lvs. flat, broad, 10–20 mm. wide, often overtopping the stem; involucral lvs. 2–5, exceeding the panicle; infl. a loose spreading compound umbel, the primary rays to 10 cm. long, scales green to brown, acute, with a prominent midrib, not awned; bristles 4, downwardly barbed, somewhat longer than ak.; stamens 2; style 2-fid; ak. whitish, 1 mm. long, ovate, lenticular, mucronate and with an obscure dorsal crest.—Along streams and on wet banks, below 7500 ft.; Coastal Sage Scrub to Montane Coniferous F.; cismontane and montane, desert edge as at Victorville; Santa Catalina Id. to Alaska, Atlantic Coast. May–Aug.

7. **S. ólneyi** Gray. Rootstocks dark, creeping, with bulbous nodes giving rise to the stout 3-angled culms 3–20 dm. tall; lf.-blades short, 2–13 cm. long; involucral lf. erect, 1–3 cm. long; infl. a capitate cluster of 5–12 spikelets; spikelets ovoid, 5–8 mm. long; scales flecked with brown, short-awned; bristles 4–6, unequal, downwardly barbed; style 3-cleft; ak. lenticular, mucronate, light brown or gray, minutely pitted, 2.5 mm. long; $n=39$ (Hicks, 1928).—Widespread in marshy places below 3000 ft.; Coastal Salt Marsh to Chaparral and Creosote Bush Scrub; San Nicolas Id.; throughout temp. N. Am. and to S. Am. June–Aug.

8. **S. robústus** Pursh. [*S. paludosus* A. Nels. *S. campestris* Britton, not Roth. *S. pacificus* Britton ex Parish.] Plate 93, Fig. B. Rootstocks slender with woody tubers; culms erect, sharply triangular, 5–15 dm. tall; lvs. mostly 4–6 mm. wide, sometimes to 12 mm.; involucral lvs. 2–5, unequal, to 30 cm. long; infl. capitate, or with one to several elongated rays; spikelets ovate, 10–25 mm. long, sometimes cylindric, to 4 cm. long; scales reddish-brown to pale straw-brown; bristles 1–6, half as long as ak.; style usually 2-fid; ak. lenticular, 3–4 mm. long; $n=53, 55, 57$ (Hicks, 1928).—Common, Coastal Salt Marsh, Freshwater Marsh, Alkali Sink, etc. below 2000 ft.; desert and cismontane; to n. Calif., most of N. Am., Argentina. April–Aug.

S. mucronàtus L., like *S. olneyi*, but with spikelets 8–10 mm. long and with horizontally rugose aks. has been reported from Santa Barbara.

9. **S. válidus** Vahl. Soft-Stem Bulrush. Resembling *S. acutus*, but the culms soft and easily compressed, pale green; spikelets ovate, 5–10 mm. long; bractlets brownish, pubescent at tip; scales glabrous or nearly so, minutely if at all spotted reddish-brown, the midrib prominent and excurrent as an awn; $n=21$ (Hicks, 1928).—Wet places below 7500 ft.; many Plant Communities; scattered in s. Calif., centering in Santa Ana R. drainage; to B.C., Atlantic Coast; trop. Am. May–Aug.

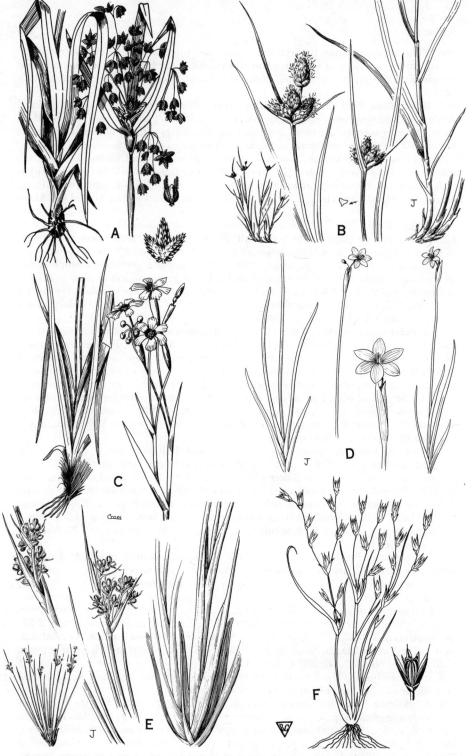

Plate 93. Fig. A, *Scirpus microcarpus*, ak. with 4 barbed bristles. Fig. B, *Scirpus robustus*. Fig. C, *Sisyrinchium bellum*. Fig. D, *Sisyrinchium elmeri*. Fig. E, *Juncus acutus* var. *sphaerocarpus*, fls. clustered. Fig. F, *Juncus bufonius*, fls. solitary.

8. Hydrocharitàceae. FROGBIT FAMILY

Dioecious or rarely monoecious or polygamous, perennial, submerged or floating, aquatic herbs. Lvs. opposite or clustered. Fls. regular, sessile or on scapelike peduncles in a spathe. Calyx usually of 3 sepals or lobes. Petals 3 or 0. Staminate fls. with 3–21 distinct or fused stamens. Pistillate fls. with 3–12 stigmas and a compound pistil fused with the tubular perianth; ovary inferior, maturing into an indehiscent submerged fr. A family of ca. 14 genera and 40 spp., widely distributed in fresh and salt waters.

1. *Elodèa* Michx. WATERWEED. ELODEA

Plants submerged forming large masses of branching elongate stems with occasional slender roots. Lvs. whorled, sometimes opposite, 1-nerved, thin, elongate, entire or minutely serrulate. Fls. 1–3 in axillary tubular spathes. Staminate fls. with 3 almost separate sepals, 3 petals, 3–9 stamens, reaching the surface or breaking off and floating. Pistillate fls. solitary, sessile, with 6-parted limb at the end of an elongate tube and with capillary style, the 3 stigmas thus at the surface. Fr. oblong or fusiform, with 1–5 seeds. A genus of perhaps a dozen spp. of temp. regions of New World. (Greek, *elodes*, marshy.)

(St. John, H. Mon. of the genus Elodea. Part I. W. N. Am. Research Studies Wash. State Univ. 30: 19–44. 1962.)

Lvs. 5–15 mm. long; fls. solitary in the spathe .... 1. *E. canadensis*
Lvs. 20–30 mm. long; fls. usually 3 in the spathe .... 2. *E. densa*

1. **E. canadénsis** Michx. [*Anacharis c.* Planch. *Philotria c.* Britton. *E. planchonii* Casp.] Stems 2–5 dm. long; lvs. narrowly oblong-ovate, 6–15 mm. long, 1.2–3 mm. wide, obtuse; sepals of ♂ fl. 4–5 mm. long, the fl. remaining attached on its long pedicel; sepals of ♀ fl. ca. 2.5 mm. long; petals 2.5–3 mm. long; fr. 10–15 mm. long; $n=12$ (Heppell, 1945), 24 (Harada, 1956).—Occasional in ponds and sluggish waters, perhaps often as an escape, as at 1350 ft., San Dimas Canyon Dam, Los Angeles Co., *Wheeler* and Dry Lake, San Bernardino Mts. at 9100 ft., *Munz*; to Canada and Atlantic Coast. June–July.

2. **E. dénsa** Planch. [*Egeria densa* St. John.] BRAZILIAN WATERWEED. Stems several dm. long, densely leafy; lvs. lance-linear, 20–30 mm. long, 3–5 mm. wide; petals 7–8 mm. long; $n=24$ (Matsuura & Suto, 1935).—Used in aquaria and sparingly natur. in sloughs and ponds at low elevs.; S. Am. July–Aug.

9. Iridàceae. IRIS FAMILY

Perennial herbs, mostly low with simple or branching stems. Lvs. mostly basal and equitant, parallel-veined, linear or sword-shaped. Fls. terminal, showy, bisexual, issuing from a spathe of herbaceous or scarious bracts. Perianth of 6 parts in 2 series, the outer sometimes calyxlike, all united into a tube that is adnate to the ovary. Stamens 3; anthers extrorse, opposite the outer perianth-segms. Pistil 1; ovary inferior, mostly 3-loculed and with axile placentation; style single; stigmas 3, sometimes expanded and petallike or divided. Fr. a few- to many-seeded loculicidal caps. Ca. 1200 spp., a cosmopolitan family with many plants used for ornamental purposes (gladioli, irises, freesias, ixias, crocuses, etc.).

Spathes 1-fld., often appearing calyxlike; infl. a panicle of many-fld. spikes; perianth with orange-red limb .... 2. *Crocosmia*
Spathes more than 1-fld.
    Infl. umbellate; perianth-segms. all alike, not more than 2 cm. long .... 6. *Sisyrinchium*
    Infl. not umbellate; perianth-segms. not all alike, or if so, more than 2 cm. long.
        Perianth with 3 erect and 3 spreading or drooping segms.; fls. not in an elongate spikelike infl. .... 5. *Iris*
        Perianth not as above; infl. spikelike.

Style branches bifid; fls. mostly less than 3.5 cm. long .... 3. *Freesia*
Style branches simple; fls. 3–7 cm. long.
    Perianth tube constricted near or below the middle into a narrow cylindrical or filiform basal part .... 1. *Chasmanthe*
    Perianth tube tapering gradually from base to throat, curved .... 4. *Gladiolus*

1. *Chasmánthe* N. E. Br.

Differing from *Gladiolus* by having the perianth tube constricted near or below the middle into a narrow or basal part. Ca. 9 spp.; African. (Greek, *chasme*, gaping, and *anthe*, flower.)

  1. **C. aethiòpica** (L.) N. E. Br. [*Antholyza a.* L.] Stems ca. 1 m. high; basal lvs. in a rosette, sword-shaped, 3–4 dm. long, 2.5 cm. wide; fls. red-yellow, 3–6 cm. long, the cylindrical part ca. ½ the whole.—Escaping from gardens, as in the vicinity of Santa Barbara. Feb.–March.

2. *Crocósmia* Planch.

Corm with reticulated tunics; lvs. many, equitant. Infl. a panicle of several spikes; spathe-valves calyxlike, notched or cut. Fl. 1 in each spathe. Perianth with tube somewhat dilated above; segms. subequal, oblong to obovate. Stamens 3, inserted at base of funnel. S. Afr. A genus of ca. 5 spp. (Greek, *crocus*, saffron, and *osme*, smell, because of odor of dried fls. immersed in water.)

  1. **C. × crocosmiflora** N. E. Br. [*Tritonia c.* Lemoine.] MONTBRETIA. Sts. to 1 m., branching; lvs. 2–4 dm. long; fls. orange-crimson, 3–5 cm. in diam.—Garden hybrid, widely cult., sometimes establishing itself as a wild plant. June–July.

3. *Freèsia* Klatt

Cormous plants with plane narrow lvs. below and showy fls. in loose secund spikes at the summit of the slender stem. Perianth tubular, funnel-shaped, the segms. ± unequal. One or more spp. from S. Afr. (E. M. *Fries*, 1795?–1876, Swedish botanist.)

  1. **F. refrácta** Klatt. Two–4 dm. tall; basal lvs. 1–1.5 dm. long; fls. solitary in the short spathes, usually ± yellow, to 3.5 cm. long, very fragrant.—Common garden plant, occasionally becoming established as about Santa Barbara. March–April.

4. *Gladìolus* L.

Plants with tunicated corms and simple ± leafy stems with large usually herbaceous spathes (bracts) each yielding 1 sessile fl. Lvs. mostly sword-shaped, broad, many-nerved, sometimes narrower. Fls. mostly several cm. across, commonly showy, with dilated funnel-shaped tube, curving upward in most spp. Perianth-segms. usually oblong, obtuse or acute, the 3 upper larger than the 3 lower. Stamens inserted below the throat. Style slender. Ca. 250 spp. from the Medit. region to S. Afr. Many in cult. (Latin, *gladiolus*, small sword.)

  1. **G. segètum** Ker. Plants ca. 1 m. high, the stem slender, several-lvd.; lvs. 1–2 dm. long, ca. 1 cm. wide; fls. several, the infl. elongating and becoming ca. 3 dm. long in fruit; outer bract of spathe 3–3.5 cm. long, inner 2–2.5 cm.; fls. purple-pink, ca. 5 cm. long; $2n = 120$ (Bamford, 1935).—Occasionally natur. in walnut groves (Orange Co.) and in fields (farther north). March–April.

5. *Ìris* L. IRIS. FLEUR-DE-LIS

Perennial herbs with creeping ± tuberous rhizomes or bulblike base. Stems erect, simple or branched, with 1–several fls. at the top and equitant mostly radical and basal lvs. Fls. in axils of bracts (or spathe) which may be opposite or alternate. Perianth of 6 clawed segms. united below into a tube, the 3 outer broad, spreading or reflexed (the sepals or falls), the 3 inner usually narrower and erect (the petals or standards). Stamens inserted at the base of the falls; anthers linear or oblong. Ovary 3-loculed; style with 3 petallike branches that arch over the stamens and bear a stigmatic surface beneath on a flat scale just below the usually 2-lobed tip (crest). Caps. oval or oblong, angled or lobed, many-seeded. Perhaps 150 spp., mostly in the N. Temp. Zone. (Greek, the rainbow, because of variegated fl.-color.)

Rhizomes ca. 5–10 mm. thick; caps. trigonal with 3 ribs or subterete; fls. 5–6 cm. long, purple to bluish-violet. Dry places in pine belt .... 2. *I. hartwegii*
Rhizomes 20–40 mm. thick; caps. hexagonal or each of the 3 angles 2-ribbed.
  Lvs. linear, generally 2–6 mm. wide. Montane native in meadows .... 3. *I. missouriensis*
  Lvs. ensiform, 15–35 mm. wide. Natur. from gardens.
    Falls bearded with multicellular hairs; fls. not wholly yellow throughout .... 1. *I. germanica*
    Falls without a beard or crest; fls. deep yellow .... 4. *I. pseudacorus*

1. **I. germánica** L. GERMAN IRIS. Rhizome stout; stem 6–9 dm. tall, bearing a 2-fld. terminal head and 2 usually 1-fld. stalked branches; spathe-valves oblong-lanceolate, tinged purple; lvs. 3–4 dm. long, 2.5–4 cm. wide; fls. lilac and lilac-purple varying to white, the obovate falls 2.5–4 cm. broad, the beard yellow; standards the same size, usually lighter-colored. This sp. yields many garden forms. European in origin. Occasional escape, as near Santa Barbara. Spring.

2. **I. hartwègii** Baker subsp. **austràlis** (Parish) Lenz. [*I. h.* var. *a.* Parish. *I. tenax* var. *a.* Foster.] Rhizome 6–10 mm. thick; lvs. 3–8 mm. wide, to 4 or 5 dm. long, usually pinkish at base; fl.-stem simple, 1–4 dm. long, with 1–several sheathing lvs. free ca. ½ their length, 2-fld.; outer spathe-bract 6–9 mm. wide, 7–12.5 cm. long; perianth-tube 7–13 mm. long; fls. purple to bluish-violet, the sepals 5–6 cm. long, the petals the same; style-branches 2–3.2 cm. long; style-crests 1.1–1.5 cm. long; $2n=40$ (Lenz, 1950).— Dry woods, 5000–7500 ft.; Yellow Pine F.; San Bernardino and to a lesser extent in San Gabriel and San Jacinto mts. May–June.

3. **I. missouriénsis** Nutt. Rhizome 2–3 cm. in diameter, clothed with dark remnants of old lvs.; lvs. rather light green, glaucous, to ca. 4.5 dm. long, 3–8 mm. wide, sometimes purplish at base; fl.-stem rather slender, 2–5 dm. high, usually branched, with 1–2 fls. in a cluster; spathe-bracts opposite, lanceolate to ovate, 4–7 cm. long; pedicel to 20 cm. long; perianth-tube funnelform, to ca. 1 cm. long; sepals ca. 6 cm. long, 2 cm. wide, largely pale lilac to whitish with lilac-purple veins; petals shorter, 1 cm. wide; style-branches to 2.5 cm. long; style-crests ca. 8 mm. long, incised; caps. 3–5 cm. long, oblong, trigonal; $2n=38$ (Foster, 1937).—Meadows, 4400–8400 ft.; largely Yellow Pine F.; mts. of San Diego Co., San Bernardino Mts., Mt. Pinos; to B.C., Rocky Mts., Coahuila. May–June.

4. **I. pseudàcorus** L. Erect, glabrous, rather glaucous, 4–15 dm. high; rhizome often 3–4 cm. in diam.; lvs. 15–25 mm. wide, ca. as long as the compressed terete scape; fls. 8–10 cm. diam., yellow, the outer segms. often purple-veined with an orange spot near the base; caps. elliptic, apiculate.—Occasional as garden escape in wet places; Ventura and Santa Barbara cos. and farther n.; native of Medit. region. April–May.

6. *Sisyrínchium* L.

Tufted perennials or less frequently annuals. Stems slender, compressed, ± winged. Lvs. narrow, grasslike. Fls. ephemeral, opening in sun, umbellate, subtended by a spathe of 2 bracts. Perianth blue, violet, purplish or yellow, sometimes white, the segms. all alike. Fils. ± united. Caps. globose; seeds subglobose or ovoid, smooth or pitted. A fairly large genus; N. Am. and W. I. (Name used by Theophrastus for a plant related to *Iris*.)

Plants annual .... 5. *S. minus*
Plants perennial.

Fls. yellow .... 2. *S. elmeri*
Fls. mostly blue to purplish-red.
    Stem bearing 1 or more lf.-bearing nodes, each node with 2 or more peduncles . 1. *S. bellum*
    Stem leafless, simple, with a single terminal spathe.
        Pedicels pubescent; perianth 9–10 mm. long; caps. not brownish.
        Kern Co. n., in ± alkaline spots .... 3. *S. halophilum*
        Pedicels glabrous; perianth 10–15 mm. long; caps. brownish. High mt. meadows
                                                                                                          4. *S. idahoense*

    1. **S. béllum** Wats. [*Bermudiana b.* Greene. *S. maritimum* Heller.] BLUE-EYED GRASS. Plate 93, Fig. C. Tufted, from 1–4 (–6) dm. high, green to glaucescent; stems 1–4 mm. wide, stout or slender, firm-margined to winged, smooth or denticulate on edges; lvs. mostly basal, soft to firm, shorter than to almost as tall as stems, mostly 2–4 (–6) mm. wide; cauline lvs. 1–3, each commonly bearing in its axil 2–4 peduncles 1–15 cm. long; spathe 2–6 mm. wide when pressed, often ± purplish, the bracts unequal to subequal, the outer mostly 2–4.5 cm., the inner 1.5–2.5 cm. long; pedicels ± exserted, slender, glabrous; perianth-segms. blue, violet, lilac, rarely white, emarginate and often aristulate, 12–15 or more mm. long; caps. dark or pale brown, 2–7 mm. long; seeds 1–few in a locule, dark, pitted; $2n=32$ (Bowden, 1945).—Widely distributed, the more typical form largely in open grassy places, below 3000 ft.; many Plant Communities near the coast. Santa Catalina, Santa Rosa, Santa Cruz, San Miguel ids. March–May.

    A variable complex in which a no. of names have been proposed: (1.) *S. eastwoodiae* Bickn. Slender with narrow lvs. and short spathes, tending to be rather tall and grasslike. —Montane Coniferous F., s. Calif. (2.) *S. funereum* Bickn. Stems rather tall, pale with very little anthocyanin; fls. 12–14 mm. long.—In ± alkaline places; Alkali Sink, Death V. and Mojave Desert. (3.) *S. hesperium* Bickn. Slender, near typical *S. bellum*, but with shorter spathe and smaller fls.; grassy slopes away from the coast; Coastal Sage Scrub, etc.; San Luis Obispo Co. to L. Calif.

    2. **S. élmeri** Greene. [*Hydastylus e.* Bickn. *H. rivularis* Bickn.] YELLOW-EYED-GRASS. Plate 93, Fig. D. Slender, mostly less than 2 dm. high, the stems 1–2 mm. wide; lvs. ca. 1–3 mm. wide, half as long as stems; outer bract of spathe 1.2–3.5 cm. long, scarcely longer than inner; perianth-segms. 8–12 mm. long, orange-yellow, with 5 dark veins, obtusish; caps. ca. 6–7 mm. long.—Boggy and wet places, 6500–7500 ft.; Yellow Pine F.; San Bernardino Mts.; Sierra Nevada. July–Aug.

    3. **S. halophilum** Greene. With the habit of *S. idahoense*; outer bract of spathe 1.5–2 cm. long, broadly scarious-margined; inner somewhat shorter; pedicels ± glandular-pubescent; perianth violet-purple, 9–10 mm. long, segms. prominently aristulate at apex; caps. 3–4 mm. high, pubescent, pale grayish-green; seeds rugulose-pitted.—In ± alkaline meadows and wet places, 4000–7000 ft.; Yellow Pine F., Foothill Wd.; Piute and Tehachapi mts., Kern Co.; to Mono Co. and w. Nev. May–June.

    4. **S. idahoénse** Bickn. [*S. oreophilum* Bickn.] Plants pale green, glaucous, 1–4.5 dm. high; stems mostly leafless, slender, 1–4 dm. high, 1–2 mm. wide, winged; lvs. ca. half as long as stems, basal, 1–3 mm. wide; spathe 1, narrow, green or faintly purplish; outer bract 2–6 cm. long, the inner somewhat shorter; pedicels glabrous, 1.5–3 cm. long; perianth-segms. 10–15 mm. long, mucronulate-aristulate; caps. 3–6 mm. diam., brown, sparsely puberulent; seeds black, rugulose.—Wet meadows, 7400–9000 ft.; Montane Coniferous F.; San Jacinto and San Bernardino mts.; Sierra Nevada to Wash. July–Aug.

    5. **S. mìnus** Engelm. & Gray. Small annual, the plants ca. 1 dm. high; fls. lavender-pink to purple-rose, white with yellow eye, or all yellow.—Taken in 1944 in grassy field, Los Angeles, by *F. W. Gould*. Native of Texas, La. Apr.–May.

10. Juncàceae. RUSH FAMILY

    Perennial or sometimes annual herbs, usually of moist places. Stems tufted or from creeping or erect rootstocks, mostly simple, terete or compressed. Lvs. alternate, sheathing, grasslike, flat to terete. Infl. usually compound or decompound, paniculate, corymbose or umbelloid, rarely reduced to 1 fl. Fls. borne singly and each subtended by a bractlet, or in headlike or spikelike clusters and not individually subtended; infl. usually subtended by 1 or more bracts. Perianth small, regular, usually 6-parted into 3 outer and 3 inner glumaceous segms. Stamens usually 3 or 6; anthers adnate, introrse, 2-celled. Ovary superior, 3- or 1-celled, with 3–many ovules, 1 style, 3 filiform stigmas. Fr. a loculicidal caps., 3–many-seeded; seeds small, round to elongate, with fleshy endosperm. Eight genera, perhaps 300 spp., widely distributed.

Lf.-sheaths open; lvs. stiff (terete to flat); stems usually with spongy pith .... 1. *Juncus*
Lf.-sheaths closed; lvs. soft (flat); stems hollow .... 2. *Luzula*

1. *Júncus* L. RUSH. WIRE-GRASS

Stems leafy or scapose, the sheaths with free margins, the blades stiff, terete, flat, channeled, or compressed and gladiate. Infl. paniculate, corymbose, capitate or 1-fld. Fls. greenish to brownish or purplish. Stamens 3 or 6. Ovary 1-celled or 3-celled by intrusion of the parietal placentae. Seeds several to many, usually reticulate or ribbed. A genus of over 200 spp., mostly temp. Often growing in meadows and swales mixed with grasses and sedges and of some economic importance for hay and pasture. (Latin name for rush, perhaps from *jungere*, to bind, the stems used for binding.)

1. Plants annual.
   2. The plants mostly 10–25 cm. high with short flat, sometimes involute lvs.
      3. Caps. oblong, 3–4.5 mm. long; perianth 4–6 mm. long .... 4. *J. bufonius*
      3. Caps. subglobose, 2–3 mm. long; perianth 3–4 mm. long .... 23. *J. sphaerocarpus*
   2. The plants mostly 0.5–10 cm. high, with ± setaceous lf.-blades.
      4. Heads often 2–several-fld.; bracts 2, subequal, 1–1.5 mm. long.
         5. Anthers longer than fils.; plants 3–10 cm. high; style 1.5–3 mm. long . 27. *J. triformis*
         5. Anthers shorter than fils.; plants 0.5–4 cm. high; style 0–1 mm. long .. 10. *J. kelloggii*
      4. Heads 1-fld.; bract 1 or 0, or if 2, these unequal and shorter.
         6. Perianth 1.5–2 mm. long; caps. shorter than perianth .... 3. *J. bryoides*
         6. Perianth 2–3 mm. long; caps. longer than perianth .... 9. *J. hemiendytus*
1. Plants perennial.
   7. Infl. seemingly lateral, the lowest bract terete and exactly like a continuation of the stem.
      8. Fls. inserted in headlike clusters which are arranged in a panicle; bracts but no bractlets present.
         9. Perianth-segms. 2–3 mm. long, the outer obtuse; caps. longer than the perianth
            1. *J. acutus*
         9. Perianth-segms. 4–5 mm. long, the outer acute; caps. ca. as long as the perianth 5. *J. cooperi*
      8. Fls. inserted singly on the racemose branches of the panicle, each with a pair of bractlets at the base.
         10. The fls. 1–3; tufted subalpine plants 1–3 dm. high; seeds conspicuously tailed
             18. *J. parryi*
         10. The fls. many; plants usually from creeping rootstocks, taller; seeds not tailed.
            11. Lf.-blades usually well developed on upper basal lf.-sheaths; stems compressed
                14. *J. mexicanus*
            11. Lf.-blades wanting or rudimentary; stems terete.
               12. Perianth 2–2.5 mm. long; anthers not longer than the fils.
                  13. Stamens 3; caps. obovoid, not apiculate .... 8. *J. effusus*
                  13. Stamens 6; caps. subglobose, apiculate .... 19. *J. patens*
               12. Perianth 3.5–5 mm. long.
                  14. Stems smooth or irregularly ridged when dry (diagonal sections of stem showing 1–2 rows of vascular bundles and no fascicles of subepidermal strengthening cells); caps. acute .... 2. *J. balticus*
                  14. Stems finely and evenly ridged (with 3–4 rows of vascular bundles and evident fascicles of subepidermal strengthening cells); caps. obtuse ... 25. *J. textilis*
   7. Infl. usually seemingly terminal, the lowest bract not exactly like a continuation of the stem, or if so, channeled along inner side.
      15. Lf.-blades transversely flattened (inserted with the flat surface facing the stem), not septate.

16. Fls. inserted singly and each subtended by 2 bractlets; lvs. with scarious elongate auricles; perianth 3-4 mm. long .... 24. *J. tenuis*
16. Fls. inserted in true heads which are subtended by bracts, but the fls. not by bractlets.
17. Perianth 5-6 mm. long.
18. Auricles not evident; panicle of 2-6 heads; lvs. 3-7 mm. wide
16. *J. orthophyllus*
18. Auricles evident; panicle of 8-25 heads; lvs. narrower ...... 12. *J. macrophyllus*
17. Perianth 2-2.5 mm. long; auricles not evident; heads few .... 6. *J. covillei*
15. Lf.-blades terete or equitant, not transversely flattened; lvs. with transverse ribbing due to internal septa.
19. The lf.-blades subterete or terete, not equitant, the septa usually complete.
20. Epidermis of stems and lvs. conspicuously transversely rugose under a lens
21. *J. rugulosus*
20. Epidermis smooth.
21. Anthers definitely shorter than the fils.; perianth 4-5 mm. long; caps. subulate; heads many-fld. .... 26. *J. torreyi*
21. Anthers equal to or longer than the fils.
22. Caps. narrowed into a long beak and well exserted when mature.
23. Auricles less than 1 mm. long; perianth 3-4 mm. long. Rare. . 15. *J. nodosus*
23. Auricles mostly 4-6 mm. long; perianth 2.5-3 mm. long. Well distributed
7. *J. dubius*
22. Caps. abruptly contracted and not well exserted; auricles 1-3 mm. long; perianth 3-3.5 mm. long .... 13. *J. mertensianus*
19. Lf.-blades flattened with the edge toward the stem, equitant; septa incomplete.
24. Perianth 4.5-5 mm. long; caps. acute below the beak and almost as long as the perianth .... 20. *J. phaeocephalus*
24. Perianth 3-4 mm. long.
25. Auricles small, but definite; perianth-segms. somewhat unequal, exceeding caps.
22. *J. saximontanus*
25. Auricles none or obscure; perianth-segms. subequal.
26. Style at least as long as ovary at anthesis; anthers mostly longer than fils.
27. Caps. shorter than perianth, obtuse below the beak; perianth purple-brown
11. *J. macrandrus*
27. Caps. exceeding perianth, tapering into a beak; perianth greenish or with brownish tinge .... 17. *J. oxymeris*
26. Style short; anthers much shorter than fils. .... 28. *J. xiphioides*

1. **J. acùtus** L. var. **sphaerocárpus** Engelm. Plate 93, Fig. E. Perennial, in large tufts, 6-12 dm. high, with stout, rigid, pungent stems; lvs. all basal, terete, nearly as long as stems, with inflated brownish sheaths; auricles from scarcely developed to several mm. long, cartilaginous; lowest bract of infl. foliose, 5-15 cm. long, spinescent; infl. paniculate, with unequal branches, 5-20 cm. long; fls. 2-4 in small clusters; perianth-segms. pale brown, 2-4 mm. long, shining, indurate; anthers 6; caps. subglobose, ca. 5 mm. long; seeds finely reticulate, obovoid.—Moist saline places; Coastal Salt Marsh, San Luis Obispo Co. to L. Calif.; alkaline seeps, Alkali Sink, Colo. Desert; Ariz. May-June.

2. **J. bálticus** Willd. [*J. b.* var. *eremicus* Jeps. *J. breweri* Engelm.] Rootstock stout, creeping; stems terete, wiry, grasslike, 3-8 dm. tall, rather slender; sheaths basal, rather loose, strawcolor to brown, without blades except sometimes with a filament-like rudiment; panicle dense to diffuse, many-fld., 2-7 cm. high, the bract 3-20 cm. long; perianth-segms. purplish-brown, often with greenish center, 3.5-5 mm. long, acuminate, scarious-margined; anthers 6, much longer than fils.; caps. ovoid, brownish, acute, mucronulate; seed many-striate; $2n=40$ (Löve & Löve, 1956).—Moist places, mostly below 5000 ft.; many Plant Communities; well distributed; N. Am. and Old World. May-Aug.

Var. **montànus** Engelm. Stems 1-2 dm. high, slender, wiry; infl. compact.—Montane Coniferous F.; San Bernardino Mts.; to Alaska, Rocky Mts. July-Aug.

3. **J. bryoìdes** F. J. Herm. Annual, 0.5-1.5 cm. high; lf.-blades 1-3 mm. long, setaceous-caniculate or triquetrous; peduncles filiform, 1-25; head 0.8-1.2 cm. across, 1-fld.; bracts 2, ovate to lanceolate, 0.5-0.9 mm. long, hyaline; perianth-segms. appressed,

somewhat unequal, 1.5–2 mm. long, elliptic-oblong, red at center, incurved at tips; anthers 3, ca. half as long as fils.; caps. elliptic-oblong to subspherical, obtuse, shorter than perianth; style none; seed smooth.—Occasional, moist sandy places, 4000–6000 ft.; Montane Coniferous F.; San Diego Co. n.; to Sierra Nevada; Utah. May–Aug.

4. **J. bufònius** L. TOAD RUSH. Plate 93, Fig. F. Tufted annual, branching from the base; stems slender, 3–20 (–30) cm. high; lvs. 1–3 on a stem, short, flat, to 1 mm. wide, often involute; auricles low, rounded, hyaline; infl. cymose, forming from $\frac{1}{3}$–$\frac{1}{2}$ of the plant, the lower nodes with lvs. having short blades and conspicuous sheaths; fls. borne singly on the branches; perianth-segms. light green, 4–6 mm. long, scarious-margined, lanceolate; anthers usually 6, ca. as long as the fils.; caps. oblong, shorter than the perianth, obtuse, mucronate; seeds minutely reticulate; $2n = 60$, 120 (Wulff, 1937).—Common in moist, especially open places, below 8000 ft.; many Plant Communities; mainland and insular, occasional on the desert; cosmopolitan except polar regions and tropics. April–Sept.

Var. **congéstus** Wahl. Fls. congested in small headlike clusters; inner perianth-segms. long-pointed.—Largely halophytic, near the coast, occasional in the interior. April–Aug.

Var. **halophìlus** Buch. & Fern. Fls. ± congested; inner perianth-segms. obtuse.—Rare, brackish spots, Coastal Salt Marsh; to Atlantic Coast.

5. **J. coóperi** Engelm. Perennial, in large tufts 6–12 dm. tall, with stout, rigid, pungent stems; lvs. all basal, terete, stiff, spinescent, much shorter than the stems; auricles 0; lowest bract of infl. spinescent, 6–10 cm. long; infl. paniculate, with uneven branches 2–10 cm. long; fls. 2–several in a cluster; perianth-segms. lanceolate, greenish or straw-color, 4–6 mm. long, acutish to acuminate; anthers 6; caps. narrowly ovoid, ca. as long as perianth; seeds reticulate, short-appendaged.—Alkali Sink, below 2000 ft.; Mojave and Colo. deserts; Nev. April–May.

6. **J. covíllei** Piper. [*J. falcatus* var. *paniculatus* Engelm.] With creeping rootstocks; stems in large tufts, slender, 1–2.5 dm. high; lvs. grasslike, mostly basal, flat, 2–3 mm. wide, ca. as long as stems; cauline lvs. 1–2; auricles 0; heads few, in small panicles, commonly 3–5-fld.; bracts and peduncles roughened; perianth-segms. 2–3.5 mm. long, ovate, obtuse, subequal, the outer pointed, the inner rounded at the tips, hyaline-margined, brown, or the cent. portion green; stamens 6; caps. oblong-ovoid, brown, depressed at the obtuse apex, exceeding perianth; seeds minutely reticulate, ribbed.—Moist sandy banks, 6500–7800 ft.; Montane Coniferous F.; San Bernardino and San Gabriel mts.; to Wash. June–July.

7. **J. dùbius** Engelm. Much like *J. rugulosus* vegetatively, but lacking the transverse ridges on the epidermis, the rather slender stems 5–10 dm. high; rigid lvs. 1–4 dm. long, conspicuously septate; perianth brownish, 2.5–3 mm. long; anthers longer than fils.—Moist places below 4000 ft.; Coastal Sage Scrub, Chaparral, San Diego Co.; Creosote Bush Scrub, Death V.; to Ore., L. Calif. Apr.–Aug.

8. **J. effùsus** L. var. **brúnneus** Engelm. Rootstocks stout; stems in dense tufts, slender, 1–2 mm. in diam., 6–12 dm. tall; sheaths loose, membranaceous, dull, red-brown at base, the inner pale and somewhat greenish toward the rounded summit, mostly 10–15 cm. long; infl. 1–3 cm. in diam.; perianth-segms. 2.5–3 mm. long, green at center and bordered by 2 dark brown bands extending almost to the margin; caps. ca. as long as perianth, brownish, obtuse to somewhat retuse, slightly apiculate; seeds reticulate in ca. 16 rows.—Damp places near the coast; Coastal Salt Marsh, Coastal Strand; Santa Barbara Co.; to B.C. June–July.

Var. **pacíficus** Fern. & Wieg. Stems 2–3.5 mm. in diam. at base; lf.-sheaths chocolate-brown, dull, 5–15 cm. long, subtruncate or emarginate at the top, with strongly converging veins and edges overlapping nearly or quite to the summit, often with terminal filamentous black-rudiment; bract of infl. 6–20 cm. long; infl. lax, fastigiate, 2.5–15 cm. long; perianth-segms. 2.5–3.5 mm. long, subequal, rather soft, pale greenish-brown, with pale brownish membranous margins; anthers 3, shorter than fils.—Rather common in moist places, below 8000 ft.; many Plant Communities; well distributed except on the deserts; L. Calif. to B.C. June–Aug. Santa Rosa Id.

9. **J. hemiendỳtus** F. J. Herm. Annual, 1–2.3 cm. high; lf.-blades 3–9 mm. long, linear-setaceous; peduncles 3–40, fairly stout; heads 1.7–2.5 mm. across, 1-fld.; bracts 1–2, elliptic, 0.5–1.25 mm. long or obsolete, usually blunt, appressed; perianth-segms. erect or ascending, 2–3 mm. long, linear-lanceolate, subequal, with broad hyaline margin and reddish midrib; stamens 3; caps. longer than perianth, narrowly oblong; style very short; seed smooth, apiculate.—Moist mud flats, at ca. 7500 ft.; Montane Coniferous F.; San Bernardino Mts., e. San Gabriel Mts.; Sierra Nevada to Wash., Nev. June–July.

10. **J. kellóggii** Engelm. Annual, 1.5–4 cm. tall; lf.-blades 5–20 mm. long, linear-setaceous; peduncles 5–many; heads 2–5 mm. in diam., 1–3-fld.; bracts 2, ovate, ca. 1.5 mm. long, hyaline, often tinged red; fls. subsessile; perianth-segms. 4 or 6, erect to appressed, 2.5–3.5 mm. long, subequal, lanceolate, ± acute and acicular, reddish with green opaque midribs; stamens 3; caps. oblong to elliptic-ovoid, blunt to emarginate, brown to reddish; style becoming a beak on the caps.; seeds ribbed lengthwise and lineolate crosswise.—Moist banks and muddy places, below 8000 ft.; many Plant Communities; scattered through cismontane and montane s. Calif.; to B.C. May–July.

11. **J. macrándrus** Cov. Much like *J. phaeocephalus* var. *paniculatus*, the lvs. 2–3 mm. wide; panicle diffuse; heads many, few-fld.; perianth 3–3.5 mm. long, brownish; caps. obtuse below the beak, definitely shorter than the perianth.—Meadows and springy places, 6500–9000 ft.; Montane Coniferous F.; San Jacinto and San Bernardino mts.; Sierra Nevada. July–Aug.

12. **J. macrophýllus** Cov. Tufted perennials, 2–9 dm. high, with stiff subterete stems; lvs. pale green, somewhat channeled; the basal striate, from half to as long as stems, 1.5–3 mm. wide; cauline lvs. 1–3, the blades mostly 8–15 cm. long; auricles 1.5–3 mm. long; infl. loosely paniculate, with 8–25 heads; lowest sheath foliaceous, shorter than infl.; heads 3–5-fld.; perianth-segms. green with reddish or brownish tinge, 5–6 mm. long, ovate, hyaline-margined, the inner distinctly longer than the outer; anthers 6, much longer than fils.; caps. short-obovoid, short-beaked, not nearly as long as perianth; seeds ca. 20-ribbed.—Wet banks and slopes, to 8500 ft.; largely Chaparral, Montane Coniferous F.; L. Calif. to Ventura Co., across the desert to Ariz. May–Aug.

13. **J. mertensiànus** Bong. var. **duránii** (Ewan) F. J. Herm. [*J. d.* Ewan.] Rootstock short, subvertical; stems tufted, capillary, 1–2 dm. high; lvs. compressed, subfiliform, 7–15 cm. long; auricles round to pointed, ca. 1 mm. long; bract of infl. subulate; heads usually solitary, 7–9 mm. in diam., flattened-hemispherical, 5–15-fld.; perianth-segms. pale brown, lance-linear, 3 mm. long, hyaline-margined; stamens 6; anthers usually much shorter than fils.; caps. obtuse-obovoid, strongly beaked, ca. as long as perianth; seeds brown, narrow-lanceolate, obscurely longitudinally lineolate.—Occasional in wet places, 6000–8300 ft.; Montane Coniferous F.; San Gabriel, San Bernardino and San Jacinto mts. July–Aug.

Subsp. **grácilis** (Engelm.) F. J. Herm. [*J. phaeocephalus* var. *g.* Engelm. *J. nevadensis* Wats.] Heads several to many, few-fld. (12 or fewer), usually dark brown; perianth-segms. usually stiffish; bracts not spathaceous; anthers longer than fils.—Moist banks, 6000–8300 ft.; Montane Coniferous F.; San Jacinto and San Bernardino mts.; Sierra Nevada to B.C., Rocky Mts. July–Aug.

Subsp. **mertensiànus** has heads usually solitary, purplish-black, the perianth-segms. flaccid; bracts spathaceous and occurs from Sierra Nevada n. and e.

14. **J. mexicànus** Willd. [*J. balticus* var. *m.* Kuntze.] Much like *J. balticus* and differing primarily in having well developed blades on the upper lf.-sheaths and stems compressed, frequently twisted; auricles short, rounded, somewhat cartilaginous; perianth-segms. greenish or strawcolor; seeds irregularly reticulate. Moist, often subalkaline spots, sea-level to ca. 12,000 ft.; many Plant Communities, from cent. Calif. s. to San Diego Co., White Mts., to Tex., Mex. May–Aug.

15. **J. nodòsus** L. The slender creeping rootstocks with tuberlike thickenings from which arise the slender stems, these 1–4 dm. high, solitary; lvs. erect, terete, septate, the cauline 2–4 in no.; auricles less than 1 mm. long; infl. open, of 1–15 heads; heads 6–30-fld.; perianth-segms. 3–4 mm. long, light brown to greenish, plainly striate; stamens 6;

caps. 3-sided, longer than perianth; seeds reticulate in 20–30 longitudinal rows.—Wild Rose Spring, Panamint Mts. at ca. 6500 ft.; to B.C. Atlantic Coast. July–Sept.

16. **J. orthophýllus** Cov. [*J. longistylis* var. *latifolius* Engelm.] Rootstocks creeping, stout; stems compressed, 2–4 dm. high, leafless or with 1–3 lvs.; lvs. grasslike. mostly basal, short to almost as long as the stem, 2–6 mm. wide; auricles not well developed; heads 1–several, mostly 6–10-fld., 8–15 mm. across; perianth-segms. minutely roughened, brown with green midrib, lance-ovate, with scarious brown margins, 5–6 mm. long, subulate-tipped, the outer shorter than the inner; anthers 6, longer than the fils.; caps. obtuse, slightly shorter than perianth; seeds short-apiculate.—Wet places, 6000–7500 ft.; Montane Coniferous F. San Bernardino Mts., Mt. Pinos; Sierra Nevada to Wash. July–Aug.

17. **J. oxýmeris** Engelm. Rootstocks creeping, stout; stems slender, 3–6 dm. high, compressed; lvs. equitant, flat, with obscure auricles, the blades 5–20 cm. long, 3–6 mm. wide; panicles well developed. compound. the sheaths short; heads many, few-fld.; perianth-segms. subequal, 3–4 mm. long, acuminate and with broad scarious margins; anthers 6, equal to or longer than fils.; caps. exceeding perianth, tapering into a slender beak; seeds reticulate.—Occasional, wet places, below 7000 ft.; Coastal Sage Scrub to Yellow Pine F.; San Diego Co. n.; to Wash. June–Aug.

18. **J. párryi** Engelm. Plate 94, Fig. A. Cespitose, mostly 1–3 dm. high, with very slender stems; sheaths slightly brownish, 1–3 cm. long; the uppermost bearing a blade; lf.-blade grooved at base, terete above, 3–6 (–8) cm. long, slender; auricles low, rounded, membranous; lowest bract of infl. 1.5–5 cm. long; fls. 1–3, inserted singly and subtended by 2 membranous bractlets; perianth-segms. lanceolate, acuminate, 5–7 mm. long, mostly tinged brown, sometimes with green in the center, with broad scarious margins, the outer segms. somewhat the longer; anthers 6, much longer than fils.; caps. narrow-oblong, usually longer than perianth; seeds finely striate, long-tailed at both ends.— Rather dry rocky places, above 9000 ft.; Subalpine F.; San Bernardino Mts.; White Mts.; Sierra Nevada to B.C., Rocky Mts. July–Aug.

19. **J. pàtens** E. Mey. Tufted perennial from short stout rootstocks; stems blue-green, slender, terete, 4–8 dm. high; basal sheaths with or without awnlike blade rudiments, brown; cauline lvs. 0; infl. usually many-fld., open-paniculate to dense, 2.5–7 cm. long, the basal bract 5–20 cm. long; perianth-segms. lanceolate, subequal, acuminate, light brown with green midrib, 2.5–3 mm. long; anthers 6, ca. as long as fils.; caps. subglobose, slightly angled, almost as long as perianth, obtuse and apiculate; seeds obscurely and irregularly reticulate.—Moist places, such as stream banks and meadows, below 5000 ft.; San Clemente Id.; Ventura Co. n.; Santa Rosa and Santa Cruz ids.; to Ore. June–July. Reported from L. Calif. April–July.

20. **J. phaeocéphalus** Engelm. var. **paniculàtus** Engelm. Rootstocks stout, creeping; stems flat, 3–9 dm. high; lvs. compressed, equitant, without auricles, the blades 6–20 cm. long, 1.5–4 mm. wide, often with complete septa; infl. of many few-fld. heads in a loose panicle, the heads mostly 5–10 mm. in diam.; perianth 4–5 mm. long, the segms. tending to be pale reddish-brown or dark-tinged at the tips; anthers 6, longer than the fils.; caps. oblong-ovoid, abruptly acuminate, prominently beaked; seeds reticulate, subtruncate at the apex.—Stream banks and meadows below 9000 ft.; several Plant Communities; San Diego Co. n. to Napa Co. May–Aug.

Var. **phaeocéphalus** [*J. p.* var. *glomeratus* Engelm.] Stems 1–5 dm. high; infl. of 1 to several heads, these 1–1.5 cm. in diam., many-fld.; perianth 4–5 mm. long, dark brown or purplish-brown.—Damp places near the coast; mostly Coastal Strand, Coastal Sage Scrub; Los Angeles Co. and Santa Rosa Id.; to Ore. May–July.

21. **J. rugulòsus** Engelm. Rootstocks stout, creeping, with rather close, single, erect, fairly stout stems 4–10 dm. tall; stems, lvs. and sheaths minutely transverse-rugose; basal sheaths bladeless; cauline lvs. 2–4, terete, 1–4 dm. long, septate; auricles commonly 4–6 mm. high; panicle diffuse, 1–2.5 dm. long, many-headed; heads 4–8-fld.; perianth-segms. light greenish-brown, 2.5–3 mm. long, lance-acuminate, subequal, hyaline-margined to near apex; anthers 6, equaling fils.; caps. narrow, beaked, longer than perianth; seeds reticulate, apiculate at each end.—Frequent in wet places below 6500

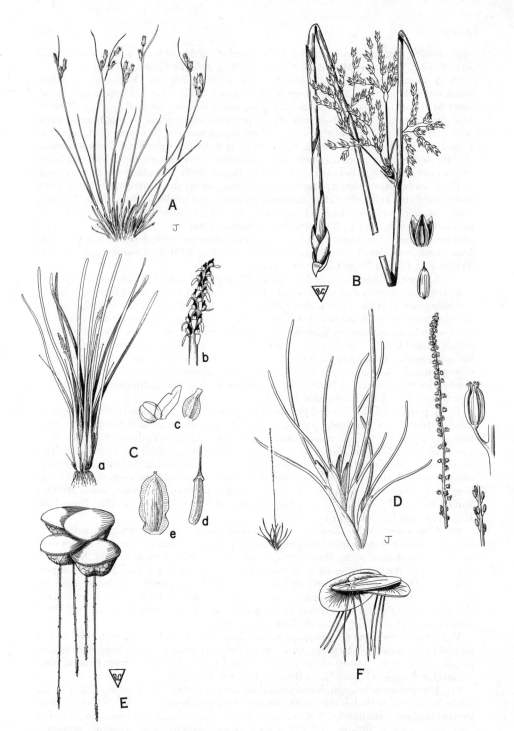

Plate 94. Fig. A, *Juncus parryi*, tufted subalpine plant. Fig. B, *Juncus textilis*, with creeping rootstock. Fig. C, *Lilaea scilloides*, each axil with spike of perfect fls. above and a basal ♀ fl. (see also *d*), *e*, *f* r. from spike. Fig. D, *Triglochin maritimum*, stigmas plumose, carpels 3–6. Fig. E, *Lemna gibba*, swollen frond with 1 rootlet. Fig. F, *Spirodela polyrhiza*, frond several-nerved, with several rootlets.

ft.; many Plant Communities; cismontane and montane s. Calif.; to San Luis Obispo Co. April–July.

22. **J. saximontànus** A. Nels. f. **brunnéscens** (Rydb.) F. J. Herm. [*J. b.* Rydb.] Rootstocks rather stout, creeping; stems flat, 4–6 dm. high; lf.-blades equitant, flat, 1–2.5 dm. long, 2–5 mm. wide, with small but definite auricles; panicle open, the lowest bract rather short; heads many, 5 mm. in diam., 5- to 12-fld.; perianth-segms. light brown, ca. 3 mm. long, lanceolate, slightly unequal; anthers 6, shorter than the fils.; caps. brown, oblong, slightly shorter than the perianth, obtuse below the beak; seeds reticulate.—Wet places, 6500–7500 ft.; Montane Coniferous F.; Santa Rosa, San Bernardino and San Gabriel mts.; Ariz. to Colo., Mex. July–Aug.

f. **saximontànus**, with few heads 7–10 mm. thick, 15–25-fld., has been taken at 6800 ft., Big Meadows, San Bernardino Mts.; Sierra Nevada to B.C., Rocky Mts.

23. **J. sphaerocárpus** Nees. Annual, with habit of *J. bufonius*, tending to be smaller; perianth 3–4 mm. long; anthers shorter than fils.; caps. subglobose to broadly ovoid; $2n = 36$ (Snogerup, 1958).—Occasional on mud flats and in damp places below 7500 ft.; several Plant Communities; throughout cismontane and montane s. Calif.; to Ore., Ida., Ariz. Old World. April–Aug.

24. **J. ténuis** Willd. var. **congéstus** Engelm. [*J. t.* var. *occidentalis* Cov. *J. o.* Wieg.] Tufted perennial with stiff stems and lvs., the former 2–6 dm. high, bright green; lvs. basal, mostly less than half as long; sheaths loose, auricles mostly less than 1 mm. long; infl. subcapitate to open, the bracts 2–3, much exceeding infl.; perianth-segms. 4–5 mm. long, with brown on each side of the central green and on to the hyaline margins; anthers 6, shorter than fils.; caps. oblong-ovoid, ca. ¾ as long as perianth; seeds obscurely oblong-reticulate.—Wet places at low elevs. near the Coast, Santa Barbara Co. n.; to Ore. May–July.

Var. **ténuis**. Stems slender, flexible; lvs. half as long or longer; auricles 1–3 mm. long; perianth-segms. green with white scarious margins.—Occasional in damp grassy places; Yellow Pine F.; San Bernardino and San Gabriel mts.; Sierra Nevada to Wash., Atlantic Coast. July–Aug.

25. **J. téxtilis** Buch. [*J. lesueurii* var. *elatus* Wats.] Plate 94, Fig. B. Rootstock stout, creeping; stems stout, terete, stiff, 1–2 m. high, 3–5 mm. thick, pale and distinctly striate and under a lens with low flat ridges; lf.-sheaths brown, basal, bladeless, the upper sometimes with a filamentose rudiment; lowest bract of infl. 5–15 cm. long, spinescent; infl. lax, many-fld., 5–15 cm. high; perianth-segms. ca. 4 mm. long, lanceolate, acuminate, subequal or the inner slightly shorter, greenish or pale brownish, scarious-margined; anthers 6, much longer than the fils.; caps. oblong-ovoid, obtuse, brownish, beaked, equal to or shorter than the perianth; seeds finely reticulate.—Wet places, below 6500 ft.; many Plant Communities; Orange and Riverside cos. to Ventura and Santa Barbara cos. May–June.

26. **J. tórreyi** Cov. [*J. nodosus* var. *megacephalus* Torr.] Rootstock slender, creeping, with tuberlike thickenings giving rise to stout stems 4–10 dm. high; lvs. septate, terete, divaricate, 2–5 mm. in diam., the upper as long as or longer than the infl.; auricles 1–2 mm. high; infl. congested, 1–20-headed, surpassed by lowest bract; heads 10–15 mm. in diam., round, many-fld.; perianth-segms. 4–5 mm. long, brownish to greenish, long-subulate, the outer longer than the inner; anthers 6, shorter than fils.; caps. subulate, slightly exceeding perianth; seeds reticulate in ca. 20 longitudinal rows.—Wet places, sea level to ca. 6000 ft.; from the coast to the mts. and sometimes in the desert; many Plant Communities; e. of the Sierra Nevada to Wash. and the Atlantic Coast. July–Sept.

27. **J. trifórmis** Engelm. [*J. t.* var. *stylosus* Engelm.] Cespitose annual, 5–12 cm. high; lf.-blades 1–2 cm. long, setaceous; peduncles 3–12, slender; heads 5–9 mm. in diam., 4–7-fld.; bracts ca. 1.2 mm. long, hyaline, ovate, blunt, spreading; pedicels ca. 1 mm. long; perianth-segms. appressed-ascending, 2.5–3.5 mm. long, slightly unequal, lance-linear, membranaceous, brownish-red with hyaline margins and green midribs, acute to acuminate; anthers 3, twice as long as fils.; caps. shorter than perianth, oblong; style 2–2.5 mm. long; seeds longitudinally ribbed.—Occasional, moist open places below

5000 ft.; Chaparral, etc.; possibly San Diego Co., San Bernardino Co.; to n. Calif. May–July.

28. **J. xiphioìdes** E. Mey. [*J. x.* vars. *auratus* and *littoralis* Engelm.] Rootstocks creeping, rather stout; stems flat, 5–9 dm. high; lvs. equitant, the blades flat, 1–4 dm. long, 3–12 mm. wide; auricles 0; bracts less than half as long as infl.; heads many, in a compound panicle, 3–20-fld.; perianth-segms. equal, lance-acuminate, light brown, 3–3.5 mm. long, narrow and revealing the caps.; anthers 6, a little shorter than fils.; caps. slightly exceeding perianth, oblong, gradually contracted below the beak; seeds reticulate.—Well distributed in wet places, sea-level to ca. 7000 ft., cismontane and montane s. Calif., occasional on the deserts; many Plant Communities; to Ore., Ariz., L. Calif. May–Oct.

2. *Lúzula* DC. Wood Rush

Perennial herbs with slender hollow simple stems. Lvs. flat, grasslike, soft, the sheaths with united margins. Infl. umbellate, paniculate or congested, bracteolate. Perianth-segms. 6. Stamens 6 in our spp. Ovary 1-celled, the 3 ovules inserted basally. Seeds 1–3, indistinctly reticulate, sometimes carunculate at base or apex. Spp. ca. 80, of temp. and colder regions, mostly in the N. Hemis. (Latin, *lucus*, a wood or thicket, in allusion to the habitat.)

Lvs. channeled; infl. dense, nodding, spicate .... 2. *L. spicata*
Lvs. flat.
  Perianth 2.5–3.5 mm. long, ca. as long as caps. Montane .... 1. *L. comosa*
  Perianth mostly 3.5–6 mm. long, much exceeding caps. Insular .... 3. *L. subsessilis*

1. **L. comòsa** E. Mey. [*L. multiflora* var. *c.* St. John. *Juncoides c.* Sheld. *L. m.* subsp. *c.* Hult.] Tufted, 1–6 dm. tall; lvs. thin, light green, 5–15 cm. long, 3–6 mm. wide, sparsely long-pilose especially at junction of sheath and blade; infl. umbellate, the rays 0.5–5 cm. long; spikes subglobose to short-cylindric, ca. 5–12 mm. long, 5–7 mm thick; lowest bract foliose, 2–5 cm. long; bractlets hyaline, subentire to lacerate, ± ciliate especially in terminal half; perianth 2.5–3.5 mm. long, light brown with broad hyaline margins; anthers equal to or longer than fils.; caps. light brown, equal to slightly shorter than perianth; seeds red-brown, 1.5 mm. long; $2n=24$ (Nordenskiold, 1951).—Dry meadows and open woods, mostly Montane Coniferous F., 3700–9100 ft.; mts. San Diego Co., San Jacinto Mts., San Bernardino Mts.; to Alaska, Rocky Mts. May–July.

2. **L. spicàta** (L.) DC. [*Juncus s.* L. *Juncoides s.* Kuntze. *L. s.* var. *nova* Smiley.] Densely tufted, 1–3.5 dm. tall; lvs. mostly basal, 3–15 cm. long, 1–4 mm. wide; infl. nodding, spikelike, often interrupted, 1–2.5 cm. long; bracts pilose-fringed; perianth 2–2.5 mm. long, brown with hyaline margins; seeds ca. 1 mm. long.—Open gravelly places, very high elevs., White Mts.; Sierra Nevada to Alaska, Atlantic Coast; Eurasia. July–Aug.

3. **L. subséssilis** (Wats.) Buch. [*L. comosa* var. *s.* Wats. *L. c.* var. *macrantha* Wats. *L. campestris* var. *m.* Fern. & Wieg.] Tufted, 1–3 dm. high; lvs. dull green, mostly basal, loosely long-ciliate, 5–15 cm. long, 3–7 mm. wide; infl. mostly umbellate, sometimes subcapitate; heads usually separate, globose to subcylindric; lowest bracts foliaceous, from shorter than to exceeding infl.; bractlets ± lacerate-ciliate, brownish with hyaline margin, 3–6 mm. long; anthers ca. as long as fils.; caps. greenish, much shorter than perianth; seeds ca. 1.25 mm. long; $2n=12$ (Nordenskiold, 1951).—Grassy and wooded places, below 4000 ft.; Chaparral, Closed cone Pine F., etc.; Santa Cruz and Santa Rosa ids., Santa Ynez Mts.; to B.C. March–May.

11. Juncaginàceae (Scheuchzeriaceae). Arrow-Grass Family

Annual or perennial marsh herbs, with rhizomes and scapes. Lvs. mostly basal, linear, sheathing at the base. Fls. small, in terminal bractless spikes or racemes or some borne singly in the axils of lvs.; fls. perfect or imperfect, anemophilous. Perianth-segms. 3 or 6, in 1 or 2 series, greenish or reddish, or perianth none. Stamens 6, 4 or 3 or 1; anthers subsessile, 2-celled, extrorse. Carpels 6, 4 or 3 or 1, superior, free or connate; style short to absent; ovule 1, basal. Fr. cylindrical to obovoid; seeds without endosperm. Several genera, mostly in cooler parts of both hemis.

Perianth present; stamens 3 or 6 .... 2. *Triglochin*
Perianth absent; stamen 1 .... 1. *Lilaea*

1. *Lilaèa* Humb. & Bonpl.

Acaulescent annuals with fibrous roots and basal spongy linear lvs. with sheathing base. Infls. of 2 types, some of the fls. basal and some in spikes on slender scapes. Basal fls. ♀, borne singly in the axils of the lvs., consisting of a single sessile, erect carpel with a long filiform style and hemispheric oblique penicillate floating stigma; fr. indehiscent, divergent, longitudinally 25–30-ribbed, obliquely beaked by the persistent sharply recurved base of the style. Fls. of the scape usually perfect, tiny, irregularly embedded in the fleshy axis of the spike, consisting of a single caducous bract and a single anther opening by lateral slits and a sessile closely appressed pistil; some spikes wholly ♀ and ebracteate; fr. similarly ribbed, but with 2–4 ribs drawn into narrow undulate wings. One sp. (Named for the French botanist, Alire Raffeneau-*Delile*, 1778–1850, who wrote on the plants of Egypt.)

1. **L. scilloìdes** (Poir.) Haum. [*Phalangium s.* Poir. *Lilaea subulata* Humb. & Bonpl.] Flowering Quillwort. Plate 94, Fig. C. Tufted; lvs. 5–20 (–35) cm. long, 1–5 mm. wide, subulate-tipped, the hyaline sheaths 3–10 cm. long; basal fls. with styles 6–20 cm. long; scapes 6–20 cm. long, the short dense erect spike borne at the water level; fl. spike 4–7 mm. long; fr. in the lf.-axils 5–7 mm. long, light brown; fr. in the spike more tapering, green; $2n=12$ (Larsen, 1966).—In muddy and marshy places, sometimes brackish, below 5000 ft.; Freshwater Marsh; scattered in s. Calif. w. of the mts.; to B.C., Rocky Mts., Mex.; S. Am. March–Oct.

2. *Triglóchin* L. Arrow Grass

Lvs. fleshy, ligulate at junction of sheath and the semiterete or compressed blade. Scapes long, naked, smooth. Perianth-segms. concave. Carpels united, 1-celled; style short; stigmas as many as carpels, plumose. Fr. a cluster of 3–6 1-seeded carpels, separating when mature from base up and dehiscing. Seeds compressed or angular. Spp. ca. 12, temp. and cooler parts of both hemis. (Greek, *tri*, three, and *glochis*, a point, referring to the fr. of some spp.)

Carpels and stigmas 6; fr. oblong or ovoid.
    Ligules 2-parted to base; blade nearly terete .... 1. *T. concinnum*
    Ligules simple, entire or emarginate; blade obcompressed .... 2. *T. maritimum*
Carpels and stigmas 3; fr. nearly globose .... 3. *T. striatum*

1. **T. concínnum** Davy var. **concínnum.** From slender elongate rootstocks; lvs. subterete, 1–2 mm. wide, with bases of old lvs. evanescent; ligules 2-parted; scapes slender, 1–4 dm. high; racemes 5–15 cm. long; pedicels 2–3 mm. long; perianth-segms. 6; carpels and stigmas 6; fr. oblong-ovoid, 3–4 mm. long, shallowly grooved; $2n=48$ (Löve & Löve, 1958).—Coastal Salt Marsh; L. Calif. to B.C. March–Aug.

Var. **débile** (Jones) J. T. Howell. [*T. maritimum* var. *d.* Jones. *T. d.* Löve & Löve.] Rootstock fibrous-coated; scapes 1.5–6 dm. high; $2n=96$ (Löve & Löve, 1958).—Wet saline places especially near hot springs, below 6500 ft.; Creosote Bush Scrub, to Pinyon-Juniper Wd.; Mojave Desert; to Modoc Co., N. Dak., Colo. May–Oct.

2. **T. marítimum** L. Plate 94, Fig. D. Densely tufted from short thick rootstocks; lvs. strongly obcompressed, ca. 2 mm. wide, with indurate-corky persistent bases; ligules entire or emarginate; scapes 3–7 dm. high; racemes 1–4 dm. long; pedicels ascending, 2–3 mm. long in anthesis, 4–5 mm. in fr.; perianth-segms. 6; carpels and stigmas 6; fr. oblong-ovoid, 3–4 mm. long, each carpel with 2 dorsal winged angles and groove

between; $2n = 12$ to 144 (Löve & Löve, 1958).—Coastal Salt Marsh, San Diego and Los Angeles cos.; alkaline flats and boggy places below 7500 ft.; Montane Coniferous F.; San Bernardino Mts.; Sierra Nevada to Alaska, Atlantic Coast, Eurasia. April–Aug.

3. **T. striàtum** R. & P. [*T. triandrum* Michx.] Scapes 1 or 2, ± angular, 1.5–2 dm. high; lvs. ca. as long as scapes, 1.5–2 mm. wide, ligules not divided; racemes 3–12 cm. long; pedicels 1–2 mm. long, not elongating in fr.; perianth-segms. 3; carpels and stigmas 3; fr. subglobose, 2 mm. thick; carpels 3-ribbed on back.—Coastal Salt Marsh; Ventura and Santa Barbara cos. to n. Calif.; e. U.S., Mex., S. Am. May–Sept.

12. Lemnàceae. Duckweed Family

Minute floating perennial aquatics, consisting of a fleshy or membranaceous, loosely cellular, thallus-like stem or frond, rootless or with 1 or more simple rootlets. Lvs. lacking. New stems produced from 2 lateral vegetative pouches or 1 terminal, the new frond being attached to the old by a short slender stalk and usually soon separating. Fls. very rare, the infl. consisting of 1 ♀ and 1 or 2 ♂ fls., borne in a saclike spathe in a flowering pouch on the edge of the upper surface of the frond. Perianth 0. Staminate fl. consists of 1 stamen, with 2–4 pollen sacs. Pistillate fl. of 1 flasklike pistil, with 1–several ovules. Fr. a 1–6-seeded utricle. Ca. 4 genera and 30 spp.

(Daubs, E. H. A monograph of Lemnaceae. Ill. Biol. Mon. 34: 1–118. 1965.)
Rootlets present.
  Roots solitary on each plant; plants usually less than 5 mm. long .... 1. *Lemna*
  Roots usually 2 or more on each plant; mature plants at least 5 mm. long ...... 2. *Spirodela*
Rootlets absent .... 3. *Wolffiella*

1. *Lémna* L. Duckweed

Fronds flattened, 1–5-nerved, each with 1 root and 2 vegetative pouches on the margin near the base. Fls. 3, borne in the flowering pouch, 2 ♂, each of 1 stamen, and 1 ♀, consisting of a single naked pistil. Ovary 1-celled, 1–several-ovuled. Fr. ribbed. Spp. 9; all continents. (Possibly Greek, *limnos*, lake, referring to living in swamps.)

Fronds oblong, long-stalked, 6–12 mm. long, mostly submerged .... 4. *L. trisulca*
Fronds oblong-ovate to elliptical, not stalked, 2–5 mm. long, floating.
  Dorsal surface flat, smooth, with no prominent protuberances; 1-veined or veinless.
    Fronds 1.5–4 mm. long, oval, symmetrical, seldom more than 2 remaining attached
                                                                        2. *L. minima*
    Fronds 2.5–5 mm. long, narrowly elliptical, often 8–10 attached .... 5. *L. valdiviana*
  Dorsal surface with ± prominent protuberances, indistinctly to prominently 3-veined.
    Ventral surface of frond flat to slightly convex, but not inflated.
      Dorsal surface mottled yellow-green; apex asymmetrical; air spaces prominent. 1. *L. gibba*
      Dorsal surface dark green; frond apex symmetrical; air spaces not prominent . 3. *L. minor*
    Ventral surface of frond noticeably convex; air spaces inflated; both surfaces showing red-purple coloring .... 1. *L. gibba*

1. **L. gíbba** L. Plate 94, Fig. E. Fronds rounded to obtuse, 1–4 in a group, 2–5 mm. long, 2–4 mm. wide, usually 3–5-nerved, convex above, pale beneath and ± gibbous; $2n = 64$ (Blackburn, 1963).—Fairly common in pools below 7000 ft.; many Plant Communities; San Diego Co. n.; to Ore., N. Dak., Tex., Mex., Eurasia.

2. **L. mínima** Phil. Fronds solitary or in 2's or 4's, oblong-elliptical, 1.5–4 mm. long, 1–1.5 mm. wide, 1-nerved or nerveless, with row of papillae along nerve on upper surface, flat or slightly convex beneath.—Occasional but well distributed, pools below 5000 ft.; several Plant Communities; San Diego Co. to desert edge in San Bernardino Co., n. to Trinity Co.; to Ore., Rocky Mts., e. U.S.; S. Am.

3. **L. mìnor** L. Fronds 1 to few together, round to round-obovate, thickish, with convex, green or purplish surfaces, 2–4 mm. long, ± faintly 3-nerved and sometimes with upper row of papillae, the apical one prominent; $2n = 40$ (Blackburn, 1933).—Stagnant

pools and quiet water below 6500 ft.; several Plant Communities; San Diego Co., n. to Wash.; all continents except S. Am.

4. **L. trisúlca** L. Fronds oblong to almost lanceolate, 6–10 mm. long, 2–3 mm. wide; remaining connected and forming chainlike colonies, dentate toward tips, obscurely 3-nerved; usually submerged, with or without rootlets; $2n = 44$ (Blackburn, 1933).—Occasional in cold springs, streams and ponds below 7500 ft., several Plant Communities; San Bernardino Mts. and neighborhood as Mentone, Victorville; e. of Sierra Nevada to B.C., Atlantic Coast; all continents.

5. **L. valdiviàna** Phil. [*L. v.* var. *abbreviata* Hegelm. *L. minor* var. *cyclostasia* Ell. *L. c.* auth.] Fronds thin, 1–2–8 in a group, oblong, subfalcate, asymmetrical, 2.5–4 mm. long, 1–1.5 mm. wide, without papillae, obscurely 1-veined.—Quiet water, below 7000 ft.; several Plant Communities; San Diego Co., n. to cent. Calif.; Atlantic Coast, S. Am.

2. *Spirodèla* Schleid. GREATER DUCKWEED

Fronds flattened, discoid, 5–15-nerved, 2–several-rooted. Vegetative pouches 2, triangular, opening as clefts in either margin of the basal portion of the frond. Two ♂ fls. and 1 ♀ borne in a flowering pouch and subtended by a saclike spathe. Fils. curving upward from the margin of the frond; anthers 2-celled. Pistil 1, naked, 2-ovuled. Fr. round-lenticular, with wing-margin. Five spp., from all continents. (Greek, *speira*, a cord, and *delos*, evident, referring to the roots.)

1. **S. polyrhìza** (L.) Schleid. [*Lemna p.* L.] Plate 94, Fig. F. Fronds solitary or in groups of 2–5, 3–8 mm. long, 3–4.5 mm. wide, flat and green above, subconvex and purplish beneath, usually sterile, with 4–several roots; $2n = 40$ (Blackburn, 1933).—Occasional in quiet water, below 5000 ft., as at Victorville; more numerous in San Joaquin V.; to B.C., Atlantic Coast; cosmopolitan.

3. *Wolffiélla* Hegelm.

Fronds minute, rootless, sickle-shaped or strap-shaped, thin, asymmetrical, curved, brown-punctate on both surfaces. Vegetative pouch 1, triangular, opening as a cleft in the basal margin of the frond. Fls. 2 in a pouch, the ♂ a single stamen, the ♀ a single pistil. Six spp., subtrop. and trop. (Diminutive of *Wolffia*, a genus named for J. F. Wolff, German botanist, 1778–1806.)

1. **W. oblónga** (Phil.) Hegelm. [*Lemna o.* Phil.] Fronds 1–2, rarely 3, tapering from the obliquely rounded base to the slightly narrower bluntly rounded apex, 2–4 mm. long, 0.5–1 mm. wide.—Pools and slow water, Santa Ana River system at low elevs., Orange and San Bernardino cos.; Mex. to S. Am.

13. **Liliàceae.** LILY FAMILY

Leafy-stemmed or scapose perennial herbs, sometimes somewhat woody or climbing; from bulbs, corms or rootstocks which may be somewhat tuberous. Lvs. mostly alternate, sometimes opposite or whorled, sometimes all basal, broad or grasslike, parallel-veined, the veins often connected by cross veinlets. Fls. perfect or imperfect, mostly showy and with colored parts, sometimes small and greenish, but then usually many in racemes, spikes or panicles. Perianth usually of 6 distinct parts, sometimes gamophyllous and 6-lobed, rarely 4-merous or 3-merous, often all parts petaloid. Stamens 6, sometimes 4 or 3, inserted at or on the base of the perianth-segms. or borne on the perianth-tube; anthers usually 2-celled, rarely 1-celled. Ovary mostly superior or nearly so, usually 3-loculed, sometimes 1–2-celled; ovules few to many in each locule and with axile placentation; styles 1 or 3; stigma capitate to lobed. Fr. a caps. or a berry, mostly several- to many-seeded. A large family, widely distributed, most abundant in temp. and subtrop. regions. Many important as food, for ornamental use, as source of drugs, fibers, etc.

1. Plant not bulbous, producing a rootstock or thickened tuberous ± branched underground parts.
    2. Green "foliage" represented by needlelike or flattened branchlets borne in the axils of scalelike lvs. .... 1. *Asparagus*
    2. Green foliage of true lvs., not borne in the axils of scales.
        3. Lvs. all basal, fleshy; root system of several tubers; stems 2–6 dm. high. Occasional introduction .... 2. *Asphodelus*
        3. Lvs. not all basal, but well distributed on stems. Native.
            4. Panicles mostly 2–6 dm. long; fr. a caps. .... 9. *Veratrum*
            4. Panicles or racemes less than 1 dm. long; fr. a berry .... 8. *Smilacina*
1. Plant with a tunicated or scaly bulb or a corm.
    5. The stems leafy, sometimes branched above.
        6. The styles 3 and distinct; perianth-segms. with a prominent gland ..... 10. *Zigadenus*
        6. Style 1, ± lobed at summit.
            7. Perianth-segms. unlike, the outer ± sepallike, the inner petaloid .... 3. *Calochortus*
            7. Perianth-segms. alike, petaloid.
                8. The perianth-segms. 15–30 mm. long, purplish or brownish .... 5. *Fritillaria*
                8. The perianth-segms. 60–90 mm. long, orange or yellow .... 7. *Lilium*
    5. The stems scapelike, with lvs. mostly at the base.
        9. Perianth-segms. united at base into a tube, the funnelform perianth 4–6 cm. long
        6. *Hesperocallis*
        9. Perianth-segms. not united at base into a tube, the spreading perianth 0.5–3 cm. long
        4. *Chlorogalum*

1. Aspáragus L. Asparagus

Perennial herbs, with thick matted rootstocks or tuberous roots. Stems much branched, ending in filiform or flattened green branchlets borne in the axils of scalelike lvs. Fls. small, greenish-yellow, usually in racemes or umbels. Perianth-segms. alike. Stamens 6; anthers versatile. Ovary 3-celled, 2- or few-ovuled; stigmas 3. Fr. a globose berry. Ca. 150 spp.; Old World. (The ancient Greek name.)

1. **A. officinàlis** L. GARDEN ASPARAGUS. Young stems simple, stout, edible, later branching and becoming 1–2 m. high; perianth bell-shaped, ca. 6 mm. long; berry red (the plants largely dioecious); $n = 10$ (Nagao, 1938), 10, 20 (Zilm, 1966).—Escape from gardens and fields, especially in low subsaline places; native of Eu. May–June. Santa Catalina Id.

Two other spp. occasionally become established, at least temporarily: **A. asparagoìdes** Wight (SMILAX of florists), a vine with ovate "lvs." 2–2.5 cm. long, shining, stiff; and **A. spréngeri** Regel, roots with short white tubers, stems scrambling or drooping and 1–2 m. long, "lvs." flat, linear, ca. 2–2.5 cm. long, 3–8 together. Both of these spp. have been taken in the Ventura–Santa Barbara region, *Pollard*.

2. Asphodèlus L. Asphodel

Root system of several tubers; stems 2–6 dm. high, glabrous; lvs. tufted, radical, spirally arranged. Fls. in open panicles, white to pinkish, with jointed pedicels, the perianth-segms. with a single colored midvein. Caps. subglobose. (Greek name of unknown meaning.)

1. **A. fistulòsus** L. Lvs. straight, semiterete; perianth 8–12 mm. long; caps. 4–6 mm. long.—Recently collected as an adventive, in San Diego, w. Riverside, Santa Barbara cos.; native of Medit. region. Spring.

3. Calochórtus Pursh. MARIPOSA-LILY. STAR-TULIP. BUTTERFLY-TULIP

From tunicated bulbs, with membranous or fibrous-reticulate coats. Stems scapiform or leafy, simple or branched, frequently bulbiferous in the axils of lowest lf. or lvs., sometimes of upper lvs. Lvs. usually linear, the basal solitary, often very large, the cauline reduced up the stem. Infl. with cent. axis or subumbellate by its suppression, the bracts

Calochortus

usually as long as pedicels and opposite them. Fls. conspicuous, globose to open-campanulate, erect or nodding, white, yellow, red, lavender, purple, bluish or brownish, often variously tinged. Perianth-segms. well differentiated, the sepals ovate to lanceolate, ± colored, usually naked; petals usually larger and broader, cuneate or clawed, usually bearded on inner face, variously spotted and patterned and bearded, with a gland near the base. Stamens inserted on the base of the segms.; fils. usually dilated; anthers basifixed, linear to oblong. Ovary 3-loculed, with many ovules in 2 rows in each locule; stigmas 3, sessile, persistent. Caps. linear to orbicular, 3-angled to -winged, septicidal, erect or nodding. Seeds irregular or flattened, usually with hexagonally reticulate coats. Ca. 60 spp. of temp. w. N. Am.; many of great beauty and horticultural interest. The bulbs of many were eaten by the Indians. (Greek, *kalos*, beautiful, and *chortus*, grass, referring to the fls. and lvs.)

(Ownbey, M. A. A monograph of the genus Calochortus. Ann. Mo. Bot. Gard. 27: 371–580. 1949.)
1. Fls. globose to closed-campanulate, nodding, 2–2.5 cm. long; caps. nodding ..... 1. *C. albus*
1. Fls. bowl-shaped to broadly campanulate, erect or spreading.
  2. Surface of gland naked or nearly so, although the bordering membrane may be fringed and extend over the gland.
    3. Petals rarely fimbriate, glabrous at apex, pinkish or purplish .... 12. *C. plummerae*
    3. Petals fimbriate, bearded nearly or quite to the apex, orange to brown or purplish
        17. *C. weedii*
  2. Surface of gland densely hairy, often with peculiar thickened hairs.
    4. Glands not depressed or but slightly so and not surrounded by a membrane.
      5. Infl. with a distinct axis, even though its internodes be short.
        6. Glands covered with linear hairs or processes.
          7. Petals with dark blotch near gland; stem erect.
            8. Processes on gland dark; petal-blotch purplish; caps. oblong, obtuse. Orange Co. to Santa Barbara Co. .... 2. *C. catalinae*
            8. Processes on gland yellow; petal-blotch red-brown; caps. linear, acute. Mts. of San Diego Co. ....,,..................... 5. *C. dunnii*
          7. Petals without dark blotch; stems sinuous. Deserts .... 6. *C. flexuosus*
        6. Glands with clavate or fungoid processes.
          9. Gland-hairs or -processes clavate; anthers white .... 11. *C. palmeri*
          9. Gland-hairs or -processes fungoid with stellate tips, or lacking; anthers purplish or blue .... 13. *C. splendens*
      5. Infl. subumbellate.
        10. Gland oblong, with slender gland-processes. Mojave Desert .... 14. *C. striatus*
        10. Gland not oblong, with short thick processes.
          11. The gland quadrate, each petal with a dark red blotch .... 16. *C. venustus*
          11. The gland not quadrate, petals with or without cent. blotch.
            12. Gland shaped like an inverted V; petals white to yellowish to lavender. Mainland .... 15. *C. superbus*
            12. Gland lunate; petals deep yellow. Insular .... 9. *C. luteus*
    4. Glands ± depressed, surrounded by a membrane.
      13. Fls. white to purplish.
        14. Petals broadly obovate, not spotted. Panamint Mts. .... 10. *C. nuttallii*
        14. Petals cuneate-obovate, sometimes spotted above the gland. Tehachapi Mts. to Laguna Mts. .... 7. *C. invenustus*
      13. Fls. red to yellow.
        15. Hairs on face of petal not enlarged distally. San Bernardino Co. to L. Calif.
          4. *C. concolor*
        15. Hairs on face of petal distally enlarged.
          16. Glands with much-branched fungoid processes; petal-hairs well represented. Cismontane .... 3. *C. clavatus*
          16. Glands with simple or somewhat branched processes; petal-hairs few. Deserts
          8. *C. kennedyi*

1. **C. álbus** Dougl. ex Benth. [*Cyclobothra a.* Benth. *Cyclobothra paniculata* Lindl. *Calochortus lanternus* Davids.] FAIRY LANTERN. GLOBE-LILY. Plate 95, Fig. A. Stem rather

Plate 95. Fig. A, *Calochortus albus*. Fig. B, *Calochortus invenustus*, petal-gland small, surrounded by a fringed membrane. Fig. C, *Calochortus splendens*, glands with fungoid processes. Fig. D, *Chlorogalum pomeridianum*. Fig. E, *Smilacina stellata*. Fig. F, *Fritillaria biflora*.

slender, erect, 2–8 dm. tall, branched; basal lf. 3–7 dm. long, 1–5 cm. wide; cauline lvs. 2–6, lanceolate to linear, 0.5–2.5 dm. long; bracts often paired, lanceolate, 1.5–5 cm. long; fls. white, globose to globose-campanulate, nodding; sepals 1–1.5 cm. long; petals elliptic or wider, ciliate and with slender hairs above the gland, 2–2.5 cm. long; gland depressed, with several transverse, fringed membranes; fils. 4–5 mm. long; anthers 4 mm. long; caps. elliptic-oblong, 3-winged, nodding, 2.5–4 cm. long; $n=10$ (Beal & Ownbey, 1943).—Shaded, often rocky places in woods or brush, below 5000 ft.; Chaparral, Foothill Wd., Yellow Pine F.; San Diego Co. to n. Calif. Santa Barbara Ids. April–June.

2. **C. catalìnae** Wats. [*Mariposa c.* Hoov. *C. lyonii* Wats.] Stem erect, somewhat zigzag, 2–6 dm. high, usually branched above, bulbiferous near base; lvs. linear, 1–2.5 dm. long, 3–6 mm. wide, the basal usually withered at anthesis; bracts 2–10 cm. long; fls. 1–several, bowl-shaped, white tinged with lilac or light purple, with purple spot near base of each sepal and petal; sepals acuminate, 2–3 cm. long; petals obovate, cuneate, rounded and obtuse, 2–5 cm. long, naked except for few hairs near base; gland not depressed, oblong, densely covered with slender processes; fils. 8–10 mm. long; anthers lilac, oblong, 4–5 mm. long; caps. narrow-oblong, obtuse, erect, 2–5 cm. long, 8–10 mm. wide; $n=7$ (Beal & Ownbey, 1943).—Heavy soil, on open grassy slopes and openings in brush, below 2000 ft.; V. Grassland, Chaparral; San Diego Co. to San Luis Obispo Co. Santa Catalina Id., Santa Rosa Id., Santa Cruz Id.

3. **C. clavàtus** Wats. ssp. **clavàtus**. [*Mariposa c.* Hoov.] Stem coarse, zigzag, rarely bulbiferous, 5–10 dm. high; lower lvs. 1–2 dm. long, linear; bracts 4–8 cm. long; fls. 1–6, subumbellate, erect, cup-shaped, deep yellow; sepals lance-ovate, 2.5–3.5 cm. long; petals broadly cuneate-obovate, with clavate hairs about the gland, 3.5–5 cm. long; gland circular, deeply impressed, surrounded by fringed membrane and densely covered with short processes with branched fungoid tips; fils. ca. 1 cm. long; anthers deep purple; caps. lance-linear, acuminate, angled, 6–9 cm. long.—Largely on soil of serpentine origin, San Luis Obispo to Santa Barbara Co. April–June.

Var. **grácilis** Ownbey. Stem slender, 2–3 dm. high; petals sparsely bearded, 3–4 cm. long.—Canyons below 2500 ft.; Chaparral; s. base of San Gabriel Mts.

Ssp. **pállidus** (Hoov.) Munz. [*Mariposa clavata* var. *pallida* Hoov.] Petals light yellow; anthers yellow to pale or medium purple.—San Joaquin and Stanislaus cos. to n. Los Angeles and Kern cos.

4. **C. concólor** (Baker) Purdy. [*C. luteus* var. *c.* Baker.] GOLDENBOWL MARIPOSA. Stem rather stout, erect, sparingly branched, rarely bulbiferous, 3–6 dm. high; lvs. glaucous, linear, becoming involute, 1–1.5 dm. long; bracts 4–8 cm. long; fls. 1–4, subumbellate, campanulate, erect, yellow within, often tinged purple in drying, usually with dark red blotch near base of each sepal and petal; sepals lance-ovate, 2.5–3 cm. long; petals cuneate-obovate, with few long yellow hairs near the gland, 3–5 cm. long; gland usually small, roundish, depressed, surrounded by fringed membrane, covered with slender unbranched processes; anthers yellowish; caps. lance-linear, acuminate, erect, 5–8 cm. long. —Dry slopes, 2000–7500 ft.; Chaparral, Yellow Pine F., s. face of San Bernardino Mts. to n. L. Calif. May–July.

5. **C. dúnnii** Purdy. [*C. palmeri* var. *d.* Jeps. & Ames.] Stem slender, erect, usually branched, 2–6 dm. high; lvs. 1–2 dm. long, 6–10 mm. wide, channeled; bracts 1–2 cm. long; fls. open-campanulate, erect, white or flushed with pink, with red-brown spot above the gland; sepals lance-ovate, 1.5–2 cm. long; petals obovate, cuneate, usually rounded above, 2–3 cm. long, with yellowish hairs near the gland; gland rounded, not depressed, covered with linear yellow processes; fils. 5–6 mm. long; anthers white, acute, 4–5 mm. long; caps. linear, erect, 2.5–3 cm. long.—Dry stony ridges, 4500–5000 ft.; Chaparral, Yellow Pine F.; Julian and Otay Mt. regions, San Diego Co.; L. Calif. May–June.

6. **C. flexuòsus** Wats. Stem branched, usually sinuous and intertwined with other plants, or straggling over the ground, 2–4 dm. long; lvs. linear, 1–2 dm. long; bracts 1–3 cm. long; fls. 1–4, campanulate, erect, white with lilac tinge and having purple spot with yellow band on each segm.; sepals lanceolate to lance-ovate, 2–2.5 cm. long; petals obovate, cuneate, rounded above, with few short hairs near the gland, 3–4 cm. long; gland

not depressed, transverse-lunate or wider, with dense short processes; fils. 6–10 mm. long; anthers oblong, 5–7 mm. long; caps. lanceolate, erect, 2.5–3.5 cm. long, 8–10 mm. thick; $n=7$ (Beal & Ownbey, 1943).—Dry stony slopes, desert hills and mesas, 2000–5000 ft.; Creosote Bush Scrub, Sagebrush Scrub; e. Mojave Desert s. to Chuckwalla Mts., Riverside Co.; to sw. Colo. April–May.

7. **C. invenústus** Greene. [*C. nuttallii* var. *australis* Munz.] Plate 95, Fig. B. Slender, erect, mostly simple, bulbiferous, 1.5–5 dm. tall; lvs. linear, 1–2 dm. long, 2–4 mm. wide, becoming involute; bracts 2–5 cm. long; fls. 1–5, subumbellate, erect, campanulate, white or dull lavender to purplish, sometimes with purplish spot below the gland; sepals lance-ovate, 1.5–2.5 cm. long; petals cuneate-obovate, obtuse to apiculate, with few short hairs near the gland, 2–3.5 cm. long; gland small, circular, slightly depressed, surrounded by a fringed membrane, densely covered with short distally branched processes; fils. 6–7 mm. long; anthers purplish or yellowish, 7–8 mm. long; caps. lance-linear, acute, 5–7 cm. long; $2n=7$ pairs (Raven, Kyhos & Hill, 1965).—Dry, mostly granitic soil, 4500–9500 ft.; Montane Coniferous F.; Laguna Mts. to Tehachapi Mts.; to cent. Calif. May–Aug.

8. **C. kénnedyi** Porter. Erect, rather simple, 1–2 (–5) dm. high, sometimes twisted, rarely bulbiferous; lower lvs. glaucous, linear, channeled, 1–2 dm. long, cauline reduced; bracts 2–4 cm. long; fls. 1–6, subumbellate, vermilion to orange, campanulate; often with brown-purple spots near the base of each segm.; sepals ovate, 1.5–2.5 cm. long; petals cuneate-obovate, 2.5–5 cm. long, with few slightly enlarged hairs near the gland; gland round, depressed, surrounded by a fringed membrane, densely covered with simple or cleft processes; fils. 4–5 mm. long; anthers purplish, 5–8 mm. long; caps. lance-linear, longitudinally striped, 4–6 cm. long; $n=8$ (Beal & Ownbey, 1943).—Heavy soil of open or brushy flats, 2000–6500 ft.; Creosote Bush Scrub, Joshua Tree Wd., Pinyon-Juniper Wd.; deserts of Inyo, Kern, Ventura and San Bernardino cos.; to Nev., Ariz. April–June. Plants from below 6000 ft. and from the Panamint Mts. to Cajon Pass have vermilion fls., while those in the e. Mojave have largely orange, but those from higher elevs. in the Panamint, Clark and Providence mts. have yellow fls. and are the var. **múnzii** Jeps.

9. **C. lùteus** Dougl. ex Lindl. [*C. l.* var. *citrinus* of Calif. refs. *Mariposa l.* Hoov.] Erect, slender, bulbiferous, 2–5 dm. high; lower lvs. 1–2 dm. long; bracts 1.5–8 cm. long; fls. 1–4, subumbellate, campanulate, deep yellow, the petals usually penciled below with red-brown lines, often with a median red-brown blotch; sepals lance-oblong, 2–3 cm. long; petals cuneate-obovate, with few slender hairs near the gland, 2.5–3.5 cm. long; gland not depressed, transverse, sublunate, with matted, short, hairlike processes; anthers 4–6 mm. long; caps. 3.5–6 cm. long; $n=7$, 10 (Beal & Ownbey, 1943).—Usually in heavy soil below 2000 ft.; Santa Cruz Id.; to n. Calif. April–June.

10. **C. nuttállii** Torr. var. **panaminténsis** Ownbey. Stem slender, 4–6 dm. tall; lower lvs. 1–1.5 dm. long, becoming involute; bracts 2–4 cm. long; fls. 1–4, subumbellate, campanulate, white tinged with lilac; sepals lanceolate, 1.5–2.5 cm. long; petals without spot above the gland; gland circular, depressed, surrounded by a conspicuous fringed membrane and densely covered with short simple or distally branched processes; anthers bluish or reddish; caps. 3–5.5 cm. long.—Dry slopes, 7500–10,500 ft.; Pinyon-Juniper Wd.; Panamint Mts., Inyo Co. June–July.

11. **C. pálmeri** Wats. var. **pálmeri**. [*C. splendens* var. *montanus* Purdy. *C. invenustus* var. *m.* Parish. *C. m.* Davids. *C. paludicola* Davids. *C. palmeri* var. *p.* Jeps. & Ames.] Erect, bulbiferous, often branched, slender, 2–6 dm. high; basal lvs. 1–2 dm. long; fls. open-campanulate, erect, white or flushed with pink, with red-brown spot above the gland; sepals lance-ovate, 1.5–2 cm. long; petals obovate, cuneate, usually rounded above, 2–3 cm. long, with yellowish hairs near the gland; gland rounded, not depressed, covered with linear yellow processes; fils. 5–6 mm. long; anthers white, 5–7 mm. long; caps. 4.5–5 cm. long; $n=7$ (Beal & Ownbey, 1943).—Meadows and places moist in early spring, 3500–6500 ft.; Chaparral, Yellow Pine F.; San Bernardino Mts. to Tehachapi Mts.; e. San Luis Obispo Co. May–July.

C. pálmeri var. **múnzii** Ownbey. Stem not bulbiferous at base; pedicels paired;

bracts opposite; glands purple or none.—Yellow Pine F., San Jacinto Mts. May–July.

12. **C. plúmmerae** Greene. [*C. weedii* var. *purpurascens* Wats.] Rather slender, usually branched, not bulbiferous, 3–6 dm. high; basal lf. 2–4 dm. long, 10–15 mm. wide, usually withered at anthesis; cauline lvs. reduced upward; bracts like upper lvs.; fls. 2, sometimes 4, broadly campanulate, erect, pink to rose, drying purplish; sepals long-acuminate, 3.5–4 cm. long; petals broadly cuneate-obovate, erose-dentate, rarely fimbriate, conspicuously bearded with long yellow hairs in transverse band; gland circular, slightly depressed, nearly naked, bordered with ring of orange hairlike processes; fils. 9–11 mm. long; anthers lance-linear, 10–14 mm. long; caps. 4–8 cm. long; $n=9$ (Beal & Ownbey, 1943). —Dry rocky places, often in brush, below 5000 ft.; Coastal Sage Scrub to Yellow Pine F.; Santa Monica Mts. to San Jacinto Mts. May–July.

13. **C. spléndens** Dougl. ex Benth. [*C. s.* vars. *atroviolaceus, major* and *ruber* Purdy. *C. davidsonianus* Abrams. *Mariposa s.* Hoov.] Plate 95, Fig. C. Erect, branched, rarely bulbiferous, 2–6 dm. high; basal lvs. 1–1.5 dm. long, 5–6 mm. wide; bracts 2–5 cm. long; fls. erect, campanulate, narrow at base, deep lilac, often with purple spot on each sepal and sometimes each petal; sepals acuminate, 2–2.5 cm. long; petals obovate, cuneate, rounded and irregular above, 2.5–5 cm. long, with scattered hairs below the middle; gland not depressed, usually with many branched, fungoid processes; fils. 7–8 mm. long; anthers dark, 5–7 mm. long; caps. erect, 5–7 cm. long, 5–8 mm. thick; $n=7$ (Beal & Ownbey, 1943); $n=14$ for *davidsonianus* (Beal & Ownbey, 1943).—Dry slopes in heavy or granitic soil, up to 7000 (–8500) ft.; Chaparral, V. Grassland, Yellow Pine F.; L. Calif. to cent. Calif. Coast Ranges. Santa Catalina Id. May–June.

14. **C. striàtus** Parish. Erect, not bulbiferous, 1–4.5 dm. high; basal lvs. 1–2 dm. long, 6–8 mm. wide; bracts linear, 1–2.5 cm. long; fls. subumbellate, campanulate, with narrow base, lavender with purple veins; sepals lanceolate, 1.5–2 cm. long, acuminate; petals obovate, cuneate, rounded and erose above, 2–2.5 cm. long, sparsely hairy about the gland; gland not depressed, oblong, covered with linear processes; fils. 5–7 mm. long; anthers 4–6 mm. long; caps. erect, 4.5–5 cm. long.—Alkaline meadows and springy places, 2500–4500 ft.; Creosote Bush Scrub; Mojave Desert at n. base of San Bernardino and San Gabriel mts.; to Las Vegas, Nev. April–June.

15. **C. supérbus** Purdy ex J. T. Howell. [*C. venustus* var. *citrinus* Baker as to type. *C. pratensis* Hoov.] Erect, bulbiferous, 4–6 dm. high; lvs. linear, 1.5–2.5 dm. long, 4–6 mm. wide; bracts 2–8 cm. long; fls. 1–3, subumbellate, erect, campanulate, white to yellowish or lavender, usually penciled with purple toward the base and each segm. with a median brown or purple blotch surrounded by a zone of bright yellow; sepals 2–3.5 cm. long; petals obovate, cuneate, rounded, obtuse, with few short hairs near the gland; gland not depressed, linear, ± Λ-shaped, densely covered with short hairlike processes; anthers 8–10 mm. long; caps. 5–6 cm. long, erect; $n=6, 7$ (Beal & Ownbey, 1943).—Open or wooded places below 5000 ft.; S. Oak Wd., etc.; Palomar Mts., San Diego Co.; Kern Co. n. to n. Calif. May–July.

16. **C. venústus** Dougl. ex Benth. Erect, stiff, 1–6 dm. high, bulbiferous; basal lvs. 1–2 dm. long, 2–5 mm. wide; bracts 2–8 cm. long; fls. 1–3, erect, campanulate, white to yellow, purple or dark red, each petal with dark red median blotch and often with a second paler blotch above the first; sepals curled back at tip, 2.5–3 cm. long; petals 3–4.5 cm. long, scattered-hairy on lower part; gland not depressed, subquadrate, covered with short processes; fils. 7–10 mm. long; anthers 7–10 mm. long; caps. linear, 5–6 cm. long, 5 mm. thick; $n=7$ (Beal & Ownbey, 1943).—Light soil, often decomposed granite, 1000–8000 ft.; V. Grassland, Foothill Wd., Yellow Pine F.; n. Los Angeles Co. to Mt. Pinos region and Kern Co. to cent. Calif. Apr.–July.

17. **C. weèdii** Wood var. **intermèdius** Ownbey. Stem slender, usually branched, not bulbiferous, 3–8 dm. high; basal lf. 2–4 dm. long, 10–15 mm. wide, withering before anthesis; fls. 2, sometimes 3–4, open-campanulate, erect, 2.5–3 cm. long; petals broadly cuneate-obovate, purplish, rounded, fringed with dark or yellow hairs, bearded on most of inner face with long yellow hairs; gland round, slightly depressed, almost naked, bordered with ring of long yellow hairs; fils. 8–12 mm. long; anthers 10–14 mm. long;

caps. erect, 4–5 cm. long.—Hills below 2000 ft.; Coastal Sage Scrub, V. Grassland; Orange Co. June–July.

Var. **véstus** Purdy. [*C. w.* var. *purpurascens* auth.] Petals subtruncate, red-brown to purplish, fringed with brown hairs; $n=9$ (Beal & Ownbey, 1943).—Dry slopes, below 2500 ft.; largely Chaparral; Ventura Co. to Monterey Co. July–Aug.

Var. **weèdii.** [*C. luteus* var. *w.* Baker. *C. citrinus* Baker.] Petals orange-yellow, flecked and often margined with red-brown, bearded on much of inner face with long yellow hairs; $n=9$ (Beal & Ownbey, 1943).—Below 5000 ft.; Chaparral; hills from Santa Ana Mts., Orange Co. to L. Calif. May–July.

4. *Chlorógalum* Kunth. SOAP PLANT. AMOLE

Herbs with fibrously or membranously coated bulbs and tall almost leafless infl. Basal lvs. several, tufted, linear; cauline reduced, the uppermost bracteose. Pedicels jointed at apex. Perianth white to pink or blue, the segms. 6, distinct, linear to oblong, persistent, twisted together above the caps., spreading. Stamens 6; fils. filiform; anthers versatile. Style long-filiform, slightly 3-cleft at apex. Caps. subglobose, loculicidal, 3-valved. Seeds black, rounded, 1–2 in a locule. Spp. 5, mostly found in Calif. (Greek, *chloros*, green, and *gala*, milk or juice.)

(Hoover, R. A. Monograph of the genus Chlorogalum. Madroño 5: 137–146. 1940.)

Fls. vespertine; style not exceeding perianth; perianth-segms. 8–30 mm. long
                                                                                                                                 2. *C. pomeridianum*

Fls. diurnal; style exceeding perianth; perianth-segms. 5–8 mm. long . . . . 1. *C. parviflorum*

1. **C. parviflòrum** Wats. [*Laothoe p.* Greene.] Bulb 4–7 cm. long, with brown membranous coats; lvs. 1–2 dm. long, 3–9 mm. wide, undulate; stem 3–8 dm. tall; pedicels 2–8 mm. long; fls. white to pink, with dark midvein, 7–8 mm. long, sub-rotate; stamens 3–4 mm. long; style 7–9 mm. long; caps. ca. 4 mm. long.—Dry open places below 2000 ft.; Coastal Sage Scrub, V. Grassland; e. Los Angeles Co. to Riverside and San Diego cos.; L. Calif. May–June.

2. **C. pomeridiànum** (DC.) Kunth. [*Scilla p.* DC. *Laothoe p.* Raf.] Plate 95, Fig. D. Bulb 7–15 cm. long, heavily coated with persistent dark brown fibers of old coats; lvs. 2–7 dm. long, 6–25 mm. wide, very wavy; stem glaucous, stout, 6–25 dm. tall, freely branching above; pedicels slender, 5–30 mm. long; perianth-segms. linear, white with green or purple midrib, 15–23 mm. long, spreading and recurved at anthesis; stamens ca. $\frac{2}{3}$ as long; style 10–15 mm. long; caps. 5–7 mm. long; $n=18$ (Cave, 1949).—Dry open hills and plains, below 5000 ft.; chiefly V. Grassland, Coastal Sage Scrub; San Diego Co. to s. Ore. The bulbs were roasted and eaten by the Indians. Uncooked they have lather-producing properties in water. Crushed material was used in streams to stupefy fish. May–July.

5. *Fritillària* L. FRITILLARY

Bulb of 1 or more fleshy scales, with or without rice-grain bulblets. Stem simple, erect. Lvs. alternate or whorled, sessile, linear to almost ovate. Fls. 1–several in a terminal raceme; perianth campanulate to funnelform, deciduous, of 6 distinct segms. in 2 series; segms. with ± evident gland or nectary above the base. Stamens 6, included, inserted on base of segms.; fils. slender; anthers extrorse, ± versatile. Ovary sessile or subsessile; style 1, entire or trifid; caps. membranaceous, 6-angled or -winged, 3-valved, loculicidal; seeds many, flat, in 2 rows in each locule, brownish. Ca. 100 spp., of N. Temp. Zone; a few of horticultural value. (Latin, *fritillus*, a dicebox, because of the shape of the caps.)

(Beetle, D. E. Monograph of the N. Am. spp. of Fritillaria. Madroño 7: 133–158. 1944.)

Lvs. on lower part of stem just above ground; fls. not mottled . . . . 1. *F. biflora*
Lvs. on upper stem, the lower part above the ground naked; fls. mottled.
    Perianth-segms. mostly 2–4 cm. long; lvs. 5–30 mm. wide. From below 2500 ft. 2. *F. lanceolata*
    Perianth-segms. 1.5–2 cm. long; lvs. 2–7 mm. wide. From 6000–10,500 ft. . . 3. *F. pinetorum*

1. **F. biflòra** Lindl. [*F. b.* vars. *ineziana* and *inflexa* Jeps. *F. succulenta* Elmer.] CHOCOLATE-LILY. MISSION BELLS. Plate 95, Fig. F. Bulb 15–20 mm. thick, of a few fleshy scales; stems 1.5–4 dm. high; lvs. 3–7, alternate, oblong to lance-ovate, 5–12 cm. long; fls. 1–7, nodding, dark brown or greenish-purple, campanulate; perianth-segms. lance-oblong, 2–3.5 cm. long, 5–12 mm. wide; gland appearing as a green band; stamens 8–10 mm. long; style deeply cleft; caps. 15–25 mm. long, not winged.—Heavy soil in grassy places below 3000 ft.; V. Grassland; San Diego and Riverside cos. n.; to Mendocino Co. Feb.–June.

2. **F. lanceolàta** Pursh. [*F. mutica* L. *F. ojaiensis* Davids.] CHECKER-LILY. Bulbs of few scales and many rice-grain bulblets; stem 3–10 dm. long; lvs. in several whorls of 3–5 on upper stem, ovate to linear-lanceolate, 4–16 cm. long; fls. 1–several, nodding, deeply bowl-shaped, brown purple mottled with yellow, to pale greenish-yellow and faintly mottled with purple; perianth-segms. 1–4 cm. long, ovate to oblong; gland yellow-green with purple dots; stamens half as long as segms.; style deeply cleft; caps. 1.5–2.5 cm. long, broadly winged; $n=12, 18, 24$ (Beetle, 1944).—Mostly below 2500 ft.; Chaparral, etc.; Ventura Co. n.; to B.C., Ida. Feb.–May.

3. **F. pinetòrum** Davids. [*F. atropurpurea* var. *p.* Jtn.] Bulb 1–2 cm. thick, with many rice-grain bulblets; stem glaucous, 1–4 dm. high, somewhat fistulous; lvs. 12–20, glaucous, somewhat whorled, linear, 5–15 cm. long; fls. 3–9, erect or nearly so, purplish, mottled with greenish-yellow; segms. 14–19 mm. long, 2–6 mm. wide; gland indefinite; style cleft to near base; caps. 12–15 mm. long, angled, with short hornlike process at base and summit of each angle.—Somewhat shaded granitic slopes, 6000–10,500 ft.; Montane Coniferous F.; San Bernardino Mts. to Mt. Pinos and cent. Sierra Nevada. May–July.

6. *Hesperocállis* Gray. DESERT-LILY

Bulb tunicated, deep set; stems stout, straight, simple, leafy at base; infl. a terminal elongate raceme. Lvs. linear, mostly strongly crisped on the white margins. Bracts subtending the white fls. scarious, conspicuous. Pedicels jointed at apex. Perianth united at middle into a tube, the segms. 6, spatulate, withering-persistent, 5–7-nerved. Stamens 6, inserted on the throat; fils. filiform; anthers linear, versatile. Ovary oblong, sessile, 3-loculed; style white, filiform, persistent, equaling segms.; stigma discoid. Caps. deeply 3-lobed, subglobose, loculicidal. Seeds many, black, flattened. One sp., of the deserts of Calif. and Ariz. Desert Indians used the bulbs for food. (Greek, *hesperos*, western, and *kallos*, beauty.)

1. **H. undulàta** Gray. Bulb ovoid, 4–6 cm. long; stem 3–20 dm. high; basal lvs. 2–5 dm. long, 8–15 mm. wide, blue-green; cauline reduced, few; raceme 1–3 dm. long; bracts broadly ovate, 1–1.5 cm. long; pedicels ca. 1 cm. long; perianth-tube 1.5–2 cm. long, the segms. 3–4 cm. long, 6–10 mm. wide, divergent, white within with a silvery-greenish lineate band on the backs; anthers golden, 7 mm. long; caps. 12–16 mm. long; seeds ca. 5 mm. long and wide; $2n=48$ (Cave, 1948).—Locally common on dry sandy flats and gentle slopes below 2500 ft.; Creosote Bush Scrub; Mojave Desert e. of Yermo and on Colo. Desert; w. Ariz. March–May.

7. *Lílium* L. LILY

Perennial herbs with scaly bulbs or scaly rootstocks. Stems simple, tall and leafy. Lvs. narrow, sessile, alternate to whorled. Fls. large, showy, solitary to many in a terminal raceme. Perianth funnelform or campanulate, deciduous, the segms. 6, spreading or recurved, each with a nectar-bearing gland near the base. Stamens 6; fils. slender, ca. $\frac{2}{3}$ as long as the perianth-segms.; anthers linear, versatile. Ovary 3-loculed; style long, deciduous, with 3-lobed stigma. Caps. loculicidal; seeds many, flat, horizontal, in 2 rows in each locule. A genus of ca. 100 spp. of the temp. N. Hemis. Of great horticultural interest. (Greek, *lirion*, the classical name.)

Lvs. in large part in conspicuous whorls.
Fls. orange-red. Plants of dry places .... 1. *F. humboldtii*
Fls. yellow. Plants of wet places .... 2. *F. pardalinum*
Lvs. scattered or only the lowest whorled .... 3. *F. parryi*

1. **F. humbòldtii** Roezl & Leichtl. var. **bloomeriànum** (Kell.) Jeps. [*L. b.* Kell. *L. ocellatum* var. *b.* Beane. *L. fairchildii* Jones.] HUMBOLDT LILY. Incorrectly called Tiger Lily. Bulbs ovoid, oblique, 5–15 cm. long, the scales not jointed; stems 9–15 dm. high; lvs. in 4–8 whorls of 10–20 each, oblanceolate, mostly bright green, 9–12 cm. long, 0.5–1.5 cm. wide; fls. nodding, few to many, orange-yellow, with dark spots usually red on their margins, the segms. 7–9 cm. long, revolute to near the base; stamens and style to ca. 5 cm. long; caps. obovoid, acutely angled, 2.5–5 cm. long.—Open flats, 4000–5500 ft.; Chaparral, Yellow Pine F.; mts. of e. San Diego Co. June–July.

Var. **ocellàtum** (Kell.) Elwes. [*L. bloomerianum* var. *o.* Kell. *L. o.* Beane. *L. humboldtii* var. *magnificum* Purdy.] Scales purplish; stems to 3.5 m. high; fls. orange-red with the maroon spots or blotches margined with red.—Gravelly soil, gulleys and canyons below 3000 ft.; Chaparral, S. Oak Wd.; Santa Barbara Co. to San Jacinto Mts. and Santa Ana Mts. Santa Cruz Id. June–July.

2. **F. pardalìnum** Kell. LEOPARD LILY. PANTHER LILY. Bulbs branching-rhizomatous, 6–10 cm. long; stems stout, 1–2.5 m. high; lvs. linear to lanceolate, 1–2 dm. long, 6–25 mm. wide, in 3–4 whorls of 9–15 and some scattered; fls. nodding, often fragrant, 1–several, yellow with some red to dark red, with maroon spots, the segms. recurved to the middle or below, 5–8 cm. long, 12–18 mm. wide; anthers 10–15 mm. long; caps. oblong, 3 cm. long; $n=12$ (Sansome & La Cour, 1934).—Forming colonies in springy places and on stream banks below ca. 6000 ft.; Yellow Pine F.; mts. of e. San Diego Co., Santa Barbara and Kern cos.; to n. Calif. May–July.

3. **F. párryi** Wats. LEMON LILY. Bulb rhizomatous, 2.5–3 cm. long, the scales jointed, 12–20 mm. long; stems slender, 6–15 dm. high; lvs. scattered, sometimes the lower whorled, linear-oblanceolate or lanceolate, 8–15 cm. long, 6–15 mm. wide; fls. 1–several, horizontal, fragrant, clear lemon-yellow, sometimes with maroon spots, trumpet-shaped, the segms. recurved or spreading in upper third, 6–10 cm. long, 8–12 mm. wide; anthers brown, 6–8 mm. long; caps. narrow-oblong, 3.5–5 cm. long; $n=12$ (Stewart, 1947).—Springy places and wet banks, 4000–9000 ft.; Montane Coniferous F.; San Gabriel Mts. to San Diego Co.; Ariz. Plants from the central San Gabriel Mts. tend to have wider lvs., to 4 cm. wide, and are the var. **késsleri** Davids. July–Aug.

8. *Smilacìna* Desf. FALSE SOLOMON'S-SEAL

Perennial herbs with creeping rootstocks. Stem simple, scaly below, leafy above. Lvs. alternate, short-petioled or sessile, lanceolate to ovate. Fls. small, usually white, spreading, in a terminal panicle or raceme; perianth of 6 equal segms. Stamens 6, inserted at base of segms.; fils. slender; anthers small, ovate. Ovary subglobose, 3-loculed, with 2 ovules in each locule; style columnar; stigma trifid to 3-grooved. Fr. a berry; seeds subglobose. Ca. 25 spp., N. Am. and Asia. (Diminutive of *Smilax.*)

(Galway, D. H. The N. Am. spp. of Smilacina. Am. Midl. Nat. 33: 644–666. 1945.)

Fls. panicled, numerous .... 1. *S. racemosa*
Fls. racemose, few to several .... 2. *S. stellata*

1. **S. racemòsa** (L.) Desf. var. **amplexicáule** (Nutt.) Wats. [*S. a.* Nutt. ex Baker. *Vagnera a.* Greene. *Unifolium a.* Greene. *S. a.* var. *glabra* Macbr. *S. racemosa* var. *g.* St. John.] Rootstock stout; stem erect, 3–9 dm. high, subglabrous to ± pubescent especially above; lvs. several, ovate to lance-oblong, acute or short-acuminate, clasping and sessile or with a short dilated clasping petiole, broadest near the base; peduncle naked or with 1–2 bracts; infl. 3–18 cm. long, many-fld.; perianth-segms. 1–2 mm. long; stamens 1.5–3 mm. long; berry ca. 5 mm. long, mostly red or with small purple spots; seeds ca. 4 mm. long; $n=18$ (Rattenbury, 1948).—Shaded woods below 8500 ft.; Montane Coniferous

F.; San Jacinto and San Bernardino mts.; Sierra Nevada and Coast Ranges n. to B.C., Rocky Mts. June–July. Some plants are quite glabrous in upper parts and slightly glaucous and constitute the var. *glabra*.

2. **S. stellàta** (L.) Desf. [*Convallaria s.* L. *Vagnera s.* Morong. *Vagnera liliacea* Rydb.] Plate 95, Fig. E. Rootstock slender; stem 3–6 dm. high, rather strict, glabrous to puberulent; lvs. oblong-lanceolate to lance-ovate, acuminate, 5–15 cm. long, 15–30 mm. wide, puberulent beneath, sessile or somewhat clasping; raceme sessile or short-peduncled, 3–15-fld.; pedicels 5–15 mm. long; perianth-segms. oblong, obtuse, 5–7 mm. long; berry red-purple, becoming black, 7–10 mm. long; $n=18$ (Stenar, 1935).—Wet places often in brush, below 8500 ft.; many Plant Communities; Cuyamaca Mts. and San Jacinto Mts. n. to B.C. and e. to Atlantic Coast. April–June.

9. *Veràtrum* L. False-Hellebore. Corn-Lily

Tall stout leafy herbs from short thick rootstocks. Lvs. broad, clasping, strongly veined and plaited. Fls. greenish-white to purple, rather large, bisexual, unisexual or together on same plant, on short pedicels, in large terminal panicles. Perianth-segms. 6, oblong-ovate, obtuse, glandless or nearly so, sometimes adnate to the base of the ovary. Stamens 6, opposite the segms. and free from them; fils. filiform; anthers cordate, their sacs confluent, ovoid. Ovary 3-loculed; styles 3, persistent. Caps. septicidal, 3-lobed, each locule several-seeded. Seeds compressed, the narrow body surrounded by a broad margin or wing. Ca. 12–14 spp. of N. Temp. Zone. Plants often reported as poisonous to stock. (Ancient name of hellebore.)

1. **V. califórnicum** Durand. Often called Skunk-Cabbage, incorrectly so. Plants 1–2 m. tall, tomentose above; lvs. ovate to broadly elliptic, 2.5–4 dm. long, 1–2 dm. wide, loose-pubescent beneath, the uppermost lanceolate; panicles 2–5 dm. long, short-tomentose; pedicels 2–6 mm. long; perianth-segms. dull white, oblong-ovate, 9–14 mm. long, with a Y-shaped green gland near the base; stamens 6–8 mm. long; caps. 2–3.5 cm. long; seeds 10–12 mm. long, strongly winged; $n=16$ (Cave, 1966).—Wet meadows and banks below 11,000 ft.; Montane Coniferous F.; from Palomar Mts. n.; to Wash., Rocky Mts.; L. Calif. July–Aug.

10. *Zigadènus* Michx. Zygadene

With tunicated bulbs in our spp. Stem simple or branched in the infl., leafy below. Lvs. linear, glabrous. Fls. greenish or yellowish-white, in terminal racemes or panicles, perfect or unisexual, subtended by acuminate bracts. Perianth withering-persistent, free or adnate to lower part of ovary; segms. ovate to lanceolate, with 2 or in our spp. 1 gland just above the base. Stamens 6, free from the segms.; anthers reniform, explanate after dehiscence. Ovary 3-loculed; styles 3, persistent; caps. 3-lobed, loculicidal. Seeds many, subrhomboid. Ca. 18 spp., N. Am. and N. Asia. At least some spp. are poisonous to cattle and sheep. (Greek, *zugon*, yoke, and *aden*, gland, referring to the paired glands of the first known sp.)

Stamens definitely longer than perianth in mature fls.; fls. normally racemose .. 3. *Z. venenosus*
Stamens not exceeding perianth.
    Perianth 5–7 mm. long; lvs. 5–10 mm. wide. Deserts .... 1. *Z. brevibracteatus*
    Perianth 10–12 mm. long; lvs. 8–25 mm. wide. Cismontane .... 2. *Z. fremontii*

1. **Z. brevibracteàtus** (Jones) Hall. [*Z. fremontii* var. *b.* Jones.] Bulbs 2–4.5 cm. long; stems slender, 3–5 dm. high, glabrous; basal lvs. 1.5–3 dm. long, 5–10 mm. wide, slightly scabrous on margins, folded; infl. usually paniculate, 1–2.5 dm. long; pedicels spreading, 1–3 cm. long; perianth-segms. 5–7 mm. long, yellowish, elliptic, the outer broader; stamens 4–6 mm. long; caps. oblong, 1.5 cm. long.—Sandy flats and mesas, 2000–5000 ft.; Creosote Bush Scrub, Joshua Tree Wd.; w. and s. Mojave Desert and Whitewater Creek, Colo. Desert; San Luis Obispo Co. April–May. *Z. exaltatus* Eastw. with wider longer lvs. is reported from the Piute Mts., Kern Co., *Twisselmann*. Sierra Nevada.

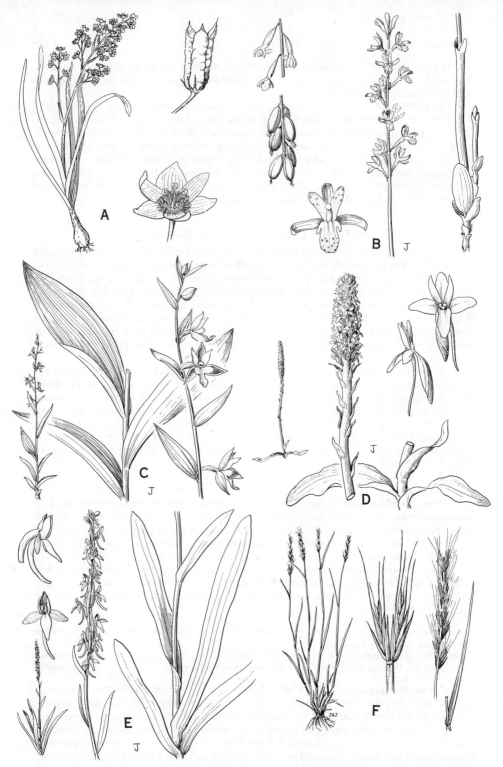

Plate 96. Fig. A, *Zigadenus fremontii*. Fig. B, *Corallorrhiza maculata*. Fig. C, *Epipactis gigantea*. Fig. D, *Habenaria elegans*. Fig. E, *Habenaria leucostachys*. Fig. F, *Agropyron trachycaulum*.

2. **Z. fremóntii** Torr. var. **fremóntii**. [*Anticlea f.* Torr. *Toxicoscordion f.* Rydb.] STAR-LILY. Plate 96, Fig. A. Bulb 3–6 cm. long; stems 3–10 dm. high, smooth; basal lvs. 2–6 dm. long, 8–25 mm. wide, folded and arched, with scabrous margins; infl. racemose or paniculate, 1–4 dm. long; bracts long-acuminate; pedicels spreading, to 4 cm. long in fr.; perianth-segms. yellowish-white, 10–12 mm. long, lance-ovate, gland greenish-yellow, the sepals subsessile; stamens 5–7 mm. long; caps. 1.5–3 cm. long; $n=11$ (Miller, 1931).—Dry grassy or bushy slopes, below 3500 ft.; V. Grassland, Coastal Sage Scrub, Chaparral; San Diego Co. and L. Calif. to n. Calif. March–May. Anacapa, Santa Cruz and Santa Rosa ids. A form from Santa Ynez Mts., with more rotate perianth-segms. and green gland has been described by Jepson as var. *inezianus*.

Var. **minor** (H. & A.) Jeps. [*Z. chloranthus* var. *m.* H. & A.] Plants 1–2 dm. high; racemes corymbose; fls. white with greenish glands.—Open moist fields near the coast; San Diego; cent. and n. Calif.

3. **Z. venenòsus** Wats. [*Toxicoscordion v.* Rydb. *Z. diegensis* Davids.] DEATH-CAMAS. Bulbs 1.5–2.5 cm. long, with dark outer coats; stems 2.5–6 dm. high, glabrous, slender; basal lvs. 1.5–3 dm. long, 4–10 mm. wide; lowest cauline lvs. with scarious sheaths; raceme usually simple, 5–20 cm. long; bracts membranous, lanceolate to setosely acuminate; pedicels ascending in fr., 1–1.5 (–2) cm. long; perianth-segms. whitish, ovate, subcordate at base, 3–4 mm. long, the gland with a well defined toothed upper margin; stamens ca. 5 mm. long; caps. cylindric, 1–1.5 cm. long.—Moist grassy places, below 8000 ft.; Coastal Sage Scrub, San Diego; Yellow Pine F., Kern Co.; to B.C.; Utah, L. Calif. May–June.

14. Najadàceae. WATER-NYMPH FAMILY

Slender branching submerged annuals with fibrous roots. Lvs. small, subopposite or verticillate, sessile with sheathing base, linear, entire or toothed. Fls. unisexual, minute, axillary, solitary, sessile or pedicelled. Staminate with 1 stamen and 2-lipped perianth, the latter entire or 4-lobed. Pistillate fl. usually naked, with a single 1-ovuled carpel. Fr. sessile, indehiscent, ellipsoidal, usually embraced by the lf.-sheath. One genus of fresh or brackish water, widely distributed in temp. and warm parts of the world.

1. Nàjas L.

A genus of ca. 35 spp. (Greek, *Naias*, a water-nymph.)

(Clausen, R. T. Studies in the genus Najas. Rhodora 38: 333–344. 1936.)

Lvs. coarsely and definitely toothed; internodes and backs of lvs. often spiny; plants dioecious
    2. *N. marina*
Lvs. almost entire, finely but remotely spined; plants monoecious .... 1. *N. guadalupensis*

1. **N. guadalupénsis** (Spreng.) Morong. [*Caulinia g.* Spreng.] Stems 3–6 dm. long, unarmed; lvs. 1–2.5 cm. long, 0.5 mm. wide, unarmed or remotely finely spined; ♂ fls. 2–3 mm. long; anthers 4-celled; ♀ fls. 2–3 mm. long; stigmas 2–3; fr. dull, finely reticulate under a lens, 2 mm. long; $2n = 12, 36, 48, 54, 60$ (Chase, 1947).—Wet places in much of Calif. at low elevs.; to Ore., Atlantic Coast; Eu. July–Sept.

2. **N. marìna** L. [*N. m.* var. *californica* Rendle.] Stems rather stout, often prickly, 1–4 dm. long; lvs. stiff, 1–4.5 cm. long, 1–3 mm. wide, with coarse teeth on margins to 1 mm. long and often spiny on the back; sheath almost entire; ♂ fls. 3–4 mm. long; anthers 4-celled; ♀ fl. 3–4 mm. long; stigmas 3; frs. 4–5 mm. long, finely reticulate; $n=6$ (Lewitzky, 1931).—Occasional in quiet fresh water, like Laguna Dam, Imperial Co., Ojai, Ventura Co ; L. Calif. to Atlantic Coast; Eurasia, Afr. July–Sept.

15. Orchidàceae. ORCHID FAMILY

Ours perennial terrestrial herbs from a short or elongate, rarely coralloid, rhizome, with fleshy to fibrous roots. Lvs. sheathing, often reduced to scales. Fls. in ours perfect, irregular, bracted, solitary or in spikes or racemes. Sepals 3, alike. Petals 3: 2 alike, usually resembling the sepals, the third forming a lip which is sometimes saccate and often spurred. Fils. 1 or 2, united with the style to form the column; the perfect anther 1 (2 in *Cypripedium*), on the apex of the column and above or behind the sticky stigmas. Pollen mostly in masses (pollinia), which in ours are usually 2 or 4. Ovary inferior, mostly elongated and twisted, 3-celled to 1-celled, forming a 3-valved caps. with very numerous minute seeds. Several hundred genera and over 15,000 spp.; most abundant in the trop., where largely epiphytic with aerial roots.

(Correll, D. S. Native orchids of N. Am. Chronica Botanica Co., 1950, pp. 1–399.)
Lvs. reduced and scalelike; plant saprophytic, not green.
    Plant white; perianth 12–15 mm. long .... 2. *Eburophyton*
    Plant brown, yellow or purple; perianth mostly shorter .... 1. *Corallorhiza*
Lvs. foliaceous; plant green.
    Fls. with a distinct slender spur which is 4–12 mm. long .... 4. *Habenaria*
    Fls. spurless, though sometimes slightly saccate at the base.
        Petals and sepals 10–18 mm. long .... 3. *Epipactis*
        Petals and sepals 2–5 mm. long.
            Fls. in a 3-ranked twisted spike; lvs. several .... 7. *Spiranthes*
            Fls. in racemes which are not twisted; lvs. 1 or 2.
                The lvs. 2, subopposite .... 5. *Listera*
                The lf. 1 .... 6. *Malaxis*

1. *Corallorhìza* Chat. CORAL ROOT

Saprophytic scapose herbs with rather short rhizomes which are much branched and coralloid. Stem brown or yellow to purple. Lvs. reduced to several sheathing scales. Infl. a loose terminal raceme. Fls. brown, yellowish or purplish, sometimes with some white or green. Sepals subequal, ascending spreading or connivent, the lateral united at base to form a short spurlike structure ± adnate to ovary. Petals ca. equal to sepals, 1–3-nerved. Lip 1–3-ridged, simple to 3-lobed. Column compressed; anthers terminal; pollinia 4, waxy, free. Caps. pendent, ovoid to ellipsoid. Genus of ca. 12 spp., N. and Cent. Am.; 1 in Eurasia. (Greek, *korallion*, coral, and *rhiza*, root.)

1. **C. maculàta** Raf. [*C. multiflora* Nutt.] Plate 96, Fig. B. Stems mostly 2–5 dm. tall, stout; raceme 5–15 cm. long; sepals and petals brownish-purple, 6–8 mm. long; lip whitish, spotted with purple, deeply 3-lobed, the middle lobe squarish, the lateral small; column yellow with magenta spots; caps. smooth, ca. 2 cm. long, compressed; $2n=42$ (Love & Simon, 1968).—Montane Coniferous F., 5000–9500 ft.; San Diego Co. n. to B.C., to Nfld., N. Car., Guatemala. June–Aug.

2. *Eburophyton* Heller. PHANTOM ORCHID

Saprophytic perennial herb, with erect stem from a branched creeping rootstock with fleshy roots. Lvs. reduced to long sheathing bracts. Fls. spicate. Lateral sepals keeled, spreading, subconcave; upper sepal and petals erect and somewhat connivent. Lip free, concave, the base saccate, the middle articulate. Column slender, with 1 stipitate anther; pollinia 2. Monotypic genus of w. U.S. (Latin, *ebur*, ivory, and *phyton*, plant, because not green.)

1. **E. aústinae** (Gray) Heller. [*Chloroea a.* Gray. *Cephalanthera a.* Heller.] Plant white, 2–5 dm. high; lvs. 2–4 cm. long; raceme 5–15 cm. long; perianth white, with yellowish tinge, 12–15 mm. long; lip shorter, 3–5-nerved; column 4 mm. long.—Dry woods below 6000 ft.; Montane Coniferous F.; Crestline, San Bernardino Mts.; Sierra Nevada and Coast Ranges to Wash., Ida. May.

3. *Epipáctis* Sw. STREAM ORCHIS. HELLEBORINE

Stem simple, leafy, from a short creeping rootstock. Lvs. lance-linear to suborbicular, plicate-venose. Fls. racemose; fl.-bracts foliaceous. Sepals and petals almost equal, spreading. Lip saccate at base, flattened ab ve, constricted at middle. Column short, with 1 sessile 2-celled anther and broad truncate stigma. Pollinia 4, mealy-granulose. Caps. ellipsoid, pendent. Ca. 20 spp. of temp. Eurasia and N. Am. (Greek, *epipegnuo*, the name for hellebore and referring to a milk-curdling property ascribed to some spp.)

1. **E. gigántea** Dougl. ex Hook. [*Peramium g.* Coult. *Serapias g.* A. A. Eat. *Helleborine g.* Druce.] Plate 96, Fig. C. Stout, 3–9 dm. tall, somewhat pubescent; lower lvs. ovate, 5–15 cm. long, upper lanceolate and gradually reduced; fls. 3–15, on pedicels 4–8 mm. long; sepals 12–18 mm. long, greenish, deeply concave; petals shorter, purplish or reddish; lip ovate-lanceolate, strongly veined and marked with purple or red, unequally 3-lobed; caps. 2–2.5 cm. long.—Moist stream banks and springy places, below 7500 ft.; many Plant Communities; L. Calif. to B.C.; occasional on deserts. May–Aug.

4. *Habenària* Willd. REIN ORCHID

Perennial terrestrial herbs with fleshy tuberlike roots. Stems erect, simple, ± leafy, at least at the base. Lvs. essentially sessile, with the basal part sheathing. Fls. in ours small, in terminal spike or raceme. Sepals free, similar or dissimilar. Petals slightly smaller, free, erect; lip entire in ours, with a basal scrotiform sac or spur. Anther-cells 2, separate, divergent; pollen granular. Caps. narrow-cylindric to ellipsoid. Ca. 500 spp. of warmer parts of the world. (Latin, *habena*, rein of a horse, because of the shape of the spur in some spp.)

Sepals 1-nerved; lvs. usually near base of stem and withering before anthesis.
  Spur at least twice as long as lip; raceme densely fld., 1.5–2.5 cm. in diam. .... 1. *H. elegans*
  Spur not much longer than the lip; raceme loosely fld., usually not more than 1 cm. in diam.
     4. *H. unalascensis*
Sepals 3-nerved; lvs. usually scattered and green at anthesis.
  Fls. white; lip abruptly dilated at base .... 2. *H. leucostachys*
  Fls. green; lip not prominently dilated at base .... 3. *H. sparsiflora*

1. **H. elegáns** (Lindl.) Boland. [*Platanthera e.* Lindl. *Piperia e.* Rydb. *H. michaelii* Greene. *H. longispicata* Parish. *H. elegans* var. *elata* Jeps.] Plate 96, Fig. D. Much like *H. unalascensis*, but the cylindrical spike up to 6 dm. long, rather densely fld., 1.5–2 cm. in diam.; fls. greenish-white; sepals 4–5 mm. long; lip 5 mm. long; spur 10–18 mm. long, filiform; $2n = 21$ pairs (Raven, Kyhos & Hill, 1965).—Mostly in dry woods below 8000 ft.; several Plant Communities; Cuyamaca Mts. n. to B.C. Santa Rosa and Santa Cruz ids. May–Aug.

2. **H. leucostàchys** (Lindl.) Wats. [*Platanthera l.* Lindl. *H. dilatata* var. *l.* Ames. *Limnorchis l.* Rydb.] Plate 96, Fig. E. Stem stout, 3–9 dm. high; lower lvs. oblanceolate, 1–2 dm. long, the upper gradually reduced, lanceolate, acute; spike rather densely fld., 1–3 dm. long; flowers white, perianth 6–8 mm. long; upper sepal ovate, 5 mm. long, lateral lanceolate, 7–8 mm. long; lip lanceolate, rhombic at base, 8 mm. long; spur narrow, 10–12 mm. long; caps. 12–16 mm. long; $n = 21$ (Raven, Kyhos & Hill, 1965).—Wet and springy places, 4500–10,000 ft.; Montane Coniferous F.; from San Diego Co. to Sierra Nevada, White Mts., Panamint Mts.; n. to B.C., Mont. May–Aug.

3. **H. sparsiflòra** Wats. [*Limnorchis s.* Rydb. *L. laxiflora* Rydb. *H. l.* Parish. *H. leucostachys* var. *virida* Jeps.] Stem 3–6 dm. high, leafy; lower lvs. oblanceolate, obtuse, 1–2 dm. long, upper reduced, acuminate; spike slender, sparsely fld., 1–3 dm. long; fls. greenish, 10–12 mm. long; upper sepal ovate, 4–5 mm. long; lip linear to lance-linear, 6–8 mm. long; spur 6–8 mm. long; caps. ca. 12–14 mm. long.—Mostly along streams or in boggy places, 4000–9000 ft.; Montane Coniferous F.; San Diego Co. to Sierra Nevada, etc. to Wash., Rocky Mts. June–Aug.

4. **H. unalascénsis** (Spreng.) Wats. [*Spiranthes u.* Spreng. *Piperia u.* Rydb. *H. cooperi* Wats.] Erect, leafy only near the base, 3–6 (–9) dm. high; lvs. oblanceolate to oblong-lanceolate, acute, 1–1.5 dm. long, withering by flowering time; spike lax, 1–3.5 dm. long, many-fld.; fls. greenish, the perianth 3–4 mm. long; upper sepals ovate, acute, the lateral sepals and petals oblong-lanceolate; lip ovate, rounded at apex, 4 mm. long; spur clavate,

ca. as long as the lip or exceeding it, but not longer than the ovary; caps. 10–15 mm. long.—Dry to moist soil, flats and slopes below 8000 ft.; many Plant Communities; L. Calif. to Alaska, Rocky Mts. April–Aug.

5. *Lístera* R. Br. TWAYBLADE

Low perennial, from fibrous roots and bearing at the middle of the stem a pair of nearly opposite sessile lvs. Fls. small, greenish, in a loose terminal raceme. Sepals and petals almost alike, spreading or reflexed; lip mostly drooping, longer than the sepals, 2-lobed or 2-cleft at the summit. Column slender, with 1 posterior anther; pollinia 2, powdery. Caps. small, pedicelled. Ca. 20 spp., in temp. and colder parts of N. Hemis. (Named for M. *Lister*, 17th-century English naturalist.)

1. **L. convallarioìdes** (Sw.) Torr. [*Epipactis c.* Sw. *Ophrys c.* W. Wight.] Slender, 1–3 dm. high, pubescent above the lvs.; lvs. oval or roundish, 3–5 cm. long; fls. 6–12, greenish; sepals and petals linear to linear-lanceolate, 3–4 mm. long; lip 9–11 mm. long, narrowly cuneate-obovate, retuse, with rounded lobes, on each side of base a short triangular claw; $n=18$ (Taylor & Brockman, 1966).—Springy shaded places at ca. 5000–7800 ft.; Montane Coniferous F.; San Jacinto Mts., San Bernardino Mts.; Sierra Nevada to Alaska, Atlantic Coast. June–Aug.

6. *Maláxis* Soland. ex Sw. ADDER'S MOUTH

Small plants with simple stems from cormlike base. Lf. 1 in ours and ovate. Fls. minute, in terminal raceme. Sepals spreading, free or the lateral connate, oblong in ours. Petals narrower, spreading. Lip often auriculate, entire or lobed, concave, spurless. Column short, terete, 2-toothed at apex in ours; anther terminal; pollinia 4, waxy. Caps. small, ovoid to subglobose. A genus of ca. 150 spp. of Eurasia, Oceania, N. Am. (Greek, *malakos*, a softening, perhaps in reference to tender nature of plant.)

1. **M. brachypòda** (Gray) Fern. [*Microstylis b.* Gray. *Microstylis monophyllus* auth. not Lindl.] Scape slender, 5–12 cm. high; lf. 3–5 cm. long, with sheathing petiole-like base; raceme spicate; bracts minute; pedicels 2–3 mm. long; sepals and petals greenish, 2–3 mm. long; lip rounded at base, with long slender tip; caps. 3–4 mm. long.—Rare, silty humps in wet meadows, 7300–8700 ft.; Montane Coniferous F.; Tahquitz V., San Jacinto Mts.; S. Fork of Santa Ana, San Bernardino Mts.; Rocky Mts. to Atlantic Coast. July–Aug.

7. *Spiránthes* Rich. LADIES' TRESSES

Roots clustered, tuberous in ours. Stems leafy only below, bracted above. Fls. white, spurless, in a twisted spike. Sepals and petals narrow, erect or somewhat connivent into a hood above; lip oblong, clawed or sessile, the base broad and embracing the column. Column short; anther 1, 2-celled, with 2 powdery pollinia. Caps. erect, ovoid or oblong. A genus of ca. 200 spp., widely distributed. (Greek, *speira*, spiral, and *anthos*, fl., alluding to the spiral character of the twisted infl.)

1. **S. porrifòlia** Lindl. [*Gyrostachys p.* Kuntze. *Orchiastrum p.* Greene. *Obidium p.* Rydb. *S. romanzoffiana* var. *p.* Ames and Correll.] Stout, glabrous, 1–5 dm. high; lower lvs. 3–5, linear to lance-linear, 0.5–3 dm. long, upper bracteate; perianth-segms. narrow, creamy-yellowish; 6–8 mm. long; lip lance-oblong, scarcely or not constricted below the apex and with 2 nipple-like callosities at the base.—Wet springy places below 8000 ft.; Montane Coniferous F.; San Bernardino Mts., Tulare Co. to Wash., Rocky Mts. July–Aug.

16. **Poàceae** (Gramineae). GRASS FAMILY (Plate 97, Fig. A.)

Herbs, rarely shrubs or trees, usually with hollow stems (*culms*) solid at the nodes, and with 2-ranked, parallel-veined lvs. consisting of 2 parts; the *sheath* and the *blade*, the former enveloping the culm with the margins overlapping or rarely grown together. At

Poaceae

the junction of the sheath and blade, on the inside, is a membranous hyaline or hairy appendage, the *ligule*. Fls. perfect, sometimes unisexual, minute, without a distinct perianth, arranged in spikelets consisting of a shortened axis, the *rachilla*, and 2–many distichous bracts, the 2 lowest, *glumes*, being empty; rarely 1 or both of them obsolete; in the axil of each succeeding bract, *lemma*, is borne a single floret, subtended and usually enveloped by a 1–2-nerved bract, *palea*, with its back to the rachilla; at the base of the fl. between it and the lemma, are usually 2 very small hyaline scales, *lodicules*. Stamens usually 3, with delicate fils. and 2-celled versatile anthers. Pistil 1, with a 1-celled, 1-ovuled ovary, usually 2 styles and plumose stigmas. Fr. a caryopsis with starchy endosperm and usually enclosed at maturity in the lemma and palea, free or adnate to the latter. The lemma with its palea and fl. constitute a floret. Spikelets arranged in spikes, racemes or panicles with bractless branches. Ca. 600 genera and 7500 spp.; from all parts of the world.

(Hitchcock, A. S. Manual of the grasses of the U.S. ed. 2, revised by Agnes Chase. U.S. Dept. Agric., Misc. Pub. 200: 1–1051. 1951. Gould, F. W. Grass systematics, 1–382. 1968.)

KEYS TO GENERA

1. Spikelets with the glumes persistent, the rachilla articulated above them, 1–many-fld.; upper lemmas frequently empty; rachilla often prolonged beyond the upper lemma.
   2. The spikelets borne in an open or spikelike raceme or panicle, usually upon distinct pedicels.
      3. Spikelets 1-fld.
         4. The spikelets with 2 sterile or ♂ lemmas below the fertile lemma; palea 1-nerved
            Group I
         4. The spikelets without sterile lemmas below the fertile lemma; palea 2-nerved
            Group II
      3. Spikelets 2–many-fld.
         5. Lemma usually shorter than the empty glumes; the awn dorsal and usually bent
            Group III
         5. Lemma usually longer than the empty glumes; the awn terminal and straight or none (rarely dorsal in *Bromus*) .... *Group IV*
   2. The spikelets borne in 2 rows, sessile or nearly so.
      6. Spikelets on 1 side of the continuous axis, forming 1-sided spikes; spikes usually more than 1 .... *Group V*
      6. Spikelets alternately on opposite sides of the axis which is often articulated; spike terminal, single .... *Group VI*
1. Spikelets falling from the pedicels entire, articulate below the glumes, naked or enclosed in bristles or burlike invols., 1-fld., or, if 2-fld., the lower fl. ♂; no lemmas empty; rachilla not extending beyond the upper lemma.
   7. The spikelets very flat, strongly laterally compressed, paniculate; glumes reduced or wanting; lemma and palea subequal, both keeled .... *Group VII*
   7. The spikelets mostly not strongly flattened; glumes, or at least 1, usually developed.
      8. Lemma and palea hyaline, thin, much more delicate in texture than the glumes.
         9. Spikelets in pairs, 1 sessile, the other pedicellate .... *Group VIII*
         9. Spikelets not in pairs .... *Group II*
      8. Lemma, at least that of the perfect fl., similar in texture to the glumes, or thicker and firmer, never hyaline and thin.
         10. The lemma and palea membranous, the 1st glume usually the larger .... *Group IX*
         10. The lemma and palea chartaceous to coriaceous, very different in color and appearance from the glumes .... *Group X*

Group I

1. Lower florets not reduced to small scalelike lemmas.
   2. Lower florets consisting of awned hairy sterile lemmas exceeding the fertile floret
      6. *Anthoxanthum*
   2. Lower florets consisting of awnless glabrous sterile lemmas enclosing the fertile floret
      33. *Ehrharta*
1. Lower florets reduced to small awnless lemmas much smaller than the fertile florets
   65. *Phalaris*

Group II

1. Rachilla articulate below the glumes, these falling with the spikelet.
    2. Glumes long-awned .... 69. *Polypogon*
    2. Glumes awnless.
        3. Glumes united toward the base, ciliate on the keel; infl. not capitate and bracteate
            .... 4. *Alopecurus*
        3. Glumes not united, glabrous; infl. capitate, in the axils of broad bracts .... 21. *Crypsis*
1. Rachilla articulate above the glumes.
    4. Lemma indurate when mature and very closely embracing the grain; callus usually well developed and bearded.
        5. Awn 3-branched, the lateral branches sometimes obsolete, then no line of demarcation between lemma and awn .... 7. *Aristida*
        5. Awn simple, a line of demarcation between the awn and lemma.
            6. The awn twisted and bent, persistent, several to many times longer than the fr. 83. *Stipa*
            6. The awn not twisted, deciduous, rarely more than 3-4 times as long as the plump fr.
                .... 60. *Oryzopsis*
    4. Lemma usually hyaline or membranaceous at maturity; callus not well developed.
        7. Glumes usually longer than the lemma (sometimes not much longer in *Agrostis*).
            8. The glumes compressed-carinate, stiff-ciliate on keel; panicle dense, cylindric or ellipsoid .... 66. *Phleum*
            8. The glumes not compressed-carinate, not ciliate.
                9. Glumes saccate at base; lemma long-awned; panicle contracted, shining
                    .... 41. *Gastridium*
                9. Glumes not saccate at base; lemma awned or not; panicle open or contracted.
                    10. Floret bearing a tuft of hairs at the base from the short callus; palea well developed, the rachilla prolonged behind the palea as a hairy bristle . 17. *Calamagrostis*
                    10. Floret without hairs at base or with short hairs; palea usually small or obsolete
                        .... 2. *Agrostis*
        7. Glumes not longer than the lemma, usually shorter (except sometimes for the awn-tips).
            11. Lemma awned from the tip or mucronate, usually 3- to 5-nerved .. 58. *Muhlenbergia*
            11. Lemma awnless or awned from the back.
                12. Floret without hairs at base.
                    13. Fr. at maturity falling from the lemma and palea; seed loose in the pericarp; lemma 1-nerved; panicle open or contracted .... 81. *Sporobolus*
                    13. Fr. not falling from the lemma and palea; but permanently enclosed in them; seed adnate to the pericarp; panicle spikelike.
                        14. Panicle short, partly enclosed in the sheath; annual .... 21. *Crypsis*
                        14. Panicle elongate; perennial .... 58. *Muhlenbergia*
                12. Floret with a tuft of hairs at the base from the short callus; lemma and palea chartaceous, awnless .... 4a. *Ammophila*

Group III

1. Florets 2, the lower perfect, awnless, the upper ♂, awned .... 46. *Holcus*
1. Florets 2 or more, all alike except the reduced upper ones.
    2. Articulation below the glumes, the spikelets falling entire.
        3. Lemmas, at least the upper, with a conspicuous bent awn; glumes nearly alike
            .... 86. *Trisetum*
        3. Lemmas awnless; second glume much wider than the first .... 80. *Sphenopholis*
    2. Articulation above the glumes, the glumes similar in shape.
        4. Lemmas bifid at apex, awned or mucronate between the lobes; spikelets several-fld.
            5. Awns conspicuous, flat, bent; spikelets 1 cm. or more long .... 27. *Danthonia*
            5. Awns minute to obsolete; spikelets not more than 5 mm. long .... 72. *Schismus*
        4. Lemmas toothed but not bifid and awned or mucronate between the lobes.
            6. Glumes 2-3.5 cm. long, 7-9-nerved; spikelets 2-fld., or with a rudimentary 3d one, pendulous; annual .... 9. *Avena*
            6. Glumes not more than 1 cm. long, 1-5-nerved; spikelets not pendulous; mostly perennial.
                7. Lemmas keeled, the awn when present from above the middle.
                    8. Rachilla-joints very short, glabrous or short-pubescent; lemmas awnless or with a straight awn from a toothed apex .... 50. *Koeleria*

8. Rachilla-joints slender, villous; lemmas with a dorsal geniculate awn in most spp.
.... 86. *Trisetum*
7. Lemmas convex, awned from below the middle.
9. Rachilla prolonged from behind the upper floret; lemmas truncate and erose-dentate at summit .... 28. *Deschampsia*
9. Rachilla not prolonged; lemmas tapering into 2 slender teeth .... 3. *Aira*

Group IV

1. Lemmas divided at top into 5 or more awns or awnlike teeth or lobes.
   2. Spikelets several-fld.; awnlike lobes 5 .... 59. *Orcuttia*
   2. Spikelets 3-fld.; awns 9, plumose .... 36. *Enneapogon*
1. Lemma awnless or 1- or 3-awned.
   3. Plants tall stout reeds with large plumelike panicles; lemmas or rachilla with silky hairs as long as the lemma.
      4. Lvs. crowded at the base of the culms .... 20. *Cortaderia*
      4. Lvs. distributed along the culms.
         5. Lemmas naked; rachilla hairy .... 67. *Phragmites*
         5. Lemmas hairy; rachilla naked .... 8. *Arundo*
   3. Plants low or to ca. 1.5 m. high, not reedlike.
      6. Plants dioecious, perennial.
         7. The plants densely tufted, erect from short rhizomes; lemmas scabrous. Of dry mountain slopes .... 43. *Hesperochloa*
         7. The plants not densely tufted, spreading by stolons or creeping rhizomes; lemmas glabrous. Of saline or alkaline places.
            8. Spikelets obscure, scarcely differentiated from the short crowded rigid lvs.
               .... 56. *Monanthochloe*
            8. Spikelets in narrow, simple, exserted panicles .... 31. *Distichlis*
      6. Plants not dioecious, except in some spp. of *Poa* (these with villous lemmas).
         9. Spikelets of 2 forms, sterile and fertile intermixed; panicle dense, ± secund.
            10. Fertile spikelets 2–3-fld.; sterile spikelets with numerous rigid awn-tipped lemmas; panicle dense, spikelike .... 24. *Cynosurus*
            10. Fertile spikelets with 1 perfect floret, long-awned; sterile spikelets with many, obtuse, sterile lemmas; panicle branches short, nodding ... 51. *Lamarckia*
         9. Spikelets all alike in the same infl.
            11. Lemmas 3-nerved, the nerves prominent.
               12. Infl. a few-fld. woolly capitate panicle overtopped by the lvs.; lemmas toothed or cleft. Low desert plants.
                  13. Lemmas cleft either side of the midverve to near the base, the lower 2 sterile, the 3d floret fertile, the 4th reduced to a 3-awned rudiment . 11. *Blepharidachne*
                  13. Lemmas 2-lobed, but not deeply cleft, all fertile but the uppermost
                     .... 39. *Erioneuron*
               12. Infl. an exserted open or spikelike panicle.
                  14. Lemmas pubescent on the nerves or callus, obtuse .... 85. *Tridens*
                  14. Lemmas not pubescent on the nerves or callus, but sometimes on the internerves, acute or acuminate.
                     15. Glumes 6–7 mm. long, unequal; infl. a dense panicle, ± secund
                        .... 22. *Cutandia*
                     15. Glumes 0.5–4 mm. long, subequal or unequal.
                        16. Glumes longer than the lemmas; lateral nerves of lemma marginal, the internerves pubescent .... 30. *Dissanthelium*
                        16. Glumes shorter than the lemmas; lateral nerves of the lemma not marginal, the internerves glabrous .... 37. *Eragrostis*
            11. Lemmas 5- to many-nerved, the nerves sometimes obscure.
               17. Lemmas as broad as long, with outspread margins; florets closely imbricate, horizontally spreading .... 15. *Briza*
               17. Lemmas longer than broad, the margins clasping the palea; florets not horizontally spreading.
                  18. The lemmas keeled on the back (sometimes somewhat rounded in *Poa*).
                     19. Spikelets strongly compressed, crowded in 1-sided clusters at the ends of stiff naked panicle-branches .... 25. *Dactylis*
                     19. Spikelets not strongly compressed, not crowded in 1-sided clusters.

                20. Lemmas awned from a minutely bifid apex, or nearly awnless; spikelets large .... 16. *Bromus*
                20. Lemmas awnless; spikelets small .... 68. *Poa*
            18. The lemmas rounded on the back.
                21. Glumes papery; lemmas firm, strongly nerved, scarious-margined; upper florets sterile, often reduced to rudiments infolded by the broad upper lemmas; spikelets tawny or purplish .... 55. *Melica*
                21. Glumes not papery; upper florets like the others.
                    22. Nerves of lemma parallel, not converging at top, or but slightly so.
                        23. Sheaths of lvs. usually connate; nerves of upper empty glumes single; styles present .... 42. *Glyceria*
                        23. Sheaths of lvs. open; nerves of upper empty glumes 3 . 70. *Puccinellia*
                    22. Nerves of lemma converging toward the summit, the lemmas narrowed at apex.
                        24. Lemmas awned or awn-tipped from a minutely bifid apex (rarely awnless); palea adhering to the caryopsis.
                            25. Spikelets in open to contracted panicles; stigmas borne at the sides of the summit of ovary .... 16. *Bromus*
                            25. Spikelets nearly sessile in strict racemes; stigmas terminal on the ovary
                                                                                14. *Brachypodium*
                        24. Lemmas entire, pointed, awnless or awned from the tip.
                            26. Spikelets awned except in a few perennial spp.; lemmas pointed
                                                                                40. *Festuca*
                            26. Spikelets awnless.
                                27. Second glumes 5–11-nerved; spikelets ca. 1 cm. or more long; lemmas broad. Sand dunes e. of Inyo Mts. .... 84. *Swallenia*
                                27. Second glumes 1–3-nerved; spikelets smaller.
                                    28. Spikelets on slender pedicels in compound panicles; perennials
                                                                                68. *Poa*
                                    28. Spikelets on short thick pedicels in simple dense panicles; annual
                                                                                73. *Scleropoa*

Group V

1. Spikelets with more than 1 perfect floret.
    2. Spikes many, slender, racemose on an elongate axis .... 53. *Leptochloa*
    2. Spikes few, digitate or nearly so.
        3. Rachis of spike not prolonged beyond the spikelets .... 34. *Eleusine*
        3. Rachis of spike extending beyond the spikelets .... 26. *Dactyloctenium*
1. Spikelets with only 1 perfect floret, though sometimes with additional imperfect florets above or below.
    4. The spikelets without additional modified florets, the rachilla sometimes prolonged.
        5. Rachilla articulate below the glumes, the spikelets falling entire; spikes not digitate, but arranged on an elongate axis .... 79. *Spartina*
        5. Rachilla jointed above the glumes; spikes digitate .... 23. *Cynodon*
    4. The spikelets with 1 or more modified florets above the perfect one.
        6. Spikes digitate or nearly so .... 19. *Chloris*
        6. Spikes racemose along a main axis .... 13. *Bouteloua*

Group VI

1. Spikelets solitary at each node of the rachis (rarely 2 in spp. of *Andropogon*, but never throughout).
    2. The spikelets 1-fld., sunken in hollows in the rachis; spikes slender, cylindric; plants low annuals.
        3. Lemmas with awns; florets lateral to the rachis. Montane .... 74. *Scribneria*
        3. Lemmas not awned; florets dorsiventral to the rachis. Coastal marshes.
            4. First glume wanting; spikes 1–2 dm. long .... 57. *Monerma*
            4. First glume present; spikes 0.7–1 dm. long .... 62. *Parapholis*
    2. The spikelets 2–several-fld., not sunken in the rachis.
        5. Spikelets placed edgewise to the rachis, the lateral ones with a single glume . 54. *Lolium*
        5. Spikelets placed flatwise to the rachis.
            6. Plants perennial .... 1. *Agropyron*
            6. Plants annual.

7. Glumes ovate, 3-nerved .... 87. *Triticum*
7. Glumes subulate, 1-nerved .... 75. *Secale*
1. Spikelets normally more than 1 at each node of rachis.
   8. Spikelets 3 at each rachis-node, 1-fld., the lateral pair pedicelled, usually reduced to awns
      .... 47. *Hordeum*
   8. Spikelets 2 or more (sometimes 1 in *Elymus*) at each node of rachis, alike, 2–6-fld.
      9. Rachis continuous, disarticulating only tardily, and that rarely; glumes not greatly elongate .... 35. *Elymus*
      9. Rachis articulating into joints at maturity; glumes usually setaceous and greatly elongate
         .... 77. *Sitanion*

Group VII

One genus .... 52. *Leersia*

Group VIII

1. Spikelets all alike, fertile, surrounded by copious soft hairs; infl. a narrow panicle . 49. *Imperata*
1. Spikelets unlike, the sessile perfect, the pedicellate ♂ or neuter.
   2. Racemes of several joints.
      3. Fertile spikelet with a hairy-pointed callus, formed of the attached supporting rachis joint or pedicel; awns strong, brown .... 44. *Heteropogon*
      3. Fertile spikelet without a callus, the rachis disarticulating below the spikelet; awns slender.
         4. Lower pair of spikelets like the others of the raceme.
            5. Pedicels and rachis internodes without a cent. groove or membranous area
               .... 5. *Andropogon*
            5. Pedicels and upper rachis usually with a cent. groove or membranous area
               .... 12. *Bothriochloa*
         4. Lower pair of spikelets sterile, awnless; racemes in pairs on slender flexuous peduncles
            .... 48. *Hyparrhenia*
   2. Racemes reduced to 1 or few joints, these in a compound panicle .... 78. *Sorghum*

Group IX

One genus in our flora .... 46. *Hilaria*

Group X

1. Spikelets sunken in the cavities of the flattened corky rachis; coarse creeping grass escaping from cult. .... 82. *Stenotaphrum*
1. Spikelets not sunken in the rachis.
   2. The spikelets subtended or surrounded by many distinct or ± connate bristles forming an invol.
      3. Bristles persistent, the spikelets deciduous .... 76. *Setaria*
      3. Bristles falling with the spikelets at maturity.
         4. The bristles not united at the base, slender, often plumose .... 64. *Pennisetum*
         4. The bristles united into a burlike invol., the bristles retrorsely barbed ... 18. *Cenchrus*
   2. The spikelets not subtended by bristles.
      5. Glumes awned or mucronate; apex of palea not enclosed by the lemma.
         6. Infl. paniculate; spikelets silky .... 71. *Rhynchelytrum*
         6. Infl. of unilateral racemes along a common axis; spikelets not silky ... 32. *Echinochloa*
      5. Glumes awnless; apex of palea usually enclosed by the lemma.
         7. Spikelets in panicles .... 61. *Panicum*
         7. Spikelets in 1-sided spikelike racemes.
            8. First glume and the rachilla joint forming a swollen ringlike callus below the spikelet; racemes several along the main axis .... 38. *Eriochloa*
            8. First glume present or lacking, not forming a ringlike callus; racemes aggregate at the summit of the culm.
               9. Racemes slender, 3–12.
                  10. Fr. flexible; 1st glumes reduced but present .... 29. *Digitaria*
                  10. Fr. rigid; 1st glume wanting .... 10. *Axonopus*
               9. Racemes stout, in pairs .... 63. *Paspalum*

1. Agropyron Gaertn. WHEAT GRASS

Perennials, often with creeping rhizomes; culms usually erect. Infl. a bilateral spike with spikelets solitary, infrequently paired, at the nodes of a continuous or disarticulating rachis. Glumes broad or narrow, mostly few- to several-nerved, awn-tipped or awnless. Lemmas firm, faintly 5- or 7-nerved, rounded on the back, awned or awnless. Palea membranous, ca. as long as lemma. Ca. 60 spp. of temp. regions. (Greek, *agrios*, wild, and *puros*, wheat, the 2 original spp. being weeds in wheat.)

(Gould, F. W. Nomenclatorial changes in Elymus with a key to the Californian spp. Madroño 9: 120–128. 1947.)

Lemmas awned; plants typically without creeping rhizomes.
    Awns of lemmas curving outward at maturity, 15–25 mm. long .... 4. *A. scribneri*
    Awns of lemmas straight, 1–8 mm. long .... 2. *A. parishii*
Lemmas awnless or nearly so.
    Plants without creeping rhizomes.
        Spikelets not much compressed; blades flat or loosely involute.
            Nodes of culms glabrous; spikelets 3–5-fld. .... 5. *A. trachycaulum*
            Nodes of culms pubescent; spikelets 6–8-fld. .... 2. *A. parishii*
        Spikelets much compressed, shorter than the internodes; glumes obtuse or truncate; blades stiff, inrolled, glaucous .... 1. *A. elongatum*
    Plants with creeping rhizomes; blades mostly involute, stiff, ± glaucous ...... 3. *A. riparium*

1. **A. elongàtum** (Host) Beauv. [*Triticum e.* Host.] TALL WHEATGRASS. Not creeping, 3–10 dm. tall, glabrous; lvs. glaucous, inrolled, stiff; spike elongate, very lax; spikelets spaced, compressed, oval, 4–8-fld.; glumes obtuse or truncate, 7- to 9-nerved.—Occasionally established as in Chaparral, Murrieta, Riverside Co. and in Kern Co., as well as farther n. Native of Old World. Spring.

2. **A. paríshii** Scribn. & Sm. [*Elymus stebbinsii* Gould.] Culms 7–12 dm. tall, pubescent at nodes, blades flat or loosely involute, 2–4 mm. wide; spike slender, nodding, 10–25 cm. long; spikelets 4–7-fld., ca. 2 cm. long, scaberulous on rachillae; glumes 3–5-nerved, 10–15 mm. long, acute; lemmas acute or short-awned (1–8 mm.); palea as long as lemma, obtuse.—Dry slopes below 5000 ft.; Chaparral, Yellow Pine F.; Cuyamaca Mts. to cent. Calif. Most of our plants have glabrous nodes and are the var. *laeve* Scribn. & Sm.

3. **A. ripàrium** Scribn. & Sm. [*Elymus rydbergii* Gould.] With vigorous rhizomes; culms 4–8 dm. high; blades mostly involute, stiff, mostly glaucous; spike 6–12 cm. long; spikelets quite imbricate, 4–8-fld., 1–1.5 cm. long; lemmas glabrous or somewhat pubescent on edges of lower part; $n=14$ (Hartung, 1946), $2n=42$ (Tateoka, 1956).—Dry or moist places, collected years ago near Riverside; to Wash., Alta., N. Dak.

4. **A. scríbneri** Vasey. [*Elymus s.* Jones.] Culms tufted, prostrate or decumbent-spreading, 2–4 dm. long; blades flat or loosely involute, 1–3 mm. wide; spike long-exserted, dense, 3–7 cm. long, with glabrous internodes, disarticulating readily; spikelets 3–5-fld.; glumes narrow, the 1 obscurely nerved, the other distinctly 2–3-nerved, awned; lemmas with strong divergent awn 15–25 mm. long; palea slightly exceeding body of lemma, 2-toothed; $2n=28$ (Tateoka, 1956; Collins, 1966).—Rocky slopes above 10,000 ft.; Alpine Fell fields; White Mts.; to Rocky Mts.

5. **A. trachycaùlum** (Link) Malte. [*Triticum t.* Link. *Elymus pauciflorus* Gould, not Lam. *A. tenerum* Vasey. *A. pauciflorum* Hitchc.] Plate 96, Fig. F. Without creeping rhizomes; culms tufted, erect, 5–10 dm. tall, mostly with glabrous sheaths; blades mostly 2–4 mm. wide; spike slender, 10–25 cm. long; spikelets remote to imbricate; glumes and lemmas awnless or nearly so; $n=14$ (Peto, 1929).—Moist to dry and rocky places, below 11,000 ft.; Montane Coniferous F., Chaparral; San Jacinto Mts. and Orange Co. n.; White Mts., to Alaska and Atlantic Coast, Mex. June–Aug.

2. Agróstis L. BENTGRASS

Annual or perennial grasses, delicate to rather coarse, with mostly flattish lvs. and open to contracted panicles of small spikelets. Spikelets 1-fld., disarticulating above the glumes, the rachilla usually not prolonged. Glumes subequal, acute, acuminate or occasionally awned, usually scabrous on the keel, sometimes also on the back. Lemma obtuse, usually shorter and thinner than the glumes, mostly 3-nerved, awnless or dorsally 3-awned. Palea usually shorter than lemma, 2-nerved or more often small and nerveless. Ca. 125 spp. in temp. and colder regions, many of importance for forage and lawns. (Greek name of a grass.)

1. Plants annual, of vernal pools.
   1a. Lemma 4-toothed at apex, the awn 3.5–4 mm. long .... 6. *A. microphylla*
   1a. Lemma deeply bifid, the awn 5–6 mm. long .... 9. *A. tandilensis*
1. Plant perennial.
   2. Palea evident, 2-nerved, at least half as long as lemma.
      3. Branches of panicle or some of them floriferous from the base; ligule 3–6 mm. long
                                                                                                 10. *A. stolonifera*
      3. Branches of panicle naked at the base; ligule 1–2 mm. long .... 11. *A. tenuis*
   2. Palea obsolete or a minute nerveless scale sometimes up to 0.5 mm. long.
      4. Plants spreading by creeping rhizomes, these short in *A. lepida*.
         5. Plants tufted, alpine; rhizome short .... 5. *A. lepida*
         5. Plants mostly from below 7500 ft.; rhizome long and slender .... 2. *A. diegoensis*
      4. Plants without rhizomes; stolons sometimes developed.
         6. Panicle narrow, contracted, at least some of the lower branches bearing spikelets from the base.
            7. Panicle loose, the branches whorled, not densely fld. at base; lemma-awn twisted, geniculate. Rare .... 1. *A. ampla*
            7. Panicle dense to loose, the branches crowded and densely fld. at base; lemma awnless or awned. Widespread .... 3. *A. exarata*
         6. Panicle open, sometimes diffuse; branches very slender, the lower not bearing spikelets at base.
            8. Panicle very diffuse, the branches capillary, not flexuous, the spikelets arranged at the ends .... 8. *A. scabra*
            8. Panicle open, but not diffuse, the branches branching at or below the middle.
               9. Spikelets ca. 2 mm. long; plants 1–3 dm. high .... 4. *A. idahoensis*
               9. Spikelets 2.5–3 mm. long; plants 6–9 dm. high .... 7. *A. oregonensis*

1. **A. ámpla** Hitchc. Tufted perennial, 3–6 dm. tall; blades 3–5 mm. wide, scabrous on margins and minutely so on nerves; panicle generally well exserted, ca. 1 dm. long, the lowest fascicle somewhat remote; glumes subequal, 2.8–3 mm. long including the 1 mm. awn, scaberulous; lemma 1.5–2 mm. long, bifid at apex, the awn 3 mm. long, geniculate near the middle; lemma obsolete.—Damp places below 5000 ft.; Chaparral; San Gabriel Mts., Santa Ynez Mts.; cent. Calif. to Wash. June–July.

2. **A. diegoénsis** Vasey. Rhizomes creeping; culms erect, 4–10 dm. tall; blades flat, lax, 2–6 mm. wide; ligule 2–3 mm. long; panicle narrow, open, 10–15 cm. long, with ascending branches, some of which are naked below; spikelets having glumes 2.5–3 mm. long, the lemma a little shorter, awned or awnless; palea obsolete; $n=21$ (Stebbins & Love, 1941).—Meadows and woods below 7500 ft.; Chaparral, Yellow Pine F.; San Diego Co. n.; to B.C., Rocky Mts. April–Aug. Most ids.

3. **A. exaràta** Trin. Plate 97, Fig. B. Mostly tufted, 2–12 dm. high, slender to stoutish; sheaths smooth to ± scabrous; ligule 4–6 mm. long; blades flat, 2–10 mm. wide, mostly scabrous; panicle narrow, rather open to dense and interrupted, 0.5–3 dm. long; glumes subequal, 2.5–4 mm. long, acuminate to awn-tipped, scabrous on keel; lemmas 1.7–2 mm. long; palea minute; $n=21$ (Stebbins & Love, 1941).—Moist open places below 8000 ft.; many Plant Communities; through much of cismontane Calif. to Alaska, Nebr., Texas, Santa Rosa and Santa Cruz ids. June–Aug.

4. **A. idahoénsis** Nash. [*A. tenuis* Vasey, not Sibth. *A. tenuiculmis* Nash.] Tufted, mostly 1–3 dm. tall; blades flat, 1–3 mm. wide; panicle 5–10 cm. long, with capillary

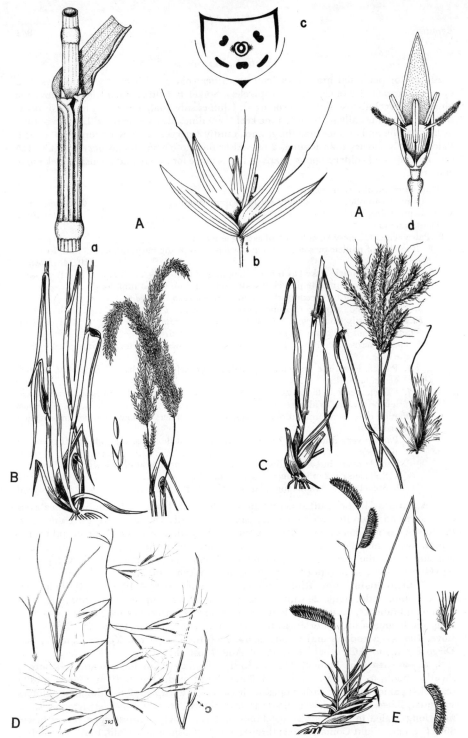

Plate 97. Fig. A, Grass features. *a*, solid nodes with intervening hollow node and ligule at junction of lf.-blade and lf.-sheath, *v*, spikelet with 2 outer glumes subtending 3 florets, the lower right floret showing a lemma awned from its back, a palea above and emerging stamen between, *d*, diagram of a floret with palea in background, 3 stamens, ovary with 2 feathery stigmas, 2 basal lodicules. Fig. B, *Agrostis exarata*. Fig. C, *Bothriochloa barbinodis*. Fig. D, *Aristida divaricata*. Fig. E, *Bouteloua gracilis*.

flexuous branches; spikelets 1.5–2 mm. long; lemma ca. 1.3 mm. long, awnless.—Mountain meadows, mostly 6000–9000 ft.; Montane Coniferous F.; San Jacinto and San Bernardino mts.; Sierra Nevada to Wash., Mont., New Mex. July–Aug.

5. **A. lépida** Hitchc. Tufted perennial, 2–3.5 dm. tall, with short rhizomes; lvs. mostly basal, the blades firm, flat or folded, 1–1.5 mm. long; ligule to 4 mm. long; panicle purple, erect, 1–1.5 dm. long, the lower branches spreading, to 5 cm. long; glumes 3 mm. long, smooth; lemma 2 mm. long; palea obsolete.—Moist places; Lodgepole F.; San Bernardino Mts.; Sierra Nevada. July–Aug.

6. **A. microphýlla** Steud. Loosely tufted annual with erect culms 1–3.5 dm. high, slender, smooth; stem-lvs. 2–3; sheaths smoothish; blades scabrid, 2–5 cm. long, flat or loosely involute, 1.5–3 mm. wide; panicle spikelike, 2–9 cm. long; glumes awned, 3–4 mm. long, aristate, green to purplish, subequal; lemma 2 mm. long, 4-toothed at apex, the awn from above the middle, geniculate below, hispidulous, 3.5–4 mm. long; palea none; $n=28$ (DeLisle, 1965).—Vernal pools, near the coast, San Diego, to Humboldt Co. and in L. Calif. Santa Rosa Id. May–July.

7. **A. oregonénsis** Vasey. [*A. hallii* var. *californica* Vasey.] Culms 6–9 dm. high; blades 2–4 mm. wide; ligule 1–2 mm. long; panicle oblong, open, 1–3 dm. long, the branches whorled, stiffish, ascending, up to 5 or 10 cm. long and branching above the middle; glumes 2.5–3 mm. long; lemmas 1.5 mm. long, awnless; palea ca. 0.5 mm. long.—Wet places at ca. 7500 ft.; Montane Coniferous F.; San Jacinto and San Bernardino mts.; Sierra Nevada to B.C., Rocky Mts. June–Aug.

8. **A. scàbra** Willd. [*A. hiemalis* Calif. refs.] Erect, tufted, 3–9 dm. tall; ligule 2–5 mm. long; blades flat, 1–3 mm. wide, 8–20 cm. long, scabrous; panicle 1.5–2.5 dm. long, with distant brittle scabrous spreading or drooping branches; spikelets 2–2.7 mm. long, loosely placed at ends of branches; glumes unequal, acuminate, scabrous on keels; lemma 1.5–1.7 mm. long, awnless; $2n=42$ (Bowden, 1960).—Meadows and moist places, 5000–9000 ft.; Montane Coniferous F.; San Jacinto Mts. n.; to Alaska, Nfld. July–Sept.

A. semiverticillàta (Forsk.) C. Chr. with contracted panicles and scabrous glumes, reported from S. Rosa and S. Cruz ids.

9. **A. tandilénsis** (Ktze.) Parodi. [*A. kennedyana* Beetle.] Much like *A. microphylla*; panicle dense, 1–4 cm. long; 1st glume 4 mm. long, 2d glume 3 mm. long; lemma deeply bifid, loosely pilose, the body 1.5–2 mm. long, the terminal teeth 1 mm., the awn 5–6 mm. long.—Vernal pools, San Diego, La Jolla; Solano Co.; Argentine. April.

10. **A. stolonífera** L. var. **màjor** (Gaud.) Farwell. [*A. alba* var. *m.* Gaud.] REDTOP. Perennial, forming turf, with elongate stolons and erect leafy sterile shoots; culms 5–12 dm. tall; blades 3–8 mm. wide; panicle ovoid to ellipsoid, purplish to green, rather lax, 1–3 dm. long, the branches spreading at anthesis, more erect in fr.; spikelets 2–3.5 mm. long; glumes subequal, usually scabrous on keel; lemma rarely awned; $n=14$ (Sokolovskaja, 1938), 21 (Stuckey & Banfield, 1946).—Escape in moist places, below 7500 ft.; many Plant Communities; introd. from Eu. Catalina Id. June–Sept.

11. **A. ténuis** Sibth. COLONIAL BENT. Tufted, slender, 2–4 dm. tall, with short stolons but not creeping rhizomes; ligule 1–2 mm. long; blades 1–3 mm. wide; panicle 5–10 cm. long, open, delicate, the branches slender, naked below; spikelets not crowded, 2–3 mm. long; lemma nearly equal to glumes, awnless; $n=14$ (Avdulov, 1931), $2n=28, 32, 34$ (Bowden, 1960).—Occasional escape from cult. as at Loma Linda, San Bernardino Co.; more common farther n.; native of Eu. July–Sept.

3. *Aìra* L. HAIRGRASS

Rather delicate annuals with narrow or open panicles of small spikelets. Lf.-blades lax, subfiliform. Spikelets 2-fld., disarticulating above the glumes, the rachilla not prolonged. Glumes boat-shaped, subequal, 1-nerved or obscurely 3-nerved, acute, membranaceous or subscarious. Lemmas firm, rounded on back, tapering into 2 slender teeth and usually with a slender geniculate twisted exserted awn from below the middle of the back. Ca. 10 spp., from s. Eu. (Old Greek name for a grass.)

1. **A. caryophýllea** L. [*Aspris c.* Nash.] Culms 1–few, slender, erect, 1–3 dm. tall; panicle open, the spikelets silvery, shining, 3 mm. long, clustered toward ends of spreading capillary branches; lemma of both florets with geniculate awn 4 mm. long, the apical teeth setaceous; $2n=14$, 28 (Böcher & Larsen, 1958).—Occasional in open places below 5000 ft.; many Plant Communities; much of cismontane Calif.; introd. from Eu. April–June.

4. *Alopécurus* L. FOXTAIL

Tufted annuals and perennials, a few rhizomatous, with flat blades and dense, cylindrical, spikelike panicles of 1-fld. spikelets. These disarticulating below the glumes, strongly compressed laterally. Glumes equal, usually united at base, ciliate on keel. Lemma ca. as long as glumes, 5-nerved, obtuse, the margins united at base, with a slender dorsal awn from below the middle which is included or 2–3 times as long as spikelet. Palea wanting. Ca. 25 spp. of temp. N. Hemis. (Greek, *alopex*, fox, and *oura*, tail, referring to the cylindrical panicle.)

Plants perennial; spikelets 2–2.5 mm. long.
   Awn scarcely exceeding the glumes .... 1. *A. aequalis*
   Awn exserted 2–5 mm. .... 2. *A. geniculatus*
Plants annual; spikelets 3–3.5 mm. long .... 3. *A. howellii*

1. **A. aequàlis** Sobol. Perennial, usually glaucous, 2–7 dm. tall; blades 1–4 mm. wide; panicle slender, 2–7 cm. long, whitish-drab to mouse-color, 3–5.5 mm. thick; spikelets 2 mm. long; glumes silky with long-ciliate keels; awn of lemma scarcely exserted; anthers 0.6–1 mm. long; $n=7$ (Avdulov, 1931), $2n=7$ pairs (Anderson, 1965).—Moist places, 5000–9200 ft.; Montane Coniferous F.; mts. from San Diego Co. n.; to Alaska, Atlantic Coast; Eurasia. May–July.

2. **A. geniculàtus** L. Near to *A. aequalis*, but with more decumbent culms rooting at the nodes; awn exserted 2–3 mm.; spikelets 2.5 mm. long, with dark purple tip; anthers ca. 1.5 mm. long; $n=14$ (Avdulov, 1928).—Wet places with *A. aequalis* n. to B.C., Atlantic Coast; Eurasia. June–July.

3. **A. howéllii** Vasey. [*A. californicus* Vasey.] Annual, 1.5–3 dm. tall, ± geniculate at lower nodes; sheaths inflated; panicle oblong to linear, 2–6 cm. long, 4–7 mm. thick; glumes 3–3.5 mm. long, ciliate on keel, appressed-pilose on lateral nerves; awn attached toward base of lemma, exserted 3–5 mm.; anthers orange, ca. 1 mm. long.—Wet places, such as drying mud flats; at low elevs.; Chaparral, Coastal Sage Scrub, etc.; San Diego Co. n. to Ore. March–June.

4a. *Ammóphila* Host. BEACHGRASS

Coarse tough perennials with scaly tough rhizomes and pale dense spikelike panicles. Spikelets 1-fld., compressed, the rachilla disarticulating above the glumes and produced beyond the palea as a short hairy bristle. Glumes subequal, chartaceous. Lemma similar and a little shorter, with bearded callus. Ca. 3 spp., on sandy coasts of Eu. and N. Am. (Greek, *ammos*, sand, and *philos*, loving.)

1. **A. arenària** (L.) Link. [*Arundo a.* L.] Stout, 6–10 dm. tall; blades very long, soon involute; ligule 10–30 mm. long; panicle 1–3 dm. long; glumes 8–10 mm. long, scabrous; $n=7$, 14 (Westergaard, 1941).—Introd. in dune areas along the coast, as at Carpenteria; native of Eu. May–Aug.

5. *Andropògon* L. BEARDGRASS

Cespitose perennials with usually stiffly erect stems, rounded or flattened and keeled sheaths, and flat or folded blades. Ligule membranous. Flowering culm from branched and broomlike to simple, each culm or branch culminated by an infl. of 2 to several racemose branches. Spikelets in pairs at each node of jointed rachis, the sessile spikelet fertile. Pediceled spikelet from absent to well developed. Glumes large, firm, awnless. Lemmas of sterile and fertile florets membranous. Ca. 150 spp. of warmer regions. (Greek, *aner*, (andr-) man, and *pogon*, beard, because of the hairy pedicels of the ♂ or sterile spikelets.)

Racemes solitary on each peduncle; rachis-joints oblique and hollow at summit . 1. *A. cirratus*
Racemes 2–many on each peduncle .... 2. *A. glomeratus*

1. **A. cirràtus** Hack. Plants pale, glaucous to purplish, 4–7 dm. high; blades 1–4 mm. wide; racemes exserted, 3–6 cm. long; sessile spikelet 8–9 mm. long, the awn 5–10 mm. long; pedicellate spikelet scarcely reduced, awnless; $n = 10$ (Brown, 1950).—Once reported from Jamacha, e. San Diego Co.; Ariz. to Tex., n. Mex.

2. **A. glomeràtus** (Walt.) BSP. [*Cinna g.* Walt.] Tufted, leafy, 5–10 dm. tall, the stems corymbosely branched above; sheaths and blades scabrous; ligules villous; blades 1–5 dm. long, 3–6 mm. wide; infl. compound, 1–3 dm. long, the racemes paired, sessile spikelet 5 mm. long; awn 1–2 cm. long; $n = 10$ (Brown, 1950).—Occasional in wet places below 2000 ft.; Coastal Sage Scrub, Chaparral, etc.; cismontane; Creosote Bush Scrub, Death V.; cent. Calif.; to Ky., W. I. Cent. Am. Sept.–March.

6. Anthoxánthum L. VERNALGRASS

Sweet smelling annual or perennial grasses with flat blades and spikelike panicles. Spikelets with 1 perfect fl. and 2 sterile reduced florets below, these falling away with the fertile one. Glumes unequal, acute or mucronate; sterile lemmas short, awned from the back; fertile lemma awnless, shorter than the sterile. Palea 1-nerved, rounded on back, enclosed in lemma. Ca. 4 spp. of Eurasia. (Greek, *anthos*, fl., and *xanthos*, yellow.)

1. **A. odoràtum** L. Tufted, erect, 3–6 dm. tall; blades 2–5 mm. wide; panicle brownish-yellow, long-exserted, 2–6 cm. long; spikelets 8–10 mm. long; 1st sterile lemma ca. 2 mm. long, brown, smooth, shining; $2n = 20$ (Aabergi, 1966).—Occasional escape from cult.; introd. from Eurasia. May–June.

7. Arístida L. TRIPLE-AWNED GRASS

Low to moderately tall annuals and perennials, these lacking rhizomes or stolons. Blades narrow, usually involute. Infl. an open or contracted panicle of usually large, 1-fld. spikelets. Disarticulation above the glumes, which are thin, lanceolate, with a strong cent. nerve and sometimes 2 lateral nerves. Lemma indurate, terete, 3-nerved, tapering to an awn-column bearing usually 3 stiff awns. Ca. 200 spp., in warmer regions. (Latin, *arista*, awn.)

1. Neck of fr. jointed at base. Deserts .... 2. *A. californica*
1. Neck of fr. not jointed at base.
  2. Lateral awns wanting or reduced to mere points .... 8. *A. orcuttiana*
  2. Lateral awns evident.
    3. Plants annual; awns 10–15 mm. long .... 1. *A. adscensionis*
    3. Plants perennial.
      4. Glumes ca. equal.
        5. Panicle open with spreading branches naked at the base.
          6. Summit of lemma narrowed into a twisted neck 2–5 mm. long .. 3. *A. divaricata*
          6. Summit of lemma somewhat narrowed but not twisted .... 6. *A. hamulosa*
        5. Panicle narrow, the branches not horizontally spreading .... 9. *A. parishii*
      4. Glumes definitely unequal.
        7 Lemma tapering into a slender ± twisted beak 5–6 mm. long; awns 1.5–2.5 cm. long, widely spreading .... 5. *A. glauca*
        7. Lemma beakless or only short-beaked.
          8. Panicle rather loose and nodding, with slender flexuous branches . 10. *A. purpurea*
          8. Panicle erect, stiff with mostly appressed branches.
            9. Panicle more than 1.5 dm. long, the branches several-fld.; awns ca. 2 cm. long
                                                                      11. *A. wrightii*

9. Panicle mostly less than 1.5 dm. long, the branches few-fld.; awns 2–several cm. long.
10. Lemma gradually narrowed above, scaberulous on upper half; lvs. mostly in a short curly cluster at base of plant .... 4. *A. fendleriana*
10. Lemma scarcely narrowed above, scaberulous only at tip; lvs. not conspicuously basal .... 7. *A. longiseta*

1. **A. adscensiònis** L. [*A. bromoides* HBK.] Annual, branched at base, often with purplish tinge, erect or spreading, 1–8 dm. tall; blades 1.5–5 cm. long, usually involute; panicle narrow, rather dense, 3–10 cm. long; 1st glume 5–7 mm. long; the 2d 8–10 mm.; lemma 6–9 mm. long, compressed toward the scarcely beaked top; awns ca. equal, mostly 1–1.5 cm. long; $n=11$ (Avdulov, 1931), $2n=22$ (Gould, 1966).—Dry open places and rocky hills below 3500 ft.; Creosote Bush Scrub, Coastal Sage Scrub, etc.; e. San Luis Obispo Co. and w. Santa Barbara Co. to the deserts and San Diego Co.; to Tex., Kans.; Argentine. Feb.–June. Santa Cruz, Santa Rosa, Santa Catalina and San Clemente ids.

2. **A. califórnica** Thurb. [*A. jonesii* Vasey. *A. c.* var. *fugitiva* Vasey.] Tufted perennial, the wiry culms 1–3 dm. tall; sheaths pubescent at base and collar; blades mostly involute; panicles loose, 2–5 cm. long, few-branched; glumes 1-nerved, awnless, 8 and 12 mm. long; lemma 5–7 mm., scaberulous toward summit; neck jointed at base, spirally twisted, 15–20 mm. long; awns equal, spreading horizontally, 2.5–3 cm. long.—Occasional in dry sandy places below 2000 ft.; Creosote Bush Scrub; both deserts; to Son. April–May.

3. **A. divaricàta** Humb. & Bonpl. ex Willd. Plate 97, Fig. D. Erect or prostrate-spreading perennial, usually 3–6 dm. long; blades flat or loosely involute, to 3 mm. wide and 15 cm. long; panicle large, diffuse, 1–3 dm. long with spreading or reflexed branches naked below; glumes subequal, 10 mm. long; lemmas 10 mm. long, the twisted beak 2–5 mm. long; awns subequal, 1–1.5 cm. long; $n=11$ (Gould, 1958).—Occasional on dry slopes; Coastal Sage Scrub, etc.; San Diego to San Bernardino V. and San Joaquin V.; to Tex., Oaxaca. May–July.

4. **A. fendleriàna** Steud. Densely cespitose erect perennial, 1.5–3 dm. high; blades forming a basal curly tuft, 3–5 cm. long; panicle narrow, 5–10 cm. long, with a few loosely arranged appressed spikelets; glumes unequal, smooth, awnless, 1-nerved, the 1st ca. 6 mm., the 2d 10–15 mm. long; lemma ca. 18 mm. long; awns 2–5 cm. long, spreading; $n=33$ (Gould, 1958).—Rare, dry slopes, 4000–6000 ft.; Pinyon-Juniper Wd.; e. Mojave Desert and Santa Rosa Mts.; to N. Dak., Tex., Mex. May–July.

5. **A. glaúca** (Nees) Walp. [*Chaetaria g.* Nees.] Erect perennial, 2–4 dm. tall; blades involute, curved or flexuous, 5–10 cm. long, ca. 1 mm. thick; panicle narrow, erect, few-fld., 7–15 cm. long; 1st glume 5–8 mm. long, 2d ca. twice as long; lemma 10–12 mm. long, with a slender, ± twisted beak 5–6 mm. long; awns equal, widely divergent, 1.5–2.5 cm. long; $2n=44$ (Gould, 1966).—Rare in dry places below 4000 ft.; Creosote Bush Scrub; both deserts; to Utah, Tex., cent. Mex. March–May.

6. **A. hamulòsa** Henr. Like *A. divaricata*; lemma somewhat narrowed at summit, not twisted; cent. awn somewhat longer than the lateral ones; $n=22$ (Stebbins and Löve, 1941).—About same range as *A. divaricata*; to w. Tex., Guatemala. May–Nov.

7. **A. longisèta** Steud. var. **robústa** Merr. Much like *A. fendleriana*, 3–5 dm. tall, the blades not in conspicuous basal tufts; panicle stiff, 5–9 cm. long; awns mostly 4–5 cm. long; $n=11$ (Gould, 1958).—Dry slopes, Creosote Bush Scrub; e. San Diego Co., Chuckwalla Mts.; to Wash., Minn. March–May.

8. **A. orcuttiàna** Vasey. [*A. schiedeana* Calif. refs.] Erect perennial, the culms scaberulous, 3–6 dm. tall; sheaths scaberulous, villous at throat; blades flat or involute, 1.5–3 mm. wide, the lower 1–2 dm. long; panicles open, 1–3 dm. long, nodding or drooping; glumes subequal, 10–15 mm. long; lemma 8–10 mm. long, narrowed into a scabrous twisted column 10–17 mm. long; cent. awn divergent, 5–10 mm. long, lateral awns up to 1 or 2 mm. long.—Reported from near San Diego; to Tex., Mex. Sept.

9. **A. paríshii** Hitchc. Tufted perennial, 2–5 dm. high; blades ± involute, sometimes

Aristida

flat, 1–2.5 dm. long, 1–2 mm. wide; panicle narrow, 1–3 dm. long, with appressed or ascending branches; glumes short-awned, the 1st 12 mm., the 2d 13–14 mm. long; lemma ca. 12 mm. long, with a short, ± twisted beak; awn subequal, ca. 2.5 cm. long, divergent.—Dry or rocky slopes, mostly below 4000 ft.; Coastal Sage Scrub, Chaparral, Creosote Bush Scrub; interior cismontane s. Calif., deserts; to Nev., Ariz. April–June.

10. **A. purpùrea** Nutt. Tufted perennial, 3–5 dm. tall; blades usually involute, 1–1.5 mm. wide when unrolled, 5–12 cm. long; panicle narrow, nodding, rather lax, usually purplish, 1–2 dm. long, with capillary branches; 1st glume 6–8 mm. long, 2d ca. 12–15 mm.; lemma ca. 10 mm. long, scarcely beaked, tuberculate-scabrous in lines upward; awns subequal, spreading, 3–5 cm. long; $n=11$ (Gould, 1958).—Occasional; Coastal Sage Scrub, Creosote Bush Scrub; cismontane and desert s. Calif.; to Ark., Kans., Tex., n. Mex. May–July.

11. **A. wrìghtii** Nash. Erect, cespitose perennial, 3–5 dm. tall; sheaths villous at throat and often with hairy line across the collar; blades involute, 1–2 dm. long, scabrous; panicle erect, narrow, 1.5–2 dm. long, with appressed branches; glumes unequal, 1-nerved, acuminate, the 1st 6 mm. long, the 2d ca. 12 mm.; lemma 10–12 mm. long, scaberulous above; awns subequal, ca. 2 cm. long, divergent; $n=33$ (Gould, 1958).—Occasional on stony slopes below 5000 ft.; largely Creosote Bush Scrub, Joshua Tree Wd.; Colo. and Mojave deserts; near Hemet, Riverside Co.; to Okla., Tex., cent. Mex. March–June, Sept.

8. *Arúndo* L. GIANT REED

Tall perennial reeds with broad blades and large plumelike terminal panicles. Spikelets several-fld., the florets successively smaller; rachilla glabrous, disarticulating above the glumes and between the florets; glumes ± unequal, membranaceous, 3-nerved, narrow, slender-pointed, ca. as long as the spikelet; lemmas thin, 3-nerved, long-pilose, narrowed upward, the nerves ending in slender teeth, the middle one becoming an awn. Ca. 6 spp. of warmer parts of Old World. (Ancient Latin name.)

1. **A. donáx** L. Culms stout, to 6 or 7 m. high; blades to 6 cm. wide; panicle 3–6 dm. long; spikelets ca. 12 mm. long; $n=55$ (Hunter, 1934).—Moist places like ditches, seeps, etc.; desert and cismontane Calif.; introd. from Eu. March–Sept.

9. *Avèna* L. OAT

Annuals, low to rather tall, with narrow to open mostly rather few-fld. panicles of rather large spikelets. Spikelets 2–3-fld., the rachilla bearded, articulating above the glumes and between the florets. Glumes subequal, membranaceous or papery, several-nerved, longer than the lower floret, usually exceeding the upper floret. Lemmas indurate except toward the summit, 5–9-nerved, bidentate at apex, with a dorsal bent and twisted awn. Ca. 10–15 spp. of temp. Eurasia. (The ancient Latin name.)

Teeth of lemmas awned or setaceous; pedicels capillary .... 1. *A. barbata*
Teeth of lemmas acute, not setaceous; pedicels stoutish.
   Lemmas pubescent with long brown hairs; spikelets usually 3-fld. .... 2. *A. fatua*
   Lemmas subglabrous; spikelets mostly 2-fld. .... 3. *A. sativa*

1. **A. barbàta** Brot. SLENDER WILD OAT. Culms slender, 3–6 dm. tall; blades flat, commonly 3–7 mm. wide; panicle lax; spikelets on curved capillary pedicels; lemma with stiff red hairs, the teeth ending in slender setae 4 mm. long; $n=14$ (Huskins, 1927).—Common weed in waste places and on open slopes, largely replacing native grasses; native of Old World. March–June. On most of our islands.

2. **A. fátua** L. WILD OAT. Culms stout, to 1.9 m. tall; blades 4–8 mm. wide; panicle loose, open, with horizontal branches; spikelets usually 3-fld.; florets readily falling from glumes; glumes ca. 2.5 cm. long, the rachilla and lower part of lemma with long stiff, mostly brownish hairs; lemmas nerved above, ca. 2 cm. long, with acuminate teeth; awn stout, geniculate, twisted below, 3–4 cm. long; $n=21$ (Philp, 1933).—Waste and

cult. places as a common weed; introd. from Eu. Reported from San Clemente, Santa Catalina and Santa Rosa ids. April–June.

3. **A. satìva** L. CULTIVATED OAT. Like *A. fatua*; spikelets mostly 2-fld.; the florets not separating readily from the glumes; lemma glabrous; awns straight, often wanting; $n=21$ (Emme, 1930).—Sometimes escaping from cult.; introd. from Eu. April–June.

10. *Axónopus* Beauv.

Stoloniferous or tufted perennials. Blades usually flat or folded. Racemes slender, spikelike, digitate or racemose along the main axis. Spikelets depressed-biconvex, oblong, usually obtuse, solitary, subsessile, alternate, in 2 rows on one side of a 3-angled rachis. First glume wanting; 2d glume and sterile lemma equal; fertile lemma and palea indurate, the lemma oblong-elliptic, usually obtuse. Ca. 70 spp. of warmer parts of Am. (Greek, *axon*, axis and *pous*, foot.)

1. **A. compréssus** (Sw.) Beauv. Stoloniferous; culms compressed, 1.5–5 dm. long; blades 8–25 cm. long, 8–12 mm. wide; raceme-spikes 2–5, mostly 4–8 cm. long; spikelets 2.2–2.5 (–2.8) mm. long, pilose.—Reported as occasional lawn weed, although sometimes planted for lawns. Se. U.S. to S. Am.

11. *Blepharidáchne* Hack.

Low tufted annuals and perennials, with short, congested, few-fld. panicles, these not or only slightly exserted above the subtending lvs. Spikelets compressed, rather few, 4-fld., the rachilla disarticulating above the glumes, but not between the florets. Glumes subequal, compressed, 1-nerved, thin, smooth; lemmas 3-nerved, deeply 3-lobed, the nerves extending into awns, conspicuously ciliate, 1st and 2d florets sterile, 3d fertile, 4th reduced to a 3-awned rudiment. Three spp. of U.S. and Argentina. (Greek, *blepharis*, eyelash, and *achne*, chaff, because of the ciliate lemma.)

1. **B. kíngii** (Wats.) Hack. [*Eremochloe k.* Wats.] Looking like *Tridens pulchellus*, but not rooting at upper nodes; culms 3–8 cm. long; sheaths with broad hyaline margins; blades involute, curved, stiff, 1–3 cm. long; panicle subcapitate, 1–2 cm. long, often purplish; glumes ca. 7–8 mm. long, acuminate, exceeding florets; sterile lemmas ca. 6 mm. long, long-ciliate on margins, pilose at base, cleft almost to middle, awn-tipped; fertile lemmas similar.—Dry rocky places, 5000–7000 ft.; Pinyon-Juniper Wd.; Panamint, Grapevine and Inyo-White mts., Inyo Co.; to Utah. May.

12. *Bothriochlòa* Ktze.

Cespitose perennials with erect or decumbent-spreading culms and flat blades. Infl. a terminal panicle, the spikelets on few to several primary spicate branches, these sometimes rebranched. Internodes of pedicels and upper rachis-internodes with a cent. groove or membranous area. Sessile spikelets of a pair fertile and awned, ± triangular in outline, the 1st glume dorsally flattened, the 2d with a median keel. Pediceled spikelet ♂ or sterile, usually well developed. Disarticulation in the rachis, the sessile spikelet falling attached to a pedicel and section of the rachis. Ca. 30 spp. of warmer parts of the world. (Greek, *bothrion*, a sucker, and *chloa*, grass.)

1. **B. barbinòdis** (Lag.) Herter. [*Andropogon b.* Lag. *A. saccharoides* of most Calif. refs.] Plate 97, Fig. C. Tufted, 6–13 dm. high, often branching below, the nodes glabrous to hispid; blades commonly glaucous, subglabrous, 3–6 mm. wide; panicle long-exserted, silvery-white, silky, dense, oblong, 7–15 cm. long; racemes 2–4 cm. long; rachis-joints and pedicels silky; spikelets 4 mm. long, the awn geniculate, twisted below, 1–1.5 cm. long; pedicellate spikelet reduced; $n=90$ (Gould, 1953).—Dry slopes below 4000 ft.; Coastal Sage Scrub, Chaparral, Joshua Tree Wd., etc.; cismontane s. Calif. and desert (Joshua Tree Nat. Mon., Providence Mts.); Catalina Id.; to Okla., Tex., Mex. Most months.

13. Bouteloùa Lag. GRAMA GRASS

Annuals and perennials, mostly tufted. Spikelets few to many, 1–2-fld., crowded in 2 rows and forming few to many, 1-sided, ± curved, sessile spikes; rachis usually conspicuously prolonged beyond the spikelets. Lower fls. perfect, the upper often ♂; glumes 2, narrow, acute, unequal, keeled. Lemma usually thinner, broader, 3-nerved, the nerves extending into short awns or mucros. Palea 2-nerved, sometimes 2-awned. Ca. 50 spp., mostly N. Am. Valuable range grasses. (C. *Boutelou*, 1774–1842, Spanish horticulturist.)

Spikelets not pectinately arranged, the spikes falling entire at maturity.
  Plants annual .... 1. *B. aristidoides*
  Plants perennial .... 3. *B. curtipendula*
Spikelets pectinately arranged at maturity, the spikes persistent, the florets falling from the persistent glumes.
  Plants annual .... 2. *B. barbata*
  Plants perennial.
    The plants decumbent or stoloniferous; culms white-woolly .... 4. *B. eriopoda*
    The plants erect or nearly so; culms not lanate.
      Spikes usually 2 .... 5. *B. gracilis*
      Spikes usually 4 or more .... 6. *B. trifida*

1. **B. aristidoìdes** (HBK.) Griseb. [*Dinebra a.* HBK.] Erect or spreading annual, the culms slender, 1.5–4 dm. long; sheaths and blades smooth, the latter rather small and few; spikes mostly 8–14, pedunculate, reflexed, readily falling, the base of the rachis forming a sharp bearded point; spikelets 2–4, narrow, appressed; glumes narrow, acuminate, the 1st ca. 2 mm. long, the 2d almost 4 mm.; lemma strigose on nerves, the lateral nerves ending in awned teeth; rudimentary floret a pilose stipe and 3 awns.—Occasional in dry sandy places, below 5500 ft.; Creosote Bush Scrub on Colo. Desert and Joshua Tree Wd. and Pinyon-Juniper Wd., s. and e. Mojave Desert, lower edge of Yellow Pine F., San Jacinto Mts.; to Nev., Tex., n. Mex. April–Sept.

2. **B. barbàta** Lag. Tufted annual with prostrate or spreading culms 1–3 dm. long; sheaths and blades glabrous; blades 1–1.5 mm. wide, 1–4 cm. long; spikes 4–7, 1–1.5 cm. long; spikelets 25–40, 2.5–4 mm. long, almost as wide; 2d glume twice the first; fertile lemma pilose at least along the sides, the awns from minute to as long as body; rudiment ± bearded at base, cleft, awned, a 2d rudiment not awned; $2n=20, 40$ (Gould, 1966).—Gravelly and sandy washes and flats, below 5000 ft.; Creosote Bush Scrub to Pinyon-Juniper Wd.; Colo. and e. Mojave deserts; to Colo., Tex., Mex. June–Oct.

3. **B. curtipéndula** (Michx.) Torr. [*Chloris c.* Michx.] Perennial, with scaly rhizomes; culms tufted, 3–8 dm. tall; blades flat or subinvolute, 3–4 mm. wide; spikes 35–50, purplish, 1–2 cm. long, spreading or twisted to 1 side of the slender axis; spikelets 5–8, appressed or ascending, 6–10 mm. long; glumes unequal, the 1st 4–5 mm. long, the 2d 7 mm. long; fertile lemma acute, mucronate, 3-nerved, 3-toothed, rudiment with 3 awns; $2n=28, 35, 40, 42, 45, 56, 70, 98$ (Fults, 1942), to 103 (Gould & Kapadia, 1964).—Dry rocky slopes, 4000–6000 ft.; Pinyon-Juniper Wd., etc.; Santa Rosa Mts. (Riverside Co.), Little San Bernardino and other ranges to the east (San Bernardino Co.); to Atlantic Coast, Mex.; Argentine. May–Aug. Gould and Kapadia refer Calif. material to var. *caespitosa* Gould & Kapalia as being more stiffly erect than in var. *curtipendula*.

B. radicòsa (Fourn.) Griffiths, with spikes mostly 7–12 and 20–25 mm. long, each with 8–11 spikelets was reported years ago from the Colo. Desert. New Mex., Mex.

4. **B. erìópoda** (Torr.) Torr. [*Chondrosium e.* Torr.] Tufted perennial with swollen bases and wiry white-woolly culms 4–6 dm. long; blades 1–1.5 mm. wide; spikes 3–8, loosely ascending, 2–3 cm. long; spikelets 12–20, narrow, 7–10 mm. long; fertile lemma acuminate, awned; rudiment slender, cleft, the lobes awned; $n=14$ (Brown, 1950).—Dry slopes, 3500–6000 ft.; Joshua Tree Wd., Pinyon-Juniper Wd.; New York Mts., Clark Mt., etc., e. Mojave Desert; to Wis., Manitoba, Tex., Mex. May–Aug.

5. **B. grácilis** (Willd. ex HBK.) Lag. [*Chondrosium g.* HBK.] Plate 97, Fig. E. Densely tufted perennial, 1–4 dm. tall; sheaths and blades glabrous, the latter flat or loosely

involute, 1–2 mm. wide; spikes usually 2, 2.5–5 cm. long, curved-spreading; spikelets many, ca. 5 mm. long; fertile lemma pilose, slender-awned; rudiment densely bearded at summit of rachilla, cleft to base, with rounded lobes and slender awns; $2n = 20, 28, 35, 40, 42, 61, 62, 77, 84$ (Fults, 1942; Snyder & Harlan, 1953), 20, 40, 60 (Gould, 1965).—Dry places, 5500–8500 ft.; Montane Coniferous F.; San Bernardino Mts.; New York Mts., Clark Mt.; to Wis., Manitoba, Tex., Mex. May–Aug.

B. hirsùta Lag. resembling *B. gracilis*, but with the rachilla prolonged beyond the spikelets as a naked point and with tuberculate glumes, was reported years ago from Jamacha, s. San Diego Co. Its range is to the east.

B. rothróckii Vasey was once reported from Jamacha, San Diego Co.; its range is to the east of s. Calif. It differs from *B. trifida* in having culms 2.5–5 dm. high; spikes 2.5–3 cm. long; spikelets 40–50.

6. **B. trífida** Thurb. Tufted perennial, 1–2 dm. high; blades mostly 1–2 cm. long; spikes 3–7, ascending, 1–2 cm. long; spikelets ca. 12, purplish, 7–10 mm. long; fertile lemma pubescent toward base, cleft over halfway, the awns 5 mm. long, winged toward base; rudiment cleft to base, awned; $2n = 20$ (Gould, 1965, 1966).—Found in Providence Mts. at 4000 ft. and in Death V.; to Nev., Tex., n. Mex. Creosote Bush Scrub. May–Sept.

14. *Brachypòdium* Beauv.

Annuals and perennials with short-pediceled or subsessile spikelets borne singly at the nodes of a stiffly erect, spicate raceme. Fls. several to many, the rachilla disarticulating above the glumes and between the florets. Glumes unequal, sharp-pointed, 5- and 7-nerved; lemmas firm, rounded or ± flattened on the back, 7-nerved, acuminate, awned or mucronate; palea as long as the body of the lemma, concave, the keels pectinate-ciliate. Ca. 15 spp., mostly in temp. regions. (Greek, *brachys*, short, and *podion*, foot, because of the short pedicels.)

1. **B. distàchyon** (L.) Beauv. [*Bromus distachyos* L.] Branched and geniculate at base, 1.5–3 dm. high, with pubescent nodes; sheaths and blades sparsely pilose to subglabrous; ligule 1.5–2 mm. long, pubescent; blades flat, 3–4 mm. wide; raceme strict, the segms. of the axis alternately concave; spikelets 1–5, ± imbricate, 2–3.5 cm. long, 5–6 mm. wide; awns slender, erect, 1–2 cm. additional; lemmas scabrous, 8–9 mm. long; $n = 15$ (Avdulov, 1931).—Becoming established occasionally as at Santa Catalina Id., near Torrey Pines Park; n. Calif.; native of Eurasia. April–May.

15. *Brìza* L. QUAKINGGRASS

Annuals or perennials, with erect culms, flat blades and open panicles with broad spikelets on capillary pedicels. Spikelets several-fld., often cordate, the florets crowded and horizontally spreading, the rachilla disarticulating above the glumes and between florets. Glumes subequal, broad, papery, with scarious margins; lemmas papery, several-nerved; palea much shorter than the lemma. Ca. 20 spp. widely scattered. (Greek, *briza*, a kind of grain.)

1. **B. mìnor** L. Annual, 1–5 dm. high; ligule of upper lf. ca. 5 mm. long, acute; blades 2–10 mm. wide; panicle erect, its slender branches spreading; spikelets pendent, 3–6-fld., pale or plum-colored, broadly cordate, 3–4 mm. long; lemmas strongly ventricose below; $n = 5$ (Avdulov, 1931), $n = 7$ (Gould, 1958).—Occasional as weed in s. Calif. (San Diego R., Camarillo); more common n.; native of Eu. April–July.

16. *Bròmus* L. BROMEGRASS

Low or rather tall annuals or perennials with closed sheaths, usually flat blades and open or contracted panicles of large spikelets. These several- to many-fld., the rachilla disarticulating above the glumes and between the florets; glumes unequal, acute, the 1st 1–3-nerved, the 2d usually 3–5-nerved; lemmas convex on the back or keeled, 5–9-nerved, 2-toothed, awned from between the teeth or awnless; palea usually shorter than

Bromus

the lemma, ciliate on the keels. Ca. 100 spp. of temp. regions. (Ancient Greek name for the Oat.)

(Wagnon, H. Keith. A revision of the genus Bromus, section Bromopsis, of N. Am. Brittonia 7: 415–480. 1952.)
1. Plants annual, largely of introd. weedy type.
    2. Spikelets strongly flattened, the lemmas compressed-keeled and with terminal teeth not more than 0.5 mm. long.
        3. Lemmas awnless or the awn less than 3 mm. long.
            4. Lemmas usually with 13 veins .... 25. *B. willdenovii*
            4. Lemmas usually with 9 veins .... 24. *B. unioloides*
        3. Lemmas with awns 7–15 mm. long.
            5. Spikelets 5–7-fld.; 2d glume almost equal to the lowest lemma ...... 2. *B. arizonicus*
            5. Spikelets 6–10-fld.; 2d glume shorter than the lowest lemma.
                6. Panicle with spreading or drooping branches; lemma 1.6–2 mm. wide. Native
                    4. *B. carinatus*
                6. Panicle with stiff ascending branches; lemma 2.5 mm. wide. Introd. 21. *B. stamineus*
    2. Spikelets terete or somewhat flattened, the lemmas not compressed-keeled, their terminal teeth mostly 0.6–5 mm. long.
        7. Awn usually geniculate, twisted at the base; lemma-teeth aristate .... 23. *B. trinii*
        7. Awn straight or spreading, sometimes minute, not twisted at base.
            8. Lemmas broad, rounded apically, not acuminate, the teeth mostly less than 1 mm. long; 1st lemma 3–5-nerved.
                9. Panicles compact, the spikelets mostly longer than the pedicels; lemmas membranaceous or scarious, prominently nerved, unequal, the lower longer than the upper
                    15. *B. mollis*
                9. Panicles more open, the spikelets mostly not longer than the pedicels; lemmas ± coriaceous, not prominently veined.
                    10. Awns straight; lemmas not conspicuously inrolled at maturity and not exposing rachilla to view.
                        11. The lemmas ca. equal; awn 10–16 mm. long; glumes pilose .. 1. *B. arenarius*
                        11. The lemmas unequal, the lower much longer than the upper; awn 4–8 mm. long; glumes glabrous .... 6. *B. commutatus*
                    10. Awns flexuous, usually ± divergent; lemmas inrolled at maturity and exposing to view the rachilla .... 10. *B. japonicus*
            8. Lemmas narrow, elongate, tapering at the tip, the teeth 2–5 mm. long; 1st glume 1-nerved.
                12. Panicle erect, contracted; awn 1–2 cm. long.
                    13. Culms pubescent below the dense panicle; sheaths pubescent. Common weed
                        20. *B. rubens*
                    13. Culms glabrous below the slightly open panicle; sheaths mostly smooth. Occasional .... 12. *B. madritensis*
                12. Panicle open, with spreading or drooping branches.
                    14. Awn 12–14 mm. long; lemmas ca. 1 cm. long; spikelets 1–2 cm. long
                        22. *B. tectorum*
                    14. Awns 35–50 mm. long; lemmas 2.5–3 cm. long; spikelets 2.5–4 cm. long
                        7. *B. diandrus*
1. Plants perennial, largely native members of our Plant Communities.
    15. First glume 3–5-nerved.
        16. Spikelets strongly flattened, the lemmas compressed-keeled.
            17. Blades canescent, densely short-pilose, 2–5 mm. wide, often involute; blades 1–3 mm. wide .... 3. *B. breviaristatus*
            17. Blades not canescent, glabrous to puberulent or sparsely pilose; blades mostly 4–12 mm. wide.
                18. Awns 2–3 mm. long; spikelets 3–4 cm. long; panicles strict, the short branches erect. Immediate coast .... 14. *B. maritimus*
                18. Awns 4–15 mm. long.
                    19. The awns 7–15 mm. long; spikelets 2–3 cm. long.
                        20. Panicles with spreading or drooping branches; lemma 1.6–2 mm. wide. Native
                            4. *B. carinatus*
                        20. Panicle with stiff ascending branches; lemma 2.5 mm. wide, more conspicuously white-margined. Introd. .... 21. *B. stamineus*

19. The awns 4–7 mm. long; spikelets 2.5–3.5 cm. long .... 13. *B. marginatus*
16. Spikelets terete before anthesis or somewhat flattened, but the lemmas not compressed-keeled.
21. Second glume 5-nerved.
22. Ligule of the culm-lvs. 2–4 mm. long; blades glabrous; glumes glabrous; Cismontane and montane .... 11. *B. laevipes*
22. Ligule of culm-lvs. 1 mm. long or less; blades mostly pilose; glumes usually with scattered hairs. Coast Ranges .... 18. *B. pseudolaevipes*
21. Second glume 3-nerved.
23. Spikelets 2–4 cm. long; sheaths retrorsely pubescent.
24. Panicle 1–1.5 dm. long; erect or pyramidal; awn 5–8 mm. long. Montane Coniferous F. .... 16. *B. orcuttianus*
24. Panicle 1.5–2 dm. long, with slender drooping branches; awn 3–6 mm. long. Below Yellow Pine F. .... 8. *B. grandis*
23. Spikelets 1.3–1.5 cm. long; sheaths mostly glabrous .... 17. *B. porteri*
15. First glume 1-nerved.
25. Creeping rhizomes present; lemma awnless or with an awn up to 3 mm. long; lvs. usually glabrous. Introd. .... 9. *B. inermis*
25. Creeping rhizomes wanting; lemma awned.
26. Anthers 1–1.3 mm. long; blades pubescent above. Moist places ...... 5. *B. ciliatus*
26. Anthers 2–3.5 mm. long; blades glabrous above. Dry places ...... 19. *B. richardsonii*

1. **B. arenàrius** Labill. AUSTRALIAN CHESS. Annual, with slender culms 2–4 dm. high; sheaths and blades pilose; blades 3–6 mm. wide; panicle pyramidal, open, nodding, 1–1.5 dm. long, the spreading branches and pedicels sinuously curved; spikelets 1–2 cm. long, ca. 5–9-fld.; glumes densely pilose, acute, scarious-margined, the 1st narrower, 3-nerved, 8 mm. long, the 2d 7-nerved, 10 mm. long; lemmas densely pilose, 7-nerved, 10 mm. long; awn staright, 10–16 mm. long.—Dry places below 6000 ft.; many Plant Communities; widely scattered over Calif., cismontane and desert; introd. from Australia. April–July.

2. **B. arizónicus** (Shear) Steb. [*B. carinatus* var. *a.* Shear.] Like *B. carinatus*, but mostly shorter, stiff, erect, rather narrow; spikelets 5–7-fld.; glumes subequal, the upper ca. as long as the lowest lemma; lemmas hirsute toward the margin, sometimes with some hairs on the back, the apical teeth 0.7–2 mm. long; $2n = 42$ (Stebbins et al., 1944).—Dry open places mostly below 2000 ft.; Creosote Bush Scrub, Chaparral, Coastal Sage Scrub, V. Grassland; L. Calif. to cent. Calif., Ariz. Santa Catalina Id. March–May.

3. **B. breviaristàtus** Buckl. [*B. subvelutinus* Shear. *B. carinatus* var. *linearis* Shear.] Erect tufted perennial 2.5–5 dm. tall; sheaths canescent to densely retrorse-pilose; blades narrow, becoming involute, mostly erect or ascending, canescent and pilose, mostly 1–3 mm. wide; panicle narrow, erect, 5–15 cm. long, with short appressed branches often bearing only 1 spikelet; spikelets 2–3 cm. long; glumes puberulent, the 1st 3–5-nerved, 8–10 mm. long, the 2d 7-nerved, 10–12 mm. long; lemmas appressed-puberulent, 12–14 mm. long; awn 3–10 mm. long; $n = 28$ (Staehlin, 1929).—Occasional in dry places, mostly 4000–8000 ft.; Montane Coniferous F.; San Jacinto Mts., Mt. Pinos, n. to B.C., Nev. May–July.

4. **B. carinàtus** H. & A. CALIFORNIA BROME. Erect annual or biennial, 5–10 (–12) dm. tall; sheaths scabrous to sparsely pilose; panicle 1.5–3 dm. long, with spreading or drooping branches; spikelets 2–3 cm. long (without the awns), mostly 6–10-fld., the florets 5-nerved, 10–15 mm. long; awns 7–15 mm. long; palea acuminate, nearly equaling the lemma; $2n = 56$ (Gould, 1966).—Frequent in dry open places below 7500 ft.; many Plant Communities; cismontane and less so in desert; s. Calif. On our islands. To L. Calif., B.C., Ida., New Mex. April–Aug. Variable; plants with smooth sheaths have been called var. *californicus* Shear and those with spikelets 3–4 mm. long, var. *hookerianus* (Thurb.) Shear.

5. **B. ciliàtus** L. FRINGED BROME. Perennial, the culms slender, 5–14 dm. tall, glabrous or retrorsely pubescent, the uppermost segms. always pubescent; sheaths pilose to glabrous; blades 5–10 mm. wide, pubescent to pilose above, mostly glabrous beneath; panicle 1–2 dm. long, open, the branches ascending to drooping, up to 1.5 dm. long;

spikelets 1.5–2.5 cm. long, 4–9-fld.; glumes glabrous, the 1st 1–3-nerved, the 2d 3-nerved; lemmas 10–12 mm. long, pubescent on margins of lower part, glabrous across the back, 5–7-nerved; awn 3–5 mm. long; $2n=14$ (Wagnon, 1952).—About meadows, 7000–8000 ft., Montane Coniferous F.; San Bernardino Mts. (S. Fork of Santa Ana R.); Sierra Nevada; to Alaska, Atlantic Coast, L. Calif. July–Aug.

6. **B. commutàtus** Schrad. HAIRY CHESS. Annual, 5–6 dm. high; sheaths retrorse-pilose; blades ± pubescent; spikelets 15–20 mm. long, 5–8-fld.; glumes narrow, the upper lanceolate; lemmas with an obtuse angle on the margin just above the middle, not strongly inrolled in fr.; awn straight, less than 1 cm. long; florets imbricate in fr., leaving no spaces at their base; $n=28$ (Nielsen, 1939).—Occasional weed in fields in s. Calif., as in Santa Barbara, Tehachapi Mts., more common in n. Calif.; native of Eu. May–July.

7. **B. diándrus** Roth. [*B. rigidus* Am. auth. *B. maximus* Desf., not Gilib.] RIPGUTGRASS. Annual, the culms 4–7 dm. tall; sheaths and blades pilose, the latter 2–10 mm. wide; panicle open or rather compact, rather few-fld., 6–12 cm. long, the lower branches mostly 1–2 cm. long; spikelets 3–4 cm. long, usually 5–7-fld.; glumes smooth, acuminate, the 1st 1-nerved, 16–20 mm. long, the 2d 3-nerved, 25–30 mm. long; lemmas 5-nerved, 25–30 mm. long, 2-toothed; awns stout, 3.5–5 cm. long; $2n=28$ (Stebbins & Love, 1941).—Common weed in waste places, fields, at low elevs., cismontane and insular; introd. from Eu. April–June. Plants with a more open panicle and lower branches up to 1 dm. long, have been designated var. *gussonei* (Parl.) Coss. & Durieu.

8. **B. grándis** (Shear) Hitchc. in Jeps. [*B. orcuttianus* var. *g.* Shear *B. porteri* var. *assimilis* Davy.] Plate 98, Fig. A. Perennial, with culms 9–14 dm. tall; nodes 3–6, retrorsely pubescent; blades 1.5–3 dm. long, 5–12 mm. wide, densely short-pubescent on both surfaces, sometimes not on upper; panicle 1.5–2 dm. long, open, with slender drooping branches; spikelets 2.5–3.5 cm. long, 7–9-fld.; glumes pubescent, the 1st mostly 3-nerved, the 2d 3–5-nerved; lemmas 11–14 mm. long, pubescent over the backs and on the margins, 7-nerved; awns 3–6 mm. long; $2n=14$ (Stebbins & Love, 1941).—Dry open or wooded slopes below 8000 ft.; Montane Coniferous F. to Coastal Sage Scrub; San Diego Co. n.; to cent. Calif. April–July.

9. **B. inérmis** Leyss. SMOOTH BROME. Perennial with creeping rhizomes; culms 5–14 dm. tall, smooth; sheaths smooth; blades 1–3 dm. long, glabrous, 5–15 mm. wide; panicle 1–2 dm. long, erect, open; spikelets 2–3 cm. long, 8–10-fld.; glumes glabrous, subulate, the 1st 6–8 mm. long, mostly 1-nerved, the 2d 7–10 mm. long, 3-nerved; lemmas 9–13 mm. long, glabrous to scabrous or sparsely puberulent on margins; $2n=56$ (Hill & Meyers, 1948).—Occasional in waste places, about meadows, etc., sea-level to 6500 ft.; native of Eurasia. May–Aug.

10. **B. japónicus** Thunb. JAPANESE CHESS. Annual, the culms erect or geniculate at base, 4–7 dm. tall; sheaths and blades pilose, the latter 2–7 mm. wide; panicle 1–2 dm. long, broadly pyramidal, diffuse, ± drooping; spikelets 2–2.5 cm. long; 1st glume 3-nerved, acute, 4–6 mm. long, the 2d 5-nerved, obtuse, 6–8 mm. long; lemmas glabrous, 7–9 mm. long, firm, obscurely 9-nerved, the margins ± inrolled at maturity; awn 8–10 mm. long, somewhat twisted and strongly divaricate at maturity; $n=7$ (Avdulov, 1928). —Occasional weed in waste places, as at San Bernardino; introd. from Eu. May–July.

B. arvénsis L. To be watched for; resembling *B. japonicus*, but with thinner, less turgid spikelets; hyaline margins of lemmas ending in prolonged acute teeth; awn straight, 7–10 mm. long; $n=7$ (Avdulov, 1928).—Native of Eu.

11. **B. laèvipes** Shear. Perennial, the culms 5–15 dm. tall, often with a decumbent base and rooting at lower nodes; sheaths and blades glabrous, the latter 4–10 mm. wide; panicle broad, lax, drooping, 1–2 dm. long; spikelets 2.5–3.5 cm. long, 5–11-fld.; glumes glabrous, smooth, the 1st 3-nerved, 6–9 mm. long, the 2d 5-nerved, 10–12 mm. long; lemmas 12–15 mm. long, densely pubescent on the margins, unevenly sparsely pubescent across the back or just on the lower half, 7-nerved; awn 4–6 mm. long; $2n=14$ (Wagnon, 1952).—Shaded stream banks and brushy slopes, below 8600 ft.; many Plant Communities; cismontane and montane Calif.; to Wash. May–Aug.

Plate 98. Fig. A, *Bromus grandis*. Fig. B, *Cenchrus incertus*, with spiny invol. Fig. C, *Cynodon dactylon*. Fig. D, *Deschampsia danthonioides*. Fig. E, *Distichlis spicata*. Fig. F, *Echinochloa colonum*.

12. **B. madriténsis** L. Resembling *B. rubens*, but the culms glabrous below the dense panicles; sheaths mostly smooth; panicles 5–10 cm. long, oblong-ovoid in outline; glumes 9–12 and 14–16 mm. long; lemmas narrow, narrow-lanceolate, 14–18 mm. long, the teeth 2–3 mm. long; $n=14$ (Cugnac & Simonet, 1941).—Occasional in open ground and waste places, as in Santa Monica Mts., and Santa Cruz Id.; introd. from Eu. April–June.

13. **B. marginàtus** Nees. Perennial, rather stout, 6–12 dm. high; sheaths pilose; blades ± pilose, flat, 4–8 (–12) mm. wide; panicles erect, rather narrow, 1–2 dm. long, the lower branches erect or ± spreading; spikelets 2.5–3.5 cm. long, closely 7–8 fld.; glumes broad, scabrous or scabrous-pubescent, the 1st subacute, 3–5-nerved, 7–9 mm. long, the 2d obtuse, 5–7-nerved, 9–11 mm. long; lemmas subcoriaceous, pubescent, 11–14 mm. long; awns mostly 4–7 mm. long; $n=21$ (Neilsen & Humphrey, 1937).—Dry open places below 9000 ft.; many Plant Communities; much of cismontane and montane Calif., Channel Ids., occasional on the deserts; to B.C., S. Dak., New Mex. April–July. Variable and intergrading with *B. carinatus*.

14. **B. marítimus** (Piper) Hitchc. [*B. marginatus* ssp. *m.* Piper.] Perennial with stout culms 3–7 dm. tall, ± geniculate at base with numerous basal leafy shoots; sheaths smooth or scaberulous; blades mostly 6–8 mm. wide, scabrous; panicles mostly 1–2 dm. long, strict, the short branches erect; spikelets 3–4 cm. long, 4–6 mm. wide, the awns 2–3 mm. long; $2n=56$ (Schulz-Schaeff. & Mark, 1957).—Sandy places, San Miguel and Anacapa ids.; San Luis Obispo Co. to Ore. April–July.

15. **B. móllis** L. [*B. hordeaceus* Calif. refs.] SOFT CHESS. Annual, softly pubescent throughout; culms 2–8 dm. tall; sheaths retrorse-pubescent; blades mostly 2–5 mm. wide; panicle contracted, erect, 5–10 cm. long or smaller; spikelets 15–20 mm. long; glumes broad, obtuse, coarsely pilose, or scabrous-pubescent, the 1st 3–5-nerved, 4–6 mm. long, the 2d 5–7-nerved, 7–8 mm. long; lemmas broad, 7-nerved, obtuse, coarsely pilose or scabrous-pubescent, bidentate, 8–9 mm. long, hyaline on margin; awn stoutish, 6–9 mm. long; $2n=28$ (Stebbins & Love, 1941).—Common weed in waste places, cismontane s. Calif., Santa Catalina and San Clemente ids.; native of Eu. April–July.

B. mollifórmis Lloyd. Near to *B. mollis*, mostly 1–2 dm. tall; lower sheaths feltypubescent, upper glabrous; panicle 2–4 cm. long; spikelets compressed; lemma ca. half as wide as long, the awn ± divaricate.—Reported from s. Calif.; introd. from Eu.

16. **B. orcuttiànus** Vasey. Perennial, the culms erect, 7–15 dm. tall, with 2–3 retrorsely pubescent nodes; sheaths pilose or ± velvety; blades glabrous, with finely scabrous margins, 6–12 mm. wide, sometimes glaucous; panicle 1–1.5 dm. long, erect; spikelets 2–4 cm. long, 5–7 (–11)-fld., subterete; glumes glabrous to pubescent, the 1st 5–8 mm. long, 1–3-nerved, the 2d 8–10 mm. long, 3-nerved; lemmas 10–16 mm. long, shortpubescent to glabrous across the back, short-pubescent to scabrous on margins, 3-, 5-, or 7-nerved; awn 5–8 mm. long; $2n=14$ (Wagnon, 1952).—Meadows, woods and rocky places, 3000–8000 ft.; largely Montane Coniferous F.; mts. from San Diego Co. to Wash. A form with blades soft-pubescent above and beneath; glumes and lemmas pubescent, occurring in mts. from Riverside Co. n. is var. *hallii* Hitchc. in Jeps.

17. **B. pòrteri** (Coult.) Nash. [*B. kalmii* var. *p.* Coult. *B. anomalus* Calif. refs.] Perennial, 3–8 dm. tall, often tufted; nodes glabrous or retrorse-pubescent; sheaths mostly glabrous; blades 1–2.5 dm. long, 2–5 mm. wide, usually erect and glabrous; panicle 7–15 cm. long, drooping, the axis mostly puberulent, branches ascending, arcuate; spikelets 13–15 mm. long, 5–11-fld.; glumes pubescent, the 1st 5–7 mm. long, subulate or with acute tip, 3-nerved, the 2d 6–10 mm. long, 3-nerved; lemmas 8–13 mm. long, pubescent on margins and across back, rarely on margins only, 7-nerved; awn 1.5–3 mm. long; $2n=14$ (Wagnon, 1952).—Rare, at high elevs.; San Bernardino Mts., InyoWhite Mts.; to Canada, Rocky Mts., New Mex. July–Aug.

18. **B. pseudolaévipes** Wagnon. Perennial, 6–12 dm. high; sheaths glabrous to pilose; blades 3–9 mm. wide, glabrous or pilose (frequently on margins only); panicle 1–2 dm. long, erect to nodding; open with ascending to spreading branches; spikelets 1.5–3.5 cm. long, 4–10 fld.; glumes mostly pubescent, the 1st 4–6 mm. long, 3-nerved, the 2d mostly 5-nerved, 6.5–8 mm. long; lemmas 10–12 mm. long, pubescent over the

back, sometimes on margins only, acute to obtuse; awns 3–5 mm. long; $2n=14$ (Wagnon, 1952).—Dry often shaded places below 3500 ft.; Coastal Sage Scrub, Chaparral, Foothill Wd.; San Diego and Kern cos. n. to San Francisco Bay; Catalina, Santa Rosa and Santa Cruz ids. April–June.

19. **B. richardsònii** Link. Perennial, 5–9 dm. tall, with glabrous nodes; sheaths glabrous or pilose; blades mostly glabrous, 1–2.5 dm. long, 3–7 mm. wide; panicle open, 1–2.5 dm. long, the branches ascending to spreading or drooping, up to 1.4 dm. long; spikelets 1.7–3.3 cm. long, 6–10-fld.; glumes glabrous, the 1st 8–10 mm. long, 1- or rarely 3-nerved, the 2d 9–12 mm. long, 3-nerved; lemmas 10–14 mm. long, pubescent on margins of lower half, glabrous across the back or ± pubescent on nerves; $2n=28$ (Elliott, 1949).—Dry open places, 4000 ft. and above; largely Pinyon-Juniper Wd., Bristle-cone Pine F.; San Bernardino and Inyo cos.; to B.C., S. Dak., Tex. June–Aug.

20. **B. rùbens** L. FOXTAIL CHESS. Annual, 1.5–4 dm. tall, puberulent below the panicle; sheaths and blades pubescent, the latter mostly 1.5–5 mm. wide; panicle erect, ovoid, compact, reddish, 2–7 cm. long; spikelets 7–11-fld., ca. 2.5 cm. long; glumes narrow, acuminate, pubescent or sometimes smooth, the 1st 1-nerved, 7–9 mm. long, the 2d 3-nerved, 10–12 mm. long; lemmas 5-nerved, lanceolate, acute, 12–16 mm. long, ending in 2 long hyaline teeth; awn 18–22 mm. long; $2n=$ca. 28 (Reese, 1957).—Common troublesome weed in waste and cult. ground at low elevs., deserts, cismontane and insular areas; introd. from Eu. March–June.

21. **B. stamíneus** Desv. in Gay. Resembling *B. carinatus*; lf.-blades ± pilose, 2–7 mm. wide, scabrous; panicle 1–2 dm. long, with stiff ascending branches; spikelets 1.8–2.4 cm. long, greenish, 4–6-fld.; glumes lance-ovate, acuminate, pubescent, the lower 7–8 mm. long (without awnlike tip), 5–7-nerved, the upper longer, 9-nerved; lemmas 10–12 mm. long, keeled, broadest above the middle, weakly 9-nerved, pubescent, with prominent whitish or slightly purple hyaline margins; awn 8–10 mm. long.—Santa Cruz Id.; weed in cent. Calif.; introd. from S. Am. May–July.

22. **B. tectòrum** L. CHEAT GRASS or DOWNY BROME. Annual, culms slender, 3–6 dm. high; sheaths and blades pubescent, the latter mostly 1.5–4 mm. wide; panicle broad, drooping, 5–12 cm. long, with slender reddish branches; spikelets nodding, 1–2 cm. long; glumes villous, the 1st 1-nerved, 4–6 mm. long, the 2d 3-nerved, 8–10 mm. long; lemmas lanceolate, villous, 5-nerved, ca. 1 cm. long, the teeth 2–3 mm. long; awn 12–14 mm. long; $n=7$ (Cugnac & Simonet, 1941).—Common weed, especially above 3000 ft., desert and cismontane slopes and mts.; introd. from Eu. May–June. A form with glabrous spikelets is var. *glabratus* Spenner.

23. **B. trìnii** Desv. in Gay. Erect annual, the culms rather slender, 3–6 dm. tall; sheaths and blades ± pilose, the latter mostly 3–8 mm. wide; panicle 1–2 dm. long, rather dense; spikelets 5–7-fld., 1.5–2 cm. long; glumes lanceolate, acuminate, smooth, the 1st 8–10 mm. long, the 2d broader, 12–16 mm. long; lemmas sparsely coarse-pubescent, 5-nerved, 12–14 mm. long, acuminate, the teeth narrow, 2 mm. long; awn 1.5–2 cm. long, twisted below, bent below the middle and strongly divaricate at maturity.— Dry plains and slopes below 5000 ft.; many Plant Communities, deserts and cismontane; San Clemente and Santa Catalina ids.; to Ore., Colo., L. Calif., Chile. March–May. A form with larger spikelets and 7-nerved lemmas occurs in the Panamint Mts. and is var. *excelsus* Shear.

24. **B. unioloìdes** (Willd.) HBK. [*B. haenkeanus* Calif. refs. *Festuca u.* Willd.] Resembling *B. willdenovii*; panicle 1–1.3 dm. long, the branches stiff, in 2's or 3's, bearing 1–3 sessile spikelets; spikelets much compressed, 2–3-fld., 1–1.3 cm. long; glumes 5–7-nerved; lemmas 10–12 mm. long, 9–11-nerved, finely scabrous-pubescent; awn very short or none.—Occasional as a weed below 4000 ft., both desert and cismontane Calif. from S. Am.

25. **B. willdenòvii** Kunth. [*B. catharticus* Vahl.] RESCUEGRASS. Annual or biennial; culms erect or spreading, 6–10 dm. tall; sheaths pilose or glabrous; blades glabrous or sparsely pilose, ca. 3–6 mm. wide; panicle open (narrow in depauperate plants), 1–2 dm. long, the branches naked at the base; spikelets 2–3 cm. long, 5–9 mm. broad, 6–12-

fld.; glumes smooth, the 1st 5-nerved, 7–10 mm. long, the 2d 7-nerved, 10–12 mm. long; lemmas acute, subcoriaceous, glabrous or scabrous, 12–16 mm. long, ca. 13-nerved; awn 0–2.5 mm. long; palea $\frac{1}{2}$–$\frac{3}{4}$ as long as lemma; $2n=42$ (Stebbins & Tobgy, 1944).— Occasional weed at low elevs.; apparently native of S. Am. April–Nov.

17. Calamagróstis Adans. REEDGRASS

Perennial usually fairly tall grasses, mostly with creeping rhizomes. Spikelets 1-fld., small, in open or more frequently narrow sometimes spikelike panicles. Rachilla disarticulating above the glumes, prolonged above the palea as a short often hairy bristle. Glumes subequal, acute or acuminate. Lemma shorter, usually more delicate than glumes, usually 5-nerved, the midvein exserted as an awn, the callus with a tuft of long hairs. Over 100 spp. of cool and temp. regions. (Greek, *calamos*, reed, and *agrostis*, a kind of grass.)

(Stebbins, G. L. Jr. A revision of some N. Am. spp. of Calamagrostis. Rhodora 32: 35–57. 1930.)
Awn longer than the glumes. Alpine .... 2. *C. purpurascens*
Awn not or scarcely longer than the glumes. From lower elevs.
    Sheaths ± pubescent on the collar .... 3. *C. rubescens*
    Sheaths glabrous on the collar .... 1. *C. densa*

1. **C. dénsa** Vasey. Rhizomes short, stout; culms stout, densely tufted, 1 m. or more tall; sheaths slightly scabrous; ligule 3–5 mm. long; blades flat or ± involute, scabrous, 1.5–2.5 dm. long; panicle spicate, pale, 1–1.5 dm. long; glumes 4.5–5 mm. long, scaberulous; lemma 3.5–4 mm. long, the awns bent, ca. as long as lemma, ± exserted at one side; callus-hairs ca. 1 mm. long; $n=14$ (Nygren, 1954).—Dry hills at 3000–4000 ft., e. San Diego Co. in Chaparral, Yellow Pine F. June–July.

2. **C. purpuráscens** R. Br. in Richards. Tufted, erect, 4–6 (–10) dm. high; sheaths mostly scabrous; blades 2–4 mm. wide, rather thick, scabrous, ± involute; panicle dense, somewhat anthocyanous, spikelike, 5–12 cm. long; glumes 6–8 mm. long, scabrous; lemma nearly as long, the apex with 4 setaceous teeth, awn from near the base, exserted ca. 2 mm.; callus-hairs rather short; $n=20, 24, 27, 28$, etc (Nygren, 1954), $2n=42$ (Taylor & Brockman, 1966).—Rocky places, 9500–13,000 ft.; Subalpine F., Alpine Fellfields; White Mts., Sierra Nevada to Alaska, Quebec. July–Sept.

3. **C. rubéscens** Buckl. Rhizomes creeping; culms tufted, slender, 6–10 dm. long; sheaths smooth, ± pubescent on collar; blades erect, 2–4 mm. wide, flat or subinvolute; panicle spikelike or ± interrupted, pale or purplish, 6–15 cm. long; glumes 4–5 mm. long; lemma pale, thin, ca. 4 mm. long, smooth, the geniculate awn from near the base, 1–2 mm. long above the bend, exserted from side of glumes; callus-hairs short, scant; $n=28$ (Nygren, 1954).—Below 2500 ft.; Chaparral, etc.; Santa Cruz Id.; San Luis Obispo Co., Monterey Co. to B.C.; Colo. June–Sept.

18. Cénchrus L. SANDBUR. BURGRASS

Ours low branching annuals with flat blades and simple racemes of spiny burs terminating the culm and branches. Spikelets 1-fld., few in a cluster, acuminate, subtended by a short-pediceled ovoid or globular invol. of rigid connate spines which is deciduous at maturity. Glumes shorter than the lemma. Ca. 25 spp. of warmer regions. (Greek, *kegchros*, a kind of millet.)

(DeLisle, D. G. Taxonomy and distribution of the genus Cenchrus. Iowa State J. Sci. 37: 259–351. 1963.)
Invol. with a ring of slender bristles at base; spikelets usually 4 in a bur .... 1. *C. echinatus*
Invol. with no ring of slender bristles at base; spikelets usually 2 in each bur.
    Spines broader at base, less than 45 in number, 2–5 mm. long .... 2. *C. incertus*
    Spines slender, usually more than 50 in number, 3.5–7 mm. long .... 3. *C. longispinus*

1. **C. echinàtus** L. Culms compressed, usually geniculate, 2.5–6 dm. long; blades 3–8 mm. wide, pilose on upper surface near base; raceme 3–10 cm. long; burs 4–7 mm.

long, equally broad or broader, pubescent, the lobes of the invol. bent inward but not interlocking; $2n=68$ (Tateoka, 1955), 70 (Gould, 1965).—Known from Imperial and San Diego cos. as a weed; to S. Car. and S. Am. Oct.

2. **C. incértus** M. A. Curtis [*C. pauciflorus* Benth.] Plate 98, Fig. B. Sometimes forming large mats, the spreading culms 2–9 dm. long, rather stout; blades 2–7 mm. wide; raceme usually 3–8 cm. long, the burs somewhat crowded, mostly 4–6 mm. long and wide, pubescent; spines spreading or reflexed, some of the upper 4–5 mm. long, usually villous at base; $n=18$ (Brown, 1948), $2n=34$ (Tateoka, 1955).—Sandy places as a weed, as at Daggett, San Bernardino Co., in Santa Barbara Co., etc.; Cent. V.; to Ore., Atlantic Coast, S. Am. July–Sept.

3. **C. longispìnus** (Hack. in Kneucker) Fern. [*C. echinatus* f. *l.* Hack. in Kneucker.] Forming large clumps with many branches; culms terete, 1–9 dm. tall; sheaths pilose on margins and at throat; ligule a rim of ciliate hairs 0.7–1.7 mm. long; blades 6–18 cm. long; infl. compact, 4–10 cm. long; burs ± globose, 8–12 mm. long; spines slender, 3.5–7 mm. long; spikelets 2–3 in a bur, 6–8 mm. long; $n=17$ (DeLisle, 1963).—Reported several s. Calif. counties, desert and cismontane as a weed; also farther n. Native of e. U.S.

19. *Chlòris* Sw. FINGERGRASS

Tufted perennials, sometimes annuals, with flat or folded scabrous blades and 2 to several, sometimes showy and feathery spikes aggregated in ± digitate fashion. Spikelets with 1 perfect floret, sessile, in 2 rows along one side of a continuous rachis, the rachilla disarticulating above the glumes, produced beyond the perfect floret and with 1 or more empty florets. Glumes ± unequal, the 1st shorter, narrow, acute; lemma keeled, usually broad, 1–5-nerved, often awned from between the short apical teeth; awn slender or reduced to a mucro. Ca. 70 spp. of warmer parts of the world. (*Chloris*, goddess of fls.)

Lemmas firm, dark brown, awnless or mucronate .... 1. *C. distichophylla*
Lemmas pale or fuscous, distinctly awned.
    Plant producing long stout stolons; awns 1–5 mm. long .... 2. *C. gayana*
    Plant not producing stolons, erect or decumbent and rooting at the nodes; awns 5–10 mm. long
        3. *C. virgata*

1. **C. distichophýlla** Lag. Culms 3–9 dm. tall, tufted, leafy; blades mostly 7–17 cm. long, 3–7 mm. wide; ligule a dense ring of hairs less than 0.5 mm. long; spikes commonly 8–15, ± brownish, usually 6–8 cm. long, closely aggregate, ascending to drooping; spikelets 2-fld., ca. 2.3 cm. long; glumes ca. 1 mm. long, minutely scabrous; lemmas awnless, brown, the lower ca. 2 mm. long, lanceolate, villous on margins; $n=10$ (Krishnaswamy, 1940), $2n=40$ (Huynh, 1965).—Once reported as escape near San Diego; native of S. Am. June–Sept.

2. **C. gayàna** Kunth. RHODESGRASS. Culms 1–1.5 m. tall, with long stout, leafy stolons; blades 5–40 cm. long, 3–5 mm. wide; spikes several to many, erect or ascending, 5–10 cm. long; spikelets pale tawny, crowded; lemma 3 mm. long, hispid on margin near summit, the awn 1–5 mm. long; $n=10$ (Brown, 1950), $2n=40$ (Tateoka, 1965).—Locally common as a weed, escaping from cult. for forage; introd. from Africa. Aug.–Sept.

3. **C. virgàta** Sw. [*C. elegans* HBK.] Annual, 4–6 (–8) dm. high; blades flat, 7–40 cm. long, 2–7 mm. wide; spikes several, 2–8 cm. long, whitish or tawny, feathery or silky; spikelets crowded, 2.5–4 mm. long; glumes 1-nerved, the 1st 1.5–2 mm., the 2d 3–3.5 mm. long; lemma 3 mm. long, long-ciliate; the awn 5–10 mm. long; $n=10$ (Moffett & Hurcombe, 1949).—Occasional in waste places, including those of the deserts; cent. Calif.; to Nebr., Tex., and farther e.; trop. Am. April–Sept.

20. *Cortadèria* Stapf. PAMPASGRASS

Large tussock grasses; lvs. largely near the base; blades narrow, attenuate, usually serrulate on edges; panicle large, terminal, plumelike. Spikelets several-fld.; internodes of rachilla jointed, the lower part glabrous, the upper bearded; florets stipitate. Glumes exceeding lower florets. Plants dioecious, the ♂ spikelets covered with long hairs. Ca. 5 spp. of S. Am. (Native Argentine name, *cortadera*, cutting, because of lf.-margins.)

1. **C. atacaménsis** (Phil.) Pilger. [*Gynerium a.* Phil. *C. selloana* of Calif. auth.] Perennial, 2–6 m. tall; lvs. 3–9 mm. wide, with tufted hairs at the throat of the sheath; ♀ panicle white to pink, 3–9 dm. long, silky-hairy; spikelets 10–15 mm. long.—An escape, especially along coastal bluffs, Ventura Co. to n. Calif.; commonly cult.; native of Chile. Late summer.

21. Crýpsis Ait.

Spreading annual with headlike infl. in axils of a pair of broad spathes that are enlarged sheaths with short rigid blades. Spikelets 1-fld., disarticulating below the glumes. Glumes subequal, narrow, acute. Lemma broad, thin, 1-nerved; palea like lemma, splitting between the nerves. Fr. readily falling from the lemma and palea, the seed free from the thin pericarp. Seven spp. of Medit. region. (Greek, *krupsis*, concealment, because of partly hidden infl.)

1. **C. niliàca** Fig. & DeNot. Prostrate, much branched, forming mats to 3 dm. in diam.; sheaths tuberculate, bearded at summit; blades flat with involute apex, spreading, 2–5 cm. long, readily deciduous from sheaths; glume ca. 3 mm. long.—Becoming quite widely spread on mud flats, sand bars, etc.; several Plant Communities; from San Diego to n. Calif. June–Sept.

22. Cutándia Willk.

Annuals, erect or decumbent, many-stemmed. Lvs. narrow, plane or involute. Infl. a dense panicle, ± secund, with very short branches bearing few spikelets. These narrow, 2–13-fld., the rachilla articulate between the florets, glabrous. Glumes narrow, rigid, unequal, 1–3-nerved. Lemmas prominently 3-nerved, entire, or 2-dentate, keeled, mucronulate or rarely aristate. Paleae shorter than the lemmas, narrow, 2-keeled. Caryopsis oblong. Ca. 6 spp. of Medit. region.

1. **C. memphítica** (Spreng.) Richt. [*Dactylis m.* Spreng.] Stems several from base, spreading, 1 1.5 dm. long, glabrous; sheaths and lf.-blades glabrous, the latter 2–4 mm. wide, flat, 4–7 cm. long; panicles few-fld., the branches flat, scabrous, zigzag; spikelets on short pedicels, 2–3-fld., 8–9 mm. long; glumes ca. 6 and 7 mm. long, keeled; lemmas 5–7 mm. long.—Sandy soils, nursery at Devil's Canyon, San Bernardino Mts.; introd. from Medit. May.

23. Cýnodon Rich.

Creeping perennials with stolons or rhizomes; blades short; spikes several, slender, digitate at summit of erect culms. Spikelets 1-fld., awnless, sessile in 2 rows along 1 side of a slender continuous rachis and closely appressed; rachilla disarticulating above glumes. Glumes narrow, acuminate, 1-nerved, subequal, shorter than floret. Lemma firm, compressed, pubescent on keel, 3-nerved, the lateral nerves near the margin. Ca. 10 spp. of warmer regions. (Greek, *kuon*, dog, and *odous*, tooth, because of hard scales on rhizomes.)

Lf.-blade with 5 primary nerves; spikes 4–5 .... 1. *C. dactylon*
Lf.-blade with 3 primary nerves; spikes usually 2 .... 2. *C. transvaalensis*

1. **C. dáctylon** (L.) Pers. [*Panicum d.* L.] BERMUDAGRASS. Plate 98, Fig. C. Culms flattened, wiry, glabrous, from tough woody scaly rhizomes, the flowering culms erect, 1–4 dm. long; ligule a conspicuous ring of white hairs; blades flat, glabrous or pilose on upper surface, mostly 1–3 cm. long; spikes 2–5 cm. long; spikelets imbricate, 2 mm. long, the acute lemma boat-shaped; $2n = 36, 36 + 1\ \text{B}, 36 + 2\ \text{B}$ (Gould, 1966).—Common weed forming very tough sod in waste and low places (fields, lawns, orchards, etc.) through much of state; to Ore., s. states; warm regions of both hemis. Sometimes used as a lawngrass, but mostly white and unsightly in winter. Also, but incorrectly, called Devil's Grass.

2. **C. transvaalénsis** Davy. Extensively creeping, the blades mostly not more than

1 mm. wide; spikes 1–3, the spikelets narrower and glumes shorter than in the preceding sp.; $n=10$ (Hurcombe, 1947).—Being used as lawn grass and escaping, as at Bard, Imperial Co.; introd. from Eu.

24. *Cynosùrus* L. DOGTAIL

Annuals or perennials with flat narrow blades and dense terminal spikelike panicles. Spikelets dimorphous, the terminal terete fertile one of each fascicle 2–3-fld., sessile, almost concealed by the modified lower sterile ones which are reduced to a rigid fanlike distichous group of narrow glumes and pointed or awned lemmas. Fertile spikelet with 2 rigid glumes; lemmas longer, terete, 3-keeled, firm, mucronate; paleas with 2 ciliate keels. Ca. 4 spp. of the Medit. region. Greek, *cynos*, of a dog, and *oura*, tail.)

Plants annual; panicles subcapitate; awns evident .... 2. *C. echinatus*
Plants perennial; panicle narrow, spikelike; awns inconspicuous .... 1. *C. cristatus*

1. C. cristàtus L. Erect perennial, 2–8 dm. tall; basal lvs. long, soft, 1–3 mm. wide; panicle strict, 1-sided, 2–8 cm. long, 0.5–1 cm. thick; pairs of spikelets ca. 5 mm. long; awns of lemmas mostly not more than 1 mm. long; $n=7$ (Avdulov, 1931).—Occasionally reported, as from Los Angeles, San Francisco, etc.; introd. from Eu.

2. C. echinàtus L. Annual, 2–4 dm. tall; blades short, soft, 1–3 mm. wide; panicle bristly, 1–4 cm. long, 2–3 cm. thick; pairs of spikelets 8–10 mm. long; awns of lemmas 6–12 mm. long; $n=7$ (Avdulov, 1928).—Common locally in fields, waste places, etc. as introd. from Eu. May–July.

25. *Dáctylis* L. ORCHARDGRASS

Coarse tufted perennials with flat blades and glomerate panicles. Spikelets 2–6-fld., compressed, subsessile, strongly overlapping, the florets perfect or the upper ♂. Glumes unequal, ciliate or scabrous on the sharp keel, acute or mucronate. Lemmas 5-nerved, hispid or ciliate-keeled, with short awn-points; paleae slightly shorter than lemmas. Ca. 3 spp. of Old World. (Greek, *dactylos*, a finger.)

1. D. glomeràta L. Glaucous, scabrous, 4–15 dm. tall; lvs. 7–30 cm. long, 2–6 mm. wide; panicle 1–2.5 dm. long, with ascending branches that become erect in fr.; lemmas 5–6 mm. long, pointed; $n=7$ (Felfoeldy, 1949), 14 or 21 (Skovsted, 1939).—Natur. ± sparingly but widely in cismontane Calif.; native of Eu. An important hay grass. May–Aug.

26. *Dactyloctènium* Willd.

Annuals or perennials, with flat blades and 2–several short thick spikes, digitate and widely spreading at summit of culms. Spikelets 3–5-fld., compressed, sessile, closely imbricate, in 2 rows along 1 side of narrow flat rachis; rachilla disarticulating above 1st glume and between florets. Glumes ± unequal, broad, 1-nerved, the 1st persistent upon rachis, 2d mucronate or short-awned below tip, deciduous. Lemmas firm, broad, keeled, acuminate or short-awned, 3-nerved, the lateral nerves faint; palea ca. as long as lemma. Ca. 3 spp. of warm regions. (Greek, *dactulos*, finger, and *ktenion*, comb, because of pectinate arrangement of spikelets.)

1. D. aegýptium (L.) Beauv. [*Cynosurus a.* L.] Culms rooting at nodes, compressed, 2–4 (–8) dm. long, forming mats; blades flat, ciliate; spikes 1–5 cm. long; $n=18$ (Moffett & Hurcombe, 1949), $2n=36$ (Gould, 1936), $2n=40$ (Tateoka, 1965).—Taken at Calexico, Imperial Co. in 1968 and at Bonsall, San Diego Co. in 1965; introd. from Old World trop. Summer.

27. *Danthònia* Lam. & DC. OATGRASS

Tufted perennial with few-fld. open or spikelike panicles of rather large spikelets. These several-fld., the rachilla disarticulating above the glumes and between the florets. Glumes subequal, broad, papery, acute, mostly exceeding the upper floret. Lemmas rounded on back, obscurely several-nerved, bifid at tip, the teeth mostly slender-awned and with a stout twisted geniculate awn arising from between the teeth. Over 100 spp., of temp. regions. (Named for E. *Danthoine*, French botanist of early 18th century.)

Lemmas glabrous on back, pilose only on the margin.
    The panicle of few to several spikelets .... 1. *D. californica*
    The panicle usually of a single spikelet .... 3. *D. unispicata*
Lemmas pilose on the back of the base and on the margins .... 2. *D. pilosa*

    1. D. califórnica Bol. var. **americàna** (Scribn.) Hitchc. [*D. a.* Scribn.] Three to 8 dm. tall, ± spreading-pilose; sheaths pilose, the blades 1–2 dm. long, often flat; panicle mostly with 2–5 spikelets, the pedicels 1–2 cm. long; glumes 15–20 mm. long; lemmas (excluding awns) 8–10 mm. long, glabrous on back, the teeth long-aristate; awns having a terminal segm. 5–10 mm. long.—On dry rocky places, 6000–8000 ft.; Montane Coniferous F.; Cuyamaca and San Bernardino mts.; Sierra Nevada to B.C., Rocky Mts. May–July.

    2. D. pilòsa R. Br. Tufted, 3–6 dm. tall, loosely pilose; panicle narrow, several-fld.; spikelets ca. 6-fld.; glumes 13–14 mm. long; lemma pilose at base and on margin, the teeth with awns 6–8 mm. long, cent. awn 12–15 mm. long; $n=24$ (Calder, 1937).—Occasionally introd. from Santa Barbara n. along the coast; native of Australia. May–July.

    3. D. unispicàta (Thurb.) Munro ex Macoun. [*D. californica* var. *u.* Thurb.] Culms 1–2 dm. long, in spreading tufts; sheaths and blades pilose to glabrous; panicle mostly of 1 spikelet, this ca. 12–15 mm. long excluding the awns; lemma usually glabrous above the hairy callus, 5–7 mm. long, gradually acuminate into the awns; $n=18$ (Stebbins & Love, 1941).—Rocky places at 4500–5000 ft., Chaparral, Yellow Pine F.; Cuyamaca Lake, San Diego Co.; Sierra Nevada to B.C., Rocky Mts. May–June.

28. *Deschámpsia* Beauv. HAIRGRASS

Annual or perennial grasses with flat or involute lvs. and open or contracted panicles. Spikelets 2 fld., disarticulating above the glumes and between the florets, the hairy rachilla extended beyond the florets or rarely terminated by a ♂ one. Glumes subequal, keeled, acute, membranous, shining. Lemmas thin, truncate, 2–4-toothed at summit, bearded at base, with a slender awn from or below the middle, the awn straight, bent or twisted. Palea narrow, 2-nerved. A genus of ca. 40 spp. from cold and temp. areas. (Named for J. C. Loiseleur-*Deslongchamps*, 1774–1849, French botanist.)

(Lawrence, W. E. Some ecotypic relations of D. caespitosa. Am. J. Bot. 32: 298–314. 1945.)
Plants annual; spikelets 4–8 mm. long; panicle open .... 2. *D. danthonioides*
Plants perennial.
    Panicle open, nodding, 1–2 dm. long .... 1. *D. cespitosa*
    Panicle narrow, elongate, almost spicate, 1–3 dm. long .... 3. *D. elongata*

    1. D. cespitòsa (L.) Beauv. [*Aira c.* L.] Densely tufted perennial, 6–12 dm. tall; lvs. principally basal, flat or folded, 1.5–4 mm. wide, short or elongate; panicle open, nodding, 1–2 dm. long, the branches capillary, scabrous, the branchlets bearing spikelets toward their tips; spikelets 4–5 mm. long, green to darkly anthocyanous, the florets distant; glumes 1-nerved or the 2d obscurely 3-nerved, acute, ca. as long as the florets; lemmas smooth; awn from near the base, straight or somewhat geniculate, usually somewhat exserted; $n=13$ (Lawrence, 1945), $2n=26$ (Taylor & Brockman, 1966).—Wet meadows, at 6000–8500 ft.; Montane Coniferous F.; San Bernardino Mts., Mt. Pinos; to Alaska, Atlantic Coast, Eurasia. July–Aug.

    2. D. danthonioìdes (Trin.) Munro ex Benth. [*Aira d.* Trin. *D. calycina* Presl.] Plate 98, Fig. D. Culms rather few, slender, 1–5 dm. tall; blades few, 2–8 cm. long, ca. 1 mm. wide; panicle open, 5–12 cm. long, the branchlets with few spikelets near the ends; glumes 6–8 mm. long, acuminate, 3-nerved, lemmas smooth, 2–3 mm. long, truncate, the geniculate awns 4–6 mm. long.—Moist places, 5000–7500 ft., largely Montane Coniferous F.; San Diego Co. n.; to Alaska, Chile. June–Aug. A rather poorly defined form is var. *gracilis* (Vasey) Munz. [*D. g.* Vasey.] Glumes 4 mm. long. On mud flats after winter pools; low elevs.; Coastal Sage Scrub, V. Grassland, Chaparral; cismontane s. Calif.; cent. Calif.

3. **D. elongàta** (Hook.) Munro ex Benth. [*Aira e.* Hook. *D. e.* vars. *ciliata* and *tenuis* Vasey.] Tufted perennial, the culms slender, erect, 3–10 dm. high; blades flat, soft, 1–1.5 mm. wide; panicle 1–3 dm. long, almost spicate; pedicels short, appressed; glumes 3-nerved, 4–6 mm. long; lemmas 2–3 mm. long, like those of *D. danthonioides*, the awns straighter, to ca. 4 mm. long; $2n = 26$ (Bowden, 1960).—Wet places, 5000–8000 ft.; Montane Coniferous F.; San Diego, Riverside and San Bernardino cos.; to Alaska, Wyo., Ariz., Mex., Chile. May–Aug.

29. *Digitària* Heister. CRABGRASS

Annual or perennial, erect to prostrate, often weedy. Racemes slender, digitate or approximate on a short axis. Spikelets in 2's or 3's, rarely 1, subsessile or short-pedicelled, alternate in 2 rows on 1 side of a 3-angled rachis; spikelets lanceolate or elliptic. First glume minute or none; 2d glume equal to or shorter than sterile lemma. Fertile lemma cartilaginous, with pale hyaline margins. Ca. 300 spp. (including *Trichachne*) of warmer regions. (Latin, *digitus*, finger, because of the infl.)

Sheaths glabrous; spikelets 2 mm. long; fertile lemma brown . . . . 1. *D. ischaemum*
Sheaths pilose or villous; spikelets 2.5–3.5 mm. long; fertile lemma pale . . . . . . 2. *D. sanguinalis*

1. **D. ischaèmum** (Schreb.) Schreb. ex Muhl. [*Panicum i.* Schreb.] Glabrous, the culms 2–4 dm. long, prostrate to ascending; lvs. 3–6 mm. wide; racemes 1–6, commonly purplish, 1–9 cm. long; glume and sterile lemma equal, short-villous between the nerves, as long as brown lemma; $n = 18$ (Brown, 1948).—Occasionally natur. at low elevs., as a weed, cismontane Calif.; native of Eu. Sept.–Nov.

2. **D. sanguinàlis** (L.) Scop. [*Panicum s.* L. *Syntherisma s.* Dulac.] Annual, usually much branched at base, often purplish, the culms 1–9 (–15) dm. long, with ascending flowering shoots; blades 5–10 mm. wide, pubescent to scaberulous; racemes few to several, 5–15 cm. long, digitate; spikelets ca. 3 mm. long; 1st glume minute, 2d ca. 1.5 mm. long, ciliate; sterile lemma nerved, the lateral internerves strigose; fertile lemma pale; $n = 18$ (Avdulov, 1931).—Common weed in lawns, gardens, etc., below 8000 ft.; several Plant Communities; native of Eu. June–Sept.

30. *Dissanthèlium* Trin.

Annual or perennial grasses with flat blades and narrow panicles. Spikelets mostly 2-fld., the florets distant; rachilla slender, disarticulating above the glumes and between the florets. Glumes subequal, acuminate, membranaceous to papery, exceeding the lower floret, the 1st 1-nerved, the 2d 3-nerved. Lemmas strongly compressed, acute, awnless, 3-nerved. Paleae somewhat shorter than the lemmas. A genus of 17 spp., mostly S. Am. (Greek, *dissos*, double, and *anthelion*, a small fl.)

1. **D. califórnicum** (Nutt.) Benth. [*Stenochloa c.* Nutt.] Annual, ca. 3 dm. tall, ± decumbent or spreading; blades flat, 1–1.5 dm. long, 2–4 mm. wide; panicle 1–1.5 dm. long, narrow, but loose, with fascicles of ascending flexuous branches, some of which are floriferous to their base; glumes 3–4 mm. long, subequal; lemmas pubescent, ca. 2 mm. long.—Open places, Coastal Sage Scrub; Santa Catalina and San Clemente ids.; ids. of L. Calif. Not collected in Calif. for many years.

31. *Dístichlis* Raf. SALTGRASS

Low dioecious perennials with wiry culms from strong creeping or deeply running rhizomes. Ligule short and evenly serrulate; lf.-blades 2-ranked, flat or ± involute.

Staminate infl. a dense spicate panicle exceeding the blades; ♀ equal to or shorter than blades. Spikelets few- to many-fld. Glumes unequal, broad, 3–7-nerved. Lemmas closely to loosely imbricate, 9- to 11-nerved, coriaceous. Paleae usually a little shorter than lemmas, 2-keeled, serrate on keels, often with a few long hairs on the back. Caryopsis brown. Ca. 4 spp., salt- or alkali-tolerant, of temp. N. and S. Am. and Sudan. (Greek, *distichos*, 2-ranked, in reference to the lvs.)

(Beetle, A. A. The N. Am. variations of D. spicata. Bull. Torr. Bot. Club 70: 638–650. 1943.)
The grass genus D. Revista Agr. Agron. 22: 86–94. 1955.)

1. **D. spicàta** (L.) Greene var. **spicàta** [*Uniola s.* L. *D. s.* var. *stolonifera* Beetle.] Plate 98, Fig. E. Culms 2–3 dm. tall, often prostrate, with a strong tendency to form stolons; blades erect, 1–2 dm. long, the upper exceeding the ♀ panicle and often equaling the ♂; the former green or purplish, club-shaped, 1.5–5 cm. long, often 2 cm. thick, of 8–35 crowded spikelets; these 5–9-fld., ca. 1 cm. long, 4 mm. broad; lower glume 2.5 mm. long, upper 3.5 mm.; lemmas 5 mm. long, faintly nerved; palea broadly winged below, with hyaline margins, serrate on keels above; caryopsis ca. 2 mm. long; ♂ infl. of 6–20 spikelets, these 7–10-fld.; glumes 3 and 3.5 mm. long; $n=20$ (Stebbins & Love, 1941).—Coastal Salt Marsh, mainland and islands to Ore., Gulf of Mex., Atlantic Coast.

Var. **strícta** (Torr.) Beetle. [*Uniola s.* Torr. *D. spicata* var. *divaricata* Beetle, var. *nana* Beetle.] Panicle of approximate but scarcely congested spikelets, the pedicels easily visible; spikelets with 3–14 florets.—Alkaline soils, at elevs. to 7000 ft.; many Plant Communities; away from the coast, cismontane and desert s. Calif. to Mont., Ia., Tex.

32. *Echinochlòa* Beauv.

Ours coarse annuals with compressed sheaths, long flat blades and terminal panicles of stout short densely flowered 1-sided racemes. Spikelets 1-fld., sometimes with a fl. below the terminal perfect one, almost sessile. Glumes unequal, spiny-hispid, mucronate. Sterile lemma similar and awned from apex or mucronate only, inclosing a hyaline palea. Fertile lemma and palea chartaceous, acuminate; margins of the glumes inrolled at summit, where the palea is not included. Ca. 20 spp. of warm regions. (Greek, *echinos*, hedgehog, and *chloa*, grass, referring to the echinate spikelets.)

(Gould, F. W. and M. A. Ali and D. E. Fairbrothers. A revision of Echinochloa in the United States. Am. Midl. Naturalist 87: 36–59. 1972.)
Leaf sheaths, at least the lowermost, hirsute or hispid, usually with papilla-based hairs
    2. *E. crusgalli oryzicola*
Leaf sheaths glabrous.
  Spikelets with at least some awns 16 mm. or more long .... 2. *E. crusgalli crusgalli*
  Spikelets awnless or with awns 15 mm. or less long.
    Primary infl. branches simple, usually 2 cm. or less long, with small (2.5–3 mm. long), awnless spikelets in 4 regular rows; palea of lower floret well developed; hairs of the infl. branches and spikelets not papilla-based .... 1. *E. colonum*
    Primary infl. branches, at least the lower, usually rebranched, the lower branches commonly more than 2 cm. long; spikelets variable in size, awned or awnless, in regular rows or not, palea of lower floret present or absent, when present then papilla-based hairs present on the infl. branches and often on the spikelets.
      Palea of lower floret absent or vestigial and much less than half as long as the lemma; papilla-based hairs not present on the panicle-branches .... 3. *E. crus-pavonis macera*
      Palea of lower floret present, more than half as long as the lemma; papilla-based hairs present or absent on the panicle branches.
        Lemma of fertile floret narrowly ovate or ellipsoid; panicle typically large, long, densely flowered, with numerous branches and a slender, often curved or drooping main axis
          3. *E. crus-pavonis crus-pavonis*
        Lemma of fertile floret broadly ovate or ellipsoid; panicle variable in size, with a stiffly erect main axis and few to numerous branches.
          Shiny coriaceous apex of lemma of fertile floret narrowly acute or acuminate, without a line of hairs and with gradual transition to a membranous, usually stiff and mucro-

nate tip; hairs of the panicle branches absent or shorter than the spikelets (excluding the awns) .... 4. *E. muricata*
Shiny coriaceous apex of lemma of fertile floret obtuse or broadly acute, with a line of minute hairs and a sharply differentiated, withering membranous tip; hairs of the panicle branches, at least some, as long as or longer than the spikelets (excluding the awns) .... 2. *E. crusgalli*

1. **E. colònum** (L.) Link. [*Panicum c.* L.] JUNGLE-RICE. Plate 98, Fig. F. Annual with slender, weak, freely branching culms; lvs. glabrous, without ligules, the blades thin and flat, 3–6 (–9) mm. broad; infl. short, few-fld., with usually 3–7 unbranched primary branches, these 1–2 (–3) cm. long; nodes of the main infl. axis and branches glabrous or with a few hairs, these never papilla-based; spikelets 2.5–3 mm. long, awnless, usually inconspicuously pubescent with short fine hairs; palea of the lower floret well developed; lemma of the fertile floret elliptic, with a sharply differentiated membranous withering tip; $2n = 54$.—Widely scattered as a weed in desert and cismontane s. Calif.; native in tropical and subtropical regions. July–Oct.

2. **E. crusgálli** (L.) Beauv. var. **crusgálli**. [*Panicum c.* L.] BARNYARD GRASS. Annual, with stiffly erect or decumbent-spreading stems, these most 3–10 dm. long; culms glabrous, the nodes slightly swollen; sheaths glabrous; blades mostly 0.3–3 cm. broad; panicles mostly 10–25 cm. long, with usually 5–25 appressed or spreading branches, the longer branches rebranched, these with stout often papilla-based setae; spikelets broadly ovate or ellipsoid, the second glume and lemma of the lower floret variously scabrous and hispid; lemma of the lower floret awnless or awned; palea of lower floret well developed, more than half as long as lemma; $2n = 54$.—A weed in waste places and damp cult. ground, mostly below 2000 ft.; Santa Catalina Id.; Old World. July–Oct.

Var. **frumentàcea** (Roxb.) Wight. [*Panicum f.* Roxb.] Panicle branches erect-appressed or slightly spreading, densely flowered with closely imbricated plump awnless usually grayish purple spikelets.—Occasional weed, as in Orange Co.; from Asia.

Var. **oryzícola** (Vasing) Ohwi. [*E. o.* Vasing in Komarov, Fl. U.R.S.S. 2: 33. 1934.] Lower sheaths hispid on the upper margins and sometimes on the collar or if sheaths glabrous, then blades usually with a few ciliate hairs on the lower margins; blades stiffly erect, folded and acuminate.—Occasional as weed in s. Calif.; introd. from Asia.

3. **E. crus-pavònis** (HBK.) Schult. var. **crus-pavònis**. [*Oplismenus crus-pavonis* HBK.] Plants to 1.5 m. tall; the robust culm with many nodes and glabrous sheath, no ligule; blade 1–2.5 cm. broad, with minutely serrate margin; panicle 1–3 dm. long, densely fld. with small densely fld. short-awned or awnless spikelets; setae absent to prominent on nodes of main panicle axis; spikelets 2.8–3.1 mm. long to base of awn; lower florets sterile with well-developed palea more than half as long as lemma; lemma of fertile florets grayish, narrow elliptic, the coriaceous apex acute or obtuse; $2n = 36$.—Occasional weed; to S. Am.

Var. **mácera** (Wiegand) Gould. [*E. zelayensis* (HBK.) Schult. var. *macera* Wiegand.] Palea of sterile floret absent or vestigial, much less than half as long as lemma.—Scattered weed; to Kansas and Mex.

4. **E. muricàta** (P. Beauv.) Fern. [*Panicum m.* Michx.] Coarse succulent annual, with erect to spreading bases; culm nodes glabrous; sheaths glabrous; ligules absent; blades glabrous, mostly 1–5 cm. long and 0.8–30 mm. broad, with scabrous margins; axes and branches of panicle glabrous or variously hairy; spikelets commonly purplish, broadly ovate, awned or awnless, 3.5 or more mm. long to base of awn; lower floret sterile; lemma of fertile floret narrowed abruptly to an acute non-withering tip; $2n = 36$.—Apparently occurring in s. Calif. as the var. **microstáchya** Wiegand, with spikelets less than 3.5 mm. long; lemma of lower floret awnless or with an awn to 6 or more mm. long; to Quebec, Wash., Mex.

33. Ehrhárta Thunb.

Annual or perennial, erect or spreading, with flat blades and narrow panicles. Spikelets laterally compressed with 1 fertile floret and 2 large sterile lemmas below enclosing the fertile floret; rachilla disarticulating above the glumes, the fertile floret and sterile lemmas falling together. Glumes ovate, obscurely keeled; sterile lemmas indurate, compressed, 3- or 5-nerved; fertile lemma indurate, ovate, obtuse, 5-nerved. A small genus. (Named for F. Ehrhart.)

1. **E. calycina** Sm. Spikelets 7–8 mm. long; sterile lemmas thinly silky-villous; fertile lemma silky on the nerves.—Reported from Ventura Co., *Pollard*; introd. from S. Afr.

2. **E. erécta** Lam. Spikelets 3–3.5 mm. long; sterile lemmas awnless, the first smooth, the 2d cross-wrinkled; natur., Santa Barbara.

34. *Eleusìne* Gaertn. GOOSEGRASS

Annual grasses with 2 to several stout spikes, digitate at the summit of the culms, sometimes with 1 or 2 a short distance below. Spikelets few- to several-fld., compressed, sessile, closely imbricate, in 2 rows along 1 side of a broad rachis; rachilla disarticulating above the glumes and between the florets. Glumes unequal, broad, acute, 1-nerved, shorter than 1st lemma. Lemmas acute, with 3 strong green nerves close together, forming a keel. Ca. 6 spp. of E. Hemis. (Named for *Eleusis*, a Greek town.)

1. **E. índica** (L.) Gaertn. [*Cynosurus i.* L.] Culms flattened, decumbent, 3–5 dm. long; sheaths loose, overlapping, compressed; blades flat or folded, 3–8 mm. wide; spikes mostly 2–6, rarely 1, 4–15 cm. long; $n=9$ (Avdulov, 1931), $2n=18$ (Singh & Godward, 1960).—Occasional, mostly as a street weed; introd. from Eu. July–Oct.

35. *Élymus* L. RYEGRASS

Annual or perennial with flat or rarely convolute blades and cylindric spikes. Spikelets sessile, 2–6-fld., sessile in 2's or rarely 3's at each node of a continuous rachis, the rachilla disarticulating above the glumes and between the florets. Glumes equal, rigid, narrow, sometimes even awnlike. Lemmas rounded on back or subterete, obscurely 5 nerved, acute or usually awned from tip. Ca. 50 spp. of n. temp. regions. (Greek, *elumos*, ancient name for a grain.)

(Gould, F. W. Nomenclatorial changes in E. with a key to the Calif. spp. Madroño 9: 120–128. 1947.)

1. Lemmas awned, the awns mostly 10–30 mm. long.
    2. Awns of lemmas curving outward at maturity .... 3. *E. glaucus jepsonii*
    2. Awns of lemmas ± straight, not curving outward.
        3. Rachis not disarticulating at maturity; glumes 3–5-nerved; culms in small clusters
            3. *E. glaucus glaucus*
        3. Rachis disarticulating at maturity, glumes 1–3-nerved; culms in dense clumps
            4. *E. macounii*
1. Lemmas awnless or the awns not more than 6 mm. long.
    4. Glumes broadly lanceolate, strongly 3–9-nerved .... 3. *E. glaucus virescens*
    4. Glumes subulate, or if lanceolate, inconspicuously nerved.
        5. Spikelets 6–40 at a node; culms mostly 6–10 mm. in diam.; blades 15–35 mm. wide
            2. *E. condensatus*
        5. Spikelets mostly 1–6 at a node; culms mostly less than 6 mm. in diam.; blades 3–15 mm. wide.
            6. Spikelets mostly more than 1 at a node.
                7. Culm-nodes or their vicinity with fine usually dense pubescence; plants mostly without rhizomes .... 1. *E. cinereus*
                7. Culm-nodes glabrous; plants with rhizomes .... 6. *E. triticoides*
            6. Spikelets mostly solitary at the nodes.
                8. Plants with rhizomes .... 6. *E. triticoides*
                8. Plants without rhizomes .... 5. *E. salinus*

1. **E. cinèreus** Scribn. & Merr. [*E. condensatus* var. *pubens* Piper.] Less robust than *E. condensatus*, 0.6–2 m. tall, typically without rhizomes, harsh-puberulent especially at nodes; sheaths and blades pubescent or glabrous, the latter mostly less than 1.5 cm. wide;

spikes 1–2.5 dm. long, mostly not branched; glumes and lemmas like those of *E. condensatus*, but the lemmas ± pubescent; $n=14, 28$ (Gould, 1945), $2n=28$ (Gould, 1966).—Dry places below 10,500 ft.; largely Sagebrush Scrub, Pinyon-Juniper Wd., etc.; deserts from San Bernardino Co. n.; to B.C., Saskatchewan, New Mex. June–Aug.

2. **E. condensàtus** Presl. Rhizomes short, thick; culms 1.5–3.5 m. high, mostly in dense clumps; blades firm, strongly nerved, flat, 1.5–3 cm. wide, glabrous or pubescent; spikes erect, dense, 1.5–5 dm. long, mostly ± compound; spikelets often in 3's, or 5's, 10–15 mm. long, 3–6-fld.; glumes subulate or flat and narrow, usually 1-nerved or nerveless, ca. as long as 1st lemma; lemmas glabrous to ± strigose, short-awned or acute, hyaline-margined; $n=14, 28$ (Gould, 1945), $2n=56$ (Skalinska et al., 1966).—Dry places below 7000 ft.; Coastal Sage Scrub, Chaparral, S. Oak Wd., cismontane s. Calif. to cent. Calif., L. Calif. Channel Ids. In Joshua Tree Wd., etc., mts. of n. Mojave Desert. June–Aug.

3. **E. gláucus** Buckl. ssp. **glaucus**. [*E. villosus* var. *glabriusculus* Torr. *E. g.* var. *breviaristatus* Davy. *E. g.* var. *maximus* Davy. *E. angustifolius* Davy and var. *caespitosus* Davy.] Plate 99, Fig. A. Culms tufted, 6–12 dm. tall; sheaths smooth or scabrous; blades flat, mostly 8–15 mm. wide, mostly scabrous, sometimes involute; spike long-exserted, 0.5–2 dm. long; spikelets 10–12 mm. long; glumes ca. as long, strongly 2–5-nerved, acuminate or awn-pointed; lemmas awned, the awn 1–2 times the body-length, erect to spreading; $n=14$ (Hartung, 1946).—Grassy and wooded places below 7500 ft.; many Plant Communities; cismontane and occasionally desert s. Calif.; to Alaska, Ontario, Ark., New Mex. June–Aug. Santa Catalina, Santa Cruz ids.

Ssp. **jepsònii** (Davy) Gould. [*E. g.* var. *j.* Davy.] Sheaths and blades ± pubescent.—Largely Montane Coniferous F.; mts. from San Diego and Riverside cos. n.; to B.C., Mont.

Ssp. **viréscens** (Piper) Gould. [*E. v.* Piper. *E. pubescens* Davy.] Lemmas awnless or the awns to 4 mm. long; $n=14$ (Hartung, 1946).—Grassy and wooded places, below 7500 ft.; many Plant Communities; cismontane, San Diego Co. to Alaska.

4. **E. macòunii** Vasey. Densely tufted, the culms slender, 5–10 dm. long; blades erect, subinvolute, mostly scabrous, 2–5 mm. wide; spike slender, 4–12 cm. long, tardily disarticulating; spikelets imbricate, mostly 2-fld., ca. 10 mm. long without the awns; glumes very narrow, ca. 1 cm. long, scabrous, short-awned; lemmas scabrous toward the tip, with slender awns 1–2 cm. long.—Meadows and open places, Chaparral, near Murrieta, Riverside Co.; n. Calif. to B.C., Minn. Material referred here is supposed to represent hybrids between *Agropyron* & *Hordeum*.

E. pacíficus Gould, like **E. salìnus,** but only 1–2 dm. high, spikes scarcely exserted.—San Miguel, Santa Rosa ids.

5. **E. salìnus** Jones. Tufted, erect, 3–8 dm. high, purplish-brown at base, sheaths scabrous; blades firm, involute, mostly scabrous; spike slender, 5–12 cm. long; spikelets 1–1.5 cm. long, mostly solitary; glumes subulate, 4–8 mm. long; lemmas ca. 10 mm. long, awnless or nearly so, glabrous or scabrous, rarely strigose, obscurely nerved.—Rocky slopes, 4500–6500 ft.; Pinyon-Juniper Wd.; Providence, New York and Clark mts. of e. San Bernardino Co.; to Rocky Mts. May–June.

6. **E. triticoìdes** Buckl. [*E. condensatus* var. *t.* Thurb. *E. orcuttianus* Vasey.] Rhizomes extensively creeping; culms mostly glaucous, 2–3.5 mm. in diam., 6–12 dm. tall, mostly in large masses; blades mostly 2–6 mm. wide, flat or soon involute; spike erect, slender to ± dense, sometimes compound, 1–2 dm. long; spikelets 12–20 mm. long; glumes subulate or narrow, firm, 0–1–3-nerved, 8–14 mm. long, awn-tipped; lemmas 6–10 mm. long, brownish to purplish, glabrous, awn-tipped; $n=14, 21$ (Gould, 1945).—Moist and alkaline places below 7500 ft.; many Plant Communities, ± throughout the state; to Wash., Mont., Tex., L. Calif. June–July. San Miguel Id., Santa Cruz Id., Santa Catalina Id.

36. *Enneapògon* Desv. ex Beauv.

Slender tufted perennials with numerous culms and flat to subinvolute lvs. Panicle spikelike, gray-green, feathery. Spikelets 3-fld., the 1st floret fertile, the 2d smaller and sterile, the 3d rudimentary. Glumes strongly 7-nerved, longer than the body of the lem-

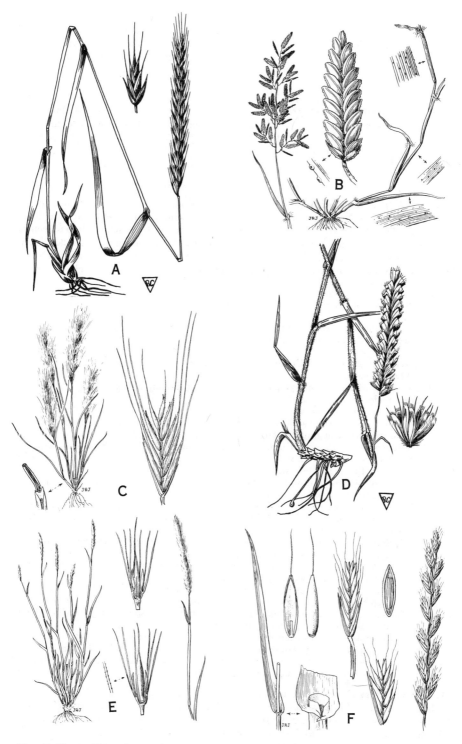

Plate 99. Fig. A, *Elymus glaucus*. Fig. B, *Eragrostis cilianensis*. Fig. C, *Festuca octoflora*. Fig. D, *Hilaria rigida*. Fig. E, *Hordeum brachyantherum*. Fig. F, *Lolium perenne* ssp. *multiflorum*.

mas. Lemmas rounded on the back, firm, the truncate apex with 9 plumose equal awns; palea slightly longer than body of lemma. Ca. 35 spp., mostly of Old World. (*Ennea*, nine, and *pogon*, beard, because of the 9 awns on the lemma.)

(Chase, A. in Madroño 7: 187–189. 1946.)

1. **E. desváuxii** Beauv. [*Pappophorum wrightii* Wats.] Culms slender, 2–4 dm. long, the nodes pubescent; blades ca. 1 mm. wide; panicle 2–5 cm. long; lowest lemma 4–5 mm. long; $2n=20$ (Gould, 1966).—Dry limestone slopes, 5000–6000 ft.; Pinyon-Juniper Wd.; Clark Mts., e. San Bernardino Co.; to Utah, Tex., Mex. Aug.–Sept.

37. *Eragróstis* Beauv. LOVEGRASS

Annuals or perennials with loose or dense terminal panicles. Spikelets strongly compressed, few–many-fld., the uppermost florets sterile; rachilla articulated, but sometimes not disjointed until after the fall of the glumes and lemmas with the seed. Glumes keeled, somewhat unequal, much shorter than the spikelets, the 1st 1-nerved, the 2d rarely 3-nerved. Lemmas 3-nerved, broad, keeled, acute or acuminate; paleae shorter than their lemmas or ca. as long. Ca. 250 spp., in trop. and temp. regions. (Greek, *eros*, love, and *agrostis*, a grass.)

1. Plants perennial; spikelets gray-green, 7–11-fld. .... 3. *E. curvula*
1. Plants annual.
    2. Plants glandular or warty along the lf.-margins, keel of lemmas, or on panicle branches.
        3. Spikelets 2.5–3 mm. wide; lemmas ca. 2.5 mm. long; glands prominent on keels of lemmas
        .... 2. *E. cilianensis*
        3. Spikelets ca. 2 mm. wide; lemmas ca. 2 mm. long; glands mostly on panicle branches
        and lvs. .... 11. *E. poaeoides*
    2. Plants not glandular on lemmas, panicle-branches, or lf.-margins, sometimes so on sheaths and near nodes.
        4. Spikelets ca. 1 mm. wide, slender, linear.
            5. Spikelets 3–5 mm. long; lemmas 1–1.5 mm. long; pedicels mostly longer than the spikelets; surface of grain smooth .... 10. *E. pilosa*
            5. Spikelets 5–7 mm. long; lemmas ca. 2 mm. long; pedicels mostly shorter than the spikelets; surface of grain reticulate .... 8. *E. orcuttiana*
        4. Spikelets ca. 1.5 mm. wide or wider, linear to ovoid.
            6. Plants creeping, rooting at the nodes to form mats .... 5. *E. hypnoides*
            6. Plants not creeping or forming mats.
                7. Panicle narrow, with ascending branches bearing spikelets almost to their bases; spikelets linear, mostly 12–15-fld. and ca. 1 cm. long .... 1. *E. barrelieri*
                7. Panicle open, ± diffuse.
                    8. Some or all of the spikelets on pedicels 1–5 mm. long; spikelets linear at maturity, appressed along primary panicle-branches.
                        9. Primary branches of panicle simple or the lower with a branchlet bearing 2–3 spikelets; culms slender, mostly less than 3 dm. high .... 9. *E. pectinacea*
                        9. Primary branches of panicle usually bearing branchlets with few to several spikelets; culms rather robust, over 3 dm. tall .... 4. *E. diffusa*
                    8. Some or all of the spikelets on pedicels 5–8 mm. long; spikelets ovate to linear, not appressed along the primary panicle branches.
                        10. Spikelets narrow-lanceolate, 1.8–2 mm. broad, lead-color, mostly 8–12-fld.; panicle with ascending branches .... 7. *E. neomexicana*
                        10. Spikelets ovate-lanceolate, 2–2.5 mm. broad, straw-color or purplish, 3–9-fld.; panicle open, with spreading branches .... 6. *E. mexicana*

1. **E. barreliéri** Daveau. Annual, with culms erect or spreading at base, 2–5 dm. tall, branched below; sheaths pilose at summit; blades flat, 2–4 mm. wide; panicle erect, open but narrow, 8–15 cm. long, with ascending branches or stiffly spreading, few-fld. and bearing spikelets nearly to the base, glabrous in axils; spikelets linear, mostly 12–15-fld., ca. 1 cm. long, 1.5 mm. wide; lemmas ca. 2 mm. long; $2n=60$ (Singh & Godward, 1960).—Waste places, Ventura and Santa Barbara cos.; to Colo., Kans. June–Sept.

Eragrostis

2. **E. cilianénsis** (All.) E. Mosher. [*E. megastachya* (Koel.) Link.] STINKGRASS. Plate 99, Fig. B. Weedy strong-scented annual with spreading to ascending culms 2–5 dm. long; with a ring of glands below the nodes; lf.-blades 3–8 mm. wide, with wartlike glands on the margins; panicles 0.5–2 dm. long, greenish lead-color, densely-fld.; spikelets 10–40-fld., 5–17 mm. long, 2.5–3 mm. wide, with closely imbricated florets; pedicels and keels of glumes and lemmas ± glandular; lemmas ca. 2.5 mm. long, with prominent lateral nerves; $n=10$ (Avdulov, 1928).—Waste places, below 6000 ft., cismontane and sometimes desert areas; through most of the U.S.; introd. from Eu. June–Dec.

3. **E. cúrvula** (Schrad.) Nees. [*Poa c.* Schrad.] Densely tufted erect perennial 6–12 dm. tall; sheaths keeled, glabrous or sparsely hispid, the lower hairy toward their base; blades involute, finely pointed, scabrous; panicle 2–3 dm. long, the branches ascending, naked at base; spikelets 7–11-fld., gray-green, 8–10 mm. long; lemmas ca. 2.5 mm. long, with prominent nerves; $2n=40$ (de Wet, 1958).—Reported from San Diego, *Fuller* and cent. Calif; s. states; introd. from Africa. Aug.–Oct.

4. **E. diffùsa** Buckl. Annual, with culms erect from a spreading base, 3–5 dm. high; sheaths somewhat pilose above; blades flat, 2–8 cm. long, 1–3 mm. wide; panicle open, diffuse, 7–20 cm. long, the primary branches with appressed secondary branchlets bearing few to several spikelets; spikelets 5–many-fld.; glumes 1-nerved, the 2d 1.5 mm. long; lemmas ca. 1.5 mm. long; $2n=60$ (Gould, 1965).—Weed in waste places, Imperial Co. and cismontane; to Rocky Mts., Tex., Mex. April–Nov.

5. **E. hypnoìdes** (Lam.) BSP. [*Poa h.* Lam.] Branching creeping annual rooting at the nodes and forming mats to several dm. across; lf.-blades scabrous or pubescent on upper side, 1–4 cm. long; panicle simple or nearly so, usually open, 1–6 cm. long; spikelets linear-lanceolate, 10–35-fld., mostly 5–15 mm. long; lemmas thin, acuminate, 1.5–2 mm. long; paleae half as long.—Shore of pond, Santa Monica Mts., Los Angeles Co.; cent. Calif. to Wash., Atlantic Coast, S. Am. Sept.

6. **E. mexicàna** (Hornem.) Link. [*Poa m.* Hornem.] Like *E. neomexicana*, but lower, the culms often simple; lvs. pale, 1–5 mm. broad; panicle 4–20 cm. long, with spreading branches; spikelets mostly not more than 7-fld., 2–2.5 mm. broad; lemmas 1.8–2.25 mm. long; $n=30$ (Avdulov, 1928).—Reported from waste places; to Tex., Mex. July–Oct.

7. **E. neomexicàna** Vasey. Rather stout annual, darkish-green, 5–10 dm. tall; sheaths glabrous, with pilose throats; blades flat, 3–8 mm. broad; panicle 2–4 dm. long, with ascending branches and ascending spikelets, mostly overlapping or imbricate; spikelets dark gray-green, ± ovate, 8–12-fld., 5–8 mm. long, 1.8–2 mm. wide; lemmas 2–2.3 mm. long; $n=28$ (Singh & Godward, 1960).—Waste places, desert and cismontane, s. Calif.; to Atlantic Coast; introd. from Eu. May–Sept.

8. **E. orcuttiàna** Vasey. Annual, the culms ascending from a ± decumbent base, 6–10 dm. tall; sheaths glabrous; blades flat, 2–6 mm. wide; panicle open, 1.5–3 dm. long, the branches, branchlets and pedicels spreading, slender, flexuous, with glabrous axils; spikelets linear, 6–10-fld., 5–7 mm. long, ca. 1 mm. wide; 2d glume 1–1.5 mm. long; lemmas ca. 1.8 mm. long, ± loosely arranged.—Moist fields and waste places, many Plant Communities; cismontane Calif., especially n.; to Ore., Rocky Mts. May–Oct.

9. **E. pectinàcea** (Michx.) Nees. [*Poa p.* Michx.] Like *E. pilosa*, the panicle-branches bearing spikelets from near the base; spikelets lance-ovate, 1.3–1.7 mm. wide, mostly more than 5 mm. long, often 10–15-fld.; lemmas usually more than 1.5 mm. long; $2n=20$, 30 (Gould, 1958).—San Bernardino, *E. Johnson*, 1944; through most of the U.S. July–Oct.

10. **E. pilòsa** (L.) Beauv. [*Poa p.* L.] Annual, the culms slender, erect or with decumbent base, 1–5 dm. tall; blades flat, 1–3 mm. wide; panicle open, 0.5–2 dm. long, the capillary branches flexuous, ascending or spreading, bearing spikelets in their distal half or more; spikelets lance-linear, ca. 1 mm. wide, 3–9-fld., 3–5 mm. long, the spreading pedicels mostly longer; glumes glabrous, 0.5 and 1 mm. long; lemmas inconspicuously nerved, 1.2–1.5 mm. long.—Occasional as a waif in waste places; introd. from Eu. July–Sept.

11. **E. poaeoìdes** Beauv. ex R. & S. Resembling *E. cilianensis*; smaller, more slender; panicles open; spikelets 1.5–2 mm. wide, 8–20-fld., the florets less densely imbricated, the

bases or rachilla-joints visible; lemmas less than 2 mm. long.—Occasional in waste places in s. Calif. as at San Jacinto; sparingly introd. from Eu. June–Sept.

38. Eriochlòa HBK. CUPGRASS

Cespitose annuals and perennials. Blades flat, mostly thin. Ligule a ring of soft hairs. Infl. a loosely contracted panicle, of several to many appressed or spreading racemes, usually approximate along a common axis. Spikelets in 1's or 2's, short-pediceled or subsessile, in 2 rows on 1 side of a narrow rachis, the back of the fertile lemma turned from the rachis; lower rachilla-joint thickened, forming a ± ringlike callus below the 2d glume, the 1st glume minute and sheathing this. Lemma usually inclosing a hyaline palea, or sometimes a ♂ fl. Fertile lemma indurate, minutely papillose-rugose, mucronate or awned, the awns often readily deciduous. Ca. 25 spp. of warmer regions. (Greek, *erion*, wool, and *chloa*, grass, because of pubescent spikelets.)

Spikelets, including awns, 7–10 mm. long .... 1. *E. aristata*
Spikelets not more than 6 mm. long, awnless or awn-tipped.
    Blades pubescent; fr. with an awn ca. 1 mm. long .... 2. *E. contracta*
    Blades glabrous; fr. apiculate .... 3. *E. gracilis*

1. **E. aristàta** Vasey. Annual, 3–8 dm. tall; blades 10–12 mm. wide; racemes several, overlapping, 3–4 cm. long, the rachis pilose; spikelets ca. 5 mm. long, the glume and sterile lemma awned; fr. 3.5 mm. long, apiculate.—Colo. R. bottom near Palo Verde; Yuma, Ariz. June–Nov.

2. **E. contrácta** Hitchc. Annual, 3–7 dm. tall; blades pubescent, 3–5 mm. wide; panicle subcylindric, to ca. 1.5 dm. long; spikelets 3.5–4 mm. long, excluding the 1 mm. awn; fr. 2–2.5 mm. long; $2n = 36$ (Gould, 1965).—Occasional, as in Imperial Co., Santa Barbara; to Nebr., La.

3. **E. grácilis** (Fourn.) Hitchc. [*Helopus g.* Fourn. *E. lemmonii* var. *g.* Gould.] Annual, 4–10 dm. tall; blades flat, glabrous, 5–10 mm. wide; racemes 2–4 cm. long; spikelets 4–5 mm. long, sparsely strigose, acuminate or short-awned; fr. ca. 3 mm. long, apiculate; $2n = 36$ (Gould, 1966).—Irrigated fields, etc.; Imperial, Riverside and San Bernardino cos.; to Okla., Tex. Aug.–Sept.

39. Erioneùron Nash.

Low, tufted, mostly stoloniferous perennial with narrow, often involute, cartilaginous-margined lvs. Infl. a short, usually capitate raceme or panicle of several-fld. spikelets, the disarticulation above the glumes and between the florets. Glumes large, membranous, subequal, 1-nerved. Lemmas broad, 3-nerved, long-hairy along the nerves at least below, the midnerve short-awned, each lateral nerve often prolonged as a short mucro. Palea slightly shorter than the lemma, ciliate on the keels, long-hairy on the lower part. Grain glossy and translucent. Five spp. of sw. U.S. and adjacent Mex. (Greek, *erion*, wool, and *neuron*, nerve.)

(Takeoka, T. A biosystematic study of Tridens. Am. J. Bot. 48: 565–573. 1961.)
Infl. leafy; plants low, creeping .... 2. *E. pulchellum*
Infl. not leafy; plants not creeping .... 1. *E. pilosum*

1. **E. pilòsum** (Buckl.) Nash. [*Uralepis p.* Buckl. *Tridens p.* Hitchc.] Densely tufted, 1–3 dm. high; sheaths pilose at throat; blades 1–1.5 mm. wide, flat or folded, ± pilose, in a dense basal cluster; panicle well exserted, ovoid, 1–2 cm. long, pale or purplish; spikelets 3–10, short-pediceled, 6–12-fld., 1–1.5 cm. long, compressed; glumes ca. ⅔ as long as lower florets; lemmas ca. 6 mm. long, pilose below and on margin, acute, with awns 1–2 mm. long; $2n = 16$ (Tateoka, 1961).—Dry slopes, 5000–6000 ft., Pinyon-Juniper Wd.; Clark Mt. and Mescal Range, e. San Bernardino Co.; to Kans., Tex., Mex. May–June.

2. **E. pulchéllum** (HBK.) Tateoka. [*Triodia p.* HBK. *Tricuspis p.* Torr. *Tridens p.*

Hitchc.] FLUFFGRASS. Tufted, mostly 5–12 cm. high, the culms slender, with a long internode bending over and rooting at the tips to form other culms; blades in fascicles, involute, sharp-pointed, scabrous, curved, mostly 2–4 cm. long; infl. a capitate panicle in the fascicle, mostly with 1–5 subsessile, white, woolly spikelets; glumes glabrous, subequal, acuminate, 6–8 mm. long, awn-pointed; lemmas 4 mm. long, long-pilose below, cleft ca. halfway, the awn scarcely exceeding the obtuse lobes; $2n = 16$ (Tateoka, 1961).—Dry rocky places, below 7000 ft.; Creosote Bush Scrub, Joshua Tree Wd.; both deserts; to Tex., Mex. Feb.–May.

40. *Festùca* L. FESCUE

Annuals or perennials, with spikelets in narrow or open panicles. Spikelets few- to several-fld., the rachilla disarticulating above the glumes and between the florets, the uppermost floret reduced. Glumes narrow, membranous, acute, unequal, the 1st sometimes very small; lemmas rounded on the back, membranous or indurate, 5-nerved, the nerves often obscure, acute to obtusish, awned from the tip or rarely from a bifid apex, sometimes awnless. More than 100 spp. of temp. and cool regions, many being important forage grasses. (Ancient Latin name for some grass.)

(Piper, C. V. N. Am. spp. of Festuca. Contr. U.S. Nat. Herb. 10: 1–48. 1906. St. Yves, A. Contribution a l'étude des Festuca. Candollea 2: 229–316. 1925.)
1. Plants annual; florets often cleistogamous, with 1 or 3 anthers (Vulpia).
   2. Spikelets densely 5–12-fld.; lemmas without scarious margin . . . . 13. *F. octoflora*
   2. Spikelets loosely 1–5-fld.; lemmas with narrow scarious margin.
      3. Infl. narrow, branches ascending or appressed-ascending.
         4. Lemmas conspicuously long-ciliate toward apex . . . . 9. *F. megalura*
         4. Lemmas not ciliate.
            5. Lower glume ⅔ to ¾ as long as the 2d . . . . 6. *F. dertonensis*
            5. Lower glume less than half as long as the 2d . . . . 11. *F. myuros*
      3. Infl. broader, the principal branches spreading.
         6. Spikelets glabrous or ± scabrous, but not with long well developed hairs.
            7. The spikelets usually 3–5-fld.; only the main branches of the infl. divergent
               14. *F. pacifica*
            7. The spikelets mostly 1–2-fld.; all branches of infl. divergent or reflexed
               16. *F. reflexa*
         6. Spikelets pubescent, with long well developed hairs.
            8. Pedicels appressed or slightly spreading; lower branches of panicle usually spreading or reflexed.
               9. Lemmas glabrous; glumes pubescent . . . . 5. *F. confusa*
               9. Lemmas hirsute; glumes glabrous or pubescent . . . . 8. *F. grqvi*
            8. Pedicels and panicle-branches all finally spreading or reflexed.
               10. Glumes glabrous; lemmas pubescent . . . . 10. *F. microstachys*
               10. Glumes pubescent; lemmas pubescent . . . . 7. *F. eastwoodae*
1. Plants perennial; florets opening; anthers mostly 3 (Festuca).
   11. Awns scarcely if at all evident, less than 1 mm. long.
      12. Lf.-blades flat, 4–8 mm. wide.
         13. Panicle 1.5–3.5 dm. long; lemmas 7–10 mm. long . . . . 2. *F. arundinacea*
         13. Panicle 1–2 dm. long; lemmas 5–7 mm. long . . . . 15. *F. pratensis*
      12. Lf.-blades narrower, involute. E. Mojave Desert . . . . 1. *F. arizonica*
   11. Awns well developed, 2–many mm. long.
      14. Collar and mouth of sheath villous . . . . 4. *F. californica*
      14. Collar and mouth of sheath not villous.
         15. Awn of lemma longer than the body; ovary pubescent at the summit
            12. *F. occidentalis*
         15. Awn of lemma shorter than the body; ovary glabrous at the summit.
            16. Culms loosely tufted and decumbent at the base; blades ± involute; sheaths redbrown with conspicuous pale nerves . . . . 17. *F. rubra*
            16. Culms closely tufted, not decumbent; blades filiform; sheaths not red-brown at base
               3. *F. brachyphylla*

1. **F. arizónica** Vasey. [*F. ovina a.* Hack.] Densely tufted perennial; culms 3–10 dm. tall; lvs. stiff, quite glaucous, mostly basal, scabrous, 1.5–3 dm. long; panicle narrow, 1–2 dm. long, the branches ascending or appressed, scabrous; spikelets mostly 5–7-fld., scabrous; glumes ca. 4 and 5–6 mm. long; lemmas 5–6 mm. long, awnless or nearly so.—Dry slopes at 5500–6000 ft.; Pinyon-Juniper Wd.; Clark Mt., e. Mojave Desert; to Ariz., Colo. Sept.–Oct.

2. **F. arundinàcea** Schreb. [*F. elatior* var. *a.* Wimm.] Resembling *F. pratensis*, but taller, 5–18 dm. long, more robust; blades 2.5–7 dm. long; panicles 1.5–5 dm. long, with more branches and spikelets; spikelets 8–12 mm. long, 4–8-fld., mostly violet-tinged; lemmas 7–10 mm. long; $n=14$ (Staehlin, 1929), $2n=42$ (Malik & Thomas, 1966).—Occasional introd. from Eu. in scattered locations. Santa Catalina Id. May–June. Widely used as a forage crop.

3. **F. brachyphýlla** Schult. [*F. ovina* ssp. *b.* Piper. *F. o.* ssp. *saximontana* var. *purpusiana* St. Yves.] Tufted perennial, with erect very slender culms 1–1.5 dm. high; blades smooth, filiform, soft, angled in drying, numerous, 2–6 cm. long; panicle short, narrow, few-fld., 2–5 cm. long; glumes and lemmas broad; spikelets 2–5-fld.; lemmas 3–3.5 mm. long; awn 1–3 mm. long; $2n=42$ (Jorgensen et al., 1958), 28, 42 (Mosquin & Hayley, 1966).—Dry, ± rocky places, 9400–11,500 ft.; Subalpine F., etc.; San Bernardino Mts.; to 14,000 ft., White Mts.; to Arctic regions, Rocky Mts., Vt. July–Sept.

4. **F. califórnica** Vasey var. **paríshii** (Piper) Hitchc. [*F. aristulata* var. *p.* Piper. *F. p.* Hitchc.] Perennial, the culms rather stout, 4–6 dm. high; sheaths pubescent; blades flat or involute when dry, 1.5–2.5 dm. long, firm; panicle loose, few-branched, ca. 1 dm. long; spikelets compressed, ca. 5-fld.; glumes 6–8 mm. long, with a scabrous keel; lemmas 8–10 mm. long, the awn 3–4 mm. long.—Dry places, 2500–6500 ft.; Chaparral, Yellow Pine F.; San Bernardino Mts. May–July.

5. **F. confùsa** Piper. Resembling *F. pacifica*; sheaths retrorsely pilose; foliage pubescent; spikelets 2–3-fld.; glumes hirsute; lemmas glabrous.—Occasional in dry places below 5000 ft.; Coastal Sage Scrub to Yellow Pine F.; San Diego Co. to Wash. March–June.

6. **F. dertonénsis** (All.) Asch. & Graebn. [*Bromus d.* All. *Vulpia d.* Volk. *F. bromoides* auth.] Similar to *F. megalura*; panicle 5–10 cm. long; 1st glume 4 mm. long, 2d 6 mm. long; lemma 7–8 mm. long, not ciliate; awn 10–12 mm. long.—Widely scattered in cismontane s. Calif. to desert edge; Coastal Sage Scrub, Chaparral, etc.; Santa Catalina Id. Introd. from Eu. April–June.

7. **F. eastwoódae** Piper. Resembling *F. reflexa*; panicle-axis and branches pubescent; glumes hirsute; lemmas hirsute; awns 3–10 mm. long.—Occasional in open places in Chaparral, etc. from Santa Ynez Mts. n. to Ore. April–June.

8. **F. gràyi** (Abrams) Piper. [*F. microstachys grayi* Abrams.] Like *F. pacifica*; often somewhat stouter; sheaths and sometimes blades pubescent; glumes glabrous to ± villous; lemmas pubescent, puberulent or ± villous.—Frequent below 6000 ft.; in open places; V. Grassland, Coastal Sage Scrub, Chaparral, etc.; cismontane s. Calif. to Ore. March–June.

9. **F. megalùra** Nutt. [*Vulpia m.* Rydb.] FOXTAIL FESCUE. Simple or tufted annual, glabrous, 2–6 dm. tall; sheaths and blades smooth; the blades flat or involute; panicle narrow, 7–20 cm. long, with appressed branches; spikelets 4–5-fld., 8–11 mm. long; glumes glabrous, very unequal, the 1st less than 2 mm. long, the 2d 4–5 mm. long; lemma obscurely 5-nerved, 4–6 mm. long, scabrous, ciliate on upper half; awn scabrous, ca. 8–15 mm. long.—Common in dry open places, below 7000 ft.; many Plant Communities; cismontane s. Calif. to B.C., Mont., L. Calif., S. Am. On most of our ids. April–June.

10. **F. microstàchys** Nutt. [*Vulpia m.* Munro ex Benth.] Like *F. reflexa*; glumes glabrous; lemmas pubescent.—Rare in open places below 5500 ft., as at Kenworthy, San Jacinto Mts.; Panamint Mts.; to Wash. April–June.

11. **F. myùros** L. [*Vulpia m.* Rydb.] Like *F. megalura*; panicle commonly somewhat shorter; 1st glume a little shorter; lemmas not ciliate; $n=7$ (Avdulov, 1928), 21 (Tanji,

1925).—Occasional but widely scattered as introd. from Eu. Santa Catalina Id. March–May.

12. **F. occidentàlis** Hook. Perennial, tufted, erect, 4–10 dm. tall; sheaths smooth; blades many, largely basal, narrow-involute; panicle 7–20 cm. long, often drooping above; spikelets 3–5-fld., 6–10 mm. long, on slender pedicels; lemmas rather thin, 5–6 mm. long; awn slender, 6–10 mm. long.—Dry wooded banks and slopes, Santa Barbara Co. n. to B.C., Mich. May–Aug.

13. **F. octoflòra** Walt. [*Vulpia o.* Rydb.] SIX-WEEKS FESCUE. Plate 99, Fig. C. Tufted annual, 0.5–4 dm. tall, glabrous or puberulent; blades involute, 2–8 cm. long; panicle erect, racemiform, 2–10 cm. long; spikelets 5–10 mm. long, densely 5–13-fld.; glumes subulate-lanceolate, the 1st 1-nerved, 3 mm. long, the 2d 3-nerved, 4 mm. long; lemmas lanceolate, convex, firm, glabrous or scabrous, 4–5 mm. long; awn scabrous, 2–4 mm. long.—Common in dry open places below 5000 ft.; many Plant Communities, cismontane Calif.; to Wash., Atlantic Coast. April–June. San Clemente Id., Santa Catalina Id.; Santa Rosa Id. A form with pubescent lemmas is the var. *hirtella* Piper and according to Twisselmann is more common on the deserts, but does occur toward the coast as on Santa Cruz Id. April–June.

14. **F. pacífica** Piper [*Vulpia p.* Rydb.] Erect or geniculate at base, 2–5 dm. tall, quite glabrous; blades soft, loosely involute; panicle 5–12 cm. long, the lower branches solitary, somewhat distant, spreading, subsecund, 1–3 cm. long; spikelets 3–6-fld., on appressed pedicels; glumes glabrous, the 1st 1-nerved, 4 mm. long, the 2d 3-nerved, 5 mm. long; lemmas lanceolate, mostly scabrous, 6–7 mm. long, with an awn 10–15 mm. long.—Dry open places below 5000 ft.; many Plant Communities; cismontane, L. Calif. to B.C.; Ariz. March–June. San Clemente Id., Santa Catalina Id., Anacapa Id.

15. **F. praténsis** Huds. [*F. elatior* Am. auth.] MEADOW FESCUE. Perennial, with stout culms 3–12 dm. long; sheaths smooth; blades flat, mostly 4–6 mm. wide, 1–2 dm. long; panicles erect or with nodding summit, subsimple or much branched, 1–2 dm. long, the branches bearing spikelets nearly to the base; spikelets mostly 6–8-fld., 8–12 mm. long; glumes lanceolate, 3 and 4 mm. long; lemmas oblong-lanceolate, coriaceous, 5–7 mm. long, scarious at apex, rarely short-awned; $2n = 14$ (Bowden, 1960).—Occasional in meadows and waste places as an escape from cult.; introd. from Eu. May–July.

16. **F. refléxa** Buckl. [*Vulpia r.* Rydb.] Annual, the culms 2–4 dm. tall; sheaths smooth or pubescent; blades narrow, flat or ± involute, 2–10 cm. long; panicle 5–12 cm. long, branches and spikelets divaricate in maturity; spikelets mostly 1–3-fld., 5–7 mm. long, the 1st glume 2–4 mm. long, the 2d 4–5 mm. long; lemmas glabrous or scaberulous, 5–6 mm. long; awns 5–8 mm. long.—Dry open places, below 5000 ft.; many Plant Communities; cismontane Calif. to Wash., Utah, Ariz. April–June. San Clemente and Santa Catalina ids.

17. **F. rùbra** L. RED FESCUE. Perennial with loosely tufted culms bent or decumbent at the ± reddish base, 4–10 dm. high; lower sheaths brown, thin, fibrillose; blades smooth, soft, ± involute; panicle 3–20 cm. long, mostly narrow with erect branches; spikelets 4–6-fld., pale, often with purplish tinge, mostly 7–8 mm. long; glumes smooth, the lower 3–4 mm., the upper 5–7 mm. long; lemmas 5–7 mm. long, smooth or scabrous upward; awn scabrous, 2–4 mm. long; $n = 7$ (Staehlin, 1929), $2n = 42$ (Jørgensen et al., 1958).—Meadows and moist places at 4000–8500 ft., Montane Coniferous F., San Bernardino and San Gabriel mts.; persistent lawn grass as at Santa Barbara; through cooler parts of N. Am. and Eu. May–July.

41. *Gastrídium* Beauv. NITGRASS

Annuals with flat blades and pale shining spicate panicles. Spikelets 1-fld., the rachilla disarticulating above the glumes. Glumes unequal, enlarged or swollen at base; lemmas much shorter than glumes, hyaline, broad, truncate, awned or awnless; palea ca. as long as lemma. Two spp. of Medit. region. (Greek, *gastridion*, small pouch because of the subsaccate glumes.)

1. **G. ventricòsum** (Gouan) Schinz & Thell. [*Agrostis v.* Gouan.] Culms 1–5 dm. tall; panicle dense, 3–8 cm. long; lf.-blades 2–12 cm. long; glumes 3 mm. long, the 2d ca. ¼ shorter than the 1st, long-pointed; lemma globular, pubescent, with geniculate awn 5 mm. long; $n=7$ (Rutland, 1941).—Weed in dry open and in waste places, most of cismontane Calif. below 4000 ft.; introd. from Eu. May–Sept. San Clemente, Santa Catalina, Santa Cruz and Santa Rosa ids.

42. *Glycèria* R. Br.

Aquatic or marsh perennials with creeping rhizomes and flat blades. Lf.-sheaths connate. Panicles open or contracted. Spikelets few- to many-fld., subterete, the rachilla disarticulating above the glumes and between the florets. Glumes unequal, 1-nerved. Lemmas broad, convex on the back, firm, usually obtuse, scarious at apex, 5–9-nerved, the usually prominent nerves parallel. Ca. 35 spp. in temp. regions. (Greek, *glukeros*, sweet, the seed of some spp. being so.)

1. **G. elàta** (Nash) Hitchc. [*Panicularia e.* Nash.] Culms 10–18 dm. tall, ± succulent; plants dark green; blades flat, thin, 6–12 mm. wide; panicle diffuse, 1.5–3 dm. long, the branches spreading or the lower deflexed; spikelets 6–8-fld., 3–6 mm. long; glumes ca. 1 and 1.5 mm. long; lemmas 2–3 mm. long; $n=10$ (Church, 1949).—Wet places, 5000–7500 ft.; Yellow Pine F.; San Bernardino Mts., San Jacinto Mts.; Sierra Nevada and N. Coast Ranges to B.C., Mont., Colo., New Mex. July–Aug.

43. *Hesperochlòa* Rydb.

Densely tufted, rhizomatous, dioecious perennial. Lvs. firm, narrow, flat or loosely involute. Panicles narrow, erect. Spikelets 3–5-fld., the rachilla disarticulating above the glumes and between the florets. Glumes subequal or unequal, shorter than the 1st lemma, lanceolate, acute, the 1st 1-nerved, the 2d 3-nerved. Lemmas rounded on the back, acute or acuminate, awnless, 5-nerved; palea as long as lemma, scabrous-ciliate on the keels; stigmas sessile, long, slender. One sp. (Greek, *esperis*, western, and *chloa*, grass.)

1. **H. kíngii** (Wats.) Rydb. [*Poa k.* Wats. *Festuca k.* Cassidy. *F. confinis* Vasey. *Leucopoa k.* Weber.] Clumps large, dense erect, 5–8 dm. tall; sheaths smooth, striate; blades firm, 3–6 mm. wide, 5–30 cm. long; panicle 7–20 cm. long, with short appressed branches, the ♂ infl. denser, with somewhat larger spikelets than the ♀; spikelets 7–12 mm. long; glumes thin, the 1st 3–4 mm., the 2d 5–6 mm. long; lemmas 5–8 mm. long, acute to acuminate, scabrous.—Dry slopes, 6000–12,000 ft.; Montane Coniferous F., Bristlecone Pine F.; San Bernardino Mts., Inyo-White Mts.; Sierra Nevada to Ore., Rocky Mts., Nebr. June–Aug.

44. *Heteropògon* Pers.

Annual or perennial, with flat or folded blades and usually solitary terminal racemes. Lower few pairs of spikelets alike, ♂, awnless; remaining sessile spikelets fertile, long-awned; pedicellate spikelets ♂ like lower ones; rachis continuous below, bearing fertile spikelets above, disarticulating at base of each joint, the joint forming a sharp barbed callus below fertile spikelet; glumes of fertile spikelet dark brown, the 1st enclosing the 2d; lemmas hyaline, fertile one with a long stout twisted geniculate awn. Ca. 8 spp. of warmer regions. (Greek, *heteros*, different, and *pogon*, beard, referring to the awnless ♂ and awned ♀ spikelets.)

1. **H. contórtus** (L.) Beauv. ex Roem. & Schult. [*Andropogon c.* L.] Perennial, tufted, 2–8 dm. tall, glabrous, with a few flowering branches at upper nodes; blades 5–15 cm. long, 3–7 mm. wide, scabrous; raceme usually long-exserted, 4–7 cm. long; 1st glume hirsute with spreading hairs; awns 5–12 cm. long, hirsute; spikelet ca. 1 cm. long.—Dehesa School, San Diego Co.. *Gander.*, n. Imperial Co., *Wheeler*; Ariz. to Texas and s. March–April.

45. Hilària HBK. GALLETA

Stiff perennial grasses with solid culms and narrow blades. Spikelets sessile, in groups of 3, appressed to the axis, in terminal spikes. The spikelet-groups falling from the axis entire, the cent. one fertile, mostly 1-fld., the 2 lateral spikelets ♂, 2-fld. Glumes coriaceous, those of the 3 spikelets forming a false invol. sometimes connate at base, ± asymmetric, usually bearing an awn on 1 side from ca. the middle. Lemma and palea hyaline, subequal. Seven spp., sw. U.S. to n. S. Am. (Auguste St. Hilaire, French naturalist.)

Stems with a felty pubescence. Widespread on deserts .... 3. H. rigida
Stems not felty pubescent.
  Cluster of spikelets flabellate; outer glumes of lateral spikelets broadest toward the summit. Se. Colo. Desert .... 1. H. belangeri
  Cluster of spikelets not flabellate; glumes of lateral spikelets narrowed toward summit. E. Mojave Desert .... 2. H. jamesii

1. **H. belángeri** (Steud.) Nash. [*Anthephora b.* Steud.] Tufted, often sending out slender stolons; culms erect, 1–3 dm. tall, villous at nodes; blades 1–2 mm. wide, flat, curly; spike ca. 2–3 cm. long, mostly with 4–8 clusters of spikelets; lateral spikelets attenuate at base, the glumes united below, firm, scabrous; fertile spikelet usually shorter than the sterile, rounded at base, glumes with deeply lobed thinner upper part, the midnerves extending into awns mostly exceeding ♂ spikelets; lemma compressed, narrowed above, awnless.—Reported from se. Calif.; to Ariz., Sonora.

2. **H. jàmesii** (Torr.) Benth. [*Pleuraphis j.* Torr.] Culms erect, 1.5–3 dm. high, slender, from decumbent bases and tough scaly rhizomes; sheaths glabrous or slightly scabrous, sparingly villous about the ligule; blades mostly 2–5 cm. long, 2–4 mm. wide, soon involute; spikelets 6–8 mm. long, villous at base; $2n=36$ (D. E. Anderson, 1965).—Dry slopes, 4000–7500 ft.; Joshua Tree Wd., Pinyon-Juniper Wd.; Mid Hills near New York Mts., Clark Mts., mts. about Death V., Argus Mts., White Mts.; to Wyo., Tex. May–June.

3. **H. rígida** (Thurb.) Benth. ex Scribn. [*Pleuraphis r.* Thurb.] Plate 99, Fig. D. Forming large open clumps with woody rhizomes; culms rigid, 5–8 dm. tall, leafy; sheaths often glabrate; blades spreading, stiff, involute, 2–6 cm. long; spikes peduncled, 4–8 cm. long; spikelets ca. 8 mm. long, woolly-ciliate, several-awned; lemma 3-nerved, villous on back, enclosing the palea and a rudimentary 2d floret in the cent. spikelet; lateral spikelets with a 2d floret like the 1st.—Common in sandy places below 4000 ft.; Creosote Bush Scrub, Joshua Tree Wd.; both deserts; to Utah, Ariz., Son. Feb.–June.

46. Hólcus L. VELVETGRASS

Perennials with flat blades and contracted panicles. Spikelets 2-fld., the pedicel disarticulating below the glumes, the rachilla curved and ± elongate below the 1st floret, not prolonged beyond the 2d. Glumes subequal, longer than the 2 florets; 1st floret perfect, the lemma awnless; 2d ♂, the lemma with a short awn on the back. Ca. 8 spp., native to Eu. and Afr. (Old Latin name for a grain.)

1. **H. lanàtus** L. [*Notholcus l.* Nash.] Grayish, velvety-pubescent, erect, 3–10 (–20) dm. tall; blades 4–8 mm. wide; panicle 8–15 cm. long, contracted, pale, tinged with purple; spikelets 4 mm. long; glumes villous, the 2d broader than the 1st, 3-nerved; lemmas smooth, shining, the 2d with a hooked awn; $n=7$ (Avdulov, 1928).—Abundantly escaped at 4000–7000 ft.; mts. from San Diego Co. n.; infrequent at lower elevs., as near Santa Barbara; more common in n. Calif.; native of Eu. June–Aug.

47. Hórdeum L. BARLEY

Annual and perennial grasses with flat blades and dense bristly spikes, disarticulating at the base of the rachis segm., this remaining as a stipe below the attached 3 spikelets. Spikelets 1-fld., 3 at each node, one or all of the spikelets fertile and according to the number that are fertile, producing spikes with 6, 4 or 2 rows. Glumes very narrow, like an invol. subtending the 3 spikelets. Lemmas awned, rounded on the back. Ca. 25 spp., of temp. regions. (Ancient Latin name for barley.)

(Covas, G. Observations on the N. Am. ssp. of Hordeum. Madroño 10: 1–21. 1949.)
1. Plants perennial.
   2. Glumes and awns 1.8–2.8 cm. long .... 7. *H. jubatum*
   2. Glumes and awns less than 1.8 cm. long.
      3. Blades pubescent, 1.5–5 mm. wide; anthers usually 1.5–3 mm. long; glumes of cent. spikelet 1.5–2.5 times as long as palea .... 3. *H. californicum*
      3. Blades glabrous, 3–9 mm. wide; anthers mostly 1–1.5 mm. long; glumes of cent. spikelet often scarcely longer than palea .... 2. *H. brachyantherum*
1. Plants annual.
   4. Glumes of cent. spikelet and the inner ones of the lateral spikelets with ciliate margins.
      5. Spike with 6–8 spikelets to each cm. of the rachis; stamens of cent. florets included at anthesis .... 6. *H. glaucum*
      5. Spike with 3–5 spikelets to each cm. of rachis; stamens of cent. florets exserted at anthesis
                                                                                                                        8. *H. leporinum*
   4. Glumes not ciliate.
      6. Auricles long; rachis continuous; all 3 spikelets sessile, fertile .... 10. *H. vulgare*
      6. Auricles obsolete; rachis articulate; lateral spikelets pedicellate, usually not fertile.
         7. Inner glumes of lateral spikelets strongly broadened, 0.6–1.8 mm. wide . 9. *H. pusillum*
         7. Inner glumes and outer linear-subulate, less than 0.6 mm. wide.
            8. Spike ovate to ovate-oblong, usually less than 5 cm. long; awns and glumes strongly spreading at maturity; bases of glumes of lateral spikelets prominent above the pedicel
                                                                                                     5. *H. geniculatum*
            8. Spike linear-oblong, usually over 5 cm. long; awns and glumes suberect; bases of glumes of lateral spikelets not prominent above the pedicel.
               9. Cent. spikelet 13–22 mm. long, including awn; pedicels of lateral spikelets almost straight; lateral florets with acute but awnless lemmas .... 4. *H. depressum*
               9. Cent. spikelet 26–32 mm. long, including awn; pedicels of lateral spikelets curved; lateral florets with acuminate very short awned apex .... 1. *H. arizonicum*

1. **H. arizónicum** Covas. [*H. adscendens* Hitchc. not HBK.] Annual, 2–6 dm. tall; lower sheaths pubescent, upper glabrous; blades 3–5 mm. wide, sparsely pubescent; spike erect, 3–12 cm. long; floret of cent. spikelets 8–9 mm. long, the awn 15–22 mm. long, the glumes slightly shorter; glumes of lateral florets nearly as long, one slightly dilated, the floret reduced to a short-awned lemma; $n=21$? (Covas, 1949).—Irrigated places, Imperial V.; to Ariz.

2. **H. brachyántherum** Nevski. [*H. nodosum*, Calif. refs., in part.] Plate 99, Fig. E. Tufted perennial, 2–7 dm. tall; blades 3–8 mm. wide, mostly glabrous; spike 8–10 cm. long, sometimes purplish; floret of cent. spikelet 7–10 mm. long, the awn ca. 1 cm., the glumes slightly shorter; glumes of lateral spikelets usually unequal, somewhat shorter, the floret ♂ to reduced and neuter, with awn 2–5 mm. long; $n=14$ (Covas & Stebbins, 1949).—Moist places, below 11,000 ft.; many Plant Communities; cismontane and montane Calif.; to Alaska, Labrador, New Mex. May–Aug.

3. **H. califórnicum** Covas & Steb. [*H. nodosum*, Calif. refs., in part.] Perennial, 2–6.5 dm. tall; blades 1.5–5 mm. wide, usually pubescent above and beneath; spike linear-oblong, green or purplish, 2.5–8 cm. long; rachis articulate; cent. spikelet sessile, 18–22 mm. long, including awns; glumes setaceous, scabrous, 8–17 mm. long; lemma usually glabrous, scabrous toward apex, tapering into awn 7–15 mm. long; palea 5.5–9.5 mm. long, acuminate; lateral spikelets pedicellate; glumes setaceous, floret mostly neuter; lemmas commonly subulate; $n=7$ (Covas & Stebbins, 1949).—Grassy and brushy places below 8500 ft.; many Plant Communities, cismontane and montane s. Calif.; to Ore. Anacapa, Santa Rosa, San Miguel ids. April–Aug.

4. **H. depréssum** (Scribn. & Sm.) Rydb. [*H. nodosum* var. *d.* Scribn. & Sm.] Annual, the culms geniculate at base, 0.5–4.5 dm. high; upper sheaths often inflated; blade pubescent, 2–4 mm. wide; spike erect, 4–7 cm. long; fl. of cent. spikelet 7–8 mm. long, subterete, the awn ca. 10 mm. long; awns of glumes and glumes of lateral spikelets sub-

equal, ca. 2 cm. long; $n=14$ (Covas, 1949).—Moist, ± alkaline places, below 5000 ft.; many Plant Communities; to B.C. April–May.

5. **H. geniculàtum** Allioni. [*H. hystrix* Roth. *H. gussoneanum* Parl.] Annual, the culms spreading at base, 1.5–4 dm. long; sheaths and blades ± pubescent; spike erect, 1.5–3 cm. long, the rachis usually not breaking easily; glumes setaceous, rigid, ca. 12 mm. long; lemma of cent. spikelet 5 mm. long, with a somewhat longer awn than the glumes; fls. of lateral spikelets reduced, short-awned; $n=7$ (Covas, 1949).—Alkaline or waste places, at scattered locations at low elevs.; to B.C., Atlantic Coast; native of Eu. April–June. Santa Catalina, San Clemente, Santa Cruz and Santa Rosa ids.

6. **H. gláucum** Steud. [*H. stebbinsii* Covas. *H. murinum* Calif. refs. in part.] Annual, 1–5 dm. tall; lvs. glaucous, with smooth sheaths; blades mostly sparsely pubescent, 2.5–7 mm. wide, auricled at base; spike ovate-oblong, 4–9 cm. long, dense; cent. spikelet 16–36 mm. long (including awn), sessile; glumes linear-lanceolate, 3-nerved, 12–22 mm. long, ciliate on both margins; floret pedicelled; lemma glabrous, the awn 7–25 mm. long; lateral spikelets on slender pedicels, ciliate inside, the florets usually neuter; $n=7$ (Covas, 1949).—Weed in waste places and fields below 5000 ft.; many Plant Communities; s. Calif. to Wash., Okla. Native of Eu. April–May. San Clemente, Santa Catalina, Anacapa, Santa Cruz, and San Miguel ids.

7. **H. jubàtum** L. FOXTAIL. Tufted perennial, 3–6 dm. tall; blades 2–5 mm. wide, scabrous; spike nodding, 5–10 cm. long, ca. as wide, pale, soft; lateral spikelets reduced to 1–3 spreading awns; glumes of perfect spikelet awnlike, 2.5–6 cm. long, spreading; lemma 6–8 mm. long, with awn equaling glumes; $n=14$ (Stebbins & Love, 1941).—Common in open moist and waste places, often a bad weed in ± alkaline pastures and meadows below 10,000 ft.; many Plant Communities; here and there throughout the state; to Alaska, Atlantic Coast, Mex. May–July.

8. **H. leporìnum** Link. [*H. murinum* Calif. refs. in part.] Spreading annual; sheaths glabrous; blades pilose to glabrous, with well developed auricle; spike 5–9 cm. long, often partly closed by the inflated uppermost sheath; glumes of cent. spikelet lanceolate, 3-nerved, long-ciliate on both margins, the nerves scabrous, the awn 2–2.5 cm. long; lateral spikelets usually ♂, the glumes much shorter, unlike; lemma broad, 10–20 mm. long, the awn 2–4 cm. long; floret 1–1.2 cm. long, the awn 2–4 cm. long; $n=14$ (Covas, 1949).—Weed in waste places, fields, etc. below 5000 ft.; cismontane and Death V. San Clemente Id., Santa Catalina Id., Santa Rosa Id. Native of Eu. April–June.

9. **H. pusíllum** Nutt. Annual, 1–3.5 dm. tall; blades erect, flat; spike erect, 2–7 cm. long; 1st glume of lateral spikelets and both glumes of cent. dilated above the base, with a slender awn 8–15 mm. long; lemma of cent. spikelet awned, smooth, of lateral spikelets awn-pointed; $n=7$ (Kihara, 1924), $2n=14$ (Rajhatky & Symko, 1966).—In open ± alkaline places below 5000 ft.; cismontane s. Calif. to Wash., Atlantic Coast, Argentina. April–May. San Clemente, Santa Catalina, and Santa Cruz ids.

10. **H. vulgàre** L. COMMON BARLEY. Erect annual, 6–12 dm. tall; blades 5–15 mm. wide; spike 2–10 cm. long excluding awns; the 3 spikelets sessile; glumes divergent at base, narrow, nerveless, stout-awned; awn of lemma straight, erect, mostly 10–15 cm. long; $n=7$ (Kihara, 1924).—Sometimes found in waste places and fields as an escape from cult.; native of Old World. BEARDLESS BARLEY (*H. v.* var. *trifurcatum* [Schlecht.] Alef) is also met. April–July.

48. *Hyparrhènia* Anderss. ex Stapf

Tall perennials, the pairs of racemes and their spathes ± crowded, forming a large elongate infl. Spikelets in pairs, those of the lower pairs alike, sterile and awnless; fertile spikelets 1–few in each raceme, terete or flattened on the back, the base usually elongate into a sharp callus. Fertile lemma with a strong geniculate awn. Ca. 70 spp. of the Old World tropics.

1. **H. hírta** (L.) Stapf. [*Andropogon h.* L.] To ca. 1 m. tall; blades to 3 mm. wide, ± involute, flexuous; racemes whitish or grayish, silky-villous.—Sometimes in cult. and reported from Los Angeles, *Raven*.

49. Imperàta Cyrill.

Erect rather coarse perennials with leafy stems and narrow silky panicles. Spikelets all alike, paired, awnless, usually pedicellate on a slender continuous rachis, surrounded by long silky hairs. Glumes subequal, membranaceous. Sterile lemma, fertile lemma and palea thin and hyaline. Ca. 7 spp. of warm regions. (F. *Imperato*, Italian naturalist.)

1. **I. brevifòlia** Vasey. [*I. hookeri* Rupr.] SATINTAIL. Culms 1–1.5 m. tall from scaly rhizomes; ligule villous; lvs. 1–5 dm. long, 8–15 mm. wide, glabrous; uppermost lvs. reduced; panicle spicate, 1.5–2.5 dm. long, somewhat tawny or pinkish, soft-silky; spikelets 3 mm. long, with hairs twice as long.—Rare, moist places; Chaparral, Coastal Sage Scrub, Creosote Bush Scrub, etc.; scattered locations in s. Calif.; to Utah, Mex. Sept.–May.

50. Koelèria Pers.

Tufted perennials or annuals with narrow blades and long shining spikelike panicles. Spikelets 2–4-fld., compressed, the rachilla disarticulating above the glumes and between the florets. Glumes mostly subequal in length, dissimilar in shape, the 1st narrow, 1-nerved, the 2d wider, broader above the middle, 3–5-nerved. Lemmas ± scarious, shining, the lowermost slightly longer than the glumes, faintly 5-nerved, pointed, the awn if present from just below the apex. Ca. 20 spp. of temp. regions. (Named for G. L. *Koeler*, an early grass student.)

1. **K. macrántha** (Ledeb.) Spreng. [*K. cristata* Am. auth.] JUNEGRASS. Culms erect, 3–6 dm. tall, leafy at base; sheaths pubescent, at least the lower; blades flat or involute, 1–3 mm. wide, glabrous or especially the lower pubescent; panicle erect, dense, often lobed or interrupted below, 4–14 cm. long; spikelets mostly 4–5 mm. long; glumes and lemmas scaberulous, 3–4 mm. long; $n=14$ (Stebbins & Love, 1941).—Dryish open places below 11,500 ft.; many Plant Communities, cismontane and montane and mts. of Mojave Desert; to B.C., Atlantic Coast; Old World. Santa Cruz and Santa Rosa ids. May–July.

51. Lamárckia Moench. GOLDENTOP

Low annual, erect, with flat blades and oblong, 1-sided compact panicles of crowded fascicled spikelets, the fertile being hidden (except the awns) by the many sterile ones; fascicles falling entire. Fertile spikelet terminal, 1-fld., the floret on a slender stipe, a rudimentary floret on a long rachilla-joint; both awned, with narrow acuminate or short-awned glumes; lemma broader, scarcely nerved, with a delicate awn just below the apex. Sterile spikelets linear, 1–3 in a fascicle; glumes 2, narrow, 1-nerved; lemmas many, empty, awnless, obtuse, a reduced spikelet borne on the pedicel with each sterile one. One sp. of s. Eu. (J. B. *Lamarck*, French naturalist.)

1. **L. aúrea** (L.) Moench. [*Cynosurus a.* L. *Achyrodes a.* Kuntze.] Erect, or ± decumbent at very base, 1–3.5 dm. tall; lvs. smooth, ligule prominent; panicle 2–7 cm. long, 1–2 cm. wide, shining, golden to purplish; pedicels fascicled, pubescent, drooping or spreading; fertile spikelet ca. 2 mm. long, the sterile 6–8 mm. long; glumes hyaline, 2 mm. long; $n=7$ (Avdulov, 1931).—Common weed in waste places at low altitudes, cismontane including islands; native of Medit. region. Feb.–May.

52. Leérsia Sw. RICE CUTGRASS

Perennial grasses, usually with creeping rhizomes. Blades flat, scabrous. Panicles mostly open. Spikelets 1-fld., strongly laterally compressed, disarticulating from the pedicel; glumes none. Lemma chartaceous, broad, oblong to oval, boat-shaped, usually 5-nerved, the lateral pair of nerves near the margins, often hispid-ciliate; palea as long as lemma, narrower, usually 3-nerved; stamens 6 or fewer. Ca. 10 spp. of trop. and temp. N. Am.; 1 sp. in Eurasia. (J. D. *Leers*, 1727–1774, German botanist.)

1. **L. oryzoìdes** (L.) Sw. [*Phalaris o.* L. *Homalocenchrus o.* Poll.] Ascending or sprawling, the culms 1–1.5 m. long, terete, from slender elongate rhizomes; lvs. scabrous, 8–10 mm.

wide; panicles terminal and axillary, 1–2 dm. long, with flexuous finally spreading branches; spikelets oblong, 4–6 mm. long, 1.5–2 mm. wide; sparsely hispidulous, with bristly ciliate keels; axillary panicle reduced, partly included in the sheaths and with cleistogamous spikelets; $n=24$ (Ramanujam, 1938).—Marshes, stream-banks, etc., about Riverside and San Bernardino; Inyo Co. to B.C., e. U.S.; Eu. Aug.–Oct.

53. Leptochlòa Beauv. SPRANGLETOP

Annuals or perennials with flat lvs. and simple elongate spikes or racemes borne on a common axis forming a long or sometimes shorter panicle. Spikelets 2- to several-fld., sessile or short-pediceled, along 1 side of a slender rachis; rachilla disarticulating above the glumes and between the florets. Glumes subequal to unequal, awnless or mucronate, 1-nerved, usually shorter than 1st lemma. Lemmas obtuse or acute, sometimes 2-toothed and mucronate or short-awned from between the teeth, 3-nerved, the nerves sometimes pubescent. Ca. 70 spp. of warmer regions. (Greek, *leptos*, slender, and *chloa*, grass, because of the slender spikes.)

Sheaths papillose-hispid; glumes exceeding 1st floret .... 2. *L. filiformis*
Sheaths smooth; glumes shorter than 1st floret.
    Lemmas awned; spikelets 7–12 mm. long .... 1. *L. fascicularis*
    Lemmas awnless; spikelets 5–7 mm. long .... 3. *L. uninervia*

1. **L. fasciculàris** (Lam.) Gray. [*Festuca f.* Lam.] Smooth annual, 3–10 dm. tall; blades flat to loosely involute; panicles ± included, mostly 1–2 dm. long, with numerous racemes 7–12 cm. long; spikelets 3–4 mm. long, the glumes much shorter than the florets; lemmas short-awned; $2n=40$ (Gould, 1966).—Usually in moist sometimes alkaline places, Creosote Bush Scrub, Valley Grassland, Coastal Sage Scrub, etc.; Imperial V., Death V., cismontane s. Calif.; to Wash., Atlantic Coast. June–Oct.

2. **L. filifórmis** (Lam.) Beauv. [*Festuca f.* Lam.] Annual, often ± reddish or purple, 3–10 dm. high; sheaths sparsely papillose-hairy; spikes 20–40, each 3–15 cm. long, spreading-ascending; spikelets 3 mm. long; glumes mucronate, nearly equaling the 3 or 4 small awnless florets; spikelets 1–2 mm. long; $n=10$ (Brown, 1950).—Moist depressions; Imperial and Coachella valleys; to Atlantic Coast, trop. Am. July–Sept.

3. **L. uninérvia** (Presl) Hitchc. & Chase. [*Megastachya u.* Presl.] Resembling *L. fascicularis*; panicle more dense; glumes obtuse; lemmas not awned, merely apiculate; $n=10$ (Gould, 1958).—Moist ± alkaline places, many Plant Communities; desert and cismontane, mostly below 2000 ft., occasional to 7000; to Ore., Atlantic Coast, S. Am. March–Dec.

54. Lòlium L. RYEGRASS. DARNEL

Annuals and short-lived perennials with simple erect stems and flat blades. Spikes terminal, flat. Spikelets several-fld., solitary, sessile, placed edgewise to the continuous rachis, one edge fitting the alternate notches; rachilla disarticulating above the glumes and between the florets. First glume wanting except on the terminal spikelet, the 2d outward, 3–5-nerved, equaling or exceeding the 2d floret. Lemmas convex, 5–7-nerved, obtuse, acute or awned. Ca. 10 spp. in temp. parts of Eurasia. (Ancient Latin name.)

Glume equaling or exceeding the spikelet .... 2. *L. temulentum*
Glume shorter than the spikelet.
    Perennial; lvs. seldom exceeding 3 mm. in width; lemma awnless .... 1. *L. perenne*
    Annual or biennial; lvs. up to 10 mm. wide; lemma usually awned .. 1. *L. perenne multiflorum*

1. **L. perénne** L. ssp. **multiflòrum** (Lam.) Husnot. ITALIAN RYEGRASS. [*L. m.* Lam.] Plate 99, Fig. F. Annual or biennial, with lvs. inrolled when young; infl. to 20 cm. long; spikelets 10–20 mm. long, of 8–14 florets; lemma broader and softer than in ssp. *perenne*, usually awned; $2n=14$ (Malik & Thomas, 1966).—Adventive and widely planted as a lawn grass, after burns, etc. Apparently with many strains and horticultural forms and

intergrading with ssp. *perenne*. Vasek & Ferguson (Madroño 17: 79–82. 1963) studying awned and awnless plants in various populations, found separation into two spp. by the traditional characters was impossible. Clapham, Tutin & Warburg, Fl. Brit. Isles, ed. 2, p. 1130, 1962, use the treatment which I here follow.

L. perénne L. ssp. **perénne**. ENGLISH RYEGRASS. A wiry perennial 3–5 dm. tall; lvs. folded when young, up to 3 mm. wide; sheaths smooth; infl. 8–15 cm. long; spikelets 5–15 mm. long, mostly of 8–11 florets; glumes of terminal spikelet subequal; upper glume of lateral spikelets usually equaling 1st floret, sometimes longer, but always shorter than spikelet; lemma linear-lanceolate, with hyaline margins in upper part; palea almost equaling lemma; $2n=14$ (Malik & Thomas, 1966).—At scattered locations, escaping from lawns and persisting for a time after planting on burns, etc.; native of Eu. May–Sept.

2. **L. temuléntum** L. DARNEL. Annual, 6–9 dm. tall; blades mostly 3–8 mm. wide; spike stiff; glumes ca. 2.5 cm. long, at least as long as the 5–7-fld. spikelet; florets plump, the lemmas to 8 mm. long, obtuse, with an awn 6–12 mm. long; $n=7$ (Jenkin & Thomas, 1938).—Common in waste places and fields, cismontane; Santa Catalina and San Clemente ids. Introd. from Eu. April–June.

55. *Mélica* L. MELIC

Perennial, the culms frequently bulbous at base; sheaths closed; blades flat. Panicle simple or compound, narrow to spreading; spikelets 1–6-fld., articulation above or below the glumes; terminal floret or florets sterile, similar to the fertile, or reduced. Glumes less firm than lemmas, with papery or hyaline margins and apices, not keeled, obtuse or acute, 3–5-nerved. Lemmas firm, not keeled, with hyaline apices and upper margins, usually 7-nerved, awned or not; palea usually ¾ the lemma. Ca. 60 spp. of cooler parts of the world. (*Melica*, an Italian name for a kind of sorghum, probably from the sweet juice [*mel*, honey].)

(Boyle, W. S. A cyto-taxonomic study of the N. Am. spp. of Melica. Madroño 8: 1–26. 1945)
Articulation above the glumes and between the florets; spikelets ascending to erect.
Lemmas awned. San Bernardino Mts. .... 1. *M. aristata*
Lemmas not awned.
  Spikelets 8–15 mm. long; fertile florets 2 to several.
    Spikelets silvery white; glumes ca. as long as the spikelet; plant 1–2 m. tall, somewhat woody
      3. *M. frutescens*
    Spikelets tawny to purplish; glumes shorter than the spikelet; plant 0.6–1.3 m. high, herbaceous .... 2. *M. californica*
  Spikelets 4–6 mm. long; fertile florets 1 or rarely 2 .... 4. *M. imperfecta*
Articulation below the glumes; spikelet falling entire at maturity, reflexed ...... 5. *M. stricta*

1. **M. aristàta** Thurb. ex Bol. [*Bromelica a.* Farw.] Culms 6–12 dm. long, not bulbous at base; blades 3–6 mm. wide, often pubescent; panicle 1–2.3 dm. long, mostly narrow; spikelets 1–2 cm. long, 2–3-fld.; glumes obtuse to subacute, 7–11 and 7–12 mm. long; lemmas awned, the lowest 8–13 mm. long, the awns 5–12 mm. long; $n=9$ (Stebbins & Love, 1941).—Dry open woods, at 7000–8000 ft.; Montane Coniferous F.; San Bernardino Mts.; Sierra Nevada to Wash. June–Aug.

2. **M. califórnica** Scribn. Culms 6–13 dm. high, ± enlarged below, but not definitely bulbous, densely tufted; blades 2–5 mm. wide; ligule 2–5 mm. long; panicle 4–30 cm. long, very narrow, mostly dense, but interrupted below; spikelets 5–15 mm. long, 2–5-fld., chaffy; glumes subequal, blunt, scarcely equaling last floret; lemma obtuse, acute or emarginate, the lowest 5–9 mm. long; rudiment usually blunt or obovoid, not exserted; $n=9$ (Stebbins & Love, 1941).—Dry rocky exposed slopes below 4000 ft.; Chaparral, Foothill Wd., etc.; Ventura Co. and Tehachapi Mts. n. April–May.

3. **M. frutéscens** Scribn. Culms 1–2 m. tall, stout; blades 2–4 mm. wide; panicle 1–4 dm. long, narrow, dense, pale and shining to purple-tinged; spikelets 1.2–1.8 cm. long, 3–6-fld.; glumes papery, 7–12 and 9–15 mm. long, prominently 5-nerved; lemmas ± obtuse, papery-scarious above, the first 8–11 mm. long, 7-nerved; $n=9$ (Boyle, 1945).—

Plate 100. Fig. A, *Melica imperfecta*. Fig. B, *Monanthochloe littoralis*. Fig. C, *Muhlenbergia microsperma*, spikelet with long pedicel and awn. Fig. D, *Oryzopsis hymenoides*. Fig. E, *Panicum pacificum*. Fig. F, *Parapholis incurva*.

Dry slopes below 5000 ft.; Creosote Bush Scrub to Pinyon-Juniper Wd.; deserts; and Coastal Sage Scrub, S. Oak Wd., etc.; L. Calif. thru interior cismontane Calif. to San Luis Obispo Co.; Ariz. March–May.

4. **M. imperfécta** Trin. [*M. colpodioides* Nees. *M. panicoides* & *poaeoides* Nutt. *M. parishii* Vasey.] Plate 100, Fig. A. Culms erect, 3–11 dm. tall; blades 1–6 mm. wide; panicle 5–36 cm. long, narrow or spreading, the branches often fascicled; spikelets 4–7 mm. long, usually 1-, occasionally 2-fld.; glumes obtuse to acutish; lemmas not pubescent above, acute to obtuse, 3–7 mm. long; $n = 9$ (Boyle, 1945).—Common on dry open often rocky slopes below 6500 ft.; Coastal Sage Scrub, Chaparral, S. Oak Wd., Yellow Pine F., etc.; L. Calif. through cismontane Calif. to cent. Calif.; occasional, Creosote Bush Scrub to Pinyon-Juniper Wd.; Mojave Desert. April–May. Santa Catalina and San Clemente ids. Plants with lower panicle branches spreading or reflexed, and pubescent lf.-blades have been called var. *refracta* Thurb. in Wats.; those with similar panicles and glabrous lf.-blades, var. *flexuosa* Bol.; and those less than 3 dm. tall, glabrous and with very narrow lf.-blades, var. *minor* Scribn.

5. **M. strícta** Bol. Culms 2–5 (–8) dm. high, densely tufted, purplish near base where thickened; blades 2–5 mm. wide; panicle 3–20 cm. long, narrow, simple, with appressed branches; spikelets 6–23 mm. long, 2–5-fld.; glumes acute to emarginate, 6–16 and 6–18 mm. long; lemmas obtuse to acute, the 1st 8–16 mm. long; anthers 1–2 mm. long; $n = 9$ (Boyle, 1945).—Dry rocky places, 4000 to 11,600 ft.; Sagebrush Scrub to Subalpine F., Bristle-cone F.; San Bernardino and San Gabriel mts. to Inyo-White Mts. and Sierra Nevada; to Ore., Utah. June–Aug. Var. *albicaulis* Boyle with paler lower sheath; anthers 2–3 mm. long; glumes less acute, ranges in mts. from Ventura Co. to San Bernardino Co.

56. *Monanthochlòe* Engelm. SHOREGRASS

Spreading wiry-stemmed perennial with clusters of short subulate lvs. Plants dioecious; spikelets 3–5-fld., only the lower fertile, borne in the axils of fascicled lvs., the uppermost florets rudimentary, the rachilla disarticulating slowly in ♀ spikelets. Glumes wanting. Lemmas rounded on back, convolute, narrowed above, 3-nerved, membranous. Paleae narrow, 2-nerved. One sp. (Greek, *monos*, one, and *anthos*, fl.)

1. **M. littoràlis** Engelm. Plate 100, Fig. B. Low, extensively creeping, with short erect branches; lf.-blades 5–10 mm. long, falcate, in ± remote clusters; spikelets 1–few, scarcely evident, partly surrounded by lf.-bases.—Coastal Salt Marsh, Santa Barbara to L. Calif.; to Fla., Mex., Cuba. Santa Catalina and Santa Rosa ids. May–June.

57. *Monérma* Beauv. THINTAIL

Low annuals with slender cylindric spikes. Spikelets 1-fld., embedded in the hard articulate rachis and falling attached to the joints. First glume wanting except on the last spikelet; 2d glume even with the surface of the rachis, hardened, nerved, acuminate, longer than the rachis-joint. Lemma with its back to the rachis, hyaline, shorter than the glume, 3-nerved. Palea slightly shorter than the lemma, hyaline. Three spp. (Greek, *monos*, one, and *erma*, support, in reference to the single spike.)

1. **M. cylíndrica** (Willd.) Coss. & Dur. [*Rottboellia c.* Willd. *Lepturus c.* Trin.] Branching, spreading, the stems 1–5 dm. long; lvs. narrow; spikes 1–2 dm. long; glume 6 mm. long, acuminate; lemmas 5 mm. long; rachis disarticulating at maturity; $n = 13$ (Avdulov, 1931), $n = 26$ (Hunter, 1934).—Coastal Salt Marsh; San Diego to Colusa Co.; introd. from Eu. May–July.

58. *Muhlenbérgia* Schreb.

Perennial or annual, tufted or with scaly rhizomes, the culms simple or much branching. Lvs. flat or involute. Infl. narrow, sometimes spikelike, sometimes an open panicle. Spikelets 1-fld., the rachilla disarticulating above the glumes. Glumes usually shorter than lemma, obtuse to acuminate or awned, keeled or convex on back, the 1st sometimes obsolete. Lemma firm-membranaceous, 3-nerved, with a very short callus, rarely long-pilose, usually minutely so, the apex acute, awned from the tip or just below it, or from between the short lobes, sometimes mucronate. Ca. 125 spp., mostly in Mex. and sw. U.S. (Named for G. H. E. *Muhlenberg*, 1753–1915, Am. botanist.)

1. Plants annual.
   2. Lemma awned; panicle open, with spreading branches .... 6. *M. microsperma*
   2. Lemma awnless.
      3. Pedicels capillary, divergent, 3–10 mm. long; glumes pubescent ..... 7. *M. minutissima*
      3. Pedicels appressed, to ca. 1 mm. long; glumes glabrous .... 5. *M. filiformis*
1. Plants perennial.
   4. Rhizomes present, usually prominent, scaly, creeping, often branching.
      5. Lf.-blades 0.5–2 mm. wide, mostly short and involute.
         6. Panicles open, the pedicels capillary, 2–15 mm. long .... 3. *M. asperifolia*
         6. Panicles narrow, ± condensed, the pedicels stouter, appressed, shorter.
            7. Lemma and palea glabrous.
               8. Culms smooth, widely creeping, the blades fine, recurved-spreading; ligules ca. 1 mm. broad .... 11. *M. utilis*
               8. Culms nodulose-roughened, erect or decumbent at base, sometimes spreading but not widely creeping; blades ascending; ligules 2–3 mm. long.
                  9. Plants mostly 1–5 dm. tall; panicle 2–10 cm. long .... 9. *M. richardsonis*
                  9. Plants 0.5–1.5 dm. tall; panicle 1–3 cm. long .... 5. *M. filiformis*
            7. Lemma and palea pilose or villous on lower half .... 2. *M. arsenei*
      5. Lf.-blades flat, at least some of them 3–5 or more mm. wide.
         10. Hairs at base of floret copious, as long as the lemma-body .... 1. *M. andina*
         10. Hairs at base of floret inconspicuous, not more than half as long as lemma
                                                                4. *M. californica*
   4. Rhizomes not developed, the culms tufted.
      11. Panicles 5–10 cm. long, open with spreading branches .... 8. *M. porteri*
      11. Panicles 25–50 cm. long, slender, mostly spicate .... 10. *M. rigens*

1. **M. andina** (Nutt.) Hitchc. [*Calamagrostis a.* Nutt. *M. comata* Thurb.] Perennial with numerous scaly rhizomes; culms erect or ± spreading, 5–10 dm. long, pubescent at the nodes; sheaths smooth or scaberulous; ligule 1 mm. long, short-ciliate; blades flat, 2–6 mm. wide, scabrous; panicle narrow, spicate, ± interrupted, 7–15 cm. long, grayish or purplish; glumes narrow, 1-nerved, 3–4 mm. long; lemma 3 mm. long, the awn 4–8 mm. long, the basal hairs almost as long as lemma-body; $n = 10$ (Brittonia 17: 108. 1965).—Open flats, 6500–10,000 ft.; Montane Coniferous F.; San Bernardino Mts., Panamint Mts.; Sierra Nevada; to B.C., Atlantic Coast. July–Sept.

2. **M. arsènei** Hitchc. Tufted perennial, 1–4 dm. high with wiry culms; lvs. near the base, the blades slender, involute, 1–3 cm. long; panicle narrow, rather loose, purplish, 2–10 cm. long, with ascending branches; spikelets 4–5 mm. long without the glumes, these shorter, acute, awnless; lemma with a flexuous awn 6–10 mm. long.—Dry limestone slopes, 5000–6000 ft.; Pinyon-Juniper Wd.; Clark Mts., e. Mojave Desert; to Utah, New Mex. Aug.–Sept.

3. **M. asperifòlia** (Nees & Mey.) Parodi. [*Vilfa a.* Nees. & Mey. *Sporobolus a.* Nees & Mey.] SCRATCHGRASS. Pale or glaucous perennial with creeping rhizomes; culms 3–6 dm. long, ascending; sheaths smooth; ligule minute, erose-toothed; blades 2–5 cm. long, 2 mm. wide, scabrous; panicles diffuse, 1–1.5 dm. long and ca. as wide, the branches scabrous; spikelets 1.5 mm. long; glumes slightly unequal, almost as long as spikelets; lemma minutely mucronate; $n = 10$ (Pohl & Mitchell, 1965).—Occasional in wet muddy, often subalkaline places below 7000 ft.; many Plant Communities; s. and interior cismontane and desert Calif.; to B.C., Atlantic Coast, s. S. Am. July–Oct.

4. **M. califórnica** Vasey. Pale leafy perennial, ± creeping at base; culms ascending, ± woody below, 3–6 dm. long; sheaths scaberulous; blades flat, 3–6 mm. wide, scabrous, 5–15 cm. long; panicle narrow, dense, interrupted, 7–15 cm. long; glumes narrow, 3–4 mm. long, pointed; lemma 3 mm. long, with an awn 1–2 mm. long; $n = 40$ (Brittonia 17: 110. 1965).—Occasional in wet places up to 7000 ft.; Coastal Sage Scrub, Chaparral,

Yellow Pine F., cismontane especially about San Bernardino V., to edge of desert. July–Sept.

5. **M. filifórmis** (Thurb.) Rydb. [*Vilfa depauperata f.* Thurb. *V. gracillima* Thurb.] Apparently annual or perennial, loosely tufted, with fibrous roots and somewhat spreading base; culms filiform, 0.5–2 (–3) dm. high; ligule ca. 2 mm. long; blades flat, 1–3 cm. long; panicles narrow, interrupted, 1–3 cm. long, few-fld.; glumes ovate, 1 mm. long, abruptly narrowed at apex; lemma lanceolate, acute, 2 mm. long, minutely pubescent.— Open moist places, mostly 5000–11,000 ft.; Montane Coniferous F.; San Jacinto and San Bernardino mts.; Sierra Nevada to B.C., S. Dak., New Mex. June–Aug.

6. **M. microspérma** (DC.) Kunth. [*Trichochloa m.* DC. *M. purpurea* Nutt.] Plate 100, Fig. C. Branching usually purplish annual with spreading culms 1–3 dm. long; sheaths smooth or scaberulous; ligule 1 mm. long; blades 2–5 cm. long, flat, 1 mm. wide; panicles narrow, lax, 2–10 cm. long; glumes ovate, 1-nerved, unequal, the 2d longer, 1 mm. long; lemma narrow, 3-nerved, 3 mm. long; awn capillary, 10–15 mm. long.—Common in dry open ± disturbed places at low elevs.; V. Grassland, Coastal Sage Scrub, Creosote Bush Scrub, etc.; L. Calif. to cent. Calif.; Nev., Ariz. San Clemente, Santa Catalina, Anacapa, Santa Rosa and Santa Cruz ids. March–May. Cleistogamous spikelets develop at the base of the lower sheaths.

7. **M. minutíssima** (Steud.) Swall. [*Agrostis m.* Steud. *M. confusa* Calif. refs. *Sporobolus microspermus* Calif. refs.] Erect to spreading annual, 5–30 cm. long; blades flat, ca. 1 mm. wide; panicle $\frac{1}{2}$–$\frac{3}{4}$ the entire plant, the pedicels slender, ascending; spikelets 1.2–1.5 mm. long, the glumes $\frac{1}{2}$–$\frac{2}{3}$ as long, minutely pilose; lemmas minutely silky-pubescent along the edges and midvein; $2n = 60$ (Gould, 1966).—Moist sandy places, 4000–7500 ft.; Pinyon-Juniper Wd., Sagebrush Scrub, Yellow Pine F.; San Jacinto and San Bernardino mts.; Sierra Nevada, etc. to Wash., Mont., Tex., Mex. July–Oct.

8. **M. pórteri** Scribn. Perennial, with culms ± woody at base, widely spreading or ascending through bushes, wiry, freely branched, 3–8 dm. long; sheaths smooth, spreading away from the branches; blades mostly ca. 1 mm. wide, flat, 2–8 cm. long, early deciduous; panicle open, 5–10 cm. long, the branches slender, widely spreading, few-fld.; glumes narrow, acuminate, 2 mm. long; lemma pilose, 3–4 mm. long, with awn 6–12 mm. long.—Dry brushy slopes, 3000–5000 ft.; Shadscale Scrub, Joshua Tree Wd., w. Colo. Desert, Mojave Desert from Twentynine Palms e.; to w. Tex., Colo. June–Oct.

9. **M. richardsònis** (Trin.) Rydb. [*Vilfa r.* Trin. *V. squarrosa* Trin. *M. s.* Rydb.] Perennial with creeping rhizomes, the culms wiry, 1–5 dm. long, smooth; sheaths smooth; ligules 1–2 mm. long; blades 2–5 cm. long, flat or involute; panicle narrow, interrupted, or sometimes rather close and spikelike, 2–10 cm. long; spikelets 2–3 mm. long, the glumes ca. half as long; lemmas lanceolate, mucronate, smooth; $n = 20$ (Stebbins & Love, 1941).—Dry or open moist ground, 5000–11,000 ft.; Santa Rosa, San Jacinto, San Bernardino, San Gabriel, Mt. Pinos and White mts.; to Wash., Atlantic Coast., L. Calif. June–Aug. The lower more rhizomatous form, more decumbent and with smaller fls. is *M. squarròsa* and the taller more slender and erect form is *M. richardsonis*.

10. **M. rìgens** (Benth.) Hitchc. [*Epicampes r.* Bent.] Culms tufted, erect, 7–15 dm. tall, rather slender; sheaths smooth or ± scabrous, covering the nodes; blades scabrous, 1.5–2.5 dm. long, involute, with long slender point; panicle slender, mostly spicate, 2–6 dm. long; glumes 2–3 mm. long, scarcely keeled, acute to obtuse; lemma scaberulous, sparsely pilose below, slightly exceeding glumes, 3-nerved upward, awnless; $n = 20$ (Stebbins & Love, 1941).—Dry or damp places below 7000 ft.; V. Grassland, Chaparral, Yellow Pine F.; L. Calif. to San Gabriel and San Bernardino mts. and Little San Bernardino Mts.; to n. Calif. June–Sept.

11. **M. ùtilis** (Torr.) Hitchc. [*Vilfa u.* Torr.] Perennial with widely creeping rhizomes; culms decumbent, widely spreading, with fine lvs., the blades less than 1 mm. wide, soon involute, ca. 2–3 cm. long; panicle narrow, ca. 1–4 cm. long, interrupted; spikelets ca. 2 mm. long, the glumes scarcely half as long; lemma pale, apiculate; $2n = 20$ (Gould, 1966).—Occasional in wet places where it forms mats; Coastal Sage Scrub, Chaparral,

etc.; w. Riverside and w. San Bernardino cos., Ventura Co., Inyo Co.; to Nev., Mex. Oct.–March.

59. Orcúttia Vasey

Low annuals with short blades and with spikes or spicate racemes. Spikelets rather large, subsessile, several-fld., the upper florets reduced, the rachilla continuous. Glumes subequal, shorter than lemmas, broad, irregularly 2–5-toothed, many-nerved, the nerves extending into teeth. Lemmas firm, prominently 13- or 15-nerved, the broad summit toothed; palea broad, as long as lemma. Five spp. of Calif. and L. Calif. (Named for C. R. *Orcutt*, 1864–1929, California collector.)

(Hoover, R. F. The genus Orcuttia. Bull. Torrey Bot. Club 68: 149–156. 1941.)

1. **O. califórnica** Vasey. Sparingly to moderately pilose, the culms 5–15 cm. long; sheaths loose; blades 2–4 cm. long; raceme loose below, dense upward, 2–5 cm. long; spikelets alternate on opposite sides of the axis, 8–12 mm. long, pilose; glumes sharply toothed, 2–4 mm. long; lemmas 4–5 mm. long, deeply toothed at tip.—Drying mud flats; V. Grassland; s. Western Ave., Los Angeles; Murrieta Hot Springs, w. Riverside Co.; n. L. Calif. May–June.

60. Oryzópsis Michx. RICEGRASS

Cespitose perennials, with flat or involute blades and narrow or open panicles. Spikelets 1-fld., disarticulating above the glumes. Glumes subequal, obtuse to acuminate. Lemma indurate, ca. as long as glumes, broad, subterete, usually pubescent, with a short blunt callus and a short deciduous sometimes bent and twisted awn; palea enclosed by the edges of the lemma. Ca. 20 spp., N. Temp. (Greek, *oruza*, rice, and *opsis*, appearance.)

Lemmas smooth, rarely pubescent.
    Blades flat, 5–10 mm. wide; spikelets ca. 3 mm. long .... 3. *O. miliacea*
    Blades ± involute, less than 2 mm. wide; spikelets 3–4 mm. long .... 2. *O. micrantha*
Lemmas pubescent.
    Branches of panicle and capillary pedicels divaricately spreading .... 1. *O. hymenoides*
    Branches of panicle erect or appressed-ascending.
        Awn 12 mm. long; culms 3–6 dm. tall .... See 1. *O. bloomeri*
        Awn 6 mm. or less long; culms 1.5–3 dm. tall .... 4. *O. webberi*

1. **O. hymenoìdes** (R. & S.) Ricker. [*Stipa h.* R. & S. *O. cuspidata* Benth. *O. membranacea* Vasey.] Plate 100, Fig. D. Cespitose, 3–6 dm. tall; sheaths smooth or scaberulous; ligule 5–6 mm. long; blades slender, involute, almost as long as culms; panicle 8–15 cm. long, with slender spreading dichotomous branches and capillary pedicels; glumes ca. 6–7 mm. long, puberulent, 3-nerved, awn-pointed; lemma turgid, 3 mm. long, densely long-pilose with white hairs 3 mm. long, the awn ca. 4 mm. long, readily deciduous; $n=24$ (Johnson & Rogler, 1943).—Common in dry sandy places below 11,000 ft.; Creosote Bush Scrub to Bristle-cone Pine F., and Lodgepole F.; both deserts; n. to B.C., Manitoba, Tex., n. Mex. April–July. Apparently hybridizing with a number of spp. of *Stipa*, some of the hybrids having ± appressed branches in the infl. and awns to 12 mm. long; they vary in the indumentum of the awn from scabrous to plumose and of the sheath from glabrous to villous; they are mostly quite sterile. One of them was described as *O. bloomeri* (Bol.) Ricker from Mono Co. and has been shown by Johnson to have as one probable parent *Stipa occidentalis*; others have *S. elmeri*, *S. thurberiana*, or *S. californica*.

2. **O. micrántha** (Trin. & Rupr.) Thurb. [*Urachne m.* Trin. & Rupr.] Densely tufted, slender, erect, 3–7 dm. tall; ligule ca. 1 mm. long; blades flat or involute, 0.5–2 mm. wide; panicle open, 10–15 cm. long, the branches distant, 2–5 cm. long; spreading to reflexed, with appressed spikelets toward the ends; glumes 3–4 mm. long; lemma elliptic, mostly glabrous, 2 2.5 mm. long, yellow or brown, with a straight awn 5–10 mm. long; $n=11$ (Johnson, 1945).—Dry limestone crevices, etc., 6000–8800 ft.; Pinyon-Juniper Wd.; Clark, Kingston and White mts., e. Mojave Desert; to Sask., N. Dak., New Mex. June–Sept.

3. **O. miliàcea** (L.) Benth. [*Agrostis m.* L.] Culms erect from a decumbent base, 6–15 dm. long; ligule ca. 2 mm. long; blades flat, 5–10 mm. wide; panicle 1.5–3 dm. long, loose, the branches spreading with numerous short-pediceled spikelets beyond the middle; glumes acuminate, 3 mm. long; lemmas smooth, 2 mm. long, with a straight awn ca. 4 mm. long; $n=12$ (Avdulov, 1928).—In widely separated waste places at low elevs., San Diego Co. n.; introd. from Medit. region. April–Sept.

4. **O. wébberi** (Thurb.) Benth. ex Vasey. [*Eriocoma w.* Thurb.] Erect, cespitose, densely tufted, 1.5–3 dm. high; blades filiform, involute, scabrous; panicle narrow, 2.5–5 cm. long, with appressed branches; glumes ca. 8 mm. long, narrow; lemma 6 mm. long, densely long-pilose, the awn up to 6 mm. long, not twisted.—Rocky slopes at 5000–10,000 ft.; Pinyon-Juniper Wd. to Bristle-cone Pine F.; Inyo-White Mts.; to Ida., Colo., Nev. June–July.

61. *Pánicum* L.

Annual or perennial grasses of various habit. Spikelets ± compressed dorsi-ventrally, in open or compact panicles, sometimes racemes. Glumes 2, green, nerved, mostly unequal, the 1st often minute, the 2d typically equaling the sterile lemma, the latter simulating a 3rd glume, bearing in its axil a membranaceous or hyaline palea. Fertile lemma and palea chartaceous-indurate, mostly obtuse, the nerves obsolete, the margins inrolled over an inclosed palea of the same texture. A genus of ca. 600 spp. of both hemis., some in cult. (*Panicum*, an old Latin name for common millet.)

1. Plants perennial.
    2. Spikelets 6–7 mm. long .... 8. *P. urvilleanum*
    2. Spikelets less than 4 mm. long.
        3. Vernal form with lf.-blades pubescent above.
            4. Upper surfaces of lf.-blades strigose; autumnal form erect or ascending
                                                                            4. *P. huachucae*
            4. Upper surfaces of lf.-blades pilose; autumnal form decumbent-spreading
                                                                            7. *P. pacificum*
        3. Vernal form with lf.-blades glabrous above .... 6. *P. occidentale*
1. Plants annual.
    5. First glume not more than ¼ as long as spikelet, truncate or broadly triangular; sheaths smooth .... 2. *P. dichotomiflorum*
    5. First glume as much as half as long as spikelet, acute or acuminate; sheaths hispid.
        6. Panicle drooping; spikelets 4.5–5 mm. long .... 5. *P. miliaceum*
        6. Panicle erect; spikelets not more than 4 mm. long.
            7. The panicle more than half the length of the entire plant .... 1. *P. capillare*
            7. The panicle not more than one-third the length of the entire plant ... 3. *P. hirticaule*

1. **P. capillàre** L. var. **occidentàle** Rydb. [*P. barbipulvinatum* Nash.] Erect annual, 2–8 dm. tall, papillose-hispid to subglabrous, usually with short flowering branches at base; sheaths hispid; blades 1–2.5 dm. long, 5–15 mm. wide, pubescent; panicles diffuse, densely fld., often half the height of the plant, exserted; spikelets 2.5–4 mm. long, attenuate at tip, subsessile along the ultimate branchlets; fr. 1.7–1.8 mm. long; $2n=18$ (Spellenberg, 1967).—Waste places at scattered localities below 5000 ft.; to B.C.; Atlantic Coast. July–Sept.

2. **P. dichotomiflòrum** Michx. Much-branched annual from a geniculate base, mostly glabrous except for a ring of hairs at the ligule and the sparse pilosity on upper surface of blades; culms 5–10 dm. long; blades 1–5 dm. long, 3–20 mm. wide; panicles terminal and axillary, mostly included at base, 1–4 dm. long; spikelets 2–3 mm. long; $n=27$ (Gould, 1958).—Moist places as a weed, as at Loma Linda, San Bernardino Co. and Santa Barbara; cent. Calif.; e. U.S. July–Oct.

3. **P. hirticáule** Presl. Erect, 2–6 dm. tall, papillose-hispid especially on the sheaths; blades 5–15 cm. long, 4–12 mm. wide; panicle 7–15 cm. long; spikelets ca. 3 mm. long, red-brown; 1st glume ½–¾ as long, acuminate; 2d glume and sterile lemma acuminate.— Open sandy places, Creosote Bush Scrub; Colo. Desert; to Tex., S. Am. July–Oct.

4. **P. huachùcae** Ashe. Vernal form typically stiff, upright, with copious spreading

papillose pubescence throughout; nodes bearded; blades firm, erect or ascending, 4–8 cm. long, 6–8 mm. wide, short-pilose on upper surface; panicle 4–7 cm. long, pilose; spikelets 1.6–1.8 mm. long, obovoid, turgid, pubescent; autumnal form erect, with fascicled branches and crowded ascending blades; $n=9$ (Brown, 1949).—Reported from Lytle Creek C., San Gabriel Mts. and from San Bernardino Mts.; Ariz. e. June–Aug.

5. **P. miliàceum** L. BROOM-CORN MILLET. Culms stout, erect, 2–10 dm. tall; sheaths papillose-hispid; lvs. 1–2.5 dm. long, 1–2.5 cm. wide; panicles dense, drooping at maturity, 1–3 dm. long, rather compact, the branches many, very scabrous, bearing spikelets near their ends; spikelets 4.5–5 mm. long, ovoid, acuminate, many-nerved; fr. 3 mm. long, straw-color to red brown; $n=18$ (Avdulov, 1931), $2n=36$ (Singh & Godward, 1960).—Occasional escape from cult.; introd. from Old World. Aug.–Oct.

6. **P. occidentàle** Scribn. Vernal form yellowish-green, 2.5–5 dm. tall, the culms leafy to spreading, pilose, with short-bearded nodes; sheaths sparsely papillose-pubescent; ligule 3–4 mm. long; blades 4–8 cm. long, 5–7 mm. wide, strigose beneath; panicle 4–7 cm. long; spikelets 1.8 mm. long, pubescent; autumnal form branching from the lower nodes.—Moist and peaty places, below 8000 ft.; many Plant Communities; scattered stations in much of Calif. June–Aug.

7. **P. pacíficum** Hitchc. & Chase. Plate 100, Fig. E. Vernal form light green, tufted, spreading or ascending, 3–6 dm. tall; nodes pilose; sheaths papillose-pilose; blades erect or ascending, 5–10 cm. long, 5–8 mm. wide, papillose-pilose on upper surface; panicles 5–10 cm. long; spikelets ca. 2 mm. long, papillose-pubescent; 1st glume $\frac{1}{4}$–$\frac{1}{3}$ the length of the spikelet, truncate; autumnal phase prostrate-spreading.—Moist places below 4500 ft.; Chaparral, etc.; mts. of San Diego Co., San Jacinto, San Bernardino and San Gabriel mts.; to B.C., Mont., Ariz. May–July.

8. **P. urvilleànum** Kunth. [*P. u.* var. *longiglume* Scribn.] Robust perennial with creeping rhizomes and solitary or tufted culms 5–10 dm. tall; nodes densely bearded; sheaths overlapping, densely retrorse-villous; blades elongate, 2–5 dm. long, 4–8 mm. wide; panicle 2–3 dm. long, rather narrow; spikelets 6–7 mm. long, densely villous; 1st glume acuminate, $\frac{2}{3}$ as long as spikelet; $n=18$ (Nuñez, 1946).—Occasional in sandy places, Creosote Bush Scrub, both deserts; V. Grassland, San Jacinto, to Ariz., S. Am. March–May.

62. *Parápholis* C. E. Hubb.

Low annuals with slender cylindric spikes. Spikelets 1–2-fld., embedded in the cylindric articulate rachis and falling attached to the joints. Glumes 2, placed in front of the spikelet and enclosing it, coriaceous, 5-nerved, pointed. Lemma with its back to the rachis, smaller than the glumes, hyaline, 1-nerved. Ca. 4 spp. of temp. parts of Old World. (Greek, *para*, beside, and *pholis*, scale, because of the 2 glumes side by side.)

1. **P. incúrva** (L.) C. E. Hubb. [*Aegilops i.* L. *Pholiurus i.* Schinz & Thell.] SICKLE-GRASS. Plate 100, Fig. F. Tufted, decumbent below, 1–2 dm. high; blades narrow; spike 7–10 cm. long, curved; spikelets 7 mm. long, acuminate; $n=18$ (Avdulov, 1931).—Coastal Salt Marsh, Coastal Strand; San Diego to Ore., e. Coast; native of Eu. April–June.

63. *Paspàlum* L.

Annual and perennial grasses with 1–many racemes digitate or racemose at the summit of the culm and branches. Spikelets 1-fld., planoconvex, subsessile, solitary or in 2's, in 2 rows on 1 side of a continuous narrow or dilated rachis. The back of the fertile lemma toward the rachis; 1st glume mostly lacking; 2d glume and sterile lemma subequal; lemma and palea chartaceously indurate, with inrolled margins. Ca. 400 spp. of warmer regions. (Greek, *paspalos*, a kind of millet.)

Racemes a pair, at the summit of the culms .... 2. *P. distichum*
Racemes several to many, forming a panicle.
    The racemes laxly spreading; spikelets 3.5 mm. long .... 1. *P. dilatatum*
    The racemes suberect; spikelets 2 mm. long .... 3. *P. urvillei*

1. **P. dilatàtum** Poir. Culms stout, 5–15 dm. tall, growing in clumps, glabrous throughout except the spikelets; lvs. 5–10 mm. wide, elongate; racemes 4–10, densely fld., 5–10 cm. long, somewhat spreading; spikelets in pairs, ovate, 3–3.5 mm. long; glume and sterile lemma long-ciliate; $n=20$ (Brown, 1948), $2n=40$ (deWet & Anderson, 1956).—Roadsides, ditches and waste places, at low elevs., cismontane and to desert edge as at Victorville; Santa Catalina Id.; to Ore., Atlantic Coast; introd. from S. Am. May–Nov.

2. **P. dístichum** L. KNOTGRASS. Creeping and rooting at the nodes, freely branching, the flowering shoots ascending, 2–6 dm. high; blades thin, 4–10 mm. wide, 5–15 cm. long; panicles terminal, the racemes mostly 2, usually erect, 2–7 cm. long; rachis 2–3 mm. wide; spikelets solitary, ovate, sparsely pubescent, 2.5–4 mm. long; $n=20$ (Brown, 1948), $2n=60$ (deWit, 1958).—Along the coast and in interior ditches; V. Grassland, Coastal Sage Scrub, Coastal Salt Marsh, etc.; cismontane; to Wash., Atlantic Coast; S. Am. Santa Catalina Id., Santa Cruz Id. June–Oct.

3. **P. urvíllei** Steud. [*P. larrangai* Arech.] Culms in large clumps, erect, 1–2 m. tall; lower sheaths hirsute; blades elongate, 3–15 mm. wide; panicle erect, 1–4 dm. long, of ca. 12–20 ascending racemes 7–14 cm. long; spikelets 2.2–2.7 mm. long, ovate, pointed, fringed with long white hairs, the glume silky; $n=20$ (Burton, 1940).—Rare as introd.; native of S. Am. April–July.

64. *Pennisètum* L. Rich.

Annual or perennial grasses, often branched, usually with flat blades and dense spikelike panicles. Spikelets solitary or in groups of 2 or 3, surrounded by an invol. of bristles (sterile branchlets), these united only at very base, often plumose, deciduous with the spikelets. First glume shorter than the spikelet, sometimes obsolete, 2d glume shorter than or equaling the sterile lemma. Fertile lemma chartaceous, smooth, thin-margined, enclosing the palea. A genus of ca. 80 spp. of warmer parts of both hemis. Some like *P. glaucum* (pearl barley) grown for food, other for ornament. (Latin, *penna*, feather, and *seta*, a bristle, because of the plumose bristles of some spp.)

Culms extensively creeping; spikelets few, hidden in the upper sheath .... 1. *P. clandestinum*
Culms not creeping; panicle exserted.
    Panicle oval, tawny, 3–10 cm. long .... 3. *P. villosum*
    Panicle elongate, purple or rosy, 15–35 cm. long .... 2. *P. setaceum*

1. **P. clandestìnum** Hochst ex Chiov. KIKUYU GRASS. Low-growing, stoloniferous and rhizomatous grass, perennial, the stolons with short internodes; sheaths broad, inflated; blades narrow; infl. of 2–4 spikelets almost entirely enclosed in the upper sheath of the short culm.—Sometimes planted for lawns, but in our area easily becoming a dangerous weed in orchards and gardens. Introd. from Africa.

2. **P. setàceum** (Forsk.) Chiov. [*Phalaris s.* Forst. *Pennisetum ruppelii* Steud.] FOUNTAIN GRASS. Tufted perennial, ca. 1 m. tall; blades scabrous, numerous, mostly 2–3 mm. wide; panicle 1.5–3.5 dm. long, mostly pink or purple, the fascicles peduncled, rather loosely arranged, with 1–3 spikelets; bristles plumose toward base, unequal, the longer 3–4 cm. long; $2n=27$ (Avdulov, 1931).—Escaping from cult. and establishing itself at scattered localities from San Diego Co. to Ventura Co.; introd. from Africa. July–Oct.

3. **P. villòsum** R. Br. [*Cenchrus longisetus* M.C. Jtn.] FEATHERTOP. Tufted perennial, 3–6 dm. tall, pubescent below the panicle; blades 3–5 mm. wide; panicle dense, feathery; spikelets 1–4 in a fascicle; fascicles short-peduncled, with a tuft of white hairs at base of peduncle; bristles many, spreading, the inner very plumose, the longer 4–5 cm. long; $2n=45$ (Avdulov, 1931).—Cult. for ornament and escaping in sandy places, Ventura Co.; introd. from Africa. June–Aug.

65. *Phálaris* L. CANARY GRASS

Annuals or perennials, with many flat blades and dense spikelike panicles. Spikelets 1-fld., laterally flattened and with 2 sterile lemmas below the fertile floret. Glumes equal, boat-shaped, often winged on the keel. Sterile lemmas reduced to small usually minute scales; fertile lemma coriaceous, shorter than glumes, enclosing the faintly 2-nerved palea. Ca. 15 spp. in temp. Eu. and Am. (An ancient Greek name for a grass.)

(Anderson, D. E. Taxonomy and distribution of the genus Phalaris. Iowa State J. Sci. 36: 1–96. 1961.)
1. Spikelets in groups of 7, 1 fertile surrounded by 6 sterile, the group falling entire . 7. *P. paradoxa*
1. Spikelets all alike, not falling entire in groups.
   2. Plants perennial; sterile lemma solitary, ca. 1.5 mm. long .... 2. *P. aquatica*
   2. Plants annual.
      3. Glumes broadly winged; panicle ovate to short-oblong.
         4. Sterile lemma solitary; fertile lemma 3 mm. long .... 6. *P. minor*
         4. Sterile lemmas 2; fertile lemma 5–6 mm. long .... 3. *P. canariensis*
      3. Glumes wingless or nearly so; panicles linear or oblong.
         5. Glumes wingless, acuminate; fertile lemma turgid, the apex smooth, acuminate
                                                                                                      5. *P. lemmonii*
         5. Glumes narrowly winged toward the summit, abruptly pointed or acute; fertile lemmas less turgid, villous to the acute apex.
            6. Panicle subcylindric, mostly 6–15 cm. long .... 1. *P. angusta*
            6. Panicle tapering to each end, mostly 2–6 cm. long .... 4. *P. caroliniana*

    1. **P. angústa** Nees ex Trin. Annual, 1–1.5 m. tall; panicle subcylindric, 6–15 cm. long, ca. 8 mm. thick; glumes 3.5–4 mm. long, narrow, abruptly pointed, the keel scabrous and narrowly winged upward; fertile lemma ovate-lanceolate, acute; sterile lemmas ca. $\frac{1}{3}$ as long; $n=7$ (Saura, 1943), $2n=14$ (Ambastha, 1956).—Uncommon, wet places, as in Fallbrook, San Diego Co.; cent. Calif. to La., S. Am. May–June.

    2. **P. aquática** L. [*P. stenoptera* Hack. *P. tuberosa* var. *s.* Hitchc.] HARDING GRASS. Perennial with a loose branching rhizomatous base; culms stout, to 1.5 m. tall; panicle narrow, not branched, 5–15 cm. long; glumes 5–6 mm. long, scabrous on keel; fertile lemma 4 mm. long, ovate-lanceolate, strigose; sterile lemma usually 1, ca. $\frac{1}{3}$ as long as fertile; $n=14$ (Nielsen & Humphrey, 1937).—Occasional escape in wet places, Santa Catalina Id., Ventura Co., etc. From Old World. May–Sept.

    3. **P. canariénsis** L. Erect annual, 3–6 dm. high; the dense broad panicle 1.5–4 cm. long; spikelets broad, pale with green stripes; glumes 7–8 mm. long, abruptly pointed, the green keel prominently pale-winged; fertile lemma 5–6 mm. long, densely strigose; sterile lemmas at least $\frac{1}{2}$ as long as fertile.—Occasional weed in waste places below 4000 ft., San Diego Co. n.; Santa Catalina Id.; introd. from Medit. region. April–July.

    4. **P. caroliniàna** Walt. Erect annual, 3–6 dm. tall; panicle oblong, 2–6 cm. long; glumes 5–6 mm. long, oblong, abruptly narrowed to an acute apex, scabrous on keel; fertile lemma lanceolate, acute, strigose, 3.5–4 mm. long; sterile lemmas 1–2 mm. long; $2n=14$ (Ambastha, 1956; Gould, 1966).—Uncommon in old fields and waste places, Riverside Co., Ventura Co. n.; introd. from e. U.S. April–May.

    5. **P. lemmònii** Vasey. Erect annual, 3–9 dm. tall; lvs. 3–9 mm. wide; panicles 5–15 cm. long, subcylindric or ± lobed below, often purplish; glumes ca. 5 mm. long, narrow, scabrous, acuminate, not winged on keel; fertile lemma lance-ovate, acuminate, 3.5–4 mm. long, brown when mature, strigose except at the acuminate tip; sterile lemmas less than $\frac{1}{3}$ as long; $n=7$ (Parthasarathy, 1938).—Moist places below 2000 ft.; Coastal Sage Scrub, V. Grassland, etc.; San Diego Co. to n. Calif. San Clemente Id. April–June.

    6. **P. mìnor** Retz. Erect annual 3–9 dm. tall; lvs. 4–15 mm. wide; panicle ovate-oblong, mostly 2–5 cm. long; spikelets narrow, the glumes oblong, 4–6 mm. long, strongly winged on keel, ± green-striped; fertile lemma ovate, acute, villous, ca. 3 mm. long; sterile lemma 1, ca. 1 mm. long; $n=14$ (Avdulov, 1931).—Frequent weed in waste and disturbed places, many Plant Communities, cismontane; to Ore. Introd. from Medit. region. On most of our ids. April–July.

    7. **P. paradóxa** L. Tufted annual, 3–6 dm. tall; panicle dense, 2–6 cm. long, oblong, with narrowed base where often enclosed in the enlarged upper sheath; spikelets in groups of 6 or 7, the cent. one fertile, with subulate-acuminate glumes bearing a tooth-

like wing near the middle of the keel, the others sterile; fertile lemma 3 mm. long, with a few hairs near summit; sterile lemmas obsolete; $n=7$ (Parthasarathy, 1938), $2n=14$ (Ambastha, 1956).—Occasional as weed from Imperial Co. n. through cismontane Calif. San Clemente Id. Introd. from Medit. region. May–Aug.

66. *Phlèum* L.

Tufted annuals and perennials. Lvs. with flat blades and membranous ligules to ca. 6 mm. long. Panicle short, cylindrical, densely contracted. Spikelets 1-fld., disarticulating above the glumes. Glumes equal, laterally flattened, mostly 3-nerved, broad, abruptly narrowed at apex into a mucro or a short awn. Lemma membranous, 3–7-nerved, broad and blunt, awnless, shorter than glumes. Palea membranous, narrow, ca. as long as lemma. Ca. 10 spp. of temp. regions. (Greek, *phleos*, old name for marsh reed.)

Panicle ovoid, not more than twice as long as thick; awn of glumes 2 mm. long ... 1. *P. alpinum*
Panicle long-cylindrical, awn of glumes 1 mm. long .... 2. *P. pratense*

1. **P. alpìnum** L. MOUNTAIN TIMOTHY. Culms 2–5 dm. high, from a decumbent tufted base; perennial; blades mostly 4–6 mm. wide; panicle mostly 1.5–2.5 cm. long; glumes ca. 5 mm. long, hispid-ciliate on the keel, the awns ca. 2 mm. long; $n=7, 14$ (Gregor & Sansome, 1930), $2n=28$ (Taylor, 1967).—Wet meadows, etc. 7000–8500 ft.; Montane Coniferous F.; San Jacinto and San Bernardino mts.; to Alaska, Atlantic Coast, Eurasia, S. Am. July–Aug.

2. **P. praténse** L. TIMOTHY. Culms 4–10 dm. high, forming large perennial clumps; blades mostly 5–8 mm. wide; panicles 3–12 cm. long, with crowded spreading spikelets; glumes ca. 3.5 mm. long, truncate with a stout awn 1 mm. long, pectinate-ciliate on keel; $n=21$ (Gregor & Sansome, 1930).—Escaping from cult. in some places, mostly below 5000 ft.; introd. from Eurasia. May–June.

67. *Phragmìtes* Trin. REED

Tall, coarse, rhizomatous and stoloniferous grasses with broad lvs. and large plumose panicles. Spikelets 3–7-fld., the rachilla clothed with long silky hairs disarticulating above the glumes and at the base of each segm. between the florets, the lowest floret ♂ or neuter; glumes 3-nerved or the upper 5-nerved, acute, lanceolate, unequal, the 2d shorter than the florets. Lemmas narrow, long-acuminate, glabrous, 3-nerved, the florets successively smaller, the summits of all subequal. Ca. 3 spp., 1 of Asia, 1 of Argentina, 1 cosmopolitan. (Greek, *phragma*, fence, i.e., hedgelike.)

1. **P. austràlis** (Cav.) Trinius ex Steudel. [*P. communis* Trin. var. *berlandieri* (Fourn.) Fern. *P. b.* Fourn.] COMMON REED. Culms stout, 2–4 m. high, from long creeping rhizomes; blades 2–6 dm. long, 1–5 cm. wide; panicle tawny, 1–3 dm. long, densely fld.; spikelets 12–15 mm. long; $n=24$ or 48 (Avdulov, 1931).—Forming canelike thickets in wet places below 5000 ft.; edge of Alkali Sink, Creosote Bush Scrub, deserts; and in scattered localities, many Plant Communities, cismontane Calif.; to Atlantic Coast and Mex. July–Nov.

68. *Pòa* L. BLUEGRASS

Annual or usually perennial. Blades relatively narrow, flat, folded or involute, ending in a boat-shaped tip. Panicles open or contracted. Spikelets 2–8-fld., the rachilla disarticulating above the glumes and between the florets, the uppermost floret often rudimentary. Glumes acute, keeled, the 1st 1-nerved, the 2d large and usually 3-nerved. Lemmas somewhat keeled, acute or sometimes rounded, often scarious apically, awnless, 5-nerved (rarely 3-nerved). Palea nearly equaling the lemma. Ca. 250 spp. in all temp. and cool regions. (Ancient Greek, *poa*, grass or fodder.)

1. Plants annual.
   2. Lemmas pubescent, but not webbed at base. Cismontane weed .... 1. *P. annua*

2. Lemmas glabrous or with a tuft of cobwebby hairs at base. Native.
   3. Panicle at length open, the elongated branches not floriferous to base. Cismontane
   .... 4. *P. bolanderi*
   3. Panicle narrow, the short appressed branches floriferous to base. Deserts . 3. *P. bigelovii*
1. Plants perennial.
   4. Creeping rhizomes present.
      5. Plants dioecious, the sexes morphologically similar. Sandbinder of the immediate coast
         .... 8. *P. douglasii*
      5. Plants not dioecious, the florets perfect (all ♀ in *P. nervosa*).
         6. Panicle open, pyramidal, its elongated lower branches lax and floriferous only in outer half.
            7. Lemmas glabrous or pubescent on nerves but not webbed at base; florets all ♀. High montane .... 12. *P. nervosa*
            7. Lemmas with a tuft of cobwebby hairs at base; florets perfect. Ubiquitous except on deserts .... 15. *P. pratensis*
         6. Panicle more contracted, oblong, its short lower branches not lax; lemmas not webbed at base, except sometimes vestigiously so in *P. compressa*.
            8. Culms conspicuously flattened, 2-edged; panicle 4–7 mm. long .... 6. *P. compressa*
            8. Culms terete or slightly flattened; panicle 8–15 cm. long ...... 2. *P. atropurpurea*
   4. Creeping rhizomes wanting.
      9. Florets mostly converted into dark purple bulblets; culm with bulblike base 5. *P. bulbosa*
      9. Florets normal, green; culms not bulblike at base.
         10. Lemmas with a tuft of cobwebby hairs at base .... 14. *P. palustris*
         10. Lemmas not webbed at base; all bunchgrasses.
            11. Lemmas glabrous or scabrid, but not pubescent or puberulent on lower part of nerves.
               12. Spikelets compressed, the lemmas keeled. White Mts. .... 7. *P. cusickii*
               12. Spikelets little compressed, elongated, the lemmas rounded dorsally, the keel obscure.
                  13. Ligule long, acuminate or sharply acute .... 13. *P. nevadensis*
                  13. Ligule short, rounded or obtuse .... 11. *P. juncifolia*
            11. Lemmas more or less hairy on back, keel, or nerves at least toward base, sometimes obscurely so (occasionally glabrous in *P. scabrella*).
               14. Spikelets distinctly compressed; lemmas keeled.
                  15. Spikelets 5–7-fld., 6–10 mm. long, pale or greenish, the bracts with prominent hyaline margin. Mid-altitudes .... 9. *P. fendleriana*
                  15. Spikelets 2–4-fld., 2.5–5 mm. long, purplish, the bracts only moderately hyaline-margined. Alpine, White Mts. .... 16. *P. rupicola*
               14. Spikelets little compressed, elongate; lemmas rounded on back, crisp-puberulent toward base, the keel obscure.
                  16. Spring-flowering. Spikelets 6–10 mm. long. Common, largely below 7000 ft.
                     .... 17. *P. scabrella*
                  16. Summer-flowering. Spikelets 5–7 mm. long. Occasional, 7000–12,000 ft.
                     .... 10. *P. incurva*

1. **P. ánnua** L. WINTERGRASS. Tufted winter annual, bright green, glabrous; culms many, often decumbent at base, 5–30 cm. long, leafy; blades lax, flat, up to 4 mm. wide; ligule 1.5–2 mm. long; panicle pyramidal, open, 3–8 cm. long; spikelets 3–6-fld., 3–6 mm. long; glumes unequal; lemmas prominently to obscurely pilose along lower third of the 5 distinct nerves, not webbed at base, the keel often ciliate almost throughout; $2n=28$ (Avdulov, 1928), $2n=14$ (Hovin, 1958).—Common weed of gardens, lawns, orchards, etc., widely distributed in cismontane Calif. and in mts. to 7500 ft.; many Plant Communities; to Alaska, Atlantic, trop. Am. Introd. from Eu. Jan.–July.

2. **P. atropurpùrea** Scribn. Rhizomatous perennial; culms 3–4.5 dm. high, leafy in lower third only; lvs. many, ± tightly folded, firm, smooth, the margin finely scabrid, 8–15 cm. long; ligule up to 2.5 mm. long, usually shorter; panicle spikelike, dense, 2.5–5 cm. long, 1–1.5 cm. wide, purplish; spikelets tightly 3–5-fld., 3.5–5 mm. long; glumes broad, chartaceous, 1.5 and 2 mm. long; lemmas 2.5–3 mm. long, entirely smooth, without web, faintly nerved.—Meadows and grassy slopes, 6000–7500 ft.; Montane Coniferous F.; San Bernardino Mts. May–June.

3. **P. bigelòvii** Vasey & Scribn. [*P. annua* var. *stricta* Vasey.] Tufted light green winter annual, with erect leafy culms 1–3 (–5) dm. long; blades soft, flat, rather short, 1–4 mm. wide; ligule 2–4 mm. long, laciniate; panicle narrow, interrupted, 5–15 cm. long, the branches short, appressed; spikelets 3–5-fld., 4–6 mm. long, broad; glumes firm, 3-nerved, conspicuously white-margined, the keel scabrous; lemmas 2.5–4 mm. long, webbed at base, conspicuously white-villous on keel and lateral nerves; anthers 0.5–0.7 mm. long.—Infrequent in desert canyons or open ground, often in protecting shade of rocks, up to 4000 ft.; Creosote Bush Scrub; w. borders of Colo. Desert and on Mojave Desert at Red Rock Canyon, Panamint Mts., Grapevine Mts., Kingston Mts., etc.; to Okla., Tex., n. Mex. March–May.

4. **P. bolánderi** Vasey. [*P. howellii* var. *chandleri* Davy. *P. b.* ssp. *c.* Piper.] Slender annual, culms 2–6 dm. high; sheaths glabrous; blades relatively short and broad, abruptly acute; spikelets usually 2–3-fld.; glumes relatively broad; lemmas glabrous except for web, scabridulous on keel and rarely on sides, the intermediate nerves very faint; $2n=28$ (Stebbins, 1943).—Occasional in Red Fir F., San Jacinto Mts.; more common further n.; to Wash. July–Aug.

Ssp. **howéllii** (Vasey & Scribn.) Keck. [*P. h.* Vasey & Scribn. *P. b.* var. *h.* Jones.] Culms 2–9 dm. high; sheaths glabrous or sometimes scaberulous; blades relatively elongated, gradually acuminate; spikelets 2–5-fld.; glumes narrow; lemmas pubescent on the lower part, sometimes obscurely so, the 5 nerves all distinct.—Shaded, rather open wooded slopes, below 5000 ft.; Chaparral, Yellow Pine F., etc.; San Diego Co. to Santa Barbara Co.; to Vancouver Id. April–June.

5. **P. bulbosa** L. Tufted, 2–5 dm. tall, the culms somewhat bulbous at base; sheaths smooth; blades of innovations usually folded, of culms usually flat, 1–2 mm. wide, smooth or with scabrous margin; ligule 2–3 mm. long; panicle lax but narrow, up to 8 cm. long, the branches ascending and scabrous; spikelets mostly proliferous, the 4–6 florets converted into bulblets with dark purple base and prolonged foliaceous tip to the lemma; unaltered lemmas 2.5 mm. long, sericeous on keel and marginal nerves, webbed at base; $2n=28$ (Armstrong, 1937) and 42.—Pastures and disturbed land, below 5000 ft.; Chaparral, Southern Oak Wd., etc.; scattered localities as near Murrieta, Riverside Co. and Tehachapi Mts.; to B.C. and the Atlantic; introd. from Eu. April–July.

6. **P. compréssa** L. CANADIAN BLUEGRASS. Rhizomatous; culms rather scattered, decumbent and ± flattened at base, erect, wiry, leafy, 2–5 dm. high; lvs. bluish-green, short, with rather flat narrow blades and short blunt ligules less than 1 m. long; panicle narrow, rather dense, 4–7 cm. long; spikelets crowded on the short branches, ovate-oblong, usually 4–6-fld.; glumes 2 mm. long, 3-nerved, acute; lemmas 2.5 mm. long, obtuse, usually purplish near tips, the keel and marginal nerves sericeous, toward base the web scant or lacking, the intermediate nerves faint; $2n=42$ (Avdulov, 1930), but races have been found with $2n=14, 35, 49, 50$ and 56.—Reported as "rare in s. Calif."; occurs in Greenhorn Mts., Kern Co. and n. to Alaska, also in Eurasia and to Atlantic Coast. May–July.

7. **P. cusíckii** Vasey. [*P. filifolia* Vasey. *P. capillarifolia* Scribn. & Will.] Densely tufted, the erect culms 1.5–5 dm. tall, with ca. 2 short lvs. near the base; blades very many, filiform, erect, short or to 25 cm. long, less than 1 mm. wide, ± scabrous; ligule 1.5–2.5 mm. long, acute; panicle usually contracted, not very dense, in some forms rather open, usually pallid and shining, 3–7 cm. long; spikelets 2–5-fld., 5–7.5 mm. long in ours; glumes rather broad, unequal, somewhat shorter than 1st floret; lemmas 4–5.5 mm. long, smooth or scabrous, rather thin, often sharply acute; $2n=42$ (Hartung, 1946).—Reported from White Mts.; Sierra Nevada to B.C., Rocky Mts. May–July.

8. **P. douglásii** Nees. [*Brizopyrum d.* H. & A. *P. californica* Steud.] Plate 101, Fig. A. Dioecious tufted perennial, spreading by deep-seated rhizomes and aerial leafy runners up to 1 or 2 m. long; culms rather stout, 2–4 dm. high; sheaths scarious, loose; blades firm, involute, glaucescent, densely puberulent ventrally, often equaling the culm; ligule mostly less than 1 mm. long, truncate, fimbriate; panicle ovate to ovate-oblong, 2–4 (–6) cm. long, pale and tawny; spikelets 3–9-fld., 6–9 mm. long; glumes broad,

Plate 101. Fig. A, *Poa douglasii*. Fig. B, *Poa rupicola*. Fig. C, *Poa scabrella*. Fig. D, *Setaria geniculata*. Fig. E, *Spartina foliosa*. Fig. F, *Stipa lepida*.

chartaceous, shiny, ciliolate on margin and hispidulous on keel, 5–6.8 mm. long; lemmas 3-nerved, the nerves pilose below, the base feebly webbed; $2n=28$ (Hartung, 1946).—Coastal Strand; San Miguel and Santa Rosa ids.; Monterey Co. to n. Calif. March–July.

9. **P. fendleriàna** (Steud.) Vasey. [*Eragrostis f.* Steud. *Panicularia f.* Ktze. *Atropis f.* Beal. *P. longiligula* Scribn. & Will. *P. f.* var. *l.* Gould.] MUTTONGRASS. Incompletely dioecious; tufted, the many culms erect, 2–6 dm. high, scabrid upward; basal lvs. flat, folded or involute, glaucescent, stiffish, 1–2 (–3.5) dm. long, those of the culm lvs. mostly short and appressed or obsolete; ligule from less than 1 mm. to 12 mm. long, truncate to acuminate, scarcely correlated with the blade characters; panicle oblong, contracted, pale, 3–10 cm. long, 1–2 cm. wide; lemmas 3–5 mm. long, broad and rounded, villous on keel and marginal nerves, usually smooth between.—Rocky mesas, open slopes or in woods, 3000–10,100 ft.; Pinyon-Juniper Wd., Yellow Pine F. to Lodgepole F.; Santa Rosa and San Jacinto mts., San Bernardino Mts., mts. of e. Mojave Desert; to n. Calif., B.C., S. Dak., n. Mex. May–July.

10. **P. incúrva** Scribn. & Will. Densely tufted; culms slender, to 4 dm. high; lvs. mostly basal, flat or folded, sometimes involute; mostly 1 mm. wide; ligule 2–4 mm. long; panicle narrow, rather dense, 5–8 (–10) cm. long; spikelets 2–4-fld., 5–7 mm. long; lemmas often scabrid apically, 3.5–5 mm. long; anthers 1.5–2.5 mm. long; $2n=90, 93, 94, 99, 105–106$ (Hartung, 1946).—Exposed open places, at 9800 ft., Sugarloaf Mt., San Bernardino Mts.; summit of Mt. Pinos; at 10,850 ft., Inyo Mts.; to n. Calif. July–Aug.

11. **P. juncifòlia** Scribn. [*P. brachyglossa* Piper. *P. fendleriana* var. *j.* Jones.] Densely tufted, light green or glaucescent, the culms slender, erect, to 10 dm. high; lvs. numerous, largely basal, blades mostly tightly involute, 1–2 dm. long, less than 2 mm. wide; ligule mostly 1–2 mm. long, obtuse; panicle narrow, elongate, strict; spikelets 3–6-fld., 7–10 mm. long; glumes broad; lemmas 3–5 mm. long, scabrous apically or throughout, like the glumes not prominently margined, obtuse or rounded; $2n=63, 78, 84$ (Hartung, 1946).—Occasional in or about alkaline meadows, 2300–3600 ft.; Creosote Bush Scrub, etc.; Antelope V., Rabbit Springs, Box "S" Springs (all on s. Mojave Desert); to Wash., Rocky Mts. June–July.

12. **P. nervòsa** (Hook.) Vasey. [*Festuca n.* Hook. *P. olneyi* Piper.] Strongly rhizomatous, tufted; culms erect, 3–6 dm. high; lower sheaths usually retrorsely pubescent and purple; upper sheaths smooth or scabrous; culm-blades usually flat, rather short, to 3 mm. wide, smooth or ventrally puberulent, those of the innovations folded, narrower; ligule 0.5–2 mm. long, rounded or truncate, puberulent; panicle open, heavy, 5–10 cm. long, the branches filiform; spikelets rather loosely 3–8-fld., 5–9 mm. long, commonly purplish; florets always ♀ in Calif. plants; glumes broad; lemmas 3.5–5 mm. long, scabrous on the nerves or throughout, not webbed; $2n=56, 63, 70$ (Hartung, 1946).—Gravelly open soils, 4500–7500 ft.; Yellow Pine F.; Cuyamaca Mts., San Bernardino Mts., Mt. Pinos; to B.C. Rocky Mts. June–Aug.

13. **P. nevadénsis** Vasey ex Scribn. [*Atropis pauciflora* Thurb. *P. p.* Benth. ex Vasey. *Panicularia thurberiana* Ktze. *Poa t.* Vasey. *P. limosa* Scribn. & Will. *Atropis n.* Beal.] NEVADA BLUEGRASS. Tufted, the many erect culms 5–10 dm. high, leafy throughout; lvs. numerous, usually elongate, to 3 dm. long; sheaths and blades often scabrous; blades flat or folded or involute, bright green or pale, 1–3.5 mm. wide, often subcapillary and stiffish; ligule of upper lvs. 3–6 mm. long, acute or acuminate; panicle narrow, up to 2.5 dm. long, pale, rather loose, the branches appressed; spikelets 2–5 (–7)-fld., 4–8 mm. long; glumes medium wide, the 2d ca. equaling the 1st floret; lemmas 3.5–5 mm. long, scabrous apically or throughout, only slightly anthocyanous, usually obtusish; $2n=63$ (Hartung, 1946).—Dry rocky places and flats, mostly below 10,000 ft.; Chaparral, Yellow Pine F., to Bristle-cone Pine F.; w. Riverside Co.; San Bernardino Co. (Oak Glen, Rabbit Springs), Inyo Co. (Death V. region, Inyo-White Mts., Panamint Mts.), Kern Co. (Piute Mts.); to Wash., Rocky Mts. June–July.

14. **P. palústris** L. FOWL BLUEGRASS. Loosely tufted, the culms decumbent and rooting at the purplish base, 3–12 dm. high; sheaths smooth or scaberulous; blades 1–3 mm. wide, often scabrous; ligule 3–5 mm. long, obtuse; panicle open, slightly nodding,

to 2.5 dm. long, the branches naked below; spikelets 2–4-fld., 3–4.5 mm. long; glumes unequal; lemmas bronze-tipped, 2.5–3 mm. long, villous on keel and marginal nerves, webbed at base; $2n=28$ (Armstrong, 1937).—Moist places, as at Santa Ana R. (San Bernardino Mts.), Big Pines (e. San Gabriel Mts.), Inyo Co. n.; circumpolar. July–Aug.

15. **P. praténsis** L. KENTUCKY BLUEGRASS. Rhizomatous, forming dense sods; culms tufted, 3–10 dm. high; innovations numerous; sheaths smooth; blades green or glaucescent, flat or folded, mostly 2–3 mm. wide, the basal up to 3 dm. long, soft, smooth or the margin scabrid; ligule mostly 1 mm. long, truncate; panicle open, pyramidal, 5–15 cm. long; spikelets 3–5-fld., 4–6 mm. long, green or purplish; glumes scabrous on keel; lemmas 2–3 mm. long, obtuse or acute, webbed at base, sericeous on keel and marginal nerves; $2n=49$ to 84 inclusive in w. N. Am. (Hartung, 1946).—Meadows, stream banks, etc. below 13,000 ft.; many Plant Communities; widespread in cismontane and montane s. Calif.; throughout n. N. Am. and Eurasia. May–Aug. Material escaped from lawns and fields may well be European in origin, montane material undoubtedly native. May–Aug.

16. **P. rupícola** Nash ex Rydb. [*P. rupestris* Vasey, not With.] TIMBERLINE BLUEGRASS. Plate 101, Fig. B. Densely tufted, the culms crowded, stiffly erect, 1–2 dm. high; lvs. sparse, usually smooth, the blades involute, stiffish, short; ligule 0.7–2 mm. long, obtuse, laciniate; panicle slender, compact, usually purplish, 1.5–5 cm. long, rarely over 1 cm. wide; spikelets 1–4-fld., 3–5 mm. long; glumes broad, apically scabrid, often rather prominently 3-nerved; lemmas silky on midrib and lateral nerves but lacking web on base.—Rocky screes and ridges, alpine, White Mts.; Sierra Nevada, to Rocky Mts. July–Aug.

17. **P. scabrélla** (Thurb.) Benth. ex Vasey. [*P. tenuifolia* Nutt. ex Buckl. not A. Rich. *Atropis californica* Munro ex Gray. and *A. c.* in Wats. *A. s.* Thurb. in Wats. *P. californica* Scribn. not Steud. *P. orcuttiana* Vasey. *P. buckleyana* Nash. *P. capillaris* Scribn. *P. nudata* Scribn.] Plate 101, Fig. C. Green, small to moderate tufts, the culms slender, erect, smooth to scabrid, 4–10 dm. high; lvs. largely basal, soft, slender, the sheaths smooth or scabrid; ligule 3–7 mm. long; acuminate; panicle usually narrow and contracted, sometimes open, 5–15 cm. long; spikelets 3–7-fld., 6–10 mm. long; lemmas 3–5 mm. long, from obviously to obscurely puberulent toward base, or sometimes glabrous on Mojave Desert; $2n=63, 66, 84, 86$ (Hartung, 1946).—Common on hillsides below 5000 ft., many Plant Communities, cismontane s. Calif.; up to 10,000 ft. in mts. of n. Mojave Desert, largely above Creosote Bush Scrub; n. to Wash., s. to L. Calif. Feb.–June. On most of the islands off the coast.

69. *Polypògon* Desf. BEARDGRASS

Low to moderately tall annuals and perennials, with weak, decumbent-erect culms, these often rooting at lower nodes. Blades thin, flat. Infl. a dense contracted panicle of small 1-fld. spikelets, these articulating below the glumes and falling entire. Glumes subequal, 1-nerved, awned from tip or awnless. Lemma broad, smooth, shining, mostly 5-nerved, shorter than glumes, awnless or with a delicate awn from the apex. Ca. 10 spp. of temp. regions. (Greek, *polus*, many, and *pogon*, beard, because of the bristly infl.)

Glumes awned.
    Awns of glumes 6–8 mm. long; plants annual .... 3. *P. monspeliensis*
    Awns of glumes 3–5 mm. long; plants perennial.
        Ligule mostly not more than 2 mm. long, at least as wide as long; awns mostly delicate, flexuous .... 1. *P. australis*
        Ligule 2–5 mm. long, definitely narrower than long; awns stiff and straight . 2. *P. interruptus*
Glumes not awned; panicle dense with glomerules .... 4. *P. semiverticillatus*

1. **P. austràlis** Brongn. Perennial, to 1 m. tall; ligule short, not longer than wide; blades commonly 5–7 mm. wide; panicle soft, lobed or interrupted, ca. 8–15 cm. long, largely with purplish awns; glumes 1.5–2 mm. long, hispidulous, the awns flexuous, delicate, 4–6 mm. long; lemma ca. $\frac{2}{3}$ as long as glumes, the awn 3 mm. long.—Found in waste low places, from Orange and Riverside cos. n.; native of s. S. Am. Summer.

2. **P. interrúptus** HBK. [*P. lutosus* Calif. refs.] Tufted perennial, the culms geniculate below, 3–8 dm. tall; ligule 2–5 mm. long; not nearly so wide; blades mostly 4–6 mm. wide; panicle oblong, 5–15 cm. long, ± interrupted or lobed; glumes 2–3 mm. long, scabrous, the awns 3–5 mm. long; lemma smooth, shining, 1 mm. long, minutely toothed at the truncate apex, its awn ca. 2 mm. long.—A weed grass in wet and waste places through much of the state; native of Eu. May–Aug. Santa Cruz, San Miguel and Santa Catalina ids.

3. **P. monspeliénsis** (L.) Desf. [*Alopecurus m.* L.] Erect annual or decumbent at base, the culms 1.5–5 dm. long; ligule 5–6 mm. long; blades mostly 4–6 mm. wide; panicle 2–15 cm. long, 1–2 cm. thick, tawny yellow when mature; glumes ca. 2 mm. long, hispidulous, the awns 6–8 mm. long; lemma smooth, shining, ca. half as long as glumes, the awn slightly exceeding them; $n=14$ (Avdulov, 1931).—Common weed in low waste places; introd. from Eu. April–Aug.

4. **P. semiverticillàtus** (Forsk.) Hylander. [*Phalaris s.* Forsk. *Agrostis s.* C. Chr. *A. verticillata* Vill.] From decumbent at base to creeping rooting; blades firm, short to elongate; panicle contracted, 3–10 cm. long, densely fld., lobed, with short whorled branches; spikelets usually falling entire; glumes equal, 2 mm. long, obtuse, scabrous; lemmas 1 mm. long, awnless, truncate and toothed; palea nearly as long; $n=14$ (Avdulov, 1931).—Weed along ditches and in moist places, widely distributed in state; introd. from Eu. Santa Catalina Id. May–June.

70. **Puccinéllia** Parl. ALKALI GRASS

Low mostly pale annuals or perennials, tufted. Lf.-sheaths open. Panicles narrow to open. Spikelets several-fld., subterete, the rachilla disarticulating above the glumes and between the florets; glumes shorter than the 1st lemma, the 1st mostly 1-nerved, the 2d 3-nerved. Lemmas usually firm, rounded on the back, obtuse or acute, rarely acuminate, usually scarious and often erose at apex, 5-nerved, the nerves mostly obscure or indistinct, sometimes conspicuous. Styles absent. Ca. 30 spp., mostly of ± saline habitats; N. Hemis. (Named for B. *Puccinelli*, Italian botanist.)

Plants annual, not more than 2 dm. tall .... 3. *P. parishii*
Plants perennial, larger
    Lemmas narrowed into an obtuse apex, ca. 3 mm. long; panicle branches usually spreading. Native .... 2. *P. nuttalliana*
    Lemmas broad, obtuse or truncate, not narrowed above, ca. 2 mm. long; panicle branches usually reflexed. Introd. .... 1. *P. distans*

1. **P. distáns** (L.) Parl. [*Poa d.* L.] Perennial, with tufted culms; spreading at base or erect, 2–4 dm. tall; blades flat or ± involute, 2–4 mm. wide, 2–10 cm. long; panicle pyramidal, loose, 5–15 cm. long, the branches rather distant, fascicled, the lower spreading to reflexed, naked in their lower half; spikelets 4–6-fld., 4–5 mm. long; glumes 1 and 2 mm. long; lemmas rather thin, obtuse or truncate, ca. 2 mm. long, with few short hairs at base; $n=7, 21$ (Avdulov, 1931).—Moist, ± alkaline places, as occasional introd.; from Eu. June–Sept.

2. **P. nuttalliàna** Schult. [*P. airoides* (Nutt.) Wats. & Coult.] Tufted perennial, the culms slender, mostly erect, rather stiff and firm at base, 3–6 dm. long; blades 1–3 mm. wide, flat to ± involute; panicle pyramidal, open, mostly 1–2 dm. long, the branches scabrous, distant, spreading, naked near the base, to 10 cm. long; spikelets 3–6-fld., 4–7 mm. long, with rather distant florets; glumes 1.5 and 2 mm. long, the 2d 3-nerved; lemmas 2–3 mm. long, narrowed into an obtuse apex; $n=28$ (Church, 1949).—Alkaline places, 6000–7000 ft.; Yellow Pine F.; Bear L., Baldwin L., San Bernardino Mts.; to B.C., Wis., New Mex. June–Sept.

3. **P. parishii** Hitchc. Tufted glabrous annual, the culms 0.3–2 dm. tall; blades flat to subinvolute, scarcely 1 mm. wide; panicle narrow or wider, 1–8 cm. long; spikelets 3–6-fld., 3–5 mm. long; glumes ca. 1.5 and 2 mm. long; lemmas ca. 2 mm. long, obtuse to truncate, scarious and ± erose at apex; pubescent on nerves nearly to tip; $n=7$

(Church, 1949).—Alkaline seeps, at 2900 ft.; Joshua Tree Wd.; Rabbit Springs. Mojave Desert; Ariz. April–May.

71. Rhynchelytrum Nees

Annuals or perennials, with rather open panicles of silky spikelets, these on short capillary pedicels. First glume minute, villous; 2d glume and sterile lemma equal, gibbous below, raised on a stipe above the 1st glume, emarginate, short-awned, covered with long silky hairs except toward the slightly spreading apex; palea well developed. Lemma cartilaginous, boat-shaped. A genus of ca. 35 spp. (Greek, *rhynchos*, beak, and *elytron*, scale, referring to the beaked 2d glume and sterile lemma.)

    1. **R. ròseum** (Nees) Stapf & Hubb. [*Tricholaena r.* Nees. *R. repens* C. E. Hubb.] NATAL GRASS. Perennial, ca. 1 m. tall; blades flat, 2–5 mm. long; panicle rosy to pink, 1–1.5 dm. long, with slender ascending branches; spikelets 5 mm. long. Escape from cult. and established at Santa Barbara, at La Mesa, San Diego Co., *Fuller*; native of S. Afr.

72. Schismus Beauv.

Low tufted annuals with filiform blades and small panicles, the slender pedicels finally disarticulating at base and falling with the spikelet or with the glumes. Spikelets several-fld., the rachilla disarticulating above the glumes and between the florets. Glumes subequal, exceeding the 1st floret, white-margined. Lemmas broad, rounded on back, several-nerved, pilose along lower part of edge, hyaline at tip, 2-toothed. Palea broad, hyaline, nerved at margin. Ca. 5 spp. of Old World. (Greek, *schismos*, splitting, because of the 2-toothed lemmas.)

Glumes 5–6 mm. long; lemma 2.5–3 mm. long, with 2 acute terminal lobes; palea acute, shorter than lemma .... 1. *S. arabicus*
Glumes 4–5 mm. long; lemmas ca. 2 mm. long, rounded and emarginate at tip; palea rounded, as long as lemma .... 2. *S. barbatus*

    1. **S. arábicus** Nees. Like the next, but spikelets slightly larger, 5–7-fld.; lemmas 2.5–3 mm. long, more pilose on margins and back, with 2 acute lobes at tip; palea acute. —Open places, w. Mojave Desert; Catalina Id.; native of sw. Asia. March–May.

    2. **S. barbàtus** (L.) Thell. [*Festuca b.* L.] Culms slender, 0.5–3 dm. long; blades mostly less than 10 cm. long; panicle narrow to oval, 1–5 cm. long, mostly dense, pale or purplish; spikelets ca. 5-fld.; glumes 5–7-nerved, shorter than spikelets, acute; lemmas 9-nerved, the margins appressed-pilose on lower part, teeth minute; $n=6$ (Gould, 1958). —Rather common in widely scattered waste places, desert and cismontane; native of Old World. March–April.

73. Scleropòa Griseb.

Annuals with slightly branched 1-sided panicles. Spikelets several-fld., linear, ± compressed, the rachilla thick, disarticulating above the glumes and between the florets, remaining as a minute stipe to the floret above. Glumes unequal, short, acutish, strongly nerved, the 1st 1-nerved, the 2d 3-nerved. Lemmas subterete, obscurely 5-nerved, obtuse, slightly scarious at apex. A small genus of the Old World. (Greek, *skleros*, hard, and *poa*, grass, because of the stiff panicle.)

    1. **S. rígida** (L.) Griseb. [*Poa r.* L. *Catapodium r.* C. E. Hubb.] Culms spreading to erect, 1–3 dm. tall; blades flat, 1–2 mm. wide; panicles stiff, narrow, 5–10 cm. long, dense, the branches short, bearing spikelets from near the base; pedicels thick, ± spreading; spikelets 4–10-fld., 5–8 mm. long; glumes ca. 1.5 and 2.5 mm. long; lemmas ca. 2.5 mm. long, glabrous, awnless; $n=7$ (Avdulov, 1931).—Becoming locally a fairly frequent weed from Santa Ynez Mts., Santa Barbara Co. n. May–Sept.

74. Scribnèria Hack.

Low annual with short narrow blades and linear cylindrical spikes. Spikelets 1-fld., solitary, laterally compressed, appressed flatwise against the ± thickened continuous rachis, the rachilla disarticulating above the glumes. These equal, firm, narrow, pointed, keeled on outer nerves, the 1st 2-nerved, the 2d 4-nerved. Lemma shorter than glumes, membranaceous, faintly nerved, short-bifid at tip, the midvein ending in a slender awn. Palea ca. as long as lemma. (Named for F. Lamson-*Scribner*, student of grasses.)

1. **S. bolánderi** (Thurb.) Hack. [*Lepturus b.* Thurb.] Culms 5–30 cm. long, tufted, ascending to erect; lvs. few, subfiliform; ligule ca. 3 mm. long; spike 2–11 cm. long, ca. 1 mm. thick; spikelets ca. 7 mm. long; lemmas pubescent near base, the erect awn 2–4 mm. long; $2n=46$ (Stebbins & Major, 1965).—Santa Ynez Mts. n. to Siskiyou Co., Calif. March–June.

75. Secàle L. RYE

Annuals with flat blades and slender spikes. Spikelets 2-fld., the glumes subulate, 1-nerved, the spikelets alternating on a long zigzag rachis. Lemmas keeled and long-awned. Five spp. of Eurasia. (Old Latin name for a grain.)

1. **S. cereàle** L. Tufted annual, 1–1.5 m. tall, blue-green; blades to ca. 12 mm. broad, long-pointed; spike slender, nodding, 7–15 cm. long; spikelets with 2 fertile fls. and sometimes a 3d sterile one; lemmas narrow; $2n=14$ (Bowden, 1966).—Cult. and sometimes escaping in waste places and fields; native of sw. Asia. May–Aug.

76. Setària Beauv. BRISTLEGRASS

Annual or perennial grasses with flat lvs. and cylindrical spikelike or looser panicles. Spikelets as in *Panicum*, but surrounded by few or many persistent awnlike branches which spring from the rachis below the articulation of the spikelets. Ca. 125 spp., of warmer regions, particularly Africa. (Latin, *seta*, bristle.)

(Rominger, J. M. Taxonomy of Setaria in N. Am. Ill. Biol. Mon. 29: 1–132. 1962.)
Bristles below each spikelet at least more than 5.
    Plants annual; spikelets 3 mm. long; lower floret ♂, the palea well developed . 3. *S. lutescens*
    Plants perennial; spikelets 2–3 mm. long.
        Panicle 4–9 cm. long, the bristles yellow or purple, 3–18 mm. long; fr. not rugose
        2. *S. geniculata*
        Panicle 8–15 cm. long, the bristles orange to purple, 3–6 mm. long; fr. finely rugose
        4. *S. sphacelata*
Bristles below each spikelet 1 or 3.
    The bristles ± retrorsely scabrous; panicles cylindrical; spikelets 2 mm. long, green
    5. *S. verticillata*
    The bristles not retrorsely scabrous.
        Upper surface of lf.-blades pilose or strigose; spikelets 2.5–3 mm. long; panicles at maturity nodding from near base .... 1. *S. faberi*
        Upper surface of lf.-blades scabrous; spikelets 1.8–2.2 mm. long; panicles at maturity nodding from apex .... 6. *S. viridis*

1. **S. fàberi** Herm. Annual, 5–20 dm. tall; lf.-blades scabrous and soft hairy on upper surface; panicles arching and drooping from near base, 6–20 cm. long; spikelets 2.5–3 mm. long; subtended by 1–6 bristles, each ca. 1 cm. long.—Reported as adventive in Los Angeles, *Fuller* and in cent. Calif.

2. **S. geniculàta** (Lam.) Beauv. [*Panicum g.* Lam. *Chaetochloa g.* Millsp. & Chase.] Plate 101, Fig. D. Perennial, ± cespitose, 3–10 dm. high, the culms slender, compressed, often geniculate at base; sheaths overlapping; blades 1–3 dm. long, 3–7 mm. wide; panicle 4–9 cm. long, 4–8 mm. thick; bristles 5–10, yellowish, 5–10 mm. long, upwardly scabrous; spikelets 2 mm. long; fr. undulate-rugose; $2=36$ (Kishimoto, 1938), $2n=36$, 72 (Gould, 1965).—Dry or moist open places at low elevs. in most of cismontane s. Calif.; Coastal Sage Scrub, V. Grassland, etc.; to Atlantic Coast, S. Am. May–Sept.

Setaria

3. **S. lutéscens** (Weigel) Hubbard. [*Chaetochloa l.* Stuntz. *S. glauca* Calif. refs.] Annual with culms branching at base, compressed, spreading or erect, 3–6 dm. long; lvs. flat, with a spiral twist, glaucous, 3–10 mm. wide; panicle 2–8 cm. long, dense, ca. 1 cm. thick; bristles 3–8 mm. long, upwardly scabrous; spikelets 3 mm. long, 1st glume half, 2d ⅔ as long as the striate undulate-rugose fertile lemma; $n=18$ (Avdulov, 1931), $2n=36$ (Singh & Godward, 1960).—Weed in waste places, fields, etc., desert and cismontane; native of Old World. June–Oct.

4. **S. sphacelàta** (Schumacher) Stapf & C. E. Hubb. [*Panicum s.* Schumacher.] Tufted perennial, 5–15 dm. tall, the culms flattened; blades 4–10 mm. wide; panicle dense, 8–15 cm. long; bristles 5 or more, 3–6 mm. long.—Occasional escape in wet places, as near Orange; introd. from Afr.

5. **S. verticillàta** (L.) Beauv. [*Panicum v.* L. *Chaetochloa v.* Scribn.] Annual, 3–6 dm. tall, tufted; lvs. scabrous, 5–10 mm. wide; panicle erect but not stiff, ± cylindrical, 5–15 cm. long, 7–15 mm. thick; bristles solitary, downwardly barbed, 3–6 mm. long; spikelets 2–2.5 mm. long; 1st glume ⅓, 2d as long as sterile lemma; fertile lemma obscurely transverse-rugose; $n=9$ (Krishnaswamy, 1935), 18 (Avdulov, 1931).—Occasional weed, as at Indio, Riverside, Upland, Santa Barbara; introd. from Eu. May–July.

6. **S. víridis** (L.) Beauv. [*Panicum v.* L. *Chaetochloa v.* Scribn.] Annual, branched at base, 1–4 dm. high; lvs. not twisted, scabrous on margins, to ca. 1 cm. wide; panicles 2–5 cm. long, green or purple, densely fld.; bristles 7–12 mm. long; spikelets 2–2.5 mm. long; $n=9$ (Tateoka, 1954).—Occasional weed, but quite widely spread, cismontane and irrigated desert valleys; introd. from Eu. June–Aug.

77. *Sitànion* Raf. SQUIRRELTAIL

Erect cespitose perennials with bristly spikes. Spikelets usually 2 at each node of a readily disarticulating rachis. Spikelets usually 2–4-fld. Glumes narrow, usually selaceous, 1–3-nerved, the midnerve extended into a long stout flexuous awn. Palea well developed, firm, the 2 keels serrulate. Four spp. of w. U.S. (Greek, *sitos*, grain for food.)

(Wilson, F. D. Revision of Sitanion. Brittonia 15: 303–323. 1963.)

Lowermost floret of 1 or both spikelets at each rachis node sterile and reduced to a subulate structure resembling extra glume segms.
    Glumes entire or bifid .... 2. *S. hystrix*
    Glumes 3–many-cleft .... 3. *S. jubatum*
Lowermost floret fertile, not reduced.
    Glumes subulate, entire; awns of the glumes exceeding those of the lemmas . 4. *S. longifolium*
    Glumes usually lanceolate, entire or 2-several-cleft; awns of the lemmas exceeding those of the glumes .... 1. "*S.*" *hansenii*

1. **"S." hansènii** (Scribn.) J. G. Sm. [*Elymus h.* Scribn. *S. anomalum* J. G. Sm.] Culms smooth, 5–10 dm. high; sheaths glabrous; blades flat, scabrous, 2–8 mm. wide; spike 8–20 cm. long; glumes 2–3-nerved, long-awned; lower lemmas ca. 8 mm. long, with mature awns divergent and 4–5 cm. long; $n=14$ (Stebbins & Love, 1941).—Dry open often rocky places below 13,000 ft.; many Plant Communities; widely spread in Calif., but especially in mts. of Mojave Desert and n. to Wash., Wyo. June–Aug. Most plants referred here are apparently sterile hybrids between *Elymus glaucus* and *Sitanion* spp.

2. **S. hýstrix** (Nutt.) J. G. Sm. var. **hýstrix**. [*Aegilops h.* Nutt. *S. minus* J. G. Sm. *S. glabrum* J. G. Sm.] Plants loosely cespitose, 1.5–3.4 dm. tall; culms slender, erect to spreading; sheaths glabrous to densely pubescent; blades flat to involute, 1–3 (–5) mm. wide, green or glaucous; spikes 4–12 cm. long; spikelets usually 2 at a node, few–severalfld.; glumes subulate or broader, 1–3-nerved (at least 1 glume of each node 2-cleft), with awns 3.5–8.5 cm. long; lemmas 8–10 mm. long, the awns shorter than those of glumes; $n=14$, $2n=28$.—Dry rocky or open places below 13,000 ft.; many Plant Communities; Riverside Co., Los Angeles Co., more common Sierra Nevada etc. n. to B.C. S. Dak. April–May.

Var. **califórnicum** (J. G. Sm.) F. D. Wilson. [*S. c.* J. G. Sm.] Glumes entire; awns of the lemmas exceeding those of the glumes.—Mts. of the e. Mojave Desert to Inyo-White Mts.; then n. and e.

3. **S. jubàtum** J. G. Sm. [*S. multisetum* J. G. Sm. *S. breviaristatum* J. G. Sm.] Plants 2-6 dm. high; sheaths smooth, villous or scabrous; blades flat or involute, 1-4 mm. wide; spike dense, 3-10 cm. long, bushy; each lobe of the glumes with an awn 3-8 cm. long; lemmas 8-10 mm. long, smooth or scabrous toward the tip, long-awned; $n=14$ (Stebbins & Love, 1941).—Rocky or brushy slopes mostly below 8000 ft.; many Plant Communities; cismontane and desert s. Calif.; to Wash., Ida. e. May-July.

4. **S. longifòlium** J. G. Sm. Plants 2.5-6 dm. tall, with erect to spreading culms; sheaths striate to indistinctly veined, mostly glabrous; blades mostly flat, 2-5 mm. wide; spikes 7-15 cm. long, loose; spikelets usually 2 at each node, few-several-fld., terminal florets mostly sterile; glumes entire, usually 1-nerved, subulate, the awns 5-11 cm. long; lemmas 7-12 mm. long, obscurely 3-5-nerved, the awn 5-10 cm. long.—At elevations below 10,000 ft., from flats in the Mojave Desert and hills and mts. from San Diego Co. through cent. Mojave Desert to Inyo Co., n. to e. Ore., e. to Rocky Mts. and plains. May-Aug.

78. Sórghum Moench

Tallish annuals or perennials, with flat blades and terminal panicles of 1- to 5-jointed tardily disarticulating racemes. Spikelets in pairs, 1 sessile and fertile, the other pedicellate, sterile but well developed, usually ♂. Ca. 35 spp. of Old World, some widely cult. (*Sorgho*, the Italian name.)

Plants perennial; panicle openly branched .... 1. *S. halepense*
Plants annual; panicle compactly branched .... 2. *S. bicolor*

1. **S. halepénse** (L.) Pers. [*Holcus h.* L.] JOHNSON GRASS. Rhizomes heavy, scaly, extensively creeping; stems erect, glabrous, coarse, 5-15 dm. tall; ligules ciliate; blades 1-5 dm. long, 1-2 cm. wide; panicles 1-3 dm. long, purplish; fertile spikelets 5 mm. long, with pubescent glumes; awn 1 cm. long, deciduous; $n=10$, 20 (Janaki Ammal, 1945), 13 (Raman et al., 1966).—Common in low wet and waste places, almost wherever irrigation water overflows, as at orchard edges, etc.; native of Old World. May-Aug.

2. **S. bícolor** Pers. (L.) Moench. [*S. vulgare* Pers.] SORGHUM. Annual, more robust; panicle more compact; $n=10$ (Kuwada, 1915).—Occasional escape from cult. Known in many forms in agriculture and often quite cornlike.

T. C. Fuller reports 3 other spp. collected by him near the Bard Experiment Station, Imperial Co. and sometimes becoming established elsewhere:

S. lanceolàtum Stapf. Robust annual to 1.5 m. tall; blades 3-6 dm. long, 2-3.5 cm. wide; panicle 2.5-4 dm. long with ascending branches; rachis joints and pedicels ciliate; spikelets ca. 6 mm. long, silky pubescent; awn ca. 1 cm. long. From trop. Afr.

S. sudanénse (L.) Pers. Annual, 2-3 m. high; lf.-blades 1.5-3 dm. long, 8-12 mm. wide; panicle erect, loose, 1.5-3 dm. long, the branches subverticillate. Cult. as a hay and pasture grass.

S. virgàtum (Hack.) Stapf. Tall annual with a narrow slender open panicle and narrowly lanceolate green finely awned spikelets.

79. *Spartina* Schreb. CORDGRASS

Coarse perennials with strong creeping rootstocks, simple rigid culms, long tough lvs. and 2 to many (in ours) appressed spikes scattered along the main axis. Spikelets 1-fld., much flattened laterally, sessile and usually closely imbricate on one side of a continuous rachis, disarticulating below the glumes. Glumes keeled, 1-nerved, or the 2d with a 2d nerve on 1 side, acute or short-awned, the 1st shorter, the 2d often longer than the lemma. Lemma firm, keeled, the lateral nerves obscure; palea 2-nerved, keeled and flattened. Ca. 16 spp., widely distributed. (Greek, *spartine*, a cord.)

(Mobberley, D. G. Taxonomy and distribution of the genus Spartina. Iowa State Coll. J. Sci. 30: 471-574. 1956.)

Blades mostly more than 8 mm. wide; spikes closely approximate forming a cylindric infl. Coastal marshes .... 1. *S. foliosa*
Blades mostly less than 5 mm. wide; spikes distinct, appressed or spreading. Alkaline interior meadows .... 2. *S. gracilis*

1. **S. foliòsa** Trin. [*S. leiantha* Benth.] Plate 101, Fig. E. Culms stout, 3–12 dm. tall, up to 1 cm. thick at base, usually rooting at lower nodes; blades 8–12 mm. wide at base, smooth; infl. dense, 1.5–2.5 dm. long; spikes many, approximate, closely appressed, 3–8 cm. long; spikelets very flat, 9–12 mm. long; glumes firm, glabrous or hispid-ciliate on keel, acute, the 1st narrow, ½–⅔ as long as the 2d, smooth, the 2d slightly hispidulous; lemma hispidulous on sides, shorter than 2d glume; palea thin, longer than lemma; $2n=56$ (Church, 1940).—Coastal Salt Marsh; L. Calif. to n. Calif. July–Nov.

2. **S. grácilis** Trin. Culms 6–10 dm. tall, more slender; blades 3–5 mm. wide, involute in age; spikes 4–8, 2–4 cm. long, appressed; spikelets 6–8 mm. long; glumes ciliate on keel; palea as long as lemma; $2n=42$ (Church, 1940).—Alkaline places, 6000–7000 ft.; Inyo and Mono cos. n. to B.C. and e. to Kans. June–Aug.

80. *Sphenópholis* Scribn. WEDGEGRASS

Mostly slender perennials, usually with flat blades and narrow terminal panicles. Spikelets 2–3-fld., the rachilla prolonged beyond the upper floret. Glumes subequal, but unlike in shape, the 1st narrow, 1-nerved, the 2d broader, 3–5-nerved, becoming coriaceous. Lemmas firm, scarcely nerved, awnless or with an awn from just below the apex, the 1st a little shorter or a little longer than the 2d glume. Palea hyaline, exposed. Four spp., of New World. (Greek, *sphen*, wedge, and *pholis*, horny scale, referring to hard obovate 2d glume.)

(Erdman, K. S. Taxonomy of the genus Sphenopholis. Iowa State J. Sci. 39: 289–336. 1965.)

1. **S. obtusàta** (Michx.) Scribn. [*Aira o.* Michx. *S. o.* var. *lobata* Scribn.] Culms tufted, erect, 3–6 dm. tall; sheaths glabrous to pubescent; blades 1–3 dm. long, 2–5 mm. wide, glabrous to pubescent; panicle spikelike to interrupted and lobed, 5–20 cm. long; spikelets 2.5–3.5 mm. long, the 2 florets very close together; 2d glume very broad, subcucullate, ± inflated at maturity, 5-nerved, scabrous; lemmas minutely papillose, the 1st ca. 2.5 mm. long; $2n=14$ (Erdman, 1965).—Damp and open places, as about San Bernardino and at 6500 ft. on the S. Fork of the Santa Ana R.; cent. Calif. to B.C., Atlantic Coast, W.I., Mex. April–July.

81. *Sporóbolus* R. Br. DROPSEED

Annuals or perennials with flat or involute lvs. and narrow or spreading panicles. Spikelets 1-fld., the rachilla disarticulating above the glumes. Glumes 1-nerved, usually unequal, the 2d often as long as the spikelet. Lemma membranous, 1-nerved, awnless. Palea usually prominent and as long as or longer than the lemma. Caryopsis falling readily from the spikelet at maturity. Ca. 100 spp. of warmer regions. (Greek, *spora*, seed, and *ballein*, to throw.)

Glumes subequal, much shorter than lemma. Introd. weed .... 5. *S. poiretii*
Glumes unequal, or if equal, as long as lemma. Natives.
    Panicle spikelike; spikelets 2.5 mm. long .... 2. *S. contractus*
    Panicle ± open, not spikelike; spikelets 1.5–2 mm. long.
        Sheath with tuft of hairs at throat; glumes scabrous on keel.
            Panicle ± included in the sheath, the branches ascending .... 3. *S. cryptandrus*
            Panicle exserted, the branches spreading .... 4. *S. flexuous*
        Sheath naked or sparingly ciliate at the throat; glumes glabrous .... 1. *S. airoides*

1. **S. airoìdes** (Torr.) Torr. [*Agrostis a.* Torr.] Perennial with large dense tufts 3–10 dm. tall; sheaths pilose at throat; ligule pilose; blades involute to flat, 1–3 mm. wide, 5–35 cm. long; panicle diffuse, 1–4 dm. long; spikelets distal on the branches, 1.5–2 mm.

long, obtuse; glumes nerved, unequal, acute, glabrous, the 1st ca. half as long as spikelet; $n=63$ (Stebbins & Love, 1941), $2n=$ ca. 80 (Gould, 1966).—Moist alkaline places below 5000 ft.; Alkali Sink, Coastal Sage Scrub, etc.; cismontane and desert s. Calif.; to Wash., S. Dak., Tex., Mex. April–Oct.

Var. **wrightii** (Munro ex Scribn.) Gould. [*S. w.* Munro.] Panicle relatively dense, 3–6 dm. long, the many short crowded densely fld. branches floriferous nearly to the base; $n=18$ (Brown, 1950).—Alkaline places, Colo. River bottom, Whipple Mts.; to Tex., Okla., Mex. April–May.

2. **S. contráctus** Hitchc. [*S. strictus* Merr., not Franch.] Tufted perennial 6–10 dm. tall; blades 3–5 mm. wide, flat or involute, 10–15 cm. long; panicles spicate, 1.5–3 (–5) dm. long; spikelets 2.5 mm. long; 1st glume ⅓ as long as spikelet; 2d glume, lemma, and palea subequal. Occasional in dry places below 6000 ft.; Creosote Bush Scrub to Pinyon-Juniper Wd.; Colo. and e. Mojave deserts; to Colo., w. Tex., Son. Sept.–Oct.

3. **S. cryptándrus** (Torr.) Gray. [*Agrostis c.* Torr.] Tufted perennial 3–8 dm. tall; sheath with conspicuous tuft of white hairs at summit; blades flat, 2–5 mm. wide, ± involute when dry, fine-pointed; panicle 5–20 cm. long, the branches ascending; spikelets ca. 2 mm. long; glumes subequal, ca. half as long as spikelet; lemma acutish; $n=18$ (Brown, 1950), $n=19$, 36 (Gould, 1958), $2n=36$ (Gould, Bowden).—Dry rocky places, 3500–8200 ft.; Joshua Tree Wd., Pinyon-Juniper Wd.; San Jacinto and San Bernardino mts. to N.Y., Clark, and Inyo-White mts.; to Atlantic Coast. Occasional weed in San Bernardino V. May–Aug.

4. **S. flexuòsus** (Thurb.) Rydb. [*Vilfa cryptandra* var. *f.* Thurb.] Resembling *S. cryptandrus* but with more open, often elongate panicles with slender spreading or drooping flexuous loosely fld. branches.—Occasional on deserts below 4000 ft.; largely Creosote Bush Scrub; w. Colo. Desert, Mojave Desert from Little San Bernardino Mts. to Death V. area; to Utah, w. Tex., n. Mex. Sept.–Oct.

5. **S. poirétii** (R. & S.) Hitchc. [*Axonopus p.* R. & S.] Erect perennial, ± tufted, 3–10 dm. tall; blades flat to subinvolute, 2–5 mm. wide, fine-pointed; panicle mostly spicate, 1–4 dm. long; spikelets ca. 2 mm. long; glumes subequal, ca. half as long as spikelet; lemma acutish; $2n=36$ (Gould, 1966).—Occasional weed in lawns and near cult. Native e. states and S. Am. July–Sept.

82. Stenotáphrum Trin. St. Augustine Grass

Creeping stoloniferous perennials with short flowering culms, rather broad and short obtuse blades, and terminal and axillary racemes. Spikelets embedded in one side of an enlarged and flattened corky rachis tardily disarticulating toward the tip at maturity, the spikelets remaining attached to the joints. First glume small, 2d glume and sterile lemma subequal, the latter with a palea or a ♂ fl. Fertile lemma chartaceous. Three spp. of warm regions. (Greek, *stenos*, narrow, and *taphros*, trench, because of the cavities in the rachis.)

1. **S. secundàtum** (Walt.) Kuntze. [*Ischaemum s.* Walt.] Rather coarse, bright green, the culms compressed, branched, with flowering stems 1–3 dm. tall; blades 5–15 cm. long, 4–10 mm. wide; racemes 5–10 cm. long; spikelets in 1's or 2's, 4–5 mm. long; $n=10$ (Brown, 1948).—Escape in low waste places from cult. in lawns, as at Santa Barbara; native se. U.S. and trop. Am. July–Sept.

83. Stìpa L. Speargrass. Needlegrass

Tufted perennials with involute lvs. and terminal panicles of 1-fld. spikelets which are articulate above the glumes, the articulation oblique, leaving a bearded sharp-pointed callus at the base of the floret. Glumes membranaceous, often papery, narrow, acute to acuminate or aristate. Lemma narrow, terete, firm or indurate, strongly convolute, terminating in a prominent awn, the junction of the body and awn evident, the awn twisted below, geniculate, mostly persistent; palea enclosed in the lemma. Ca. 150 spp. of temp. regions. (Greek, *stupe*, tow, alluding to the feathery awns of the type sp.)

Stipa

(Dedecca, D. M. Studies on the Calif. spp. of Stipa. Madroño 12: 129–139. 1954.)
1. Panicle broad, open, loose, with spreading branches; lemma at least 4 times as long as palea.
    2. Lemma less than 7 mm. long; awn up to 4 cm. long .... 10. *S. lepida*
    2. Lemma more than 7 mm. long; awn much more than 4 cm. long.
        3. Lemma slender, cylindrical; middle culm lvs. 1.2–2.4 mm. broad; foliage usually glaucous; awn slender, flexuous beyond the 2d bend, mostly 9–12 times as long as lemma
            .... 4. *S. cernua*
        3. Lemma fusiform; middle culm lvs. 2.4–6 mm. broad; foliage usually green; awn stout, stiff, mostly 7–9 times as long as lemma .... 14. *S. pulchra*
1. Panicle narrow, slender or contracted; lemma less than 4 times as long as palea.
    4. Lemma densely appressed-villous, with hairs 3–4 mm. long and rising above the summit like a pappus-crown.
        5. Lemma ca. 8 mm. long; awn 4–5 cm. long .... 6. *S. coronata*
        5. Lemma up to 6 mm. long; awn ca. 2 cm. long .... 13. *S. pinetorum*
    4. Lemma sparsely strigose to subglabrous, never long-villous.
        6. Awns pubescent, most commonly plumose.
            7. First segm. of the once-geniculate awn strongly plumose, the hairs 5–8 mm. long
                15. *S. speciosa*
            7. First segm. of the awn conspicuously pubescent, the hairs not more than 2 mm. long.
                8. Ligule hyaline, 3–6 mm. long .... 16. *S. thurberiana*
                8. Ligule opaque, less than 3 mm. long.
                    9. Palea 4–5 mm. long .... 9. *S. latiglumis*
                    9. Palea not more than 3 mm. long.
                        10. Sheaths glabrous.
                            11. Hairs on upper part of lemma longer than those below; awn with rather short hairs .... 3. *S. californica*
                            11. Hairs of lemma all short; awn with rather long hairs ... 12. *S. occidentalis*
                        10. Sheaths pubescent .... 8. *S. elmeri*
        6. Awns scabrous to subglabrous, rarely appressed-hispid, not plumose.
            12. Mature lemma brownish.
                13. Lemma 8–12 mm. long, ± sparsely pubescent; awn 10–15 cm. long. Common native .... 5. *S. comata*
                13. Lemma 3.5–6 mm. long, pubescent in lines; awn 1.1–1.8 cm. long. Rare introduction .... 2. *S. brachychaeta*
            12. Mature lemma yellowish, not more than 8 mm. long.
                14. Palea nearly as long as lemma; hairs at summit of lemma longer than those of body
                    11. *S. lettermannii*
                14. Palea ½–¼ as long as lemma; lemma short-hirsute throughout.
                    15. Culms densely pubescent below the nodes; palea 3–4 mm. long . 7. *S. diegoensis*
                    15. Culms glabrous below the nodes; palea not more than 2.5 mm. long.
                        16. Awn 4–6 mm. long, obscurely geniculate, with flexuous terminal segm.; lemma short-hirsute except near the glabrous tip. .... 1. *S. arida*
                        16. Awn usually 3–5 cm. long, twice geniculate; lemma hirsute throughout
                            17. *S. williamsii*

1. **S. árida** Jones. Culms 3–8 dm. tall; blades 1–2 dm. long, 1–2 mm. wide, scabrous, flat or involute; panicle 1–1.5 dm. long, narrow, pale or silvery; glumes 8–12 mm. long; lemma 4–5 mm. long, strigose below, scaberulous upward; awn 4–6 cm. long, loosely twisted for 1–2 cm., flexuous beyond.—Dry probably limestone slopes, 4000–5700 ft.; Pinyon-Juniper Wd.; Funeral Mts. and Clark Mt. of e. Mojave Desert; to Colo., Tex. May–June.

2. **S. brachychaèta** Godr. Densely cespitose perennial to 1 m. tall; blades firm, flat or loosely involute; panicle narrow, open, the few spikelets on slender pedicels; glumes 6–8 mm. long; lemma 3.5–6 mm. long, brown, pubescent in lines; awn 1.1–1.8 cm. long. —Reported from near Camarillo, Ventura Co.; *Fuller*; native of Argentina.

3. **S. califórnica** Merr. & Davy. Culms 6–15 dm. high, glabrous or with pubescent nodes; sheaths glabrous or the lower puberulent; ligule firm, 1–2 mm. long; blades 1–4 mm. wide, flat (later ± involute), 1–2 dm. long; panicle 1.5–3 dm. long or more, slender, pale; glumes pale, hyaline, glabrous, 3-nerved, ca. 12 mm. long; lemma 6–8 mm. long, sparsely villous, the uppermost hairs ca. 1.5 mm. long; awn 2.5–3.5 cm. long, twice bent,

the 1st and 2d segms. plumose; $n=18$ (Stebbins & Love, 1941).—Dry open places, 4500–10,400 ft.; Sagebrush Scrub to Bristle-cone Pine F.; White Mts.; Yellow Pine F.; San Bernardino Mts.; to Wash., Ida. May–Aug.

4. **S. cérnua** Steb. & Love. Culms mostly 6–9 dm. tall, in rather large clumps; basal lvs. many, narrow, glaucous, cauline, 1.2–2.4 mm. wide; panicle open with slender flexuous branches; glumes acuminate, the 1st 12–19 mm. long, the 2d shorter; lemma 5–10 mm. long, papillose, pilose below, and on nerves; awn 6–11 cm. long, twice bent, the terminal segm. flexuous, scabrous or basally short-pubescent; $n=35$ (Stebbins & Love, 1941).—Dry slopes below 4500 ft.; Coastal Sage Scrub, S. Oak Wd., etc.; San Diego Co. to n. Calif. Santa Catalina and Santa Cruz ids. April–May.

5. **S. comàta** Trin. & Rupr. Culms glabrous, 3–6 dm. high; sheaths glabrous, naked at throat; ligule thin, 3–4 mm. long; blades 1–3 dm. long, 1–2 mm. wide, flat or involute; panicle narrow, 1–2 dm. long, commonly included at base; glumes 15–20 mm. long, with ± hyaline attenuate tips; lemma 8–12 mm. long, pale or ± brownish, somewhat sparsely pubescent; awn 10–15 cm. long, indistinctly bent twice, slender, loosely twisted below, flexuous above, often deciduous.—Dry places, 5000–8600 ft.; largely Pinyon-Juniper Wd., Yellow Pine F., Lodgepole F.; San Bernardino Mts., mts. of n. and ne. Mojave Desert; to Alaska, Ind., Tex. June–July.

6. **S. coronàta** Thurb. in Wats. Culms very stout, 1–2 m. tall, 4–6 mm. thick at base, erect, glaucous; sheaths glabrous with pubescent throat; blades flat, 3–6 dm. long, 5–10 mm. wide, involute toward tip; panicle narrow, pale or purplish, ± nodding, 2–5 dm. long; glumes unequal, the 1st 15–20 mm. long, the 2d 13–16 mm. long, both 5-nerved; lemma 9 mm. long, appressed-villous; awn 3.5–5 cm. long, scabrous, twice-bent, ca. 1 cm. long, the 2d a little shorter; $n=20$ (Stebbins & Love, 1941).—Dry slopes, usually below 5500 ft.; Coastal Sage Scrub, Chaparral; L. Calif. to Napa Co. April–June.

Var. **depauperàta** (Jones) Hitchc. [*S. parishii* var. *d.* Jones. *S. parishii* Vasey.] Culms usually 3–6 dm. tall; panicle 1–1.5 dm. long, dense; awns ca. 2.5 cm. long, once geniculate.—Dry rocky slopes, mostly 3000–9000 ft.; Pinyon-Juniper Wd., etc.; desert slopes of mts. from e. San Diego Co. to San Bernardino Mts. and to Inyo Co.; then to Utah, Ariz. May–Aug.

7. **S. diegoénsis** Swall. Culms 7–10 dm. tall, scaberulous, pubescent; ligule 1–2 mm. long; blades 1.5–4 dm. long, 2–4 mm. wide, flat or involute; panicle 1.5–3 dm. long, dense, narrow; glumes acuminate, the 1st 9–10 mm. long, 1-nerved, the 2d 8–9 mm. long, 3-nerved; lemma ca. 7 mm. long, with summit hairs 1–2 mm. long; awn 2–3 cm. long, twice bent, scabrous.—Along vernal stream at 800 ft.; Chaparral; Jamul, e. San Diego Co., n. L. Calif. May–June.

8. **S. élmeri** Piper & Brodie ex Scribn. Culms erect, puberulent, 4–8 dm. tall; sheaths pubescent; ligule very short; blades 1.5–3 dm. long, 2–4 mm. wide, flat or later involute, pubescent on upper surface; panicle narrow, 1.5–3.5 dm. long, rather loose; glumes 12–14 mm. long, long-acuminate, hyaline above; lemma ca. 7 mm. long, strigose; awn 4–5 cm. long, twice bent into subequal segms., the 1st and 2d finely plumose; $n=18$ (Stebbins & Love, 1941).—Dry gravelly and rocky places, 7000–11,200 ft.; Montane Coniferous F.; San Jacinto and San Bernardino mts.; Sierra Nevada to Wash., Ida. June–Aug.

9. **S. latiglùmis** Swall. Culms slender, 5–10 dm. tall, strigose below; at least lower sheaths pubescent; blades flat or ± involute, pilose on upper surface; ligule 1–4 mm. long; panicle narrow, 1.5–3 dm. long, loosely fld.; glumes subequal, firm, acute to acuminate, 3-nerved, ± purplish, 1.3–1.5 cm. long; lemma densely pubescent, 8–9 mm. long, the awns twice bent, 3.5–4.5 cm. long, plumose on 1st and 2d segms.; $n=35$ (Dedecca, 1954).—Dry slopes, 4000–6000 ft.; Yellow Pine F.; San Jacinto Mts.; Sierra Nevada. June–July.

10. **S. lépida** Hitchc. [*S. eminens*, Calif. refs.] Plate 101, Fig. F. Slender, puberulent below the nodes, 6–10 dm. tall; sheaths smooth, rarely puberulent, ± villous at throat; ligule very short; blades 1–3 dm. long, flat, 2–4 mm. wide, pubescent on basal part of upper surface; panicle rather loose, open, 1.5–2 dm. long, with slender distal branches;

glumes 3-nerved, smooth, acuminate, the 1st 6–10 mm. long, the 2d ca. 4–8 mm. long; lemma ca. 6 mm. long, sparingly villous; awn indistinctly twice bent, 2.5–4 cm. long, scabrous; $n=17$ (Stebbins & Love, 1941).—Dry slopes below 4000 ft.; Chaparral, Coastal Sage Scrub, etc.; L. Calif. to n. Calif. Channel Ids. March–May.

11. **S. lettermánnii** Vasey. Culms 2–3 dm. tall, in large tufts; blades filiform, strongly involute, 2–3 dm. tall; panicle slender, narrow, 1–1.5 dm. long, glumes equal, 7–9 mm. long; lemma 4.5–6 mm. long, the apical hairs ca. 1.5 mm. long, those on the body slightly shorter; palea 3.5–4.5 mm. long; awn glabrous on the 2 lower segms., ± scaberulous on the 3d; $n=16$ (Johnson, 1962).—At 6500–8000 ft.; Yellow Pine F.; San Bernardino Mts. June–July.

12. **S. occidentàlis** Thurb. [*S. stricta* var. *sparsiflora* Vasey. *S. o.* var. *montana* Merr. & Davy.] Culms slender, 1.5–4 dm. tall; sheaths glabrous to pubescent; ligule ca. 1 mm. long; blades 1–2 dm. long, 1–2 mm. wide, mostly involute, white-puberulent on upper surface; panicle 1–2 dm. long, lax; glumes ca. 12 mm. long, hyaline at the pointed tips; lemma pale brown, ca. 7 mm. long, sparsely strigose; awns 3–4 cm. long, twice bent, plumose, the hairs on the 2 lower segms. ca. 1 mm. long, on the upper short; $n=18$ (Stebbins & Love, 1941).—Dry open woods, 5000–11,000 ft.; Yellow Pine F.; Bristle-cone F., etc.; San Jacinto, San Bernardino and San Gabriel mts., Panamint and Inyo-White mts.; to Wash., Ariz. June–Aug.

13. **S. pinetòrum** Jones. In large tufts, 3–5 dm. tall; ligule to ca. 0.5 mm. long; lvs. mostly basal, the blades involute-filiform, 5–12 cm. long, slightly scabrous; panicle narrow, 8–10 cm. long; glumes ca. 9 mm. long; lemma narrowly fusiform, ca. 5 mm. long, with a conspicuous tuft of hairs above and with 2 hyaline teeth 1 mm. long at apex; awn ca. 2 cm. long, twice bent, subglabrous.—Dry rocky places, 7000–12,000 ft.; Pinyon-Juniper Wd., Bristle-cone Pine F.; Panamint and Inyo-White mts.; to Mont., Ida., Colo. June–Aug.

14. **S. púlchra** Hitchc. [*S. setigera* Calif. refs.] Culms 6–10 dm. tall; blades flat or involute, 2.5–6 mm. wide, deep green; panicle nodding, loose, ca. 1.5–2 dm. long, with slender spreading branches; lower glume 15–26 mm. long, the 2d slightly shorter, 3–5-nerved; lemma 7.5–13 mm. long, fusiform, pubescent throughout or at base and on nerves to middle or top; awn 6–9 cm. long, short-pubescent to 2d bend, the 1st segms. 1.5–2 cm. long, the 2d shorter, the 3d 4–6 cm.; $n=32$ (Stebbins & Love, 1941).—Dry slopes below 5000 ft.; Chaparral, Coastal Sage Scrub, etc.; L. Calif. to n. Calif.; Channel Ids. March–May.

15. **S. speciòsa** Trin. & Rupr. Culms many, erect, 3–6 dm. high; sheaths firm, the lowermost pubescent, the throat short-villous; ligule short; blades slender, coriaceous, 2–4 dm. long, involute-filiform, mostly basal; panicle dense, narrow, 1–1.5 dm. long; glumes subequal, pale, papery, ca. 15 mm. long, the 1st 3-nerved, the 2d 5-nerved; lemma ca. 8 mm. long, densely long-pilose on lower half or more, the hairs 6–8 mm. long; awn with 1 sharp bend, the 1st section 1.5–2 cm. long, long-pilose, the upper section scabrous, ca. 2.5 cm. long; $n=30$ (Stebbins & Love, 1941).—Dry rocky places, below 6500 ft.; Creosote Bush Scrub, Joshua Tree Wd., etc. on both deserts, occasional in Chaparral of interior cismontane valleys; n. to cent. Calif.; to Colo., Ariz., S. Am. April–June.

16. **S. thurberiàna** Piper. Culms 1.5–5 dm. tall; sheaths smooth or ± scabrous; ligule 3–6 mm. long; blades mostly basal, 1.5–2.5 dm. long, involute-filiform, scabrous to soft-pubescent, flexuous; panicle 8–15 cm. long, strigose, the awn 4–5 cm. long, twice bent, the 1st and 2d segms. plumose with hairs 1–2 mm. long; $n=17$ (Stebbins & Love, 1941).—Dry open woods, 5000–8500 ft.; Pinyon-Juniper Wd.; Inyo-White Mts.; to n. Calif. June–July.

17. **S. williámsii** Scribn. Erect, 6–10 dm. tall, puberulent especially about the nodes; sheaths puberulent; ligule very short; blades ± puberulent, 1–3 mm. wide, 1–2 dm. long; panicle narrow, 1.5–2 dm. long; glumes thin, subequal, 1 cm. long; lemma ca. 7 mm. long; awn usually 3–5 cm. long.—Dry places, at 8000 ft.; Montane Coniferous F.; S. Fork, San Bernardino Mts.; n. Calif. to Wash., Rocky Mts. June–July.

84. Swallènia Soderstrom & Decker

Stiff perennial, freely branched, from a long thick scaly rhizome with woolly nodes; flowering culms erect or ascending, 2.5–3.5 dm. tall, ridged, glabrous except for the puberulent summit. Lvs. distant, stiff, harsh, the sheaths villous on the margin near the summit; ligule a ring of hairs; blades 5–10 cm. long, 3–5 mm. wide, with pungent apex. Panicles narrow, simple, 5–10 cm. long, with pubescent compressed branches. Spikelets several-fld., the glumes and lemmas persistent on the short-jointed rachilla; glumes subequal, broad, 7–11-nerved, 9–14 mm. long; lemmas rounded on back, imbricate, thin, 5–7-nerved, 7–9 mm. long, densely villous on margins of lower part; palea equal to lemma. Monotypic. (Named for J. R. *Swallen*, American grass student.)

1. **S. alexándrae** (Swallen) Soderstrom & Decker. [*Ectosperma a.* Swallen.] Forming extensive masses on sand dunes, at 3000–3500 ft.; Creosote Bush Scrub; s. end of Eureka V. and mouth of Marble Canyon, e. side of Inyo Mts., Inyo Co. April–June.

85. Trìdens R. & S.

Low or moderately tall perennials with blades usually flat and panicles open or contracted. Spikelets several-fld., disarticulating above the glumes and between the florets. Glumes mostly thin and membranous, subequal, the 1st 1-nerved, the 2d 1–3-nerved, rarely 5-nerved. Lemmas broad, thin, 3-nerved, short-hairy on the nerves below, rounded on the back. Palea slightly shorter than the lemma, strongly 2-nerved. Ca. 16 spp. of N. and S. Am. (Latin, *tria*, thrice, and *dens*, tooth, referring to the lemma.)

1. **T. mùticus** (Torr.) Nash. [*Tricuspis m.* Torr. *Triodia m.* Scribn.] Culms slender, densely tufted, 2–5 dm. tall; sheaths and blades scaberulous, the former loosely pilose, especially at summit; blades ± involute, 1–3 mm. wide; panicle 4–14 cm. long, interrupted; spikelets 6–8-fld., ca. 1 cm. long, pale to purplish, subterete; glumes scaberulous, 5–6 mm. long, 1-nerved; lemmas ca. 5 mm. long, thin, obtuse, pilose on nerves near their base, ± purplish; $2n = 40$ (Tateoka, 1961).—Rocky places, often on limestone, 3000–6000 ft.; Creosote Bush Scrub, Pinyon-Juniper Wd.; deserts from Chuckawalla and Eagle mts. n. through Providence Mts. and Clark Mt. to Sierra Nevada; Utah, Mex. Spring and Autumn.

86. Trisètum Pers.

Tufted perennials with flat blades and open or contracted shining panicles. Spikelets usually 2-fld., sometimes 3–5-fld., the rachilla prolonged beyond the upper floret, usually villous. Glumes ± unequal, acute, the 2d usually exceeding the 1st floret. Lemmas usually short-bearded at base, 2-toothed at apex, the teeth often awned, usually bearing from the back a straight or bent awn. Ca. 75 spp. of colder and temp. regions. (Latin, *tri*, three, and *seta*, bristle, because of the awn and 2 teeth of lemma.)

1. **T. spicàtum** (L.) Richt. [*Aira s.* L.] Culms slender, erect, densely tufted, 1–4 dm. high, glabrous or puberulent; sheaths and usually the blades puberulent; panicle dense, almost spikelike, pale or dark purple, 5–15 cm. long; spikelets 4–6 mm. long; 1st glume 1-nerved, 2d broader, acute, 3-nerved; lemmas scaberulous, 5 mm. long, the 1st exceeding the glumes, with setaceous teeth; awn attached ca. $\frac{1}{3}$ way below the tip, 5–6 mm. long, geniculate, exserted; $2n = 28$ (Tateoka, 1954).—At 7000–13,700 ft.; Montane Coniferous F., San Bernardino Mts. and San Jacinto Mts.; Bristle-cone Pine F.; White Mts.; to the Arctic, also in Antartic. July–Aug.

87. Tríticum L.

Annuals with flat blades and thick spikes. Spikelets 2–5-fld., solitary, placed flatwise at each joint of the rachis, the rachis often disarticulating above the glumes and between the florets. Glumes rigid, keeled, 3–several-nerved, the apex abruptly mucronate or toothed or awned. Lemmas broad, keeled, many-nerved, pointed or awned. Ca. 30 spp. of Medit. region and w. Asia. (Latin name for wheat.)

1. **T. aestìvum** L. WHEAT. Six–10 dm. tall; blades 1–2 cm. wide; spike mostly 5–12 cm. long; rachis-internodes 3–6 mm. long; spikelets broad; glumes usually keeled toward 1 side; 2n=42 (Chapman & Riley, 1966).—Commonly cult. and often escaping in waste places; native of Old World. April–July.

17. Pontederiàceae. PICKEREL-WEED FAMILY

Perennial aquatic or bog plants. Lvs. petioled, with rounded to elongate blades. Fls. perfect, 1 to many and spicate, regular to irregular, without bracts, subtended by leaflike spathes. Perianth with a tube, 6-lobed or -parted, petaloid, free from the ovary. Stamens 3 or 6, inserted on the perianth; anthers 2-celled, introrse. Ovary superior, 1- or 3-celled; style 1; stigma 3-lobed or 6-toothed. Fr. a caps. or ak.; endosperm copious. Six genera and ca. 20 spp., mostly in warmer parts of the world.

1. *Eichhórnia* Kunth. WATER-HYACINTH

Floating or rooting at nodes on mud. Lvs. floating or emersed, obovate, cordate to lanceolate. Fls. spicate, rarely paniculate; perianth funnelform with long or short tube. Stamens 6, unequally inserted, some of them exserted. Ovary sessile, 3-celled, many-ovuled; style filiform. Six spp. in tropics. (Named for J. A. F. *Eichhorn*, 1779–1856, German statesman.)

1. **E. cràssipes** (Mart.) Solms. [*Pontederia c.* Mart. *Piaropus c.* Britton.] Lvs. 1–12 cm. wide, ovate to rounded, slightly scabrous above; petioles inflated at base; scape 1–4 dm. high, sheathed near middle; fls. many, showy; perianth ca. 5 cm. long, 6-lobed, violet, the upper lobe enlarged and with patch of blue having yellow center; $n=16$ (Taylor, 1925), $2n=32$ (Briggs, 1962).—Occasionally natur. in sloughs and ponds, Santa Ana R. system and at Ramona and San Ysidro, San Diego Co.; native of trop. Am., where it may be a serious pest interfering with navigation. June–Oct.

18. Potamogetonaceae. PONDWEED FAMILY

Aquatic herbs, entirely submerged or with some floating lvs. Lvs. threadlike or grass-like or some broader and floating, commonly with sheathing base or stipules. Fls. small, floating or submersed or barely raised above the water-surface, in spikes or heads, perfect, with or without a calyx of 4 distinct sepals. Petals none. Stamens 4 or 2; anthers extrorse. Carpels 4 or more, sessile or stipitate, indehiscent, the seeds solitary, without endosperm. A small family, widely distributed.

Sepals 4, distinct; stamens 4; carpels sessile .... 1. *Potamogeton*
Sepals none; stamens 2; carpels stipitate in fr. .... 2. *Ruppia*

1. *Potamogèton* L. PONDWEED

Herbs of ponds and streams, with jointed mostly rooting stems and 2-ranked lvs., which are usually alternate or imperfectly opposite, the submerged lvs. often pellucid, the floating ones often dilated and firmer in texture. Stipules sheathing. Fls. in spikes which are sheathed by the stipules in the bud, later mostly raised on a peduncle to the surface of the water. Fls. perfect, small. Perianth of 4 free, rounded, short-clawed, valvate segms. Stamens 4, inserted on the claws; anthers 2-celled, sessile, extrorse. Carpels 4, free, sessile, 1-celled and 1-ovuled; stigmas on short styles or sessile. Fr. drupelike when fresh, ± compressed, with bony endocarp; seed 1, without endosperm. A widely distributed genus of perhaps 70 spp., many of which are important as food for wild fowl and various mammals. In some cases the entire plant, in others the starchy rhizomes, are eaten. (Greek, *potamos*, river, and *geiton*, a neighbor, because of the habitat.)

(Fernald, M. L. The linear-leaved N. Am. spp. of Potamogeton, section Axillares. Mem. Am. Acad. Arts & Sci. 17: pt. 1: 1932. Ogden, E. C. The broad-leaved spp. of P. of N. Am. n. of Mex. Rhodora 45: 57–104, 119–163, 171–213, 1943.)

1. Lvs. all submerged and similar.
   2. The lvs. linear.
      3. Stipules fused with base of lf. to form a sheath at least 1 cm. long.
         4. The lvs. fascicled terminally, 2–4 mm. wide. Little Lake, Inyo Co. ... 5. *P. latifolius*
         4. The lvs. generally distributed, scarcely 1 mm. wide. Widespread. ..... 8. *P. pectinatus*
      3. Stipules free from the lvs. or nearly so.
         5. The lvs. without basal glands; nerves 3–5 .... 3. *P. foliosus*
         5. The lvs. with a pair of basal glands; nerves mostly 3 .... 9. *P. pusillus*
   2. The lvs. broad (lanceolate to elliptic or ovate), never linear.
      6. Margins of lvs. serrulate throughout .... 1. *P. crispus*
      6. Margins of lvs. serrulate at tip only .... 4. *P. illinoensis*
1. Lvs. of 2 kinds, floating (broad and coriaceous) and submerged (broad or narrow, but thin).
   7. Submerged lvs. linear or threadlike, not differentiated into petiole and blade; floating lvs. elliptical, subcordate at base .... 6. *P. natans*
   7. Submerged lvs. linear to lanceolate or ovate, with true blades.
      8. The submerged lvs. ovate to lanceolate, more than 4 mm. wide.
         9. Blades of submerged lvs. petioled and not sharp-pointed at apex .... 7. *P. nodosus*
         9. Blades of submerged lvs. sessile, sharp-pointed .... 4. *P. illinoensis*
      8. The submerged lvs. filiform to linear, 0.5–1.5 mm. wide .... 2. *P. diversifolius*

1. **P. críspus** L. Stems compressed, branching; lvs. all submerged, sessile, half-clasping, oblong, 2–6 cm. long, 5–12 mm. wide, obtuse, crisped and serrulate, with prominent midrib and at least 2 fine, lateral veins; stipules scarious, deciduous, splitting; propagating buds prominent in axils of decayed lvs.; spikes cylindric, 1–1.8 cm. long; fr. ovoid, 3-keeled, almost 3 mm. long, with a beak ca. 2 mm. long; $n=26$ (Scheerer, 1939). —Natur. in Santa Ana R. system, Mohave R., Santa Catalina Id.; cent. Calif., etc.; to Atlantic Coast. Native of Eu. July–Sept.

2. **P. diversifòlius** Raf. [*P. dimorphus* of Calif. refs.] Stems freely branched, flattened to subterete, 2–10 dm. long; submerged lvs. narrowly linear, acute, 0.5–1.5 mm. wide, 2.5–7 cm. long, faintly 3-nerved; their stipules 6–10 mm. long, fused ca. half their length; floating lvs. elliptic to narrowly obovate, the blades 1–4 cm. long, 3–20 mm. wide, 7–15-nerved, on usually shorter petioles; their stipules usually becoming fibrous; submerged spikes subglobose, on peduncles to 5 mm. long, emersed spikes cylindric, 5–20 mm. long, on peduncles 5–15 mm. long; frs. suborbicular, 1–1.5 mm. long, the beak obtuse.— Lake Surprise, San Jacinto Mts. at 9000 ft.; cent. Calif. n. and e. to Atlantic Coast, Mex.

3. **P. foliòsus** Raf. [*P. f.* var. *californicus* Morong. *P. c.* Piper.] Stems flattened, leafy, slender, freely branched, 2–10 dm. long; winter-buds on very short branches; lvs. submerged, numerous, linear, 3–10 cm. long, 1.5–2.5 mm. wide, 3–5-nerved, without basal glands; the midrib with 1–3 rows of loosely cellular tissue on each side near the base; stipules at first connate, later free, 15–25 mm. long, early deciduous; peduncles 5–20 mm. long; spikes many, few-fld. and short; frs. broadly ovoid, pitted, 2–2.5 mm. long, 3-keeled, the middle keel winged and denticulate; $2n=28$ (Stern, 1961).—Hard or brackish water below 5000 ft.; well distributed in cismontane Calif. and in Mohave R. drainage; to B.C., Atlantic Coast, Cent. Am. July–Oct.

4. **P. illinoénsis** Morong. [*P. lucens* auth., not L.] Stem simple or branched, 1.5–5 mm. thick, 3–6 dm. long; submerged lvs. elliptic to lanceolate, often somewhat arcuate, the blades 5–20 cm. long, 1.5–4 cm. wide, sessile or with petioles to 4 cm. long; stipules persistent, obtuse, 2.5–8 cm. long, 0.5–1.2 cm. wide, 15–35-nerved; floating lvs. absent, or, if present, subcoriaceous, elliptic to oblong, the blades 4–12 cm. long, 2–6.5 cm. wide, cuneate or rounded at base, 13–29-nerved; stipules broader than for submerged lvs.; peduncles ± thickened, 4–15 cm. long; spikes cylindric, 8–15-whorled, dense and 3–6 cm. long in fr.; frs. obovoid to orbicular, 2.5–3.5 mm. long, prominently and acutely keeled, the beak deltoid, 0.5 mm. long; $n=52$ (Stern, 1961).—Below 5000 ft., San Diego Co. to B.C., Atlantic Coast, L. Calif. June–Aug.

5. **P. latifòlius** (J. W. Robbins) Morong. [*P. pectinatus* var. *l.* J. W. Robbins.] Stem

1–2 mm. thick, whitish, freely branched above, 2–6 dm. long; lvs. all submerged, fascicled terminally, linear, obtuse or short-apiculate, 2–8 cm. long, 2–4 mm. wide, 3–5-nerved, with many cross-veins; sheaths loose, swollen, 1–2.5 cm. long, much thicker than stem; peduncles 2–10 cm. long; spikes 1–3 cm. long, interrupted; fr. obovoid, 3 mm. long, 2 mm. wide, the style slender, incurved, forming a beak.—In slightly alkaline ponds, such as Little Lake, Inyo Co.; to Ore., Nev. July–Aug.

6. **P. nàtans** L. Stems simple or nearly so, terete, 1–2 mm. thick; submerged lvs. coriaceous, linear, 10–20 cm. long, 1–2 mm. wide, obscurely nerved; stipules clasping the stem, whitish, fibrous, persistent, linear to lanceolate, 4–9 cm. long, 2-keeled; floating lvs. ovate to oblong-ovate, usually subcordate at base, the blades 4–9 cm. long, 2.5–6 cm. wide, 23–37-nerved; stipules broader than those of submerged lvs.; peduncles scarcely thickened, 3–8 cm. long; spikes 8–14-whorled, 3–5 cm. long in fr.; frs. obovoid, 3.5–5 mm. long, rounded on back, beak short and broad; $n=21$ (Stern, 1961).—At 4000–7500 ft.; San Bernardino Mts.; Coast Ranges and Sierra Nevada to Alaska, Atlantic Coast; Eurasia. June–Aug.

7. **P. nodòsus** Poir. [*P. americanus* Cham. & Schlecht. *P. lonchites* auth. *P. fluitans* auth.] Plate 102, Fig. A. Stems terete (often flat when pressed), branched, 1–1.5 mm. thick, 1–2 m. long; submerged lvs. thin, 7–15-nerved, lance-linear to lance-elliptic, the blades 9–20 cm. long, 1–3.5 cm. wide, gradually tapering into long petioles; stipules brownish, linear, 3–6 (–9) cm. long; floating lvs. coriaceous, elliptical, cuneate or rounded at base, 5–9 cm. long, 2–4 cm. wide, 13–21-nerved; stipules 3–6 cm. long, linear, somewhat keeled; peduncles thicker than stem, 3–15 cm. long; spikes 10–15-whorled, 3–6 cm. long in fr.; frs. obovoid, 3.5–4 mm. long, prominently keeled, brown or reddish when mature, with erect linear beak to 1 mm. long.—Ponds and streams, from sea-level to 7400 ft.; from San Diego Co. n.; to B.C., Atlantic Coast, S. Am. May–Aug.

8. **P. pectinàtus** L. Plate 102, Fig. B. Rootstock creeping, branched, slender, with terminal thickened tubers; stems filiform, ca. 1 mm. thick, much-branched, 5–20 dm. long; lvs. submerged, setaceous, mostly 1-nerved, attenuate at apex, 3–15 cm. long, scarcely 1 mm. wide; sheaths 2–5 cm. long, only slightly thicker than stem, often bleached; peduncles filiform, 5–25 mm. long; whorls 2–6, widely spaced, fr. obliquely obovoid, 2.5–4 mm. long, 2–3 mm. wide, the style slender, incurved, forming a beak; $2n=78$ (Harada, 1942), $2n=42$ (Misra, 1966).—Brackish or fresh water, widely distributed below 7000 ft.; to B.C., Atlantic Coast; temp. and trop. region. The small tubers important for wild ducks. May–July.

9. **P. pusíllus** L. [*P. panormitanus* Biv.] Stems filiform, branching, 3–10 dm. long; lvs. submerged, linear, pointed to rounded at tip, 2–6 cm. long, 0.5–3 mm. wide, with a translucent gland on each side of the base; stipules hyaline, 6–15 mm. long, free from the lvs. except at very base, edges connate so as to form a tube; peduncles filiform, 6–8 cm. long; spikes 6–12 mm. long, few-whorled, interrupted; fr. obliquely obovoid, 2–3 mm. long, rounded on back, obscurely keeled; $2n=26$ (Harada, 1956).—Ponds and slow streams below 8000 ft.; occasional, San Diego Co. n.; to Alaska, Atlantic Coast; Eurasia. May–June. Variable as to lf.-width.

2. Rùppia L. Ditch-Grass

Herbs of saline water, with filiform forking stems and almost capillary alternate lvs., abruptly sheathing at the base. Fls. 2, approximate on a slender spadix, which is at first inclosed in the sheathing spathelike base of a lf., consisting of 2 single stamens, each with 2 anther-sacs, and 4 small sessile carpels with single ovules. Stigma sessile, depressed. Frs. small, obliquely ovoid, pointed drupes, each on a slender stalk that develops after flowering, the spadix then raised on a filiform peduncle. Fls. raised to surface of water at anthesis. Apparently 2 spp. (Named for H. B. *Rupp*, an 18th century German botanist.)

(Gamerro, J. C. Observaciones sobre la biología floral y morfología de la Potamogetonacea *Ruppia cirrhosa* (Petag.) Grande. Darwiniana 14: 575–608. 1968.)

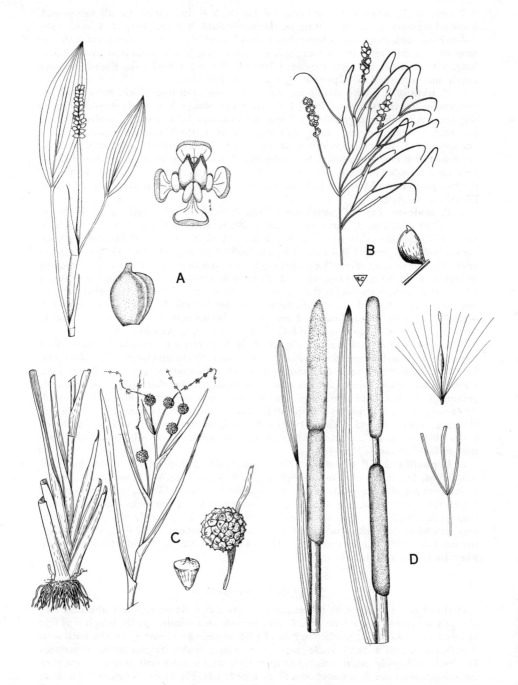

Plate 102. Fig. A, *Potamogeton nodosus*, fl. with 4 perianth-segms., 4 stamens, 4 carpels. Fig. B, *Potamogeton pectinatus*, whorls of fls. 2–6. Fig. C, *Sparganium eurycarpum*, ♂ fl.-heads above. ♀ beneath and also drawn separately. Fig. D, *Typha* sp., upper spike ♂, lower, ♀ lower right fl. with 3 stamens, ♀ fl. with stipitate ovary and basal hairs.

Peduncles 3–30 cm. long, spiraling or flexuous; lvs. acute .... 1. *R. cirrhosa*
Peduncles 0.2–2.5 cm. long, not spiraling; lvs. obtusish .... 2. *R. maritima*

1. **R. cirrhòsa** (Petag.) Grande. [*Buccaferrea c.* Petag. *Ruppia spiralis* L. ex Dumort.] Much like *R. maritima*, but the peduncles becoming 5–30 cm. long in fr. and spiraling or flexuous; lvs. acute, often ca. 1 mm. wide; $2n=40$ (Reese, 1962).—Mostly in salty ponds below 4000 ft. and of the interior (Death V., Little Lake, Castaic Lake, Palmdale, Salton Sea), more coastal from Santa Barbara Co. n.; cosmopolitan.

2. **R. marítima** L. [*R. m.* var. *intermedia* Asch. & Graebn. and var. *rostrata* Ag. *R. rostellata* Koch.] Stems branching, 6–10 dm. long, with internodes 2–8 cm. long; lvs. threadlike, 2–10 cm. long, 0.3–0.5 mm. wide, blunt at apex, with basal stipular sheath 6–10 mm. long; peduncles 2–15 mm. long; stipes of carpels 1–2.5 cm. long in fr.; fr. ovoid, ca. 2 mm. long; $2n=20$ (Reese, 1962).—Coastal Salt Marsh of the mainland and Santa Catalina and Santa Cruz ids.; widely distributed over the world. Variable, with a number of named varieties.

19. Sparganiàceae. BUR-REED FAMILY

Perennial herbs from a creeping rhizome. Stems simple or branched, leafy. Lvs. elongate, alternate, sheathing at base, floating and flaccid or erect and stiffer. Fls. unisexual, densely crowded in globose heads on the upper stem and branches, the ♂ heads above the ♀. Perianth of a few chaffy elongate scales. Staminate fls. of 3 or more stamens; fils. free or partly united; anthers oblong, basifixed. Pistillate fls. with a sessile mostly 1-celled ovary narrowed at base; style simple or forked; ovule 1. Frs. indehiscent, crowded, narrowed at base, nutlike. Seed with mealy endosperm. One genus.

1. *Spargànium* L. BUR-REED

Characters of the family. Ca. 15 spp. of aquatic habitats, temp. and cold regions of N. Hemis. and Australasia. (Greek, *sparganion*, a swaddling band, referring to the long narrow lvs.)

Stigmas 2; mature aks. rather truncate at apex .... 2. *S. eurycarpum*
Stigma 1; mature aks. tapered at apex.
    Lvs. 1.5–4 mm. wide, floating, the middle and upper enlarged and sheathing at the base
                                                                                             1. *S. angustifolium*
    Lvs. 5–12 mm. wide, usually erect, scarcely inflated at the base ..... 3. *S. multipedunculatum*

1. **S. angustifòlium** Michx. [*S. simplex* var. *a.* Torr.] Stems floating, slender, 2–10 dm. long; lvs. flaccid, largely floating, usually longer than stem, 2–4 (–5) mm. wide, with the nerves 0.2–0.8 mm. apart, rounded on the back, the upper lvs. with enlarged sheathing bases; infl. simple; ♂ heads 2–6; ♀ 2–4, the lower on supra-axillary peduncles, 1.2–2 cm. thick when mature; aks. fusiform, 5–6 mm. long, stipitate, the beak ca. 1 mm. long; $2n=30$ (Löve & Löve, 1942).—Ponds and slowly moving water; Montane Coniferous F.; San Bernardino Mts.; Sierra Nevada to Alaska, Atlantic Coast. July–Aug.

2. **S. eurycárpum** Engelm. [*S. greenei* Morong. *S. e.* var. *g.* Graebn. *S. californicum* Greene.] Plate 102, Fig. C. Stems 5–25 dm. high, erect; lvs. equaling or shorter than the stem, flat, somewhat keeled beneath, 7–12 mm. wide; infl. branched; ♂ heads 5–many; ♀ heads 1–4 on a branch, sessile or peduncled, 2–3 cm. in diam. when mature; perianth-scales almost as long as aks.; aks. sessile, obpyramidal, subtruncate at summit, 6–9 mm. long, the beak ending in 2 stigmas.—Swamps and along streams at low elevs., cismontane and montane; L. Calif. to B.C., Atlantic Coast. April–Aug. The material near the coast seems to have shorter frs. (5–8 mm. long) and more rounded at the summit and is the basis for the var. *greenei*.

3. **S. multipedunculàtum** (Morong) Rydb. [*S. simplex* var. *m.* Morong. *S. s.* auth., not Huds.] Stems 4–10 dm. high, erect; lvs. flat, scarcely inflated at the base, commonly exceeding the stem, 5–12 mm. wide, with nerves 0.8–2 mm. apart; infl. simple; ♂ heads

4–8; ♀ heads 2–5, the lower usually supra-axillary and peduncled, 18–25 mm. in diam. when mature; aks. fusiform, stipitate, 4–5 mm. long, the beak ca. 1.5 mm. long, with 1 stigma.—Swamps and ponds below 8000 ft.; Bluff Lake, San Bernardino Mts.; Sierra Nevada to Alaska, Atlantic Coast. June–Aug.

20. Typhàceae. Cat-Tail Family

Tall perennial herbs from creeping rhizomes. Stems simple, submerged at base, cylindrical, jointless. Lvs. mostly radical, elongate-linear, alternate, rather thick and spongy. Fls. unisexual, anemophilous, very numerous, crowded in a terminal elongate spike, the ♂ above, the ♀ below. Perianth of many slender, jointed threads. Staminate fls. with 2–5 stamens; fils. free or connate; anthers linear, basifixed. Fertile ♀ fls. with 1-celled stipitate ovary, narrowed into a style with narrow or ligulate stigma. Many ♀ fls. sterile and with abortive terminal ovaries and perianth hairs beneath. Fr. dry, tardily dehiscent. Seed striate, with mealy endosperm. One genus.

1. *Týpha* L. Cat-Tail (Plate 102, Fig. D)

Characters of the family. Ca. 10 spp. of fresh water and marshy places, temp. and trop. regions. (The ancient Greek name.)

(Smith, S. Galen. Am. Midl. Nat. 78: 257–287, 1967.)

Lvs. mostly 5 mm. wide, strongly convex on back, dark green; ♀ spikes much overtopped by lvs. and with bracts very dark brown, opaque and firm; lf.-sheaths auriculate ... 1. *T. angustifolia*
Lvs. 6–15 mm. wide; ♀ spikes not or moderately overtopped by lvs. and with bracts light brown and translucent or lacking.
    Lf.-blades moderately convex on back, light yellow-green; bracts of ♂ spikes cuneate, laciniate, brown; bracts of ♀ spikes ovate and apiculate; interval between ♂ and ♀ parts of spikes 1–4 cm. long .... 2. *T. domingensis*
    Lf.-blades flat, light green; bracts of ♂ spikes simple, hairlike, white; bracts of ♀ spikes none; no interval between ♂ and ♀ portions of spikes .... 3. *T. latifolia*

1. **T. angustifòlia** L. Stems 1–1.5 m. tall; lvs. 7–13, much exceeding ♀ spikes, sometimes by half their length, usually ca. 5 mm. wide, strongly convex on the back, dark green, the sheaths usually auriculate at their summit; ♀ portion of mature spike 8–20 cm. long, 1.3–2 cm. thick, uniform, dark-brown or red-brown in color, becoming greenish brown as stigmas wear off and finally mottled dark brown and buff; bracts spatulate, blunt, dark brown, opaque and firm; stigmas dark brown, linear, not fleshy; interval between ♂ and ♀ parts of spike variable, but usually ca. twice the diam. of the ♀ spike; ♂ spike with simple or forked, linear, brown bracts and light lemon-yellow 1-celled pollen; $n=15$ (S. Galen Smith).—Mostly in subalkaline water at low elevs.; Freshwater Marsh; to Atlantic Coast; Eurasia. June–July.

2. **T. domingénsis** Pers. [*T. truxillensis* HBK. *T. bracteata* Greene. *T. angustifolia* many refs.] Plant 2–3 m. tall; lvs. 6–9, equaling or slightly exceeding ♀ spikes, 6–12 mm. wide, moderately convex on back, light yellowish green, the sheaths tapering into the blade; mature ♀ spike 15–25 cm. long, 1.5–2.2 cm. thick, uniform, light cinnamon-brown, becoming buffy or grayish; bracts light brown, translucent and spongy; stigmas light brown, linear, not fleshy; interval between ♀ and ♂ spikes usually ca. the diam. of the former; ♂ spikes with cuneate laciniate brown bracts and golden-yellow 1-celled pollen. —Widely distributed in subsaline habitats below 5000 ft.; Freshwater Marsh; L. Calif. to Ore.; s. Atlantic Coast, S. Am.; Eu. June–July. Santa Catalina and San Clemente ids.

3. **T. latifòlia** L. Plant 1–2.5 m. high; lvs. 12–16, moderately exceeding the ♀ spikes, 8–15 mm. wide, nearly flat, light green; the sheaths tapering into the blade or truncate; mature ♀ spike 10–18 cm. long, 1.8–3 cm. thick, often thickened upward, dark greenbrown to red-brown, becoming whitish as stigmas wear off; bracts none; stigmas medium to dark brown, lance-ovate, fleshy; no interval between ♀ and ♂ spikes; ♂ spikes with

simple, hairlike, white bracts and deep orange-yellow, 4-celled pollen; $2n=30$ (Harada, 1947).—Freshwater Marsh; throughout Calif., below 5000 ft.; Channel Ids.; to Alaska, Atlantic Coast, Eu. June–July.

T. × gláuca Godron. Under this name have been included many plants which seem to combine characters of the other spp. and represent hybrids between *T. angustifolia* and *T. latifolia.* According to Mason (A flora of the marshes of Calif., p. 43. 1957), the one outstanding character of these hybrids is the presence of yellow pith in the stem immediately above the rhizome.

21. Zannichelliàceae. HORNED PONDWEED FAMILY

Submerged aquatic monoecious or dioecious herbs with slender creeping rhizome. Stems capillary, branched. Lvs. alternate or opposite or crowded at nodes, linear, sheathing at base, the sheaths usually ligulate. Fls. minute, axillary, solitary or cymose. Perianth of 3 small free scales or lacking. Stamens 3, 2, or 1; anthers 2–1-celled, opening lengthwise. Carpels free, 1–9; style short or long, simple or 2–4-lobed; stigma capitate, peltate or spathulate. Fruiting carpels indehiscent, 1-seeded, sessile or stipitate. Seed lacking endosperm. Ca. 6 genera, widely distributed, in fresh, brackish or salt water.

Plants dioecious; style shorter than the stigma; pollen filiform. Salton Sea ...... 1. *Halodule*
Plants monoecious; style longer than stigma; pollen globose. Widespread ..... 2. *Zannichellia*

1. Halódule Endl.

Habit of *Zannichellia*; dioecious; style shorter than stigma; pollen filiform; carpels 2 and united, or 1. Fr. a nutlet. Two spp.

1. **H. wrìghtii** Aschers. [*Diplanthera w.* Aschers.] Lvs. bicuspidate at the semi-lunate apex; anthers ca. 6 mm. long; mature fr. black.—Introd. intentionally into the Salton Sea from Texas and now well established. Native in se. states and W. Indies.

2. Zannichéllia L. HORNED PONDWEED

Lvs. opposite, filiform but flat, 1-nerved. Staminate and ♀ fls. in the same axil, enclosed in the bud by a hyaline spathelike envelope; ♂ solitary, without a perianth; stamen 1. Pistillate fls. 2–5 (usually 4) almost sessile, the carpels in a hyaline invol., each with short style and peltate stigma. Frs. flattish, usually toothed down one side. One or 2 spp. of fresh or brackish waters; cosmopolitan. (G. G. *Zannichelli*, 1662–1729, Italian botanist.)

1. **Z. palústris** L. Plate 103, Fig. A. Stems 3–5 dm. long; lvs. 2–8 cm. long; frs. 2–2.5 mm. long, short-stipitate; $2n=12$ (Reese, 1957).—Pools and slow streams, below 7000 ft.; desert, cismontane and insular; cosmopolitan. March–Nov.

22. Zosteràceae. EEL-GRASS FAMILY

Submerged marine perennials with creeping rhizomes. Stems flattened, slender, simple or branched. Lvs. in 2 rows, linear, sheathing at base. Plants monoecious or dioecious, the fls. arranged on 1 side of a flattened axis or spadix, at first enclosed in the lf.-sheath; bracts lacking. Perianth absent or represented by a row of bractlike lobes on each side of the axis. Staminate fls. reduced to a single anther, arranged alternately in 2 rows on the spadix. Pistillate fl. an ovary with 2 stigmas and 1 ovule. Fr. indehiscent or bursting irregularly. Seed without endosperm. Family of 2 widely distributed genera.

Plants dioecious; fr. cordate at base; lvs. 1.5–4 mm. wide .... 1. *Phyllospadix*
Plants monoecious; fr. ovoid; lvs. mostly 6–12 mm. wide .... 2. *Zostera*

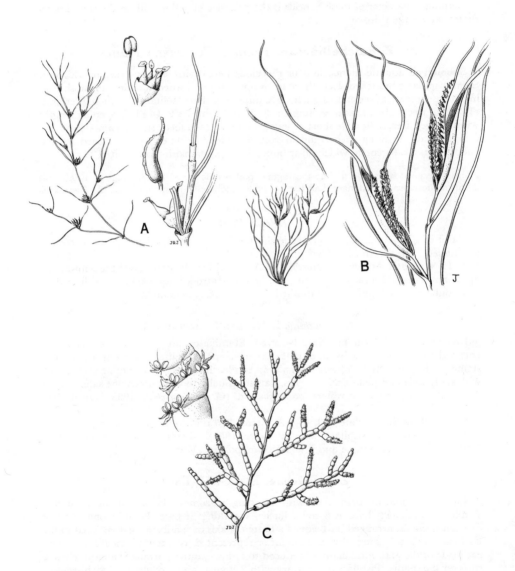

Plate 103. Fig. A, *Zannichellia palustris*, ♂ fl. with 1 stamen, ♀ fls. 4. Fig. B, *Phyllospadix torreyi*, with spathe around 2 or 3 spadices or spikes. Fig. C, *Allenrolfea occidentalis*, of the Chenopodiaceae, jointed stems and alternate branches.

1. Phyllospádix Hook. SURF-GRASS

Rootstocks thickened. Plants dioecious, with the infl. at or near the summit of the slender stems, the spikes (spadices) solitary or 2 or 3 within a spathe. Staminate fls. in 2 rows, many, each of 1 sessile stamen. Pistillate of simple sessile ovaries; style short, with 2 filiform stigmas. Fr. beaked, cordate-sagittate. Seed globose, not ridged. Few spp. of the n. Pacific. (Greek, *phullon*, lf. and *spadix*, spicate infl.)

Pistillate spadix usually 1, on flowering stems 1–2 dm. long .... 1. *P. scouleri*
Pistillate spadices 2–6, borne on flowering stems 3–10 dm. long .... 2. *P. torreyi*

1. **P. scoùleri** Hook. Stems 1–2 (–4) dm. long; lvs. flat, mostly 2–4 mm. wide; ♂ spadix ca. 6 cm. long; ♀ 2–5 cm. long; fr. ca. 3.5 mm. long.—On surf-beaten rocky shores, San Diego to B.C.; Japan. May–July. Santa Catalina Id., San Nicolas Id., Santa Barbara Id.

2. **P. tórreyi** Wats. Plate 103, Fig. B. Stems flat, 3–10 dm. long; lvs. linear, from flat to subterete, ca. 1.5 mm. wide, 1–2 m. long; spadices cauline, in 2–3 pairs, ♀ usually ca. 5 cm. long, ♂ shorter; fr. flask-shaped, 2–3 mm. long, beaked by the persistent style, cordate-sagittate at base.—Usually below low-tide level in quiet waters; L. Calif. to Ore. May–Nov. San Clemente, Santa Catalina, Santa Rosa ids. May–Nov.

2. Zóstera L. EEL-GRASS

Rootstocks slender. Monoecious; lvs. to 2 m. long. Fls. imperfect, the ♂ and ♀ alternating, in 2 rows in a 1-sided spike. Anther 1-celled; pollen threadlike. Style elongate; stigmas capillary. Mature carpels flasklike, beaked, forming a utricle. Seed usually ribbed. A genus of ca. 5 spp. of both hemis. Of great importance as food for waterfowl. (Greek, *zoster*, a girdle or band, because of the ribbonlike lvs.)

1. **Z. marìna** L. [*Z. m.* var. *latifolia* Morong. *Z. l.* Morong. *Z. pacifica* Wats. *Z. oregona* Wats.] Stems branched, 1–3 m. long; lvs. 3–15 dm. long, (2–) 6–12 mm. wide, with 3–7 main nerves; spadix 2.5–6 cm. long, bearing ca. 10–20 of each kind of fls.; anthers escaping from the spathe and discharging pollen into water; stigmas protruding through the spathe and dropping off before anthers of same spadix open; seed 20-ribbed, ca. 3 mm. long; $n=6$ (Wulff, 1937).—Shallow water of bays, usually in muddy bottoms in the intertidal zone; San Diego to Alaska; Atlantic Coast; Eurasia. June–Sept. There has been some tendency to refer the Pacific Coast plants to var. *latifolia*, with lvs. 6–12 mm. wide and seeds not ribbed, while var. *marina* would refer to Atlantic plants, but the differences seem untenable.

ADDITIONS AND CORRECTIONS

Dicotyledones

18. Brassicàceae

29. *Rorippa* Scop.

R. L. Stuckey (Sida 4: 279–426. 1972) has a revision of *Rorippa* for North America, which he confines to the yellow-flowered species. The white-flowered Water Cress is not included, but probably would be put in the genus *Nasturtium*. The following key for southern California spp. is modeled on Stuckey (distribution is difficult to give for our area since Stuckey does not give definite localities, but maps only):

1. Petals 1.2–3.5 mm. long; oldest siliques on lower portion of the terminal raceme
                                                         3. *R. palustris* ssp. *occidentalis* var. *occidentalis*
1. Petals usually less than 1.2 mm. long; oldest siliques on lower axillary racemes or siliques nearly equal in development at corresponding points on terminal and lateral racemes.
   2. Replum margin minutely hirsute and the valves readily dehiscent, or if replum margin glabrous, the valves not readily dehiscent; petals spreading between the sepals in living plants.
      3. Plants prostrate; lvs. usually oblanceolate, pinnatifid to midrib with narrow lobes; siliques mostly 4–8.3 mm. long; replum margin minutely hirsute
                                                 1. *R. curvisiliqua* var. *orientalis*
      3. Plants erect; lvs. spatulate, unlobed or with wide lobes; siliques 8–17 mm. long; replum margin glabrous .... 1. *R. curvisiliqua* var. *occidentalis*
   2. Replum margin glabrous; valves readily dehiscent; petals erect in living plants.
      4. Siliques short- to elongate-cylindrical; replum oblong to triangular in outline.
         5. Plants mostly shorter than 3 dm.; cauline lvs. usually fewer than 10; silique valves minutely papillose.
            6. Siliques rough with minute papillae, tapering to apex, slightly curved, not constricted at center; lvs. lyrate-divided, with entire margin .... 5. *R. tenerrima*
            6. Siliques glabrous, scarcely tapering to apex, straight, constricted at center; lvs. pinnate-divided to midrib, the margins with angular teeth .... 6. *R. truncata*
         5. Plants mostly taller than 3 dm.; cauline lvs. usually more than 10; silique-valves glabrous .... 2. *R. intermedia*
      4. Siliques globose; replum circular in outline .... 4. *R. sphaerocarpa*

1. **R. curvisilíqua** (Hook.) Bessey ex Britton is represented in southern California by: (1) var. **orientàlis** Stuckey and apparently occurs at Little Lake, Inyo Co., in San Bernardino Mts. and Palomar Mts.

(2) var. **occidentàlis** (Greene) Stuckey. [*Nasturtium occidentale* Greene.] in wet places at low elevs., Los Angeles and Orange cos. and in the pine belt of the San Bernardino Mts.

2. **R. intermèdia** (O. Ktze.) Stuckey. [*Cardamine palustris mexicana intermedia* O. Ktze.] may have been collected at Cuyamaca Lake, San Diego Co.; it ranges south to Costa Rica.

3. **R. palústris** (L.) Besser subsp. **occidentàlis** (Wats. in Gray) Abrams var. **occidentàlis** is apparently the common *Rorippa* in so. Calif., being the same as given in this Flora.

4. **R. sphaerocárpa** (Gray) Britton [*Nasturtium s.* Gray.] is occasional in the San Bernardino Mts. as on Fish Creek at 7200 ft.

5. **R. tenérrima** Greene seems to be in the mts. of San Diego Co.

6. **R. truncàta** (Jeps.) Stuckey. [*R. sinuata* (Nutt.) Greene var. *truncata* Jeps.] is found between 5500 ft. and 8000 ft., San Gabriel and San Bernardino mts. It has largely been called *R. obtusa* by auth.

40. Ericàceae

4a. *Ornithostáphylos* Small

Shrubs with erect stiff branched stems. Lvs. opposite, numerous, persistent, narrow, thick, entire, revolute, short-petioled. Fls. many, paniculate. Calyx persistent, mostly 5-lobed. Corolla globular-urceolate, mostly 5-lobed. Stamens mostly 10, included; fils. unappendaged, pubescent; anther-sacs awned. Ovary depressed, seated on a thick disk. Drupe dry, with thin pulp.

1. **O. oppositifòlia** (Parry) Small. [*Arctostaphylos oppositifolia* Parry.] Lvs. 2.5–8 cm. long, ± tomentulose beneath; panicles partly drooping; calyx ca. 2.5 mm. wide, with deltoid lobes; corolla 3–3.5 mm. long; drupes 4–6 mm. long. This common native in n. L. Calif. has recently been collected in chaparral at n. end of Tijuana Hills, 3 miles w. of San Ysidro, San Diego Co. by *R. Moran*.

GLOSSARY

Compiled by David D. Keck

Abaxial. Located on the side away from the axis.
Abortive. Imperfect or barren.
Abrupt. Terminating suddenly; not tapering.
Acaulescent. Stemless or essentially so. (Cf. *Caulescent.*)
Accessory. Additional to the usual number of organs.
Accrescent. Increasing in size with age, as often with the calyx after flowering.
Accumbent. Lying against something, as cotyledons against the radicle.
Acerose. Needle-shaped, as the leaves of pines.
Acicular. Needlelike.
Actinomorphic. Exhibiting radial symmetry, as a regular flower.
Acumen. A tapering point.
Acuminate. Gradually tapering to a short point. (Cf. *Acute, Attenuate.*)
Acute. Sharp-pointed, but less tapering than acuminate.
Adaxial. Located on the side nearest the axis.
Adnate. Grown together with an unlike part, as the calyx-tube with an inferior ovary, or an anther by its whole length with the filament.
Adventitious, adventive. Out of the usual place; introduced but not yet naturalized.
Aestivation. The arrangement of the parts in a flower bud.
Aggregate. Collected into dense clusters or tufts. *Aggregate fruit:* one formed by the clustering together of pistils that were distinct in the flower, as blackberry.
Akene (achene). A small, dry, hard, indehiscent, 1-seeded fruit.
Alliaceous. Onionlike, usually in respect to odor.
Alpine. Strictly applicable to plants growing above timber line.
Alternate. Any arrangement of parts along the axis other than opposite or whorled; situated regularly between other organs, as stamens alternate with petals.
Alveolate. Honeycombed; with deep angular cavities (alveoli) separated by thin partitions; faveolate.
Ament. Catkin.
Amplexicaul. Clasping the stem, as the base of certain leaves.
Ampliate. Enlarged; dilated.
Anastomosing. Netted; particularly applied to veins so connected by cross veins as to form a network.

Anatropous. An inverted and straight ovule, with the micropyle next to the hilum.
Ancipital. Two-edged, as certain flattened stems.
Androecium. The whole set of stamens.
Androgynous. Having staminate and pistillate flowers in the same inflorescence, or in *Carex* in the same spikelet, the former above the latter.
Anemophilous. Wind-pollinated.
Annual. Of one year's or season's duration from seed to maturity and death. *Winter annual:* a plant from autumn-germinating seed that fruits in the following spring.
Annular. Circular; in the form of a ring, or marked transversely by rings.
Annulus. A ring-shaped part or organ, such as surrounds the sporangium in some ferns.
Anterior. On the front side; in the flower the side away from the axis and toward the subtending bract.
Anther. The pollen-bearing part of the stamen.
Antheridium. The male sexual organ of ferns, etc., analogous to the anther.
Antheriferous. Anther-bearing.
Anthesis. Strictly, the time of expansion of the flower, but also used for the period during which the flower is open and functional.
Anthocarp. A structure in which the fruit proper is united with the perianth or receptacle.
Anthocyanous. Showing anthocyanin in the herbage, a class of soluble glucoside pigments producing reddish or purplish coloring.
Antipetalous. Opposite the petals. *Antisepalous:* opposite the sepals.
Antrorse. Directed upward or forward.
Apetalous. Without petals.
Aphyllopodic. Without leaves at the base.
Aphyllous. Leafless.
Apical. Situated at the tip.
Apiculate. Terminated abruptly in a little point.
Apomictic. Producing seed without any form of fertilization or sexual union.
Appendage. Any attached supplementary or secondary part.
Appressed. Pressed flat against another organ.
Approximate. Near together.
Arachnoid. Cobwebby; of soft and slender entangled hairs.
Arborescent. Treelike in tendency.
Archegonium. The female sexual organ of ferns, etc., analogous to the pistil.

[1019]

Arcuate. Moderately curved, as if bent like a bow.

Areola (-ae). A little area defined on a surface, as the angular space between vein reticulations. In Asteraceae the circle at the summit of the akene where sat the corolla. *Areolate:* with areolae; reticulate.

Aril. A process of the placenta adhering about the hilum of a seed. *Arillate:* with an aril.

Aristate. Awn- or bristle-tipped, as the floral bracts of barley. *Aristulate:* bearing a short awn.

Articulate. Jointed; having a place for natural separation with a clean-cut scar.

Ascending. Rising obliquely or curving upward.

Asepalous. Without sepals.

Asperous. Rough to the touch; scabrid.

Assurgent. Ascending, rising.

Attenuate. Slenderly tapering or prolonged; more gradual than acuminate.

Auricle. An ear-shaped appendage. *Auriculate:* bearing auricles.

Awn. A terminal slender bristle on an organ.

Axil. Upper angle formed by a leaf or branch with the stem.

Axile. Belonging to, or situated in, the axis.

Axillary. In an axil.

Axis. The central stem along which parts or organs are arranged; the central line of any organ.

Baccate. Berrylike and pulpy.

Banner. Upper petal of a papilionaceous flower.

Barbate. Bearded with long stiff hairs.

Barbed. Bearing sharp rigid reflexed points like the barb of a fish-hook.

Barbellate. With short, usually stiff hairs. *Barbellulate:* the diminutive.

Barbulate. Finely bearded.

Basal. Relating to, or situated at, the base.

Basifixed. Attached by the base.

Beak. A prolonged firm tip, particularly of a seed or fruit. *Beaked:* ending in a beak.

Bearded. Bearing long stiff hairs.

Berry. A pulpy indehiscent fruit with no true stone, as the tomato.

Bi- or Bis-. Latin prefix signifying two, twice, or doubly.

Bidentate. Having two teeth.

Biennial. Of two years' duration from seed to maturity and death.

Bifid. Two-cleft to about the middle.

Bifurcate. Two-forked or -pronged.

Bilabiate. Two-lipped (calyx or corolla).

Bilocular. Two-celled.

Binate. In pairs.

Bipartite. Divided into two parts almost to the base; two-parted.

Bipinnate. Doubly or twice pinnate.

Bipinnatifid. Twice pinnately cleft.

Bisected. Completely divided into two parts.

Bladdery. Thin and inflated.

Blade. The expanded part of a leaf or petal.

Bole. The trunk or stem of a tree.

Bract. A reduced leaf subtending a flower, usually associated with an inflorescence. *Bracteate:* with bracts.

Bractlet. A secondary bract borne on a pedicel instead of subtending it; sepaloid organs subtending the sepals in many Rosaceae. *Bracteolate:* with bractlets.

Bristly. Bearing stiff hairs.

Bud. An undeveloped stem, leaf, or flower. Buds are often enclosed by reduced or specialized leaves termed bud-scales.

Bulb. An underground leaf-bud with thickened scales or coats like the onion.

Bulbel. Daughter bulbs arising around the mother bulb.

Bulblet. A small bulb, especially one borne aerially as in a leaf axil or in the inflorescence.

Bullate. Blistered or puckered.

Caducous. Falling off very early or prematurely.

Calcarate. Spurred.

Callosity. A hardened thickening.

Callus. A callosity; the thickened extension at the base of the lemma in some grasses. *Callose:* bearing callosities.

Calyculate. Bearing bracts around the calyx imitating an outer calyx; said of the short bracts imitating an outer involucre in some Asteraceae.

Calyptra. A lid or hood.

Calyx. The external, usually green, whorl of a flower, contrasted with the inner showy corolla.

Calyx-lobe. In a gamosepalous calyx, the free projecting parts.

Campanulate. Bell-shaped.

Campylotropous (ovule or seed). So curved as to bring apex and base nearly together.

Canaliculate. Longitudinally channeled or grooved.

Canescent. Covered with grayish-white or hoary fine hairs.

Capillary. Hairlike; exceedingly slender.

Capitate. Head-shaped; aggregated into very dense clusters or heads.

Capitulum. See *Head.*

Capsule. A dry dehiscent fruit composed of more than one carpel.

Carinate. Keeled; with a sharp longitudinal ridge.

Carpel. A simple pistil, or one of the modified leaves forming a compound pistil.

Carpophore. A prolongation of the floral axis between the carpels, as that which supports the pendulous fruit of the Apiaceae.

Cartilaginous. Like cartilage in texture; tough.

Caruncle. An excrescense or outgrowth at or near the hilum of certain seeds.

Caryopsis. The grain or fruit of grasses.

Glossary

Castaneous. Chestnut-colored; dark brown.
Catkin. A scaly deciduous spike; ament.
Caudate. Bearing a tail or slender taillike appendage.
Caudex. The woody base of an otherwise herbaceous perennial.
Caulescent. With an obvious leafy stem; plants with radical leaves and flowers on a scape are called acaulescent.
Cauline. Belonging to the stem.
Cell. A cavity of an anther containing the pollen, or of an ovary containing the ovules. *Cellular:* made up of cells or marked off so as to resemble cells.
Cenospecies. All the ecospecies so related that they may exchange genes among themselves to a limited extent through hybridization.
Centripetal. Growing from without toward the center; an indeterminate inflorescence.
Cernuous. Nodding; drooping.
Cespitose. In little tufts or dense clumps; said of low plants of turfy habit.
Chaff. Thin dry scales.
Channeled. Deeply grooved longitudinally, like a gutter.
Chaparral. A xerophytic formation of dense, impenetrable thickets composed of stiff or thorny, mostly small-leaved, evergreen shrubs.
Chartaceous. With the texture of writing paper.
Ciliate. Fringed with hairs on the margin. *Ciliolate:* the diminutive.
Cincinnus (-i). A curl; used in the plural here for the branches of a unilateral scorpioid cyme.
Cinereous. Ash-colored; light gray.
Circinate. Coiled from the top downward with the apex as a center.
Circumpolar. Occurring around the pole, as of arctic plants mostly confined to far northern latitudes.
Circumscissile. Dehiscing by a transverse line around the fruit or anther, the top falling as a lid.
Cismontane. This side of the mountains, or west of the main Sierran crest, as opposed to the deserts.
Clasping. Leaf partly or wholly surrounding the stem; amplexicaul.
Clavate. Club-shaped; gradually thickened toward the apex from a slender base. *Clavellate:* the diminutive.
Claw. The narrow petiolelike base of some petals and sepals.
Cleft. Cut about halfway to the midrib.
Cleistogamous. Small flowers self-fertilizing without opening; usually additional to the ordinary flower and inconspicuous, as in some violets.
Coalescent. Said of organs of one kind that have grown together.

Cochleate. Coiled like a snail shell.
Coherent. Congenitally united with another organ of the same kind (*coalescent*), or of another kind (*adnate*).
Collar. Outer side of the grass leaf at the junction of sheath and blade.
Columella. The persistent axis of certain capsules.
Column. Body formed by union of stamens and pistil in orchids, or of stamens in mallows and milkweeds.
Coma. A tuft of hairs, particularly on a seed. *Comose:* furnished with a coma.
Commissure. The face by which two carpels cohere, as in Apiaceae.
Complanate. Flattened.
Complete. Having all the parts belonging to it, as a flower with sepals, petals, stamens, and pistils.
Complicate. Folded together.
Compound. Having two or more similar parts in one organ. *Compound leaf:* one with two or more separate leaflets. *Compound pistil:* having two or more carpels united.
Compressed. Flattened laterally.
Concave. Hollow.
Concolor. Of uniform color.
Conduplicate. Folded together lengthwise, as the leaves of many grasses.
Confluent. Blending of one part into another.
Congested. Crowded together.
Conglomerate. Densely clustered.
Conic. Cone-shaped, with the point of attachment at the broad base.
Conjugate. Joined in pairs.
Connate. Congenitally united, as similar organs joined as one. *Connate-perfoliate:* united at base in pairs around the supporting axis.
Connective. Portion of the filament connecting the two cells of an anther.
Connivent. Converging or coming together, but not organically united.
Constricted. Tightened or drawn together.
Continuous. Not articulated (grass rhachis) or interrupted.
Contorted. Twisted, bent, or distorted.
Contracted. Narrowed in a particular place, or shortened; the opposite of open or spreading (inflorescence).
Convex. Rounded on the surface.
Convolute. Rolled up longitudinally. Said of blades or floral envelopes in the bud when one edge is outside and the other inside.
Coralloid. Corallike.
Cordate. Heart-shaped with the notch at the base and ovate in general outline.
Coriaceous. Leathery in texture; tough.
Corm. A short, bulblike, underground stem, as the "bulb" of gladiolus.
Corneous. Of the texture of horn.
Corniculate. Bearing little horns or hornlike processes.

Corolla. The inner perianth of a flower, composed of colored petals, which may be almost wholly united.

Corona. A crown; the crownlike cup found at the orifice of the corolla-tube in *Narcissus.*

Coroniform. Crown-shaped.

Corrugated. Wrinkled; folded.

Cortex. Rind or bark.

Corymb. A flat-topped or convex racemose flower-cluster, the lower or outer pedicels longer, their flowers opening first. *Corymbose:* in corymbs.

Costa. A rib. *Costate:* ribbed; having longitudinal elevations.

Cotyledon. The primary leaf or leaves of the embryo.

Crateriform. Shallowly cup-shaped.

Creeping. Spreading over or beneath the ground and rooting at the nodes.

Crenate. Having the margin cut with rounded teeth; scalloped. *Crenulate:* the diminutive.

Crested. Having a crest, elevated appendage, or ridge on the summit of an organ.

Crinite. Bearded with long and weak hairs.

Crispate. Crisped or curled.

Crisped. Irregularly curled (said of hairs or leaf-margins).

Cristate. Crested or tufted.

Crown. The persistent base of an herbaceous perennial; the top of a tree; a circle of appendages on the throat of a corolla, etc.

Cruciferous. Cross-bearing; a flower with four petals placed opposite each other at right angles.

Crustaceous. Of brittle texture.

Cucullate. Hooded, or hood-shaped.

Cucullus. A hoodlike process on some seeds; cf. caruncle.

Culm. The type of hollow or pithy slender stem found in grasses and sedges.

Cuneate, cuneiform. Wedge-shaped; triangular, with the narrow part at point of attachment.

Cupulate. Cup-shaped, as the cup (involucre) of the acorn.

Cuspidate. Tipped with a cusp, or sharp, short, rigid point.

Cylindraceous. Somewhat or nearly cylindrical.

Cyme. A flat-topped or convex paniculate flower-cluster, with central flowers opening first. *Cymose:* arranged in cymes. *Cymule:* a small or few-flowered cyme.

Deciduous. Falling off, as petals fall after flowering, or leaves of nonevergreen trees in autumn.

Declined. Curved downward.

Decompound. More than once divided or compounded.

Decumbent. Lying down, but with the tip ascending.

Decurrent. Extending down the stem below the insertion; said of leaves or ligules.

Decussate. Opposite pairs (usually leaves) alternating at right angles with those above or below.

Deflexed. Turned abruptly downward.

Dehiscent. Opening spontaneously when ripe to discharge the contents, as an anther or seed vessel.

Deliquescent. Dissolving or melting away.

Deltoid. Equilaterally triangular.

Dendritic. Treelike, as the branching hairs of some Brassicaceae.

Dentate. Having the margin cut with sharp salient teeth not directed forward. *Denticulate:* slightly and finely toothed.

Depauperate. Dwarf, starved.

Depressed. Low, as if flattened from above.

Dextrorse. Turned to the right.

Diadelphous. Stamens united by their filaments into two sets.

Diandrous. Having two stamens.

Diaphanous. Transparent.

Dicarpellary. Having two carpels.

Dichotomous. Repeatedly forking in pairs.

Didymous. Twin; found in pairs.

Didynamous. With four stamens in two pairs of unequal length, as in most Lamiaceae.

Diffuse. Scattered; widely spread.

Digitate. Fingered; shaped as an open hand; compound with the members arising from one point.

Dilated. Flattened and broadened, as an expanded filament.

Dimorphic, dimorphous. Having two forms, as flowers with short stamens and long styles or long stamens and short styles.

Dioecious. Having staminate and pistillate flowers on different plants.

Diploid. Having two basic chromosome sets (twice the number in normal, haploid gametes).

Disarticulating. Separating joint from joint at maturity.

Discoid. Disklike. In the Asteraceae, a head without ray-florets.

Discrete. Separate; not coalescent.

Disk. A fleshy development of the receptacle about the base of the ovary. In Asteraceae, the tubular flowers (disk-florets) of the head as distinct from the ray.

Dissected. Deeply divided into numerous fine segments.

Distal. Opposite the point of attachment; apical; away from the axis.

Distichous. In two vertical rows or ranks.

Distinct. Separate; not united with parts in the same circle. Cf. *Free.*

Divaricate. Widely divergent.

Divergent. Extending away from each other by degrees.

Divided. Separated to the base.

Dolabriform. Axe-shaped or hatchet-shaped.

Glossary

Dorsal. Pertaining to the back; the surface turned away from the axis.
Dorsifixed. Attached to the back.
Dorsiventral. Having distinct dorsal and ventral surfaces.
Downy. Closely covered with very short and weak soft hairs.
Drooping. Erectish at base but bending downward above, as the branches of a grass panicle.
Drupe. A fleshy one-seeded indehiscent fruit containing a stone with a kernel; a stonefruit such as a plum.
E-, Ex-. Latin prefix meaning without, out of, from.
Ebracteate. Without bracts.
Echinate. Prickly, a hedgehog.
Ecospecies. All individuals so related that they are able to exchange genes freely without loss of fertility or vigor in the offspring.
Ecotype. Those individuals that are fitted to survive in only one kind of environment occupied by the species.
Edaphic. Pertaining to, or influenced by, soil conditions.
Ellipsoid. An elliptic solid.
Elliptic. In the form of a flattened circle more than twice as long as broad.
Emarginate. With a small notch at the apex.
Embryo. The incipient plantlet in the seed.
Emersed, emergent. Raised above the water.
Enation. An outgrowth on the surface of an organ.
Endocarp. The inner layer of the pericarp.
Endogenous. Forming new tissue within.
Endosperm. The nutritive tissue surrounding the embryo of a seed and formed within the embryo-sac.
Ensiform. Sword-shaped, as the leaves of *Iris*.
Entire. Undivided; the margin continuous, not incised or toothed.
Entomophilous. Insect-pollinated.
Epappose. Without pappus.
Ephemeral. Lasting for a day or less.
Epicarp. The outer layer of the pericarp.
Epigynous. Borne on the ovary; said of floral parts when the ovary is wholly inferior.
Equitant. Astride, as if riding, such as the leaves of an Iris.
Erect. Upright in relation to the ground, or sometimes perpendicular to the surface of attachment. A lip of a corolla is erect when in line with the tube.
Erose. Irregularly toothed as if gnawed.
Estipulate. Without stipules.
Evanescent. Quickly disappearing.
Evergreen. Remaining green through the winter.
Exalbuminous. Without endosperm, referring to seeds.
Excurrent. Projecting beyond the edge, as the midrib of a mucronate leaf.

Exocarp. The outer layer of the pericarp.
Exogenous. Forming new tissue outside the old.
Explanate. Spread out flat.
Exserted. Protruding, as stamens projecting beyond the corolla; not included.
Extrorse. Facing outward from the axis, as the dehiscence of an anther.
Falcate. Sickle-shaped.
Farinaceous. Containing starch; mealy in texture.
Farinose. Covered with a meallike powder.
Fascicle. A close cluster or bundle of flowers, leaves, stems, or roots. *Fasciculate:* in a fascicle.
Fastigiate. Clustered, parallel, erect branches.
Faveolate, favose. Honeycombed, as the receptacle in many Asteraceae; alveolate.
Fenestrate. With transparent areas or windowlike openings.
Ferruginous. Rust-colored.
Fertile. Said of pollen-bearing stamens and seed-bearing fruits.
Fetid. Disagreeably odorous.
Fibrillose. Furnished with little fibers.
-fid. A suffix meaning deeply cut.
Filament. A thread, especially the stalk of an anther.
Filiform. Threadlike.
Fimbriate. Fringed (with longer or coarser hairs as compared with ciliate). *Fimbrillate:* the diminutive.
Fistular, fistulous. Hollow throughout, as the leaf of an onion.
Flabellate, flabelliform. Fan-shaped; broadly wedge-shaped.
Flaccid. Weak, limp, soft or flabby.
Flagellate. With very slender runners.
Fleshy. Thick and juicy; succulent.
Flexuous. Zigzag.
Floccose. With *flocs* or tufts of soft woolly hair. *Flocculent:* the diminutive.
Floral tube. A more or less elongate tube consisting of perianth or other floral parts.
Floret. The individual flower of the Asteraceae and Poaceae; a small flower of a dense cluster.
Floriferous. Bearing or producing flowers.
Foliaceous. Leaflike; said especially of sepals or bracts that in texture or appearance resemble leaves.
Foliolate. Having leaflets.
Foliose. Having numerous leaves.
Follicle. A dry, monocarpellary fruit, opening only on the ventral suture, as in the larkspur.
Foveate. Pitted. *Foveolate:* the diminutive.
Free. Not joined to other organs; the reverse of adnate.
Frond. Leaf of a fern.
Fruit. The ripened pistil with all its accessory parts.
Frutescent, fruticose. Shrubby or bushy in the

sense of being woody. Fruticulose: applied to a little shrub.
Fugacious. Perishing very early.
Fulvous. Tawny; dull yellow.
Funnelform. Gradually widening upwards, like a funnel.
Furcate. Forked.
Furfuraceous. Covered with branlike scales; scurfy.
Fuscous. Grayish-brown.
Fusiform. Spindle-shaped; thickest near the middle and tapering toward each end.
Galea. The helmetlike upper lip in certain bilabiate corollas. *Galeate:* having a galea.
Gametophyte. The sexual form of the plant (as in ferns) contrasted with the sporophyte or asexual form.
Gamopetalous. Corolla with petals united. Same as sympetalous and monopetalous.
Gamophyllous. Composed of coalescent leaves or leaflike organs.
Gamosepalous. Calyx with sepals united.
Geminate. In pairs, twin, binate.
Genetic. That which is inherited.
Geniculate. Bent abruptly, as a knee.
Gibbous. Swollen on one side; ventricose.
Glabrous. Without hairs; incorrectly used in the sense of smooth, the antonym of rough. *Glabrate:* Almost glabrous; tending to be glabrous. *Glabrescent:* becoming glabrous.
Gland. A depression, protuberance, or appendage on the surface of an organ, which secretes a usually sticky fluid. *Glandular:* bearing glands or glandlike.
Glaucous. Covered or whitened with a bloom, as a cabbage leaf. *Glaucescent:* becoming glaucous.
Globose. Spherical or rounded. *Globular:* somewhat or nearly globose.
Glochid. A barbed hair or bristle. *Glochidiate:* barbed at the tip, as a bristle.
Glomerate. Densely compacted in clusters or heads. *Glomerulate:* arranged in small clusters.
Glomerule. A compact capitate cyme.
Glumes. The pair of bracts at the base of a grass spikelet. *Glumaceous:* resembling glumes.
Glutinous. With a gluey exudation.
Graduate. Marked with small regular distances.
Granular, granulose. Covered with very small grains or granules; minutely mealy.
Gregarious. Growing in groups or colonies.
Gynaecandrous. Having staminate and pistillate flowers in the same spikelet, the latter above the former.
Gynandrous. Stamens adnate to the pistil.
Gynobase. An elongation or dilation of the receptacle to support the carpels or nutlets, as in many borages.
Gynoecium. The pistils collectively of a flower.

Gynophore. The prolonged stipe of a pistil, as in *Cleome.*
Habit. General appearance of a plant.
Habitat. The normal situation in which a plant lives.
Halophyte. A plant of salty or alkaline soils.
Hastate. Halberd-shaped; of the shape of an arrowhead but with the basal lobes turned outward.
Haustoria. The suckerlike attachment organs of parasites like *Cuscuta.*
Head. A dense globular cluster of sessile or subsessile flowers arising essentially from the same point on the peduncle; capitulum.
Hemispheric. Shaped like half a sphere.
Herb. A plant without persistent woody stem, at least above ground.
Herbaceous. Pertaining to an herb; opposed to woody; having the texture or color of a foliage leaf; dying to the ground each year.
Herbage. Collectively, the green parts of a plant.
Heterogamous. Producing two or more kinds of flowers.
Heteromorphic. Of more than one kind of form.
Heterosporous. Having spores of two sizes or shapes.
Hexamerous. Having the floral whorls composed of six members.
Hexaploid. Having six basic chromosome sets (six times the gametic number).
Hilum. The scar at the point of attachment of an ovule or seed.
Hirsute. Rough with coarse or shaggy hairs. *Hirsutulous:* the diminutive.
Hirtellous. Minutely hirsute.
Hispid. Rough with stiff or bristly hairs. *Hispidulous:* the diminutive.
Hoary. Covered with white down.
Holosericeous. Covered with fine and silky hairs.
Holotype. The one specimen on which a species or other taxon is based.
Homogamous. A head or cluster with flowers alike throughout.
Homonym. In nomenclature, a name rejected because it duplicates a name previously and validly published for a group of the same rank based on a different type.
Homosporous. With spores all alike in size and shape.
Host. A plant which nourishes a parasite.
Humistrate. Spread over the surface of the ground.
Hyaline. Colorless or translucent, transparent.
Hybrid. A cross between two species.
Hydrophyte. Partially or wholly immersed water plant.
Hypanthium. A cup-shaped enlargement of the receptacle on which the calyx, corolla, and often the stamens are inserted; in perigyny the "calyx-tube."
Hypogynous. Borne on the receptacle below

Glossary

or free from the pistil; said of petals or stamens.
Imbricate. Overlapping as shingles on a roof.
Immersed. Growing under water.
Imparipinnate. Odd-pinnate, having a terminal leaflet.
Incised. Cut rather deeply and sharply.
Included. Not protruding beyond the surrounding organ or envelope.
Incumbent. Lying upon anything; said of cotyledons when the back of one rests against the stalk of the embryo.
Incurved. Bending inwards.
Indefinite. Relating to number, inconstant or very numerous.
Indehiscent. Not splitting open, as an akene.
Indeterminate. Not terminated absolutely, as a raceme.
Indigenous. Native to the country.
Indument. Any hairy covering or pubescence.
Induplicate. Valvate aestivation in which the margins of the leaves are bent or folded inward.
Indurate. Hard, hardened.
Indusium. In ferns, the epidermal outgrowth that covers or invests the sorus.
Inequilateral. Unequal-sided.
Inferior. Lower or beneath. *Inferior ovary:* one that is adnate to the hypanthium and situated below the calyx-lobes.
Inflated. Blown up; bladdery.
Inflexed. Turned abruptly inward.
Inflorescence. The flower-cluster of a plant, or, more correctly, the disposition of the flowers on an axis.
Infundibuliform. Funnelform.
Innate. Borne on the apex of the support; in an anther the antithesis of adnate.
Innocuous. Harmless, hence unarmed or spineless.
Innovation. In Gramineae a new sterile basal shoot; an offshoot from a stem.
Insectivorous. Consuming insects, i.e. by dissolving out the organic parts.
Inserted. Attached to or growing upon.
Insertion. The place or mode of attachment of an organ to its support.
Intercostal. Situated between ribs or costae.
Internerves. Spaces between the nerves.
Internode. The portion of stem between two nodes.
Interrupted. Not continuous or regular.
Interval. Space between ridges.
Introrse. Turned inward, towards the axis.
Intruded. Projecting inward or forward.
Inverted. Upside down; turned over.
Involucel. A secondary involucre, as the bracts subtending the secondary umbels in the Apiaceae.
Involucrate. Having an involucre.
Involucre A whorl of bracts (phyllaries) subtending a flower cluster, as in the heads of Asteraceae.
Involute. With the edges rolled inward, i.e., toward the upper side.
Irregular. Showing a lack of uniformity; asymmetric, as a zygomorphic flower.
Isotype. A specimen of the type collection other than the holotype.
Joint. The node of a grass culm; an articulation.
Keel. A prominent dorsal ridge, analogous to the keel of a boat; the two lower and united petals of a papilionaceous corolla.
Labellum. A lip; the odd petal of an orchid.
Labiate. Lipped; a member of the Labiatae.
Lacerate. Appearing irregularly torn or cleft.
Laciniate. Cut into narrow lobes or segments.
Lamellate. Made up of thin plates.
Lamina. The blade or expanded part of a leaf, petal, etc.
Lanate, lanose. Woolly; densely clothed with long entangled hairs. *Lanulose:* short-woolly.
Lanceolate. Lance-shaped; much longer than broad, tapering from below the middle to the apex and (more abruptly) to the base.
Lanuginous. Cottony or woolly; lanate.
Lateral. At or on the side.
Lax. Loose, distant.
Leaflet. A segment of a compound leaf.
Legume. A superior 1-celled fruit of a simple pistil usually dehiscent into two valves, having the seeds attached along the ventral suture; a leguminous plant.
Lemma. In grasses, the lower of the two bracts immediately enclosing the floret.
Lenticels. Corky spots on young bark, arising in relation to epidermal stomates.
Lenticular. Lens-shaped.
Lepidote. Covered with small scurfy scales; scurfy; furfuraceous.
Ligneous. Woody.
Ligule. The strap-shaped part of a ray corolla in Compositae; the thin, collarlike appendage on the inside of the blade at the junction with the sheath in grasses. *Ligulate:* provided with a ligule; strap-shaped or tongue-shaped.
Limb. A border; in particular, the expanded portion of a gamopetalous corolla or a gamosepalous calyx.
Linear. Resembling a line; long and narrow, of uniform width, as the leaf-blade of grasses.
Lineate. Marked with parallel lines.
Lingulate. Tongue-shaped.
Lip. One of the two divisions of a bilabiate corolla or calyx, hence an upper lip and a lower lip, although one lip may be wanting; the upper lip of orchids, by a twist of the ovary, usually appears to be the lower.
Littoral. Of a shore, particularly of the seashore.
Lobe. A division or segment of an organ, usually

rounded or obtuse; cut less than halfway to the midrib (of a leaf). *Lobed:* bearing lobes.
Locular. Having cells or loculi, as a bilocular pistil or anther is two-celled.
Loculicidal. Dehiscent longitudinally through the middle of the back of a pericarp, between the partitions into the cavity.
Lodicules. The 2 or 3 minute hyaline scales at the base of the stamens in grasses, representing the perianth.
Loment. A legume made up of 1-seeded joints.
Lunate. Crescent-shaped.
Lyrate. Lyre-shaped; pinnatifid, with the terminal lobe considerably larger than the others.
Macro-. Greek prefix meaning large or, more properly, long.
Macrosporangium. The organ in which macrospores are produced.
Macrospore. The larger of the two kinds of spores in Selaginellaceae, etc.
Macrosporophyll. The modified leaf that bears the macrosporangium.
Maculate. Spotted or blotched.
Malpighiaceous hairs. Straight appressed hairs attached by the middle and tapering to the free tips.
Mammillate. Having nipples.
Marcescent. When withered not falling off, especially leaves and corollas.
Marginate. Distinctly margined.
Maritime. Of the seacoast.
Medial. Of the middle.
Medullary. Pertaining to the pith.
Megaspore. A synonym of macrospore.
Membranaceous, membranous. Of the nature of a membrane; thin, soft and pliable.
-merous. Greek suffix, having parts, as pentamerous or 5-merous, having 5 parts.
Mesocarp. The middle layer or coat of a fruit.
Mesophyte. A plant that grows under medium moisture conditions.
Micro-. Greek prefix meaning small.
Micropyle. The minute orifice in the integuments of an ovule through which the pollen tube enters the seed cavity.
Microsporangium. The organ in which microspores are produced.
Microspore. The smaller of the two kinds of spores in Selaginellaceae, etc.
Microsporophyll. The modified leaf that bears the microsporangium.
Midrib. The central rib of a leaf or other organ.
Monadelphous. Stamens united by their filaments into a tube surrounding the gyncecium, as in malvaceous flowers.
Moniliform. Resembling a string of beads.
Monocephalous. Bearing only one head.
Monoecious. Having staminate and pistillate flowers on the same plant but not perfect ones.

Monotypic. Having a single type or representative, as a genus with only one species.
Montane. Pertaining to mountains.
Mucro. A small and short abrupt tip of an organ, as the projection of the midrib of a leaf. *Mucronate:* with a mucro. *Mucronulate:* minutely mucronate.
Multi-. Latin prefix for many.
Multicipital. Descriptive of a crown of roots or a caudex from which several stems arise.
Multifid. Cleft into many narrow lobes or segments.
Muricate. Rough with short and firm sharp excrescences. *Muriculate:* the diminutive.
Muticous. Pointless, blunt, awnless.
Naked. With a usual covering wanting, as a flower destitute of a perianth.
Napiform. Turnip-shaped.
Nascent. In the act of being formed.
Navicular. Cymbiform; boat-shaped; shaped like the bow of a canoe.
Nectariferous. Nectar-bearing; having a *nectary*, or an organ which secretes nectar.
Nerve. A simple vein or slender rib of a leaf or bract. *Nervelet:* an ultimate branch of a nerve.
Neutral. Devoid of functional stamens or gynoecium.
Node. The joint of a stem; the point of insertion of a leaf or leaves.
Nodding. Hanging down. Cf. *Pendent.*
Nodose. Furnished with knots or knobby nodes. *Nodulose:* the diminutive.
Nut. A hard-shelled and one-seeded indehiscent fruit derived from a simple or a compound ovary.
Nutlet. Diminutive of nut; applied to any small and dry nutlike fruit or seed. Thicker-walled than an akene.
Ob-. Latin prefix signifying the reverse or contrariwise.
Obcompressed. Flattened the other way, anteroposteriorly instead of laterally, opposite of the usual way.
Obconic. Inversely conical, with the point of attachment at the small end.
Obcordate. Inversely cordate, the notch at the apex.
Oblanceolate. Inversely lanceolate.
Oblique. Of unequal sides (as in leaves), slanting.
Oblong. Much longer than broad with nearly parallel sides.
Obovate. Inversely ovate.
Obovoid. Inversely ovoid.
Obsolete. Rudimentary or not evident; applied to an organ that is almost entirely suppressed: vestigial.
Obtuse. Blunt or rounded at the end.
Ochroleucous. Yellowish-white.

Glossary

Ocrea. A sheath around the stem derived from the leaf-stipules; used chiefly in the Polygonaceae.
Offsets. Short basal lateral shoots from which new plants can develop.
Operculum. A lid. *Operculate:* furnished with a lid.
Opposite. Set against, as leaves when two at a node; one part before another, as a stamen in front of a petal.
Orbicular, orbiculate. Approximately circular in outline.
Orifice. The mouthlike opening of a tubular corolla at the junction of limb and throat or tube.
Orthotropous (ovule or seed). Erect, straight, with the micropyle at the apex and the hilum at the base.
Oval. Broadly elliptic.
Ovary. The part of the pistil that contains the ovules.
Ovate. With the outline of a hen's egg in longitudinal section, the broader end downward.
Ovoid. Solid ovate or solid oval.
Ovule. The megasporangium of a seed plant; the body in the ovary which becomes a seed. *Ovulate, ovuliferous:* bearing ovules.
Palate. In personate corollas, the projecting part of the lower lip, which closes the throat, as in the snapdragon.
Palea. One of the chafflike scales on the receptacle of many Asteraceae; the inner bract of a grass floret, often partly invested by the lemma. *Paleaceous:* chaffy; composed of small membranaceous scales.
Palmate. Hand-shaped with the fingers spread; in a leaf, having the lobes or divisions radiating from a common point.
Palmatifid. Cleft so as to resemble the outstretched fingers of the hand. *Palmatisect:* palmately divided.
Palustrine. Of or pertaining to marshes.
Panduriform. Fiddle-shaped; obovate and with a contraction on each side.
Panicle. A compound racemose inflorescence. *Paniculate:* borne in a panicle.
Pannose. Feltlike in texture or appearance.
Papilionaceous. Applied to the butterfly-like corolla of the pea, with banner, wings, and keel.
Papillate, papillose. Bearing minute conical processes or papillae.
Pappose. Pappus-bearing.
Pappus. The modified calyx-limb in Asteraceae, consisting of a crown of bristles or scales on the summit of the akene.
Parietal. Attached to the wall of the ovary, instead of to the axis; said of ovules or a placenta.
Paripinnate. See Pinnate.

Parted. Deeply cleft nearly to the base.
Pectinate. With narrow closely set divisions like the teeth of a comb.
Pedate. Palmate, with the lateral lobes 2-cleft; said of leaves.
Pedicel. The stalk of a single flower in a flower-cluster or of a spikelet in grasses. *Pedicellate:* having a pedicel, as opposed to sessile.
Peduncle. The general term for the stalk of a flower or a cluster of flowers. *Pedunculate:* having a peduncle.
Pellicle. A thin skin or filmy covering.
Pellucid. Transparent, clear.
Peltate. Shield-shaped; a flat body having the stalk attached to the lower surface instead of at the base or margin.
Pendent, pendulous. Suspended or hanging, as an ovule that hangs from the side of the locule; nodding.
Penicillate. Ending in a tuft of fine hairs.
Pepo. A gourd fruit, with hard rind, one-loculed, many seeded.
Perennial. Lasting from year to year.
Perfect. A flower having both stamens and pistils.
Perfoliate. With the leaf entirely surrounding the stem.
Perianth. The floral envelopes collectively; usually used when calyx and corolla are not clearly differentiated.
Pericarp. The ripened walls of the ovary, referring to a fruit.
Perigynium. The inflated saclike organ surrounding the pistil in *Carex.*
Perigynous. Borne around the ovary in contrast to beneath it, as the stamens and corolla are inserted on the floral-tube.
Perisperm. The nutritive tissue surrounding the embryo of the seed and formed outside the embryo-sac.
Persistent. Remaining attached, as a calyx on the fruit.
Personate. A bilabiate corolla having the throat nearly closed by a prominent palate.
Petal. One of the leaves of a corolla, usually colored. *Petaloid:* having the aspect of or colored as petals.
Petiole. A leaf-stalk. *Petiolate:* having a petiole.
Petiolule. The stalk of a leaflet.
Phyllary. An individual bract of the involucre of an Asteraceae.
Phyllotaxy. Leaf arrangement.
Phyllode. A dilated petiole serving as a leaf-blade.
Phyllopodic. With a leafy base.
Phylogeny. The race history of a plant or natural group; relationship by descent.
Phylum. A primary division of the plant kingdom.
Pilose. Bearing soft and straight spreading hairs. *Pilosulous:* the diminutive.

Pinna. A leaflet or primary division of a pinnate leaf.

Pinnate. A compound leaf, having the leaflets arranged on each side of a common petiole; featherlike. *Odd-pinnate:* pinnate with a single terminal leaflet (imparipinnate). *Abruptly pinnate:* pinnate without an odd terminal leaflet (paripinnate).

Pinnatifid. Pinnately cleft into narrow lobes not reaching to the midrib.

Pinnatisect. Pinnately divided to the midrib.

Pinnule. A division of a pinna.

Pistil. The ovule-bearing organ of a flower, consisting of stigma and ovary, usually with a style between; gynoecium.

Pistillate. Provided with pistils and without stamens; female.

Pitted. Having little depressions or pits: foveate.

Placenta. The ovule-bearing surface in the ovary.

Placentation. The arrangement or orientation of the placentas.

Plane. Surface flat and even, not curved.

Plicate. Plaited; folded as a fan.

Plumose. Feathery; having fine hairs on each side as a plume.

Pod. Any dry dehiscent fruit; specifically a legume.

Pollen. The male fecundating spores found in the anther.

Pollinia. The pollen-masses of the orchids and milkweeds.

Polygamous. Bearing unisexual and bisexual flowers on the same plant.

Polymorphous. Of various forms.

Polyploid. Having a chromosome number that is a multiple of a basic number for a group of forms. *Polyploid series:* examples—$2n = 8$ (diploid), 16 (tetraploid), 24 (hexaploid), 32 (octoploid), etc. *Polyploid complex:* intimately related members of a polyploid series.

Polystichous. In several vertical rows or ranks.

Pome. An applelike fruit.

Porrect. Directed outward and forward; vertical to the substratum.

Posterior. On the side toward the axis; the upper side of the flower.

Praemorse. As it were, bitten off; said of roots.

Precocious. Flowering before the appearance of the leaves.

Prickle. Sharp outgrowth of the bark or epidermis. *Prickly:* armed with prickles, as the rose.

Procumbent. Trailing on the ground, but not rooting.

Proliferous. Bearing offsets, bulbils, or other vegetative progeny; abnormal or redundant development, as when a leafy shoot develops from a flower part.

Prostrate. Lying flat upon the ground.

Proterandrous. Shedding the pollen before the stigma of the flower is receptive. *Proterogynous:* having the stigma receptive before the stamens of the flower mature.

Proximal. Nearest the axis or base, as contrasted with distal.

Pruinose. Covered with a coarse waxy powder, more pronounced than when glaucous.

Puberulent. Minutely pubescent.

Pubescent. Covered with short soft hairs; downy.

Pulverulent. Dusted as with fine powder.

Pulvinate. Cushion-shaped.

Pulvinus. A swelling close under the insertion of a leaf; the swollen base of a petiole.

Punctate. Dotted with punctures or with translucent pitted glands or with colored dots. *Puncticulate:* the diminutive.

Pungent. Ending in a rigid, sharp point or prickle; acrid (to the taste or smell).

Pustulate. Bearing irregular blisterlike swellings or *pustules*, mostly at the bases of hairs.

Pyriform. Pear-shaped.

Quadrate. Square.

Quadri-. Latin prefix signifying four.

Raceme. A simple, elongated, indeterminate inflorescence with each flower subequally pedicelled. *Racemose:* of the nature of a raceme or in racemes.

Rachilla. A small rachis, specifically the axis of a grass spikelet.

Rachis. The axis of a spike or raceme, or of a compound leaf.

Radiate. Spreading from a common center; bearing rays.

Radical. Belonging to or proceeding from the root.

Radicle. That portion of the embryo below the cotyledons.

Rameal. Belonging to a branch.

Ramose. Branching or branchy. *Ramulose:* with many branchlets.

Raphe. The ridge connecting the two ends of an anatropous ovule.

Ray. A primary branch of an umbel; the ligule of a ray-floret in Asteraceae, the ray-florets being marginal and differentiated from the disk-florets.

Receptacle. That portion of the floral axis upon which the flower parts are borne, or, in Asteraceae, that which bears the florets in the head.

Reclined, reclinate. Turned downward, with the tip resting on the ground.

Recurved. Bent backwards.

Reflexed. Abruptly bent downward.

Regular. Said of a flower having radial symmetry, with the parts in each series alike.

Relict. A localized plant left over from an earlier geological period.

Remote. Distantly spaced.

Reniform. Kidney-shaped.

Glossary

Repand. With an undulating margin, less strongly wavy than sinuate.
Repent. Creeping (prostrate and rooting).
Replicate. Folded backward.
Replum. The septum of certain pods that persist after the valves have fallen, as in the fruit of Brassicaceae.
Resiniferous. Producing resin.
Resupinate. Upside down; inverted by the twisting of the pedicel, as the flowers of orchids.
Reticulate. With a network; net-veined.
Retrorse. Bent backward or downward.
Retuse. Notched shallowly at a rounded apex.
Revolute. Rolled backward from both margins, i.e., toward the underside.
Rhizome. An underground stem or rootstock, with scales at the nodes and producing leafy shoots on the upper side and roots on the lower side. *Rhizomatous:* having a rhizome.
Rhombic. Somewhat diamond-shaped. *Rhomboidal.* A solid with a rhombic outline.
Rib. The primary vein of a leaf, or a ridge on a fruit. *Ribbed:* with prominent ribs.
Ringent. Gaping, as the mouth of an open-throated bilabiate corolla.
Rootstock. See *Rhizome.*
Rosette. A crowded cluster of radiating leaves appearing to rise from the ground.
Rostrate. Beaked. *Rostellate:* the diminutive.
Rosulate. In the form of a rosette.
Rotate. Wheel-shaped; said of a sympetalous corolla with obsolete tube and with a flat and circular limb.
Rotund. Rounded in outline.
Rubellous. Reddish. *Rubescent:* turning red.
Ruderal. Weedy; growing in waste places.
Rudimentary. Imperfectly developed; vestigial.
Rufous. Reddish-brown.
Ruga(-e). A wrinkle or fold.
Rugose. Wrinkled. *Rugulose:* the diminutive.
Runcinate. Sharply pinnatifid or incised, the lobes pointing downward.
Runner. A slender trailing stem rooting at the nodes or end; a very slender stolon.
Sac. The cavity of an anther.
Saccate. Furnished with a sac or pouch.
Sagittate. Shaped as an arrowhead, with the basal lobes turned downward. Cf. *Hastate.*
Salverform. A corolla with slender tube abruptly expanding into a flat limb.
Samara. An indehiscent winged fruit.
Saprophyte. A plant living on dead organic matter and hence without chlorophyll.
Sarmentose. Bearing long slender prostrate runners.
Scabrous. Rough to the touch, owing to the structure of the epidermis or to the presence of short stiff hairs. *Scabrid:* somewhat rough. *Scaberulous:* minutely roughened.
Scalariform. Ladderlike.

Scale. Any thin scarious bract; usually a vestigial leaf. *Scaly:* squamose; scarious.
Scandent. Climbing.
Scape. A leafless peduncle rising from the ground in acaulescent plants. *Scapiform, scapose:* resembling or bearing a scape.
Scarious. Thin, dry, and membranaceous, not green.
Scorpioid. Said of a unilateral inflorescence circinately coiled in the bud.
Scrobiculate. Marked by minute or shallow depressions.
Scurfy. Clothed with small branlike scales; furfuraceous.
Secund. Arranged on one side only; unilateral.
Seed. The ripened ovule.
Seep(s). A moist spot where underground water comes to or near the surface.
Segment. A division or part of a leaf or other organ that is cleft or divided but not truly compound.
Sepal. A leaf or segment of the calyx. *Sepaloid:* sepallike.
Septicidal. Dehiscence of a capsule through the septa and between the locules.
Septum. A partition between cavities. *Septate:* divided by partitions or septa.
Seriate. Disposed in series or rows.
Sericeous. Silky; clothed with appressed fine and straight hairs.
Serrate. Saw-toothed, the sharp teeth pointing forward. *Serrulate:* finely serrate.
Sessile. Attached directly by the base, not stalked, as a leaf without a petiole.
Seta. A bristle, or a rigid, sharp-pointed, bristlelike organ. *Setaceous:* bristly or bristlelike. *Setigerous:* bristle-bearing.
Setose. Clothed with bristles. *Setulose:* bearing minute bristles.
Several. Fewer than *many,* perhaps 6 to 8 or 10.
Sheath. The tubular basal part of the leaf that encloses the stem, as in grasses and sedges. *Sheathing:* enclosed as by a sheath, vaginate.
Shrub. A woody plant of smaller proportions than a tree, which usually produces several branches from the base.
Sigmoid. Doubly curved, like the letter S.
Silique. A narrow many-seeded capsule of the Cruciferae, with 2 valves splitting from the bottom and leaving the placentae with the false partition (replum) between them. *Silicle:* a short silique, not much longer than wide.
Silky. See sericeous.
Simple. Unbranched, as a stem or hair; uncompounded, as a leaf; single, as a pistil of one carpel.
Sinistrorse. Directed toward the left.
Sinuate. With a strongly wavy margin. Cf. *Repand.*

Sinus. The cleft or recess between two lobes of an expanded organ such as a leaf.

Smooth. Not rough to the touch. Cf. *glabrous,* without hairs, which may be either smooth or scabrous.

Solitary. Borne singly.

Sordid. Of a dull or dirty hue.

Sorus (plural, *sori*). A fruit-dot, or -cluster on the back of the fronds of ferns.

Spadix. A spike on a succulent axis enveloped in a spathe.

Spathe. A broad sheathing bract enclosing a spadix, as in the calla. *Spathaceous:* resembling or having a spathe.

Spatulate. Like a spatula, a knife rounded above and gradually narrowed to the base.

Spicate. Having the form of or arranged in a spike. *Spiciform:* spikelike.

Spike. An elongated rachis of sessile flowers or spikelets.

Spikelet. A secondary spike; the ultimate flower-cluster in grasses, consisting of two glumes and one or more florets, and in sedges.

Spine. A sharp-pointed, stiff, woody body, arising from below the epidermis; commonly the counterpart of a leaf or stipule. Cf. *Prickle. Spinescent:* more or less spiny, spine-tipped. *Spinose:* bearing spines. *Spinulose:* bearing diminutive spines.

Sporangium. A spore case or sac.

Spore. The reproductive body of pteridophytes and lower plants, analogous to the seed.

Sporocarp. A receptacle containing sporangia or spores.

Sporophyll. A spore-bearing leaf.

Sporophyte. The asexual or diploid generation of ferns and their allies, the fern plant itself.

Spreading. Diverging almost to the horizontal; nearly prostrate. *Spreading hairs:* not at all appressed, but erect. *Spreading lower lip:* diverging from the main axis of the flower.

Spur. A slender, saclike, nectariferous process from a petal or sepal.

Squama. A scale, usually the homologue of a leaf. *Squamella:* Diminutive squama, applied to some types of pappus in Compositae. *Squamellate:* like a little scale. *Squamose, squamate:* covered with scales; scaly. *Squamulose:* provided with small scales.

Squarrose. Spreading rigidly at right angles or more, as the tips of bracts.

Stamen. The male organ of the flower, which bears the pollen.

Staminate. Having stamens but not pistils; said of a flower or plant that is male, hence not seed-bearing.

Staminiferous. Stamen-bearing.

Staminode. A sterile stamen (lacking an anther), or what corresponds to a stamen.

Stellate. Star-shaped. *Stellulate:* resembling a little star or stars.

Sterile. Infertile or barren, as a stamen lacking an anther, a flower wanting a pistil, a seed without an embryo, etc.

Stigma. The receptive part of the pistil on which the pollen germinates. *Stigmatic:* pertaining to the stigma.

Stipe. The leaf-stalk of a fern; the stalk beneath an ovary. *Stipitate:* having a stipe or stalk, as an elevated gland.

Stipule. One of the pair of usually foliaceous appendages found at the base of the petiole in many plants. *Stipulate:* possessing stipules.

Stolon. A modified stem bending over and rooting at the tip; or creeping and rooting at the nodes; or a horizontal stem that gives rise to a new plant at its tip. Cf. *Runner,* a very slender stolon, and *Rhizome,* a subterranean stem. *Stoloniferous:* having stolons.

Stomate. A breathing pore or aperture in the epidermis.

Stomium. Line of dehiscence in a fern sporangium.

Stone. The bony endocarp of a drupe.

Stramineous. Straw-colored.

Striate. Marked with fine longitudinal lines or furrows.

Strict. Very straight and upright, not at all lax or spreading.

Strigose. Clothed with sharp and stiff appressed straight hairs. *Strigillose* or *strigulose:* minutely strigose.

Strobilus. Conelike aggregation of sporophylls.

Strophiole. An appendage at the hilum of certain seeds.

Style. The contracted portion of the pistil between the ovary and the stigma. *Style-branches* may be only in part stigmatic, the remainder then being *appendage.*

Stylopodium. An enlargement or disklike expansion at base of the style, as in Umbelliferae.

Sub-. Latin prefix meaning somewhat, almost, of inferior rank, beneath.

Subtend. To be below and close to, as the leaf subtends the shoot borne in its axil.

Subulate. Awl-shaped.

Succulent. Juicy; fleshy and soft.

Suffrutescent. Obscurely shrubby; very little woody, but not necessarily low.

Suffruticose. Woody but very low; diminutively shrubby. Cf. *Fruticulose.*

Sulcate. Longitudinally grooved, furrowed, or channeled.

Sulcus (-*i*). A furrow or groove.

Superior. Growing above, as an ovary that is free from the other floral organs.

Suture. The line of dehiscence of fruits or anthers; the line of a natural union or division between coherent parts.

Swale. A moist meadowy area.

Symmetrical. Said of a flower having the same number of parts in each circle.

Glossary

Sympatric. Growing together with, or having the same range as.
Sympetalous. With petals united in a one-piece corolla; gamopetalous.
Sympodium. An apparent main axis formed of successive secondary axes, each of which represents one fork of a dichotomy, the other being much weaker or entirely suppressed. *Sympodial:* of the nature of a sympodium.
Syn-. Greek prefix meaning united.
Synonym. A systematic name, as for a species, that was superfluous when published, or for some other reason is rejected in favor of another.
Taproot. A primary stout vertical root giving off small laterals but not dividing.
Taxon (plural, *taxa*). Any taxonomic unit, as an order, genus, variety, etc.
Tawny. Dull brownish-yellow; fulvous.
Tendril. A slender, coiling or twining organ by which a climbing plant grasps its support.
Terete. Cylindrical; round in cross section.
Ternary. Consisting of threes; trimerous.
Ternate. In threes, as a leaf consisting of three leaflets.
Tesselated. The surface marked by checkered work, either as depressions or color patterns.
Testa. The outer seed-coat.
Tetradynamous. Having four long and two short stamens.
Tetragonal, tetragonous. Four-angled.
Tetramerous. Having the floral organs in fours or multiples of four.
Tetraploid. Having four basic chromosome sets (four times the gametic number).
Thorn. See *Spine,* but technically a sharp-pointed stiff woody body derived from a modified branch.
Throat. The orifice of a gamopetalous corolla; the expanded portion between the limb and the tube proper.
Thyrse, thyrsus. A compact, ovate panicle; strictly, with main axis indeterminate, but with other axes cymose. *Thyrsoid:* like a thyrse.
Tomentose. With tomentum; covered with a rather short, densely matted, soft white wool. *Tomentulose:* the diminutive. *Tomentum:* a covering of such densely matted woolly hairs.
Tooth. Any small marginal lobe. *Toothed:* dentate.
Torose. Cylindrical with alternate swellings and constrictions, as a rhizome. *Torulose:* the diminutive.
Tortile. Twisted or twining.
Tortuous. Bent or twisted in different directions.
Torus. See *Receptacle.*
Tri-. A Greek and Latin prefix signifying three, thrice, or triply.
Triandrous. Having three stamens.
Trichotomous. Three-forked.
Trifid. Three-cleft to about the middle.
Trigonous. Three-angled, the faces between plane.
Trimerous. Having the parts in threes.
Tripartite. Three-parted.
Triquetrous. Three-edged, the faces between concave.
Triternate. Thrice ternate.
Truncate. As if cut off squarely at the end.
Tube. The narrow basal portion of a gamopetalous corolla or a gamosepalous calyx.
Tuber. A thickened, solid and short underground stem, with many buds. *Tuberous:* bearing a tuber; resembling a tuber.
Tubercle. A small tuberlike prominence or nodule; the persistent base of the style in some Cyperaceae. *Tubercled, tuberculate:* beset with tubercles or warty excrescences; verrucose.
Tuberiferous. Bearing tubers.
Tubular. Shaped like a hollow cylinder.
Tunicate, tunicated. Having coats or tunics, as a bulb.
Turbinate. Top-shaped; inversely conical.
Turgid. Swollen; inflated.
Umbel. A flat or convex flower-cluster in which the pedicels arise from a common point, like rays of an umbrella. *Umbellate:* borne in an umbel. *Umbellet:* a secondary umbel. *Umbelliferous:* bearing umbels. *Umbelliform:* umbel shaped.
Umbilicus. A navel; the hilum of a seed. *Umbilicate:* depressed in the center.
Umbonate. Bearing an umbo or boss or conical projection in the center, as the scale of a pine cone.
Uncinate. Hooked at the tip.
Undershrub. A very low shrub.
Undulate. Wavy; repand; with less pronounced "waves" than sinuate.
Unguiculate. Contracted at base into a claw.
Unilateral. One-sided, or turned to one side of an axis: secund.
Unilocular. Having one locule or cell.
Uniseriate. Arranged in one horizontal row.
Unisexual. Flowers having only stamens or pistils; of one sex.
Urceolate. Urn-shaped or pitcherlike, contracted at the mouth.
Utricle. A small, bladdery one-seeded fruit. *Utricular:* having little bladders; inflated.
Vaginate. Loosely surrounded by a sheath.
Valve. One of the segments into which a dehiscent capsule or legume separates. *Valvate:* opening as if by valves, as most capsules and some anthers; in aestivation, meeting at the edges without overlapping.
Vein. A vascular bundle of a leaf or other flat

organ. Cf. *Nerve*. *Veinlet:* one of the ultimate branches of a vein.

Velum. The membranous indusium in *Isoetes*.

Velutinous. Velvety; covered with a fine and dense silky pubescence.

Venation. The arrangement of the veins of a leaf; nervation; veining. *Venose:* veiny; abounding in veins. *Venulose:* abounding in veinlets.

Venter. Belly; under part.

Ventral. Relating to the inward face of an organ, in relation to the axis; anterior; front; opposed to dorsal.

Ventricose. Inflated or swelling out on one side or unequally; gibbous.

Vermicular, vermiform. Worm-shaped or worm-like.

Vermiculate. Marked with tortuous impressions, as if worm-eaten.

Vernation. The arrangement of foliage leaves within the bud.

Vernicose. As if varnished.

Verrucose. Warty; covered with wartlike excrescenses; tuberculate.

Versatile. An anther attached near the middle and capable of swinging freely on the filament.

Verticil. A whorl, or circular arrangement of similar parts about the same point on an axis. *Verticillate:* whorled.

Verticillaster. A false whorl, composed of a pair of nearly sessile cymes in the axils of opposite leaves, as in many mints.

Vesicle. A little bladder or air cavity. *Vesicular:* pertaining to, or having the form of, a vesicle.

Vespertine. Blossoming in the evening.

Vestigial. Reduced to a vestige or trace of a part or organ once more perfectly developed.

Villous. Bearing long and soft and not matted hairs; shaggy.

Virgate. Wand-shaped; slender, straight and erect.

Viscid, viscous. Sticky; glutinous. *Viscidulous:* slightly viscid.

Whorl. A ring of similar organs radiating from a node; verticil.

Wing. A thin and usually dry extension bordering an organ; a lateral petal of a papilionaceous flower. *Winged:* bearing a wing; alate.

Woolly. Having long, soft, entangled hairs; lanate. Cf. *Tomentose*.

Xerophyte. A drought-resistant or desert plant.

Zygomorphic, zygomorphous. Bilaterally symmetrical; that which can be bisected by only one plane into similar halves.

INDEX TO
COMMON AND SCIENTIFIC NAMES

Common names that are hyphenated or consist of more than one word are to be sought under the last word. For example, oaks such as Blue Oak, Canyon Oak and Poison-Oak are in alphabetical order under Oak. In the case of scientific names, roman type is used for species or varieties that are maintained, while italics are used for synonyms.

Abies, 37
  concolor, 4, 38
  macrocarpa, 42
Abronia, 578
  alba, 580
  angulata, 579
  aurita, 580
  covillei, 579
  exalata, 579
  latifolia, 578
  maritima, 579
  micrantha, 579
  nana covillei, 579
  pinetorum, 580
  platyphylla, 580
  pogonantha, 579
  turbinata, 579
  umbellata, 579
  variabilis, 580
  villosa
    aurita, 580
    villosa, 580
Abutilon
  newberryi, 564
  palmeri, 562
  parvulum, 562
  theophrasti, 562
Acacia, 420
  Blackwood, 421
  decurrens, 421
    mollis, 421
  farnesiana, 421
  greggii, 421
  longifolia, 421
  melanoxylon, 421
  mollissima, 421
  retinodes, 421
Acalypha californica, 406
Acamptopappus
  shockleyi, 105
  sphaerocephalus, 105
Acanthaceae, 53, 54
Acanthogonum
  corrugatum, 672
  polygonoides

longispinum, 672
rigidum, 674
Acanthomintha, 523
  ilicifolia, 524
  obovata, 524
Acanthoscyphus abramsii, 700
Acarphaea artemisiaefolia, 136
Acer
  bernardinum, 54
  dactylophyllum, 54
  diffusum, 54
  glabrum diffusum, 54
  leptodactylon, 54
  macrophyllum, 54
  negundo californicum, 54
  politum, 54
Aceraceae, 46, 49, 54
Acerates asperula, 92
  capricornu occidentalis, 92
Achillea
  arenicola, 106
  borealis californica, 106
  californica, 106
  lanulosa, 106
  millefolium, 106
    arenicola, 106
    californica, 106
    lanulosa, 106
    millefolium, 107
    pacifica, 107
  pacifica, 107
Acleisanthes longiflora, 580
Achyrachaena mollis, 107
Achyranthes lanuginosa, 63
Achyrodes aureum, 978
Achyronychia
  cooperi, 335
  rixfordii, 342
Acnida tamariscana, 63
Acourtia microcephala, 214
Acrolasia
  desertorum, 551
  montana, 555
  tridentata, 558
  viridescens, 551

Actaea
  arguta, 720
  rubra arguta, 720
Actinea acaulis arizonica, 194
Actinolepis
  lanosa, 164
  lemmonii, 233
  multicaulis, 164
  pringlei, 164
Adder's Mouth, 934
Adelia
  neomexicana, 586
  parvifolia, 586
Adenostegia
  canescens, 801
  eremica, 802
  helleri, 801
  maritima, 801
Adenostoma, 4, 741
  fasciculatum, 741
    obtusifolium, 741
  sparsifolium, 741
Adiantum
  capillus-veneris, 25
  emarginatum, 25
  jordani, 25
  modestum, 25
  pedatum aleuticum, 25
Adolphia californica, 732
Aegilops
  hystrix, 999
  incurvus, 987
Aegochloa
  atractyloides, 664
  intertexta, 666
  pubescens, 666
Aesculus californica, 495
Agarista calliopsidea, 150
Agastache urticifolia, 524
Agavaceae, 863
Agave
  americana, 864
  consociata, 864
  deserti, 864
  nevadensis, 864

Agave (cont.)
  shawii, 864
  utahensis nevadensis, 864
Ageratina
  adenophora, 107
  herbacea, 107
Ageratum
  lineare, 214
  wrightii, 237
Agianthus bernardinus, 303
Agoseris
  glauca monticola, 108
  grandiflora, 108
  heterophylla, 108
  monticola, 108
  retrorsa, 108
Agrimonia gryposepala, 741
Agrimony, 741
Agropyron, 940
  elongatum, 940
  parishii, 940
  pauciflorum, 940
  riparium, 940
  scribneri, 940
  tenerum, 940
  trachycaulum, 940
Agrostis, 941
  airoides, 1001
  alba major, 943
  ampla, 941
  cryptandra, 1002
  diegoensis, 941
  exarata, 941
  hallii californica, 943
  hiemalis, 943
  idahoensis, 941
  kennedyana, 943
  lepida, 943
  microphylla, 943
  minutissima, 984
  miliacea, 986
  oregonensis, 943
  scabra, 943
  semiverticillata, 943, 996
  stolonifera major, 943
  tandilensis, 943
  tenuiculmis, 941
  tenuis, 943
  ventricosa, 974
  verticillata, 996
Ailanthus
  altissima, 829
  glandulosa, 829
Aira, 943
  caespitosa, 961
  caryophyllea, 944
  danthonioides, 961
  elongata, 962
  obtusata, 1001
  spicata, 1006
Aizoaceae, 47, 49, 55

Alcaea rosea, 562
Alchemilla, 741
  cuneifolia, 742
  occidentalis, 742
Alcina perfoliata, 210
Alder, White, 245
Aleuritopteris californica, 29
Alfalfa, 463
Alhagi camelorum, 422
Aligera
  insignis, 849
  grayii, 849
Alisma
  berteroi, 870
  plantago-aquaticum, 868
    americanum, 868
    michalettii, 868
    triviale, 868
Alismataceae, 868
Allenrolfea, 5, 352
  occidentalis, 352
Allionia
  comata, 585
  incarnata, 580
  linearis, 585
  nyctaginea, 585
  pumila, 585
Allium, 871
  acuminatum bidwelliae, 873
  amplectens, 872
  atrorubens, 873
    inyonis, 873
  attenuifolium, 872
    monospermum, 872
  austinae, 873
  bidwelliae, 873
  bisceptrum, 873
  bullardii, 873
  burlewii, 873
  campanulatum, 873
  crispum, 873
  cristatum, 876
  croceum, 877
  davisiae, 874
  decipiens, 873
  fimbriatum, 874
    denticulatum, 874
    diabolense, 874
    fimbriatum, 874
    mohavense, 874
    munzii, 875
    parryi, 875
  haematochiton, 875
  howellii
    clokeyi, 875
    howellii, 875
    hyalinum, 875
    praecox, 876
    inyonis, 873
    kessleri, 875
    lacunosum, 875

  maritimum, 881
  marvinii, 875
  mohavense, 874
  monospermum, 872
  monticola, 875
    keckii, 875
  montigenum, 876
    neapolitanum, 876
    nevadense cristatum, 876
    occidentale, 872
    parishii, 876
      keckii, 876
      parryi, 875
      parvum jacintense, 876
      peirsonii, 875
    peninsulare, 876
      crispum, 873
    praecox, 876
    pseudobulbiferum, 874
    tenellum, 873
    tribracteatum, 876
Allocarya
  acanthocarpa, 264
  bracteata, 265
  cooperi, 266
  hispidula, 265
  leptoclada, 266
  salsa, 266
  undulata, 267
Allophyllum, 639
  divaricatum, 640
  gilioides, 640
  glutinosum, 640
  integrifolium, 640
  violaceum, 641
Allosurus mucronatus, 31
All-thorn, 521
Almond, Desert, 757
Alnus rhombifoa, 245
Alopecurus
  aequalis, 944
  geniculatus, 944
  californicus, 944
  howellii, 944
  monspeliensis, 996
Aloysia wrightii, 851
Alsine
  bocconei, 345
  crispa, 348
  graminea, 348
  jamesiana, 348
  longipes, 348
  media, 348
  rubella, 339
Alternanthera
  lanuginosa, 63
  peploides, 60
  philoxeroides, 60
Althaea rosea, 562
Alum-Root, 779
Alyssum maritimum, 296

Index

Sweet-, 296
Amaranth, 60
Amaranthaceae, 58
Amaranthus
  albomarginatus, 61
  albus, 61
  arenicola, 61
  blitoides, 61
  californicus, 61
  carneus, 61
  caudatus, 62
  cruentus, 62
  deflexus, 62
  fimbriatus, 62
  graecizans, 61
  hybridus, 62
  palmeri, 62
  powellii, 62
  pringlei, 62
  retroflexus, 63
  spinosus, 63
  tamariscanus, 63
  watsonii, 63
Amarella californica, 489
Amaryllidaceae, 871
Amauria dissecta, 126
Amblogyna fimbriata, 62
Amblyopappus
  neomexicanus, 126
  pusillus, 109
Ambrosia, 109
  acanthicarpa, 109
  ambrosioides, 110
  bipinnatifida, 110
  californica, 112
  chamissonis, 115
  chenopodiifolia, 110
  confertiflora, 110
  dumosa, 4, 110
  eriocentra, 112
  ilicifolia, 112
  psilostachya californica, 112
  pumila, 112
Amelanchier
  alnifolia cuyamacensis, 742
  covillei, 742
  gracilis, 742
  utahensis, 742
Amellus villosus, 141
Ammannia coccinea, 560
Ammi
  majus, 69
  visnaga, 69
Ammobroma sonorae, 545
Ammophila arenaria, 944
Ammoselinum
  giganteum, 69
  occidentale, 69
Amole, 926
Amorpha
  californica, 422

fruticosa occidentalis, 422
  occidentalis, 422
Amphiachyris
  fremontii, 112
  spinosus, 112
Amphipappus
  fremontii, 112
  fremontii spinosus, 112
  spinosus, 112
Amsinckia, 247
  carnosa, 249
  douglasiana, 248
  eastwoodae, 248
  gloriosa, 248
  intermedia, 248
    eastwoodae, 248
  lemmonii, 248
  macrantha, 248
  maritima, 248
  menziesii, 248
  parviflora, 248
  pustulata, 249
  retrorsa, 248
  spectabilis, 248
  nicolai, 249
  tessellata, 249
  vernicosa, 249
Amsonia
  brevifolia, 87
  tomentosa, 88
  tomentosa, 88
Anabasis glomeratus, 365
Anacardiaceae, 49, 64
Anacharis canadensis, 905
Anacyclus australis, 153
Anagallis, 715
  arvensis, 717
  minimus, 717
Ananthrix decumbens, 92
Anaphalis
  margaritacea, 6, 113
  occidentalis, 113
Ancistrocarphus filagineus, 232
Andropogon, 944
  barbinodis, 948
  cirratus, 945
  contortus, 974
  glomeratus, 945
  hirtus, 977
  saccharoides, 948
Androsace
  acuta, 717
  elongata acuta, 717
  septentrionalis
    subumbellata, 717
Androstephium brevifolium, 877
Anemone tuberosa, 720
Anemopsis californica, 770
Anethum
  foeniculum, 76

graveolens, 70
Angelica
  californica, 70
  kingii, 70
  lineariloba, 70
    culbertsonii, 70
  tomentosa, 70
Angiospermae, 45
Anisocoma acaulis, 113
Anogra californica, 614
Antennaria, 113
  alpina media, 114
  corymbosa, 114
  dimorpha, 114
  dioica
    corymbosa, 114
    marginata, 114
  marginata, 114
  media, 114
  rosea, 114
Anthemis cotula, 114
Anthephora belangeri, 975
Antheropeas rubellum, 165
Antholyza aethiopica, 906
Anthoxanthum odoratum, 945
Anticlea fremontii, 931
Antirrhinum
  confertiflorum, 816
  cooperi, 791
  coulterianum, 791
    orcuttianum, 791
  dalmaticum, 805
  elatine, 804
  filipes, 791
  glandulosum, 793
  hookerianum, 791
  kelloggii, 791
  kingii, 793
  majus, 793
  maurandioides, 806
  multiflorum, 793
  nuttallianum, 793
    pusillum, 793
  purpureum, 805
  spurium, 804
  strictum, 791
Apache-Plume, 746
Aphanisma blitoides, 352
Aphora serrata, 409
Aphyllon cooperi, 618
Apiaceae, 50, 67
Apiastrum angustifolium, 71
Apium
  graveolens, 71
  leptophyllum, 71
Aplopappus, see Haplopappus
Aplopappus
  sphaerocephalus, 105
  tortifolius, 204
Apocynaceae, 52, 87

Apocynum
  androsaemifolium, 88
  cannabinum glaberrimum, 88
  medium floribundum, 88
  floridum, 88
  lividum, 88
  pumilum pumilum, 88
    rhomboideum, 89
  rhomboideum, 89
  sibiricum salignum, 90
  salignum, 90
Apple
  Squaw-, 750
  Thorn-, 830
Aptenia cordifolia, 55
Aquilegia, 720
  eximia, 721
  formosa
    hypolasia, 721
    pauciflora, 721
    truncata, 721
  hypolasia, 721
  mohavensis, 721
  pauciflora, 721
  shockleyi, 721
  truncata, 721
Arabis, 270
  arcuata perennans, 273
  breweri, 271
    pecuniaria, 271
  deserti, 300
  dispar, 271
  filifolia, 300
  glabra, 271
  glaucovalvula, 272
  hirsuta glabrata, 272
  hoffmannii, 272
  holboellii
    pinetorum, 272
    retrofracta, 272
  inyoensis, 272
  johnstonii, 273
  longirostris, 303
  maxima hoffmannii, 272
  parishii, 273
  perennans, 273
  perfoliata, 272
  pinetorum, 272
  platysperma, 273
  pulchra, 273
    gracilis, 275
    munciensis, 275
  pycnocarpa glabrata, 272
  rectissima, 275
  repanda, 275
  retrofracta, 272
  shockleyi, 275
  sparsiflora
    arcuata, 275
    californica, 275

virginica, 301
Araceae, 882
Aralia californica, 91
Araliaceae, 50, 90
Araujia sericofera, 91
Arbutus menziesii, 398
Arceuthobium, 856
  abietinum, 857
  californicum, 857
  campylopodum *cryptopodum*, 857
  cyanocarpum, 857
  divaricatum, 857
  douglasii abietinum, 857
  occidentale, 857
Arctium lappa, 115
Arctomecon merriamii, 623
Arctostaphylos, 4, 398
  bicolor, 405
  catalinae, 399
  confertifolia, 399
  diversifolia, 402
  drupacea, 401
  glandulosa, 399
    adamsii, 400
    crassifolia, 400
    mollis, 400
    glandulosa, 399
  glauca, 400
    eremicola, 400
    puberula, 400
  insularis, 400
  oppositifolia, 1018
  otayensis, 400
  parryana, 400
    pinetorum, 400
  patula platyphylla, 400
  pinetorum, 400
  pringlei drupacea, 401
  pungens, 401
    platyphylla, 400
  refugioensis, 401
  subcordata, 401
  tomentosa
    insulicola, 401
    subcordata, 401
    viridissima, 401
Arctotis
  fastuosa, 115
  grandis, 115
  stoechadifolia grandis, 115
Arecaceae, 882
Arenaria, 335
  aculeata, 337
  brevifolia californica, 337
  californica, 337
  confusa, 337
  congesta
    charlestonensis, 337
    subcongesta, 337
  douglasii, 337

  fendleri
    glabrescens, 337
    subcongesta, 337
  kingii
    compacta, 338
    glabrescens, 337
  macradenia, 338
    arcuifolia, 338
    ferrisiae, 338
    kuschei, 338
    parishiorum, 338
  macrophylla, 338
  macrotheca, 347
  nuttallii gracilis, 338
  paludicola, 338
  pusilla, 338
    diffusa, 338
  rubella, 339
  rubra, 347
    marina, 347
    saxosa, 337
  serpyllifolia, 339
  ursina, 339
Argemone
  corymbosa, 624
  munita
    argentea, 624
    munita, 624
    rotundata, 624
    robusta, 624
  rotundata, 624
Argythamnia
  adenophora, 408
  californica, 408
  clariana, 408
  lanceolata, 409
  neomexicana, 409
  serrata, 409
Aristida, 945
  adscensionis, 946
  bromoides, 946
  californica, 946
    fugitiva, 946
  divaricata, 946
  fendleriana, 946
  glauca, 946
  hamulosa, 946
  jonesii, 946
  longiseta robusta, 946
  orcuttiana, 946
  parishii, 946
  purpurea, 947
  schiedeana, 946
  wrightii, 947
Armeria
  maritima californica, 637
Arnica
  bernardina, 116
  chamissonis foliosa, 115
  discoidea, 116
  foliosa, 115

Index

mollis, 116
Aromia tenuifolia, 109
Arrowhead, 870
Arrowscale, 357
Arrowweed, 219
Artemisia, 4, 116
  albula, 117
  arbuscula nova, 118
  biennis, 117
  bigelovii, 117
  californica, 117
    insularis, 118
  cana, 4
  douglasiana, 117
  chamissoniana saxatilis, 118
  dracunculus, 117
  dracunculoides, 117
  glauca, 117
  gnaphalodes, 118
  incompta, 118
  ludoviciana
    albula, 117
    incompta, 118
    ludoviciana, 118
  matricarioides, 210
  nesiotica, 118
  norvegica saxatilis, 118
  nova, 4, 118
  palmeri, 118
  parishii, 119
  rothrockii, 118
  spinescens, 4, 119
  tridentata, 4, 119
    angustifolia, 119
    nova, 118
    parishii, 119
    rothrockii, 118
    tridentata, 119
  vulgaris californicaa, 117
    heterophylla, 117
Artemisiastrum palmeri, 118
Arthrocenemum subterminale, 368
Arundo
  arenaria, 944
  donax, 947
  selloana, 959
Asarina
  antirrhiniflora, 806
  filipes, 791
  petrophila, 806
Asclepiadaceae, 52, 91
Asclepias, 91
  albicans, 92
  asperula, 92
  californica, 92
  eriocarpa, 92
    microcarpa, 93
  erosa, 93
    obtusata, 93
  fascicularis, 93
  fremontii, 92

mexicana, 93
nyctaginifolia, 93
rothrockii, 93
subulata, 93
vestita, 93
  parishii, 93
Asclepiodora decumbens, 92
Ash, 586
  Arizona, 588
  Flowering, 588
Asparagus, 92
  asparagoides, 920
  Garden, 920
  officinalis, 920
  sprengeri, 920
Aspen, 771
  Quaking, 772
Asphodel, 920
Asphodelus fistulosus, 920
Aspidiaceae, 17
Aspidium aculeatum
    scopulinum, 21
  argutum, 19
  filix-mas, 19
  munitum, 19
    imbricans, 21
  puberulum, 21
Aspidotis californica 26
Aspleniaceae, 22
Asplenium
  trichomanes vespertinum, 22
  vespertinum, 22
Aspris caryophyllea, 944
Astephanus utahensis, 94
Aster, 119
  abatus, 204
  adscendens, 121
  alpigenus andersonii, 121
  andersonii, 121
  bernardinus, 121
  canescens, 203
  carnosus, 122
  chilensis, 122
    adscendens, 121
  cognatus, 204
  defoliatus, 121
  Desert, 204
  deserticola, 121
  eatonii, 122
  ericaefolius, 202
  ericoides, 202
  exilis, 122
  foliaceus eatonii, 122
  frondosus, 122
  Golden-, 140
  greatai, 122
  hesperius, 122
  intricatus, 122
  leucanthemifolius, 204
  menziesii, 122
  occidentalis, 123

orcuttii, 204
pauciflorus, 123
radulinus, 123
scopulorum, 123
spinosus, 123
standleyi, 204
tephrodes, 204
tortifolius, 204
Asteraceae, 54, 95
Astragalus, 422
  acutirostris, 426
  agninus, 431
  albens, 426
  antisellii phoxus, 436
  aridus, 426
  atratus mensanus, 426
  bernardinus, 426
  bicristatus, 426
  brauntonii, 426
  calycosus, 426
  casei, 427
  catalinensis, 428
  chuckwallae, 433
  cimae
    cimae, 427
    sufflatus, 427
  coccineus, 427
  crotalariae, 427
  curtipes, 427
  deanei, 427
  deserticola, 433
  didymocarpus
    didymocarpus, 428
    dispermus, 428
  dispermus, 428
  douglasii
    douglasii, 428
    megalophysus, 428
    parishii, 428
    perstrictus, 428
  filipes, 428
  fremontii, 431
  funereus, 428
  gambellianus, 428
  gaviotus, 436
  gilmanii, 428
  hasseanus, 436
  hemigyrus, 432
  hookerianus
    pinosus, 436
    whitneyi, 436
  hornii, 429
  insularis harwoodii, 429
  inyoensis, 429
  jaegerianus, 429
  kentrophyta
    elatus, 429
    implexus, 429
  layneae, 429
  lentiginosus, 431
    albifolius, 431

Astragalus (cont.)
  lentiginosus (cont.)
    antonius, 431
    borreganus, 431
    coachellae, 431
    coulteri, 431
    fremontii, 431
    idriensis, 431
    micans, 431
    nigricalycis, 431
    semotus, 431
    sierrae, 431
    variabilis, 431
  leucolobus, 432
  magdalenae peirsonii, 432
  malacus, 432
  marcus-jonesii, 435
  metanus, 434
  miguelensis, 432
  minthorniae villosus, 432
  mohavensis
    hemigyrus, 432
    mohavensis, 432
  nevinii, 432
  newberryi, 432
  nigrescens, 428
  nutans, 433
  nuttallianus
    cedrosensis, 433
    imperfectus, 433
  oocarpus, 433
  oophorus, 433
  pachypus
    jaegeri, 433
    pachypus, 433
  palmeri, 434
  panamintensis, 434
  parishii, 428
  peirsonii, 432
  platytropis, 434
  pomonensis, 434
  preussii
    laxiflorus, 434
    preussii, 434
  purshii
    lectulus, 434
    gaviotus, 434
    leucolobus, 432
    longilobus, 434
    ordensis, 434
    tinctus, 434
  pycnostachyus
    lanosissimus, 435
  remulcus, 435
  sabulonum, 435
  serenoi, 435
  tehachapensis, 431
  tejonensis, 428
  tener titi, 435
  tephrodes remulcus, 435
  tidestromii, 435

  traskiae, 435
  tricarinatus, 435
  trichopodus
    antisellii, 436
    leucopsis, 436
    lonchus, 436
    triflorus morans, 428
    trichopodus, 436
    vaseyi, 434
      johnstonii, 434
    whitneyi, 436
    wootonii, 436
  Ataenia gairdneri, 83
  Athyrium, 18
    alpestre cyclosorum, 18
    americanum, 18
    filix-femina californicum, 18
      cyclosorum, 18
      sitchense, 18
  Athysanus pusillus, 275
  Atrichoseris platyphylla, 123
  Atriplex, 5, 352
    argentea expansa, 354
    californica, 354
    canescens
      canescens, 354
      linearis, 354
    confertifolia, 4, 354
    coronata notatior, 354
    coulteri, 355
    decumbens, 358
    depressa, 357
    elegans fasciculata, 355
    expansa, 354
      mohavensis, 354
    fasciculata, 355
    hastata, 357
    hymenelytra, 355
    johnstonii, 355
    lentiformis
      breweri, 355
      lentiformis, 355
    leucophylla, 355
    linearis, 354
    microcarpa, 355
    nummularia, 355
    pacifica, 355
    parishii, 357
    parryi, 357
    patula hastata, 357
    phyllostegia, 357
    polycarpa, 357
    rosea, 357
    saltonensis, 355
    semibaccata, 358
    serenana, 358
      davidsonii, 358
    sordida, 354
    spinifera, 358
    torreyi, 358
    truncata, 358

  vesicaria, 358
  watsonii, 358
Atropis
  californica, 995
  fendleriana, 994
  nevadensis, 994
  pauciflora, 994
  scabrella, 995
Audibertia
  capitata, 538
  clevelandii, 536
  dorrii, 537
  grandiflora, 539
  humilis, 539
  nivea, 538
  pachystachya, 538
  stachyoides revoluta, 536
  vaseyi, 539
Audibertiella argentea, 537
Aulospermum
  aboriginum, 74
  panamintense, 74
Avena
  barbata, 947
  fatua, 947
  sativa, 948
Avens, 746
Axonopus
  compressus, 948
  poirettii, 1002
Ayenia
  californica, 841
  compacta, 841
  pusilla, 841
Azalea
  occidentalis, 404
  Western, 404
Azolla filiculoides, 35

Baby Blue-Eyes, 501
Baccharis, 123
  brachyphylla, 124
  Broom, 125
  coerulescens, 124
  consanguinea, 125
  douglasii, 124
  emoryi, 124
  glutinosa, 124
  haenkei, 124
  pilularis consanguinea, 125
  plummerae, 125
  sarathroides, 125
  sergiloides, 125
  viminea, 124
Baeria
  californica, 197
  chrysostoma, 197
  gracilis, 197
  hirsutula, 197
  microglossa, 197
  minor, 197

Index 1039

parishii, 197
Bahia, 125
  artemisiaefolia, 165
  confertiflora, 163
  dissecta, 126
  neomexicana, 126
Baileya
  multiradiata, 126
    nudicaulis, 126
    pleniradiata, 126
  nervosa, 126
  pauciradiata, 126
  pleniradiata, 126
Balardia platensis, 347
Balsaminaceae, 52, 241
Balsamorrhiza deltoidea, 127
Banalia occidentalis, 367
Barbarea
  americana, 276
  orthoceras, 276
    dolichocarpa, 276
Barberry, 242
Barley, 975
  Common, 977
  Beardless, 977
Bartonia
  albicaulis, 552
  laevicaulis, 554
  micrantha, 554
Basellaceae, 49, 242
Bassia hyssopifolia, 359
Batidaceae, 46, 47, 242
Batis
  californica, 242
  maritima, 242
  vermiculata, 369
Bay, California, 544
Bayonet, Spanish, 866
Beak, Birds', 800
Bean
  Castor, 417
  Horse-, 477
  Screw-, 466
Bean-Caper, 862
Beard
  Crown-, 239
  Goat's, 236
  Hawk's-, 153
Beard-Tongue, 818
Bebbia
  aspera, 127
  juncea, 127
Beckwithia andersonii, 728
Bedstraw, 762
  Sweet-Scented, 769
Beeplant, Rocky Mountain, 329
Beet
  Garden, 359
  Sugar, 359
Bell, Wild Canterbury, 512

Bellis perennis, 127
Bells
  Mission, 927
  Whispering, 496
Beloperone californica, 54
Berberidaceae, 50, 242
Berberis, 242
  amplectens, 243
  dictyota, 243
  fremontii, 243
  haematocarpa, 243
  higginsae, 245
  nevinii, 245
    haematocarpa, 243
  pinnata insularis, 245
  pinnata pinnata, 245
Bergerocactus emoryi, 310
Bergia texana, 396
Bermuda-Buttercup, 622
Bermudiana bella, 908
Bernardia
  incana, 406
  myricifolia, 406
Berry
  Bane, 720
  Bar-, 242
  Black, 759
    California, 760
    Cut-Leaf, 760
  Blue, 404
  Buffalo, 396
  Christmas, 747
  Coffee, 738
  Dew, 760
  Elder, 333
  Goose-, 784
  Hack-, 845
  Himalaya, 760
  Huckle, 404
  Lemonade-, 64
  Mul-, 576
  Poke, 632
  Rasp-, 759
    Western, 760
  Red, 739
  Service, 742
  Snow, 333
  Straw-, 746
  Thimble-, 760
  Twin, 332
Berula erecta, 71
Bessera brevifolia, 877
Bet, Bouncing, 341
Beta vulgaris, 359
Betula
  fontinalis, 246
  occidentalis, 246
Betulaceae, 46, 245
Bicuspidaria tricuspis, 558
Bidens
  californica, 128

  frondosa, 128
  laevis, 128
  pilosa, 128
Bigelovia
  acradenia, 176
  albida, 145
  brachylepis, 179
  cooperi, 177
  depressa, 145
  douglasii, stenophylla, 145
  furfuracea, 181
  intricata, 122
  mohavensis, 144
  paniculata, 144
  parishii, 179
  teretifolia, 144
  veneta sedoides, 181
  viscidiflora, 145
Bignonia linearis, 246
Bignoniaceae, 53, 246
Big-Root, 393
Bindweed,
  Black, 703
Birch, Water, 246
Bird's Beak, 800
Biscutella californica, 288
Bitterroot, 714
Bladderpod,
  Double, 297
Bladderwort, 547
Blechnaceae, 22
Blennosperma
  californicum, 128
  nanum, 128
Blepharidachne kingii, 948
Blepharipappus
  glandulosa, 199
  platyglossa, 199
Blite, Sea-, 369
Blitum
  californicum, 361
  nuttallianum, 366
Bloomeria
  clevelandii, 879
  crocea
    crocea, 877
    montana, 877
  gracilis, 877
  maritima, 881
  montana, 877
Blow-Wives, 107
Bluecurls, 543
  Woolly, 544
Blue Dicks, 879
Bluegrass,
  Fowl, 994
  Kentucky, 995
  Timberline, 995
Boerhaavia
  annulata, 582
  caribaea, 582

Boerhaavia (*cont.*)
  coccinea, 582
  coulteri, 582
  erecta intermedia, 582
  hirsuta, 582
  intermedia, 582
  triquetra, 583
  wrightii, 588
Boisduvalia
  densiflora, 590
  glabella, 590
  stricta, 590
Bolelia cuspidata, 323
Borage, 249
Boraginaceae, 52, 246
Borago officinalis, 249
Boschniaka strobilacea, 617
Bothriochloa barbinodis, 948
Botrychium
  lunaria, 34
    minganense, 34
  simplex, 34
    compositum, 34
Boussingaultia
  gracilis pseudo-baselloides, 242
Bouteloua, 949
  aristidoides, 949
  barbata, 949
  curtipendula, 949
    caespitosa, 949
  eriopoda, 949
  gracilis, 949
  hirsuta, 950
  radicosa, 949
  rothrockii, 950
  trifida, 950
Bower, Virgin's, 721
Bowlesia
  incana, 72
  lobata, 72
  septentrionalis, 72
Box Elder, 55
Box-Thorn, 831
Boykinia
  elata, 778
  rotundifolia, 778
Brachyactis frondosus, 122
Brachypodium distachyon, 950
Brachyris microcephala, 174
Brandegea bigelovii, 391
Brass-Buttons, 153
Brassica, 276
  campestris, 277
  eruca, 289
  geniculata, 276
  hirta, 277
  juncea, 277
  kaber, 277
  napus, 277

  nigra, 277
  oleracea, 277
  orientalis, 285
  rapa sylvestris, 277
  tournefortii, 278
Brassicaceae, 50, 51, 267
Brayulinea densa, 63
Breweria minima, 375
Brickellia, 128
  arguta, 129
  atractyloides arguta, 129
  californica, 129
    desertorum, 129
  desertorum, 129
  frutescens, 131
  incana, 131
  knappiana, 131
  linifolia, 131
  longifolia, 131
  microphylla, 131
  mohavensis, 131
  multiflora, 131
  nevinii, 131
  oblongifolia linifolia, 131
  watsonii, 131
Bride, Mourning-, 395
Briza minor, 950
Brizopyrum douglasii, 992
Brodiaea, 877
  candida, 882
  capitata, 878
    pauciflora, 878
  clementina, 881
  coronaria kernensis, 878
  dudleyi, 881
  filifolia, 878
    orcuttii, 878
  jolonensis, 878
  kinkiensis, 878
  laxa, 882
  lutea scabra, 882
  maritima, 881
  orcuttii, 878
  scabra, 882
  terrestris kernensis, 878
Brome, 950
  California, 952
  Downy, 956
  Fringed, 952
  Smooth, 953
Bromelica aristata, 980
Bromus
  anomalus, 955
  arenarius, 952
  arizonicus, 952
  arvensis, 953
  breviaristatus, 952
  carinatus, 952
    arizonicus, 952
    linearis, 952
  catharticus, 956

  ciliatus, 952
  commutatus, 953
  dertonensis, 972
  diandrus, 953
  distachyos, 950
  grandis, 953
  haenkeanus, 956
  hordeaceus, 955
  inermis, 953
  japonicus, 953
  kalmii porteri, 955
  laevipes, 953
  madritensis, 955
  marginatus, 955
  maritimus, 955
  maximus, 953
  molliformis, 955
  mollis, 955
  orcuttianus, 955
    grandis, 953
  porteri, 955
    assimilis, 953
  pseudolaevipes, 955
  richardsonii, 956
  rigidus, 953
  rubens, 956
  stamineus, 956
  subvelutinus, 952
  tectorum, 956
  trinii, 956
  unioloides, 956
  willdenovii, 956
Brooklime, 828
Broom, 438
  French, 438
  Scale-, 200
  Spanish, 470
  Turpentine-, 771
Broom-Rape, 617
Brush
  Buck, 733
  Coyote, 125
  Deer, 734
  Paint-, 793
  Rabbit, 4, 141
  Sage-, 116
  Turpentine, 178
Bryanthus breweri, 402
Buckeye, 495
Buckthorn, 738
Buckwheat
  California, 684
  Wild, 677
Bud, Scale, 113
Buddleja utahensis, 560
Buffalo-Bur, 840
Bugler, Scarlet, 821
Bugloss, 262
  Viper's, 258
Bugseed, 364
Bulbostylis

annua, 220
californica, 129
microphylla, 131
Bulrush
　Hard-Stem, 902
　Soft-Stem, 903
Bupleurum subovatum, 72
Burdock, 115
Bur Head, 868
　Buffalo, 840
　Sand, 109
Burnet, 761
Burrielia
　lanosa, 164
　microglossa, 197
Bursera microphylla, 307
Burseraceae, 50, 307
Bush
　Antelope, 758
　Black-, 745
　Brittle-, 158
　Burning, 349
　Chaff, 112
　Creosote, 860
　Fern, 745
　Flannel, 841
　Indigo, 438
　Iodine, 352
　Pine-, 179
　Ribbon, 741
　Salt-, 352
　Silk-Tassel, 484
　Snow, 733
　Snowdrop, 843
　Squaw, 66
　Sugar, 4, 66
　Sweet, 127
　Wishbone, 584
Bushrue, 771
Butter-and-Eggs, 806, 817
Buttercup, 727
　Bermuda, 622
Butterfly-Tulip, 920
Button
　Bachelor's, 134
　Ranger's, 86
Buttons, Brass, 153
Buxaceae, 46, 308
Buxus chinensis, 308

Cabbage, 277
　Skunk-, 929
　Squaw-, 283
Cactaceae, 49, 308
Cactus
　ficus-indica, 317
　radiosus alversonii, 311
Cactus,
　Barrel, 312
　Beavertail, 315
　Devil's, 320

Foxtail, 311
Giant, 310
Grizzly-Bear, 317
Hedgehog, 311
Mound, 312
Pencil, 320
Cakile
　californica, 278
　edentula californica, 278
　maritima, 278
Calabazilla, 392
Calais
　douglasii, 211
　linearifolia, 212
　platycarpha, 211
　tenella, 212
Calamagrostis
　andina, 983
　densa, 957
　purpurascens, 957
　rubescens, 957
Calamintha
　chandleri, 539
　ilicifolia, 524
Calandrinia, 708
　ambigua, 709
　breweri, 709
　ciliata menziesii, 709
　maritima, 709
　nevadensis, 713
Calendula
　arvensis, 132
　officinalis, 132
California-Bay, 544
Calla aethiopica, 882
Calliandra eriophylla, 436
Callichroa platyglossa, 199
Calligonum canescens, 354
Calliprora scabra, 882
Callisteris arizonica, 654
Callitrichaceae, 47, 321
Callitriche, 321
　bolanderi, 322
　heterophylla bolanderi, 322
　longipedunculata, 322
　marginata, 322
　　longipedunculata, 322
　nuttallii, 322
　peploides semialata, 322
　palustris, 322
　verna, 322
Calocedrus decurrens, 4, 36
Calochortus, 920
　albus, 921
　catalinae, 923
　citrinus, 926
　clavatus
　　clavatus, 923
　　gracilis, 923
　　pallidus, 923
　　concolor, 923

davidsonianus, 925
dunnii, 923
flexuosus, 923
invenustus, 924
montanus, 924
kennedyi, 924
　munzii, 924
lanternus, 921
luteus
　citrinus, 926
　concolor, 923
lyonii, 923
montanus, 924
nuttallii
　australis, 924
　panamintensis, 924
palmeri
　dunnii, 923
　munzii, 924
　palmeri, 924
　paludicola, 924
paludicola, 924
plummerae, 925
pratensis, 925
splendens
　atroviolaceus, 925
　major, 925
　montanus, 924
striatus, 925
superbus, 925
venustus, 925
　citrinus, 925
weedii
　purpurascens, 925
　intermedius, 925
　vestus, 926
　weedii, 926
Calothyrsus californica, 495
Caltrop, 862
Calycadenia tenella, 132
Calycoseris, 132
　parryi, 133
　wrightii, 133
　californica, 133
Calyptridium, 709
　monandrum, 711
　parryi
　　nevadense, 711
　　parryi, 711
　pygmaeum, 711
　umbellatum, 711
Calystegia
　fulcrata, 373
　　tomentella, 374
　　deltoidea, 374
　longipes, 373
　macrostegia
　　arida, 373
　　cyclostegia, 373, 374
　　intermedia, 373, 374
　　longiloba, 373, 374

Calystegia (cont.)
  macrostegia (cont.)
    macrostegia, 373, 374
    tenuifolia, 373, 374
    malacophylla pedicellata, 374
  peirsonii, 374
  purpurata, 374
  sepium, 374
    binghamiae, 374
    soldanella, 374
Camas, Death-, 931
Camelina
  microcarpa, 279
  sativa, 279
Camissonia, 590
  arenaria, 592
  bistorta, 592
  boothii
    alyssoides, 593
    condensata, 594
    decorticans, 593
    desertorum, 594
    intermedia, 594
    inyoensis, 594
    rutila, 594
  brevipes
    brevipes, 594
    pallidula, 594
  californica, 594
  campestris, 595
  cardiophylla
    cardiophylla, 595
    robusta, 595
  chamaenerioides, 595
  cheiranthifolia
    cheiranthifolia, 595
    suffruticosa, 595
  claviformis
    aurantiaca, 596
    funerea, 596
    integrior, 596
    lancifolia, 596
    peirsonii, 596
    yumae, 597
  confusa, 597
  graciliflora, 597
  guadalupensis clementina, 597
  heterochroma, 597
  hirtella, 597
  ignota, 597
  intermedia, 598
  kernensis
    gilmanii, 598
    kernensis, 598
  lewisii, 598
  micrantha, 598
  munzii, 598
  pallida
    hallii, 598

    pallida, 598
  palmeri, 599
  pterosperma, 599
  pubens, 599
  pusilla, 599
  refracta, 599
  robusta, 599
  strigulosa, 599
  walkeri tortilis, 600
Campanula *biflora*, 328
Campanulaceae, 53, 322
Campion, 342
  Bladder, 343
Camus,
  Death-, 931
Canaigre, 707
Canbya candida, 625
Canchalagua, 486
Candle,
  Desert, 283
  Our Lord's, 867
Candlewood, 483
Cannabis sativa, 575
Caper, Bean-, 862
Capnorea leporina, 499
Capparaceae, 50, 51, 328
Capparidaceae, synonym of Capparaceae
Caprifoliaceae, 53, 331
Capsella
  bursa-pastoris, 279
  procumbens, 291
Cardamine, 279
  breweri, 279
  californica, 279
  filifolia, 300
  gambelii, 281
  menziesii, 286
  oligosperma, 281
  palustris mexicana intermedia, 1017
  virginica, 301
Cardaria
  draba, 281
  pubescens elongata, 281
Cardionema ramosissimum, 339
Cardoon, 154
Carduus, 133
  californicus, 147
  marianus, 227
  nutans, 133
  occidentalis, 147
  pycnocephalus, 133
  tenuiflorus, 133
  undulatus, 148
  vulgaris, 148
Carex, 884
  abramsii, 891
  abrupta, 887
  adusta congesta, 892

  aequa, 894
  alma, 887
  arenicola pansa, 892
  athrostachya, 887
  aurea, 887
    celsa, 890
  auriculata, 893
  austromontana, 893
  barbarae, 887
  bernardina, 888
  bifida, 894
  bishalii, 893
  bolanderi, 887
  brevipes, 888
  brongniartii densa, 888
  cephalophora, 888
  chrysoleuca, 888
  comosa, 888
  deflexa boottii, 888
  densa, 888
  diandra, 888
  disperma, 888
  douglasii, 888
    brunnea, 893
  echinata ormantha, 892
  eleocharis, 889
  exserta, 889
  festivella, 889
  feta, 889
    multa, 889
  filifolia erostrata, 889
  fracta, 889
  globosa, 889
  gracilior, 890
  hassei, 890
  haydeniana, 890
  heteroneura, 890
  hoodii, 890
    nervosa, 890
  illota, 890
  integra, 890
  jacintoensis, 892
  jonesii, 891
  lacunarum, 887
  lanuginosa, 891
  lemmonii, 891
  mariposana, 891
  microptera, 891
  molesta, 891
  montereyensis, 891
  monticola, 895
  multicaulis, 892
  multicostata, 892
  nebrascensis, 892
  nubicola, 890
  occidentalis, 892
  ormantha, 892
  pachycarpa, 892
  pansa, 892
  praegracilis, 893
  quadrifolia, 890

Index 1043

caeca, 890
lenis, 890
rosea texensis, 895
rossii, 893
rostrata, 893
sartwelliana, 893
schottii, 893
senta, 893
serratodens, 894
simulata, 894
spissa, 894
stenoptera, 894
straminea mixta, 889
subbracteata, 894
subfusca, 894
subnigricans, 894
teneraeformis, 894
texensis, 895
triquetra, 895
tumulicola, 895
usta, 893
utriculata globosa, 893
vernacula, 895
vitrea, 887
wilkesii, 887
yosemitana, 893
Carnegiea gigantea, 310
Caryopitys edulis, 39
monophylla, 40
Carphephorus junceus, 127
Carpobrotus
aequilaterus, 56
chilensis, 56
edulis, 56
Carrot, Wild, 75
Carthamus
baeticus, 133
nitidus, 133
tinctorius, 134
Caryophyllaceae, 47, 50, 334
Cascara, 738
Cassia
armata, 437
covesii, 437
tomentosa, 437
Castanea sempervirens, 478
Castanopsis sempervirens, 478
Castela emoryi, 829
Castilleja, 793
affinis
affinis, 795
contentiosa, 795
insularis, 795
minor, 797
anacapensis, 797
angustifolia, 795
breweri, 795
californica
chromosa, 795
cinerea, 795
clokeyi, 797

disticha, 795
douglasii, 795
ewanii, 797
foliolosa, 795
gleasonii, 796
grisea, 796
gyroloba, 796
hoffmannii, 796
hololeuca, 796
jepsonii, 796
latifolia insularis, 795
linariaefolia, 796
martinii, 796
clokeyi, 797
ewanii, 797
miniata, 797
minor, 797
mollis, 797
nana, 797
plagiotoma, 797
roseana, 796
stenantha, 797
wightii, 797
Castor-Bean, 417
Catapodium rigidum, 997
Catclaw, 421
Catchfly, 342
Catnip, 530
Cat-Tail, 1012
Caucalis microcarpa, 72
Caulanthus, 281
amplexicaulis, 282
cooperi, 282
coulteri, 282
crassicaulis, 283
major, 283
glaucus, 283
hallii, 283
heterophyllus, 304
inflatus, 283
major, 283
pilosus, 285
simulans, 285
stenocarpus, 285
Caulinia guadalupensis, 931
Ceanothus, 4, 732
arboreus, 733
austromontanus, 734
cordulatus, 733
crassifolius, 733
planus, 733
cuneatus, 733
cyaneus, 734
divaricatus, 734
foliosus, 734
greggii
perplexans, 734
vestitus, 734
insularis, 735
integerrimus, 734
puberulus, 734

leucodermis, 734
macrocarpus, 735
megacarpus
insularis, 735
megacarpus, 735
oliganthus, 735
palmeri, 735
papillosus roweanus, 735
perplexans, 734
puberulus, 734
sorediatus, 735
spinosus, 735
tomentosus olivaceus, 737
verrucosus, 737
Cedar,
Incense-, 35
Pigmy-, 218
Celastraceae, 51, 349
Celery, 71
Wild-, 71
Celtis
douglasii, 845
reticulata, 845
Cenchrus
echinatus, 957
longispinus, 958
incertus, 958
longispinus, 958
longisetus, 988
pauciflorus, 958
Centaurea
calcitrapa, 134
cineraria, 134
cyanus, 134
diluta, 135
iberica, 135
melitensis, 135
moschata, 135
muricata, 135
repens, 135
solstitalis, 135
Centaurium
calycosum, 486
exaltatum, 486
venustum, 486
Centaury, 486
Centromadia
fitchii, 186
pungens, 187
Centrostegia
leptoceras, 673
thurberi, 676
Centunculus minimus, 717
Century Plant, 864
Cephalanthera austinae, 932
Cerastium
glomeratum, 339
viscosum, 339
vulgatum, 340

Cerasus
  demissa, 757
  emarginata, 757
  ilicifolia, 757
Ceratophyllaceae, 46, 350
Ceratophyllum
  apiculatum, 351
  demersum, 351
Cercidium
  floridum, 437
  microphyllum, 437
Cercis
  occidentalis, 438
  nephrophylla, 438
Cercocarpus, 4, 742
  alnifolius, 743
  betuloides, 743
    blancheae, 743
    traskae, 743
  intricatus, 743
  ledifolius, 4, 743
  minutiflorus, 743
  traskae, 743
Cereus
  coccineus melanacanthus, 312
  emoryi, 310
  engelmannii, 311
  gigantea, 310
  mojavensis, 312
  munzii, 312
Ceropteris
  triangularis, 32
  viscosa, 32
Chaenactis, 135
  artemisiaefolia, 136
  attenuata, 136
  carphoclinia, 136
    attenuata, 136
    peirsonii, 136
  denudata, 137
  douglasii, 136
    achilleaefolia, 137
    rubricaulis, 137
  filifolia, 138
  fremontii, 137
  glabriuscula, 137
    curta, 137
    denudata, 137
    glabriuscula, 137
    lanosa, 137
    orcuttiana, 138
    tenuifolia, 138
  heterocarpha curta, 137
  lanosa, 137
  latifolia, 138
  macrantha, 138
  orcuttiana, 138
  panamintensis, 137
  peirsonii, 136
  santolinoides, 138
    indurata, 138

  stevioides, 138
  tenuifolia orcuttiana, 138
    xantiana, 138
    integrifolia, 138
Chaetadelphia wheeleri, 139
Chaetaria glauca, 946
Chaetochloa
  geniculata, 998
  lutescens, 999
  verticillata, 999
  viridis, 999
Chaetopappa
  alsinoides, 139
  aurea, 139
  lyonii, 139
Chamaebatia australis, 743
Chamaebataria millefolium, 745
Chamaesaracha coronopus, 830
Chamise, 741
Charlock, 277
  Jointed, 297
Chamaesyce
  albomarginata, 411
  fendleri, 412
  maculata, 413
  melanadenia, 413
  micromera, 413
  ocellata, 413
  parishii, 414
  parryi, 414
  pediculifera, 414
  polycarpa, 414
  prostrata, 415
  revoluta, 415
  serpyllifolia, 415
  setiloba, 415
  supina, 415
  vallis-mortae, 417
Chasmanthe aethiopica, 906
Checker, 570
Cheeses, 568
Cheilanthes
  californica, 26
  clevelandii, 26
  cooperae, 26
  covillei, 26
  feei, 26
  fibrillosa, 28
  gracilis, 26
  intertexta, 26
  jonesii, 29
  lanuginosa fibrillosa, 28
  newberryi, 29
  parishii, 28
  parryi, 29
  sinuata cochisensis, 29
  viscida, 28
  wootoni, 28

Cheiranthus
  capitatus, 290
  cheirii, 290
  incanus, 297
  suffrutescens, 290
Chelidonium glaucium, 628
Chenopodium, 359
  album, 360
  ambrosioides, 360
    anthelminticum, 361
  atrovirens, 361
  berlandieri
    sinuatum, 361
    zschackei, 361
  botrys, 361
  californicum, 361
  carinatum, 364
  chenopodioides, 361
  dessicatum
    dessicatum, 361
    leptophylloides, 361
  fremontii, 361
    incanum, 363
  fruticosum, 370
  gigantospermum, 363
  glaucophyllum, 364
  glaucum salinum, 363
  humile, 361
  inamoenum, 363
  incanum, 363
  incognitum, 363
  leptophyllum, 363
    farinosum, 363
  missouriense, 363
  multifidum, 363
  murale, 364
    farinosum, 363
  nevadense, 364
  pratericola dessicatum, 361
  pumilio, 364
  rubrum, 364
  salinum, 363
  spinosum, 365
  strictum glaucophyllum, 364
  zschackei, 361
Chenopodiaceae, 46, 47, 351
Cherry,
  Bitter, 757
  Catalina, 757
  Ground-, 836
  Western Choke-, 757
Chess,
  Australian, 952
  Foxtail, 956
  Hairy, 953
  Japanese, 953
  Soft, 955
Chestnut, Horse-, 495
Chia, 536

Chickweed,
  Common, 348
  Indian, 57
  Mouse-Ear, 339
Chicory, 145
Chilianthus oleaceus, 560
Chilopsis
  linearis, 246
  saligna, 246
Chimaphila, 401
  menziesii, 402
  occidentalis, 402
  umbellata occidentalis, 402
Chinese Houses, 799
Chinquapin, Bush, 478
Chloris
  curtipendula, 949
  distichophylla, 958
  elegans, 958
  gayana, 958
  virgata, 958
Chloroea austinae, 932
Chlorogalum
  parviflorum, 926
  pomeridianum, 926
Cholla, 315
  Golden, 316
  Jumping, 316
  Silver, 316
  Valley, 318
Chondrosium
  gracile, 949
  eriopodum, 949
Chorizanthe, 670
  brevicornu
    brevicornu, 671
    spathulata, 672
  californica, 672
    suksdorfii, 672
  chrysacantha, 674
  clevelandii, 672
  coriacea, 672
  corrugata, 672
  discolor, 674
  fernandina, 673
  fimbriata, 672
    laciniata, 673
  insularis, 676
  lastarriaea, 672
  leptoceras, 673
  leptochloa, 673
  membranacea, 673
  nudicaulis, 674
  orcuttiana, 673
  parryi, 673
    fernandina, 673
  perfoliata, 673
  polygonoides longispina, 673
  procumbens, 674
    albiflora, 674

rigida, 674
spathulata, 672
spinosa, 674
staticoides
  brevispina, 674
  chrysacantha, 674
  elata, 674
  latiloba, 674
  staticoides, 674
  thurberi, 676
  uncinata, 674
  watsonii, 676
  wheeleri, 676
xanti
  leucotheca, 676
  xanti, 676
Chrysanthemum
  carinatum, 140
  coronarium, 140
  Garland, 140
  leucanthemum, 140
  nanum, 128
  parthenium, 140
  Tricolor, 140
Chrysocoma
  nauseosa, 143
  pumila, 145, 218
Chrysolepis sempervirens, 478
Chrysopsis, 140
  acaulis, 176
  breweri, 141
  echioides, 141
  fastigiata, 141
  sessiliflora, 141
  villosa
    echioides, 141
    fastigiata, 141
    hispida, 141
    sessiliflora, 141
  wrightii, 141
Chrysothamnus, 4, 141
  albidus, 142
  asper, 144
  consimilis, 144
  depressus, 142
  gnaphalodes, 144
  gramineus, 216
  mohavensis, 144
  nauseosus, 142
    bernardinus, 142
    consimilis, 144
    gnaphalodes, 144
    hololeucus, 144
    leiospermus, 142
    mohavensis, 144
    viridulus, 144
  parryi
    asper, 144
    imulus, 144
    teretifolius, 144
  viscidiflorus

pumilus, 145
puberulus, 145
stenophyllus, 145
viscidiflorus, 145
Chuparosa, 54
Chylismia
  arenaria, 592
  heterochroma, 597
  lancifolia, 596
  scapoidea
    aurantiaca, 596
    purpurascens, 596
Cicely, Sweet-, 82
Cicendia exaltata, 486
Cichorium intybus, 145
Cicuta douglasii, 73
Cinna glomerata, 945
Cinquefoil, 752
Circaea
  alpina pacifica, 600
  pacifica, 600
Cirsium, 145
  arvense, 146
  brevistylum, 146
  californicum, 147
    bernardinum, 147
  congdonii, 147
  coulteri, 147
  drummondii, 147
  edule, 146
  lanceolatum, 148
  mohavense, 147
  neomexicanum, 147
  nidulum, 147
  occidentale, 147
  ochrocentrum, 148
  proteanum, 148
  tioganum, 148
  undulatum, 148
  vulgare, 148
Cistaceae, 48, 370
Cistus
  villosus, 370
    tauricus, 370
    undulatus, 370
Citron, 391
Citrullus
  lanatus
    citroides, 391
    lanatus, 391
  vulgaris, 391
Cladium
  californicum, 895
  mariscus californicum, 895
Cladothrix oblongifolia, 64
Clarkia, 600
  cylindrica, 601
  deflexa, 601
  delicata, 601
  dudleyana, 603
  elegans, 604

Clarkia (cont.)
  epilobioides, 603
  prostrata, 603
  purpurea
    quadrivulnera, 603
    viminea, 603
  rhomboidea, 603
  similis, 604
  speciosa, polyantha, 604
  unguiculata, 604
  xantiana, 604
Claytonia
  ambigua, 709
  chamissoi, 713
  lanceolata peirsonii, 712
  linearis, 714
  perfoliata, 712
    depressa, 712
      parviflora, 712
        depressa, 712
      perfoliata, 712
      utahensis, 712
  spathulata
    exigua, 713
    spathulata, 712
    tenuifolia, 713
    viridis, 713
  tenuifolia, 713
Cleavers, 762
Clematis, 721
  lasiantha, 722
  ligusticifolia, 722
    californica, 722
    brevifolia, 722
  pauciflora, 722
Cleome
  isomeris, 330
  lutea, 329
  pinnata, 302
  serrulata, 329
  sparsifolia, 329
Cleomella
  brevipes, 329
  mojavensis, 330
  obtusifolia, 329
    florifera, 329
    jonesii, 329
    pubescens, 330
  parviflora, 330
  plocasperma
    mojavensis, 330
    stricta, 330
Clintonia pulchella, 323
Clistoyucca arborescens, 866
Clotbur, Spiny, 241
Clover, 470
  Alsike, 473
  Bur-, 463
  Hop, 475
  Jackass-, 331
  Owl's-, 817

Prairie-, 465
Red, 475
Sweet-, 464
White, 475
Club-moss, Little, 14
Clusiaceae, synonym of
  Hypericaceae, 519
Clypeola maritima, 296
Cneoridium dumosum, 771
Cnicus
  benedictus, 148
  nidulus, 147
  tioganus, 148
Cockle, 349
Cocklebur, 241
Cogswellia
  chandleri, 80
  dasycarpa, 78
  lucida, 79
  hassei, 79
  nevadensis pseudorientalis, 80
  utriculata, 80
  vaseyi, 80
Coinogyne carnosa, 196
Coldenia, 249
  brevicalyx, 250
  canescens, 250
    pulchella, 250
    subnuda, 250
  nuttallii, 250
  palmeri, 250
  plicata, 250
Coleogyne ramosissima, 4, 745
Coleosanthus
  californicus, 129
  desertorum, 129
  frutescens, 131
  knappianus, 131
  linifolius, 131
  longifolius, 131
  microphyllus, 131
  nevinii, 131
Collinsia
  bartsiaefolia davidsonii, 798
  bicolor, 799
    austromontana, 799
  callosa, 798
  childii, 799
  concolor, 799
  davidsonii, 798
  heterophylla
    austromontana, 799
    heterophylla, 799
  parryi, 799
  parviflora, 799
  torreyi wrightii, 799
  wrightii, 799
Collomia
  gilioides, 640
  glutinosa, 640
  grandiflora, 641

linearis, 641
Colubrina
  californica, 737
  texensis californica, 737
Columbine, 720
Comandra
  pallida, 776
  umbellata californica, 776
Comarostaphylis
  diversifolia, 402
  planifolia, 402
Commelinaceae, 883
Compositae, synonym of
  Asteraceae, 95
Condalia
  globosa pubescens, 737
  parryi, 738
Condaliopsis
  lycioides canescens, 738
  parryi, 738
Coniferae, 35
Coniferales, 35
Coniothele californica, 128
Conium maculatum, 73
Conringia orientalis, 285
Convallaria stellata, 929
Convolvulaceae, 52, 53, 371
Convolvulus, 1
  althaeoides, 375
  aridus
    intermedius, 374
    longilobus, 374
    tenuifolius, 374
  arvensis, 375
  binghamiae, 374
  cyclostegius, 374
  deltoides, 374
  fulcratus, 373
  longipes, 373
  luteolus
    fulcratus, 373
    purpuratus, 374
  macrostegius, 373
  malacophyllus, 374
  peirsonii, 374
  pentaploides, 375
  purpureus, 379
  sepium americanus, 374
  simulans, 375
  soldanella, 374
  tomentellus, 374
  villosus pedicellatus, 374
Conyza
  bonariensis, 149
  canadensis, 149
  coulteri, 149
  purpurascens, 219
Conyzella coulteri, 149
Copperleaf, California, 406
Coral Root, 932

Corallorhiza
  maculata, 932
  multiflora, 932
Cordylanthus, 800
  bernardinus, 801
  canescens, 801
  ferrisianus, 801
  filifolius, 801
    brevibracteatus, 802
  helleri, 801
  littoralis platycephalus, 801
  maritimus, 801
  nevinii, 802
  parviflorus, 802
  platycephalus, 801
  ramosus eremicus, 802
  rigidus brevibracteatus, 802
  tecopensis, 802
Coreopsis, 149
  bigelovii, 150
  californica, 150
  calliopsidea, 150
  douglasii, 150
  gigantea, 150
  lanceolata, 150
  maritima, 151
Corethrogyne
  brevicula, 152
  detonsa, 177
  filaginifolia
    bernardina, 151
    brevicula, 152
    glomerata, 152
    incana, 152
    latifolia, 152
    linifolia, 152
    pacifica, 152
    peirsonii, 152
    pinetorum, 152
    robusta, 152
    rigida, 152
    sessilis, 152
    virgata, 152
    viscidula, 152
  flagellaris, 152
  incana, 152
  sessilis, 152
  virgata, 152
    bernardina, 151
  viscidula, 152
Corispermum hyssopifolium, 364
Cornaceae, 50, 379
Cornflower, 134
Cornus
  californica, 382
  glabrata, 381
  nuttallii, 381
  occidentalis, 381
  sericea occidentalis, 381
  stolonifera, 381

Coronopus didymus, 285
Cortaderia, 958
  atacamensis, 959
  selloana, 959
Coryphantha, 310
  desertii, 311
  rosea, 311
  vivipara
    alversonii, 310
    desertii, 311
    rosea, 311
Cotoneaster pannosa, 745
Cotton, Lavender-, 222
Cottonwood, 771
  Black, 772
Cotula, 152
  australis, 153
  coronopifolia, 153
Cotyledon, 382
  caespitosa, 385
  nudicaulis, 386
  orbiculata, 382
  saxosa, 387
  viscida, 388
Cowania
  mexicana
    dubia, 746
    stansburiana, 745
Cranesbill, 491
Crassula
  aquatica, 382
  erecta, 382
Crassulaceae, 49, 52, 382
Crepis
  acuminata, 153
    intermedia, 153
  intermedia, 153
  nana, 154
  occidentalis
    occidentalis, 154
    pumila, 154
  pumila, 154
  vesicaria taraxacifolia, 154
Cress, Hoary, 281
  Penny-, 305
  Rock-, 270
  Shield-, 294
  Wart-, 285
  Water-, 298
  Winter-, 276
Cressa
  minima, 376
  truxillensis
    minima, 376
    vallicola, 375
  vallicola, 375
Crinitaria viscidiflora, 145
Crocosmia crocosmiflora, 906
Crossosoma
  bigelovii, 389
  californicum, 389

Crossomataceae, 48, 389
Crossostephium insulare, 118
Croton, 406
  arenicola, 408
  californicus
    californicus, 408
    mohavensis, 408
    tenuis, 408
  tenuis, 408
  wigginsii, 408
Crowfoot, 727
Crown-Beard, 239
Crownscale, 354
Cruciferae, synonym of Brassicaceae
Crunocallis chamissoi, 713
Cryophytum
  crystallinum, 56
  nodiflorum, 57
Crypsis niliaca, 959
Cryptantha, 250
  abramsii, 253
  affinis, 252
  ambigua echinella, 254
  angustifolia, 252
  barbigera, 253
  brandegei, 253
  circumscissa, 253
    hispida, 253
  clevelandii, 253
    florosa, 253
  confertiflora, 253
  corollata, 253
  costata, 254
  decipiens, 254
    corollata, 254
  dumetorum, 254
  echinella, 254
  flaccida, 254
  flavoculata, 254
  ganderi, 254
  geminata, 252
  gracilis, 255
  hispidissima, 253
  hoffmannii, 255
  holoptera, 255
  inaequata, 255
  intermedia, 255
    dumetorum, 254
    johnstonii, 255
    rigida, 257
  jamesii abortiva, 255
  maritima, 256
  micrantha
    lepida, 256
    micrantha, 256
  micromeres, 256
  microstachys, 256
  muricata
    clokeyi, 257
    denticulata, 257

Cryptantha (cont.)
  muricata (cont.)
    jonesii, 256
    muricata, 256
  nevadensis
    nevadensis, 257
    rigida, 257
  nubigena, 257
  pterocarya, 257
    cycloptera, 257
    purpusii, 257
  racemosa, 257
    lignosa, 257
  recurvata, 257
  roosiorum, 258
  seorsa, 254
  similis, 258
  simulans, 258
  traskiae, 258
  tumulosa, 258
  utahensis, 258
  virginensis, 258
Cryptogramma
  crispa acrostichoides, 28
Cucumber, Wild, 392
Cucumis
  dudain, 391
  melo dudain, 391
  myriocarpus, 391
Cucurbita
  californica, 392
  digitata, 392
  foetidissima, 392
  palmata, 392
  pepo, 392
Cucurbitaceae, 49, 54, 389
Cupressaceae, 35
Cupressus
  arizonica stephensonii, 36
  forbesii, 36
  stephensonii, 36
Cups, Cream, 629
Currant, 784
  Chaparral, 787
  Golden, 785
  Sierra, 788
  Squaw, 785
  White-flowered, 787
Cuscuta, 376
  arvensis, 377
  californica, 376
    apiculata, 377
    papillosa, 377
  campestris, 377
  ceanothi, 377
  denticulata, 377
  indecora, 377
  nevadensis, 377
  obtusiflora glandulosa, 377
  occidentalis, 377
  pentagona, 378

  salina, 378
    apoda, 377
    major, 378
  suaveolens, 378
  subinclusa, 377
  suksdorfii subpedicellata, 378
  veatchii apoda, 377
Cuscutaceae included in Convolvulaceae, 371
Cutandia memphitica, 959
Cycladenia
  humilis venusta, 90
  venusta, 90
Cyclobothra alba, 921
Cycloloma atriplicifolium, 365
Cymopterus, 73
  aboriginum, 74
  cinerarius, 74
  deserticola, 74
  gilmanii, 74
  montanus purpurascens, 74
  multinervatus, 74
  nevadensis, 84
  panamintensis
    acutifolius, 74
    panamintensis, 74
  petraeus, 84
  purpurascens, 74
Cynanchum utahense, 94
Cynara cardunculus, 154
Cynodon
  dactylon, 959
  transvaalensis, 959
Cynoglossum penicillatum, 263
Cynosurus
  aegyptium, 960
  aureus, 978
  cristatus, 960
  echinatus, 960
  indicus, 965
Cyperaceae, 883
Cyperus, 895
  acuminatus, 896
  alternifolius, 896
  aristatus, 896
    inflexus, 896
  brevifolius, 897
  bromoides, 898
  californicus, 898
  diandrus capitatus, 897
  difformis, 897
  eragrostis, 897
  erythrorhizos, 897
  esculentus, 897
    hermannii, 897
  ferax, 898
  inflexus, 896
  laevigatus, 897
  lateriflorus, 897
  melanostachyus, 898

niger
  capitatus, 897
  castaneus, 898
  rivularis, 898
  occidentalis, 897
  odoratus, 898
  parishii, 898
  rotundus, 898
  serrulatus, 897
  unioloides, 898
  vegetus, 897
Cypress
  Cuyamaca, 36
  Summer-, 366
  Tecate, 36
Cyrtomium falcatum, 18
Cystopteris fragilis, 18
Cytisus
  linifolius, 438
  monspessulanus, 438

Dactylis
  memphitica, 959
  glomerata, 960
Dactyloctenium aegyptium, 960
Daisy,
  African, 115
  English, 127
  Ground-, 236
  Ox-Eye, 140
  Seaside-, 161
Dalea, 438
  arborescens, 439
  californica, 439
  emoryi, 439
  fremontii
    fremontii, 439
    minutifolia, 439
    saundersii, 441
  johnsonii, 439
  mollis, 441
  mollissima, 441
  parryi, 441
  polyadenia, 441
  schottii, 441
  spinosa, 441
Damasonium californicum, 868
Dandelion, 234
  Common, 234
  Desert-, 208
  Mountain, 108
  Red-seeded, 234
Danthonia, 960
  americana, 961
  californica
    americana, 961
    unispicata, 961
  pilosa, 961
  unispicata, 961

Darnel, 979
Darwinia exaltata, 470
Dasylirion bigelovii, 865
Datisca glomerata, 393
Datiscaceae, 48, 393
Datura
　discolor, 830
　inoxia, 831
　meteloides, 830
　stramonium, 831
　　tatula, 831
　tatula, 831
　wrightii, 831
Daucus
　brachiatus, 72
　carota, 75
　pusillus, 75
　visnaga, 69
Delphinium, 4, 722
　amabile, 724
　　pallidum, 723
　cardinale, 723
　glaucum, 723
　hesperium cuyamacae, 723
　inopinum, 723
　kinkiense, 723
　parishii, 724
　　inopinum, 723
　　pallidum, 723
　　purpureum, 724
　　subglobosum, 724
　parryi, 724
　　montanum, 724
　patens
　　hepaticoideum, 724
　　montanum, 724
　subglobosum, 724
　variegatum thornei, 724
Dendromecon
　harfordii, 625
　　rhamnoides, 625
　rigida, 625
　　harfordii, 625
　　rhamnoides, 625
Dentaria
　californica, 279
　integrifolia californica, 279
Deschampsia
　caespitosa, 961
　calycina, 961
　danthonioides, 961
　　gracilis, 962
　elongata, 962
　　ciliata, 962
　　tenuis, 962
　gracilis, 962
Descurainia, 285
　californica, 286
　obtusa adenophora, 286
　pinnata
　　glabra, 286

halictorum, 286
menziesii, 286
richardsonii
　incisa, 286
　viscosa, 287
　sophia, 287
Deweya
　arguta, 86
　hartwegii, 86
　vestita, 82
Dicentra, 625
　chrysantha, 626
　ochroleuca, 626
Dichaeta
　tenella, 197
　uliginosa, 197
Dichelostemma
　capitata, 878
　insularis, 878
　pauciflora, 879
　pulchella, 878
　　pauciflora, 879
Dichondra, 378
　donnelliana, 379
　occidentalis, 379
　repens, 379
Diclytra
　chrysantha, 626
　ochroleuca, 626
Dicoria, 154
　canescens canescens, 156
　　clarkae, 156
　　hispidula, 156
　clarkae, 156
　　hispidula, 156
Dicotyledones, 45
Dietaria gracilis, 178
Digitalis purpurea, 802
Digitaria
　ischaemum, 962
　sanguinalis, 962
Dill, 70
Dimorphotheca
　ecklonis, 156
　sinuata, 156
Dinebra aristidoides, 949
Diotis lanata, 365
Diphyma crassifolium, 56
Diplacus
　aridus, 809
　australis, 809
　calycinus, 813
　clevelandii, 810
　longiflorus, 813
　parviflorus, 811
　puniceus, 814
Diplanthera wrightii, 1013
Diplopappus
　ericoides, 177
　hispidus, 141
Diplostephium canum, 176

Diplotaxis, 287
　muralis, 287
　tenuifolia, 287
Dipsacaceae, 53, 395
Dipsacus sativus, 395
Dipterostemon
　capitatum, 878
　insulare, 878
　pauciflorum, 879
Disphyma crassifolium, 56
Dissanthelium californicum, 962
Distasis concinnus, 162
Ditaxis
　adenophora, 408
　californica, 408
　lanceolata, 409
　neomexicana, 409
　sericophylla, 409
　serrata, 409
Distichlis, 5, 962
　spicata
　　divaricata, 963
　　nana, 963
　　spicata, 963
　　stricta, 963
　　stolonifera, 963
Dithyraea, 287
　californica, 288
　maritima, 288
Dock, 705
　Curly, 706
　Fiddle, 707
Dodder, 376
Dodecatheon, 717
　alpinum majus, 718
　clevelandii
　　clevelandii, 718
　　insulare, 718
　　sanctarum, 718
　hendersonii, 718
　redolens, 718
　sanctarum, 718
Dogbane, 88
Dogtail, 960
Dogwood,
　American, 381
　Brown, 381
　Mountain, 381
Dondia
　californica, 369
　depressa, 369
　torreyana, 370
Donia
　ciliata, 176
　squarrosa, 173
Downingia
　concolor brevior, 323
　cuspidata, 323
　immaculata, 323
　pulchella, 323

Draba, 288
  corrugata
    corrugata, 288
    saxosa, 288
  crockeri, 289
  cuneifolia
    cuneifolia, 288
    integrifolia, 288
  douglasii crockeri, 289
  integrifolia, 288
  reptans stellifera, 289
  saxosa, 288
  sonorae, 288
  stenoloba nana, 289
  verna, 289
  vestita, 288
Dropseed, 1001
Drosanthemum
  floribundum, 56
  speciosum, 56
Drymocallis glandulosa, 753
Dryopteris
  arguta, 19
  feei, 26
  filix-mas, 19
  puberula, 21
  rigida arguta, 19
Dudleya, 383
  abramsii, 384
  aloides, 387
  arizonica, 384
  attenuata orcuttii, 385
  bernardina, 385
  blochmanae
    blochmanae, 385
    brevifolia, 385
    insularis, 385
  brauntonii, 386
  caespitosa, 385
  candelabrum, 385
  congesta, 386
  cymosa
    marcescens, 385
    minor, 385
    ovatifolia, 385
  delicata, 387
  densiflora, 386
  echeverioides, 386
  elongata, 386
  edulis, 386
  farinosa, 386
  granliflora, 387
  greenei, 386
  hallii, 386
  hassei, 386
  hoffmannii, 386
  lanceolata, 386
  lurida, 386
  minor, 385
  multicaulis, 386
  nesiotica, 387

  ovatifolia, 385
  parishii, 386
  parva, 387
  pulverulenta, 387
  pumila, 385
  regalis, 386
  robusta, 386
  saxosa
    aloides, 387
    saxosa, 387
  stolonifera, 387
  tenuis, 384
  traskae, 388
  variegata, 388
  virens, 388
  viscida, 388
Duckweed,
  Greater, 919
Duster, Fairy, 436
Dyssodia, 156
  cooperi, 157
  papposa, 157
  pentachaeta
    belenidium, 157
    porophylloides, 157
  thurberi, 157

Ear,
  Cat's, 194
  Hare's, 285
  Lion's, 525
Ear-drops, Golden, 626
Eburophyton austinae, 932
Echeveria
  abramsii, 384
  lagunensis, 384
  lanceolata, 386
  monicae, 386
  palensis, 385
  pulverulenta, 387
Echidiocarya
  californica, 265
  ursina, 265
Echinocactus, 311
  acanthodes, 312
  johnsonii, 314
  lecontei, 312
  polyancistrus, 321
  polycephalus, 311
  viridescens, 312
Echinocereus
  emoryi, 310
  engelmannii, 311
    acicularis, 311
    armatus, 312
    chrysocentrus, 312
    engelmannii, 311
    munzii, 312
  triglochidiatus, 312
    melanacanthus, 312
    mojavensis, 312

Echinochloa, 963
  colonum, 964
  crusgalli
    crusgalli, 964
    frumentacea, 964
    oryzicola, 964
  crus-pavonis, 964
    macera, 964
  muricata, 964
    microstachya, 964
    zelayensis, 964
    macera, 964
Echinocystis
  horrida, 393
  macrocarpa, 393
  parviflora, 391
Echinodorus, 868
  berteroi, 870
  cordifolius, 870
  rostratus, 870
Echinomastus johnsonii, 314
Echinopsilon hyssopifolia, 359
Echium, 258
  menziesii, 248
  plantagineum, 259
Eclipta alba, 157
Ectosperma alexandrae, 1006
Edwinia californica, 781
Egeria densa, 905
Ehrharta, 964
  calycina, 965
  erecta, 965
Eichhornia crassipes, 1007
Elaeagnaceae, 46, 395
Elaeagnus angustifolia, 395
Elaphrium microphyllum, 307
Elaterium bigelovii, 391
Elatinaceae, 49, 396
Elatine
  brachysperma, 396
  californica, 396
  chilensis, 397
  gracilis, 397
  rubella, 397
  triandra, 397
Eleocharis, 1, 898
  acicularis, 899
    bella, 899
    occidentalis, 899
    radicans, 900
  arenicola, 900
  atropurpurea, 899
  bella, 899
  bernardina, 899
  caribaea, 900
  coloradoensis, 899
  disciformis, 900
  geniculata, 900
  macrostachya, 900
  mamillata, 900
  montana, 900

Index

montevidensis, 900
  parishii, 900
palustris, 900
parishii, 900
parvula coloradoensis, 899
pauciflora, 900
  bernardina, 899
quinqueflora, 900
radicans, 900
rostellata, 900
  occidentalis, 900
Elephant's-Head, Little, 818
Ellisia
  aurita, 517
  chrysanthemifolia, 498
  membranacea, 517
  micrantha, 498
  racemosa, 518
  torreyi, 498
Eleusine indica, 965
Elm, 845
  Chinese, 846
  English, 846
  Siberian, 846
  Smooth-leaved, 846
Elodea
  canadensis, 905
  densa, 905
  planchonii, 905
Elymus, 965
  angustifolius, 966
    caespitosus, 966
  cinereus, 965
  condensatus, 966
    pubens, 965
    triticoides, 966
  glaucus
    breviaristatus, 966
    glaucus, 966
    jepsonii, 966
    maximus, 966
    virescens, 966
  hansenii, 999
  macounii, 966
  orcuttianus, 966
  pacificus, 966
  pauciflorus, 940
  pubescens, 966
  rydbergii, 940
  salinus, 966
  scribneri, 940
  stebbinsii, 940
  trachycaulum, 940
  triticoides, 966
  villosus glabriusculus, 966
  virescens, 966
Elytrospermum californicum, 902
Emex spinosa, 676
Emmenanthe pendulifera, 496
Emplectocladus faciculatus, 757
Encelia, 4, 157

actoni, 158
californica, 158
eriocephala, 169
farinosa, 158
  phenicodonta, 158
frutescens, 158
  actoni, 158
  nudicaulis, 159
  virginensis
    actoni, 158
    virginensis, 158
viscida, 169
Enceliopsis
  argophylla grandiflora, 159
  covillei, 159
  grandiflora, 159
  nudicaulis, 159
Encina, 479
Enneapogon desvauxii, 968
Ephedra
  aspera, 43
  californica, 43
    funerea, 44
  clokeyi, 44
  fasciculata, 44
    clokeyi, 44
    funerea, 44
    nevadensis, 44
      aspera, 43
    viridis, 44
  trifurca, 44
  viridis, 44
Ephedraceae, 43
Ephedrales, 43
Epicampes rigens, 984
Epilobium
  adenocaulon
    adenocaulon, 605
    holosericeum, 605
    parishii, 605
  americanum, 606
  angustifolium circumvagum, 605
  brevistylum, 606
    ursinum, 606
  californicum, 605
    parishii, 605
  ciliatum, 606
  exaltatum, 606
  glaberrimum, 606
    fastigiatum, 606
  holosericeum, 605
  lactiflorum, 606
  oregonense, 606
  paniculatum, 607
    adenocladon, 607
    laevicaule, 607
    subulatum, 607
  perplexans, 606
  ursinum, 606
Epipactis, 932

convallarioides, 934
gigantea, 933
Equisetaceae, 15
Equisetae, 15
Equisetales, 15
Equisetum
  arvense, 15
  ferrisii, 16
  fontinale, 16
  funstoni, 16
  hyemale, 15
    affine × *laevigatum*, 16
    californicum, 15
    intermedium, 16
    robustum, 15
  kansanum, 16
  laevigatum, 16
  maximum, 16
  praealtum, 15
  robustum affine, 15
  telmateia, 16
    braunii, 16
Eragrostis, 968
  barrelieri, 968
  cilianensis, 969
  curvula, 969
  diffusa, 969
  fendleriana, 994
  hypnoides, 969
  megastachya, 969
  mexicana, 969
  neomexicana, 969
  orcuttiana, 969
  pectinacea, 969
  pilosa, 969
  poaeoides, 969
Eremalche
  exilis, 563
  parryi, 563
  rotundifolia, 563
Eremiastrum
  bellioides, 213
  orcuttii, 213
Eremocarpus setigerus, 409
Eremochloe kingii, 948
Eremonanus mohavensis, 164
Eriastrum, 641
  densifolium
    austromontanum, 642
    mohavense, 642
    sanctorum, 642
  diffusum, 642
  eremicum, 642
  filifolium, 643
  pluriflorum
    pluriflorum, 643
    sherman-hoytae, 643
  sapphirinum
    ambiguum, 643
    dasyanthum, 643
    gymnocephalum, 643

Eriastrum (cont.)
  sapphirinum (cont.)
    sapphirinum, 643
    sparsiflorum, 643
    wilcoxii, 644
Ericaceae, 51, 52, 53, 397
Ericameria
  arborescens, 176
  brachylepis, 179
  cuneata, 177
  ericoides, 177
  laricifolia, 178
  monactis, 177
  paniculata, 144
  parishii, 179
  pinifolia, 179
Erigeron, 159
  andersonii, 121
  aphanactis, 160
    congestus, 160
  argentatus, 160
  bonariensis, 149
  breweri
    breweri, 160
    jacinteus, 160
    porphyreticus, 160
  californicus, 161
  canadensis, 149
  clokeyi, 160
  compactus, 161
  compositus
    discoideus, 161
    glabratus, 161
  concinnus aphanactis, 160
  congestus, 160
  covillei, 161
  discoidea, 149
  divergens, 161
  foliosus
    covillei, 161
    foliosus, 161
    stenophyllus, 161
  glaucus, 161
  jacinteus, 160
  linifolius, 149
  lonchophyllus, 162
  multifidus, 161
  parishii, 162
  philadelphicus, 162
  porphyreticus, 160
  pumilus concinnoides, 162
  racemosus, 162
  sanctarum, 162
  stenophyllus, 161
  tenuissimus, 161
  trifidus, 161
  uncialis, 162
  utahensis, 162
Eriochloa
  aristata, 970
  contracta, 970

  gracilis, 970
  lemmonii gracilis, 970
Eriocoma webberi, 986
Eriodictyon, 497
  angustifolium, 497
  californicum australe lanatum, 498
  capitatum, 497
  crassifolium, 497
    denudatum, 497
    nigrescens, 497
  parryi, 518
  traskiae
    smithii, 498
    traskiae, 498
  trichocalyx
    lanatum, 498
    trichocalyx, 498
Eriogonella
  leptoceras, 673
  membranacea, 673
  spinosa, 674
Eriogonum, 677
  anemophilum, 695
  agninum, 695
  angulosum, 681
    gracillimum, 686
    maculatum, 690
  apiculatum, 681
  arborescens, 681
  aridum, 697
  aureum ambiguum, 691
  baileyi, 681
    brachyanthum, 682
    tomentosum, 685, 693
  baratum, 683
  bifurcatum, 681
  bloomeri, 695
  brachyanthum, 682
  brachypodum, 682
  breedlovei, 682
  caespitosum, 682
  carneum, 686
  cernuum, 682
    tenue, 682
  cinereum, 683
  citharaeforme, 683
    agninum, 695
  crocatum, 683
  davidsonii, 683
  deductum, 692
  deflexum
    baratum, 683
    brachypodum, 682
    deflexum, 683
    insigne, 688
  deserticola, 684
  effusum foliosum, 691
  elatum
    glabrescens, 684
    villosum, 684

  elongatum, 684
  eremicola, 684
  esmeraldense, 684
  fasciculatum, 4, 684
    fasciculatum, 684
    flavoviride, 685
    foliolosum, 685
    polifolium, 685
  foliosum, 685
  giganteum
    compactum, 685
    formosum, 685
    giganteum, 685
  gilmanii, 685
  glandulosum, 686
  glaucum, 688
  gracile, 686
    citharaeforme, 688
  gracilipes, 686
  gracillimum, 686
  grande
    grande, 686
    rubescens, 686
    timorum, 686
  heermannii
    argense, 687
    floccosum, 687
    heermannii, 687
    humilius, 687
    sulcatum, 687
  hirtiflorum, 687
  hoffmannii
    hoffmannii, 687
    robustius, 687
  hookeri, 688
  inerme hispidulum, 688
  inflatum
    contiguum, 688
    deflatum, 688
  insigne, 688
  intrafractum, 688
  kearneyi monoense, 690
  kennedyi
    alpigenum, 690
    austromontanum, 690
    gracilipes, 686
    kennedyi, 690
    purpusii, 690
  latifolium pauciflorum, 692
  luteolum, 698
  maculatum, 690
  mensicola, 693
  microthecum
    ambiguum, 691
    corymbosoides, 691
    foliosum, 691
    johnstonii, 691
    lapidicola, 691
    laxiflorum, 691
    panamintense, 692
  mohavense, 692

molestum, 683
nidularium, 692
nodosum, 698
　　jaegeri, 694
nudum
　　deductum, 692
　　pauciflorum, 692
　　pubiflorum, 692
　　saxicola, 692
ochrocephalum
　　agnellum, 695
　　breedlovei, 682
ordii, 692
ovalifolium
　　ovalifolium, 692
　　vineum, 693
palmerianum, 693
panamintense
　　mensicola, 693
　　panamintense, 693
parishii, 693
parryi, 682
parvifolium, 693
　　paynei, 694
plumatella, 694
　　jaegeri, 694
　　palmeri, 693
polifolium, 685
purpusii, 690
pusillum, 694
racemosum, 694
reniforme, 694
rixfordii, 694
rosense, 695
roseum, 695
rubescens, 686
rupinum, 695
saxatile, 695
saxicola, 692
shockleyi, 695
spergulinum reddingianum, 695
sulcatum, 687
　　argense, 687
thomasii, 695
thurberi, 696
trachygonum, 698
trichopes, 696
trichopodum, 696
umbellatum, 696
　　aridum, 697
　　dicrocephalum, 697
　　ferrissii, 697
　　minus, 697
　　munzii, 697
　　subaridum, 697
　　versicolor, 697
vimineum
　　citharaeforme, 683
　　davidsonii, 683
　　gracile, 686

polygonoides
　　vineum, 693
　　virgatum, 695
　　viridescens, 697
　　wrightii
　　　　membranaceum, 697
　　　　nodosum, 698
　　　　subscaposum, 698
　　　　trachygonum, 698
　　　　wrightii, 697
Erioneuron
　　pilosum, 970
　　pulchellum, 970
Eriophyllum
　　ambiguum, 163
　　confertiflorum
　　　　confertiflorum, 163
　　　　discoideum, 163
　　　　laxiflorum, 164
　　　　tridactylum, 163
　　　　trifidum, 164
　　lanatum
　　　　hallii, 164
　　　　obovatum, 164
　　lanosum, 164
　　mohavense, 164
　　multicaule, 164
　　nevinii, 164
　　obovatum, 164
　　pringlei, 164
　　staechadifolium
　　　　artemisiaefolium, 165
　　　　depressum, 165
　　　　wallacei, 165
　　　　rubellum, 165
Eritrichium
　　angustifolium, 253
　　barbigerum, 253
　　canescens arizonicum, 264
　　cooperi, 266
　　holopterum, 255
　　intermedium, 255
　　micranthum lepidum, 256
　　micranthum micranthum, 256
　　micromeres, 256
　　nothofulvum, 266
　　pterocaryum, 257
　　racemosum, 258
　　torreyi, 267
Erodium
　　botrys, 490
　　cicutarium, 490
　　cygnorum, 491
　　macrophyllum, 491
　　moschatum, 491
　　obtusiplicatum, 491
　　texanum, 491
Eruca sativa, 289
Ervum lens, 466

Eryngium
　　aristulatum, 75
　　　　aristulatum, 75
　　　　parishii, 75
　　armatum, 76
　　jepsoni, 75
　　parishii, 75
　　petiolatum, 76
Erysimum, 289
　　ammophilum, 290
　　asperum, 290
　　capitatum, 290
　　cheirii, 290
　　insulare, 290
　　officinale, 301
　　suffrutescens, 290
Erythraea
　　calycosa, 486
　　venusta, 486
Eschscholzia, 626
　　arvensis dilatata, 627
　　caespitosa
　　　　kernensis, 626
　　　　caespitosa, 626
　　californica, 627
　　　　crocea, 627
　　　　douglasii, 627
　　　　maritima, 627
　　　　peninsularis, 627
　　covillei, 627
　　douglasii, 627
　　elegans, 628
　　glyptosperma, 627
　　mexicana, 627
　　minutiflora, 627
　　　　darwinensis, 627
　　parishii, 628
　　ramosa, 628
　　wrigleyana, 628
Eucalyptus
　　camaldulensis, 577
　　globulus, 577
Euclisia amplexicaulis, 282
Eucnide
　　rupestris, 550
　　urens, 550
Eucrypta
　　chrysanthemifolia
　　　　bipinnatifida, 498
　　　　chrysanthemifolia, 498
　　micrantha, 498
Eulobus californica, 594
Eulophus
　　parishii, 83
　　pringlei, 83
Eunanus
　　bigelovii, 809
　　breweri, 810
Euonymus
　　occidentalis parishii, 349
　　parishii, 349

Eupatorium
  adenophorum, 107
  ageratifolium herbaceum, 107
  herbaceum, 107
  pasadenense, 107
Euphorbia, 409
  abramsiana, 411
  albomarginata, 411
  arenicola, 413
  arizonica, 411
  crenulata, 412
  dictyosperma, 415
  eremica, 414
  eriantha, 412
  exstipulata, 412
  fendleri, 412
  helioscopia, 412
  incisa, 412
  lathyris, 412
  maculata, 413
  melanadenia, 413
  micromera, 413
  misera, 413
  nutans, 413
  ocellata
    arenicola, 413
    ocellata, 413
  palmeri, 413
  parishii, 414
  parryi, 414
  patellifera, 414
  pediculifera, 414
  peplus, 414
  platysperma, 414
  polycarpa
    hirtella, 415
    polycarpa, 414
  preslii, 413
  prostrata, 415
  revoluta, 415
  schizoloba, 412
  serpens, 415
  serpyllifolia, 415
    hirtula, 415
  setiloba, 415
  spathulata, 415
  supina, 415
  vallis-mortae, 417
Euphorbiaceae, 46, 405
Euploca convolvulacea californica, 259
Eurotia, 4, 365
  lanata, 365
    subspinosa, 365
Eurytera lucida, 79
Eustoma exaltatum, 486
Euthamia occidentalis, 228
Eutoca aretioides, 500
  brachyloba, 506
  grandiflora, 510
  viscida, 517

Evax
  acaulis, 165
  caulescens sparsiflora, 165
  sparsiflora, 165
Everlasting, 169
  Pearly, 113

Fabaceae, 48, 51, 52, 419
Fagaceae, 46, 478
Fagonia
  californica
    glutinosa, 860
    laevis, 859
  laevis, 859
  pachyacantha, 860
Fallugia paradoxa, 4, 746
Family,
  Acanthus, 54
  Agave, 863
  Amaranth, 58
  Amaryllis, 871
  Arrow-Grass, 916
  Arum, 882
  Barberry, 242
  Basella, 242
  Batis, 242
  Beech, 478
  Bellflower, 322
  Bignonia, 246
  Birch, 245
  Bladderwort, 545
  Borage, 246
  Box, 308
  Broom-Rape, 617
  Buckeye, 495
  Buckthorn, 731
  Buckwheat, 670
  Bur-Reed, 1011
  Cacao, 841
  Cactus, 308
  Caltrop, 859
  Caper, 328
  Carpetweed, 55
  Carrot, 67
  Cat-Tail, 1012
  Crossosoma, 389
  Crowfoot, 719
  Cypress, 35
  Datisca, 393
  Dogbane, 87
  Dogwood, 379
  Duckweed, 918
  Eel-Grass, 1013
  Elm, 845
  Evening-Primrose, 589
  False Mermaid, 547
  Figwort, 789
  Flax, 547
  Four-O'clock, 577
  Frankenia, 484
  Frogbit, 905

Gentian, 485
Geranium, 489
Ginseng, 90
Goosefoot, 351
Gourd, 389
Grape, 858
Grass, 934
Heath, 397
Honeysuckle, 331
Horned Pondweed, 1013
Hornwort, 350
Iris, 905
Jewel-Weed, 241
Junco, 521
Krameria, 522
Laurel, 544
Leadwort, 636
Lily, 919
Lizard-Tail, 776
Logania, 559
Loosestrife, 560
Madder, 761
Mallow, 561
Maple, 54
Mare's-Tail, 495
Martynia, 574
Mignonette, 730
Milkweed, 91
Milkwort, 669
Mint, 522
Mistletoe, 856
Morning-Glory, 371
Mulberry, 575
Mustard, 267
Myoporum, 576
Myrtle, 577
Nettle, 846
Nightshade, 829
Ocotillo, 483
Oleaster, 395
Olive, 586
Orchid, 931
Oxalis, 620
Palm, 882
Pea, 419
Peony, 622
Phlox, 637
Pickerel-Weed, 1007
Pine, 37
Pink, 334
Pittosporum, 632
Plantain, 633
Pokeweed, 632
Pondweed, 1007
Poppy, 622
Primrose, 715
Purslane, 708
Quassia, 828
Rafflesia, 719
Rock-Rose, 370
Rose, 739

Rue, 770
Rush, 908
St. John's-Wort, 519
Sandalwood, 776
Saxifrage, 778
Sedge, 883
Silk-Tassel, 484
Spiderwort, 883
Spurge, 405
Staff-Tree, 349
Stick-Leaf, 549
Stonecrop, 382
Storax, 843
Sumac, 64
Sunflower, 95
Sycamore, 636
Tamarisk, 843
Teasel, 395
Torchwood, 307
Tropaeolum, 845
Valerian, 847
Vervain, 850
Violet, 852
Walnut, 521
Waterleaf, 495
Water-Milfoil, 493
Water-Nymph, 931
Water-Plantain, 868
Water-Starwort, 321
Waterwort, 396
Wax-Myrtle, 576
Willow, 771
Wood-Sorrel, 620
Fat
  Mule, 124
  Winter, 365
Felwort, 489
Fendlerella utahensis, 779
Fennel, Sweet, 76
Fenzlia dianthiflora, 661
Fern,
  Adder's-tongue, 34
  Bird's Foot, 31
  Bracken, 32
  Brake, 32
  Brittle, 18
  Chain, 23
  Coffee, 31
  Cotton, 29
  Duckweed, 35
  Five-finger, 25
  Goldenback, 32
  Grape, 34
  Holly, 18
  Lace, 26
  Ladder-brake, 32
  Lady, 18
  Lip, 25
  Maidenhair, 25
  Male, 6, 19
  Parsley, 28

Parry Cloak, 29
Polypody, 23
Shield, 19
Silverback, 32
Sword, 19
Venus-hair, 25
Wood, 19
Ferns, 17
Ferocactus
  acanthodes
    acanthodes, 312
    lecontei, 312
  johnsonii, 314
  viridescens, 312
Ferula caruifolia, 78
Fescue, 971
Festuca, 971
  aristulata parishii, 972
  arizonica, 972
  arundinacea, 972
  barbata, 997
  brachyphylla, 972
  bromoides, 972
  californica parishii, 972
  confinis, 974
  confusa, 972
  dertonensis, 972
  eastwoodae, 972
  elatior arundinacea, 972
  fascicularis, 979
  filiformis, 979
  grayi, 972
  kingii, 974
  megalura, 972
  microstachys, 972
  grayi, 972
  myuros, 972
  nervosa, 994
  occidentalis, 973
  octoflora, 973
  ovina
    arizonica, 972
    brachyphylla, 972
    saximontana, 972
  pacifica, 973
  parishii, 972
  pratensis, 973
  reflexa, 973
  rubra, 973
  unioloides, 956
Ficus
  carica, 575
  pseudocarica, 575
Fiddleneck, 247
Fig,
  Common, 575
  Hottentot-, 56
  Indian, 317
  Sea-, 56
Figwort, 825
Filago, 165

arizonica, 166
californica, 166
depressa, 166
gallica, 166
Filaree, 490
Filicales, 17
Filicae, 16
Filix fragilis, 18
Fimbristylis thermalis, 901
Finger, Five-, 25, 752
Fir, White, 38
Fivespot, Desert, 563
Flaveria trinervia, 166
Flax, 548
  Common, 549
  False-, 278
Fleabane, 159
  Marsh, 219
Fleece, Golden, 176
Fleur-de-Lis, 906
Flower,
  Bladder-, 91
  Blanket, 166
  Ghost, 816
  Monkey-, 806
  Paper-, 221
  Popcorn, 266
  Star, 879
  Sun, 182
  Wall, 289
  Wind, 720
Flower-of-an-Hour, 562
Fluellin, 801
Foam, Meadow-, 547
Foeniculum vulgare, 76
Food, Sand, 545
Forestiera neomexicana, 586
Forsellesia, 349
  nevadensis, 350
  pungens glabra, 350
  stipulifera, 350
Fouquieria splendens, 4, 483
Fouquieriaceae, 52, 483
Four-O'Clock, 584
Foxglove, 802
Foxtail, 977
Fragaria
  californica, 746
  vesca californica, 746
Frankenia
  grandifolia, 484
    campestris, 484
  palmeri, 484
Frankeniaceae, 50, 484
Franseria
  acanthicarpa, 109
  albicaulis, 110
  bipinnatifida dubia, 110
  bipinnatisecta, 110
  californica, 109
  chamisonnis bipinnatisecta, 110

Franseria (cont.)
  chenopodiifolia, 110
  confertiflora, 110
  cuneifolia, 110
  dumosa, 110
  eriocentra, 112
  ilicifolia, 112
  palmeri, 109
  pumila, 112
Frasera
  albomarginata, 487
  neglecta, 487
  parryi, 487
  puberulenta, 487
Fraxinus, 586
  anomala, 588
  coriacea, 588
  dipetala, 588
  velutina
    coriacea, 588
    velutina, 588
Freesia refracta, 906
Fremontia, 841
Fremontia
  californica, 841
    diegensis, 842
  mexicana, 842
  vermicularis, 369
Fremontodendron
  californicum, 841
  mexicanum, 842
Fritillaria, 926
  atropurpurea pinetorum, 927
  biflora, 927
    ineziana, 927
    inflexa, 927
  lanceolata, 927
  mutica, 927
  ojaiensis, 927
  pinetorum, 927
  succulenta, 927
Fritillary, 926
Fuchsia, California-, 616
Fumaria
  officinalis, 628
  parviflora, 628
Fumitory, 628
Funastrum hirtellum, 94
Furze, 476

Gaertneria
  ambrosioides, 110
  bipinnatifida, 110
  chenopodiifolia, 110
  confertiflora, 110
  dumosa, 110
  eriocentra, 112
Gaillardia, 166
  aristata, 167
  pulchella, 167
Galinsoga parviflora, 167

Galium, 762
  andrewsii
    andrewsii, 763
    gatense, 764
    intermedium, 764
  angustifolium
    angustifolium, 764
    borregoense, 764
    diffusum, 764
    foliosum, 764
    gabrielense, 764
    gracillimum, 764
    jacinticum, 765
    nudicaule, 765
    pinetorum, 768
    subglabrum, 768
  aparine, 765
  argense, 765
  bifolium, 765
  bolanderi, 765
  buxifolium, 766
  californicum
    flaccidum, 765
    miguelense, 765
    primum, 765
  catalinense
    buxifolium, 766
    catalinense, 766
  cliftonsmithii, 766
  flaccidum, 765
  gabrielense, 764
  grande, 766
  hallii, 766
  hilendiae
    carneum, 766
    hilendiae, 766
    kingstonense, 766
  hypotrichium tomentellum, 768
  jepsonii, 768
  johnstonii, 768
  matthewsii, 768
  miguelense, 765
  multiflorum, 768
    parvifolium, 769
  munzii, 768
    carneum, 766
  nuttallii, 768
    cliftonsmithii, 766
    insulare, 769
    nuttallii, 768
    ovalifolium, 769
  parishii, 769
  parisiense, 769
  pubens grande, 766
  spurium echinospermum, 769
  stellatum eremicum, 769
  trifidum
    pacificum, 769
    pusillum, 769
    subbiflorum, 769

  triflorum, 769
  wrightii, 770
Galvezia speciosa, 803
Gambelia speciosa, 803
Garrya
  elliptica, 485
  flavescens pallida, 485
  fremontii, 485
  pallida 485
  veatchii, 485
Garryaceae, 46, 484
Gasoul
  crystallinum, 56
  nodiflorum, 57
Gastridium, 973
  ventricosum, 974
Gaura
  coccinea, 607
  decorticans, 593
  heterandra, 611
  odorata, 607
  parviflora, 608
    lachnocarpa, 608
  sinuata, 608
Gayophytum, 608
  caesium, 611
  decipiens, 610
  diffusum
    diffusum, 610
    parviflorum, 610
    villosum, 610
  helleri, 611
  heterozygum, 610
  humile, 610
  nuttallii, 610
  oligospermum, 611
  racemosum, 611
  ramosissimum, 611
  strictum, 590
Gazania longiscapa, 167
Genista linifolia, 438
Gentian, 489
  Catchfly-, 486
  Green, 487
  Moss, 489
  Sierra, 489
Gentiana
  acuta, 489
  amarella, 489
  fremontii, 489
  holopetala, 489
  serrata helopetala, 489
  simplex, 489
Gentianaceae, 53, 485
Geraea, 167
  canescens, 169
  viscida, 169
Geraniaceae, 51, 489
Geranium, 491, 493
  botrys, 490
  californicum, 492

Index

carolinianum, 492
cicutarium, 490
concinnum, 492
dissectum, 492
Fish, 493
grossularioides, 493
Horseshoe, 493
leucanthum, 492
molle, 492
moschatum, 491
retrorsum, 492
richardsonii, 492
zonale, 493
Germander, 543
Geum macrophyllum, 746
Gilia, 644
  abrotanifolia, 648
  achilleaefolia
    achilleaefolia, 646
    multicaulis, 647
  aliquanta
    aliquanta, 647
    breviloba, 647
  angelensis, 647
  arenicola, 659
  aurea, 659
  australis, 647
  austrooccidentalis, 647
  bella, 660
  bigelovii, 660
  brecciarum
    brecciarum, 647
    argusana, 648
    neglecta, 648
  brevicula, 660
  californica, 657
    glandulosa, 657
  cana
    bernardina, 648
    speciformis, 648
    speciosa, 648
    triceps, 648
  capillaris, 648
  capitata abrotanifolia, 648
  caruifolia, 648
  ciliata, 660
  clivorum, 649
  clokeyi, 649
  collina grantii, 653
  concinna, 661
  davyi, 650
  demissa, 661
  densifolia, 642
    austromontana, 642
    mohavensis, 642
    sanctora, 642
  depressa, 654
  dianthiflora, 661
  dianthoides, 661
    farinosa, 661
  dichotoma, 661

integra, 661
diegensis, 649
divaricata, 640
  volcanica, 640
filifolia, 643
  diffusa, 642
filiformis, 649
flaviflora, 655
floccosa ambigua, 643
floribunda, 662
gilioides, 640
  integrifolia, 640
gilmanii, 652
gracilis, 663
grandiflora, 641
hallii, 657
hoffmanii, 653
hutchinsifolia, 649
inconspicua
  diegensis, 649
  sinuata, 652
  sinuata deserti, 647
interior, 649
jacens, 649
jaegeri, 657
jonesii, 661
latiflora
  caruifolia, 648
  cosana, 650
  cuyamensis, 650
  davyi, 650
  elongata, 650
  excellens, 650
  exilis, 652
  latiflora, 650
  latifolia, 650
lemmonii, 662
leptantha
  leptantha, 650
  pinetorum, 651
  purpusii, 651
  vivida, 652
leptomeria, 651
  rubella, 649
linearis, 641
luteola, 659
maculata, 662
malior, 651
micrantha, 659
micromeria, 651
minor, 651
modocensis, 651
montana, 654
multicaulis, 647
  clivorum, 649
nevinii, 652
nudata, 652
nuttallii, 662
ochroleuca
  bizonata, 652
  ochroleuca, 652

exilis, 652
transmontana, 653
vivida, 652
opthalmoides, 652
parryae, 663
pharnaceoides, 662
pluriflora, 643
polycladon, 654
prostrata, 666
pungens pulchriflora, 657
pusilla, 663
pygmaea, 663
ripleyi, 652
sapphirina, 643
scopulorum, 652
setosissima punctata, 655
sherman-hoytae, 643
sinuata, 652
sparsiflora, 643
splendens
  australis, 647
  grantii, 653
  splendens, 653
stellata, 653
tenuiflora
  altissima, 653
  caruifolia, 648
  hoffmannii, 653
  latiflora excellens, 650
  speciosa, 648
  triceps, 648
tetrabreccia, 653
transmontana, 653
traskiae, 640
violacea, 641
wilcoxii, 644
Gill-Over-the-Ground, 524
Gilmania luteola, 698
Githopsis
  diffusa, 323
  filicaulis, 324
  gilioides, 323
  rariflora, 324
  specularioides
    candida, 324
    specularioides, 324
Gladiolus segetum, 906
Glasswort, 367
Glaucium flavum, 628
Glecoma hederacea, 524
Glossopetalon
  nevadensis, 350
  stipulifera, 350
Glyceria elata, 974
Glycyrrhiza
  glabra, 442
  lepidota glutinosa, 442
Glyptopleura
  marginata, 169
  setulosa, 169
  setulosa, 169

Gnaphalium, 169
  albidum, 171
  beneolens, 170
  bicolor, 170
  californicum, 170
  chilense, 170
    confertifolium, 171
  decurrens californicum, 170
  dimorphum, 114
  leucocephalum, 171
  luteo-album, 171
  margaritaceum, 113
  microcephalum, 171
  palustre, 171
  peregrinum, 171
  purpureum, 171
  ramosissimum, 171
  spathulatum, 171
  thermale, 171
  ustulatum, 171
  wrightii, 172
Gnetae, 43
Goatnut, 308
Godetia
  bottae cylindrica, 601
  cylindrica, 601
  deflexa, 601
  delicata, 601
  dudleyana, 603
  epilobioides, 603
  quadrivulnera, 603
  viminea, 603
Goldenhead, 105
Goldenrod, 228
  California, 228
Goldentop, 978
Goldfield, 196
Gomphocarpus
  tomentosus, 92
    xanti, 92
  torreyi, 92
Gonolobus
  californicus, 94
  hastulatus, 94
  parvifolius, 94
Gooseberry, 784
  Bitter, 785
  Fuchsia-flowered, 788
  Sierra, 788
Goosefoot, 359
Gormania anomala, 389
Gorse, 476
Gourd, 392
Gramineae, 934
Grape, 859
Grass
  Alkali, 996
  Arrow-, 917
  Barnyard, 964
  Beach, 944
  Beard, 944, 995

Bent, 941
  Colonial, 943
  Bermuda, 959
  Blue-eyed, 908
  Blue, 990, 994
    Canadian, 992
    Kentucky, 995
    Nevada, 994
    Timberline, 995
  Bristle, 998
  Brome, 950
    California, 952
    Smooth, 953
  Bur, 957
  Canary, 988
  Cheat, 956
  Cord, 1000
  Crab, 962
  Cup, 970
  Cut, Rice, 978
  Ditch-, 1009
  Eel-, 1015
  Feathertop, 988
  Fescue, 971
    Foxtail, 972
    Meadow, 973
    Red, 973
    Six-Weeks, 973
  Finger, 958
  Fluff, 971
  Foxtail, 944, 977
  Galleta, 975
  Goldentop, 978
  Goose, 965
  Grama, 949
  Hair, 943, 961
  Harding, 989
  Johnson, 1000
  June, 978
  Kikuyu, 988
  Knot, 988
  Love, 968
  Mutton, 994
  Natal, 996
  Needle, 1002
  Nit, 973
  Nut-, 897
  Oat, 947, 960
  -of-Parnassus, 783
  Orchard, 960
  Pampas, 958
  Pepper, 291
  Quaking, 950
  Reed, 957, 990
  Rescue, 956
  Rhodes, 958
  Rib-, 635
  Rice, 985
    Jungle-, 964
  Ripgut, 953
  Rye, 965, 979, 998

St. Augustine's, 1002
  Salt, 5, 962
  Saw-, 895
  Shore, 982
  Sickle, 987
  Spear, 1002
  Stink, 969
  Surf-, 1015
  Triple-Awned, 945
  Velvet, 975
  Vernal, 945
  Wedge, 1001
  Wheat, 940, 1007
    Tall, 940
  Winter, 991
  Wire-, 909
  Yellow-eyed, 908
Grayia spinosa, 4, 365
Greasewood, 369, 741
Greenocharis circumscissa, 253
Grindelia
  aphanactis, 172
  camporum, 172
  hallii, 172
  hirsutula, 173
  latifolia, 173
  robusta, 173
    bracteosa, 173
    latifolia, 173
    rigida, 172
  squarrosa, 173
Gromwell, 261
Grossularia
  lasiantha, 787
  velutina, 788
Groundsel, 223
  Common, 227
Guizotia abyssinica, 173
Gum, Blue, 577
Gum-Plant, 172
Gutierrezia, 173
  bracteata, 174
  californica bracteata, 174
  lucida, 174
  microcephala, 174
  sarothrae, 4, 174
Gymnogramme triangularis, 32
  viscosa, 32
Gymnolomia nevadensis, 240
Gymnopteris triangularis, 32
Gynerium atacamense, 959
Gyrostachys porrifolia, 934

Habenaria
  cooperi, 933
  dilatata leucostachys, 933
  elegans, 933
    elata, 933
  laxiflora, 933
  leucostachys, 933
    virida, 933

longispicata, 933
  michaelii, 933
  sparsiflora, 933
  unalascensis, 933
Halimolobos diffusa jaegeri, 291
Halliophytum hallii, 418
Halodule wrightii, 1013
Halogeton glomeratus, 365
Haloragaceae, 47, 49, 493
Halostachys occidentalis, 352
Haplopappus, 4, 174
  acaulis, 176
  acradenius
    acradenius, 176
    eremophilus, 176
  arborescens, 176
  brachylepis, 179
  brickellioides, 176
  canus, 176
  ciliatus, 176
  cooperi, 177
  cuneatus, 177
    spathulatus, 177
  detonsus, 177
  ericoides, 177
    blakei, 177
    ericoides, 177
  gilmanii, 177
  gooddingii, 177
  gracilis, 178
  interior, 178
  junceus, 178
  laricifolius, 178
  linearifolius, 178
  macronema, 178
  monactis, 177
  palmeri
    pachylepis, 178
    palmeri, 179
  parishii, 179
  pinifolius, 179
  propinquus, 179
  racemosus
    glomeratus, 179
    sessiflorus, 179
  squarrosus
    grindelioides, 179
    obtusus, 181
  uniflorus gossypinus, 181
  venetus furfuraceus, 181
    oxyphyllus, 181
    sedoides, 181
    vernonioides, 181
Harpagonella palmeri, 259
Hartmannia
  fasciculata, 186
  pungens, 187
  rosea, 615
  speciosa, 615
Hasseanthus

  elongatus, 386
  kessleri, 385
  multicaulis, 386
  nesioticus, 387
  variegatus, 388
Hawkbit, 200
Hawksbeard, 153
Hazardia
  caoa, 176
  obtusa, 181
Head
  Arrow-, 870
  Bur, 868
  Snake's, 270
Heads, White, 86
  Nigger-, 311
Heather, Mountain-, 402
Hecastocleis shockleyi, 182
Hedeoma
  incanum, 532
  namum californicum, 524
  thymoides, 524
Hedera helix, 91
Hedynois cretica, 182
Helenium
  bigelovii, 182
  puberulum, 182
Heleocharts
  leptos, 899
    johnstonii, 899
Helianthella nudicaulis, 159
Helianthemum
  greenei, 370
  occidentale, 370
  scoparium, 371
    aldersonii, 371
    scoparium, 371
    vulgare, 371
Helianthus, 182
  annuus
    jaegeri, 183
    lenticularis, 183
  californicus, 183
  ciliaris, 183
  gracilentus, 183
  jaegeri, 183
  lenticularis, 183
  niveus canescens, 184
    tephrodes, 184
  nuttallii
    nuttallii, 184
    parishii, 184
  oliveri, 184
  parishii, 184
  petiolaris, 184
    canescens, 184
Heliotrope, 259
  Wild-, 509
Heliotropium
  amplexicaule, 259
  californicum, 259

convolvulaceum californicum, 259
curassavicum oculatum, 261
  oculatum, 261
Hellebore, False-, 929
Helleborine giganteum, 933
Helopus gracilis, 970
Helosciadium californicum, 80
Helxine soleirolii, 847
Hemicarpha
  micrantha minor, 901
  occidentalis, 901
Hemizonella minima, 206
Hemizonia, 184
  arida, 185
  australis, 185
  clementina, 185
  conjugens, 185
  fasciculata, 186
  fitchii, 186
  floribunda, 186
  glomerata, 186
  heermannii, 189
  kelloggii, 186
  laevis, 186
  minima, 206
  minthornii, 186
  mohavensis, 186
  pallida, 187
  paniculata, 187
    increscens, 187
    pungens, 187
    ramosissima, 187
    tenella, 132
    wheeleri, 205
    wrightii, 186
Hemlock
  Poison-, 73
  Water-, 72
Hemp, 575
  Colorado-River-, 470
  Indian, 88
Henbit, 525
Heracleum
  lanatum, 76
  maximum, 76
  sphondylium montanum, 76
Herb
  Cow-, 349
  Willow-, 604
Hermidium alipes, 583
Herniaria cinerea, 340
Herpestis pilosa, 814
Herrea elongata, 57
Hesperocallis undulata, 927
Hesperochiron
  californicus, 499
  watsonianus, 499
Hesperochloa kingii, 974
Hesperocnide tenella, 846
Hesperocordium maritimum, 881

Hesperolinon micranthum, 548
Hesperonia cedrosensis, 584
Hesperoyucca whipplei, 867
Heterocodon rariflorum, 324
Heterogaura
　californica, 611
　heterandra, 611
Heteromeles arbutifolia, 4, 747
　macrocarpa, 747
Heteropogon contortus, 974
Heterotheca, 187
　breweri, 141
　echioides, 141
　floribunda, 188
　grandiflora, 188
　scabra, 188
　subaxillaris, 188
Heuchera, 779
　abramsii, 780
　alpestris, 780
　brevistaminea, 780
　duranii, 780
　elegans, 780
　hirsutissima, 780
　leptomeria peninsularis, 780
　maxima, 780
　parishii, 780
　rubescens pachypoda, 781
Heyderia decurrens, 36
Hibiscus
　denudatus, 563
　trionum, 563
Hieracium
　albiflorum, 188
　argutum
　　argutum, 188
　　parishii, 188
　grinnellii, 188
　horridum, 189
　parishii, 188
Hilaria, 975
　belangeri, 975
　jamesii, 975
　rigida, 975
Hippocastanaceae, 50, 495
Hippuridaceae, 47, 495
Hippuris vulgaris, 495
Hirschfeldia adpressa, 276
Hoffmannseggia
　densiflora, 442
　falcata, 442
　microphylla, 442
　stricta, 442
Hofmeisteria pluriseta, 219
Holcus
　halepensis, 1000
　lanatus, 975
Hollyhock, Desert-, 562
Holly,
　Desert-, 355

Summer-, 402
Holocantha emoryi, 829
Holocarpha
　heermannii, 189
　virgata
　　elongata, 189
Holodiscus
　boursieri, 747
　discolor
　　discolor, 747
　　franciscanus, 748
　microphyllus, 748
　　microphyllus, 748
　　sericeus, 748
Homalocenchrus oryzoides, 978
Homopappus glomeratus, 179
Honeysuckle, 331
Hookera
　capitata, 878
　filifolia, 878
　laxa, 882
　multipedunculata, 878
　orcuttii, 878
　pulchella, 878
　parviflora, 878
　pauciflora, 879
Hops, 576
Hordeum, 975
　adscendens, 976
　arizonicum, 976
　brachyantherum, 976
　californicum, 976
　depressum, 976
　geniculatum, 977
　glaucum, 977
　gussoneanum, 977
　hystrix, 977
　jubatum, 977
　leporinum, 977
　murinum, 977
　nodosum, 976
　pusillum, 977
　stebbinsii, 977
　vulgare, 977
Horebound, 527
　Water-, 526
Horkelia
　argyrocoma, 749
　bolanderi
　　clevelandii, 748
　　parryi, 748
　cuneata, 748
　　puberula, 749
　elata, 749
　hispidula, 749
　truncata, 749
　wilderae, 749
Hornwort, 351
Horsetail, 15
Horsfordia
　alata, 564

　newberryi, 564
Hosackia
　americana
　　argophylla, 445
　　argyraea, 446
　　crassifolia, 446
　　grandiflora, 446
　　haydonii, 447
　　heermannii, 447
　　humilis, 449
　　oblongifolia, 449
　　ornithopus, 446
　　procumbens, 449
　　purshiana, 449
　　rigida, 449
　　scoparia, 449
　　strigosa, 450
　　subpinnata, 450
　　tomentella, 450
　　venusta, 445
Houses, Chinese, 799
Huegelia
　densifolia, 642
　eremica, 642
Hulsea, 189
　algida, 190
　californica, 190
　　inyoensis, 192
　heterochroma, 190
　mexicana, 190
　vestita
　　callicarpha, 192
　　inyoensis, 192
　　parryi, 192
　　pygmaea, 192
　　vestita, 190
Humulus lupulus, 576
Hutchinsia procumbens, 291
Hyacinth
　Water-, 1007
　Wild-, 879
Hydastylus
　elmeri, 908
　rivularis, 908
Hydrocharitaceae, 905
Hydrocotyle, 76
　cuneata, 77
　polystachya triradiata, 77
　ranunculoides, 77
　sibthorpioides, 77
　umbellata, 77
　verticillata
　　triradiata, 77
　　verticillata, 77
Hydrophyllaceae, 53, 495
Hymenatherum thurberi, 157
Hymenoclea
　monogyra, 192
　pentalepis, 192
　salsola
　　pentalepis, 192

salsola, 192
Hymenopappus, 192
  douglasii, 136
  eriopodus, 193
  filifolius
    eriopodus, 193
    lugens, 193
    megacephalus, 193
    nanus, 193
    nanus, 193
    wrightii, 193
    viscidulus, 193
Hymenophysa pubescens, 281
Hymenothrix
  loomisii, 193
  wrightii, 193
Hymenoxys
  acaulis arizonica, 194
  chrysanthemoides
    excurrens, 194
  cooperi
    canescens, 194
    cooperi, 194
  odorata, 194
Hyparrhenia hirta, 977
Hypericaceae, 48, 519
Hypericum
  anagalloides, 519
  canariense, 519
  formosum scouleri, 519
  perforatum, 519
  scouleri, 519
Hypochoeris, 194
  glabra, 195
  radicans, 195
Hypolepis californica, 29
Hyptis emoryi, 525

Ibicella lutea, 574
Illecebrum peploides, 60
  repens, 60
  densum, 63
Impatiens balfourii, 242
Imperata
  brevifolia, 978
  hookeri, 978
Incienso, 4, 158
Indigo, False, 422
Infantea chilensis, 109
Innocence, 799
Inula
  ericoides, 202
  scabra, 188
  subaxillaris, 188
Inyonia dysodioides, 218
Ipheion uniflorum, 879
Ipomoea
  hederacea, 379
  mutabilis, 379
  nil, 379
  purpurea, 379
  triloba, 379
Ipomopsis, 653
  aggregata arizonica, 654
  congesta montana, 654
  depressa, 654
  polycladon, 654
  tenuifolia, 654
Iridaceae, 905
Iris, 906
  German, 907
  germanica, 907
  hartwegii australis, 907
  missouriensis, 907
  pseudacorus, 907
  tenax australis, 907
Ironwood,
  Catalina, 750
  Desert-, 464
Ischaemum secundatum, 1002
Isocoma
  acradenia, 176
  eremophila, 176
  oxyphylla, 181
  vernonioides, 181
Isoetaceae, 12
Isoetales, 12
Isoetes, 12
  bolanderi, 12
  californica, 12
  howellii, 12
  melanopoda californica, 12
  nuttallii orcuttii, 12
  orcuttii, 12
Isolepis
  carinatus, 903
  koilolepis, 903
  pygmaea californica, 903
  subsquarrosa minor, 901
Isomeris
  angustata, 330
  arborea
    angustata, 330
    arborea, 330
    globosa, 330
    insularis, 330
  globosa, 330
Isopyrum occidentale, 725
Iva
  acerosa, 195
  axillaris
    pubescens, 195
    robustior, 195
  hayesiana, 195
Ivesia
  argyrocoma, 749
  callida, 749
  jaegeri, 750
  santolinoides, 750
Ivy,
  English, 91
  German, 226
  Ground-, 524

Jamesia americana
  californica, 781
Jaumea carnosa, 196
Jepsonia
  malvaefolia, 781
  parryi, 781
Jew, Wandering, 883
Johny-Jump-Up, 854
Jojoba, 308
Juglandaceae, 45, 521
Juglans
  californica, 521
  regia, 521
Juncaceae, 908
Juncaginaceae, 916
Juncoides
  comosa, 916
  spicata, 916
Juncus, 904
  acutus sphaerocarpus, 910
  balticus, 910
    eremicus, 910
    mexicanus, 912
    montanus, 910
  breweri, 910
  brunnescens, 915
  bryoides, 910
  bufonius, 911
    congestus, 911
    halophilus, 911
  cooperi, 911
  covillei, 911
  dubius, 911
  durani, 912
  effusus
    brunneus, 911
    pacificus, 911
  falcatus paniculatus, 911
  hemiendytus, 911
  kelloggii, 912
  lesueurii elatus, 915
  longistylis latifolius, 913
  macrandrus, 912
  macrophyllus, 912
  mertensianus
    duranii, 912
    gracilis, 912
    mertensianus, 912
    mexicanus, 912
  nevadensis, 912
  nodosus, 912
    megacephalus, 915
  occidentalis, 915
  orthophyllus, 913
  oxymeris, 913
  parryi, 913
  patens, 913
  phaeocephalus
    glomeratus, 913

Juncus (cont.)
  phaecephalus (cont.)
    gracilis, 912
    paniculatus, 913
    phaeocephalus, 913
    rugulosus, 913
  saximontanus
    brunnescens, 915
    saximontanus, 915
  sphaerocarpus, 915
  tenuis
    congestus, 915
    occidentalis, 915
    tenuis, 915
  textilis, 915
  torreyi, 915
  triformis, 915
    stylosus, 915
  xiphioides, 916
    auratus, 916
    littoralis, 916
Juniper, 36
Juniperus
  californica, 4, 37
  occidentalis australis, 37
  osteosperma, 4, 37
  utahensis, 37
Jussiaea
  grandiflora, 612
  natans stipitata, 612
  peploides, 612
  repens californica, 612
  uruguayensis, 612

Kallstroemia
  californica, 860
  grandiflora, 860
  parvifolia, 860
Keckia antirrhinoides, 803
Keckiella
  antirrhinoides
    antirrhinoides, 803
    microphylla, 803
  breviflora, 803
  cordifolia, 804
  rothrockii
    jacintensis, 804
    rothrockii, 804
  ternata
    septentrionalis, 804
    ternata, 804
Kelloggia galioides, 770
Kentranthus ruber, 849
Kentrophyllum baeticum, 133
Kicksia
  elatine, 804
  spuria, 804
Knawel, 342
Kochia, 366
  americana, 366
    vestita, 366

  californica, 366
  scoparia, 366
    subvillosa, 396
Koeberlinia
  spinosa, 521
    tenuissima, 521
Koeberliniaceae, 48, 51, 521
Koeleria
  aristata, 978
  macrantha, 978
Kopsiopsis strobilacea, 617
Krameria
  canescens, 522
  grayi, 522
  imparata, 522
  parvifolia
    glandulosa, 522
    imparata, 522
Krameriaceae, 50, 521
Krynitzkia
  affinis, 252
  barbigera inops, 257
  decipiens, 254
  denticulata, 257
  dumetorum, 254
  glomerata virginensis, 258
  jonesii, 256
  maritima, 256
  microstachys, 256
  trachycarpa, 267
  utahensis, 258
Kyllinga brevifolia, 897

Lace,
  Queen Anne's, 75
  St. Catherine's, 685
Lactuca
  ludoviciana, 196
  scariola, 196
  serriola integrata, 196
Lagophylla
  minima, 196
  ramosissima, 196
Lamarckia aurea, 978
Lamiaceae, 53, 522
Lamium amplexicaule, 525
Lampranthus coccineus, 57
Langloisia
  matthewsii, 655
  punctata, 655
  schottii, 655
  setosissima, 655
Lantana
  camara, 850
  montevidensis, 850
Lantern, Fairy-, 921
Laothoe
  parviflora, 926
  pomeridiana, 926
Laphamia
  inyoensis, 216

  megalocephala, 216
  villosa, 216
Lappula
  echinata, 261
  redowskii, 261
    desertorum, 261
Larkspur, Scarlet, 723
Larrea, 4, 860
  divaricata, 860
  tridentata, 860
Lasthenia, 196
  ambigua, 163
  chrysostoma, 197
  coronaria, 197
  glabrata coulteri, 197
  microglossa, 197
  minor, 197
Lastrea augescens, 21
Lathyrus, 442
  alefeldii, 443
  brownii, 443
  hitchcockianus, 443
  laetiflorus
    alefeldii, 443
    barbarae, 443
    laetiflorus, 443
  latifolius, 443
  odoratus, 443
  pauciflorus
    brownii, 443
  puberulus, 444
  splendens, 443
  strictus, 443
  tingitanus, 444
  vestitus puberulus, 444
  violaceus barbarae, 443
Lauraceae, 46, 544
Laurel,
  California-, 544
  Cherry-, 755
Lavatera
  arborea, 564
  assurgentifolia, 564
  cretica, 566
Lavauxia primiveris, 615
Lavender,
  Desert-, 525
  Sea-, 637
Layia
  glandulosa
    glandulosa, 199
    lutea, 199
  heterotricha, 199
  platyglossa
    campestris, 199
    breviseta, 199
    platyglossa, 199
  ziegleri, 200
Leaf,
  Arrow, 219
  Velvet, 562

Index

Leersia oryzoides, 978
Leguminosae, synonym of
 Fabaceae, 419
Lemmonia californica, 499
Lemna gibba, 918
 minima, 918
 minor, 918
  cyclostasa, 919
  oblonga, 919
  polyrhiza, 919
  trisulca, 919
  valdiviana, 919
   abbreviata, 919
Lemnaceae, 918
Lennoaceae, 52, 545
Lens culinaris, 466
Lentibulariaceae, 53, 545
Lentil, 466
Leonotis leonurus, 525
Leontodon
 laevigatum, 234
 leysseri, 200
Lepechinia
 calycina, 526
 cardiophylla, 526
 fragrans, 526
 ganderi, 526
Lepidium, 291
 bernardinum, 295
 densiflorum, 292
 dictyotum, 292
  acutidens, 292
  dictyotum, 292
 didymus, 285
 draba, 281
 flavum, 292
  felipense, 292
 fremontii, 292
 intermedium pubescens, 295
 lasiocarpum, 294
  georginum, 294
 latipes, 294
 montanum cinereum, 294
 nitidum, 294
 oblongum, 294
 perfoliatum, 294
 pinnatifidum, 295
 procumbens, 291
 ramosissimum, 295
 strictum, 295
 thurberi, 295
 virginicum
  pubescens, 295
  robinsonii, 295
Lepidospartum
 latisquamum, 200
 squamatum, 200
  palmeri, 201
  striatum, 200
Lepidostephanus madioides, 107
Leptochloa

 fascicularis, 979
 filiformis, 979
 univervia, 979
Leptodactylon
 californicum
  californicum, 657
  glandulosum, 657
 jaegeri, 657
 melingii, 662
 nuttallii, 662
 pungens
  hallii, 657
  pulchriflorum, 657
Leptoseris sonchoides, 208
Leptosiphon
 bicolor, 660
 luteus, 659
Leptosyne
 californica, 150
 douglasii, 150
 gigantea, 150
Leptotaenia
 californica, 78
 multifida, 78
Lepturus
 bolanderi, 998
 cylindricus
Lesquerella
 bernardina, 296
 gordonii sessilis, 296
 kingii
  bernardina, 296
  cordiformis, 296
  kingii, 296
  palmeri, 296
  wardii, 296
Lessingia
 germanorum
  glandulifera, 201
  lemmonii, 201
  peirsonii, 201
  ramulosissima, 201
  tomentosa, 201
 glandulifera
  glandulifera, 201
  tomentosa, 201
 heterochroma, 202
 lemmonii, 201
  lemmonii, 201
  peirsonii, 201
  ramulosissima, 201
 tenuis, 202
Lettuce, Wild, 196
 Miner's-, 712
 Water-, 882
Leucanthemum vulgare, 140
Leucelene ericoides, 202
Leucopoa kingii, 974
Leucopsis spinosa, 123
Leucoseris
 saxatilis, 208

 tenuifolius, 208
Lewisia
 brachycalyx, 713
 minor, 714
 nevadensis, 713
 pygmaea, 714
 rediviva minor, 714
Libocedrus decurrens, 36
Licorice, 442
Lilac, California, 732
Lilaea
 scilloides, 917
 subulata, 917
Liliaceae, 919
Lily, 927
 Calla-, 882
 Checker-, 927
 Chinese-, 881
 Chocolate-, 927
 Corn-, 929
 Desert-, 927
 Globe-, 920
 Humboldt, 928
 Lemon, 923
 Leopard, 928
 Mariposa-, 920
 Panther, 928
 Star-, 931
Lilium, 927
 bloomerianum, 928
  ocellatum, 928
 fairchildii, 928
 humboldtii
  bloomerianum, 928
  magnificum, 928
  ocellatum, 928
  ocellatum bloomerianum, 928
 pardalinum, 928
 parryi, 928
 kessleri, 928
Limnanthaceae, 51, 547
Limnanthes
 gracilis parishii, 547
 versicolor parishii, 547
Limnorchis
 leucostachys, 933
 sparsiflora, 933
Limonium, 1, 637
 californicum, 637
  mexicanum, 637
 perezii, 637
 sinuatum, 637
Limosella acaulis, 805
Linaceae, 50, 547
Linanthastrum nuttallii, 662
Linanthus
 androsaceus
  luteolus, 659
  luteus, 659
  micranthus, 659
 arenicola, 659

Linanthus (cont.)
aureus, 659
  decorus, 660
bellus, 660
bicolor, 660
bigelovii, 660
breviculus, 660
ciliatus, 660
concinnus, 661
demissus, 661
dianthiflorus
  dianthiflorus, 661
  farinosus, 661
dichotomus, 661
jonesii, 661
killipii, 662
lemmonii, 662
liniflorus pharnaceoides, 662
maculatus, 662
melingii, 662
nudatus, 662
nuttallii
  floribundus, 662
  nuttallii, 662
orcuttii pacificus, 663
pacificus, 663
parryae, 663
pygmaeus, 663
Linaria, 805
bipartita, 805
canadensis texana, 805
dalmatica, 805
elatine, 804
maroccana, 805
purpurea, 805
spuria, 804
texana, 805
vulgaris, 806
Linosyris
arborescens, 176
depressa, 142
obtecta, 201
palmeri, 201
squamata, 200
teretifolia, 144
viscidiflora puberula, 145
Linum
grandiflorum, 548
lewisii, 548
perenne lewisii, 548
puberulum, 548
rigidum puberulum, 548
usitatissimum, 549
Lippia
Garden, 850
incisa, 850
lanceolata, 85
nodiflora
  canescens, 851
  rosea, 850
wrightii, 851

Liquorice, 441
Listera convallarioides, 934
Lithocarpus densiflorus, 479
Lithophragma, 782
affine
  affine, 782
  mixtum, 782
bolanderi, 782
cymbalaria, 782
heterophyllum, 782
maximum, 783
parviflorum, 783
scabrellum peirsonii, 782
tenellum, 783
tripartitum, 782
Lithospermum, 261
circumscissum, 253
incisum, 262
Live-Forever, 383
Loasaceae, 49, 50, 549
Loasella rupestris, 549
Lobelia
cardinalis
  graminea, 324
  multiflora, 325
  pseudosplendens, 324
dunnii serrata, 325
graminea, 324
Scarlet, 324
Lobularia maritima, 296
Locust, 469
Loeflingia
pusilla, 340
ramosissima, 339
squarrosa, 340
Loeselia
matthewsii, 655
tenuifolia, 654
Loganiaceae, 53, 559
Lolium
multiflorum, 979
perenne, 979
  multiflorum, 979
  perenne, 980
temulentum, 980
Lomatium, 77
californicum, 78
caruifolium, 78
dasycarpum, 78
  tomentosum, 78
dissectum, multifidum, 78
fimbriatum, 79
foeniculaceum
inyoense, 79
insulare, 79
inyoense, 79
lucidum, 79
mohavense, 79
longilobum, 79
nevadense
  holopterum, 80

parishii, 79
pseudorientale, 80
parryi, 80
tomentosum, 78
utriculatum, 80
vaseyi, 80
Lonicera, 331
catalinensis, 332
denudata, 332
hispidula
  californica, 332
  vacillans, 332
interrupta, 332
involucrata, 332
  ledebourii, 332
japonica, 332
johnstonii, 332
subspicata
  denudata, 332
  johnstonii, 332
  subspicata, 332
Lophanthus urticifolius, 524
Lophotocarpus calycina, 870
Lotus, 444
argensis, 449
argophyllus
  adsurgens, 446
  argenteus, 446
  argophyllus, 445
  decorus, 446
  hancockii, 446
  niveus, 446
  ornithopus, 446
argyraeus
  argyraeus, 446
  multicaulis, 446
corniculatus, 446
crassifolius, 446
davidsonii, 446
eriophorus, 447
grandiflorus, 446
hamatus, 447
haydonii, 447
heermannii, 447
humistratus, 447
leucophyllus, 449
micranthus, 447
nevadensis, 447
nuttallianus, 447
oblongifolius, 449
procumbens, 449
purshianus, 449
rigidus, 449
rubellus, 450
salsuginosus
  brevivexillus, 449
  salsuginosus, 449
scoparius, 449
  brevialatus, 450
  dendroideus, 449
  scoparius, 449

traskae, 450
spencerae, 447
strigosus, 450
  hirtellus, 450
subpinnatus, 450
tomentellus, 450
wrangelianus, 450
wrightii multicaulis, 446
Lousewort, 818
Ludwigia, 611
  natans stipitata, 612
  palustris pacifica, 612
  peploides, 612
  repens stipitata, 612
  uruguayensis, 612
Lupine, 450
  False, 470
Lupinus, 450
  adsurgens, 453
  affinis, 462
  agardhianus, 453
  albifrons
    albifrons, 453
    douglasii, 453
    eminens, 453
    medius, 457
  aliclementinus
  alpestris, 453
  andersonii, 454
  arboreus, 454
  arbustus montanus, 454
  argenteus tenellus, 454
  arizonicus, 454
  austromontanus, 457
  benthamii, 454
  bernardinus, 461
  bicolor
    marginatus, 454
    microphyllus, 455
    umbellatus, 455
  brevicaulis, 455
  brevior, 455
  breweri
    breweri, 455
    bryoides, 455
    clokeyanus, 455
    grandiflorus, 455
  bridgesii, 459
  brittonii, 453
  caespitosus, 455
  campbellae bernardinus, 455
  candidissimus, 461
  caudatus, 455
  chamissonis, 456
    longifolius, 460
  concinnus
    agardhianus, 453
    arizonicus, 454
    concinnus, 460
    optatus, 460
    orcuttii, 460

confertus, 456
cystoides, 459
densiflorus
  austrocollium, 456
  glareosus, 456
  lacteus, 456
  palustris, 456
desertorum, 461
elatus, 457
eminens, 453
excubitus
  austromontanus, 457
  excubitus, 457
  hallii, 457
  johnstonii, 457
  medius, 457
flavoculatus, 457
formosus
  bridgesii, 459
  formosus, 457
  hyacinthinus, 459
funstonianus, 453
glareosus, 456
guadalupensis, 459
hallii, 457
hirsutissimus, 459
holmgrenanus, 459
horizontalis platypetalus, 459
hyacinthinus, 459
intermontanus, 461
jaegerianus, 461
keckianus, 461
kerrii, 460
lacteus, 456
latifolius
  latifolius, 459
  parishii, 460
laxiflorus montanus, 454
longifolius, 460
luteolus, 460
magnificus, 460
  glareola, 460
micranthus, 461
mollisifolius, 460
moranii, 459
munzii, 453
nanus
  latifolius, 460
  nanus, 460
  nevadensis, 460
  odoratus, 460
  pilosellus, 460
pallidus, 461
palmeri, 461
palustris, 456
parishii
pasadenensis, 457
paynei, 457
peirsonii, 461
polycarpus, 461

polyphyllus
  bernardinus, 461
  superbus, 461
pusillus intermontanus, 461
rivularis, 459
ruber, 462
shockleyi, 462
sparsiflorus
  arizonicus, 454
  brevior, 455
  inopinatus, 462
  mohavensis, 462
  sparsiflorus, 462
  stiversii, 462
  subhirsutus, 462
  subvexus, 462
  succulentus, 462
  superbus, 461
  tenellus, 454
  truncatus, 463
Luzula
  campestris macrantha, 916
  comosa, 916
    macrantha, 916
    subsessilis, 916
  multiflora comosa, 916
  spicata, 916
    nova, 916
  subsessilis, 916
Lycium, 4, 831
  andersonii, 831
  barbarum, 832
  brevipes
    brevipes, 832
    hassei, 832
  californicum, 832
  cooperi, 832
  fremontii, 832
  halimifolium, 832
  hassei, 832
  pallidum oligospermum, 832
  parishii, 834
  torreyi, 834
  verrucosum, 834
Lycopersicum esculentum, 834
Lycopodiae, 11
Lycopsida, 11
Lycopsis arvensis, 262
Lycopus
  americanus, 527
  lucidus, 527
Lygodesmia
  exigua, 202
  spinosa, 202
Lyonothamnus
  floribundus, 750
    asplenifolius, 750
    floribundus, 750
Lyrocarpa
  coulteri palmeri, 296
  palmeri, 296

Lythraceae, 51, 560
Lythrum
　adsurgens, 561
　californicum, 561
　hyssopifolia, 561

Machaeranthera
　arida, 203
　canescens
　　canescens, 203
　　ziegleri, 204
　cognata, 204
　lagunensis, 204
　leucanthemifolia, 204
　orcuttii, 204
　tephrodes, 204
　tortifolia, 204
Machaerocarpus californicus, 868
Macronema discoideum, 178
Macrorhyncus
　heterophyllus, 108
　retrorsus, 108
Madaria elegans, 205
Madaroglossa
　heterotricha, 199
　elegans, 199
Madder, Field, 770
Madia
　densifolia, 205
　dissitiflora, 206
　elegans
　　densifolia, 205
　　elegans, 205
　　wheeleri, 205
　exigua, 205
　glomerata, 205
　gracilis, 206
　minima, 206
　sativa, 206
Madrone, 398
Maguey, 864
Mahogany, Mountain-, 743
Mahonia
　amplectens, 243
　dictyota, 243
　higginsae, 245
　nevinii, 245
　pinnata, 245
Maids, Red, 709
Malacomeris incanus, 208
Malacothamnus
　aboriginum, 566
　clementinus, 567
　davidsonii, 567
　densiflorus
　　densiflorus, 567
　　viscidus, 567
　fasciculatus
　　catalinensis, 568
　　fasciculatus, 567
　　laxiflorus, 568

　nesioticus, 568
　nuttallii, 568
　marrubioides, 568
　orbiculatus, 568
　parishii, 568
Malacothrix, 206
　altissima, 208
　blairii, 213
　californica, 207
　californica glabrata, 208
　clevelandii, 207
　coulteri, 207
　　cognata, 207
　floccifera, 207
　foliosa, 207
　glabrata, 208
　implicata, 208
　incana, 208
　obtusa, 207
　saxatilis
　　altissima, 208
　　implicata, 208
　　saxatilis, 208
　　tenuifolia, 208
　　tenuissima, 208
　similis, 208
　sonchoides, 208
　stebbinsii, 210
Malaxis brachypoda, 934
Mallow, 568
　Desert-, 572
　Globe-, 571
　Indian-, 562
　Rose-, 562
　Tree-, 564
Malosma laurina, 66
Malperia tenuis, 210
Malva
　caroliniana, 569
　coccinea, 607
　crispa, 569
　fasciculata, 567
　mauritiana, 569
　nicaeensis, 569
　parviflora, 569
　Rosa, 564
　sylvestris, 569
　verticillata, 569
Malvaceae, 48, 52, 561
Malvastrum
　aboriginum, 566
　catalinense, 568
　clementinum, 567
　coulteri, 573
　davidsonii, 567
　densiflorum, 567
　exile, 563
　gabrielense, 568
　marrubioides, 568
　nesioticum, 568
　nuttallii, 568

　orbiculatum, 568
　parishii, 568
　rotundifolium, 563
　thurberi laxiflorum, 568
　viscidum, 567
Mammillaria, 313
　alversonii, 311
　arizonica, 311
　desertii, 311
　dioica, 313
　　incerta, 313
　grahamii, 313
　microcarpa, 313
　radiosa alversonii, 311
　tetrancistra, 313
Man, Old, 317
Manzanita, 398
Maple,
　Big-leaf, 54
　Canyon, 54
　Flowering-, 562
Marah
　fabaceus agrestis, 393
　horridus, 393
　macrocarpus, 393
Marigold, 233
　Bur-, 128
　Cape-, 156
　Desert, 126
　Field, 132
　French, 233
　Pot-, 132
Marijuana, 575
Mariposa
　catalinae, 923
　clavata, 923
　　pallida, 923
　Goldenbowl, 923
　lutea, 924
　splendens, 925
Mariscus californicus, 895
Marrubium vulgare, 527
Marsilea vestita, 33
Marsileales, 33
Martynia
　althaeifolia, 575
　lutea, 574
　parviflora, 575
Martyniaceae, 53, 574
Matelea parvifolia, 94
Matthiola incana, 297
Matricaria
　matricarioides, 210
　occidentalis, 210
　suaveolens, 210
Maurandya
　antirrhiniflora, 806
　petrophylla, 806
Meconella
　californica denticulata, 629
　denticulata, 629

Meconopsis heterophylla, 632
Medicago
  hispida, 463
  lupulina cupaniana, 463
  minima, 463
  polymorpha, 463
    brevispina, 463
  sativa, 463
Medick, Black, 463
Megastachya uninervia, 979
Melampodium perfoliatum, 210
Melic, 980
Melica, 980
  aristata, 980
  californica, 980
  colpodioides, 982
  frutescens, 980
  imperfecta, 982
  panicoides, 982
  parishii, 982
  poaeoides, 982
  stricta, 982
Melilotus
  albus, 464
  indicus, 464
  officinalis, 464
Melon, 392
  Water, 391
Mengea californica, 61
Menodora
  scabra, 588
  laevis, 588
  scoparia, 589
  spinescens, 589
    mohavensis, 589
Mentha, 527
  arvensis
    arvensis, 527
    villosa, 528
  canadensis villosa, 528
  citrata, 528
  longifolia, 528
  piperita, 528
  pulegium, 528
  rotundifolia, 528
  silvestris, 528
  spicata, 528
    rotundifolia, 528
Mentzelia, 550
  affinis, 551
  albicaulis, 552
    jonesii, 554
  californica, 552
  congesta, 552
  desertorum, 552
  dispersa obtusa, 552
  eremophila, 553
  gracilenta, 553
  hirsutissima stenophylla, 553
  involucrata, 553
    involucrata, 553

  megalantha, 554
  jonesii, 554
  laevicaulis, 554
  lindleyi
    eremophila, 553
    longiloba, 554
    micrantha, 554
    mojavensis, 554
    montana, 555
    nitens, 555
    obscura, 555
    oreophila, 555
    peirsonii, 553
    pinetorum, 557
    pterosperma, 557
    puberula, 557
    ravenii, 557
    reflexa, 557
    stenophylla, 553
    tricuspis, 558
      brevicornuta, 558
    tridentata, 558
    urens, 550
    veatchiana, 558
Merimea texana, 396
Mesembryanthemum
  aequilaterum, 56
  chilense, 56
  coccineum, 57
  cordifolium, 55
  crassifolium, 56
  crystallinum, 56
  edule, 56
  elongatum, 57
  floribundum, 56
  nodiflorum, 57
  pugioniforme, 57
  speciosum, 56
Mesquite, 5, 466
Micropus californicus, 211
Microseris
  douglasii
    douglasii, 211
    platycarpha, 211
    tenella, 212
  elegans, 212
  heterocarpa, 212
  linearifolia, 212
  sylvatica, 212
Microsteris
  gracilis
    gracilis, 663
    humilis, 663
  humilis, 663
Microstylis
  brachypodus, 934
  monophyllus, 934
Mignonette, 731
Milfoil, Water-, 493
Milkwort, 669
Milla

  capitata
    pauciflora, 879
    laxa, 882
Miller, Dusty, 134
Millet Broom-Corn, 987
Mimetanthe pilosa, 814
Mimosa farnesiana, 421
Mimulus, 806
  androsaceus, 809
  aridus, 809
  aurantiacus australis, 809
  bigelovii
    bigelovii, 809
    cuspidatus, 809
    panamintensis, 809
  brandegei, 809
  brevipes, 810
  breweri, 810
  cardinalis, 810
  clevelandii, 810
  congdonii, 810
  densus, 810
  diffusus, 811
  exiguus, 811
  flemingii, 811
  floribundus, 811
  fremontii, 811
  guttatus
    grandis, 812
    guttatus, 811
    litoralis, 813
  inconspicuus latidens, 813
  johnstonii, 812
  latidens, 813
  longiflorus
    calycinus, 813
    longiflorus, 812
    rutilus, 813
  mohavensis, 813
  montioides, 813
  moschatus, 813
    longiflorus, 814
    sessiliflorus, 814
  nasutus, 814
  palmeri, 814
    androsaceus, 809
  parishii, 814
  pilosellus, 814
  pilosus, 814
  primuloides
    pilosellus, 814
    primuloides, 814
  puniceus, 814
  purpureus, 815
  rubellus, 815
  rupicola, 815
  suksdorfii, 815
  tilingii
    corallinus, 815
    tilingii, 815
    traskiae, 815

Mint,
　Bergamot, 528
　Horse, 528
　Mountain-, 533
　Pepper, 528
　Spear, 528
　Thorn, 523
Mirabilis, 583
　aspera, 584
　bigelovii
　　aspera, 584
　　bigelovii, 584
　　retrorsa, 584
　californica
　　californica, 584
　　cedrosensis, 584
　　cordifolia, 584
　froebelii
　　froebelii, 584
　　glabrata, 584
　jalapa, 584
　tenuiloba, 584
Mistletoe, 858
Modiola caroliniana, 569
Mohavea
　breviflora, 816
　confertifolia, 816
Mollugo
　cerviana, 57
　tetraphylla, 341
　verticillata, 57
Molly, Green-, 366
Monanthochloe littoralis, 982
Monarch-of-the-Veld, 239
Monarda pectinata, 528
Monardella, 529
　australis, 531
　breweri, 529
　candicans exilis, 529
　cinerea, 529
　epilobioides, 530
　exilis, 529
　hypoleuca
　　hypoleuca, 530
　　lanata, 530
　lanata, 530
　lanceolata, 530
　　glandulifera, 530
　　microcephala, 530
　linoides
　　linoides, 530
　　oblonga, 530
　　stricta, 530
　　viminea, 530
　macrantha
　　arida, 531
　　hallii, 531
　　macrantha, 530
　nana
　　arida, 531
　　leptosiphon, 531

　　tenuiflora, 531
　　nana, 531
　　oblonga, 530
　　odoratissima australis, 531
　　pringlei, 531
　　robisonii, 531
　　saxicola, 531
　　tenuiflora, 531
　　villosa leptosiphon, 531
　　viminea, 530
　　viridis saxicola, 531
Monerma cylindrica, 982
Monocotyledones, 862
Monolepis
　minor, 197
　nuttalliana, 366
　spathulata, 367
Monolopia lanceolata, 212
Monoptilon
　bellidiforme, 213
　bellioides, 213
Montbretia, 906
Montia, 714
　chamissoi, 713
　fontana
　　amporitana, 714
　　variabilis, 714
　funstonii, 714
　hallii, 714
　linearis, 714
　parviflora, 712
　perfoliata, 712
　　depressa, 712
　　utahensis, 712
　spathulata, 712
　viridis, 713
Montiastrum lineare, 714
Moraceae, 47, 575
Morning-Glory, 371, 375, 379
　Beach, 374
Mortonia
　scabrella utahensis, 350
　utahensis, 350
Morus alba, 576
Moss, Spike-, 14
Mourning-Bride, 395
Mouse-Tail, 725
Mucronea
　californica, 672
　perfoliata, 673
Mudwort, 805
Muhlenbergia, 982
　andina, 983
　arsenei, 983
　asperifolia, 983
　californica, 983
　comata, 983
　confusa, 984
　filiformis, 984
　microsperma, 984
　minutissima, 984

　porteri, 984
　purpurea, 984
　richardsonis, 984
　rigens, 984
　utilis, 984
Muilla
　clevelandii, 879
　coronata, 879
　maritima, 881
　serotina, 881
　tenuis, 881
Mullein, 827
Moth, 827
　Turkey-, 409
Munzothamnus blairii, 213
Mustard, 276
　Black, 277
　Field, 277
　Hedge-, 301
　Tansy-, 285
　Tower-, 271
　Tumble-, 301
　White, 277
Myagrum
　rugosum, 298
　sativum, 278
Myoporaceae, 52, 576
Myoporum laetum, 576
Myosotis
　flaccida, 254
　muricata, 256
　redowskii, 261
　tenella, 267
Myosurus
　aristatus, 725
　cupulatus, 725
　minimus
　　apus, 725
　　filiformis, 727
Myrica californica, 576
Myricaceae, 46, 576
Myriophyllum exalbescens, 495
Myriopteris gracilis, 26
Myrtaceae, 46, 577
Myrtle, Wax-, 576
Myzorrhiza corymbosa, 619

Najadaceae, 931
Najas
　guadalupensis, 931
　marina, 931
　　californica, 931
Nama, 499
　aretioides, 500
　biflorum spathulatum, 500
　demissum
　　demissum, 500
　　desertii, 500
　depressum, 500
　hispidum spathulatum, 500
　revolutum, 500

Index 1069

parryi, 518
pusillum, 500
roctrockii, 501
stenocarpum, 501
Narcissus tazetta, 881
Nasturtium, 845
  gambelii, 281
  Garden, 845
  occidentale, 1017
  officinale, 298
  obtusum, 298
  sinuatum, 300
  sphaerocarpum, 1017
Navarretia
  atractyloides, 664
  densifolia jacumbana, 642
  hamata, 664
    foliacea, 666
    leptantha, 666
  intertexta, 666
  latifolia, 650
  mitracarpa, 666
  peninsularis, 666
  prostrata, 666
  pubescens, 666
  schottii, 655
  setiloba, 666
  setosissima, 655
  tagetina, 666
  virgata
    dasyantha, 643
    gymnocephala, 643
Needle, Shepherd's, 85
Needles, Spanish, 214
Negundo californicum, 54
Neillia monogyna alternans, 752
Nemacladus
  adenophorus, 326
  capillaris, 325
  glanduliferus
    glanduliferus, 326
    orientalis, 326
  gracilis, 326
  longiflorus
    breviflorus, 326
    longiflorus, 326
  pinnatifidus, 326
  ramosissimus, 326
    gracilis, 326
    pinnatifidus, 326
  rigidus, 326
    capillaris, 325
    rubescens, 326
  rubescens, 326
  sigmoideus, 328
  tenuissimus, 326
Nemacaulis
  denudata
    denudata, 698
    gracilis, 698
    nuttallii, 698

Nemophila, 501
  aurita, 517
  insignis, 501
  integrifolia, 501
  menziesii, 501
    integrifolia, 501
    annulata, 502
    incana, 501
    rotata, 502
    racemosa, 518
    sepulta, 502
Neolloydia johnsonii, 314
Neomammillaria
  dioica, 313
  microcarpa, 313
  tetrancistra, 313
Nepeta, 531
  cataria, 532
Nettle, 847
  Dwarf, 847
  Hedge, 540
  Rock, 550
  Silverleaf, 839
Nicandra physalodes, 834
Nicolletia occidentalis, 213
Nicotiana, 834
  attenuata, 835
  bigelovii, 835
    bigelovii, 835
    wallacei, 835
  clevelandii, 835
  glauca, 835
  trigonophylla, 835
Nightshade, 838
  Enchanter's, 600
Ninebark, 752
Nitrophila
  mohavensis, 367
  occidentalis, 367
Nolina, 865
  bigelovii, 865
    parryi, 865
  interrata, 865
  parryi, 865
    wolfii, 865
  wolfii, 865
Notholaena
  californica, 29
  californica nigrescens, 29
  cochisensis, 29
  cretacea, 29
  jonesii, 29
  newberryi, 29
  parryi, 29
  sinuata cochisensis, 29
Notholcus lanatus, 975
Nothoscordum maritimum, 881
Nut-Grass, 897
  Purple, 895
Nyctaginaceae, 47, 52, 577

Oak, 4, 479
  Blue, 480
  California Black, 481
  Canyon, 480
  Coast Live, 479
  Engelmann, 480
  Interior Live, 483
  Island, 481
  Jerusalem-, 361
  Maul, 480
  Oracle, 481
  Poison-, 67
  Scrub, 4, 480
  Valley, 481
Oat, 947
  Cultivated, 948
  Slender Wild, 947
Obidium, porrifolium, 934
Obione
  confertifolia, 354
  coulteri, 355
  hymenelytra, 355
  lentiformis, 355
  leucophylla, 355
  microcarpa, 355
  phyllostegia, 357
  polycarpa, 357
  torreyi, 358
  truncata, 358
Ocotillo, 4
Oenanthe
  californica, 80
  sarmentosa, 80
Oenothera, 612
  alyssoides, 593
  arguta, 615
  avita
    avita, 613
    eurekensis, 613
  berlandieri, 615
  bistorta, 592
    veitchiana, 593
  boothii
    alyssoides, 593
    condensata, 594
    decorticans, 593
    desertorum, 594
    intermedia, 594
    inyoensis, 594
  brevipes, 594
    pallidula, 594
  bufonis, 615
  caespitosa
    crinita, 613
    marginata, 613
  californica, 611
  campestris, 595
  cardiophylla
    longituba, 592
    robusta, 595
  splendens, 592

Oenothera (*cont.*)
  chamaenerioides, 595
  cheiranthifolia, 595
    suffruticosa, 595
  claviformis, 596
    aurantiaca, 596
    funerea, 596
    integrior, 596
    peirsonii, 596
    yumae, 597
  clutei, 614
  contorta
    flexuosa, 599
    strigulosa, 599
  decorticans, 593
  delessertiana, 615
  deltoides, 614
    cineracea, 614
    eurekensis, 613
  densiflora, 590
  dentata
    campestris, 595
    gilmanii, 598
    johnstonii, 595
  epilobioides, 603
  glabella, 590
  graciliflora, 597
  guadalupensis clementina, 597
  hallii, 598
  heterochroma, 597
    monensis, 597
  heterophylla, 592
  hirta, 598
  hirtella, 597
  hookeri
    angustifolia, 614
    grisea, 614
    venusta, 614
  ignota, 597
  kernensis, 598
    gilmanii, 598
    mojavensis, 598
  laciniata, 614
  leptocarpa, 594
  longissima clutei, 614
  marginata, 613
  micrantha
    jonesii, 597
    ignota, 597
  munzii, 598
  nitida, 595
  palmeri, 599
  primiveris
    bufonis, 615
    caulescens, 615
    primiveris, 615
  pterosperma, 599
  quadrivulnera, 803
  refracta, 599
  rosea, 615
  rutila, 594

  scapoidea
    aurantiaca, 596
    purpurascens, 596
    tortilis, 600
  sinuata, 614
  speciosa, 615
    childsii, 615
  strigulosa pubens, 599
  stricta, 615
  venusta, 614
    grisea, 614
  viminea, 603
Oleaceae, 46, 50, 53, 586
Oleaster, 395
Oligomeris linifolia, 730
Olive
  Desert-, 586
  Russian, 395
Olneya tesota, 464
Onagraceae, 47, 50, 589
Onion, Wild, 871
Ophioglossaceae, 33
Ophioglossales, 33
Ophioglossum
  californicum, 34
  lusitanicum californicum, 34
Ophrys convallarioides, 934
Oplismenus crus-pavonis, 964
Opuntia, 314
  acanthocarpa
    coloradensis, 315
    ganderi, 315
    major, 315
  basilaris, 315
    basilaris, 315
    brachyclada, 316
    ramosa, 316
    treleasei, 316
  bernardina, 318
  bigelovii
    bigelovii, 316
    hoffmannii, 316
  brachyclada, 316
  chlorotica, 316
  "demissa", 316
  discata, 318
  echinocarpa
    echinocarpa, 316
    major, 315
    parkeri, 316
    wolfii, 316
  engelmannii littoralis, 317
  erinacea
    erinacea, 317
    ursina, 317
    utahensis, 317
    xanthostemma, 317
  ficus-indica, 317
  fosbergii, 316
  littoralis, 317
    austrocalifornica, 317

    martiniana, 317
    piercei, 317
    vaseyi, 318
  macrocentra martiniana, 317
  macrorhiza, 318
  mesacantha vaseyi, 318
  missouriensis rufispina, 320
  mojavensis, 320
  munzii, 318
  "occidentalis", 318
  oricola, 318
  parishii, 320
  parryi
    parryi, 318
    serpentina, 318
  phaeacantha
    discata, 318
    major, 320
    piercei, 317
  polyacantha rufispina, 320
  prolifera, 320
  ramosissima, 320
  rhodantha, 317
  sphaerocarpa utahensis, 317
  stanlyi parishii, 320
  tessellata, 320
  tomentosa, 320
  treleasei, 316
  ursina, 317
  wigginsii, 321
Orange, Mock-, 783
Orchiastrum porrifolium, 934
Orchid, Phantom, 932
  Rein, 933
Orchidaceae, 931
Orchis
  Stream, 932
Orcuttia californica, 985
Oreocarya
  abortiva, 255
  confertiflora, 253
  flavoculata, 254
  hoffmannii, 255
  lutea, 253
  nubigena, 257
  tumulosa, 258
Oreonana vestita, 82
Ornithostaphylos oppositifolia, 1018
Orobanchaceae, 52, 617
Orobanche, 617
  bulbosa, 618
  californica
    californica, 620
    claremontensis, 620˚
    feudgei, 618
    grandis, 618
    parishii, 619
  cooperi, 618
  corymbosa, 619
  fasciculata, 619

grayana
  feudgei, 618
  vallicola
  ludoviciana cooperi, 618
    latiloba, 618
    valida, 619
  multiflora arenosa, 619
  parishii
    brachyloba, 619
    parishii, 619
  tuberosa, 618
  uniflora occidentalis, 619
  valida, 619
  vallicola, 620
Orthocarpus, 816
  attenuatus, 817
  densiflorus
    densiflorus, 817
    gracilis, 817
  erianthus, 817
  gracilis, 817
  hispidus, 817
  lasiorhynchus, 817
  purpurascens
    ornatus, 818
    pallidus, 818
    purpurascens, 817
Oryzopsis
  bloomeri, 985
  cuspidata, 985
  hymenoides, 985
  membranacea, 985
  micrantha, 985
  miliacea, 986
  webberi, 986
Osmadenia tenella, 132
Osmorhiza
  brachypoda, 82
    fraterna, 82
  chilensis, 82
  nuda, 82
Osmunda lunaria, 34
Ourisia californica, 499
Oxalidaceae, 49, 51, 52, 620
Oxalis
  albicans
    californica, 620
    pilosa, 620
  cernua, 622
  corniculata, 620
  pes-caprae, 622
  pilosa, 620
Oxybaphus
  coccineus, 585
  comatus, 585
  froebelii, 584
  linearis, 585
  nyctagineus, 585
  pumilus, 585
Oxypolis occidentalis, 82
Oxyria digyna, 699

Oxystylis lutea, 331
Oxytenia acerosa, 195
Oxytheca
  caryophylloides, 699
  dendroidea, 699
  emarginata, 699
  luteola, 700
  parishii
    abramsii, 700
    parishii, 700
  perfoliata, 700
  reddingiana, 695
  trilobata, 700
  watsonii, 700
Oxytropis
  oreophila, 464
  viscida, 465

Pachylophis
  crinitus, 613
  marginatus, 613
Pachypodium integrifolium, 304
Paeonia
  brownii californica, 622
  californica, 622
Paeoniaceae, 48, 622
Palafoxia
  linearis, 214
    arenicola, 214
    gigantea, 214
    linearis, 214
Palm, Fan, 883
Palmae, synonym of
  Arecaceae, 882
Palmerella debilis serrata, 325
Palo Verde, 437
  Mexican, 465
Panicularia
  elata, 974
  fendleriana, 994
  thurberiana, 994
Panicum
  barbipulvinatum, 986
  capillare occidentale, 986
  colonum, 964
  crusgalli, 964
  dactylon, 959
  dichotomiflorum, 986
  frumentaceum, 964
  geniculatum, 998
  hirticaule, 986
  huachucae, 986
  ischaemum, 962
  miliaceum, 987
  muricatum, 964
  occidentale, 987
  pacificum, 987
  sanguinale, 962
  sphacelatum, 999
  urvilleanum, 987
  longiglume, 987

verticillatum, 999
viride, 999
Pansy, Wild, 856
Papaver
  californicum, 629
  hybridum, 629
  somniferum, 629
Papaveraceae, 48, 51, 52, 622
Paper-White, 881
Pappophorum wrightii, 968
Parapholis incurva, 987
Parietaria
  debilis, 846
  floridana, 846
Parishella californica, 328
Parkinsonia
  aculeata, 465
  florida, 437
  microphylla, 437
  torreyana, 437
Parnassia
  californica, 783
  cirrata, 783
  palustris californica, 783
Parosela
  arborescens, 439
  californica, 439
    simplifolia, 439
  emoryi, 439
  fremontii, 439
  johnsonii minutifolia, 439
  mollis, 441
  mollissima, 441
  neglecta, 439
  parryi, 441
  polyadenia, 441
  schottii, 441
  spinosa, 441
Parsley, 83
  Hedge-, 87
  Marsh-, 71
Parsnip, 83
  Cow-, 76
  Water-, 71
Parthenopsis maritimus, 239
Pasania densiflora, 479
Paspalum, 987
  dilatatum, 988
  distichum, 988
  larrangai, 988
  urvillei, 988
Pastinaca sativa, 83
Paws, Pussy, 711
Pea, 442
  Chaparral, 465
  Everlasting, 443
  Garden, 466
  Sweet, 443
  Tangier, 443
Peach, 757
  Desert, 755

Pear, 758
  Pancake-, 316
  Prickly-, 314
Pearlwort, 341
Pectis papposa, 214
Pectocarva, 262
  gracilis platycarpa, 263
  heterocarpa, 262
  linearis ferocula, 262
  penicillata, 263
    heterocarpa, 262
    platycarpa, 263
    recurvata, 263
    setosa, 263
      aptera, 263
      holoptera, 263
Pedicularis
  attolens, 818
  densiflora, 818
  semibarbata, 818
Pelargonium
  grossularioides, 493
  hortorum, 493
  zonale, 493
Pellaea
  andromedaefolia
    andromedaefolia, 31
      rubens, 31
      pubescens, 31
  breweri, 31
  bella, 31
  compacta, 31
  longimucronata, 31
  mucronata, 31
    californica, 31
  ornithopus, 31
  rafaelensis, 31
  wrightiana
    californica, 31
    compacta, 31
    longimucronata, 31
Pellitory, 846
Pennisetum
  clandestinum, 988
  ruppelii, 988
  setaceum, 988
  villosum, 988
Penny, Tinker's, 519
Pennyroyal, 528
  Mock, 524
Pennywort, Marsh, 76
Pentacaena ramosissima, 339
Pentachaeta
  alsinoides, 139
  aurea, 139
  lyonii, 139
Penstemon, 818
  albomarginatus, 820
  ambiguus thurberi, 825
  antirrhinoides, 803
  austinii, 822

barbatus labrosus, 822
breviflorus, 803
bridgesii, 820
caesius, 820
calcareus, 821
californicus, 821
centranthifolius, 821
clevelandii
  clevelandii, 821
  connatus, 821
  mohavensis, 821
confusus patens, 821
cordifolius, 804
desertorum, 821
eatonii, 821
  undosus, 822
floridus
  austinii, 822
  floridus, 822
fruticiformis
  amargosae, 822
  fruticiformis, 822
  incertus, 822
glaber, 824
  utahensis, 824
grinnellii
  grinnellii, 822
  scrophularioides, 822
heterophyllus
  australis, 822
  heterophyllus, 822
  incertus, 822
jacintensis, 804
labrosus, 822
laetus, 824
linarioides californicus, 821
micranthus, 803
monoensis, 824
munzii, 822
parishii, 824
pseudospectabilis, 824
pumilus thompsoniae, 825
rothrockii, 804
scapoides, 824
scrophularioides, 822
speciosus, 824
spectabilis, 825
  subvicosus, 825
stephensii, 825
ternatus, 804
  septentrionalis, 804
thompsoniae, 825
thurberi, 825
utahensis, 825
Peramium giganteum, 933
Peraphyllum ramosissimum, 750
Perezia microcephala, 214
Pericome caudata, 215
Perideridia, 83
  gairdneri, 83

parishii parishii, 83
  latifolia, 83
  pringlei, 83
Perityle
  emoryi, 215
  inyoensis, 216
  megalocephala
    megalocephala, 216
    oligophylla, 216
  villosa, 216
Periwinkle, 90
Petalonyx, 558
  gilmanii, 559
  linearis, 559
  nitidus, 559
  thurberi
    gilmanii, 559
    thurberi, 559
Petalostemon searlsiae, 465
Petradoria
  discoidea, 216
  pumila, 218
Petrophytum caespitosum, 752
Petunia
  parviflora, 835
  violacea, 836
Peucedanum
  argense, 79
  dasycarpum, 78
  euryptera, 79
  hassei, 79
  insulare, 79
  mohavense, 79
  parishii, 79
  parryi, 80
  tomentosum, 78
  utriculatum, 80
  vaseyi, 80
Peucephyllum schottii, 218
Phaca
  crotalariae, 427
  davidsonii, 434
  deanei, 427
  douglasii, 428
  lanosissima, 435
  leucopsis, 436
  trichopoda, 436
Phacelia, 502
  affinis, 505
  amabilis, 505
  ambigua, 508
  anelsonii, 505
  austromontana, 505
  barnebyana, 505
  bicolor, 505
  brachyloba, 506
  californica
    calycosa, 511
    jacintensis, 512
  calthifolia, 506

Index

campanularia
  campanularia, 506
  vasiformis, 506
cicutaria
  cicutaria, 506
  hispida, 506
ciliata, 506
coerulea, 508
crenulata
  ambigua, 508
  crenulata, 508
  funerea, 508
cryptantha, 508
  derivata, 508
curvipes, 508
  davidsonii, 508
davidsonii, 508
distans, 509
divaricata insularis, 509
douglasii
  douglasii, 509
  cryptantha, 509
egena, 509
floribunda, 509
fremontii, 509
frigida dasyphylla, 510
geraniifolia, 514
gooddingii, 514
grandiflora, 510
grisea, 510
gymnoclada, 510
hallii, 509
heterophylla virgata, 510
hispida, 506
  heliophila, 517
humilis
  dudleyi, 511
  humilis, 510
  humilis lobata, 505
imbricata
  bernardina, 511
  imbricata, 511
  patula, 511
insularis, 509
irritans, 510
ivesiana, 511
  pediculoides, 514
keckii, 516
lemmonii, 511
lobata, 505
longipes, 511
lyonii, 512
magellanica
  barbata egena, 509
  patula, 511
micrantha, 498
minor, 512
minutiflora, 512
mohavensis, 512
mustellina, 512
mutabilis, 512

nashiana, 513
neglecta, 513
nemoralis, 513
novenmillensis, 513
oreopola
  oreopola, 513
  simulans, 513
pachyphylla, 513
parishii, 513
parryi, 514
pedicellata, 514
pediculoides, 514
peirsoniana, 514
perityloides
  jaegeri, 514
  perityloides, 514
pulchella gooddingii, 514
ramosissima
  austrolitoralis, 516
  suffrutescens, 516
  valida, 516
  rotundifolia, 516
suaveolens keckii, 516
tanacetifolia, 516
vallis-mortae
  heliophila, 517
  vallis-mortae, 516
virgata bernardina, 511
viscida, 517
Phalangium scilloides, 917
Phalaris, 988
  angusta, 989
  aquatica, 989
  canariensis, 989
  caroliniana, 989
  lemmonii, 989
  minor, 989
  oryzoides, 978
  paradoxa, 989
  semiverticillata, 996
  setacea, 988
  stenoptera, 989
  tuberosa stenoptera, 989
Pharnaceum cervianum, 57
Phegopteris alpestris americana, 18
Phellosperma tetrancistra, 313
Phellopterus
  multinervatus, 74
  purpurascens, 74
Philadelphus, 783
  microphyllus
    pumilus, 784
    stramineus, 784
  pumilus, 784
  stramineus, 784
Philibertella
  hartwegii, 94
  hirtella, 94
Philotria canadensis, 905
Phleum

alpinum, 990
pratense, 990
Phlomis leonurus, 525
Phlox
  austromontana, 667
  bernardina, 668
  covillei, 667
  diffusa, 667
    subcarinata, 668
  dolichantha, 668
  longifolia compacta, 668
  Prickly, 657
  superba, 668
  stansburyi superba, 668
  viridis compacta, 668
Pholisma
  arenarium, 545
  paniculatum, 545
Pholistoma
  auritum, 517
  membranaceum, 517
  racemosum, 518
Pholiurus incurvus, 987
Phoradendron
  bolleanum
    densum, 858
    pauciflorum, 858
  californicum, 858
  juniperinum
    juniperinum, 858
    libocedri, 858
    ligatum, 858
    tomentosum, 858
    villosum, 858
Photinia
  arbutifolia, 747
  cerina, 747
  macrocarpa, 747
Phragmites
  australis, 990
  berlandieri, 990
  communis berlandieri, 990
Phyla
  incisa, 850
  lanceolata, 850
Phyllodoce breweri, 402
Phyllogonum luteolum, 698
Phyllospadix, 1015
  scouleri, 1015
  torreyi, 1015
Physalis, 836
  acutifolia, 836
  angulata lanceifolia, 836
  crassifolia
    crassifolia, 837
    versicolor, 837
  fendleri cordifolia, 837
  greenei, 837
  hederaefolia cordifolia, 837
  hirsuta integrifolia, 837
  ixocarpa, 837

Physalis (*cont.*)
  lanceifolia, 836
  pensylvanica cinerascens, 837
  philadelphica, 837
  pubescens
    grisea, 837
    integrifolia, 837
  versicolor, 837
  viscosa cinerascens
    wrightii, 836
Physaria
  chambersii, 297
  cordiformis, 296
Physocarpus
  alternans, 752
    annulatus, 752
    panamintensis, 752
Phytolacca
  americana, 632
  decandra, 632
Phytolaccaceae, 47, 632
Piaropus crassipes, 1007
Picea
  concolor, 38
  lowiana, 38
Pickeringia
  montana, 465
    tomentosa, 466
Picris echioides, 218
Pill-wort, 33
Pilostyles thurberi, 719
Pilularia americana, 33
Pimpernel, 717
  Water-, 719
Pimpinella parishii, 83
Pinaceae, 37·
Pine
  Bishop, 40
  Bristlecone, 4, 40
  Coulter, 39
  Digger, 42
  Jeffrey, 39
  Knobcone, 39
  Limber, 4, 39
  Lodgepole, 4, 40
  Pinyon, 39
    One-leaved, 40
    Four-leaved, 40
  Ponderosa, 40
  Santa Cruz Island, 42
  Sugar, 39
  Tamarac, 40
  Torrey, 42
  Yellow, 40
Pinedrops, 403
Pink Ground-, 661
Pinus
  aristata, 40
  attenuata, 39
  balfouriana aristata, 40
  benthamiana, 40

  californica, 39
  cembroides
    edulis, 39
    monophylla, 40
    parryana, 40
  contorta murrayana, 40
  coulteri, 4, 39
  deflexa, 39
  edgariana, 40
  edulis, 39
    monophylla, 40
  flexilis, 4, 39
  fremontiana, 40
  jeffreyi, 4, 39
  lambertiana, 39
  longaeva, 4, 40
  lophosperma, 42
  macrocarpa, 39
  monophylla, 4, 40
  muricata, 40
  murrayana, 4, 40
  parryana, 40
  ponderosa, 4, 40
  quadrifolia, 40
  remorata, 42
  sabiniana, 42
  torreyana, 42
  tuberculata, 39
Pinyon
  Four-leaved, 40
  One-leaved, 40
Piperia
  elegans, 933
  unalascensis, 933
Pipsissewa, 401
Pistia stratiotes, 882
Pisum
  arvense, 466
  sativum, 466
Pitavia dumosa, 771
Pittosporaceae, 51, 632
Pittosporum, 632
  tobira, 633
  undulatum, 633
Pityrogramma, 31
  triangularis triangularis, 32
    maxoni, 32
    viridis, 32
    viscosa, 32
Plagiobothrys, 263
  acanthocarpus, 264
  arizonicus, 264
  bracteatus, 265
  californicus
    californicus, 265
    fulvescens, 265
    gracilis, 265
    ursinus, 265
  canescens, 265
  hispidulus, 265
  jonesii, 266

  leptocladus, 266
  nothofulvus, 266
  parishii, 266
  salsus, 266
  tenellus, 267
  torreyi, 267
  trachycarpus, 267
  undulatus, 267
Plant,
  Century, 864
  Gum-, 172
  Ice-, 56
  Oyster, 236
  Sandpaper, 558
  Snow, 404
  Soap, 926
  Umbrella-, 896
  Unicorn-, 574
Plantaginaceae, 52, 633
Plantago, 633
  aristata, 634
  bigelowii
    bigelowii, 634
    californica, 634
  californica, 634
  coronopus, 634
  erecta
    erecta, 634
    rigidior, 634
  galeottiana, 634
  hirtella galeottiana, 634
  hookeriana californica, 634
  indica, 635
  insularis, 635
  juncoides californica, 635
  lanceolata, 635
  major, 635
  maritima californica, 635
  oblonga, 635
  parishii, 634
  purshii oblonga, 635
  pusilla, 635
  rhodosperma, 636
  virginica, 636
Plantain, 633
  Common, 635
  English, 635
  Water-, 870
Platanaceae, 46, 51, 636
Plants, Flowering, 45
Platanthera
  elegans, 933
  leucostachys, 933
Platanus
  californica, 636
  racemosa, 636
Platyloma bella, 31
Platystemon, 629
  californicus
    californicus, 631
    ciliatus, 631

Index

crinitus, 631
horridulus, 631
nutans, 631
ornithopus, 631
horridulus, 631
Plectritis
brachystemon, 849
ciliosa insignis, 849
congesta brachystemon, 849
macrocera, 849
grayii, 849
Pleuraphis
jamesii, 975
rigida, 975
Pleurocoronis pluriseta, 219
Pluchea
borealis, 219
camphorata, 219
purpurascens, 219
sericea, 219
Plumbaginaceae, 49, 52, 636
Plume,
Apache-, 746
Prince's, 302
Poa, 990
annua, 991
stricta, 992
atropurpurea, 991
bigelovii, 992
bolanderi, 992
howellii, 992
brachyglossa, 994
buckleyana, 995
bulbosa, 992
californica, 992, 995
capillarifolia, 992
capillaris, 995
compressa, 992
curvula, 969
cusickii, 992
distans, 996
douglasii, 992
fendleriana, 994
juncifolia, 994
longiligula, 994
filifolia, 992
howellii, 992
chandleri, 992
hypnoides, 969
incurva, 994
juncifolia, 994
kingii, 974
limosa, 994
longiligula, 994
mexicana, 969
nervosa, 994
nevadensis, 994
nudata, 995
olneyi, 994
orcuttiana, 995
palustris, 994

pauciflora, 994
pectinacea, 969
pilosa, 969
pratensis, 995
rigida, 997
rupestris, 995
rupicola, 995
scabrella, 995
tenuifolia, 995
Poaceae, 934
Pod,
Bladder-, 295, 330
Fringe-, 306
Lace-, 306
Spectacle-, 287
Podistera nevadensis, 84
Pogogyne
abramsii, 532
nudiuscula, 532
Poinsettia eriantha, 412
Polemoniaceae, 53, 637
Polemonium
caeruleum amygdalinum, 668
micranthum, 668
occidentale, 668
Poliomintha incana, 532
Polyanthus narcissus, 881
Polycarpon
depressum, 340
tetraphyllum, 341
Polygala
acanthoclada, 669
californica, 669
cornuta
fishiae, 669
pollardii, 669
fishiae, 669
subspinosa heterorhyncha, 670
Polygyalaceae, 50, 52, 669
Polygonaceae, 46, 47, 670
Polygonum, 700
amphibium
emersum, 703
stipulaceum, 701
arenastrum, 703
argyrocoleon, 703
aviculare, 703
baldschuanicum, 703
bistortoides, 703
coccineum, 703
convolvulus, 703
douglasii
douglasii, 703
johnstonii, 703
emersum, 703
fusiforme, 704
hartwrightii, 701
hydropiperoides
asperifolium, 704

kelloggii, 704
lapathifolium, 704
natans, 701
parryi, 704
pensylvanicum, 704
persicaria, 704
punctatum, 704
ramosissimum, 704
sawatchense, 703
Polymnia abyssinica, 173
Polypappus sericeus, 219
Polypodiaceae, 23
Polypodium
acuminatum, 21
australe, 23
californicum, 23
kaulfussii, 23
carnosum, 24
falcatum, 18
filix-mas, 19
fragile, 18
hesperium, 23
intermedium, 23
pachyphyllum, 24
scouleri, 24
vulgare
columbianum, 23
intermedium, 23
Polypogon
australis, 995
interruptus, 996
lutosus, 996
monspeliensis, 996
semiverticillatus, 996
Polystichum
mohrioides scopulinum, 21
munitum
curtum, 21
imbricans, 21
munitum, 19
scopulinum, 21
Pontederia crassipes, 1007
Pontederiaceae, 1007
Poplar, 771
Silver, 772
Poppy
Bear, 623
California, 626
Matilija, 631
Opium, 629
Prickly, 623
Sea, 628
Tree, 625
Populus, 771
alba, 772
fremontii
arizonica, 772
fremontii, 772
macdougallii, 772
pubescens, 772
macdougallii, 772

Populus (cont.)
  tremuloides, 772
  trichocarpa
    ingrata, 772
    trichocarpa, 772
Porophyllum gracile, 219
Portulaca
  mundula, 715
  oleracea, 715
Portulacaceae, 48, 49, 51, 708
Potamogeton, 1007
  americanus, 1009
  californicus, 1008
  crispus, 1008
  dimorphus, 1008
  diversifolius, 1008
  fluitans, 1009
  foliosus, 1008
    californicus, 1008
  illinoensis, 1008
  latifolius, 1008
  lonchites, 1009
  lucens, 1008
  natans, 1009
  nodosus, 1009
  panormitanus, 1009
  pectinatus, 1009
    latifolius, 1008
  pusillus, 1009
Potamogetonaceae, 1007
Potato, Tule-, 870
Potentilla, 752
  anserina, 753
    grandis, 753
    sericea, 753
  argentina, 753
  argyrocoma, 749
  biennis, 753
  callida, 749
  clevelandii, 748
  egedei grandis, 753
  elata, 749
  glandulosa
    ewanii, 754
    glandulosa, 753
    hansenii, 754
    nevadensis, 754
    reflexa, 754
  gracilis nuttallii, 754
  hansenii, 754
  hispidula, 749
  jaegeri, 750
  lindleyi, 748
  millegrana, 754
  monspeliensis, 754
  multijuga, 754
  norvegica monspeliensis, 754
  nuttallii, 754
  patellifera, 754
  pectinisecta, 754

  puberula, 749
  rivalis millegrana, 754
  santolinoides, 750
  saxosa, 754
  sibbaldii, 761
  truncata, 749
  wheeleri
    rimicola, 755
    wheeleri, 755
  wilderae, 749
  wrangelliana glandulosa, 753
  Poteridium occidentale, 761
  Prenanthes pauciflora, 231
Pride of California, 443
Primulaceae, 52, 715
Pritchardia
  filifera, 883
  filamentosa, 883
Proboscidea, 574
  althaeifolia, 575
  parviflora, 575
Prosopis, 5, 466
  chilensis, 466
  glandulosa, torreyana, 466
  julifera, 466
  odorata, 466
  pubescens, 466
  strombulifera, 466
  velutina, 468
Prunella, 532
  pennsylvanica lanceolata, 533
  vulgaris
    lanceolata, 533
    vulgaris, 533
Prunus
  andersonii, 755
  caroliniana, 755
  emarginata, 757
  eriogyna, 757
  fasciculata, 757
  fremontii, 757
  ilicifolia, 757
  lyonii, 757
  mahaleb, 757
  persica, 757
  virginiana demissa, 757
Psathyrotes
  annua, 220
  incisa, 237
  ramosissima, 220
  schottii, 218
Pseudotsuga
  douglasii macrocarpa, 42
  macrocarpa, 42
Psilactis coulteri, 203
Psilocarphus
  brevissimus, 220
  globiferus, 220
  tenellus, 220
Psilostrophe cooperi, 221
Psoralea

  bituminosa, 468
  californica, 468
  castorea, 468
  macrostachya, 469
  mephitica, 469
  orbicularis, 469
  physodes, 469
  rigida, 469
Pteridaceae, 24
Pteridium
  aquilinum
    lanuginosum, 32
    pubescens, 32
Pteris vittata, 32
Pteropsida, 16, 35, 43
Pterospora andromedea, 403
Pterostegia drymarioides, 705
Pteryxia petraea, 84
Ptilomeris
  anthemoides, 197
  coronaria, 197
Ptiloria
  cichoriacea, 230
  exigua, 231
  pauciflora, 231
  pleurocarpa, 231
  tenuifolia, 231
  virgata, 231
Puccinellia
  airoides, 996
  distans, 996
  nuttalliana, 996
  parishii, 996
Puccoon, 261
Pugiopappus
  bigelovii, 150
  calliopsidea, 150
Pulicaria hispanica, 221
Pumpkin, Field, 392
Purpusia saxosa, 758
Purse, Shepherd's-, 279
Purshia glandulosa, 4, 758
Purslane, 715
  Horse-, 58
  Sea-, 58
Pussytoes, 113
Pycnanthemum californicum, 533
Pyrola
  aphylla, 404
  asarifolia, 403
    incarnata, 403
  californica, 403
  dentata integra, 404
  menziesii, 402
  minor, 403
  occidentalis, 402
  picta, 403
    aphylla, 404
    integra, 404
  secunda, 404

Index

Pyrolaceae, included in
    Ericaceae, 397
Pyrrocoma
    gossypina, 181
    grindelioides, 179
    sessiliflora, 179
Pyrus communis, 758

Quamoclidium
    froebellii glabratum, 584
Quarters, Lamb's, 360
Quercus, 4, 479
    agrifolia
        agrifolia, 479
        oxyadenia, 479
    alvordiana, 483
    chrysolepis, 480
    densiflora, 478
    douglasii, 480
    dumosa, 480
        kinselae, 480
        turbinella, 483
    dunnii, 480
    engelmannii, 480
    ganderi, 480
    garryana semota, 481
    grandidentata, 481
    hindsii, 481
    kelloggii, 4, 481
    lobata, 481
    macdonaldii, 481
    morehus, 481
    palmeri, 480
    tomentella, 481
    townei, 480
    turbinella
        californica, 483
        turbinella, 483
    wislizenii
        frutescens, 483
        parvula, 483
        wislizenii, 483
Quillwort, Flowering, 917

Radish, Wild, 297
Rafflesiaceae, 719
Rafinesquia
    californica, 221
    neomexicana, 222
Ragweed, Western, 112
Ragwort, 223
Raillardella
    argentea, 222
Raillardia argentea, 222
Raimannia laciniata, 614
Ramtilla, 173
Ranunculaceae, 45, 47, 719
Ranunculus, 4, 727
    alismellus, 727
    alismaefolius alismellus, 727
    andersonii, 728

aquatilis
    capillaceus, 728
    hispidulus, 728
californicus
    austromontanus, 728
    californicus, 728
    cuneatus, 728
    canus ludovicianus, 728
    cymbalaria saximontanus,
        728
    eisenii, 729
    eschscholtzii oxynotus, 729
    flammula ovalis, 729
    hebecarpus, 729
    ludovicianus, 728
    occidentalis eisenii, 729
    oxynotus, 729
    uncinatus, 729
Rape, 277
    Broom-, 617
Raphanus
    raphanistrum, 297
    sativum, 297
Rapistrum rugosum, 298
Redbud, 438
Redscale, 357
Reed
    Bur-, 1011
    Common, 990
    Giant, 947
Reseda
    alba, 731
    lutea, 731
Resedaceae, 48, 50, 730
Rhagadiolus cretica, 182
Rhamnaceae, 46, 51, 731
Rhamnus, 4, 738
    californica
        californica, 738
        cuspidata, 738
        tomentella, 739
        ursina, 739
    crocea, 739
        ilicifolia, 739
        insularis, 739
    cuspidata, 738
    ilicifolia, 739
    pilosa, 739
    pirifolia, 739
    tomentella, 739
    ursina, 739
Rhododendron occidentale,
    404
Rhubarb, Wild-, 707
Rhus
    diversiloba, 67
    integrifolia, 64
    laurina, 66
    ovata, 66
    trilobata pilosissima, 66
        anisophylla, 66

quinata, 66
Rhynchelytrum
    roseum, 997
    repens, 997
Ribes, 784
    amarum, 785
    aureum
        aureum, 785
        gracillimum, 785
    californicum hesperium, 785
    canthariforme, 785
    cereum, 785
    divaricatum parishii, 787
    gracillimum, 785
    hesperium, 785
    indecorum, 787
    lasianthum, 787
    malvaceum
        clementinum, 787
        malvaceum, 787
        viridifolium, 787
    menziesii, 787
        thacherianum, 788
    montigenum, 788
    nevadense, 788
        jaegeri, 788
        parishii, 787
    quercetorum, 788
    roezlii, 788
    speciosum, 788
    thacherianum, 788
    velutinum, 788
        glanduliferum, 789
    viburnifolium, 789
Rice, Jungle-, 964
Ricinus communis, 417
Riddellia cooperi, 221
Rigiopappus leptocladus, 222
Robinia pseudo-acacia, 469
Roble, 479
Rocket,
    Garden-, 289
    London-, 301
    Sand-, 287
    Sea-, 278
    Wall-, 287
Romero, 544
Romneya
    coulteri, 631
        trichocalyx, 631
    trichocalyx, 631
Root,
    Alum-, 779
    Balsam, 127
    Bitter-, 714
    Coral, 932
    Snake, 84
    Squaw, 83
Rorippa
    curvisiliqua, 298, 1017
    orientalis, 1017

Rorippa (cont.)
  curvisiliqua (cont.)
    occidentalis, 1017
    intermedia, 1017
  islandica occidentalis, 300, 1017
  nasturtium-aquaticum, 298
  obtusa, 298
  palustris occidentalis, 300, 1017
  sinuata, 300
  sphaerocarpa, 1017
  tenerrima, 1017
  truncata, 1017
Rosa, 758
  californica, 759
  gratissima, 759
  gymnocarpa, 759
  mohavensis, 759
  woodsii
    ultramontana, 759
    glabrata, 759
    gratissima, 759
Rosaceae, 46, 47, 49, 51, 739
Rose, 758
  Rock-, 370
  Wood, 759
Rosemary, Marsh-, 637
Rottboellia cylindrica, 982
Roubieva multifida, 363
Rubiaceae, 53, 761
Rubus, 759
  almus, 760
  ganderi, 760
  glaucifolius ganderi, 760
  laciniatus, 760
  leucodermis bernardinus, 760
  nutkanus, 760
  parviflorus, 760
  procerus, 760
  ursinus, 760
Rue, 771
  Bush-, 771
  Meadow-, 729
Rugging, Turkish, 674
Rumex, 705
  acetosella, 706
  angiocarpus, 706
  californicus, 706
  conglomeratus, 706
  crassus, 706
  crispus, 706
  digynus, 699
  fueginus, 707
  hymenosepalus, 707
  kerneri, 707
  lacustris, 707
  obtusifolius agrestis, 707
  persicarioides, 707
  pulcher, 707

  salicifolius, 708
    ecallosus, 706
    triangulivalvis, 708
    spinosus, 676
    triangulivalvis, 708
    violascens, 708
Ruppia, 1009
  cirrhosa, 1011
  maritima, 1011
    intermedia, 1011
    rostrata, 1011
    rostellata, 1011
    spiralis, 1011
Rush, 909
  Scouring-, 15
  Spike-, 898
  Toad, 911
  Wood, 916
Ruta chalepensis, 771
Rutaceae, 770
Rye, 998
Ryegrass,
  English, 980
  Italian, 979

Sabina
  californica, 37
  osteosperma, 37
Safflower, 134
Sage, 533
  Black, 538
  Bladder-, 533
  Bur-, 110
  Hop-, 365
  Pitcher, 526
  Thistle, 536
  White, 536
Sagebrush, 116
  Basin, 119
  Coastal, 117
Sagina
  apetala barbata, 341
  linnaei, 341
  occidentalis, 341
  saginoides hesperia, 341
St. John's Wort, 519
Sagittaria
  arifolia, 870
  calycina, 870
  cuneata, 870
  greggii, 870
  latifolia, 870
  longiloba, 870
  montevidensis calycina, 870
  sanfordii, 870
  variabilis, 870
Sahuaro, 310
Salazaria mexicana, 4, 533
Salicaceae, 46, 771
Salicornia, 1, 5, 367
  ambigua, 368

  bigelovii, 368
  depressa, 368
  europaea, 368
  pacifica, 368
  rubra, 368
  subterminalis, 368
  utahensis, 368
  virginica, 368
Salix, 773
  babylonica, 774
  caudata bryantiana, 774
  exigua, 774
    stenophylla, 774
  geyeriana, 774
  gooddingii
    gooddingii, 774
    variabilis, 774
  hindsiana
    hindsiana, 774
    leucodendroides, 774
    parishiana, 774
  laevigata
    araquipa, 775
    laevigata, 774
  lancifolia, 775
  lasiandra
    abramsii, 775
    lancifolia, 775
    lasiandra, 775
  lasiolepis
    bracelinae, 775
    lasiolepis, 775
  lemmonii, 775
  lutea watsonii, 775
  macrostachya leucodendroides, 774
  melanopsis bolanderiana, 775
  parishiana, 774
  scouleriana, 776
Salpichroa, 837
  origanifolia, 838
  rhomboidea, 838
Salsify, 236
Salsola
  depressa, 369
  hyssopifolia, 359
  iberica, 368
  kali
    ruthenica, 368
    tebuifolia, 368
  paulsenii, 368
  scoparia, 366
Saltbush, 352
  Australian, 358
Saltwort, 242
Salvia, 4, 533
  apiana, 536
    compacta, 536
  bernardina, 536
  brandegei, 536

carduacea, 536
carnosa
　gilmanii, 537
　pilosa, 537
clevelandii, 536
columbariae
　columbariae, 536
　ziegleri, 536
compacta, 536
dorrii
　argentea, 537
　dorrii, 537
　gilmanii, 537
　eremostachya, 537
　funerea, 537
　greatai, 537
　leucophylla, 538
　mellifera, 538
　　jonesii, 538
　mohavensis, 538
　munzii, 538
　pachyphylla, 538
　pachystachya, 538
　palmeri, 538
　pratensis, 539
　sonomensis, 539
　spathacea, 539
　vaseyi, 539
Salviniaceae, 34
Salviniales, 34
Sambucus
　caerulea, 333
　glauca, 333
　mexicana, 333
　microbotrys, 333
Samolus
　floribundus, 719
　parviflorus, 719
Samphire, 367
Sandwort, 335
Sanguisorba
　annua, 761
　minor, 761
　occidentalis, 761
Sanicle,
　Poison, 85
　Purple, 85
Sanicula, 84
　arguta, 84
　bipinnata, 85
　bipinnatifida, 85
　crassicaulis, 85
　graveolens, 85
　hoffmannii, 85
　menziesii, 85
　nevadensis, 85
　pinnatifida, 85
　tuberosa, 85
Santalaceae, 776
Santolina chamaecyparissus, 222

Sanvitalia, 222
　abertii, 223
　procumbens, 223
Saponaria
　officinalis, 341
　vaccaria, 349
Sarcobatus vermiculatus, 369
Sarcodes sanguinea, 404
Sarcostemma
　cynanchoides hartwegii, 94
　heterophyllum, 94
　　hirtellum, 95
　hirtellum, 95
Sarratia berlandieri, 62
Satintail, 978
Satureja
　chandleri, 539
　douglasii, 539
Saururaceae, 47, 776
Saxifraga
　arguta, 789
　californica, 789
　elata, 778
　malvaefolia, 781
　neonuttalliana, 781
　odontoloma, 789
　parryi, 781
Saxifragaceae, 48, 49, 50, 778
Saxifrage, 789
Scabiosa atropurpurea, 395
Scandix pecten-veneris, 85
Scheuchzeriaceae, synonym of Juncaginaceae, 916
Schinus molle, 66
Schismus
　arabicus, 997
　barbatus, 997
Schkuhria neomexicana, 126
Schmaltzia
　anisophylla, 66
　ovata, 66
　integrifolia, 64
　malacophylla, 66
　ovata, 66
　　traskiae, 66
　quinata, 66
Schoenus nigricans, 901
Scilla pomeridiana, 926
Scirpus, 1, 902
　acicularis, 899
　acutus, 902
　　occidentalis, 902
　americanus, 902
　　polyphyllus, 902
　atropurpureus, 899
　californicus, 902
　cernuus californicus, 903
　campestris, 903
　coloradoensis, 899
　geniculatus, 900
　koilolepis, 903

microcarpus, 903
mucronatus, 903
olneyi, 903
pacificus, 903
paludosus, 903
pauciflorus, 900
quinqueflorus, 900
robustus, 903
rostellatus, 900
validus, 903
Scleranthus annuus, 342
Sclerocactus polyancistrus, 321
Sclerocarpus
　exigua, 205
　gracilis, 206
Scleropoa rigida, 997
Scopulophila
　nitrophiloides, 342
　rixfordii, 342
Scorzonella sylvatica, 212
Scribneria bolanderi, 998
Scrophularia, 825
　californica
　　desertorum, 826
　　floribunda, 826
　desertorum, 826
　villosa, 826
Scrophulariaceae, 53, 789
Scutellaria
　austinae, 540
　bolanderi austromontana, 540
　lateriflora, 540
　tuberosa, 540
Secale cereale, 998
Sedge, 884
　Black, 901
　Umbrella-, 895
Sedum
　album, 388
　blochmanae, 385
　edule, 386
　niveum, 388
　oblongorhizum, 386
　sanctae-monicae, 386
　spathulifolium anomalum, 389
　variegatum, 388
Selaginella
　asprella, 14
　bigelovii, 14
　bryoides, 14
　cinerascens, 14
　eremophila, 14
　leucobryoides, 14
　parishii, 14
　watsonii, 15
Selaginellaceae, 12
Selaginellales, 11, 12
Selfheal, 532

Selinocarpus diffusus, 586
Selinum
  eryngifolium, 86
  kingii, 70
Senecio, 223
  aphanactis, 224
  astephanus, 224
  bernardinus, 224
  breweri, 224
  californicus, 224
  canus, 224
  douglasii
    douglasii, 226
    monoensis, 226
  floccifera, 207
  fremontii occidentalis, 226
  ilecetorum, 224
  ionophyllus, 224
    bernardinus, 224
    sparsilobatus, 226
  lyonii, 226
  mikanioides, 226
  mohavensis, 226
  monoensis, 226
  multilobatus, 227
  occidentalis, 226
  serra, 227
    sanctus, 227
  spartioides, 227
  stygius, 227
  triangularis, 227
  vulgaris, 227
Senkenbergia coulteri, 582
Senna, 437
Serapias gigantea, 933
Sericotheca
  boursieri, 747
  franciscana, 748
  microphylla, 748
Serratula arvensis, 146
Sesbania
  exaltata, 470
  macrocarpa, 470
Sesuvium
  sessile, 58
  verrucosum, 58
Setaria
  faberi, 998
  geniculata, 998
  lutescens, 999
  sphacelata, 999
  verticillata, 999
  viridis, 999
Shank, Red, 741
Shepherdia argentea, 396
Sherardia arvensis, 770
Shinleaf, 403
Sibara
  deserti, 300
  filifolia, 300
  rosulata, 301

virginica, 301
Sibbaldia procumbens, 761
Sida
  alata, 564
  hederacea, 569
  leprosa hederacea, 569
  malvaeflora, 570
Sidalcea
  hickmannii parishii, 570
  malvaeflora
    californica, 571
    dolosa, 571
    malvaeflora, 570
    neomexicana thurberi, 571
    pedata, 571
Sideranthus gooddingii, 177
  gracilis, 178
  junceus, 178
Sieversia paradoxa, 746
Silene, 342
  andersonii, 344
  anglica, 343
  antirrhina, 343
  bernardina, 343
  conoidea, 343
  cucubalus, 343
  dichotoma, 343
  dorrii, 344
  gallica, 343
  inflata, 343
  laciniata
    augustifolia, 344
    latifolia, 344
    major, 343
  latifolia, 343
    dichotoma, 343
  lemmonii, 344
  menziesii dorrii, 344
  montana bernardina, 343
  multinervia, 344
  parishii
    latifolia, 344
    parishii, 344
    platyota, 345
    viscida, 344
  platyota, 345
  verecunda
    andersonii, 344
    eglandulosa, 345
    platyota, 345
Silybum marianum, 227
Simaroubaceae, 48, 828
Simmondsia
  chinensis, 308
  californica, 308
Simsia
  canescens, 169
  frutescens, 158
Sinapis
  alba, 277
  geniculata, 276

  incana, 276
  juncea, 277
  kaber, 277
  nigra, 277
Sisymbrium
  altissimum, 301
  curvisiliquum, 298
  diffusum jaegeri, 291
  incisum, 286
  irio, 306
  murale, 287
  nasturtium-aquaticum, 298
  officinale, 301
  orientale, 302
  sophia, 287
  tenuifolium, 287
Sisyrinchium, 907
  bellum, 908
  eastwoodiae, 908
  elmeri, 908
  funereum, 908
  halophilum, 908
  hesperium, 908
  idahoense, 908
  maritimum, 908
  minus, 908
  oreophilum, 908
Sitanion
  anomalum, 999
  breviaristatum, 1000
  californicum, 999
  glabrum, 999
  hansenii, 999
  hystrix
    californicum, 999
    hystrix, 999
  jubatum, 1000
  longifolium, 1000
  minus, 999
  multisetum, 1000
Sium
  douglasii, 73
  erectum, 71
Skullcap, 540
Smartweed, 700
  Water, 701
Smelowskia californica, 286
Smilacina
  amplexicaulis, 928
  glabra, 928
  racemosa
    amplexicaulis, 928
    glabra, 928
  stellata, 929
Smilax, 920
Snapdragon, Common, 793
Snow, Evening, 661
Solanaceae, 53, 829
Solanum, 838
  arborescens, 839
  californicum, 840

Index

clokeyi, 839
coronopus, 830
douglasii, 839
elaeagnifolium, 839
lanceolatum, 839
lycopersicum, 834
marginatum, 839
nigrum, 839
nodiflorum, 839
parishii, 840
rostratum, 840
sarrachoides, 840
tenuilobatum, 840
triflorum, 840
umbelliferum
  glabrescens, 840
  incanum, 840
wallacei, 840
xanti
  glabresceus, 840
  hoffmannii, 840
  intermedium, 841
  montanum, 841
  wallacei, 840
  xanti, 840
Soleirola soleirolii, 847
Solidago
  altissima, 228
  californica, 228
  confinis, 228
  guiradonis spectabilis, 229
  occidentalis, 228
  pumila, 218
  sarothrae, 174
  spectabilis, 229
Soliva
  daucifolia, 229
  sessilis, 229
Sollya fusiformis, 633
Solomon's-Seal, False, 928
Sonchus, 229
  arvensis, 230
  asper, 230
  oleraceus, 230
  tenerrimus, 230
Sophia
  adenophora, 286
  glabra, 286
  halictorum, 286
  menziesii, 286
  viscosa, 287
Sorghum
  bicolor, 1000
  halepense, 1000
  lanceolatum, 1000
  sudanense, 1000
  virgatum, 1000
  vulgare, 1000
Sorrel, 705
  Mountain, 698

Sheep, 706
Wood-, 620
Sparganiaceae, 1011
Sparganium
  angustifolium, 1011
  californicum, 1011
  eurycarpum, 1011
  greenei, 1011
  greenei, 1011
  multipedunculatum, 1011
  simplex
    angustifolium, 1011
    multipedunculatum, 1011
Spartina, 1000
  foliosa, 1001
  gracilis, 1001
  leiantha, 1001
Spartium junceum, 470
Spear, Ithuriel's, 882
Specularia biflora, 328
Speedwell, 827
Spergula
  arvensis, 345
  villosa, 347
Spergularia
  atrosperma, 345
  bocconii, 345
  macrotheca
    leucantha, 347
    macrotheca, 347
  marina, 347
  platensis, 347
  rubra, 347
  villosa, 347
Sphacele
  calycina, 526
  fragrans, 526
Sphaeralcea, 571
  ambigua
    ambigua, 572
    keckii, 572
    monticola, 572
    rosacea, 572
    rugosa, 572
  angustifolia cuspidata, 572
  arida, 573
  coulteri, 573
  cuspidata, 572
  emoryi
    arida, 573
    emoryi, 573
    nevadensis, 573
    variabilis, 573
  eremicola, 574
  orcuttii, 573
  parvifolia, 573
  pulchella, 572
  rosacea, 572
  rusbyi eremicola, 574
  variabilis, 573

Sphaerostigma
  chamaenerioides, 595
  cheiranthifolium, 595
  hirtellum, 597
  hirtum, 598
  pallidum, 598
  pubens, 599
  strigulosum, 599
Sphenopholis
  obtusata, 1001
    lobata, 1001
Sphenopsida, 15
Sphenosciadium capitellatum, 86
Spikenard, 90
Spinach
  New-Zealand-, 58
  Sea-, 58
Spiraea
  boursieri, 747
  caespitosa, 752
  discolor, 74
  millefolia, 745
  Rock-, 750
Spiranthes
  porrifolia, 934
  romanzoffiana porrifolia, 934
  unalascensis, 933
Spirodela polyrhiza, 919
Spirostachys occidentalis, 352
Sporobolus
  airoides, 1001
    wrightii, 1002
  asperifolius, 983
  contractus, 1002
  cryptandrus, 1002
  flexuosus, 1002
  microspermus, 984
  poiretii, 1002
  strictus, 1002
Spraguea umbellata, 719
Sprangletop, 979
Spray, Ocean, 747
Spruce,
  Big-Cone, 42
Spurge, 409
  Caper, 412
  Petty, 414
Spurrey, 345
  Sand-, 345
Squirreltail, 999
Stachys, 540
  ajugoides, 541
  albens, 541
  bullata, 541
  californica, 541
  quercetorum, 541
  rigida
    rigida, 541
    quercetorum, 541

Stanleya
　elata, 302
　pinnata
　　inyoensis, 302
　　pinnata, 302
　pinnatifida, 302
Star
　Blazing, 550
　Desert, 213
　Shooting, 717
　Woodland-, 782
Starwort, 348
Statice
　californica, 637
　perezii, 637
　sinuata, 637
Stellaria
　crispa, 348
　graminea, 348
　jamesiana, 348
　longipes, 348
　media, 348
　nitens, 348
Stemodia durantifolia, 826
Stenochloa
　californica, 972
Stenotaphrum secundatum, 1002
Stenotus linearifolius, 178
Stephanomeria
　blairii, 213
　cichoriacea, 230
　cinerea, 231
　exigua, 231
　　deanei, 231
　　pentachaeta, 231
　myrioclada, 231
　parryi, 231
　pauciflora, 231
　　parishii, 231
　runcinata, 231
　virgata, 231
Sterculiaceae, 46, 841
Stevia linearis, 214
Stickseed, 261
Stick-Tight, 128
Stillingia
　annua, 418
　linearifolia, 417
　paucidentata, 417
　spinulosa, 418
Stipa, 1002
　arida, 1003
　brachychaeta, 1003
　californica, 1003
　cernua, 1004
　comata, 1004
　coronata, 1004
　　depauperata, 1004
　diegoensis, 1004
　elmeri, 1004

eminens, 1004
hymenoides, 985
latiglumis, 1004
lepida, 1004
lettermannii, 1005
occidentalis, 1005
　montana, 1004
parishii, 1004
　depauperata, 1004
pinetorum, 1005
pulchra, 1005
setigera, 1005
speciosa, 1005
stricta sparsiflora, 1005
thurberiana, 1005
williamsii, 1005
Stock, 297
Stonecrop, 388
Stone-Fruits, 755
Storax, 843
Storksbill, 490
Streptanthella
　longirostris, 303
　　derelicta, 303
Streptanthus
　amplexicaulis, 282
　arcuatus, 275
　bernardinus, 303
　campestris, 303
　cordatus, 340
　coulteri, 282
　crassicaulis, 283
　glaucus, 283
　hallii, 283
　heterophyllus, 304
　inflatus, 283
　insignis, 304
　major, 283
　pilosus, 285
　simulans, 285
Strobus lambertiana, 39
Stylocline
　acaulis, 165
　filaginea, 232
　gnaphalioides, 232
　micropodoides, 232
Stylomecon heterophylla, 632
Stylopappus grandiflora, 108
Stylophyllum
　densiflorum, 386
　hassei, 386
　orcutti, 385
　traskae, 388
　virens, 388
　viscidum, 388
Styphonia integrifolia, 64
Styracaceae, 53, 843
Styrax
　californica fulvescens, 843
　officinalis fulvescens, 843
Suaeda, 1, 5, 369

californica, 369
depressa
　depressa, 369
　erecta, 369
fruticosa, 370
minutiflora, 369
torreyana, 370
　ramosissima, 370
Sultan, Sweet, 135
Sumac, Laurel, 66
Sunflower, Desert-, 169
Svida, glabrata, 381
Swallenia alexandrae, 1006
Sweet, Desert, 745
Sycamore, 4, 636
Sympetaleia rupestris, 550
Symphoricarpos
　longiflorus, 333
　mollis, 334
　parishii, 334
　parvifolius, 334
Syntherisma sanguinalis, 962
Syntrichopappus
　fremontii, 232
　lemmonii, 233
Syringa, 783
Syrmatium
　argophyllum, 445
　davidsonii, 446
　dendroideum, 449
　glabrum, 449
　nevadense, 447
　niveum, 446
　traskae, 450

Tagetes
　minuta, 233
　papposa, 157
　patula, 233
Tail,
　Cat-, 1012
　Dog, 960
　Mare's-, 495
　Mouse, 725
　Satin, 978
　Squirrel, 999
　Thin, 982
Talinum
　menziesii, 709
　pygmaeum, 714
Tamaricaceae, 49, 843
Tamarisk, 5, 843
Tamarix, 843
　africana, 844
　aphylla, 844
　aralensis, 844
　chinensis, 844
　gallica, 844
　parviflora, 844
　ramosissima, 844
Tanacetum canum, 233

Index

Taraxacum
  californicum, 234
  ceratophorum bernardinum, 234
  erythrospermum, 234
  laevigatum, 234
  officinale, 234
  vulgare, 234
Taraxia
  graciliflora, 597
  palmeri, 599
Tauschia
  arguta, 86
  hartwegii, 86
  parishii, 86
Tea,
  Mexican, 360
  Mormon, 43
Teasel, 395
Tellima
  heterophylla, 782
  parviflora, 783
Tetracoccus
  dioicus, 418
  hallii, 418
  ilicifolius, 418
Tetradymia, 234
  argyraea, 235
  axillaris, 235
  canescens, 235
  comosa, 235
  glabrata, 235
  ramosissima, 220
  spinosa, longissima, 235
  stenolepis, 236
Tetragonia
  expansa, 58
  tetragonioides, 58
Tetraneuris arizonica, 194
Tetranthera californica, 544
Teucrium
  cubense depressum, 543
  depressum, 543
  glandulosum, 543
Thalictrum, 729
  fendleri, 730
  polycarpum, 730
  sparsiflorum, 730
    nevadense, 730
Thamnosma montana, 771
Thelypodium
  affine, 304
  cooperi, 282
  deserti, 300
  integrifolium, 304
  jaegeri, 305
  lasiophyllum
    lasiophyllum, 305
    utahense, 305
  stenopetalum, 305
    utahense, 305
Thelypteris

acuminata, 21
augescens, 21
puberula, 21
Thermopsis
  californica, 470
  macrophylla
    macrophylla, 470
    semota, 470
Thistle, 145
  Barneby's, 135
  Blessed, 148
  Bull, 148
  Canada, 146
  Italian, 133
  Milk, 227
  Musk, 133
  Plumeless, 133
  Russian-, 368
  Sow, 229
  Star, 134
Thlaspi
  arvense, 305
  bursa-pastoris, 279
Thorn,
  Box, 381
  Buck, 738
  Camel, 422
  Cotton-, 235
  Crucifixion, 829
Thrift, 637
Thrincia leysseri, 200
Thuja decurrens, 36
Thumb, Lady's, 704
Thymus
  chamissonis, 539
  douglasii, 539
Thysanocarpus
  conchuliferus, 307
  crenatus, 307
  curvipes
    curvipes, 306
    elegans, 306
    eradiatus, 306
    desertorum, 307
    elegans, 306
  laciniatus
    conchuliferus, 307
    crenatus, 307
    desertorum, 307
    hitchcockii, 307
    ramosus, 307
    rigidus, 307
  pusillus, 276
Tickseed, 149
Ticks, Beggar-, 128
Tidestromia
  lanuginosa, 63
  oblongifolia, 64
Tillaea
  aquatica, 382
  erecta, 382

Timothy, 990
  Mountain, 990
Tips, Tidy, 199
Tiquilia
  brevifolia plicata, 250
Tissa
  clevelandii, 347
  leucantha, 347
    glabra, 347
  macrotheca scariosa, 347
  marina, 347
  rubra, 347
Tithymalus
  crenulatus, 412
  dictyospermus, 415
  helioscopia, 412
  lathyris, 412
  palmeri, 413
  peplus, 414
  schizolobus, 412
Toadflax, 805
  Bastard, 776
Tobacco, 834
  Tree, 835
Tocalote, 135
Tomatillo, 837
Tomato, 834
Tongue,
  Beard, 818
  Ox, 218
Tordylium nodosum, 87
Torilis nodosa, 87
Tornillo, 466
Townsendia scapigera, 236
Toxicodendron
  comarophyllum, 67
  diversilobum, 67
  isophyllum, 67
  radicans
    diversilobum, 67
Toxicoscordion
  fremontii, 931
  venenosus, 931
Toyon, 747
Tracheophyta, 11
Tradescantia fluminensis, 883
Tragia
  ramosa, 419
  stylaris, 419
Tragopogon porrifolium, 236
Tree,
  Elephant, 307
  Gum, 577
  Joshua, 866
  Judas, 438
  of Heaven, 829
  Pepper-, 66
  Plane-, 636
  Smoke, 441
Trefoil, Bird's Foot, 444
Tresses, Ladies', 934

Trianthemum portulacastrum, 58
Tribulus terrestris, 862
Tricardia watsonii, 518
Tricerastes glomerata, 393
Trichochloa microsperma, 984
Trichocoronis wrightii, 237
Tricholaena rosea, 997
Trichoptilium incisum, 237
Trichostema
  austromontanum, 543
    compactum, 543
  lanatum, 544
    denudatum, 544
  lanceolatum, 544
  micranthum, 544
  parishii, 544
Tricuspis
  muticus, 1006
  pulchellus, 970
Tridens
  muticus, 1006
  pilosus, 970
  pulchellus, 970
Trifolium, 470
  aciculare, 476
  albopurpureum, 471
  amplectens, 472
  barbigerum, 472
  bifidum decipiens, 472
  catalinae, 473
  ciliolatum, 472
  cyathiferum, 472
  depauperatum, 472
  dubium, 472
  fucatum
    fucatum, 472
    gambelii, 473
  gambelii, 473
  gracilentum, 473
    inconspicuum, 473
  grantianum, 475
  hirtum, 473
  hybridum, 473
  insularum, 471
  involucratum, 476
  longipes atrorubens, 473
  macraei, 473
  microcephalum, 473
  microdon pilosus, 473
  monanthum
    grantianum, 475
    monanthum, 475
  obtusiflorum, 475
  palmeri, 475
  pratense, 475
  procumbens, 475
  repens, 475
  rusbyi atrorubens, 473
  traskae, 471
  tridentatum

    aciculare, 476
    tridentatum, 475
    variegatum, 476
    wormskioldii, 476
Triglochin
  concinnum
    concinnum, 917
    debile, 917
    maritimum, 917
  debile, 917
  striatum, 918
  triandrum, 918
Triodanis biflora, 328
Triodia
  mutica, 1006
  pulchella, 970
Tripolium frondosum, 122
  occidentale, 123
Trisetum spicatum, 1006
Triteleia
  candida, 882
  clementina, 881
  dudleyi, 881
  laxa, 882
  scabra, 882
  uniflora, 879
Triticum
  aestivum, 1007
  elongatum, 940
  trachycaulum, 940
Tritonia crocosmiflora, 906
Trixis californica, 237
Tropaeolaceae, 52, 845
Tropaeolum majus, 845
Tropidocarpum
  dubium, 307
  gracile, 307
    dubium, 307
Troximum
  grandiflorum, 108
  heterophyllum, 108
  plebeium, 108
  retrorsum, 108
Trumpet, Desert, 688
Tuckermannia
  gigantea, 150
  maritima, 151
Tule, 902
Tulip, Star-, 920
Turricula parryi, 518
Turritis
  glabra, 272
  lasiophylla, 305
Twayblade, 934
Typha, 1, 1012
  angustifolia, 1012
  bracteata, 1012
  domingensis, 1012
  glauca, 1013
  latifolia, 1012
  truxillensis, 1012

Typhaceae, 1012

Ulex europaeus, 476
Ulmaceae, 46, 845
Ulmus, 845
  chinensis, 846
  minor, 846
  parvifolia, 846
  procera, 846
  pumila, 846
Umbelliferae, synonym of Apiaceae, 67
Umbellularia californica, 544
Unifolium amplexicaule, 928
Uniola
  spicata, 963
  stricta, 963
Urachne micrantha, 985
Uralepis pilosa, 970
Uropappus heterocarpa, 212
Urtica
  holosericea, 847
  urens, 847
Urticaceae, 47, 846
Utricularia vulgaris, 547

Vaccaria
  pyramidata, 349
  segetalis, 349
Vaccinium ovatum, 404
Vagnera
  amplexicaulis, 928
  liliacea, 928
  stellata, 928
Valerian, Red, 849
Valerianaceae, 53, 847
Velaea
  arguta, 86
  parishii, 86
Venegasia, 237
  carpesioides, 239
  deltoides, 239
Venidium fastuosum, 239
Veratrum californicum, 929
Verbascum
  blattaria, 827
  thapsus, 827
  virgatum, 827
Verbena
  bracteata, 851
  bracteosa, 851
  gooddingii, 851
  lasiostachys, 851
    abramsii, 852
  menthaefolia, 852
  prostrata, 851
  robusta, 852
  Sand-, 578
  scabra, 852
  tenuisecta, 852
Verbenaceae, 53, 850

Index

Verbesina
  alba, 157
  dissita, 239
  encelioides, exauriculata, 239
Veronica, 827
  americana, 828
  anagallis-aquatica, 828
  arvensis, 828
  comosa, 828
  peregrina xalapensis, 828
  persica, 828
  scutellata, 828
  serpyllifolia, 828
  xalapensis, 828
Vervain, 851
Vesicaria kingii, 296
Vetch, 476
  Common, 477
  Hungarian, 477
  Milk, 422
  Spring, 478
  Winter, 478
Vicia, 476
  americana, 477
    linearis, 477
  angustifolia, 477
  atropurpurea, 477
  benghalensis, 477
  californica, 477
  dasycarpa, 477
  exigua, 477
  faba, 477
  oregana, 477
  pannonica, 477
  sativa, 478
  truncata, 477
  villosa, 478
Viguiera, 239
  deltoidea parishii, 240
  laciniata, 240
  multiflora nevadensis, 240
  reticulata, 240
  parishii, 240
  tephrodes, 184
Vilfa
  asperifolia, 983
  depauperata, 984
  flexuosa, 1002
  gracillima, 984
  richardsonis, 984
  squarrosa, 984
  utilis, 984
Villanova chrysanthemoides, 126
Vinca major, 90
Vincetoxicum hastulatum, 94
Vine
  Lily-of-the-Valley, 838
  Madeira, 242

Matrimony, 832
Mignonette, 242
Puncture, 862
Viola, 852
  adunca, 853
  arvensis, 853
  aurea mohavensis, 853
  blanda, 854
  douglasii, 854
  lobata, 854
    integrifolia, 854
  macloskeyi, 854
  nephrophylla, 854
  pedunculata, 854
  purpurea
    mesophyta, 856
    purpurea, 854
    xerophyta, 856
  quercetorum, 856
  sheltonii, 856
  tricolor, 856
Violaceae, 51, 852
Violet, 852
Viscaceae, 46, 856
Vitaceae, 49, 858
Vitis girdiana, 859
Vulpia
  dertonensis, 972
  megalura, 972
  microstachys, 972
  myurus, 972
  octoflora, 972
  pacifica, 972
  reflexa, 972

Walnut, 4
  English, 521
Wapato, 870
Warrior, Indian, 818
Washingtonia
  brachypoda, 82
  brevipes, 82
  filifera, 883
  robusta, 883
Waterwort, 396
Wattle,
  Black, 421
  Golden, 421
  Green, 421
  Silver, 421
Wedelia incarnata, 580
Wedeliella incarnata, 580
Wedgescale, 358
Weed,
  Alkali, 375
  Alligator, 60
  Arrow-, 219
  Bind-, 375
  Bishop's, 69
  Blue, 183
  Burro-, 4, 110

Cheese, 569
Chinch, 214
Cud, 169
Devil-, 123
  Mexican, 123
Dove, 409
Duck, 918
Fire, 605
Gum, 206
Hawk, 188
Horse, 149
Jimson, 830
Klamath, 519
Knap, Russian, 135
Knot, 700
Loco, 422, 464
Match, 173
May, 114
Milk, 91
Pickle, 367
Pig, 359
  Winged, 364
Pineapple, 210
Poke, 632
Pond, 1007
  Horned, 1013
Poverty, 195
Rattle, 422
Rattlesnake, 75, 411
Rosin, 132
Seep, 369
Silver, 753
Smart, 700
Sneeze, 182
Spike, 186, 187
Stink, 329
Tar, 184, 189, 205
  Chile, 206
Telegraph, 187
Tobacco-, 123
Tumble-, 61
Vinegar, 544
Wart, 412
Water, 905
  Brazilian, 905
  Squaw, 125
Wheat, 1007
Whipplea utahensis, 779
Whitetop, 281
Whitlavia minor, 512
Wigandia caracasana macrophylla, 519
Willow,
  Arroyo, 775
  Desert-, 246
  Narrow-Leaf, 774
  Sandbar, 774
  Seep-, 124
  Weeping, 774
Windmills, 580
Wintergreen, 403

Wislizenia
    californica, 331
    refracta, 331
Wives, Blow-, 107
Wolffiella oblonga, 919
Woodsia
    oregana, 22
    plummerae, 22
    scopulina, 22
Woodwardia
    chamissoi, 23
    fimbriata, 23
    radicans americana, 23
Wormwood, 116
Wyethia, 240
    coriacea, 241
    ovata, 241

Xanthium
    acutum, 241
    californicum, 241
    campestre, 241
    canadense, 241
    italicum, 241
    palustre, 241
    spinosum, 241
    strumarium
        canadense, 241
        strumarium, 241
Xanthocephalum sarothrae, 174
Xanthoxalis californica, 620
Ximensia exauriculata, 239
Xylococcus bicolor, 405
Xylorhiza
    cognata, 204
    orcuttii, 204

tortifolius, 204
Xylosteon involucrata, 332
Xylothermia
    montana, 465
    tomentosa, 466

Yarrow, 105
    Golden-, 163
Yerba Buena
    de la Rabia, 580
    Santa, 497
Yerba Mansa, 776
Yucca, 866
    arborescens, 866
    baccata
        baccata, 866
        vespertina, 866
    brevifolia, 4, 866
        brevifolia, 866
        herbertii, 866
        jaegeriana, 866
        wolfei, 866
    californica, 867
    draconis arborescens, 866
    graminifolia, 867
    Mohave, 867
    mohavensis, 867
    ortgesiana, 867
    schidigera, 4, 867
    whipplei, 867
        caespitosa, 867
        intermedia, 867
        parishii, 867
        percursa, 867

Zannichellia palustris, 1013
Zannichelliaceae, 1013

Zantedeschia aethiopica, 882
Zappania nodiflora rosea, 850
Zauschneria
    californica, 616
        angustifolia, 616
        latifolia, 616
        mexicana, 616
    cana, 617
    garrettii, 617
    latifolia, 616
        johnstonii, 616
        viscosa, 616
    mexicana, 616
    villosa, 616
Zigadenus
    brevibracteatus, 929
    chloranthus, 931
    diegensis, 931
    exaltatus, 929
    fremontii
        brevibracteatus, 929
        fremontii, 931
        minor, 931
    venenosus, 931
Zonanthemis clementina, 185
Zostera
    latifolia, 1015
    marina, 1015
        latifolia, 1015
        oregona, 1015
        pacifica, 1015
Zosteraceae, 1013
Zygadene, 929
Zygophyllaceae, 50, 859
Zygophyllum
    fabago brachycarpum, 862